# 33rd Annual International Pittsburgh Coal Conference (IPCC 2016)

## Coal – Energy, Environment and Sustainable Development

Cape Town, South Africa
8-12 August 2016

Volume 1 of 2

ISBN: 978-1-5108-5366-9

**Printed from e-media with permission by:**

Curran Associates, Inc.
57 Morehouse Lane
Red Hook, NY 12571

**Some format issues inherent in the e-media version may also appear in this print version.**



Printed by Curran Associates, Inc. (2018)



International Pittsburgh Coal Conference
University of Pittsburgh
Swanson School of Engineering
3700 O'Hara St.
940 Benedum Hall
Pittsburgh, Pennsylvania 15261
USA

Phone: (412) 624-7440
Fax: (412) 624-1480

ipcc@pitt.edu

**Additional copies of this publication are available from:**

Curran Associates, Inc.
57 Morehouse Lane
Red Hook, NY 12571 USA
Phone: 845-758-0400
Fax: 845-758-2633
Email: curran@proceedings.com
Web: www.proceedings.com

# TABLE OF CONTENTS

## VOLUME 1

### PLENARY SPEAKERS



### ORAL SESSIONS

### SESSION 1: CLEAN COAL DEMONSTRATION AND COMMERCIAL PROJECTS - 1



### SESSION 2: COMBUSTION TECHNOLOGIES - 1



### SESSION 3: CARBON MANAGEMENT - 1

## SESSION 12: COAL MINING AND BENEFICIATION - 2



## SESSION 13: CLEAN COAL DEMONSTRATION AND COMMERCIAL PROJECTS - 3



## SESSION 14: COMBUSTION TECHNOLOGIES - 3



## SESSION 15: CARBON MANAGEMENT - 3

## SESSION 32: COAL SCIENCE - 6



## SESSION 33: GASIFICATION TECHNOLOGIES: GENERAL - 3



## SESSION 34: CLEAN COAL AND GAS TO FUELS: PROCESSING



## SESSION 35: GASIFICATION TECHNOLOGIES: MODELING - 2

## POSTER SESSION 5: COAL SCIENCE

THIRTY - THIRD ANNUAL

# INTERNATIONAL PITTSBURGH COAL CONFERENCE

University of Pittsburgh - Swanson School of Engineering

# FINAL PROGRAM

## COAL - ENERGY, ENVIRONMENT AND SUSTAINABLE DEVELOPMENT

August 8 - 12, 2016
Cape Town, SOUTH AFRICA
Cape Town International Convention Center

**University of Pittsburgh**
**Swanson School of Engineering**

NORTH-WEST UNIVERSITY ®
YUNIBESITI YA BOKONE-BOPHIRIMA
NOORDWES-UNIVERSITEIT

AFRICARY
African Carbon Energy

# General Information

**WELCOME!**

On behalf of the Conference Advisory Board, Conference Committees, and the University of Pittsburgh we welcome you to the Thirty-ThirdAnnual International Pittsburgh Coal Conference held August 8 - 12, 2016 at the Cape Town International Convention Center in Cape Town, South Africa. The Conference is hosted by the University of Pittsburgh.

The theme of this year's conference is "Coal - Energy, Environment and Sustainable Development" which covers a wide spectrum of important topics on coal technology, synfuel and environmental issues. The topics cover energy and environmental issues and technologies related to coal and its byproducts. Over 200 technical papers and posters will be presented throughout the conference. The Poster Sessions will be held on Wednesday, August 10 from 17:30 - 19:30. For detailed information on technical sessions, papers and speakers, please turn to page 6 in the Technical Program.

The invited Plenary Speakers include: Grace M. Bochenek, Director, National Energy Technology Laboratory, Department of Energy, USA; Wang Hongqiao, Vice President, China National Coal Association, CHINA; Robert Jeffrey, Managing Director and Senior Economist, Econometrix, SOUTH AFRICA and Rosemary Falcon, Professor, School of Chemical and Metallurgical Engineering, University of the Witwatersrand, SOUTH AFRICA; Aleksander Sobolewski, Institute For Chemical Processing of Coal, POLAND; Xiangzeng Wang, Senior Vice President, Shaanxi Yanchang Petroleum Group Company, CHINA; Paul de Mattos, Managing Director, Clarke Energy, SOUTH AFRICA; Satya P. Chauhan, Senior Program Director, Battelle Columbus Operations, USA; and Johan Brand, Managing Director, Africary, SOUTH AFRICA.

We express our sincere gratitude to the contributors for their support and involvement, to all the authors and co-authors of the technical papers and to all the members of the Program Committee, Awards Committee, International Committee and Membership Committee. Special thanks go to all session chairs, speakers and international delegates for their contributions to the 2016 technical program.

As the chair of the Advisory Board of the Conference, I deeply appreciate your participation and interest in this year's Conference and we invite you to join us next year for the Thirty-Fourth Annual International Pittsburgh Coal Conference, which will be held in Pittsburgh, PA from September 5 - 8, 2017.

Sincerely,

Richard Winschel, Chair
CONSOL Energy Inc., Pittsburgh, PA, USA

## CONFERENCE REGISTRATION

On-Site Registration will begin Tuesday, August 9, from 15:00 - 18:00 and continues Wednesday, Thursday, from 7:00 - 15:30, and Friday from 7:00 until 10:00.

**Please check in even if you have Pre-Registered!**

## PITT AWARD

***The Award for Innovation in Coal Conversion*** was founded by the Chemical and Petroleum Engineering Department, University of Pittsburgh in 1983 with industrial support. Since 1992, it has been fully funded by CONSOL Energy Inc.

## The International Pittsburgh Coal Conference

EXECUTIVE DIRECTOR:
Dr. Badie I. Morsi

CONFERENCE ORGANIZER:
Mrs. Heidi M. Peck

University of Pittsburgh
Swanson School of Engineering
940 Benedum Hall
Pittsburgh, PA 15261 USA
Tel: +1-412-624-7440
FAX: +1-412-624-1480
Email: ipcc@pitt.edu
www.pccpitt.org

## CONFERENCE OVERVIEW

### MONDAY, AUGUST 8, 2016

| | |
|---|---|
| Technical Tour - Johannesburg | 06:45 - 17:30 |
| Technical Tour - Cape Town | 07:15 - 13:00 |

### TUESDAY, AUGUST 9, 2016

| | |
|---|---|
| Vineyard Tour, Cape Town | 08:15 - 13:00 |
| Registration | 15:00 - 18:00 |
| Welcome Cocktail Reception | 17:30 - 19:30 |

### WEDNESDAY, AUGUST 10, 2016

| | |
|---|---|
| Registration | 07:00 - 15:30 |
| Opening Session | 08:30 - 08:45 |
| Plenary Session – 1 | 08:45 - 10:30 |
| Concurrent Tech. Sessions | 10:50 - 12:30 |
| Conference Lunch | 12:30 - 13:30 |
| Concurrent Tech. Sessions | 13:30 - 17:10 |
| Poster Session | 17:30 - 19:30 |

### THURSDAY, AUGUST 11, 2016

| | |
|---|---|
| Registration | 07:00 - 15:30 |
| Opening Session | 08:30 - 08:45 |
| Plenary Session – 2 | 08:45 - 10:30 |
| Concurrent Tech. Sessions | 10:50 - 12:30 |
| Conference Lunch | 12:30 - 13:30 |
| Concurrent Tech. Sessions | 13:30 - 17:10 |
| Gala Dinner | 18:30 - 21:30 |

### FRIDAY, AUGUST 12, 2016

| | |
|---|---|
| Registration | 07:00 - 10:00 |
| Opening Session | 08:30 - 08:45 |
| Plenary Session – 3 | 08:45 - 10:00 |
| Concurrent Tech. Sessions | 10:20 - 12:00 |
| Awards Lunch | 12:30 -13:30 |
| Advisory Board Meeting | 17:00 -19:00 |

## PLENARY SPEAKERS

### WEDNESDAY, AUGUST 10, 2016

**Dr. Grace M. Bochenek**
Director, National Energy Technology Laboratory, Department of Energy, USA
**"Defining the Future – The Imperative of Coal Research"**

**Mr. Wang Hongqiao**
Vice President
China National Coal Association, CHINA
**"Current Situation and Prospect of Coal Industry in China"**

**Mr. Robert Jeffrey**
Managing Director and Senior Economist
Econometrix, SOUTH AFRICA
**"Energy and its Effect on Economic Growth - The Role of Coal and Gas in the Future Economic Environment"**

### THURSDAY, AUGUST 11, 2016

**Dr. Aleksander Sobolewski**
Institute For Chemical Processing of Coal, POLAND
**"Coal Gasification in Poland - Perspectives and Key Driving Forces"**

**Dr. Xiangzeng Wang**
Senior Vice President
Shaanxi Yanchang Petroleum Group Company, CHINA
**"Cleaner Coal and Greener Oil Production: An Integrated CCUS Approach by the Yanchang Petroleum Group"**

**Mr. Paul de Mattos**
Managing Director, Clarke Energy, SOUTH AFRICA
**"The Future of Electricity Production from Coal and Gas"**

### FRIDAY, AUGUST 12, 2016

**Dr. Satya P. Chauhan**
Senior Program Director
Battelle Columbus Operations, USA
**"Scale-Up of Battelle's Direct Coal-To-Liquids (CTL) Process for Jet Fuel Using Biomass-Derived Solvents"**

**Mr. Johan Brand**
Managing Director, Africary, SOUTH AFRICA
**"Environment Coal, Energy, and CCS Potential in South Africa - A Review"**

# Technical Program Schedule

| | | | | | | |
|---|---|---|---|---|---|---|
| **Monday, August 8, 2016** | | | | | | |
| 6:45-17:30 | Technical Tour – Eskom Lethabo and Sasol Secunda Plant Bus Tour, Gauteng (Johannesburg) | | | | | |
| 7:15-13:00 | Technical Tour – Koeberg Power Plant, Cape Town | | | | | |
| **Tuesday, August 9, 2016** | | | | | | |
| 8:15-13:00 | Vineyard Tour, Cape Town | | | | | |
| 15:00-18:00 | Registration - Outside Ballroom | | | | | |
| 17:30-19:30 | Welcome Cocktail Reception - Clivia Conservatory & Jasminum Restaurant | | | | | |
| **Wednesday, August 10, 2016** | | | | | | |
| 7:00-15:30 | Registration - Outside Ballroom | | | | | |
| 8:30-8:45 | Opening Session – West Ballroom | | | | | |
| 8:45-10:30 | Plenary Session – West Ballroom | | | | | |
| 10:30-10:50 | **Morning Tea** | | | | | |
| ROOM | **West Ballroom** | **Room 1.41 and 1.42** | **Room 1.43** | **Room 1.44** | **Room 1.61** | **Room 1.62** |
| | Session 1 | Session 2 | Session 3 | Session 4 | Session 5 | Session 6 |
| 10:50-12:30 | Clean Coal Demonstration and Commercial Projects - 1 | Combustion Technologies 1 | Carbon Management - 1 | Coal Science - 1 | Gasification Technologies: Gas Cleanup | Coal Mining and Beneficiation - 1 |
| 12:30-13:30 | **Conference Lunch - East Ballroom** | | | | | |
| | Session 7 | Session 8 | Session 9 | Session 10 | Session 11 | Session 12 |
| 13:30-15:10 | Clean Coal Demonstration and Commercial Projects - 2 | Combustion Technologies - 2 | Carbon Management - 2 | Coal Science - 2 | Gasification Technologies: UCG - 1 | Coal Mining and Beneficiation - 2 |
| 15:10-15:30 | **Afternoon Tea** | | | | | |
| | Session 13 | Session 14 | Session 15 | Session 16 | Session 17 | Session 18 |
| 15:30-17:10 | Clean Coal Demonstration and Commercial Projects - 3 | Combustion Technologies - 3 | Carbon Management - 3 | Coal Science - 3 | Gasification Technologies: General - 1 | Coal Mining and Beneficiation - 3 |
| 17:30-19:30 | **Poster Session with Wine Tasting - East Ballroom** | | | | | |
| **Thursday, August 11, 2016** | | | | | | |
| 7:00-15:30 | Registration - Outside Ballroom | | | | | |
| 8:30-8:45 | Opening Session – West Ballroom | | | | | |
| 8:45-10:30 | Plenary Session – West Ballroom | | | | | |
| 10:30-10:50 | **Morning Tea** | | | | | |
| ROOM | **West Ballroom** | **Room 1.41 and 1.42** | **Room 1.43** | **Room 1.44** | **Room 1.61** | **Room 1.62** |
| | Session 19 | Session 20 | Session 21 | Session 22 | Session 23 | |
| 10:50-12:30 | Clean Coal and Gas to Fuels: Catalytic Study - 1 | Sustainability and Environment - 1 | Shale and Coal Bed Gas - 1 | Coal Science - 4 | Gasification Technologies: General - 2 | |
| 12:30-13:30 | **Conference Lunch - East Ballroom** | | | | | |
| | Session 24 | Session 25 | Session 26 | Session 27 | Session 28 | |
| 13:30-15:10 | Clean Coal and Gas to Fuels: New Technologies | Sustainability and Environment - 2 | Gasification Technologies: UCG - 2 | Coal Science - 5 | Gasification Technologies: Modeling - 1 | |
| 15:10-15:30 | **Afternoon Tea** | | | | | |
| | Session 29 | Session 30 | Session 31 | Session 32 | Session 33 | |
| 15:30-17:10 | Clean Coal and Gas to Fuels: Catalytic Study - 2 | Sustainability and Environment - 3 | Gasification Technologies: General - 4 | Coal Science - 6 | Gasification Technologies: General - 3 | |
| 18:30-21:30 | **Gala Dinner - East Ballroom** | | | | | |
| **Friday, August 12, 2016** | | | | | | |
| 7:00-10:00 | Registration - Outside Ballroom | | | | | |
| 8:30-8:45 | Opening Session – West Ballroom | | | | | |
| 8:45-10:00 | Plenary Session – West Ballroom | | | | | |
| 10:00-10:20 | **Morning Tea** | | | | | |
| ROOM | **West Ballroom** | **Room 1.41 and 1.42** | **Room 1.43** | **Room 1.44** | **Room 1.61** | **Room 1.62** |
| | Session 34 | Session 35 | Session 36 | | | |
| 10:20-12:00 | Clean Coal and Gas to Fuels: Processing | Gasification Technologies: Modeling - 2 | Gasification Technologies: General - 5 | | | |
| 12:30-13:30 | **Awards Luncheon - East Ballroom** | | | | | |
| 17:00-19:00 | **Advisory Board Meeting - Cullinan Hotel** | | | | | |

# LOCATION GUIDE

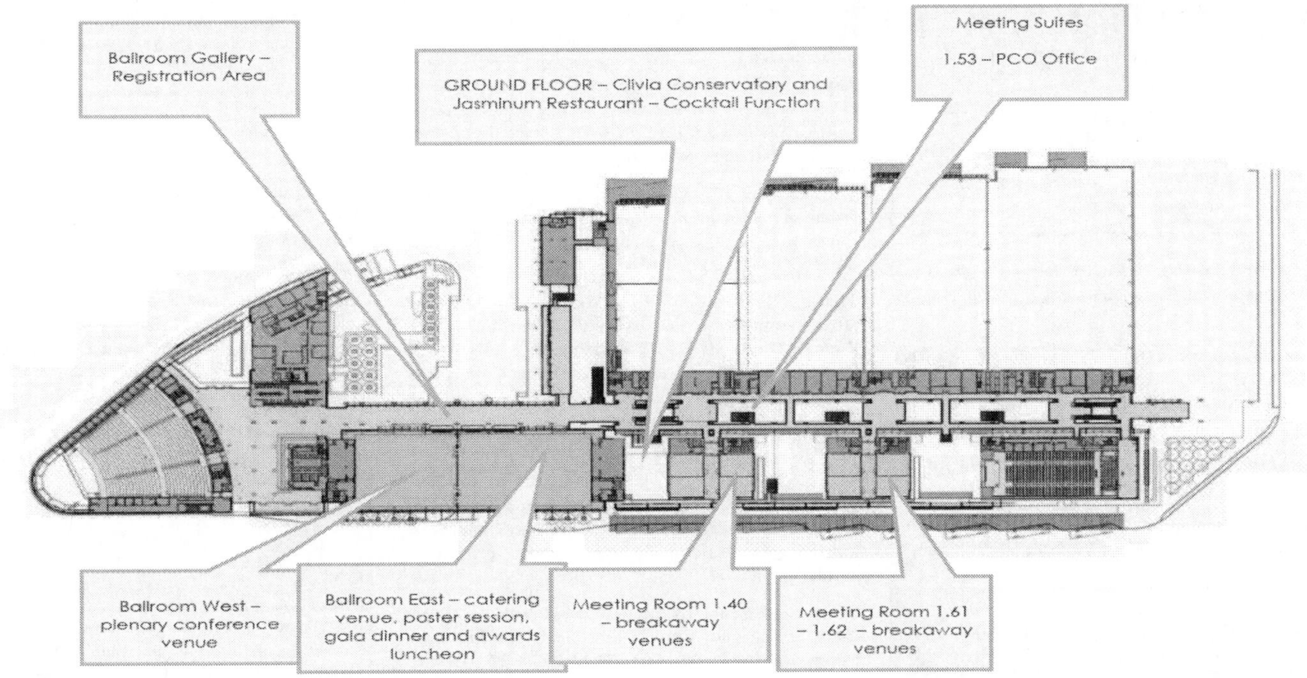


First Floor

**SPONSORS**

John Wiley & Sons Inc.

NWU
(North West University)

Multotec

**EXHIBITORS**

Scantech

University of Kentucky - Center for Applied Energy Research

**SESSION MEETING ROOMS**

**West Ballroom**
Sessions 1, 7, 13, 19, 24, 29, 34

**Room 1.41 and 1.42**
Sessions 2, 8, 14, 20, 25, 30, 35

**Room 1.43**
Sessions 3, 9, 15, 21, 26, 31, 36

**Room 1.44**
Sessions 4, 10, 16, 22, 27, 32

**Room 1.61**
Sessions 5, 11, 17, 23, 28, 33

**Room 1.62**
Sessions 6, 12, 18

**Registration**
Foyer outside Ballroom

**Tuesday Welcome Cocktail Reception**
Clivia Conservatory & Jasminum Restaurant

**Opening Session**
West Ballroom

**Plenary Session**
West Ballroom

**Conference Luncheons**
East Ballroom

**Poster Presentations**
East Ballroom

**Exhibits**
Ballroom

# TECHNICAL PROGRAM

## ORAL SESSIONS
## Wednesday, August 10

### SESSION 1
### CLEAN COAL DEMONSTRATION AND COMMERCIAL PROJECTS - 1
*John Bunt and Thomas Sarkus*

**The Value of Fuel Flexible Circulating Fluidized Bed Technology**, Robert Giglio, Antonio Castilla Oliver, Amec Foster Wheeler, USA

**Application Update of OMB CWS Gasification Process**, Guangsuo Yu, Qinghua Guo, Xueli Chen, Yifei Wang, Zhenghua Dai, Haifeng Liu, Fuchen Wang, Xin Gong, ECUST, CHINA

**Ohio Valley University's Alternative Clean Energy Facility,** Marcus Wiley, TCG Global, LLC, USA

**SES SGT Commercial & Operation Plant Updates**, Francis Lau, Synthesis Energy Systems, USA

### SESSION 2
### COMBUSTION TECHNOLOGIES - 1
*Mohsen Alimandegari and Johan van Dyk*

**CFD Simulation of an Industrial Wet Flue Gas Desulfurization Spray Tower: A Comprehensive Model with Special Attention Devoted to the Modeling of Absorption and Chemical Reactions,** Raymond Everson, Arif Arif, Hein W.J.P. Neomagus, Dawie Branken, North-West University, SOUTH AFRICA

**Optimization of Coal and Combustion Air Distribution for 500 MW Tangential Coal-Fired Plant**, Kitae Kang, Korea Southern Power (KOSPO) CO., Ltd, KOREA

**NLP/MINLP for Modeling and Optimization of Lignite Pyrolysis Process**, Wengying Li, Yi Huang, Qun Yi, Minhui Gong, Guosheng Wu, Taiyuan University of Technology, CHINA

### SESSION 3
### CARBON MANAGEMENT - 1
*Esta van Heerden and John Duddy*

**Status of the Carbon Capture International Test Center Network**, Roxann Laird, Southern Company Services, USA

**Status of Technology Development at the National Carbon Capture Center**, Roxann Laird, Southern Company Services, USA

**Performing a Push-Pull Test in a Shallow Aquifer to Assess Impacts of $CO_2$ and Impurities Leakage on Freshwater Quality**, Stéphane Lafortune, Philippe Gombert, Zbigniew Pokryszka, INERIS, FRANCE; Maria del Camino Martin Torre, Universidad de Cantabria, SPAIN

**Geological Carbon Sequestration and its Influence on Subsurface Microbial Diversity and Metabolic Carbon Cycling**, Mariana Erasmus, Esta van Heerden, University of the Free State, SOUTH AFRICA; Tullis Onstott, Princeton University; Mary DeFlaun, Geosyntec Consultants; Tommy Phelps, University of Tennessee, USA

### SESSION 4
### COAL SCIENCE - 1
*Frans Waanders and Daniel Roberts*

**Physical and Chemical Characteristics of Coke by using Ash-Free Coal as Binder**, Gyeong-Min Kim, Kevin Yohanes Lisandy, Jin-Ho Kim, Chung-Hwan Jeon, Pusan National University, KOREA

**The Influence of $K_2CO_3$ on the Swelling and Caking Behavior of Large South African Coal Particles**, S. Coetzee, H.W.J.P Neomagus, J.R. Bunt, North-West University, SOUTH AFRICA

**Time of Flight Secondary Ion Mass Spectrometry and High-Resolution Transmission Electron Microscopy/ Energy Dispersive Spectroscopy: A Study of the Distribution of Hazardous Elements in Coal Ashes**, Diego de Medeiros, Marcos L.S. Oliveira, Bianca D. Lima, Luis F. O. Silva, Centro Universitário La Salle, BRAZIL; Frans Waanders, North West University, SOUTH AFRICA

**Investigation of Solvent Crystallization Process of Anthracene**, Wenying Li, Xiaoxiao Ding, Cuiping Ye, Mingming Fan, Hai Mu, Jie Feng, Taiyuan University of Technology, CHINA

**Pyrolysis Product Evaluation and Prediction From Coals of Different Rank**, Gustav Pretorius, J.R. Bunt, North-West University, SOUTH AFRICA; M. Gräbner, Air Liquide, GERMANY

### SESSION 5
### GASIFICATION TECHNOLOGIES: GAS CLEANUP
*Shehzaad Kauchali and Francis Lau*

**Improving Response Rates of Acid Gas Absorber Columns,** Chris Higman, Higman Consulting GmbH; Bernd Epple, Christain Heinze, TU Darmstadt, GERMANY; Jose Marasigan, EPRI, USA

**Thermodynamic Possibilities and Constraints for the Regeneration of Spent Solid $H_2S$ Sorbents Based on Ce-, La-, and Zn-Oxides,** Karel Svoboda, Miloslav Hartman, Institute of Chemical Process Fundamentals of the CAS; Jindřich Leitner, Jiří Brynda, Siarhei Skoblia, University of Chemistry and Technology Prague, CZECH REPUBLIC; Yau-Pin Chyou, Yi-Shun Chen, Institute of Nuclear Energy Research, TAIWAN

**Particle Fragmentation Behavior in a Bench Scale Entrained-Flow Coal-Water Slurry Gasifier**, Qinghua Guo, Guangsuo Yu, Yifei Wang, Xin Gong, East China University of Science and Technology, CHINA

### SESSION 6
### COAL MINING AND BENEFICIATION - 1
*Tarunjit Butalia*

**The Product Yield Problem and the History and Future of Washability Analysis**, Herman Dorland, Nicola Wagner, Xitshembiso Makukule, University of Johannesburg, SOUTH AFRICA

**Estimation of Moisture Dynamics within a Composite Coal Stockpile**, Kyle Enslin, Chris van Alphen, Eskom; Walter Schmitz, University of Witwatersrand, SOUTH AFRICA

**Seal Design Investigations for Omerler Underground Coal Mine in Turkey**, C. Atilla Ozturk, Abdullah Fisne, Istanbul Technical University; Mustafa Ozdingis, Sahika Yurek, Kıvanc Het, Turkish Coal Enterprises, TURKEY

### SESSION 7
### CLEAN COAL DEMONSTRATION AND COMMERCIAL PROJECTS - 2
*John Bunt and Thomas Sarkus*

**The Future of Coal**, Farid A. Malik, Forman Christian College (American University), PAKISTAN

**Status Update on the Kemper County IGCC Project**, Diane Revay Madden, DOE/NETL; Tim Pinkston, Southern Company, USA

**ITEA CCS Flameless Pressurized Oxy-Combustion: The Strategic Role of Intermediate Scale (50 MWth) Demonstration Units, Tailored on Local Quality and Low Rank Coals in Each Major Geographic Area, to Mature Technology in View of Industrial Exploitation**, Massimo Malavasi, ITEA S.p.a., ITALY

**Shenhua Coal Refining Technology Development**, Cliff Y. Guo, National Institute of Clean-and-low-carbon Energy, CHINA

### SESSION 8
### COMBUSTION TECHNOLOGIES - 2
*Mohsen Alimandegari and Johan van Dyk*

**Fuel Additive Efficient Coal Combustion Technology,** Marc Ottolini, International Innovative Technologies Ltd.; William Nimmo, Syed Sheraz Daood, University of Sheffield, UNITED KINGDOM

**Impact of Biomass Co-Firing with Unreactive Coals in Pulverised Fuel,** C. Bridge, C. Snape, University of Nottingham; C. J. Waldron, GE Power, UNITED KINGDOM

**Effects of Temperature on Physico-Chemical Characteristics Evolution of Rice Husk Char under Reburning Conditions**, Jiateng Shi, Xinyi Yin, Ping Lu, Nanjing Normal University, CHINA

**Low $NO_X$ Coal Burner Temperature Profile Evaluation,** Dewan Smit, Sandile Peta, Eskom; Walter Schmitz, Reshendren Naidoo, University of Witwatersrand, SOUTH AFRICA

### SESSION 9
### CARBON MANAGEMENT - 2
*Esta van Heerden and John Duddy*

**Acid Gas Control by Dry Sorbent Injection in Air and Oxy-Fuel Combustion,** Reinhold Spörl, Reyhane Youssefi, Jörg Maier, Günter Scheffknecht, Institute of Combustion and Power Plant Technology - IFK, University of Stuttgart, GERMANY

**Introduction to the Wyoming Integrated Test Center**, Jason Begger, Wyoming Infrastructure Authority, USA

**Adsorption Characteristics of Bituminous Coal Seams in Zonguldak Basin**, Olgun Esen, Samet Can Ozer, Abdullah Fisne, Istanbul Technical University, TURKEY

### SESSION 10
### COAL SCIENCE - 2
*Nikki Wagner and Daniel Roberts*

**Determination of Rare Earth Elements (REE's) in Coals and their Associated Ash Products from Selected Localities in South Africa**, Matiane R. Arnold, Nikki Wagner, University of Johannesburg, SOUTH AFRICA

**Rare Earth Elements from Coal and Coal By-Products**, Mary Anne Alvin, DOE/NETL, USA

**Multifaceted Nanominerals and Ultrafine Particles in Phosphogypsum: A By-Product of Coal Cleaning Rejects**, Felipe B. de Leão, Centro Universitário La Salle and Universidade Federal do Rio Grande do Sul; Silvio R. Taffarel, Bianca D. Lima, Luis F. O. Silva, Centro Universitário La Salle; Carlos H. Sampaio, Universidade Federal do Rio Grande do Sul, BRAZIL; Frans Waanders, North West University, SOUTH AFRICA

**Briquetting of Disk-Pelletized Fine Coal**, Darrell Taulbee, Robert Hodgen, University of Kentucky Center for Applied Energy Research, USA

**SESSION 11**
**GASIFICATION TECHNOLOGIES: UCG - 1**
*Shehzaad Kauchali and Francis Lau*

**UCG – Unlocking Southern-Africa's Coal Potential: Update on the Africary's Theunissen Underground Coal Gasification (TUCG) Project in South-Africa,** JF Brand, JC van Dyk, Africary, SOUTH AFRICA

**The Queensland Underground Coal Gasification Pilots 2008-2016 – Evidence Based Assessment and Lessons in Government and Public Relations,** Cliff Mallett, Carbon Energy Limited, AUSTRALIA

**Leaching of Metals From Post-Underground Coal Gasification Cavity Residues,** Krzysztof Stańczyk, Aleksandra Strugała-Wilczek, Central Mining Institute, POLAND

**Coupled Thermo-Chemical-Mechanical Modelling of Cavity Growth and Farfield Geomechanical Assessment in Underground Coal Gasification**, Sevket Durucan, Ana Korre, Epameinondas Andrianopoulos, Guangyao Si, Ji-Quan Shi, Royal School of Mines, Imperial College London; Sebastian Franzsen, Seamwell International Ltd., UNITED KINGDOM

**SESSION 12**
**COAL MINING AND BENEFICIATION - 2**
*Tarunjit Butalia and Christien Strydom*

**Avdan Lignite Mine Open Pit Planning and Design Studies for Thermal Power Plant**, C. Atilla Ozturk, Murat Ozkan, Mohammad Lashgari, Selamet Ercelebi, Istanbul Technical University, TURKEY

**Modelling a Multi-Layer Coal Mine Deposit: A Case Study of Avdan Lignite Site in Denizli, Turkey**, Mohammad Lashgari, Murat Ozkan, C. Atilla Ozturk, Selamet Ercelebi, Istanbul Technical University; Mustafa Darici, Anadolu Mining Co., TURKEY

**Mineral Matter in Coals From the Soutpansberg Coalfield, South Africa**, Maseda Mphaphuli, Nicola Jane Wagner, University of Johannesburg; John Sparrow, Coal of Africa Limited, SOUTH AFRICA

**Assessing Risks Related to Stimulation of Firedamp Drainage Boreholes in Order to Recommend Prevention Measures and Monitoring Methods to Help Early Risk Management During Design of Future Drainage Projects**, Stéphane Lafortune, Régis Farret, Candice Lagny, Zbigniew Pokryszka, INERIS, FRANCE

**SESSION 13**
**CLEAN COAL DEMONSTRATION AND COMMERCIAL PROJECTS - 3**
*John Bunt and Thomas Sarkus*

**Illinois Industrial Carbon Capture and Storage Project**, Scott McDonald, Archer Daniels Midland Company, USA

**Process Optimization and Solvent Reclaiming Enhanced the Operation and Production of an Existing 800 Tonnes Per Day Plant Capturing $CO_2$ From Flue Gases of a Coal-Fired Power Plant**, Ahmed Aboudheir, Lionel Kambeitz, HTC Purenergy, CANADA

**Optimization of an Existing 130 Tonnes Per Day Plant Capturing $CO_2$ From Flue Gases of a Coal-Fired Power Plant**, Ahmed Aboudheir, Lionel Kambeitz, HTC Purenergy, CANADA

**The U.S. Department of Energy's Current Portfolio of Major Fossil Energy Demonstration Projects**, Thomas A. Sarkus, Diane Madden, Theodore J. McMahon, Anthony Zinn, Sai Gollakota, Venkat Venkataraman, Nelson Rekos, DOE/NETL, USA

**SESSION 14**
**COMBUSTION TECHNOLOGIES - 3**
*Mohsen Alimandegari and Daniel Roberts*

**Vapor Phase Chemical Looping Combustion of Coal**, Tomasz Wiltowski, Debalina Dasgupta, Kanchan Mondal, Quan Zhang, Southern Illinois University, USA

**Copper-Based Chemical Looping with Oxygen Uncoupling: Process Development, Reactor Scale-Up, Pilot-Scale Studies and System Modeling**, Kevin J. Whitty, Matthew A. Hamilton, Kirsten M. Merritt, Kyle O'Malley, Shakar Ali, Emir Rahislic, Andrew R. Fry, JoAnn S. Lighty, The University of Utah, USA

**A Study on the Mineral Liberation and Transformations During Pulverised Bituminous Coal Combustion**, Laurence Silvester, Cheng-gong Sun, Colin Snape, Hao Liu, Ofonime Udoudo, University of Nottingham; William Livingston, Doosan Babcock Energy Limited, UNITED KINGDOM

**Combustion Properties of Bamboo When Blended with a South African Coal**, O. Makwarela, S O Bada, R Falcon, University of the Witwatersrand, SOUTH AFRICA

**An Experimental Study of Nitrogen Oxides Formation of Pulverized Coal in $O_2/N_2$ and $O_2/CO_2$ Atmospheres**, Jeong-Woo Kim, Hong-Min Cho, Lim Ho, Chung-Hwan Jeon, Pusan National University, KOREA

**SESSION 15**
**CARBON MANAGEMENT - 3**
*Esta van Heerden and John Duddy*

**Monolithic Amine-Grafted Sorbents for Post-Combustion $CO_2$ Capture**, Steven S. C. Chuang, Jie Yu, Long Zhang, Yuxin Zhai, Hailiang Jin, Sihan Wang, The University of Akron, USA

**Mathematical Modeling of Adsorption Behaviour of SOD-ZMOF/Chitosan Adsorbent During Post-Combustion $CO_2$ Capture**, Kelvin Odafe Yoro, Muofhe Singo, Michael O. Daramola, Jean L. Mulopo, University of the Witwatersrand, SOUTH AFRICA

**Synthesis of Chitosan-Impregnated SOD-Zeolite Metal Organic Frameworks (SOD-ZMOF) for $CO_2$ Capture: A Preliminary Investigation**, Muofhe Singo, Michael O. Daramola, University of the Witwatersrand, SOUTH AFRICA

**Nanocomposite Architecture Hydroxy Sodalite/α-Alumina Membrane for $CO_2$ Capture**, O. Oloye, Michael O. Daramola, University of the Witwatersrand, SOUTH AFRICA

**SESSION 16**
**COAL SCIENCE - 3**
*Christien Strydom and Nikki Wagner*

**Prediction of Spontaneous Combustion in Coal by Use of Thermogravimetry**, Zama A Mthabela, Nicola J Wagner, Nthabiseng Maledi, University of Witwatersrand, SOUTH AFRICA

**Mechanical and Thermal Characterization of Briquetted High Ash South African Coal Fines Using Low Density Polyethylene as the Binding Agent**, Thabo Mthombo, Christien Strydom, John Bunt, Hein Neomagus, North-West University, SOUTH AFRICA

**Evaluation of a Domestic-Use Low-Smoke Fuel Produced via the Pyrolysis of Lump Coal**, LP Sumbane, John R Bunt, Hein WJP Neomagus, Stuart SJ Picketh, Thys Kuhn, North-West University, SOUTH AFRICA

**Research of the Masking Effect of Surfactants on O-Containing Functional Groups in Low-Rank Coals**, Geng Jianchun, China University of Mining and Technology, CHINA

**SESSION 17**
**GASIFICATION TECHNOLOGIES: GENERAL - 1**
*Shehzaad Kauchali and Francis Lau*

**Soot Gasification in Syngas Atmosphere**, G. Worringer, Linde AG - Engineering Division; R. Reimert, Karlsruhe Institute of Technology (KIT), GERMANY

**Comparison Study on Gasification of Long-Flame Briquette Coal with $CO_2$ and $H_2O$**, Lingmei Zhou, Mo Chu, Zhibing Chang, Fang Xu, Chengliang Hao, China University of Mining and Technology, CHINA

**Experiments of Coal Partial Gasification on a Circulating Fluidized Bed**, Chao Ye, Qinhui Wang, Zhongyang Luo, Mengxiang Fang, State Key Laboratory of Clean Energy Utilization, Zhejiang University, CHINA

**Catalytic Reduction Mechanism Investigation of $K_2CO_3$ Modified Iron-Based Oxygen Carrier by Carbon**, Zhongliang Yu, Shunxuan Hu, Chunyu Li, Tao Liu, Xing Zhou, Yitian Fang, Jiantao Zhao, Institute of Coal Chemistry, CHINA

**SESSION 18**
**COAL MINING AND BENEFICIATION - 3**
*Tarunjit Butalia*

**Development of a Condition Monitoring Philosophy for a Pulverised Fuel Vertical Spindle Mill**, Andre Govender, Eskom; Walter Schmitz, Reshendren Naidoo, University of the Witwatersrand, SOUTH AFRICA

**Measurement and Prediction of Coal Flow Rates in Pulverised Fuel Pipes**, Dheshan Naran, Bonny Nyangwa, Eskom; Walter Schmitz, University of Witwatersrand, SOUTH AFRICA

**Applications of Online Coal Analysers in South Africa**, Michael Edwards, Scantech International Pty Ltd, AUSTRALIA

# ORAL SESSIONS
## Thursday, August 11, 2016

### SESSION 19
### CLEAN COAL AND GAS TO FUELS: CATALYTIC STUDY - 1
*Somayeh Farzad and Wenhua Li*

**Fischer-Tropsch Synthesis in the Presence of Solvent Using Novel Carbon-Oxide Composite Supported Cobalt and Iron Catalysts**, Atsushi Ishihara, Yusuke Maejima, Hiroyuki Nasu, Tadanori Hashimoto, Mie University, JAPAN

**Carbon from Pine Cones as a Support for High Temperature Fischer-Tropsch Synthesis,** Eric van Steen, Pierre Cilliers, Doreen Nabaho, Chelsea Tucker, University of Cape Town; Wanda Bout, Augustine Ofomaja, Elvera Viljoen, Vaal University of Technology, SOUTH AFRICA

**Bifunctional Hydrocracking: Is the Well-Established 'Classical' Fluid Phase Diffusion Mechanism a Delusion?**, Walter Böhringer, Peter Wynne, Roald Brosius, Stephen Roberts, Jack CQ Fletcher, University of Cape Town, SOUTH AFRICA

**Catalyst Deactivation During Hydroprocessing of Coal-Derived Liquids at Pilot Plant Scale**, Rui Wang, Zhuihua Han, Linge Ma, Jingyun Chen, Xuelei Duan, Chenjing Qian, Yinhua Long, Xiaofen Guo, National Institute of Clean-and-Low-Carbon Energy, CHINA

### SESSION 20
### SUSTAINABILITY AND ENVIRONMENT - 1
*Frans Waanders and Joseph Bittinger*

**Hazardous Particles from Coal Fires from Santa Catarina, South Brazil**, Diego de Medeiros, Marcos L.S. Oliveira, Fagner T.C. Sá, Silvio R. Taffarel, Luis F. O. Silva, Centro Universitário La Salle, BRAZIL; Frans Waanders, North West University, SOUTH AFRICA

**An Investigation on Incorporation of a Local Thermal Power Station Boiler Ash as a Raw Material in Clinker Manufacture**, Nonhlanhla G. Mguni, Marcia R. Nyika, Liberty L. Mguni, National University of Science and Technology, ZIMBABWE

**Environmental and Economic Evaluation of the Production of Organic Platform Chemicals from Different Feedstocks**, Florian Keller, Christian Wolfersdorf, Bernd Meyer, IEC, TU Bergakademie Freiberg; Lutz Schiffer, Saxon Academy of Sciences in Leipzig, GERMANY

**Process Simulation and Assessment of Coke Oven Gas to Methanol Integrated with $CO_2$ Recycle**, Wenying Li, Minhui Gong, Qun Yi, Yi Huang, Jie Feng, Guosheng Wu, Taiyuan University of Technology, CHINA

### SESSION 21
### SHALE AND COAL BED GAS
*Frans Waanders and Leslie Ruppert*

**Status of Worldwide CMM and CBM Activity Including Sub-Saharan Africa**, Jonathan Kelafant, Advanced Resources International, Inc.; Felicia Ruiz, U.S. Environmental Protection Agency, USA

**Consideration of a Possibility of Enhanced Coalbed Methane Recovery by Injecting Air Into Coal Seam**, Noritaka Shimatani, Yuichi Sugai, Kyuro Sasaki, Kyushu University, JAPAN

**The Role of the Residual Bitumen of Mature Lacustrine Shale to the Gas Storage Capacity: A Case Study of the Triassic Yanchang Shale, Ordos Basin, China**, Jianfa Chen, Feiran Chen, Zhenxue Jiang, Zhilong Huang, Guoheng Liu, Fengyang Xiong, China University of Petroleum, CHINA

**Analyses of Adsorbed and Free Gas Production Using Methane Carbon Isotope Fractionation During Canister Desorption of Shale Gas**, Yong Ma, Ningning Zhong, China University of Petroleum, CHINA; Zhejun Pan, CSIRO Energy Flagship, AUSTRALIA

**Anisotropic Permeability and its Relationship with Pore/Fracture Structure of Longmaxi Shales in the Upper Yangtze Region, China**, Ningning Zhong, Yong Ma, China University of Petroleum, CHINA

### SESSION 22
### COAL SCIENCE - 4
*Nikki Wagner and Joseph Bittinger*

**Desulphurization of South African Coal Using Low Power Microwave Energy**, Herman Potgieter, University of the Witwatersrand and Manchester Metropolitan University; Sanja Potgieter-Vermaak, Manchester Metropolitan University, UNITED KINGDOM; Waseela Mohammed, University of the Witwatersrand, SOUTH AFRICA

**Application of Coal Petrography in Understanding the Variation of Coal Composition in the Coalfields of Botswana**, Ndivhuho Nendouvhada, Nicola Wagner, University of Johannesburg, SOUTH AFRICA; Alan Golding, Analytika Holdings, BOTSWANA

### SESSION 23
### GASIFICATION TECHNOLOGIES: GENERAL - 2
*Daniel Roberts and Aleksander Sobolewski*

**Char Gasification Kinetics in Mixtures of $CO_2$ and $H_2O$: Understanding the Roles of Actives Sites and Catalysis,** Daniel Roberts, Mark Kochanek, David Harris, CSIRO, AUSTRALIA

**Green Coal Development for Application in Fixed-Bed Catalytic Gasification,** John R Bunt, Sanette Marx, Frans B Waanders, North-West University, SOUTH AFRICA

**Experimental Investigation on Char Gasification Kinetics at Elevated Pressures**, Sascha Russig, Victor Gonzalez, Mark Zink, Steffen Krzack, Stefan Guhl, Bernd Meyer, TU Bergakademie Freiberg, GERMANY

**Ash Agglomeration in the Modified Internal Circulating COORVED Coal Gasifier**, Martin Schurz, Alexander Laugwitz, Steffen Krzack, Bernd Meyer, TU Bergakademie Freiberg, GERMANY

### SESSION 24
### CLEAN COAL AND GAS TO FUELS: NEW TECHNOLOGIES
*Somayeh Farzad and Wenhua Li*

**Supercritical $CO_2$ Cycles for Power Production**, Mike Holmes, Joshua Stanislowski, Jason Laumb, Energy & Environmental Research Center, University of North Dakota; Xijia Lu, Brock Forrest, Mike McGroddy, 8 Rivers Capital, LL, USA

**Integration of Gasification and Fuel Cells: Interaction Between the Anode and Contaminants in the Syngas**, Marcos Millan, Esther Lorente, Paul Boldrin, Nigel Brandon, Imperial College, UNITED KINGDOM

**Liquefaction of South African Bituminous Coal Using Phenol as a Solvent**, Zach Sehume, C.A. Strydom, J.R. Bunt, North-West University, SOUTH AFRICA; H.H. Schobert, Penn State University, USA

**Effect of Direct Addition of Alcohols to an Oligomerisation Reaction System**, Mekia Makhubu, Klaus Moller, University of Cape Town; Masikana Mdleleni, PetroSA Synthetic Fuels Innovation Center (PSFIC), University of Western Cape, SOUTH AFRICA

### SESSION 25
### SUSTAINABILITY AND ENVIRONMENT - 2
*Frans Waanders and Joseph Bittinger*

**Environmental Pollution and Nano-Mineralogical Characterization of Coal, Overburden and Sediment from Assam (India) Coal Mining Acid Drainage,** Frans B. Waanders, North West University, SOUTH AFRICA; Silvio R. Taffarel, Diego de Medeiros, Cesar M. N. L. Cutruneo, Luis F.O. Silva, Centro Universitário La Salle, BRAZIL; Madhulika Dutta, CNB College and CSIR-North East Institute of Science & Technology; Binoy K Saikia, CSIR-North East Institute of Science & Technology, INDIA

**Bench-Scale Testing of Stabilized Flue Gas Desulfurization Materials to Treat Acid Mine Drainage,** Tarunjit S. Butalia, The Ohio State University, USA

**Orbitrap Analysis of Fullerene on Surface Soils and River Sediments from Santa Catarina (Brazil) Coal Area,** Diego de Medeiros, Marcos L.S. Oliveira, Silvio R. Taffarel, Luis F. O. Silva, Centro Universitário La Salle, BRAZIL

**Fireside Corrosion Monitoring, Control and Prevention in a 600 MWe Pulverized Fuel Boiler,** Otilia Mwape, University of the Witwatersrand, SOUTH AFRICA

### SESSION 26
### GASIFICATION TECHNOLOGIES: UCG - 2
*Johan Brand and Thomas Sarkus*

**Can UCG Give Coal a New Lease on Life in 21st Century?,** Michael S. Blinderman, Ergo Exergy Technologies Inc., CANADA; Simon S. Maev, Laurus Energy, Inc., USA

**Life Cycle Sustainability Assessment of Integrated UCG-CCS Systems, Comparison with Clean Coal Alternatives,** Anna Korre, Sevket Durucan, Zhenggang Nie, Royal School of Mines, Imperial College London; Sebastian Franzsen, Seamwell International Ltd., UNITED KINGDOM

**Ex-Situ Simulation of High Pressure Underground Gasification of Hard Coal and Lignite,** Krzysztof Stańczyk, Krzysztof Kapusta, Marian Wiatowski, Central Mining Institute, POLAND; Sevket Durucan, Imperial College London, UNITED KINGDOM

**Experimental and Computational Studies on Underground Gasification of Turkish Lignite,** Mesut Gür, Nurdil Eskin, Hasancan Okutan, Ahmet Arısoy, Erhan Böke, Ünal Altıntaş, Oğuz Büyükşirin, E. Deniz Canbaz, Olgu Yıldırım, Istanbul Technical University, TURKEY

### SESSION 27
### COAL SCIENCE - 5
*John Bunt and Leslie Ruppert*

**Trace Elements Associated with the Upper Coal Zones of the Springbok Flats Basin, South Africa,** Valerie Nxumalo, Nicola Wagner, Jan Kramers, Clarisa Vorster, Bruce Cairncross, University of Johannesburg; John Paul Hunt, Council for Geoscience and University of Johannesburg; Wojciech Przybylowicz, iThemba LABS, SOUTH AFRICA

**Organic and Inorganic Characterisation of Different Coals by Single Particle Analysis,** Sanja Potgieter-Vermaak, Manchester Metropolitan University and University of the Witwatersrand; Nthabiseng Maledi, Herman Potgieter, University of the Witwatersrand; Nicola Wagner, University of Johannesburg, SOUTH AFRICA; Robert Sparkes, Manchester Metropolitan University, UNITED KINGDOM

**Coal Cleaning Rejects and Sulphide Oxidation by Fenton's Reaction,** Diego de Medeiros, Anderson S. Pires, Marcos L.S. Oliveira, Bianca D. Lima, Luis F. O. Silva, Centro Universitário La Salle, BRAZIL; Frans Waanders, North West University, SOUTH AFRICA

**Petroleum Coke-Based Activated Carbon Modified by Binary Metal Oxides for Elemental Mercury Removal in Flue Gas,** Yang Li, Ning Zhang, Lijun Jin, Haoquan Hu, Dalian University of Technology, CHINA

**SESSION 28**
**GASIFICATION TECHNOLOGIES: MODELING - 1**
*Daniel Roberts and Aleksander Sobolewski*

**Effect of Alkali Species in Australian Coals on Slag Composition and Viscosity,** Alexander Ilyushechkin, Daniel Roberts, David Harris, CSIRO, AUSTRALIA

**Numerical Study of the Hitecom Reactor for In-Situ Measurement of Single Char Particle Conversion,** Fengbo An, Andreas Richter, Felix Küster, Stefan Guhl, Bernd Meyer, Technische Universität Bergakademie Freiberg, GERMANY

**CFD-Based Accelerated Gasification Technology Development,** Andreas Richter, Philip Rößger, Thomas Förster, Peter Seifert, Bernd Meyer, TU Bergakademie Freiberg, GERMANY

**Syngas-Based Annex Concepts and $CO_2$-Based Power-To-X Concepts Within Pulverized Coal Combustion Power Plants,** Christian Wolfersdorf, Clemens Forman, Bernd Meyer, IEC, TU Bergakademie Freiberg, GERMANY

**SESSION 29**
**CLEAN COAL AND GAS TO FUELS: CATALYTIC STUDY - 2**
*Somayeh Farzad and Wenhua Li*

**Spontaneously Activatable Iron-Based Catalysts in Fischer-Tropsch Synthesis: From Lab-Scale Study to Pilot-Scale Demonstration,** Dong Hyun Chun, Geun Bae Rhim, Ji Chan Park, Ho-Tae Lee, Jung-Il Yang, Sungjun Hong, Heondo Jung, Heon Jung, Korea Institute of Energy Research, KOREA

**Single Pass Fischer-Tropsch Synthesis Over Iron-Based Catalysts,** Eric van Steen, Michael Claeys, Klaus P. Möller, Doreen Nabaho, University of Cape Town, SOUTH AFRICA

**Low-Temperature Methanation of Syngas in Slurry Reactor Over NiO@SiO$_2$ Core-Shell Catalyst,** Yizhuo Han, Junfeng Zhang, Hui Wang, Institute of Coal Chemistry, Chinese Academy of Sciences, CHINA

**SESSION 30**
**SUSTAINABILITY AND ENVIRONMENT - 3**
*Frans Waanders and Joseph Bittinger*

**American Electric Power's Experience with Mercury and Air Toxics Standards (MATS) Compliance,** Joseph E. Bittinger, Dean Bell, Jason Baker, Douglass Ritzenthaler, American Electric Power, USA

**Future of Coal Utilization in the US and China,** Qingyun Sun, Richard Bajura, West Virginia University, USA; Min Zhou, Xianliang Meng, China University of Mining &Technology, CHINA

**Carbon Tax in South Africa: Sustainability Challenges,** L. Landu, Council for Mineral Technology (MINTEK); R Minnitt, R Falcon, The University of the Witwatersrand, SOUTH AFRICA

**The Application of Coal Quality Database Over Time and Space Across the Mpumalanga Coalfields,** M. P. Kekana, N. J. Wagner, University of Johannesburg, SOUTH AFRICA

**SESSION 31**
**GASIFICATION TECHNOLOGIES: GENERAL - 4**
*Johan Brand and Thomas Sarkus*

**Kinetic Studies for $CO_2$ Gasification of HOK Cokes Using Thermogravimetric Analyzers at High Pressures,** Ziad Abosteif, Stefan Guhl, TU Bergakadmie Freiberg, GERMANY

**Study on the Effect of Particle Size on Gasification Characteristics of Different Ranked Chars Under High Temperature,** Xin Gong, XiaoPeng Zou, Haifeng Lu, Qinghua Guo, East China University of Science and Technology, CHINA

**Experimental Study of Particle Evolution Characteristics at Different Region in an Opposed Multi-Burner Gasifier,** Qing Zhang, Yan Gong, Qinghua Guo, Guangsuo Yu, East China University of Science and Technology, CHINA

**Corrosion Testing of Steel for Production and Injection Well Applications in a UCG Process,** JC van Dyk, African Carbon Energy and North-West University; J Brand, African Carbon Energy; FB Waanders, AC van Wyk, North-West University; C Kahts, Aqua Alpha, SOUTH AFRICA

**SESSION 32**
**COAL SCIENCE - 6**
*John Bunt and Leslie Ruppert*

**Property of Extract Prepared by Hydrothermal Extraction of Brown Coal,** Masato Morimoto, Qingxin Zheng, Toshimasa Takanohashi, Research Institute of Energy Frontier, National Institute of Advanced Industrial Science and Technology, JAPAN

**Process Characteristics Investigation of Shendong Coal Co-Pyrolysis through Solid Heat Carrier,** Wenying Li, Baofu Li, Xiaohong Li, Jie Feng, Taiyuan University of Technology, CHINA

**The Prediction of Pyrolysis Products in Grading Lignite Pyrolysis Process,** Wenying Li, Lan Yi, Jie Feng, Yuhong Qin, Yuhong Qin, CHINA

**Influence of Oxygen Group and Coal Structure on the Migration of Biomass Alkalis During Coal/ Biomass Co-Pyrolysis,** Wengying Li, Rong Zhu, Yuncai Song, Laisong Wang, Jie Feng, Taiyuan University of Technology, CHINA

**Heat Release Characteristics of the Oxidation of Aliphatic Hydrocarbon Groups and its Effects on the Spontaneous Combustion of Coal,** Tao Xu, Jianghong Liu, Jinhui Wang, Yu Jiao, Yutao Kang, Xiaobei Yin, Qimiao Xie, Xianbin Li, Shanghai Maritime University; Haihui Xin, Deming Wang, China University of Mining and Technology, CHINA

**SESSION 33**
**GASIFICATION TECHNOLOGIES: GENERAL - 3**
*Christien Strydom and Aleksander Sobolewski*

**$CO_2$ and $H_2O$ Gasification Under Chemically and Diffusion Controlled Conditions,** Thobias Kreitzberg, Benjamin Gövert, Stefan Pielsticker, Reinhold Kneer, Institute of Heat & Mass Transfer (WSA), RWTH Aachen University, GERMANY

**New Approach in Gasifying Reactive Low Rank Coals Under Non/Partial Slagging Entrained-Bed Conditions: 20 TPD Pilot Test Results,** Yongseung Yun, Jin-Wook Lee, Seok Woo Chung, Institute for Advanced Engineering; Young-Tai Choi, Korea Institute of Industrial Technology, KOREA

**Influence of Quench Water Temperature and pH Value Variations on Selected Trace Components in Autothermal Gas Reforming Processes,** Ibrahim Kolawole Muritala, Technische Universität Bergakademie Freiberg, GERMANY

**Influences of the Cross Angle on the Granular Jet in Annular Gas Stream with Thick Wall of Inner Channel,** Chenchen Fang, Haifeng Liu, East China University of Science and Technology, CHINA

## ORAL SESSIONS
## Friday, August 12, 2016

**SESSION 34**
**CLEAN COAL AND GAS TO FUELS: PROCESSING**
*Somayeh Farzad and Wenhua Li*

**Methane Quality, Processing and Hydrogen Production,** Klaus Rademann, Thomas Ryll, Humboldt-Universität zu Berlin, GERMANY

**The Development of Highly Anisotropic Coke from an Aliphatic Fischer-Tropsch Waxy Oil Residue,** JG Clark, Sasol Performance Chemicals, SOUTH AFRICA

**The Conversion of FT Olefins: Multiphase Reactor Simulation Study,** Klaus Möller, University of Cape Town; Bongani Mthombeni-Möller, Intens Engineering Solutions; Masikana Mdleleni, PetroSA Synthetic Fuels Innovation Centre (PSFIC), University of the Western Cape; Cyril Knottenbelt, PetroSA, SOUTH AFRICA

**Coal to Liquids by Oxidative Cracking in Near-Critical and Super-Critical Water,** Marcos Millan, Pedro Arcelus-Arrillaga, Klaus Hellgardt, Imperial College, UNITED KINGDOM

**Optimization of Configurational Alternatives for a Gas to Liquids Process Based on Fischer-Tropsch Technology,** Ashcaan Khazali, Klaus Möller, Eric van Steen, University of Cape Town, SOUTH AFRICA

**SESSION 35**
**GASIFICATION TECHNOLOGIES: MODELING - 2**
*Johan van Dyk and Aleksander Sobolewski*

**Syngas Production from Brazilian Coal Chars in the Presence of Steam,** Humberto Jorge José, Regina de Fátima P. M. Moreira, Gabriela Carvalho Collazzo, Michele Di Domenico, Federal University of Santa Catarina, BRAZIL

**Evaluation of $CO_2$ Gasification Reactivity of Biomass and Coal Chars by Physicochemical Properties,** Zhiqing

Wang, Jiejie Huang, Yitian Fang, Jiantao Zhao, Libo Dong, State Key Laboratory of Coal Conversion, Institute of Coal Chemistry, Chinese Academy of Sciences, CHINA

**The CUMT International Research Centre for Underground Coal Gasification**, Cliff Mallett, Carbon Energy Limited, AUSTRALIA

**SESSION 36**
**GASIFICATION TECHNOLOGIES: GENERAL - 5**
*Johan Brand and Francis Lau*

**Online Detection and Retention of Trace Contaminants in Reducing Gasification Atmosphere**, Philip Edinger, École Polytechnique Fédérale de Lausanne (EPFL) and Paul Scherrer Institut (PSI); Mohamed Tarik, Paul Scherrer Institut (PSI); Christian Ludwig, École Polytechnique Fédérale, SWITZERLAND

**Interaction Between Rice Straw and Jincheng Anthracite Single Particle During Ashing Process**, Wenying Li, Zibing Zhao, QingQing Han, Chengcheng Zhu, Yuhong Qin, Jie Feng, Taiyuan University of Technology, CHINA

**Catalytic Gasification with Carbon Dioxide Characteristics of Alkali and Alkaline Earth Metals Catalysts in the Fixed-bed Reactor**, Yifei Wang, Jilin Li, Guangsuo Yu, Fuchen Wang, East China University of Science and Technology, CHINA

**Oxygen Co-Gasification of Coal and Rice Straw in Drop Tube Furnace**, Xueli Chen, Jianye Tang, Qinghua Guo, Guangsuo Yu, East China University of Science and Technology, CHINA

# POSTER SESSIONS
## Wednesday, August 10, 2016

**POSTER SESSION 1**
**COMBUSTION**

**Influence of Combustion Temperature and Oxygen Concentration in Combustion Medium on Emissions of $NO_x$, $N_2O$ and $SO_2$**, Karel Svoboda, Jaroslav Moško, Tomáš Durda, Michael Pohořelý, Boleslav Zach, Michal Šyc, Aneta Krausová, Michal Jeremiáš, Institute of Chemical Process Fundamentals, Academy of Science of the Czech Republic, CZECH REPUBLIC

**POSTER SESSION 2**
**CLEAN COAL AND GAS TO FUELS**

**Methanol Synthesis by Catalytic Hydrogenation and Photoelectrochemical Reduction of $CO_2$**, Alberto Pettinau, Mauro Mureddu, Francesca Ferrara, Sotacarbo S.p.A., ITALY

**Effect of Sparger Design and Internals Configurations on the Hydrodynamics and Overall Performance of a Slurry Bubble Column Reactor for Fischer-Tropsch Synthesis**, Omar M. Basha, Badie I. Morsi, University of Pittsburgh, USA

**Isotope Analysis for Understanding the Hydrogen Transfer Mechanism in Direct Liquefaction of Bulianta Coal**, Haoquan Hu, Ben Niu, Zhiwei Shi, Lijun Jin, Yang Li, Dalian University of Technology, CHINA

**Design ZSM-22 Crystal Morphology with Urea and its Catalytic Property**, Debao Li, Pengyu Niu, Litao Jia, Bo Hou, Institute of Coal Chemistry Chinese Academy of Science, CHINA

**POSTER SESSION 3**
**SUSTAINABILITY AND ENVIRONMENT**

**Geochemical Mobility of Potential Hazardous Elements of Coal Cleaning Rejects**, Diego de Medeiros, Marcos L.S. Oliveira, Bianca D. Lima, Luis F. O. Silva, Centro Universitário La Salle, BRAZIL; Frans Waanders, North West University, SOUTH AFRICA

**Occurrence of the Hazardous Elements in Agricultural Soils Surrounding a Coal Power Plant Complex From Santa Catarina (Brazil)**, Diego de Medeiros, Marcos L.S. Oliveira, Fagner T.C. Sá, Luis F. O. Silva, Centro Universitário La Salle, BRAZIL; Frans Waanders, North West University, SOUTH AFRICA

**A Density Functional Theory Study on Acetic Acid Decomposition on Ni (111)**, Wenying Li, Yanxiong Ran, Zhenyi Du, Jie Feng, Taiyuan University of Technology, CHINA

**POSTER SESSION 4**
**CARBON MANAGEMENT**

**Pretreatment of Flue Gas Using Polymeric Membrane for $CO_2$ Capture**, Soohyun Choi, Ilhyun Baek, Korea Institute of Energy Research, KOREA

**Experimental Study on $CO_2$ Capture From Simulated Flue Gas by Using Aqueous Ammonia with External Magnetic Field**, Qi Zhang, Jiayi Wu, Ping Lu, Nanjing Normal University, CHINA

**POSTER SESSION 5**
**COAL SCIENCE**

**Coal Cleaning Residues and Fe-Ultrafine Particles and Fe-Minerals Consequences**, Diego de Medeiros, Felipe B. de Leão, Fagner T.C. de Sá, Luis F. O. Silva, Centro Universitário La Salle, BRAZIL; Frans Waanders, North West University, SOUTH AFRICA

**Zinc Occurrence in Coal Power Plant Burning Combinations of Coal and Tires**, Diego de Medeiros, Fernanda R. Silveira, Silvio R. Taffarel, Luis F. O. Silva, Centro Universitário La Salle, BRAZIL; Frans Waanders, North West University, SOUTH AFRICA; James C. Hower, University of Kentucky Center for Applied Energy Research, USA

**Hard Carbon From Coal Heavy Organics as Anode Material for Li-Ion Batteries**, Ying Zhou, Yulong Li, Ruifeng Liu, Jun Guan, Peng Wan, Lei He, Weiran Cao, Dongju Ma, Dalian University of Technology, CHINA

**The Relationship between Oxygen Transfer and Pyrolysis Conditions**, Wenying Li, Chaopeng Wang, Huijun Huang, Xiaohong Li, Jie Feng, Taiyuan University of Technology, CHINA

# ADVISORY BOARD



**2016 Proceedings CD**

Please note the 2016 Proceedings CD will be mailed to you after the conference for quality purposes.

# ANNOUNCING:

# Thirty-fourth Annual International Pittsburgh Coal Conference

**SHERATON PITTSBURGH HOTEL AT STATION SQUARE**
**PITTSBURGH, PA, USA**
**SEPTEMBER 5 - 8, 2017**

PCC

Abstracts must be submitted by March 1, 2017.
Please forward paper title, intended topic area, authors, affiliations, contact information with valid email address and a one-page abstract to:

Conference Secretary
ipcc@pitt.edu

**Please visit the PCC WEBSITE: www.pccpitt.org**

# Defining the Future
## The Imperative of Coal Research

NATIONAL ENERGY TECHNOLOGY LABORATORY

Grace M. Bochenek, Ph.D., Director

## NETL is…

NATIONAL ENERGY TECHNOLOGY LABORATORY

…the only U.S. National Laboratory devoted to Fossil Energy Technology Discovery, Development and Deployment.

# NETL's Mission

NATIONAL ENERGY TECHNOLOGY LABORATORY

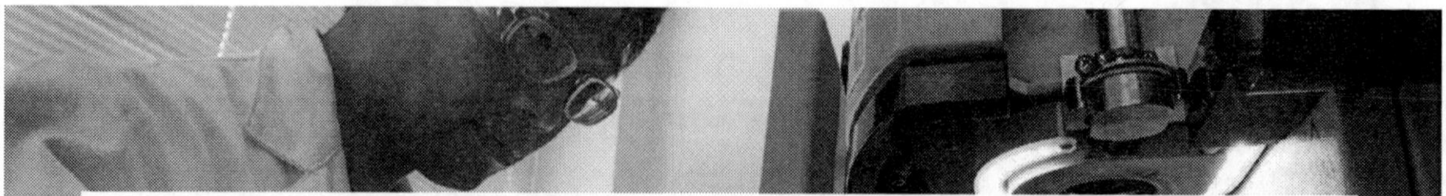

To discover, integrate, and mature technology solutions to enhance the Nation's energy foundation and protect the environment for future generations

**EFFECTIVE RESOURCE DEVELOPMENT**

Developing technologies that improve the effectiveness and economics of producing our fossil energy resources

**EFFICIENT ENERGY CONVERSION**

Discovering cleaner, cheaper, and more efficient energy conversion technologies for the production of high-value energy commodities

**ENVIRONMENTAL SUSTAINABILITY**

Accelerating the development of transformative and enabling solutions to protect our air, land, and water for future generations

U.S. DEPARTMENT OF ENERGY 3

# Energy Poverty

NATIONAL ENERGY TECHNOLOGY LABORATORY

**3.5 BILLION people** lack proper access to electricity

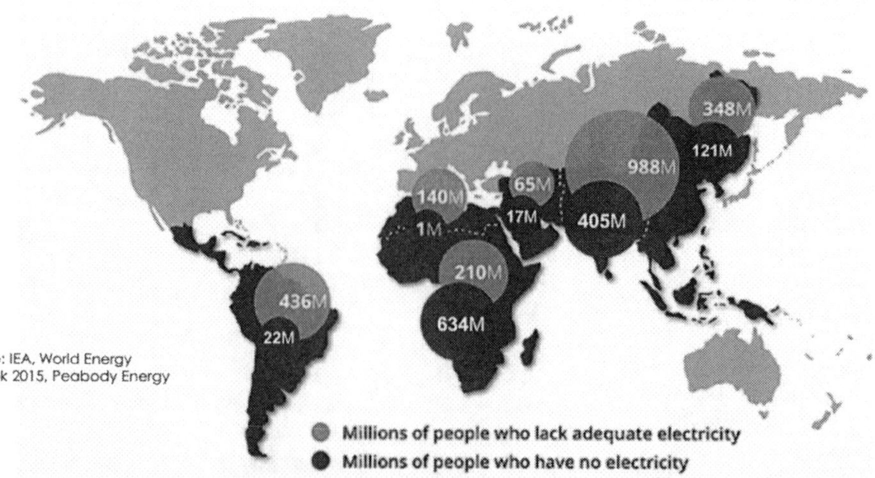


Source: IEA, World Energy Outlook 2015, Peabody Energy

U.S. DEPARTMENT OF ENERGY 4

# Global Drivers

NATIONAL ENERGY TECHNOLOGY LABORATORY

NATIONAL GEOGRAPHIC MAGAZINE

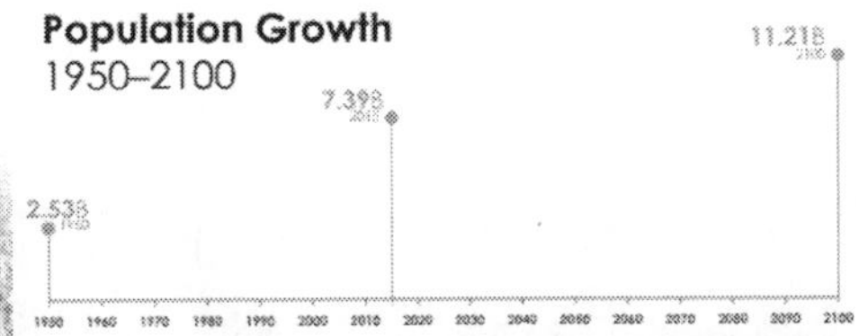


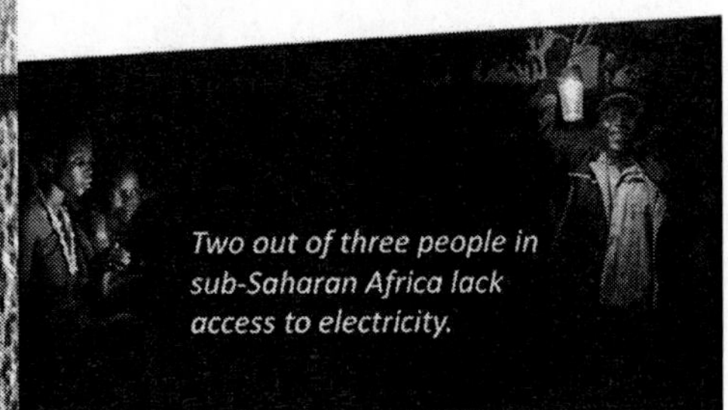




# Electricity Production from Coal Sources

NATIONAL ENERGY TECHNOLOGY LABORATORY

## Coal Use is Critical to Global Energy Requirements

| | |
|---|---|
| South Africa | 94% |
| Mongolia | 93% |
| Poland | 85% |
| Kazakhstan | 81% |
| China | 75% |
| India | 73% |
| Australia | 65% |
| Israel | 54% |
| Czech Republic | 51% |
| Germany | 47% |
| Ukraine | 42% |
| Korea, Rep. | 41% |
| Denmark | 41% |
| United States | 40% |
| **Sub-Saharan Africa** | **54%** |
| **World** | **41%** |

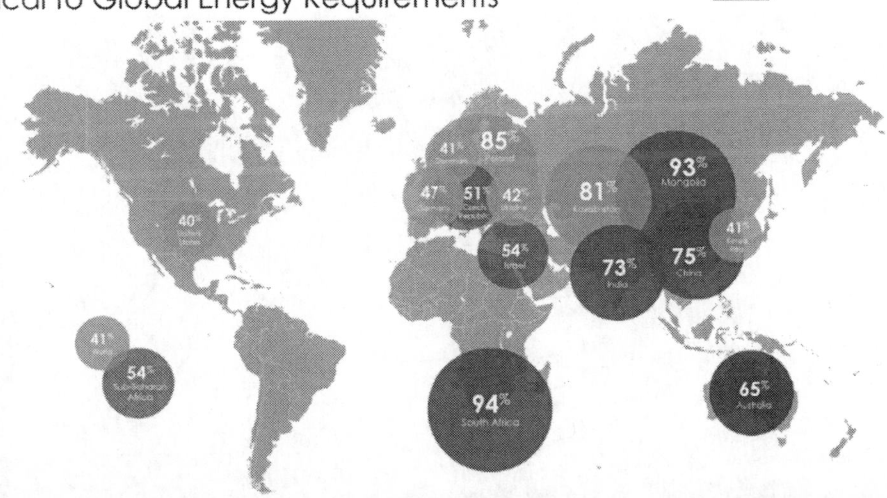

# The World and U.S. Energy Future



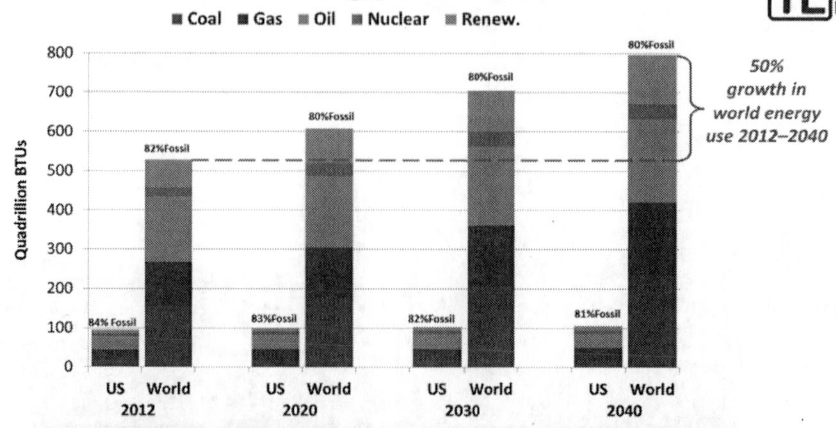


*≥80% Fossil Energy Today AND Tomorrow*
*Dominated by Global Growth*



# The Imperative of Coal Research

- Coal R&D will define the future of coal use
- Without innovative coal R&D, global energy needs will not be met
- Coal R&D leads to sustainable solutions

## R&D Success is Foundation of Current Coal Demand

NATIONAL ENERGY TECHNOLOGY LABORATORY

Mercury Removal from Flue Gas

Mining Safety/Mechanization

Emissions Controls for Acid Rain Gases

U.S. DEPARTMENT OF ENERGY

9

## Continued Success Requires Vision, Courage and Perseverance

NATIONAL ENERGY TECHNOLOGY LABORATORY

- We must continuously fill the pipeline with innovative technology
- We must take on large investments as R&D moves forward
- We must accelerate maturation of technologies

U.S. DEPARTMENT OF ENERGY

10

# Filling the Pipeline

NATIONAL ENERGY TECHNOLOGY LABORATORY

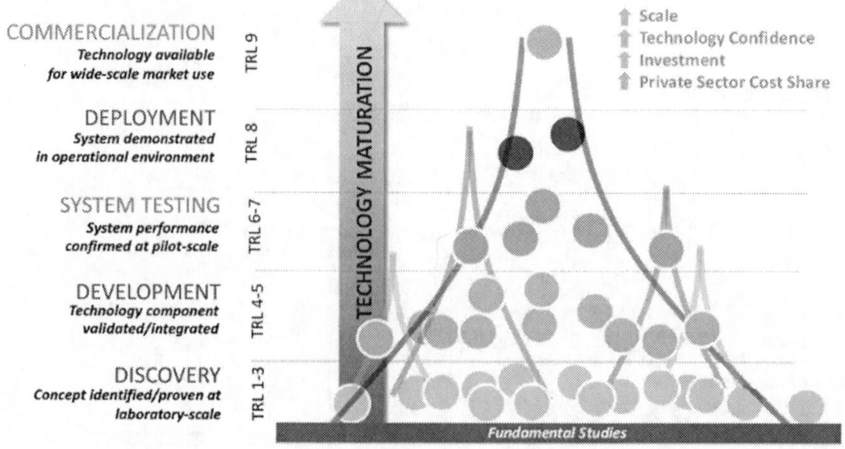




# RD³ – Research Development, Demonstration, Deployment

NATIONAL ENERGY TECHNOLOGY LABORATORY

Foundation to NETL's Mission Success

Advanced Combustion Turbine - Airfoil Design

*From Concept to Market Readiness*

In 2013, Siemens facility opens in Charlottesville, VA for commercial production of airfoil ceramic cores for gas turbine blades and vanes using MikroSystems TOMO℠ technology

**For decades, NETL's mission-unique resources and facilities have been used to develop and nurture technology from concept to market readiness**

Commercial sales generated in technologies supported by FE/NETL's Demonstration Program total in excess of $62 billion in environmental and advanced power generation industries

# Development of the TRIG™ for Power and Chemical Production

TRIG™ Leverages Long History of KBR Fluid Catalytic Cracking (FCC) Expertise

| 1980's | 1990 | 1996 | 1996 | USA 2016 |
|---|---|---|---|---|
| Design Based on FCC Technology | Pilot Plant Tech Center | Grand Forks, ND 2,600 Hours Test Run | PSDF at Wilsonville, AL. | TRIG™ In Kemper County, MS, USA |



# Simulation-Based Engineering

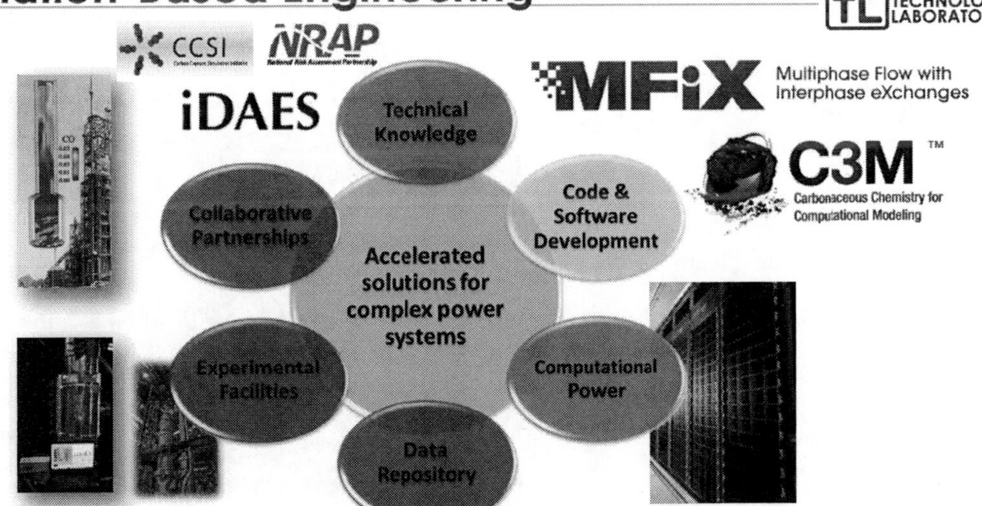

## Designing Advanced Energy Systems

Simulation-Based Engineering

**Benefits**

Accelerate innovation in order to:

- **Enable COP21 goals**
- **Halve technology maturation cost and deployment time**
- **Increase energy sector efficiency by 50%**

***Identifying technology solutions in the context of the full energy portfolio***

***Framework applicable to any multi-dimensional & complex challenge***

15

U.S. DEPARTMENT OF ENERGY | NATIONAL ENERGY TECHNOLOGY LABORATORY

## Simulation-Based Engineering of Energy Devices

NATIONAL ENERGY TECHNOLOGY LABORATORY

Foster device-scale innovation in energy transformation technology, pushing the limits of thermodynamics ... and imagination

**Benefits**

- **Pushing thermodynamic limits**
- **Radically new device designs**
- **Reducing time for optimization**
- **Reducing the technical and financial risk of maturing technology**

***Identifying device solutions in context of process needs***

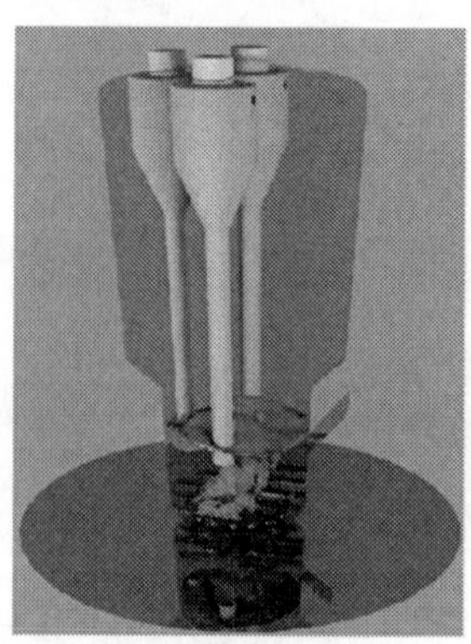

U.S. DEPARTMENT OF ENERGY

NETL R&D Initiatives are Defining the Future
NATIONAL ENERGY TECHNOLOGY LABORATORY
Accelerated technology development
Improved $CO_2$ removal costs
Expanded coal markets
Innovative coal technologies
Post-Combustion Test Facility, National Carbon Capture Center, Wilsonville, AL
U.S. DEPARTMENT OF ENERGY
17

Solutions for Today
Options for Tomorrow
For More Information, Contact NETL
www.netl.doe.gov
NATIONAL ENERGY TECHNOLOGY LABORATORY
U.S. DEPARTMENT OF ENERGY

# Current Situation and Prospect of Coal Industry in China

Mr. Wang Hongqiao

Vice President of China National Coal Associati

Augu 6

**China National Coal Association**

On behalf of China National Coal Association, I would like to extend heartfelt congratulations on the opening of Thirty-Third International Pittsburgh Coal Conference, and sincere gratitude to the organizing committee.

China National Coal Association

Over the past ten years, remarkable changes have taken place in the international coal industry.

Great progress and development has been made in aspects of coal science and technology, work safety, occupational health, clean utilization, etc.

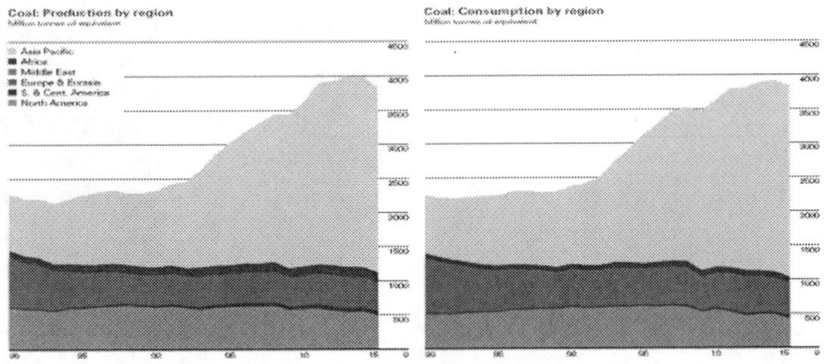


BP Statistical Review of World Energy June 2016

China National Coal Association

Coal accounts for 72% in primary energy production and 64% in consumption in China.

As the world's largest coal producing and consuming country, China's coal production and consumption account for 47% and 50%, respectively, of the world, increased by 16% and 18% compared with 2002.

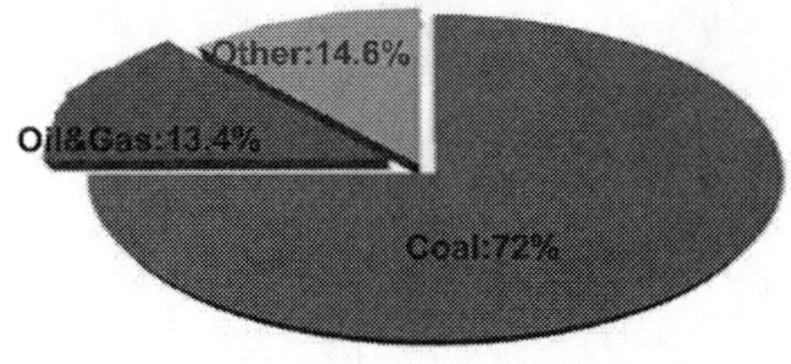


China Energy Production Pie Chart 2015

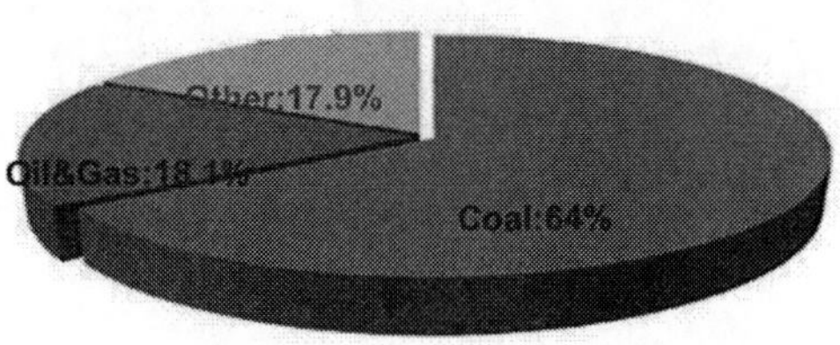


China Energy Consumption Pie Chart 2015

**China National Coal Association**

### 1. Development and construction of coal resources

**Special drilling technology and equipment such as freezing shaft sinking, borehole mining and grouting have made breakthroughs, providing technical support for the construction of large-scale modern coal mine.**

Freezing shaft sinking

**China National Coal Association**

### 2. Safe and high-efficient coal mining

**China has successfully developed fully mechanized top-coal caving mining technology at ultra-thick coal seams, partly realized unmanned operation and intelligent production.**

Complete set of intelligent mining system

The fully mechanized top-coal caving mining technology at ultra-thick coal seams

China National Coal Association

Coal to MEG

### 3. Clean and high-efficient utilization of Coal

**Nearly 90 million kilowatts of large coal-fired units have now adopted ultra low emission technology; Modern coal chemical technology and processes such as coal liquefaction, coal to olefin, coal to Monoethylene glycol (MEG) , coal to methyl ether have made breakthroughs.**

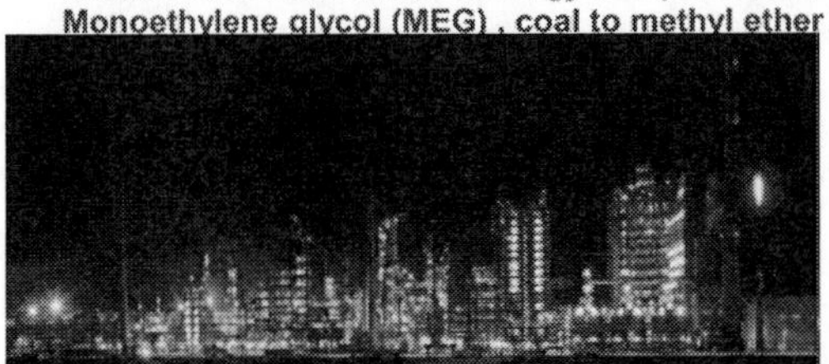

Direct Coal Liquefaction of Shenhua Group

Indirect Coal Liquefaction of Yankuang Group

China National Coal Association

Tashan Coal Mine of Datong Coal Mine Group

### 4. Ecological mining area construction

**Many mining areas including Tashan Coal Mine of Datong Coal Mine Group, Shanxi Coking Coal Group have built large Circular Economy Parks with comprehensive utilization of resources and environmental protection.**

Circular Economy Park of Shanxi Coking Coal Group

**China National Coal Association**

## 5. Resolving problem of overcapacity, achieving transformation and upgrading

**The Chinese government attaches great importance to the healthy development of coal industry.**

**China has promulgated a series of policies and measures to promote coal industry to resolve the problem of overcapacity and achieve turnaround development.**

**China National Coal Association**

## 6. Ultra-low emission of coal-fired power generation

**In order to control air pollution caused by coal-fired power, China not only carries out the strictest emission standard of air pollutants for thermal power plants, but also has made provisions that the emission of all thermal power plant must implement emission standard for natural gas power plants by 2020.**

Ultra-low emission of thermal power plant in China

China National Coal Association

7. Promoting international exchanges and cooperation, improving level of opening up

**We will promote the cooperation of capacity, technology and equipment between China and coal producing countries of the One Belt One Road region, expanding the international market space.**

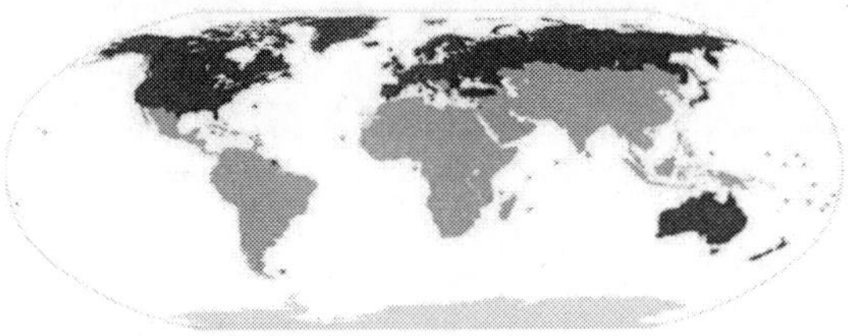

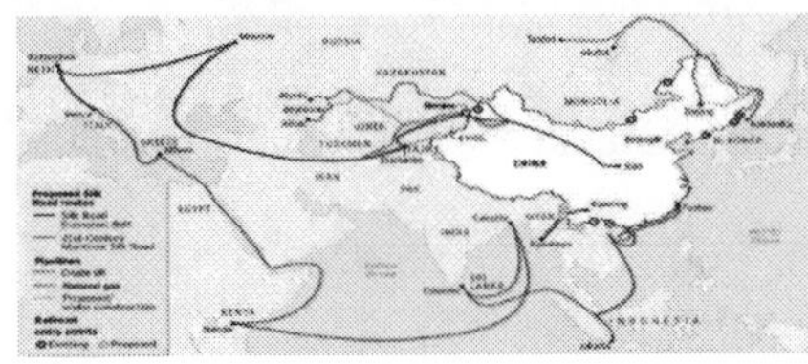

China National Coal Association

World Coal Association Technology Committee Workshop 2016

Mr. Liang Jiakun, Chairman of World Coal Association Technology Committee

Pesident Wang Xianzheng give a speech at workshop

## China National Coal Association

International Cooperation

Cooperation between US Mine Safety and Health Administration and China National Coal Association

MOU signed with Brazilian Coal Association

## China National Coal Association

International Coal Summit 2015

The 16th China International Coal & Mining Expo

**China National Coal Association hopes to strengthen cooperation with the organizing committee of Pittsburgh coal conference, government, NGOs, and enterprises.**

Thank You!

# ENERGY AND ITS EFFECT ON ECONOMIC GROWTH:

## The Role of Coal in the Future Economic Environment of Southern Africa

[1]R Jeffrey and [2]R Falcon
[1]Econometrix
[2]University of the Witwatersrand

PCC Conference 2016

# CONTENTS

- Introduction
- Major Trends in Future African Economic Growth
- Electricity and Economic growth
- Comparative costs of electricity
- Challenges in Southern Africa
- Environmental issues - IPCC, COP21 and Carbon Tax
- Solutions - Clean Coal Technologies
- Conclusion

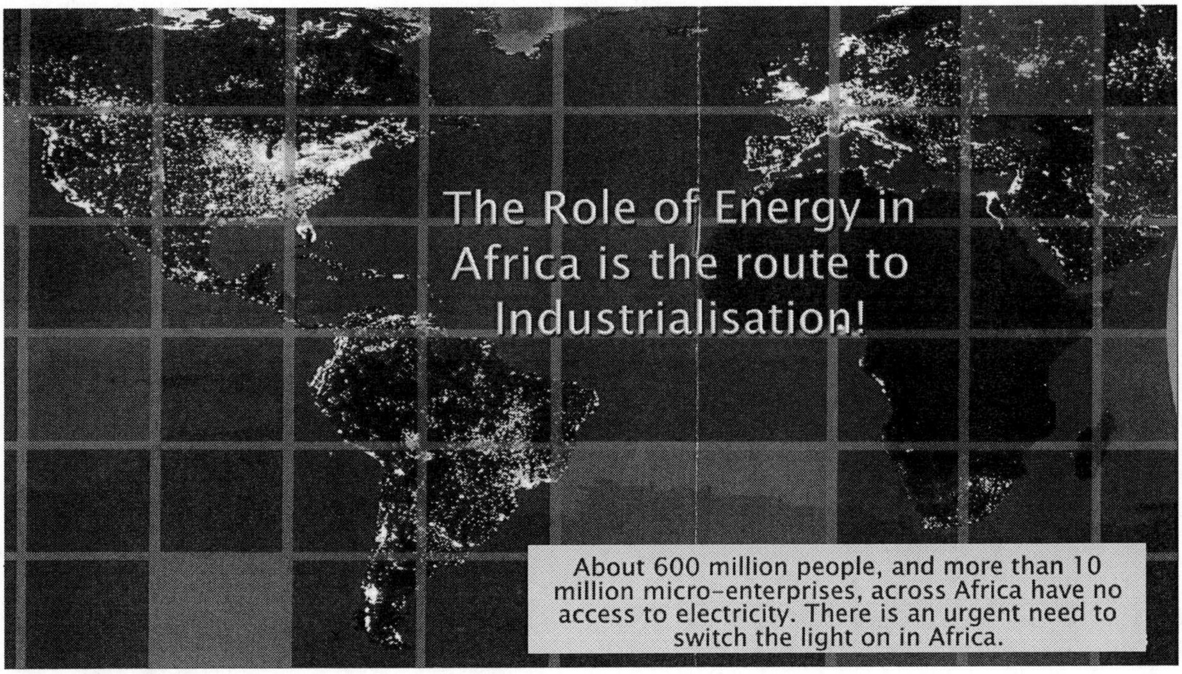
The Role of Energy in Africa is the route to Industrialisation!
About 600 million people, and more than 10 million micro-enterprises, across Africa have no access to electricity. There is an urgent need to switch the light on in Africa.

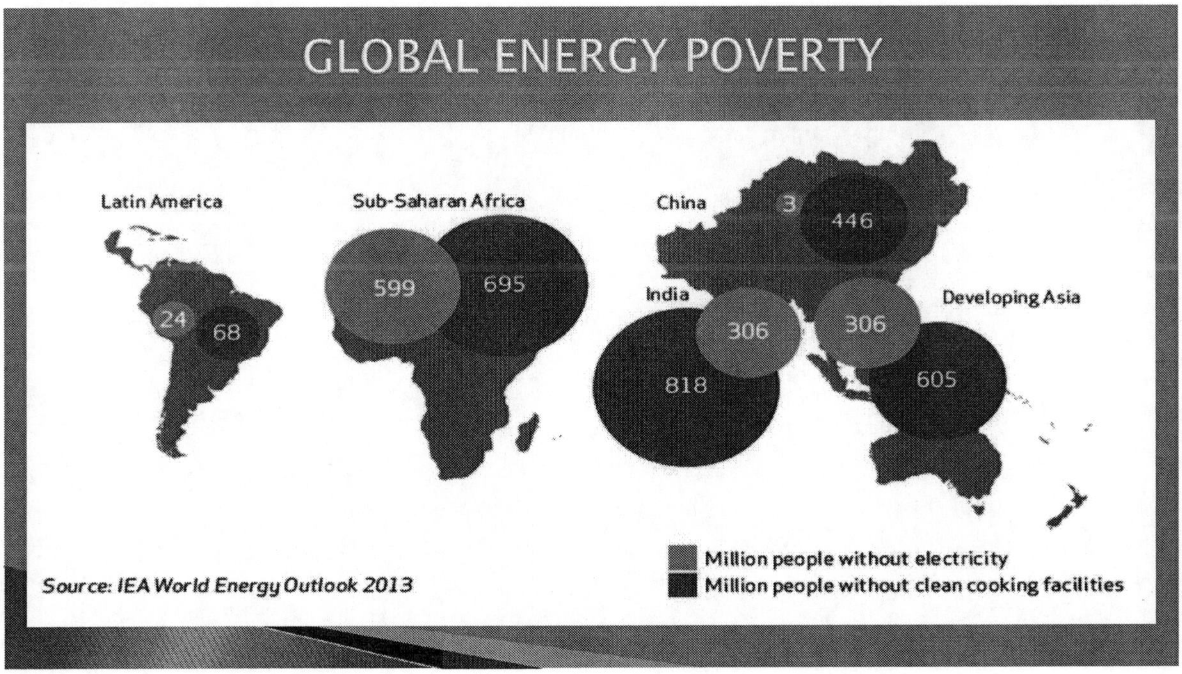
GLOBAL ENERGY POVERTY
Latin America
24
68
Sub-Saharan Africa
599
695
China
3
446
India
306
818
Developing Asia
306
605
Million people without electricity
Million people without clean cooking facilities
Source: IEA World Energy Outlook 2013

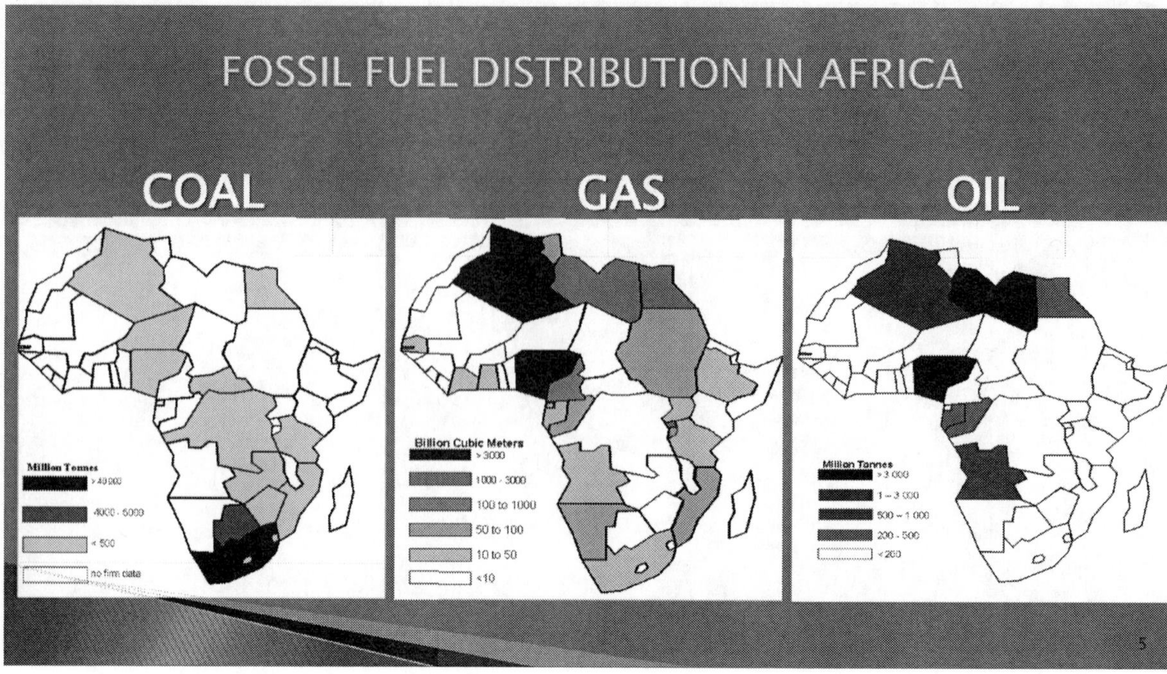

FOSSIL FUEL DISTRIBUTION IN AFRICA
COAL
GAS
OIL
Million Tonnes
> 40 000
4000 - 6000
< 500
no firm data
Billion Cubic Meters
> 3000
1000 - 3000
100 to 1000
50 to 100
10 to 50
<10
Million Tonnes
> 3 000
1 – 3 000
500 – 1 000
200 - 500
< 200
5


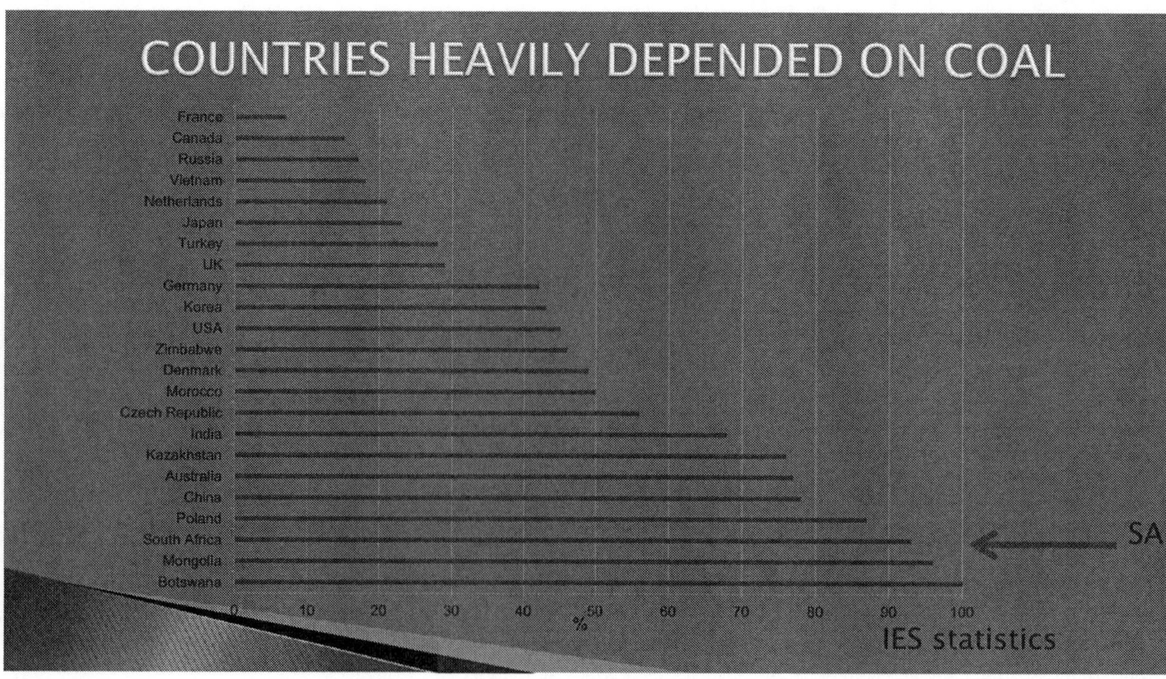

COUNTRIES HEAVILY DEPENDED ON COAL
France
Canada
Russia
Vietnam
Netherlands
Japan
Turkey
UK
Germany
Korea
USA
Zimbabwe
Denmark
Morocco
Czech Republic
India
Kazakhstan
Australia
China
Poland
South Africa
Mongolia
Botswana
0 10 20 30 40 50 60 70 80 90 100
%
SA
IES statistics

# PREDICTED LIFE SPAN OF PROVEN COAL RESERVES

- **GLOBAL PREDICTION**: **118 years** (including hard and brown coal); [**Gas 59** years and **Oil 46** years] Source: World Coal Association/IEA 2011 report on 2010 data
- **SADC PREDICTION**: **Likely to be 200 years** under current usage, subject to exploration, market and logistics
- **SOUTH AFRICA**: **120 years** (hard coal) under current production – NB: an updated Coal Resource and Reserve Assessment us currently being undertaken.

7

# CONTENTS

Introduction
Major Trends affecting Future African Economic Growth
Electricity and Economic growth
Comparative costs of electricity
Challenges in Southern Africa
Environmental issues – IPCC, COP21 and Carbon Tax
Solutions Clean Coal Technologies
Conclusion

# INFRASTRUCTURE IN SUB-SAHARAN AFRICA'S LESS DEVELOPED COUNTRIES

| Low income countries | | |
|---|---|---|
| | Sub-Saharan Africa | Rest of world |
| **Total Road density** *(km/km²)* | 137 | **211** |
| **Paved road density** *(km/km²)* | 31 | **134** |
| **Power generation capacity** *(mw/pop in millions)* | 37 | **326** |
| **Electricity access** *(% of population)* | 16 | **41** |
| **Access to reliable water** *(% of pop)* | 60 | **72** |
| **Access to sanitation** *(% of population)* | 34 | **51** |

# POOR GOODS PRODUCING SECTOR RESULTS IN INCREASING IMPORTS AND INSUFFICIENT EXPORTS - AND LOW GDP

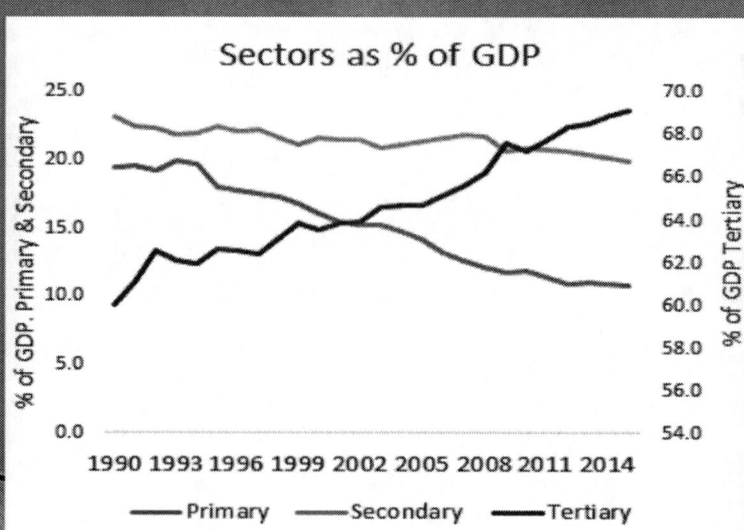


- **PRIMARY SECTOR** - goods produced
- **SECONDARY SECTOR** – goods exported
- **TERTIARY SECTOR** – goods
- imported

# WORLD POPULATION (Millions)

| | Population | | 2050 Population as a Multiple of 2013 |
|---|---|---|---|
| | Mid-2013 | Mid-2050 | |
| Sub-Saharan Africa | 926 | 2185 | 2.4 |
| Africa | 1100 | 2431 | 2.2 |
| Northern Africa | 208 | 316 | 1.5 |
| World | 7137 | 9727 | 1.4 |
| United States | 316 | 400 | 1.3 |
| Latin America & the Caribbean | 606 | 780 | 1.3 |
| Asia (excl. China) | 2945 | 3970 | 1.3 |
| China | 1357 | 1314 | 1.0 |
| Western Europe | 190 | 199 | 1.0 |
| Southern Europe | 153 | 146 | 1.0 |
| Eastern Europe | 295 | 260 | 0.9 |

# LOW GDP GROWTH MEANS LOW EMPLOYMENT GROWTH

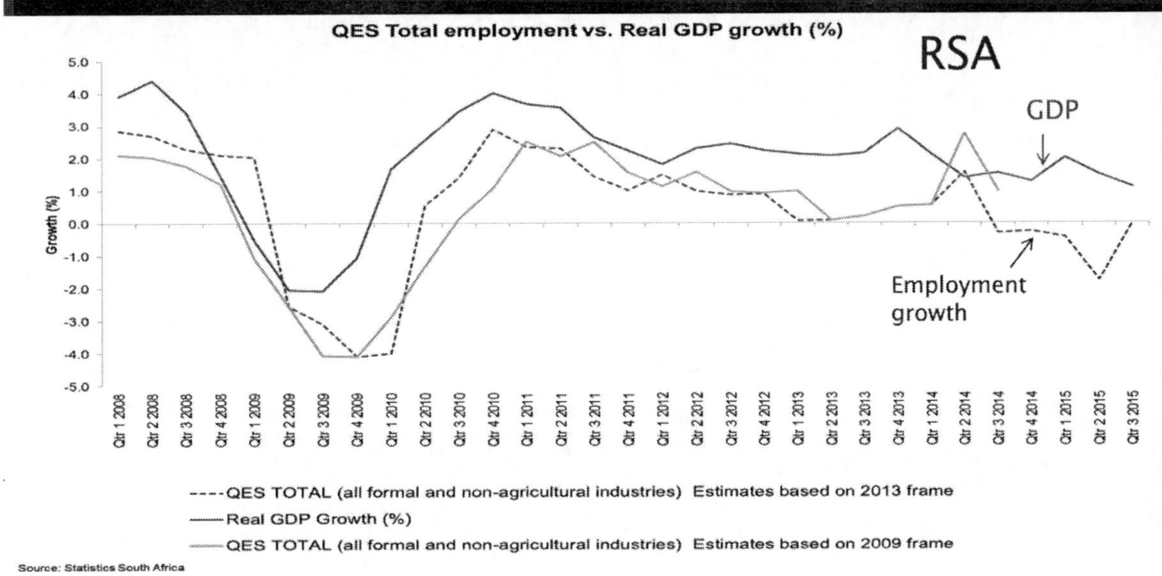

# RSA - RISING UNEMPLOYMENT RATE

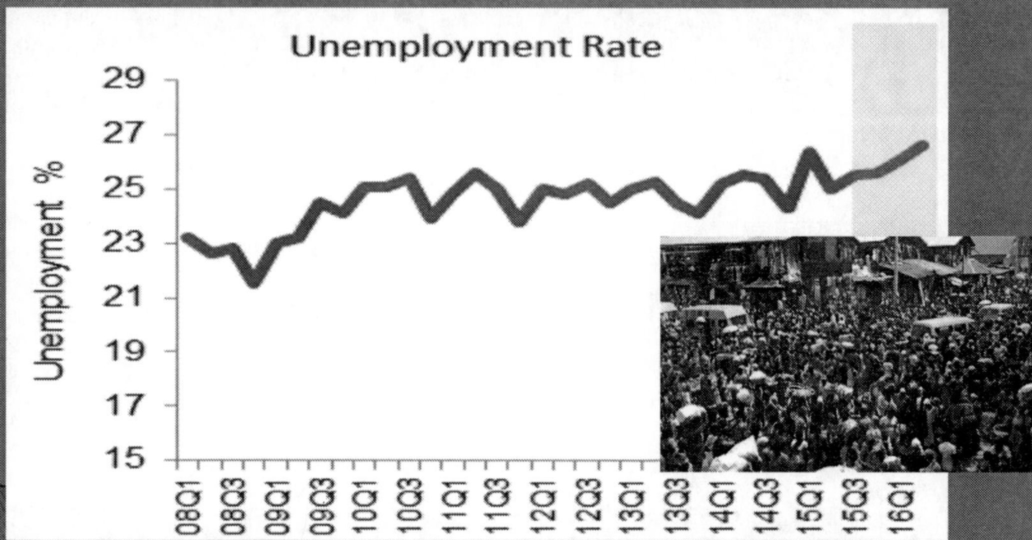


# KEY ISSUES FACING EMERGING ECONOMIES

- Rising unemployment, increasing social issues, and slow growth,
- There are three major policy objectives.
  - POVERTY ALLEVIATION,
  - REDUCING INEQUALITY AND
  - RAISING STANDARDS OF LIVING
- Economic growth is of paramount importance in the goods-producing sectors.
- Electricity growth is essential to support economic growth
- Emerging economies require security of supply of electricity at the lowest possible cost.

## FUTURE LOW GROWTH MEANS RISING UNEMPLOYMENT

| Population GDP growth and unemployment | | | |
|---|---|---|---|
| Current Statistics | 1995 | 2015 | 2030 |
| Population | 45m | 54m | 69m |
| Unemployment | 3.7m | 8.4m | ? |
| GDP Growth 2008-2015 | | 1.9% pa | ? |
| Current growth rate less than | | 2% pa | 2.5% pa ? |
| Jobs created 2008- 2015 QLF | | 1.2million | 2.5 million |
| Jobs required by 2030 | | | 16 million |

R Jeffrey 2016

## CONTENTS

# ELECTRICITY AND ECONOMIC DEVELOPMENT

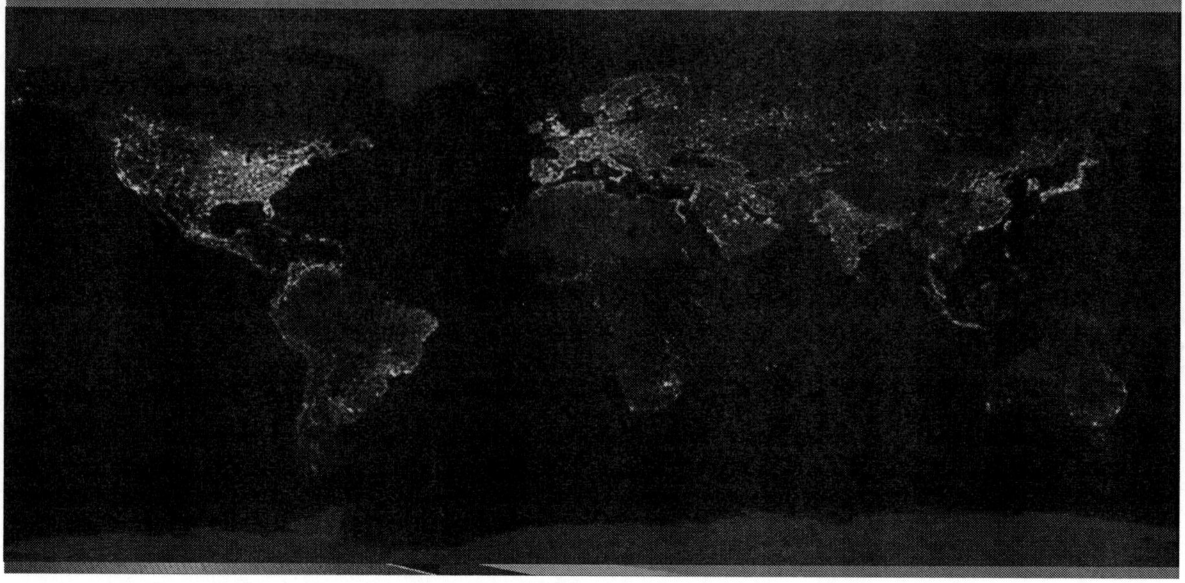

# ELECTRICITY AND PRODUCTION

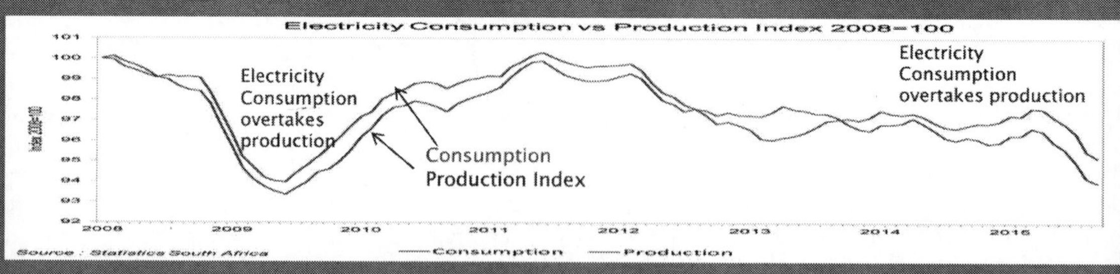


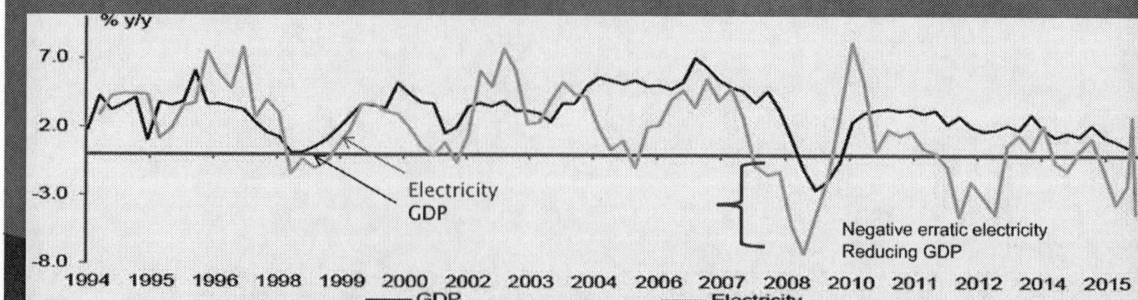


Stats SA

## THE ESSENTIAL ECONOMIC TRUTHS

- Economic growth is essential for raising standards of living, reducing unemployment and reducing inequality
- But economic growth can only take place with secure and increasing electricity production
- Low electricity growth leads to low economic growth

## CONTENTS

Introduction

Major Trends affecting Future African Economic Growth

Electricity and Economic growth

Comparative costs of electricity

Challenges in Southern Africa

Environmental issues - IPCC, COP21 and Carbon Tax

Solutions Clean Coal Technologies

Conclusion

ALL POWER SOURCES HAVE COSTS AND BENEFITS
©Photography by Scott Highton All rights reserved

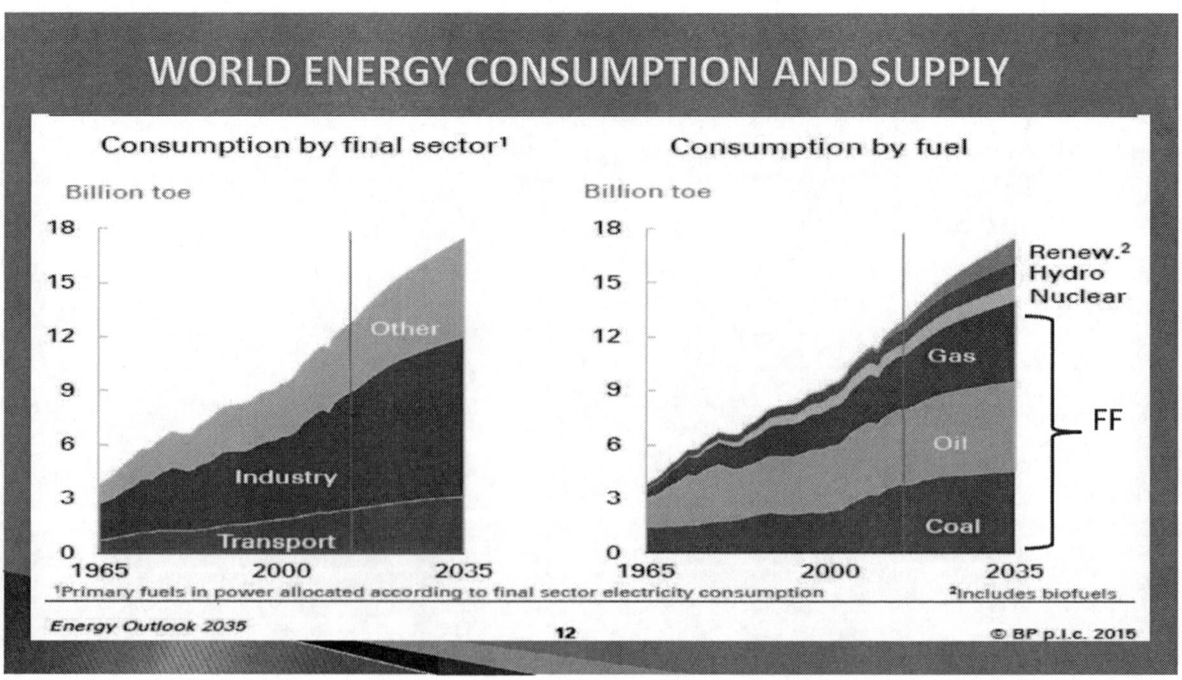
WORLD ENERGY CONSUMPTION AND SUPPLY
Consumption by final sector¹
Billion toe
Other
Industry
Transport
Consumption by fuel
Billion toe
Renew.²
Hydro
Nuclear
Gas
Oil
Coal
FF
¹Primary fuels in power allocated according to final sector electricity consumption
²Includes biofuels
Energy Outlook 2035
12
© BP p.l.c. 2015

# RELATIVE LEVELISED COST OF ELECTRICITY (LCOE)

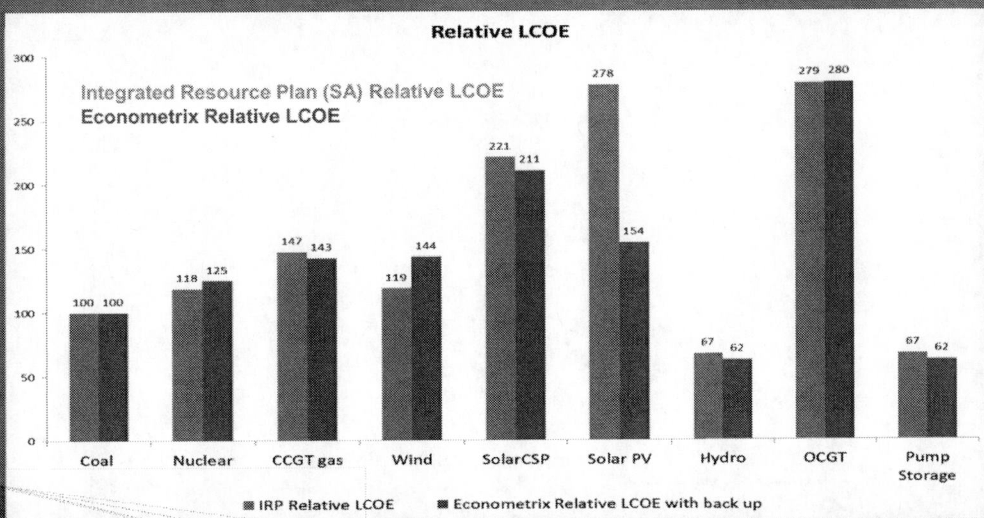


Econometrix 2016

# RENEWABLES

- Renewables are **intermittent and variable** increase these costs
  - They only supply power approximately **31% of the time**.
  - Require **100% back up**

- Until **storage costs** come down, these are unlikely to change significantly in the near future.

- Wind and solar CSP are not suitable for **base load power. They are too variable and expensive**

## REAL COSTS OF RENEWABLES

- Renewables being intermittent and variable increase these costs and ONLY supply power 31% of the time
- Relative Real overnight capital costs (R in millions) are therefore
  - Wind at 31%: 200
  - Solar at 31%: 712
  - Nuclear at 92%: 244
  - Coal at 80%: 125
- Wind and solar are significantly more expensive than coal and nuclear particularly when back up power costs are added
- **Latest data from overseas estimate that the LCOE of wind is 48% higher than for coal**

R Jeffrey, Econometrix '16

## RENEWABLES SECURITY OF SUPPLY AND COST

- There is increasing evidence that renewables (wind and Solar CSP) are not suitable for mass power in industrialised or industrialising countries
- ENERGIEWENDE IN GERMANY have driven up electricity prices to unacceptable and uneconomic levels
- THE GERMAN GOVERNMENT have capped wind and are in the process of withdrawing subsidies.
- SOUTH AUSTRALIA'S Lessons for the World:- Wind Power The Fastest Route to Social & Economic Disaster
- SOUTH AUSTRALIA'S Intermittent & unreliable renewable energy is leading to the deindustrialisation and reduction in its citizens' living standards.

## ENERGY POVERTY - NEW TO INDUSTRIALISED COUNTRIES

- **54 million** people in Europe are affected by **rising energy poverty, around 1.5 million** people could be dying prematurely
- Average real electricity prices for households and industries have increased **by 29%** between 2005 and 2011.
- **UK household** electricity prices have increased of over **80%** since 2005.
- The number of households affected by **energy poverty** in the UK, increased from **2 million to 5 million**.
- Green-energy policies in **California** have resulted in as many as **15 percent of households** falling into **energy poverty**.
- In the EU coal-fired power plants generate electricity at **half the price** of wind turbines and a **quarter of the price** of solar PVs

## CONTENTS

Introduction

Major Trends affecting Future African Economic Growth

Electricity and Economic growth

Comparative costs of electricity

Challenges to the use of coal in Southern Africa

Environmental issues - IPCC, COP21, Carbon Tax

Solutions - Clean Coal Technologies

Conclusion

## COP 21: PARIS OUTCOME

- The outcome of COP21 was an excellent agreement. What was important was not what was agreed but was not agreed.
- The agreement gave a set of sound long term global objectives. It asked for no commitments by any country.
  - Major countries either could not get agreement internally
  - Countries not going to make commitments were ASEAN countries, China, Russia, India, Vietnam, Korea and Poland
  - They were embarking on major expansions of coal and fossil fuels
- In summary, countries were expected to do what was in their **best economic interests**

## CARBON DIOXIDE EMITTERS

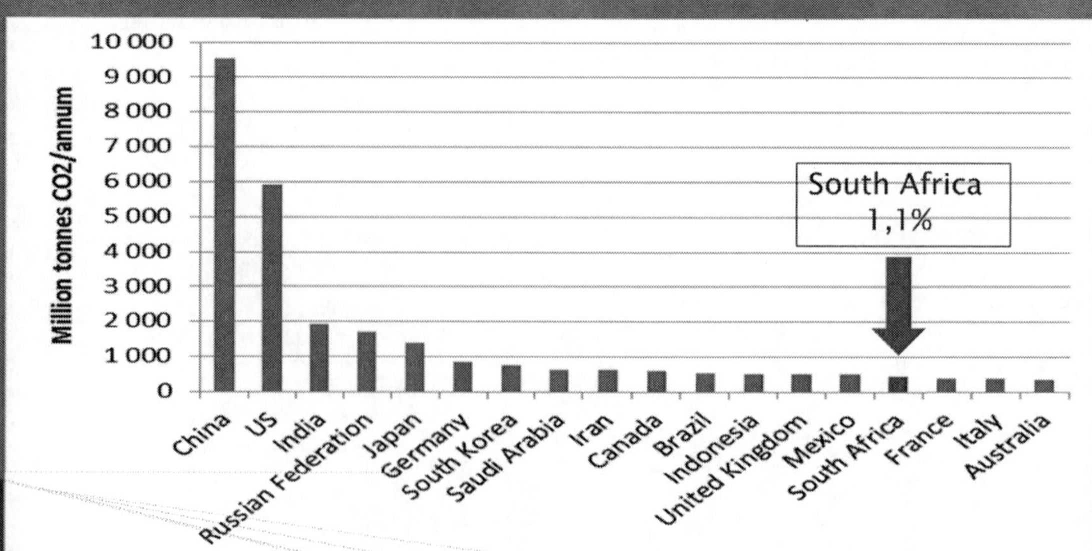

## CO2 EMISSIONS IN SOUTH AFRICA

| | $CO_2$ conc | Mt $CO_2$/a |
|---|---|---|
| Coal to Liquids | 85% | 21,7 |
| Power Generation | 8-12% | 224,6 |
| Other industries | 8-30 | 31,8 |

31

## REDUCING EMISSIONS - THE ROLE OF TECHNOLOGY

Raising global average **efficiency** of coal plants from **34% to 40%** with off-the-shelf cleaner coal technology

- Would save **2 Gigatonnes of CO2.**
- More than the annual $CO_2$ emissions of India- 4th largest emitter

Initiatives needed to **cut 2 Gigatonnes of CO2** emissions

- Run **EU Emissions** trading scheme for **53 years**
- Run the **Kyoto Protocol 3X** over
- Multiply the world's current **solar capacity 195X**

## SOUTH AFRICA'S PROPOSED FUTURE CLEAN ENERGY FUTURE

- At the 2009 UN climate change talks, South Africa **undertook to reduce domestic greenhouse gas emissions trajectory by 34% by 2020, and by 42% by 2025**, subject to adequate financial and technical support
- South Africa introduced the **SA Renewables Initiative in October 2011** – obtaining funds for climate change adaptation, energy-saving equipment, moderating the spike in the price of electricity generated from renewable sources
- South Africa introduced the **Integrated Resource Plan (IRP) in 2010** – the **IRP2010**, mixed sources of energy

33

## LOW CARBON ECONOMY LEADING TO A REDUCTION IN $CO_2$ EMISSIONS IN SA - IRP2010 [Integrated Resource Plan 2010)

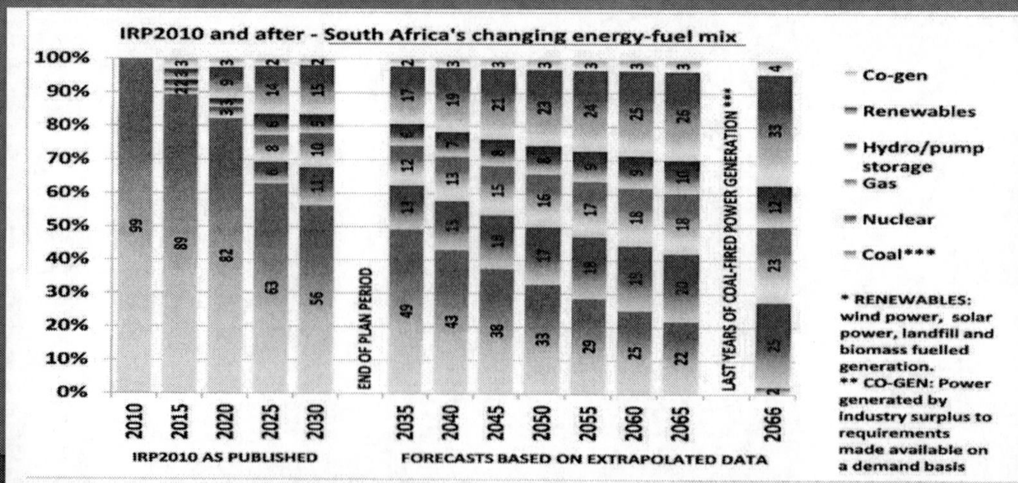


34

IRP2010/12

## LOW CARBON ECONOMY LEADING TO A REDUCTION IN $CO_2$ EMISSIONS IN SA - IRP2010 [Integrated Resource Plan 2010)

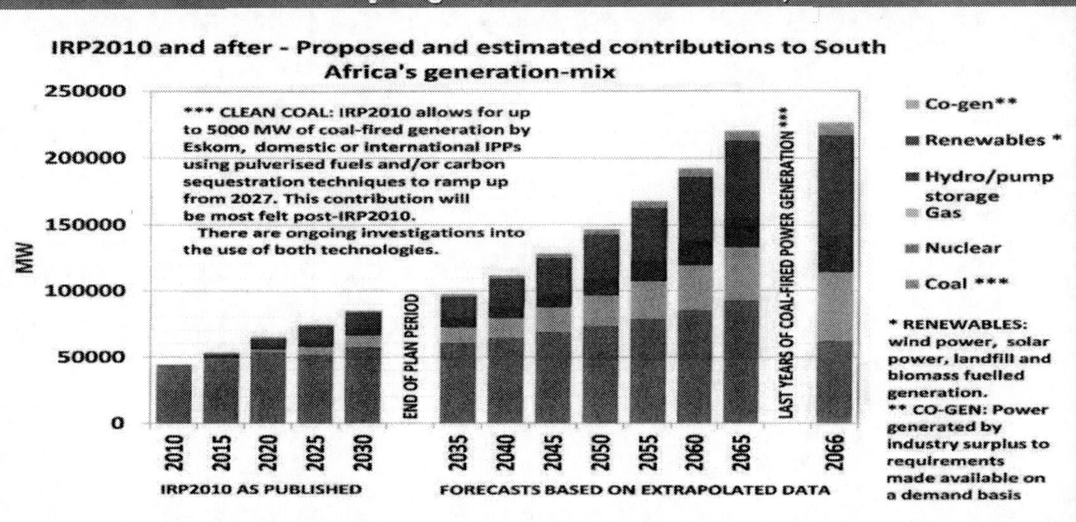


IRP2010/12

35

# CONTENTS

- Introduction
- Major Trends affecting Future African Economic Growth
- Electricity and Economic growth
- Comparative costs of electricity
- Challenges in Southern Africa
- Environmental issues - IPCC, COP21 and Carbon Tax
- Solutions - Clean Coal Technologies
- Conclusion

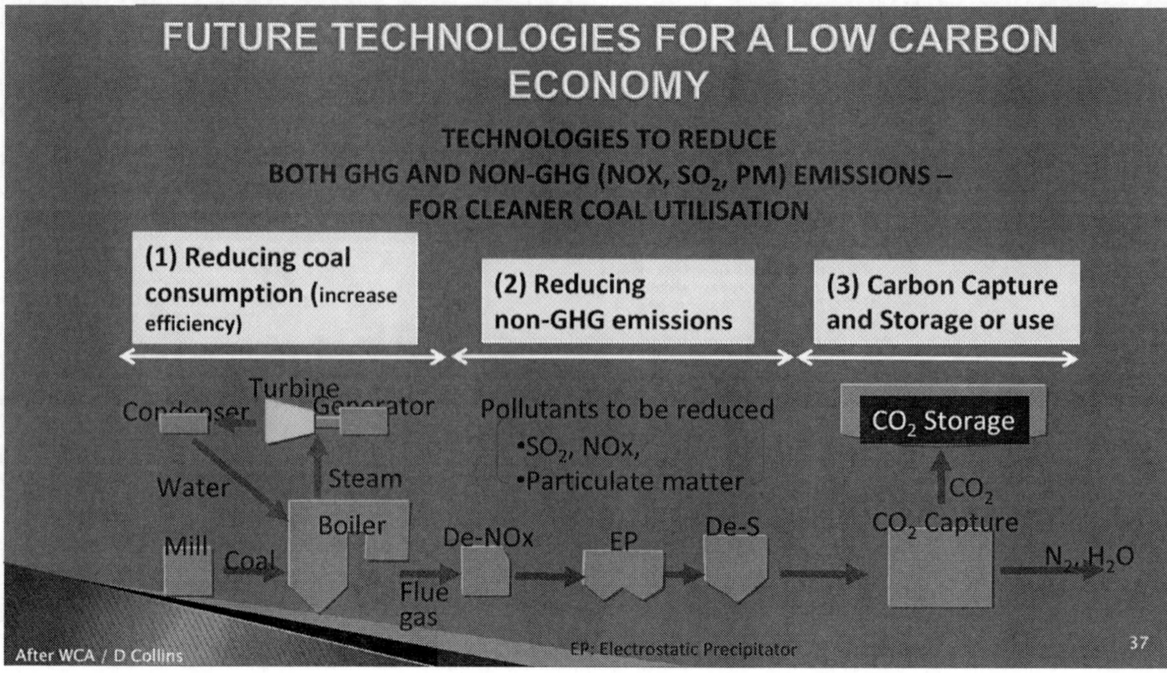
FUTURE TECHNOLOGIES FOR A LOW CARBON ECONOMY
TECHNOLOGIES TO REDUCE
BOTH GHG AND NON-GHG (NOX, $SO_2$, PM) EMISSIONS –
FOR CLEANER COAL UTILISATION
(1) Reducing coal consumption (increase efficiency)
(2) Reducing non-GHG emissions
(3) Carbon Capture and Storage or use
Turbine
Generator
Condenser
Water
Steam
Boiler
Mill
Coal
Pollutants to be reduced
•$SO_2$, NOx,
•Particulate matter
De-NOx
EP
De-S
Flue gas
$CO_2$ Storage
$CO_2$
$CO_2$ Capture
$N_2$, $H_2O$
After WCA / D Collins
EP: Electrostatic Precipitator
37

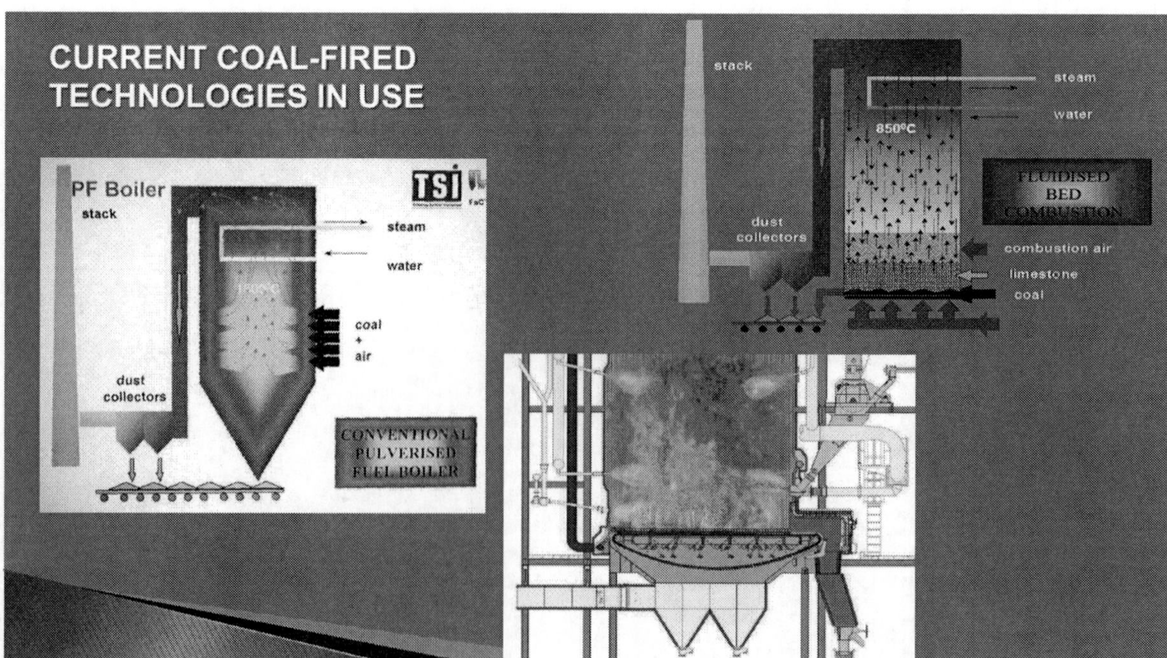
CURRENT COAL-FIRED TECHNOLOGIES IN USE
PF Boiler
stack
steam
water
coal
+
air
dust collectors
CONVENTIONAL PULVERISED FUEL BOILER
stack
steam
water
850ºC
FLUIDISED BED COMBUSTION
dust collectors
combustion air
limestone
coal

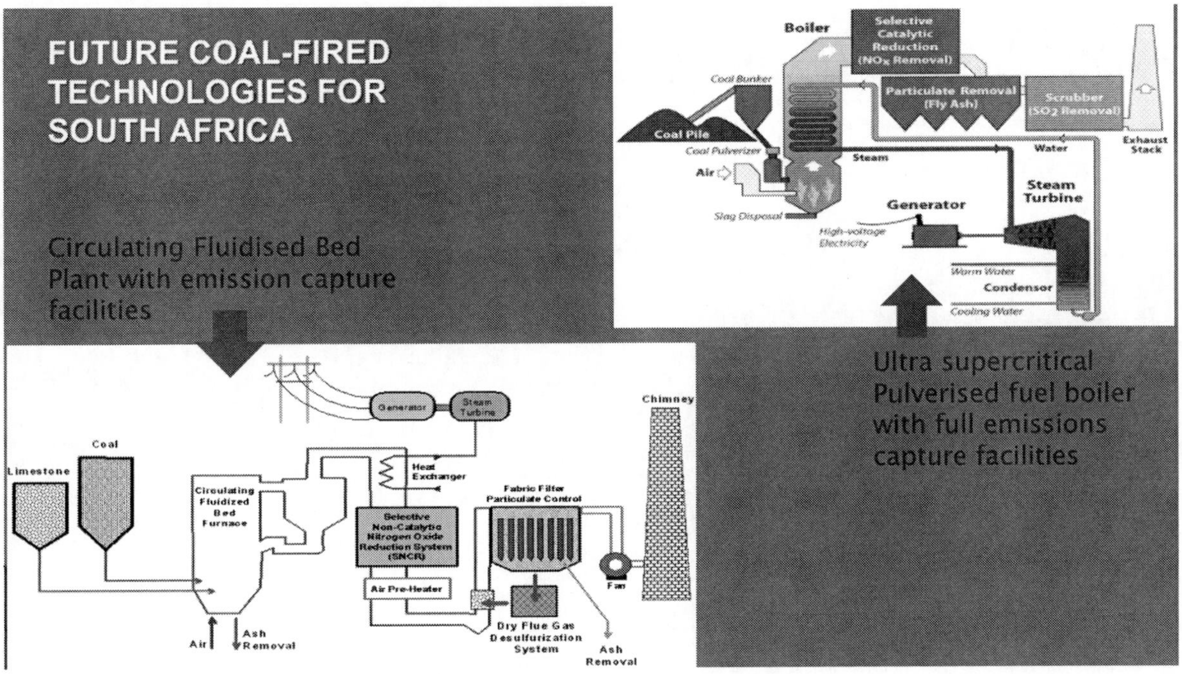
FUTURE COAL-FIRED TECHNOLOGIES FOR SOUTH AFRICA
Circulating Fluidised Bed Plant with emission capture facilities
Ultra supercritical Pulverised fuel boiler with full emissions capture facilities
Boiler
Selective Catalytic Reduction (NOx Removal)
Coal Bunker
Particulate Removal (Fly Ash)
Scrubber (SO2 Removal)
Coal Pile
Coal Pulverizer
Exhaust Stack
Steam
Water
Air
Steam Turbine
Generator
Slag Disposal
High-voltage Electricity
Warm Water
Condensor
Cooling Water
Generator
Steam Turbine
Chimney
Coal
Limestone
Heat Exchanger
Circulating Fluidized Bed Furnace
Fabric Filter Particulate Control
Selective Non-Catalytic Nitrogen Oxide Reduction System (SNCR)
Air Pre-Heater
Fan
Air
Ash Removal
Dry Flue Gas Desulfurization System
Ash Removal

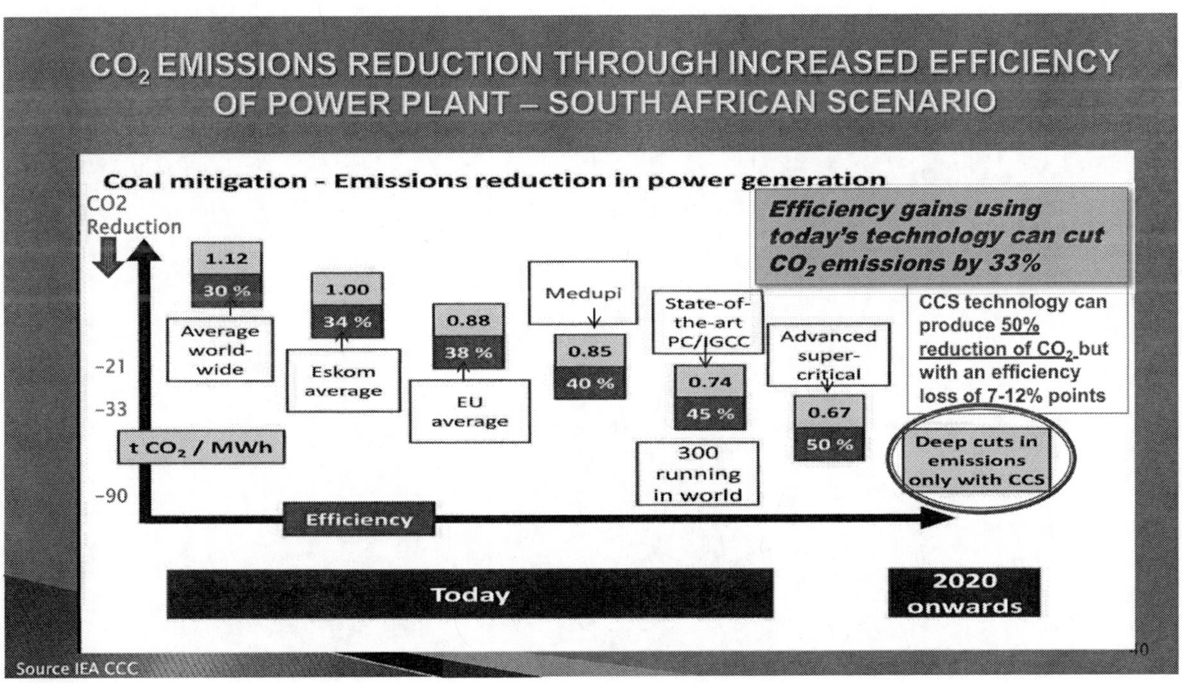
CO2 EMISSIONS REDUCTION THROUGH INCREASED EFFICIENCY OF POWER PLANT – SOUTH AFRICAN SCENARIO
Coal mitigation - Emissions reduction in power generation
CO2 Reduction
Efficiency gains using today's technology can cut CO2 emissions by 33%
1.12
30 %
Average world-wide
1.00
34 %
Eskom average
0.88
38 %
EU average
Medupi
0.85
40 %
State-of-the-art PC/IGCC
0.74
45 %
300 running in world
Advanced super-critical
0.67
50 %
CCS technology can produce 50% reduction of CO2 but with an efficiency loss of 7-12% points
Deep cuts in emissions only with CCS
-21
-33
-90
t CO2 / MWh
Efficiency
Today
2020 onwards
Source IEA CCC

## CLEANER AIR AND REDUCING POLLUTION

**Eskom's drive for cleaner air**

Our efforts to improve air quality focused on reducing emissions of particulates or ash. The relative particulate emissions from coal-fired power stations have reduced by more than 90% over the last 35 years, as can be seen in the graph below.

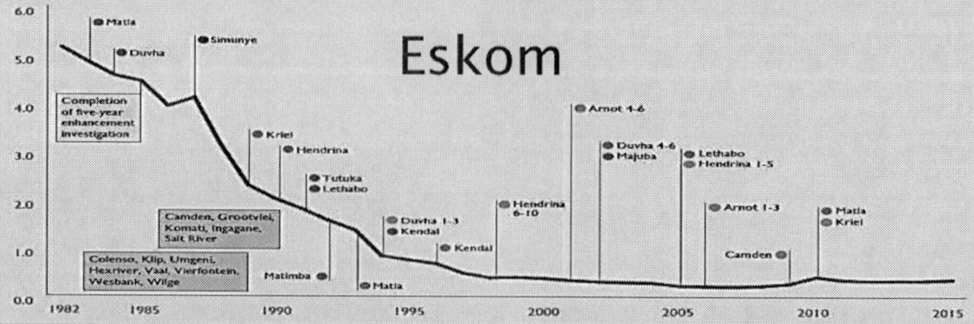


**Relative particulate emissions, kg/MWhSO**

Stations commissioned — Stations decommissioned — $SO_3$ plant installed — Fabric filter plant retrofits installed

Reduction in relative particulate minimum emissions from Eskom's power stations from 1982 to the present, due to the decommissioning of older power stations and commissioning of more efficient technology.

## FACTORS AFFECTING EFFICIENCY OF SOUTHERN AFRICAN COALS

## COAL-BEARING KAROO BASINS OF SOUTHERN AFRICA (

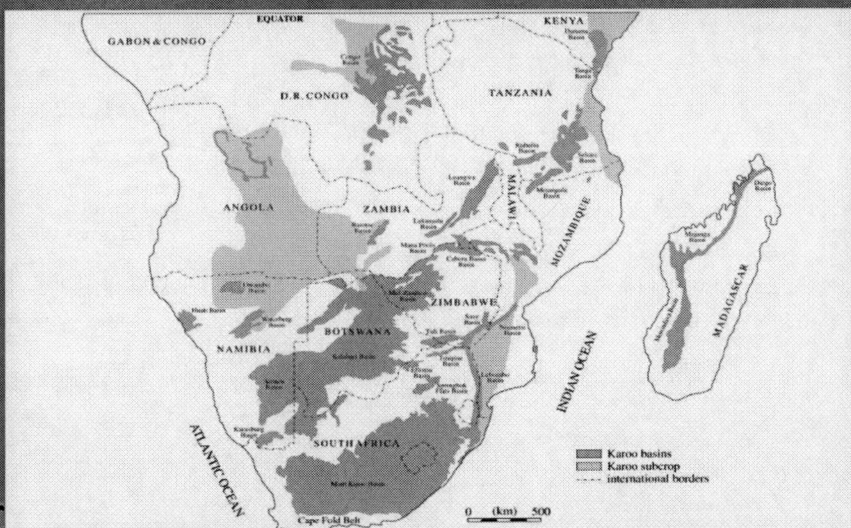


AFTER CATUNEANU *ET AL.*, 2005)

43

## DIFFERENCES BETWEEN NORTHERN HEMISPHERE (LAURASIAN) COALS AND SOUTHERN HEMISPHERE (GONDWANALAND) COALS

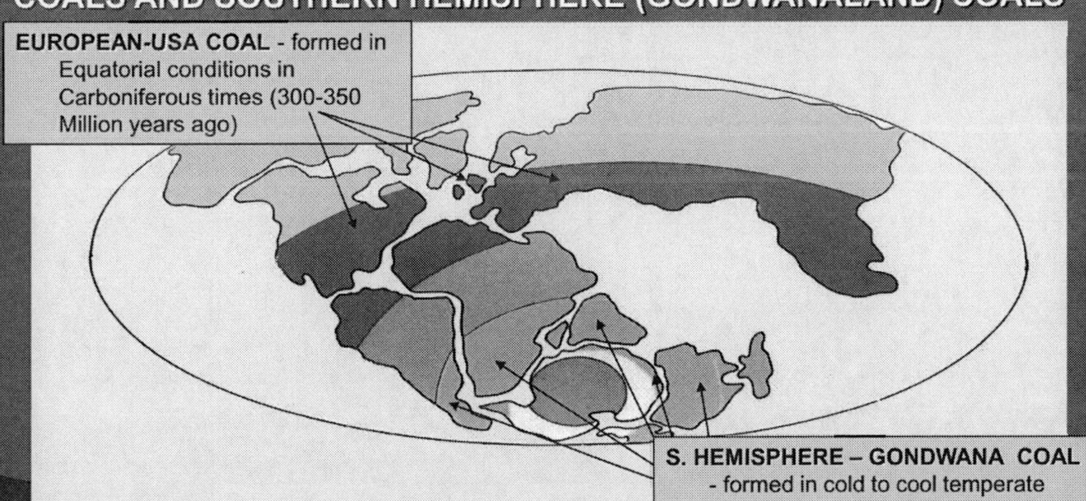


44

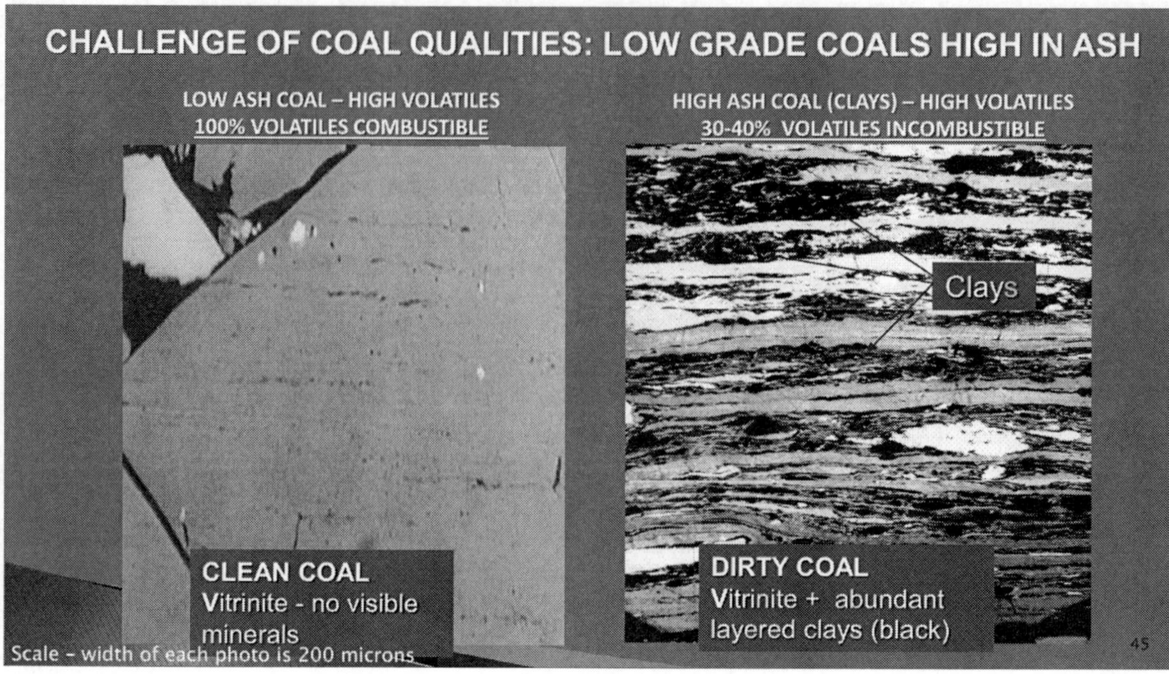
CHALLENGE OF COAL QUALITIES: LOW GRADE COALS HIGH IN ASH
LOW ASH COAL – HIGH VOLATILES
100% VOLATILES COMBUSTIBLE
HIGH ASH COAL (CLAYS) – HIGH VOLATILES
30-40% VOLATILES INCOMBUSTIBLE
Clays
CLEAN COAL
Vitrinite - no visible minerals
DIRTY COAL
Vitrinite + abundant layered clays (black)
Scale – width of each photo is 200 microns
45

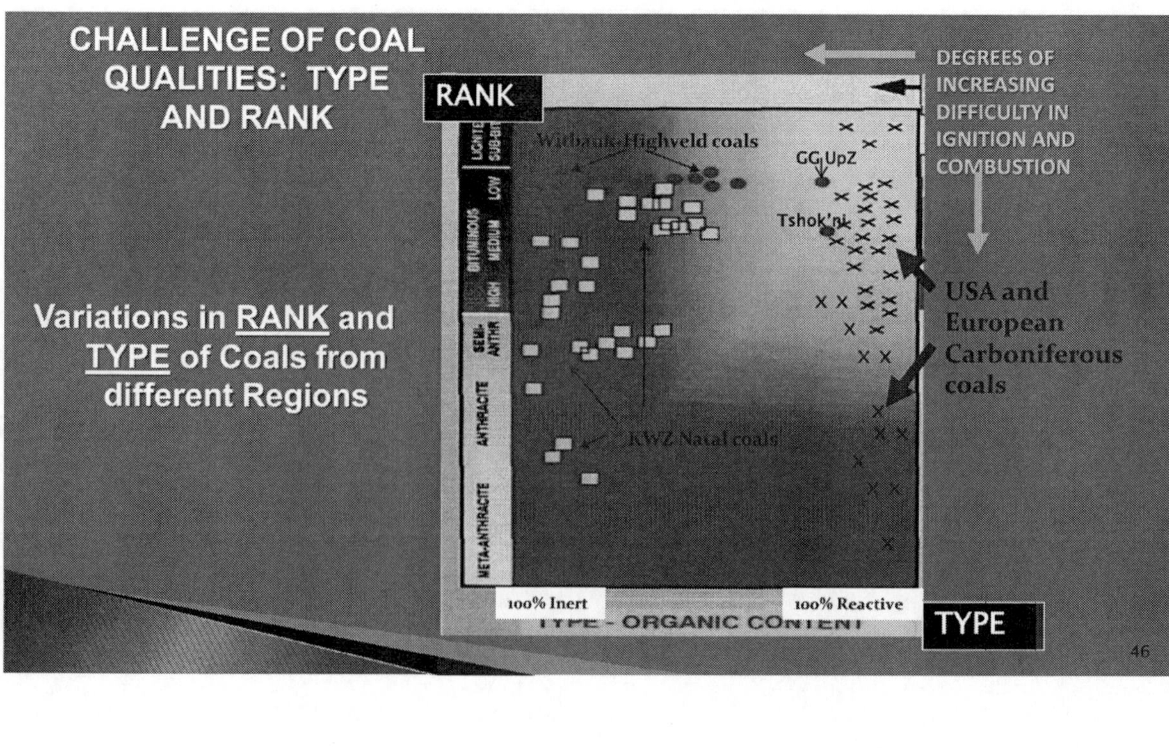
CHALLENGE OF COAL QUALITIES: TYPE AND RANK
Variations in RANK and TYPE of Coals from different Regions
RANK
DEGREES OF INCREASING DIFFICULTY IN IGNITION AND COMBUSTION
Witbank-Highveld coals
GG UpZ
Tshok'ni
KWZ Natal coals
USA and European Carboniferous coals
BITUMINOUS
HIGH
MEDIUM
LOW
SEMI-ANTHR
ANTHRACITE
META-ANTHRACITE
100% Inert
100% Reactive
TYPE - ORGANIC CONTENT
TYPE
46

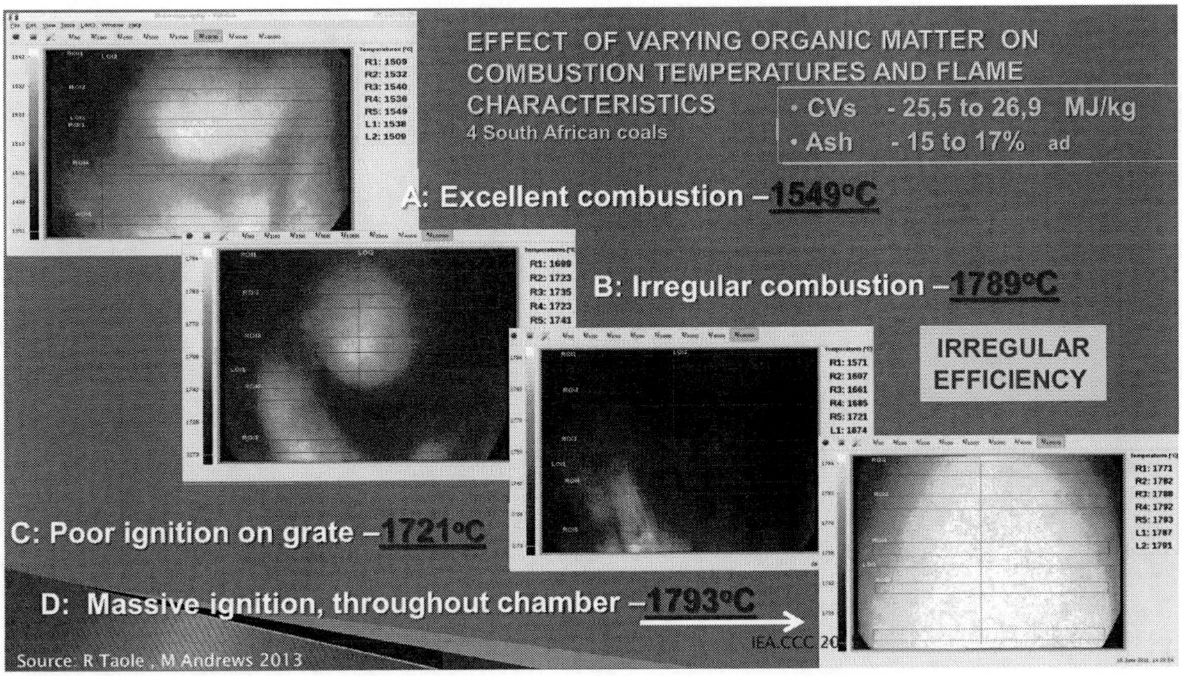


# CLEAN COAL TECHNOLOGIES IN SOUTHERN AFRICA

- Coal beneficiation of low grade coals – dry systems
- Subcritical pulverised combustion - low NOx burners , dry cooling
- Ultra-supercritical pulverised combustion +FGD – two new power stations
- Co-firing of coal and biomass – currently in agricultural industry
- Co-generation – use of industrial waste heat
- Underground coal gasification – successful demonstration

**Future systems**

- Circulating Fluidised bed boilers - in-bed reduction of SOx and NOx
- Hybridisation – solar + coal-fired power

# THE VALUE OF COAL in SOUTH AFRICA

- **SOUTH AFRICA**
  - 7th largest producer of coal in the world
  - 7th largest exporter of coal
- **COAL IN SA ACCOUNTS FOR**
  - 1st highest foreign exchange earnings in the country
  - 1st highest minerals income earner , beating gold , pgms
  - 95% of SA energy production
  - >90% of carbon reductants in the metallurgical industry
  - >30% of liquid fuel, petrol and diesel requirements
  - >200 major chemicals for 1000s of carbon-based products

49

# THE TRUE VALUE OF COAL

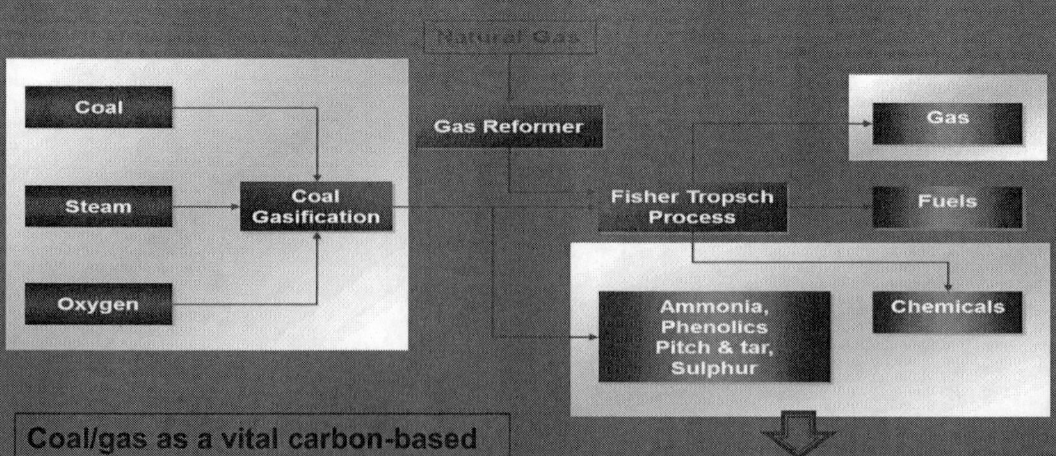


Coal/gas as a vital carbon-based chemical feedstock

Petrol, diesel, liquid fuels,
Explosives, fertilisers, plastics
Food, materials, cosmetics, sulphur
Tars, pitch, bitumen, petcoke +

# CONTENTS



## THE NECESSITY FOR FOSSIL FUELS

- India China Bangladesh Indonesia Russia have major coal and Fossil Fuel expansion programmes
- COP 21 deliberately requires no commitments so that countries can decide to do what is in their best interests
- **Unless emerging countries use Fossil Fuels they will heavily prejudice their future growth and result in increased unemployment and poverty**
- **This will detrimentally effect the economic political and social structures of many emerging nations**

R Jeffrey 2016

# CONCLUSIONS

- **Energy, Electricity leading to increased GDP and Employment Growth are the keys to Southern Africa's -**
  - Future Economic, Social and Political Prosperity
  - Sustainability,
  - Stability and
  - Individual Economic Freedom

**Where the country and the region's coal reserves play a vital part.**

54

## ACKNOWLEDGEMENTS

The authors wishes to acknowledge the contributions made by associated colleagues in the research reviewed above.

This research was undertaken under the auspices of Econometrix and the SARChI Chair of Clean Coal Technology funded by the National Foundation for Research (NRF) of South Africa.

University of the Witwatersrand,
Johannesburg, South Africa

Econometrix
Johannesburg


55

INSTITUTE FOR CHEMICAL PROCESSING OF COAL

1955-2016

33rd International Pittsburgh Coal Conference
August 8-12, 2016 - Cape Town, South Africa

# Coal gasification in Poland - perspectives and key driving forces

**Aleksander Sobolewski**

## Scope of presentation

1. Energy balance of Poland : coal vs. natural gas
2. Coal gasification
   - Technology overview
   - Gasification R&D in Poland
3. Strategy for coal gasification development in Poland
4. Economy and profitability
5. Actions taken in Poland 2015/2016
6. Conclusions

## Background : Poland 2016

1. population: 38 million
2. history : coal-based economy
3. 85% of electricity from coal
4. a lot of old coal-fired power plants
5. excellent R&D coal processing
6. European environmental protection regulations

7. coal overproduction : 8 MTPY
   - export ?
   - shut down of some coal mines ?
   - alternative usage !



## Energy policy in Poland

MINISTRY OF ENERGY

**Krzysztof Tchórzewski**
Minister of Energy

**Grzegorz Tobiszowski**
Deputy Minister of Energy, Government Plenipotentiary, Restructuring of coal mining

*„The basis for covering the demand for electricity will be the use of energy generated in domestic sources, primarily based on energy resources, mainly coal."*

Krzysztof Tchórzewski, IV National Energy and Economic Summit, 20.04.2016

*„The role of carbon energy sector in Poland must be seen as an opportunity for our country to be the European leader in the development of clean coal technologies."*

Krzysztof Tchórzewski, II Congress of Polish Electrical Engineering 11.04.2016

# The energy balance of Poland

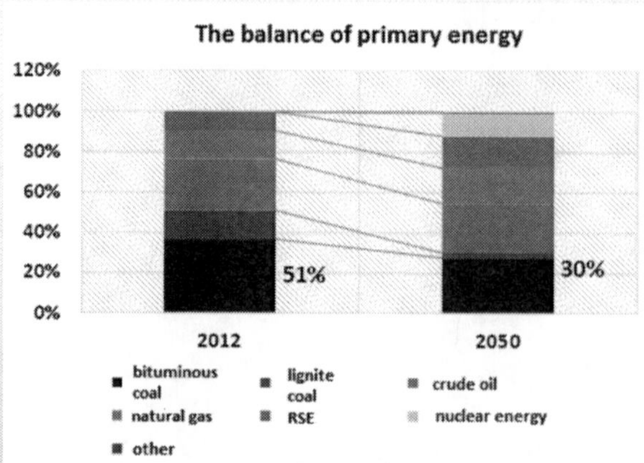


Power generation

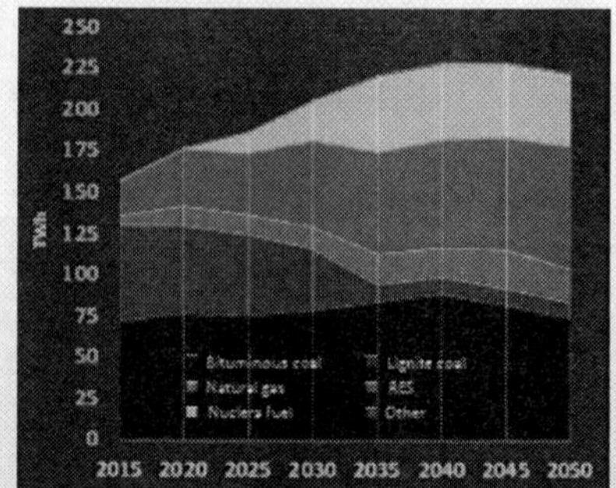


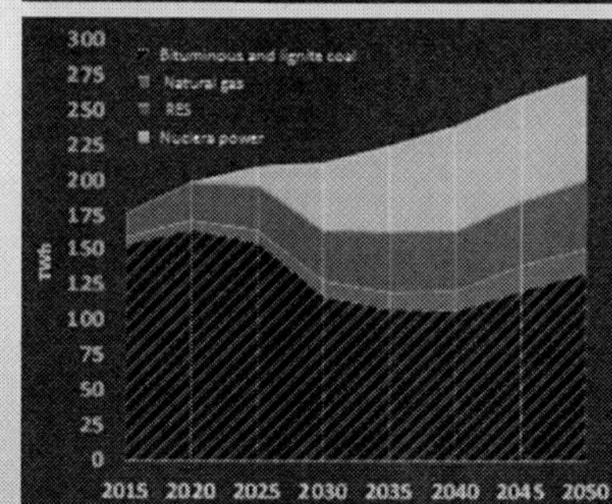


Conclusions from the analysis for Polish Energy Policy until 2050; Annex 2 .: the Polish Energy Policy until 2050; The project version 02.



# Natural gas sources for Polish economy

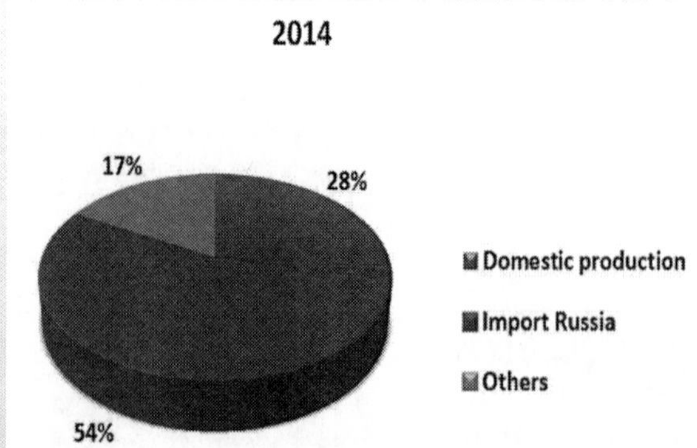


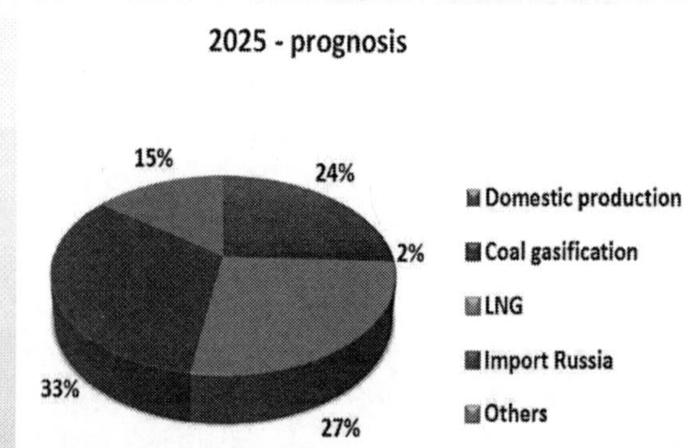


*REPORT ON THE RESULTS OF MONITORING GAS FUEL SUPPLY SECURITY from the period from 1 January 2014 till 31 December 2014The Ministry of Economy 2015*

*The forecast on the basis of the Polish Energy Policy until 2050 and taking into account LNG terminal (amount of gas production and import from Germany and the Czech Republic as in 2014), Coal gasification: 1 million t/year.*

# Diversification of natural gas supply

*Conclusions from the analysis for the purpose of the Energy Policy of Poland until 2050; Annex 2 .: the Energy Policy of Poland until 2050; The project version 02.*

**... 3) Coal will remain the basis for energy security of Poland in the foreseeable period (ie. 2050), but its share will decrease ...**

**Despite that, the need of NG source diversification is indicated repeatedly in the new Energy Policy of Poland (2050), one can not find in document that:**
**exist the possibility of its efficient substitution by gasification of domestic coal.**



# Coal gasification technology:
## Advantages

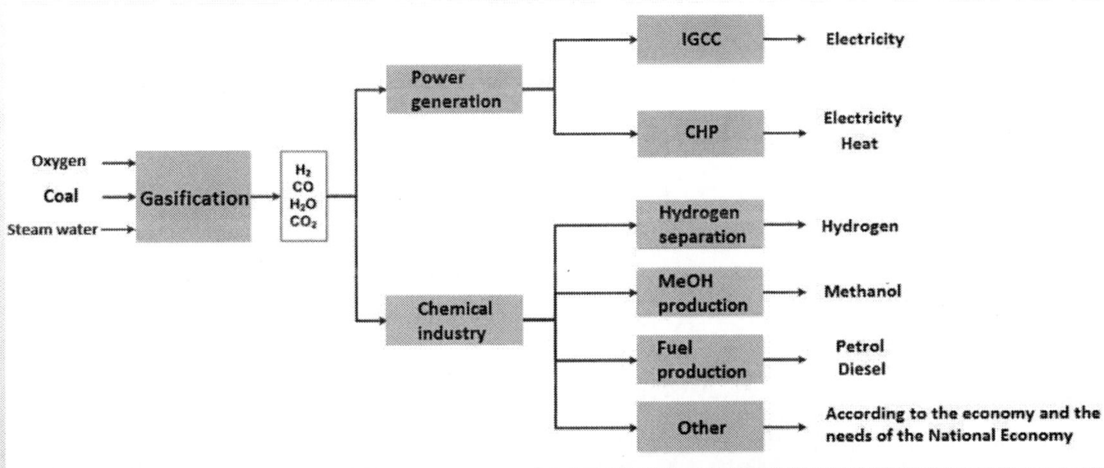


- Zero emission power generation – CCS/current CCU model
- Substitution liquid and gaseous fuel – chemical synthesis
- Polygeneration systems (chemical and energy)

# Coal gasification technology:
## Commercial technology suppliers

- Entrained flow technologies
  - Water slurry and dry feeding
- Fluidized bed technologies
- Fixed bed technologies

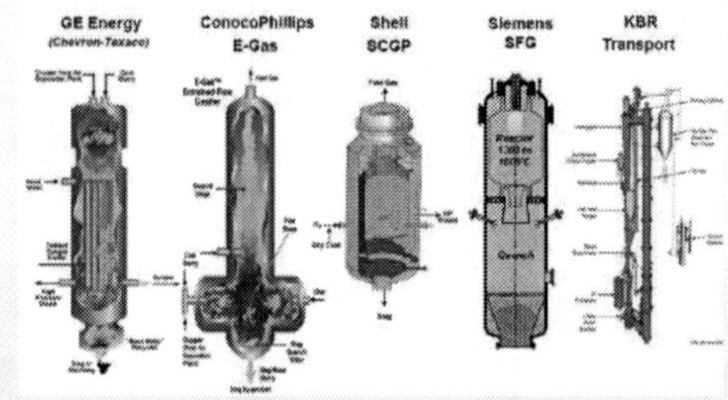


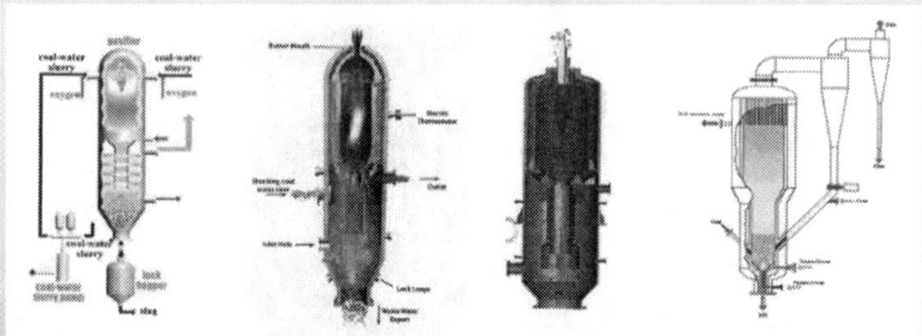

- East China University of Science and Technology **(OMB: Opposed Multi-Burner Gasifier)**
- Northwest Research Institute **(MCSG: The Multi-Component Slurry Reactor )**
- Aerospace Science and Technology Corporation **(HT-L: Pressurized, Down-Flow, Entrained Reactor)**
- Institute of Coal Chemistry **(AFB: Ash Agglomerated, Fluidized-Bed Reactor)**

9 ...beyond standards! ICHPI INSTITUTE FOR CHEMICAL PROCESSING OF COAL

# Coal gasification technology:
## State of development

Coal gasification:

- 25% of global ammonia production
- 30% of global methanol production

Production directions – example of China

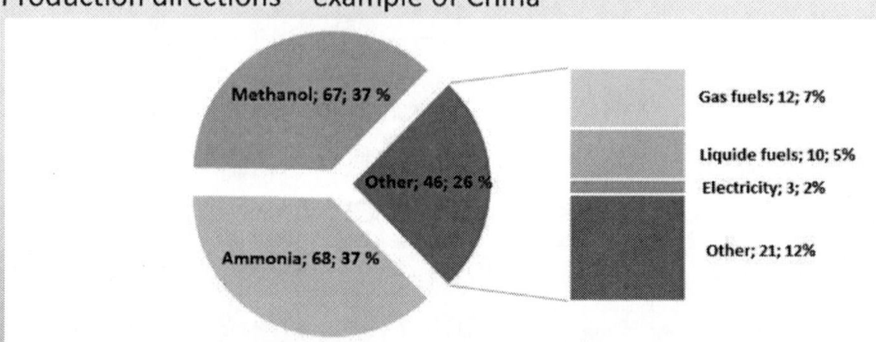


10 ...beyond standards! ICHPI INSTITUTE FOR CHEMICAL PROCESSING OF COAL

# Coal gasification: R&D in Poland: 40 years of IChPW experience

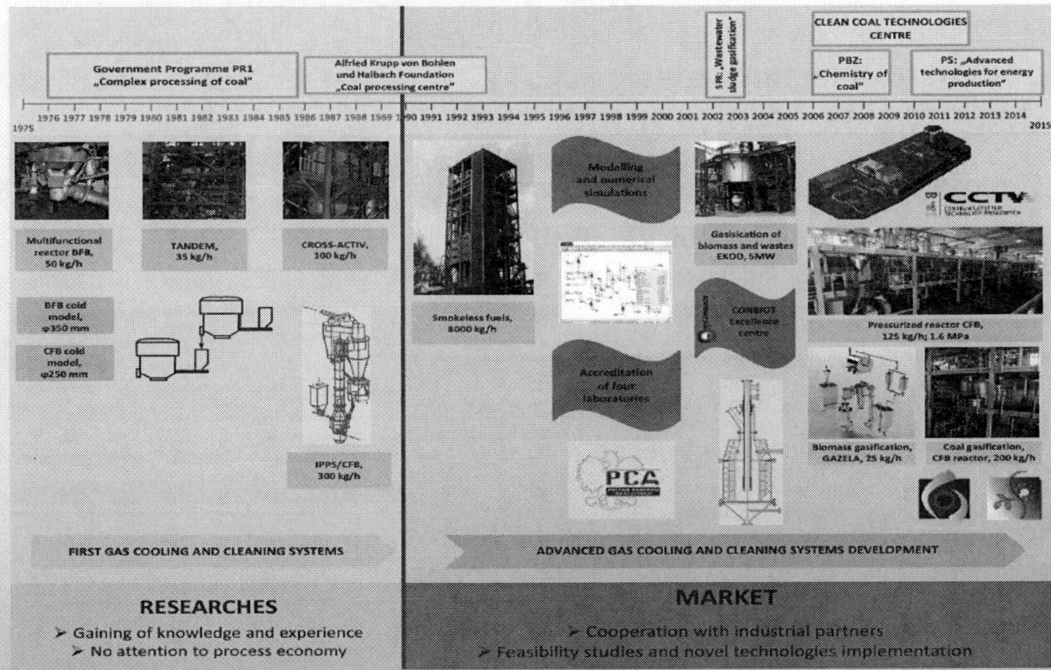


11 ▶ ...beyond standards! INSTITUTE FOR CHEMICAL PROCESSING OF COAL

# $CO_2$ enhanced coal gasification in CFB reactor

## Technology developed by IChPW

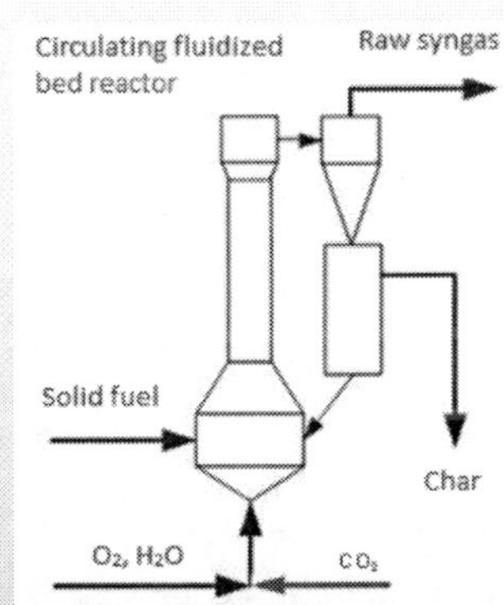

**Results:**

- **Increasing the efficiency of the process and the improvement of economic factors** of production of syngas.
- **Reducing the consumption of fossil fuel** with a constant production of syngas.
- **Reduction of oxygen consumption** (use of $CO_2$ as a carrier of oxygen).
- **Utilization of $CO_2$,** which is produced during the conversion of fossil fuels.

**The level of technological readiness: 6**

▶ 12 ...beyond standards! INSTITUTE FOR CHEMICAL PROCESSING OF COAL

## Strategy for coal gasification development in Poland

- **Commercial technologies: short term perspective 2020**
- **Developing technologies: medium term perspective 2030**

(Results of Task No.3: „ Development of coal gasification technology for highly efficient production of fuel and electricity")

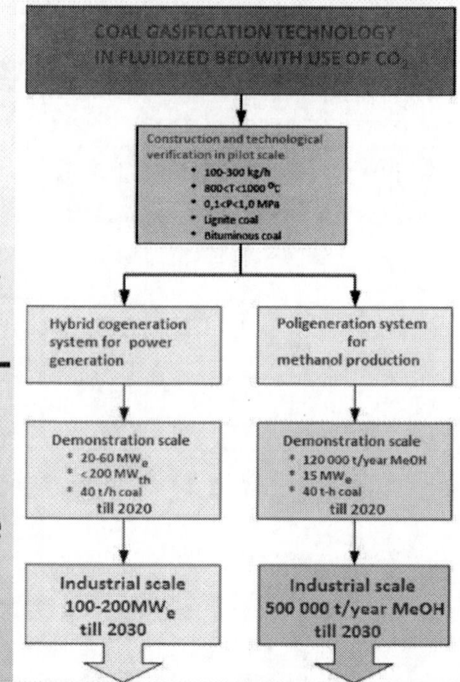


- **Polygeneration systems** integrated with coal gasification
- **Substitution of natural gas** to produce **hydrogen** for **ammonia** synthesis
- **Methanol** synthesis and **olefin** production

▶ 13 ...beyond standards! INSTITUTE FOR CHEMICAL PROCESSING OF COAL

## Strategy for coal gasification development in Poland

**Development variant/direction** | **Possible date of implementation (the first commercial property)** / **Scale of single system/ yearly coal consumption** | **Technology source and development stage**

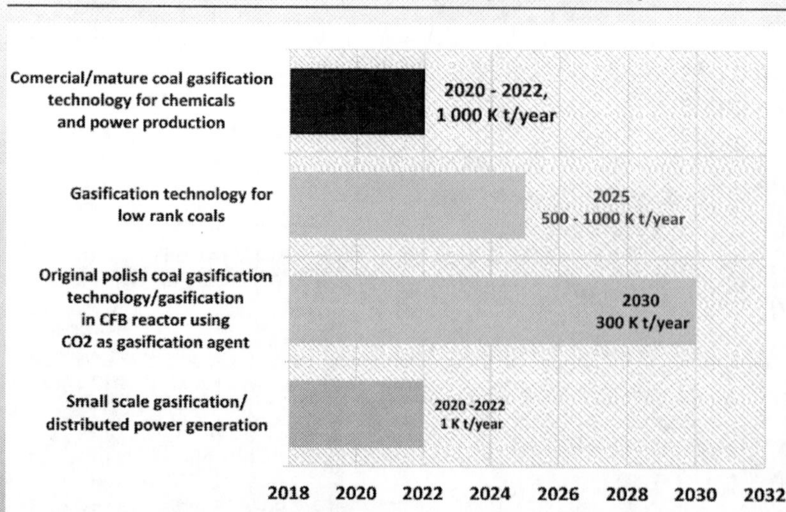


License purchase
Mature Technology

Technological partnership
Requires demo scale operation
Common technology

The level of technological readiness: 6
Requires demo scale operation
Original polish technology

Different technology.
The level of technological readiness: 3-6

▶ 14 ...beyond standards! INSTITUTE FOR CHEMICAL PROCESSING OF COAL

## Economic analysis – assumptions (IChPW, Poland 2015)

- **2 coals:**
  - Bituminous coal – ZW Janina
  - Lignite coal – KWB Bełchatów
- **Entrained flow reactor - dry feeding**
  - One reactor (1000 K t/y)
  - Shell, Siemens
- **CCS ready system** ($CO_2$ capture and compression)
- **Products: H2 /MeOH/ Propylene**
- **The scale of production can meet the needs of the polish industry in:**
  - 17% in the case of propylene
  - 100 % in the case of methanol

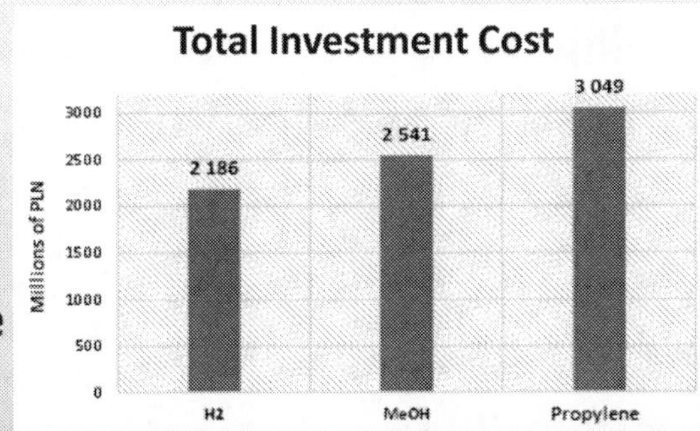


▶ 15 ...beyond standards! INSTITUTE FOR CHEMICAL PROCESSING OF COAL

## Economic analysis– results (IChPW, Poland 2015)

- **Coal cost:**
  - Bituminous coal: 2,4 €/GJ
  - Lignite coal: 1,7 € /GJ
- **Sales prices of the products:**
  - Hydrogen: 1,55 € /kg (260 € /K $m^3$)
  - Methanol: 354 €/t (average market price 2014-2015: 363 €/t; max: 450 €/t)
  - Propylene: 1330 €/t

1 € = 4,2 PLN

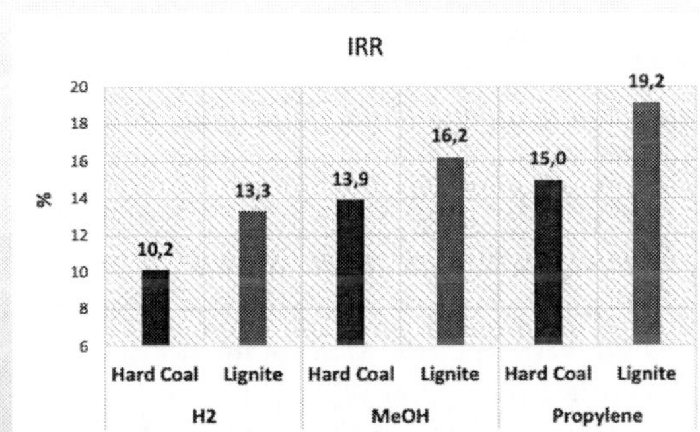


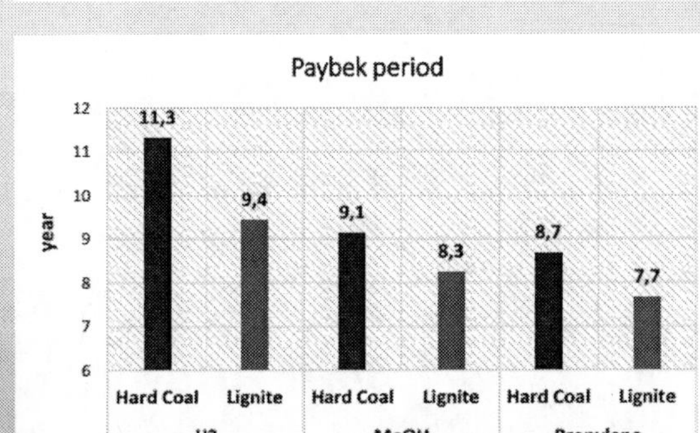


▶ 16 ...beyond standards! INSTITUTE FOR CHEMICAL PROCESSING OF COAL

## Possible substitution of natural gas

☑ **To meet the needs of domestic nitrogen plants (2,5 billions m$^3$ of natural gas) 7 – 8 millions tons of coal** must be gasified (depending on the fuel quality and gasification technology)

☑ **To produce 500 000 t/year of methanol** about **1 million tons of coal** is needed, what is equivalent of approx. **400 million m$^3$ of natural gas** consumption.



## Actions taken in Poland 2015/2016

**Consortium:**
**Grupa Azoty SA, TAURON SA, KGHM SA**

➲ „Verification of Assumptions for the Feasibility Study polygeneration power plant"
- Development of pre-feasibility study for different configuration of chemical and power production systems integrated with coal gasification.
- Contractor: **IChPW, Zabrze**
- **The results of this work were the basis for next actions**

➲ **Feasibility Study of Polygeneration Power Plant**
- **Complete Feasibility Study** for different chemical production systems integrated with coal gasification.
- **Location: Grupa Azoty, Zakłady Azotowe Kędzierzyn S.A.**
- Contractor: **Amec Foster Wheeler Italiana**

## Actions taken in Poland 2015/2016

### Polska Grupa Energetyczna SA

(the biggest power producer in Poland)

- **Formation of the Working Group for the coal gasification technology**
- **The report – the review of coal gasification technology**
  - **Objective:** To analyze the possibility of developing gasification technology in Poland with the use of bituminous and lignite coal deposits
  - Analysis of the available and world leading commercial gasification technologies
  - Study visits, expert meetings - including direct meetings with gasification technology suppliers.

▶ 19 ...beyond standards! INSTITUTE FOR CHEMICAL PROCESSING OF COAL

## Summary

### Results of works done in 2010 – 2015 :

- ✓ **Detailed analysis of the potential for coal gasification in Poland, well-established team of experts and researchers to promote the industrial implementation**
- ✓ **Qualitative and quantitative analysis of usefulness the polish coals for gasification**
- ✓ **Knowledge of the existing coal gasification technology (commercial and being under development)**
- ✓ **Collected and analyzed detailed information on the investment and operating costs, economic efficiency of different production directions and internal and external markets for products.**

▶ 20 ...beyond standards! INSTITUTE FOR CHEMICAL PROCESSING OF COAL

## Summary

**Poland is ready to take the strategic decision about construction and start-up the first industrial coal gasification plant**

▶ 21 ...beyond standards! ICHPW INSTITUTE FOR CHEMICAL PROCESSING OF COAL

## Poland – future for coal gasification

MINISTRY OF ENERGY

**Krzysztof Tchórzewski**
Minister of Energy

**Grzegorz Tobiszowski**
Deputy Minister of Energy,
Government Plenipotentiary.
Restructuring of coal mining

*(...) Coal gasification seems to be very promising type of clean coal technology. I see the potential in the development of $CO_2$ processing technology, in management of this gas in carbochemistry and production of fuels, including biofuels „*

Grzegorz Tobiszowski, Conference: "Clean coal technologies in the context of achieving the objectives of the National Energy Policy", 07.04. 2016, Warsaw.

*„We need to combine high efficiency, flexibility and environmental parameters with the real economic effect (…). It will also be important alternative possibilities of coal use, like gasification."*

Michał Kurtyka, Undersecretary of State,
"EU Energy Policy - pillars and the prospect of development" Conference, 25.04. 2016

▶ ...beyond standards! ICHPW INSTITUTE FOR CHEMICAL PROCESSING OF COAL

## Summary

### Key driving forces :

- ✓ availability and quality of coal
- ✓ lack of natural gas
- ✓ well-known technology
- ✓ educated staff and R&D
- ✓ negative trade balance for chemicals

**....and**

# We are still waiting!



# Thank you for your attention !

# 清洁煤与油田绿色开发：
# 延长石油集团CCUS一体化项目
# Clean Coal and Green Development of Oilfield: The Integrated CCUS Project of Yanchang Petroleum Group


报告人：王香增 博士　高级副总裁

Speaker: Dr. Xiangzeng Wang　Senior Vice President

中国　陕西延长石油（集团）有限责任公司

Shaanxi Yanchang Petroleum (Group) Co., Ltd, China


2016年8月11日 11th August, 2016

延长石油 YANCHANG PETROLEUM 1905

## 引　言
## Introduction

- 随着全球工业的快速发展，气候变化受到各国政府的高度重视.
- As the rapid development of global industry, climate change is getting rising concerns.
- $CO_2$捕集、利用与封存技术（CCUS）可有效降低碳排放量，是能源企业积极应对全球气候变化，实现碳减排的有效途径之一。
- As an important tool to reduce climate impact, $CO_2$ Capture Utilizing and Storage (CCUS) is a effective way that fossil energy business players take.
- 煤炭相关产业是当前二氧化碳排放的主要来源，开展CCUS项目将有助于推动煤炭的清洁化利用。
- Coal based industry is one major source of $CO_2$, CCUS project is founded to utilize coal in a cleaner way.
- 延长石油集团发挥自身优势，打造了全球首个集煤化工$CO_2$捕集、特低渗油田$CO_2$驱油与封存为一体的 CCUS项目，在技术创新方面积累了一些经验，愿与全球煤炭行业专家分享。
- With its core competencies, Yanchang Petroleum Group has built the very first integrated CCUS project, which combined coal chemical $CO_2$ capture and low permeability reservoirs $CO_2$ EOR and storage together, and our experiences in technology innovation will be shared here.

# 内容提要
# Table of Contents



## （一）延长石油集团概况
## Overview of Yanchang Petroleum Group

- 延长石油地处中国鄂尔多斯盆地，是中国陆上发现最早的油田，成立于1905年；
- Yanchang Petroleum was founded in 1905, and it is located in Ordos basin, which is known as the first onshore oil field of China.
- 2015年原油产量1241万吨，连续9年稳产千万吨以上；
- In 2015, total oil production is approximate 100MMbbl, and this capacity has been maintained for 9 years.
- 业务涵盖油气勘探开发、煤矿开采、油气煤综合化工及科技研发等；
- Its business includes oil and gas E&D, coal mining, chemical industry, and R&D etc.
- 世界企业“500强”最新排名325位。
- Rank 325 in Global Fortune Top 500 (latest).

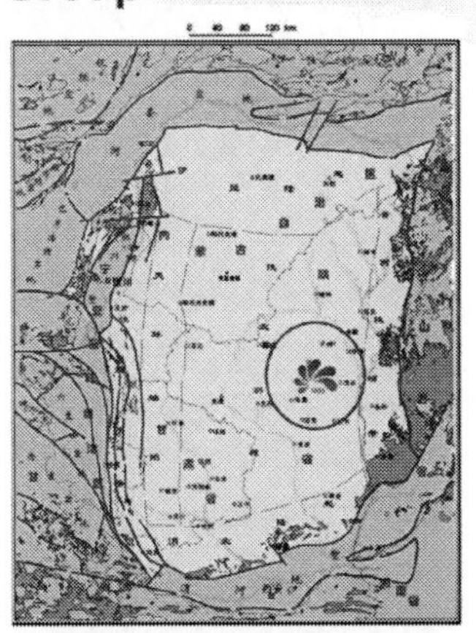

油田开发 Oilfield

综合化工 chemical industry

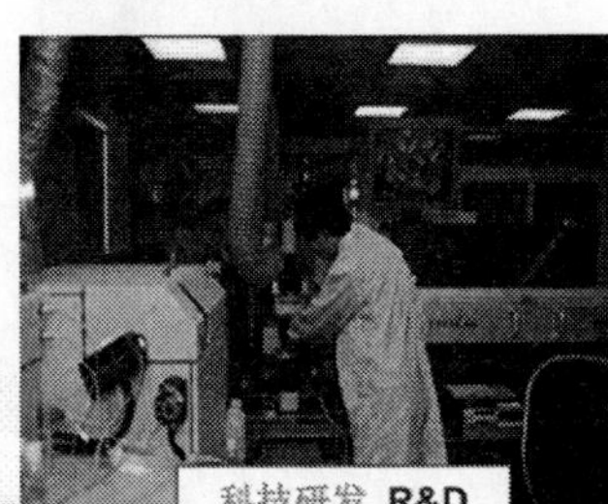

科技研发 R&D

## （一）延长石油集团概况

## Overview of Yanchang Petroleum Group

延长石油拥有多个煤矿、煤化工企业和油气田，可自主开展全流程一体化CCUS项目

Yanchang petroleum owns several coal mines ,coal-chemical plants and oil-gas fields, and can independently carry out the whole process of CCUS projects.

煤矿 **Coal Mine**

煤化工企业 **Coal-Chemical Plants**

特低渗油田 **Oil Fields with Low Permeability**

## （二）延长石油开展CCUS项目的有利条件

## Core competencies of Yanchang Petroleum to have CCUS

### 1、具有充足的碳源和较低的捕集、输送成本

Abundant carbon supply and low cost in $CO_2$ capture and transportation.

- **碳源充足**：相对集中的煤化工企业提供了充足的碳源；
- **Abundant carbon supply from** relatively concentrated coal chemical plants.
- **捕集成本低**：煤化工企业产生的$CO_2$浓度高，捕集成本约20美元/吨；
- **Low cost** of high concentration $CO_2$ from coal chemical plants **(approx. 20 USD per ton)**
- **输送成本低**：在产油田与煤化工企业处于同一区域，输送距离小于150公里
- **Low pipeline cost**, less than 150 km between coal chemical plants and oil fields.

碳捕集装置 **Equipment of $CO_2$ capture**

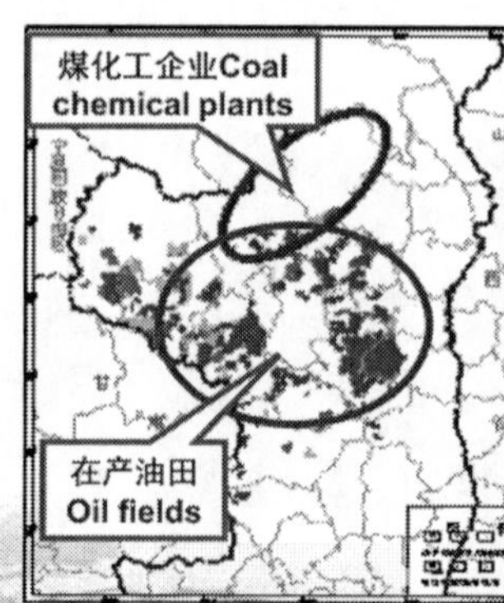

# （二）延长石油开展CCUS项目的有利条件

## 延长石油 Core competencies of Yanchang Petroleum to have CCUS

### 2、有大量低渗透油藏适合开展$CO_2$驱油
Large amount of oil reservoirs suits $CO_2$ flooding.

- **有助于油田长期稳产**：延长油田属于特低渗透油藏，产量递减较快，储层特征和温压条件适合开展$CO_2$驱油，可大幅度提高采收率；
- Yanchang oilfields are low permeability reservoirs, reservoir characteristics, formation temperature, and pressure are ideal to launch $CO_2$ flooding, which greatly improve the recovery factor.
- **有利于节约水资源**：陕北地区水资源匮乏，用$CO_2$驱油代替注水开发可节约水资源。
- Shanbei region suffers water shortage, so $CO_2$ flooding saves more water than water flooding.

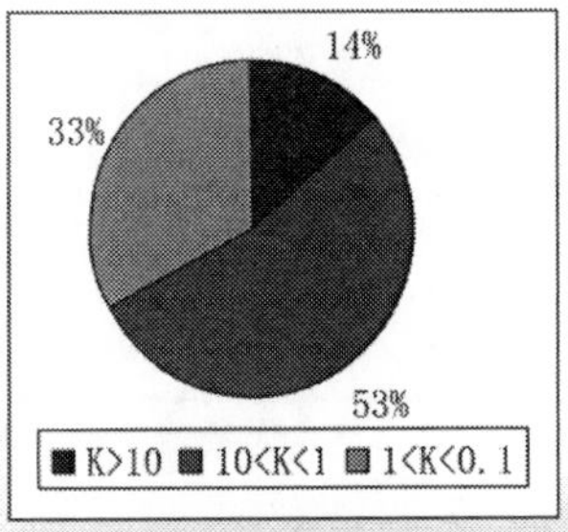


**不同渗透率级别储层储量分布图**
**Reserve Distribution for low, ultra-low, permeability reservoirs**

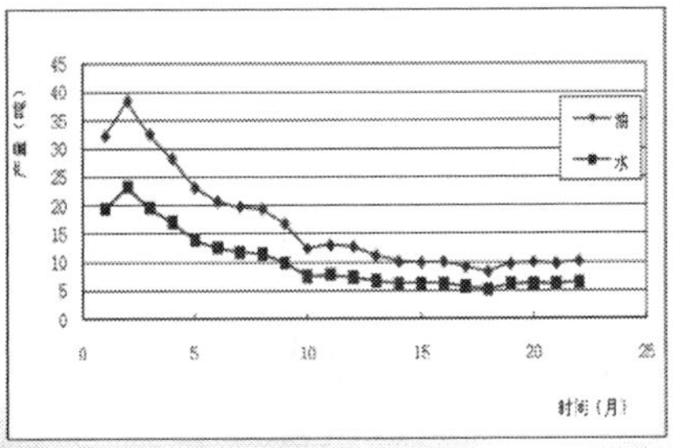


**单井月产量递减较快**
**Well production rate change**

**地表干旱缺水**
**Droughty land**

# （二）延长石油开展CCUS项目的有利条件

## 延长石油 Core competencies of Yanchang Petroleum to have CCUS

### 3、地层条件有利于$CO_2$长期封存
Geological condition benefits long-term $CO_2$ storage

- **封存条件好**：陕北斜坡构造简单，地层稳定，断层不发育，$CO_2$封存安全可靠；
- Shanbei Slope has simple structure, stable formation, and fault development, so $CO_2$ sequestration is secured and reliable.
- **封存空间大**：有大量需要提高采收率的油藏和盐水层，适宜于$CO_2$封存。
- There is plenty of low permeability reservoirs and saltwater layers to store $CO_2$.

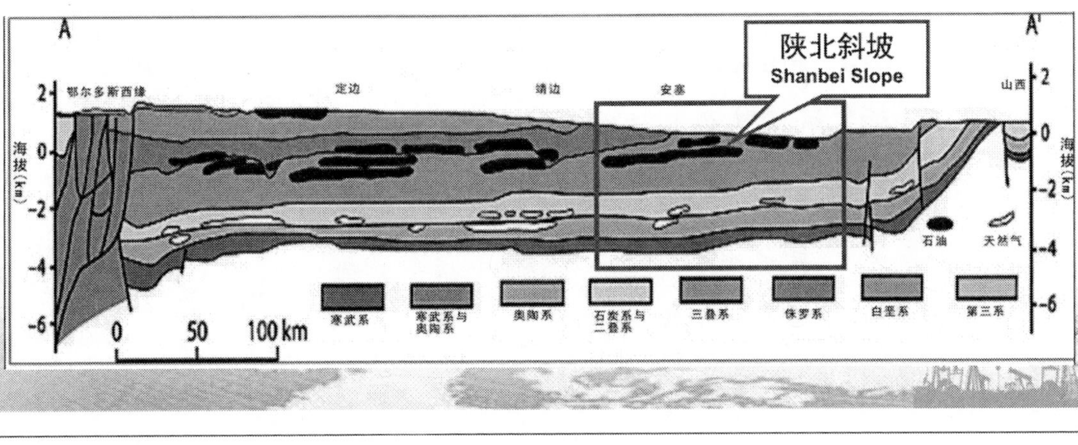

## （三）延长石油CCUS项目现状

## Current Status of Yanchang CCUS Project

**历时8年，投资近5000万美元。**

Lasts 8 years & invested almost 50 million USD.

- 形成了碳捕集、利用与封存技术系列。
- Developed a series of CCUS technologies.
- 建成13万吨/年$CO_2$捕集装置。
- Established $CO_2$ capture plant with 130K ton per year capacity.
- 建成两个$CO_2$-EOR与封存先导试验区。
- Built two pilot test zones for $CO_2$-EOR and storage.
- 开展了非常规油气井$CO_2$压裂矿场试验。
- Developed $CO_2$ fracturing technologies and conducted tests in unconventional oil & gas wells.

通过开展CCUS项目，将**"碳捕集—提高油田采收率—碳封存—碳减排"**融为一体，是延长石油**实现低碳、可持续发展**的必然选择。

CCUS project integrates **$CO_2$ capture, enhanced oil recovery rate**, $CO_2$ storage and reduced $CO_2$ emission together, and it will help Yanchang Petroleum.to realize **low carbon and sustainable development**

## 内容提要

## Table of Contents

## （一）形成煤化工低成本$CO_2$捕集技术
## Developed low-cost $CO_2$ capture technology from coal chemical plant.

**1、 实现高浓度$CO_2$低成本捕集。**

$CO_2$ Capture with High Concentration and Low-cost.

- 2012年延长石油榆林煤化公司建成5万吨/年的$CO_2$捕集装置，采用**低温甲醇洗技术**，捕集成本仅20美元/吨。
- In 2012, Yanchang Yulin Coal-Chemical Company completed the capture plant with 50K tons p.a capacity by **Rectisol process**. The capture cost is only $20 per ton.

## （一）形成煤化工低成本$CO_2$捕集技术

## Developed low-cost coal chemical plant $CO_2$ capture technology

**2、针对低浓度$CO_2$，采用氨吸收技术捕集，已完成工艺包开发和中试装置建设。**

Industrialized process package and pilot plants have been developed for $CO_2$ capture with low concentration.

**多功能中试装置multifunctional pilot plants**

$CO_2$捕集工艺与仿真模拟 **$CO_2$ capture and analog simulation**

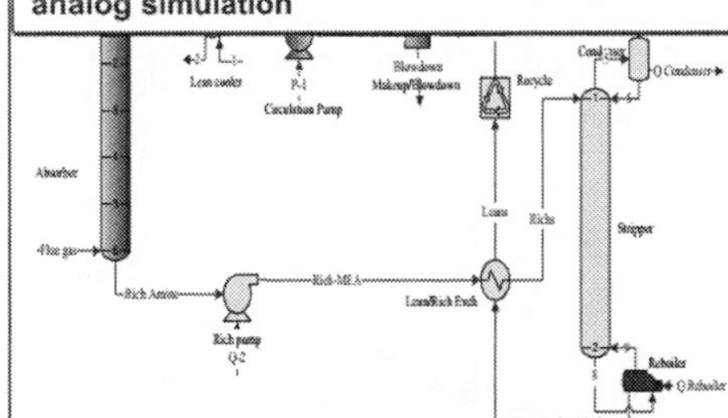


**$CO_2$捕集能力近期将达到50万吨/年**

**$CO_2$ capture capacity will reach 500K tons per year recently.**

## （二）形成特低渗油藏$CO_2$驱油与埋存技术

## Developed $CO_2$-EOR and storage technology in ultra-low reservoir

### 1、建立了完整的$CO_2$驱油室内实验评价方法

Completed the experimental evaluation method of $CO_2$-EOR.

- 微观和宏观有机结合，开展了$CO_2$驱油室内实验评价，对**$CO_2$驱油机理及驱替特征**进行研究
- Integrated micro and macro analysis methods are applied, the experimental evaluation for mechanism and characteristics of $CO_2$-EOR has been done

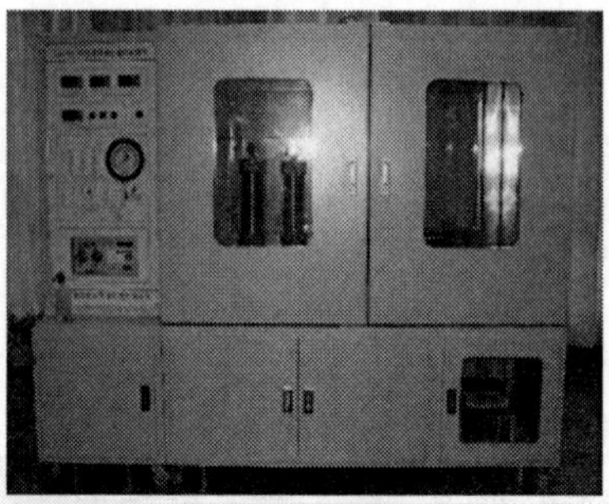

**自动岩心流动实验仪**
**Experiment device of core displacement**

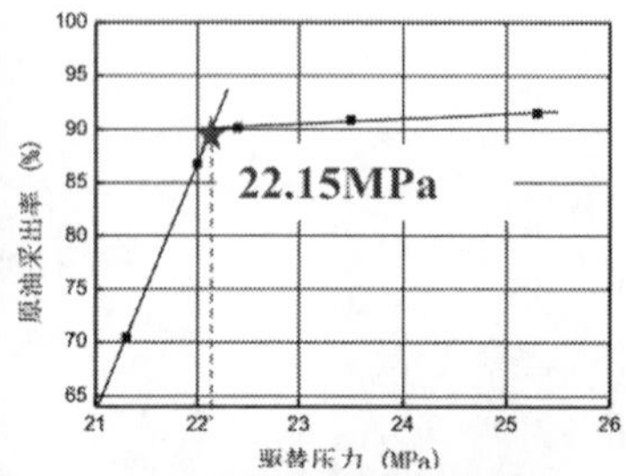


**靖边乔家洼长$_6$油层最小混相压力**
**Miscible pressure of Jinbian oilfield**

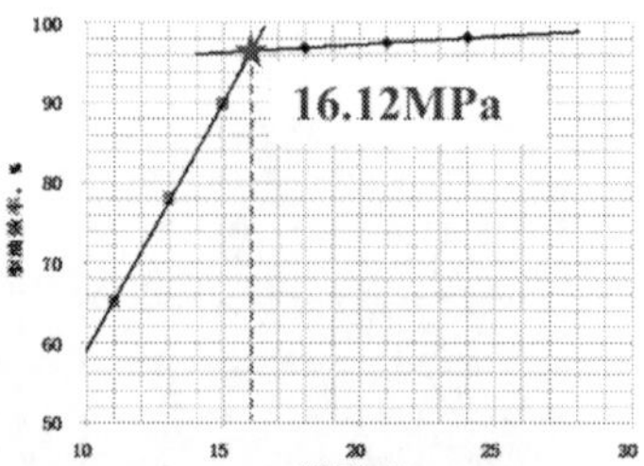


**吴起油沟长$_6$油层最小混相压力**
**Miscible pressure of Wuqi oilfield**

## （二）形成特低渗油藏$CO_2$驱油与埋存技术

## Developed $CO_2$-EOR and storage technology in ultra-low reservoir

模拟研究不同$CO_2$注入方式的采油效果，优选水气交替注入（WAG）

Determine the $CO_2$ injection mode by indoor simulation, WAG is the optimum way.

- 连续气注入最终采出程度为**33.73%**，提高采出程度**8.41%**；
- For continuous gas injection, ultimate recovery rate is 33.73%, increased by 8.41%.
- 水气交替注入（WAG）最终采出程度为**44.7%**，提高采出程度**20.95%**，较连续气驱的采出程度提高了**12.54%**。
- For WAG, ultimate recovery rate is 44.7%, increased by 20.95%, exceed continuous gas injection by 12.54%.

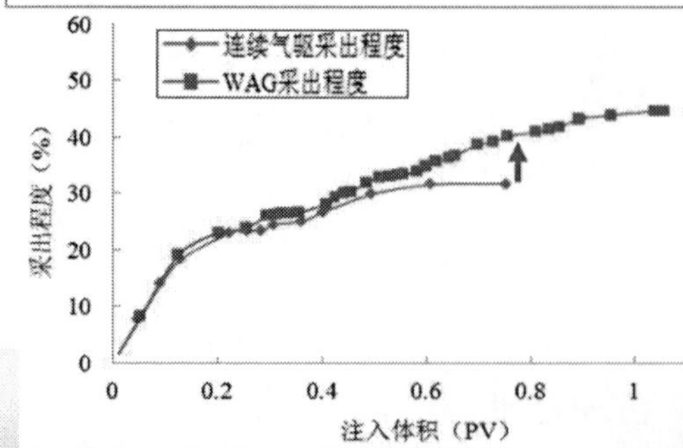


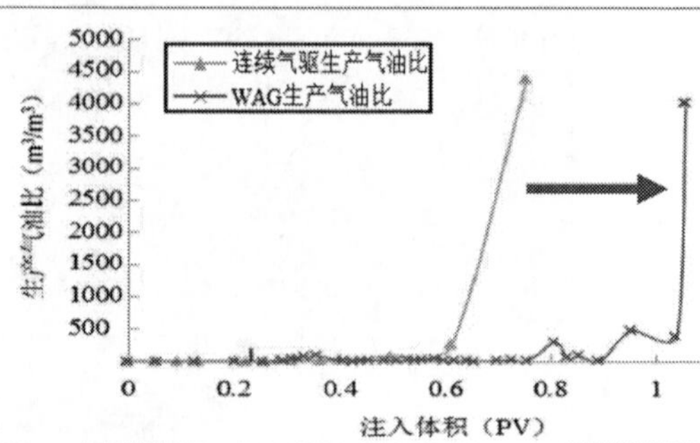


**WAG与连续气驱驱油动态曲线**
**The dynamic curve of WAG and Continous Gas Injection**

## （二）形成特低渗油藏$CO_2$驱油与埋存技术

## Developed $CO_2$-EOR and storage technology in ultra-low reservoir

### 2、优化了$CO_2$驱油油藏工程设计

Optimized reservoir engineering design for $CO_2$-EOR

- 通过油藏精细地质研究，优化了**注采井网、注气方式、注气压力**和**初期开发指标**等关键参数。
- Optimized parameters, such as injection-production pattern, gas injection method, gas injection pressure and initial development parameters by detailed geological studies.

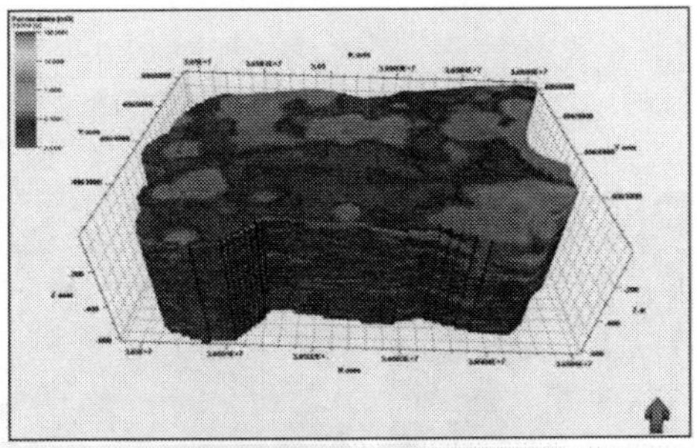

**油藏渗透率三维模型**
**3D model of reservoir permeability**

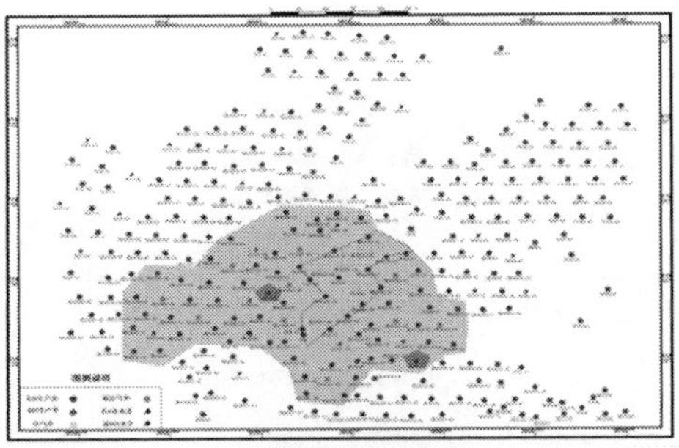

**$CO_2$驱注采井网**
**$CO_2$ injection-production pattern**

## （二）形成特低渗油藏$CO_2$驱油与埋存技术

## Developed $CO_2$-EOR and storage technology in ultra-low reservoir

### 3、形成了注采工艺配套技术及防腐技术

Formed the related techniques of $CO_2$ injection and oil production, and the anti-corrosion technologies.

优化了注采管柱设计，优选了**"普通碳钢+缓蚀剂"**防腐配套工艺，现场应用效果良好。

Injection & production string design has been optimized by a number of tests, and "ordinary steel+ inhibiter" has been selected as the best anti-corrosion approach. It performs excellent in field application.

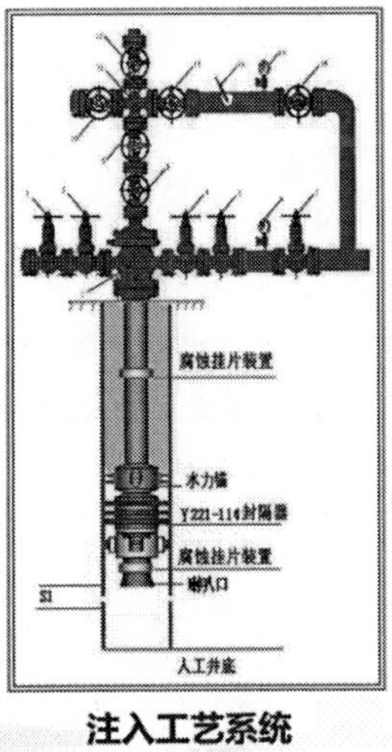


**注入工艺系统**
**Injection system**

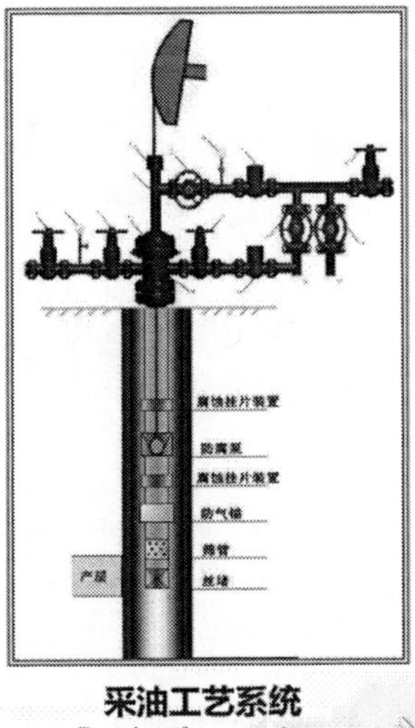

**采油工艺系统**
**Production system**

## （二）形成特低渗油藏$CO_2$驱油与埋存技术

## Developed $CO_2$-EOR and storage technology in ultra-low reservoir

### 4、初步形成$CO_2$安全监测技术

Developed $CO_2$ monitoring technology prototype.

- 从油藏监测、地表监测、大气监测三个角度，形成了**$CO_2$地质埋存安全风险评价和预警方法。**
- Has set up $CO_2$ storage risk assessment and warning technology in three aspects: reservoir monitoring, surface monitoring and atmospheric monitoring.
- 首次创立了利用**$^{13}$C同位素**监测$CO_2$前缘流动位置和近地表$CO_2$泄露的监测方法。
- Has initiated utilizing $^{13}$C isotop in $CO_2$ flow front and surface leakage detection.

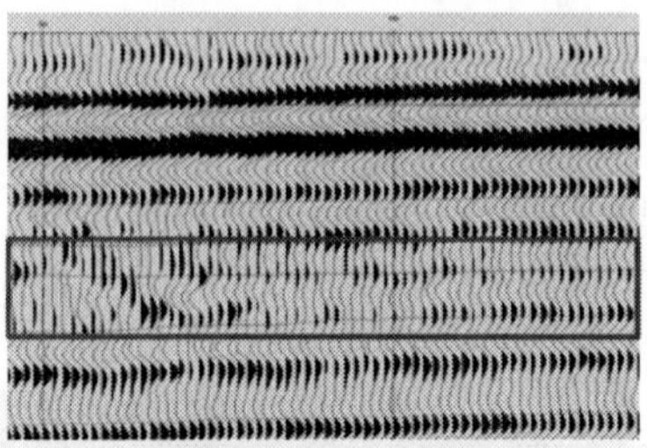

**油藏监测 ——四维地震**
**Reservoir monitoring —4D Seismic**

**地表监测——植被监测**
**surface monitoring —Vegetation monitoring**

**大气监测——采出气组分分析**
**Atmospheric monitoring —Gas sample testing**


## （二）形成特低渗油藏$CO_2$驱油与埋存技术

## Developed $CO_2$-EOR and storage technology in ultra-low reservoir


### 5、开展了$CO_2$再捕集与循环利用技术研究

Carried out research on $CO_2$ recapture and recycling technologies .

根据气油比高低，开展了**低能耗$CO_2$胺吸收法**和**油气分离压缩直接回注**室内研究和技术开发，实现了再捕集和循环利用。

The low energy consumption amine absorption and gas-oil separation and recycling method can be used respectively for high and low gas-oil ratio reservoirs, which can realize the recapture and recycling.

**低能耗$CO_2$胺吸收法**
**The equipment of amine absorption**

## （二）形成特低渗油藏$CO_2$驱油与埋存技术

## Developed $CO_2$-EOR and storage technology in ultra-low reservoir

**现场应用效果:**

Field application performance:

- 建成两个试验区，目前10口注气井，累计注入7万吨液态$CO_2$，单井产量提高50%，预计可提高原油采收率8%，实现动态封存$CO_2$ 7万吨。
- Two pilot test zones with 10 gas-injecting wells have been built. About 70K tons of $CO_2$ have been injected and sealed, the current performance of EOR meet the expectation.

驱油现场

$CO_2$ flooding site

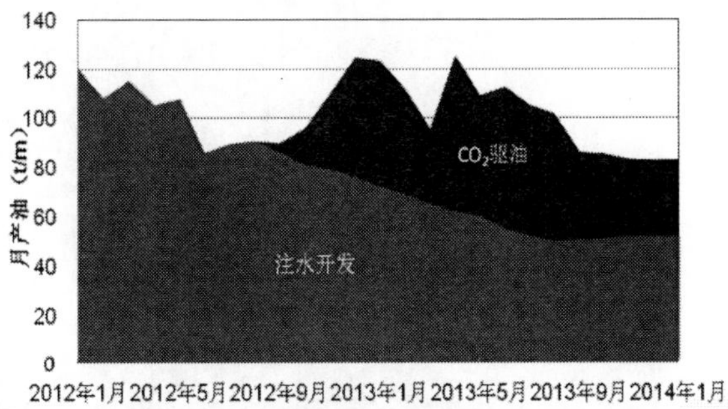


一线受效井产量变化曲线

Production change of Affected well

## （三）形成非常规油气藏$CO_2$压裂技术

## Developed unconventional hydrocarbon reservoirs $CO_2$ fracturing technology

针对非常规油气储层特征，开发了$CO_2$增能压裂技术、VES-$CO_2$泡沫压裂技术、纯液态$CO_2$压裂技术

$CO_2$ energized, VES-$CO_2$ foam and pure liquid $CO_2$ fracturing technologies have been developed for each type of unconventional oil & gas reservoirs.

1）提高了单井产量。实施致密气井$CO_2$压裂12井次，平均单井产量提高30%以上；实施致密油井$CO_2$压裂12井次，产量提高2.8倍，返排率提高70%。

Tight gas average well output has been increased by over 30%. Tight oil average well has been 2.8 timed, and flowback rate has been increased by 70% after 12 times $CO_2$ fracturing.

2）提高了返排效率和速度。在页岩气井实施$CO_2$压裂47井次，返排率提高35%，排液周期缩短1/3。

Shale gas flowback rate has been increased by 35%, the fluid flowback period has been shortened by 1/3 after 47 times $CO_2$ fracturing in shale gas wells.

压裂现场

Fracturing Site

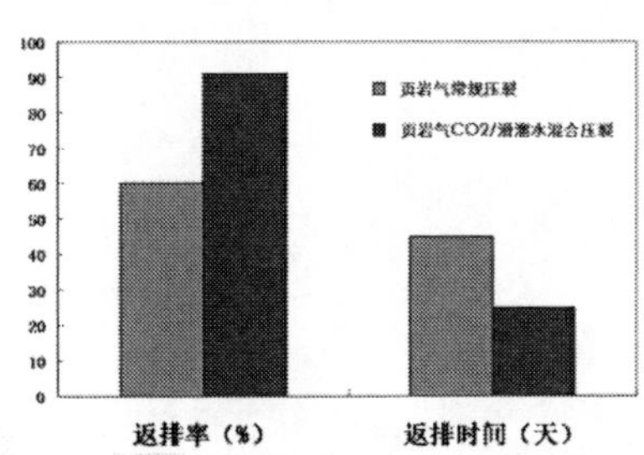


页岩气井$CO_2$压裂返排率对比

Comparison of the fluid flowback rate.

## （三）形成非常规油气藏$CO_2$压裂技术

## Developed unconventional hydrocarbon reservoirs $CO_2$ fracturing technology

2015年10月19日，自主设计的**无增稠干法二氧化碳压裂技术在云页4井**成功实施，对于陕北缺水地区实现页岩气开发与环境保护具有重要推广意义。

On 19th October, 2015, the first Chinese pure $CO_2$ fracturing was carried out in YY4 well, which was designed and operated by Yanchang Petroleum independently. It means significantly to environmental protection and shale gas development in Shanbei water shortage regions.

## （四）取得的科技成果

## S&T Achievements

在国内外期刊发表学术论文**41**篇（国外期刊**8**篇），获国家专利**12**项，相关成果荣获科技奖励**8**项。

Until so far, Yanchang has got many S&T achievements in the field of CCUS, including 41 academic papers, 12 patents, and 8 S&T awards.

**学术论文**
**Academic papers**

**发明专利**
**Patents**

**科技成果奖励**
**S&T Awards**

## （五）获得广泛国际认可
## International Recognition

- **中-澳碳捕集、利用与封存（CCUS）一体化国际合作示范项目：**获得技术与资金支持；
- Sino-Australia International Cooperation on CCUS: Acquired technologies and funds support;
- **中-美气候变化工作组合作项目：** 与美国西弗吉尼亚大学、怀俄明大学等共同开展合作研究。
- US-China Climate Change Working Group Cooperation Project: signed the memorandum with West Virginia University and University of Wyoming, aiming at the study of Yanchang CCUS project.

**与澳大利亚全球碳捕集与封存研究院签署合作协议**
**Cooperation with GCCSI**

**与西弗吉尼亚大学、怀俄明大学签署合作备忘录**
**Cooperation with WVU and UW**

## （五）获得广泛国际认可
## International Recognition

■ 十分高兴看到延长CCUS项目取得如此大的进展，感谢您的报告。我们赞成您的远景目标，这将有助于将延长项目建设成**世界一流的CCUS 项目**。

-----胡里奥.弗里德曼博士
前美国能源部助理部长

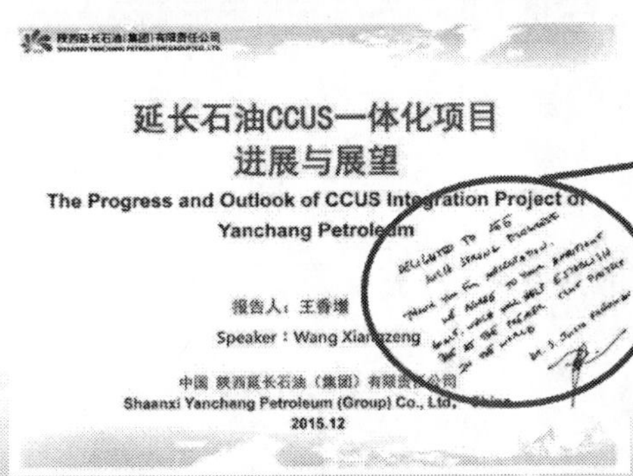


Delighted to see such strong progress, thank you for presentation. We agree to your ambitious goals, which will help establish this as **the premier CCUS project** in the world.

-----Dr. S. Julio. Friedmann

## （五）获得广泛国际认可
## International Recognition

- 2015年6月，靖边CCS项目获得全球碳封存领导人论坛（CSLF）认证，成为**发展中国家第一个**得到国际认证的CCS项目。
- In June 2015, Jingbian CCS Project was recognized as the first internationally certified CCS project from a developing country by Carbon Sequestration Leadership Forum.

**认证会现场（加拿大里贾纳）**
**CSLF meeting at Regina**

Carbon Sequestration Leadership Forum
www.cslforum.org

CERTIFICATE OF RECOGNITION

The Carbon Sequestration Leadership Forum is pleased to welcome the **Jingbian CCS Project** into its portfolio of recognized projects.

Richard Lynch
Richard Lynch
CSLF Secretariat

**靖边CCS项目认证书**
**Certificate of Recognition from CSLF**


## （五）获得广泛国际认可
## International Recognition


- 2015年9月，习近平主席访美期间，在《中美元首气候变化联合声明》中，提到**“关于2014年中美气候变化联合声明中所提的碳捕集、利用和封存项目，两国已选定由陕西延长石油公司运行的位于中国陕西省延安－榆林地区的项目场址。”**
- In September 2015, “U.S.-China Joint Presidential Statement on Climate Change” mentioned that “On the CCUS project agreed to in the 2014 Joint Announcement, the two countries have identified the project site in Yan’an-Yulin, Shan’xi Province, China, operated by Shaan’xi Yanchang Petroleum”

U.S.-China Joint Presidential Statement on Climate Change

13. Building on the robust bilateral cooperation initiatives that support the achievement of ambitious domestic actions, the two sides commit to further deepen and enhance these efforts through the U.S.-China Climate Change Working Group (CCWG), the premier mechanism for facilitating constructive U.S.-China dialogue and cooperation on climate change. The two sides have made concrete progress in each of the initiatives, including heavy-duty and other vehicles, smart grids, carbon capture, utilization, and storage (CCUS), energy efficiency in buildings and industry, collecting and managing greenhouse gas emissions data, climate change and forests, industrial boilers efficiency and fuel switching, and climate-smart/low-carbon cities, and will continue to work together on green ports and vessels and zero emission vehicles, as well as the enhanced policy dialogue and cooperation on HFCs. Furthermore, a new Domestic Policy Dialogue was established this year to share information on respective domestic actions. The two sides will continue to devote significant effort and resources to the existing initiatives. On the CCUS project agreed to in the 2014 Joint Announcement, the two countries have identified the project site in Yan'an-Yulin, Shan'xi Province, China, operated by Shan'xi Yanchang Petroleum. The two sides will continue to collaborate to demonstrate the utilization of CO2 for enhanced water recovery.

# 内容提要
# Table of Contents



## 挑战一：全球首个由企业独立承担的CCUS全流程一体化项目，无成熟经验可借鉴。

Challenge 1: This project is the very first integrated CCUS project operated by just one company alone, so there is no reference to learn from.

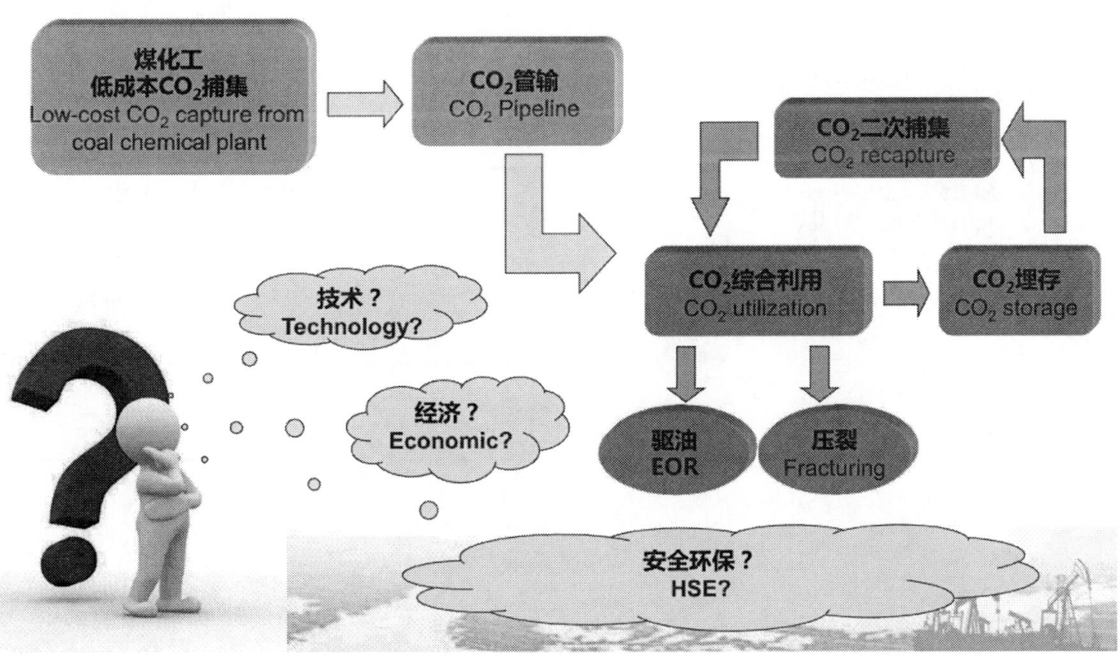

## 挑战二：项目区地形复杂，$CO_2$输送管网建设难度大。

## Challenge 2: Terrain brings more difficulties to $CO_2$ pipeline construction

$CO_2$输送管网在温压控制、计量方式、增压站建设等方面与北美CCUS项目差异较大，需克服复杂地形带来的影响。

Yanchang $CO_2$ pipeline have to conquer geographic difficulties that result in temperature and pressure control, measurement and construction difficulties compared with North America CCUS projects.

延长石油CCUS项目区地形
——沟壑纵横
**Terrain of Yanchang project region**

北美CCUS项目区地形
——平原为主
**Terrain of America CCUS Region**

## 挑战三：油田地层压力低于最小混相压力，难以实现$CO_2$混相驱油。

## Challenge 3: Formation pressure of oilfield lower than minimum miscibility pressure makes $CO_2$ flooding miscibility harder

地层压力与埋深交汇图
**Formation pressure and depth graph**

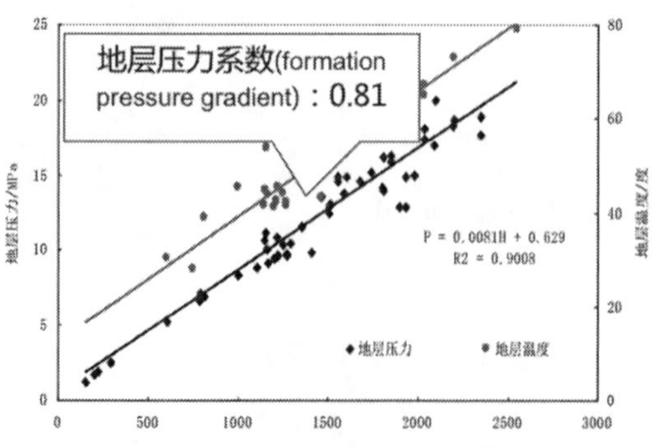


原始地层压力与最小混相压力对比表
**Comparision of IFP and MMP**

| 项目区 Region | 原始地层压力 IFP (MPa) | 最小混相压力 MMP (MPa) |
|---|---|---|
| 靖边项目区 Jinbian | 12.1 | 22.15 |
| 吴起项目区 Wuqi | 15 | 16.12 |

## 内容提要
## Table of Contents



## 项目具有良好的发展前景
## Project future is prospected to be bright

- **充足的碳源**：项目区二氧化碳年排放量超过1亿吨。
- **Abundant carbon resources:** annual $CO_2$ production is more than 100 million tons in the project area.
- **油气田封存潜力**：延长石油有12.5亿吨石油地质储量适合$CO_2$驱油与封存，可封存$CO_2$6.5亿吨
- **Storage capability of oil and gas fields:** Yanchang has 1.25 billion tons of OOIP for $CO_2$ flooding and storage. It can store 650million $CO_2$.
- **盐水层封存潜力**：奥陶系的盐水层可封存$CO_2$数十亿吨，其它地层封存量可达数百亿吨
- **Storage capability of saline water layers:** Ordovician saline water layers can store billions tons of $CO_2$. Other formations can store tens billions tons of $CO_2$.

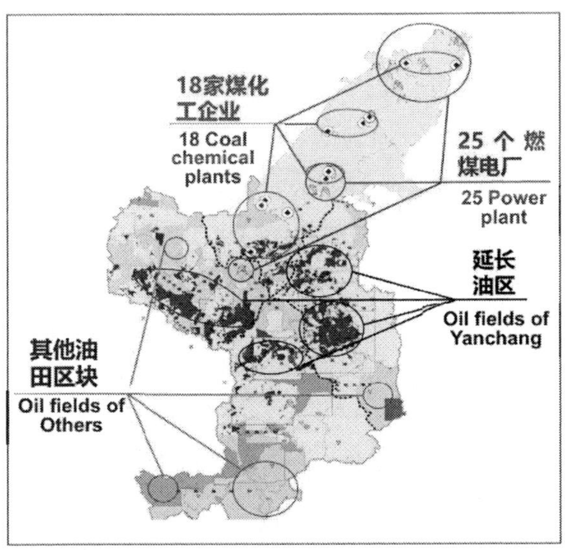


陕北油区、煤化工、燃煤电厂分布图
The map of Oil field, Coal chemical and Power plant

延长石油
YANCHANG PETROLEUM

## 制定了中、长期发展规划
## Development strategy has been done

**近期目标**：2017年建成36万吨/年CCUS示范区；
**Short-term goal:** By the end of 2017, finish construction of integrated CCUS demonstration zone with 360K ton/year capacity
**中期目标**：2020年建成100万吨/年CCUS示范工程；
**Medium-term target:** By the end of 2020, finish construction of CCUS demonstration project with 1 million tons/year capacity;
**远期目标**： 2030年建成400万吨/年能力CCUS示范基地。
**Long-term goals:** By 2030, finish construction of CCUS demonstration base with 4 million tons/year capacity.

### 最终目标：Ultimate goal:

形成完整的煤化工$CO_2$捕集、驱油、封存技术体系与标准，探索CCUS商业运作模式，为中国碳减排和$CO_2$资源化利用起到示范引领作用。
Complete formatting CCUS technology system and standard, explore CCUS commercial operating mode, and provide relevant experiences for other domestic regions.

延长石油
YANCHANG PETROLEUM

## 结束语
## Finally

二氧化碳捕集、利用与封存（CCUS）是能源企业积极应对气候变化，实现碳减排和循环发展的有效途径。将煤化工与碳捕集相结合，将驱油与碳封存相结合，可以降低成本，保障CCUS项目可持续发展。延长石油愿与国际同行进一步加强交流与合作，积极倡导低碳理念，为全球化石能源的高效、清洁、低碳利用做出积极的贡献。

For energy enterprises，CCUS is an positive response towards climate change, and an effective way to achieve low $CO_2$ emission and recycle. Combined coal chemical industry and carbon capture together, displacement and storage together, cost will be remarkably down, and CCUS projects will be sustainably developed. Yanchang Group is open to knowledge exchanges and work collaborations with worldwide researchers on advocating low carbon style, making contribution to utilizing fossil energy in a more efficient, cleaner way with lower $CO_2$ emission.

谢谢关注

Thanks for Your Attention

Clarke
Energy®

Engineer - Install - Maintain

Distributor & Service Provider
Reciprocating engines

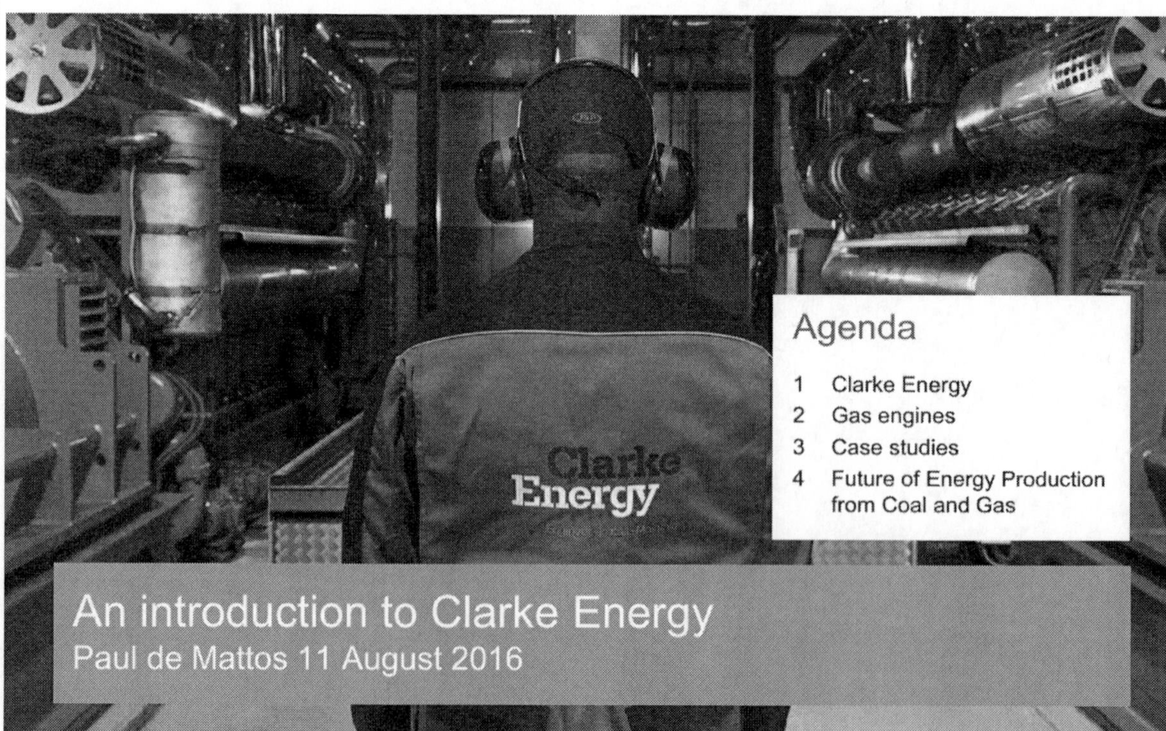

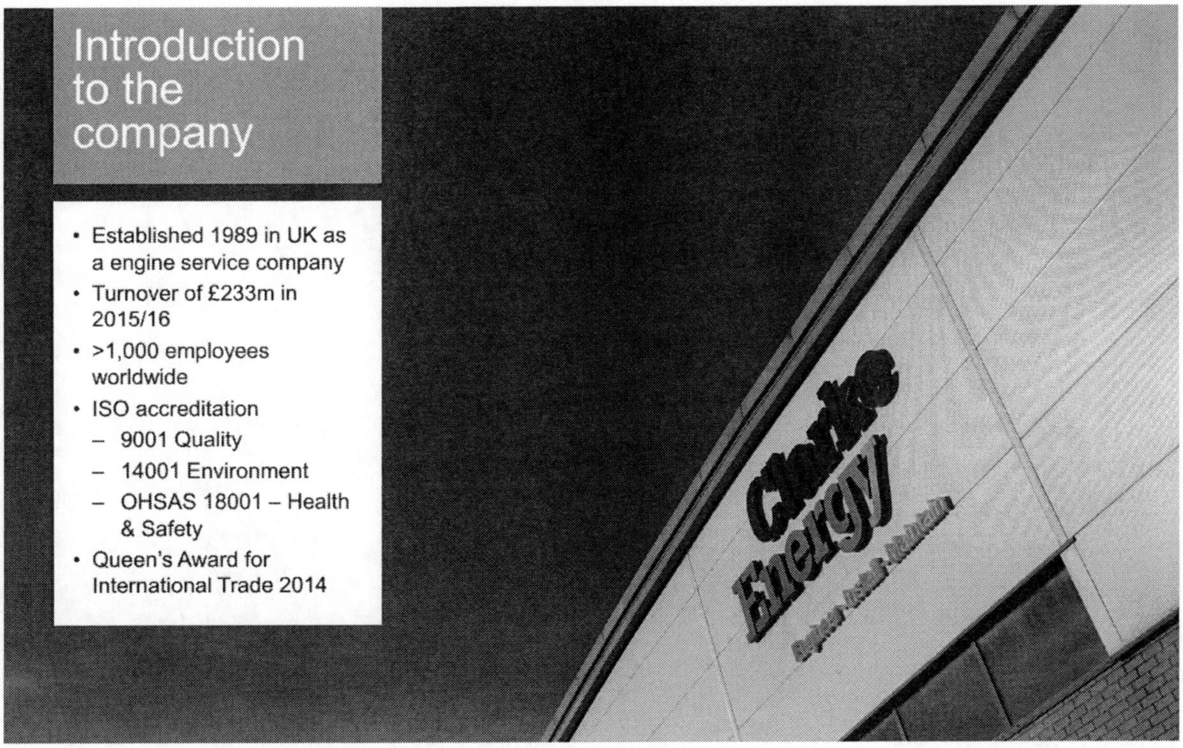
Introduction to the company
• Established 1989 in UK as a engine service company
• Turnover of £233m in 2015/16
• >1,000 employees worldwide
• ISO accreditation
– 9001 Quality
– 14001 Environment
– OHSAS 18001 – Health & Safety
• Queen's Award for International Trade 2014
Clarke Energy

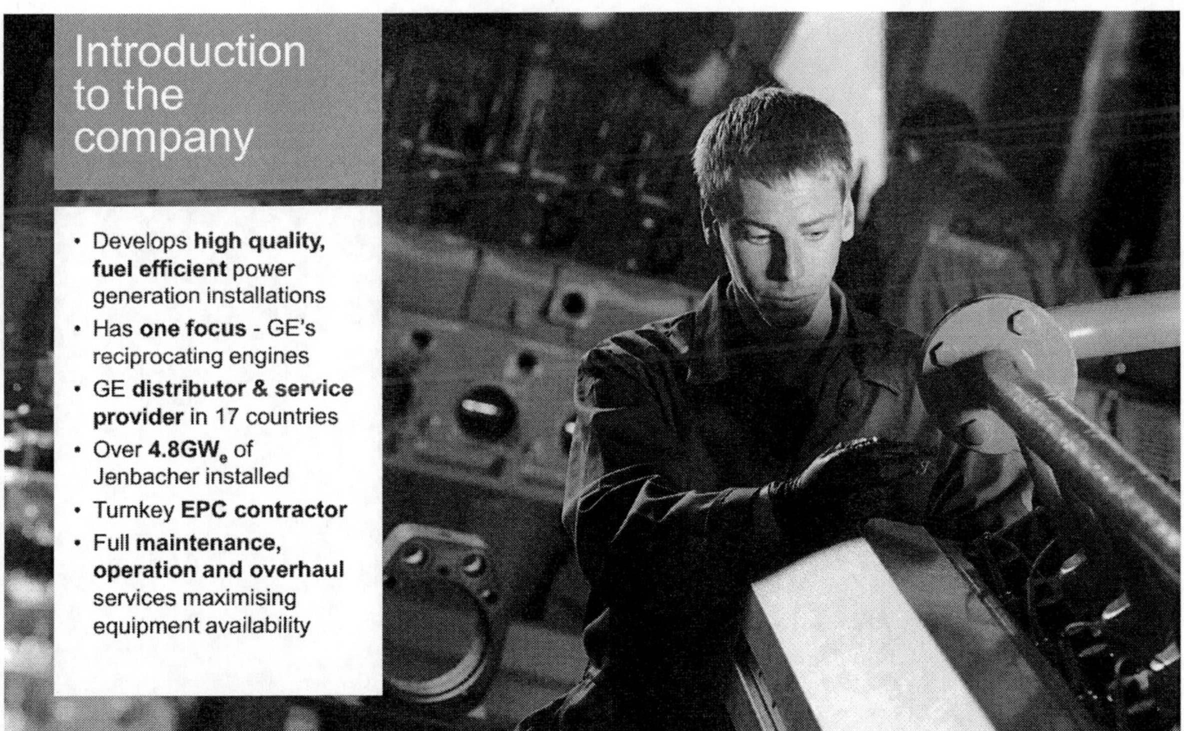
Introduction to the company
• Develops high quality, fuel efficient power generation installations
• Has one focus - GE's reciprocating engines
• GE distributor & service provider in 17 countries
• Over $4.8GW_e$ of Jenbacher installed
• Turnkey EPC contractor
• Full maintenance, operation and overhaul services maximising equipment availability

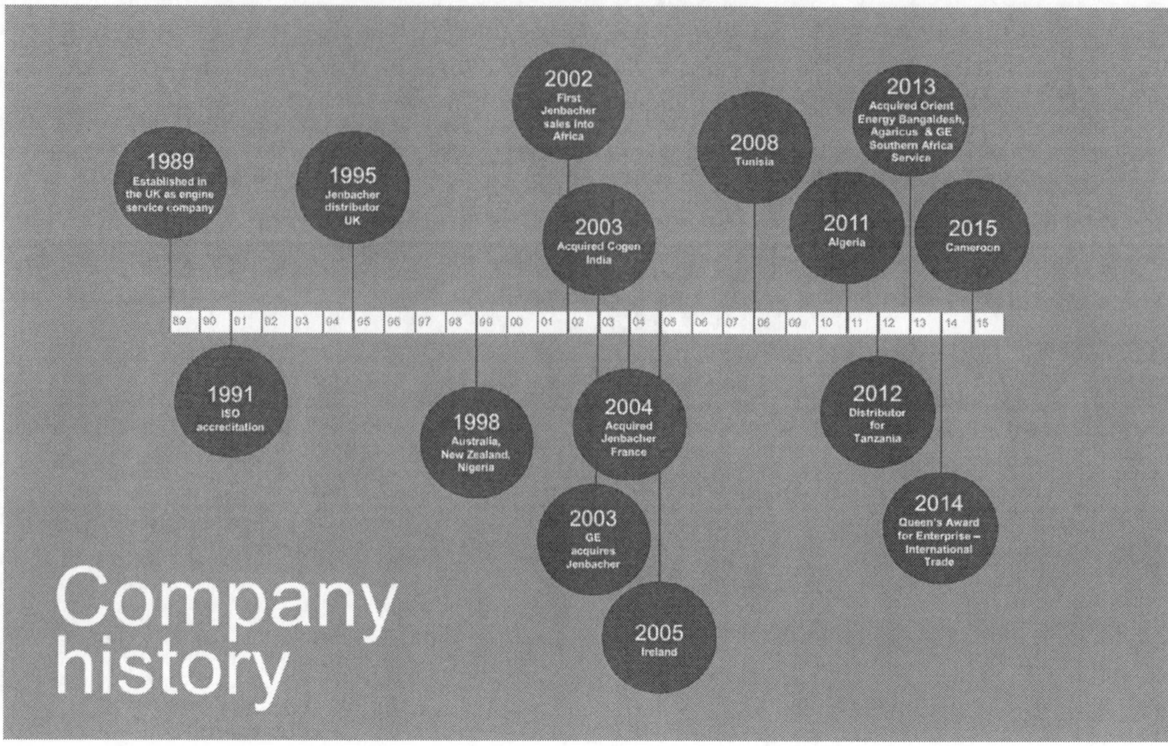
1989
Established in the UK as engine service company
1991
ISO accreditation
1995
Jenbacher distributor UK
1998
Australia, New Zealand, Nigeria
2002
First Jenbacher sales into Africa
2003
Acquired Cogen India
2003
GE acquires Jenbacher
2004
Acquired Jenbacher France
2005
Ireland
2008
Tunisia
2011
Algeria
2012
Distributor for Tanzania
2013
Acquired Orient Energy Bangaldesh, Agaricus & GE Southern Africa Service
2014
Queen's Award for Enterprise – International Trade
2015
Cameroon
89 90 91 92 93 94 95 96 97 98 99 00 01 02 03 04 05 06 07 08 09 10 11 12 13 14 15
Company history

Territories & Offices

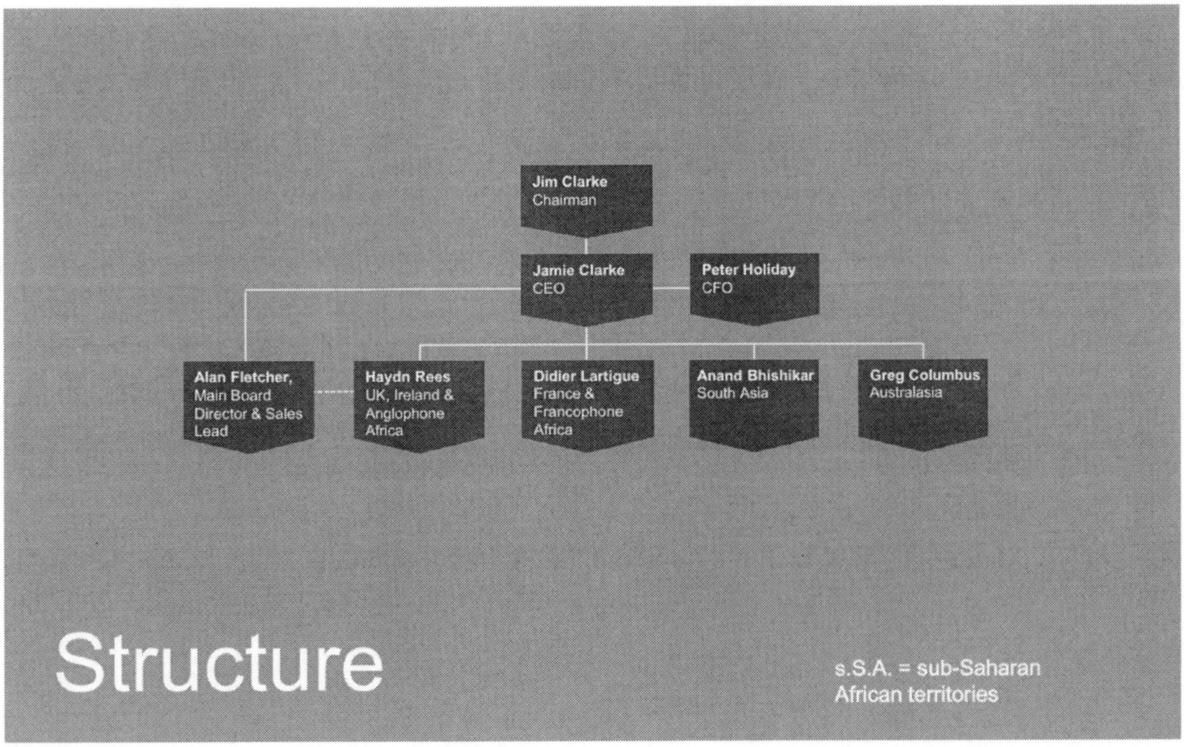
Jim Clarke
Chairman
Jamie Clarke
CEO
Peter Holiday
CFO
Alan Fletcher,
Main Board
Director & Sales
Lead
Haydn Rees
UK, Ireland &
Anglophone
Africa
Didier Lartigue
France &
Francophone
Africa
Anand Bhishikar
South Asia
Greg Columbus
Australasia
Structure
s.S.A. = sub-Saharan
African territories

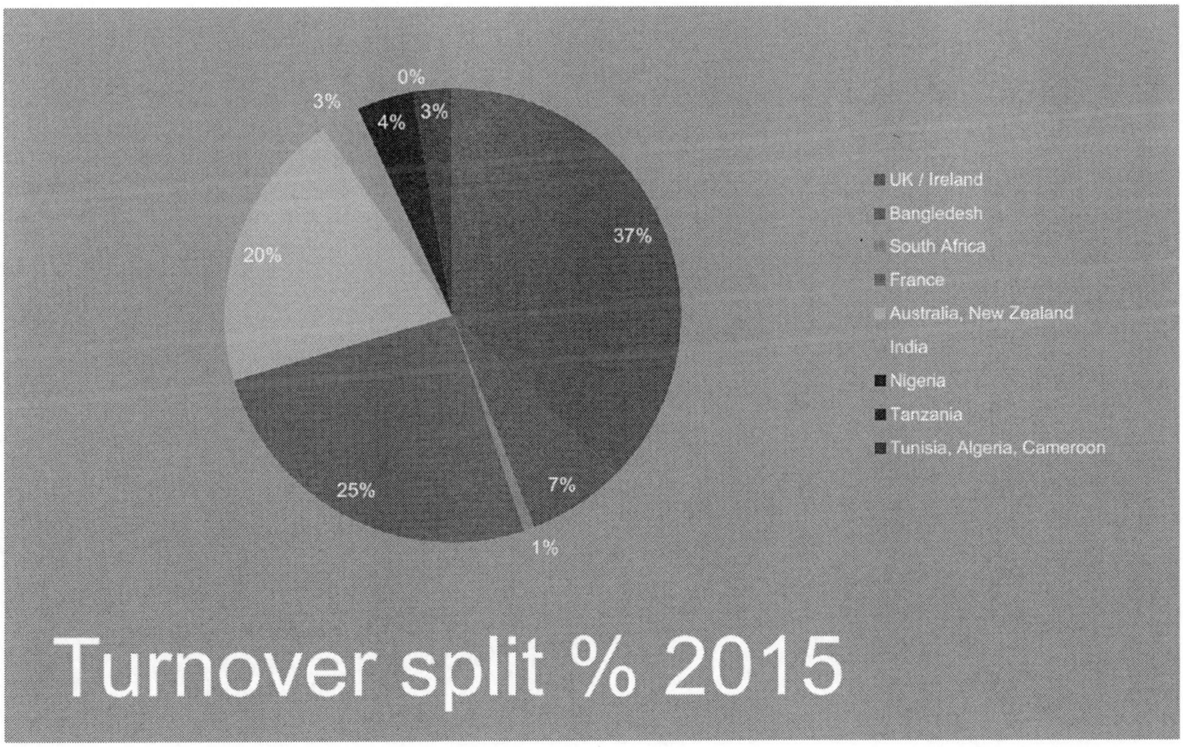
0%
3%
4%
3%
37%
20%
25%
7%
1%
UK / Ireland
Bangledesh
South Africa
France
Australia, New Zealand
India
Nigeria
Tanzania
Tunisia, Algeria, Cameroon
Turnover split % 2015

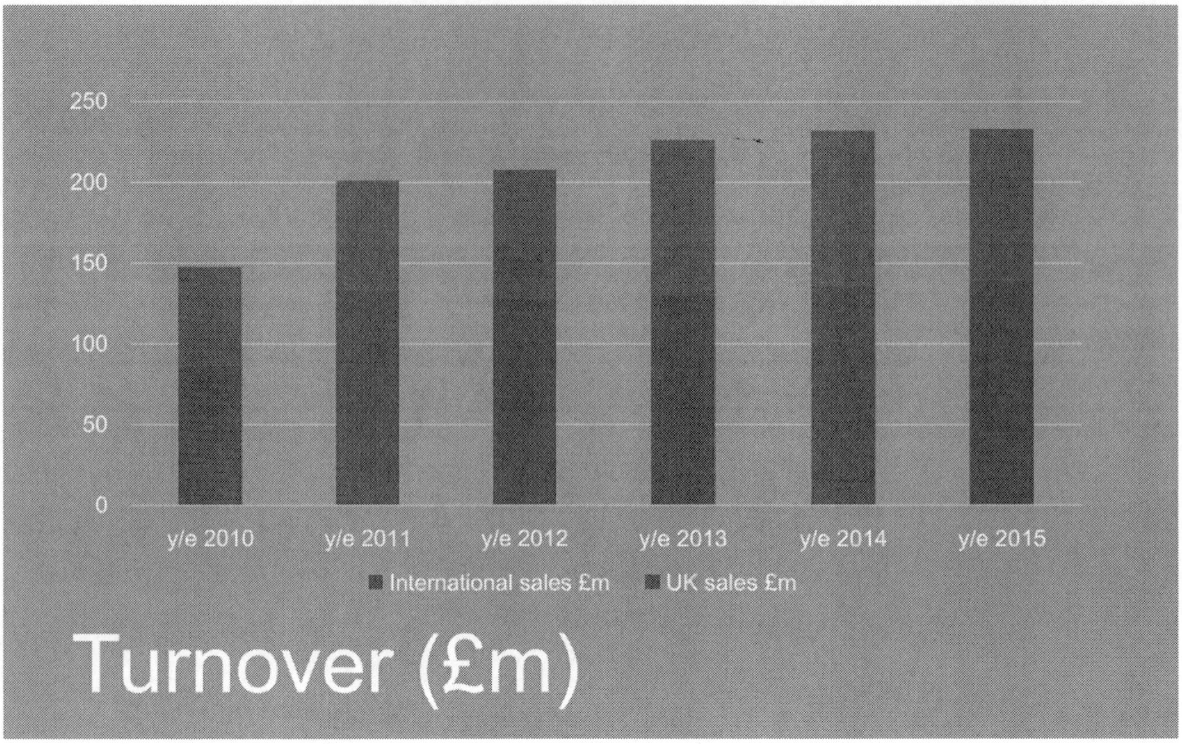
250
200
150
100
50
0
y/e 2010
y/e 2011
y/e 2012
y/e 2013
y/e 2014
y/e 2015
International sales £m
UK sales £m
Turnover (£m)

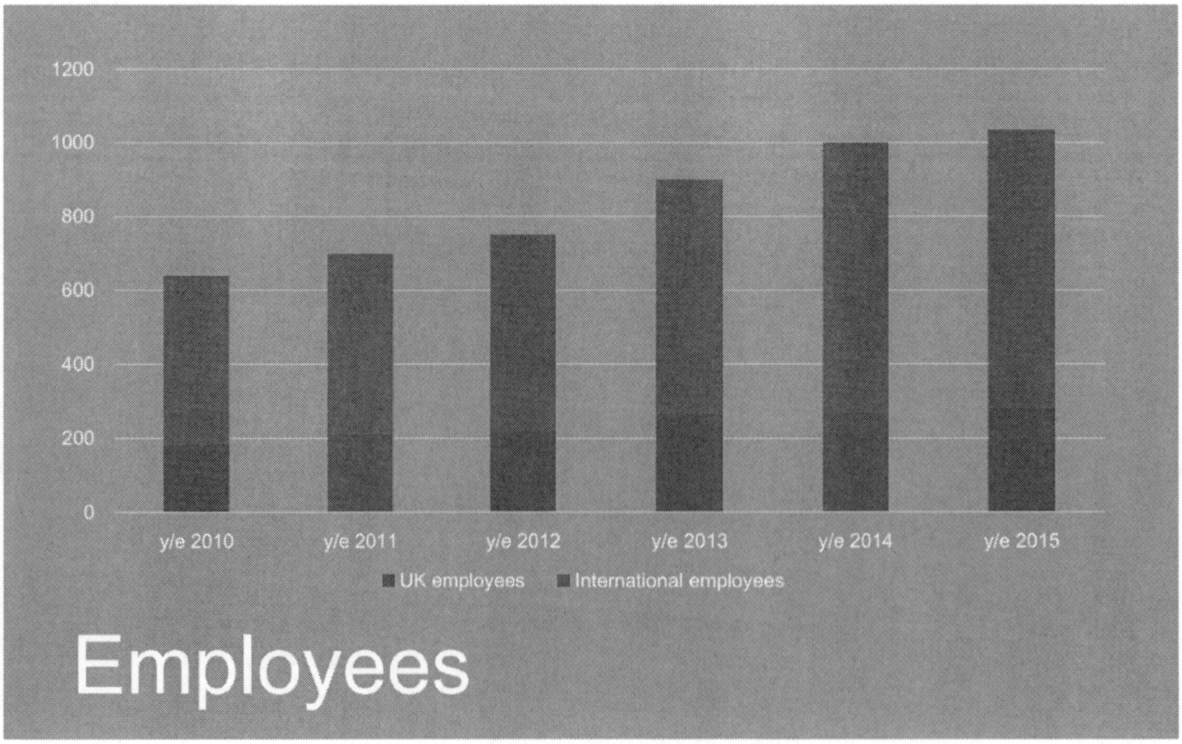
1200
1000
800
600
400
200
0
y/e 2010
y/e 2011
y/e 2012
y/e 2013
y/e 2014
y/e 2015
UK employees
International employees
Employees

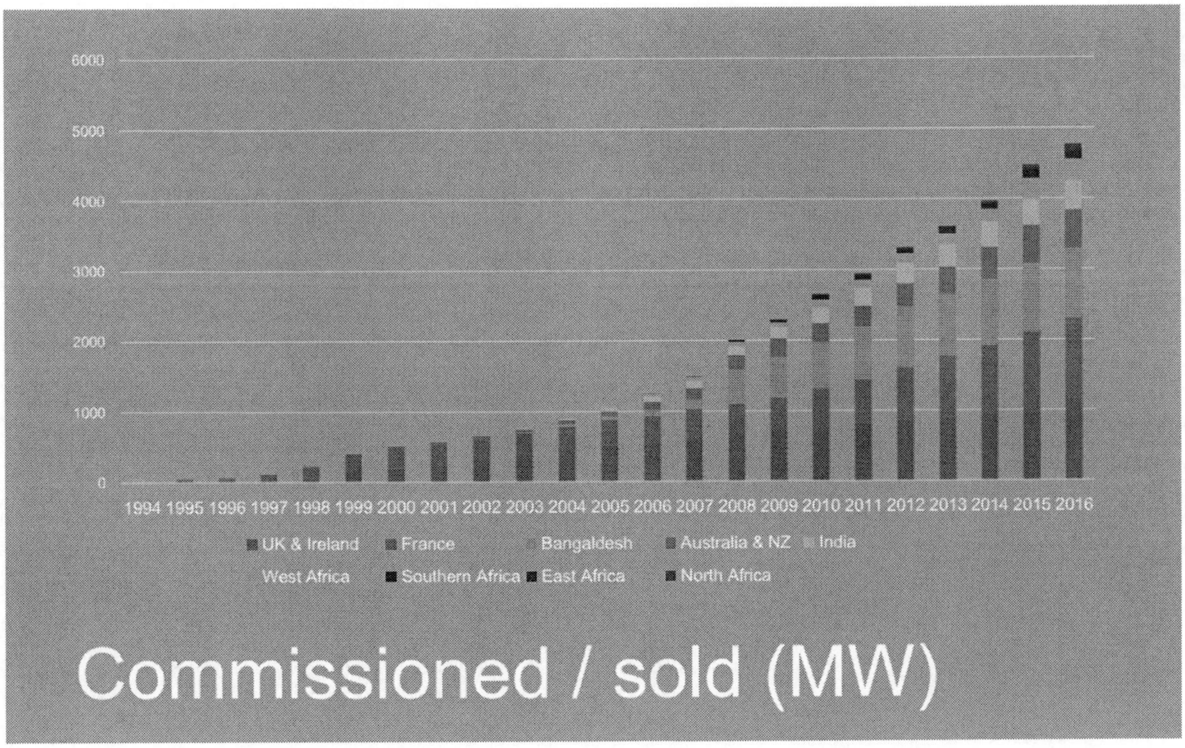
6000
5000
4000
3000
2000
1000
0
1994 1995 1996 1997 1998 1999 2000 2001 2002 2003 2004 2005 2006 2007 2008 2009 2010 2011 2012 2013 2014 2015 2016
UK & Ireland
France
Bangaldesh
Australia & NZ
India
West Africa
Southern Africa
East Africa
North Africa
Commissioned / sold (MW)

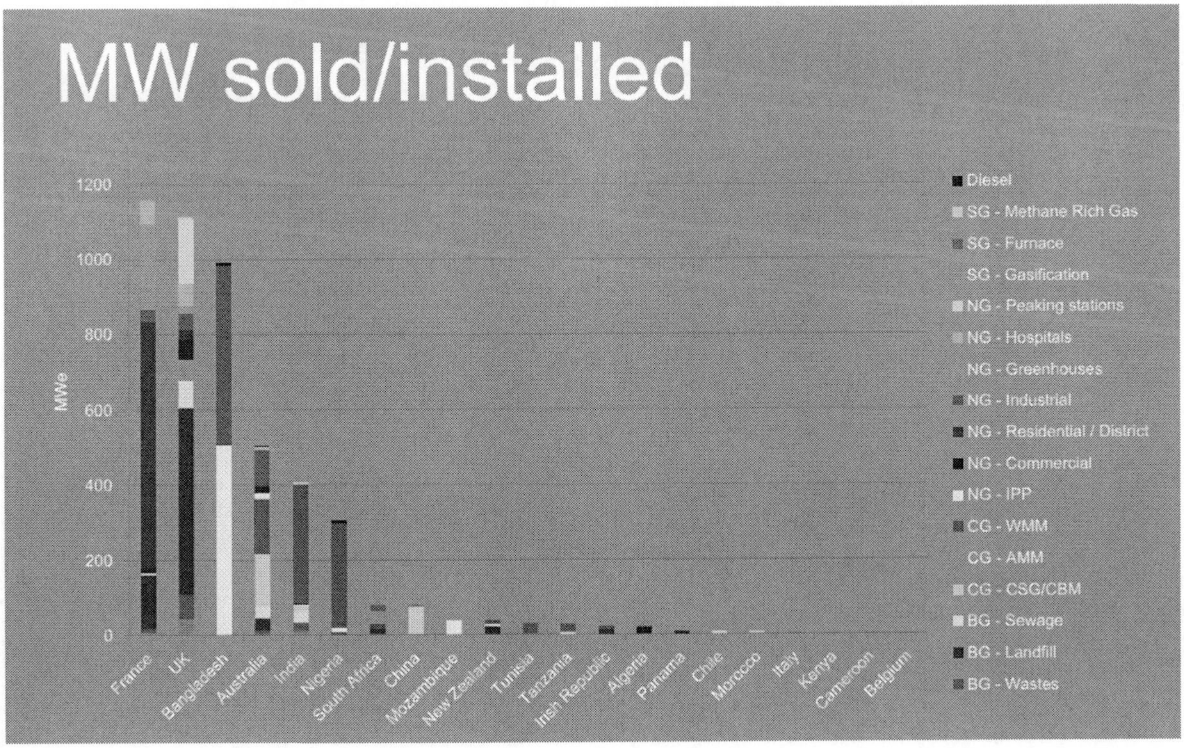
MW sold/installed
1200
1000
800
600
400
200
0
MWe
France
UK
Bangladesh
Australia
India
Nigeria
South Africa
China
Mozambique
New Zealand
Tunisia
Tanzania
Irish Republic
Algeria
Panama
Chile
Morocco
Italy
Kenya
Cameroon
Belgium
Diesel
SG - Methane Rich Gas
SG - Furnace
SG - Gasification
NG - Peaking stations
NG - Hospitals
NG - Greenhouses
NG - Industrial
NG - Residential / District
NG - Commercial
NG - IPP
CG - WMM
CG - AMM
CG - CSG/CBM
BG - Sewage
BG - Landfill
BG - Wastes

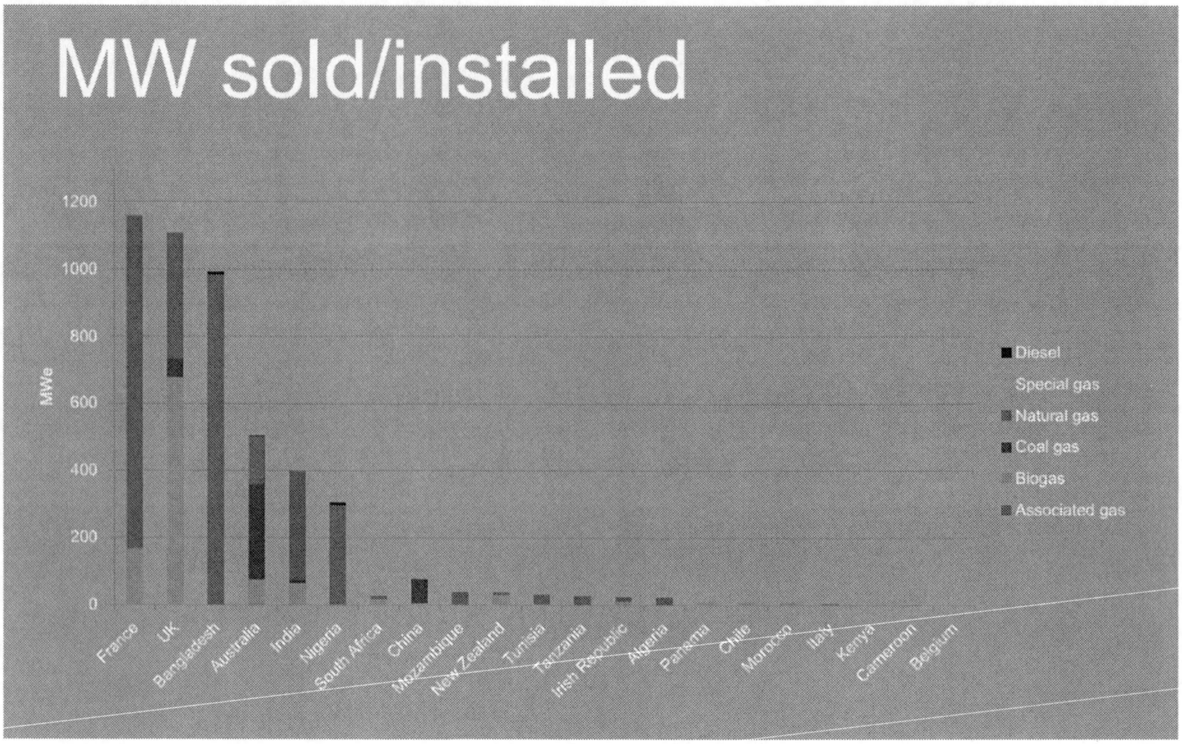
MW sold/installed
MWe
1200
1000
800
600
400
200
0
France
UK
Bangladesh
Australia
India
Nigeria
South Africa
China
Mozambique
New Zealand
Tunisia
Tanzania
Irish Republic
Algeria
Panama
Chile
Morocco
Italy
Kenya
Cameroon
Belgium
Diesel
Special gas
Natural gas
Coal gas
Biogas
Associated gas

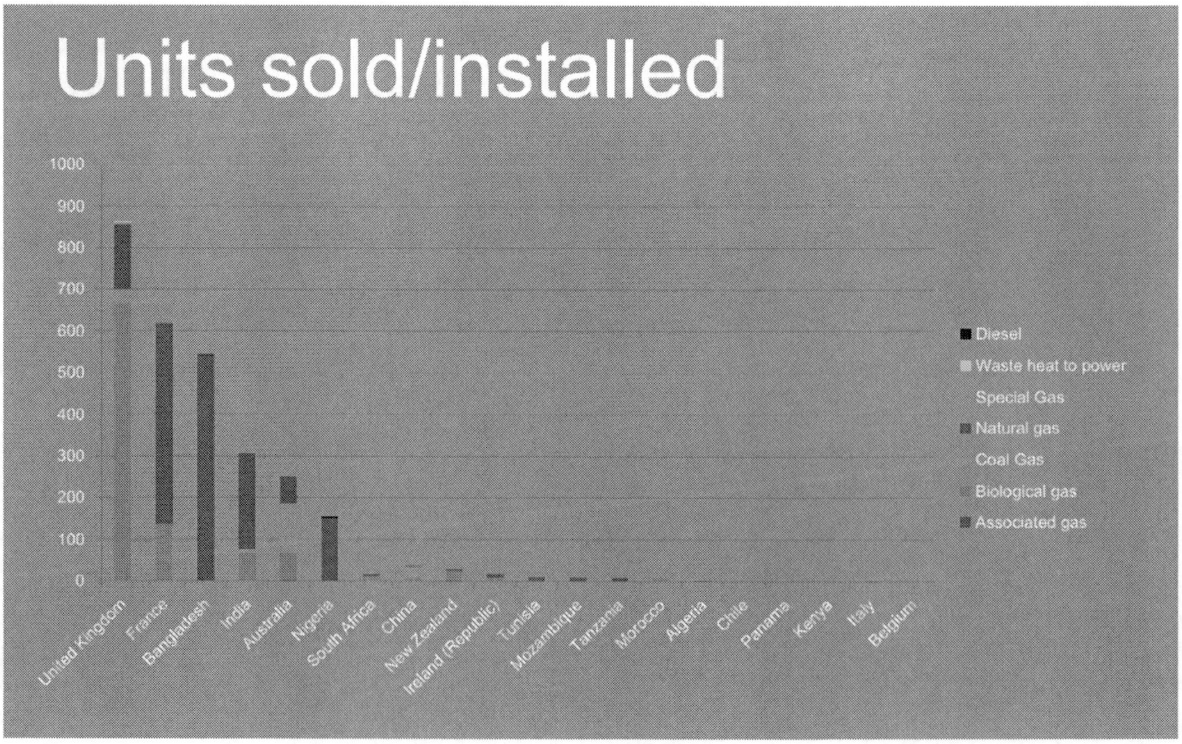
Units sold/installed
1000
900
800
700
600
500
400
300
200
100
0
United Kingdom
France
Bangladesh
India
Australia
Nigeria
South Africa
China
New Zealand
Ireland (Republic)
Tunisia
Mozambique
Tanzania
Morocco
Algeria
Chile
Panama
Kenya
Italy
Belgium
Diesel
Waste heat to power
Special Gas
Natural gas
Coal Gas
Biological gas
Associated gas

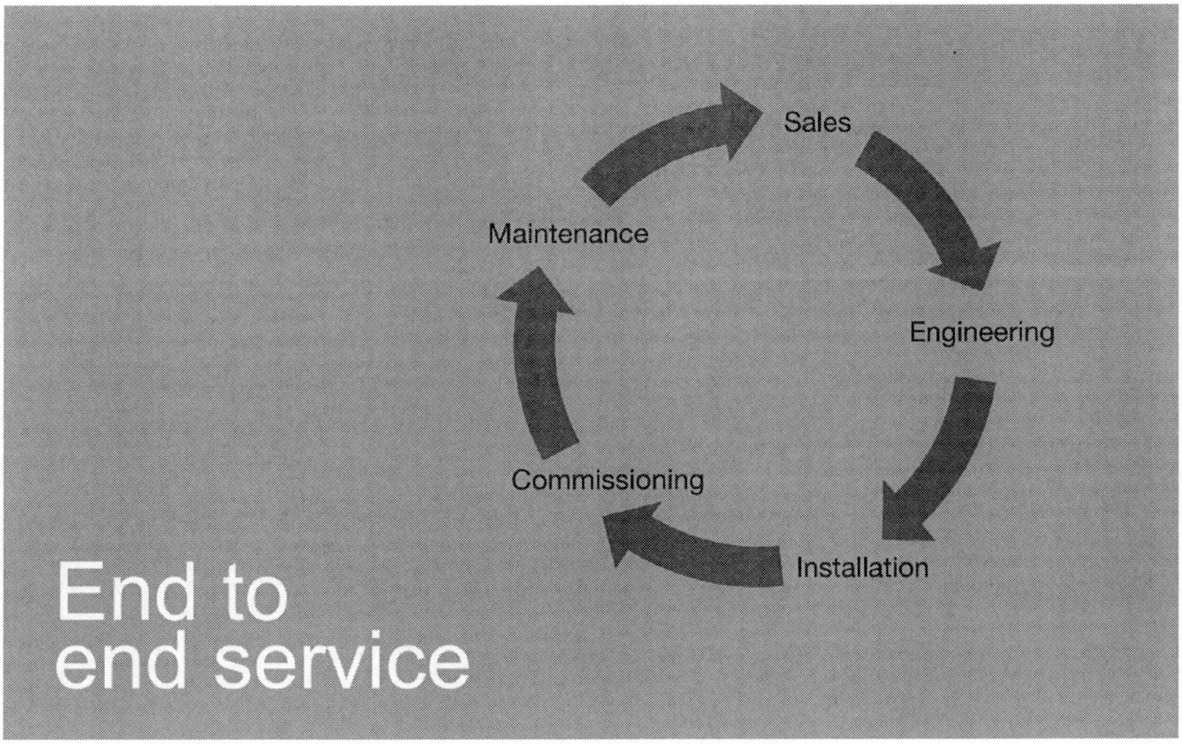
Sales
Engineering
Installation
Commissioning
Maintenance
End to
end service

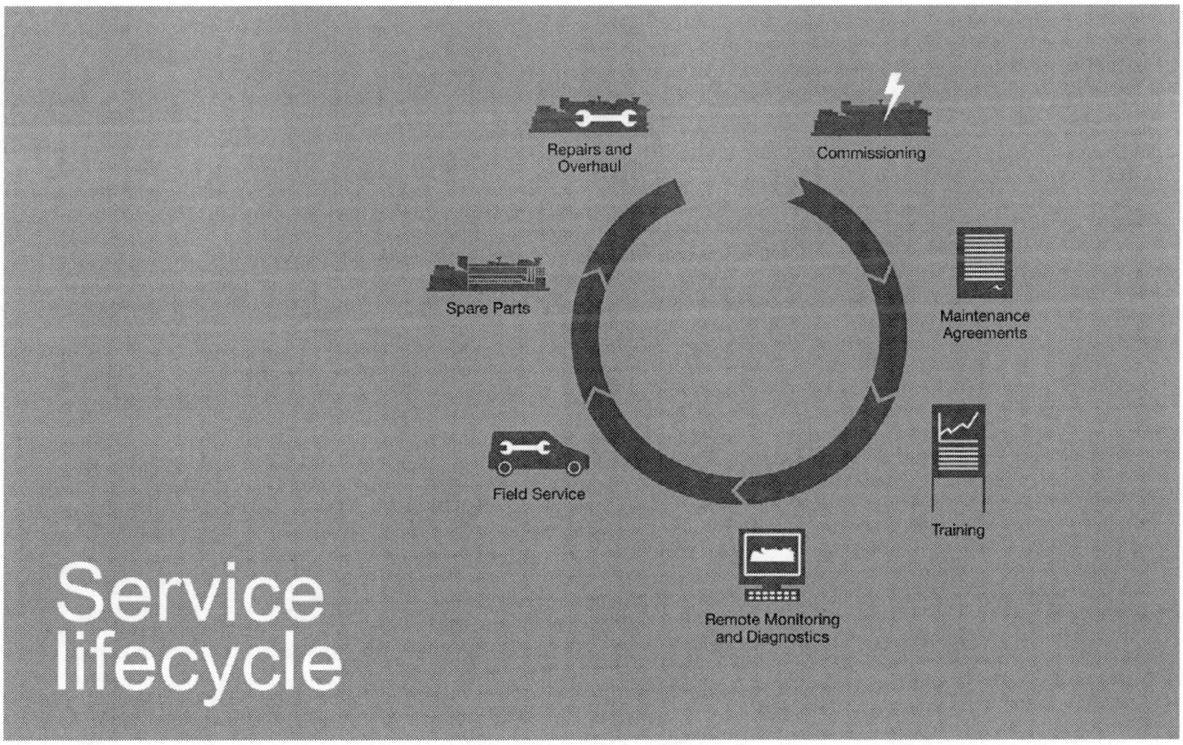
Repairs and
Overhaul
Commissioning
Maintenance
Agreements
Training
Remote Monitoring
and Diagnostics
Field Service
Spare Parts
Service
lifecycle

Maximising equipment availability
• In-country commissioning support
• Maintenance agreements supported by field-service engineers
• Site-based operation and maintenance
• GE certified trainers
• Remote monitoring and diagnostics capability
• In-country GE approved spare parts inventory

Clarke Energy
Local, Equipped Field engineers

Workshop
facilities

Clarke
Energy
Product quality
focused
Engine design
Engine development
Engine manufacture
Parts production
Customer project
focused
Installation design
Project management
Installation works
Commissioning
Maintenance
Parts stockholding
Partnership

# Jenbacher Gas Engines

**Designed solely for operation on various gases**
**Power range 0.3-9.5MWe**

# GE's Reciprocating Engines

0.1 – 5.0 MW | 5 – 25 MW

**Gas Products**
**Power Generation**

**Mechanical Drives**

**Diesel Products**

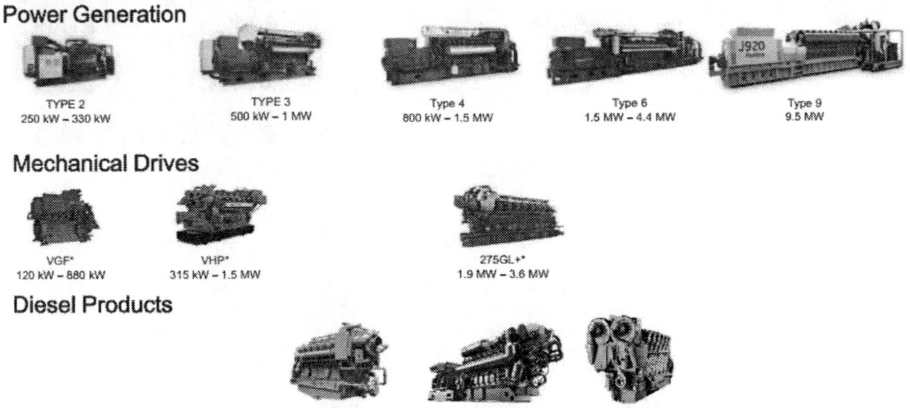

Power Generation: TYPE 2, 250 kW – 330 kW; TYPE 3, 500 kW – 1 MW; Type 4, 800 kW – 1.5 MW; Type 6, 1.5 MW – 4.4 MW; Type 9, 9.5 MW

Mechanical Drives: VGF*, 120 kW – 880 kW; VHP*, 315 kW – 1.5 MW; 275GL+*, 1.9 MW – 3.6 MW

Diesel Products: 228, 1.5 – 3.2 MW; 616, 2.6 MW; 250, 1.5 – 4.5MW

Distributor & Service Provider
Reciprocating engines

# Fuel Flexibility

# Flexible delivery model

1. Genset
- Gas engine
- Generator
- Base frame
- Fuel gas train
- Genset control panel
- Design and Documentation
- Options

2. Power module
- Item 1 plus
- Ventilation system
- Acoustic/weatherproof enclosure
- Silencer
- Radiators
- Fire and gas detection
- Electrical and I&C
- Further options
- Co/tri-generation

3. Power plant
- Item 2 plus
- Civil work
- Gas conditioning plant
- Lube oil system
- HV/LV switch room, control room, electrical sub-stations , metering and reticulation
- Black-start diesel generators
- Storage / workshops
- Further options

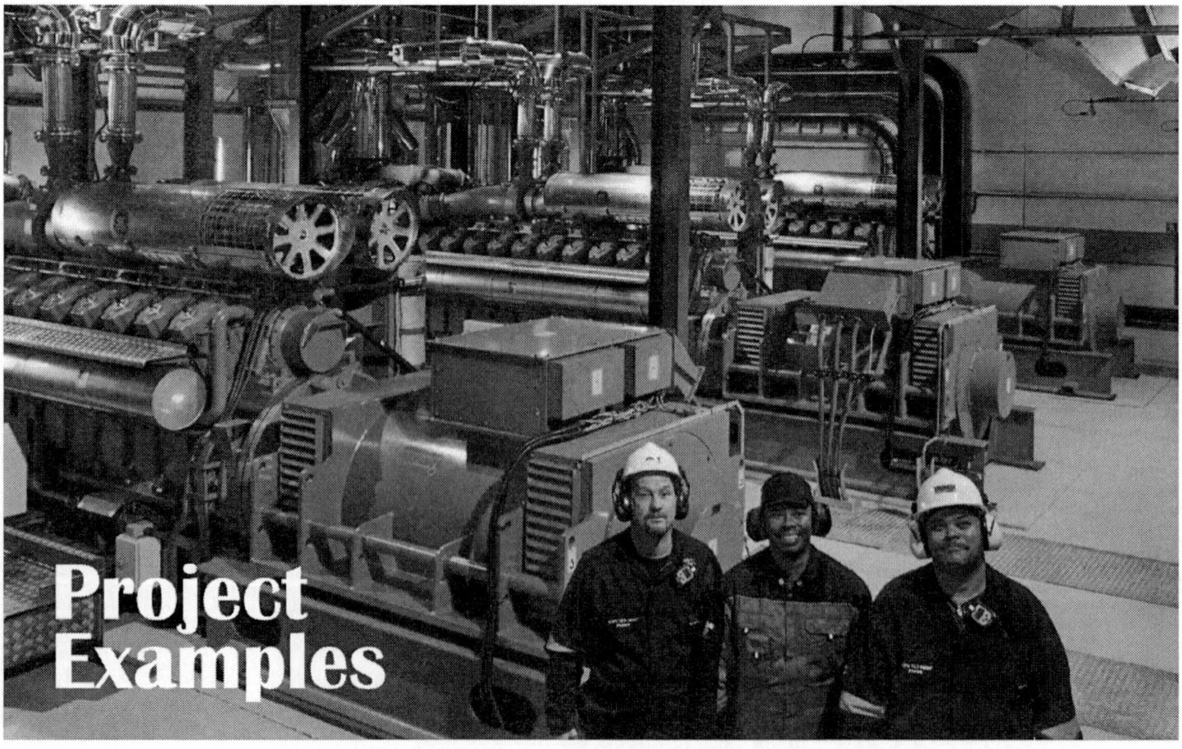

The Shard
London, England, Natural gas-fuelled CHP for a commercial building, Largest residential tower block in European Union, 1 x JMS416, 1.2$MW_e$

St Thomas' Hospital
London, England, Natural gas fuelled CHP (electricity, hot water and steam), 1 x JMS620, 3$MW_e$, Saving £1m fuel costs and 5,500t $CO_2$ per year

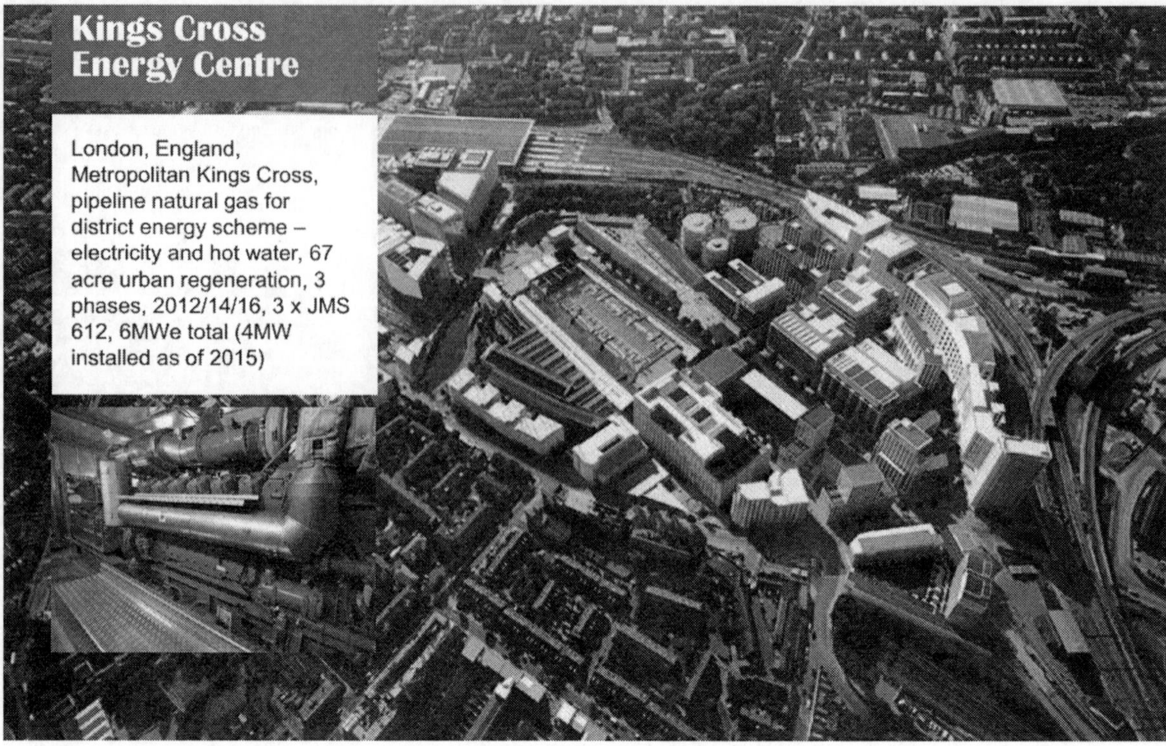
Kings Cross
Energy Centre
London, England, Metropolitan Kings Cross, pipeline natural gas for district energy scheme – electricity and hot water, 67 acre urban regeneration, 3 phases, 2012/14/16, 3 x JMS 612, 6MWe total (4MW installed as of 2015)

Reliance Industries Ltd.
Naroda, Gujarat, India
Pipeline natural gas
Combined heat and power plant providing steam and electricity to a textile facility
7 x JMS620, 21.3 $MW_e$

Flour Mills of Nigeria
Lagos, Nigeria, pipeline natural gas, captive power plant for flour mill, 11 x JMS620, 33MW$_e$

Tamar Energy
Biogas from organic wastes
5 sites in England, CHP
Farleigh, 2 x JGMC420, 3MW$_e$
Halstead, 2 x JMS416, 2.2MW$_e$
Hoddeston, 2 x JGMC420, 2.8MW$_e$
Holbeach Hurn 1 x JGMC420, 1.4MW$_e$
Retford, JGMC420, 1.5MW$_e$
10.9MW$_e$ total output

Elgin
Fruit
Grabouw, Western Cape,
South Africa
Biogas from fruit pulp
Western Cape, South Africa
1 x JMS312, 0.5MWe

Tronox
Namakwa
Sands
Western Cape, South Africa
Waste furnace gases
Combined heat and power
8 x JGS620, 13.6$MW_e$

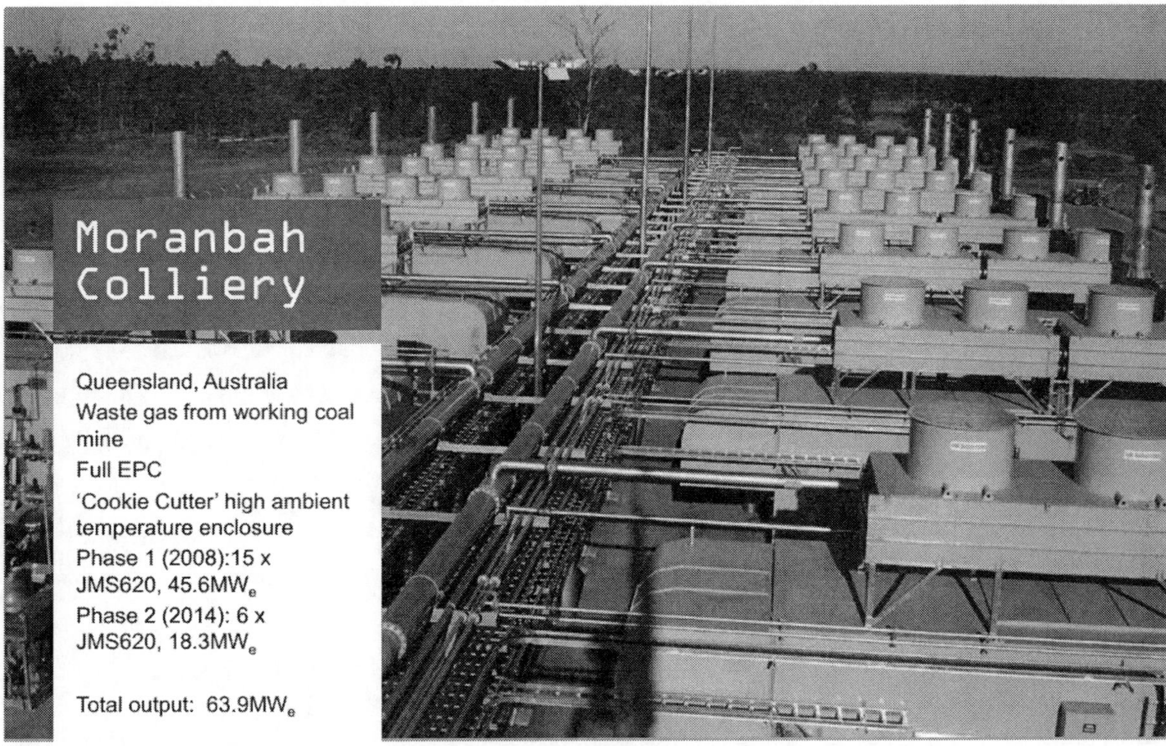
Moranbah Colliery
Queensland, Australia
Waste gas from working coal mine
Full EPC
'Cookie Cutter' high ambient temperature enclosure
Phase 1 (2008):15 x JMS620, 45.6MW$_e$
Phase 2 (2014): 6 x JMS620, 18.3MW$_e$
Total output: 63.9MW$_e$

APLNG* Projects
Queensland, Australia
Eurombah (2014)
Natural gas fuelled cogeneration (electricity and cooling)
Full EPC
9 x JMS620
27.4MW$_e$
Drop-over enclosures, load banks heat recovery for absorption chillers
*Australia Pacific Liquefied Natural Gas

## Future of Electricity Production from Coal and Gas

- Emphasis on Southern Africa
- Ever growing demand for power as economies grow.

- **Coal Conventional**
- Coal likely to around for many years to come with ~67 billion tonnes of reserves in SA alone.
- Also Botswana, Zimbabwe, Mozambique, Tanzania, Zambia and Malawi having substantial reserves.
- Likely to remain 1$^{o}$ energy source in southern Africa for many decades.
- Mining to conventional thermal power stations in SA,

## Future of Electricity Production from Coal and Gas

- Mozambique Moatize,
- Tanzania Mbeya and Mchuchuma,
- Botswana Morupule.
- Malawi Salima
- Zambia Maamba
- Both conventional and supercritical steam plant.
- SA has large discard dumps at relatively low cost that can be used in circulating fluidized bed boilers.
- Coal still the cheapest energy source in terms of R/GJ. Based on $30/te ZAR 450/te this equates to $1.15/GJ or ZAR 17/GJ. By comparison current commercial pipeline NG is ZAR105-120/GJ and CNG ZAR 180-240/GJ.
- The technology mature and robust.
- But High Carbon Foot print and environmental cost.

## Future of Electricity Production from Coal and Gas

**• Coal Unconventional**

- Mining and thermal power stations have a high environmental impact. This cost is generally not included in energy charges to consumers. Certainly not in SA.
- Hence "clean coal"/gasification technologies particularly UCG
- Many reserves are unsuitable for either surface or underground mining for different reasons.
- Location, depth, energy value, geology etc. However these may be amenable for recovery by UCG.
- Ash and some N & S remain underground. However ground water contamination must be prevented by good site selection and management.
- UCG can be used in conventional Rankine steam cycle at lower cost with little clean up
- UCG can be used in higher efficiency gas engines/gas turbines only after substantial clean up. SA has many years experience and a large pool of experience in gas cleaning after some 60 plus years of Fischer Tropsch coal to liquid plant operation at SASOL, 1, 2 and 3.

# Future of Electricity Production from Coal and Gas

- **Coal Unconventional**

- This gas cleaning/conditioning is aimed at providing:
- Steady gas pressure and flow
- Steady gas composition in terms of $H_2$, CO, $CO_2$, $O_2$ and $N_2$.
- Steady LHV
- Free from water particulates $NH_3$ and S compounds
- This is exactly what the Fischer-Tropsch process used by SASOL requires and there is vast experience in SA of achieving the required gas requirements for engines and turbines.
- Need removal of water, S N and particulates for use in gas engines/gas turbines. TUCG 60 MW project awaiting Gas IPP RFQ issue.
- Little or no surface impact by comparison to strip mining and ash disposal.
- Similar C footprint to coal but difficult to measure.
- Surface gasification can provide the same opportunities as UCG but with all the environmental problems of conventional coal mining and thermal power stations.

# Future of Electricity Production from Coal and Gas

- **CBM and CMM**

- Southern African coals not particularly gassy. ~5m3/ te vs ~30m3/te in Australia and elsewhere.
- Majority of SA coal is produced from Surface mining hence ther is no CMM potential.
- Where there is underground mining it is generally shallow <100m below ground. Some methane produced in long wall mining from goaf collapse but production is too variable to be of any use.
- However potential in various areas.
- SA Volksrust, Springbok Flats, Free State, Waterberg
- Botswana Central and North East
- Zimbabwe North West Hwange Lupane, and Zambesi Valley
- Mozambique Zambesi Valley Tete, Moatize
- Zambia Maamba
- Several projects in SA and Botswana coming close to commercial production in SE Mpumalanga and W Free State. Potential likely to be in the 100's of MW not GW's

## Future of Electricity Production from Coal and Gas

**• FUTURE**

- Fuel in the form of Coal, CBM/CMM and UCG is available
- Technology is available in the form of Conventional, CFB and supercritical power stations
- Gas conditioning/cleaning capability
- Boilers, Steam Turbines, Gas Engines and Gas Turbines
- Power distribution system but this requires expansion and rehabilitation.
- Interlinked Southern African Power Pool

## Future of Electricity Production from Coal and Gas

**• Economics**

- Growing demand from business and domestic markets as GDP increases
- Improved tariff from an IPP point of view
- Hopeful return to increased growth in Global GDP

**• What we have**

- Slow cumbersome government controlled bidding programme
- Sometimes uneconomic tariff caps
- Current Refusal of Eskom to sign further PPA's with IPP's

## Future of Electricity Production from Coal and Gas

**• What is Not Available**

- Clear well defined regulatory space
- Even playing field between IPP's and State Owned Utility companies which operate pretty much as a monopoly in most of Southern Africa in both generation and distribution
- Separation of generation and distribution activity
- Power Trading platform on willing buyer willing seller basis
- Fair simple generation licensing

**• What do we need**

- All of the above
- Plus FDI and FDI friendly legislation

## Future of Electricity Production from Coal and Gas

**• A light on the Horizon maybe**

- Current approaches to uses of power by others are
- Own consumption grid parallel
- Net metering
- Wheeling
- Island mode and across the fence
- Different approach is the Aggregator Model

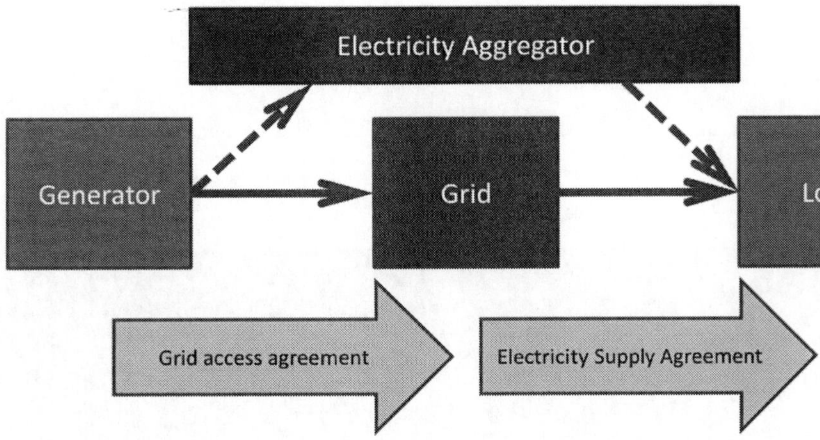


## Electrcity aggregator model

- An electricity aggregator operates a "virtual power station"
- It enters into:
  - power purchase agreements with numerous generators
  - electricity supply agreements with numerous off takers
  - use-of-systems agreements with the grid
- Agreements can be long term basis or shorter term. Can trade on an exchange

Clarke Energy
Engineer - Install - Maintain
Clarke Energy South Africa (Pty) Ltd
Unit 2
105 EP Malan Road
Pomona
Kempton Park
www.clarke-energy.com/south-africa

Dr. Satya Chauhan, Battelle
Email: chauhan@battelle.org • 614.424.4812
August 8 - 12, 2016

# SCALE-UP OF BATTELLE'S DIRECT COAL-TO-LIQUIDS (CTL) PROCESS FOR JET FUEL USING BIOMASS-DERIVED SOLVENTS

**33rd International Pittsburgh Coal Conference, Cape Town, South Africa**

1

Battelle
*The Business of Innovation*

## Outline

- Background on jet fuel from non-petroleum sources
- Indirect vs. direct liquefaction of coal-to-liquids (CTL) technologies for jet fuel
- Battelle's novel approach to improve direct CTL for jet fuel using biomass-derived solvents
- Scale-up of Battelle's new direct CTL process

2

Battelle
*The Business of Innovation*

# Need for Alternative Feedstocks for Jet Fuels

- Increasing demand and limited supply of petroleum crude
  - 500 million barrels per year U.S. demand for jet fuel
- Secure supply of jet fuel is a priority of the United States
  - Large U.S. coal reserves available and easily accessible
  - Coal can be easily stored and stocks can be drawn on in emergencies
- Greenhouse gas (GHG) emission reduction requirements are driving consideration of biomass as a feedstock
  - Due to the dispersed nature and limited renewal rate of biomass, options are being considered for hybrid processes using coal plus biomass





# Advanced Technologies for Alternative Jet Fuels

- Fischer-Tropsch (FT) technology
  - Gasification of coal, biomass, or natural gas, followed by FT synthesis and upgrading
  - FT Synthetic Paraffinic Kerosene (FT-SPK) and FT Synthetic Paraffinic Kerosene with Aromatics (FT-SKA) approved by FAA
  - Example Process: Battelle-developed microchannel FT, being commercialized by Velocys
- Lipids-based fuels
  - Deoxygenation and upgrading of edible and non-edible oils (lipids); Hydroprocessed Esters and Fatty Acids (HEFA) approved by FAA; paraffinic
  - Example Process: Catalytic Hydrothermolysis (CH) of lipids to introduce aromatics; under development by ARA (ReadiJet)
- Carbohydrate-based fuels
  - Conversion to sugars followed by fermentation and then conversion to jet fuels; Direct Sugar to Hydrocarbons (DSHC) and Alcohol to Jet (ATJ) approved by FAA



Battelle
The Business of Innovation

# Alternative Jet Fuels Testing

- Battelle has been working with the U.S. Air Force and UDRI for the last 10 years to generate small quantities of alternative jet fuels for characterization
- Battelle has designed, built, and operated a 1 barrel/day facility called Assured Aerospace Fuels Research Facility (AAFRF); tested on FT and bio-based intermediates

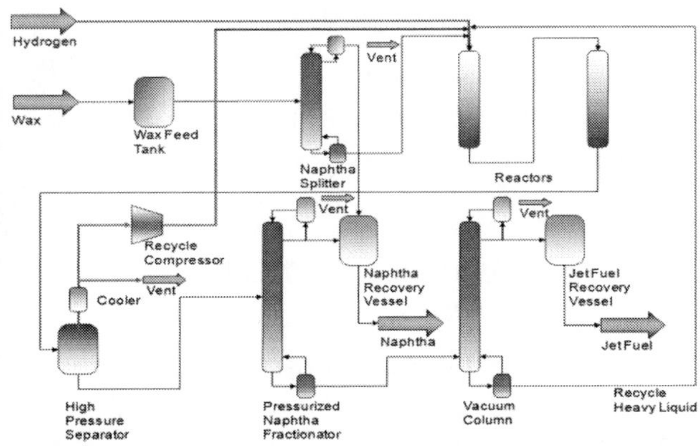


5

Battelle
*The Business of Innovation*

# AAFRF SPK Produced from FT Wax

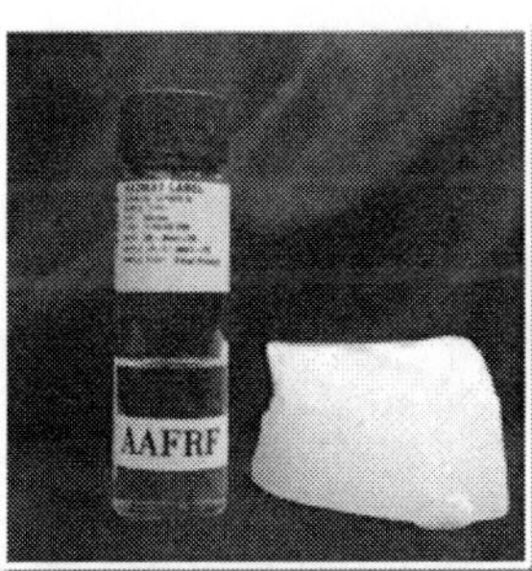


AAFRF SPK properties are nearly identical to other jet fuels

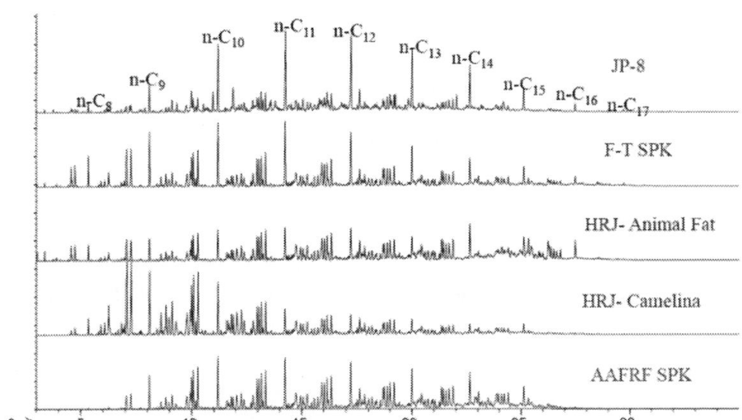


6

Battelle
*The Business of Innovation*

# Some Shortcomings of Alternative Jet Fuels Technology

- Most products are paraffinic in composition, but a minimum of 8% aromatics are currently required for Jet-A fuels
  - The jet fuel fraction is $C_8$ to $C_{16}$ paraffins; blended with alternative sources of aromatics
- Most technologies produce only about 30% in jet fuel range; rest is diesel remnants or naptha

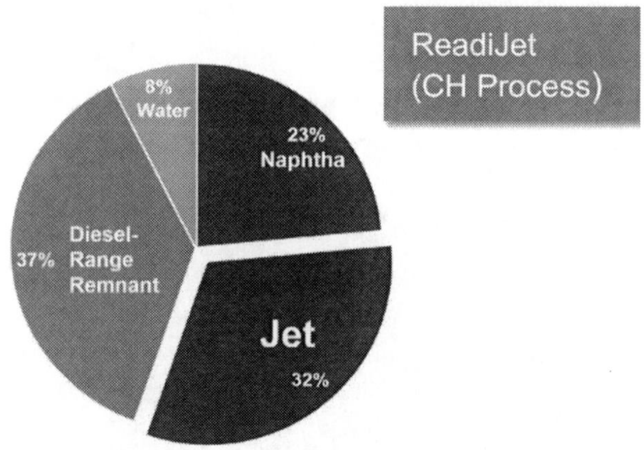


7

Battelle
*The Business of Innovation*

# Direct Coal-to Liquids (CTL) for Jet Fuel

- Direct CTL Technology Options
  - Conventional, three-phase hydroliquefaction
  - Two-stage liquefaction using hydrogen-donor solvents
  - Coal-tar distillate/petroleum cycle-oil co-processing (PennState)
- Common concerns of direct CTL
  - While more thermally efficient than indirect CTL, current direct CTL is still inefficient, costly, and generates excessive amounts of GHG emissions
  - Provides far more aromatics than the minimum 8% desired in jet fuel
  - Complicated design and operation to maintain the proper solvent properties and balance
  - Requires molecular hydrogen ($H_2$); uses it inefficiently
  - Process is economical at larger (>5,000 tons/day of coal) scale

8

Battelle
*The Business of Innovation*

# A New Technology for Direct Coal Liquefaction

- Battelle concept for using biomass-derived solvents for coal liquefaction
  - Provides much more hydrogen transfer than ~0.3% required to dissolve coal (*Chauhan, et.al., "Short Residence Time Coal Liquefaction, ACS Symposium Series, No. 139; 1980*)
- Radically different from previously researched technologies
  - Solvent Refined Coal II process using coal-derived solvents
  - Exxon Donor Solvent process using tetralin as hydrogen donor solvent
  - PennState effort using cyclic petroleum-derived fractions as solvent
- Once-through use of the hydrogen-donor, biomass-derived solvent eliminates need to regenerate and recycle solvent
- Need for molecular $H_2$ for coal liquefaction completely eliminated; reduces pressure and eliminates need for on-site $H_2$ generation

9

Battelle
The Business of Innovation

# New CTL Technology

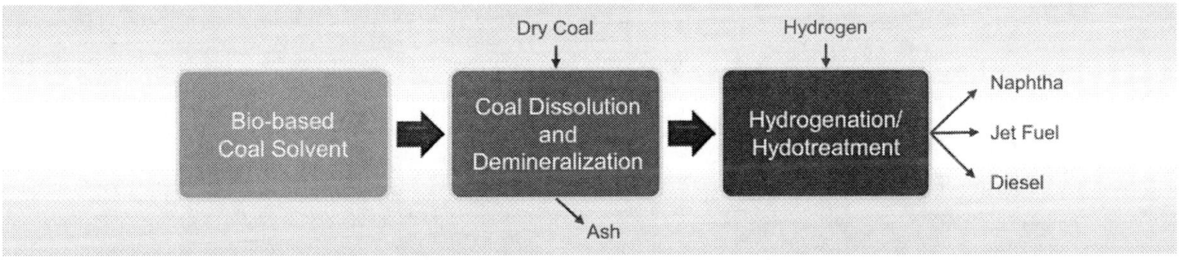


- Straightforward integration of proven subsystems with novel liquefaction chemistry
- Significant reduction in the capital and operating costs due to mild operating conditions (500 vs. 2500 psi)
- Elimination of CCS at coal liquefaction site and minimization of CCS at the coal-derived-syncrude-refining site due to reduced $H_2$ demand
- Meet jet fuel specifications with minimal blending with petroleum-based components

10

Battelle
The Business of Innovation

# Bench-Scale Testing

- Three coals
  - West Virginia (WV) Coal; High Volatile A, Bituminous
  - Ohio (OH) Coal; High Volatile A, Bituminous
  - Wyoming (WY) Coal; Sub-bituminous
- Biomass-Derived Solvents (BS)
  - 40 proprietary solvents tested and compared with tetralin
  - Some in native form; others pre-treated to improve performance
- Liquefaction Testing
  - 0.5L autoclave
  - Microreactor (at PennState)
- Catalytic Upgrading
  - Two-stage catalytic upgrading
  - Multiple microreactors (~20 g catalyst)
  - GC-MS, GCXGC, and ultimate analysis

# Typical Batch Liquefaction Processing Conditions

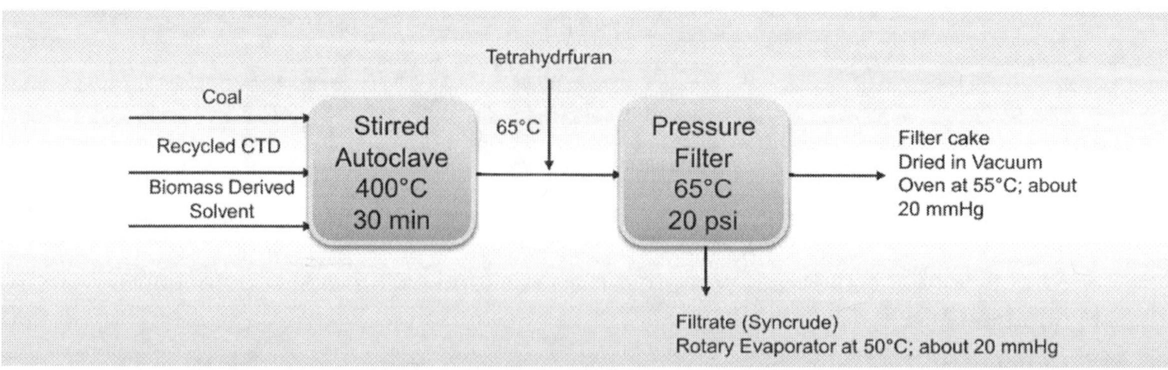


- Run in 0.5L Batch System; Coal Tar Distillate (CTD) as recycle solvent
- Percent liquefied coal calculated based on dried solid; some additionally confirmed by ash analysis
- Syncrude quality checked by measuring viscosity at 50°C

# Batch Autoclave Coal Liquefaction Screening Results on WV Coal

| Solvent | Solvent/Coal wt% | Coal Solubilized % MAF Coal |
|---|---|---|
| Tetralin | 60.2 | 86.6 |
| Bio-solvent #2 (BS-2) | 60.1 | 53.1 |
| BS-3 | 60.1 | 86.7 ✓ |
| BS-9 | 60.1 | 85.5 ✓ |
| BS-10 | 60.0 | 69.8 |
| BS-14 | 60.1 | 56.9 |
| BS-15 | 60.0 | 87.9 ✓ |

✓ *Exceeded 80% solubility target; typical H/C ranged from 0.86 in coal to 1.08 in syncrude*





# Batch Autoclave Coal Liquefaction Screening Results for Ohio Coal

| Biomass-Derived Solvent | Solvent/Coal wt% | Coal Solubilized %MAF Coal THF Soluble | Syncrude Viscosity @ 50°C cp |
|---|---|---|---|
| Tetralin | 60.0 | 84.9 | 325 |
| BS-12 | 60.3 | 88.3 | 96.6 |
| BS-15 | 60.4 | 86.2 | 9.2 |
| BS-15A | 24.0 | 85.1 | 20.0 |
| BS-19 | 60.0 | 84.5 | 190 |
| BS-19A | 23.9 | 82.0 | 50.0 |
| BS-19B | 18.0 | 82.2 | 31.7 |
| BS-23 | 60.3 | 85.2 | 1632 |
| BS-25 | 60.2 | 83.6 | 1330 |
| BS-27B | 24.1 | 88.2 | 403 |
| BS-32 | 11.9 | 85.5 | 658 |
| BS-40D | 24.0 | 92.1 | 639 |

# Solubility Verses Hydrogen Available for Transfer

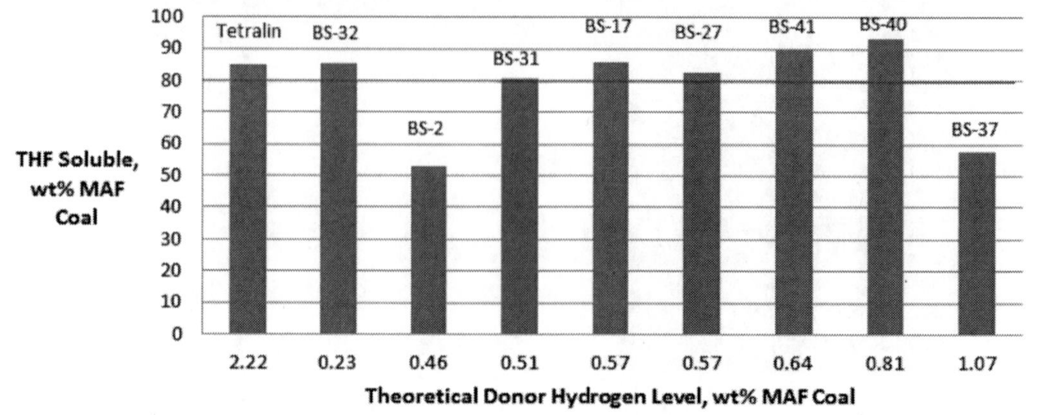


15

Battelle
*The Business of Innovation*

# Bench-Scale Two-Stage Hydrotreatment/Hydrogenation

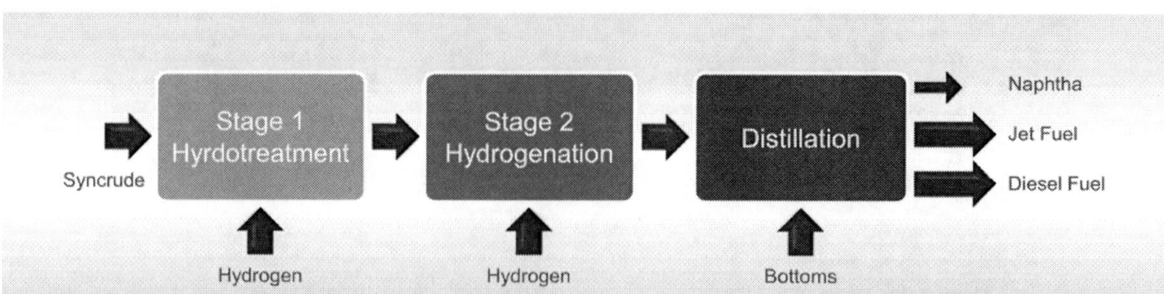


- Stage 1: Heteroatom removal
- Stage 2: Hydrogenation/hydrocracking
- Distillation: Separation into distillate fractions; not performed at bench-scale

16

Battelle
*The Business of Innovation*

# Stage 1 Hydrotreatment

- Objective: Reduce N, S, and O content (LHSV = 0.15 $hr^{-1}$)
  - Achieved 99.7 and 99.9% HDN and HDS after Stage 1
  - Residual oxygen reduced to below analysis limit

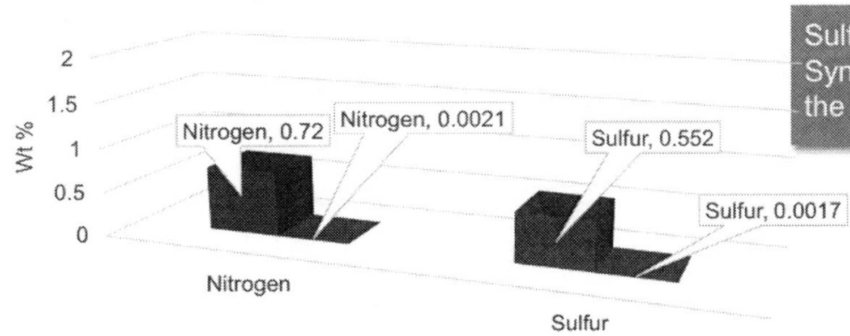


Sulfur in hydrotreated Syncrude is far below the Jet A 0.3% S limit

| | Nitrogen | Sulfur |
|---|---|---|
| ■ Before | 0.72 | 0.552 |
| ■ After | 0.0021 | 0.0017 |

■ Before ■ After



Battelle
The Business of Innovation

# Syncrude Feed and Bench-Scale Upgrading Products

Hydrotreatment
O, N, S removal

Hydrogenation

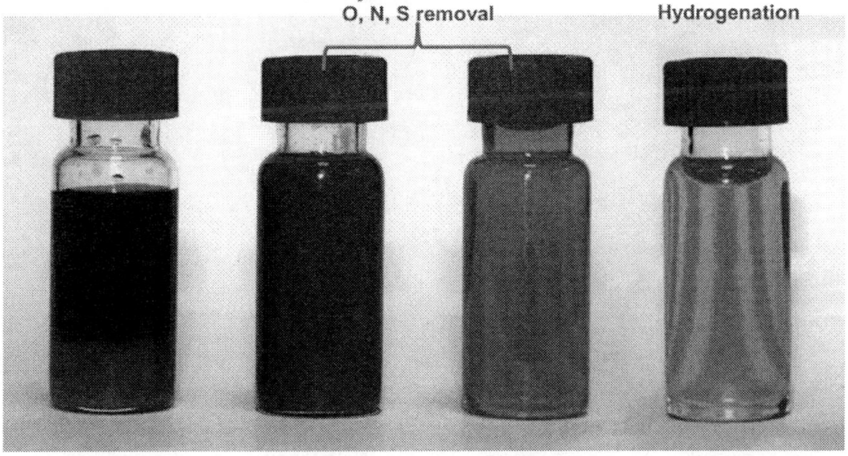

| Syncrude | Stage 1a | Stage 1b | Stage 2 |
|---|---|---|---|
| 1.13 g/mL | 380°C, 1250psi | 380°C, 1250psi | 240°C, 1250 psi |
| | 0.97 g/mL | 0.93 g/mL | 0.93 g/mL |



Battelle
The Business of Innovation

## Simulated Distillation of Feed and Bench-Scale Upgrading Products

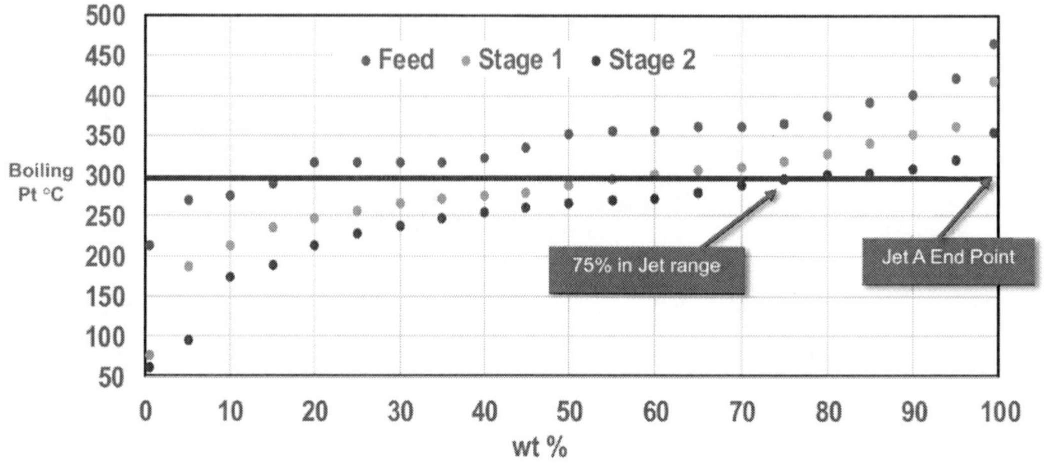






## Liquefaction Scaled Up

- Process scaled up at Quantex 1 ton/day facility (6 tests)
- 400-410°C; 500 psig pressure; 30-minute treatment time
- Product slurry centrifuged; centrate split in a single-stage evaporator at ~380°C

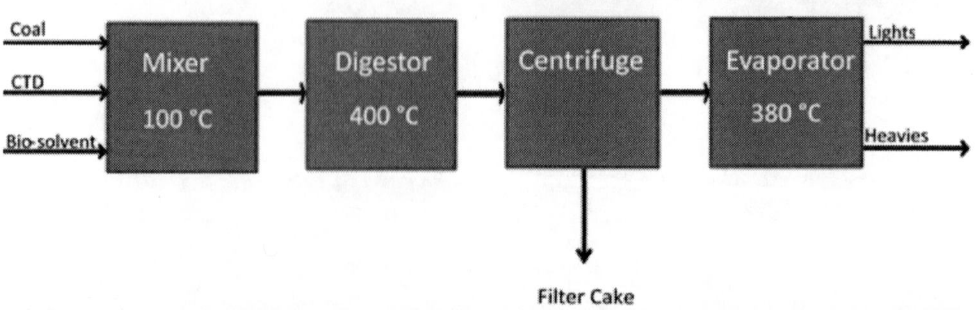

# Pilot-Plant Liquefaction Results

- THF-soluble yield of 75-85 wt.% of MAF coal; somewhat lower than bench test yield values
- The product slurry was easy to centrifuge, yielding over 40 wt% solid in centrifuge cake
- The centrate had a viscosity of ~60 cp at 50°C, compared to ~100-260 cp in corresponding bench test
  - This is considerably lower than the 325 cp viscosity for bench test with twice as much tetralin
  - The ash content was reduced to <0.07 wt.%
- Two batches of syncrude were distilled to 500°C cut for scale-up of upgrading process
  - ~80 wt.% was <500°C boiling point



Battelle
*The Business of Innovation*

# Scale-Up of Hydrotreating/Hydrogenation of Syncrude

- Scaled-up at Intertek's 20 L/day facility
- Two 150 hr. continuous runs using 2-stage catalytic upgrading

Intertek's 20 L/day Continuous Syncrude Hydrotreatment System



Battelle
*The Business of Innovation*

# Residual "S" and "N" at Different LHSV in Lab & Pilot Scale Reactors

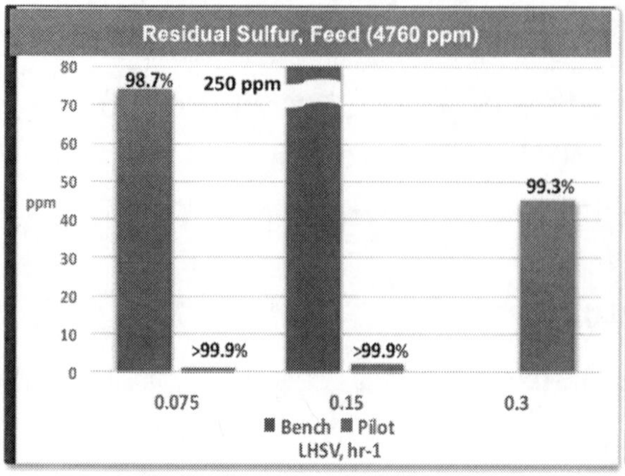


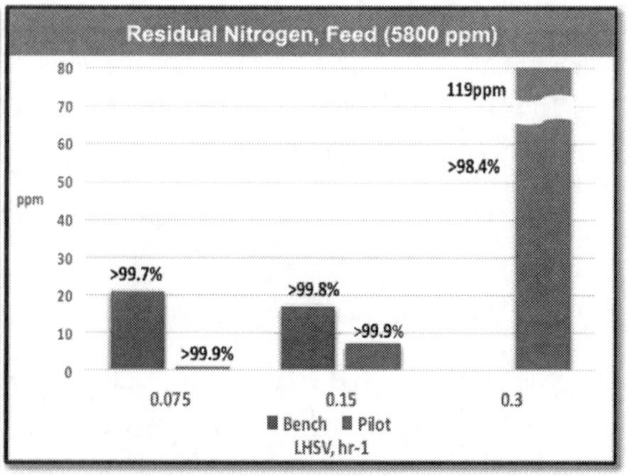


23

Battelle
The Business of Innovation

# Stage1 & 2 Composition by GCxGC-MS

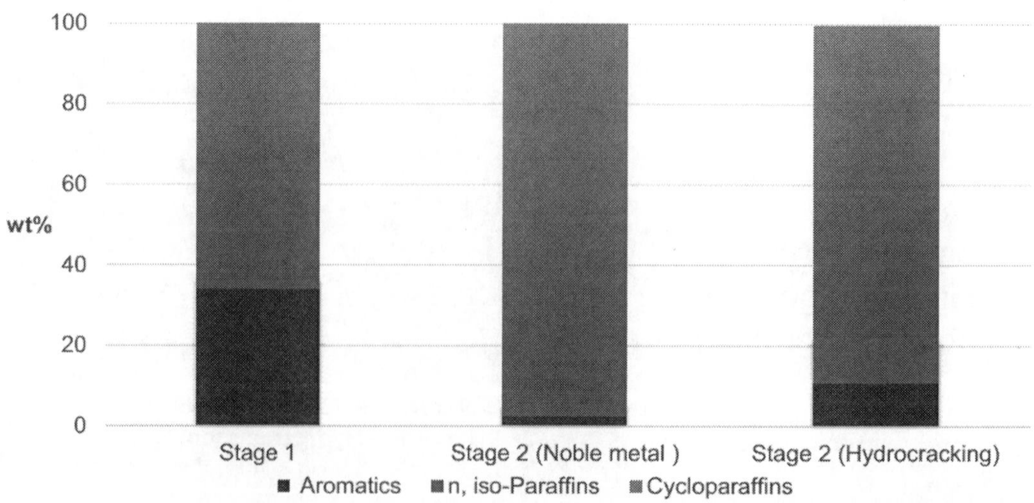


24

Battelle
The Business of Innovation

# Syncrude Feed and Pilot-Scale Upgrading Products

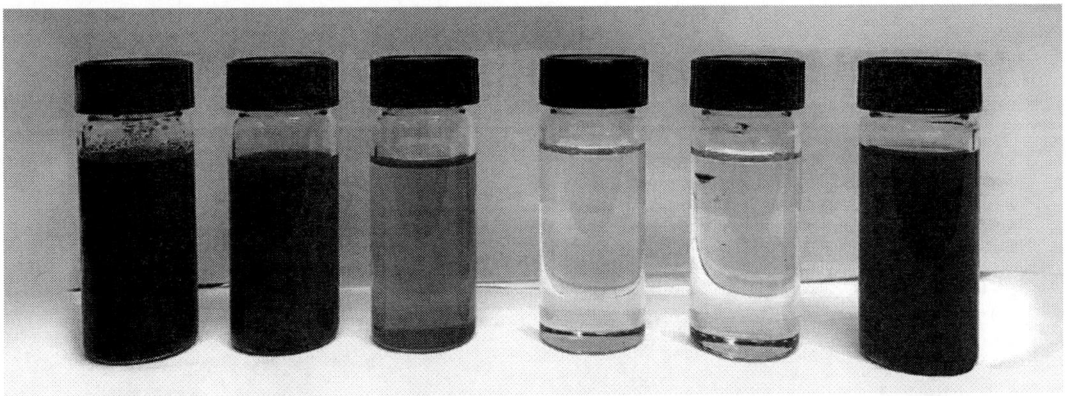

| Syncrude | Stage 1 | Stage 2 | Jet fuel | Diesel | Fraction |
|---|---|---|---|---|---|
| 1.13 g/mL | 380°C, 1250psi 0.97 g/mL | 240°C, 1250psi 0.93 g/mL | 0.89 g/mL | 0.91 g/mL | >300°C 0.91 g/mL |

25

Battelle
The Business of Innovation

# Simulated Distillation of Feed and Pilot-Scale Upgrading Products (Run 1)

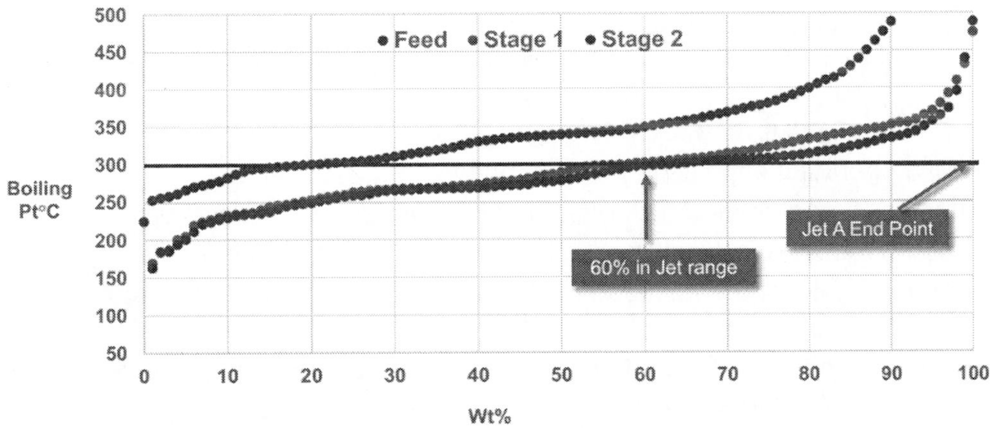


26

Battelle
The Business of Innovation

# Simulated Distillation of Feed and Pilot-Scale Upgrading Products (Run 2)

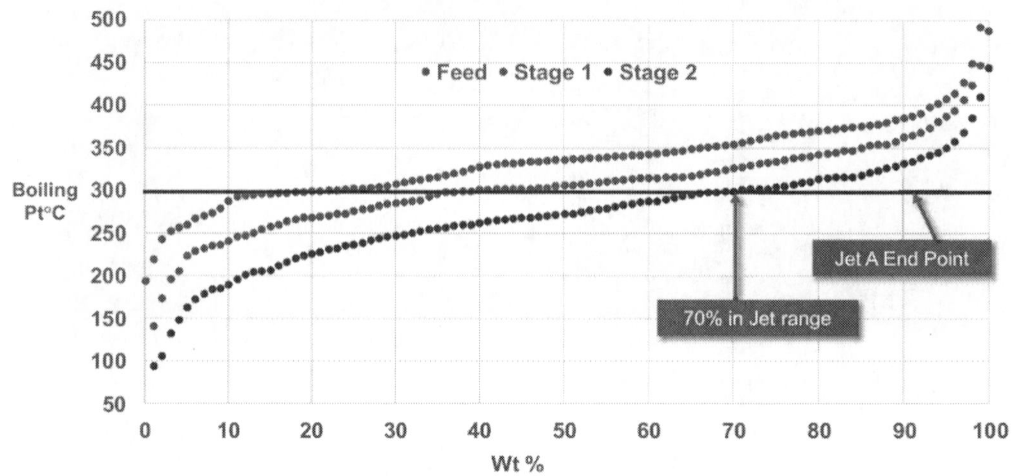






# Biomass-Derived CTL Process Benefits

- Blend of biomass-derived solvents to
  - Provide effective hydrogen transfer
  - Solubilize products of bond cleavage in coal
  - Improve slurry transport
  - Provide high H/C precursors for jet fuel and diesel fuel
- Lower Capital and Operating Costs
  - Mild operating conditions for liquefaction
  - Order of magnitude smaller liquefaction plants (<1000 tons/day)
  - No $H_2$ required for coal liquefaction, eliminating need for on-site generation
  - Approximately 30% increase in H/C ratio during liquefaction without using $H_2$
  - Reduced complexity due to once-through solvent use and no catalyst required for liquefaction
  - Carbon Capture & Storage (CCS) avoided at liquefaction plant and much reduced at syncrude upgrading site due to reduced $H_2$ required

## Conclusions

- The Battelle CTL process employing once-through use of biomass-derived coal solvent is a unique approach
  - Elimination of H2 and regeneration of solvent for liquefaction greatly simplifies the process
  - Use of biomass-derived solvents provides a synergistic effect during liquefaction, which also reduces the GHG footprint
- Various bituminous and subbituminous coals were successfully converted to syncrude using one of many identified biomass-derived solvents without the use of H2 or catalyst
- A 2-stage, catalytic hydrotreatment/hydrogenation can convert 60-70% of the CTL syncrude to jet fuel and ~100% to diesel fuel
  - More than 99.9% of nitrogen and sulfur is removed, and oxygen is reduced below detection limit
- Both coal liquefaction and syncrude hydrotreating/hydrogenation sub-systems have been successfully scaled from Technology Readiness Level (TRL) 2-3 to TRL 5 (1 ton/day coal liquefaction and 20 L/day upgrading)

29

Battelle
The Business of Innovation

## Acknowledgements

- Funding Agencies: U.S. DOE/NETL (Jason Lewis, Technical Lead) and State of Ohio (OCDO/ODSA; Gregory Payne, Technical Lead)
- Battelle: Daniel B. Garbark; Rachid Taha; Grady Marcum
- UDRI: John J. Graham; Matt DeWitt
- PennState: Caroline B. Clifford; Ron Wincek
- Quantex: Elliott B. Kennel; Gilbert Chalifoux
- ARA: Ed Coppola; Sanjay Nana
- Intertek: John Brophy; Chad Chrostowski

U.S. DOE/NETL

30

Battelle
The Business of Innovation

*For More Information Contact:*

Dr. Satya P. Chauhan, 614-424-4812, Chauhan@Battelle.org

Battelle

*The* Business *of* Innovation

800.201.2011 | solutions@battelle.org | www.battelle.org

31

# ENERGY AND CCS POTENTIAL IN SOUTH AFRICA – A REVIEW

AFRICARY

African Carbon Energy

JC van Dyk[1,*], JF Brand[2] and E Roberg[3]

[1]African Carbon Energy, South-Africa, +27825500473, email: johan.vandyk@africary.com
[2]African Carbon Energy, South-Africa, +27824971995, email: johan.brand@africary.com
[3] African Carbon Energy, South-Africa, +27826515138, email: einar.roberg@africary.com

## Roadmap of presentation

AFRICARY

- ❑ Background on South-Africa
- ❑ Coal – primary energy source
  - ❑ *Coalfields*
  - ❑ *Role of coal in current economy*
  - ❑ *Power and liquid fuels production from coal*
- ❑ Energy and coal related challenges
- ❑ Energy efficiency improvement
  - ❑ *Clean Coal Technologies*
  - ❑ *UCG*
- ❑ Carbon capture and storage options
  - ❑ *Sequestration options*
- ❑ Coal Policy, Government and Legislation
- ❑ *Future developments and R&D needs*

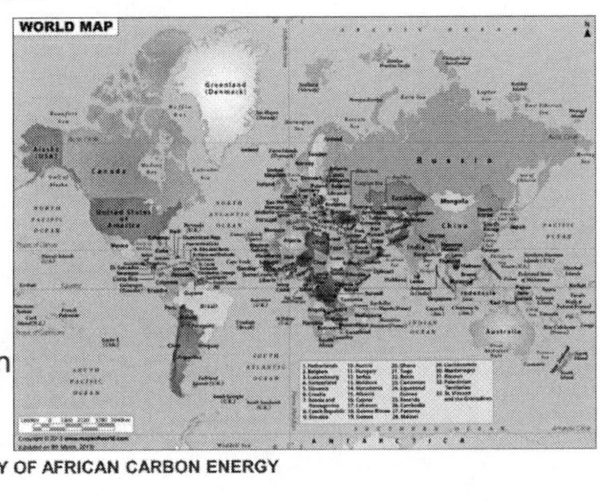

# Background on South Africa

AFRICARY

- 2 798 km's of coastline along the southern Atlantic and Indian Oceans.
- $25^{th}$-largest country in the world by land area and >55 million people.
- Today about 80% South Africans are of Sub-Saharan African ancestry, speaking different dialects, nine of which have official status, with the remaining population consists of European (white), Asian (Indian), and coloured ancestry - referred to as the "Rainbow Nation“.
- Upper-middle income economy by the World Bank. $3^{rd}$ largest economy in Africa and is the $34^{th}$-largest in the world.
- **>25% of the population unemployed and living on <US$1.25 a day.**
- **World's largest and oldest commercial coal to liquids (CTL) plant.**
- **In Top20 of the most carbon-intensive economies in the world.**



# Background on South Africa (cont.)

AFRICARY

Extended unemployment 36.3% (real 26.7%)

Scholars, pensioners, home workers, etc.

**Unemployment by race**

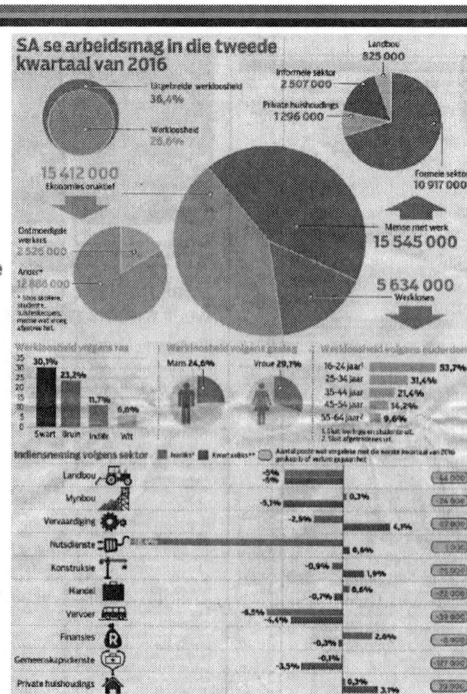


Qualified people unemployed >5M

Unemployment by age

Decline in mining sector 5.1% -24 000 employees

# Background on South Africa (cont.)

AFRICARY

**MINING – A VOLATILE SECTOR**
FROM A 6.8% GROWTH IN LABOUR MARKET
TO 18,1% DOWN IN CONTRIBUTION TO SA'S
ECONOMY (Q4 2015 TO Q1 2016)

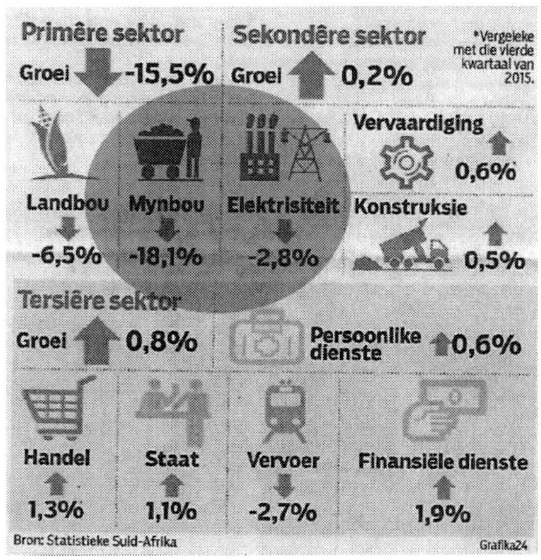




# Coal – primary energy source

AFRICARY

- First mined on a commercial basis in South Africa in 1857.
- Currently 6th largest coal producer in the world.
- Primary energy source for domestic power generation – and set to dominate the energy mix for the foreseeable future.
- **Eskom (http://www.eskom.co.za), the state-owned national electricity supply utility, has generated >96% of the country's electricity (1923).**
- **Sasol - 160 000bbl/day coal-to-liquids plant**
- Coal is furthermore used extensively in the metallurgical industry (titanium, ferrochrome, ferromanganese and steel industries).
- The largest energy resource was and remains coal; the country is endowed with estimated coal reserves sufficient to cater for both domestic and export demand for 200 years.

# Coalfields

AFRICARY

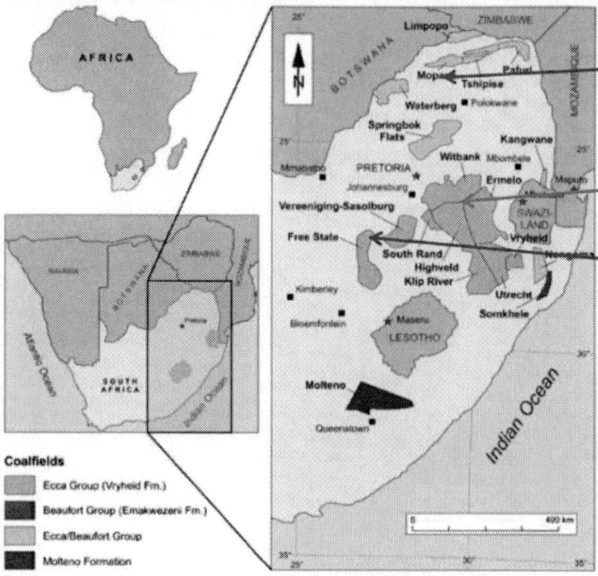


- ❑ Occur in the rock strata of the Main Karoo Basin (MKB) and its associated sub-basins.
- ❑ The MKB forms part of a major series of Gondwana basins.



# Coalfields (cont.)

AFRICARY

- ❑ Estimates for **economical recoverable reserves range from 9 - 59 billion** tons - latest estimates by the Minerals Bureau suggests 33 billion ton.
- ❑ As much as 70% of that coal is located in the Waterberg, Witbank, and Highveld coalfields, with smaller percentages in the Ermelo, Free State and Springbok Flats coalfields.
- ❑ 96% of the reserves are bituminous coal, 2% metallurgical and the remainder mainly anthracite.
- ❑ Coal reserves as referring ‘only to that portion of the total coal resources of which the nature and distribution have been fairly well established and which is at present economically recoverable or borders on economic recoverability’.

# Coalfields (cont.)

AFRICARY

- New extraction technologies such as Underground Coal Gasification, as well as suitable uses and markets for low-grade, high-ash coal, are required before the country can utilize its vast coal resources to full potential.
- Total coal output for 2012, was 285 million tons (Mt), and ranked fifth as a hard coal exported at 67 MT in 2009 – current output similar but not verified

| PRODUCERS | MT / annum (2009[1] and 2012) | EXPORTERS | MT / annum |
|---|---|---|---|
| China | 2971 / 4017 | Australia | 262 |
| USA | 919 / 1106 | Indonesia | 230 |
| India | 526 / 649 | Russia | 116 |
| Australia | 335 / 463 | Colombia | 70 |
| Indonesia | 263 / 488 | South Africa | 67 |
| South Africa | 247 / 285 | USA | 53 |
| Russia | 229 / 390 | Canada | 28 |
| Kazakhstan | 96 / 138 | Vietnam | 26 |
| Poland | 78 / 158 | China | 23 |
| Colombia | 73 / 98 | Kazakhstan | 23 |
| WORLD | 5990 / 8687 | World | 944 |

# Role of coal in current economy

AFRICARY

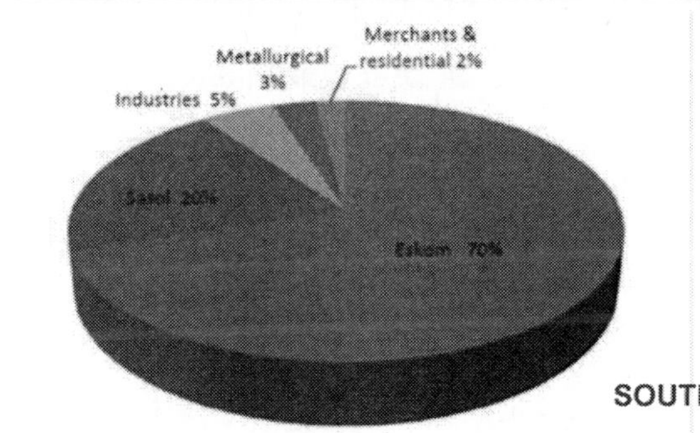


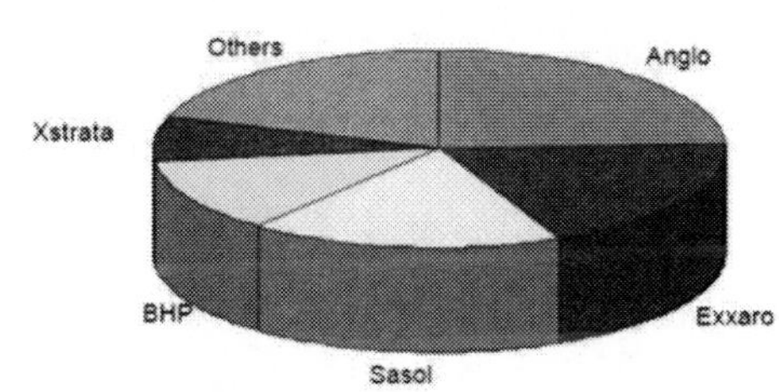


SOUTH AFRICAN COMMODITY SUMMARY FOR TOTAL SALES IN RANDS (ZAR)

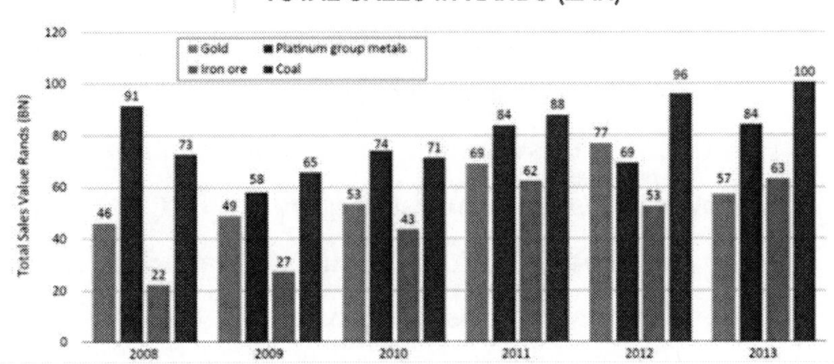

# Role of coal in current economy (cont.)

AFRICARY

*>8 GW of new coal capacity is required to achieve the higher levels of economic growth required by the National Development Plan.*

*Coal is the only source of local energy that can provide sufficient security for base load electricity supply for this to be a realistic, achievable goal.*

FEBRUARY 2015

Powering Africa

Antonio Castellano, Adam Kendall, Mikhail Nikomarov, and Tarryn Swemmer

There is a direct correlation between economic growth and electricity supply. If sub-Saharan Africa is to fulfill its promise, it needs power—and lots of it.

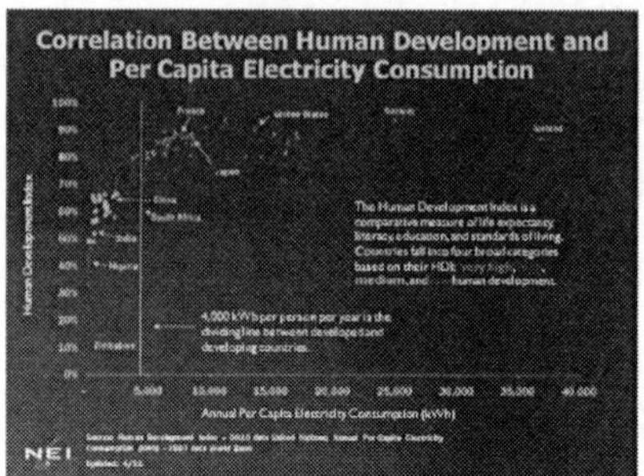


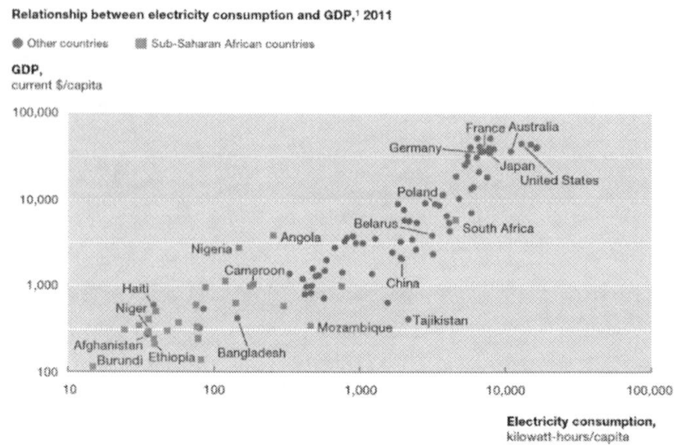




# Power and liquid fuels production

AFRICARY

- ❑ Eskom operates a fleet of 23 power stations with a total nominal capacity of 42 090MW, comprising 35 721MW of coal-fired stations, 1 860MW of nuclear power, 2 409MW of gas-fired, 600MW hydro and 1 400MW pumped storage stations, as well as the recently commissioned 100MW wind farm .
- ❑ Two new coal-fired power stations, Medupi and Kusile (similar supercritical coal-fired power stations with a planned efficiency of 38% compared to 34% of the subcritical technologies in the older stations), are currently in construction or commissioning.
- ❑ **Integrated Resources Plan for Electricity (IRP) envisages a further 6250 MW of new coal being required by 2030. It is believed this will be totally inadequate if South Africa is to achieve higher rates of economic and industrial growth.**

# Power and liquid fuels production (cont.)

AFRICARY

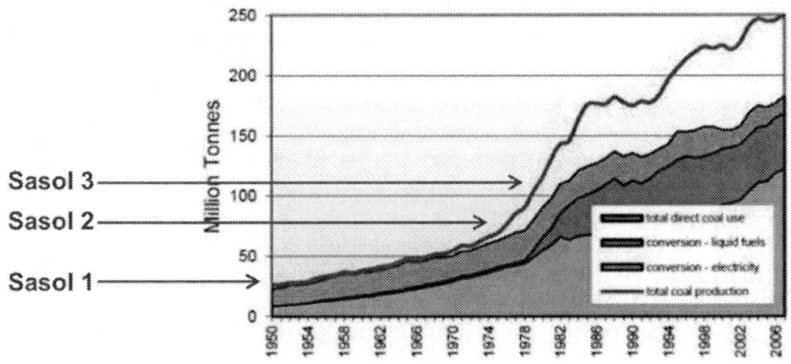


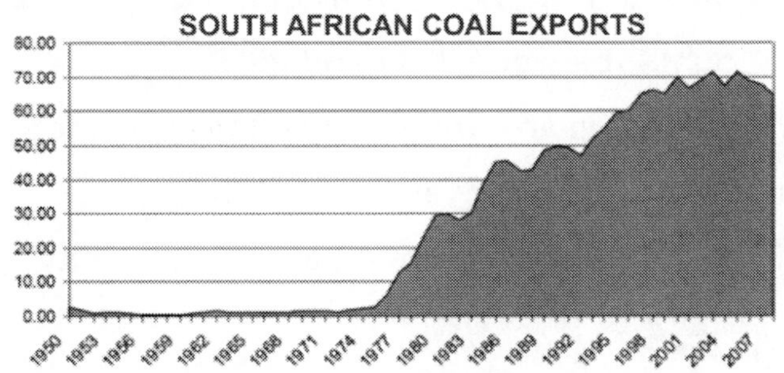




# Power and liquid fuels production (cont.)

AFRICARY

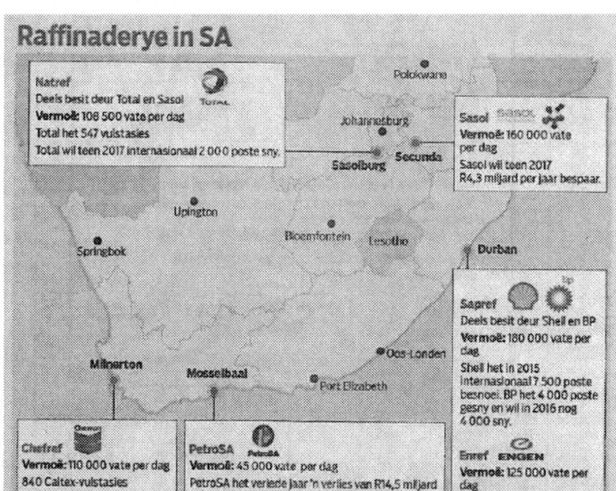


Fuel from coal <22%

| OWNERS | CAPACITY (bbls/ day) |
|---|---|
| NATREF (Total + Sasol) | 108 5000 |
| SASOL | >160 000 |
| SAPREF (Shell + BP) | 180 000 |
| PETROSA | 45 000 |
| CHEFREF (Chevron + Caltex) | 110 000 |

# Power and liquid fuels production (cont.)

AFRICARY

- ❑ The IRP envisages an additional 9600 MW of new nuclear power by 2023 – questionable given the accepted lead time for nuclear power is 14 years.
- ❑ Coal is a more reliable and lower-cost for base load option in the short to medium term – (also lowest capital cost)
- ❑ **South Africa is facing a dramatic power crisis largely caused by the lack of sufficient coal for power generation over the next five years. The size of the coal shortfall ranges from 60 to 120 Mt and is expected to impact the country between 2014 and 2019.**
- ❑ **South Africa's biggest near-term energy challenge, therefore, is:**
  - ❑ **To secure Eskom supply over the next five years.**
  - ❑ **Ageing Eskom's Power Plants – replacing capacity**



# Power and liquid fuels production (cont.)

AFRICARY

THE CONCERNING FUTURE OF SA'S POWER PLANTS

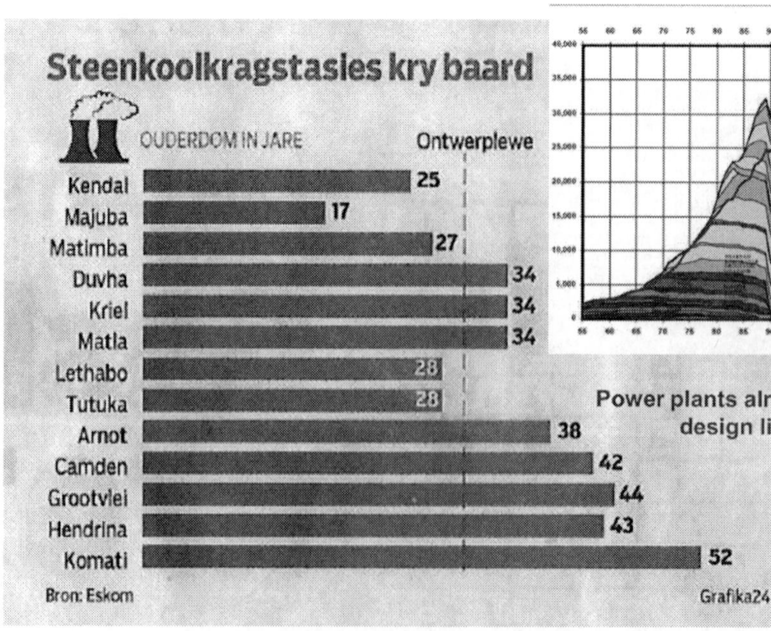


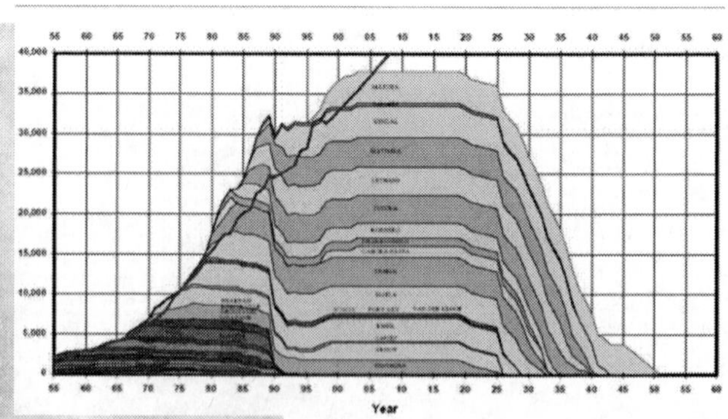


Power plants already past design life

# Energy and coal related challenges

AFRICARY

Coal is the heart of the energy in South Africa as the primary source, but the future of coal in South Africa also faces many challenges, not the least of which is a lack of a clear strategy for the future.

The National Planning Commission (NPC) noted: ***"While most of South Africa's energy comes from coal, it is striking that government has no integrated coal policy. South Africa ranks fifth internationally as a coal producer and exporter, yet government has no clear export strategy."***

*There is also no integrated development of mining, rail and port infrastructure to facilitate either exports or anticipated increases in local production and consumption, within acceptable environmental constraints*".



# Energy and coal related challenges - most critical challenges

AFRICARY

1. ROM coal qualities are **higher in ash content** (lower grade) with better quality coal already been extracted. Higher grade coal still available but mostly exported. Beneficiation is therefore mandatory to provide products for local usage and export market purposes.
2. **Shortage of water** - this is a limiting factor in the complete value chain from coal to product. Key performance target reduce to 1.20 L/kWh by 2017.
3. Coalfields of the future will require considerable infrastructure and development. **Logistics** for the transport of coal to and from remote coalfields export terminal are constrained by railage limitations.

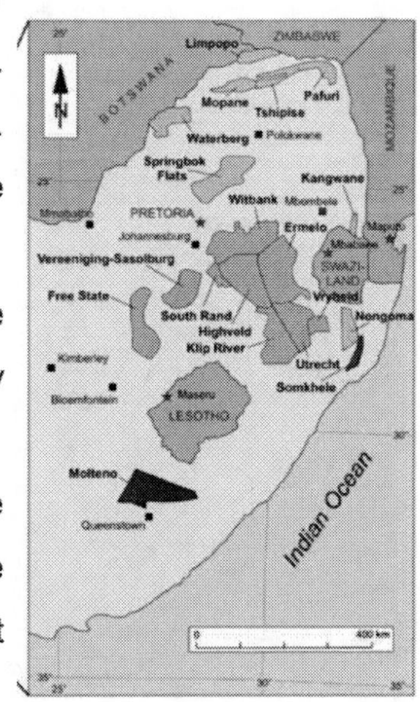

# Energy and coal related challenges - most critical challenges

AFRICARY

4. Ever-increasing tonnages of **discard stockpiles**. Currently >2 billion tons are stored in stockpiles, with >60 million tons accumulating each year.
5. **Acid mine drainage**. Conventional mining breaks up the rock mass, allowing free access of water and sulphuric acid producing reactions between iron sulphide (pyrite) and acidifies soil and rivers. Threat of acid mine drainage from defunct mines remains an ever-present concern.
6. The storage of $CO_2$, produced predominantly from the power stations boilers and petrochemical gasifiers, may be limited by a **lack of underground storage capacity** in South Africa.



# Energy and coal related challenges - in numbers

AFRICARY

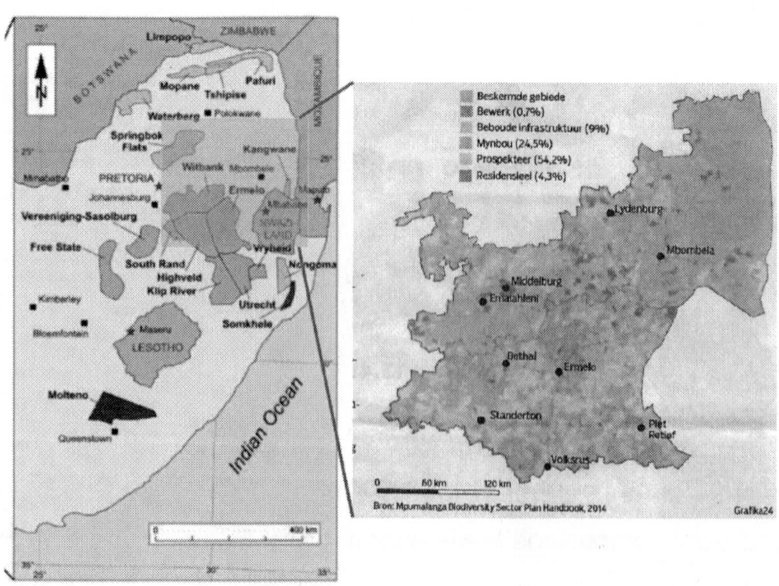


- ±5 000 trucks on road daily
- Transport responsible for 49% of carbon emissions
- 2015 – 239 active mines and 788 abandoned mines with only 5 government safety officers for this area
- Only 2 officers from Department of Water
- Illegal mines received licenses, not adhering to mining regulation and safety but still in operation
- Conventional mining threatened food security

# Energy efficiency improvement (the policy / dream)

AFRICARY

| Policy / Action | Cumulative emissions | Period | Annual emissions* |
|---|---|---|---|
| Montreal protocol | 135.0bn | 1989-2013 | 5.6bn |
| Hydropower worldwide | 2.8bn | 2010 | 2.8bn |
| Nuclear power worldwide | 2.2bn | 2010 | 2.2bn |
| **Increase average global efficiency of coal-fired power plants to 40%** | | | 2bn |
| China one-child policy | 1.3bn | 2005 | 1.3bn |
| Other renewables worldwide | 600m | 2010 | 600m |
| US vehicle emissions & fuel economy standards* | 6.0bn | 2012-2025 | 460m |
| Brazil forest preservation | 3.2bn | 2005-2013 | 400m |
| India land-use change | 177m | 2007 | 177m |
| Clean Development Mechanism | 1.5bn | 2004-2014 | 150m |
| US building & appliances codes | 3.0bn | 2008-2030 | 136m |
| China SOE efficiency targets | 1.9bn | 2005-2020 | 126m |
| Collapse of USSR | 709m | 1992-1998 | 118m |
| Global Environment Facility | 2.3bn | 1991-2014 | 100m |
| EU energy efficiency | 230m | 2008-2012 | 58m |
| US vehicle emissions & fuel economy standards† | 270m | 2014-2018 | 54m |
| EU renewables | 117m | 2008-2012 | 29m |
| US building codes (2013) | 230m | 2014-2030 | 10m |
| US appliances (2013) | 158m | 2014-2030 | 10m |
| Clean technology fund | 1.7bn | project lifetime | na |
| EU vehicle emission standards | 140m | 2020 | na |

**Increased efficiency to 40% (2bn annually)**

* Annual emissions are cumulative emissions divided by the relevant period.
The estimate for the current emissions avoided under the Montreal protocol is eight billion tonnes of $CO_2e$.
The annual figure for the collapse of the USSR refers to the years 1992-1998. *Cars and light trucks † Heavy trucks



# Energy efficiency improvement (the policy / dream)

AFRICARY

- ❑ Divestment from coal does not recognise the reality of growing energy demand. The 21st century has been dominated by coal. Coal's share of global primary energy consumption in 2013 reached 30.1% (3% growth).
- ❑ IEA predicted that half of the on-grid electricity needed to meet the "energy" demand would need to come from coal. Coal is an easily accessible, reliable source and the most affordable base load energy option.
- ❑ The role of technology, and specifically the application of efficient and clean fossil fuel technologies, is the key to divestment. **Cleaner coal technologies urgently needed as coal demand is going to continue.**
- ❑ Raising the global average efficiency of coal plants from 34% to only 40% with the off-the-shelf technology available today would save 2 Gt of $CO_2$.

COPYRIGHT RESERVED – PROPERTY OF AFRICAN CARBON ENERGY

## Clean Coal Technologies

AFRICARY

Examples where South Africa resulted world-class coal R&D achievements:

- ❑ **Acid mine drainage** - water can be used for agriculture, with fly ash from power stations used as a filter in the upgrading/treatment process.
- ❑ **Advanced beneficiation** on fine coal, which provides cleaner coal that can improve plant efficiencies and reduce $CO_2$ emissions.
- ❑ **Dry coal technologies** to reduce the amount of water.
- ❑ **Upgrading and use of discards** as a source of energy in fluidized bed combustion and gasification processes.
- ❑ **Ash utilization** and upgrading for use in cement manufacture, building and road materials.
- ❑ Cleaner and more environmentally friendly processes such as **UCG** is being developed.



## Clean Coal Technologies (cont.)

AFRICARY

- ❑ **Super-critical pulverised fuel combustion** - increasing the steam temperature and pressure in conventional pulverised fuel plants. The current efficiency of Eskom's plants is between 33% and 36%. Super-critical design can achieve improvements of 2,5%, resulting in less coal utilised and lower emissions.

- ❑ **Fluidised Bed Combustion (FBC)**, with the addition of a sorbent to the coal that will mitigate environmental impacts - the sorbent "scrubs" the gases that are released. Advantage to burn extremely low grade coal and using significantly less water in the control of emissions, but currently not economically viable for South Africa. Adding to the challenges of FBC is that limestone and dolomite, are not widely available.

# Underground Coal Gasification

## the obvious solution for clean coal

AFRICARY

- 25% less $CO_2$ per MWh and in large-scale combined-cycle mode can reach energy efficiencies of up to 58% versus 35% in boilers.
- No particulate emissions or ash handling and little or no leaching of trace elements from ash when operated correctly.
- Less sulphur and heavy metals are released or emitted.
- Monetize economically unmineable coal. (<26% of SA coal reserves are economically / technically recoverable with conventional mining.)
- Deployment can create new high-value jobs.
- UCG projects can be located in economically depressed areas.
- 90% less water usage
- No chemicals are used - only air and water are required.
- Fracking is not required and no drilling chemicals are injected to create the boreholes.



# Carbon capture and storage options

AFRICARY

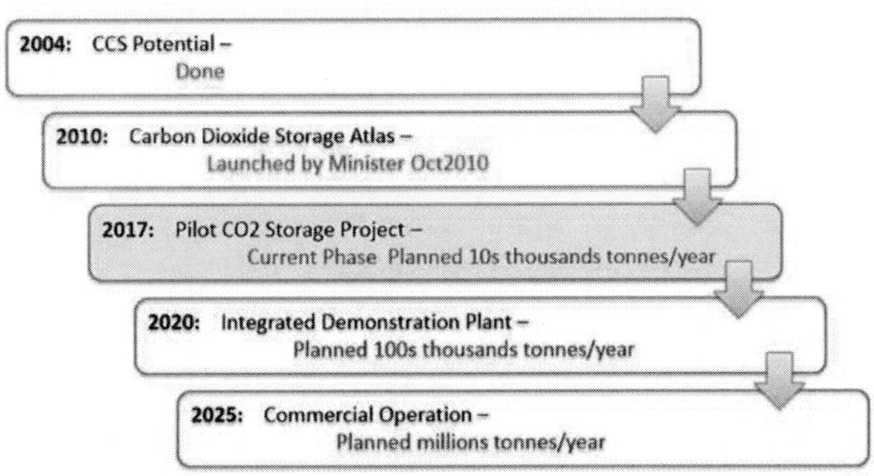

# Carbon capture and storage options (cont.)

AFRICARY

**150 G tonne theoretical potential**

Orange Basin: Area of previous natural gas exploration, however the gas finds were too small to commercialise.

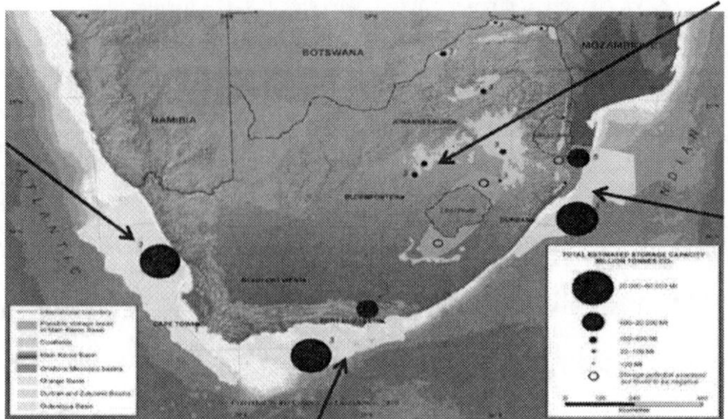


Karoo Basin: On-shore and near the majority of the coal-based electricity generation stations and synfuels production plants.

Durban/Zululand Basin

Outeniqua Basin + Algoa Basin. The offshore area includes the only gas/petroleum producing wells.

Unmineable Coal Seams: Such storage may render those coal seams 'sterile' as potential exploitation with future technology is likely to release the $CO_2$.



# Carbon capture and storage options (Biological sequestration)

AFRICARY

Medium to long term storage of carbon, derived from the atmosphere, in living and dead plants and organic soil matter.

The Net Primary Productivity in the biological arena is relatively small and implies that the carbon sequestration potential is also small.

| BIOME | Delta (gC/m²) | Period (yr) | Area (Mha) | Total (gC (Mt)) | Rate (gC/y (Mt/y)) |
|---|---|---|---|---|---|
| Savannas | 7000 | 30 | 40 | 2800 | 93 |
| Karoo | 1000 | 30 | 38 | 380 | 13 |
| Thicket | 8000 | 30 | 2.5 | 200 | 7 |
| Low tillage farming | 2400 | 30 | 10.6 | 254 | 8 |
| Grassland to forest | 8000 | 30 | 1.8 | 144 | 5 |
| TOTAL | | | | 3778 | 126 |

COPYRIGHT RESERVED – PROPERTY OF AFRICAN CARBON ENERGY

# Carbon capture and storage options (Geological sequestration)

AFRICARY

Direct or indirect sequestration where $CO_2$ is stored in underground formations (depleted oil / gas reservoirs, non-mineable coal seams and saline reservoirs)

- **Oil and gas reservoirs**

  Recently become an oil and gas producing country – mainly offshore. Production figures of 1.4 billion $m^3/y$, may result in 0.7 billion $m^3/y$ $CO_2$ being able to be sequestered*. (*Extracted gas calculated at normal pressure and the sequestered $CO_2$ at 80 bar pressure)

- **Deep saline formations**

  Vryheid formation (average thickness of 350m over an area of $1km^2$ and only 2% of its volume, is ±7 million $m^3$ assuming all volume pore-space could be used. This equates to 183 750 million $m^3$ of storage capacity, which would be sufficient to store all SA's $CO_2$ emissions for 500 year.

  Katberg Formation can add another 16 000 million $m^3$ to the potential



# Carbon capture and storage options (Summary of SA's potential)

AFRICARY

| POTENTIAL SINK | Tonnage (million t/yr) | Duration (years) | Comments |
|---|---|---|---|
| Afforestation<br>Reduced tillage<br>Savanna thickening | 3.9<br>0.4<br>7.9 | 20<br>20<br>20 | An effort is required to store $CO_2$ in "perpetuity" |
| Gas reservoirs | 1 | Long | Enhanced gas recovery is possible |
| Mines | >10 | Site specific | More study is required |
| Vryheid Formation<br>Katberg Formation | >18 375<br>>1 600 | Very long<br>Very long | Poor porosity and permeability, more study is required |
| Coalbed methane | Small | Long | May enhance methane recovery |
| Chemical capture | 1-5 | Indefinite | Large volumes of reactive material is needed |
| Deep ocean | Unlimited | Hundred of years | Ocean sequestration poorly understood |
| Ocean fertilization | Not known | Not known | Study required |

**From the latest results published, it is estimated that of the >400 million tonnes of annual $CO_2$ emitted, ±60% can be sequestered**

COPYRIGHT RESERVED – PROPERTY OF AFRICAN CARBON ENERGY

# Coal Policy

AFRICARY

- ❑ **No approved explicit coal policy**, although draft regulation and working legislation are in progress. Focus on programmes to increase household access to energy, with little attention to security of supply.

- ❑ Eskom initiated a forum for government departments and related to coal supply challenges and take the lead on regulation support:
  - ❑ Concerns around the quality, security and price of its coal supplies.
  - ❑ **South African Coal Roadmap** - it aims to facilitate policy, planning and strategy together.



# Coal Policy (cont.)

AFRICARY

- ❑ Energy Policy White Paper published in 1998:
  - ❑ *Continued deregulation*
  - ❑ *Maintaining a coal resource database*
  - ❑ *Promotion of low-smoke coal for households*
  - ❑ *Use of discard coal*
  - ❑ *Promotion of end-use efficiency*
  - ❑ *Clean Coal Technologies*
  - ❑ *Use of coal-bed methane*

22 Mei 2016 Rapport

**Klimaat gaan miljarde kos**

Yolandi Groenewald

**Nuwe taks vir $CO_2$ 'stink'**

Produksie van hernubare energie dalk ondersteun

- ❑ A number of key policy questions remain unanswered - the most important is the answer around carbon tax and future exposure to international climate change commitments aimed at reducing greenhouse gas emissions.

## Legislative Framework

AFRICARY

- Effort to increase ownership and participation by black South Africans in the economy and in the mining sector.
- Overhaul of the licensing regime for prospecting and mining - dramatic increase in black-economic empowerment BEE.
- Minerals and Petroleum Resources Development (MPRD) Act No. 28 of 2002 - previously owned by the land-owner, the legislation transferred ownership of mineral rights to the state, which is now empowered to grant, control, administer or refuse prospecting or mining rights.
- Policy and legislative initiatives have had a positive effect on black-economic empowerment, but overall impact on the mining sector has been more mixed. Investment and production of most commodities in South Africa has stagnated or even declined over the past decade.



## Market drivers

AFRICARY

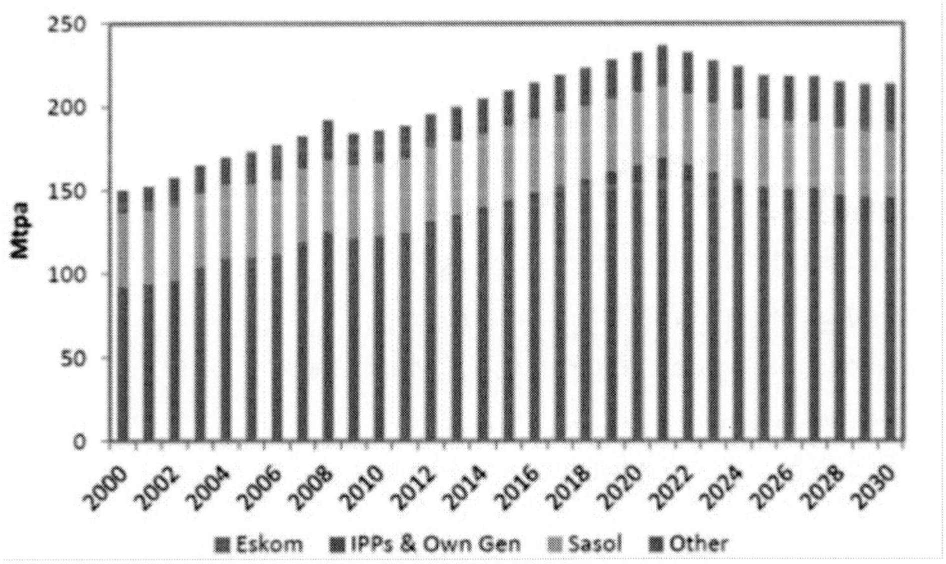

# Future developments and R&D needs

AFRICARY

- <ZAR1 billion spent as gross expenditure on R&D on energy-related topics
- Investments in R&D and Clean Coal Technologies not sufficiently funded
- Very little / insufficient attention is given to research into shale gas
- Research in renewable energy is growing, albeit at a pace lower than required to meet the national targets and expectations
- Human capital development, as well as R&D in nuclear energy is diminishing and the country is losing critical skills in this field
- Energy efficiency efforts towards long-term planning
- Energy-related Masters and Doctoral degree output is not growing
- Coordination and cooperation in energy and energy-related R&D is insufficient, resulting in overlaps / duplication and gaps in terms of national priorities



# Future developments and R&D needs (cont.)

AFRICARY

- A number of research strengths have been identified. South African National Energy Development Institute (SANEDI) is supporting a significant and well-coordinated suite of energy projects, aligned with national priorities. The following institutes have well coordinated research programs:
- Eskom (fossil fuels and renewable energy)
- Sasol (fossil fuel and Fischer-Tropsch)
- NWU (coal, renewable energy and energy efficiency)
- SU (renewable energy)
- and other such as UCT, UWC, WITS and UP.

# Future developments and R&D needs (cont.)

AFRICARY

- Coordination – government departments with an energy budget establish a formal coordination mechanism, with a mandate to steer, plan and coordinate energy R&D funded with public money and eliminating gaps.
- Funding – a more substantial portion of the national R&D vote be allocated to energy research. 1.5% of the fiscal appropriation is proposed for R&D support, with a higher proportion towards energy.
- Human capital development – More research chairs and Centres of Excellence be established and funded.
- Coal, Gas, Renewable Energy, Nuclear Energy, Energy efficiency, Energy economy and policy



# Conclusions

AFRICARY

- Energy is vital to our daily lives – produce food, fuel transport and power communication channels.
- As global population rises (including South Africa), more people are moving out of poverty by gaining access to energy.
- Human capital, innovation and technology have to be aligned to power the future and limit our impact on the environment.
- Growth in SA will be primarily driven by domestic coal power.
- Environmental constraints will more than likely imply that total coal usage will reach a plateau in the next decade and penalised with Carbon Tax.
- New coal mines will have to be developed as existing mines exhaust reserves (i.e. Highveld area), together with Eskom, Sasol and new IPP's must expand alternative production.

# Conclusions

AFRICARY

- ❑ The future sustainability of energy in South Africa—whether based upon coal, gas, nuclear, or renewables—will depend on the ability of the government of the day to attract and fund large infrastructure projects.
- ❑ But, The Consumer, who ultimately foots the bill, will be the deciding factor in determining what the energy mix of the future looks like.

***"The future will depend on the vision of those yet to come, for the road is long and the plans of action are more than one generation in duration.***
***One thing is certain, however: The value of coal as an energy and highly prized carbon-based chemical source can never be underestimated.***
***Coal is arguably the most important commodity of the South African economy."***



# References

AFRICARY


1. Carbon Capture and Storage in South Africa, S Mwaksonda and H Winkler, p95-109, Chapter 6: Growing in the Greenhouse: Protecting the climate by putting development first.
2. www.cornerstonemag.net, The benefits and challenges associated with coal in South Africa, R Jeffrey, R Falcon and A Kinghom, p. 66-70.
3. The Potential for Sequatration of Carbon Dioxide in South Africa, Carbon Capture and Storage in South Africa, S Hietkamp, A Engelbrecht, B Scholes and A Golding, CSIR MSM, PO Box 395, Pretoria, 0001, shietkam@csir.co.za
4. Integrated Assessment of Carbon Capture and Storage (CCS) in South Africa's Power Sector, Peter Viebahn, Daniel Vallentin and Samuel Höller, Energies 2015, 8(12), p. 14380-14406.
5. https://en.wikipedia.org/wiki/South_Africa
6. Program on Energy and Sustainable Development, Working Paper #100, January 2011, The future of South African coal: Market, Investment and Policy challenges, A Eberhard, Stanford University, 2011
7. International Journal of Coal Geology 132 (2014) 170–254, South Africa's coalfields — A 2014 perspective P. John Hancox, Annette E. Götz
8. The Journal of The South African Institute of Mining and Metallurgy FEBRUARY 2005 95, Characterization of the coal resources of South Africa by L.S. Jeffrey*
9. BREDELL, J.H. South African coal resources explained and analysed. Geological Survey of South Africa. 1987. pp. 1987–0154.
10. https://www.eia.gov/cfapps/ipdbproject/iedindex3.cfm?tid=1&pid=7&aid=1&cid=regions&syid=2012
11. WWW.ESKOM.CO.ZA
12. Beeld Sake, Tuesday 26 January 2016
13. The South African Coal Roadmap. (2013, July). I Hall (Ed). Fossil Fuel Foundation, www.fossilfuel.co.za/initiatives/2013/SACRMRoadmap
14. WORLD COAL ASSOCIATION, COAL MATTERS – Divestment and the future role of coal, 2014, info@worldcoal.org
15. At the United National Framework Convention for Climate Change (UNFCCC)
16. The study was undertaken by the CSIR under contract to the DoE
17. http://www.sacccs.org.za/wp-content/uploads/2010/11/Atlas.pdf
18. The Benefits and Challenges Associated With Coal in South Africa, By Rob Jeffrey, MD, Econometrix Rosemary Falcon, SARChI Chair of Clean Coal Technology, University of the Witwatersrand, Johannesburg, Andrew Kinghorn, Executive Director, Fossil Fuel Foundation, Society and Culture, p.66
19. Dr. Tony Surridge, SANEDI (South African National Energy Development Institute) and South African Centre for Carbon Capture & Storage, SANEDI (South African National Energy Development Institute), +27 (11) 038 4348, tonys@sanedi.org.za
20. The Potential for Sequatration of Carbon Dioxide in South Africa, Carbon Capture and Storage in South Africa, S Hietkamp, A Engelbrecht, B Scholes and A Golding, CSIR MSM, PO Box 395, Pretoria, 0001, shietkam@csir.co.za
21. Geological Carbon Sequastration and its influence on subsurface microbial diversity and metabolic carbon cycling, PhD Thesis, University of the Free State, South Africa, M Erasmus, August 2015.
22. Integrated Assessment of Carbon Capture and Storage (CCS) in South Africa's Power Sector, Peter Viebahn, Daniel Vallentin and Samuel Höller, Energies 2015, 8(12), p. 14380-14406.
23. Review of South African environmental regulatory requirements relevant to injection and storage of CO2 for the Test Injection Project, IMBEWU Sustainability Legal Specialists (Pty) Ltd, Contributor/editor - Gillian Arenstein, 2012
24. The State of Energy Research in South Africa, Academy of Science of South Africa, August 2014.


COPYRIGHT RESERVED – PROPERTY OF AFRICAN CARBON ENERGY

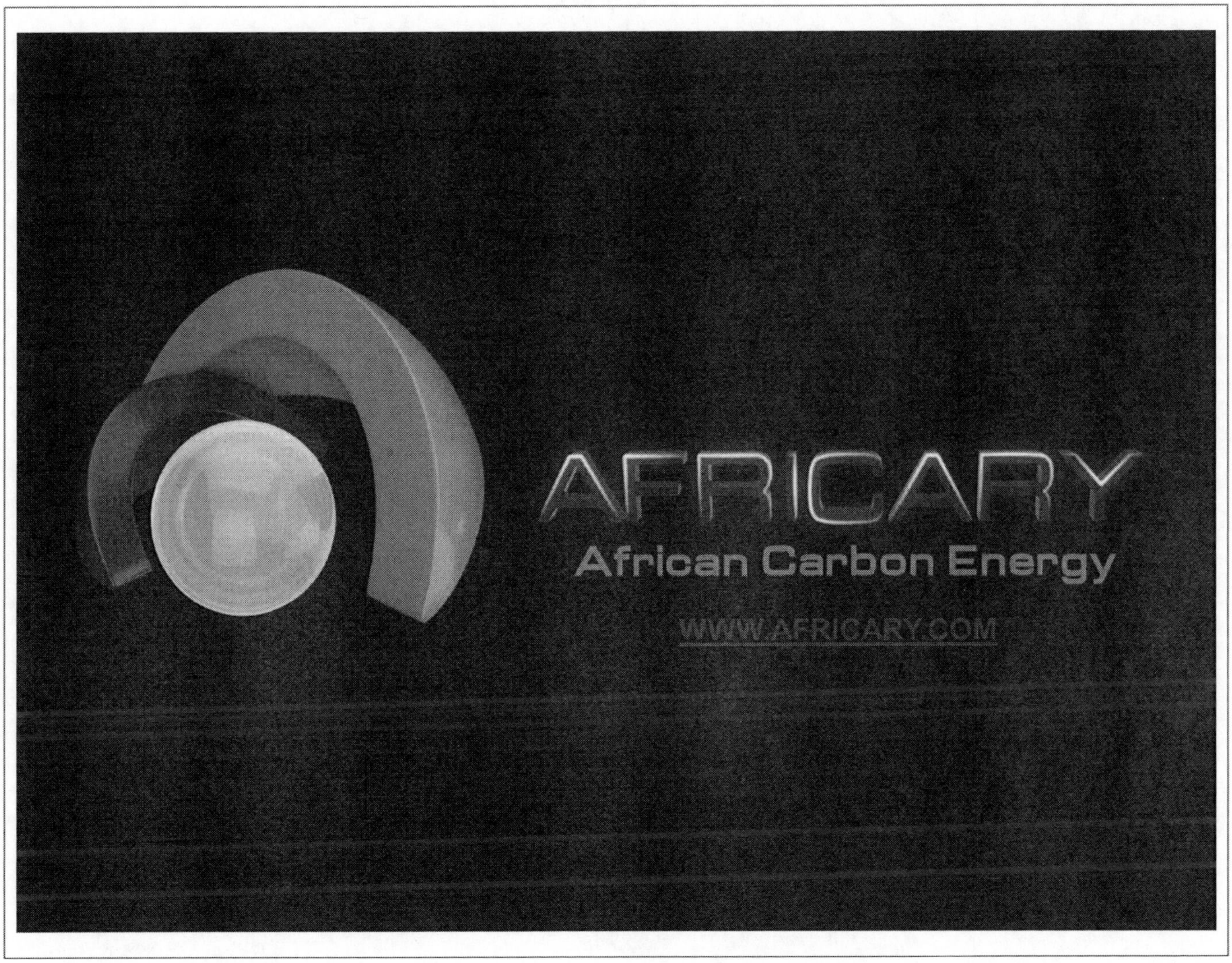
AFRICARY
African Carbon Energy
WWW.AFRICARY.COM

# The Value of Fuel Flexible Circulating Fluidized Bed Technology

**Robert Giglio**
Amec Foster Wheeler
Hampton, New Jersey, USA

**Antonio Castilla Oliver**
Amec Foster Wheeler
La Roza, Madrid, Spain



**Presented at:**
33rd International Pittsburgh Coal Conference
Cape Town, South Africa
Aug 8-12, 2016

# The Value of Fuel Flexible Circulating Fluidized Bed Technology

Robert Giglio, VP of Strategy and Business Development
Amec Foster Wheeler's Global Power Group

Antonio Castilla Oliver, Regional Sales VP for South Europe,
Middle East, Africa, Latin America,
Amec Foster Wheeler's Global Power Group

***Abstract***

The use of Circulating Fluidized Bed (CFB) combustion technology is growing strongly in large utility power plant applications. It has now reached sizes and efficiency levels never imagined by many executives of utility power companies.

This paper will explore the market drivers and technology advancements which allowed CFB boiler technology to grow from small-scale industrial applications to large ultra-supercritical power plants in less than 20 years.

The CFB's fuel flexibility is the primary driver behind the decisions of many power company executives for selecting CFB boiler technology over conventional PC technology. This is due to a growing uncertainty in the global coal market and a trend of declining coal quality vs. attractive price discounts.

However, fuel flexibility means more than just being able to burn a wide range of coals and biomasses. It also means that plant reliability, maintenance, ease of operation, and stack emissions are not significantly impacted over the wider fuel range. Design features of Amec Foster Wheeler's CFBs will be reviewed and compared to conventional PC boiler technology showing how they contribute to high reliability, low maintenance, and cost effective designs.

This paper will also provide an updated review of some of the largest and most advanced CFB plants around the world. Examples would be Tauron's 460 MWe supercritical CFB Lagisza power plant in Bedzin, Poland successfully running since 2009 and the 2,200 MWe ultra-supercritical CFB plant in Samcheok, South Korea, scheduled to come on-line in 2016.

Finally, this paper provides a view for the value that CFB technology can bring to Africa as a valuable part of South Africa's IPP market.

***Why is CFB Technology Growing So Strongly in the Utility Power Sector?***

Circulating Fluidized Bed (CFB) combustion technology has been around for over 40 years, and over the last 10 years its use in large utility power plants has grown strongly. It has now reached sizes and efficiency levels never imagined; drawing attention from utility power companies around the world.

Figure 1. Lagisza CFB Power Plant located in Bedzin, Poland

Figure 1 shows the most advanced CFB power plant operating in the world today, located at Tauron's Lagisza power plant in Bedzin, Poland which has been successfully running since 2009. At the heart of the plant is a single 460 MWe supercritical Amec Foster Wheeler CFB featuring many unique first-of-a-kind design features and a very impressive net plant LHV efficiency of 43.3%. But, its most profound feature is that this plant meets its permitted stack emissions without SCR or FGD equipment, saving Tauron over $100 million in its construction cost.

An even more impressive example is the 2,200 MWe Green Power Plant under commissioning in Samcheok, South Korea, shown in figure 2. The Samcheok plant has four larger 550 MWe Amec Foster Wheeler CFBs utilizing ultra-supercritical steam conditions (257 barg, 603/603°C). The Samcheok plant will meet even tighter stack emissions without using FGD equipment, saving Korea's Southern Power Company (KOSPO) over $250 million in construction cost. These CFBs are now the most advanced units in the world.

Figure 2. The 2,200 MWe Advanced Green Power CFB Plant in Samcheok, South Korea

## *How did the CFB go from Small Industrial Units to a 2,200 MWe Ultra-Supercritical Power Plant?*

*The answer is just two words:* ***Fuel Flexibility****.*

CFBs were originally developed as a solution for industrial facilities that had a need for steam and power, combined with growing supplies of waste bark, wood, plastic, cardboard, paper, sludge's and other waste products. Over the last 20 years, Amec Foster Wheeler has broadened the CFB's fuels range and unit size so much that many power companies took notice and saw the CFB as a way to produce low cost power from low quality fuels such as brown coals, lignites, and waste coals, as well as, high-energy, hard-to-burn fuels like anthracite and petcoke.

Conventional PC boilers have trouble accepting these off-spec fuels due to their narrow fuel specs typically calling for heating values above 5500 kcal/kg, ash and moisture levels below 30%-35%, and volatility's above 20%. However, this is not the case for CFB technology due to its capability of burning both the worst and best coals and lignites with heating values ranging from 1000 to 8500 kcal/kg, ash and moisture levels as high as 60%, and volatilities down to 5%, as shown in figure 3.

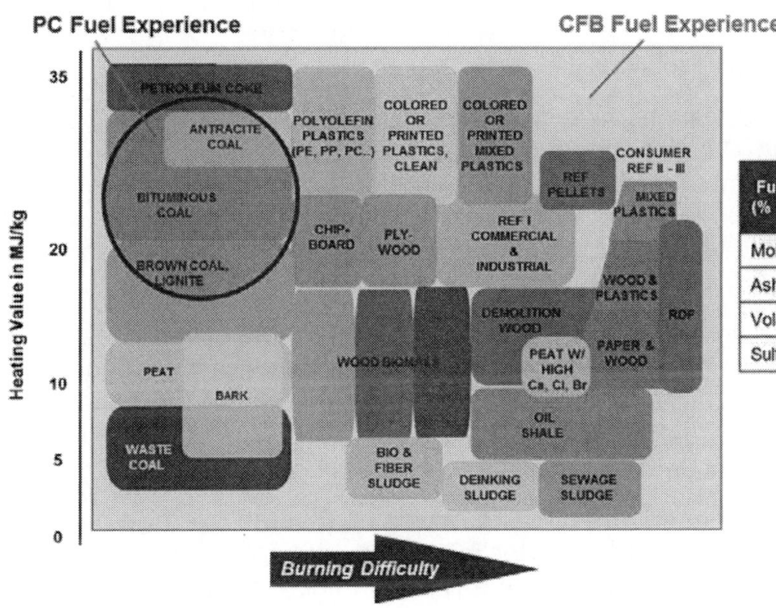


**GPG Fuel Range Experience**

| Fuel Parameter (% As Received) | PC | CFB |
|---|---|---|
| Moisture | 5% - 35% | 5% - 60% |
| Ash | 2% - 30% | 0.1% - 60% |
| Volatile Matter | 15% - 45% | 5% - 50% |
| Sulfur | 0.1% - 5% | 0% - 8% |

Figure 3. Fuel experience of Amec Foster Wheeler CFBs

## *Meeting a Market Need*

Large, utility scale CFB power plant references are growing just in time to meet the challenge of declining quality in internationally traded coals and to support the growing use of economical, low-quality and hard-to-burn domestic fuels in many countries.

Since 2005, Indonesian coal exports have grown faster than coal exports from all other countries combined, nearly quadrupling to over 400 million metric tons over the last few years (see figure 4) and projections show Indonesia maintaining its dominant position in the world's global coal market over the long term.

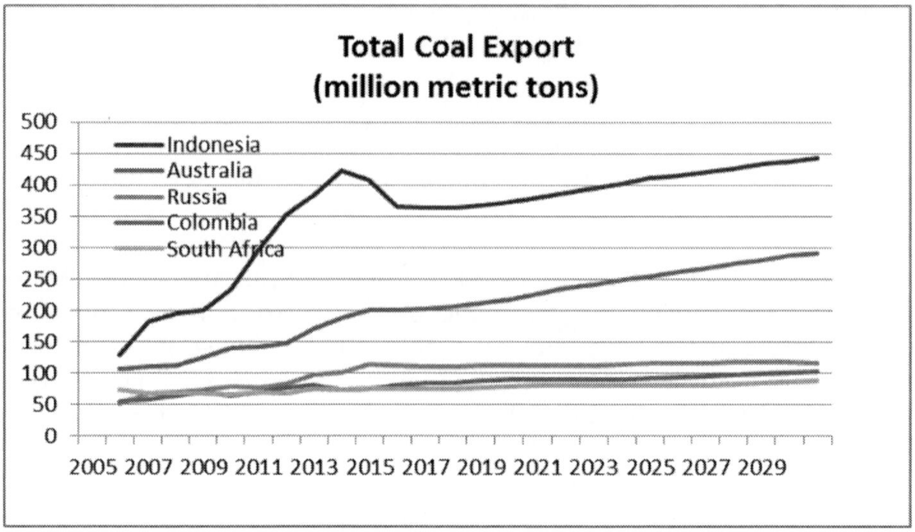


Figure 4. Global coal exports. Source: IHS

However, today about 50% of Indonesian coal exports consists of high moisture sub-bituminous coals with gross-as-received (GAR) higher heating values ranging between 3900-4200 kcal/kg. Further, the best quality Indonesian coal reserves are expected to produce coals with average heating values no greater than 5200 kcal/kg (with economically washing levels). These heating values are well below the 6000 kcal/kg benchmark used in the international coal market for the last 50 years

The primary driver for the ballooning of Indonesian coal trading is simple economics. Even after accounting for the difference in heating value, on a comparative energy basis, Indonesian coals offer 10-20% discounts as compared to premium coals, which goes right to the bottom line of a power plant's balance sheet. Since fuel makes up 75%-85% of the operating cost of a large power plant, the economic benefits of using low quality fuels are tough to ignore.

However, there is another trend that needs mentioning, which is, that the average heating value of Indonesian export coals have been steadily declining and that this trend is expected to continue, as shown in figure 5. This happens in many coal mines as they age and mining operations move to lower quality coal seams. To save cost and reduce coal waste pile-up, instead of increasing coal washing to maintain heating value, miners will sell the higher ash fuels to the market at very attractive discounts.

We can also see the growing use of low quality fuels in several domestic markets as well, where low quality coals and lignites play a major role in power production. For example, 38% of Turkey's solid fuel power is produced from low quality lignite, sub-bituminous and waste coals, while in Germany and the US this number grows to 45% and 49% respectively. Today, nearly all new build coal projects in Turkey are planned around the use of low quality Turkish lignite.

In addition to Turkey, use of low rank coals, lignites and petcokes for power production are growing in India, China, Indonesia, Philippines, Australia, and South Africa driven by better economics, lower fuel price volatility and higher fuel supply security. Further, the use of biomass

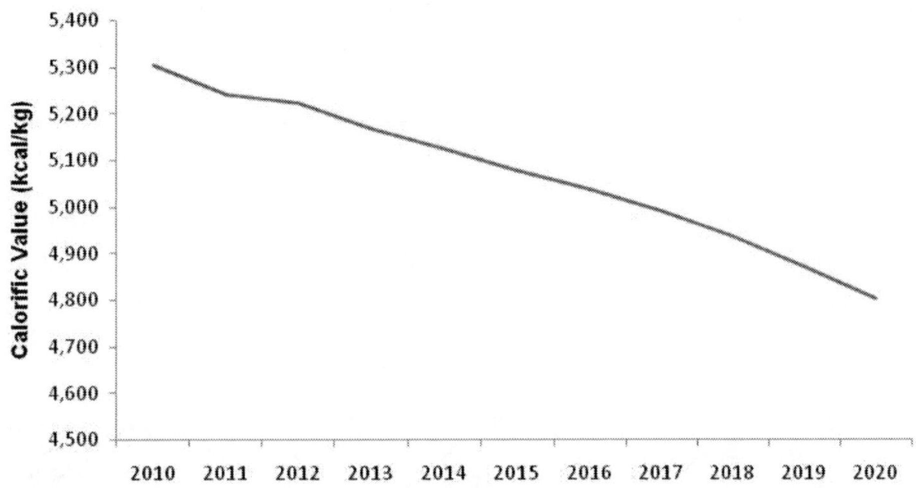


Figure 5. Average Gross Heating Value of Indonesian coal exports. Source: Marketing, Sales and Logistics Analyst, Banpu PCL

for power production is growing in many countries to reduce net carbon emissions and meet future global warming targets.

This trend is not expected to change anytime soon. Instead it looks to be a permanent shift toward a more flexible coal price vs. quality market, where buyers and sellers will trade coal quality for price, very similar to many other commodity and finished good markets.

These expanding low quality solid fuel markets across the globe have dramatically increased the value of fuel flexibility for large scale power plants and have been the primary driver behind the large CFB power plants coming on-line over the last 20 years.

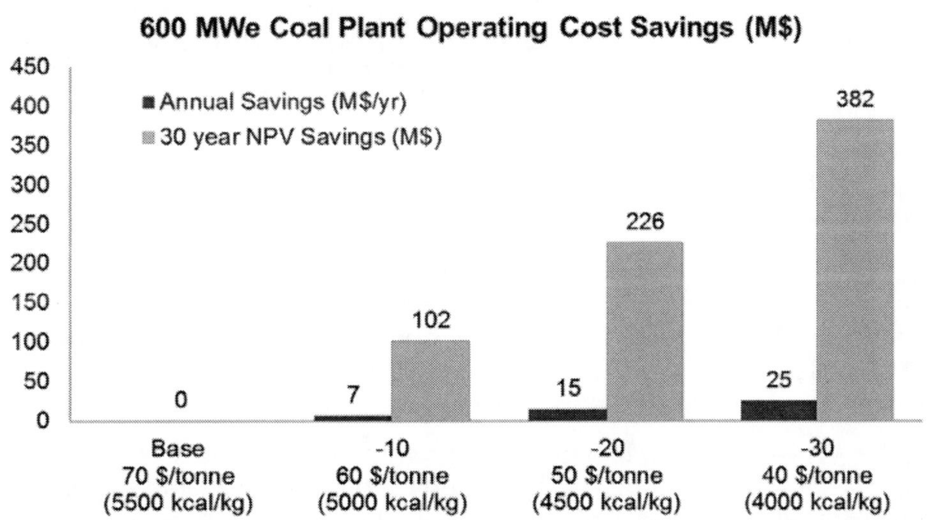


Figure 6. Operating cost saving for 600 MWe supercritical coal plant with 90% annual capacity factor firing 70 $/tonne (5500 kcal/kg) coal as base fuel. NPV based on 30 year term and 5% discount rate.

Fuel flexibility can add up to huge operating cost savings for a power plant over its typical 30 year life. Figure 6 show the savings in plant operating cost that can be realized for a 600 MWe coal plant by buying discounted lower quality fuels. Whereas, a power plant with a narrow fuel spec will be at a competitive disadvantage, resulting in limited utilization and financial return.

In addition to fuel flexibility, the CFB also provides emission flexibility. It can achieve low air emissions without post combustion SCR and flue gas desulfurization saving over 100 M$ on plant construction and operating cost over life of a 600 MWe power plant. This flexibility is important since emission regulations are continually tightening in nearly all countries.

To sum up the CFB's value as compared to PC, table 1 adds up both the capital and operating cost savings for each of the fuel and emission issues discussed above. The results show that the CFB can deliver over four times the value as compared to PC boiler technology.

| Technology Value Points (M$) | PC | CFB | Comment |
|---|---|---|---|
| **Fuel Flexibility** | | **156** | NPV of fuel cost savings for CFB assuming average 10 $/tonne fuel cost discount over 30 year plant life due to ability to fire lower quality fuels. Based on 50/40 $/tonne (4500/4000 kcal/kg) coals |
| **Boiler cost savings** | **60** | | NPV of PC boiler assuming 100 $/KWe installed boiler cost discount for a PC without SCR as compared to CFB |
| **SCR cost savings** | | **36** | NPV cost of adding SCR to PC boiler since it is not needed in CFB assuming a 40 $/KWe installed SCR cost and 30 year NPV of operating expense to meet IFC DA stack NOx emissions of 200 mg/Nm3 |
| **Post boiler FGD savings** | | **39** | NPV of avoiding FGD for CFB assuming a 60 $/KWe installed seawater scrubber cost and 30 year NPV of operating expense to meet IFC DA stack SO2 emissions level of 400 mg/Nm3 |
| **Biomass co-firing** | | **35** | NPV of future CO2 carbon credits. Assume 10% biomass co-firing in CFB and average 10 $/tonne for future CO2 credit value over 30 year plant life. |
| **Total Value (M$)** | **60** | **266** | |

Table 1. Value of 600 MWe CFB vs. PC plant. NPV based on 30 year term and 5% discount rate.

## ***Fuel Flexibility is No Good without Plant Reliability***

Plants with Amec Foster Wheeler CFBs have demonstrated plant availabilities well above conventional PC boilers over a wider fuel range. Figure 7 shows the results from a study comparing PC plant availabilities to plants utilizing Amec Foster Wheeler CFBs.

The Amec Foster Wheeler CFB plant availability values were based on client supplied data reported over the 2000-2008 period. The PC values are based on client supplied data reported in VGB's PowerTech 2012 Report, TW103Ve, over the 2002-2011 period for units located in Europe. PC plants co-firing multiple fuels (anthracite, brown coals, lignite, etc.) do not exist and therefore availability data for them are not shown in the chart.

As shown by figure 7, the plants with Amec Foster Wheeler CFBs had about a 5% higher availability than the PC plants. This higher availability difference was maintained for even

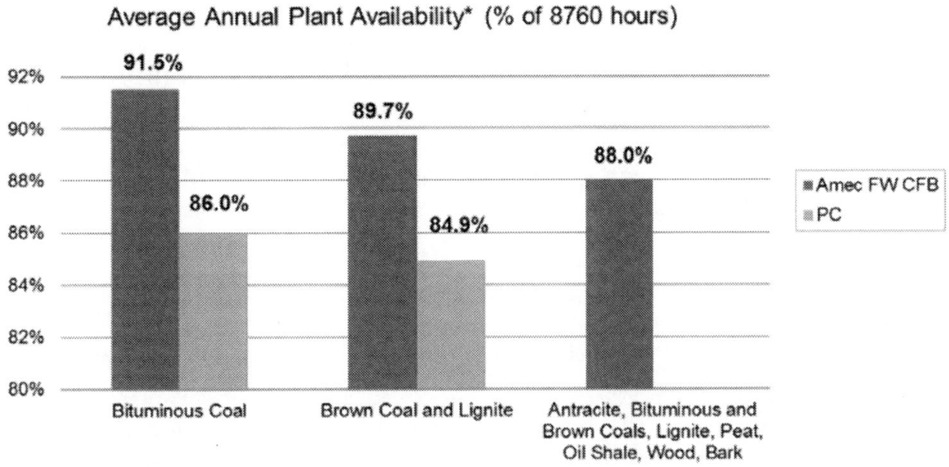


Figure 7. Results from Reliability study of PC vs. CFB Power Plants. *Availability means total time plant is available to run accounting for both planned and unplanned downtime.

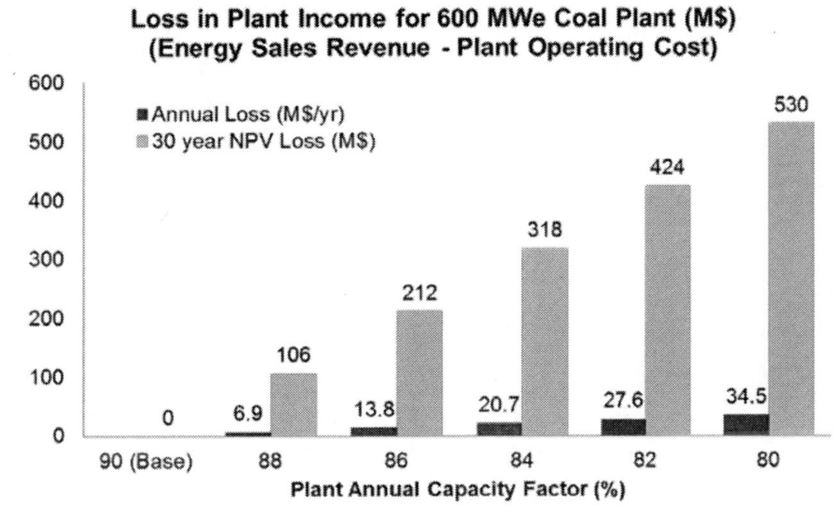


Figure 8. Impact of plant utilization factor on annual plant net income of 600 MWe supercritical coal plant firing 50 $/tonne (4500 kcal/kg) Indonesian coal, selling power at 100 $/MWh at a base 90% capacity factor. Plant Capacity Factor = Actual electricity produced by plant/ possible maximum (600 MWe x 8760 hrs).NPV based on 30 year term and 5% discount rate.

brown coals and lignites which can translate into over a $250 million NPV gain in net income over the life of a 600 MWe power plant, as shown in figure 8.

## ***<u>CFB's Fuel Flexibility comes from its Unique Combustion Process</u>***

The CFB's fuel flexibility is rooted in its unique flameless, low-temperature combustion process. As shown in figure 9, unlike conventional pulverized coal (PC) or oil/gas boilers, instead of an open flame, circulating solids are used to achieve high combustion and heat transfer efficiency to burn a wide range of fuels. The fuel's ash does not melt or soften which allows the CFB to avoid the fouling and corrosion problems encountered in conventional boilers with an open flame.

From an environmental aspect, the low temperature CFB combustion process minimizes NOx formation and allows limestone to be fed directly into the furnace to capture SOx as the fuel burns. In most cases, a SCR and FGD are not needed, dramatically reducing the plant's installed operating cost and water consumption while improving plant reliability and efficiency.

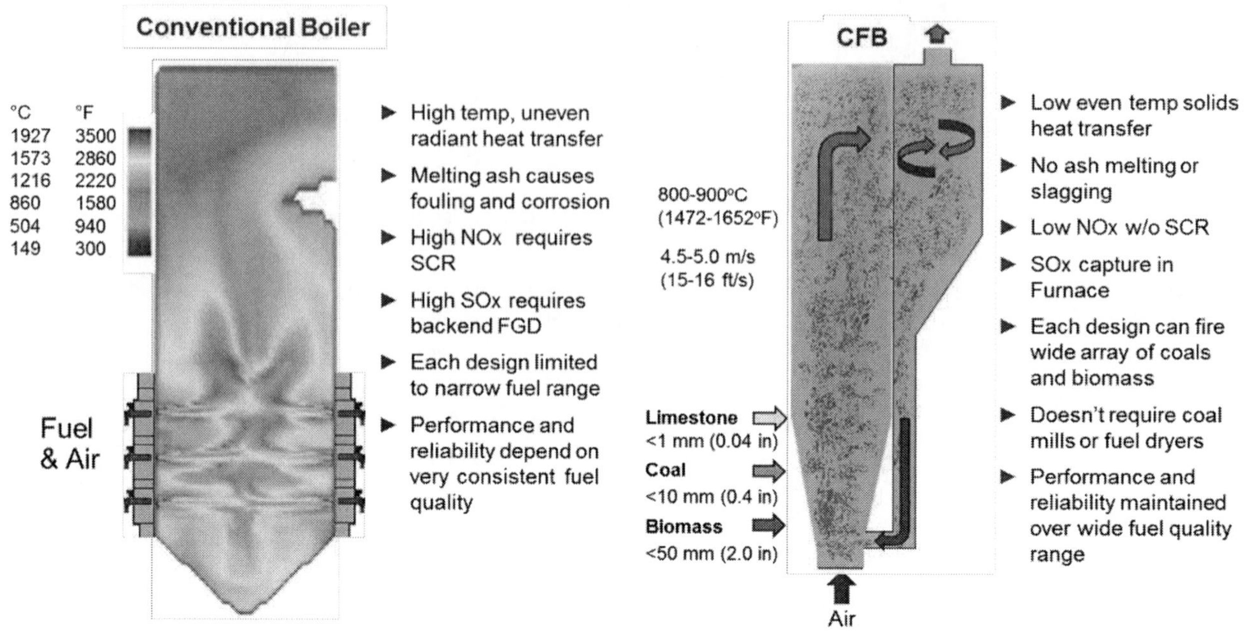


Figure 9. Comparison of PC vs. CFB combustion process

### ***Furnace Size vs. Fuel Quality***

Since the fuel's ash doesn't soften or melt in a CFB, the size of the furnace doesn't grow as much as PC boilers when firing lower quality fuels. As can be seen in figure 10, in order to control fouling, slagging and corrosion, PC furnace height increases by 45% and its footprint increases by over 60% when firing low quality fuels like high sodium lignite, whereas, the CFB boiler height increases by only 8% and its footprint increases by only 20%. This results in a smaller and lower cost CFB boiler as compared to the PC boiler for lower quality fuels.

Further, unlike a PC, a CFB doesn't need soot blowers to control the build-up of deposits and slag in the furnace since the circulating solids themselves keep the furnace walls, panels and steam coils clean for the most efficient heat transfer.

### ***Superheater and Reheater Design Considerations***

Another very important feature of the CFB involves the final superheat and reheat steam coils. These coils operate at the highest metal temperatures in the boiler making them the most vulnerable to corrosion and fouling attack. This vulnerability increases significantly for supercritical boilers with high steam temperatures.

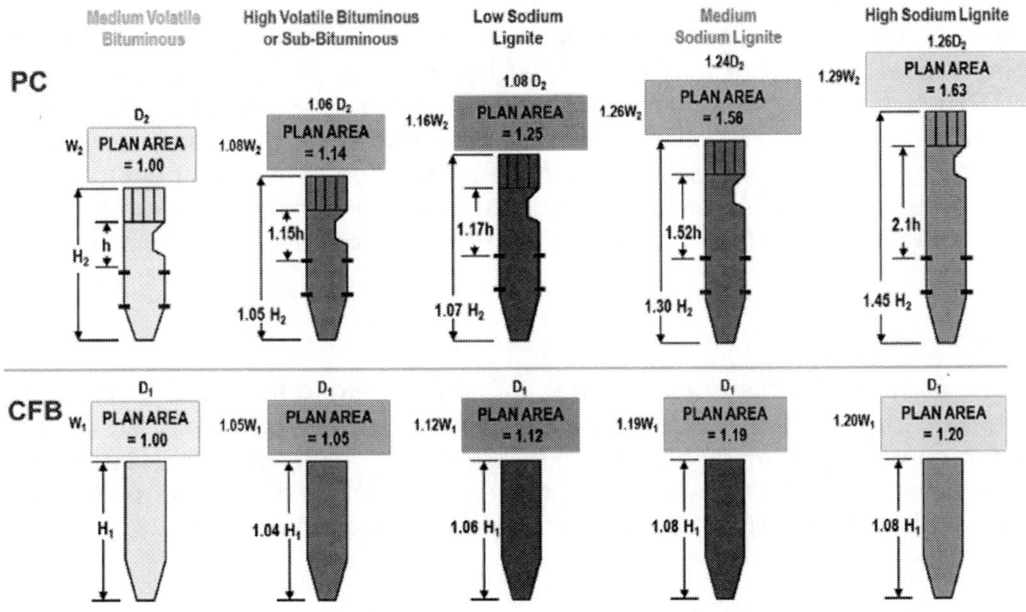


Figure 10. PC vs. CFB - Impact on furnace size as fuel quality degrades

In a PC or oil/gas boiler, these coils are hung from the furnace ceiling and are directly exposed to the slagging ash and corrosive gases (sodium and potassium chlorides) in the hot furnace flue gas. To cope with this undesirable situation, boiler designers use costly high grade alloys

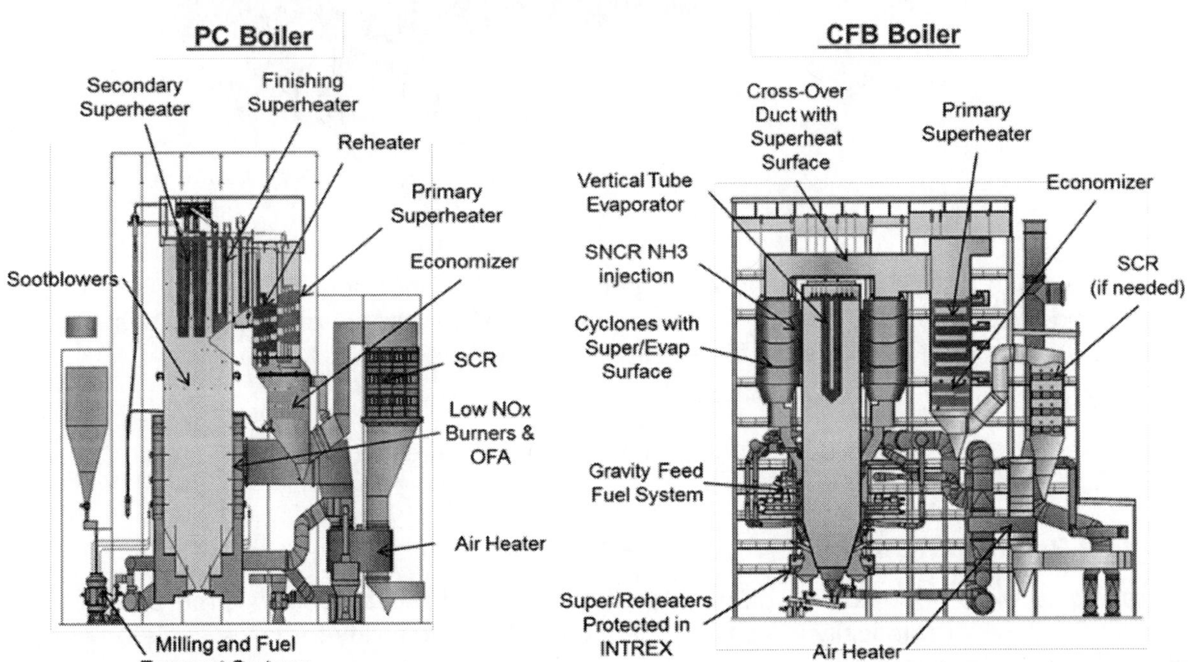


and recommend a high level of cleaning and maintenance for these coils.

Figure 11.Typical fuel feed system for Amec Foster Wheeler CFBs

As shown in figure 11, this design weakness is avoided in Amec Foster Wheeler's CFBs by submerging these coils in hot solids, fluidized by clean air in heat exchanger compartments called INTREXs, protecting them from the corrosive flue gas. The bubbling solids efficiently conduct their heat to the steam contained in the coils and since the solids never melt or soften, fouling and corrosion of these coils are minimal. Further, due to the high heat transfer rate of the solids (via conduction heat transfer), the final superheat and reheat coil sizes are many times smaller than the pendent and convective coils in PCs saving more capital and operating cost.

### ***Fuel Delievery System***

One last important design feature is the fuel delivery system to the boiler. A PC boiler requires coal pulverizes to grind the coal to a powder which is then pneumatically transported and distributed in coal pipes to many burners. For low quality, high ash fuels, like brown coals and lignites, the initial cost, maintenance and power consumption of the fuel pulverizers increase dramatically while the reliability of the entire fuel delivery system declines.

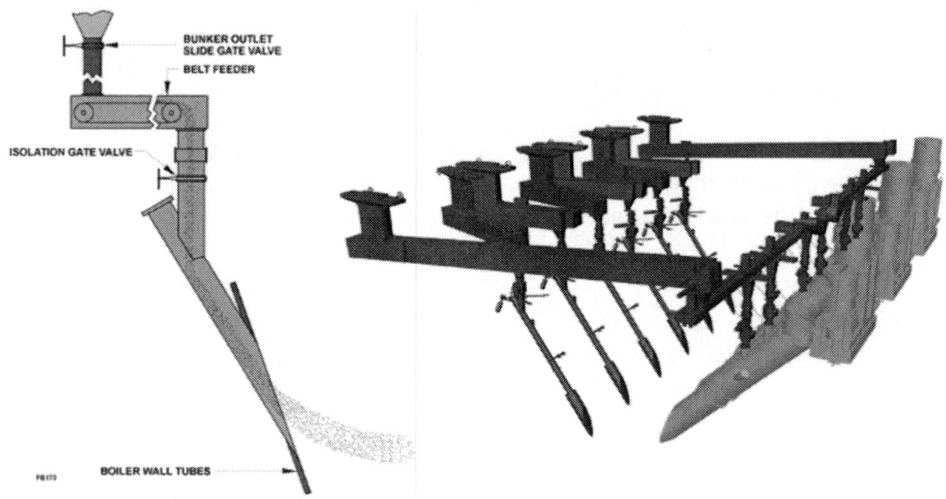


For a CFB, these issues are avoided since pulverizers are not needed. The fuel is only coarsly crushed and fed to the CFB directly from the fuel silos via a simple gravity feed system. Figure 12 shows a typical fuel feed system for an Amec Foster Wheeler CFB.

Figure 12. PC vs. CFB boiler design feature comparison

### ***Important Points to Remember***

The traditional fixed 6000 kcal/kg global steam coal market has moved to a more flexible "price vs. quality" market dramatically increasing the value of fuel flexibility for large scale power plants and has been the primary driver behind the large CFB power plants coming on-line over the last 10 years.

PC power plants with tight coal specifications will have a limited ability to capitalize of the growing supply of discounted coals, domestic lignites, waste coals and petroleum cokes. These plants will have a choice to stay within the premium steam coal market or venture into the broader market and trade fuel price discounts for lower plant output, reduced reliability and higher plant maintenance cost.

This is not to say that new PC boiler power plants can't be designed to burn these low rank fuels, because they can. The point for consideration is that once a PC is designed for a specific low rank or hard-to-burn fuel, it is difficult to burn other fuels without negatively impacting plant performance, reliability and maintenance cost.

However, power plants utilizing CFB boiler technology allows power generators to fully capitalize on fuel cost savings by accessing the full range of discount coals (even for ultra-supercritical designs); buying fuels for maximum economic benefit with little impact to their plant's performance, maintenance and reliability.

For countries with domestic low rank coals, lignites, waste coals or hard-to-burn fuels, like anthracite and petcoke, the CFB opens the door to affordable and secure power over the long term while lessening the risk of future carbon regulation due to the CFB's ability to utilize biomass and other carbon neutral fuels.

THIRTY - THIRD - ANNUAL
INTERNATIONAL PITTSBURGH COAL CONFERENCE

# Application Update of OMB CWS Gasification Process

Guangsuo Yu, ECUST

August 8 - 12, 2016

Cape Town, SOUTH AFRICA

## Team

- ICCT (Institute of Clean Coal Technology) of ECUST, since 1995
- ~100 people (19 faculty members including 7 Prof., 5 Associate Prof. , and about ~80 graduate students)
- Key Lab of Coal Gasification and Energy Chemical Engineering of Ministry of Education (MOE) was constituted in 2006.
- ICCT was honored by Innovative Research Team of MOE and Innovative Research Team of MOST.

# Research Fields

**Research on the carbonaceous fuel entrained-flow gasification reaction, flow mechanism, heat and mass transfer process**

- **Multiphase turbulent reactive flow and impinging streams**
- **Liquid atomization and particle dispersion**
- **Slurry ability and rheological characteristics of liquid fuel**
- **Hopper discharge and dense phase pneumatic conveying of pulverized coal**
- **Process and mechanism of gasification reaction**
- **Flame temperature distribution and chemiluminescent emission**
- **Slag physicochemical properties and slag layer thermal stress characteristics**
- **Gas-liquid-solid multi-phase heat and mass transfer and separation**
- **Multiphase flow and heat transfer in Radiant Syngas Cooler**
- **Process simulation of entrained-flow coal gasification system**
- **Biomass pyrolysis and gasification**

# Patents

- **>100 granted Chinese patents**

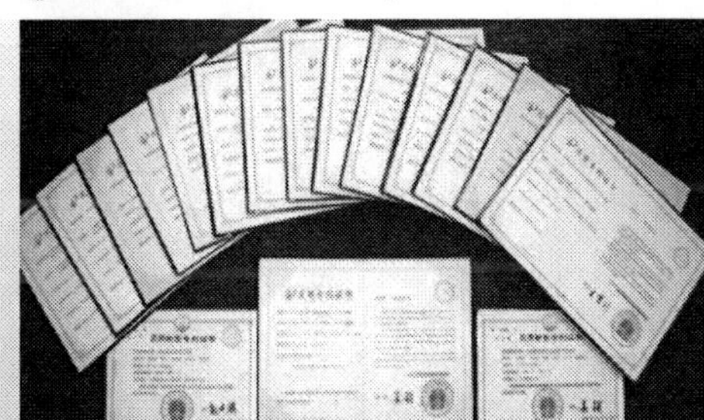

| 申请号 | 专利名称 |
|---|---|
| ZL95111750.5 | 带有旋流器的三通道组合式水煤浆气化喷嘴 |
| ZL98110616.1 | 多喷嘴对置式水煤浆或煤粉气化炉及其应用 |
| ZL00111437.9 | 煤基两段组合式气化工艺及其装置 |
| ZL01112701.5 | 碳氢化合物为原料煤气生产装置中的含渣废水热回收方法 |
| ZL01112880.1 | 一种复合床高温煤气冷却洗涤设备及其工业应用 |
| ZL01112700.7 | 以含碳氢化合物为原料气流床生产煤气的初步净化装置 |
| ZL01112702.3 | 一种高温煤气冷却洗涤设备及其工业应用 |
| ZL02111205.3 | 干煤粉加压气化过程中干煤粉的加料方法与装置 |
| ZL02151143.8 | 带有内分布器的三喷口天然气部分氧化制合成气烧嘴 |
| ZL03115227.9 | 三相气流床生产甲醇的方法和装置 |
| ZL03115820.X | 一种以合成气为原料循环气流床生产二甲醚的方法和装置 |
| ZL02151238.8 | 多通道天然气部分氧化制合成气烧嘴 |
| ZL01126381.4 | 高温气体脱氯剂及其制备方法 |
| ZL02139941.7 | 煤灰型碱金属蒸汽吸附剂 |
| ZL02139942.5 | 一种吸附碱金属蒸汽的吸附剂 |
| ZL00119799.1 | 一种在三相淤浆床中合成二甲醚的方法 |
| ZL02136724.8 | 一种在组合床中催化合成二甲醚的方法 |
| ZL03110108386.X | 能够调节控制和均匀分配煤粉流量的装置和操作方法 |
| ZL94214422.8 | 双通道内旋组合式喷嘴 （共同专利权人） |
| ZL97235458.1 | 五通道负荷可调式水煤浆气化喷嘴 |
| ZL01210097.8 | 顶部设有喷嘴的多喷嘴对置式水煤浆或煤粉气化炉 |
| ZL01239008.9 | 以含碳氢化合物为原料气流床生产煤气的排渣装置 |
| ZL01239007.0 | 气流床生产煤气装置中回收含渣废水热量的热水塔 |
| ZL02216472.3 | 一种用于煤粉加压气化装置控制粉煤加入量的气密式加料器 |
| ZL02266704.0 | 一种射流型洗涤冷却水均布器 |
| ZL200320108896.2 | 设有定位块的洗涤冷却水均布器 |
| ZL200420041534.0 | 气相介质气化喷嘴 （共同专利权人） |
| ZL200420041535.5 | 气液混合相气化喷嘴 （共同专利权人） |
| ZL200420041457.0 | 气化喷嘴的滑动密封装置 （共同设计人） |
| ZL200420114032.6 | 一种粉煤部分氧化制备合成气喷嘴 |
| ZL200420114033.0 | 加压气化反应器 |
| …… | |

- **2 granted US patents**

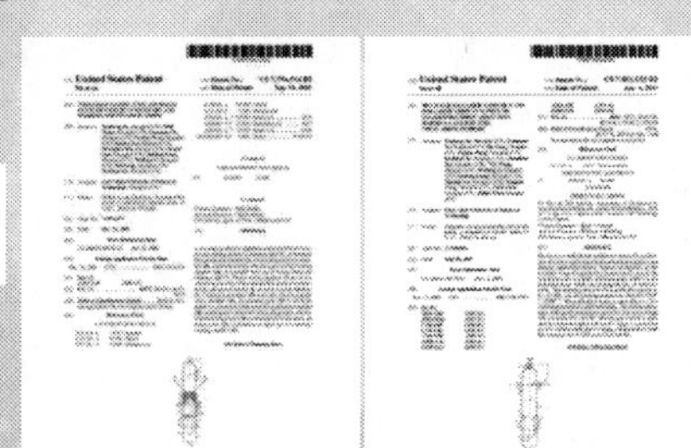

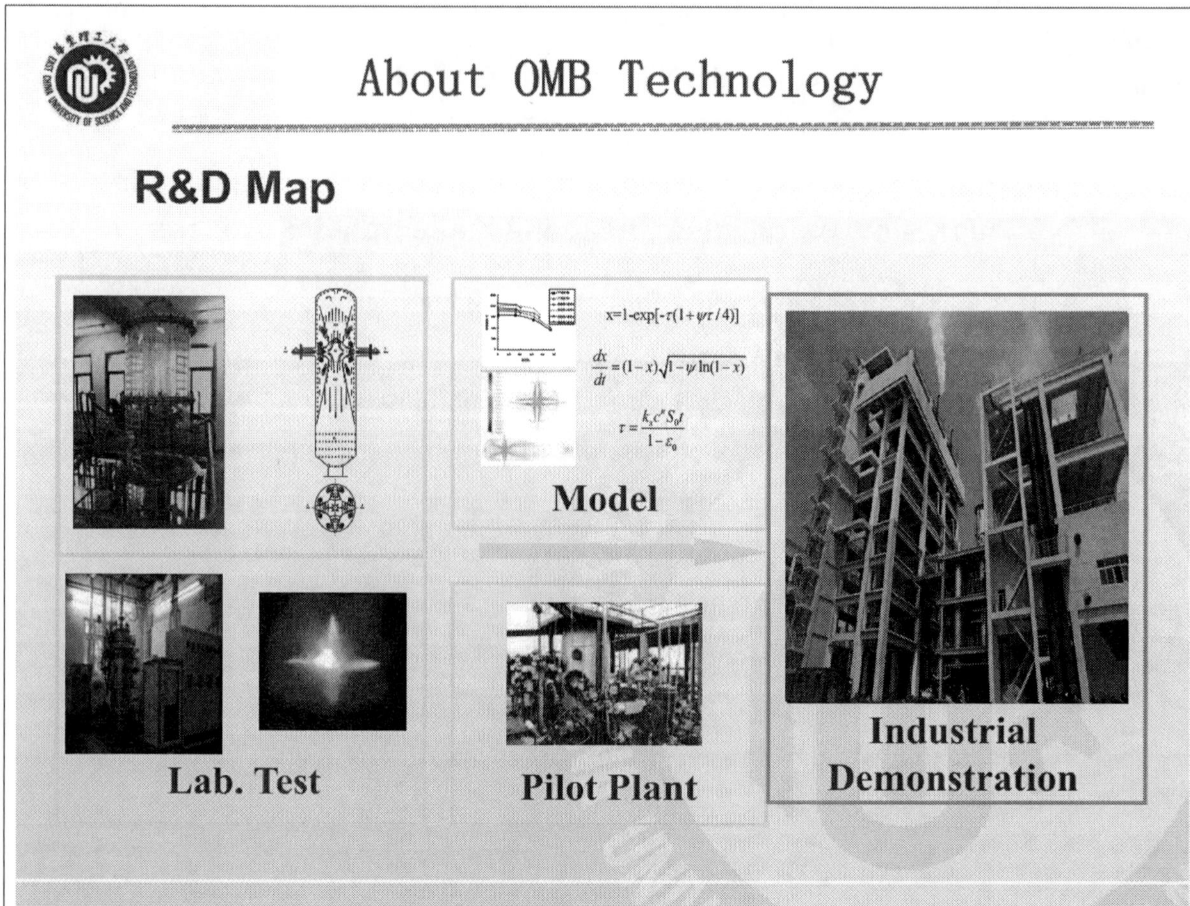
About OMB Technology
R&D Map
$x = 1 - \exp[-\tau(1 + \psi\tau/4)]$
$\frac{dx}{dt} = (1-x)\sqrt{1-\psi\ln(1-x)}$
$\tau = \frac{k_s c^n S_0 t}{1-\varepsilon_0}$
Model
Lab. Test
Pilot Plant
Industrial Demonstration

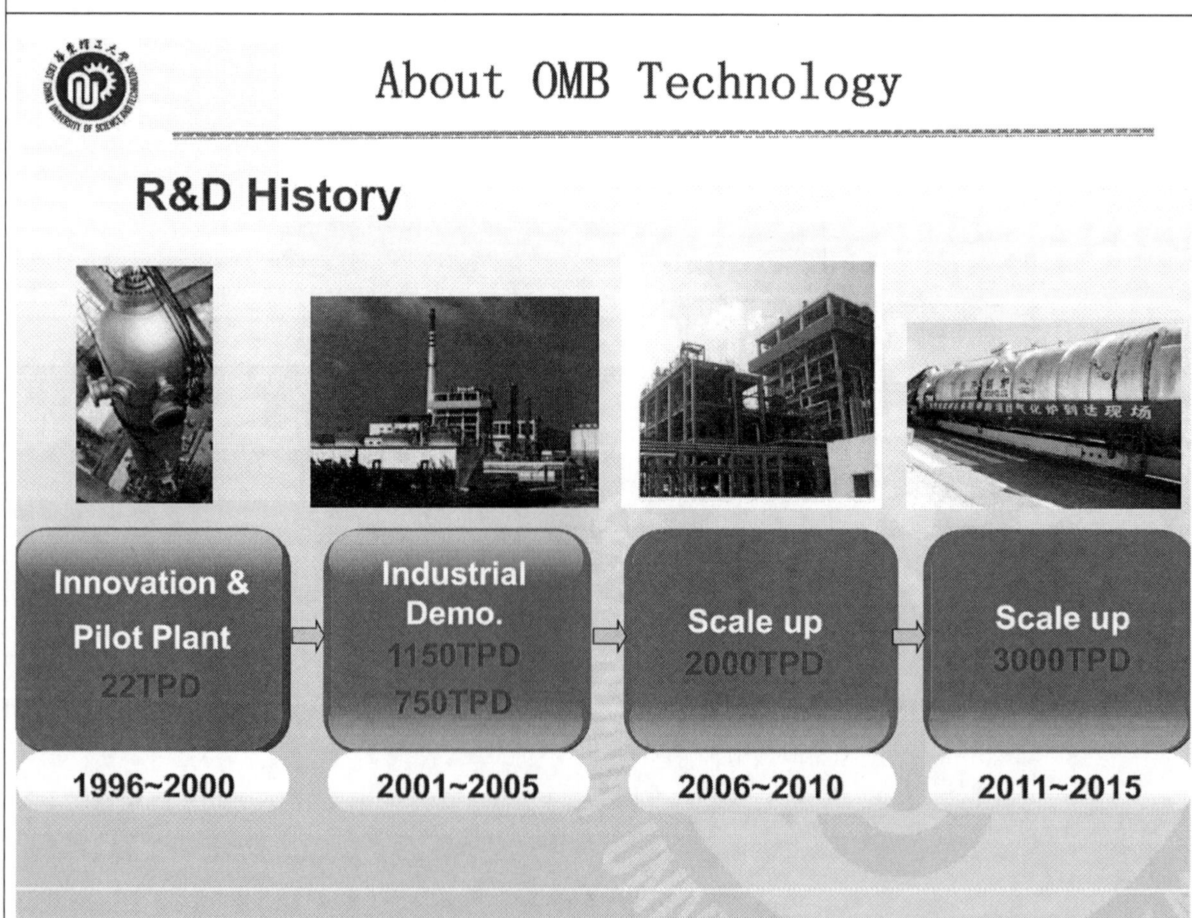
About OMB Technology
R&D History
Innovation & Pilot Plant
22TPD
Industrial Demo.
1150TPD
750TPD
Scale up
2000TPD
Scale up
3000TPD
1996~2000
2001~2005
2006~2010
2011~2015

# OMB Gasification Technology

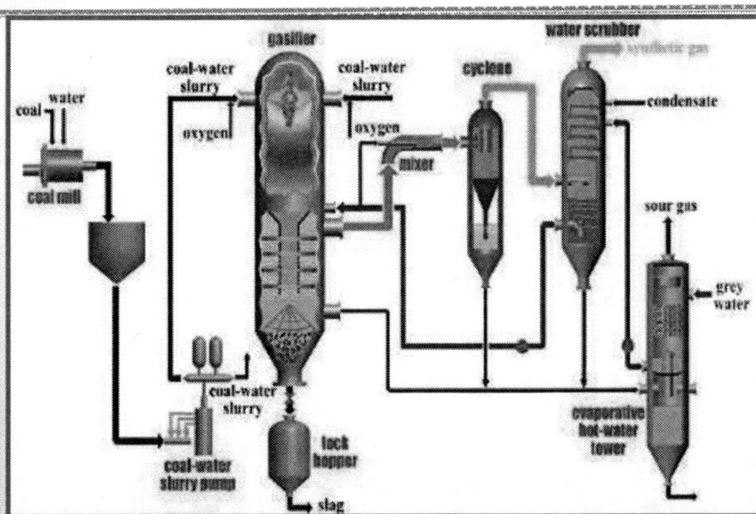

# Technology Features

- **High carbon conversion and low oxygen/coal consumptions:** opposed burner and better mixing process
- **Easy to scale up (2000-3000TPD):** multi-burner
- **High efficiency of syngas primary purification section:** low process pressure drop and low fly-ash content in syngas (1mg/Nm$^3$ syngas) using "multi-stages" method
- **High heat recover efficiency and stability in slag water treatment section:** direct heat exchange between grey water and vapor from flash of slag water

## Industrial Applications

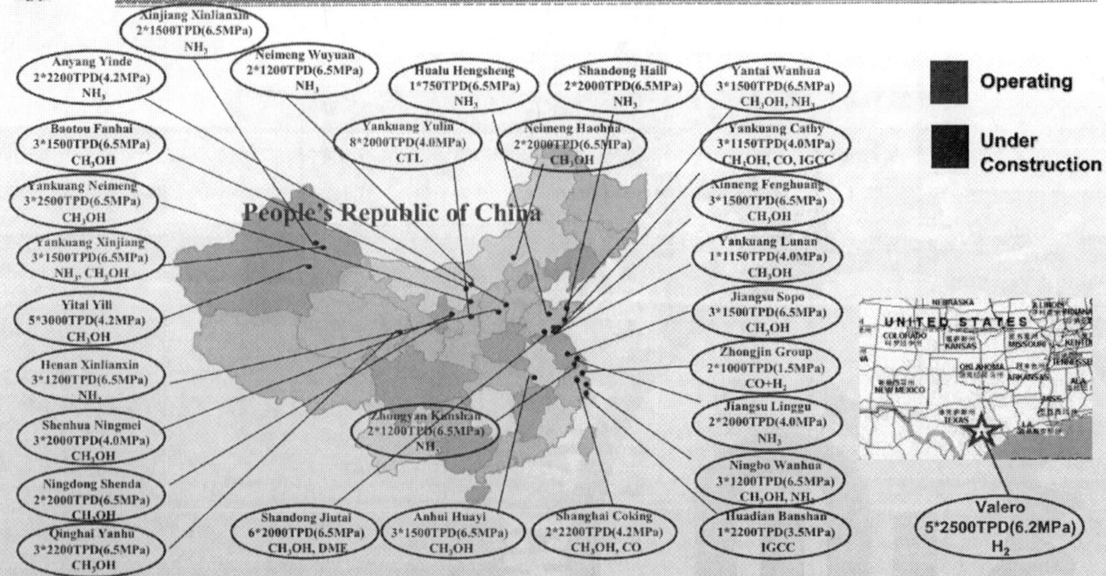


- 45 projects (122 gasifiers)
- The total capacity of coal for all projects is ~140,000 TPD
- The capacity of single gasifier ranges from 750 to 3250

## Industrial Applications

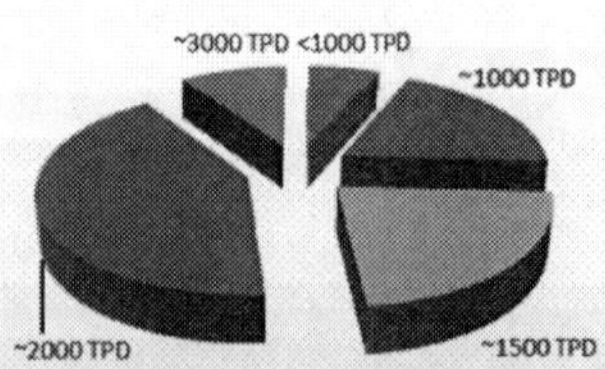


Total Projects = 43

| Des. Cap. Per Gasifier | Num. of Projects | Num. of Gasifiers | Num. of Projects in Operating | Num. of Gasifiers in Operating |
|---|---|---|---|---|
| <1000 TPD | 2 | 3 | 2 | 3 |
| ~1000TPD | 9 | 14 | 3 | 8 |
| ~1500 TPD | 11 | 44 | 7 | 19 |
| ~2000 TPD | 15 | 43 | 8 | 22 |
| 2500~3000 TPD | 6 | 14 | 2 | 5 |
| *Sums* | *43* | *118* | *22* | *57* |

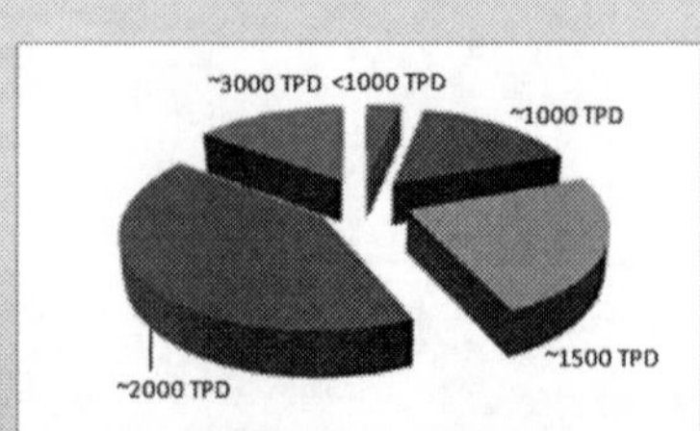


Total Gasifiers = 118

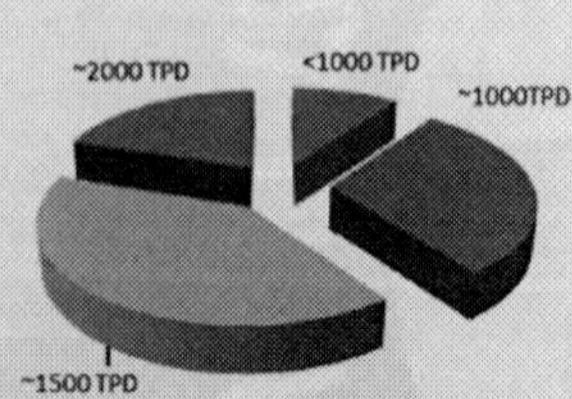


Projects in O.P. = 21

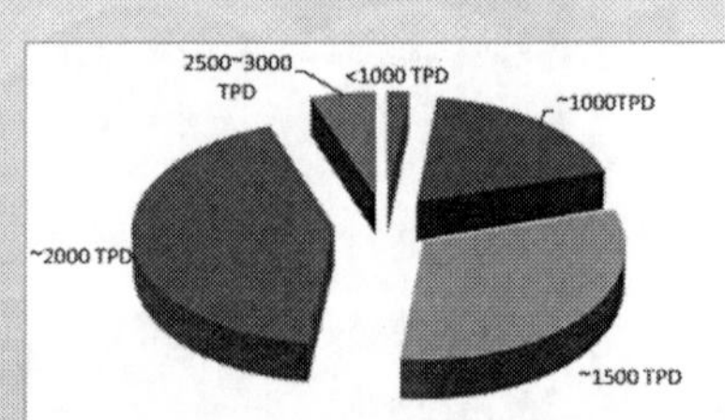


Gasifiers in O.P. = 57

# Industrial Applications

- Company: Hengli Petrochemical(Dalian) Refinery Co. Ltd.
- Gasifiers: 4+2
- Pressure: 6.5MPa(G)
- Products: Hydrogen & gas
- Syngas ($CO+H_2$) flowrate: 460000Nm$^3$/h
- Syngas ($CO+H_2$) flowrate(will be modified): 800000Nm$^3$/h
- PDP contract: June 2014
- Commencement ceremony: Dec. 9, 2015
- The largest hydrogen & gas production plant in China

# Industrial Applications

In Operation (22 projects, 57 gasifiers)

## Industrial Applications-Jiangsu Sopo, 1500TPD

Gasifiers: 2+1
Pressure: 6.5MPa(G)
Startup: Sep. 8, 2009

In 2013
Operation time: 8756h
Availability: 99.9%

In 2015
The availibility of double-gasifier is 98.5%.

From Jan. 2013 to June 10, 2014, the continuous operation time for the whole plant is 511d.
From 2015 to 2016, the continuous operation time is 404d(shutdown due to the failure of external power grid).

# Industrial Applications-Jiangsu Linggu, 2000TPD

Gasifiers: 2+1

Pressure: 4.0MPa(G)

Startup: June 15, 2009

Availability

2011: 97.8% (357d)

2012: 96.4% (352d)

2013: 97.2% (355d)

2014: 94.8% (346d)

2015: 93.7% (342d)

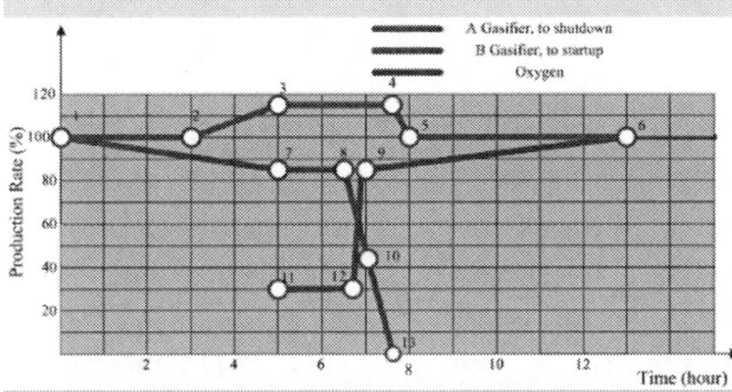


| Feed | Shenhua coal | |
|---|---|---|
| | Designed | Performance |
| Carbon conversion % | 98 | 99.2 |
| Specific oxygen consumption $Nm^3/1000Nm^3(CO+H_2)$ | 390 | 352 |
| Specific coal consumption $kg/1000Nm^3(CO+H_2)$ | 590 | 568 |
| Syngas component v% | 81 | 82.9 |
| Cold gas efficiency % | 74 | 75 |

# Industrial Applications-Shanghai Coking, 2200TPD

- Gasifiers: 1+1
- Pressure: 4.2MPa(G)
- Startup: March 31, 2013
- Carbon conversion: 98.9%
- Operation capacity: $120m^3/h$ CWS

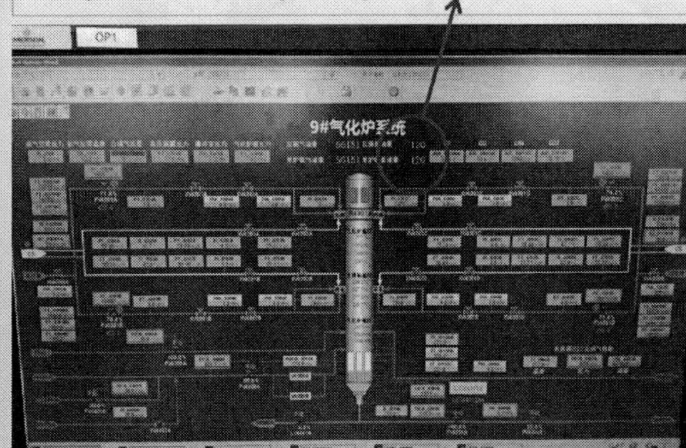

## Industrial Applications-Inn. Mon. Rongxin, 3000TPD

- Gasifiers: 2+1
- Pressure: 6.5MPa(G)
- Startup: June 24, 2014
- Carbon conversion: ~99.5%
- Operation capacity: ~134m³/h CWS
- Performance test: Sept. 23-26, 2015

| Items | Guarantee | Performance |
|---|---|---|
| Carbon Conversion % | >98 | 99.63 |
| Oxygen Consumption $Nm^3/kNm^3(CO+H_2)$ | <400 | 388.3 |
| Coal Consumption $kg/kNm^3(CO+H_2)$ | <600 | 559.5 |
| $CO+H_2$ in dry syngas vol% | >80 | 80.98 |

## Industrial Applications-Shanxi Future Ener., 2000TPD

- Gasifiers: 6+2
- Pressure: 4.0MPa(G)
- Startup: July 31, 2015
- Product: F-T synthesis of 1Mt/a

- First application in large scale (1Mt/a) CTL
- Largest OMB gasification plant in operation

# Industrial Applications-Ningbo Zhongjin, 850TPD

- Gasifiers: 1+1
- Pressure: 1.5MPa(G)
- Startup: May 24, 2015
- Raw material: Coal+petcoke
- Product: Gas

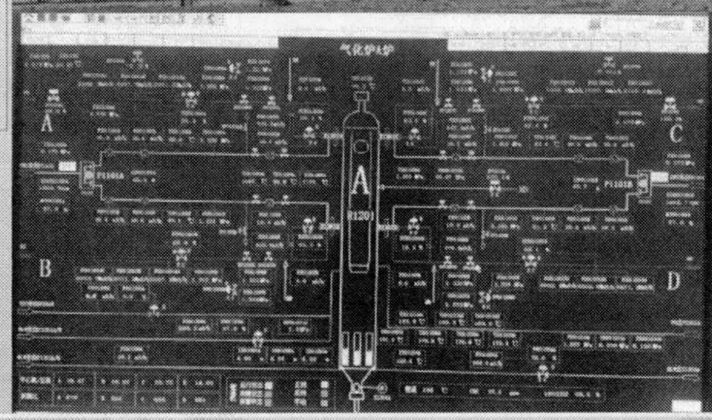

# Industrial Applications

## □ Burner of OMB gasifier (YCCC)

| Year | Average lifespan (d) | Longest lifespan (d) | Shortest lifespan (d) | Shutdown number due to burner error |
|---|---|---|---|---|
| 2005 | 17 | 31 | 15 | - |
| 2006 | 28 | 68 | 4 | 6 |
| 2007 | 29 | 72 | 13 | 9 |
| 2008 | 47 | 100 | 20 | 5 |
| 2009 | 65 | 104 | 47 | 1 |
| 2010 | 93 | 152 | 53 | 0 |
| 2011 | 89 | 139 | 41 | 0 |

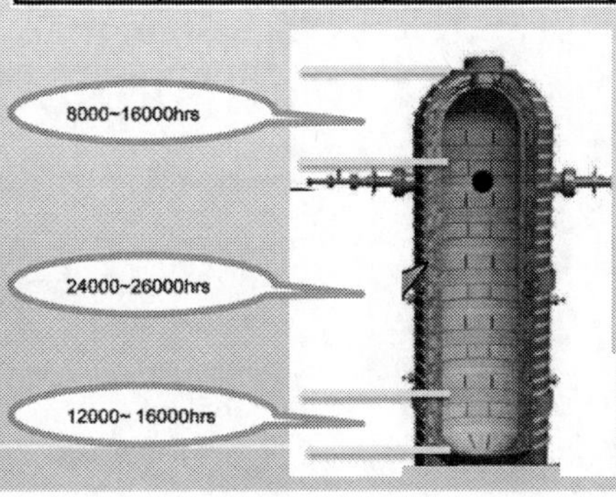


## □ The life of dome refractory for OMB gasifier is about 15000hrs.

# Excellent Performance

## High carbon conversion

| Company | Capacity TPD O.P. + S.P. | Operation pressure MPa | Carbon conversion | Remarks |
|---|---|---|---|---|
| Hualu Hengsheng | 750 1 | 6.5 | 98.3% | Reported |
| Yangkuang Cathay | 1150 2+1 | 4.0 | 98.8% | 168h performance test Dec. 11-18, 2005 |
| Xinneng Fenghuang | 1500 2+1 | 6.5 | 99.2% | 72h performance test Oct. 13-16, 2010 |
| Shenhua Ningmei | 2000 2+1 | 4.0 | 98.9% | 72h performance test Sep. 24-27, 2010 |
| Jiangsuo Linggu | 2000 1+1 | 4.0 | 99.2% | 72h performance test Nov. 25-28, 2011 |
| Anhui Huayi | 1500 2+1 | 6.5 | 98.94% | 72h performance test Aug. 20-24, 2012 |
| Shanghai Coking | 2200 1+1 | 4.2 | 98.9% | Reported |
| Anyang Yingde | 2200 1+1 | 4.2 | 99.3% | 72h performance test Oct. 29-31, 2014 |
| Henan Xinlianxin | 1200 2+1 | 6.5 | 98.7% | 72h performance test Jan. 12-14, 2015 |
| Wanhua Chemical | 1500 2+1 | 6.5 | 99.38% | 72h performance test Aug. 17-19, 2015 |
| Inner Mongonia | 3000 2+1 | 6.5 | 99.63% | 72h performance test Sep. 23-26, 2015 |

# OMB Conference

# 中国石油和化学工业联合会文件

## China Petroleum and Chemical Industry Federation File

中石化联煤发（2016）128号

CPCIF Announcement (2016) No.128

**Pre-Advice of the 5th International Conference of Opposed Multi-Burner (OMB) Coal Gasification Technology Promotion and Application**

**Nov. 2016, Nanjing, China**

# Industrial Partners

兖矿集团 WWW.YKJT.CN

sopo 索普

WANHUA 万华

HUAYI 华谊集团

盈德气体

YCCC

兖矿鲁南化肥厂

江苏灵谷化工有限公司 JIANGSU LINGGU CHEMICAL CO.,LTD.

上海焦化有限公司 Shanghai Coking & Chemical Corporation

ENN 新奥

神华宁夏煤业集团

山东联盟化工集团

恒力石化

DME 久泰能源

华鲁恒升 HUALU-HENGSHENG

中国华电集团公司 CHINA HUADIAN CORPORATION

河南心连心化肥有限公司 HENAN XINLIANXIN FERTILISER LIMITED COMPANY

SINOPEC 中国石油化工股份有限公司 CHINA PETROLEUM & CHEMICAL CORPORATION

中国海洋石油总公司 CHINA NATIONAL OFFSHORE OIL CORP.

湖北宜化

VALERO

TCC 中国天辰工程有限公司 CHINA TIANCHEN ENGINEERING CORPORATION

HUALU 华陆工程科技有限责任公司 HUALU ENGINEERING & TECHNOLOGY CO.,LTD （化学工业部第六设计院）

东华工程科技股份有限公司 EAST CHINA ENGINEERING SCIENCE AND TECHNOLOGY CO., LTD.

# International Activities

## THIRTY - THIRD - ANNUAL INTERNATIONAL PITTSBURGH COAL CONFERENCE

# Thanks !

**http://icct.ecust.edu.cn**

Alternative Clean Energy
OHIO VALLEY UNIVERSITY

OHIO VALLEY
UNIVERSITY

# Ohio Valley University's Alternative Clean Energy Facility

TCG GLOBAL

**Marcus Wiley, P.E.**
**Chairman - TCG Global, LLC**

THIRTY - THIRD ANNUAL
INTERNATIONAL PITTSBURGH
COAL CONFERENCE
University of Pittsburgh - Swanson School of Engineering
Cape Town, South Africa - August 10, 2016



## Mark Wiley
## Chairman, TCG Global, LLC

- A registered professional engineer with 45 years of experience in engineering and management with an emphasis on coal
- Worked for (at the time the second largest US coal company) Consolidation Coal (now Consol Energy) – in mine exploration, planning, permitting, front line supervision, construction, mine superintendent and general mine management in operations located in four major U.S. coal fields
- Co-inventor and co-patent holder for a revolutionary gasification process and personally led the engineering, procurement and construction of a commercial scale reference plant based on this technology
- Part owner and President of two different coal companies with mines in Oklahoma, Arkansas and West Virginia
- Holds a BS in Mine Engineering from New Mexico Institute of Mining & Technology
- Taught for 4 years and is a PhD Candidate at Colorado School of Mines
- Currently serves on the Oklahoma Christian University Board of Trustees

ive Clean Energy
ALLEY UNIVERSITY

# TCG Global, LLC

- TCG Global was formed in 2010 to further the commercialization of an innovative gasification technology
- TCG Global, LLC utilizes patented technology to design, build, own, operate, and market gasification plants
- Patents of Thermo Technologies, LLC are exclusively licensed to Thermo Conversions, LLC and TCG operates using this technology
- Ohio Valley University has an operating permit for a plant from Thermo Conversions

OVU-A

ive Clean Energy
ALLEY UNIVERSITY

# US Coal Map

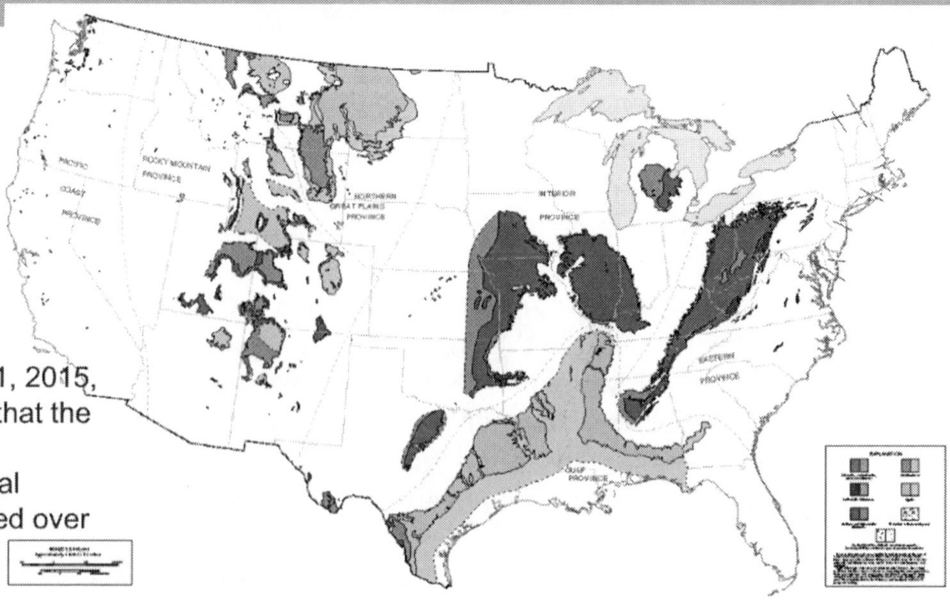


As of January 1, 2015, EIA estimated that the remaining U.S. recoverable coal Reserves totaled over

**255 billion short tons**

http://pubs.usgs.gov/of/1996/of96-092/other_files/us_coal.pdf - www.eia.gov/**coal/reserves/**

tive Clean Energy
VALLEY UNIVERSITY

# West Virginia Coal Map

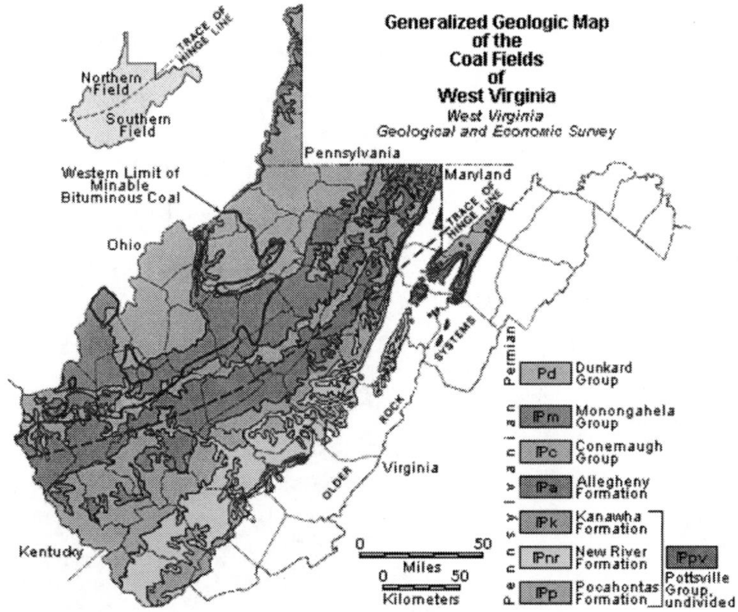


2009 Estimated Recoverable Resources equals **51 billion tons**

http://wvges.wvnet.edu/
http://www.wvminesafety.org/PDFs/resources2009.pdf

OVU-A

tive Clean Energy
VALLEY UNIVERSITY

# Ohio Valley University

OHIO VALLEY
UNIVERSITY

**Chartered in 1958 as a private, independent, regional, faith-proclaiming, liberal arts university**

**Annually recognized as one of the nation's best comprehensive baccalaureate universities in *U.S. News & World Report's America's Best Colleges Guide***

atïve Clean Energy
O VALLEY UNIVERSITY

# OVU Location

**Located in Vienna, Wood County, West Virginia in the Ohio River Valley, a community of approximately 80,000**

**OVU is situated above the valley on 266 wooded acres**

OVU-

atïve Clean Energy
O VALLEY UNIVERSITY

# Energy Management and Production Engineering

Ohio Valley University's Bachelor of Science degree in Energy Management & Production Engineering (EMPE) is designed to provide students with entry-level capabilities to manage operations within the energy industry.

https://ovu.edu/images/academics/empe/2015%20EMPE%20Program%20Guide.pdf

ative Clean Energy
VALLEY UNIVERSITY

# Alternative Clean Energy Project

- Development of a fully commercial scale coal to liquids facility in Vienna/Parkersburg, WV owned by OVU-ACE Foundation
- Establish a training center and expand the curriculum to include operation and maintenance of a clean energy production facility
- Revenue from operations net of debt service would be used to fund OVU and other Universities' general expenses thus lowering costs for students
  - Goal is to have students graduate with zero debt
  - Work study and intern opportunities for students
- Future college of engineering in alternative energy
- Host site for multiple projects demonstrating $CO_2$ utilization

OVU-A

ative Clean Energy
VALLEY UNIVERSITY

# Why ACE at OVU?

- Leadership
  - Education and business expertise
- Place where innovation happens
  - Stretch goal to graduate students debt free
- No faculty, administration, trustee conflicts
- Fit with the University's Strategic Plan
- Agility
  - "Turn on a dime"

tive Clean Energy
VALLEY UNIVERSITY

# OVU-ACE Project Objectives

- Demonstrate conversion of coal into clean energy products
- Support OVU operating budget and act as a model for other institutions
- Serve as an educational resource for Energy Management & Production Engineering degree program
- Host site for other clean energy technology demonstrations
- Research site for other coal related by-products such as recovery of rare earth elements from coal ash

OVU-A

tive Clean Energy
VALLEY UNIVERSITY

# Why Coal?

- Abundant domestic and world wide energy resource
- Coal pricing is relatively stable compared to other forms of energy
- Supports an industry vital to West Virginia
- Compelling economics when converting to clean energy products
- Coal has highest energy per unit of volume of any fossil fuel
- Technology exists to utilize coal with potentially zero Carbon Dioxide, Sulfur, NOx and Mercury emissions

# OVU-ACE Facility Concept

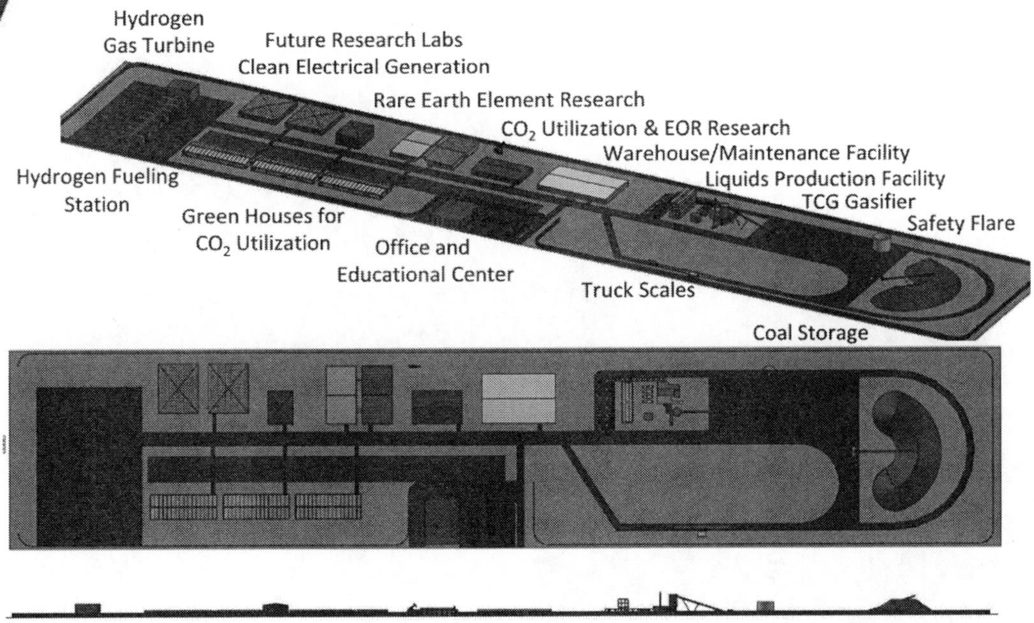


# Office and Educational Center

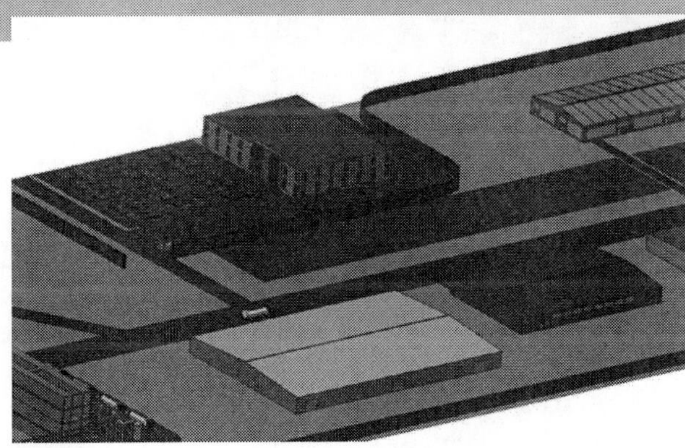

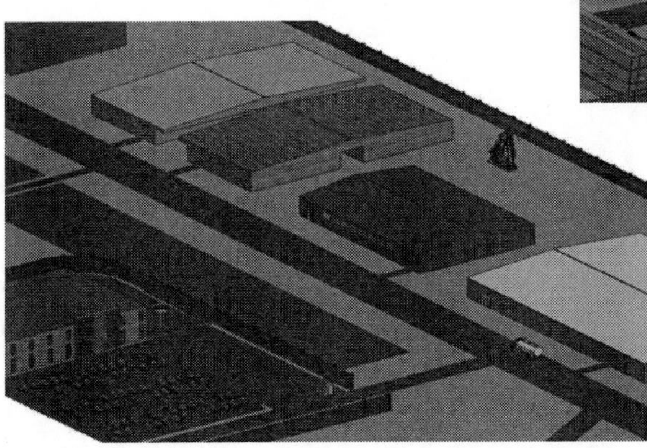

# Liquids Production Facility with TCG Gasifier

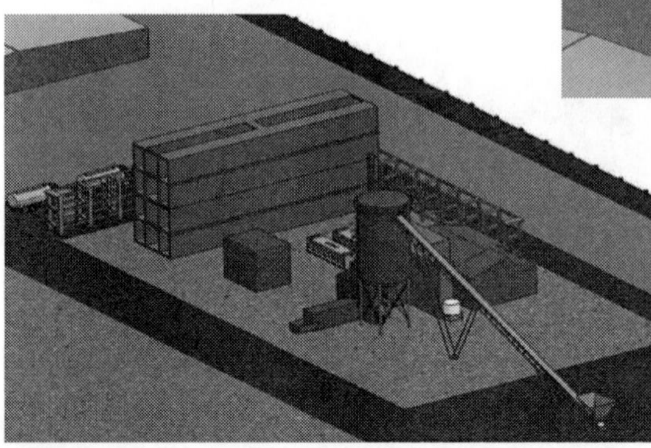

# Coal Storage

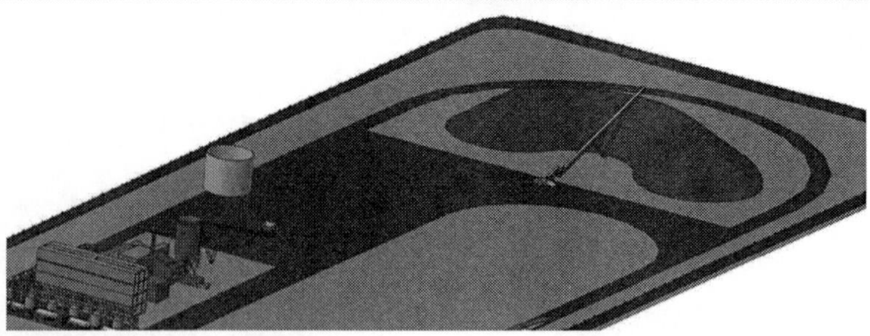

native Clean Energy
O VALLEY UNIVERSITY

# Future Research Labs and $CO_2$ Utilization Green Houses

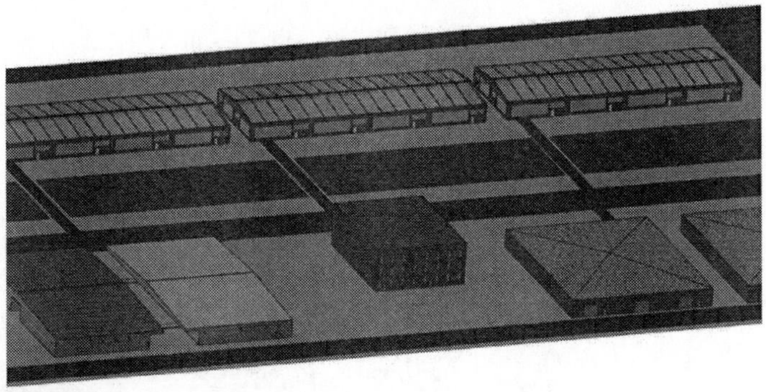

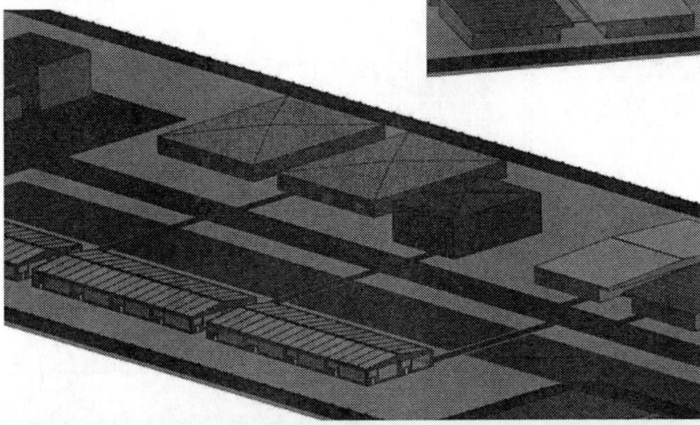

OVU-

native Clean Energy
O VALLEY UNIVERSITY

# Hydrogen Fueling Station and Hydrogen Turbine

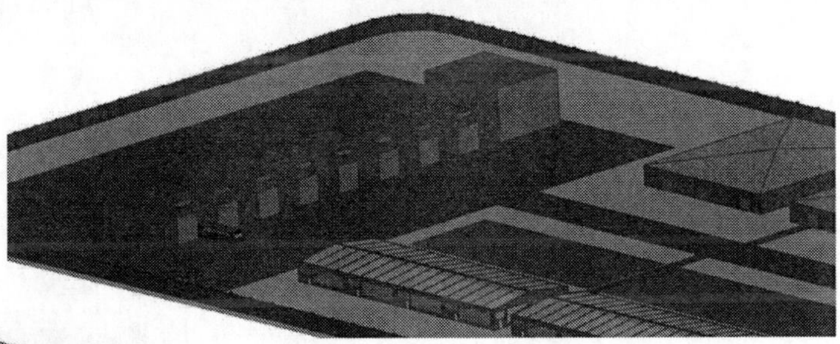

# Technology Platform

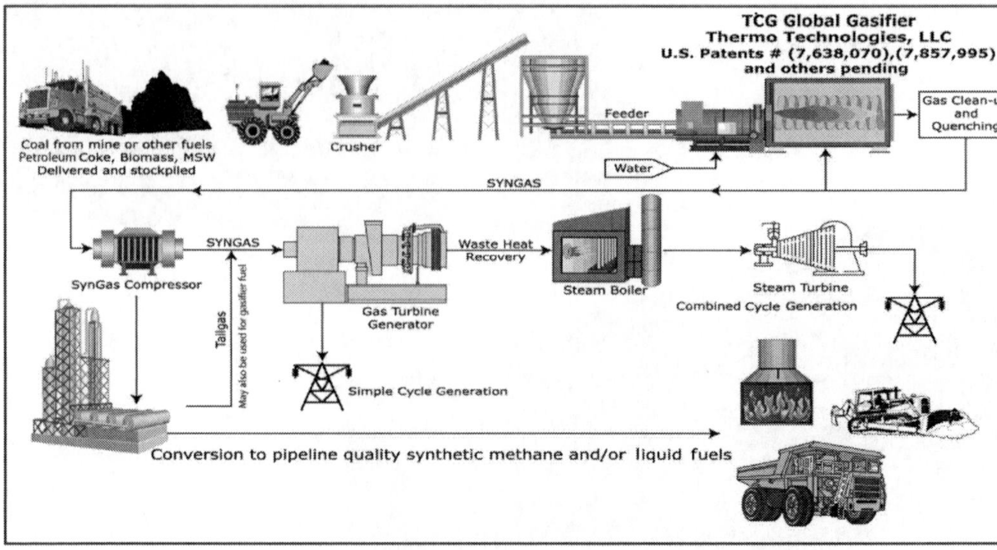


OVU-AC

# Product Options

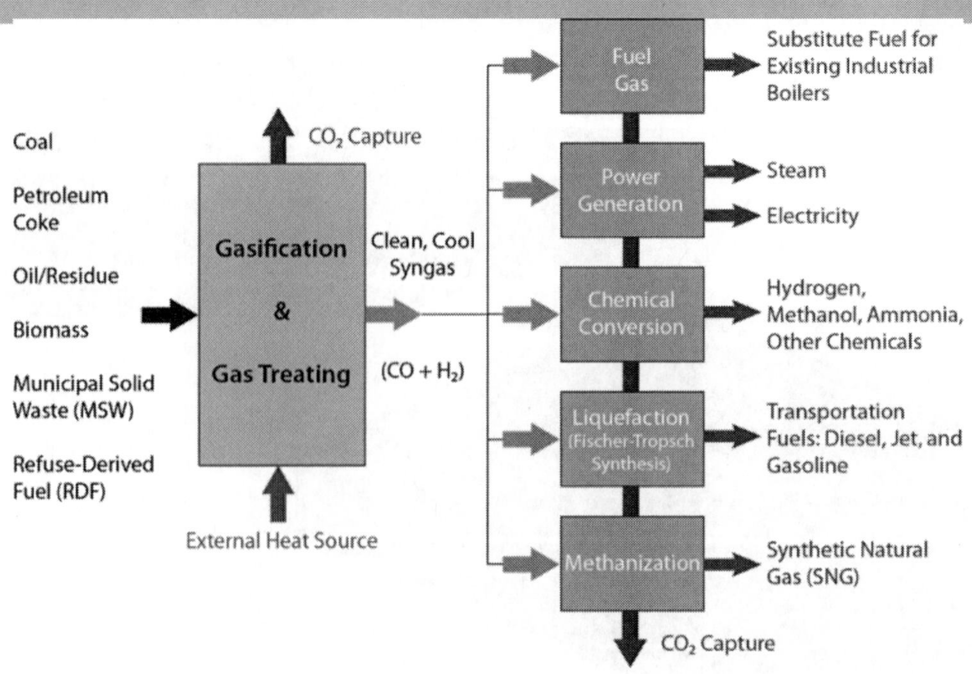

# Production Products

- The OVU-ACE facility can produce a variety of products from the coal derived clean syngas
  - Diesel
  - Mixed Alcohols
    - Methanol
    - Ethanol
    - Propanol
  - Fischer Tropsch waxes
  - Electricity
  - Carbon Dioxide
  - Hydrogen
- Rare Earth Elements can be recovered from the coal ash

OVU-AC

# Efficient Design for Liquids Fuel Production

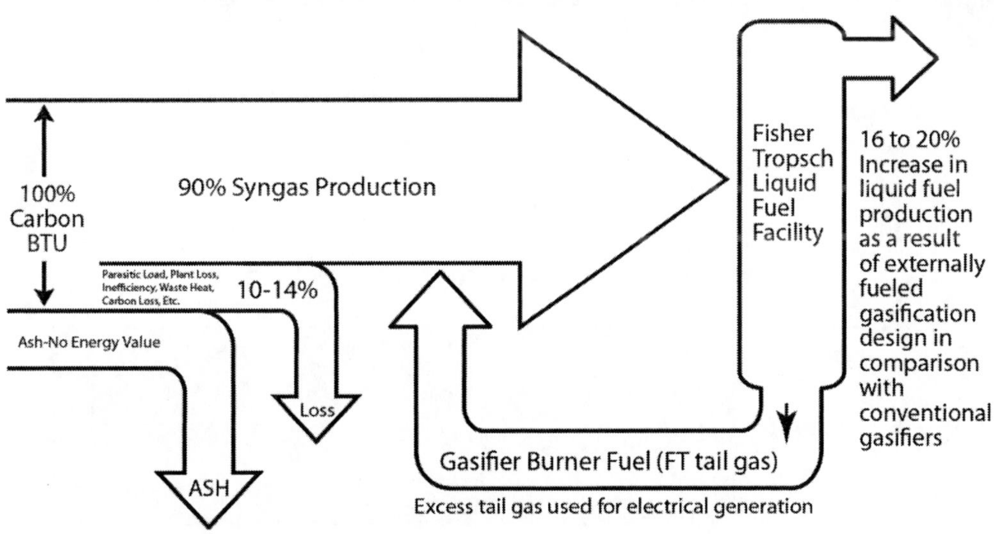

rnative Clean Energy
HIO VALLEY UNIVERSITY

# Clean Emissions & $CO_2$ Capture

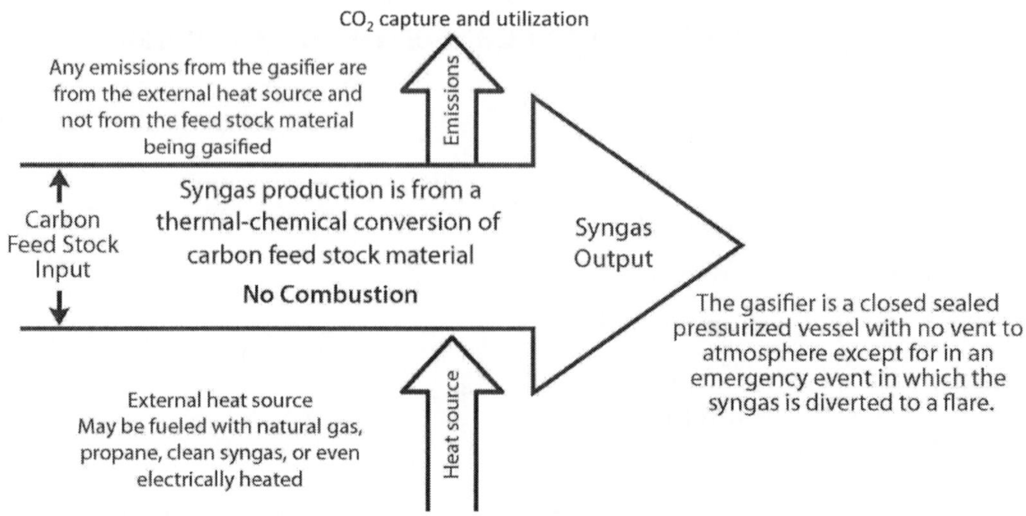


OVU

Alternative Clean Energy
OHIO VALLEY UNIVERSITY

# WV Geological & Economic Survey

**WVGES has offered to assist OVU-ACE in the development of several options for CO2 sequestration in the Vienna/Parkersburg area, and perhaps even a utilization (enhanced oil recovery) project in the depleted Parkersburg oil field.**

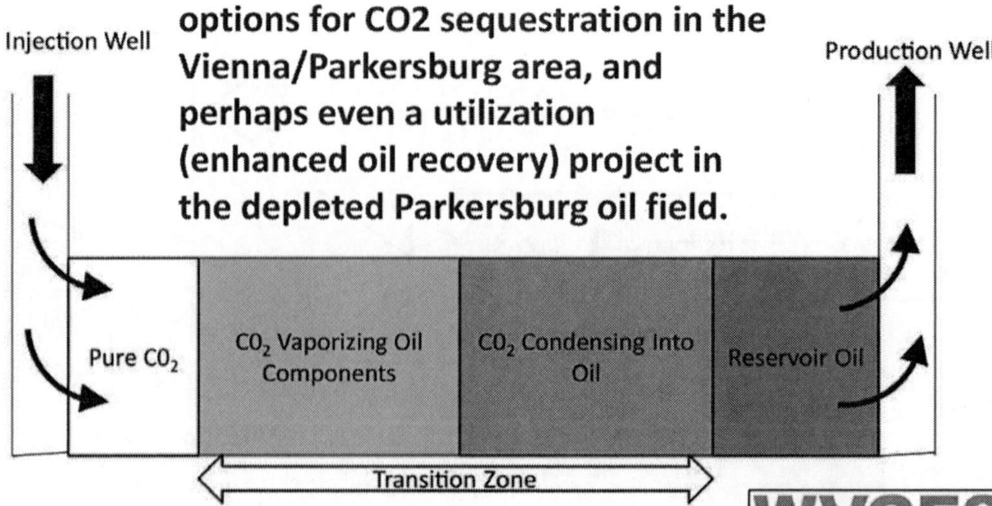


WVGES
GEOLOGY UNDERLIES IT ALL

OVU-ACE , LLC

Alternative Clean Energy
OHIO VALLEY UNIVERSITY

# $CO_2$ Enhanced Oil Recovery

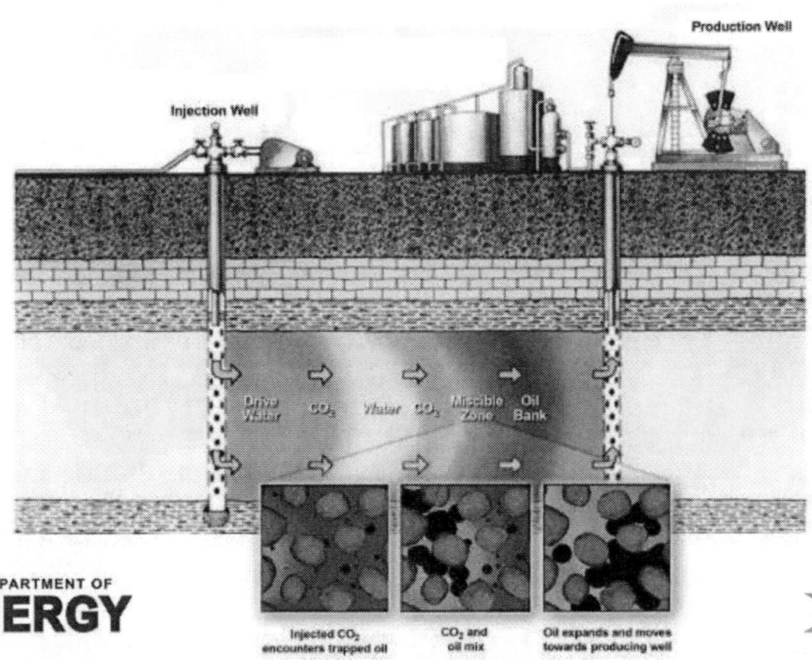


U.S. DEPARTMENT OF ENERGY

NETL

Source: http://energy.gov/fe/science-innovation/oil-gas-research/enhanced-oil-recovery

OVU

Alternative Clean Energy
OHIO VALLEY UNIVERSITY

# WV Geological & Economic Survey

**WVGES would serve as Technical Advisor to Ohio Valley University for improved characterization of Rare-Earth/Strategic Element composition of WV Coals. Staff from the Coal Program would serve as advisory scientists and will provide guidance pertaining to:**

- General coal geology of West Virginia
- Location/characteristics of key mines/seams
- Relative concentration of REEs in coals and associated coal waste (fly ash)
- Integration of legacy data with newly-collected x-ray fluorescence (XRF) analyses

**This guidance would include compilation of relevant scientific literature, feedback pertaining to curriculum design, and hands-on, inquiry-based demonstrations**

WVGES
GEOLOGY UNDERLIES IT ALL

OVU-ACE , LLC

# Rare Earth Elements

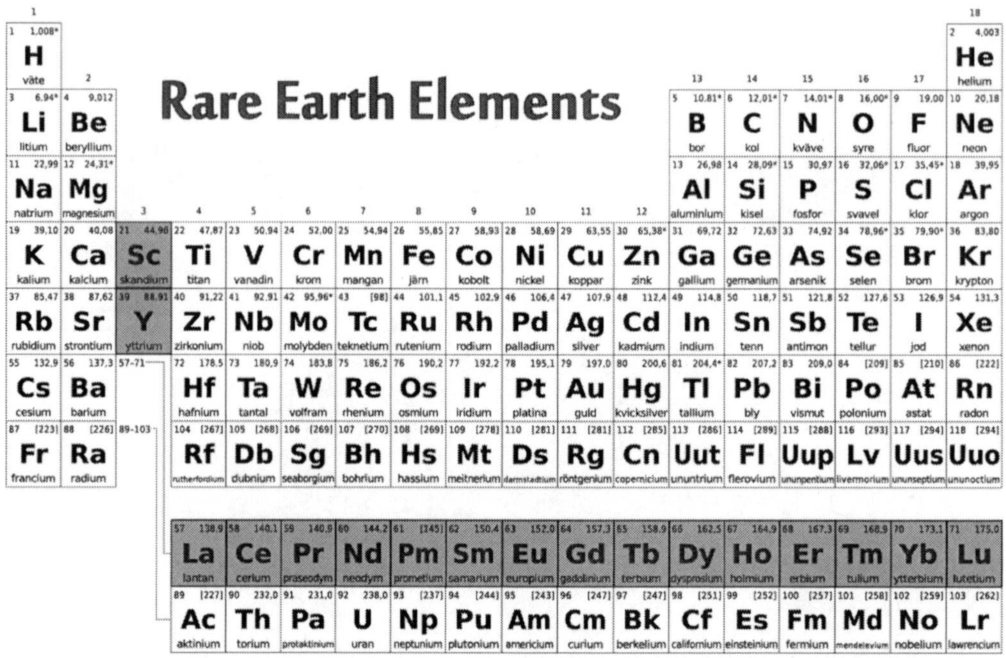


OVU-AC

# Rare Earth Elements

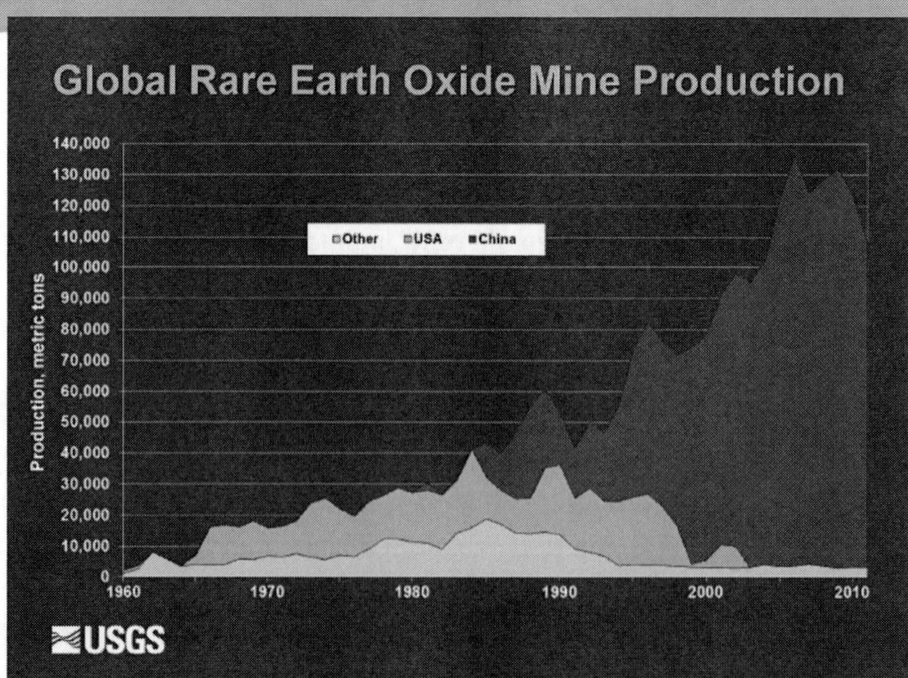

# WV Geological & Economic Survey

**Development and Implementation of a Training Curriculum for Rare Earth Processing of West Virginia Coals**

**Project Goals:**

- workforce development related to next-gen coal technologies
- filling a critical gap in development of a domestic source of strategic minerals
- promotion of green technologies through processing of materials traditionally thought of as waste products

WVGES
GEOLOGY UNDERLIES IT ALL

OVU-ACE , LLC

# Project Status

- ✓ **OVU retained AAIG and North Slope Capital for Overall Project Funding**
- ✓ **Secured additional development funding**
- ✓ **Site acquisition documents executed**
- ✓ **Preliminary engineering and budgeting completed**
- ✓ **Supply and Offtake agreements underway**
- ✓ **Permitting implemented**
- **Final engineering design**
- **Financial close**
- **Permit issuance**
- **Construction**
- **Operation**

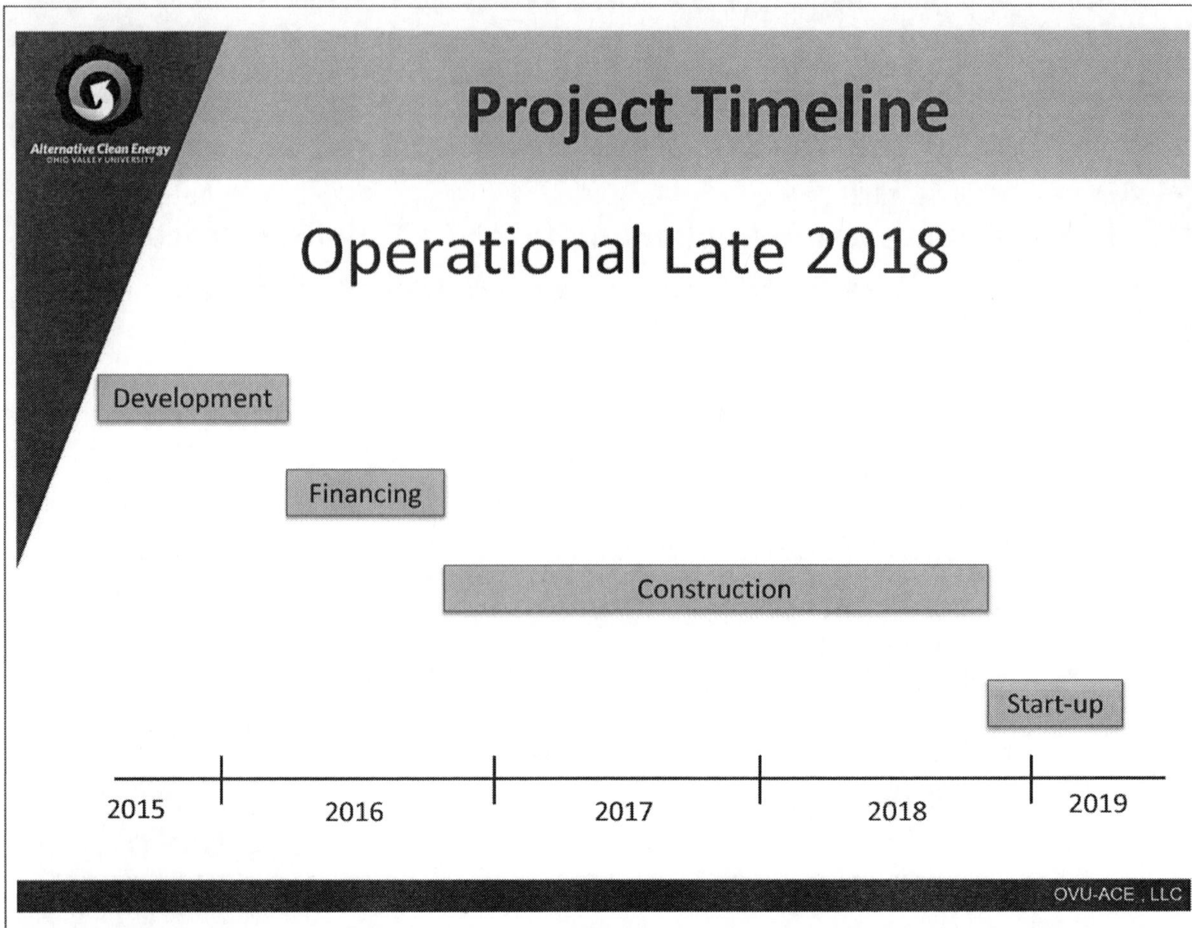
Alternative Clean Energy
OHIO VALLEY UNIVERSITY
Project Timeline
Operational Late 2018
Development
Financing
Construction
Start-up
2015
2016
2017
2018
2019
OVU-ACE , LLC

Alternative Clean Energy
OHIO VALLEY UNIVERSITY
Supporting and Contributing Organizations
OHIO VALLEY UNIVERSITY
WVGES
GEOLOGY UNDERLIES IT ALL
CITY OF PARKERSBURG
North Slope Capital Advisors
TCG GLOBAL
CITY OF VIENNA
All American Investment Group
LOCKTON
noble group
FLUOR
OVU-ACE , LLC

Alternative Clean Energy
OHIO VALLEY UNIVERSITY

# Questions?

OHIO VALLEY UNIVERSITY

WVGES GEOLOGY UNDERLIES IT ALL

CITY OF VIENNA

North Slope Capital Advisors

TCG GLOBAL

CITY OF PARKERSBURG

All American Investment Group

LOCKTON

noble group

FLUOR

OVU-ACE , LLC

# OVU-ACE Team

## Jeff Dimick – OVU-ACE President & OVU EVP

Jeff is an experienced executive that has overseen and successfully completed numerous complex projects. He currently manages OVU's operations and represents OVU with OVU President Harold Shank. Prior to joining OVU, Jeff worked for 27 years at The Boeing Company (and its predecessor Hughes Aircraft). At Boeing, he filled roles of increasing responsibility, culminating as a Program Manager in the Advanced Systems group. Senior management recognized Jeff as a 'fixer,' who was brought in to get projects back on track. In addition, he has shown an entrepreneurial streak and skills by founding and managing One World Post, a video post production company in Los Angeles. Jeff holds a Bachelors of Science in Mathematics & Computer Science from Oklahoma Christian University and a Masters of Science in Pure Mathematics from Oklahoma State University.

Currently serves on the Oklahoma Christian University Board of Trustees

e Clean Energy
LEY UNIVERSITY

# OVU-ACE Team

## Bob Winland – OVU-ACE Board & Advisor

Bob's background is in Petroleum Engineering with an emphasis on project development and economics. He recently concluded a 36-year career with Marathon Oil Company, working on projects ranging from simple, low-cost onshore developments through complex, multi-billion-dollar International offshore developments. Projects for which he provided assistance include: reservoir engineering management for the Sakhalin project in Russia, the country's first offshore development; exploration and conceptual development economics for a large exploration program offshore Angola; and economic modeling for Marathon's $5 billion entry into the Eagle Ford project in Texas. Bob holds an Associate of Science degree from Ohio Valley University (and is on the Board of Trustees) and a Bachelor of Science degree in Petroleum Engineering from Marietta (OH) College

OVU-AC

e Clean Energy
LEY UNIVERSITY

# OVU-ACE Team

## Dan Blair – OVU-ACE Board & OVU Dean of College of Business

For over 30 years, Dr. Dan. Blair has been leading performance improvement initiatives. He began his work life as a summer-time clerk typist for the Department of the Army, and then served in the United States Air Force. For two decades, Dan dedicated his talents to the Hewlett-Packard Company, where he led several important initiatives, including the development of HP's first worldwide training management system. At OVU, Dan established the Office of Institutional Effectiveness and currently serves as the Dean of the College of Business. He holds Bachelor of Science and Master of Arts degrees, and a doctorate from the University of Southern California

# City of Vienna

- Supports OVU-ACE project
- Founded in 1795
- Once home to many industrial sites
- Now home to retail trade for surrounding area
- Regional population ~200,000 residents
- Population: 12,000 residents
- A debt-free city
- Low crime rate
- Government: Mayor & Council
- Available skilled workforce

# City of Parkersburg

- Supports OVU-ACE project
- Named Parkersburg in 1810
- Once home to many industrial sites
- Now home to retail trade for surrounding area
- Regional population ~200,000 residents
- Population: 31,000 residents
- Low crime rate
- Government: Mayor & Council
- Available skilled workforce

rnative Clean Energy
HIO VALLEY UNIVERSITY

# All American Investment Group

- Provides exceptional service to an existing client base
- Provides innovative and unique debt structures for the taxable and tax-exempt fixed income market place
- The firm obtains its transactions through four primary sources:
  - direct relationships with issuers
  - established vendor programs
  - relationships with financial advisors
  - relationships with sales consultants

All American
Investment Group

OVU

Alternative Clean Energy
OHIO VALLEY UNIVERSITY

# North Slope Capital Advisors

- North Slope works closely with a University to identify and prioritize goals such as savings targets, affordability of new debt, debt policies and interest rate management, future capital planning and debt issuances, current debt portfolio reviews, and financial analysis/credit rating strategies.
- Financial product evaluation
- Structuring advice
- Bond sale price guidance
- Deal review

North Slope
Capital Advisors

OVU-ACE , LLC

ative Clean Energy
O VALLEY UNIVERSITY

# Fluor

- Fluor is one of the world's largest publicly traded engineering, procurement, construction (EPC), maintenance, and project management companies. Fluor works with governments and Clients in diverse industries around the world to design, construct, and maintain complex and challenging capital projects

- Clients depend on the expertise of Fluor's 59,000 employees operating globally to deliver capital projects safely, on schedule, within budget, and with the quality they expect

FLUOR®

OVU-

ative Clean Energy
O VALLEY UNIVERSITY

# Noble Group & Revelation Energy

- Noble Americas and Revelation Energy have agreed to provide a long term coal supply contract for the OVU-ACE project
- Noble Group manages a portfolio of global supply chains across a range of industrial and energy products. They market, process, finance and transport key commodities, connecting low-cost producing regions with high-demand growth markets
- Ranked number 77 in the 2015 Fortune Global 500.
- Revelation Energy, LLC is a West Virginia Coal Company

noble group

# Lockton

- Lockton provides a variety of insurance products through companies that they represent
- Lockton is the world's largest privately owned insurance brokerage firm
- Clients most frequently describe Lockton as team members who make their businesses better
- Energy, innovation and deep expertise fuel their focus on solving clients' problems and achieving real results
- A full process warrantee insurance wrap is available to support the financing of this project

LOCKTON

OVU-ACE , LLC

# WV Geological & Economic Survey

WVGES will also serve as a liaison between OVU and other agencies/institutions with current research projects dedicated to industrial processing of REEs

Final Deliverable= Technical Training Manual designed for use by any interested stakeholder or educational institution wishing to investigate rare-earth technologies

WVGES
GEOLOGY UNDERLIES IT ALL

OVU-ACE , LLC

# TCG Proven Technology

Alternative Clean Energy
OHIO VALLEY UNIVERSITY

## University of Toledo plant constructed in 2007

**Testbed for DOE study of Biomass-to-Diesel processes**

OVU-ACE , LLC

# Third Party Report on Syngas Quality

Renewable Energy Testing Center

Accelerating the Marketplace for Renewable Technologies

RETC

US Army Contract W15QKN-05-D-0030
Task 5 RETC, WBS # 4.6.0

Final Technical Report

Technikon # 1602-460 NA

December 2009
*(Revised for public distribution - February 2010)*

Operated by
TECHNIKON
5301 Price Avenue
McClellan, CA 95652
916-929-8001 (p)
916-929-8020 (f)
www.technikon.us

Funded through the
Department of Defense

**Methods, Assumptions and Procedures**

**The primary goal of the RETC subtask 1.1.1 effort is to determine the primary and trace constituents in syngas generated from a commercial scale thermochemical conversion (TCC) system. This 300 dry ton per day (dtpd) system was developed and built by Thermo Conversions, LLC (T.CON) and is currently being operated by Red Lion Bioenergy (RLB) at the University of Toledo.**

**Results and Discussion**

**The results of multiple test runs show that the technical approach being utilized by Red Lion Bio-Energy is producing a higher quality syngas than any historically documented gasification system**

(Page 14)

# Conclusions of US Army Report

- The testing of the Red Lion Bio-Energy gasification system has produced promising results
- This design shows significant improvement over the previous generations of gasification systems
- Testing results support that the pyrolysis/steam reforming [illegible] variety of biomass resources that has an ideal composition for the production of fuels and the co-production of electricity:
  - High quality syngas can be generated from a wide variety of feedstocks
  - The ratio of $H_2/CO$ is in the ideal range of 1.8 to 2.5 for diesel and alcohol fuel production
  - The concentrations of contaminants in the syngas are much lower than the concentration of contaminants generated from thermochemical systems that use oxygen or air. This significantly reduces the cost of syngas purification systems.
    - Several contaminants level specifications have been met (page 16)

# Synthesis Energy Systems, Inc.

GROWTH WITH BLUE SKIES

Francis Lau

33rd International PCC

August 10, 2016

Cape Town, South Africa

# Forward Looking Statements

This presentation includes "forward-looking statements" within the meaning of Section 27A of the Securities Act of 1933, as amended, and Section 21E of the Securities Exchange Act of 1934, as amended. All statements other than statements of historical fact are forward-looking statements. Forward-looking statements are subject to certain risks, trends and uncertainties that could cause actual results to differ materially from those projected. Among those risks, trends and uncertainties are the development stage of the operations of SES, its estimate of the sufficiency of existing capital sources, its ability to successfully develop its licensing business, its ability to raise additional capital to fund cash requirements for future investments and operations including its China platform initiative, its ability to reduce operating costs, the limited history and viability of its technology, commodity prices and the availability and terms of financing opportunities, its results of operations in foreign countries, its ability to diversify, its ability to complete the restructuring of the ZZ joint venture and production of methanol, its ability to obtain the necessary approvals and permits for future projects, the estimated timetables for achieving mechanical completion and commencing commercial operations for the Yima project as well as the ability of the Yima project to produce revenues and earnings, the sufficiency of internal controls and procedures, its ability to grow its business and generate revenues and earnings as a result of its proposed China and India platform initiatives, and its ability to develop its power business unit and marketing arrangement with GE and its other business verticals, steel and renewables. Although SES believes that in making such forward-looking statements its expectations are based upon reasonable assumptions, such statements may be influenced by factors that could cause actual outcomes and results to be materially different from those projected. SES cannot assure you that the assumptions upon which these statements are based will prove to have been correct.

## SES – A Global Energy And Gasification Technology Company

SYMX
NASDAQ®
LISTED

- Publicly listed energy technology company headquarter in Houston, Texas
- Asia Pacific headquarter in Shanghai with Engineering and Technology expertise
- Exclusive global license to U-GAS® gasification technology from the Gas Technology Institute (GTI®). U-GAS® is a commercially proven gasification technology under SES
- SES advanced the technology by improving fuel utilization, plant efficiency, and cost competitiveness through an expanding I.P. library—forming SES Gasification Technology (SGT).

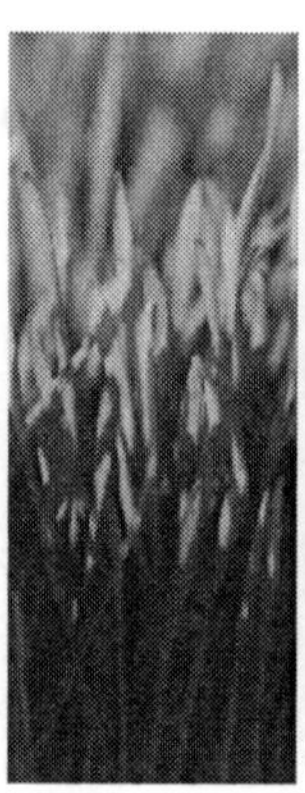

# SES SGT Advanced Gasification Technology

## SES Gasification Technology (SGT) Platform - U-GAS®

- FUEL FLEXIBLE – ALL RANKS OF COAL, BIOMASS & WASTES
- SINGLE FLUIDIZED BED GASIFIER: SIMPLE OPERATION AND LOW-COST DESIGN
- HIGH COLD GAS EFFICIENCY WITH HIGH CARBON CONVERSION
- LOW WATER USAGE AND ENVIRONMENTALLY FRIENDLY

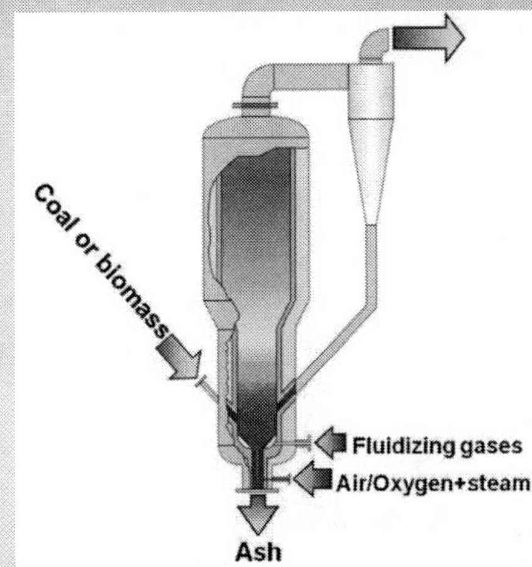


5

## SES Gasification Technology (SGT) Island

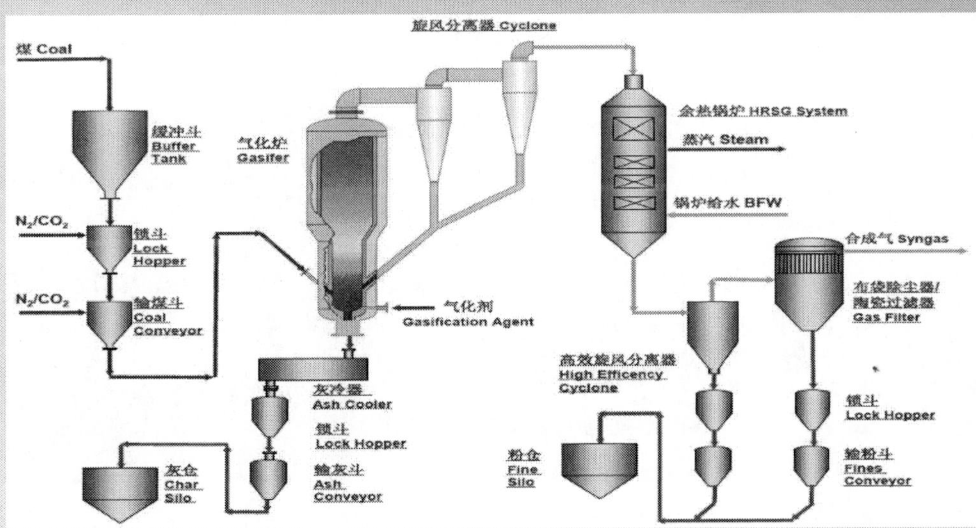


6

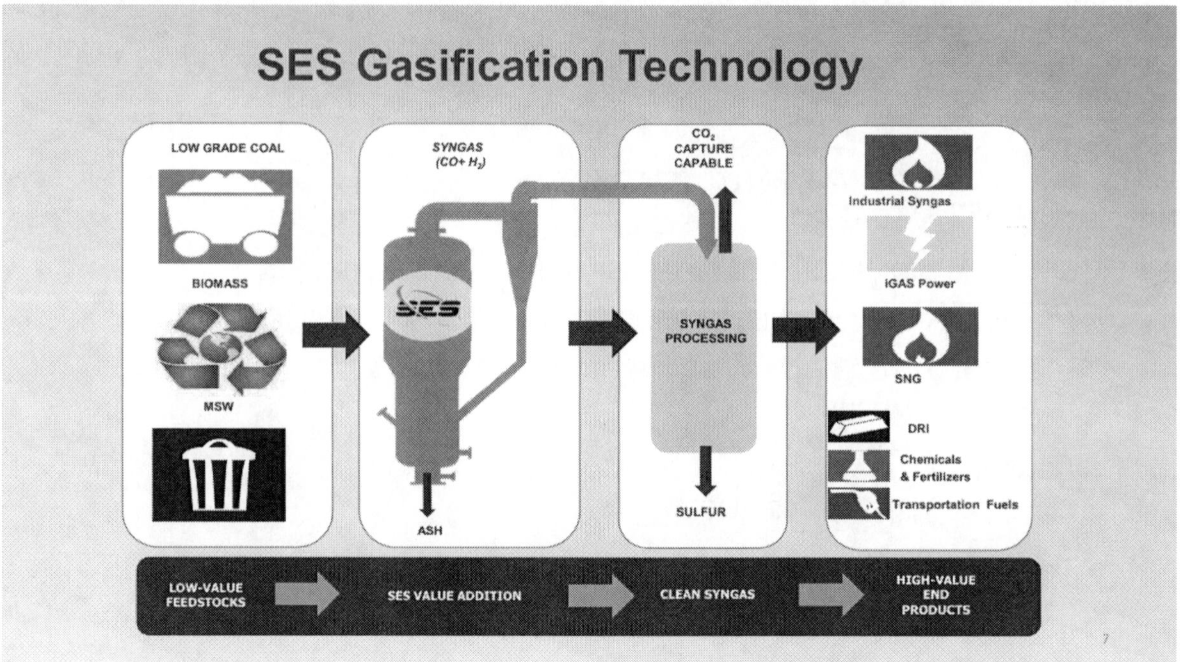
SES Gasification Technology
LOW GRADE COAL
BIOMASS
MSW
SYNGAS
(CO+ H2)
ASH
CO2
CAPTURE
CAPABLE
SYNGAS
PROCESSING
SULFUR
Industrial Syngas
IGAS Power
SNG
DRI
Chemicals
& Fertilizers
Transportation Fuels
LOW-VALUE
FEEDSTOCKS
SES VALUE ADDITION
CLEAN SYNGAS
HIGH-VALUE
END
PRODUCTS
7

SES Gasification
Technology
Development &
Commercialization

8

## Technology from the USA and Commercialized in China

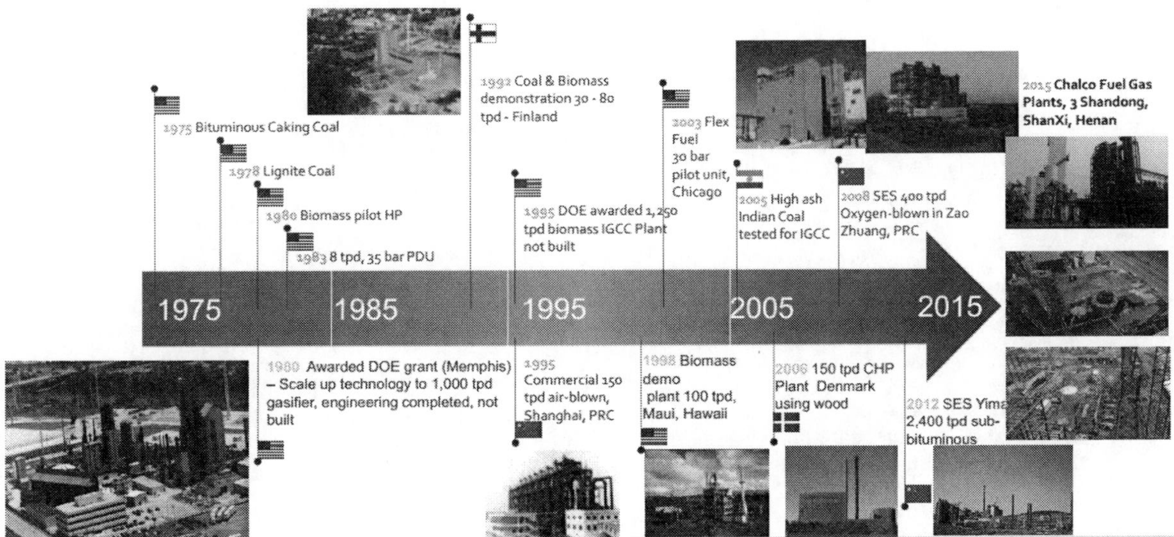


CONFIDENTIAL 9

## GTI Energy Development Center

Gas Technology Institute ("GTI") Energy Development Center
(1965 – 1996)

10

## Feedstocks gasified

**Bituminous and Sub-bituminous Coals**
- Western Kentucky No. 9, washed & ROM
- Western Kentucky No. 9 and 11
- Illinois No. 6, Peabody No. 10 and Crown III
- Pittsburgh No. 8, Champion and Ireland
- Australian, Bayswater No. 2, Sydney Basin
- Polish, Silesia
- French, Merlebach - ROM
- Utah - ROM
- Colombian
- Chinese, Shen Fu, ZaoZhuang, Shen Mu,Henan, Shanxi
- Indian, North Karanpura, washed and ROM
- Ambre Subbituminous coal

**Low Rank Coals**
- Montana Rosebud, Colstrip Wyoming, Big Horn, North Dakota lignite, Freedom
- Saskatchewan Lignite
- Inner Mongolia lignite

**Coke Char, Peat, Wastes**
- Metallurgical Coke, U.S., China, Poland
- Western Kentucky No. 9 coal char
- Illinois No. 6 coal char
- Finnish Peat, Viidansuo and Savaloneva
- Automobile Shredder Residue
- Oil shale
- Refuse derived Fuel (Sorted MSW)

**Biomass**
- Finnish waste wood and pulp mill waste
- Danish Willow
- Danish Straw
- Pelletized alfalfa stems
- Pelletized waste wood
- Bagasse
- Rice straw
- Agri-wastes
- Chicken litter

## SES SGT U-GAS Commercial Operation Experiences

SES' ASSETS IN CHINA

12

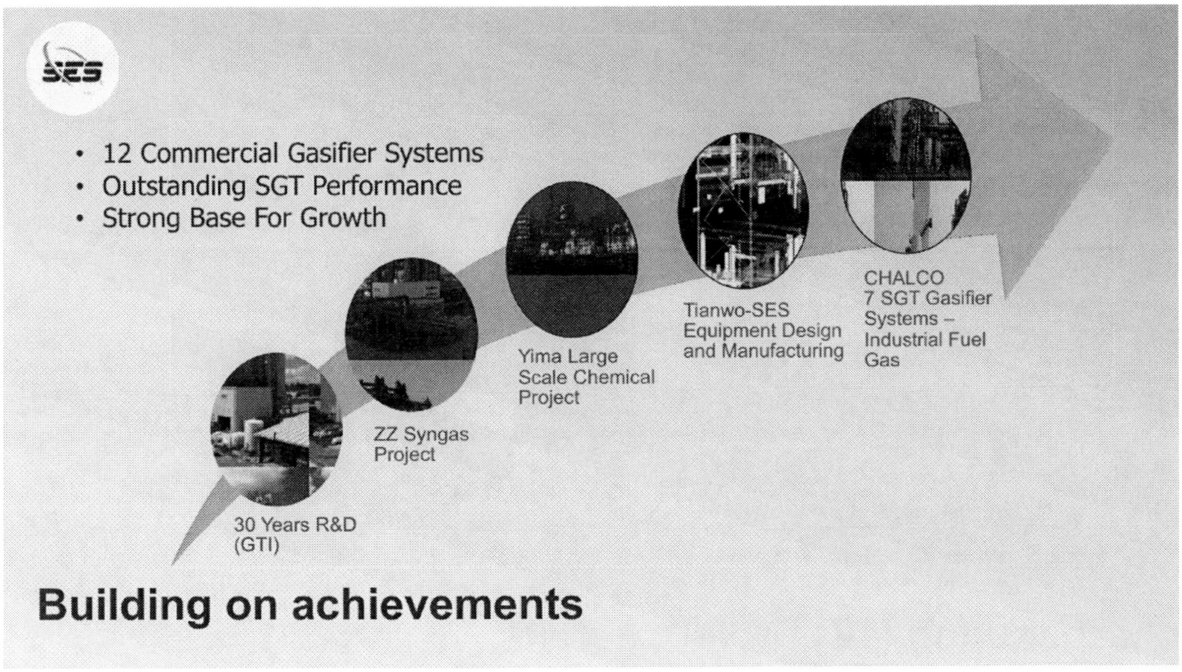

SES
• 12 Commercial Gasifier Systems
• Outstanding SGT Performance
• Strong Base For Growth
30 Years R&D (GTI)
ZZ Syngas Project
Yima Large Scale Chemical Project
Tianwo-SES Equipment Design and Manufacturing
CHALCO 7 SGT Gasifier Systems – Industrial Fuel Gas
Building on achievements


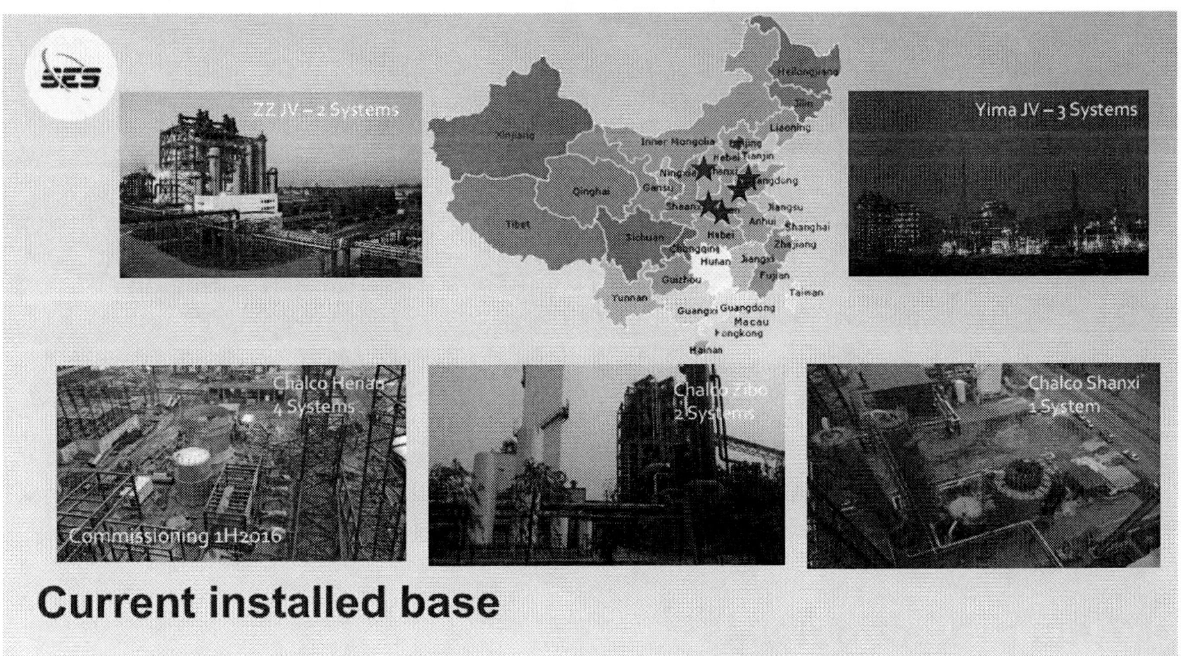

SES
ZZ JV – 2 Systems
Yima JV – 3 Systems
Xinjiang
Tibet
Qinghai
Sichuan
Yunnan
Chalco Henan 4 Systems
Commissioning 1H2016
Chalco Zibo 2 Systems
Chalco Shanxi 1 System
Current installed base

## Zaozhuang (ZZ) project details

- **Design Details**
  - 2 Trains
  - 20,000 nm3/hr syngas
  - 400 MTPD coal
  - Methanol Production
  - 2.2 Bar Operating Pressure
- **Schedule**
  - Construction Complete 12/07; First Syngas 1/08; First Methanol 3/08
- **Maximum Carbon Conversion: >99%**
- **High Cold Gas Efficiency: 85 %**
- **Demonstrated stable operations**
  - Successfully operated on over 20 coals; demonstrated flexibility of SES Gasification Technology at commercial scale

## Yima project details

- **Design Details**
  - 3 Trains, 2 x 1 Operation
  - 90,000 nm3/hr syngas
  - 1,200 MTPD coal per gasifier
  - 300,000 tonne/yr Methanol Production
  - 10 Bar Gasifier Operating Pressure
- **Schedule**
  - 12/12 First Methanol Production
- **Highest Carbon Conversion: >99%**
  - 85% CGE
  - Achieved All Performance targets
- **Improved Technology Demonstrations**
  - Higher Pressure; Ash Cooling & discharge (L-valve); Cyclone Fines Capture System Design; Fines recycle; Simplified startup burner design

- **Design Details**
  - 2 Trains
  - 82,000 nm3/hr industrial fuel gas
  - 460 MTPD coal per gasifier
  - ~3 Bar Operating Pressure
- **Schedule**
  - Project Announced 12/14; Commissioning Started 6/15; First Syngas 7/15
- **Highest Carbon Conversion: >99%**
- **Demonstrated stable operations and delivered spec fuel gas**
- **Improvements: feed coal and bottom char cooling and discharge arrangements**

## Chalco Shandong project details

- **Design Details**
  - 1 Train
  - 28,000 nm3/hr industrial fuel gas
  - 360 MTPD
  - 2 Bar Operating Pressure
- **Schedule**
  - Project Announced 12/14
- **Current Status**
  - In Start-Up

## Chalco Shanxi project details

- **Design Details**
  - 4 Trains
  - 120,000 nm3/hr industrial fuel gas
  - 305 MTPD coal per gasifier
  - 2.5 Bar Operating Pressure
- **Schedule**
  - Project Announced 12/14
- **Current Status**
  - In Start-Up

## Chalco Henan project details

SES founded in **2003**, based on **U-GAS®** developed by GTI for over **30 years.**

By **2016, SES** has **12** commercial scale **SGT** gasifiers in **China.**

Over **99%** **Carbon Conversion** - **Cold Gas Efficiency** demonstrated at over **85%.**

**100%** **Syngas Availability** for **2 years** during methanol production in **ZZ plant**

First **SGT** gasifier in Shandong Project produced syngas **6 months** after start of construction,

**2 SGT** gasifiers save China Aluminum Corp in Shandong **50K USD** per day in NG fuel costs

**SES** has invested **90M USD** in projects using **SGT gasifiers** so far

SGT's **XL 3000** Fluidized bed gasification design up to **55 bars** pressure and **1000°C**

## SGT by the numbers…

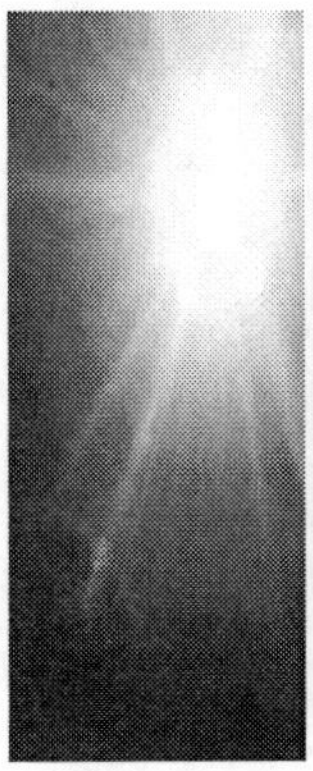

## SES Market Focus

- China, India, Indonesia, Brazil, Australia, Turkey, Pakistan
- Power
- DRI Steel

GROWTH WITH BLUE SKIES

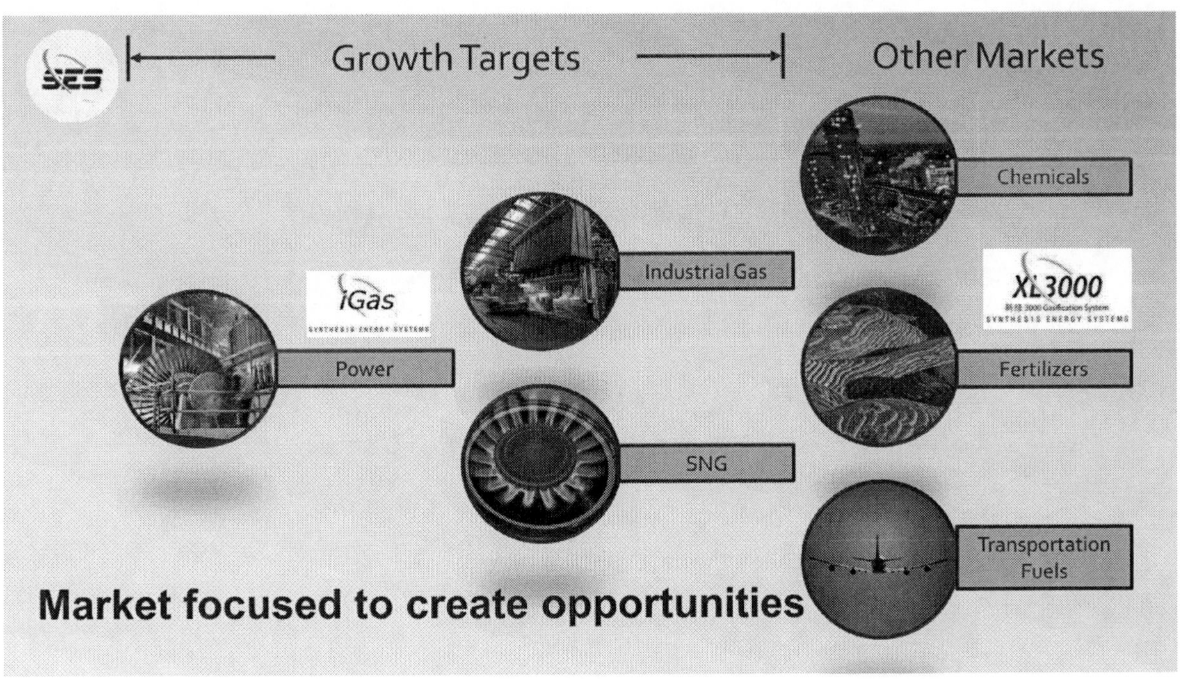

## SES Activities

- Joint Venture Operations
  - **Tianwo-SES JV** – First projects(3) in producing industrial fuel gases
  - **Yima JV** – Syngas to Methanol operating plant
  - **ZZ New Gas Company JV** – Repurpose to other end products under evaluation
- Two new projects under development:
  - **Industrial fuel gas** for metals industry
  - **Hydrogen for refineries** in industrial parks
- Markets and Global Partnerships:
  - **GE Packaged Power** – Distributed Power
  - **Midrex Technologies**, a subsidiary of Kobe Steel – DRI Steel

23

## Distributed Power

**SES and GE : Mini-IGCC Distributed Power**

- Demand for clean power is on the rise
- Global co-marketing partner: General Electric's Packaged Power Division
- **GE-SES Distributed Power Generation Plants to combine SES Gasification Technology and GE model LM2500+G4 aero-derivative gas turbines** and related power generation equipment, to produce clean iGas
- Initial regional targets: China, Pakistan, and India

24

## Direct Reduced Iron

**SES and MIDREX :DRI Steel from Low-quality Coal**

- Cleaner technology replacing coal-fired blast furnace for steelmaking
- **Global exclusive partner, Midrex Technologies, a subsidiary of Kobe Steel, is world leader with 63% market share**
- DRI steel annual growth projected at ~5MTPY through 2030
- Typical China plant size: 1MM TPY, requiring two gasification systems
- Initial target markets: India, China

July 22, 2012 Footer text here 25

# Thank You

## Synthesis Energy Systems, Inc.

GROWTH WITH BLUE SKIES

**NASDAQ:SYMX**
**synthesisenergy.com**

**2016 International Pittsburgh Coal Conference**
**Cape Town, SOUTH AFRICA**
**August 8 – 12, 2016**

**Abstract Submission**

**PROGRAM TOPIC:**
**FLUE GAS CLEAN UP AND ASH CHEMISTRY**

**Exact Title of Paper:**

# CFD SIMULATION OF AN INDUSTRIAL WET FLUE GAS DESULFURIZATION SPRAY TOWER: A COMPREHENSIVE MODEL WITH SPECIAL ATTTENTION DEVOTED TO THE MODELING OF ABSORPTION AND CHEMICAL REACTIONS

***Presenting Author***
***Professor Raymond Everson***
***School of Chemical and Minerals Engineering, North-West University, Potchefstroom Campus, Private Bag X6001, Potchefstroom 2520, South Africa***
***E-mail: ray.everson@nwu.ac.za, Tel: +2718 299 1986***

*Co-Authors*
*Mr Arif Arif*
*School of Chemical and Minerals Engineering, North-West University, Potchefstroom Campus, Private Bag X6001, Potchefstroom 2520, South Africa*
*E-mail: arif.afrif@nwu.ac.za, Tel: +2718 299 1991*

*Prof. Hein W.J.P. Neomagus*
*School of Chemical and Minerals Engineering, North-West University, Potchefstroom Campus, Private Bag X6001, Potchefstroom 2520, South Africa*
*E-mail: hein.Neomagus@nwu.ac.za, Tel: +2718 299 1991*

*Dr Dawie Branken*
*School of Chemical and Minerals Engineering, North-West University, Potchefstroom Campus, Private Bag X6001, Potchefstroom 2520, South Africa*
*E-mail: dawie.branken@nwu.ac.za, Tel: +2718 299 1376*

**Abstract:**
A complete CFD model for an industrial scale flue gas desulphurisation (WFGD) spray tower was developed and validated which included the hydrodynamics of the gas and droplet phases as well as the mass and heat transfer between the phases and the chemical reactions inside the slurry droplets. The model is capable of estimating the desulfurization efficiency, pH, enhancement factor and the concentration of chemical species in the slurry droplets.The mass transfer is based on the two film theory and the reactions in the droplets are assumed all to be at equilibrium. A non-isothermal operation was modeled with all parameters adjusted accordingly. The model was validated against plant results obtained from an industrial WFGD absorber which included the prediction of flue gas velocity, temperature and desulfurization efficiency. The dependence of the desulfurization efficiency on the enhancement factor and slurry pH obtained from the chemical model was found to be well aligned with the plant tests and literature results.

**2016 International Pittsburgh Coal Conference**
**Cape Town, SOUTH AFRICA**
**August 8 – 12, 2016**

**Abstract Submission**

**PROGRAM TOPIC: COMBUSTION TECHNOLOGIES**

**Exact Title of Paper:**

# OPTIMIZATION OF COAL AND COMBUSTION AIR DISTRIBUTION FOR 500MW TANGENTIAL COAL-FIRED PLANT

*Presenting Author: Kitae Kang, Senior Researcher, Korea Southern Power (KOSPO) CO., Ltd*
*Hadong Thermal Power Site Division, Combustion Technology Center,*
*509 Gyeongjesaneop-ro, Geumseong-Myun, Hadong-Gun, REPUBLIC OF KOREA*

*Contact Information [Email: kkang@kospo.co.kr, Phone: 82-70-7713-1664, Fax: 82-70-7713-1960]*

**Abstract:**

Recently aggressive reductions of greenhouse gas in coal-fired power plant emissions are mandated. So the electric utility industry has been moving toward improvement of plant efficiency. As the most cost effective means of improving efficiency in coal-fired power plant, the combustion modifications should be reviewed as the first step in any strategy.

The purpose of this project is to increase the plant efficiency through the optimization of coal and combustion air distribution in boiler.

The investigation of optimal operating condition through combustion test in fully operating boiler can be extremely time-consuming and expensive. The approach in this project was to use CFD to quantify the impacts of variations of burner fuel flow and secondary air flow on boiler operating parameters. And then based on CFD result, detailed optimum operating condition was derived by performing the combustion test.

Firstly, the impacts of variations of two operating parameters such as burner fuel flow and secondary air flow distribution were investigated using CFD with focusing on unburned carbon and NOx emissions. The following general conclusions were drawn from CFD. Unburned carbon reduced when secondary air flow to row C, D burner was increased by 3%. The added air is pulled from the other four rows proportional to the baseline secondary air flows of each burner. And unburned carbon also reduced when coal flow to row C burner was decreased by 15%. The removed fuel is the distributed between the other four rows proportional to the baseline fuel flows of each burner.

These results mean that pulverized coal particles from row C burner contribute most to unburned carbon.
First coal from the bottom burners(row A,B) experiences the most residence time for burnout. And coal from the middle burner row tends to pass through the central fuel rich core of fireball, seeing very low concentrations of $O_2$.
However coal particles from top burner row circulate at larger radius, in closer proximity to combustion air introduced through the row F, OFA, seeing high concentration of $O_2$. As a result, even though the coal from row D, E burner experience lower residence times in the furnace, they experience higher oxygen concentrations, critical for complete burnout. And secondary air distribution has larger impact on unburned carbon than fuel distribution.

So combustion test was performed to derive the optimum operating conditions only about secondary air distribution especially air flow to row C and D, focusing on unburned carbon and NOx, furnace gas temperature, heat absorption rate. In the test, optimum rate of increase in secondary air flow to row C&D was investigated according to OFA (over fire air) damper position 10, 26, 53% (average of OFA #1~4 ports damper position). The added secondary air flow to row C&D is pulled from other four rows. As a result of combustion test, the optimum rate of increase in secondary air flow to row C&D is each 1.8/ 6.1/ 14.4% at OFA damper position 10/ 26/ 53%. The optimum rate of increase in secondary air flow to row C&D is proportional to OFA damper position. At OFA damper position 10, 26%, unburned carbon decreases and NOx increases by increasing secondary air flow to row C&D. But at OFA damper position 53%, both unburned carbon and NOx decrease. Also it is possible to increase plant efficiency (0.13%p) and decrease cost of fuel($330thousand/year) by optimization of secondary air distribution.

# 1. Introduction

Recently reductions of greenhouse gas in coal-fired power plant emissions are very important. So the electric utility industry has been moving toward improvement of plant efficiency. As the most cost effective means of improving efficiency in coal-fired power plant, the combustion modifications should be reviewed as the first step in any strategy.

The purpose of this project is to increase the plant efficiency through the optimization of coal and combustion air distribution in boiler. The investigation of optimal operating condition through combustion test in fully operating boiler can be extremely time-consuming and expensive. The approach in this project was to use CFD to quantify the impacts of variations of burner fuel flow and secondary air flow on boiler operating parameters such as NOx and unburned carbon in fly ash(UBC). And then based on CFD result, detailed optimum operating condition was derived by performing the combustion test.

# 2. Combustion Simulation

## 2.1 Modeling method

In this research, we utilized three dimensional, multiphase, turbulent reacting flow commercial code of Fluent to perform the combustion simulation. And we selected Tangentially coal fired boiler for 3D modeling. This boiler has one PM(Pollution Minimum) burner in each 4 corners and 6 elevations, so total are 24set of PM burner for low NOx. And one PM burner is divided into concentric coal burner(rich coal) and weak coal burner(lean coal). The coal flow ratio of concentric coal burner and weak coal burner is about 7.5 : 2.5. Over fire air port is located upper region of burner F for low NOx. Figure.1 shows the configuration of the firing system of the 500MW T-Fired boiler.

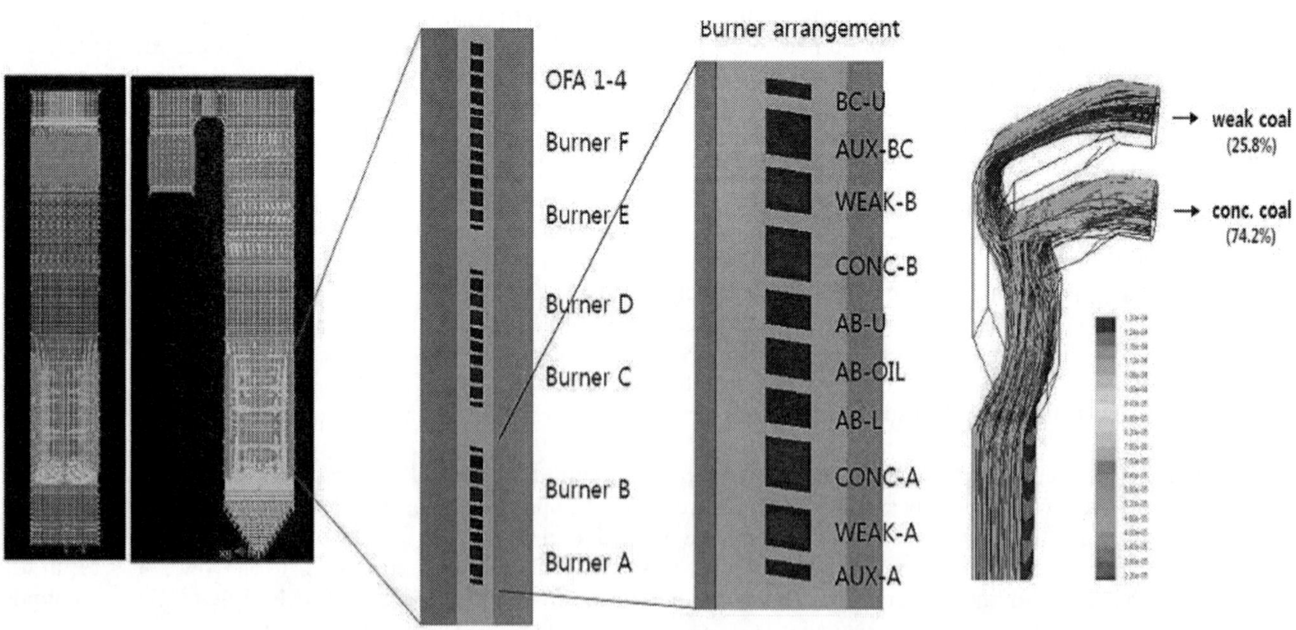


Figure 1 Configuration of the Firing System for the 500MW T-Fired Boiler

2.2 Calculation Condition

Table.1 lists calculation conditions of base case. Coal flow and coal calorific value, total air flow is calculated based on the gross power out, 525MW. Primary air temperature is 60℃ and secondary air(combustion air) is 305℃. Coal injected to 5 burner, A-E. Over fire air damper position is 10%. Table.2 lists the coal analysis value.

Table 1 Base Conditions

| Coal Flow (ton/hr) | Air flow (ton/hr) | | | Air Temperature (℃) | | Number of mills | OFA Damper Position(%) |
|---|---|---|---|---|---|---|---|
| | Primary Air | Secondary Air | Total | Primary Air | Secondary Air | | |
| 209 | 397 | 1105 | 1502 | 60 | 305 | 5EA (#A~E) | 10 |

Table 2 Coal Analysis used for simulation

| Items | | Test Coal |
|---|---|---|
| Coal Calorific value (As received base) | | 5621 |
| Proximate Analysis | Fixed carbon | 44.98 |
| | V.M | 33.31 |
| | Ash | 8.73 |
| | Moisture | 12.98 |
| | Sum | 100 |
| Ultimate Analysis | C | 68.24 |
| | H | 4.78 |
| | O | 12.85 |
| | N | 2.06 |
| | S | 0.93 |
| | Ash | 11.14 |
| | Sum | 100 |

2.1.1 Combustion air row biasing

Summary baseline and parametric of combustion air row biasing conditions is presented in Table 3. Secondary air damper increase 10% each elevation for case 1-3 and increase 20% each elevation for case 4-6. Total secondary air flow is same for baseline and case1-6. The added secondary air is pulled from the other 2 rows proportional to the baseline secondary air flows of each burner.

Table 3 Summary baseline and parametric of combustion air row biasing conditions

| Items | | Base | Case 1 | Case 2 | Case 3 | Case 4 | Case 5 | Case 6 |
|---|---|---|---|---|---|---|---|---|
| Secondary Air Damper Bias | EL.EF | 1 | 1 | 1 | 1.1 | 1 | 1 | 1.2 |
| | EL.CD | 1 | 1 | 1.1 | 1 | 1 | 1.2 | 1 |
| | EL.AB | 1 | 1.1 | 1 | 1 | 1.2 | 1 | 1 |

2.1.2 Coal flow row biasing

Summary baseline and parametric of coal flow row biasing conditions is presented in Table.4. Coal flow decrease 15% each burner in case 1-5. The removed fuel is the distributed between the other four rows proportional to the baseline fuel flows of each burner.

Table 4 Coal flow row biasing conditions

| Items | | Base | | Case 1 | | Case 2 | | Case 3 | | Case 4 | | Case 5 | |
|---|---|---|---|---|---|---|---|---|---|---|---|---|---|
| | | Bias | Coal | Bias | Coal | Bias | Coal | Bias | Coal | Bias | Coal | Bias | Coal |
| Bias & Coal Flow (T/H) | E | 1 | 41.84 | 1 | 43.41 | 1 | 43.41 | 1 | 43.41 | 1 | 43.41 | 0.85 | 35.57 |
| | D | 1 | 41.84 | 1 | 43.41 | 1 | 43.41 | 1 | 43.41 | 0.85 | 35.57 | 1 | 43.41 |
| | C | 1 | 41.84 | 1 | 43.41 | 1 | 43.41 | 0.85 | 35.57 | 1 | 43.41 | 1 | 43.41 |
| | B | 1 | 41.84 | 1 | 43.41 | 0.85 | 35.57 | 1 | 43.41 | 1 | 43.41 | 1 | 43.41 |
| | A | 1 | 41.84 | 0.85 | 35.57 | 1 | 43.41 | 1 | 43.41 | 1 | 43.41 | 1 | 43.41 |

2.3 Results and discussion

Firstly, the impacts of variations of two operating parameters such as secondary air flow distribution and burner fuel flow were investigated using CFD with focusing on unburned carbon and NOx emissions.

2.3.1 Combustion air row biasing

Unburned carbon reduced the most in the case2 from 14.75% to 5.59% when secondary air damper to row C, D burner was increased by 10%, but NOx concentration increased from 129ppm to 135ppm. Figure.2 shows that O2 in region of burner A-B, B-C reduced and O2 in region of burner C-D, burner E-Over fire air port increased more than baseline condition when secondary air damper to row C, D burner was increased by 10%, Case2. Figure.3 shows average gas temperature as function of boiler height in case of Base and Case2.

Table 5 Calculation results of air row biasing

| Items | | Calculation Result | | | | | | |
|---|---|---|---|---|---|---|---|---|
| | | Base | Case 1 | Case 2 | Case 3 | Case 4 | Case 5 | Case 6 |
| UBC | % | 14.75 | 12.18 | 5.59 | 6.66 | 6.29 | 9.19 | 20.41 |
| NOx | ppm | 129 | 148 | 135 | 126 | 136 | 144 | 121 |

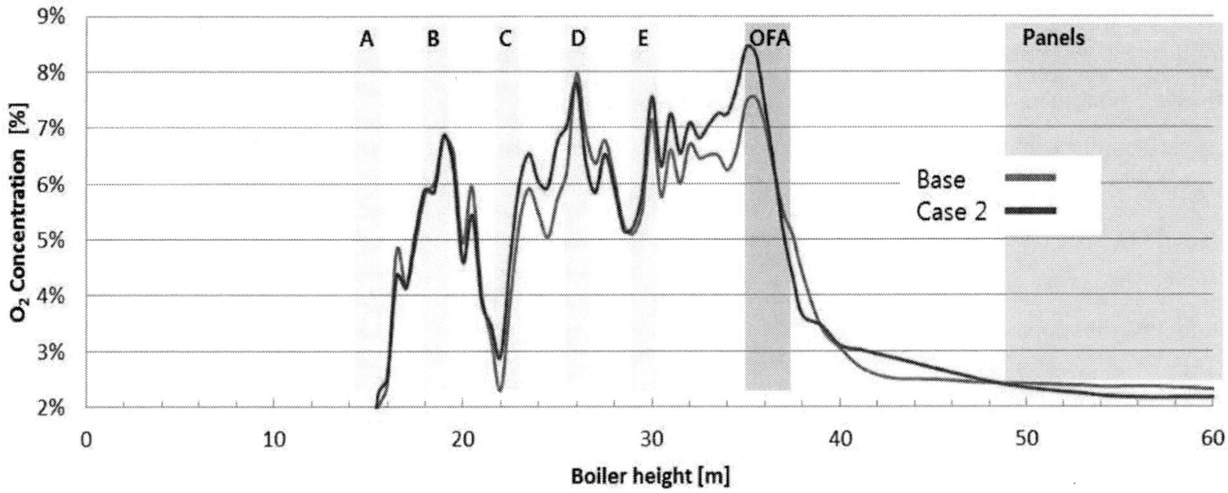


Figure 2 Average $O_2$ concentration as function of furnace height

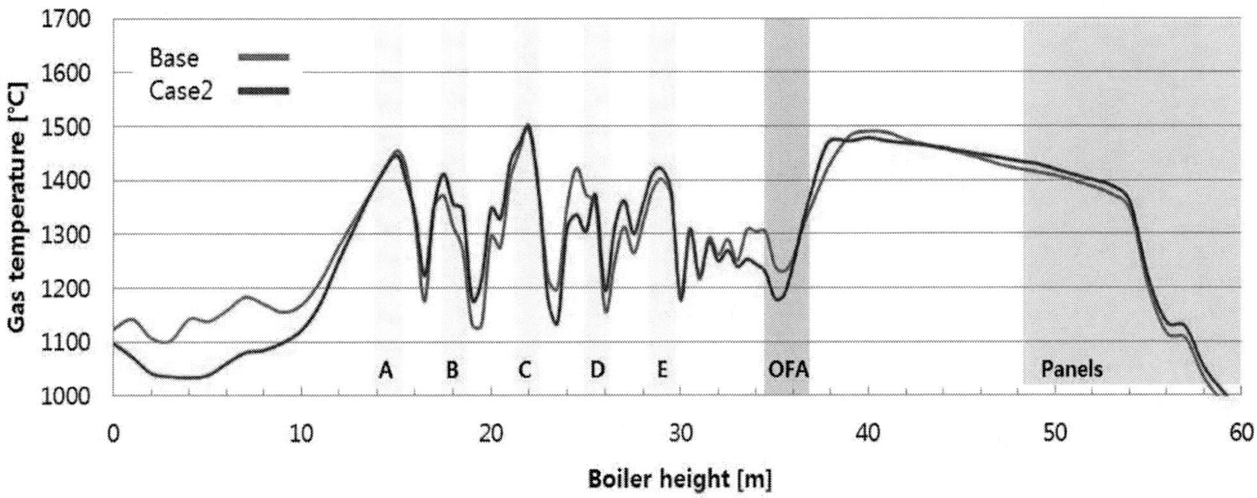


Figure 3 Average gas temperature as function of furnace height

2.3.2 Coal flow row biasing

Unburned carbon also reduced the most in the case.3 when coal flow to row C burner was decreased by 15% but NOx concentration increased from 128ppm to 133ppm.

Table 6 Calculation results of coal flow biasing

| Items | | Calculation Result | | | | | |
|---|---|---|---|---|---|---|---|
| | | Base | Case 1 | Case 2 | Case 3 | Case 4 | Case 5 |
| UBC | % | 7.17 | 12.96 | 7.70 | 6.95 | 8.23 | 17.69 |
| NOx | ppm | 128 | 136 | 134 | 133 | 138 | 118 |

2.4 Conclusion of Simulation

These results mean that pulverized coal particles from row C burner contribute most to unburned carbon.
First coal from the bottom burners(row A, B) experiences the most residence time for burnout. And coal from the middle burner row tends to pass through the central fuel rich core of fireball, seeing very low concentrations of O2. However coal particles from top burner row circulate at larger radius, in closer proximity to combustion air introduced through the row F, OFA, seeing high concentration of O2. As a result, even though the coal from row D, E burner experience lower residence times in the furnace, they experience higher oxygen concentrations, critical for complete burnout. And secondary air distribution has larger impact on unburned carbon than fuel distribution.

Table 7 Distribution $O_2$ in furnace and coal particle trajectory each burner

| O2 Distribution | A Burner | B Burner | C Burner | D Burner | E Burner |
|---|---|---|---|---|---|
| 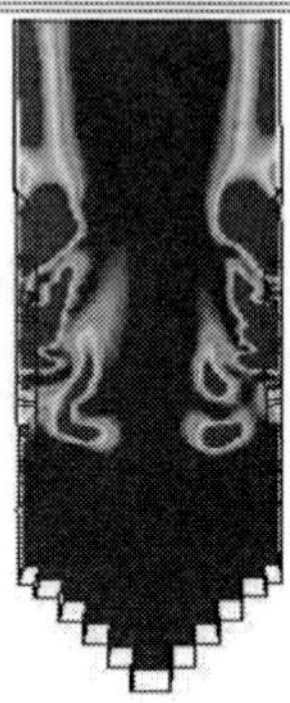 | 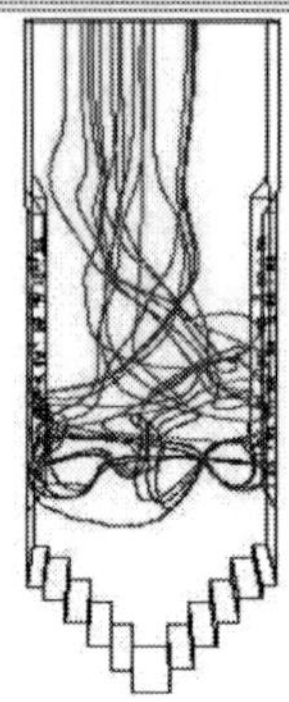 | 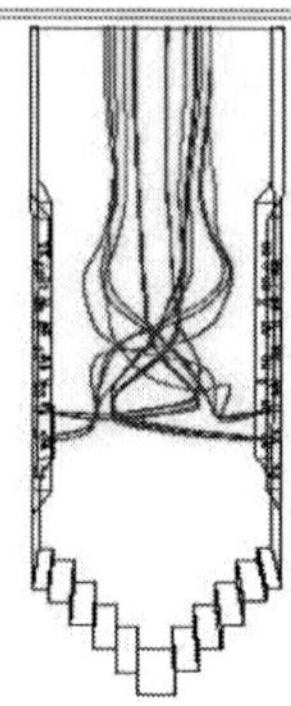 | 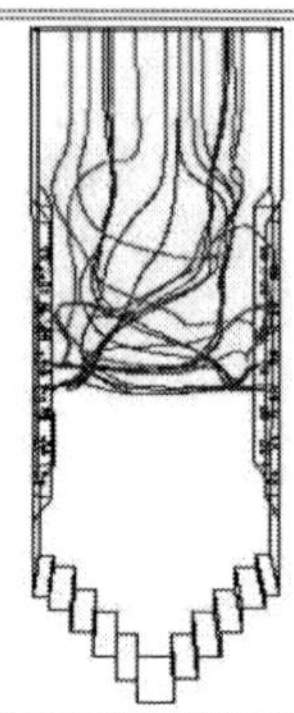 | 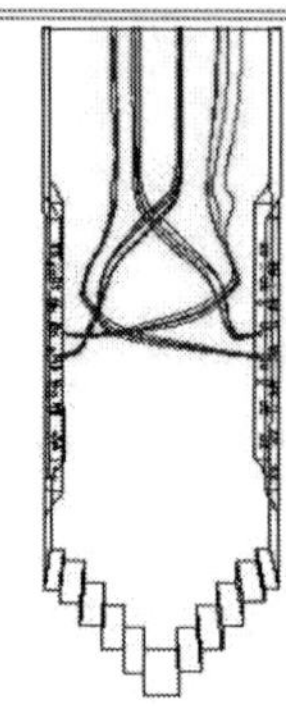 | 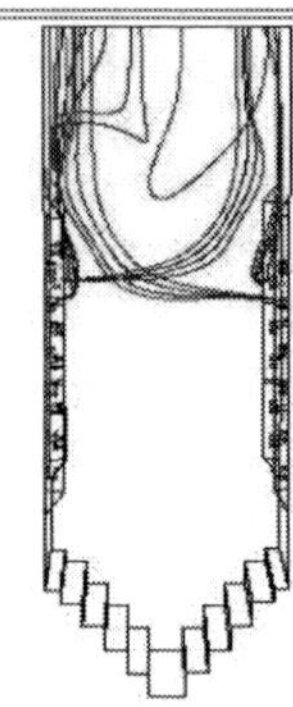 |

# 3. Combustion Test

3.1 Condition of test

So combustion test was performed to derive the optimum operating conditions only about secondary air distribution especially for air flow to row C and D according to Over fire air damper, focusing on unburned carbon and NOx, furnace exit gas temperature.

In the test, optimum rate of increase in secondary air flow to row C&D (EL.CD) was investigated according to OFA (over fire air) damper position 10, 26, 53% (average of OFA #1~4 ports damper position). The added secondary air flow to row C&D is pulled from other four rows. Table 8, 9 described the operating conditions for baseline and parametric.

Table 8 Operating conditions for Baseline Conditions

| Items | Contents |
|---|---|
| Gross Power Output | Automatic Generation Control, 510~515 MW |
| Test Coal | 2~3 Blended coals, 5600kcal/kg |
| Coal Flow | 200~220ton/hr |
| Air Flow | 340Nm3/s |
| OFA Damper Position | 10, 25, 53% |

Table 9 Parametric matrix

| Items | | Increasement of EL.CD Secondary Air Damper (%P) | | | |
|---|---|---|---|---|---|
| | | Base | Case 1 | Case 2 | Case 3 |
| OFA Damper Position | 10% | 0 | 3↑ | 4↑ | 5↑ |
| | 25% | 0 | 3↑ | 5↑ | 7↑ |
| | 53% | 0 | 4↑ | 7↑ | 10↑ |

3.2 Results

Relative change of unburned carbon in fly ash is seen Figure.5. At OFA 10%, case1 which is the increasement 3%P of EL.CD damper position is the best condition for reducing UBC. And at OFA 25%, case3 which is the increasement 7%P of EL.CD damper position is the best. Finally at OFA 53%, case3 which is the increasement 10%P of EL.CD is the best condition. Usually UBC is proportional to OFA Damper opening. But if secondary air flow to row C,D increase at optimum condition with opening the OFA damper, unburned carbon at OFA 53% is similar with unburned carbon with OFA 10%, seen in Figure.6.

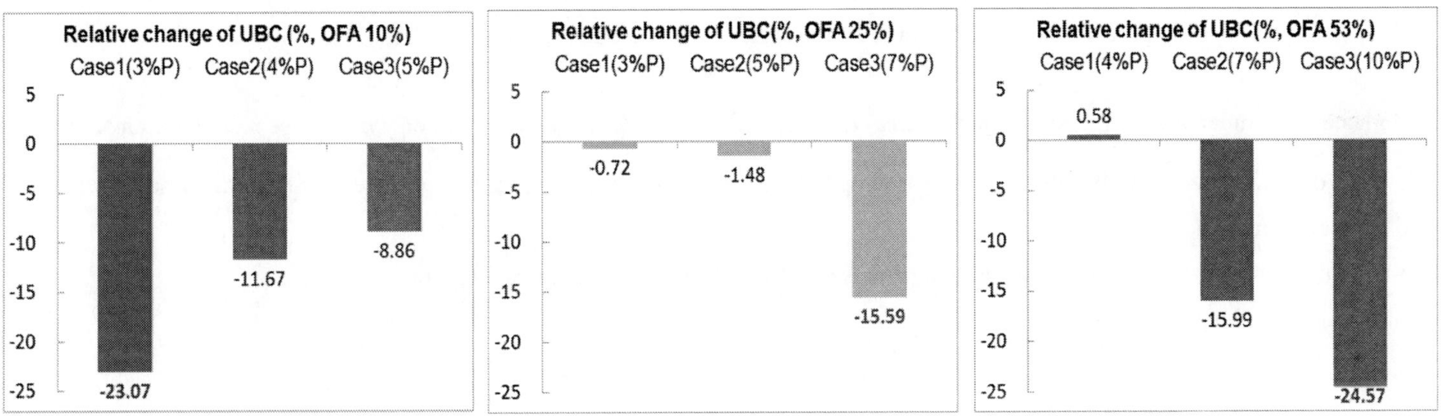


Figure 4 Relative change of Unburned Carbon according to OFA damper position and EL.CD air flow

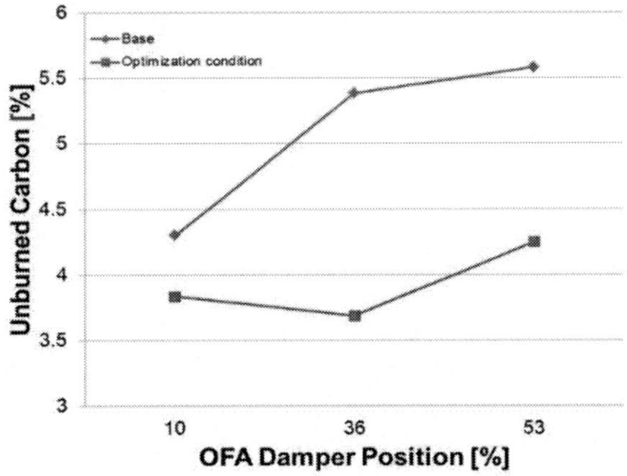


Figure 5 Comparison of trends of unburned carbon between Baseline and Optimum condition according to OFA damper position

Change of NOx concentration at SCR inlet and unburned carbon is described in Figure7. NOx is inversely proportional to unburned carbon in case of OFA 10, 25%. But at OFA 53% with Case2(7%P) and Case3(10%P), both NOx and unburned carbon reduced. Because at the region of EL.AB and EL.EF, reducing condition is more stronger than before.

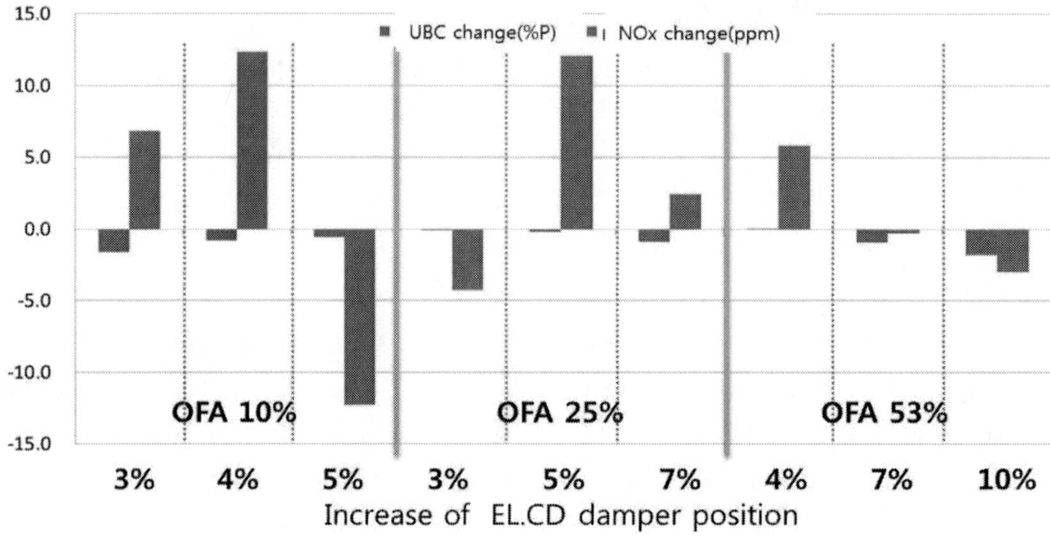


Figure 6 Change of Unburned carbon and NOx according to OFA and EL.CD air flow

Furnace gas temperature(at Primary Reheater area) increased Max. 10℃ (592 → 602℃) when we increase secondary air flow to EL.CD at optimum condition. But furnace exit gas temperature(at Gas air heater inlet) didn't increase. It means that heat absorption and plant efficiency increase. Figure

- UBC decrease → Furnace Gas Temperature increase → Heat absorption and steam temperature increase → Plant Efficiency increase

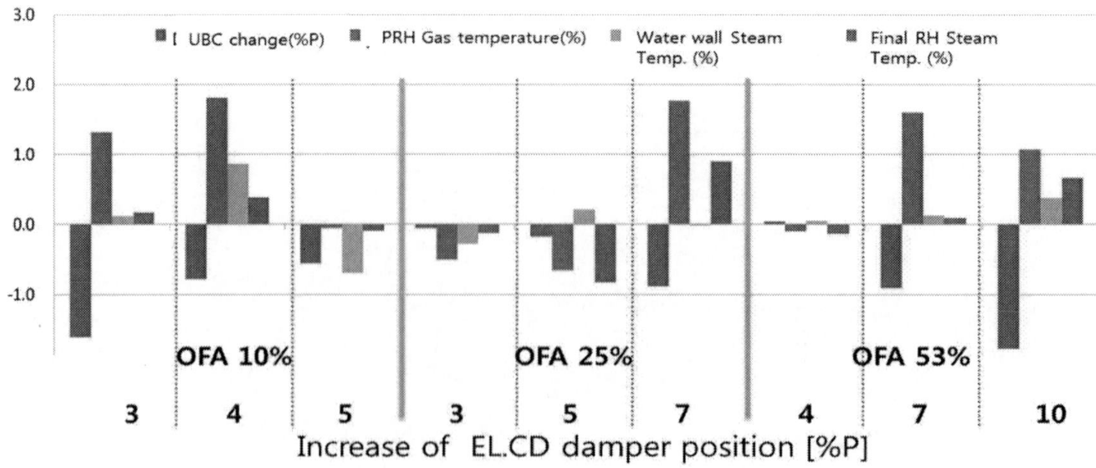


Figure 7 Relative change of UBC, Gas temperature, steam temperature according to OFA damper position and EL.CD air flow

# 4. Conclusion

As a result of combustion test, the optimum rate of increase in secondary air flow to row C&D is each 1.8/ 6.1/ 14.4% at OFA damper position 10/ 26/ 53%. The optimum rate of increase in secondary air flow to row C&D is proportional to OFA damper position. At OFA damper position 10, 26%, unburned carbon decreases and NOx increases by increasing secondary air flow to row C&D. But at OFA damper position 53%, both unburned carbon and NOx decrease. Also it is possible to increase plant efficiency (0.13%p) and decrease cost of fuel($330thousand/year) by optimization of secondary air distribution.

Table 10 Analysis of Economic effect

| OFA damper | Increase of EL. CD damper | UBC [%] | Final RH Steam Temp. [℃] | RH Spary [t/h] | Unit Efficiency [%] | Cost Savings [$/year] |
|---|---|---|---|---|---|---|
| Close 10% | Baseline | 6.76 | 539.71 | 6.73 | 0.099↑ ~ 0.135↑ | 330,000 |
| | Case1 (3%P) | 5.14 | 540.63 | 6.86 | | |
| | Change | 1.62↓ | 0.92↑ | 0.13↑ | | |
| Open 25% | Baseline | 5.69 | 534.20 | 1.48 | | |
| | Case3 (7%P) | 4.80 | 539.02 | 4.40 | | |
| | Change | 0.89↓ | 4.81↑ | 2.92↑ | | |
| Open 53% | Baseline | 7.22 | 531.66 | 1.29 | | |
| | Case3 (10%P) | 5.44 | 535.18 | 1.30 | | |
| | Change | 1.77↓ | 3.52↑ | 0.02↑ | | |

# NLP/MINLP for Modeling and Optimization of Lignite Pyrolysis Process

*Yi Huang, Qun Yi*, Minhui Gong, Guosheng Wu, Wenying Li**

*1Key Laboratory of Coal Science and Technology (Taiyuan University of Technology), Ministry of Education and Shanxi Province, Taiyuan 030024, P. R. China*

*2Training Base of State Key Laboratory of Coal Science and Technology Jointly Constructed by Shanxi Province and Ministry of Science and Technology, Taiyuan 030024, P. R. China.*

** Corresponding author. E-mail: ying@tyut.edu.cn*

**Abstract:** In this work, a Nonlinear Programming (NLP) model was established to describe the process of low rank coal (LRC) pyrolysis, and the pyrolysis model was validated through comparing with the pilot or semi-industry technologies such as Toscoal process, Lurgi-Ruhrgas (LR) process, Char Oil Energy Development (COED) technology, Dalian University of Technology (DG) process, and China Coal Research Institute (CCRI) Multistage Rotary Furnace (MRF) technology. All of the results showed a good agreement with the actual production values, suggesting that the model can be regarded as a universal one for simulating LRC pyrolysis. On the basis of this model, the product distribution and energy penalty can be obtained from different coal and pyrolysis technologies. Furthermore, in order to analyze each process comprehensively, we took the Hulunbuir lignite as an example to evaluate the five pyrolysis processes, respectively and comparatively.

***Keywords:*** low rank coal (LRC), pyrolysis process modeling, Nonlinear Programming (NLP)

## 1. Introduction

Low rank coal (LRC) pyrolysis based on polygeneration system is an efficient and clean way for LRC utilization. The polygeneration system integrated with different configurations and processes, which can realize reasonable conversion of element and cascade utilization of energy, thus， improve the energy efficiency and economic benefits of the whole system and reduce the emission of pollutants. Many studies focused on LRC utilization efficiency. Yi Qun [1] proposed a lignite pyrolysis by solid heat carrier coupled with gasification system, and optimized the energy efficient of the system. Chen [2] investigated the LRC pyrolysis system and analyzed the energy consumption in each unit, integrating the heat of the system. In order to utilize the LRC reasonably in the LRC pyrolysis technology, many researches [3-5] focused on the technology of improving the tar yield. Pyrolysis process is core unit of the lignite pyrolysis based on polygeneration system, and the main products from lignite pyrolysis are semi-coke, tar and gas. Besides, the distribution of these products, especially the tar components distribution, has great influence on downstream processes.

Consequently, some researches [6-8] considered a non-linear programming (NLP) modeling to estimate the products distribution. The research concentrated on the characteristics of different pyrolysis processes and the distribution of the products in different coals is beneficial to the efficient and clean utilization of the low rank coals. There are several low rank coal pyrolysis (LRCP) processes available, three of which are the most popular LRCP technologies based on [9]: solid heat carrier, moving bed process, and fluidized bed, such as Toscoal process [10], Lurgi-Ruhrgas (LR) process [11], Char Oil Energy Development (COED) technology [12], Dalian University of Technology (DG) process [13], and China Coal Research Institute (CCRI) Multistage Rotary Furnace (MRF)technology [14].

Table 1 Characteristics of LRC pyrolysis processes

| Process | Developer | Heat carrier | Furnace type | Temperature/℃ | Pressure/atm | Coal hold–up time |
|---|---|---|---|---|---|---|
| Toscoal[15] | Toaco | Eramicite | Moving bed | 425-540 | 1 | <5 min |
| LR[11] | Lurgi-Rahrgas | Semi-coke | Moving bed | 450-600 | 1 | <20 sec |
| DG[13] | DUT | Semi-coke | Moving bed | 450-650 | 1 | 0.5 min |
| COED[15] | FMC Corp. | Flue gas | Fluidized bed | 288-816 | 1.5 | 1-4 hr |
| MRF[14] | CCRI | Flue gas | MRF | 500-700 | 1 | 22-25 min |

During the pyrolysis process, the pyrolysis technology and the coal species have the influence on the distribution of the products, which is a result of different coal ranks, types of furnace, and heat carrier used in pyrolysis process.

Meanwhile, when it comes to a certain area of coal, a pyrolysis process applied in polygeneration system has the effect on the economy, environment, element utilization efficiency, and energy efficiency of the system. Thus, it is advantageous to use the LRC to study different kinds of LRC, hence make use of different pyrolysis processes and perform the different results. This different results are mainly of the water consumption, energy consumption, and product distribution in the pyrolysis process. Due to the different characteristics of various pyrolysis processes, the pyrolysis process can be evaluated by the aforementioned performance of the pyrolysis process. Water consumption, energy efficiency, exergy evaluation, $CO_2$ emission, and economy were generally taken into account to assess the system.

## 2. Mathematical Model

In this work, the LRC pyrolysis process was simulated to LRC as feedstock in the process, in which process coal was converted into semi-coke, tar, pyrolysis-gas, and pyrolysis-water.

$$\text{LRC} \xrightarrow{\textit{Technology}} \text{Semi-coke} + \text{Coal-tar} + \text{Pyrolysis-gas} + \text{Pyrolysis-}H_2O \quad (1)$$

### 2.1 Process Description

Coal pyrolysis is the first step to fulfill the staged conversion of LRC resources through the coal-pyrolysis based on pylygeneration system. In a coal pyrolysis process, coal was at first feed into a pre drying unit, and then mixed with heat carrier in the pyrolysis furnace. Attributing to the differences between the types of furnace and the heat carrier materials, different volatile releasing rates and second reactions were in the process.

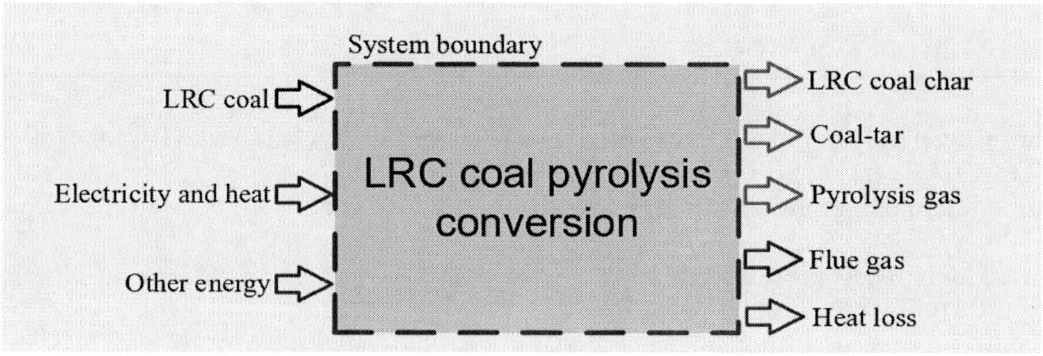


Fig.1 The input and output of LRC coal pyrolysis conversion process

In the LRC pyrolysis process model, LRC through a pyrolysis process presented different product distribution, energy consumption, water consumption, and $CO_2$ emission.

### 2.2 Construction the LRC pyrolysis model

Many factors have the influence on the overall chemical reactions of coal pyrolysis based on low rank coal pyrolysis technology. A model of the kinetic with coal pyrolysis process based on the coal molecular structure will be the challenged. In this section, the statistics and stoichiometric method was employed, referred to the literature of different pyrolysis process product distributions. A NLP model was constructed to predict the LRC pyrolysis product distribution based on different pyrolysis processes. In this paper, five LRC pyrolysis process were taken into consideration, i.e.TOSCOAL, LR, DG, COED, and MRF process respectively. The coal species used in this five pyrolysis process are shown in Table 2.

Table 2 Coal species proximate analysis of different LRC pyrolysis process

| Pyrolysis process | Coal | Proximate analysis wt%, daf | |
|---|---|---|---|
| | | V | FC |
| TOSCOAL[10] | WYODAK | 47.4 | 52.6 |
| | BIG HORN | 44.3 | 55.7 |
| | BIG SKY | 47.5 | 52.5 |
| | SAVAGE | 42.1 | 57.9 |
| LR[11] | LUKAVAC | 55.3 | 44.7 |
| | PROSPER | 40.5 | 59.5 |
| | NORMANBY | 50.9 | 49.1 |
| DG[16] | GDY | 47.0 | 53.0 |
| | GMZ | 48.7 | 51.3 |
| | GLC | 41.5 | 58.5 |
| | GYX | 49.5 | 50.5 |
| COED[17] | W.KY | 41.3 | 58.7 |
| | Illinois No. 6 | 46.6 | 53.4 |
| | Illinois coal | 42.2 | 57.8 |
| MRF[14] | ELKOL | 43.2 | 56.8 |
| | HM | 46.2 | 53.8 |
| | CY | 41.8 | 58.2 |
| | BN | 45.8 | 54.2 |

NLP pyrolysis production distribution model was based on the balance of elements and energy, and pilot plant data of distribution of pyrolysis products in five pyrolysis processes. The pyrolysis product distribution was determined by the pyrolysis technology, and feedstock's characteristics. The product distribution boundary in pyrolysis process is shown in Table 3 and Table 4.

Table 3 LRC pyrolysis product data from different process

| | Toscoal[10] | | LR[11] | | DG[16] | | COED[17] | | MRF[14] | |
|---|---|---|---|---|---|---|---|---|---|---|
| | .lo | .up | .lo | .up | .lo | .up | .lo | .up | .lo | .up |
| Product | (Basis: LRC, wt%, daf) | | | | | | | | | |
| Semi-coke | 73.5 | 77.7 | 60.7 | 67.8 | 60.5 | 69.2 | 59.7 | 61.2 | 66.5 | 69.7 |
| Tar | 13.2 | 16.8 | 9.8 | 12.1 | 7.4 | 7.9 | 13.1 | 23.2 | 4.9 | 7.1 |
| Pyrolysis-gas | 8.8 | 10.4 | 12.9 | 14.5 | 19.4 | 23.2 | 11.5 | 17.8 | 21.6 | 24.4 |
| Pyrolysis-water | 1.0 | 2.0 | 9.8 | 13.5 | 4.0 | 6.0 | 3.1 | 5.9 | 3.8 | 4.6 |

| | | | | | | | | | | |
|---|---|---|---|---|---|---|---|---|---|---|
| Semi-coke Ultimate analysis | | | | | | | | | (Basis: LRC, wt%, daf) | |
| C | 84.62 | 97.89 | 83.59 | 87.42 | 80.61 | 91.27 | 93.73 | 94.79 | 84.27 | 87.39 |
| H | 2.93 | 3.57 | 3.11 | 3.65 | 0.97 | 1.45 | 0.23 | 0.53 | 2.09 | 4.61 |
| O | 8.98 | 9.07 | 6.29 | 6.79 | 12.06 | 16.51 | 1.75 | 2.92 | 7.94 | 8.45 |
| S | 0.28 | 0.39 | 0.54 | 1.03 | 0.16 | 0.51 | 2.25 | 4.63 | 0.19 | 0.43 |
| N | 0.09 | 0.13 | 0.21 | 1.07 | 1.07 | 1.35 | 0.86 | 1.23 | 0.94 | 1.54 |
| Pyrolysis-gas component distribution | | | | | | | | | (Basis: molar %) | |
| CO | 12.7 | 17.5 | 10.8 | 12.7 | 13.6 | 15.9 | 18.3 | 21.7 | 14.9 | 17.0 |
| $CO_2$ | 41.5 | 47.9 | 43.9 | 54.0 | 19.0 | 23.4 | 32.4 | 40.7 | 9.7 | 11.7 |
| $CH_4$ | 19.8 | 26.0 | 21.0 | 25.4 | 22.1 | 27.2 | 11.0 | 13.1 | 34.4 | 41.3 |
| $H_2$ | 4.7 | 6.5 | 12.0 | 14.7 | 28.6 | 35.2 | 27.9 | 35.6 | 22.3 | 26.9 |
| $SO_2$ | 0.7 | 0.9 | 0.8 | 1.0 | 0.5 | 0.6 | 0.3 | 0.3 | 0.2 | 0.2 |
| $C_2H_6$ | 5.3 | 7.4 | 1.8 | 2.1 | 6.3 | 7.6 | 1.0 | 1.7 | 5.3 | 6.3 |
| $C_3H_8$ | 3.8 | 5.3 | 1.6 | 2.0 | 2.2 | 2.7 | 0.4 | 0.4 | 5.5 | 6.8 |
| $H_2S$ | 0.3 | 1.1 | 0.5 | - | - | - | - | - | - | - |
| $N_2$ | - | - | - | - | - | - | - | - | - | - |
| $NH_3$ | - | - | - | - | - | - | - | - | - | - |

Table 4 Coal tar fraction distribution in different pyrolysis process

Unit: ℃

| | Toscoal | | LR | | DG | | COED | | MRF | |
|---|---|---|---|---|---|---|---|---|---|---|
| wt% | .lo | .up | .lo | .up | .lo | .up | .lo | .up | .lo | .up |
| IBP | - | - | - | - | 95 | 105 | 64 | 92 | - | - |
| 2.5 | 199 | 212 | - | - | - | - | - | - | 161 | 170 |
| 5 | - | - | 140 | 143 | 160 | 176 | 119 | 171 | - | - |
| 10 | 207 | 254 | 160 | 165 | 190 | 201 | 142 | 211 | 183 | 192 |
| 15 | - | - | 180 | 183 | - | - | - | - | - | - |
| 20 | 235 | 302 | - | - | 210 | 214 | 184 | 252 | 204 | 209 |
| 30 | 285 | 341 | 203 | 217 | 230 | 236 | 216 | 266 | 226 | 213 |
| 40 | 338 | 377 | - | - | 255 | 260 | 256 | 271 | 232 | 249 |
| 50 | 385 | 407 | 213 | 226 | 290 | 302 | 281 | 335 | 271 | 280 |
| 60 | - | - | 237 | 320 | 320 | 335 | 317 | 369 | 287 | 312 |
| 70 | 413 | 416 | 260 | 342 | - | - | 347 | 402 | 340 | 360 |
| 80 | - | - | - | - | 335 | 347 | 379 | 428 | - | - |
| 90 | 429 | 438 | 280 | 357 | - | - | 417 | 460 | - | - |
| 95 | - | - | - | - | - | - | 440 | 488 | - | - |
| FBP | - | - | - | - | - | - | 451 | 510 | - | - |

Note: IBP = initial boiling point FBP = final boiling point

The coal tar composition is very complicated, which contains thousands of kinds of organic compounds, however, at present, only 500 species are identified. Low temperature coal tar are brown, have smaller density, which contain 2% - 10% alkane, 3% - 5% alkenes, and 15% - 25% aromatic hydrocarbons, the phenolic content up to 20% - 35%, almost 10 times higher than that of high temperature coal tar. Low temperature coal tar was mainly composed of three parts composition, i.e. low boiling phenol, C3-C4 alkyl phenol, naphthol and alkyl naphthols, respectively [18]. According to the characteristics of each coal tar fractions distribution in different pyrolysis processed showed in Table 4, higher contents of coal tar components were selected as model compounds, as shown in Table 5.

Table 5 Model compounds of the coal tar fractions

| Species name /℃ | <170 | 170-210 | 210-270 | 270-300 | 300-340 | >340 |
|---|---|---|---|---|---|---|
| Phenols | Phenol | Xylenol | Tert-Butylphenol | | | |
| Alkane | | | | | | |
| Olefins | Octane | Dodecane | Cetane | | | |
| Cycloalkanes | | | | | | |
| Arene | Methylbenzene | Tetrahydronaphthalene | Naphthalene | | phenanthrene | |
| Oxy-compound | n-pentanoic acid | | Methylbenzoic acid | | | |
| Nitro-compound | Indole | | Quinoline | | | |
| Heterocyclic compound | Thiophene | | Methylbenzothiophene | | | |
| Pitch | | | | | | Pyrene |

Therefore, thirty substances represent the pyrolysis products in the pyrolysis. pyrolysis-gas = {CO, $CO_2$, $CH_4$, $H_2$, $SO_2$, $C_2H_6$, $C_3H_8$, $H_2S$, $N_2$, $NH_3$}; tar = {$C_4H_4S$, $C_7H_8$, $C_8H_{18}$, $C_6H_6O$, $C_5H_{10}O_2$, $C_{10}H_{12}$, $C_{12}H_{26}$, $C_{10}H_8$, $C_8H_6S$, $C_8H_{10}O$, $C_9H_7N$, $C_{10}H_{14}O$, $C_8H_7N$, $C_8H_8O_2$, $C_{16}H_{34}$, $C_{12}H_{10}$, $C_{14}H_{10}$, $C_{16}H_{10}$}. The properties of the pyrolysis-gas and tar components are shown in Table 6.

Table 6 Pyrolysis product model compounds and their properties

| Name | Formula | Mol. wt. | BP/℃ | H/C | Exergy (MJ/kg) | LHV/25 ℃(MJ/kg) |
|---|---|---|---|---|---|---|
| Thiophene | $C_4H_4S$ | 84.14 | 84 | 1 | 33.31 | 46.20 |
| Toluene | $C_7H_8$ | 92.14 | 111 | 1.14 | 42.44 | 40.94 |
| n-Octane | $C_8H_{18}$ | 114.22 | 126 | 2.25 | 47.49 | 44.79 |
| Phenol | $C_6H_6O$ | 94.11 | 182 | 1 | 33.09 | 47.83 |
| n-Pentanoic acid | $C_5H_{10}O_2$ | 102.13 | 186 | 2 | 29.02 | 44.08 |
| 1,2,3,4-Tetrahydronaphthalene | $C_{10}H_{12}$ | 132.2 | 208 | 1.2 | 42.72 | 43.35 |

| | | | | | | |
|---|---|---|---|---|---|---|
| n-Dodecane | $C_{12}H_{26}$ | 170.33 | 217 | 2.17 | 47.14 | 37.26 |
| Naphthalene | $C_{10}H_8$ | 128.18 | 217.9 | 0.8 | 33.91 | 44.72 |
| Benzothiophe | $C_8H_6S$ | 134.2 | 220 | 0.75 | 35.47 | 42.71 |
| 2,5-Xylenol | $C_8H_{10}O$ | 122.16 | 225 | 1.25 | 36.15 | 46.92 |
| Quinoline | $C_9H_7N$ | 129.16 | 238 | 0.78 | 36.07 | 44.38 |
| p-tert-Butylphenol | $C_{10}H_{14}O$ | 150.22 | 238 | 1.4 | 38.07 | 42.25 |
| Indole | $C_8H_7N$ | 117.15 | 254 | 0.88 | 36.05 | 43.67 |
| o-Toluic acid | $C_8H_8O_2$ | 136.15 | 259 | 1 | 29.78 | 46.61 |
| n-Hexadecane | $C_{16}H_{34}$ | 226.44 | 287 | 2.13 | 46.96 | 28.02 |
| Anthracene | $C_{14}H_{10}$ | 178.22 | 340 | 0.71 | 40.25 | 41.15 |
| Acenaphthene | $C_{12}H_{10}$ | 154.21 | 279 | 1.2 | 27.93 | 35.61 |
| Phenanthrene | $C_{16}H_{10}$ | 202.26 | 394 | 0.63 | 39.78 | 31.38 |
| Carbon monoxide | $CO$ | 28.01 | - | - | 9.81 | 10.96 |
| Carbon dioxide | $CO_2$ | 44.01 | - | - | 0.45 | - |
| Methane | $CH_4$ | 16.04 | - | - | 50.72 | 50.03 |
| Hydrogen | $H_2$ | 2.02 | - | - | 116.88 | 119.91 |
| Sulfur dioxide | $SO_2$ | 64.06 | - | - | 4.85 | - |
| ethane | $C_2H_6$ | 30.07 | - | - | 49.73 | 47.52 |
| Propane | $C_3H_8$ | 44.10 | - | - | 48.78 | 46.34 |
| Hydrogen sulfur | $H_2S$ | 34.08 | - | - | 23.82 | 15.20 |
| Nitrogen | $N_2$ | 28.01 | - | - | 0.02 | - |
| Ammonia | $NH_3$ | 17.03 | - | - | 19.84 | 18.60 |

Not all kinds of coal vs. the pyrolysis process type data of the pyrolysis products distribution have been done. In this part, a NLP LRC pyrolysis model was established according to the aforementioned pyrolysis model compounds, which varied within lower and upper bounds in each pyrolysis process. In this model, the feedstocks are LRC, and fixed carbon (FC), the content varying from 45% to 60% in dry ash-free basis (daf). The assumptions for the coal pyrolysis model were presented below.

i. The elements composition of LRC are C, H, O, N, S, and on a daf basis.

$i_{coal}$ = {C, H, O, N, S}

ii. The pyrolysis products are semi-coke, pyrolysis-gas, tar, and pyrolysis water. The coposition of the semi-coke, pyrolysis-gas, tar are in the following set.

$i_{char} = \{C, H, O, N, S\}$;

$j_{gas} = \{CO, CO_2, CH_4, H_2, SO_2, C_2H_6, C_3H_8, H_2S, N_2, NH_3\}$;

$k_{tar} = \{C_4H_4S, C_7H_8, C_8H_{18}, C_6H_6O, C_5H_{10}O_2, C_{10}H_{12}, C_{12}H_{26}, C_{10}H_8, C_8H_6S, C_8H_{10}O, C_9H_7N, C_{10}H_{14}O, C_8H_7N, C_8H_8O_2, C_{16}H_{34}, C_{12}H_{10}, C_{14}H_{10}, C_{16}H_{10}\}$;

iii. In the pyrolysis process, 40% of the N elements are transferred to the $N_2$, and 50% of the S elements are transferred to the $H_2S$ [19].

iv. The ash of coal is not involved in the coal pyrolysis reaction, which only has thermal effect on the whole process.

v. In this model, the range of coal is only the low rank coal, and the content of fixed carbon is from 60% to 45% (daf).

The optimization model are listed as following:

Element balance:

$$\begin{cases} Mcoal * [w(i) - pchar * m(i)] = \sum_{pyrp,gas} n(i) * mw(i) \\ \quad + \sum_{pyrp,tar} m(i) * \frac{mw(i)}{M(i)}, \quad i \in (C, H, O, N, S) \\ 0.5 * Mcoal * w(S) = M(S) * n(H_2S) \\ 0.4 * Mcoal * w(N) = M(N) * n(N_2) \end{cases} \tag{2}$$

Each elements in char distribution:

$$\begin{cases} \frac{m_{pyrp,\ i}}{m_{pyrp,\ i+1}} - \frac{\dot{m}_{pyrp,\ i}}{\dot{m}_{pyrp,\ i+1}} \geq -R_i \\ \frac{m_{pyrp,\ i}}{m_{pyrp,\ i+1}} - \frac{\dot{m}_{pyrp,\ i}}{\dot{m}_{pyrp,\ i+1}} \leq R_i \end{cases} \quad \forall i \in char \tag{3}$$

Prolysis-gas distribution:

$$\begin{cases} \frac{n_{pyrp,\ j}}{n_{pyrp,\ j+1}} - \frac{\dot{n}_{pyrp,\ j}}{\dot{n}_{pyrp,\ j+1}} \geq -R_j \\ \frac{n_{pyrp,\ j}}{n_{pyrp,\ j+1}} - \frac{\dot{n}_{pyrp,\ j}}{\dot{n}_{pyrp,\ j+1}} \leq R_j \end{cases} \quad \forall j \in gas \tag{4}$$

Tar component distribution:

$$\begin{cases} \dfrac{m_{pyrp,\ k}}{m_{pyrp,\ k+1}} - \dfrac{m^*_{pyrp,\ k}}{m^*_{pyrp,\ k+1}} \geq -\mathrm{R}_k \\ \\ \dfrac{m_{pyrp,\ k}}{m_{pyrp,\ k+1}} - \dfrac{m^*_{pyrp,\ k}}{m^*_{pyrp,\ k+1}} \leq \ \mathrm{R}_k \end{cases} \quad \forall\ \mathrm{k} \in \mathrm{tar} \tag{5}$$

Pyrolysis process three product distribution:

$$\begin{cases} \dfrac{\sum_{i\in char} m_{pyrp,\ i}}{\sum_{j\in gas} m_{pyrp,\ j}} - \dfrac{\sum_{i\in char} m^*_{pyrp,\ i}}{\sum_{j\in gas} m^*_{pyrp,\ j}} \geq -S_1 \\ \dfrac{\sum_{j\in char} m_{pyrp,\ i}}{\sum_{k\in gas} m_{pyrp,\ j}} - \dfrac{\sum_{j\in char} m^*_{pyrp,\ i}}{\sum_{k\in gas} m^*_{pyrp,\ j}} \leq \ \mathrm{S}_1 \end{cases} \tag{6}$$

$$\begin{cases} \dfrac{\sum_{i\in gas} m_{pyrp,\ j}}{\sum_{j\in tar} m_{pyrp,\ k}} - \dfrac{\sum_{i\in gas} m^*_{pyrp,\ j}}{\sum_{j\in tar} m^*_{pyrp,\ k}} \geq -\mathrm{S}_2 \\ \dfrac{\sum_{j\in gas} m_{pyrp,\ j}}{\sum_{k\in tar} m_{pyrp,\ k}} - \dfrac{\sum_{j\in gas} m^*_{pyrp,\ j}}{\sum_{k\in tar} m^*_{pyrp,\ k}} \leq \ \mathrm{S}_2 \end{cases} \tag{7}$$

$$\begin{cases} \dfrac{\sum_{i\in tar} m_{pyrp,\ k}}{m_{char} + m_{gas} + m_{tar} + m_{water}} - \dfrac{\sum_{i\in tar} m^*_{pyrp,\ k}}{m^*_{char} + m^*_{gas} + m^*_{tar} + m^*_{water}} \geq -\mathrm{S}_3 \\ \dfrac{\sum_{i\in tar} m_{pyrp,\ k}}{m_{char} + m_{gas} + m_{tar} + m_{water}} - \dfrac{\sum_{i\in tar} m^*_{pyrp,\ k}}{m^*_{char} + m^*_{gas} + m^*_{tar} + m^*_{water}} \leq \ \mathrm{S}_3 \end{cases} \tag{8}$$

Model constraints:

$$\begin{cases} R_{i,\ j,\ k} > 0, \ \ S_{1,\ 2,\ 3} > 0 \\ \\ m_{pyrp,\ i,\ k} > 0, \quad n_{pyrp,\ j} > 0 \end{cases} \quad \forall\, i \in char,\ \ j \in gas, k \in tar \tag{9}$$

$$\begin{cases} n^*_{pyrp,j,lo} \leq n_{pyrp,j} \leq n^*_{pyrp,j,up} \\ m^*_{pyrp,i,k,lo} \leq m_{pyrp,i,k} \leq m^*_{pyrp,i,k,up} \end{cases} \quad \forall\, i \in char,\ \ j \in gas, k \in tar \tag{10}$$

Objective function:

$$\mathrm{Min} \sum_{i=1}^{5} \sum_{j=1}^{7} \sum_{k=1}^{17} \sum_{h=1}^{3} \left(\alpha R_i + \beta R_j + \gamma R_k + \delta S_h\right) \tag{11}$$

The variable m*, n* denote the reference mass flow rate and molar flow rate just seen Tables

3-4, and the variable m and n denote the expected mass and molar flow rate. The objective function is minimizing the positive slack variables, called R and S. α, β, γ, δ, represent the correction factor, taking account of this model, the variable of the three component are more important. In this model, we choose α= β= γ=1, δ=10.

## 3. Results and discussions

### 3.1 Case study

In this paper, the Hulunbuir lignite was selected as a feedstock in the pyrolysis process, the proximate and ultimate analysis seen Table 7. This coal are suitable for the pyrolysis process model.

Table 7 Proximate and ultimate analysis of the Hulunbuir lignite

| | proximate analysis, wt/% | | | | ultimate analysis, wt/% | | | | |
|---|---|---|---|---|---|---|---|---|---|
| | M | A | V | FC | C | H | O* | N | S |
| ad | 9.69 | 11.01 | 33.6 | 45.7 | 58.56 | 3.51 | 15.83 | 1.02 | 0.38 |
| daf | - | - | 42.4 | 57.6 | 73.85 | 4.42 | 19.96 | 1.29 | 4.79 |

Note: *- By difference.

The calculation of Hulunbuir coal used as feedstock in the model, solving the model using the GAMS 23.5, the variable $n_{pyrp}$ and $m_{pyrp}$ are listed in Table 8 (transformed into mass flow rate).

Table 8 Pyrolysis product in each pyrolysis process

(Basis: coal 100g/s)

| | Species | | Pyrolysis product in each process (unit: g/s) | | | | |
|---|---|---|---|---|---|---|---|
| | Name | Index | Toscoal | LR | DG | COED | MRF |
| product (coal.d) | Pyrolysis water | $H_2O$ | 3.69 | 7.71 | 6.71 | 6.06 | 4.08 |
| | Semi-coke | PCHAR | 71.75 | 63.78 | 63.47 | 58.79 | 66.13 |
| | Pyrolysis gas | PGAS | 9.97 | 13.64 | 22.24 | 17.27 | 23.06 |
| | Coal tar | PTAR | 14.59 | 14.86 | 7.57 | 17.88 | 6.74 |
| coal-tar (coal.daf) | Thiophene | $C_4H_4S$ | 0.448 | 0.092 | 0.071 | 0.131 | 0.016 |
| | Toluene | $C_7H_8$ | 1.225 | 0.485 | 0.437 | 0.232 | 0.137 |
| | n-Octane | $C_8H_{18}$ | 0.274 | 0.802 | 0.519 | 0.505 | 0.214 |
| | Phenol | $C_6H_6O$ | 1.906 | 0.479 | 0.961 | 1.270 | 0.446 |
| | n-Pentanoic acid | $C_5H_{10}O_2$ | 0.925 | 0.371 | 0.206 | 1.915 | 0.264 |
| | 1,2,3,4-Tetrahydronaphthalene | $C_{10}H_{12}$ | 1.075 | 1.281 | 0.623 | 1.768 | 0.441 |
| | n-Dodecane | $C_{12}H_{26}$ | 2.724 | 1.382 | 0.650 | 0.997 | 0.256 |
| | Naphthalene | $C_{10}H_8$ | 2.357 | 2.049 | 0.532 | 0.888 | 0.271 |

| | | | | | | | |
|---|---|---|---|---|---|---|---|
| | Benzothiophe | $C_8H_6S$ | 0.179 | 0.244 | 0.151 | 0.465 | 0.422 |
| | 2,5-Xylenol | $C_8H_{10}O$ | 0.114 | 1.524 | 0.398 | 0.904 | 0.161 |
| | Quinoline | $C_9H_7N$ | 0.585 | 0.188 | 0.072 | 0.308 | 0.339 |
| | p-tert-Butylphenol | $C_{10}H_{14}O$ | 1.821 | 1.637 | 0.480 | 1.148 | 0.543 |
| | Indole | $C_8H_7N$ | 0.008 | 0.369 | 0.105 | 0.140 | 0.104 |
| | o-Toluic acid | $C_8H_8O_2$ | 0.073 | 1.121 | 0.496 | 0.536 | 0.556 |
| | n-Hexadecane | $C_{16}H_{34}$ | 0.513 | 1.536 | 0.254 | 2.920 | 0.563 |
| | Acenaphthene | $C_{12}H_{10}$ | 0.534 | 0.299 | 0.372 | 0.792 | 0.565 |
| | Phenanthrene | $C_{14}H_{10}$ | 0.285 | 0.669 | 0.790 | 1.830 | 0.427 |
| | Pyrene | $C_{16}H_{10}$ | 0.754 | 1.273 | 0.783 | 0.652 | 1.376 |
| | Carbon monoxide | CO | 1.310 | 1.446 | 3.693 | 3.647 | 4.903 |
| | Carbon dioxide | $CO_2$ | 5.698 | 9.105 | 8.106 | 10.699 | 5.633 |
| | Methane | $CH_4$ | 0.815 | 1.369 | 4.230 | 1.255 | 6.038 |
| | Hydrogen | $H_2$ | 0.028 | 0.098 | 0.688 | 0.429 | 0.494 |
| pyrolysis-gas (coal.daf) | Sulfur dioxide | $SO_2$ | 0.121 | 0.208 | 0.311 | 0.115 | 0.070 |
| | ethane | $C_2H_6$ | 0.439 | 0.220 | 2.215 | 0.323 | 1.744 |
| | Propane | $C_3H_8$ | 0.465 | 0.305 | 1.154 | 0.105 | 2.798 |
| | Hydrogen sulfur | $H_2S$ | 0.093 | 0.001 | 0.003 | 0.002 | 0.004 |
| | Nitrogen | $N_2$ | 0.492 | 0.535 | 0.788 | 0.669 | 0.858 |
| | Ammonia | $NH_3$ | 0.511 | 0.353 | 1.057 | 0.020 | 0.988 |
| | Carbon | C | 54.02 | 57.21 | 57.95 | 55.08 | 56.74 |
| | Hydrogen | H | 2.08 | 1.98 | 0.95 | 0.29 | 1.52 |
| semi-coke (coal.daf) | Oxygen | O | 7.25 | 4.34 | 7.68 | 1.70 | 5.62 |
| | Sulfur | S | 0.06 | 0.45 | 0.70 | 0.71 | 0.60 |
| | Nitrogen | N | 0.29 | 0.32 | 0.19 | 1.29 | 0.20 |

The products of the five pyrolysis processes are shown in figure 3. Comparing the coal pyrolysis product distribution in five pyrolysis processes, the LR process got the most yield of pyrolysis water, Toscoal process had the maximum semi-coke, DG and MRF process had the more pyrolysis gas and the less coal tar.

### 3.2 Sustainability evaluation of coal pyrolysis process

Coal-based polygeneration assessment are generally contains economic analysis, energy efficiency, and environment evaluation [20].

#### 3.2.1 Energy efficiency and exergy efficiency analysis

Energy efficiency and exergy efficiency are the general evaluation parameter for the chemical power system, the formulation is presented as follows, shown in Eq.16.

$$\eta_e = \frac{E_o}{E_{in}} \times 100\% \tag{12}$$

$$\eta_{ex} = \frac{EX_o}{EX_{in}} \times 100\% \tag{13}$$

Where $\eta_e$ and $\eta_{ex}$ denotes the energy efficiency and exergy efficiency of the process. Eo and Ein are the total energy input and output based on low heat value (LHV), respectively. EXo and EXin represent the total output exergy and input exergy of the system. The energy and exergy input contains the heat consumption, electricity consumption and coal as feedstock. The energy output contains the three products, i.e. semi-coke, coal-tar, and pyrolysis-gas, respectively. The energy and exergy of flue gas and pyrolysis water can be ignored.

Table 9 The energy and exergy of the coal and pyrolysis product

Unit: MJ/kg

| Process | Coal | | Semi-coke | | Coal-tar | | Pyrolysis-gas | |
|---|---|---|---|---|---|---|---|---|
| | energy | exergy | energy | exergy | energy | exergy | energy | exergy |
| Toscoal | 25.19 | 26.10 | 28.53 | 29.50 | 41.67 | 38.47 | 7.67 | 11.78 |
| LR | 25.19 | 26.10 | 30.05 | 31.02 | 40.81 | 38.94 | 4.99 | 9.75 |
| DG | 25.19 | 26.10 | 26.89 | 28.00 | 44.39 | 38.36 | 14.80 | 23.55 |
| COED | 25.19 | 26.10 | 29.17 | 30.45 | 39.79 | 38.63 | 4.45 | 10.23 |
| MRF | 25.19 | 26.10 | 28.70 | 29.72 | 39.59 | 37.62 | 21.12 | 27.96 |

The energy and exergy of the pyrolysis product obtained by Eq.14.

$$E_p = m\% \cdot \sum_i E_{pyrp} \qquad p \in \text{char, gas, tar} \tag{14}$$

where E represent the energy or exergy, m% represent the mass percent of each component in coal-tar and pyrolysis gas. The energy and exergy of component are shown in Table 6.

### 3.2.2 Economy assessment

The economy of the system is the important concern at the design/operation level. The maximum profit and the minimum total capital cost are generally taken as the optimization objective when design a chemical and power system [21]. The plant life of 20 years was considered in this study. The economies of scale was referred to production scope exponential method [22], see

Eq.19.

$$I_{new} = I_0 \cdot (\frac{Q_{new}}{Q_0})^n \tag{15}$$

$$NPV = \sum_t \frac{NCF(t) \cdot y(t)}{(1+DR)^{n(t)}} \tag{16}$$

$$\sum_{t=0}^{n} NFC(t) \cdot (1+IRR)^{-t} = 0 \tag{17}$$

Where $I_{new}$ and $I_0$ represent the total capital cost of the studied and reference, and $Q_{new}$ and $Q_0$ are of the scale of the studied and reference, and n is the scale factor. NPV and IRR usually as an economic indicator to evaluate the chemical and energy project, usually to choose the process have the maximum NPV and IRR. Besides, the IRR indicator was also used to evaluate the feasibility of a design [23]. In this study, discount ratio was 12%, if the IRR< 12%, this design is not economically feasible, and will not bring economic benefits.

Table 10 Assumption for the economic analysis

| parameters | value |
|---|---|
| plant availability, $10^4$ t/y | 100 |
| depreciation life, y | 10 |
| design time, y | 1 |
| construction life, y | 1 |
| plant life, y | 20 |
| discount ratio, % | 12 |
| depreciation rate, % | 8 |
| tax rate, % | 25 |
| standard coal price, CNY/t | 380 |
| coal as feedstock price, CNY/t | 200 |
| electricity price, CNY/kWh | 1 |
| semi-coke price, CNY/t | 300 |
| pyrolysis gas price, CNY/$Nm^3$ | 0.8 |

### 3.2.3 Environment analysis

As the global warning and water source shortage phenomenon become an upgrading problem, carbon dioxide ($CO_2$) as a greenhouse gas (GHG) emission is the urgent problem, it is necessary to discuss the $CO_2$ emission in a system. The carbon emission over the operation can be obtained from the heat energy resource supply and consumption of electricity, which were calculated by as follows:

$$\textbf{GHG emission} = \sum_f fuel(f) \cdot c \qquad f \in \text{coal, electricity} \tag{18}$$

Here, GHG emission denote the $CO_2$ emission, f denote the set of energy type, and c is the amount of carbon dioxide emission when consumed per unit of energy. In the pyrolysis process, the $CO_2$ emission comes from the electricity consumption and heat energy supply. In general, 1 tone standard coal combustion lead to 2.6 tone CO2 emission, and can produce electricity 2860 kWh.

The major water consumption in the coal pyrolysis is in the coke quenching unit, and cooling water of the system. The consumption of water is different in each coal pyrolysis process, the water consumption per tons coal is calculated by Eq.19.

$$\textbf{Water consumption} = \frac{\sum fresh\ water\ consumed}{\sum mass\ of\ coal\ feedin} \tag{19}$$

### 3.2.4 Comprehensive analysis of coal pyrolysis process

A comprehensive analysis of five coal pyrolysis processes were described in Table 11. the capital cost and the variable cost of each pyrolysis process are based on the literature, and Yulin semi-coke industry information network service platform [24]. These five kinds of coal pyrolysis processes are only semi industrialization, so we chose the annual coal handling capacity of 1 million to deal with.

Table 11 Performance of each pyrolysis process

Scale: 1 million coal per year

| Item | Toscoal | LR | DG | COED | MRF |
|---|---|---|---|---|---|
| **Reference scale and investment** | | | | | |
| base scale $10^4$ t/y | 31 | 60 | 60 | 100 | 10 |
| base cost (millions) | 107 | 190 | 190 | 350 | 30 |
| scaling factor | 0.67 | 0.67 | 0.67 | 1 | 0.67 |
| **Investment** | | | | | |
| the total capital cost (million CNY) | 300 | 268 | 268 | 350 | 140 |
| O&M cost (million CNY/y) | 263 | 284 | 261 | 264 | 309 |
| labour No. | 120 | 100 | 100 | 120 | 100 |
| labour cost (million CNY/y) | 10 | 8 | 8 | 10 | 6 |

| | | | | | |
|---|---|---|---|---|---|
| coal cost (million CNY/y) | 200 | 200 | 200 | 200 | 200 |
| the total variable cost (million CNY/y) | 17 | 13 | 13 | 12 | 8 |
| **Consumption** | | | | | |
| heat energy (MJ/t-coal) | 1613 | 1173 | 1104 | 1537 | 2430 |
| fresh water (t/t-coal) | 0.40 | 0.35 | 0.15 | 0.15 | 0.16 |
| electricity (kWh/t-coal) | 25 | 11 | 25 | 20 | 33 |
| **Performance** | | | | | |
| semi-coke production ($10^4$ t/y) | 68 | 62 | 61 | 58 | 63 |
| tar production ($10^4$ t/y) | 12 | 12 | 6 | 14 | 6 |
| pyrolysis-gas production ($10^4$ t/y) | 8 | 11 | 18 | 14 | 18 |
| $CO_2$ emission ( t/t-LRC) | 0.017 | 0.016 | 0.025 | 0.022 | 0.042 |
| energy efficiency $\eta_e$ (%) | 91.33 | 89.47 | 81.71 | 85.22 | 87.21 |
| exergy efficiency $\eta_{ex}$ (%) | 90.51 | 89.72 | 85.73 | 87.31 | 90.59 |
| IRR (%) | 23.54 | 24.08 | 22.64 | 24.55 | 22.75 |
| NPV(million CNY) | 388 | 350 | 324 | 485 | 172 |

The semi-coke yield of five processes are probably about 600 thousand tons. COED process had the highest total capital cost, and MRF process had the highest energy consumption in five processes, mainly because of its furnace type consuming a large amount of electricity. The COED process had the highest IRR, reaching to 35.14%, and had the maximum NPV, namely, 485 million CNY.

Using the radar chart to comprehensively analyze the five kinds of pyrolysis processes. There are six indicators to be analyzed, water consumption, energy efficiency, exergy efficiency, GHG emission, NPV and IRR respectively. In order to normalize the indicator, a relative value of indicator was calculated by Eq.20.

$$Y_{a,b} = \frac{y_{a,b} - W\{a\}}{W\{a\} - B\{a\}} \quad (20)$$

a∈{water consumption, energy efficiency, exergy efficiency, GHG emission, NPV, IRR }

b∈{Toscoal, LR, DG, COED, MRF}

The $Y_{a,b}$ denotes the a indicator relative value of b pyrolysis process, it value range in [0,1]. $y_{a,b}$ represent the actual value of an indicator in b process. W {a} and B {a} represent the worst value and the best value in sets a or b respectively. The result are shown in figure 2.

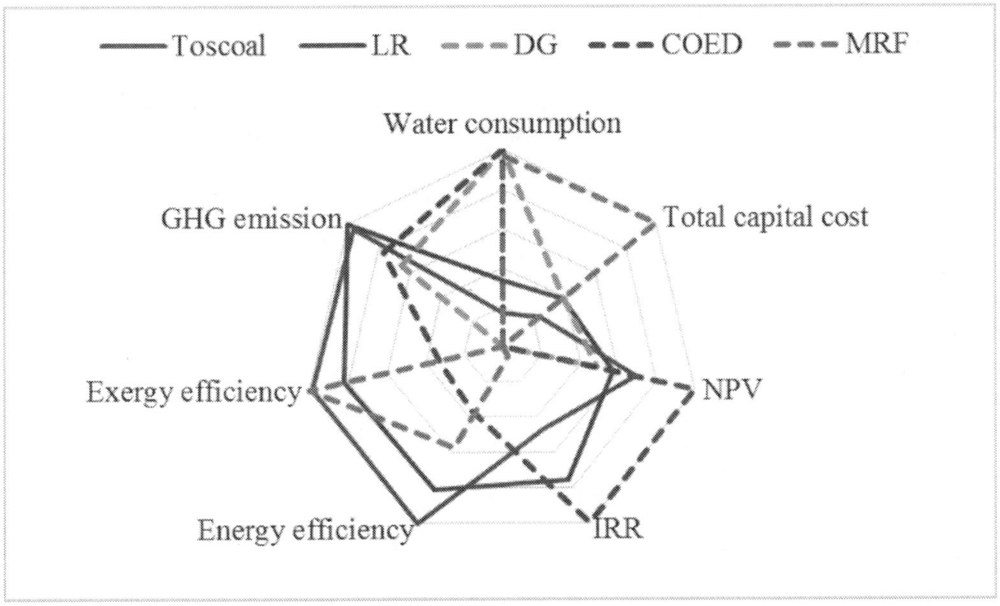


Fig.2 Sustainability evaluation of coal pyrolysis process

In the design level, if we consider the economic indicator as the chief concern, the COED process was favorable. If we took the energy efficiency or exergy efficiency as main targets, Toscoal and LR were preferred. If the design requirements of the lower water consumption was the optimized target, DG technology could be better. If less fixed investment was required, the MRF process was the best choice.

## 4. Conclusions

Based on the NLP modeling method, the pyrolysis product distribution model of five kinds of LRC pyrolysis processes was established. Taking Hulun Buir the lignite as an example, the five models were calculated respectively by using the sustainability approach to evaluate these five processes, the results showed that the TOSCOAL and LR processes could be well performed in terms of energy efficiency. COED process was prominent in economic performance, but the fixed investment was relatively large. Fixed investment of MRF process was the least, however, the process of power consumption was large, indicating higher operating costs, and poor performance in terms of energy efficiency. DG process was advantageous in term of water consumption

## Acknowledgements

The author would like to thank the financial support from the National Natural Science Foundation of China (51404164, 51276120, U1361202).

## References

[1] Yi Q, Feng J, Lu B, Deng J, Yu C, Li W. Energy Evaluation for Lignite Pyrolysis by Solid Heat Carrier Coupled with Gasification. Energy & Fuels. 2013;27(8):4523-33.

[2] Chen X, Zheng D, Guo J, Liu J, Ji P. Energy analysis for low-rank coal based process system to co-produce semicoke, syngas and light oil. Energy. 2013;52:279-88.

[3] Liang L, Huang W, Gao F, Hao X, Zhang Z, Zhang Q, et al. Mild catalytic depolymerization of low rank coals: a novel way to increase tar yield. RSC Adv. 2015;5(4):2493-503.

[4] Fan S, Yuan X, Zhao L, Xu L-H, Kang T-J, Kim H-T. Experimental and kinetic study of catalytic steam gasification of low rank coal with an environmentally friendly, inexpensive composite K2CO3–eggshell derived CaO catalyst. Fuel. 2016;165:397-404.

[5] Fan Y, Yi Q, Li F, Yi L, Li W. Tech-economic assessment of a coproduction system integrated with lignite pyrolysis and Fischer–Tropsch synthesis to produce liquid fuels and electricity. Fuel Processing Technology. 2015;135:180-6.

[6] He C, Feng X, Chu KH. Process modeling and thermodynamic analysis of Lurgi fixed-bed coal gasifier in an SNG plant. Applied Energy. 2013;111:742-57.

[7] He C, Feng X, Chu KH, Li A, Liu Y. Industrial-scale Fixed-bed Coal Gasification: Modeling, Simulation and Thermodynamic Analysis. Chinese Journal of Chemical Engineering. 2014;22(5):522-30.

[8] Baliban RC, Elia JA, Floudas CA. Toward novel hybrid biomass, coal, and natural gas processes for satisfying current transportation fuel demands, 1: Process alternatives, gasification modeling, process simulation, and economic analysis. Industrial & Engineering Chemistry Research. 2010;49(16):7343-70.

[9] Yang Q, Yang S, Qian Y, Kraslawski A. Application of House of Quality in evaluation of low rank coal pyrolysis polygeneration technologies. Energy Conversion and Management.

2015;99:231-41.

[10] Atwood MT, Schulman BL. The TOSCOAL process-pyrolysis of western coals and lignites for char and oil production. Preprints of Papers American Chemical Society Division of Fuel Chemical. 1977;22(2):233-52.

[11] Guo S. Coal and oil shale retorting technology——LR solid heat carrier fast pyrolysis method. Coal Chemical Industry, (China). 1986;4:008.

[12] Research FCC, Center D, Eddinger RT. Char Oil Energy Development: Final Technical Report: FMC Corporation, Chemical Research and Development Center, 1965.

[13] Guo S, Luo C, Han Z. Lignite Retorting Using Solid Heat Carrier. Fuel Science and Technology International. 1990;8(1):39-49.

[14] Ma G, Dai H. Experimental study on pyrolysis of Shenmu coal in mutistage rotary furnace. Coal Science and Technology, (China). 1994;22(11):37-9.

[15] Sliepcevich C, Bodle W, Conn A, Katzer J, Neskora D, Mullowney J, et al. Assessment of Technology for the Liquefaction of Coal. Nat Acad Sci# FE/1216-3. 1977.

[16] Guo S, Luo C, Han Z. Lignite retorting using solid heat carrier. Petroleum Science and Technology. 1990;8(1):39-49.

[17] Jones JF, Schmid MR, Sacks ME, Chen Y-C, Gray CA. Char oil energy development. FMC Corp., Princeton, NJ (USA); 1967.

[18] Guo S. Coal chemical technology. Beijing: Chemical Industry Press (China), 2006.

[19] He C, You F, Feng X. A novel hybrid feedstock to liquids and electricity process: Process modeling and exergoeconomic life cycle optimization. AIChE Journal. 2014;60(11):3739-53.

[20] Liu P, Pistikopoulos EN, Li Z. A mixed-integer optimization approach for polygeneration energy systems design. Computers & Chemical Engineering. 2009;33(3):759-68.

[21] Swanson RM, Platon A, Satrio JA, Brown RC. Techno-economic analysis of biomass-to-liquids production based on gasification. Fuel. 2010;89:S11-S9.

[22] Karellas S, Braimakis K. Energy–exergy analysis and economic investigation of a cogeneration and trigeneration ORC–VCC hybrid system utilizing biomass fuel and solar power. Energy Conversion and Management. 2016;107:103-13.

[23] http://www.ylltxx.com/show/showCokeDetails?cokeId=10. 2016/6/13.

# INTERNATIONAL TEST CENTER NETWORK


**Roxann Laird**

***Director, National Carbon Capture Center***
***Southern Company, USA***


U.S. DEPARTMENT OF ENERGY

SOUTHERN COMPANY

NETL

# Mission

**Share $CO_2$ Capture Knowledge**
to encourage global collaboration and accelerate technology development of cost effective $CO_2$ capture processes

INTERNATIONAL TEST CENTER NETWORK

# CC ITCN Objectives

- Share lessons learned and results from parallel activities
- Provide enhanced technical learning and confidence
- Increase insight and awareness of different technologies that may reduce risks and increase investments in CO2 capture technology
- Enhance public awareness and acceptance of the technology
- Work with technology developers to scale up tests and operate under different conditions

INTERNATIONAL TEST CENTER NETWORK

# Focus

Technical and business development collaboration

Enhance awareness of network

Increase network membership

INTERNATIONAL TEST CENTER NETWORK

# Members

INTERNATIONAL TEST CENTER NETWORK

# Key Accomplishments

- Established Network Charter and Demonstrated relationship and contractual model
- Drafted Technical Report
- Served as a model for creation of CSLF's Large-Scale Saline Storage Project Network
- Successful transition to Leadership by the United States
- Increased public awareness of capture R&D
- Supported progressive scale up of capture technology to larger demonstrations
- Technical Sharing Lessons Learned
    - measuring emissions and amine degradation products from post-combustion capture
    - aerosols and mist formation and mitigation and how to measure mist particle size and distributions

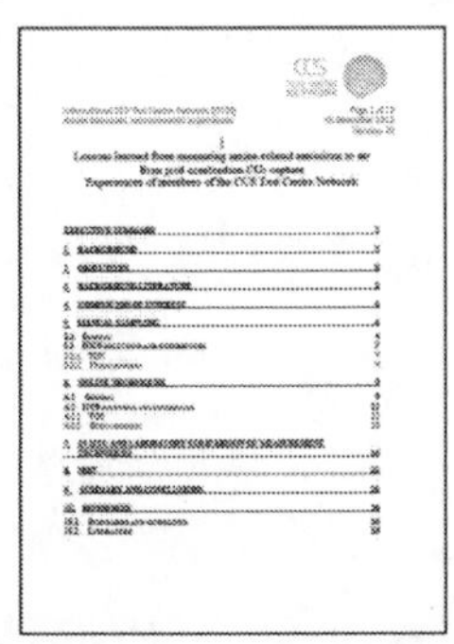

INTERNATIONAL TEST CENTER NETWORK

# Possible Future Technical Collaboration

- Evaluation of Flue Gas from natural gas sources and the benefits of reducing oxygen concentration
- Consider alternative base runs to MEA solvent such as a mixture of solvents or other public domain solvents.
- How to support model development and advanced simulations with a focus on reducing capital cost and opex?
- Data and equipment configuration required for input
  - Is it possible to participate with US DOE CCSI?
  - How data is collected and used in simulations? Accuracy of data?
  - Parametric tests that inform simulation.
- Addition of on-line monitoring of liquid absorbents.

INTERNATIONAL TEST CENTER NETWORK

INTERNATIONAL TEST CENTER NETWORK

# Acknowledgements

U.S. DEPARTMENT OF ENERGY

SOUTHERN COMPANY

NETL

# Project Status

*Roxann Laird, Director*

U.S. DEPARTMENT OF ENERGY

SOUTHERN COMPANY

NETL

NATIONAL CARBON CAPTURE CENTER

# Project Facts

- Located in Wilsonville, Alabama
- Sponsored by the U.S. Department of Energy and its National Energy Technology Laboratory
- Partners include the Electric Power Research Institute and leaders in the power and coal industries

U.S. DEPARTMENT OF ENERGY

NETL

EPRI | ELECTRIC POWER RESEARCH INSTITUTE

AEP AMERICAN® ELECTRIC POWER

SOUTHERN COMPANY

CLOUD PEAK ENERGY®

DUKE ENERGY.

Luminant

NATIONAL CARBON CAPTURE CENTER

# Mission and Core Values

Offering a **world-class neutral** test facility and a highly specialized staff, to **accelerate the commercialization** of advanced technologies and enable coal based power plants to achieve **near-zero emissions (low cost $CO_2$)**.

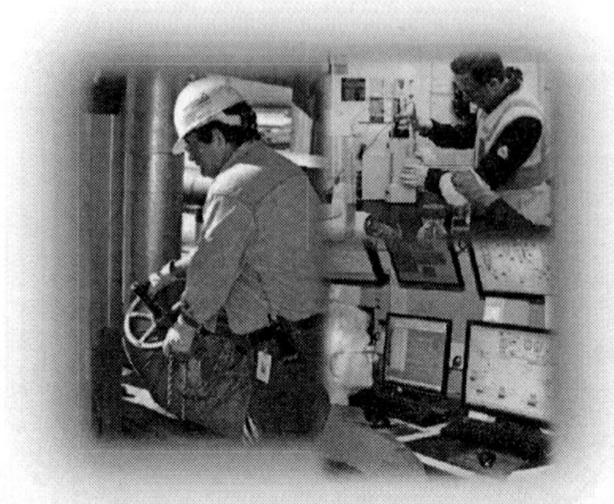

**Unquestionable Trust**
**Superior Performance**
**Total Commitment**

NATIONAL CARBON CAPTURE CENTER

# Test Sites

NATIONAL CARBON CAPTURE CENTER

# Accomplishments

- 20 years with no lost time accidents
- 80,000 test hours for technology developers from U.S. and six other countries since 2008 founding of NCCC
- First coal-derived gas testing of solid oxide fuel cells and certain solvents, membranes, enzymes, etc.
- On-site scale-ups and process enhancements for ten technologies; scale-ups for testing at larger sites for five solvents; scale-up to commercial operation for one solvent
- Full compliance with environmental and government regulations, including on-time submittal and publication of technical reports

NATIONAL CARBON CAPTURE CENTER

# What NCCC Provides

- A cost-efficient test site for numerous technology developers
- Real industrial conditions with coal-derived flue gas and syngas
- Capability for testing at multiple scales and for on-site scale-ups
- Expert staff for support of design, installation, and testing
- High quality data acquisition and gas/liquid sampling and analysis

NATIONAL CARBON CAPTURE CENTER

# Technology Development Process

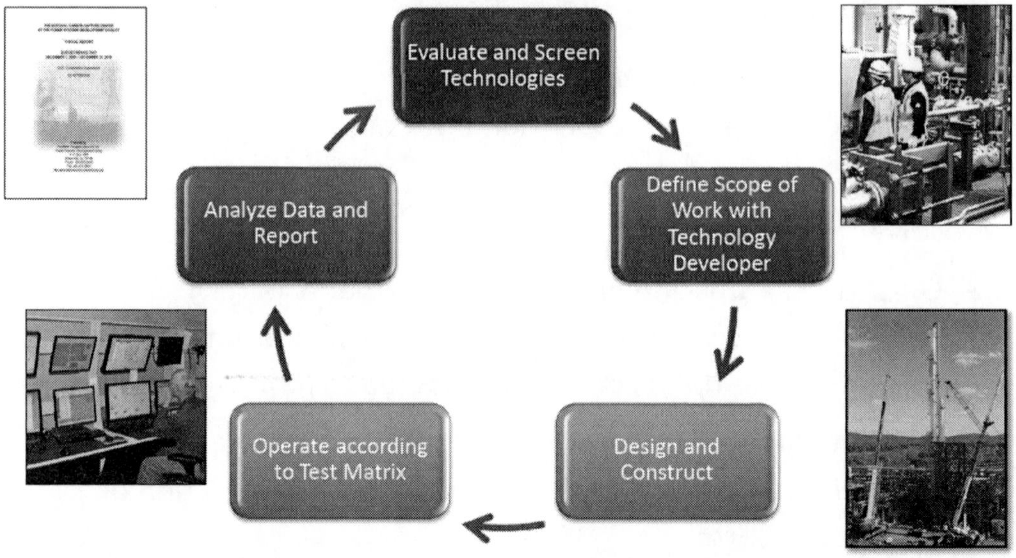




# PC4 Bench-Scale

- Simultaneous operation of up to five developers' test units
- Slipstream Solvent Test Unit (SSTU) for solvents in early development
- SSTU also used for solvent emissions studies and emission mitigation processes
- Flue gas/utilities and gas analysis systems operating independently of PC4 pilot-scale area

# PC4 Pilot-Scale

- Simultaneous operation of developer test units and Pilot Solvent Test Unit (PSTU)
- PSTU offers flexible operation to match developers' planned commercial configuration
- PSTU also supports solvent emissions and degradation studies

NATIONAL CARBON CAPTURE CENTER

# Post-Combustion Accomplishments

- PC4 operation supported over 40,000 hours of technology testing
  - Over 6,000 hours under natural gas conditions
  - More than 20 developer projects completed
  - Tested enzymes, membranes, sorbents, solvents, and associated systems
  - Continued relationship with technology developers to achieve scale-ups and process enhancements
- PSTU operation for more than 13,000 hours
  - Demonstrated near 100% mass and energy balance closures
  - Supported commercial developers and DOE Carbon Capture Simulation Initiative
  - Several solvents progressed to further testing at other facilities
- Facility construction and upgrades
  - PC4 constructed in under three years
  - Plant capacity more than doubled from 12,000 to 30,000 lb/hr flue gas
  - Added systems (SSTU, air dilution, etc.) and enhanced instrumentation, sampling methods, and analysis systems

NATIONAL CARBON CAPTURE CENTER

# Pre-Combustion/Gasification Infrastructure

- Coal-derived syngas from the Transport Gasifier train, in operation since 1996
- Operations conducted for nominally three 750-hour runs per year
- Syngas Conditioning Unit (SCU) with slipstreams at 5 to 500 lb/hr and syngas pre-treatments and process conditions tailored for each developer's needs
- Pilot unit for operation at 1,000 lb/hr
- State-of-the-art gas analysis and data acquisition
- On-site labs for sample analysis and processing

NATIONAL CARBON CAPTURE CENTER

# Gasification & Pre-Combustion Accomplishments

- Gasifier operation supported over 38,000 hours of technology testing
  - Biomass gasification in air- and oxygen-blown operation
  - Sensors: Tunable Diode Laser, particulate monitor, thermowells, coal feeder instrumentation
  - Catalysts: Fischer-Tropsch, water-gas shift, and COS hydrolysis
  - Sorbents: trace metals, $CO_2$, ammonia
  - Membranes: hydrogen and $CO_2$
  - Advanced processes: ammonium carbonate/bicarbonate solvent, syngas chemical looping, pressure-swing adsorption, pressure-swing Claus
  - Fuel cells
- Additional operation with $CO_2$ solvents, both on-line and off-line
- Achieved scale-ups and process intensification for several technologies including membranes, catalysts, sorbents
- Major upgrades to infrastructure such as new SCU control room and gas analysis building

MTR

NATIONAL CARBON CAPTURE CENTER

# Current Post Combustion Tests

Pilot-scale
- GE w/ PSTU
- Linde/BASF @PB2
- Air Liquide @PB3

Bench-scale
- DOE Membrane
- Akermin (PH2)
- SSTU
  - Baseline/emission
  - Green Technology: additive to amine
  - Cansolv

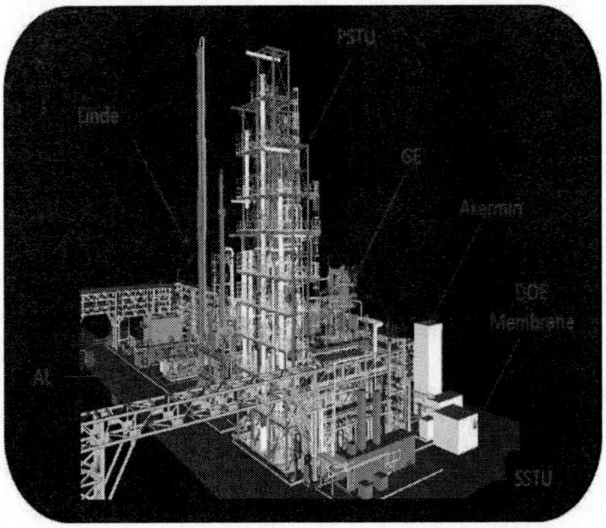


NATIONAL CARBON CAPTURE CENTER

# International Collaboration

**Carbon capture R&D is an international issue that requires international collaboration**

- Support DOE goal of international cooperation
  - Multiple paths for involvement; partners, developers, network members and workshops
  - Ease of collaboration since intellectual property is not being shared
- Broad effort in China, India, Middle East, Korea, Japan, EU, Australia, Canada

Aker CleanCarbon · CHIYODA · MITSUBISHI HEAVY INDUSTRIES, LTD · BASF The Chemical Company · Linde · The University of Edinburgh · HITACHI Inspire the Next · Cansolv · CCS Carbon Clean Solutions

NATIONAL CARBON CAPTURE CENTER

# International Test Center Network

**Share $CO_2$ Capture Knowledge to encourage global collaboration and accelerate technology development of cost effective $CO_2$ capture processes**

# Successful Testing and Partnerships

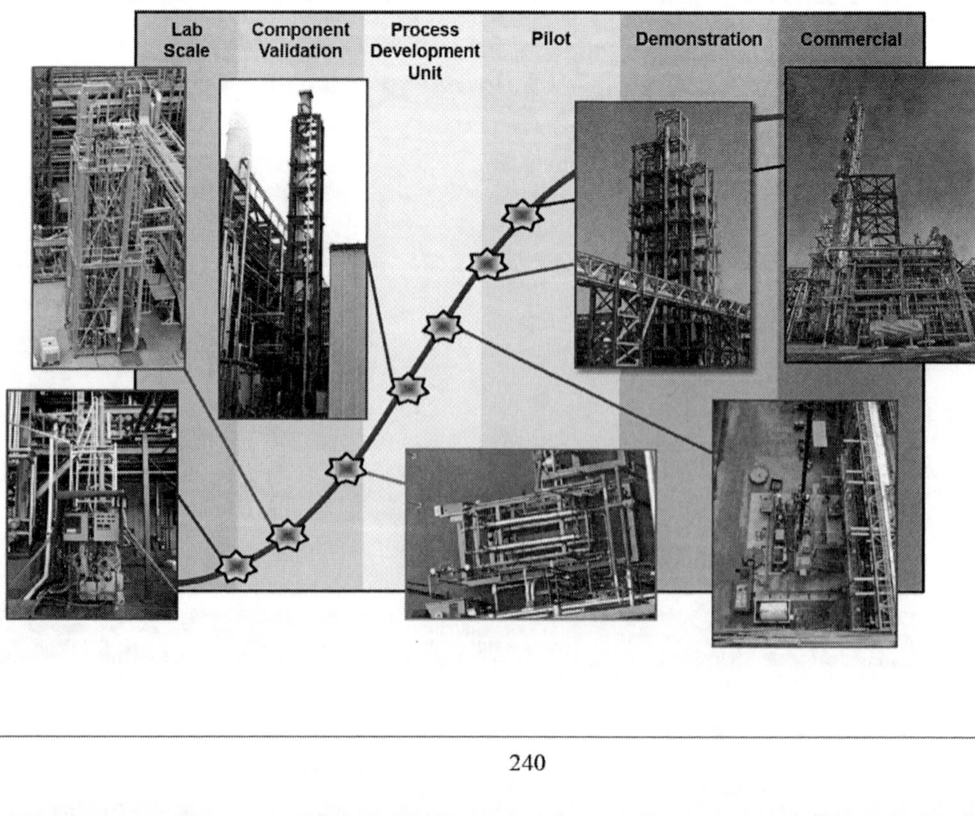


National Carbon Capture Center

# Acknowledgements

U.S. DEPARTMENT OF ENERGY

NETL

SOUTHERN COMPANY

EPRI | ELECTRIC POWER RESEARCH INSTITUTE

Luminant

DUKE ENERGY.

AEP AMERICAN® ELECTRIC POWER

CLOUD PEAK ENERGY®

More information: *www.nationalcarboncapturecenter.com*

NATIONAL CARBON CAPTURE CENTER

# $CO_2$QUEST UE FP7 research programme

## Performing a push-pull test in a shallow aquifer to assess impacts of $CO_2$ and impurities leakage on freshwater quality


Philippe GOMBERT, Stéphane LAFORTUNE and Zbigniew POKRYSZKA
**INERIS, France**
Maria del Camino MARTIN TORRE
**Universidad de Cantabria, Spain**



**33d international Pittsburgh coal conference (Cape Town, South Africa, 8-12 August 2016)**
Session 3 / Carbon management – 1 / 10 August 2016 / 11:30-11:50


UCL Imperial College London UPPSALA UNIVERSITET EWRE BGR OCAS INERIS UNIVERSITY OF LEEDS

## INERIS in brief

**INERIS = French national competence centre for industrial safety and environmental protection.**

**Created in 1990, INERIS is a public institution under the aegis of the French ministry for ecology, whose role is to assess and avert accidental and chronic risks to people, workers and the environment.**

**Our area of expertise in ground and underground risks management:**

- Rock stability and modelling long-term evolution.
- Prevention plans and risk assessment cartography above natural voids and quarries/mines.
- Monitoring of engineering structures.
- Fluid migration, geochemical and thermo-mechanical modellings.
- *In situ* experiments and lab tests.

INERIS

A budget of 80 M€

A staff of ~600, including 335 R&D engineers and 15 professors

More than 90 ISI articles per year

25 000 m² of laboratories and large test facilities

More than 1 000 French and international customers per year

$CO_2$QUEST 2

## $CO_2$QUEST in brief

**Research project for the study of the impact of the quality of $CO_2$ on storage and transport.**

**CO2QUEST focusses on the development of state-of-the-art mathematical models and of large scale experiments to identify impacts of $CO_2$ mixtures on different parts of the CCS chain such as:**

- Pipeline pressure drop and compressor power requirement.
- Pipeline propensity to ductile and brittle facture propagation.
- Corrosion.
- Geochemical interactions within the storage site (here).

**Results will be recommendations for the safe design and economic operation of CCS.**

Coordinated by University College London (UCL)

10 industrial and academic partners from 8 countries in Europe and China

Funded by the UE 7th framework programme for research, technological development and demonstration

UE funding: 4M€

Started in 2013 for 3 years

$CO_2$QUEST 3

## $CO_2$ leakage in a shallow & freshwater aquifer

**Simulation of a leakage from a deep storage site to a shallow aquifer through a <u>push-pull test</u> performed on an experimental site in the Paris region.**

**Objectives:**

- To investigate the impact of industrial grade $CO_2$ + impurities (here **Fe**, **Mn**, **As**, **Cu**, **Zn** and **Li**) on freshwater aquifers.
- To recommend appropriate leakage monitoring methods.

**Results:**

- Both *in situ* monitoring & modelling.

3 tanks of $CO_2$ saturated brine + impurities

**Injection mixture:**
- **3 m$^3$ of water**
- **90 kg of natural (deicing) salt**
- **6 kg of $CO_2$ ⇨ -1.8 pH (7.1 to 5.3)**
- **8 g of metallic impurities + tracers**

$CO_2$QUEST 4

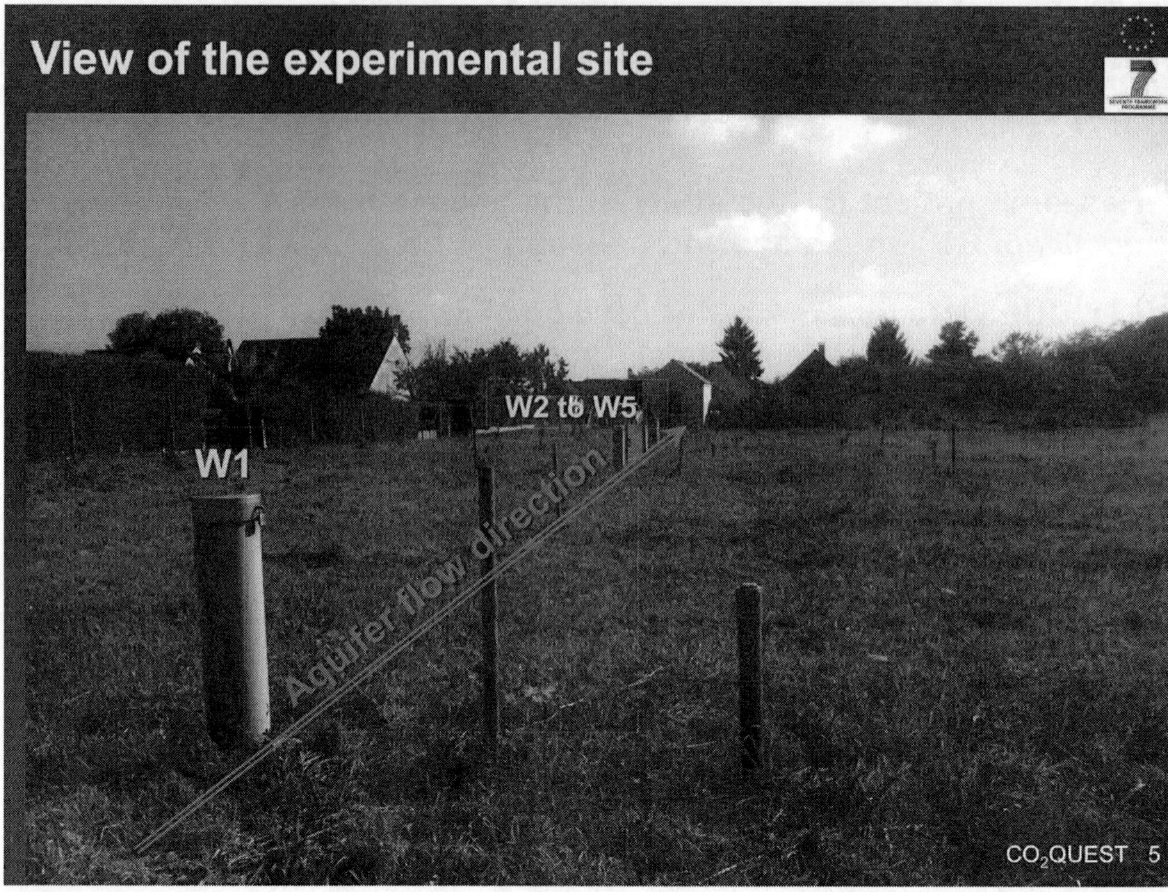
View of the experimental site
W1
W2 to W5
Aquifer flow direction
CO2QUEST 5

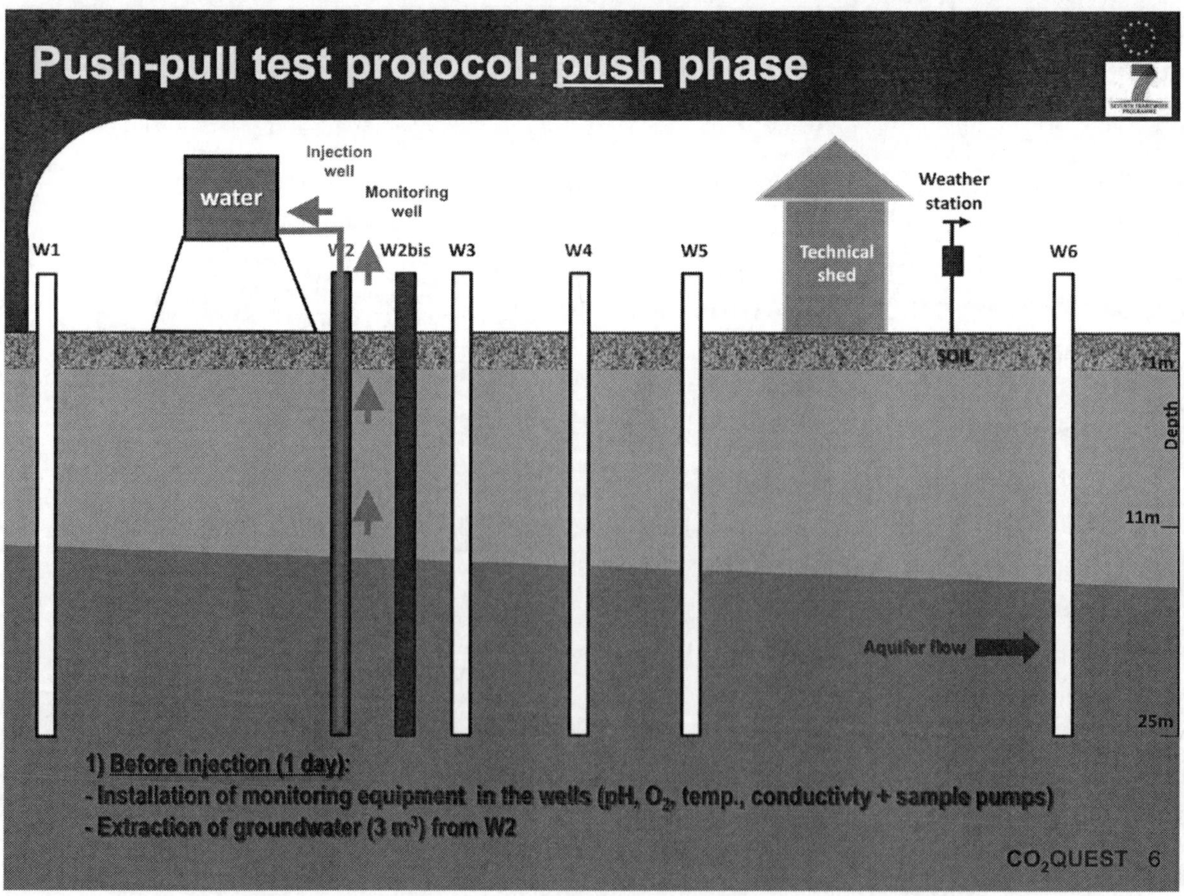
Push-pull test protocol: push phase
Injection well
Monitoring well
water
Weather station
Technical shed
W1
W2
W2bis
W3
W4
W5
W6
SOIL
1m
Depth
11m
25m
Aquifer flow
1) Before injection (1 day):
- Installation of monitoring equipment in the wells (pH, O2, temp., conductivty + sample pumps)
- Extraction of groundwater (3 m3) from W2
CO2QUEST 6

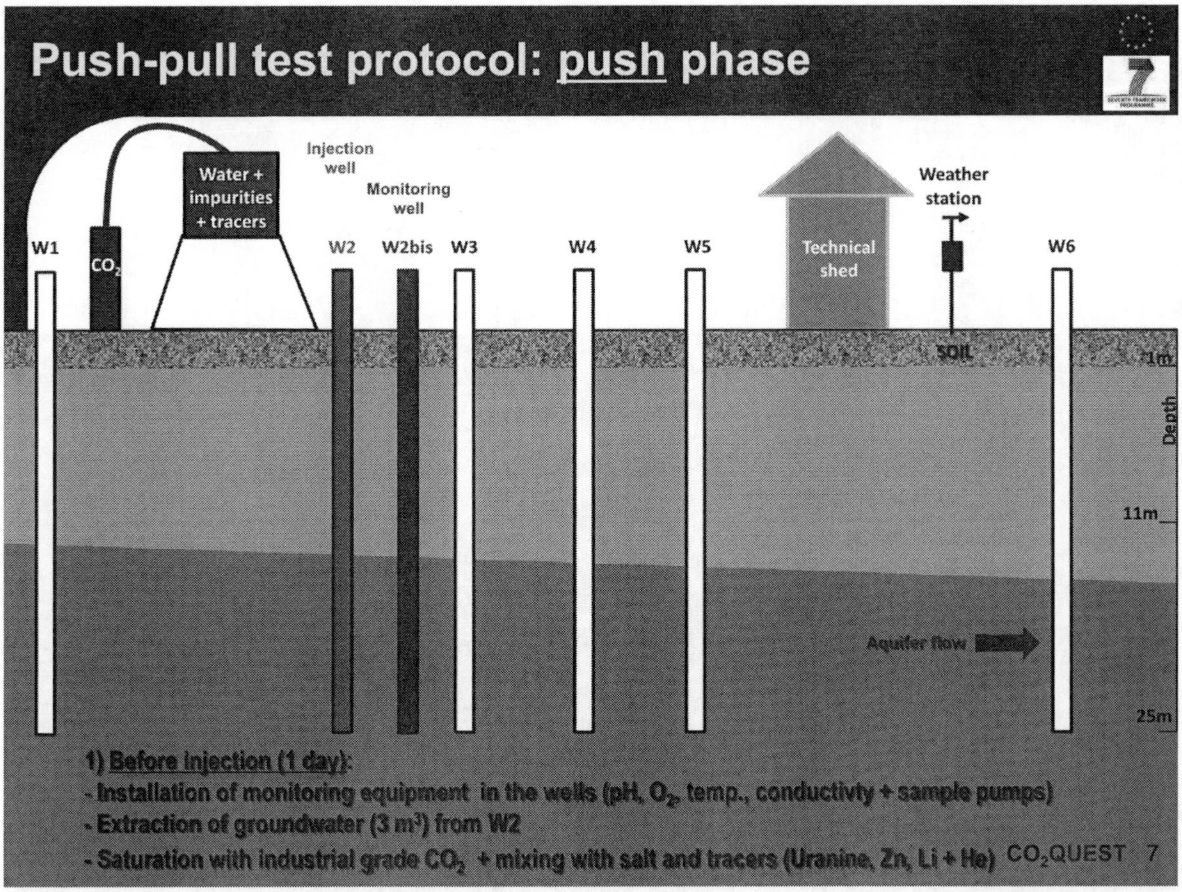
Push-pull test protocol: push phase
Injection well
Monitoring well
Water + impurities + tracers
CO2
W1
W2
W2bis
W3
W4
W5
Technical shed
Weather station
W6
SOIL
1m
Depth
11m
Aquifer flow
25m
1) Before injection (1 day):
- Installation of monitoring equipment in the wells (pH, O2, temp., conductivty + sample pumps)
- Extraction of groundwater (3 m3) from W2
- Saturation with industrial grade CO2 + mixing with salt and tracers (Uranine, Zn, Li + He)
CO2QUEST 7

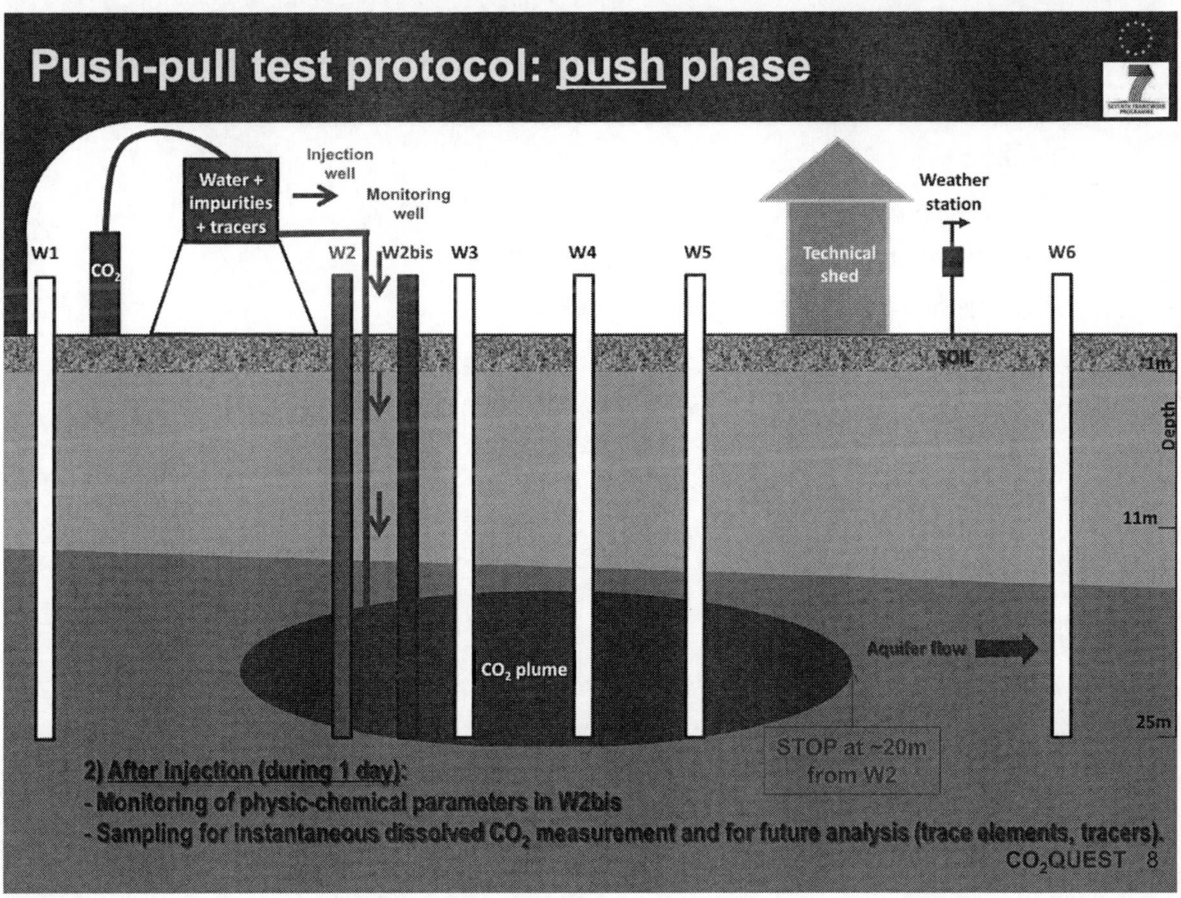
Push-pull test protocol: push phase
Injection well
Monitoring well
Water + impurities + tracers
CO2
W1
W2
W2bis
W3
W4
W5
Technical shed
Weather station
W6
SOIL
1m
Depth
11m
CO2 plume
Aquifer flow
25m
STOP at ~20m from W2
2) After injection (during 1 day):
- Monitoring of physic-chemical parameters in W2bis
- Sampling for instantaneous dissolved CO2 measurement and for future analysis (trace elements, tracers).
CO2QUEST 8

## Push-pull test protocol: pull phase

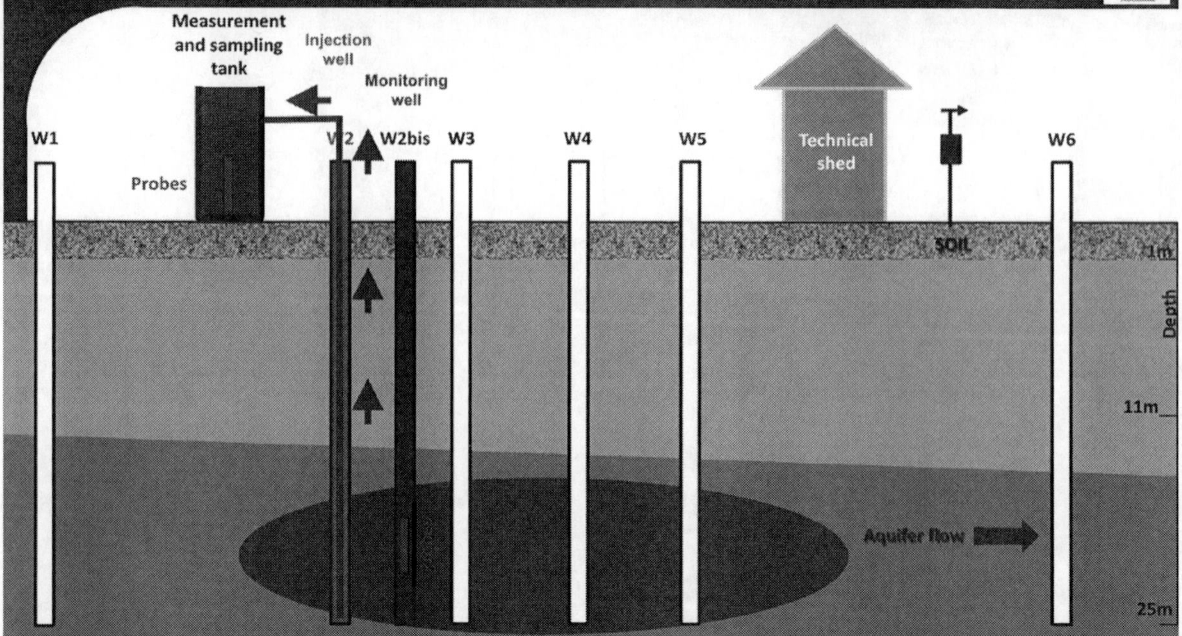


3) After 3 days of water-rock interaction :

- Installation of monitoring equipment in W2bis (pH, T, conductivty, automatic sampler)
- Extraction of groundwater (580 m³ ie 193 times the injected vol.) from W2.

$CO_2$QUEST 9

## Push phase results at monitoring well W2bis

**Results:**

- $CO_2$ peak = 3.3 $10^{-3}$ mol/L (background =1.5 $10^{-3}$ mol/L)
- Alcalinity peak = 9 mg/L (background = 5 mg/L)
- Uranine 600 µg/L (background = 0 µg/L)

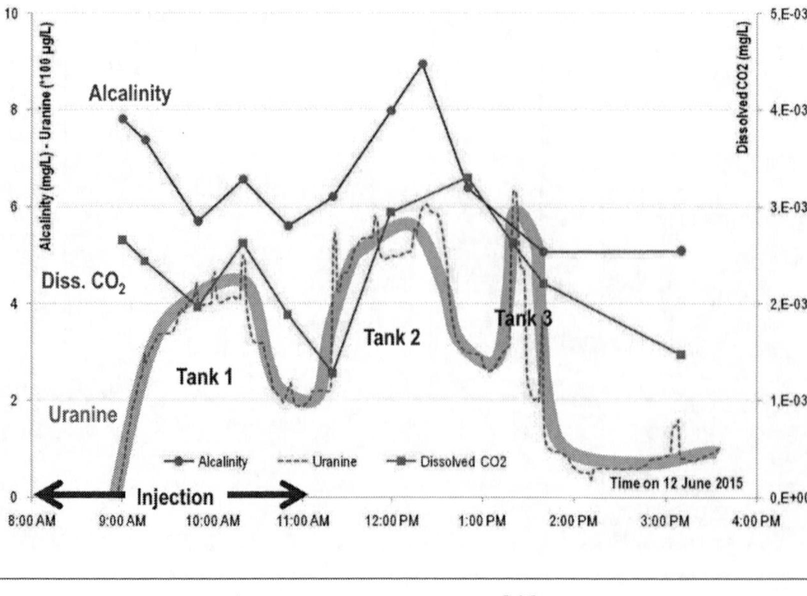


Evolution of dissolved $CO_2$, alcalinity and uranine (tracer) during the push phase in the W2bis well

$CO_2$QUEST 10

# Pull phase results at monitoring well W2bis

## Evolution of pH and alcalinity compared to Li ⇨ restitution

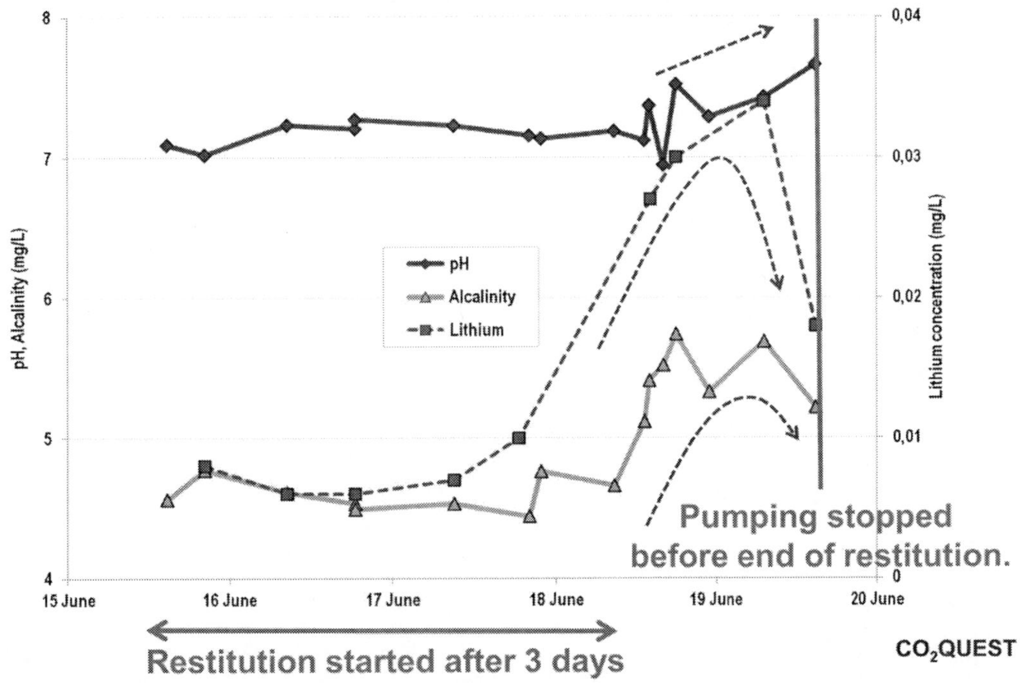


CO$_2$QUEST 11

# Conclusion of the *in situ* experiment

## Behaviour of metallic impurities at monitoring well W2bis

| Element | Initial concentration in injected brine (in mg/L) | Peak in the W2BIS (in mg/L) | Ratio Peak/Initial Concentration |
|---|---|---|---|
| Fe | 0.01 | 0 | 0 |
| Mn | 0.01 | 0 | 0 |
| As | 0.028 | 0.008 | 0.29 |
| Cu | 0.06 | 0 | 0 |
| Zn | 1 | 0.53 | 0.53 |
| Li | 0.633 | 0.248 | 0.39 |
| Uranine | 1 | 0.632 | 0.63 |

←**During push phase:**

**40-60% restitution of tracers**

**30-50% restitution of As, Zn**

no restitution of Fe, Mn, Cu

| Element | Initial concentration in injected brine (in mg/L) | Peak in the W2BIS (in mg/L) | Ratio Peak/Initial Concentration |
|---|---|---|---|
| Fe | 0.01 | 0 | 0 |
| Mn | 0.01 | 0 | 0 |
| As | 0.028 | 0 | 0 |
| Cu | 0.06 | 0 | 0 |
| Zn | 1 | 0 | 0 |
| Li | 0.633 | 0.034 | 0.054 |
| Uranine | 1 | 0.036 | 0.036 |

←**During pull phase:**

**4-6% restitution of tracers**

no other restitution (incl. no metals: Fe, Mn, Cu + As & Zn)

↓

***adsorption, dilution < DL?.***

CO$_2$QUEST 12

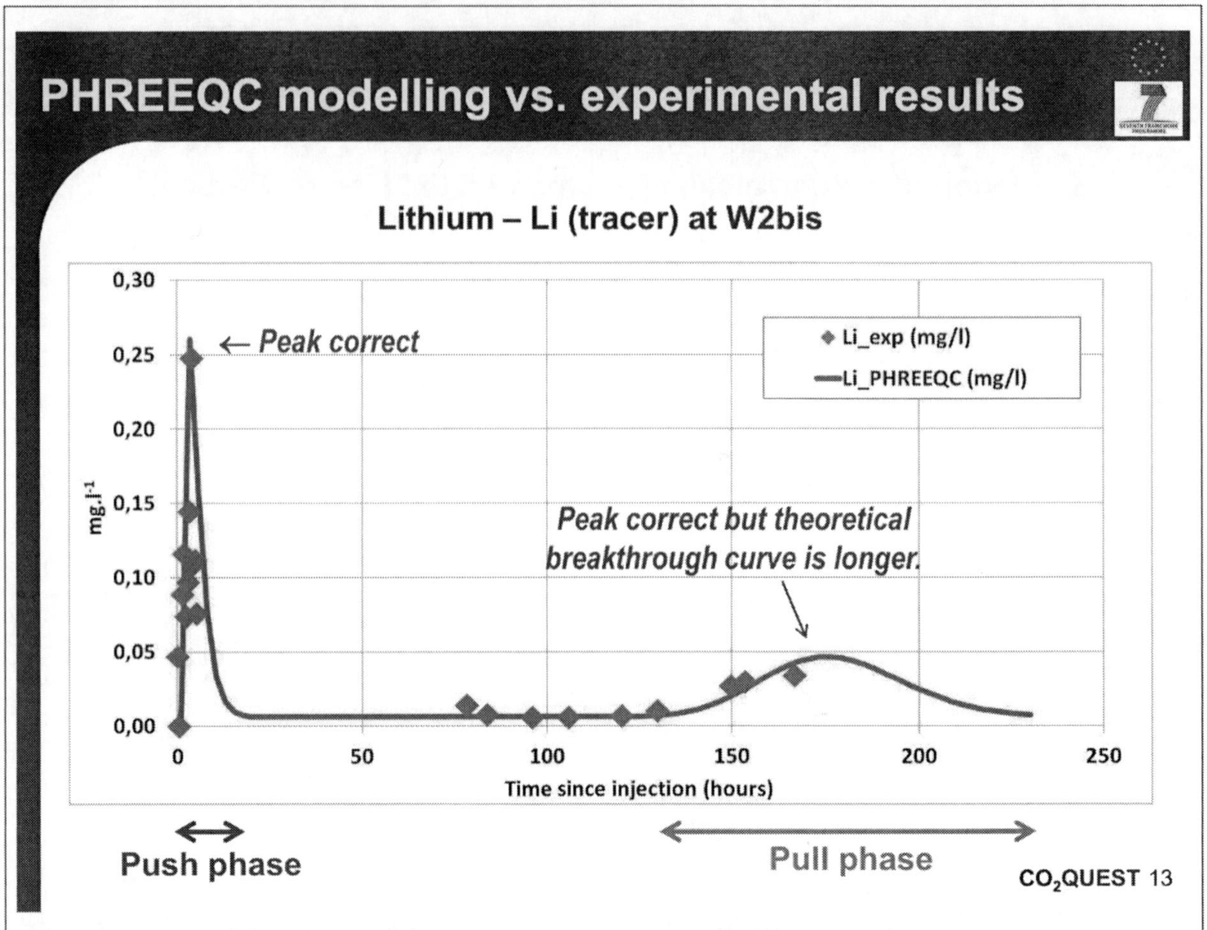
PHREEQC modelling vs. experimental results
Lithium – Li (tracer) at W2bis
0,30
0,25
0,20
0,15
0,10
0,05
0,00
mg.l-1
0
50
100
150
200
250
Time since injection (hours)
Li_exp (mg/l)
Li_PHREEQC (mg/l)
← Peak correct
Peak correct but theoretical breakthrough curve is longer.
Push phase
Pull phase
CO2QUEST 13

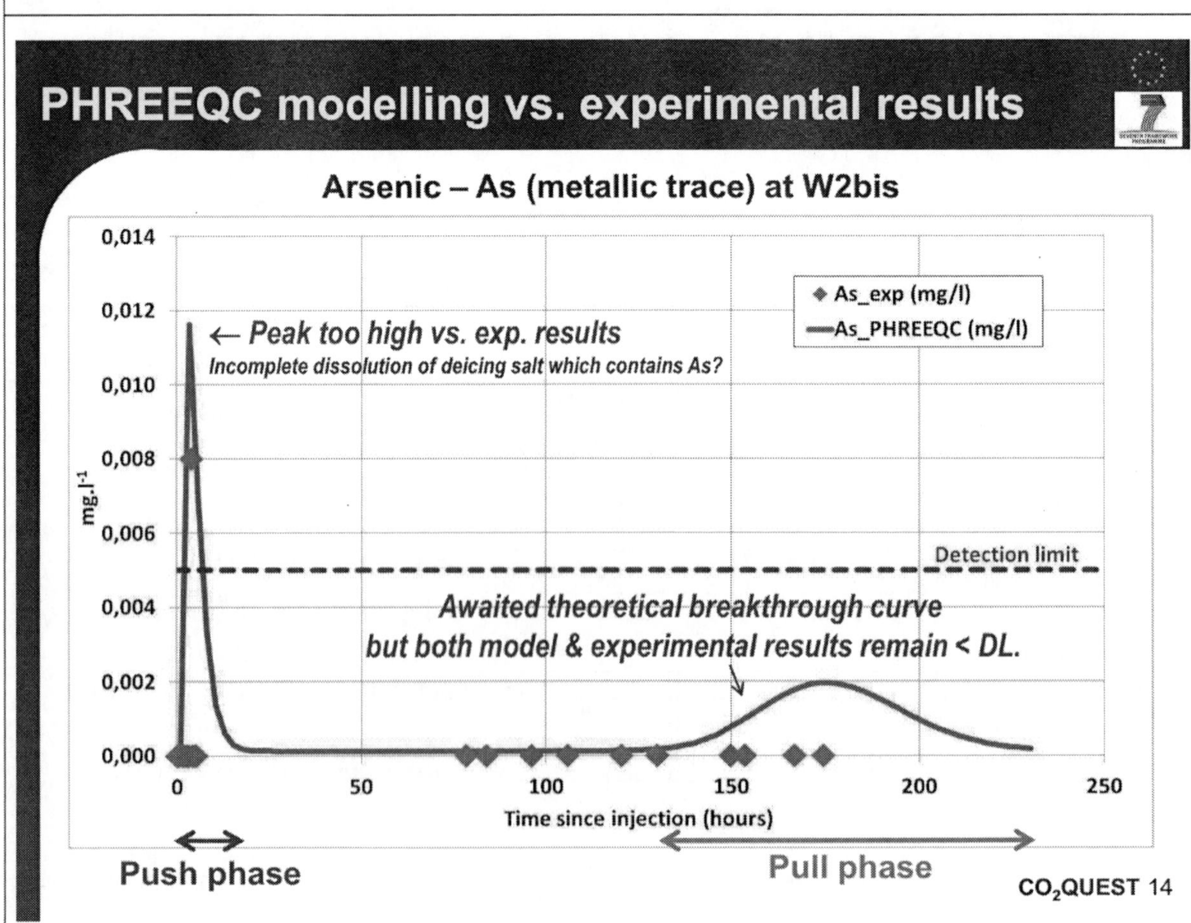
PHREEQC modelling vs. experimental results
Arsenic – As (metallic trace) at W2bis
0,014
0,012
0,010
0,008
0,006
0,004
0,002
0,000
mg.l-1
0
50
100
150
200
250
Time since injection (hours)
As_exp (mg/l)
As_PHREEQC (mg/l)
← Peak too high vs. exp. results
Incomplete dissolution of deicing salt which contains As?
Detection limit
Awaited theoretical breakthrough curve
but both model & experimental results remain < DL.
Push phase
Pull phase
CO2QUEST 14

## Modelling vs. experimental results

### Zinc – Zn (metallic trace element) at W2bis

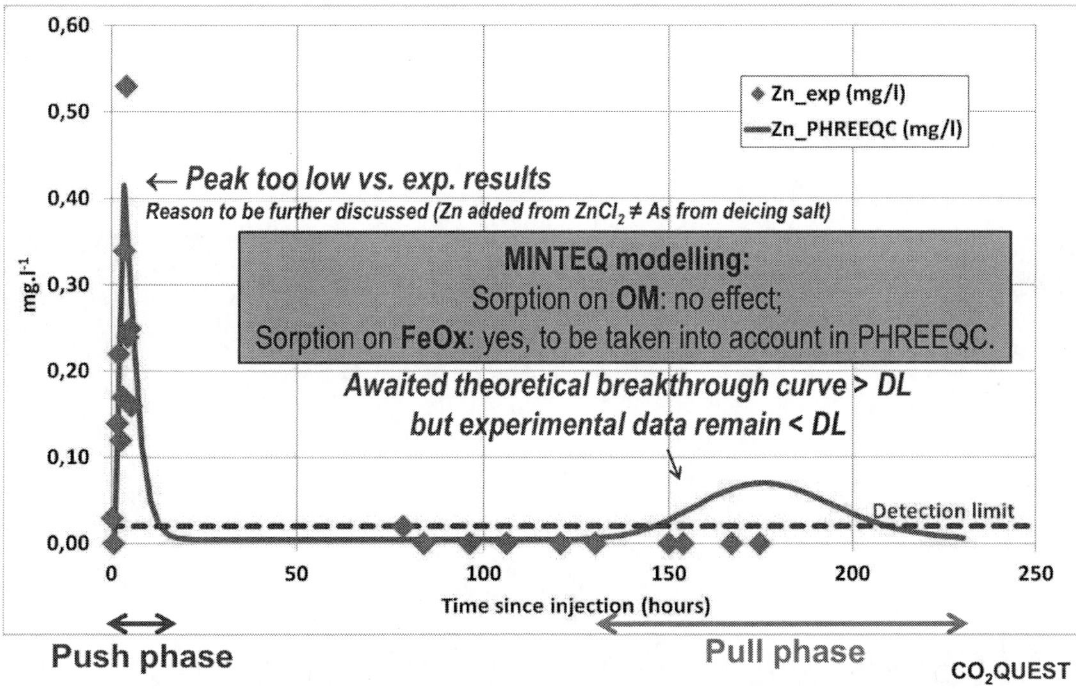


$CO_2$QUEST 15

## Conclusions and future works

➊ Preparation of artificial impure $CO_2$ by dissolving metallic impurities (Fe, Mn, As, Cu, Zn) in brine (30 g/L) + tracers (Uranine, Li)

➋ Injection of $CO_2$ saturated brine + impurities in W2 = "push phase"

➌ Pumping of ~200 times of injected volume in W2 during 4 days = "pull phase"

➍ Monitoring of $CO_2$ plume at W2bis (5 m downstream) during pull phase:
⇨ observed restitution of $CO_2$, tracers but no metals

➎ Geochemical modelling: sorption on FeOx has to to taken into account.
A workshop dedicated to intercomparison of modelling approaches is planned end-2016: participants welcome.

**⇨ No significant impact from the studied impurities in the push-pull test conditions.**

$CO_2$QUEST 16

# Acknowledgements and disclaimer

# Merci.
# Thank you for your attention!

**Philippe GOMBERT & Stéphane LAFORTUNE**
**philippe.gombert@ineris.fr**
**stephane.lafortune@ineris.fr**

The research leading to the results described in this presentation has received funding from the European Union 7th Framework Programme FP7-ENERGY-2012-1-2STAGE under grant agreement number 309102.

The presentation reflects only the authors' views and the European Union is not liable for any use that may be made of the information contained therein.

**$CO_2$QUEST** 17

**2016 International Pittsburgh Coal Conference**
**Cape Town, SOUTH AFRICA**
**August 8 – 12, 2016**

**Abstract Submission**

**PROGRAM TOPIC:**
**CARBON MANAGEMENT**

**Exact Title of Paper:**

# Geological Carbon Sequestration and its Influence on Subsurface Microbial Diversity and Metabolic Carbon Cycling

***Mariana Erasmus, Dr, University of the Free State, Bloemfontein, Free State, RSA***
***ErasM@ufs.ac.za, +27 51 401 7578***

*Tullis Onstott, Prof, Princeton University, Princeton, New Jersey, USA*
*tullis@Princeton.EDU, +1 609 258 4101*

*Mary DeFlaun, Prof, Geosyntec Consultants, Ewing, New Jersey, USA*
*MDeFlaun@geosyntec.com, +1 609 895 1400*

*Tommy Phelps, Prof, University of Tennessee, Knoxville, Tennessee, USA*
*tphelps@utk.edu, +1 865 974 8031*

*Esta van Heerden, Prof, University of the Free State, Bloemfontein, Free State, RSA*
*vheerde@ufs.ac.za, +27 51 401 2472*

**Abstract:**
As an attempt to manage carbon resources more efficiently, South Africa will be storing anthropogenic $CO_2$ in geological formations. However, within these deep subsurface environments, it is well documented that diverse microbial communities can be found and the metabolic activities of these microorganisms are closely tied to the geochemical and mineralogical processes. Therefore, to define possible influences that the $CO_2$ could have on the diversity and ultimately carbon cycling, the interaction was studied between the sequestered $CO_2$ and the subsurface biome, as well as the geological environments. To define the current activity of the biomes *in situ*, the biogeochemistry for a study site, similar to proposed sites for carbon sequestration in South Africa, was characterized comprehensively using sampling techniques that were designed to maintain subsurface conditions with preservative addition for reliable analyses, including stable isotopic analysis to define current biogenic impacts. The diversity of a subsurface biome from the study site was characterized using DGGE and NGS analyses and was found to contain microorganisms belonging to the Archaeal, Bacterial, and Eukaryal domains, but was dominated specifically by Proteobacteria. A novel, continuous, high-pressure bioreactor was designed and used to evaluate the influence of the stored $CO_2$ on the subsurface microbial diversity and metabolic carbon cycling by simulating carbon sequestration conditions.

Due to the extreme conditions of the deep subsurface, most of these microorganisms have adapted to survive in extreme energy limiting conditions while lowering their metabolic activities. The research indicated that even though in most of the conditions tested the microorganisms were not able to multiply or to produce detectable amounts of expected metabolic products, they surprisingly remained viable, most likely due to the presence of biofilm formations that allowed for protection and resistance in these harsh conditions. However, a considerable decrease in the diversity was observed while the transcriptome data revealed that the selected subsurface metabolic activities like sulphur, nitrogen, and carbon were active. Diversity analyses of the bioreactor confirmed that microorganisms, utilizing these metabolisms, remained viable after six weeks under carbon sequestration conditions. These conditions, mainly due to the acidic environment created by the dissolved $CO_2$, allowed for significant leaching of metals from the sandstone, especially aluminium and iron, creating even more stressful environments, thus restricting the growth of the biome. This is the first study in South Africa that demonstrates the impacts carbon sequestration could have on the biogenic carbon cycling with additional considerations to changes in biogeochemical conditions to be considered in the ventures to capture and store carbon in terrestrial environments.

# Physical and Chemical Characteristics of Coke by using Ash-Free Coal as Binder

Gyeong-Min Kim [a,1], Jin-Ho Kim [a], Kevin Yohanes Lisandy [a], Chung-Hwan Jeon [a,b,*]

[a] School of Mechanical Engineering, Pusan National University, Busan 609-735, Republic of Korea

[b] Pusan Clean Coal Center, Pusan National University, Busan 609-735, Republic of Korea

## Abstract

Coke strength can be increased by adding an ash-free coal (AFC) binder. The effect of adding an AFC binder during the development of coke on the physical and chemical properties of coke was investigated to understand the mechanism of increase in coke strength. In this study, two AFC samples extracted from an Indonesian low rank coal were used to understand the effect of the binder. Further, the addition of a torrefied fuel produced from biomass along with the coal binders was considered to realize a reduction in $CO_2$ emission in the steel-making plants. The interface between the base coal and AFC was observed through a scanning electron microscope (SEM). Moreover, the coke strength was investigated through indirect tensile strength test and $^1$H-NMR, $^{13}$C-NMR analyses. To investigate the mechanism of increase in the coke strength, thermal coal (which has no free swelling index) used at power plants was bound with the AFC, and SEM analysis was performed. The results show that the indirect tensile strength between the molecular structures, determined through $^1$H-NMR and $^{13}$C-NMR analyses of the cokes, is proportionate to the average number of aromatic rings.

*Keywords:* Coke; Ash-free Coal; Coal binder; Coke strength

* Corresponding author. Tel.: +82-51-510-3051; fax: +82-51-510-5236.
E-mail address: Energy_min@pusan.ac.kr (G.-M. Kim), jinolfc@pusan.ac.kr (J.-H. Kim), Kevinlie23@pusan.ac.kr (K.Y. Lisandy), chjeon@pusan.ac.kr (C.-H. Jeon).

[1] First author

## 1. Introduction

Coal is widely distributed around the world. Half of it consists of sub-bituminous coal and lignite, and the amount of caking-coal suitable for manufacturing coke is limited. In addition, it is essential to use caking coals to produce high-quality coke. However, caking coal is expensive because the amount of its reserves is small. Therefore, a technology must be devised that does not reduce the quality and increases the blending ratio of the noncaking and/or slightly caking coals. Generally, coke strength decreases with increase in the ratio of noncaking and/or slightly caking coals; however, it is possible to recover or improve the strength by adding ash-free coal (AFC) produced by extracting solvents.

Shishido et al.[1)] reported that the higher the bending ratio of noncaking and/or slightly caking coals, the more the blending effect on the coke strength improved. Thus, adding hyper coal (HPC) manufactured using a similar method to AFC from Japan is effective in improving the strength of coke blended with noncaking and/or slightly caking coals. Takanohashi et al.[2)] reported that coke strength increased by producing HPC from a caking coal and adding it to blending coals, including 35 mass% of noncaking and/or slightly caking coals.

In addition, AFC is expected to be produced from cost-effective low-rank coals. Koyano et al.[3)] reported that the addition of large peri-condensed aromatic-ring compounds to a coal blend markedly enhanced the resulting coke strength. The addition of various polyaromatic hydrocarbons to coals as caking additives increases the coke strength when more than four aromatic rings are present, and the effect enhances with increase in the number of aromatic rings. Moreover, comparing between the polyaromatics with the same ring size, the coke strength improved. It was previously reported that the average number of aromatic rings in low-rank coals is approximately 1–2.[4)] If the number of aromatic rings in AFC produced from low-rank coals is nearly equal (1–2 rings), the improvement of coke strength may not be expected, although it should not be easily compared with model compounds because their polarity or molecular weight are different. Moreover, as low-rank coals contain a large amount of oxygen functional groups, the plasticity of blending coals may significantly decrease. The details on the molecular structure of AFC and the mechanism of its effect have not been clarified so far. In addition, Sakimoto et al.[5] measured that HPC produced from various ranks of coal to evaluate the effect of the chemical structure of HPC on the coke strength, and showed that the coke strength increased with the number of aromatic rings of HPC. The objective of this study is to compare the characteristics of HPC produced in Japan and AFC produced in Korea.

In the current study, AFC produced from low-rank coals was structurally analyzed to clarify its molecular structure and characteristics. Next, this AFC was added to a slightly caking coal to investigate the effect on coke strength and its relationship with molecular structure.

## 2. Experimental methods

### *2.1. Coal sample preparation*

**Table 1**. Particle size distribution of coal for coke-making processes

| Sample | Particle size | No additive | 3% AFC | 3% AFC + 3% Torrefied Fuel |
|---|---|---|---|---|
| Coking coal A | 0.38-1.0mm | 33 mass% | 33 mass% | 33 mass% |
| | 0.15-0.38mm | 33 mass% | 33 mass% | 33 mass% |
| | Under 0.15mm | 34 mass% | 31 mass% | 28 mass% |
| AFC | Under 0.15mm | 0 mass% | 3 mass% | 3 mass% |
| Torrefied Fuel | 0.3-0.5mm | 0 mass% | 0 mass% | 3 mass% |

An Indonesian low-rank coal KCH sub-bituminous was used to produce AFCs through solvent extraction: Hexane and Oven coals. In addition, Coking coal A was used as a standard coal for manufacturing coke. Another sub-bituminous coal, KPU from Indonesia was selected as a representative of base coal. The KPU coal does not have any thermoplastic properties, and can thus be used to investigate the principle of strength improvement by AFC binder. The samples are ground to three particle size distributions: under 0.15, 0.15–0.38, and 0.38–1.00 mm. Table 1 shows the mass contents for the size distributions. Torrefaction of wood pallets was heated electric furnace. A nitrogen flow of 2 L $min^{-1}$ was used for torrefaction process. An Wood pallet sample was charged into the reactor at room temperature, and the reactor was heated at 10 °C $min^{-1}$ under atmospheric pressure. Both clean wood chips and forest residue chips were torrefied at 225 °C at 30 min residence time. The torrefaction residence time of 30 min was chosen to be optimal for torrefaction reaction from preliminary studies.

*2.2. Sample Analysis*

**Table 2.** Proximate and elemental analyses and free swelling index of coal

| Coal Properties | | Coking Coal A | KCH-Hexane | KCH-Oven | Torrefied Fuel |
|---|---|---|---|---|---|
| Proximate Analysis, (db, wt%) | Moi. | 2.46 | 0.43 | 0.31 | 4.07 |
| | VM | 34.64 | 77.16 | 91.61 | 68.11 |
| | FC | 54.47 | 22.30 | 8.04 | 26.41 |
| | Ash | 8.43 | 0.11 | 0.04 | 1.42 |
| Elemental Analysis, (db, wt%) | C | - | 78.58 | 82.09 | 54.66 |
| | H | - | 8.37 | 9.10 | 5.95 |
| | N | - | 0.34 | 0.24 | 0.18 |
| | O | - | 12.44 | 8.41 | 39.21 |
| | S | - | 0.27 | 0.16 | 0.00 |
| F.S.I | | 6.1 | >4 | >4 | - |

Approximately 5 g of the sample was analyzed using a thermogravimetric analyzer (TGA 701, LECO Co., USA) and the American society of testing materials (ASTM) proximate analysis standard, while an ultimate analysis was conducted through a commercial device (Leco-TruSpec Micro CHNS). The free swelling index (FSI) was determined according to ASTM D 720 method. Table 2 presents the characteristics of coal.

### *2.3. Solvent extraction*

The AFC process was modeled by the KIER (Korea Institute of Energy Research) using a pilot-scale apparatus.[6)] Actually, to produce AFC, 1-methylnaphthalene, nonpolar (1-MN) was used as a solvent to extract the hydrocarbon contents from coal. The coal and solvent were mixed at 0.1 Mpa and 110 °C. The extraction column pressure is approximately 2.99 Mpa and 350 °C, and the particle diameter of the extracted residual coal was approximately 80 μm.

### *2.4. Solid state $^{1}$H-NMR Measurement*

The hydrogen-type distribution of the AFC samples and the aromaticity (fa) were measured using solid state $^{1}$H-NMR (400 MHz Avance II + Brunker solid-state NMR; KBSI Seoul Western Center). The hydrogen were classified into four types: an aromatic hydrogen (Har, 6.0–10.0 ppm) and three aliphatic hydrogen; hydrogen attached on carbon at alpha position (Ha, 2.0–5.0 ppm), at beta position (Hb, 1.1–2.0 ppm), and at gamma position (Hr, 0.2–1.1 ppm).

Further, the chemical shift is based on that of a standard tetramethylsilane (TMS).

### *2.5. Solid state $^{13}$C-NMR Measurement*

The oxygen-type distribution of the AFC samples and the aromaticity (fa) were measured using solid-state $^{13}$C-NMR (400 MHz Avance II + Brunker solid-state NMR; KBSI Seoul Western Center). The classification of the carbon types from 13C-NMR spectrum was determined by referring to relevant literature[7)]: aliphatic carbon (Cali, 0–93 ppm) and aromatic carbon (Car, 93–171 ppm).

In this study, it was assumed that three oxygen types are present in the NMR spectrum: the ether, phenolic hydroxyl group (148–171 ppm), and carboxyl group (171–200 ppm). The ether type was calculated by removing the phenolic hydroxyl and carboxyl oxygen contents from the total oxygen content.

*2.6. Coke Strength Measurement*

The coke samples were manufactured using raw coking coal by adding AFCs (Hexane and Oven) and torrefied fuel (originated from Wood Pallet). The added ratio of AFCs and that of the torrefied fuel are 3 mass%.

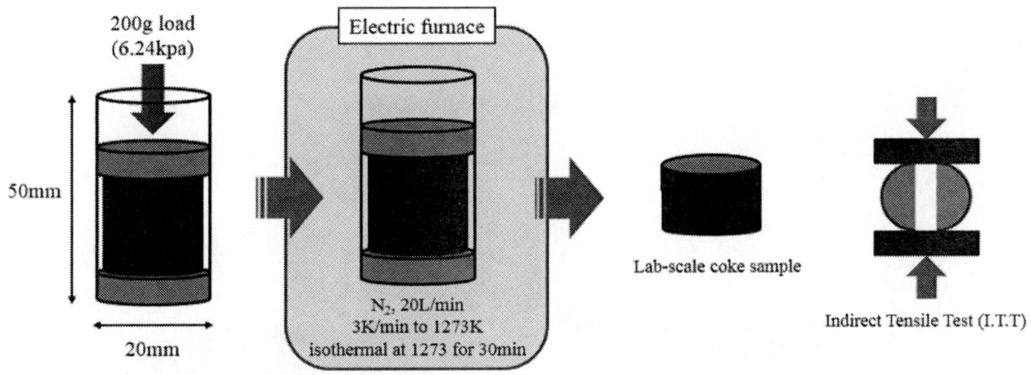


**Fig. 1** Schematic of lab-scale coke-making apparatus and indirect tensile test

Fig. 1 shows the schematic of the lab-scale coke-making apparatus and strength measurement. Approximately 5 g of blended coal sample was charged into the vessel. The bulk density of blended coal was approximately 800 kg/m$^3$. A load of 200 g was weighted onto the packed sample until its carbonization with the nitrogen environment. The furnace was heated from room temperature to 1273 K at a heating rate of 3 K/min. Once the temperature reached 1273 K, the sample was held for 30 min, and then cooled down at 473–573 K by using nitrogen gas. Moreover, the coke samples were produced using the same method.

The compressive load of the coke samples was measured using the universal testing machine (KSU-10M; 10 kN). In addition, the indirect tensile strength of coke was evaluated using the indirect tensile test method developed by Miyagawa et al., and the equation is as follows.[8)]

$$S = \frac{4 \cdot P}{\pi \cdot r \cdot l} \quad \cdots\cdots (1)$$

where $S$ is the indirect tensile strength of coke (Mpa), $P$ is the failure load of coke (N), $r$ is the coke sample radius (mm), and $l$ is the thickness of the sample (mm).

## 3. Results and discussion

### *3.1. Analysis of sample morphology by using SEM image*

**Fig. 2** Analysis of sample morphology by using SEM image a) KPU + hexane, b) KPU + Oven

Fig. 2 shows the SEM image of carbonized KPU coal particles. KPU coal has no FSI, and therefore cannot be bonded. In addition, it is possible to observe the role of the AFC with the binding principle. Fig. 2 (a) and (b) show the SEM images of KPU coal carbonized with Hexane (AFC) and Oven (AFC), respectively. The KPU coal particles are bound together by the AFC binder, which coats and fills the void space of the KPU coal.

### *3.2. $^{13}C$ NMR results*

#### *3.2.1. $^{1}H$-NMR results*

**Table 3** Distribution of hydrogen type and number of hydrogen types per 100 carbons

| Sample | Hydrogen fraction [%] | | | | Number of Hydrogen [-/100 Carbons] | | | |
|---|---|---|---|---|---|---|---|---|
| | Har | Ha | Hb | Hr | Har | Ha | Hb | Hr |
| Hexane | 38.5 | 28.0 | 16.4 | 17.2 | 49.2 | 35.7 | 20.9 | 22.0 |
| Oven | 38.0 | 28.8 | 16.2 | 17.0 | 50.5 | 38.3 | 21.6 | 22.6 |

Table 3 shows the distribution of hydrogen types and the number of hydrogen types per 100 carbons. A similar trend is observed for both of the AFCs in the hydrogen contents, even though their extraction methods are different. The Hexane coal has a slightly higher aromatic hydrogen (Har) amount of 0.5%; however, when considering the number of hydrogen types per 100 carbons, the amount of aromatic hydrogen content is higher: 1.3.

#### *3.2.1. $^{13}C$ NMR results*

Fig. 3 shows the solid-state $^{13}C$-NMR spectra of KCH raw coal and AFC coals (Hexane and Oven) extracted from KCH raw coal. The results were obtained through a standardized method using TMS as an internal standard (0 ppm).

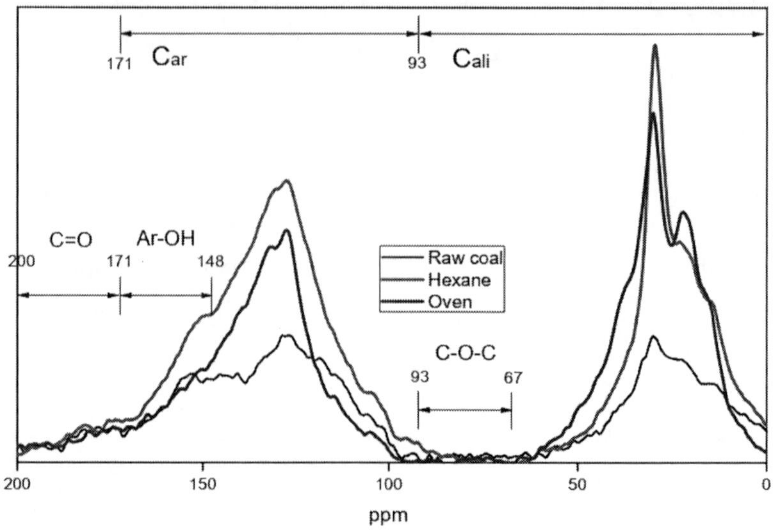


**Fig. 3** Solid-state $^{13}C$-NMR spectra of KCH-Hexane and KCH-Oven coals

**Table 4** Distribution of oxygen type and number of oxygen types per 100 carbons

| Sample | Oxygen fraction [%] | | | Number of Oxygen [-/100 Carbons] | | |
|---|---|---|---|---|---|---|
| | COOH | Ar-OH | -O-(diff.) | COOH | Ar-OH | -O-(diff.) |
| Hexane | 24.0 | 66.6 | 9.4 | 2.9 | 7.9 | 1.1 |
| Oven | 30.0 | 60.1 | 9.8 | 2.3 | 4.6 | 0.8 |

Fig. 3 shows the solid-state $^{13}C$-NMR spectra of Hexane and Oven coals. The results were standardized to be in the same spectrum area. Table 4 shows the distribution of oxygen types and the number of oxygen types per 100 carbons. The number of oxygen per 100 carbons in carboxyl group of Hexane is more than in Oven coal by 0.6. The phenolic hydroxyl group of Hexane has much more contents than that of Oven by 3.3.

*3.3. Bulk density and tensile strength of the cokes*

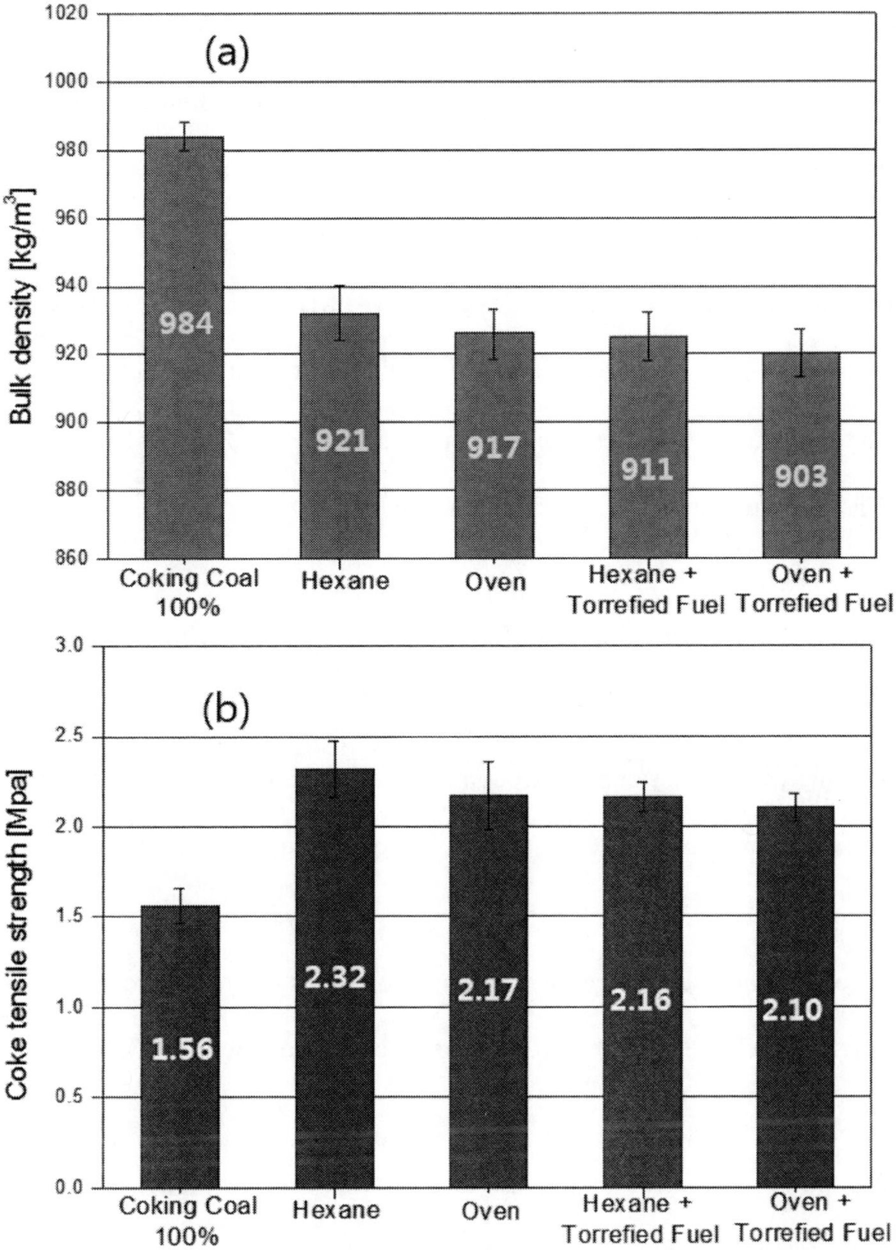


**Fig. 4** (a) The bulk density of the coke and (b) the coke tensile strength

Fig. 4 shows the bulk density and tensile strength of coke. The bulk density decreased with the addition of AFC and torrefied fuel. This is because the densities of AFC and torrefied fuel are lighter than coking coal density. The indirect tensile strength of the coke with the additives and torrefied fuel increased from 100% coking coal.

## 4. Conclusions

Coke strength was increased by adding an ash-free coal (AFC) binder. The effect of adding an AFC binder during the development of coke on the physical and chemical properties of coke was successfully investigated. The interface between the base coal and AFC was observed through a scanning electron microscope (SEM) and coke strength was investigated through indirect tensile strength test. The results show that the indirect tensile strength between the molecular structures, determined through 1H-NMR and 13C-NMR analyses of the cokes, is proportionate to the fraction of aromatic.

1. Coke strength can be successfully increased by adding an AFC binder.

2. The analysis of the samples through thermal coal and AFC morphology shows that the wetted area is an important factor for increasing the physical strength.

3. In spite of the slight decrease in densities of cokes because of low-density AFCs and torrefied fuel, the coke strength increased approximately over 35%.

4. The $^{13}C$ NMR analysis shows that the fraction of the aromatic carbon related with coke strength and the coke strength between the molecular structures is proportionate to the average number of aromatic rings.

**Acknowledgments**

This work was supported by the Power Generation & Electricity Delivery Core Technology Program of the Korea Institute of Energy Technology Evaluation and Planning (KETEP), and granted financial resource by the Ministry of Trade, Industry & Energy, Republic of Korea (No. 20131010101810)

**References**

[1] T. Shishido, K. Sakai, N. Okuyama, M. Hamaguchi and N. Komatsu, *Kobe Steel Eng. Rep.*, 60 (2010), 62.

[2] T. Takanohashi, T. Shishido and I. sato, *Energ. Fuel.*, 22 (2008), 1779

[3] K. Koyano, T. Takanohashi and I. Saito, *Energ. Fuel.*, 25 (2011), 2565

[4] J. P. Mathews and A. L. Chafee, *Fuel*, 96 (2012), 1.

[5] N. Sakimoto, T. Takanohashi, A. Matsumura, S. Sato and I. Saito, *Fuel Process. Technol*, 86 (2004), 61

[6] S. H. Park., S. W. Chung., S. K. Lee., H. K. Choi. And S. H. Lee., *Appl. Therm. Eng.*, 89 (2015), 843

[7] T. Yoshida, *Sekiyu Gakkaishi*, 35 (1992), 1.

[8] T. Miyagawa and I. Fujishima, J. Fuel Soc. Jpn., 54 (1975), 983

# The influence of $K_2CO_3$ on the swelling and caking behaviour of large South African coal particles

**Dr. Sansha Coetzee**
Prof. Hein W.J.P. Neomagus
Prof. John R. Bunt

**8-12 August 2016**

NORTH-WEST UNIVERSITY ®
YUNIBESITI YA BOKONE-BOPHIRIMA
NOORDWES-UNIVERSITEIT

**University of Pittsburgh**
**Swanson School of Engineering**

AFRICARY
African Carbon Energy

---

## Background

- Coals may undergo physical transformations, such as fragmentation or plastic deformation (softening and swelling), during heating[1-3]
- Coal swelling and caking are undesired occurrence in fixed-bed and fluidised-bed operations, and influence particle size, density, porosity and char reactivity, and also cause channelling[4]
- Various methods have been investigated to reduce and/or eliminate swelling and caking of coal[5-8]:
  - Pre-oxidation
  - Slow heating rate
  - Additive addition such as alkali and alkaline earth metal salts
- Until recently, studies investigating swelling reduction have mainly focused on powdered coal samples (< 500 µm)

[1] HABERMEHL *et al.* 1981. Chemistry of Coal Utilisation. New York; Wiley-Interscience. p317-368. [2] DAKIČ *et al.* 1989. *Fuel*, 68:911-916.
[3] BREWER. 1945. Agglutinating, Agglomerating, and Swelling Properties of Coals. New York: John Wiley & Sons Inc. P160-309.
[4] SHENG & AZEVEDO. 2000. *Proceedings of the Combustion Institute*, 28:2225-2232. [5] CREWE et al. 1975. *Fuel*, 54:20-23.
[6] FORNEY *et al.* 1964. *I&EC Product Research and Development*, 3:48-53. [7] BEXLEY *et al.* 1986. *Fuel*, 65:47-53.
[8] PATRICK & SHAW. 1972. *Fuel*, 51:69-75.

2

# Background

- A recent study investigated the influence of potassium carbonate ($K_2CO_3$) on the degree of swelling and caking of large coal particles (5, 10, 20 mm)[9,10,11]
- Three South African coals with varying swelling characteristics were selected, and experimental conditions included constant impregnation concentration (5 M $K_2CO_3$) and constant devolatilisation heating rate (7 °C/min) [9,10]
- Results showed that $K_2CO_3$-impregnation reduced the swelling and caking behaviour of a moderately swelling coal (Waterberg coal) most effectively[9,10]
- Investigate the influence of different $K_2CO_3$ concentrations, and varying devolatilisation heating rates

[9] COETZEE *et al.* 2015. *Energy & Fuels*, 29:4255-4263. [10] COETZEE *et al.* 2014. *Fuel*, 29:6197-6205. [11] COETZEE *et al.* 2014. *Fuel*, 136:79-88.

3

# Aim and objectives

- The aim of this study is to quantify the influence of different $K_2CO_3$ concentrations and varying devolatilisation heating rates on the swelling and caking of large Grootegeluk (GG) coal particles
- **Objectives:**
  - Quantify the swelling behaviour of 20 mm GG particles as a function of $K_2CO_3$ concentration and heating rate, using volumetric swelling ratios ($SR_V$)
  - Quantify the caking propensity of 10 mm GG particles as a function of $K_2CO_3$ concentration and heating rate, using bridging neck size measurements

4

# Experimental

## Experimental outline

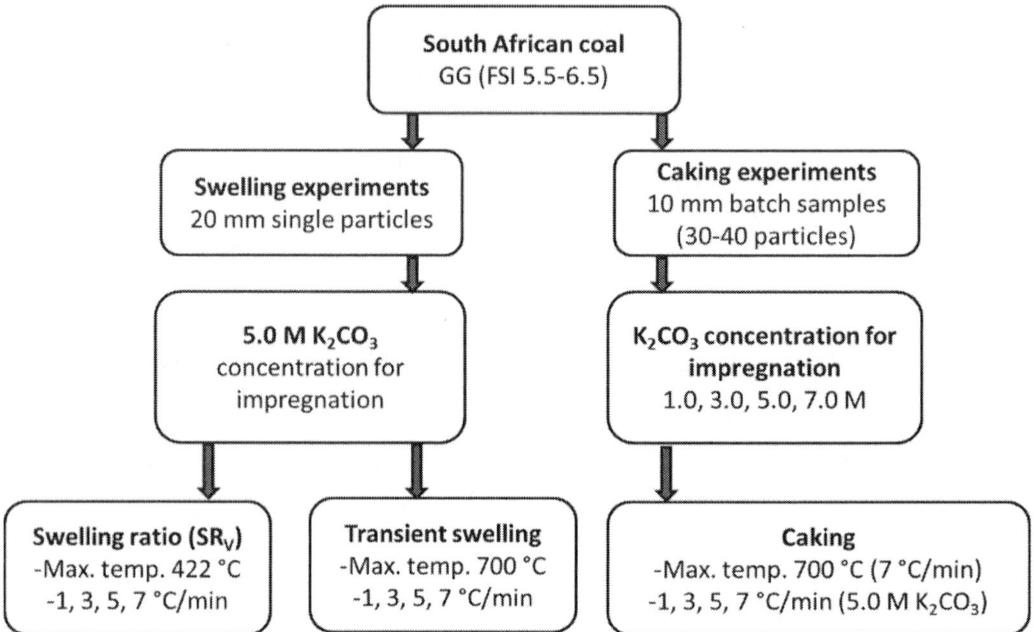


5

# Experimental

## Devolatilisation

- Swelling and caking experiments were conducted in a Carbolite Biomass CAF furnace, equipped with a camera

| | Operating conditions |
|---|---|
| Initial temperature (°C) | 100 |
| Heating rate (°C/min) | Specified |
| Maximum temperature (°C) | Specified |
| Gas atmosphere | Nitrogen (6 L/min) |

- Images were acquired every 1 °C temperature increase

6

# Experimental

**Swelling experiments**

- 2D interpretation of images acquired during devolatilisation using ImageJ – qualitative

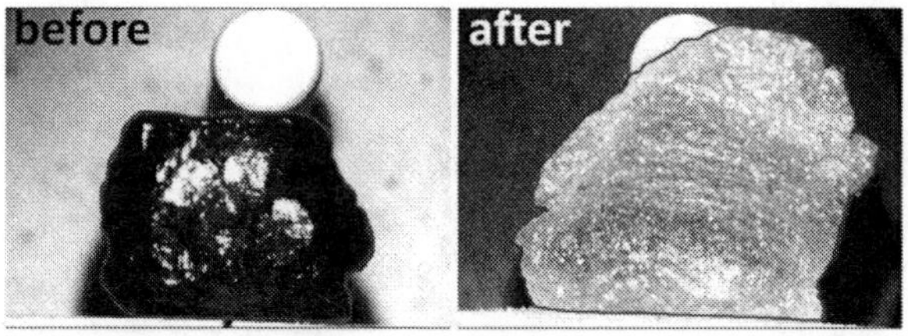


•Image analysis swelling ratio (based on area)

$$SR_{IA} = \frac{A_f}{A_i}$$

•Mercury submersion swelling ratio (based on volume)

$$SR_V = \frac{V_f}{V_i}$$

7

# Experimental

**Caking experiments - Bridging neck size measurements**

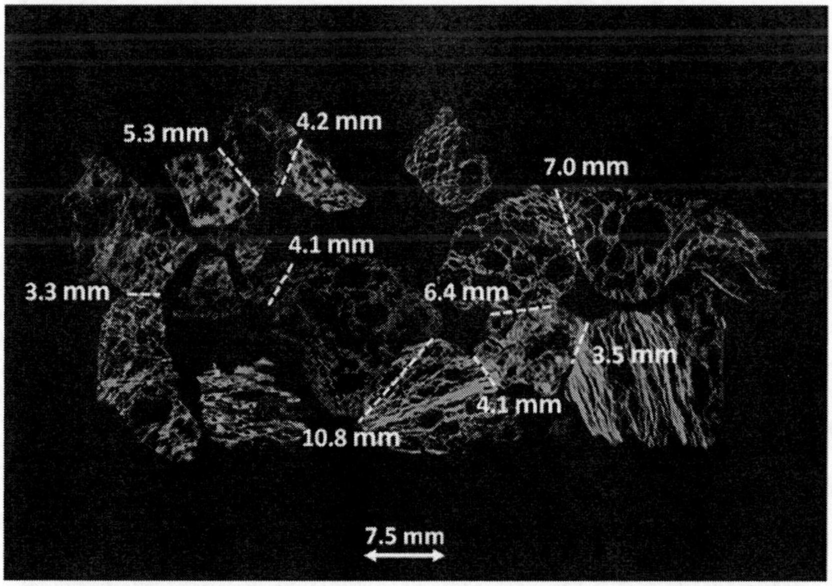


8

# Results

## Coal Characterisation

| Analyses | | | GG 10 mm | GG 20 mm |
|---|---|---|---|---|
| Proximate analysis | Inherent moisture (adb) | wt% | 2.5 | 2.5 |
| | Ash content (adb) | wt% | 13.0 | 13.9 |
| | Volatile matter (adb) | wt% | 36.6 | 36.9 |
| | Fixed carbon (adb) | wt% | 47.9 | 46.7 |
| Ultimate analysis | Carbon (adb) | wt% | 68.5 | 67.8 |
| | Hydrogen (adb) | wt% | 4.7 | 4.8 |
| | Nitrogen (adb) | wt% | 1.5 | 1.4 |
| | Oxygen (adb) [by difference] | wt% | 8.5 | 8.1 |
| | FSI (Free Swelling Index) | | 5.5 | 6.5 |
| Gieseler fluidity | Initial softening temperature | °C | 402 | 400 |
| | Maximum fluid temperature | °C | 433 | 430 |
| | Solidification temperature | °C | 451 | 457 |
| | Maximum fluidity | ddpm | 41 | 50 |
| Dilatation | Softening temperature | °C | 363 | 371 |
| | Temperature of maximum contraction | °C | 415 | 418 |
| | Temperature of maximum dilatation | °C | 441 | 435 |
| | Maximum contraction | % | 36 | 33 |
| | Maximum dilatation | % | 19 | 14 |

9

# Results

## Swelling & fluidity-pulverised (-212 μm) coal

| Analysis | GG6.5 | | |
|---|---|---|---|
| | Raw | 5 wt.%* | 10 wt.%* |
| *Free swelling index, FSI* | 6.5 | 1.0 | 0 |
| *Gieseler fluidity (°C)* | | | |
| Initial softening temp. | 400 | 422 | - |
| Maximum fluid temp. | 430 | 431 | - |
| Solidification temp. | 457 | 440 | - |
| Maximum fluidity (DDPM) | 50 | 2 | 0 |
| *Dilatation (°C)* | | | |
| Softening temp. | 371 | 415 | - |
| Temp. of max. contraction | 418 | 500 | - |
| Temp. of maximum dilatation | 435 | - | - |
| Max. contraction (%) | 33 | 10 | 0 |
| Max. dilatation (%) | 14 | - | 0 |

*-$K_2CO_3$

10

## Results

**Swelling - Visual observations 20 mm (devolatilisation at 422 °C)**

**GG Raw**

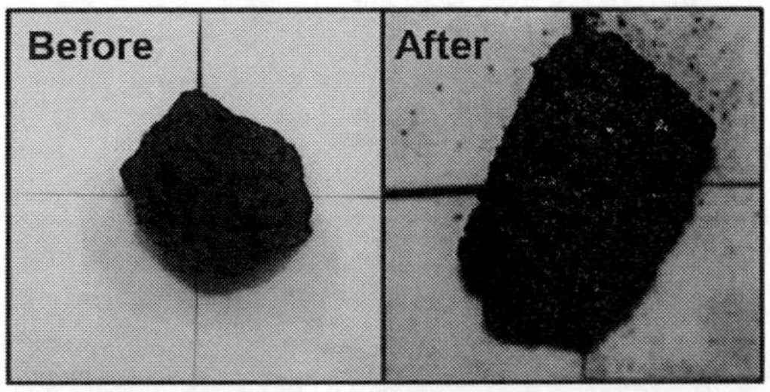


**GG Impregnated**

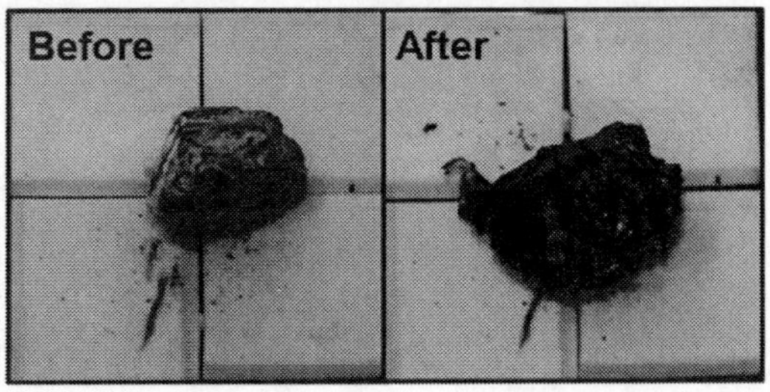


11

## Results

**Quantification of swelling (20 mm) - $SR_V$**

| Heating rate (°C/min) | $SR_V$ raw coal | $SR_V$ coal with $K_2CO_3$ | % Reduction in swelling |
|---|---|---|---|
| 1 | 2.0 ± 0.2 | 2.0 ± 0.1 | none |
| 3 | 2.8 ± 0.4 | 2.2 ± 0.2 | moderate |
| 5 | 3.1 ± 0.2 | 2.5 ± 0.3 | moderate |
| 7 | 3.2 ± 0.4 | 2.8 ± 0.2 | moderate |

moderate <30% reduction in swelling

12

## Results

**Transient swelling during devolatilisation**

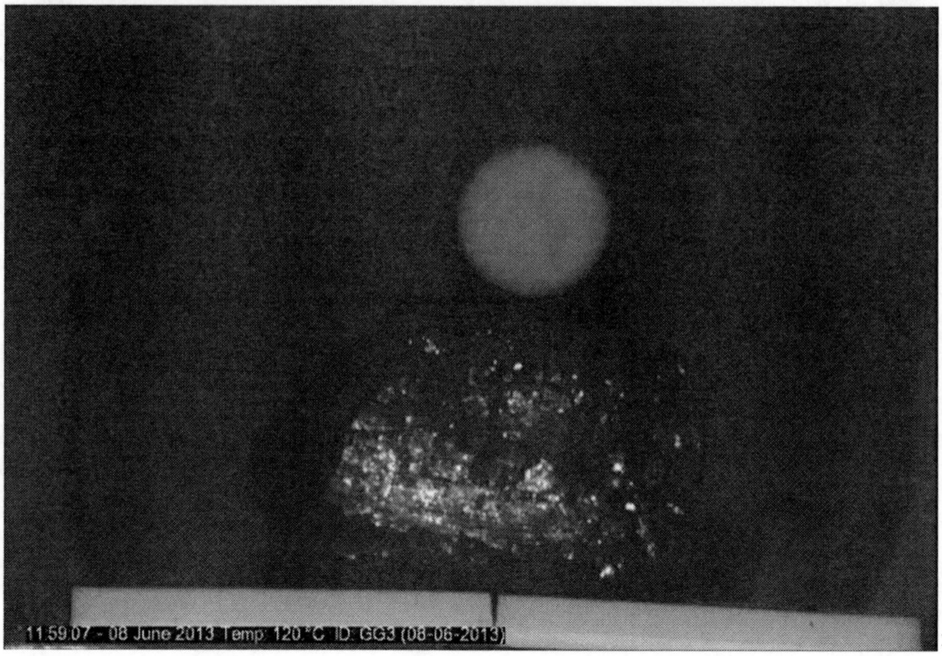


3

## Results

**Transient swelling behaviour (20 mm) – 1 °C/min**

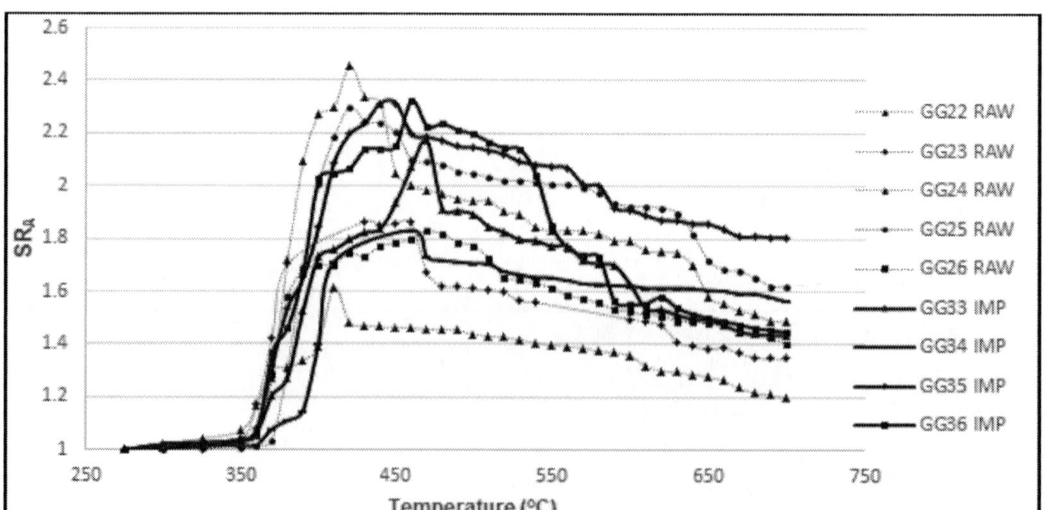


14

# Results

**Caking - Visual observations 10 mm (700 °C, 5.0 M $K_2CO_3$)**

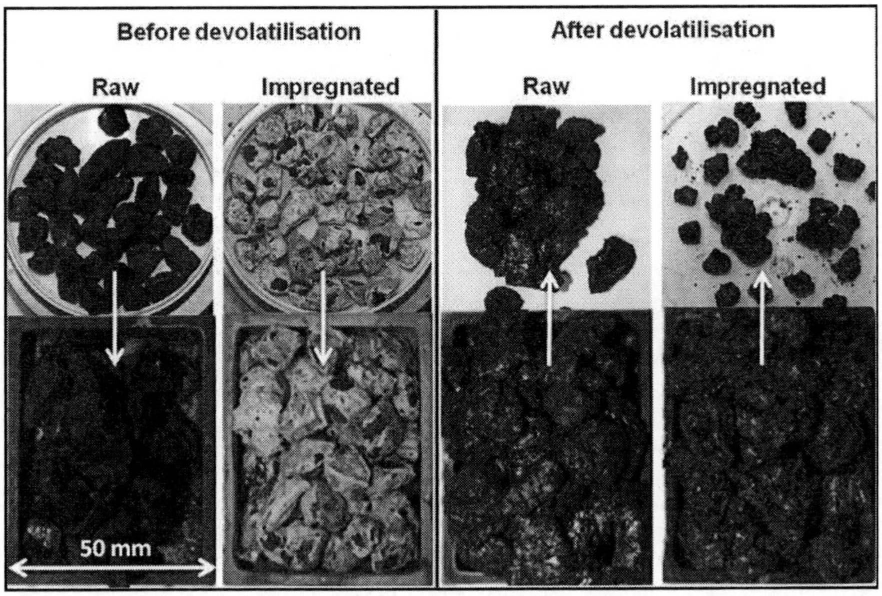


15

# Results

**Quantification of caking (10 mm) – Bridging neck sizes**

| [$K_2CO_3$] (M) (7 °C/min) | Avg. bridging neck size (mm) |
|---|---|
| Raw | 5.42 |
| 1.0 | 3.35 |
| 3.0 | 3.59 |
| 5.0 | 3.58 |
| 7.0 | 3.47 |
| **Heating rates (°C/min) (5.0 M $K_2CO_3$)** | |
| 1.0 | 1.39 |
| 3.0 | 2.16 |
| 5.0 | 2.78 |
| 7.0 | 3.58 |

16

## Results

### Quantification of caking (10 mm) – Bridging neck sizes

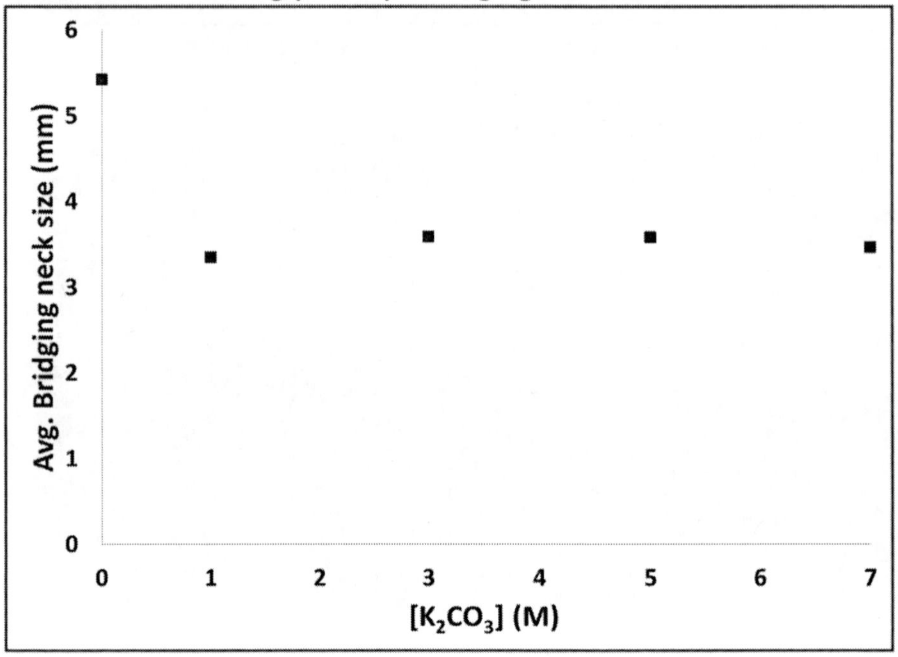


17

## Results

### Quantification of caking (10 mm) – Bridging neck sizes

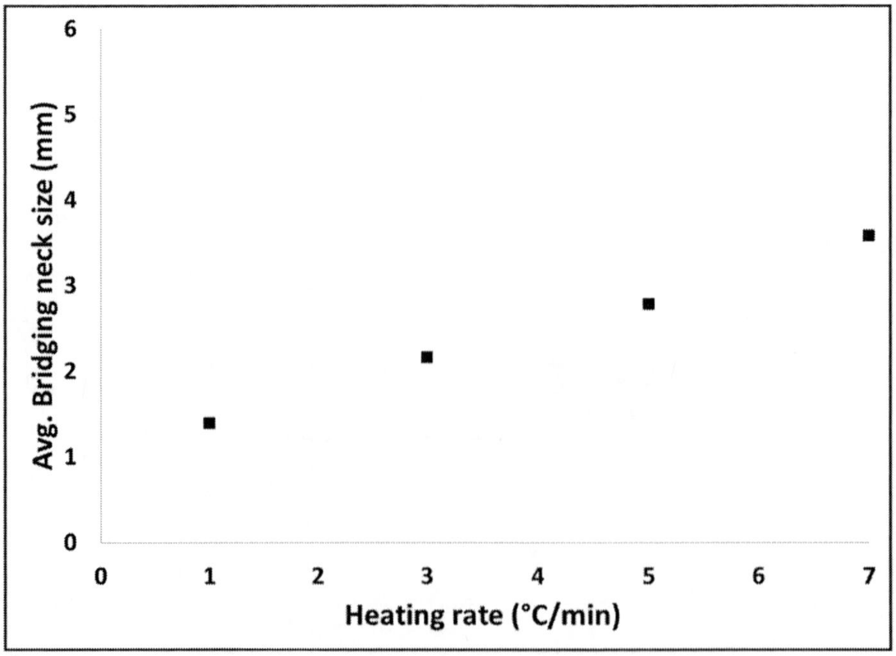


18

## Conclusions

- Mercury submersion and bridging neck size measurements can be used for the quantification of swelling and caking of large coal particles
- $K_2CO_3$-impregnation reduced the swelling of 20 mm GG particles by 20-30 % at higher heating rates (3, 5, 7 °C/min), while no difference was observed at 1 °C/min
- $SR_V$ increased as the devolatilisation heating rate increased
- $K_2CO_3$-impregnation reduced caking propensity of 10 mm GG by up to 40 %
- Degree of caking reduction was not influenced by $K_2CO_3$ concentration, and results showed a 1.0 M solution to be sufficient
- The caking propensity of impregnated (5.0 M $K_2CO_3$) samples was directly proportional to the devolatilisation heating rate
- For reduction of both the swelling and caking propensity of GG particles, it was found that the heating rate has a more significant influence compared to additive addition

19

## Acknowledgement


- Final year students for acquisition of results:
  - Joani van Rooyen
  - Jeanique van der Westhuizen
  - Janine Naudé
  - Xoliswa Ngonyama

NRF National Research Foundation | RISA Research and Innovation Support and Advancement

This work is based on the research supported by the South African Research Chairs Initiative of the Department of Science and Technology and National Research Foundation of South Africa.

Any opinion, finding or conclusion or recommendation expressed in this material is that of the author(s) and the NRF does not accept any liability in this regard.


20

**Thank you**

21

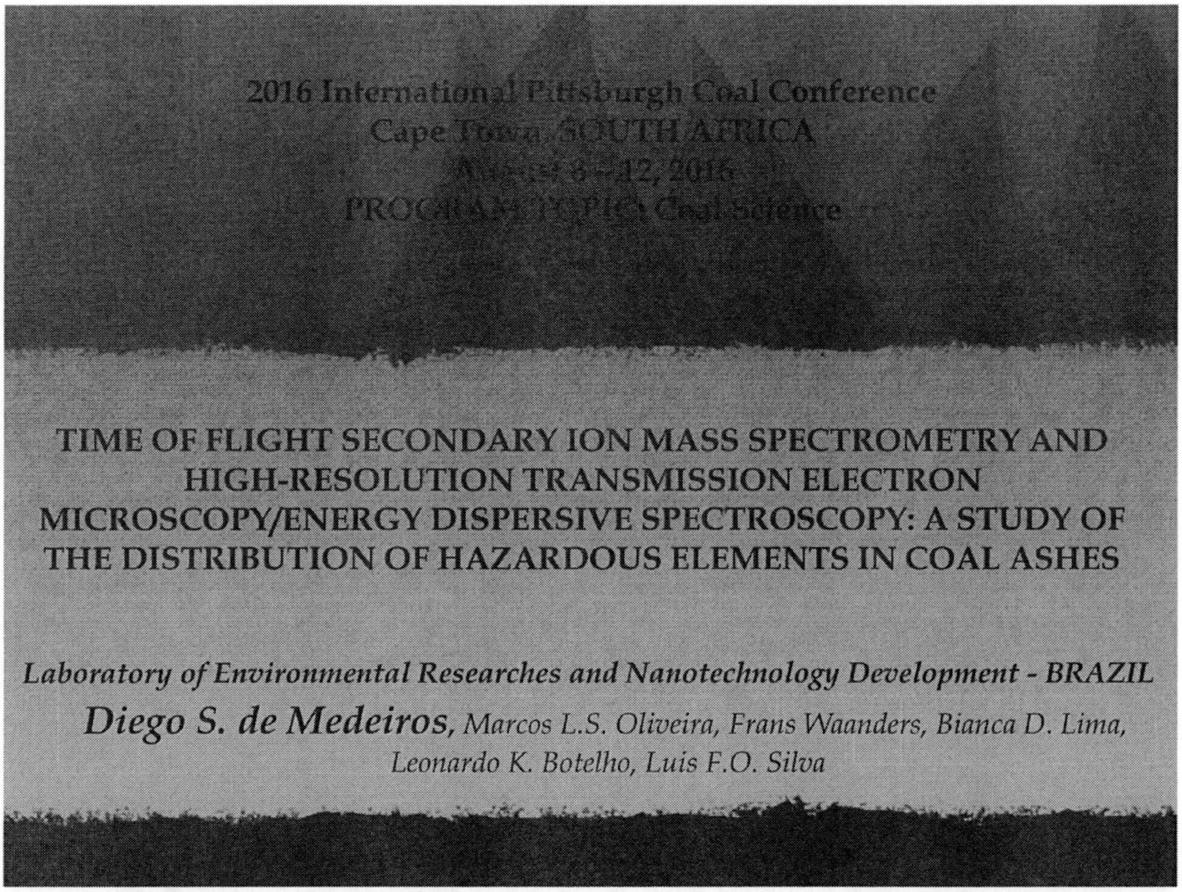
2016 International Pittsburgh Coal Conference
Cape Town, SOUTH AFRICA
August 8 – 12, 2016
PROGRAM TOPIC: Coal Science
TIME OF FLIGHT SECONDARY ION MASS SPECTROMETRY AND HIGH-RESOLUTION TRANSMISSION ELECTRON MICROSCOPY/ENERGY DISPERSIVE SPECTROSCOPY: A STUDY OF THE DISTRIBUTION OF HAZARDOUS ELEMENTS IN COAL ASHES
Laboratory of Environmental Researches and Nanotechnology Development - BRAZIL
Diego S. de Medeiros, Marcos L.S. Oliveira, Frans Waanders, Bianca D. Lima, Leonardo K. Botelho, Luis F.O. Silva

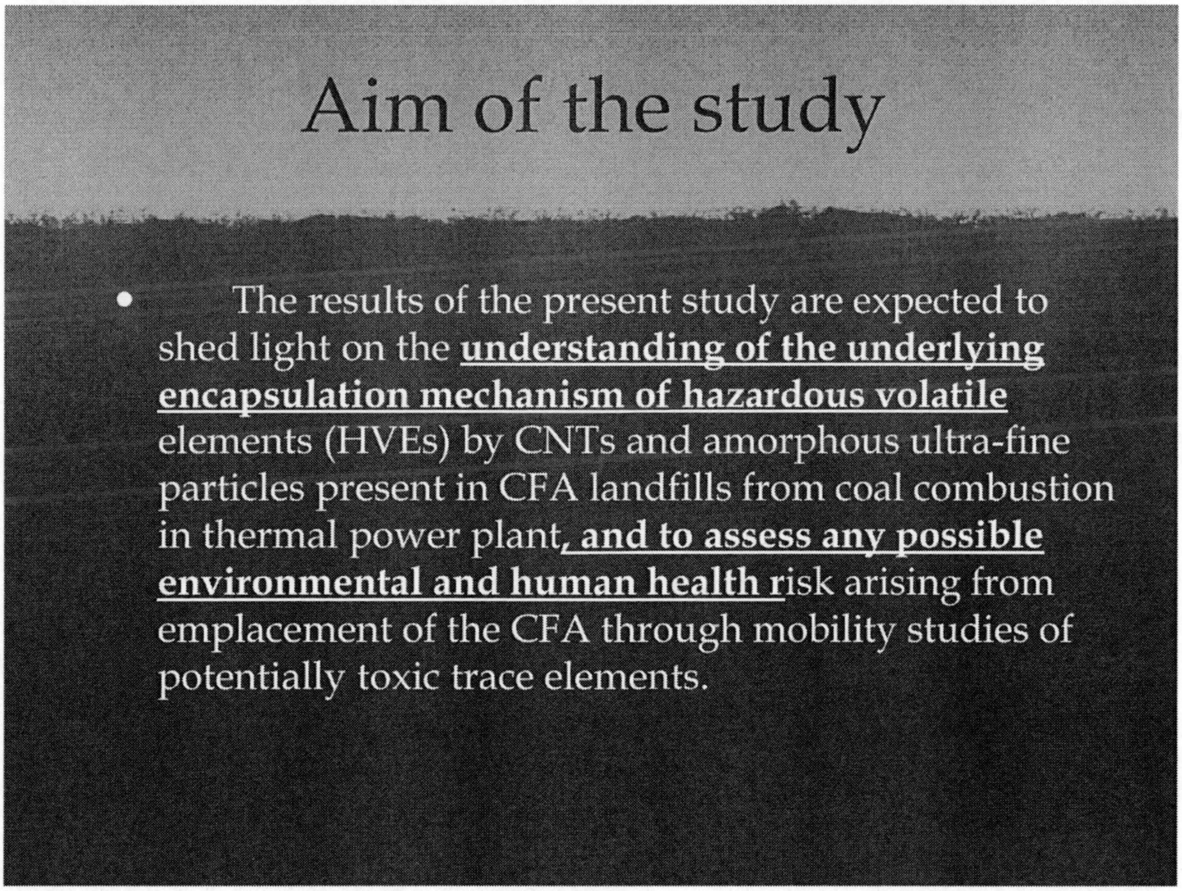
Aim of the study
• The results of the present study are expected to shed light on the understanding of the underlying encapsulation mechanism of hazardous volatile elements (HVEs) by CNTs and amorphous ultra-fine particles present in CFA landfills from coal combustion in thermal power plant, and to assess any possible environmental and human health risk arising from emplacement of the CFA through mobility studies of potentially toxic trace elements.

# Study Area

The Tapada do Outeiro thermal power plant operation started in 1959 and was completely deactivated in 2004. The anthracite exploited in the Douro Coalfield was the most important fuel supply of the thermal power plant and was co-fired with fuel oil (due to the hard ignition of this high-rank coal) at 1300-1400 °C, from the beginning of its activity until 1997. From 1997 until 2004, fuel oil was the only energy source used. The thermal power plant is located in Gondomar (Porto) close to the Douro Coalfield area. It was constructed in the watershed of the Crestuma-Lever dam of the Douro River, whose proximity facilitated the coal transportation from the mining centers .

Thermal power plant and fly ash landfills location.

# Sampling Method

- For this study a total of six CFAs were collected from two landfills (landfill 1 - FA 3, FA 5 and FA 6; landfill 2 - FA 13, FA 15 and FA 16).

# *Analytical procedures*

- The characterization of carbonaceous nanoparticles and ultra-fine particles included field-emission scanning electron microscopy (FE-SEM), high resolution transmission electron microscopy (HR-TEM) and time of flight secondary ion mass spectrometry (ToF-SIMS).
- The determination of some HVEs of CFA particles was achieved with EDS coupled to FE-SEM and HR-TEM.
- To support the data obtained by this advanced techniques the analysis of the CFA samples also include the determination of mercury (Hg), sulfur (S), carbon (C) and nitrogen (N) contents.
- The micron to nano-particles characterization includes the energy dispersive X-ray spectroscopy (EDS), coupled with FE-SEM and HR-TEM

# Results and discussion

According with further FE-SEM/EDS and HR-TEM/EDS analysis samples are, mostly, composed of: (i) larger size (≈100 µm) thick and thin walled cenospheres,

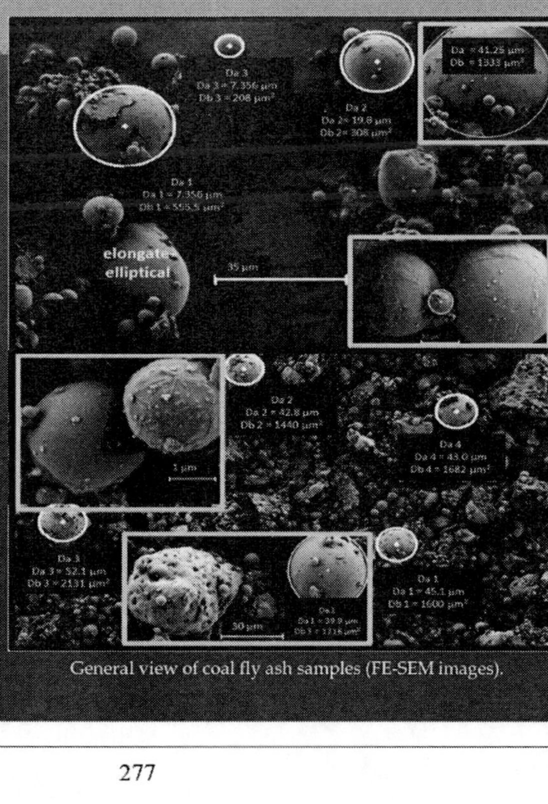


General view of coal fly ash samples (FE-SEM images).

# Results and discussion

The table presents the C, S, N and Hg contents in the bulk fly ash samples. The C content of the fly ash samples is relatively high, varying between 8.24 wt. % and 16.00 wt. %. The N varies between 0.01 wt. % and 0.06 wt. % while the S content ranges from 0.06 wt. % to 0.12 wt. %.

| Sample | HVEs % | Hg mg/kg | C % | N % | S % |
|---|---|---|---|---|---|
| FA 3 | 1.10 | 0.96 | 8.24 | 0.01 | 0.06 |
| FA 5 | 2.08 | 1.66 | 15.70 | 0.02 | 0.12 |
| FA 6 | 1.54 | 0.89 | 12.30 | 0.02 | 0.10 |
| FA 13 | 1.62 | 0.49 | 12.00 | 0.06 | 0.10 |
| FA 15 | 1.84 | 0.44 | 16.00 | 0.04 | 0.10 |
| FA 16 | 1.86 | 0.76 | 11.30 | 0.01 | 0.10 |

Hazardous volatile elements (determined by EDS coupled with FE-SEM and HR-TEM for 106 nanoparticles between 0.1-100nm), Hg, C, N and S contents in fly ash samples

# Results and discussion

Hower et al. (2010) demonstrated that the Hg concentrations in coal ash are largely a function of the (i) amount of carbon, (ii) the flue-gas temperature at the point of collection, (iii) the composition of the flue gas, and (iv) the type of CFA carbon, including variations dependent on the rank of the feed coal.

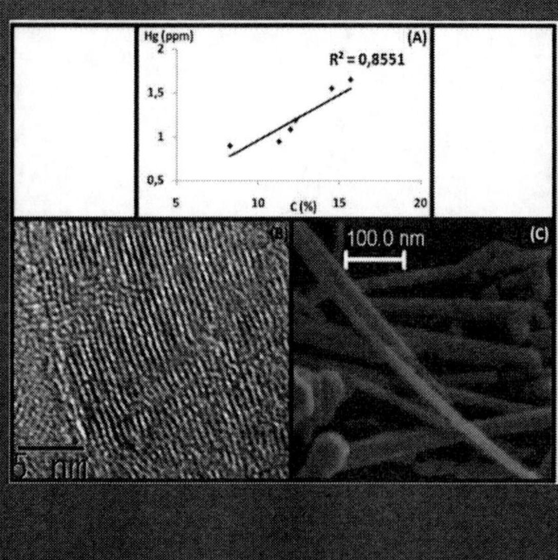


(A)Linear correlation between Hg and C contents in fly ash samples;

(B)TEM image of fragmented MWCNTs magnification containing Hg nanoparticles;

(C)TEM image showing interaction between MWCNTs and amorphous nanoparticles containing Hg and Se.

# Results and discussion

The CFAs samples present of a wide range of organic and inorganic compounds which were determined by chemical maps obtained in ToF-SIMS analysis. Unlike fullerene-rich CFAs, the fly ash samples analyzed in this work do not reveal the presence of fullerenes, that illustrates that the detected ions do not correspond to fullerene mass. In this study the ToF-SIMS results demonstrate the presence of Pb in the CFAs samples in elemental form,

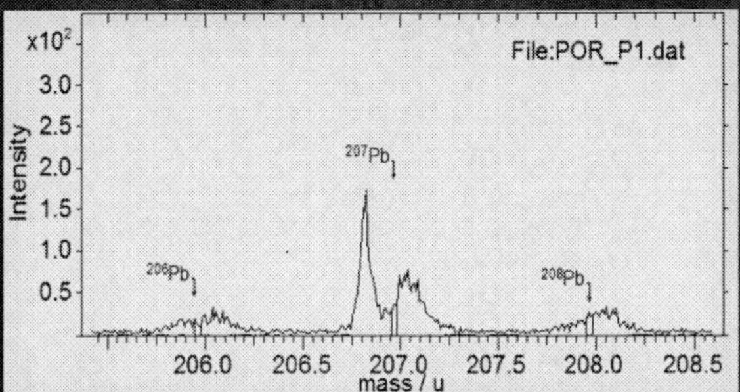


In this work the presence of abundant MWCNTs and amorphous carbon particles, both containing HVEs, was evidenced by the FE-SEM/EDS and HR-TEM/EDS analysis

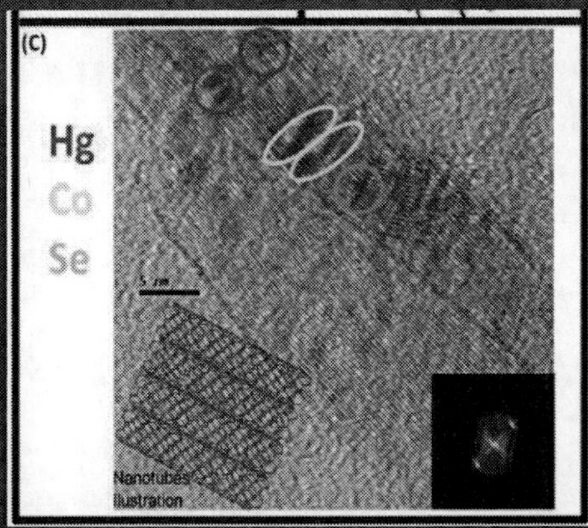


(C) Mixed SWCNTs and MWCNTs encapsulating hazardous elements detected by EDS analysis (HR-TEM image and FFT).

# Conclusion

- The FE-SEM/EDSand HR-TEM/EDS analysis reveals that, generally, the anthracite CFA samples include carbonaceous, glassy and metallic solid spheres with some containing mixed amorphous/crystalline phases.
- The studied samples comprise: (i) larger size (≈100 µm) thick and thin walled cenospheres, and dense spheroidal particles often containing embedded quartz; (ii) irregular shaped agglomerates of small size glassy spheres, quartz and iron-rich phases; (iii) ultrafine size (≈ 100nm) glassy spheres dense or having thick walls; (iv) carbonaceous nanospheres (< 50nm) contain several hazardous volatile elements such as As, Hg, Pb, Se and nanoparticles with multiple nanominerals assemblages.
- The EDS analysis coupled with the FE-SEM and HR-TEM observations of the CFAs particles with 100 to 0.1 nm demonstrates that these materials contain a small but significant proportion of encapsulated HVEs.
- The correlation between Hg and C contents in the fly ash samples indicates that Hg capture is function of the C content.
- The presence of abundant MWCNTs and amorphous carbon particles, both containing HVEs, was also evidenced by the FE-SEM/EDS and HR-TEM/EDS analysis.
- Fly ash samples with wide range of organic and inorganic compounds were determined by chemical maps obtained in ToF-SIMS analysis.

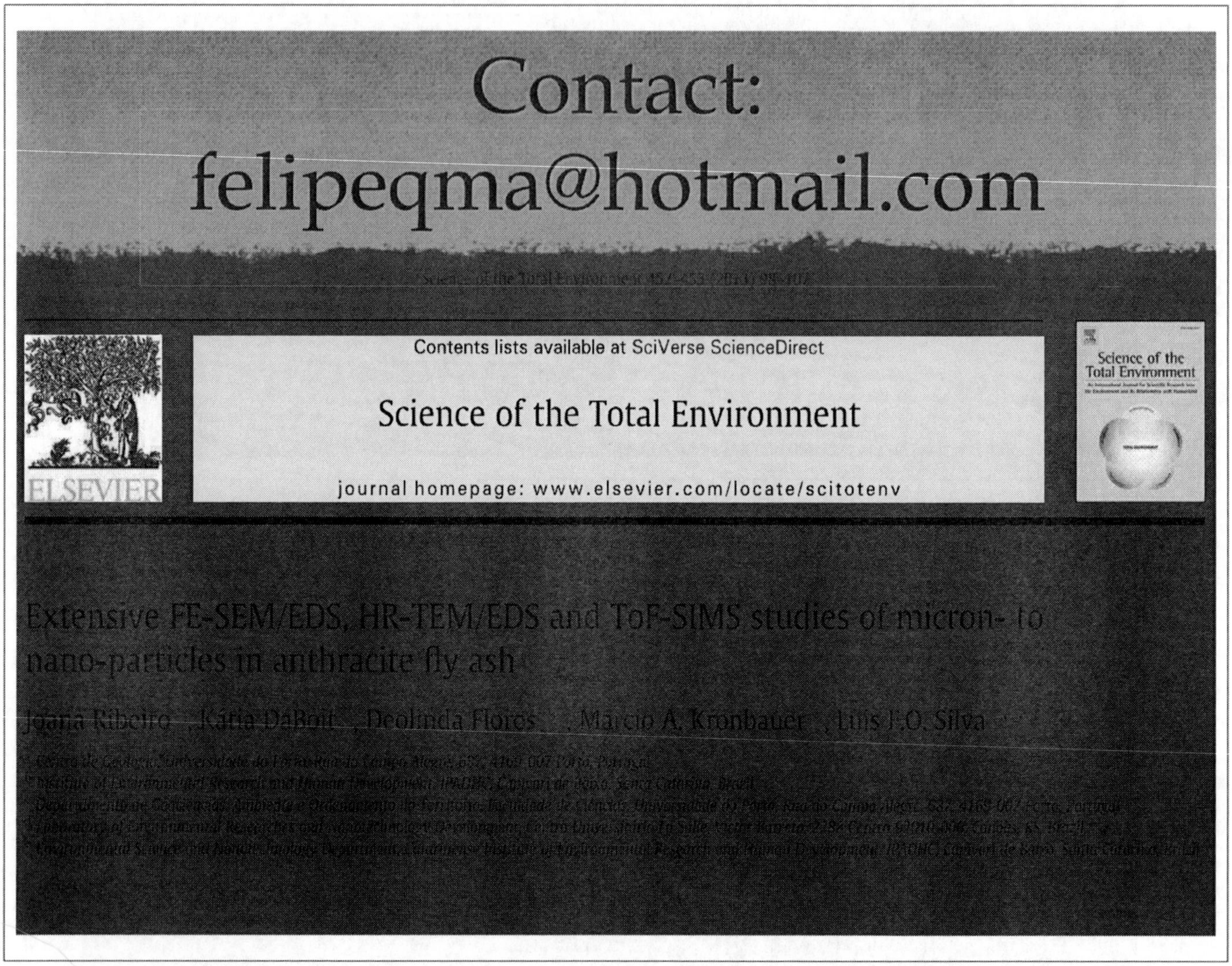
Contact:
felipeqma@hotmail.com
Science of the Total Environment 452–453 (2013) 98–107
Contents lists available at SciVerse ScienceDirect
Science of the Total Environment
journal homepage: www.elsevier.com/locate/scitotenv
ELSEVIER
Science of the Total Environment
Extensive FE-SEM/EDS, HR-TEM/EDS and ToF-SIMS studies of micron- to nano-particles in anthracite fly ash
Joana Ribeiro, Kátia DaBoit, Deolinda Flores, Marcio A. Kronbauer, Luis F.O. Silva

# Investigation of Solvent Crystallization Process of Anthracene

*Xiaoxiao Ding[1,2], Cuiping Ye[1,2,3], Wenying Li[1,2*], Mingming Fan[1,2], Hai Mu[1,2], Jie Feng[1,2]*

*1.Key Laboratory of Coal Science and Technology (Taiyuan University of Technology), Ministry of Education and Shanxi province, Taiyuan 030024, PR China*

*2.Training Base of State Key Laboratory of Coal Science and Technology Jointly Constructed by Shanxi Province and Ministry of Science and Technology, Taiyuan 030024, PR China*

*3.College of Environmental Science and Technology;*

* *Corresponding author. E-mail: ying@tyut.edu.cn*

**Abstract**: Anthracene is one of the most value-added components of coal tar, which is an irreplaceable chemical raw material in dye, pharmaceutical, and photoelectric industries. Because of its low cost and energy consumption, solvent crystallization is the most widely used in large scale to refine anthracene from crude anthracene. To obtain perfect anthracene crystal with higher purity, the effects of cooling mode, stirring rate, and solvent on the crystal size, crystal morphology, and purity were investigated in the solvent crystallization process in the point of crystal nucleation and solvent-solute intermolecular interaction. Powder X-ray diffraction (XRD) and scanning electron microscope (SEM) were used to explore the morphology and crystal size of anthracene. The purity was measured by gas chromatography (GC). The perfect crystal was obtained by using N,N-dimethylformamide (DMF) with the initial temperature of 353.15 K at the stirring rate of 450 r/min, when the cooling curve was controlled by a forced circulating system.

***Keywords:*** solvent crystallization; anthracene; solvent; cooling mode

# 1.Introduction

Anthracene is one of the most valued by-products from crude anthracene, which is a distillate primarily containing heavy components in coal tar. The content of anthracene in coal tar and crude anthracene are up to 1.8 wt% and 36 wt%[1], respectively. Because of its special polycyclic structure, antharcene and its derivatives show lots of unique photoelectric properties and biological activities, which are widely used in photoelectric material[2], pharmaceuticals[3], pesticides[4], dyes[5], fuels, and many other fields. The separation and refining of anthracene from coal tar not only makes better use of coal tar, but also reduces the pollutant emission. In recent years, the demand for anthracene exceeds the supply, especially in OLED industry nbecause of the form of high-quality single crystals and thin films they perform[6, 7]. To meet the demand of market, many refining methods have been applied, the most common methods are solvent crystallization, distillation, and chemical method. Many new method including emulsion membrane[8], supercritical fluid extraction (SFE)[9] and other processes are also developed. Amidst these methods, solvent crystallization is most widely used in a large scale in China due to its low cost and low energy consumption with a simple operation process. The key of the solvent crystallization process is the highly efficient solvent selection. Based on the common principle "dissolve like dissolve", progresses were made, however, the fact that when the product purity was up to 98.93 wt%[10], it was too expensive to improve the purity further at the expense of plenty of solvent and multi-stage operation was still unsatisfactory in market.

Solvent crystallization is a common method to separate and purify solid products in industry, which not only involves heat and mass transfer, but also surface reaction. In the solvent crystallization process, solvent properties and trace impurity are factors of greater significance determining the final product quality, except for the temperature for crystallization and filtration, liquid-solid ratio, and solvent dosage. Besides the purity and yield, the crystallization process also need to control the particle size, size distribution, morphology and other physical properties of crystal in order to acquire desired crystal products. Furthermore, there are close relationship between purity of final product and crystal habits[11, 12]. The crystal habits are not only determined by the internal structure, but largely influenced by external factors such as temperature, solvent, stirring, and impurity which could exert dramatic effects on crystal nucleation and crystal growth, and hence the crystal morphology[13]. To control crystal habits presents a major challenge since they considerably influence the downstream processing factors such as filtration, drying, transportation, storage and so forth. In the research of separating anthracene from crude anthracene, much effort has been directed successfully to improve the purity and yield of the product. Many researchers have conducted massive experiments in terms of intermolecular interaction and predicted the solvent by many models in order to screen the best solvent[10, 14-16]. However, the fact that the similarity of anthracene and other components (mainly carbazole) in chemical structure in crude anthracene have slight difference in physico-chemical properties, and they are inclined to form solid solution[17] in the separating process, makes it a rather tough task to get high purity anthracene from crude anthracene. Present researches have screened solvents with high selectivity on the basis of crystallization thermodynamic, however, the theoretical foundation is weak and unclear. There are little researches which explore the effects of solvent on the crystallization nucleation and growth, hence influence the purity, morphology, size, and other properties of final crystal products. However, a good understanding of the interaction of solvent and solute, and the effect of solvents have on the crystallization and final products will to a large extend make a benefit of selecting proper solvents.

In this paper, the proper crystallization conditions mainly containing cooling mode and stirring in DMF were determined. Afterwards, we measured the solubility of anthracene in four organic solvents and performed crystallization experiments in the above solvents in the proper cooling mode and stirring speed to screen the optimal solvent for anthracene crystallization and explore the effects of solvent on the crystal habits and purity of anthracene in the aspect of crystal nucleation and solute-solvent intermolecular interaction. The anthracene crystals with desired crystal habit and high purity were obtained.

## 2. Experimental Section

### 2.1. Materials

Anthracene (analytical reagent, Aladdin) were further purified recrystallization with DMF, and the purity was above 99 wt%. The standards used in the experiment were anthracene (99.4 wt%) and 9-fluorenone (98 wt%, internal standard substance) purchased from West Chester, PA. The analytical reagent grade N,N-dimethylformamide (DMF), xylene (DMB), tetrachloroethylene (TCE), and diethylene glycol dimethyl ether (DGDE) were used without further purification. Acetonitrile (HPLC, Dikama Technology) was used to dissolve the samples for gas chromatography analysis.

### 2.2. Analysis Apparatus and Conditions

**GC**. The weight composition of various samples were determined by Shimadzu GC-2014 (Japan), equipped with an Rtx®-5 (USA) capillary column (0.32 mm×30 m×0.25 μL), wall-coated with 5 % diphenyl / 95 % dimethyl polysiloxane. A flame ionization detector was used as the detector. The temperature program was as follows: 443.15 K for 1 min, then increased by 3.5 K/min to 473.15 K, held for 1 min, then increased by 10 K/min to 523.15 K. The injection volume was 0.2 μL. Nitrogen was used as the carrier gas with column flow 1 mL/min. Split ratio was 30:1. The internal standard curves method with 9-fluorenone as the internal standard was used for quantitative analysis. The calibration curve is shown in Figure 1.

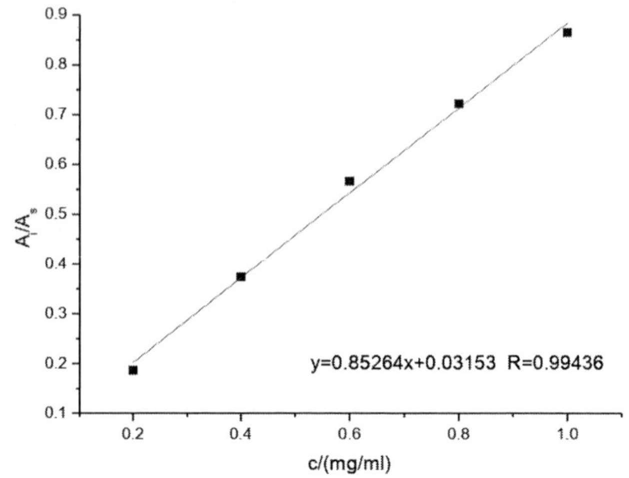


Figure 1. Calibration curve of anthracene using 9-fluorenone as the internal standard substance

**XRD**. The diffraction patterns were obtained by using Cu Kα radiation on a Rigaku Ultima IV X-ray diffractometer (Japan). Crystal samples were analyzed over a diffraction-angle (2θ) range of 5° to 60° with a scanning step of 4°/min, at a step size of 0.02°, a voltage of 40 kV, and a current of 40 mA. 50 mg sample weighed by a analytical balance (0.1 mg, Mettler) was directly tabletted on a glass holder because the samples were all in powder form.

**SEM**. The images of crystal samples were obtained by Hitachi TM 3000 desktop scanning electron microscopy when the accelerating voltage was 5 kV. The samples were mounted on a brass specimen club with double-sides adhesive tape and made electrically conductive by coating with gold under vacuum in an argon atmosphere before observation.

### 2.3. Crystallization Procedures

**Solubility Determination.** The solubility of anthracene was measured by a self-made equilibrium equipment based on the solid-liquid equilibrium theory. Solvent and excess solute was placed in the apparatus mentioned above and allowed to equilibrate in a constant temperature which was controlled by a water thermostat bath with agitation for three days. The solutions were equilibrated at a higher temperature and measured repetitively after a minimum of one additional day. Then the upper portion in different solvents was transferred by a pipettor, diluted with acetonitrile. Subsequently, temperature was lowered at a interval of 5 K. The samples were mixed with a internal standard of 9-fluorenone in a certain ratio before analyzed by gas chromatography. The measurements are repeated three times, and the average is considered to be the solubility.

**Solvent Crystallization.** Experiments were conducted in a 300 mL jacketed glass crystallizer. Saturated solutions were prepared in a beaker by dissolving an appropriate amount of anthracene in DMF, xylene (DMB), tetrachloroethylene (TCE), and diethylene glycol dimethyl ether (DGDE), respectively at certain temperature with agitation rate of 300 r/min, 450 r/min, and 600 r/min by a magnetic stirrer in a thermostatic water bath. After anthracene crystals disappeared, additional 30 min was needed to make sure the complete dissolution. Then, the solutions were filtered at the high temperature and transferred immediately to a jacketed glass vessel, and cooled to 298.15 K in room temperature by a cooling curve controlled by thermostatic water bath and a temperature-controlled water bath with a forced circulating system (0.1 K). Finally the crystals solid was isolated from the suspensions quickly, afterwards the crystals were collected and dried at room temperature in air. Then the composition of crystals was determined by GC, and the crystal structure and crystal size and morphology were characterized by XRD and SEM, respectively. To collect the crystal nucleation data indirectly, the induction time which included a period between the on set of opacity and the final cooling temperature was also measured in the solvent crystallization process.

## 3. Results and Discussion

### 3.1. Effect of Cooling Mode on Crystallization

Different cooling conditions could exert a remarkable influence on the quality of crystal product. In crystallization industry, a controlled cooling mode is preferred, because it can control the supersaturation within metastable zone, in which the crystal growth rate predominates the secondary nucleation rate to get narrow CSD with large mean size of final crystals, which obviously facilitates the downstream process[18-20]. The cooling curves obtained in different conditions were presented in Figure 2. The cooling rate of both curves gradually decreased, especially when the temperature reached 320.15 K. Moreover, the cooling rate in a forced circulating system was obviously larger than thermostatic bath.

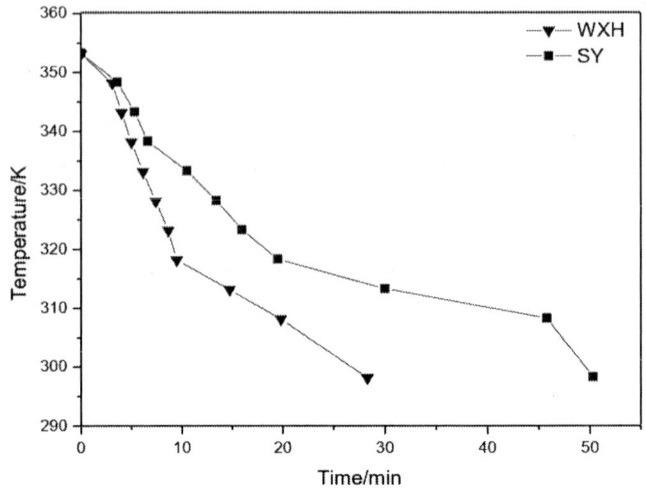


Figure 2.Cooling curves controlled by a force circulating system (WXH) and by a thermostat bath system (SY)

The XRD results and SEM images of crystals obtained in different cooling modes were shown in Figure 3 and Figure 4. There were little changes of the main peak positions and intensities, implying that the crystals obtained in different mode were in similar crystal structure, all of the crystal were in plate-like hexagon and belonged to monoclinic system. The results obtained from the SEM images indicated that anthracene crystal grown under the cooling curve controlled by a forced circulating system exhibited mean size and intact morphology.

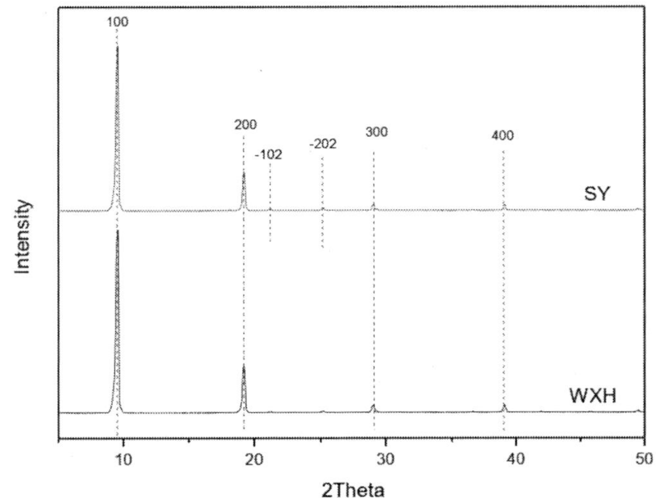


Figure 3. XRD patterns of anthracene crystals obtained in different cooling modes

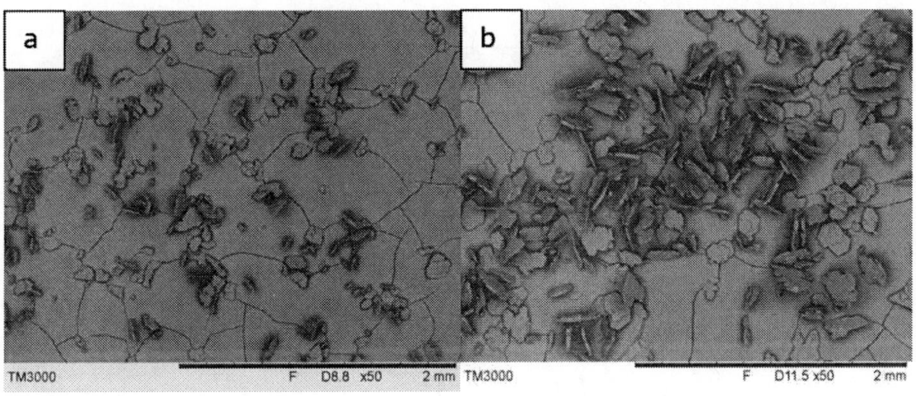


Figure 4. SEM images (×50) of anthracene crystals obtained in different cooling modes. a) cooling in a thermostatic water bath (SY), b) cooling controlled by a forced circulating system (WXH).

Crystal products in desired quality are usually produced in the metastable zone which is a virtual zone between solubility curve and supersaturation curve. The metastable zone width is proportional to cooling rate[18]. The favorable crystals obtained in the forced circulating system was attributed to the bigger cooling rate, which widened the metastable zone for the growth of anthracene crystals. The crystals grown in thermostatic water bath was distinctly smaller in size than that in forced circulating system due to a relatively small metastable zone with the lower cooling rate. The crystals grown in the controlled curves were uniform in size, despite of the different size. The anthracene crystals was distributed in a disordered form in size and agglomerated seriously in room temperature, which was mainly due to different local supersaturation caused by the uneven heating. In high supersaturation level, the crystals readily grew in a larger size, while in a lower level, the crystals grew slower or did not grow at all. The nonuniform and small crystals crystallized from solution are detrimental to the downstream process, such as filtration, drying, storage and etc. In consequence, the following experiments were all operated in the forced circulating system.

### 3.2. Effect of Stirring on Crystallization

Vigorous, efficient stirring is indispensable to optimize the contact between the phases in order to minimize diffusion resistances in solution crystallization. Crystals always heavily agglomerate in a crystallization system without agitation. Agglomeration usually occurs in small particles and a nonuniform local supersaturation[13].

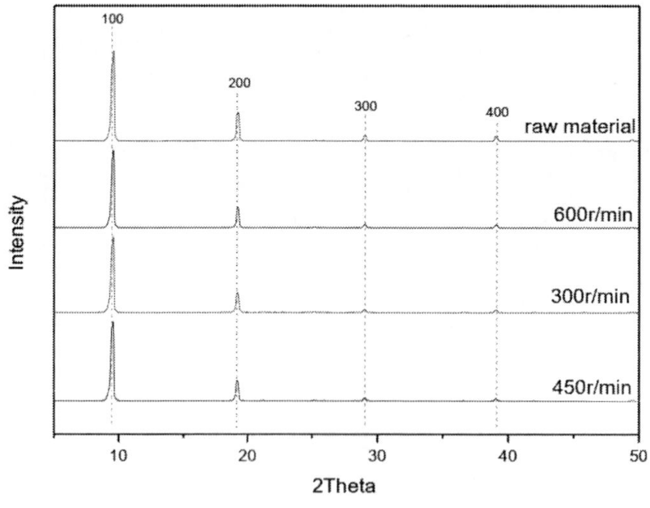


Figure 5. XRD patterns of anthracene crystals obtained in different stirring speeds

The results of SEM and XRD of anthracene crystals obtained in different stirring rates were shown in Figure 5 and Figure 6. The sharp peaks were given in the diffratograms of all the samples in different stirring rates, indicating no polymorphic modifications. The crystal images obtained from the scanning electron microscope (SEM) showed considerably different crystal habits. Crystals grown at an agitation of 300 r/min were in small and uneven size. However, when the agitation speed was up to 450 r/min and 600 r/min, the crystal size distribution was obviously improved compared with 300 r/min. at the same time, the particles tended to be smaller and the uniformity of sizes was worse with the increasing of agitation speed.

When it comes to agitation, we should mention the fluid force field which is closely related to stirring in the crystallizer, and could have a notable effect on the crystal nucleation and crystal growth[13, 21]. As shown in Figure 6, the increasing speeds

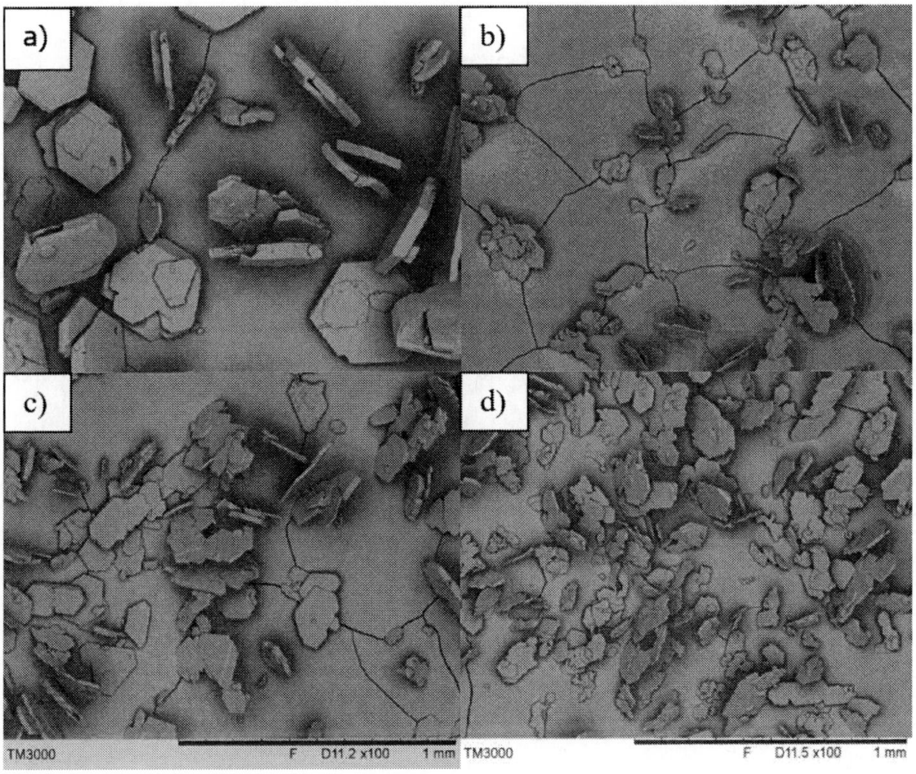


Figure 6. SEM images of anthracene crystals obtained in different stirring speeds (×100). a) raw material, b) 300r/min, c) 450r/min, d) 600r/min

made a full mixture in the system, hence reduced the uneven level local supersaturation and crystal nucleation, the crystals tended to grow in larger size. Especially when the stirring speed was of 450 r/min, the crystals in uniform size and intact shape were obtained. Taking the product uniformity and size distribution into consideration, the other experiments were performed at the agitation speed of 450 r/min.

### 3.3. Solvent Selection

Solvent is a crucial issue in solvent crystallization which has the important effect on the solubility[22, 23], crystallization kinetic[24], and the crystal habit[25]. The crystal morphology varied with different solvents could be ascribed to the effects of solvents on the relative growth rates of the individual crystal surfaces[26-29].

The temperature dependence of solubility is shown in Table 1. the solubility increased significantly with increasing temperature. When purifying the targeted crystals, the relative solubility of targeted substance over impurity is significant. The relative solubility is bigger, the easier the operation[22]. The solubility in four organic solvents is in the decreasing order of DMF,DMB,DGDE,TCE, indicating that there was a higher supersaturation which was the driving force for crystal nucleation and growth.

Crystallization induction time is inversely proportional to crystal nucleation rate, therefore, researchers usually measure the induction time to estimate the nucleation rate qualitatively[24, 28]. The crystallization induction time in different solvents were presented in Table 2. Induction time was in the increasing sequence: DGDE, DMB, DMF, TCE. Therefore, the biggest nucleation rate was obtained in DGDE. When the DGDE was taken as solvent, there was burst nucleation in which the majority of anthracene crystals had a tendency to nucleate not to grow, hence the crystals size was smaller but uniform.

Table 1. Solubility of anthracene in four organic solvents in mole fraction ($\times 10^3$)

| Temperature(K) | TCE | DMB | DMF | DGDE |
|---|---|---|---|---|
| 303.15 | 8.05 | 9.69 | 11.32 | 8.71 |
| 313.15 | 11.27 | 15.56 | 15.66 | 12.06 |
| 323.15 | 15.99 | 20.72 | 20.88 | 16.72 |
| 333.15 | 21.64 | 31.24 | 28.42 | 20.84 |
| 343.15 | 30.54 | 40.1 | 40.72 | 38.36 |

Table 2. Crystallization induction time in four organic solvents

| Solvent Series | Induction Time(min) |
|---|---|
| DMB | 11 |
| TCE | 24 |
| DGDE | 10 |
| DMF | 13 |

The effect of solvents on crystal habits of anthracene was shown in Figure 7 and Figure 8. The crystal X-diffraction patterns from DMF, DMB, and TCE showed no change in peaks position but some change in peak intensities, which indicated that the crystals grown from different solvents were identical in crystal structure, and the difference between the peak intensities could be attributed to different morphology and size of crystals prepared in different experiments. The main appearance faces of anthracene crystals grown from DMF, DMB, and TCE are (100), (200), (300), and (400). However, more main crystal faces were observed for crystal from DGDE, i.e. (100), (200), (300), (400), and (-202), which implied that crystal structure changed when crystallization was performed in DGDE solution. The SEM images of crystals produced from different solvents showed in Figure 8 presented different morphologies and sizes, which corresponded to the results revealed in XRD. It could be seen that the crystal habit of anthracene from DMF and TCE was plate-like hexagon, cracked anomalous plate in DMB, while the crystals performed an obviously different morphology from other solvents in DGDE, it is in scattered prismatic shape. The crystal size derived from different solvents decreased in the following sequence: TCE>DMF>DMB>DGDE. The lowest solubility of anthracene in TCE of the four solvents usually generates a lower supersaturation. Thus, the crystal nucleation may not happen easily, the small amount of nucleated crystals took a tendency to grow, crystals in this condition were usually thicker plate-like hexagon than that in DMF, which corresponded to the results in Table 2 and Figure 8.

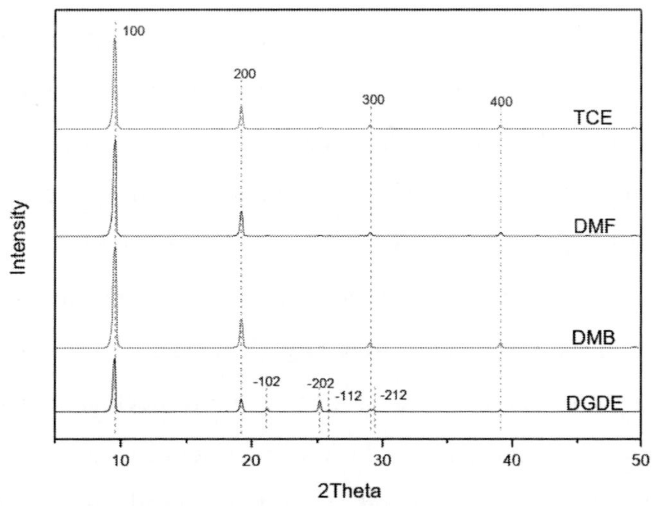


Figure 7. XRD patterns of anthracene crystals obtained in four organic solvents

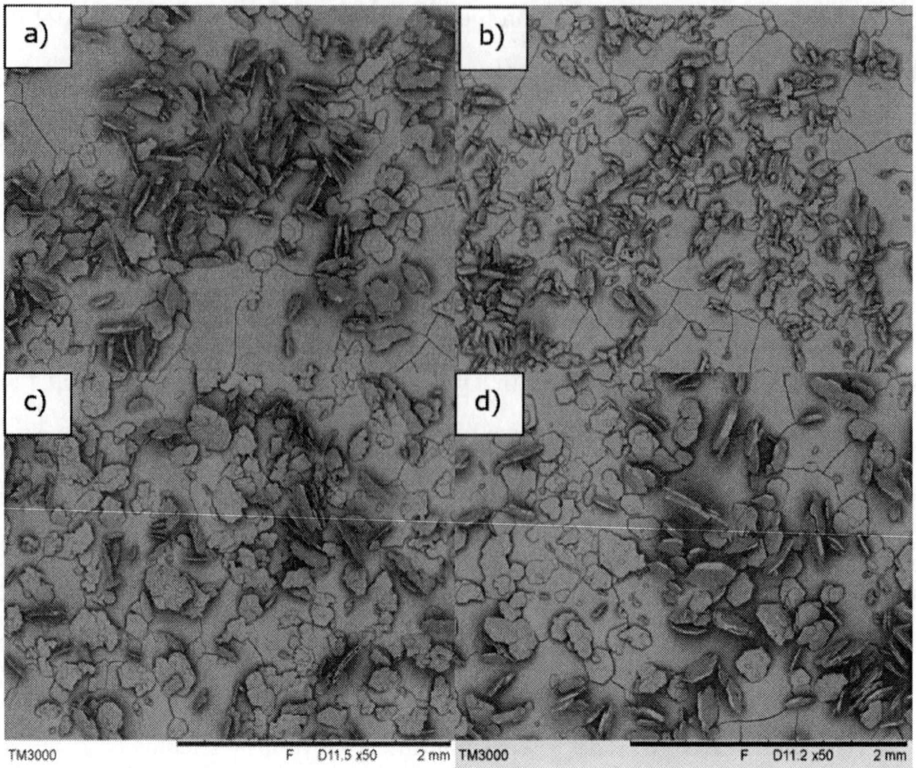


Figure 8. SEM images (×50) of anthracene crystals obtained in four organic solvents. a) DMF, b) DGDE, c)DMB, d)TCE.

The influences of solvent on the crystal habits and the solubility of solute in different solvents can be explained by the molecular structure of solvent and anthracene as well as the interaction between solvent and anthracene molecules at given crystal face. Because of the symmetric three ring condensed aromatic structure with greater intermolecular forces, the interaction between anthracene and solvent molecules is mainly due to intermolecular van der Waals forces. Li[30] et al have proved that there were no hydrogen bond formed between DMF and anthracene. The four solvents are all aprotic solvents, but the relative polarity of them are different, the polarity decreases in the order: DMF>DGDE>DMB>TCE. The interaction between solvent and solute is usually assumed as incorporation mechanism usually caused by intermolecular force which often blocks the incorporation of crystal molecules to some degree and hence slow the growth of certain crystal faces down. The crystals finally performed different morphologies. When the DGDE was used as solvents, the growth of different crystal faces were slowed down, especially the surface (-202), thus presented a different crystal habit from other solvents. However, the crystals obtained from DGDE was in smaller size and scattered prismatic shape, which was detrimental in the filtration and storage. Meanwhile, when DMF was used as solvent, the intact crystal was gained.

## 4. Conclusion

Cooling mode, stirring rate, and solvent had little effects on the crystal structure of anthracene, but influenced the anthracene crystal habits significantly, when DMF, DMB, TCE were taken as solvent, the crystals all in plate-like hexagon, only differing in crystal size and integrities, in the same time, a different crystals in small and prismatic shape were obtained from DGDE. In addition, the perfect crystals in good shape were obtained by in DMF with the initial temperature of 353.15 K at an agitation rate of 450 r/min, when the crystallization temperature was controlled by a forced circulating system.


## ACKNOWLEDGEMENTS

This work is financially supported by the National Nature Science Foundation of China (U1361202 and 51276120).


## References


1.Xiao R. Coal Tar Chemical Engineering [M]. Metallurgical Industry Press. 2002.

2.Fukagawa H, Shimizu T, Ohbe N, Tokito S, Tokumaru K, Fujikake H. Anthracene derivatives as efficient emitting hosts for blue organic light-emitting diodes utilizing triplet–triplet annihilation[J]. Organic Electronics. 2012;13(7):1197-203.

3.Kraicheva I, Vodenicharova E, Shenkov S, Tashev E, Tosheva T, Tsacheva I, et al. Synthesis, characterization, antitumor activity and safety testing of novel polyphosphoesters bearing anthracene-derived aminophosphonate units[J]. Bioorg Med Chem. 2014;22(2):874-82.

4.Lin H, Zhang Y-W, Zheng L-H, Meng X-Y, Bao Y-L, Wu Y, et al. Anthracene and Anthraquinone Derivatives from the Stem Bark of Juglans mandshuricaMaxim.[J]. HelveticaChimica Acta. 2011;94.

5.Ooyama Y, Aoyama S, Furue K, Uenaka K, Ohshita J. Fluorescence sensor for water based on PET (photo-induced electron transfer): Anthracene-bis(aminomethyl)phenylboronic acid ester[J]. Dyes and Pigments. 2015;123:248-53.

6.Jang YJ, Lim BT, Yoon SB, Choi HJ, Ha JU, Chung DS, et al. A small molecule composed of anthracene and thienothiophene devised for high-performance optoelectronic applications[J]. Dyes and Pigments. 2015;120:30-6.

7.Seo J-A, Lee C-W, Gong M-S. Spirobenzofluorene linked anthracene derivatives: Synthesis and application in blue fluorescent host materials[J]. Dyes and Pigments. 2013;96(1):211-9.

8.Xiaopeng Z, Xiaoyue L, Ruili G. Study on Separation of the Mixture of Anthracene，Phenanthrene and Carbazole by Emulsion Liquid Membrane[J]. Journal of Shihezi University(Natural Science). 2012;30(2).

9.Liou Y, Chang CJ. Separation of Anthracene from Crude Anthracene Using Gas Antisolvent Recrystallization[J]. Separation Science and Technology. 1992;27(10):1277-89.

10.Wu T. PROCESS IMPROVEMENT OF REFINING ANTHRACENE AND CARBAZOLE FROM CRUDE ANTHRACENE BY SOLVENT CRYSTALLIZATION METHOD[D]. Taiyuan University of Technology. Taiyuan. 2013.
11.Koay GFL, Chuah T-G, Zainal-Abidin S, Ahmad S, Choong TSY. Solvent crystallization of palm based dihydroxystearic acid with isopropyl alcohol: Effects of solvent quantity and concentration on particle size distribution, crystal habit and morphology, and resultant crystal purity[J]. Industrial Crops and Products. 2011;34(1):1135-40.
12.Raditladi ZS, Ngila JC, Kgarebe BV. Effect of habit modifiers on the morphology and purity of soda ash[J]. Journal of Crystal Growth. 2003;257(3–4):344-9.
13.Beckmann W. Crystallization-Basic Concepts and_Industrial Applications[M]. 2013.
14.Ye C, Zheng H, Wu T, Fan M, Feng J, Li W. Optimization of solvent crystallization process in obtaining high purity anthracene and carbazole from crude anthracene[J]. AIChE Journal. 2014;60(1):275-81.
15.Zheng H. Investigation on Purification of High Purity Carbazole from Anthracene Reside by Solvent Crystallization Method[D]. Taiyuan University of Technology. Taiyuan. 2012.
16.Ye C-P, Ding X-X, Li W-Y, Wu T-T, Fan M-M, Feng J. Highly Efficient Solvent Screening for Separating Carbazole from Crude Anthracene[J]. Energy & Fuels. 2016;30(4):3529-34.
17.Robinson PM, Scott HG. The Anthracene-Carbazole System[J]. Molecular Crystals. 1969;5(4):387-404.
18.Doki N, Kubota∗ N, Sato A, Yokota M. Effect of cooling mode on product crystal size in seeded batch crystallization of potassium alum[J]. Chemical Engineering Journal. 2001;81:313-6.
19.Hojjati H, Rohani S. Cooling and seeding effect on supersaturation and final crystal size distribution (CSD) of ammonium sulphate in a batch crystallizer[J]. Chemical Engineering and Processing: Process Intensification. 2005;44(9):949-57.
20.Yang G, Kubota N, Sha Z, Louhi-Kultanen M, Wang J. Crystal Shape Control by Manipulating Supersaturation in Batch Cooling Crystallization[J]. CRYSTAL GROWTH&DESIGN. 2006;6(12):2799-803.
21.Wang J. Crystallization[M]. Tianjin University. 1998.
22.Givand JC, Teja AS, Rousseau RW. Effect of Relative Solubility on Amino Acid Crystal Purity[J]. separation. 2001;47(12).
23.T.Matsumotoa, M.Okazakib, M.Inouec, Y.Hamadaa, M.Tairaa, J.Takahashia. Crystallinity and solubility characteristics of hydroxyapatite adsorbed amino acid[J]. Biomaterials. 2002;23:2241-7.
24.Du W, Yin Q, Gong J, Bao Y, Zhang X, Sun X, et al. Effects of Solvent on Polymorph Formation and Nucleation of Prasugrel Hydrochloride[J]. Crystal Growth & Design. 2014;14(9):4519-25.
25.Chen J, Wang J, Ulrich J, Yin Q, Xue L. Effect of Solvent on the Crystal Structure and Habit of Hydrocortisone[J]. Crystal Growth & Design. 2008;8(5):1490-4.
26.Hod I, Mastai Y, Medina DD. Effect of solvents on the growth morphology of dl-alanine crystals[J]. CrystEngComm. 2011;13(2):502-9.
27.Lahav M, Leiserowitz L. The e!ect of solvent on crystal growth and morphology[J]. Chemical Engineering Science. 2001;56:2245-53.
28.Liu J, Chang Z, He Y, Yang M, Dang J. Solvent effects on the crystallization of avermectin B1a[J]. Journal of Crystal Growth. 2007;307(1):131-6.
29.Yang X, Qian G, Zhang X, Duan X, Zhou X. Effects of Solvent and Impurities on Crystal Morphology of Zinc Lactate Trihydrate[J]. Chinese Journal of Chemical Engineering. 2014;22(2):221-6.
30.Li W, Zheng H, Ye C, Wu T, Fan M, Feng J. Effect of the Intermolecular Hydrogen Bond between Carbazole andN,N-Dimethylformamide/Isopropanolamine on the Solubility of Carbazole[J]. Energy & Fuels. 2012;26(10):6316-22.

# Pyrolysis product evaluation and prediction from coals of different rank

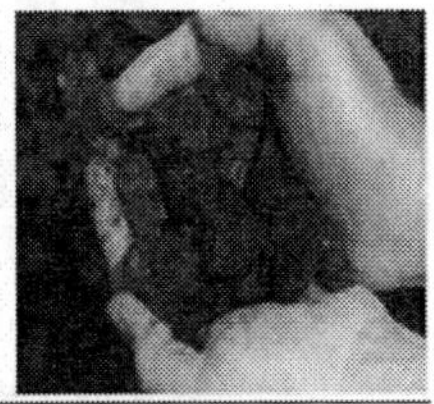


*by*
*G.N. Pretorius*

*Supervisors:*
*Prof. J.R. Bunt (NWU, South Africa)*
*Prof. R.C. Everson (NWU, South Africa)*
*Prof. H.W.J.P. Neomagus (NWU, South Africa)*
*Dr. M. Gräbner (Air Liquide R&D, Germany)*

NORTH-WEST UNIVERSITY
YUNIBESITI YA BOKONE-BOPHIRIMA
NOORDWES-UNIVERSITEIT
POTCHEFSTROOMKAMPUS

AIR LIQUIDE


CARBON

1

# Background and Motivation

- Coal gasification applications[1,3]:
  - Production of synthesis gas for production of methanol, ammonia, SNG, FT liquids etc.
  - Gaseous fuel for electricity generation.
- Fixed bed gasifier with four reaction zones [2,3]:
  - Drying
  - **Devolatilization**
  - Reduction
  - Combustion

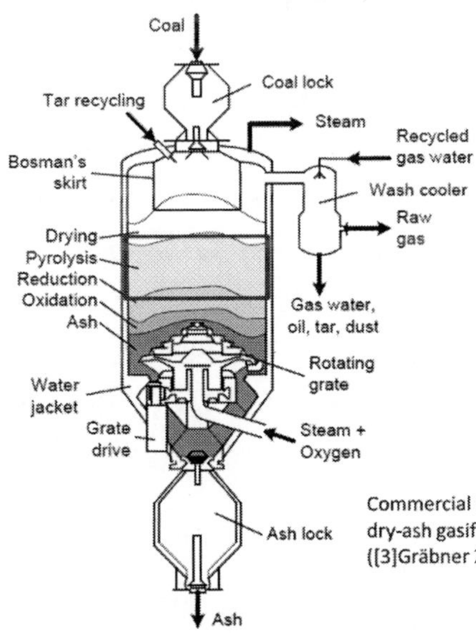


Commercial fixed-bed dry-ash gasifier. ([3]Gräbner 2014, p.246)


[1] Bell, D.A., Towler, B.F. & Fan, M. 2011. Coal gasification and its applications. Amsterdam: Elsevier. 416p.
[2] Bunt, J.R.. & Waanders, F.B. 2008. Identification of the reaction zones occurring in a commercial-scale Sasol–Lurgi FBDB gasifier. Fuel, 87:1814-1823
[3] Gräbner, M. 2014. Industrial Coal Gasification Processes Covering Baseline and High-ash Coal, Weinheim: Wiley-VCH.


NORTH-WEST UNIVERSITY
YUNIBESITI YA BOKONE-BOPHIRIMA
NOORDWES-UNIVERSITEIT
POTCHEFSTROOMKAMPUS

CARBON

2

# Background and Motivation

- Coal -> Char(s) + Volatiles(g)
- Main pyrolysis products: char, water, gas, tar [5].
- Pyrolysis product yields dependent on [5]:
  - Coal rank
  - Operating conditions
- Pyrolysis models in literature -> **FLASHCHAIN** [7]

NORTH-WEST UNIVERSITY
YUNIBESITI YA BOKONE-BOPHIRIMA
NOORDWES-UNIVERSITEIT
POTCHEFSTROOMKAMPUS

[4] CASAL, M.D., DÍEZ, M.A., ALVAREZ, R. & BARRIOCANAL, C. 2008. Primary tar of different coking coal ranks. *International Journal of Coal Geology*, 76:237-242.
[6] SOLOMON, P.R., FLETCHER, T.H. & PUGMIRE, R.J. 1993. Progress in coal pyrolysis. *Fuel*, 72:587-597.
[7] NIKSA, S. & KERSTEIN, A.R. 1991a. FLASHCHAIN Theory for Rapid Coal Devolatilization Kinetics. 1. Formulation. *Energy & Fuels*, 5:647-665.
[8] SCHOBERT, H.H. & SONG, C. 2002. Chemicals and materials from coal in the 21st century. *Fuel*, 81:15-32.

3

CARBON

# Background and Motivation

- Tars and oils are saleable products . Prediction of their yield and quality from a coal sample can have high impact on business case.
- Predicting pyrolysis gas yields (e.g. $CH_4$) from a coal sample can help to design downstream processes more accurately.
- Existing models are developed for flash pyrolysis.
- Testing a variety of very different coals can allow to setup a validated model serving as coal-sample-based indicator for tar/oil and gas yields.

NORTH-WEST UNIVERSITY
YUNIBESITI YA BOKONE-BOPHIRIMA
NOORDWES-UNIVERSITEIT
POTCHEFSTROOMKAMPUS

4

CARBON

# Research aim and objectives

## Aim

Test **models** for **five** different coals of different rank in order to **predict** gas, tar, char and water **yields** and the **liquid tar composition** when undergoing pyrolysis at **low heating rates** (8 °C/min).

## Objectives

- Characterisation of five different coals.
- Use a modified Fischer Assay setup to investigate gas, tar, char and water yields.
- Prediction modelling (FLASHCHAIN) based on coal properties to:
  - predict tar, char, gas and water yields;
  - Predict liquid tar elemental composition.

NORTH-WEST UNIVERSITY
YUNIBESITI YA BOKONE-BOPHIRIMA
NOORDWES-UNIVERSITEIT
POTCHEFSTROOMKAMPUS

5

CARBON

# Coal characterisation

| | A | B | C | D | E |
|---|---|---|---|---|---|
| | | **Proximate analysis (air-dried)** | | | |
| **wt% Inherent moisture content** | 30.1 | 8.5 | 12.8 | 5.2 | 7.3 |
| **wt% Ash content** | 6.6 | 26.0 | 10.7 | 27.9 | 15.2 |
| **wt% Volatile Matter** | 29.4 | 30.0 | 28.5 | 24.5 | 27.5 |
| **wt% Fixed carbon *(by difference)*** | 33.9 | 35.5 | 48.0 | 42.4 | 50.1 |
| **Gross Calorific value (MJ/kg)** | 17.1 | 20.5 | 22.3 | 20.0 | 23.7 |
| | | **Ultimate analyses (dry, ash free)** | | | |
| **wt% Carbon Content** | 66.55 | 74.18 | 74.06 | 75.40 | 76.80 |
| **wt% Hydrogen Content** | 4.79 | 5.10 | 3.92 | 4.13 | 4.33 |
| **wt% Nitrogen Content** | 1.04 | 1.80 | 0.83 | 1.96 | 1.29 |
| **wt% Total sulphur** | 0.92 | 0.93 | 0.43 | 1.63 | 0.97 |
| **wt% Oxygen Content (difference)** | 26.72 | 18.01 | 20.77 | 16.91 | 16.62 |

NORTH-WEST UNIVERSITY
YUNIBESITI YA BOKONE-BOPHIRIMA
NOORDWES-UNIVERSITEIT
POTCHEFSTROOMKAMPUS

6

CARBON

# Experimental setup

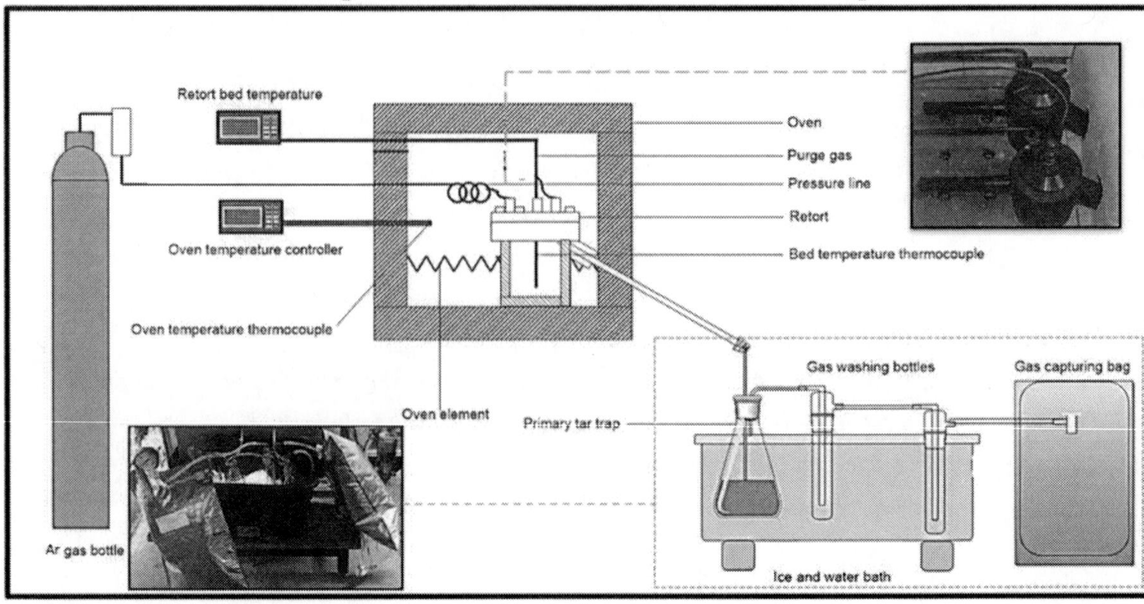


ROETS, L. 2014. *The effect of mineral addition on the pyrolysis products derived from typical Highveld coal.*

NORTH-WEST UNIVERSITY
YUNIBESITI YA BOKONE-BOPHIRIMA
NOORDWES-UNIVERSITEIT
POTCHEFSTROOMKAMPUS

CARBON

7

# Research methodology

5 coals of **different rank**
90% <1mm, not more than 50%<0.2mm

Fischer Assay
Heating rate: 8 °C/min
Argon atmosphere
Atmospheric pressure
Sample mass: 50g

520 °C
720 °C
920 °C

Char yield

Tar yield

Water yield

Gas yield

NORTH-WEST UNIVERSITY
YUNIBESITI YA BOKONE-BOPHIRIMA
NOORDWES-UNIVERSITEIT
POTCHEFSTROOMKAMPUS

CARBON

8

# Fischer Assay yields

Char | Tar
Water | Gas

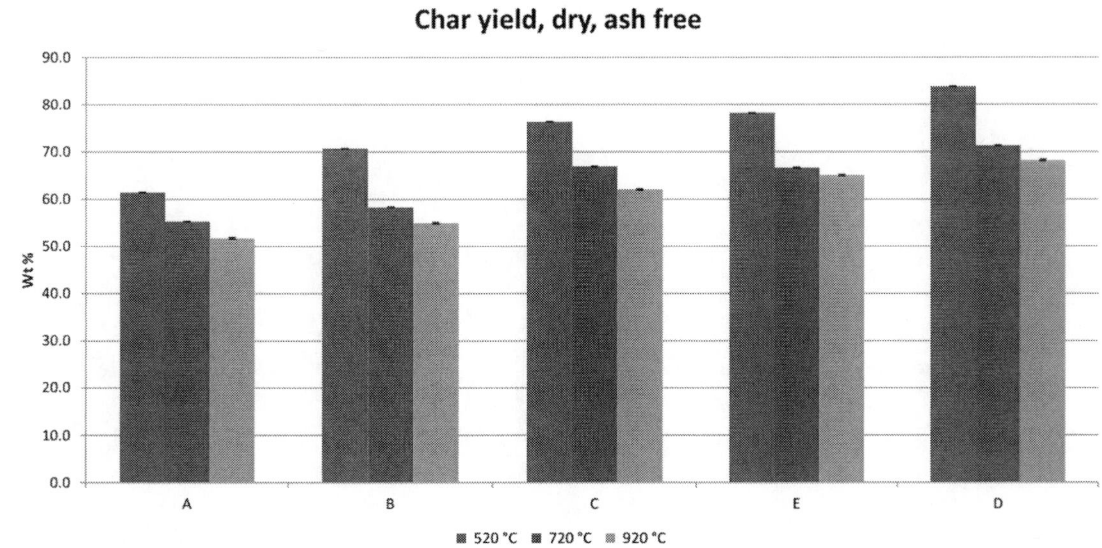


NORTH-WEST UNIVERSITY
YUNIBESITI YA BOKONE-BOPHIRIMA
NOORDWES-UNIVERSITEIT
POTCHEFSTROOMKAMPUS

CARBON

9

# Fischer Assay yields

Char | Tar
Water | Gas

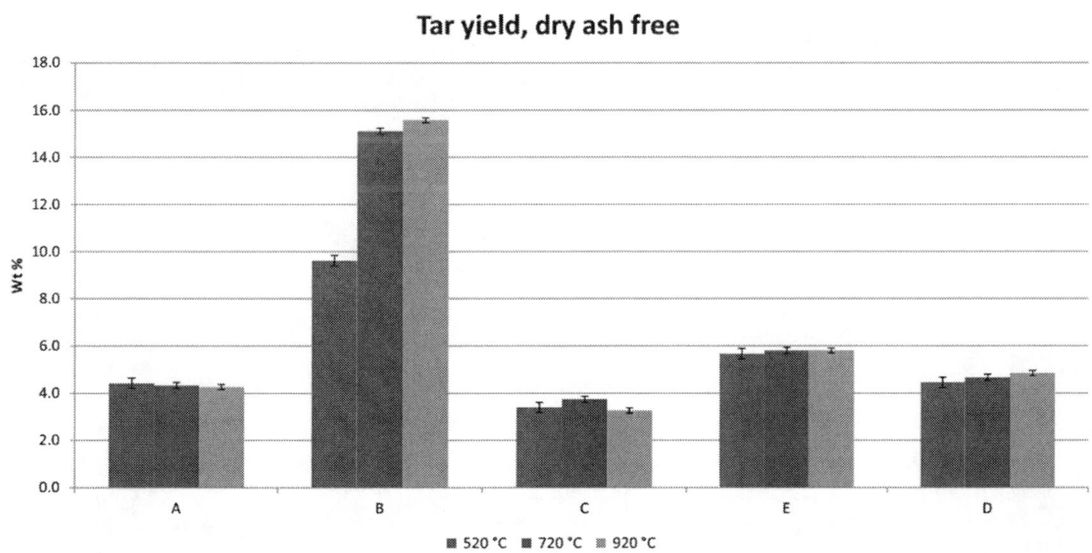


NORTH-WEST UNIVERSITY
YUNIBESITI YA BOKONE-BOPHIRIMA
NOORDWES-UNIVERSITEIT
POTCHEFSTROOMKAMPUS

CARBON

10

# Fischer Assay yields

Char | Tar
Water | Gas

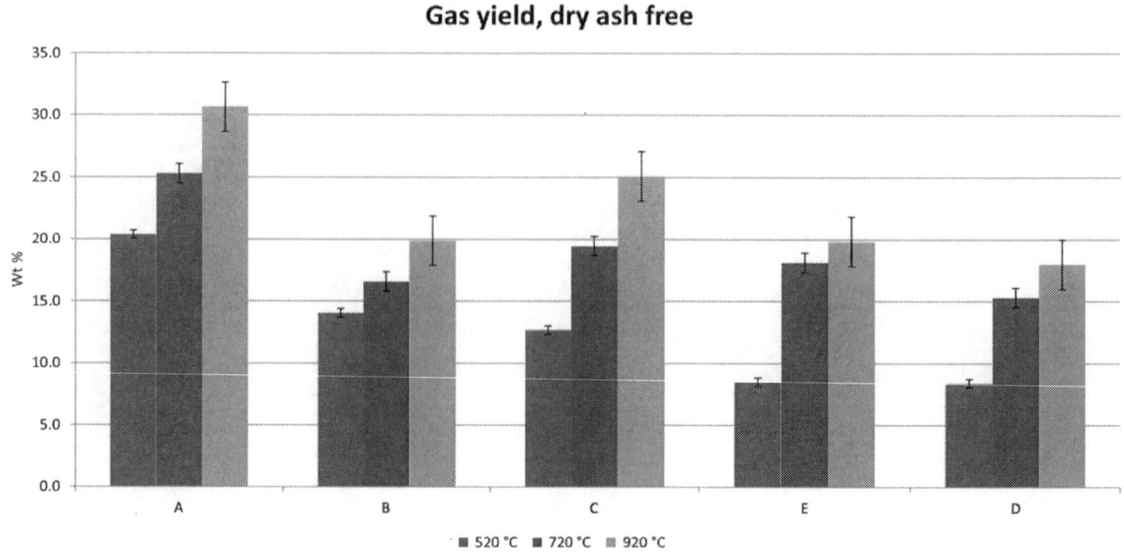


NORTH-WEST UNIVERSITY
YUNIBESITI YA BOKONE-BOPHIRIMA
NOORDWES-UNIVERSITEIT
POTCHEFSTROOMKAMPUS

CARBON

11

# Research methodology

Char

**Proximate analysis**
Fixed carbon
Ash content
Moisture content
Volatile matter

**Fischer Assay**

Ul

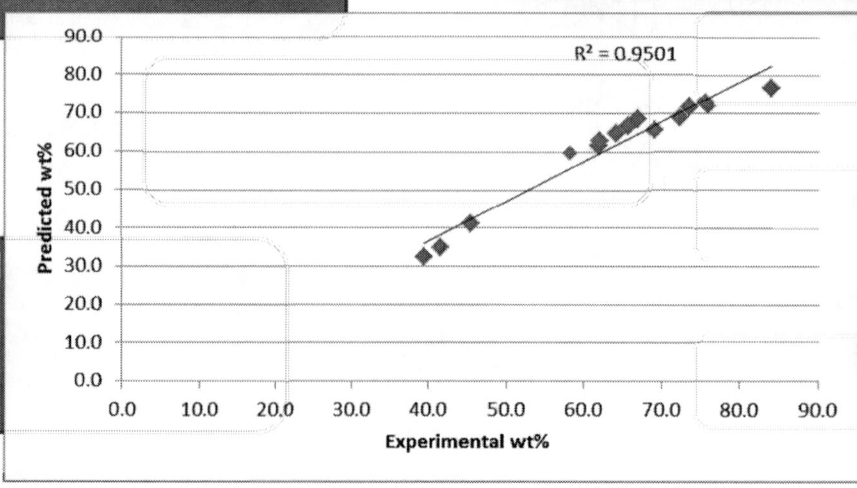


NORTH-WEST UNIVERSITY
YUNIBESITI YA BOKONE-BOPHIRIMA
NOORDWES-UNIVERSITEIT
POTCHEFSTROOMKAMPUS

[9] NIKSA, S. & KERSTEIN, A.R. 1991a. FLASHCHAIN Theory for Rapid Coal Devolatilization Kinetics. 1. Formulation. *Energy & Fuels*, 5:647-665.

12

CARBON

# Total volatiles yield

Tar

Gas

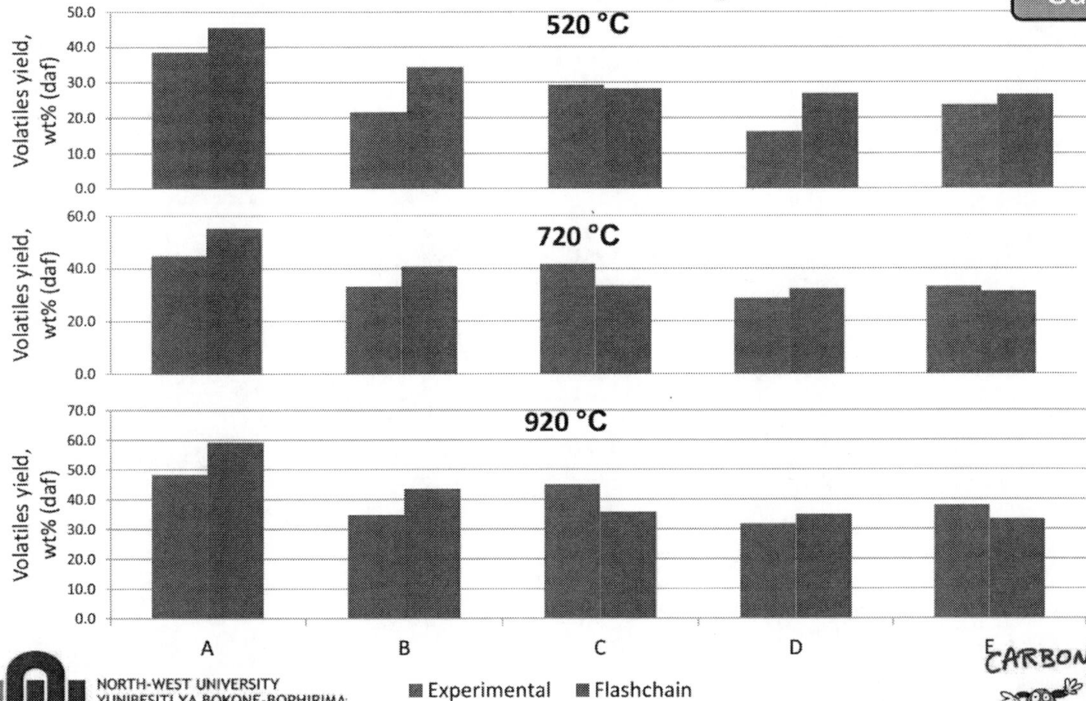


NORTH-WEST UNIVERSITY
YUNIBESITI YA BOKONE-BOPHIRIMA
NOORDWES-UNIVERSITEIT
POTCHEFSTROOMKAMPUS

CARBON

13

# Tar CHO

Tar

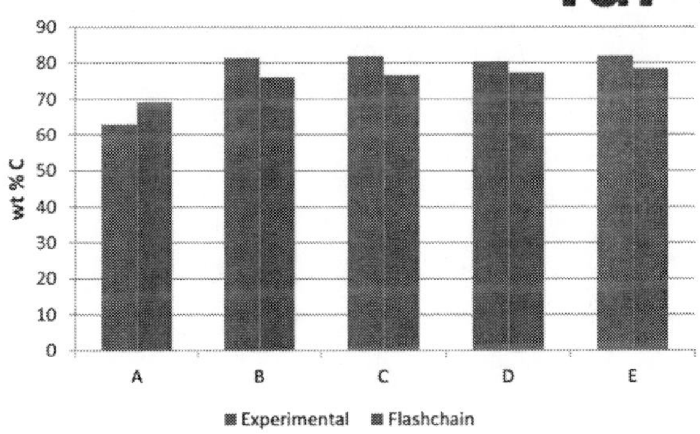


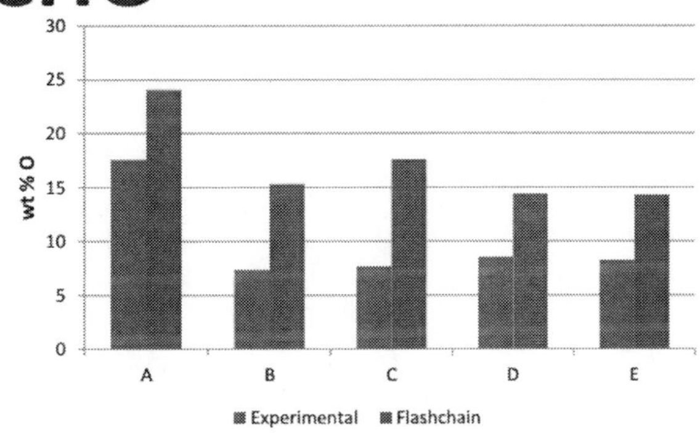


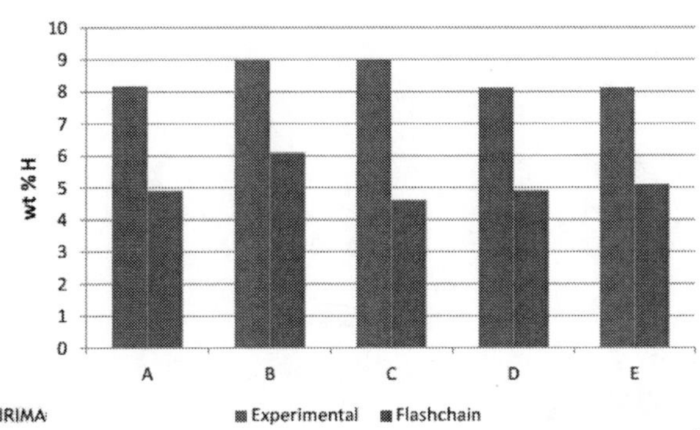


NORTH-WEST UNIVERSITY
YUNIBESITI YA BOKONE-BOPHIRIMA
NOORDWES-UNIVERSITEIT
POTCHEFSTROOMKAMPUS

CARBON

14

Tar

# Tar SEC-UV

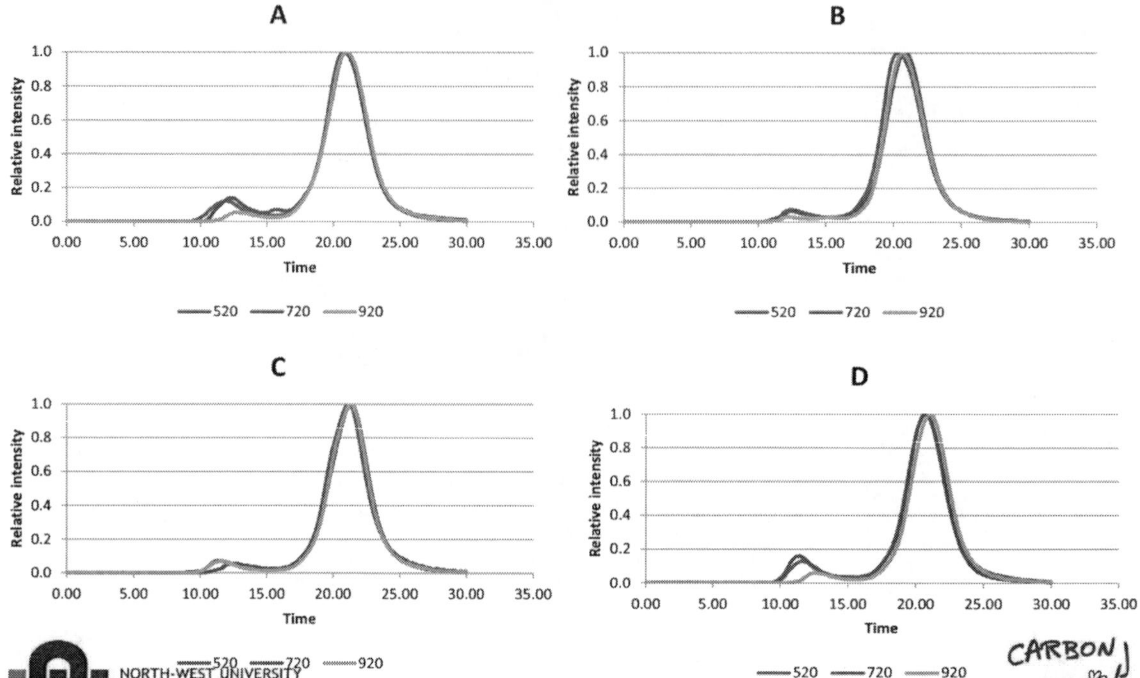


NORTH-WEST UNIVERSITY
YUNIBESITI YA BOKONE-BOPHIRIMA
NOORDWES-UNIVERSITEIT
POTCHEFSTROOMKAMPUS

CARBON

15

Tar

# Tar SEC-UV

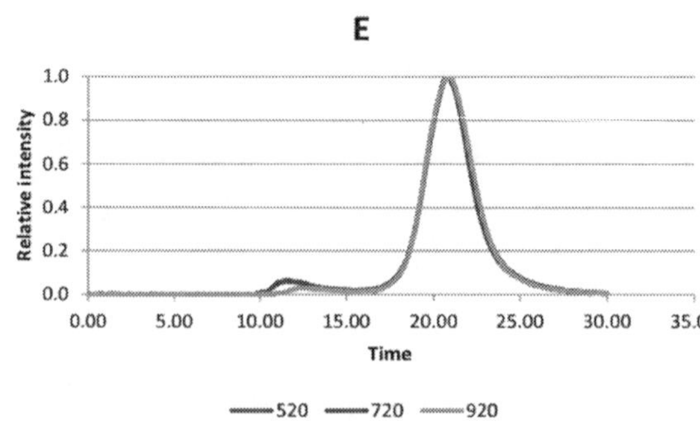



NORTH-WEST UNIVERSITY
YUNIBESITI YA BOKONE-BOPHIRIMA
NOORDWES-UNIVERSITEIT
POTCHEFSTROOMKAMPUS



CARBON


16

Tar

# Tar SimDis

**Simulated Distillation Tar/Oil Samples**

- **Sample Analysed** (Boiling point < 450°C)
  - Heavy naphtha (Boiling point < 205°C)
  - Kerosene (Boiling point 205°C < bp < 260°C)
  - Diesel (Boiling point 260°C < bp < 340°C)
  - Gas oil (Boiling point 340°C < bp < 425°C)
  - Residue fraction (Boiling point 425°C < bp < 450°C)
- **Sample not Analysed** (Boiling point > 450°C)

NORTH-WEST UNIVERSITY
YUNIBESITI YA BOKONE-BOPHIRIMA
NOORDWES-UNIVERSITEIT
POTCHEFSTROOMKAMPUS

CARBON

17

Tar

# Tar SimDis

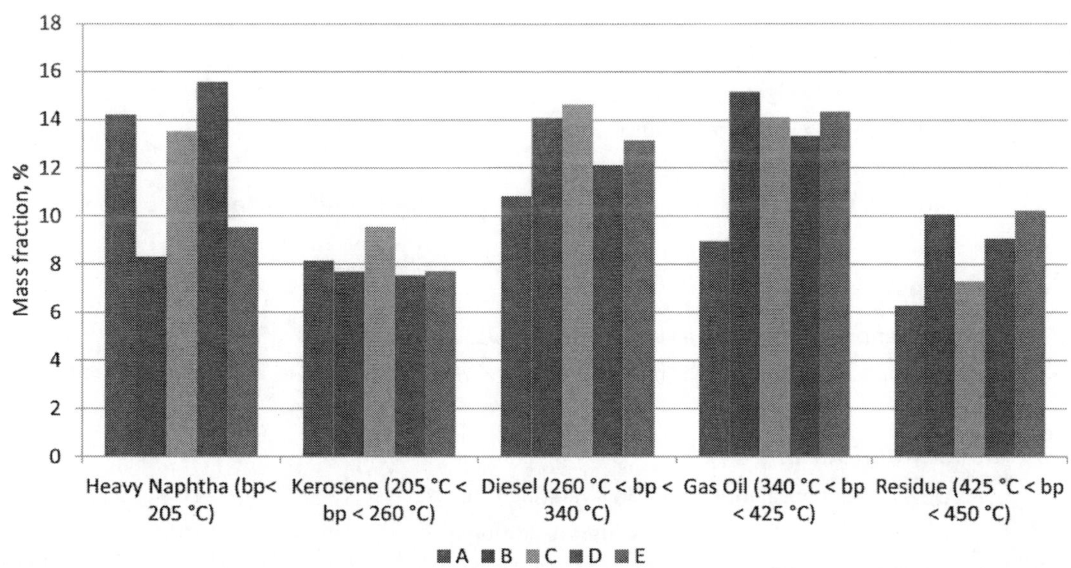


NORTH-WEST UNIVERSITY
YUNIBESITI YA BOKONE-BOPHIRIMA
NOORDWES-UNIVERSITEIT
POTCHEFSTROOMKAMPUS

CARBON

18

Gas

# Gas composition

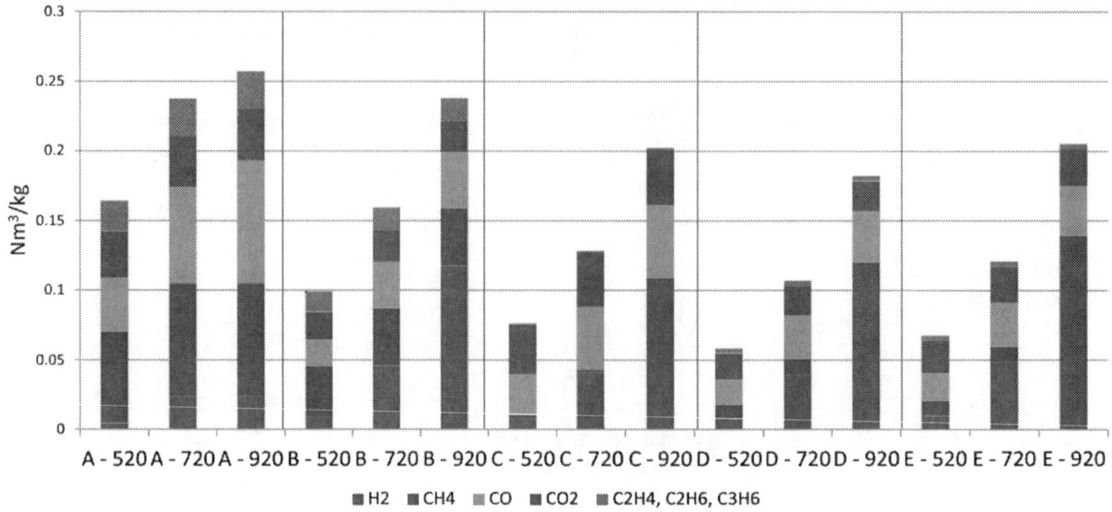


NORTH-WEST UNIVERSITY
YUNIBESITI YA BOKONE-BOPHIRIMA
NOORDWES-UNIVERSITEIT
POTCHEFSTROOMKAMPUS

19

CARBON

# Concluding remarks

- Experimental
  - Higher rank -> higher char yields.
  - Higher rank -> lower gas yields.
  - Temperatures above 720 °C -> mainly $H_2$ release.
  - Slow heating rates -> Pyrolysis temperature has minimal effect on tar composition.
- Modelling
  - Char and total volatiles -> FLASHCHAIN proves to be a good early indicator
  - Product compositions -> Some limitations -> Mineral effects & slow heating rate
- Recommendations
  - Develop statistical models -> Slow heating rate
    -> Mineral additions
  - Combine statistical model with FLASHCHAIN model.

NORTH-WEST UNIVERSITY
YUNIBESITI YA BOKONE-BOPHIRIMA
NOORDWES-UNIVERSITEIT
POTCHEFSTROOMKAMPUS

20

CARBON

# Acknowledgements


- Thanks to Air Liquide R&D, Germany, for financially supporting this M-ENG investigation.

- This work is based on the research supported by the South African Research Chairs Initiative of the Department of Science and Technology and National Research Foundation of South Africa (Chair Grant No. 86880, UID85643, UID85632). Any opinion, finding or conclusion or recommendation expressed in this material is that of the author(s) and the NRF does not accept any liability in this regard.


NORTH-WEST UNIVERSITY
YUNIBESITI YA BOKONE-BOPHIRIMA
NOORDWES-UNIVERSITEIT
POTCHEFSTROOMKAMPUS

AIR LIQUIDE

21

CARBON

Thank You


NORTH-WEST UNIVERSITY
YUNIBESITI YA BOKONE-BOPHIRIMA
NOORDWES-UNIVERSITEIT
POTCHEFSTROOMKAMPUS

AIR LIQUIDE


22

CARBON

# Improving Response Rates of Acid Gas Absorber Columns

Pittsburgh Coal Conference
Cape Town, 10th August, 2016

C. Heinze, TU Darmstadt; C. Higman, Higman Consulting GmbH;
J. Marasigan, EPRI; B. Epple, TU Darmstadt



## Overview

- **Background/Motivation**
  - Analysis of bottleneck for IGCC flexibility
  - Hypothesis for ramping limitations in AGR
- **Project Scope**
  - Static model of typical absorber at different loads
  - Dynamic model confirming observed effects
  - Development of "Rapid Response Absorber" in dynamic model
- **Results**
- **Outlook**

## Background

- Renewable sources (wind, solar) can vary in output from 0% to 100% over a short period – other generating resources must cycle to accommodate these variations.
- EPRI-Report examined design options for enhancing IGCC Flexible Operations Performance and Economics to match Rapid Response Gas Turbines.
- The analysis showed that the Acid Gas Removal (AGR) unit is the current bottleneck to increasing ramp rates.

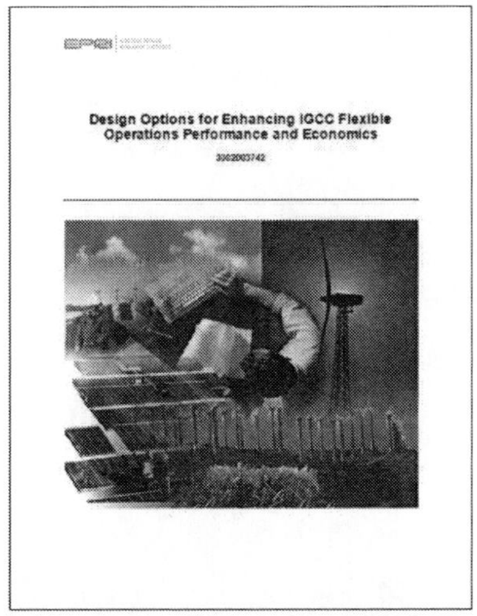
Design Options for Enhancing IGCC Flexible Operations Performance and Economics



## Main IGCC Components

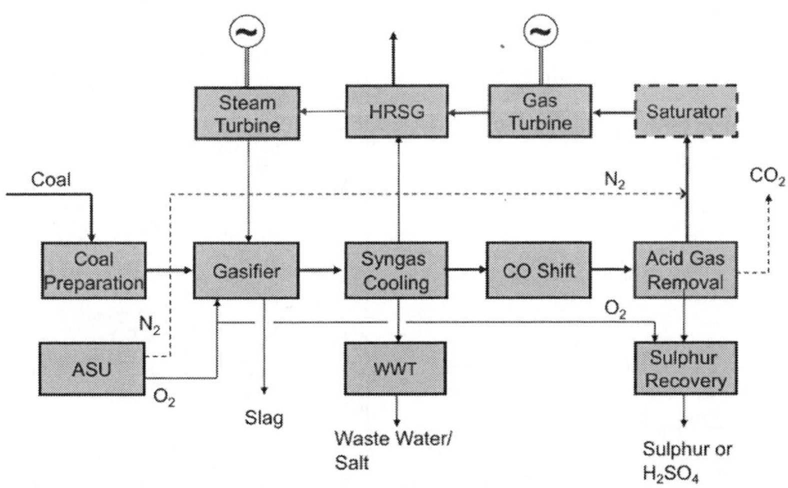

## Functionality of AGR

- Absorbs acid gases ($H_2S$ and $CO_2$) in a chemical or physical solvent.
- Regenerate solvent by heating and/or flashing.
- Details can be more complicated, particularly if selectivity (separation of $H_2S$ from $CO_2$) is required.
- Absorber internals can be random or structured packing or trays.

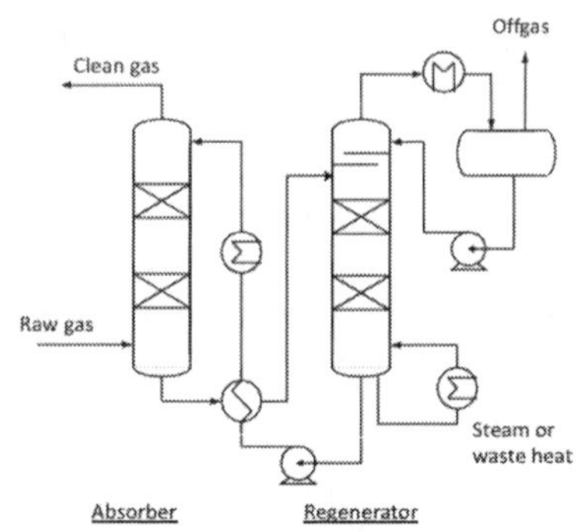




## Typical Tray Type Internals

- Solvent flows down via downcomer onto tray.
- Solvent level on tray is maintained by weir to next downcomer.
- Gas flow up through holes in tray.
- Downcomer submerged to avoid gas bypassing.
- Tray details vary from simple sieve trays (as shown) to e. g. valve trays which provide better downturn.

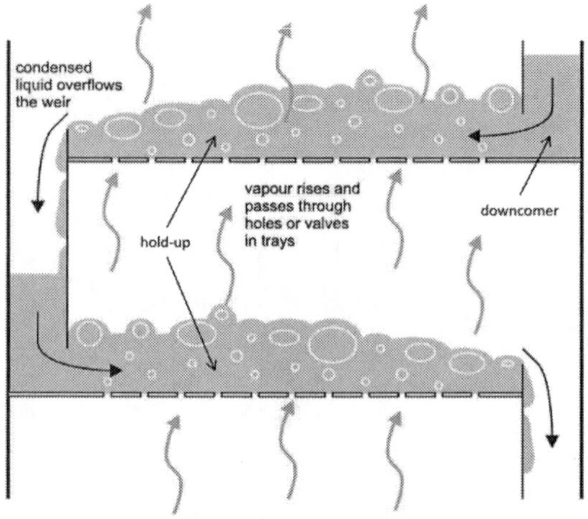

## AGR Ramping – Current Experience

- Solvent flow ratio controlled from gas flow (proportional)
- Ramp up:
  - Operators limit ramp up speed; typical limit is < 3 %/min
  - From experience faster ramps lead to sulfur-breakthrough
  - Sulfur-breakthrough even occurs with solvent flow kept at 100%
- Ramp down:
  - Experience with very fast ramp down on multi-gasifier installations exist.
  - No syngas quality issues occur, though problems have been seen in the regenerator.

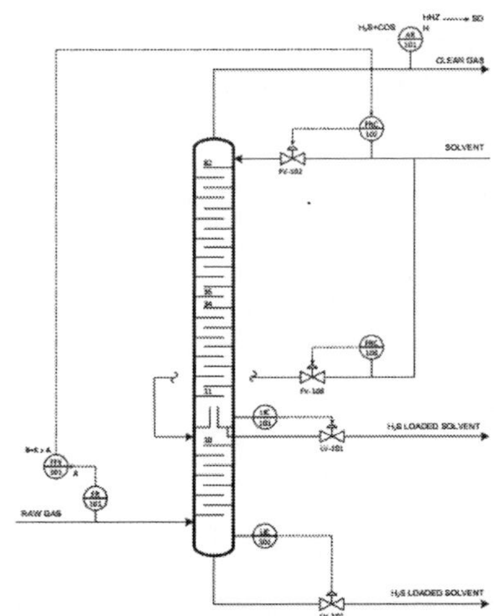




## Hypothesis

- **In the previous report, a hypothesis was developed to explain poor ramp rate capability of acid gas absorbers.**
- **The hypothesis was developed for trayed columns, but is expected to apply equally to packed columns.**
- **Hypothesis**
  1. At higher operating loads the liquid levels (hold-up) on the trays are higher than at lower loads.
  2. During ramp up transient, the additional liquid inventory must be built up over the whole column.
  3. The upper tray levels will increase first, robbing flow to lower trays, which, being starved, will become ineffective.
  4. This could be solved by additional liquid supply to lower trays.

# Steady State Modelling in ASPEN PLUS

- Absorber with physical solvent and prewash section fully modelled according to data sheet of a reference plant.
- Good agreement between supplied data and simulation results for all species.
- Correct sulfur concentration was adjusted using an appropriate tray efficiency (ca. 25% Murphree)

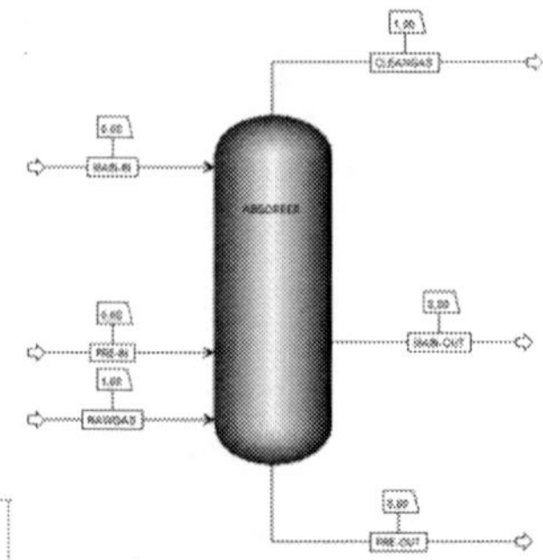

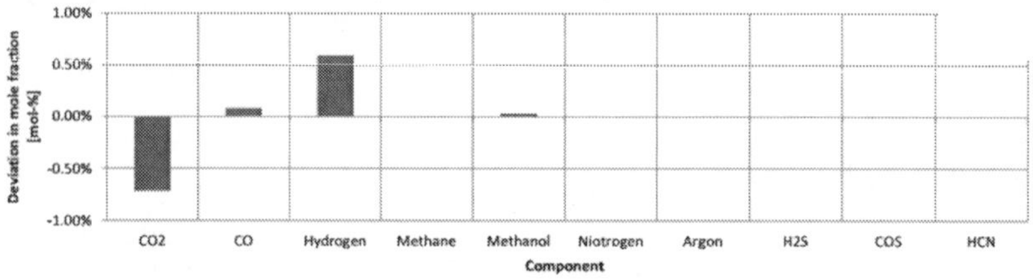




# Dynamic Modelling in ASPEN Dynamics

- Qualitatively ASPEN Dynamics can demonstrate the expected $H_2S$ concentration profile over time for a ramp of 40% in 6 min (ca. 6.7 %/min)
- Result is independent of absolute values/Murphree efficiency.

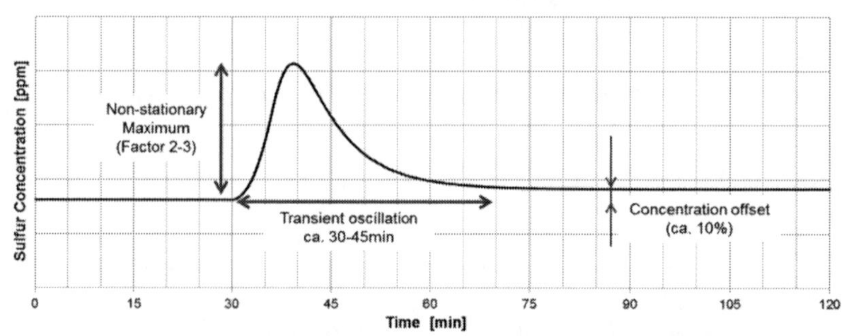

# Dynamics – Weir Equation

- When the solvent flow is kept constant, the ASPEN Dynamics results do not match the expectation. No sulfur peak occurs.
- Reason is the simple weir model in ASPEN that only considers liquid height to calculate the solvent flow rate.
- The more sophisticated Stichlmair equation also considers the effect of bubbling and foaming of solvent caused by the gas flow.
- With the Stichlmair equation the simulation matches the expectations.

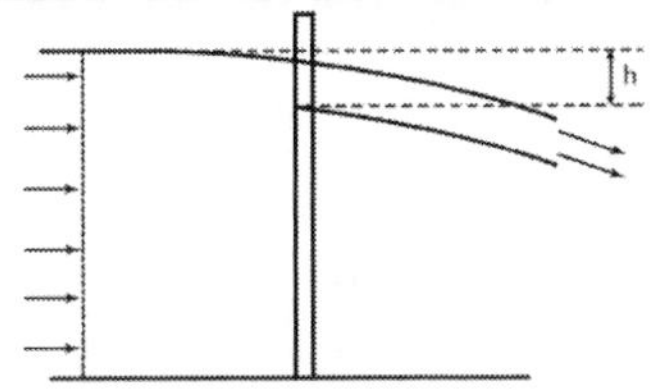


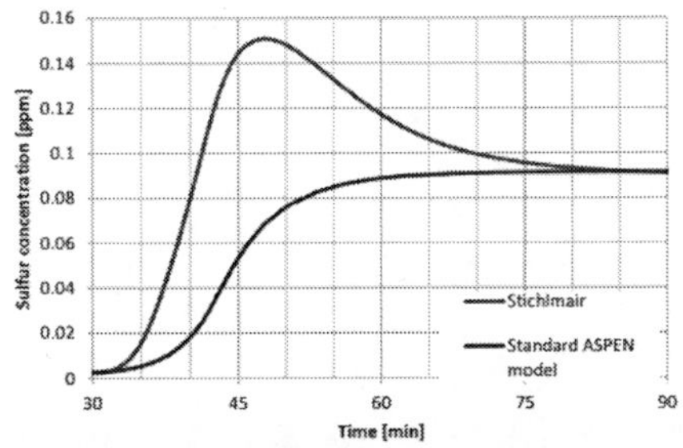




# Developing a Solution

- **Base case:**
  - Standard arrangement as used in previous simulations.
- **Rapid Response Absorber Type A:**
  - Provide additional solvent to the column and feed it into the top.
- **Rapid Response Absorber Type B:**
  - Provide additional solvent to the column and feed it into an additional inlet halfway down the column to the "starved" trays.
- **The additional solvent flow is calculated via a differential controller without any proportional gain.**

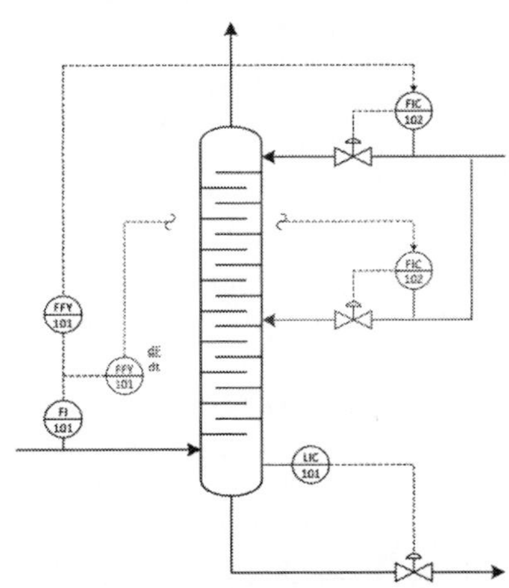

## Results – Performance of the RRA

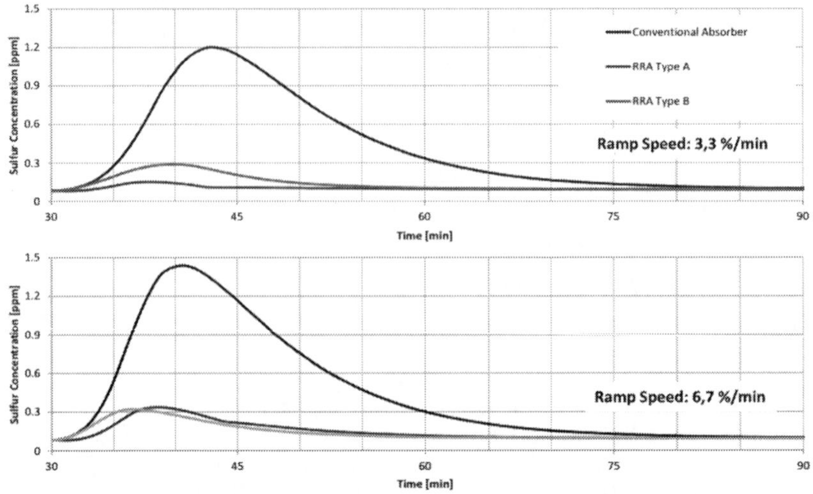




## Results – Downcomer Limitations

- **Both approaches show promising results.**
- **With faster ramps the RRA Type B is advantageous.**
  - The downcomer only allows a certain solvent flow. Higher flows will lead to flooding of the top trays.
  - Hence, even with very high feed rates at the top, the lower trays will eventually starve.
  - Then a dedicated feed for the lower trays becomes necessary.

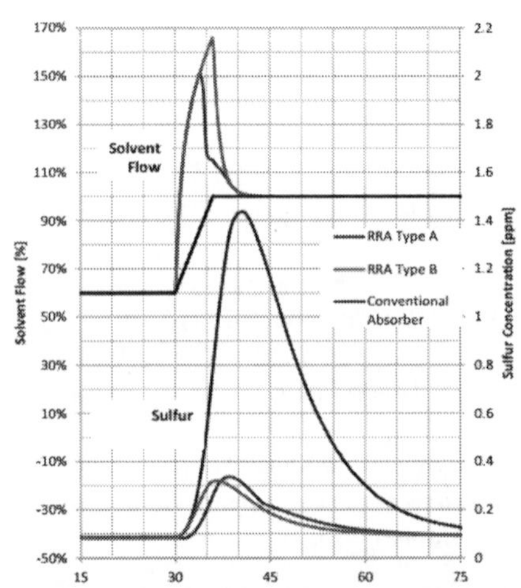

## Results – Summary

- **Dynamic simulation performed, using an improved ASPEN model, provides a better understanding of column dynamics:**
  - The different hold-up at different column loads lead to unwanted transient behavior.
  - The effect of the gas flow on the effective tray level can not be neglected.
- **The performance during transients can be improved by temporary addition of solvent and appropriate control scheme.**
  - The simulation show that a feed to the top or to the middle trays is feasible.
  - For very high rates of gas flow change the solvent needs to be fed to the lower trays. Otherwise the large amount of solvent at the top trays will lead to flooding and the bottom trays starve.
- **The absolute values for sulfur slip must be considered as provisional at this stage, since the dynamic model is not yet validated.**



## Outlook

- **Model improvements and model additions:**
  - Rate-based effects can alter the performance of trays at different levels of solvent flow.
    In this context the sufficient phase contact at low solvent supplies have to be considered as well.
  - The model will be extended with at least the desorber to embed the absorber into the whole process.

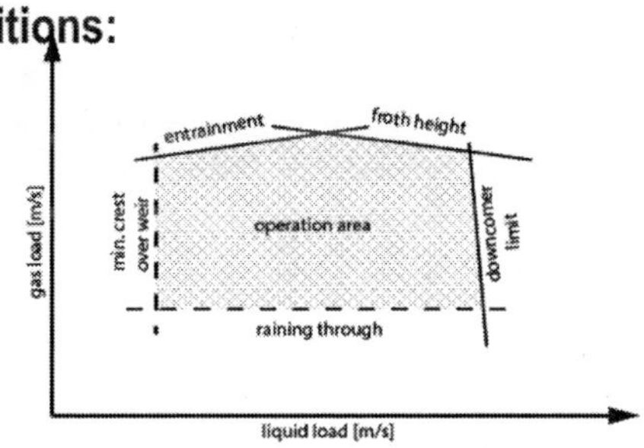


- **Validation of the ASPEN model as well as experimental data for chemical absorption**
- **Investigation of packed columns to confirm that the hypothesis applies here as well**

## Acknowledgement

The authors thank EPRI for funding this work under the TI Polaris Initiative.

# Thermodynamic possibilities and constraints for the regeneration of spent solid $H_2S$ sorbents based on Ce-, La- and Zn-oxides

Karel Svoboda[1], Jindřich Leitner[2], Jiří Brynda[1,3], Siarhei Skoblia[3], Yau-Pin Chyou[4], Yi-Shun Chen[4], Miloslav Hartman[1]

[1]Institute of Chemical Process Fundamentals of the CAS, v.v.i., Rozvojová 135, 165 02 Prague 6, Czech Republic, E-mail: svoboda@icpf.cas.cz Tel.: +420 220 390 160

[2] Dep. of Solid State Engineering, Faculty of Chemical Technology, University of Chemistry and Technology Prague, Technická 3, 166 28 Prague 6, Czech Republic, E-mail: jindrich.leitner@vscht.cz

[3] Department of Gas, Coke and Air Protection, University of Chemistry and Technology Prague, Technická 5, 166 28 Prague 6, Czech Republic, E-mail: sergej.skoblja@vscht.cz

[4] Chemistry Division, Institute of Nuclear Energy Research, No. 1000, Wenhua Rd., Jiaan Village, Longtan Township, Taoyuan County 32546, Taiwan (R.O.C.)
E-mail: ypchyou@iner.gov.tw E-mail: chenys@iner.gov.tw

## Abstract

The thermodynamic analyses are focused on suitable ways and conditions for regeneration of spent desulfurization sorbents from dry cleaning of producer gas in a temperature range 600 – 1200 K. Three sulfidic compounds (ZnS, $Ce_2O_2S$ and $La_2O_2S$) were considered for regeneration to the corresponding oxides or oxy-carbonates by mild oxidation, reaction with $SO_2$ and reaction with water vapour and $CO_2$. Zinc sulfide can be regenerated practically only by mild oxidation either at $O_2$/ZnS molar ration around 1.5 at temperatures between approx. 950 and 1050 K or at temperatures over 1200 K at higher $O_2$/ZnS molar ratios. The temperature of regeneration higher than approx. 1100 K causes sintering/changes of sorbent texture. The $Ce_2O_2S$-based spent cerium sorbents are advantageously regenerated by reaction with $SO_2$ at temperatures over 700 K (between 700 and 900 K). In this way $CeO_2$, elemental sulfur and small concentrations of $S_2O$ are theoretically produced. If the spent Ce-sorbent contains also Ce-halides (e.g. $CeCl_3$) , they are not decomposed by reaction with $SO_2$. The La-based spent sulfur sorbents are restored by reaction with $CO_2$ and $H_2O$ vapor leading to formation of $La_2O_2CO_3$ and $H_2S$. The suitable operating conditions involve temperatures between 700 and 800 K and higher operating pressures (above 1 MPa). The regeneration of $La_2O_2S$ by simple oxygen oxidation is complicated by formation of very stable oxy-sulfate $La_2O_2SO_4$.

***Keywords:*** Spent $H_2S$ sorbents; ZnS; $La_2O_2S$ and $Ce_2O_2S$; Regeneration; Thermodynamics.

## 1 Introduction

Sorbents based on ZnO, $Ce_2O_3$, and $La_2O_3$/$La_2O_2CO_3$ can be used for deep, dry, medium/high temperature desulfurization of fuel gas or syngas [1-5]. The spent solid sorbents from producer or syngas high/medium temperature desulfurization (sulfidation) processes can be regenerated generally by three methods [6-15]: mild oxidation, reactions with steam (in fact the reverse sulfidation reaction) or by reactions with $SO_2$ leading to elemental sulfur production.

The problems in application of the oxidation reactions for spent sorbents include highly exothermic reactions, necessity to use higher temperatures to overcome formation of stable sulfates (e.g. $ZnSO_4$), oxy-sulfates (e.g. $La_2O_2SO_4$) and highly thermodynamically stable basic sulfates (e.g., $ZnO*2ZnSO_4$) [16-19]. Necessity to use relatively high temperature and precisely defined oxygen/sorbent molar ratio is usually a practical obstacle for highly selective conversion to reactive, regenerated (non-sintered) oxide.

In regeneration by steam, the highest conversions to $H_2S$ are attained for the less efficient $H_2S$ sorbents [20] (e.g., $CeO_2$). More advantageous regeneration reactions with steam are those with the decreasing number of moles of gaseous products, because in such a case the higher operation pressure in $H_2S$ desorption has a beneficial effect. But, the common steam regeneration reactions of spent sulfide sorbent (back to metal oxides) are equimolar in nature. Only the formation of e.g., stable carbonates/oxy-carbonates [21-23] instead of oxides leads to decreasing number of moles of gaseous compounds.

Formation of elemental sulfur in the sorbent regeneration processes presents an advantage of versatile, easy-storable solid product. However, as it is suggested by thermodynamics [11,15], the only applicable regeneration reaction leading to elemental sulfur is reaction of $Ce_2O_2S$ with $SO_2$.

Because in literature relatively often partly incomplete and partly controversial data can be found on the regeneration reactions, chemistry, thermodynamics, and kinetics, we analyze here possible chemical products and thermodynamics for several selected regenerations of solid sorbents.

The very usual ZnO-based sulfided sorbents can be regenerated from the thermodynamic point of view practically only by mild oxidation reactions with various constraints.

For the spent Ce-based sorbents, the regeneration process based on the reaction with $SO_2$ - potentially forming elemental sulfur appears as the best suitable regeneration process. The regeneration reaction with water vapor cannot be influenced by operating pressure and all oxidation reactions of $Ce_2O_2S$ are highly exothermic with excessive heat effects.

The $La_2O_3$/$La_2O_2CO_3$ based sorbents can be regenerated [24] after sulfidation by either oxidation or by reaction with $H_2O$ vapor and $CO_2$ leading to lanthanum-oxy-carbonate ($La_2O_2CO_3$). The oxidation regeneration reactions, as the thermodynamics and TG analyses indicate [17,24], give rise to the formation of thermally very stable oxy-sulfate ($La_2O_2SO_4$) with the temperature of decomposition above approx. 900 °C (1200 K). The regeneration reaction with water vapor and $CO_2$ is enhanced by the increasing operating pressure. The formation of $La_2O_2CO_3$ is disadvantage from the point of view of efficient $H_2S$ removal, but on the other hand, significant content of $CO_2$ in fuel gas (or syngas) causes at temperatures below approx. 800 °C the formation of $La_2O_2CO_3$ together with attenuation of $H_2S$ equilibrium sorption [25].

This is the reason why we have focused in this study on thermodynamic analysis of the regeneration of ZnS by oxidation reactions, thermodynamic analysis of the $Ce_2O_2S$ regeneration by reaction with $SO_2$ and on analysis of the $La_2O_2S$ regeneration by reactions with mixtures of $H_2O$ vapor and $CO_2$.

## 2 Chemistry and thermodynamic analysis of ZnS regeneration to ZnO

The basic chemical reactions considered and arranged according the increasing oxygen/ZnS ratio are as follows:

$ZnS + 0.5\ O_2 \leftrightarrow ZnO + (1/x)S_x$ (1)
$ZnS + 1.5O_2 \leftrightarrow ZnO + SO_2$ (2)
$3ZnS + 5.5O_2 \leftrightarrow ZnO*2ZnSO_4 + SO_2$ (3)
$ZnS + 2O_2 \leftrightarrow ZnSO_4$ (4)

Other pertinent reactions of formed ZnO leading to zinc sulfate, basic-Zn-sulfate, oxidation of $SO_2$ and reactions of elemental sulfur and water vapor can be considered:

$ZnO + SO_2 + 0.5O_2 \leftrightarrow ZnSO_4$ (5)
$ZnO + 2ZnSO_4 \leftrightarrow ZnO*2ZnSO_4$ (6)
$(1/x)S_x + O_2 \leftrightarrow SO_2$ (7)
$SO_2 + 0.5O_2 \leftrightarrow SO_3$ (8)
$(3/x)S_x + 2H_2O \leftrightarrow 2H_2S + SO_2$ (9)

According to the thermodynamic analysis (values of $\Delta G_r$) for the reactions 3 and 4, at temperatures over approx. 900 K, the reaction 3 (formation of $ZnO*2ZnSO_4$) prevails (i.e. it has lower $\Delta G_r$ values). Similarly at temperatures above 900 K reaction 6 prevails in comparison with reaction 5.

Because the system is relatively complicated with several possible reactions and by-products, we have used the minimization of Gibbs energy [26-28] for the temperature range 900 – 1200 K with three input (initial) values of $O_2/ZnS$ molar ratio. The molar ratio 1 means the conditions between reactions 1 and 2 (possible formation of $SO_2$, $S_x$, ZnO, and also possible incomplete course of reaction 2). The input molar ratio $O_2/ZnS = 1.6$ means the conditions between reactions 2 and 3. The input molar ratio $O_2/ZnS > 3$ means a relatively higher surplus of oxygen with possible formation of $SO_3$. To see influence of occurrence of water vapor, we also considered its small concentrations in all cases. The results are given in **Table 1**, **Table 2,** and **Table 3**. Thermodynamic data (the Gibbs energy of formation, $\Delta G_f$) have been taken from the respected literature sources [29,30].

Remaining concentrations of oxygen in gas (at molar ratio $ZnS/O_2 = 1$) have been found to be very low (e.g., at the temperature 1200 K the $O_2$ concentration is only 2.22 E-10 mol. %, at lower temperature it is even lower). Presence of water vapor causes, in a small extent, the formation of $H_2S$ and hydrogen. As it was proved experimentally [31], $H_2O$ can also react directly with ZnS to form $SO_2$ and $H_2$ at conditions with insufficient presence of oxygen.

**Table 1:** Composition of gases and solid products in equilibrium (after the minimization of Gibbs energy for the system) . Input conditions: 1 mole ZnS, 100 moles of gas (1 mole $O_2$, 0.5 mole $H_2O$, balance $N_2$).

| T (K) | $SO_2$ (mol %) | $S_2$ (mol %) | $H_2$ (mol %) | $H_2S$ (mol %) | Zn(g) (mol %) | ZnO(s) (ZnS) (mol) |
|---|---|---|---|---|---|---|
| 900 | 0.6689 | 1.4 E-05 | 1.45 E-04 | 2.58 E-05 | 2.15 E-07 | 0.6668 (0.3332) |
| 1000 | 0.6690 | 8.14 E-05 | 5.45 E-04 | 7.04 E-05 | 1.68 E-05 | 0.667 (0.333) |
| 1050 | 0.6691 | 1.73 E-04 | 9.62 E-04 | 1.08 E-04 | 1.08 E-04 | 0.6673 (0.3326) |
| 1100 | 0.6693 | 3.43 E-04 | 1.61 E-03 | 1.6 E-04 | 5.83 E-04 | 0.6675 (0.3319) |
| 1200 | 0.6727 | 1.14 E-03 | 3.4 E-03 | 3.17 E-04 | 1.10 E-02 | 0.6626 (0.3263) |

At higher temperatures, relatively significant concentrations of SO and $S_2O$ were indicated in gas phase (at 1100 K 6.6 E-06 mol. % $S_2O$ and 1.02 E-04 mol. % SO). Zinc volatility increases with the increasing temperature and is particularly important at T > 1100 K. Volatility of zinc is negligibly influenced by the presence of small concentrations of water vapor in gas. In the range 0.5 < $O_2$/ZnS ratio < 1.5 only the reaction 2 clearly prevails with incomplete ZnS regeneration.

**Table 2:** Composition of gases and solid products in equilibrium (after the minimization of Gibbs energy for the system). Input conditions: 1 mole ZnS, 100 moles of gas ($O_2$:1.6 moles, $H_2O$: 0.5 moles, balance $N_2$).

| T (K) | $O_2$ (mol %) | $SO_2$ (mol %) | $SO_3$ (mol %) | $H_2SO_4$ (mol %) | ZnO (mol) | ZnO*2$ZnSO_4$ (mol) |
|---|---|---|---|---|---|---|
| 900 | 1.06 E-08 | 0.8065 | 6.03 E-05 | 3.87 E-09 | 0.7001 | 0.0997 |
| 1000 | 1.21 E-04 | 0.8067 | 1.75 E-03 | 3.31E-08 | 0.703 | 0.09901 |
| 1050 | 6.3 E-03 | 0.8188 | .7.32 E-03 | 8.29 E-08 | 0.73 | 0.09012 |
| 1070 | 0.0254 | 0.856 | .0.0125 | 1.17 E-07 | 0.794 | 0.0686 |
| 1100 | 0.09044 | 0.985 | .0.0201 | 1.43 E-07 | 1 | 0 |
| 1200 | 0.0962 | 0.9964 | .0.0087 | 2.76 E-08 | 1 | 0 |

At the temperature 1200 K the concentration of Zn(g) in gas phase is only 5.3 E-06 mol. %. $H_2SO_4$ formation is caused by the presence of water vapor in gas. At conditions between the reactions 2 and 3 (1.5 < $O_2$/ZnS < 1.833 = 5.5/3), there is no ZnS present in the solid products, but with increasing $O_2$/ZnS molar ratio, the amount of ZnO*2$ZnSO_4$ in solid products increases steeply. The temperature of ZnO*2$ZnSO_4$ decomposition (disappearing from the solid products) increases with the increasing $O_2$/ZnS ratio (between 1.5 and 1.83).

**Table 3:** Composition of gases and solid products in equilibrium (after the minimization of Gibbs energy for the system) . Input conditions : 1 mole ZnS, 100 moles of gas ($O_2$: 5 moles, $H_2O$: 0.5 moles, balance $N_2$).

| T (K) | $O_2$ (mol %) | $SO_2$ (mol %) | $SO_3$ (mol %) | $H_2SO_4$ (mol %) | ZnO (mol) | ZnO*2$ZnSO_4$ (mol) |
|---|---|---|---|---|---|---|
| 900 | 3.123 | 0.148 | 0.1903 | 1.23 E-05 | 0 | 0.3333 |
| 1000 | 3.172 | 0.2507 | 0.0879 | 1.68 E-06 | 0 | 0.3333 |
| 1050 | 3.187 | 0.2818 | 0.05672 | 6.47 E-07 | 0 | 0.3333 |
| 1100 | 3.197 | 0.3018 | 0.0367 | 2.63 E-07 | 0 | 0.3333 |
| 1200 | 3.475 | 0.95 | 0.05 | 1.59 E-07 | 1 | 0 |

At an $O_2$/ZnS ratio higher than 1.83, there is only formation of ZnO*2$ZnSO_4$ at temperatures between approx. 900 K and 1150 K. It means that for the complete decomposition of this basic Zn-sulfate, temperatures over approx. 1150 K are needed. With the increasing $O_2$/ZnS molar ratio (over approx. 3), $SO_3$ concentrations are increasingly important. Concentrations of Zn(g) in gas phase are negligible.

The overall „map“ of ZnS regeneration (regions in a graphics with $O_2$/ZnS molar ratio vs. temperature) is shown in **Fig. 1**. Only at temperatures below 900 K and at $O_2$/ZnS input molar ratio over 1.83, the formation of $ZnSO_4$ is thermodynamically possible.

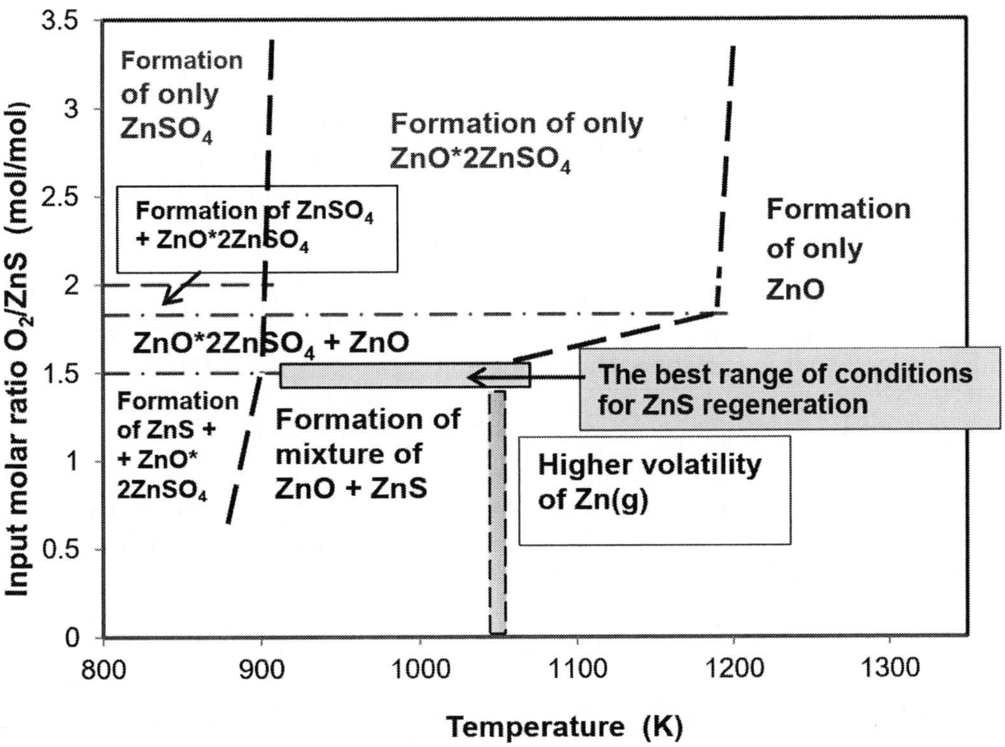


**Fig. 1** Map of products (approximate dependence on operating conditions, temperature and $O_2/ZnS$ input molar ratio) in the regeneration of ZnS to zinc oxide (gas with 1 – 5 vol. % $O_2$).

As obvious, there is only very limited range of conditions for the selective and complete regeneration of the spent ZnO-based sorbent without formation of undesired products (Zn(g) and $ZnO*2ZnSO_4$) at lower temperatures (below 1000 K – 1050 K). It requires $O_2/ZnS$ input molar ratio tightly (closely) about 1.5. At a slightly lower molar ratio there is incomplete ZnS regeneration, at a slightly higher $O_2/ZnS$ ratio than 1.5, the formation of stable $ZnO*2ZnSO_4$ starts. Complete decomposition of $ZnO*2ZnSO_4$ requires high temperature over 1150 K – 1200 K with probable sintering of the sorbent material and generally with deterioration of sorption properties of the regenerated sorbent.

The presence of water vapor in gas under reducing conditions ($O_2/ZnS < 1.5$) causes the formation of small concentrations of $H_2S$ and hydrogen. Under strongly oxidizing conditions ($O_2/ZnS > 3$) the presence of water vapor has generally the only effect on the formation of vapors of sulfuric acid from $SO_3$.

The convenient conditions for ZnS regeneration are either generally at temperatures over 1200 K and at a molar ratio $O_2/ZnS > 1.6$ or at the temperatures 950 – 1050 K in a very narrow region of $O_2/ZnS \approx 1.5$. Regenerative technology based on those conditions is, however, very sensitive to changes of the operating conditions and thus is not robust.

## 3 Chemistry and thermodynamic analysis of $Ce_2O_2S$ regeneration to $Ce_xO_y$

For regeneration of $Ce_2O_2S$, the reaction with water vapor or with gaseous $SO_2$ can be used. The reaction with water vapor (reaction 10) could be used rather at temperatures above

$$Ce_2O_2S + 2H_2O \leftrightarrow 2CeO_2 + H_2S + H_2 \qquad (10)$$

800 K to produce gas with content of $H_2S$ exceeding 10 vol. % for the case of regeneration by pure (non-diluted) inlet steam. If the spent Ce-sorbent contains also $CeCl_3$ and $CeF_3$, they both can be decomposed by non-diluted steam at temperature over 900 K.

Higher concentrations of sulfur products (over 50 %) can be achieved by regeneration reaction with $SO_2$ leading to production of elemental sulfur:

$$Ce_2O_2S + SO_2 \leftrightarrow 2CeO_2 + (2/x)\, S_x \qquad (11)$$

with practically only one important side reaction:

$$Ce_2O_2S + 5SO_2 \leftrightarrow Ce_2(SO_4)_3 + (3/x)S_x \qquad (12)$$

The Gibbs energy $\Delta G_r$ for reaction 12 with formation of only gaseous sulfur products is -45.54 kJ/mol at 700 K, and -30.84 kJ/mol at 1100 K. In the temperature range between approx. 500 and 1100 K, there is a mixture of $S_x$ molecules, where $2 \leq x \leq 8$. If we consider only sulfur molecules $S_2$, $S_4$, $S_6$, and $S_8$, then the approx. dependence of sulfur molecule size (sulfur molecule speciation $S_x$) on temperature can be expressed as shown in **Fig. 2**. The sulfides ZnS and $La_2O_2S$ are unsuitable for such regeneration because of thermodynamic constraints (high positive values of $\Delta G_r$ for the regeneration reactions):

$$2ZnS + SO_2 \leftrightarrow 2ZnO + (3/x)S_x \qquad (13)$$
$$2La_2O_2S + SO_2 \leftrightarrow 2La_2O_3 + (3/x)S_x \qquad (14)$$

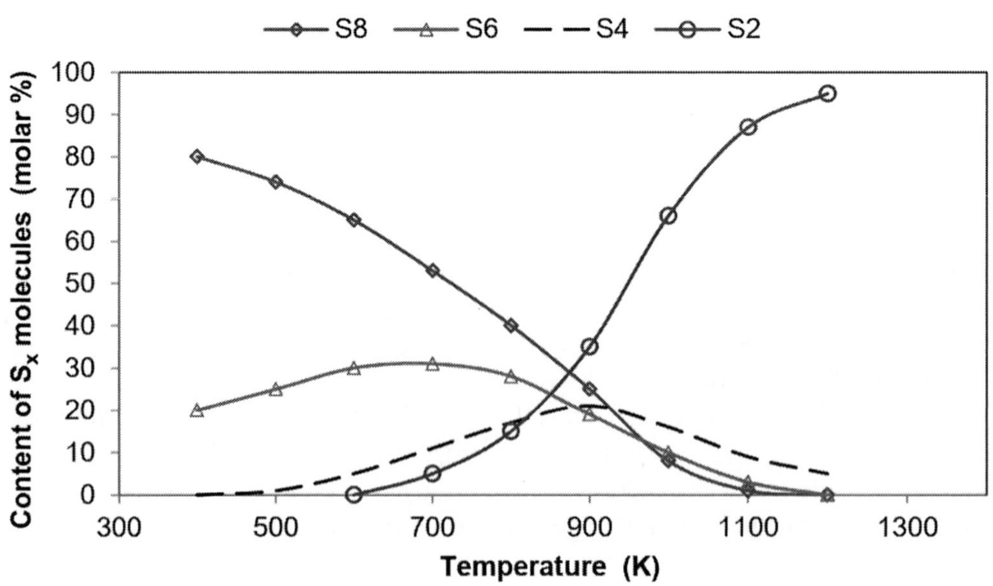


**Fig. 2:** Approximate dependence of sulfur molecule size ($S_x$) distribution in vapor phase on temperature [32,33].

Results of computation of the minimization of Gibbs energy for the system conaining $Ce_2O_2S$ with 1 vol. % and 10 vol. % $SO_2$ in gas phase are compared in **Table 4.**

**Table 4:** Equilibrium composition from the minimization of Gibbs energy for the system of $Ce_2O_2S$ regeneration with $SO_2$. Input – case A: 100 moles of gas with $SO_2$ (10 moles), 90 moles of $N_2$ and 5 moles of $Ce_2O_2S$. Case B: input 100 moles of gas with 1 mole of $SO_2$ , 99 moles of $N_2$ and 0.5 mole of $Ce_2O_2S$.

| T (K) | Case A Prevailing S -form | Case A $SO_2$ (mol) | Case A $Ce_2(SO_4)_3$ (mol) | Case B Prevailing S -form | Case B $SO_2$ (mol) | Case B $Ce_2(SO_4)_3$ (mol) |
|---|---|---|---|---|---|---|
| 700 | Mixture: $S_2$, $S_4$, $S_6$, $S_8$ | 4.999 | 0 | Mixture | 0.4998 | 0 |
| 800 | $S_2$ and $S_4$ | 4.995 | 0 | $S_2$ | 0.4996 | 0 |
| 900 | $S_2$ | 4.991 | 0 | $S_2$ | 0.4995 | 0 |
| 1000 | $S_2$ | 4.989 | 0 | $S_2$ | 0.4993 | 0 |

The solid product from the regeneration theoretically contains only $CeO_2$. Particularly at higher concentrations of $SO_2$ in gas, the formation of sulfur sub-oxide ($S_2O$) was proved. The dependence of $S_2O$ concentrations on temperature for two different input concentrations of $SO_2$ in gas phase are shown in **Fig. 3**. At temperatures over 700 K and at normal atmospheric pressure, no formation of liquid elemental sulfur and no formation of cerium(III) sulfate were thermodynamically proved (confirmed).

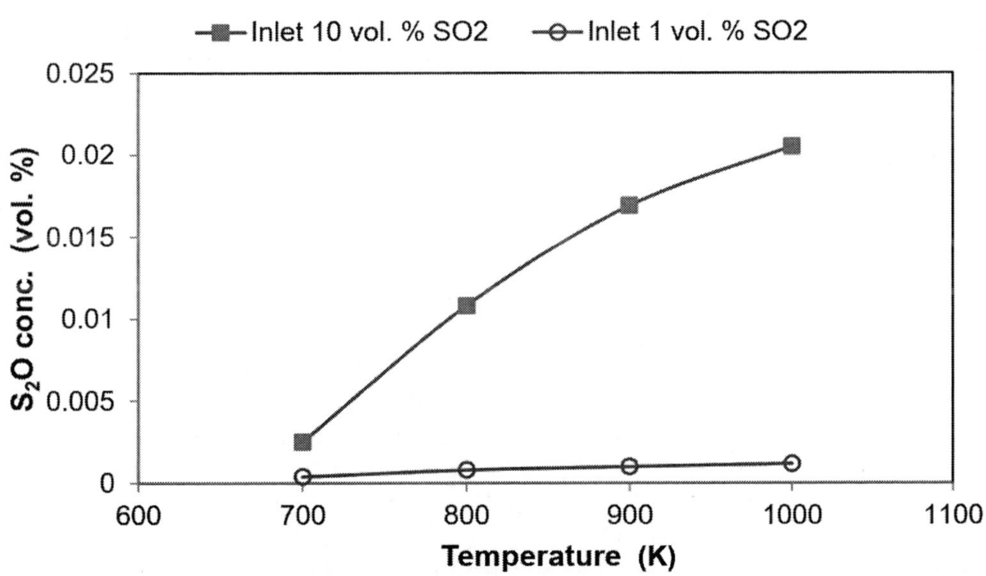


**Fig. 3**: Dependence of equilibrium $S_2O$ concentrations in gas phases on temperature in the regeneration of $Ce_2O_2S$ by $SO_2$ for 10 vol. % and 1 vol. % $SO_2$ as input concentrations in the gas.

The regeneration of $Ce_2O_2S$ by $SO_2$ is not much exothermic and the selectivity is high at temperatures over 700 K and at $SO_2/Ce_2O_2S$ input molar ratios lower than approx. 3 (no formation of Ce-sulfates). Only at high $SO_2/Ce_2O_2S$ input molar ratios, at higher operating pressures and temperatures below approx. 750 K, the formation of $Ce_2(SO_4)_3$ according to reaction 11can be significant. This kind of Ce-sorbent regeneration is relatively robust and without bigger operating problems. On the other hand, if the spent Ce-based sorbent contains cerium halides (e.g. $CeF_3$, $CeCl_3$), they are not decomposed and regenerated to $CeO_2$ by $SO_2$ [25], as thermodynamics predicts. In such a case rather regeneration of $Ce_2O_2S$ based on reaction with steam with formation of $H_2S$ and $CeO_2$ should be preferred. The cerium halides are decomposed under steam regeneration conditions.

Regeneration of ZnS and $La_2O_2S$ by this method is thermodynamically very limited and practically not feasible (**Table 5**). In the mixed sorbents containing $Ce_2O_3$, if the prevailing part of sulfur is bound with Ce, such sorbents are also suitable for regeneration by $SO_2$.

**Table 5**: Comparison of the $\Delta G_r$ data for sorbent regeneration reactions by $SO_2$ leading to formation of elemental sulfur.

| Spent sorbent | Regeneration reaction | $\Delta G_r$ (kJ/mol) 500 K | 1100 K |
|---|---|---|---|
| $Ce_2O_2S$ | $Ce_2O_2S + SO_2 = 2CeO_2 + 2/n\ S_n$ | -58.08 | -27.45 |
| $La_2O_2S$ | $2La_2O_2S + SO_2 = 2La_2O_3 + 3/n\ S_n$ | 61.36 | 165.08 |
| ZnS | $2ZnS + SO_2 = 2ZnO + 3/n\ S_n$ | 102.02 | 135.47 |

## 4 Chemistry and thermodynamic analysis of $La_2O_2S$ regeneration to $La_2O_3/La_2O_2CO_3$

The regeneration of $La_2O_2S$ by oxidation is complicated by formation of very stable oxy-sulfate $La_2O_2SO_4$ [17,19,24]. Regeneration by means of $SO_2$ is not thermodynamically preferred (**Table 5**). Therefore, for practical applications, only the regeneration based on the reaction with steam and $CO_2$ remains (reaction 15). This reaction and equilibrium $H_2S$ concentrations can be influenced by operating pressure.

$$La_2O_2S(s) + H_2O(g) + CO_2(g) \leftrightarrow La_2O_2CO_3(s) + H_2S(g) \qquad (15)$$

The results of computations of equilibrium $H_2S$ concentrations based on assumption of unit activities of solid phases [34] are shown in **Fig. 4**.

Attaining of high $H_2S$ concentrations in gas (over 5 vol. %) requires rather lower temperatures (below 800 K) and higher operating pressure (over approx. 1 MPa). By this regeneration, $La_2O_2CO_3$ is formed which is stable in producer gas up to temperatures about 1100 K. The equilibria for $H_2S$ removal by $La_2O_2CO_3$ are less favorable than in the case of $La_2O_3$, but under realistic conditions (producer gas with 10 – 30 vol. % $CO_2$) majority of $La_2O_3$ based sorbents will be anyway, at least partly, transformed into oxy-carbonate ($La_2O_2CO_3$).

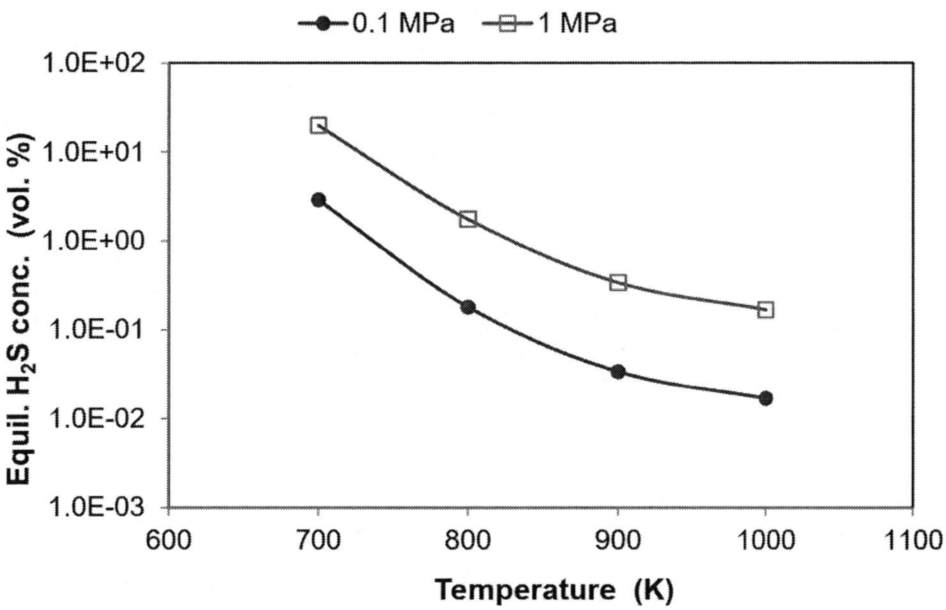


**Fig. 4**: Attainable equilibrium concentrations of $H_2S$ in the regeneration of $La_2O_2S$ by an equimolar mixture of steam and $CO_2$ (50 vol. % water vapor and 50 vol. % $CO_2$ in gas phase).

If LaOF and/or LaOCl are present together with $La_2O_2S$ in the spent sorbent [25], the regeneration process based on the reaction with $CO_2$ and steam is able to convert the La-halides into $La_2O_2CO_3$, particularly at temperatures above 900 K and at higher operating pressures.

## 5 Conclusions

Regeneration of ZnO, $Ce_2O_3/CeO_2$, and $La_2O_3/La_2O_2CO_3$-based sorbents after sulfidation can be accomplished by mild oxidation of the resulting sulfides, by reaction with $SO_2$ or by pressurized reactions with steam or mixtures of steam with $CO_2$.

The Zn-based spent sorbents should be regenerated by mild oxidation at temperatures between approx. 950 and 1050 K and at the $O_2$/ZnS molar ratio about 1.5 to avoid problems with the formation of $ZnO*2ZnSO_4$ and excessive volatilization of zinc vapors (at the $O_2$/ZnS molar ratio slightly below 1.5). Practical application of such (narrow region) regeneration conditions is, however, very difficult. Regeneration at high temperatures (above 1200 K) is a robust technological process leading to ZnO, but with inevitable consequences on textural properties of the sorbent. Regeneration with $SO_2$ has thermodynamic constraints and is, therefore, very difficult or even impossible. Regeneration by steam results in very low $H_2S$ concentrations.

The Ce-based sulfur sorbents can be advantageously regenerated by reaction with $SO_2$ leading to the formation of $CeO_2$ and vapors of elemental sulfur at temperatures over 700 K. The only (minor) side product is sulfur sub-oxide. Reaction of $Ce_2O_2S$ with steam, if considered for application, should be operated at temperatures over 800 K. Mild oxidation of $Ce_2O_2S$ could also be taken into consideration, but due to significant heat effects, the changes of texture of the sorbent are very likely. The main problem in the sulfidation-regeneration cycle of the cerium oxide-based sorbents is necessity of the consecutive reduction of less efficient $CeO_2$ to the efficient $Ce_2O_3$ form in the sorbent and difficult regeneration of cerium halides ($CeF_3$,$CeCl_3$) to cerium oxides.

The $La_2O_3$/$La_2O_2CO_3$-based spent sorbents are not suitable for the regeneration by oxidation because $La_2O_2SO_4$ is stable up to high temperatures (over 1200 K). The regeneration with the use of $SO_2$ is not thermodynamically preferred. Therefore, only one method of regeneration is recommendable – the reaction with steam and $CO_2$ under pressurized conditions. This reaction promises theoretically attainable $H_2S$ concentration in resulting gas over 5 vol. % at temperatures below 800 K and at higher operating pressures (above 1 MPa). Potential small contents of La-oxy-halides (LaOF and LaOCl) in the spent sorbent are converted into $La_2O_2CO_3$ in such a regeneration process.

***Acknowledgments***

*The authors appreciate financial support from the Grant Agency of Czech Republic (GAČR), bilateral grant project of GAČR and National Science Council (NSC) Taiwan, Registr. No. in CR: 14-09692J (Registr. No. of foreign project 102WBS0300011) and Grant No. in Taiwan: NSC 103-2923-E-042A-001 -MY3.*

**References**

1. P. J. Woolcock and R. C. Brown, *Biomass and Bioenergy*, **52**, 54 (2013).
2. P. V. Aravind and W. de Jong, *Progress in Energy and Comb. Sci.*,**38**, 737 (2012).
3. I. Valsamakis, R. Si and M. Flytzani-Stephanopoulos, *J.of Power Sources*, **195**, 2815 (2010).
4. M. Flytzani-Stephanopoulos, M. Sakbodin and Y. Wang, *Science*, **312**, 1508 (2006).
5. K. M. Dooley, V. Kalakota and S. Adusumilli, *Energy and Fuels*, **25**, 1213 (2011).
6. V. Girard, A. Baudot, D. Chiche, D. Bazer-Bachi, C. Bounie and C. Geantet, *Fuel*, **128**, 220 (2014).
7. V. Girard, D. Chiche, A. Baudot, D. Bazer-Bachi, I. Clémoncon, F. Moreau and C. Geantet, *Fuel*, **140**, 453 (2015).
8. J. M. Sánchez-Hervás, J. Otero and E. Ruiz, *Chem. Eng. Sci.*, **60**, 2977 (2005).
9. V. R. Kalakota, Sulfur removal using regeneragle sorbents of rare earth/Transition metal oxides, *Ph.D. Thesis, Dep. of Chem. Eng., Louisiana State University*, USA (2008).
10. O. A. Sofekun, A kinetic study of the sulfidation-regeneration of zinc oxide sorbent, *Ph.D. Thesis, Dep. of Chem. Eng., Iowa state University*, Iowa, USA (1996).
11. D. P. Harrison, F. R. Groves, W. N. Huang, A. Lopez-Ortiz, J. D. White, Y. Zeng and S. Zhang, Advanced sulfur control concepts for hot gas desulfurization technology, *Final Report Dep. Of Chem. Eng. Louisiana State University*, Baton Rouge LA, USA (1998).
12. Y. Feng, J. Dou, A. Tahmasebi, J. Xu, X. Li, J. Yu and F. Yin, *Energy and Fuels*, **29**, 7124 (2015).
13. J. T. Konttinen, C. P. A. Zevenhoven, K. P. Yrjas and M. Hupa, *Ind. End. Chem. Res.*,**36**, 5432 (1997).
14. J. Huang, J. Zhao, X. Wei, Y. Wang and X Bu, *Powder Technol.*, **180**, 196 (2008).
15. Y. Zeng, S. Zhang, F. R. Groves and D. P. Harrison, *Chem. Eng. Sci.*, **54**, 3007 (1999).
16. F. Jones; H. Tran; D. Lindberg; L. Zhao and M. Hupa, *Energy and Fuels*, **27**, 5663 (2013).
17. J.A. Poston, R. V. Siriwardane, E. P. Fisher and A. L. Miltz, *Applied Surface Science*, **214**, 83 (2003).

18. L. Lisi, G. Mancino and S. Cimino, *Internat. J. of Hydrogen Energy*, **40**, 2047 (2015).
19. K. H. Stern, *High temperature properties and thermal decomposition of inorganic salts with oxyanions*. London: CRC Press (2001), ISBN 0-8493-0256-0.
20. K. Svoboda, M. Hartman, J. Havlica, J. Leitner, M. Pohořelý, M. Šyc and P. C. Chen, Thermodynamics of HCl and HF Interferences in Hot Producer Gas Desulfurization by Zinc, Cerium and Lanthanum Oxide Based Sorbents. *Internat. Conf. on Chem. and Biochem. Engineering*. Proceedings - Abstracts-book, CE095, p. 202, Paris-Saint Denis,France, 20-22 July (2015).
21. K. Foge, M. Hoang and T. W. Turney, *J. of Materials Science*, **27**, 77 (1992).
22. A. N. Shirsat, M. Ali, K. N. G. Kaimal, S. R. Bharadwaj and D. Das, *Thermochim Acta*, **399**, 167 (2003).
23. B. Bakiz, F. Guinneton, M. Arab, A. Benhlhachemi, S. Villain, P. Satre and J. R. Gavarri, Carbonation and decarbonation kinetics in $La_2O_3$-$La_2O_2CO_3$ system under $CO_2$ gas flows. *Advances in Materials Science and Engineering* (2010), Article ID 360597, 6 pages, doi:10.1155/2010/360597.
24. B. J. Feist, Hot Gas Desulfurization with Gamma-Alumina Supported Lanthanum Oxide. *Ph. D. Thesis, Dep. of Chem. and Petrol. Engineering, University of Calgary*, Canada, (2014).
25. K. Svoboda, M. Hartman, J. Havlica, J. Leitner, M. Pohořelý, M. Šyc and P. C. Chen, Thermodynamics of HCl and HF Interferences in Hot Producer Gas Desulfurization by Zinc, Cerium, and Lanthanum Oxide-based Sorbents, *Chem. Papers*, submitted (2015).
26. R. Holub and P. Voňka, *Chemical Equilibrium of heterogeneous and condensed systems* (in Czech). Praha, CZ: Academia (1984).
27. D. S. Ebel, M. S. Ghiorso, R. O. Sack and L. Grossman, *J. of Computat. Chemistry*, **21**, 247 (2000).
28. P. Voňka and J. Leitner, *CALPHAD*, **19**(1), 25 (1995).
29. I. Barin and O. Knacke, *Thermochemical data of pure substances*. 3rd ed. Weinheim, VCH (1995).
30. O. Knacke, O. Kubaschewski and K. Hesselmann, *Thermochemical Properties of Inorganic Substances*. 2nd Ed. Berlin, Springer (1991).
31. E. Sasaoka, M. Hatori and N. Sada, *Ind. Eng. Chem. Res.*, **39**, 3844 (2000).
32. *Ullmann´s encyclopedia of Industrial Chemistry*, 5-th edition, (sulfur molecule speciation). VCH Verlagsgesellschaft, Germany (1987).
33. B. Meyer, *Chemical Reviews*, **76**(3), 367 (1976).
34. K. Svoboda, G. Slowinski, J. Rogut and D. Baxter, *Energy Conversion and Manag.*, **48**, 3063 (2007).

# Particle fragmentation behavior in a bench scale entrained-flow coal-water slurry gasifier

Qinghua GUO, Guangsuo YU*, Yifei WANG, Xin GONG

*Key Laboratory of Coal Gasification and Energy Chemical Engineering of Ministry of Education, Shanghai Engineering Research Center of Coal Gasification, East China University of Science and Technology, Shanghai 200237, PR China*

**Abstract:** Fragmentation behavior of coal char particles in an entrained-flow coal-water slurry gasifier is essential for the gasifier slag discharging process and the downstream syngas primary clean-up system. The phenomenon of particle fragmentation process occurring in the gasifier captured by a high speed video camera on a bench scale opposed multi-burner (OMB) gasifier illustrated that the secondary fragmentation behavior exists in the entrained-flow gasifier after the particles moving from high temperature impinging flame region to the gasification chamber outlet region. Particles larger than 200 μm in the impinging flame region have porosity structures with fragile shapes, while most of the particles smaller than 100 μm have spherical shapes. As long as the coal char particles move from the impinging flame region toward the gasification chamber outlet region, the primary fragmentation is almost finished in the impinging flame region and the secondary fragmentation is probably taken place as a result of percolative fragmentation and undergoing gasification reactions. The particle size distributions under different oxygen to carbon ratios (O/C) also indicate that there has a significant fragmentation when the particle size is larger than 200 μm.

**Keywords:** Coal gasification; Fragmentation; Coal char; Particle size distribution

* Corresponding author. Tel: +86-21-64252974. Fax: +86-21-64251312. E-mail: gsyu@ecust.edu.cn

## 1. Introduction

Entrained-flow coal gasification process is a partial oxidation process with an average particle size normally smaller than 100 μm [1]. In terms of the coal-water slurry gasification with quenching process, all the fly ash, char, slag and other granular materials generated in the gasifier will be discharged through the quench chamber by the quenching water. The ash/slag formation during entrained-flow coal gasification is one of the major factors influencing the deposition and slag discharging behaviors in the gasifiers [2-3], which may also have significant influence on the efficiency of primary syngas cleanup process and downstream processes with catalyst utilization.

Ash/slag particle formation during coal gasification is a complex process that heavily depends on gasification conditions and complicated coal char fragmentation behavior. Coal char fragmentation is an important factor that should be considered during the ash/slag formation process. Ash formation during pulverized coal combustion has been systematically studied using bituminous coal with particle sizes between 63 and 90 μm [4, 5]. Studies on the char-slag properties and the formation mechanism during the coal gasification process have been presented as well [6-8]. Several studies have experimentally quantified particle fragmentation during the coal devolatilization in the fluidized bed combustion while the operating temperature is normally below 1273 K [9, 10]. During the coal combustion and/or gasification process, the porous structure of coal char plays an important role in the coal char particle shrinkage and fragmentation [11]. E. Bar-Ziv and I.I. Kantorovich have discussed the shrinkage, fragmentation and porosity processes and properties based on the mutual effects of char porosity and reactivity [12]. Besides, theoretical analysis and simulation work have been carried out to identify the mechanisms for particle fragmentation in the high temperature entrained-flow bed reactors [13-15]. Particle size distribution affected by particle fragmentation is mainly caused by many stresses and mechanisms including pressure-induced stress, thermal stress, etc. Thermal stress theory is considered as an important factor that affects the coal particle primary fragmentation. R. Chirone and L. Massimilla [16] have presented the Weibull theory about the primary fragmentation of coal during devolatilization. The starting point for anisotropy of fragmentation is a simple monoaxial stress field, consisting of internal forces which are opposite to resistance forces of the material due to volatile pressure. No and Syred [17] have proposed that thermal stresses will produce fragmentation when particles are smaller than 1000 μm, especially when the coal particles are larger than 100 μm. Fragmentation is the main cause of the decrease of particle size in entrained-flow pulverized coal combustion process [18]. Particle collision, agglomeration and percolative fragmentation are main performances in the particle interactions. Percolative fragmentation has been investigated directly on over 200 coal and char particles in the pulverized size range [19].

However, the experimental studies on coal char fragmentation in the entrained-flow gasifier are rarely seen in the literature. In terms of the application of the entrained-flow coal-water slurry gasification, a proper understanding of the gasification behavior of coal particles, especially the particle fragmentation which may have effect on the slag discharging, is essential to the successful design and operation of a coal gasifier. The Opposed Multi-Burner (OMB) gasification technology, one of the representative entrained-flow gasification technologies, has been widely used in China [20]. In order to provide further effective technical support for the application of OMB technology, experimental work has been carried out to characterize the particulate matters and slag deposition properties in a bench-scale OMB gasifier [21, 22]. In this paper, particle samples are taken directly and subsequently Scanning Electron Microscope (SEM) and particle size distribution tests are performed under different operating conditions. Coal char particle fragmentation behavior inside the gasifier is presented as well.

## 2. Experimental work

### 2.1 Experimental setup

Experimental study was carried out in an entrained-flow OMB bench scale coal-water slurry gasifier. The schematic diagram can be seen in the literature [23, 24], feeding stock coal used for the current investigation is similar to that used in previous studies. The proximate, ultimate, sink/float, and ash analysis for the coal are presented in Table 1.

**Table 1 Properties of coal used in the experiment (dry basis, wt %)**

| | | |
|---|---|---|
| Proximate Analysis | | |
| | ash | 10.10 |
| | volatile matter | 36.96 |
| | fixed carbon | 52.94 |
| Ultimate Analysis | | |
| | carbon | 73.51 |
| | hydrogen | 4.90 |
| | nitrogen | 1.26 |
| | sulfur | 1.61 |
| | oxygen | 8.62 |
| Minerals Distribution by Sink/Float Analysis | | |
| | included | 81.44 |
| | excluded | 18.56 |
| Ash Analysis | | |
| | $SiO_2$ | 46.42 |
| | $Al_2O_3$ | 33.74 |
| | $Fe_2O_3$ | 6.78 |
| | NaO | 4.07 |
| | CaO | 3.25 |
| | $SO_3$ | 2.44 |
| | MgO | 1.09 |
| | $TiO_2$ | 1.03 |
| | Others | 1.18 |

In order to capture high temperature flame and particle images inside the gasifier, a Φ38 mm endoscope with cooling water jacket combined with the Photron high speed camera was employed and mounted on the top of the gasifier. Monochrome CMOS sensor and monochromatic filter were used in this camera to capture the radiation gray-scale images under certain shutter speeds and exposures.

The experimental conditions were conducted under different oxygen to carbon atomic ratios (O/C). During the experiments, each burner's coal-water slurry flow rate was held constant at 10 kg/h with a weight concentration of 60% coal. Oxygen flow rates varied from 3.7 $m^3$/h to 4.6 $m^3$/h under normal operating pressure, corresponding to the O/C ratios from 0.9 to 1.1, the average temperatures in the impinging flame region were about 1600 K, respectively. Meanwhile, the three corresponding temperatures in the gasification chamber outlet region were about 1320 K [23].

**2.2 Particle sampling and analysis**

The particle sampling system consists of sampling tubes with cooling water jacket followed by a filter, cooler, dryer, rotameter and vacuum pump [21, 22]. In order to get the whole size range of the particles, inner diameter of the sampling tube is 3 mm. Two sampling ports were selected for this work, 1# is located near the impinging flame region which is 100 mm below the burner plane; 2# near the gasification chamber outlet region which is 1200 mm below the burner plane. Particle size distribution and average particle size of the samples were analyzed by a Malvern Particle Size Analyzer after sampling.

Samples in the two regions exhibit different behaviors due to the coal gasification reaction mechanisms [21]. Coal-water slurry feeding in the gasifier firstly enters into the high temperature impinging flame region where the devolatilization, combustion and gasification mainly occur. Physical and chemical behaviors of coal particles in the impinging flame region are complicated due to the gasification, combustion, fragmentation, collision, shrinking or swelling which will cause the coal char particle size changes. Primary fragmentation can take place as a result of devolatilization or thermal stress in the impinging flame region where coal particles are heated and react. We assume the primary fragmentation process of the particles sampled from sampling port 1# has finished, and then the analysis along the gasifier below the impinging flame region is mainly focus on the secondary fragmentation.

## 3. Results and discussion

### 3.1 Fragmentation visualization in the gasifier

Direct optical visualization offers an important method of coal char particle fragmentation [25]. However, it is hard to record a clear process in a real gasifier due to high temperature, high moisture concentration and complicated media inside the gasifier. Occasionally, the fragmentation behavior inside the gasifier is observed and captured by the real time visualization method according to many times experiment. Using the high temperature endoscope coupled with a high speed video camera, Fig. 1 shows a particle fragmentation process inside the gasifier. Particle in the frame of 0.25 ms was fragmented into several pieces in the frame of 0.75ms. The visualized fragmentation location is probably taken place away from the impinging flame region where the primary fragmentation should be finished according to the set position of the camera focus plane.

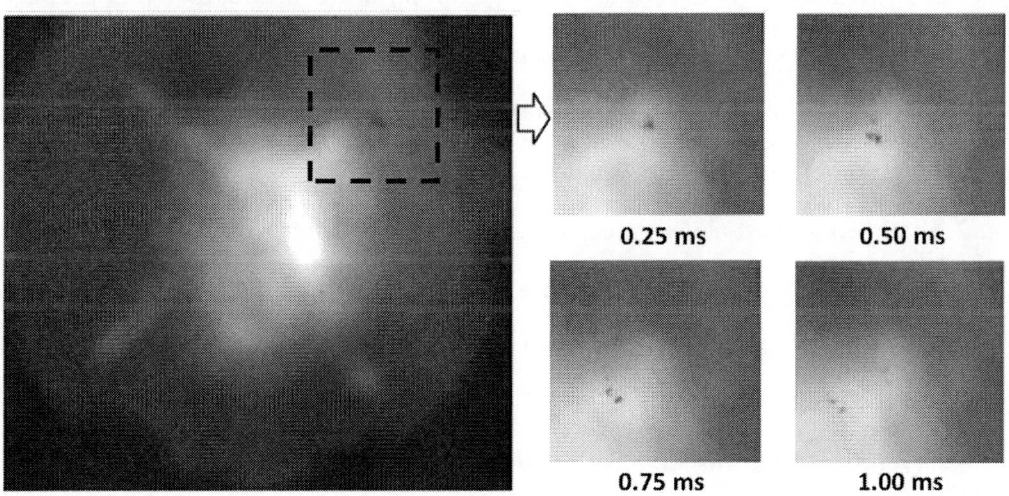


**Fig. 1 Coal char particle fragmentation process inside the gasifier**

Despite the fragmentation phenomenon can be captured inside the gasifier, it is impossible to capture the whole particle size range fragmentation behaviors due to the camera resolution and other strict limited conditions. Using visualization method cannot easily give a quantity analysis of the particle size changes either. It is necessary to study the secondary fragmentation mechanisms inside the gasifier based on the sampling analysis method.

### 3.2 Coal char fragmentation analysis

The decrease of coal char particle size can be caused by several reasons, including the mineral fragmentation, gasification-induced shrinkage, thermal stress fragmentation, percolative fragmentation, etc. Coal char particles formed during the gasification process contain a certain amount of minerals and residual carbon [26]. Consequently, coal char fragmentation is the result of mineral interaction and coal char gasification which may cause the particle size changes. The amount of the mineral matter is substantially constant, therefore only the particle size distribution changes during the mineral fragmentation and agglomeration processes. Usually, the fragmentation occurs in the excluded minerals while the agglomeration occurs with the included minerals [27].

In order to analyze the effects of included and excluded minerals on fragmentation, the low temperature ash sample which can prevent from destroying mineral structure due to the effect of high temperature preheating was prepared. Samples from different sampling ports under the oxygen/carbon atom ratio of 1.0 were prepared by the low temperature plasma ashing method and then compared to the low temperature ash particle sizes and the coal char particle sizes. As shown in Fig. 2, the average particle size decreases along the gasifier from top to the bottom while the low temperature ash particle size changes are very small. The mineral particle size decreases in the impinging flame area when particles move from the burner plane to the near plug flow region. Thus it can be concluded that in the impinging flame area the degree of fragmentation of the excluded minerals is greater than that of agglomeration of the included minerals. However, in the low temperature ash, particle size near the impinging flame region is 19.5 μm while the outlet is 16.6 μm. The mineral fragmentation contributes less to the particle size decreasing during the coal gasification process due to this small difference.

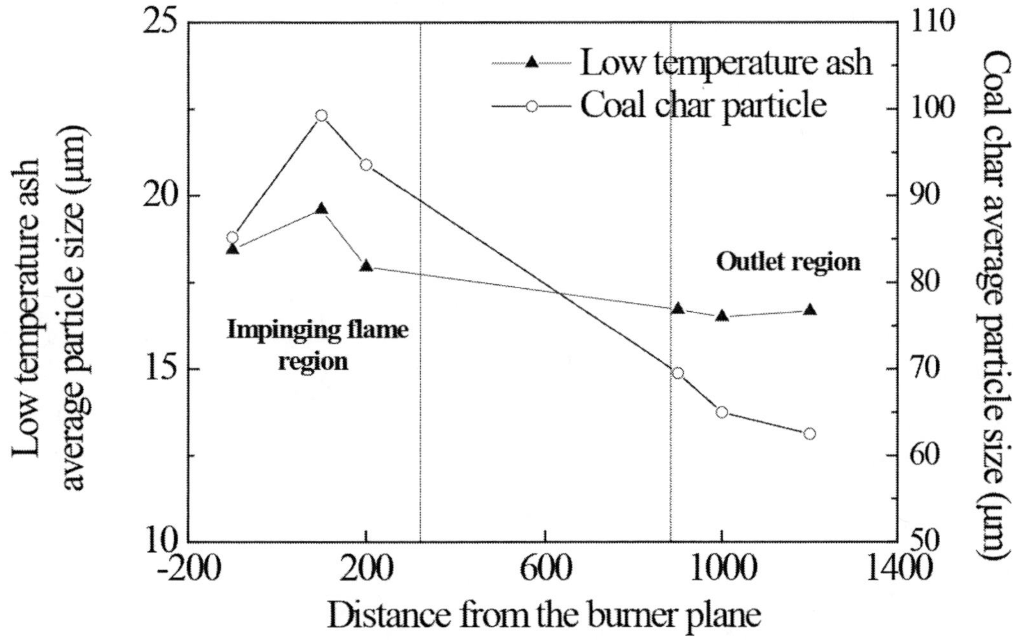


**Fig. 2 Average sizes of coal char particle and its low temperature ash**

Additionally, samples from sampling ports 1# and 2# were dried and sieved into six ranges (>200 μm, 100~200 μm, 75~100 μm, 50~75 μm, 38~50 μm and <38 μm), the weight percentage and residual carbon content with different ranges are shown in Fig.3. The residual carbon content of the particles from the impinging flame region is very high and close to the carbon content of raw coal, which is much higher than that of the particles from the gasification chamber outlet region. The smaller particle size the lower residual carbon content in particles from the gasification chamber outlet region. As long as the particles move toward the gasifier outlet region, residual carbon inside the coal char particles is further reacted. But the residual carbon gasification reaction cannot be completely finished inside the gasifier mainly due to the limited residence time. The proportion of different sizes in sampling 1# and 2# changes greatly, especially for the particle size smaller than 50 μm and greater than 100 μm. The proportion of particle sizes smaller than 50 μm increases sharply, whereas the proportion of particle sizes larger than 100 μm decreases more than half when particles move from the burner plane region to the gasification outlet region. This phenomenon is in accordance with the previous assumption that the particle primary fragmentation mainly occurs in the impinging region. So the secondary fragmentation and carbon reaction should be the main reasons leading to the particle size decrease.

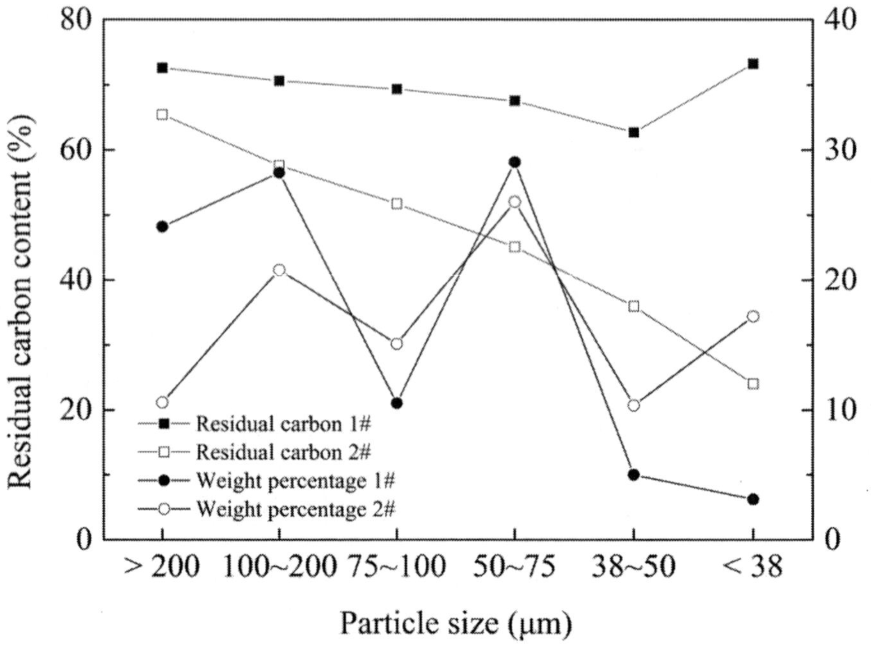


**Fig. 3 Coal char particle residual carbon content and weight percentage**

Coal char gasification reaction occurs in the gasifier with $CO_2$ and $H_2O$ leading to the particle size decreases. In order to analyze the gasification-induced shrinkage, assuming homogeneous spherical particles with a given density do not change their features when they undergo fragmentation, a simple method can be established for average diameter size based on the experimental results. When a coal char particle in the gasifier moves from position *A* to *B*, the conservation of mass is below:

$$M_a = M_b + (M_a C_a - M_b C_b)$$

where, $M_a$ and $M_b$ represent particle mass in positions *A* and *B*; $C_a$ and $C_b$ represent particle residual carbon content in positions *A* and *B*, respectively.

With the assumption that the particle density is constant, particle diameters of $D_a$ and $D_b$ are

$$D_b = D_a \left( \frac{1-C_a}{1-C_b} \right)^{1/3}$$

Here, sampling 1# is selected as position *A* with an average particle diameter 98.7 μm, and sampling 2# is selected as position *B*. Experimental results show that the residual carbon contents are 69.8% and 46.3%, respectively. The calculated value of $D_b$ is 81.5 μm which is higher than the experimental result of 61.5 μm. It can be seen that residual carbon gasification reaction is one of the reasons that causes the particle size decreasing from 98.7 μm to 81.5 μm. However, the experimental data 61.5 μm is much smaller than that of the calculated data. So, secondary fragmentation probably occurs during the particles moving process.

Figure 4 shows the particle morphologies with two different particle sizes sieved from sampling port 1#, one is larger than 200 μm, the other is smaller than 100 μm. Most of coal char particles larger than 200 μm have a porous structure with hollow spherical or ellipsoid shape which possesses the requirement of percolative fragmentation conditions, while most of the particles smaller than 100 μm are solid spherical particles. Higher porosity resistance against the thermal stress is strong but against the mechanical stress is weak. These bigger size particles are more fragile than the smaller size particles. The porous structure of coal char particles will increase the surface area and the char gasification reaction rate in the gasifier due to the high residual carbon content in the coal char particles. Water vapor and $CO_2$ in the gasifier can subsequently react with the residual carbon in the coal char resulting in a fragile particle structure and increasing the possibility of particle fragmentation. The reduced particle average size is mainly resulted from both coal char fragmentation and residual carbon gasification reaction.

Figure 5 shows the particle morphologies with two different size ranges sieved from sampling port 2#. It can be seen that in Fig.5a less particles are within the range of 100~200 μm, some of the coal char particles with hollow sphere and solid network structures in Fig.4a are broken into small piece as well as the solid network are progressively eroded until the solid phase becomes spatially discontinuous. These findings are in accordance with the Percolative fragmentation mechanism. Fine particles smaller than 38 μm with low residual carbon content are formed in the surface tension contraction of molten mineral matters released from the coal char particles. The smaller the particle size, the deeper gasification reaction occurs, corresponding to the lower residual carbon content in these spherical particles.

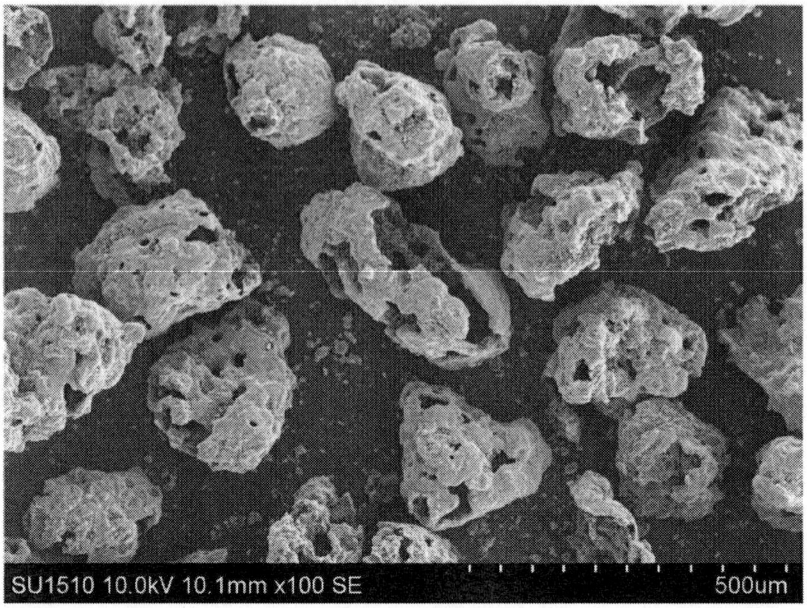


(a) >200 μm

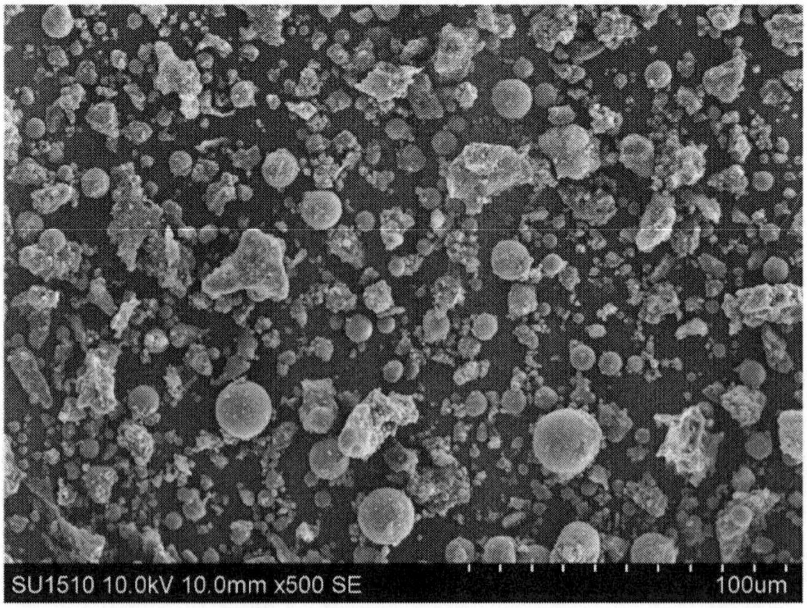


(b) <100 μm

**Fig. 4 SEM of different size particles from the impinging flame region**

(a) 100~200 μm

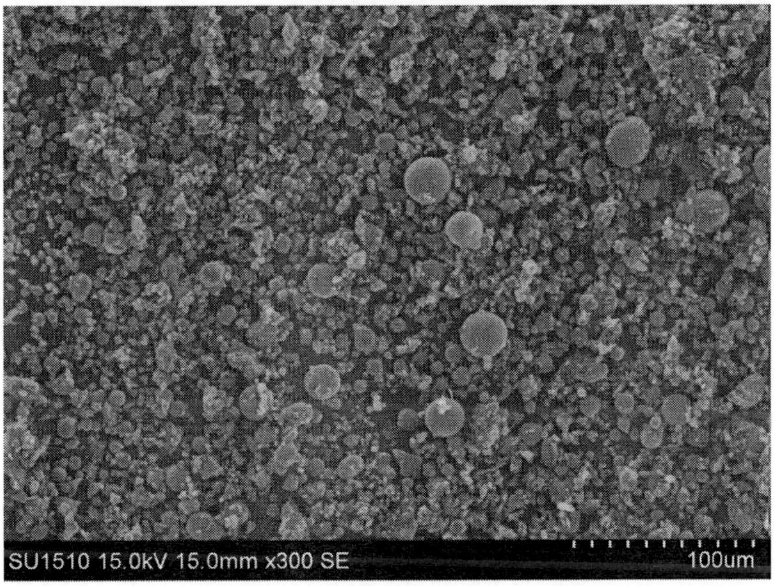


(b) <38 μm

**Fig. 5 SEM of different size particles from the gasification chamber outlet region**

### 3.3 Particle size distribution

Fragmentation has been inferred from measurements of particle size distributions. Comparison between the raw coal particle size distribution and the coal char particles from the two sampling ports under different O/C ratios are shown in Fig. 6. Coal char particles have a more wide range of particle size distribution compared with the raw coal. Without considering the impact of coal-water slurry atomization, particles swell to generate hollow sphere structures and agglomeration should be the main reasons. So, even though almost all the raw coal particle sizes are less than 200 μm in the present study, there is a big portion of sampling coal char particle sizes larger than 200 μm as well as such proportion in the high temperature flame region (+100 mm) is much higher than that in the gasification chamber outlet region (+1200 mm). Coal char particle size distribution in the impinging flame region is trimodal type with a main peak around 70 μm while in the gasification chamber outlet region it is bimodal type. Due to the coal char particles agglomeration, swelling, fragmentation, devolatilization and partial combustion occurring in the impinging flame region, particle size distribution in this region is more complicated than in the gasification chamber outlet region. It is complicated to study the coal char particles inside the impinging flame region. Particles near the impinging flame region have high residual carbon content, and the residue carbon is continuous consumed as long as the coal char particles move toward the gasification chamber outlet. With the residual carbon consumption and particle fragmentation, inherent mineral in the coal char particles will release to form as spherical particles, the main peak value of the particle size distribution decreases slightly.

Operating temperature has great impact on the coal ash fragmentation behavior [28]. Normally, the degree of particle fragmentation increases as temperature rises. Increasing the O/C ratio as well as the burner jet velocity is an effect of atomization improving, and subsequently the operation temperature rises as well. Temperature difference between the impinging flame region and the gasification chamber outlet region is around 250~400 K in terms of current operating conditions. As long as the temperature rises, coal char particles are easy to break, thus the trimodal trend becomes weakened in the impinging flame region, and the particle size distribution range becomes narrow for the particles in the gasification chamber outlet region.

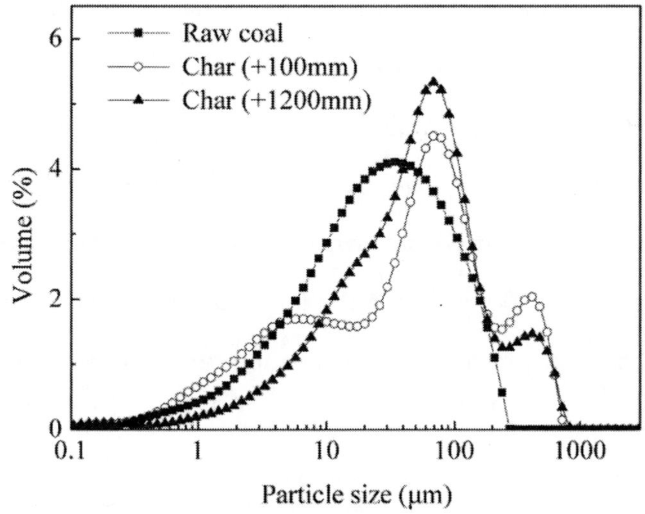


**(a) O/C=0.9**

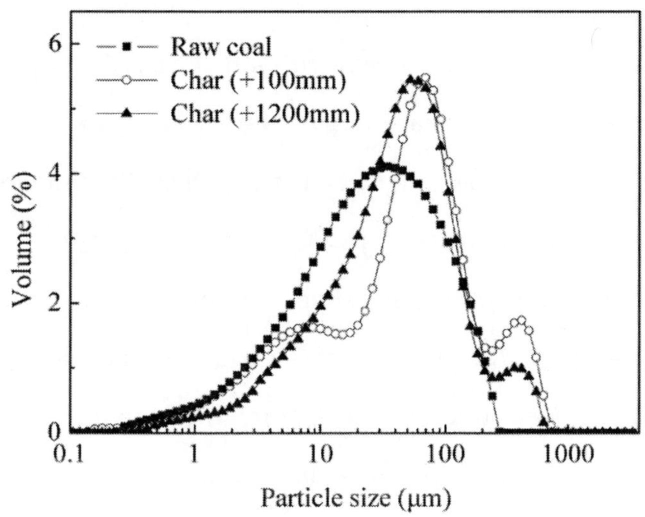


**(b) O/C=1.0**

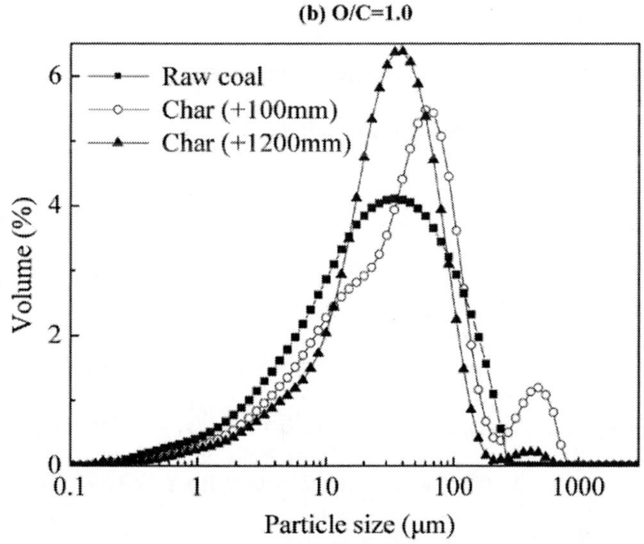


**(c) O/C=1.1**

**Fig. 6 Particle size distribution under different O/C ratios**

Experimental results also show that when particle size is larger than 200 μm, the amount of particle reduces dramatically and the fragmentation becomes strengthened while the particle moves from the burner plane downward to the gasification chamber outlet. It can be inferred that secondary fragmentation behavior causes the large particle size fraction percentage decreases rapidly when particle size is larger than 200 μm. With the increase of O/C ratios, the peak value of particles larger than 200 μm becomes weakened, and the amount of this type of particles decreases.

## 4. Conclusions

Coal char particle fragmentation is a complicated process inside the gasifier. The capture of particle fragmentation process with a high speed video camera indicates that coal char particle has secondary fragmentation behavior after devolatilization and partial oxidation in the impinging flame region where the primary fragmentation is probably finished. Particles larger than 200 μm in the impinging flame region have porosity structures with fragile shapes, while most of the particles smaller than 100 μm have spherical shapes. When the particles move from the impinging flame region to the gasification chamber outlet, percolative fragmentation as well as the residual carbon gasification reactions take the main responsibilities for the particle size decreasing. The particle size distributions under different O/C ratios indicate that there has a significant fragmentation when the particle size is larger than 200 μm. With the of O/C ratios, the peak value of particles larger than 200 μm becomes weakened.

**Acknowledgement**

This work is financially supported by the Shanghai Pujiang Program (15PJD011), the National Nature Science Foundation of China (Grant No. 21376082), and the Fundamental Research Funds for the Central Universities (222201514336).

**References:**

[1] I.S. Ruiz, J.C. Crelling, Applied coal petrology: The role of petrology in coal utilization, Academic Press, New York, 2008.

[2] P. Wang, M. Massoudi, Slag Behavior in Gasifiers Part I: Influence of Coal Properties and Gasification Conditions, Energies 6 (2013) 784-806.

[3] A. R. Mclennan, G.W. Bryant, B.R. Stanmore, T.F. Wall, Ash Formation Mechanisms during pf Combustion in Reducing Conditions, Energy Fuels 14 (2000) 150-159.

[4] H. Wu, T. Wall, G. Liu, G. Bryant. Ash Liberation from Included Minerals During Combustion of Pulverized Coal: The Relationship with Char Structure and Burnout. Energy Fuels 13 (1999) 1197-1202.

[5] H. Wu, G. Bryant, T. Wall. The Effect of Pressure on Ash Formation During Pulverized Coal Combustion. Energy Fuels 14 (2000) 745-750.

[6] J.C. Van Dyk, S.A. Benson, M.L. Laumb, B. Waanders, Coal and coal ash characteristics to understand mineral transformations and slag formation, Fuel 88 (2009) 1057-1063.

[7] K.J. Whitty, Physical phenomena of char-slag transition in pulverized coal gasification, Fuel Process. Technol. 95 (2012) 127-136.

[8] L. Sun, Y. Wu, K.J. Whitty, Ash deposition behavior during char-slag transition under simulated gasification conditions. Energy Fuels 24 (2010) 1868-1876.

[9] R. Chirone, L. Massimilla, P. Salatino, Comminution of carbons in fluidized bed combustion, Prog. Energ. Combust. 17 (1991) 297-326.

[10] B.R. Stanmore, A. Brillard, P. Gilot, L. Delfosse, Fragmentation of small coal particles under fluidized-bed combustor conditions. Twenty-Sixth Symposium (International) on Combustion/the Combustion Institute 26 (1996) 3269-3275.

[11] E. Bar-Ziv, I.I. Kantorovich. Role of Porous Structure in Char Oxidation. Appl. Therm. Eng. 18 (1998) 991-1003.

[12] E. Bar-Ziv, I.I. Kantorovich. Mutual Effects of Porosity and Reactivity in Char Oxidation. Prog. Energy Combust. Sci. 2001, 27, 667-697.

[13] T. Yamashita, Y. Fujii, Y. Morozumi, H. Aoki, T. Miura, Modeling of gasification and fragmentation behavior of char particles having complicated structures, Combust. Flame 146 (2006) 85-94.

[14] B. Kreutzkam, C. Wieland, H. Spliethoff. Improved numerical prediction of ash formation and deposition using a novel developed char fragmentation model, Fuel 98 (2012) 103-110.

[15] O. Senneca, M. Urciuolo, R. Chirone, A semidetailed model of primary fragmentation of coal, Fuel 104 (2013) 253-261.

[16] R. Chirone, L. Massimilla, The application of weibull theory to primary fragmentation of a coal during devolatilization, Power technol. 57 (1989) 197-212.

[17] S.Y. No, N. Syred, Thermal stress and pressure effects on coal particle fragmentation and burning behavior in a cyclone combustor, J. Energy Inst. 63 (1990) 195-202.

[18] Santiago J, Javier B. Study of the evolution of particle size distributions and its effects on the oxidation of pulverized coal[J]. Combustion and Flame. 2007, 151: 482-494

[19] Robert Hurt, Kevin Davis. Percolative fragmentation and spontaneous agglomeration. Combustion and Flame. 1999, 116: 662-670.

[20] Z. Zhou, Update of ECUST OMB Gasification Process, Gasification Technologies Conference, Colorado, United States, 2015.

[21] Q. Guo, Y. Gong, J. Xu, G. Yu, F. Wang, Particulate matter properties in a bench-scale entrained-flow coal-water slurry gasifier, Powder Technol. 254 (2014) 125-130.

[22] G. Yu, Q. Zhu, G. Chi, Q. Guo, Z. Zhou, Study on slag composition and flow property in a bench-scale OMB gasifier, Fuel Process. Technol. 104 (2012) 136-143.

[23] Y. Gong, Q. Guo, J. Zhang, P. Fan, Q. Liang, G. Yu, Impinging flame characteristics in an opposed multiburner gasifier, Ind. Eng. Chem. Res. 52 (2013) 3007-3018.

[24] Y. Gong , G. Yu, Q. Guo, Z. Zhou, F. Wang, Y. Liu. Experimental study of the particle deposition characteristics in an entrained flow gasifier. Chemi. Eng. Sci. 138(2015) 291-302.

[25] J. Friedemann, A. Wagner, A. Heinze, S. Krzack, B. Meyer. Direct optical observation of coal particle fragmentation behavior in a drop-tube reactor. Fuel. 166(2016) 382-391.

[26] T. Wu, M. Gong, E. Lester, F. Wang, Z. Zhou, Z. Yu, Characterisation of residual carbon from entrained-flow coal-water slurry gasifiers, Fuel 86 (2007) 972-982.

[27] A.C.D. Latorre. Ash formation under pressurized pulverized coal combustion conditions. Ph.D. Thesis, University of Connecticut, United States, 2005.

[28] G. Fix, W. Seames, M. Mann, S. Benson, D. Miller, The effect of combustion temperature on coal ash fine-fragmentation mode formation mechanisms, Fuel 113 (2013) 140-147.

# The product yield problem and the history and future of washability analysis

H.C. Dorland[1], N.J. Wagner[1], X.M. Makukule[1], Q.P. Campbell[2], M. le Roux[2], P.A. Van Vuuren[3], W.C. Venter[3], P. Erasmus[4], M.I. Dorland[5]

[1]University of Johannesburg, Department of Geology, CIMERA, South Africa

[2]North-West University, School of Chemical and Minerals Engineering, South Africa

[3]North-West University, School of Electrical, Electronic and Computer Engineering, South Africa

[4]EMT Consulting, South Africa

[5]The Workshop et al, South Africa

## Outline

- Current method
- The product yield problem
- Washability simulation

# Float and Sink

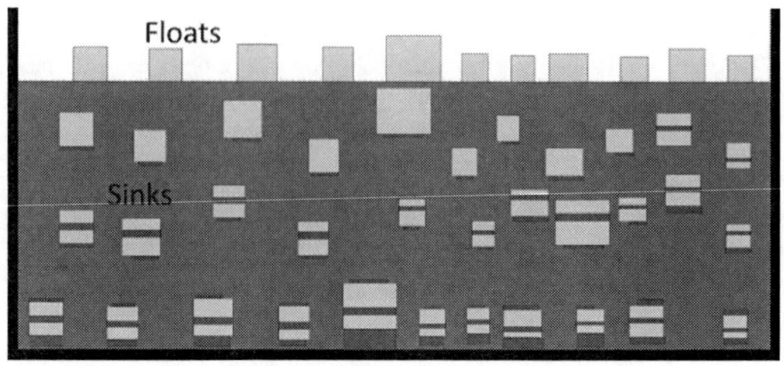


# Cumulative Washtable

| Sinks | Floats | Yield | Ash | CV | Sulphur | Moisture | Volatiles | FC |
|---|---|---|---|---|---|---|---|---|
| | | % | % | MJ/kg | % | % | % | % |
| - | 1.30 | 6.90 | 4.0 | 29.34 | 0.84 | 5.5 | 35.1 | 55.4 |
| 1.30 | 1.35 | 17.36 | 7.3 | 28.37 | 0.89 | 5.3 | 35.4 | 52.0 |
| 1.35 | 1.40 | 28.46 | 9.8 | 27.63 | 0.96 | 4.7 | 35.8 | 49.7 |
| 1.40 | 1.45 | 35.80 | 11.5 | 27.04 | 1.03 | 4.5 | 35.7 | 48.2 |
| 1.45 | 1.50 | 46.42 | 14.3 | 26.05 | 1.08 | 4.3 | 35.4 | 46.0 |
| 1.50 | 1.55 | 48.77 | 14.8 | 25.81 | 1.09 | 4.3 | 35.4 | 45.6 |
| 1.55 | 1.60 | 51.40 | 15.5 | 25.53 | 1.09 | 4.2 | 35.2 | 45.0 |
| 1.60 | 1.70 | 54.91 | 16.7 | 25.05 | 1.09 | 4.2 | 35.0 | 44.1 |
| 1.70 | 1.80 | 56.13 | 17.2 | 24.84 | 1.09 | 4.2 | 35.0 | 43.7 |
| 1.80 | 1.90 | 58.10 | 18.1 | 24.46 | 1.09 | 4.1 | 34.9 | 43.0 |
| 1.90 | 2.00 | 59.37 | 18.7 | 24.18 | 1.10 | 4.1 | 34.8 | 42.4 |
| 2.00 | 2.10 | 60.48 | 19.3 | 23.92 | 1.11 | 4.1 | 34.7 | 42.0 |
| 2.10 | 2.20 | 61.97 | 20.1 | 23.53 | 1.11 | 4.0 | 34.6 | 41.3 |
| 2.20 | - | 100.00 | 41.7 | 15.72 | 4.86 | 3.3 | 27.1 | 27.9 |

## Current Methods

- Float and sink analysis
- X-ray computed tomography

## The product yield problem

> "The Moatize Basin of Mozambique is often described as being comparable to Queensland's Bowen Basin in the 1960s, [...]. If there were any doubt about the potential, [*Company A*] moved to put that to rest with its $3.9 billion acquisition of [the mine] this year."
>
> ***The Australian Financial Review*** *17/11/2011*

## [*Company A*] reviews Mozambique as miners retreat from big plans

"[*Company A*] has begun a review of its Mozambique coal mining operations which cost it a ***$3 billion write-off***, reconsidering development plans, partners and its options for getting the coal from pit to port."

"Mozambique's infrastructure had proved more challenging than expected, [...] and ***estimates of recoverable coking coal used in steel production were lower than expected***."

**Mineweb** 22 January 2013

## Retrenchments follow as [*Company C*] closes [*Mine X*] in Limpopo

"[*Company C*] has shut down its troubled [*Mine X*] in Limpopo for the second time..."

"Also, that coal was produced at very poor operating efficiencies. [They] had to treat 162,289 tonnes of coal in 2012 to produce the 46,066 sales tonnes, which indicates a ***recovery efficiency of 28%.***"

"The obvious conclusion is that the plant is not working. ***To put it bluntly, it appears to be a dud.***"

***Business Day Live*** *13 October 2013*

NORTH-WEST UNIVERSITY
YUNIBESITI YA BOKONE-BOPHIRIMA
NOORDWES-UNIVERSITEIT

# So, what was the problem?

- These plants did not achieve the required yield
- Washabilites were done, but...
- It overestimated the potential yields
- Liberation effect was not understood / considered.

cimera

UNIVERSITY OF JOHANNESBURG

NORTH-WEST UNIVERSITY
YUNIBESITI YA BOKONE-BOPHIRIMA
NOORDWES-UNIVERSITEIT

cimera

UNIVERSITY OF JOHANNESBURG

NORTH-WEST UNIVERSITY
YUNIBESITI YA BOKONE-BOPHIRIMA
NOORDWES-UNIVERSITEIT

cimera

UNIVERSITY OF JOHANNESBURG

NORTH-WEST UNIVERSITY
YUNIBESITI YA BOKONE-BOPHIRIMA
NOORDWES-UNIVERSITEIT

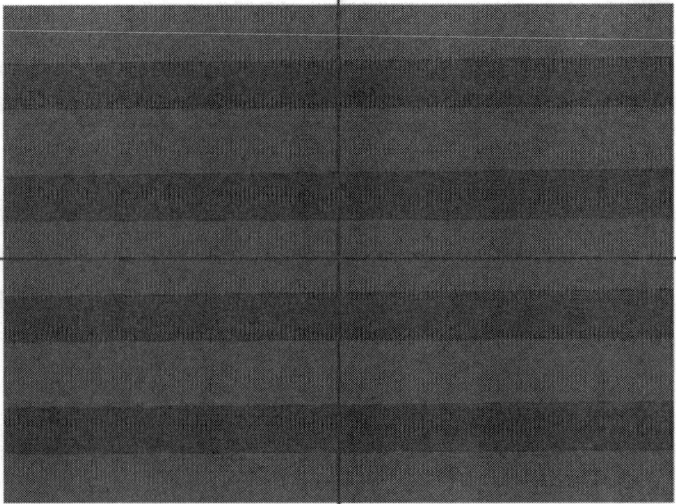

cimera

UNIVERSITY OF JOHANNESBURG

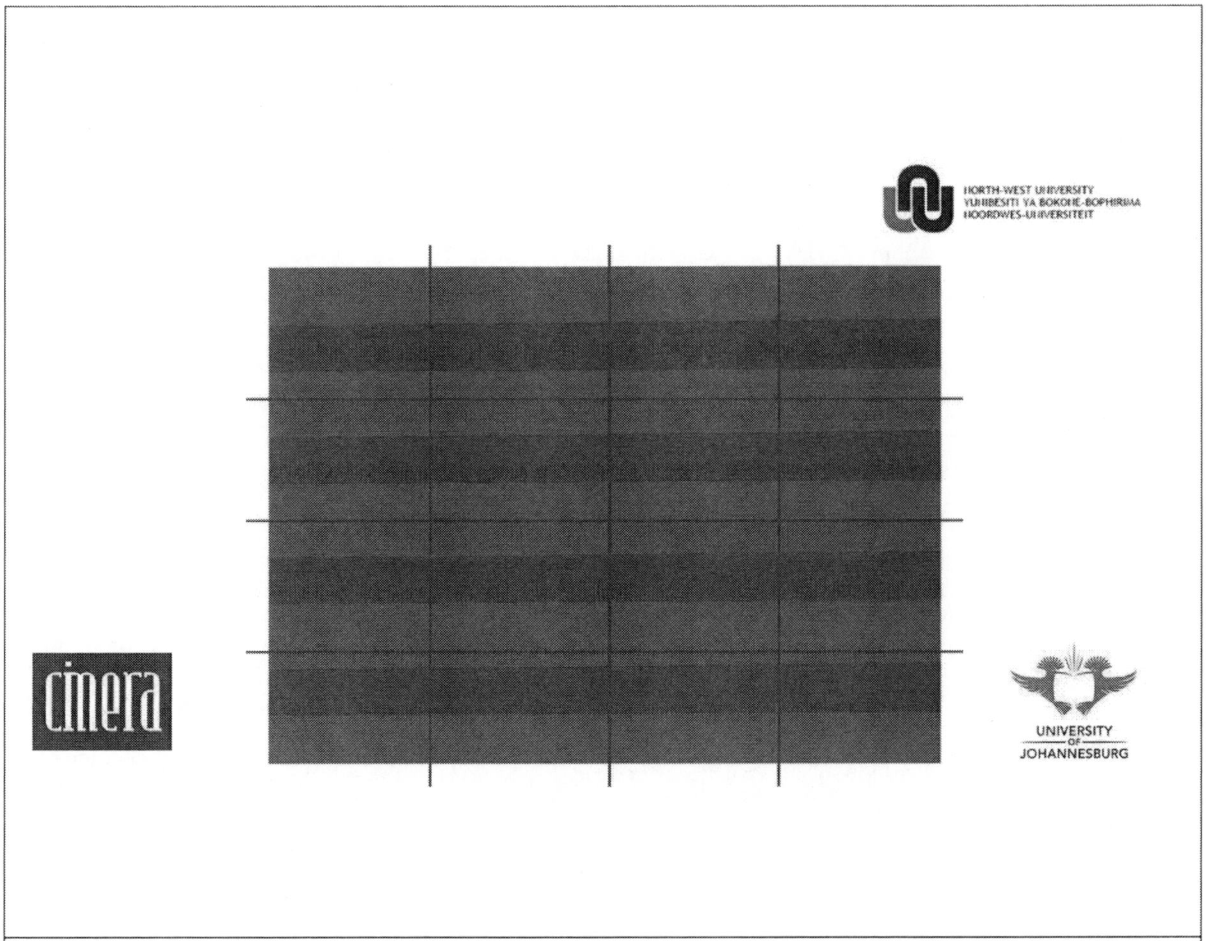
NORTH-WEST UNIVERSITY
YUNIBESITI YA BOKONE-BOPHIRIMA
NOORDWES-UNIVERSITEIT
cimera
UNIVERSITY
OF
JOHANNESBURG

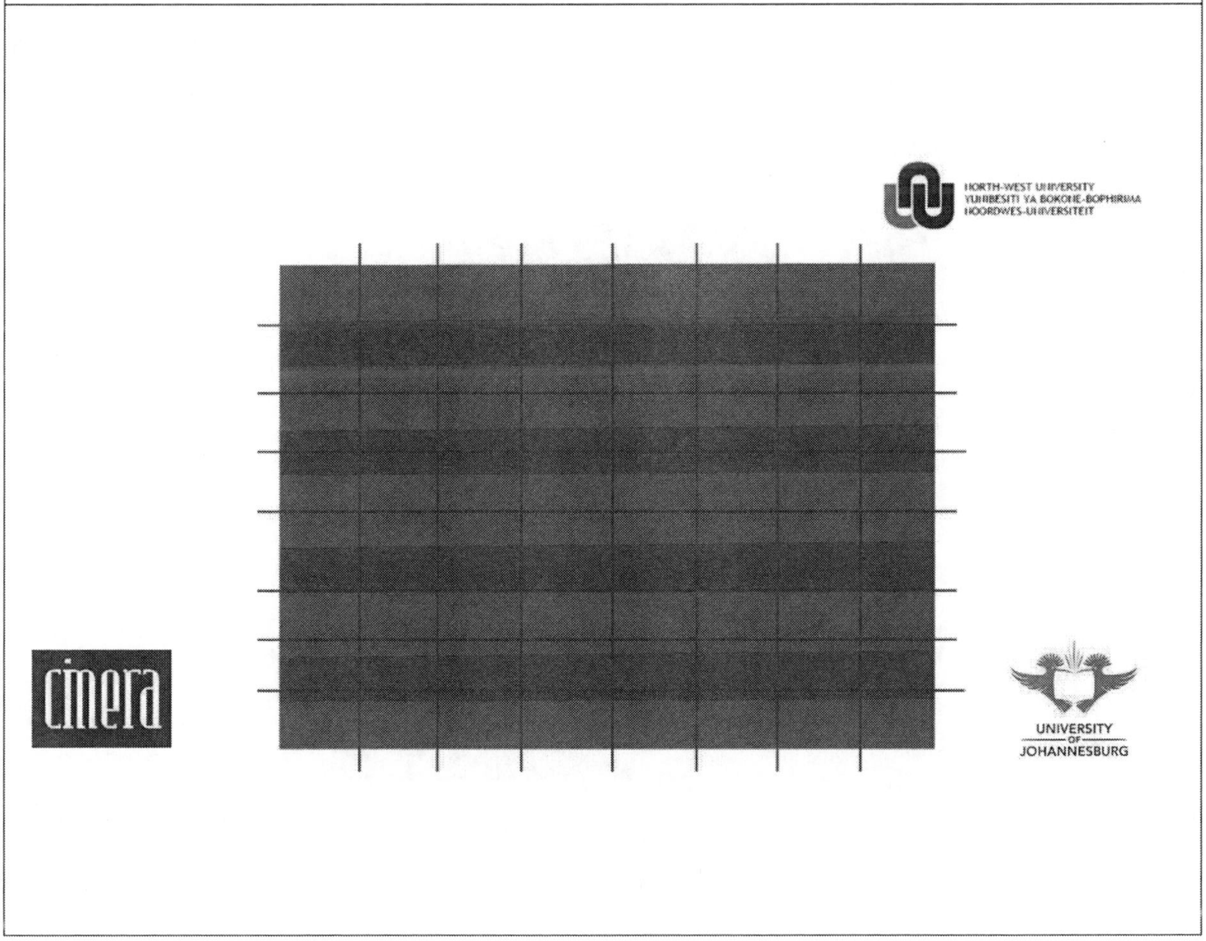
NORTH-WEST UNIVERSITY
YUNIBESITI YA BOKONE-BOPHIRIMA
NOORDWES-UNIVERSITEIT
cimera
UNIVERSITY
OF
JOHANNESBURG

NORTH-WEST UNIVERSITY
YUNIBESITI YA BOKONE-BOPHIRIMA
NOORDWES-UNIVERSITEIT

cimera

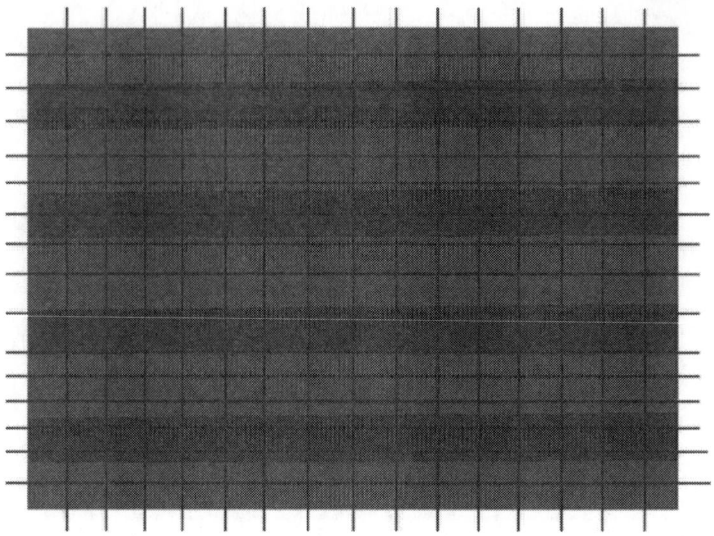

UNIVERSITY OF JOHANNESBURG

NORTH-WEST UNIVERSITY
YUNIBESITI YA BOKONE-BOPHIRIMA
NOORDWES-UNIVERSITEIT

cimera

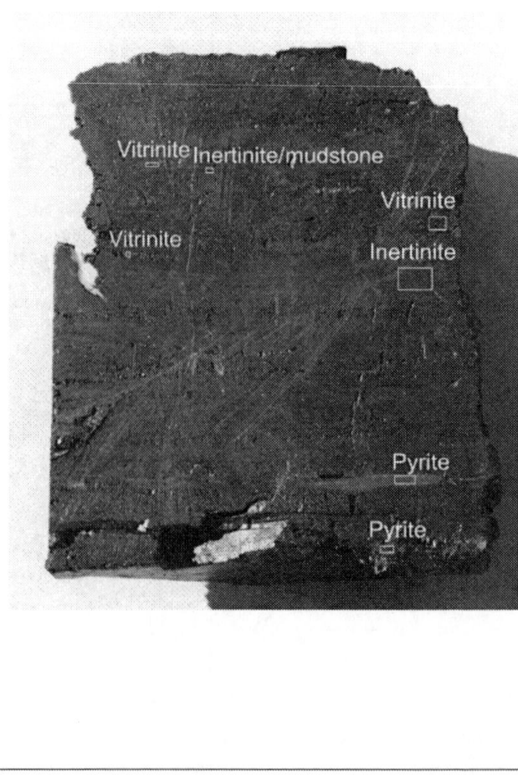


UNIVERSITY OF JOHANNESBURG

NORTH-WEST UNIVERSITY
YUNIBESITI YA BOKONE-BOPHIRIMA
NOORDWES-UNIVERSITEIT

cimera

UNIVERSITY OF JOHANNESBURG

NORTH-WEST UNIVERSITY
YUNIBESITI YA BOKONE-BOPHIRIMA
NOORDWES-UNIVERSITEIT

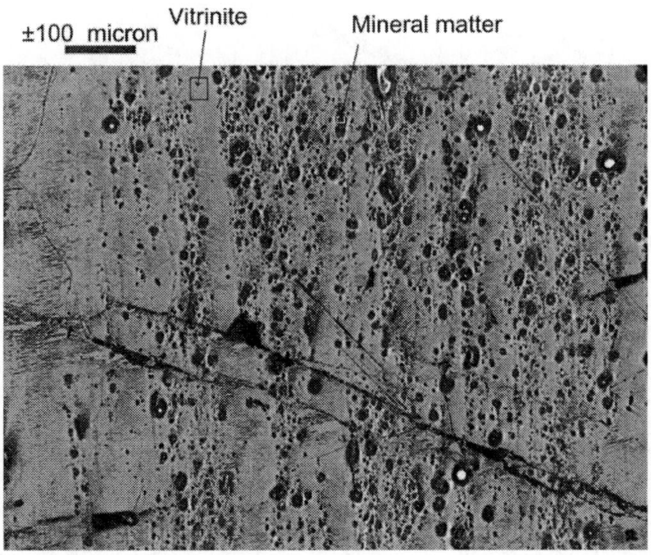


cimera

UNIVERSITY OF JOHANNESBURG

NORTH-WEST UNIVERSITY
YUNIBESITI YA BOKONE-BOPHIRIMA
NOORDWES-UNIVERSITEIT

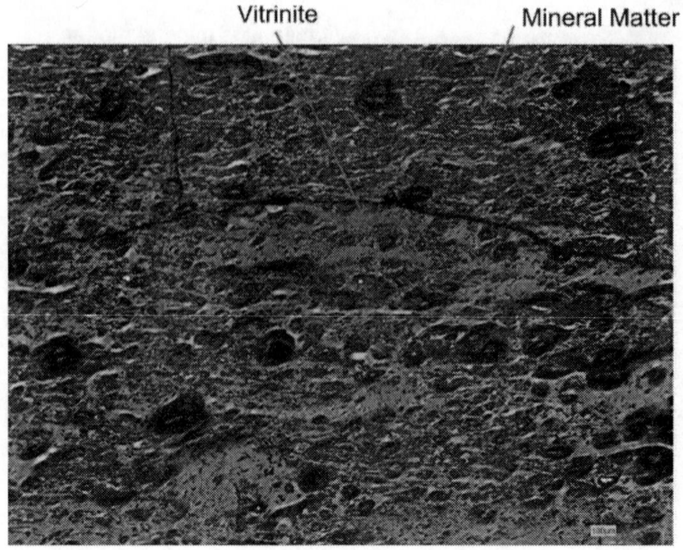


cimera

UNIVERSITY OF JOHANNESBURG

NORTH-WEST UNIVERSITY
YUNIBESITI YA BOKONE-BOPHIRIMA
NOORDWES-UNIVERSITEIT

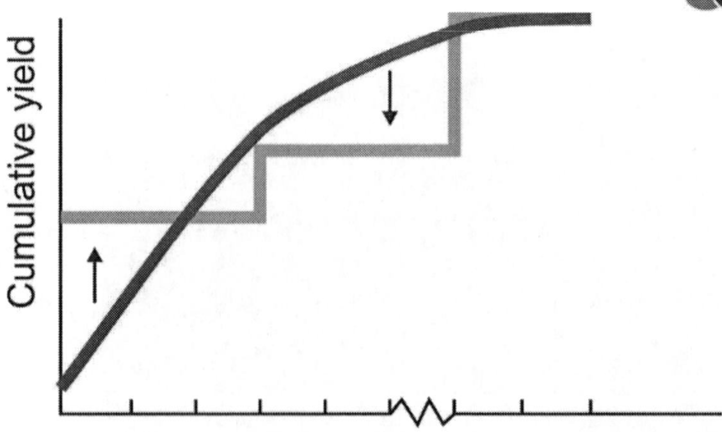


cimera

UNIVERSITY OF JOHANNESBURG

## Limitations of washability tests

- Particle breakage essentially random/complex
- Impossible to predict exactly what shape and size of particles are produced when coal is crushed
- Washability does not usually account for
  - Different top sizes
  - Different PSD
  - Different breakage patterns
- Washability behaviour is unpredictable.

## The purpose of the current work

Develop a methodology to simulate many virtual crush experiments in order to understand the washability of coal before it is crushed, and to predict the yield and quality at any top size.

NORTH-WEST UNIVERSITY
YUNIBESITI YA BOKONE-BOPHIRIMA
NOORDWES-UNIVERSITEIT

# Washability simulation

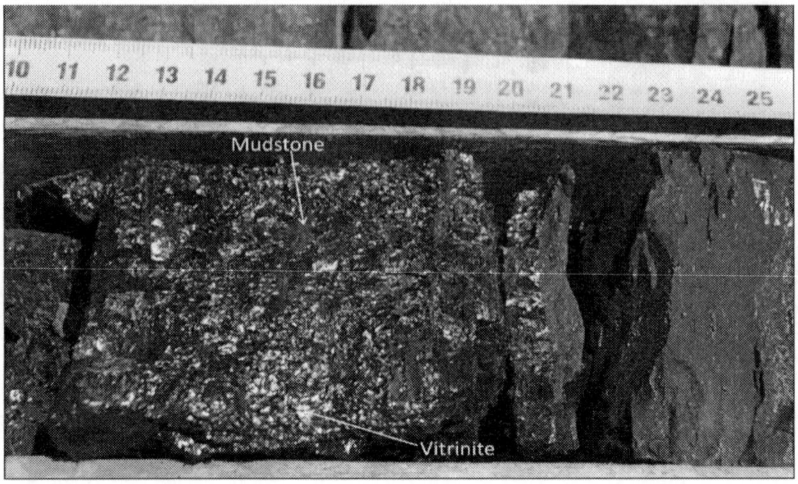


cimera

UNIVERSITY OF JOHANNESBURG

NORTH-WEST UNIVERSITY
YUNIBESITI YA BOKONE-BOPHIRIMA
NOORDWES-UNIVERSITEIT

cimera

UNIVERSITY OF JOHANNESBURG

NORTH-WEST UNIVERSITY
YUNIBESITI YA BOKONE-BOPHIRIMA
NOORDWES-UNIVERSITEIT

| Component | Density (g/cm³) | Ash (wt %) |
|---|---|---|
| Vitrinite | 1.35 | 4 |
| Mudstone | 2.54 | 98 |

cimera

UNIVERSITY OF JOHANNESBURG

NORTH-WEST UNIVERSITY
YUNIBESITI YA BOKONE-BOPHIRIMA
NOORDWES-UNIVERSITEIT

# Current Assumptions

- Crushing occurs in cubical blocks
- All breakage products are same size
- 2D surface information can be extrapolated to 3D
- No preferential breakage along interfaces / cleats
- All identified components have consistent density and ash content.

cimera

UNIVERSITY OF JOHANNESBURG

NORTH-WEST UNIVERSITY
YUNIBESITI YA BOKONE-BOPHIRIMA
NOORDWES-UNIVERSITEIT

# Washability results

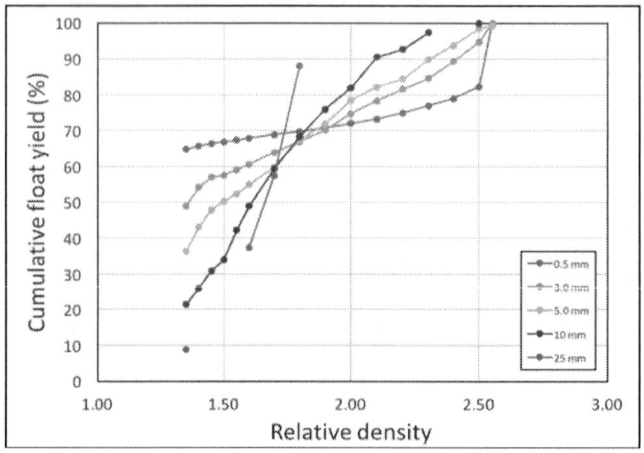


cimera

UNIVERSITY OF JOHANNESBURG

NORTH-WEST UNIVERSITY
YUNIBESITI YA BOKONE-BOPHIRIMA
NOORDWES-UNIVERSITEIT

# Washability results

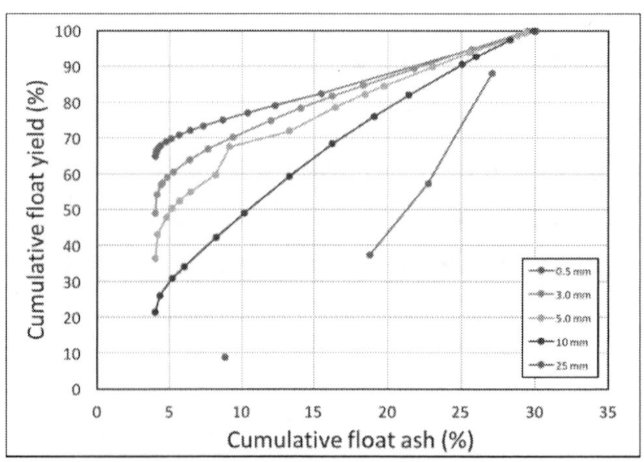


cimera

UNIVERSITY OF JOHANNESBURG

# Yield curve – 10% ash

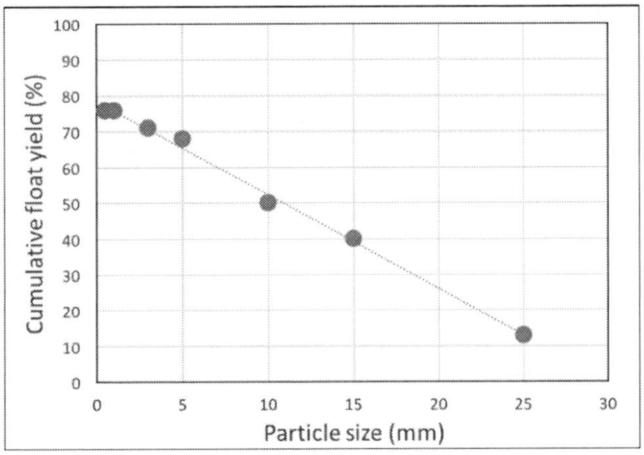


# Automated Classification

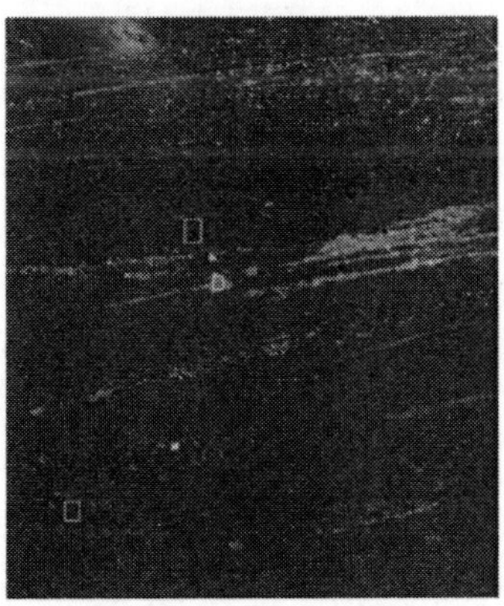

Final: blue = vitrinite, green = inertinite, purple = pyrite

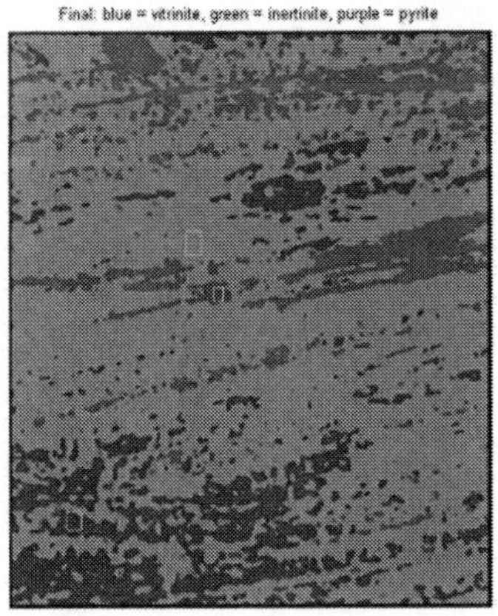

# Performance of classifier on polished block

| | | Actual | | |
|---|---|---|---|---|
| | | V | V+M | P |
| Mapped | V | 136 | 2 | 0 |
| | V+M | 87 | 558 | 35 |
| | P | 8 | 22 | 79 |

| | V | V+M | P |
|---|---|---|---|
| Actual | 231 | 583 | 114 |
| Mapped | 138 | 680 | 109 |
| Estimation (%) | 60 | 117 | 96 |

Total correct 83%

# Single 5.005 mm simulation

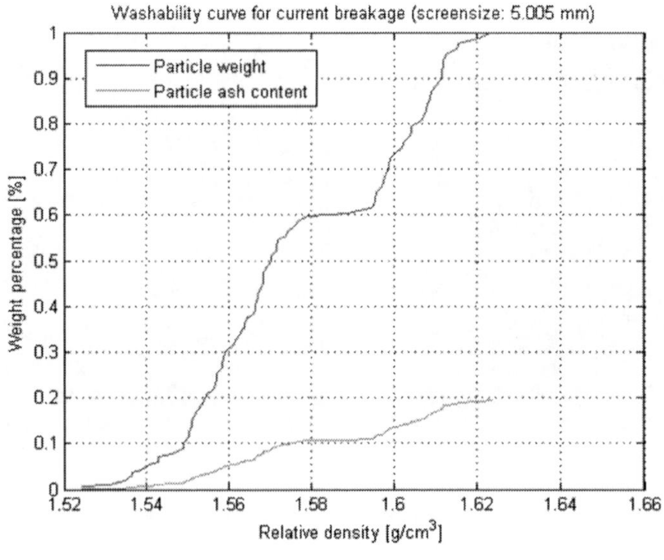

# Thousands of simulations

NORTH-WEST UNIVERSITY
YUNIBESITI YA BOKONE-BOPHIRIMA
NOORDWES-UNIVERSITEIT

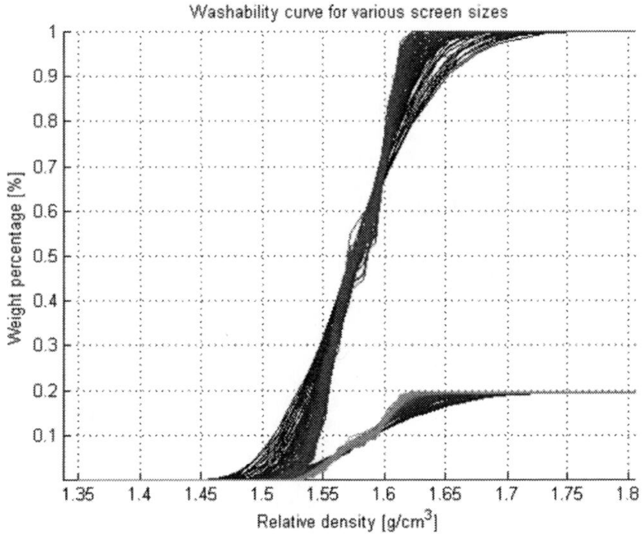


cimera

UNIVERSITY OF JOHANNESBURG

---

NORTH-WEST UNIVERSITY
YUNIBESITI YA BOKONE-BOPHIRIMA
NOORDWES-UNIVERSITEIT

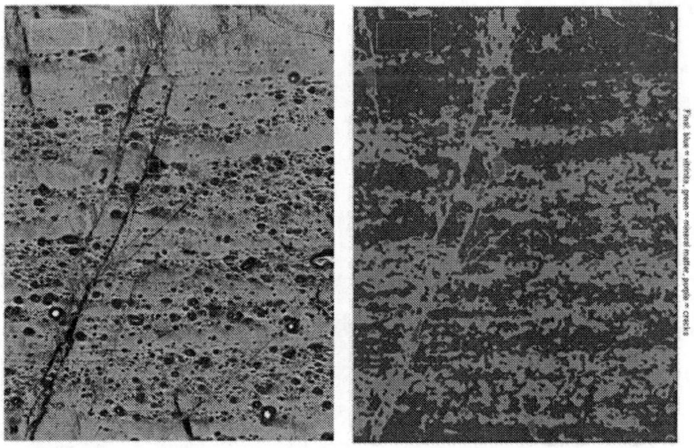


cimera

UNIVERSITY OF JOHANNESBURG

## Simulation versus traditional

- Will the coal break as in the single scenario in the laboratory?
- No subsampling
- No problems with scooping, density of baths, short cuts
- Probability of most likely washability, worst case and best case

## Conclusions

- Infinitely many washability scenarios for a specific volume of coal
- Current washability analysis give some information about only one scenario
- Great monetary losses due to not understanding the impact of particle size and shape on product yield and plant design
- Washability simulation is a method to address the above problems
- Virtual coal laboratory possible?

Thank you for listening

NORTH-WEST UNIVERSITY
YUNIBESITI YA BOKONE-BOPHIRIMA
NOORDWES-UNIVERSITEIT

COALTECH
2020

cimera

UNIVERSITY
OF
JOHANNESBURG

Eskom

# Estimation of Moisture Dynamics in a Composite Coal Stockpile


Presenter: Kyle Enslin

Academic Mentor: Walter Schmitz

Industrial Mentor: Chris van Alphen


Date: August 2016


UNIVERSITY OF THE WITWATERSRAND, JOHANNESBURG


## Problem Statement

Eskom

- Eskom needs to dewater coal as cheaply as possible however existing technologies are expensive.
- Coal stockyards are never designed as dewatering instruments however, during times of wet coal, they are utilized as such.
- General assumption is a coal stockpile will drain to a desired moisture content given enough time
- Eskom would like to know what drainage rates and equilibrium moisture contents are possible given any coal

## Research Question/Hypothesis

Eskom

- Questions:
  - What is the main control for coal drainage and equilibrium moisture content?
  - Can the equilibrium moisture content be modeled using the particle size distribution?
- Hypothesis
  - By controlling the fines content (<500micron) the drainage rate and equilibrium moisture content can be improved.



## Hypothesis

Eskom

- Inherent moisture cannot be removed by drainage
- Drainage controlled by permeability
- Permeability controlled by particle size distribution and degree of saturation (ie: moisture content)

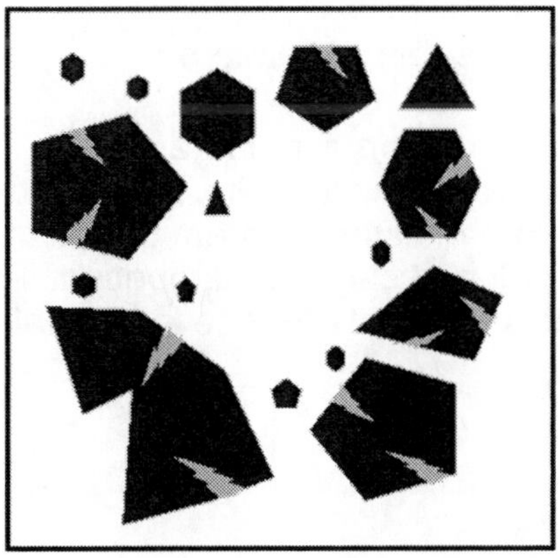

## Literature Review: Summary

Eskom

- Eskom has not standardized it's coal reporting with regard to size grading
- Industry has not clearly defined a particle size classification system (Hand, 2000; Cumming, 2003; Williams and Whitton, 2008)
- Coals behave similar to soils in terms of drainage (Williams and Whitton, 2008)
- Saturated and unsaturated soil mechanics can be applied to coals (Williams and Whitton, 2008)
- Fines, like clays and silts, greatly affect drainage (Merwe and Campbell, 2002; Cumming, 2003)
- The degree to which fines (-500micron) affect drainage and moisture content is yet to be established.



## Assumptions

Eskom

Assumptions:

- Effects of wind and sun on dewatering is negligible
- Fines defined as <500micron
- Each size fraction contributes to the total equilibrium moisture content of a composite coal stockpile. This contribution is proportional to the equilibrium moisture content of the size fraction and its weighting towards the overall particle size distribution. Thus, the total equilibrium moisture content is the sum of these individual moisture contributions.

## Model

Eskom

- The model then being validated is:

$$w_{TM}\% = \sum_{i=A}^{i} w_{Ti} \times f_i = (w_{TA} \times f_A) + (w_{TB} \times f_B) + \cdots$$

Where $w_{TM}$= Total Equilibrium Moisture Content of Composite Sample

$w_{Ti}$=Total Equilibrium Moisture Content of Individual Size Fraction

$f_i$=contribution of size fraction to overall size distribution



## Model

Eskom

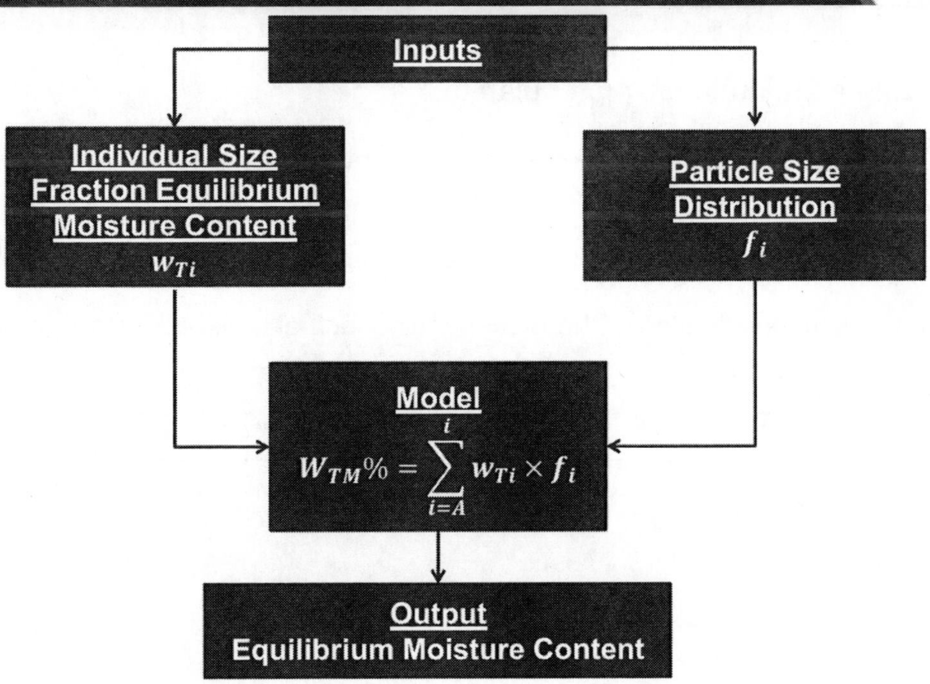

## Methodology

Eskom

- Utilizing a 25L drum I:

  1) Isolated individual size fractions (8) to establish which size fraction has the greatest affect on drainage and equilibrium moisture content. This was then used as input into the model.

| Size Fraction (mm) | | | | | | | | |
|---|---|---|---|---|---|---|---|---|
| A | B | C | D | E | F | G | H | I |
| -0.5 | +0.5-1 | +1-2 | +2-3 | +3-4.75 | +4.75-6 | +6-10 | +10-19 | +19 |

  2) Varied this size fraction (-500micron) to establish correlations between drainage rate and equilibrium moisture content

| Fines content (%) | | | | | | | | | | |
|---|---|---|---|---|---|---|---|---|---|---|
| 0 | 1 | 2 | 3 | 4 | 5 | 6 | 7 | 8 | 9 | 10 |

  3) Model outputs were then validated against actual moisture contents to determine accuracy

University of the Witwatersrand, Johannesburg    9    Eskom Power Plant Engineering Institute

## Methodology

Eskom

- Utilizing a 2m drainage pipe I plan to:

  1) See whether the results from the 25L drum experiments are constant when the experiments are scaled up.

  2) Variance of moisture content and fines content within the coal stockpile

  3) Continued validation of the model against actual moisture data

| Fines Content (%) | | | | | |
|---|---|---|---|---|---|
| 0 | 2 | 4 | 6 | 8 | 10 |

University of the Witwatersrand, Johannesburg    10    Eskom Power Plant Engineering Institute

## Testing: Equipment

Eskom

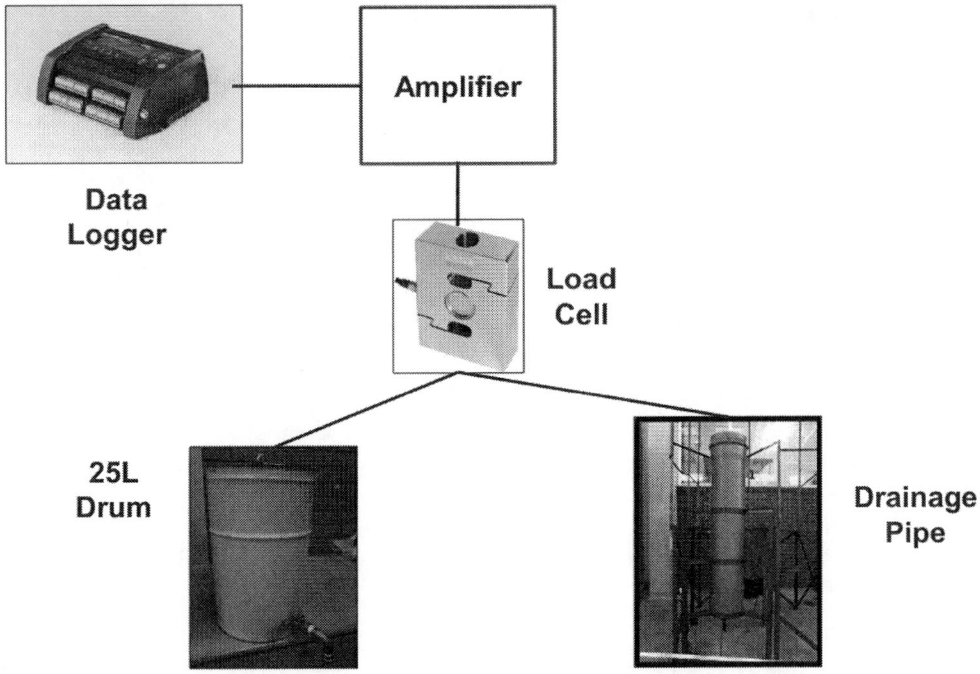


UNIVERSITY OF THE WITWATERSRAND, JOHANNESBURG

11

Eskom Power Plant Engineering Institute

## Results: 25L Drum –Individual Size Fractions

Eskom

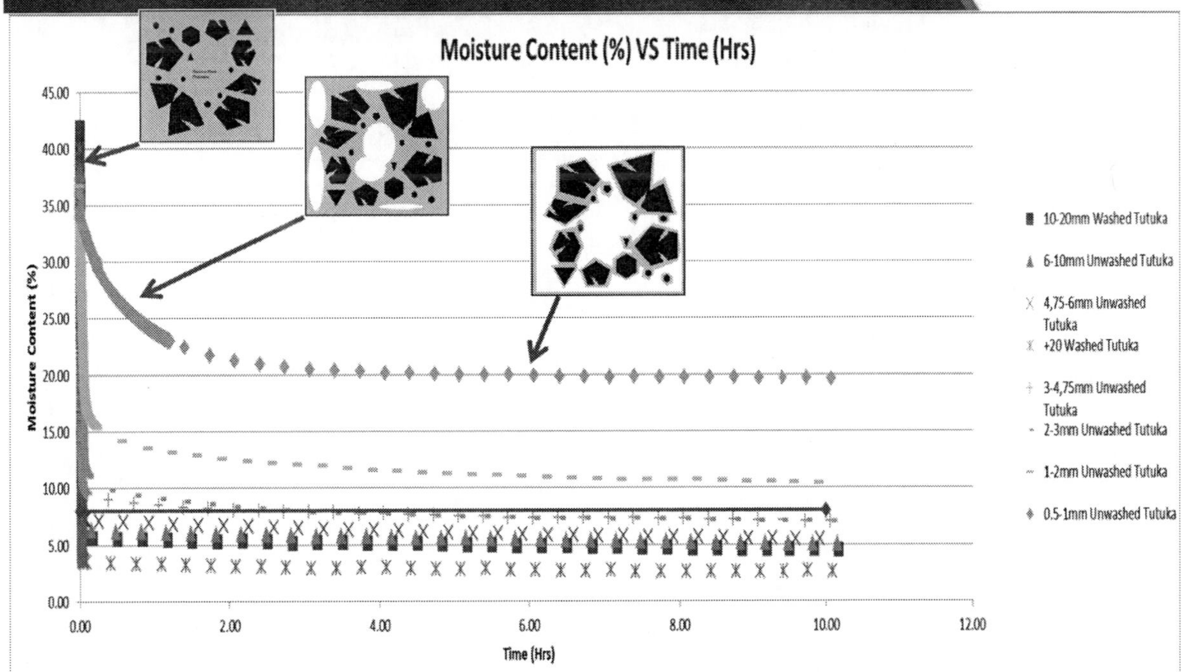


UNIVERSITY OF THE WITWATERSRAND, JOHANNESBURG

12

Eskom Power Plant Engineering Institute

# Results: 25L Drum –Individual Size Fractions

Eskom

- Equilibrium Moisture Content

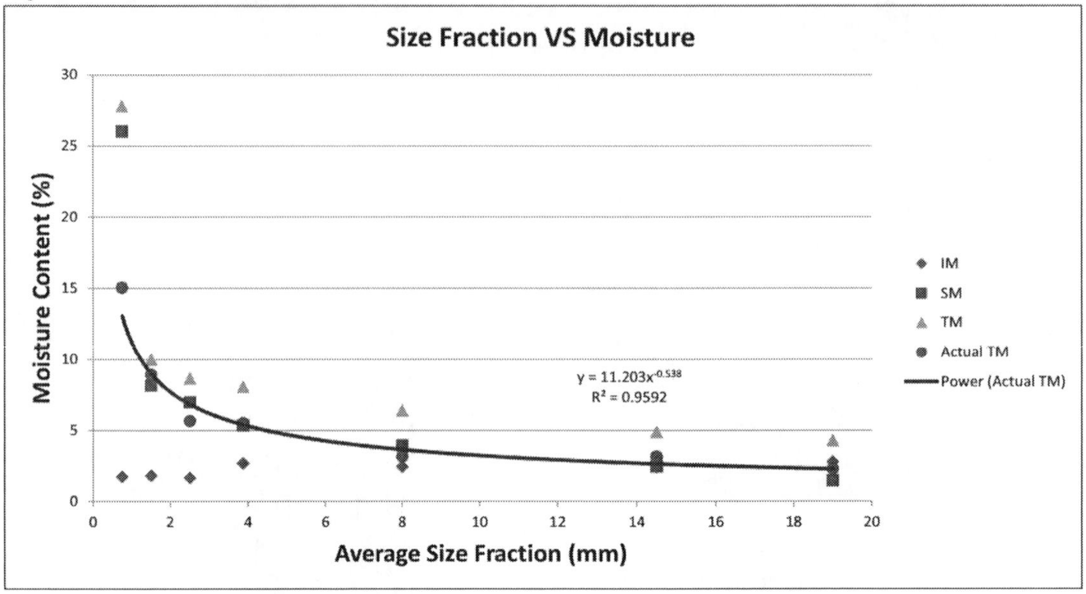




# Results: 25L Drum –Fines Content

Eskom

- Final Moisture Contents –Drum B

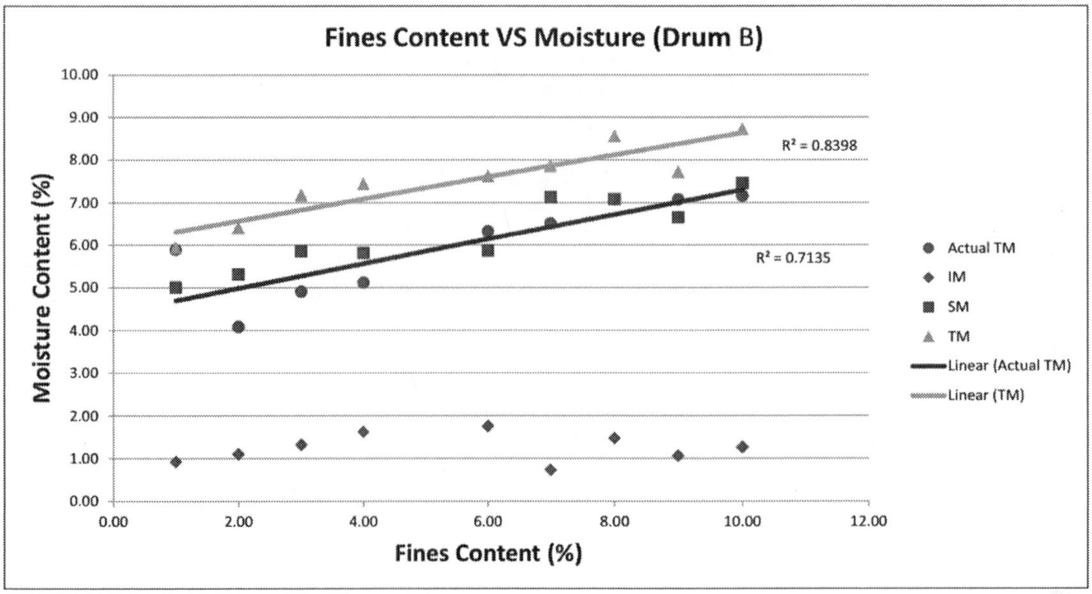

## Results: Drainage Pipe

Eskom

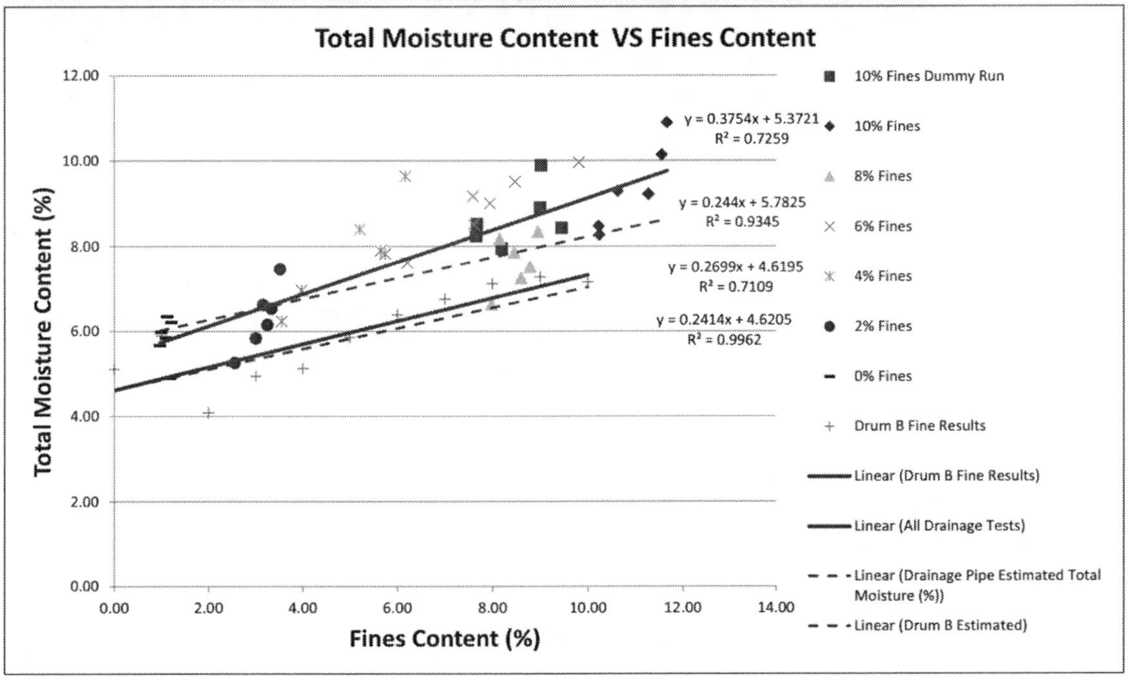




## Results: Drainage Pipe

Eskom

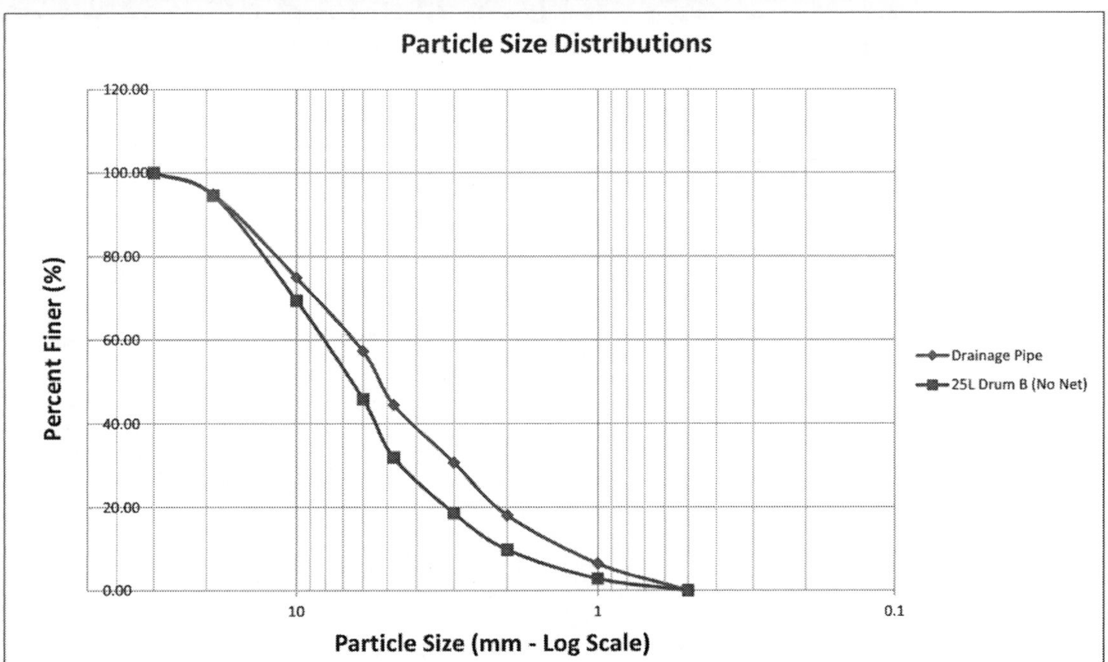

## Results: Drainage Pipe

Eskom

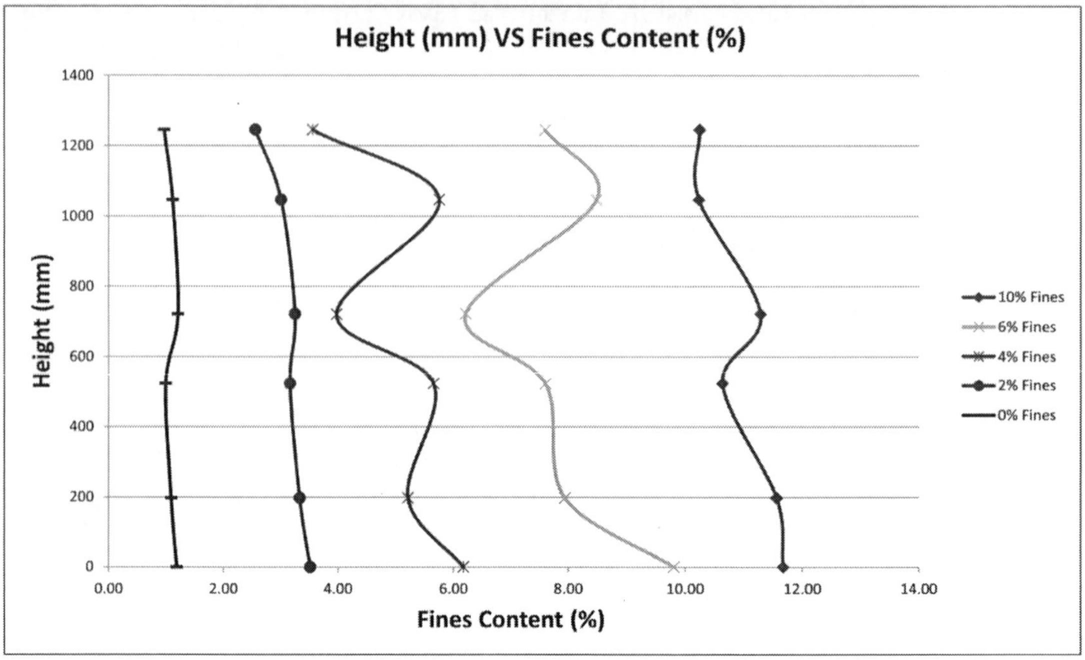


UNIVERSITY OF THE WITWATERSRAND, JOHANNESBURG

17

## Results: Drainage Pipe

Eskom

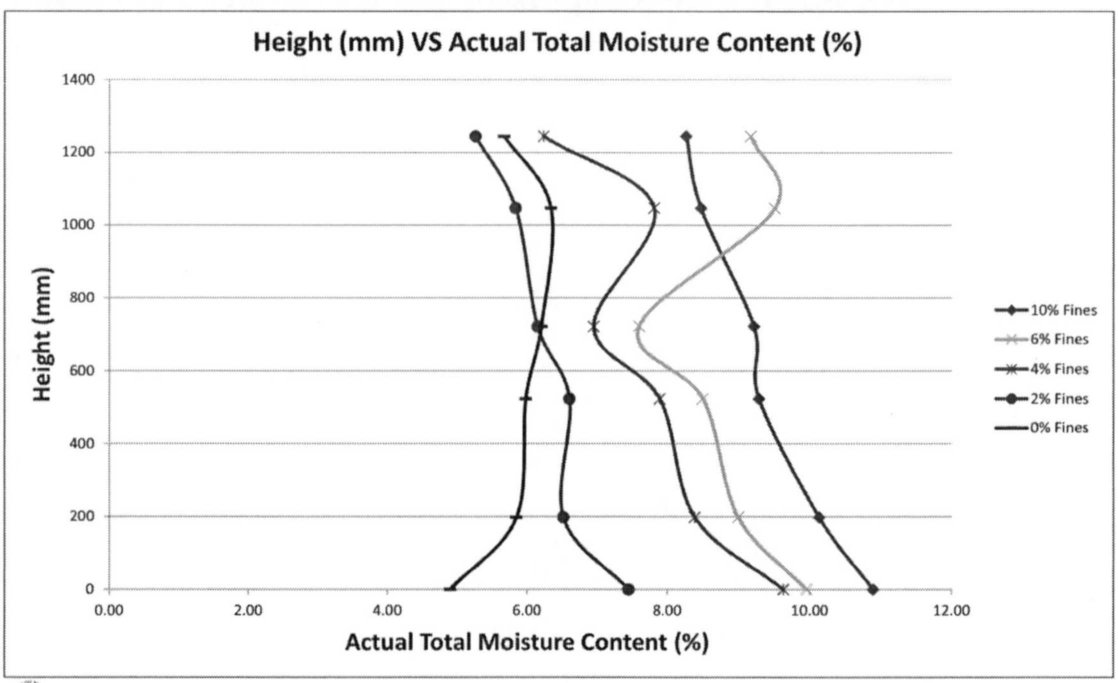


UNIVERSITY OF THE WITWATERSRAND, JOHANNESBURG

18

## Discussion and Conclusion

Eskom

- Drum Correlation Coefficient: 0.85
- Drainage Pipe Correlation Coefficient: 0.89
- 1% increase in fines due to small amount of handling
- Moisture content largely follows fines content
- The results are consistent when scaled up

UNIVERSITY OF THE WITWATERSRAND, JOHANNESBURG

19

## Discussion and Conclusion

Eskom

- Better Coal Stockpile design
- Improved coal stockpile operating due to better moisture retention understanding and aid of model
- Better foundation for coal contract negotiations by knowing what value of fines will yield a desired moisture content.
- Model does not include the effects of specific surface area, surface tension and capillary forces. It assumes the data from the individual size fraction tests already account for these effects

UNIVERSITY OF THE WITWATERSRAND, JOHANNESBURG

20

Eskom

# Thank you

CELL: 072 217 5191

E-mail: kyle.enslin@eskom.co.za

UNIVERSITY OF THE WITWATERSRAND, JOHANNESBURG

**Seal Design Investigations for Omerler Underground Coal Mine in Turkey**

*C. Atilla Ozturk[a],*
*Abdullah Fişne[a], Mustafa Özdingiş[b], Şahika Yürek[c], Kıvanç Het[d]*

[a] *Assoc. Prof. Dr. Istanbul Technical University, Faculty of Mines, Turkey, atilla.ozturk@itu.edu.tr*
[b] *Deputy General Manager, [c]Project Manager, [d]Ph.D. Mining Engineer, Turkish Coal Enterprises, Turkey*

**Abstract**

Western Lignite's Enterprises (GLI) under the branch of Turkish Coal Enterprises (TKI) located in Tuncbilek, Kutahya has been operated since 1940's. According to the current reserve estimation, the site has around 170 million tons of lignite. Annual production of the site reaches 4.2 million tons while 20% of that amount is met from underground mining. Omerler underground lignite mine produces around 1.000.000 tons of lignite from fully mechanized longwalls. Average thickness of lignite seams is 8 m to 9 m. Top caving system is used to produce 6-m thickness coal from the backside of hydraulic support system while the initial 3-m thickness of lignite produced from shearer. Seals are used to form a barrier to isolate abandoned working areas from the current production sites. 29 seals were constructed up to date for the underground mine. In this study, 29 seals were firstly classified based on construction material and cross-section area, and then classified seals were also evaluated based on seal thickness. 4 different types of seals were found in the underground mine while seal thicknesses were also changed as 45 cm, 50 cm, 70 cm, 100 cm. Reinforced concrete, brick, and aerated concrete blocks were used as construction materials. Engineering properties of construction materials as well as surrounding rock mass of the seals were also determined from the existing studies to evaluate the sufficiency of seals against collapse caused by explosion and/or in-situ stress-strain conditions. The following reasons are summarized to the requirement of re-design and evaluation of the existing permanent seals to protect the underground mine against any type of accidents. Excessive number of abandoned mining panels increases the stress over seals as well as the effect of depth that is around 250 m. The geological materials around the seals are mostly classified as poor or very poor. Space behind the seals were not filled up to 100 m for 21 number of seals and that increases the risk of gas explosion or flood. The quality control of construction materials was not documented in the mine. Because of these, permanent seals designs were evaluated and re-designed in this study to protect the underground mine. The current design procedures proposed as per USA, UK, and German systems discussed in this study were applied to evaluate the sufficiency of existing seals as well. As a result of this study, new designs of permanent seals are proposed after evaluating and re-designing the existing permanent seals of Omerler Underground Lignite Mine.

Key Words: Coal, Safety, Stability, Seal

## 1. Introduction

Coal mining is still the most important source for energy production from thermal power plant. Production of coal from underground mining methodologies is the important part of coal exploitation especially for energy production. Tuncbilek coal region operated by Western Lignite's Enterprises (GLI) under the branch of Turkish Coal Enterprises (TKI) was firstly discovered in 1906. First underground mining production was commenced in 1932 while coal production was initiated in 1962. GLI was established in 1938. The Enterprises has been also producing lignite from Omerler underground coal mine. The location of the mine can be seen in Figure 1.

The lignite region base formed by serpentinized ultrabasic rocks. The region emanated on miocene and pliocene series that are called as upper lignite (Tuncbilek basin) and lower lignite (Domanic basin). The production have been made on the Tuncbilek basin in between marl horizons. Roof and floor materials of the coal is classified as marl. The material strength ranges between 30 – 70 MPa for floor while it fluctuates between 50 and 60 MPa for roof. Main lignite seam thickness changes from 5 to 11 m including interburden clay classified as A-clay, B-clay and C-clay as well as yellow clay over the C-clay.

Current underground mine's, Omerler, lignite thickness is around 8 to 9 m with a depth of 200 – 250 m. 3 different clay layer as per interburden can be found in the Omerler lignite seam. Longwall mining method is used as underground mining method in the mine. Average length of longwalls is 90 m while width of the panels generally differ from 450 m to 600 m. Large steel hydraulic supports temporarily maintain the roof. Top coal caving method is used for the production of upper lignite in the longwall. Retreating type of production is used in the fully mechanized longwall panels. 2,8 m height coal produces by shearer and the remaining 5.2 m height of the seam producing from top coal caving. Armored face conveyors are equipped in the longwalls to haul the coal from longwall to the main haulage gallery while belts are used to transfer the coal from underground to the surface. A typical geological cross-section of Omerler underground coal mine lignite seam and Tuncbilek coal region is given in Figure 2.

Figure 1. Google Earth view of the study area.

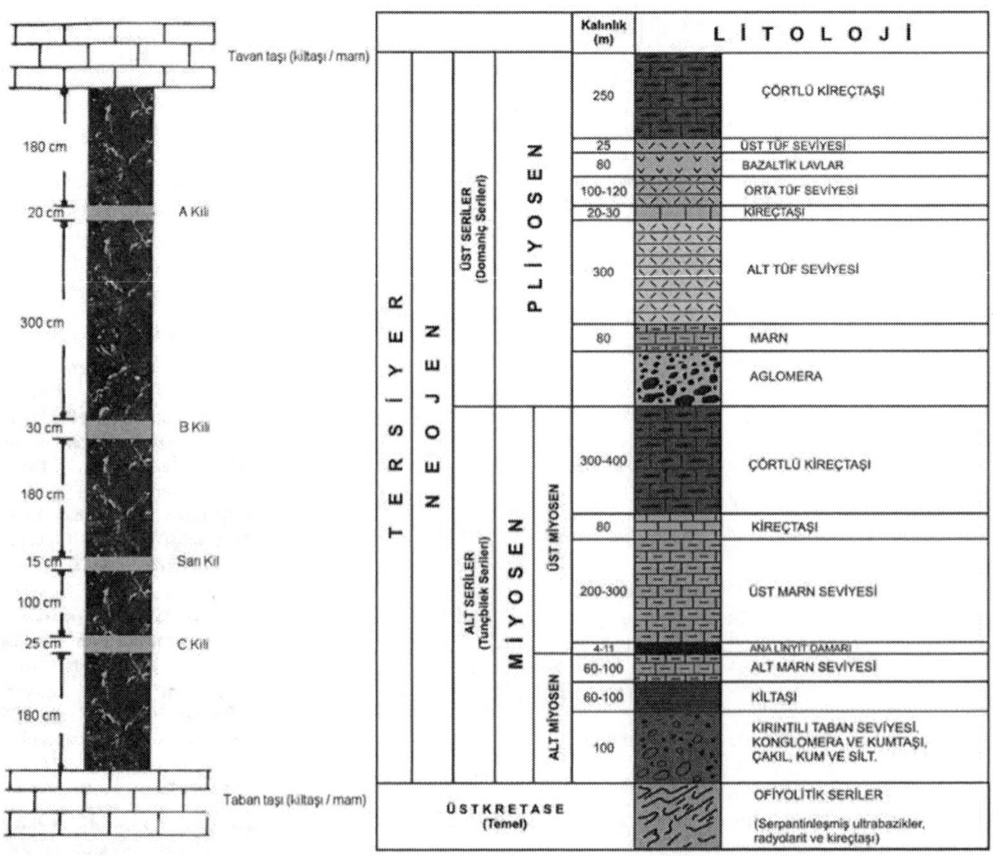


Figure 2. Typical cross-section of coal seam and Tuncbilek coal region (Başarır and Karpuz, 2004).

Abandoned panels were isolated by seals in the underground mine to protect the working areas against flood and fire. 29 seals were constructed up to date in the mine having different thickness from 45 cm to 100 cm. Reinforced concrete, brick, and aerated concrete blocks were used as construction materials for the seals. 1,000,000 ton annual production target and the excessive numbers of abandoned working areas increases the importance of seal design against any type of failure. Main objective of this paper is to present current conditions of the seals in Omerler Underground Coal Mine and evaluate the seal's stability.

## 2. Geotechnical Evaluations

Geotechnical investigations were carried out by using a report prepared by Beyhan and Yuvka (2016) for Igdekuzu underground mine adjacent to Omerler. In this report, roof, lignite, and floor were classified based on Rock Mass Rating (RMR) classification. Hoek – Brown failure criteria is also used to find out the modulus of elasticity and uniaxial compressive strength of rock masses. Table 1 and Table 2 is prepared from this research to show a brief geotechnical view of lignite and surrounding rock environment.

Table 1. Rock mass classifications based on RMR Beyhan and Yuvka (2016).

| Definition | Floor Marl | Lignite | Roof Marl |
|---|---|---|---|
| Uniaxial compressive strength, MPa | 11.00 | 5.90 | 19.80 |
| RQD, % | 63.80 | 45.00 | 47.40 |
| Joint spacing, mm | 185.00 | 54.00 | 240.00 |
| Joint continuity | >20 m | >20 m | >20 m |
| Joint aperture | None | None | None |
| Roughness | Low | Low | Low |
| Filling | None | None | None |
| Alteration | Low | Low | Low |
| Underground water condition | Dropping | Dropping | Dropping |
| Basic RMR | 45.80 | 40.60 | 44.90 |
| Joint orientation | Medium | Medium | Medium |
| Blasting Adjustment | 0.94 | 0.94 | 0.94 |
| Modified RMR | 38.40 | 33.50 | 37.50 |
| Rock Mass class | Poor | Poor | Poor |
| Cohesion, MPa | 0.10 - 0.20 | | |
| Internal friction angle, $^0$ | 15 - 25 | | |

Table 2. Rock mass properties based on Hoek-Brown Beyhan and Yuvka (2016).

| Definition | | Floor Marl | Lignite | Roof Marl |
|---|---|---|---|---|
| Hoek - Brown Constants | $m_b$ | 0.423 | 0.313 | 0.361 |
| | s | 0.0011 | 0.0003 | 0.0006 |
| | a | 0.507 | 0.513 | 0.509 |
| $E_M$, MPa | | 817.04 | 131.73 | 438.9 |
| $\sigma_{cm}$, MPa | | 0.669 | 0.318 | 0.605 |

## 3. Seals

Details about the seals are given in this section. There are 29 seals already constructed in the mine. Concrete seals have been constructed since 2010. Brick and aerated concreted were used as construction material before that date. Wooden chocks were used front and back of seals as additional supports of the roof. Pipes were installed on seal for gas and water drainage. Thickness of these current seals are 45 cm, 50 cm, 70 cm, 1.0 m, and 2.0 m. Seals were classified in the mine as per dimensions and construction material in Table 3.

Table 3. Classifications of current seals.

| Seal Type | Width (m) | Height (m) | Construction Material | Number of Seals |
|---|---|---|---|---|
| A | 2.80 | 2.37 | Brick | 7 |
| B | 2.80 | 2.37 | Concrete | 11 |
| C | 4.60 | 3.50 | Concrete | 10 |
| D | 4.60 | 3.50 | Aerated Concrete | 1 |

A simple example can be seen in Figure 3 to demonstrate the location of seals in the underground mine. Dimensions as well as as-built projects for the seals can be also seen in Figure 4. Sample photos can be also seen in Figure 5 to demonstrate the current view of the seals.

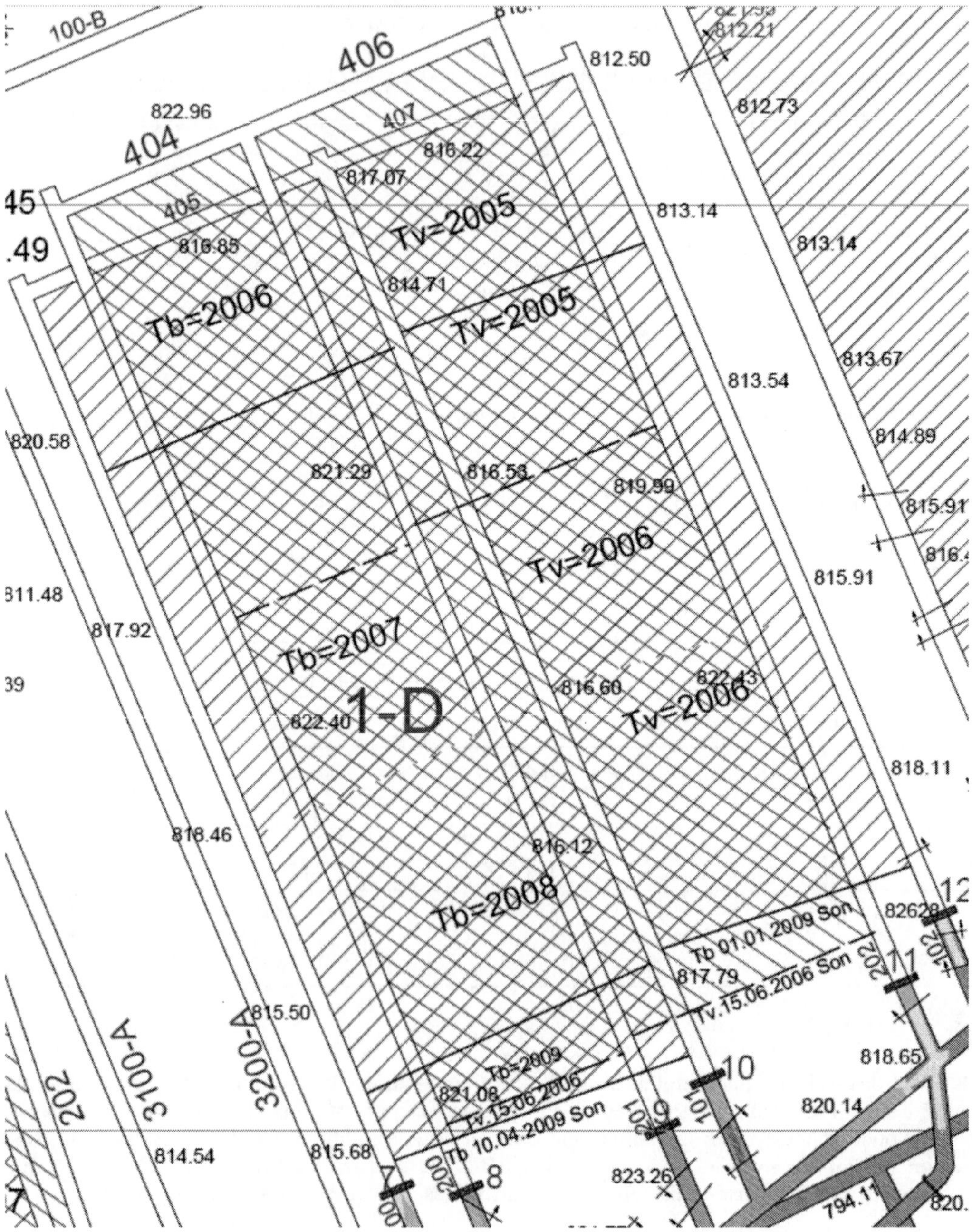


Figure 3. Seal location examples on the underground mine project.

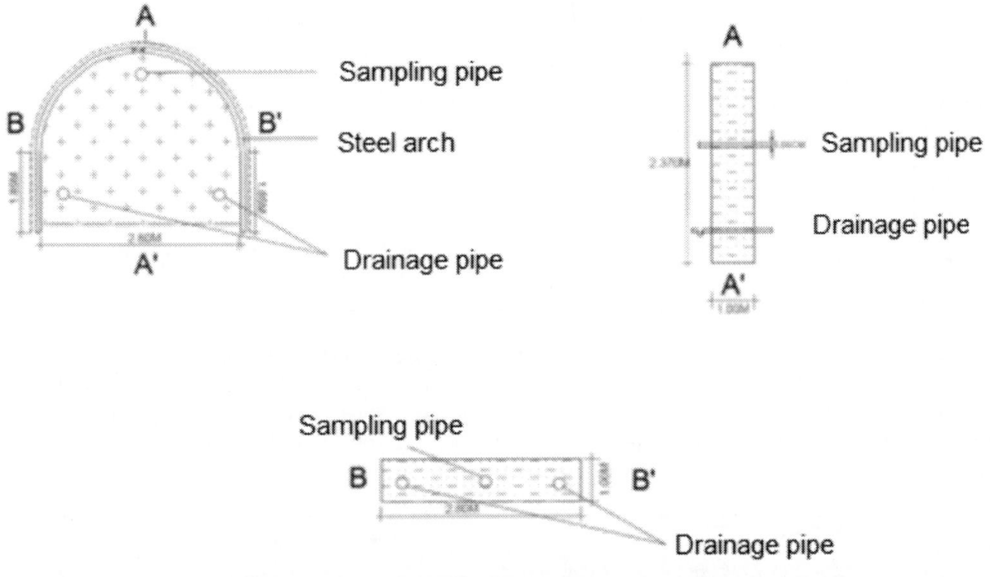

Figure 4. As-built projects for 1-m thick seal.

Material: Concrete, Height: 3.5m Width: 1.0 m

Material: Aeriated Concrete, Height: 3.5m Width: 1.0m

Material: Concrete, Height: 2.37m Width: 2.0 m

Material: Concrete, Height: 2.37m Width: 1.0 m

Figure 5. Sample photos of current seals in the mine

**4. Design and Evaluation**

Seals are used to make a barrier between abandoned panels and current main galleries and panels. Seals should stop air circulation through abandoned panels to eliminate oxidation of coal in the gob. On the other hand seals can be also used to protect the current working areas against dangerous mine gases. Added to that, seals are also important barrier especially to protect the working panels and galleries against flood risk. The gap between seal and back is also important for seal stability. In some cases this gap fills with some materials such as ashes, in some cases the gap is short due to the seal location. Residual distance between seal and back increases the importance of seal's stability. This distance in the mine changes from 10 m to 700 m. 21 number of seals have more than 100 m gap between seal and back.

Design studies mostly carried out based on countries engineering skills and investigations. That is why design implementations are generally called as English, German, Poland, Australian approaches in this study. National

Institute for Occupational Safety and Health (NIOSH) from United States of America also proposed own approaches for the design of seal (Zipf et.al., 2007).

As reported from the study of Zipf et.al. (2007), the required seal thickness against explosion can be calculated by the formula given below which is assigned as English approach (Hornsby et.al., 1985).

$t = \frac{W+H}{2} + 0.6$ *(t: seal thickness, w: width, h: height)* [1]

German approach proposed by Michelis and Kleine (1989) can be applied by the below equation.

$t = \frac{0.7\,a}{\sqrt{\sigma_e}}$ *(t: seal thickness, a: largest dimension-height or width, $\sigma_e$:flexural strength)*[2]

Niosh proposed a different graphical approaches based on full-scale tests to design thickness of seal. The system proposed different seal thickness for different construction material based on material uniaxial compressive strength. 3 different nomograms are used to withstand 3 different design pressure-time curve that are 4.4 MPa, 800 KPa, and 345 KPa. An example of 4.4 MPa nomogram can be seen in Figure 6. Further readings about American approach can be found in the study of Zipf et.al. (2007).

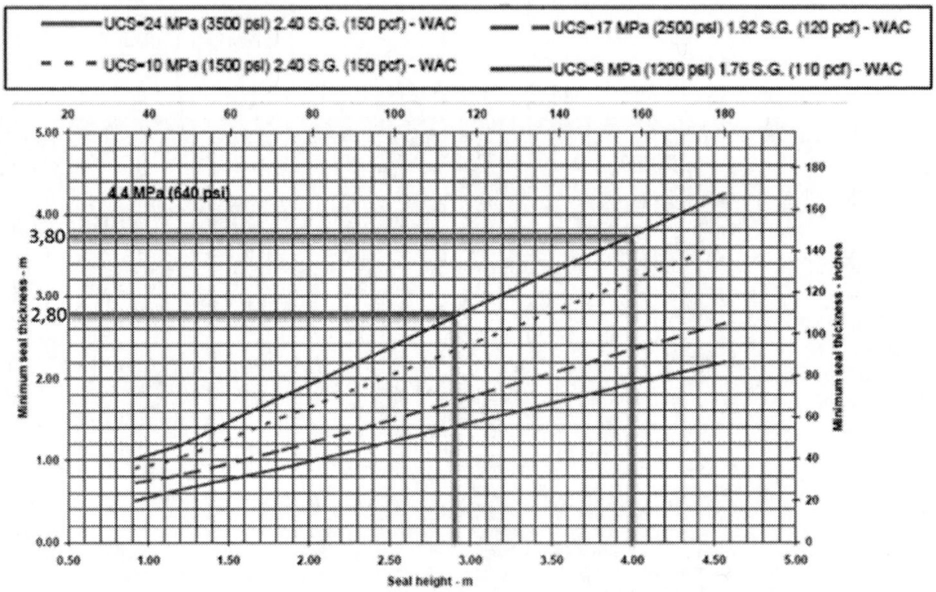


Figure 6. Seal thickness from American approach for 4.4 MPa design pressure-time curve (Zipf et.al., 2007).

These different approaches were used in this study to design the thickness of seals in Omerler Underground Mine. Due to the uncertainties based on methodologies, geology and hydrogeology, a factor of safety is applied in this study to protect the underground mine against disasters. 1.50 is selected as factor of safety. 3 different approaches were applied for 4 different type of seals classified in Table 3. The output of the study summarized in Table 4.

Table 4. Studies of seals to design thickness.

| Seal Type | Width [A] (m) | Height [B] (m) | Seal Thickness (m) England [C] | German [D] | NIOSH [E] | Design Thickness | Factor of Safety | Design Thickness (m) |
|---|---|---|---|---|---|---|---|---|
| **A** | 3.30 | 2.87 | 3.69 | 2.31 | 2.80 | **3.69** | 1.50 | **5.50** |
| **B** | 3.30 | 2.87 | 3.69 | 2.31 | 2.80 | **3.69** | 1.50 | **5.50** |
| **C** | 5.10 | 4.00 | 5.15 | 3.57 | 3.80 | **5.15** | 1.50 | **7.70** |
| **D** | 5.10 | 4.00 | 5.15 | 3.57 | 3.80 | **5.15** | 1.50 | **7.70** |

[A; B] *Extra 50 cm added to width and height due to the as built constructions,* [C]*Equation 1,* [D]*Equation 2, flexural strength is taken as 1 MPa,* [E]*Figure 6*

The results show that minimum thickness for the seals should be more than 3.7 m while maximum seal thickness can be reached up to 7.7 m. If the uncertainties regarding geology, hydrogeology and the other issues such as design methodologies will be covered, less values can be assumed as seal thickness due to decreasing amount of factor of safety.

## 5. Conclusion

This study is prepared to outline the main important steps for the design of seals using as a barrier between working and abandoned panels in underground coal mine. Omerler underground lignite mine is selected as case study area operated by Western Lignite Enterprises (GLI). 3 different approaches, English, German and American, were applied to decide the seal thickness which is an independent parameter apart from the height and width. A factor of safety is used to decrease the uncertainties based on geology, geotechnics, and hydrogeology. The increasing amount of knowledge will decrease the factor of safety as well as design thickness of seals. Further investigations should be carried out to design seal to include geological, geotechnical and hydrogeological aspects in the design studies.

## Acknowledgments

The research was a part of the industrial project supported by Turkish Coal Enterprises (TKI) Western Lignite's Enterprises (GLI). The authors gratefully acknowledge the assistance and cooperation provided by the Enterprises.

## References

Basarir, H. and Karpuz, C., 2004. *A rippability classification system for marls in lignite mines*, Engineering Geology, v. 74, p.303–318.

Beyhan, S. and Yuvka, S., 2016. *Park Teknik Elektrik San. ve Tic. A.Ş. İğdekuzu Underground Mine Seal Design for Panel E (in Turkish).* Project report, unpublished.

Hornsby, C.D., Hallam, P, Allan J.A., Walker, G., Boyle, J.C., and Criddle, S.J., 1985. *Sealing off fires underground.* Memorandum from the Committee of the Institution of Mining Engineers, U.K.

Michelis, J. and Kleine, W., 1989. *Development of components designed to resist explosion pressures of approximately 1 MPa for use in ventilation structures in underground mines.* In: Proceedings of the 23rd International Conference of Safety in Mines Research Institutes (Washington, DC, September 11–15, 1989). Pittsburgh, PA: U.S. Department of the Interior, Bureau of Mines, p. 859–857.

Zipf, R.K., Sapko, M.J. and Brune, J.F., 2007. *Explosion Pressure Design Criteria for New Seals in U.S. Coal Mines.* Information Circular 9500, NIOSH, Pittsburg, USA.

# THE FUTURE OF COAL

2016 International Pittsburgh Coal Conference
Capetown, South Africa
August 8 – 12, 2016

*Dr. Farid A. Malik*
*FCC University, Lahore – Pakistan*

*Engr. Abid Aziz*
*Pak Motors (Pvt.) Ltd., Pakistan*

1

# THE FUTURE OF COAL

# COAL WITH NO TOLL

2

# THE FUTURE OF COAL

## Table of Contents



3

## THE FUTURE OF COAL

### Abstract


- Coal has been the historic fuel for humans. The Romans started its use in 2nd century AD. For over 2000 years it met human needs from heating to cooking. It fuelled the entire Industrial Revolution in Europe. In 1880 the British stretched its use to transportation, power generation, steel making and cement production. Every year around 1000M tons of Lignite coal is mined globally.
- Environmental considerations have challenged the use of coal as a 21st century fuel. The slogan being: 'COAL WITH NO TOLL'. Mississippi Power has accepted this challenge by building the Kemper County Energy facility where state of the art Integrated Gasification Combined Cycle (IGCC) technology has been applied to produce 582 megawatt electric power, together with production of Sulphuric Acid and Anhydrous Ammonia while 65% of the emitted carbon dioxide will be captured and sold to the oil companies. Water will be recycled this being the first coal plant with almost no emission which is expected to be in service in first half of 2016. Future of coal rests on the success of such technology initiatives. From 2nd century Rome to 21st century Mississippi coal as a fuel has come of age.


4

## THE FUTURE OF COAL

- Invention of the wheel
  - Ancient Sumerians got humanity rolling in 3200 BC
  - Propulsion without wheels ?

  - Discovery
  - Earliest source of heat and light
  - Chinese started using coal 3000 years ago
  - Romans started use of coal in 2nd century AD
  - In 1880 the British stretched use of coal to transportation, power generation, steel making and cement
  - In the 21st century though in decline coal still produces more than 50% power globally

5

## THE FUTURE OF COAL

COUNTRIES WITH THE LARGEST COAL RESERVES

Figure I

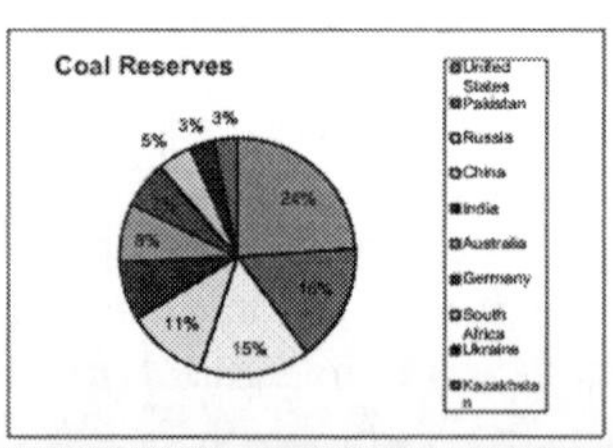


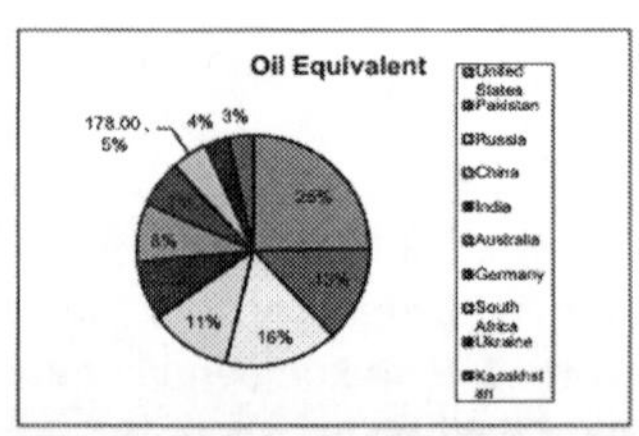


| Country | Coal Reserves Billion Tones | Coal Reserves (%) | Oil Equivalent Billion Barrels |
|---|---|---|---|
| United States | 274.00 | 24.0 | 887.00 |
| Pakistan | 184.00 | 16.1 | 480.00 |
| Russia | 173.00 | 15.1 | 560.00 |
| China | 126.00 | 11.0 | 407.00 |
| India | 93.00 | 8.1 | 301.00 |
| Australia | 90.00 | 7.9 | 291.00 |
| Germany | 73.00 | 6.4 | 236.00 |
| South Africa | 55.00 | 4.8 | 178.00 |
| Ukraine | 38.00 | 3.3 | 123.00 |
| Kazakhstan | 37.00 | 3.2 | 120.00 |

6

## THE FUTURE OF COAL

- FUTURISTIC PROJECTS
  - **LONGVIEW POWER WEST VIRGINIA (WV)**
    - Supercritical pulverized coal fired boiler technology producing 700 MW net electric power
    - Meets or exceeds Best Available Control Technology (BACT) standards by utilizing:
      - Low nitrogen oxide (Nox) burners and staged combustion
      - Selective catalytic reduction for final Nox control
      - Hydrated lime injection for acid mist control
      - > 99% removal fabric filter baghouse for particulate control
      - > 99.5% removal of wet flue gas desulfurization (FGD) for $SO_2$ control
      - Lowest $CO_2$ EMISSIONS OF ANY COAL FIRED PLANT IN (WV)

7

## THE FUTURE OF COAL

- **Kemper County Energy Facility Mississippi**

  - Integrated Gasification Combined Cycle (IGCC) technology using TRIG system producing 582 MW net electric power
  - $CO_2$ capture > 65% and sold to oil industry
  - Chemical production of sulphuric acid and anhydrous ammonia
  - Closed loop water recirculation
  - First coal facility with almost zero emissions

8

## THE FUTURE OF COAL
### How it Works

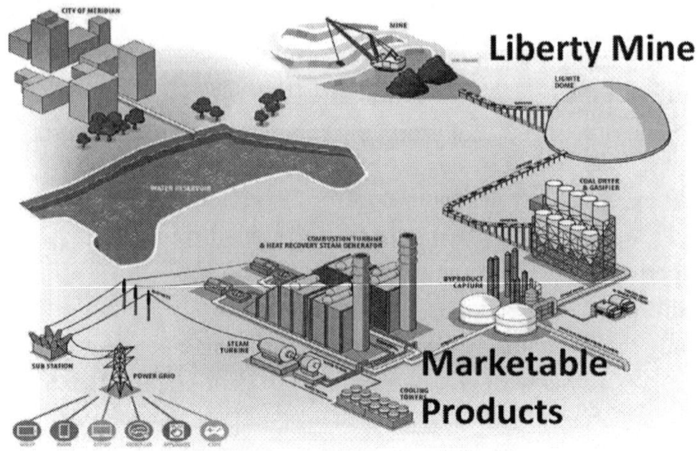

Liberty Mine

Process Water

Output

Marketable Products

## THE FUTURE OF COAL
### GASIFIER

# TRIG™

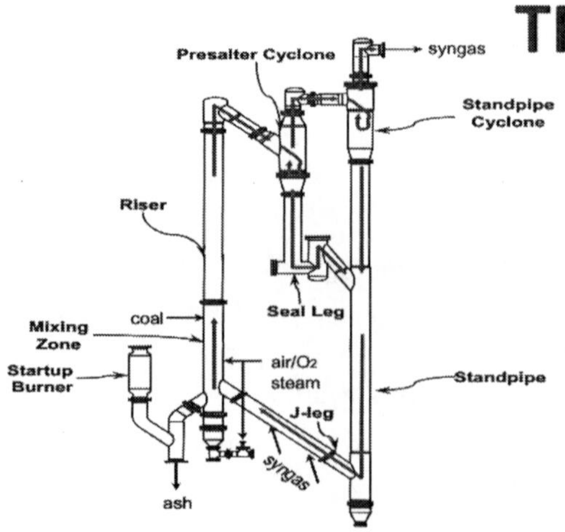


- Innovative design
- High Reliability Design
- Lower Fuel Costs
- Creates synthesis gas
- Excellent Environmental Performance

# THE FUTURE OF COAL

**SNG vs LNG**
(National Taiwan University)

- **SNG**

1. Air Separation Unit (ASU)
2. Gasification
3. Syngas Treatment (water gas – shift reaction, cooling, acid gas removal)
4. Methanation Reaction
5. Electricity Production
6. Energy Conversion Efficiency 60.38%

**Cost $10.837 (USD / GJ)**

- **LNG**

1. Liquification
2. Transportation
3. Gasification

**Cost $ 14-17 (USD / GJ)**

11

# THE FUTURE OF COAL

Conclusions:

- With close to 1200 billion tons of global coal deposits it will remain an important energy resource in the future
- Use of around 1000 million tons of Lignite coal annually calls for environmental sensitivities.
- Energy deficient nations like Pakistan have no option but to rely on their indigenous coal reserves reaching upto 30 – 40% of the mix
- SNG will continue to be a viable option for coal rich countries like Pakistan
- State of the art and environmental friendly technologies like IGCC hold promise for the future of coal
- Kemper County Energy facility will prove to be the way forward for:

**COAL WITH NO TOLL**

12

# Update on the Kemper County IGCC Project

NATIONAL ENERGY TECHNOLOGY LABORATORY

Diane Revay Madden, Project Manager
U.S. Department of Energy
National Energy Technology Laboratory

## Kemper County IGCC Project Overview

NATIONAL ENERGY TECHNOLOGY LABORATORY

- **2x1 Integrated Gasification Combined Cycle (IGCC)**
  - 2 Transport Gasifiers.
  - 2 Siemens SGT6 - 5000F CTs.
  - 1 Toshiba steam turbine.
  - 582 MW peak and 524 MW on syngas.
  - Heat rate 12,150 Btu/kWh.
  - 28.1% HHV efficiency w/ $CO_2$ control and >40% moisture coal.
  - Selexol for $H_2S$ and $CO_2$ removal.
  - >65% $CO_2$ capture (~800 lb/MWh emission rate).
  - Mine mouth lignite.

| Kemper Lignite Composition | | | | |
|---|---|---|---|---|
| | | Average | Min | Max |
| Heat Content | btu/lb | 5,290 | 4,765 | 5,870 |
| Moisture | % | 45.5 | 42 | 50 |
| Ash | % | 12.0 | 8.6 | 17 |
| Sulfur | % | 1.0 | 0.35 | 1.7 |

- **Owner and Operator: Mississippi Power**

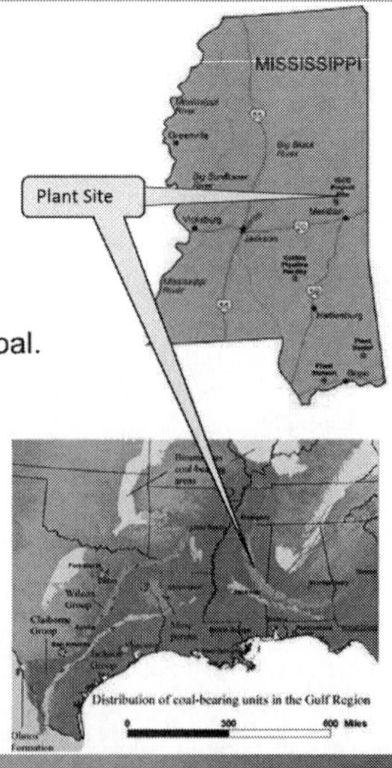

## Kemper County IGCC Project Overview

NATIONAL ENERGY TECHNOLOGY LABORATORY

- **Commercial Operating Date: 3rd Quarter 2016**
  - Combined cycle in-service August 2014.
- **Zero Liquid Discharge.**
  - Make-up water from City of Meridian treated effluent.
- **By-Products (TPY)**
  - ~3,000,000 – Carbon dioxide used for EOR
  - ~140,000 – Sulfuric acid
  - ~22,000 – Ammonia

U.S. DEPARTMENT OF ENERGY



## Kemper County IGCC Infrastructure

NATIONAL ENERGY TECHNOLOGY LABORATORY

- ~70 miles transmission
- ~ 60 miles $CO_2$ pipeline (for EOR)
- ~5 miles natural gas pipeline
- ~31,000 acre mine site
- ~2,900 acres plant site
- ~ 30 miles treated effluent line

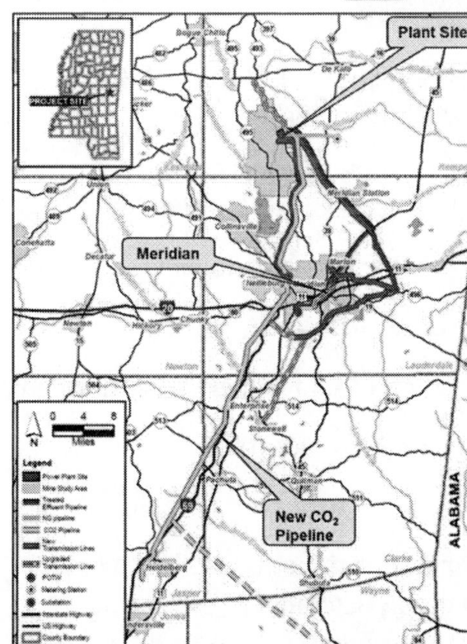


U.S. DEPARTMENT OF ENERGY

## Development of the Transport Gasifier for Power and Chemical Production

NATIONAL ENERGY TECHNOLOGY LABORATORY

***TRIG™ Leverages Long History of Fluid Catalytic Cracking (FCC) Expertise***

U.S. DEPARTMENT OF ENERGY 5

## Transport Gasifier

NATIONAL ENERGY TECHNOLOGY LABORATORY

- Moderate gas temperatures compared to other gasifiers
  - Allows for less expensive materials to be used
  - No slag—ash leaves as a powder
- Air-blown design
- Dry feed ideal for high moisture coals
- No expansion joints
- Long refractory life
- Simple construction scale-up (based on catalytic cracking experience)

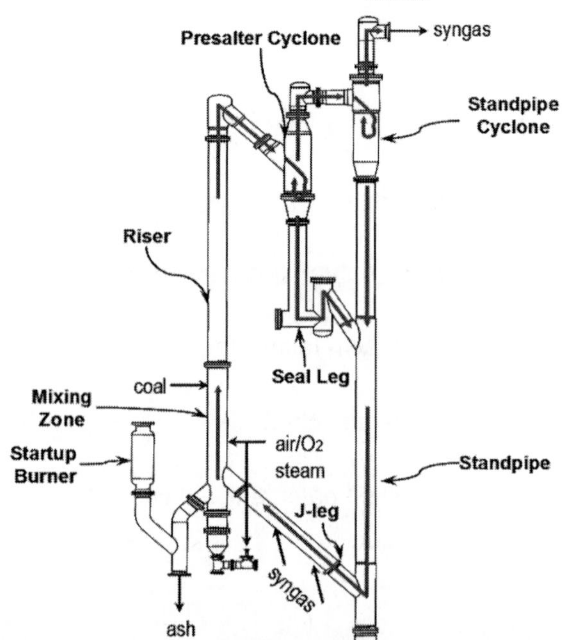


U.S. DEPARTMENT OF ENERGY 6

# Construction Update

**>99% Complete**

Construction complete on all plant equipment and piping

0.43 Project Recordable Incident Rate (7 times safer than Construction Industry average)

U.S. DEPARTMENT OF ENERGY

7

# Operator Training Simulator

- Has proven to be beneficial for procedure, DCS controls, SIS, and interface checkout.
- Numerous findings have saved countless hours that would have been spent during startup.

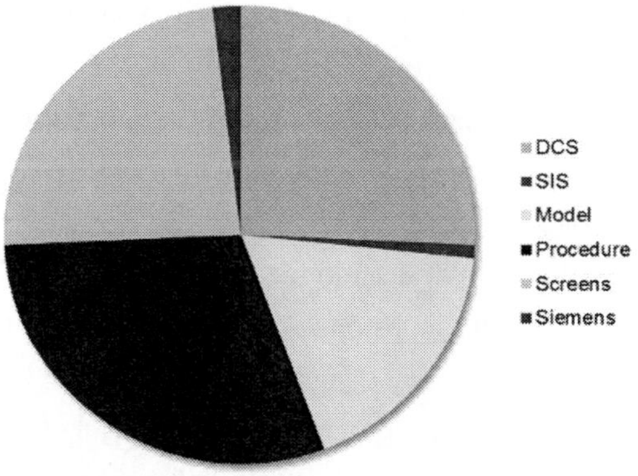


**Issues Solved Using the Simulator**

**Current Status**

- Delivered to site on December 23rd, 2015.
- Site Acceptance Test (SAT) completed in three weeks with only minor issues.
- Plant operations and training staff ran the simulator during the SAT with very little engineering or vendor support.
- Currently used for operator training.

U.S. DEPARTMENT OF ENERGY

## Gasifier Testing Update

NATIONAL ENERGY TECHNOLOGY LABORATORY

- Successful solids circulation tests on both trains.
- Excellent solids circulation rates achieved.
- Gasifier B heat-up to full operating temperature, and refractory cure-out complete.
- Successful operation of Gasifier B startup burners and diesel injection heat-up system.
- Coal feed/syngas production on Gasifier B began July 14, 2016 at 3:10 PM.

Pressurized Transport Gasifier
RX1002 RX2002

U.S. DEPARTMENT OF ENERGY

## Current Status

NATIONAL ENERGY TECHNOLOGY LABORATORY

- Commissioning activities essentially complete.
  - Essentially all rotating equipment tested.
  - All reactors loaded with catalysts.
  - Selexol unit filled with solvent.
  - Refrigeration system being charged.
  - Pressure checks complete.
  - Virtually all systems tested.
- Next steps—continue normal startup.
  - Complete Gasifier B controls tuning.
  - Water-gas-shift catalyst sulfiding.
  - Syngas to carbon capture unit.
  - Syngas to turbine.

U.S. DEPARTMENT OF ENERGY

U.S. DEPARTMENT OF ENERGY

# Commercial Operations

NATIONAL ENERGY TECHNOLOGY LABORATORY

- **Combined Cycle placed in service: August 9, 2014**
- **EFOR: about 1 percent. (Four times better than typical combined cycle of similar size).**

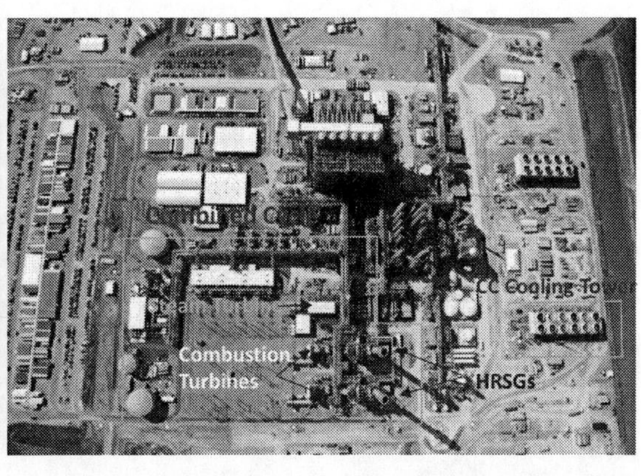




U.S. DEPARTMENT OF ENERGY

12

# Liberty Mine

NATIONAL ENERGY TECHNOLOGY LABORATORY

# First Syngas Produced July 14, 2016

NATIONAL ENERGY TECHNOLOGY LABORATORY

# For Additional Information…

NATIONAL ENERGY TECHNOLOGY LABORATORY

- **Office of Fossil Energy:** **www.fe.doe.gov**
- **NETL:** **www.netl.doe.gov**
- **Mississippi Power Company:** **www.mississippipower.com/about-energy/plants/kemper-county-energy-facility/home.cshtml**

# Flameless Pressurized Oxy-Combustion Technology

**The Strategic Role of Intermediate Scale (50 MWth) large pilot/demonstration Units in Each Major Geographic Area**

**tailored on Local, Quality and Low Rank, Coals to Mature Technology in View of Industrial Exploitation**

PCC 2016
Cape Town - August 8-12 20156

2

Content

## Itea Flameless Combustion 3

FPOC technology

FPOC enjoys «flameless» fundamental performances

**Traditional "flame" combustion**

**"chaotic"**

**non locally controllable**

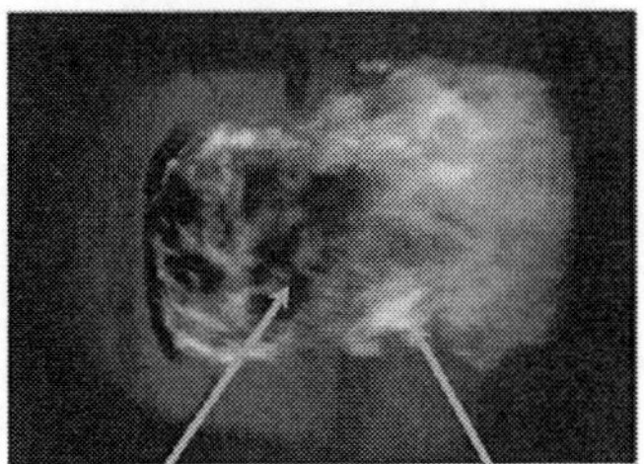

**Cool Zone**

**Flame Front: Peak Temperatures**

**Itea «Flameless combustion »**

***MILD, flameless, « volume combustion»***

**High Uniform Temperature**

NO NASTY COMBUSTION RESIDUES IN FUMES

## The product of the combustion: Vitrified Slag 4

FPOC technology

FPOC enjoys «flameless» fundamental performances

HEAVY AND FLY ASHES MELTING

Ashes quantitative melting and coalescence is unique to Isotherm process.

- Quantitative Melting
- Liquid phase triggers slag coalescence
- Molten slag drained at combustor bottom

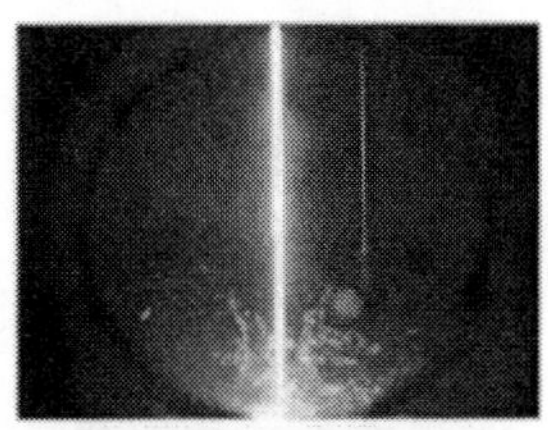

- Molten slag quenched in a water bath
- Vitreous Granular
- Zero Residual Carbon
- Impervious to Heavy Metal Leaching
- Fully inert

## FPOC Industrial Waste Treatment: Overall Emissions 5

| Flue Gas Emissions | | |
|---|---|---|
| Noxious Gas: | EU 2010/75 | Isotherm PWR® at stack |
| CO | 50 mg/m³, peak value 200 | < 1 mg/m³ (1) |
| NOx | 200 mg/m³, peak value 400 | **< 100 mg/m³** (1) |
| SOx | 50 mg/m³, peak value 200 | **< 10 mg/m³** |
| TOC | 10 mg/m³, peak value 20 | **<0.05 mg/m³** (1) |
| HCl | 10 mg/m³, peak value 60 | < 0.1 mg/m³ |
| PAH | 0,1 mg/m³ | **<0.0001mg/m³** (1) |
| Dust (total) | 10 mg/m³, peak value 30 | < 1 mg/m³ |
| **PM 2.5** | **Not yet regulated (Industrial avg. 1.000 – 5.000 µg/m³)** | **<10 µg/m³** |
| Dioxin, Furans | 0,1 ng/m³ | **<0.0001 ng/m³** (1) |
| Heavy metals | 0,5 mg/m³ | < 0.1 mg/m³ |
| **SOOT / Organic Carbon** | **Not yet regulated** | **Zero** (1) |

(1) Performances already attained at combustor outlet.

## FPOC Block Diagram 6

FPOC technology

FPOC «flameless» performs clean fumes since combustor outlet

simple process, efficient, compact, easy to run

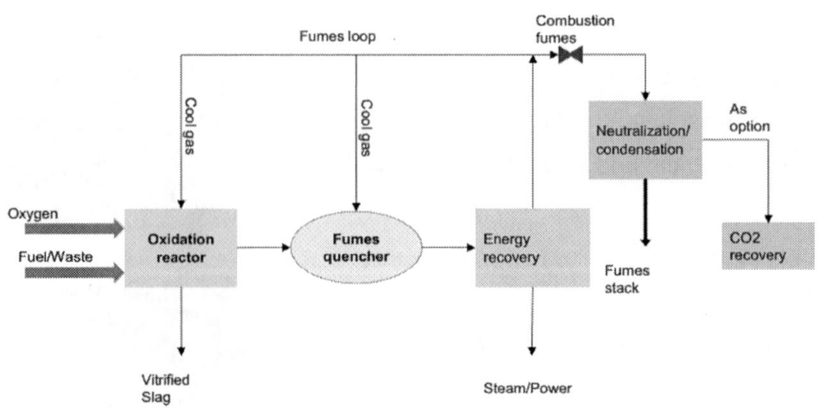

## FPOC application development pilot 7

*Flameless Pressurized Oxy-combustion*

**5 MWth Pilot Unit - Aerial View**
**in operation since 2004**

**120 ft X 60 ft**

**Small**
**Compact**
**Simple**
**(few unit operation)**

**Fully automated**
**Easy to operate**

At Ansaldo CCA Test Rig Gioia del Colle (BA) Italy

## FPOC strategy for industrial exploitation

recently

FPOC is gaining momentum

It is driving the attention of

Governmental Institutions
Investors

as the sole technical solution for coal CCS
at attractive LCOE

* FPOC is the candidate of election for CCS at Sulcis (Sardinia – Italy)
  First step: 50 MWth large pilot

* FPOC got FOA award pre-approval for 50 MWth large pilot feasibility study in USA

WHY FPOC ?
WHY 50 MWth large pilot ?

## WHY FPOC



Original Theoretical Studies Projections

LCOE Table 2 COE projections (different studies) (2)

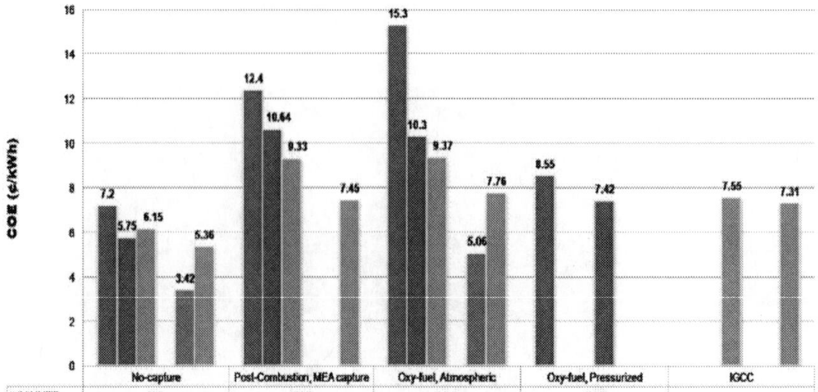


| | No-capture | Post-Combustion, MEA capture | Oxy-fuel, Atmospheric | Oxy-fuel, Pressurized | IGCC |
|---|---|---|---|---|---|
| CANMET | 7.2 | 12.4 | 15.3 | 8.55 | |
| NETL | 5.75 | 10.64 | 10.3 | | |
| Xu, B., et al. | 6.15 | 9.33 | 9.37 | | 7.55 |
| Proposed | | | | 7.42 | |
| Xiong, J., et al. | 3.42 | | 5.06 | | |
| Davison, J. | 5.36 | [illegible] | 7.76 | | 7.31 |

**Fig. 3. Cost of Electricity**

(2) Proceedings of ASME IMECE 2010– J. Hong, A. Ghoniem, R. Field, M. Gazzino

## WHY FPOC



Original CCS Theoretical Studies Projections
and Demo Projects FID road-block

The majority of the launched CCS DEMO projects collected engineered figures which did not pass the financial decision point, because of:

1 significantly lower efficiency, and much higher CAPEX vs theoretical projections

2 to high capital coupled with first-in-kind risk

3 Uncertain regulation scenario (timing, entity), and deep negative economics at the verge

2 & 3 they have practically made DEMO projects non-bankable

## WHY FPOC



FPOC 350 MWe Detailed Engineering Study

FPOC experimental development/validation for coal firing
at 5 MWth FPOC Itea pilot
extended to a broad spectrum of different type/rank of coal

it gave so promising results that:

ENEL European Utility Co-leader
in cooperation with
ITEA FPOC pioneer

decided to straightforward perform

FPOC 350 MWe DEMO unit
FEED and Detailed Engineering Study in 2010-2012

- Almost 90% based on vendor data and performance
- and, Relying on ENEL decades of experience in engineering innovative technologies

to ground development strategy with reliable economic figures

## WHY FPOC



FPOC 350 MWe Detailed Engineering Study

Table 3 FPOC competitive positioning – base line SCPC

| Parameter | unit | Non-CCS | CCS | | | | | |
|---|---|---|---|---|---|---|---|---|
| | | | Post capture | Oxy-combustion | | | | |
| | | SC base line | SC + amine | IGCC (1) | Atmospheric Oxy-Comb (1) | FPOC Flameless Pressurized Oxy-comb (1) | | |
| Power in | MWth | 1345 | 1880 | 1770 | 1760 | 1520 | | |
| Gross Power | MWe | 580 | 661 | | 786 | 723 | | |
| Net Power | MWe | 550 | 550 | 550 | 550 | 550 | | |
| Efficiency | % | 40.9 | 29.3 | 31.0 | 31.3 | 36.2 | | |
| Capital | M€ | 936 | 1693 | 2403 | 2350 | 1690 | | |
| CAPEX | €/kWenet | 1700 | 3078 | 4370 | 4270 | 3700 | | |
| LCOE | €/MWh | 76 | 133 | 140 | 137 | 104 | | |
| LCOE with low rank coal (-35% coal cost) | | | | | | 94 (2) | | |

Note: Location in Europe - Quality coal - CO2 storage not included
Elaborated according to std methods (DOE-NETL)
(1) Recent advance 0.16 kWh/kg O2 applied to all Oxy-combustion technologies
(2) FPOC only performs low rank coals, with efficiency and Opex almost equivalent to quality coals

## WHY FPOC – WHY 50 MWth large pilot



FPOC is the sole technology
enjoying
Engineered Figures close to Theoretical Projections

Efficiency CAPEX OPEX LCOE
top of competitive engineered figures vs other candidates theoretical projections
and, pretty close to DOE target (+30% vs SCPC) for CCS LCOE

however
detailed engineering study teaches
the direct jump from ITEA 5 MWth pilot to industrial
causes significant economics drawbacks

- Huge overdesign
- Lack of OEM experience in innovative, non-conventional unit operations
- Capital expensive, parasitic power demanding, complex, lego-type combinations of conventional guaranteed equipment
- The direct jump makes the industrial initiative non-bankable

if an intermediate (e.g. 50 MWth) development step is not materialized

## WHY 50 MWth large pilot



FPOC palrge pilot potential impact

Table 4 Expected Competitive Positioning Improvement

| Parameter | unit | Non-CCS | CCS | | | | | |
|---|---|---|---|---|---|---|---|---|
| | | | Post capture | Oxy-combustion | | | | |
| | | SC base line | SC + amine | IGCC (1) | Atmospheric Oxy-Comb (1) | FPOC Flameless Pressurized Oxy-comb (1) | FPOC optimized (post 50 MWt pilot) (1) | |
| Power in | MWth | 1345 | 1880 | 1770 | 1760 | 1520 | 1425 | |
| Gross Power | MWe | 580 | 661 | | 786 | 723 | 681 | |
| Net Power | MWe | 550 | 550 | 550 | 550 | 550 | 550 | |
| Efficiency | % | 40.9 | 29.3 | 31.0 | 31.3 | 36.2 | 38.6 | |
| Capital | M€ | 936 | 1693 | 2403 | 2350 | 1690 | 1390 | |
| CAPEX | €/kWenet | 1700 | 3078 | 4370 | 4270 | 3700 | 2710 | |
| LCOE | €/MWh | 76 | 133 | 140 | 137 | 104 | 91 | |
| LCOE with low rank coal (-35% coal cost) | | | | | | 94 (2) | 82 (2) | |

Note: Location in Europe - Quality coal - $CO_2$ storage not included
Elaborated according to std methods (DOE-NETL)
(1) Recent Oxy-Manufacturer advance : 0.16 kWh/kg $O_2$ ; applied to all Oxy-combustion technologies
(2) FPOC only performs low rank coals, with efficiency and Opex almost equivalent to quality coals

## FPOC 50 MWth large pilot at Sulcis (Sardinia) 15

Sulcis Project Background

SOTACARBO
SOCIETÀ TECNOLOGIE AVANZATE CARBONE S.P.A.

SOTACARBO ENEA
$CO_2$ TECHNOLOGY CENTRE - SULCIS

A CONSORTIUM BETWEEN SOTACARBO AND ENEA

is managing

THE CONSTITUTION of a CENTER OF EXCELLENCE for CLEAN ENERGY TECHNOLOGY

based on

PRESSURIZED OXY-COMBUTION TECHNOLOGY

contributed by

ITEA Spa (Sofinter Group)

It will materialize the more promising R&D Project for Coal CCS Technology

that ITALY contributes to Environment SAVING

## FPOC 50 MWth large pilot at Sulcis (Sardinia) 16

Sulcis Project Background

CCS FLAMELESS PRESSURIZED OXY-COMBUTION
R&D PROJECTS AND RELEVANT GRANTS

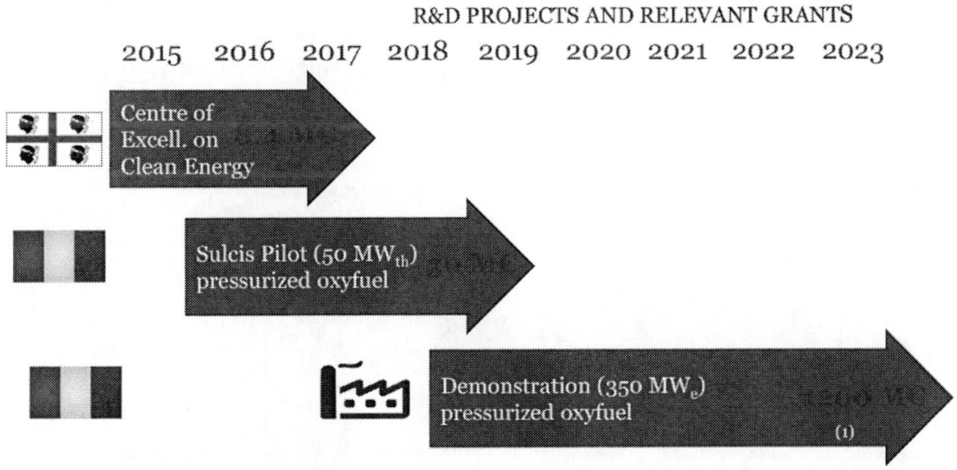


**Italian law 21 Feb. 2014, N. 9:**

**Incentive** for covering CCS extra costs: 30 €/MWh; maximum: 2.100 GWh/yr.

(1) **60 M€/y = 1.2 Billion € in 20 years**

## FPOC 50 MWth large pilot at Sulcis (Sardinia)

17

Sulcis Project

Project Leader: Sotacarbo-ENEA
Technology: Flameless Pressurized Oxy-Combustion contributed by ITEA Spa

Location: CO2 Technology Center – Sulcis (Carbonia- Sardinia) at Sotacarbo Research Center

Capacity: 48 MWth
Firing loop
Fuels: high ash content coals
high Sulfur coals
O&G tars
co-firing with municipal wastes fractions
Specific ancillary features: SO2 conversion to commercial sulfuric acid
CO2 for CCUS preliminary tests

Estimated cost
(including running cost and 20% contingency): ~50 M €

## FPOC 50 MWth large pilot at Sulcis (Sardinia)

18

Sulcis Project

Technical Target: The pilot will validate many of the advanced solutions not considered in the 350 MWe detailed engineering design

**an example: 48 MWth large pilot design evolution 2010-2015**

**2010**

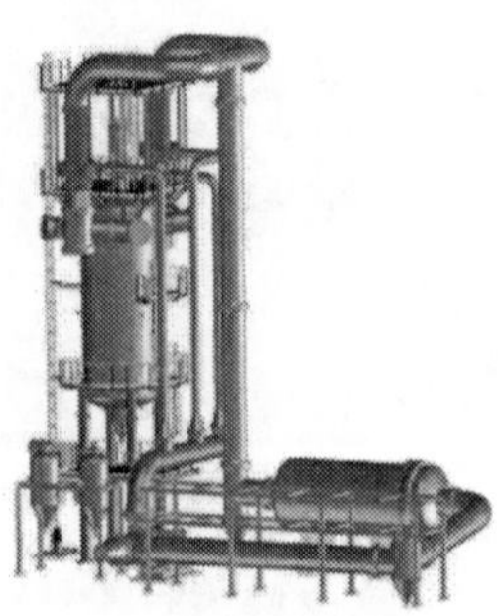

**2015**

Sulcis (Sardinia) 50 MWth large scale pilot

## FPOC 50 MWth large pilot at Sulcis (Sardinia) 19

Sulcis Project

**Project Plot Plan**
+/- 6 mth

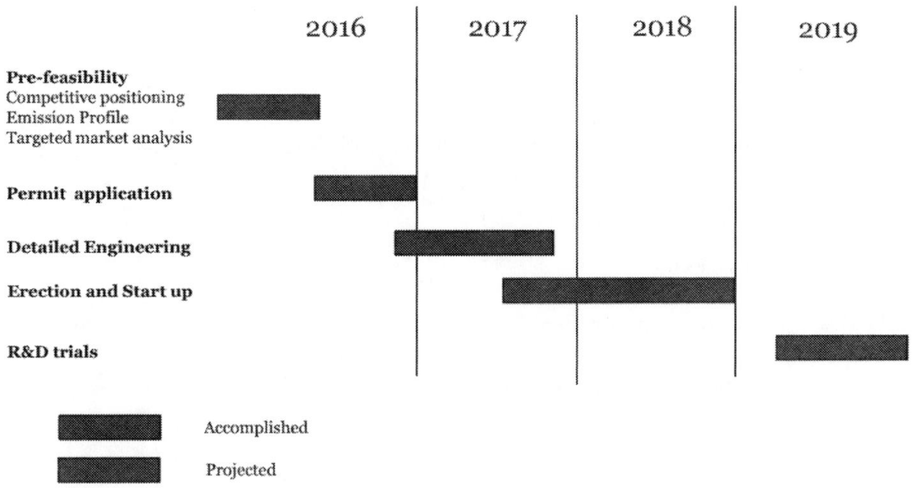


## FPOC 50 MWth large pilot in USA 20

Funding Opportunity Number: DE-FOA-0001459

Pre-feasibility CCS large pilot Funding
Reserved to advanced Technologies: < 30% LCOE increase vs SCPC
Capacity: > 20 MWe large pilot

**FPO Project filed**

| | | |
|---|---|---|
| <u>Project Leader</u>: | SwRI | (USA) Joshua Schmitt |
| <u>Partners</u>: | EPRI | (USA) Andrew Maxson |
| | GE Global Research | (USA) Douglas Hofer |
| | Jacobs | (USA) Stephen Scott |
| | ITEA | (Italy) Massimo Malavasi |
| | PRA | (UK) Peter Reineck |

With ITEA contributing Flameless process basics

Project already got pre-approval

Expected project startup – September 2016

## FPOC 50 MWth large pilot in USA

21

FPO Project Profile

Location: to be defined in the feasibility study

Firing: 50 MWth flameless firing loop
Fuels: US coals
PRB
Lignite and virgin lignite
Specific ancillary features: pressurized hot fumes turboexpansion feasibility study

Technical Targets
- Validation of scale up rules
- Scale up validation for local fuels
- Detailed design of non conventional equipment and systems and develop relevant vendor data and performance

## FPOC: READY-FOR-CCS solution

22

FPOC easy retrofit for brindging to CCS world

FPOC «Ready-For-CCS» solution

To respond to Uncertain Scenario issue

Flameless Fundamentals

Clean Fumes

Pressurized (12 barg)

High $CO_2$ concentration

=

it means ease in $CO_2$
Compression-liquefaction
for EOR/sequestration

they allow to

in the interim
increase cycle efficiency
via
hot fumes turbo-expansion
in non-CCS set up

later on
complement partially/totally
depending on regulation
with limited capital
to CCS set up

## FPOC: READY-FOR-CCS solution



FPOC technology further development step

Table 5 Resulting Economic Profile: CCS and Ready For CCS

| Parameter | unit | Non-CCS | CCS | | | | | CCS ready |
|---|---|---|---|---|---|---|---|---|
| | | | Post capture | Oxy-combustion | | | | |
| | | SC base line | SC + amine | IGCC (1) | Atmospheric Oxy-Comb (1) | FPOC Flameless Pressurized Oxy-comb (1) | FPOC optimized (post 50 MWt pilot) (1) | FPOC CCS ready (1) |
| Power in | MWth | 1345 | 1880 | 1770 | 1760 | 1520 | 1425 | 1425 |
| Gross Power | MWe | 580 | 661 | | 786 | 723 | 681 | 681 |
| Net Power | MWe | 550 | 550 | 550 | 550 | 550 | 550 | 613 |
| Efficiency | % | 40.9 | 29.3 | 31.0 | 31.3 | 36.2 | 38.6 | 42.9 |
| Capital | M€ | 936 | 1693 | 2403 | 2350 | 1690 | 1390 | 1210 |
| CAPEX | €/kWenet | 1700 | 3078 | 4370 | 4270 | 3700 | 2710 | 1974 |
| LCOE | €/MWh | 76 | 133 | 140 | 137 | 104 | 91 | 75 |
| LCOE with low rank coal (-35% coal cost) | | | | | | 94 (2) | 82 (2) | 68 (2) |

Note: Location in Europe - Quality coal - $CO_2$ storage not included
Elaborated according to std methods (DOE-NETL)
(1) Recent Oxy-Manufacturer advance: 0.16 kWh/kg $O_2$ ; applied to all Oxy-combustion technologies
(2) FPOC only performs low rank coals, with efficiency and Opex almost equivalent to quality coals

## ITEA's FPOC



FPOC

a call for partners
around a competitive technology development

for commercial equipment/systems development synergistic with FPOC potential

for joining technology final development step around large pilot initiatives

Acknowledgements
ITEA Spa acknowledges the tangible contribution of:
- ENEL, R&D and Engineering
- MIT, Mechanical Engineering Department and Energy Initiative Department
to Pressurized Oxy-Combustion coal application experimental development,
and to Pressurized Oxy-Coal Basic Process and Economics analysis



# THANK YOU FOR YOUR ATTENTION

Massimo Pietro Malavasi

ITEA Spa

massimo.malavasi@iteaspa.com

**2016 International Pittsburgh Coal Conference**
**Beijing, CHINA**
**August 8 – 12, 2016**

**PROGRAM TOPIC:**
**CLEAN COAL DEMONSTRATION AND COMMERCIAL PROJECTS**

# EXACT TITLE OF PAPER:
# SHENHUA COAL REFINING TECHNOLOGY DEVELOPMENT

*Presenting Author (Cliff Y. Guo, Director of Coal Upgrading R&D Center, National Institute of Clean-and-low-carbon Energy, P.O. Box 001 Shenhua NICE, Future Science & Technology City, Changping District, Beijing 102209, CHINA)*
*Contact Information: guoyi@nicenergy.com, Tel:86-10-57339619, Fax:86-10-57339649-9619*

**Abstract:**
Low rank coal (LRC) accounts for over 40% of the total coal reserves in China and US. The high efficiency and clean utilization of LRC plays an important role to meet the strong energy demand, reduce SOx/NOx/Hg and PM2.5 emissions. It is very important for China to develop advanced LRC utilization technologies. Various LRC upgrading technologies have been studied and developed over past decades, mainly in the directions of drying upgrading, pyrolysis polygeneration and gasification. But as today, there are no mature technologies for the efficient utilization of LRC in commercial scale. Driven by the strong energy demand and great market potentials, National Institute of Clean-and-low-carbon Energy (NICE) developed advanced Coal Refining (CoalRef®) technology. The CoalRef technology aims at efficiently produce high value liquid and solid products from LRC by using the principle of Oil Refining. The CoalRef technology is the integration of drying, pyrolysis, quenching/cooling, deactivation process, and further processing of gas/liquid/solid products to clean fuel and chemicals. The upgraded coal is not only stable and has much higher heating value than the raw coals, but also has much lower sulfur and Hg content, and can be used for gasification and combustion in a much cleaner way. The valuable liquid products can be used directly or upgraded to high value fuels and chemicals.

The pilot plant was built in ShaanXi Yulin of China to demo. the latest Coal Refining technology and ready for the test this fall. The successful demo. of this technology will have significant impact on the clean and efficient utilization of LRC in China and world.

international
INNOVATIVE
technologies

# FUEL ADDITIVE EFFICIENT COAL COMBUSTION

Marc Ottolini
CEO
10 August 2016

William Nimmo & Syed Sheraz Daood,
Energy Engineering Group, University of Sheffield, UK

**2016 Pittsburgh Coal Conference**

Cape Town, SOUTH AFRICA

---

international
INNOVATIVE
technologies

## Silanite™ Fuel Enrichment additive

Patented fuel additive system for coal-fired power stations

Multiple benefits:

3.4% volume equivalent additive yields:

- 1.1% increase in boiler efficiency
- 8-10% $NO_x$ reduction
- About 20% reduction in dust and particulate emissions
- Significantly better quality fly ash
- > 40% corrosion reduction

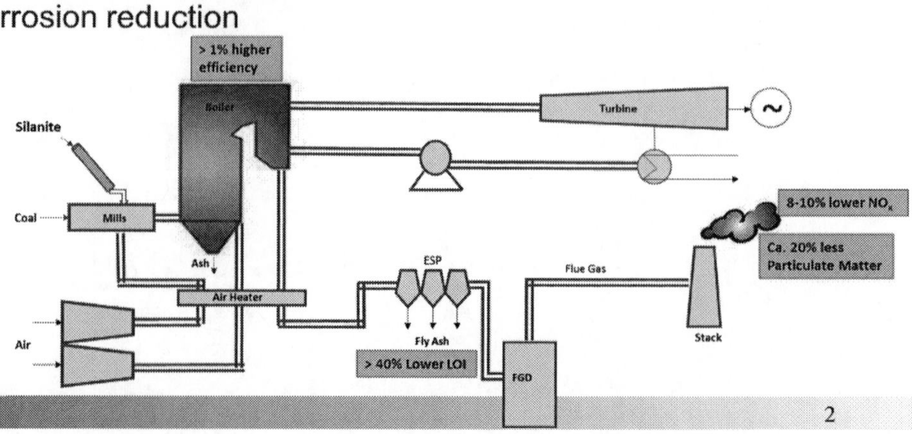

## Fundamental science



$$C_nH_m + O_2 + 4Fe_2O_3 + 2Al_2SiO_5 \rightarrow C + 2CO + CO_2 + C_xH_y + Fe_2SiO_4 + 2Al_2O_3 + SiO_2 + 2Fe_3O_4$$
$$or\ C_nH_m + O_2 + 2Fe_3O_4 + 4Al_2SiO_5 \rightarrow C + 2CO + CO_2 + C_xH_y + 3Fe_2SiO_4 + 4Al_2O_3 + SiO_2$$
$$or\ C_nH_m + O_2 + Fe_2O_3 + SiO_2 \rightarrow C + CO + CO_2 + C_xH_y + Fe_2SiO_4$$

A

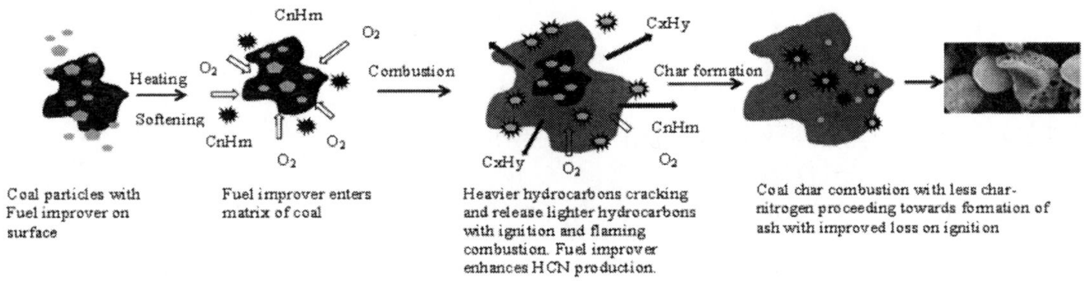


B

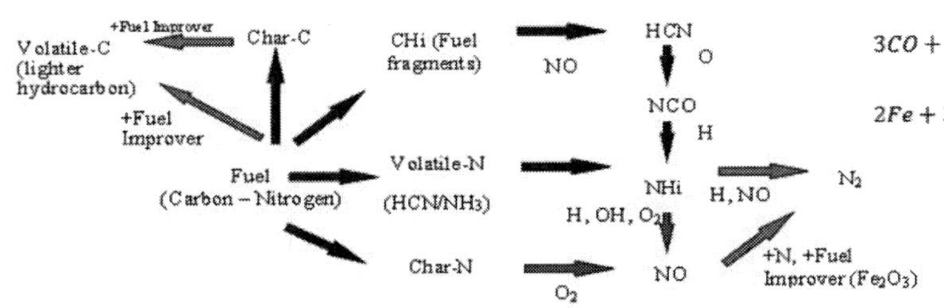


$$3CO + Fe_2O_3 \rightarrow 3CO_2 + 2Fe$$
$$2Fe + 3\,NO \rightarrow \frac{3}{2}N_2 + Fe_2O_3$$

 3

## Silanite™ Fuel Enrichment: Gas/Oil ratio

international
INNOVATIVE
technologies

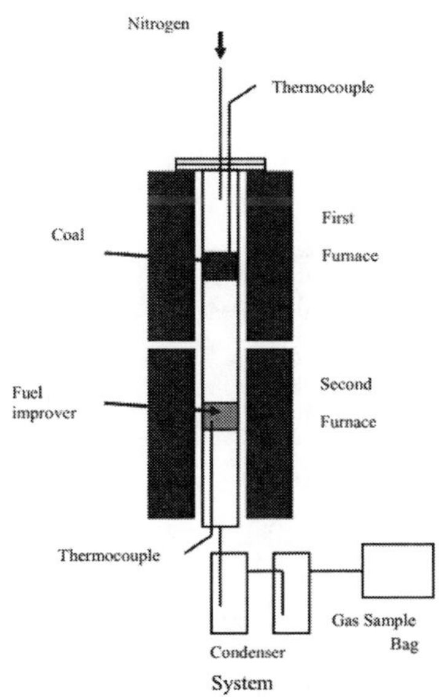


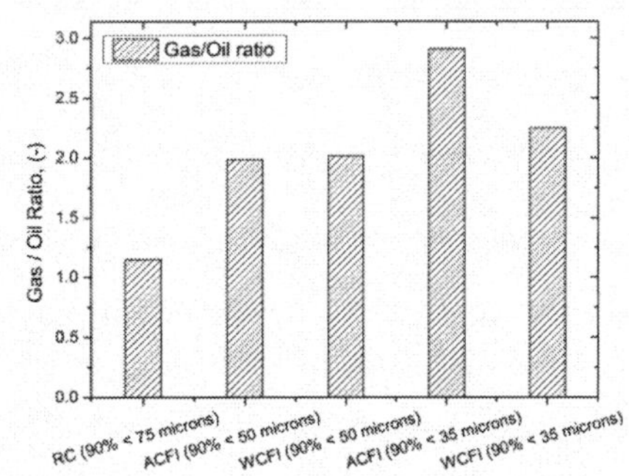


| Gaseous compositions | RC d(0.9) < 75 μm | ACFI d(0.9)<35 μm | WCFI d(0.9)<35 μm |
|---|---|---|---|
| CO | 16.50 | 18.80 | 18.85 |
| $H_2$ | 58.42 | 55.67 | 56.55 |
| $CO_2$ | 3.54 | 4.38 | 3.99 |
| $CH_4$ | 19.64 | 19.15 | 20.61 |
| $C_2H_4$ | 0.0184 | 0.0018 | 0.002 |
| $C_2H_6$ | 0.0004 | 0.0001 | 0.0001 |

 4

## Fundamental science

The effects of Silanite™ are based on 2 key mechanisms:

- Additive particles interact with coal particles causing heavier hydrocarbons to crack into lighter hydrocarbons
  - ➡ increased gas yield
  - ➡ more complete combustion
- Iron oxide in fuel additive acts as catalyst causing additional $NO_x$ reduction reactions




## Silanite™ additive: full-scale boiler

SembCorp Utilities Ltd: Test coal boiler

- 260-280 ton/h max. steam operating flow
- 170 ton/hr minimum steam flow
- Furnace dimensions = 29m x 7.5 m x 7.6 m
- Throat diameter for burners: 650mm, 12 burners placed in 3 landings
- Forced-induced draft fans = 2 each; Electrostatic precipitator; Doosan ball/ ring type coal mills; Main steam output splits into HP and IP steams.
- Plant related data recorded by Delta V control system

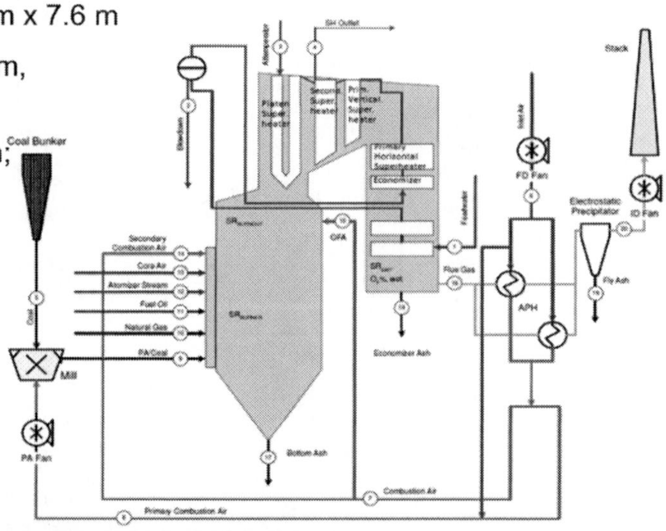

## Silanite™ additive: Test plan

Test Regimes- Description of activities

| Time Regimes | Description |
|---|---|
| **7:30 to 10:20** | Coal baseline |
| **10:20** | Silanite additive injection started |
| **10:20 to 11:00** | Transition from coal to additive plus coal |
| **10:40 to 15:00** | Additive + coal steady state |
| **14:44 to 19:00** | Coal mill loads on both B and C mill started to increase |
| **15:00** | Additive injection stopped |
| **15:00 to 15:30** | Transition from additive plus coal back to coal only |
| **15:30 to 16:30** | Coal ending baseline; not steady state as the coal mill load kept on increasing due to customer demand. |




## Silanite™ additive: NO reduction

**8.4% reduction in NOx for 3.4 vol. % Silanite Injection**

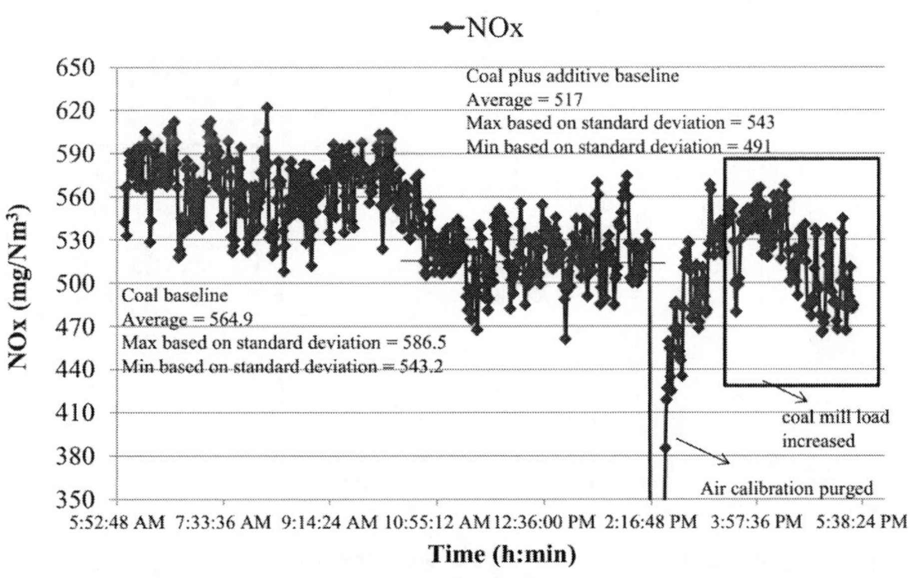

## Silanite™ additive: Flame zone temperature



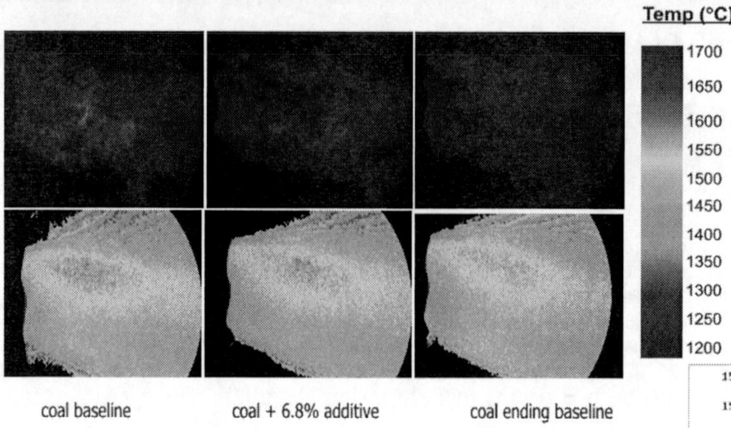


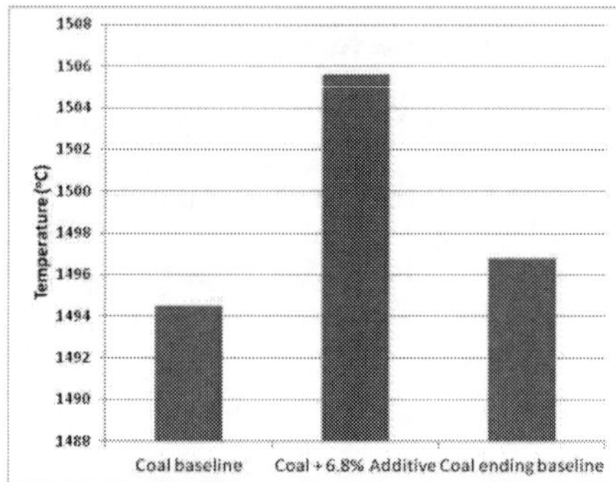


 

## Silanite™ additive: Loss on Ignition

international
INNOVATIVE
technologies

> 40% reduction in LOI

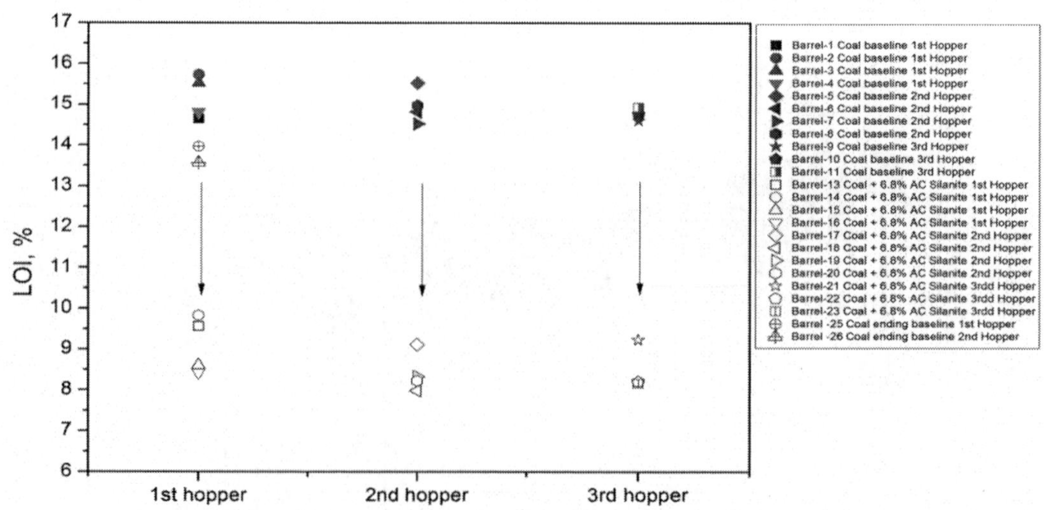


Fly ash sampling points from the back of the boiler

 

## Silanite™ additive: Boiler efficiency

international INNOVATIVE technologies

Net boiler efficiency according to BS EN 12952-15:2003

Net improvement boiler efficiency > 1%

The main influential factor for the higher net boiler efficiency is the unburned combustion content in the fly ash with and without the additive.

$$\eta\,(N)B = \frac{Q_N}{Q_{(N)Ztot}}$$

$$Q_N = m_{ST}(h_{ST} - h_{FW}) + m_{SS}\,(h_{FW} - h_{SS})$$

$$Q_{(N)\,Ztot} = m_F\,H_{(N)tot} + Q_{(N)Z}$$

Where

$$Q_{(N)Z} = P_M + P$$

$$H_{(N)tot} = (H_{(N)} + C_F(t_F - t_r))\,/\,(1 - l_u) + \mu_{As}h_{(N)AS} + \mu_A C_{pA}(t_A - t_r)$$

$$l_u = \frac{\gamma_{Ash}\,(1-v)}{1-\gamma_{Ash} - \gamma_{H2O}}\left[\frac{u_{SL}}{1-u_{SL}}\eta_{SL} + \frac{u_{FA}}{1-u_{FA}}\eta_{FA}\right]$$

(Based on estimated ash collection efficiency)

The above $Q_{(N)Ztot}$ equation gets simplified into the following equation when running on coal;

$$Q_{(N)}\,Ztot = m_F\left[(H_{(N)} + C_F(t_F - t_r))\,/\left[1 - \frac{\gamma_{Ash}\,(1-v)}{1-\gamma_{Ash} - \gamma_{H2O}}\left(\frac{u_{SL}}{1-u_{SL}}\eta_{SL} + \frac{u_{FA}}{1-u_{FA}}\eta_{FA}\right)\right] + \mu_A C_{pA}(t_A - t_r)\right] + P_M + P$$



## Silanite™ additive: improved dust collection in ESP

international INNOVATIVE technologies

17%; 18%; 17% less dust concentration compared to coal baseline on E, C&W legs

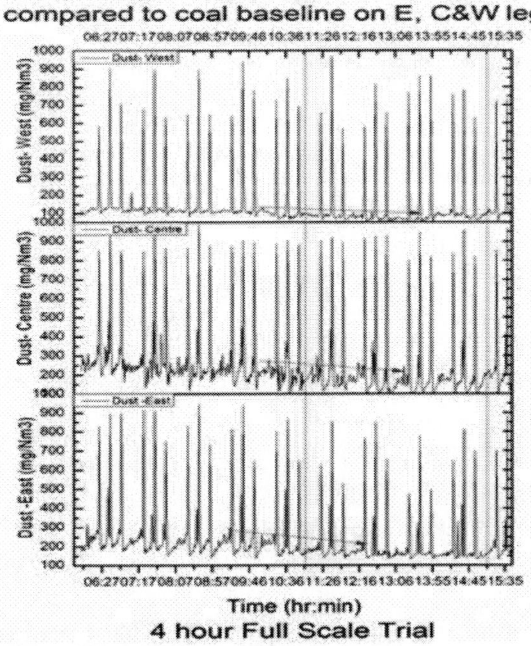


4 hour Full Scale Trial

## Silanite™ additive: Risk assessment

- Risk assessment conducted in 2014 with inputs from Sembcorp, Parsons Brinckerhoff and University of Leeds

### Erosion

Risk areas:

- Pulverizer mill
- PF tubes
- Boiler tubes

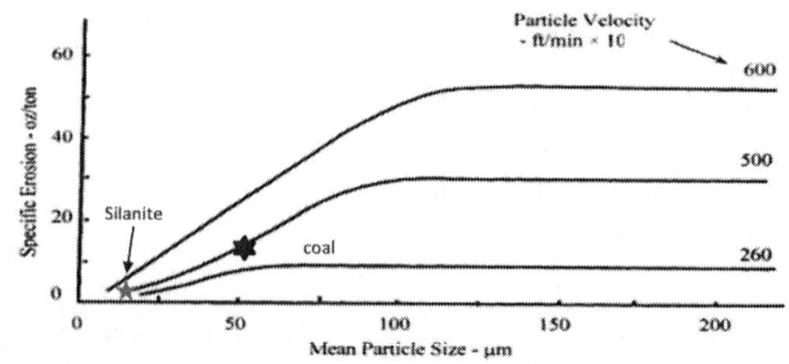


| Data: PSD | $d_{90}$ | $d_{50}$ | Actual Specific Density $\varphi_{p\,(approx)}$ |
|---|---|---|---|
| Ex-mill coal | 180µm | 54µm | 2 g/cm$^3$ |
| Silanite™ | 47µm | 17µm | 4 g/cm$^3$ |

| | $d_{90}$ | $d_{50}$ |
|---|---|---|
| Base line fly ash | 508 µ | 21 µ |
| Fly ash with 6.8% additive | 335 µ | 11 µ |

Silanite™ particles and Silanite™ fly ash too small to have material effect

 13

## Silanite™ additive: Risk assessment (2)

### Slagging & Fouling

With additive:

- Slagging index *Rs* moved from 0.08 to 0.19
  *Rs* < 0.6 low slagging propensity
- Fouling index *Fu* moved from 0.43 to 0.87
  *Fu* < 0.6 low fouling propensity
- $K_2O$ concentration lower
- Higher fly ash softening temperature

Conclusion: Effect of Silanite™ additive insignificant

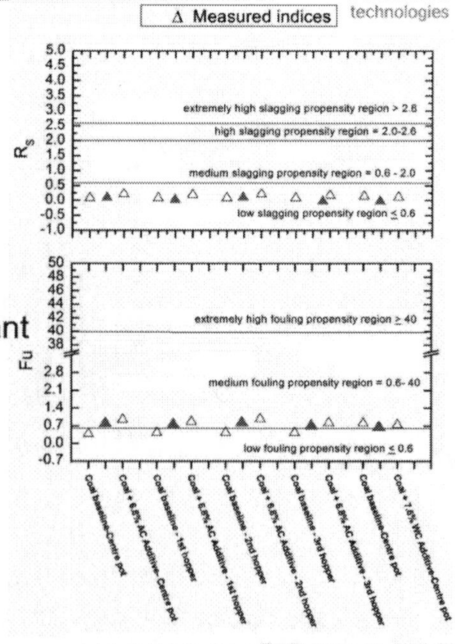


### Corrosion

- Electro-chemical corrosion happens in presence of liquid phase
- Silanite™ fly ash has higher fly ash softening temperature, thus reducing corrosion effect
- Test results indicate Silanite™ reduces corrosion

Conclusion: Silanite™ additive has no detrimental effect on boiler installation

 14

## Silanite™ additive: Ash fusibility temperatures

international INNOVATIVE technologies

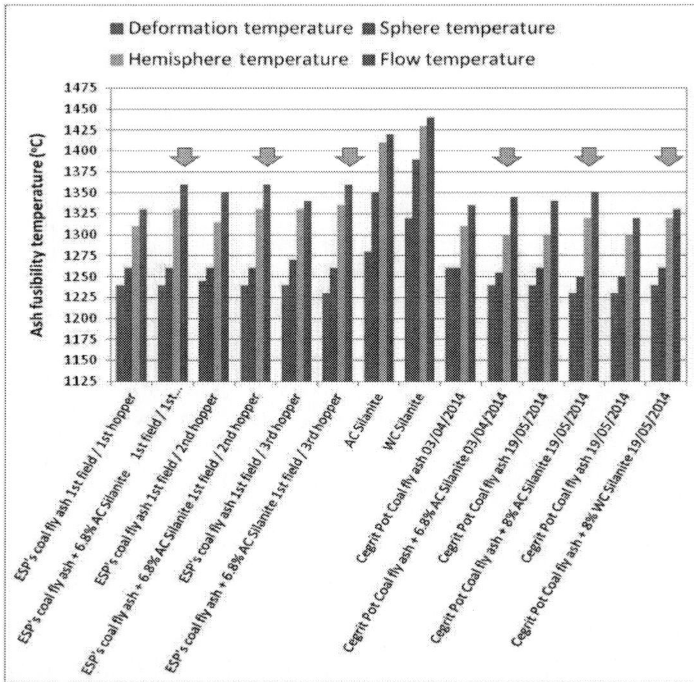


 15

## Silanite™ additive: Corrosion test facility

international INNOVATIVE technologies

- 6 silicon carbide electrically heated elements
- A gas mixture skid supplying simulated combustion gas products ($O_2$: 3-4%, $CO_2$: 14-16%, HCl: 0.03-0.05%, $SO_2$: 0.13-0.15%, $N_2$: 75-78%, $H_2O$: 6-8%)
- T22 ferritic steel alloy (Cr: 2.25%, Mo: 1%) coupons coated with the coal fly ash and Silanite mixed fly ash
- Exposure time: 1000 hrs

 16

## Silanite™: Corrosion test results: 40 – 50% reduction

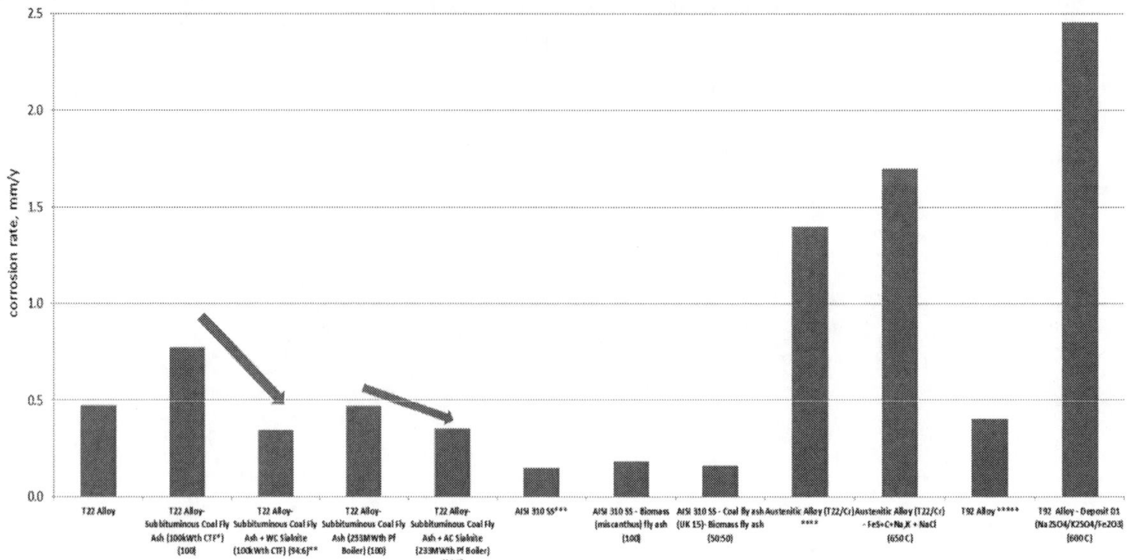


* Combustion test facility

** (94:6) Coal fly ash to additive weight ratio in coating

*** Onaivi Daniel A and Bill Nimmo. Fireside corrosion in oxy fuel combustion of coal/biomass. Intl J.of Engineering Research & Tech., 2016, Vol. 5, , Pages 400-404.

**** K. Natesan. High- temperature corrosion in power generating systems. Energy Technology Division, Argonne National Laboratory, Argonne, illinois.

***** T Hussain., A.U.Syed., N.J. Simms. Trends in fireside corrosion damage to superheaters in air and oxy-firing of coal/biomass. Fuel 2013, Vol 113, Pages 787-797.

 17

## Silanite™ additive: SEM / EDS images

international INNOVATIVE technologies

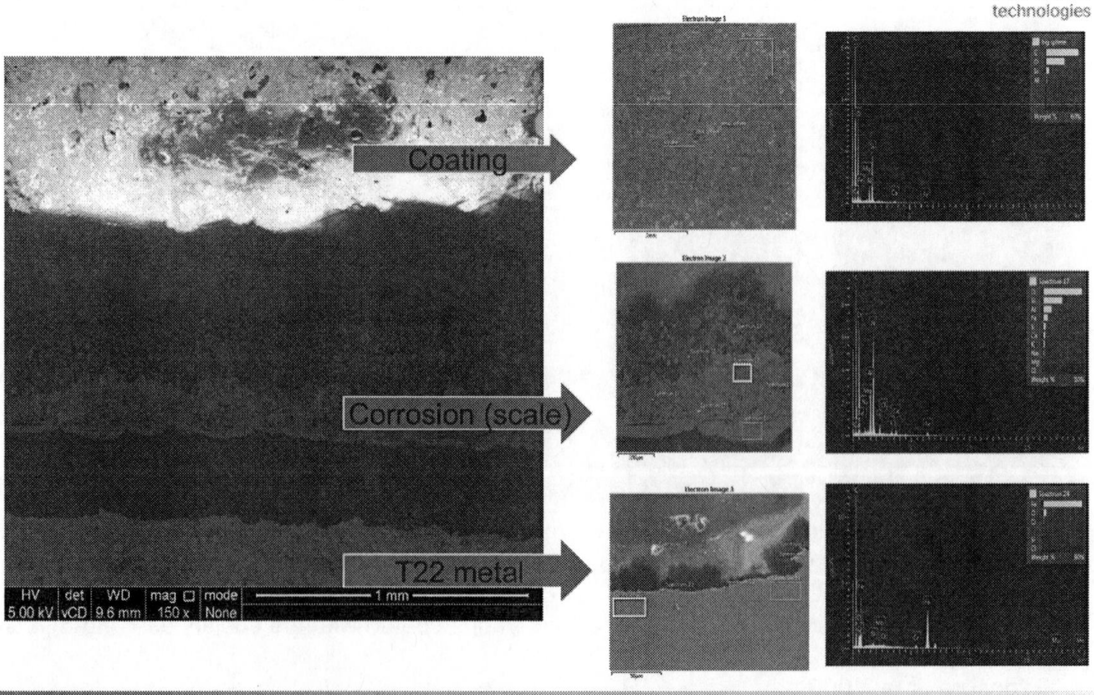


 18

# Silanite™ Additive: finer PSD, higher iron

international INNOVATIVE technologies

Silanite™ additive creates higher quality fly ash:

- > 40% reduction of PSD

| | $d_{90}$ | $d_{50}$ |
|---|---|---|
| Base line fly ash | 508 μ | 21 μ |
| Fly ash with 6.8% additive | 335 μ | 11 μ |

➡ Enhanced suitability for use in concrete

| | Coal fly ash | Coal + 6.8% Silanite fly ash |
|---|---|---|
| | % | % |
| **$Na_2O$** | 0.371 | 0.513 |
| MgO | 2.125 | 2.002 |
| **$Al_2O_3$** | 23.317 | 17.970 |
| **$SiO_2$** | 56.047 | 46.210 |
| $P_2O_5$ | 0.590 | 0.517 |
| $SO_3$ | 1.156 | 1.137 |
| Cl | 0.031 | 0.007 |
| **$K_2O$** | 2.183 | 1.645 |
| **CaO** | 6.174 | 4.871 |
| TiO2 | 1.029 | 0.751 |
| $V_2O_5$ | 0.018 | 0.012 |
| $Cr_2O_3$ | 0.016 | 0.025 |
| MnO | 0.048 | 0.058 |
| **$Fe_2O_3$** | 6.503 | 23.148 |
| CoO | 0.010 | 0.023 |
| NiO | 0.011 | 0.016 |
| CuO | 0.011 | 0.275 |
| ZnO | 0.023 | 0.349 |
| $ZrO_2$ | 0.049 | 0.037 |
| SrO | 0.253 | 0.172 |
| Mo | 0.001 | 0.103 |
| $SnO_2$ | 0.001 | 0.025 |
| Sb | 0.001 | 0.009 |
| PbO | 0.007 | 0.088 |





# Silanite™ + SNCR: enabling 70 - 90% AFR reduction

international INNOVATIVE technologies

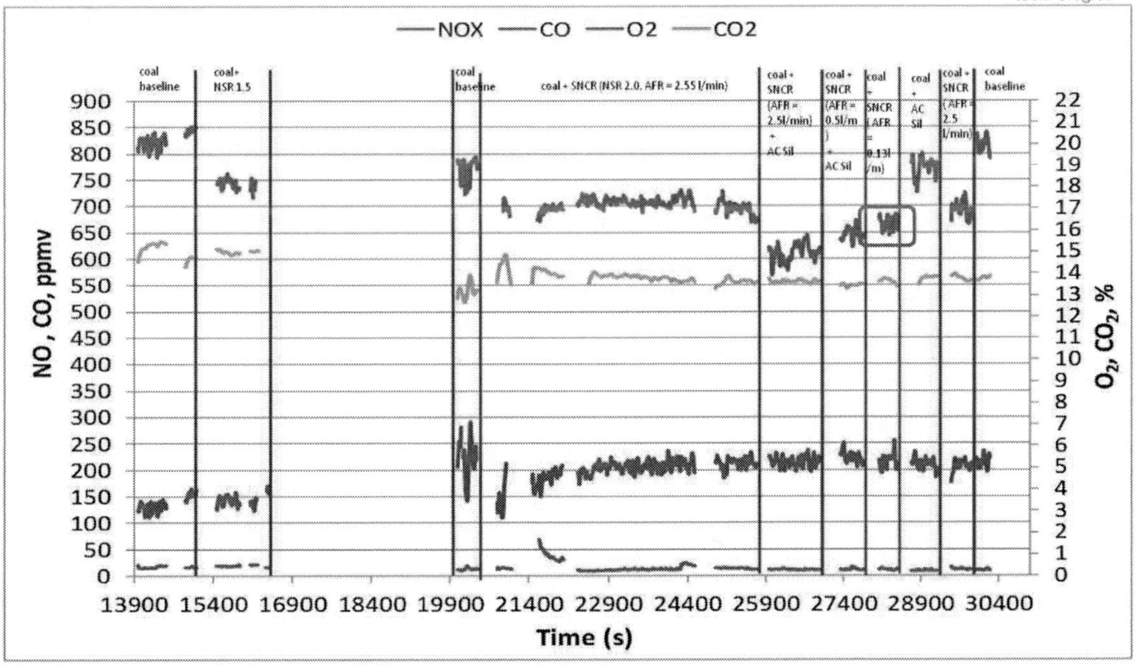

## Silanite™ + SNCR: supplementary effect

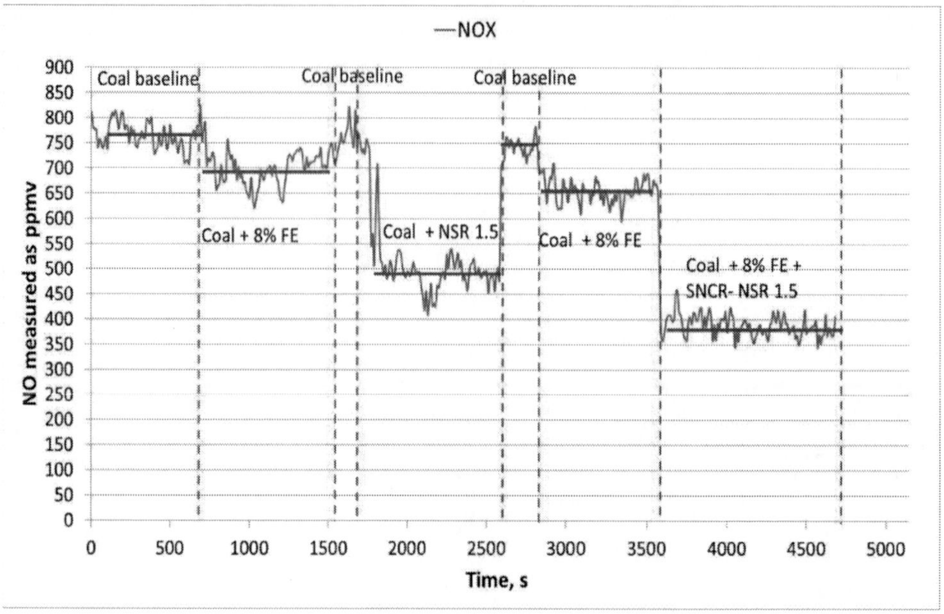





## Silanite™ Fuel Enrichment additive: summary

Multiple benefits:
3.4% volume equivalent additive yields:

- 1.1% increase in boiler efficiency
- 8-10% $NO_x$ reduction + advantageous interaction with SNCR
- About 20% reduction in dust and particulate emissions
- Significantly better quality fly ash
- > 40% corrosion reduction

- Easy to retrofit without need for outage

The University of Nottingham
UNITED KINGDOM · CHINA · MALAYSIA

Loughborough University

UNIVERSITY OF BIRMINGHAM

Research Councils UK Energy For a Low Carbon Future

# Biomass Co-firing to Improve the Burn-out of Unreactive Coals in Pulverised Fuel Combustion

Chris Bridge 10/08/2016

Efficient Fossil Energy Technologies
Engineering Doctorate (EngD) Centre

Biomass and Fossil Fuel Research Alliance BF2RA

The University of Nottingham
UNITED KINGDOM · CHINA · MALAYSIA

## Outline

- Introduction & Legacy Work
- Heating Rate Study
- DTF Devolatisation
- DTF Char Reactivity
- Summary

Efficient Fossil Energy Technologies
Engineering Doctorate (EngD) Centre

Biomass and Fossil Fuel Research Alliance BF2RA

The University of Nottingham
UNITED KINGDOM · CHINA · MALAYSIA

# Introduction

- Project Objectives
  - Understand how biomass can improve the combustion performance of unreactive bituminous coals during co-firing
  - Develop a quantitative assessment of biomass to coal ratio and the biomass species' that maximise beneficial effects
- Why?
  - Co-firing has become uneconomic in UK, however it is widely used around the world
  - Previous work observed catalytic effects when biomass ash was added to coals
  - Synergistic effects not agreed upon in literature
  - Benefits of catalytic effect due to alkali and alkaline metals include:
    - Increased volatile yield and improved char burn-out
    - Leading to decreased emissions


1. Quyn, D. M., Wu, H., Li, C-Z., *Volatilisation and catalytic effects of alkali and alkaline earth metallic species during the pyrolysis and gasification of Victorian brown coal*, Fuel, 2002
2. C. A. Querini, L. M. C., M. A. Ulla, E. E. Miro *Applied Catalysis B; Enviromental* 1999, *20*, 165.
3. H. Haykiri-Acma, S. Y. *Fuel* 2007, *86*, 373.


***Efficient Fossil Energy Technologies Engineering Doctorate (EngD) Centre***

Biomass and Fossil Fuel Research Alliance BF2RA

The University of Nottingham
UNITED KINGDOM · CHINA · MALAYSIA

# Legacy Work

Drop Tube Furnace (DTF) -200 ms

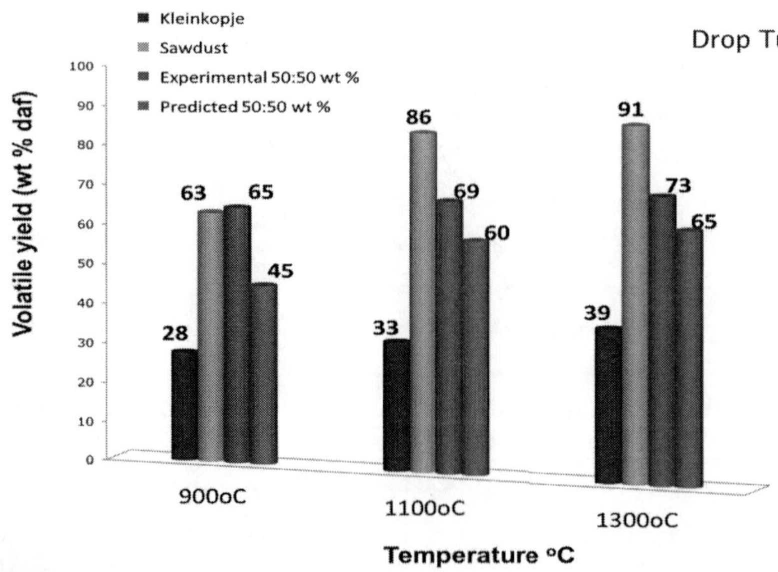


Farrow, T. S., *Work conducted at the University of Nottingham*, 2014

***Efficient Fossil Energy Technologies Engineering Doctorate (EngD) Centre***

Biomass and Fossil Fuel Research Alliance BF2RA

# Legacy Work

The University of Nottingham

UNITED KINGDOM · CHINA · MALAYSIA

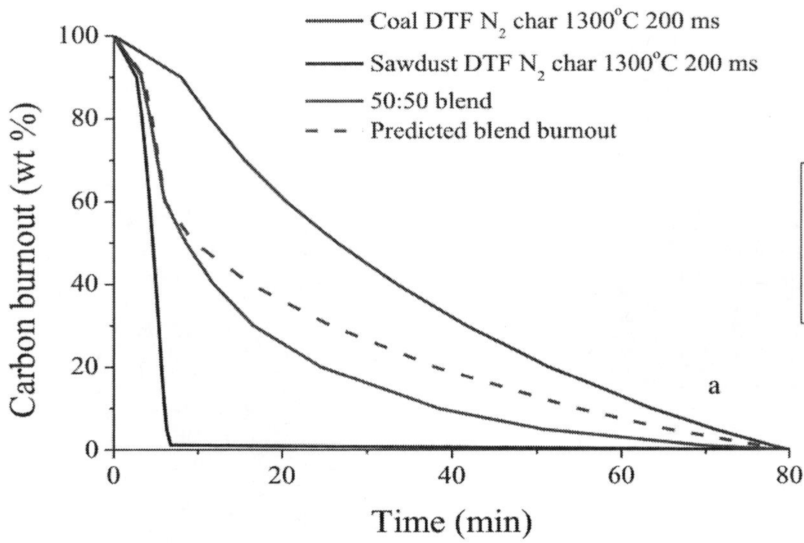


Char produced from DTF experiment on previous slide with TGA burn-out conducted

Farrow, T. S., *Work conducted at the University of Nottingham*, 2014

***Efficient Fossil Energy Technologies Engineering Doctorate (EngD) Centre***

Biomass and Fossil Fuel Research Alliance BF2RA

# Fuels

The University of Nottingham

UNITED KINGDOM · CHINA · MALAYSIA

Biomass

- Miscanthus
- Olive Cake
- Raw Bagasse
- Eucalyptus
- Commercial Wood Pellet
- Steam Exploded Fuel Pellet

Coal

- South African

***Efficient Fossil Energy Technologies Engineering Doctorate (EngD) Centre***

Biomass and Fossil Fuel Research Alliance BF2RA

The University of Nottingham
UNITED KINGDOM · CHINA · MALAYSIA

# Heating Ramp Rate Study
## Thermogravimetric Analysis (TGA)

TA SDT Q600
50 °C/min

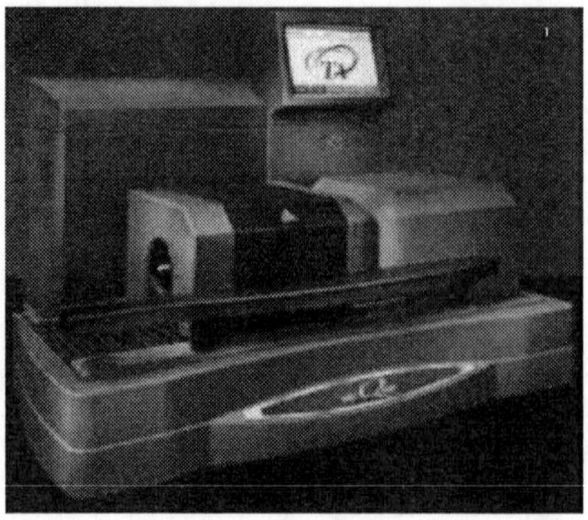

TA Discovery TGA
500 °C/min
>2000 °C/min

1. Bargal Analytical Instruments, Product Catalogue
2. Labwrench, Product Catalogue

*Efficient Fossil Energy Technologies Engineering Doctorate (EngD) Centre*
Biomass and Fossil Fuel Research Alliance BF2RA

The University of Nottingham
UNITED KINGDOM · CHINA · MALAYSIA

# Heating Proximate Analysis Method

**Ramp to 110°C** → **Ramp to 850°C** → **Switch gas** → **Cool down phase**

- Ramp to 110°C
  - 50°C/min
  - Nitrogen
  - Isotherm 5 mins
- Ramp to 850°C
  - 50°C/min
  - Nitrogen
  - Isotherm 15 mins
- Switch gas
  - From Nitrogen
  - To Air
  - Isotherm 30 mins
- Cool down phase
  - Cooled at 50°C/min

Sample weight 10 mg

3 experimental runs per blend

Predictions assumed an additive relationship

*Efficient Fossil Energy Technologies Engineering Doctorate (EngD) Centre*
Biomass and Fossil Fuel Research Alliance BF2RA

The University of Nottingham

UNITED KINGDOM · CHINA · MALAYSIA

# Slow Heating Ramp Rate

## TGA

**Volatile Matter of South African Coal Blended with Biomass at 50°C/min**

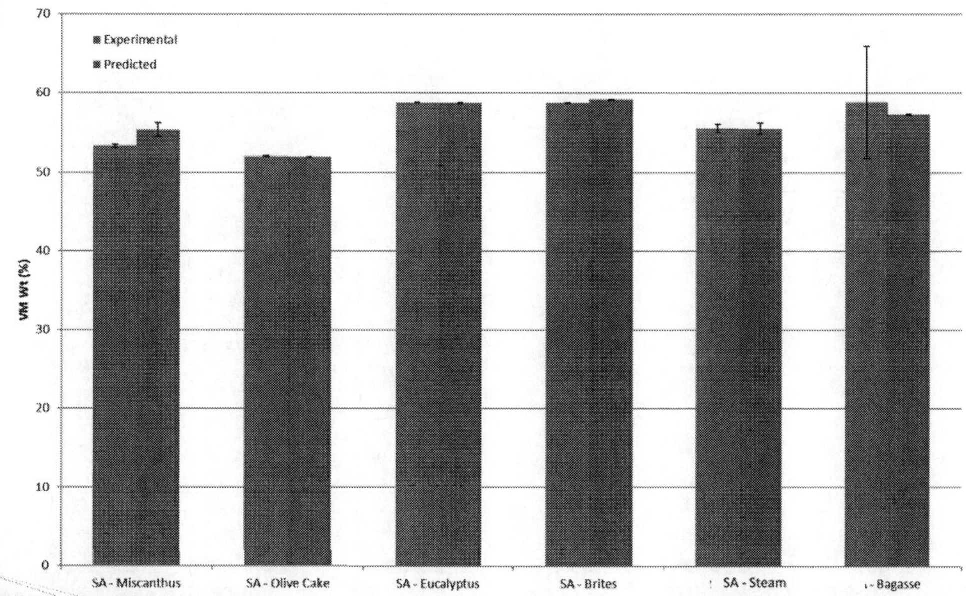


***Efficient Fossil Energy Technologies Engineering Doctorate (EngD) Centre***

Biomass and Fossil Fuel Research Alliance BF2RA

The University of Nottingham

UNITED KINGDOM · CHINA · MALAYSIA

# Intermediate Heating Ramp Rate

## TGA

**Volatile Matter of South African Coal Blended with Biomass at 500°C/min**

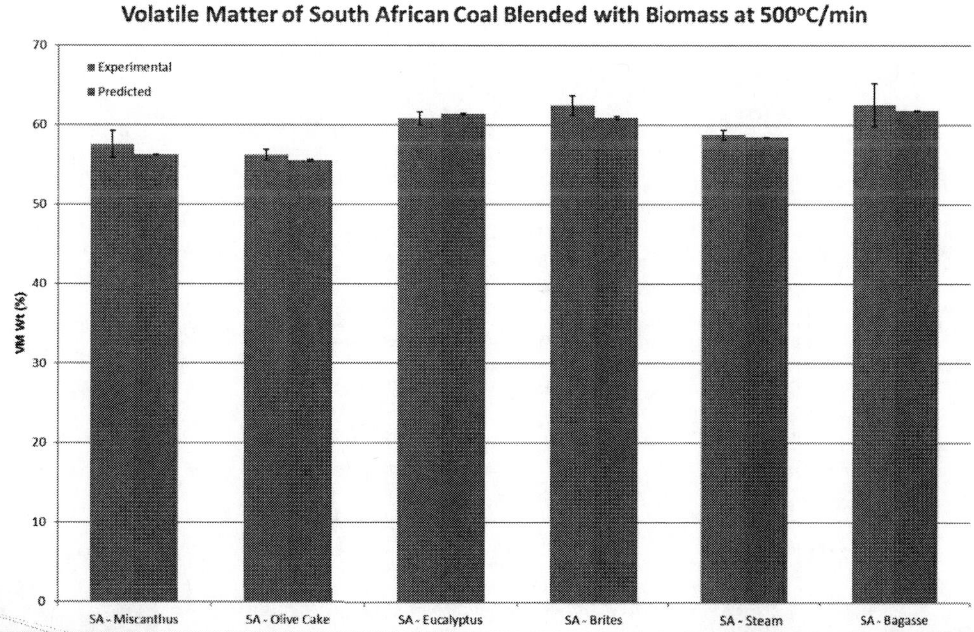


***Efficient Fossil Energy Technologies Engineering Doctorate (EngD) Centre***

Biomass and Fossil Fuel Research Alliance BF2RA

The University of Nottingham
UNITED KINGDOM · CHINA · MALAYSIA

# Fast Heating Ramp Rate

## TGA

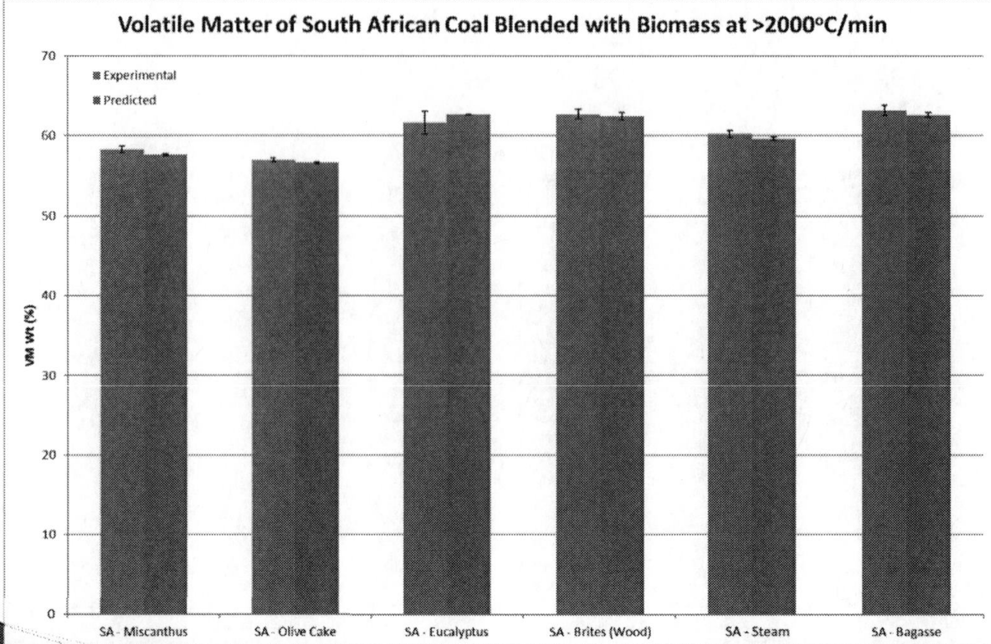


***Efficient Fossil Energy Technologies Engineering Doctorate (EngD) Centre***

Biomass and Fossil Fuel Research Alliance BF2RA

The University of Nottingham
UNITED KINGDOM · CHINA · MALAYSIA

# Devolatisation

## Drop Tube Furnace (DTF)

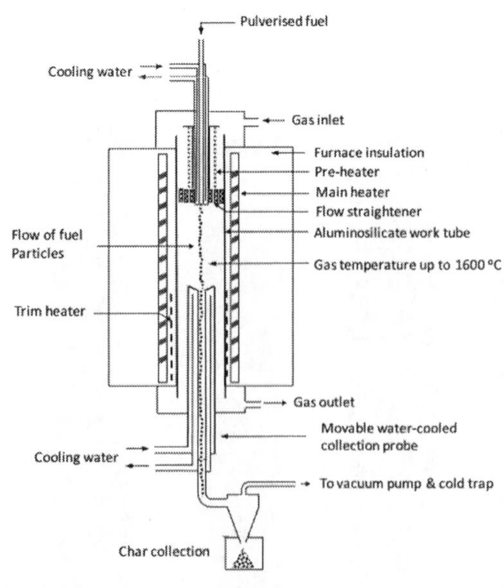


$$V_{DTF} = \frac{10^4 \times (Ash_{Char} - Ash_{Raw\ Fuel})}{Ash_{Char} \times (100 - Ash_{Raw\ Fuel})} \quad (\%\ daf)$$

Farrow, T. S, *PhD Thesis, University of Nottingham*, 2013

***Efficient Fossil Energy Technologies Engineering Doctorate (EngD) Centre***

Biomass and Fossil Fuel Research Alliance BF2RA

The University of Nottingham
UNITED KINGDOM · CHINA · MALAYSIA

# Devolatisation

## Drop Tube Furnace (DTF)

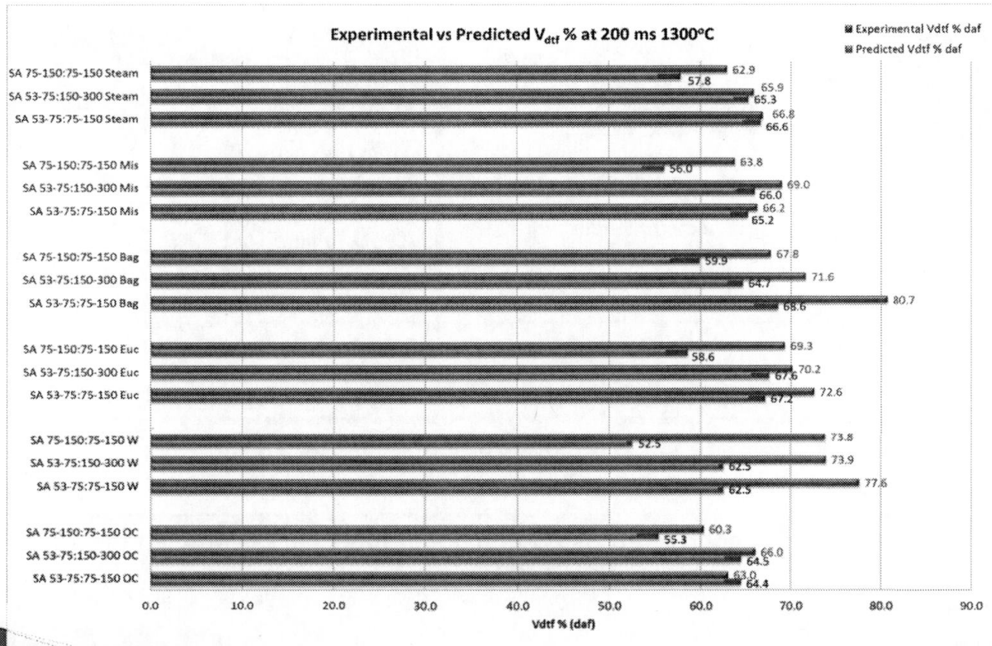


***Efficient Fossil Energy Technologies***
***Engineering Doctorate (EngD) Centre***

Biomass and Fossil Fuel Research Alliance BF2RA

The University of Nottingham
UNITED KINGDOM · CHINA · MALAYSIA

# Devolatisation

## Drop Tube Furnace (DTF)

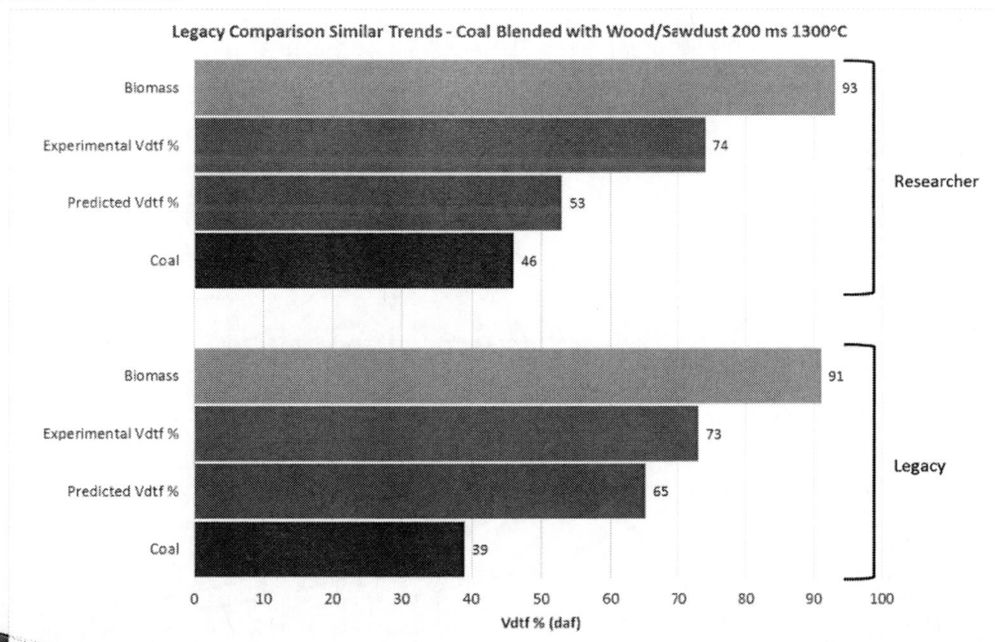


***Efficient Fossil Energy Technologies***
***Engineering Doctorate (EngD) Centre***

Biomass and Fossil Fuel Research Alliance BF2RA

The University of Nottingham
UNITED KINGDOM · CHINA · MALAYSIA

# Char Reactivity

## Thermogravimetric Analysis (TGA)

**Ramp to 110°C**
- 50°C/min
- Nitrogen
- Isotherm 5 mins

**Ramp to 525°C**
- 500°C/min
- Nitrogen
- Isotherm 5 mins

**Switch gas**
- From Nitrogen
- To Air
- Isotherm 180/240 mins

**Cool down phase**
- Cooled at 50°C/min

Sample weight ~10 mg

3 experimental runs per sample

Simulations conducted using individual fuel chars produced in DTF

*Efficient Fossil Energy Technologies Engineering Doctorate (EngD) Centre*

Biomass and Fossil Fuel Research Alliance BF2RA

The University of Nottingham
UNITED KINGDOM · CHINA · MALAYSIA

# Char Reactivity

## Thermogravimetric Analysis (TGA)

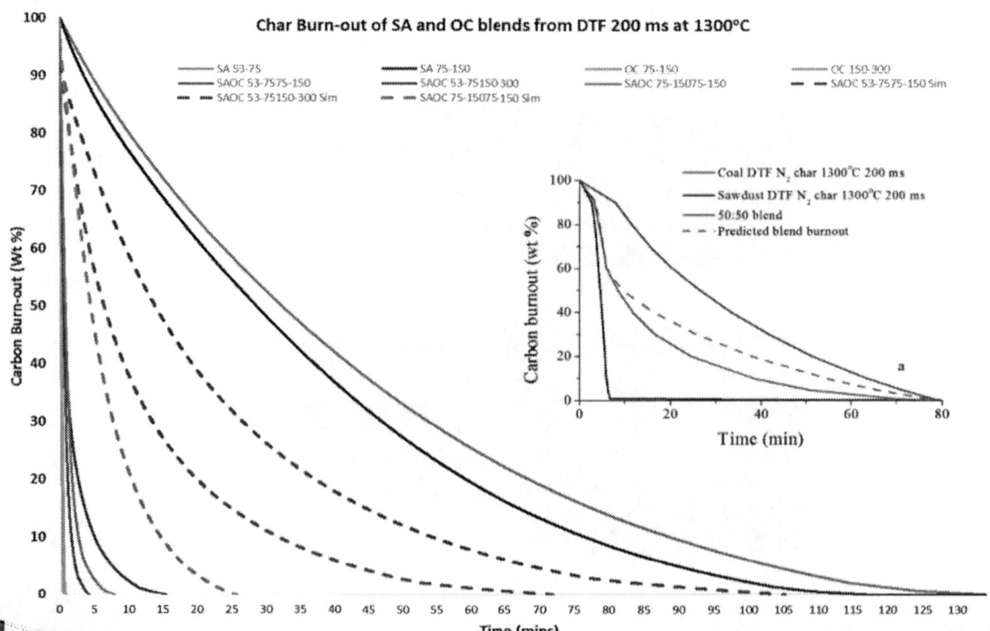


*Efficient Fossil Energy Technologies Engineering Doctorate (EngD) Centre*

Biomass and Fossil Fuel Research Alliance BF2RA

The University of Nottingham
UNITED KINGDOM · CHINA · MALAYSIA

# Research Summary

**Heating Ramp Rate Analysis (TGA)**

- Completion of heating ramp rate experiments both individual and blends at various ramp rates
- Results within experimental error therefore no synergistic effects observed

**Devolatisation (DTF)**

- Investigation conducted with various biomasses blended with South African coal at 200 ms
- Synergistic effects observed with varying degrees of impact
- Comparison with legacy work shows agreement with woody biomass
- Cause – increased steam oxidation due to minerals being present

**DTF Char Reactivity (TGA)**

- DTF char burn-out conducted using TGA
- Synergistic effect observed with a significant decrease in burn-out time across all particle size blends
- Biomass with higher ash content shows greater impact than low ash biomass used in legacy work

*Efficient Fossil Energy Technologies Engineering Doctorate (EngD) Centre*
Biomass and Fossil Fuel Research Alliance BF2RA

The University of Nottingham
UNITED KINGDOM · CHINA · MALAYSIA

# Further Work

**Devolatisation (DTF)**

- Further testing being carried out to investigate the impact of varying residence time
- Investigation into blending Welsh coal with biomasses underway
- Testing of acid washed biomass samples will be conducted

**DTF Char Reactivity (TGA)**

- TGA analysis underway on all chars produced from devolatisation testing
- Kinetics to be investigated

**Char Analysis**

- Surface analysis – SEM and BET
- Chemical analysis – EA and XRF

*Efficient Fossil Energy Technologies Engineering Doctorate (EngD) Centre*
Biomass and Fossil Fuel Research Alliance BF2RA

The University of Nottingham
UNITED KINGDOM · CHINA · MALAYSIA

# Acknowledgements

| | |
|---|---|
| Academic Supervisors | Prof. Colin Snape |
| | Dr. Cheng-gong Sun |
| | Dr. Hao Liu |
| Industrial Supervisor | Dr. Dave Waldron |

BF2RA (Biomass and Fossil Fuels Research Alliance)

EPSRC

***Efficient Fossil Energy Technologies Engineering Doctorate (EngD) Centre***

Biomass and Fossil Fuel Research Alliance BF2RA

The University of Nottingham
UNITED KINGDOM · CHINA · MALAYSIA

# The End

## Thank you for your attention

Are there any questions?

***Efficient Fossil Energy Technologies Engineering Doctorate (EngD) Centre***

Biomass and Fossil Fuel Research Alliance BF2RA

# Effects of Temperature on Physico-chemical Characteristics Evolution of Rice Husk Char under Reburning Conditions

SHI Jiateng, YIN Xinyi, Lu Ping *

School of Energy & Mechanical Engineering, Nanjing Normal University, Nanjing 210042, P. R. China

**Abstract:** Different kinds of rice husk char were prepared in an entrained flow reactor under reburning conditions. The physico-chemical properties of pyrolyzed biomass chars were characterized by scanning electron microscopy-energy dispersive spectrometer(SEM-EDS), specific surface area and pore size analyzer, X-ray diffraction(XRD), X-ray fluorescence spectrometer(XRF), and Fourier transform infra-red(FTIR) spectroscopy, respectively. The effects of temperature on evolution of physico-chemical structures of bio-cokes were discussed. The results indicated that the specific surface area and pore volume increases significantly with increasing the reaction temperature of 850-1150°C. The elements in the chars are mainly K, Na, Ca, Mg, Cl and Si which exist as $SiO_2$, $K_2SO_4$ and silicates. The functional groups in the chars decrease significantly with increasing the reaction temperature.

**Keywords:** reburning; rice husk; rice husk char; reaction temperature; physico-chemical characteristics

## 1. Introduction

Nitrogen oxides ($NO_x$) and mercury are the major pollutants emitted from coal-fired power station, which leads to acid precipitation, photochemical smog, and heavy metal migration and transformation that impose a significant threat to the environment and human health. The combination technology of sorbent injection and flue gas conditioning has been considered as a promising mercury emission control technology [1, 2]. The research focus is how to produce a cost-efficiency sorbent and to form the suitable flue gas components during fuel combustion process. Biomass reburning has been considered as a promising $NO_x$ control technology, due to biomass behaving renewable nature, $CO_2$-neutral character, and relative less contents of sulfur and toxic heavy metals (such as mercury) [3-5]. Furthermore, biomass with relative abundant contents of AAEMs (Alkali and alkali earth metals, mainly K, Na, Ca, and Mg) and chlorine (Cl) could form reasonable concentrations of AAEMs and Cl in the flue gas in biomass reburning process, which is benefit to promote NO reduction and mercury catalytic oxidation, and finally to achieve simultaneous control of $NO_x$ and mercury in conventional air pollution control devices (APCDs). However, too much release of AAEMs and chlorine species bounded in biomass during the thermos-chemical conversion process may result in slagging and/or fouling problems for the operation of boiler heating surface areas [4-5].

The physico-chemical characteristics evolution of biomass under reburning condition are affected by my factors, such as biomass species, temperature, particle size, residence time, etc. Biagini[6] found temperature has a significant effect on the physico-chemical characteristics evolution, and demonstrated high temperature is benefit to the formation of pore structure of biomass char. Luan[7] found that the specific surface area and pore volume of biomass chars were increased with the increase of pyrolysis temperature under the condition of fast pyrolysis. Chen[8] studied the effects of temperature on characteristics of biomass under the condition of fast pyrolysis and found that the hole collapses and hybridization appear at 1000°C on the surface of rice husk char and its main microelements are C, O, Si and K. Feng[9] prepared the biomass chars in the one-dimensional entrained flow reactor in $N_2$ atmosphere and found micro-chemical structures of oxygen, sugar and aliphatic functional groups gradually decreased with increasing the pyrolysis temperature. Yang[10] prepared palm oil shell char in a fixed bed at pyrolysis temperature of 300 - 1000°C and found the main functional group of char are C=O, C-C, C-H, C-O and OH. Xiao [11] produced biomass ash at different temperatures and found the contents of AAEMs and chlorine in the char decreased with increasing temperature, which mainly formed sulphates, silicate and chloride. Some studies on chlorine in biomass ash indicate that organic chlorine released out with volatile matter in low temperature range (200-400°C) and inorganic chlorine mainly converted into part of the biomass ash and released at higher temperature [12-15]. To our knowledge, there are fewer studies on the physico-chemical characteristics evolution of biomass under reburning process.

In this paper, rice husk (RH) were pyrolyzed in an entrained flow reactor under reburning condition. The physico-chemical properties of pyrolyzed biomass (biomass char) were characterized by scanning electron microscopy - energy dispersive spectrometer (SEM-EDS), specific surface area and pore size analyzer, X-ray diffraction (XRD), X-ray fluorescence spectrometer (XRF) and Fourier transform infra-red (FTIR) spectroscopy to understand the physico-chemical characteristics evolution.

## 2. Experimental

### 2.1 Raw materials

Rice husk (RH) was collected from Gongyi, Henan, China. Biomass sample was dried in oven at 45°C for 18 h and then was ground and sieved less than 450 μm. The proximate, ultimate analysis and lower heating value (LHV) of tested samples on air dry basis are listed in Table1.

**Table 1 Proximate analysis, ultimate analysis and low heating value of rice husk**

| Proximate analysis Wt /% | | | | Ultimate analysis Wt/% | | | | | $Q_{net,ar}$ (MJ/kg) |
|---|---|---|---|---|---|---|---|---|---|
| $V_{ad}$ | $FC_{ad}$ | $M_{ad}$ | $A_{ad}$ | $C_{ad}$ | $H_{ad}$ | $N_{ad}$ | $S_{ad}$ | $O_{ad}$ | |
| 3.32 | 68.86 | 18.30 | 9.52 | 42.44 | 4.91 | 37.49 | 1.77 | 0.55 | 16.20 |

### 2.2 Experimental setup and methods

The experiments was carried out in an entrained flow reactor (EFR) (shown in Fig.1). It consists of a high temperature drop tube furnace (DTF), a mini-screw feeder, a simulated flue gas supplying system, a temperature measurement and control system, a water vapor generation system and a flue gas analysis system, etc. The DTF is heated by electric heating elements and its temperature ($T_p$) can be controlled upto 1350°C by an automatic programmed temperature controller. A quartz tube with $\Phi$45 mm in inner diameter (ID) and 1800 mm in length is placed in center of the DTF as a reactor. The stoichiometric ratio in the reburning zone (SR2) is kept at 0.9, which is determined by oxygen content in the simulated flue gas and the feeding of biomass.

The pyrolysis procedure for each sample is as follows: 1) heat the entrained flow reactor from room temperature to the set pyrolysis temperature; 2) switch on the circuit of simulated flue gas and regulate the flue gas composition (shown in Table 2); 3) feed the biomass sample into the reactor through the mini-screw feeder. The residence time in the reactor ($\tau$=1.0s) of tested biomass was determined through the total flow rate of simulated flue gas. The rice husk char prepared at different temperatures are called as RH850, RH950, RH1050 and RH1150, respectively.

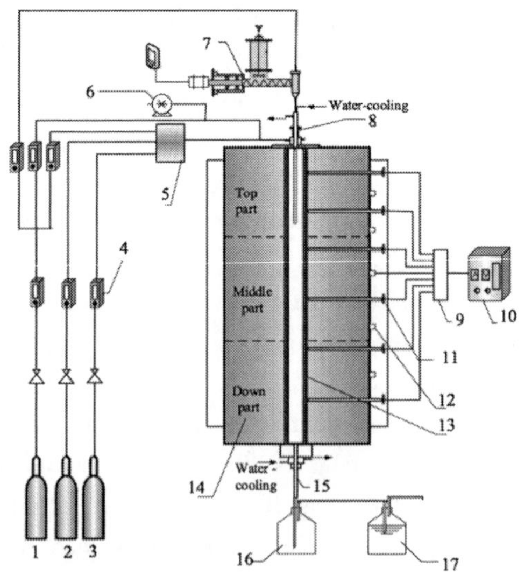


**Fig. 1 Schematic of the entrained flow reactor**

1-3-cylinders of $N_2$, $O_2$, $CO_2$; 4-mass flowmeter; 5-mixing container; 6-syringe pump; 7-mini-screw feeder; 8-water-cooling injector; 9-switcher; 10-controller; 11-heating elements; 12-thermocouples; 13-quartz tube; 14-high temperature DTF; 15-ID fan; 16- char collection ; 17- Flue gas cleaning

**Table 2 The simulated flue gas components**

| $CO_2$ | $O_2$ | $H_2O$(g) | $N_2$ |
|---|---|---|---|
| % | % | % | % |
| 12 | 0-4 | 5 | balance gas |
| 99.9 | 99.995 | DI water | 99.999 |

## 3. Results and discussion

### 3.1 Evolution of pore structure

SEM photos of RH850, RH950, RH1050 and RH1150 enlarged 2000 times are shown in Fig.2. the surface of RH850 char is smooth with a small amount of raised parts and few obvious holes can be observed. The surface of RH950 char becomes uneven and there are much more raised parts compared with RH850. Several small holes appear at the surface of RH1050 char and the surface becomes rough. The amount of small holes increase significantly at the surface of RH1150 and some holes start to collapse. Rice husk is a variety of complex macromolecular organic compounds, and cellulose, hemicellulose and lignin are the main composition. Carbon skeleton were damaged at different degrees with the decompositon of the macromolecular substances, and resulting in many macropores, mesopores and micropores formed with a release of a large amount of volatile compounds.

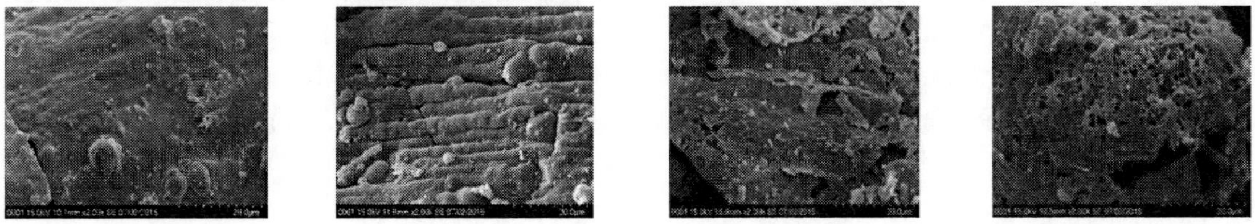

(a) RH850 (b) RH950 (c) RH1050 (d) RH1150

**Fig. 2 SEM photos of the RH-chars**

According to classification provided by the IUPAC (International Union of pure and applied chemistry), pore is classified as micropores (pore size is less than 2 nm), mesopores (pore size ranges at 2-50 nm) and macropores (pore size is large than 50 nm). Table 3 shows the pore structure parameters of RH chars at different temperature. It shows that the specific surface area increases significantly while temperature increases from 850°C to 1150°C, and the specific surface area ($S_{BE}$T, $S_{D\text{-}R}$) of RH1150 is 8 times larger than that of RH850. Micropore volume ($V_M$) and the total pore volume ($V_T$) also increase with increasing temperature and the micro-porosity ($X = V_M/V_T$) ranges from 49.88% to 75.41%. The average pore size($D_a$) is in the range 2.76-5.19 nm. The pore structure mainly affected by temperatute and heating rate under the pyrolysis temperature of 1300°C, the thermal stress became larger and more cracks emerged at high temperature, meanwhile, the devolatilization and escape velocity of volatile compounds increases so that the particle porosity developed rapidly[16].

**Table 3 Pore structure parameters of the pyrolyzed biomass chars**

| Char sample | $S_{BET}$ | $S_{D\text{-}R}$ | $V_M$ | $V_T$ | $X$ | $D_a$ |
|---|---|---|---|---|---|---|
| | ($m^2/g$) | $m^2/g$ | $mm^3/g$ | $mm^3/g$ | % | nm |
| RH850 | 15.47 | 22.74 | 10.3 | 20.05 | 49.88 | 5.19 |
| RH950 | 45.01 | 65.88 | 20.1 | 37.60 | 53.19 | 3.34 |
| RH1050 | 112.56 | 164.75 | 60.1 | 79.57 | 75.41 | 3.14 |
| RH1150 | 119.63 | 175.27 | 60.7 | 82.54 | 72.69 | 2.76 |

Fig.3 shows the pore size distributions of rice husk chars at different temperature. The beginning portion of pore size distribution curve raised due to the contribution of micropore and the peak emerged in the range of mesopore, the pore volume of macropore can be ignored because it is very small. The high contents of inorganic matter and the existence of ash both hinder the develop of pore structure, but the network framework of rice husk formed during the pyrolysis process remaining stable with the releasing of organic matter, the micropores expanded and macropores collapsed with temperature increased so that micropore decreased and mesopore

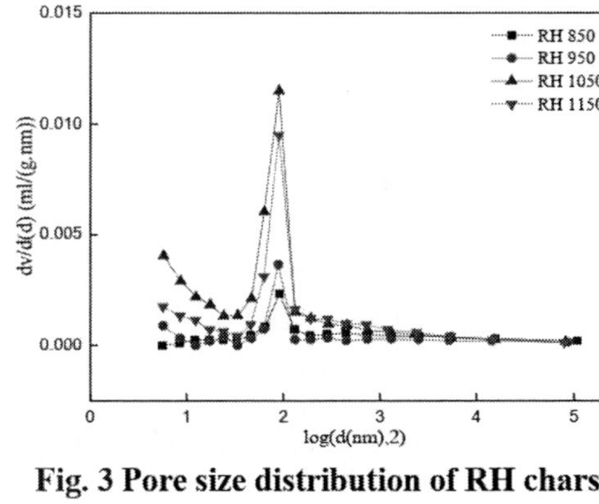


**Fig. 3 Pore size distribution of RH chars**

increased. At the temperature of 1150°C, the amount of micropore and mesopore both decreases but the specific surface didn't change significantly compared with that of 1050°C. The results indicate that reburning temperature has important effects on the surface morphology and pore structure. Increasing the reaction temperature is beneficial to the formation of micropore and mesopore but pore may collapse and transform at high temperature.

### 3.2 Evolution of chemical composition

The results of XRF analysis are shown in Table 4. The results indicate that rice husk chars mainly compose of inorganic substances formed by Si, Al, Mg, Fe, Ca, K and Na. The content of $SiO_2$ increased with increasing the reaction temperature. Generally, the content of $SiO_2$ is the most and the content of $K_2O$ is second-most while the content of $Na_2O$ is the least. During the process of reburning, the volatile elements (K, Na) couldn't volatilize completely and parts of them still remained in the char. The results combined with Table 5 indicate that the evaporation of K is more than that of Na because Na may react with $SiO_2$ to make silicate which decreased the releasing of Na[17]. There were still some AAEMS remained in the chars. Zygarlicke [18] holds the view that Ca may react with $SiO_2$ to make calcium silicate at the temperature 1100-1700 K and the existence of K can catalyze the reaction.

**Table 4 XRF analysis results of RH chars**

| Char sample | $SiO_2$ % | $Na_2O$ % | $K_2O$ % | CaO % | MgO % | $Al_2O_3$ % | $Fe_2O_3$ % | $P_2O_5$ % | $SO_3$ % |
|---|---|---|---|---|---|---|---|---|---|
| RH850 | 38.30 | 0.58 | 30.17 | 7.97 | 6.61 | 2.01 | 1.38 | 8.02 | 4.95 |
| RH950 | 42.31 | 0.50 | 27.11 | 8.52 | 6.25 | 2.23 | 1.54 | 7.50 | 4.06 |
| RH1050 | 40.29 | 0.53 | 27.16 | 9.16 | 6.56 | 2.40 | 1.57 | 7.82 | 4.52 |
| RH1150 | 51.76 | 0.49 | 24.91 | 6.18 | 4.42 | 1.95 | 1.27 | 5.52 | 3.73 |

**Table 5 ICP results of RH chars**

| Char sample | Na % | K % | Ca % | Mg % | Cl % | Fe % | Al % | Zn μg/g | Mn μg/g |
|---|---|---|---|---|---|---|---|---|---|
| RH850 | 1.75 | 10.14 | 4.50 | 1.65 | 1.15 | 0.86 | 0.84 | 306.11 | 561.99 |
| RH950 | 1.32 | 5.66 | 2.66 | 1.22 | 0.63 | 0.58 | 0.63 | 141.67 | 424.35 |
| RH1050 | 1.24 | 7.08 | 2.53 | 1.18 | 0.54 | 0.50 | 0.56 | 158.37 | 399.02 |
| RH1150 | 1.10 | 5.88 | 2.38 | 0.97 | 0.31 | 0.46 | 0.48 | 263.13 | 328.91 |

The XRD curves of the rice husk chars are shown in Fig.4. It shows that the crystalline are $SiO_2$, $K_2SO_4$, $Na_2Si_2O_5$, $CaSiO_3$ and $MgSiO_3$, the main crystalline is $SiO_2$ and the other crystalline silicates is made from AAEMS and $Si_xO_y$. Some AAEMS still remained in the chars in the form of silicates with temperature increased. It can be observed that the diffraction peak didn't change significantly at different temperature and the KCl diffraction peak couldn't be found in the two XRD curves because KCl may fully volatilize in the form of gaseous above the temperature of 850°C. The result is consistent with Mi's conclusion[19]. Yu[20]concluded that the residual chlorine in the chars may release at higher temperature and rice husk with high sulfur content could lead to a large amount of $K_2SO_4$ remained in the chars[21].

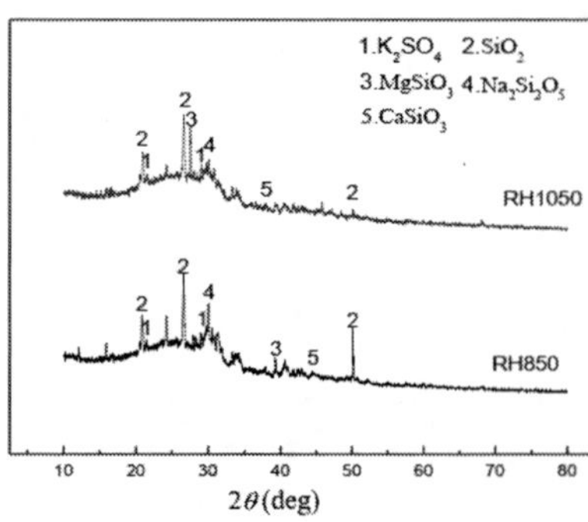


Fig. 4 XRD curves of the RH chars

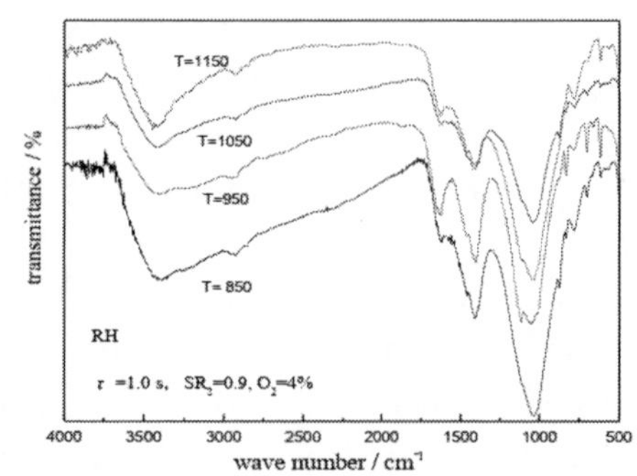


Fig. 5 FTIR curves of the RH chars

### 3.3 Evolution of functional groups

The organic functional groups broke, condensated and restructured with pyrolysis temperature increased. The organic compounds decomposed, volatilized and released in the form of CO, $CO_2$, $CH_4$ and other small molecular gas so that the species and quantity of organic functional groups also changed[22,23]. Fig.5 shows the FTIR curves of the rice husk chars. It shows that the main organic functional groups are -OH (hydroxyl), C-H (alkane, alkenyl and aromatic ring), C=O(carbonyl group), C-O (alcohol, ether) and the absorption peak area of each functional group also changes with increasing temperature. The infrared absorption peak area of -OH gradually decreases at 1425 $cm^{-1}$ because -OH breaks and restructures at the temperature of 850 -1150°C, and it finally releases in the form of $H_2O$ during the pyrolysis process, which leads to the decreasing of the amount of -OH. Also, the bending vibration peak of C-H at 1510 $cm^{-1}$ and the stretching vibration absorption peak of C=C at 873 $cm^{-1}$ decreases with temperature increased because C-H and C=C breaks and forms gaseous hydrocarbon (such as $CH_4$, $C_2H_4$). The stretching vibration absorption peak of C-O bond in alcohol, phenol, ether or ester appears at 1035 $cm^{-1}$ and it can be determined by the existence of phenolic hydroxyl groups on the surface of the chars. Similarly, the absorption peak decreases as the reaction temperature increasing from 850°C to 1050°C. Therefore, it can be concluded that high temperature is not conducive to the existence of organic functional groups in the chars.

## 4. Conclusions

(1) Reburning temperature had important effects on the surface morphology and pore structure. Increasing the reaction temperature is beneficial to the formation of micropore and mesopore, however, pores may collapse and transform at high temperature. The specific surface area and pore volume increases significantly with increasing the reaction temperature of 850-1150°C.

(2) Rice husk char mainly composes of inorganic substances formed by Si, Al, Mg, Fe, Ca, K, and Na. The evaporation of K is more than that of Na. The main crystalline are $SiO_2$, $K_2SO_4$ and silicate made by AAEMS and $Si_xO_y$.

(3) The main organic functional groups are -OH (hydroxyl), C-H (alkane, alkenyl and aromatic ring), C=O (carbonyl group), and C-O (alcohol, ether). High temperature is not conducive to the existence of organic functional groups in the chars.

## Acknowledgments

The financial supports of National Natural Science Foundation of China (51476079) and Jiangsu Province Prospective Co-operation Research Project of China (BY2015001-01) are gratefully acknowledged.

## References

[1] Yang H, Xu Z, Fan M, Bland AE, Judkins RR. Adsorbents for capturing mercury in coal-fired boiler flue gas [J]. Journal of Hazardous Materials, 2007, 146(1-2): 1-11.

[2] Smoot LD, Hill SC, Xu H. $NO_x$ control through reburning. Progress in Energy and Combustion Science, 1998, 24(4): 385-408.

[3] Lu P, Wang Y, Huang Z, Lu F, Liu Y. Study on NO reduction and its heterogeneous mechanism through biomass reburning in an entrained flow reactor. Energy Fuels, 2011, 25(7): 2956-2962.

[4] Sami M, Annamalai K, Wooldridge M. Cofiring of coal and biomass fuel blends. Progress in Energy and Combustion Science, 2001, 27(2): 171-214.

[5] Demirbas A. Combustion characteristics of different biomass fuels. Progress in Energy and Combustion Science, 2004, 30(2): 219-230.

[6] Biagini E, Cioni M, Tognotti L. Development and characterization of a lab-scale entrained flow reactor for testing biomass fuels. Fuel, 2005, 84(12–13): 1524-1534.

[7] Luan J, Sun R, Lu J, Yao N, Wu S. Experimental studies on reburning with biomass for $NO_x$ reduction. Proceedings of the CSEE, 2008, 28(14): 73-79.

[8] Chen Y, Zhu B, Wang X, et al. Evolutoion of char's physical and chemical structure during biomass

pyrolysis. Acta Energiae Solaris Sinica, 2012(8): 1267-1272.

[9] Feng D. Rsearch on physical and chemical characteristics of biomass pyrolysis char. Harbing: School of energy science and engineering, Harbin Institute of Technology, 2014.

[10] Yang H, Chen H, Chen Y, et al. Investigation on the forming property of palm oil shell char physicochemical structure during biomass pyrolysis. Proceedings of the CSEE, 2008, 28(32): 106-110.

[11] Xiao R, Chen X, Wang F, Yu G, et al. The physicochemical properties od different biomass ash. Acta Energiae Solaris Sinica, 2011, 32(3):364-369.

[12] Du S, Chen H, Yang H, et al. Transformation behavior of F and Cl during biomass pyrolysis. Proceedings of the CSEE, 2010, 30(14): 115-120.

[13] Jensen PA, Frandsen FJ, Dam-Johansen K, et al. Experimental investigation of the transformation and release of potassium and chorine during straw pyrolysis. Energy Fuels, 2000, 14(7):1280-1285.

[14] Sippula O, Lind T, Jokiniemi J. Effects of chorine and sulphur on particle formation in wood combustion performed in a laboratory scale reactor. Fuel, 2008, 87(12): 2425-2436.

[15] Chen W. Effects of pyrolysis conditions on char structure and its intrinsic gasification reactivity. Taiyuan: College of Chemistry and Chemical Engineering, Taiyuan University of Technology, 2014.

[16] Zhou J, Zhang H, Lu J, Yue G. Effect of pyrolysis temperature on porous structure of anthracite chars produced at high temperatures. Journal of Fuel Chemistry and Technology, 2007, 35(2): 155-159.

[17] Brinsmead KH, Kear RW. Study of the kinetics of the mechanism of solid state reactions at increasing temperature. Fuel, 1956, 35: 84-93.

[18] Zygarlicke CJ, Steadman EN, Benson SA. Ag-glomeration phenomena in fluidized bed combustion of straw. Prog. Energy Combust Sci., 1990, 16:195-204.

[19] Mi T, Chen H, Wu Z, et al. Chemistry characteristic study on biomass ash. Acta Energiae Solaris Sinica, 2004, 25(2): 236-241.

[20] Yu M, Dong Y, Wang P, Ma C. Progress of effects of chloride on mercury removal for coal-fired flue gas. Chemical Industry and Engineering Progress, 2012, 31(7): 1610-1614.

[21] Du S. Fundamental study on transformation behavior of inorganic components during thermochemical conversion of biomass and ash fusion characteristics. Wuhan: School of Energy and Power Engineering, Huazhong University of Science and Technology, 2014.

[22] Fu P, Hu S, Xiang J, et al. Evolution of pore structure of biomass particles during pyrolysis. CIESC Journal, 2009, 60(7): 1793-1799.

[23] Demirbas A. Mechanisms of liquefaction and pyrolysis reactions of biomass. Energy Conversion and Management, 2000, 41(6): 633-646.

Eskom

# Low NOx Coal Burner Temperature Profile Evaluation

Presenter: Dewan Smit

Academic Mentor: Walter Schmitz

Academic Co-mentor: Reshendren Naidoo

Industrial Mentor: Sandile Peta

Date: July 2016

## Contents

Eskom

1. Subject Background
2. Research Objectives
3. Devolatilization Kinetics
4. Char Combustion Kinetics
5. Full Scale Single Burner Combustion CFD Model

## Subject Background

Eskom

- Stringent emissions legislation has led to increased demand for cleaner and more efficient coal combustion technologies.
- A primary technology for the reduction of emissions of oxides of nitrogen (NOx) is the installation of low NOx coal combustion burners (least costs involve).
- Extensive research to improve on the combustion performance of low NOx coal burners has been extensively undertaken and is ongoing.
- Computational fluid dynamics (CFD) is a definite tool to aid with analysis, development and design of pulverized coal combustion burners as it offers a substantially lower cost solution and more importantly zero risk to production loss.

3

## Subject Background

Eskom

- Low NOx coal burners reduce NOx emissions by staging the mixing of coal and air.
- This results in a fuel rich region for char combustion, longer flames and lower peak flame temperature.
- The aerodynamics plays a major role in the chemical reactions (and therefore the pollutant formation) in the flame

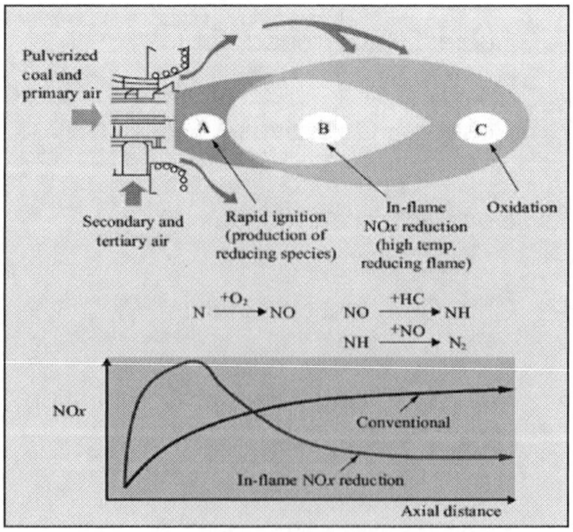


4

## Subject Background

Eskom

- Comprehensive investigation in the form of a combustion reliability project showed the air flow to the burners at full load is lower than design.
- Consequently the power plant is experiencing:
  - Unstable coal combustion particularly at lower loads when the pf and air flow are even further reduced.
  - Very poor turndown capability (Can't operate steadily below 140-150 MW).
- Therefore the need for a study to be done to determine the combustion characteristics of the specific LNASB with specific reference to ignition point and temperature profile during current operating conditions.

5

## Research Objectives

Eskom

- Perform in-flame temperature measurements, at the power plant itself, on a single LNASB to determine the flame temperature profile at current operating conditions under controlled air and coal supply.

- Determine devolatilization and char single rate reaction kinetics of the specific coal through drop tube furnace (DTF) experiments to better characterise the coal for computational fluid dynamics (CFD) modelling.

- Develop a complete combustion CFD model of an industrial scale single LNASB.

6

# Research Method

Eskom

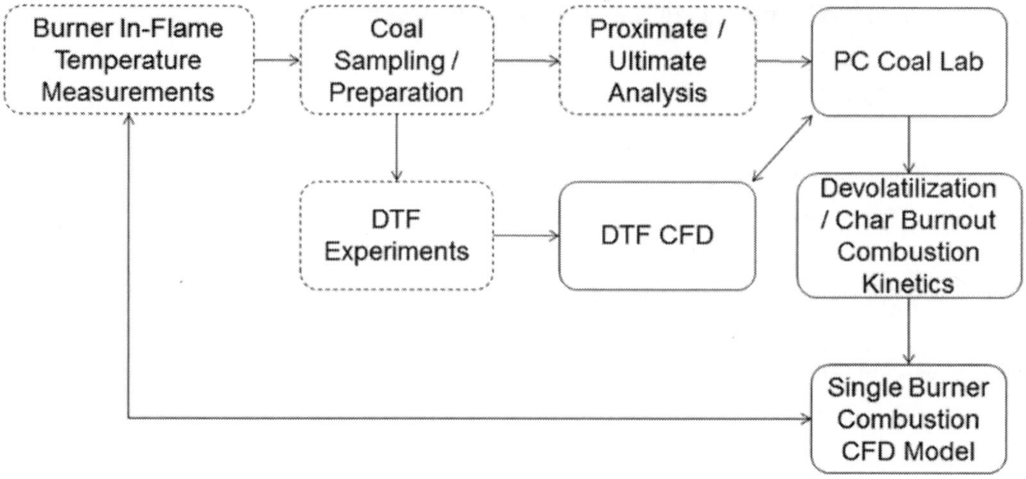


7

# Combustion Kinetics: DTF Experimental / Numerical Setup

Eskom

**DTF Experimental Setup**

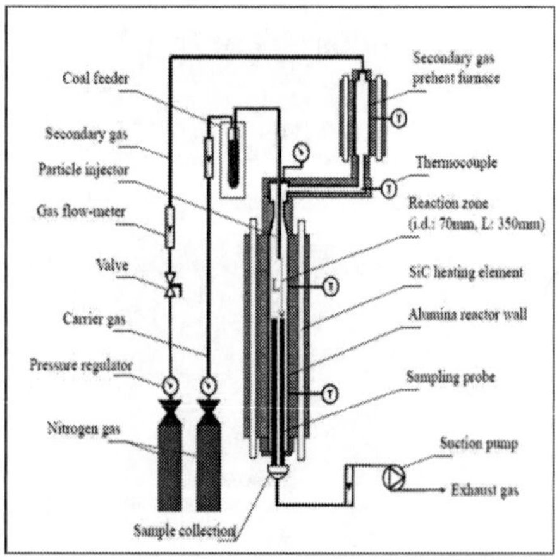


Peta, S. (2014). Ignition characteristics of low volatile quality high ash coal. MSc. Dissertation. Johannesburg: University of Witwatersrand.

**PC Coal Lab Input**

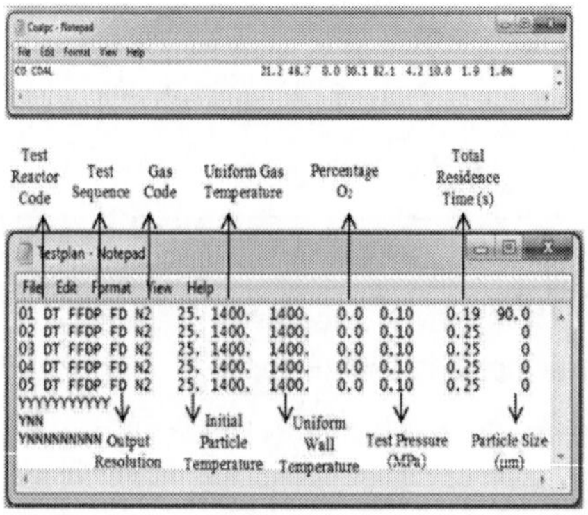


**DTF CFD Geometry**

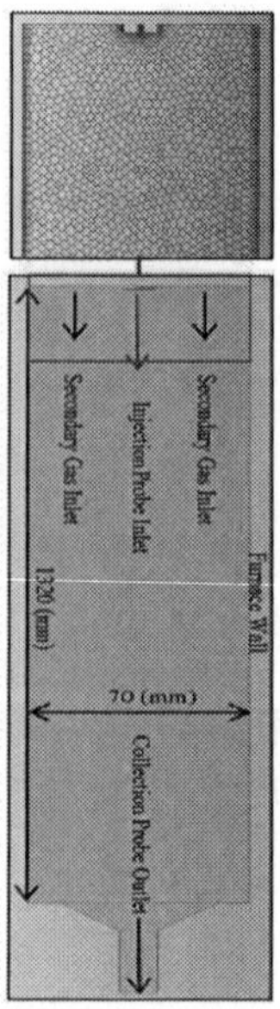


8

## Arrhenius Single First Order Reaction Kinetics Calculation

Eskom

$$k_s = A\exp\left(-\frac{Ea}{RT}\right) = \frac{\frac{dV(t)}{dt}}{V^{Ult} - V(t)}$$

| | | |
|---|---|---|
| $k_s$ | - | Arrhenius rate constant. |
| A | - | Pseudo-frequency factor (1/s). |
| $E_a$ | - | Apparent Activation Energy (J/Kmol). |
| R | - | Universal gas constant (J/mol.K). |
| T | - | Particle temperature (K). |
| $V_{(t)}$ | - | Instantaneous yield. |
| $V_{ult}$ | - | Ultimate yield. |

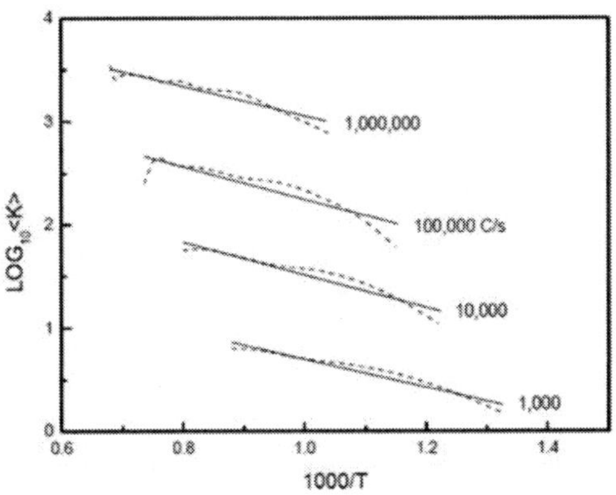


Niksa Energy Associates, (1997-2015). User Guide and Tutorial, PC Coal Lab, Version 4.3.

9

## Arrhenius Single First Order Reaction Kinetics Calculation (Numerically)

Eskom

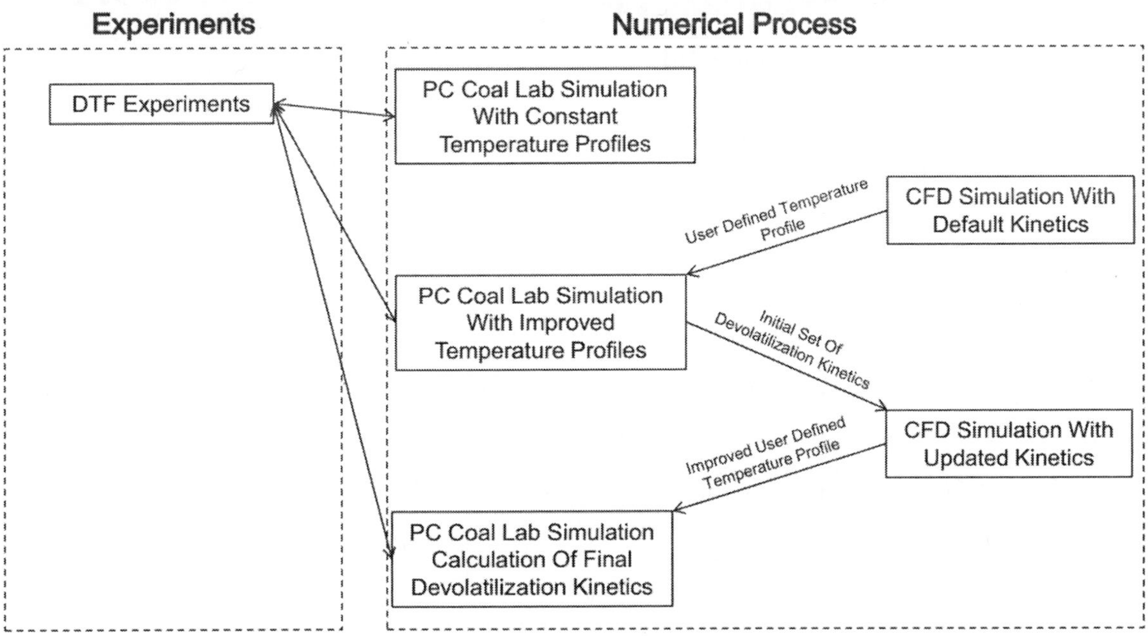


10

# Development of Numerical Char Burnout Result (@1400°C) – Final Char Burnout Result

Eskom

**PC Coal Lab Result After User Defined Temperature Profiles Used**

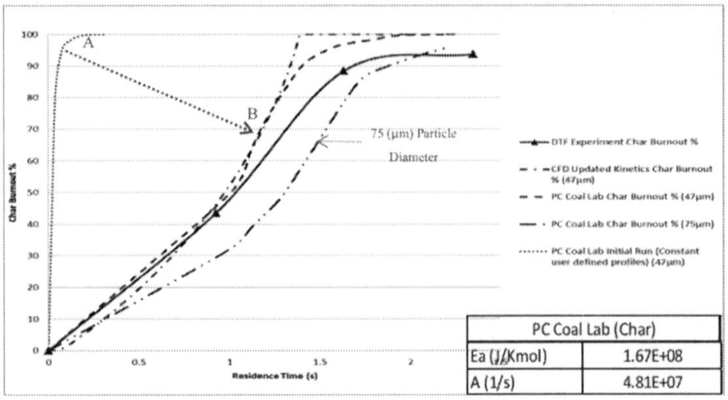


| PC Coal Lab (Char) | |
|---|---|
| Ea (J/Kmol) | 1.67E+08 |
| A (1/s) | 4.81E+07 |

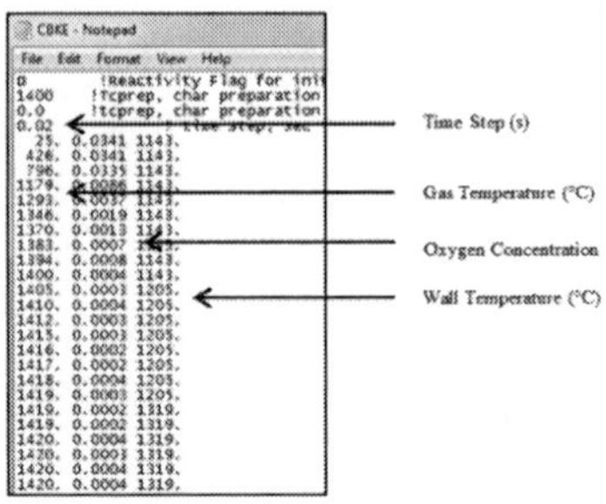


**PC Coal Lab User Defined Profiles**

**DTF Temperature Contour**

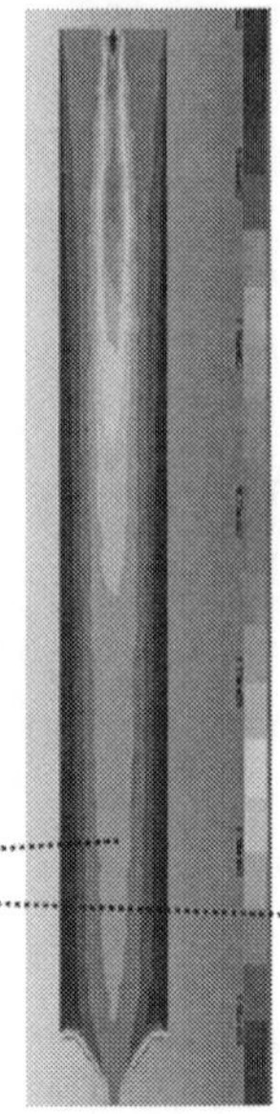

**Char Burnout Oxygen Concentration Contour**

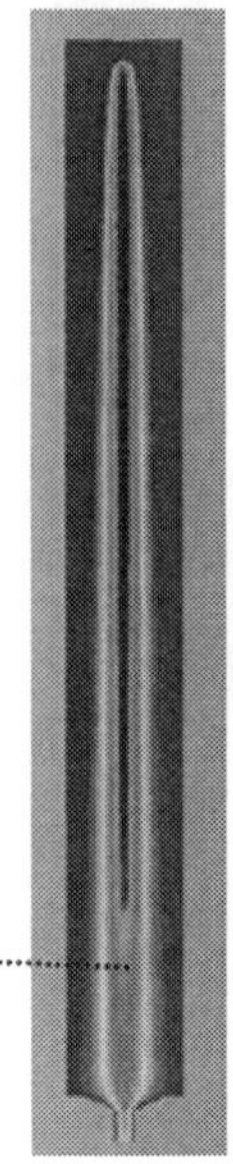

11

# Full Scale Single Burner Combustion CFD

Eskom

Two case studies investigated:

1. Effect of kinetic input on simulation result.
   - Diffusion Limited
   - Diffusion Limited / Default Kinetics
   - Diffusion Limited / Updated Kinetics

2. Effect of a reduced amount of air through the burner.
   - Aerodynamic result
   - Effect on temperature distribution

12

## Full Scale Single Burner Combustion CFD – Models Used

Eskom

- Turbulence: Standard K-epsilon, Enhanced wall treatment
- Radiation: Discrete Ordinates (Particle radiation included – Physical models)

  Weighted-sum-of-gray-gases model (WSGGM) – For radiative properties of the gas
- Species Models: Species Transport

  Turbulence-Chemistry interaction – Finite-Rate/Eddy-Dissapation

  Diffusion limited / Kinetics
- Discrete Phase: Particle surface injection (Lagrangian)

  Particles injected at each DPM iteration 34 600

13

## Full Scale Single Burner Combustion CFD - Geometry

Eskom

**Geometry Single Low NOx Burner**

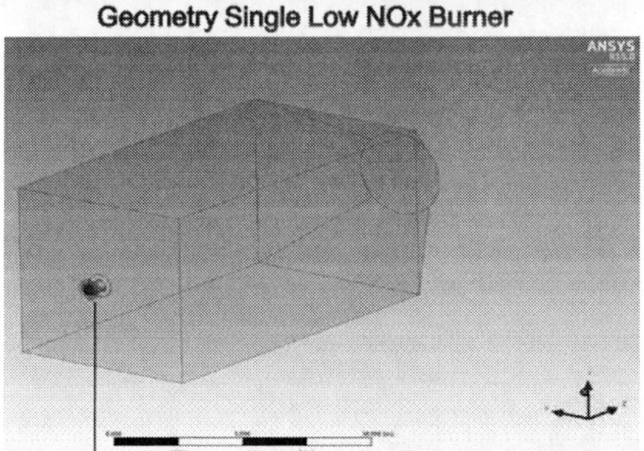


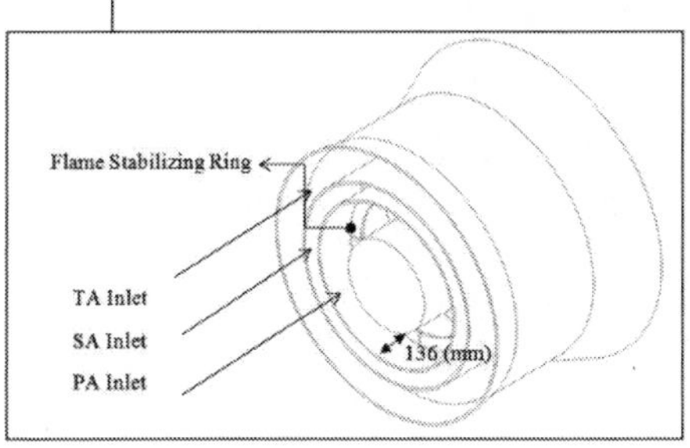


**Geometry – Burner Pair**

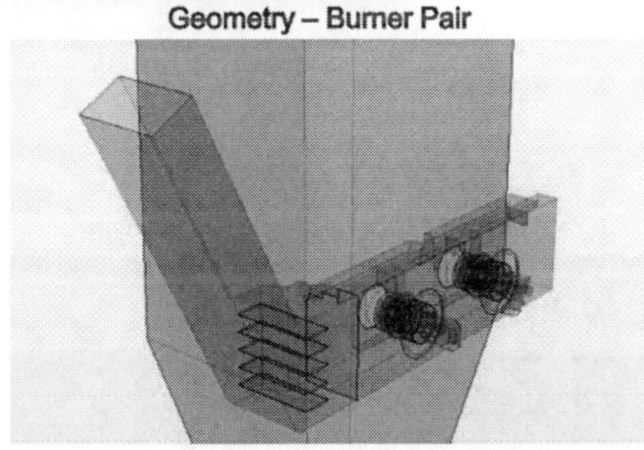

Naidoo, R., Schmitz, W. (2010). Low NOx burner flow characterisation using CFD. Eskom Technical Report.

**Velocity Profile Full Geometry**

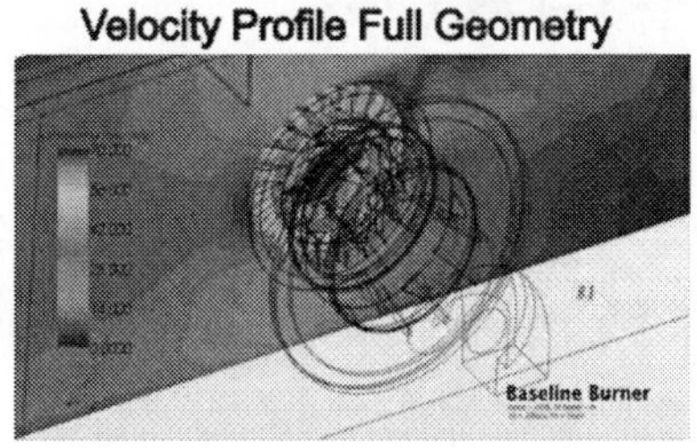


Naidoo, R., Schmitz, W. (2010). Low NOx burner flow characterisation using CFD. Eskom Technical Report.

14

# Full Scale Single Burner Combustion CFD – Kinetic Input Result

Eskom

**Kinetic comparison through centre-line**

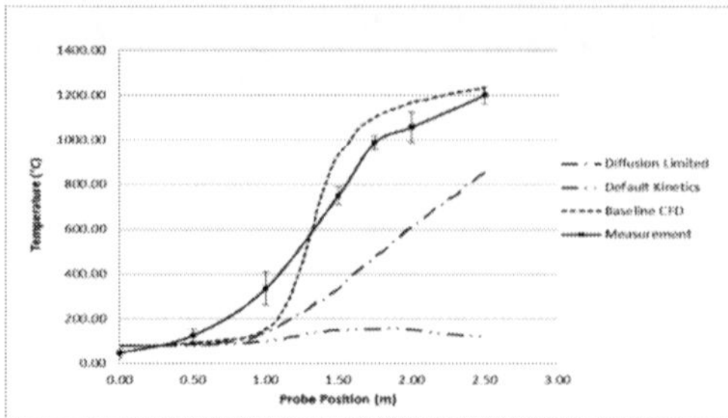


**Kinetic comparison through bottom inspection port**

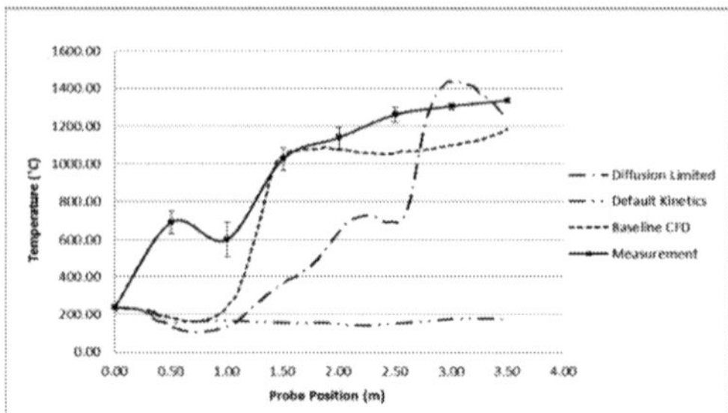


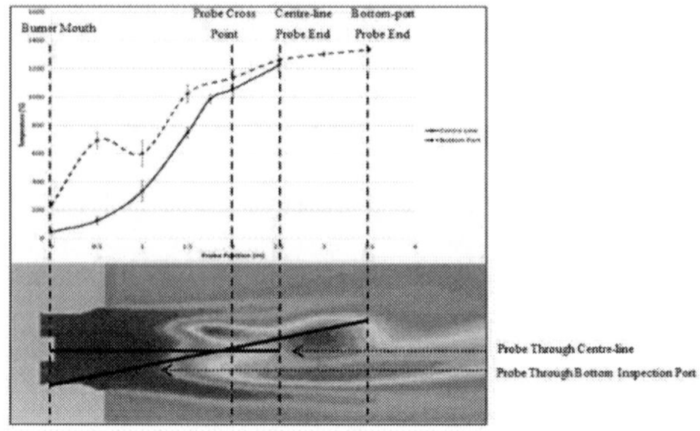


15

# Full Scale Single Burner Combustion CFD – Air Flow & Swirl Effect

Eskom

**Velocity Contours (m/s)** **Temperature Contours (°C)**

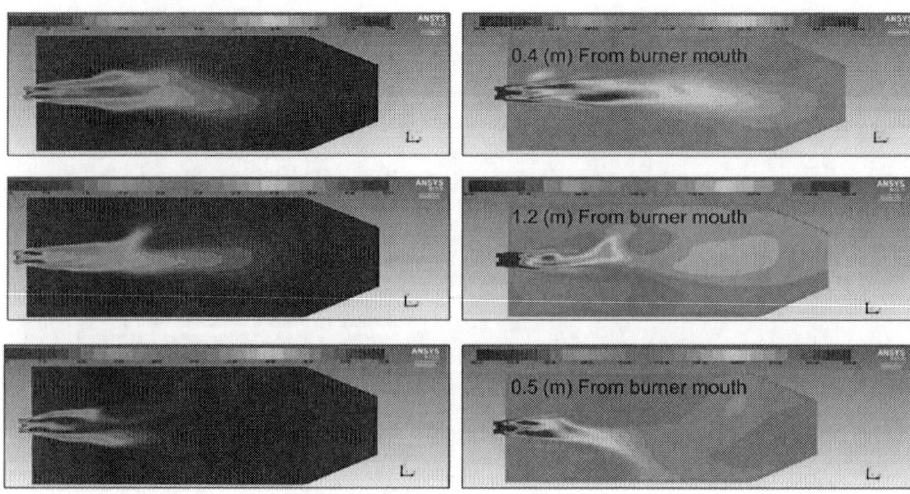


Design Flows

Baseline Simulation

Baseline Simulation & Increased Swirl Number

16

## Conclusion

Eskom

- The temperature profile of an industrial low NOx coal burner was evaluated through a combination of experimental and numerical techniques.
- To improve on the burner CFD model results the coal devolatilization and char burnout combustion kinetics had to be determined.
- The method used in calculating the devolatilization and char single rate kinetic parameters through CFD and PC coal lab was found to be effective and accurate.
- The single burner CFD model temperature results compared well with the in-flame temperature measurements once the calculated coal combustion kinetic parameters were defined.
- The single burner CFD model showed that during reduced flow conditions a reduced amount of swirl is created compared to the design flow simulation.
- With the baseline flow and an increased amount of swirl the flame offset from the burner mouth can be decreased from 1.2 (m) to approximately 0.5 (m) away.

17

Eskom

# Thank you

Eskom | Powering your world

**Institute of Combustion and Power Plant Technology**
**Prof. Dr. techn. G. Scheffknecht**

Universität Stuttgart ifk

## 33rd Annual International Pittsburgh Coal Conference

Cape Town, South Africa

Wednesday, 10th August 2016

# Acid Gas Control by Dry Sorbent Injection in Air and Oxy-Fuel Combustion

by
Reinhold Spörl, Reyhane Youssefi, Jörg Maier, Günter Scheffknecht

Contacts:
*Reinhold.Spoerl@ifk.uni-stuttgart.de*
*Reyhane.Youssefi@ifk.uni-stuttgart.de*
*Joerg.Maier@ifk.uni-stuttgart.de*

## Outline

ifk

1. Introduction
2. Background & Motivation
3. Methodology
4. Experimental results
5. Summary and outlook

2

## Oxy-fuel Combustion Systems

- Practical **oxy-fuel** configuration for low S & Cl fuels

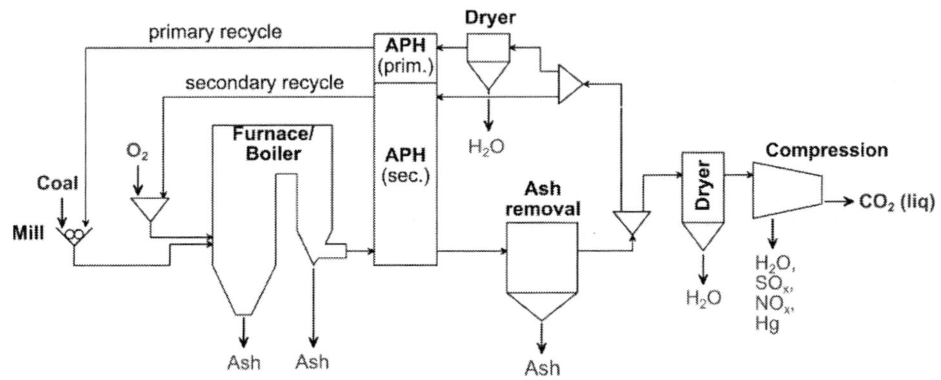


**Problems:** - Excessive $SO_x$ (& HCl) levels with most coals

- $SO_x$ (& HCl) control within recycle to reduce corrosion
- Final acid gas removal required before $CO_2$ liquefaction

3

## Acid Gas Removal within Oxy-Fuel Recycle

Dry Sorbent Injection (DSI) [e.g. $CaCO_3$ or $Ca(OH)_2$]:

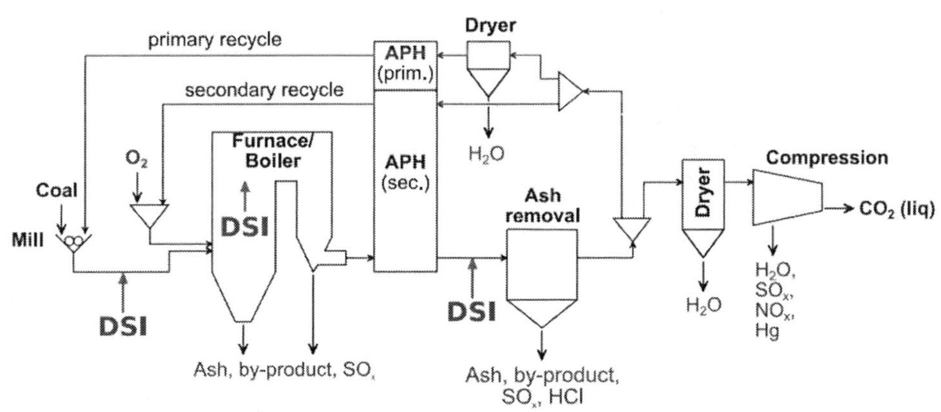


**Benefits/drawbacks:** + Low costs (no scrubber required)

+ No excessive gas cooling
- Dilution of ash

4

## Outline

1. Introduction
2. Background & Motivation
3. Methodology
4. Experimental results
5. Summary and outlook

5

## Acid Gases in Oxy-Fuel Combustion

- Air-fired combustion:

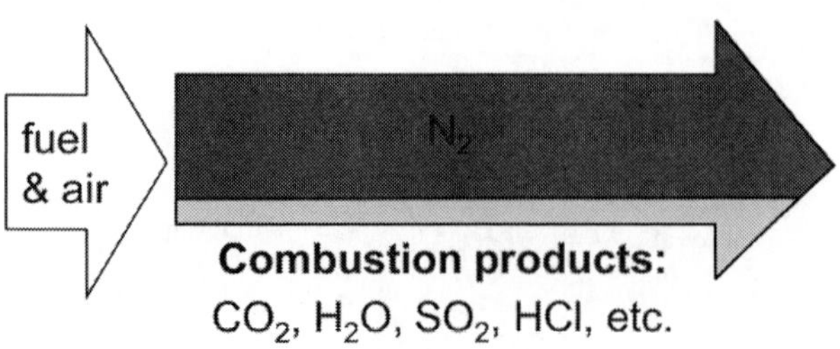


6

## Acid Gases in Oxy-Fuel Combustion

ifk

- Oxy-fuel combustion:

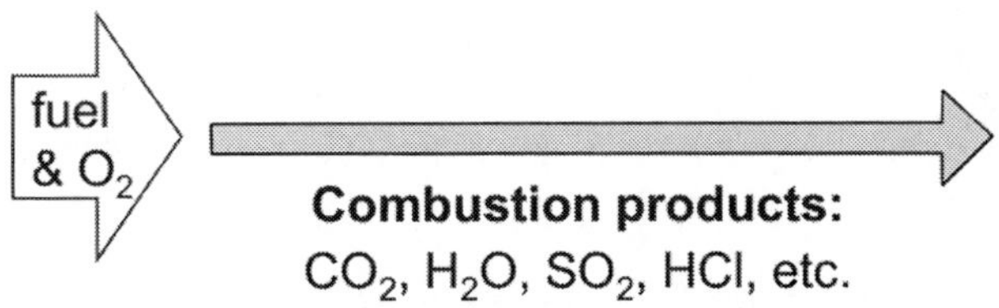


**Exclusion of airborne $N_2$:**

- Generation of $CO_2$ rich gas for **sequestration/utilization**
- **Increase of acid gas** concentrations (factor 4-5)

  (Recycle combustion does not alter gas concentrations)

**High impurity level beneficial for abatement reactions:**

- $Ca(OH)_2 + \mathbf{SO_2} \rightarrow CaSO_3 + H_2O$ *(low temperature)*
- $Ca(OH)_2 + \mathbf{SO_2} + {}^1/_2\,O_2 \rightarrow CaSO_4 + H_2O$ *(high temperature)*
- $CaCO_3 \quad + \mathbf{SO_2} \; + {}^1/_2\,O_2 \rightarrow CaSO_4$ *(high temperature)*

7

## Acid Gases in Oxy-Fuel Combustion

ifk

- Oxy-fuel recycle combustion:

**Flue gas recirculation to control:**

- Temperatures
- Gas flows

➢ HEAT TRANSFER

8

## Acid Gases in Oxy-Fuel Recycle Combustion

ifk

- Recycle combustion:

**Acid gases ($SO_2$, HCl):**

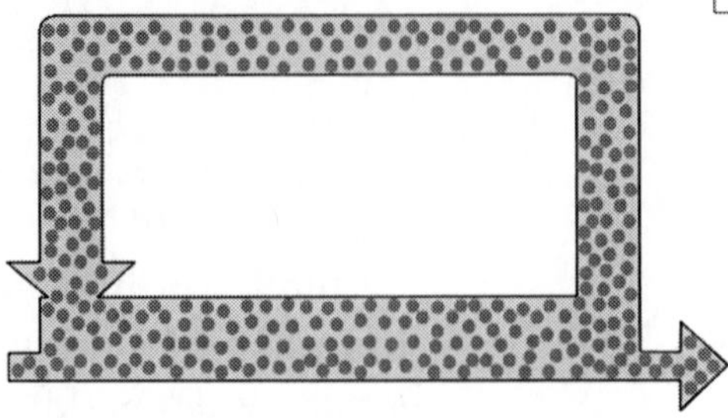

9

## Acid Gas Removal Recycle Combustion

ifk

- Recycle combustion:

**Acid gases ($SO_2$, HCl):**

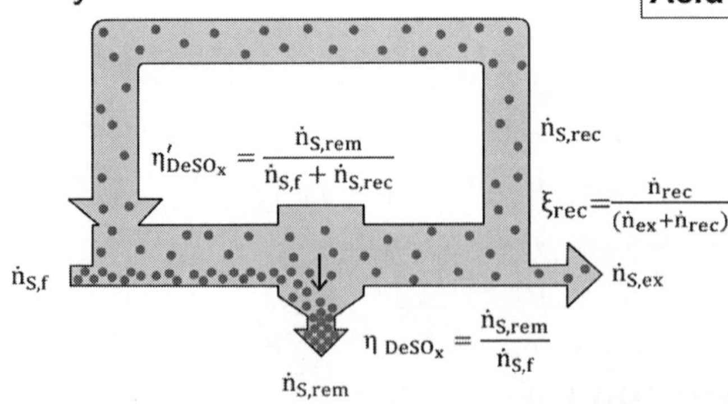


**Effect of flue gas recirculation on acid gas removal:**

- Acid gas removal lowers initial impurity levels (negative)
- Practical residence time of impurities is extended (beneficial)
- Both effects balance each other to some degree

10

## Motivation

ifk

**Open Questions:**

- Impact of oxy-fuel conditions on DSI performance:
  - High acid gas levels (e.g.: $SO_x$, HCl)
  - High $CO_2$ levels
  - Changed temperature-residence-time profile
  - Flue gas recirculation

**Goal of this study:**

- Identify limitations and advantages for DSI technology in oxy-fuel
- Demonstrate DSI applicability/performance under oxy-fuel conditions

11

## Outline

ifk

1. Introduction
2. Background & Motivation
3. Methodology
4. Experimental results
5. Summary and outlook

12

## Fuel and Sorbents

- Fuel for DSI tests: **high S and Cl US coal**

| | NCV | $H_2O$ | Ash | V | $C_{fix}$ | C | S | H | N | Cl | O |
|---|---|---|---|---|---|---|---|---|---|---|---|
| | *[MJ/kg, raw]* | *[%, raw]* | *[%, wf]* | *[%, waf]* | *[%, waf]* | *[%, waf]* | *[%, waf]* | *[%, waf]* | *[%, waf]* | *[%, waf]* | *[%, waf]* |
| **US2.5** | 30.5 | 1.6 | 9.43 | 38.7 | 61.3 | 81.9 | **2.67** | 5.2 | 1.6 | **0.30** | 8.33 |

- acid gas levels (w/o DSI):

| scale | mode | $SO_2$ | HCl *[ppm]* | $SO_3$ |
|---|---|---|---|---|
| **500 kW** | **air** | 1500 | 210 | 9 |
| | **oxy-fuel** | 4800 | 560 | 18 |
| **20 kW** | **air** | 1600 | 195 | - |
| | **$CO_2/O_2$** | 4650 | 375 | 17 |

- Sorbents:
  - ➢ Hydrated lime ($Ca(OH)_2$): Sorbacal® SPS [Lhoist]
  - ➢ Finely milled limestone ($CaCO_3$) [Lhoist]

13

## 20kW Electrically Heated Combustion Test Rig

**Sorbent injection locations:**

- Co-Injection (> 1300°C)
- In-furnace (~ 1100°C)

**Combustion conditions:**

- Air & $CO_2/O_2$
- $T_{furnace}$:1350°C -1100°C
- $\lambda \approx 1.1\text{-}1.15$
- $\dot{V}_{flue\ gas} = 11.5\ m^3(STP)/h$

**Gas sampling:**

- After filter

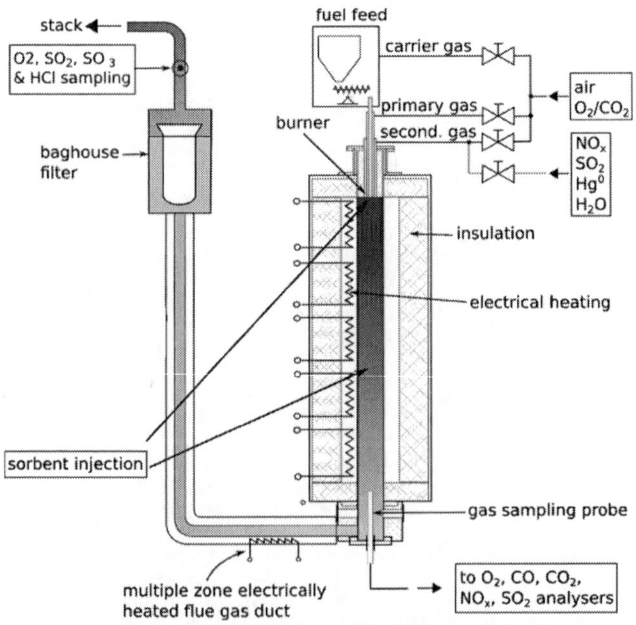


**For additional 20kW DSI results see 5th Oxyfuel Combustion Research Network Meeting, Wuhan, China, 2015:**
http://www.ieaghg.org/docs/General_Docs/5oxy%20presentations/Session%207A/7A-05%20-%20R.%20Spoorl%20(IFK%20Stuttgart%20University).pdf

14

## 500kW Air/Oxy-Fuel Pilot Rig

**Combustion conditions:**

- Air & Oxy-Fuel
- $\lambda \approx 1.14\text{-}1.17$

**Sorbent injection:**

- Loss-in-weight control
- Venturi dispersion system with air/$CO_2$ as feed gas
- Co-injection (with fuel)
- In-furnace (~ 1100°C)
- Low temperature (~ 220 °C)

**Gas sampling:**

- $SO_2$ (NDIR), HCl (FTIR), $SO_3$ (controlled condensation)
- Downstream ESP

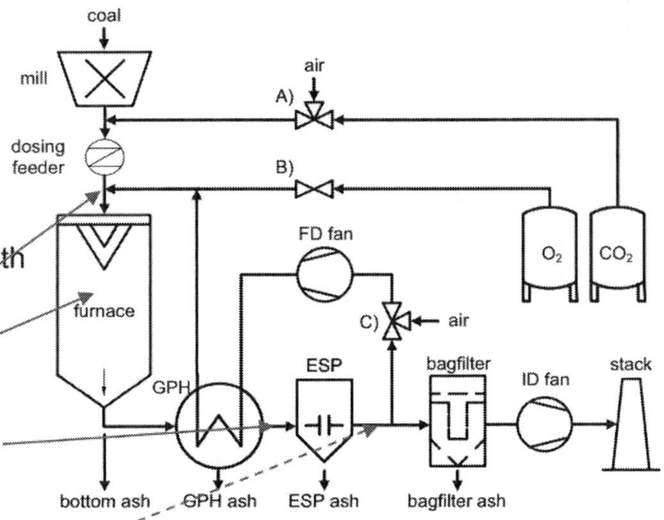


15

## Outline

1. Introduction
2. Background & Motivation
3. Methodology
4. Experimental results
5. Summary and outlook

16

## Results: Example of DSI test

### Oxy-fired, co-injection of limestone at <u>500 kW</u> test rig

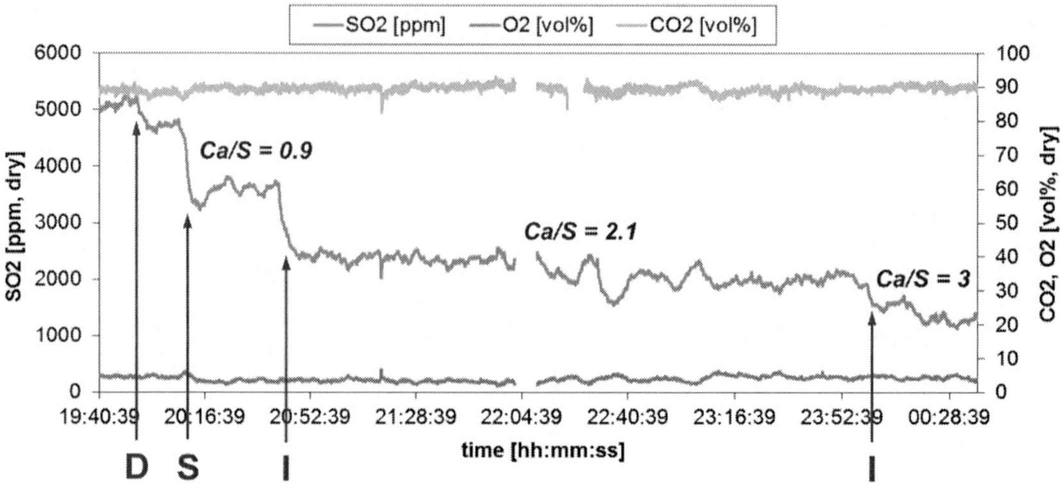


**Explanation:** **D** dosing gas on; **S** sorbent on;
**I** injection increased;

17

## 20 kW: Sorbent Performance – Air vs. $CO_2/O_2$

### <u>$CaCO_3$ & $Ca(OH)_2$ Co-/In-Furnace Injection</u>: (<u>Target – $SO_2$</u>)

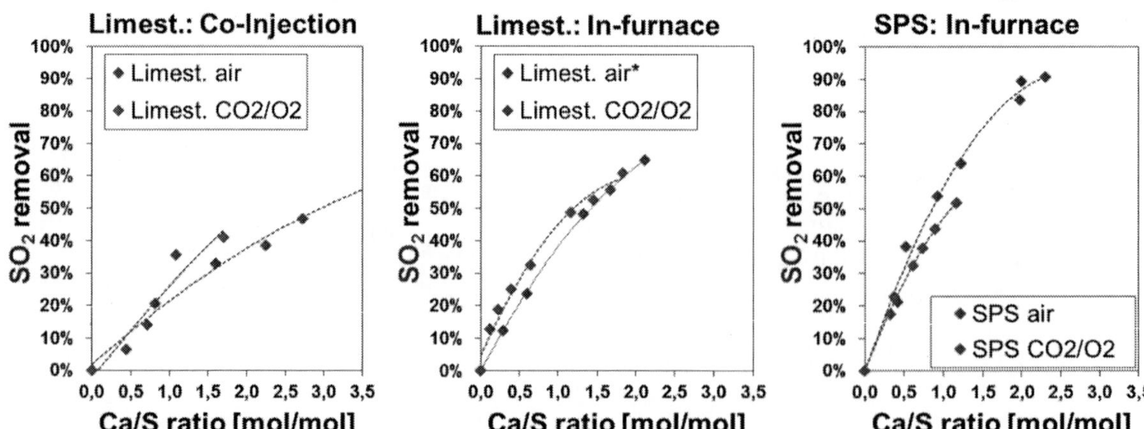


* air Ca/S corrected due to an offset of the sorbent balance

- Similar $SO_2$ removal in air and $CO_2/O_2$ atmosphere
- In-furnace injection yields higher removal than co-injection (sintering ?)
- $Ca(OH)_2$ performs better than $CaCO_3$ (up to 90% at Ca/S = 2 in air mode)

18

## Air: Sorbent Performance – <u>20 vs. 500 kW scale</u>

ifk

### <u>$CaCO_3$ & $Ca(OH)_2$ Co-/In-Furnace Injection</u>: (<u>Target – $SO_2$</u>)

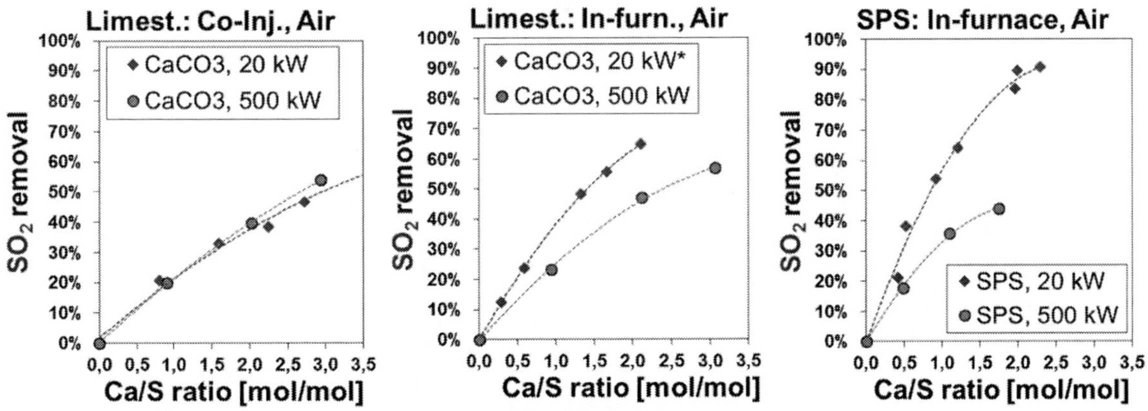


** air Ca/S corrected due to an offset of the sorbent balance*

- Co-injection performance at both scales similar
- In-furnace injection at 500 kW scale lower: <u>likely imperfect sorbent dispersion</u>
- Oxy-fuel and $CO_2/O_2$ tests at both scales are not directly comparable (different Ca/S definitions)

19

## 500 kW: Sorbent Performance – <u>Air vs. Oxy-Fuel</u>

ifk

### <u>$CaCO_3$ & SPS Co- & In-Furnace Injection</u>: (<u>Target – $SO_2$</u>)

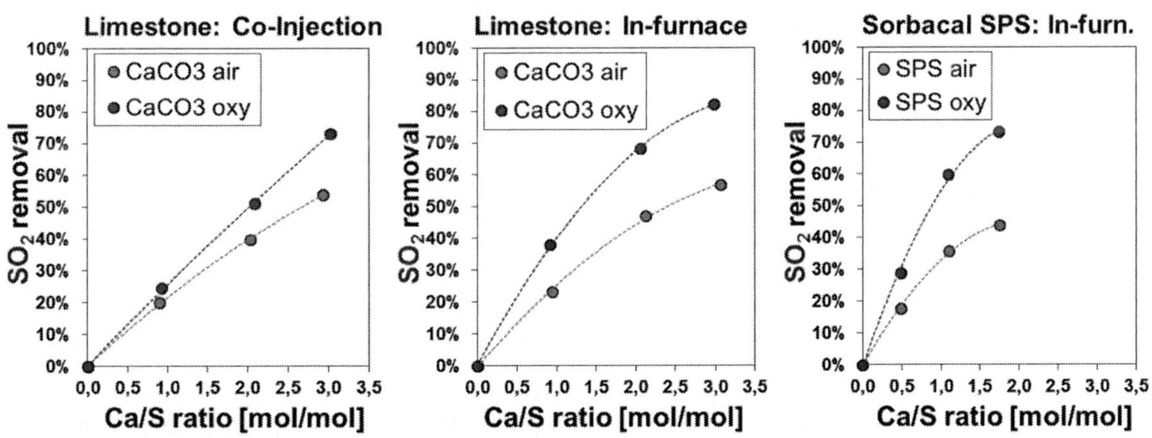


- $SO_2$ removal performance in oxy-fuel is better than in air mode (19-29 %-points)
- In-furnace injection yields higher removal than co-injection (sintering ?)
- High removal efficiencies with $CaCO_3$: 68% at Ca/S = 2.1 in oxy mode
- Even better removal with SPS ($Ca(OH_2)$: 73% at Ca/S = 1.7 in oxy mode

20

## 500 kW: Sorbent Performance – Air vs. Oxy-Fuel

ifk

### In-furnace Injection of SPS: (Co-benefit HCl & Target – $SO_2$ )

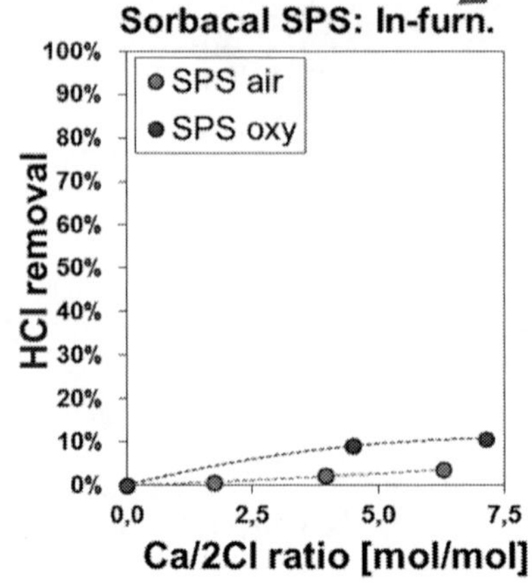


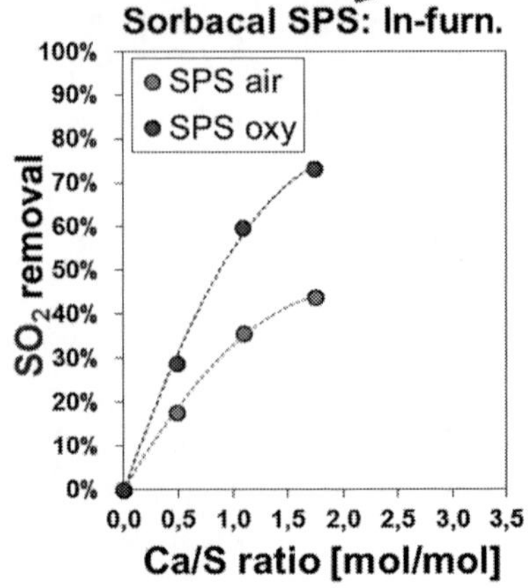


- HCl removal performance in oxy-fuel is better than in air mode
- $SO_2$ capture consumes majority of sorbent > low HCl reduction

21

## 500 kW: Sorbent Performance – Air vs. Oxy-Fuel

ifk

### Injection of SPS upstream ESP: (Target – HCl & Co-benefit $SO_2$ )

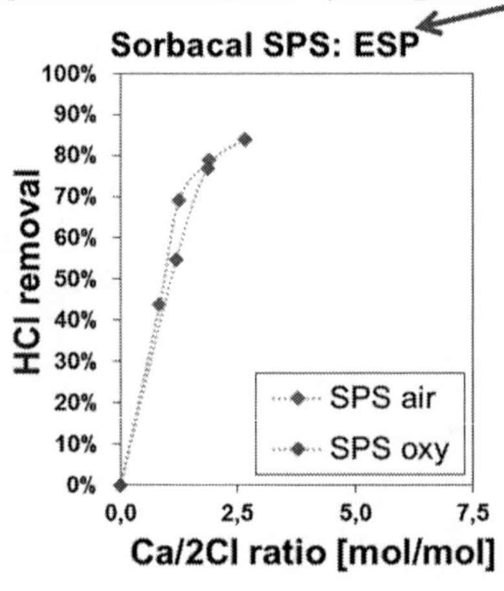


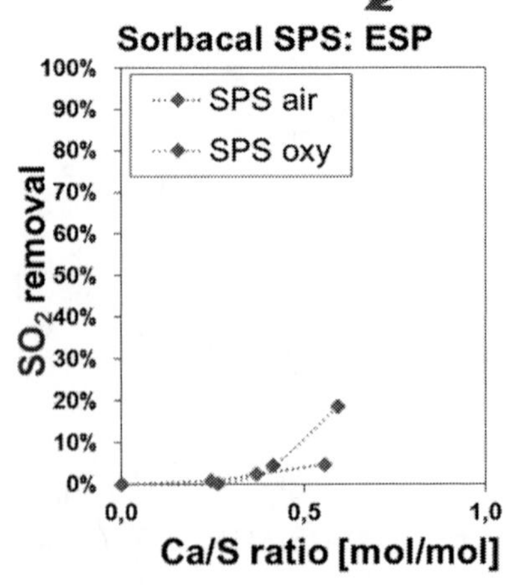


- Similar and efficient HCl removal performance in oxy-fuel and air mode
- HCl capture consumes majority of sorbent > low $SO_2$ reduction
- Increased $SO_2$ removal with excess sorbent in respect to HCl

22

## 500 kW: Sorbent Performance – $SO_3$

**Injection into furnace & upstream ESP**: (Target: $SO_2$ or HCl)

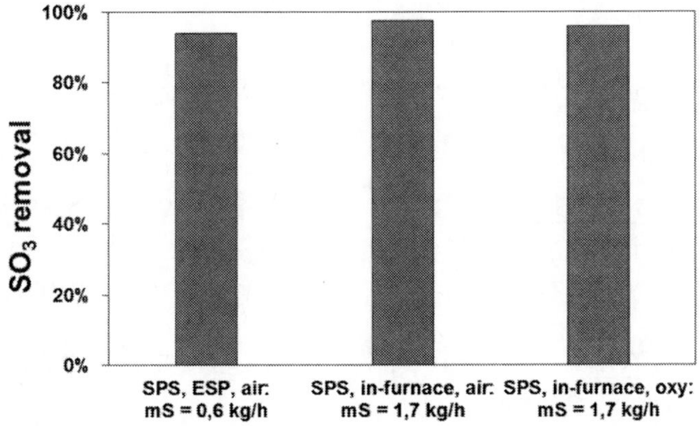


*mS: mass flow of sorbent*

- Efficient $SO_3$ removal (94-98%) as co-benefit to SPS injection for $SO_2$ and HCl control
- Similar $SO_3$ reduction for in-furnace SPS injection in air & oxy mode
- Injection upstream ESP yields high $SO_3$ removal at moderate Ca/S

23

## Outline

1. Introduction
2. Background
3. Motivation
4. Methodology
5. Experimental results
6. Summary and outlook

24

## Summary and Outlook

ifk

**DSI for acid gas control is highly promising for the oxy-fuel recycle process:**

- Efficient removal of HCl in air & oxy-fuel with $Ca(OH)_2$ injection upstream ESP (additionally $SO_2$ removal with excess sorbent)
- Efficient removal of $SO_2$ in air & oxy-fuel with co-injection and in-furnace injection of $CaCO_3$ and $Ca(OH)_2$
(oxy-fuel: $CaCO_3$: 68% at Ca/S = 2.1; $Ca(OH)_2$: 73% at Ca/S = 1.7)
- Efficient $SO_3$ removal (94-98%) as co-benefit to $Ca(OH)_2$ injection in air and oxy-fuel operation

- Limitation: - Feeding of $Ca(OH)_2$ with $CO_2$ is problematic
  - No problems observed with limestone

25

## Summary and Outlook

ifk

- Clear $SO_2$ removal performance gain (19-29 %-points) for sorbent co-injection and in-furnace injection under oxy-fuel recycle conditions.

**Outlook:**

- DSI tests (June 2016) at CIUDEN's 20MW oxy-fuel plant performed
- More details on air/oxy-fuel DSI tests will be published in future
- Final public project workshop on 2nd generation biomass co-firing and dry sorbent technologies: 22nd/23rd Nov. 2016, IFK, Stuttgart
For more details contact: *aaron.fuller@ifk.uni-Stuttgart.de*

26

Uniper Technologies Limited

ifk

# Thank you for your attention!


**ACKNOWLEDGMENTS**

The research leading to these results has received funding from the European Union's Research Fund for Coal and Steel (RFCS) research programme under grant agreement num. RFCR-CT-2013-00010 (RFCS research project BiOxySorb: http://bioxysorb.eu-projects.de/). The authors gratefully acknowledge this financial contribution and the support by advices and expertise of the BiOxySorb project partners Fundación Ciudad de la Energía, Uniper Technologies Limited, Lhoist Recherche et Développement SA and Gestamp Biomass Solutions. The authors also thank all colleagues of IFK's department "Firing Systems" who contributed to this work and in particular S. Pek, B. Ebner, M. Faulhaber, M. Pagano and T. Wagner, as well as W. Ross and his team of IFK's ´Laboratory for Fuels, Ashes and Slag´ for their support in the performed experiments.


27

## Normalized Stoic. Ratios & DSI Performance

ifk

- Injection stoichiometry determines acid gas removal efficiency:
  *(i.e.: How many moles of sorbent per mole of $SO_2$/$SO_3$/HCl are injected?)*

$$\frac{Ca}{S} = \frac{\dot{m}_{sorbent}/W_{sorbent}}{y_{SO_2} \times V_{FG}/V_{mol}}; \quad \frac{Ca}{2Cl} = \frac{\dot{m}_{sorbent}/W_{sorbent}}{2 \times y_{HCl} \times V_{FG}/V_{mol}}; \quad \frac{Na}{Cl} = \frac{\dot{m}_{sorbent}/W_{sorbent}}{y_{HCl} \times V_{FG}/V_{mol}};$$

*Where:*

$\dot{m}_{sorbent}$= mass flow sorbent [g/h], $W_{sorbent}$= mol. weight sorbent [g/mol], $y_{SO_2}$= $SO_2$ conc. [ppm],

$y_{HCl}$= HCl conc. [ppm], $V_{FG}$= flue gas flowrate [m³ (STP)/h], $V_{mol}$= molar volume [m³ (STP)/mol].

- Percentaged acid gas removal: $\eta_i$

$$\eta_i = \frac{y_i^0 - y_i^{sorbent}}{y_i^0} \times 100\%$$

*Where:*

$y_i^0$= conc. of component i without sorbent injection,

$y_i^{sorbent}$= equilibrated conc. of component i during sorbent injection,

$i$= acid gas component: $SO_2$, $SO_3$, HCl.

28

## Acid Gas Removal Recycle Combustion

ifk

- Recycle combustion:

**Acid gases ($SO_2$, HCl):**

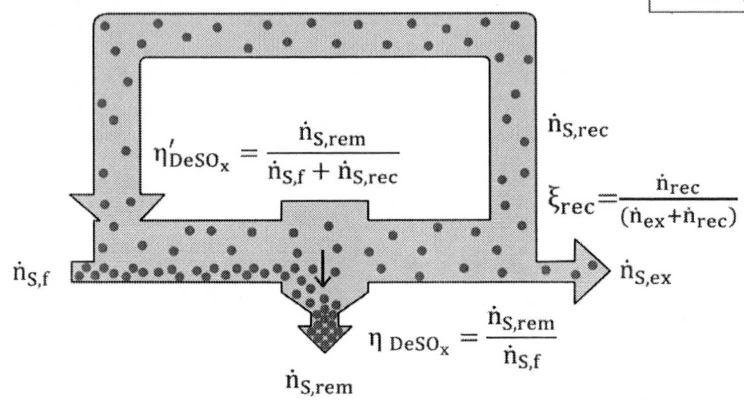


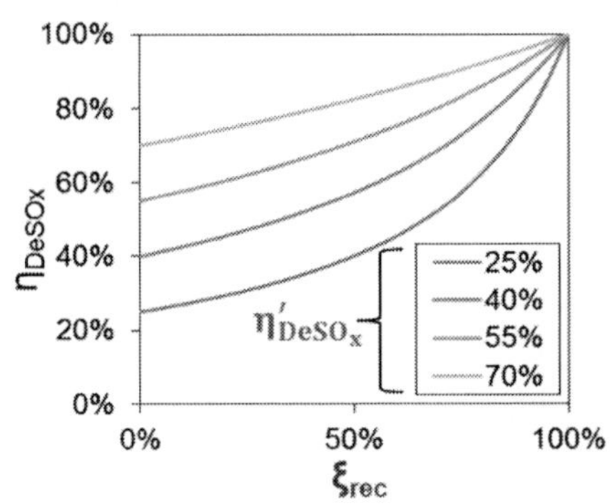


**Effect of flue gas recirculation on acid gas removal:**

$$\eta_{DeSO_x} = \frac{\eta'_{DeSO_x}}{1 - (1 - \eta'_{DeSO_x})\xi_{rec}}$$

29

ITC
WYOMING
INTEGRATED
TEST CENTER

# Powering the Nation, Fueling Innovation

Who?
Why?
How?
What?
Where?
When?

ITC
WYOMING
INTEGRATED
TEST CENTER

# Wyoming Infrastructure Authority

The WIA is an instrumentality of the State of Wyoming and was created in 2004 by the Wyoming State Legislature.

**Our Mission:** The WIA will diversify and expand the state's economy by adding value to Wyoming's energy resources and infrastructure for the benefit of Wyoming and the region.

**Our Vision**: The WIA will advance Wyoming's Energy Strategy by promoting the value of Wyoming's energy resources; supporting the necessary infrastructure; enhancing resource development and operation; and ensuring a credible and objective voice for Wyoming.

Within the statutory purview of the WIA is "advanced coal technology and advanced energy technology facilities. The facilities and related supporting infrastructure may include all facilities, structures and properties incidental and necessary or useful in the production or transmission of energy". In addition, effective July 1, 2014, the responsibilities of the WIA include coal exports pursuant to HB 147 which was enacted in the 2014 Legislative Session.

In addition, the WIA has $1 billion in bonding authority to finance transmission and generation and coal export infrastructure.

ITC
WYOMING
INTEGRATED
TEST CENTER

# Who?

State of Wyoming - $15 million
Basin Electric – Host at Dry Fork Station
Tri-State G&T - $5 million
National Rural Electric Cooperatives Association - $1 million
XPRIZE Foundation – tenant
Wyoming Infrastructure Authority – Managing Entity

ITC
WYOMING
INTEGRATED
TEST CENTER

BASIN ELECTRIC POWER COOPERATIVE
A Touchstone Energy Cooperative

nrg | cosia CARBON XPRIZE

TRI-STATE
Generation and Transmission Association, Inc.
A Touchstone Energy Cooperative

National Rural Electric Cooperative Association
A Touchstone Energy Cooperative

# Why is coal important to Wyoming?

- Jobs
- School construction funding
- About $1 billion annually in direct tax payments to Wyoming
- Affordable energy and grid reliability

Meadowlark Elementary – Laramie #1 Jan. 2016

ITC
WYOMING
INTEGRATED
TEST CENTER

# Why do we need take the lead?

- The Federal Government is not prioritizing coal research
- The Administration recommends restructuring the National Energy Technology Laboratory (NETL). There will no longer be an NETL In-House R&D program included in the Coal CCS & Power Systems budget.
- Puts Wyoming is in a much stronger position to request resources

| DOE Program (dollars in millions) | FY 2015 Omnibus | FY 2016 Omnibus | FY17 Request | Increase or Decrease Over Enacted | |
|---|---|---|---|---|---|
| | | | | $ | % |
| Fossil Energy* | 560,587 | 632,000 | 600,000 | -32,000 | -5.1 |
| Energy Efficiency and Renewable Energy | 1,914,195 | 2,069,194 | 4,233,400 | +2,164,206 | +104.6 |
| Electricity Delivery & Reliability | 146,975 | 206,000 | 262,300 | +56,300 | +27.3 |
| Nuclear | 833,379 | 986,161 | 993,896 | +7,735 | +0.8 |
| Science | 5,067,738 | 5,347,000 | 5,672,069 | +325,069 | +6.1 |
| ARPA-e | 279,982 | 291,000 | 500,000 | +209,000 | +71.8 |

* When the $240 million in CCPI funding is applied to the program, it leaves a balance of $360 million for the program, which represents a 43% cut in funding from FY16 enacted levels, which does not appear to be consistent with increasing funding by 20% as agreed to under the Mission Innovation initiative.

ITC
WYOMING
INTEGRATED
TEST CENTER

# The Goals

How to keep coal viable and an important part of Wyoming's economy while addressing societal concerns over carbon emissions?

Can we turn a liability into an asset?

Let's position Wyoming as a leader on solutions!

ITC
WYOMING
INTEGRATED
TEST CENTER

# How? Technology development works

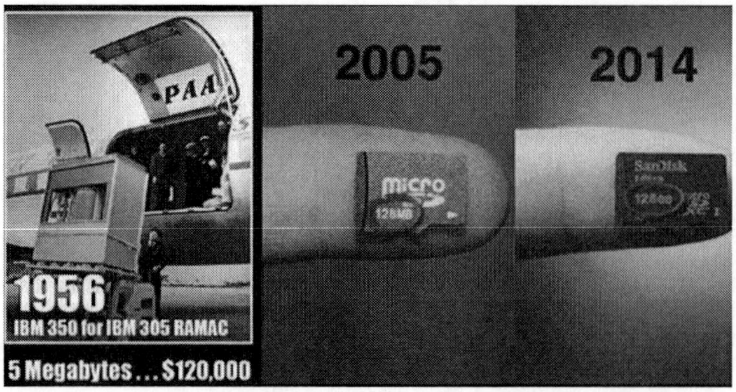


ITC
WYOMING
INTEGRATED
TEST CENTER

# Technology development works

ITC
WYOMING
INTEGRATED
TEST CENTER

# Emissions Technology Works

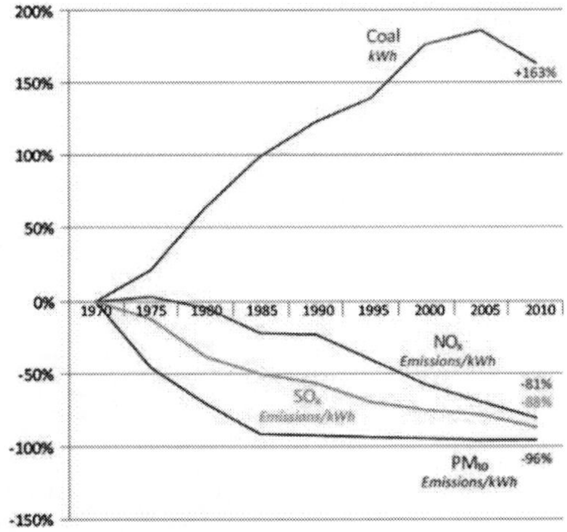


ITC
WYOMING
INTEGRATED
TEST CENTER

# What? The Integrated Test Center

- The Wyoming Legislature appropriated $15 million in 2014
- ITC Working group – Governor, UW, industry, R&D
- Tri-State Generation and Transmission matched $5 million
- The National Rural Electric Cooperatives are providing $1 million
- Basin Electric Cooperative is providing the host site
- Rocky Mountain Power and Black Hills Corp. provided technical expertise
- XPRIZE Foundation is the first tenant
- $12 million capital construction, remainder for operations

ITC
WYOMING
INTEGRATED
TEST CENTER

# What exactly is the ITC?

- At completion, it will be one of only six similar facilities in the world
- Supplied with 20 MW of flue gas from the Dry Fork Power Station
- Simple design keeps costs low, provides flexibility for researchers and quick turnaround time
- Designed for maximum flexibility and scalability for testing – not a laboratory

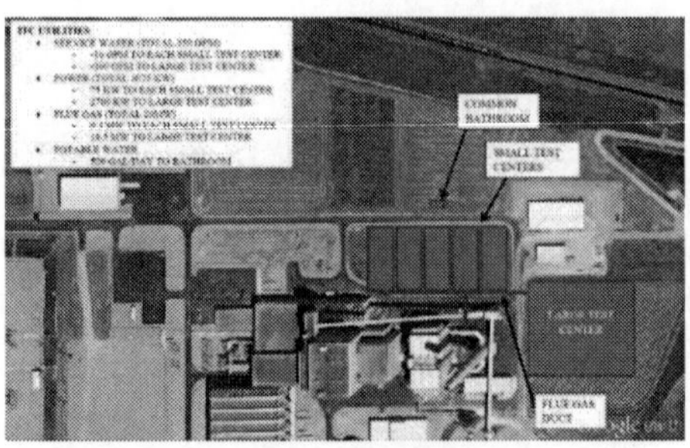

Credit: Basin Electric Cooperative

ITC
WYOMING
INTEGRATED
TEST CENTER

# Small Test Bays

Site of XPRIZE competition

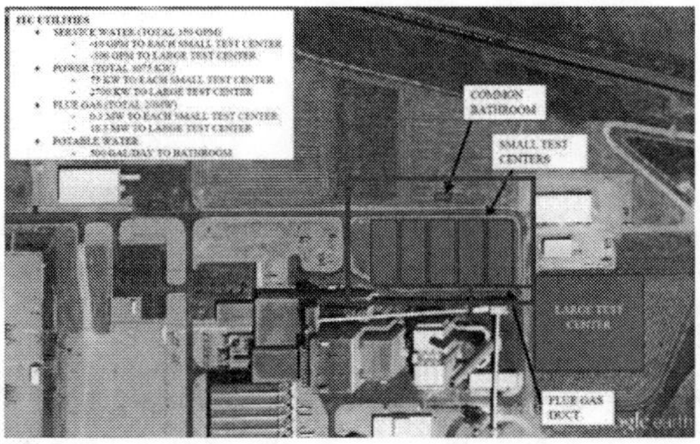


Credit: Basin Electric Cooperative

ITC
WYOMING
INTEGRATED
TEST CENTER

# XPRIZE Foundation

nrg | cosia CARBON XPRIZE

"The $20M NRG COSIA Carbon XPRIZE will challenge the world to reimagine what we can do with $CO_2$emissions by incentivizing and accelerating the development of technologies that convert $CO_2$ into valuable products."

- In September 2015 announced Carbon XPRIZE competition with NRG and COSIA
- Two tracts
  - Natural gas
  - Coal
- Three rounds of competition per tract
  - 1 – Technical Papers – up to 15 teams will move on
  - 2 – Small scale, laboratory testing – up to 5 teams will share $2.5 million prize
  - 3 – Real world, larger scale – grand prize of $7.5 million

XPRIZE is a tenant and at the completion of the competition, the space will be available to new testers.

ITC
WYOMING
INTEGRATED
TEST CENTER

# Large Test Bay

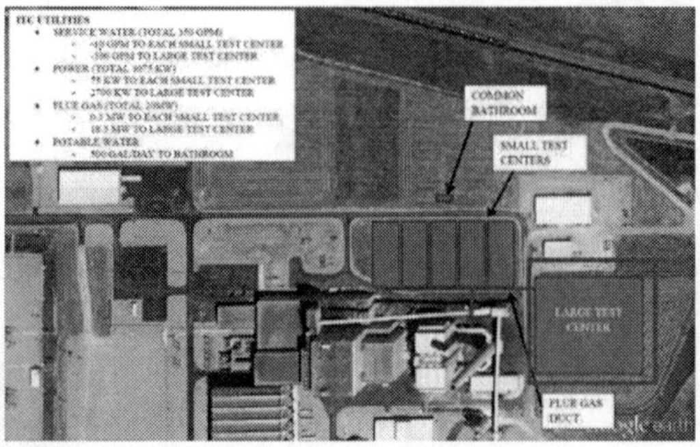


Credit: Basin Electric Cooperative

ITC
WYOMING
INTEGRATED
TEST CENTER

# When? Major Milestones

- March 2014 – Legislative Authorization
- Spring 2014 – Creation of ITC Working Group
- Summer 2015 – Feasibility Study Completed
- Summer 2015 – Governor Mead authorizes ITC construction based upon legislative milestone completion
- August 2015 – WIA appointed to lead ITC construction
- September 2015 – MOU between UW-SER and WIA executed
- October 2015 – MOU between XPRIZE and WIA executed
- October 2015 – Basin Electric orders damper and steel for first phase of construction
- November 2015 – MOU between Tri-State G&T and WIA executed
- December 2015 – Communications and PR development begins
- February 2016 – Launch website – www.wyomingitc.org
- March – April 2016 – Installation of the damper
- April 2016 – MOU between WIA and NRECA Executed
- April 2016 – Official Groundbreaking
- *Summer 2016 – Completion of MOU with Basic Electric*
- *Fall 2015-Summer 2016 – Final engineering completed*
- *Summer 2016 – Finalization of Operating Entity*
- *Spring-Fall 2016 – Construction of test bays, utilities and flue gas distribution network*
- *Winter 2017 – Testing and Commissioning*
- *May 2017 – Completion and ready for testing!*
- *Fall 2017 – XPRIZE begins on-site testing*

ITC
WYOMING
INTEGRATED
TEST CENTER

# Damper Installation Mar-April 2016

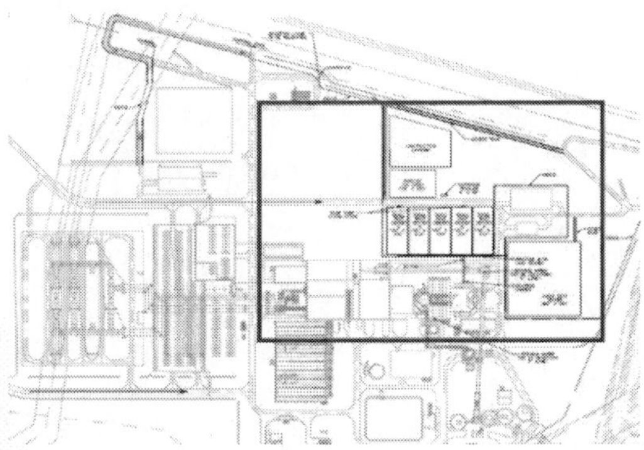

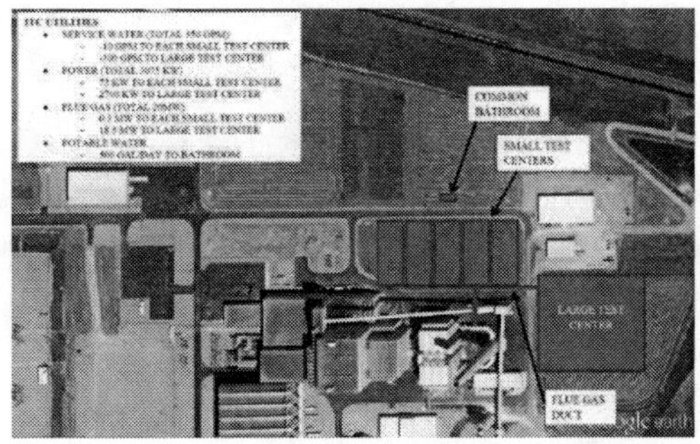


Credit: Basin Electric Cooperative

ITC
WYOMING
INTEGRATED
TEST CENTER

# Damper Installation Mar-April 2016

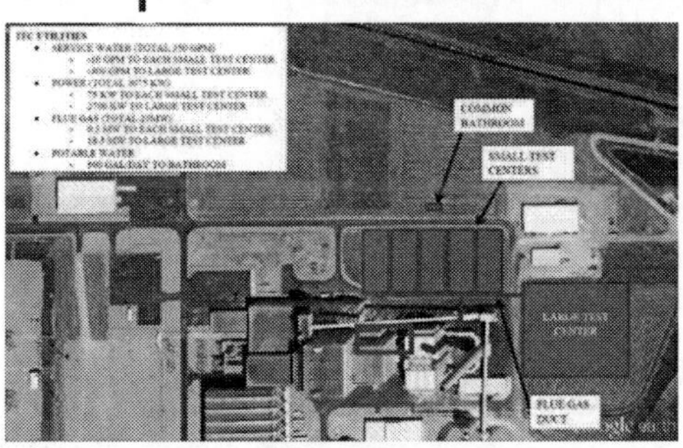


Credit: Basin Electric Cooperative

**TIE-IN DUCTWORK STUB – MODEL SHOT**

ITC
WYOMING
INTEGRATED
TEST CENTER

# Damper Installation Mar-April 2016

ITC
WYOMING
INTEGRATED
TEST CENTER

# Damper Installation Mar-April 2016

ITC
WYOMING
INTEGRATED
TEST CENTER

# Damper Installation Mar-April 2016

ITC
WYOMING
INTEGRATED
TEST CENTER

# Next Steps

- Explore opportunities to leverage additional funds
  - Have reached out to Department of Energy
  - Held discussions with other coal producing states
- Open dialog with international organizations to promote and advertise ITC
  - The International Carbon Capture Centre Network
    - Scheduled to deliver presentation for membership Q4 2016
  - World Coal Association
    - R&D MOU

Construction is ahead of schedule, on budget!

ITC
WYOMING
INTEGRATED
TEST CENTER

WORLD COAL ASSOCIATION

U.S. DEPARTMENT OF ENERGY

CCS TEST CENTRE NETWORK

# Questions?

Jason Begger
Executive Director
Wyoming Infrastructure Authority
jason.begger@wyo.gov
(307) 256-0098

www.wyomingitc.org

ITC
WYOMING
INTEGRATED
TEST CENTER

Wyoming
INFRASTRUCTURE AUTHORITY

**2016 International Pittsburgh Coal Conference**
**Cape Town, SOUTH AFRICA**
**August 8 – 12, 2016**

**Manuscript Submission**

**PROGRAM TOPIC:**
**8. COAL MINING**

# ADSORPTION CHARACTERISTICS OF BITUMINOUS COAL SEAMS IN ZONGULDAK BASIN

***Olgun ESEN, Research Assistant, Istanbul Technical University Mining Engineering Dept. 34469 Maslak-Istanbul, TURKEY***
***Contact Information: esenolgun@itu.edu.tr, +902122857364, +902122856131***

*Samet Can ÖZER, Research Assistant, Istanbul Technical University Mining Engineering Dept. 34469 Maslak-Istanbul, TURKEY*
*Contact Information: canozer@itu.edu.tr, +902122857444, +902122856131*

*Abdullah FIŞNE, PhD, Assistant Professor Dr., ITU Mining Engineering Department, 34469 Maslak Istanbul, TURKEY*
*Contact Information: fisnea@itu.edu.tr, +902122856363, +902122856131*

**Abstract:**

Zonguldak Coal Basin, located on the Black Sea coast, is the main part of the Upper Carboniferous bituminous coal basin of Turkey. Much of the bituminous coal mining has thus been concentrated in the Zonguldak Basin. The estimated coal reserve, carried to 1200 m under sea level, is about $1.32x10^9$ tons of coal in this basin. 57 coal seams have been identified and 38 are workable with their thickness varying from 0.7 - 10 m and dips from 0 - 90°. The most important coal seams are the Cay and Acilik seams having average thickness of 6 and 4.5 m, respectively. Furthermore, the Sulu, Hacimemis, Domuzcu and Buyuk seams having an average thickness ranging from 1.5 m to 2.5 m are the other coal seams which are regarded as important for coal production. The aim of this study is to determine the gas adsorption characteristics of coal seams in the Zonguldak Basin. A number of samples were collected from 8 different seams. The adsorption characteristics of samples were measured and attempts made to correlate the results with different coal properties determined with the help of proximate analyses. Methane adsorption isotherms were measured at constant temperature of 25°C at different gas pressures up to 50 bar and the results were discussed in relation to the gas capacities of coal seams. Langmuir equation was used to interpret the adsorption isotherms. Results of the interpretation of the adsorption isotherms indicate that Langmuir (monolayer) adsorption capacity of coal seams varies between 10.3 $m^3/t$ and 18.2 $m^3/t$. Although the respective data are highly scattered, a certain correlation could be defined between different coal properties and the adsorption capacity. The most apparent correlation is observed between the ash content and the gas adsorption capacity. The trends between Langmuir adsorption capacity and ash content are generally in accordance with the trends observed in other coal basins.

## INTRODUCTION

The release of methane during coal mining has often been a factor limiting production, in some cases it has been known to stop coal production completely. Because of highly disturbed geological and tectonic conditions, methane emission is an important problem in Zonguldak coalfield, Turkey. A knowledge of the gas content and rate of methane released from a given coal seam is important for assessing the potential danger of emissions during mining operations. It is recognized that most methane present in virgin coal occurs in the adsorbed state (Yalcin and Durucan, 1991a-b).

Coal is one of the richest known sources of and reservoirs for major accumulations of natural gas (mainly methane). Gas is retained in coals in several forms: (1) as adsorbed molecules within micropores (2 nm in diameter) and/or included in the molecular structure of the coals; (2) as trapped gas within the matrix porosity; (3) as free gas within meso pores and macro pores of coal, in cleats and fractures; and (4) as dissolved gas in water within the coal fractures. The primary mechanism of methane storage in a coal bed is the sorption within molecules and/or in the molecular structure of coals. Sorption of methane in coals is mainly controlled among other primary controls such as type, rank and moisture content also by different coal properties such as density, pore volume, pore size distribution and surface area. These parameters can be determined by density measurements, mercury porosimeter and gas adsorption measurements (Gurdal and Yalcin, 2000).

The purpose of this study here is to determine the gas adsorption capacity of coal seams in Zonguldak Basin and to investigate the controlling parameters in terms of adsorption characterization. Thus, this controlling parameters influence on gas adsorption in coal structure are also investigated. Furthermore chemical analyses of coal samples were carried out on the purpose of characterization and classification of coal seams to explain the adsorption characteristics of bituminous coal seams in the Zonguldak basin.

## EXPERIMENTAL

The gas adsorption capacity (gas storage capacity) of coal seams are determined by the adsorption (sorption) isotherms (Kim, 1977; Diamond et al, 1976; McFall et al., 1986). Gas is being held by a physical and chemical forces in coal pores when a porous material is under-pressured in a sealed system, is described as Adsorption. The amount of gas sorbed by solid material surface depend on amount of solid material, temperature of the condition and pressure. It is known whether the gas being adsorbed by solid surface at constant temperature, the adsorbed gas amount will only be changed by pressure. Also this change is become the function of pressure. In this case, adsorption isotherm is explained by the plotted graph which shows the relationship between pressure and the adsorbed amount of gas (Gurdal, 1998). The gas content–pressure relation for a given coal cannot be predicted from coal properties nor determined by wireline logs and must, therefore, be measured in the laboratory (Seidle, 2011).

A sorption isotherm, or simply an isotherm, for a given coal is measured in a laboratory by re-adsorbing a gas onto the coal sample or charging a coal sample with gas and measuring the gas content as pressure is reduced. At low pressures, the amount of sorbed gas is the proportional of the gas pressure. Hardly, gas adsorption comes to a stopping point whether the gas pressure is increased. The result is the inner surface of coal could be hold maximum gas capacity. The coal sample gas adsorption capacities were determined by the high-pressured sorption test apparatus which was owned by Hard Coal Agency of Turkey in Zonguldak. The sorption device is shown in Figure 1 below.

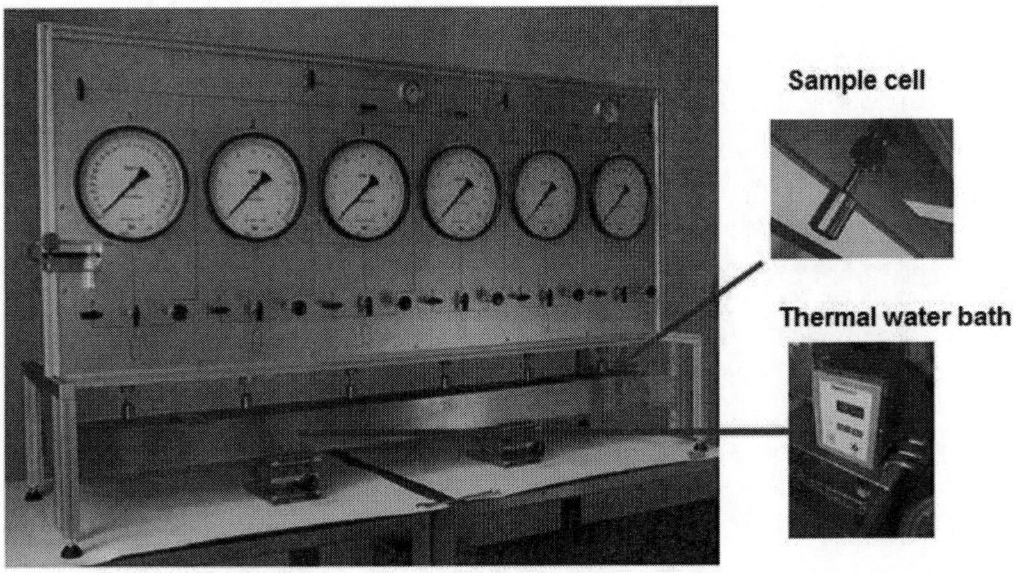


Figure 1. High-pressured sorption test apparatus.

A coal sample was weighted to 30 g and divided in to 6 sample cells as 5 grams of coal sample in each cell and sieved the grain sizes between +400 μm – 630 μm. Then the pre-sorption was performed on each sample cell during 4 days. The pre-sorption process involves 20 bar for the first three sample cell and the other cells were under pressured on 50 bar. The injected gas was methane. After the methane pre-sorption was performed at desired pressures, methane gas was sorbed in to sample cells at 0 bar, 4 bar, 8 bar, 16 bar, 32 bar, 40 bar next two days. Then amount of gas adsorbed was determined by the following procedures below.

Tight containers are connected to the analyzer for measuring the methane concentration which sample emitted into the container. Figure 2 shows the gas analyzer used for gas concentration measurements. This process is then repeated the next day and samples are ground using a sealed vibrating cup mill. A burette is connected to the mill to determine the volume of released gas, as shown in Figure 3. Then the amount of gas adsorbed was measured and determined.

Figure 2. Methane gas analyzer for measuring the methane concentration.

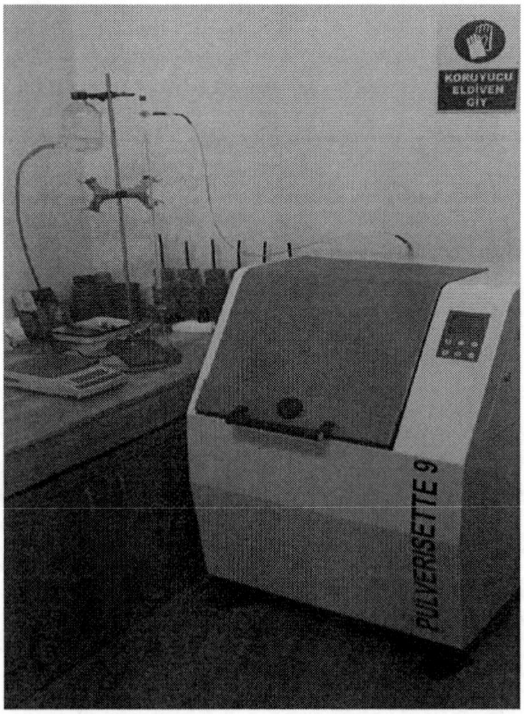


Figure 3. Determination of released gas by ring mill.

The gas adsorption capacities of the coal seams might be determined by the adsorption isotherms which are the plotted explanation of sorption experiment results. Equations from various sorption models have been employed to describe experimental coal gas sorption data. Among the most popular is the Langmuir Isotherm equation which was developed by Langmuir (1918).

The amount of adsorbed gas which is sorbed in microporous materials can be explained and described by the Type-1 Langmuir Isotherm. The Langmuir Equation has most common usage in various coal rank studies and is given in Equation 1.

$$V = V_L \frac{P}{P + P_L} \qquad (1)$$

Where,

| | |
|---|---|
| V | : Amount of gas adsorbed, $cm^3/g$, |
| $V_L$ | : Langmuir volume constant, $cm^3/g$, |
| P | : Pressure, MPa, |
| $P_L$ | : Langmuir pressure constant, Mpa. |

As pressure increases without bound, Equation (1) indicates gas content approaches $V_L$, the Langmuir volume constant. In a physical sense, the Langmuir volume constant represents the maximum amount of gas that can be sorbed onto the given sample at infinite pressure. When pressure equals the Langmuir pressure constant, $P_L$, gas content calculated from Equation (1) is ½ $V_L$. Thus, the Langmuir pressure constant represents the half-saturation pressure, that is, the pressure at which the coal holds one-half the maximum gas volume (Seidle, 2011).

It has been accepted when using Langmuir model that adsorbed gas to be ideal, the adsorption limited to a single layer, solid (coal) was homogeneous the energy distribution on the surface and where there is interaction between the adsorbed molecules (Marsh, 2001).

In this study, Langmuir Model was applied to determine the gas adsorption capacities of coal seams by the provided adsorption values. Determination of Langmuir constants from a given set of sorption data has traditionally been done by fitting the data to a rewritten version of equation that is shown in Equation 2.

$$\frac{P}{V} = \frac{P}{V_L} + \frac{P_L}{V_L} \qquad (2)$$

Linearized plot will be provided by the plotted graph of P versus P/V. The Langmuir volume constant, $V_L$, is the reciprocal of the slope. The pressure constant, $P_L$, is determined by dividing the intercept by the slope. After the determination of Langmuir constants, the sorption isotherms of coal seams were plotted to the following process.

**RESULTS**

This study aims to determine the gas adsorption capacities and controlling parameters on adsorption of coal seams lying on the Zonguldak coal basin. For the determination of gas adsorption capacities of coal seams, studies were performed by the high-pressured sorption test apparatus owned by Hard Coal Agency of Turkey (TTK) Occupational Safety and Education Office in Zonguldak. 19 coal samples were taken from Acılık, Çay, Sulu, Büyük, Hacımemiş, Kurul, Unudulmuş, Domuzcu, Piç, Taşbaca and Hacıpetro Seams and were analyzed for sorption tests.

Coal samples were sieved between +400 μm – 630 μm. All analyses were made at constant temperature as 27 °C. Pre-sorption tests of the coal samples performed for 4 days on each 5 g coal sample which were sorbed in 6 sample cells. Total sample weight was 30 grams for 6 days experiment. Methane injection was applied for the first three sample cells at 20 bar when other 3 sample cells have 50 bar. The saturated sample cells by the pre-sorption tests, the desired pressure was performed on coal sample cells at 0 bar, 4 bar, 8 bar, 16 bar, 32 bar and 40 bar by the following 2 days. Gas measurement process which was explained previously on experimental in this study, was made for sorbed gas amount determination.

Gas adsorption capacities of coal seams can be calculated by the assessment of adsorption isotherms which are the explanation of experimental sorption data set. The calculations were made by mathematical models to describe the isotherms. In this study, the Langmuir Model was applied and used to describe the sorption data of the coal seams. Provided sorption data results are given in Table 1.

In Zonguldak coal basin, number of 11 coal seams' gas adsorption capacities vary in original (as received) between 9,13 $m^3/t$ and 12,06 $m^3/t$, in dry and ash basis between 10,27 $m^3/t$ and 18,15 $m^3/t$. All these minimum and maximum adsorption capacity values are owned by Hacımemiş and Büyük coal seams respectively.

Table 1. The adsorption capacities of the coal seams in Zonguldak coal basin.

| **Drilling Code / Colliery / Coal Seam** | **Langmuir Constants (as received basis)** | | **Langmuir Constants (dry and ash free basis)** | |
|---|---|---|---|---|
| | **$V_L$ ($m^3$/t)** | **$P_L$ (bar)** | **$V_L$ ($m^3$/t)** | **$P_L$ (bar)** |
| 20140327 Kozlu Acılık | 11.88 | 9.03 | 12.85 | 8.90 |
| 20140903 Kilimli Acılık | 12.06 | 7.36 | 12.79 | 7.38 |
| 20141209 Kilimli Acılık | 11.81 | 10.66 | 12.64 | 10.64 |
| 20140318 Kozlu Çay | 15.82 | 11.67 | 17.48 | 11.52 |
| 20140424 Kozlu Sulu | 14.25 | 9.44 | 16.75 | 9.32 |
| 20141124 Kozlu Sulu | 11.99 | 9.94 | 12.63 | 9.90 |
| 20140703 Kozlu Büyük | 16.98 | 19.40 | 18.15 | 19.45 |
| 20140813 Armutçuk Büyük | 12.52 | 10.35 | 13.40 | 10.32 |
| 20140821 Armutçuk Büyük | 10.71 | 9.36 | 11.33 | 9.48 |
| 20141023 Kozlu Hacımemiş | 13.19 | 16.94 | 14.08 | 16.92 |
| 20141030 Kozlu Hacımemiş | 13.35 | 7.54 | 14.08 | 7.48 |
| 20141030 Kozlu Hacımemiş | 11.75 | 10.26 | 14.39 | 10.21 |
| 20141030 Kozlu Hacımemiş | 9.13 | 8.62 | 10.27 | 8.87 |
| 20140611 Gelik Kurul | 13.35 | 9.21 | 15.13 | 8.92 |
| 20140829 Kilimli Unudulmuş | 15.38 | 15.54 | 16.72 | 15.54 |
| 20140618 Kozlu Domuzcu | 14.25 | 8.09 | 15.36 | 8.03 |
| 20150317 Kozlu Piç | 11.66 | 8.79 | 12.27 | 8.76 |
| 20150330 Kozlu Taşbaca | 12.95 | 3.96 | 14.01 | 3.97 |
| 20140116 Kozlu Hacıpetro | 14.08 | 10.33 | 15.80 | 10.33 |

Effect of moisture, ash, volatile matter and fixed carbon on coal gas adsorption capacity of the coal samples were also investigated.

Moisture effect on gas adsorption capacity of coal samples are shown in Figure 4. Due to various moisture contents of the coal samples, effect of moisture on gas adsorption capacities have not been completely determined. Thus, there are quite a few researches exists related with effect of moisture content on gas adsorption capacity (Joubert et al., 1973; Levine, 1993). These studies show that increase in moisture content up to a critical value such as %2 causes to one – three times decrease in gas adsorption capacity of coals. Above the critical moisture content, the effect of moisture on gas adsorption capacity is disappeared.

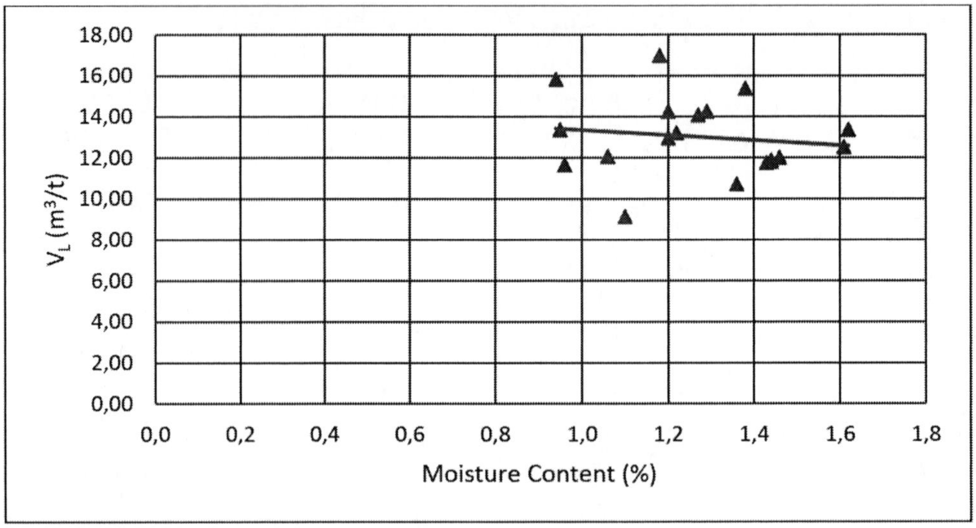


Figure 4. Relationship between gas adsorption capacity and moisture content of coals.

Ash, like moisture, can decrease the adsorption capacity of coal (Figure 5). Ash content in coal causes a negative effect on adsorption capacity due to gas is held on the organic matter surface.

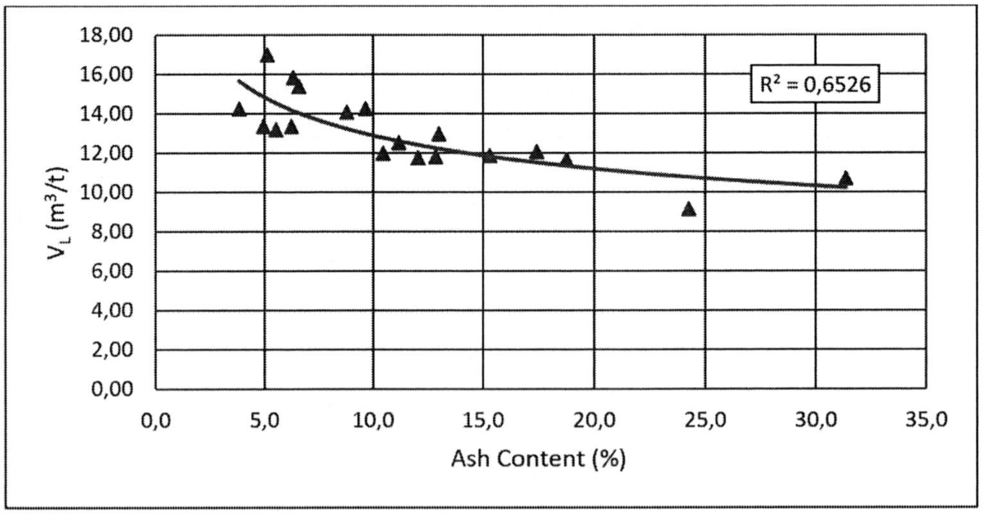


Figure 5. Relationship between gas adsorption capacity and ash content of coals.

Volatile matter content is a maturity parameter which effects on gas generation and gas adsorption capacity of coals. Gas generation in coals especially begins while volatile matter content exceeds %37 and increases with maturity of coals. Gas is especially held by micropores in low maturity and high volatile matter content coals. Gas adsorption capacity decreases when volatile matter decrease first, and volatile matter decreasing from %15 then increases (Gurdal, 1998). For the reason of decrease is macropores and mesopores are deformed under pressure. The relationship between volatile matter and gas adsorption capacities of coals is given in Figure 6.

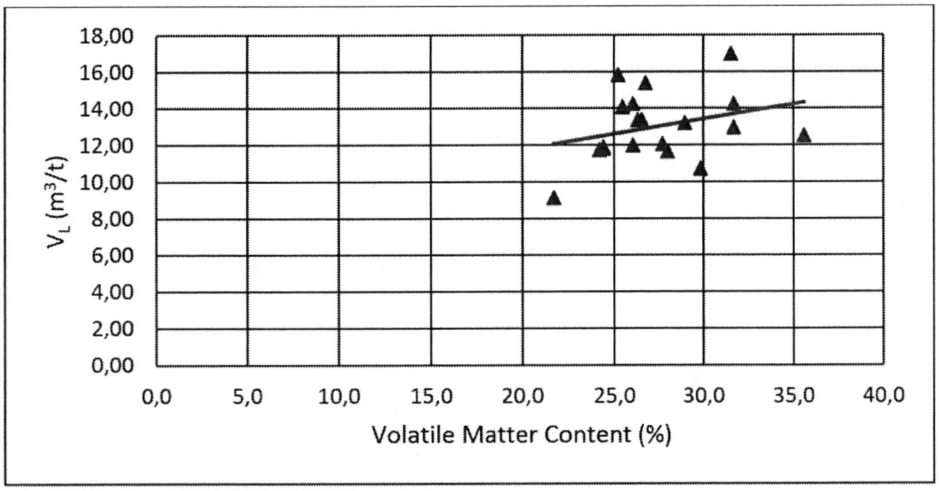


Figure 6. Relationship between volatile matter and gas adsorption capacity of coals.

Volatile matter contents of coal samples vary between % 20 – 35. This narrow range makes the distribution scattered. Low correlation values gives the minimum gas adsorption capacity when increase of volatile matter. Effect of fixed carbon on gas adsorption capacity of coals is shown in the Figure 7. Scattered distribution of values are observed during assessment of fixed carbon and gas adsorption capacity of coals. The expectation was increasing in fixed carbon makes the adsorption capacity higher, however, effect of other parameters this expectation would not be made. Although the scattered distribution, it can be said that there exists a positive correlation between this two parameters. Moreover, this study shows the same fixed carbon value coals gives different gas adsorption capacities. This type of scattering is determined by different researches (Mahajan, 1991).

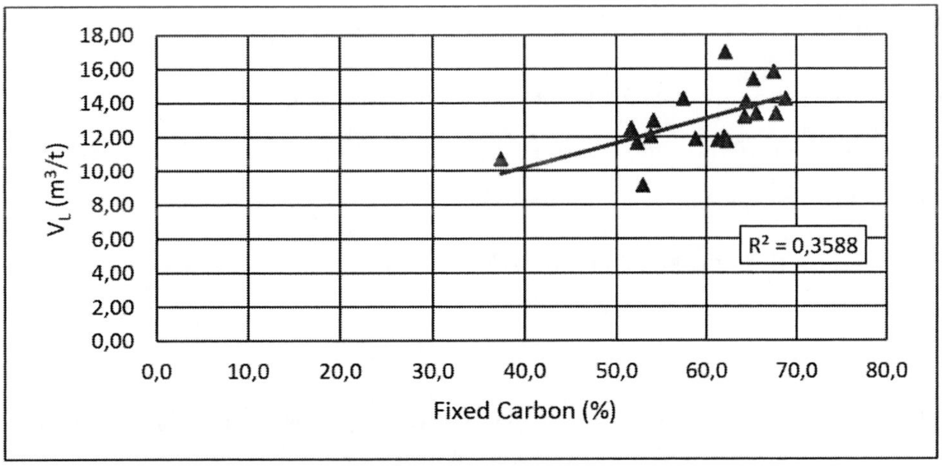


Figure 7. Relationship between fixed carbon and gas adsorption capacity of coals.

## CONCLUSIONS

This paper involves the gas adsorption capacities and coal reservoir properties of coal seams which were determined by indirect methods associate with Langmuir isotherms in the Zonguldak coal basin. Sorption analyses were carried out on the 19 coal samples taken from Acılık, Çay, Sulu, Büyük, Hacımemiş, Kurul, Unudulmuş, Domuzcu, Piç, Taşbaca and Hacıpetro coal seams. Gas adsorption capacities of 11 different coal seams at as received basis vary between 9.13 $m^3/t$ and 12.06 $m^3/t$, and dry-ash free basis vary between 10.27 $m^3/t$ and 18.15 $m^3/t$. As a result of the sorption isotherms, Hacımemiş seam has the minimum adsorption capacity and Buyuk seam has the maximum adsorption capacity all over the basin.

It was investigated the influence of moisture, ash, volatile matter and fixed carbon content of the coal seams on the coal adsorption capacity. The moisture effect on adsorption capacity couldn't be completely determined due to moisture contents of coals varied between much closer values. Thus, increase in ash content up to %15, the adsorption capacity of coals has an adverse effect and was considered a decrease in capacities, however, has a good exponential correlation. Volatile matter content of the coal samples varied between %20 – 35. This range is also very closer results a scattered distribution on adsorption capacity data on the plotted graph. Depend on the increase of fixed carbon content, nevertheless adsorption capacities of the coal samples must be increased. However, it is understood that due to possibility of external factors, adsorption capacities were decreased when fixed carbon is increased. So, it can be said that there is a good correlation between fixed carbon content and adsorption capacities of the coal samples.

## ACKNOWLEDGEMENTS

This study was carried out as part of Scientific and Technological Research Council of Turkey (TUBITAK) founded project "Determination of the Gas Contents and Gas Desorption Characteristics of the Coal Seams in Zonguldak Hardcoal Basin", Grant No. 113M460. The authors gratefully acknowledge to the Scientific and Technological Research Council of Turkey (TUBITAK) for funding this research.

## REFERENCES

**Diamond, W. P., Murrie, G. W., McCulloch, C. M. 1976.** "Methane Gas Content of the Mary Lee Group of Coalbeds, Jefferson, Tuscaloosa, and Walker Counties" Ala: US Department of the Interior, Bureau of Mines.

**Gurdal, G. 1998.** Controlling parameters of coal seam gas storage in Zonguldak coal basin. PhD Dissertation, Istanbul Technical University, Faculty of Mines, Istanbul, Turkey.

**Gurdal, G., Yalçın, M. N. 2000.** "Gas adsorption capacity of Carboniferous coals in the Zonguldak Basin (NW Turkey) and its controlling factors", Fuel, 79(15), 1913-1924.

**Joubert, J. L., Grein, C. T., Bienstock, D. 1973.** "Sorption of Methane in Moist Coal", Fuel, 52, 181-185.

**Kim, A. G. 1977.** "Estimating methane content of bituminous coalbeds from adsorption data", Department of the Interior, Bureau of Mines.

**Langmuir, I. 1918.** "The Constitution and Fundamental Properties of Solids and Liquids", J Am Chem Soc, 12, 2062-2076.

**Levine, J. R. 1993.** Coalification: The Evolution of Coal as Source Rock and Reservoir Rock for Oil nad Gas. Law, B. E., Rice, D.D. (Ed), Hydrocarbons from Coal AAPG Studies in Geology, 28, 39-77.

**Mahajan O.P. 1991.** CO surface area of coals: the 25-year paradox. Carbon 29, 735-742.

**Marsh, H. 2001.** Activated carbon compendium. A collection of papers from the journal carbon 1996-2000, Gulf Professional Publishing.

**McFall, K. S., Wicks, D. E., Kuuskraa, V. A., Lewin. 1986.** A geologic assessment of natural gas from coal seams in the Warrior Basin, Alabama. Gas Research Institute.

**Seidle, J. 2011.** Fundamentals of Coalbed Methane Reservoir Engineering, PennWell, Oklahoma, USA.

**Yalçın, E., Durucan, Ş. 1991a.** "Methane Desorption Characteristics of Zonguldak Coals", Mining Science and Technology, 13, 207-214.

**Yalçın, E., Durucan, Ş. 1991b.** "Methane Capacities of Zonguldak Coals and the Factors Affecting Methane Adsorption", Mining Science and Technology, 13, 215-222.

# Determination of Rare Earth Elements (REE's) in coal and ash products in South Africa.

By

Matiane Arnold R.

Supervisor: Prof Nikki Wagner

MSc research project

PCC: 10-August-2016

UNIVERSITY OF JOHANNESBURG

Introduction | Approach | Results | The End!

## Introduction

UNIVERSITY OF JOHANNESBURG

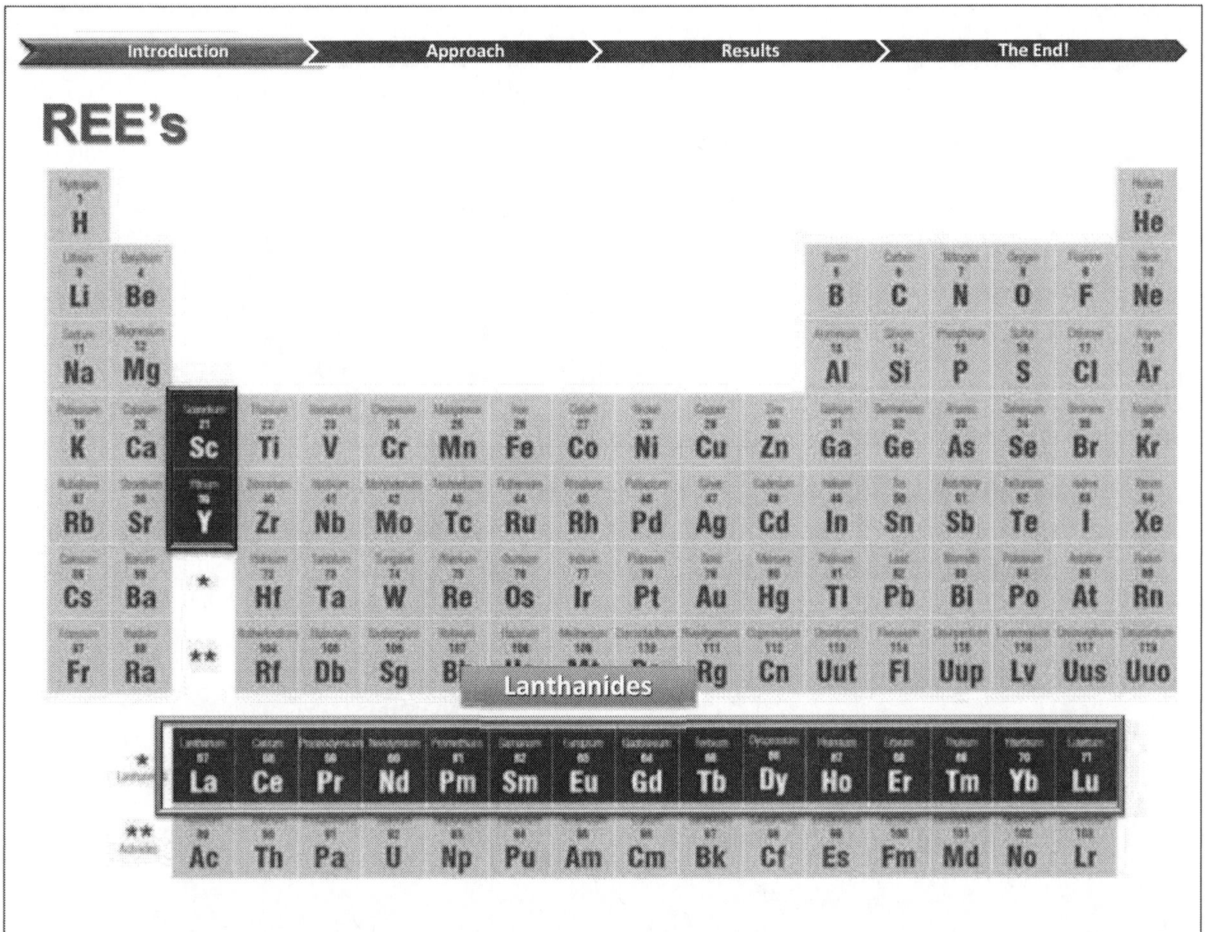

Introduction
Approach
Results
The End!
REE's
H He
Li Be B C N O F Ne
Na Mg Al Si P S Cl Ar
K Ca Sc Ti V Cr Mn Fe Co Ni Cu Zn Ga Ge As Se Br Kr
Rb Sr Y Zr Nb Mo Tc Ru Rh Pd Ag Cd In Sn Sb Te I Xe
Cs Ba * Hf Ta W Re Os Ir Pt Au Hg Tl Pb Bi Po At Rn
Fr Ra ** Rf Db Sg Rg Cn Uut Fl Uup Lv Uus Uuo
Lanthanides
* La Ce Pr Nd Pm Sm Eu Gd Tb Dy Ho Er Tm Yb Lu
** Ac Th Pa U Np Pu Am Cm Bk Cf Es Fm Md No Lr


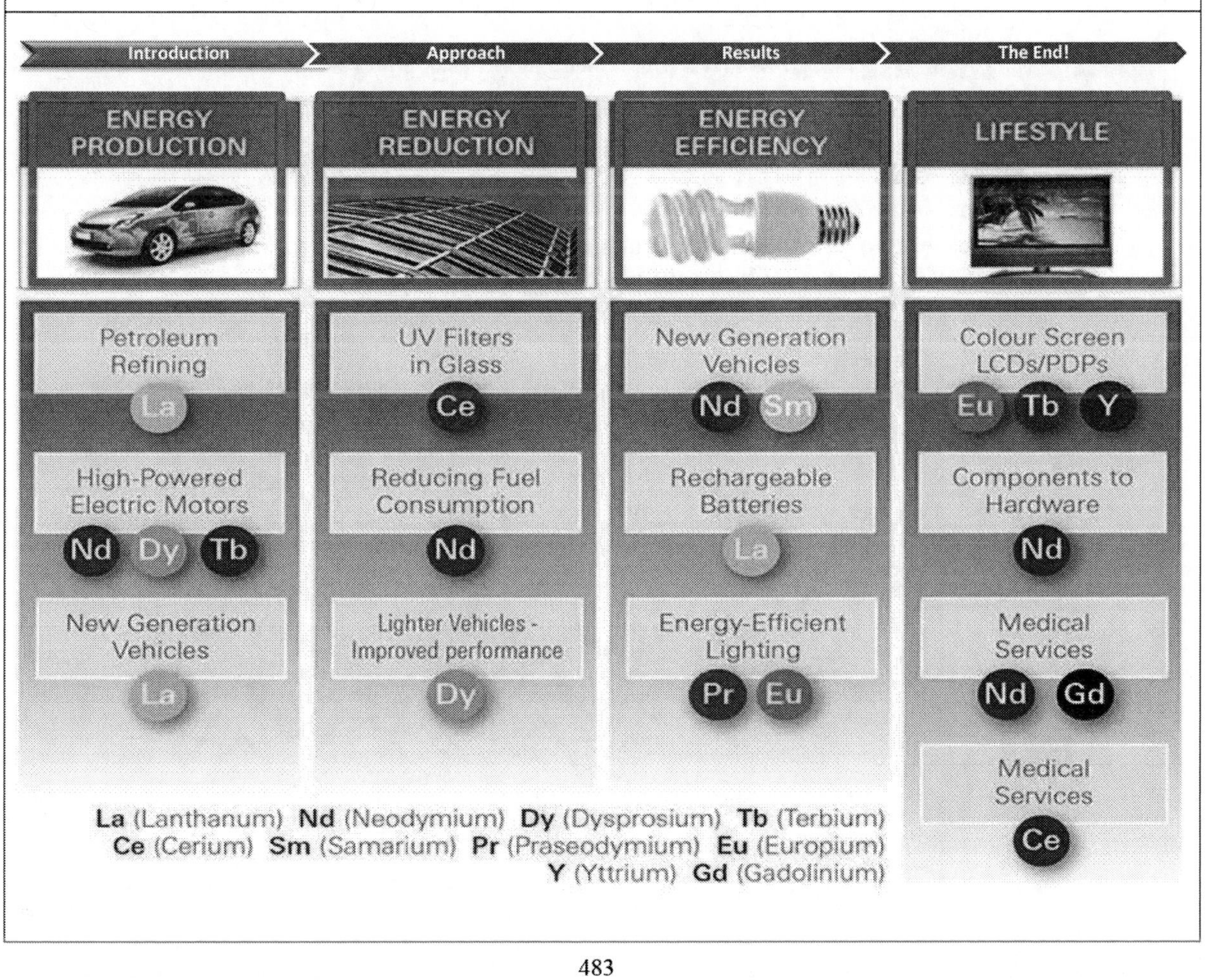

Introduction
Approach
Results
The End!
ENERGY PRODUCTION
ENERGY REDUCTION
ENERGY EFFICIENCY
LIFESTYLE
Petroleum Refining
La
High-Powered Electric Motors
Nd Dy Tb
New Generation Vehicles
La
UV Filters in Glass
Ce
Reducing Fuel Consumption
Nd
Lighter Vehicles - Improved performance
Dy
New Generation Vehicles
Nd Sm
Rechargeable Batteries
La
Energy-Efficient Lighting
Pr Eu
Colour Screen LCDs/PDPs
Eu Tb Y
Components to Hardware
Nd
Medical Services
Nd Gd
Medical Services
Ce
La (Lanthanum) Nd (Neodymium) Dy (Dysprosium) Tb (Terbium)
Ce (Cerium) Sm (Samarium) Pr (Praseodymium) Eu (Europium)
Y (Yttrium) Gd (Gadolinium)

Introduction > Approach > Results > The End!

# Why This Project?

- **Background**
- **Justification**

UNIVERSITY OF JOHANNESBURG

Introduction > Approach > Results > The End!

- REE's deposits:
  - Carbonatite, alkaline igneous rocks, ion-adsorption clay (Chinese ore), monazite-xenotime-bearing placer deposits.
  - There is a search for new sources.
- **Comparison of two commercial sources of REE's to crustal average and to REE's associated with major coal formations (Ekmann, 2012).**

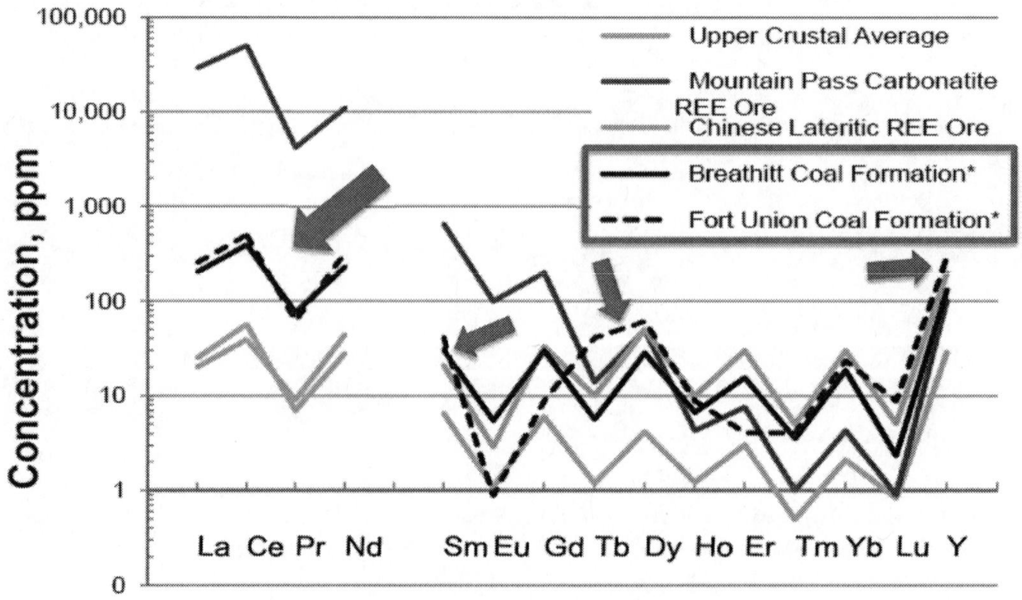

Introduction | Approach | Results | The End!

- Since the late 1980's, China has provided 85-95% of the world's REE's;
  - 2013: 85% of REE's production took place in China.
  - China controls 48% of global REE's supply (Taggart, 2015).

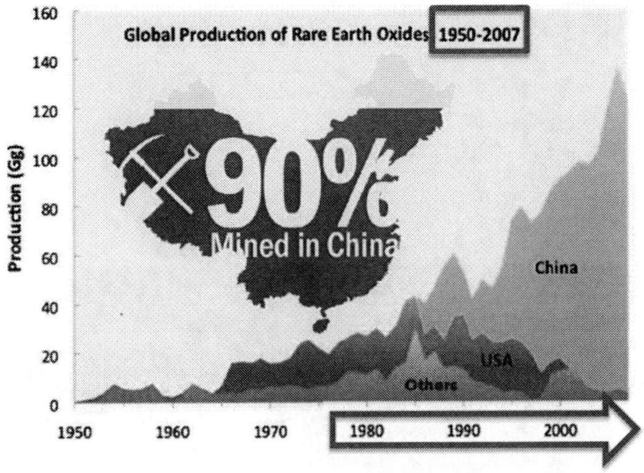


- 2010: China announced their intention to reduce REE's exports;
  - REE's use increased substantially (Taggart, 2015).

Introduction | Approach | Results | The End!

## Potential economic value of coal ash

- Selected set of ash samples: **Europe, US, Mexico and Spain** (Lucinda, 2015).
- Coal deposits: **US, Russia, China, and Middle East** (Lucinda, 2015).
- (**US$/tonne** and **£/tonne**).

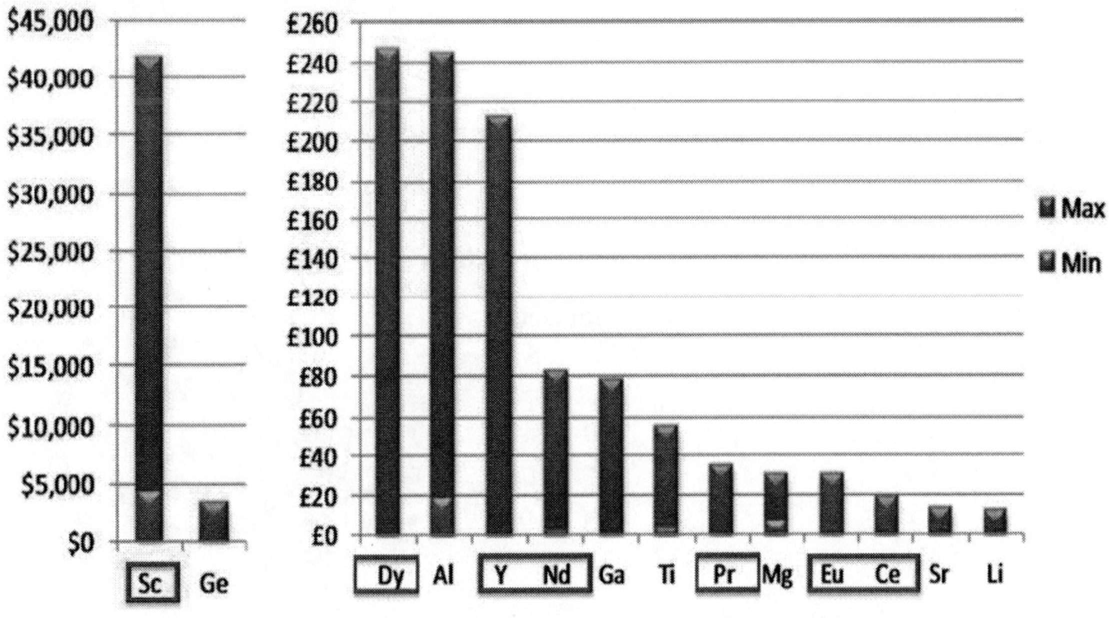

Introduction | Approach | Results | The End!

# Approach

- **Aim and Objectives**
- **Sampling Localities**
- **Techniques**

UNIVERSITY OF JOHANNESBURG

Introduction | Approach | Results | The End!

- ## Main Aim

- The aim of the study is to examine the association of Rare Earth Elements (REE's) with coal and coal ash from selected locations in South Africa.

- **Objectives**:

➢ **To identify the best techniques that can be used to analyse REE's in coal and coal ash.**

➢ To determine the occurrence, concentration and distribution of REE's in coal and coal ash from selected localities.

## Sampling Localities

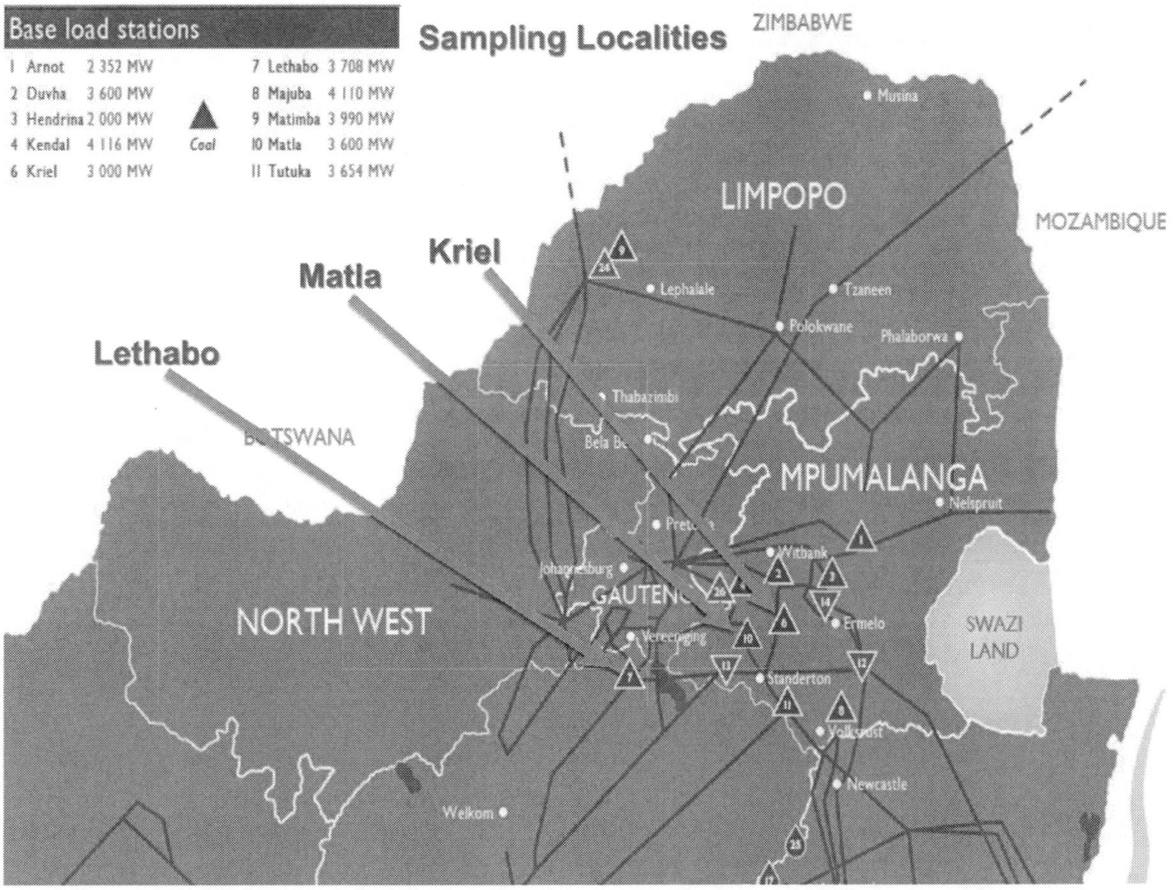


Introduction > Approach > Results > The End!

## Coal Ash Samples

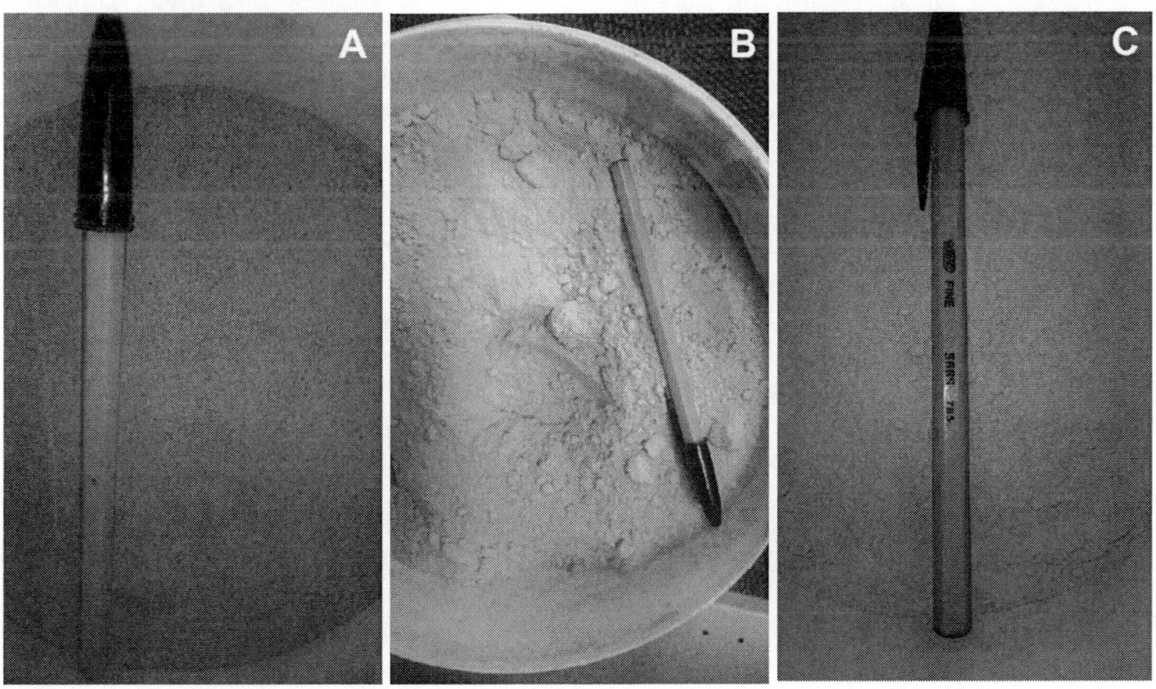

# Techniques

- Common: XRD, SEM, ICP-MS, ICP-OES, XRF and INAA (Instrumental Neutron Activation Analysis).
- Proximate analysis.
- Petrographic analyses.
- XRD.
- SEM.
- ICP Techniques (ICP-MS and ICP-OES);
  - Whole rock digestion ($HNO_3$, HCl, & HF).
  - Sodium Peroxide fusion ($Na_2O_2$ & $HNO_3$)
- INAA.

Introduction Approach Results The End!

- **Basic Principle of INAA**

Neutron flux
↓
Radioactive nuclides
↓
Radioactive nuclides decay
↓
Emission of gamma rays
↓
Energy characterisation for each nuclide

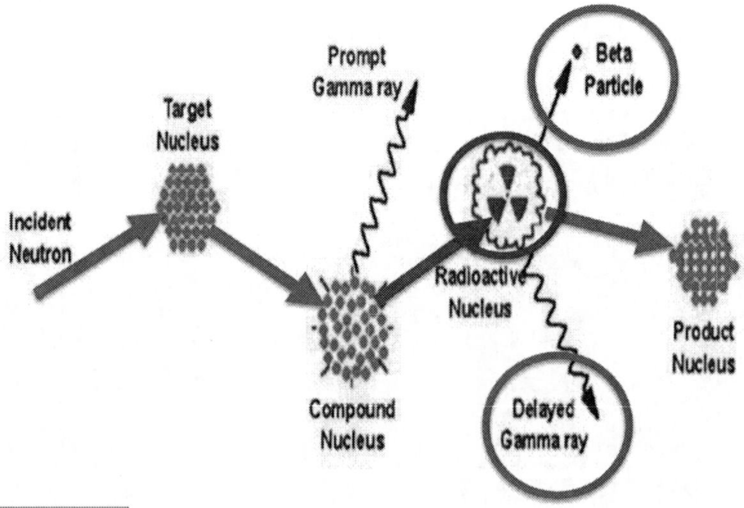

# Results

- **Characterization**
- **Concentration**

UNIVERSITY OF JOHANNESBURG

Introduction | Approach | Results | The End!

## Characterization of coal and coal ash

- **Proximate analysis:**
  - Total moisture content.
  - Volatile matter.
  - Fixed carbon.
  - ✓ **Ash content;**
    - ➢ **Coal (Lethabo = 34%).**
    - ➢ **Coal ash (94-95%).**
- **Petrographic analyses (coal only):**
  - ➢ Maceral count (**High inertinite**).
  - ➢ Reflectance of vitrinite (**Bituminous**).
- **XRD**
  - ➢ No REE's minerals

Introduction | Approach | Results | The End!

## SEM

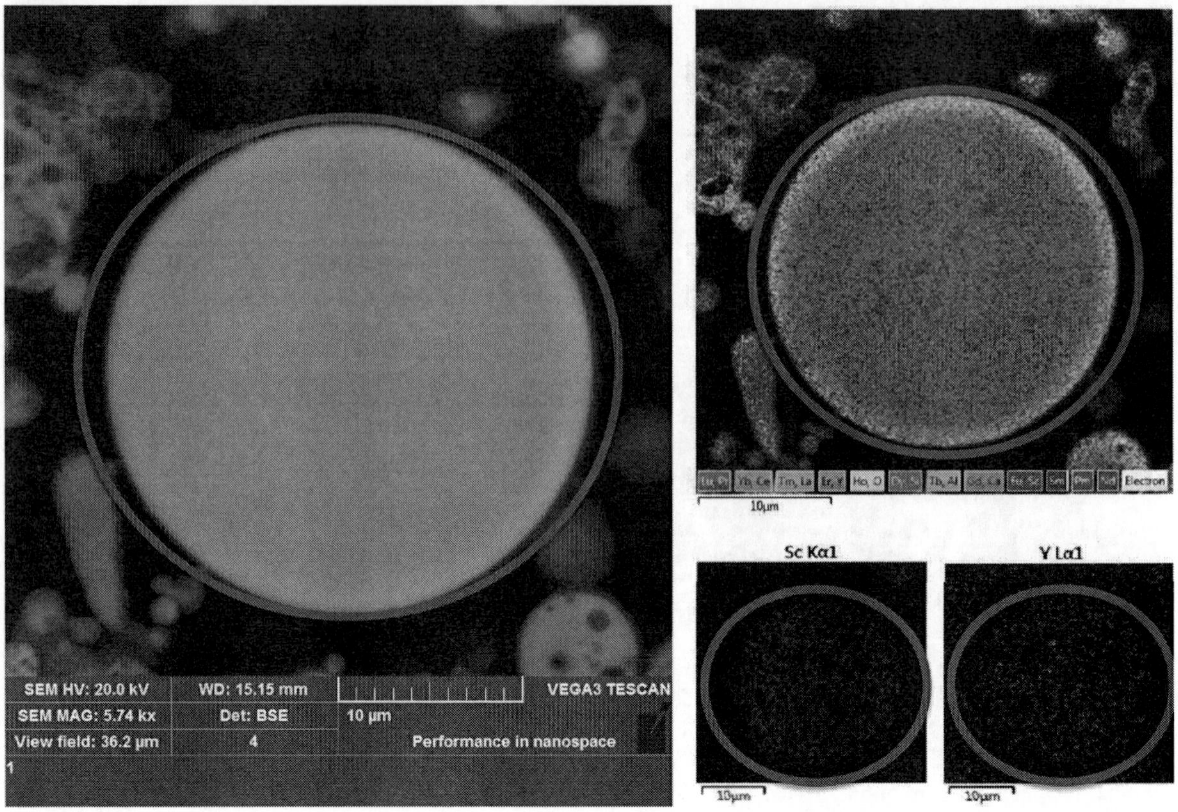


Introduction | Approach | Results | The End!

## SEM cont…

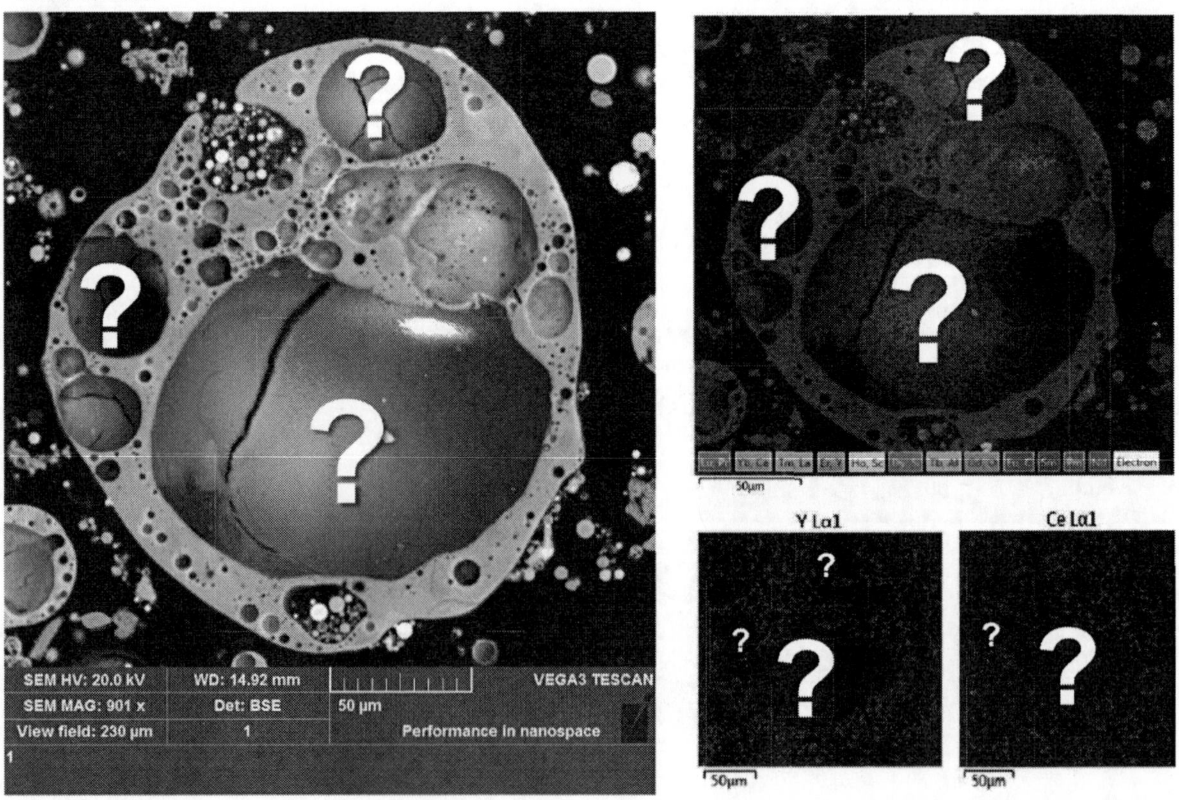

Introduction > Approach > Results > The End!

## REE's Concentration by ICP-MS

- SARM 18, 19 & 20.
- Two methods & 5 trials:
  - UJ, WITS, & SGS.
- Sodium Peroxide fusion = Relatively High recovery
- Multi-Acid digestion = Relatively low recovery

- **Reference material (Recovery %)**

| | SARM 20 | |
|---|---|---|
| | Multi-Acid digestion ($HNO_3$, HCL, & HF) | Sodium Peroxide fusion ($Na_2O_2$ & $HNO_3$) |
| Ce | 32,4 | 89,4 |
| La | 23,0 | 96,3 |
| Sc | 163,9 | 0,0 |
| Sm | 12,9 | 96,8 |
| Eu | | 101,0 |
| Tb | | 101,1 |
| Y | 3,3 | 86,9 |
| Yb | | 105,0 |

Introduction > Approach > Results > The End!

## Coal: REE's Concentration by ICP-MS cont...

- Only sodium peroxide fusion (sintering) results;
  - UJ & SGS.
  - Good recovery.
- Sc, Lu, La, & Tm ?
- **95 ppm < TREE < 149 ppm.**

| Sample I.D. | Techniques | Rare Earth Elements (REE's) - ppm | | | | | | | | | | | | | | | | |
|---|---|---|---|---|---|---|---|---|---|---|---|---|---|---|---|---|---|---|
| | | Sc | Y | Lu | La | Ce | Pr | Nd | Sm | Eu | Gd | Tb | Dy | Ho | Er | Tm | Yb | TREE's |
| Kriel O/C | UJ | _ | 29,64 | 0,00 | 0,00 | 43,14 | 3,91 | 15,14 | 3,02 | 0,31 | 2,83 | 0,00 | 1,95 | 0,32 | 1,1 | 0,00 | 1,08 | **102,44** |
| | SGS | _ | 15,8 | 0,15 | 20,40 | 45,10 | 5,05 | 18,70 | 4,00 | 0,66 | 3,49 | 0,50 | 3,02 | 0,56 | 1,6 | 0,16 | 1,60 | **120,80** |
| Kriel U/G | UJ | _ | 29,90 | 0,00 | 0,00 | 38,74 | 3,43 | 13,59 | 2,83 | 0,27 | 2,50 | 0,00 | 1,71 | 0,28 | 1,0 | 0,00 | 0,98 | **95,22** |
| | SGS | _ | 15,9 | 0,10 | 16,40 | 43,00 | 4,69 | 17,40 | 3,70 | 0,72 | 3,37 | 0,44 | 2,69 | 0,46 | 1,4 | 0,10 | 1,50 | **111,93** |
| Matla | UJ | _ | 4,69 | 0,00 | 0,00 | 56,02 | 5,06 | 18,70 | 3,80 | 0,44 | 3,53 | 0,06 | 2,62 | 0,44 | 1,4 | 0,00 | 1,37 | **138,17** |
| | SGS | _ | 23,1 | 0,17 | 19,70 | 53,80 | 5,87 | 20,70 | 4,10 | 0,78 | 4,14 | 0,50 | 3,44 | 0,60 | 1,9 | 0,18 | 2,10 | **141,08** |
| Lethabo | UJ | _ | 5,18 | 0,00 | 0,00 | 57,81 | 5,27 | 19,10 | 3,87 | 0,44 | 3,62 | 0,06 | 2,58 | 0,43 | 1,4 | 0,00 | 1,37 | **141,17** |
| | SGS | _ | 23,8 | 0,16 | 21,70 | 56,80 | 6,11 | 22,50 | 4,60 | 0,84 | 4,32 | 0,49 | 3,46 | 0,63 | 1,9 | 0,19 | 2,00 | **149,58** |

## Coal Ash: REE's Concentration by ICP-MS cont...

- Sc, Lu, & La?
- **402 ppm < TREE < 598 ppm:**
  - **TREE >= The average global basis {445 ppm} (Kertis and Yudowich, 2009).**

| Sample I.D. | Techniques | Rare Earth Elements (REE's) - ppm | | | | | | | | | | | | | | | |
|---|---|---|---|---|---|---|---|---|---|---|---|---|---|---|---|---|---|
| | | Sc | Y | Lu | La | Ce | Pr | Nd | Sm | Eu | Gd | Tb | Dy | Ho | Er | Tm | Yb | **TREE's** |
| Kriel Bot. Ash | UJ | _ | 94,13 | 0,00 | 0,00 | 177,14 | 18,10 | 66,27 | 12,49 | 2,23 | 12,44 | 1,36 | 9,47 | 1,82 | 5,17 | 0,54 | 4,90 | **406,07** |
| | SGS | _ | 62,80 | 1,06 | 93,00 | 188,00 | 20,80 | 74,00 | 15,80 | 3,39 | 14,20 | 2,33 | 11,90 | 2,58 | 6,80 | 1,18 | 6,40 | **504,24** |
| Kriel Class N | UJ | _ | 23,12 | 0,00 | 0,00 | 215,27 | 21,38 | 75,47 | 13,78 | 2,40 | 14,02 | 1,54 | 10,44 | 2,04 | 5,80 | 0,64 | 5,36 | **491,26** |
| | SGS | _ | 77,00 | 1,06 | 112,00 | 222,00 | 24,20 | 86,50 | 18,40 | 3,52 | 16,80 | 2,47 | 14,20 | 2,87 | 7,92 | 1,19 | 7,80 | **597,93** |
| Kriel Class S | UJ | _ | 06,52 | 0,00 | 0,00 | 202,52 | 20,32 | 72,68 | 13,43 | 2,34 | 13,24 | 1,47 | 10,18 | 1,96 | 5,54 | 0,60 | 5,33 | **456,14** |
| | SGS | _ | 69,20 | 0,96 | 108,00 | 208,00 | 22,50 | 80,30 | 17,00 | 3,32 | 15,50 | 2,48 | 13,30 | 2,61 | 7,70 | 1,13 | 7,20 | **559,20** |
| Lethabo PozzFill | UJ | _ | 00,70 | 0,00 | 0,00 | 176,37 | 17,55 | 62,37 | 11,44 | 1,95 | 11,32 | 1,21 | 8,76 | 1,64 | 4,70 | 0,47 | 4,33 | **402,80** |
| | SGS | _ | 61,10 | 0,76 | 93,30 | 184,00 | 19,60 | 69,70 | 15,10 | 2,73 | 13,50 | 2,17 | 11,60 | 2,31 | 6,37 | 0,87 | 5,80 | **488,91** |
| Lethabo DuraPozz | UJ | _ | 05,41 | 0,00 | 0,00 | 191,32 | 19,49 | 70,06 | 13,15 | 2,19 | 12,73 | 1,44 | 9,89 | 1,90 | 5,38 | 0,56 | 4,99 | **438,49** |
| | SGS | _ | 65,30 | 0,86 | 104,00 | 200,00 | 21,50 | 78,30 | 16,40 | 3,01 | 15,60 | 2,55 | 13,10 | 2,49 | 7,29 | 0,94 | 6,60 | **537,94** |
| Lethabo SuperPozz | UJ | _ | 99,76 | 0,00 | 0,00 | 188,63 | 19,48 | 70,45 | 13,16 | 2,21 | 12,55 | 1,43 | 9,97 | 1,92 | 5,48 | 0,57 | 5,08 | **430,69** |
| | SGS | _ | 64,60 | 0,83 | 102,00 | 198,00 | 21,30 | 78,30 | 16,40 | 2,77 | 14,60 | 2,44 | 13,40 | 2,52 | 7,38 | 0,93 | 6,80 | **532,27** |
| Matla Unclassified | UJ | _ | 20,49 | 0,00 | 0,00 | 240,28 | 23,77 | 84,27 | 15,32 | 2,70 | 15,44 | 1,76 | 11,60 | 2,26 | 6,45 | 0,71 | 5,94 | **531,00** |
| | SGS | _ | 74,60 | 0,95 | 110,00 | 222,00 | 23,80 | 84,10 | 17,80 | 3,44 | 16,40 | 2,51 | 13,80 | 2,74 | 7,63 | 1,06 | 6,70 | **587,53** |
| Matla Classified | UJ | _ | 17,99 | 0,00 | 0,00 | 210,82 | 21,26 | 73,56 | 13,47 | 2,45 | 13,46 | 1,50 | 10,31 | 1,99 | 5,66 | 0,59 | 5,17 | **478,24** |
| | SGS | _ | 71,50 | 0,94 | 119,00 | 225,00 | 23,80 | 84,60 | 17,90 | 3,44 | 16,80 | 2,76 | 14,40 | 2,78 | 7,70 | 1,08 | 7,00 | **598,70** |



# REE's Concentration by INAA

- **6 REE's.**
- **310 ppm < TREE < 420 ppm.**

| Sample I.D | | REE's - ppm | | | | | | |
|---|---|---|---|---|---|---|---|---|
| | | Sc | La | Ce | Eu | Sm | Tb | **Total REE's** |
| Kriel | Class N | 25,40 | 110,00 | 222,00 | 3,22 | 16,50 | 1,90 | **379,02** |
| | Class S | 24,20 | 102,00 | 219,00 | 2,63 | 15,90 | 1,90 | **365,63** |
| Matla | Unclassified | 30,70 | 133,00 | 234,00 | 3,20 | 17,40 | 2,33 | **420,63** |
| | Classified | 21,30 | 105,00 | 220,00 | 3,16 | 16,30 | 1,81 | **367,57** |
| Lethabo | Pozzfill | 23,00 | 85,40 | 184,00 | 2,56 | 13,50 | 1,54 | **310,00** |

Introduction | Approach | Results | The End!

## ICP-MS AND INAA

- Comparison of ICP-MS and INAA results;
  - ➢ SGS.
- Only for 6 REE's;
  - ➢ **TREE's.**

| Sample I.D. | Techniques | Rare Earth Elements - ppm | | | | | | |
|---|---|---|---|---|---|---|---|---|
| | | Sc | La | Ce | Eu | Sm | Tb | Total REE's |
| Kriel Class N | ICP-MS | _ | 122,00 | 222,00 | 3,52 | 18,40 | 2,47 | **368,39** |
| | INAA | 25,40 | 110,00 | 222,00 | 3,22 | 16,50 | 1,90 | **379,02** |
| Kriel Class S | ICP-MS | _ | 108,00 | 208,00 | 3,32 | 17,00 | 2,48 | **338,80** |
| | INAA | 24,20 | 102,00 | 219,00 | 2,63 | 15,90 | 1,90 | **365,63** |
| Lethabo PozzFill | ICP-MS | _ | 93,30 | 184,00 | 2,73 | 15,10 | 2,17 | **297,30** |
| | INAA | 30,70 | 133,00 | 234,00 | 3,20 | 17,40 | 2,33 | **420,63** |
| Matla Unclassified | ICP-MS | _ | 110,00 | 222,00 | 3,44 | 17,80 | 2,51 | **355,75** |
| | INAA | 21,30 | 105,00 | 220,00 | 3,16 | 16,30 | 1,81 | **367,57** |
| Matla Classified | ICP-MS | _ | 119,00 | 225,00 | 3,44 | 17,60 | 2,76 | **367,80** |
| | INAA | 23,00 | 85,40 | 184,00 | 2,56 | 13,50 | 1,54 | **310,00** |

Introduction | Approach | Results | The End!

# Conclusion

- **REE's occurrence, distribution, and concentration in coal and ash:**
  - ➢ There's a correlation between REE's and presence of aluminium and silicon oxides.
  - ➢ REE's concentration: ash > coal.
- **Techniques:**
  - ➢ Sodium peroxide is more effective than multi acid digestion.
  - ➢ Single technique cannot be reliable.
  - ➢ ICP-MS correlate with INAA.

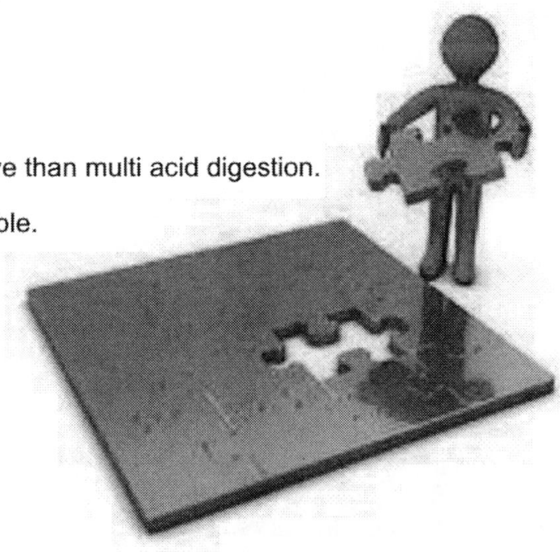

Introduction > Approach > Results > The End!

# Acknowledgement

Eskom


- I would like to express my sincere appreciation to CIMERA for funding this project.
- I would also like to thank Prof R.A. Kruger for his recommendation on sampling areas and his guidance.
- Special thanks to Eskom, Ulula and Ash Resources (Pty) Ltd for supplying us with samples.
- And also not forgetting Eve (spectrum) for helping with sample digestion and analysis.
- **This work is based on the research supported wholly by NRF.**


cimera DST-NRF Centre of Excellence for Integrated Mineral and Energy Resource Analysis

necsa We're in your world

UNIVERSITY OF JOHANNESBURG

Introduction > Approach > Results > The End!

# THE END

- **Thank you for listening and now if there are any questions, I would be pleased to answer them....!**

cimera DST-NRF Centre of Excellence for Integrated Mineral and Energy Resource Analysis

UNIVERSITY OF JOHANNESBURG

# Rare Earth Elements from Coal & Coal By-Products

NATIONAL ENERGY TECHNOLOGY LABORATORY

Thomas J. Tarka, PE

International Pittsburgh Coal Conference

August 9, 2016

## NETL is…

NATIONAL ENERGY TECHNOLOGY LABORATORY

1,400 R&D projects/50 states
1,400 employees
5 labs across U.S.
FY16 budget $868M

U.S. DEPARTMENT OF ENERGY

2

## NETL's Mission

NATIONAL ENERGY TECHNOLOGY LABORATORY

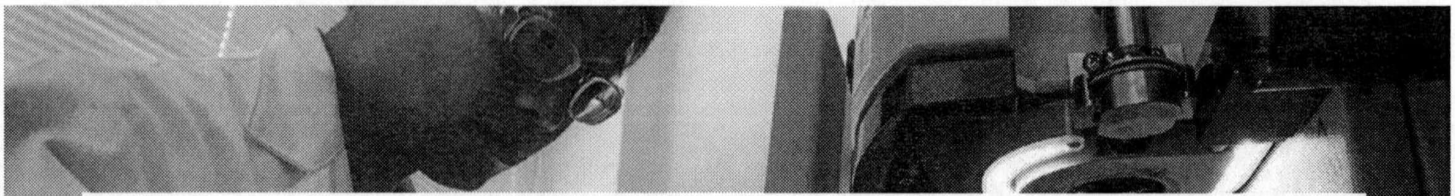

To discover, integrate, and mature technology solutions to enhance the Nation's energy foundation and protect the environment for future generations

**EFFECTIVE RESOURCE DEVELOPMENT**

Developing technologies that improve the effectiveness and economics of producing our fossil energy resources

**EFFICIENT ENERGY CONVERSION**

Discovering cleaner, cheaper, and more efficient energy conversion technologies for the production of high-value energy commodities

**ENVIRONMENTAL SUSTAINABILITY**

Accelerating the development of transformative and enabling solutions to protect our air, land, and water for future generations

U.S. DEPARTMENT OF ENERGY 3

## Rare Earth Elements

NATIONAL ENERGY TECHNOLOGY LABORATORY

15 lanthanides, 2 transition elements

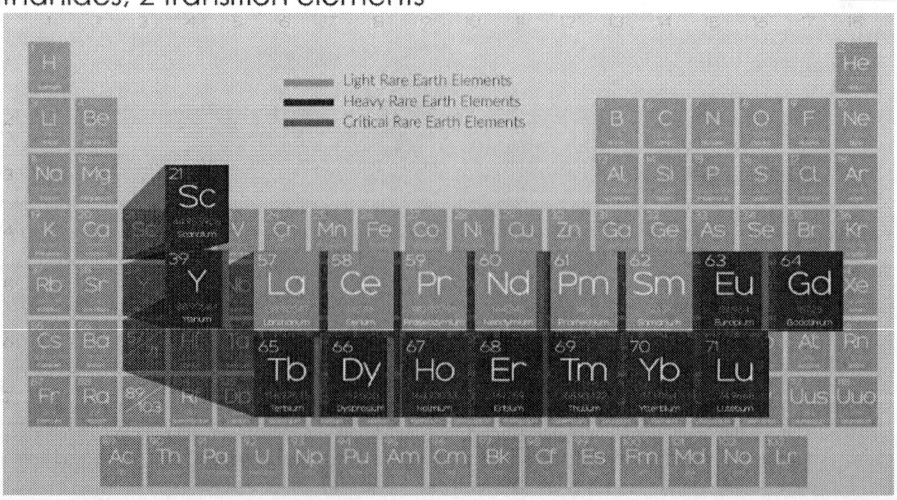


U.S. DEPARTMENT OF ENERGY 4

# Rare Earth Elements

Applications Overview

NATIONAL ENERGY TECHNOLOGY LABORATORY

- **Primary Application Sectors**
  - Magnetics
  - Phosphors
  - Ceramics
  - Metal Alloys
  - Catalysts
  - Polishing Powders
  - Glass Additives

- **Defense Applications**
  - Satellite Communications
  - Guidance Systems
  - Aircraft Structures
  - Fly-by-wire
  - Smart Missiles

**Examples of REE Applications**

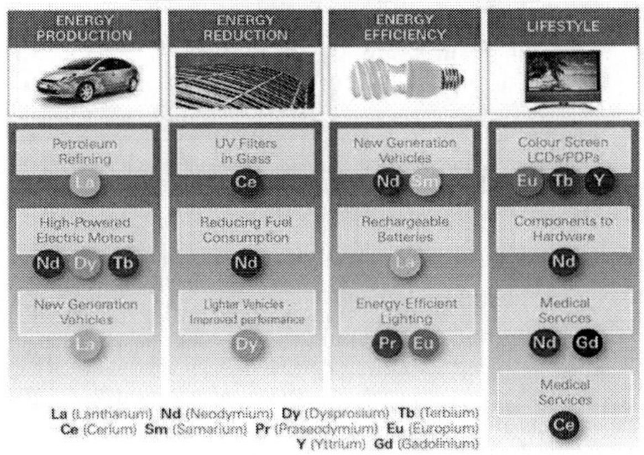




# Rare Earth Elements

Market Overview

NATIONAL ENERGY TECHNOLOGY LABORATORY

- **Prices appear stable after drop in 2011-2012**
  - Oxide prices remain above $1,500 per tonne, range to ~$50,000

- **Market remains complicated**
  - Risks of substitution
  - Current oversupply

- **Magnets appear to dominate value opportunity**
  - 75-80% of market value

**Recent REE Market Value Evaluations**

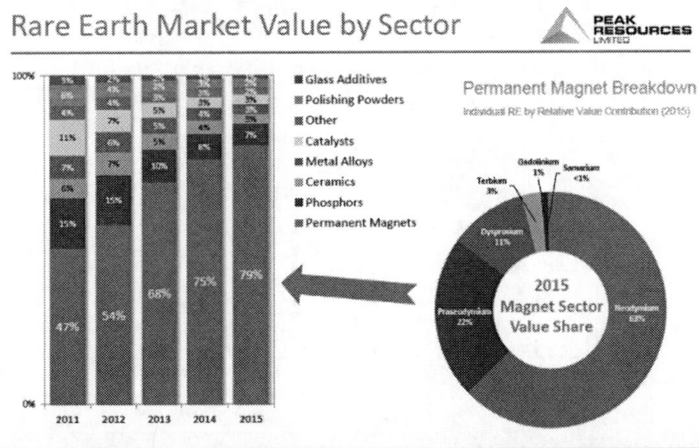

## Rare Earth Elements in Coal

Where are Rare Earth Elements found in coal?

- **Coal & Surrounding Strata**
  - Coal type and seam
  - Over- and Underburden types
  - Wide variations in element distribution, even in a single coal bed
- **Coal By-products**
  - Combustion residuals
  - Mining waste & tailings
  - Mine drainage sludge

**Examples of REE Containing Coal Materials**

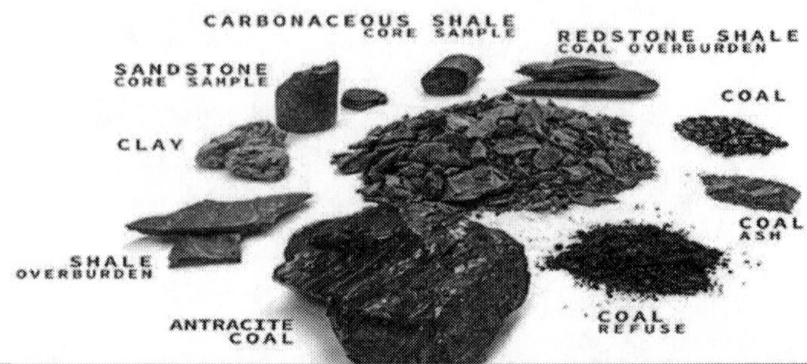




## Rare Earth Elements in Coal

Opportunities for REE Recovery Span Coal Value Chain

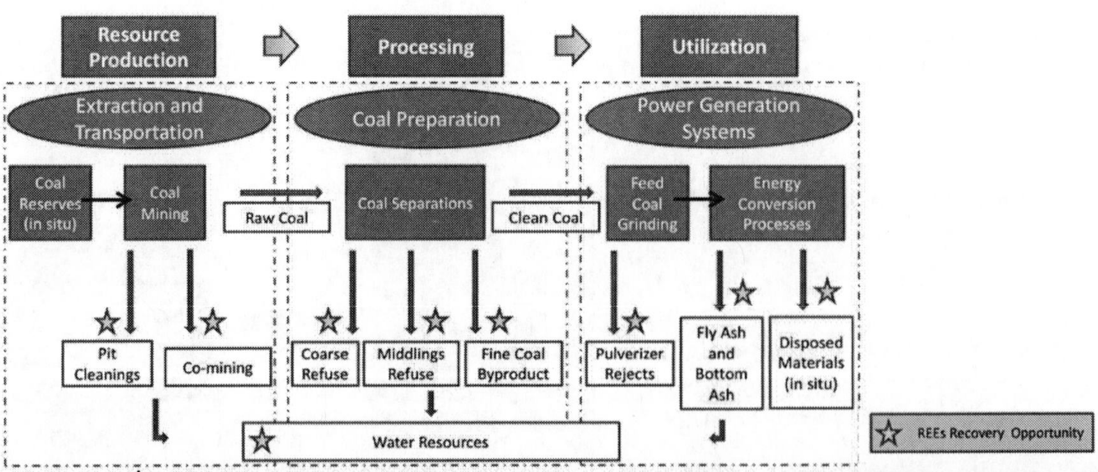

# Opportunities & Challenges

Rare Earth Elements from Coal and Coal By-Products

NATIONAL ENERGY TECHNOLOGY LABORATORY

| + | − |
| --- | --- |
| Significant volumes of low concentration REEs in potentially no or low cost feedstocks | Concentrations vary greatly depending on specific source |
| Other valuable materials can be co-produced, improving system economics | Perceived economic & environmental challenges in processing |
| Environmental remediation benefits may improve economics | Opaque & Volatile Market |

U.S. DEPARTMENT OF ENERGY 9

# Rare Earth Elements in Coal

Environmental Remediation & Beneficiation Opportunities

NATIONAL ENERGY TECHNOLOGY LABORATORY

- **Targets**
  - Coal Combustion By-products: Fly-Ash in Ponds
  - Fines, tailings, and gob piles
  - Acid mine drainage
- **Opportunity**
  - Revenue stream from products reduces remediation cost
  - Recover heating value of waste coal while adding environmental benefit
  - Obviate or reduce future waste streams
  - High visibility of ash containment ponds can provide driver
- **Barriers**
  - Liability for legacy mining operations
  - Perceived (?) potential environmental damage from disturbing gob or waste

U.S. DEPARTMENT OF ENERGY 10

# NETL Coal Rare Earth Element Program

Mission and Vision

NATIONAL ENERGY TECHNOLOGY LABORATORY

**Mission:** Economic recovery of rare earth elements from coal and coal by-product streams to ensure a sustainable domestic supply of critical materials.

**Objective:** Advance existing and/or develop new technologies to improve the economics and reduce the environmental impact of a coal-based REE value chain.

| Improve Knowledge | Develop Capabilities | Deliver Solutions | Leverage Value |
|---|---|---|---|
| • Market Analysis<br>• Techno-Economics<br>• Systems Engineering<br>• Sampling and Analysis<br>• Characterization | • Advanced Separations Expertise<br>• Bench-Scale/Pilot-Scale Testing & Demonstration Facilities<br>• Coal and Coal By-Products Chemistry<br>• Simulations and Modeling | • Separations Technologies &<br>• Separations Facilities<br>• Sensor Monitoring & Control<br>• Advanced Characterization Techniques<br>• Training for Next Generation Scientists & Engineers | • New Businesses/Spin-Off Companies<br>• Technology Transfer to OtherTechnical Areas<br>• Supply of REEs: Commercial & Defense Products/Equipment |

**Goals**

- ✓ *Develop production technology that can* ***economically deployed****, enabling additional domestic REE supply*
- ✓ ***Reduce the environmental impact*** *of coal and REE production, through advances in production from coal & coal by-products*
- ✓ *Deliver advanced technologies that can be developed, manufactured, and* ***deployed within the United States***

*Execute through NETL partnerships with Industry, Universities, National Labs*



# NETL Coal Rare Earth Element Program

Rare Earth Elements from Coal and Coal By-Products

NATIONAL ENERGY TECHNOLOGY LABORATORY

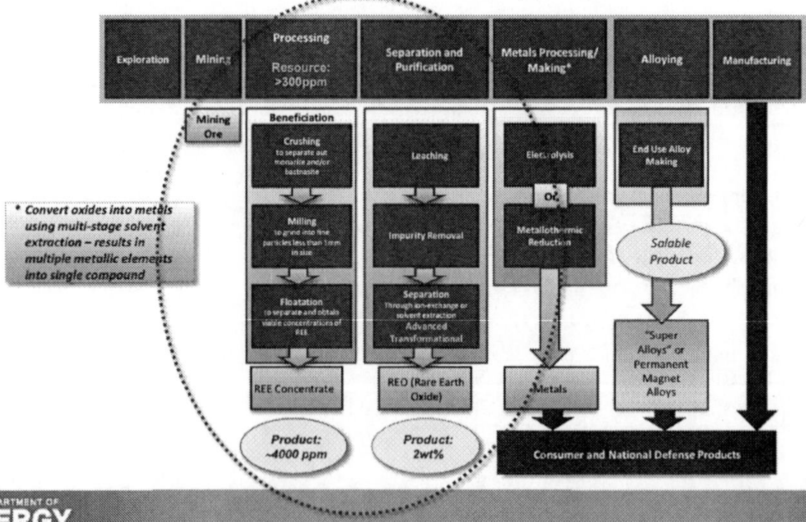

## NETL R&IC Research Focus Areas

Rare Earth Elements from Coal and Coal By-Products

NATIONAL ENERGY TECHNOLOGY LABORATORY

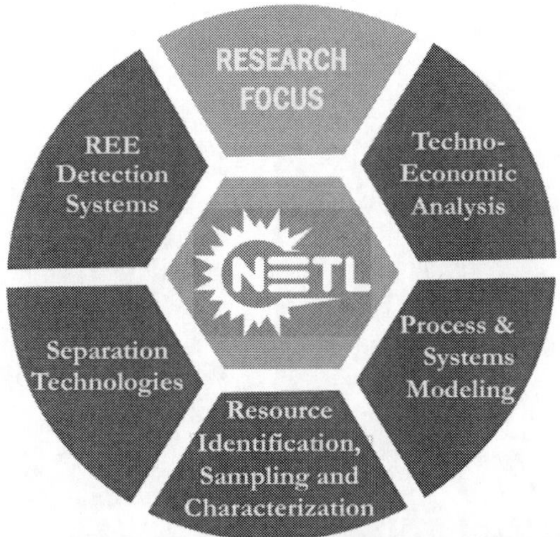

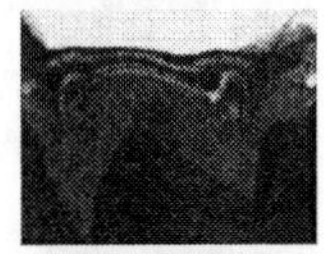

U.S. DEPARTMENT OF ENERGY

13

## NETL R&IC Research Focus Areas

Rare Earth Elements from Coal and Coal By-Products

NATIONAL ENERGY TECHNOLOGY LABORATORY

- Grinding, Size, Density Separation
- Froth Flotation
- Fine Grinding (5-10μm)
- Reactive Grinding/High Intensity Grinding
- High Gradient Wet Magnetic Separation

- Thermal Extraction
- Sequential Extraction
- Photophoretic Separation
- Ammonium Sulfate (Ion Exchange Clays)
- Electrochemical
- $CO_2$ Saturation Brine
- Sequential Extraction
- Deep Eutectic Solvents & ILs
- Alkaline/Acid Roasting and Reactive Ball Milling

- Sorbents: Polymer & Organic Matrices on Clay Inorganic Supports
- Sorbents: Organic Matrices on Clays/ Inorganics
- Sorbents: Inorganic Clay Matrices
- Sorbents: Layered Double Hydroxides

- Optical Sensors
- LIBS
- Laser

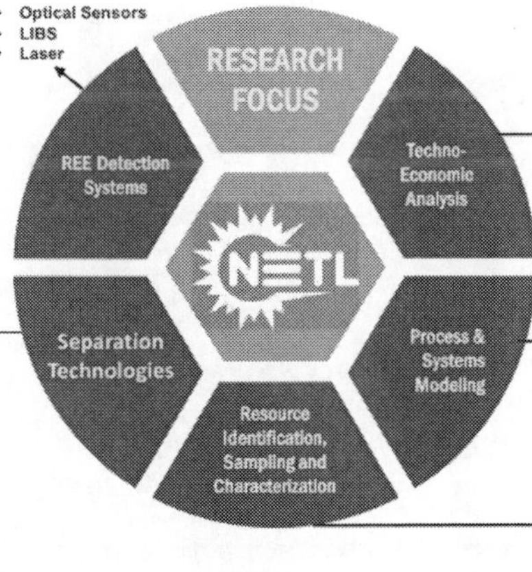


- Development of Market Awareness
- Development of Strategic Processing Cases
- Assessment of Domestic REE Production
- Assessment of FY16 FWP/FOA REE Recovery Pathways

- Modeling of Ion Exchange REE Separation from Clays

- Field Sampling XRF
- ICP-MS
- XRD & SEM
- Laser Ablation MS
- Synchrotron
- High Temperature Confocal Scanning Laser Microscopy
- Optical Sensors

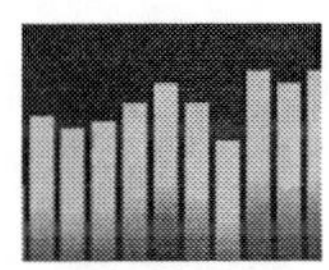
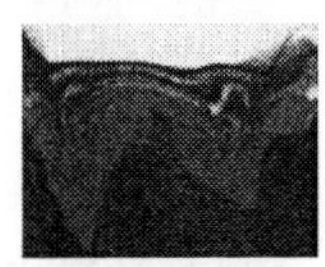

U.S. DEPARTMENT OF ENERGY

14

# NETL R&IC Research Focus Areas

NATIONAL ENERGY TECHNOLOGY LABORATORY

## Accomplishments

**Resource Identification**

~450 Appalachian Basin Field Samples Acquired. REE Concentration ~600 ppm

**Advanced Analytical Characterization**

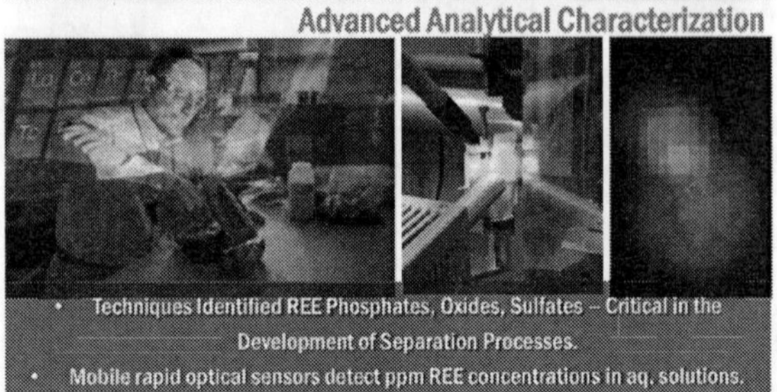

- Techniques Identified REE Phosphates, Oxides, Sulfates – Critical in the Development of Separation Processes.
- Mobile rapid optical sensors detect ppm REE concentrations in aq. solutions.

**Separations Technology Development**

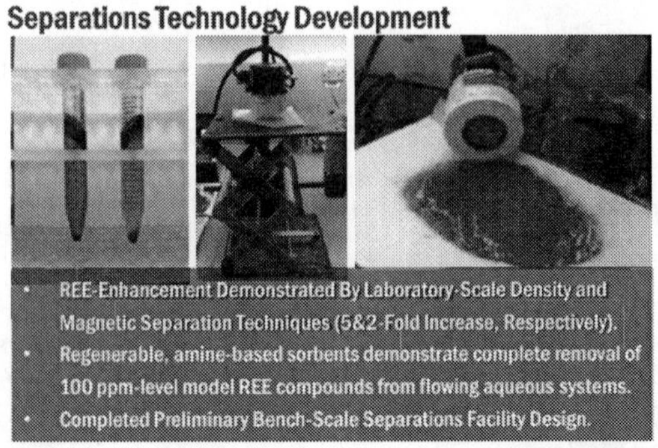

- REE-Enhancement Demonstrated By Laboratory-Scale Density and Magnetic Separation Techniques (5&2-Fold Increase, Respectively).
- Regenerable, amine-based sorbents demonstrate complete removal of 100 ppm-level model REE compounds from flowing aqueous systems.
- Completed Preliminary Bench-Scale Separations Facility Design.

U.S. DEPARTMENT OF ENERGY



# NETL R&IC Research Focus Areas

NATIONAL ENERGY TECHNOLOGY LABORATORY

## Additional Accomplishments

- 9 Recipients selected for cooperative research
- 23 Conference Presentations
- 7 Journal Publications
- 3 Conference Sessions Organized

- DOE Led Research Team Makes Significant Rare Earth Discovery – Ion Exchange Process
- EDX database upgrade
- Launch of the REE website
- Patent NOIs Being Considered for Submission 3
- Received Geological Society of America Best Paper Award

U.S. DEPARTMENT OF ENERGY

## NETL R&IC Research Focus Areas

NATIONAL ENERGY TECHNOLOGY LABORATORY

Nine Techno-Economic Analysis Funding Awards

Bench & Pilot-Scale REE Separation Projects

SOUTHERN RESEARCH · UNIVERSITY OF KENTUCKY · UNIVERSITY OF WYOMING · UNIVERSITY OF NORTH DAKOTA · PHYSICAL SCIENCES INC. · DUKE UNIVERSITY · Battelle · West Virginia University · Tusaar

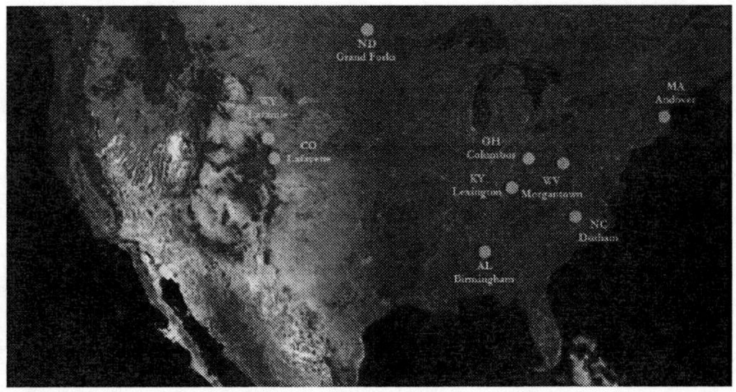


Five Bench-Scale & Four Pilot-Scale Separation Processes Awarded
March 1, 2016, for Design, Techno-Economic Analysis & Performance Feasibility

U.S. DEPARTMENT OF ENERGY

U.S. DEPARTMENT OF ENERGY

# 2016 International Pittsburgh Coal Conference
# Cape Town, SOUTH AFRICA
# August 8 - 12, 2016

**Coal Science**

## Multifaceted nanominerals and ultrafine particles in phosphogypsum: A by-product of coal cleaning rejects


***Felipe B. de Leão***[a,c]; *Silvio R. Taffarel*[a]; *Frans Waanders*[b]; *Carlos H. Sampaio*[c]; *Bianca D. Lima*[a]; *Luis F. O. Silva*[a]

[a] *Laboratory of Environmental Researches and Nanotechnology Development, Centro Universitário La Salle, Mestrado em Avaliação de Impactos Ambientais em Mineração. Victor Barreto, 2288 Centro 92010-000, Canoas, RS, Brazil.*

[b] *School of Chemical and Minerals Engineering North West University (Potchefstroom campus) Potchefstroom 2531, South Africa*

[c] *Universidade Federal do Rio Grande do Sul, Escola de Engenharia, Departamento de Metalurgia, Centro de Tecnologia, Av. Bento Gonc¸alves, 9500. Bairro Agronomia. CEP: 91501-970. Porto Alegre – RS, Brasil*



**Abstract**

Phosphogypsum ($CaSO_4{\cdot}2H_2O$), a by-product of phosphate rock processing, contains a large amount of impurities such $P_2O_5$, F, radioactive elements, organic substances, secondary nanominerals, and ultrafine particles (UFP) enriched in metals and metalloids. In this study, we examined phosphogypsum (PG) collected from abandoned fertilizer industry facility in south Brazil (Santa Catarina state). The fragile nature of nanominerals and UFP assemblages from fertilizer industry systems required novel techniques and experimental approaches. The investigation of the geochemistry of complex nanominerals and UFP assemblages was a prerequisite to accurately assess the environmental and human health risks of contaminants and cost-effective chemical and biogeological remediation strategies. Particular emphasis was placed on the study and characterization of the complex mixed nanominerals and UFP containing potentially toxic elements. Nanometer-sized phases in PG were characterised using energy-dispersive X-ray spectrometer (EDS), field-emission scanning electron microscope (FE-SEM), and high-resolution transmission electron microscopy (HR-TEM) images. The chemical composition and possible correlations with morphology of nanominerals and UFP, as well as aspects of nanominerals and UFP, were discussed in the context of human health exposure, as well as in relation to management of the nanominerals and UFP in PG environments.

# DRYING AND BRIQUETTING OF DISK-PELLETIZED FINE COAL


Darrell Taulbee, Robert Hodgen

University of Kentucky Center for Applied Energy Research


UK Center for Applied Energy Research

caer.uky.edu

## University of Kentucky
## Center for Applied Energy Research

- **Located in Lexington, KY**
  - **~150 employees**
- **Mission**
  - **Research and Development**
  - **Service to Industry**
  - **Teaching and Instruction**
- **Major Research areas**
  - ► Carbon Capture
  - ► Carbon Materials
  - ► Biofuels
  - ► Fischer Tropsch
  - ► Clean fuels and chemicals
  - ►Electrochemical power sources
  - ► Coal combustion byproducts

UK Center for Applied Energy Research

PCC16-Cape town

caer.uky.edu

# Engineered Fuels Group

- Initial work began in 2002
- Objective: Produce a high-quality steam or stoker fuel from waste coal fines or waste coal fines co-briquetted with biomass

# Waste Coal Fines

- **Pros**
  - Technologies to recover and clean waste fines are available
  - Waste coal fines are abundant
    - Estimates of 2-3 Billion tons in gob piles and impoundments in the USA
    - Mostly higher-value bituminous fines

- **Cons**
  - High moisture
    - Difficult to transport
      - Freezes in winter; hangs in chutes, rail cars, hoppers, etc.
    - Difficult and expensive to dry
    - Reduces heating value
  - Fugitive dust problems if dried

# Why Make Briquettes/Pellets from Wet Coal Fines

- **Advantages**
  - Flowable
  - Higher net heating value after drying
  - Can be transported, stored, conveyed, and burned in existing facilities
  - Suppressed dust emissions
  - Can easily incorporate additives
    - Lime for sulfur/arsenic capture
    - Biomass as renewable energy source
- **Obstacles**
  - Cost!!!
    - Capital and operational costs
    - Binder costs
    - Energy costs (drying)
  - ***Available drying technology***

# Drying Wet Coal Fines

- ❑ Fluidized bed historically used to dry wet coal fines
  - New FB dryers effectively prohibited in US
    - ➢ Fire/Explosion hazards
    - ➢ Dust emissions

30 kW micro-wave drier

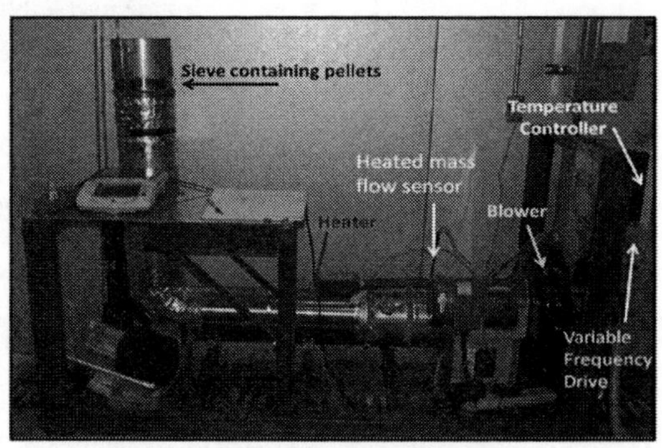


65 kW forced-air thermal drier

# Alternate Approach

## Sequential Pelletization - Conveyed Bed Drying - Briquetting

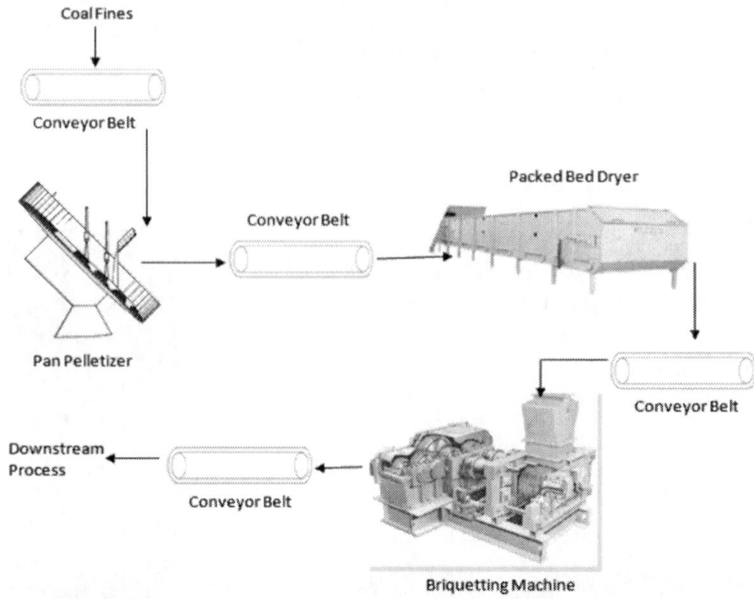




# Pan (Disk) Pelletization

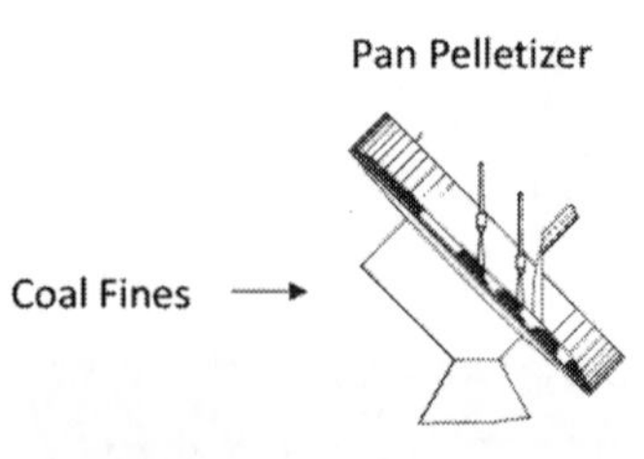

- ➢ Form spherical pellets by tumbling fines in an inclined pan
  - ✓ Requires a wet feedstock → well suited for wet coal fines
  - ✓ Low energy cost and equipment erosion rates
  - ✓ High throughput
  - X Pellets have low to moderate durability

# Thermal Drying in Conveyed Bed

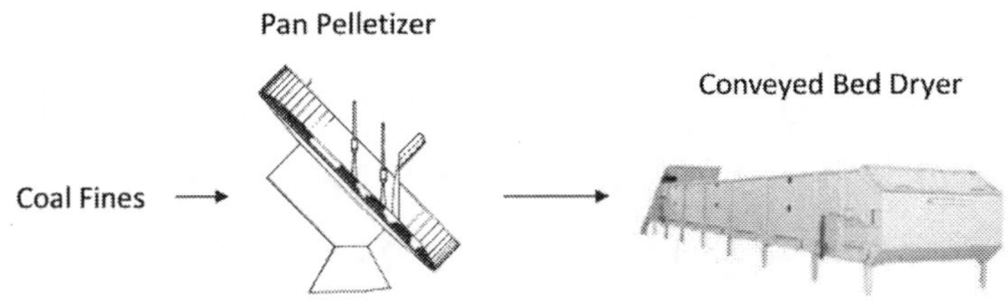

- Hot flue gas can be passed through a packed bed of wet coal pellets
- Advantages compared to a fluidized bed dryer:
  - ✓ Lessens fire/explosion hazards and lowers dust emissions
  - ✓ Simpler to operate
  - ✓ Operating permits can still be obtained in USA
  - ✓ Dried pellets are highly flowable
    - Won't bridge in hoppers or hang in chutes



# Roll Briquetting

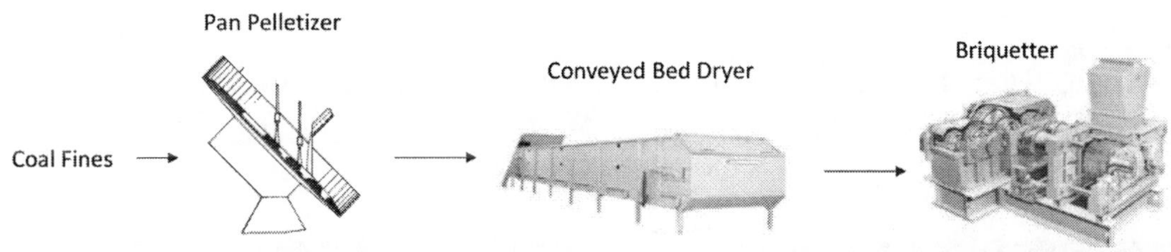


- ✓ High throughput rates
- ✓ Low to moderate energy costs
- ✓ Low to moderate erosion rates
- ✓ Well suited for coal/biomass blends

X Often requires a binder ($$$)

## Briquettes from Thermally-Dried Coal Pellets

- Dry pellets flow readily and feed consistently to the briquetter
  - ✓ Fewer run stoppages/more consistent product
- Briquettes typically have acceptable green strengths (no curing needed)
- **Briquettes made with hot pellets tend to be more durable than briquettes made with an equivalent feedstock briquetted at ambient temperature**

## Thermal Drying/Briquetting of Coal Pellets: Lab Studies

- Prepared pellets in either a 40 or a 90-cm diameter pan pelletizer then dried in a 65 kW forced-air thermal dryer

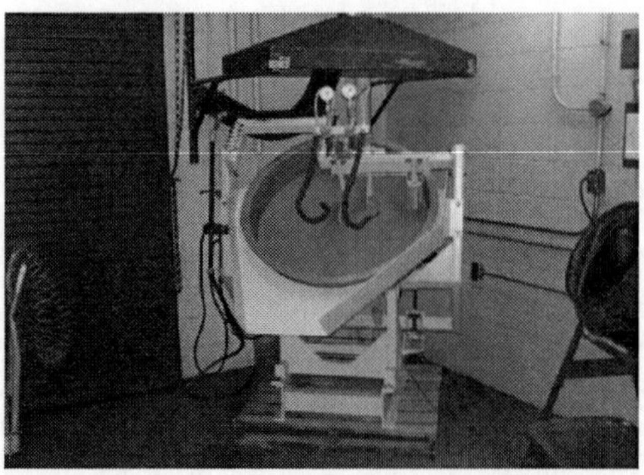

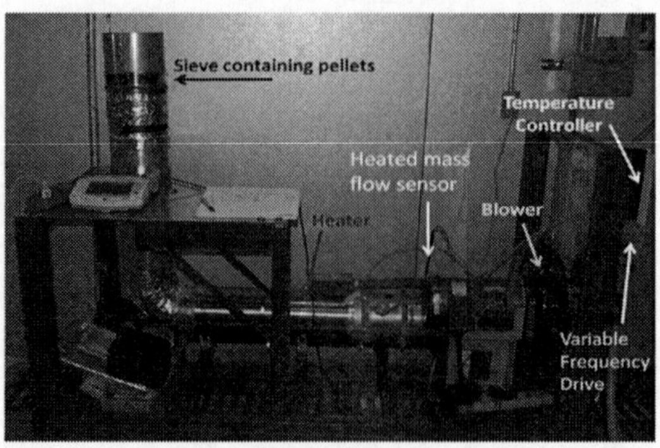

# Parametric Studies

| Parameter | Parameter set points | | |
|---|---|---|---|
| Air Velocity (m/s) | 1 | 2 | 3 |
| Drying time (min) | 5 | 10 | 15 |
| Pellet bed Depth (cm) | 5 | 10 | 15 |
| Drying Temperature ( °C) | 80 | 140 | 200 |

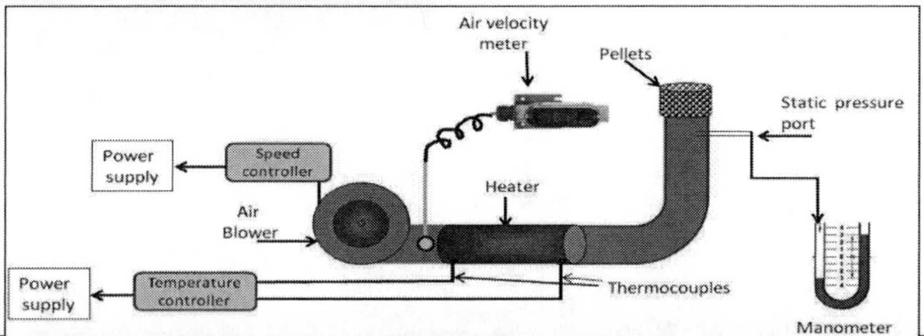




# Energy Requirements for Reduction of Pellet Moisture as a Function of Drying Time, Bed Depth, and Gas Velocity

1 m/s flue-gas velocity

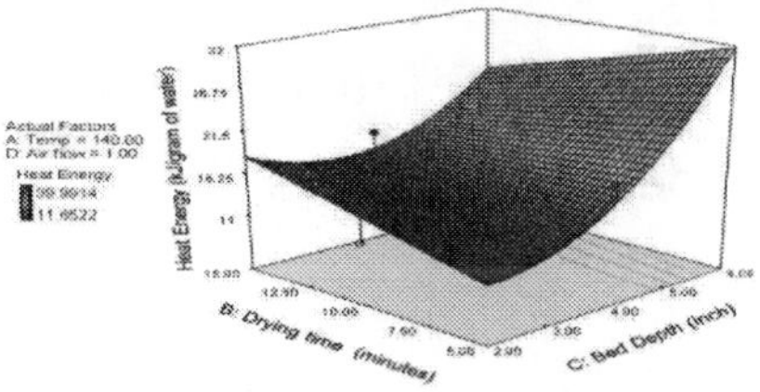


(a)

kJ/g water
Flue-gas Temperature = 140 C

2 m/s flue-gas velocity

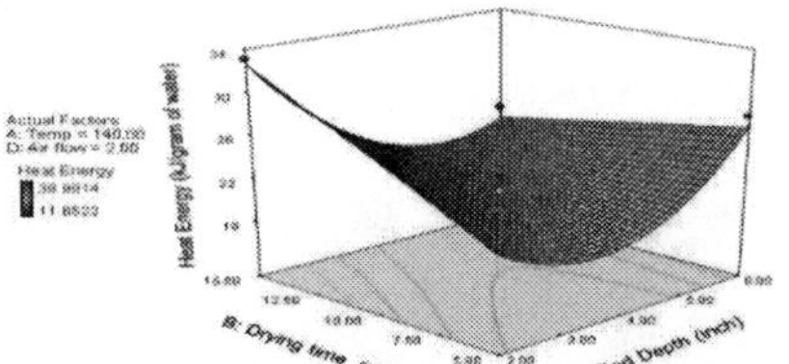


(b)

3 m/s flue-gas velocity

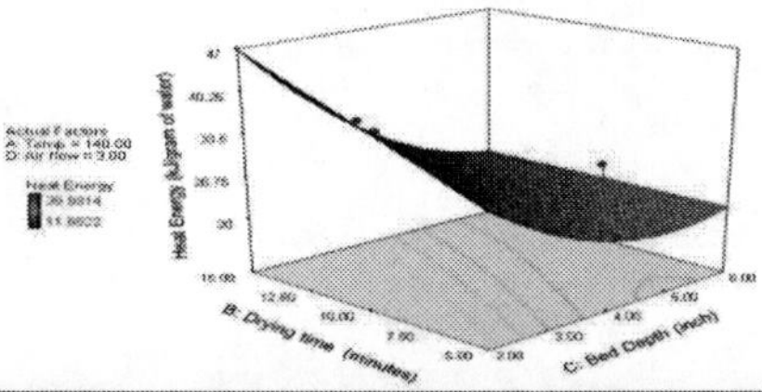


(c)

# Model Predictions of Energy Input per Gram of Water Removed

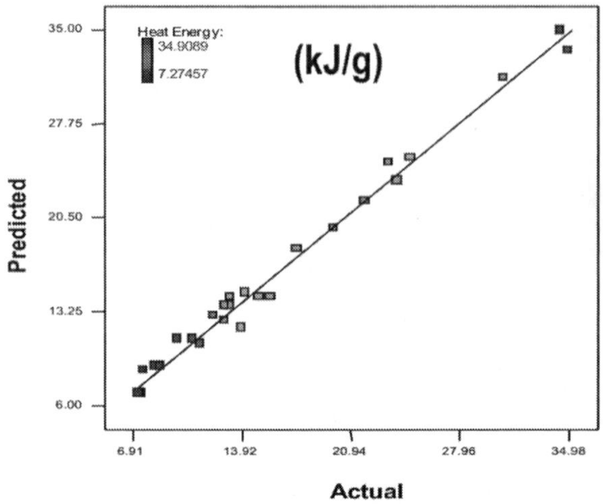


- ❑ Models incorporate:
  - ➢ Drying gas temperature
  - ➢ Drying gas velocity
  - ➢ Drying time
  - ➢ Bed depth
  - ➢ Significant cross terms



# Briquetting Questions?

- ➢ Does pelletization result in stronger coal briquettes
  - ➢ More uniform dispersion of binder?
  - ➢ More suitable feed properties?
- ➢ Does briquetting at an elevated temperature produce stronger briquettes?

# Briquetter Feedstocks

| Feedstock | Briquetting Temperature |
|---|---|
| 1. Air-dried coal fines | Ambient |
| 2. Microwave-dried fines | Elevated |
| 3. Air-dried coal pellets | Ambient |
| 4a. Thermally-dried coal pellets | Ambient |
| 4b. Thermally-dried coal pellets | Elevated |

# Briquettes from Air Dried Fines

Air Dry

## Briquettes from Air Dried Pellets

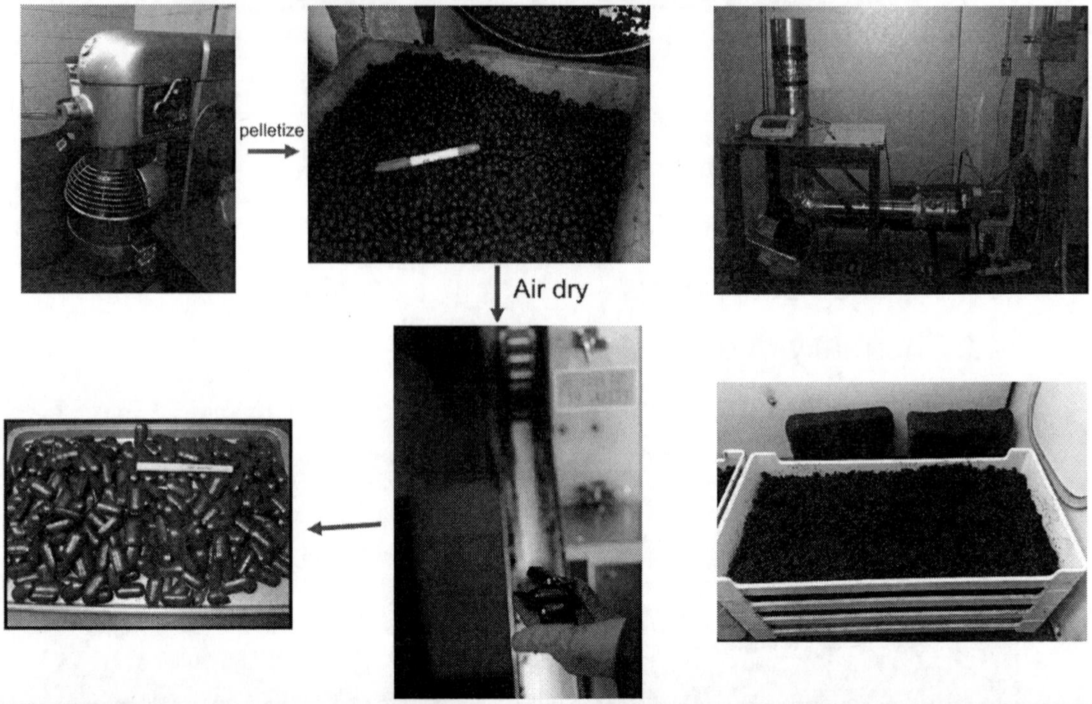


Center for Applied Energy Research

PCC16-Cape town

caer.uky.edu

## Thermally-Dried Pellets Briquetted Hot vs Ambient

Center for Applied Energy Research

PCC16-Cape town

caer.uky.edu

## Mivrowave-Dried Fines Briquetted Hot vs Ambient

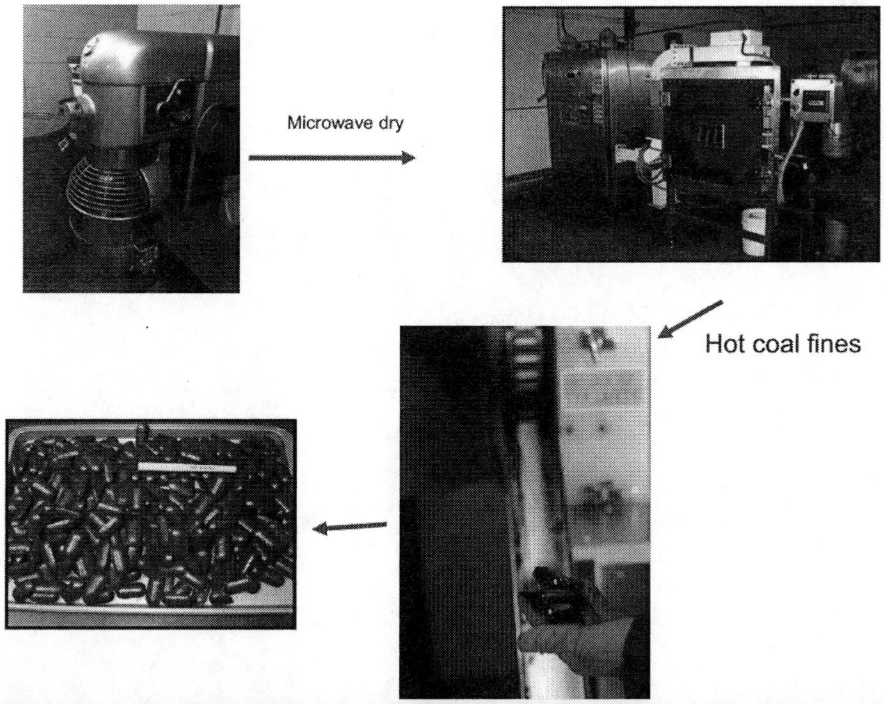


Center for Applied Energy Research PCC16-Cape town caer.uky.edu

## Fine Coal and Biomass Samples

- **Coals (Bituminous)**
  - PCC SB – Bituminous screen-bowl fines from Appalachian coal field
  - RV SB – Bituminous screen-bowl fines from Illinois Basin
  - MRC FL – Bituminous froth flotation fines from Appalachian coal field

- **Biomass (-6mm)**
  - Wheat straw (WS)
  - Corn stover (CS)
  - Miscanthus (MIS)
  - Biochar (bioCh)
  - Kiln-dried sawdust (SD)

UK UNIVERSITY OF KENTUCKY see blue.

# Study Coals: Chemistry

| Sample ID | Description | Ash % | Moisture (%) | Volatile Matter (%) | Fixed C (%) | Calorific value (kJ/kg-GCV) |
|---|---|---|---|---|---|---|
| PCC SB | Perry Co Screen Bowl | 6.86 | 1.03 | 34.60 | 57.51 | 31720 |
| RV SB | River View Screen Bowl | 13.42 | 1.00 | 32.06 | 53.52 | 26242 |
| MRC FL | McClure River Flotation | 11.00 | 0.87 | 27.83 | 60.30 | 31657 |

| | C (%) | H (%) | N (%) | $S_{tot}$ (%) | O (%) |
|---|---|---|---|---|---|
| PCC SB | 77.97 | 5.24 | 1.50 | 0.75 | 7.68 |
| RV SB | 67.18 | 4.10 | 1.41 | 2.70 | 11.19 |
| MRC FL | 77.98 | 4.90 | 1.41 | 0.89 | 3.82 |

Center for Applied Energy Research caer.uky.edu

# Study Coals: Particle-Size Distributions

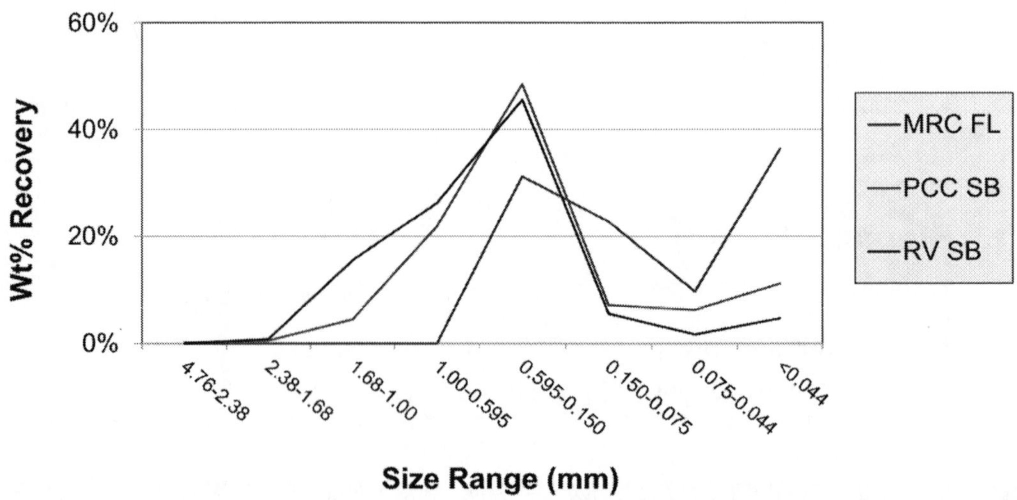


Center for Applied Energy Research PCC16-Cape town caer.uky.edu

# Compressive Strengths - PCC

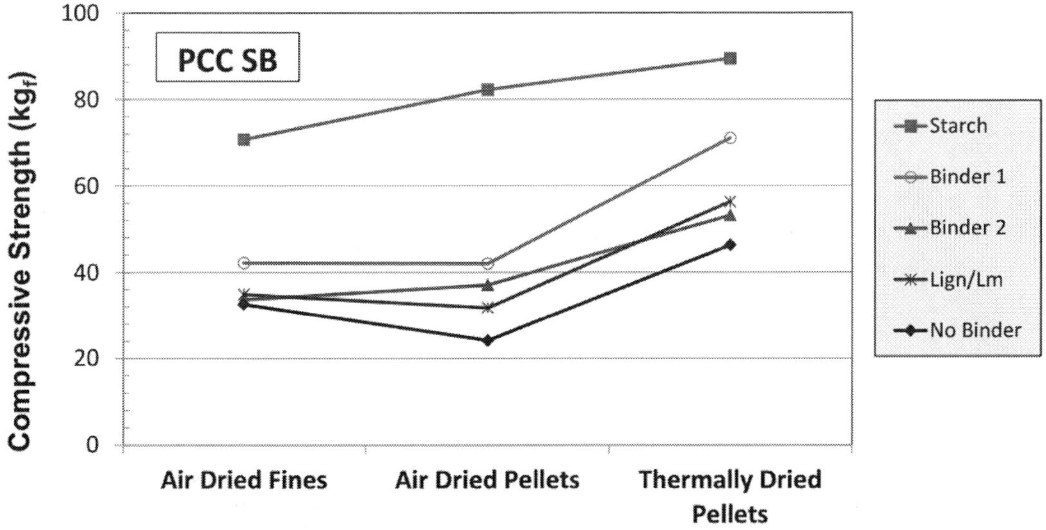


Center for Applied Energy Research

PCC16-Cape town

caer.uky.edu

# Compressive Strengths - RVSB

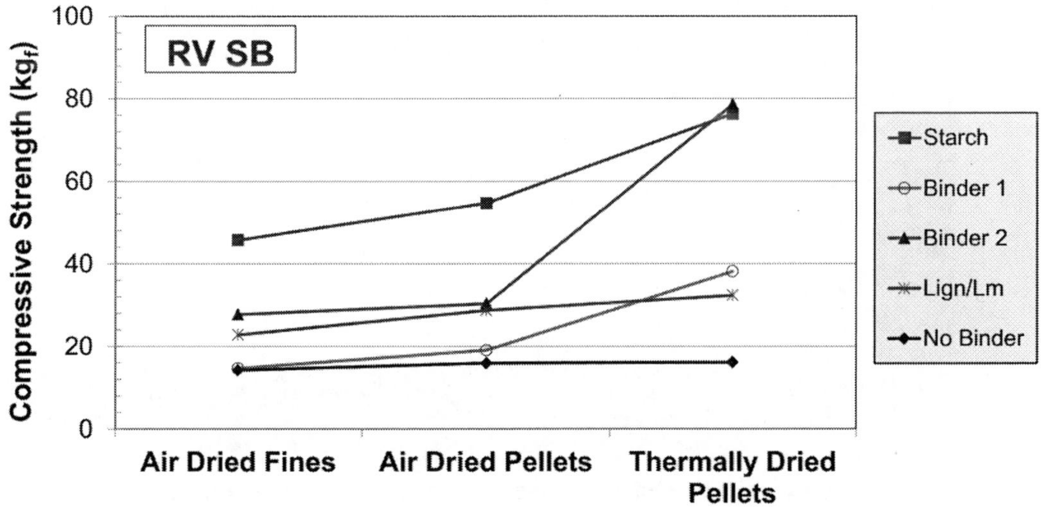


Center for Applied Energy Research

PCC16-Cape town

caer.uky.edu

# Cured Compressive Strengths - MRC

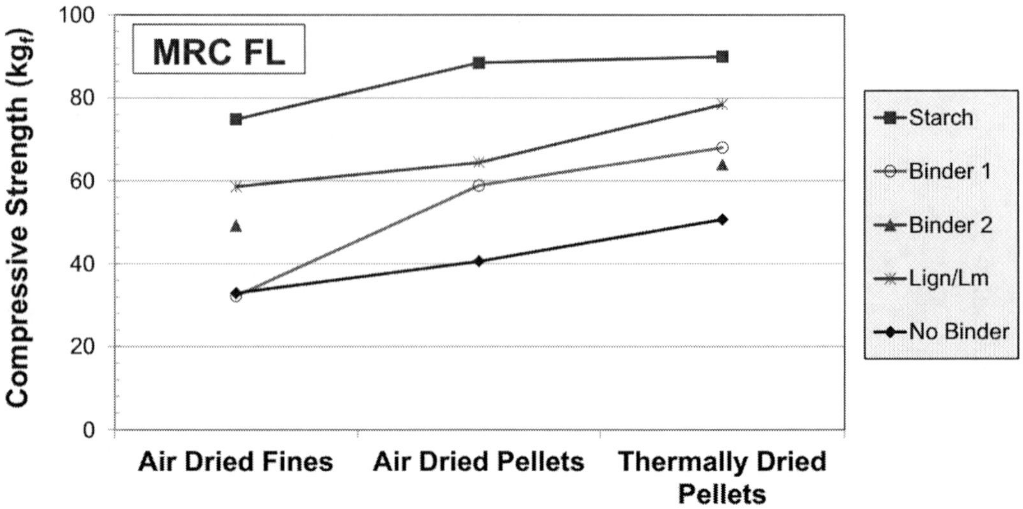


Center for Applied Energy Research PCC16-Cape town caer.uky.edu

# Average Attrition Index as a Function of Binder Type

| Binder | Air-dried fines[a] | Air-dried pellets[a] | Thermally-dried pellets[b] |
|---|---|---|---|
| ID | Attrition index (AI) | Attrition index (AI) | Attrition index (AI) |
| Binderless | 0.53 | 0.54 | 0.60 |
| Starch | 0.85 | 0.87 | 0.87 |
| Binder 1 | 0.53 | 0.61 | 0.76 |
| Binder 2 | 0.62 | 0.63 | 0.83 |
| Lignin/lime | 0.70 | 0.70 | 0.74 |
| Averages | 0.65 | 0.67 | 0.76 |
| [a] briquetted at ambient temperature | | | |
| [b] briquetted hot | | | |

Attrition Index = Fraction of briquettes remaining as + 6mm diameter particles after tumbling for 5 min at 40 rpm in a 40-cm-diameter cylinder equipped with 3 lifters

Center for Applied Energy Research PCC16-Cape town caer.uky.edu

## Comparison of Briquettes Made with Thermally-Dried Pellets Briquetted while Hot or after Cooling

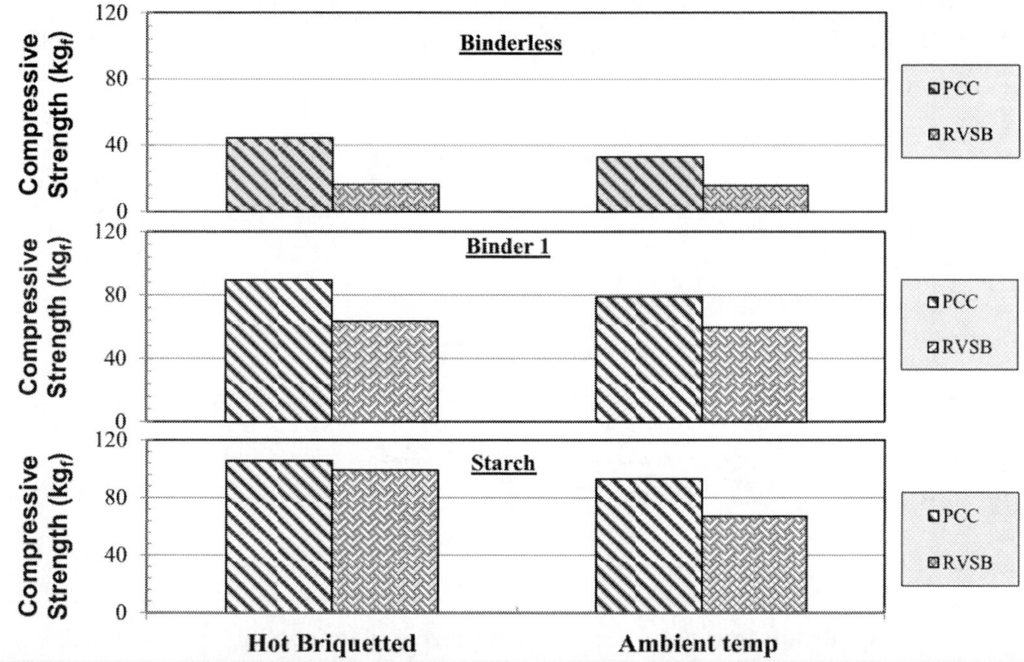


- Pellets were thermally dried then half were briquetted while hot with remainder briquetted after cooling

## Thermally-Dried Coal Pellets Briquetted with 10-Wt% Biomass Briquettes Produced at Elevated versus Ambient Temperature

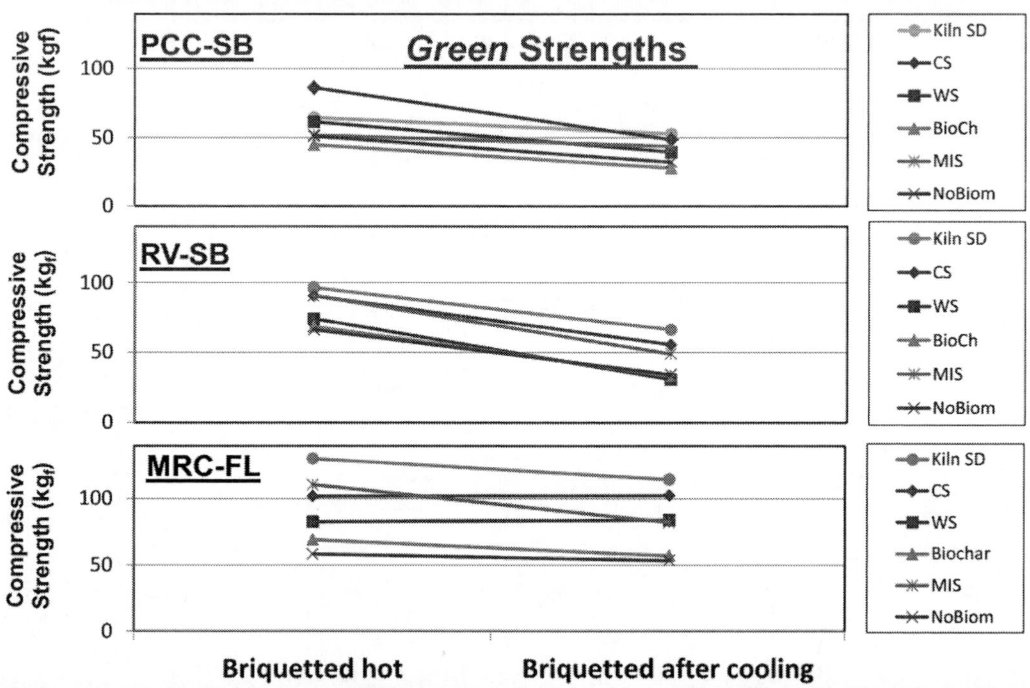


- Pellets made with starch binder and thermally dried. One half of each pellet sample was briquetted hot with remainder briquetted after pellets cooled

## Thermally-Dried Coal Pellets Briquetted with 10-Wt% Biomass Briquettes Produced at Elevated versus Ambient Temperature

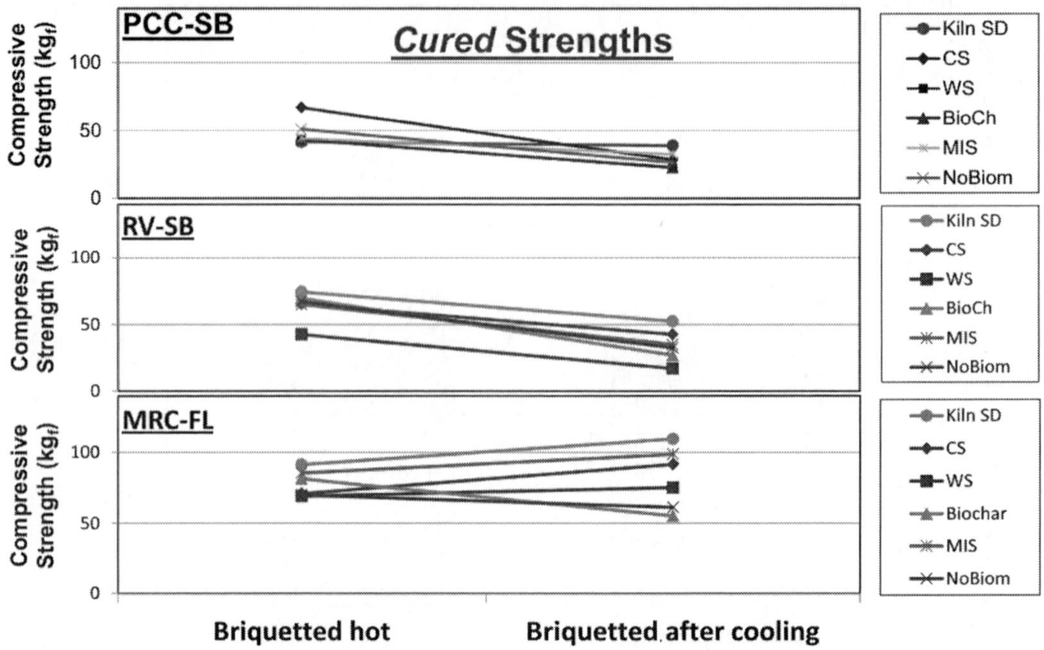


- Pellets made with starch binder and thermally dried then half of each pellet sample was briquetted hot with remainder briquetted after pellets cooled

## Thermally-Dried Coal Pellets Briquetted with 10-Wt% Biomass Briquettes Produced at Elevated versus Ambient Temperature

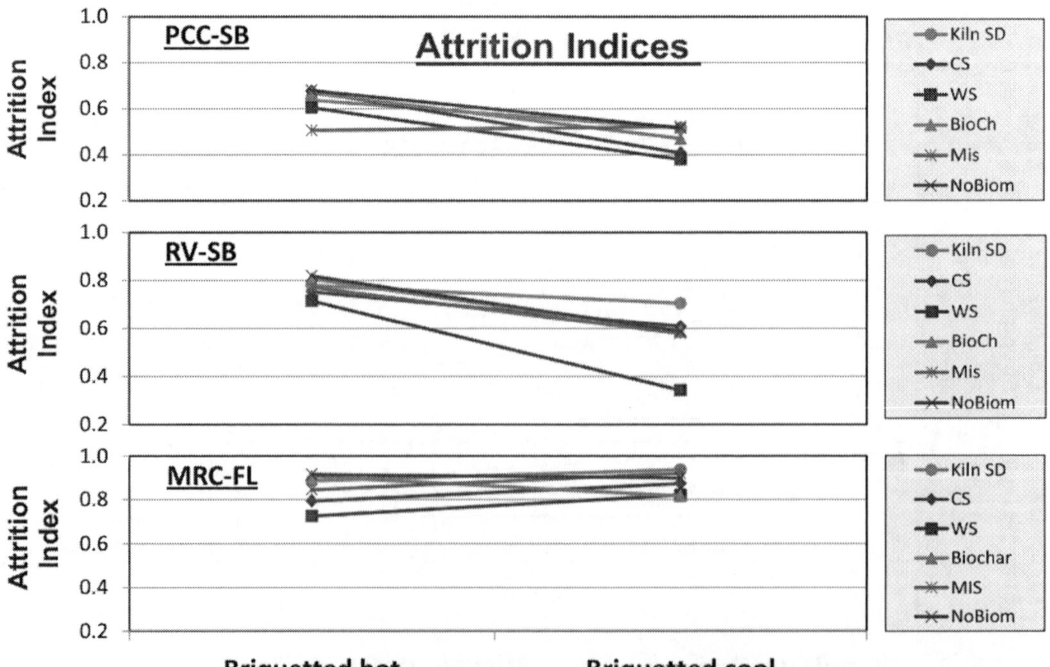


- Pellets made with starch binder and thermally dried then half of each pellet sample was briquetted hot with remainder briquetted after pellets cooled

## Compressive Strengths: Air-Dried Fines Briquetted at Ambient Temp vs Microwave-Dried Fines Briquetted Hot

RVSB Coal Fines

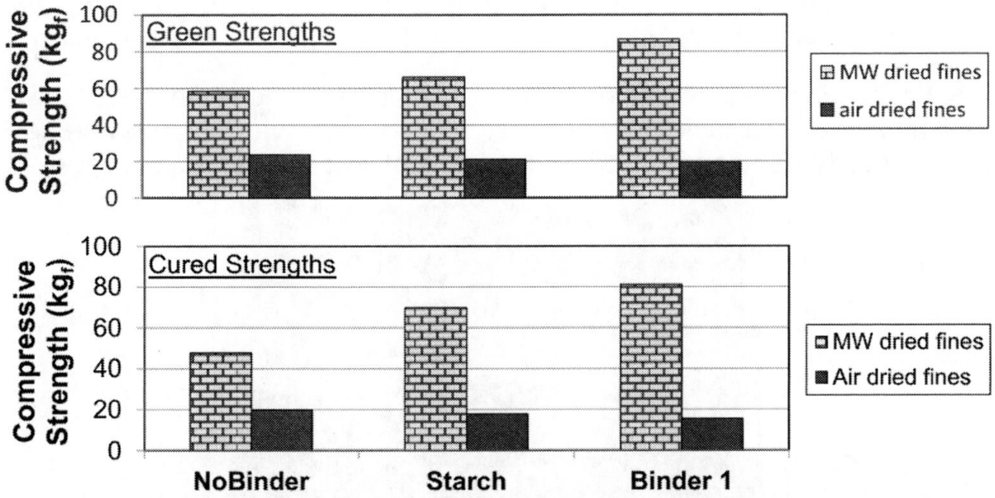


Center for Applied Energy Research PCC16-Cape town caer.uky.edu

## Attrition Indices: Air-Dried Fines Briquetted at Ambient Temp vs Microwave-Dried Fines Briquetted Hot

RVSB Coal Fines

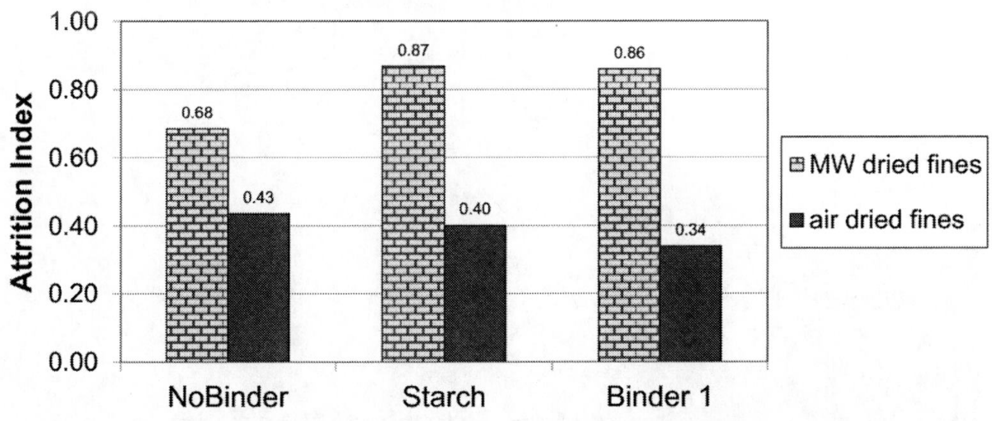


Center for Applied Energy Research PCC16-Cape town caer.uky.edu

## Summary

- Conveyed bed drying of fine coal pellets formed in a pan pelletizer is a viable alternative to fluidized bed drying
- Pelletizing fine coal prior to briquetting results in a small improvement in briquette properties, which can be attributed to either more uniform dispersion of binder or a feedstock that can be fed more consistently
- Briquetting coal pellets or coal fines at an elevated temperature, with or without biomass, provides a significant improvement in briquette durability compared to briquetting at ambient temperature

Center for Applied Energy Research

caer.uky.edu

## Questions?

Center for Applied Energy Research

PCC16-Cape town

caer.uky.edu

# Acknowledgements

❑ Financial support:

- U.S. Department of Energy National Energy Technology Laboratory
- Center for Advanced Separation Science administered by Virginia Tech
- Kentucky Department of Energy Development and Independence

Center for Applied Energy Research PCC16-Cape town caer.uky.edu

# Briquetted Fuel Sales Price Required to Provide a 10% Rate of Return on Investment

- Plant capacity → 50 tph/338,040 tpy
- 10 year plant life
- \$11.6M capital investment
- \$6.98M annual operating cost
  - Operating cost includes a binder addition rate of \$15/ton

$$\frac{\left(\$11{,}669{,}679 \times \frac{10\% \times (1+10\%)^{10}}{(1+10\%)^{10}-1} + \$6{,}980{,}401\right)/year}{338{,}040 tons/year}$$

$$= \mathbf{\$26.27/ton}$$

Center for Applied Energy Research caer.uky.edu

# Model Predictions of Moisture Reduction

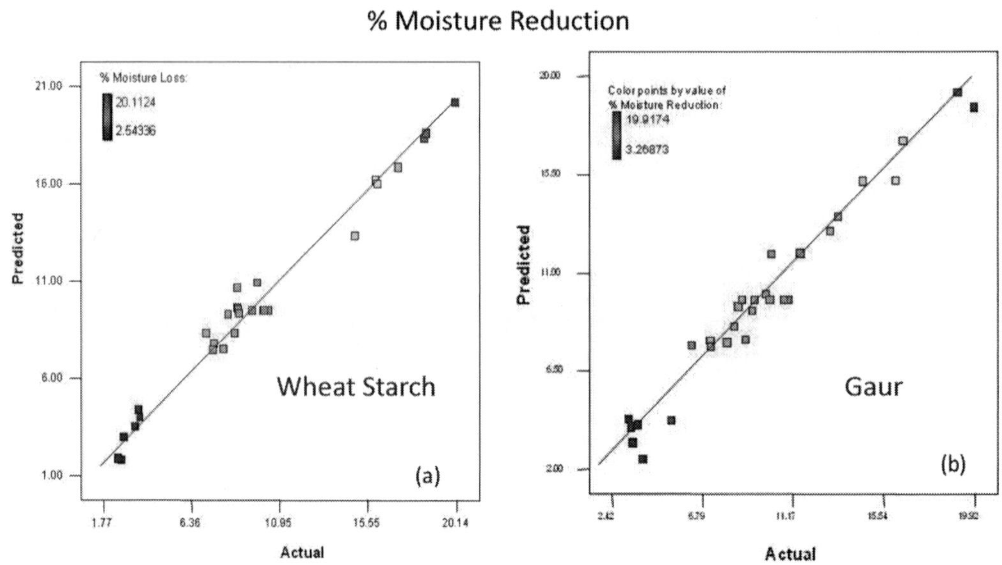


Center for Applied Energy Research caer.uky.edu

# Attrition Indices

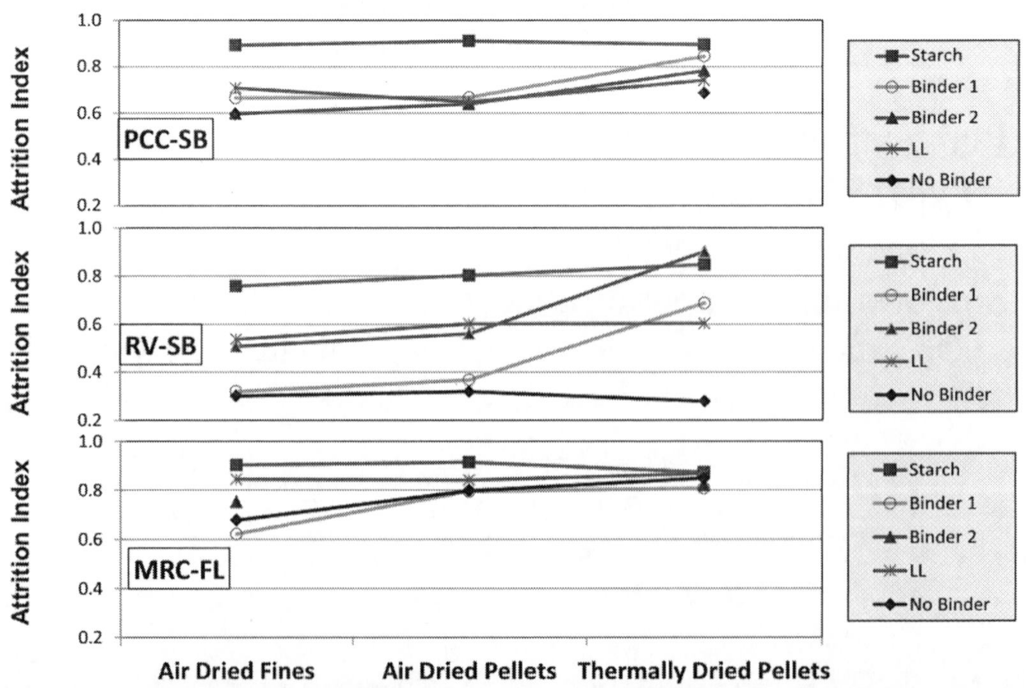


Center for Applied Energy Research PCC16-Cape town caer.uky.edu

| | | Compressive Strengths | | | Moisture Degradation | | | |
|---|---|---|---|---|---|---|---|---|
| | Briquette Moisture | 30-min kgf | 24-hr kgf | 7-day kgf | Attrition Index | Ave Briq Wt (g) | Wet Str kgf | Water Uptake |
| **PCC-Air Dried Fines** | | | | | | | | |
| Run #1 | 2.1% | 68 | 75 | 71 | 0.89 | 9.72 | — | — |
| Run #2 | 2.3% | 67 | 70 | 70 | 0.90 | 9.56 | — | — |
| Average | 2.2% | 67 | 73 | 71 | 0.89 | 9.64 | — | — |
| **PCC-Air Dried Pellets** | | | | | | | | |
| Run #1 | 2.0% | 83 | 89 | 85 | 0.91 | 10.10 | — | — |
| Run #2 | 2.1% | 74 | 82 | 79 | 0.91 | 9.83 | — | — |
| Average | 2.0% | 79 | 86 | 82 | 0.91 | 9.97 | — | — |
| **PCC-Thermally Dried Pellets** | | | | | | | | |
| Run #1 | 1.2% | 98 | 96 | 92 | 0.89 | 9.39 | 19 | 14.0% |
| Run #2 | 1.5% | 89 | 90 | 87 | 0.90 | 9.37 | 18 | 13.8% |
| Average | 1.4% | 94 | 93 | 89 | 0.90 | 9.38 | 18 | 13.9% |
| **RVSB-Air Dried Fines** | | | | | | | | |
| Run #1 | 2.7% | 51 | 52 | 51 | 0.79 | 9.14 | 1 | 33.8% |
| Run #2 | 2.5% | 47 | 46 | 40 | 0.73 | 9.15 | 3 | 33.8% |
| Average | 2.6% | 49 | 49 | 46 | 0.76 | 9.15 | 2 | 33.8% |
| **RVSB-Air Dried Pellets** | | | | | | | | |
| Run #1 | 3.5% | 57 | 57 | 57 | 0.81 | 9.60 | 4 | 31.7% |
| Run #2 | 3.5% | 59 | 53 | 52 | 0.80 | 9.57 | 5 | 30.2% |
| Average | 3.5% | 58 | 55 | 55 | 0.80 | 9.59 | 4 | 31.0% |
| **RVSB-Thermally Dried Pellets** | | | | | | | | |
| Run #1 | 1.9% | 75 | 75 | 75 | 0.84 | 9.17 | 35 | 14.5% |
| Run #2 | 2.3% | 83 | 75 | 78 | 0.85 | 8.97 | 21 | 17.6% |



# Roll Briquetting

- ✓ High throughput rates
- ✓ Low to moderate energy costs
- ✓ Low to moderate erosion rates
- ✓ Well suited for coal/biomass blends
- X Often requires a binder ($$$)

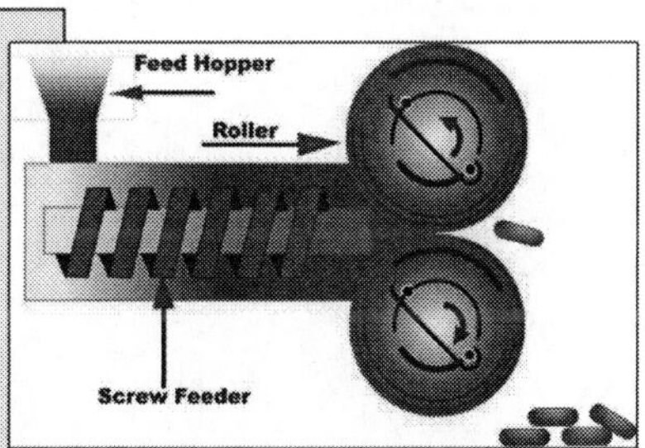

Questions?
UK CAER Energy Campus

# UCG – UNLOCKING SOUTHERN-AFRICA'S COAL POTENTIAL

## UPDATE ON THE AFRICARY's THEUNISSEN UNDERGROUND COAL GASIFICATION (TUCG) PROJECT IN SOUTH-AFRICA

JF Brand*, johan.brand@africary.com, +27 82 497 1955(Presenting author)
JC van Dyk, johan.vandyk@africary.com, +27 82 550 0473

African Carbon Energy (Africary) is putting South Africa at the forefront of Underground Coal Gasification (UCG) by developing a one billion-tonne coal deposit at Theunissen in the Free State to fill a vital gap in the country's electricity supply. Working to register as an Independent Power Producer (IPP) with the Department of Energy, Africary is championing a technique that had been commonplace several decades ago, and one which holds several advantages over conventional coal mining which has now become the mainstream.

Why UCG? Underground Coal Gasification is a mining method that extracts previously stranded coal reserves in-situ, through a gasification process by constructing at least two horizontal boreholes, where one borehole is used to inject air/oxygen to convert the coal into syngas and another to bring the syngas to surface. It is an economically and environmentally viable option for mining deep reserves (200 metres or deeper), and with two thirds of the planet's coal unable to be mined through conventional techniques it is the most promising technology for the future.

Africary Holdings (Pty) Ltd (Africary), is planning to establish an underground coal gasification (UCG) project in the Free State, South-Africa. The company has acquired prospecting rights for coal over an area of more than 300 square kilometres and has identified potential target areas for the establishment of a UCG project at a site located about 24 km north-north-west of Theunissen and 27 km south-south-west of Welkom in the Free State Province. The company has expertise in mining, gasification and power generation project development, specifically based on UCG and conventional coal mining.

The average annual selling price of local electricity has risen >60% from 2010 to 2015 and is set to rise further in the years to come, making power generation from gas produced by UCG competitive. The current project will involve gasification of ±5 million tons of in-situ coal over 20 years under an area of about 150 hectares and will generate >50 MW (nett) clean electricity to the grid from an installed gross capacity of ±77 MW (or ±31t/hr coal flow, ±2400 kmol/h wet gas and ±53kNm$^3$/h equivalent or >80 000kg/hr). The UCG process will operate with an average of 35% oxygen enriched air as agent feed to produce a clean Syngas with a heating value of >7MJ/m$^3$n at gas-engine intake.

Africary may become the world's first commercial UCG operator to supply electricity for base load power supply. The product of this application will transform the face of coal mining and electricity production in South-Africa to new future perspectives and heights. The project will be able to compete with coal-fired power generation without the need for any rebates, and with the increase in tariffs makes electricity generation with UCG cost competitive.

Once UCG is seen to be commercial acceptable, opportunities will arise for poly-generation, using the syngas to make diesel, fuels or chemicals. UCG not only turned out to be safer, but also more economical and more efficient.

2016 International Pittsburgh Coal Conference
Cape Town, SOUTH AFRICA
August 8-12, 2016

# THE QUEENSLAND UNDERGROUND COAL GASIFICATION PILOTS 2008-2016 –EVIDENCE BASED ASSESSMENT AND LESSONS IN GOVERNMENT AND PUBLIC RELATIONS

*Cliff Mallett, Technical Director, Carbon Energy Limited, 301 Coronation Drive, Milton, Queensland,AUSTRALIA*
cmallett@carbonenergy.com.au *phone +61419753719*

## Introduction

UCG trials commenced in Queensland in 1999 when Linc Energy ignited a vertical well design at Chinchilla and produced syngas for several years. The principal funder was a Queensland Government owned power utility company CS Energy. Interest in UCG world wide was growing and by 2006, two other UCG companies were also active. Cougar Energy was founded by the originator of Linc Energy after control of the Linc Energy company passed to other investors, and Carbon Energy was a commercial spin out company of the CSIRO, the Australian government's national scientific research organisation, using new UCG designs developed over a ten year research program.

## Permitting processes

Permits to operate were issued under mining legislation as underground coal gasification was considered a coal mining activity, and this also regulated the health and safety of the operations. Site functions were further segregated as all underground activity was regulated by the coal mining section of the department, and surface facilities by the petroleum section within the department. In addition environmental permits to operate were issued by a different government department under environmental legislation. A new category of resource extraction was created to describe the production of UCG syngas in Queensland, "Mineral F". As the concept of underground coal gasification was new and little known, obligations under these permits were initially minimal. The first permits imposed only limited obligations for monitoring and reporting environmental information, and as awareness of potential issues grew with government regulators, subsequent permits imposed increasingly stringent requirements on impacts and reporting. Cougar Energy, as the last company to obtain a permit was burdened with the strictest requirements, including some which were in practice impossible to meet, and indicated that the regulators did not clearly understand the physical processes within and around underground gasifier cavities.

At the same time as these exploration and mining licences over coal were being granted for UCG, the coal seam gas (or coalbed methane) industry was starting up in the same coal basins. Coal seam gas operates under petroleum exploration and development licences, and these were being granted over the same areas as the coal exploration and mining licences. The coal seam gas industry expanded rapidly and was promising investment into Queensland of more than $60 billion dollars. They saw the UCG permits as competing for access to their coal gas and launched a major lobbying effort with the government to stop the UCG industry.

In 2009 the Queensland government effectively imposed a moratorium on UCG development by initiating a program of UCG Pilots to demonstrate the environmental safety of the process, before they would establish a policy for commercial UCG operation. An Independent Scientific Panel of three was appointed by the government to develop a Pilot review process and report to government on the outcome. Only companies already involved in UCG sites were included and other companies planning to commence UCG were prevented from further work until the government determined a long term policy on commercial development of UCG.

The three companies with advanced UCG activities at that time were given permission to operate Pilot operations. All companies held multiple exploration areas where UCG might be possible, but government deadlines forced the selection of sites already investigated, rather than allowing time for exploration to identify the best UCG sites. As a result Linc Energy began operations on seams at 130m depth (where they had commenced in 1999), Cougar Energy at 160m depth, and Carbon Energy at 200m depth. This proved very significant in the subsequent UCG Pilot process.

## UCG Technologies

A variety of UCG gasifier designs were used in the Pilots. Linc Energy commenced operation well before the Pilot scheme was instigated using vertical wells as commonly employed in the former USSR. Subsequently they tried a four modules of different designs, in time order, vertical wells combustion linked, vertical wells borehole linked, in-seam parallel CRIP (borehole layout similar to Carbon Energy's Keyseam* design), and finally, a linear CRIP module. Modules 2-5 operated at different times between 2008 and 2014 and the company was active on the site until it went into liquidation in 2016. Cougar Energy used the same technology provider and vertical well design as employed at the 1999 Linc site. Carbon Energy used a new UCG design (Keyseam*) developed by CSIRO, based on the Rocky Mountain trials in Wyoming and extensive research into conventional longwall mining.

UCG gasifier design needs to provide for continuous gas production in commercial operations. The gasifier vessel is a cavity created in coal by gasification, and there has to be access to supply the oxidant gas and to remove product gas. Gasification occurs on a coal face where hot gases can interact with coal and groundwater. So any UCG design must simultaneously feed oxidant gas, maintain a gasification face and capture product gas in a continuous process. Early methods did this by establishing fixed injection and extraction points in a coal seam, linking them by some method, and then gasifying coal between the two points which created a cavity which gradually expanded as the coal was consumed. Individual cavities were limited to around 15m diameter before the gasification efficiencies declined and a new gasification module needed to be established. The Linear CRIP design developed by Lawrence Livermore Laboratories strung these individual gasifier modules along an in-seam borehole, by episodic retractions of the injection point. The Rocky Mountain trial demonstrated a different type of gasifier, where the gasification face developed between two in-seam boreholes, drilled at an angle to each other and both intersecting a vertical well for ignition of the coal. One horizontal well was for injection and the other one for syngas removal. There is only one gasification face, located between the injection borehole and the product extraction borehole. To maintain gasification, the injection point must be continually retracted to align it with the fresh coal face. This became the basis of the Keyseam* Parallel CRIP method used by Carbon Energy and the design concept was also applied in China with their Tunnel gasifiers within conventional coal mines. Hence the Queensland UCG Pilots provided a comprehensive spread of UCG borehole designs.

Government relations

Each of the Pilots adopted individual approaches to project development and dealing with regulators, local governments, other land users and the community. Linc Energy combined UCG pilot production simultaneously with the development of Gas to Liquids technology and built a significant Fisher Tropsch experimental plant which was fed from the UCG pilot operations. The Pilot was promoted as an alternative liquid fuels project and adopted a high public profile in the business press and media. The effective moratorium on UCG imposed by the Pilot program was a significant impediment to Linc's Coal-to-Liquids business plan and relations between the government and the company were strained from the start. As mentioned above, Linc's legal environmental reporting obligations were the first issued and were minimal. In the confrontational environment that developed, the company adhered strictly to the limited reporting requirements in their agreement, which became a major point of frustration with regulators. The other reporting requirement was for the Independent Scientific Panel, that were charged with assessing the environmental performance of the Pilots. This panel took a long time to determine what information it needed, but under the confidentiality conditions associated with the ISP team, Linc provided comprehensive descriptions of their operations.

Carbon Energy was the second UCG Pilot to obtain its environmental conditions which while they were more comprehensive than those of Linc Energy, conditions were consistent with other mining and chemical industry projects in the State. With their research background, Carbon Energy staff were used to comprehensive monitoring and analysis activities associated with operations and consistently provided the government monthly reports on all environmental matters and welcomed them to the site. This continues up to today. They believed that the more the regulators knew and understood about underground gasification, the fewer would be their concerns. The company also worked cooperatively with the Independent Scientific Panel.

Cougar Energy had protracted negotiations with the government on their environmental obligations, and confronted with the choice of accepting unworkable terms or very significant delays to starting, they opted to accept the conditions demanded. They ignited coal in March 2010 and "shortly after" the well blocked. In May and June 2010 Cougar detected benzene and toluene in two groundwater monitoring wells which was re-confirmed by the analytical laboratory on 7th June and reported to the government on 30th June. They had agreed to never allowing any contaminants to escape from the gasifier into surrounding groundwater. In July 2011 the government forced the shut down of the Cougar Pilot. Later in 2011 Cougar initiated legal action against the government seeking $34 million compensation, but dropped the action in July 2013, when financial statements provided to the Australian stock exchange suggested they lacked the funds for an extended court action. They then pleaded guilty to three non-compliance issues alleged by the government, and in September 2013 the company was fined for: not constructing the production well according to Environmental Approval requirements, contaminating groundwater, and not notifying the government of monitoring well results in a "reasonable" timeframe. Defence submissions that the levels of contamination in groundwater were so low to be of no danger (for example lower than air contaminant levels in the courthouse carpark) were not considered relevant.

Community relations

The activities of Linc Energy and Carbon Energy were initially generally welcomed. Linc Energy employed a significant local workforce and the councils and community appreciated the potential local industry promised for the region. Their work sites were relatively small and remote. As the UCG companies were listed on the Australian share market, many 'dads and mums' bought shares and, for example Carbon Energy had several thousand share holdings by such small investors. At the same time the coal seam gas industry started exploration and development, and their heavy handed approach to land entry created a major community backlash which developed a sensitivity for all energy industry developments. Green activist groups developed and began targeting specific proposals. They had never had much common ground with the rural community, but now they exploited the concerns landholders had over the over-riding rights of entry that mineral licence holders had to explore and develop underground resources. A "Lock the Gates" campaign blossomed and the press had a field day. High profile campaigns were launched against specific mining sites on the basis that they were destroying "prime agricultural land". Promises to restore land after mining, both economically and technically feasible, were ridiculed.

Cougar Energy was located only 11km from Kingaroy, a significant rural town surrounded by productive volcanic soils. A coal mine and power station had been established in the region in the 1980s and exploration to extend the open cut coal mines was active. A well organised activist group had emerged locally to fight the take over of farms for new coal mines. Unfortunately the site selected by Cougar Energy was in the middle of this contested ground. A perfect storm erupted when an analytical laboratory made a mistake with a water monitoring result for the Cougar site. Cougar staff asked for a check to be made, but word got out to the community, and a local independent politician facing an election and seeking publicity, questioned the minister about the result at public parliamentary inquiry. By the time the laboratory confirmed that the result was a mistake several days later, the activists had called community meetings which were told the town's water supply had been poisoned, the government had put a hold on Cougar Energy operations, prohibited farmers in a 20km radius from taking any water from bores, forced Cougar Energy to pay for the delivery of water to all farmers affected by the ban, and there was a major controversy raging over all UCG projects. The government put a hold on activities at Carbon Energy and Linc Energy Pilots, and began a search for something they could use as the reason for their bans. In Carbon Energy's case they prosecuted the company in relation to a process water spill of a year previous, that had been cleaned up, and the government subsequently notified, and site inspected without comment, however the company was subsequently forced to conduct over a $1 million of site investigations that showed zero contamination present. Never the less, the company was prosecuted for not supplying a report on the incident which had no environmental impact, in an appropriate timeframe, and the Pilot was delayed for over a year. The social concerns in the Kingaroy area were so severe that the government investigated the Cougar Energy site, accumulating evidence that would justify them closing the site completely, which they did in January 2011. As Cougar Energy had dropped its legal action against the government and pleaded guilty to the three Environmental Authority non-compliances in 2013, there was little evidence released of what had actually happened at the site.

Independent Scientific Panel (ISP)

The UCG Pilot program was to be reviewed by a panel of independent experts selected by the government and panel members decided how they would like each UCG site to report. None of the experts had direct experience with UCG but all worked in some aspect of the energy industry. The chairman was at the time the director of a major research centre funded by the coal seam gas industry. The process was interrupted by the hiatus associated with the events at Kingaroy, and Cougar Energy withdrew from the Pilot process in 2011. The remaining two companies provided comprehensive reports in mid-2012 on their activities up to that time, which were reviewed by the independent scientific panel. The panel was assisted by the Queensland Government Chief Scientist, in developing policy options in relation to the technical issues, and the ISP report was released in July 2013. In essence it reported that both pilots had commissioned and operated the syngas production phases in a satisfactory manner, but recommended that as neither had demonstrated decommissioning and rehabilitation planning, the moratorium on commercial UCG should be extended until this was done. However, in the time taken for the review to be presented, both sites had decommissioned multiple UCG modules, and Carbon Energy provided reports on Decommissioning (August 2014) and a Rehabilitation Plan (October 2014) as requested. No formal response has yet been provided to Carbon Energy with respect to these reports. Linc Energy has not submitted comparable reports and in late 2013 relocated from Australia to Singapore, citing Queensland government actions an important factor in the move.

Government Responses

Following the Kingaroy incident a subsequent report presented to the Queensland Parliament by the State Ombudsman severely criticised the oversight of the UCG trials by the government's environment department. This initiated a period of intense scrutiny of activities by departmental officers, far in excess of the attention provided to comparable industries. The "letter of the law" was applied to all conditions in the environmental approvals, and compliance became the focus rather than environmental harm management. In this confrontational environment Carbon Energy attempted to cooperate and work with departmental officers, but Linc Energy, which still had relatively lighter obligations, only provided what was specified. It was a frustration for government officers that Carbon Energy was providing detailed monthly reports on monitoring of the whole site, and Linc Energy provided minimal data. When a member of the public, identity protected by 'whistleblower' legislation, made a complaint about historical activities on the Linc Energy site, the government was empowered to investigate Linc Energy. What followed was a truly remarkable legal and investigative process that grew into the largest environmental prosecution case ever undertaken in Queensland, with a legal team of over fifteen and expenditure of at least $15 million. A preliminary hearing determined that the case should go to court but the company went into liquidation before court actions could commence. Once again accusations of environmental harm from UCG have escaped rigorous court examination, and the government and activists groups use alleged contamination as evidence of contamination. Within days of the Linc Energy liquidation, the Queensland government announced a ban on UCG in Queensland, providing no basis other than having considered the risks and benefits.

GIG

# LEACHING OF METALS FROM POST-UNDERGROUND COAL GASIFICATION CAVITY RESIDUES

*Krzysztof Stańczyk and Aleksandra Strugała-Wilczek*
*Central Mining Institute, Katowice, POLAND*

2016 International Pittsburgh Coal Conference
Cape Town, SOUTH AFRICA
August 8 – 12, 2016

## Plan of presentation

- Background of the study
- The aim of the research
- Methods
- Results and discussion
- Conclusions

2

## Background

Underground coal gasification process is accompanied by the generation of organic and inorganic waste therefore the threat of ground water pollution is the dominant environmental concern during and after UCG technology implementation. The potential environmental problems have been reported in the 1980s and 1990s as well as recently by Queensland Government, Australia. In the post-reaction cavity (formed as a result of the underground coal gasification) a part of the solid products remain as ash residue or carbon chars. Particularly little is known about inorganic pollutants, so research on metal elution from the cavity residues seems to be necessary for understanding a possible water contamination.

3

## Background

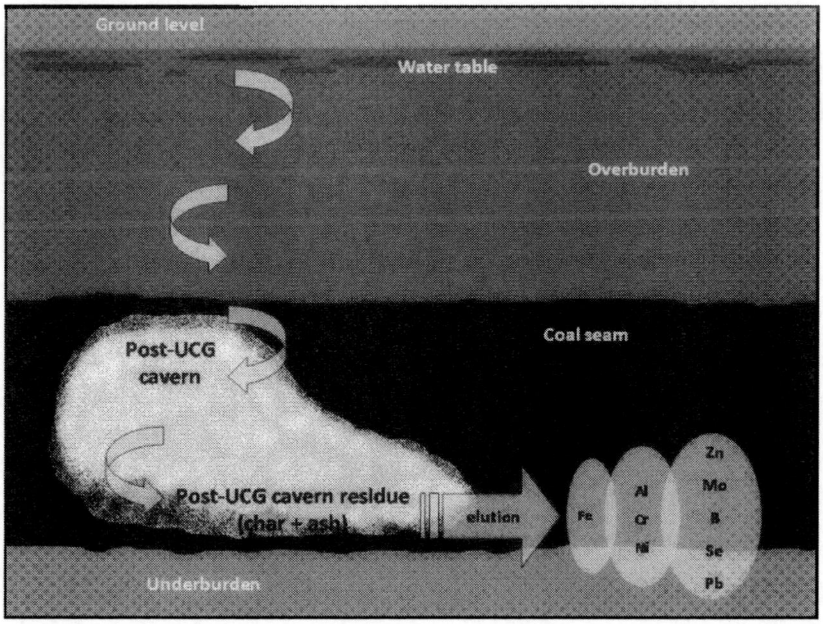


4

## Background

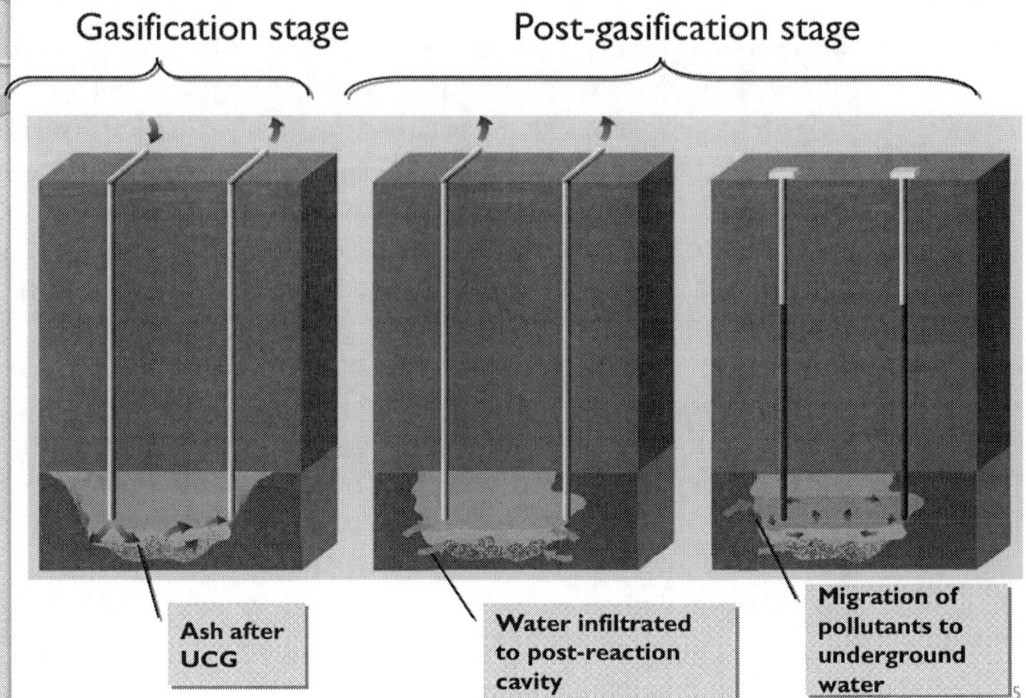


5

## The aim of the study

- The aim of this study was to examine whether water entering the post-reaction area after underground coal gasification (UCG) leaches metals from the post-processing waste in the cavity to such a degree that it can pose a threat to the environment.

6

## Abstract

The work presents a comparison of chosen metal elution from cavern residue after underground coal gasification and from the ash obtained during combustion of the same samples of coal in a boiler. The influence of coal rank (hard coal and lignite are compared) on metal leaching behaviour is demonstrated. Also the influence of the range of water mineralization on metal elution is shown.

7

Exemplary samples of test materials (for hard coal Bobrek and lignite Bełchatów)

**Coal and lignite**

**Char from the UCG process**

**Ash from the UCG process**

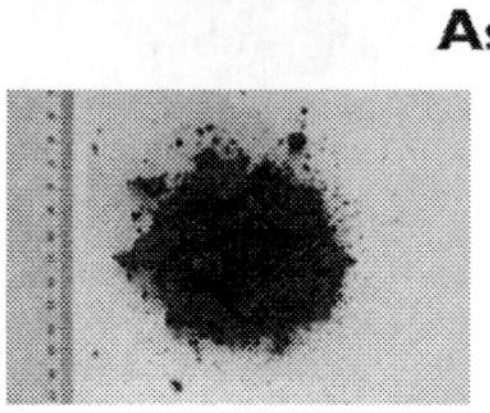

8

## Classification of major elements and trace elements due to their behaviour during the underground gasification and combustion of coals

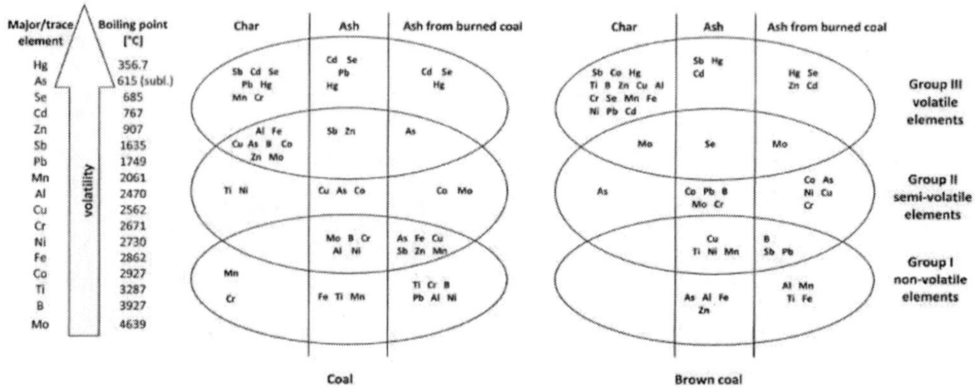


A. Strugała-Wilczek, K. Stańczyk, Comparison of metal elution from cavern residue after underground coal gasification and from ash obtained during coal combustion, ***Fuel***, 2015, 158, 733–743

9

## A distribution of the elements in terms of mobility for elution into the water phase

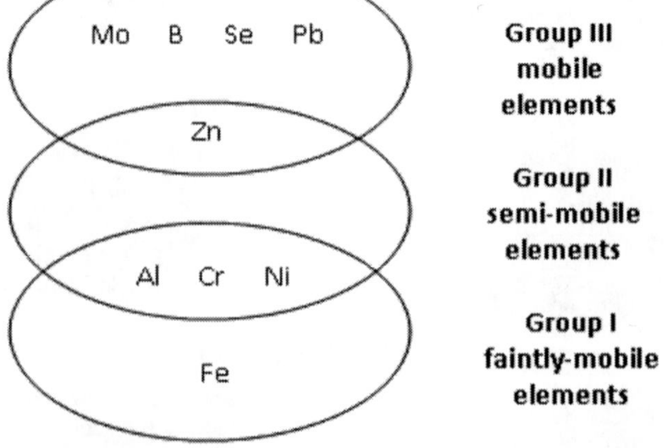


A. Strugała-Wilczek, K. Stańczyk, Comparison of metal elution from cavern residue after underground coal gasification and from ash obtained during coal combustion, ***Fuel***, 2015, 158, 733–743

10

## Mo content in the water extracts in relation to the hydro-geochemical background values of groundwater

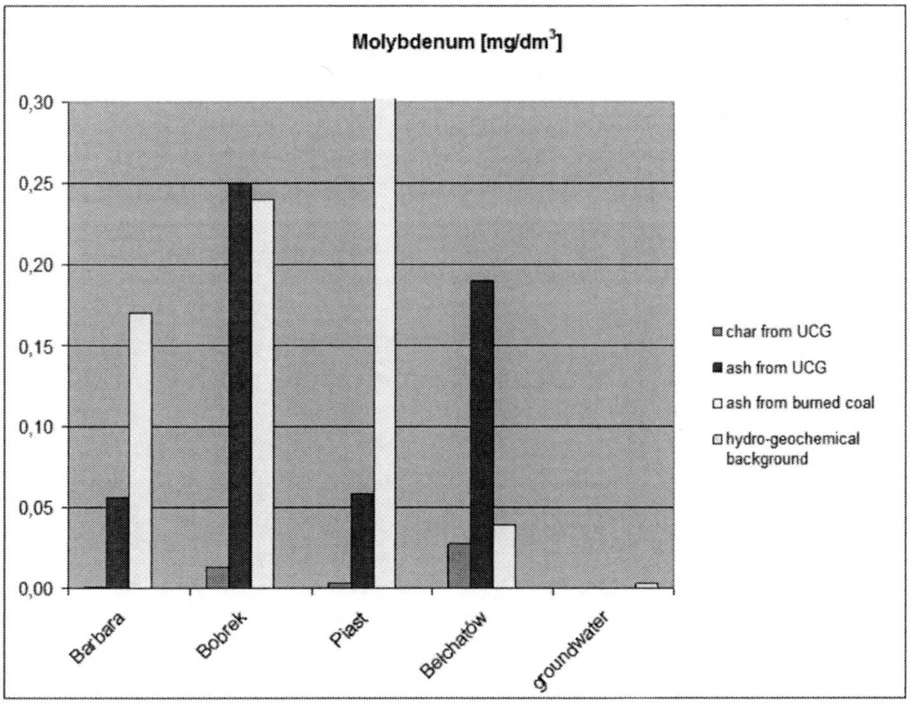


11

## Pb content in the water extracts in relation to the hydro-geochemical background values of groundwater

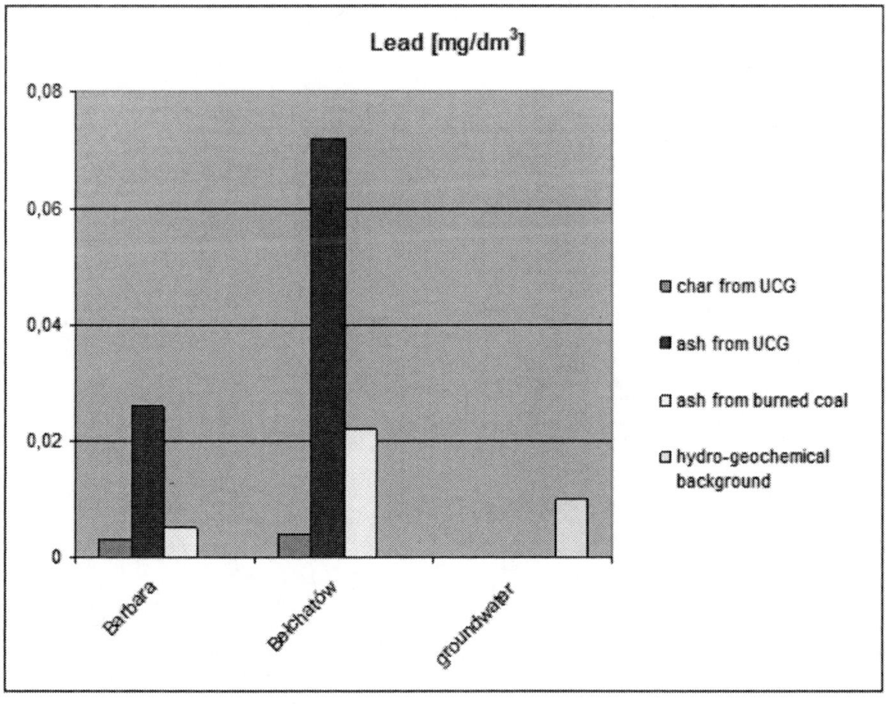


12

# Se content in the water extracts in relation to the hydro-geochemical background values of groundwater

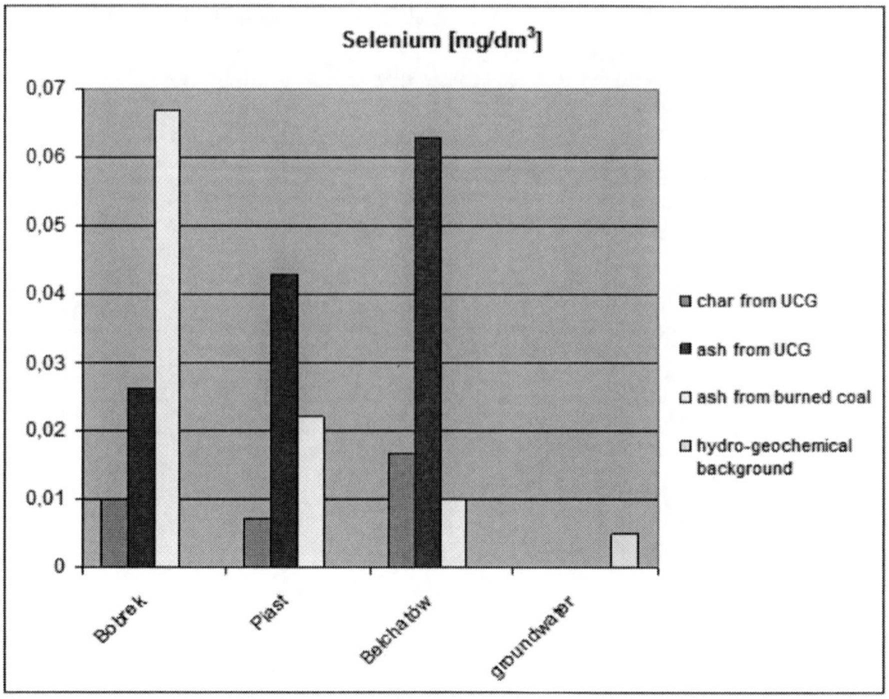


13

# Classification of major and trace elements due to their behaviour during the underground gasification of hard coal and lignite

| Major/ trace element | Boiling point [°C] |
|---|---|
| Hg | 356.7 |
| As | 615 (subl.) |
| Se | 685 |
| Cd | 767 |
| Zn | 907 |
| Sb | 1635 |
| Pb | 1749 |
| Mn | 2061 |
| Al | 2470 |
| Cu | 2562 |
| Cr | 2671 |
| Ni | 2730 |
| Fe | 2862 |
| Co | 2927 |
| Ti | 3287 |
| B | 3927 |
| Mo | 4639 |

volatility

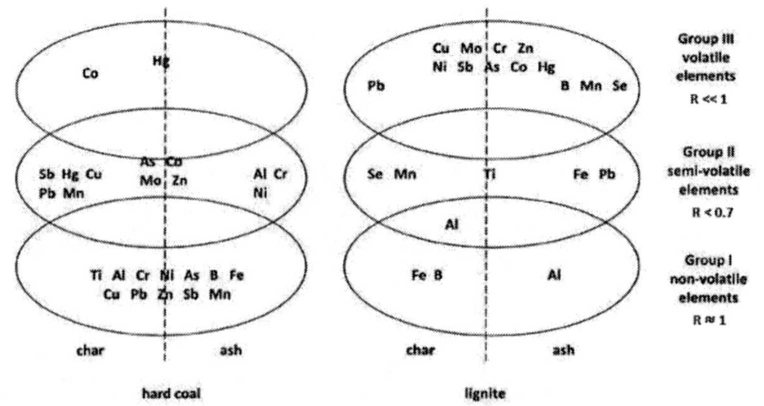


A. Strugała-Wilczek, K. Stańczyk, Leaching behaviour of metals from post-underground coal gasification cavity residues in water differing in mineralization, ***Fuel***, 2016, 173, 106-114

14

## Distribution of the elements in terms of mobility for elution into the water phase (UCG of hard coal and lignite)

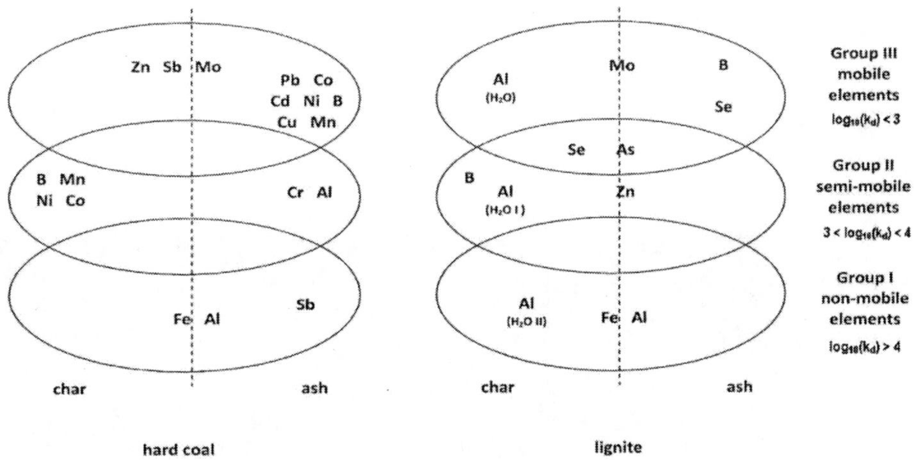


A. Strugała-Wilczek, K. Stańczyk, Leaching behaviour of metals from post-underground coal gasification cavity residues in water differing in mineralization, ***Fuel***, 2016, 173, 106-114

15

## Conclusions

- The majority of examined metals showed a tendency to elute to a higher extent from ash residue than from chars. Metals present in the ash derived from burned coal exhibited a greater affinity for the aqueous phase than the metals present in the ash residue from UCG
- The study allowed to classify the metals into three groups in terms of volatility (the most volatile metals are Hg, Se and Cd)
- Based on the estimated distribution coefficients, metals were classified into groups according to their mobility during their elution with water
- A higher relation to the water phase is shown by the residue of gasified lignite compared to residues from hard coal

16

Thank you for your kind attention

17

Imperial College
London

# Coupled Thermo-Chemical-Mechanical Modelling of Cavity Growth and Farfield Geomechanical Assessment in Underground Coal Gasification

Sevket Durucan
A. Korre, E. Andrianopoulos, G. Si, R. Govindan, J-Q. Shi, S. Franzsen*

mERG
Minerals, Energy and Environmental Engineering Research
Department of Earth Science and Engineering
Royal School of Mines
Prince Consort Road
London, SW7 2AZ, UK

Seamwell INTERNATIONAL
Seamwell International Ltd., 3 Shepherd Street, London, W1J 7H, UK

## Outline

- Thermo-Chemical-Mechanical Modelling
  - Subsurface gasifier layout
  - Chemical process model
  - Geomechanical process model
  - Comparative assessment of cavity growth, syngas composition and heating value for two lignites and a bituminous coal resource at depths of 400 and 600 m, using oxygen as the reagent
- Far field geomechanical modelling
  - Geological model
  - Geomechanical model
  - Optimisation of UCG panel and pillar sizes
  - Vertical displacement profiles over UCG panels

## Thermo-Chemical-Mechanical Modelling



## Thermo-Chemical-Mechanical Modelling of Cavity Growth

### Modelling framework

Schematic of the coupling process between the Aspen Plus™ and the FLAC3D™ software tools

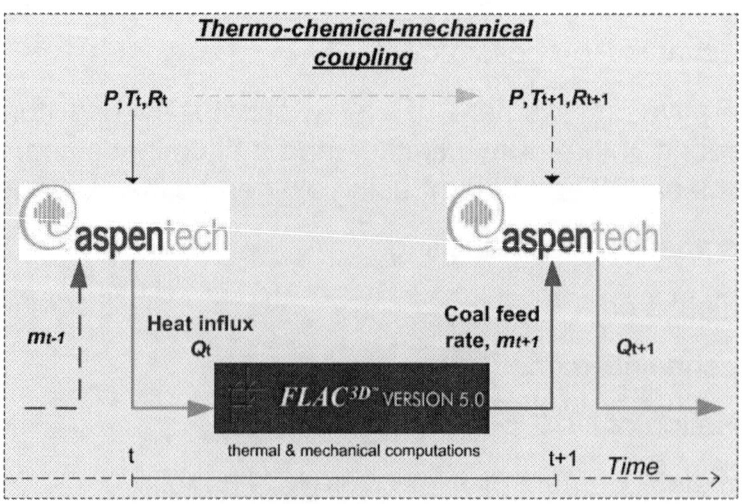

## Subsurface gasifier layout

Continuous Retracting Injection Point (CRIP) UCG method with linked vertical wells

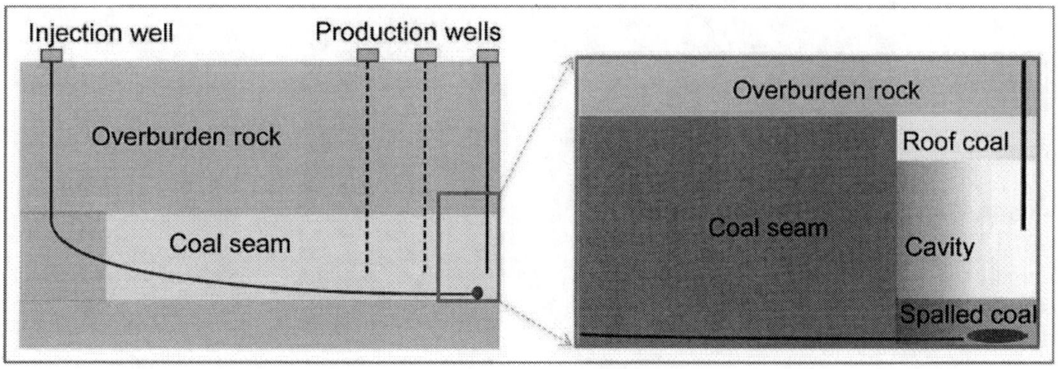






## Chemical process model

Aspen Plus$^{TM}$ chemical process model developed for the simulation of CRIP subsurface layout.

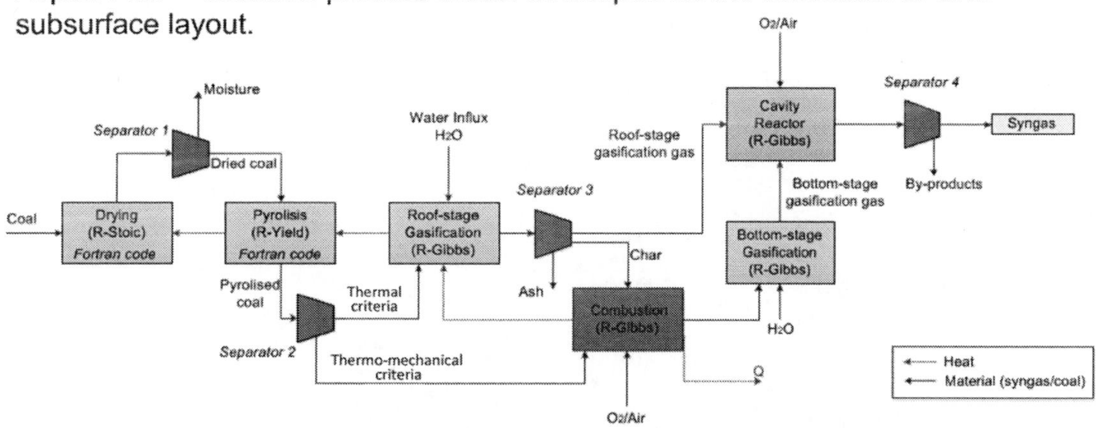


- The drying and pyrolysis stage models in Fortran in order to regulate the model performance
- Thermodynamic, mass and heat transfer modelling under equilibrium conditions to calculate the amount of gases produced
- Thermo-mechanically induced stresses and the new rock temperatures in FLAC3D.
- An additional module to calculate the convective heat transfer between the product gas and the cavity boundaries, as well as calculating the $O_2$ and $H_2O$ injection rates

# UCG model domain developed in FLAC3D™

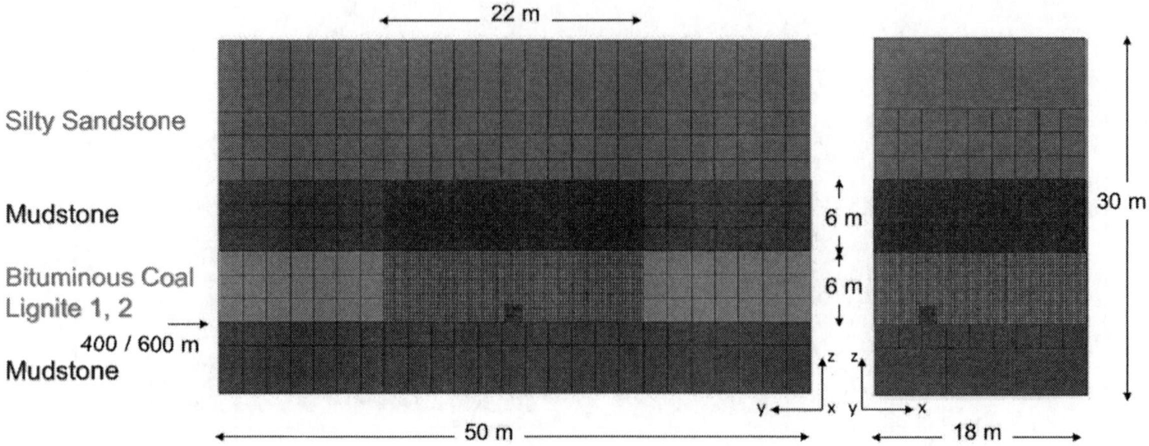






# Lignite, bituminous coal and rock properties

| | Lignite 1 | Lignite 2 | Bituminous Coal |
|---|---|---|---|
| *Proximate analysis (% w/w, dry basis)* | | | |
| Fixed carbon | 29.71 | 38.75 | 62.74 |
| Volatile material | 64.43 | 26.39 | 32.58 |
| Ash(g) | 5.86 | 34.86 | 4.68 |
| *Ultimate analysis (% w/w, dry basis)* | | | |
| Carbon | 56.10 | 49.29 | 79.51 |
| Hydrogen | 5.25 | 2.85 | 4.27 |
| Nitrogen | 0.72 | 0.89 | 1.26 |
| Sulphur | 0.74 | 0.13 | 0.31 |
| Ash | 5.86 | 34.86 | 4.68 |
| Oxygen | 31.33 | 11.98 | 9.97 |
| High Heating Value (dry basis, kJ/kg) | 19.73 | 16.17 | 30.97 |
| Moisture (% w/w) | 31.00 | 28.00 | 4.93 |

| | Unit | Mudstone | Silty Sandstone | Bituminous Coal | Lignite 1, 2 |
|---|---|---|---|---|---|
| *Mechanical Parameters* | | | | | |
| Young's modulus (E) | GPa | 3.60 | 30 | 2.30 | 0.85 |
| Tensile strength ($\sigma_i$) | MPa | 1.8 | 4.8 | 1.3 | 0.86 |
| Friction angle (g) | ° | 30 | 40 | 28 | 22 |
| Cohesion (c) | MPa | 5.20 | 18.68 | 4 | 2.90 |
| Poisson's ratio ($\nu$) | - | 0.35 | 0.2 | 0.34 | 0.42 |
| Density ($\rho$) | kg/m³ | 2,530 | 2,540 | 1,342 | 1,390 |
| *Thermal parameters* | | | | | |
| Linear thermal expansion coefficient (a) | °C$^{-1}$ | $0.873\times10^{-5}$ | $0.55\times10^{-5}$ | $0.65\times10^{-5}$ | |
| Specific heat ($C_p$) | J/(kg °C) | 985 | 852 | 1,066 | |
| Thermal conductivity ($\lambda$) | W/(m°C) | 1.76 | 2.4 | 0.28 | |

## Model results: Cavity growth along the x-z plane

Lignite 1, Lignite 2, Bituminous Coal seam depth 400 m at base

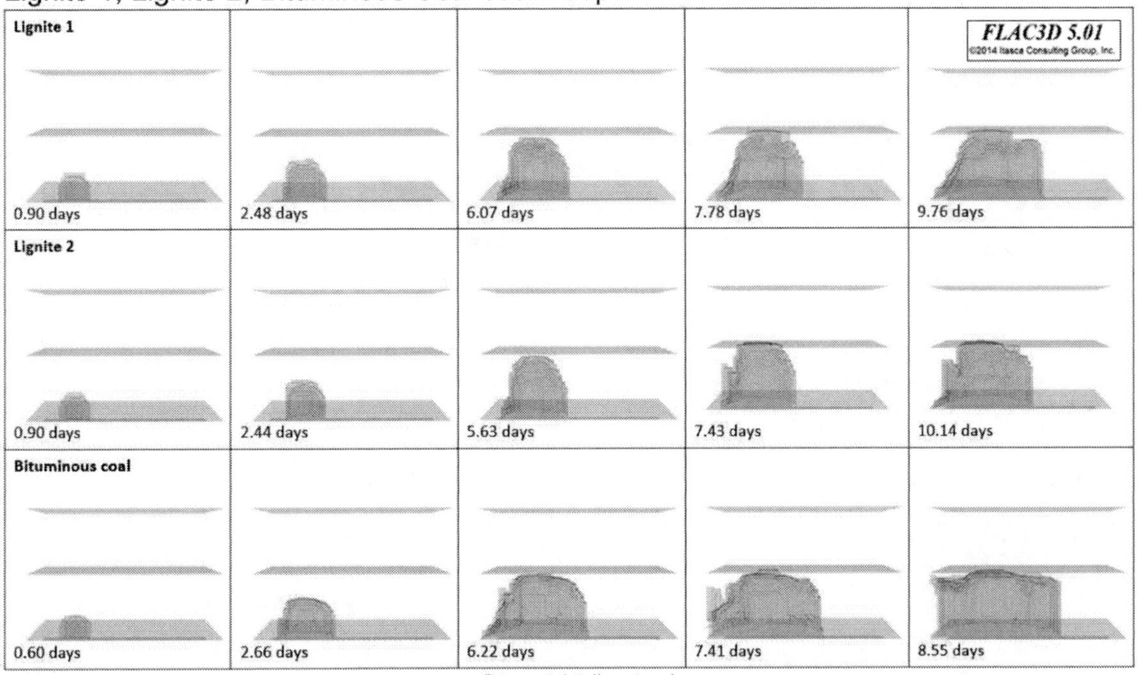




## Model results: Cavity growth and temperature profiles

Lignite 1 at a coal seam depth of 600 m at base

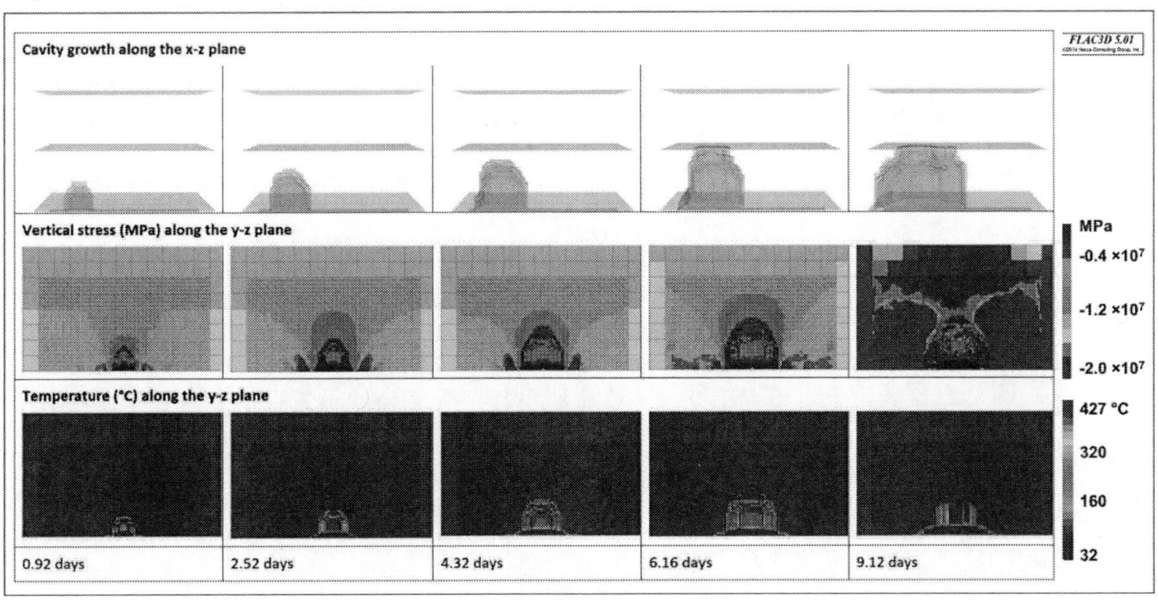

# Model results: Cavity dimensions over time

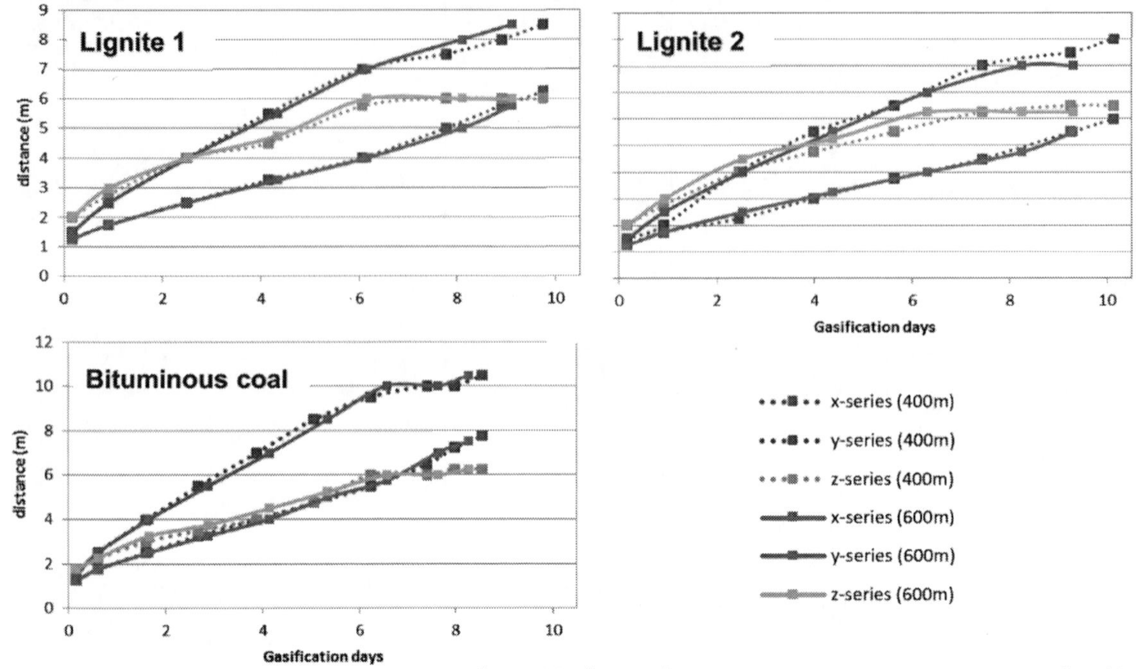






# Model results:
# Gas composition and heating value evolution

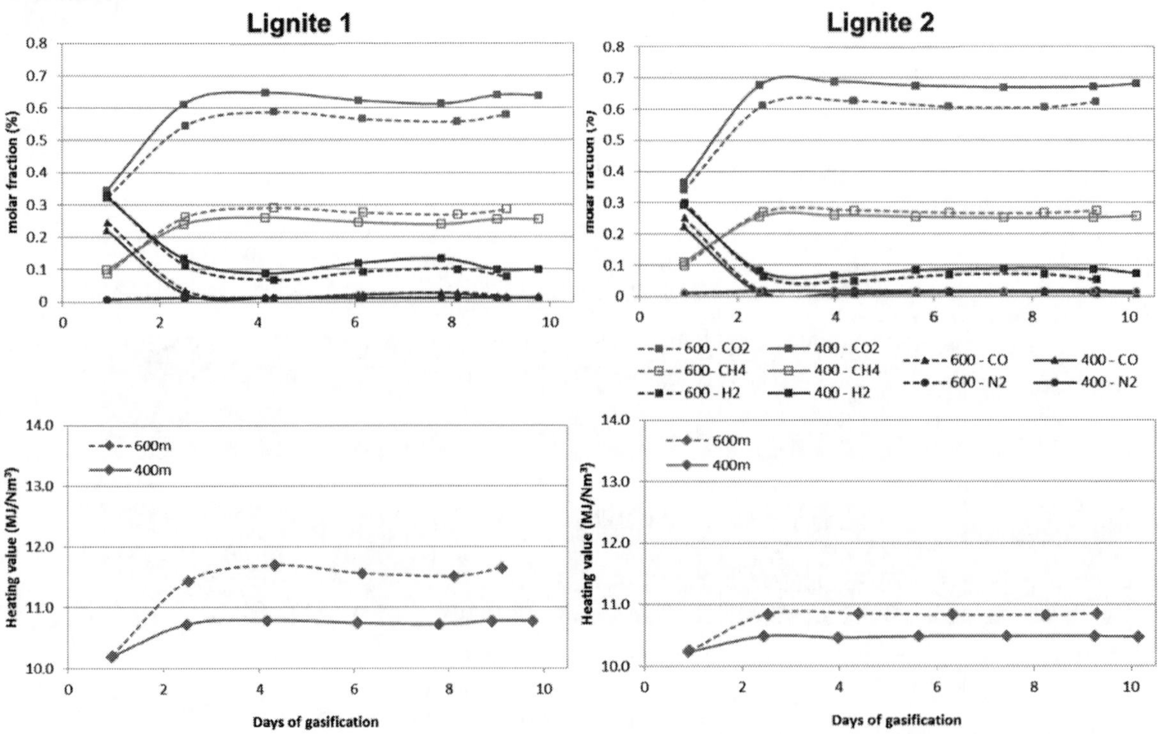

# Model results:
# Gas composition and heating value evolution

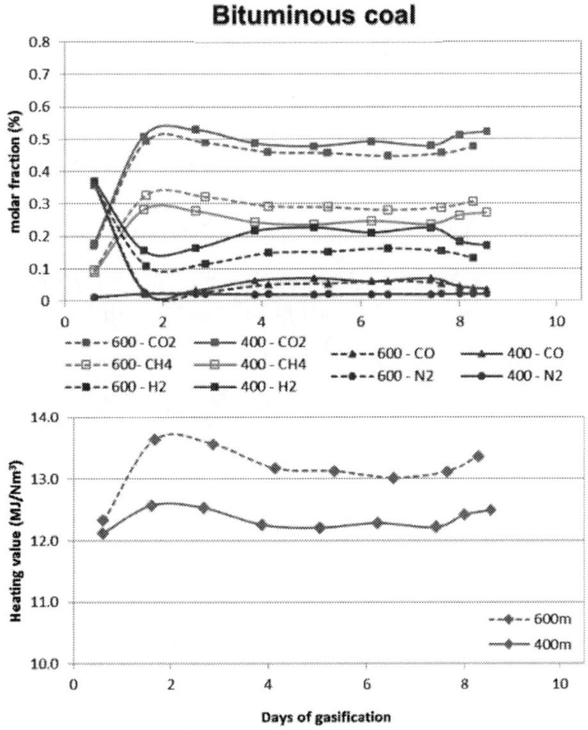


Page 13

---

## Far field geomechanical modelling

Page 14

## Geological model development

A coal field and surrounding geosphere model with an areal coverage of 7km × 5.5km was developed in ISATIS®.

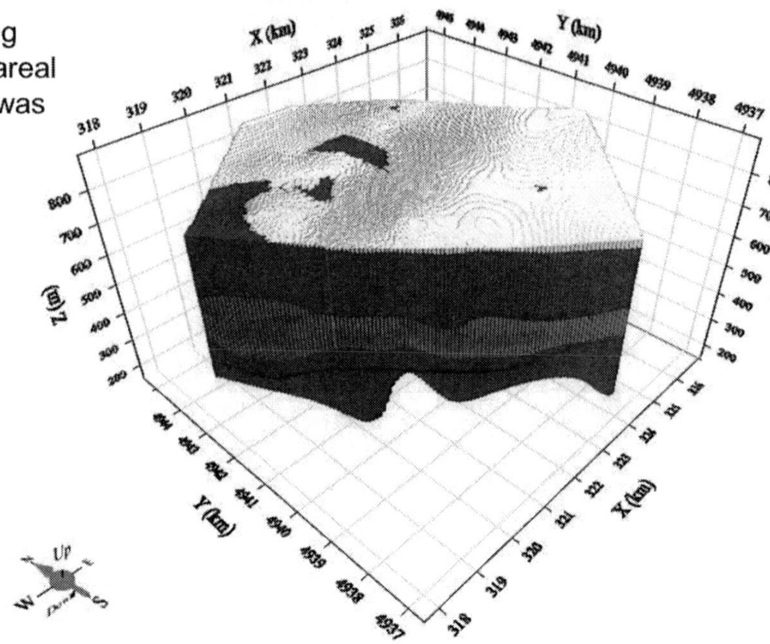


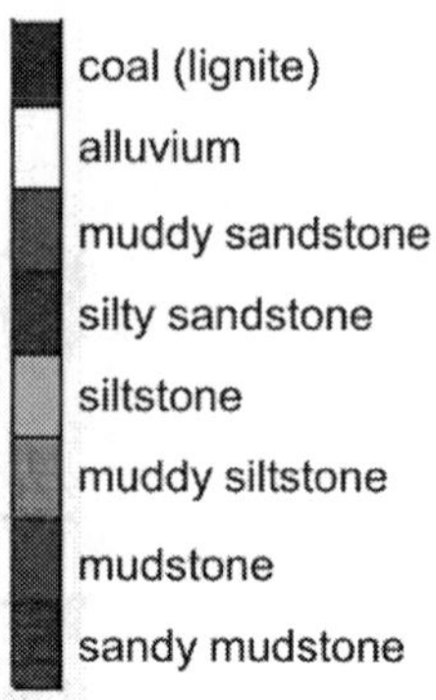






## Geological model development

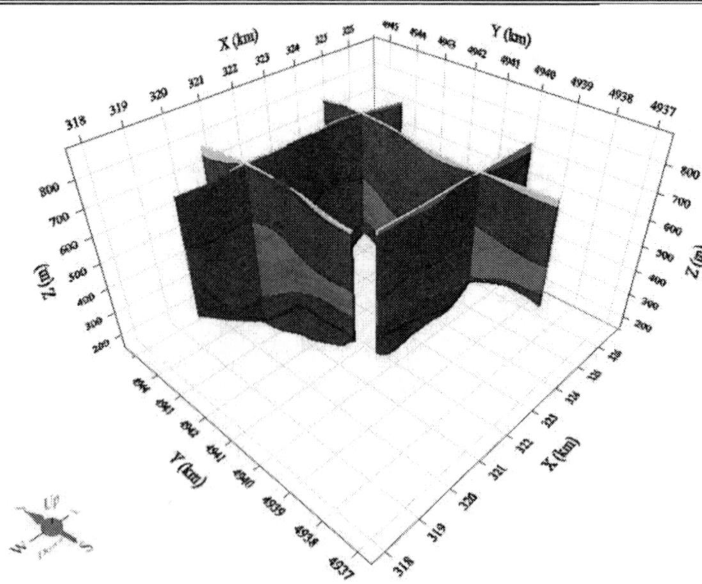


Three distinct geological formations, representing depositional environments varying from mudstone, shale, siltstone, and sandstone-type rocks and alluvium close to the surface.

## Geological model development

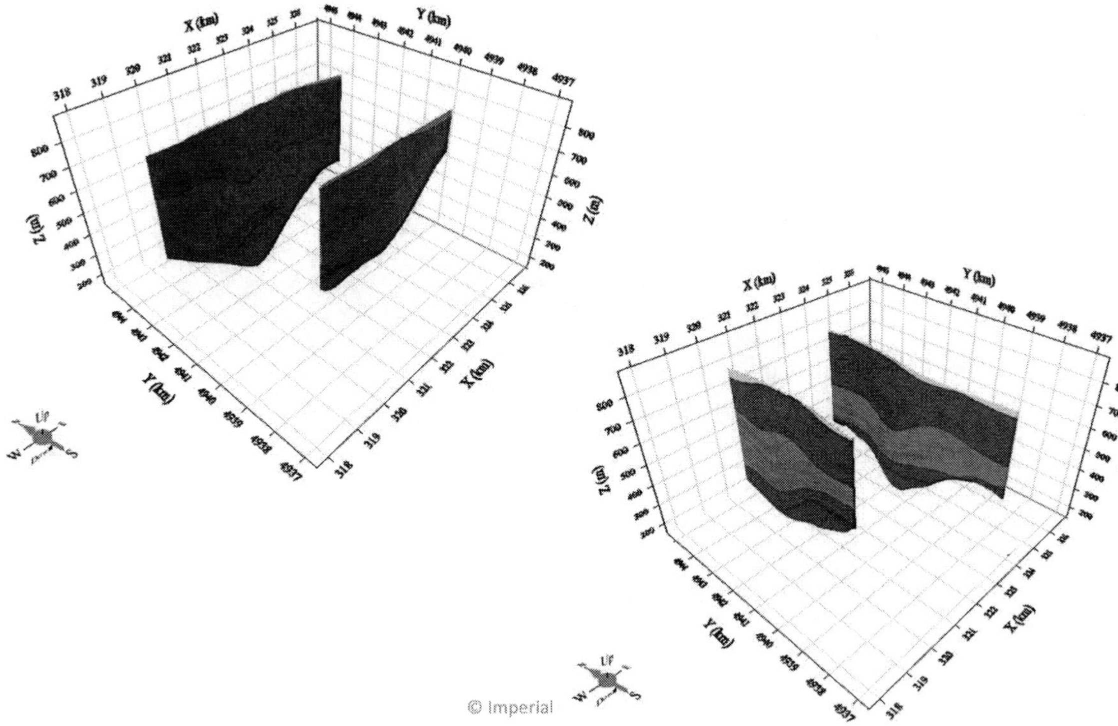


## Geological model development

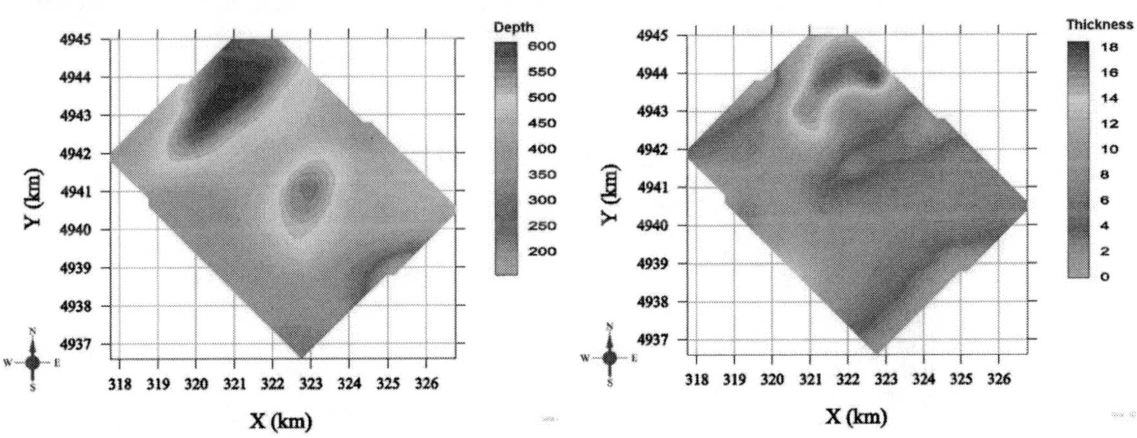


- The bottom-most layer primarily containing mudstone rocks hosts the target coal seam having thicknesses ranging between 0 m to 18 m, and depths ranging between 150 m to 600 m.
- The coal field model has a grid resolution of 50 m × 50 m × 1 m.
- The total tonnage of coal represented in the model is 415 Mt, of which approximately 372 Mt of coal correspond to seam thicknesses > 5 m.

## Geomechanical model set-up

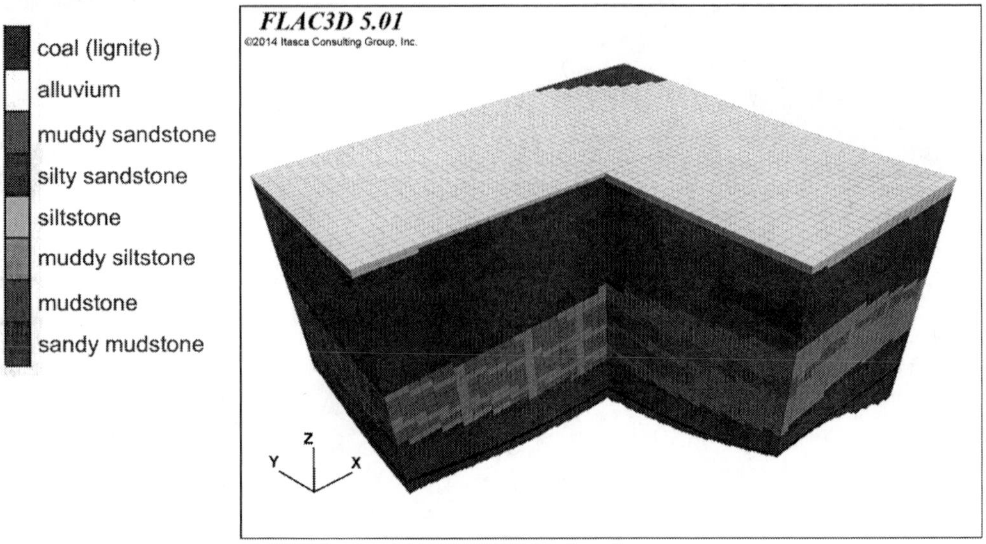


Based on the field-scale geological model, two contrasting rock types were distributed into each formation to represent a more realistic lithology in the overburden.





## Geomechanical model development

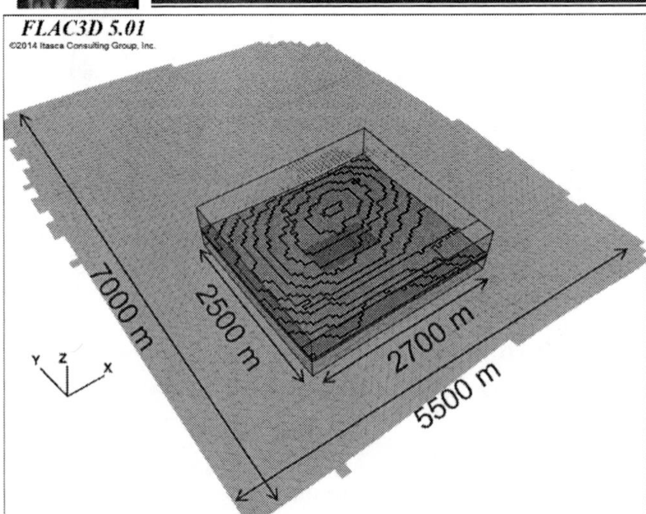


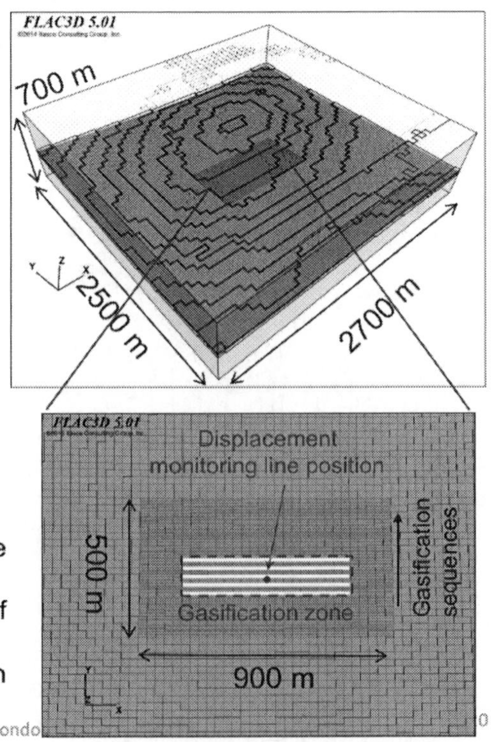


- A refined area was pre-defined at the centre of the model domain.
- 5 UCG panels: Each 600 m long with a number of panel and pillar sizes modelled.
- Vertical displacement along the centre line of each gasification zone was monitored.

## Geomechanical model set-up

| Layer | Constitutive Model | Young's modulus (GPa) | Poisson's ratio (-) | Internal frictional angle (°) | Cohesion (MPa) | Tensile cut-off (MPa) |
|---|---|---|---|---|---|---|
| Coal | M-C | 0.71 | 0.40 | 22 | 2.90 | 0.86 |
| Alluvium | Elastic | $1.03 \times 10^{-3}$ | 0.29 | -- | -- | -- |
| Muddy sandstone | Elastic | 21.30 | 0.15 | -- | -- | -- |
| Silty sandstone | Elastic | 30.00 | 0.10 | -- | -- | -- |
| Siltstone | M-C | 12.90 | 0.20 | 40 | 18.68 | 4.80 |
| Muddy siltstone | M-C | 5.90 | 0.25 | 35 | 12.45 | 2.50 |
| Mudstone | M-C | 3.60 | 0.35 | 30 | 5.20 | 1.80 |
| Sandy mudstone | M-C | 4.80 | 0.30 | 25 | 7.40 | 2.30 |

- Boundary conditions: sides roller; base fixed
- *In-situ* vertical stress was assumed to be induced by the weight of the overburden.





## Optimisation of UCG panel sizes

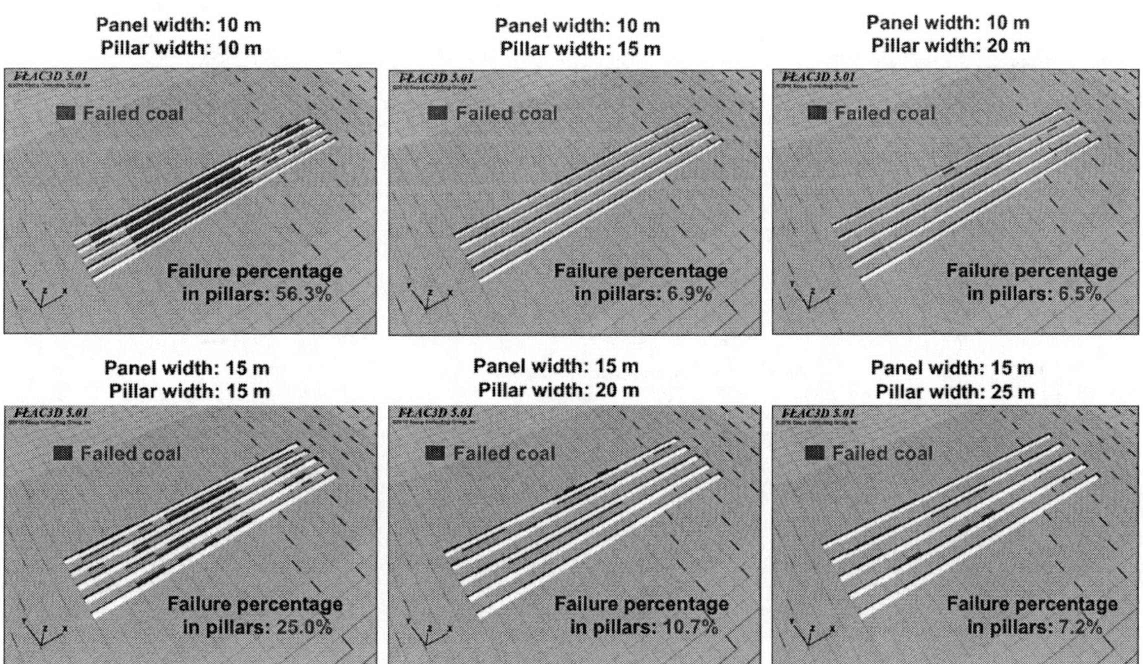

# Optimisation of UCG panel sizes

Panel width: 30 m
Pillar width: 15 m

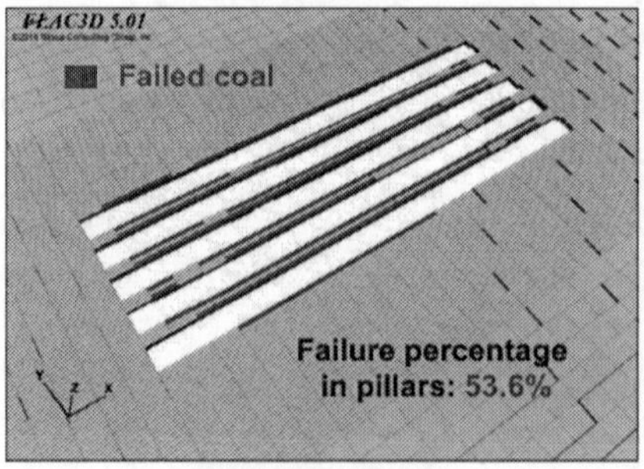


Panel width: 30 m
Pillar width: 20 m

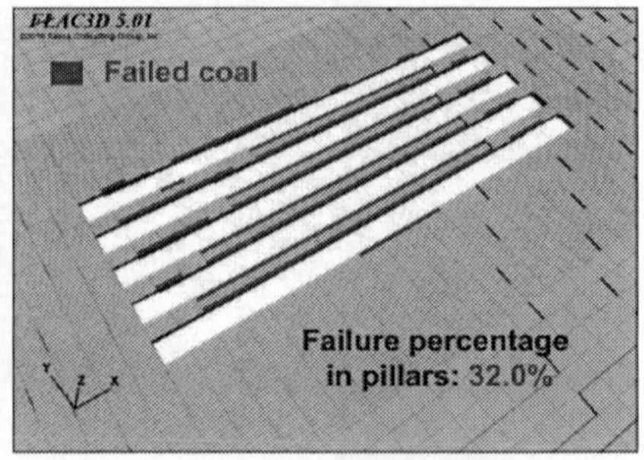


Panel width: 30 m
Pillar width: 25 m

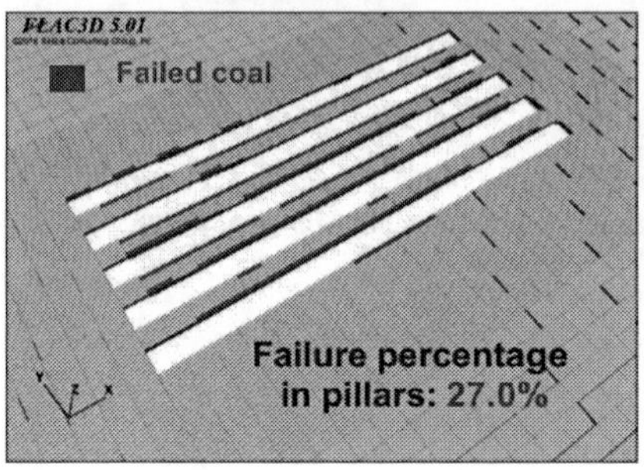


Panel width: 30 m
Pillar width: 30 m

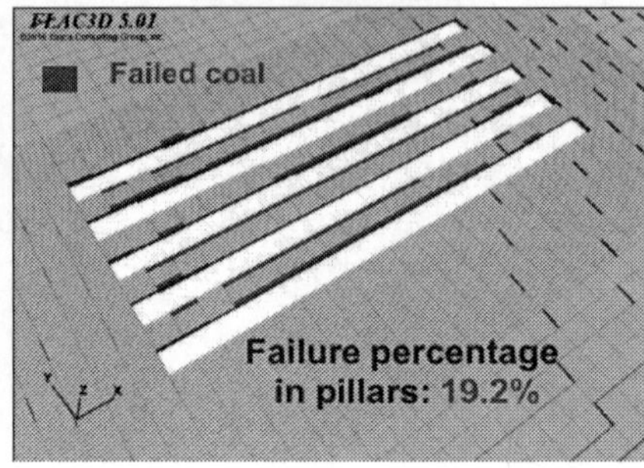


23

# Vertical displacement profiles over UCG panels

Panel width: 10 m
Pillar width: 10 m

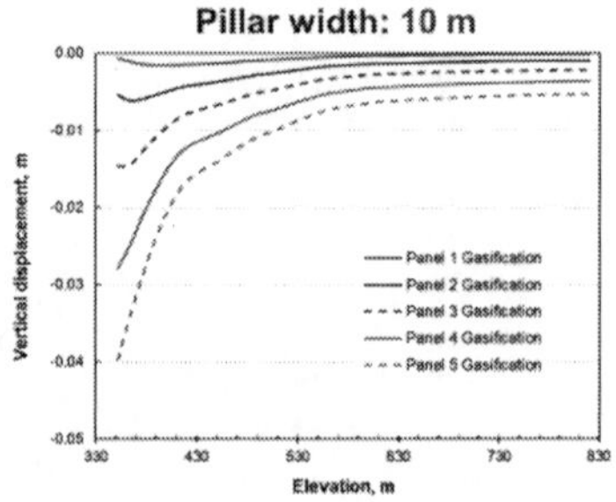


Panel width: 10 m
Pillar width: 15 m

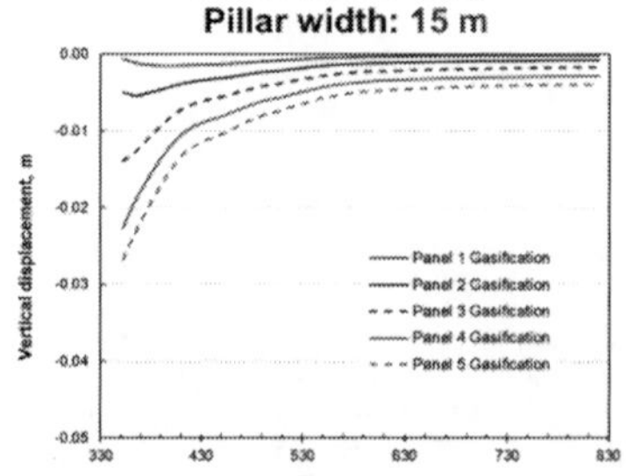


Panel width: 10 m
Pillar width: 20 m

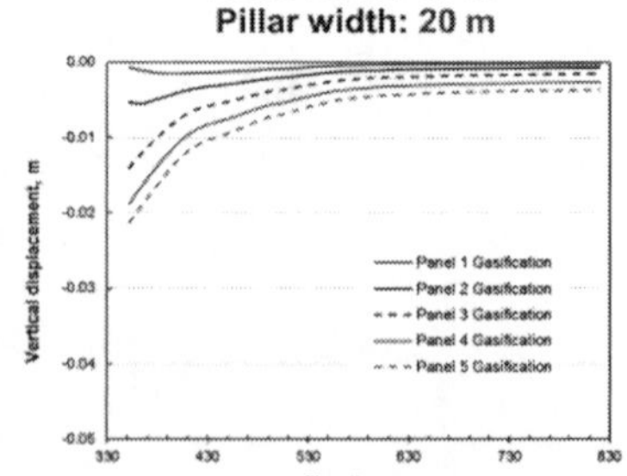


Panel width: 15 m
Pillar width: 15 m

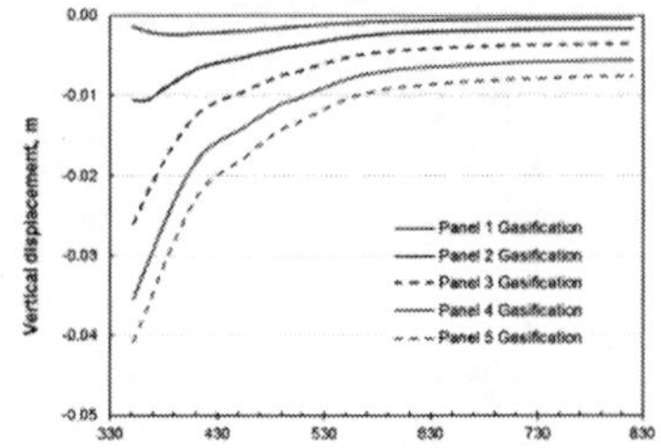


Panel width: 15 m
Pillar width: 20 m

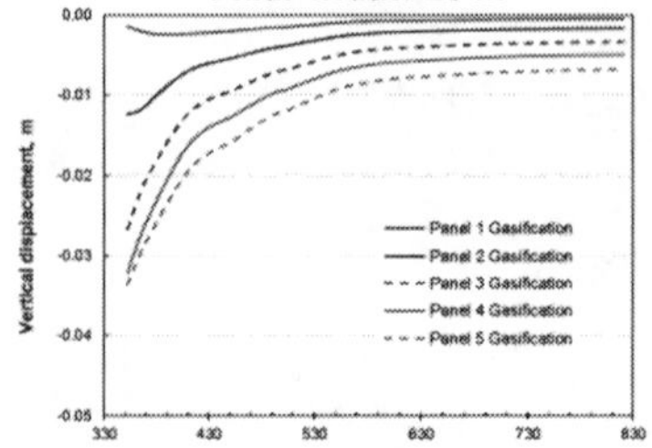


Panel width: 15 m
Pillar width: 25 m

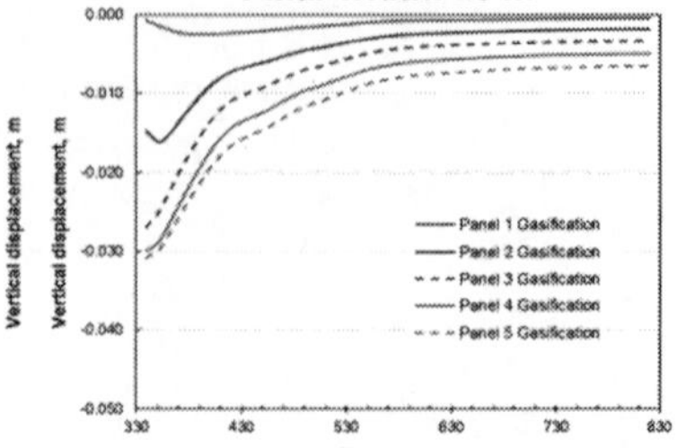




Page 24

## Vertical displacement profiles over UCG panels

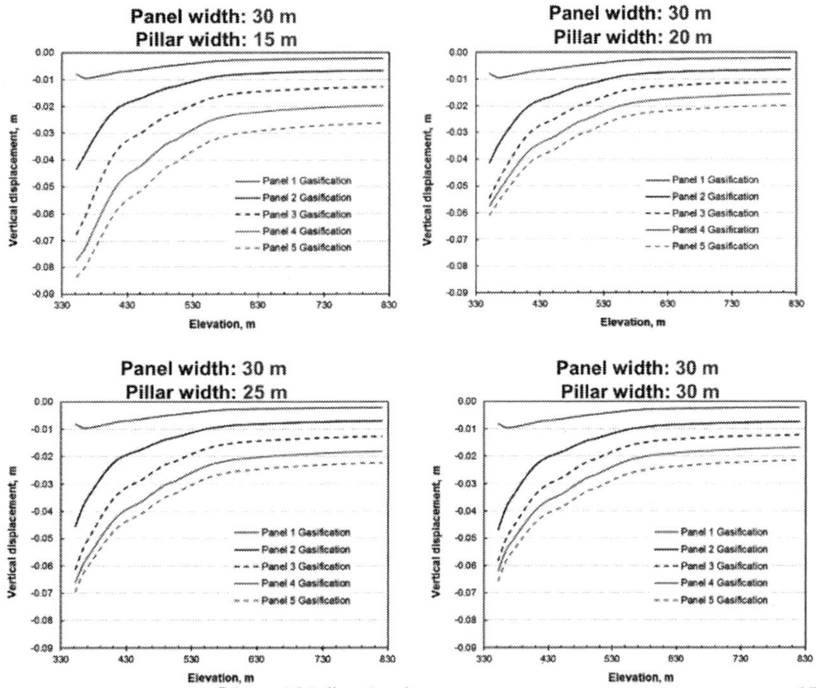






## Vertical displacement profiles over UCG panels

Comparison of different UCG panel and pillar size performance (vertical displacement after full field gasification):

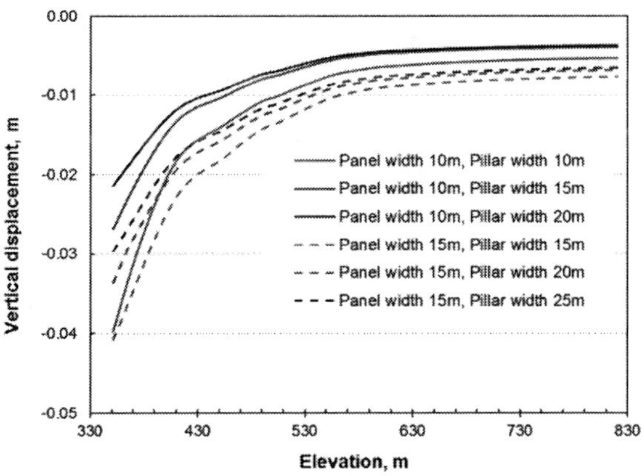


Narrow panel sizes

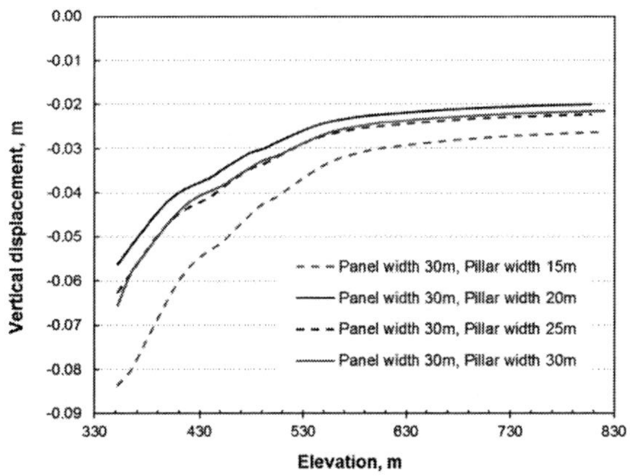


Wider panel sizes

In general, vertical surface displacement decreases with increased pillar sizes, however, there is an optimum pillar size beyond which there may not be further reduction in surface subsidence.

# Acknowledgements

**This project is receiving funding from the European Union's Seventh Framework Programme for research, technological development and demonstration under Grant Agreement No. 608517 ("Technology Options for Coupled Underground Coal Gasification and $CO_2$ Capture and Storage")**

# Avdan Lignite Mine Open Pit Planning and Design Studies for Thermal Power Plant

*C. Atilla Ozturk**,
*Murat Ozkan, Mohammad Lashgari, Selamet Ercelebi*

**Assoc. Prof. Dr. Istanbul Technical University, Faculty of Mines, Turkey, atilla.ozturk@itu.edu.tr*

**Abstract**

Sustainability of a thermal power plant shall be conditioned by a well-planned coal mine production. According to thermal power plant capacity, annual production target should be decided based on the quality of coal deposit. Average lower calorific value of coal deposit can be used to calculate the annual production target. Once the target is resolved, mine planning and design studies can be initiated. First selection must be performed between open pit and underground mining production. In this study, Avdan lignite site located in the western part of Denizli, Turkey was selected as a case study area. Average lower calorific value is modelled based on the geostatistical applications and determined as 1,719 kcal/kg while resource and reserve calculations were also performed by using integrated mining software and found as 30.5 MTon and 26.4 MTon, successively. The capacity of the thermal power plant is decided as 150 MWe, so annual lignite production target is selected as 1.5 MTon/year based on the thermal plan capacity assumptions that will be discussed in this paper in detail. Open pit mine is selected as production method due to the depth and economic reasons. Firstly geotechnical and slope stability investigations were applied at the site to design bench and overall slope geometrical properties such as bench height, width, angle, and overall slope angle related with the production depth. Once the stability studies are completed, planning and design studies then can be commenced. Optimization should be performed to select the most suitable place for initial production location minimizing the stripping ratio especially in the early years of the production to reduce the duration for returning on investment earlier. Design and planning studies although it is involuntarily is like a trial-and-error method to lower the stripping ratio that is the most important factor for unit cost of a production as well as investment amount. After a series of trial-and-error studies based on optimization techniques, open pit mine planning projects were performed annually for the first 5 years of the production, and the rest of the production planning projects were performed based on 5 years' production target. According to the reserve and production target, open pit mine will serve to the thermal power plant nearly for 18 years. Final open pit limit and design was also performed in the study as a main output of this research. Target and designed production amount as well as annual and cumulative stripping amount were determined according to all open pit mine planning and design studies. Finally, 26.4 MTons of coal will be produced after stripping 380 $Mm^3$ overburden and 50 $Mm^3$ interburden materials in 18 years that makes this project will be important open pit mine project in the western part of Turkey to supply coal for the 150 MWe capacity thermal power plant.

Key Words: Coal, Energy, Mine Planning and Design, Stability

## 1. Introduction

Coal increases its materiality due to the increasing growing and production rate for Turkish Economy as an important energy source. Thermal power plants will use to produce energy while coal is the main raw material for those plants. Production of coal from local sources is of the essence for the sustainability of energy production from the plants. Feasibility of thermal power plants is other critical issue for the sustainability, and the plants are generally constructed at least for 25-year energy production.

150 MWe-capacity thermal plant will be constructed in the city of Denizli, Turkey. The location is in the western part of Turkey as can be seen in the Figure 1. The plant will become neighbors to a coal deposit called as Avdan Lignite Mine Site. Data for coal quality were determined from 103 boreholes that 98 of them cut the coal zone. Lower calorific value, total sulphur, humidity, ash, volatile matter, and density were investigated to understand the quality of coal at the site. Two main coal structures were determined in the site called as Sekkoy and Yenidere based on the geology. Reserve calculations as well as quality distribution maps were obtained from integrated mining software, Micromine. Coal site topography and boreholes can be also seen in Figure 1. Details about the coal quality and reserve estimation is given in this study, as well.

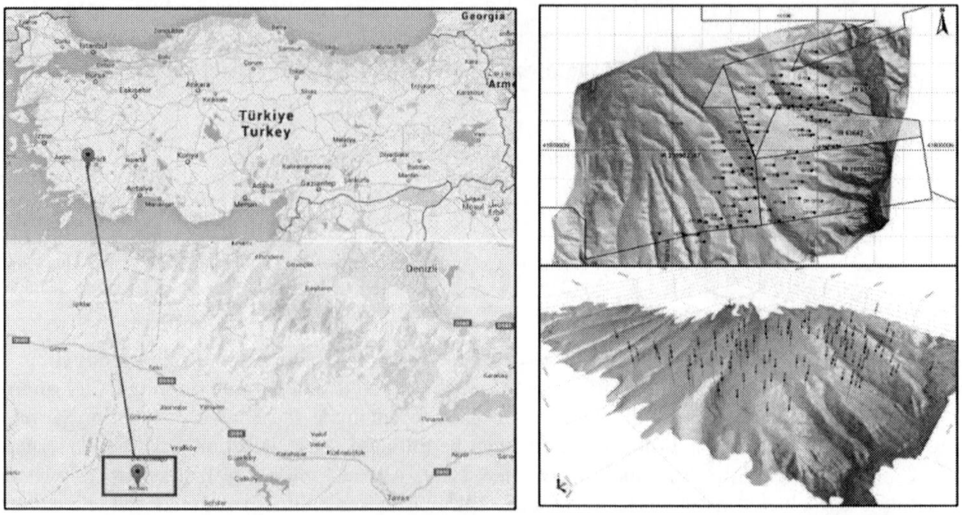


Figure 1. Google Earth view, topography, and boreholes of the study area.

Annual coal production target must be decided according to the capacity of thermal power plant as well as the average lower calorific value of coal. Following assumptions for the plant are used to calculate the annual coal production target. Annual working hours of the plant is 7500. Loss of production is calculated as 5% for coal production. Average lower calorific value is 1719 kcal/kg for Sekkoy and Yenidere coal zones. Production yield of the plant is 40%. Based on these general assumptions, annul target of the coal mine was decided as 1.5 million ton.

In this study, the procedure applied to plan and design open pit lignite mine in the Avdan coal region is given and the outputs of the planning and design studies are shared.

## 2. Brief Geology

The following information on the geology can be used to understand the geological structures in the coal mine site.

*i) Strata (Tm)*

The basis of the area is formed by Paleozoic - Mesozoic quartz , marble, slate, limestone, creates, radiolarite and ultramafic rocks.

*ii) Neogene Sediments*

Tertiary old rocks overlie on strata rocks. It is divided into two groups. They are Akçay Group (Karadere, Mortuma, Yenidere, Kale and Künar Formation passed through it and Muğla Group (Turgut, Sekköy, Yatağan and Milet Formation).

- Akçay Group: License area and its surroundings are represented Yenidere Formation.

Yenidere Formation (Ty): On Oligosen old Mortuma Formation there are angular unconformities. Thickness is 250 - 1450 m. There are three member divisions which are Esenkaya (Tye), Ortaköy (Tyo) and Karakaya (Tyk).

Esenkaya is consisting of large conglomerate, mudstone, sandstone and siltsone and has a thickness of 25-250m.

Ortaköy is consisting of sandstone, siltstone, claystone and conglomerate. Thickness in the central part of it, is up to 2.90 m has reached lignite containing lignite horizons. The thickness of this member varies between125-600 m.

Karakaya; is consisting of claystone and siltstone and totally has a thickness between 70 - 100 m.

Yenidere Formation is consisting of braided meandering rivers, coastal plains, lagoons, alluvium, and sand which are deposited in alluvial environments. Age is the Lower Miocene.

- Muğla Group: There unconformities overlaying on Akçay Group. Licence area and its surroundings are can be presented by Sekköy ve Yatağan formations.

Sekköy Formation (Tms): Sekköy Formation near Avdan and Omurca exist upon Yenidere formation as unconformities and has a thickness of 50 - 150 m. Yellowish white clay - Micritic - clastic limestone, gray siltstone, is composed of tuffs and tuff. From formation base there are lignite seams with the thickness of the 0.05 – 2.20 m. Sekköy Formation was deposited in swamps and lake environments.

Yatağan Formation (Tmy): is upon Sekköy with unconformities. Composed of Conglomerate, mudstone, sandstone, tuff ve tuffite. Thickness is around 250 m. It is deposited in alluvial environment.

Milet Formation (Tmm) is upon Yatağan with unconformities. Licensed area and surroundings constitutes of plain and high plateaus. White yellowish beige colored limestone and the thickness of the unit consisting of clay and limestone is about 100 to 120 m. The age of the formation of sediments of the lake environment is Pliocene.

*iii) Quaternary*

In river beds gravel, sand, materials consisting of silt and clay have been seen. In Figure 2, license area and the surroundings geology is shown. There are numerous studies have been carried out since 1869. Further readings can be found in the literature such as Tchihatcheff (1869), Philippson (1915), Altınlı (1955), Şimşek (1982 ), Hakyemez (1989), Sözbilir et al. (2000), Alçiçek et al. (2000), Bayhan-Hasdiğen (2003).

Figure 2. Geological map of the study area.

## 3. Reserve Estimation

Coal at Avdan site is a part of delta formation, so it is not possible to correlate single or multi-layer correlated coal seams. Because of that, coal zones instead of coal seams are defined for Sekkoy and Yenidere formation, separately. Coal zone boundaries are determined based on international standards. Coal thickness's threshold value for production is taken as 40 cm while lower calorific value is taken as 950 kcal/kg. Solid models for Sekkoy and Yenidere coal zones were determined from the borehole database, and geostatistical approaches used to obtain the coal quality distribution maps. Solid 3D models can be seen in Figure 3.

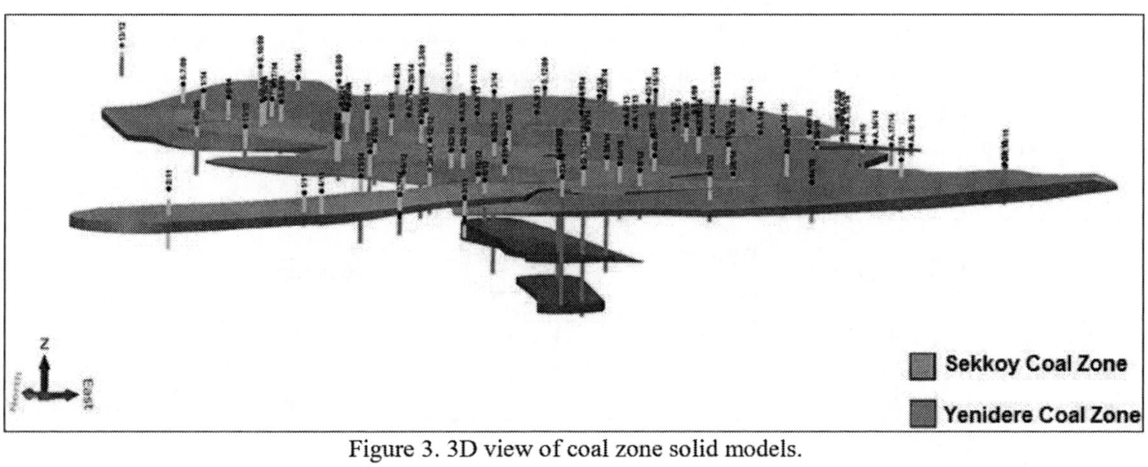


Figure 3. 3D view of coal zone solid models.

Coal thickness, lower calorific value, ash, sulphur, humidity, and volatile matter distribution maps are than determined based on geostatistical approaches. Lower calorific value distribution maps for Sekkoy and Yenidere coal zones are given in Figure 4 and 5 as per samples of these studies.

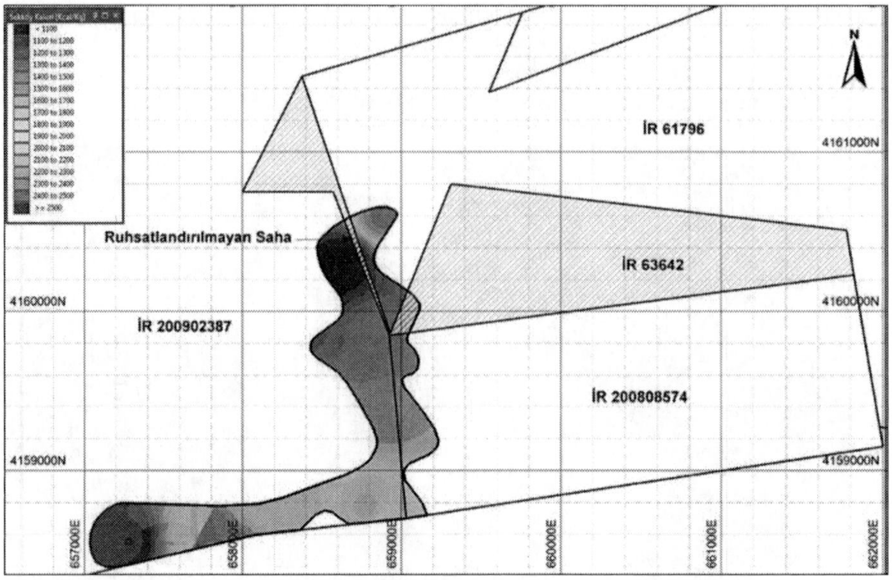


Figure 4. Sekkoy coal zone lower calorific value distribution map.

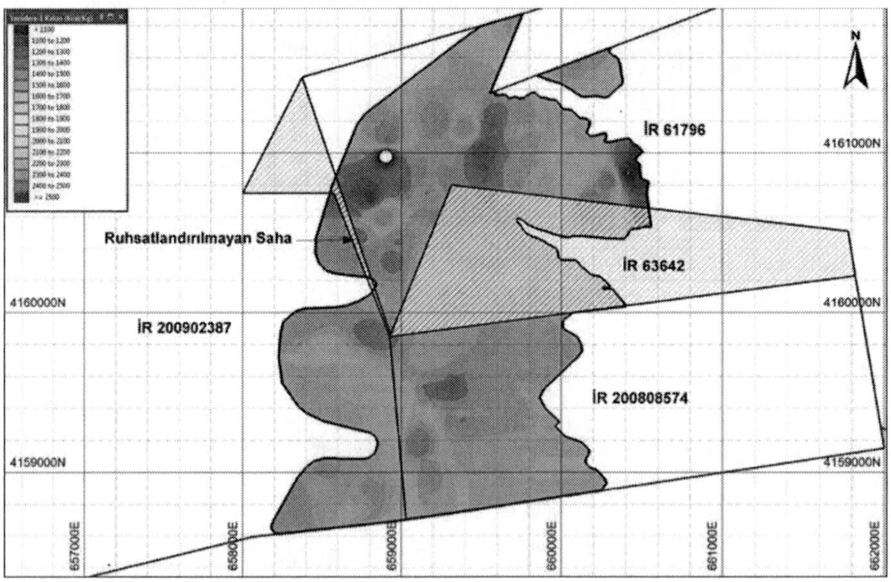


Figure 5. Yenidere coal zone lower calorific value distribution map.

After all of the reserve and quality estimation studies following summary table is determined to show the reserve and the average lower calorific value at Avdan coal site.

Table 1. Reserve calculation results.

| Coal Zone | Reserve (Ton) | Density (gr/cm$^3$) | Average Calorific Value (kcal/kg) | Breakdown |
|---|---|---|---|---|
| Sekkoy | 4,978,236 | 1.54 | 1,465 | 16% |
| Yenidere | 25,494,492 | 1.48 | 1,768 | 84% |
| Total | 30,472,728 | | | 100% |

### 4. Open Pit Mine Planning and Design

Planning and design of an open pit mine has a multi layer work-flow and a simplified example is given in Figure 6. After determining geological aspects as well as reserve and resource modelling of coal site, 3 boreholes were drilled to obtain rock and soil types samples required for geotechnical analyses. Geotechnical studies were then performed to obtain mechanical and physical properties of rock samples. Hence, slope stability investigations could be initiated. Stabile bench designs were achieved from the stability analyses performed based on limit equilibrium analysis method. Summaries complied from geotechnical and slope stability studies ca be found in Table 2, Table 3, and Table 4, successively.

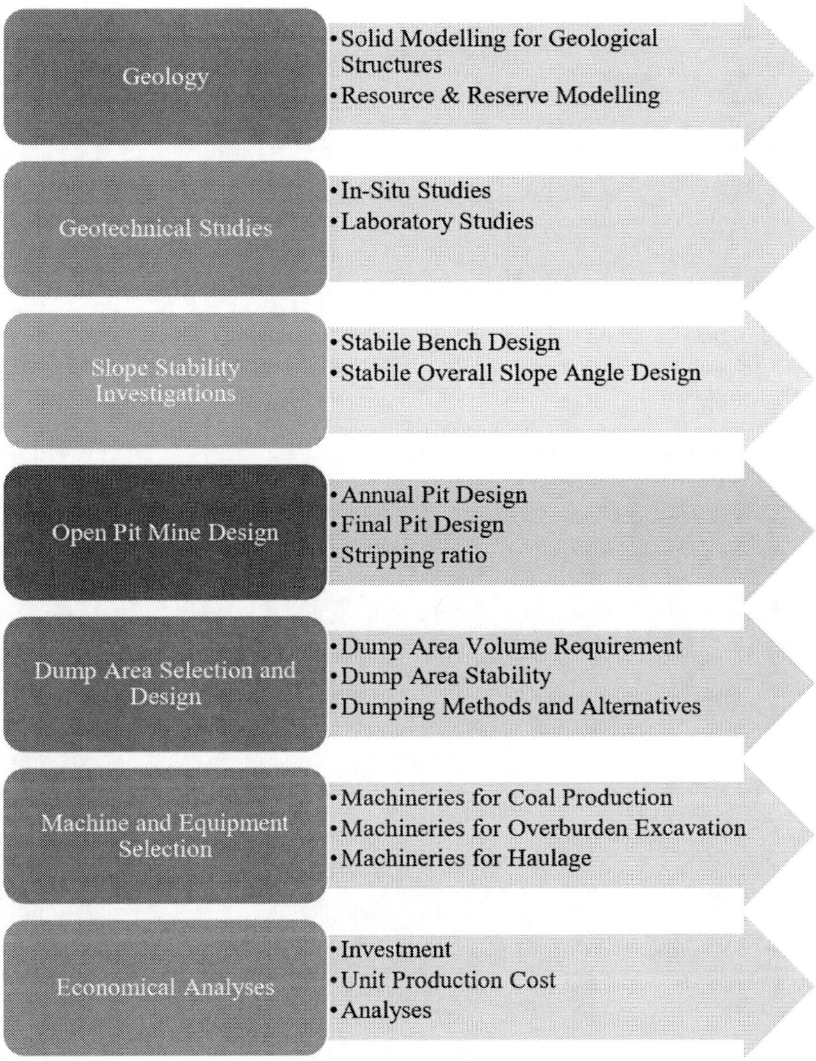


Figure 6. Work-flow diagram applied in this project.

Table 2. Summary table from geotechnical studies.

| Formation | Rock Type | Cohesion (kPa) | Internal Friction Angle ($^0$) | Unit Volume Weight (kN/m$^3$) |
|---|---|---|---|---|
| Sekköy | Claystone-Siltstone | 15 | 25.77 | 19.78 |
| | Marl - Limestone | 380 | 42.61 | 14.77 |
| Yenidere | Claystone-Siltstone | 25 | 23.73 | 19.52 |
| | Sandstone - Marl | 160 | 16.91 | 18.49 |
| | Lignite | 220 | 29.93 | 12.93 |
| Yenidere | Conglomerate | 1050 | 57.50 | 20.00 |

Table 3. Summary table from bench design studies.

| Formation | Rock Type | Angle ($^0$) | Height (m) |
|---|---|---|---|
| Sekköy | Claystone-Siltstone | 60 | 5 |
| | Marl - Limestone | 75 | 15 |
| Yenidere | Claystone-Siltstone | 75 | 7 |
| | Sandstone - Marl | 75 | 15 |
| | Lignite | 75 | 15 |

Table 4. Summary table from overall slope angle studies.

| Total Height (m) | Overall Slope Angle ($^0$) | | | |
|---|---|---|---|---|
| | Sekköy Formation Max. | Yenidere Formation Max. | Average Structures Developing Slopes | Average Structures Final Slopes |
| 130 | 17 | 17 | 17 | 17 |
| 100 | 18 | 17 | 18 | 18 |
| 50 | 21 | 22 | 22 | 22 |
| 25 | 26 | 34 | 32 | 28 |

Once benches were designed based on stability and the annual production amount set as a target, open pit mine design studies could be initiated. First 5 year of the production designed as per annually while the remaining production schedule designed as per 5-year periods. Final pit design devised for the production of the remaining coal in the coal site that will be the 18$^{th}$ production year. Designs for the open pit mine which were carried out by using integrated mining software can be found in Figure 7.

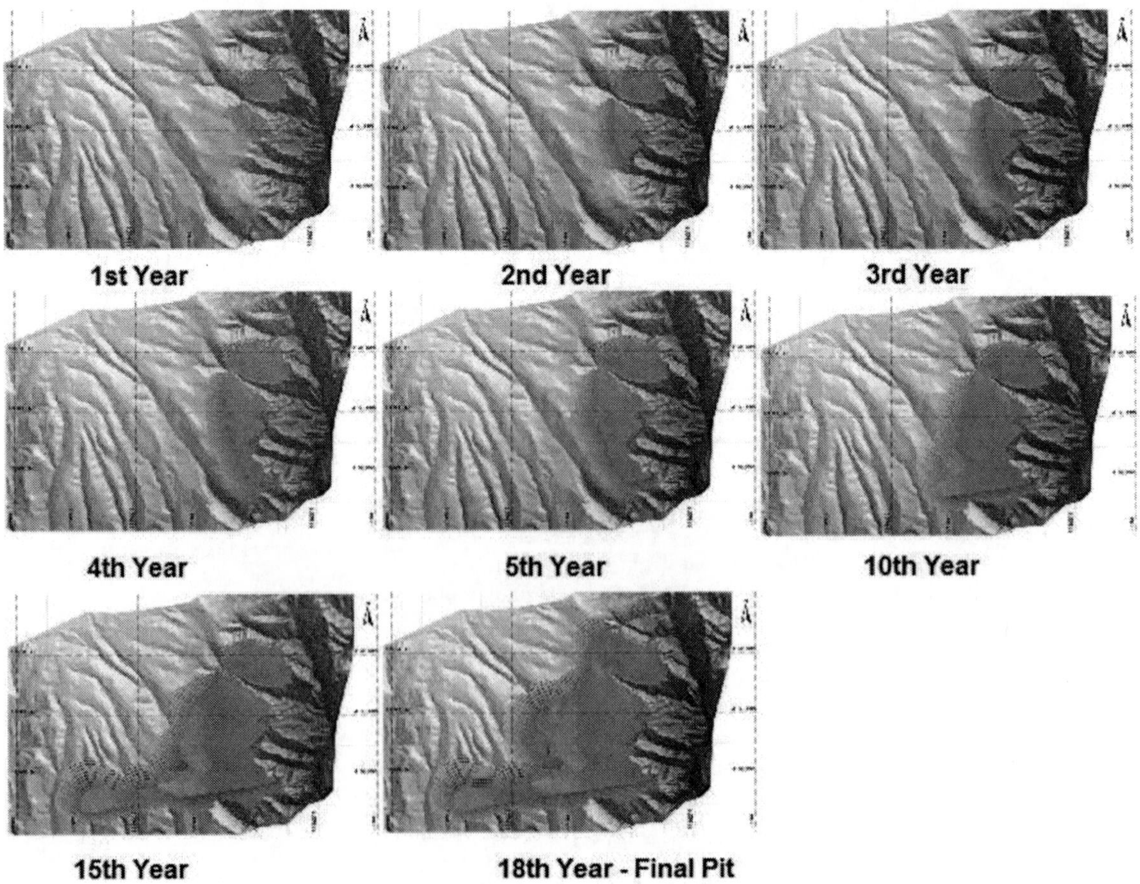


Figure 7. Open pit mine designs.

Table 5 prepared based on the data gathered from the open pit design studies. In order to eliminate the difficulties of high investment, optimization was applied to obtain lowest stripping ratio in the beginning of the project.

Table 5. Cumulative coal production and overburden and interburden removal amount from designs.

| Year | Coal Production (MTon) | | Cumulative Over-Interburden Removal ($Mm^3$) | | | Stripping Ratio ($m^3$/ton) |
|---|---|---|---|---|---|---|
| | Target Production | Design Production | Overburden | Interburden | Total | |
| 1 | 1.5 | 1.5 | 15.7 | 1.7 | 17.4 | 11.6 |
| 2 | 3.0 | 3.0 | 35.0 | 4.2 | 39.2 | 13.0 |
| 3 | 4.5 | 4.5 | 54.6 | 8.2 | 62.7 | 13.9 |
| 4 | 6.0 | 6.0 | 80.0 | 11.4 | 91.4 | 15.2 |
| 5 | 7.5 | 7.7 | 97.5 | 15.7 | 113.2 | 14.8 |
| 6-10 | 15.0 | 14.9 | 189.7 | 30.3 | 219.9 | 14.7 |
| 11-15 | 22.5 | 22.1 | 309.2 | 42.6 | 351.8 | 15.9 |
| 16-18 | 26.5 | 26.5 | 378.6 | 53.3 | 431.9 | 16.3 |

Dump area design is also a part of open pit planning and design studies. A waste dump area for the overburden and interburden excavated material was designed for the first 5 year of the production. After 5 years, the mine will use internal dump area instead of outside dump area. Design of waste dump area can be seen in Figure 8.

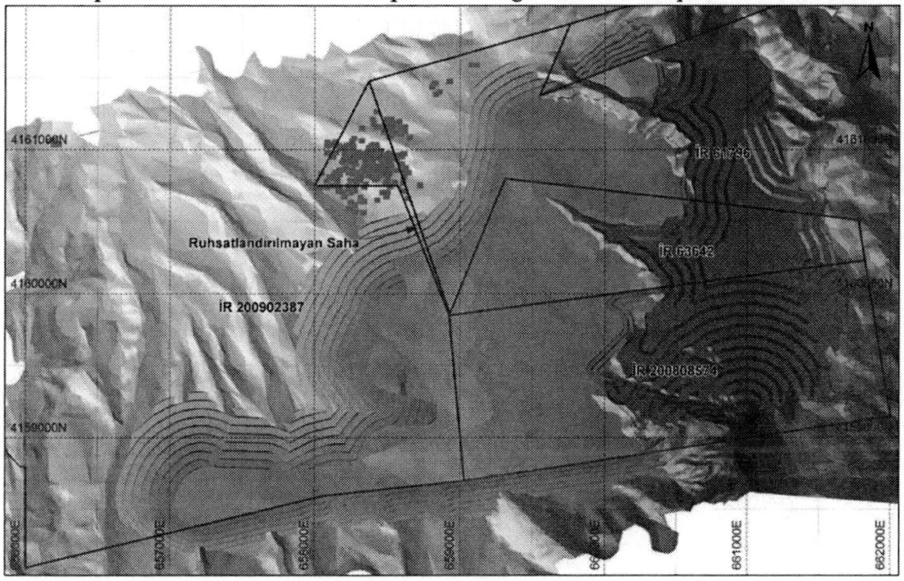


Figure 8. Dump area design.

## 5. Results

An open pit mine design and planning project phases was outlined in this research. A coal mine as a case study area was projected located in Avdan village of Denizli City, Turkey. Annual production target decided as per coal quality and the capacity of thermal power plant adjacent to the coal mine. Resource and reserve modelling was the first stage as well as geology investigations to understand the quality and quantity of coal. After that 1.5 MTon annual coal productions set as a target. According to the target, open pit mine project was initiated. Geotechnical and slope stability investigations were carried out to construct stabile benches in the open pit. Once stability investigation completed, then open pit mine planning and design works could be started. Most important aim of this study is to lower the stripping ratio especially for the first production years to lower the initial investments of the project. The procedure applied in this study will be used as a common procedure for the scientist and engineers especially working for open pit mining design works.

**Acknowledgments**

The research was a part of the industrial project supported by Avdan Mining and Energy Company (Avdan Madencilik Enerji San. ve Tic. A.Ş.). The authors gratefully acknowledge the assistance and cooperation provided by the Company.

**References**

Alcicek, M., Kazanci, N. and Ozkul, M. 2000. *Cameli Acipayam (Denizli, GB Turkiye) Neojen havzasi ve tortul dolgusu*. 53. Turkiye Jeoloji Kurultayi, 201-203 (in Turkish).

Altınlı, I.E. 1955. *Denizli güneyinin jeolojik incelemesi*. İst. Üniv. Fen Fak.Mecm., B, XX, 1 - 2, 1 - 47 (in Turkish).

Hakyemez, Y. 1989. *Kale-Kurbalik (GB Denizli) bolgesindeki Senozoyik yasli cokel kayalarin jeolojisi ve stratigrafisi*. MTA Dergisi, 109, 9-21 (in Turkish).

Philippson, A. 1915. *Reisen und Forschungen im vvestlichen Kleinasien, (Heft V)*. Petermanns Mitt., Ergansungsband, 39.

Simsek, S. 1982. *Denizli- Saraykoy- Buldan alaninin jeolojisi ve jeotermal enerji olanaklari*. Ist. Uni.Yerbilimleri Dergisi, 1, 145-162 (in Turkish).

Sozbilir, H., Akgun, F. and Akkiraz, S. 2000. *Uyumsuzluklarla sinirlendirilmis Tersiyer yasli tortul istiflerin Denizli ve Izmir arasindaki stratigrafisi*. 53. Turkiye Jeoloji Kurultayi, 179-182 (in Turkish).

Hasdigen, S. and Bayhan, E. 2003. *Kale (GB Denizli) bolgesindeki Tersiyer yasli kayaclarin kil sedimantolojisi*. Yerbilimleri, 27, 47-58 (in Turkish).

Tchihatcheff, P. de. 1869. *Asie mineure*, Geologie III.

# Modelling a Multi-Layer Coal Mine Deposit: A Case Study of Avdan Lignite Site in Denizli, Turkey

Mining Eng. Mohammad Lashgari*,

Murat Ozkan[B], Atilla Ozturk[C], Selamet Ercelebi[D], Mustafa Darici[E]

* Mining Eng., Istanbul Technical University, Turkey, lashgari@itu.edu.tr

**Abstract**

Coal is one of the most important sources for energy production in all over the world. Sustainability of energy production from a thermal power plant should be correlated with reserve quality and quantity of coal deposit. Reserve quantity and quality must be determined from innovative approaches to decrease risk arising from ore body modelling. Geological formation of the coal deposit can be assumed as the initial stage for coal deposit research. Coal deposit can be found as vein, lake, or delta types based on the location, geological background and evolution. Understanding the geological background of coal deposit is useful for estimating possible development of coal deposit based on the location. Vein type coal deposits are the easiest to model due to the continuity while delta types are the most complex structures. Interburden defined as the material separating coal seams within strata is also continuous for vein type coals while several interburdens can be found randomly in coal band alternation for delta type deposit. Different veins in a single coal deposit can be correlated based on whole coal dating however it is hard to correlate the several different coal bands in a single delta type coal deposit. Avdan lignite mine deposit is located in the western part of Turkey. This deposit is important to supply coal for a 150 MWe capacity thermal power plant located close to the mine site nearly for 25 years. The lignite mine will be operated by open pit mining based on the feasibility project carried out for thermal power plant. The deposit is delta type that was a challenging study for modelling ore body. A database was formed from 103 drill logs in the study area for modelling the coal deposit. Maximum distance between the drill logs is prevented to exceed more than 250 m to obtain coal deposit reserve classification as per measured coal resource. Border of the coal deposit was defined from the database based on the geological structure. Cut-off values for minimum coal thickness and lowest calorific value are selected as 40 cm and 950 kcal/kg during the coal deposit modelling. These values were selected as per production method as well as thermal power plant technology. Afterwards, database was used to model the coal structures in the lignite zones. Two main coal structures that are named as Sekkoy and Yenidere Coal Zones were derived from the investigations. Due to the impossibility of correlations among the coal bands, coal zones were defined for these two main coal structures. Coal zone definitions that are the most innovative aspect of this research would be useful to determine the quantity of the coal reserve for the different coal zones in the study area. Hence, although it was not possible to correlate coal bands from one drill log to the other, it is possible to design the coal zones as well as the coal thickness that must be required for reserve estimation. On the other hand, the quality of the deposit based on calorific value, humidity, ash, and sulphur was also determined for each coal zone from geostatistical research. Thereby, quality distribution maps as well as calorific value and reserve estimation graph were also designed for the study area. Micromine software were used to model delta type coal deposit as well as geostatistical applications to obtain quality distribution maps.

Keywords: Multi-Layer, Coal, Micromine, Coal Zone, Lignite, Reserve Estimation, Seam Correlation

## 1. Introduction

Every mineral project is based on a geological deposit. Every effort must be made to observe, record objectively, and describe all geologic features that may affect the characteristic, geometry and variability of the study area. Thus, all the lithological, mineralogical, alteration, structural, and rock competency data must be collected by means of which the geologist would be able to subdivide the study area in key subunits in order to improve description, correlation, projection, and understand the genesis of the deposit (SME, 1992). Project evaluation is the process of identifying the economic feasibility which determines that capital investment is worthy of application or not. Mineral reserve assessments always have been critical for mine planning and feasibility evaluation. Complex mineral projects often have many uncertainties caused by nature of geologic reserve estimations; severe problems in forecasting commodity prices and production cost; long evaluation process during which economic changes may happen; uncertainty in regulatory; local economy changes and neighboring countries economic and social status. Increasing demand for energy beside exemption from oil market have been a concentrating area for many countries. Turkey currently has about 8.4 Gt. Of lignite reserves -and most were formed in intermontane basins (for example, in Soma and Cayirhan)(A.I. Karayigitand *et al* 1998) -, of which 3.9 Gt. are exploitable. Most of the known lignite deposits in Turkey have low calorific valuesand high ash, moisture, and total sulphur contents. Almost 80% of the total reserves have calorific values below10.46 MJ.kg –1. The lignites having calorific values lower than 10.46 MJ.kg–1are generally consumed in power plants, while those having calorific values higher than stated value are directed towards household and industrial uses. The majority of Turkish lignite deposits are worked as open-pit mines, though there are also some underground operations. Miocene, Pliocene, and Quaternary coals are formed in limnic and limnic fluvial environments with some volcanogenic intercalations. The total extent of the coal-bearing Miocene (835.1 km2) and Pliocene (526.9 km2) deposits is 1,362 km2 (Tuncali *et al* 2002). Moreover, from a strategical point of view taking advantage of domestic energy potentials makes the small and even low quality coal resources feasible. In the studied area, stratified geologic formations have created multi-layer coal deposit consisted of seams with thickness from 2 cm to 390 cm. This structure can cause the 3D correlation modelling of the reserve be a demanding and even impossible task. In order to stablish a practical protocol, the minimum extractable thickness is set to 40 cm and the figures lower than that have assumed as non-extractable so the interburden.

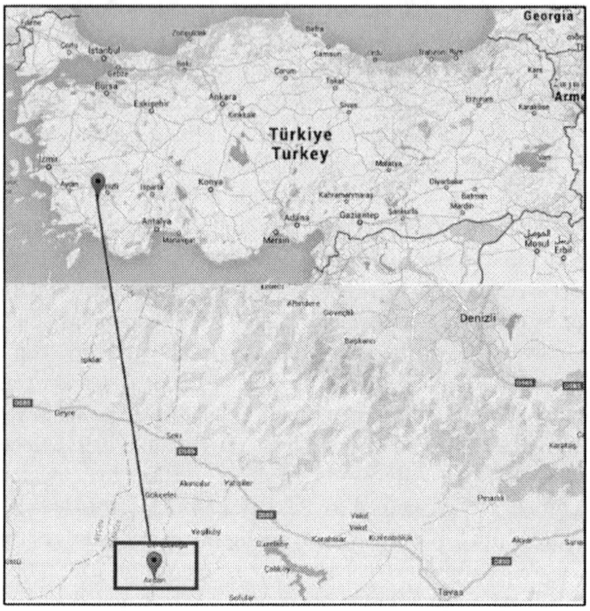


Fig.1. Study Area Location

## 2. Geology of the Site

The study of the geology of the region is known to begin in 1869. These studies have been continued until today and are summarized as follows by Geological Engineer, Mustafa Darici. License area which included the district of Denizli Tavas as the first study trip notes were taken into consideration by Tchihatcheff (1869). The structure of the area is formed in three geological eras; First, Paleozoic - Mesozoic quartz, marble, slate, limestone, creates, radiolarite and ultramafic rocks is forming the strata; Second, Neogene sediments such as tertiary old rocks overlie on strata rocks which are divided into two groups; They are Akçay Group (Yenidere) and Muğla Group (Sekköy); And third, Quaternary, in which on the river beds gravel, sand, materials consisting of silt and clay have been seen.

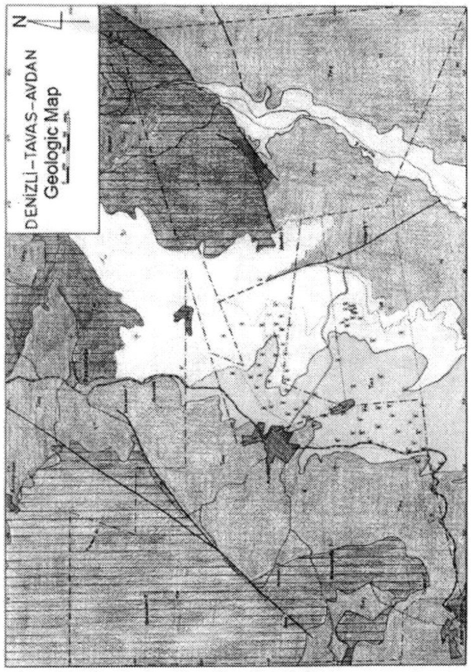


a) Study Area Geologic Map

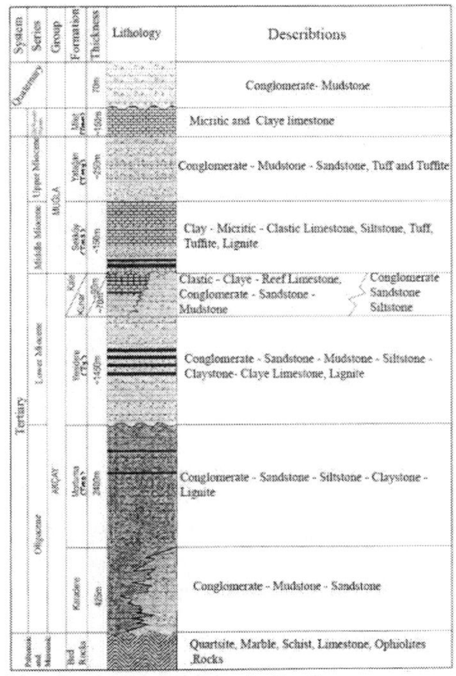


b) Study Area Geologic Section

Fig.2. License Area Borders and the Geologic Configuration (Hakyemez, 1989)

## 3. Quantity Investigations

In order to obtain detailed information from the geometry and quality of coal bed extensive drilling has taken place and the collar coordinates has been surveyed covering approximately 4.6 km$^2$ of exploration area. Apparently there has not been a square drilling pattern and as the geology logging of samples shows drilling pattern had been defined by results of previous borehole.

### 3.1. Borehole Information

In total, 103 boreholes have been drilled of which 98 boreholes cut coal. The dip of all the boreholes is -90 degrees and average distance between collars is 353 m. Maximum depth of boreholes 334m and minimum depth is 25m. The elevation range of the boreholes collars is 213 m and the topography and geology show the delta type basin is forming the coal bed.

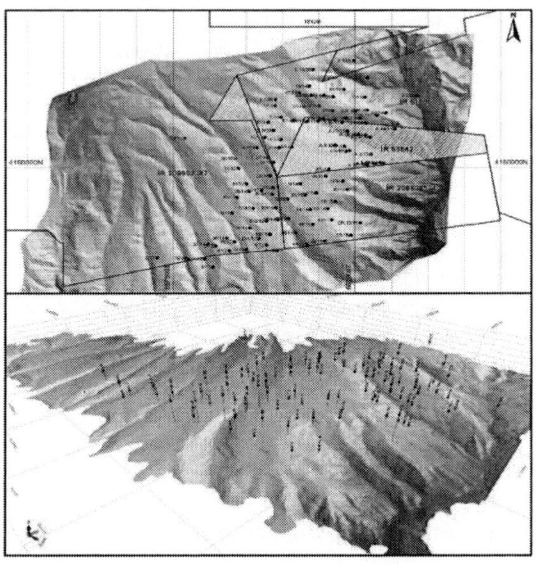

Fig.3. Study Area Topography and Borehole Locations

### 3.2. Formations

Overbuden, coal, and interburden are located in geologic formations which are Yatağan, Sekköy and Yenidere. Coal is mainly formed in Sekkoy and Yenidere formations and in both formations there are several coal seams including interburdens.

#### 3.2.1. Sekköy

In the Sekkoy formation coal seams are located between the depth of 21 m and 94.4 m. It should be mentioned that this wide range originated by the delta shaped topography and the overburden thickness is directly affected by the topography and so on the coal seams are almost tabluar. Moreover, in this formation the minmum and maximum coal seam thickness is 0.15 m and 3.90 m, respectively. In addition, the interburden thickness changes between 0.15 m 11.00 m. Interburden is consisted of Marl, limestone and clay thoroughly.

#### 3.2.2. Yenidere

Exactly beneath the Sekkoy formation Yenidere is lying. The same as the Sekkoy coal seams these of Yenidere are tabular with no sensible dip and surprisingly the minimum and maximum of coal seam thickness are 0.15 m and 3.80 m respectively and the thickness interburdens lying between coal seams are changing from 0.10 m and 35.13 m. Yenidere interburdens are mainly consisted of Siltstone, Mudstone, Clay, Sandstone and Marl. In fig. 4, several coal seams available in the borehole Sj.34 is shown as an example.

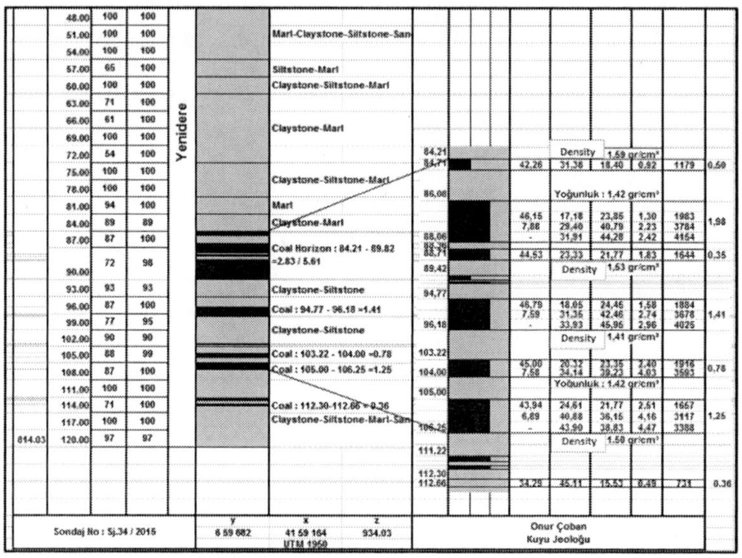


Fig.4. Borehole Sj.34 Log

### 3.3. Coal Zone Determination

Determination of a correlation between each pair of boreholes in a multi seam reserve is a demanding task. Furthermore, in a complicated field, sometimes it is impossible to do so. In the studied area through the Sekkoy and Yenidere formations maximum numbers of coal seams are 12 and 8, respectively. Due to mining machinery capabilities for minimum seam thickness to excavate seams with the thickness of less than 40 cm were waved. Moreover, due high probability of dilution because of thin seams the layers with calorific value less than 950 Kcal/Kg were ignored. In order to be able to determine a correlation and form a unique solid model for each formation, all the coal seams including the interburdens have been taken into solid model as the formation coal zone model. Theses coal zones named based on the formation they belong to and Sekkoy and Yenodere coal zones were modeled.

### 3.4. Solid Models

There are conventional methods to create a 3D model of ore body. In this study in order to acquire unity for the shape of coal zones and the actual coal seams thickness model, gridding method is applied to estimate the RL (Reduced Level) by means X and Y coordinates and using the known RL values available in the boreholes to create two surfaces of roof and floor of the coal zones. Grid is a regular network formed by triangle, square or rectangular generally (Kaan Erarslan ,2012) and in the approach triangular grid is used. The gridding model best correspondent to a lignite ore body geometry from the geological aspect of view was determine to be basic general linear model, which obviously can present the linear behavior of changes of stratified reserves and simultaneously can solve the complexity of a 3D estimation method. The Linear model is special case of power model the type

$$\gamma(h) = C_0 + p.h^x \qquad \text{(f.1)}$$

Where $C_0$is the nugget effect; $p$ is a constant (slope in the linear model); $h$ is lag; and $x$ is the exponent of $h$ such that $0 \leq x \leq 2$ ($x = 1$, in linear case).

The basic general linear model parameters best fit for the study area are, $x = 1.5$, $C_0 = 0$, $p = 0.015$ and the search ellipsoid radius was 700 m and had four sectors and in each sector maximum points were 2.

Having created the roof and floor of coal zones they unified to build up the solid model.

As it is shown in the fig. 5 that would be easy to estimate that interburdens are included in the coal zone and the volume of ore solid model was exaggerating too high. To correct the deviation, another solid model of total thickness of coal seams in each borehole was created. The "coal thickness solid model" was allocated exactly in the middle of coal zone model. In order to be form the exact parallel surfaces of roof and floor for the coal thickness model the same estimating method and parameters was precisely used. Therefore, the coal zone model can serve as a location model and simultaneously the coal thickness model can represent the real quantity specification of total coal seams.

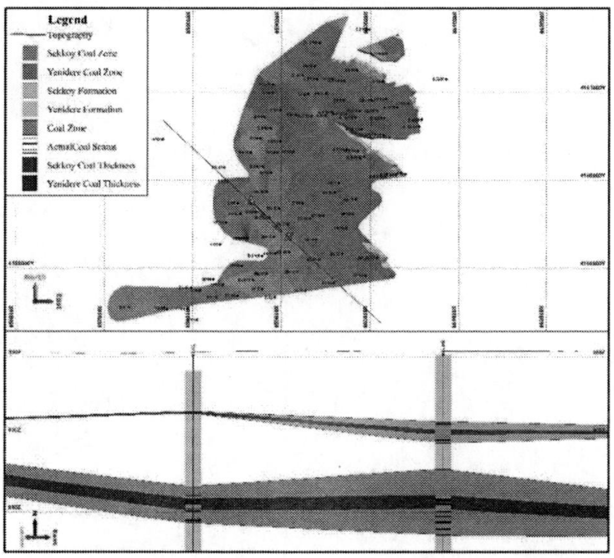


Fig.5. Coal Thickness Model in the Middle of Coal Zone Model

## 4. Quality Investigations

A block model with the parent block size of $10 \times 10 \times 4\, m$ $(X \times Y \times RL)$ and sub-block factor of 10 for each direction was created to be fit to coal thickness solid model. In order to have a realistic quality features of the ore body and use them for pit optimization process, estimation for calorific value, ash, moisture, Sulphur and volatile matter was taken into action.

### 4.1. Geostatistical Background

Choosing between the Ordinary kriging (OK) and the Inverse Distance Weighted has always been a controversial problem. Some studies show that results derived by OK indicate lower errors in comparison with raw data (Shahab Shahbeik *et al* ,2013). OK, a common spatial estimator, is used to map the general trend of spatial distribution of coal properties. Investigating the isotropic properties of the ore body in various directions including the downhole, it was obviously determined that there is isotropic distribution for the coal properties. It can be inferred that there is an acceptable correlation between the spatial modeling method and the coal quality estimating model when there is not much erosion and secondary distortion on the coal ore body (Mohamad N. Heriawan *et al*, 2008). The same kriging model was used to estimate the coal quality distribution to evaluate the block properties.

## 4.2. **Distribution Maps**

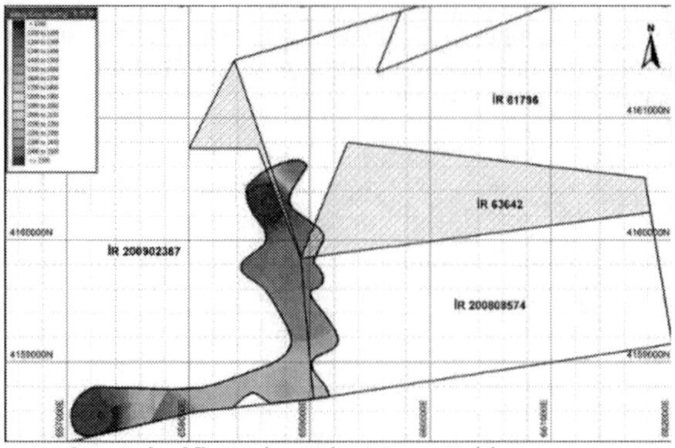


a) Calorific Value Dis. Map - Sekkoy

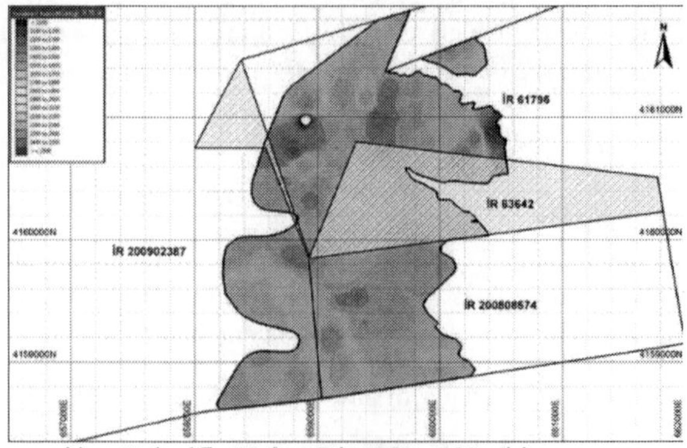


b) Calorific Value Dis. Map - Yenidere

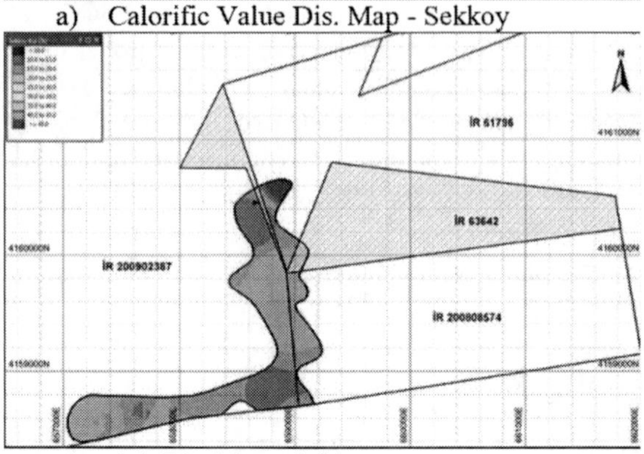


c) Ash Dis. Map - Sekkoy

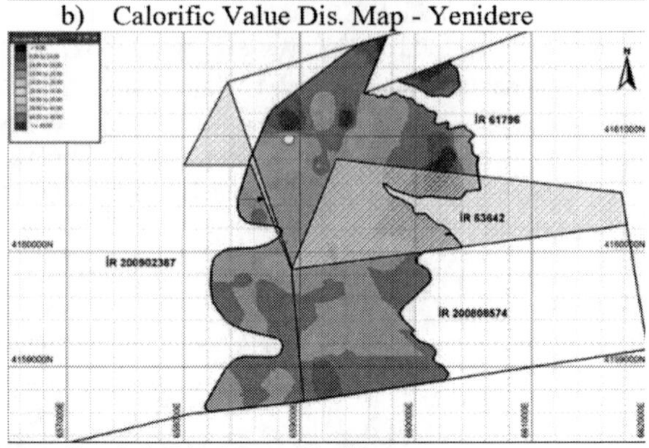


d) Ash Dis. Map - Yenidere

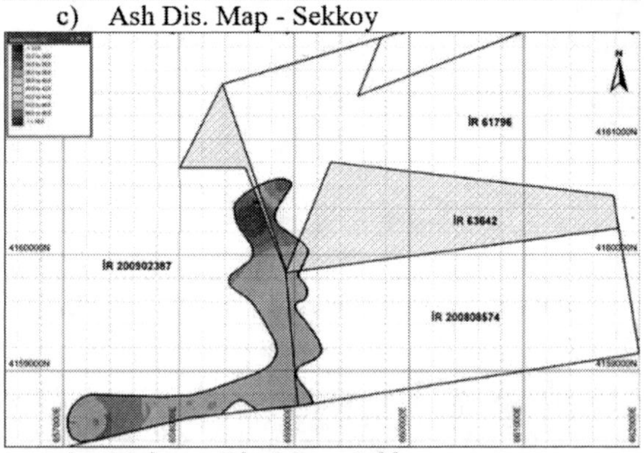


e) Moisture Dis. Map - Sekkoy

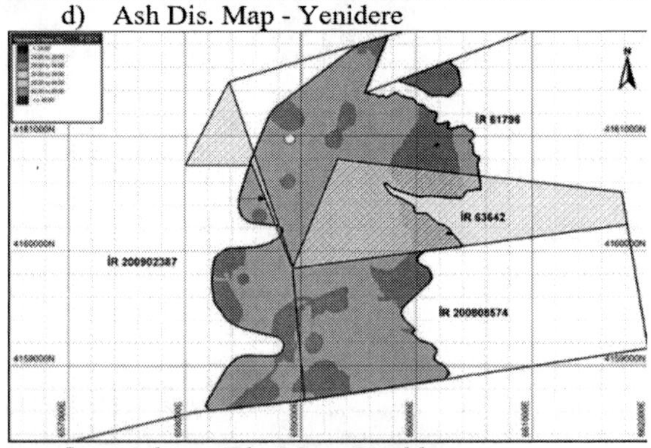


f) Moisture Dis. Map - Yenidere

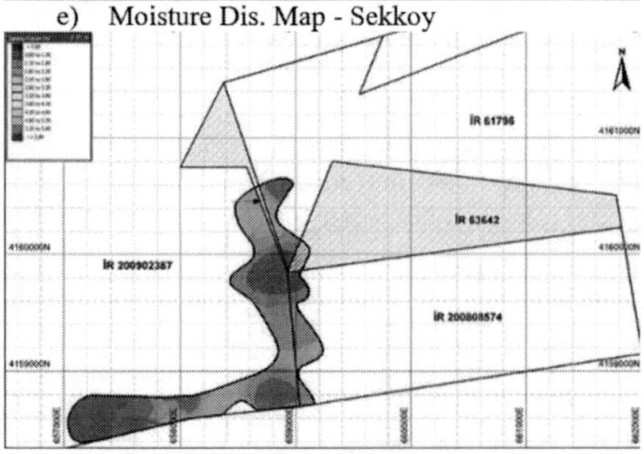


g) Sulphur Dis. Map - Sekkoy

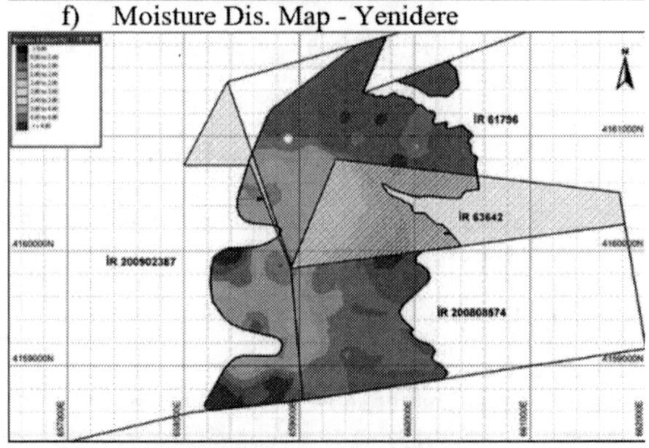


h) Sulphur Dis. Map - Yenidere

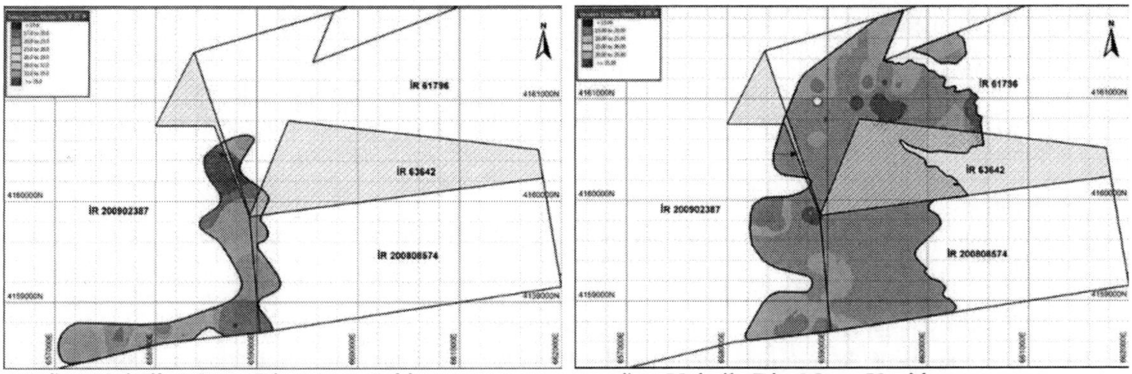


i) Volatile Matter Dis. Map - Sekkoy j) Volatile Dis. Map - Yenidere

Fig.6. Coal Quality Distribution Maps

### 4.3. Statistics of Quality Parameters

Feeding the extracted coal to thermal power plant make the calorific value the most important parameter for these coal miners. In order to have a criterion for a feasibility decision to know the whole reserve quality grade – ton graph has always been noteworthy.

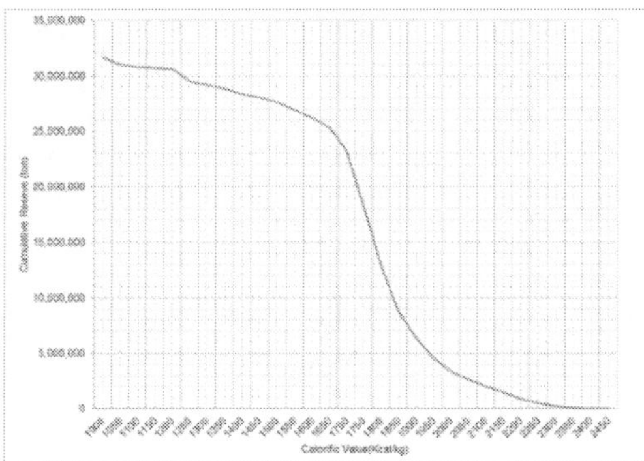

Fig.7. Grade (C.V.) – Ton Graph

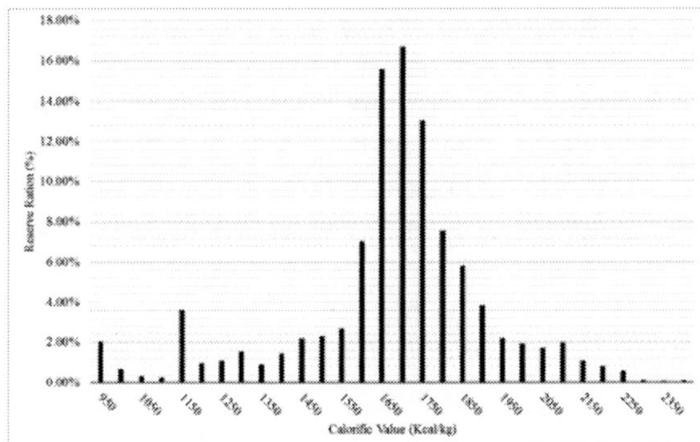


Fig.8. C.V. – Reserve Ratio Graph

That would be useful any time to have the average quality of ore reserve. The following table is the average of coal quality block parameters.

Table 1. Reserve Quality Average Parameters

| Coal Zone | Reserve (MTon) | Density (gr/cm³) | C.V. (Kcal/kg) | Moisture (%) | Ash (%) | V.M. (%) | Sulphur(%) |
|---|---|---|---|---|---|---|---|
| **Sekkoy** | 5.0 | 1.5 | 1.5 | 39.6 | 28.3 | 26.5 | 2.7 |
| **Yenidere** | 25.5 | 1.5 | 1.8 | 42.5 | 24.0 | 23.6 | 2.3 |

## 5. Results and discussions

In this study a practical method for modelling a multi-layer coal reserve is used. In traditional polygonal method determination of a correlation between each pair of seams found in a borehole is inevitable. In addition to be a demanding task of what stated, restricting a block model to what would be created may cause deviations in volume and mass calculations. In the approach discussed in this study it should be noted that the pit optimization results may have some deviations originated from the interburden volume and exact location because automatically it is assumed that the any identical point the interburden volume in the coal zone in upper and lower part are exactly equal.

## 6. Acknowledgments

The research was a part of the industrial project supported by Avdan Mining and Energy Company (Avdan Madencilik Enerji San. ve Tic. A.Ş.). The authors gratefully acknowledge the assistance and cooperation provided by the Company.

## 7. References

1. Sinclair, Alastair J.,Blackwell, Garston H. (2002), Applied Mineral Inventory Estimation, Cambridge University Press
2. Mohamad Nur Heriawan, Katsuaki Koike (2008), Uncertainty assessment of coal tonnage by spatial modeling of seam distribution and coal quality, International Journal of Coal Geology
3. Kaan Erarslan (2012). Computer Aided Ore Body Modelling and Mine Valuation, Earth Sciences, Dr. Imran Ahmad Dar (Ed.), InTech, DOI: 10.5772/26020
4. Hakyemez, (1989), Kale-Kurbalik (GB Denizli) bolgesindeki Senozoyik yasli cokel kayalarin jeolojisi ve stratigrafisi. MTA Dergisi, 109, 9-21 (in Turkish).
5. Tchihatcheff, P. de. 1869. Asie mineure, Geologie III.
6. Karayigit, F. Akgun, R.A. Gayer, A. Temel, (1998), Quality, Palynology, and Palaeoenvironmental interpretation of the Ilgin lignite
7. Tuncali E, Çiftci B, Yavuz N, Toprak S, Köker A, Gencer Z, Ayçık H, Şahin N (2002). Chemical and Technological Properties of Turkish Tertiary Coals. 1st ed. Ankara, Turkey: General Directorate of Mineral Research and Exploration
8. Howard L. Hartman, 1992, Mining Engineering Handbook, SME
9. Shahab Shahbeik, P. Afzal, P. Moarefvand, M. Qumarsy (2013), Comparison between ordinary kriging (OK) and inverse distance weighted (IDW) based on estimation error. Case study: Dardevey iron ore deposit, NE Iran, Saudi Society for Geosciences

# Mineral matter in coals from the Soutpansberg Coalfield, South Africa

MSc Project

By Maseda Mphaphuli
Supervisor: Professor. N Wagner
Co-supervisor: Mr J Sparrow
UNIVERSITY OF JOHANNESBURG

## Overview of research project

The main aim:

- To understand coal quality and petrographic variations on either side of the Tshipise fault

The objectives of the project will include:

- To determine the impact of the Tshipise fault on the coking properties if any
- To attempt to reconstruct the depositional environment of Soutpansberg Coalfield

UNIVERSITY OF JOHANNESBURG

## Study Area of Soutpansberg coalfield (SC)

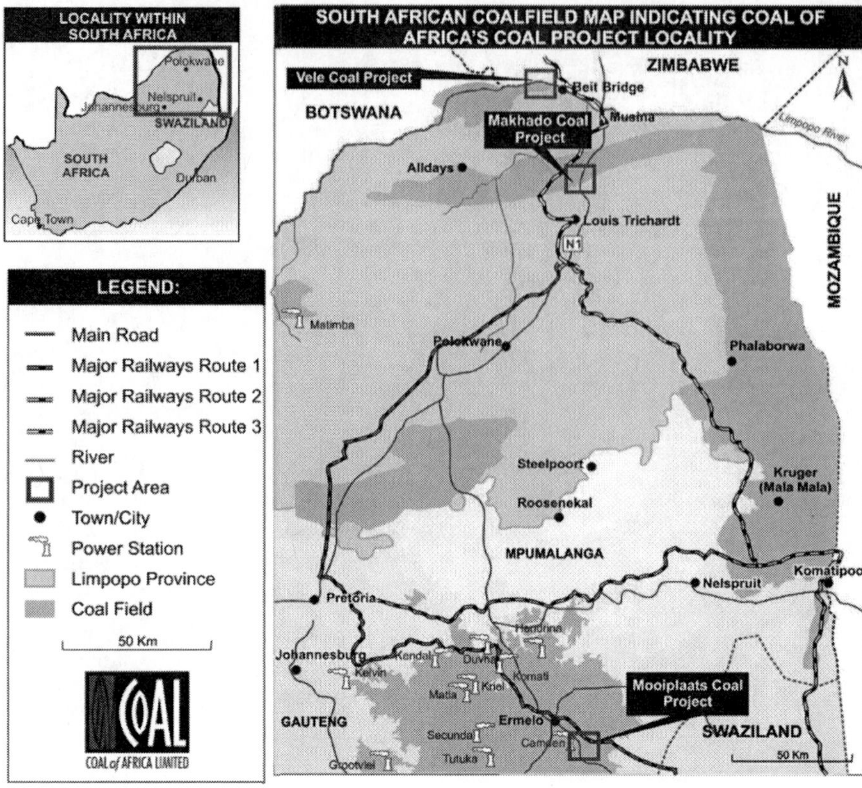


## Overview of presentation

- Why study mineral matter
- Study area
- Methods
- Results
- Conclusion
- Acknowledgement

UNIVERSITY OF JOHANNESBURG

# Why study mineral matter

- Mineral matter = inorganic material
- Dilutant to coal products.
- Aims characterize mineral matter in SC
- Sedimentological and geochemical history of peat and coal forming environments
- Assess how different coals behave when undergoing utilization processes

# Methods

- S
- S
- C
- F
- X
- S
- E

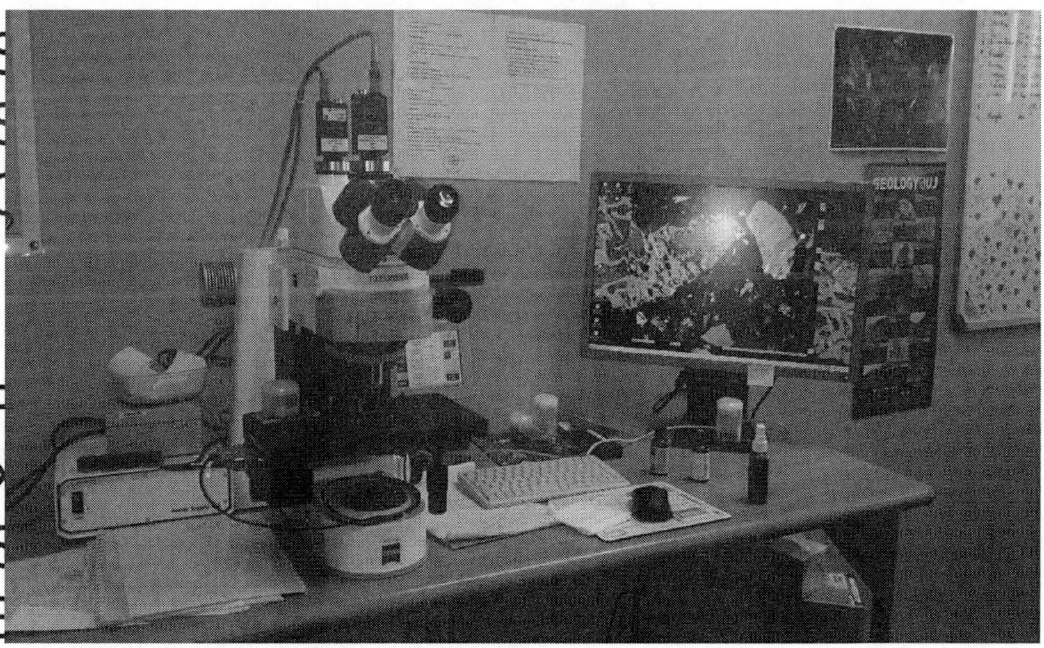

Zeiss Axio Imager M2 microscope on Hilgers

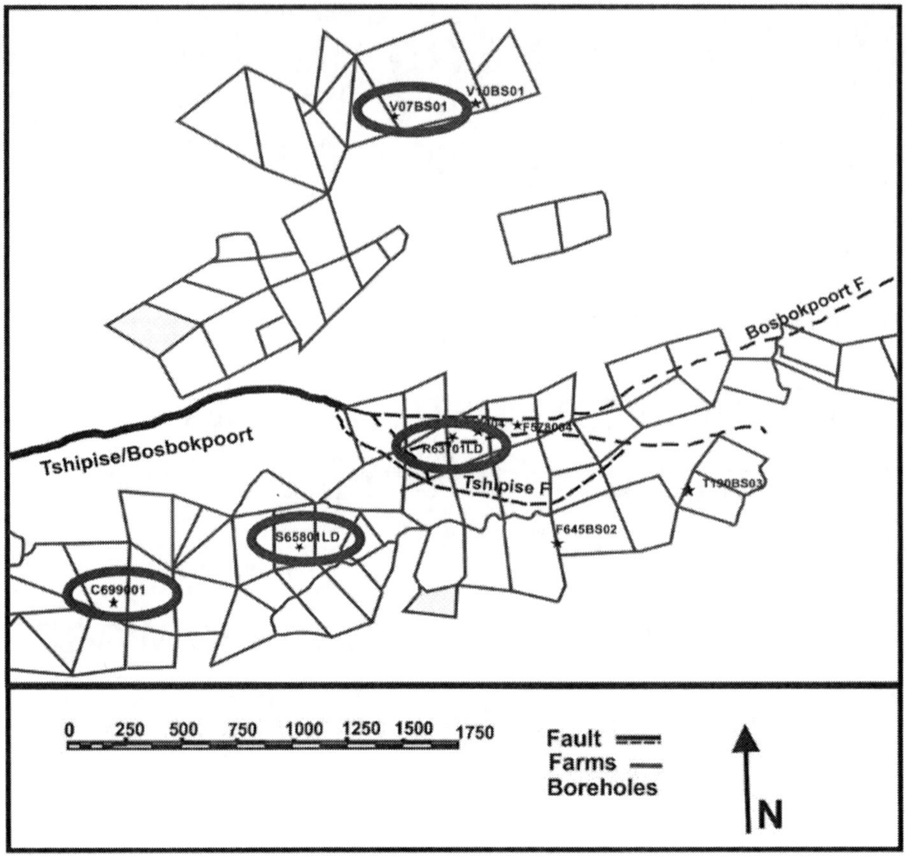


# Results

| | Ash | MC | VM | FC | CV | TS |
|---|---|---|---|---|---|---|
| VB01 | 34.92 | 0.60 | 25.97 | 38.51 | 20.34 | 9.49 |
| VB02 | 36.60 | 1.26 | 24.60 | 37.55 | 25.42 | 2.79 |
| VB03 | 34.99 | 0.56 | 21.75 | 42.70 | 22.06 | 3.93 |
| 15147 | 28.99 | 2.82 | 20.92 | 47.26 | 22.69 | 0.93 |
| 15148 | 34.59 | 0.87 | 24.71 | 39.84 | 18.54 | 1.16 |
| 15149 | 16.02 | 0.67 | 21.20 | 62.11 | 28.24 | 0.88 |
| 15143 | 36.17 | 0.64 | 22.41 | 40.78 | 17.07 | 2.16 |
| 15144 | 20.06 | 1.01 | 27.55 | 51.38 | 26.87 | 1.45 |
| 15145 | 42.87 | 0.55 | 24.57 | 32.01 | 11.51 | 0.47 |
| 15152 | 30.20 | 0.70 | 19.93 | 49.17 | 22.28 | 1.40 |
| 15153 | 26.11 | 0.89 | 26.65 | 46.35 | 23.91 | 0.67 |

# Results

| Sample no: | MACERAL GROUP | | | | (RoV%) |
|---|---|---|---|---|---|
| | Vitrinite | Inertinite | Liptinite | MM | |
| VB01 | 32.8 | 3.5 | 0.9 | 62.8 | 0.77 |
| VB02 | 67.7 | 6.6 | 0.9 | 24.8 | 0.78 |
| VB03 | 67.8 | 3.2 | 1.7 | 26.7 | 0.76 |
| 15147 | 64 | 10.6 | 1.4 | 24 | 0.90 |
| 15148 | 41.9 | 15.3 | 1.2 | 41.6 | 0.92 |
| 15149 | 69 | 22.8 | 0.2 | 8 | 0.94 |
| 15143 | 34.6 | 8.8 | 3.6 | 53 | 0.71 |
| 15144 | 75.3 | 9.6 | 3.3 | 11.8 | 0.70 |
| 15145 | 9.8 | 63.8 | 0.7 | 25.7 | 0.79 |
| 15152 | 42.8 | 17.8 | 1.6 | 37.8 | 0.89 |
| 15153 | 64.8 | 10.5 | 0.2 | 24.5 | 0.95 |

# Mineral matter

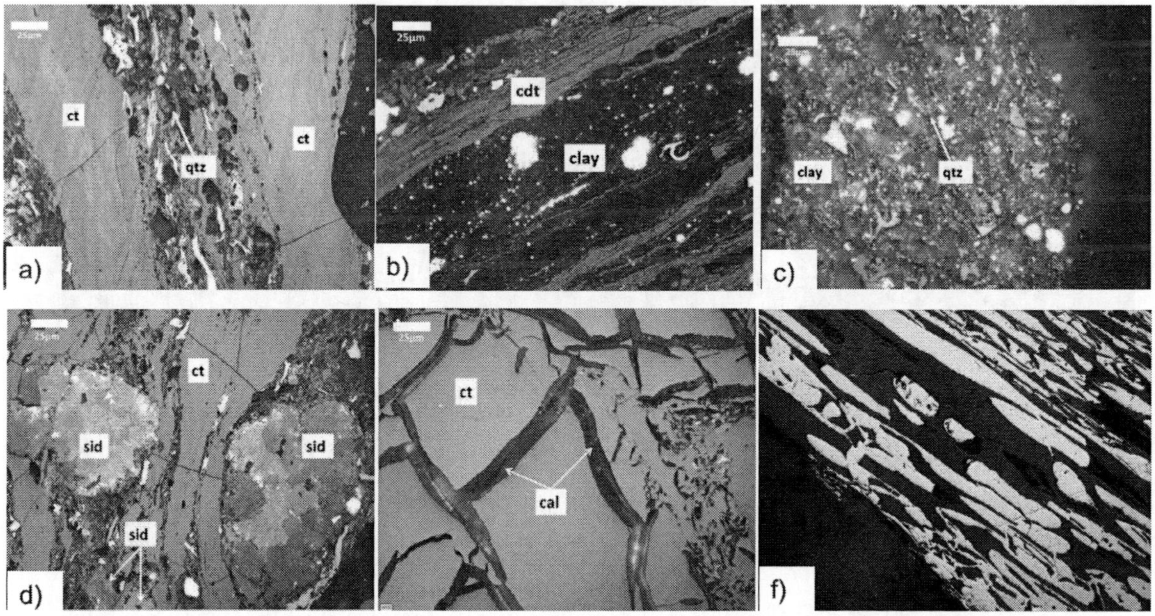


a) Quartz grains in a vitrinite grain, b) clay minerals with pyrite, c) clay and quartz d) carbonate mineral, e) calcite mineral in vitrinite maceral, and f) pyrite filling cell and pore structures.

# XRD results

| Compound name | Chemical Formula |
|---|---|
| Calcium magnesium (Dolomite) | $CaMg(CO_3)_2$ |
| Iron sulphide (Pyrite) | $FeS_2$ |
| Silicon oxide (Quartz) | $SiO_4$ |
| Aluminium silicate ( Kaolinite) | $Al_2Si_2O_5(OH)_4$ |
| Iron carbonate (Siderite) | $FeCO_3$ |
| Calcium, iron, magnesium, manganese carbonate (Ankerite) | $Ca(Fe,Mg,Mn)(CO_3)_2$ |

# Variation in pyrite

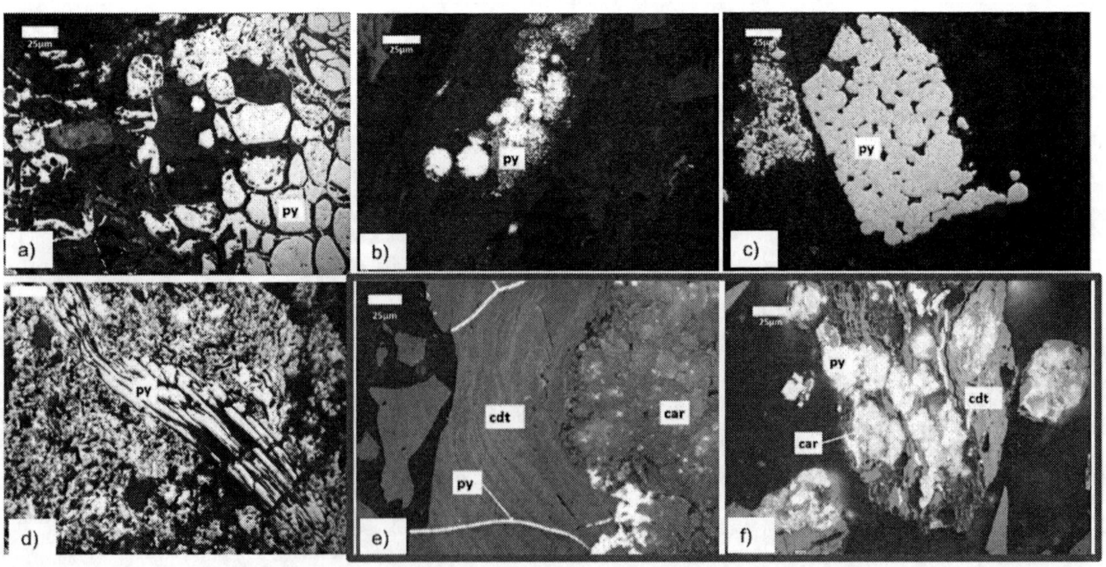


Syngenetic and epigenetic pyrite a) syngenetic pyrite as cell structure is still preserved b) and c) Framboidal pyrite d) massive pyrite e) pyrite infilling cracks and f) replacement of carbonates

# Botryoidal pyrite

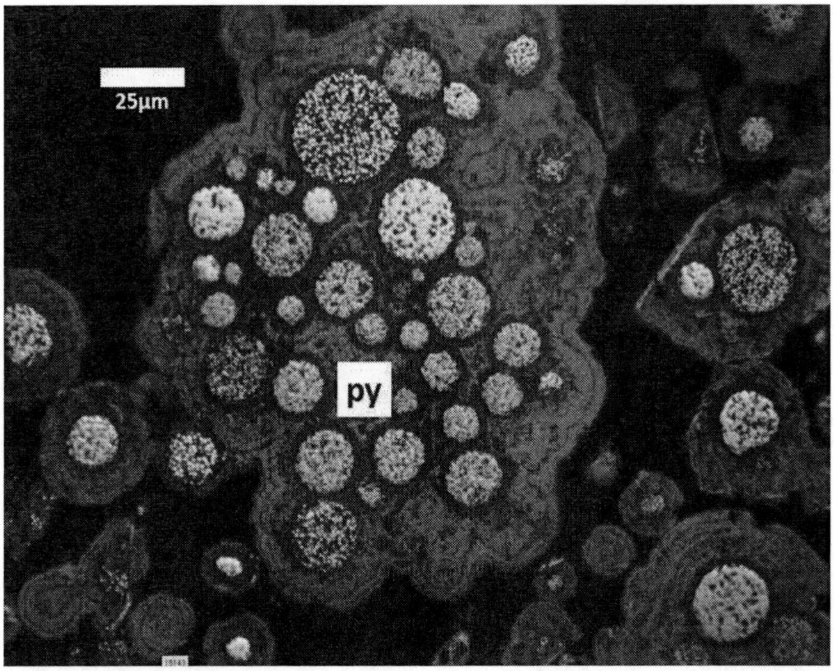


Petrographic microscope

# EMPA results for pyrite

| Sample no | 15143 | 15143 | VB01 | VB01 | VB01 |
|---|---|---|---|---|---|
| | overgrowth | core | cell filling | crack filling | massive pyrite |
| | | | | | |
| S | 43.46 | 53.19 | 53.4 | 52.88 | 53.34 |
| Mn | 0 | 0 | 0 | 0 | 0.01 |
| Fe | 35.36 | 46.45 | 45.96 | 45.07 | 46.08 |
| Co | 0.57 | 0.02 | 0.01 | 0.01 | 0.01 |
| Ni | 1.51 | 0 | 0 | 0 | 0 |
| Cu | 0 | 0.04 | 0 | 0.02 | 0 |
| Zn | 0 | 0 | 0 | 0 | 0.01 |
| As | 1.39 | 0.01 | 0.01 | 0.01 | 0 |
| Se | 0.19 | 0 | 0 | 0 | 0 |
| Pb | 1.4 | 0.02 | 0.01 | 0.05 | 0 |
| Total | 83.89 | 99.72 | 99.4 | 98.04 | 99.46 |

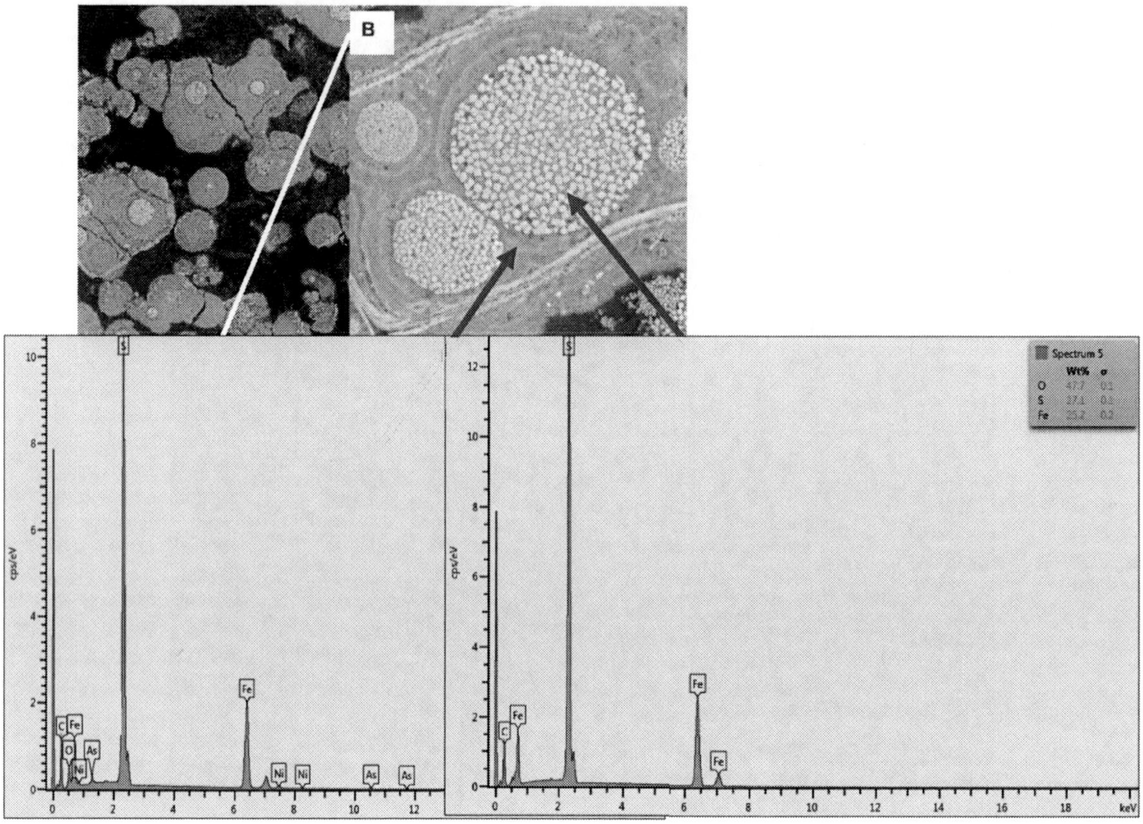


# Variation in Carbonates

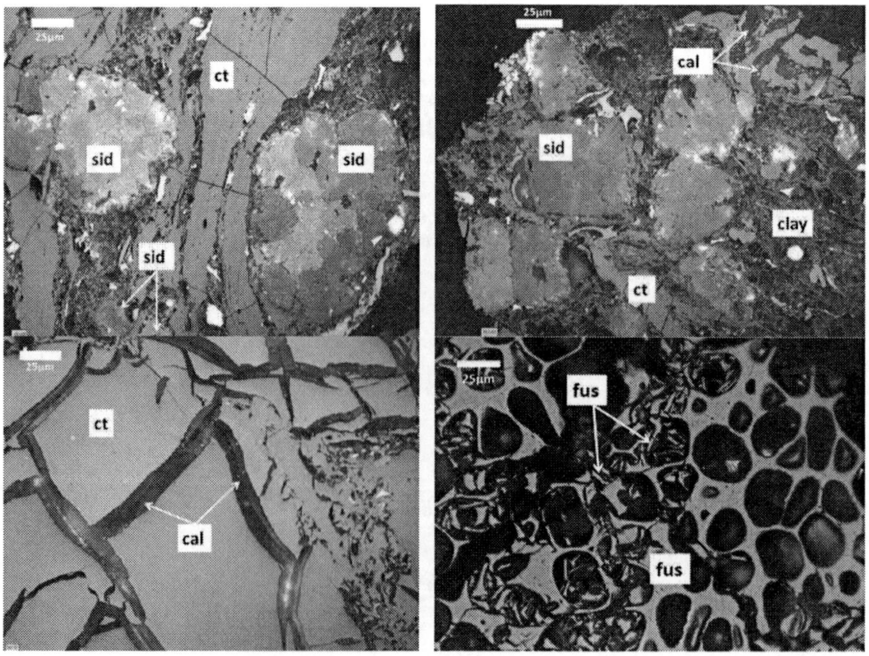

# EMPA results for carbonates

| Sample ID | 15148 | 15148 | 15148 | 15136 | 15143 |
|---|---|---|---|---|---|
| | Siderite T | Siderite N | Siderite N | Calcite | cleats (bp) |
| CaO | 4.18 | 2.78 | 0.96 | 51.1 | 30.73 |
| MgO | 2.66 | 2.05 | 0.31 | 1.95 | 13.34 |
| SrO | 0.01 | 0 | 0.01 | 0.09 | 0.09 |
| SiO2 | 0.29 | 0.04 | 0.09 | 0.05 | 0.24 |
| Al2O3 | 0.08 | 0.05 | 0.03 | 0 | 0.13 |
| FeO | 51.89 | 53.72 | 58.59 | 3.68 | 9.42 |
| MnO | 1.22 | 1.53 | 1.42 | 0.15 | 0.15 |
| BaO | 0.03 | 0.04 | 0.04 | 0.06 | 0.09 |
| ZnO | 0 | 0.01 | 0.04 | 0.03 | 0.04 |
| CO2 | 39.61 | 39.78 | 38.5 | 42.87 | 45.77 |
| Total | 99.97 | 99.99 | 100 | 99.99 | 100 |

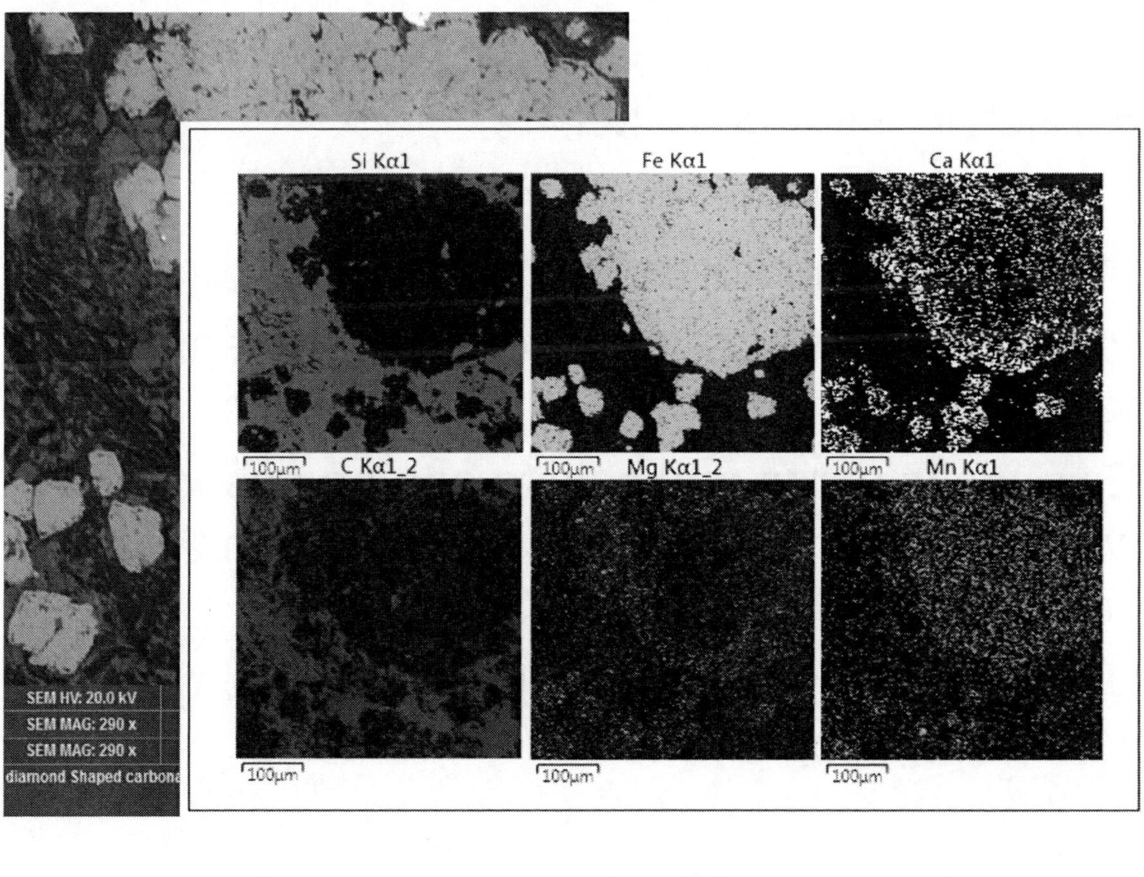

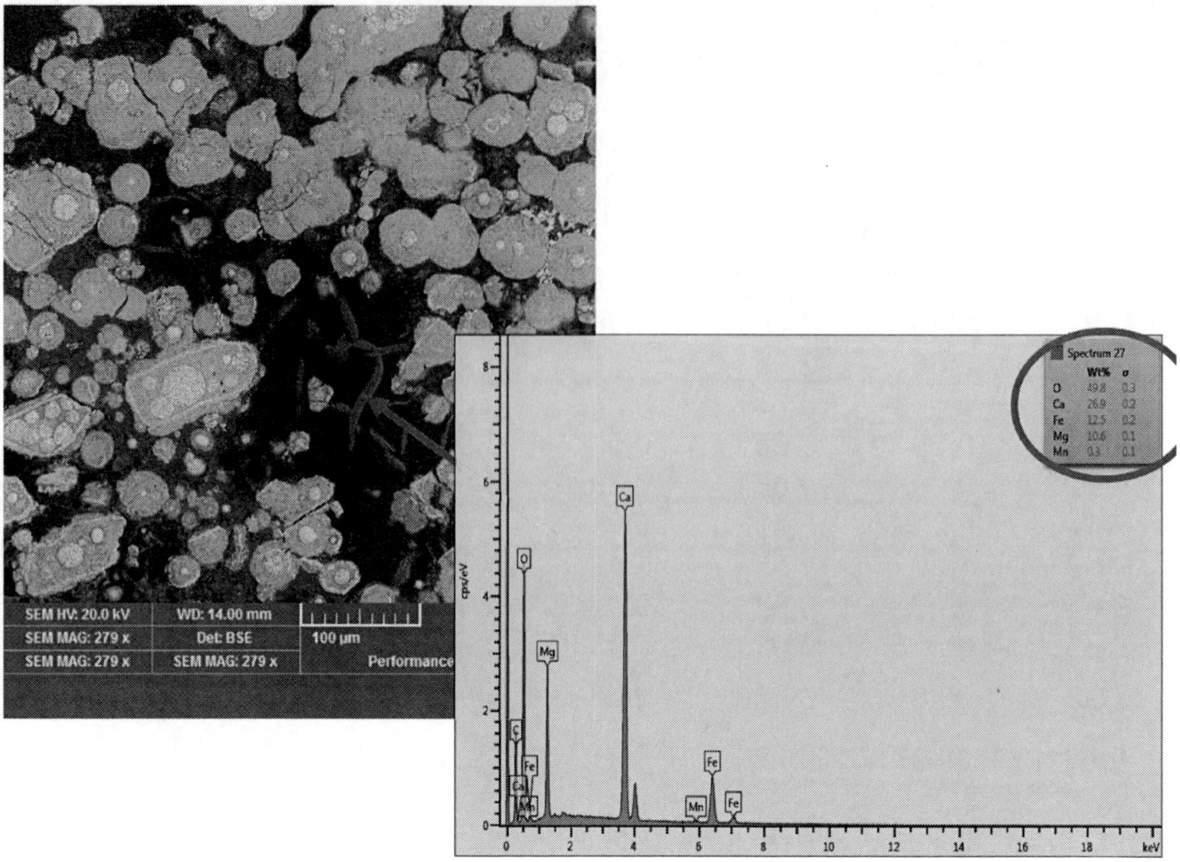


# Conclusion

- SC rich in mineral matter
- Silicates, sulphides and carbonates
- Syngenetic and epigenetic mode of occurrences
- Possible marine depositional environment for samples rich in syngenetic pyrite
- Epigenetic pyrite and calcite can be associated with hydrothermal fluids associated with the Tshipise fault
- Data provided can help with utilization process when designing plants

## Acknowledgements

- This work is based on research supported wholly/in part by the National Research Foundation of South Africa (Grant number 103466)
- Coal Of Africa, John Sparrow and Nthabiseng Masunyane, for providing the samples and support
- Eskom Tertiary Education Support Programme (TESP)
- Zama Mthabela, WITS

UNIVERSITY OF JOHANNESBURG

COAL of AFRICA LIMITED

NRF National Research Foundation


## Thank you

UNIVERSITY OF JOHANNESBURG

**2016 International Pittsburgh Coal Conference**
**Cape Town, SOUTH AFRICA**
**August 8 – 12, 2016**

**Abstract Submission**

**PROGRAM TOPIC:**
**8. COAL MINING (GAS MANAGEMENT IN COAL MINES)**

**Exact Title of Paper:**

# ASSESSING RISKS RELATED TO STIMULATION OF FIREDAMP DRAINAGE BOREHOLES IN ORDER TO RECOMMEND PREVENTION MEASURES AND MONITORING METHODS TO HELP EARLY RISK MANAGEMENT DURING DESIGN OF FUTURE DRAINAGE PROJECTS

***Presenting Author:***
***Stéphane LAFORTUNE***
***INERIS, French national competence centre for industrial environment and risks***
***Parc technologique ALATA, B.P. 2, 60 550 Verneuil-en-Halatte, FRANCE***
***stephane.lafortune@ineris.fr, phone: +33 (0)3 44 55 67 91, fax: +33 (0)3 44 55 67 00***

*Co-Authors:*
*Régis FARRET, Candice LAGNY, Zbigniew POKRYSZKA*
*INERIS, French national competence centre for industrial environment and risks*
*Parc technologique ALATA, B.P. 2, 60 550 Verneuil-en-Halatte, FRANCE*
*firstname.name@ineris.fr, phone: +33 (0)3 44 55 66 77, fax: +33 (0)3 44 55 67 00*

**Abstract:**

Coal mining in Europe is extending to deeper and deeper levels, facing ever increasing coal seam methane contents and much higher methane emissions at production districts. Methane emissions raise a serious safety problem but also limit production advancement rates.

Current drainage and pre-drainage methods applied in European coal mines are not sufficiently effective to face growing risks related to methane emissions. Therefore, in order to improve safety and increase production rates, it is essential to develop new gas control and drainage techniques.

The Research fund for coal and steel of the European commission funds the GASDRAIN project between July 2014 and June 2018. The main objective of this project is to investigate stimulation techniques to improve drainage efficiency by increasing the permeability in the vicinity of firedamp drainage boreholes. As all stimulation techniques cannot be appropriate to all coal seams and mining conditions, 4 techniques are investigated by GASDRAIN partners. They are: hydraulic fracturing, open or cased hole cavitation, high pressure water jet slotting and use of explosives.

Application of these stimulation techniques induces risks during stimulation process but also later during mining of stimulated/fractured coal seams or rock strata. For instance, self-heating of coal and spontaneous combustion are possible risks due to penetration of oxygenated air within coal seams or rock strata through fractures created by stimulation.

Thus, a detailed risk analysis related to the use of the 4 stimulation techniques has thus been carried out. Firstly, all relevant incidents associated with stimulation of firedamp drainage boreholes have been identified through a bibliographical review and by collecting feedback from partners of the project through a questionnaire. Then, a core set of relevant risk scenarios has been defined and was represented with a bow-tie diagram (or event tree). Best practice measures to prevent and mitigate the risks (i.e. safety barriers) have also been selected.

Authors will detail results from the detailed risk analysis (relevant incidents, bow-tie diagram) and present recommendations on prevention measures or monitoring methods which can help early risk management during design of future drainage projects in European coal mines.

# Illinois Industrial Carbon Capture & Storage Project

## *Eliminating $CO_2$ Emissions from the Production of Bio Fuels - A 'Green' Carbon Process*

ADM

**Pittsburgh Coal Conference**

**August 10, 2016**

**Scott McDonald**

**Biofuels Development Director**

**scott.mcdonald@adm.com**


ILLINOIS
INDUSTRIAL CARBON
CAPTURE & STORAGE

---

ADM

# Acknowledgements

ILLINOIS
INDUSTRIAL CARBON
CAPTURE & STORAGE


- **The Industrial Carbon Capture and Storage (ICCS) project is administered by the U.S. Department of Energy's Office of Fossil Energy and managed by the National Energy Technology Laboratory (award number DE-FE-0001547) and by a cost share agreement with the Archer Daniels Midland Company, University of Illinois through the Illinois State Geological Survey, Schlumberger Carbon Services, and Richland Community College.**
- **The Midwest Geological Sequestration Consortium is funded by the U.S. Department of Energy through the National Energy Technology Laboratory via the Regional Carbon Sequestration Partnership Program (contract number DE-FC26-05NT42588) and by a cost share agreement with the Illinois Department of Commerce and Economic Opportunity, Office of Coal Development through the Illinois Clean Coal Institute.**
- **The Midwest Geological Sequestration Consortium (MGSC) is a collaboration led by the geological surveys of Illinois, Indiana, and Kentucky**


ARCHER DANIELS MIDLAND COMPANY

# Illinois Basin Decatur Project (IBDP)

ILLINOIS
INDUSTRIAL CARBON
CAPTURE & STORAGE

**Program Objective**

**Large scale geologic test to inject 1.0 million tons of $CO_2$ over a three year period (1,000 MT/day).**

**Project Team Members**

ADM  Schlumberger Carbon Services  NETL

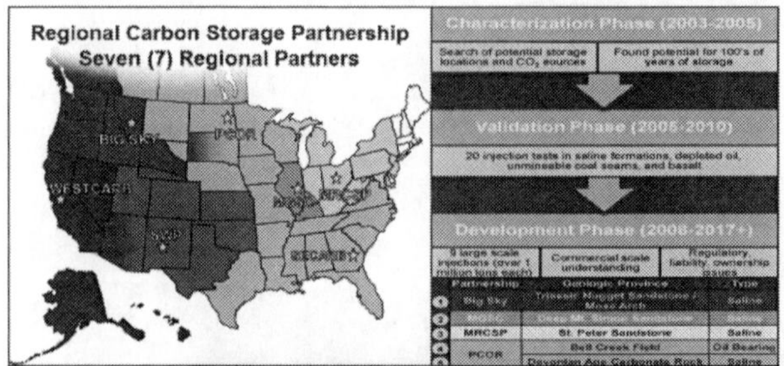


**Knowledge Base**

- Site Geological Characterization
- Risk Assessment & Reservoir Modeling
- Engineering Design & MVA

***Nov 2014-Completed goal of injecting and storing 1,000,000 tons of $CO_2$.***

ARCHER DANIELS MIDLAND COMPANY

# Illinois Industrial CCS Project (IL-ICCS)

ILLINOIS
INDUSTRIAL CARBON
CAPTURE & STORAGE

**Program Objectives**

- Target & Demonstrate Advanced CCS Technologies at Industrial Scale Facilities
- Inject and Store One Million Tons of $CO_2$ Annually (3,000 tons/day)

**Project Team Members**

ADM  Schlumberger Carbon Services  Richland Community College  NETL

**Knowledge Base**

- Site Geological Characterization
- Risk Assessment & Reservoir Modeling
- Engineering Design & MVA
- Education and Public Outreach

***Study the interaction between the $CO_2$ plumes from two injection wells within the same formation.***

ARCHER DANIELS MIDLAND COMPANY

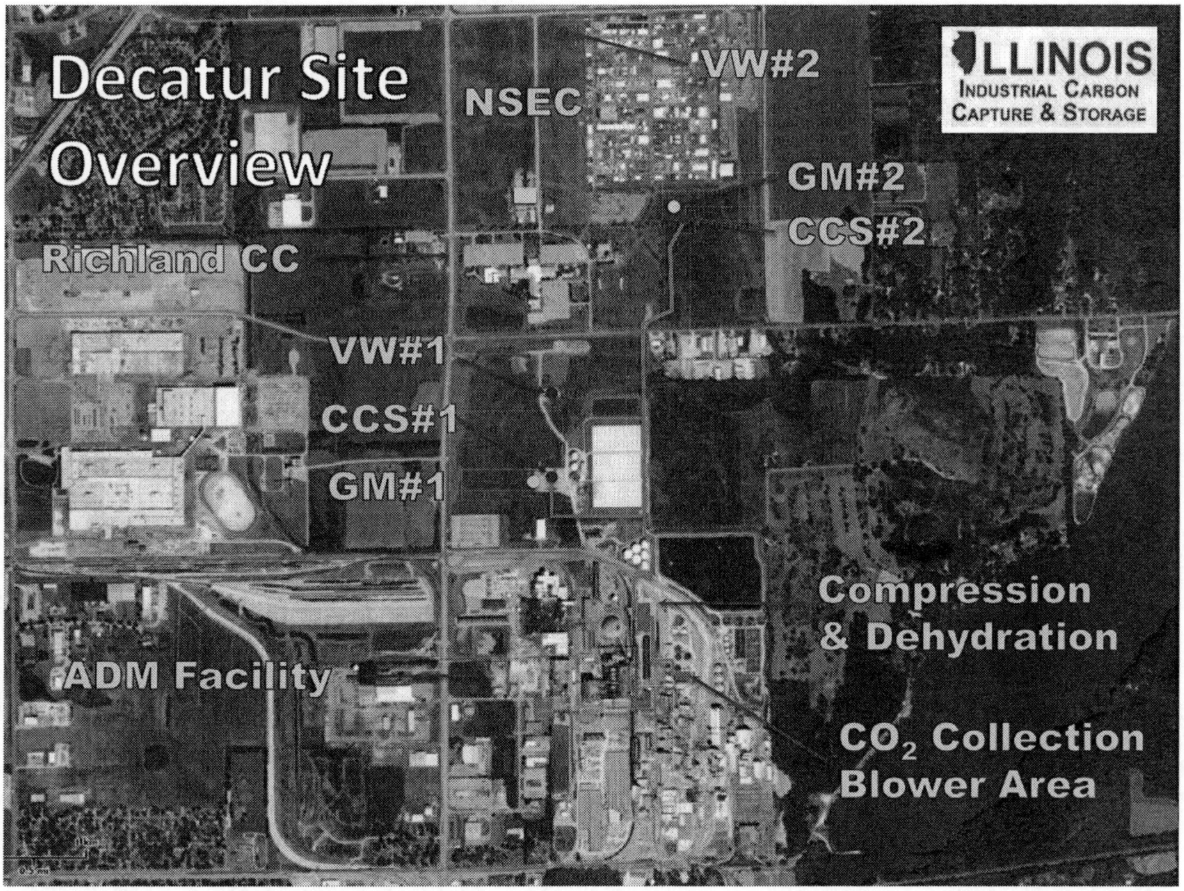
Decatur Site Overview
ILLINOIS
INDUSTRIAL CARBON CAPTURE & STORAGE
NSEC
VW#2
GM#2
CCS#2
Richland CC
VW#1
CCS#1
GM#1
Compression & Dehydration
ADM Facility
$CO_2$ Collection Blower Area

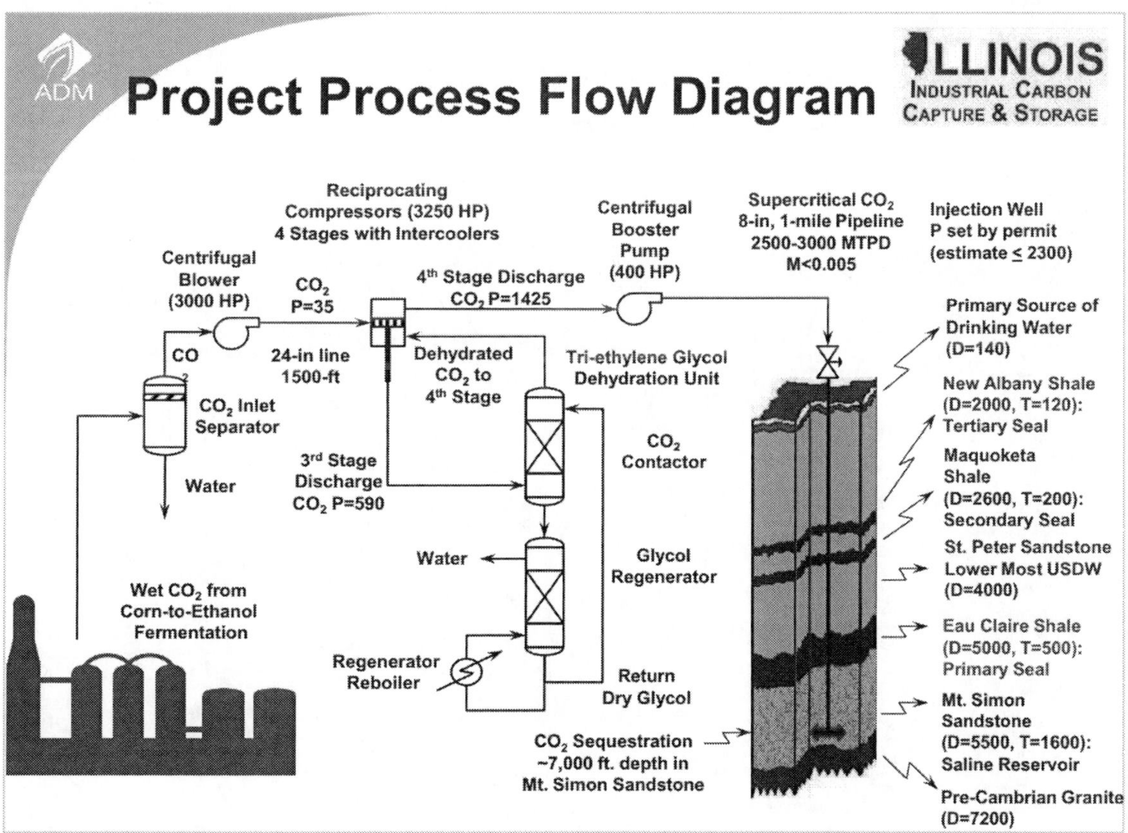
ADM
Project Process Flow Diagram
ILLINOIS
INDUSTRIAL CARBON CAPTURE & STORAGE
Reciprocating Compressors (3250 HP) 4 Stages with Intercoolers
Centrifugal Booster Pump (400 HP)
Supercritical $CO_2$ 8-in, 1-mile Pipeline 2500-3000 MTPD M<0.005
Injection Well P set by permit (estimate ≤ 2300)
Centrifugal Blower (3000 HP)
$CO_2$ P=35
4th Stage Discharge $CO_2$ P=1425
CO2
24-in line 1500-ft
Dehydrated $CO_2$ to 4th Stage
Tri-ethylene Glycol Dehydration Unit
$CO_2$ Inlet Separator
Water
3rd Stage Discharge $CO_2$ P=590
$CO_2$ Contactor
Water
Glycol Regenerator
Wet $CO_2$ from Corn-to-Ethanol Fermentation
Regenerator Reboiler
Return Dry Glycol
$CO_2$ Sequestration ~7,000 ft. depth in Mt. Simon Sandstone
Primary Source of Drinking Water (D=140)
New Albany Shale (D=2000, T=120): Tertiary Seal
Maquoketa Shale (D=2600, T=200): Secondary Seal
St. Peter Sandstone Lower Most USDW (D=4000)
Eau Claire Shale (D=5000, T=500): Primary Seal
Mt. Simon Sandstone (D=5500, T=1600): Saline Reservoir
Pre-Cambrian Granite (D=7200)
ARCHER DANIELS MIDLAND COMPANY

ADM

# Site Selection

## *Regional Geologic Characterization*

ILLINOIS
INDUSTRIAL CARBON
CAPTURE & STORAGE

- **Cratonic basin**
- **60,000 square mile area**
- **Structurally complex to the south with faulting and seismicity**
- **ADM Decatur facility is located near the center of this geologic formation**
- **Estimated $CO_2$ storage capacity between 27 to 109 billion metric tons**

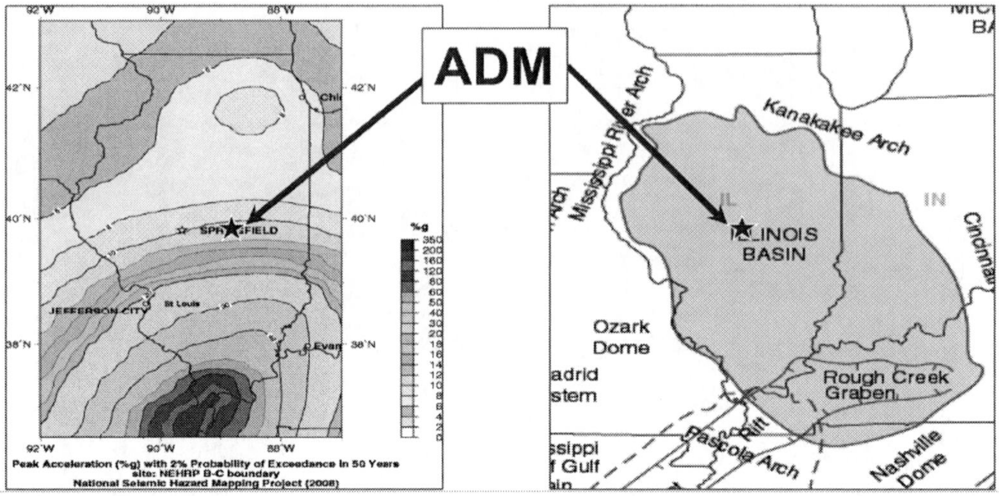


ARCHER DANIELS MIDLAND COMPANY

ADM

# Site Characterization

## *Seismic Acquisition*

ILLINOIS
INDUSTRIAL CARBON
CAPTURE & STORAGE

ARCHER DANIELS MIDLAND COMPANY

ADM
Test Well Construction
ILLINOIS
Industrial Carbon
Capture & Storage
Deep Monitoring & Geophysical Wells
ARCHER DANIELS MIDLAND COMPANY

ADM
Coring and Well Logging
ILLINOIS
Industrial Carbon
Capture & Storage
Core Samples and Well Logging
ARCHER DANIELS MIDLAND COMPANY

ADM

# Geophysical Model Development

ILLINOIS
INDUSTRIAL CARBON
CAPTURE & STORAGE

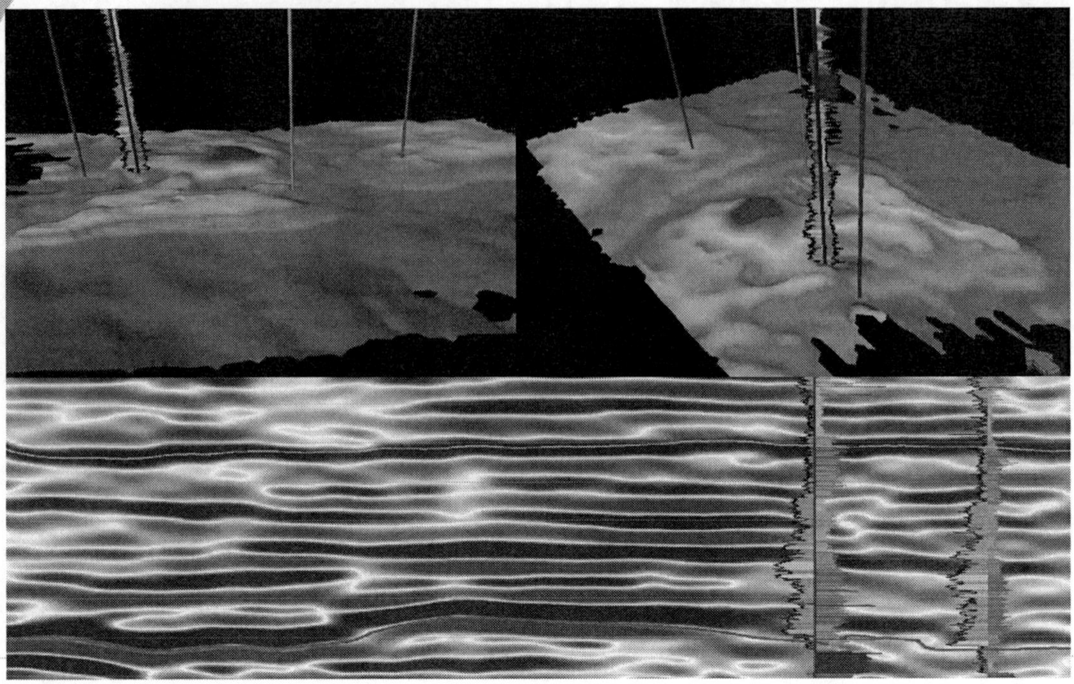

ADM

# Site Permitting

*USEPA: UIC Class VI Permit*

***1st UIC Class VI Permit Application Reviewed by the USEPA***

- **UIC Class VI permit application submitted on July 25, 2011.**
- **US EPA Region V issued a draft permit on April 15, 2014.**
- **Public hearing conducted on May 21, 2014.**
- **Public Comment Period concluded May 31, 2014.**
- **Final permit issued December 28, 2014.**

ARCHER DANIELS MIDLAND COMPANY

ADM

# Geophysical Model Framework

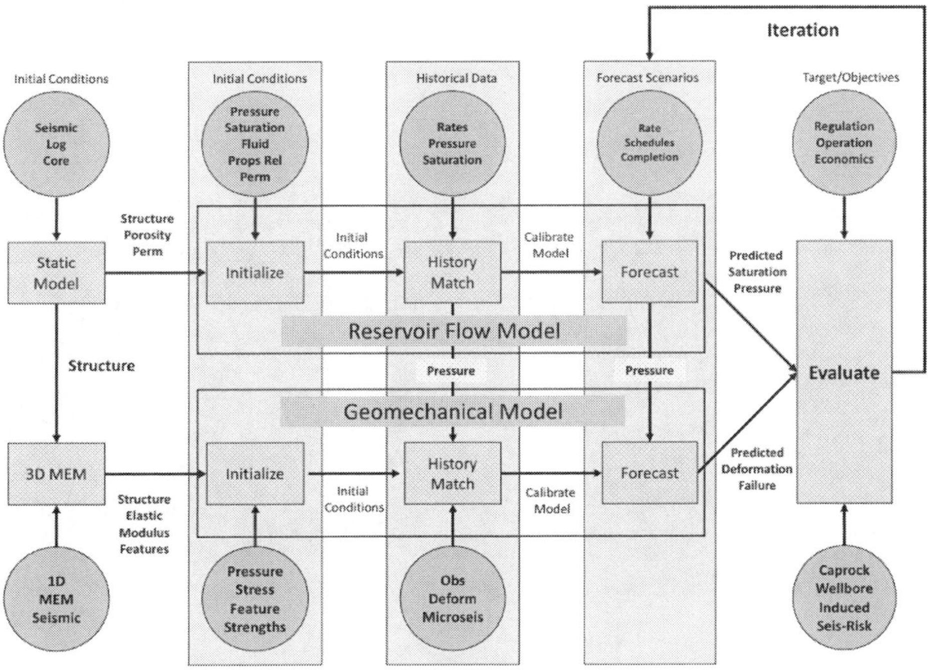


ARCHER DANIELS MIDLAND COMPANY

ADM

ILLINOIS
INDUSTRIAL CARBON
CAPTURE & STORAGE

# Geophysical Modeling

## *$CO_2$ Plume Position - 2012*

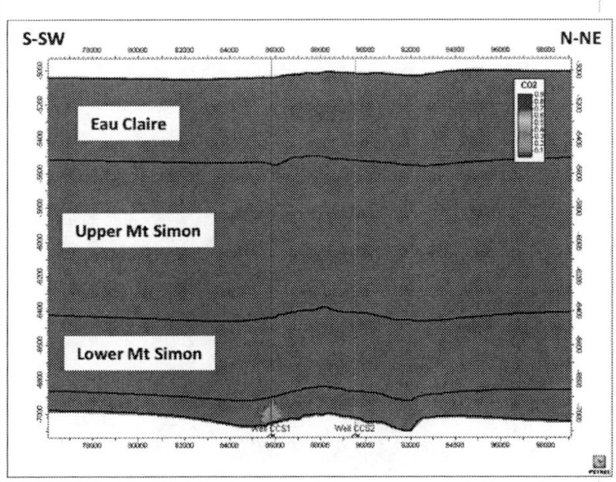


ARCHER DANIELS MIDLAND COMPANY

ADM

# Geophysical Modeling

## *$CO_2$ Plume Position - 2013*

ILLINOIS
Industrial Carbon
Capture & Storage

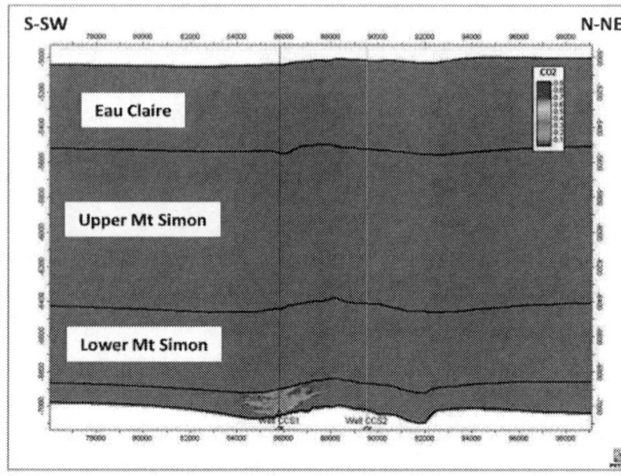




ADM

# Geophysical Modeling

## *$CO_2$ Plume Position - 2014*

ILLINOIS
Industrial Carbon
Capture & Storage

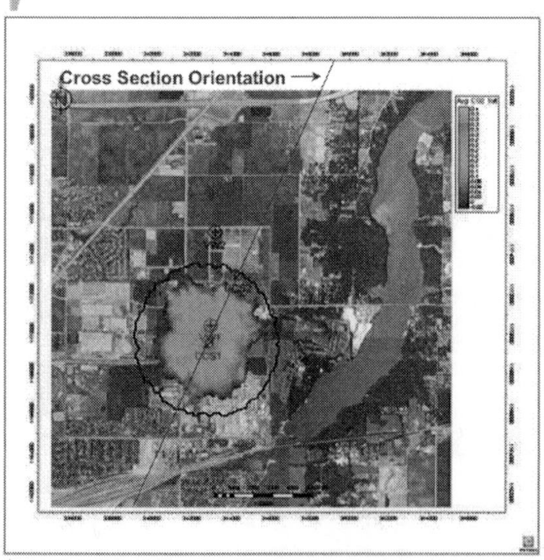


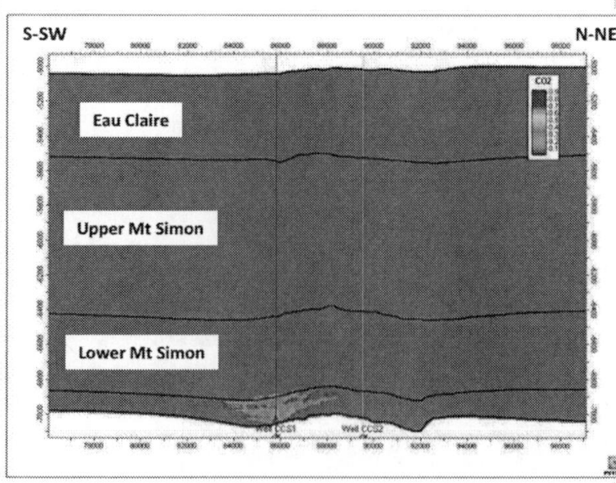

ADM

ILLINOIS
INDUSTRIAL CARBON
CAPTURE & STORAGE

# Geophysical Modeling

## *$CO_2$ Plume Position - 2015*

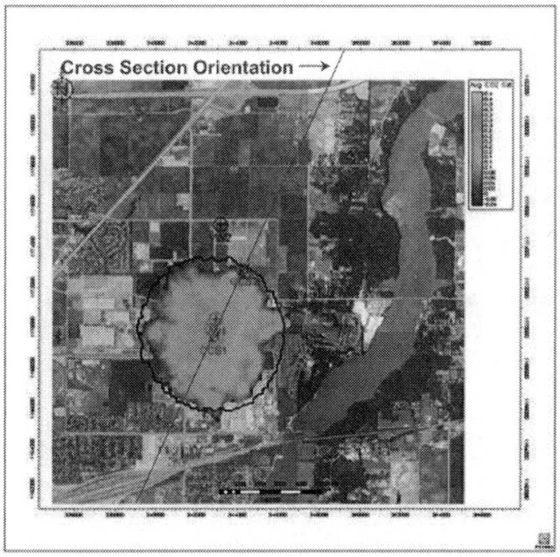


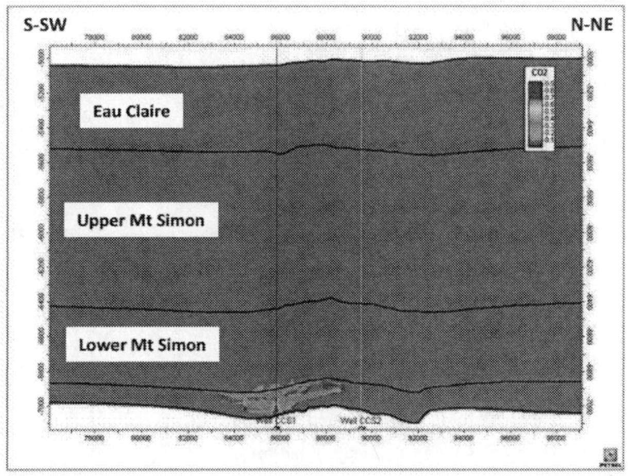




ADM

ILLINOIS
INDUSTRIAL CARBON
CAPTURE & STORAGE

# Geophysical Modeling

## *$CO_2$ Plume Position - 2016*

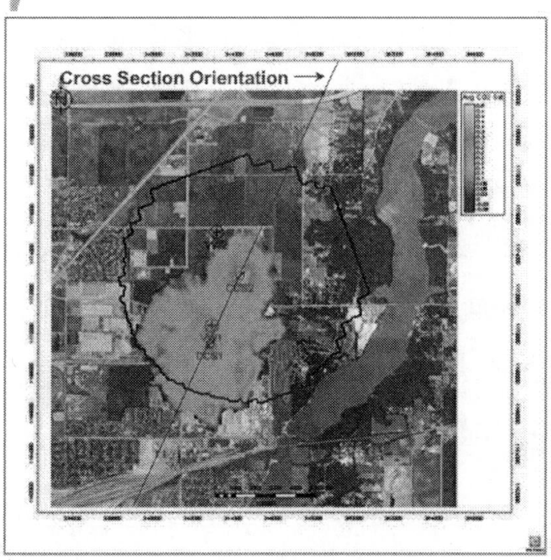


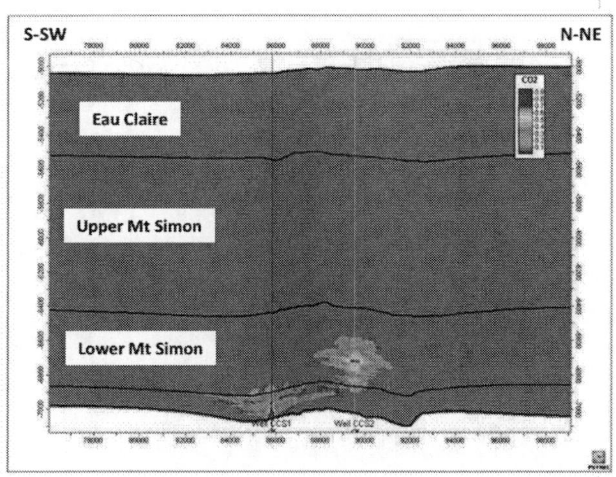


ARCHER DANIELS MIDLAND COMPANY

ADM

ILLINOIS
INDUSTRIAL CARBON
CAPTURE & STORAGE

# Geophysical Modeling

## *$CO_2$ Plume Position - 2017*

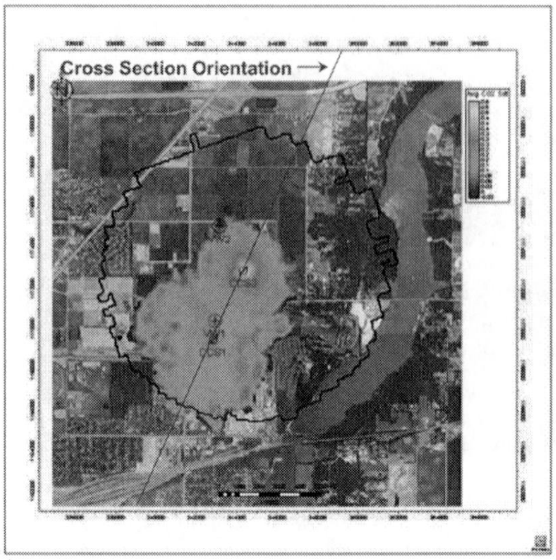


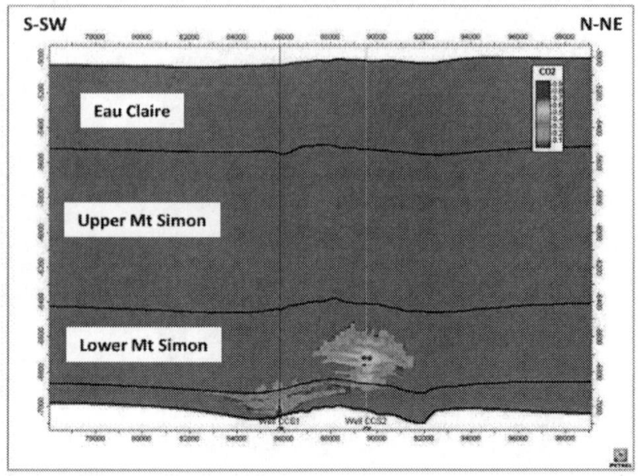


ARCHER DANIELS MIDLAND COMPANY

ADM

ILLINOIS
INDUSTRIAL CARBON
CAPTURE & STORAGE

# Geophysical Modeling

## *$CO_2$ Plume Position - 2018*

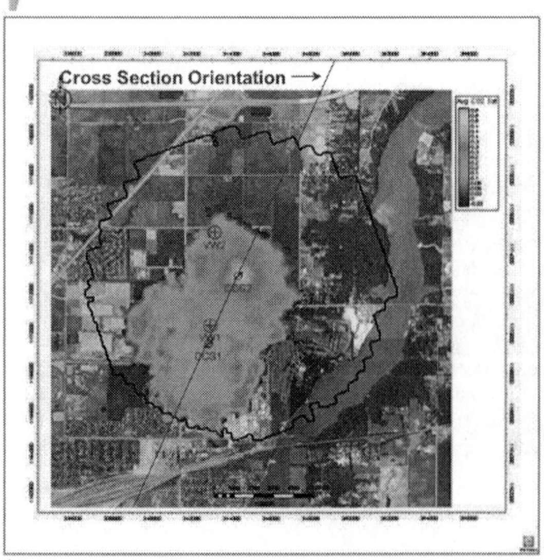


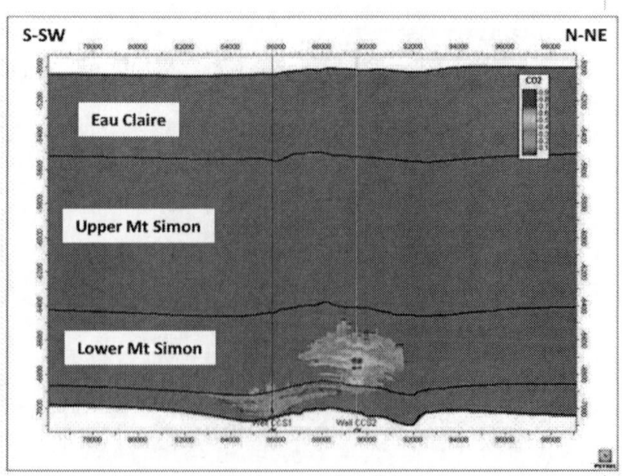


ARCHER DANIELS MIDLAND COMPANY

ADM

# Geophysical Modeling

## *$CO_2$ Plume Position - 2019*

ILLINOIS
INDUSTRIAL CARBON
CAPTURE & STORAGE

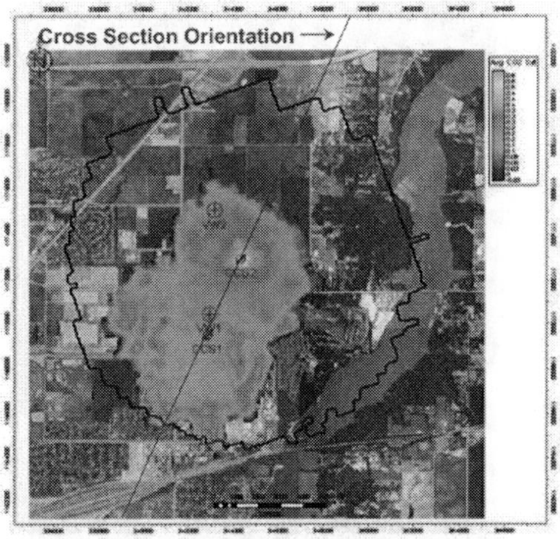


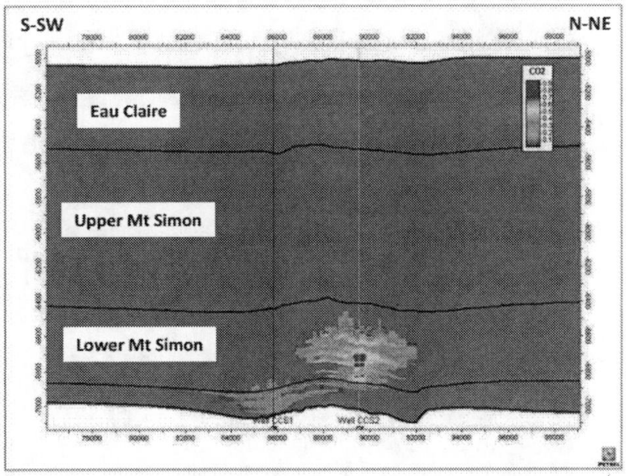


ARCHER DANIELS MIDLAND COMPANY

ADM

# Geophysical Modeling

## *$CO_2$ Plume Position - 2020*

ILLINOIS
INDUSTRIAL CARBON
CAPTURE & STORAGE

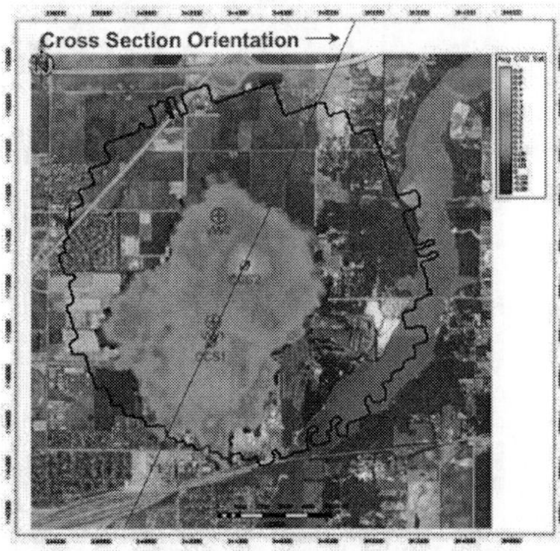


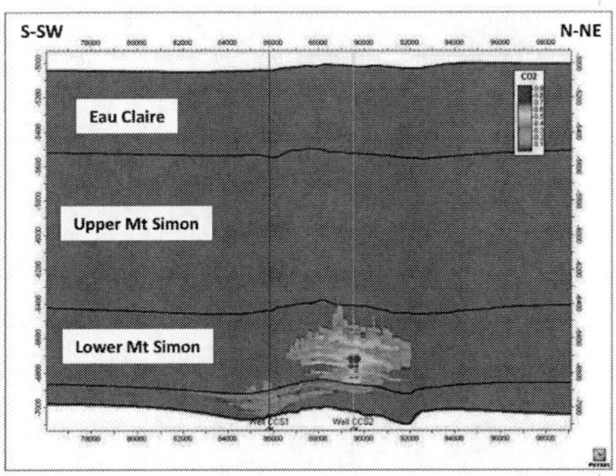


ARCHER DANIELS MIDLAND COMPANY

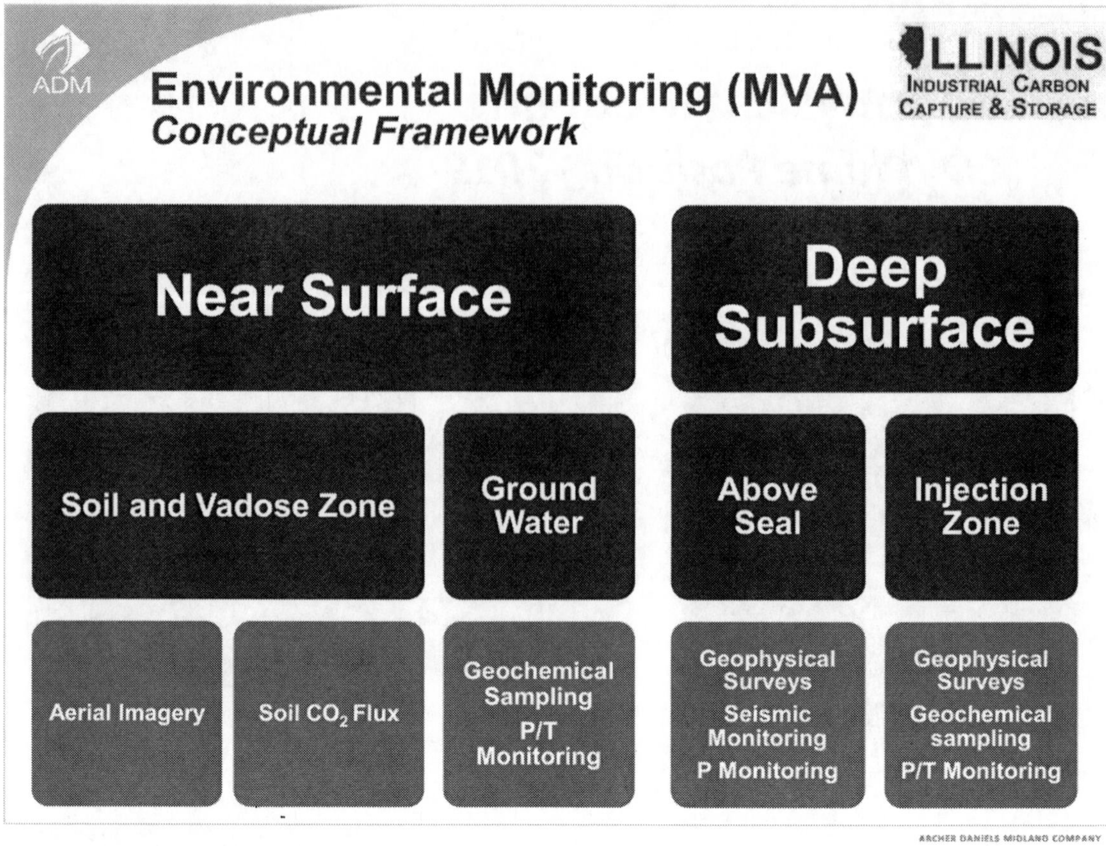
ADM
ILLINOIS
INDUSTRIAL CARBON
CAPTURE & STORAGE
Environmental Monitoring (MVA)
Conceptual Framework
Near Surface
Deep Subsurface
Soil and Vadose Zone
Ground Water
Above Seal
Injection Zone
Aerial Imagery
Soil $CO_2$ Flux
Geochemical Sampling
P/T Monitoring
Geophysical Surveys
Seismic Monitoring
P Monitoring
Geophysical Surveys
Geochemical sampling
P/T Monitoring
ARCHER DANIELS MIDLAND COMPANY

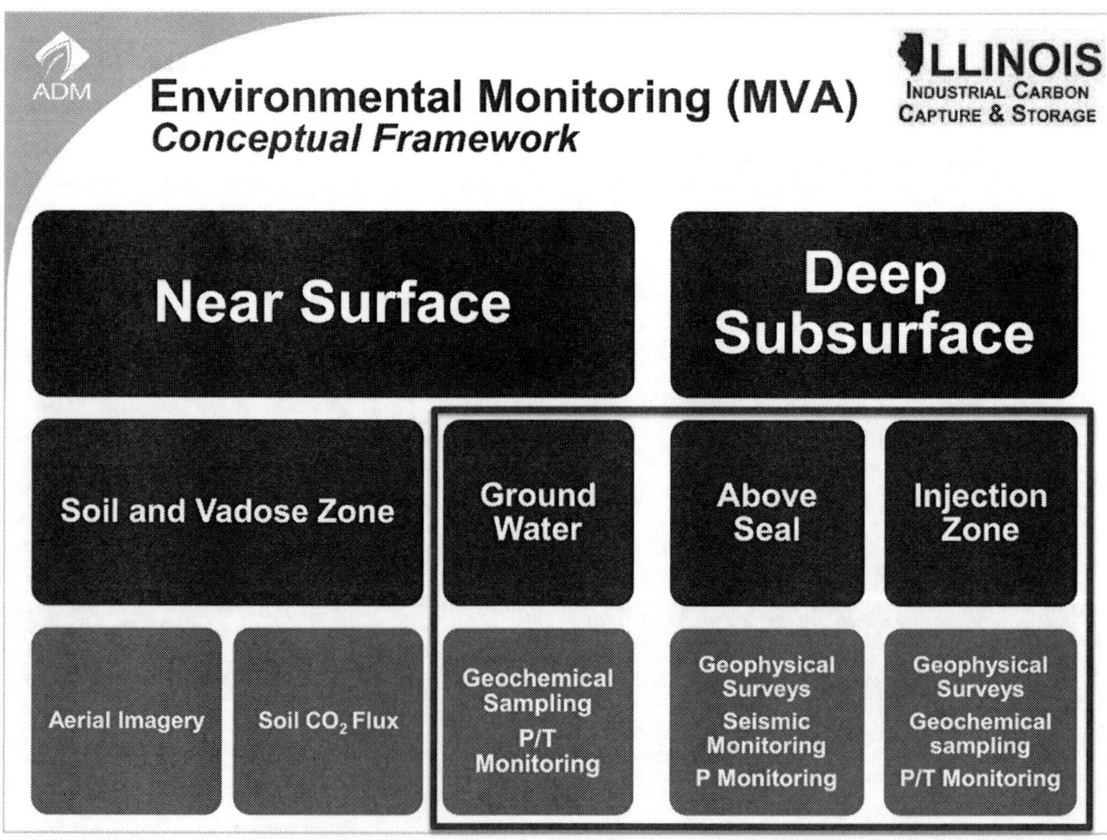
ADM
ILLINOIS
INDUSTRIAL CARBON
CAPTURE & STORAGE
Environmental Monitoring (MVA)
Conceptual Framework
Near Surface
Deep Subsurface
Soil and Vadose Zone
Ground Water
Above Seal
Injection Zone
Aerial Imagery
Soil $CO_2$ Flux
Geochemical Sampling
P/T Monitoring
Geophysical Surveys
Seismic Monitoring
P Monitoring
Geophysical Surveys
Geochemical sampling
P/T Monitoring
ARCHER DANIELS MIDLAND COMPANY

# Environmental Monitoring
## *Near Surface Monitoring*

ILLINOIS
INDUSTRIAL CARBON
CAPTURE & STORAGE

- **Near infrared aerial imagery will be used to evaluate plant stress**
- **Soil resistivity characterized shallow depths for identification of optimum GWM locations**
- **GWM for baseline conditions and operational surveillance**
- **Surface soil $CO_2$ flux monitoring**

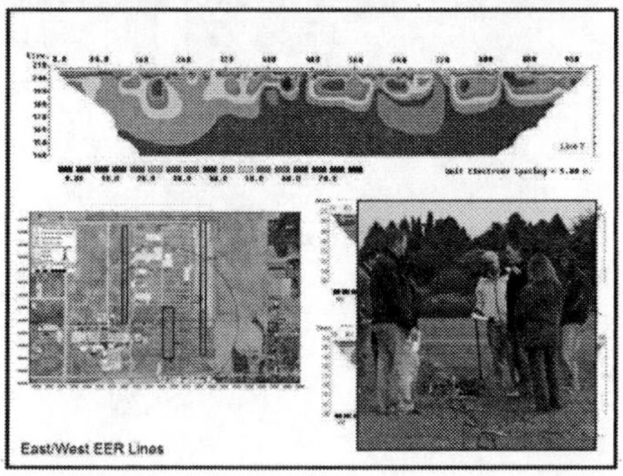

# Environmental Monitoring
## *Deep Subsurface Monitoring*

ILLINOIS
INDUSTRIAL CARBON
CAPTURE & STORAGE

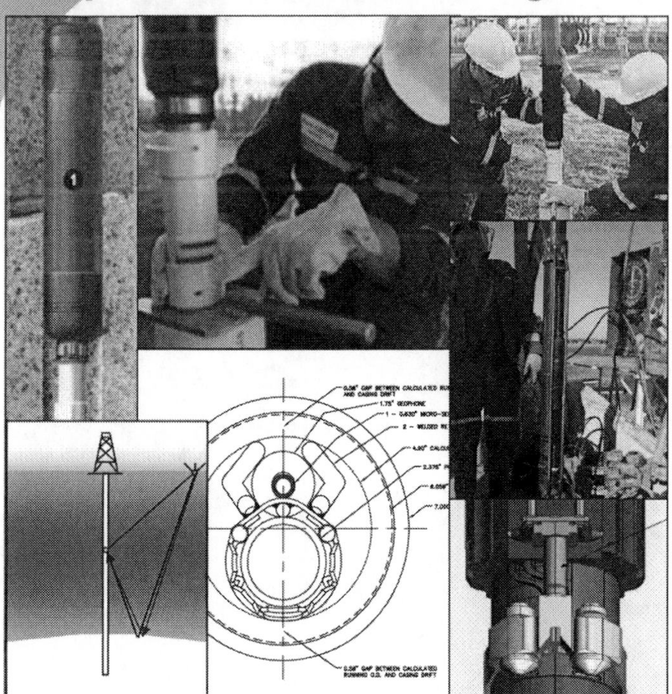

- **Deep P/T monitoring**
- **Distributed Temp Sensor**
- **Distributed Vibration Sensor**
- **Multi-level fluid sampling ports**
- **P/T monitoring above the seal formation**
- **Deep geophones arrays allowing vertical seismic profiling (VSP)**
- **Pulse neutron well logging**

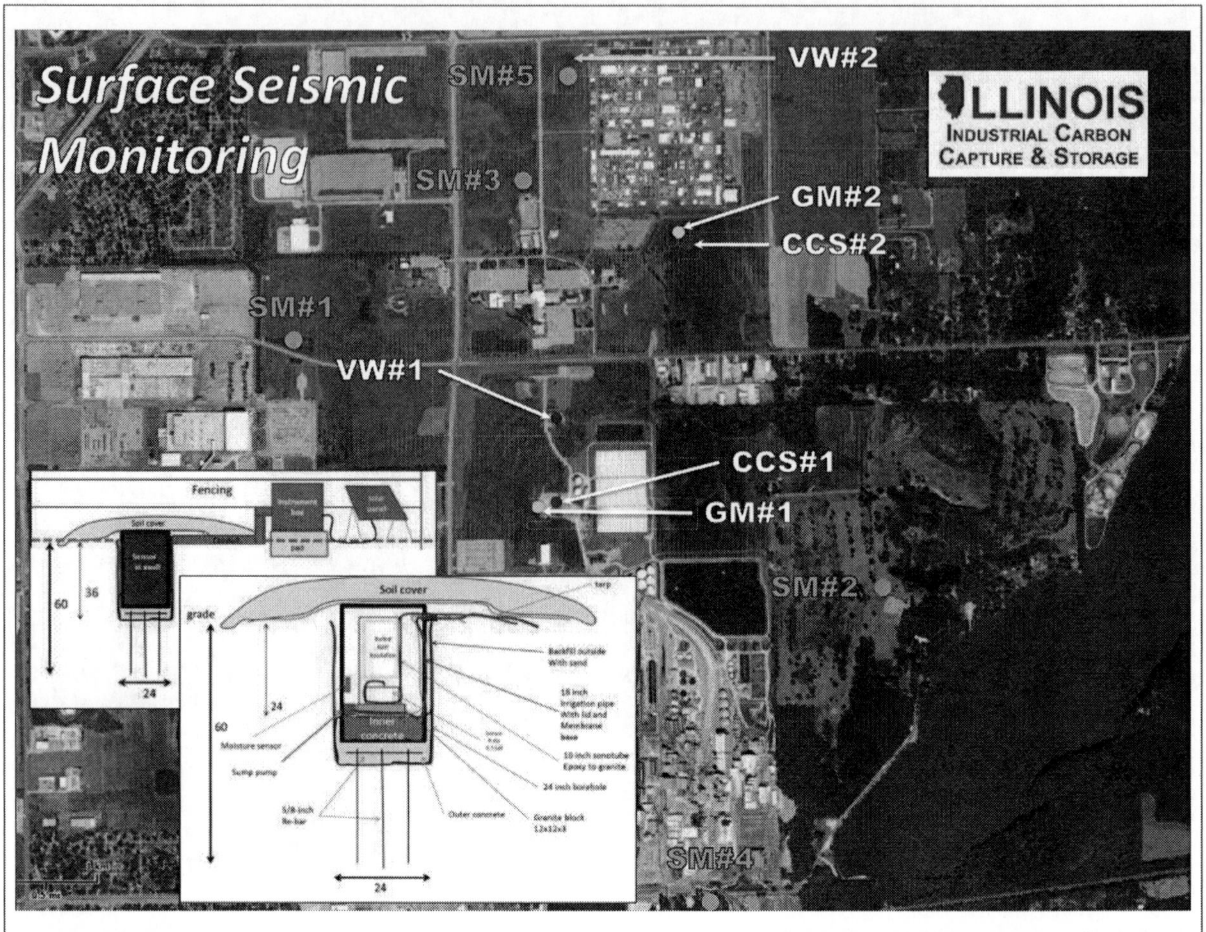
Surface Seismic Monitoring
ILLINOIS
INDUSTRIAL CARBON CAPTURE & STORAGE
VW#2
SM#5
SM#3
GM#2
CCS#2
SM#1
VW#1
CCS#1
GM#1
SM#2
SM#4
Fencing
Soil cover
60
36
24
Soil cover
grade
60
24
24
Moisture sensor
Sump pump
Inner concrete
Outer concrete
Backfill outside With sand
24 inch borehole

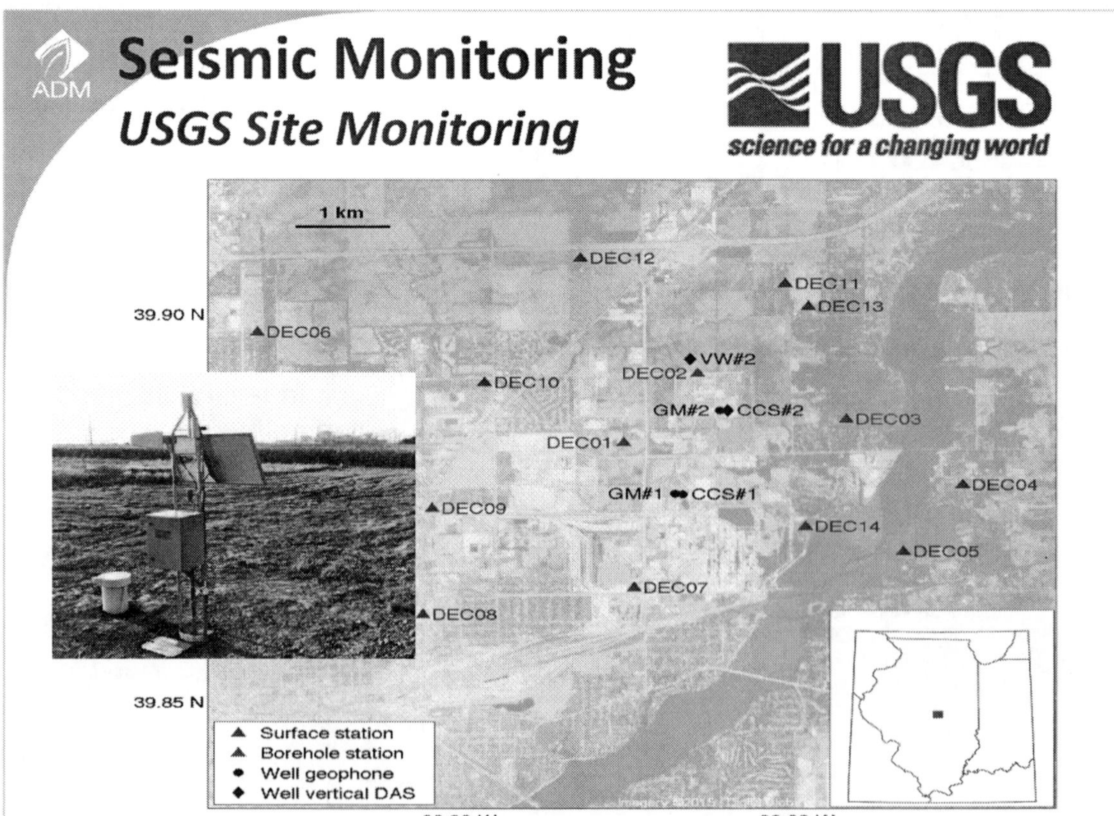
ADM
Seismic Monitoring
USGS Site Monitoring
USGS
science for a changing world
1 km
DEC12
DEC11
DEC13
39.90 N
DEC06
VW#2
DEC02
DEC10
GM#2
CCS#2
DEC03
DEC01
DEC04
GM#1
CCS#1
DEC09
DEC14
DEC05
DEC07
DEC08
39.85 N
Surface station
Borehole station
Well geophone
Well vertical DAS
88.92 W
88.88 W

ADM

**ILLINOIS**
INDUSTRIAL CARBON
CAPTURE & STORAGE

# Deep Seismic Monitoring

- 3 Geophone Arrays in CCS1
- 24 Active Geophones in GM1
- 5 Level Geophone Array in VW2
- 7 Borehole Stations (500 ft)
- 14 Surface Stations

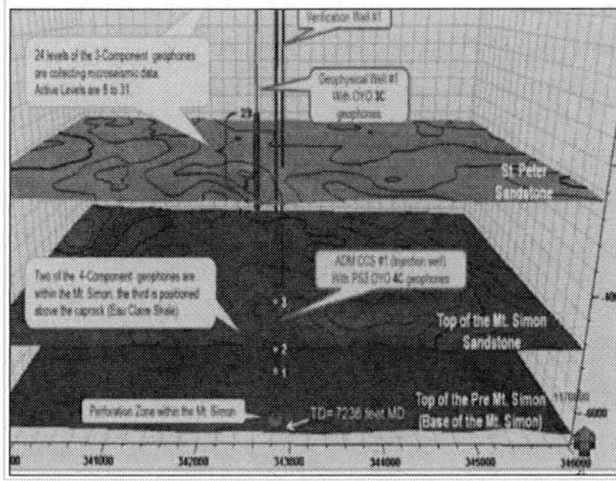


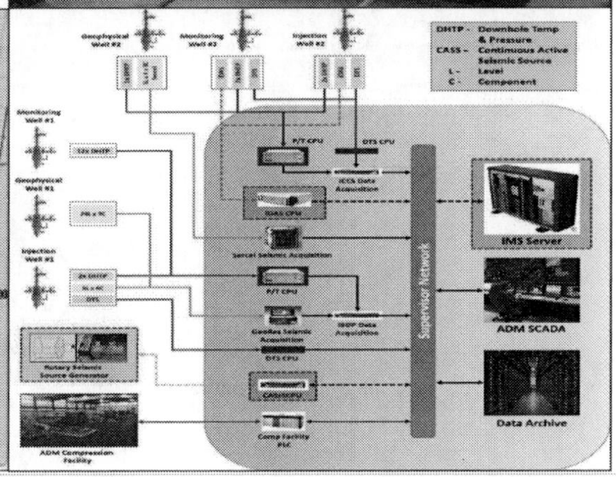


ARCHER DANIELS MIDLAND COMPANY

ADM

**ILLINOIS**
INDUSTRIAL CARBON
CAPTURE & STORAGE

# Seismic Monitoring Data

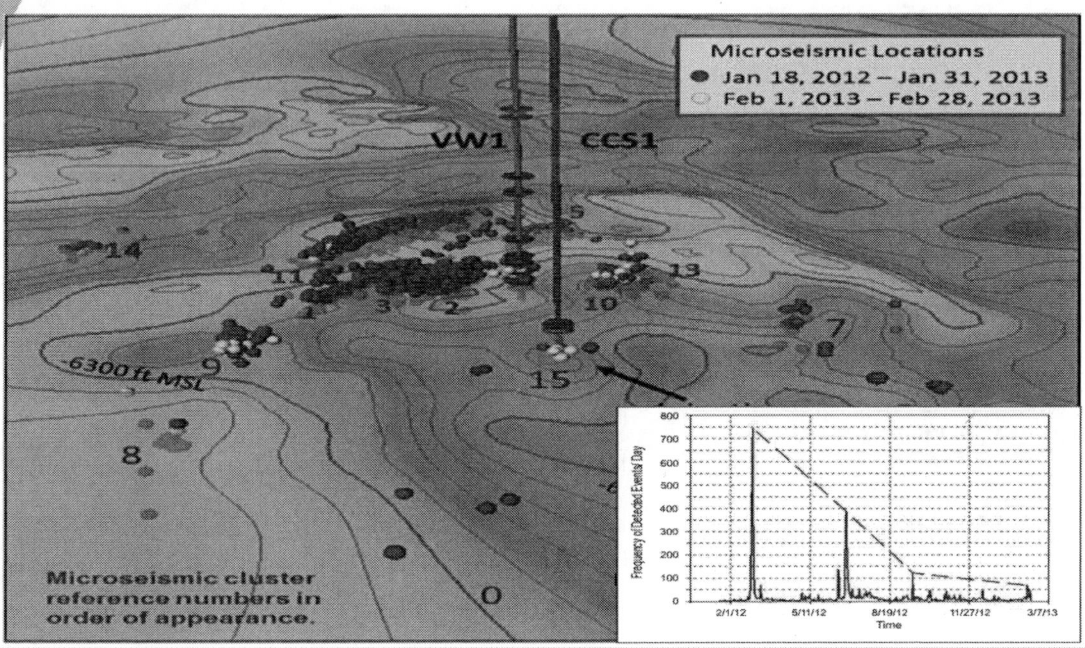


ARCHER DANIELS MIDLAND COMPANY

# Time Lapse VSP Surveys

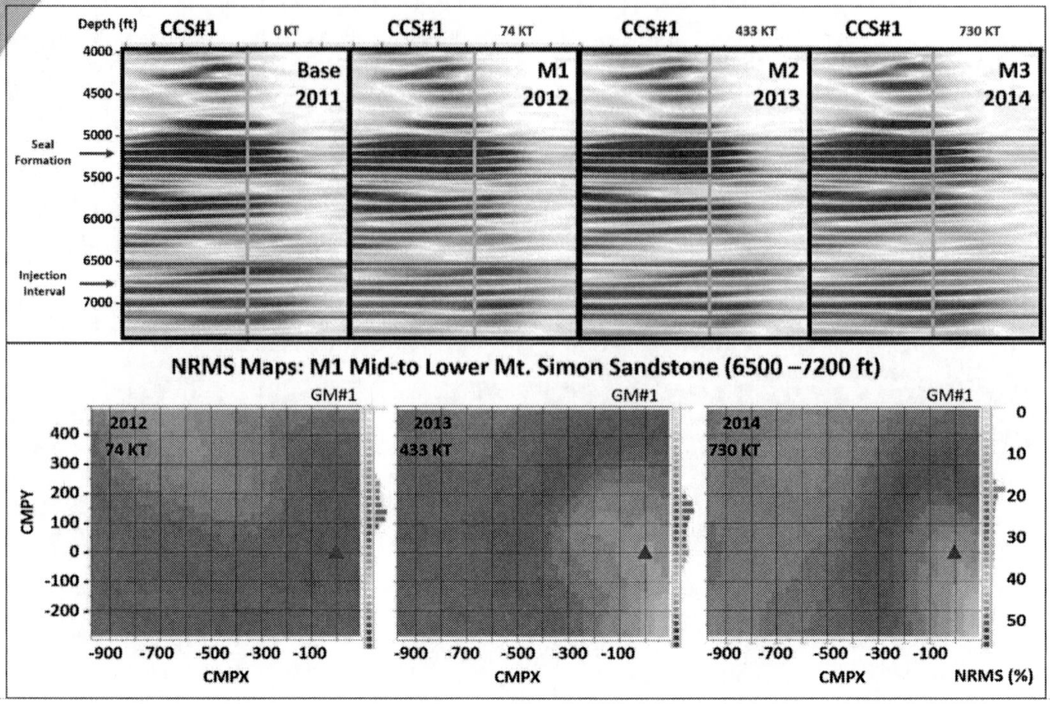




# $CO_2$ Based Chemicals

- **Carbonates**
  - Glycerol Carbonate
  - Propylene Carbonate
  - Dimethyl Carbonate
- **Fertilizers**
- **Alcohols**
- **Fuels**
- **Acids**
- **Others**

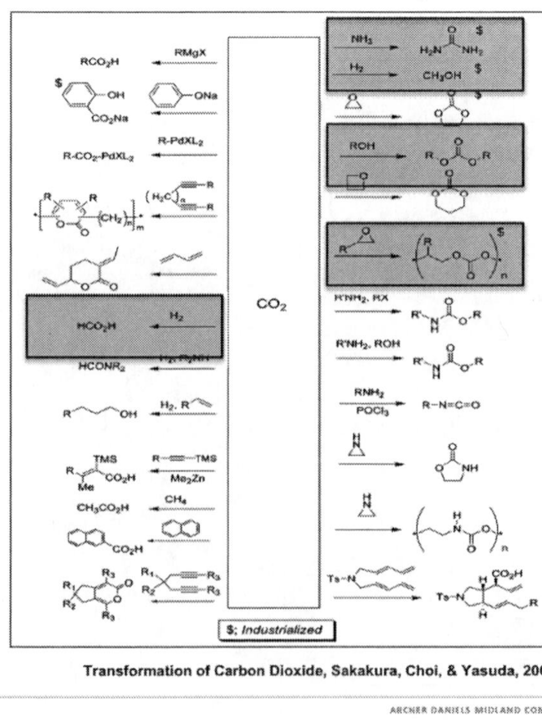


Transformation of Carbon Dioxide, Sakakura, Choi, & Yasuda, 2007

# Enhanced Oil Production

http://www.youtube.com/watch?v=azLVjYij5U4

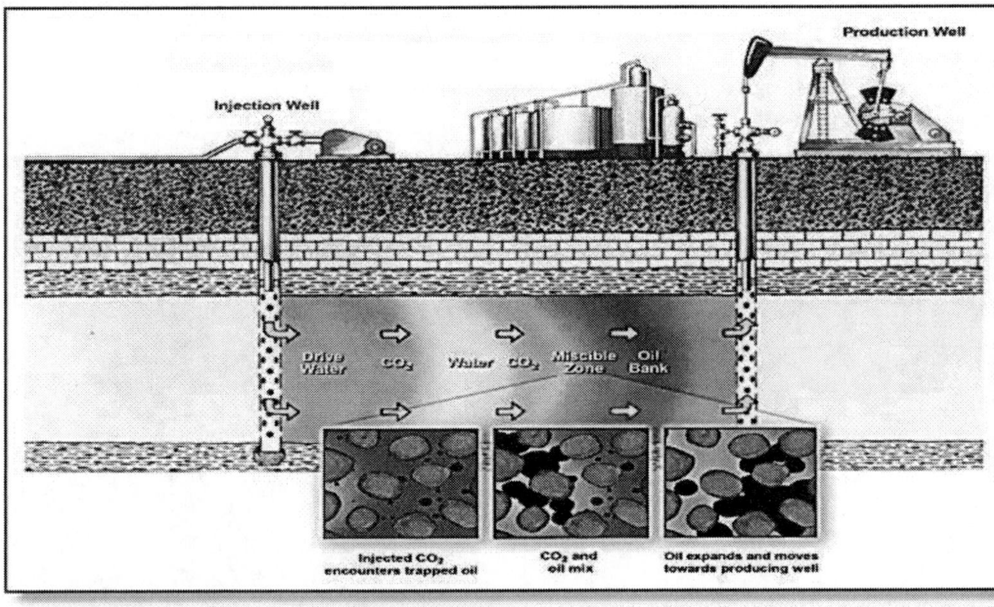


*Cross-section illustrating how carbon dioxide and water can be used to flush residual oil from a subsurface rock formation between wells*



# Illinois Basin Potential

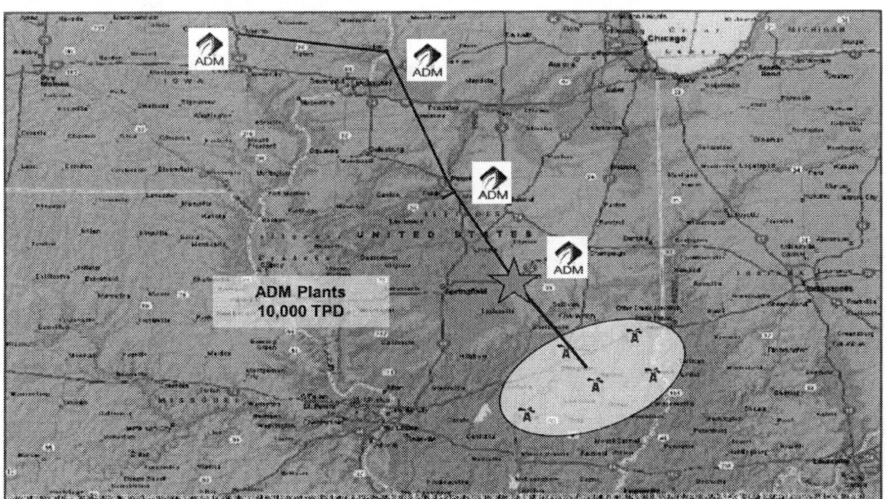


- Illinois Basin - Oil Producing Region
- Est. Recoverable Oil = 700 million bbls[(1)]
- Est. $CO_2$ Requirements = 150 million tons

(1) BASIN ORIENTED STRATEGIES FOR CO2 ENHANCED OIL RECOVERY: ILLINOIS AND MICHIGAN BASIN OF ILLINOIS, INDIANA, KENTUCKY AND MICHIGAN; Advanced Resources International, February 2006

# Intelligent Monitoring System (IMS)

ADM

ILLINOIS INDUSTRIAL CARBON CAPTURE & STORAGE

**Program Objectives**

- Develop and validate software tools that advance CCS-specific IMS by enabling access, integration and analysis of real-time surface and subsurface data for decision-making and automation of process
- Demonstrate integration of system components to validate feasibility of real-world application to CCS.

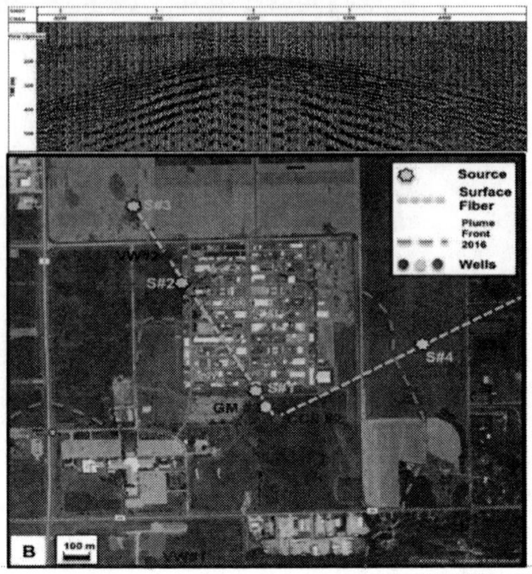


**Project Team Members**

ADM, Berkeley Lab, Silixa, NETL, USGS science for a changing world, Richland Community College, Schlumberger Carbon Services, Illinois

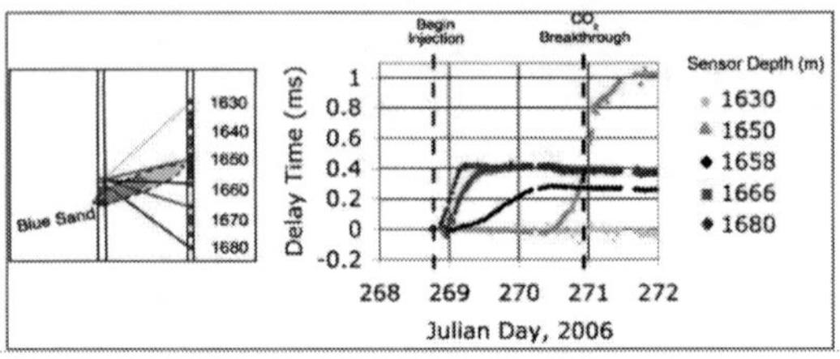




# Project Background

- Under the UIC Class VI permit ADM is required to monitor the plume and demonstrate plume stabilization prior to site closure.
- Initial technology choice was to conduct 3-D time lapse surface seismic surveys.
- Per the permit, ADM is required to conduct three 3D surface seismic surveys; one (1) during injection operation and two (2) surveys during the post injection survey site closure period.

# Project Background

- Based on the results of the vertical seismic profiling (VSP) and time lapse 3D seismic survey, visualization for the CO2 is very limited.
- No effect is seen when examining the amplitude data.
- Limited visualization is seen when interpreting the RMS data.

37

# VSP Results - *Amplitude Data*

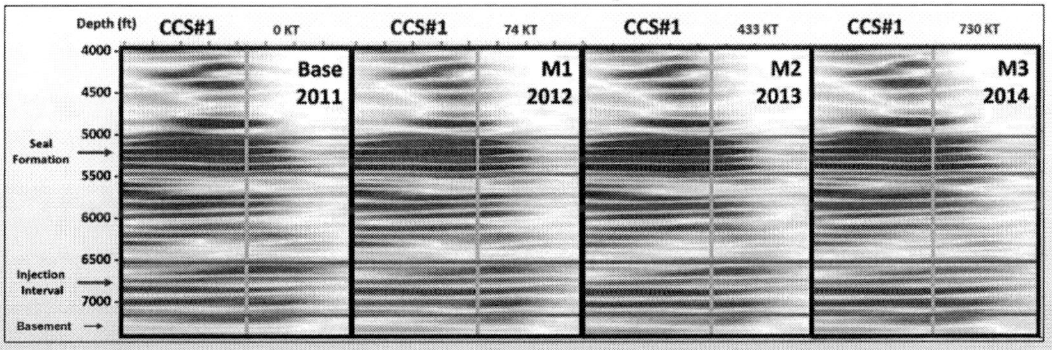


- **One should see increasing contrast (darker reflection in the seal and basement formations.**
- **The three time lapse images look virtually identical, indicating no change in the formation properties and no resolution of the $CO_2$ plume.**

38

# VSP Results – *RMS Data*

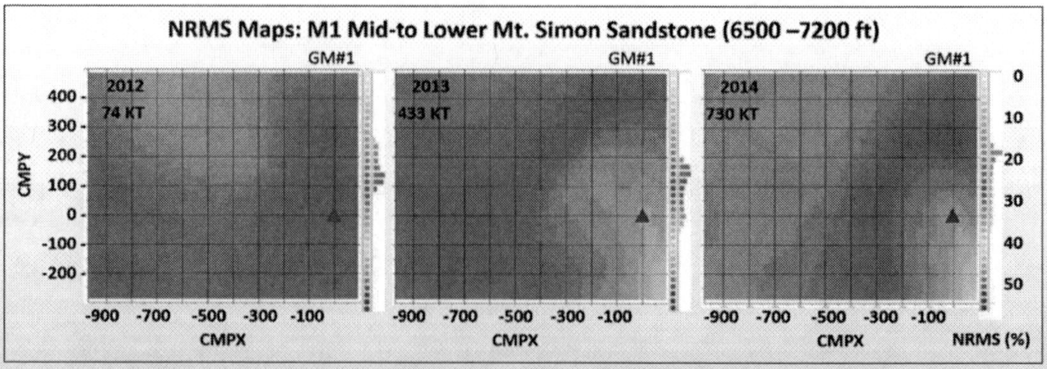


- **Examining a vertical volume of data, one can discern discrete changes in the formation properties (fuzzy horizontal plume).**
- **But comparing 2013 versus 2014 there are inconsistencies. The 2013 plume appears larger than the 2014 plume (impossible).**

39

# IMS Project Benefits

- **Reduce overall storage cost.**
- **Increase monitoring sensitivity.**
- **Supplant expensive seismic surveys that appear to have limited benefits.**
- **Increase monitoring reliability by using an integrated system.**
- **Optimize operation and maintenance activities.**
- **Reduce project risk during and after the injection of $CO_2$.**
- **Offset ~$6.0 million in UIC Class VI monitoring activities specifically 3D seismic surveys.**

40

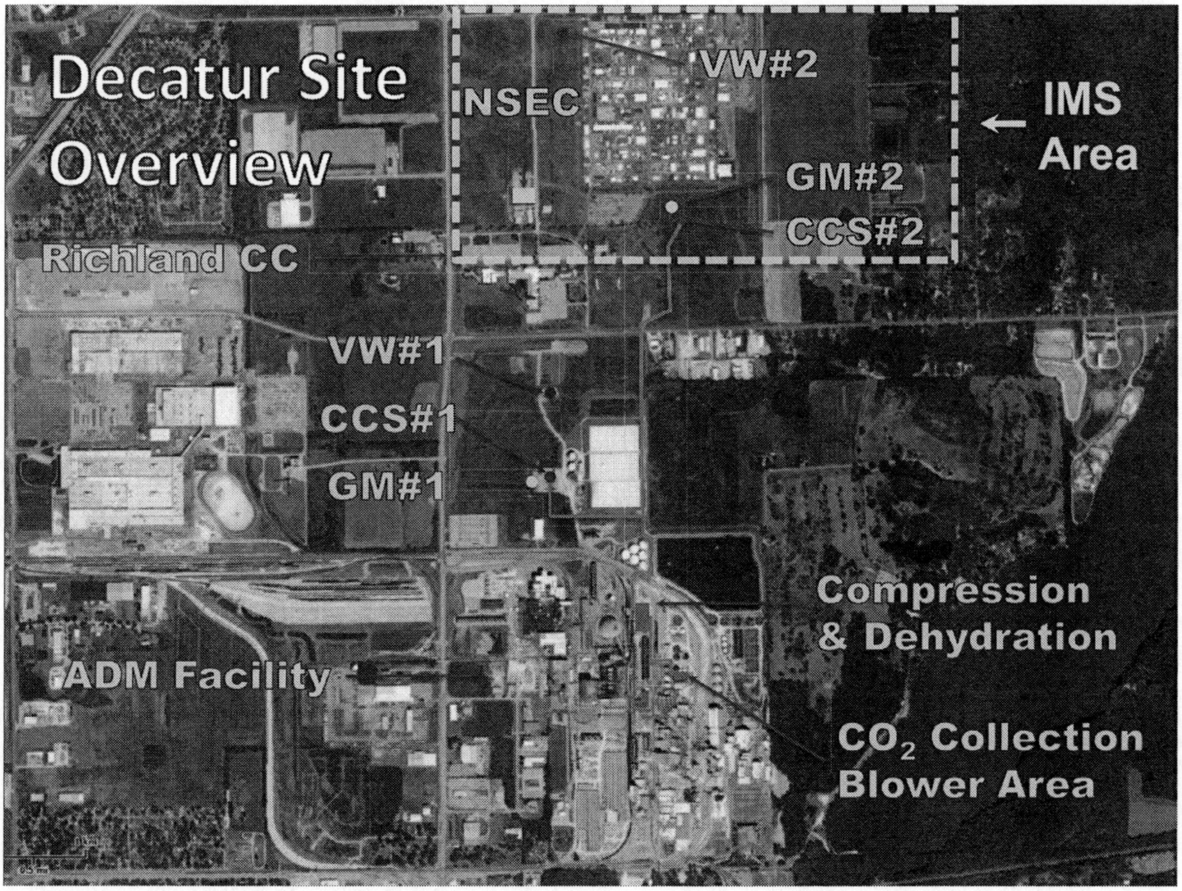


# USGS Seismic Monitoring

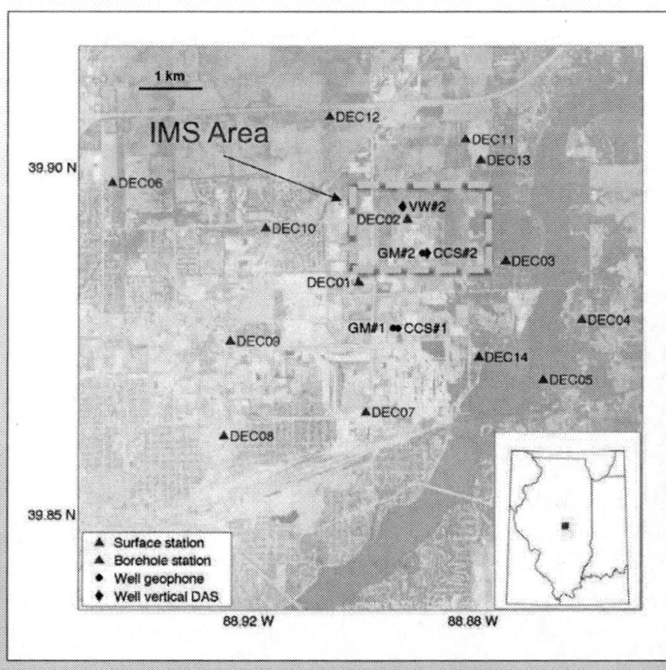


42

# Methodology

The proposed ICCS IMS program is transformative. It integrates into a single real-time processing framework the following emergent technologies.

- DAS seismic imaging (VSP & surface reflection)
- Permanent rotary surface source
- Hybrid geophone/DAS microseismic array
- DTS well integrity monitoring
- Integration into the existing Decatur Monitoring system.

43

# Distributed Acoustic Sensing (DAS)

Example from PTRC Aquistore
DAS Baseline 3D-VSP

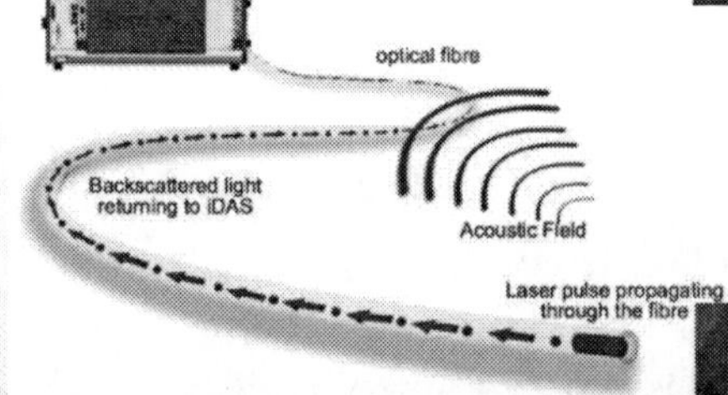

DAS VSP is becoming accepted technology.

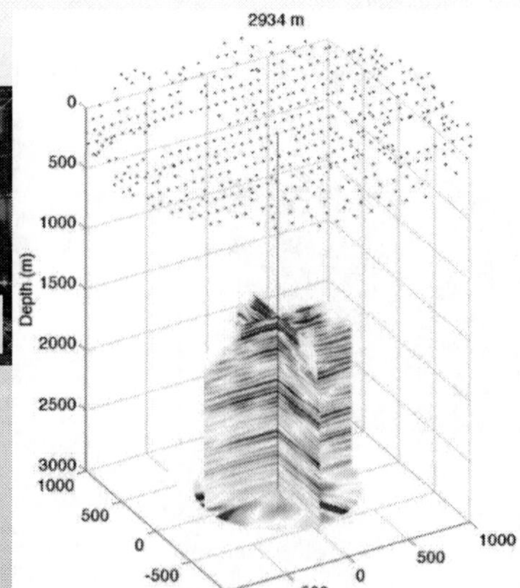

# Existing vs. New Technology

**Convention Seismic**

**DAS Seismic**

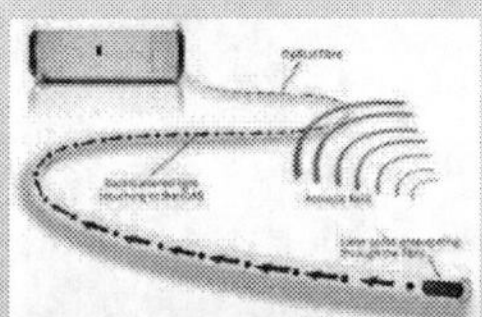

**Seismic surveys are considered the backbone technique for $CO_2$ storage monitoring programs.**

**Stringing thousands of cables and running thumper trucks every few years can test the limits of good neighbors. Costs are high.**

**Permanent reservoir monitoring offers a way to obtain higher quality information with minimal intrusion into surrounding lands –**

- **DAS provides high spatial and temporal resolution.**
- **Installation can be in horizontal directionally drilled boreholes beneath bodies of water, existing infrastructure.**
- **Excitation of DAS cables can be achieved through permanent fixed rotary sources for continuous monitoring.**

45

# Technologies Developed under DOE Core R&D are incorporated into the IMS

Rotary Orbital Source

Broadside Sensitive Cable

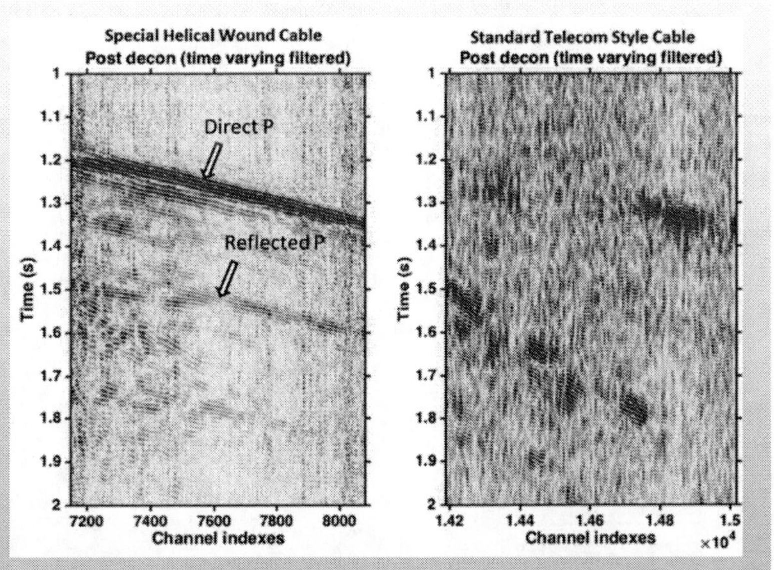


46

# IMS Equipment Layout

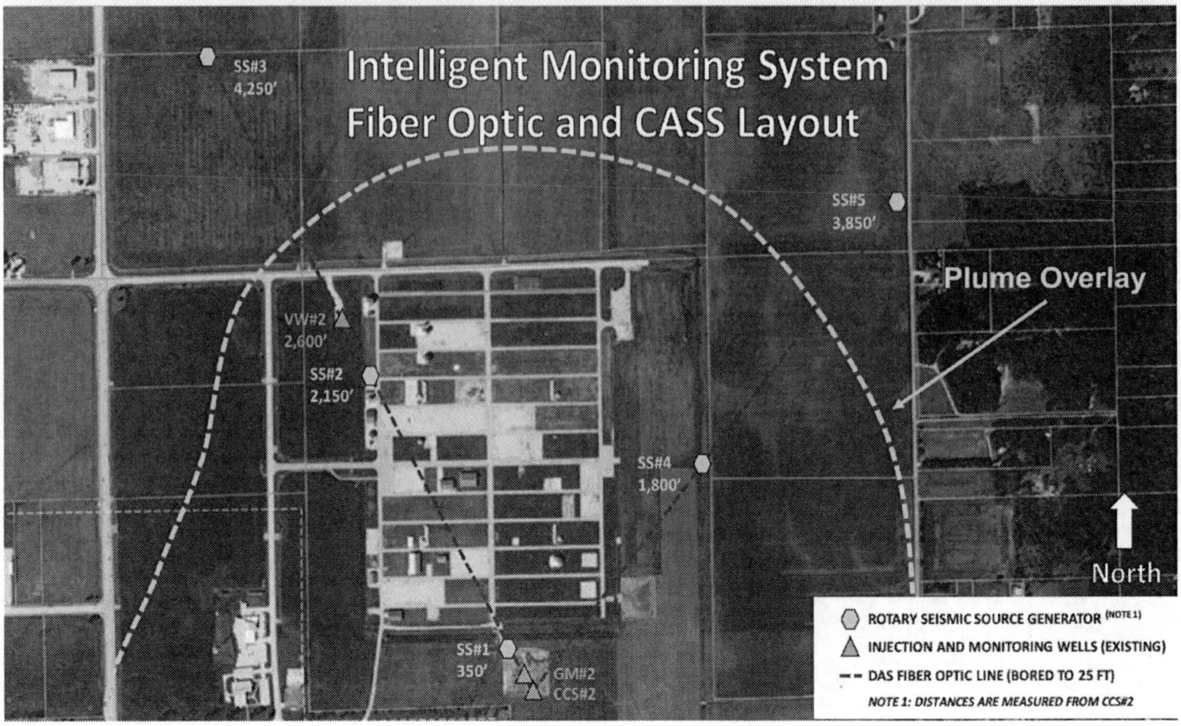


# IMS Data Acquisition Architecture

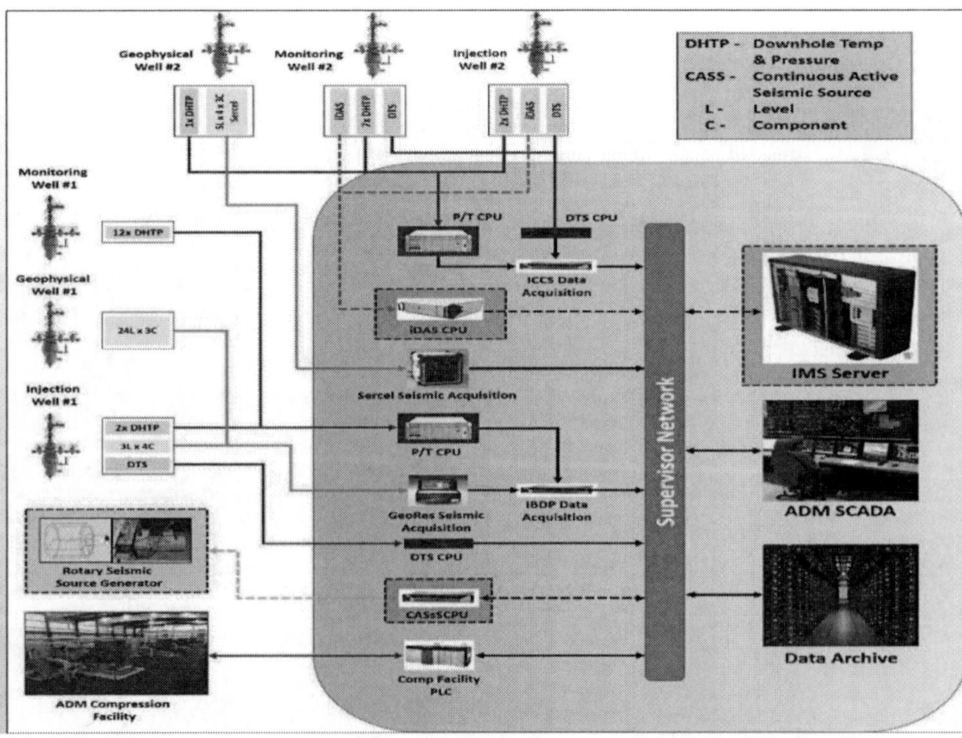


48

# IMS Project Timeline/Milestones

- Project began with scoping and DOE approval activities in October 2015.
- Project will be funded by the DOE for work performed from January 2016.

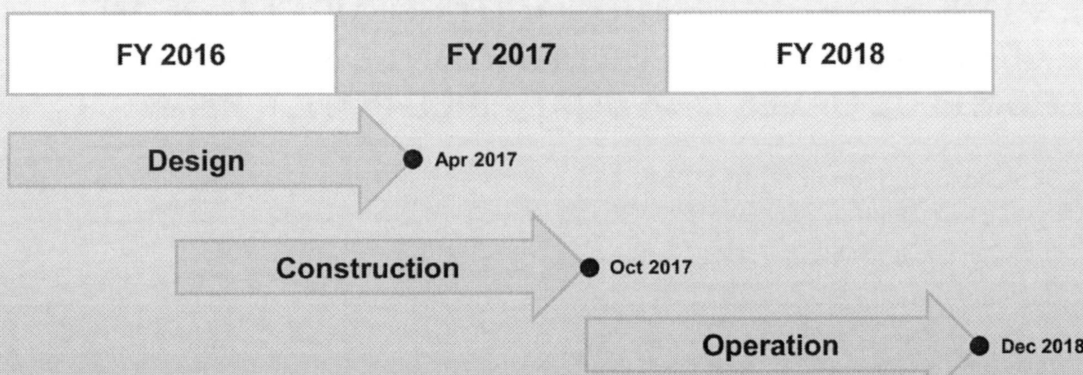


49

# Thank You!

ADM

ILLINOIS
INDUSTRIAL CARBON
CAPTURE & STORAGE

## Industrial Carbon Capture and Storage Project:

- U.S. Department of Energy Award No. DE-FE-0001547
- Administered by the DOE's Office of Fossil Energy
- Managed by the National Energy Technology Laboratory

## Cost Share Agreements:

- Archer Daniels Midland Company
- University of Illinois through the Illinois State Geological Survey
- Schlumberger Carbon Services
- Richland Community College

## Project Team Members Contacts:

- Dr. Sai Gollakota (NETL-DOE) Sai.Gollakota@NETL.DOE.GOV
- Dr. Sallie Greenberg, (ISGS) greenberg@isgs.illinois.edu
- John Medler, (Schlumberger Carbon Services) jmedler@slb.com
- Dr. Douglas Brauer (RCC) dbrauer@richland.edu



ARCHER DANIELS MIDLAND COMPANY

# PROCESS OPTIMIZATION AND SOLVENT RECLAIMING ENHANCED THE OPERATION AND PRODUCTION OF AN EXISTING 800 TONNES PER DAY PLANT CAPTURING $CO_2$ FROM FLUE GASES OF A COAL-FIRED POWER PLANT

**Dr. Ahmed Aboudheir**, PhD., P.Eng.
President, Delta Purification
Chief Technology Officer, HTC Purenergy
Regina, SK, Canada

**2016 International Pittsburgh Coal Conference**
**Cape Town, SOUTH AFRICA**
**August 8 – 12, 2016**



# Searles Valley Minerals, SVM 800 TPD $CO_2$ Capture Plant

- **Two Trains:**
- **Absorption Columns:14.5 ft ID x high 119 ft**
- **Stripper Columns: 12.5 ft ID x 77 ft high**
- **$CO_2$ Use: Raw material for $Na_2CO_3$**

# Delta Reclaimer Integration at the SVM $CO_2$ Capture Plant

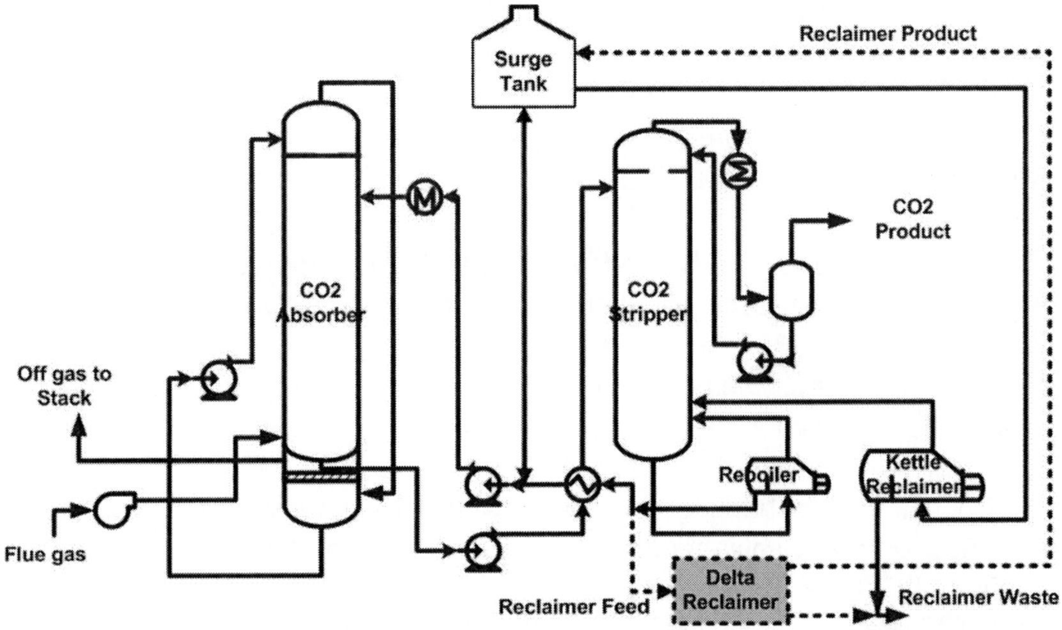


Searles Valley Minerals

HTC Purenergy purenergy solutions

delta purification reclaim recycle reuse



# SVM $CO_2$ Capture Plant; Main Columns

Plant data

| Process equipment | Absorber | | Stripper | |
|---|---|---|---|---|
| Number of Unit | 2 | | 2 | |
| Unit operator | Absorber | Wash section | Stripper | Reflux Section |
| Packing Type and Size | 2 in pall ring plastic | Tray | Tray | Tray |
| Column Diameter, ft | 14.5 | 14.5 | 12.5 | 12.5 |
| Packing Height, ft/Number of Tray | 46 | 2 Tray | 20 Tray | 2 Tray |
| Number of packed bed section | 2 | 1 | - | - |

No DCC, But water sprayed on blower blades to cool the flue gas temperature up to 136 F

HTC Purenergy purenergy solutions

delta purification reclaim recycle reuse

# Plant Design, Troubleshooting, and Optimization using **PDOEngine™**

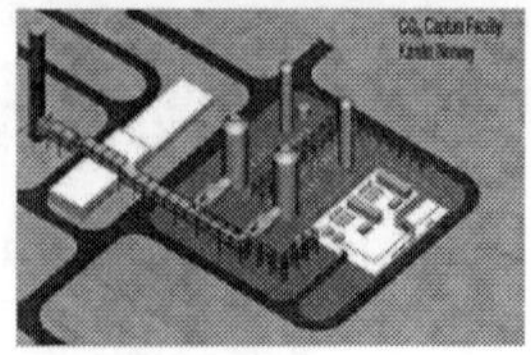

- PDOEngine™, is the HTC design and diagnostic tools to troubleshoot and optimize the operation of existing acid gas treatment plants in the chemical, petrochemical, oil, and gas industries.
- The PDOEngine™ is based on proprietary rigorous models/software and accumulative experience to perform the following activities:
  - Producing comprehensive material, momentum and energy balances.
  - Optimum allocation of scarce resources
  - Sensitivity analysis
  - Process troubleshooting, debottlenecking, and optimization
  - Single or mixed solvents recommendation to meet the production and cleanup targets.
  - Solvent diagnostic and recommendation for suitable reclaiming process.

HTC Purenergy purenergy solutions | delta purification reclaim recycle reuse



# Rate Based Model Validation

- Comparison Performance Parameters:
  - Production capacity, solvent circulation rate, rich loading, rich solvent concentration, steam consumptions, temperatures of the main streams, flow rate of main streams, and composition of main streams

| | Plant Name | Production Capacity | Solvent | Plant Configuration | Average Absolute Deviation, % |
|---|---|---|---|---|---|
| Pilot Plant | ERTF (UK) | 1 TPD | MEA | TKO™ | 12.2% |
| | | | RS-2™ | | 10.6% |
| Commercial Plants | AES (USA) | 130 TPD | MEA | Standard | 8.0% |
| | SVM (USA) | 800 TPD | MEA | Standard | 6.7% |

HTC Purenergy purenergy solutions | delta purification reclaim recycle reuse

# SVM, Measured and Predicted Performance, AAD% ± 6.7

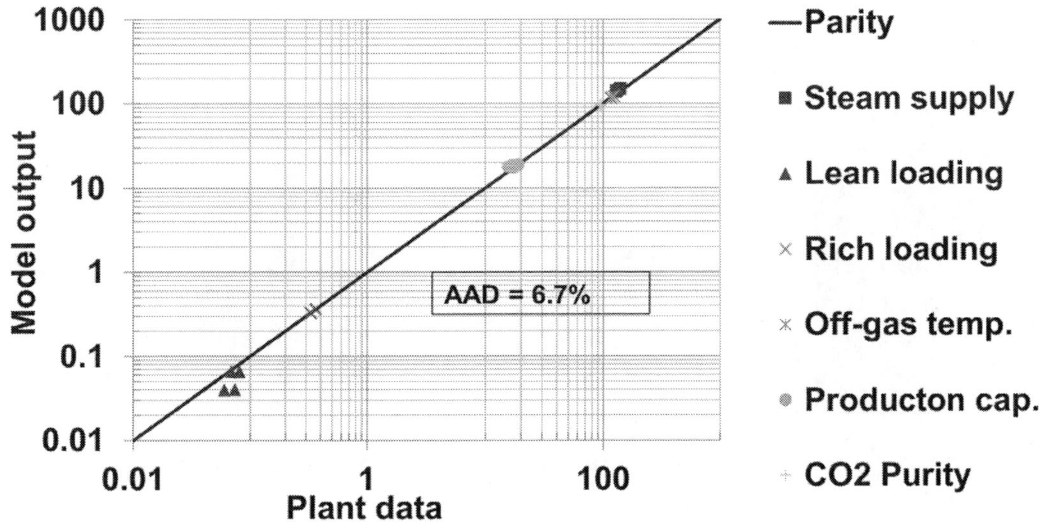




# SVM $CO_2$ Capture Plant Optimization

Parameters

- Effect of Solvent Concentration
- Effect of L/G Ratio (Solvent Circulation Rate)
- Effect of Flue Gas Inlet Temperature
- Effect of Packing Change

Objectives

- Reduce steam consumption for current capture capacity
- Increase capture capacity with available resources
- Improve performance by minimal revamp

August 10, 2016 2016 PCC, Cape Town, South Africa, August 8-12, 2016 9

# Effect of Solvent Concentration

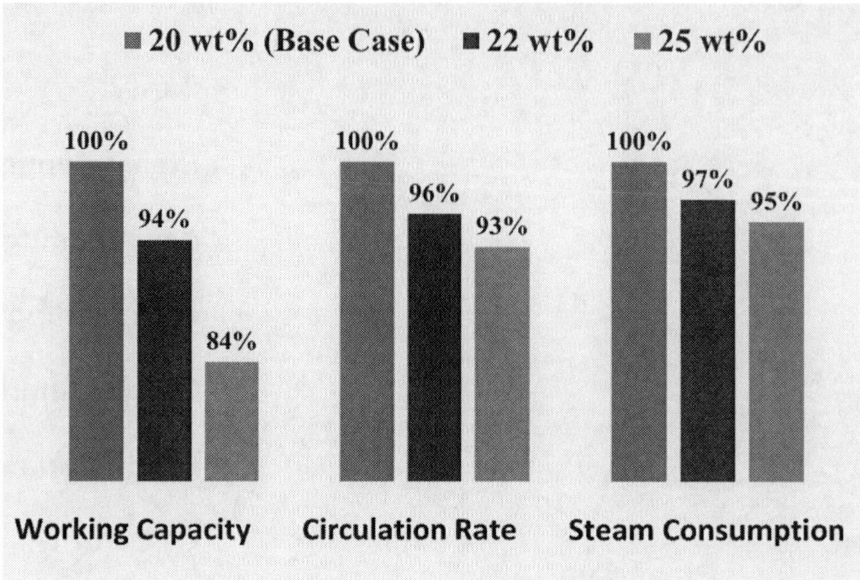


August 10, 2016 2016 PCC, Cape Town, South Africa, August 8-12, 2016 10

# Effect of L/G Ratio (Solvent Circ. Rate)

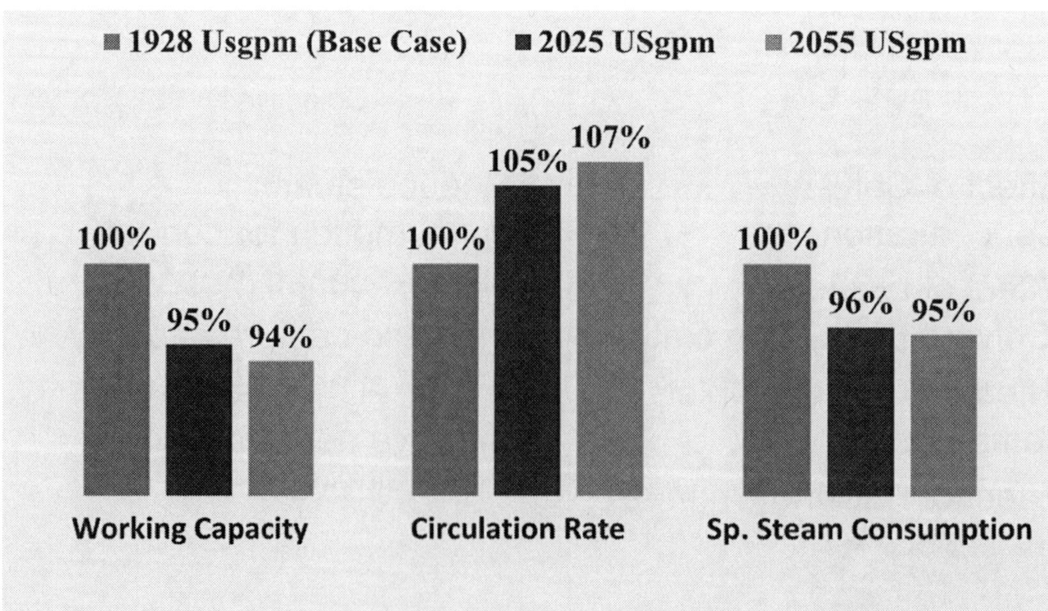

# Effect of Flue Gas Inlet Temperature

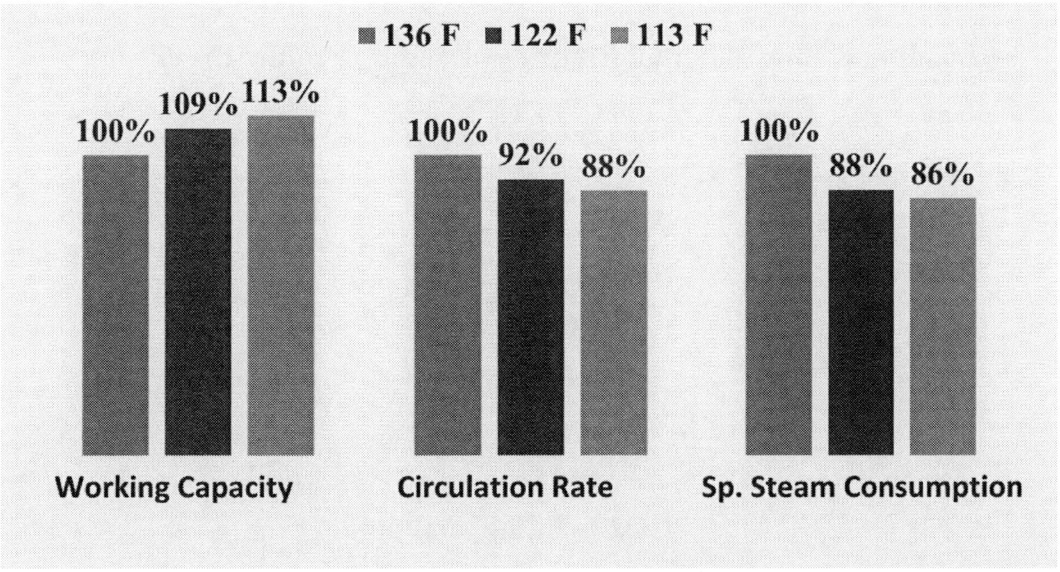


HTC Purenergy purenergy solutions | delta purification reclaim recycle reuse



# Effect of Packing Change

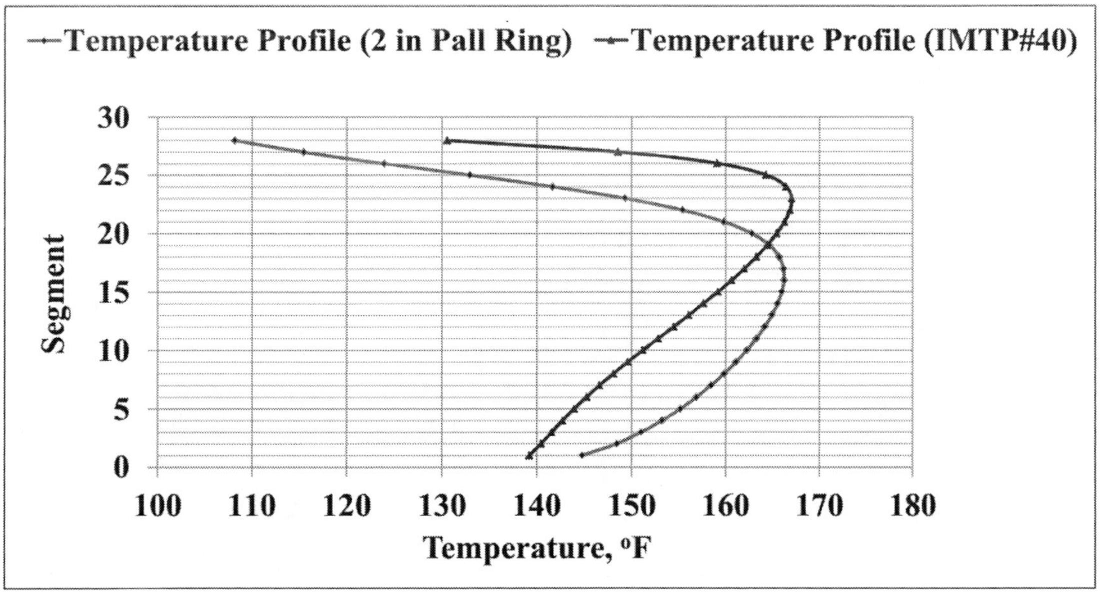


HTC Purenergy purenergy solutions | delta purification reclaim recycle reuse

# Effect of Packing Change

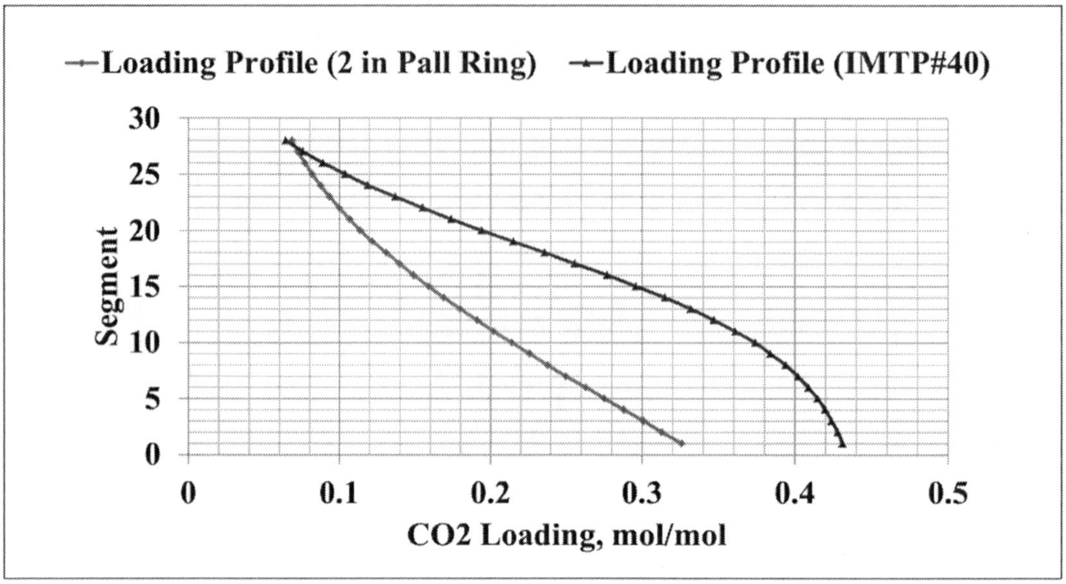




# Effect of Packing Change

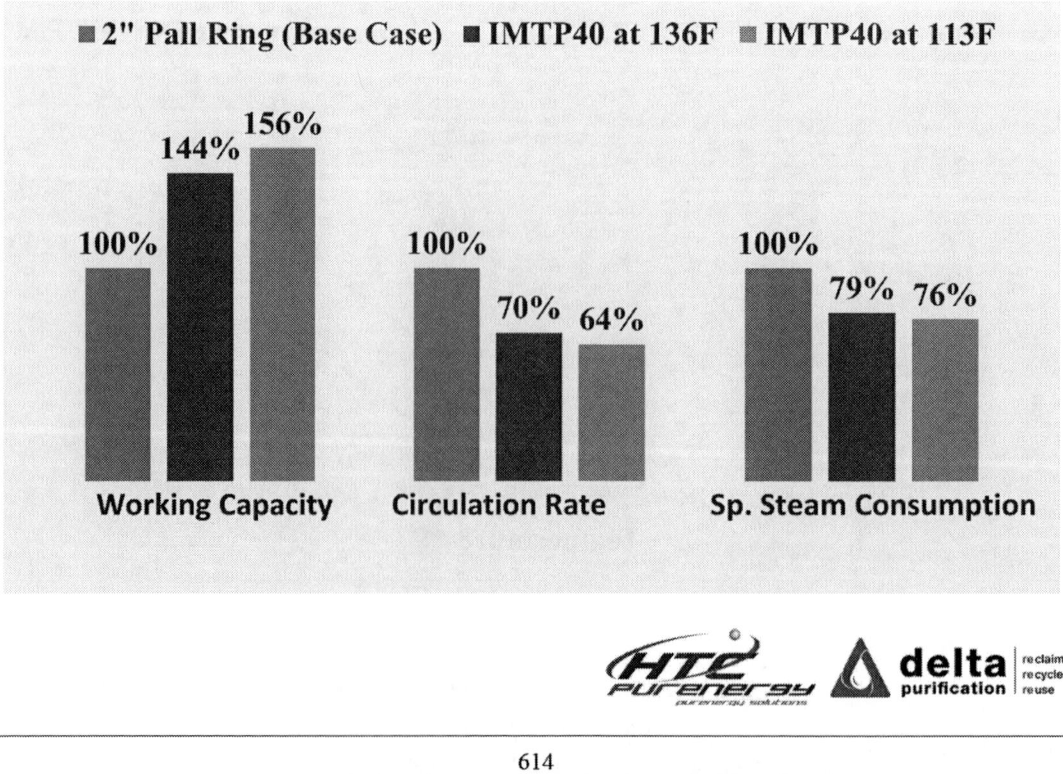

# Summary of Results

| Scenario | Solvent circulation | Reboiler steam consumption | Solvent Working Capacity |
|---|---|---|---|
| Increasing the solvent concentration | 7% ↓ | 5% ↓ | 16% ↓ |
| Relaxing the solvent loading approach | 7% ↑ | 5% ↓ | 6% ↓ |
| Reducing the flue gas temperature | 12% ↓ | 14% ↓ | 13% ↑ |
| Changing the existing packing | 30% ↓ | 21% ↓ | 56% ↑ |

HTC Purenergy — delta purification (reclaim recycle reuse)



# Recommendations

- Implement the optimum operating parameters
- IMTP # 40 or equivalent is recommended
- Efficiency of Liquid Distributor/redistributor is to be investigated further
- Flue gas cooling below 115 F is recommended for improved plant performance

HTC Purenergy — delta purification (reclaim recycle reuse)

# Delta Technology for Cleanup Degraded Solvents and Glycols

HTC Purenergy purenergy solutions | delta purification reclaim recycle reuse



# Reclaimer

- Reclaimer unit is designed to remove from the solvent:
  - high-boiling degradation products,
  - ionic species (heat stable salts, HSS),
  - impurities and fine suspended solids.
- The type of degradation products is a function of:
  - the impurities in the gaseous streams,
  - the solvent type,
  - the operating conditions of the acid gas absorption plant.
- The type of reclaiming processes
  - Ion Exchange; capable of removing HSS only
  - Electro-dialysis; capable of removing HSS only
  - **Thermal distillation,** capable of removing of HSS, degradation products, and solids

HTC Purenergy purenergy solutions | delta purification reclaim recycle reuse

# Types of Thermal Reclaimers in Use

- Kettle reclaimer for primary amines; MEA and DGA solvents. Operated at the stripper pressure in order to return the reclaimer product vapour to the stripper. Main disadvantages:
  - Extensive Energy use
  - Large volume of waste, which include solvent
- Reclaimer unit operated under reduced pressure for secondary and tertiary amines; DEA, DIPA and MDEA. Main disadvantage:
  - Amines reclaimed on a contract basis on-site using mobile units or off-site at high cost
  - Use of complex design units (High CAPEX & OPEX)
  - Require highly experienced operators

HTE Purenergy purenergy solutions delta purification reclaim recycle reuse



# Potential Problems from Accumulation of Degradation Products in Solvents

- Reduction in solvent absorption capacity
- Change of solvent physical properties leads to change the solvent characteristics and performance.
- Foaming and increased solvent foaming tendency in absorber/stripper reduce the contact area between gas and liquid phases.
- Increasing solvent corrosivity due to accumulation of corrosive species, such as Heat Stable Salts.
- Degradation products and heat stable salts lead to increase the replacement/change-out of mechanical and activated carbon filters and solvents makeup, which will increase the operating and maintenance costs.

HTE Purenergy purenergy solutions delta purification reclaim recycle reuse

# Delta Reclaimer™ Advantages Compared to Existing Technologies

- Less capital cost
- Less operating cost
- Less waste for disposal
- Less utility consumptions
- Simple to operate and maintain
- Higher recovery rate for solvent
- Reclaims single, mixed, and formulated solvents

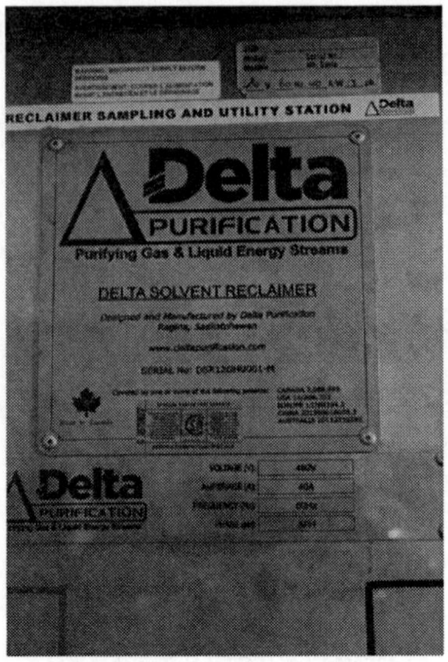




# Delta Reclaimer™ Process Diagram; Patented Technology

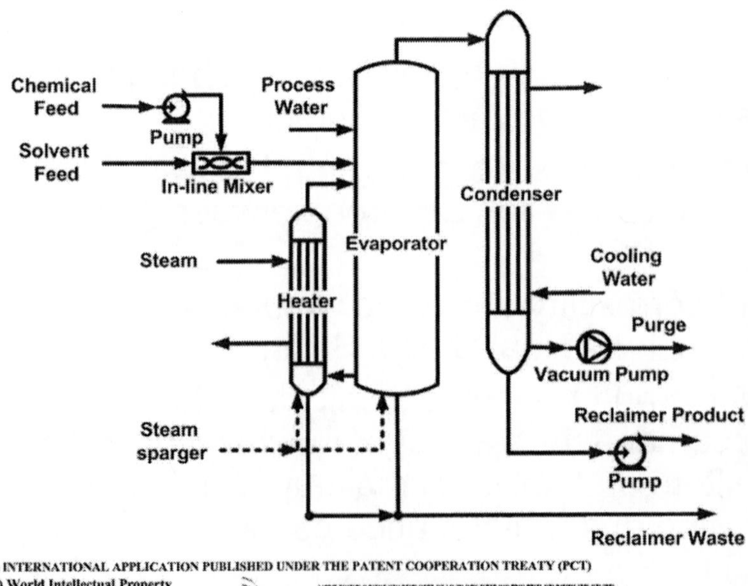

(12) INTERNATIONAL APPLICATION PUBLISHED UNDER THE PATENT COOPERATION TREATY (PCT)

(19) World Intellectual Property Organization
International Bureau

(43) International Publication Date
3 October 2013 (03.10.2013)

WIPO | PCT

(10) International Publication Number
WO 2013/143001 A1

# Delta Reclaimer Main Operating Parameters

- Saturated low pressure steam at 40 psig
- Reclaimer operating at an absolute pressure of 190 mmHg (Torr)
- Reclaimer operating temperature 180 to 205°F
- Cooling water at 75 °F
- Chemical injection
  - 28 wt% Soda ash or
  - 50 wt% Caustic soda

HTE Purenergy purenergy solutions — delta purification reclaim recycle reuse



# Samples of the Main Process Streams

- Colours of the Samples from right to left; Reclaimer Feed, Reclaimer Product, and Process Fluid

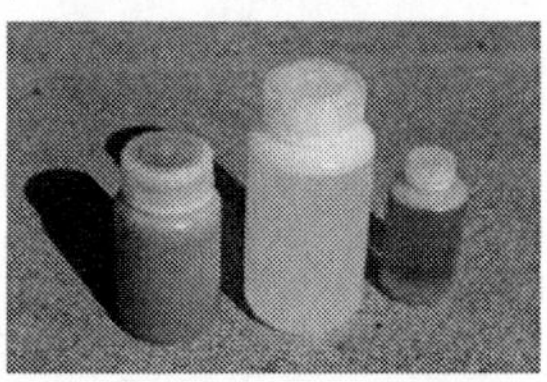

- Collected concentrated Waste for Disposal on a Continues Basis

HTE Purenergy purenergy solutions — delta purification reclaim recycle reuse

# Solvent recovery rate and energy consumption

| Process Data | Run No. 1 | Run No. 2 | Run No. 3 |
|---|---|---|---|
| Reclaimer Operating Pressure, mmHg | 190 | 190 | 190 |
| Reclaimer Operating Temperature, °F | 185 ±5 | 185 ±5 | 200 ±5 |
| Chemical Injection for neutralization | 28 wt% $Na_2CO_3$ | 28 wt% $Na_2CO_3$ | 50 wt% NaOH |
| Average total impurities, wt% | 1.45 | 5.80 | 8.50 |
| Total Solvent Recovery, % | 98% | 95% | 91% |
| lb saturated steam (40 psig) per lb product | 0.97 | 1.1 | 0.95 |



# Run # 1: Sample analysis

| Process Streams | | Feed | Product |
|---|---|---|---|
| MEA Concentration, wt% | | 17.51 | 17 |
| $CO_2$ loading, mol/mol | | 0.085 | 0.128 |
| Specific gravity | | 1.0393 | 1.013 |
| Refractive Index at 20 °C | | 1.3636 | 1.352 |
| pH with $CO_2$ | | 10.4 | 10.65 |
| pH without $CO_2$ | | 10.91 | 11.39 |
| Heat stable salts (HSS), wt% as MEA | | 1.24 | 0.15 |
| Organic HSS | T. Formate, ppm | 846 | 13 |
| | T. Acetate, ppm | 793 | 9 |
| | T. Glycolate, ppm | 289 | 78 |
| | T. Oxalate, ppm | 372 | 78 |
| Inorganic HSS | Sulfate, ppm | 8,656 | 25 |
| | Chloride, ppm | 6,452 | 16 |
| | Nitrite, ppm | 68 | 0 |
| Degradation Products | N-(hydroxyethyl)-glycine, ppm | 3,874 | 0 |
| | N-(N-hydroxyethyl-ethylamine)-ethylenediamine, ppm | 15 | 0 |
| Heavy metals | Fe, ppm | 5 | 0.3 |

# Run # 2: Sample analysis

| Process Streams | | Feed | Product |
|---|---|---|---|
| MEA Concentration, wt% | | 16.75 | 17.04 |
| $CO_2$ loading, mol/mol | | 0.100 | 0.120 |
| Specific gravity | | 1.0485 | 1.0145 |
| Refractive Index at 20 °C | | 1.3673 | 1.3493 |
| pH with $CO_2$ | | 10.42 | 10.57 |
| pH without $CO_2$ | | 10.91 | 11.25 |
| Heat stable salts (HSS), wt% as MEA | | 1.18 | 0.23 |
| Organic HSS | T. Formate, ppm | 915 | 0 |
| | T. Acetate, ppm | 917 | 35 |
| | T. Glycolate, ppm | 424 | 11 |
| | T. Oxalate, ppm | 356 | 185 |
| Inorganic HSS | Sulfate, ppm | 10,575 | 76 |
| | Chloride, ppm | 10,263 | 23 |
| | Nitrite, ppm | 61 | 0 |
| Degradation Products | N-(hydroxyethyl)-glycine, ppm | 4,771 | 16 |
| | N-(N-hydroxyethyl-ethylamine)-ethylenediamine, ppm | 7 | - |
| Heavy metals | Fe, ppm | 7 | 0.1 |



# Run # 3: Sample analysis

| Process Streams | | Feed | Product |
|---|---|---|---|
| MEA Concentration, wt% | | 17.25 | 16.87 |
| $CO_2$ loading, mol/mol | | 0.070 | 0.055 |
| Specific gravity | | 1.0705 | 1.023 |
| Refractive Index at 20 °C | | 1.3701 | 1.3492 |
| pH with $CO_2$ | | 10.51 | 10.94 |
| pH without $CO_2$ | | - | 11.17 |
| Heat stable salts (HSS), wt% as MEA | | 1.105 | 0.1 |
| Organic HSS | T. Formate, ppm | 644 | 30 |
| | T. Acetate, ppm | 673 | 9 |
| | T. Glycolate, ppm | 477 | 141 |
| | T. Oxalate, ppm | 284 | 94 |
| Inorganic HSS | Sulfate, ppm | 11,183 | 23 |
| | Chloride, ppm | 17,193 | 21 |
| | Nitrite, ppm | 56 | 0 |
| Degradation Products | N-(hydroxyethyl)-glycine, ppm | 4,291 | 0 |
| | N-(N-hydroxyethyl-ethylamine)-ethylenediamine, ppm | - | - |
| Heavy metals | Fe, ppm | 4.8 | 0.1 |

# Heat stable salts of feed and product streams

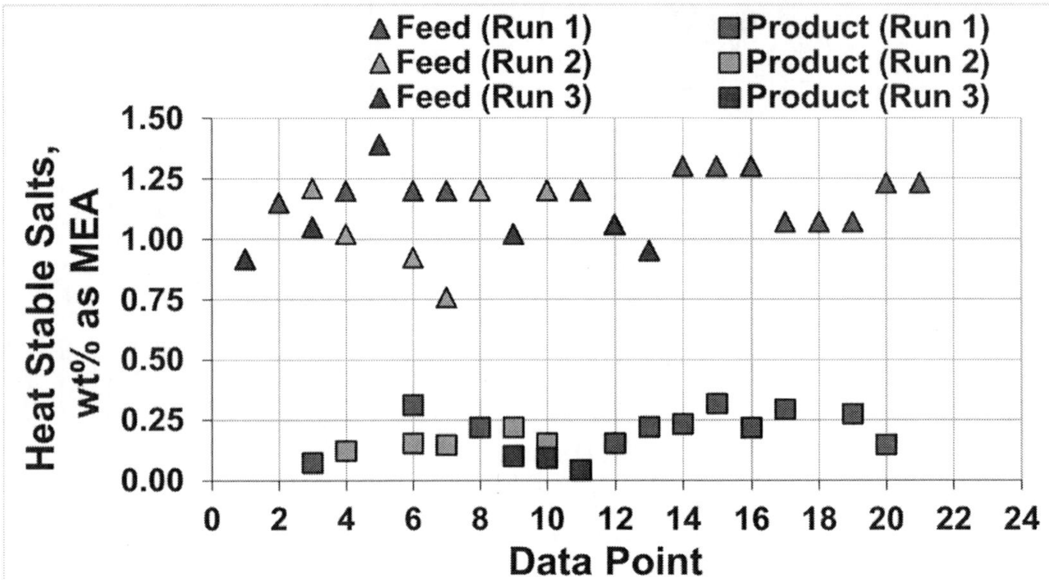




# Run 2: Impurities concentration in main streams

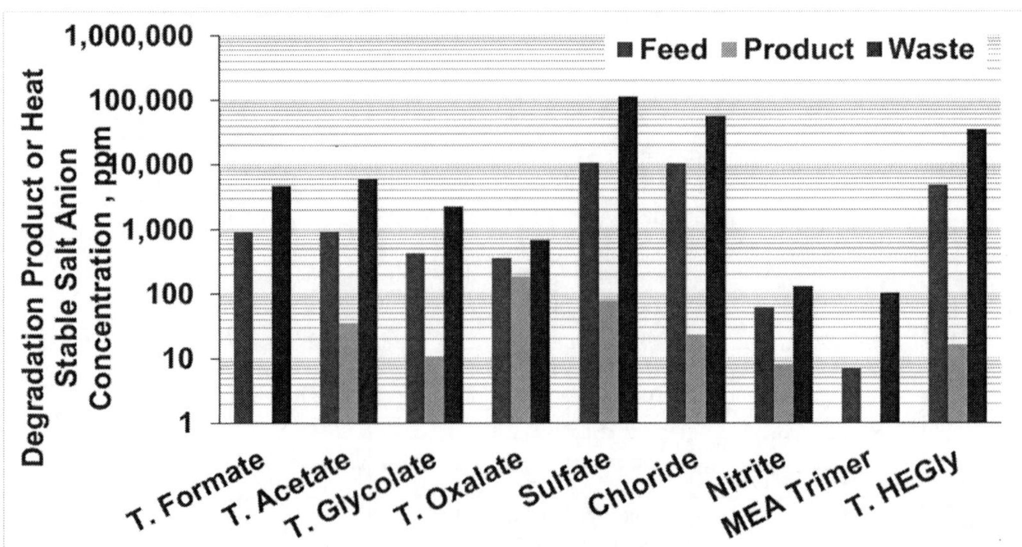


MEA Trimer = N-(N-hydroxyethyl-ethylamine)-ethylenediamine
HEGly = N-(hydroxyethyl)-glycine

# Conclusion & Findings

- Excessive contaminated solvent has been restore to almost its original purity: colorless, 17 wt% MEA & minimum impurities.
- Low Operating Cost;
  - 98% recovery rate for solvents with 1.5 wt% contaminants &
  - 91% recovery rate for solvents with 8.5 wt% contaminants
  - Energy consumption 1 lb steam (40 psig) per 1 lb product
  - Easy to operate and maintain
  - Minimum attention from the plant operator.
- Minimum waste for disposal, concentrated waste was collected periodically while the reclaimer process was in continuous operation.
- Low Capital Cost; simple process configuration and few unit operations
- Reclaims single, mixed, and formulated solvents
- Assist the CO2 Capture Plants to meet the production capacity and cleanup targets at minimum operating cost, emission to atmosphere, and waste for disposal

HTC Purenergy delta purification reclaim recycle reuse

# THANK YOU …

**For more information:**
Ahmed Aboudheir, Ph.D, P.Eng.
002 – 2305 Victoria Ave
Regina, SK, S4P 0S7, Canada
aaboudheir@htcenergy.com
+1-306-352-6132
+1-306-501-8227

HTC Purenergy delta purification reclaim recycle reuse

# OPTIMIZATION OF AN EXISTING 130 TONNES PER DAY PLANT CAPTURING $CO_2$ FROM FLUE GASES OF A COAL-FIRED POWER PLANT

**Dr. Ahmed Aboudheir**, PhD, P.Eng.
President, Delta Purification
Chief Technology Officer, HTC Purenergy
Regina, SK, Canada

**2016 International Pittsburgh Coal Conference**
**Cape Town, SOUTH AFRICA**
**August 8 – 12, 2016**

HTC Purenergy purenergy solutions | delta purification reclaim recycle reuse



# AES $CO_2$ Capture Plant, Slipstream Coal Flue Gas

| Plant Description | |
|---|---|
| Production Capacity | 130 TPD |
| Plant in Operation | 15 years |
| Flue Gas Source | Coal-fired Boiler |
| Solvent type | 16 wt% MEA |
| Solvent rate | 153 $m^3$/hr |
| Process Configuration | Two Blower, SO2 Scrubber, One Absorber, One Stripper |
| Blower Operating Capacity | 60% of Design Capacity |
| $CO_2$ End Use | Food and Beverages |

HTC Purenergy purenergy solutions | delta purification reclaim recycle reuse

# Process Flow Diagram

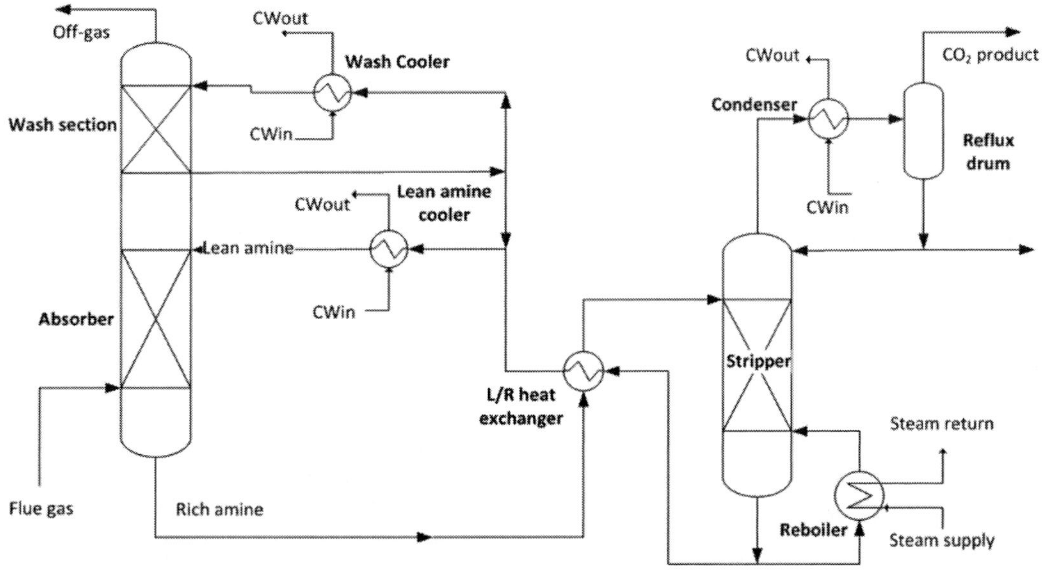




# AES $CO_2$ Capture Plant; Main Columns

| Process Equipment | Absorber | | Stripper | |
|---|---|---|---|---|
| Number of unit | 1 | | 1 | |
| Unit Operator | Absorber | Wash Section | Stripper | Reflux Section |
| Packing/Tray type | 2 inch Polypropylene Ring | 1 inch Polypropylene Ring | 22 Valve Trays | 4 Valve Trays |
| Column Diameter, m | 2.90 | 2.90 | 2.60 | 2.60 |
| Packing Height, m | 6.85 | 3.00 | - | - |
| # of Packed Bed Section | 2 | 1 | - | - |

# Plant Design, Troubleshooting, and Optimization using **PDOEngine™**

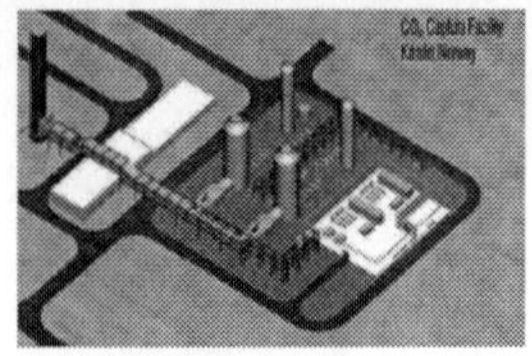

- PDOEngine™, is the HTC design and diagnostic tools to troubleshoot and optimize the operation of existing acid gas treatment plants in the chemical, petrochemical, oil, and gas industries.
- The PDOEngine™ is based on proprietary rigorous models/software and accumulative experience to perform the following activities:
  - Producing comprehensive material, momentum and energy balances.
  - Optimum allocation of scarce resources
  - Sensitivity analysis
  - Process troubleshooting, debottlenecking, and optimization
  - Single or mixed solvents recommendation to meet the production and cleanup targets.
  - Solvent diagnostic and recommendation for suitable reclaiming process.



# Rate Based Model Validation

- Comparison Performance Parameters:
  - Production capacity, solvent circulation rate, rich loading, rich solvent concentration, steam consumptions, temperatures of the main streams, flow rate of main streams, and composition of main streams

| | **Plant Name** | **Production Capacity** | **Solvent** | **Plant Configuration** | **Average Absolute Deviation, %** |
|---|---|---|---|---|---|
| **Pilot Plant** | ERTF (UK) | 1 TPD | MEA | TKO™ | 12.2% |
| | | | RS-2™ | | 10.6% |
| **Commercial Plants** | AES (USA) | 130 TPD | MEA | Standard | 8.0% |
| | SVM (USA) | 800 TPD | MEA | Standard | 6.7% |

# Plant Data Versus Predicted Data, ±8% average absolute deviation

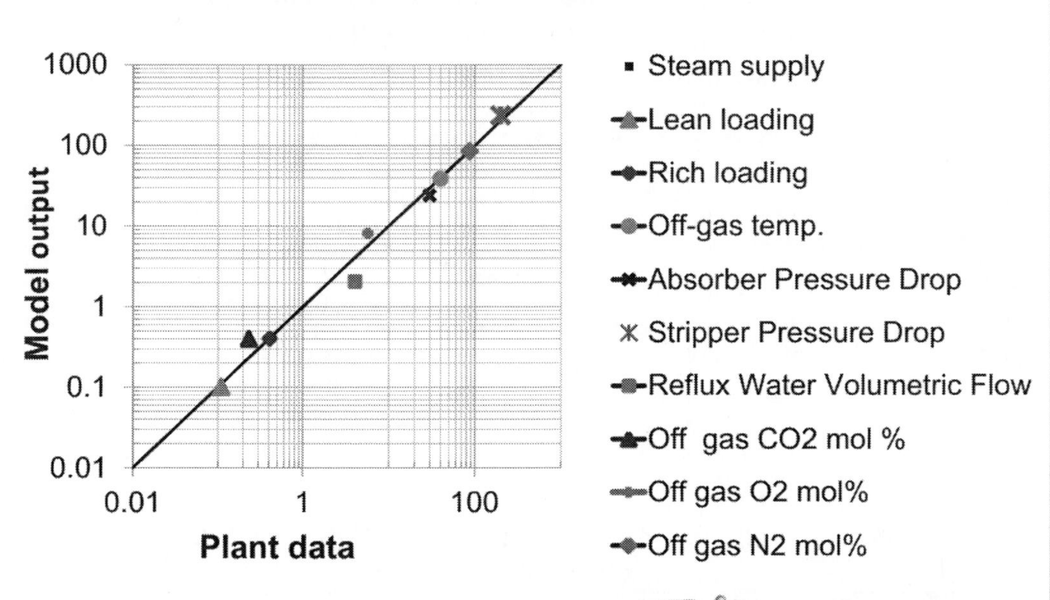


HTE purenergy purenergy solutions
delta purification reclaim recycle reuse



# Optimization with no Capital Expenditure Required

- **Circulation rate is reduced by 19%**
- **Steam consumption reduced by 21%**

| Operation Case | Lean Loading | Rich Loading | Solvent Rate | Steam Flow | Steam Consumption |
|---|---|---|---|---|---|
| | mol/mol | mol/mol | $m^3/hr$ | kg/hr | kg/kg |
| Base | 0.101 | 0.402 | 159.0 | 17,240 | 3.17 |
| Case 1 | 0.173 | 0.461 | 159.0 | 14,970 | 2.86 |
| Case 2 | 0.106 | 0.467 | 128.3 | 13,610 | 2.59 |

HTE purenergy purenergy solutions
delta purification reclaim recycle reuse

## Increase in Production by Increasing Gas Flow Rate

- **The flue gas flow rate can be increased up to 30%**
- **Production capacity will be increased by 28 TPD**
- **Unit steam consumption will be reduced from 3.17 to 2.61 kg/kg**

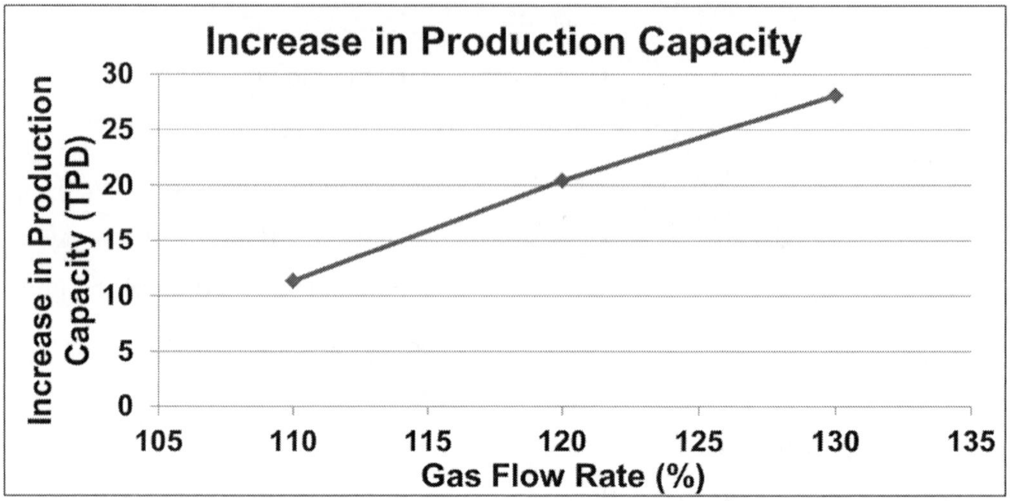


HTE Purenergy delta purification reclaim recycle reuse



# Solvent Concentration

Change in Solvent Concentration reduces solvent rate by 40% from the base case and by 25% from the optimized case

| Case | Solvent Rate | Steam Flow | Lean Loading | Rich Loading | Working Capacity |
|---|---|---|---|---|---|
| | $m^3$/hr | kg/hr | mol/mol | mol/mol | mol/mol |
| 16 wt% (Base Case) | **159** | 17,240 | 0.101 | 0.402 | 0.301 |
| 16 wt% (Optimized ) | **128.3** | 13,610 | 0.106 | 0.467 | 0.361 |
| 20 wt% (Optimized) | **96.53** | 13,610 | 0.100 | 0.479 | 0.379 |

HTE Purenergy delta purification reclaim recycle reuse

August 10, 2016 | 2016 PCC, Cape Town, South Africa, August 8-12, 2016 | 11

# Optimization with Additional Capital Expenditure

- Production can be increased by increasing the inlet flue gas rate and changing the packing to IMTP; Solvent Rate 159 $m^3$/hr & Steam rate 17,240 kg/hr.

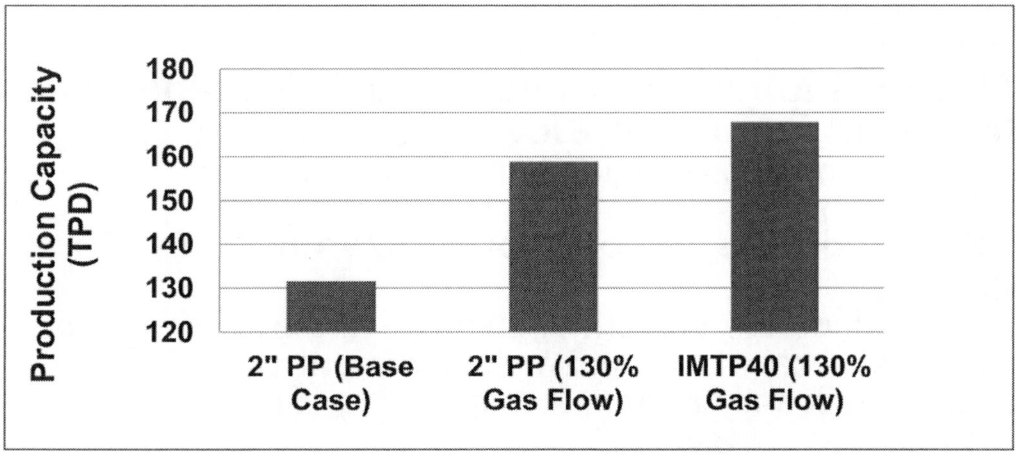


HTE purenergy · delta purification reclaim recycle reuse

August 10, 2016 | 2016 PCC, Cape Town, South Africa, August 8-12, 2016 | 12

# Optimization with Additional Capital Expenditure

- Specific steam consumption can be decreased by increasing inlet flue gas rate and changing the packing to IMTP; Solvent Rate 159 $m^3$/hr & Steam rate 17,240 kg/hr

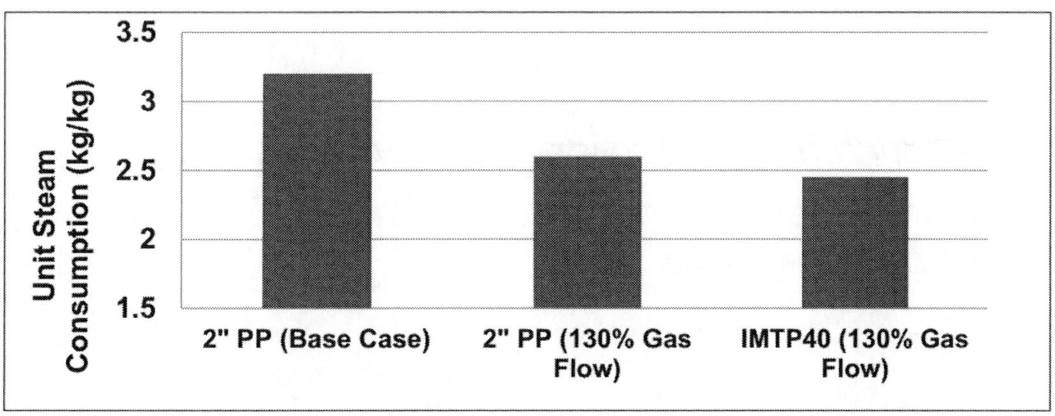


HTE purenergy · delta purification reclaim recycle reuse

## Recommendations with No Capital Expenditure Required

- **Operate at optimum solvent rate is 128.3 m³/hr instead of 159 m³/hr .**
  - will reduce the specific steam consumption from 3.17 to 2.59 kg/kg and the power requirement of re-circulating pumps by ~19%.
- **Increasing gas flow rate up to 30% within the hydraulic capacity & Blower Capacity**
  - will Increase production capacity by 28 TPD.
- **Changing the solvent concentration from 16 to 20 wt%**
  - will reduce the solvent circulation rate by ~40% from the base case and steam requirement by 21%. This will significantly reduce the operating expenses of the plant.

HTE Purenergy purenergy solutions — delta purification reclaim recycle reuse



## Recommendation with Capital Expenditure

- Changing absorber column packing from 2 inch pall ring to high performance IMTP40 and increasing flue gas rate by 30%
  - will increase production capacity by 38 TPD.
  - will decrease specific steam consumption from 3.17 kg/kg to 2.46 kg/kg

HTE Purenergy purenergy solutions — delta purification reclaim recycle reuse

August 10, 2016 2016 PCC, Cape Town, South Africa, August 8-12, 2016 **15**

# THANK YOU …

**For more information:**
Ahmed Aboudheir, Ph.D, P.Eng.
002 – 2305 Victoria Ave
Regina, SK, S4P 0S7, Canada
aaboudheir@htcenergy.com
+1-306-352-6132
+1-306-501-8227

HTC Purenergy — purenergy solutions
delta purification — reclaim recycle reuse

# Major CCS Demonstration Projects

NATIONAL ENERGY TECHNOLOGY LABORATORY

Project Locations & Cost Share

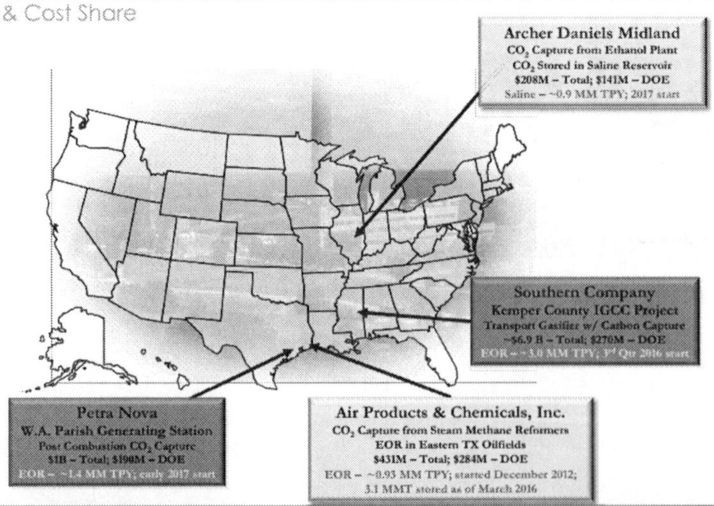


U.S. DEPARTMENT OF ENERGY 2

# Petra Nova – NRG W.A. Parish

NATIONAL ENERGY TECHNOLOGY LABORATORY

Advanced Post Combustion $CO_2$ Capture

- **Thompsons, TX (near Houston)**
- **240 MWe slipstream at NRG Energy's W.A. Parish power plant (originally 60 MWe)**
- **Fuel: PRB sub-bituminous coal**
- **90% $CO_2$ capture (KM CDR Process®) 1,400,000 tonnes $CO_2$/year**
- **EOR: Hilcorp West Ranch oil field**
- **Total Project Cost: ~\$1 billion DOE Share: \$190 million**

Key Dates

- Project Awarded: May 2010
- Air Permit: December 2012
- NEPA Record of Decision: May 2013
- Financial Close: July 2014
- Construction: March 2014 (LNTP); July 2014 (NTP)
- Operation: January 2017

Status

- All major foundations completed
- Construction of all major equipment complete: quencher, absorber, regenerator, HRSG, cooling tower
- Final construction activities on-going: installation of pipe, heat tracing, & lighting, and final grading and paving
- Commissioning activities initiated
- Pipeline mechanically complete, hydro-test completed
- Overall EPC effort: 98.5% complete (6/30/16)
- Construction: 96.5% complete (6/30/16)

U.S. DEPARTMENT OF ENERGY 3

# Petra Nova – NRG W.A. Parish

NATIONAL ENERGY TECHNOLOGY LABORATORY

Partners and Transaction Structure

**Project Ownership Structure**

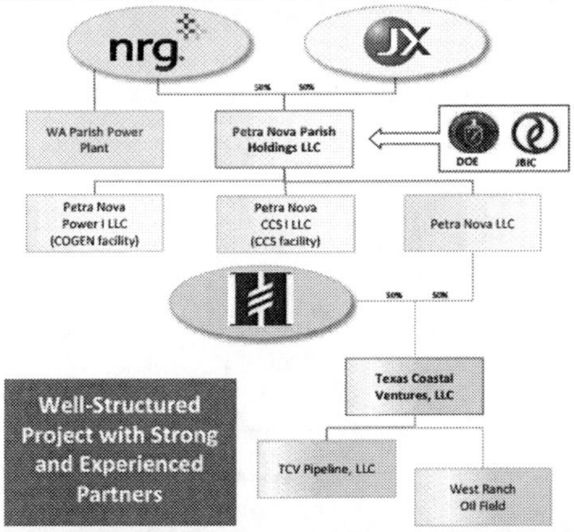


**Well-Structured Project with Strong and Experienced Partners**

**Partner Summary**

**Hilcorp Energy**
- One of the largest privately-held oil and natural gas E&P companies in the US
- Strong track record of implementing new production techniques into mature reservoirs
- Specialized team that has extensive experience implementing $CO_2$ floods

**JX Nippon Oil & Gas Exploration**
- Currently conducting oil and natural gas business in 14 countries
- Parent company, JX Holdings, is a leading integrated energy, resources, and materials company

**Japan Bank for Int'l Cooperation**
- Policy-based financial institution
- Wholly-owned by the Japanese government
- Over 25,000 loan and equity commitments valued at over \$479 billion

NEXI

**U.S. DOE**
- Awarded \$190 MM grant
- Funded through Clean Coal Power Initiative

U.S. DEPARTMENT OF ENERGY 4

# NRG W.A. Parish Generating Station

NATIONAL ENERGY TECHNOLOGY LABORATORY

Petra Nova $CO_2$ Capture Project

# Petra Nova – NRG W.A. Parish

NATIONAL ENERGY TECHNOLOGY LABORATORY

Overview

- The project will capture $CO_2$ and send it to nearby oil fields for carbon dioxide enhanced oil recovery ($CO_2$-EOR), which is a long-proven method of producing stranded oil.
- DOE's goals are to demonstrate integration of carbon capture at an existing plant, protocols for $CO_2$ monitoring to ensure permanence of sequestration, identify areas for cost reduction.
- Design will utilize a stand-alone technology natural gas cogen to simplify retrofit, provide flexibility in matching parasitic energy demands, without disrupt existing plant performance.

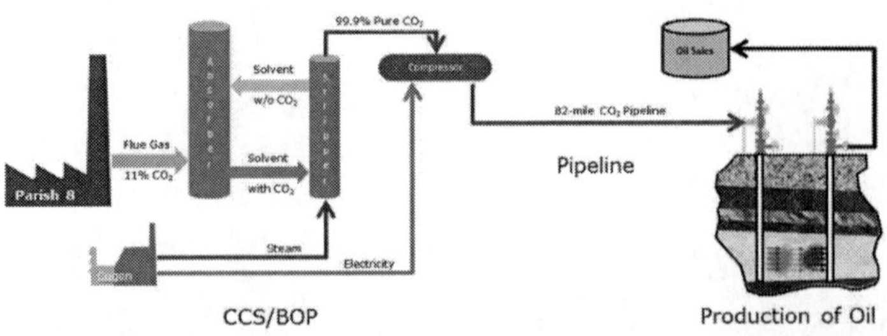

## Petra Nova – NRG W.A. Parish

NATIONAL ENERGY TECHNOLOGY LABORATORY

U.S. 500 Metric TPD Coal Fired CCS Demo Project

Successful demonstration of the SoCo 500 Metric TPD Integrated CCS Demo Project has lead to MHI being able to provide guarantees for coal flue gas CDR plants

SOUTHERN COMPANY

U.S. DEPARTMENT OF ENERGY

7

## Petra Nova – NRG W.A. Parish

NATIONAL ENERGY TECHNOLOGY LABORATORY

MHI Commercial $CO_2$ Capture Plants

World leading LARGE-SCALE post combustion $CO_2$ capture technology licensor, with:

**11 COMMERCIAL PLANTS IN OPERATION**

U.S. DEPARTMENT OF ENERGY

8

## Petra Nova Construction

NATIONAL ENERGY TECHNOLOGY LABORATORY

Moving/Lifting Module A2 (Absorber)

## Setting Top Section of Absorber

NATIONAL ENERGY TECHNOLOGY LABORATORY

# Absorber Column

# Regenerator – Shipped from Korea to Texas

26' Diameter & 180' Length

Regenerator Loaded for Shipment (Sep 2015)

Offload at the Peirce Landing Jetty (Nov 2015)

Heavy Haul to Parish (Dec 2015)

Dressed out at site (Jan 2016)

# Regenerator 500 Ton Lift

January 30, 2016

U.S. DEPARTMENT OF ENERGY

13

# Petra Nova Construction

27,000 HP 8-stage $CO_2$ Compressor

U.S. DEPARTMENT OF ENERGY

14

# Advanced Post Combustion $CO_2$

NATIONAL ENERGY TECHNOLOGY LABORATORY

Capture with Enhanced Oil Recovery

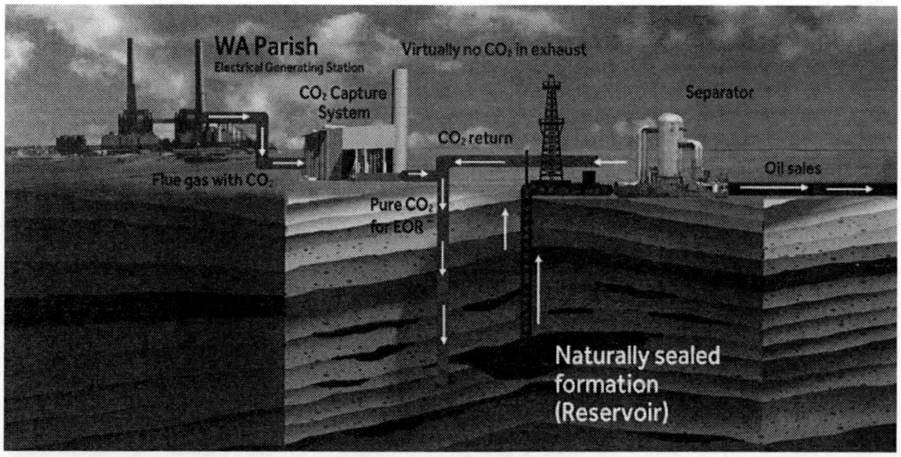




# Petra Nova – NRG W.A. Parish

NATIONAL ENERGY TECHNOLOGY LABORATORY

$CO_2$-EOR Fundamentals

- Large scale $CO_2$-EOR has been proven successfully since the early 1970s to produce otherwise unrecoverable oil from mature fields (tertiary recovery).
- It's a closed-loop system where the $CO_2$ produced with the oil is separated and re-injected.
- 1 ton of $CO_2$ can produce up to 4 bbls of oil.
- 60MM bbls are estimated to be recoverable at West Ranch

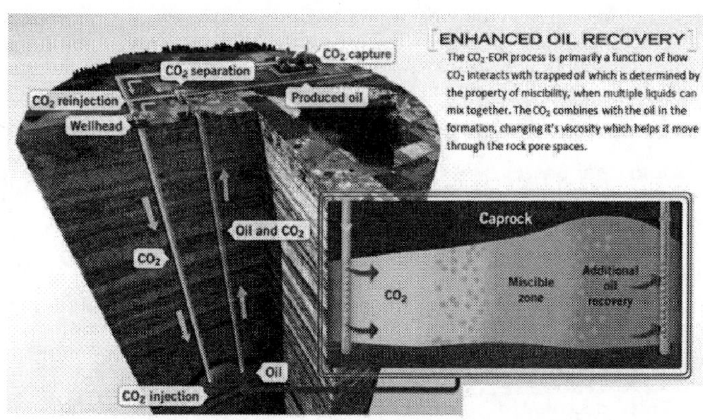

# Southern Company Services, Inc.

NATIONAL ENERGY TECHNOLOGY LABORATORY

Kemper County Advanced IGCC with $CO_2$ Capture

- **Kemper County, MS**
- **582 MWe (net) with duct firing; 2 TRIG™ gasifiers, 2 Siemens combustion turbines, 1 Toshiba steam turbine**
- **Fuel: Mississippi lignite**
- **67+% $CO_2$ capture (Selexol® process); 3,000,000 tons $CO_2$/year**
- **EOR: Denbury Onshore LLC**
- **Total DOE CCPI Project: $2.01 B; DOE Share: $407 MM (20%)**
- **DOE Share: ~$ 407 MM**
- **Total estimated project cost: ~$ 6.99 B**

Status
- Lignite mine in commercial operation: June 2013
- Combined cycle commercial operation on natural gas: Aug. 2014
- Gasifier "First Fire" of natural gas startup pilot burners: Mar 2015
- First lignite feed to gasifier: July 2016

Key Dates
- Project Awarded: Jan. 30, 2006
- Project moved to MS: Dec. 5, 2008
- NEPA Record of Decision: Aug. 19, 2010
- Initiate excavation work: Sept. 27, 2010
- Operations: 3rd Qtr 2016

U.S. DEPARTMENT OF ENERGY 17

# Air Products & Chemicals, Inc.

NATIONAL ENERGY TECHNOLOGY LABORATORY

Steam Methane Reforming with $CO_2$ Capture

- Port Arthur, TX (Hydrogen plant at Valero Refinery)
- 90%+ $CO_2$ capture (Vacuum Swing Adsorption) from 2 steam-methane reformers (SMRs) yielding ≈925,000 tonnes $CO_2$/year
- ≈30 MWe cogeneration unit to supply makeup steam to SMRs and operate VSA and compression equipment
- $CO_2$ to Denbury "Green" pipeline for EOR in Texas at West Hastings oil field
- Total Project: $431 MM; DOE Share: $284 MM (66%)

Key Dates
- Phase 2 Awarded: June 15, 2010
- FEED completed: Nov. 2010
- Permit By Rule (PBR) and Standard Air Permits issued: May 2011
- NEPA FONSI: July 2011
- Construction started: Aug. 2011
- Operation started: Dec. 2012

Status
- PA-1 initiated operation: March 3, 2013
- PA-2 initiated operation: Dec. 16, 2012
- Full capacity achieved: April 2013
- Has operated at >100% of design when necessary
- 1 MM tonnes CO2 delivered as of 4/24/14
- 2 MM tonnes CO2 delivered as of 5/15/15
- 3 MM tonnes CO2 delivered as of 6/30/16

U.S. DEPARTMENT OF ENERGY 18

## Air Products & Chemicals, Inc.

NATIONAL ENERGY TECHNOLOGY LABORATORY

Port Arthur 2

U.S. DEPARTMENT OF ENERGY

19

## APCI Port Arthur 1 Overview

NATIONAL ENERGY TECHNOLOGY LABORATORY

U.S. DEPARTMENT OF ENERGY

20

## APCI Port Arthur 2 Overview

NATIONAL ENERGY TECHNOLOGY LABORATORY

U.S. DEPARTMENT OF ENERGY 21

## Port Arthur (Air Products)

NATIONAL ENERGY TECHNOLOGY LABORATORY

Key Project Components

- Vacuum swing adsorption (VSA) system
  - Proven technology, but previously never used for $CO_2$ capture at the project scale
- $CO_2$ Compressor
  - 8 stages
  - Export pressure over 2,000 psig (≈140 bar)
  - Currently largest single active train $CO_2$ compressor in operation in US
- 13-mile (21 km) $CO_2$ Pipeline connecting to Denbury's "Green" 300+ Mile (≈500 km) $CO_2$ Pipeline

U.S. DEPARTMENT OF ENERGY 22

# Port Arthur (Air Products)

Vacuum Swing Adsorption (VSA) Process

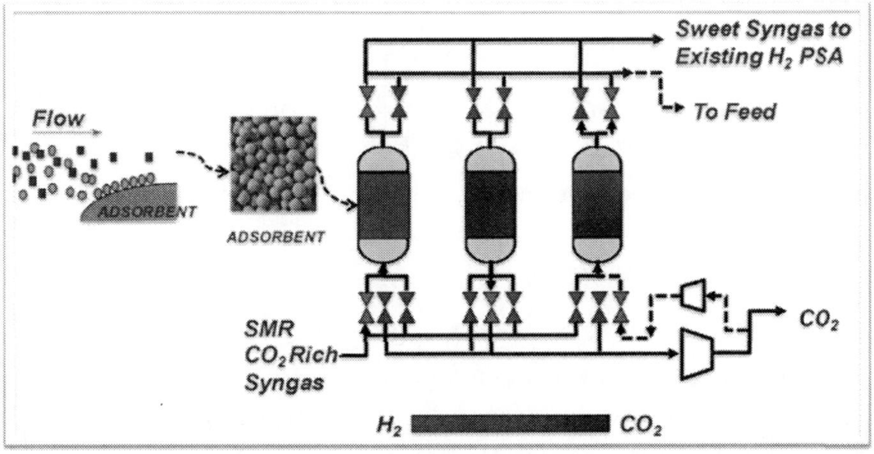




# Archer Daniels Midland Company

$CO_2$ Capture from Biofuel Plant

- Decatur, IL
- $CO_2$ (>99% purity) is a by-product from production of fuel-grade ethanol via anaerobic fermentation
- Up to 90% $CO_2$ capture, dehydration (via tri-ethylene glycol) & compression
- ~900,000 tonnes $CO_2$ /year
- Sequestration in Mt. Simon Sandstone saline fm.
- Total Project: $208 MM; DOE Share: $141 MM (68%)

Key Dates
- Phase 2 Awarded: June 15, 2010
- FEED Completed: April 2011
- NEPA FONSI: April 2011
- Construction started: May 2011
- UIC Class VI Injection Well Permit: Sept. 2014; UIC Class VI Operating Permit: Dec 2016
- Sequestration start: 1st Qtr 2017

Status
- Construction >99% complete July 2016
- Two monitoring wells drilled: Nov. 2012
- New Hans substation energized: Nov. 2014
- Commissioning compression and dehydration system completed: Sept. 2015
- Injection well drilled and completed: Sept. 2015

## Lessons Learned – Major Demonstrations

NATIONAL ENERGY TECHNOLOGY LABORATORY

- A multi-award, multi-FOA 'down-select' approach may be optimal:
  - Pre-FEED awards → fewer FEED awards → fewer construction/operation awards
  - Multiple, discrete, sequential awards allow work scope and budgets to be well-defined and better managed
  - Allows DOE to respond to changes in market/business environment
  - Each step might be full-and-open competition FOA soliciting proposals from existing projects plus new ones (or ones unsuccessful in an earlier round)
  - Allow low recipient cost shares for the pre-FEED and FEED awards
- Encourage recipients to identify and address fundamental issues related to CCS deployment in their state or region
  - UIC Class VI permits
  - Whether or not the state asks for primacy from the U.S. EPA
  - Pore space ownership
  - Premium electricity rates or line charges to recover CCS costs
  - Unique tax treatments to help the business case for commercial CCS projects
  - State or other entity assuming risk of long-term storage liability for CCS projects
- Incorporation of $CO_2$ storage alternatives to facilitate projects proceeding into construction while UIC Class VI permits are being processed



## For Additional Information

NATIONAL ENERGY TECHNOLOGY LABORATORY

Office of Fossil Energy
www.fe.doe.gov

NETL
www.netl.doe.gov

Solutions for Today
Options for Tomorrow
For More Information, Contact NETL
www.netl.doe.gov
NATIONAL
ENERGY
TECHNOLOGY
LABORATORY
U.S. DEPARTMENT OF
ENERGY

# Vapor Phase Chemical Looping Combustion of Coal

**Tomasz Wiltowski**[+], ***Debalina Dasgupta***[+], ***Quan Zhang, Kanchan Mondal***

**[+]Advanced Coal and Energy Research Center**
**Mechanical Engineering and Energy Processes**
**Southern Illinois University, Carbondale**

**33[rd] International Pittsburgh Coal Conference**
**Cape Town, SA**
**August, 2016**

SIU Southern Illinois University
CARBONDALE

# What is CLC??

$(2n+0.5\ m)\ MeO_{x-1} + (n+0.25\ m)\ O_2 \rightarrow (2n+0.5\ m)\ MeO_x$

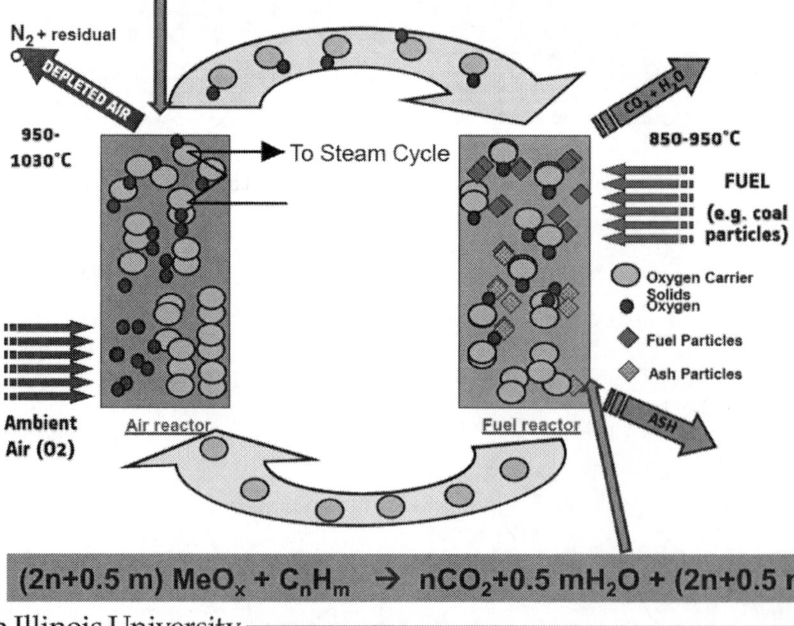


$(2n+0.5\ m)\ MeO_x + C_nH_m \rightarrow nCO_2 + 0.5\ mH_2O + (2n+0.5\ m)\ MeO_{x-1}$

SIU Southern Illinois University
CARBONDALE

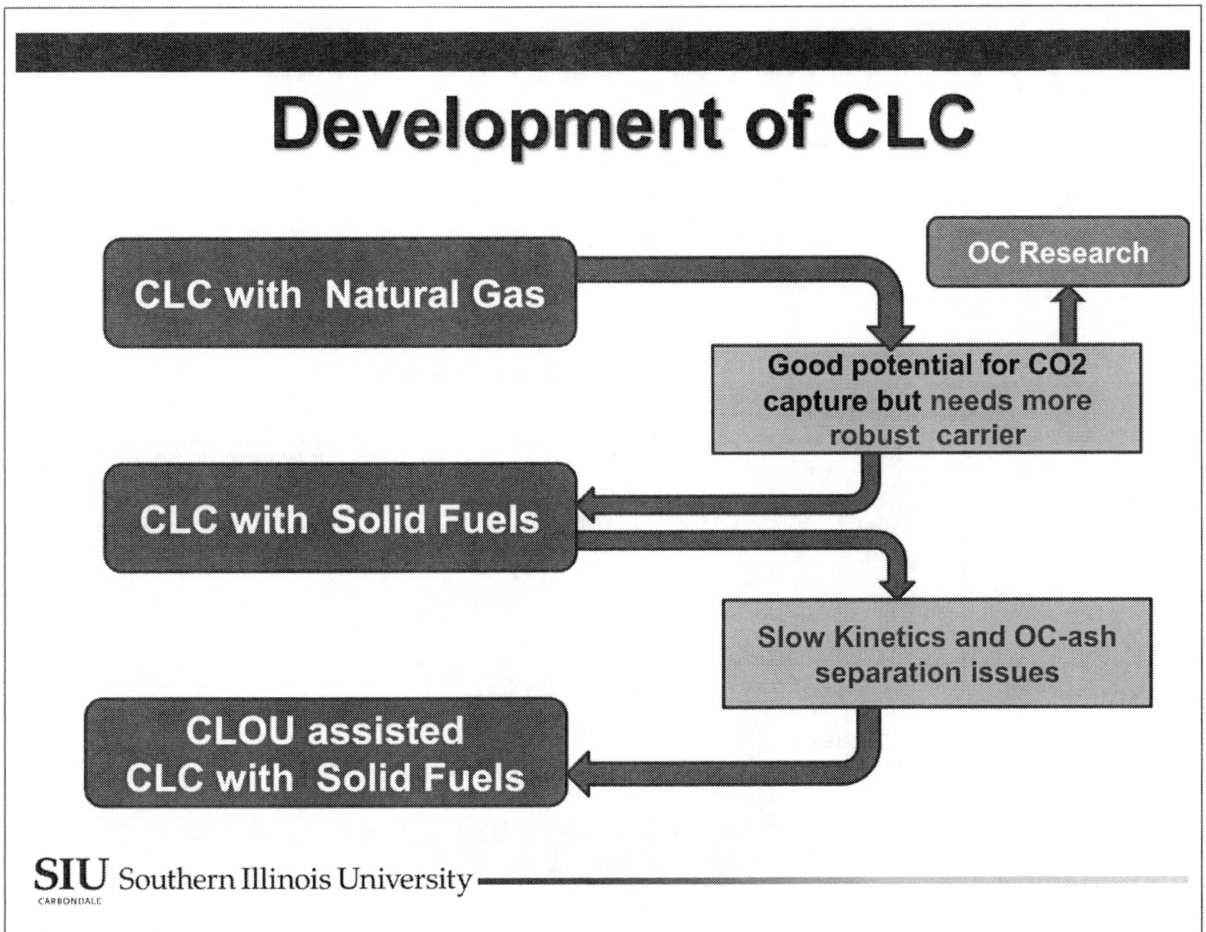
Development of CLC
CLC with Natural Gas
OC Research
Good potential for CO2 capture but needs more robust carrier
CLC with Solid Fuels
Slow Kinetics and OC-ash separation issues
CLOU assisted CLC with Solid Fuels
SIU Southern Illinois University
CARBONDALE

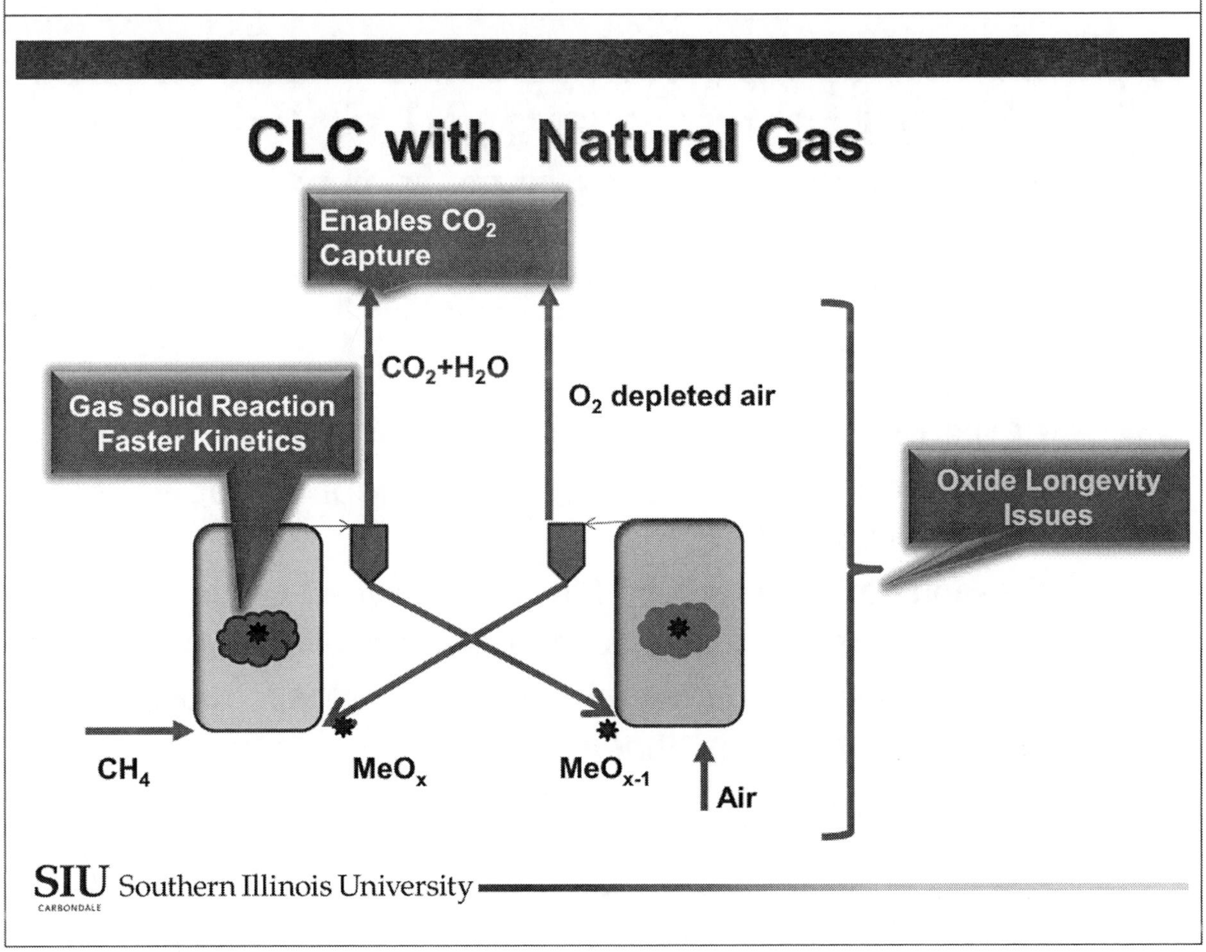
CLC with Natural Gas
Enables $CO_2$ Capture
Gas Solid Reaction Faster Kinetics
$CO_2$+$H_2O$
$O_2$ depleted air
Oxide Longevity Issues
$CH_4$
$MeO_x$
$MeO_{x-1}$
Air
SIU Southern Illinois University
CARBONDALE

# Coal Direct CLC

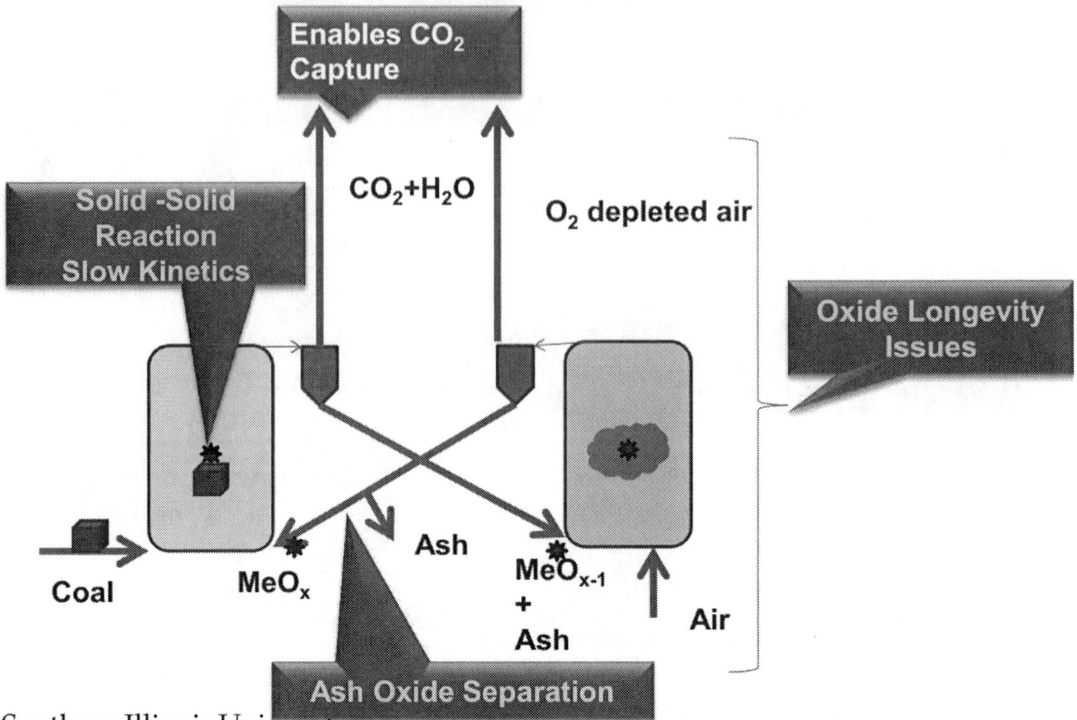


SIU Southern Illinois University CARBONDALE

# Need to have a fluid solid reaction to improve kinetics

- Since the fuel is solid in CDCL, the oxidant needs to be in fluid form
- However, the oxygen carrier cannot be gaseous after oxidation to avoid gas-gas separation
- Current technology uses CLOU concept to apply to this approach
  - Gaseous oxygen released from the oxygen carrier
  - Oxygen carrier remains solid at all times
  - Actual kinetics is dependent on the $O_2$ release rates
  - Actual conversion is dependent on $O_2$ capacity

SIU Southern Illinois University CARBONDALE

# Coal Direct CLOU based CLC

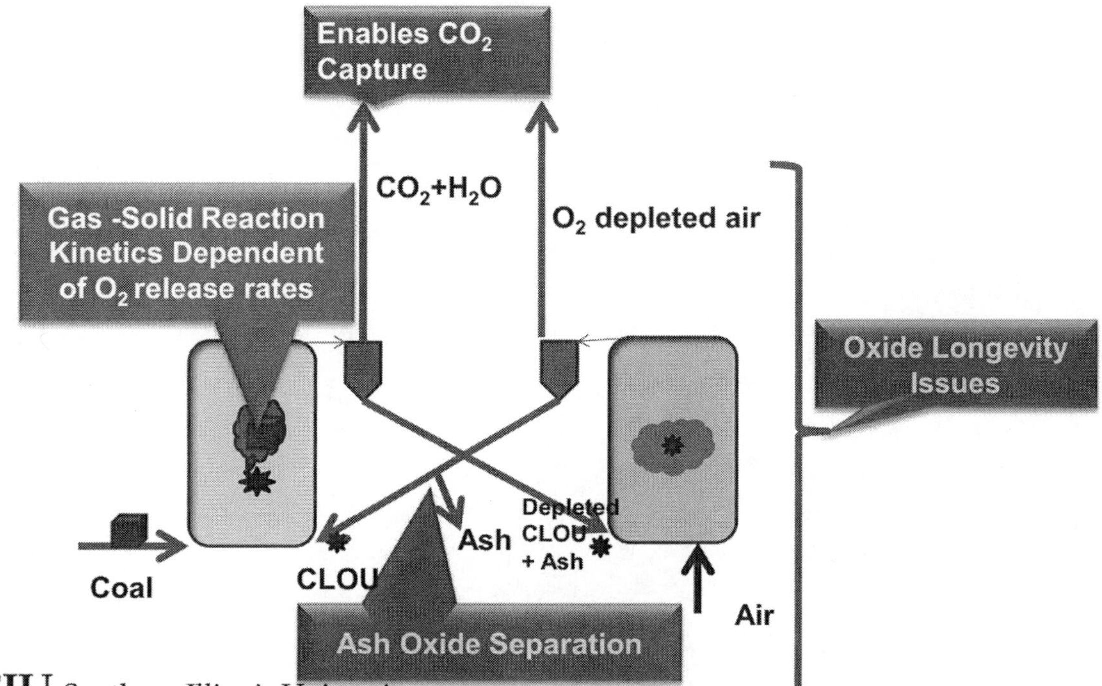


SIU Southern Illinois University
CARBONDALE

# What if the $MeO_x$ was vapor??

- Gas – solid reaction
  - As in conventional CLC
  - Conversion rates limited by reaction rates and not $O_2$ release rates
- Ash – Oxide separation can be made easier if the fully oxidized form could be melted or vaporized at temperatures below ash fusion.
- Oxide longevity issues may be eliminated

**Optimal :reduced form of the oxide is solid in the fuel reactor**

SIU Southern Illinois University
CARBONDALE

## Vapor Phase Coal Direct CLC

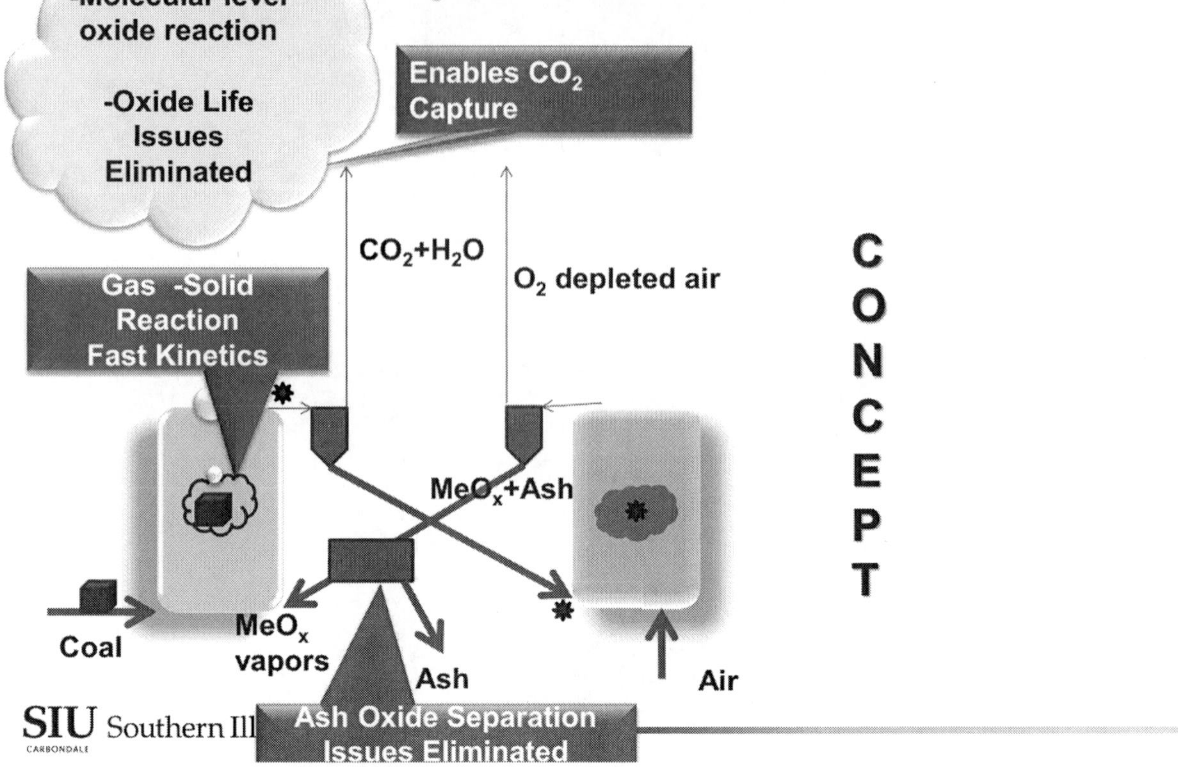




# Salient Features and Associated Benefits

- Active oxide is Vapor at Operation temperatures
  - Enables gas –solid reactions
- Reduced oxide is solid at operational temperatures
  - Reduced oxides separates from the gas by precipitation
  - Reduced oxide converts to acitve form in air reactor
  - Subsequent heat exchange vaporizes active oxide and
  - No solid-solid separation required
- Molecular gas reacts with solid and reduced oxide sublimes to molecular gas upon oxidation
  - Oxide deactivation is not an issue
- Simpler Design as with conventional CLC with natural Gas

# Additional Benefits

- Smaller Reactor due to fast kinetics
- Raw Material loss reduced during separation
- Elimination of solid solid separator
- Lowered makeup oxide requirement
- Fuel Flexible
  - Can also be used with natural gas
    - Gas-gas reaction with reduced oxide precipitated

SIU Southern Illinois University
CARBONDALE

# Work Completed to Date

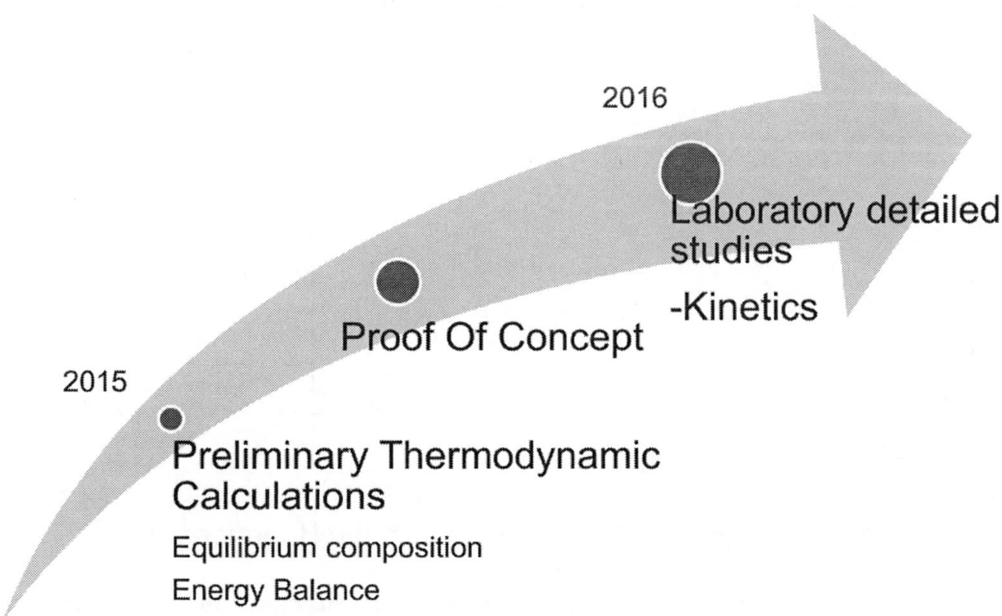


SIU Southern Illinois University
CARBONDALE

# Preliminary Equilibrium Calculations

## Sub-Stoichiometric amounts of $MeO_x$ to C

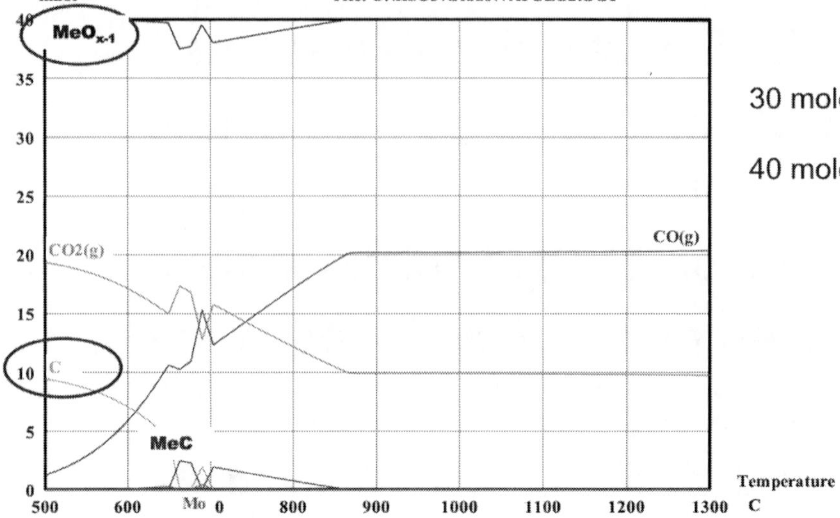


30 moles of C

40 moles of of $MeO_x$

**All $MeO_x$ converted to solid reduced forms and all carbon converted by 700 °C**

**CO predominates over $CO_2$ when complete carbon conversion is achieved**

SIU Southern Illinois University
CARBONDALE

# Preliminary Equilibrium Calculations

## Stoichiometric amounts of $MeO_x$ to C

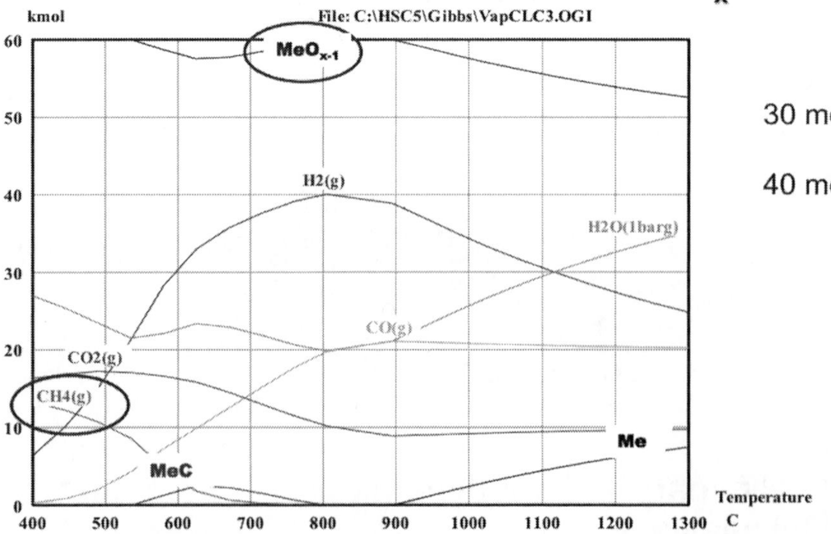


30 moles of C

40 moles of of $MeO_3$

**Further reduction of $MeO_{x-1}$ to Me with the $H_2$ generated**

**Syngas produced when complete methane conversion is achieved**

SIU Southern Illinois University
CARBONDALE

# Energy Balance

- Enthalpys of the reactions in fuel reactor
  - $2MeO_x(g) + C = 2MeO_{x-1} + CO_2$
  - > 800 kJ produced /mol C

- Enthalpys of the reactions in air reactor
  - $2MeO_{x-1} + O_2 = 2MeO_x\ (g)$
  - 400 kJ needed when the two moles of $MeO_2$ is oxidized and vaporized

Net heat Produced is 400 kJ/mol C
Primary heat generator will be the fuel reactor

SIU Southern Illinois University
CARBONDALE

# Experimental Setup

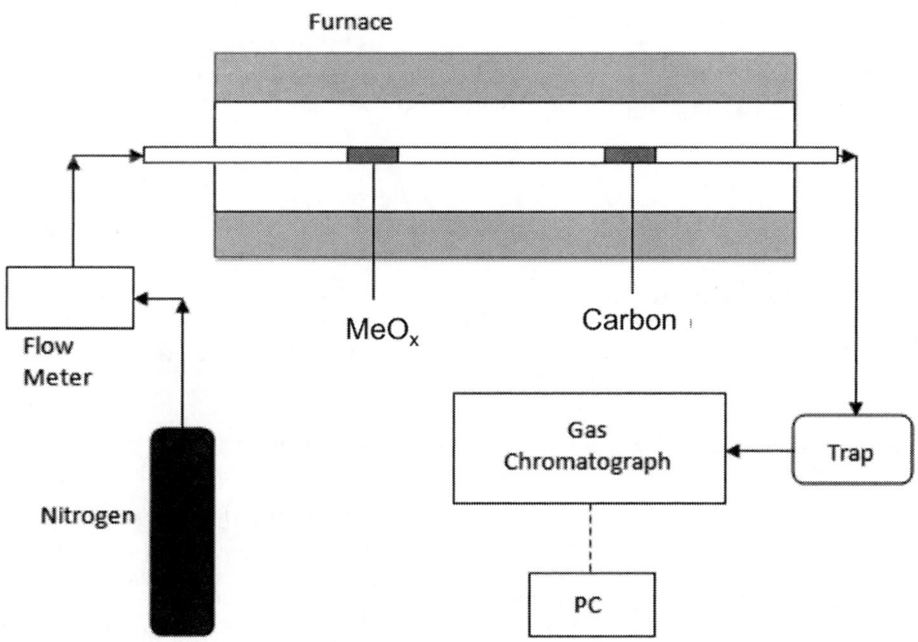


SIU Southern Illinois University
CARBONDALE

# Preliminary Parametric Effects

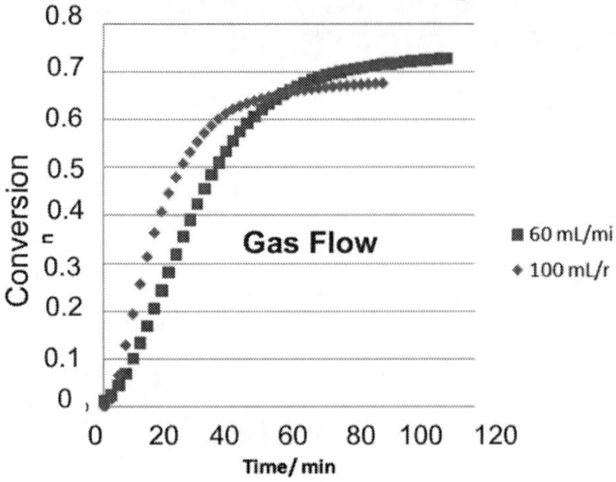


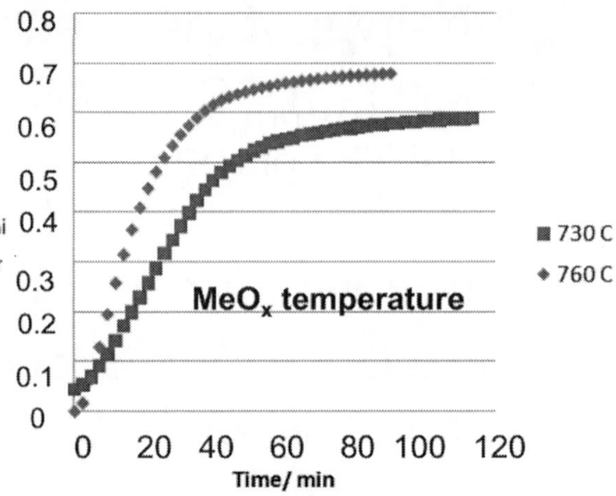


Improves Rate

May Impact Conversion
- may be due to $MeO_{x-1}$ coating

Improves rate and conversion



# Preliminary Parametric Effects

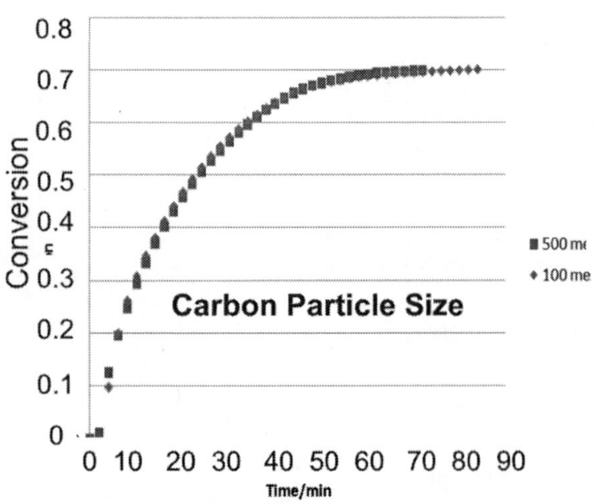


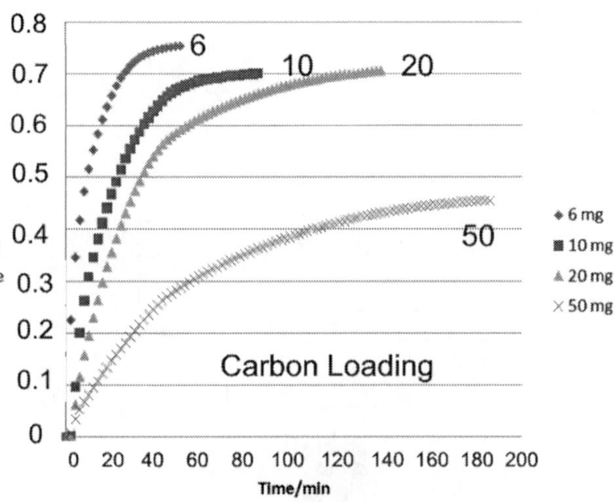


No Impact

-not limited by intra particle diffusion?

Lower Carbon Loading
increases rate and conversion

- Extra particle diffusion limitation??

# Kinetic Studies

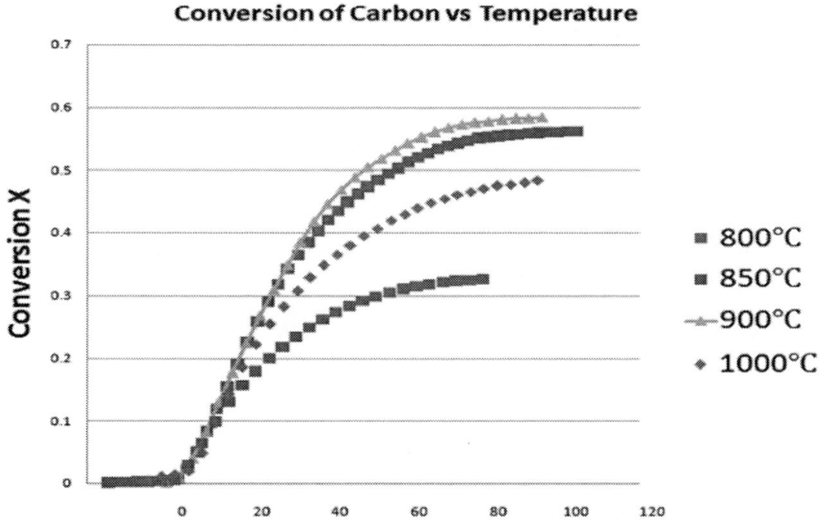


Note at 1000 C, final conversion lowers
- reaction kinetics vs diffusion through solid $MeO_{x-1}$ deposits???



# Kinetic Studies

Approach $-[\ln(1-X)]^{1/n} = k_{app} * t$

$k = A*\exp\{-Ea/RT\}$

Initial Assumption **n=1**

Conditions $MeO_x$ temp: 730 -810 C,
Loading 6 to 50 mg
Carbon Temperature 800 – 1000 C

- Activation energy varied with $MeO_x$ zone temperature
- Activation energy ranged from 41.57 to 174.59 kJ/mol
- Activation energy also varied with loading.
- Activation energy ranged from 99 to 116 kJ/mol

Model needs to be revisited to separate the kinetics zone and mass transport limitation regime (due to $MeO_{x-1}$ layering)

# Alternate Kinetic Model - Reduction

$C + (a+2b)\ MeO_x \rightarrow aCO + bCO_2 + (a+2b)\ MeO_{x-1}$

$CO + MeO_x \rightarrow CO_2 + MeO_{x-1}$

$C + CO_2 \rightarrow 2CO$

$[C] + [CO_2] + [CO] = constant = [C]_{initial}$

Work under progress

SIU Southern Illinois University
CARBONDALE

# Conclusions

**Vapor Phase CLC is feasible and has the potential to solve Solid Fuel based CLC issues**

- Candidate oxides identified
  - MeOx in vapor form in fuel reactor
  - $MeO_{x-1}$ is solid
  - Oxides show favorable equilibrium conditions and energy balance
- Candidate oxides tested
  - Activated Char and not coal were tested to evaluate solid char effects rather than the just the volatiles in coal
  - Vapors oxidize the carbon
  - Conversion seems to be limited after rapid oxidation – probably due to pore blockage by the condensing vapors
- Preliminary parametric evaluation completed
- Kinetic modeling underway
  - Rudimentary 1st order kinetics show reaction rate limitation during the initial period
  - Activation energy calculated

SIU Southern Illinois University
CARBONDALE

SIU Southern Illinois University
CARBONDALE

THE INSTITUTE FOR CLEAN AND SECURE ENERGY

•155 South 1452 East Room 380 •Salt Lake City, Utah 84112 •1-801-585-1233

# Copper-Based Chemical Looping with Oxygen Uncoupling: Process Development, Reactor Scale-Up, Pilot-Scale Studies and System Modeling

Kevin J. Whitty,
Matthew A. Hamilton, Kirsten M. Merritt, Kyle O'Malley,
Shakar Ali, Emir Rahislic, Steve Stafsholt,
Andrew R. Fry, JoAnn S. Lighty

Department of Chemical Engineering
Institute for Clean and Secure Energy
The University of Utah
Salt Lake City, Utah, USA

Contact: kevin.whitty@utah.edu

Thirty-Third Annual Pittsburgh Coal Conference
Cape Town, South Africa
August 8-12 2016

## Outline

- Technology background
  - Chemical looping combustion
  - CLOU
- CLOU process development and scale-up
  - Oxygen carrier production
  - Fundamentals of reactions and rates
  - Reactor design and operation
  - Process modeling and reactor simulation
- Future plans

# Technology Background: Chemical Looping Combustion

- CLC achieves *in situ* air separation by using a metal to transport oxygen from *air reactor* to *fuel reactor*
- Fuel (e.g. natural gas, coal) fed to fuel reactor is indirectly combusted by oxygen on oxidized metal
- Metal returns to reduced state in fuel reactor and "loops" back to air reactor
- Overall balance same as for conventional combustion
- Economic evaluations indicate CLC yields lowest COE of any $CO_2$-capture technology

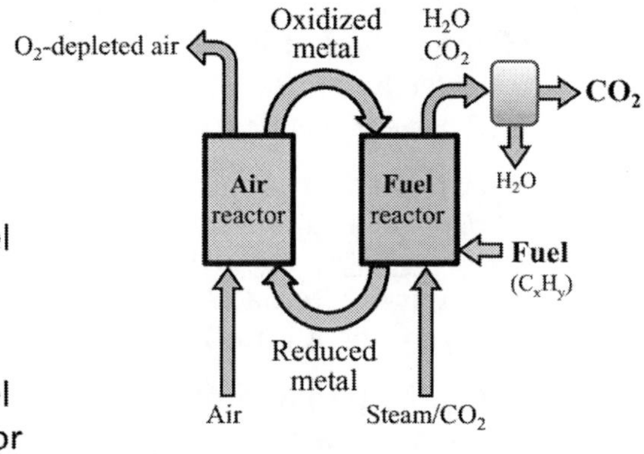


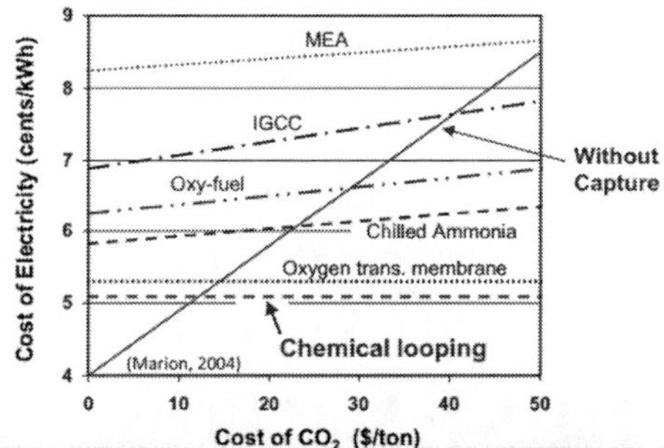


THE INSTITUTE FOR CLEAN AND SECURE ENERGY 3

# Technology Background: Chemical Looping with Oxygen Uncoupling (CLOU)

$$Cu_2O(s) + \tfrac{1}{2} O_2(g) \rightleftarrows 2\,CuO(s)$$

- Copper is one of few metals for which oxidation equilibrium ($Cu_2O$/CuO) lies within CLC operating temperatures.
- $Cu_2O$ is oxidized in air reactor
- CuO spontaneously releases $O_2$ in fuel reactor due to low $O_2$ partial pressure
- Released $O_2$ reacts with solid coal char, converting more than 50x faster than with non-CLOU oxygen carriers

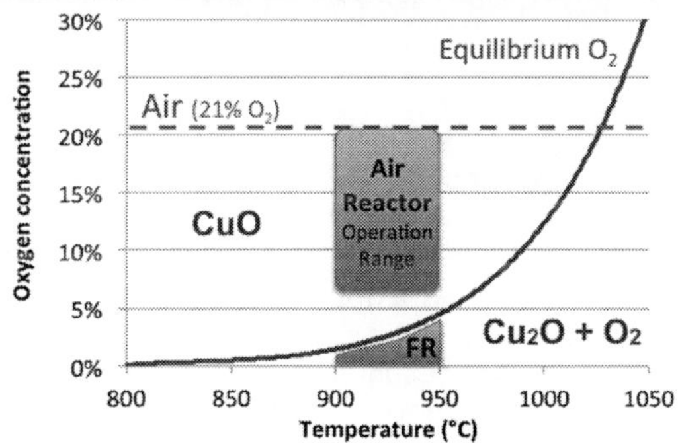


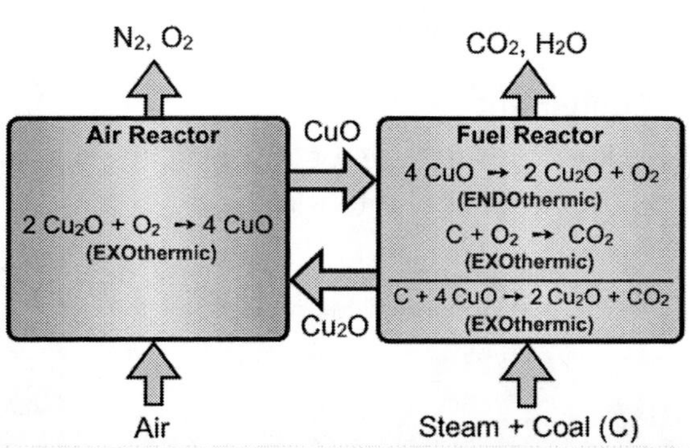


THE INSTITUTE FOR CLEAN AND SECURE ENERGY 4

# CLOU Process Research and Development

- **CLC intensively researched worldwide**
  - University of Utah researching since 2007
  - 10 projects totaling more than $5 million
- **Oxygen carrier development and production**
  - Focus on <u>inexpensive</u> copper-based carriers with <u>scalable</u> production
  - Dozens of alternatives tested
  - Current focus is CuO-on-SiC
- **Reactor design and operation**
  - Fundamental studies of CLOU reaction kinetics
  - Lab-scale experiments of coal conversion
  - Design and initial construction of 200 kW PDU
- **Process modeling and reactor simulation**
  - Aspen Plus modeling of CLC system
  - Barracuda VR® modeling of integrated fluidized bed system

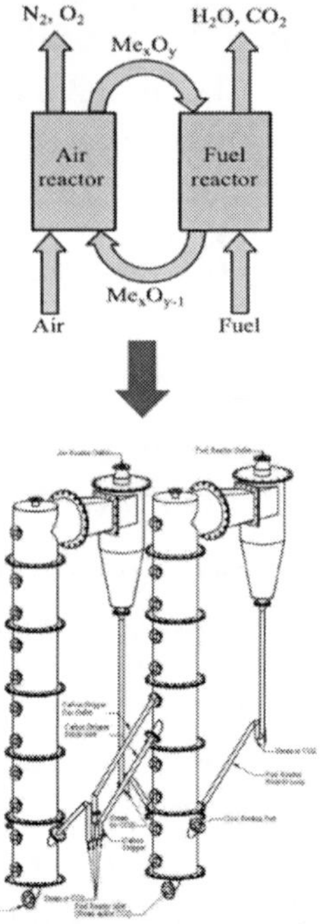




# Scale-up of CLOU Oxygen Carrier Production

- **Procedure**
  - Incipient wetness with copper nitrate
  - Current support: SiC
    - Best based on previous screening
  - Research focuses on identifying optimum production "recipe"
    - $Cu(NO_3)_2$ concentration
    - Number of additions
    - Calcining time
    - Solvent
    - Support pretreat
- **Equipment**
  - Rotary evaporator for screening
  - Three scales of rotary kiln
    - 1 kg lab scale
    - 10 kg bench scale
    - 100 kg pilot-scale production kiln

| System | Type | Capacity | Heating | Max T | Length | Diam |
|---|---|---|---|---|---|---|
| RV-1 | Rotary evap | 1 kg | Water bath | 95°C | n/a | 0.15 m |
| RK-1 | Rotary kiln | 1 kg | Elec Inductive | 800°C | 0.15 m | 0.1 m |
| RK-10 | Rotary kiln | 10 kg | Elec radiative | 350°C | 0.8 m | 0.2 m |
| RK-100 | Rotary kiln | 100 kg | Natural gas | 500°C | 1.4 m | 0.4 m |

RK-1 lab-scale induction kiln

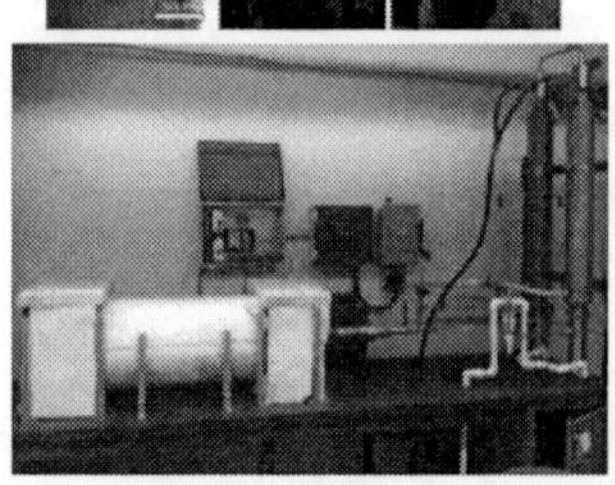

RK-10 bench-scale rotary kiln

RK-100 oxygen carrier production kiln

# Scale-up of CLOU Oxygen Carrier Production (2)

- **Characterization**
  - TGA for oxygen loading/rates
  - BET for available surface area
  - SEM for morphology
  - Crush strength
  - Lab-scale fluidized bed for long-term performance in a cycling fluidized bed reactor
- **Current Status**
  - Over 35 different carriers produced and characterized
  - Test batch of 50 kg CuO-on-SiC successfully produced in RK-100 pilot kiln

TGA for oxygen capacity and rate tests

CuO-on-SiC oxygen carrier

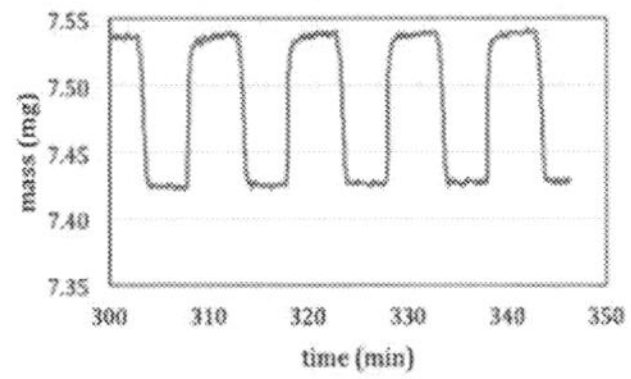


Oxidation and reduction of Cu-based carrier

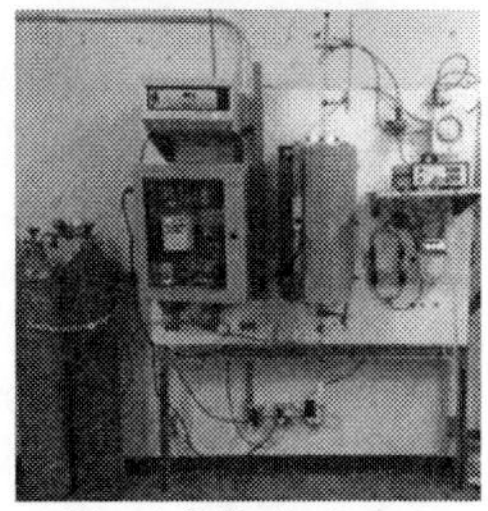

Lab-scale fluidized bed system

BET surface area analyzer

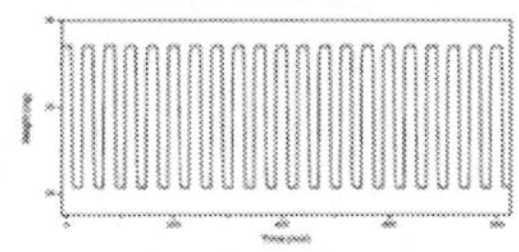

Sustained cyclability of carrier



# Process and Reactor Design: Underlying Chemistry and Rates

- Goal is to develop universal rate expressions suitable for incorporation into system models.
  - For oxidation: $rate = A\exp\left(\frac{E_a}{RT}\right)\left[P_{O2} - P_{O2,eq}\right]^{\alpha}\left[Cu_2O\right]^{\beta}$
  - For reduction: $rate = A\exp\left(\frac{E_a}{RT}\right)\left[P_{O2,eq} - P_{O2}\right]^{\alpha}\left[CuO\right]^{\beta}$
- Range of interest for CLOU rates
  - X = fraction of Cu as CuO, with remainder as $Cu_2O$
  - PDU design assumption: Carrier cycling between X = 0.75 exiting air reactor and X = 0.30 exiting fuel reactor

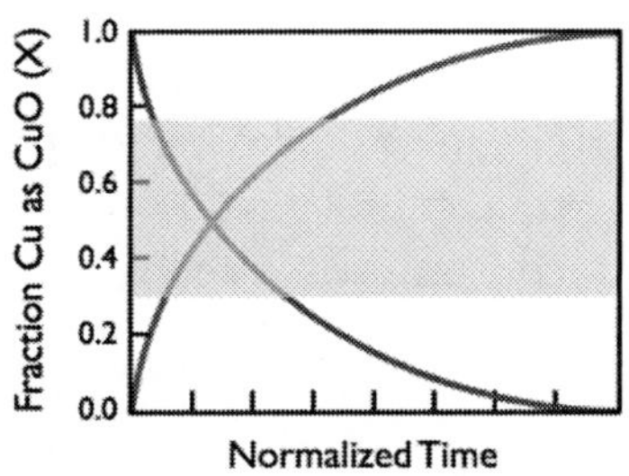


Oxidation: $2Cu_2O + O_2 \rightarrow 4CuO$

Reduction: $4CuO \rightarrow 2Cu_2O + O_2$

# Oxidation of $Cu_2O$ to CuO

- Oxidation experiments present interesting challenge
  - Driving force for oxidation decreases with temperature
  - Fundamental chemical rate increases with temperature ($E_a$)
  - Possible grain boundary sintering may also contribute to reduced rate at high temperature
- Resulting "oxidation rate peak" observed by many groups
- Deciphering true kinetics is challenging

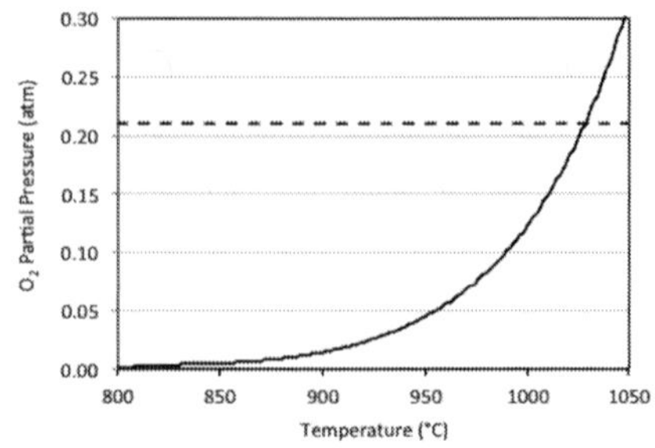


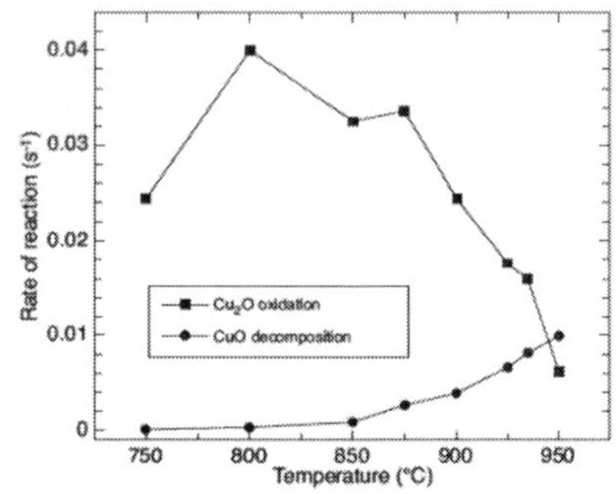




# Oxidation: Isolating Influences of $O_2$ Driving Force and Temperature

- Oxidation rate affected by
  - Temperature (fundamental kinetics)
  - $O_2$ driving force, which depends on temperature
- Series of experiments to isolate these influences
  - Hold T constant while changing $O_2$ driving force
  - Hold $O_2$ driving force constant while changing T
- Rate evaluation allowed distinction between rate influences

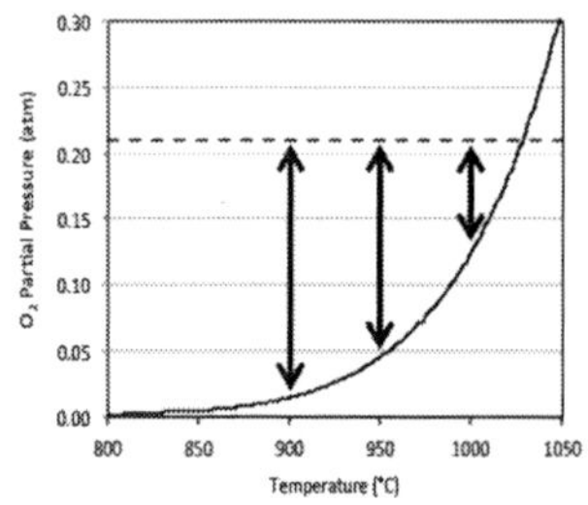


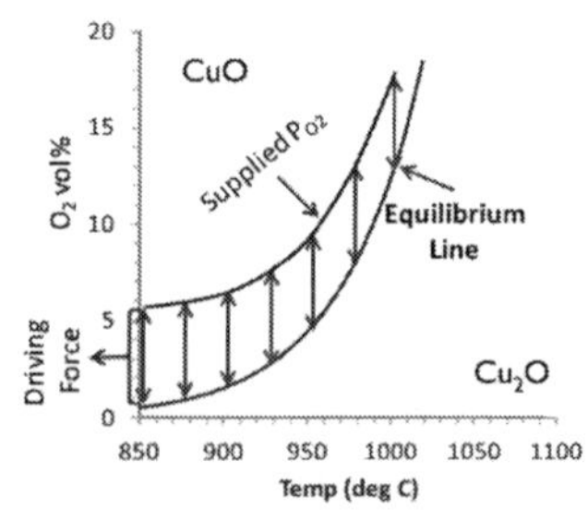


| Experiment number | Temperature (°C) | Equilibrium $O_2$ part. press. (atm) | Supplied $O_2$ part. press. (atm) | $O_2$ "driving force" (atm) |
|---|---|---|---|---|
| 1 | 876 | 0.010 | 0.050 | 0.040 |
| 2 | 962 | 0.060 | 0.100 | 0.040 |
| 3 | 994 | 0.110 | 0.150 | 0.040 |
| 4 | 1017 | 0.170 | 0.210 | 0.040 |

| Experiment number | Temperature (°C) | Equilibrium $O_2$ part. press. (atm) | Supplied $O_2$ part. press. (atm) | $O_2$ "driving force" (atm) |
|---|---|---|---|---|
| 1 | 850 | 0.005 | 0.050 | 0.045 |
| 2 | 850 | 0.005 | 0.100 | 0.095 |
| 3 | 850 | 0.005 | 0.150 | 0.145 |
| 4 | 850 | 0.005 | 0.210 | 0.205 |

# Measured Oxidation Rates

- Range of experimental conditions
  - Temperature
  - Reacting gas composition
- Four types of carrier materials
  - Various production techniques
  - Various CuO loadings

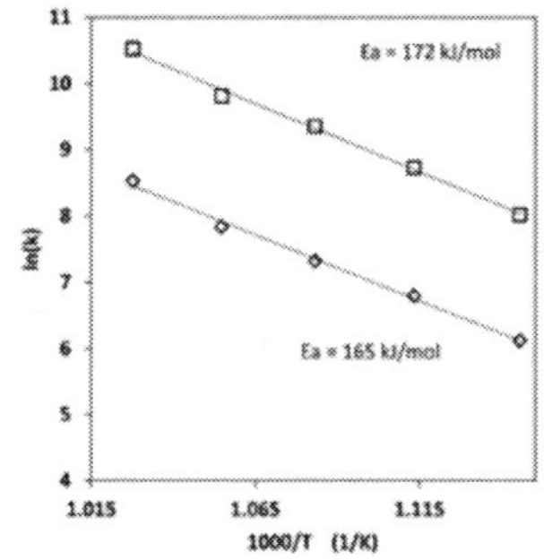


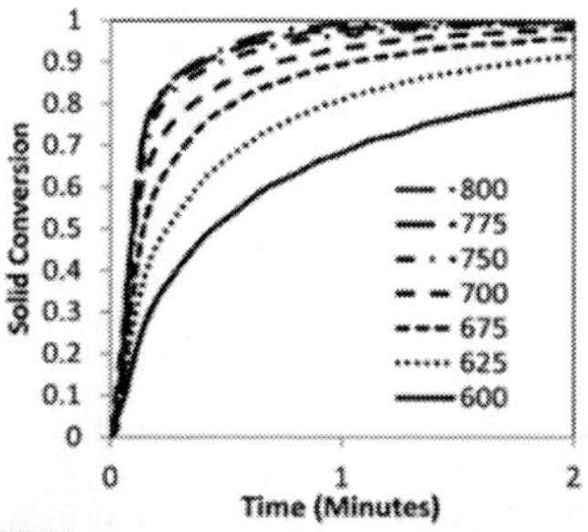


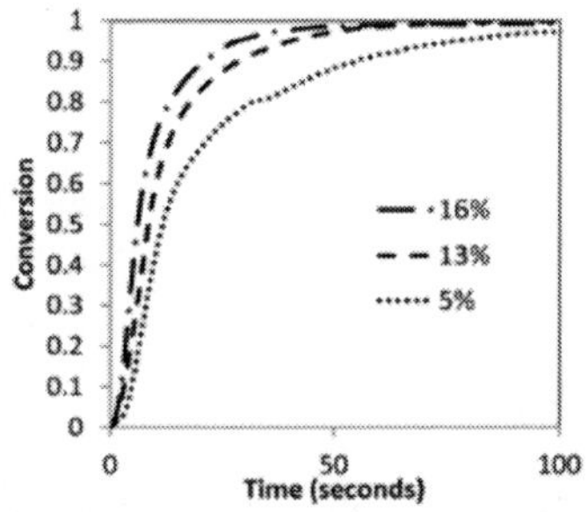


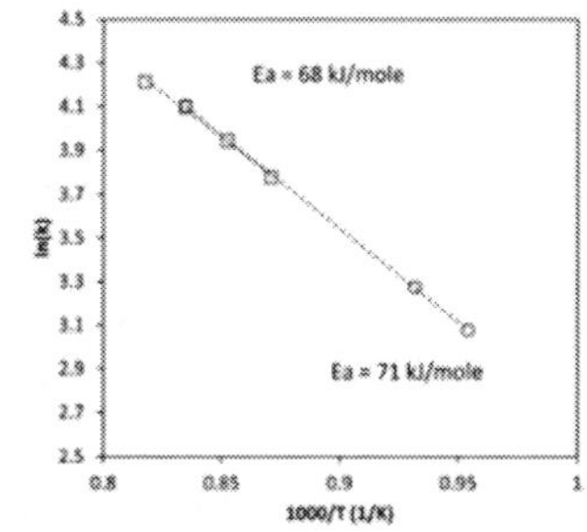




# Modeling of Oxidation Rates

- Mechanism determined to be more challenging than simple reversible reaction kinetics
- Two regimes of reaction behavior identified
  - Low temperature, non-CLOU region
    - Best described by pore blocking kinetic mechanism

$$\frac{dX}{dt} = k\frac{1}{\exp\left(\frac{X}{\lambda}\right)}[P_{O_2}^{\beta} - P_{O_2,eq}^{\beta}]$$

  - High temperature CLOU region
    - Activation energy must be separated into thermodynamic and kinetic barriers
    - Best described by nucleation and growth mechanism

$$[-\ln(1 - X)]^{1/n} = k_{app} \times t$$

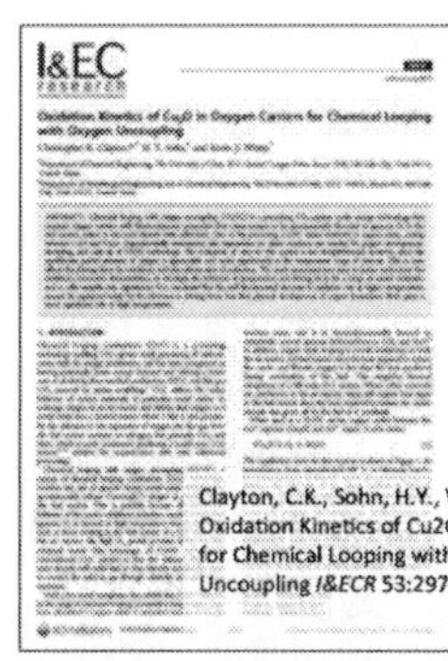

Clayton, C.K., Sohn, H.Y., Whitty, K.J. Oxidation Kinetics of Cu2O in Oxygen Carriers for Chemical Looping with Oxygen Uncoupling *I&ECR* 53:2976-2986 (2013).

# Measurement of CuO Reduction Rates

- Similar to oxidation studies
- Range of conditions
  - Temperature
  - Gas composition
- Challenge of having absolutely zero $O_2$ in gas phase
- Reaction order in CuO = 0
- Apparent activation energy 274 kJ/mol

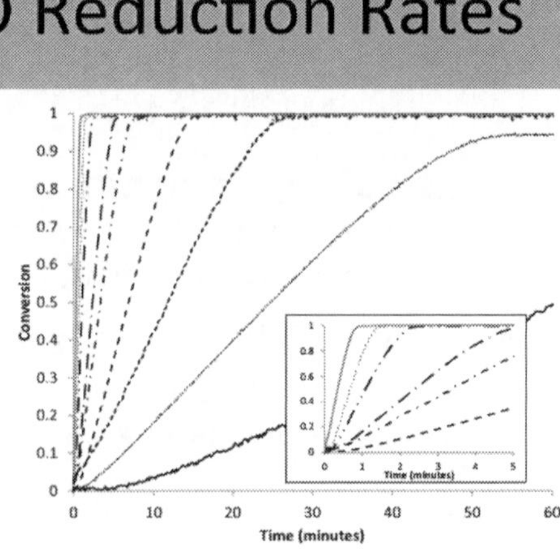


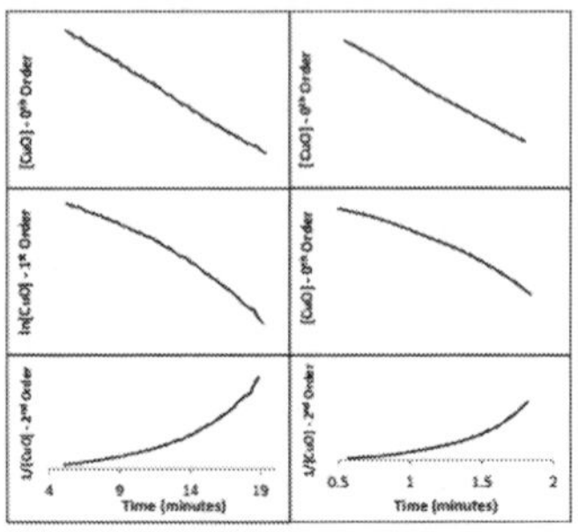


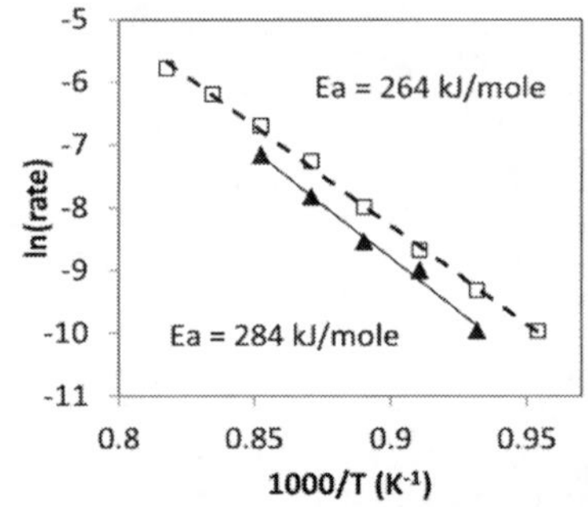




# Modeling of Carrier Reduction Rates

- Any oxygen in gas phase reduces driving force for reduction
- Used similar methodology to deciphering specific influences for oxidation
  - Vary $(p_{O2,eq} - p_{O2})$ at constant temperature
  - Hold $(p_{O2,eq} - p_{O2})$ constant at various temperatures
- Could decipher constants in rate expression
- Universal rate expression:

$$\text{rate}\left(\frac{g}{s}\frac{1}{g_{Cu}}\right) = 3.90 \times 10^{-4} \exp\left(-\frac{62,000}{R \times T}\right)[6.057 \times 10^{-11}\exp^{0.02146\times(T-273)} - P_{O2}]$$

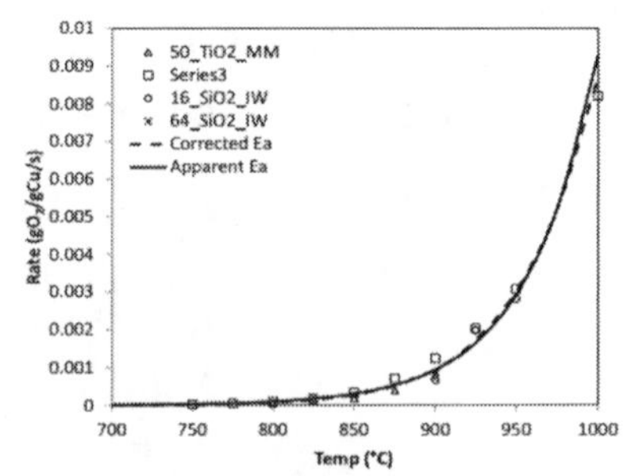

Clayton, C.K., Whitty, K.J., Measurement and Modeling of Decomposition Kinetics for Copper-Oxide Based Chemical Looping with Oxygen Uncoupling, *Applied Energy* 116:416-423 (2013).

# 10 $kW_{th}$ Bench-Scale Chemical Looping System

- Design
  - Electrically heated
  - 1.5 kg/hr coal feed rate
  - Bubbling bed fuel reactor
    - 10 cm diameter
  - Bubbling bed air reactor
    - 15 cm diameter
  - Riser to lift particles to cyclone
  - Filters to capture carryover particles
  - Induced draft fans for pressure balance

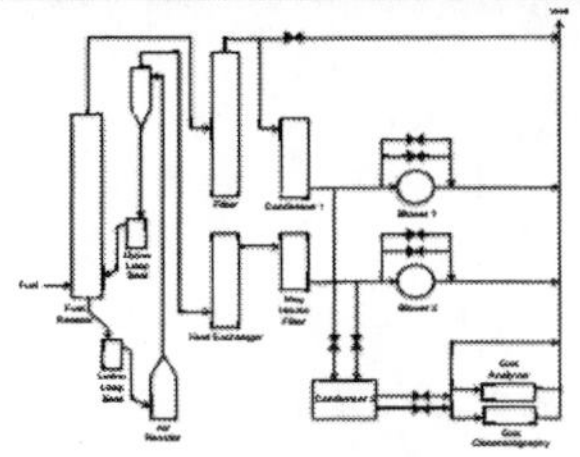

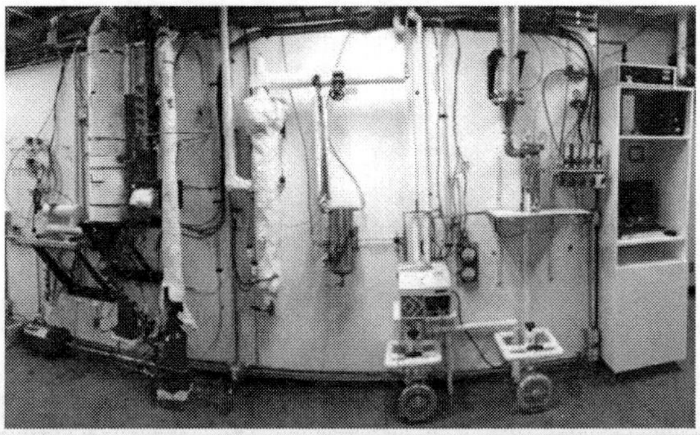

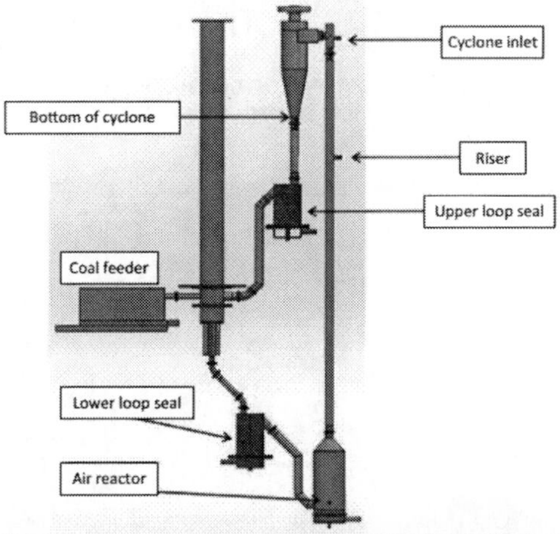




# 200 $kW_{th}$ Process Development Unit

- **PDU Design**
  - Two interconnected CFBs
    - approx 5 m tall
    - 0.3 m diam
  - Refractory-lined
  - Electric + gas air/steam preheat
  - Bag house filters
  - Operation on natural gas or coal
  - Approx. 200 kg bed inventory
- **Status**
  - Construction complete
    - Some rebuilds/repairs were necessary
  - Function of subsystems has been confirmed
  - Oxygen carrier circulation rates to 12 tons/hr ilmenite achieved
  - CLC testing with gas in progress; coal starting soon

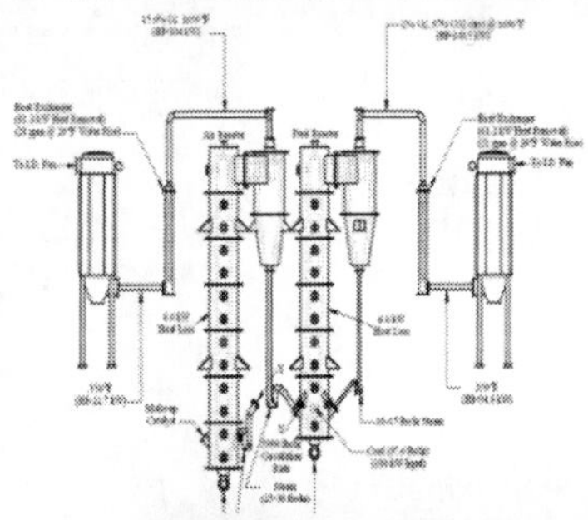

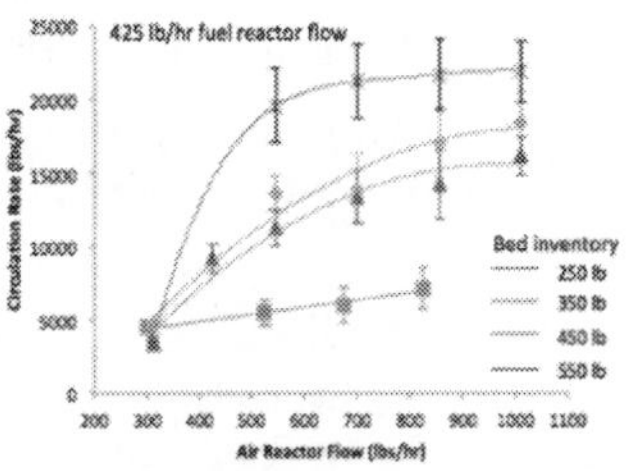

# Chemical Looping Process Modeling

- Process models
  - Aspen-based for whole process
  - Excel or Matlab for reactors
- Overall energy balance
- Energy distribution between reactors
- Evaluation of preheating needs for air reactor and fuel reactor
- Assessment for both UofU PDU and extension to industrial-scale systems

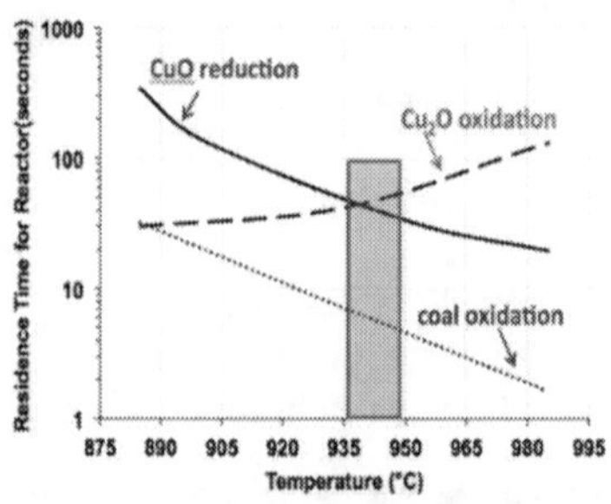


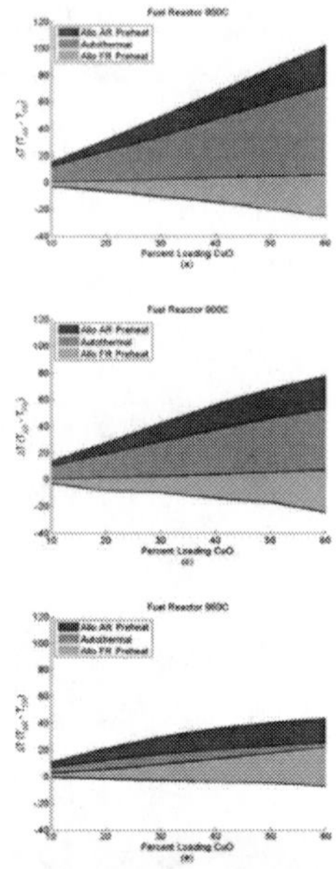

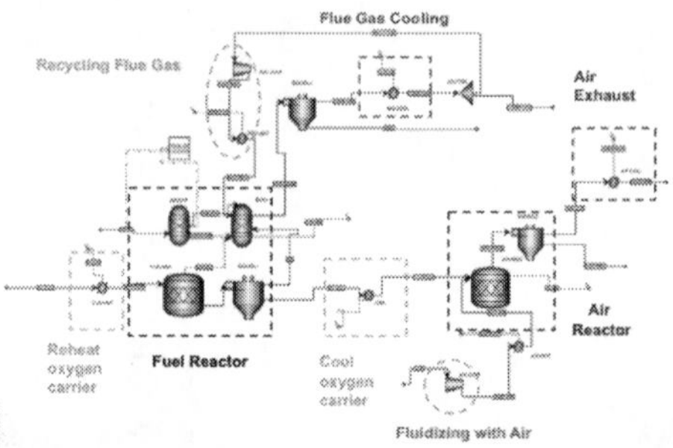


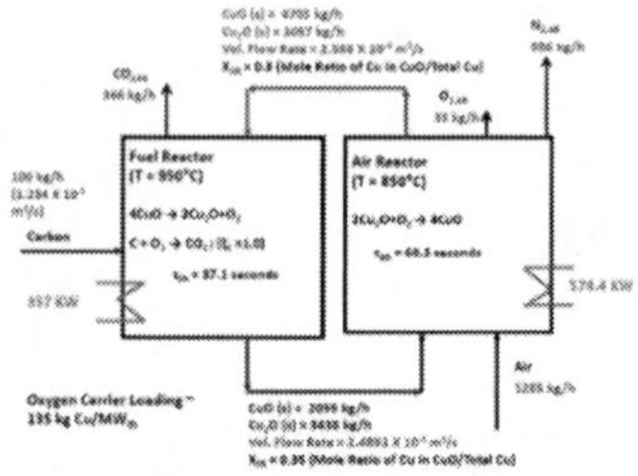



# Progress and Current Status: Chemical Looping Reactor Simulation

- Using CPFD Barracuda VR®
- Models of both 10 kW bench-scale and 200 kW pilot-scale reactors
- Simulations include
  - hydrodynamics
  - heat transfer
  - Oxygen carrier chemistry/kinetics
  - Coal combustion chemistry/kinetics
- Plexiglas cold-flow model of UofU PDU to help validation
- Understanding from simulations has been valuable in starting up and interpreting behavior of pilot-scale system

Cold-flow model of UofU PDU

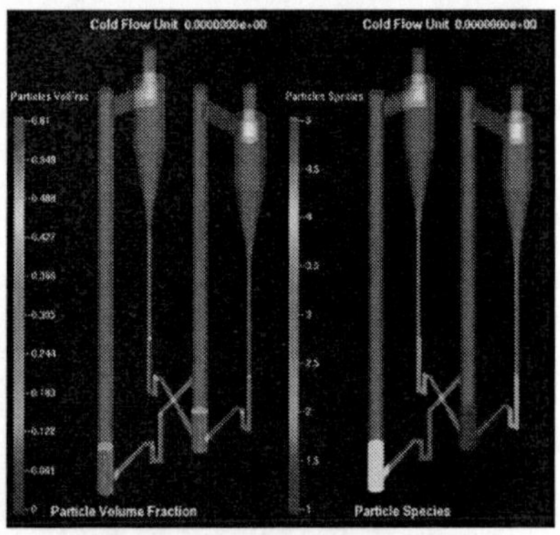


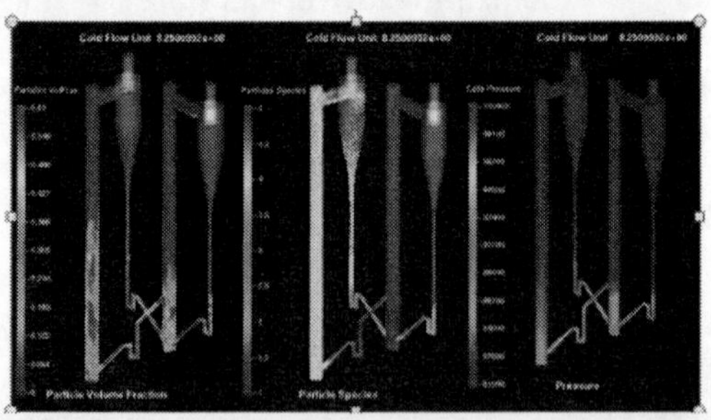

# Future Plans

- Near-term
  - Produce CuO-based CLOU carrier for PDU testing
    - Initial batch targets 20% CuO to ensure no agglomeration
    - Eventually target 40-45% CuO to increase load
  - Parametric testing of PDU with CuO (CLOU) carrier and coal
    - Vary coal, coal particle size, air reactor flow rate (circulation rate),
    - Measure $CO_2$ capture, $CO_2$ purity, fuel conversion, overall performance
    - Design, install and test carbon stripper to improve coal conversion and $CO_2$ capture.
  - Advance computational simulation
    - Validate simulation of PDU with operational data
    - Simulate larger (e.g. 10 and 100 MW) reactors
- Long-term
  - Continued operation and experience with PDU
  - Pursue opportunities for larger pilot (3-10 MW) system



# Acknowledgements

- Institute for Clean & Secure Energy
- Dana Overacker, Denis Balic
- University of Wyoming School of Energy Resources
- This material is based in part upon work supported by the U.S. Department of Energy under award number DE-NT0005105 and DE-FE0025076.

**2016 International Pittsburgh Coal Conference**
**Cape Town, SOUTH AFRICA**
**August 8 – 12, 2016**

**Abstract Submission**

**Combustion Technologies**

# A study on the mineral liberation and transformations during pulverised bituminous coal combustion

***(Laurence Silvester, Mr, University of Nottingham, Faculty of Engineering, Jubilee Campus, Triumph Road, Nottingham, NG7 2TU, United Kingdom)***
***Contact Information: enxls2@nottingham.ac.uk , +447944232320***

*(Cheng-gong Sun, Dr., University of Nottingham, Faculty of Engineering, Jubilee Campus, Triumph Road, Nottingham, NG7 2TU, United Kingdom)*
*Contact Information: Cheng-gong.Sun@nottingham.ac.uk , +441157484577*

*(Colin Snape, Prof., University of Nottingham, Faculty of Engineering, Jubilee Campus, Triumph Road, Nottingham, NG7 2TU, United Kingdom)*
*Contact Information: colin.snape@nottingham.ac.uk , +441159514166*

*(Hao Liu, Prof., University of Nottingham, Faculty of Engineering, Jubilee Campus, Triumph Road, Nottingham, NG7 2TU, United Kingdom)*
*Contact Information: liu.hao@nottingham.ac.uk , +441158467674*

*(William Livingston, Dr., Doosan Babcock Energy Limited, Technology Centre, High Street, Renfrew, PA4 8UW*
*Contact Information: bill.livingston@doosan.com , +441418864141*

*(Ofonime Udoudo, Dr., University of Nottingham, University Park, Faculty of Engineering, L4, Nottingham, NG7 2RD, United Kingdom)*
*Contact Information: ofonime.udoudo@nottingham.ac.uk , +441159514080*

Ash deposition is a major cause of reduced boiler efficiency and unscheduled shutdowns from induced blockage, corrosion and erosion in pulverised coal fired boilers[1, 2]. A detailed understanding of the ash formation mechanism due to the inorganic minerals in coal is the first step in predicting the ash deposition behaviour on convective heat transfer surfaces. This study investigates the mineral liberation and transformations of six pulverised bituminous coals during devolatilisation and subsequent combustion at 1450°C in a laboratory-scale drop tube furnace (DTF). The mineral properties of the coals and derived chars and ashes will be examined with mineral liberation analysis (MLA) to identify mineral transformation, morphological and particle liberation changes. X-ray diffraction (XRD) will be used to qualitatively validate the coal mineral phases. The physical changes due to devolatilisation and combustion will be determined through digital image analysis for particle size distribution and shape factor data. The combustion residues obtained from DTF tests and those collected from the convective boiler sections of four power plants will be compared in terms of their mineral distribution and physical properties, particularly the presence of particle agglomerates. Preliminary results suggest that the coals contain the major minerals quartz, kaolinite, illite and pyrite with minor minerals montmorillonite and calcite. Electron micrograph images of the power plant fly ashes show large agglomerating particles, with one fly ash showing unburned char particles, indicating an oxygen deficient boiler environment.

References:

1. Ian Barnes. Slagging and fouling in coal-fired boilers. IEACCC/147, London, IEA Coal Research, 2009.
2. 2. Couch G R. Ash management in coal-fired power plants. IEACCC/118, London, IEA Clean Coal Centre, 2006.

**Clean Coal Technology Research**

- *Beneficiation*
- *Characterisation*
- *Utilization*
- *Environment*

# Combustion Properties of Bamboo when Blended with a South African coal

Olive Makwarela
*School of Chemical and Metallurgical Engineering, University of the Witwatersrand*

*Co-Investigators:*
Dr Samson Bada and Prof Rosemary Falcon

**Pittsburgh Coal Conference, August 8 - 12, 2016 - Cape Town, SOUTH AFRICA**

1

## Presentation Outline

- Background
- Relevance of this work
- Aim
- Objectives
- Experimental work
- Results
- Conclusions
- Recommendations
- Acknowledgements

2

# Background

- RSA is economically vulnerable to climate change.
- Currently more than 90% of RSA's electricity is from coal combustion.
- Coal is a fossil fuel that emits greenhouse gases during combustion (Eskom emissions=233Mt $CO_2$).
- $CO_2$ emissions can be reduced by 50% by co-firing coal with biomass (1:1).
- Existing power stations and new IPPs can be used for co-firing with minor modifications.
- Biomass has poor physical and fuel properties.

3

# Coal fired power station

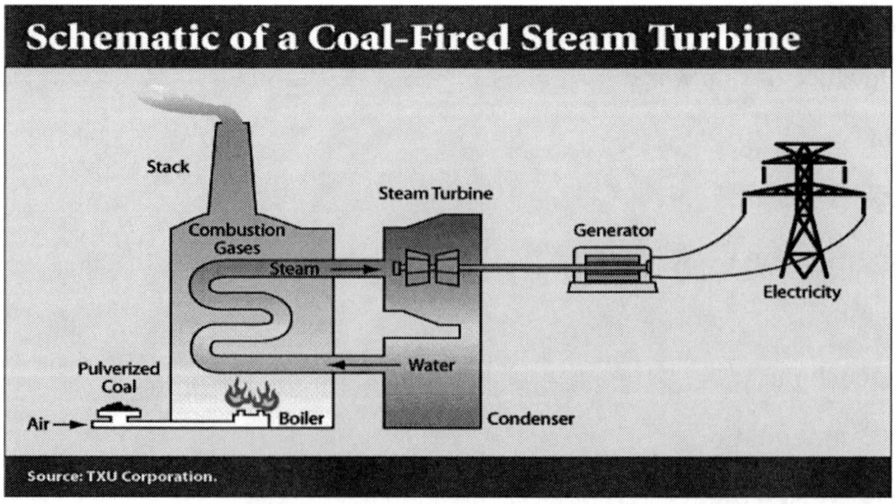


Coal power plant schematic (Combs et al., 2008)

4

## Relevance of this work

The relevance of this work to South Africa is as follows:

- It is vital for SA to undertake clean coal technologies i.e. To reduce $CO_2$ emissions when burning coal for heat and power generation. One key method is to introduce co-firing of biomass with coal as biomass is considered carbon neutral.
- To follow this route, it is necessary to select the most appropriate form of biomass, and to understand how to handle it and how it combusts.
- This can be done by studying the combustion of biomass with coal - in varying proportions and under different conditions – in order to establish the most optimal conditions for power generation.

5

## Research Aim

The aim of this research work is to study the physical properties of raw, torrefied and carbonised bamboo to establish the potential of these materials as suitable biomass sources for co-firing with coal.

6

## Objectives

Given that grinding and comminution are known to be difficult when handling fibrous forms of biomass, and that different parts of plants combust in different ways, the objectives in this research are the following:

- To investigate the impact of raw and thermal treatment on the grindability of the materials, using the Hardgrove Grindability Index (HGI).
- To investigate the impact of thermal treatment on the fuel properties of bamboo.
- To determine the combustibility of the materials and thereafter the compatibility when co-fired with coal using TGA.
- To compare different ages of Bamboo in order to establish the optimum growth phase for maximum end use application.

7

## Experimental Work

**1. Samples**

- 1.1. Coals – A, B, C, D and E.
- 1.2. Bamboo, *Bambusa balcooa* (1, 3 and 4 years + old) from Western Cape.

**2. Thermal treatment**

- 2.1. Torrefaction, 250°C and 280°C.
- 2.2. Low temperature carbonization, 350°C and 400°C.
- 2.3. 3°C/min, $N_2$ atmosphere, Residence time (RT)=40min.

**3. Sample characterization**

- 3.1. Proximate and ultimate analyses.
- 3.2. Calorific value (CV)

**4. Grindability test**

- 4.1. Retsch PM 100 ball mill
- 4.2. Calibration curve of HGI vs mass passing 75μm
- 4.3. Determination of HGI for both raw and thermally treated *bambusa balcooa* (1, 3 and 4 years +)

**5. Thermo-gravimetric analysis**

- 5.1. Combustion in air, rate=10°C/min to T= 850°C
- 5.2. Raw, treated bamboo and coal (particle size -212 μm)
- 5.3. Sample size of 100 mg
- 5.4. Co-firing bamboo and coal ( 75%, 50% and 25% bamboo).

8

# Images of bamboo samples

Photographic image of raw bamboo

Photographic image of treated bamboo

9

# Physico-chemical properties of 1, 3 and 4 year old Bamboo

**Table 1: Ultimate analyses of untreated 1, 3 and 4yr Bambusa balcooa**

| Sample ID | *Ultimate, Wt.% as received or air dried* | | | | |
|---|---|---|---|---|---|
| | N | C | H | S | O |
| 1 year old bamboo | 0.88 | 44.91 | 6.17 | 0.07 | 39.07 |
| 3 year old bamboo | 0.41 | 46.76 | 6.22 | 0.02 | 37.03 |
| 4 years old bamboo | 0.32 | 46.52 | 6.26 | 0.03 | 37.25 |

N: Nitrogen; C: Total carbon; H: Hydrogen; S:Total sulphur; and O: Oxygen by difference [100-(M+Ash+S+H+C+N)]

# Physico-chemical properties of coal, raw, thermally treated 1, 3 & 4 year old Bamboo

**Table 2: Fuel properties of raw and thermally treated *bambusa balcooa***

| Sample ID | *As received proximate (Wb.%)* | | | | *Dry proximate (Db.%)* | | | CV Mj/kg |
|---|---|---|---|---|---|---|---|---|
| | M | FC | VM | Ash | FC | VM | Ash | |
| Raw 1 year | 7.97 | 16.59 | 74.58 | 0.86 | 18.03 | 81.04 | 0.93 | 18.53 |
| 1 year T250 | 1.26 | 45.58 | 50.72 | 2.44 | 46.16 | 51.36 | 2.47 | 25.94 |
| 1 year T280 | 0.92 | 50.21 | 46.41 | 2.47 | 50.67 | 46.84 | 2.49 | 27.78 |
| 1 year C350 | 2.52 | 66.91 | 27.02 | 3.56 | 68.64 | 27.72 | 3.65 | 28.51 |
| 1 year C400 | 1.99 | 68.55 | 26.88 | 2.58 | 69.94 | 27.43 | 2.63 | ↓28.54 |
| Raw 3 year | 7.01 | 16.97 | 73.64 | 2.37 | 18.25 | 79.20 | 2.55 | 17.10 |
| 3 year T250 | 3.05 | 52.32 | 38.33 | 6.30 | 53.97 | 39.54 | 6.50 | 27.34 |
| 3 year T280 | 3.18 | 64.60 | 25.10 | 7.13 | 66.71 | 25.92 | 7.36 | 28.39 |
| 3 year C350 | 3.30 | 66.34 | 22.13 | 8.25 | 68.60 | 22.88 | 8.53 | 29.76 |
| 3 year C400 | 4.14 | 68.46 | 18.90 | 8.50 | 71.42 | 19.72 | 8.87 | 30.17 |
| Raw 4 year | 8.09 | 17.69 | 72.81 | 1.41 | 19.25 | 79.22 | 1.53 | 17.63 |
| 4 year T250 | 1.41 | 57.76 | 37.25 | 3.59 | 58.58 | 37.78 | 3.64 | 25.46 |
| 4 year T280 | 1.62 | 62.61 | 31.63 | 4.15 | 63.64 | 32.15 | 4.21 | 26.38 |
| 4 year C350 | 2.07 | 71.62 | 21.62 | 4.69 | 73.14 | 22.08 | 4.78 | 29.16 |
| 4 year C400 | 2.62 | 74.59 | 17.75 | 5.04 | **76.60** | 18.23 | 5.18 | **30.24** |
| Coal | 6.28 | 29.64 | 19.36 | 44.72 | 31.63 | 20.66 | 47.72 | **12.70** |

M: Moisture; FC: Fixed carbon; VM: Volatile matter; Wb: Wet basis and Db: Dry basis

# Energy characteristic of the raw and thermally treated Bamboo

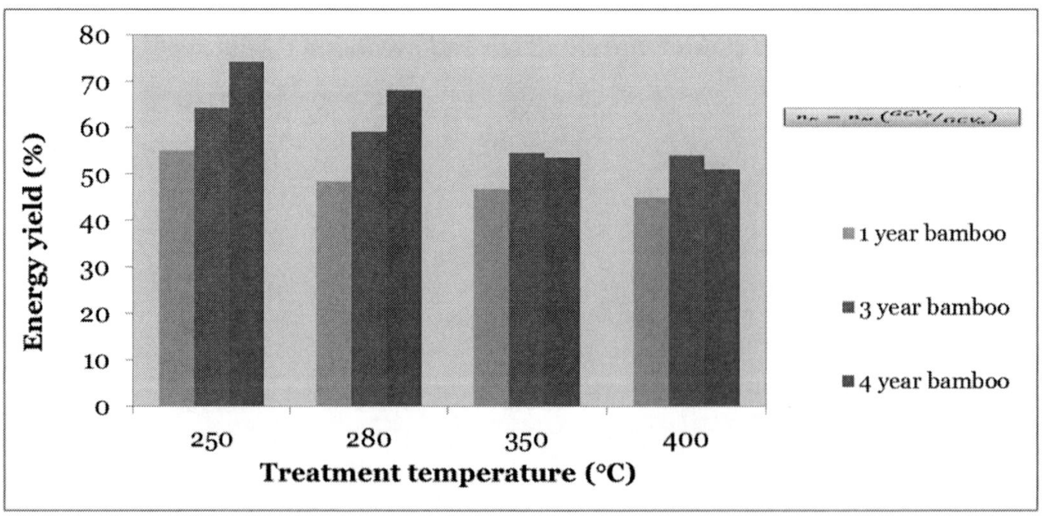


12

# Fixed Carbon and Volatile matter content of the thermally treated Bamboo

A typical power station coal FC = 30 % (db) or more

VM of about 30 % (db) required for stable flame in the boiler (PF)

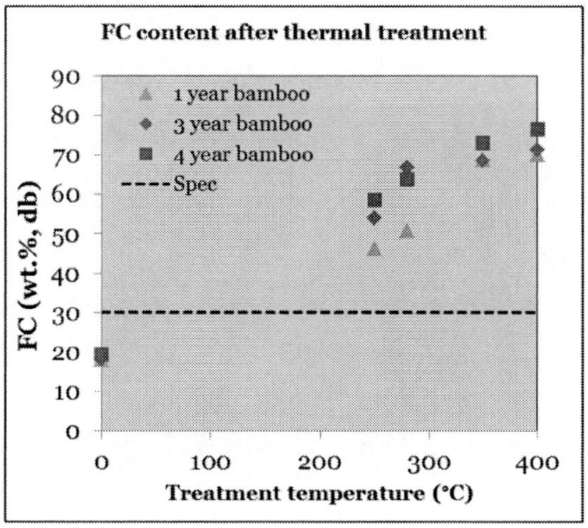


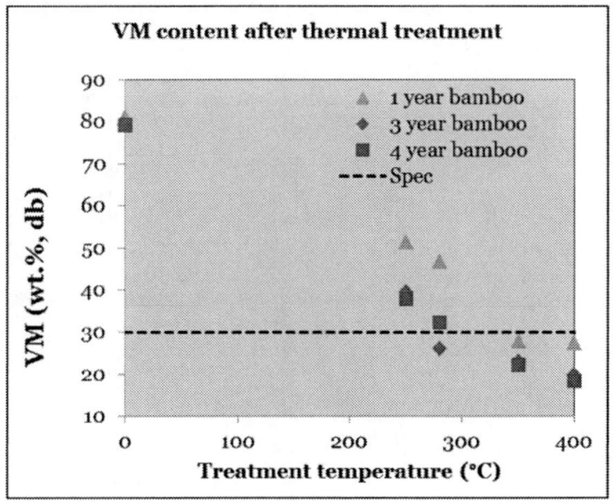


13

# Fuel Ratio of 1, 3 and 4yr old *Bambusa Balcooa* and Coal

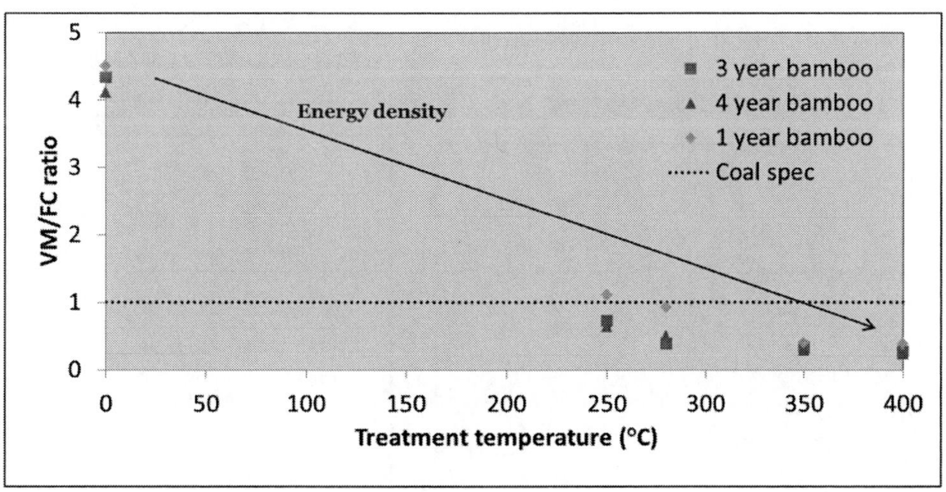


14

## Grindability test on four different coal

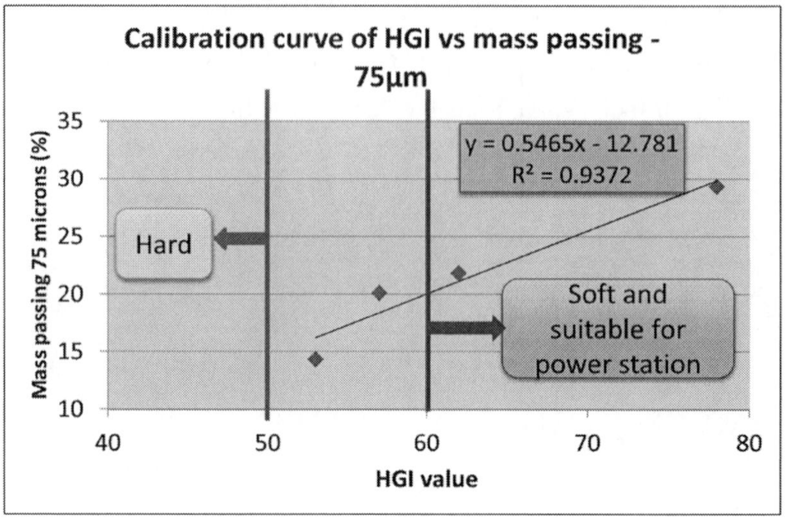


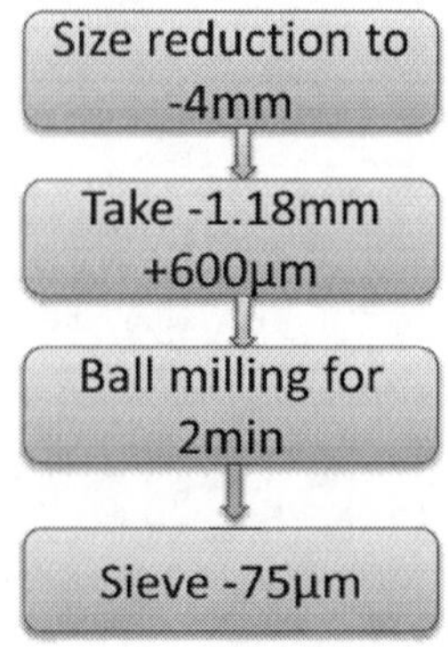


$$HGI_B = \frac{Mass_{-75\mu m} + 12.781}{0.5465}$$

## HGI of raw 1, 3, 4 year old and thermally treat *Bambusa Balcooa*

For Power station, HGI ≥ 60 desired

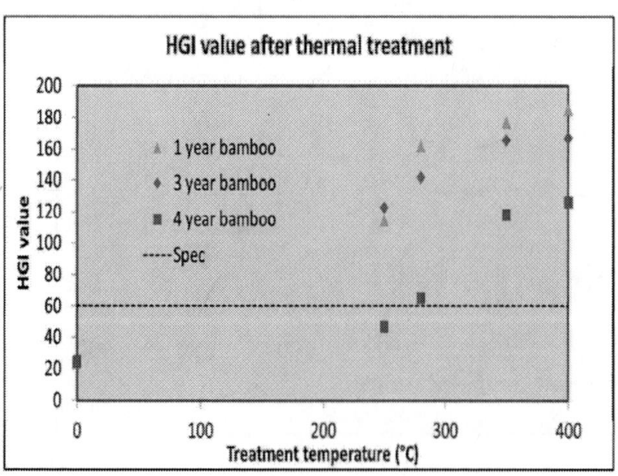


**Table 8: Equivalent HGI values for the raw and thermally treated bamboo**

| Sample | Raw | | T250 | | T280 | | C350 | | C400 | |
|---|---|---|---|---|---|---|---|---|---|---|
| | m | HGI | m | HGI | M | HGI | m | HGI | m | HGI |
| 1 year | 0.47 | **24** | 49.70 | 114 | 75.63 | 162 | 83.88 | 177 | 88.50 | **185** |
| 3 year | 0.71 | **25** | 54.27 | 123 | 64.41 | 141 | 77.61 | 165 | 78.44 | **167** |
| 4 year | 0.88 | **25** | 12.72 | 47 | 22.90 | 65 | 51.64 | 118 | 56.18 | **126** |

HGI: Hardgrove Grindability Index, m: mass of particles passing 75 µm

# Particle size distribution of the milled reference coal and all *Bambusa Balcooa samples*

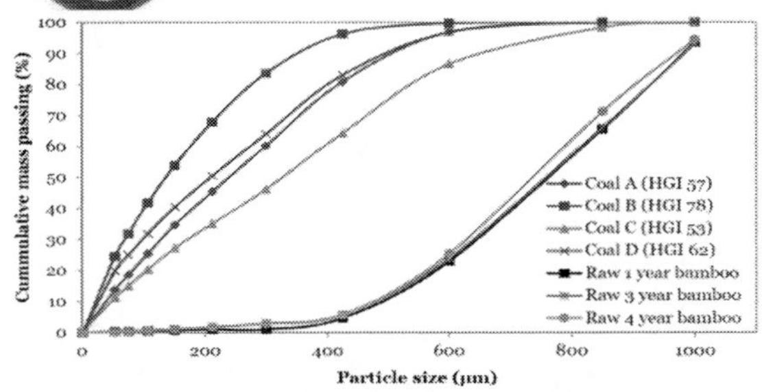


PSD for the reference coals and raw Bamboo samples

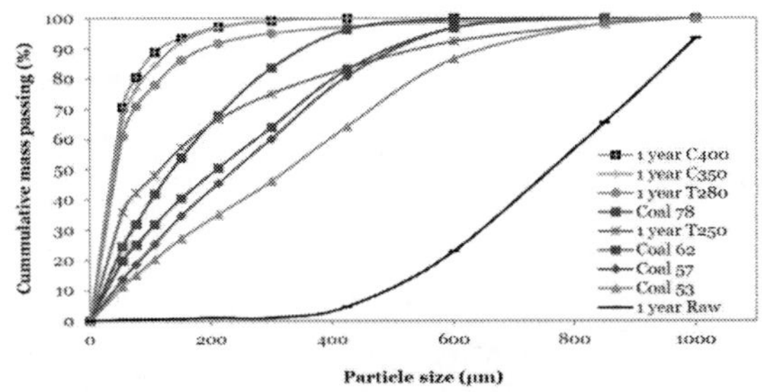


PSD for the reference coals and 1 yr old Bamboo samples

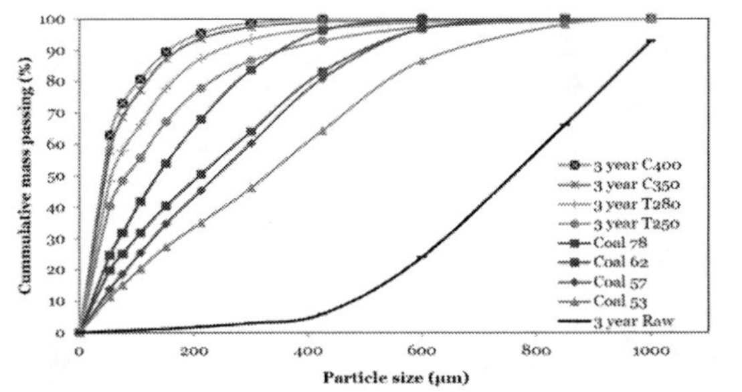


PSD for the reference coals and 3 yr old Bamboo samples

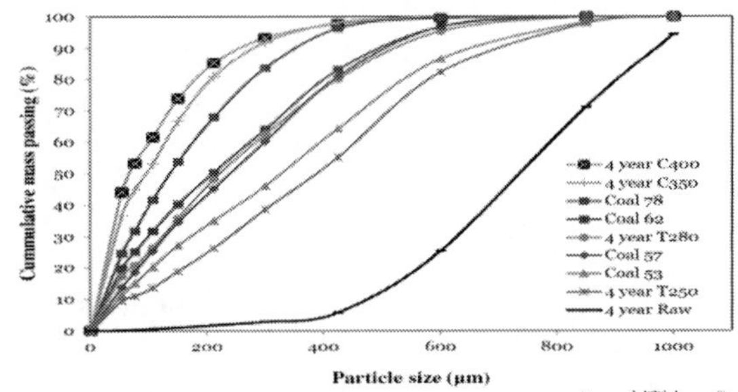


PSD for the reference coals and 4 yr old Bamboo samples

# Combustion of the raw bamboo samples and coal

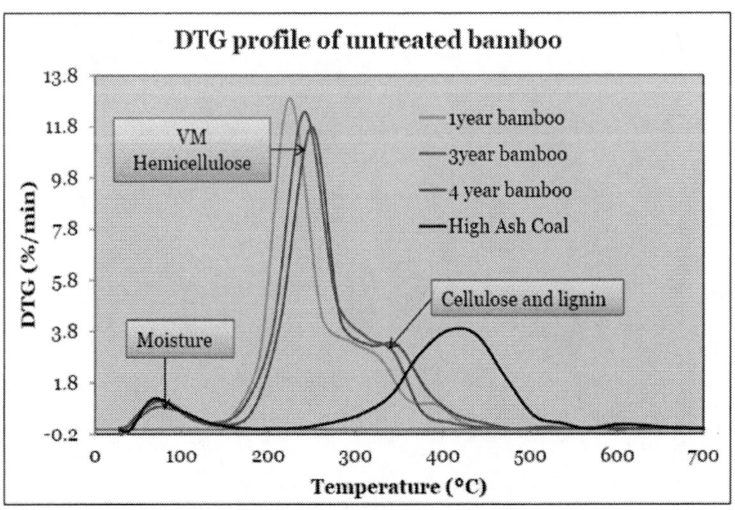


**Figure. DTG curves for 1, 3 and 4 year *Bambusa balcooa with* coal.**

**Observations;**

(i) The DTG profile presents two different peaks for the bamboo samples

(ii) The shoulders observed within 210 – 300 °C could be attributed to the decomposition of hemicellulose & some cellulose.

(iii) At temperature >300 °C, lignin is being degraded.

(iv) The ignition of the bamboos occurs at the lower temp region

(v) The *B. bamboos* has the highest peak/reactivity compared to all fuels in Fig 4b

(vi) The coal was seen with the highest burnout temperature

18

# Combustion of the all bamboo samples and coal

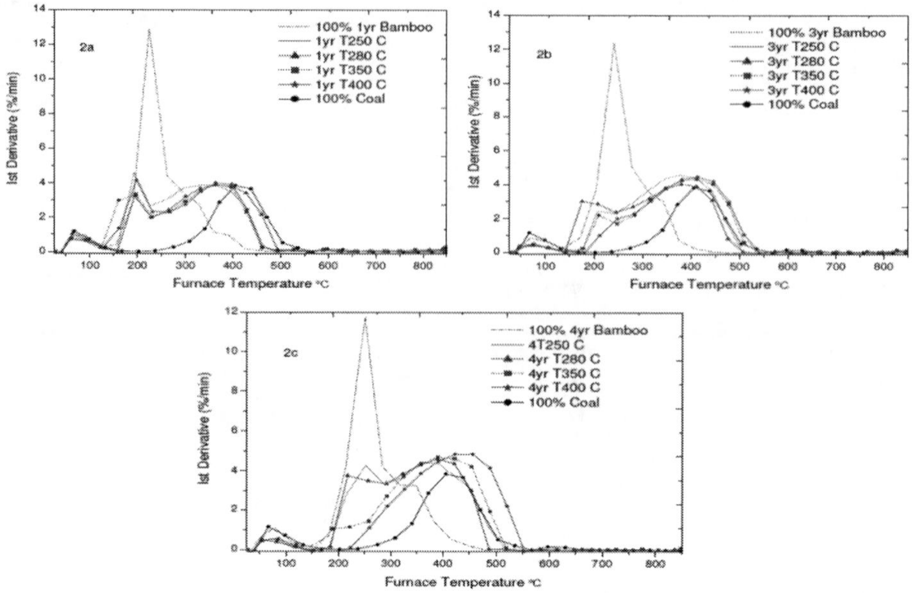


**DTG curves for raw, thermally treated bamboo and coal (a:1yr; b:3yr and c:4yr)**

19

# Co-combustion profiles of all raw bamboo (1,3 & 4 yr) samples and coal

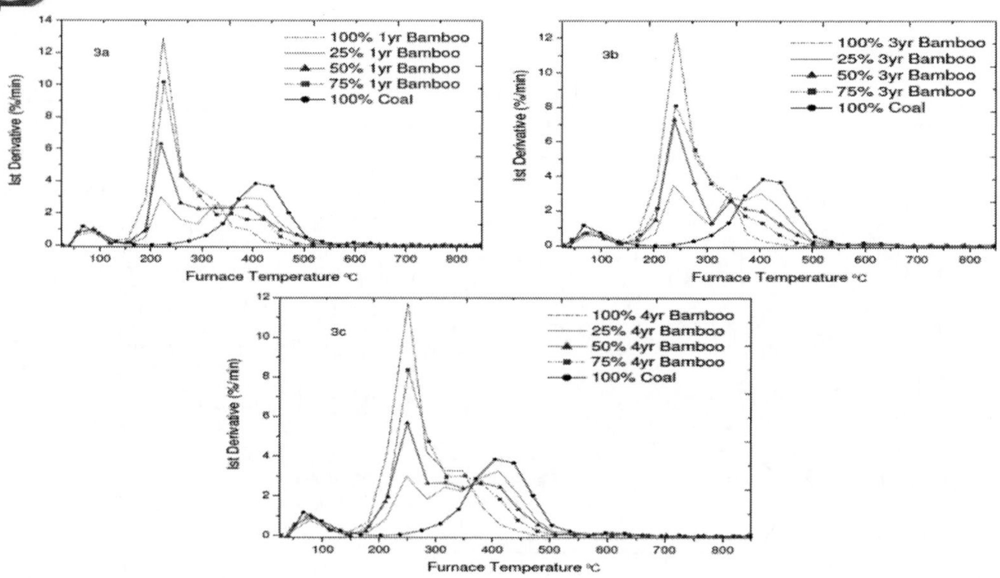


**DTG curves of co-combustion of all raw bamboo/coal (3a:1yr; 3b:3yr and 3c:4yr)**

20

# Co-combustion profiles of 1 yr thermally treated bamboo/coal

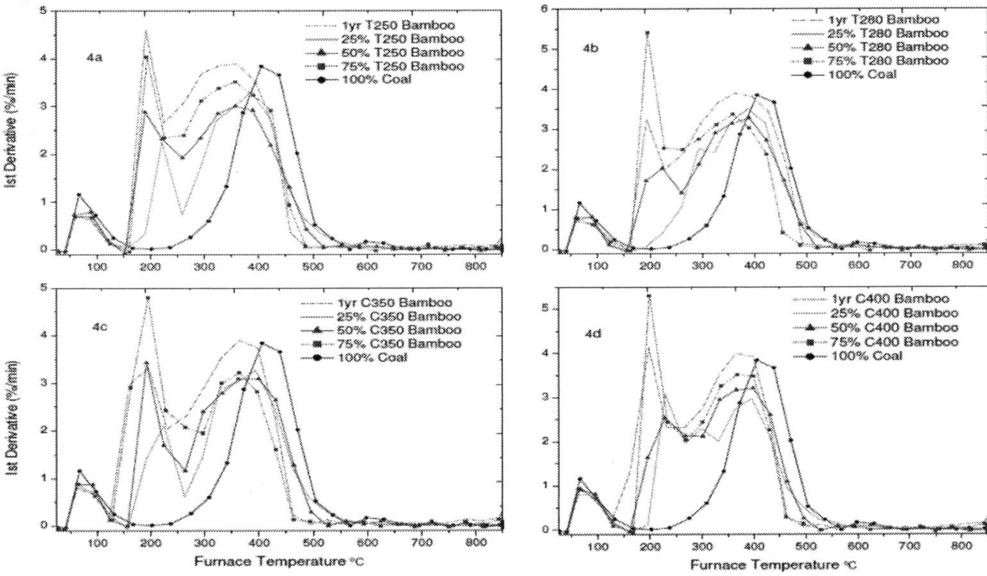


**DTG curves of co-fired thermal treated 1 year old bamboo/coal (4a; 4b; 4c and 4d)**

21

# Co-combustion profiles of 3 year low carbonized bamboo/coal

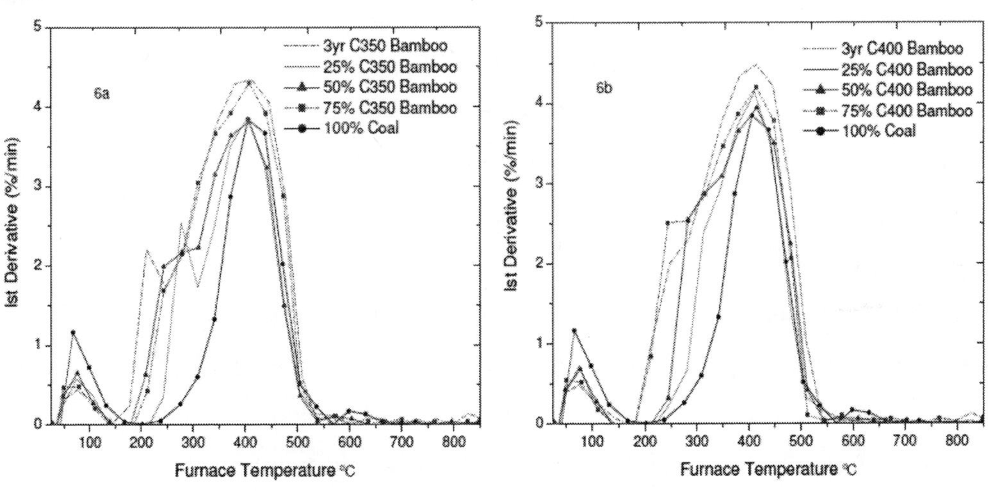


**DTG curves for co-combustion of 3 year C350 & C400/coal**

22

## Co-combustion profiles of 4 year low carbonized bamboo/coal

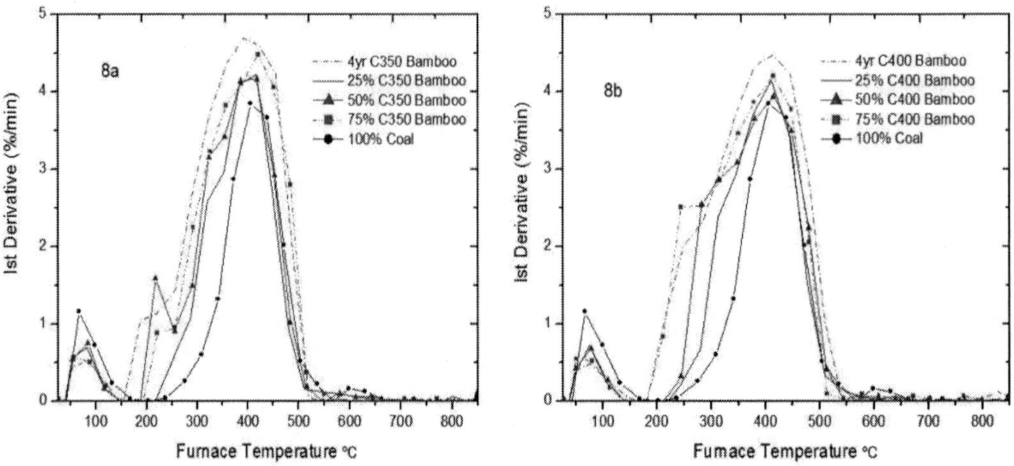


**DTG curves for co-combustion of 4 year C350 & C400/coal**

23

## Conclusions

- With heat treatment, the ***grindability*** of bamboo was improved, the younger bamboo having better (higher) HGI values compared to the older plants.
- **Energy yield** was highest when 4 year old bamboo had been treated at torrefaction temperatures (68% and 74%).
- **Fuel properties** (FC content) of bamboo were improved by all levels of thermal treatment and highest (76.6%) at carbonisation temperatures.
- ***Combustion properties*** (i.e. higher reactivity, lower burning temperatures) were produced when bamboo plant material was blended with coal.
- In summary, the results indicate that the 4 year old bamboo carbonized at 400 $^{o}C$ has the closest ***burning compatibility*** with coal.

24

## Recommendations

Investigations yet to be undertaken:

- Pilot scale growth patterns of selected bamboo
- Life cycle analysis of rehabilitation sites using bamboo as the biofuel crop of choice
- Life cycle analysis of the co-firing plant
- Techno-economics studies

25

## Acknowledgements


- I would like to thank my supervisors:

  Dr Samson Bada

  Prof Rosemary Falcon
- Funding for the project from NRF


26

# THANK YOU

# QUESTIONS.......?

27

# An Experimental Study of Nitrogen Oxides Formation of Pulverized Coal in $O_2/N_2$ and $O_2/CO_2$ atmosphere

Jeong-Woo Kim [a,1], Hong-Min Cho [a], Lim Ho [a], Chung-Hwan Jeon [a,b,*]

[a] School of Mechanical Engineering, Pusan National University, Busan 609-735, Republic of Korea

[b] Pusan Clean Coal Center, Pusan National University, Busan 609-735, Republic of Korea

**Abstract**

The effort to reduce emission of greenhouse gas is an irreversible global trend. Oxy-fuel combustion is a promising technology that can be applied to coal fired power plants. As regulations and standards of air pollutant emissions are being tightened, nitrogen oxides (NOx) exhaust remains an important fator for the choice and application of oxy-fuel combustion technology. Several researchers investigating the difference in NOx formation under oxy-fuel and conventional conditions reported that the tendency of nitrogen release in the pyrolysis process under the $CO_2$ condition was dissimilar from that of $N_2$. The purpose of this study is to understand the effect of nitrogen release on NOx formation under the oxy-fuel condition. A thermogravimetric analyzer and drop tube furnace were used for the observation of devolatilization and gasification characteristics in $CO_2$ and $N_2$ atmosphere at increasing temperatures. Combustion experiments to measure NOx concentration under $O_2$-$CO_2$ and $O_2$-$N_2$ atmospheres were conducted in a drop tube furnace by converting the temperature and stoichiometric ratio. The results of the TGA experiments show that water evaporation and devolatilization occurred, and that the trends of these processes were similar. However, in case of the $CO_2$ at temperatures greater than 800℃, the rate of the weight loss started to

* Corresponding author. Tel.: +82-51-510-1379; fax: +82-51-510-5236.
E-mail address: jeongwoo_kim@pusan.ac.kr (J.-W. Kim), hongmincho@pusan.ac.kr (H.-M. Choi), holim@pusan.ac.kr(H, Lim) , chjeon@pusan.ac.kr (C.-H. Jeon).

[1] First author

increase, whereas the weight loss under the $N_2$ atmosphere, was uniform as gasification occurred. This tendency can also be seen in the results of the experiments conducted by DTF. The amount of apparent volatile yields increased as the temperature in the reactor increased. The results show that the apparent volatile yield under the $CO_2$ atmosphere exceeds that under $N_2$ at approximately 1300℃. In the combustion experiments, the measured NOx concentration tended to increase as the temperature and stoichiometric ratio of $O_2$ increased under both $O_2/CO_2$ and $O_2/N_2$ atmosphere. However, the total amount of NOx under the $O_2$-$CO_2$ atmosphere under all conditions was lower than the $N_2$-$CO_2$ atmosphere.

*Keywords:* Oxy-fuel, NOx, UBC, fuel reburning

## 1. Introduction

Oxy fuel combustion, which is under investigation and is currently commercialized for coal power plants, is a promising technology to reduce greenhouse gas emissions. By using recycled flue gas, high purity $CO_2$ can be obtained for sequestration without stripping the $CO_2$ from the gas stream [1,2]. In addition, flue gas recycling implies new options for the control of nitrogen oxides(NOx)[1]. However, the limitation of condensed NOx which remains in liquefied $CO_2$ has not been established and remains an important issue in oxy fuel combustion[3]. $CO_2$, as the main component in gas for the combustion of fuel, affects the heat transfer, flame ignition, coal burnout and the composition of radicals, which is significant in NOx formation and reduction. Although a number of studies have been conducted on NOx formation and reduction mechanism using various reactors in $O_2/N_2$ and $O_2/CO_2$ atmospheres[4-6], it is fair to say that there are more disagreements than agreements[8]. R. K. Ranthnam et al.[7] and Taniguchi et al.[4] reported that the char + $CO_2$ gasification reaction is promoted in the oxygen deficient atmospheres of $N_2$ and $CO_2$, such as in the later stage of burnout and this CO formation is associated with the non-hydrocarbon reburning process which may remove 20-30% of the nitric oxide[8].

This study is based on measurements of CO and NOx in $O_2/N_2$ and $O_2/CO_2$ atmosphere at various temperature and stoichiometric ratios(SR). The objective of the present study is to evaluate experimentally the ability of CO to reduce nitric oxide by a non-hydrocarbon reburning process. A Thermogravimetric analyzer and drop tube furnace were used under $N_2$ and $CO_2$ atmosphere to study of the formation of CO. Combustion experiments using DTF at various SR's and temperatures conducted to observe the relationship between NOx and CO. The reburning effect of CO on NOx reduction is discussed.

## 2. Experiment

### Devolatilization experiments.

The purpose of devolatilization in this study is to observe the trend of pyrolysis and gasification, respectively, under the oxygen deficient atmospheres (later stage of combustion) $N_2$ and $CO_2$. Devolatilization experiments were conducted using TGA-DSC(LECO Q-600) and DTF under the conditions shown in Table 1. Bituminous glencore imported from Indonesia, was selected and 75-90 μm cut coal particles were used in the experiments. The proximate and ultimate analyses of coal are shown in Table 2. This study adopts the non-isothermal method in which the fuel sample is heated at a constant heating rate. The raw coal particles were used in TGA and DTF to study the pyrolysis and gasification, respectively. Devolatilization using DTF was also conducted by increasing the temperature from 1173 to 1773K. The residence time varied from 1.2 s to 2 s, and the coal was fed at a rate of 1g/min.

### 2.2 Drop tube furnace experiments

The experiments were conducted in a Two staged Drop Tube Furnace (Two staged DTF) which was designed by the Pusan Clean Coal Center. Two staged DTF is comprised of an injection, reaction, and collection part. In the injection part, a pneumatic vibration feeder was used to ensure coal feeding at a set rate. $N_2$ were used as a carrier gas to transport coal particles, floating in a glass test tube, to the reactor. The coal particle and gas mixture of nitrogen and oxygen supplied using MFCs were injected into the upper part of the furnace. In additions, staged gas was supplied at the connection of the upper and lower part of the furnace. In the reaction part, two alumina tubes were connected in series. These alumina tubes were heated by electric heaters, which could reach approximately 1,850 ℃ and could be controlled separately for each tubes. The temperature in close proximity to the heater was uniform, and the temperature profile is shown in Fig. 1

*2.2.1 Experimental procedure*

Single stage combustion (one reaction zone) was conducted at various SRs and temperatures in a mixture of $O_2/N_2$ (air) and $O_2/CO_2$(oxy-fuel) atmosphere to observe the effect of SR and temperature on unburned carbon(UBC) and NOx emission characteristics. Experimental conditions were set for have SRs of 0.6, 0.8, 1.0 and 1.2 and oxygen concentration of approximately 9, 12, 15 and 18% at 1,100 ℃, 1,300℃ and 1,500℃. The total feeding rate maintained at approxamately1g/min. The gas extracted at the exit of the furnace was analyzed. Continuous analyzer MK 9000(Ecom-rbr) was used to measure $O_2$, $CO_2$, CO and NOx. The UBC was calculated using the ash tracer method[9].

The residence time was in the range of 1.5 -2.0s , set according to temperature. The total flow rate of the gas mixture was 10NL/min. The experimental conditions are shown in Table 3.

## 3. Results and discussion

*3.1 TGA and DTF devolatilization*

TGA and DTF devolatilization experiments were conducted to observe the coal burnout in oxygen deficient atmosphere. The results of the TGA devolatilization (Fig.2) shows that the normalized mass loss of the coal particles is similar in $N_2$ and $CO_2$ atmospheres below 850 ℃.

Vaporization and pyrolysis were observed at approximately 100 and 300 ℃, respectively. However, the mass loss in $CO_2$ stared to increase significantly above 850 ℃. This may attributed th the char+$CO_2$ gasification reaction, as previously observed by Varhegyi et al[10] DTF devolatilization was carried out in the range of 900-1,500 ℃ and the result is shown in Fig. 3. The mass loss under in $N_2$ and $CO_2$ atmospheres was similar below 1,200 ℃. Then, the mass loss started to increase significantly in $CO_2$ above 1,200 ℃. where that of $N_2$ was maintained at proximately 45%. Mass loss in $CO_2$ was also caused by gasification[7]. These results indicate that coal burnout could be enhanced by char + $CO_2$ (from coal combustion) gasification reaction when oxygen is insufficient. Moreover, in oxy-fuel atmospheres, gasification could be the main reason of coal burnout.

*3.2 Coal burnout*

Single stage combustion (one reaction zone) experiments were conducted under the conditions shown in Table 1 and the results are shown in Fig. 4 and Fig. 5. The results indicates that coal burnout increased with the increase in SR in the $O_2/N_2$ and $O_2/CO_2$ atmospheres. The increase in oxygen concentration leads to increased coal burnout in both atmospheres. Coal burnout in the fuel rich atmosphere at SRs is lower than 1.0 was increased by increasing the temperature, even if there was insufficient oxygen as the temperature increased in the $O_2/N_2$ and $O_2/CO_2$ atmospheres. Furthermore, the amount of coal burnout in the fuel rich atmosphere was higher in $O_2/CO_2$ than in $O_2/N_2$. The oxygen concentration at the furnace exit when the coal was burned at 1,100 ℃ with an SR of 1.2 was approximately 5%, and the measured oxygen concentration decreased with increasing temperature and SR. Ranthnam et al.[7] researched coal burnout at various oxygen concentrations in $CO_2$ and $N_2$, and observed enhanced coal burnout by gasification at an oxygen concentration of 5% in $CO_2$.

Similar results were observed in this study.

### *3.3 Effect of coal gasification on NOx formation*

Nitric oxide can be removed by the reaction with hydrocarbon-derived radicals, in adittion, nitrogen containing species are efficient in reducing NO[8]. However, as reported by Mackrory and Tree, $NH_3$ and HCN profiles show the same trend with NOx, where both $NH_3$ and HCN are reduced to a low levels in the reducing zone[3]. With the limitation of NOx reduction by HCN and $NH_3$, the rapid NOx reduction trend with increasing CO concentration shown in Fig. 6 can not be explained. The flue gas composition measured by continuous gas analyzer is shown in Fig. 6. After coal was injected, the $O_2$ concentration started to decreased and the amount of NOx and $CO_2$ increased. However, at low $O_2$ concentration, NOx and $CO_2$ decreased with the CO created by the gasification reaction in the later stage of burnout. CO gasification has an effect on NOx reduction either directly or indirectly. Glarborg et al.[8] reported that non-hydrocarbons such as CO and $H_2$ may remove 20-30%of the nitric oxide. This reburning process with CO in the later stage of coal burnout decreased the NOx concentration, and is expressed as

$$CO + NOx \rightleftharpoons CO2 + N \quad (1)$$

The measured CO concentration at each conditions are shown in Fig. 7 and Fig. 8. CO significantly increased by decreasing SR. The CO concentration at each conditions in $O_2/CO_2$ is higher than that in $O_2/N_2$ as CO formation enhanced under the oxygen insufficient and high temperature atmosphere.

The measured NOx concentration at each condition was converted to a nitrogen conversion ratio (CR).

$$CR = \frac{C_{NO}}{\dfrac{2.24 \times 10^{-2} \times \dfrac{FN}{1.4 \times 10^{-2}}}{V_{dry}}} \quad (2)$$

CR represents the degree of conversion of fuel N to NOx. where $C_{NO}$ is the NO concentration at the furnace exit and $V_{dry}$ the flow rate of dry air for the unit coal feed rate($Nm^3/kg$). The denominator of equation 2 represents the NO concentration when all fuel N is converted into NOx. CR in $O_2/N_2$ and $O_2/CO_2$ are shown in Fig. 9 and Fig. 10, respectively. The measured CO concentration at SRs greater than 1.0 were almost under 500ppm in $O_2/N_2$, and similar trends are observed for $O_2/CO_2$. CO started to increase with decreasing SR and increasing temperature. The CO concentration was significantly higher in fuel rich conditions (SR<1.0) under the $O_2/CO_2$ rather than $O_2/N_2$ atmosphere.

When both the temperature and SR decreased the CO concentration increased in fuel rich atmosphere (SR<1.0), and even the CO concentration in the $O_2/CO_2$ atmosphere were higher than CO in $O_2/N_2$ at each condition. The $CR_{O_2/CO_2}$ at each conditions were lower than $CR_{O_2/N_2}$ as high CO concentration promoted CO reburning which removes NOx. The trend of CR is inversely proported to CO formation. However at sufficiently low CR values, the effect of temperature on CR is decreased because it is inverse reaction of equation (1). In addition, it is observed the $CR_{O_2/CO_2}$ under fuel rich atmosphere s is less sensitive SR than $CR_{O_2/N_2}$.

## 4. Conclusions

The following conclusions can be taken from this study.

1. TGA and DTF devolatilization experiments were conducted. CO was produced by the char + $CO_2$ reaction. The CO formation was enhanced at temperatures above above 850 ℃. When sufficient heat was supplied (above 1,200 ℃) to coal particles under the $CO_2$ atmosphere in DTF, the mass loss was promoted by the char + $CO_2$ gasification reaction.
2. Coal burnout increased with increasing stoichiometric ratio and temperature for the experiments conducted in the DTF. The burnout in fuel-rich atmosphere was

enhanced by gasification at high temperatures. Furthermore, this gasification was promoted in the $O_2/CO_2$ atmosphere.

3. The measured NOx concentrations were converted to CR's which decreased at low stoichiometric ratios and lower temperature. CR is inverse proportion to CO formation as the reburning process by CO removes NOx. $CR_{O_2/N_2}$ is more sensitive to $CR_{O_2/CO_2}$ .

**References**


1. Buhre, B. J. P., et al. "Oxy-fuel combustion technology for coal-fired power generation." Progress in energy and combustion science 31.4 (2005): 283-307.
2. Sangras R. The 29th international technical conference on coal utilization and fuel systems, Clearwater, FL; April 18-22 2004.
3. Mackrory, Andrew J., and Dale R. Tree. "Measurement of nitrogen evolution in a staged oxy-combustion coal flame." Fuel 93 (2012): 298-304.
4. Taniguchi, Masayuki, et al. "A role of hydrocarbon reaction for NOx formation and reduction in fuel-rich pulverized coal combustion." Combustion and flame 157.8 (2010): 1456-1466.
5. Houshfar, Ehsan, et al. "Effect of excess air ratio and temperature on NOx emission from grate combustion of biomass in the staged air combustion scenario." Energy & Fuels 25.10 (2011): 4643-4654.
6. Okazaki, K., and T. Ando. "NO x reduction mechanism in coal combustion with recycled CO 2." Energy 22.2 (1997): 207-215.
7. Rathnam, Renu Kumar, et al. "Differences in reactivity of pulverised coal in air (O 2/N 2) and oxy-fuel (O 2/CO 2) conditions." Fuel Processing Technology 90.6 (2009): 797-802.
8. Glarborg, Peter, et al. "Nitric oxide reduction by non-hydrocarbon fuels. Implications for reburning with gasification gases." Energy & Fuels 14.4 (2000): 828-838.

9. Lee, Byoung-Hwa, Eric G. Eddings, and Chung-Hwan Jeon. "Effect of coal blending methods with different excess oxygen on unburned carbon and NO x emissions in an entrained flow reactor." Energy & Fuels 26.11 (2012): 6803-6814.

10. Várhegyi, Gábor, et al. "Mathematical modeling of char reactivity in Ar-O2 and CO2-O2 mixtures." Energy & fuels 10.6 (1996): 1208-1214.

## List of Figures

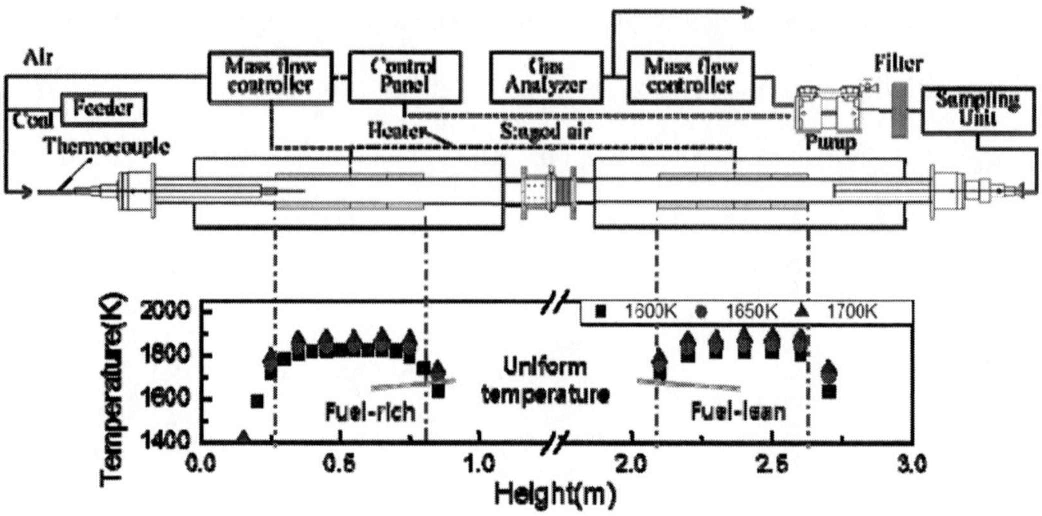


**Fig. 1. Axial temperature profile of the Two staged DTF**

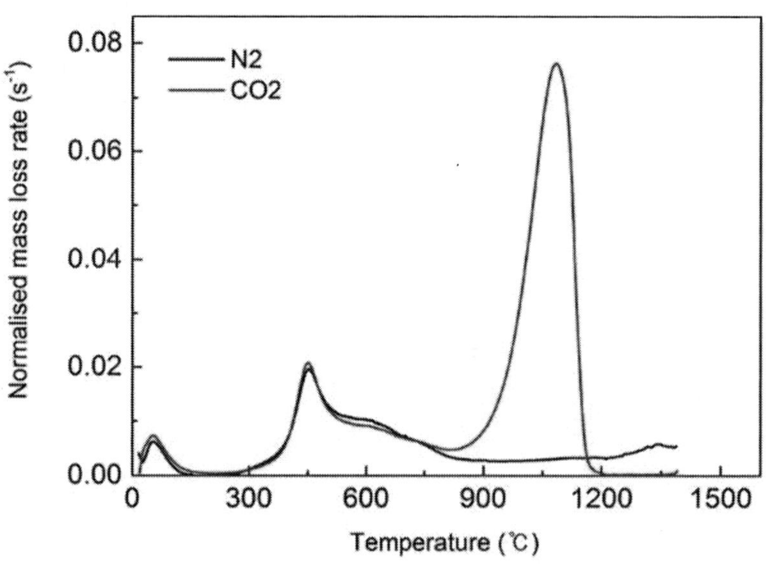


**Fig. 2. Nomalised mass loss rate measured under N2 and CO2 atmosphere.**

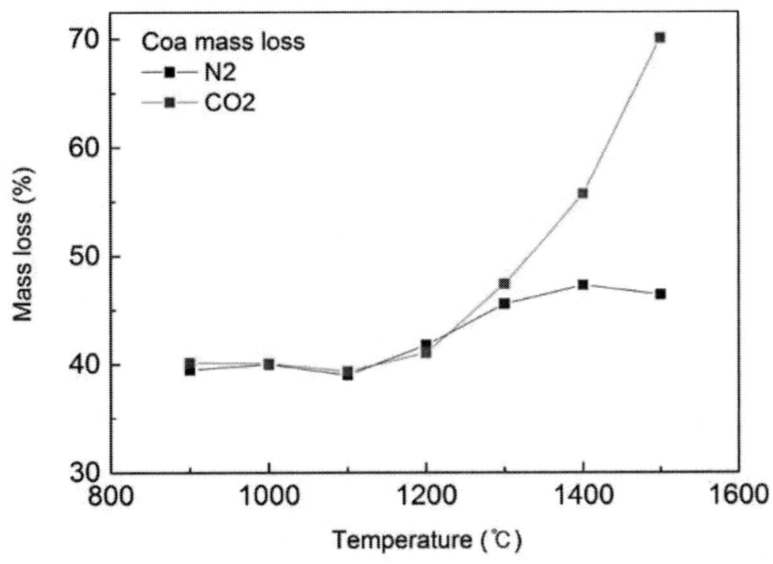


**Fig. 3. Mass loss of coal particle during pyrolysis under N2 and CO2 in DTF**

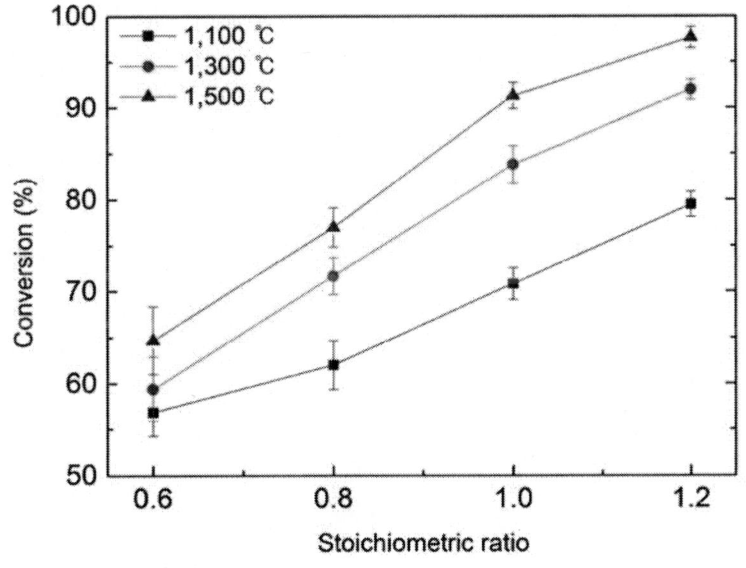


**Fig. 4. Coal conversion at each experimental condition under O2/N2 atmosphere.**

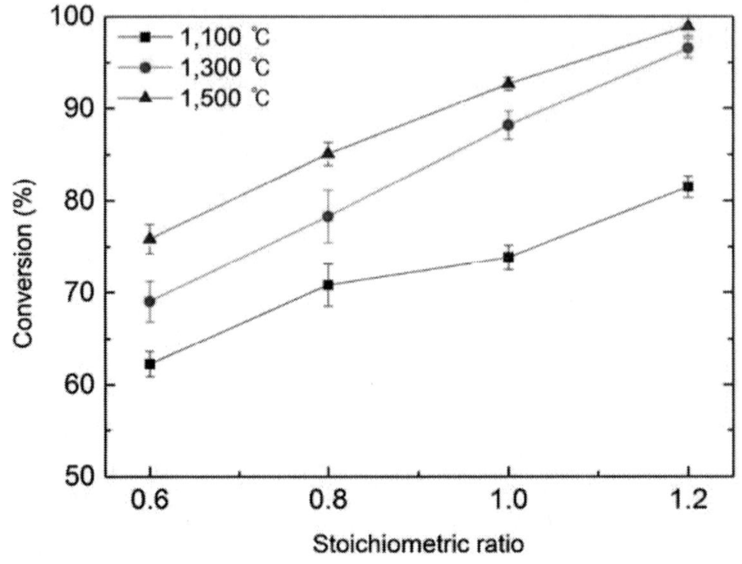


**Fig. 5. Coal conversion at each experimental condition under O2/CO2 atmosphere**

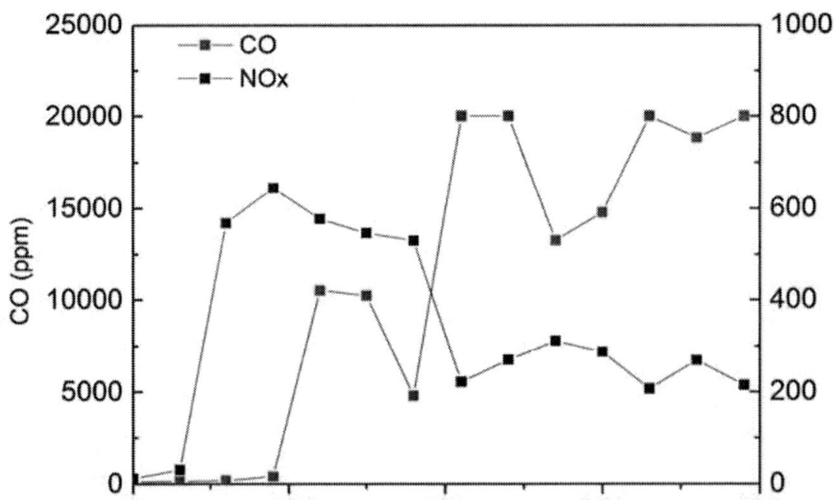


**Fig. 6. The formation and reduction trend of NOx with CO**

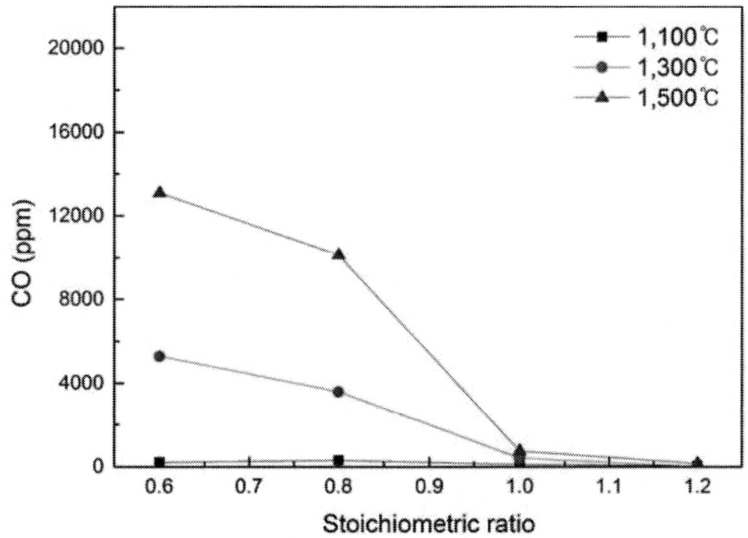


**Fig.7. CO concentration under O2/N2 atmosphere by temperature as a function of SR**

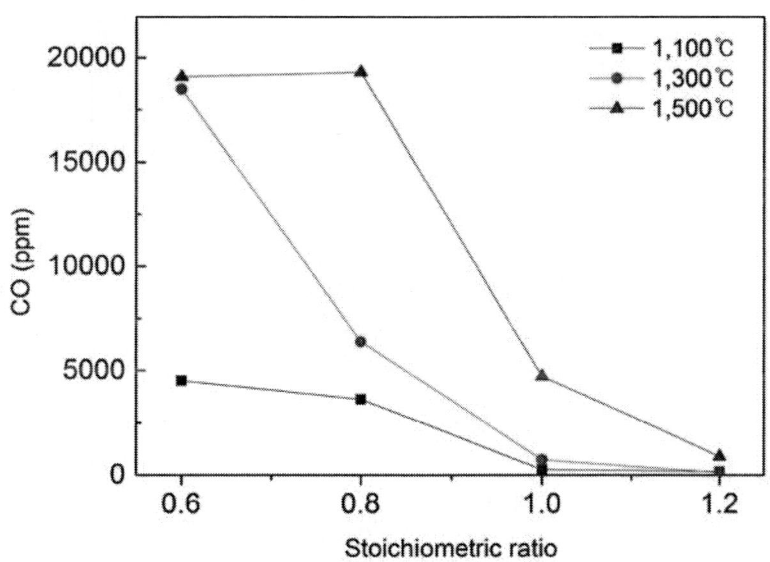


**Fig. 8. CO concentration under O2/CO2 atmosphere by temperature as a function of SR**

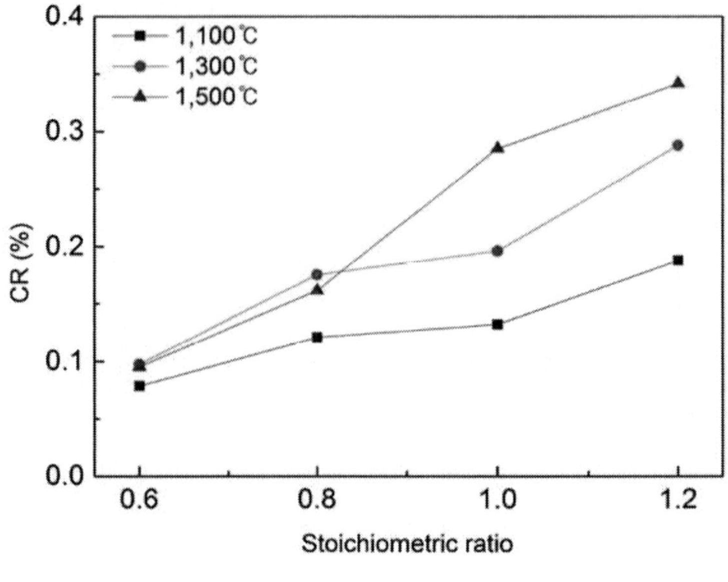


**Fig. 9. CR obtatined from experimental results under O2/N2 atmosphere.**

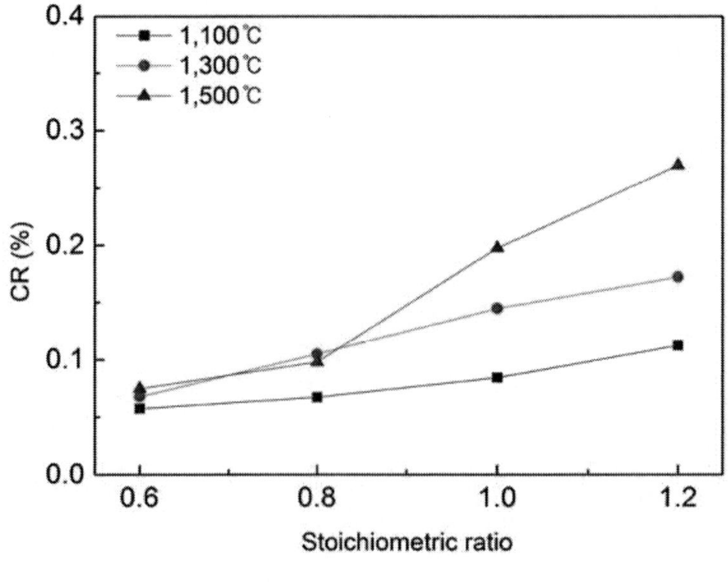


**Fig. 103. CR obtatined from experimental results under O2/CO2 atmosphere.**

**List of Tables**

**Table 1. Experimental conditions for devolatilization in TGA and DTF**

| Parameter for TGA | Condition |
|---|---|
| Sampling mass (mg) | 20 |
| Temperature range (℃) | 25 to 1400 |
| Heating rate (℃/min) | 25 |
| Gas flow rate (ml/min) | 20 |
| Atmosphere | $N_2$ and $CO_2$ |
| **Parameter for DTF** | **condition** |
| Coal feed rate (g/min) | 1 |
| Total gas flow rate (L/min) | 10 |
| Gas temperature (℃) | 900-1500 |
| Atmosphere | $N_2$ and $CO_2$ |

**Table 2. Proximate and ultimate analysis of glencore.**

| Glencore | |
|---|---|
| Proximate analysis (wt%,air dried basis) | |
| Air dried moisture | 1.23 |
| Volatile matter | 28.32 |
| Fixed carbon | 61.3 |
| Ash | 9.147 |
| Ultimate analysis (wt%,dry ash free basis) | |
| Carbon | 84.72 |
| Hydrogen | 5.4 |
| Oxygen | 7.75 |
| Nitrogen | 1.28 |
| Sulphur | 0.85 |

**Table 3. Experimental condition for combustion of DTF**

| Parameter | |
|---|---|
| Gas temperature (℃) | 1,100, 1,300, 1,500 |
| Stoichiometric ratio (oxygen concentration %) | 0.6 (9), 0.8 (12), 1.0 (15), 1.2 (18) |
| Coal feeding rate (g/min) | 1 |
| Total gas flow rate (g/min) | 10 |
| Residence time (s) | 1.5 - 2 |
| Atmosphere | $O_2/N_2$ , $O_2/CO_2$ |

# Monolithic Amine-Grafted Sorbents for Post-Combustion $CO_2$ Capture

Steven S.C. Chuang, E. Willet, J. Yu, L. Zhang, Y. Zhai, H. Jin, S. Wang

Department of Polymer Science
The University of Akron
Akron, OH 44325-3909 USA

U.S. DEPARTMENT OF ENERGY | Ohio Development Services Agency | FirstEnergy | aspen aerogels

## Introduction - $CO_2$ emission

US electricity generation by source

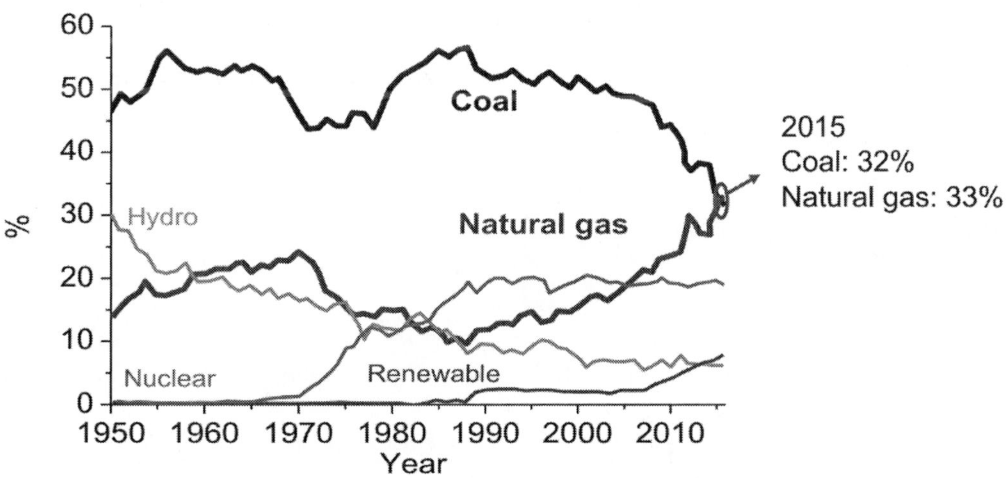


1. https://www.eia.gov , accessed by April 7th, 2016
2. Lecomte, F. et. al, $CO_2$ capture: Technologies to Reduce Greenhouse Gas Emissions. 2010
3. http://www.eia.gov/tools/faqs/faq.cfm?id=77&t=11, accessed by April 7th, 2016
4. Boden, T.A., Marland, G., and Andres, R.J. (2015).

# Introduction - $CO_2$ emission

Flue gas compositions (vol%)[2]

| | Natural gas | Coal |
|---|---|---|
| $CO_2$ | 3-5 | 12-15 |
| $H_2O$ | 4-7 | 7-11 |
| $N_2$ | 75-80 | 70-75 |
| $O_2$ | 12 | 3-5 |
| $SO_x$ (ppm) | 150-200 | 2000-3000 |

$CO_2$ emissions from U.S. electric power sector by source, 2015 [3]

| Source | Coal | Natural gas |
|---|---|---|
| Million metric tons | 1,364 | 530 |
| Share of total | 71% | 28% |


1. https://www.eia.gov , accessed by April 7th, 2016
2. Lecomte, F. et. al, $CO_2$ capture: Technologies to Reduce Greenhouse Gas Emissions. 2010
3. http://www.eia.gov/tools/faqs/faq.cfm?id=77&t=11, accessed by April 7th, 2016
4. Boden, T.A., Marland, G., and Andres, R.J. (2015).


# $CO_2$ Source and Utilization

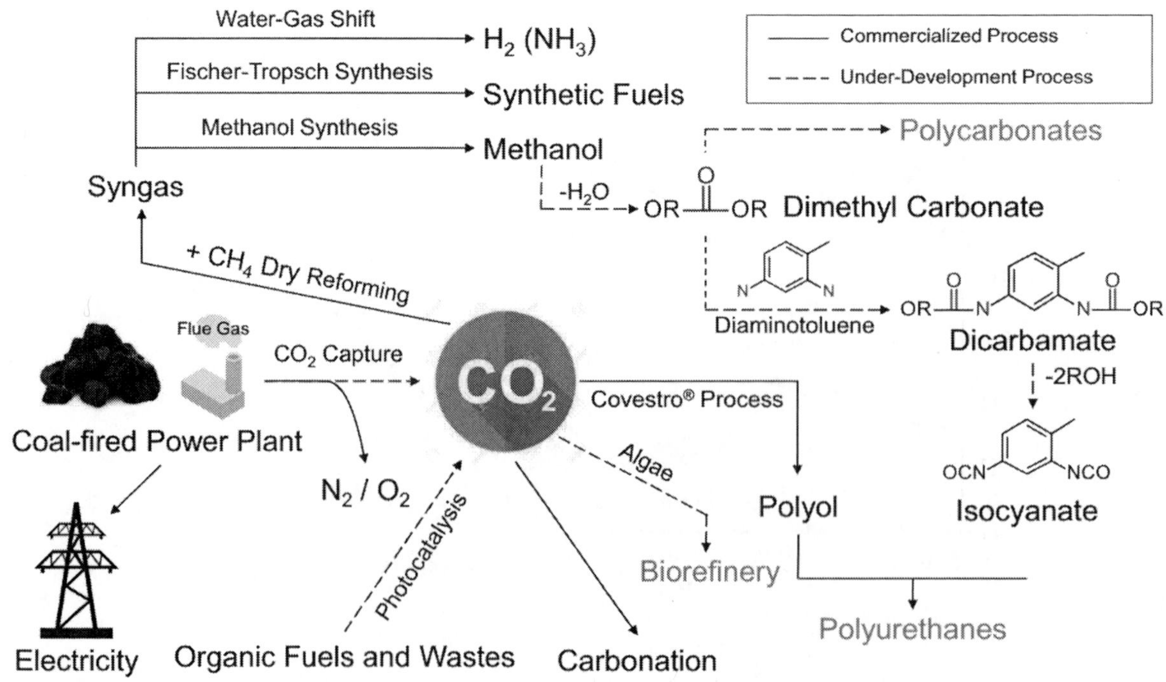

## Minimum Work

$$W_{\min} = \Delta G$$
$$\equiv \Delta H - T\Delta S$$
$$W_{\min} = -RT\left( N_1 \sum_{k=1\ldots n} X_{1,k} \ln X_{1,k} - N_2 \sum_{k=1\ldots n} X_{2,k} \ln X_{2,k} - N_3 \sum_{k=1\ldots n} X_{3,k} \ln X_{3,k} \right)$$

Kurt Zenz House et al. PNAS 2011;108:20428-20433

## Fundamental Study and Scale-up Development for $CO_2$ Capture: Metal Monolith

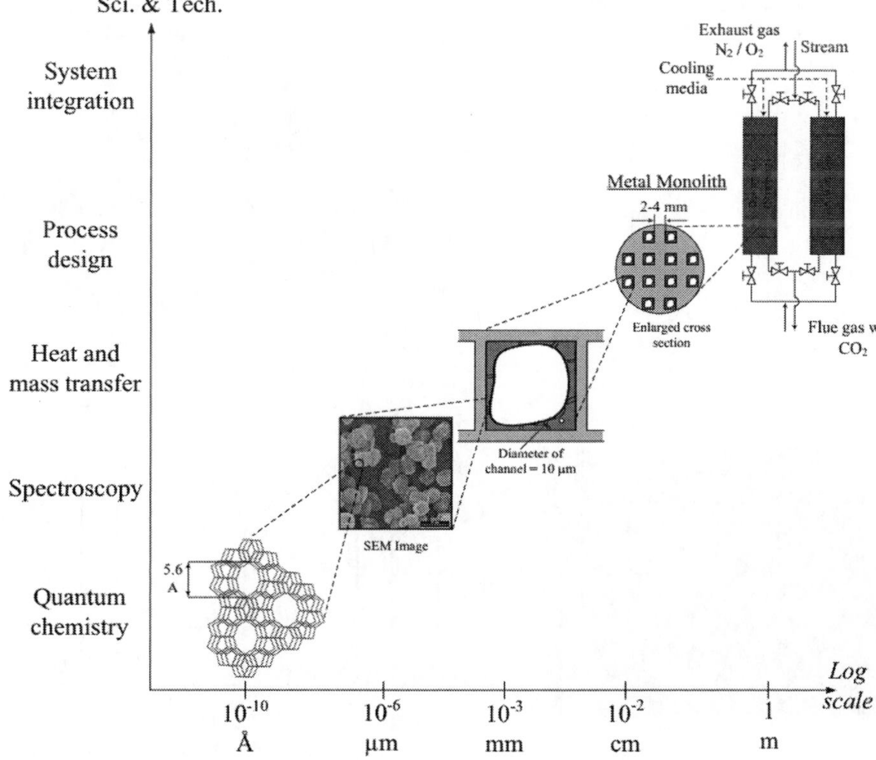

# Background – solid amine sorbent and its degradation

- **Solid amine sorbent = amine species + binder + support**
  - •Amine species = primary ($RNH_2$), secondary ($R_2NH$), and tertiary ($R_3N$)

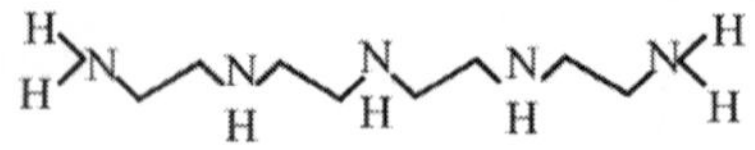

(TEPA)

- •Support = metal oxides, fiber glass, activated carbon, etc.
- •hydorphobicity/hdrophilicity, porosity, mechanical strength

($SiO_2$)

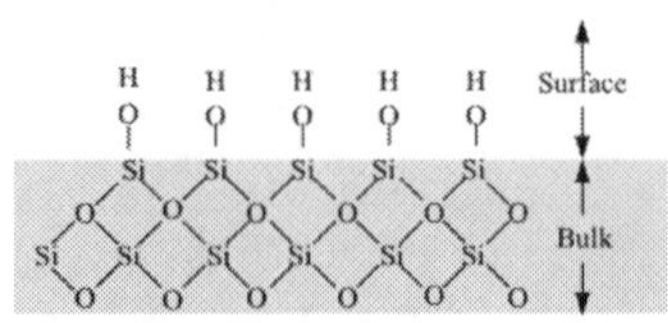


$$\mathbf{CO_2 + RNH_2 \leftrightarrow RNHCOO^{-}H^{+}}$$

$$CO_2 + 2RNH_2 \leftrightarrow RNH_3^{+} + \underset{\text{(carbamate)}}{RNHCOO^{-}}$$

$$\underset{\text{(carbamate)}}{RNHCOO^{-}} + H_2O \leftrightarrow RNH_2 + \underset{\text{(bicarbonate)}}{HCO_3^{-}}$$

7

# In Situ Infrared Spectroscopic Study

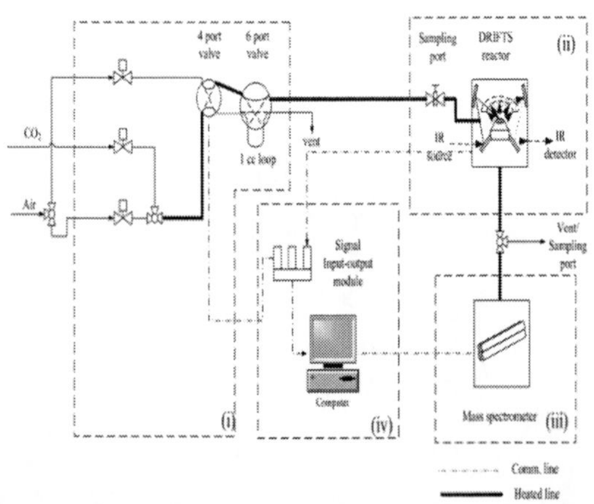


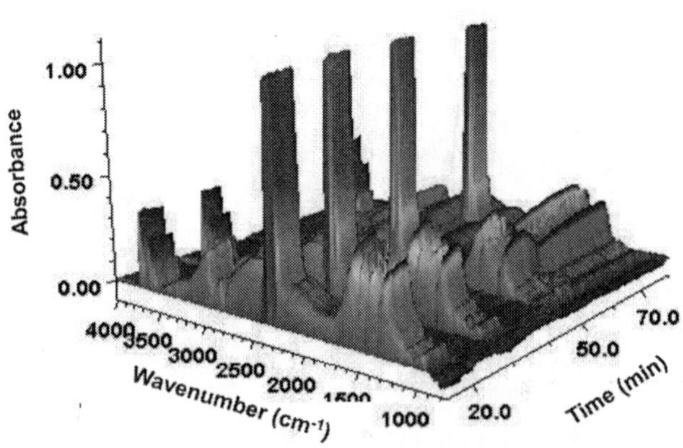


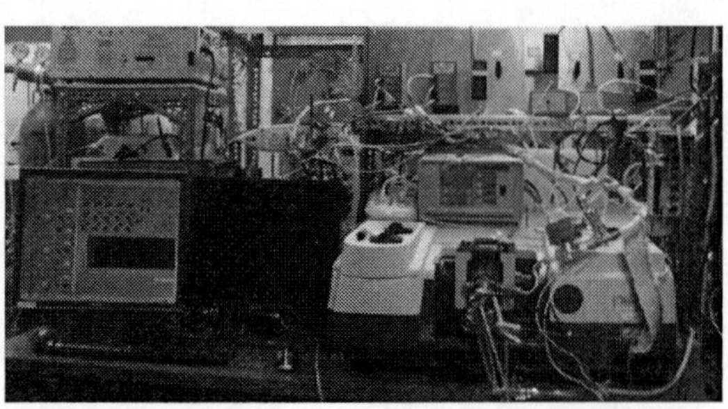

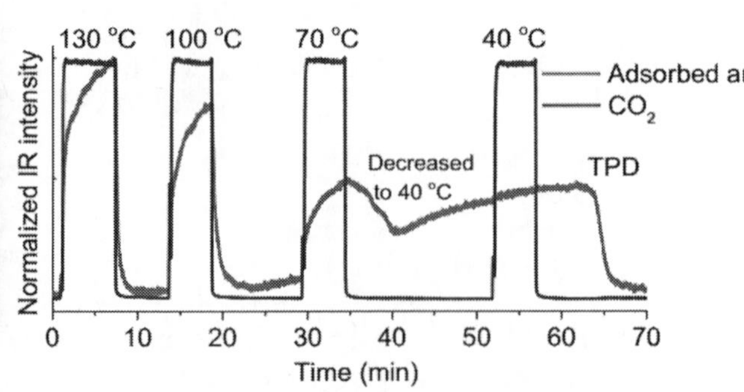

## Adsorption of $CO_2/H_2O$

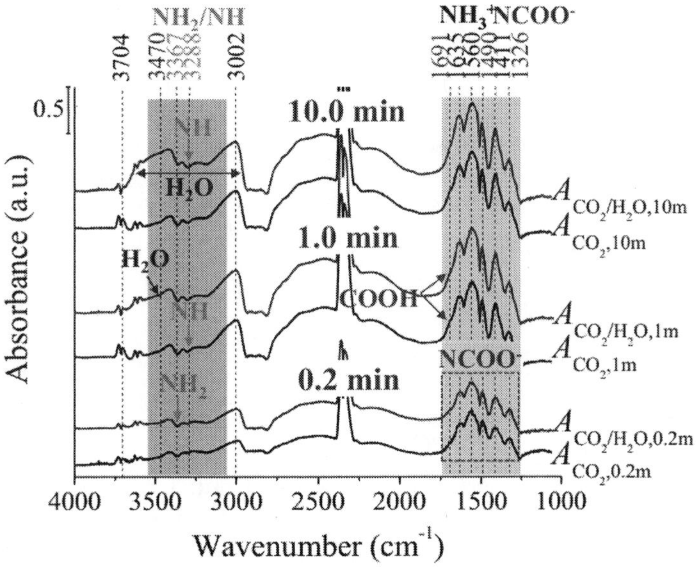


- $CO_2$ adsorbed on $RNH_2$ as carbamate and on $R_2NH$ as carbamic acid and carbamates.
- $H_2O$ exhibited better affinity to $R_2NH$ and enhanced $CO_2$ capture capacity.

## IR of Strongly and Weakly Adsorbed $CO_2$

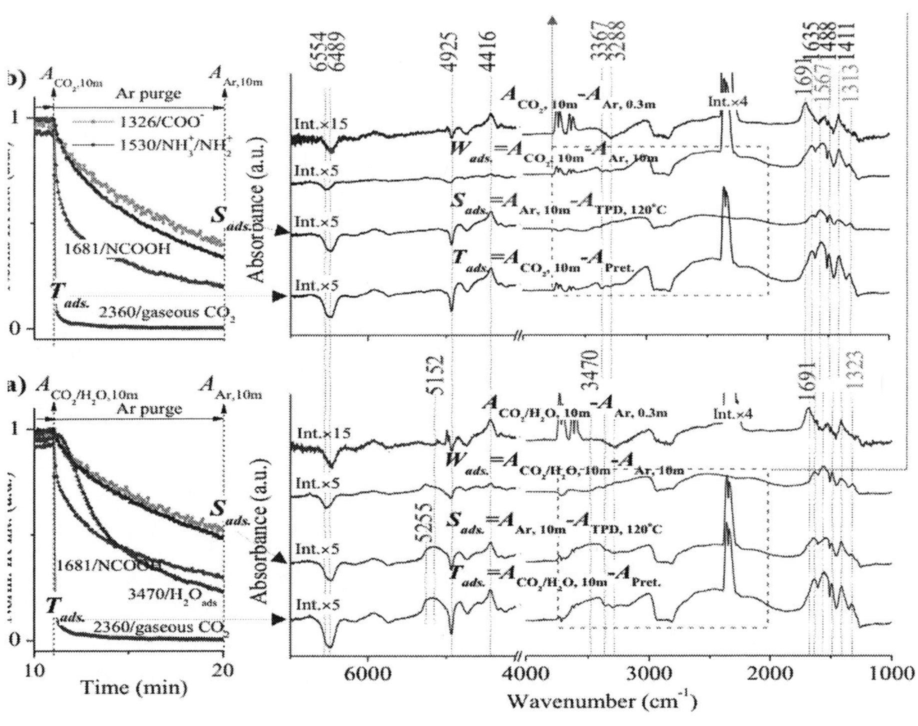

# The Role of H2O in CO2 Capture on Amines

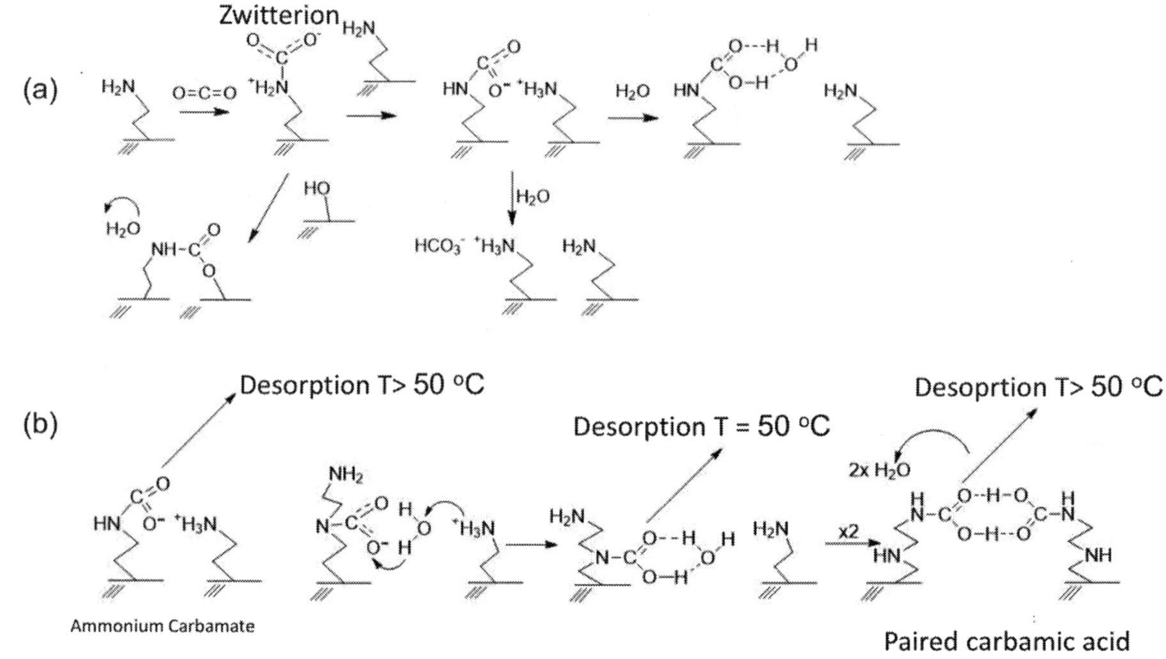


# Grafted Propylamine Model

*B3LYP/6-311+G with IEFPCM solvation*

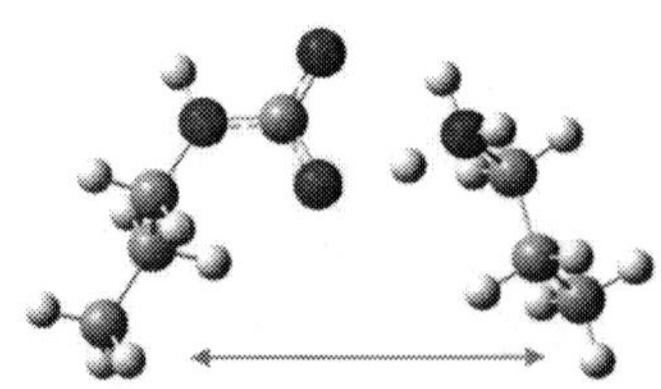

Fixed terminal carbons to adjust spacing, site density

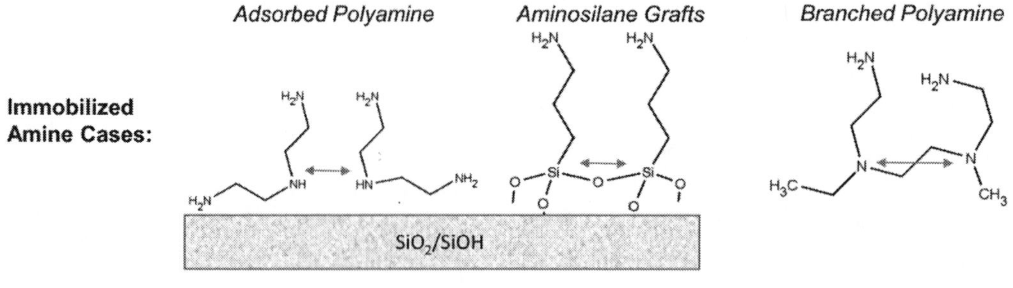

# Varying Graft Distance

*B3LYP/6-311+G with IEFPCM solvation*

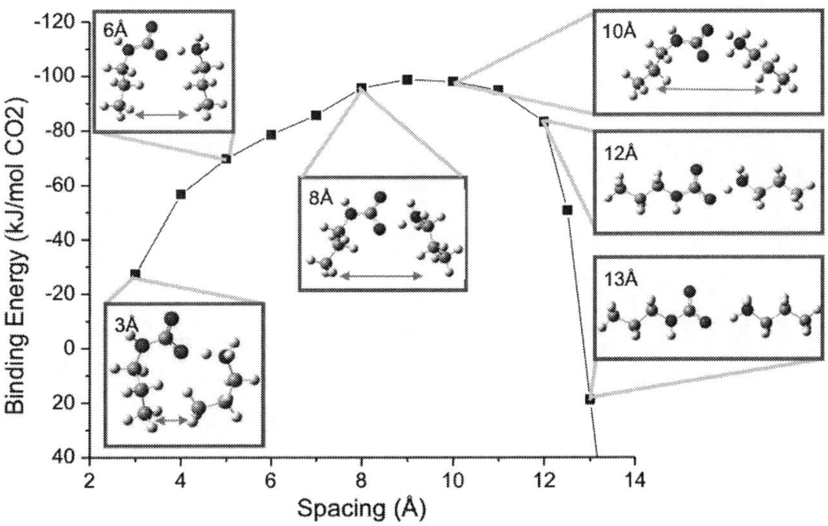


# Further Immobilization

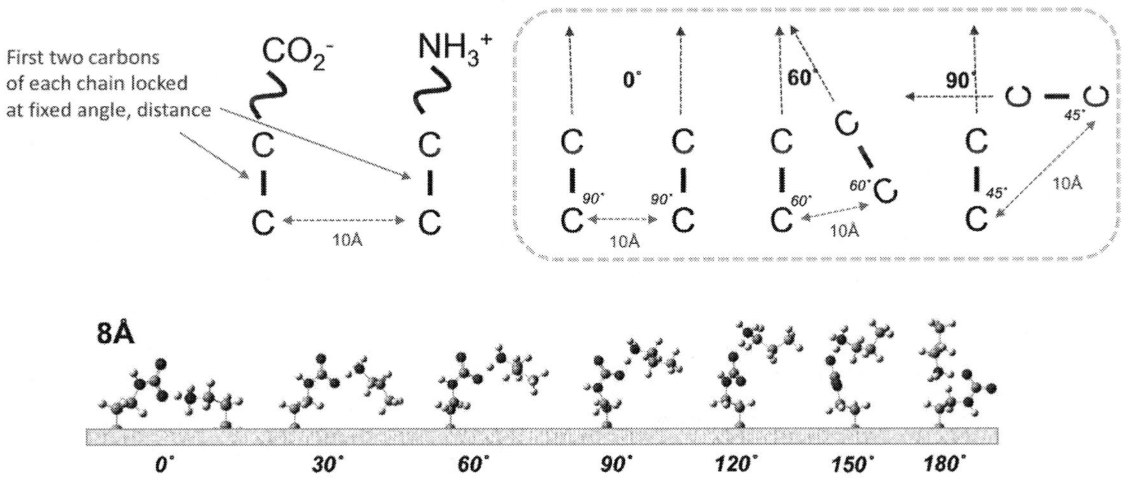


Planas, N.; Dzubak, A. L.; Poloni, R.; Gagliardi, L. *JACS* **2013**, *135*, 7402

# Orientational Effect

*B3LYP/6-311+G with IEFPCM solvation*

+0 to -111 kJ/mol by varying bond angle/distance

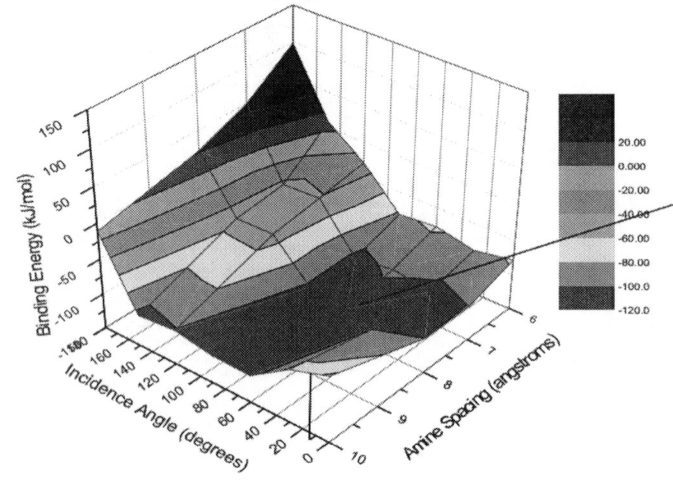


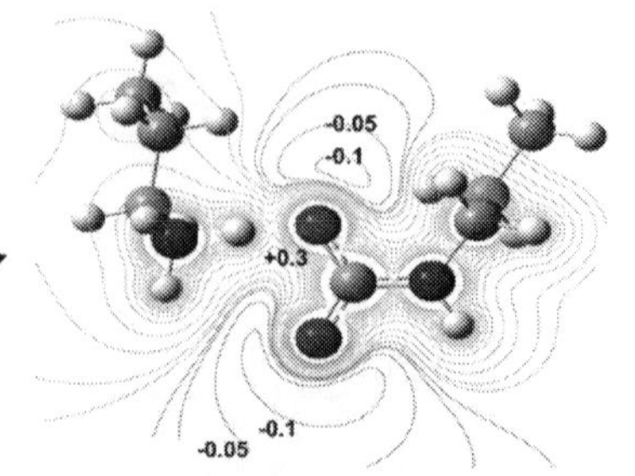


Electrostatic potential mapping on 7Å-60° pair

High/low incidence angle and spacing prevent optimal alignment of $COO^-/NH^+$ charge

# Binding Energy Results

**Table 3** Calculated binding energies for adsorbed species at the B3LYP/6-31++G* and B3LYP/6-31G optimized level.

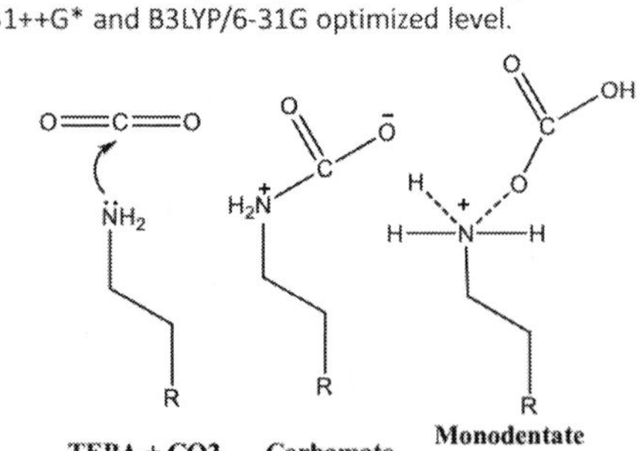

| Basis set | TEPA + CO2 (kJ/mol) | Carbamate (kJ/mol) | Monodentate bicarbonate (kJ/mol) |
|---|---|---|---|
| 6-31++G* | -23.19 | -19.93 | -54.07 |
| (6-31G) | (-45.71) | (-50.22) | (-105.49) |

Optimized Structure
B3LYP//6-31+G*

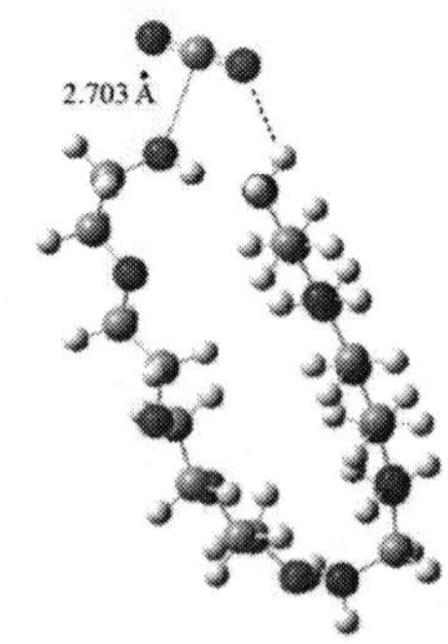


**Table 4** Calculated binding energies for adsorbed species at the B3LYP/6-31+G* and B3LYP/6-31G optimized level.

| Basis set | PEG + $CO_2$ (kJ/mol) | TEPA + $CO_2$ (kJ/mol) | TEPA +PEG + $CO_2$ (kJ/mol) |
|---|---|---|---|
| 6-31+G* | -4.81 | -23.19 | -15.07 |
| (6-31G) | (-15.05) | (-47.23) | (-21.60) |

# MS Analysis of PEG Loading Effect

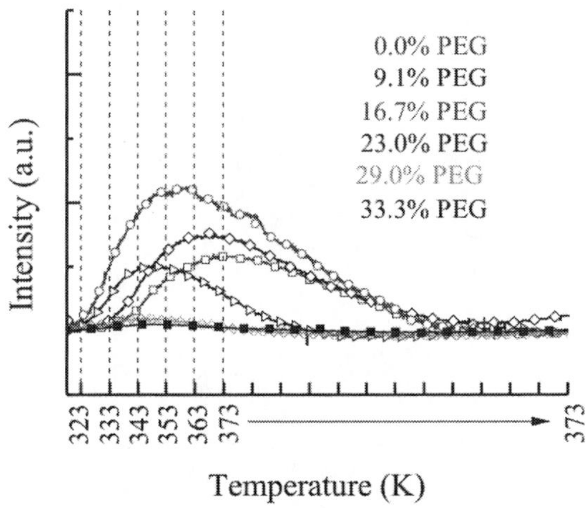


**Figure 5** MS analysis during the $CO_2$ TPD over PEG-TEPA.

**Table 2** $CO_2$ Capture on TEPA-PEG thin film and TEPA-PEG/$SiO_2$

| PEG Loading (wt %) | Thin-film $CO_2$ (μmol) | $SiO_2$ Support $CO_2$ (mmol/kg) | Temp. (K) |
|---|---|---|---|
| 0 | 6.84 | 1514 | 373 |
| 9.1 | 8.13 | 1124 | 373 |
| 16.7 | 12.83 | 2314 | 353 |
| 23 | 6.41 | 2400 | 343 |
| 29 | 0.49 | 2445 | 323 |
| 33.3 | 0.41 | 2266 | 323 |
| | ATR-IR | DRIFTS | |

# Tubular Testing System

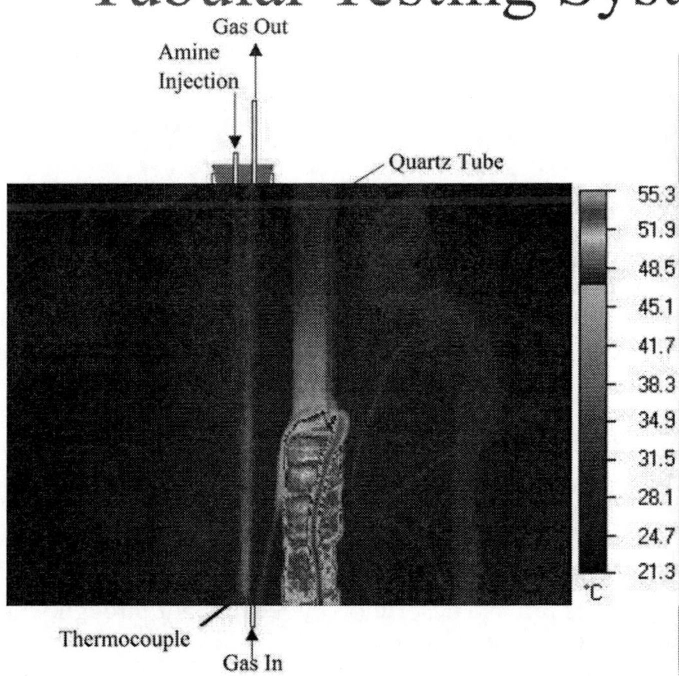


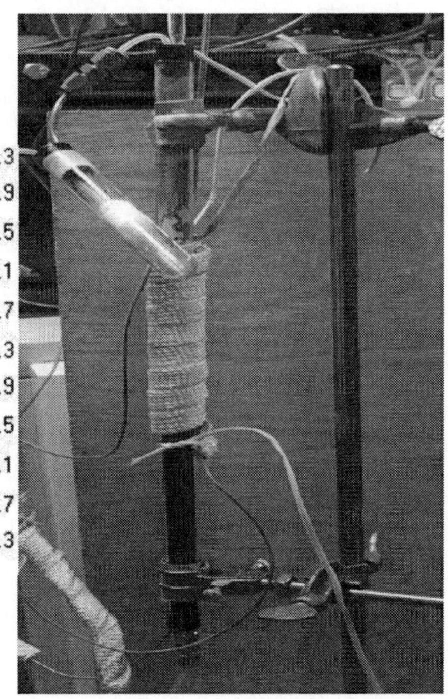

# 10 Kg Moving Bed Reactor for $CO_2$ Capture

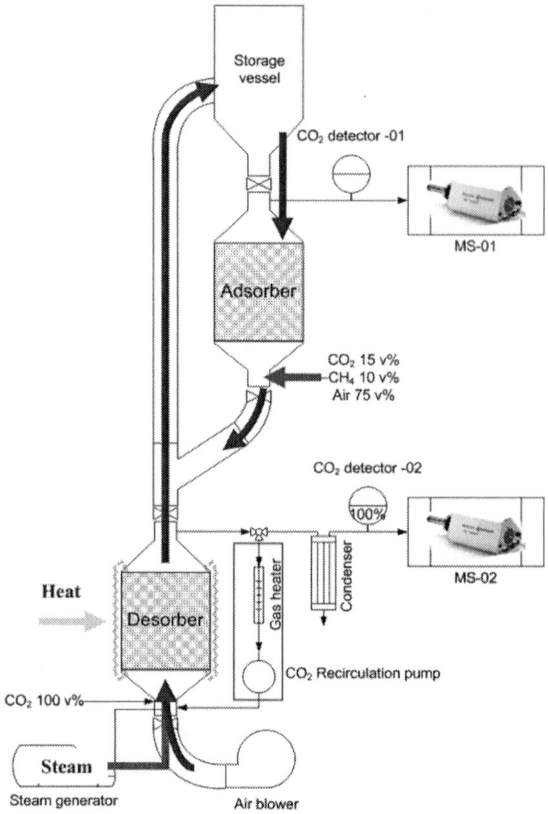


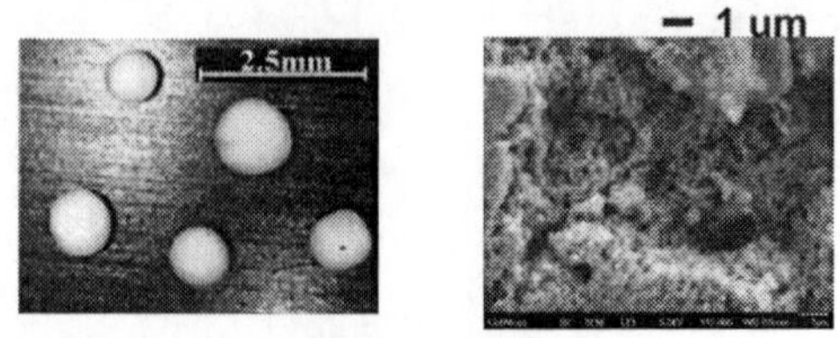

# Moving Bed Reactor: $CO_2$ purge and recycle

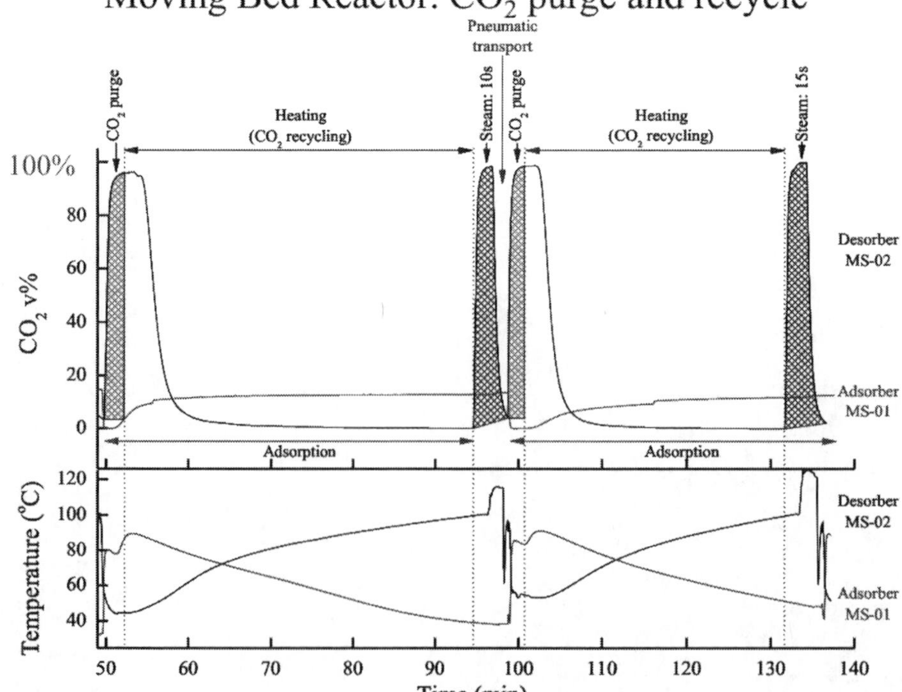


1. $CO_2$ purge protects weakly adsorbed species prior to desorption by recycling it throughout the heating stage. $CO_2$ concentration keeps high at all times.
3. $CO_2$ from purge is recuperated along with the $CO_2$ desorbed from the sorbent yielding to:

Total $CO_2$ out = Desorbed $CO_2$ + Purge $CO_2$

# $CO_2$ Capture with Radiator Monolith Adsorber

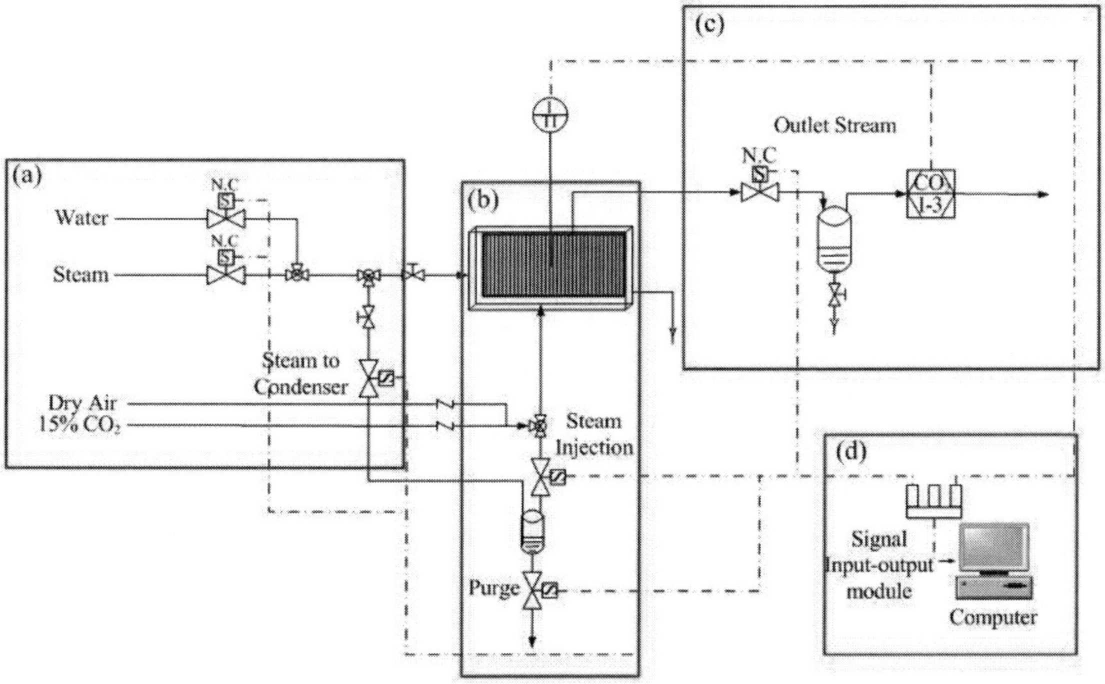




# $CO_2$ Capture with Radiator Monolith Adsorber

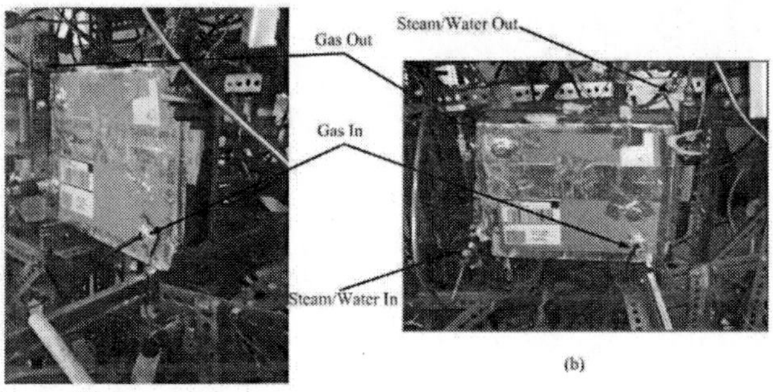

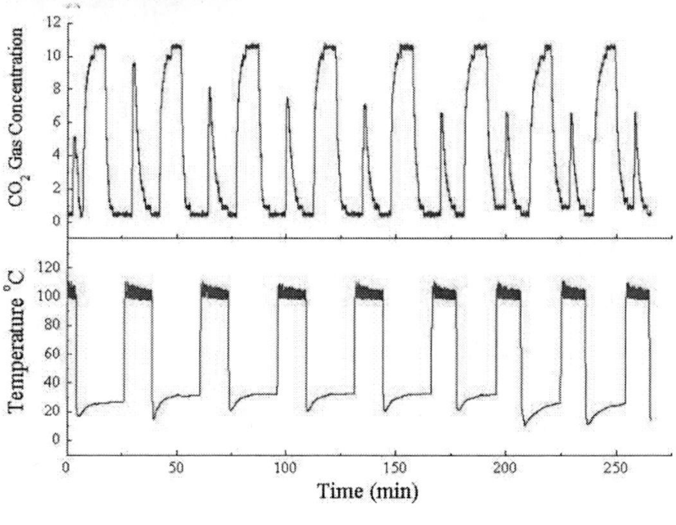


Metal monoliths provide excellent means to accelerate the heat transfer during the $CO_2$ capture process.

## $CO_2$ Capture Factors

| | Amine solid sorbent* | MEA |
|---|---|---|
| Adsorption temperature (°C) | 35-75 | 40-50[1] |
| Regeneration temperature (°C) | 100-120 | 120-140[2] |
| $CO_2$ capture capacity (mol/kg) | 2.9 at 35 °C | 4-5.5[1] |
| Cost of $CO_2$ avoidance ($/ton) | 40[3] | 116[4] |
| Heat capacity (kJ/kg K) | 3.5[2] | 4.5[2] |
| Regeneration energy (kJ/kg) | 746.4[2] | 3300[3] |
| $K_{eq}$ | 35°C: 1330**<br>55°C: 911<br>75 oC:244 | Henry's constant: 71 pa |
| Degradation | $O_2$, $CO_2$, $SO_x$/$NO_x$ poisoning, Fe/Cu from equipment | |

* Our TEPA/$SiO_2$ adsrobents

** The equilibrium constants were fitted from a dual site adsorption mechanism

$$2RNH_2 + CO_2 \rightarrow RNHCOO^- + RNH_3^+$$

$$C_{CO2^*eq} = \frac{C_T\sqrt{K_C P_{CO2}}}{1+2\sqrt{K_C P_{CO2}}}$$


1. Wang, M., et. al,. 2011, *89 (9)*, 1609-1624.
2. Yu, C.H., et. al,. 2012, 12, 745-769
3. Lohrasbi,M. et. al, Techno-economic assessment of U Akron silica-based solid sorbent for $CO_2$ capture.
4. IEAGHG. Report 2012/8. Cheltenham, UK: IEAGHG; 2012.


$$\text{Cost of } CO_2 \text{ Avoided (\$/tonne)} = \frac{(\$/kWh)_{capture} - (\$/kWh)_{reference}}{(\text{tonne } CO_2/kWh)_{reference} - (\text{tonne } CO_2/kWh)_{capture}}$$

# Summary

- Binding strength of adsorbed $CO_2$ on amine sorbents can be tuned by
  - the addition of PEG
  - control of the spacing and angles between amine functional groups.
- Rapid heating and cooling can be achieved with metal monoliths.
- Flexibility

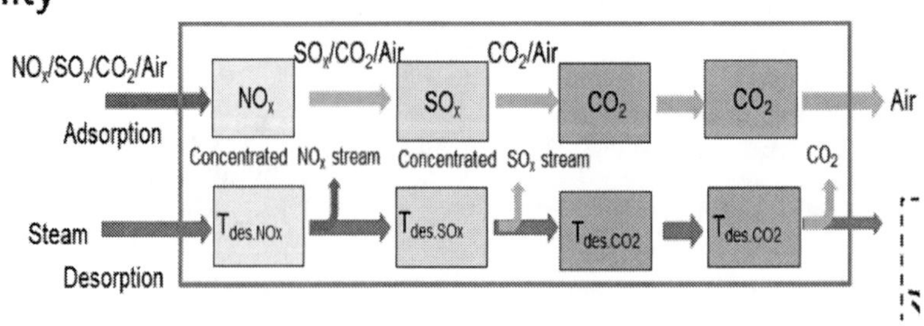


24

# Mathematical Modeling of Adsorption Behavior of Sod-ZMOF/Chitosan Adsorbent during Post-Combustion $CO_2$ Capture

Kelvin O. Yoro, Muofhe Singo, Michael O. Daramola*, Jean L. Mulopo,

School of Chemical and Metallurgical Engineering, Faculty of Engineering and the Built Environment, University of the Witwatersrand, Wits 2050, Johannesburg South Africa.

**Abstract**

Sod-ZMOF/Chitosan is a promising adsorbent for $CO_2$ capture. However, its adsorption behavior under different conditions is not sufficiently investigated and reported in literature. This study investigated the behavior of Sod-MOFs/Chitosan as an adsorbent for post combustion $CO_2$ capture via mathematical modeling. Some selected models were explored in order to describe the behavior of this adsorbent. It was found that Sod-ZMOF/Chitosan obviously showed a two-stage adsorption process with a rapid initial $CO_2$ uptake and subsequently a slow $CO_2$ adsorption process under different adsorption temperatures. The effects of temperature on the behavior of Sod-ZMOF/Chitosan during post-combustion $CO_2$ capture was investigated and corresponding breakthrough curves were obtained, presented and discussed. The kinetics of adsorption of $CO_2$ onto Sod-ZMOF/Chitosan was also studied. The Lagergen's pseudo $1^{st}$ order kinetic model accurately predicted the $CO_2$ adsorption behavior of the Sod-ZMOF/Chitosan material at every adsorption condition which indicated that the adsorption of $CO_2$ onto Sod-ZMOF/Chitosan is a physical adsorption. The experiments used in validating these models were conducted at an operating pressure of 200 KPa gas flow rate of 2.5 ml/s, mass of adsorbent was 0.1 g and a temperature of 30, 45 and 60°C. The same conditions were used for the simulation with mathematical models in the MATLAB R2014a environment in order to understand the behavior of this adsorbent. The model proposed by Bohart-Adams perfectly described the latter part of the breakthrough curve while the Lagergen's pseudo-first order kinetic model perfectly described the kinetics of adsorption of adsorbent at specified conditions.

**Keywords:** Mathematical modeling, Adsorption, Sod-ZMOFs/Chitosan, Post combustion $CO_2$ capture, Adsorption kinetics

*Corresponding author: Telephone: +27 11 717 7536 Email: michael.daramola@wits.ac.za

## 1. Introduction

The emission of $CO_2$ into the atmosphere needs to be stabilized in order to mitigate the unfettered release of greenhouse gases into the atmosphere because $CO_2$ is the main cause of global warming [1]. Capturing $CO_2$ from flue gas stream is a very important step in addressing this problem of $CO_2$ emission. The adsorption technology is envisaged to be a viable technology for $CO_2$ capture [2]. This is because, several studies conducted on $CO_2$ capture by adsorption indicates that the adsorption technique is an attractive technique for post combustion $CO_2$ capture [2]. Climate change due to emission of greenhouse gases, especially $CO_2$, has become a serious concern today. It is therefore imperative to capture $CO_2$ because it is the main cause of climate change. Consequently, materials and methods of various degrees of complexity have been developed and reported in literature as regards $CO_2$ capture. These methods include the adsorption of $CO_2$ onto materials such as zeolites and membranes as well as the chemical absorption of $CO_2$ in amine-rich solvents [3]. However, absorption technology is regarded as "mature technology" for $CO_2$ capture [4]. Nonetheless, it is plagued with some disadvantages such as its high energy requirement, corrosiveness and high financial cost which makes it too expensive to implement on an industrial scale [5]. Various $CO_2$ capture technologies have been reported in literature [8]. Littel et al [9], Chiesa and Consonni [10] reported on physical absorption of $CO_2$. Bishnoi and Rochelle [11] worked extensively on Chemical absorption of $CO_2$, Harlick and Tezel [12] as well as Cheng and Tan [13] reported $CO_2$ capture using the adsorption technology while Powelle and Qiao [14] reported on $CO_2$ capture using the membrane technology. They all concluded that these different technologies all have their disadvantages during $CO_2$ capture from power plants because of the existence of a significant mass transfer limitation in the process and the high amount of flue gas needed to be pretreated.

Among all these technologies listed above, the absorption technology using amine compounds is considered as more matured technology for $CO_2$ capture [4]. Despite the maturity level of the absorption technology, the conventional aqueous amine absorption processes has a low $CO_2$ loading, low contact area between gas and the liquid absorbent as well as a severe corrosion caused by the liquid absorbents [8]. In view of this, the solid adsorption process becomes a very good alternative in $CO_2$ capture, hence considered in this study. The preferential adsorption of $CO_2$ onto porous membranes and adsorbents is the basis for $CO_2$ capture by adsorption. Finding a very suitable adsorbent or membrane material that will sufficiently capture $CO_2$ is the first and most important step in every $CO_2$ capture process [15]

In view of this, new classes of porous materials called Sodalite-Metal Organic Frameworks (Sod-MOFs) impregnated with Chitosan for adsorption process was developed and it is envisaged to exhibit an excellent performance in capturing carbon dioxide from a flue gas mixture but its behavior during $CO_2$ capture is not widely reported. Experiments have been conducted and reported in literature on the application of Sod-ZMOFs for post-combustion $CO_2$ capture [6]. However, reports on the modeling of the adsorption behavior of Sod-ZMOF/Chitosan during $CO_2$ capture are still limited. Understanding the behavior and performance of this adsorbent, especially when impregnated with chitosan, during post combustion $CO_2$ capture is very essential. On this note, modeling will be a very useful tool to understand the behavior of the adsorbent during the $CO_2$ capture process.

This study is devoted to understanding the adsorption behavior of Sod-ZMOFs/chitosan material which is a MOF of the Sodalite topology with chitosan impregnation for post combustion $CO_2$ capture via mathematical modeling and simulations. The adsorption technology was considered in this study because of its system simplicity and its very promising low energy consumption [7]. In addition, the adsorbent considered in this study is Sodalite-Zeolite Metal Organic frameworks grafted with Chitosan (Sod-ZMOFs/Chitosan) because of its good physical properties such as porosity, high thermal stability, fast adsorption kinetics etc. The $CO_2$ adsorption capacity of Sod-ZMOFs/Chitosan was determined experimentally and compared with that obtained from the mathematical models in this study. In addition, this study also gives further insight into the effects of temperature on the adsorption capacity of Sod-MOFs/Chitosan during $CO_2$ capture.

## 2. Experimental Section

The adsorbent used in this study was Sod-ZMOFs/Chitosan and was synthesized by a colleague from modified protocol described elsewhere [16]. As earlier stated, information regarding the synthesis and characterization of sod-ZMOF/Chitosan considered in this study was reported in a different study by this research group. As such, it will not be reported again in this study. Sod-ZMOF was synthesized using the method adapted from the work reported in literature [16]. For preparation of chitosan, chitin was sieved and subjected to demineralization, deproteinization and decolonization. Chitin used in this work was produced from exoskeletons of crabs. The procedure used for impregnation of chitosan was adapted from the work reported in literature

[17] Chitosan was first added in acetic acid and stirred under heat for about 1 hour after which sod-ZMOF was added under continuous stirring for 1 hour. The resultant suspension was filtered, washed with distilled water to a neutral $P^H$ and dried in an oven at 50°C for 24 hours.

The gases used in this experimental study is a standard mixture of $CO_2/N_2$ (15 % $CO_2$ v/v balance $N_2$) supplied by Afrox South Africa. The Sod-ZMOF/chitosan adsorbent was pretreated in the adsorption column by passing in Nitrogen gas through the adsorption column at a flowrate of 1.5 ml/s and a temperature of 100°C for about one hour to remove impurities. The temperature of the adsorption bed was adjusted to 60, 45, and 30°C after the pretreatment stage under $N_2$ flow. The flue gas mixture (15 % $CO_2$/85% $N_2$ v/v) was then fed to the adsorption bed at a flow rate 2.5 ml/s and the adsorption column was by-passed in order to first determine the initial concentration of $CO_2$ from the ABB-AO2020 gas analyzer. The gas flow rate was regulated using a mass flow controller and the outlet $CO_2$ concentrations were obtained from the ABB-AO2020 gas analyzer. The experiments reported in this study were conducted at isothermal conditions.

## 3. Model Description

The mathematical models used in this study were already-existing gas-solid adsorption models derived from the mass balance around the adsorption column. Complete set of mathematical models capable of describing the $CO_2$ adsorption behavior of Sod-ZMOFs/Chitosan as an adsorbent for post combustion $CO_2$ capture based on ordinary differential and algebraic equations were studied based on the underlying physical principles governing the behavior of the gas-solid adsorption process. Kinetic data of the adsorption process is another characteristic that is very important for the simulation and design of a $CO_2$ capture system. This is because, kinetic data significantly influence the residence time required for the completion of any adsorption process as well as the unit capital cost [21]. In this study, the kinetics of adsorption of Sod-ZMOFs/Chitosan was investigated using the Lagergen's pseudo 1$^{st}$ and 2$^{nd}$ order kinetic models. Variety of gas-solid adsorption models with different level of complexities to quantitatively describe the behavior of several adsorbents during $CO_2$ adsorption process have been studied and reported in literature [22]. Due to complexities associated with the description of model parameters, this study considered a common approach which involves curve fitting of the data obtained from experiments to a number of kinetic models. The model that correlate well with the experimental data is always considered as the best model describing the behavior of the adsorbent [23].

The $CO_2$ adsorption breakthrough curves in this study were obtained by plotting the ratio of the concentration of $CO_2$ exit at time to the initial concentration of $CO_2$. The breakthrough curves were obtained at various temperatures, constant pressure and constant flow rates. The breakthrough curves in this study were described using the Bohart-Adams [24].The model results obtained from these mathematical models were compared with experimental results in order to validate the models and conclude on which of the models best describe the adsorption behavior of the sod-ZMOFs/chitosan adsorbent during post combustion $CO_2$ capture. Model parameters obtained from the solution in MATLAB R2014a described the adsorption capacity of the adsorbent and it was validated using experimental breakthrough curves.
Several kinetic models to describe the adsorption behavior of adsorbents have been reported in literature [25–29]. Amongst the reported kinetic models, the Lagergen's pseudo 1$^{st}$ order and pseudo 2$^{nd}$ order kinetic models are the most widely used because of their simplicity in describing the kinetics of adsorption as well as the interactions between the adsorbent and the adsorbate. However, in this study, the Largergen's pseudo 1$^{st}$ and 2$^{nd}$ order kinetic models (Equation (1) and Equation (2)) were used to describe the $CO_2$ adsorption behavior of Sod-ZMOF under a wide range of temperatures and constant adsorption pressures. The kinetic models used to describe the adsorption kinetics were validated using kinetic data obtained from the experiment.

$$\frac{dq}{dt} = K_f(q_e - q_t) \qquad (1)$$

$$\frac{dq}{dt} = K_s(q_e - q_t)^2 \qquad (2)$$

Where qe and qt are the equilibrium adsorption capacity and adsorption capacity at a given time in mmol/g. $K_f$ is the 1$^{st}$ order kinetic constant. Where $Ks$ is the 2$^{nd}$ order kinetic constant in $min^{-1}$. At boundary conditions $q_t$=0, t=∞, $q_t = q_e$. Equation (35) can be expressed as;

In order to determine the accuracy of each mathematical model explored in this study, an error function based on the normalized standard deviation reported in a related study [23] was applied as seen in equation (5) below;

$$\% \, Err = \sqrt{\frac{(\Sigma[(qt(exp) - qt(mod))/qt(exp))^2}{(N-1)}} \; X \; 100 \qquad (5)$$

Where % Err is the error function from the study in percentages, qt (exp) is the average amount of $CO_2$ adsorbed experimentally at a given time, qt(mod) is the amount of $CO_2$ adsorbed as predicted by the mathematical model and N is the total number of experimental runs.

## 4. Results and Discussion

In a bid to investigate the behavior of Sod-ZMOF/Chitosan using its breakthrough curves and kinetics of adsorption, the adsorbent was tested under different temperatures. Fig. 3 shows a breakthrough curve for the adsorption of $CO_2$ on Sod-ZMOF/chitosan at 30, 45 and 60°C. Fig 4 shows a breakthrough curve at 30°C fitted to the Bohart-Adams model. In Fig. 4, the Bohart-Adams model did not describe the initial part of the breakthrough curve but it clearly described the latter part of the breakthrough curve. This simply means that the thermodynamic model alone cannot sufficiently describe the behavior of Sod-ZMOF/Chitosan, hence kinetic models were also studied. Table 2 presents the model parameters obtained from the solutions in MATLAB R2014a. Bohart-Adams model also predicted a higher amount of $CO_2$ adsorbed by the material which is closer to that obtained experimentally. A maximum of 1.9 mmol of $CO_2$ per gramme adsorbent was obtained as the amount of $CO_2$ adsorbed by the adsorbent from the experiment while the 1$^{st}$ order kinetic model predicted that a maximum of 1.8 mmol $CO_2$ per gramme of adsorbent was adsorbed by the material and the Pseudo 2$^{nd}$ order predicted a maximum of 1.7 mmol of $CO_2$ per gramme of adsorbent. From the breakthrough curves in Fig. 3 and Fig. 4 and the adsorption curve in Fig. 5 , it could be inferred that the adsorption of $CO_2$ on Sod-ZMOF/Chitosan showed a two-stage adsorption process under the specified conditions with a rapid initial uptake of $CO_2$ and subsequently a slow adsorption process. This two-stage adsorption process observed in this study is similar to that observed during the adsorption of $CO_2$ on calcium oxide [30] , Lithium Zirconate [31] and porous magnesium oxide [32]. The two-stage adsorption observed in this study could be attributed to the adsorption products generated on the surface of the adsorbent and the porosity distribution within the material. The adsorption products generated on the surface of the adsorbent slowed down the diffusion of $CO_2$ into the inner part of the particle thereby lowering the rate of adsorption of $CO_2$ on this adsorbent. The plots at different temperatures all showed a non-linear relationship. The Bohart-Adams model perfectly describes the latter part of the breakthrough curve at 30°C, but does not describe sufficiently the initial part of the curve because a rectangular or step isotherm as well as gradients in axial direction was assumed for the model and these assumptions did not completely favor the adsorbent under investigation in this study. On Table 1, the adsorption temperature exhibited distinct effect on the adsorption behavior of Sod-ZMOF/chitosan. A higher concentration of $CO_2$ was adsorbed at very low temperatures. This is perfectly in agreement with literature reports that lower temperatures favor the adsorption process. From the breakthrough curve in Fig. 3, 80% vol/vol of $CO_2$ was adsorbed by the adsorbent and equilibrium was attained after 900 seconds at 30°C. At 45°C, 70% v/v $CO_2$ was adsorbed and equilibrium was attained after 750 seconds while at a maximum temperature of 60°C, a minimum of 55% vol/vol $CO_2$ was adsorbed on the surface of Sod-ZMOF/Chitosan, equilibrium faster at 700 seconds. This clearly explains that the kinetics of adsorption of CO2 onto sod-ZMOF/Chitosan is not too fast therefore it could be studied using kinetic models. From the adsorption curves reported in this study, it can be speculated that adsorption started rapidly at the initial stage after 50 seconds while the slow adsorption stage lasted for about 1000 seconds. This is due to the inter-particle diffusion resistance which could be attributed to the pore size distribution of the adsorbent. Film diffusion resistance controlled the adsorption of $CO_2$ on the sod-ZMOF/chitosan material at the initial stage but upon diffusion of the $CO_2$ on the gas film, intra-particle diffusion became a rate-limiting step in the adsorption process until adsorption equilibrium.

From the normalized standard deviation of the kinetic models in Table 1, it is clearly seen that the Lagergen's pseudo-first order model predicted the amount of $CO_2$ adsorbed with a smaller error than the Lagergen's pseudo-second order model. The pseudo 1$^{st}$ order model had a minimum and maximum percentage error of 3.72% and 5.6% respectively. While the pseudo 2$^{nd}$ order model predicted the maximum amount of $CO_2$ adsorbed on Sod-ZMOF/Chitosan with a maximum percentage error of 8.32% and a minimum error of 7.44%. These errors could be attributed to some minor experimental errors in the course of the experiment. The Lagergen's pseudo-first order model was also more consistent with the experimental results as seen in Fig 6. The pseudo-first order model best described the adsorption of $CO_2$ unto Sod-ZMOF/Chitosan material. As reported in literature [27], the pseudo-1st order model is well suited in describing the adsorption process on physical adsorbents while the pseudo-second order is best suited to describe a process involving chemical adsorption[33]. On this note, it can be speculated that physical adsorption governs the adsorption of $CO_2$ onto the Sod-ZMOF/Chitosan material in this study. However, estimation of the adsorption enthalpies will be necessary to make a logical conclusion.

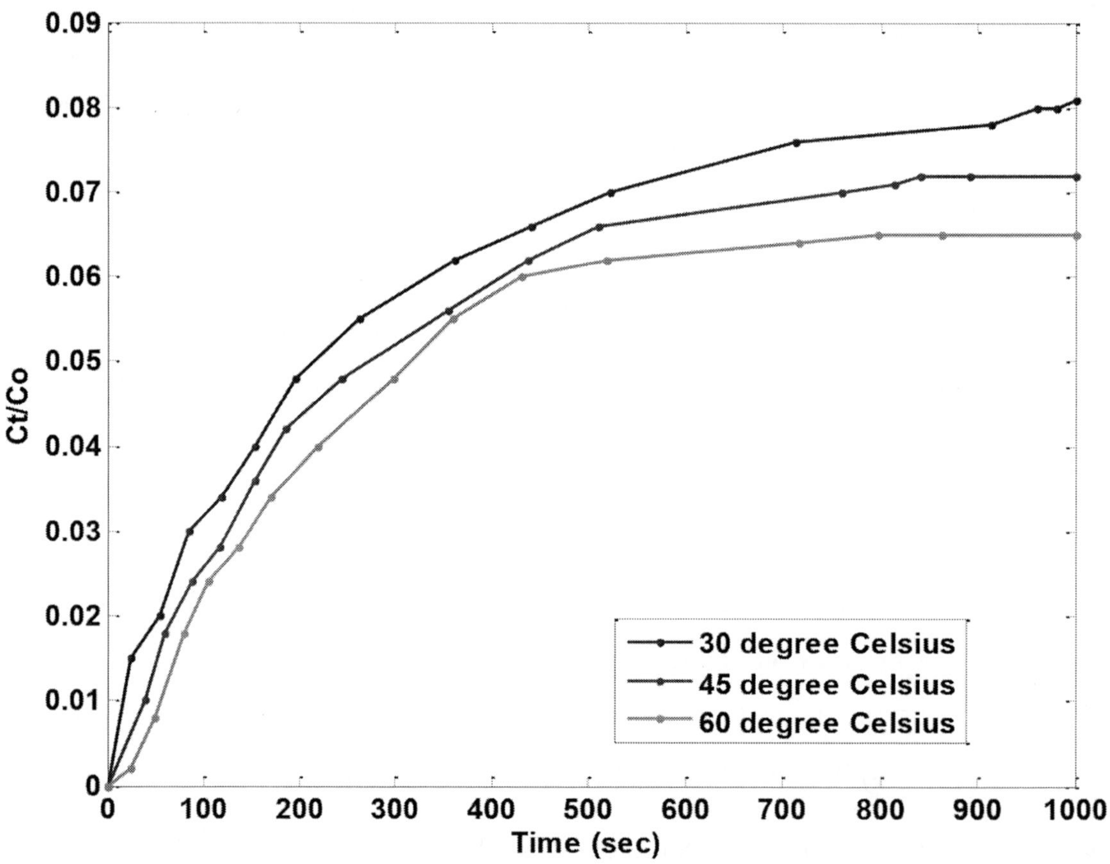


Fig 3. Experimental Breakthrough curves of 14.9% $CO_2$ balance $N_2$ on Sod-ZMOF/Chitosan at various temperatures

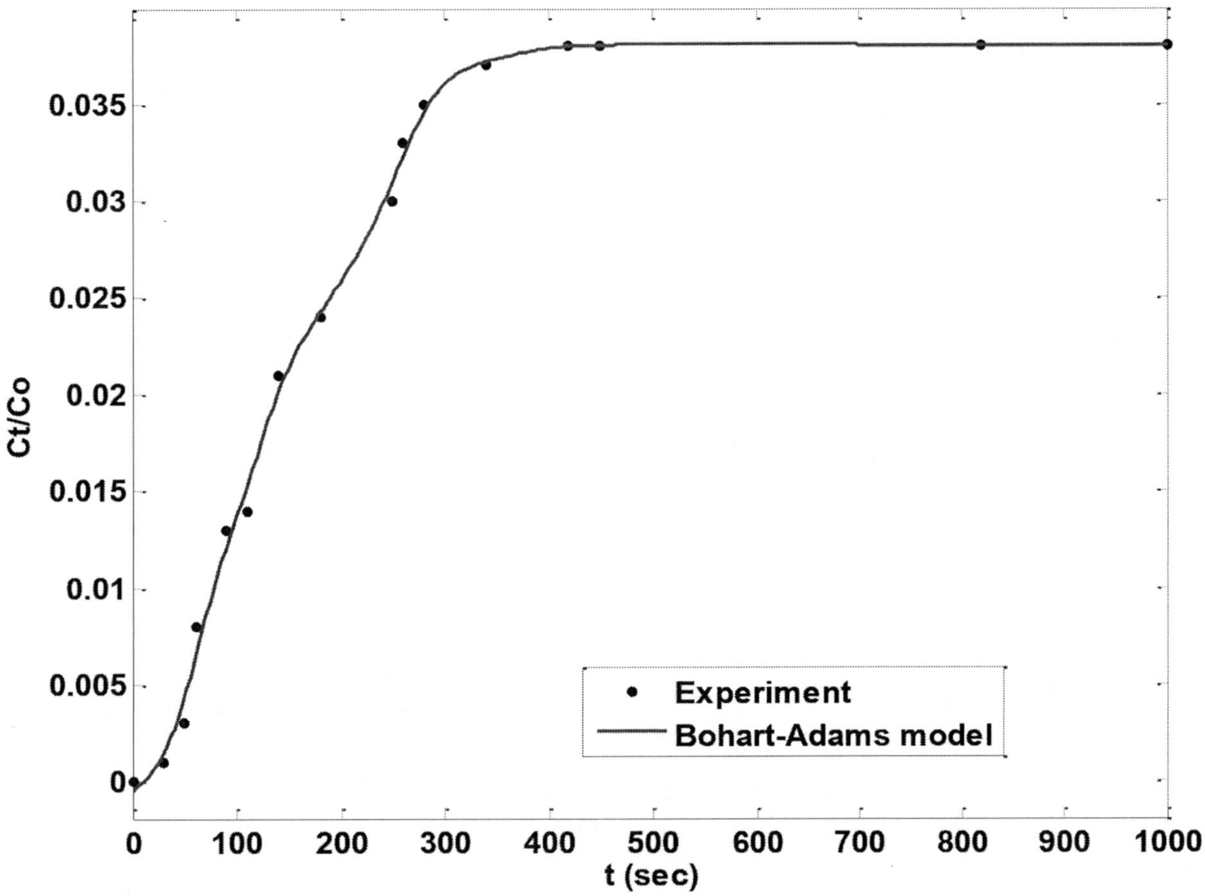


Fig 4. Experimental breakthrough curve at 30°C and corresponding fit to the Bohart-Adams model

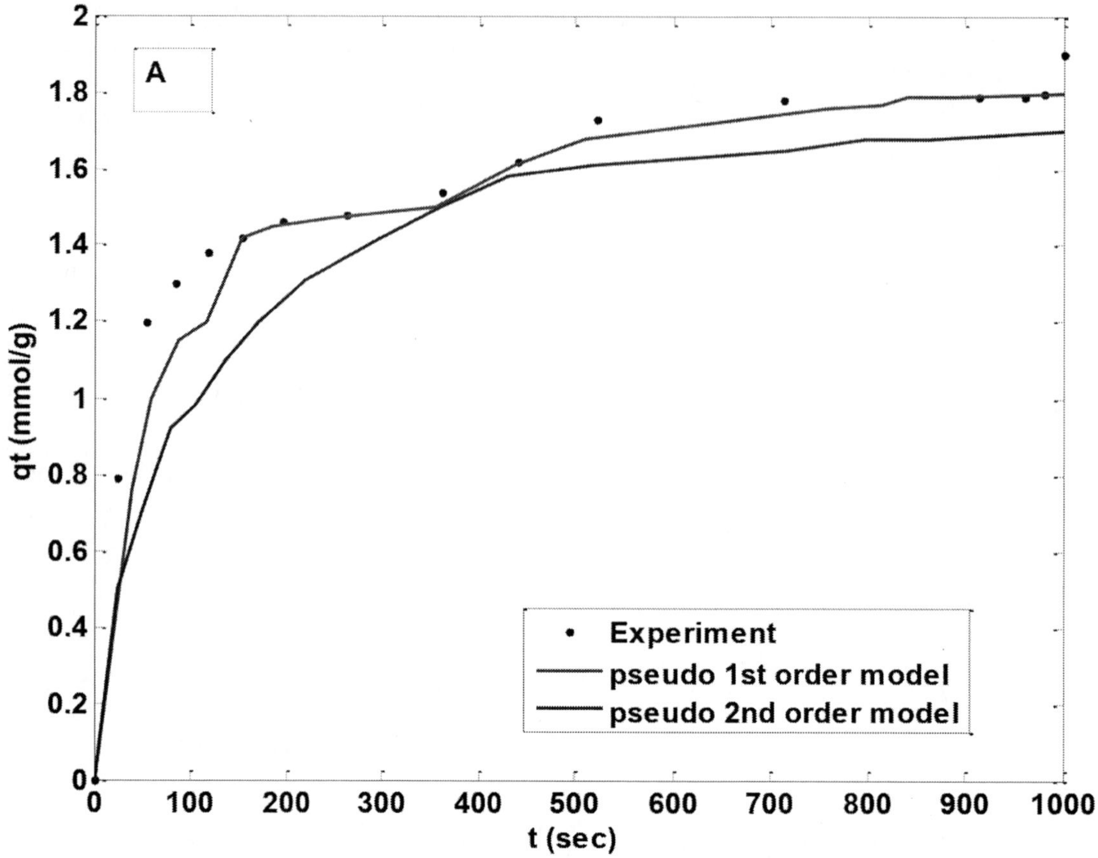
A
Experiment
pseudo 1st order model
pseudo 2nd order model
qt (mmol/g)
t (sec)
0
0.2
0.4
0.6
0.8
1
1.2
1.4
1.6
1.8
2
0
100
200
300
400
500
600
700
800
900
1000

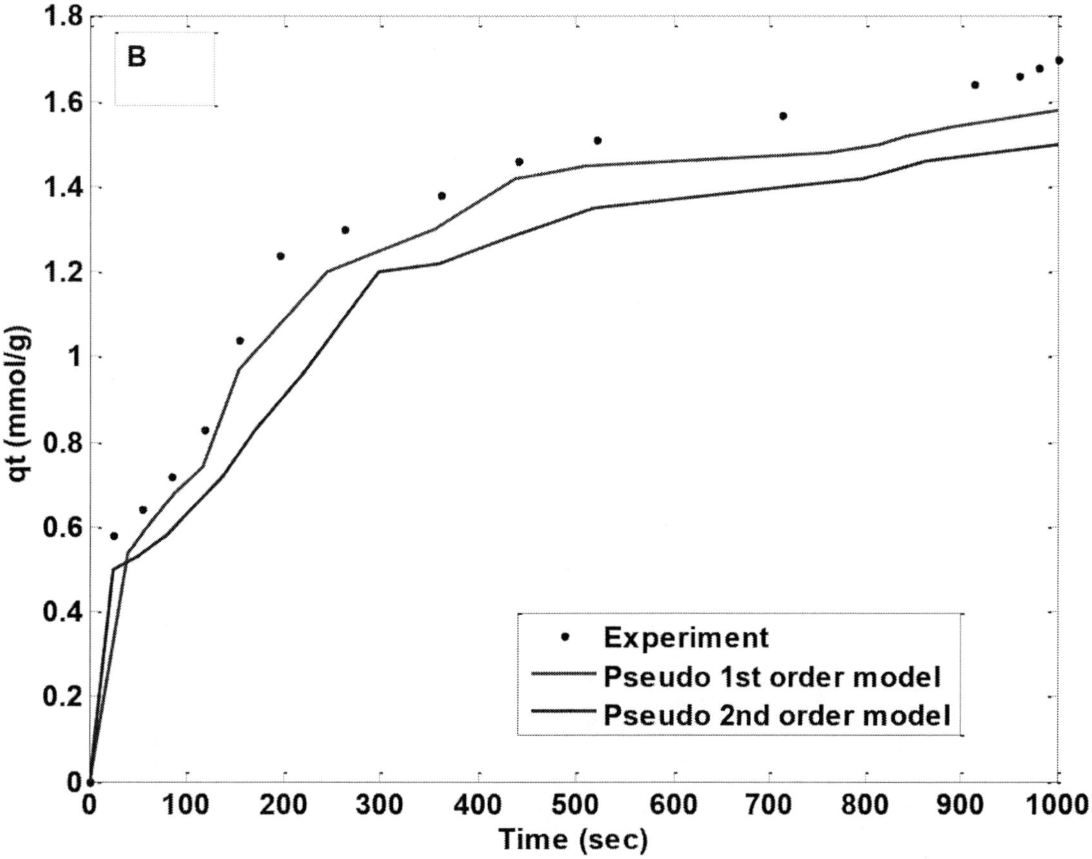
B
qt (mmol/g)
Time (sec)
Experiment
Pseudo 1st order model
Pseudo 2nd order model
1.8
1.6
1.4
1.2
1
0.8
0.6
0.4
0.2
0
0
100
200
300
400
500
600
700
800
900
1000

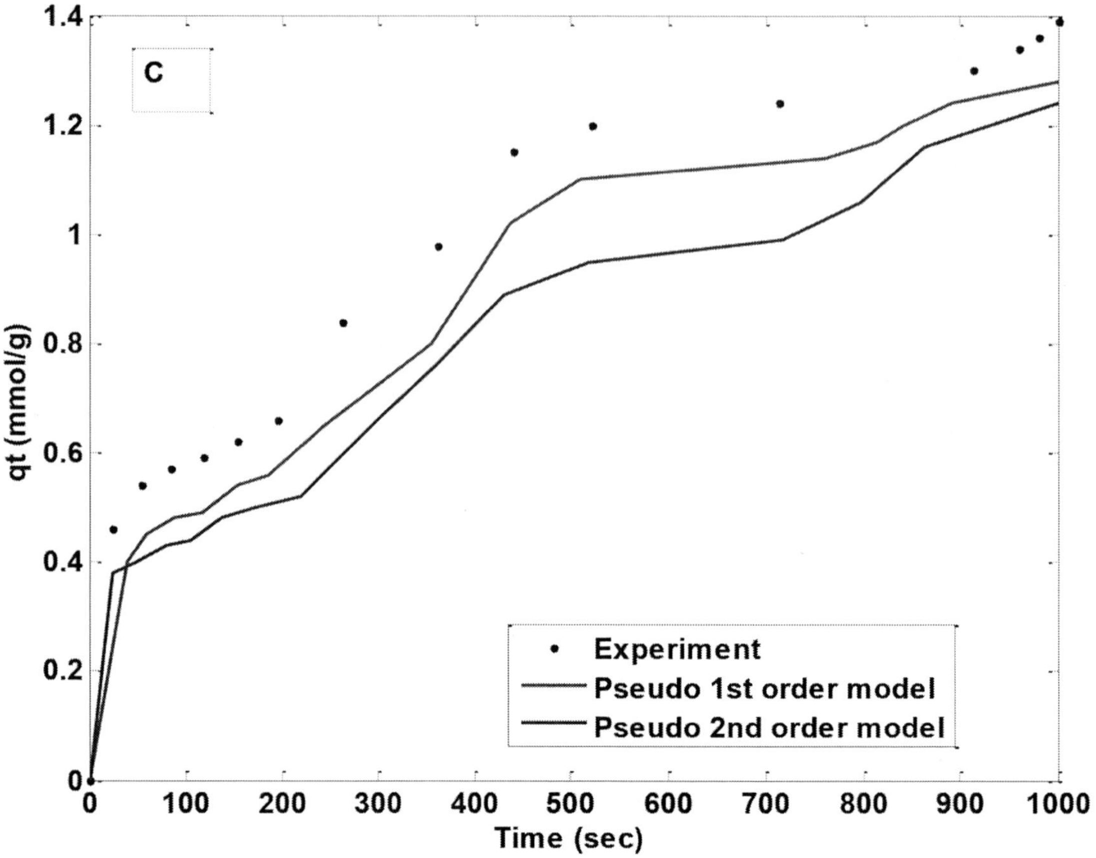


Fig 5. Experimental and predicted $CO_2$ adsorption curves of Sod-ZMOF/Chitosan material at constant pressure and different temperatures; **(A)** 2bar, 30°C, **(B)** 2bar 45°C **(C)** 2bar 60°C

Table 1. Kinetic parameters of pseudo 1$^{st}$ and 2$^{nd}$ order models under different adsorption conditions

| | | **Pseudo 1$^{st}$ order** | | | **Pseudo 2$^{nd}$ order** | |
|---|---|---|---|---|---|---|
| **Adsorption Condition** | $q_{exp}$ | $q_{cal}$ | **Δq Error (%)** | $q_{exp}$ | $q_{cal}$ | **Δq Error (%)** |
| **2bar, 30C** | 1.9 | 1.8 | 3.72 | 1.9 | 1.7 | 7.44 |
| **2bar, 45C** | 1.7 | 1.58 | 4.99 | 1.7 | 1.5 | 8.32 |
| **2bar, 60C** | 1.39 | 1.28 | 5.60 | 1.39 | 1.24 | 7.63 |

Table 2. Mathematical relations and corresponding model parameters of used models

| Models Studied | Model Parameters | | | Mathematical Equation |
|---|---|---|---|---|
| **Bohart-Adams** | ***Kb*** | ***q*** | | $\ln(C/Co) = K(Cot - q.\frac{z}{v})$ |
| | 0.184 | 1.8 | | |
| **Pseudo 1st order** | $K_f$ | | | $q_t = q_e(1\text{-}e^{-Kft})$ |
| | **30°C** | **45°C** | **60°C** | |
| | 0.07 | 0.075 | 0.08 | |
| **Pseudo 2nd order** | $K_s$ | | | $q_t = \frac{q_e^2 K_s t}{1 + q_e K_s t}$ |
| | 0.076 | 0.078 | 0.079 | |

## Conclusion

The $CO_2$ adsorption behavior of Sod-ZMOF/Chitosan was studied under various adsorption temperatures and at a constant pressure and flowrate of 2 bars and 2.5 ml $CO_2$/s, respectively. Sod-ZMOF/Chitosan exhibited a two-stage $CO_2$ adsorption process characterized by a rapid initial step and a sluggish following step. The initial step which was reported to be very fast in this study was as a result of film diffusion from the bulk gas phase to the surface of the sod-ZMOF/Chitosan material. The latter step which was reported to be slow was attributed to intra-particle diffusion resistance as a result of the narrow pores on the adsorbent which hampers on the $CO_2$ diffusion rate. This study also reveals that the sluggish diffusion of $CO_2$ onto Sod-ZMOF/Chitosan has effect on the adsorption kinetics of the Sod-ZMOF/Chitosan material. The Lagergen's pseudo-first and second order models were used to fit the experimental adsorption curves of Sod-ZMOF/Chitosan. The accuracy of the predictions by these kinetic models were tested using an error function based on the normalized standard deviation as reported in literature[27]. The Lagergen's pseudo-first perfectly described the kinetics of adsorption of $CO_2$ onto Sod-ZMOF/Chitosan at various temperatures in this study. The least error was also reported by the Lagergen's-first order kinetic model. Based on this, it can be concluded that the mechanism of adsorption of $CO_2$ onto Sod-ZMOF is governed by physical adsorption. Sod-ZMOF/Chitosan also displayed good adsorption characteristics under various temperatures, it had a relatively fast kinetics therefore it is speculated that Sod-ZMOF/Chitosan is a promising adsorbent for post-combustion $CO_2$ capture based on its general behavior in this study.

Finally, this study also confirms reports from literature that high pressure and low temperature favors adsorption process because the highest amount of $CO_2$ adsorbed by the adsorbent in this study was obtained at a high pressure of 200 kPa and at the lowest temperature of $30^{o}C$ while at a higher temperature of $60^{o}C$ in this study, a lower amount of $CO_2$ was adsorbed.

**Acknowledgement**

The authors wish to acknowledge the National Research Foundation of South Africa grant number ***94874*** and the University of the Witwatersrand, Johannesburg South Africa for their financial support.

# REFERENCES


[1] M. S. Shafeeyan, W. M. A. Wan Daud, and A. Shamiri, A review of mathematical modeling of fixed-bed columns for carbon dioxide adsorption, *Chem. Eng. Res. Des.*, vol. 92, no. 5 (2014) 961–988.

[2] Cheng-Hsiu Yu, Chih-Hung Huang, and Chung-Sung Tan, A Review of CO2 Capture by Absorption and Adsorption, *Aerosol Air Qual. Res.* 12 (2012)745–769.

[3] P. Luis, Use of monoethanolamine (MEA) for CO2 capture in a global scenario: Consequences and alternatives, *Desalination*, vol. 380 (2016) 93–99.

[4] D.M Ruthven, Principles of adsorption and adsorption processes, *John Wiley Sons*, (1984).

[5] R. Sabouni, H. Kazemian, and S. Rohani, Carbon dioxide capturing technologies: a review focusing on metal organic framework materials (MOFs), *Environ. Sci. Pollut. Res.*, vol. 21, no. 8 (2013) 5427–5449.

[6] R. Sabouni, Carbon Dioxide Adsorption by Metal Organic Frameworks (Synthesis, Testing and Modeling), *Electron. Thesis Diss. Repos* (2013).

[7] J. Zhang, P. Xiao, G. Li, and P. A. Webley, Effect of flue gas impurities on $CO_2$ capture performance from flue gas at coal-fired power stations by vacuum swing adsorption, *Energy Procedia*, vol. 1, no. 1 (2009) 1115–1122.

[8] D. Y. C. Leung, G. Caramanna, and M. M. Maroto-Valer, An overview of current status of carbon dioxide capture and storage technologies, *Renew. Sustain. Energy Rev.*, vol. 39 (2014) 426–443.

[9] R. J. Littel, G. F. Versteeg, and W. P. M. Van Swaaij, Physical absorption into non-aqueous solutions in a stirred cell reactor, *Chem. Eng. Sci.*, vol. 46, no. 12 (1991) 3308–3313.

[10] P. Chiesa and S. Consonni, Shift Reactors and Physical Absorption for Low-$CO_2$ Emission IGCCs, *J. Eng. Gas Turbines Power*, vol. 121, no. 2 (1999) 295–305.

[11] S. Bishnoi and G. T. Rochelle, Absorption of carbon dioxide into aqueous piperazine: reaction kinetics, mass transfer and solubility, *Chem. Eng. Sci.*, vol. 55, no. 22 (2000) 5531–5543.

[12] Harlick, P.J.E. and Tezel, F.H, An Experimental Adsorbent Screening Study for $CO_2$ Capture from N2, *Microporous Mesoporous Mater*, (2004) 76: 71–79.

[13] H.-H. Cheng and C.-S. Tan, Carbon dioxide capture by blended alkanolamines in rotating packed bed, *Energy Procedia*, vol. 1, no. 1 (2009) 925–932.

[14] Powell, C.E. and Qiao, G.G, Polymeric $CO_2/N_2$ Gas Separation Membranes for the Capture Carbon Dioxide from Power Plant Flue Gases," *J Membr Sci*, no. 279 (2006) 1–49.

[15] T. L. P. Dantas, F. M. T. Luna, S. Jr, I. J, A. E. B. Torres, D. Azevedo, D. C. S, A. E. Rodrigues, and R. F. P. M. Moreira, Modeling of the fixed - bed adsorption of carbon dioxide and a carbon dioxide - nitrogen mixture on zeolite 13X, *Braz. J. Chem. Eng.*, vol. 28, no. 3 (2011) 533–544.

[16] C. Chen, J. Kim, D.-A. Yang and W.-S. Ahn, Carbon dioxide adsorption over zeolite-like metal organic frameworks (ZMOFs) having a sod topology: Structure and ion-exchange effect, *Chem. Eng. J.*, vol. 168, no. 3 (2011) 1134–1139.

[17] Sun, H., Lu, L., Chen, X. & Jiang, Z., Surface modified zeolite-filled chitosan membranes for pervaporation dehydration of ethanol. Applied Surface Science, (2008) 5367–5374.

[18] Anne Elise Creamer and Bin Gao, Carbon-Based Adsorbents for Post combustion $CO_2$ Capture: A Critical Review, *Environ. Sci. Technol.* (2016).

[19] R. W. Breault and E. D. Huckaby, Parametric behavior of a $CO_2$ capture process: CFD simulation of solid-sorbent CO2 absorption in a riser reactor, *Appl. Energy*, vol. 112, (2013) 224–234.

[20] A. Ghribi and M. Chlendi, Modeling of Fixed Bed Adsorption: Application to the Adsorption of an Organic Dye, *Asian J. Text.*, vol. 1, no. 4, (2011) 161–171.

[21] R. Serna-Guerrero and A. Sayari, Modeling adsorption of CO2 on amine-functionalized mesoporous silica. 2: Kinetics and breakthrough curves, *Chem. Eng. J.*, vol. 161, no. 1–2 (2010) 182–190.

[22] Anne Elise Creamer and Bin Gao, "Carbon-Based Adsorbents for Post combustion CO2 Capture: A Critical Review," *Environ. Sci. Technol.*,( 2016).

[23] G. S. Bohart and E. Q. Adams, Some aspects of the behavior of charcoal with respect to chlorine, *J. Frankl. Inst.*, vol. 189, no. 5 (1920) 669.

[24] Lunhong Ai, Chunying Zhang, and Lanying Meng, Adsorption of Methyl Orange from Aqueous Solution on Hydrothermal Synthesized Mg-Al Layered Double Hydroxide, *J. Chem. Eng. Data*, vol. 56, (2011) 4217 – 4225.

[25] J. M. Borah, J. Sarma, and S. Mahiuddin, Adsorption comparison at the α-alumina/water interface: 3, 4-Dihydroxybenzoic acid vs. catechol," *Colloids Surf. Physicochem. Eng. Asp.*, vol. 387, no. 1–3 (2011) 50–56.

[26] C. W. Cheung, J. F. Porter, and G. Mckay, Sorption kinetic analysis for the removal of cadmium ions from effluents using bone char, *Water Res.*, vol. 35, no. 3 (2001) 605–612.

[27] S. Sen Gupta and K. G. Bhattacharyya, Kinetics of adsorption of metal ions on inorganic materials: A review, *Adv. Colloid Interface Sci.*, vol. 162, no. 1–2 (2011) 39–58.

[28] A. R. Cestari, E. F. S. Vieira, G. S. Vieira, and L. E. Almeida, The removal of anionic dyes from aqueous solutions in the presence of anionic surfactant using aminopropylsilica—A kinetic study, *J. Hazard. Mater.* vol. 138, no. 1 (2006)133–141.

[29] B. H. Hameed, D. K. Mahmoud, and A. L. Ahmad, Sorption equilibrium and kinetics of basic dye from aqueous solution using banana stalk waste, *J. Hazard. Mater.* vol. 158, no. 2–3 (2008) 499–506.

[30] J. M. Borah, J. Sarma, and S. Mahiuddin, Adsorption comparison at the α-alumina/water interface: 3,4-Dihydroxybenzoic acid vs. catechol, *Colloids Surf. Physicochem. Eng. Asp.*, vol. 387, no. 1–3 (2011)50–56.

[31] S. K. Bhatia and D. D. Perlmutter, Effect of the product layer on the kinetics of the CO2-lime reaction, *AIChE J.*, vol. 29, no. 1, (1983) 79–86.

[32] R. Xiong, J. Ida, and Y. S. Lin, Kinetics of carbon dioxide sorption on potassium-doped lithium zirconate," *Chem. Eng. Sci.*, vol. 58, no. 19, (2003)4377–4385.

[33] G. Song, X. Zhu, R. Chen, Q. Liao, Y.-D. Ding, and L. Chen, "An investigation of CO2 adsorption kinetics on porous magnesium oxide," *Chem. Eng. J.*, vol. 283 (2016) pp. 175–183.

# SYNTHESIS OF CHITOSAN-IMPREGNATED SOD-ZEOLITE METAL ORGANIC FRAMEWORKS (SOD-ZMOF) FOR CARBON DIOXIDE ($CO_2$) CAPTURE: A PRELIMINARY INVESTIGATION

Muofhe. C .Singo[1], Olugbenga O. Oluwasina[2], Michael.O.Daramola[1,*]

[1]School of Chemical and Metallurgical Engineering, Faculty of Engineering and the Built Environment, University of the Witwatersrand, Private Bag X3, Wits 2050, Johannesburg, South Africa

[2]Department of Chemistry, Federal University of Technology, Akure, Ondo State, Nigeria

*Corresponding authors: Michael.Daramola@wits.ac.za

## ABSTRACT

In this study, zeolite metal organic framework (ZMOF) with sodalite topology (sod-ZMOF) was synthesized in order to prepare sod-ZMOF-chitosan composite adsorbent for $CO_2$ capture. The chitosan was impregnated on sod-ZMOF by using acetic acid as solvent. After synthesis, characterization of the adsorbent was done using X-Ray diffraction (XRD), Thermo gravimetric analysis (TGA), Fourier Transform Infrared Radiation (FTIR) and Nitrogen physi-sorption for BET surface area, pore volume and pore size. XRD patterns showed that sod-ZMOF is crystalline and TGA analysis showed that it is thermally stable up to 250°C. FTIR analysis showed functional groups that are present in the organic linker (4.5-imidazoledicarboxylic acid) are also present in sod-ZMOF. In addition, the FTIR spectra showed that chitosan was successfully impregnated in the sod-ZMOF. After characterization, performance evaluation of sod-ZMOF, chitosan and sod-ZMOF-chitosan during $CO_2$ capture was done using a packed-bed adsorption column. A gas mixture containing 15% $CO_2$ and 85% $N_2$ was used in this study. Impregnation of chitosan in sod-ZMOF-chitosan resulted in a 12% increase in $CO_2$ adsorption capacity as compared to the ordinary sod-ZMOF.

**Keywords:** Chitosan, $CO_2$ capture, Adsorption, MOFs, Packed-bed column

## 1. INTRODUCTION

Climate change is one of the challenges that the world is facing today due to the increase in atmospheric concentration of greenhouse gases such as $CO_2$. Carbon dioxide ($CO_2$) is a gas that is naturally present in the atmosphere as part of the earth's carbon cycle. It is also produced from human activities such as burning of fossil fuels for energy production. In South Africa, about 90 % of electricity needs is produced from burning of fossil fuel [1]. In the past, the atmospheric sink was large enough to accommodate additional carbon dioxide produced by human activities. However, since the beginning of the industrial revolution, atmospheric $CO_2$ concentration has increased from 280 ppm to over 380 ppm in 2006 [2]. In 2015, the IEA reported that atmospheric $CO_2$ concentration continues to increase with an average growth of 2 ppm per year [3]. To date, atmospheric $CO_2$ concentration is slightly above 400 ppm and continues to rise [4]. If nothing is done to reduce the amount of $CO_2$ being released into the atmosphere, the upper safe limit of $CO_2$ will be reached soon. Much effort is being devoted to finding ways of reducing the amount of $CO_2$ concentrations being released into the atmosphere.

Carbon capture and storage (CCS) have been identified for reducing the amount of $CO_2$ being emitted into the atmosphere whilst ensuring the continued use of fossil fuels for energy production. CCS comprises of 3 stages; capture transportation and storage. Research shows that capture is the most expensive stage and as such capture technologies have been investigated and developed over the past decade [5].

Currently, the most common and mature process for $CO_2$ capture is amine based absorption. However, this process has several drawbacks such as the amine solvents used are corrosive, the process is energy intensive as large amount of energy are required to regenerate the amine after absorption [6]. This calls for an alternative $CO_2$ capture technique. Adsorption using solid adsorbents is a promising alternative because it utilises less energy and it is more environmentally friendly as compared to absorption [7]. Adsorption can either be physical or chemical depending on the type of bond between the adsorbent and adsorbate. For physical adsorption, there are weak van der Waal forces while in chemical adsorption there are strong chemical forces between adsorbent and adsorbate [8]. Properties of adsorbents are characterised based on their surface area, pore volume, cost, selectivity and energy of regeneration [8].

Adsorption of $CO_2$ using solid adsorbents such as activated carbon, zeolites, silica materials, and amine based adsorbents and metal organic frameworks (MOFs) have been actively investigated. MOFs are highly crystalline materials that have metal ions as centres and organic materials as linkers [9]. MOFs are advantageous over other adsorbents because of their large surface areas, high porosity and open metal sites. MOFs also require less energy for regeneration after adsorption is complete [9]. There are different types of MOFs depending on choice of building blocks (metal ion, organic linker and structure directing agent). Chen *et al* [5] investigated a type of MOF known as sod-ZMOF for $CO_2$ capture and found the $CO_2$ adsorption capacity of this material to be 53 mg $CO_2$/ g adsorbent. The material displayed a relatively high $CO_2$ adsorption capacity as compared to other types of MOFs. Additionally, properties of MOFs can be enhanced by incorporating it with other materials through grafting, impregnation and ion exchange to improve its $CO_2$ adsorption capacity [10]. Chen *et al* [11], grafted ethylenediamine (EDA) on sod-ZMOF to enhance its adsorption capacity and EDA grafted sod-ZMOF gave a 30% increase in adsorption capacity as compared to the ordinary sod-ZMOF [10]. In this study, chitosan impregnated sod-ZMOF (Sod-ZMOF-chitosan) was synthesized and evaluated for $CO_2$ capture.

Chitosan is a biocompatible, biodegradable and non-toxic polymer derived from chitin [12]. Chitin is a natural resource produced from crabs, insects and shrimps. Chitosan is a green material and is the second largest most abundant polymer on earth [12]. Properties of chitosan include high viscosity, high solubility in various media, ability to form films and capability to bind with various metal ions [12]. Chitosan being a biodegradable organic material with several anime groups depending on its degree of deacetylation can be a green material that could be used to substitute synthetic liquid amine chemical. Also, the final product; sod-ZMOF-chitosan would have very little environmental effect.

## 2. EXPERIMENTAL

### 2.1.Materials

The chemicals used in the study: 4, 5 Imidazoledicarboxylic acid, imidazole, indium (III) nitrate hydrate, hydrochloric acid, N, N-dimethylformamide ,sodium hydroxide (DMF), nitric acid, acetonitrile and acetic acid were all purchased from Sigma Aldrich South Africa. The chitin was obtained from the shell of crabs (usually a waste material from restaurants). The gas mixture, 15%-85% $CO_2$-$N_2$, 100% $CO_2$ and nitrogen baseline were obtained from Afrox, South Africa.

### 2.2.Synthesis of sod-ZMOF and sod-ZMOF-chitosan

#### 2.2.1. Synthesis of Sod-ZMOF

The method used to synthesize sod-ZMOF in this study was adapted from the work reported elsewhere [5]. About 0.84 g 4.5-imidazoledicarboxylic acid, 0.6 g indium (III) nitrate hydrate ($In(NO_3)_3.2H_2O$), 60 mL N, N-dimethylformamide (DMF), 20 mL acetonitrile ($CH_3CN$), 0.82 g imidazole in 8 mL DMF and 1.8 mL nitric acid ($HNO_3$) in 12 mL DMF were added into a 100 mL in a 100 mL flask. The solution was mixed thoroughly until a clear solution was obtained. The solution was heated up to 85°C under reflux for 12 h. after 12 h, the temperature was increased to 105°C and continued heating for 23 hours. Polyhedral crystals formed where collected by filtration; washed with methanol to remove any DMF from the pores and surface and then dried in an oven at 100 °C for 2 h. The dried poly crystals obtained were sod-ZMOF.

#### 2.2.2. Preparation of chitosan

The crabs flesh was removed, the exoskeletons thoroughly washed and sundry. The dried exoskeletons were then chopped into smaller sizes, pulverized to a particle size of < 74 nm. For preparation of chitosan, chitin was sieved and subjected to demineralization, deproteinization and decolonization.

Chitin was added to 10 % (v/v) hydrochloric acid (HCl) with a solvent to solid ratio of 1:10. The reaction was carried out at 65 °C for 2 h with continuous stirring, after which the obtained material was filtered and washed with distilled water until filtrate pH was neutral. The resulting sample was dried in an oven at 50 °C for 24 h.

The demineralized chitin was added to 4 % (w/v) sodium hydroxide (NaOH) with a solvent to solid ratio of 1:10 and heated at 65 °C for two h with continuous stirring. The resultant product was then filtered and washed with distilled water to a neutral pH. The sample was dried in an oven at 50 °C for 24 h.

About 50 g of Deproteinized chitin was added to 50% (w/v) of sodium hydroxide (NaOH) at 90 °C for 5 h with continuous stirring. The resultant product was filtered, washed with distilled water to a neutral pH and dried in an oven at 50 °C for 24 h.

### 2.2.3. Impregnation of chitosan to sod-ZMOF

The procedure used for impregnation of chitosan was adapted from the work reported elsewhere [13]. Chitosan was dissolved in 2 wt. % acetic acid and stirred at 80 °C for 1 h to obtain 2 wt. % chitosan solution. 1 g of sod-ZMOF was added into the solution under stirring for 1 h. The suspension mixture was filtered and the obtained product dried in an oven at 50 °C for about 2 h. Characterization

X-ray diffraction (XRD) was done in order to determine the crystallinity of synthesized adsorbent. The XRD patterns in this study were obtained using a cobalt (CoK∝) radiation with wavelength ($\lambda$) of 1.79$\dot{A}$. Thermo gravimetric analysis (TGA) characterization was done in order to determine the thermal stability of the sod-ZMOF. Fourier Transform Infrared Spectroscopy (FTIR) was done to check the surface chemistry of the synthesized adsorbents. $N_2$ physisorption at 77 K was carried out to determine the BET surface area, pore volume and the pore size of the synthesized adsorbent.

### 2.2.4. Evaluation of adsorbent for $CO_2$ capture

Adsorption set-up illustrated in Figure 1 was used to evaluate the performance of the synthesized adsorbents. The set-up consists of a packed-bed column connected to a $CO_2$ detector in order to read the $CO_2$ concentration before, during and after adsorption. About 0.1 g of adsorbent was packed in the column. The sample in the column was then heated from room temperature to 110°C under the continuous flow of nitrogen through the sample and held for another 30 minutes at 110°C. This was done in order to remove impurities such as moisture from the adsorbent. Then the temperature was dropped to the 30 °C and gas mixture of about 15% $CO_2$ (balance being nitrogen) at a flowrate of 2.5 ml/s and a pressure of 2 bar was passed through the sample. This was done for about 15 minutes. A gas mixture containing 15% $CO_2$ (balance nitrogen) was used to mimic the industrial flue gas, which contains about 15% $CO_2$ concentration. Adsorption using 100% $CO_2$ was also investigated for sod-ZMOF-chitosan in order to compare the adsorption behaviour of the adsorbent with pure $CO_2$ to understand the effect of $N_2$ on the adsorption capacity of the material.

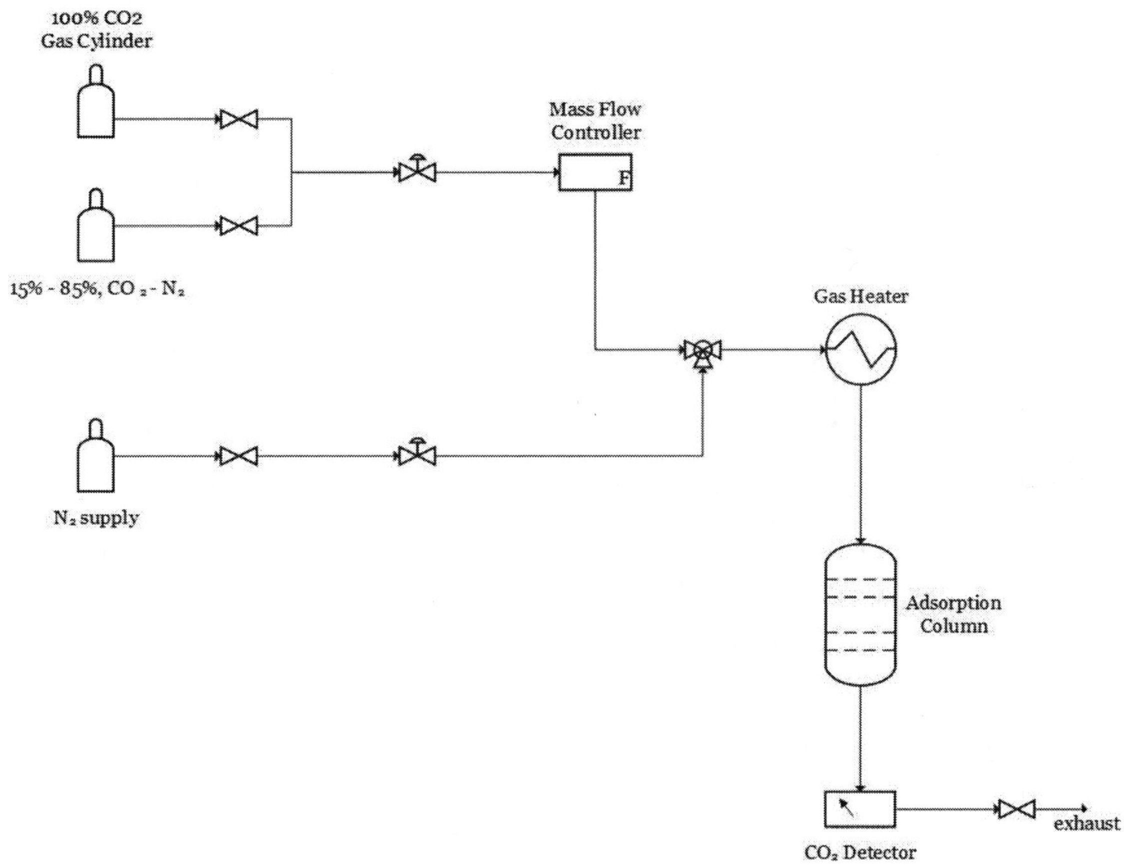


Figure 1: Experimental set-up for the CO2 adsorption experiment

## 3. RESULTS AND DISCUSSION

### 3.1. Characterization

First the synthesized sod-ZMOF characterisation results were compared to that obtained from literature [5]. This was done to confirm if the synthesized sod-ZMOF is similar to that reported in literature. Figure 2 is an XRD pattern of sod-ZMOF. XRD characterization was done in order to determine if sod-ZMOF is crystalline or amorphous. The sharpness of the XRD pattern is important in identifying the crystallinity of sod-ZMOF.

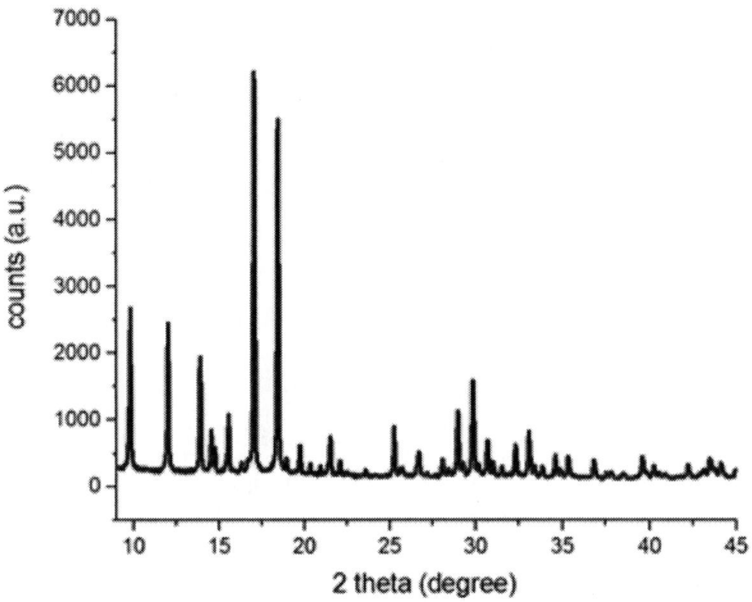


Figure 2: XRD pattern of sod-ZMOF

From literature XRD patterns were obtained using a copper (CuK∝) radiation with a wavelength of ($\lambda$) of 1.54$\dot{A}$. In order to compare experimental results to that from literature, normalization of the XRD result obtained in this study was done. At a 2-theta of between 15 and 20, The XRD pattern showed two significant sharp peaks. The sharp peaks on the XRD patterns in Figure 2 show that sod-ZMOF is a highly crystalline material. Figure 3 shows the measurements of change in weight of sod-ZMOF as a function of temperature. These measurements are used to determine thermal stability of sod-ZMOF at different temperatures.

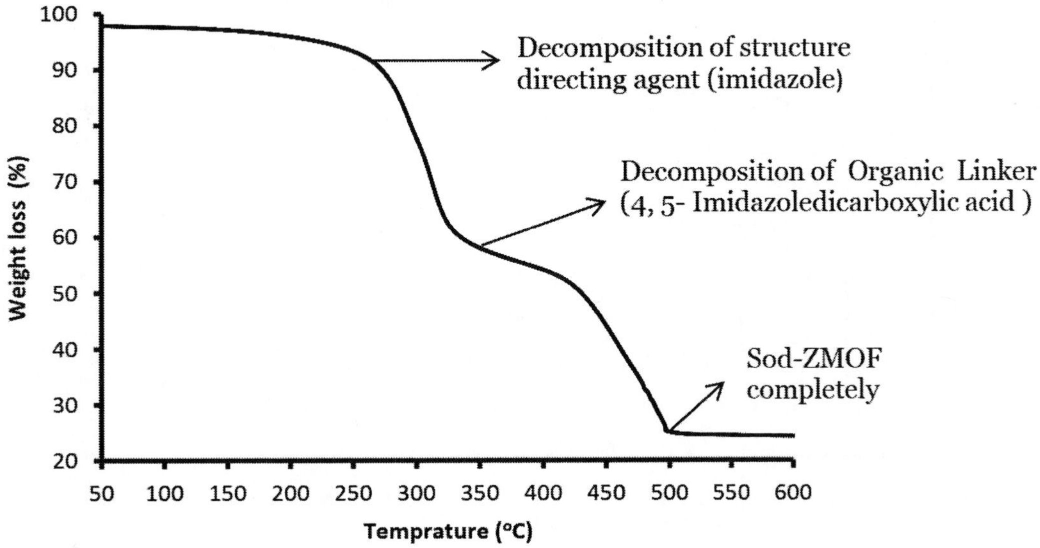


Figure 3: TGA analysis of-sod-ZMOF

The TGA results show that the thermal degradation of sod-ZMOF began at around 250 °C, although there was a small change in weight before this temperature attributable to moisture and the structure directing agent (SDA). According to [14, 15], loss in weight at this temperature is as a result of loss of structure directing agent (imidazole). The next major weight loss occurs at about 350°C, this is due to the decomposition of organic linker (4.5-imidazoledicarboxylic acid) [14, 15]. At a temperature of about 500 °C, the structure of sod-ZMOF has completely decomposed. The same results were obtained and reported elsewhere [5].

Figure 4 represent a diagram of FTIR spectra of the synthesized sod-ZMOF, chitosan and sod-ZMOF-chitosan.

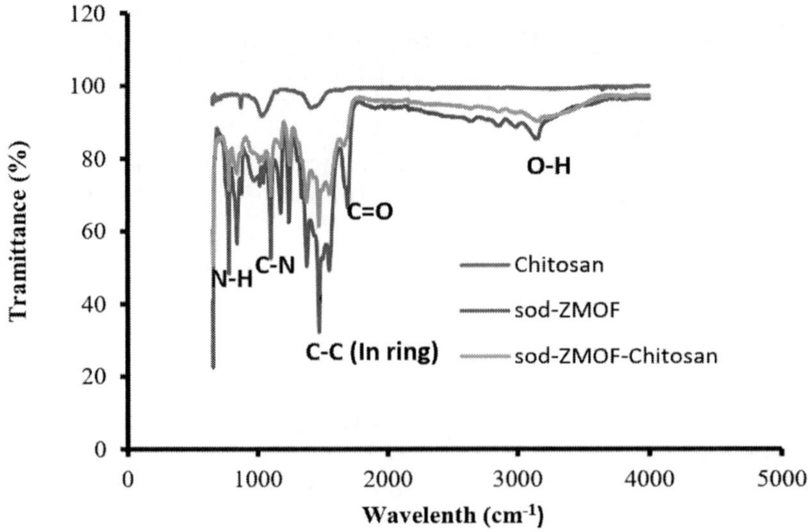


Figure 4: FTIR spectra of synthesized samples

The organic linker in sod-ZMOF is 4.5-Imidazoledicarboxylic acid and as such it is expected that sod-ZMOF and this organic linker have similar functional groups. The functional groups present in 4.5-Imidazoledicarboxylic acid are amine (N-H), carboxylic acid (O-H), aromatics (C-C (in ring)), aromatic amine (CN) and (C=O). The FTIR analysis of sod-ZMOF in Figure 4 shows the presence of (O-H, C-C (in ring), C-N and C=O, which therefore confirms the successful synthesis of sod-ZMOF. Figure 4 also shows FTIR spectrum for sod-ZMOF-chitosan. The FTIR spectrum of sod-ZMOF-chitosan showed functional groups similar to those found in sod-ZMOF. However, there is a drastic decrease of the O-H (carboxylic acid) band in the IR spectrum for sod-ZMOF-chitosan as compared to the ordinary sod-ZMOF. When Chen *et al.* [11] grafted sod-ZMOF with ethylenediamine (EDA), the FTIR also showed a decrease in O-H band after grafting. Chen *et al* [11] speculated that the decrease in O-H band could be as a result of carboxylic acids being transformed to amide groups. In this work, the decrease in the O-H band could also be due to the impregnation of chitosan which then confirms the presence of chitosan (amine group) in sod-ZMOF chitosan.

The Nitrogen physi-sorption at 77 K was conducted on the synthesised samples in order to determine the pore size, pore volume and BET surface area. Table 1 shows a summary of the results obtained in this study. From work done by Chen *et al.* [5], sod-ZMOF has a pore volume of 0.16 $cm^3/g$ and a BET surface area of 375 $cm^2/g$. In this study, sod-ZMOF had a pore volume and BET surface which were much less than that from the work done by Chen *et al.* [5]. The low pore volume and the BET surface area could be attributed to the presence of material such as Structure Directing Agent (SDA) occluded in the structure. As this study is preliminary, effort will be made to remove the material from the structure in the subsequent optimization study of the material. In addition, impregnation of chitosan on sod-ZMOF resulted in a decrease in surface area and pore volume. The decrease in surface area and pore size is expected as chitosan impregnated occupies the surface area and pore volume in sod-ZMOF. The decrease in surface area is a significant indication of the successful impregnation of chitosan. Chen et al. [11] also observed a decrease in surface area and pore volume when they grafted amine in sod-ZMOF.

Table 1: BET surface area, pore volume and pore size of sod-ZMOF and sod-ZMOF-chitosan samples

| **Sample** | **Pore Size (nm)** | **Pore Volume ($cm^3/g$)** | **Surface area ($cm^2/g$)** |
|---|---|---|---|
| Sod-ZMOF | 3.40 | 0.0445 | 52.4 |
| Chitosan | 27.0 | 0.119 | 17.6 |
| Sod-ZMOF–chitosan | 201 | 0.0265 | 0.527 |

### 3.2. Performance evaluation

$CO_2$ adsorption capacity of sod-ZMOF, chitosan and sod-ZMOF-chitosan was determined using the adsorption set-up illustrated in Figure 1. The performance evaluation of sod-ZMOF, chitosan and sod-ZMOF-chitosan was determined using a gas mixture of $CO_2$ and $N_2$ containing 15% and 85% of $CO_2$ and $N_2$, respectively. Operating conditions used were a pressure of 2 bar, inlet gas mixture flowrate of 2 ml/s and a temperature of 30°C and adsorption time of 15 minute. At these conditions, sod-ZMOF-chitosan, chitosan and sod-ZMOF- displayed a $CO_2$ adsorption capacity of 3035 mg $CO_2$/ g adsorbent, 2720 mg $CO_2$/ g adsorbent and 2720 mg $CO_2$/ g adsorbent, respectively (see Figure 5). Chitosan impregnated sod-ZMOF displayed about 12% increase in adsorption capacity. The increase in the adsorption capacity could be attributed to the role played by impregnated chitosan in the sod-ZMOF. The interaction between acidic $CO_2$ and basic impregnated chitosan facilitates $CO_2$ adsorption through formation of covalent bonds [12].

Performance evaluation of sod-ZMOF-chitosan was also investigated using 100% $CO_2$ at the same operating conditions (30°C, 2 bar and inlet flowrate of 2.5 ml/s). The $CO_2$ adsorption capacity obtained using 100% $CO_2$ feed gas after an adsorption time of 83 seconds was 2900 mg $CO_2$/ g adsorbent. At the same operating conditions, the $CO_2$ adsorption capacity obtained using 15% $CO_2$ feed gas was 233 mg $CO_2$/ g adsorbent. This is 8% of the adsorption capacity obtained when using 100% $CO_2$ feed gas (see Figure 6). This decrease in $CO_2$ adsorption capacity when gas mixture was used could be attributed to the competition between $CO_2$ and $N_2$ for the available pore volume of the material. One of the significant chemical properties of chitosan is that it contains reactive amino groups [16]. Amino groups are known to react immediately with $CO_2$ to form carbamates complexes [11]. Thus the impregnation of chitosan in sod-ZMOF resulted in an increase in $CO_2$ adsorption capacity.

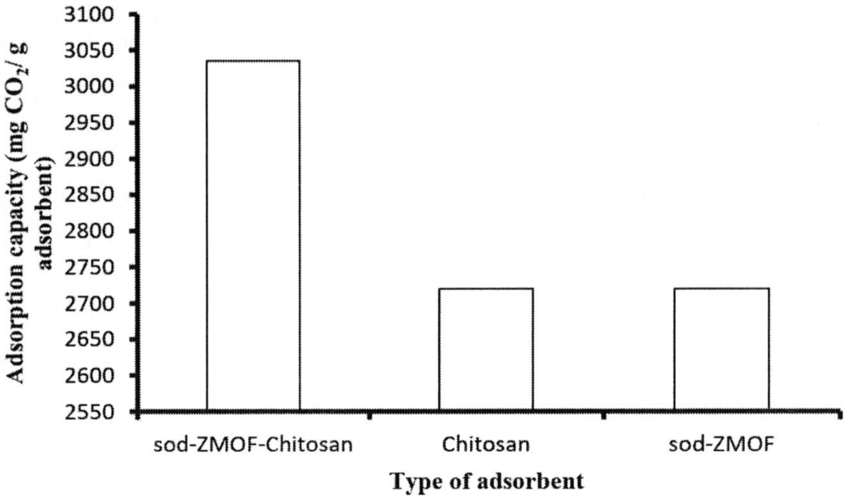


Figure 5: $CO_2$ adsorption capacity of the synthesised adsorbents using gas mixture (15% $CO_2$ and 85% $N_2$)

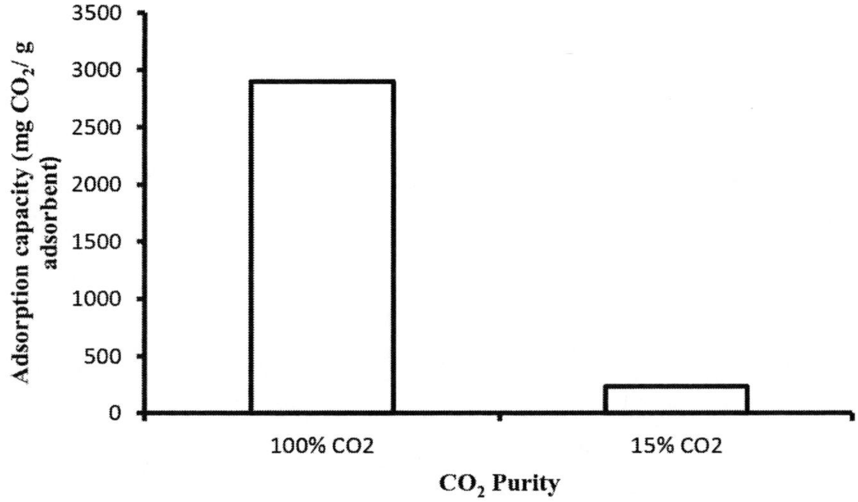


Figure 6: $CO_2$ adsorption capacity of sod-ZMOF using pure $CO_2$ (100% $CO_2$) and gas mixture (15% $CO_2$ and 85% $N_2$)

Effect of temperature on the $CO_2$ adsorption capacity of Sod-ZMOF-chitosan was further investigated by varying the temperature while keeping other parameters constant. $CO_2$ adsorption was investigated at 30°C, 45°C and 60°C . Figure 7 illustrates the adsorption kinetics of sod-ZMOF-chitosan at the 3 temperatures. $CO_2$ adsorption capacity is highest at a temperature of 30°C and lowest at a temperature of 60°C. This is because adsorption is an exothermic process; thus increasing the temperature decreases the adsorption capacity [17]. At higher temperature $CO_2$ molecules also have higher kinetic energy, hence increasing temperature decreases $CO_2$ adsorption capacity.

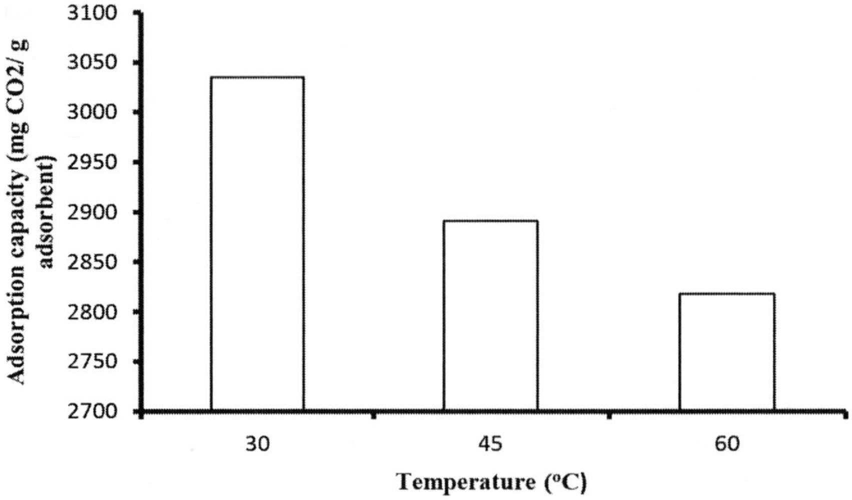


Figure 7: Sod-ZMOF-chitosan adsorption capacity at 3 different temperatures

Results of this study were compared to literature is presented in Table 2. In literature, several researchers commonly use TGA to investigate $CO_2$ adsorption capacity of their adsorbents. In this study a packed bed was used with the advantage that there is direct contact between the gas and the adsorbent as compared to the TGA. This is because with a packed-bed column, the gas flows through the adsorbent while with the TGA; the gas just flows on top of the gas and there is no direct contact. In addition, a packed bed can be up-scaled for the industrial applications. Various materials such as zeolites, activated carbon, silica materials and other types of MOFs have been investigated for $CO_2$ capture. These materials are mostly investigated using pure $CO_2$ and at high pressures and temperatures [18]. Sod-ZMOF synthesized in this work and other adsorbents reported in literature are compared in Table 2.

Table 2: Results of this study compared with literature

| Type of Adsorbent | Adsorption Capacity (mg $CO_2$/ g adsorbent) | Adsorption equipment | Adsorption conditions | Reference |
|---|---|---|---|---|
| Sod-ZMOF | 2720 | Packed bed | 15% $CO_2$, 30°C, 2 bar, 2.5 ml/s | This work |
| Sod-ZMOF | 20.14 | TGA | 15% $CO_2$, 25°C, 1 bar | [11] |
| Sod-ZMOF-chitosan | 3035 | Packed bed | 15% $CO_2$, 30°C, 2 bar, 2.5 ml/s | This work |
| ED-sod-ZMOF | 33.12 | TGA | 15% $CO_2$, 25°C, 1 bar | [11] |
| PAA | 44.2 | TGA | 100% $CO_2$, 40 °C, 1.1 bar, 1 ml/s | [7] |
| ZIF | 5.4 | TGA | 15% $CO_2$, 25°C, 1 bar | [11] |

## 4. CONCLUSIONS

Sod-ZMOF was synthesized and characterized. Characterization using XRD, FTIR and TGA revealed a synthesized structure consistent with literature. BET results however differed from those in literature. Impregnation of chitosan to sod-ZMOF showed an increase in adsorption capacity of 12% when adsorption was undertaken for 15 minutes. The time for adsorption should be increased so as to allow for equilibrium which is the current effort of the investigation. Further study will optimize the synthesis protocol of the material and also investigate effect of operating conditions in detail.


## 5. ACKNOWLEDGEMENTS

The authors acknowledge the financial assistance from the National Research Foundation (NRF) of South Africa, the University of the Witwatersrand (WITS) and ABB for the gas detector utilized in this research.

# 6. REFERENCES


1. Nyajowa, W. ., 2014. South Africa's coal demands and trends. (SHAM MEDIA corp) Accessed October 8, 2015, http://www.sham-media.net/article.php?45
2. Lackner, K., 2009. Carbon Capture: sequestration and storage.
3. International energy agency (IEA). 2015. $CO_2$ emissions from fuel combustion highlights: Accessed July 29, 2016, https://www.iea.org/publications/freepublications/publication/CO2EmissionsFromFuelCombustionHighlights2015.pdf.
4. Global monitoring division, 2016. Global greenhouse gas reference network, Trends in atmospheric carbon dioxide. Accessed July 25, 2016, http://www.esrl.noaa.gov/gmd/ccgg/trends/.
5. Chen, C., Kim, J., Yang, D. & Ahn, W., 2011. Carbon dioxide adsorption over zeolite-like metal organic frameworks (ZMOFs) having a sod topology: structure and ion exchange effect. Chemical Engineering Journal., pp. 1134-1139
6. Cummings, P., & Schneider, S. (2010). Liquids-Based Absorption. Report of carbon capture: beyond 2020, pp 9-16
7. Chitsiga, T.C., Daramola, M.O., Wagner, N., Ngoy, J., 2016 Effect of the presence of water-soluble amines on the carbon dioxide ($CO_2$) adsorption capacity of amine-grafted polysuccinimide (PSI) adsorbent during $CO_2$ capture. Energy Procedia, pp. 90-105.
8. Pires, J., Martins, F., Aliuim-Ferraz, M. & Simoes, M., 2011. Recent developments on carbon capture and storage: An overview. Chemical Engineering Research and Design, pp. 1446-1460.
9. Rowsell, J. & Yaghi, O., 2004. Metal-organic frameworks: a new class of porous materials. Microporous and Mesoporous Materials, pp. 3-14.
10. Imteaz, A. & Sung, H., 2014. Composite of metal organic framework: Preparation and Application in adsorption. Materials today, pp. 136-146.
11. Chen, C., Kim, J., Park, D. & Ahn, W., 2013. Ethylenediamine grafting on zeolite like metal organic frameworks (ZMOF) for $CO_2$ capture. Materials letters, pp. 344-347.
12. Shukla, S., Mishra, A., Arotiba, O. & Mamba, B., 2013. Chitosan-based nanomaterials: A state-of the-art review. International Journal of Biological Macromolecules, pp. 46-58.

13. Sun, H., Lu, L., Chen, X. & Jiang, Z., 2008. Surface modified zeolite-filled chitosan membranes for pervaporation dehydration of ethanol. Applied Surface Science, pp. 5367–5374.
14. Calleja, G., Botas, JA., Sa´nchez-Sa´nchez, M., Orcajo, MG., 2010. Hydrogen adsorption over Zeolite-like MOF materials modified by ion exchange.International Journal of Hydrogen Energy, pp 9916-9923
15. Ayşe, K., Çiğdem, A., Ahmet, S., Birgül, T., 2015. Sod-ZMOF/Matrimids mixed matrix membranes for $CO_2$ separation. Journal of Membrane Science , pp 81-89
16. Dutta, PK., Dutta, J., Tripathi, VS., 2004. Chitin and chitosan: Chemistry, Properties and applications. Journal of Scientific and Industrial Research, pp 20-31
17. Yu, C., Huang, C ., Tan, C., 2012. A Review of CO2 Capture by Absorption and Adsorption. Aerosol and Air Quality Research, pp 754-769
18. Lee, SY., Park, SJ. 2014 A review on solid adsorbents for carbon dioxide capture. Journal of Industrial and Engineering Chemistry, pp 1-11

# Nanocomposite Sodalite/Ceramic Membrane for Pre-combustion $CO_2$ Capture: Single Gas Permeation Measurement

*Olawale Oloye, Michael O. Daramola**

*School of Chemical and Metallurgical Engineering, Faculty of Engineering and the Built Environment, University of the Witwatersrand, Wits 2050, Johannesburg, South Africa*

**Corresponding author (e-mail: michael.daramola@wits.ac.za)*

## Abstract

The continuous anthropogenic carbon dioxide ($CO_2$) emission into the atmosphere, causing detrimental effect like global climate change, is receiving substantial attention worldwide. Solving the problem of climate change through the reduction of $CO_2$ emission constitutes a technological and scientific challenge. However, numerous technologies (absorption, adsorption and membrane technology) for $CO_2$ capture from power plants have been proposed and evaluated. Membrane technology has gained attraction among researchers and remains an active topic. In this study, nanocomposite hydroxy sodalite membrane was prepared on α-alumina support by pore plugging hydrothermal synthesis technique. The morphology and the quality of the synthesized membrane were checked with a Scanning Electron Microscope (SEM) and Basic Desorption Quality Test (BDQT), respectively. Single gas permeation test using pure component of $H_2$ and $CO_2$ was conducted on the membrane to evaluate its $H_2/CO_2$ separation performance. The single gas permeation was carried out at 298 K and at transmembrane pressure of 25 kPa. The gas permeation measurement shows that the membrane displayed $H_2$ permeance and $CO_2$ permeance of $3.9 \times 10^{-7}$ mols$^{-1}$m$^{-2}$Pa$^{-1}$ and $8.38 \times 10^{-8}$ mols$^{-1}$m$^{-2}$Pa$^{-1}$, respectively. The $H_2/CO_2$ ideal selectivity at this condition was 4.63.

***Keywords**:* Nanocomposite, Sodalite, Membrane, Gas permeation, Carbon Capture and Storage.

## 1 Introduction

Carbon footprint predominantly from combustion of carbonaceous fuels (example; natural gas and coal), which is consequential of energy generation and industrial developments, has caused environmental and climatic changes(*IEA*, 2015). Greenhouse gases (GHGs) in the atmosphere has been increasing geometrically since the advent of the industrial age. This increase, is in direct relationship with urbanization and continuous increase in energy demands(*IEA*, 2015). Several approaches are being developed to cushion the effect of the GHGs emission. Carbon capture and storage (CCS), amongst all other options, is identified to have great potentials to alleviate the aforementioned challenges. CCS, also known as carbon sequestration, which allows the continuous combustion of carbonaceous fuels (fossil fuel) with the option of capturing GHG components prior to final or after combustion(Pires et al., 2011). Three (3) predominant CCS approaches that are being employed today are; post-combustion, pre-combustion and oxy-combustion.

In pre-combustion $CO_2$ capture, the carbonaceous fuel (coal) is either gasified, partially oxidized or reformed in the presence of oxygen or air and sometimes steam to give a mixture of gases, mainly of CO and $H_2$. During Water Gas Shift reaction over a catalytic reactor, CO reacts with steam to delivers $CO_2$ and supplementary $H_2$ (Olajire, 2010). The $H_2/CO_2$ mixture could be separated and $H_2$ serves as the fuel source in a gas turbine combined-cycle plant for power generation (Daramola et al., 2016; Olajire, 2010). Moreover, if coal is used, this process is referred to as Integrated Gasification of Combined Coal Cycle (IGCC)(Barbieri et al., 2011; Olajire, 2010). The $H_2$ gas could be separated from the gas mixture using a membrane system.

Several membrane materials are being developed for pre-combustion $CO_2$ capture (microporous amorphous silica, doped silica, carbon molecular sieve (CMS) and zeolites). Sodalite (zeolite) membranes have great potential in such gas separation application owing to its molecular sieving ability for the target molecule ($H_2$) and thermal stability. Sodalite is a crystalline zeolite made of cubic array of β-cages as primary building block. It has a cage aperture in the range of 0.26 and 0.29 nm making it a potential candidate for the separation of small molecules (for example hydrogen (0.27 nm)) under certain process conditions. In addition, previous studies by Daramola et al. (2016, 2012) indicated that membrane technology offers high selectivity and high driving force with simple modular design. Some other reports made comparison of conventional separation process (absorption, adsorption and distillation) with membrane technology(Franz et al., 2014; Daramola et al., 2012; Merkel et al., 2010).They provide insight of cost and energy efficiency of membrane system over conventional techniques. Dense palladium-based membranes, which are mainly hydrogen selective, suffer from poor performance due to plasticization (sorption of condensable gases), fouling and chemical interactions(Al-Mufachi et al., 2015). Moreover, polymeric membranes have been extensively studied and are prone to degradation at elevated temperatures and harsh conditions (Daramola et al., 2012). Zeolite membranes overcome some of the aforementioned challenges. Zeolite membranes are relatively thermally stable and possess good chemical inertness. In spite of this, reproducibility of high quality zeolite membranes and high cost of membrane substrate (support) could hamper the development and application of zeolite membrane in pre-combustion carbon capture (Daramola et al., 2012; Domínguez-Domínguez et al., 2008). However, pore plugging hydrothermal (PPH) synthesis has been reported to enhance reproducibility and this has been validated by Miachon et al. (2006) for nanocomposite MFI-alumina membrane. Also, Daramola et al.(2016, 2015) reported the synthesis and morphological characterization of nanocomposite sodalite/ceramic membrane with higher degree of reproducibility.

To date, however, no studies have been reported on gas permeation for nanocomposite sodalite/ ceramic membranes. In this study, further investigations were conducted to evaluate the quality and separation performance of synthesized nanocomposite sodalite/ceramic membranes.

## 2 Experimental section

### 2.1 Materials

The supports for the zeolite membranes were asymmetrical $\alpha$-$Al_2O_3$ tubes (inner diameter / outer diameter = 7 mm / 10 mm, length 150 mm), with 10 mm enameled at both ends. Only 130 mm length of the supports were available for gas permeation with active area of 2.6 x $10^{-3}$ $m^2$. The cross-sectional layers of the substrate were of dimensions: outer layer, 1200 nm; intermediate layer, 800 nm; inner layer, 200 nm. Chemicals used for membrane synthesis, anhydrous sodium metasilicate ($Na_2SiO_3$), sodium hydroxide (NaOH) and anhydrous sodium aluminate ($NaAlO_2$), were purchased from Sigma Aldrich (Pty) SA. The de-ionized water was prepared in-house and all the chemicals were used without further purification. Graphite O-rings (for sealing of the membrane and the membrane module) were provided by BMG Selby (Pty) Ltd SA. Pure single gas components (hydrogen and carbon dioxide) of purity > 99.9 % were supplied by Afrox SA.

## 2.2. Sodalite membrane synthesis and characterization

Prior to membrane synthesis, 3 supports (substrates) used for synthesis were pretreated to remove dirt and loose particles by boiling them in deionized water for about 1 hour and subsequently dried in an oven at 100 °C for 1 hour. The hydroxy sodalite membranes were synthesized via pore-plugging hydrothermal synthesis techniques using a precursor solution of molar composition ($5SiO_2$:$Al_2O_3$:$50Na_2O$:$1005H_2O$) (Daramola et al., 2015). About 250 ml of the homogenous precursor was poured in the Teflon-lined autoclave and the 3 pre-treated selected α-$Al_2O_3$ tubes were totally immersed in the solution. The content was left for about 15 minutes to allow the penetration of precursor into the pores of the substrates and displacement of air from the pores of the supports. Experimental procedures used for the synthesis of the membrane is depicted in Figure 1. The membranes were synthesized by adopting procedures described by (Daramola et al., 2016, 2015; Li et al., 2008) using the temperature profile depicted in Figure 2. After synthesis, the sodalite/ceramic membranes and the sodalite crystals formed at the bottom of the Teflon-lined autoclave were thoroughly washed in de-ionized water until pH of the wash water was about neutral. Afterwards, the membranes and the sodalite crystals were dried in an oven at 100 °C overnight.

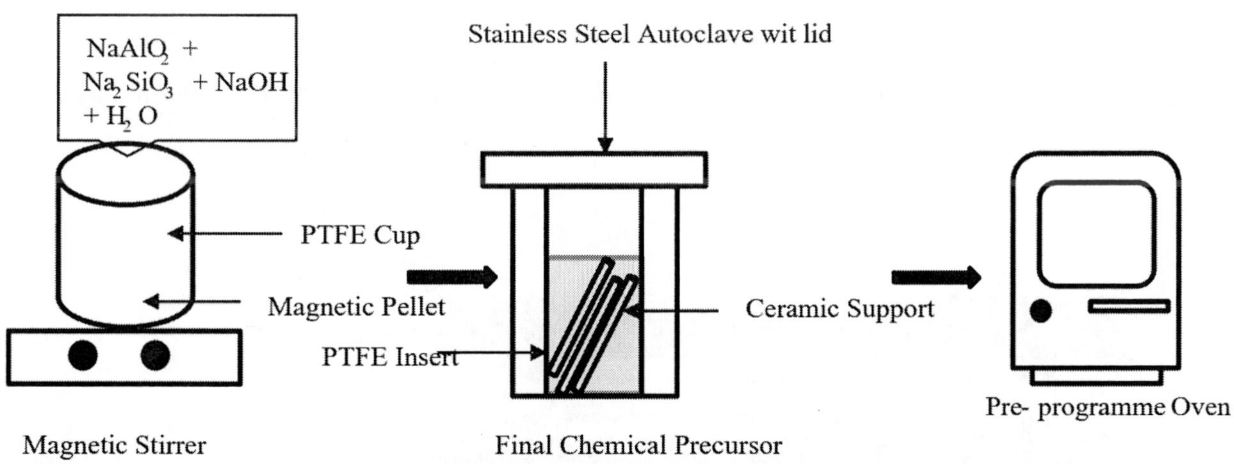


**Fig. 1** Experimental procedures for preparation of sodalite membranes(Daramola et al., 2016, 2015)

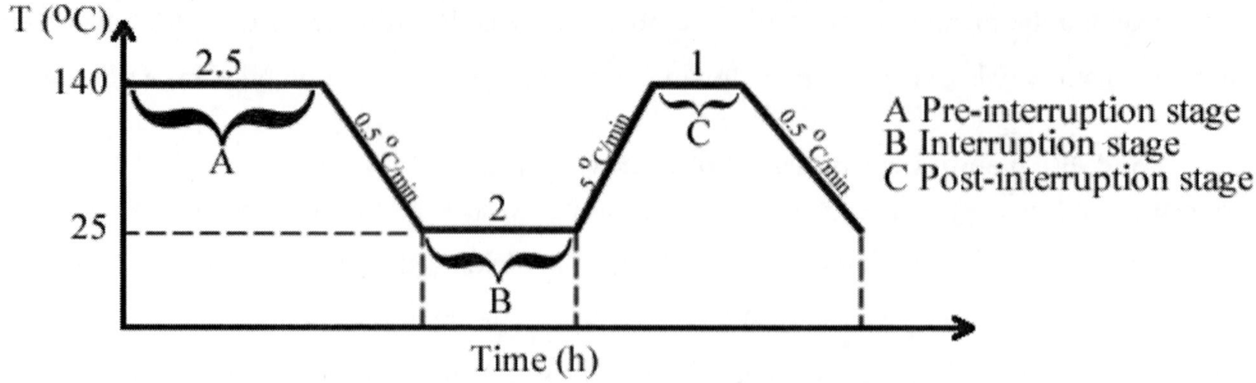


**Fig. 2** Temperature profile adapted for the pore-plugging hydrothermal synthesis

For the static characterization of the synthesized membranes, one of the membranes was cracked and scanning electron microscopy (SEM) was used to check the morphology. The SEM micrographs were obtained using Phillips XL 20 equipment. The morphology of the sodalite crystals obtained from the bottom of the Teflon-lined autoclave was checked also to confirm the crystals in the membranes.

### 2.3. Membrane quality

The quality of membranes was checked using the Basic Desorption Quality Test (BDQT)(Pachtova, 2003). Prior to the BDQT, the membranes were subjected to thermal treatment at temperature of 100 °C for time duration of 1.5 hours to remove moistures that might have been absorbed within the membrane pores. Then, the membrane was allowed to cool to ambient temperature before n-butane saturation. Figure 3 shows the membrane module employed for thermal treatment, BDQT and gas permeation experiments. Graphite cylindrical rings were used as seals between membrane and at both ends of the module to prevent gas leakage and ensure that the module was gas-tight.

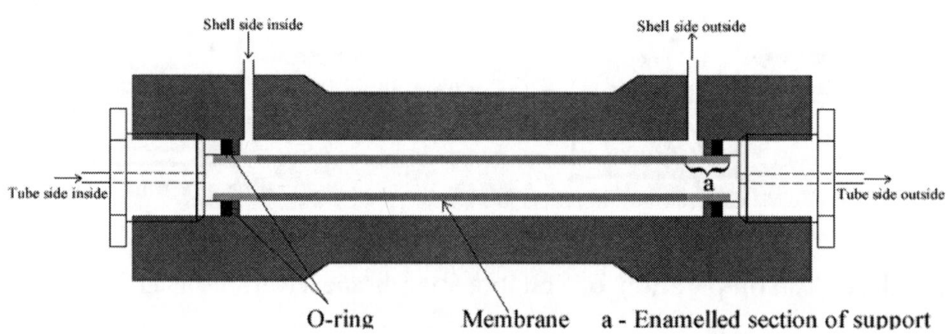


**Fig. 3** Schematic of the membrane module employed for BDQT & gas permeation experiments.

The quality of the membranes was evaluated using BDQT procedure reported by (Pachtova, 2003). Butane is known to condense into the zeolite pores, hence blocks permeation of non-condensable gases (e.g. $N_2$, $H_2$) through the pores of the membranes. However in the presence of defect or non-zeolitic pores, non-condensable gases will permeate. The synthesized sodalite/ceramic membranes were saturated with n-butane for a duration of 1.5 hours at ambient temperature and a partial pressure of 14 kPa. Hydrogen was used as the non-condensable gas to open the blocked pores of n-butane at room temperature. The permeance of hydrogen from the membrane was obtained with the use of Equation 1. For the BDQT and single gas permeation measurement of the nanocomposite sodalite/ceramic membranes, the experimental setup depicted in Figure 4 was used.

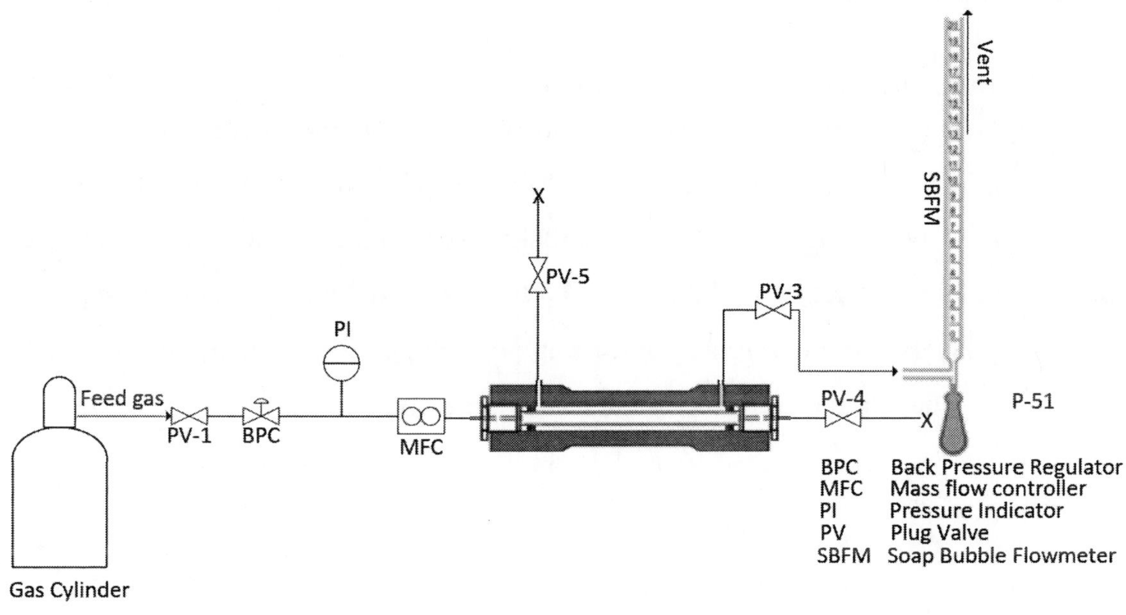


**Fig. 4** Schematic of the experimental set-up used for the BDQT and single gas permeation experiments

## 2.4. Single gas permeation measurement of $H_2$ & $CO_2$

Single gas permeation of $H_2$ and $CO_2$ were measured in dead-end mode. The permeation experiments were carried out at room temperature and at transmembrane pressure of 25 kPa. In order to perform the permeation experiment in dead-end mode, valve 4 and 5 were closed (see Figure 4). The flow rate of the permeated gas was measured using a bubble soap flowmeter. The permeance of the $H_2$ and $CO_2$ through the membrane ($\prod_i$) and the $H_2/CO_2$ ideal selectivity ($S_{H2/CO2}$), were calculated using Equation (1) and Equation (2), respectively:

$$\prod_i = \frac{V_i}{A\Delta P} \tag{1}$$

$$S_{H2/CO2} = \frac{\prod_{H2}}{\prod_{CO2}} \tag{2}$$

$\prod_i$ is the permeance of component $i$, ( $i$, is $H_2$ or $CO_2$); $V_i$ is the volumetric flowrate of the permeated gas from the membrane in mol/s; A is the effective membrane permeation area ($m^2$); $\Delta P$ is the transmembrane pressure (Pa); $\prod_{H2}$ and $\prod_{CO2}$ are the $H_2$ and $CO_2$ permeance in $mol.s^{-1}.m^{-2}.Pa^{-1}$, respectively.

## 3. Results and discussion

### 3.1. Membrane synthesis and characterization

Morphology of the surface and cross-sectional view of the synthesized nanocomposite sodalite/ceramic membranes are depicted in Figure 5 (3 layers) and Figure 6 (inner layer). Evidently, the SEM micrographs show that sodalite were indeed grown within the layers of the support. The 200 nm layer could be seen to be more plugged with sodalite/ceramic membrane than the other layers (see Figure 5). A multi-step pore plugging hydrothermal synthesis might be required to plug the 800 nm and 1200 nm layers completely.

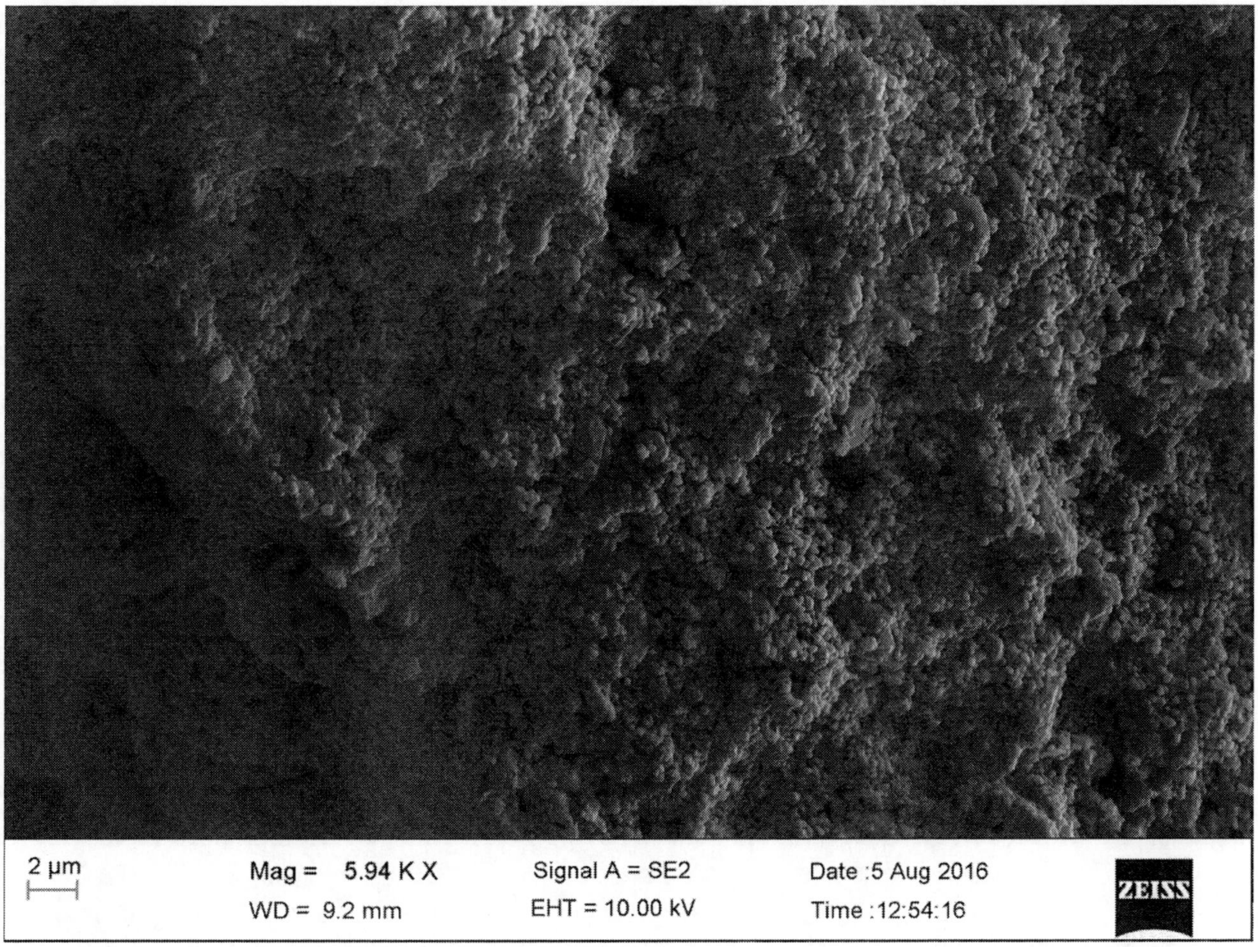


**Fig. 5** Cross section of the sodalite/ceramic membrane showing inner layers of asymmetric support

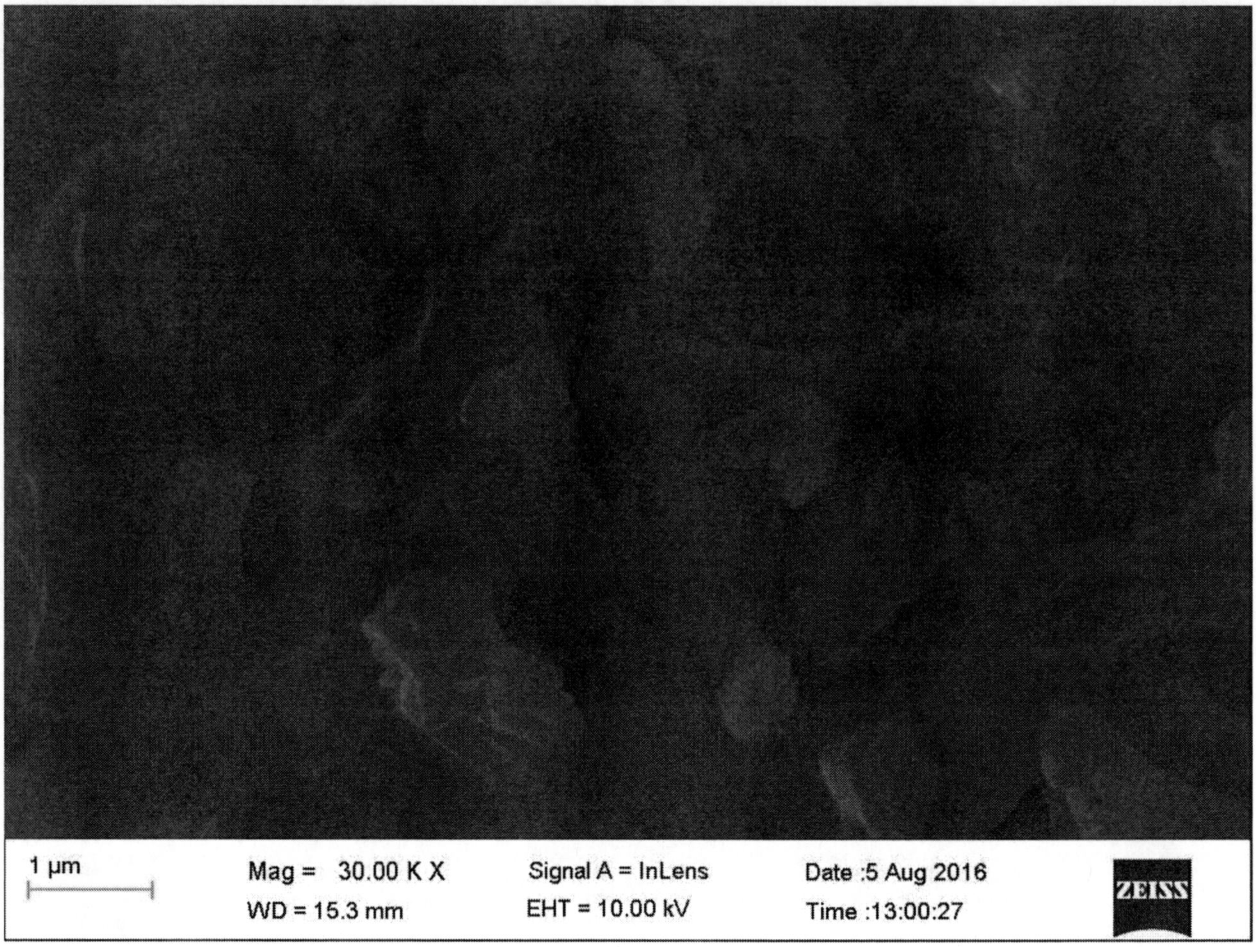


**Fig. 6** SEM micrograph showing inner surface of sodalite/ceramic membrane

### 3.2. Membrane quality

Hydrogen permeance through saturated n-butane sodalite membrane (labelled M1, M2 and M3) were calculated from experimental results, obtained from hydrogen flowrate through the three (3) membranes. The strong absorption of n-butane in the nanocomposite sodalite/ceramic pores blocks the $H_2$ permeation. Hydrogen permeation through the saturated sodalite/ceramic membranes (M2 & M3) was faster in comparison with membrane M1. Rapid hydrogen permeation through nanocomposite sodalite/ceramic membrane M2 & M3 could be due fast desorption of n-butane within the membrane pores (inter-crystalline and intra-crystalline pores). Sodalite membrane M1 shows the least permeance among the three (3) membranes as observed in Figure 7. The porosity of these membranes which corroborate with the permeation of hydrogen through the n-butane saturated nanocomposite sodalite membrane is described in this order of increase; M1 > M3 > M2. However, this indicate that sodalite membrane M1 possess the best quality and it was subsequently used for single gas permeation of $H_2$ and $CO_2$. It is to be noted that results obtained from n-butane desorption using hydrogen was reproducible. The experiment was repeated at least twice for each of the membranes.

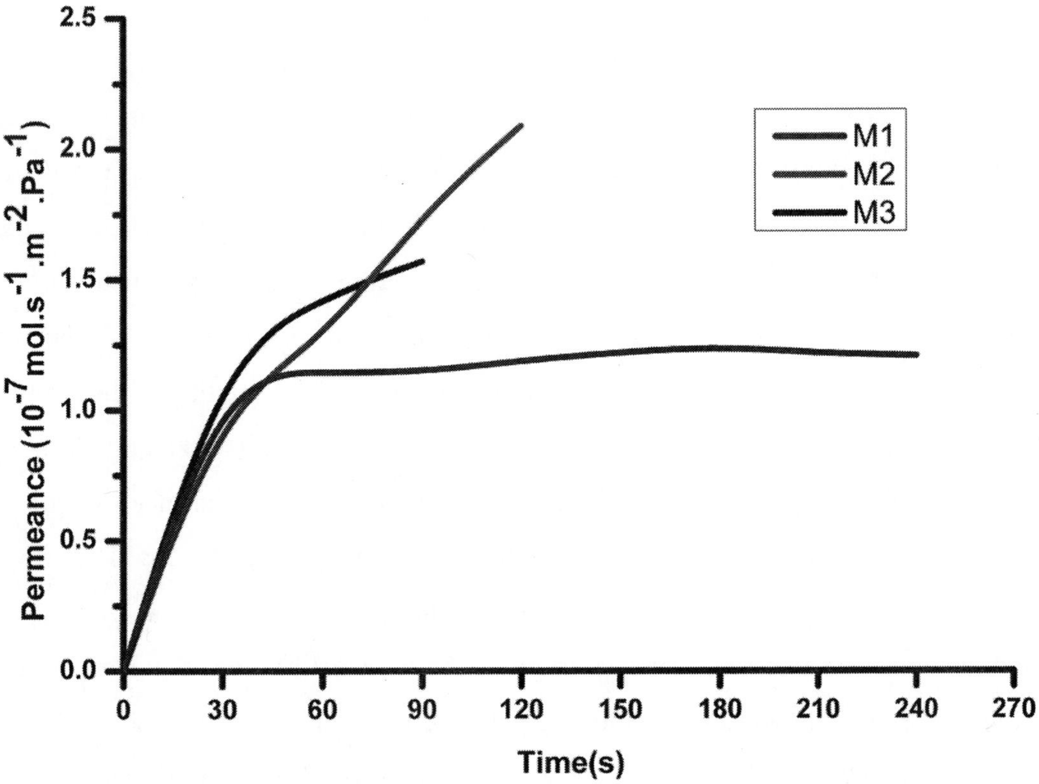


**Fig.** 7 $H_2$ permeance versus time for n-butane desorption of the three sodalite/ceramic membranes (M1, M2 & M3).

### 3.3. Single gas permeation measurement of the as-prepared nanocomposite sodalite/ceramic membrane

Further investigation was conducted in order to evaluate the separation performance of the sodalite/ceramic membrane. Single gases ($H_2$ and $CO_2$) permeation measurements were carried out at ambient temperature and transmembrane pressure of 25 kPa over a permeation period of 170 s. Results obtained indicated that $H_2$ has the highest permeance through nanocomposite sodalite/ceramic membrane, this could be due to its small kinetic diameter of 0.27 nm in comparison with $CO_2$ (0.33 nm). However, the presence of pin-holes or defects or intercrystalline pores (non-zeolitic pores) is known to impact such single gas permeation.

In principle, gas permeation through porous media is described by mechanism such as Knudsen diffusion, surface diffusion, Poiseuille flow and molecular sieving(Abas et al., 2015). Sodalite membrane behavior in gas separation could be best described by molecular sieving and Knudsen diffusion. In such mechanism, the ratio of pore diameter (sodalite crystal pore diameter) to molecular diameter (for example, hydrogen) tends to unity. This is an indication that separation is being effected by the molecular structure, as a result of interactions between permeating molecules and pore walls. In this work, hydrogen and carbon dioxide permeance through the sodalite membrane reached a value of about $3.9 \times 10^{-7}$ $mols^{-1}m^{-2}Pa^{-1}$ & $8.38 \times 10^{-8}$ $mols^{-1}m^{-2}Pa^{-1}$, respectively (see Figure 8). Experimental result obtained in this work is in close agreement to value reported for membranes evaluated for single gas permeation of hydrogen and carbon dioxide. For example, nanocomposite MFI membrane (pore size ca. 0.5 nm) was synthesized via pore-plugging hydrothermal synthesis by Alshebani et al. (2008). The quality of the synthesized MFI was evaluated using single gas of hydrogen and carbon dioxide. Also, Huang et al., (2012) prepared multilayer (three layer) NaA zeolite and evaluated it for gas permeation properties. Single gas permeation reported by Wang et al., (2015); Huang et al., (2012); Hosseini et al., (2008); Guan et al., (2003, 2001) were evaluated at higher transmembrane pressure when compared to this present work. However, this could be responsible for the ideal selectivity reported by these authors. In addition, because of the kinetic diameter of $H_2$ (0.27 nm) which is smaller than that of $CO_2$ (0.33 nm), higher permeance of $H_2$ through the defect-free nanocomposite sodalite/ceramic membrane is expected. Ideal selectivity of 4.6 was obtained for the permeance of $H_2$ and $CO_2$. This value is slightly lower than the Knudsen separation coefficient 4.7 for $H_2/CO_2$. A brief comparison of the results obtained in this study to those reported in literature is shown in Table 1.

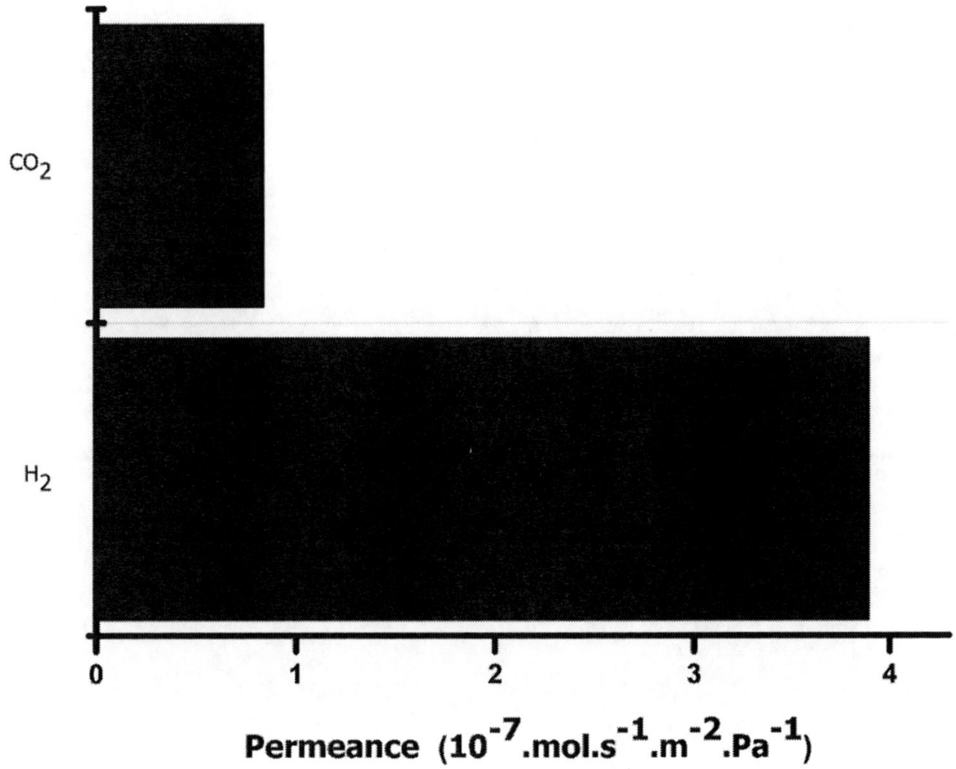


**Fig. 8** $H_2$ and $CO_2$ single gas component permeance versus time in nanocomposite sodalite/ceramic membrane

**Table 1** Comparison of permeation and selectivity result obtained and literature

| Membrane | Synthesis method | Permeation conditions | | Ideal selectivity | Permeance ($10^{-7}$ mols$^{-1}$m$^{-2}$Pa$^{-1}$) | | Reference |
|---|---|---|---|---|---|---|---|
| | | Temperature (K) | Transmembrane pressure (kPa) | ($H_2/CO_2$) | $H_2$ | $CO_2$ | |
| Hydroxy Sodalite | Pore-plugging hydrothermal (1-step pore plugging) | 298 | 25 | 4.63 | 3.89 | 0.84 | This work |
| Ion-Exchanged LTA | Direct hydrothermal | 308 | 101 | - | 0.08-0.26 | - | Guan et al., (2001) |
| Multi-Layer LTA (3- layers) | *in-situ* hydrothermal (3 step synthesis) | 373 | 101 | 12.50 | 2.10 | 0.17 | Huang et al., (2012) |
| MDES modified MFI | *in-situ* crystallization | 296-723 | - | 17.50 | 1.86 | 0.11 | Gu et al., (2008) |
| Amine modified Mg-MOF-74 | Solvothermal | 298 | 101 | 11.00 | 1.21 | 0.11 | Wang et al., (2015) |
| Matrimind /PBI | *in-situ* hydrothermal | 308 | 358-1023 | 26.20 | 10.74 | 0.41 | Hosseini et al., (2008) |
| $AlPO_4$ | Two-step hydrothermal | 310 | 101 | 23.90 | 3.36 | 0.14 | Guan et al., (2003) |
| ZIF-7 | Microwave-assisted secondary growth | 493 | 101 | 13.6 | 0.45 | 0.03 | Li et al., (2010) |
| Ni-MOF-74 | Two-stage (layer by layer seeding and secondary solvothermal crystallization) | 298 | - | 9.1 | 127 | 0.07 | Lee et al., (2012) |
| CVD modified DDR-3 | Secondary seeded growth | 298-773 | - | 5.59 | 2.18 | 0.39 | Kanezashi et al., (2008) |

## 4. Conclusion

In the present work, successfully nanocomposite sodalite/ceramic membranes were synthesized and evaluated for $H_2/CO_2$ separation using single gas permeation measurement. However the synthesized nanocomposite sodalite/ceramic membrane, which is selective to hydrogen with ideal selectivity of 4.63, is not good enough for industrial application. The SEM images revealed that the supports were not fully plugged with sodalite crystals during one-step pore plugging hydrothermal synthesis. Further research is being conducted to enhance the performance of the sodalite/ceramic membrane using a multi-step PPH. Multi-step PPH would enhance the total pore-plugging of the porous of the support (which is 200 nm in size). In addition, separation of gas mixture ($H_2/CO_2$) will be conducted to ascertain the performance of the sodalite/ceramic membrane at pre-combustion carbon dioxide capture conditions.

**Acknowledgements**

The authors acknowledge the financial support from EnPe-NORAD grant for $CO_2$ capture project.

**References**

Abas, N., Kalair, A., Khan, N., 2015. Review of fossil fuels and future energy technologies. Futures 69, 31–49.

Al-Mufachi, N.A., Rees, N.V., Steinberger-Wilkens, R., 2015. Hydrogen selective membranes: A review of palladium-based dense metal membranes. Renew. Sustain. Energy Rev. 47, 540–551.

Alshebani, A., Pera-Titus, M., Landrivon, E., Schiestel, T., Miachon, S., Dalmon, J.-A., 2008. Nanocomposite MFI – Ceramic hollow fibres: Prospects for $CO_2$ separation. Microporous Mesoporous Mater., 4th International Zeolite Membrane Meeting 115, 197–205.

Barbieri, G., Brunetti, A., Caravella, A., Drioli, E., 2011. Pd-based membrane reactors for one-stage process of water gas shift. RSC Adv. 1, 651–661.

Daramola, M., Aransiola, E., Ojumu, T., 2012. Potential Applications of Zeolite Membranes in Reaction Coupling Separation Processes. Materials 5, 2101–2136.

Daramola, M.O., Dinat, A., Hasrod, S., 2015. Synthesis and characterization of nanocomposite hydroxy-sodalite/ceramic membrane via pore-plugging hydrothermal synthesis technique. J. Membr. Sep. Technol. 4, 1-7.

Daramola, M.O., Oloye, O., Yaya, A., 2016. Nanocomposite sodalite/ceramic membrane for pre-combustion $CO_2$ capture: synthesis and morphological characterization. Int. J. Coal Sci. Technol. 1-7.

Domínguez-Domínguez, S., Berenguer-Murcia, A., Morallón, E., Linares-Solano, A., Cazorla-Amorós, D., 2008. Zeolite LTA/carbon membranes for air separation. Microporous Mesoporous Mater. 115, 51–60.

Franz, J., Maas, P., Scherer, V., 2014. Economic evaluation of pre-combustion $CO_2$-capture in IGCC power plants by porous ceramic membranes. Appl. Energy 130, 532–542.

Gu, X., Tang, Z., Dong, J., 2008. On-stream modification of MFI zeolite membranes for enhancing hydrogen separation at high temperature. Microporous Mesoporous Mater. 111, 441–448.

Guan, G., Kusakabe, K., Morooka, S., 2001. Gas Permeation Properties of Ion-Exchanged LTA-Type Zeolite Membranes. Sep. Sci. Technol. 36, 2233–2245.

Guan, G., Tanaka, T., Kusakabe, K., Sotowa, K.-I., Morooka, S., 2003. Characterization of $AlPO_4$-type molecular sieving membranes formed on a porous α-alumina tube. J. Membr. Sci. 214, 191–198.

Hosseini, S.S., Teoh, M.M., Chung, T.S., 2008. Hydrogen separation and purification in membranes of miscible polymer blends with interpenetration networks. Polymer 49, 1594–1603.

Huang, A., Wang, N., Caro, J., 2012. Synthesis of multi-layer zeolite LTA membranes with enhanced gas separation performance by using 3-aminopropyltriethoxysilane as interlayer. Microporous Mesoporous Mater. 164, 294–301.

IEA Statistics: Carbon dioxide emissions from fuel combustion (Highlights), 2015. International Energy Agency, Paris (accessed on June 15, 2016).

Kanezashi, M., O'Brien-Abraham, J., Lin, Y.S., Suzuki, K., 2008. Gas permeation through DDR-type zeolite membranes at high temperatures. AIChE J. 54, 1478–1486.

Lee, D.-J., Li, Q., Kim, H., Lee, K., 2012. Preparation of Ni-MOF-74 membrane for $CO_2$ separation by layer-by-layer seeding technique. Microporous Mesoporous Mater. 163, 169–177.

Li, Y., Liang, F., Bux, H., Yang, W., Caro, J., 2010. Zeolitic imidazolate framework ZIF-7 based molecular sieve membrane for hydrogen separation. J. Membr. Sci. 354, 48–54.

Li, Y., Pera-Titus, M., Xiong, G., Yang, W., Landrivon, E., Miachon, S., Dalmon, J.-A., 2008. Nanocomposite MFI-alumina membranes via pore-plugging synthesis: Genesis of the zeolite material. J. Membr. Sci. 325, 973–981.

Merkel, T.C., Lin, H., Wei, X., Baker, R., 2010. Power plant post-combustion carbon dioxide capture: An opportunity for membranes. J. Membr. Sci. 359, 126–139.

Miachon, S., Landrivon, E., Aouine, M., Sun, Y., Kumakiri, I., Li, Y., Prokopova, O., Guilhaume, N., Giroirfendler, A., Mozzanega, H., 2006. Nanocomposite MFI-alumina membranes via pore-plugging synthesis: Preparation and morphological characterisation. J. Membr. Sci. 281, 228–238.

Olajire, A.A., 2010. $CO_2$ capture and separation technologies for end-of-pipe applications – A review. Energy 35, 2610–2628.

Pachtova, O., 2003. Dynamic desorption of adsorbing species under cross membrane pressure difference: a new defect characterisation approach in zeolite membranes. J. Membr. Sci. 226, 101–110.

Pires, J.C.M., Martins, F.G., Alvim-Ferraz, M.C.M., Simões, M., 2011. Recent developments on carbon capture and storage: An overview. Chem. Eng. Res. Des. 89, 1446–1460.

Wang, N., Mundstock, A., Liu, Y., Huang, A., Caro, J., 2015. Amine-modified Mg-MOF-74/CPO-27-Mg membrane with enhanced $H_2/CO_2$ separation. Chem. Eng. Sci., Metal-Organic Frameworks for Emerging Chemical Technologies 124, 27–36.

**Clean Coal Technology Research**

- *Beneficiation*
- *Characterisation*
- *Utilization*
- *Environment*

# Prediction of spontaneous combustion in coal by use of thermogravimetry


Zama. A. Mthabela

*School of Chemical and Metallurgical Engineering, University of the Witwatersrand*

Dr. Nicola Wagner and Ms Nthabiseng Maledi


**33rd Annual International Pittsburgh Coal Conference**

# Table of Content

# Energy sources in SA (Eskom, 2012)

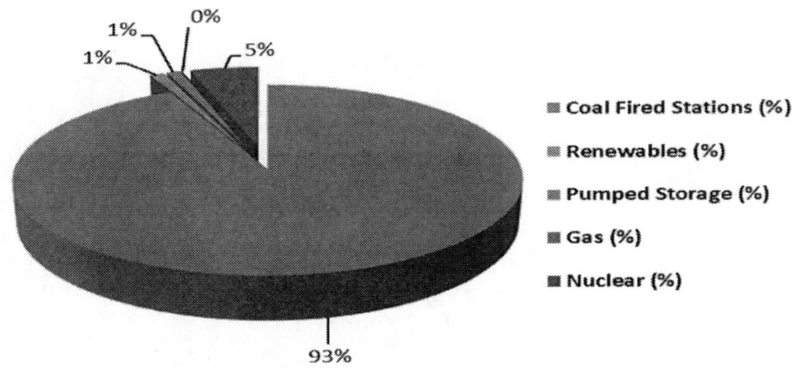


South Africa's (SA) energy is heavily dependent on coal as a source of energy (World Coal Association, 2005).
Coal mining, storage, and utilisation comes with environmental problems, one of which is spontaneous combustion.

# Problem Statement & Aim

**Problem Statement**

- Spontaneous combustion of coal creates a need to predict coal's liability and a need for a rapid and reliable analysis.

**Aim**

- To assess and validate a proposed thermogravimetric method for prediction of spontaneous combustion liability.

# Spontaneous Combustion

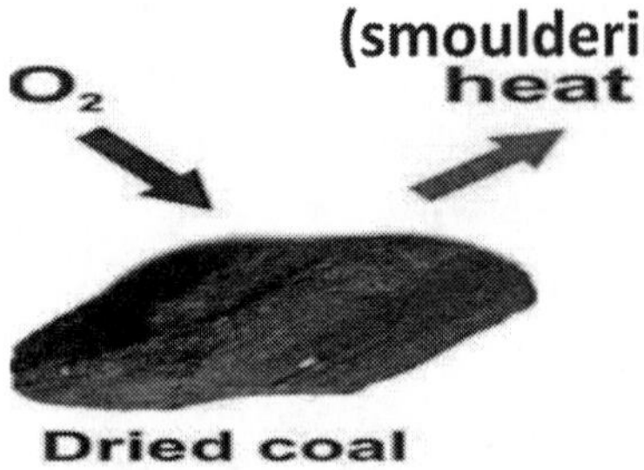


(smouldering) → self-heating

↓

Ignition point

↓

Spontaneous Combustion

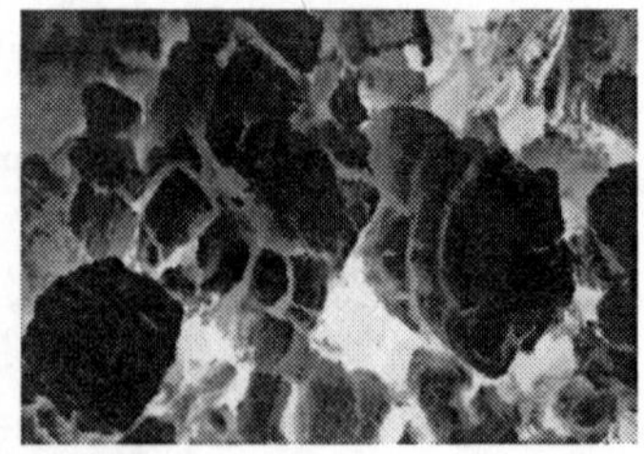

C + O2 → CO2 + Heat..........Equation 1 (Grapes, 2010)

Factors leading to spontaneous combustion:

coal rank, moisture content, pyrite content, coal particle size, porosity, maceral content, ventilation of coal, etc .

# Relevance to Industry

- Spontaneous combustion of coal is a very destructive occurrence that has severe impacts on:
- Safety – hazard to mine workers and potential loss of life
- Capital – loss of exports, equipment, possible [illegible] increased difficulty in coal [illegible]
- **Environment** – fo[illegible] gases [illegible] ($H_2$, CO, $CH_4$) and [illegible] emissions [illegible] (CO, [illegible] $NO_x$)

By products of combustion - potassium, iron and calcium sulphates - Anna coal mine dump in Germany ( F de Wit)

# Methodology

**Two spont comb liability test methods were used:**

- ❑ TGA method based on a thesis by Avila (2012).
- ❑ Smith-Glasser oxidation method (Smith and Glasser, 2005)

**Characterisation Analyses**

- Proximate Analysis
- Petrographic Analysis
- Calorific Value Analysis
- Sulphur Analysis

# Thermogravimetry

Thermogravimetric analysis is based on the measurement of mass loss of material as a function of temperature.

TGA can provide data on.... (prox, combustion reactions, melting points, ignition points....)

For spontaneous combustion liability analysis two sets of non-isothermal tests were required (Marinov ,1977; Earnest ,1988), namely :

- Ignition test (derivative weight vs time curve)
- Oxygen adsorption (weight vs temperature)

# TGA Advantage

- Static Isothermal Method,
- Adiabatic Calorimetry - R70 method,
- Crossing Point Index,
- Wits-Ehac liability Index

**Quick turnaround time of TGA**

# TGA Ignition Test

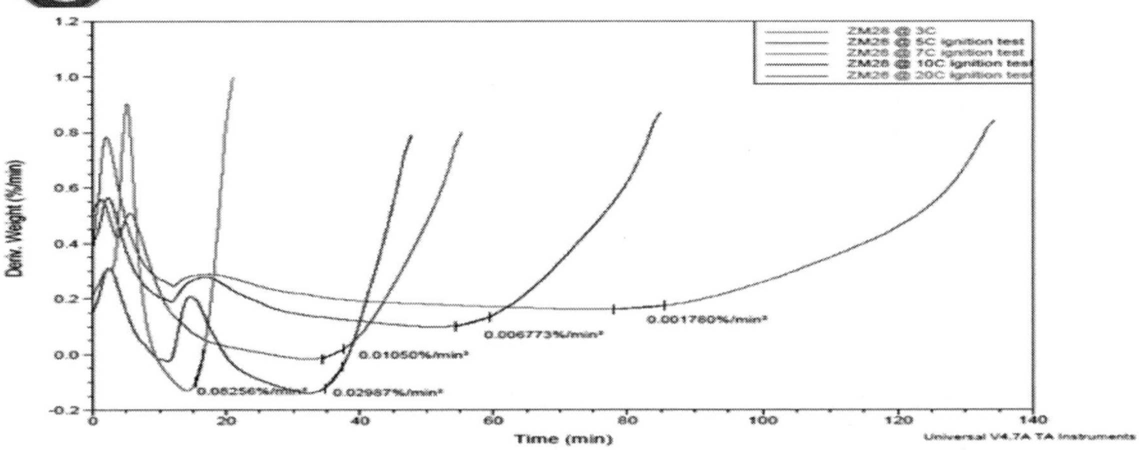


**Figure 11: ZM28 Derivative weight curve at different temperature ramps (%/min)**

The slope of the linear segment of the derivative weight curve were obtained from the 5 different heating rates. The five slopes were use to obtain the $TG_{spc}$ index (thermogravimetry spontaneous combustion index).

# $TG_{spc}$ Index

Slopes of linear segment $_{calculated}$ = $\frac{\text{slope of linear segment}}{\text{temperature ramp}}$ (wt% °C$^{-1}$ min$^{-1}$) ...............Equation 1

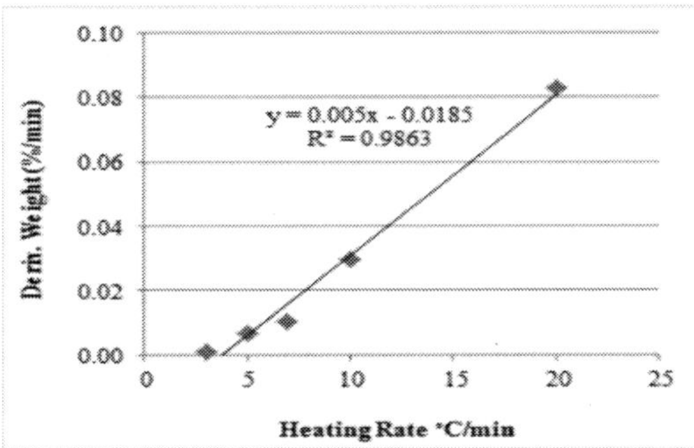


**Figure12: ZM 28 Slope of derivative weight curves (Linear segment) used to obtain used $R^2$ graph**

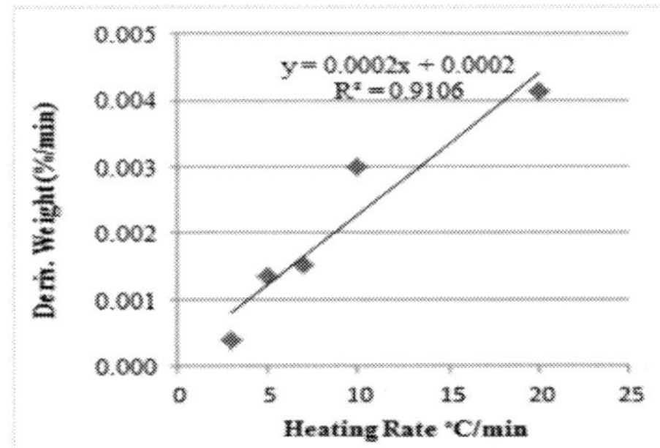


**Figure13: ZM 28 Calculated slope of derivative weight curves used to obtain $R^2_{calculated}$**

The $TG_{spc}$ Index is obtained by using the following formula:

$TG_{spc}$ Index $= R^2_{(graph)} - R^2_{(calculation)}$ (wt% °C$^{-1}$ min$^{-1}$) ...... Equation 2 (Avila, 2012)

$TG_{spc}$ Index$_{ZM28}$ = 0.0757

# $TG_{spc}$ Index Results

| $TG_{spc}$ Index | |
|---|---|
| High Reactive : | 0.070 > |
| Reactive : | 0.036 – 0.069 |
| Less Reactive : | 0.030 – 0.035 |
| Not Reactive : | 0.020 – 0.029 |
| Unreactive : | 0.00 – 0.019 |

**Table 1: Coal Reactivity Guidelines (Avila, 2012)**

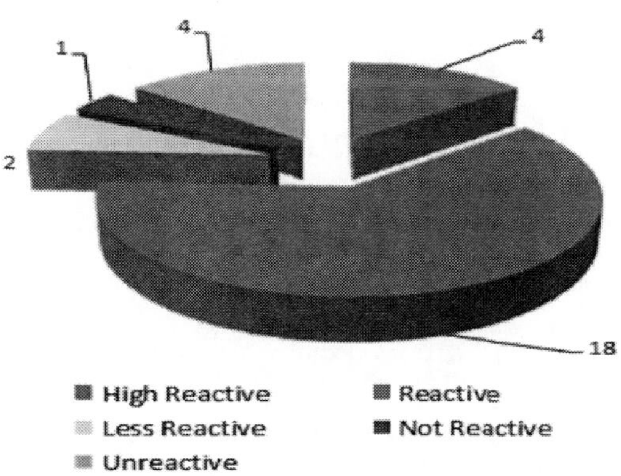


**Figure 14: Coal Sample Reactivity using the TGspc Index as a guideline set by Avila (2012)**

ZM 28 according to the $TG_{spc}$ Index guidelines is highly reactive

# TGA Oxygen Adsorption

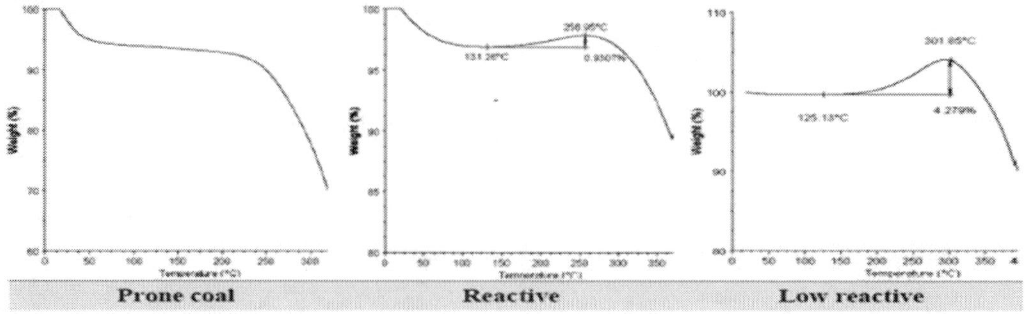


**Figure 15: Oxygen adsorption curves illustrating the weight gain (Avila, 2012)**

Results are interpreted as follows:

- Curve 1 is ruled by the desorption stage of water molecules; this coal is very reactive and most likely to spontaneously combust.
- Curve 2 is in between the desorption and adsorption stage, this coal under specific conditions, might self-heat.
- Curve 3 is ruled by the O2 adsorption, this coal will not self-heat.

# Correlation of results

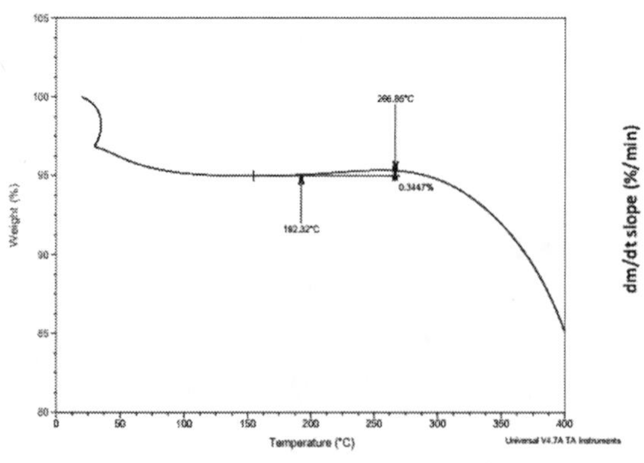


**Figure 16: ZM28 (3°C/min ) oxygen adsorption graph**

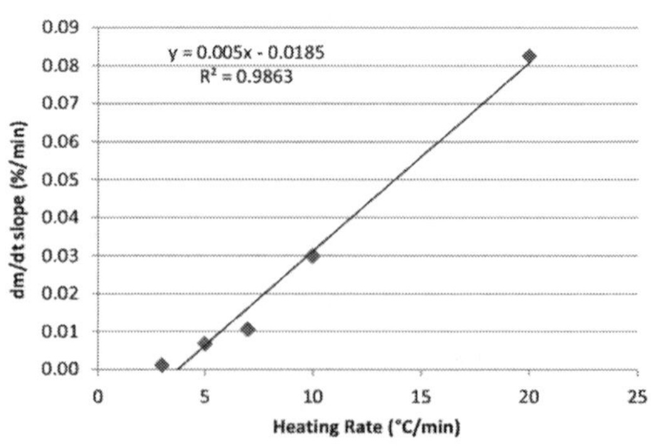


**Figure 17: ZM28 - heating ramp applied against slope of derivatives curves**

- Very little mass increase - sample has gone through desorption.
- High TGspc Index = 0.0757
- Low Peak Temperature = 266.85°C
- Very reactive - most likely to spontaneously combust.

# Correlation of results

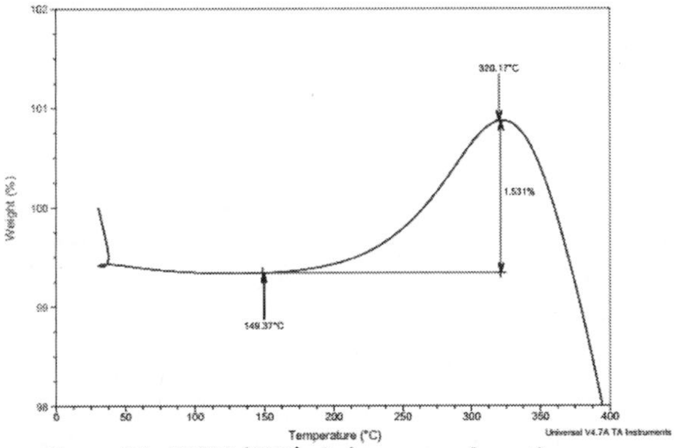


**Figure 18: ZM29 (3°C/min ) oxygen adsorption graph**

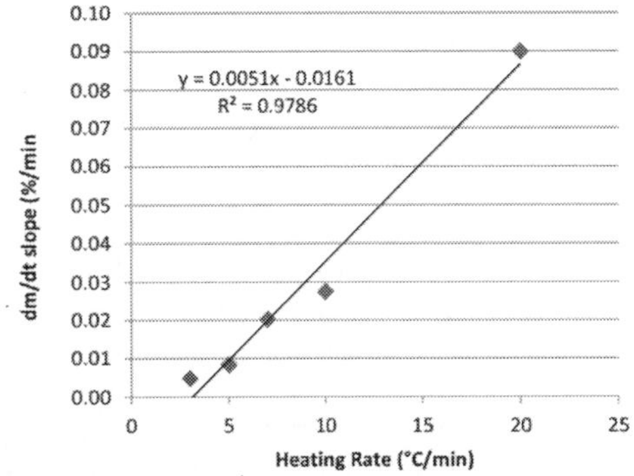


**Figure 19: ZM29 - heating ramp applied against slope of derivatives curves**

- High mass increase - sample has gone through adsorption.
- Average $TG_{spc}$ Index = 0.0525
- High Peak Temperature = 320.17C
- Low reactive Coal – least likely to spontaneously combust.

# Smith-Glasser Oxidation Test

The Static isothermal method measures the reaction of coal with $O_2$, Smith and Glasser (2005) worked on the principle that when the coal adsorbs $O_2$, the oil in a beaker is displaced due to absorption; this becomes evident by the increase in the oil reading in the pipette.

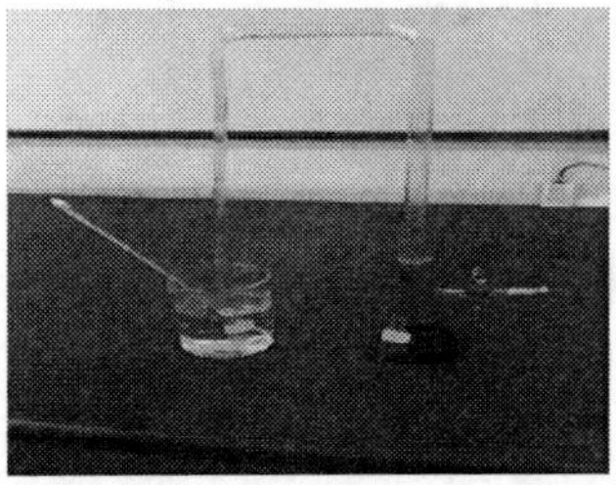

**Figure 20: Smith-Glasser test setup**

**Figure 21: Water bath placement**

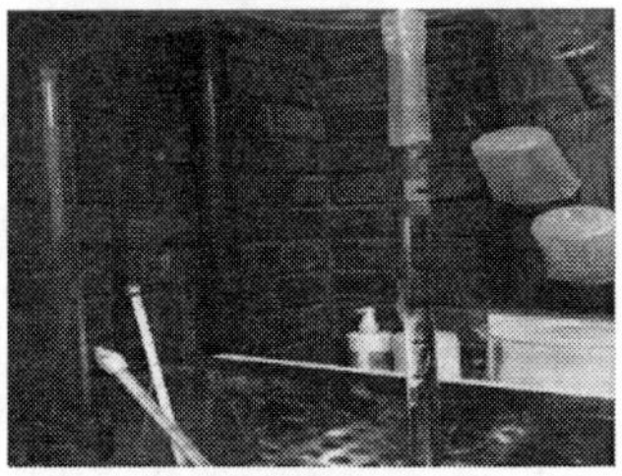

**Figure 22: Rise in oil (mm)**

**Figure 23: Water condensation**

# Overall Results Interpretation Guidelines

| | Total Reactive Macerals | Sulphur (%) | Volatile Matter (%) | CV (MJ/kg) | Mass Increase (%) | Peak Temp (°C) | TGspc Index | Highest Oil Reading (ml) |
|---|---|---|---|---|---|---|---|---|
| High Reactivity | 55> | 2.0> | 24> | 23> | 0 - 0.2 | 270< | 0.070 | 7> |
| Reactive | 35 - 54 | 1.5 - 1.9 | 20 - 23 | 17 - 22 | 0.2 - 0.6 | 271 - 280 | 0.036 - 0.069 | 1 - 6 |
| Less Reactive | 20 - 34 | 0.8 - 1.4 | 17 - 19 | 11 - 16 | 0.7 - 1.1 | 281 -300 | 0.030 - 0.035 | 0.5 - 0.9 |
| Not Reactive | 1 - 9 | 0.4 - 0.7 | 1 - 12 | 5 - 10 | 1.2 – 1.7 | 301 - 310 | 0.020 - 0.029 | 0.1 - 0.4 |
| Unreactive | 0 - 9 | 0.0 - 0.3 | 0 - 12 | 0 - 4 | 2> | 311 > | 0.000 - 0.019 | 0.0 |

**Table 3: Overall Project Guidelines**

The guidelines were assumed using the overall analytical and test results.

# Spontaneous Combustion Prediction

| Coal Samples | Spontaneous Combustion Prediction |
|---|---|
| ZM1 | Highly likely |
| ZM2 | Less likely |
| ZM3 | Likely |
| ZM4 | Likely |
| ZM5 | Highly likely |
| ZM6 | Likely |
| ZM7 | Less likely |
| ZM8 | Likely |
| ZM9 | Highly likely |
| ZM10 | Likely |
| ZM11 | Likely |
| ZM12 | Highly likely |
| ZM13 | Likely |
| ZM14 | Not likely |
| ZM15 | Likely |
| ZM16 | Less likely |
| ZM17 | Highly likely |
| ZM18 | Likely |
| ZM19 | Likely |
| ZM20 | Less likely |
| ZM21 | Less likely |
| ZM22 | Not likely |
| ZM23 | Highly likely |
| ZM24 | Likely |
| ZM25 | Highly likely |
| ZM26 | Likely |
| ZM27 | Likely |
| ZM28 | Likely |
| ZM29 | Less likely |

**Table 5: Spontaneous Combustion Prediction**

- The predictions were based on the consistency of results for each sample.
- If 3 or more components in Table 4 had the same outcome (High reactive, Reactive, Less Reactive or Not Reactive) then a conclusion was made on the sample's liability to self-heat (Table 5).

# Conclusion

- No correlation between the two methods.
- Both test methods did not fully correlate with the coal characteristics.
- The TGA method could be used, but with reservations.
- The TGA method does not fully explain the intensity of the coal reaction with $O_2$.
- The TGA method needs both the coal characterisation information and Smith-Glasser oxidation method.

# Recommendation

- It is recommended for a TGA reference standard to be made.
- It is recommended that the different types of sulphur analysis (organic, pyritic and sulphate) be performed.
- It is highly recommended that further TGA repeatability tests be performed.
- BET surface area analyses are also recommended to be performed.

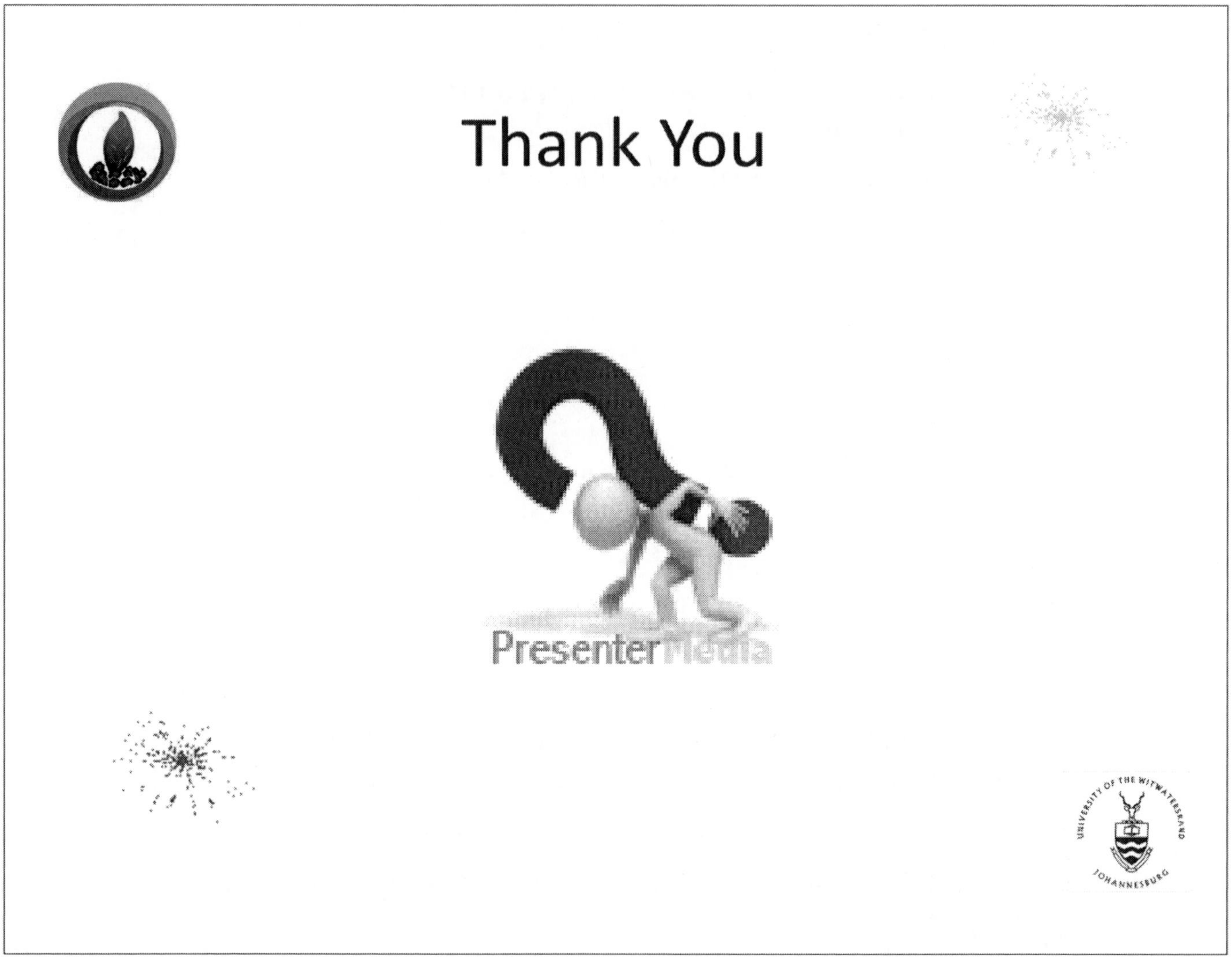
Thank You
PresenterMedia
UNIVERSITY OF THE WITWATERSRAND
JOHANNESBURG

# Mechanical and thermal characteristics of briquetted coal fines using LDPE as the binding agent


**T.S. Mthombo, C.A. Strydom, J.R . Bunt & H.W.J.P. Neomagus**



33rd PCC. 8-12th, August 2016. Cape Town, SA.



NORTH-WEST UNIVERSITY ®
YUNIBESITI YA BOKONE-BOPHIRIMA
NOORDWES-UNIVERSITEIT
POTCHEFSTROOM CAMPUS


It all starts here ®

# Presentation Outline

1

- Introduction
  - Fine coal issues
  - Waste plastic issues
  - Solution
- Experimental Methods
  - Sample preparation
  - Characterization
- Results and discussion
  - Mechanical
  - Thermal
- Conclusion

# Fine coal issues

2

- Significant growth in fine waste coal
  - improved high output mining technologies [1]

- Challenges [2,3]:
  - Environmental concerns (slime dams)
  - Land shortage
  - Transportation costs

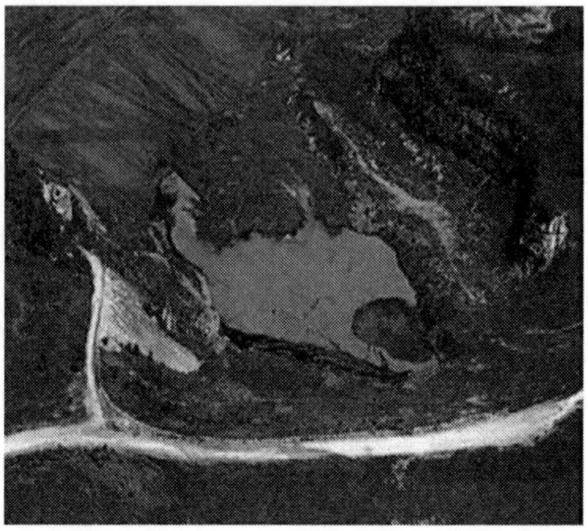

[1] Radloff, B., Kirsten, M., and Anderson, R. (2003). Wallerawang Colliery Rehabilitation: Minerals Engineering. 17: 153-157.
[2] Mangena, S.J., de Korte, G.J., McCrindle R.I., and Morgan D.L. (2004). Fuel Processing Technology, 85: 1647–1662.
[3] Mani, S., Sokhansanj, S., Bi X., and Turhollow, A., (2001). Applied Engineering Agriculture. 22: 421-426.

# Waste plastic issues

3

- Plastic SA 2015 stats:
  - 292 917 tons; a 3.5% increase from 2014
  - 15.4 % increase in operational costs [4]

- Challenges:
  - Environmental concerns (land fills)
  - Decreasing credibility of recycling processes [5]
  - Escalating operational costs

[4] Plastic South Africa 2015 recycling statistics. Press release, 30 May 2016.
[5] Municipal Solid Waste Generation, Recycling, and Disposal in the United States: Facts and Figures for 2010. EPA 2010.

# Solution

4

- Densification of fine coal
- Incorporation of plastic wastes into existing densification processes as raw materials
- Provides an attractive way to dispose off coal and also convert plastics into higher value fuels [6-8]
- Thus combining the two does not only minimize costs, but also protects the environment.

[6] Marin, N., Collura, S., Sharypov, V.I., Beregovtsova, N.G., Baryshnikov, S.V. (2002). Journal of Applied Pyrolysis 65:41–55.
[7] Vivero, L., Barriocanal, .C, Alvarez, R., Diez, M.A. (2005). Journal of Applied Pyrolysis 74:327–36.
[8] Ishaq, M., Ahmad, I., Shakirullah, M., Khan, MA., Rehman, H., Bahader, A. (2006).Energy Conversions Management 47:3216–23.

# What's new?

5

- Wide research into combustion characteristics of various fuel products derived from and/or as combination of:
  - coal
  - sludge
  - plastics
  - biomass *etc.*
- However, few reports on the use of plastics as minor components, acting as binding agents [9-11]
- AIM: To understand the interactions of plastics, acting as binders, with coal fines during briquetting: it's influence on thermal and mechanical behavior

[4] Dı´ez, M.A., Ivarez, R. A., Barriocanal, C., Canga, C.S. (2000). Eurocarbon II. 719-26.
[5] Nomura S., Kato, K., Nakagawa,T., Komaki, I., (2003). Fuel 82: 1775- 83.
[11] Sakurovs, R. Fuel (2003) 82: 1911-19.

# LDPE

6

- Main component of waste plastics (>60%), with desirable waterproofing, energy and binding characteristics .

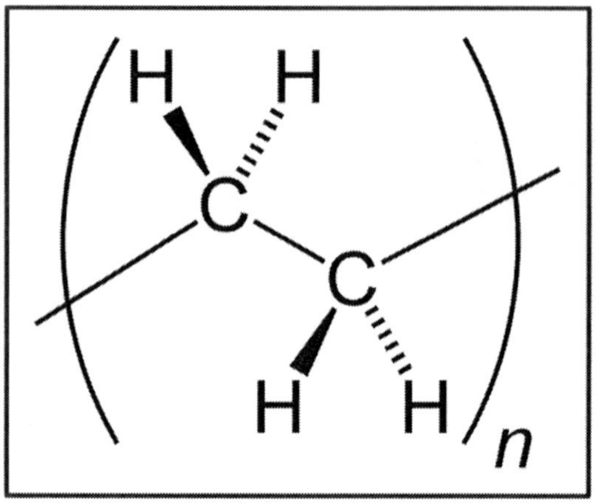

# Sample preparation

7

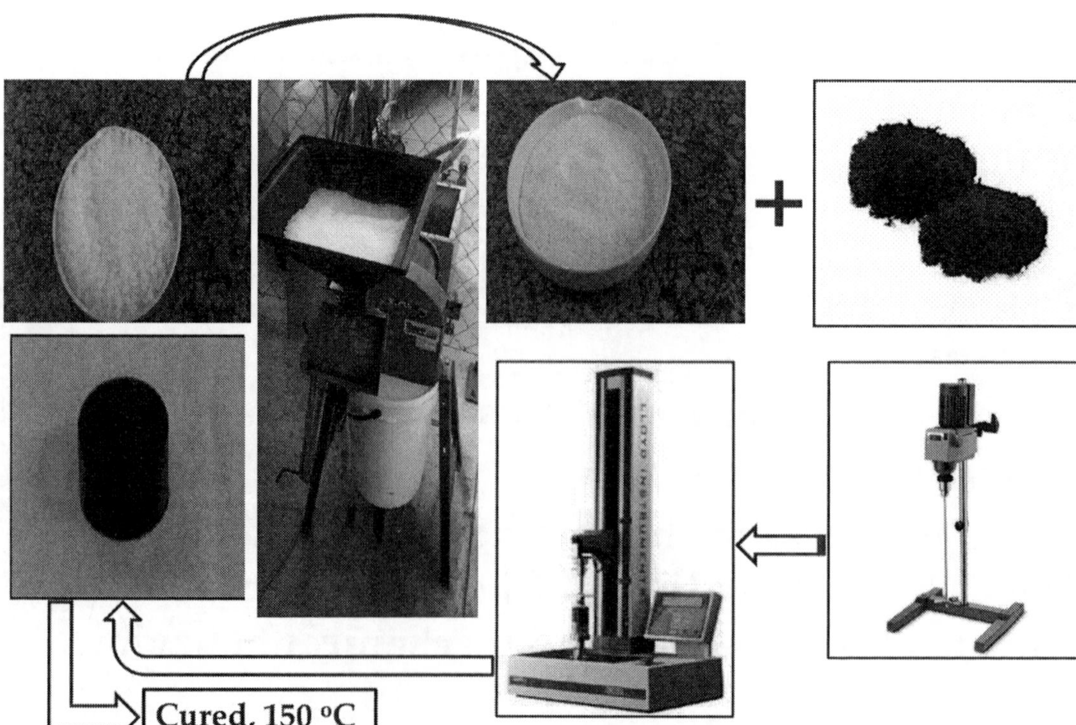

# Material properties

8

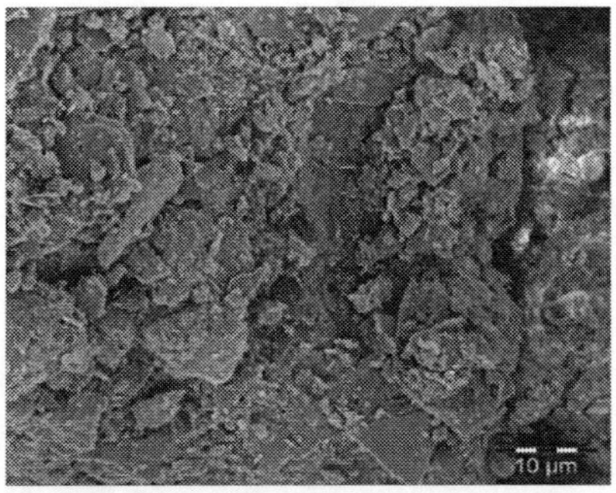


Coal fines

| Proximate (% adb)* | |
|---|---|
| Moisture | 5.3 |
| Ash | 28.3 |
| Volatile | 25.8 |
| F C | 40.6 |

*adb = air-dry basis
**dafb = dry ash-free basis

| Ultimate (% dafb)** | |
|---|---|
| C | 77.4 |
| H | 4.2 |
| N | 1.9 |
| S | 1.3 |
| O | 15.2 |

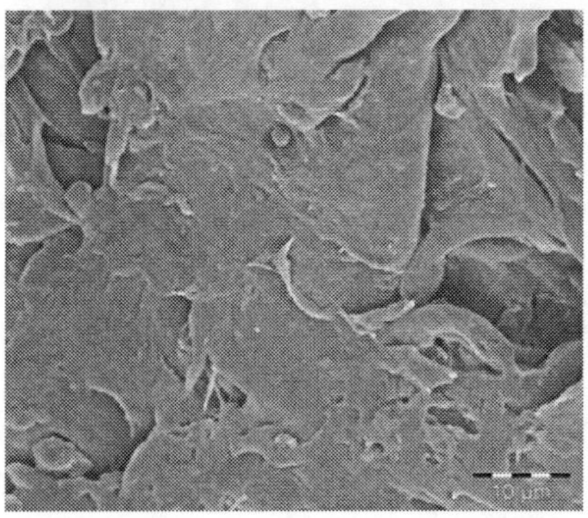


LDPE

| Proximate analysis (% dry basis) | | |
|---|---|---|
| Ash | Volatile | Fixed carbon |
| 0.3 | 99.7 | - |

| Ultimate analysis (% dry basis) | | |
|---|---|---|
| Carbon | Hydrogen | Sulphur |
| 85.5 | 14.3 | 0.2 |

# Coal/LDPE blend

9

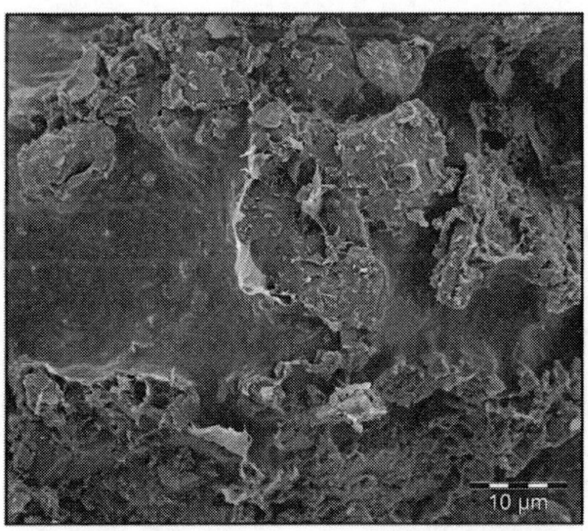


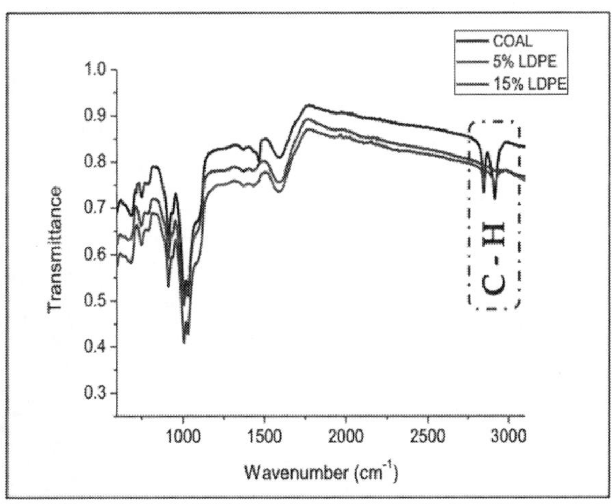


- Blends show characteristic C-H peaks in 2800-3000 cm$^{-1}$ range
- No significant changes in the peaks, save for some minor differences in peak intensities with varying coal/LDPE ratios

# Compression strength

10

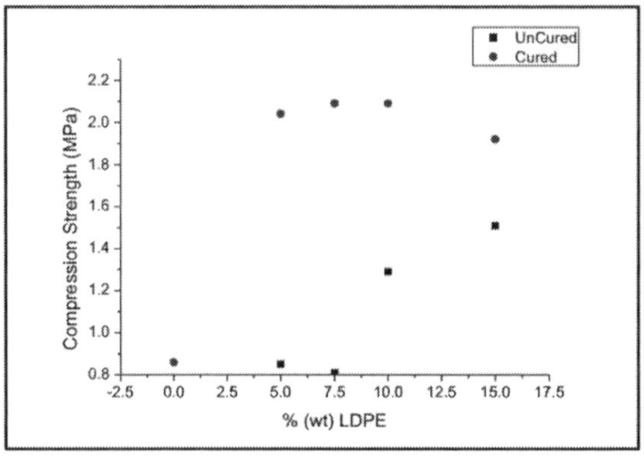


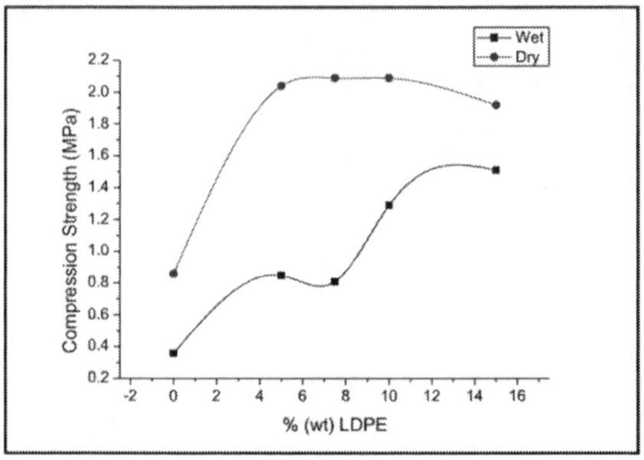


# Attrition Index

11

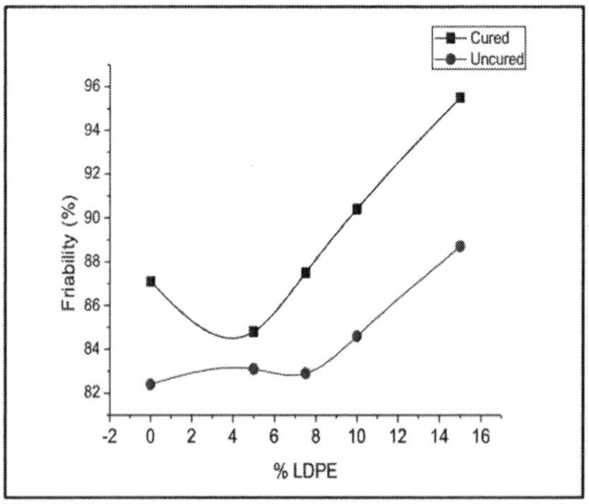

# TGA/DTG



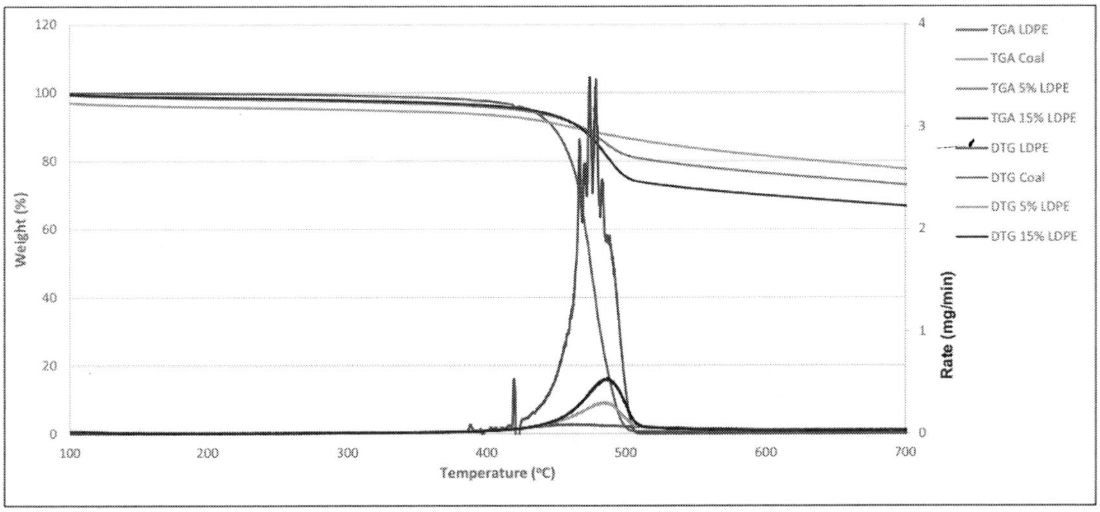


- **Multiple devolitization in LDPE in 400 – 500 °C**
- **Seemingly suppressed with blending**
- **Non-conclusive: Lower heating rate**

# Higher Heating Values



| **Plastic** | **HHV (MJ/Kg)** |
|---|---|
| PVA | 22.1 |
| PS | 38.2 |
| PP | 44.6 |
| LDPE | 46.1 |
| HDPE | 46.1 |

| **Sample** | **HHV (MJ/kg)** | |
|---|---|---|
| | Measured | Calculated |
| Coal fines | 19.6 | 19.6 |
| 5% LDPE | 21.1 | 20.9 |
| 7.5% LDPE | 21.5 | 21.6 |
| 10% LDPE | 21.9 | 22.3 |
| 15% LDPE | 24.5 | 23.6 |

# Conclusion

14

- Positive influence of LDPE on mechanical properties
  - LDPE/coal blends deform significantly before fracturing
  - 15% LDPE retain strength remarkably after exposure to water for 24h
  - Attrition data had similar correlations
- LDPE significantly improves HHV of fuel product
  - HHV of LDPE over twice that of waste coal
  - A 10% LDPE blend has an 11.7% greater HHV over the parent coal
- Addition of LDPE causes decomposition at significantly lower temperatures than those observed in the waste coal.

# Going forward...

15

- Comparative study with real municipal solid waste (MSW)
  - Effect of pretreatment (type)
- Detailed TGA/DTA profiles (at slower heating rates) to understand and solve the prediction of possible chlorinated organic compounds

# Acknowledgements


This work is based on the research financially supported by the South African Research Chairs Initiative of the Department of Science and Technology and National Research Foundation of South Africa.

TIA

NRF National Research Foundation | RISA Research and Innovation Support and Advancement

MTB

Any opinion, finding or conclusion or recommendation expressed in this material is that of the author(s) and the NRF does not accept any liability in this regard.


It all starts here

NORTH-WEST UNIVERSITY ®
YUNIBESITI YA BOKONE-BOPHIRIMA
NOORDWES-UNIVERSITEIT
POTCHEFSTROOM CAMPUS

# Thank you

# Evaluation of a Domestic-use Low-smoke Fuel Produced via the Pyrolysis of Lump Coal.

**LP Sumbane[1], M. Kühn[1], J. Bunt[1], H. Neomagus[1], S. Piketh[2]**
*[1]Coal Research Group, Unit for Energy systems, School of Chemical and Mineral Engineering, North West University, Potchefstroom, 2520, South Africa*
*[2]Climate Research Group, School of Geo and Spatial Sciences, North West University, Potchefstroom, 2520, South Africa*

33[rd] Annual International Pittsburgh Coal Conference (PCC).
August 8 – 12, 2016.
Cape Town, South Africa.

National Research Foundation
NORTH-WEST UNIVERSITY
YUNIBESITI YA BOKONE-BOPHIRIMA
NOORDWES-UNIVERSITEIT
POTCHEFSTROOM CAMPUS

2

# Introduction

## Domestic Coal Usage in South Africa

- 3 mil. Tons/yr used in ~1 million homes as primary energy source [1, 2]
- 3-5% of national coal consumption
- 20% of country's particulate emission
- Higher impact than adjacent power plants [2,3]
- Leads to a host of illnesses including respiratory infections] which costs the state ~R1.2 billion p.a. to treat [4, 5]

[1] Balmer, M. 2007. Household coal use in an urban township in South Africa. Journal of energy in Southern Africa. 18(3):27-32
[2] Sowazi, S and Maake R. 2001.Clean Energy Initiatives in South Africa: Qalabotja Micro-Scale Experiments as a Case Study. In Domestic Energy Conference. Minerals and Energy Policy Centre.
[3]Lloyd , P. 2014. Challenges in Household Energisation and the Poor. Journal of Energy in Southern Africa. Vol 25(2)
[4] Scorgie et al. 2001. Assessing the potential of low-smoke fuel in addressing air pollution and human health impacts due to coal burning emissions. Clean air journal. 0(8):6-16
[5] WHO. 2002.WHO report 2002: reducing risks, promoting healthy life. Geneva, Switzerland: World Health Organization.

4

# Introduction

## Domestic Coal Usage in South Africa

- 3 mil. Tons/yr used in ~1 million homes as primary energy source [1, 2]

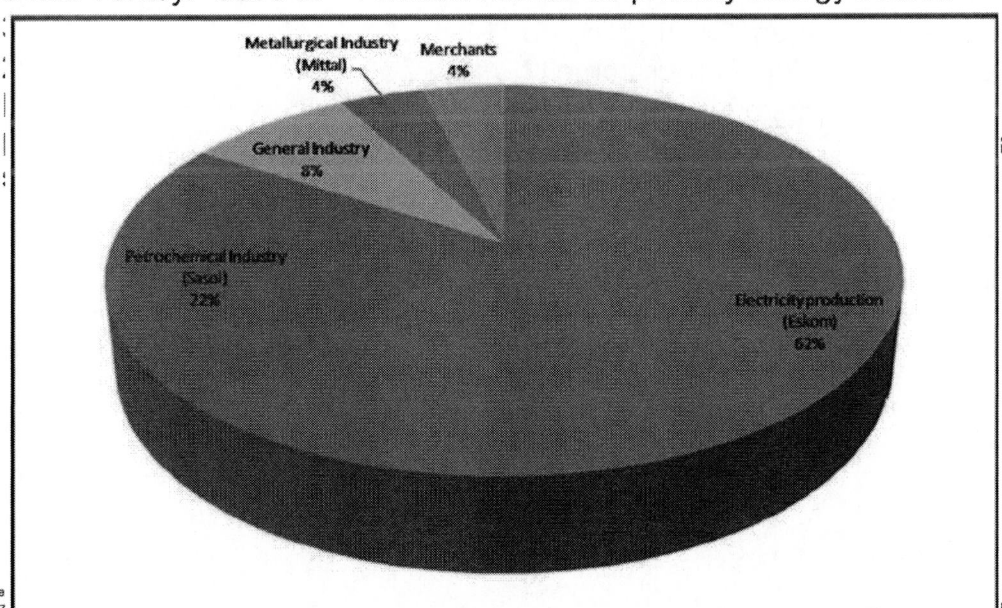


2015 Statistics from DoE website

ts the

[1] Balme
[2] Sowaz ... y Policy Centre.
[3]Lloyd ,
[4] Scorgi
[5] WHO. 2002.WHO report 2002: reducing risks, promoting healthy life. Geneva, Switzerland: World Health Organization.

4

# Introduction

## Domestic Coal Usage in South Africa

- 3 mil. Tons/yr used in ~1 million homes as primary energy source [1, 2]
- 3-5% of national coal consumption
- 20% of country's particulate emission
- Higher impact than adjacent power plants [2,3]
- Leads to a host of illnesses including respiratory infections] which costs the state ~R1.2 billion p.a. to treat [4, 5]

## There exists a need for a fuel to low-income households that

- Meets the energy needs
- is easily available and competitively priced
- Limits the exposure of users to pollutants emitted during coal combustion

[1] Balmer, M. 2007. Household coal use in an urban township in South Africa. Journal of energy in Southern Africa. 18(3):27-32
[2] Sowazi, S and Maake R. 2001.Clean Energy Initiatives in South Africa: Qalabotja Micro-Scale Experiments as a Case Study. In Domestic Energy Conference. Minerals and Energy Policy Centre.
[3]Lloyd , P. 2014. Challenges in Household Energisation and the Poor. Journal of Energy in Southern Africa. Vol 25(2)
[4] Scorgie et al. 2001. Assessing the potential of low-smoke fuel in addressing air pollution and human health impacts due to coal burning emissions. Clean air journal. 0(8):6-16
[5] WHO. 2002.WHO report 2002: reducing risks, promoting healthy life. Geneva, Switzerland: World Health Organization.

4

# Introduction

## Previous attempt at solving the pollution issue

- NOVA/Sasol innovations [6]
- Basa Njengo Magogo lighting method [7]
- DME low-smoke fuels programme [7]
- UJ /SeTAR novel stove programs [8]

## This Study

- Evaluate the technical viability of a low-smoke fuel (LSF) by:
  - Producing a LSF via pyrolysis at preset temperatures
  - Assessing the performance and pollution reduction achieved by the LSF in an everyday coal-stove
  - Comparing the performance of the LSF to the untreated coal

[6] Sasol representative. Personal communication
[7] van Niekerk & Swanepoel. 2002. An assessment of various options for improving low-cost, clean household energy. Report submitted to DME: Low-smoke fuels programmer.
[8] Ibraimo et al. Modelling of a bottom-lit down-draft (BLDD) clean-burning coal stove

5

# Experimental

National Research Foundation

NORTH-WEST UNIVERSITY
YUNIBESITI YA BOKONE-BOPHIRIMA
NOORDWES-UNIVERSITEIT
POTCHEFSTROOM CAMPUS

## Experimental Schedule

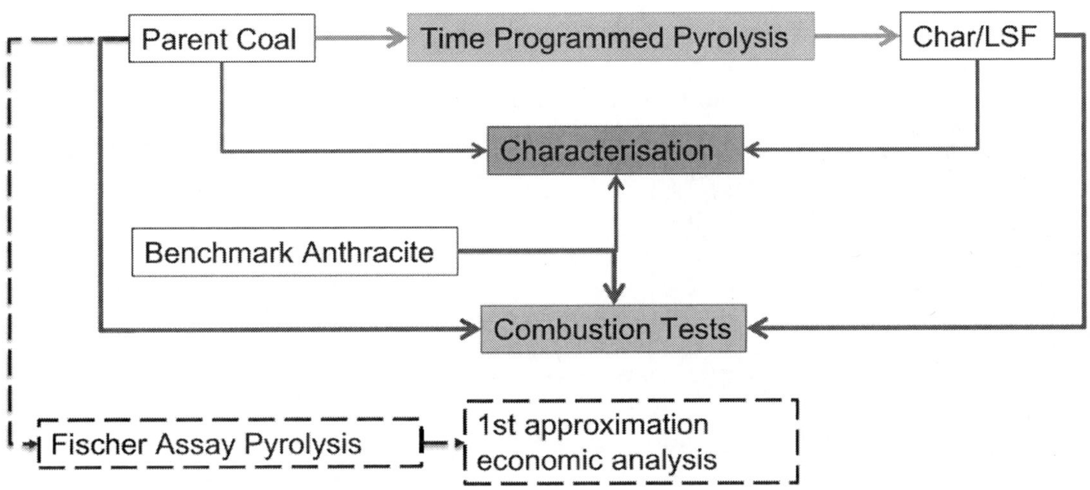


6

# Experimental

National Research Foundation

NORTH-WEST UNIVERSITY
YUNIBESITI YA BOKONE-BOPHIRIMA
NOORDWES-UNIVERSITEIT
POTCHEFSTROOM CAMPUS

Time Programmed Pyrolysis
- 450°C, 550°C, 650°C, and 750°C
- 5°C/min
- 3 LN/min sweep $N_2$
- 2 hr holding time

Tube Furnace

7

# Experimental

National Research Foundation

NORTH-WEST UNIVERSITY
YUNIBESITI YA BOKONE-BOPHIRIMA
NOORDWES-UNIVERSITEIT
POTCHEFSTROOM CAMPUS

Combustion Tests
- 2 hour runs in Union no 7 stove
- Performance Assessments
- Pollution Reduction Assessments
- Fuels were used as a composite of sizes according to the PSD of the bulk sample.

Stove and Temperature Probes

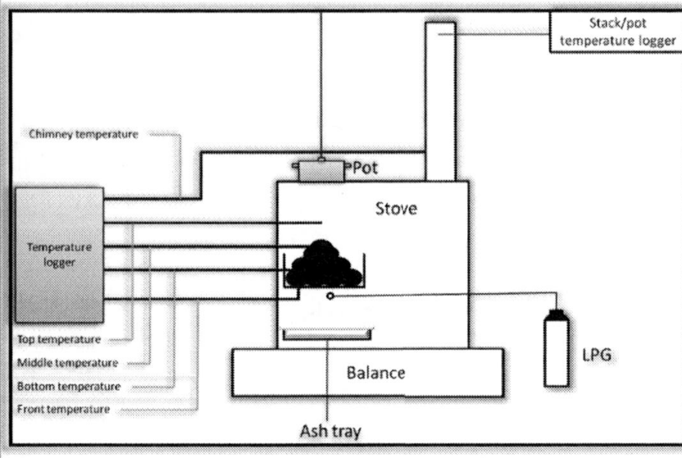


8

# Experimental

National Research Foundation

NORTH-WEST UNIVERSITY
YUNIBESITI YA BOKONE-BOPHIRIMA
NOORDWES-UNIVERSITEIT
POTCHEFSTROOM CAMPUS

Combustion Tests
- 2 hour runs in Union no 7 stove
- Performance Assessments
- Pollution Reduction Assessments
- Fuels were used as a composite of sizes according to the PSD of the bulk sample.

Chimney and Pollution Analysis Probes

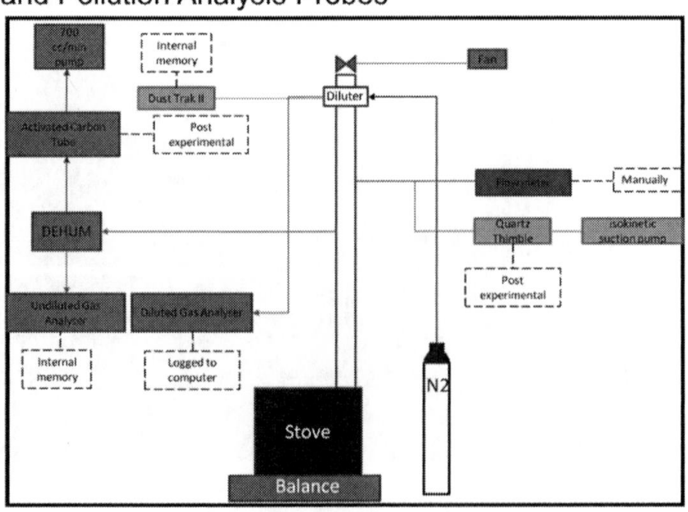


9

# Combustion Test Results

National Research Foundation NRF

NORTH-WEST UNIVERSITY
YUNIBESITI YA BOKONE-BOPHIRIMA
NOORDWES-UNIVERSITEIT
POTCHEFSTROOM CAMPUS

## Pollution Reduction Assessments

Treated fuels show a marked decrease in total **Particulate emissions**

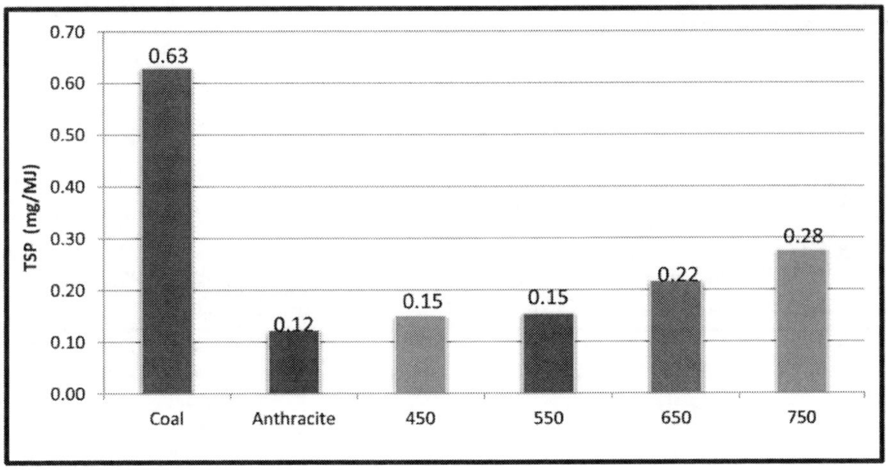


Anthracite – 59%
450-char – 71%
550-char – 75%
650-char – 72%
750-char – 71%

LSF Limit[9] : 0.07g/MJ

[9] Le Roux, L., Cilliers, K. & van Vuuren D. 2004. Low-smoke fuels: Standard Testing and Verification. CSIR report.

11


# Combustion Test Results

National Research Foundation NRF

NORTH-WEST UNIVERSITY
YUNIBESITI YA BOKONE-BOPHIRIMA
NOORDWES-UNIVERSITEIT
POTCHEFSTROOM CAMPUS

## Pollution Reduction Assessments


Treated fuels show a decrease in **$SO_2$ emissions**

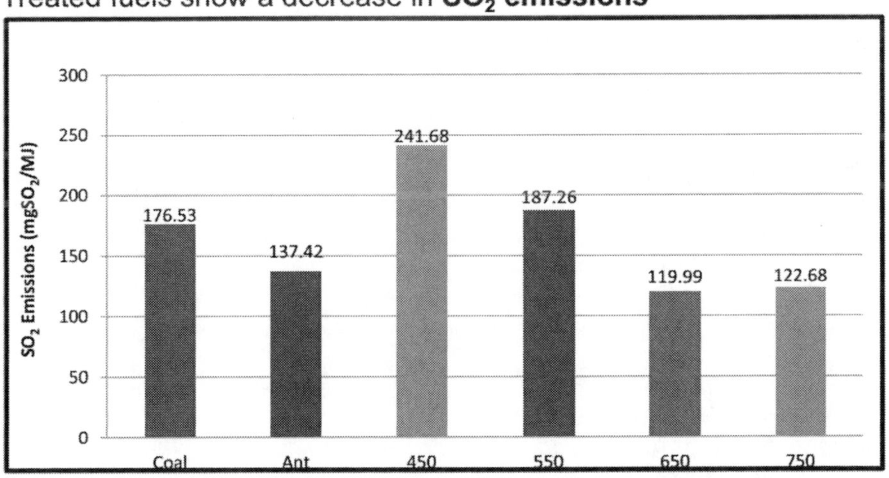


Decrease:
Anthracite – 22%
550-char – 6%
650-char – 32%
750-char – 31%

Increase
450-char – 37%

LSF Limit[9]: 300m/MJ

[9] Le Roux, L., Cilliers, K. & van Vuuren D. 2004. Low-smoke fuels: Standard Testing and Verification. CSIR report.

12

# Combustion Test Results

National Research Foundation

NORTH-WEST UNIVERSITY YUNIBESITI YA BOKONE-BOPHIRIMA NOORDWES-UNIVERSITEIT POTCHEFSTROOM CAMPUS

## Pollution Reduction Assessments

Treated fuels show a decrease in aromatic and polyatomic aromatic **volatile compounds**

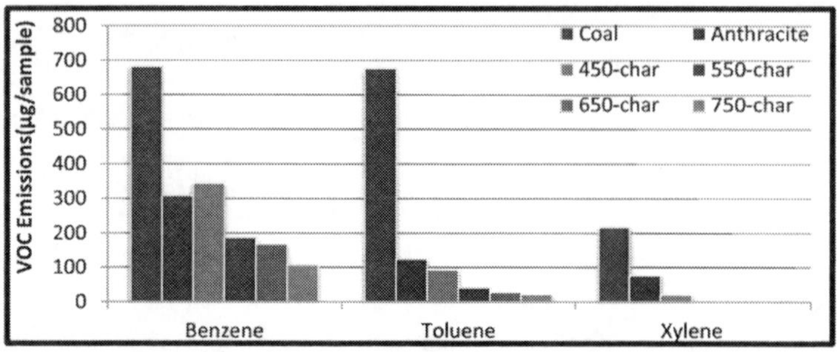


•BTX most abundant

Only PAH is naphthalene (responsible for ARIs)

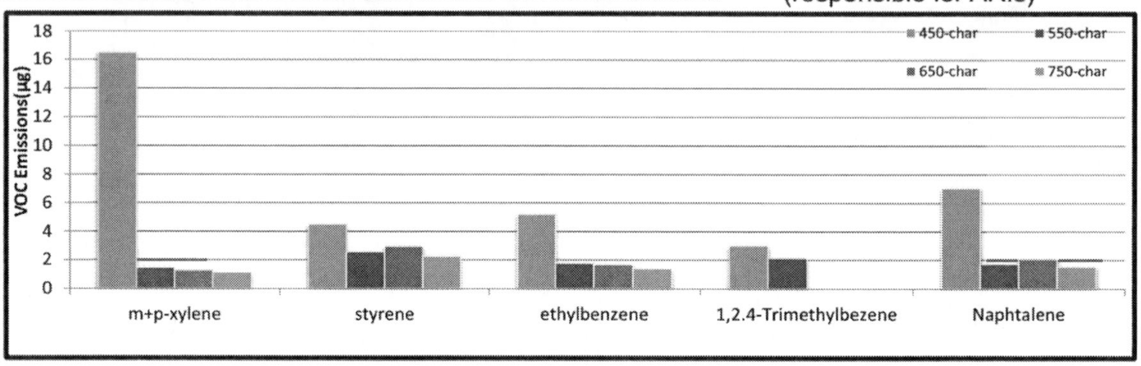


13


# Combustion Test Results

National Research Foundation

NORTH-WEST UNIVERSITY YUNIBESITI YA BOKONE-BOPHIRIMA NOORDWES-UNIVERSITEIT POTCHEFSTROOM CAMPUS


## Performance Assessments

**Water Boiling Time** increases with increasing devolatisation temperature

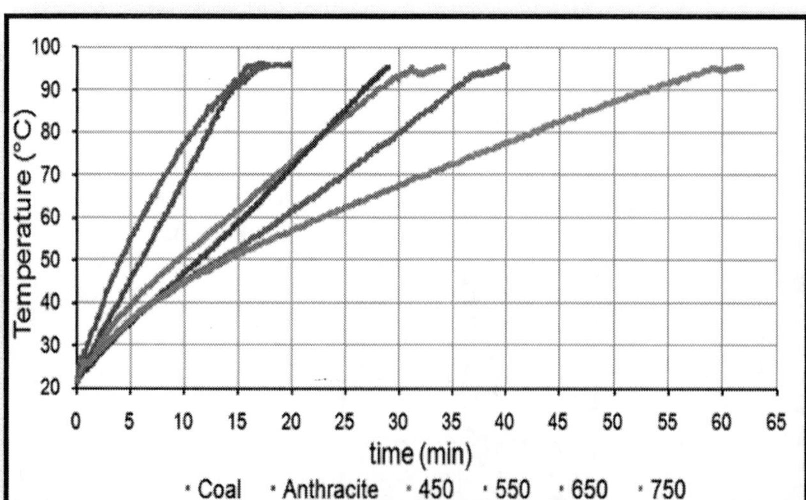


Coal – 20 min.
Anthracite – 16 min
450-char – 34 min
550-char – 28 min
650-char – 40 min
750-char – 62 min

LSF Limit[9] : 05:30 mm:ss

[9] Le Roux, L., Cilliers, K. & van Vuuren D. 2004. Low-smoke fuels: Standard Testing and Verification. CSIR report.

14

## Combustion Test Results



### Performance Assessments

**Space-Heating Time** increases with increasing devolatisation temperature

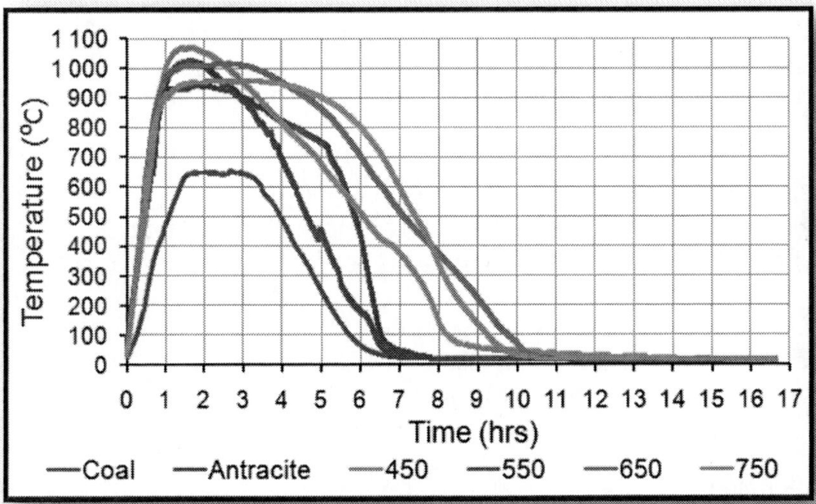


Coal – 6 hr
Anthracite – 6.5 hr
450-char – 8.5 hr
550-char – 6.5 hr
650-char – 10 hr
750-char – 10 hr

15

## Combustion Test Results

National Research Foundation NRF
NORTH-WEST UNIVERSITY
YUNIBESITI YA BOKONE-BOPHIRIMA
NOORDWES-UNIVERSITEIT
POTCHEFSTROOM CAMPUS

### Performance Assessments

**Other Performance Parameters**

| Fuel | Ignition time (mm:ss) | Max. stove top temp (°C) | Combustion efficiency (%) |
|---|---|---|---|
| Coal | 24:50 | 320 | 69.0 |
| Anthracite | 30:50 | 600 | 80.5 |
| 450-char | 30:00 | 375 | 78.6 |
| 550-char | 25:30 | 400 | 77.0 |
| 650-char | 30:00 | 380 | 67.3 |
| 750-char | 27:00 | 280 | 62.3 |

LSF limits[6]: Ignition time 45 min
Combustion Efficiency 89%

16

## Conclusion Statements

1. Comparative Performance and Pollution Reduction Assessments
   - Low-volatile fuels (650-char,750-char) perform better in pollution (PM, $SO_2$ and VOC) reduction
   - High-volatile fuels (450-char, 550-char) were best performing in terms of efficiency, boiling and ignition time.
   - The high fixed carbon content of the high-temperature chars resulted in longer space heating times

2. Selection of Optimum Production Temperature
   - A decision Matrix taking into account all the assessed parameters showed that the 550-char (*i.e* 550°C) had the best average performance.

3. The way forward
   - Address shortcomings in ignition time and combustion efficiency
   - Use various stoves/combustion vessels in order to increase efficiency

   Further off into the future
   - Pilot Tests on best performing fuel
   - Ascertain the economic implication and possibility of mass production
   - design the facility to carry out the devolatisation and make LSF available to public

16

## Acknowledgements

This work is based on the research financially supported by the South African Research Chairs Initiative of the Department of Science and Technology and National Research Foundation of South Africa.

Any opinion, finding or conclusion or recommendation expressed in this material is that of the author(s) and the NRF does not accept any liability in this regard.

It all starts here

**Thank you**

# Experimental

NRF National Research Foundation

NORTH-WEST UNIVERSITY
YUNIBESITI YA BOKONE-BOPHIRIMA
NOORDWES-UNIVERSITEIT
POTCHEFSTROOM CAMPUS

Characterisation Results (%wy daf)

| | Coal | Anthracite | 450-Char | 550-Char | 650-Char | 750-Char |
|---|---|---|---|---|---|---|
| **Proximate Analysis** | | | | | | |
| % Volatile Matter | 33.33 | 13.19 | 21.77 | 17.87 | 7.82 | 3.90 |
| %Fixed carbon | 66.67 | 86.81 | 78.23 | 82.13 | 92.18 | 96.10 |
| **Ultimate Analysis** | | | | | | |
| % Carbon Content | 76.43 | 88.79 | 81.42 | 84.01 | 90.54 | 93.18 |
| % Hydrogen Content | 3.94 | 3.51 | 3.35 | 2.37 | 1.44 | 0.53 |
| % Nitrogen Content | 1.75 | 2.59 | 1.89 | 1.89 | 1.87 | 1.70 |
| % Oxygen Content | 16.53 | 3.42 | 12.29 | 8.69 | 5.16 | 3.55 |
| % Total sulphur | 1.35 | 1.68 | 1.05 | 3.04 | 0.99 | 1.04 |
| **Gross Calorific Value (MJ/kg)** | | | | | | |
| | 29.34 | 35.25 | 31.93 | 30.90 | 33.45 | 33.81 |

A-1

# Experimental

National Research Foundation | NORTH-WEST UNIVERSITY YUNIBESITI YA BOKONE-BOPHIRIMA NOORDWES-UNIVERSITEIT POTCHEFSTROOM CAMPUS

Combustion Efficiency Formula

$$\eta_{combustion} = 100 \times \left[1 - \frac{32.8 \times (M_{ASH} \times LOI)}{M_{START} \times G_{CV}}\right]$$

where:

| | |
|---|---|
| 32.8 | the gross calorific values ($G_{CV}$) of carbon in MJ/kg |
| $G_{CV}$ | the gross calorific value of the fuel in MJ/kg |
| $M_{START}$ | is the mass of fuel at the beginning of the test (i.e. 5 kg) |
| $M_{ASH}$ | the mass of remaining ash/unburnt fuel after ~16hour of burning |
| LOI | the loss on ignition in % calculated during the ashing of a 100g batch of ash as: |

$$LOI = \frac{100[g] - ASH_{out}[g]}{100}$$

**A-1**

# Repeatability Study: Coal

National Research Foundation | NORTH-WEST UNIVERSITY YUNIBESITI YA BOKONE-BOPHIRIMA NOORDWES-UNIVERSITEIT POTCHEFSTROOM CAMPUS

Statistical analyses performed on coal results to ascertain repeatability of results.

Ignition Time

| | Run1 | Run 2 | Run 3 | Average | Std Dev. | % Error |
|---|---|---|---|---|---|---|
| Ignition time (mm:ss) | 21:00 | 27:00 | 26:30 | 24:50 | 03:17 | 48.67 |

Temperature Measurements

| | Run 1 | Run 1 | Run 1 | Average | Std Dev. | % Error |
|---|---|---|---|---|---|---|
| **Maximum Temperature (°C)** | | | | | | |
| **Fire Bottom** | 747 | 1015 | 1022 | 928 | 156.80 | 62 |
| **Fire Core** | 723 | 969 | 650 | 780.6 | 167 | 78.7 |
| **Stove Top** | 334 | 399 | 425 | 386 | 46.9 | 44 |
| **Final Temperature (°C)** | | | | | | |
| **Fire Bottom** | 633 | 875 | 936 | 814.6 | 160.30 | 72 |
| **Fire Core** | 716 | 824 | 650 | 730 | 87.8 | 44 |
| **Stove Top** | 246 | 280 | 265 | 263.7 | 46.9 | 23.7 |

**A-2**

# Repeatability Study: Coal

National Research Foundation

NORTH-WEST UNIVERSITY
YUNIBESITI YA BOKONE-BOPHIRIMA
NOORDWES-UNIVERSITEIT
POTCHEFSTROOM CAMPUS

Only trends were observed for the gaseous emissions

NO Emissions

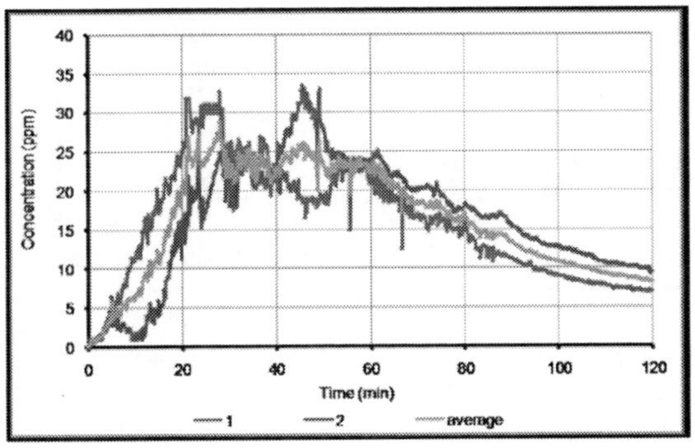


$SO_2$ Emissions

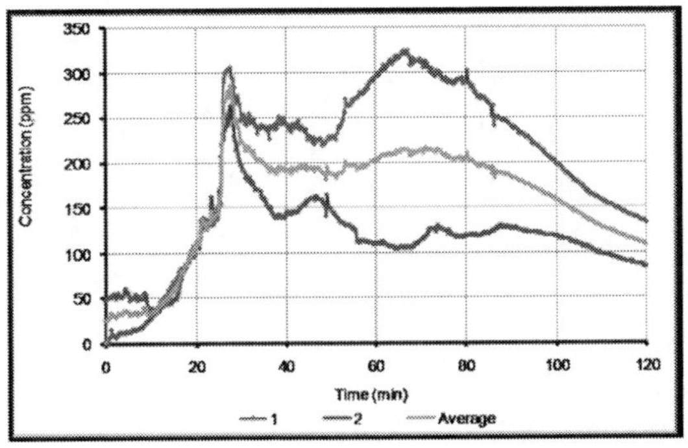


Trends were similar for the separate runs. All gaseous species were found t display realistic trends.

**A-3**


# Repeatability Study: Coal

National Research Foundation

NORTH-WEST UNIVERSITY
YUNIBESITI YA BOKONE-BOPHIRIMA
NOORDWES-UNIVERSITEIT
POTCHEFSTROOM CAMPUS

Only trends were observed for the gaseous emissions


CO Emissions

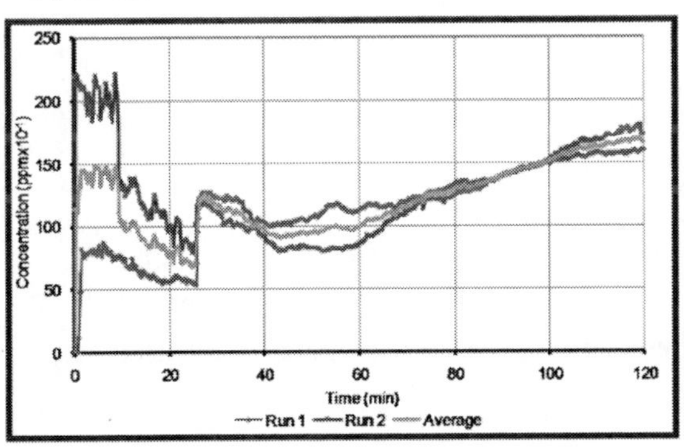


$CO_2$ Emissions

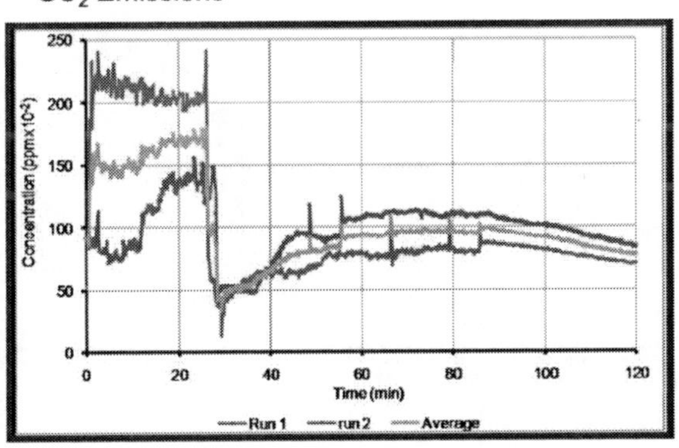



Trends were similar for the separate runs. All gaseous species were found t display realistic trends.


**A-4**

## Repeatability Study: Coal

NRF National Research Foundation — NORTH-WEST UNIVERSITY YUNIBESITI YA BOKONE-BOPHIRIMA NOORDWES-UNIVERSITEIT POTCHEFSTROOM CAMPUS

Gaseous Emissions: CO and $CO_2$

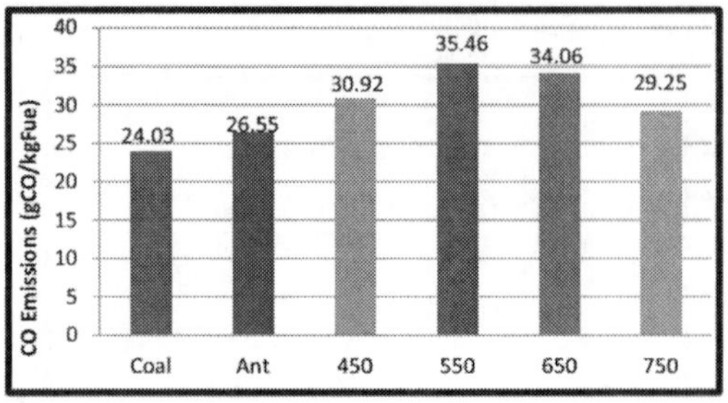


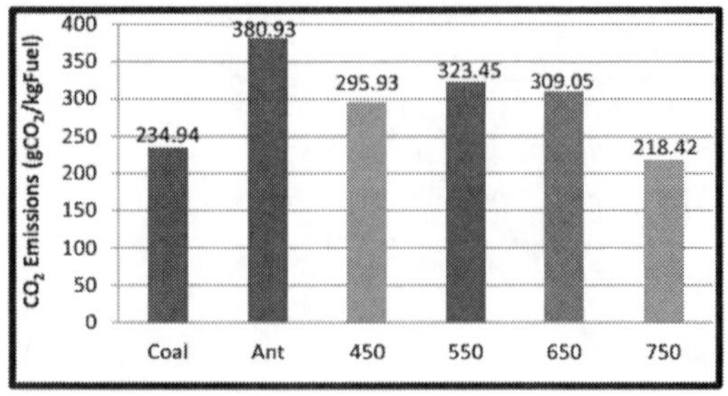


$CO:CO_2$ Emissions ratio
Anthracite – 14
450-char – 9.8
550-char – 9.1
650-char – 9
750-char – 7.5

$CO_2$ Emissions (% reduction/increase)
Anthracite – 10%
450-char – 29%
550-char – 48%
650-char – 42%
750-char – 22%

$CO:CO_2$ ratio speaks to efficiency of combustion
$CO_2$ is a greenhouse gas and a reduction in emissions is desired

**A-5**

## Repeatability Study: Coal

NRF National Research Foundation — NORTH-WEST UNIVERSITY YUNIBESITI YA BOKONE-BOPHIRIMA NOORDWES-UNIVERSITEIT POTCHEFSTROOM CAMPUS

Volatile Organic Compound Emissions: Aliphatic Compounds

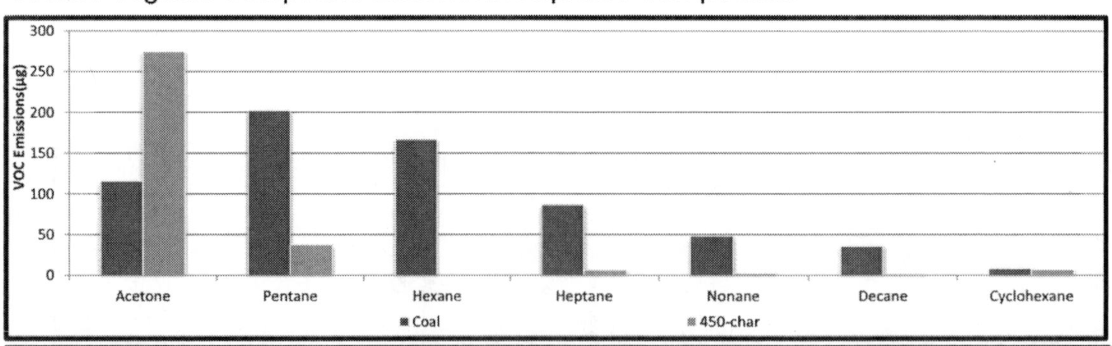


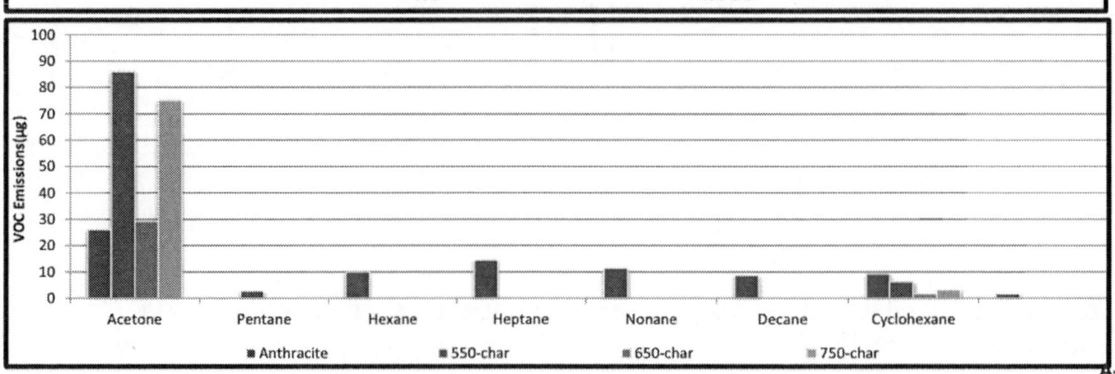


**A-11**

# Research of The Masking Effect of Surfactants On O-containing Functional Groups In Low-rank Coals

Geng Jianchun

(School of Chemical and Environmental Engineering, China University of Mining and Technology(Beijing), Beijing 100083, China)

**Abstract:** Pingshuo long flame coal and Shengli brown coal, which are quite different in metamorphic grade and are of low impurity content, were chosen as test samples in this experiment. The samples were surface modified with three kinds of surfactants, which are CTAB, SDS and Span-80. The surface modified samples were taken into floatation experiment. Masking effect of the surfactants on the main oxygen containing functional groups in coal under different modifying conditions(dosage of surfactants and stirring conditions) was researched. The relationship between floatability of coal and its oxygen containing functional group content was also researched. Results of the experiment showed that masking effect of surfactants altered as the surfactant type or modifying conditions changed. The adsorption of surfactants on coal surface could decrease the content of carboxyl groups and phenolic hydroxyl groups effectively and improve the floatability of low rank coals.

**Key words:** surfactant; surface modification; O-containing functional group; floatability

The efficiently wash and upgrading utilize of low rank coal is a hotspot of coal processing and utilization in recent years. Coal flotation is one of the most effective technologies that can make the target products separate from gangue and it is based on the differences of their surface properties. However, since there are lots of oxygen containing functional groups in low rank coals, which can absorb water molecules through hydrogen bond , and generate a stable hydrated film on the coal surface and hinder the collision and attachment between coal particles and bubbles, and make low rank coals very difficult to float[1-6]. It is concluded that carboxyl groups and phenolic hydroxyl groups are the most important functional groups that affect the wettability of coals through a series of researches on the oxygen containing functional groups and wettability of coals[7-11]. The floatability of low rank coals can get improved by surfactants through the interaction between surfactants and coal surface, which can make some oxygen containing functional groups masked[12-15]. There is no systematic research on the regularity of masking effect with different kinds of surfactants and under different surface modifying conditions(dosage of surfactants, rate and time of agitation) or on the relationship between oxygen containing functional group content and the floatability of coals. In this paper, Pingshuo long flame coal and Shengli brown coal were taken as test samples and surface modified with three kinds of surfactants respectively. The masking effect of surfactants on main oxygen containing functional groups in low rank coals were researched and the relationship between oxygen containing functional groups and the floatability of coals were explored.

## 1. Experiment

### 1.1 Samples and reagents

Pingshuo long flame coal and Shengli brown coal were chosen as test samples and both the samples were pretreated throuth float and sink experiment. The ash content of the samples were 11.95% and 5.94% respectively and were crushed to 0.15-0.074mm.

The surfactants used in the experiment were cetyl trimethyl ammonium bromide(CTAB, cationic surfactant), sodium dodecyl sulfate(SDS, anionic surfactant) and Span-80(nonionic surfactnt)[16-17].The constitutional formula of the three kinds of surfactants are shown in table 1.

Table 1 Constitutional formula of surfactants

| Surfactants | Type | Constitutional formula |
|---|---|---|
| CTAB | Cationic | 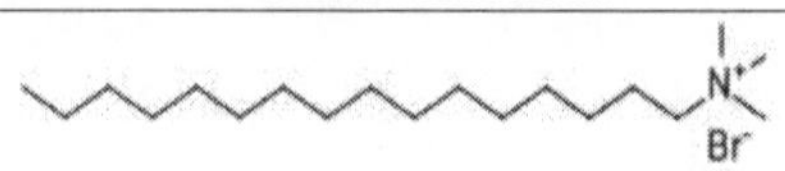 |
| SDS | Anionic |  |
| Span-80 | Nonionic | 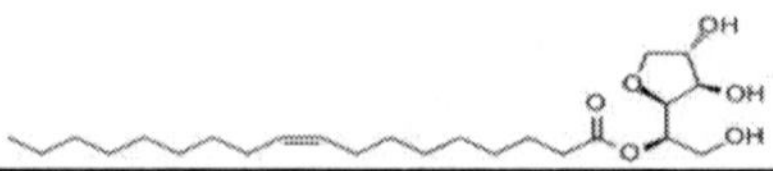 |

### 1.2 Experiment procedure

(1) Surface modifying

10 g prepared samples were taken into a beaker with 500ml water. Keep stirring until the samples were totally wetted. Surfactant was added into the slurry followed with stirring ,a certain rate of stirring were kept since the surfactant was added. The slurry was filtered and the filter cake was wash with water repeatedly. The filter cake was dried in a vacuum drying oven under 40℃ and surface modified samples were get.

(2) Determination of the oxygen containing functional group content

Oxygen containing functional group content was determined through ion exchange method in this experiment[18 19]. Based on the result of author's former research[20], ion exchange time for determining oxygen containing functional group content in Pingshuo long flame coal was chosen as 24h. Meanwhile, there are a great amount of small organic molecules in Shengli brown coal which are of lower metamorphic grade, and the small organic molecules can be easily dissolved during the ion exchange procedure and reduce the accuracy of the experiment. In order to reduce the interferences of small organic molecules, ion exchange time for Shengli brown coal was chosen as 2h.

(3) Coal floatation

Coal floatation was carried out with a modified Hallimond tube to raise the accuracy of floatation test. No collectors nor frother were added to avoid their interferences on the result. The procedure of floatation test was: take 1 g prepared samples into the Halimond tube and add a little water. Make all the samples wetted by stirring. Then fit the Halimond tube up with water. Keep the rate of stirring at 1600r/min and nitrogen was improted into the Halimond through a capillary. The floatation was finished after 5min and products were gathered and dried[21].

## 2 Effect of modifying conditions on the oxygen containing functional group content

Surfactant absorbed on the polar point of the coal surface with the polar head groups oriented towards the surface and nonpolar tails oriented to water. As some of the oxygen containing functional groups were masked by the surfactants, part of the hydrophilic coal surface get hydrophobic[22-23]. Type and dosage of surfactants and conditions for surface modifying are all important factors that affect the adsorption of surfactants.

### 2.1 Effect of the dosage of surfactants on the oxygen containing functional group content

Keep the rate of stirring at 1500 r/min and stirring for 2min with different dosage of surfactants. The content of carboxyl groups and phenolic hydroxyl groups in each sample is shown in Fig 1 and Fig 2. It can be concluded that:

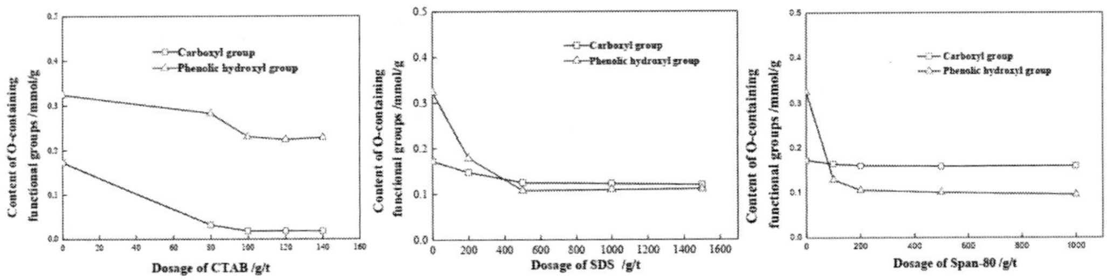


Fig.1 O-containing functional group content of Pingshuo long-flame coal with different dosage of surfactants

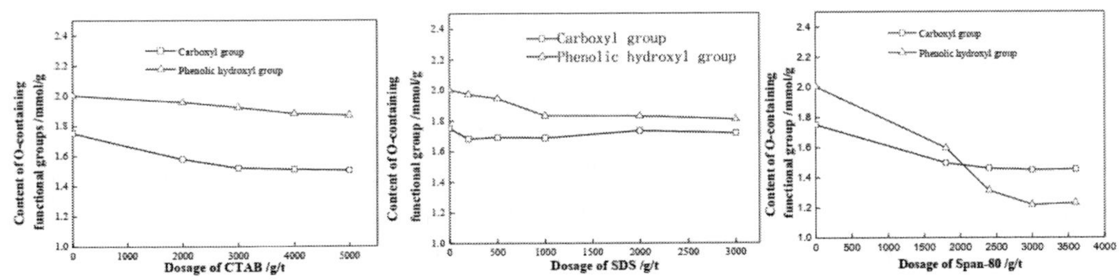


Fig.2 O-containing functional group content of Shengli brown coal with different dosage of surfactants

Table 2 Dosage of surfactants which generated the minimum carboxy group content and phenolic hydroxyl group content

| Surfactants | Dosage of surfactants corresponding to the minimum O-containing functional group content g/t | |
|---|---|---|
| | Pingshuo long flame coal | Shengli brown coal |
| CTAB | 100 | 4000 |
| SDS | 500 | 1000 |
| Span-80 | 200 | 3000 |

(1) Content of carboxyl groups and phenolic hydroxyl groups in Pingshuo long flame coal and Shengli brown coal decreased as the dosage of surfactants increased. The content decreased sharply as the dosage of surfactants is low and gradually gets gently as the dosage increased. The content of oxygen containing functional groups approached to the minimum value as the dosage of surfactant beyond a certain point . The minimum values of carboxyl group and phenolic hydroxyl group content in surface modified samples are shown in table 2.

(2) The maximum reduction of oxygen containing functional group in surface modified samples are shown in table 3. CTAB masked more carboxyl groups as it adsorbed on the coal, while SDS and Span-80 masked more phenolic hydroxyl groups. It was assumed that surfactant adsorbed on the coal surface with the polar head of it oriented to the coal surface. The dominating driving force for the adsorption of SDS and Span-80 was their hydrogen bond with the oxygen containing functional groups in coal. Since there is delocalization effect between phenolic hydroxyl group and its connected benzene ring and the conjugation lead to the decrease of the electron cloud density and increased of the positive electricity on the oxygen atom in phenolic hydroxyl group. Then the hydrogen bond between the sulfate radical of SDS or the ester group of Span-80 and phenolic hydroxyl group is enhanced and attributed to a better masking effect on phenolic hydroxyl grouos. While the he dominating driving force for the adsorption of CTAB was its electrostatic attraction with negatively charged oxygen containing functional groups, which is weakened by the delocalization effect. As a result, the adsorption of CTAB on phenolic hydroxyl groups was less[24-26].

Table 3 The minimum value of O-containing functional group content and corresponding decrement

| Surfactant | Pingshuo long falme coal | | | | Shengli brown coal | | | |
|---|---|---|---|---|---|---|---|---|
| | Carboxyl group mmol/g | Reduction mmol/g | Phenolic hydroxyl group mmol/g | Reduction mmol/g | Carboxyl group mmol/g | Reduction mmol/g | Phenolic hydroxyl group mmol/g | Reduction mmol/g |
| 无 | 0.1720 | - | 0.3240 | - | 1.7516 | - | 2.0040 | - |
| CTAB | 0.0182 | 0.1538 | 0.2602 | 0.0638 | 1.5067 | 0.2449 | 1.8708 | 0.1332 |
| SDS | 0.1250 | 0.0470 | 0.1076 | 0.2164 | 1.6874 | 0.0642 | 1.8084 | 0.1956 |
| Span-80 | 0.1598 | 0.0122 | 0.1055 | 0.2185 | 1.4512 | 0.3004 | 1.2199 | 0.7841 |

## 2.2 Effect of stirring conditions on the oxygen containing functional group content

Modify the surface of Pingshuo long flame coal and Shengli brown coal with three kinds of surfactants respectively under different stirring conditions( rate and time of stirring). The dosage of each surfactant was chosen according to table 2. Since some of the coal surface would be peeled since the strong friction and shearing action at high stirring rate or after long time of stirring, the rate of stirring was chosen as 500r/min, 1000r/min and 1500r/min, the time of stirring was chosen as 0.5min, 1min, 2min, 5min and 10min[27-29]. Result of each test is shown in Fig3 and Fig 4. It can be concluded that ,

(1) The content of oxygen containing functional group in Span-80 modified samples can get obviously decreased as the rate of stirring increased, while no similar phenomenon was found in SDS or CTAB modified samples. The reason is that Span-80 is difficult to get dissolved in water and its dispersion was accelerated remarkablely, which promoted its contact to coal particles and adsorption on coal particles. SDS and CTAB can get dissolved easily in water and their diffusion was not significantly affected by the rate of stirring.

(2) The masking action of surfactants on oxygen containing functional groups can be promoted with a longer stirring time. The dispersion of surfactants and coal particles would be promoted since the time of stirring was prolonged, and the probability of contact and adsorption was improved. While the time of stirring was prolonged, the dispersion of surfactants and particles gradually turned to a maximum level and no further improvement was found as the time of stirring prolonged more[30].

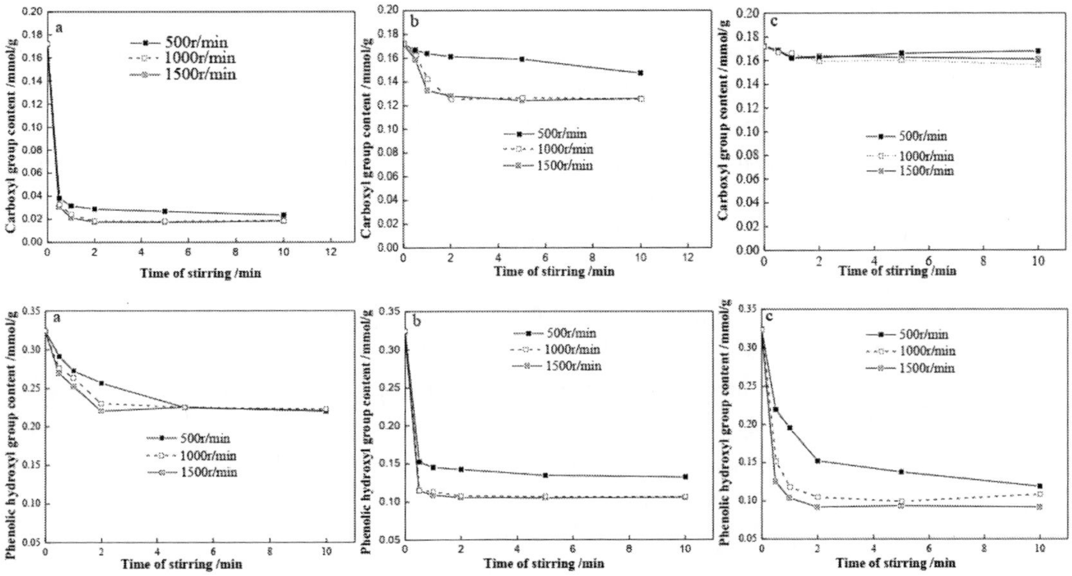


Fig.3 O-containing functional group content of Pingshuo long flame coal under different stir conditions

a-CTAB as surfactant b-SDS as surfactant c-Span-80 as surfactant

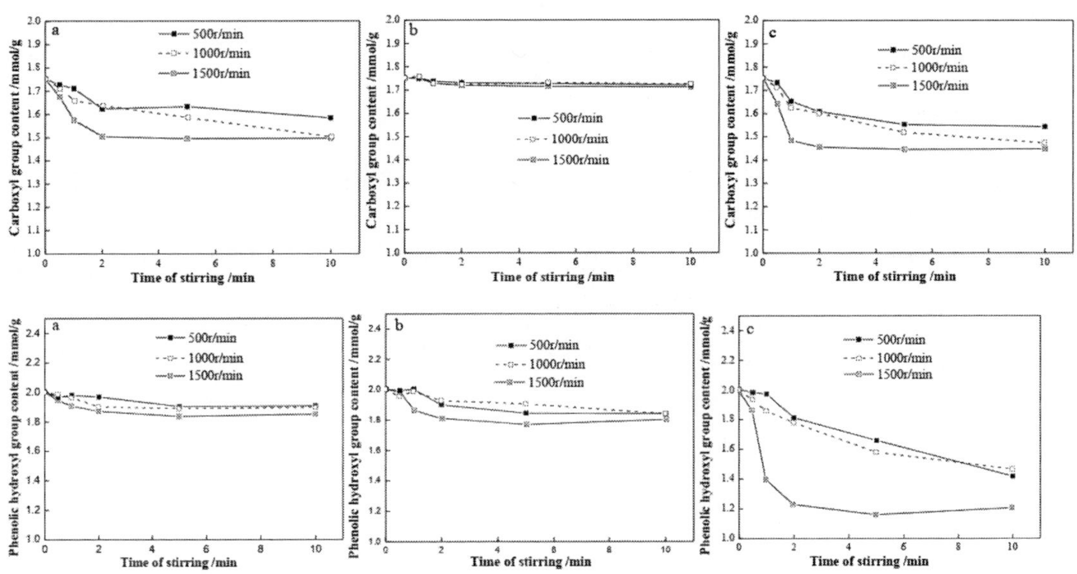


Fig.4 O-containing functional group content of Shengli brown coal under different stir conditions

a-CTAB as surfactant b-SDS as surfactant c-Span-80 as surfactant

## 3 Relationship between oxygen containing functional groups and floatability

Coal floatation test was carried out in a Hallimond tube with the surface modified samples. It should be noted that there was no difference in other characters except the content of oxygen containing functional groups between surface modified sample and its original coal. Referring to the method for evaluating floatability of bituminous coal and anthracite , the floatability of low rank coals were evaluated according to their combustible recovery. The relationship between combustible recovery and oxygen containing functional group content are shown in Fig 5 and Fig 6. The conclusion can be drawn that,

(1)No significant correlation between floatability and carboxyl group content was found for Pingshuo long flame coal, while the correlation between floatability and phenolic hydroxyl group content was obviously. The combustible recovery of Pingshuo long flame coal increased from 13.30% to 23.45% as the content of phenolic hydroxyl groups decreased. The reason is that there were much less carboxyl groups(0.1720 mmol/g) in Pingshuo long flame coal than phenolic hydroxyl groups(0.3240 mmol/g). Moreover, the variable of carboxyl group content caused by SDS or Span-80 was quite small and the corresponding change of wettability was not obviously. The carboxyl group content in CTAB modified samples distributed in a rather small range from 0.0179mmol/g to 0.038mmol/g and no corresponding change in combustible recovery could be found.

(2) The floatability of Shengli brown coal was well associated with its carboxyl group content and phenolic hydroxyl group content. As the content of carboxyl groups and phenolic hydroxyl groups decreased, the combustible recovery increased from 6.81% to 9.22%. The reason is that Shengli brown coal was of lower metamorphic grade and more oxygen containing functional groups. The reduction of any of the oxygen containing functional groups would improve the hydrophobicity of the coal significantly.

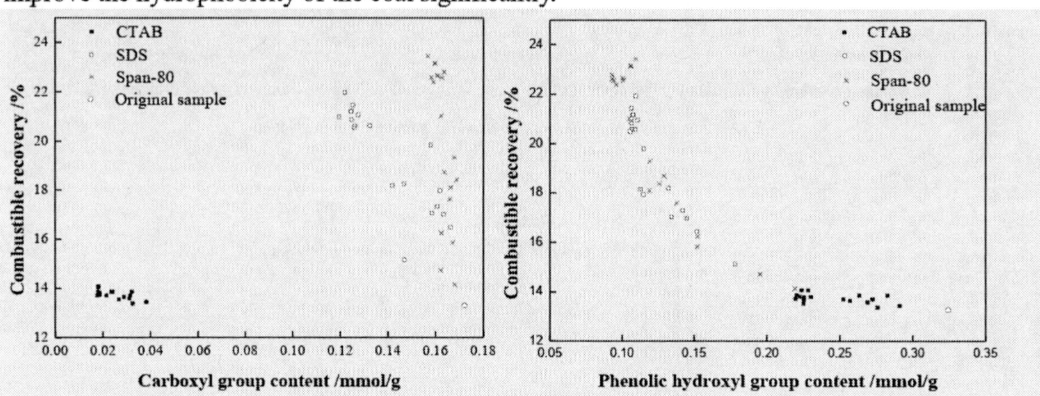


Fig.5 Combustible material recovery of Pingshuo long flame coal with different O-containing functional group content

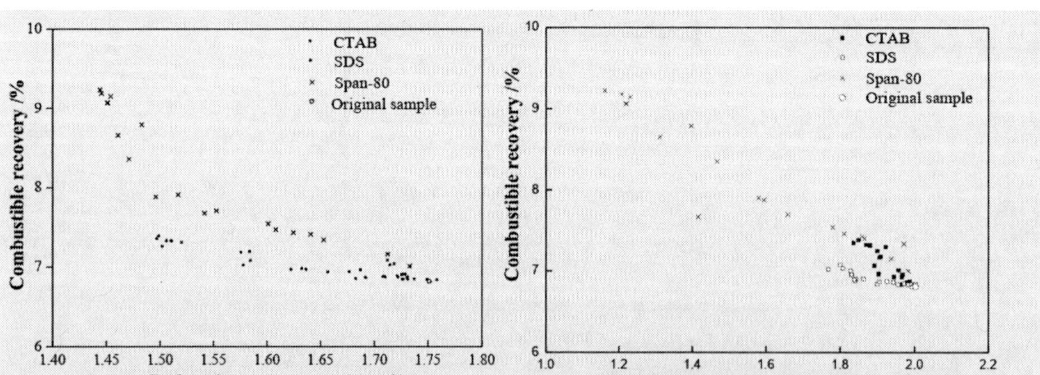


Fig.6 Combustible material recovery of Shengli brown coal with different O-containing functional group content

## 4 Conclusion

(1) Part of the oxygen containing functional groups were masked as surfactants adsorbed on the coal. Moreover, the masking effect of each type of surfactant was different. CTAB masked more carboxyl group than phenolic hydroxyl groups ,while SDS and Span-80 worked reversely.

(2) The dispersity of coal particles and surfactants could be improved as the rate of stirring increased or the time of stirring prolonged, which made the contact and adsorption of surfactant to coal particles promoted. The best modifying effect can get with a stirring rate of 1500r/min and a stirring time of 2min, and the oxygen containing functional group content get the minimum level.

(3) High oxygen containing functional group content is the main reason that lead to the poor floatability of Pingshuo long flame coal and Shengli brown coal. The combustible recovery of them could increase from 13.30% and 6.81% to 23.45% and 9.22% respectively as their surface modified with surfactants.

## References


1 Zhu Xuedong, Zhu Zibin, Han Chongjia,et al. Quantitative determination of oxygen containing functional groups in coal by FTIR spectroscopy[J].Journal of fuel chemistry and technology, 1999,(04):335-339.

2 D W Fuerstenau, John M Rosenbaum, J Laskowski. Effect of surface functional groups on the flotation of coal[J].Colloids and surfaces,1983,8:153-174.

3 Zhao Zhenguo. Principle of the application of adsorption[M].Beijing: Chemical industry press,2005:201.

4 Guo Mengxiong, Yu Qiang. Study on coal floatability[J].Journal of China coal society, 1987,03:15-22.

5 Li Chunzhu. Advances in the science of Victorian brown coal[M].Beijing: Chemical industry press,2009:16.

6 Sarikaya M. Flotation characteristics of oxidized coal[J].Fuel,1995,74:291-294.

7 Murada Teisen. Research and application of coal wettability[M].Beijing: China coal industry publishing house,1992

8 D W Fuerstenau. Oxidation phenomena in coal flotation part I. Correlation between oxygen functional group concentration, immersion wettability and salt flotation response[J].Coal preparation, 1987,4:161-182.

9 Wang Yonggang, Zhou Jianlin, Lin Xiongchao. Deposit conditions of oxygen functional groups in low rank coal and affect to surface properties[J]. Coal science and technology, 2013,(09):182-184.

10 Cheng Weimin, Xue Jiao, Zhou Gang et al. Study of coal dust wettability based on FTIR[J].Journal of China coal society, 2014,11:2256-2262.

11 Huang Xin, Zhang Shu, Lin Chaoxiong et al. Deoxygenation effect on hydraphilicity changes of Shengli lignite during pressurized pyrolysis at low temperature[J]. Journal of Fuel Chemistry and Technology, 2013,12: 1409-1414.

12 M S Jena, S K Biswal, M V Rudramuniyappa. Study on flotation characteristics of oxidized Indian high ash sub-bituminous coal[J]. International journal of mineral processing. 2008(87):42-50.

13 Hussin A M Ahmed, Jan Drzymala. Effect of flotation procedure and composition of reagents on yield of a difficult to float coal[J]. Physicochemical problems of mineral processing, 2004,38:53-63.

14 Li Xiang. Surface-coating improves the coal water slurry formation of low rank coal and biomass[D]. Zhejiang university, 2012.

15 Yakup Cebeci. The investigation of the floatability improvement of Yozgat Ayridam lignite using various collectors[J].Fuel,2002,81:281-289.

16 Bimal P Smingh. The role of surfactant adsorption in the improved dewatering of fine coal[J]. Fuel,1999,78:501-506.

17 R J Crawford, D E Mainwaring. The influence of surfactant adsorption on the surface characterization of Australian coals[J]. Fuel,2001,80:313-320.

18 Zhou Jianlin, Wang Yonggang, Huang Xin et.al. Determination of O-containing functional groups distribution in low rank coals by chemical titration[J].Journal of fuel chemistry and technology, 2013,(02):134-138.

19 D J Allardice, L M Clemow, W. R. Jackson. Determination of the acid distribution and total acidity of low rank coals and coal-derived materials by an improved barium exchange technique[J]. Fuel,2003,82:35-40.

20 Geng Jianchun, Liu Wenli, Zhao Jili. Research of test conditions of carboxyl groups in low rank coals[J]. Coal science and technology, 2014,(09):139-141.

21 Xu Shi. Research of the ore washability[M].Beijing: Metallurgical industry press, 1989:166-167.

22 Wencheng Xia, Jianguo Yang, Chuan Liang. A short review of improvement in flotation of low rank/oxidized coals by pretreatments[J].Powder technology,2013,237:1-8.

23 Han Yanna. Influence of the interaction between inorganic electrolyte and surfactant on the surface properties and floatability of low rank coal[D]. Taiyuan university of technology,2011.

24 S Paria, K C Khilar. A review on experimental studies of surfactant adsorption at the hydrophilic solid-water interface [J].Adv. Colloid Interface Sci., 2004,110:75-95.

25 R Zhang, P Somasundaran.Advances in adsorption of surfactants and their mixtures at solid/solution interfaces[J]. Adv. Colloid Interface Sci.,2006,123-126:213-229.

26 Cui Zhenggang. Fundamentals of surfactants, colloids, and interface chemistry[M].Beijing: Chemical industry press.

27 Ren Huailiang. Study on flotation technology to treat difficult floated fine coal[J].Coal science and technology, 2011,39:103-107.

28 Li Yanfeng, Zhang Xiaobo, Gui Xiahui. et al. The morphological feature of hard to float fine coal and enhancing its floatability by stirring[J]. Journal of China university of mining and technology, 2012,(06):930-935.

29 Ma Liqiang, Chen Qingru. Study status and progress on slime water slurry blending technology and equipment[J]. 2011,(04):11-16.

30 Huang Gen. Interfacial effects of flotation conditioning and process intensification[D]. China university of mining and technology.

**2016 International Pittsburgh Coal Conference**
**Cape Town, SOUTH AFRICA**
**August 8 – 12, 2016**

# SOOT GASIFICATION IN SYNGAS ATMOSPHERE

***G. Worringer,***
***Linde AG, Dr.-Carl-von-Linde-Str. 6-14, 82049 Pullach, Germany,***
***gregor.worringer@linde-le.com, Phone: +49 89 7445 – 4646, Fax: +49 89 4943***

*R. Reimert, Prof. Dr.-Ing.,*
*Karlsruhe Institute of Technology (KIT), Engler-Bunte-Institut, Division of Fuel Chemistry and Technology, Karlsruhe, Germany*
*rainer.reimert@kit.edu, Phone: +49 721 608 – 44891, Fax: +49 721 9640227*

## Abstract

An experimental set-up for measuring the gasification reactivity of soot in dependence of temperature and soot age was installed and tested. Soot was produced from a mixture of propane ($C_3H_8$) and $N_2$ by pyrolysis at a maximum temperature of 1,200 °C in a ceramic tube reactor. The soot was collected in a packed bed of ceramic particles and $CO_2$ was added directly upstream the loaded filter in order to gasify the solid carbon at temperatures of 1,100 – 1,300 °C. During experiments, besides typical soot, a graphitic carbon material was found. Formation rate of the second carbon material as well as gasification rates of both carbon materials were calculated.

## 1. Introduction

Entrained flow gasification (EFG) of liquid feedstocks is prone to soot formation due to its process conditions: low oxygen content, non-ideal gas phase mixing, temperatures of 1200°C and above, as well as elevated pressure. Soot consists of nanoscale carbon particles formed in a series of complex radical polymerization reactions and physical processes. The reactions start from a pool of molecule fragments formed during feedstock pyrolysis preceding gasification. The most common liquid feedstocks for EFG are heavy residual oils. However, gasification of low rank biomass derived fuels, e. g. bio-syncrudes originating from flash pyrolysis, comes currently into focus for subsequent syntheses of bio-based chemicals or fuels.

Soot particles are usually associated with fly ash which is removed from the raw synthesis gas to prevent deposition and fouling in piping and downstream equipment as heat exchangers, cleaning processes and chemical reactors. In commercial size plants this is done by water scrubbing. However, the produced soot water has to be cleaned with high technical effort which creates considerable capital and operational expenditures [1]. Furthermore, cooling the gas in the scrubber contradicts an energy efficient process topology characterized by raw gas cleaning at the temperature level of the subsequent process chain. By this way, costly equipment for cooling and reheating and efficiency losses could be avoided. Hot gas cleaning processes as an alternative are subjects of current R&D projects

[2], [3]. High temperature filters applied in these processes will benefit if particle loading is low already right at the gasifier's raw gas exit.

Reduction of the soot loading could be achieved by increasing the soot conversion either directly in the gasifier´s post flame region, or by gasification of the soot on the surface of the filter candles. The presented work aims at identification of the kinetic parameters of soot conversion reactions in a synthesis gas environment. Regarding soot gasification in the filter influence of soot aging as a result of tempering on its reactivity has to be considered.

In a laboratory set-up the reaction rate of fresh soot and of soot aged by tempering at gasification temperature and atmospheric pressure at the reactor outlet was measured.

## 2. Symbols and Definitions

| Symbol | Description | Definition | Unit |
|---|---|---|---|
| $A$ | Surface | $L^2$ | $m^2$ |
| $a_m$ | Mass specific surface | $A/m$ | $m^2$ / kg |
| $c_i$ | Concentration | $n_i/V$ | mol / $m^3$ |
| $d$ | Diameter | $L$ | m |
| $F$ | Force | various | N |
| $k$ | Reaction rate coefficient | $^{m}Q/f(c,A,X))$ | diverse |
| $L$ | Length | Basic | m |
| $M$ | Molar mass | $m/n$ | kg / mol |
| $m$ | Mass | Basic | kg |
| $n$ | Molecular amount of species | Basic | mol |
| $p$ | Pressure | $F/A$ | Pa |
| $^{m}Q$ | Reaction rate | $dm/dt$ | kg / s |
| $^{m}S$ | Deposition rate | $dm/dt$ | kg / s |
| $T$ | Temperature | Basic | K |
| $t$ | Time | Basic | s |
| $V$ | Volume | $L^3$ | $m^3$ |
| $X$ | Conversion | $1-m/m_0$ | - |
| $y_i$ | Mole fraction | $n_i/n$ | - |
| $Y_C$ | Yield | $\frac{\nu_j \cdot dn_i/dt}{\nu_i \cdot dn_j/dt}$ | - |

| Symbol | Description | Definition | Unit |
|---|---|---|---|
| $x_i$ | Mass fraction | $m_i/m$ | |
| $\varepsilon_H$ | Porosität | $V_H/V_{total}$ | - |
| $^m\Phi$ | Mass transport flow | $dm/dt$ | mole / s |
| $^V\Phi_N$ | Standard volume flow rate | $dV/dt$ | mole / s |
| $\eta$ | Deposition efficiency | $(c_{in}-c_{out})/c_{in}$ | - |
| $\vartheta$ | Temperature | $T + 273{,}15\ K$ | °C |
| $v_m$ | volume density | $V_{Po}/m$ | $m^3/kg$ |
| $v_{m,cum}$ | cumulated volume density | $\sum_i^N v_{m,i,Po}$ | $m^3/kg$ |
| $\rho$ | Density | $m/V$ | $kg/m^3$ |

## 3. Experimental

A flow diagram of the experimental set-up is shown in Figure 1. Its main part is a 34 mm diameter reaction tube of alumina in which gas mixtures flow from top to down. The tube is heated over a length of 2 m with three equivalent electric furnaces. In the tube's lower section a filter of randomly packed alumina particles is embedded. The filter consists of a 40 mm high support layer of 1 - 3 mm diameter particles, a 100 mm filter layer of particles smaller than 0.5 mm, and a 50 mm top layer of 0.5 - 1 mm particles. Directly upstream of the filter bed gases can be injected additionally by means of an injection tube. The flow rates of all the feed gases are controlled via mass flow controllers. The product gas components CO, $CO_2$ and $CH_4$ are measured in-line in time intervals of one second by infra-red analyzers, and $H_2$ is measured by means of a heat conductivity analyzer. Hydrocarbons were measured discontinuously via gas chromatograph (GC). The pressure drop over the filter is measured continuously by pressure transmitters, and the temperature in the filter is measured by thermocouples inserted in the injection tube.

For collecting un-treated soot samples the reactor tube is used without the filter bed, and soot is then collected in quartz glass filters downstream of the reactor tube. The filter cake is blown out discontinuously by $N_2$ and collected in the soot sampling vessel.

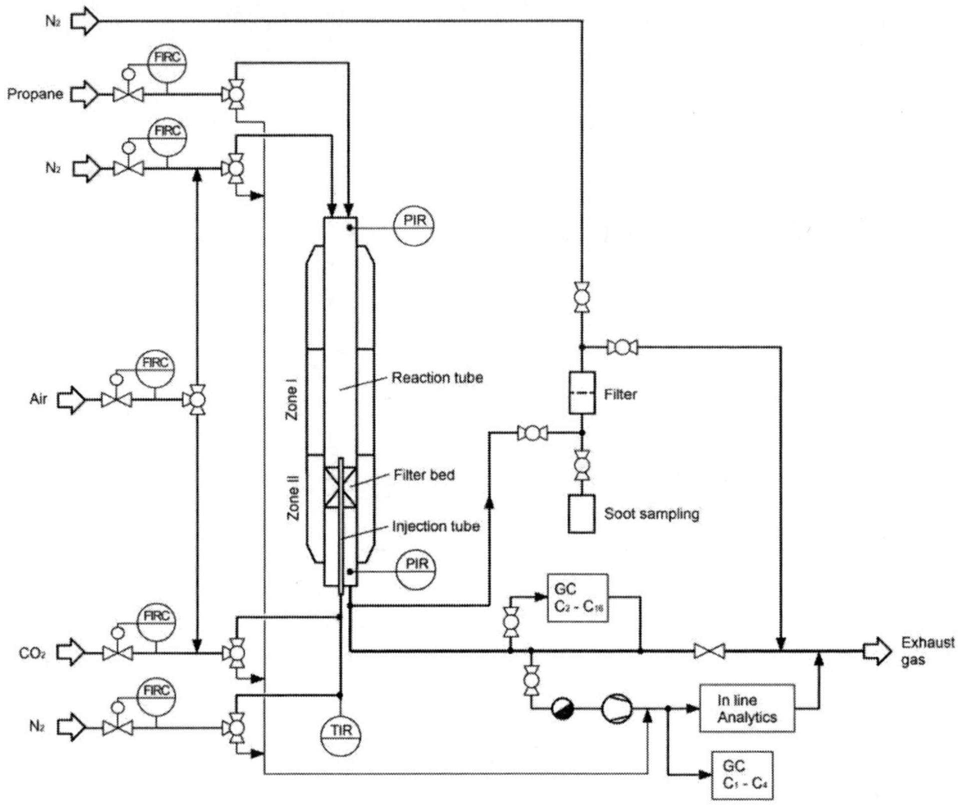


Figure 1: Flow diagram of experimental set-up

The measuring procedures of the two types of experiments performed are shown in Figure 2 for consecutive, and in Figure 3 for simultaneous gasification experiments. They differ by the reactor feeding sequences. All measurements where performed at 10 mbar above ambient pressure.

**Consecutive gasification:**

After flushing the reactor with nitrogen a mixture of 0.2 mole % propane in nitrogen is routed through the reaction tube (step b in Figure 2). Thereby soot is produced in zone I of the reactor at a maximal temperature of 1,200 °C and a residence time of 2 seconds. After 30 min of soot collection zones I and II are flushed with nitrogen for 1 min. Thereafter, $CO_2$ is injected into zone II (step d) resulting in a gas mixture of 20 mole % of $CO_2$ in $N_2$ for gasifying the soot collected in the filter. Measurements were performed at gasification temperatures of 1,100, 1,150, 1,200 1,250 and 1,300 °C, respectively. The main gasification reaction is:

$$C + CO_2 \longrightarrow 2\,CO \qquad \Delta_R H_{298K} = 173 \text{ kJ/mol} \qquad (1)$$

The whole reactor was flushed with nitrogen for 1 min (step e) before remaining carbon was quantified by combustion and concurrent $CO/CO_2$ measurement. First, air was injected through the injection tube (step f) to burn the carbon remaining in the filter, and, finally, air was fed to the reactor inlet (step g) to burn also the carbon in the reaction tube upstream of the filter.

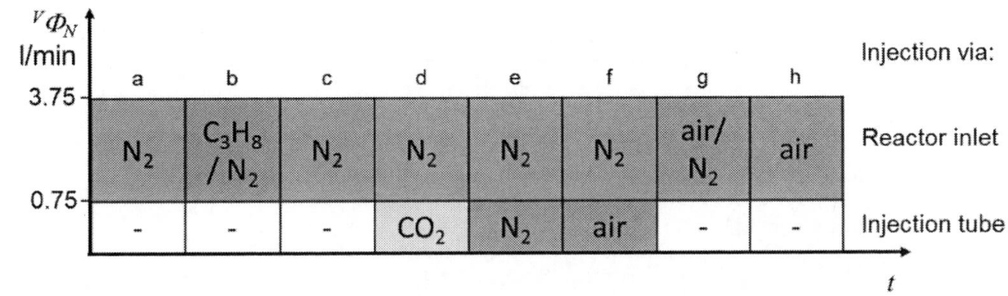


Figure 2: Sequence of consecutive gasification experiment. Molar gas concentrations: $y_{C3H8, zone\ I}$ = 0.2 %, $y_{CO2, zone\ II}$ = 20 %, $y_{air}$ = 23.8 %, rest: $N_2$.

**Simultaneous gasification:**

The simultaneous gasification experiments differ from the consecutive mainly in step b as $CO_2$ is injected additionally to the reactor gas flow above the filter. This allows that soot produced in the reactor zone I is simultaneously gasified downstream in zone II in the filter bed where reaction (1) takes place. As hydrogen is produced in zone I by propane pyrolysis, CO and $H_2O$ can be generated by the reverse water gas shift reaction (2) in zone II.

$$CO_2 + H_2 \longrightarrow CO + H_2O \qquad \Delta_R H_{298K} = 41.2 \text{ kJ/mol} \qquad (2)$$

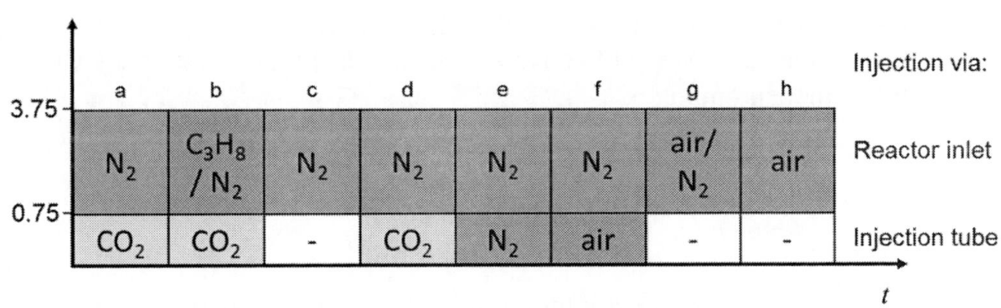


Figure 3: Sequence of simultaneous gasification experiment. Molar gas concentrations: $y_{C3H8, zone\ I}$ = 0.2 %, $y_{CO2, zone\ II}$ = 20 %, $y_{air}$ = 23.8 %, rest: $N_2$.

## Data processing:

For data evaluation the reactor is artificially divided in zone I (pyrolysis) and zone II (filter bed) as shown in Figure 4. There, all the feed streams $^m\Phi$ of $C_3H_8$, $N_2$, $CO_2$ and air are shown as well as the intermediate not measured streams from zone I to zone II. Also shown are soot formation rate $^mQ_{C,I}$, gasification rate $^mQ_{C,II}$, wall deposition rate in zone I $^mS_{C,I}$ and filter deposition rate in zone II $^mS_{C,II}$. The gasification rates were calculated for time steps of 1 s by balancing the measured inlet and outlet mass flows.

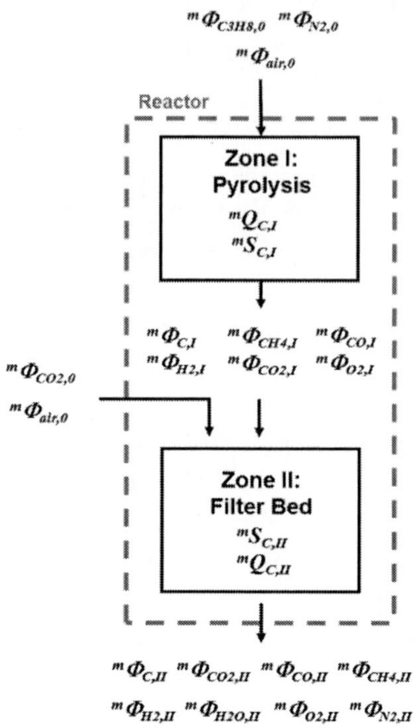


Figure 4: Component streams and reaction and deposition rates used for reactor balancing including formation of CO in zone I

For the **consecutive gasification** experiments the carbon mass in the filter $m_{c,f}$, collected during pyrolysis is determined indirectly via the CO stream during subsequent gasification (step d in Figure 2) and via $CO_2$ stream during combustion in reactor zone II (step f). Assuming stationary conditions, the soot conversion is directly proportional to the CO stream determined by in-line measurement:

$$\frac{dm_{c,f}(t)}{dt} \propto {}^m\Phi_{CO,II} \tag{3}$$

Obtaining $^mQ_{C,II}$ from **simultaneous gasification** experiments is more complex, because pyrolysis and gasification take place at the same time in zone I and II, respectively. Starting point is the soot balance of the reactor, equation (4).

$$^mQ_{C,I} = {}^mS_{C,I} + {}^mS_{C,II} + {}^mQ_{C,II} + {}^m\Phi_{C,II} \tag{4}$$

Filter soot mass $m_{c,f}$ is obtained by integration of $^mS_{C,II}$ over time. To calculate the latter from equation (4) it is necessary to make the following assumptions:

- The soot production rate is constant:

$$^mQ_{C,I} = 3 \cdot {}^m\Phi_{C3H8,0} \cdot Y_C \tag{5}$$

- Soot yield $Y_C$ (equation 6) is obtained from pyrolysis measurement under same conditions (Figure 2, step b). Soot mass $m_{c,I}$ is obtained by balancing the carbon flows during the measurement, and $m_{C3H8}$ is the propane mass which was fed to the reactor.

$$Y_C = \frac{1}{3} \cdot \frac{m_{C,I}}{m_{C3H8,0}} \tag{6}$$

- The soot stream leaving the reactor $'''\Phi_{C,II}$ is also assumed to be constant. By use of the deposition efficiencies $\eta_I$ and $\eta_{II}$ it can be written:

$$'''\Phi_{C,II} = 3 \cdot Y_C \cdot {}'''\Phi_{C3H8,0} \cdot (1-\eta_I) \cdot (1-\eta_{II}) \tag{7}$$

with wall deposition efficiency: $$\eta_I = \frac{m_{C,w}}{m_{C,I}} \tag{8}$$

and filter deposition efficiency: $$\eta_{II} = \frac{m_{C,f}}{m_{C,I} \cdot (1-\eta_I)} \tag{9}$$

- In order to determine the gasification rate $'''Q_{C,II}$ the reverse water gas shift reaction (2) and the formation of CO from reactor wall have to be taken into account. Reaction (2) is included by calculating $H_2O$ from an H-balance. Values for $''\Phi_{CO,I}$ formed in reactor zone II are taken from pyrolysis measurements under same conditions (Figure 2 step b).

$$'''Q_{C,II} = 0.5 \cdot M_C \cdot ({}'''\Phi_{CO,II} - {}'''\Phi_{H2O,II} - {}'''\Phi_{CO,I}) \tag{10}$$

Soot adhered at wall $m_{c,w}$ in zone I is determined by combustion in steps g and h in Figure 3. The soot mass produced in the experiment $m_{c,I}$ is calculated by integration of equation (5) over time.

## 5. Results

The gas analyses taken at the reactor outlet (see Figure 5, left) show that propane is converted completely to $H_2$, $CH_4$ and soot. The gas chromatographs did not detect other hydrocarbons than $CH_4$. During the first six minutes small amounts of CO were found. Possibly CO results from oxygen compounds adhered at the ceramic surface. In the course of pyrolysis the filter is loaded with soot indirectly shown by an increase of the pressure drop $\Delta p$.

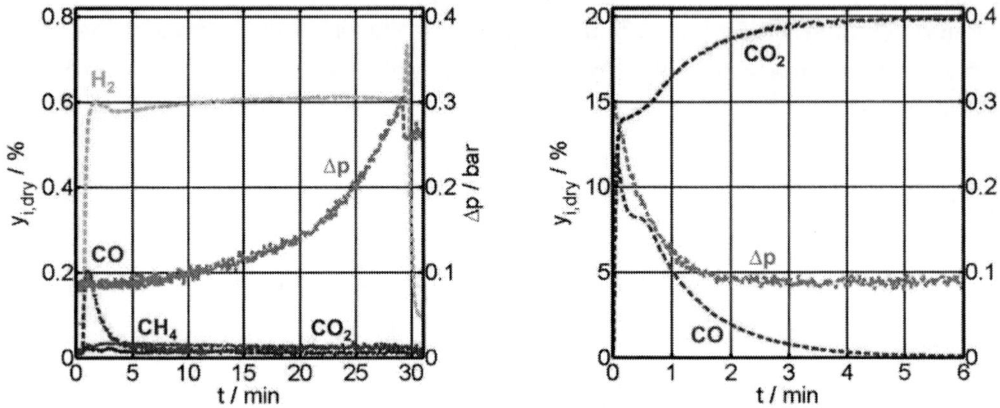


Figure 5: Composition and pressure drop during propane pyrolysis (left) and consecutive gasification (right) at 1,200°C, Feed: ${}^{V}\Phi_{N,C3H8}$ = 0.3 l/h, $y_{C3H8}$ = 0.2% und $y_{CO2}$ = 20%.

The conversion of the soot during consecutive gasification in Figure 5, right is shown by the decrease of $\Delta p$. The CO production decreases in the first minutes to a very low level of $y_{CO}$, an indication that solid carbon was still gasified. When the $CO_2$ injection was stopped after 30 min, not shown in Figure 5, $y_{CO}$ decreased to zero.

In the case of simultaneous gasification, after 45 min all detected gas components have constant concentrations (see Figure 6, left). This state of stationarity was not reached at gasification temperatures below 1,200 °C, e.g. at 1,160°C. The state of stationarity is also proven by the constant $\Delta p$. It is notable that no hydrogen was detected at the reactor outlet. At the temperature level of the experiment the reaction rate of the reverse water gas shift reaction obviously is high enough for a complete conversion of hydrogen to $H_2O$.

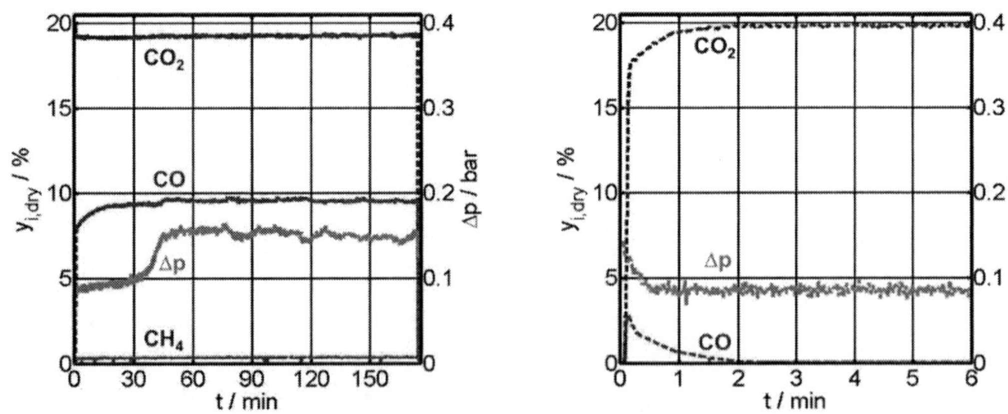


Figure 6: Composition and pressure drop during propane pyrolysis and simultaneous soot gasification (left) and subsequent consecutive gasification (right) at 1,200°C. Feed: ${}^{V}\Phi_{N,C3H8}$ = 0.3 l/h, $y_{C3H8}$ = 0.2% und $y_{CO2}$ = 20%.

The filter soot produced in this kind of experiment behaved during gasification like the one from the strict consecutive tests as a comparison of the right sides of Figure 5 and 6 shows.

However, the concentration of CO at the beginning is significantly lower. Reason might be a lower $CO_2$ residence time with respect to the soot due to a thinner soot layer in the filter. The first peak is also followed by a constant CO production on low level. The residual carbon mass in the filter was $m_{Cf}$ = 153 mg after 30 min of pyrolysis, and $m_{Cf}$ = 48 mg after 175 min of simultaneous gasification, respectively .

The conversions $X$ of the pyrolysis soot and of the residual soot after simultaneous gasification are shown in Figure 7. The fast conversion during the first minutes changes quickly to a very slow conversion. The change takes place for the pyrolysis soot at 1-$X$ = 0.1 and for the residual soot at 1-$X$ = 0.68 and 1-$X$ = 0.74 depending on duration of the previous simultaneous gasification.

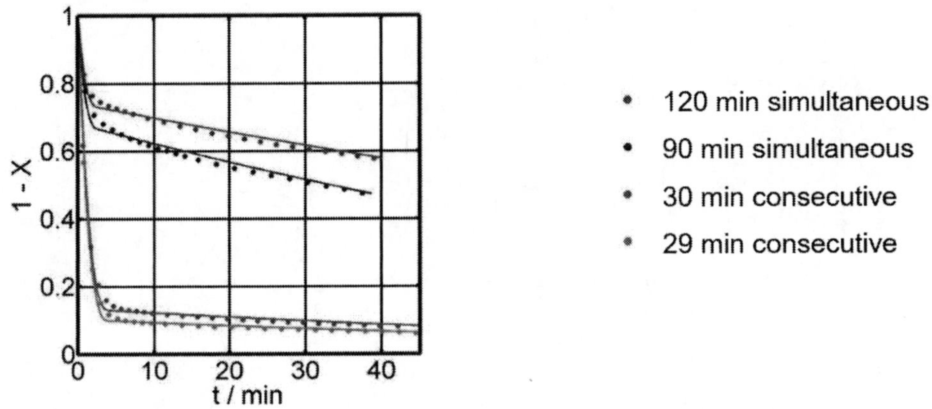


Figure 7: gasification with $CO_2$ at 1,200°C, comparison of calculation (lines) with measurements (dots). Soot from pyrolysis (left), residual soot from simultaneous gasification measurements of different duration (right).

The shape of the conversion curves can be explained with the presence of different kinds of carbon materials found in the reaction tube after cool-down under nitrogen flushing. The inner reactor wall was covered with a black soot powder, Figure 8, top. Below this layer brittle platelets with silvery color were revealed - obviously graphitic carbon, Figure 8, bottom. Images of transmission electron microscopy (TEM) of both carbon materials are also shown in Figure 8. Transmission electron microscopy (TEM) images show for the soot the typical clusters of primary particles with an average diameter of 23 nm. The primary particles are randomly structured. The graphitic carbon shows spherical and tube shaped geometries with long interference lines rich in contrast indicating a high degree of graphitization. While the spheres obviously originate from soot particles which were altered at high temperature in $CO_2$ atmosphere the nanometer sized tubes were probably formed by catalytic interaction with the reactor wall. The high degree of graphitization explains the lower activity for gasification in comparison with the soot particles.

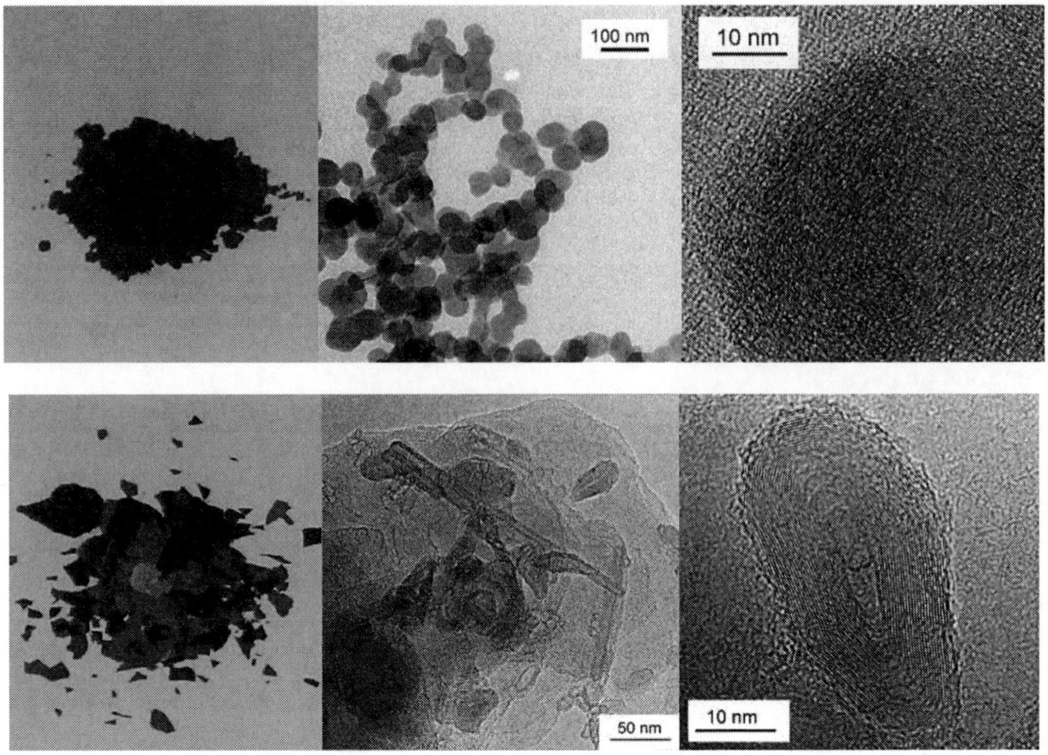


Figure 8: Photos (left) and TEM-picture of carbon materials found in reactor: soot (above) and graphitic carbon (bottom).

The results of the ultimate analyses of the soot are:

$x_{C,daf}$ = 95.3% ± 0.1, $x_{H,daf}$ = 0.7 ± 0.1, $x_{N,daf}$ = 0.3 ± 0.0%, $x_{Rest,daf}$ = 3.8 ± 0.1; neither sulphur nor ash. These data are typical for soot from combustion processes and for carbon black [5], [6].

The gas adsorption analysis revealed a moderate specific BET area of $a_{m,BET}$ = 65 $m^2/g$.

## 6. Data evaluation

To determine soot and graphitic carbon gasification reactivity (kinetic rate coefficients) the reaction scheme shown in Figure 9 is used. It reflects the finding that the fresh soot is gasified with $CO_2$ with the reaction rate $^{m}Q_1$ and in parallel forms the graphitic carbon material (graph) with reaction rate $^{m}Q_2$. The graphitic material is then gasified with $CO_2$ with reaction rate $^{m}Q_3$.

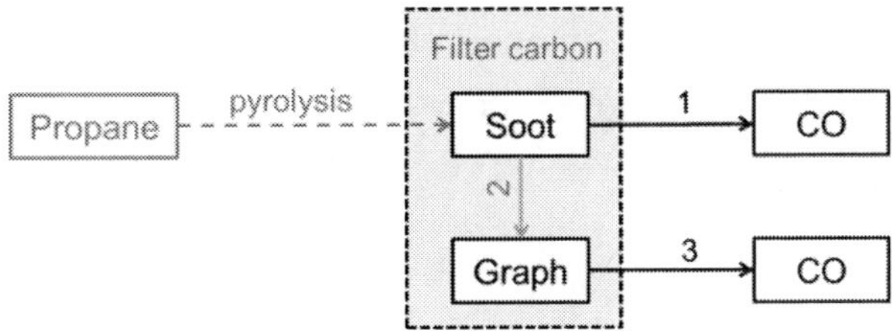


Figure 9: Reaction scheme for gasification and aging of soot. Orange colored reaction steps considered for simultaneous gasification solely.

For the rate equations of reactions 1 and 3 a reaction at the surface is assumed while reaction rate 2 is considered to be proportional to the actual mass:

$$^{m}Q_1 = -k_1 \cdot m_{soot}^{2/3} \quad (11) \qquad ^{m}Q_3 = -k_3 \cdot m_{graph}^{2/3} \quad (12)$$

$$^{m}Q_2 = k_2 \cdot m_{soot} \quad (13) \qquad m_{C,f} = m_{soot} + m_{graph} \quad (14)$$

$$^{m}Q_{C,II} = \left(^{m}Q_1 + ^{m}Q_2 + ^{m}Q_3\right) \quad (15)$$

The temperature dependences of coefficients $k_1$, $k_2$, and $k_3$ are described by the Arrhenius equation. The parameters $k_1$ and $k_3$ were determined by numerical fitting of model equations (11), (13) and (14) to the (indirectly) measured masses of filter carbon $m_{C,f}$ as a function of time obtained from consecutive gasification at temperatures of 1,100, 1,150...1,300 °C. Thereby, it was assumed that during step d graphitization reaction 2, equation (13) is much slower than gasification reactions and can therefore be ignored when determining coefficients $k_1$ and $k_3$. Some results are shown in Figure 7 exemplarily.

Coefficient $k_2$ was obtained afterwards by numerical fitting the whole set of equations (11) to (15) to the mass of filter carbon $m_{C,f}$ at the end of the simultaneous gasification experiment, detected during sequences d, f, g, h as shown in Figure 3. Thereby fitted values of $k_1$ and $k_3$ from consecutive gasification were used and $k_2$ was adjusted. In Figure 10, left side, measured and calculated values $^{m}Q_{C,II}$ are compared.

After a running time of 2 min calculated and measured values correspond fairly well. This shows that the soot gasification rates in the consecutive and in the simultaneous experiments are practically equal. In the first two minutes the measured values are above the calculated. This could be an indication that the reactivity of the fresh soot is even higher at the beginning of the experiment. However, the effect can also be related to the oxigenic influence of the reactor surface, observed during the pyrolysis measurement. Therefore, the reactivity of the soot in the very first seconds of simultaneous gasification experiment cannot be quantified at this time.

On the right side of Figure 10 the masses of soot and graphitic carbon are shown together with the experimental and calculated filter carbon mass. The soot mass remains constant after the first few seconds. The mass of graphitic carbon increases monotonously during the whole measurement but with decreasing slope. In spite of the increase of carbon mass in the filter the pressure difference remained constant over time (Figure 6). This can be explained with the differences in density of soot and graphitic carbon. Graphitic carbons exhibit a more than 30 times higher bulk density ($\rho_{Graphit}$ = ca. 1800 kg/m$^3$ [7]) than soot which accumulates voluminously from the gas phase. Hence, the volume of the graphitic carbon formed in the filter bed is small enough to leave the gas flow practically unaffected.

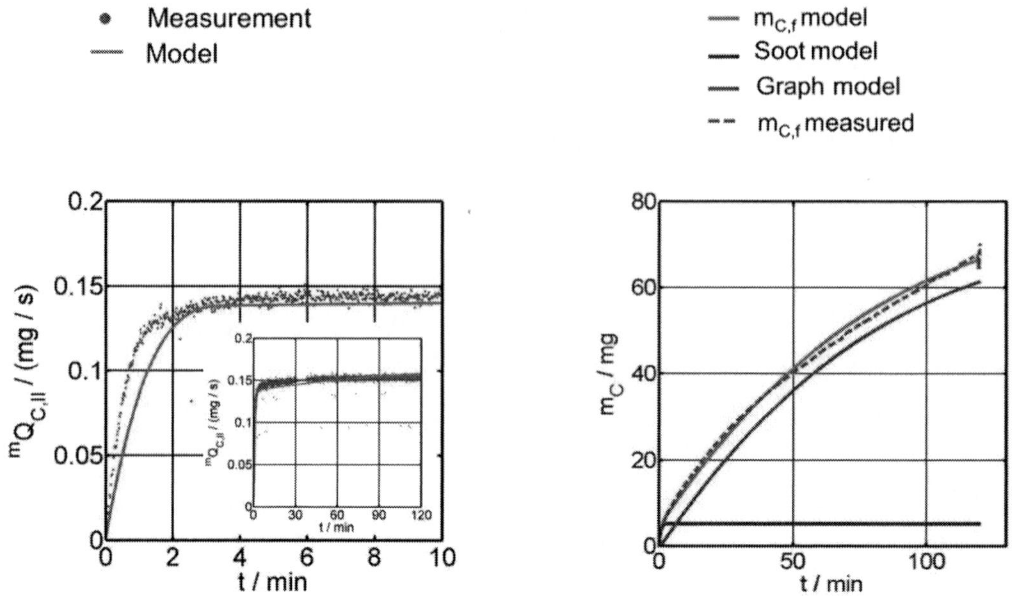


Figure 10: Comparison of measured and with equation (15) calculated gasification rates for simultaneous gasification experiment at 1,200 °C.

Figure 11 shows the logarithms of the reaction coefficients in the Arrhenius plot. Over a temperature range from 1,100 °C to 1,300 °C rate coefficients $k_1$ are practically equal for consecutive and for simultaneous gasification. Values of $k_3$ are significantly lower and exhibit larger discrepancy between both types of experiments at 1050 and 1100°C. This can be explained by uncertainties in measuring CO concentrations. During gasification of graphitic carbon at temperatures below 1200°C the measured values were below the measurement range of the analyzer For temperatures below 1200°C no stationarity has been found during simultaneous gasification, and therefore no values of $k_2$ where determined. Graphitization reaction rates $^{m}Q_{C,2}$ at the beginning of consecutive gasification measurements were lower by a factor of approximately 50 than soot gasification rates $^{m}Q_{C,1}$. Hence, ignoring reaction 2 at consecutive gasification experiments can be justified, considering the comparatively short time for full soot conversion.

Soot is a carbon material with micro- and nano-porous character. Hence its reactivity can be affected by mass transport [8]. In soot pores diffusion takes place in Knudsen regime [9]. The influence of mass transport on reaction rate can be estimated with the Wagner-Weisz-Wheeler criterion [10] showing that in the soot aggregates of the present study reaction is limited by mass transport [15].

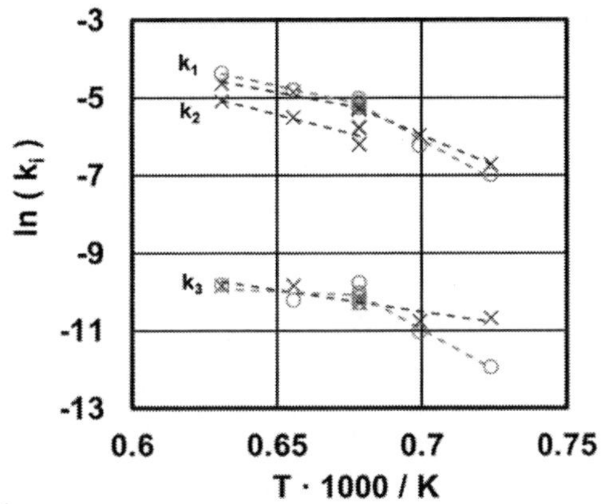


Figure 11: Reaction coefficients. Units of $k_1$ and $k_3$ in g$^{1/3}$ / s, $k_2$ in 1/s. Symbols: o consecutive, x simultaneous experiment.

## 7. Conclusion

The measurements of simultaneous and consecutive gasification of soot generated in the same reactor by pyrolysis of propane have shown that graphitic carbon material is formed on the alumina ceramic surface of the reactor and in the filter bed. Owing to its high atomic order the graphitic carbon is less reactive during gasification than the soot formed in the gas phase. The used reactor material proved not to be completely inert as in the first minutes of pyrolysis oxygen was set free resulting in some CO. In order to take into account side reactions some assumptions had to be made to calculate the reaction rates of the carbon materials captured in the filter. It is likely that also the formation of graphitic carbon is induced catalytically on the surface of the reactor material.

A reaction scheme was introduced consisting of gasification of fresh soot and in parallel conversion to graphitic carbon which is gasified, as well. Comparisons of simulation and measured results have shown that the measurement results of simultaneous and consecutive gasification could well be represented by the reaction scheme. The gasification rate of graphitic material was considerably lower than that of soot.

The gasification rate of soot was found similar in simultaneous and consecutive gasification. The soot is converted at 1200°C completely within 3 min. Hence, in the time range between a few minutes to 15 minutes of soot collection at pyrolysis condition no significant change of soot gasification rate is observed. Thus, soot aging takes place in shorter time intervals after its generation. Because of assumptions for processing the measured data it is not worthwhile to interpret the first two minutes of the simultaneous gasification experiments. For this purpose in situ techniques which enable to measure the gasification rates of soot within the gas phase are recommended. Evaluated techniques are e.g. laser light scattering methods [16], mobility analysis in different modifications [17], [18] mass spectrometry [19] or soot sampling on TEM platelets [20], [17], [21].

## 8. References

[1]. **Higman, C. und van der Burgt, M.** *Gasification.* Burlington : Butterworth Heinemann, 2009. Bd. 2. SBN-13: 978-0750685283.
[2]. **Leibold, H., Hornung, A. und Seifert, H.** HTHP syngas cleaning concept of two stage biomass gasification for FT synthesis. *Powder Technology.* 2008, 180, S. 265-270.
[3]. **Licht, W. R. und et al.** *In-situ gasification of soot contained in exothermically generated syngas stream. 7,534,276 B2* USA, 19. 05. 2009.
[4]. **Medalia, A.I., Rivin, D. und Sanders, D.R.** A comparison of carbon black with soot. *The Science of the Total Environment.* 1983, 31, S. 1-22.
[5]. **Clague, A.D.H., et al.** A comparison of diesel engine soot with carbon black. *Carbon.* 1999, 37, S. 1553-1565.
[6]. **Frohs, Wilhelm, et al.** *Ullmann - Carbon, 3. Graphite.* Weinheim : Wiley-VCH Verlag GmbH & Co. KGaA, 2010.
[7]. **Stanmore, B. R., Brilhac, J. F. und Gilot, P.** The oxidation of soot: a review of experiments, mechanisms and models. *Carbon.* 39, 2001, 2247 - 2268.
[8]. **Gilot, P., et al.** Determination of Kinetic Data for Soot Oxidation. Modeling of Competition Between Oxygen Diffusion and Reaction During Thermogravimetric Analysis. *Combustion and Flame.* 1993, 95, S. 87-100.
[9]. **Levenspiel, O.** *Chemical Reaction Engineering.* New York : John Wiley & Sons, 1999. ISBN 0-471-25424-X.
[10]. **Worringer, G.** *Reaktionskinetische Untersuchungen zum Umsatzverhalten von Ruß im Reaktionsgas der Flugstromvergasung.* Institute of Technology Karlsruhe (KIT) : Dissertation, in prep.
[11]. **Neoh, K. G., Howard, J. B. und Sarofim, A. F.** Effect of oxidation on the physical structure of soot. *Twentieth Symposium (International) on Combustion.* 1984, 20, S. 951-957.
[12]. *Size resolved soot surface oxidation kinetics.* **Lall, A. A. und Zachariah, M. R.** Villa Orlandi : Karlsruhe : KIT Scientific Publishing , 2009.
[13]. **Echavarria, C. A. und et al.** Studies of soot oxidation and fragmentation in a two-stage burner under fuel-lean and fuel-rich conditions. *Processings of the Combustion Intitute.* 2010, doi: 10.1016/j.proci.2010.06.149.
[14]. **Mätzing, H. und et al.** Detection of electrically charged soot particles in laminar premixed flames. *Combustion and Flame.* 2012, 159, S. 1082-1089.
[15]. **Vander Wal, R. L. und Tomasek, A. J.** Soot oxidation: dependence upon initial nanostructure. *Combustion and Flame.* 2003, 134, S. 1 - 9.
[16]. **Song, J. und al., et.** Diesel soot oxidation during the late combustion phase. *Combust. Flame.* 2010, doi: 10.1016/j.combustflame.2010.09.017.

# COMPARISON STUDY ON GASIFICATION OF LONG-FLAME BRIQUETTE COAL WITH $CO_2$ AND $H_2O$

*Presenting Author (Lingmei Zhou, Doctor, China University of Mining and Technology (Beijing), D11, Xueyuan Road, Haidian District, 100083, Beijing, China)*

*Contact Information: lingmeizhou@hotmail.com, 13521621519*

*Corresponding author (Mo Chu, Professor, China University of Mining and Technology (Beijing), D11, Xueyuan Road, Haidian District, 100083, Beijing, China)*

*Contact Information: cm@cumtb.edu.cn, 62331863*

*Co-Authors (Zhibing Chang, Chengliang Hao, Fang Xu, China University of Mining and Technology (Beijing), D11, Xueyuan Road, Haidian District, 100083, Beijing, China)*

**Abstract:**

Long-flame coal, as a low rank coal, is normally used for power-station and coal chemical industry. However, it is easily to produce a large number of fine particles through long-distance transport, which brings serious environmental pollution problems and also reduces the economic benefit. In the present study, a typical Chinese long-flame coal was made into three kinds of briquette coals (FH, JW and PM) as feed for gasification, by addition of different binders. Partial (one hour) and complete (8 hours) gasification experiment with three briquette coals was studied in a fixed bed reactor with $CO_2$ and $H_2O$ and compared with the raw lump coal. For one hour partial gasification experiment with $CO_2$, the carbon conversion in sequence agreed with that for $CO_2$ reactivity test of samples: raw>FH>PM>JW. However, for gasification experiment with $H_2O$, the carbon conversion in sequence was inconsistent with that for $CO_2$ reactivity test of samples: FH>PM>JW>raw. The reason for the different was illustrated in the present study, which was mainly related to fracture of samples under different atmosphere and different formula of binders. For 8 hours complete gasification experiment, the behaviors were consistent with that for partial gasification process.

**Keywords:** Long-flame coal, briquette coal, gasification

## 1. Introduction

China is the largest producer and consumer of coal all over the world, accounting for about 37% of world production of coal. The proportion of coal in China's primary energy supply mix is about 70%, which will be not fundamentally changed in the next few decades [1-2]. Among many kinds of coal resources in China, the low-level bituminous coal, such as Long-flame coal, accounts for 42.45% of proved coal resources [3], mostly in Xinjiang, Inner Mongolia and other provinces. The main characteristics of low-level bituminous coal are the "three low (low ash, sulfur and phosphorus) and three high (high calorie, volatile and chemical activity)", which is a good feed for power station and coal chemical industry. However, it owns high moisture content and is easy to be weathered and oxidized. It will produce a large number of fine particles through long-distance transport, which brings serious environmental pollution problems and also reduces the economic benefit. Therefore, it is necessary to process the Long-flame coal into high added value and easy transportation products.

Briquette technology is an effective method to process coal, especially for the fines. In China, many small companies, such as the ceramics factory, need cheap and clean combustible gas. Briquette made from Long-flame coal fines can be used as cheap feedstock for fixed-bed gasification to produce combustible gas. Much literature [4-6] focused coal gasification technology in the recent years, since it can not only enhance the carbon conversion ratio, but also can reduce the atmospheric pollution. However, little literature reported gasification behaviors of briquette coal [7,8]. It is generally recognized that the reactivity of briquette coal is less than that of raw lump coal, since the binders for producing briquette are normally carbon-free materials, which would reduce the carbon conversion and reaction rate. However, Cui et al. [8] reported that the reactivity of briquette anthracite was higher compared to the raw coal. But they did not give specific reasons for this phenomenon. Therefore, the present study focused on comparison of gasification reactivity in both $CO_2$ and steam atmosphere between briquette Long-flame coal and the raw lump coal, especially the mechanism for it.

In the present study, a typical Chinese long-flame coal was made into three kinds of briquette coals (FH, JW and PM) as feed for gasification, by addition of different binders. Gasification experiment with three briquette coals was studied in a fixed bed reactor with $CO_2$ and $H_2O$ and compared with the raw lump coal. The aim was to investigate the gasification behavior of briquette coals, especially the effects of different kinds of binders in different atmospheres. In addition, the gasification experiments for 1 hour (partial gasification process) and 8 hours (complete gasification process) were also researched, which could provide more comprehensive description of gasification process.

## 2 Materials and methods

### 2.1 Materials

Two kinds of samples were used in this study, raw lump long-flame coal (RL) and three briquette coals (FH, JW and PM) with different binders. The raw lump long-flame coal, with particle size of 20-25mm, was used as the comparative sample to illustrate the gasification reactivity of the briquette coals. The characteristics of raw coal were performed according to the related GB standards. The ultimate, proximate, ash fusion temperatures, thermal stability and the $CO_2$ gasification reactivity of raw sample are shown in Table 1. For proximate analysis, it can be seen that the RL coal had a high volatile matter (36.23 wt.%, daf), low ash (3.03 wt.%, d) and high calorific value (32.59 MJ/kg). For ultimate analysis, RL coal owned typical characteristics of low rank coal: high oxygen content, high hydrogen content and medium sulfur content. The hardness of RL is not high, and it is easy to fragmentation and pulverization. For ash composition analysis, the content of the alkali metal oxides calcium oxide (CaO) and sodium oxide ($Na_2O$) content was relatively high, resulting in low ash melting point (the Flowing Temperature was around 1130 ℃). Thermal stability of coal sample was significant for fixed bed reactor. The $BTS_{+13}$ value was as high as 78.8%, which was much higher than the demand for coal sample used in fixed bed reactor (≥60%).

**Table 1 Analysis of raw lump coal and briquette coals**

| Sample | Raw lump coal (RL) | Briquette JW | Briquette FH | Briquette PM |
|---|---|---|---|---|
| Ultimate analysis (wt.%) | | | | |

| | | | | |
|---|---|---|---|---|
| Carbon, $C_d$ | 87.16 | 73.87 | 71.44 | 71.01 |
| Hydrogen, $H_d$ | 6.00 | 5.24 | 5.01 | 4.93 |
| Nitrogen, $N_d$ | 1.36 | 1.16 | 1.10 | 1.10 |
| Sulfur, $S_{t,d}$ | 0.69 | 0.86 | 0.72 | 0.89 |
| Proximate analysis (wt.%) | | | | |
| Moisture, $M_{ad}$ | 13.70 | 6.98 | 4.70 | 4.35 |
| Ash, $A_d$ | 3.03 | 10.09 | 14.81 | 15.35 |
| Volatile matter, $V_{daf}$ | 36.23 | 35.59 | 33.88 | 33.79 |
| Fixed carbon, $FC_d$ | 61.84 | 57.90 | 56.33 | 56.05 |
| $Q_{net,d}$/ (MJ/kg) | 32.59 | 27.71 | 26.45 | 26.43 |
| $CO_2$ gasification reactivity | | | | |
| 800°C | 11.00 | 15.10 | 13.90 | 10.50 |
| 850°C | 31.40 | 27.20 | 32.60 | 23.10 |
| 900°C | 55.40 | 44.40 | 53.00 | 41.90 |
| 950°C | 81.40 | 58.90 | 68.00 | 59.40 |
| 1000°C | 89.70 | 77.30 | 80.50 | 77.60 |
| 1050°C | 94.90 | 88.70 | 87.60 | 91.20 |
| 1100°C | 98.00 | 93.80 | 97.80 | 97.20 |
| Ash fusion temperatures (℃) | | | | |
| Deforming Temperature (DT) | 1110 | 1160 | 1150 | 1160 |
| Sphere Temperature (ST) | 1120 | 1180 | 1168 | 1185 |
| Hemisphere Temperature (HT) | 1130 | 1190 | 1180 | 1190 |
| Flowing Temperature (FT) | 1130 | 1200 | 1230 | 1250 |
| Thermal stability (%) | | | | |
| $BTS_{+13}$ | 78.8 | 74.3 | 79.6 | 80.4 |
| $BTS_{-3}$ | 5.5 | 10.8 | 4.0 | 2.5 |

The three briquette coals (FH, JW and PM) were shaped under optimized conditions as Fig. 1, and the formulas of binders are showed as follows (the content of additives are not given for the confidentiality reasons):

JW briquette: Bentonite + Carboxymethyl starch + borax + 1/3 coking coal

FH briquette: Sodium humate + Bentonite + Pitch

PM briquette: Sodium humate + Bentonite + Pitch + 1/3 coking coal

**Fig.1 The shaping machine and the briquette coals**

The characteristics of raw and briquette coals compared with industry standards are shown in Table 2. It can be seen that the properties of three briquette coals meet the requirements for industry standards of fixed bed gasification.

**Table2 Characteristics of raw and briquette coals compared with industry standards**

| Samples | RL | JW | FH | PM | Coal quality demand (MT/T1010-2006) |
|---|---|---|---|---|---|
| $A_d$ (%) | 3.03 | 10.09 | 14.81 | 15.35 | Very good ≤12%<br>Good: 12-18% |
| $S_{t,d}$ (%) | 0.69 | 0.86 | 0.72 | 0.89 | ≤1% |
| $BTS_{+13}$(%) | 78.8 | 74.3 | 79.6 | 80.4 | >50% |
| $BTS_{-3}$(%) | 5.5 | 10.8 | 4.0 | 2.5 | <15% |
| $Q_{net.ar}$ (MJ/kg) | 32.59 | 27.71 | 26.45 | 26.43 | >21MJ/kg |
| ST (℃) | 1120 | 1180 | 1168 | 1185 | >1100℃ ($A_d$≤ 12%)<br>>1150℃ ($A_d$: 12-18%) |
| $CO_2$ gasification reactivity (1100℃) | 98.0 | 93.8 | 97.2 | 97.8 | ---- |

The ultimate, proximate, ash fusion temperatures, thermal stability and the $CO_2$ gasification reactivity of briquette coals are also shown in Table 1. Compared to raw lump coal, the fixed carbon, volatile matter and heating value decreased due to the addition of binders. The ash content increased, leading to higher ash fusion temperature. The $CO_2$ gasification reactivity of all briquette coals was lower, especially in the temperature range from 900 to 1000 ℃. This was consistent with the findings in literature, caused by the addition of non-carbonaceous binders. The $CO_2$ gasification reactivity of raw and briquette coals in descending order are: RL> FH> PM≥ JW, at temperature from 900 to 1000℃. At higher temperatures, the $CO_2$ gasification reactivity of samples was approximate, because the gasification reaction was affected by the diffusion. For thermal stability, it can be found that the $BTS_{+13}$ value of briquette coals was higher than that of raw lump coal, which was more suitable for fixed bed gasification process. To conclude, the gasification reactivity of briquette coals seemed to be lower than that of raw lump coal, but the thermal stability of briquette coals was higher. The gasification behavior in fixed bed reactor was studied in the present study, which was significant to guide the industrial production.

## 2.2 Methods

### 2.2.1 Gasification experiments in a fixed bed reactor

A fixed bed reactor, shown in Fig.2, is used for gasification of raw lump coal and briquette samples. It can also collect the gases, tar and char during the pyrolysis process for analysis. The briquette and raw lump coals (100 g) were firstly put into the reactor under $N_2$ atmosphere, and heated from room temperature to 500 ℃ by heating rate 10 K/min, 500 to 1000 ℃ by heating rate 5 K/min. After the temperature reached to 1000 ℃ for 30 minutes, the gas was shifted from $N_2$ to $CO_2$ (1L /min) or $H_2O$ (1.5 g/m in), to carry out the gasification experiments. Two kinds of experiments were adopted in the study: (1) Partial gasification reaction (1 hour); (2) Complete gasification reaction (8 hours). For complete gasification reaction, 100 g (about 12 briquettes) raw or briquette coals were used for steam gasification experiments, and 40 g (about 5 briquettes) raw or briquette coals were used for $CO_2$ gasification experiments, since $CO_2$ gasification reaction is slower than the steam gasification reaction. The process was assumed as almost complete when the $CO/H_2$ content in the product gas was lower than 5 vol.% for $CO_2/H_2O$ gasification experiments.

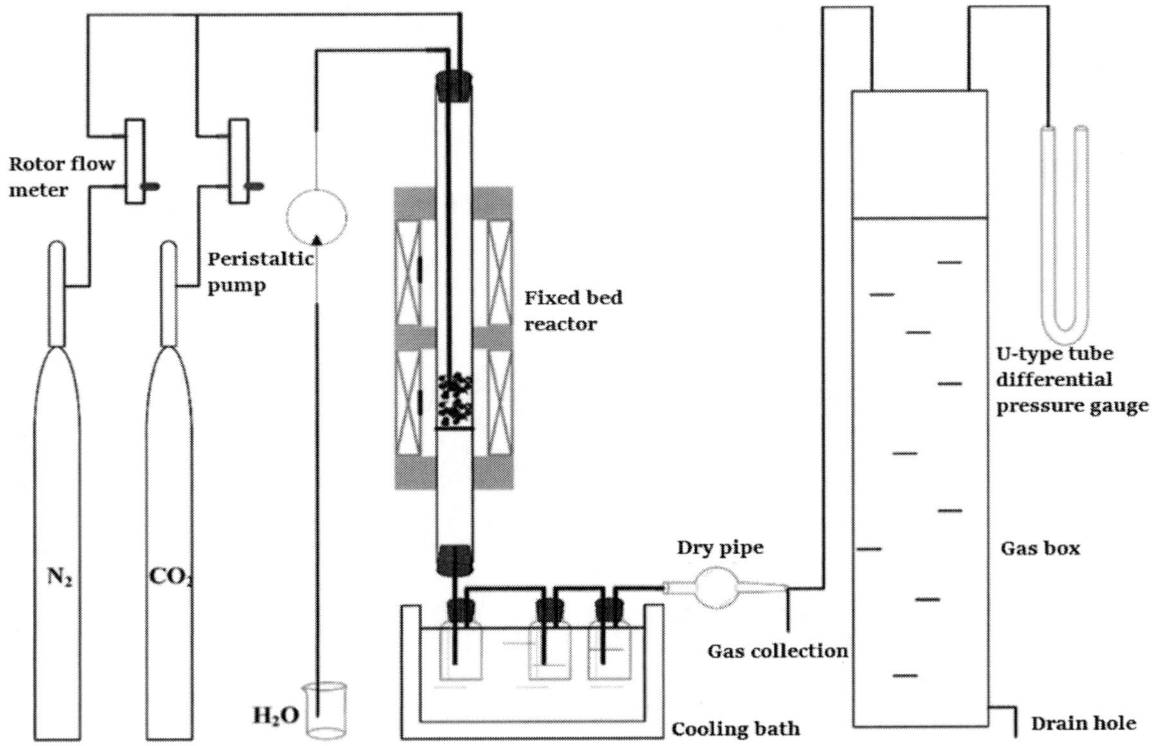


**Fig.1 Schematic of fixed bed reactor**

### 2.2.2 Experimental parameters and equations

The present study chooses four indicators to investigate the gasification behavior: carbon conversion, gas composition, gas yield and gas heating value. Each index is defined as follows:

1. Carbon conversion of char

Carbon conversion is the fraction of consumed carbon mass accounting for the total carbon mass, as calculated by Eq.1:

$$\alpha = \frac{(m_1 \times n_1 - m_2 \times n_2)}{m_1 \times n_1} \times 100\% \qquad (1)$$

where $\alpha$ is carbon conversion, $m_1$ is the coal/char mass, $m_2$ is the char/ash mass, $n_1$ is the carbon content of coal/char, $n_2$ is the carbon content of char/ash.

2. Gas composition

The main reaction during $CO_2$ gasification process is shown in Eq.2. The main reactions during steam gasification process are shown in Eq.3 and Eq.4. The main product gases during $CO_2$ gasification process are CO and $CO_2$. The main product gases during steam gasification process are $H_2$, CO and $CO_2$.

$$C + CO_2 \leftrightarrows 2CO \qquad + 172\ MJ/Kmol \qquad (2)$$

$$C + H_2O \leftrightarrows CO + H_2 \qquad + 131\ MJ/Kmol \qquad (3)$$

$$CO + H_2O \leftrightarrows CO_2 + H_2 \qquad - 41\ MJ/Kmol \qquad (4)$$

3. Gas yield

Gas yield during gasification process are calculated as Eq.5:

$$\omega = \frac{V_{H_2}+V_{CO}+V_{CO_2}}{m_1} \quad (5)$$

where $\omega$ ($Nm^3$/kg) is the gas yield during gasification process; $V_{H2}$, $V_{CO}$ and $V_{CO2}$ ($Nm^3$) are gas volumes under standard condition; $m_1$ (kg) is the weight of char.

4. Gas heating value

1) Heating value of gasification gases

For gasification process, the main gases are CO, $CO_2$ and $H_2$, so the low heating value (LHV) of product gases can be approximately calculated as Eq.6:

$$Q = 10.79 * \varphi_{H_2} + 12.64 * \varphi_{CO} \qquad (6)$$

Where Q ($MJ/Nm^3$) is the low heating value of product gases; $\varphi_{H_2}/\varphi_{CO}$ (vol.%) is the volume fraction of $H_2$/CO in the product gases.

2) Heating value of pyrolysis gases

For pyrolysis process, the gas heating value can be calculated as Eq.7 and Eq.8:

(1) High heating value:

$$Q_1=39.82\times CH_4\%+70.3\times C_2H_6\%+63.4\times C_2H_4\%+101.2\times C_3H_8\%+93.61\times C_3H_6\%+133.8\times n\text{-}C_4H_{10}\%+132.96\times i\text{-}C_4H_{10}\%+125.76\times C_4H_8\% +12.74\times H_2\%+12.64\times CO\% \qquad (7)$$

(2) Low heating value:

$$Q_2=35.88\times CH_4\%+64.35\times C_2H_6\%+59.44\times C_2H_4\%+93.18\times C_3H_8\%+87.6\times C_3H_6\%+123.56\times n\text{-}C_4H_{10}\%+122.77\times i\text{-}C_4H_{10}\%+117.61\times C_4H_8\%+10.79\times H_2\%+12.64\times CO\% \qquad (8)$$

Where $Q_1/Q_2$ (MJ/Nm3) is the low/high heating value of pyrolysis gases; X(gas)% (vol.%) is the volume fraction of gases.

## 3 Results and discussion

### 3.1 Pyrolysis of raw and briquette coals

In drying and pyrolysis stage, the moisture and volatile content evolution would lead to fracture of coal/briquette, resulting in increased bed resistance in gasifier. The fragmentation status is one of the key conditions, to keep the gasification reaction proceed smoothly. Therefore, it is significant to investigate the pyrolysis behavior of coal and briquette samples and the fragmentation status, and their impact on gasification process. The pyrolysis conditions were the same as described in the gasification process. The briquette and raw lump coals (100 g) were heated from room temperature to 500 ℃ by heating rate 10 K/min, 500 to 1000 ℃ by heating rate 5 K/min and kept at 1000 ℃ for 30 minutes.

1. The products yields

Table 3 shows the yields of pyrolysis products for raw and briquette coals. Overall, the pyrolysis products yields of raw and briquette coals were approximate. The char yields of briquette coals (70 to 71%) were slightly lower than that of raw coal (about 72%). The yields of oil (including water) products of briquette coals (8.84 to 9.78%) were slightly higher than that of raw coal (about 8.36%). The yield of gas products of JW (13.63%) was approximate to that of raw coal (13.26%), while that of FH and PM were lower. In addition, the recoveries of pyrolysis products were higher than 92%, indicating that the experimental data were reliable.

**Table 3 Products yield during pyrolysis process of raw and briquette coals**

| Sample | Products yield, d (%) | | | Recovery |
|---|---|---|---|---|
| | Char | Oil/water | Gases | (%) |
| RL | 72.41 | 8.36 | 13.26 | 94.03 |
| JW | 71.24 | 8.84 | 13.63 | 93.71 |
| FH | 70.02 | 9.64 | 12.85 | 92.51 |
| PM | 70.92 | 9.78 | 11.9 | 92.61 |

2. The gas composition

The gases were collected when the pyrolysis temperature reached to 1000 ℃ and kept for 30 minutes. The gas products were collected in the gas gathering cabinet shown in

Fig.2. The gas compositions were measured by gas chromatography and shown in Table 4. As can be seen from Table 4, the gaseous products from raw and briquette coals were mainly consisted by the following gases: methane ($CH_4$), hydrocarbons ($C_2$-$C_4$), hydrogen ($H_2$), carbon dioxide ($CO_2$) and Carbon monoxide (CO). The $H_2$ content is highest, more than 57%, followed by $CH_4$ more than 18% and the content of hydrocarbons/ CO/ $CO_2$ was less than 10%. The gases were collected at 1000 ℃, at which the devolatilization process has been almost completed. The gases were mainly from condensation reaction, which produced a large amount of hydrogen.

**Table 4 Gas composition during pyrolysis process of raw and briquette coals**

| Sample | Gas composition (vol.%) | | | | |
|---|---|---|---|---|---|
| | $H_2$ | $CH_4$ | $C_2$-$C_4$ | CO | $CO_2$ |
| RL | 57.46 | 21.31 | 3.69 | 9.5 | 8.12 |
| JW | 58.41 | 20.14 | 3.28 | 9.19 | 8.98 |
| FH | 61.47 | 18.69 | 3.5 | 8.52 | 7.81 |
| PM | 60.20 | 19.43 | 3.54 | 8.62 | 8.2 |

3. Gas heating value

Table 5 shows the yield and heating value of gas products during pyrolysis process, calculated by gas composition illustrated in Table 4, by Eq.7 and Eq.8. As can be seen from Table 5, the heating value of raw lump coal was highest, 19.89 MJ/Nm$^3$ for high heating value and 17.71 MJ/Nm$^3$ for low heating value, due to the higher hydrocarbons content ($CH_4$, $C_2$-$C_4$). In contrast, the heating value of three briquette coals was lower, but the difference is small.

**Table 5 The yield and heating value of gas products during pyrolysis process**

| Sample | HHV | LHV | Yield |
|---|---|---|---|
| | (MJ/Nm$^3$) | | (Nm$^3$/kg) |
| RL | 19.89 | 17.71 | 13.26 |
| JW | 19.17 | 17.04 | 13.63 |
| FH | 19.22 | 17.06 | 12.85 |
| PM | 19.32 | 17.16 | 11.90 |

4. Properties of chars

The weight and proximate analysis of raw and briquette chars are shown in Table 6. Compared to Table 1, the carbon content of chars increased significantly, hydrogen and sulfur content decreased, while the nitrogen content did not change obviously. Furthermore, it can be found that the hydrogen content of briquettes was lower than raw coal, while sulfur content of briquettes was higher.

**Table 6 Sample weight (g) and proximate analysis (wt.%) of raw and briquette chars**

| Sample | Weight | C | H | N | S |
|---|---|---|---|---|---|
| | g | wt.% | | | |
| RL | 63.33 | 83.94 | 0.66 | 1.14 | 0.21 |
| JW | 68.10 | 84.71 | 0.22 | 1.13 | 0.50 |
| FH | 68.22 | 83.59 | 0.24 | 1.18 | 0.25 |
| PM | 74.82 | 79.08 | 0.20 | 1.06 | 0.71 |

Figure 2 shows fragmentation status of chars after the pyrolysis process. It can be seen that the thermal stability of raw coal was worst. Many raw lump coals broke into small pieces and also a lot of cracks on the surface. In contrast, the thermal stability of three coal briquette was obviously better. The fragmentation did not occur by pyrolysis, only cracks on the surface of the particles. Among the three briquettes, the thermal stability of PM was better than that of FH and JW. The binders for PM and FH were the same, expect the addition of 1/3 coking coal into PM. Therefore, it can be deduced that the 1/3 coking coal addition can improve the thermal stability and intensity of coals, which was good to keep the stable material layer.

**Fig. 2 Fragmentation status of chars after the pyrolysis process**

### 3.2 Gasification of raw and briquette coals

### 3.2.1 Partial gasification of raw and briquette coals

In this section, the briquette and raw lump coals (100 g) were firstly pyrolyzed under the conditions as the pyrolysis process in section 3.1. After the temperature reached to 1000 ℃ for 30 minutes, the gas was shifted from $N_2$ to $CO_2$ (1L /min) or $H_2O$ (1.5 g/m in), to carry out the gasification experiments for 1 hour. By measuring the gas composition and carbon conversion rate and comparing the fragmentation state of different gasification feeds, the gasification behavior of raw lump and briquette coals were investigated in fixed bed reactor.

1. $CO_2$ gasification reaction

1）Gas composition

Table 7 shows the gas composition for partial gasification process for one hour, which was the average content of gases. As can be seen, a little hydrogen generated, compared to lots of CO (more than 70 vol.%) . In addition, the CO content of raw lump coal was highest (83.85 vol.%), while the lowest was for JW (74.26 vol.%). Since the reaction time here is one hour, only part of the carbon was gasified. This was consistent with the $CO_2$ gasification reactivity shown in Table 1. The CO content here can represent the $CO_2$ gasification reactivity of samples.

**Table 7 Gas composition during $CO_2$ gasification process (1000℃, 1 hour)**

| Sample | $H_2$ | CO | $CO_2$ |
|---|---|---|---|
| | | (vol.%) | |
| RL | 1.95 | 83.85 | 14.20 |
| JW | 1.09 | 74.26 | 24.65 |
| FH | 1.92 | 79.79 | 18.29 |
| PM | 1.69 | 76.70 | 21.61 |

Figure 3 shows the gases $H_2$, $CO_2$, CO composition with increasing reaction time during $CO_2$ partial gasification process. For $H_2$ content, the value decreased with reaction time, since the dehydrogenation reaction weaken with decreasing H percent in chars. From Fig. 3(b) and (c), it can be seen that the percentage of $CO_2$ was increasing with decreasing CO percent. From the point of view of the chemical reaction, the reaction rate would decrease after the sharp increase at the beginning. In addition, it can be seen that the CO content of raw lump coal was highest during the whole process, indicating high reactivity of raw coal.

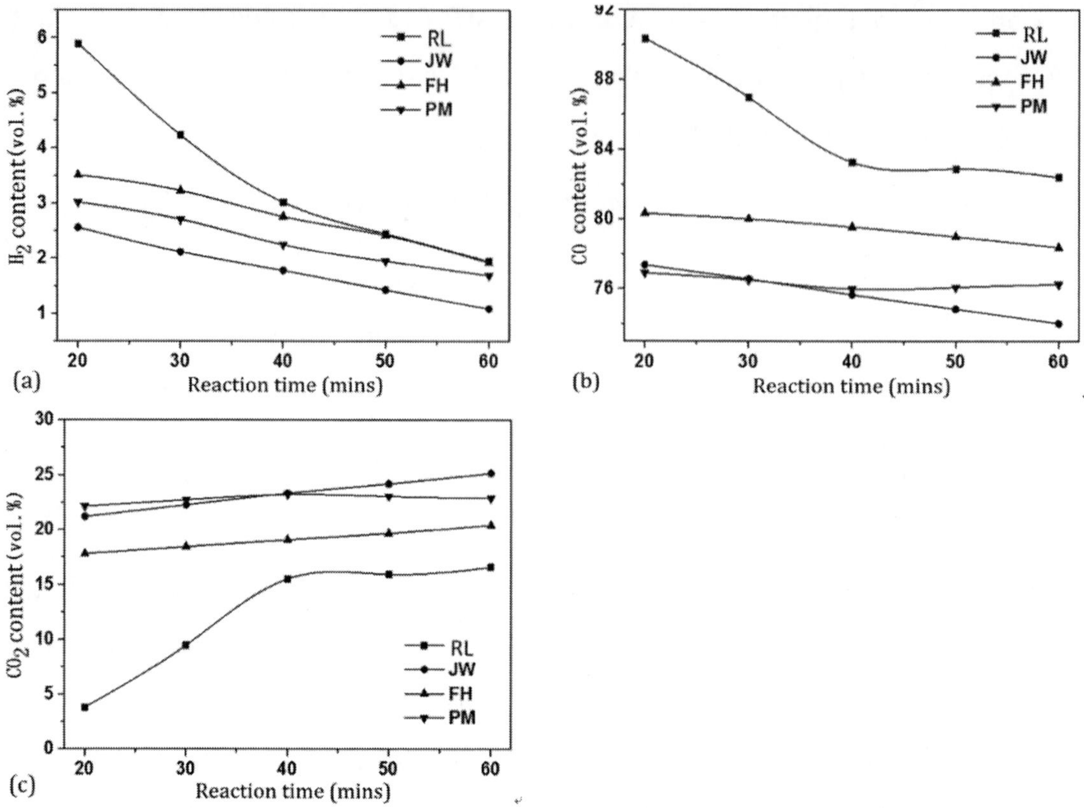


**Fig. 3 Gas composition for $CO_2$ reaction with reaction time, (a) $H_2$, (b) CO, (c) $CO_2$**

2) Gas heating value

The briquette coals in this study intended to be used in ceramics industrial to provide the fuel gas, so the gas yield and heating value were the main indicators to measure the gasification reactivity, as shown in Table 8. As can be seen, for the 1 hour gasification process, the gas yields of raw coal (0.92 $Nm^3/kg$) and FH briquette coal (0.83 $Nm^3/kg$) were higher than that of JW and PM. Taking into account the pyrolysis process, the gas yield of raw coal was slightly higher than the three briquette coals, also for the gas heating value.

**Table 8 Gas yield and heating value during $CO_2$ gasification process (1000℃, 1 hour)**

| Sample | RL | JW | FH | PM |
|---|---|---|---|---|
| Yield of gasification gas（$Nm^3/kg$） | 0.92 | 0.54 | 0.83 | 0.56 |
| LHV of gasification gas（$MJ/Nm^3$） | 10.81 | 9.5 | 10.29 | 9.88 |
| Yield of pyrolysis gas（$Nm^3/kg$） | 0.13 | 0.14 | 0.13 | 0.12 |

| | | | | |
|---|---|---|---|---|
| LHV of pyrolysis gas（MJ/Nm$^3$） | 17.71 | 17.04 | 17.06 | 17.16 |
| Yield of product gas（P+G） | 1.05 | 0.68 | 0.96 | 0.68 |
| LHV of product gas（MJ/Nm$^3$） | 11.66 | 11.05 | 11.21 | 11.16 |

3) Fragmentation status of chars

Figure 4 shows the fragmentation status of chars after 1 hour gasification process. It can be seen that, there was no significant fragmentation or chalking for briquette coals, compared to small amount of powder generated from raw coal, indicating that briquette coals can form a stable layer material during $CO_2$ gasification process, without causing serious clogging of the bed.

**Fig. 4 Fragmentation status of chars after 1 hour $CO_2$ gasification process**

4) Carbon conversion for $CO_2$ partial gasification process

After one hour of the gasification reaction, the char residues of different materials were subjected to proximate analysis, shown in Table 9. It can be found that the values for proximate analysis of raw and briquette coals were similar to each other.

**Table 9 Proximate analysis of char residues for $CO_2$ gasification process (1000℃, 1 hour)**

| Sample | $C_d$ | $H_d$ | $N_d$ | $S_{t,d}$ |
|---|---|---|---|---|
| | (%) | | | |

| | | | | |
|---|---|---|---|---|
| RL | 81.82 | 0.26 | 1.05 | 0.4 |
| JW | 82.55 | 0.19 | 1.11 | 0.37 |
| FH | 78.18 | 0.19 | 1.10 | 0.30 |
| PM | 79.17 | 0.19 | 1.05 | 0.51 |

The carbon conversion during gasification of raw and briquette coals, shown in Table 10, were calculated by Eq.1, based on the values in Table 6 and Table 9. It can be found that the carbon conversion of raw flame coal was highest, reaching to 22.02%, followed by 19.63% of FH, carbon conversion of JW was lowest, only 11.66%. The order of carbon conversion agreed with the measured $CO_2$ gasification reactivity: raw> FH> PM> JW.

**Table 10 Carbon conversion during $CO_2$ partical gasification process (1000℃, 1 hour)**

| Sample | Char | Char C% | Ash | Ash C% | Carbon consume | Carbon conversion |
|---|---|---|---|---|---|---|
| | (g) | (%) | (g) | (%) | (g) | (%) |
| RL | 64.90 | 83.94 | 45.77 | 81.82 | 17.03 | 22.02 |
| JW | 68.54 | 84.71 | 58.90 | 82.55 | 9.44 | 11.66 |
| FH | 68.12 | 83.59 | 52.37 | 78.18 | 16.00 | 19.63 |
| PM | 71.61 | 79.08 | 57.64 | 79.17 | 11.00 | 12.14 |

The lower carbon conversion of briquette coals was due to the addition of binders, which increased the ash content. In addition, the voids of briquette coal particles can be filled with binders, leading to the decrease of reaction rate. For example, the binders for FH and PM contained pitch, which would melt when heating to clog voids and channels of coal particles, leading to lower gasification rate. The lowest carbon conversion was for JW. There were two reasons for that: (1) The fragment of JW shown in Fig.4 was most serious; (2) The binders of FH and PM contained Sodium Humate, which has catalytic effects for gasification reaction.

2. $H_2O$ gasification reaction

1）Gas composition

Steam gasification reaction was more complex than $CO_2$ gasification reaction. Table 11 shows the gas compositions for partial steam gasification process in one hour, which was the average content of gases. As can be seen, the $H_2$ content among gas products was highest (50-55 vol.%), followed by CO content (35-40 vol.%), and around 10 vol.% $CO_2$. It can be deduced that the water shift reaction (Eq.4) might occur during steam gasification process.

**Table 11 Gas composition during $H_2O$ gasification process (1000℃, 1 hour)**

| Sample | $H_2$ | CO | $CO_2$ |
|---|---|---|---|
| | | (%) | |
| RL | 52.06 | 34.21 | 10.00 |
| JW | 51.78 | 36.02 | 8.24 |
| FH | 52.54 | 37.76 | 9.89 |
| PM | 52.62 | 36.64 | 11.37 |

Figure 5 shows the gases $H_2$, $CO_2$, CO composition with increasing reaction time during $H_2O$ partial gasification process. From Fig.5 (a), it can be seen that the hydrogen content maintained relatively high values for all samples, in a downward trend with increasing reaction time. From Fig.5 (b) and (c), the CO content decreased, along with the increased CO2 content, due to the occurrence of water shift reaction.

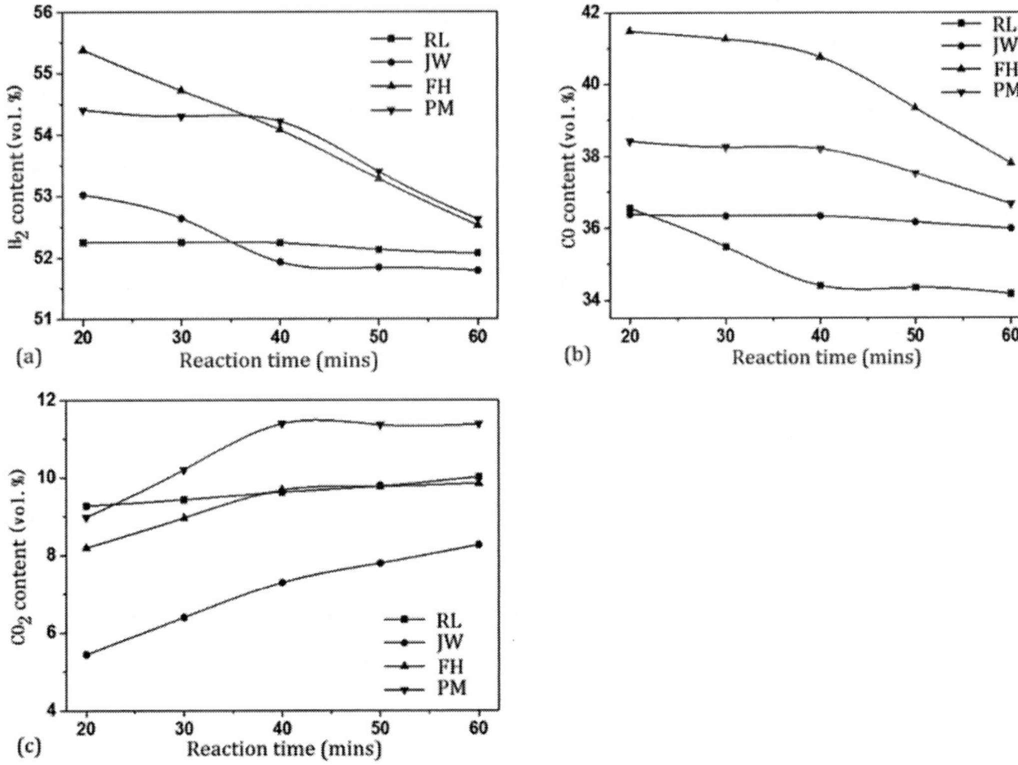

**Fig. 5 Gas composition for $H_2O$ gasification with reaction time, (a) $H_2$, (b) CO, (c) $CO_2$**

2) Gas heating value

As can be seen in Table 12, for the 1 hour steam gasification process, the gas yields of raw and three briquette coals were approximate, from 0.54-0.57Nm$^3$/kg. The lowest value was for raw lump coal, which was different compared to $CO_2$ gasification process. Taking into account the pyrolysis process, the gas yield and heating value of raw coal and three briquette coals were also approximate.

**Table 12 Gas yield and heating value during $CO_2$ gasification process (1000℃, 1 hour)**

| Sample | RL | JW | FH | PM |
|---|---|---|---|---|
| Yield of gasification gas（Nm$^3$/kg） | 0.54 | 0.56 | 0.57 | 0.56 |
| LHV of gasification gas（MJ/Nm$^3$） | 9.94 | 10.14 | 10.44 | 10.31 |
| Yield of pyrolysis gas（Nm$^3$/kg） | 0.13 | 0.14 | 0.13 | 0.12 |
| LHV of pyrolysis gas（MJ/Nm$^3$） | 17.71 | 17.04 | 17.06 | 17.16 |
| Yield of product gas（P+G） | 0.67 | 0.70 | 0.70 | 0.68 |
| LHV of product gas（MJ/Nm$^3$） | 11.45 | 11.52 | 11.67 | 11.52 |

3) Fragmentation status of chars

Figure 6 shows the fragmentation status of chars after 1 hour steam gasification process. It can be seen that, there was significant fragmentation for raw coal, and large amount of powders generated after gasification reaction. Among briquette coals, there were obvious cracks on the surface of JW, while FH and PM kept relatively good shape. Generally, it can be deduced that the briquette coals would have more stable bed layer, compared to raw lump coal.

**Fig. 6 Fragmentation status of chars after 1 hour steam gasification process**

4) Carbon conversion for steam partial gasification process

After one hour of the steam gasification reaction, the char residues of different materials were subjected to proximate analysis, shown in Table 13. As can be seen, the carbon content of raw sample was little higher than that of briquette coals, which might be caused by binders addition in briquette coals.

**Table 13 Proximate analysis of char residues for steam gasification process (1000℃, 1 hour)**

| Sample | $C_d$ | $H_d$ | $N_d$ | $S_{t,d}$ |
|---|---|---|---|---|
| | (%) | | | |
| RL | 86.59 | 0.81 | 0.97 | 0.16 |
| JW | 79.53 | 0.54 | 0.93 | 0.48 |
| FH | 78.00 | 1.48 | 0.75 | 0.31 |
| PM | 76.68 | 1.50 | 0.75 | 0.52 |

The carbon conversion during gasification of raw and briquette coals, shown in Table 14, were calculated by Eq.1, based on the values in Table 6 and Table 13. From the values, it can be seen that the highest carbon conversion, during steam gasification reaction, was for FH reaching to 26.66%, followed by PM (26.31%). The carbon conversion of JW and raw coal was relatively low, around 22.97% and 22.56%. For briquette coals, the carbon conversion in order was FH>PM>JW, which was consistent with the measured $CO_2$ gasification reactivity. However, the reactivity of raw lump decreased, even lower than the three briquettes.

**Table 14 Carbon conversion during steam partical gasification process (1000℃, 1 hour)**

| Sample | Char | Char C% | Ash | Ash C% | Carbon consume | Carbon conversion |
|---|---|---|---|---|---|---|
| | (g) | (%) | (g) | (%) | (g) | (%) |
| RL | 89.64 | 63.57 | 83.94 | 46.18 | 86.59 | 12.04 |
| JW | 96.20 | 69.27 | 84.71 | 58.31 | 79.53 | 13.48 |
| FH | 97.23 | 67.57 | 83.59 | 53.11 | 78.00 | 15.06 |
| PM | 96.50 | 70.64 | 79.08 | 53.69 | 76.68 | 14.70 |

As seen in Fig.6, after one hour steam gasification reaction, the raw lump coal showed serious fragmentation, which produced lots of powders. On one hand, the fines could cover the surface of lump coal to hinder the inner diffusion of water vapor from outside to inside of the coal. On the other hand, the fines would clog the material bed to hinder the outer diffusion, resulting in decrease of reactivity. For briquette coals, JW also cracked significantly, while FH and PM kept relatively good shape. Therefore, the carbon conversion of FH and PM were higher than that of raw and JW coals.

By comparison, it can be found that the gasification behaviors in steam and $CO_2$ atmosphere were quite different. The carbon conversion of raw lump coal after one hour gasification process with $CO_2$ was highest, while lowest with steam. The gasification reaction with steam was much intense than that with $CO_2$, resulting in much more serious fragmentation of raw lump coal. In addition, the binders of FH and PM contained sodium compounds, which would transform into ionic compound more rapidly in steam atmosphere. According to the literature [9], only compound in ionic state can catalyze the gasification reaction. These can be the reasons for the order of carbon conversion during steam gasification: FH>PM>JW>RL.

### 3.2.2 Complete gasification of raw and briquette coals

In this section, the briquette and raw lump coals were firstly pyrolyzed under the conditions as the pyrolysis process in section 3.1. After the temperature reached to 1000 ℃

for 30 minutes, the gas was shifted from $N_2$ to $CO_2$ (1L /min) or $H_2O$ (1.5 g/m in), to carry out the gasification experiments for 8 hours. For complete gasification reaction, 100 g (about 12 briquettes) raw or briquette coals were used for steam gasification experiments, and 40 g (about 5 briquettes) raw or briquette coals were used for $CO_2$ gasification experiments, since $CO_2$ gasification reaction is slower than the steam gasification reaction.

1. $CO_2$ gasification reaction

1) Gas composition

Since Eq.2 ($C+CO_2=2CO$) was the principal reaction for $CO_2$ gasification process, the CO and $CO_2$ concentration with increasing reaction rate was investigated to represent the gasification reactivity, as shown in Fig.7. When the reaction time was 1h, the CO concentration of RL was 56.7%, while the CO concentration of three briquette coals was about 45%. With the increasing reaction time, the CO concentration decreased. When reaction time was 4 hour, the concentration of RL was lower than 5%, which the reaction can be considered to be completed. However, the CO concentration at the same time of briquette coals PM and JW was around 20%, showing that the reactivity of raw lump coal was higher than that of briquettes. The behaviors during complete gasification process agreed the behaviors in partical gasification reaction well, which was raw RL>FH>PM≥ JW.

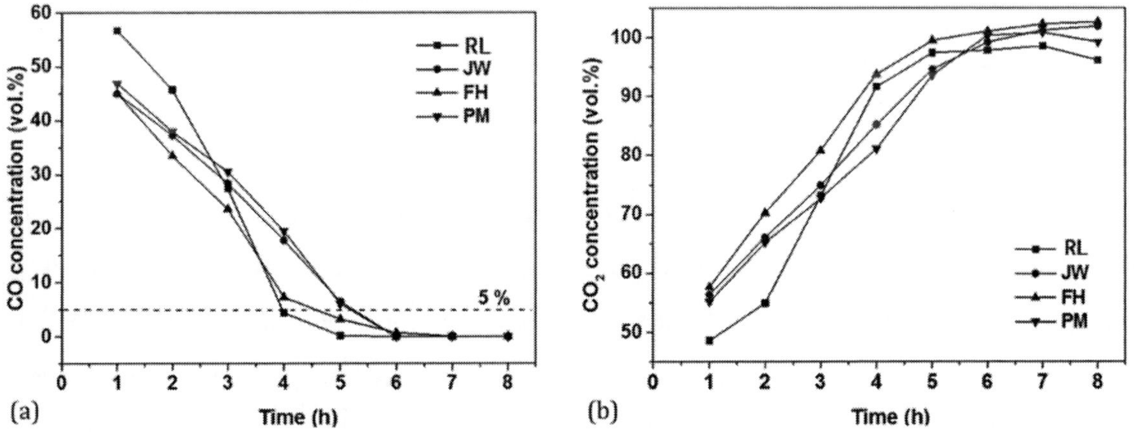


**Fig. 7 Gas composition for $CO_2$ complete gasification reaction, (a) CO, (b) $CO_2$**

2) Ash of raw and briquette coals

Figure 8 shows the ash of raw lump and three briquette coals, after 8 hours complete $CO_2$ gasification process. For raw lump coal, the ash was all in powder, except some large pieces of gangue. For the briquette coals, all the ash kept the shape as the original briquette coals, which might be caused by the effects of binders. For JW, it can be found the mild melting phenomenon occurred, which could hinder the internal and external diffusion. The binder of JW was different with that of FH and PM. It contained starch and borax, which would melt at high temperature. Therefore, the gasification reactivity of JW was lowest among the samples.

**Fig. 8 Ash of raw and briquette coals after 8 hour $CO_2$ gasification process**

2. Steam gasification reaction

1) Gas composition

Since Eq.3 ($C+H_2O=CO+H_2$) and Eq.4 ($CO+H_2O=CO_2+H_2$) were the principal reactions for steam gasification process. The CO and $H_2$ concentration with increasing reaction rate was investigated to represent the gasification reactivity, as shown in Fig.9. When the reaction time was 1h, the $H_2$ concentration of raw lump coal and briquette coals was around 40-45%. With the increasing reaction time, the $H_2$ concentration decreased. When reaction time was 6 hour, the concentrations of three briquette coals were lower than 5% and lower than 1% at 8 hour, which the reaction can be considered to be completed. However, the $H_2$ concentration at 6 hour of raw lump coal was around 21% and even 11% at 8 hour, showing that the reaction has not been finished. It can be found that the reactivity of raw lump coal was lowest. Taking the time required to complete the reaction (H2≤5%) as the measure of steam gasification reactivity, the order of which was FH> PM> JW> RL. This agreed the order for 1 hour partial gasification reaction.

For CO and $CO_2$ concentration versus reaction time, the CO concentration of briquette coals was higher than that of raw lump, while the $CO_2$ concentration was lower. It showed that the CO shift reaction occurred more significantly for raw coal steam gasification process. That was because the diffusion from outer to inner of the lump coal was more difficult, resulting higher vapor concentration to promote the CO shift reaction. By comparison, the briquette coals were shaped by coal particles smaller than 3mm. There were pores with different sizes between coal particles, leading to better diffusion of steam.

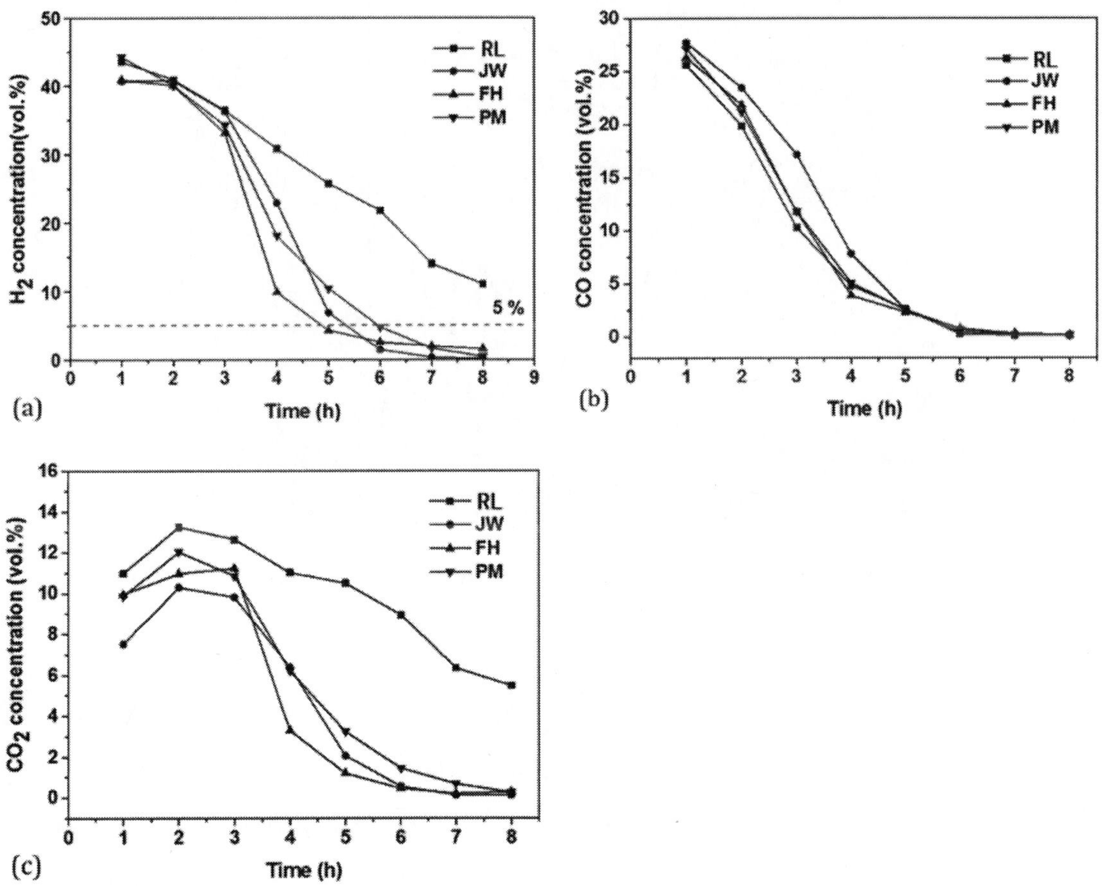


**Fig. 9 Gas composition for $CO_2$ complete gasification reaction, (a) CO, (b) $CO_2$**

2) Ash of raw and briquette coals

Figure 10 shows the comparison of ash between three briquette coals and raw lump coal. For raw lump coal, the ash was almost in powder, except some large pieces of gangue. For the briquette coals, a part of ash kept the shape as the original briquette coals, while the left part was in power. For JW, it can be found the mild melting phenomenon also occurred to hinder the diffusion of steam. In addition, it can be seen that the ash of raw lump coal contained some unreacted or not-fully-reacted particles in black, compared to grey ash for briquette coals. As shown in partial steam gasification reaction (Fig.6), there were already many fines produced after 1 hour reaction, which could block the material bed and reduce the reaction rate. This can also explain the lowest reactivity of raw lump coal during steam gasification process.

**Fig. 10 Ash of raw and briquette coals after 8 hour steam gasification process**

## 4. Conclusion

The present research mainly focused on the gasification behavior of briquette coals (addition of different binding agents) made of long-flame coal in both steam and $CO_2$ atmosphere and compared with the behavior of raw lump coal. From the discussions, it can be concluded that:

(1) One hour partial gasification reaction: For $CO_2$ gasification reaction, the raw lamp showed the highest gasification reactivity and also for the gas heating value. There was no significant fragmentation for briquette coals, compared to small amount of powders generated from raw coal. However, for steam gasification reaction, the raw lamp coal showed the lowest gasification reactivity, since the raw lump coal occurred serious fragmentation and produced lots of powders. The fines could cover the surface of lump coal to hinder the inner diffusion of water vapor from outside to inside of the particle and clog the material bed to hinder the outer diffusion, resulting in decrease of reactivity. For both $CO_2$ and steam gasification reactions, the gasification reactivity of briquette coals in order was FH>PM>JW, since the fragmentation of JW was most serious and the binders of FH and PM contained Sodium Humate (catalytic effects).

(2) Eight hours complete gasification reaction: For $CO_2$ gasification reaction, the behaviors during complete gasification process agreed the behaviors in partical gasification reaction well, which was raw RL>FH>PM≥JW. The ash of raw lump coal was all in powder, while the ash of briquette coals kept the shape as the originals. For steam gasification reaction, the order of gasification reactivity was FH> PM> JW> RL, agreed that for 1 hour partial gasification reaction. The ash of raw lump coal contained some unreacted or not-fully-reacted particles, also showing its lowest reactivity.

**Acknowledgement:**

The authors are grateful to the financial support of the National Basic Research Program of China (973, No. 2014CB744301) and the Natural Science Foundation of China (No. U1261101).

**References**

[1] Sheng Hong-zhi, Liu Dian-fu, Wei Xiao-lin. Characteristics of Semi-Cokes — the Solid Residues from Coal Partial Gasification. Journal of Combustion Science and Technology, 10 (2004), 187-191.

[2] Cui Guo-xing, Lin Ming-sui. Gasification characteristics and kinetics of briquette using humic acid binder. Journal of Fuel Chemistry and Technology, 40 (2012), 1289-1294.

[3] Wang Yong, Wang Tong, Kang Gao-feng. A potential analysis of the availability of coal resources in China. GEOLOGY IN CHINA, 36 (2009), 845-852.

[4] Lingmei Zhou, Guanjun Zhang, Martin Schurz et al. Kinetic study on CO2 gasification of brown coal and biomass chars: reaction order. Fuel, 173 (2016) , 311–319.

[5] Kajitani S, Suzuki N, Ashizawa M, et al. CO2 gasification rate analysis of coal char in entrained flow coal gasifier. Fuel, 85(2006), 163–169.

[6] H. Liu, M. Kaneko, C. Luo, et al. Effect of pyrolysis time on the gasification reactivity of char with CO2 at elevated temperatures. Fuel, 83 (2004), 1055-1061.

[7] Tian Ya-peng, Fu Sheng-shi. Improvement of Production Technology for Producing Gasification Briquette From Long Flame Coal by Luqi Gasification Furnace. Coal Technology, 28 (2009), 132-133.

[8] Cui Guo-xing, Lin Ming-sui. Gasification characteristics and kinetics of briquette using humic acid binder. Journal of Fuel Chemistry and Technology, 40 (2012), 1289-1294.

[9] R.C. Brown, Q. Liu, G. Norton. Catalytic effects observed during the co-gasification of coal and switchgrass. Biomass and Bioenergy, 18 (2000), 499-506.

# Experiments of coal partial gasification on a circulating fluidized bed

**Chao Ye, Dr, State Key Laboratory of Clean Energy Utilization, Zhejiang University, Zheda road 38, Hangzhou, Zhejiang, China**

yechao-zju@zju.edu.cn, **15700078345**

**Qinhui Wang, Dr, Professor, State Key Laboratory of Clean Energy Utilization, Zhejiang University, Zheda road 38, Hangzhou, Zhejiang, China**

qhwang@zju.edu.cn

**Zhongyang Luo, Dr, Professor, State Key Laboratory of Clean Energy Utilization, Zhejiang University, Zheda road 38, Hangzhou, Zhejiang, China**

zyluo@zju.edu.cn

**Mengxiang Fang, Dr, Professor, State Key Laboratory of Clean Energy Utilization, Zhejiang University, Zheda road 38, Hangzhou, Zhejiang, China**

**mxfang@zju.edu.cn**

## Abstract

Huating coal was partially gasified in a laboratory scale circulating fluidized bed (CFB) gasifier under a mixture of $O_2$ and steam. The CFB gasifier system consists of furnace, coal feed system, air feed system, steam generation system, gas-solid separation system, measuring system, and material returning system. The furnace surrounded with heat strips whose height was 2600mm and diameter is 120mm. The experiments were conducted at temperatures between 885°C and 970°C. The coal particle size of 0.35mm~0.9mm were chosen for experiments. The coal particle is gasified in the mixture of $CO_2$ and steam. Different oxygen-coal ratios and steam-coal ratios are studied. As the results show that higher oxygen-coal ratio lead to higher gasification temperature and carbon conversion. While slightly higher steam-coal ratio is beneficial for the quality of gas composition. The gas generated from the furnace is collected for gas composition testing by using the gas chromatography. The structure characteristics of char samples are obtained by adopting the Raman spectroscopy and FT-IR spectroscopy. The results show that the volume concentration of combustible gas and the lower heating value reach as high as 70% and 9.1MJ/Nm3, respectively. It can be obtained from the reasults of Raman spectra and FT-IR spectra that the structure of chars become higher ordered because of loss of different functional groups gradually as the gasification reactions process. The chars become less reactive as higher ordered results in less reactive.

**Key words**: Coal partial gasification, circulated fluidized bed gasified gas composition, char characteristics,

# Introduction

As one kind of clean and efficient technology, gasification is used for converting carbonaceous fuels into valuable energy products. [1] There are mainly three types of gasification technology in view of gasifier configurations and flow geometry, including entrained flow gasifiers, fluidized bed gasifiers and moving bed gasifiers. [2] Nowadays, the most widely used is entrained bed gasifier. [3] It aims at high temperature and high pressure to reach high carbon conversion. Therefore, the cost of equipment and operation is high because of extreme furnace condition. It is well known that the CFB has lower investment and higher fuel flexibility. In this paper, coal partial gasification involves of cascading utilization of coal and CFB technology is raised to increase the thermal efficiency and reduce the investment of power generation.

The cascading utilization of coal and not over-pursing high carbon conversion are core concepts of coal partial gasification. In the partial gasification, coal is grinded into particles then sent into the gasifier together with steam and oxygen. In the gasifier, the active part of coal is gasified and the syngas mainly composed of CO, $H_2$, $CO_2$, and $CH_4$ are produced. The syngas can be used for power generation or chemical product. The ungasified char is combusted in the combustor which is the primary characteristic of coal partial gasification. The heat from combustor is used for steam which enhances the steam parameters. The thermal efficiency of coal partial gasification is improved by this way. The air and steam coal partial gasification in an atmospheric fluidized bed is studied by Hongcang Zhou et.al.[4] The thermal and exergy efficiency are researched of a combined-cycle power generation based on oxygen-blown coal partial gasification using AspenPlus by Guoqiang Zhang and et. al [5] [6]The structure development of partial gasified char is discussed in a bench scale fluidized-bed using $CO_2$ by Evgeniia Komarova et. al. [7] The structures of bituminous coal under $CO_2$ and Ar in a fixed reactor is investigated by Yonghui Bai et. al. [8]

However, the experiments in above-mentioned literatures are only conducted in a single reaction atmosphere, or in a fixed reactor. In this paper, the experiments are carried out in a laboratory scale
fluidized-bed gasifier. The oxygen and steam are adopted as reactant atmosphere. The experiments are conducted under different temperature. And the effect of different oxygen coal ratio and steam coal ratio on the gas composition and char characteristic is investigated. Since the coal partial gasification technology is not employed in the practical industrial application. This paper aims at providing new data for feasibility of coal partial gasification.

# Materials and methods

**Coal and bed material preparation.** Huating coal adopted for experiments is screened into 0.35mm~0.9mm. The proximate and ultimate analysis of coal is summarized in the Table 1. The yellow sand chosen for experiments whose particle size is 0.35mm~0.45mm.

Table 1 The proximate analysis and ultimate analysis of Huating coal

| Proximate | | | | Ultimate | | | | | $Q_{net,ar}$ |
|---|---|---|---|---|---|---|---|---|---|
| Mar% | Aar% | Var% | FCar% | Car% | Har% | Nar% | Sar% | Oar% | MJ/kg |
| 7.55 | 20.94 | 25.12 | 46.39 | 61.14 | 3.18 | 1.23 | 0.56 | 12.95 | 21.93 |

**Experimental Facility.**

The schematic diagram of the circulated fluidized bed system for coal partial gasification is showed in Figure 1. The whole system is mainly consist of the fluidized bed gasifier, coal feeding system, gas supplying system, steam generation system, gas-solid separation system, Distributed Control System(DCS), induced fan, and material returning system. The furnace is made of ceramic, the diameter and height of which are 120mm and 2600mm. There is air distributor at the bottom of furnace. There are heating wires around the furnace, and the highest temperature of the heating wire can reach 880°C. The temperature can be controlled and measured by the DCS system. The air and oxygen are supplied by a centrifugal fan and the oxygen cylinder, respectively. The gas is preheated to 350°C through a air preheater before entering the furnace. The steam is generated in a steam generator, the temperature and pressure of which are 180°C and 0.4MPa. The steam is mixed with the other gas before entering the furnace. The coal particle is transported into the furnace by a screw feeder. The screw feeder is driven by an electromotor which is controlled by the DCS. There are two-stage cyclones for gas-solid separation. The char and ash from the primary cyclone is sent back to the furnace through the material returning tube. While char and ash from the secondary cyclone is collected by an ash hopper. The J valve is adopted as the material returning device.

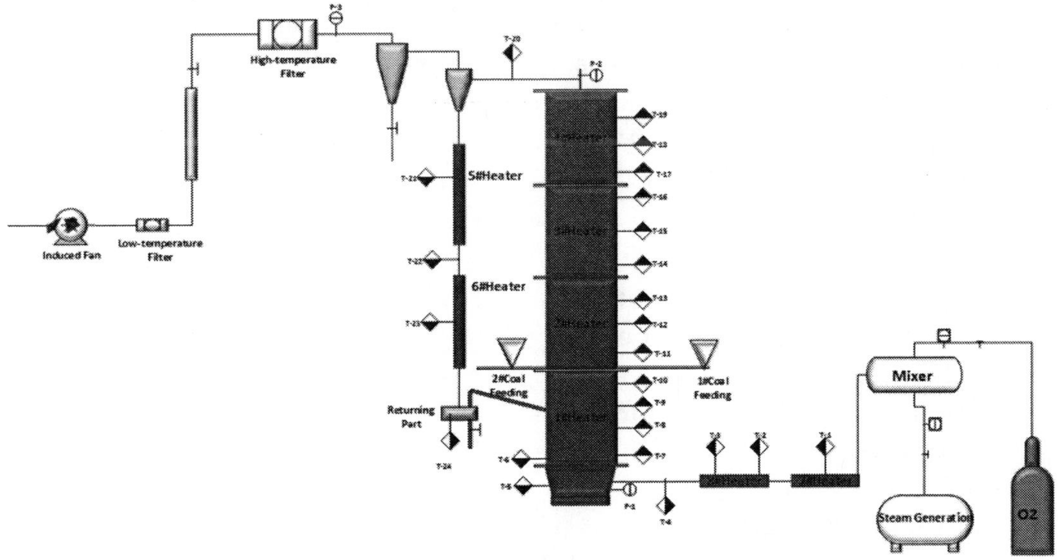


Figure 1 The schematic diagram of the circulated fluidized bed gasifier

### Experimental Procedure

The experiments are conducted as the following steps:

(1) Turn on the heating wire of the furnace, the riser and air preheater. The set temperature of these parts are 650°C, 300°C and 350 °C, respectively.

(2) Switch on the oxygen cylinder to release some gas to make sure the bed material flow in the furnace. Turn on the steam generator but make sure the valve is closed .

(3) About 750g yellow sand and 3000g coal particle are added into the furnace and coal hopper.

(4) Turn on the coal feeder and adjust the flowrate of oxygen to make sure the complete combustion of coal. When bed temperature reaches the demanded gasification temperature, open the valve of steam generator.

(5) The oxygen-coal ratio and steam-coal ratio are adjusted to the required value. After the gasification enters the steady state, the gas sample is collected at a 5min interval.

(6) After 20min, the coal feeder, heating wires and steam is stopped and the oxygen is switch to nitrogen for the furnace cooling down. After the temperature of furnace cools down to 150 °C, the char samples are collected for further analysis.

### Methods of Analysis

The gas sample is collected from the top of the furnace and is analyzed by using the a gas chromatography( Agilent 7890B GC). The char sample is collected after the furnace cools down. Then, the char sample is screened and analyzed by Raman spectroscopy (DXR) which is from the Thermofisher corporation and Fourier Transform Infrared Spectrometer (FT-IR) ( Nicolet 5700).

## Results and Discussions

**Reactions of Coal Gasification** As is well known by people that the coal is a very complex matter. Therefore the chemical reactions involved in the gasifier are very complicated. There mainly two types of reactions, including the gas-solid reaction and gas-gas reaction, which is summarized bellow.(Rui Zhang, Dong Liu, Qinhui Wang, Zhongyang Luo, Mengxiang Fang, Kefa Cen, Coal Char Gasification on a Circulating Fluidized Bed for Hydrogen Generation: Experiments and Simulation, Volume 3, Issue 10, 2015, 1059-1067) [9] All of the reactions are exothermic reactions, except for reaction 3 and reaction 4.

(1) Gas-solid Reactions:

$C+0.5O_2=CO$, $\triangle H=-110.6$ kJ/mol (1)

$C+O2=CO_2$, $\triangle H=-393.6$ kJ/mol (2)

$C+H_2O=H_2+CO_2$, $\triangle H=131.3$ kJ/mol (3)

$C+CO_2=2CO$, $\triangle H=172.5$ kJ/mol (4)

$C+2H_2=CH_4$, $\triangle H=-74.9$ kJ/mol (5)

(2) Gas-gas Reactions:

$H_2+0.5O_2=H_2O$, $\triangle H=-241.9$ kJ/mol (6)

$CO+0.5O_2=CO_2$, $\triangle H=-283$ kJ/mol (7)

$CO+H_2O=H_2+CO_2$, $\triangle H$=-41.2 kJ/mol (8)

$CO+3H_2=CH_4+H_2O$, $\triangle H$=-206.2 kJ/mol (9)

$2CO+2H_2=CH_4+CO_2$, $\triangle H$=-203.3 kJ/mol (10)

$CO_2+4H_2=CH_4+2H_2O$, $\triangle H$=-208.9 kJ/mol (11)

As two main factors, the oxygen-coal ratio and steam-coal ratio have great influence on the results of coal gasification. In the next section, the effect of oxygen-coal ratio and steam coal ratio on the temperature, gas composition and heat value of syngas is studied.

**The effects of oxygen-coal ratio on the results of coal partial gasification**

In this section, the steam-coal ratio is kept as a constant value to study the effect of oxygen-coal ratio on the results of coal partial gasification. The value of oxygen-coal ratio ranges from 0.54 to 1.15. Figure 2 shows the influence of oxygen-coal ratio gas composition. The effect of oxygen-coal ratio on the temperature, lower calorific value and gasification share. In the following part, the gasification temperature is represented by the value of T-9. As can be seen from Figure 3 that the gasification temperature increases from 955°C to 987 °C with the increase of oxygen-coal ratio which can be explained by more coal combustion. It also can be obtained from Figure 3 that the gasification share of coal increases. The oxygen-coal ratio has great effect on the gasification temperature. In case of the gas composition, the volume fraction of CO and $H_2$ decreases by ~3% and ~10%, respectively, while the volume fraction of $CO_2$ increases by ~12% as the oxygen-coal ratio increases which could be accounted for combustible gas combustion. The decrement of $H_2$ is much larger than that of CO. The reason why the volume fraction of $CO_2$ almost reaches more than 37% is that the oxygen content of coal reaches 12.95%. The volume fraction of combustible gas accounts for 64%. It is different from other three types of gas, the volume of $CH_4$ differs slightly. The lower calorific value of gas deceases from 8.4MJ/Nm$^3$ to 7 MJ/Nm$^3$ which is attributed to the variation of gas composition. It is much larger than that of the air and steam coal partial gasification.[4, 10]

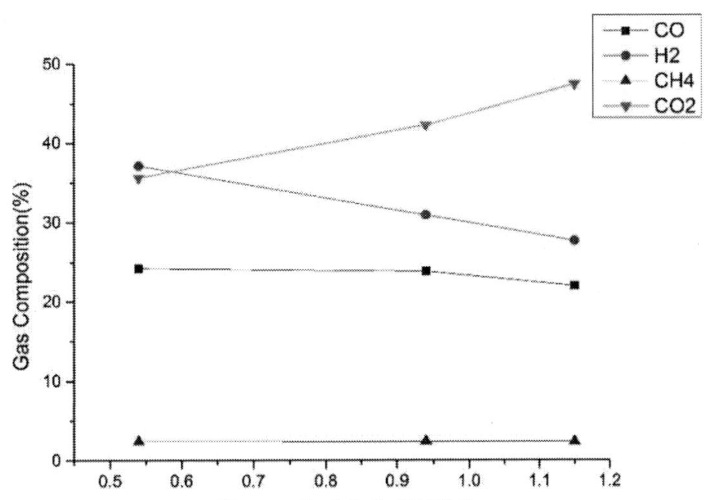


Figure 2 The effects of oxygen-coal ratio on the gas composition

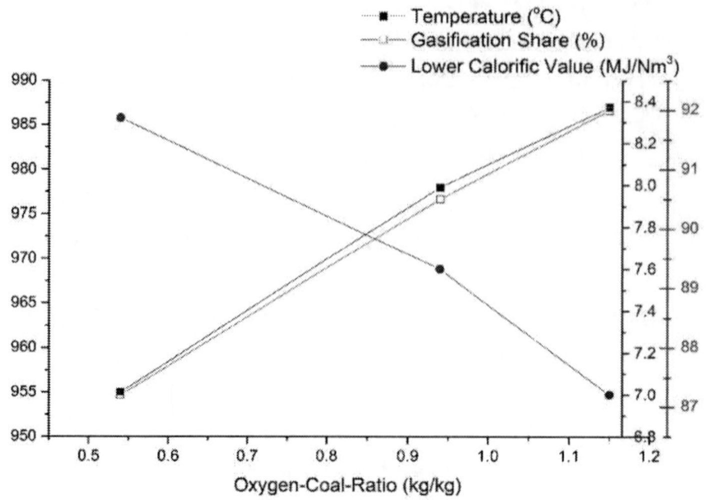


Figure 3 The effects of oxygen-coal ratio on the temperature, gasification share and lower calorific value of syngas

**The effects of steam-coal ratio on the results of coal partial gasification**

In this section, the oxygen-coal ratio is kept as a constant value to explore the effect of steam-coal ratio on the results of coal partial gasification. The effect of steam-coal ratio on the gas composition, temperature, gasification share and lower heating value are shown in the Figure 4 and Figure 5. The steam-coal ratio is increased from 0.32 to 0.51. As can be seen from Figure 4 that the volume fraction of CO and H2 increase by 2% and 9%, respectively. As more steam is added into the furnace, the increment of CO is smaller than that of $H_2$ because of reaction 3 and reaction 8. There is some CO consumed during reaction 8. The volume fraction of CO2 decreases from 40.92% to 29.93%. Therefore the volume fraction of combustible gas increase from 59% to 70% which is very considerable. It is similar to above section that the volume fraction of $CH_4$ differs slightly. It can be obtained that the gasification temperature keeps declining since excessed steam absorbs heat. The gasification share increases since more carbon reacts with the steam which increases from 88.3% to 91.2%. However, the increment of gasification share is not larger than that of last section. As the volume fraction of combustible gas increase, the maximum value of gas reaches as high as 9.1 MJ/Nm$^3$.

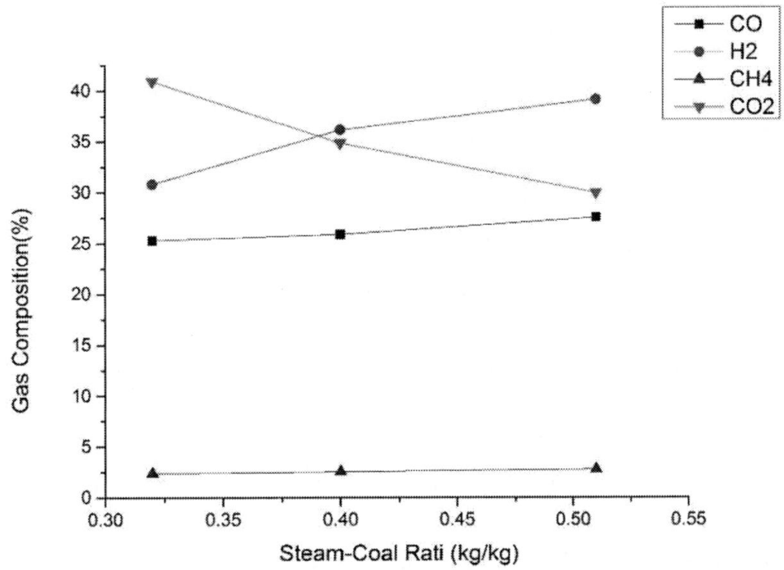


Figure 4 The effects of steam-coal ratio on the gas composition

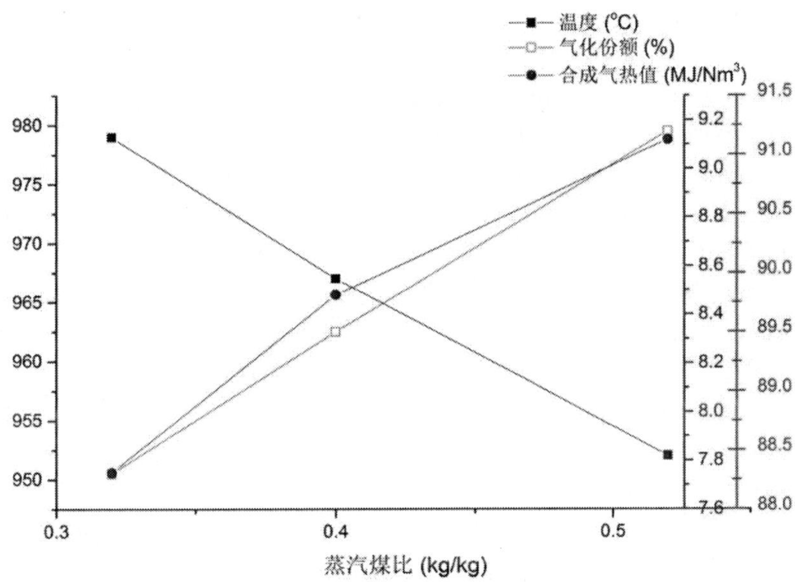


Figure 5 The effects of steam-coal ratio on the temperature, gasification share and lower calorific value of syngas

**Char characterization**

In this section, char samples under three kinds of experimental conditions are adopted for Raman spectra and FT-IR spectra texting. The experimental conditions are summarized in the Table 2. As mentioned above, more oxygen-coal ratio results in higher gasification temperature.

Table 2 Number of different chars under experimental conditions

| Experimental Conditions | Char-1 | Char-2 | Char-3 |
|---|---|---|---|
| Temperature/°C | 860 | 890 | 930 |

| | | | |
|---|---|---|---|
| Oxygen-Coal Ratio/Kg/Kg | 0.365 | 0.461 | 0.492 |
| Steam-Coal Ratio /Kg/Kg | 0.375 | 0.311 | 0.362 |

**Raman spectra acquisition of char**

The Raman spectroscopy is used extensively to explore the structure features of carbonaceous matters. [11] In this section, the graphite structure (G) and the disorder structure (D) are mainly adopted to investigate the characteristic structure of char. However, the quantitative connection between the Raman spectra and structural features is not acquired by only taking the G band and D band in to consideration.[12] There are overlapped areas for highly disordered carbonaceous material between the two bands, including $G_I$, G, $G_r$, $V_I$, D, $S_I$, S, $S_r$, R. [13, 14] [15]However, some of them are too small to observe for some kinds of chars. Since the reason stated above, it is unable to find the detailed information to study the disordered carbon material. Therefore, it is necessary to conduct the deconvolution of Raman spectra into Gaussian bands by Origin 9.[16] The detailed Raman spectra is shown in Figure 6. The texting spectra ranges from 800cm$^{-1}$ to 2200cm$^{-1}$. There are manly five bands are existing in the derived char samples. The D1 band around 1350cm$^{-1}$ is considered as the medium to large aromatic rings including 6 or more fused rings and graphitic lattice without especial order. [17] The D2 band around 1620cm$^{-1}$ is attributed to a lattice vibration involving graphene layers. [16] The D3 around 1530cm$^{-1}$ is related to the small ring system(less than 6 fused rings). [18] The D4 band around 1150 cm$^{-1}$ is considered as the mixer $sp^2$-$sp^3$ bond form at the periphery of crystallites or the C-C and C=C. [19] The G band around 1580 cm$^{-1}$ is related to the stretching vibration of the ideal graphitic lattice. [17]

The total peak area shown in Figure 7 is used as a measurement to describe the overall Raman intensity. As the literature stated that growth of aromatic rings and loss of O-containing functional groups results in decrease of Raman intensity.[14] While the O-containing functional group leads to higher Raman intensity. The sample of NO.4 in the Figure 7 represents the total peak intensity of raw coal which is much larger than that of other three samples. It indicates that the raw coal is much less ordered and contains more different functional groups. As the gasification temperature increases, the reactions during the gasification becomes much more drastic and more compositions and functional groups are involved in the reactions. While the total peak area decreases which implies that the char structure grows more ordered. It could be explained that there are more aromatic rings forms and O-containing functional group consumed as the gasification condition gets hasher and hasher.

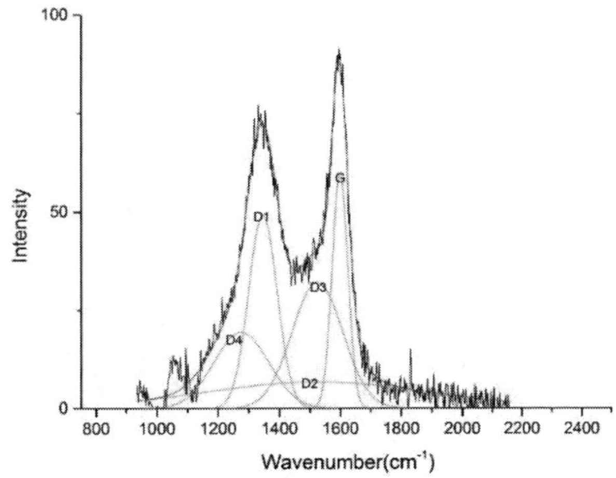


Figure 6 The deconvoluted Raman spectra of coal char

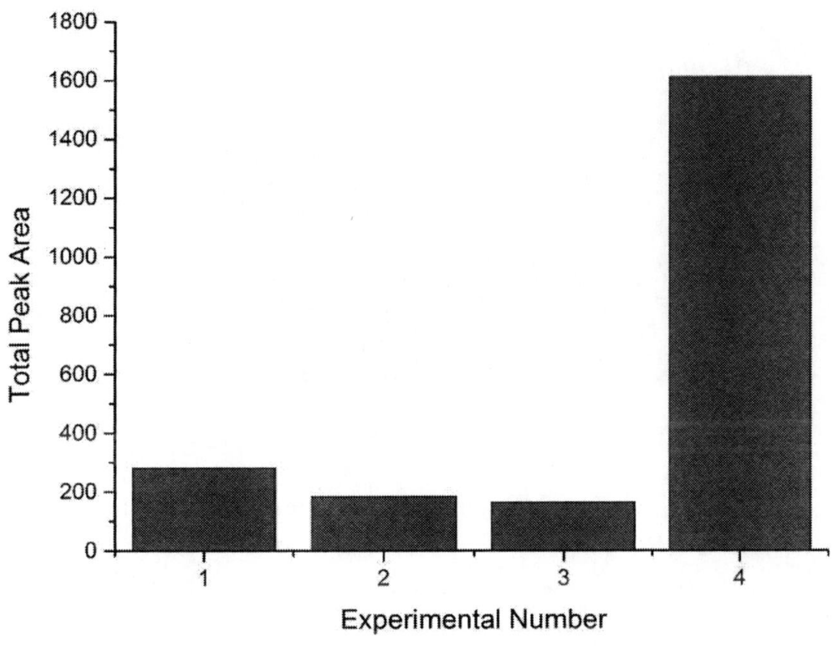


Figure 7 The total peak intensity of different chars at different conditions

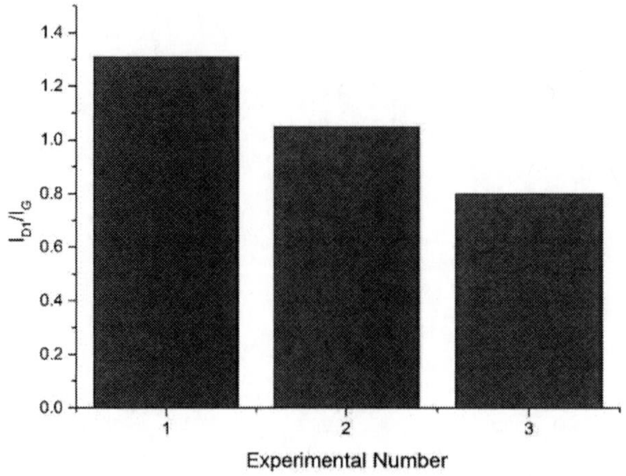


Figure 8 $I_{D1}/I_G$ of different chars at different conditions

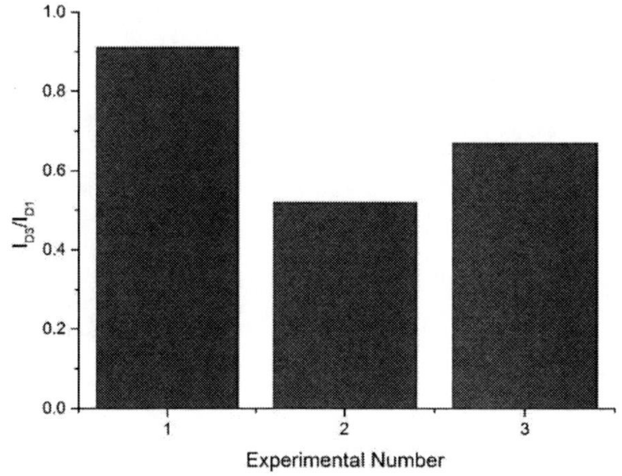


Figure 9 $I_{D3}/I_{D1}$ of different chars at different conditions

The ratio of some major bands can be used for charactering the structure of carbon crystalline. [16] The ratios of $I_{D1}$ and $I_G$, $I_{D3}$ and $I_{D1}$ are shown in the Figure 8 and Figure 9. The ratio of $I_{D1}$ and $I_G$ indicates the degree of disorder of char, while $I_{D3}/I_{D1}$ represents the ratio of small aromatic rings and large aromatic rings. [14] [13] [20] It can be seen that $I_{D1}/I_G$ decreases as the gasification temperature increases. As it is stated by other literature that the disordered carbon structure is higher reactive than the high ordered one. It is explained that as gasification proceeds the reactive part is generally consumed which is comprehensive compared to the actual situation. After the active part is consumed, the residue char can be combusted for steam generation. It is supplied the experimental experience and theoretical basis for coal partial gasification technology. It can be obtained from Figure 9 that $I_{D3}/I_{D1}$ first decreases and then increases slightly, but still is lower than NO.1. It is consistent with the previous research results. [14] When the temperature is not very high(less than 900°C) the smaller aromatic rings are converted into larger ones or consumed.[20] While the temperature is higher than 900 °C, it is possible that the bigger aromatic rings turn into smaller ones again.

**FT-IR spectra acquisition of char**

The FT-IR spectra is adopted for explore the characteristic variations of surface chemistry during the gasification process by others.[21] The char is screened into powers mixed with KBr and tableted before the sample is tested. The FT-IR spectra is between 400cm$^{-1}$ and 4000cm$^{-1}$. The band around 3430 cm$^{-1}$ is represented v(O-H) bond stretching vibrations of hydroxyl groups. [16, 20] The band around 1585 cm$^{-1}$ is attributed to v(C=C) vibrations.[22] The band around 1425 cm$^{-1}$ means the δ ( O-H) vibrations. [23] The band around 1170 cm$^{-1}$ is represents the v(C-O) vibrations of ether-type structures. [23] The band around 1090 cm$^{-1}$ is attributed to v(C-O) vibrations in secondary hydroxyl groups.[24] The bands around 875, 795, 750 and 695 cm$^{-1}$ are attributed to aromatic structures of isolated hydrogen, three adjacent-hydrogen atoms per ring, four adjacent-hydrogen atoms per ring and five adjacent-hydrogen atoms per ring, respectively.[23] The FT-IR spectra of char and coal is shown in Figure 10. It can be obtained that some bans disappear gradually as the gasification process. The peak area of the v(C=C) (~1585 cm$^{-1}$) decreases as gasification temperature increases. It could be explained that more stable structure, such as alkyl-aryl, develops from the C=C band. ([23, 25] The peak area of band (~1090 cm$^{-1}$) is the highest among all of the bands which means C-O in the form of the secondary hydroxyl groups accounts for most. The band (~1035 cm$^{-1}$) disappears as the temperature increases which results from the aromatic condensation. [23] In the general, the band (~695 cm-1) increases and becomes obvious as the the gasification temperature increases, which means that larger aromatic rings formed from the smaller aromatic rings. It is consistent with the trend of results of Raman spectra.

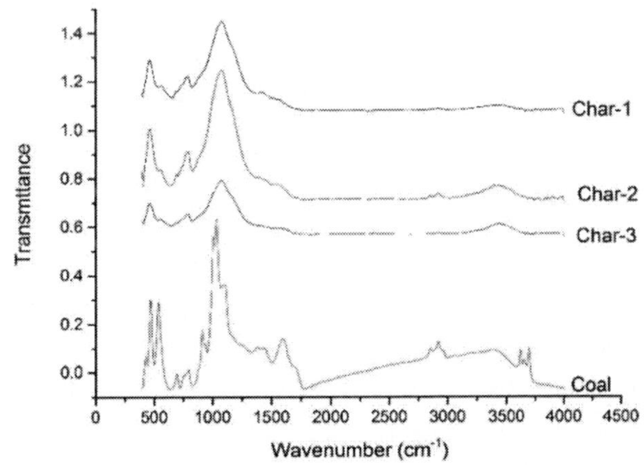


Figure 10 The FT-IR spectra of the char and coal

**Results**

The coal particle is gasified in a bench-scale circulated fluidized bed gasified under atmospheric pressure at 885oC-970oC. The effects of oxygen-coal ratio and steam-coal ratio on the gasification temperature, gas composition, the lower heating value and the gasification share are studied. The gas composition and characteristics of chars are tested by the gas chtomatography, Raman spectroscopy and FT-IR spectroscopy, respectively. The gasification temperature and gasification share increase while the volume concentration of combustible gas decreases as the oxygen-coal ratio increases from 0.54 to 1.15. As the steam-coal ratio increases from 0.32 to 0.51, the gasification temperature decreases while the volume concentration of combustible gas, the lower heating value and gasification share increase. The volume concentration of combustible gas and lower heating value can reach as high as 70% and 9.1 MJ/Nm3 which is much higher than the results from the existed literatures. It can be obtained from the reasults of Raman spectra and FT-IR spectra that the structure of chars become higher ordered because of loss of different functional groups gradually as the gasification reactions process. The chars become less reactive as higher ordered results in less reactive.

**Acknowledgements**

The work is supported by the collaboration project of the Clean Energy Research Centre On Advanced Coal Technology Consortium(CERC-ACTC, 2014DFG2060).

1. Miura, Nakagawa et al. 2004, Kajitani, Tay et al. 2013, Wang, Chen et al. 2015.
2. Minchener, A.J., *Coal gasification for advanced power generation.* Fuel, 2005. **84**:2222-2235.
3. Emun, F., et al., *Integrated gasification combined cycle (IGCC) process simulation and optimization.* Computers & Chemical Engineering, 2010. **34**:331-338.
4. Hongcang Zhou, B.J., Zhaoping Zhong, Yaji Huang, Rui Xiao, *Air and Steam Coal Partial Gasification in an Atmospheric Fluidized Bed.* Energy & fuels, 2005. **19**:1619-1623.
5. GuoQiang Zhang, Y.Y., HongGuan Jin, Gang Xu, Kai Zhang, *Proposed combined-cycle power system based on oxygen-blown coal.* Applied Energy, 2013. **102**:735-745.
6. Li, Y., et al., *Thermodynamic analysis of a coal-based polygeneration system with partial gasification.* Energy, 2014. **72**:201-214.
7. Komarova, E., S. Guhl, and B. Meyer, *Brown coal char CO2-gasification kinetics with respect to the char structure. Part I: Char structure development.* Fuel, 2015.
8. Bai, Y., et al., *Structural features and gasification reactivity of coal chars formed in Ar and CO2 atmospheres at elevated pressures.* Energy, 2014. **74**:464-470.
9. Duan, Jin et al. 2010, Emun, Gadalla et al. 2010, Emami-Taba, Irfan et al. 2013, Zhang, Wang et al. 2013.
10. Xiao, R., et al., *Air blown partial gasification of coal in a pilot plant pressurized spout-fluid bed reactor.* Fuel, 2007. **86**:1631-1640.
11. Li, Hayashi et al. 2006, Wang, Roberts et al. 2014, Zhong, Gao et al. 2015
12. Li, Hayashi et al. 2006, Tay and Li 2010, Wang, Sun et al. 2012.
13. Wang, M., et al., *Raman Spectroscopic Investigations into Links between Intrinsic Reactivity and Char Chemical Structure.* Energy & Fuels, 2014. **28**:285-290.
14. Li, X., J. Hayashi, and C. Li, *FT-Raman spectroscopic study of the evolution of char structure during the pyrolysis of a Victorian brown coal.* Fuel, 2006. **85**:1700-1707.
15. Wu, Z., et al., *Thermal Behavior and Char Structure Evolution of Bituminous Coal Blends with Edible Fungi Residue during Co-Pyrolysis.* Energy & Fuels, 2014. **28**:1792-1801.
16. Wang, B., et al., *Char Structural Evolution during Pyrolysis and Its Influence on Combustion Reactivity in Air and Oxy-Fuel Conditions.* Energy & Fuels, 2012. **26**:1565-1574.

17. Azargohar, R., et al., *Effects of temperature on the physicochemical characteristics of fast pyrolysis bio-chars derived from Canadian waste biomass.* Fuel, 2014. **125**:90-100.

18. Wang, Sun et al. 2012, Li, Zhang et al. 2014.

19. Li, T., et al., *Effects of gasification atmosphere and temperature on char structural evolution during the gasification of Collie sub-bituminous coal.* Fuel, 2014. **117**:1190-1195.

20. Tay, H.-L., et al., *Effects of gasifying agent on the evolution of char structure during the gasification of Victorian brown coal.* Fuel, 2013. **103**:22-28.

21. Pastor-Villegas, Meneses Rodríguez et al. 2007, Wang, Sun et al. 2012, Bai, Wang et al. 2013).

22. Wang, Sun et al. 2012, Meng, Yu et al. 2014).

23. Meng, F., et al., *Characteristics of Chars from Low-Temperature Pyrolysis of Lignite.* Energy & Fuels, 2014. **28**:275-284.

24. V. Gomez-Serrano 1996, Wang, Sun et al. 2012, Meng, Yu et al. 2014).

25. V. Gomez-Serrano , J.P.-V., A. Perez-Florindo,C. Duran-Valle, C. Valenzuela-Calahorro, *FT-IR study of rockrose and of char and activated carbon.* Journal of Analytical and Applied Pyrolysis, 1996. **36**:71-80.

**2016 International Pittsburgh Coal Conference**
**Cape Town, SOUTH AFRICA**
**August 8 – 12, 2016**

**Abstract Submission**

**COMBUSTION TECHNOLOGIES**

# Catalytic Reduction Mechanism Investigation of $K_2CO_3$ Modified Iron-based Oxygen Carrier by Carbon


***Zhongliang Yu, Dr., Institute of Coal Chemistry, 27# Taoyuan Road, Shanxi, CHINA***
***yuzl@sxicc.ac.cn, +8618734861008***

*Shunxuan Hu, Mr., Institute of Coal Chemistry, 27# Taoyuan Road, Shanxi, CHINA*
*hushunxuan2012@163.com, +8618734861008*

*Chunyu Li, Dr., Institute of Coal Chemistry, 27# Taoyuan Road, Shanxi, CHINA*
*licy@sxicc.ac.cn, +8613994288980*

*Tao Liu, Mr., Institute of Coal Chemistry, 27# Taoyuan Road, Shanxi, CHINA*
*Tao_08@126.com, +8618735172377*

*Xing Zhou, Mr., Institute of Coal Chemistry, 27# Taoyuan Road, Shanxi, CHINA*
*413217568@qq.com, +8615035159358*

*Yitian Fang, Prof., Institute of Coal Chemistry, 27# Taoyuan Road, Shanxi, CHINA*
*fyt@sxicc.ac.cn, +8613099003126*

*Jiantao Zhao, Prof., Institute of Coal Chemistry, 27# Taoyuan Road, Shanxi, CHINA*
*zhaojt@sxicc.ac.cn, +8613835142096*



**Abstract:**
It is almost infeasible to directly reduce iron oxide oxygen carrier only by coal char due to the slow reaction rate in the temperature ranges of chemical looping combustion (CLC). With alkali metal catalyst addition, however, this reduction can be significantly increased, which makes this direct reduction feasible in coal direct CLC. Because of the complexity of the reduction, the mechanism has not been explained in detail. This work tries to investigate this mechanism with $K_2CO_3$ decorated iron based oxygen carrier by activated carbon. The results show that K migrated from oxygen carrier to activated carbon at the initial stage of reduction. A possible migration mechanism can be summarized as following: firstly the $K_2CO_3$ on oxygen carrier decompose to $K_2O$; then the $K_2O$ melt to the contact point of iron oxide and carbon; $K_2O$ react with C to produce K, and some K volatilized and absorbed on carbon. The migrated K could then directly catalyze the C-$CO_2$ reaction. Besides the catalytic effect of K, a remarkable autocatalysis of iron oxide itself was also observed. Moreover, this autocatalysis seems to be related with the K addition. $K_2CO_3$ could also improve the reduction step of iron oxide by CO, however, this improvement is far lower than that of the whole reduction.

Eskom

# 33rd Annual International Pittsburgh Coal Conference

**Cape Town International Convention Centre**
**8 – 12 August 2016**

## *Development of a Condition Monitoring Philosophy for a Pulverised Fuel Vertical Spindle Mill*


*Presenter:* André Govender, *PrEng., MSc. Eng. (Wits)*

*Co- Authors:* *Prof. Walter Schmitz (Wits)*
*Mr. Reshendren Naidoo (Wits)*

UNIVERSITY OF THE WITWATERSRAND, JOHANNESBURG


# Introduction

Eskom Primary Energy Cost Breakdown, R Million (% of energy mix)

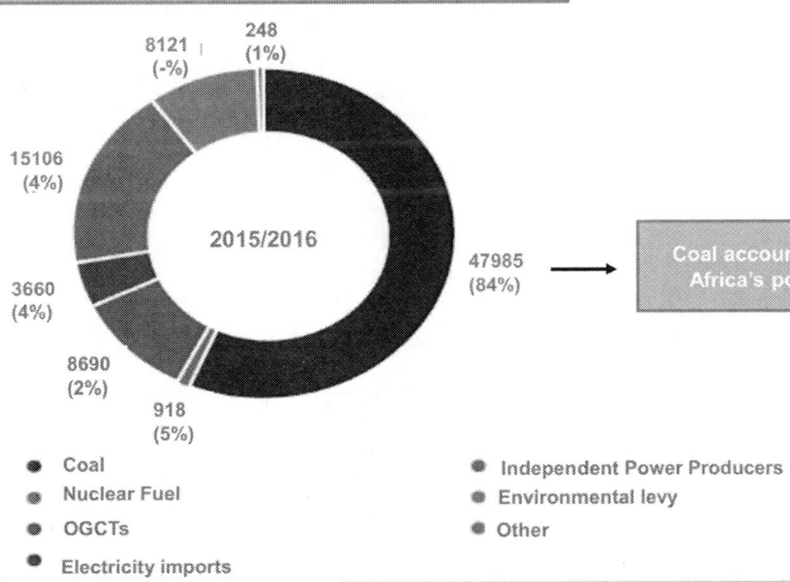


Coal accounts for 84% of South Africa's power requirements

Eskom

2016/09/15 2

# Introduction

- Combustion of coal is central to the power generation process.
- Eskom employs vertical spindle (VSM) and tube mills to grind, dry and pneumatically convey coal particles to the furnace.
- Mills operate in a hostile environment and the degradation mechanism of internal components is predominantly caused by wear and impact damage.
- Depleting coal qualities and time-based maintenance strategies.
- The lack of pulveriser condition monitoring techniques is perhaps a primary determinant of poor pulveriser performance leading to unstable and inefficient coal combustion.

Eskom

2016/09/15 3

# Vertical Spindle Mill Operation

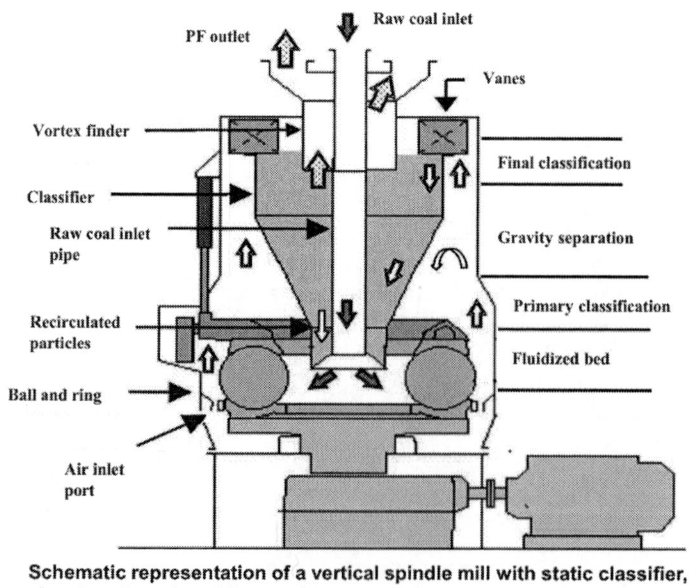


**Schematic representation of a vertical spindle mill with static classifier, showing air and coal flow paths**

**Comments**

1. Up to 6 coal mills per unit. Nominal rate of 25 tons/hr.
2. Coal ground coal from 40mm to 65-75% passing a 75 μm sieve by mass.
3. **Areas prone to wear**:
   Throat gap
   Grinding balls and rings
   Classifier cone
   Classifier vanes
   Vortex finder

Eskom

2016/09/15 4

# Research Aim & Objectives

**Aim**:

The aim of this study was to develop a Condition Monitoring Philosophy for PF VSM.

**Objectives**:

- Establish the raw coal feed rate to the VSM by developing and validating a pulveriser mass and energy balance model.
- Experimentally ascertain the static pressure profile in a VSM.
- Develop and validate a Computational Fluid Dynamic (CFD) model of the VSM to predict the static pressure profile (single phase flow).
- Define a relationship, if any, between the static pressure drop measured across critical components and the mechanical condition of the pulveriser.
- Highlight the effect of worn components on the coal particle size distribution at the pulveriser exit through CFD application.

Eskom



# Research Method

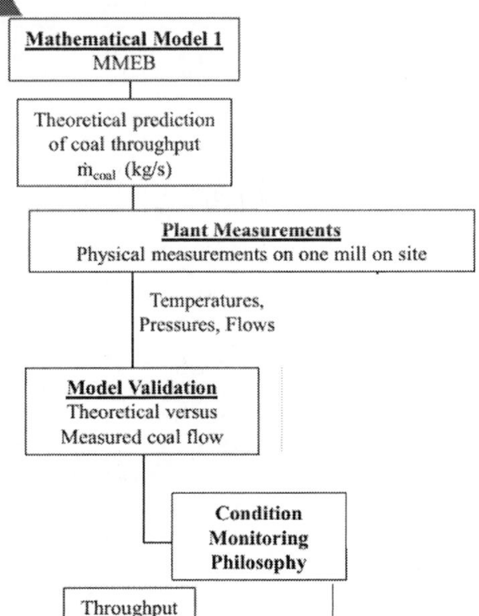


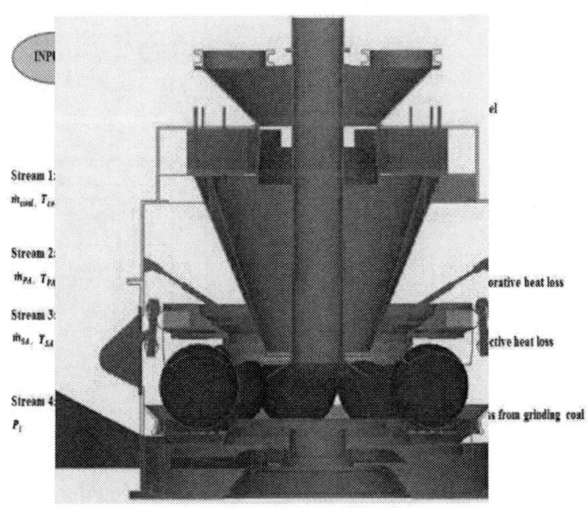


Eskom

# Development of Mathematical Model 1: MMEB

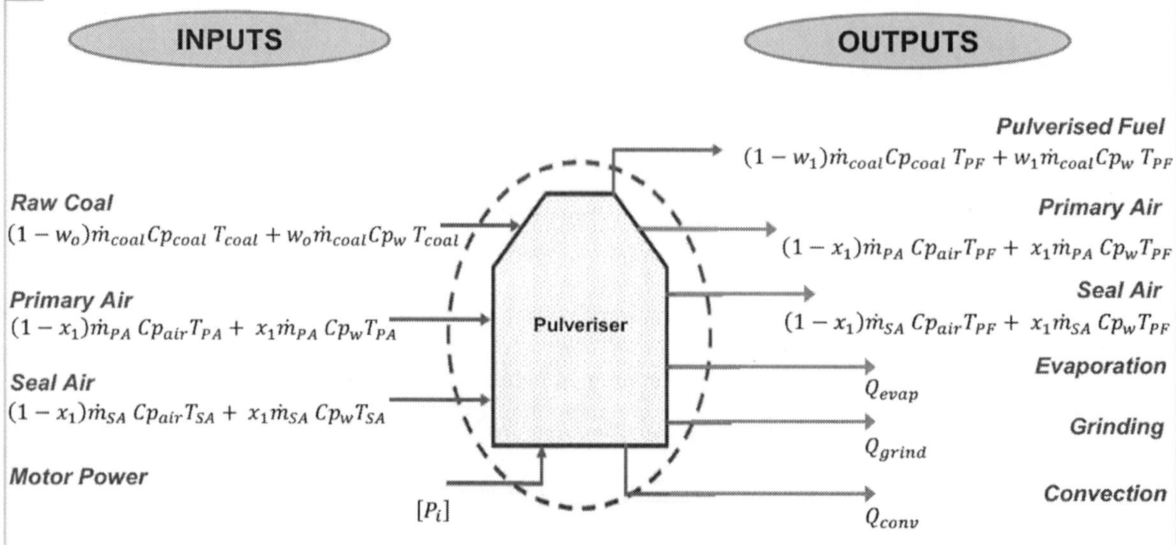


Pulveriser control volume

Eskom

2016/09/15 7

# Experiment Part 1: MMEB

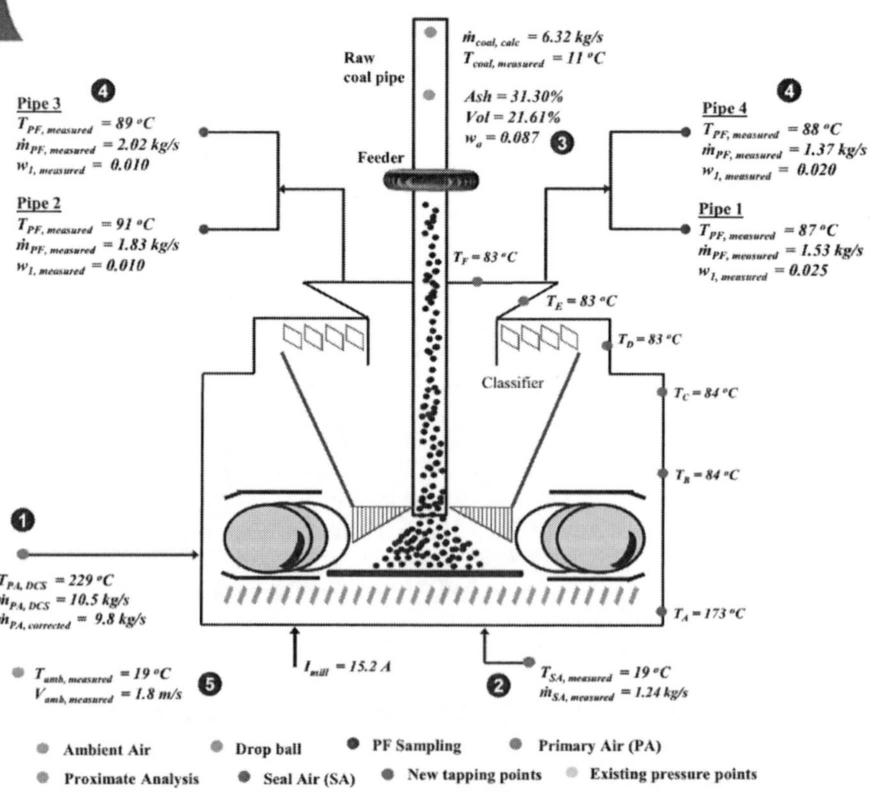


**Comments**

1. PA Flow Traverse: Quantify VSM air input
2. SA flow traverse: Quantify SA input
3. Proximate Analysis: Coal properties defined
4. Isokinetic sampling: Quantify PF - flow, temperature and moisture
5. Ambient conditions: Quantify air temperature and velocity
6. Casing temperature: Quantify convective heat loss to environment

Eskom

8

# MMEB Model Validation

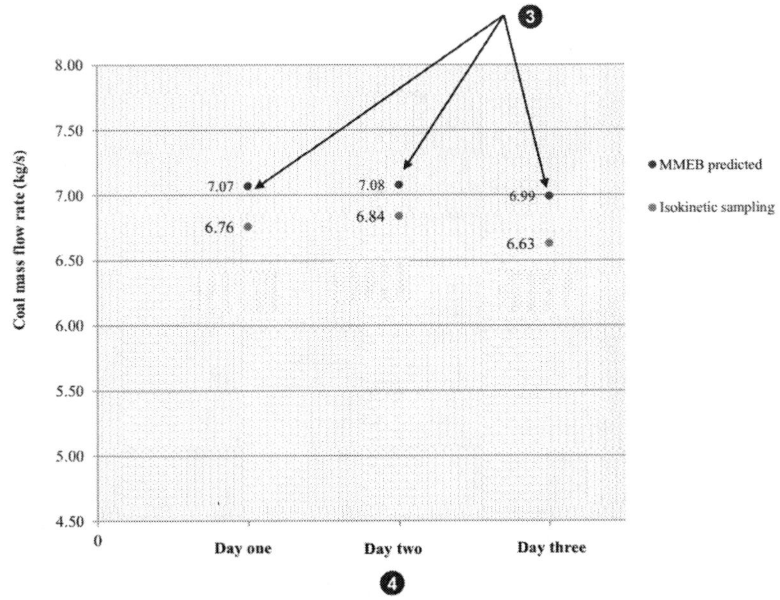


**Comments**

1. Validation of the model assessed through experimental correlation.
2. MMEB predicted results demonstrate close correlation to isokinetic experimental results (less than 5% variance)
3. MMEB results observed to be consistently higher than experimental.
4. Experiment performed in triplicate and results were reproducible, highlighting precision and accuracy of the measured outcomes.

Eskom



# MMEB Sensitivity Analysis

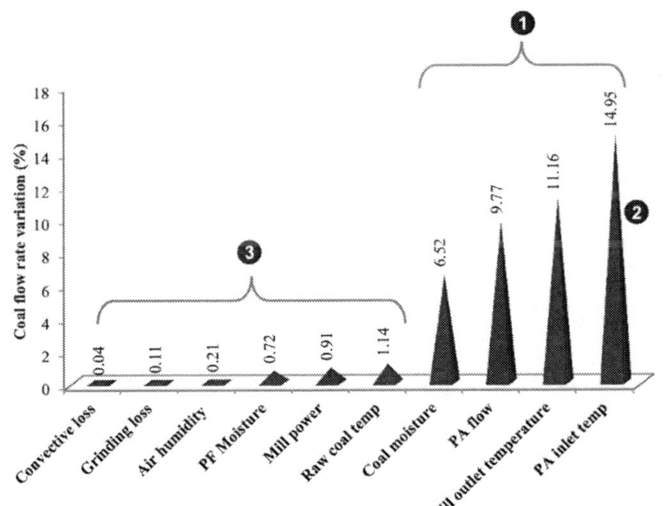


**Comments**

1. MMEB most influential parameters identified.
2. PA inlet temperature is the largest contributor to model output variability, at approx. 15%.
3. Statistically insignificant variation in coal flow for these variables.
4. Results are congruent with results from previous studies, served to verify the model developed.

Each model input parameter varied individually by 10%, whilst other variables remained constant

Eskom

# Experiment Part 2: Static Pressure Measurements

Cross section of 8.5E VSM

Top view of 8.5E VSM

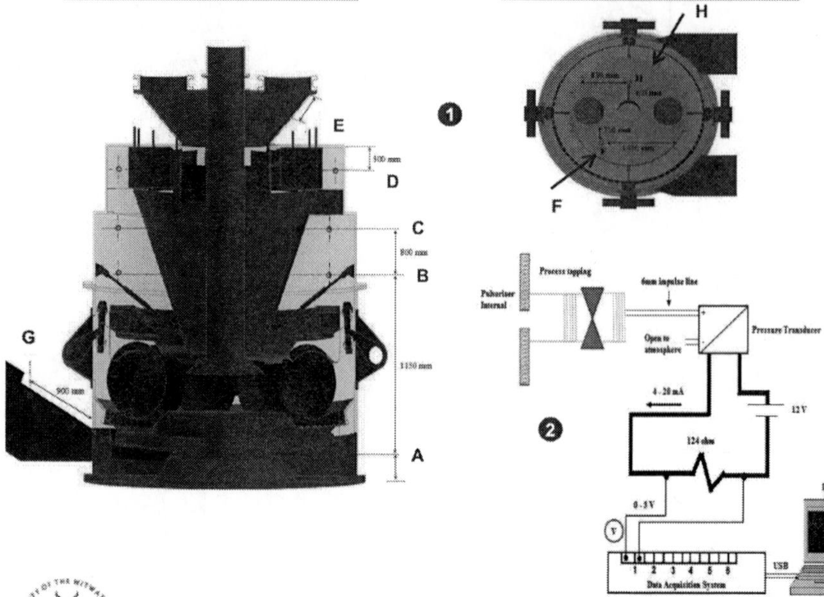


**Comments**

1. Existing pressure measurement points insufficient (G,H) – New points installed (A-F).

Eskom

2016/09/15

11

# VSM Experimental Pressure Profile

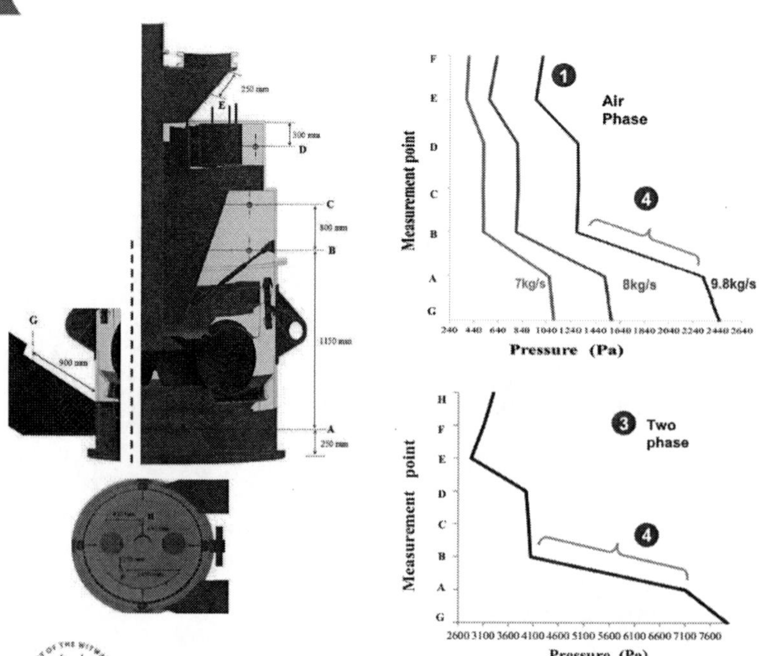


**Comments**

1. Air phase static pressure profile. Three PA flows exhibit similar profile, highlights precision and accuracy of results.
2. Highest pressure in PA duct (G), lowest in outlet plenum (E).
3. Two-phase static pressure profile, similar pattern to that exhibited by single phase flow.
4. Highest pressure drop across the throat (A-B), ~70% of total mill ΔP.

Eskom

2016/09/15

12

# Development of Model 2: CFD

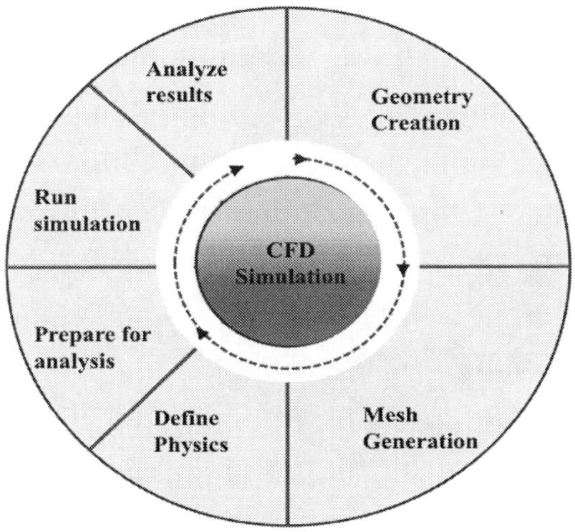


**Generic approach employed to solve problems using CFD simulation**

**Comments**

1. StarCCM+ 9.02 was employed for the study.
2. Prediction of static pressure.
3. Effect of worn components on PSD.

Benefits of CFD:

- Prediction of flow phenomenon in full scale models in a fraction of the time and cost as compared to physical prototyping
- More efficient solutions can be achieved

Eskom



# Development of Model 2: CFD

**3D view of vertical spindle mill prepared for CFD**

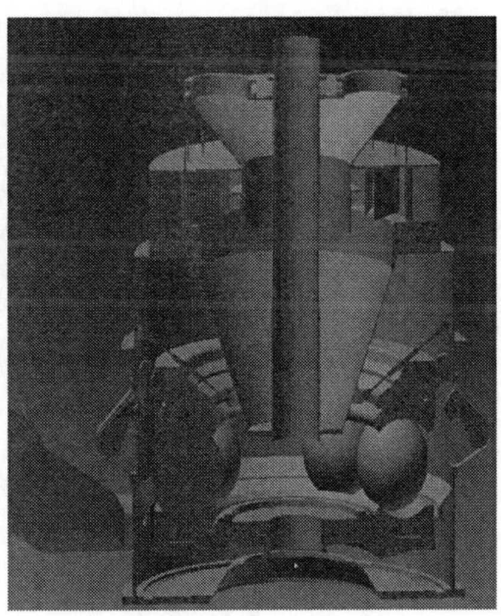

**Sectional view of VSM**

**Comments**

1. Full scale 3D model of 8.5E VSM.
2. Highly detailed model – captured internal components accurately.

Eskom

# Development of Model 2: CFD

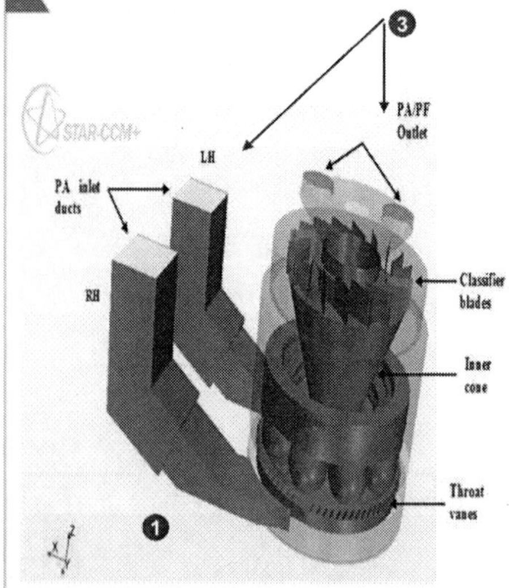


3D model of 8.5E VSM

Mesh Generated

**Comments**

1. Mesh generation:
   3.5 million cells.
   Polyhedral & Prism layers (at wall boundaries).
2. Finer cells in high velocity zones.
3. Mass flow inlet combined with pressure outlet boundaries.



# Development of Model 2: CFD

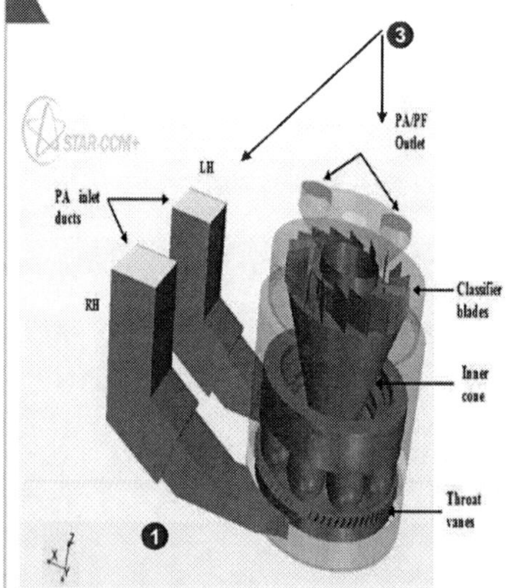


3D model of 8.5E VSM

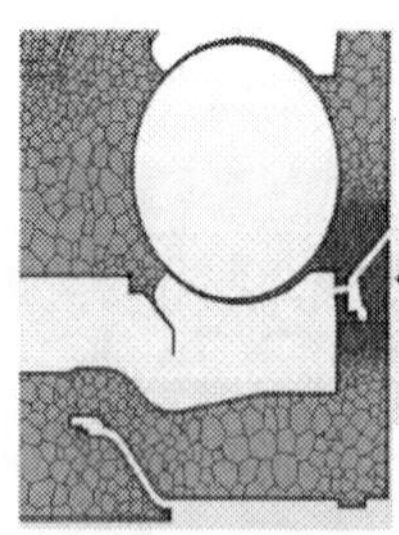

High mesh resolution in throat

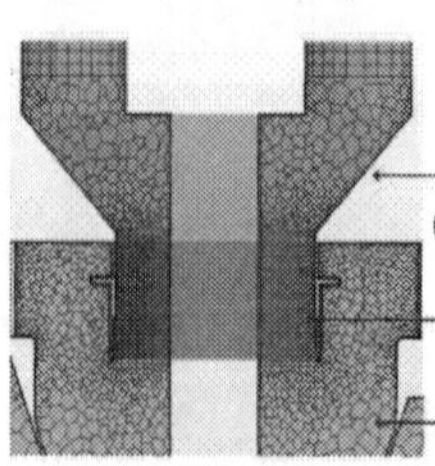

Outlet plenum

Increased cells in vortex

Coarse cells in adjacent areas

Mesh Generated

**Comments**

1. Mesh generation:
   3.5 million cells.
   Polyhedral & Prism layers (at wall boundaries).
2. Mesh Sensitivity:
   Finer cells in high velocity zones.
3. Mass flow inlet combined with pressure outlet boundaries.

# CFD Results: Velocity Profile

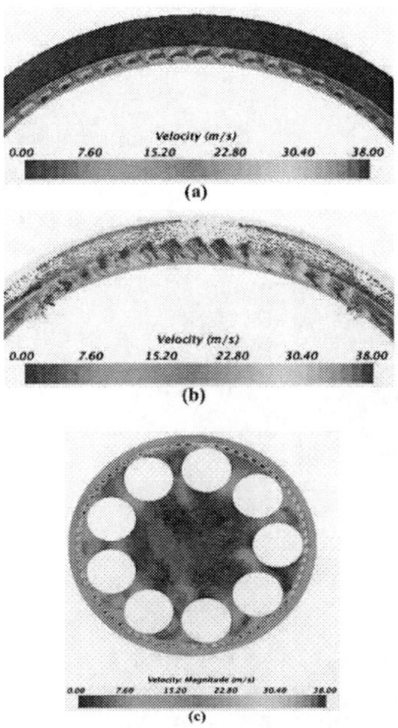


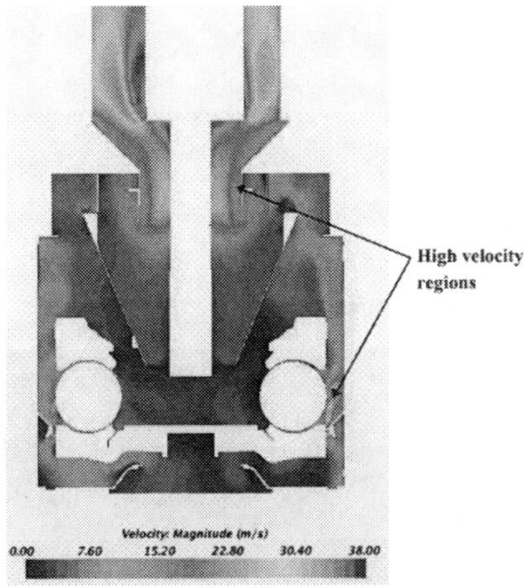


Velocity profile

**Comments**

1. Profile developed using air at ambient temperature (CAC).
2. Air velocity discernably affected by the throat gap and vortex finder.
3. Throat blades impart a swirl component.
4. Results are congruent with previous studies – served to verify the model developed.

Eskom



# CFD Results: Pressure Profile

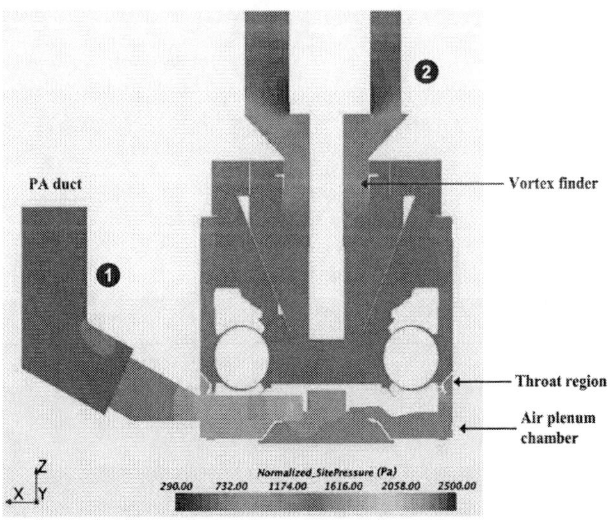


Pressure profile: Single phase

**Comments**

1. Highest pressure at PA duct/Air plenum chamber.
2. Lowest pressure at mill outlet.
3. Largest pressure drop across the throat region.

Eskom

# CFD Results and Validation (Air only)

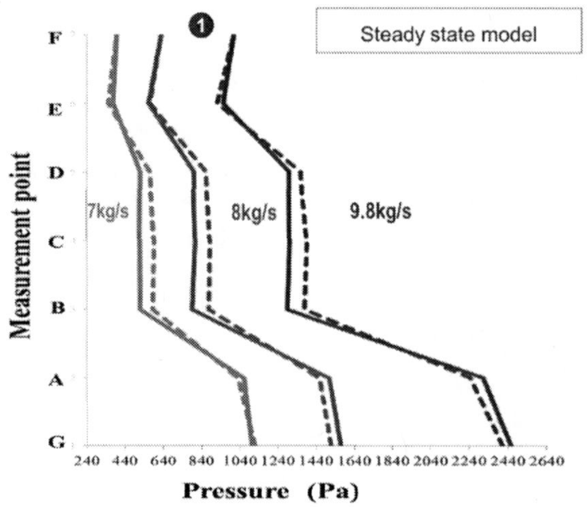


*Experimental (solid) and simulated (dashed) static pressure profile*

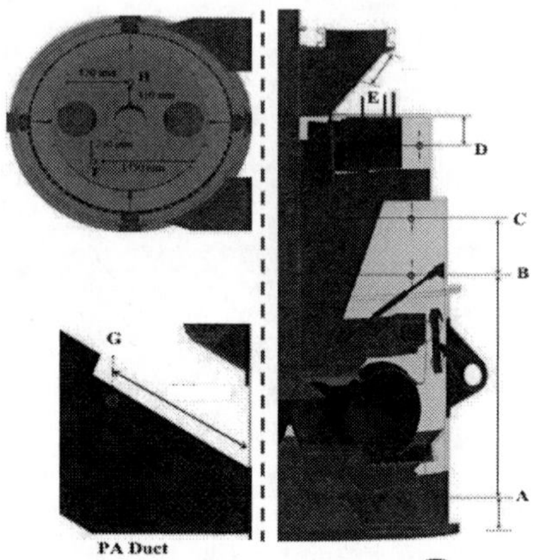


**Comments**

1. Validation assessed through experimental correlation.
2. Excellent correlation demonstrated between experimental and CFD predicted results, less than 5% variance.

Eskom



# CFD (Two-Phase: Uncoupled Particle size distribution)

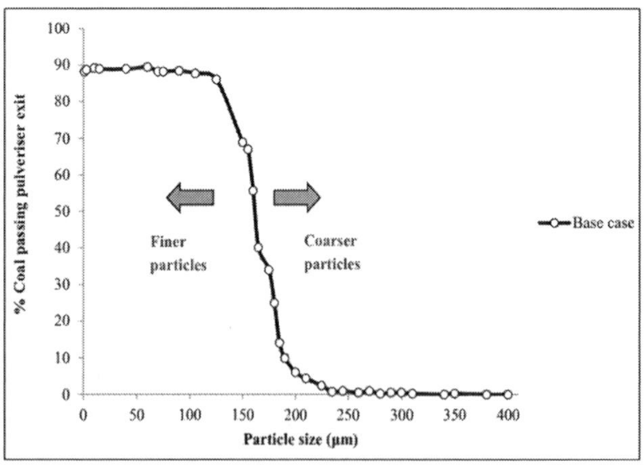


$$Coal\ passing\ (\%) = \frac{(Number\ of\ particles\ at\ exit\ )}{(Number\ of\ particles\ injected)} \times 100$$

**Comments**

1. 34 constant size injectors.
2. Particles were tracked for a residence time of 24s.
3. Higher percentage of fines at exit.

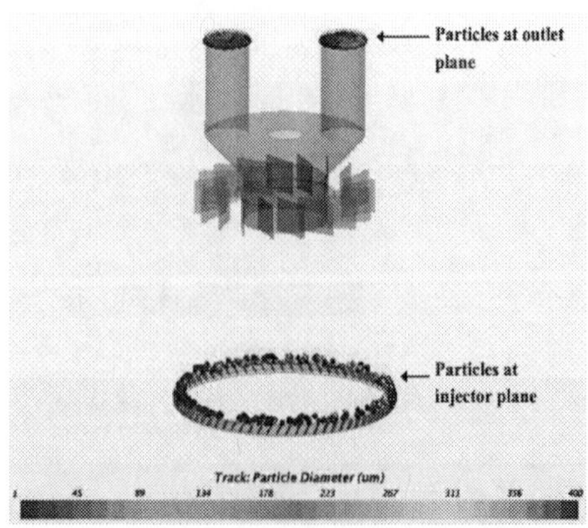

# Condition Monitoring Philosophy

- Condition monitoring approaches, employing statistical evaluation of measured outcomes, are extensively applied in industry. However, to a much less extent in coal pulverisers.
- For a successful condition monitoring system, <u>*a sensing system able to monitor a systems response to changes*</u> combined with an intelligent data analysis procedure are fundamental prerequisites.
- The single-phase CFD model developed provided a platform to assess the static pressure profile in a mill under conditions that mimic wear scenarios, typically those that cannot be tested in real life.
- Hence, candidate damage scenarios identified were simulated to assess the viability of employing the static pressure to achieve the aim of this study.



# Scenario 1: Damaged Vortex Finder

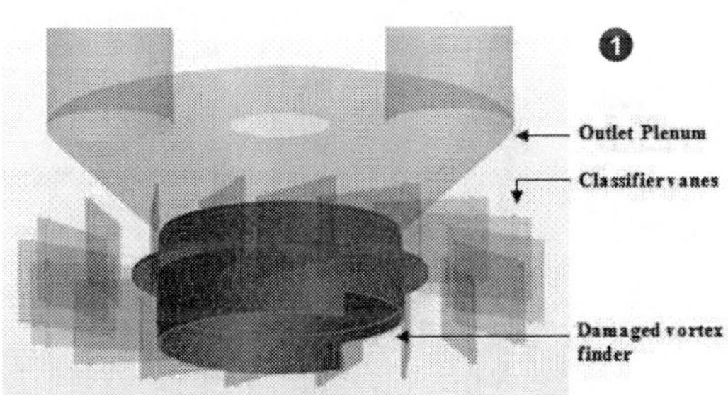


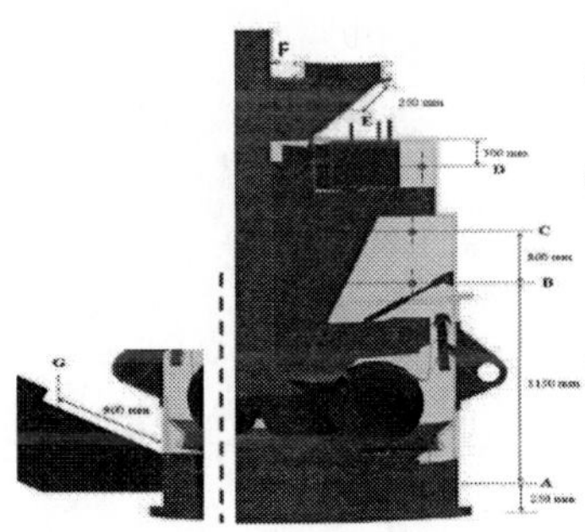


| Pressure drop | Base case (Pa) | Scenario 1 (Pa) | Variation (%) |
|---|---|---|---|
| $\Delta P_{GA}$ | 168.41 | 163 | 3.21 |
| $\Delta P_{AB}$ | 865.39 | 900.21 | 4.02 |
| $\Delta P_{BC}$ | -12.52 | -12.67 | 1.20 |
| $\Delta P_{BF}$ | 371.03 | 331.03 | 10.78 |
| $\Delta P_{CF}$ | 383.55 | 343.70 | 10.39 |
| $\Delta P_{GF}$ | 1404.83 | 1394.23 | 0.75 |

2

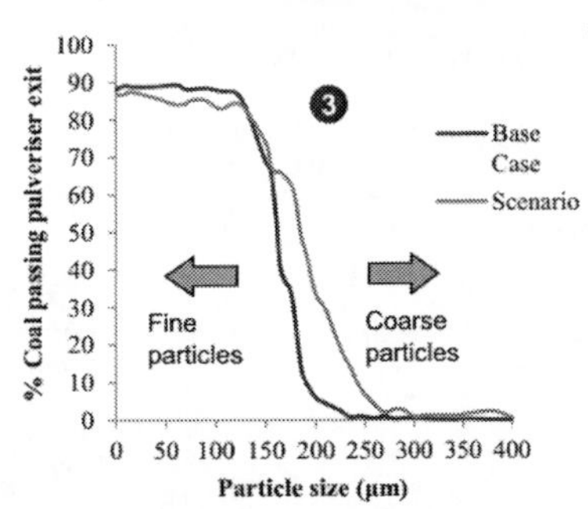


**Comments**

1. Impact damage to vortex finder.
2. Variation in static pressure across classifier region is noteworthy, pressure drop decreases.
3. Approx. 10 000 particles injected (uncoupled), larger particles exiting pulveriser with this condition.
4. No significant variation across other regions observed.

# Scenario 2: Damaged Inner Cone

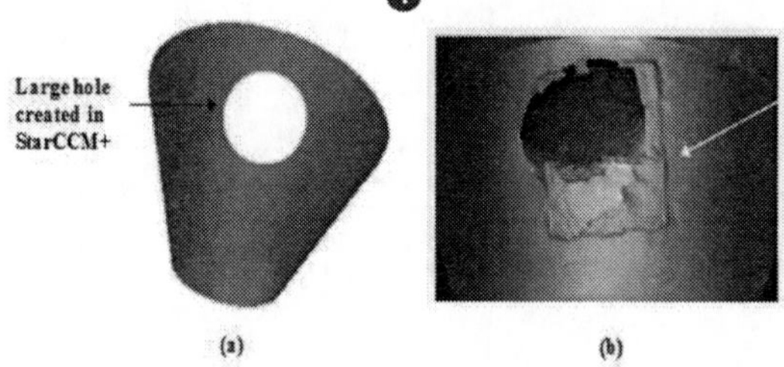


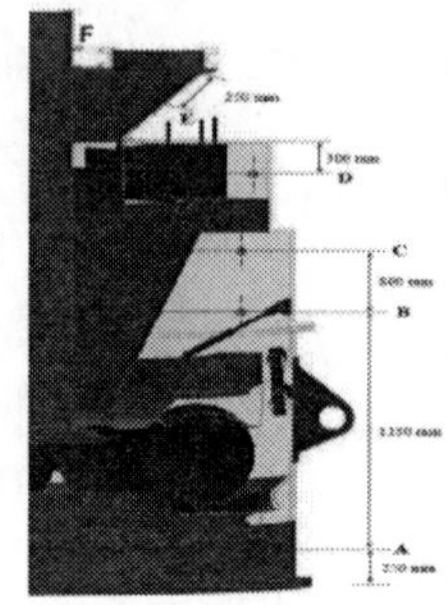

**Comments**

1. Impact damage to inner cone.
2. Variation in static pressure across classifier region is significant.
3. Additional variation highlighted across BC, directly related to worn cone.
4. Coarser particles exit the mill, perhaps due to classifier bypass.

| Pressure drop | Base case (Pa) | Scenario 2 (Pa) | Variation (%) |
|---|---|---|---|
| $\Delta P_{GA}$ | 168.41 | 162.11 | 3.74 |
| $\Delta P_{AB}$ | 865.39 | 881.57 | 1.87 |
| $\Delta P_{BC}$ | -12.52 | -8.51 | 32.06 ❸ |
| $\Delta P_{BF}$ | 371.03 | 247.48 | 33.30 ❷ |
| $\Delta P_{CF}$ | 383.55 | 255.99 | 33.26 ❷ |
| $\Delta P_{GF}$ | 1404.83 | 1291.16 | 8.09 |

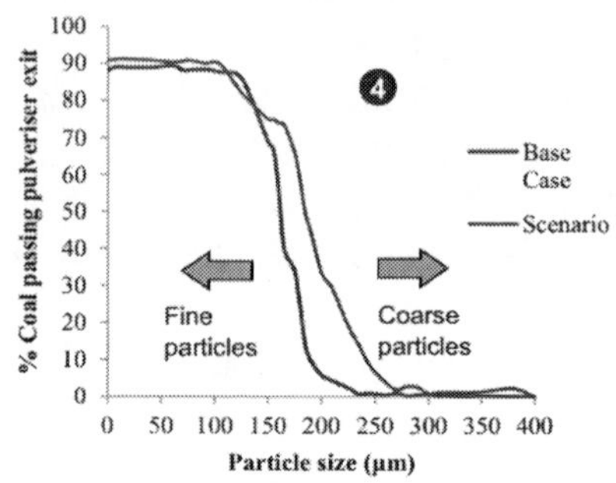




# Scenario 3: Blocked throat gap (5%)

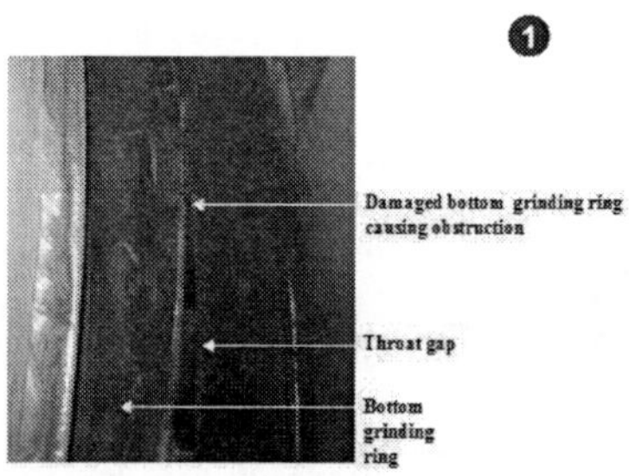


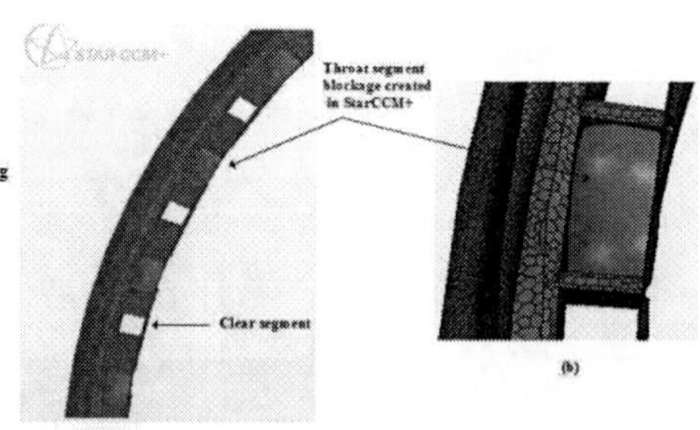


**Comments**

1. Damaged grinding ring, PA flow restriction.
2. Variation in static pressure across throat region (AB) is highlighted.
3. Higher percentage of finer particles exit the mill, perhaps due to higher PA velocity.

| Pressure drop | Base case (Pa) | Scenario 3 (Pa) | Variation (%) |
|---|---|---|---|
| $\Delta P_{GA}$ | 168.41 | 156.96 | 6.80 |
| $\Delta P_{AB}$ | 865.39 | 968.75 | 11.94 ❷ |
| $\Delta P_{BC}$ | -12.52 | -13.12 | 4.81 |
| $\Delta P_{BF}$ | 371.03 | 364.83 | 1.67 |
| $\Delta P_{CF}$ | 383.55 | 377.95 | 1.46 |
| $\Delta P_{GF}$ | 1404.83 | 1490.54 | 6.10 |

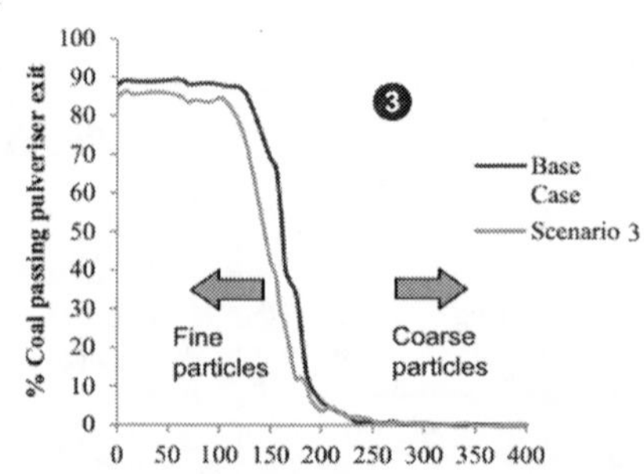

## Conclusions

- **MMEB**
  - MMEB model derived to predict the coal mass flow rate was successfully developed.
  - Experimental and model predicted values demonstrated close correlation, less than 5% variance (isokinetic sampling).
  - Sensitivity analysis highlighted significantly influential parameters on the predictive model output.
  - Application of this model:
    - Quantification of coal flow through individual mills (table feeders).
    - Enable engineers to continuously monitor the mill loading.

Eskom

2016/09/15 25

## Conclusions (Contd.)

- **Condition Monitoring Philosophy**
  - CFD model of 8.5E VSM was successfully developed.
  - Experimental and Model predicted measured outcomes displayed close correlation, less than 5% variance (Air only).
  - CFD simulated results highlighted a decrease in pressure drop across critical components as the mill component wear.
  - Additionally, CFD depicts larger particles exit the mill as the components wear.
  - Application of this model:
    - Assess the impact of mill component deterioration on mill performance.
    - Proactively address changes in mill condition, thereby improving mill output, reducing cost and mill downtime.

Eskom

2016/09/15 26

# Thank you

2016/09/15

27

# AUTHOR INDEX

# AUTHOR INDEX

# AUTHOR INDEX

# AUTHOR INDEX

# AUTHOR INDEX

**International Pittsburgh Coal Conference**
University of Pittsburgh
Swanson School of Engineering
3700 O'Hara St.
940 Benedum Hall
Pittsburgh, Pennsylvania 15261
USA

ISBN 978-1-5108-5366-9

9 781510 853669

CURRAN ASSOCIATES INC.
proceedings.com

# 33rd Annual International Pittsburgh Coal Conference (IPCC 2016)

## Coal – Energy, Environment and Sustainable Development

Cape Town, South Africa
8-12 August 2016

Volume 2 of 2

# 33rd Annual International Pittsburgh Coal Conference (IPCC 2016)

## Coal – Energy, Environment and Sustainable Development

Cape Town, South Africa
8-12 August 2016

Volume 2 of 2

ISBN: 978-1-5108-5366-9

**Printed from e-media with permission by:**

Curran Associates, Inc.
57 Morehouse Lane
Red Hook, NY 12571

**Some format issues inherent in the e-media version may also appear in this print version.**



Printed by Curran Associates, Inc. (2018)



International Pittsburgh Coal Conference
University of Pittsburgh
Swanson School of Engineering
3700 O'Hara St.
940 Benedum Hall
Pittsburgh, Pennsylvania 15261
USA

Phone: (412) 624-7440
Fax: (412) 624-1480

ipcc@pitt.edu

**Additional copies of this publication are available from:**

Curran Associates, Inc.
57 Morehouse Lane
Red Hook, NY 12571 USA
Phone: 845-758-0400
Fax: 845-758-2633
Email: curran@proceedings.com
Web: www.proceedings.com

# TABLE OF CONTENTS

## VOLUME 1

### PLENARY SPEAKERS



### ORAL SESSIONS

### SESSION 1: CLEAN COAL DEMONSTRATION AND COMMERCIAL PROJECTS - 1



### SESSION 2: COMBUSTION TECHNOLOGIES - 1



### SESSION 3: CARBON MANAGEMENT - 1

## SESSION 4: COAL SCIENCE - 1



## SESSION 5: GASIFICATION TECHNOLOGIES: GAS CLEANUP



## SESSION 6: COAL MINING AND BENEFICIATION - 1



## SESSION 7: CLEAN COAL DEMONSTRATION AND COMMERCIAL PROJECTS - 2

## SESSION 8: COMBUSTION TECHNOLOGIES - 2



## SESSION 9: CARBON MANAGEMENT - 2



## SESSION 10: COAL SCIENCE - 2



## SESSION 11: GASIFICATION TECHNOLOGIES: UCG - 1

## SESSION 12: COAL MINING AND BENEFICIATION - 2



## SESSION 13: CLEAN COAL DEMONSTRATION AND COMMERCIAL PROJECTS - 3



## SESSION 14: COMBUSTION TECHNOLOGIES - 3



## SESSION 15: CARBON MANAGEMENT - 3

## SESSION 28: GASIFICATION TECHNOLOGIES: MODELING - 1



## SESSION 29: CLEAN COAL AND GAS TO FUELS: CATALYTIC STUDY - 2



## SESSION 30: SUSTAINABILITY AND ENVIRONMENT - 3



## SESSION 31: GASIFICATION TECHNOLOGIES: GENERAL - 4

## SESSION 32: COAL SCIENCE - 6



## SESSION 33: GASIFICATION TECHNOLOGIES: GENERAL - 3



## SESSION 34: CLEAN COAL AND GAS TO FUELS: PROCESSING



## SESSION 35: GASIFICATION TECHNOLOGIES: MODELING - 2

## POSTER SESSION 5: COAL SCIENCE

Eskom

# Measurement and prediction of coal flow rates in pulverised fuel pipes

Presenter: Dheshan Naran

Academic Mentor: Walter Schmitz

Industrial Mentor: Bonny Nyangwa

Date: 10/08/2016

UNIVERSITY OF THE WITWATERSRAND, JOHANNESBURG

## Background

Eskom

(Picture: http://www.bdlive.co.za/business/mining/2016)

±120 million tons of coal per annum

85.5% of South Africa's Electricity

(Picture: theconversation.com)

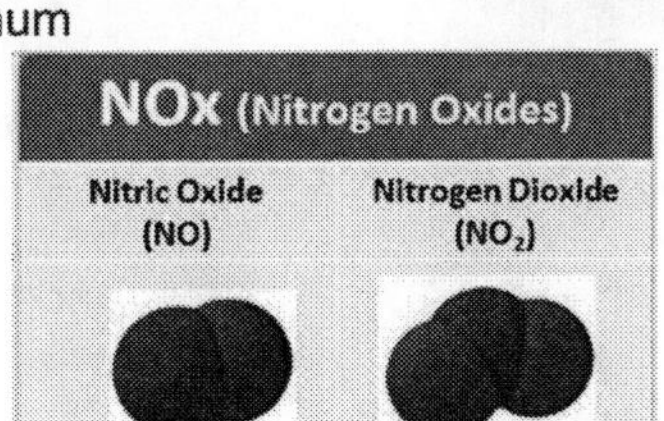


(Picture: http://www.pfonline.com/articles/nox-scrubbing-technology-breakthrough)

900kT to 1000kT per annum

UNIVERSITY OF THE WITWATERSRAND, JOHANNESBURG

## Background



Air Quality Management (AQM) Act 39, of 2004

(Picture: www.cchowardcounty.gov)

(Picture: www.detroitstoker.com)

Low NOx Burner

- The act states amongst other limits, the Nitrogen Oxide (NOx) emissions should not exceed 750mg/Nm$^3$ (@10% O2) by 2020.

UNIVERSITY OF THE WITWATERSRAND, JOHANNESBURG

## Background

Eskom

Pulverised Fuel

Balanced (mass flow and size distribution)

Pictures:
(monitored) – http://seriouslymen.com
(balanced) – www.huffingtonpost.com
(controlled) – www.123inkjets.com.au

To realise **lower NOx emissions and optimise combustion efficiency**, pulverised fuel (PF) flow must be monitored, controlled and evenly distributed (mass flow and size) between the PF pipes and burners.

UNIVERSITY OF THE WITWATERSRAND, JOHANNESBURG

## Background

Eskom

Power utilities **rely on Isokinetic PF sampling** to determine the PF (mass and size) distribution between burners.

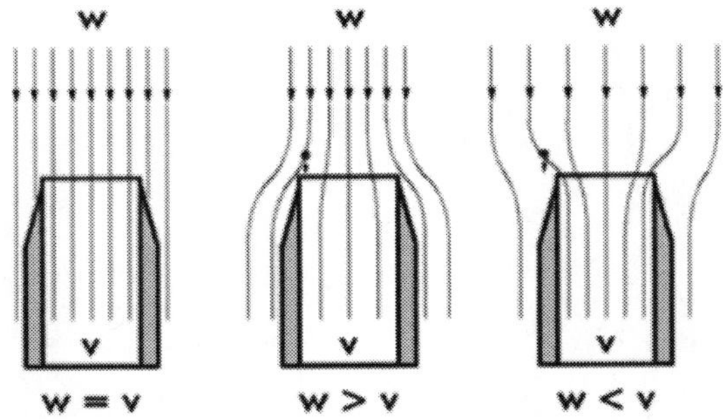


(Picture: www.photometer.com)

(Picture: www.coachtishtimes.com)

Isokinetic sampling can be defined as sampling at such a rate that the velocity and the direction of the gas entering the sampling nozzle is the same as that of the gas in the pipe/duct/stack at the same sampling point [17].

UNIVERSITY OF THE WITWATERSRAND, JOHANNESBURG

## Purpose of the Study

Eskom

(Picture: http://www.coachtishtimes.com)

The purpose of the study was to evaluate current pulverised fuel (PF) measurement technology, standards and practices and recommend a best practice for PF measurements in Power Utilities

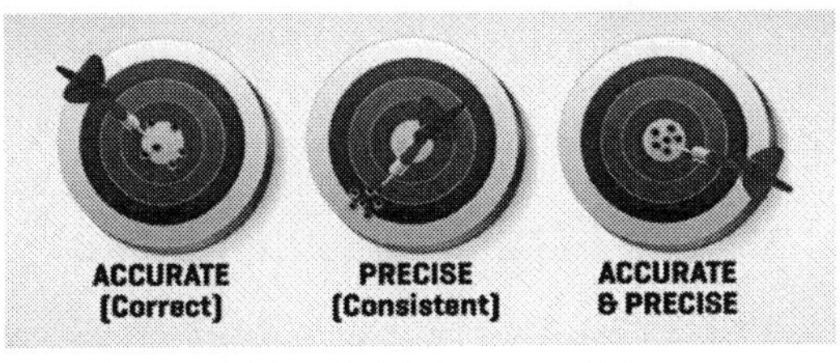


(Picture: https://www.hotdesign.com/marketing/accuracy-vs-precision)

UNIVERSITY OF THE WITWATERSRAND, JOHANNESBURG

## Sampling Equipment

Eskom

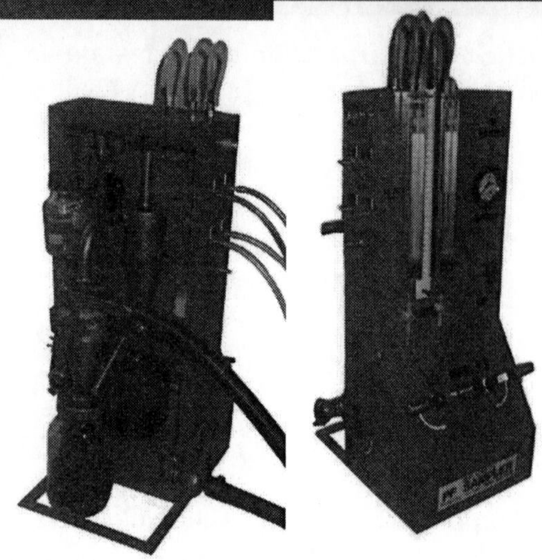

Manual Sampler

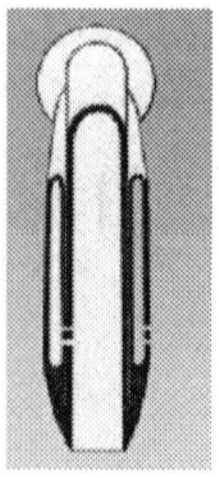

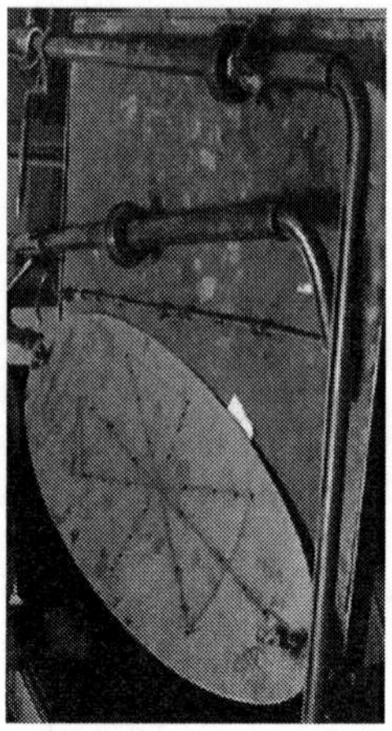

(Pictures: LC4 operating manual)



## Application of Manual Sampler

Eskom

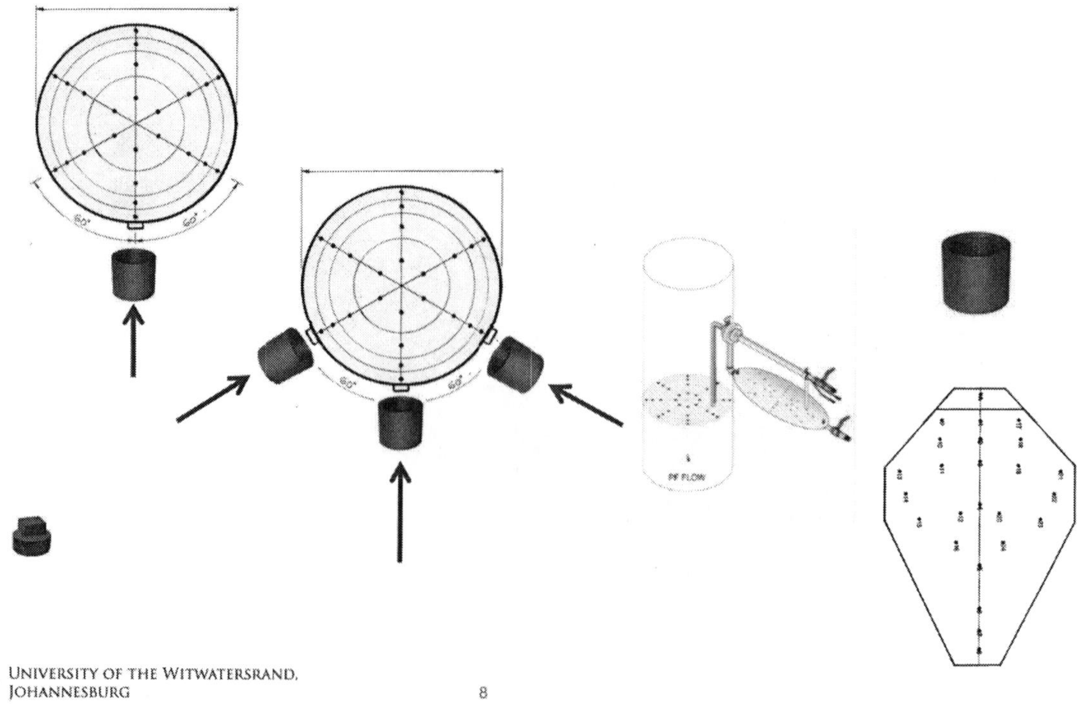

## Sampling Equipment

Eskom

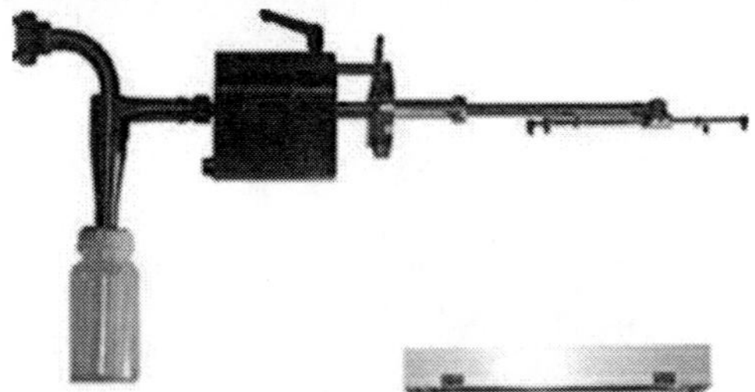

Dustless connector

Auto PF Sampler

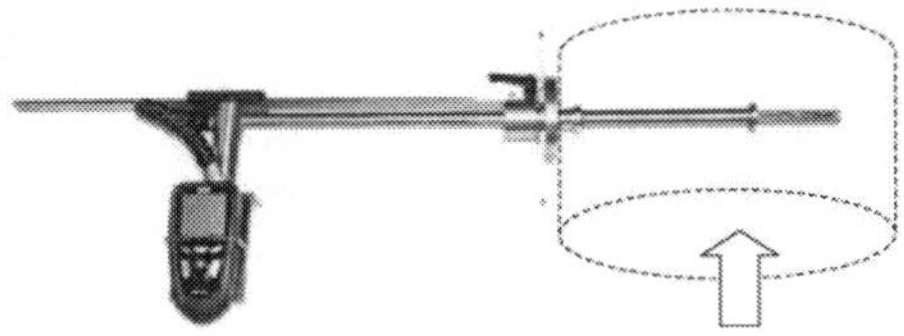

Dirty air probe with manometer

(Pictures: MW manual)

UNIVERSITY OF THE WITWATERSRAND, JOHANNESBURG

## Sampling locations

Eskom

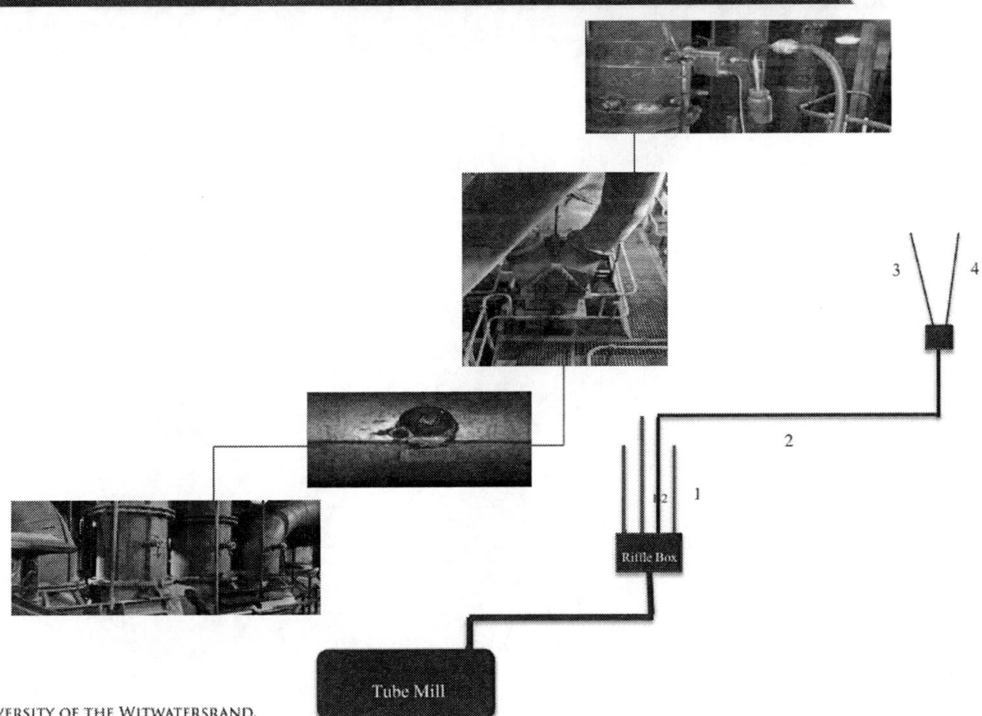


UNIVERSITY OF THE WITWATERSRAND, JOHANNESBURG

## Samples & Analysis

Eskom

- 924 PF samples were extracted
- Raw coal was also sampled

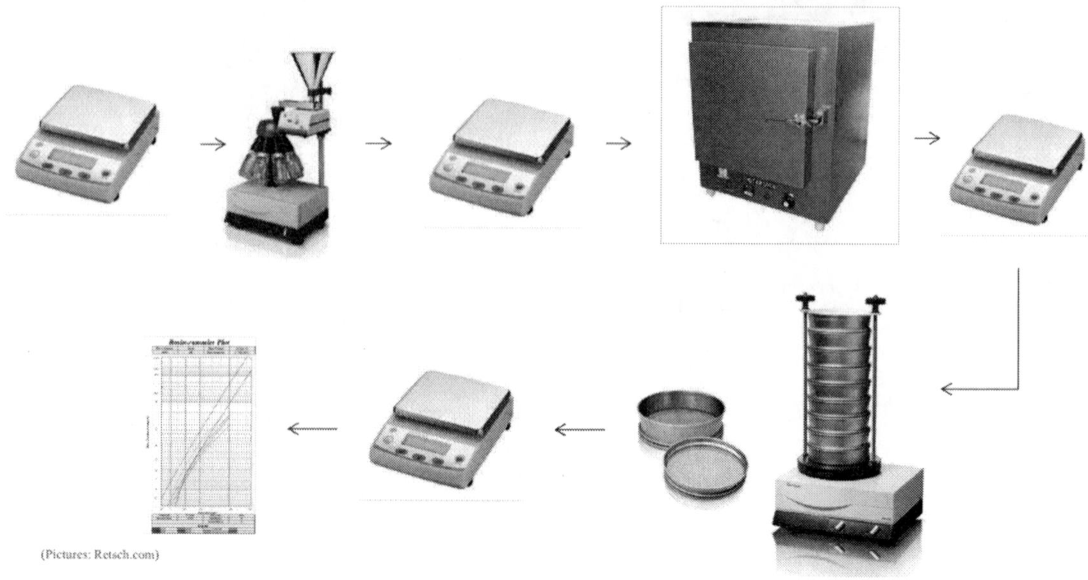

(Pictures: Retsch.com)

UNIVERSITY OF THE WITWATERSRAND, JOHANNESBURG

## Incremental Sampling

Eskom

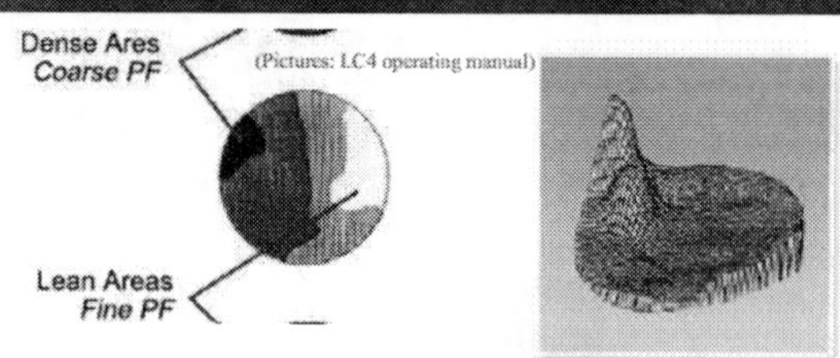


(Pictures: LC4 operating manual)

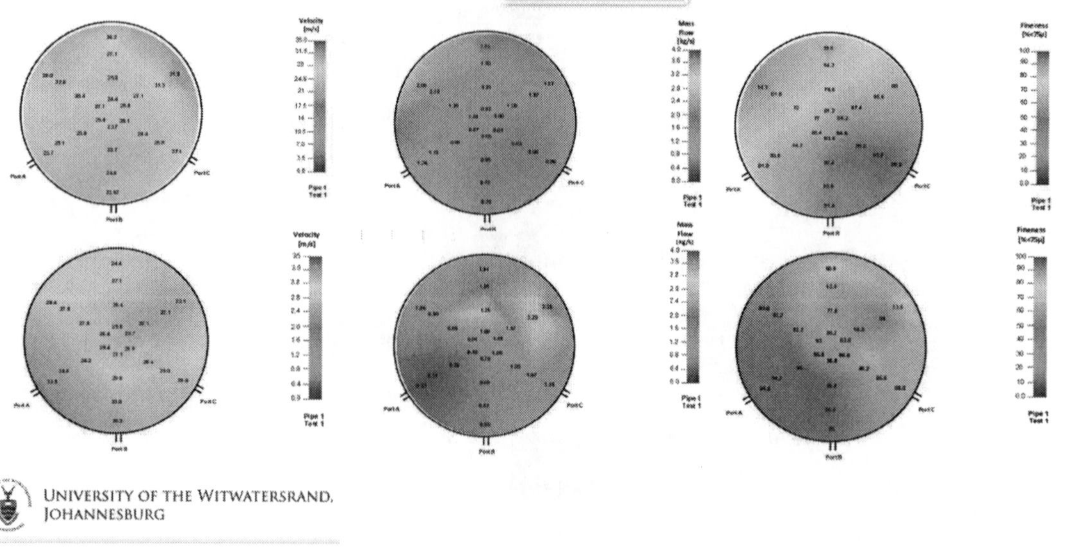

UNIVERSITY OF THE WITWATERSRAND, JOHANNESBURG

## CFD

Eskom

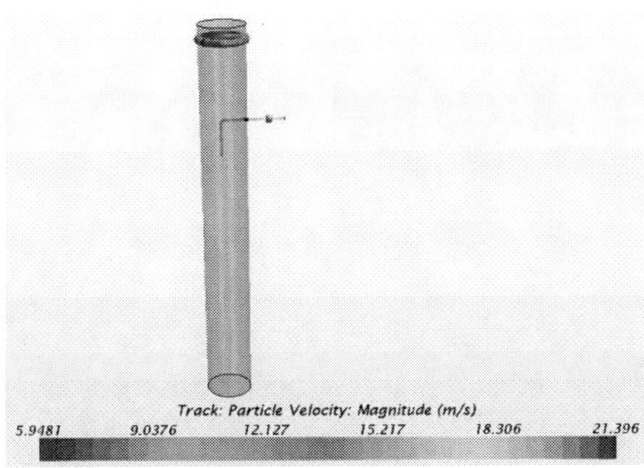


Sampling nozzle axial to flow

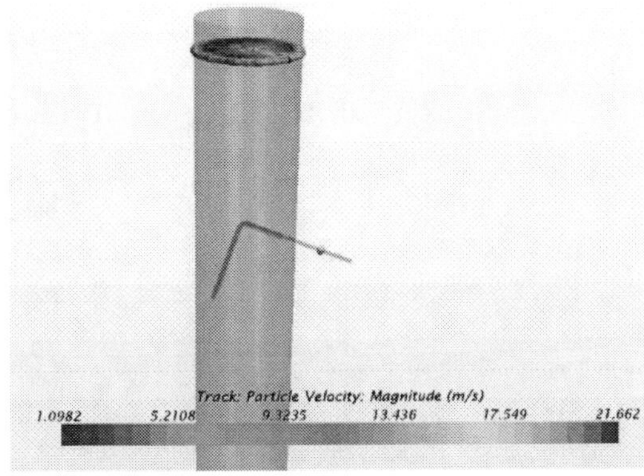


Sampling nozzle angled at 30°

Good correlation with experimental work.

- $15^0$ reduction in particles collected ( -3.5 % coal mass flow)
- $22^0$ increase in particles sampled (+18.28 % coal mass flow)
- $30^0$ increase in particles collected (+18.78 % coal mass flow)

UNIVERSITY OF THE WITWATERSRAND, JOHANNESBURG

## Repeatability

Eskom

Standard deviation employed

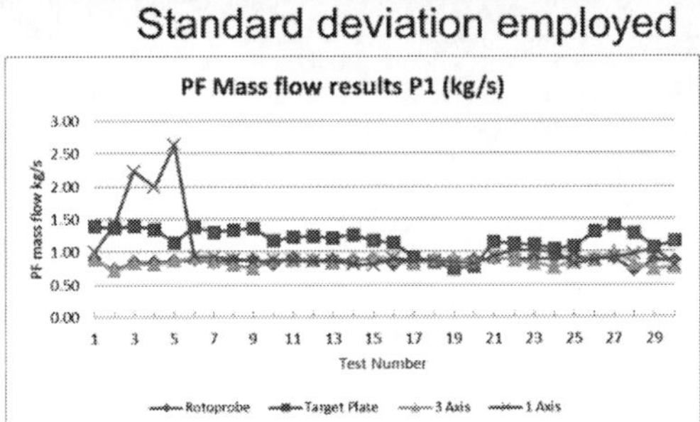


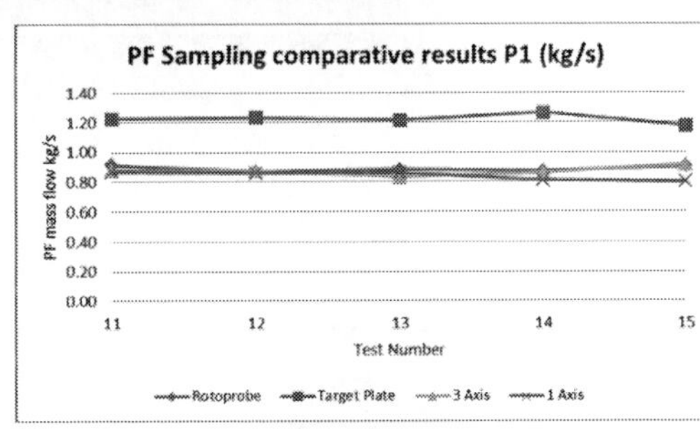


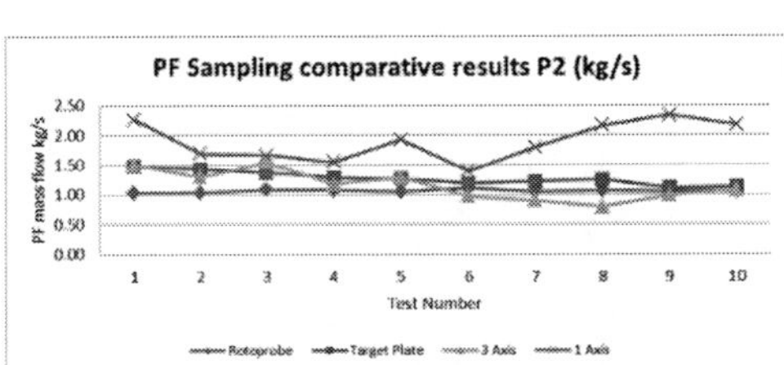


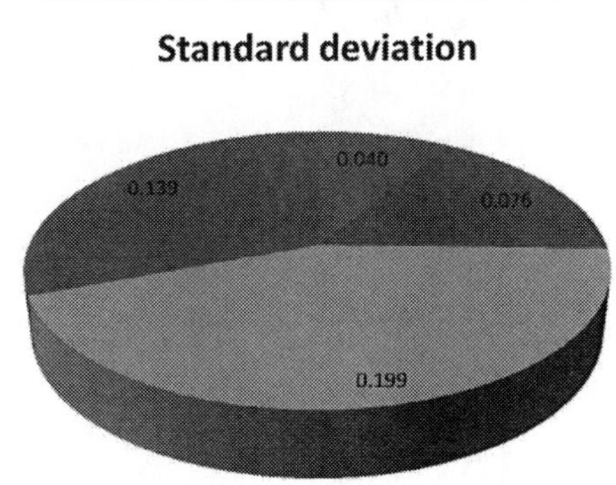





UNIVERSITY OF THE WITWATERSRAND, JOHANNESBURG

# Accuracy

**Coal Mass Balance**

Raw coal fed into the mills (kg/s) = $\sum$ of PF in pipes (kg/s)

$$MRC_d = MRC_w - (MRC_w.\%M_c)$$

$$MF_d = \frac{M_d}{n.t}\left(\frac{D}{d}\right)^2$$

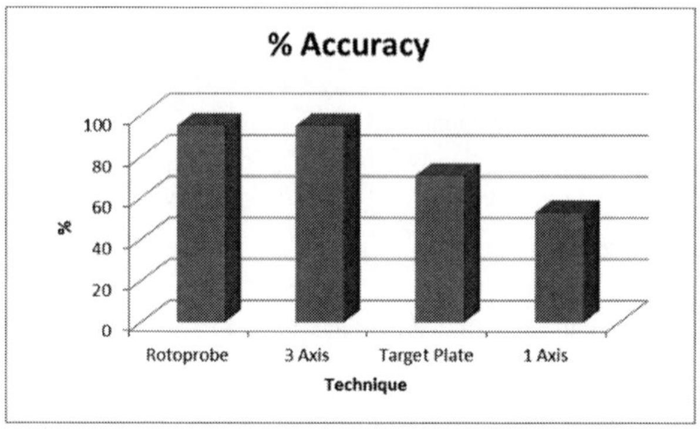




# Conclusions

| Sampling Technique | Pipe Orientation | Recovery Rate | % Error | Degree of Accuracy % | Standard Deviation | Accuracy & Repeatability |
|---|---|---|---|---|---|---|
| Rotoprobe | Vertical | 104.95 | 4.95 | 95.05 | 0.040 |  |
| 3 Axis | Vertical | 105.23 | 5.23 | 94.77 | 0.076 |  |
| Target Plate | Vertical | 129.09 | 29.09 | 70.91 | 0.199 |  |
| 1 Axis | Vertical | 146.93 | 46.93 | 53.07 | 0.139 |  |

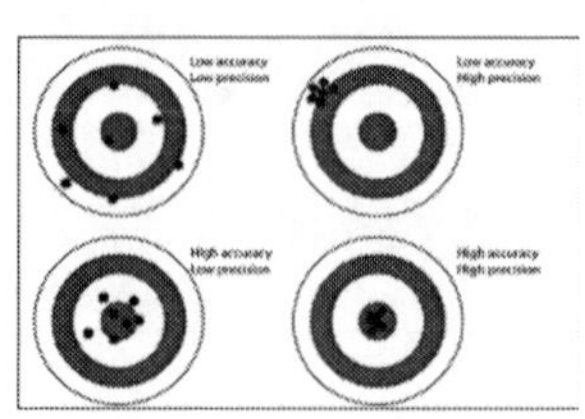

(Picture: http://blog.wheaton.com/accuracy-vs-precision-know-the-difference)

Eskom

# Thank you

UNIVERSITY OF THE WITWATERSRAND, JOHANNESBURG

## References

Eskom


1. 2014. . [ONLINE] Available at: http://www.eskom.co.za/AboutElectricity/FactsFigures/Documents/ES0003EskomsCommitmentEnvironmentRev9.pdf. [Accessed 23 September 2014].
2. 2014. . [ONLINE] Available at: http://www.utgjiu.ro/revista/ing/pdf/2008-02/36_LAZA%20IOAN.pdf. [Accessed 24 September 2014].
3. Cabrejos & Klinzing, FJ & GE , (1994). Pickup and saltation mechanisms of solid particles in horizontal pneumatic transport. In Powder Technology, 1994. 173-186.
4. David Mills Dip Tech (Eng) PhD CEng MIMechE, 2004. Pneumatic Conveying Design Guide, Second Edition. 2 Edition. Butterworth-Heinemann.
5. Edited by Clayton T. Crowe, 2005. Multiphase Flow Handbook (Mechanical and Aerospace Engineering Series). 1 Edition. CRC Press.
6. H. A. Stoess Jr., 1983. Pneumatic Conveying. 2 Edition. Wiley-Interscience.
7. G.E. Klinzing, 1997. Pneumatic Conveying of Solids: A theoretical and practical approach (Particle Technology Series). 2nd Edition. Springer.
8. 2014. [ONLINE] Available at: http://www.e-inst.com/docs/Combustion-Booklet-2013.pdf. [Accessed 22 September 2014].
9. 2014. . [ONLINE] Available at: http://www.aiche.org/sites/default/files/docs/webinars/JacobK-PneumaticConveyingPDFmin.pdf. [Accessed 22 September 2014].
10. L. M. Gomes and A. L. Amarante Mesquita. (January - March, 2014). ON THE PREDICTION OF PICKUP AND. Brazilian Journal. 31 (1), p35-46
11. Chen Xiao-ping, Zhou Yun, Liang Cai, Meng Qing-min, Lu Peng, Pu Wen-hao, Xu Pan. (2009). Effect of properties of pulverized coal on dense phase. Journal of Physics: Conference Series 147 (2009) 012060. 147 (1), 1-9.


UNIVERSITY OF THE WITWATERSRAND, JOHANNESBURG

## References

Eskom


12. D. Geldart and S. J. Ling*. (1992). Saltation velocities in high pressure conveying of fine coal. Powder Technology. 69 (1), 157-162.
13. S.S. Mallick, P.W. Wypych. (2009). Minimum transport boundaries for pneumatic conveying of powders. Powder Technology. 194 (1), 181–186.
14. Peter J. Jones and L. S. Leung. (1978). A Comparison of Correlations for Saltation Velocity in Horizontal. Ind. Eng. Chem. Process Des. 17 (4), 572-575.
15. JIA Zhi-hai, FAN Xue-liang, LI Jun-feng, CAI Xiao-shu,LIU Ji-ze, CONGXiao, JIA Jing-chen. (2009). On-line Measurement of Pneumatic Conveying of Pulverized. Journal of Physics: Conference Series . 147 (1), 1-7.
16. Zwiegelaar J, Pulverise Fuel Sample Collection and Size Grading, Eskom Standard 240-76440331
17. Woollatt G, An Introduction to Source Sampling (Presentation)
18. Steyn KH (1999), Pulverised Fuel Sampling Procedure GGP 0855 (Matimba).
19. ISO 9931 (1991), Coal-Sampling of pulverised coal conveyed by gases in direct fired coal systems.
20. BS893-1978, Methods for the measurement of the concentration of particulate matter in ducts carrying gases.
21. TRR/P89/022, van Den Boorn JMM (1989), Development and Operating Instruction for a Isokinetic Pulverised Sampler.
22. van Eeden F & Keir J (2001), Pulverised fuel measurement, Eskom report.
23. Bilirgen & Levey, H & E, 2000. Mixing and dispersion of particle ropes in lean phase pneumatic conveying. Powder Technology, 119, 134-152.
24. Archary H & Schmitz W & Jestin L (2014), Condition Monitoring of Pulverised Fuel Vertical Spindle mills, 12 International conference on Boiler Technology
25. Dense Phase Pneumatic Conveying System, Pneumatic Conveying Regimes, Solid Dense Conveying Systems, Mumbai, India. 2014. Mumbai, India. [ONLINE] Available at: http://www.macawberindia.com/product1.htm. [Accessed 01 November 2014].


UNIVERSITY OF THE WITWATERSRAND, JOHANNESBURG


## References

Eskom


Pictures


(NOx) – www.pfonline.com
Legislation – www.cchowardcounty.gov
(low NOx burner) – www.detroitstoker.com
(PF) – www.mhi-global.com
(monitored) – http://seriouslymen.com
(balanced) – www.huffingtonpost.com
(controlled) – www.123inkjets.com.au
(isokinetic) – www.photometer.com
(why) – www.en.hdyo.org
(analysis) – www.Retsch.com
(LC4 sampler) – PF sampling manual
(Incremental sampling) – PF sampling manual
(Rotoprobe) – MW manual



UNIVERSITY OF THE WITWATERSRAND, JOHANNESBURG

## Samples & Analysis

Eskom

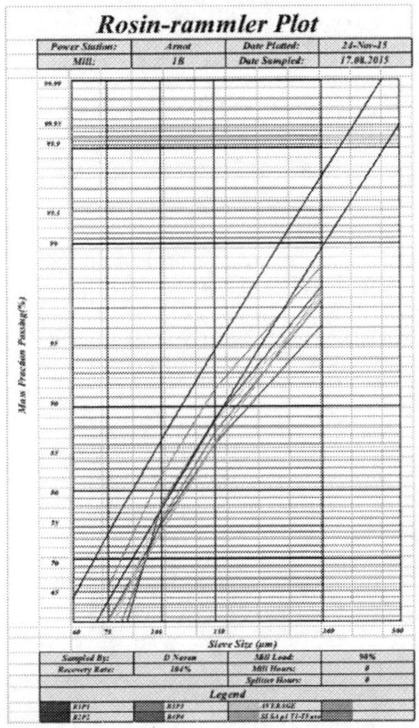


- **Standard Burner Requirements :-**
- Passing through 75µm : 65% to 75%
- Passing through 106µm : 77% to 86%
- Passing through 150µm : 88% to 94%
- Passing through 300µm: 98.8 to 99.8%

- **Low Nox Burners (with over fire air) :-**
- Passing through 75µm : 70% to 84%
- Passing through 106µm: 83% to 91.5%
- Passing through 150µm : 92% to 96.5%
- Passing through 300µm : 99.5% to 99.75%

UNIVERSITY OF THE WITWATERSRAND, JOHANNESBURG

## Samples & Analysis

Eskom

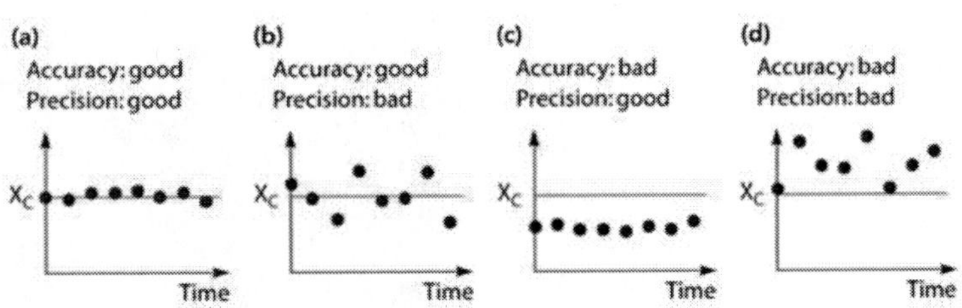


UNIVERSITY OF THE WITWATERSRAND, JOHANNESBURG

# Online Analysis in the South African Coal Industry

SCANTECH
*Process control specialists*

August 2016

**Pittsburgh Coal Conference**

# Benefits of On-Belt Real-Time Analysis

**Faster process control advice – Reduce Risk and Cost**

- Increased control over coal deliveries, reducing chance of out of spec delivery
- Quicker identification and correction of equipment problems
- Better asset utilisation
- Improve process efficiency
- Less sampling and laboratory analysis required
- Automated process eliminates human errors

SCANTECH
*Process control specialists*

August 2016

**Pittsburgh Coal Conference**

## AUSTMINE Presentation: November 2015

Mark Ryan, Business Manager – Automation: Sandvik:

Technology is a vital and integral part of the future mining process. It allows our customers to automatically capture multiple data points, use and simulate data to assist in process planning, identify bottle necks, establish solutions, confirm effectiveness of these solutions and provide an overall view of the entire mining process. Data is the key driver in achieving continuous improvements in safety, utilisation, productivity and optimisation. *Today data is generally collected manually which is subject to human error, but the main problem is that only a small number of data points are collected. More data points equals a more in-depth process understanding.*



## Applications for Coal Analysers

- Coal Monitoring at Mine
- Coal Monitoring at Load Out
- Coal Monitoring at Wash Plant
- Coal Monitoring at Power Station

SCANTECH
Process control specialists
August 2016
Pittsburgh Coal Conference

# Coal Monitoring at Mine

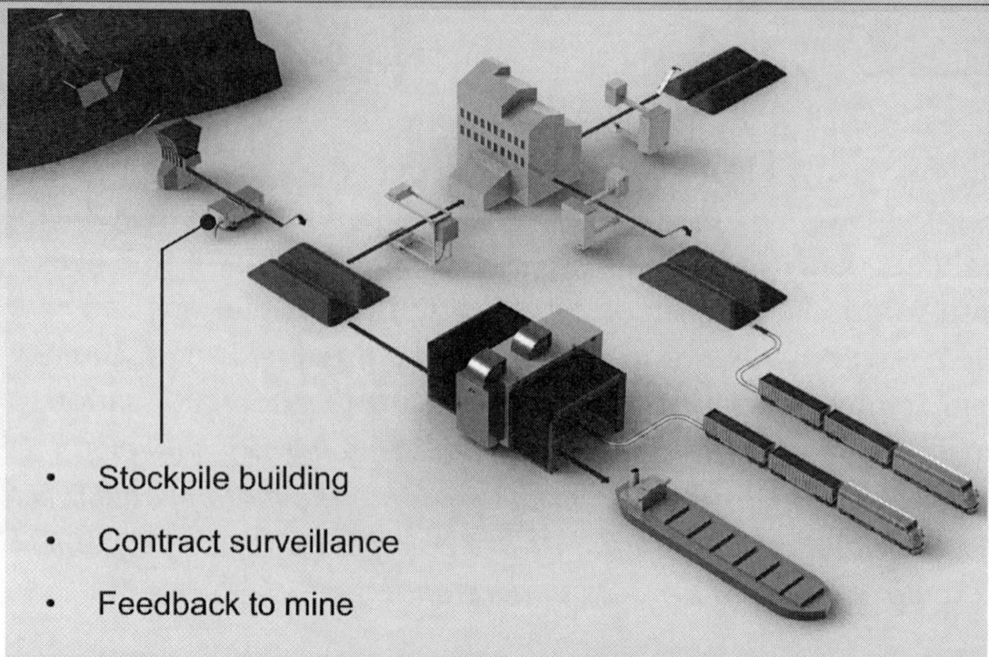

- Stockpile building
- Contract surveillance
- Feedback to mine

SCANTECH
*Process control specialists*

August 2016

**Pittsburgh Coal Conference**

# Coal Monitoring at Mine Loadout

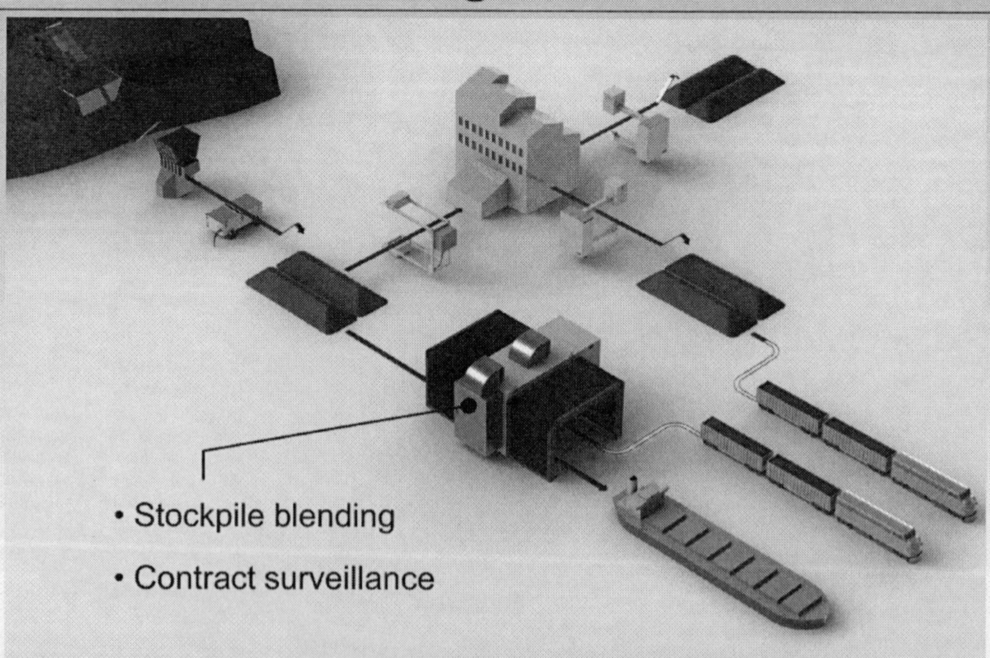

- Stockpile blending
- Contract surveillance

SCANTECH
*Process control specialists*

August 2016

**Pittsburgh Coal Conference**

## Number of Daily Rapid Analyses

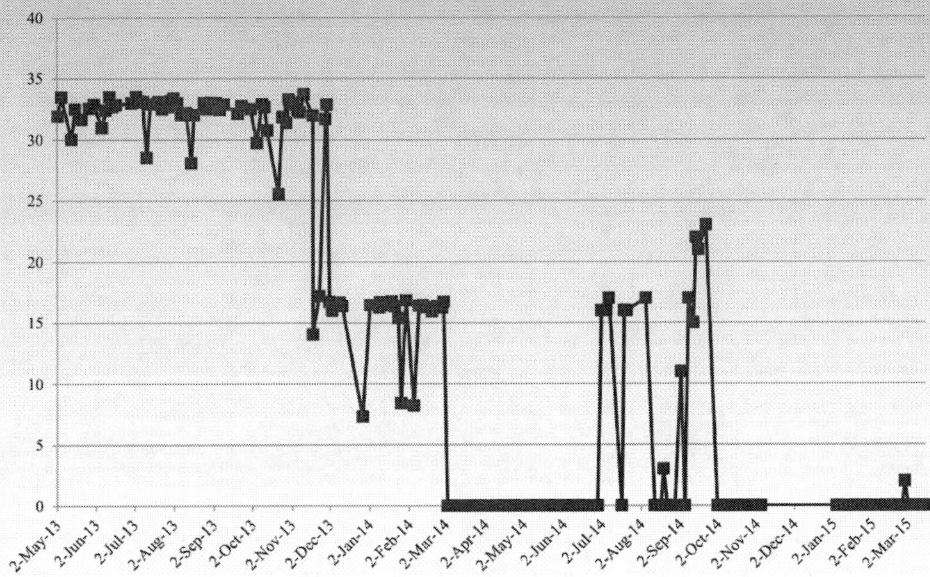


SCANTECH
*Process control specialists*

August 2016

**Pittsburgh Coal Conference**

## Number of Daily Rapid Analyses

| Reduce Rapids | 32 / day |
|---|---|
| Average Cost | $100 / rapid |
| Average Total Cost | $3,200 / day |
| Cost Reduction | Eliminate Rapids |
| Days per year | 300 |
| Total Improvement | $960,000/year |
| Cost of Analyser | $300,000 |
| Infrastructure | $300,000 |
| **Payback** | **7.5 months** |

SCANTECH
*Process control specialists*

August 2016

**Pittsburgh Coal Conference**

# Coal Monitoring at Washery

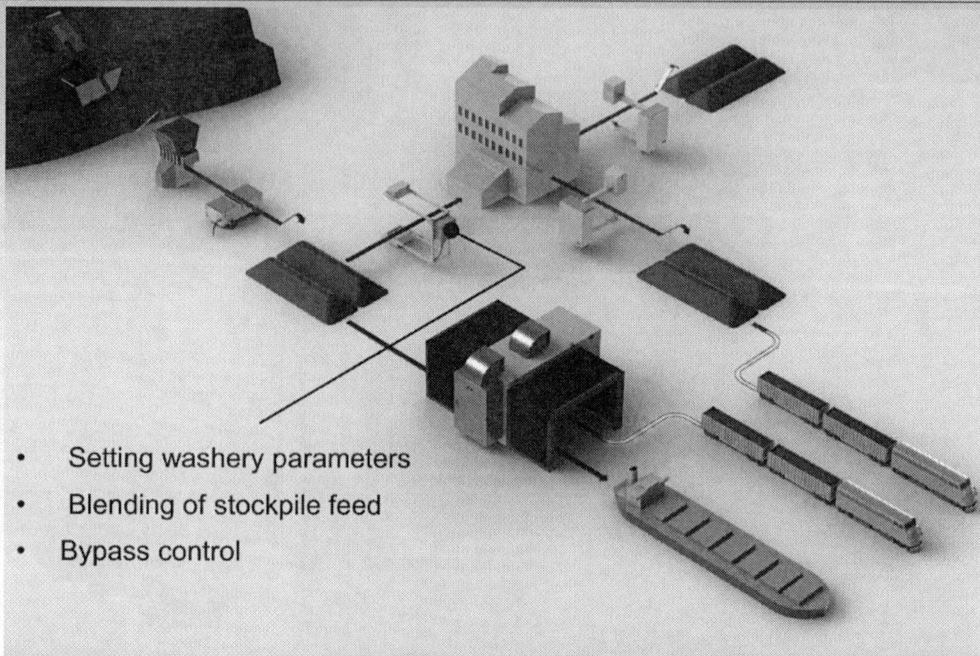

- Setting washery parameters
- Blending of stockpile feed
- Bypass control

SCANTECH
*Process control specialists*

August 2016

**Pittsburgh Coal Conference**

# CHPP Yield Variability

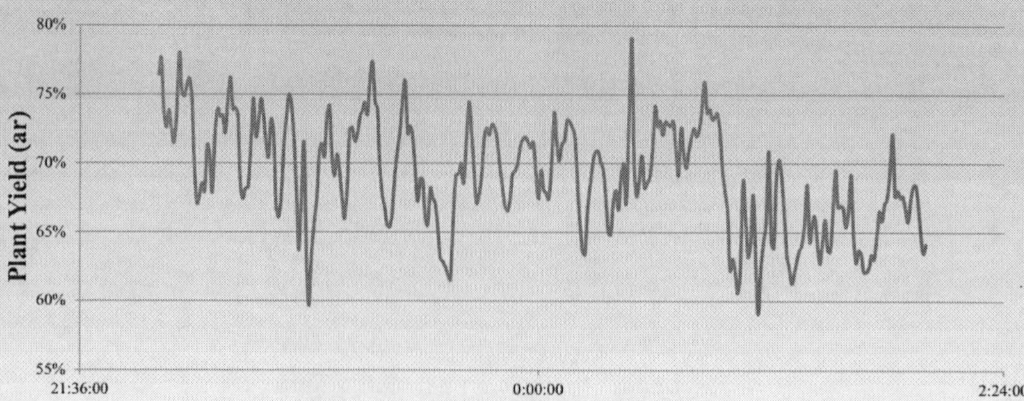



SCANTECH
*Process control specialists*

August 2016

**Pittsburgh Coal Conference**

# CHPP Yield Variability

| Increase Yield | 2% |
|---|---|
| Average Capacity | 300 t/hr |
| Improvement | 6 t/hr |
| Selling Price | \$50/t |
| Total Improvement | \$300/hr |
| Hours per day | 20 |
| Total Improvement | \$6,000/day |
| Days per year | 300 |
| Total Improvement | \$1,800,000/year |
| Cost of Analyser | \$300,000 |
| Infrastructure | \$300,000 |
| **Payback** | **4 months** |

SCANTECH
*Process control specialists*

August 2016

Pittsburgh Coal Conference

# Bypass Washing

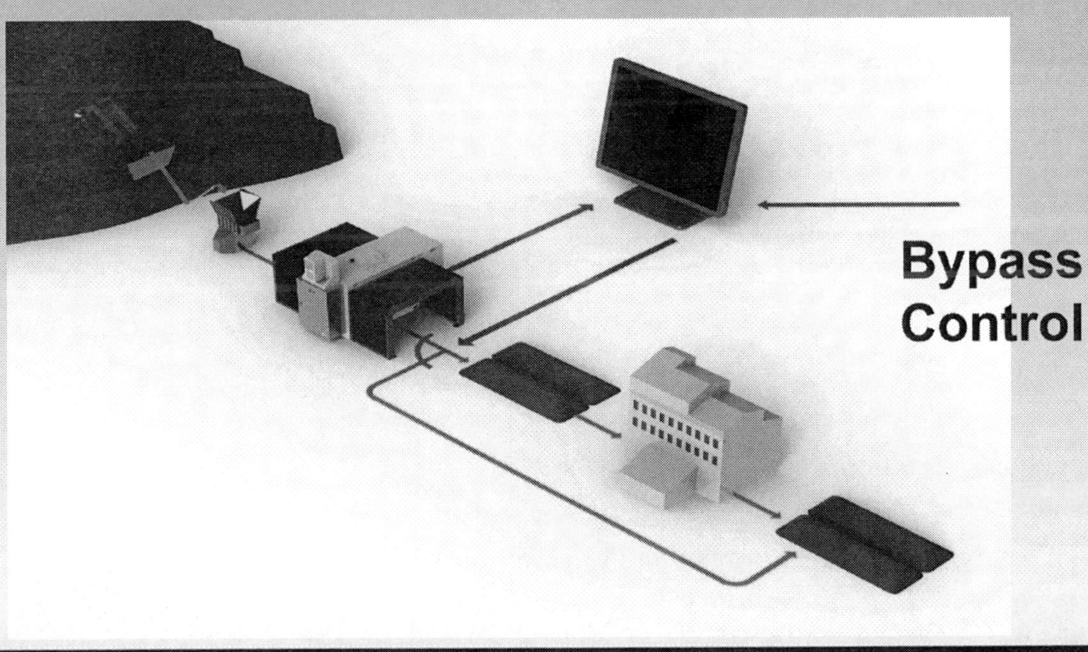


SCANTECH
*Process control specialists*

August 2016

Pittsburgh Coal Conference

# Bypass Assumptions

| | | | | | |
|---|---|---|---|---|---|
| Mining Assumptions | | | No Analyser | | With Analyser |
| | Raw Tons | | 2,000,000 | | 2,000,000 |
| | Mining cost per Tonne | $ | 10 | $ | 10 |
| | Total Mining Cost | $ | 20,000,000 | $ | 20,000,000 |
| Divert Raw coal from Prep plant | | | | | |
| | Prep Plant ByPass | | 0% | | 15% |
| | Raw tonnes bypassed | | 0 | | 300000 |
| Prep Plant Assumptions | | | | | |
| | Tonnes delivered to plant | | 2,000,000 | | 1,700,000 |
| | Recovery | | 55% | | 56% |
| | Total clean Tonnes | | 1,100,000 | | 952,000 |
| | Washiong cost per raw tonne | $ | 2.75 | $ | 2.75 |
| | Total Washing Cost | $ | 5,500,000 | $ | 4,675,000 |
| Total cost (Mining + Washing) | | $ | 25,500,000 | $ | 24,675,000 |
| Revenue Assumptions | | | | | |
| | Price per tonnes | $ | 50.00 | $ | 50.00 |
| | Total Clean Coal Tonnes sold | | 1,100,000 | | 1,252,000 |
| | Total Revenue | $ | 55,000,000 | $ | 62,600,000 |

SCANTECH
Process control specialists

August 2016

Pittsburgh Coal Conference

# Bypass Payback

| | | | | | |
|---|---|---|---|---|---|
| Profit from Operations | | | | | |
| | Total Revenue | $ | 55,000,000 | $ | 62,600,000 |
| | Total cost (Mining + Washing) | $ | 25,500,000 | $ | 24,675,000 |
| | Profit | $ | 29,500,000 | $ | 37,925,000 |
| | Profit % | | 53.64% | | 60.58% |
| ADDITIONAL PROFIT | | | | $ 8,425,000 | |
| Cost of Analyser (Including Infrastructure) | | | | $ 600,000 | |
| **Pay Back** | | | | | **0.85 Months** |

SCANTECH
Process control specialists

August 2016

Pittsburgh Coal Conference

# Coal Monitoring at Washery

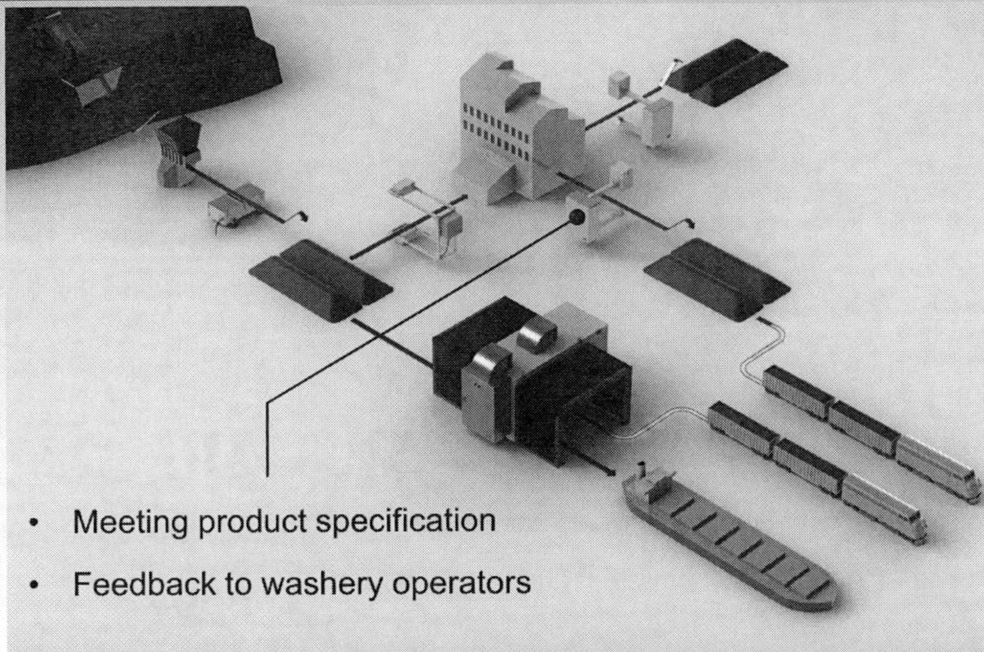

- Meeting product specification
- Feedback to washery operators

SCANTECH
*Process control specialists*

August 2016

**Pittsburgh Coal Conference**

# Coal Monitoring at Power Plant

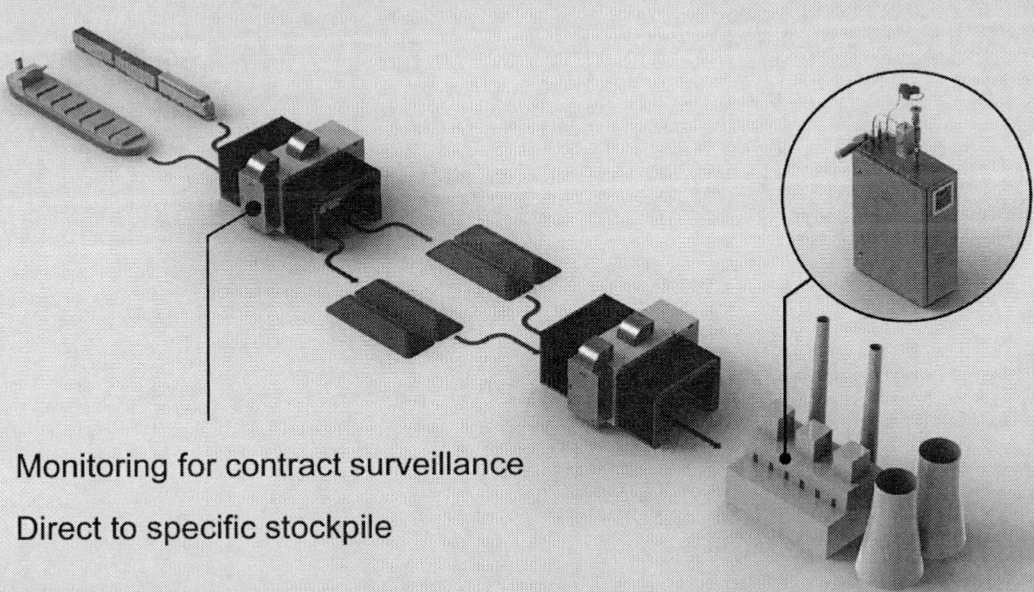

- Monitoring for contract surveillance
- Direct to specific stockpile

SCANTECH
*Process control specialists*

August 2016

**Pittsburgh Coal Conference**

# Coal Monitoring at Power Plant

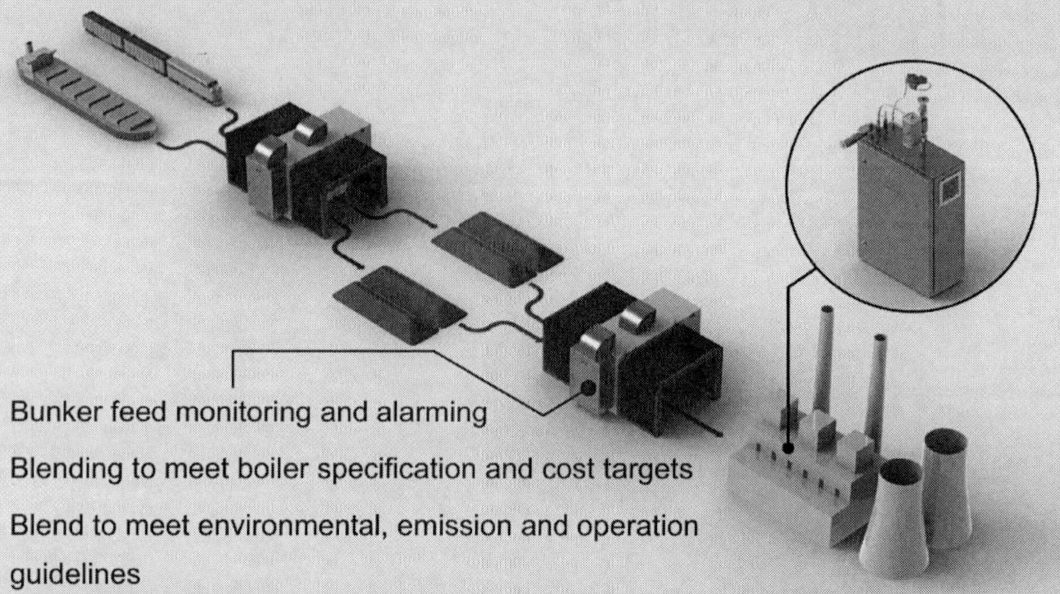

- Bunker feed monitoring and alarming
- Blending to meet boiler specification and cost targets
- Blend to meet environmental, emission and operation guidelines



# Natural Gamma Radiation Measurements

# Natural Gamma Analyser

- Measures Ash using correlations with naturally occurring K, U and Th gamma rays
- No radioactive source
- No bed depth limit
- Ideal for dilution monitoring

SCANTECH
Process control specialists

August 2016

Pittsburgh Coal Conference

# Dual Energy Transmission (DUET)

SCANTECH
Process control specialists

August 2016

Pittsburgh Coal Conference

# DUET Analyser

- Higher energy Cs-137 gammas measure density
- Lower energy Am-241 gammas measure density and ash
- Typical bed depths from 50mm to 350mm (can increase bed depth with a stronger source)
- Lump size and belt speed do not affect analysis
- Fe, Ca and Ba oxide variability can be compensated using multiple calibration curves

SCANTECH
*Process control specialists*

August 2016

**Pittsburgh Coal Conference**

# Dual Beam – Single Detector

Analysis of spectra from Am-241 and Cs-137 sources leading to measurement of ash

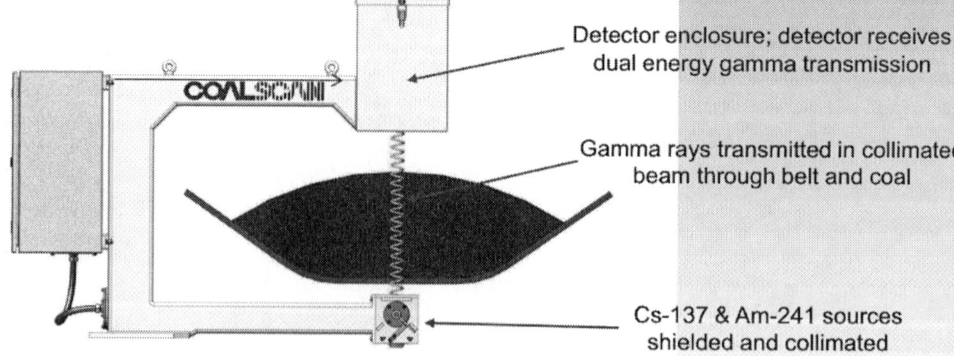


SCANTECH
*Process control specialists*

August 2016

**Pittsburgh Coal Conference**

# Moisture Analysis using Microwave Transmission

SCANTECH
Process control specialists

August 2016

Pittsburgh Coal Conference

## Microwave Moisture Monitor

- Based on Microwave Transmission technique; microwaves interact with free (surface) water molecules, causing them to oscillate
- Oscillating water molecules are measured
- Changing moisture content influences the microwave beam's attenuation and delay

SCANTECH
Process control specialists

August 2016

Pittsburgh Coal Conference

## Microwave Moisture Monitor

SCANTECH
*Process control specialists*
August 2016
**Pittsburgh Coal Conference**

# Prompt Gamma Neutron Activation Analysis (PGNAA)

SCANTECH
*Process control specialists*
August 2016
**Pittsburgh Coal Conference**

# PGNAA - How it works

- A Californium 252 source emits neutrons
- Neutrons are absorbed by elements in the coal
- Each element emits a unique gamma ray spectrum

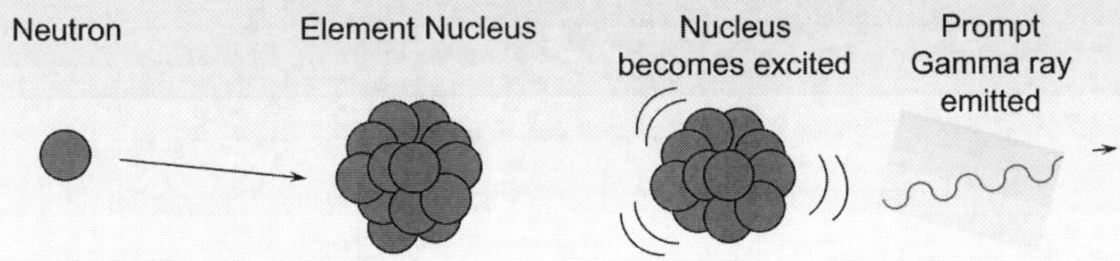




# The PGNAA Technique

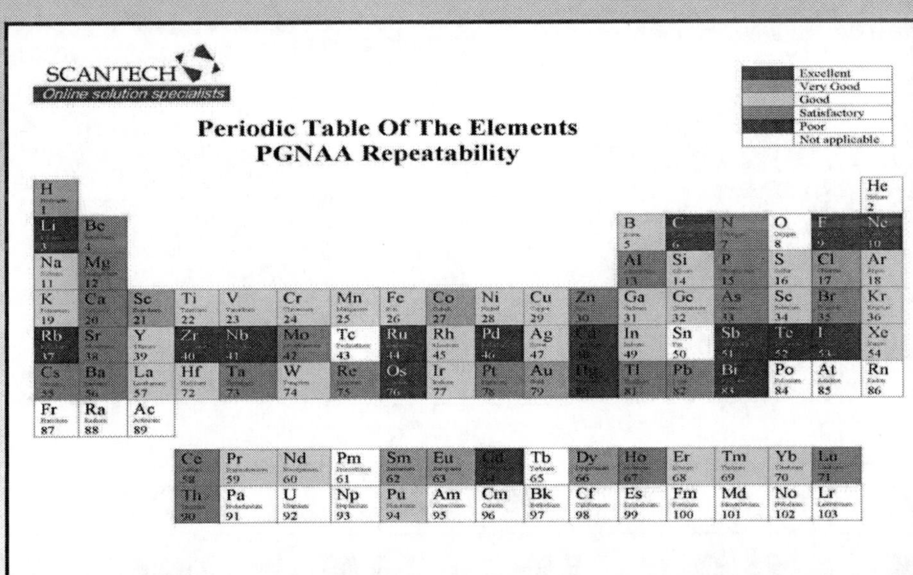

# PGNAA – Options

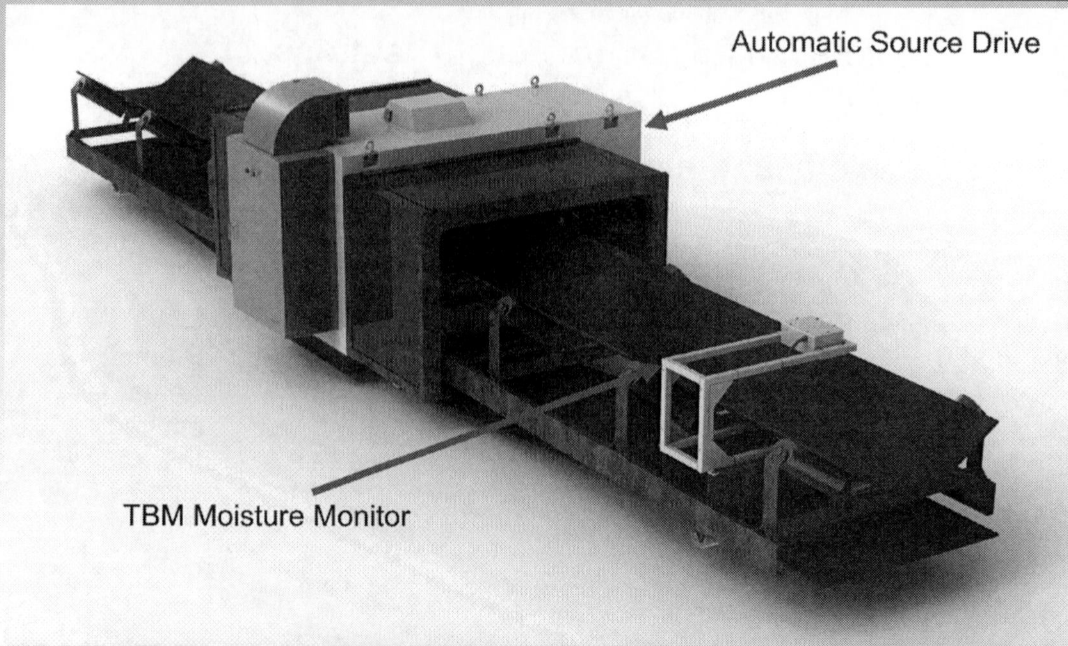




# PGNAA – How it works

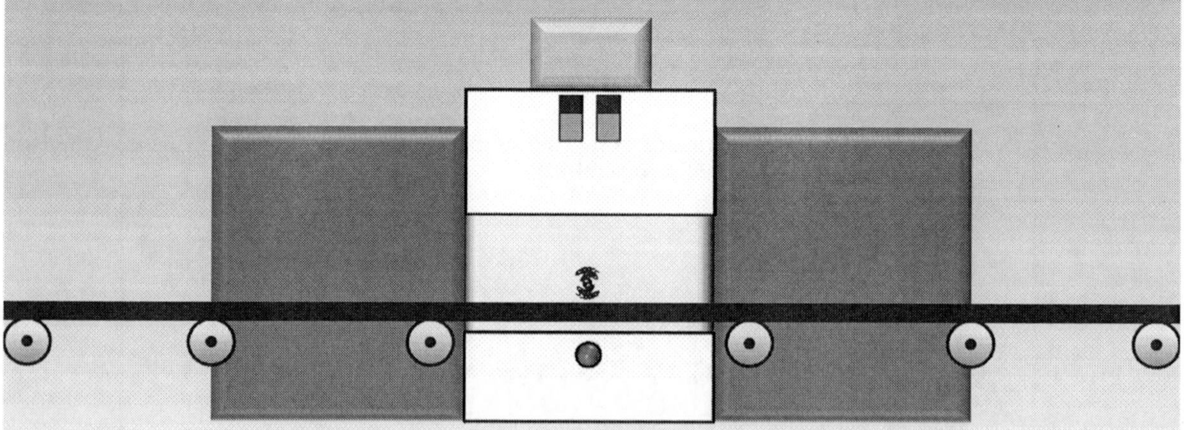

| $SiO_2$ % | $Al_2O_3$ % | Fe % | $Na_2O$ % | CaO % | TS % |
|---|---|---|---|---|---|
| 6[illegible].[illegible] | 2[illegible].[illegible] | [illegible] | 0.[illegible] | 1.[illegible] | 0.[illegible] |

## PGNAA Benefits

- Provides accurate elemental analyses such as Si, Al, Fe, Ca, Mg, Na, Ti, K, Cl, S, Mn.
- Can be combined with a microwave moisture measurement.
- Provides accurate on line analysis of coal for ash, moisture, specific energy, sulphur and oxides in ash, independent of coal type.

SCANTECH
*Process control specialists*

August 2016

**Pittsburgh Coal Conference**

## Analyser Verification

- Coal that passes through the analyser is sampled
- Coal samples are analysed in a laboratory
- Analyser results are calibrated against corresponding laboratory analyses

SCANTECH
*Process control specialists*

August 2016

**Pittsburgh Coal Conference**

## AUSTMINE Presentation: November 2015

Andrew Fisher: St Barbara:

Without doubt technology will continue to play a large part in underpinning operational capabilities, delivery and output. *Process control systems are providing real time interpretation and analysis of performance and lead to improved and informed decision making, directly correlated to cost savings*. We view this as an opportunity to expand on and realise further un-tapped benefits.

SCANTECH
*Process control specialists*

August 2016

**Pittsburgh Coal Conference**

# Fischer-Tropsch Synthesis in the Presence of Solvent Using Novel Carbon-Oxide Composite Supported Cobalt and Iron Catalysts

*Atsushi Ishihara*, Yusuke Maejima, Hiroyuki Nasu and Tadanori Hashimoto*

Division of Chemistry for Materials, Graduate School of Engineering, Mie University, 1577 Kurima Machiya-Cho, Tsu-City, Mie 5148507, Japan, Tel: +81-59-231-9434, Email: *ishihara@chem.mie-u.ac.jp

## 1. Introduction

Currently, industrial progress caused the mass consumption of resources. The reserves of carbon resources under the ground are decreasing and as a result, the drying up of carbon resources has begun. On the other hand, the use of carbon resources has caused the environmental pollution such as global warming. The exhaust of harmful matters affects human bodies badly. Therefore, Fischer-Tropsch synthesis (FTS) using synthesis gas produced by the gasification of biomass attracts much attention because the biomass is regarded as a carbon neutral renewable matter and the FTS produces completely sulfur-free fuels. Nowadays publications related to the FTS are more than hundred and not only academic community but also industrial world have noticed the importance of this technology because of the drastic increase in the price of crude oil and the increasing demand for the clean alternative fuels.[1] In this method called gas to liquid (GTL), one of the key technologies is the development of catalysts. So far various heterogeneous transition metal catalysts such as Fe, Co, Ni and Ru have been developed for this reaction. It is well-known that among them, Co based catalysts are advantageous for the practical application due to the high activity and selectivity to long chain linear hydrocarbons, high resistance toward deactivation, and lower price than Ru. Generally Co and Fe catalysts are deposited on the supports with high surface area (e.g., $SiO_2$, $Al_2O_3$, $TiO_2$ and $ZrO_2$) in order to increase the number of active Co, Fe metal species and silica-supported cobalt is one of the promising catalysts.[2,3] Further, due to the high resistance to attrition and the ability to stabilize small clusters, alumina is one of the attractive support materials for Co or Fe based FT catalysts. Although usual transition metal-supported catalysts are made by the impregnation method, the preparation of catalysts using sol-gel method has become one of promising methods.

On the way of researches to prepare catalysts using the sol-gel method, we are interested in the preparation of transition metal-supported catalysts including carbon sources because some of these we prepared were found to be active for steam reforming and hydrothermal gasification of ethanol and aqueous phenol.[4,5] Further, we are also interested in the FTS in the presence of a solvent because there are few reports utilizing such types of the reactor in the FTS. In this study, we prepared Co and Fe catalysts supported on carbon-$SiO_2$, $Al_2O_3$, $TiO_2$ and $ZrO_2$ composites using polyethylene glycol as a carbon source. Their reactivity for the FTS in the presence of decalin solvent was investigated.

## 2. Experimental

### 2.1. Preparation of carbon-oxide composite-supported cobalt and iron catalysts

Silica, alumina, titania and zirconia were used as oxide supports. Polyethylene glycol (PEG) was added as a carbon source. Starting materials for silica, alumina, titania and zirconia were tetraethyl orthosilicate (TEOS), aluminum tri-sec-butoxide, titanium isopropoxide and zirconium butoxide, respectively. Cobalt nitrate hexahydrate or iron nitrate nonahydrate was used as a cobalt or an iron source. These catalysts were prepared using the sol-gel method. The typical preparation procedure is as follows: To ethanol solution of TEOS was added aqueous solution of cobalt nitrate at 0°C over 1 h, and the mixture was stirred for 10 h. After that, the mixture was dried at 90°C for 3 h to obtain a clay-like paste. The solid obtained was calcined under nitrogen atmosphere at 500°C for 3 h. Iron catalysts were prepared similarly.

### 2.2. Characterization and Catalytic reaction

Measurements of XRD and nitrogen adsorption were performed to analyze the crystal structure and pore structure in a catalyst. The F-T reaction was performed under the conditions, gas flow 40ml/min, syngas composition $H_2$ 66%, CO 33%, Ar 1%, and reaction temperature 250-340 °C. The liquid solvent decalin was introduced with 10 cc/h. The collected liquid and gas were analyzed using gas chromatograph (GC-2010 FID, GC-2014 TCD).

## 3. Results and Discussion

### 3.1. Characterization and catalytic reactivity of cobalt catalysts

**Fig. 1** shows XRD patterns of $16Co63C21SiO_2$, $16Co63C21Al_2O_3$ and $16Co63C21ZrO_2$ catalysts before and after the reaction. $16Co63C63SiO_2$ catalysts did not show the XRD peaks, indicating that Co metal was dispersed on the support. On the other hand, $16Co63C21Al_2O_3$ catalysts exhibited peaks of $Co_2Al_2O_4$ before and after the reaction. Further, $16Co63C21ZrO_2$ catalyst exhibited both peaks of Co and $ZrO_2$ after the reaction while only cobalt metal was shown before the reaction.

**Table 1** shows the results from the $N_2$ adsorption and desorption measurement. When surface areas were compared among $16Co63C21SiO_2$, $16Co63C21Al_2O_3$ and $16Co63C21ZrO_2$, those for catalysts using $SiO_2$ and $Al_2O_3$ were more than 200m$^2$/g while that for $ZrO_2$ supported catalyst was about 120m$^2$/g. In contrast, the size of pore volume decreased in the order $16Co63C21Al_2O_3$ > $16Co63C21ZrO_2$ > $16Co63C21SiO_2$. Further, the differences between BET and BJH methods were observed for $16Co63C21SiO_2$ and $16Co63C21Al_2O_3$ while there was no difference between BET and BJH method in $16Co63C21ZrO_2$, indicating that almost all pores in $16Co63C21ZrO_2$ consist of mesopores. These results can be confirmed by the BJH plots and MP plots of catalysts shown in **Figs. 2 and 3**, respectively. These three catalysts have mesopores however $16Co63C21ZrO_2$ hardly have micropores. Further, $16Co63C21ZrO_2$ and $16Co63C21Al_2O_3$ have the wide range distribution of mesopores while $16Co63C21SiO_2$ has a sharp distribution of mesopore.

Conversions of carbon monoxide at 300°C in the FT synthesis using $16Co63C21SiO_2$, $16Co63C21Al_2O_3$ and $16Co63C21ZrO_2$ catalysts were 6.2%, 0.4 % and 17%, respectively, as shown in **Fig. 4**. When the amount of PEG carbon source added was decreased, the conversions decreased. The conversions for $16Co21C63SiO_2$ and

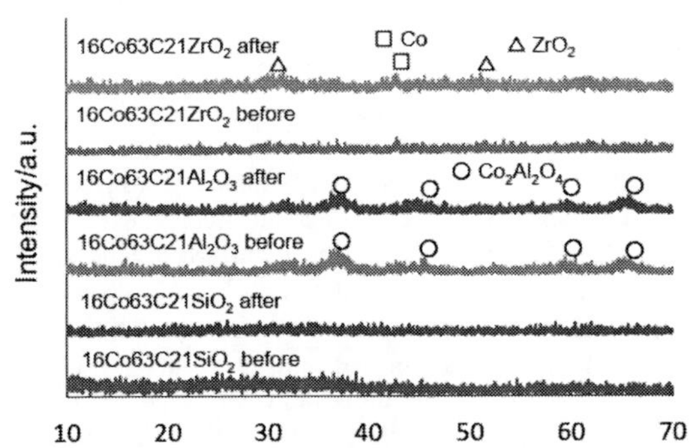


**Fig. 1** XRD patterns of $16Co63C21SiO_2$, $16Co63C21Al_2O_3$ and $16Co63C21ZrO_2$ before and after the reaction

**Table 1** $N_2$ adsorption-desorption measurement of catalysts before and after the reaction

| Sample | BET SA ($m^2/g$) | BET PV ($cm^3/g$) | BET PD (nm) | BJH SA ($m^2/g$) | BJH PV ($cm^3/g$) | BJH PD (nm) |
|---|---|---|---|---|---|---|
| $16Co63C21SiO_2$ before | 203 | 0.18 | 3.6 | 124 | 0.13 | 3.7 |
| $16Co63C21SiO_2$ after | 208 | 0.20 | 3.8 | 147 | 0.15 | 3.7 |
| $16Co63C21Al_2O_3$ before | 212 | 0.37 | 7.0 | 163 | 0.33 | 3.7 |
| $16Co63C21Al_2O_3$ after | 211 | 0.35 | 6.7 | 152 | 0.32 | 3.7 |
| $16Co63C21ZrO_2$ before | 126 | 0.23 | 7.2 | 117 | 0.21 | 3.7 |
| $16Co63C21ZrO_2$ after | 131 | 0.22 | 6.8 | 113 | 0.20 | 3.7 |

SA: Surface area, PV: Pore volume, PD: Pore diameter

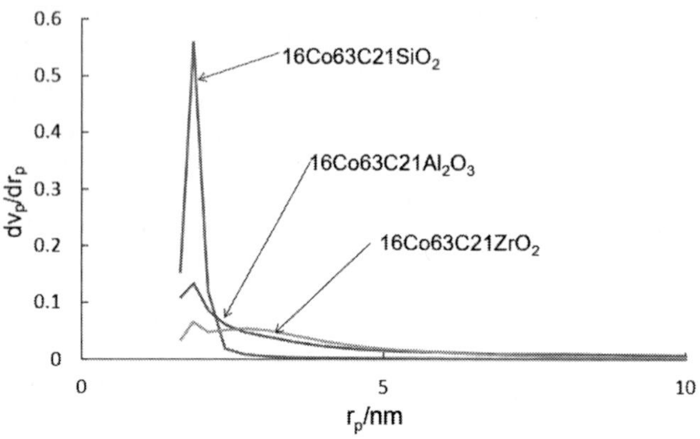


**Fig. 2** BJH plots in the nitrogen adsorption measurement of catalysts

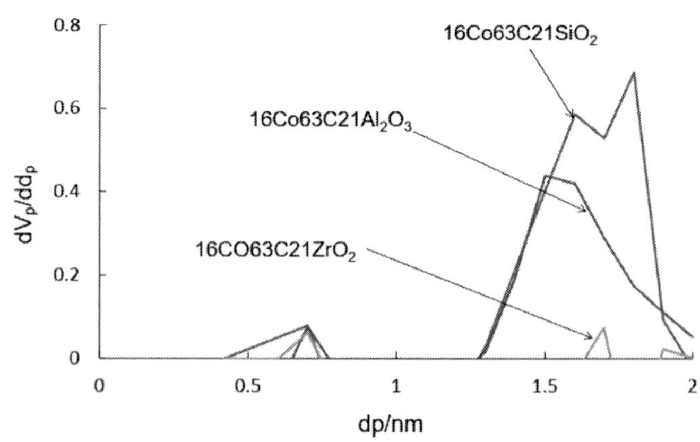


**Fig. 3** MP plots in the nitrogen adsorption measurement of catalysts

$16Co21C63ZrO_2$ at 300°C were 0.21% and 3.3%, respectively. The conversion of $16Co63C21ZrO_2$ exhibited the highest CO conversion among the catalysts examined. The result would be related to the pore size distribution of $16Co63C21ZrO_2$, which only consists of mesopores, because mesopores have advantages for the diffusion of products, reactants and a solvent into the catalyst pore. In our catalyst system, decalin is included and it is thought that the trickle bed reactor is formed and that such system would strongly be affected by the mass transfer.

**Fig. 4** also shows the carbon number distribution of products. Methane was a major product for all the catalysts examined. Products

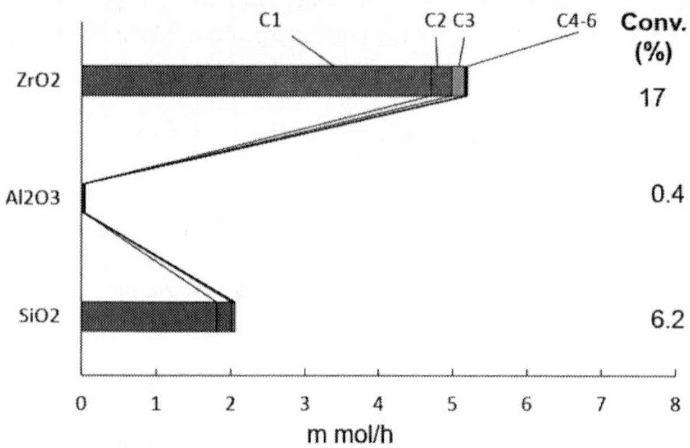


**Fig. 4** Effects of supports in cobalt catalysts on carbon number distribution of products

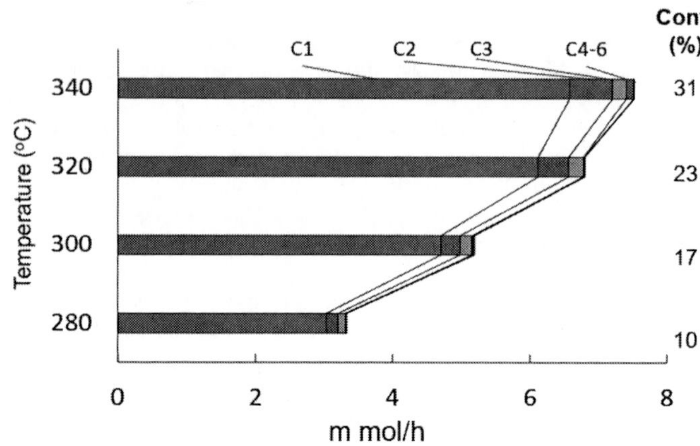


**Fig. 5** Effects of temperature on carbon number distribution of products in FT synthesis catalyzed by$16Co63C21ZrO_2$.

of $C_1$-$C_3$ were confirmed in $16Co63C21SiO_2$ while products with more than $C_4$ were not detected. $16Co63C21Al_2O_3$ catalyst only produced $C_1$ and $C_2$ compounds. In contrast, products with $C_1$-$C_6$ were observed using $16Co63C21ZrO_2$.

**Fig. 5** shows the effect of temperature on the conversion and the product distribution for FT reaction using $16Co63C21ZrO_2$ catalyst. The conversion increased with increasing temperature and reached 31% at 340°C.

### 3.2. Characterization and catalytic reactivity of iron catalysts

XRD patterns of $16Fe63C21SiO_2$, $16Fe63C21Al_2O_3$, $16Fe63C21TiO_2$ and $16Fe63C21ZrO_2$ catalysts before and after the reaction were measured (**Figs. 6 and 7**). $16Fe63C63SiO_2$ catalysts did not show the XRD peaks, indicating that iron species were dispersed on the support. On the other hand, $16Fe63C21Al_2O_3$ and $16Fe63C21TiO_2$ catalysts exhibited peaks of $Fe_3O_4$ before and after the reaction. Further, $16Fe63C21ZrO_2$ catalyst calcined at 500°C exhibited peaks of cubic $ZrO_2$ before and after the reaction while iron metals were shown in the catalysts calcined at 700°C and 900°C before and after the reaction. Surprisingly, particle sizes of irons decreased after the reactions, specifically from 42nm to 7.2 nm for the catalyst calcined at 700°C.

$N_2$ adsorption and desorption of catalysts were measured to estimate sizes of pores. When surface areas were compared among $16Fe63C21SiO_2$, $16Fe63C21Al_2O_3$, $16Fe63C21TiO_2$, and $16Fe63C$-$21ZrO_2$, those for catalysts using $SiO_2$, and $Al_2O_3$ were more than $120m^2/g$ while those for $ZrO_2$ and $TiO_2$ supported catalysts were less than $120m^2/g$. However, it is also noteworthy that the sizes of

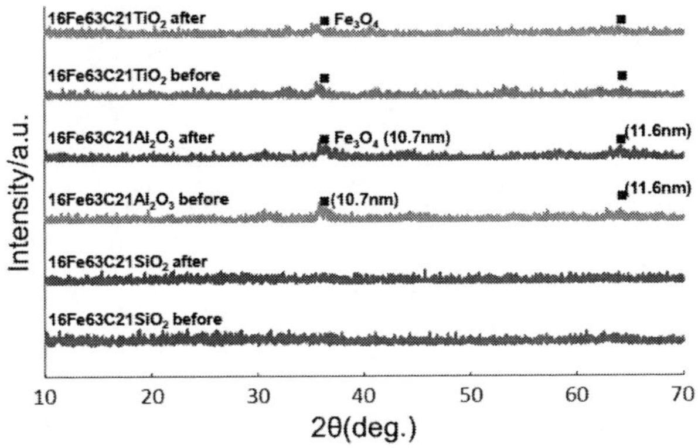


**Fig. 6** XRD patterns of $16Fe63C21TiO_2$, $16Fe63C21Al_2O_3$ and $16Fe63C21SiO_2$ before and after the reaction

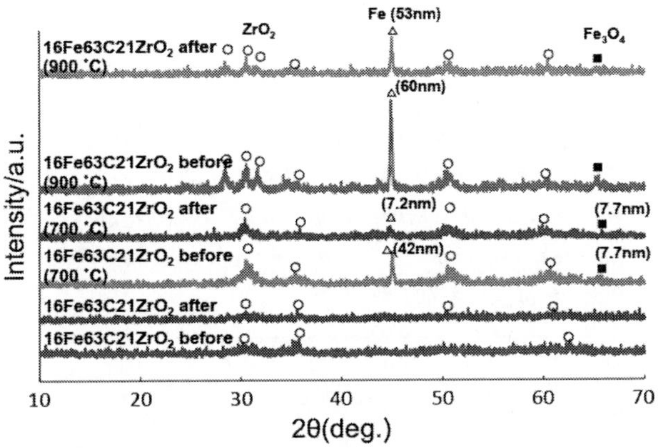


**Fig. 7** XRD patterns of $16Fe63C21ZrO_2$ calcined at 900°C, $16Fe63C21ZrO_2$ calcined at 700°C and $16Fe63C21ZrO_2$ calcined at 500°C before and after the reaction

surface area and pore volume for $ZrO_2$ supported catalysts increased with increasing calcination temperature. Further, the differences between BET and BJH methods were observed for all the iron catalysts, indicating that micropores developed when iron catalysts were prepared. This result was not consistent with that of cobalt catalysts.

Conversions of carbon monoxide at 300°C in the FT synthesis using $16Fe63C21SiO_2$, $16Fe63C21Al_2O_3$, $16Fe63C21TiO_2$ and $16Fe63C21ZrO_2$ catalysts were 0.3%, 3.3 %, 0.3% and 10%, respectively, as shown in **Fig. 8**. When the calcination temperature increased to 700°C for $16Fe63C21ZrO_2$ catalyst, the conversions increased to 55%, the highest CO conversion among the catalysts examined. The result would be related to the generation of very fine iron metal in $16Fe63C21ZrO_2$ calcined at 700°C because highly dispersed iron particles would have advantages for the increase of active sites.

**Fig. 8** also shows the carbon number distribution of products. Methane was a major product for all the catalysts examined. Products of $C_1$-$C_6$ were confirmed in $16Fe63C21ZrO_2$. When reaction temperature increased from 300°C to 320 and 340°C using $16Fe63C21ZrO_2$, conversions decreased.

**Fig. 9** shows TEM images of $16Fe63C21ZrO_2$ calcined at 700°C before and after the reaction. It was observed that particles of the catalyst congregated as reaction temperature increased to 340°C. However, very fine iron particles were still present for the catalyst after the reaction, which is consistent with the results from XRD.

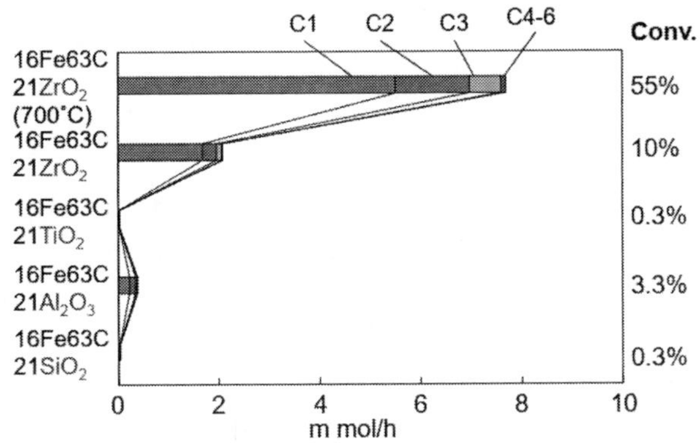


**Fig. 8** Effects of supports in iron catalysts on carbon number distribution of products

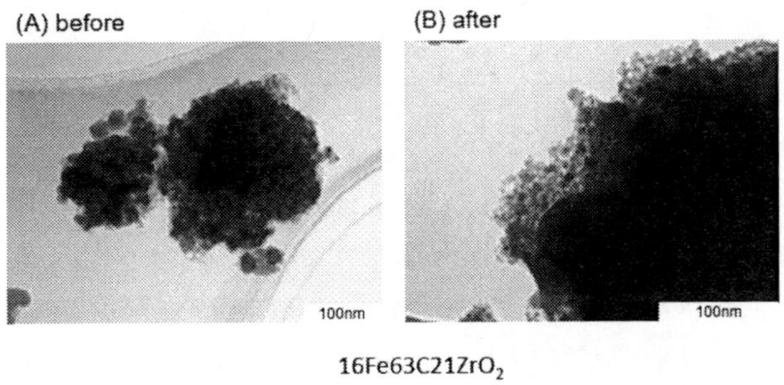


**Fig. 9** TEM images of $16Fe63C21ZrO_2$ calcined at 700°C.

## 4. Conclusions

Fischer-Tropsch synthesis in the presence of solvent was examined for novel carbon-oxide composite supported cobalt catalysts. The CO conversion increased in the order $16Co63C21Al_2O_3 < 16Co63C21SiO_2 < 16Co63C21ZrO_2$. Most of product was methane for all the cobalt catalysts. $16Co63C21ZrO_2$ catalyst showed the highest conversion 31% at 340°C probably because $16Co63C21ZrO_2$ catalyst consists of only mesopores which have advantages for mass transfer.

Novel carbon-oxide composite supported iron catalysts were prepared similarly and reactivity for Fischer-Tropsch synthesis in the presence of solvent was examined. The CO conversion increased in the order $16Fe63C21SiO_2 = 16Co63C21TiO_2 < 16Fe63C21Al_2O_3 < 16Fe63C21ZrO_2$ calcined at 500°C < $16Fe63C21ZrO_2$ calcined at 700°C. Although most of product was methane for all the iron catalysts, $16Fe63C21ZrO_2$ calcined at 700°C also produced C2-C6 hydrocarbons. The results may be related to the facts that the sizes of surface area and pore volume for $ZrO_2$ supported catalysts increased with increasing calcination temperature and that particle sizes of irons decreased after the reactions, specifically from 42nm to 7.2 nm for the catalyst calcined at 700°C.

## References

1) A. Ishihara, 1.4 Petroleum Chemistry, Annual review in J. Jpn. Energy Inst., in press (2015) (in Japanese). 2) W. Ma, G. Jacobs, P. Gao, T. Jermwongratanachai, W. D. Shafer, V. Pendyala, R. Ramana, C. H. Yen, J. L. S. Klettlinger, B. H. Davis, Appl. Catal., A: Gen., 475, 314-324 (2014). 3) A.Y. Khodakov, W. Chu, P. Fongarland, Chem. Rev. 107, 1692–1744 (2007). 4) A. Ishihara, T. Imai, T. Hashimoto, H. Nasu, Fuel Process. Technol., 136, 34-40 (2015). 5) A. Ishihara, K. Imanishi, T. Hashimoto, H. Nasu, J. Jpn. Petrol. Inst., 58 (2015) 302-311.

Pittsburgh Coal Convention

Cape Town 2016

# Carbon from Pine Cones as a Support for High Temperature Fischer-Tropsch Synthesis

Pierre Cilliers[a], Chelsea Tucker[a], Wanda Bout[b], Doreen Nabaho[a], Augustine Ofomaja[b], Elvera Viljoen[b], Eric van Steen[a,*]

*[a]Centre for Catalysis Research, Department of Chemical Engineering, University of Cape Town, Private Bag X3, Rondebosch 7701, South Africa; [b] Department of Chemistry, Vaal University of Technology, Vanderbylpark, South Africa.*

**Abstract**

*The use of carbon as a support for iron-based Fischer-Tropsch catalysts originating from pine cones was investigated for the use in the high temperature Fischer-Tropsch synthesis at 350 °C and 20 barg for a feed with $H_2$:CO of 1. Impurities present in the support do not negatively affect the catalyst performance, and an iron based catalyst supported on carbon from pine cone material, synthesised using 800 W microwave energy input for 20-minute radiation time yielded the largest $C_2$–$C_4$ olefin yield of 40 ± 1 % with a low $CH_4$ selectivity of 6.0 ± 0.1 %.*

***Keywords:*** *biomass, Fischer-Tropsch, iron, carbon, light olefins*

## Introduction

Lower or light olefins ($C_2$-$C_4$) are a platform chemical in the chemical industry due to a wide range of derivatives, which can be obtained from these compounds including plastics, drugs, textiles, cosmetics and antifreezing agents (Torres Galvis and de Jong 2013a). Their major use is as a monomer in the production of plastics, such as LDPE, HDPE and PP, which can be applied in a large variety of materials.

The majority of light olefins is nowadays produced in the oil refinery in processes such as steam cracking and catalytic cracking. These compounds can also be formed indirectly, e.g. using the methanol to olefin (MTO) process or more directly via the Fischer-Tropsch to olefin (FTO) process (Torres Galvis, et al. 2012).. The Fischer-Tropsch (FT) process is a catalytically aided hydrogenation of carbon monoxide and hydrogen (synthesis gas) into higher molecular weight hydrocarbons (Dry 2001). A benefit of the FTO process is the fact that it can utilise biomass as the source of synthesis gas.

Catalysts that can be used industrially in the FT process must exhibit sufficient CO hydrogenation activity (Torres Galvis and de Jong 2013a). According to Dry (2001), the only metals with a high enough activity are iron, cobalt, nickel and ruthenium. Iron is relatively inexpensive in comparison to the other metal catalysts and is therefore typically used for the production of light olefins at high temperatures (Hirsa M. Torres Galvis and Krijn P. de Jong 2013).

Bulk (unsupported) iron catalysts exhibit mechanical instability through chemical attrition, also caused by carbon deposition. Furthermore, they are susceptible to sintering under reaction conditions. Supported iron catalysts have been shown to have reduced sintering effects improving catalyst stability (Torres Galvis and de Jong 2013a). Activated carbon has been shown to be a suitable support due to its high surface area, weak interaction with the catalytic active phase and low production costs when obtained from natural sources (Torres Galvis and de Jong 2013a). An inexpensive source of activated carbon is biomass such as coconut shells. Pine cones can also act as a source for the synthesis of activated carbon (Duman, et al. 2009).

There are two fundamental challenges associated with using biomass-derived support material for the FTO process. Firstly, pine cones may contain impurities that could result into undesirable side effects when used as a support for Fischer-Tropsch catalysts, such as lower activity, lower selectivity towards light olefins or reduced stability. Secondly, the preparation method associated with the production of the activated carbon is an important factor in order to produce a high-surface area catalyst.

Microwave irradiation is a relatively novel alternative to heating that has been found to not only produce low cost activated carbon, but decrease the energy required to do so (Hesas, et al. 2013, Oghbaei and Mirzaee 2010). Hesas et al. (2010) investigated the effect of radiation time and power on the adsorption capacity of activated carbon produced from two natural sources, i.e. apple peel and apple pulp. Both sources were prepared in a microwave at 500 W, 700 W and 1000 W over a time period of 10 minutes, 12.5 minutes and 15 minutes. It was observed that the adsorption capacity of the obtained activated carbon increased upon increasing the energy input (power x exposure time) due to the increase in porosity and surface area. However, the study indicated that passed a specific threshold of energy, the pore volume decreased due to burn-off of carbon, which decreases the effectiveness of the activated carbon.

The aim of this research was twofold. Firstly, to determine the viability of utilising pine cone based activated carbon as a support for iron-based catalysts in relation to a sample of bulk iron. The second aim is to determine the effect of preparation conditions (in terms of energy input and radiation time) on the iron catalyst supported on activated carbon ex pine cone in a high temperature Fischer-Tropsch reactor.

## Materials & Methods

The iron based, pine cone biomass supported, catalyst samples were synthesized by contacting crushed pine cones (0.1-0.12 mm) with an aqueous $Fe(NO_3)_3.9H_2O$ (7 mM) solution in a liquid to solid ratio of 4 ml/g for 4 hrs. The excess liquid was decanted after centrifugation and the obtained solid was dried for overnight at 110 °C in a vacuum oven to a constant weight. The final catalyst sample was obtained under pyrolysis at varied microwave power input (W) and radiation time (minutes). To analyse the effect that power input and radiation time had on the catalyst selectivity and activity, three samples were tested. These samples were iron based (Fe) supported on activated carbon (AC) referred here after as Fe/AC-800 W, 20 min, Fe/AC-800 W, 60 min and Fe/AC-1200 W, 20 min which contained an iron weight composition of 24 wt.%, for comparative purposes an unsupported bulk iron catalyst ex hematite, $Fe_2O_3$, (Chonco, et al. 2013)was also tested at the same reaction conditions.

The experimental runs were performed in a Fischer-Tropsch (FT) laboratory set-up with $H_2$ and CO as the synthesis gas, $N_2$ as the internal reference; pressure controlled addition of Ar was used to regulate the pressure. The Fischer-Tropsch synthesis was performed at 350 °C and 20 barg with a $H_2$:CO ratio of 1:1. The activated carbon supported iron catalysts were tested at a space velocity of 1000 $ml_n$CO/min/$g_{Fe}$, whereas the unsupported catalyst was tested at a higher space velocity of 1400 $ml_n$ CO/min /$g_{Fe}$,. The catalyst samples were operated under these conditions for 75 hours on stream to allow for short term deactivation observation. The product flows were analysed using an online GC-TCD and offline GC-FID.

**Results and Discussion**

The composition of the supported iron catalyst in its calcined state was determined using XRF. The iron loading in the calcined catalyst was determined to be 24 wt.%. Other elements of interest found in the pine cone supported catalyst sample are indicated in Table 1. In terms of the additional elements found in the pine cones - potassium, manganese, copper and ruthenium are all documented promoters of iron catalyst Fischer-Tropsch systems (Torres Galvis and de Jong 2013a, Venter, et al. 1987, Li, et al. 2002, Farias, et al. 2011, Torres Galvis, et al. 2013b). The levels of impurities are very low indicating that their influence on activity/selectivity might be marginal

*Table 1: XRF elemental analysis of Fe/AC – 800 W, 20 min*

| *Element* | *Wt.%** | *Element* | *Wt.%** | *Element* | *Wt.%** |
|---|---|---|---|---|---|
| Iron | 23.9 | Calcium | 0.01 | Manganese | 0.01 |
| Potassium | 0.05 | Chromium | 0.01 | Cobalt | 0.05 |
| Copper | 0.01 | Ruthenium | 0.03 | Zinc | 0.01 |

**Due to the low concentrations true concentration values are difficult to quantify. The XRF results should be taken more as indication to the presence of these elements and not the true concentrations present.*

An online GC-TCD was used to analyse the entering and exit flows of CO, $H_2$, $CO_2$ and $CH_4$ in order to determine the conversion of CO achieved in each catalyst sample run. The conversion over the full run cycle can be seen in Figure 1. The catalysts reach a high level of conversion after ca. 24-40 hrs on stream. From there on, a decrease with time on line can be noted. The activity drop can be related to catalyst deactivation.

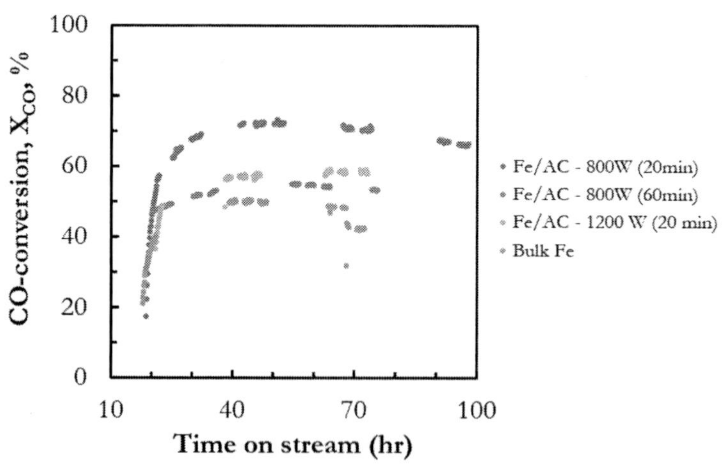


***Figure 1:*** *Activity of Fe/AC - 800 W, 20 min; 800 W, 60 min; 1200 W, 20 min (tested at a space velocity of 1250 $ml_n/g_{Fe}/min$) and bulk iron catalyst (tested at a space velocity of 1900 $ml_n/g_{Fe}/min$) as a function of time on stream*

The relative activities for Fe/AC-800 W, 20 min, Fe/AC-1200 W, 20 min and Fe/AC-800 W, 60 min had relative activities ($Activity_{sample}/Activity_{bulk}$) of 1.53, 1.10 and 0.97. The activity of Fe/AC-800 W, 20 min was shown to be consistently larger with a steady state activity after 40 hours of 10.7 ± 0.1 /g·Fe. The lowest activity was that of Fe/AC-800 W, 60 min at 6.7 ± 0.06/g·Fe similar to the bulk iron catalyst. This is predicted to be due to the pores of Fe/AC-800 W, 60 min (irradiated at 2880 kJ) having been burned off, therefore decreasing active surface area in a similar way to the activated carbon investigated by Hesas et al. (2013).

A key question for this investigation focussed on the stability of the catalyst. The first order rate constant of deactivation was calculated for each sample. Two of the pine cone supported catalysts (Fe/AC – 800 W, 20 min and Fe/AC - 800 W, 60 min) exhibited slight on-time deactivation with a coefficient of deactivation of -0.0217 and -0.0158. The Fe/AC – 1200 W, 20 min did not deactivate within the time on stream.

The bulk iron sample experienced a rapid drop in activity at approximately 68 hours on stream. This rapid drop was not seen in any of the other samples and is an indication of sintering. This is expected as bulk iron lacks support material that prevents sintering and maintains a higher level of mechanical stability within the catalyst.

### 3.2.2 Selectivity

The catalyst samples relative selectivity towards $CO_2$ and $CH_4$ and can be seen in Table 2:

*Table 2: Catalyst sample selectivities for $CO_2$ and $CH_4$ after sufficient pseudo-steady state time achieved*

| Catalyst Sample | Fe/AC – 800 W, 20 min | Fe/AC – 1200 W, 20 min | Fe/AC – 800 W, 60 min | Bulk Iron |
|---|---|---|---|---|
| $CO_2$ selectivity (75 hours) (%) | 25.5 ± 0.1 | 22.9 ± 0.2 | 35.4 ± 1.3 | 26.6 ± 0.2 |
| $CH_4$ selectivity (75 hours) (%) | 6.0 ± 0.1 | 4.7 ± 0.1 | 8.0 ± 0.2 | 6.6 ± 0.2 |

The Fe/AC – 1200 W, 20 min catalyst corresponded to the lowest relative selectivity of $CO_2$ and $CH_4$ with the Fe/AC – 800 W, 60 min catalyst corresponding to the highest relative selectivity. However, low selectivities are not the only criteria for an optimal catalyst as the Fe/AC- 800 W, 20 min sample also achieved lower selectivities than that of the bulk iron. In order to determine which preparation technique gives the optimal catalyst the relative yields of low olefin formation ($C_2$ to $C_4$) was calculated and can be seen in Figure 2 in decreasing order of yield.

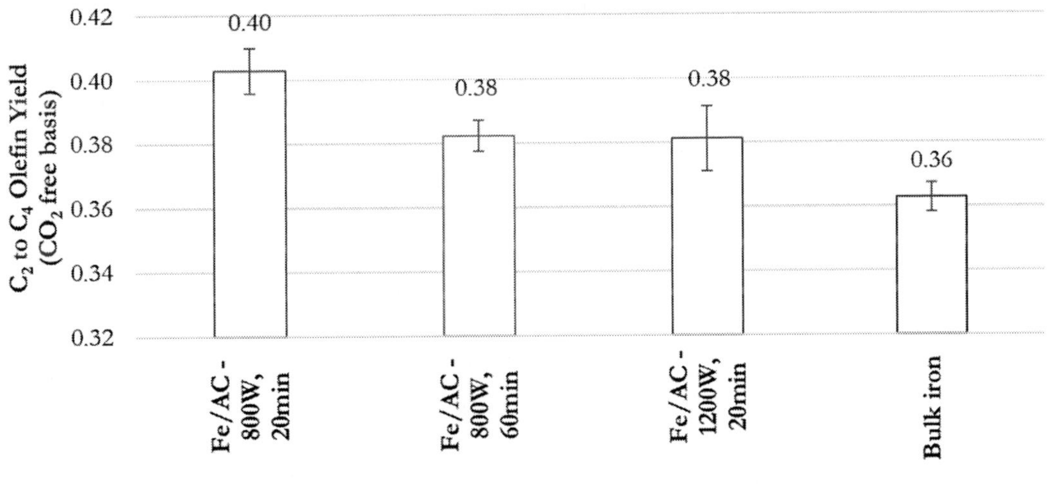


*Figure 2: Low olefin formation for Fe/AC - 800 W, 20min / 800 W, 60 min / 1200 W, 20 min and bulk iron*

As seen in Figure 2 the highest low olefin yield occurs with the Fe/AC – 800 W, 20 min sample, this corresponds to an increase in yield relative to bulk iron by 11 % and is therefore a far more desirable catalyst sample.

The key observation that was shown was that for all samples the yield of low olefins was higher than that of the bulk iron. The higher yields and low $CH_4$ and $CO_2$ selectivities shown by the supported activated carbon material, from pine cone biomass, in comparison to bulk iron is a positive indication that this material is a potentially viable source for high temperature Fischer-Tropsch reactions and should be further investigated.

**Conclusions**

In conclusion the analyses of the Fe/AC catalyst, performed with XRF, found potassium, ruthenium and copper detectable amounts which have been recorded as having a positive effect on the activity of iron-based catalysts (Li, et al. 2002, Torres Galvis, et al. 2013b). BET analyses were inconclusive at this stage due to pyrolysis oil on the catalyst surfaces from the synthesis method of the catalysts.

Relative to the bulk iron catalyst the activity for Fe/AC-800 W, 20 min, Fe/AC-1200 W, 20 min and Fe/AC-800 W, 60 min were 1.53, 1.10 and 0.97. The highest activity was that of Fe/AC-800 W, 20 min with a steady state activity after 40 hours of run of 10.7 ± 0.1 /g.Fe, this catalyst also corresponded to low $CO_2$ and $CH_4$ selectivity of 25.5 ± 0.1 % and 6.0 ± 0.1 % respectively. Due to the high activity of Fe/AC-800 W, 20 min the catalyst sample experienced the highest deactivation of the supported catalyst samples. Despite the high deactivation Fe/AC-800 W, 20 min exhibited the highest $C_2$-$C_4$ olefin yield of 40 ± 1 %, higher than the bulk iron yield of 36 ± < 1 %.

The activity and low olefin yield of Fe/AC-800 W, 20 min have exhibited desirable results which indicate that a low power input, low radiation time iron based catalyst supported on biomass material, such as that from pinecone biomass, could be used in Fischer-Tropsch reaction and should further more investigation as a potential support material at industrial scales.

**References**


Chonco, Z.H., A. Ferreira, L. Lodya, M. Claeys, E. van Steen. "Ferrites – a model structure to study group 11 promoters in Fe-based Fischer-Tropsch catalysts." *Journal of Catalysis* 308 (2013): 363-373.

Dry, Mark. E. "High quality diesel via the Fischer-Tropsch process - a review." *Journal of Chemical Technology and Biotechnology* 77 (2001): 43-50.

Duman, Gozde, Yunus Onal, Cagdas Okutucu, Sermin Onenc, and Jale Yanik. "Production of Activated Carbon from Pine Cone and Evaluation of Its Physical, Chemical and Adsorption Properties." *Energy & Fuels* 23 (2009): 2197-2204.

Farias, F. E. M., R. C. Rabelo Neto, M Baldanza, M Schmal, and F, Fernandes. "Effect of K promoter on the structure and catalytic behavior of supported iron-based catalysts in fischer-tropsch synthesis." *Brazilian Journal of Chemical Engineering* 28, no. 3 (2011).

Galvis, Hirsa M Torres, et al. "Effect of sodium and sulphur on catalystic performance of supported iron based catalysts for the Fischer-Tropsch synthesis of lower olefins." *Journal of Catalysis* 303 (2013): 22-30.

Hanekom, Anton. *Overview of the Plastics Industry in South Africa.* Johannesburg: Plastics SA, 2014.

Hesas, Roozbeh, Arash Arami-Niya, Wan Mohd Daud, and J.N Sahu. "Preparation and Characterisation of Activated Carbon from Apple Waste by Microwave-Assisted Phosphoric Acid Activation." *BioResources* 8, no. 2 (2013): 2950-2966.

Hu, Jin, Fei Yu, and Yongwu Lu. "Application of Fischer-Tropsch Synthesis in Biomass to Liquid Conversion." *Catalysts* 2 (2012): 303-326.

Li, Senzi, Sundaram Krishamoorthy, Anwu Li, George D Meitzner, and Enrique Iglesia. "Promoted Iron-Based Catalysts for the Fischer-Tropsch Synthesis: Design, Synthesis, Site Densities and Catalytic Properties." *Journal of Catalysis* 206 (2002): 202-217.

Oghbaei, Morteza, and Omid Mirzaee. "Microwave versus conventional sintering: A review of the fundementals, advantagesand applications." *Journal of Alloys and Compounds* 494 (2010): 179-189.

Torres Galvis, H. M Torres, Johannes H Bitter, Chaitanya B. Khare, Matthijs Ruitenbeek, A. Iulian Dugulan, and Krijn P de Jong. "Supported Iron Nanoparticles as Catalysts for Sustainable Production of Lower Olefins." *Science* 335 (2012): 835-838.

Torres Galvis, H.M. Torres, and Krijn P. de Jong. "Catalyst Production of Lower Olefins from Synthesis Gas: A Review." *ACS Catalysis*, 2013a: 2130-2149.

Tucker, Chelsea, Pierre Cilliers, Doreen Nabaho, and Eric van Steen. *Carbon from Biomass as a support for high temperature Fischer-Tropsch synthesis.* Cape Town: University of Cape Town, 2015.

Venter, J, M Kaminsky, G Geoffrey, and M Vannice. "Effect of Pottasium on Iron Based catalysts." *Catalysis* 103 (1987): 450-465.

CENTRE FOR CATALYSIS RESEARCH

# Bifunctional Hydrocracking:

## Is the well-established 'classical' fluid phase diffusion mechanism a delusion?


Walter Böhringer
Peter Wynne, Roald Brosius
Stephen Roberts, Jack C.Q. Fletcher

Centre for Catalysis Research
Department of Chemical Engineering
University of Cape Town
South Africa



2016 International Pittsburgh Coal Conference, August 8–12,
Cape Town Int. Convention Centre, Cape Town, South Africa


UNIVERSITY OF CAPE TOWN
CENTRE FOR CATALYSIS RESEARCH

## Bifunctional hydrocracking catalysis

**Typical feedstocks:**
Crude oil derived heavy distillates and heavier fractions,
Fischer-Tropsch wax from CTL, GTL, BTL, WTL ...

**Typical target products:**
Diesel, jet fuel and naphtha (petrol)

**Typical catalysts:**
Sulphided base metals (bifunctional)
**Pt, Pd, Ni ... (metal sites)**
**on silica-alumina, H-zeolites ... (acid sites)**

**Typical make of the latter:**
Metal co-catalyst impregnated onto acid catalyst


UNIVERSITY OF CAPE TOWN
CENTRE FOR CATALYSIS RESEARCH

## The classical 'Hop Over' hydrocracking mechanism

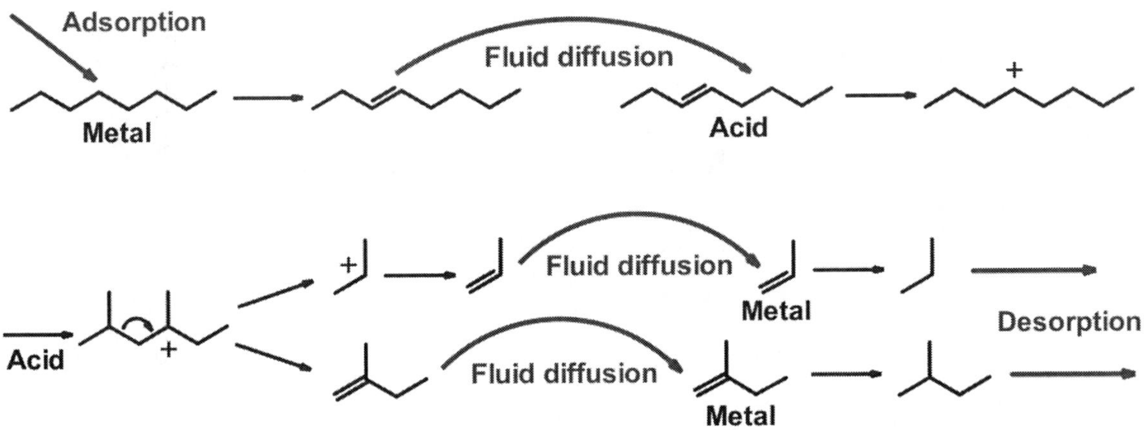


Mills, G.A., Heinemann, H., Milliken, T.H., Oblad, A.G., J. Ind. Eng. Chem. 45 (1953) 134
Weisz, P.B., Swegler, E.W., Science 126 (1957) 31
Weisz, P.B., Adv. Catal. 13 (1962) 137
Coonraadt, H.L., Garwood, W.E., Ind. Eng. Chem. Proc. Des. Dev. 3 (1964) 38

UNIVERSITY OF CAPE TOWN
CENTRE FOR CATALYSIS RESEARCH

## 'Hop Over' mechanism still dominates the recent literature

- Subjects of scientific papers are based on the 'Hop Over' mechanism
- Many papers referring to the 'Hop Over' mechanism in the introduction (such as 'widely accepted')

**Does this 'hopping over' take place indeed?**

UNIVERSITY OF CAPE TOWN
CENTRE FOR CATALYSIS RESEARCH

## Hydroisomerisation of cyclohexane in layered bed

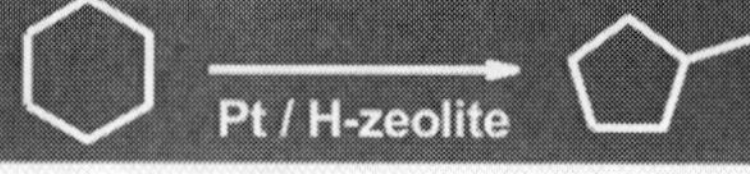


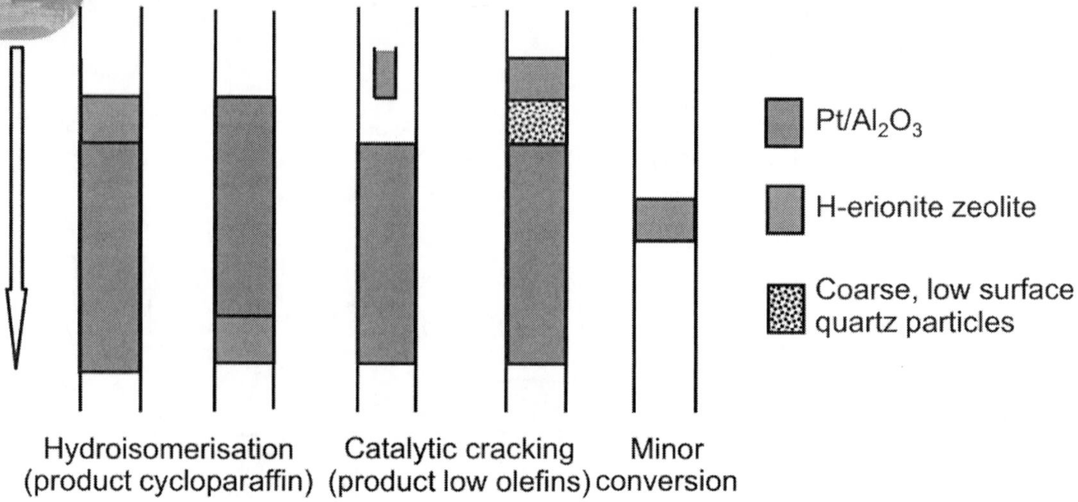


Gas phase, 1 bar $H_2$, 224°C


Roland, U., Braunschweig, T., Roessner, F., J. Mol. Catal. A: Chemical 127 (1997) 61
Steinberg, K.-H., Mroczek, U., Roessner, F., Stud. Surf. Sci. Catal. 46 (1988) 81
Roessner, F., Hagen, A., Mroczek, U., Stud. Surf. Sci. Catal. 77 (1993) 151
Roessner, F., Roland, U., J. Mol. Catal. A: Chemical 112 (1996) 401


UNIVERSITY OF CAPE TOWN
CENTRE FOR CATALYSIS RESEARCH

## Hydrocracking of n-hexadecane in layered bed

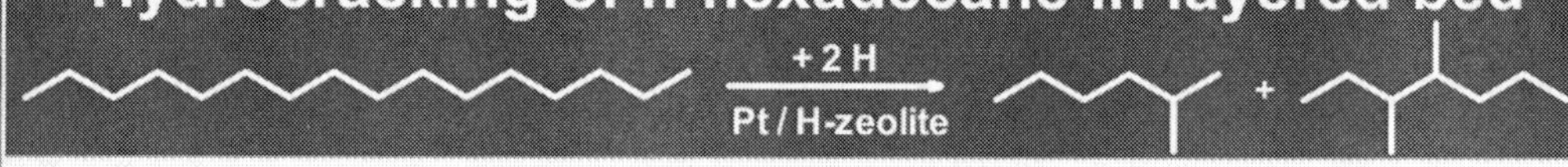


**Pd/$SiO_2$ / H-ZSM-5 catalyst**

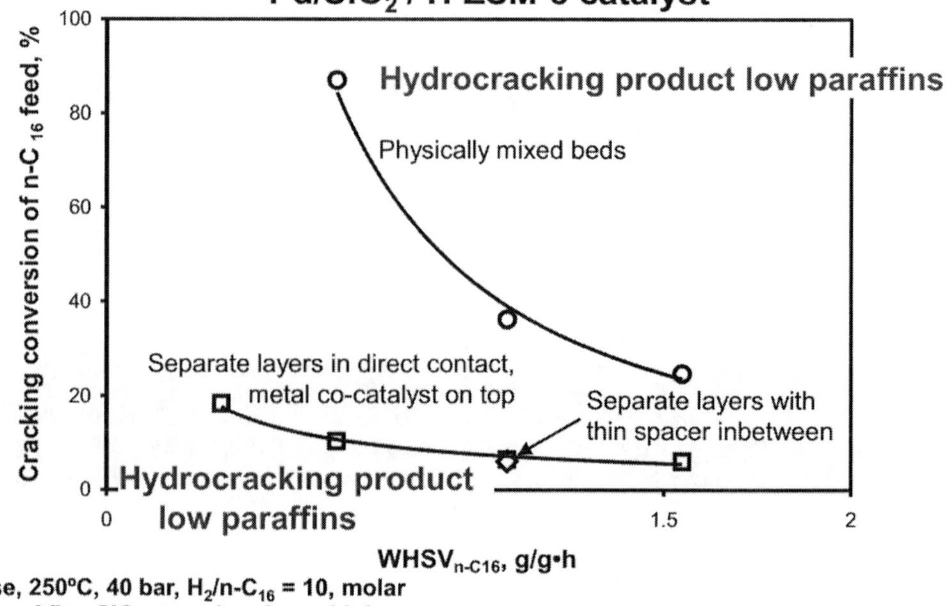


**Liquid phase, 250°C, 40 bar, $H_2$/n-$C_{16}$ = 10, molar**
**Spacer: layer of fine $SiO_2$ granules, 2 mm high**


Kukard, R., Wynne, P., Honours Theses, Centre for Catalysis Research, University of Cape Town (2006), unpublished



UNIVERSITY OF CAPE TOWN
CENTRE FOR CATALYSIS RESEARCH

## Hydrocracking of paraffin in layered bed

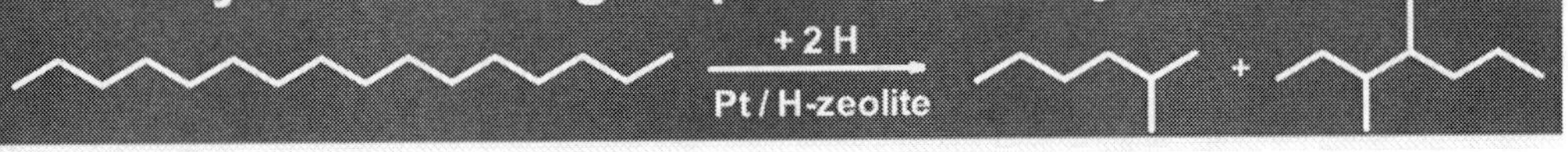


Hydroisomerising and hydrocracking require splitting off hydrogen from the feed molecules.

Hydrogen must finally be added again to saturate the isomerisation or crack product.

According to the classical mechanism, this requires 'hopping' of olefinic species between the metal and the acid sites through the fluid phase.

**How would the classical 'Hop Over' hydrocracking mechanism work with layered catalyst beds ?**

UNIVERSITY OF CAPE TOWN
CENTRE FOR CATALYSIS RESEARCH

## Classical 'Hop Over' hydrocracking mechanism in a layered catalyst bed ?

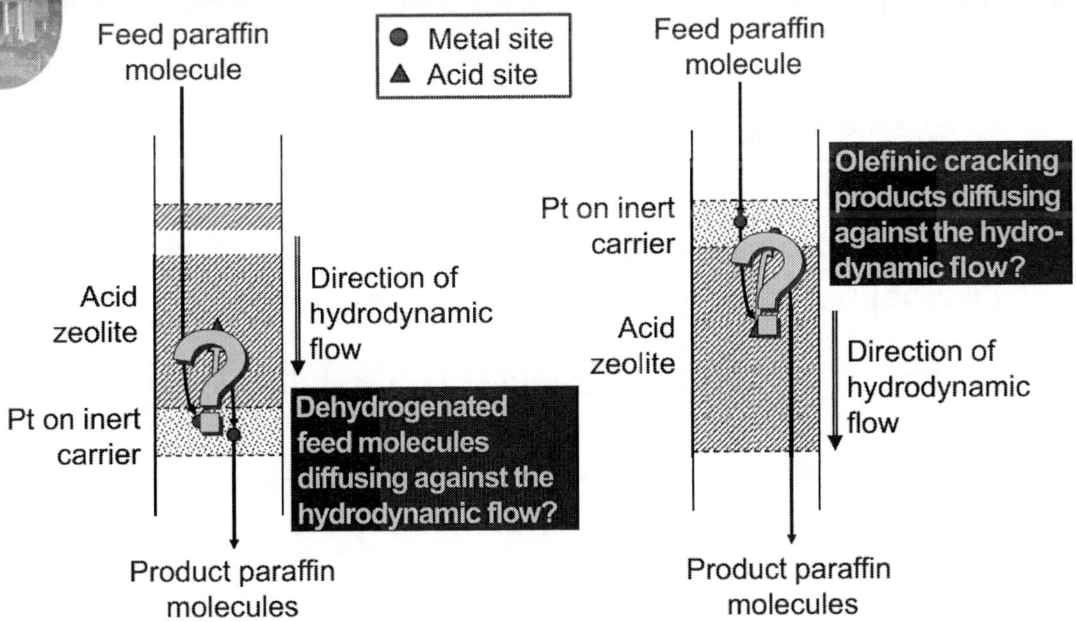


UNIVERSITY OF CAPE TOWN
CENTRE FOR CATALYSIS RESEARCH

## Classical 'Hop Over' hydrocracking or hydrogenation mechanism with metal functions shielded ?

**Catalyst functions physically shielded from each other by:**

- **Encapsulation of metal in layered silicate[1]**
- **Separation of metal from feed and acid by membrane[2]**
- **Encapsulation of metal in narrow pore zeolite[3]**

**No feed and no intermediate or product can get from noble metal sites to acid sites and *vice versa.***

**S-compounds and CO in the feed do not kill the metal co-catalyst.**

**S-compounds and CO cannot get to the metal sites[1,2,3].**

**Despite the separation of the individual sites, hydrogenation and hydrocracking still take place[1,2,3].**

[1] Metal encapsulated in layered silicate: Kuhlmann, A., Roessner, F., Schwieger, W., Gravenhorst, O., Selvam, T., Catal. Today 97 (2004) 303 and Roessner, F., Kuhlmann, A. and Schwieger, W., EP1530999 (2005).
[2] Metal shielded by ceramic membrane: Najafabadi, N.I., Chattopadhyaya, G., Smith, K.J., Appl. Catal. A: General 235 (2002) 47
[3] Metal encapsulated in zeolite A supercages: Yang, H., Chen, H., Chen, J., Omotoso, O., Ring, Z., J. Catal. 243 (2006) 36

UNIVERSITY OF CAPE TOWN
CENTRE FOR CATALYSIS RESEARCH

## Classical 'Hop Over' hydrocracking mechanism with separated catalyst beds ?

**In layered beds, the classical 'Hop Over' mechanism requires olefin diffusion against the hydrodynamic flow.**

**Such diffusion against the hydrodynamic flow will hardly occur (at least not over 'macroscopic' distances).**

**In shielded beds such diffusion is physically blocked.**

**This suggest that the 'classical' mechanism is not the process at work or at least not the only process at work.**

UNIVERSITY OF CAPE TOWN
CENTRE FOR CATALYSIS RESEARCH

## Classical 'Hop Over' hydrocracking mechanism

**Does hydrocracking according to the classical 'Hop Over' mechanism take place at all ?**

UNIVERSITY OF CAPE TOWN
CENTRE FOR CATALYSIS RESEARCH

## Classical 'Hop Over' hydrocracking mechanism in a layered catalyst bed ? - IV

**The classical 'Hop Over' mechanism:**

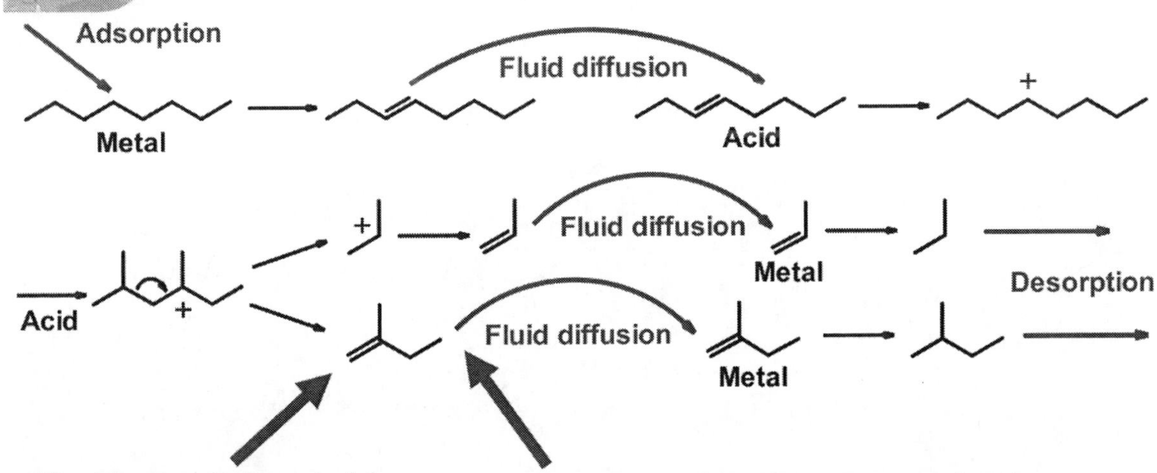


The fragments from acid catalysed bond cleavage are olefinic

The classical 'Hop Over' mechanism demands desorption of the crack olefins from the acid sites

UNIVERSITY OF CAPE TOWN
CENTRE FOR CATALYSIS RESEARCH

## Classical 'Hop Over' hydrocracking mechanism in a layered catalyst bed ? - V

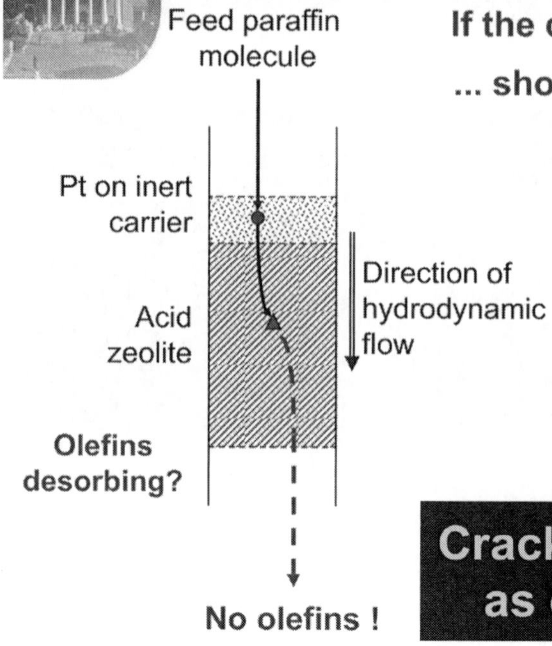


**If the classical mechanism was correct …**

**… shouldn't we see these desorbed crack olefins in the reactor effluent ?**

**But the product is paraffinic !**

**We do NOT see olefins in the reactor effluent !**

**Crack products do NOT desorb as olefins from the acid sites**

UNIVERSITY OF CAPE TOWN
CENTRE FOR CATALYSIS RESEARCH

## Olefin 'Hop Over' hydrocracking mechanism ???

**The entire bifunctional hydrocracking reaction, from the activation of the feed molecules up to the saturation of the crack products, occurs on the acid sites only, without desorption of olefinic intermediates**

UNIVERSITY OF CAPE TOWN
CENTRE FOR CATALYSIS RESEARCH

# A possible alternative: hydrogen spillover

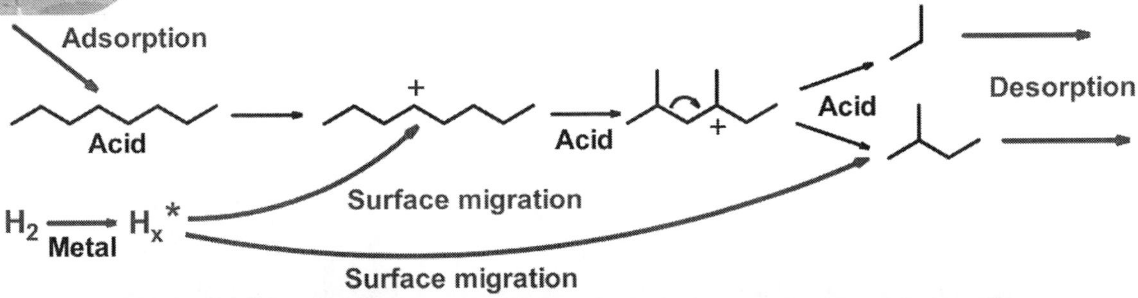


**No formation / desorption / adsorption of olefinic intermediates**

Roessner, F., in Handbook of Heterogeneous Catalysis, Wiley-VCH, Weinheim, 2nd ed., vol 3 (2008) 1574
Fujimoto, K., Stud. Surf. Sci. Catal. 127 (1999) 37
Roessner, F., Roland, U., J. Mol. Catal. A: Chemical, 112 (1996) 401
Roessner, F., Roland, U., Braunschweig, T., Faraday Trans. I, 91 (1995) 1539
Conner, W.C., jr., Chemical Industries, 31 (1988) 311
Inui, T., Funabiki, M., Suehiro, M., Sezume, T., Fraday Trans. I, 75 (1979) 787

UNIVERSITY OF CAPE TOWN
CENTRE FOR CATALYSIS RESEARCH

# The hydrogen spillover mechanism

- Activated hydrogen is produced on the metal sites.
- The activated hydrogen species migrate / spill over to the acid sites by surface diffusion.
- Surface diffusion is unaffected by the hydrodynamic flow.
- The activated hydrogen species can migrate on the surface against the direction of the hydrodynamic flow.
- Hydrogen spillover provides a constant flow of activated hydrogen – the hydrocracking reaction is a hydrogen sink!
- The entire reaction sequence takes place on the acid sites.

UNIVERSITY OF CAPE TOWN
CENTRE FOR CATALYSIS RESEARCH

## Hydrogen spillover in a layered bed

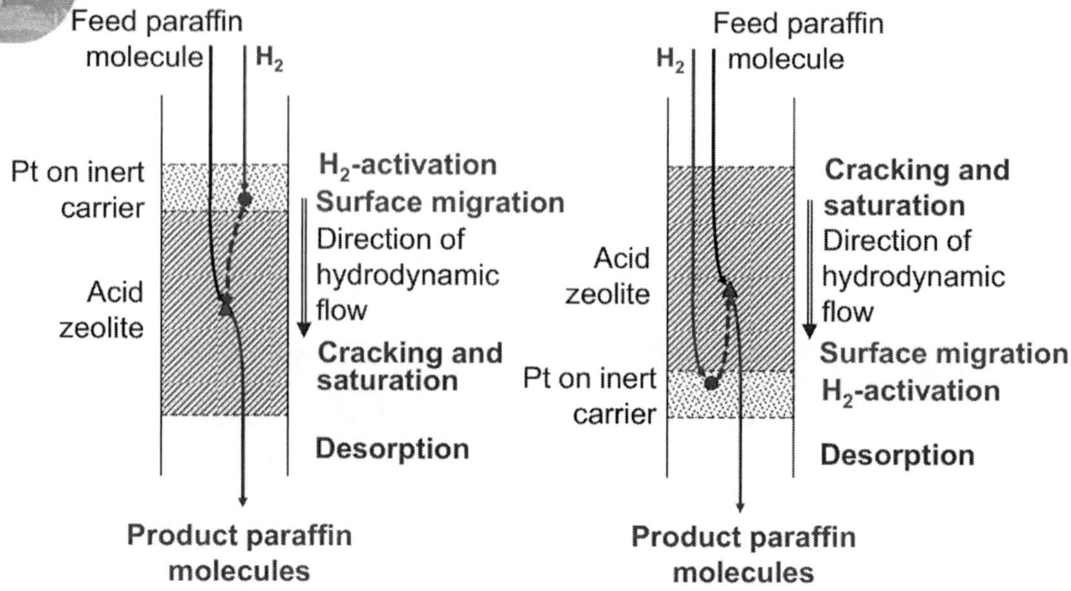


UNIVERSITY OF CAPE TOWN
CENTRE FOR CATALYSIS RESEARCH

## Hydrogen surface diffusion

Like every diffusional flux, the total flux of activated hydrogen is controlled by the following:

- Number of hydrogen activating metal sites.
- Length of the diffusional path (distance between metal and acid sites).
- Width of the diffusional path (surface diffusion).
- Properties of the surface.

**Maximum intimacy of sites would therefore be required to minimise noble metal contents.**

Hydrogen availability on the acid site is also depending on the consumption of hydrogen, that is, by the

- Metal / acid site ratio
- Reaction rate constant

⋮

UNIVERSITY OF CAPE TOWN
CENTRE FOR CATALYSIS RESEARCH

## Summary and conclusions

- Hydrogen spillover appears to be the mechanism at work in hydrocracking and hydroisomerisation.
- Activated hydrogen is produced on the metal sites and spills over to the acid sites by surface diffusion.
- The entire reaction sequence takes place on the acid sites.
- The classical mechanism that requires desorption of olefins twice and diffusion through the fluid phase has proved false.

**There is no 'olefin hopping' or 'hopping over' of olefins in hydrocracking**

UNIVERSITY OF CAPE TOWN
CENTRE FOR CATALYSIS RESEARCH

## ~~Olefin 'Hop Over' hydrocracking mechanism ???~~

**To those who still 'believe' in the 'olefin hopping' hydrocracking mechanism**

UNIVERSITY OF CAPE TOWN
CENTRE FOR CATALYSIS RESEARCH

## Critical rationalism – by Karl Popper

- Austrian / British philosopher Sir Karl Popper is regarded as one of the most important and influential philosophers of science of the 20th century.
- Sir Karl's philosophy of critical rationalism says that no number of positive outcomes of experiments can confirm a scientific theory but, logically, a single contradicting example is decisive:

  It shows the respective theory to be false.

Thornton, S., "Karl Popper", The Stanford Encyclopedia of Philosophy (Winter 2011 Edition), E.N. Zalta (ed.), http://plato.stanford.edu/archives/win2011/entries/popper/

UNIVERSITY OF CAPE TOWN
CENTRE FOR CATALYSIS RESEARCH

## The end of the classical 'Hop Over' hydroisomerisation / hydrocracking model

Following Sir Karl, the classical mechanism fails to explain experimental results and, therefore, is false.

This does not imply that the alternative must be correct, since according to Sir Karl, a theory can never be proofed right.

Nevertheless, a hydrogen spillover mechanism can explain the experimental outcome.

UNIVERSITY OF CAPE TOWN
CENTRE FOR CATALYSIS RESEARCH

## The outlook ...

**Max Planck (1858-1947), German theoretical physicist, Nobel Prize laureate in physics, 1918:**

"A new scientific truth does not triumph by convincing
its opponents and making them see the light,
but rather because its opponents eventually die,
and a new generation grows up that is familiar with it."

Science Quotes by Max Planck.
Available: http://www.todayinsci.com/P/Planck_Max/PlanckMax-Quotations.htm.

UNIVERSITY OF CAPE TOWN
CENTRE FOR CATALYSIS RESEARCH

## Acknowledgements

c*change
DST-NRF Centre of Excellence in Catalysis

University of Cape Town • iYunivesithi yaseKapa • Universiteit van Kaapstad

# Thank you

Department of
Science & Technology

NRF

UNIVERSITY OF CAPE TOWN
CENTRE FOR CATALYSIS RESEARCH

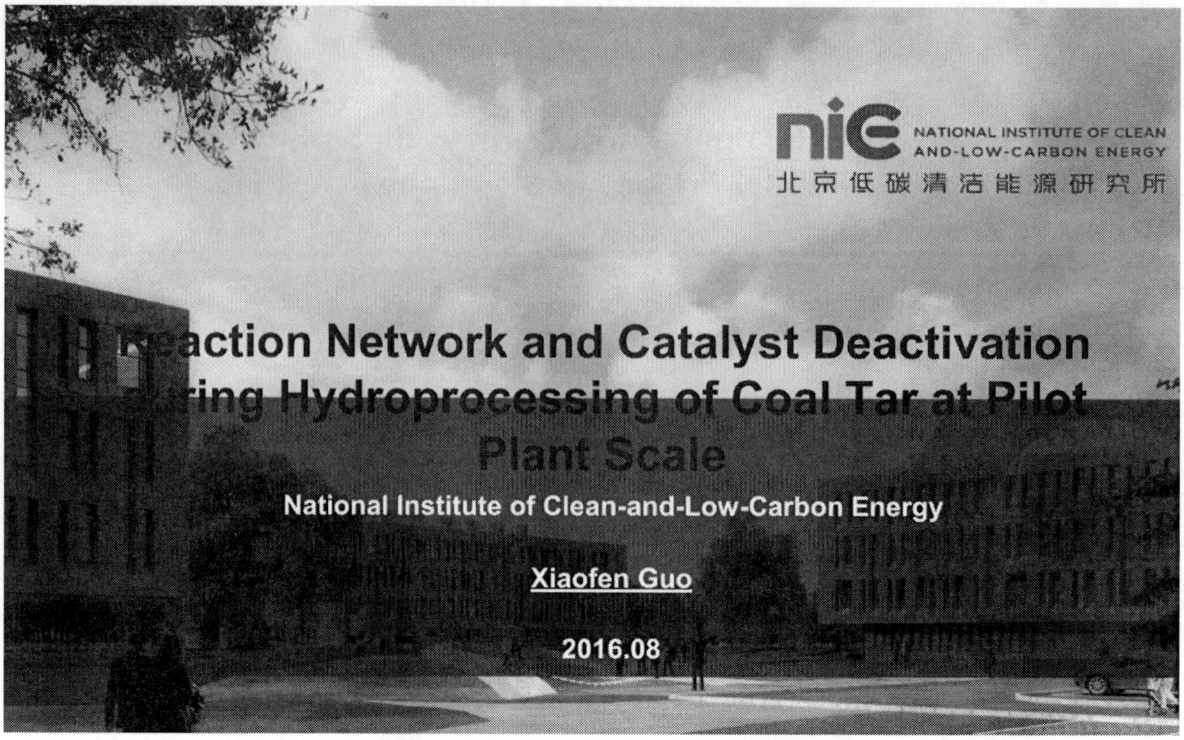


# Outline

- Background
- Results and discussion
  - Hydroprocessing performance
  - Transformation of hetero-atom compounds
  - Catalyst deactivation
- Summary
- Acknowledgement

National Institute of Clean & low-carbon Energy (NICE) www.nicenergy.com NICE Confidential

## Introduction

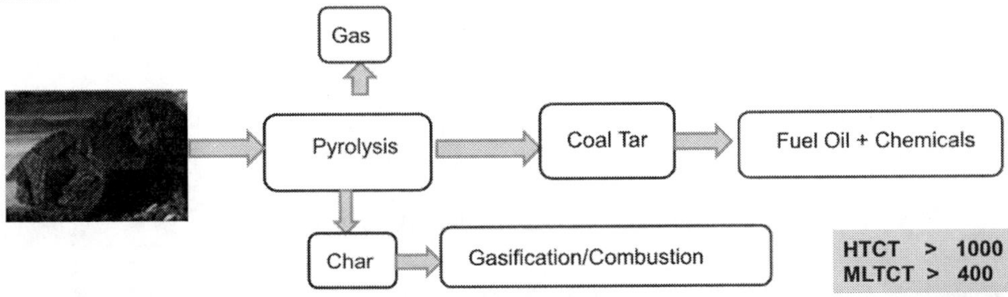


HTCT > 1000 $10^4$ t
MLTCT > 400 $10^4$ t

**The Clean Utilization Route of Low-rank Coal**

- **Traditional coal tar utilization**
  - ☐ Coal tar deep processing, chemicals extraction

  Complex route, high energy-consumption, low economic profits
  - ☐ Burn directly as fuels

  NOx and SOx emission, environmental pollution, low-value product



## Business case and Current technology

*Delayed Coking Unit (500K t/a)* from Shengmu Tanyu

*Fixed Bed Unit (120Kt/a)* from Fuyou Nengyuan

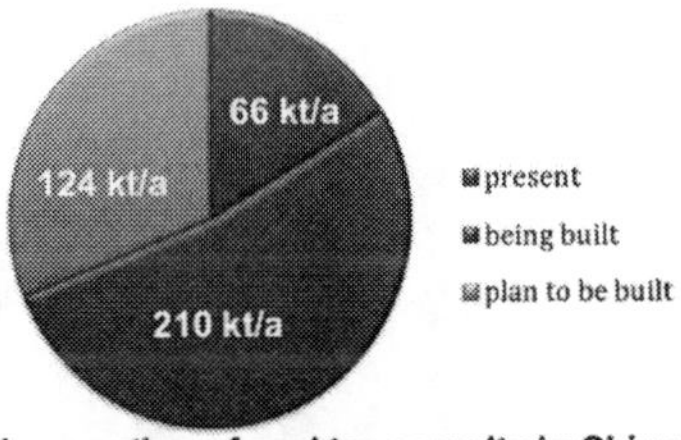


**Hydrogenation of coal tar capacity in China**

| Current upgrading technology : | | |
|---|---|---|
| TianYuan | Delayed coking+ Fixed bed HT/P hydroprocessing | Liquid yield is not high(~75%) |
| FuYou | Whole fraction HP Fixed bed hydroprocessing | High operation pressure (~16 MPa) |
| FRIPP+ BRICC | EB/Slurry bed+ Fixed bed hydroprocessing | No industrialized demonstration<br>High operation cost |

☐ Hydrogenation of coal tar to produce high-value clean transportation fuels is the most effective route of clean utilization of coal tar.

## NICE Whole Fraction Coal Tar Upgrading Research & Development Process

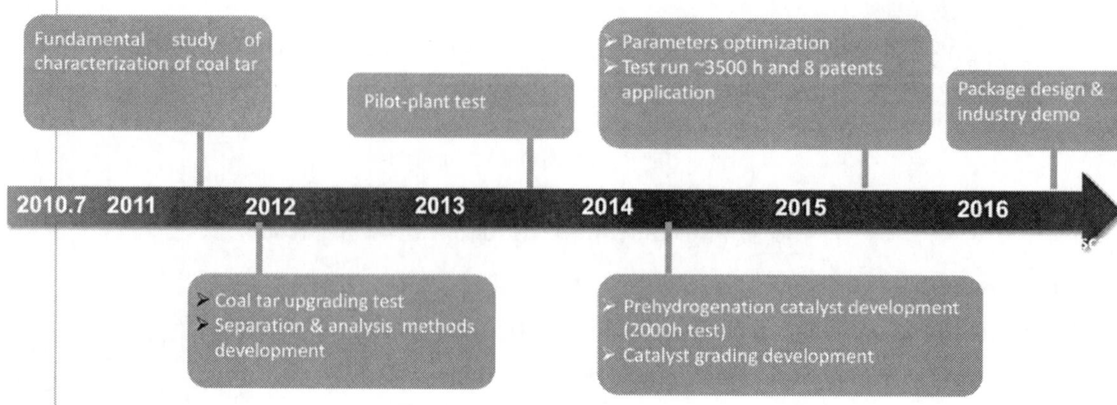


National Institute of Clean & low-carbon Energy (NICE) www.nicenergy.com NICE Confidential

## NICE Whole Fraction Coal Tar Upgrading Process

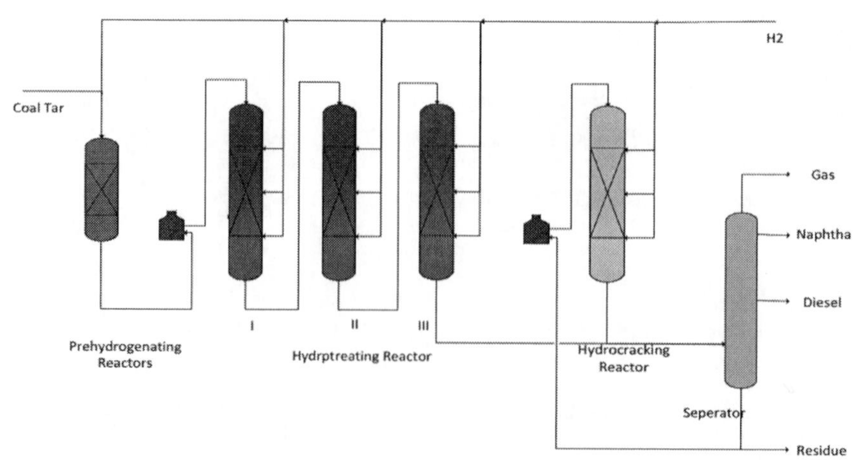


- ❑ Whole fraction feedstock provides maximum utilization of coal tar;
- ❑ Pre-hydrogenation catalyst and rational catalyst grading provide the long-term running;
- ❑ Middle pressure operation save OPEX & High liquid oil yield

National Institute of Clean & low-carbon Energy (NICE) www.nicenergy.com NICE Confidential

# Main Properties of Coal Tar and Catalysts

Main properties of middle and low-temperature coal tar (MLTCT)

| Properties | | Properties | |
|---|---|---|---|
| Density, g/cm$^3$ | 1.0752 | TI, wt% | 1.14 |
| Viscosity, 50 °C, mm$^2$/s | 218.81 | QI, wt% | 0.28 |
| C, wt% | 80.97 | Fe, mg/kg | 117 |
| H, wt% | 7.52 | Ca, mg/kg | 103 |
| O, wt% | 8.71 | Ni+V, mg/kg | <0.3 |
| S, wt% | 0.18 | Distillation range, °C | |
| N, wt% | 0.68 | IBP | 185 |
| H/C molar ratio | 1.11 | 10% | 254 |
| Aromatic carbon ratio, % | 66.61 | 50% | 385 |
| $C_7$ asphaltenes, wt% | 22.58 | 90% | 480 |

Properties and chemical compositions of catalysts

| Catalyst | Mo-Ni/ $\gamma$-$Al_2O_3$ | Mo-Ni/ $\gamma$-$Al_2O_3$ | W-Ni/ USY |
|---|---|---|---|
| Type | HDM | HDT | HDC |
| Shape | ring | lobed | bar |
| Active metal, wt% | | | |
| Mo | 5 | 10 | - |
| W | - | - | 16 |
| Ni | 1.2 | 3 | 5 |
| Physical properties | | | |
| $S_{BET}$, m$^2$/g | 106 | 215 | 324 |
| Pore volume, cm$^3$/g | 0.47 | 0.42 | 0.39 |
| Pore diameter, nm | 15.7 | 7.8 | 4.8 |



# Hydroprocessing of coal tar: 1st -HDM

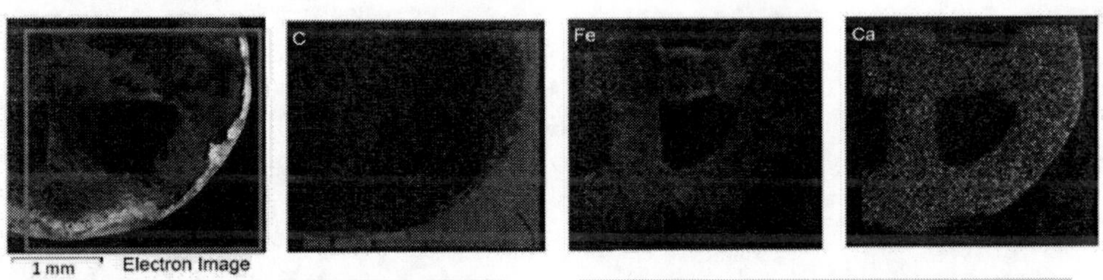


| Properties | Coal Tar | HDM |
|---|---|---|
| Density, g/cm$^3$ | 1.0752 | 1.0474 |
| H/C molar ratio | 1.11 | 1.13 |
| Aromatic carbon ratio,% | 66.61 | 64.08 |
| HDS, % | - | 15 |
| HDN, % | - | 8.5 |
| HDM, % | - | 89.5 |
| $C_7$-asphaltenes | 22.58 | 9.68 |

| HDM catalyst | Fresh | Spent |
|---|---|---|
| $S_{BET}$ (m$^2$/g) | 106 | 79 |
| Pore volume, cm$^3$/g | 0.47 | 0.34 |
| Pore diameter, nm | 17.5 | 17.4 |
| Micropore (<2 nm), % | 1.3 | - |
| Mesopore (2-50 nm), % | 63.7 | 60.5 |
| Macropore (>50 nm), % | 35 | 39.5 |

- HDM catalyst combines high capacity of metal removal and asphaltenes conversion with low HDS/HDN activities.

## Hydroprocessing Performance

### 2nd-HDT

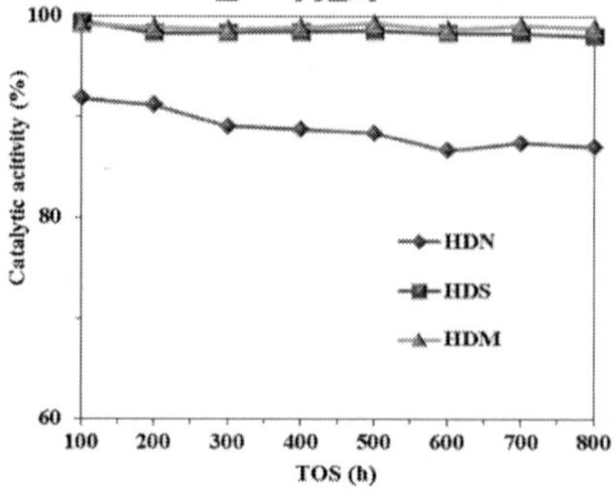


### 3rd-HDC

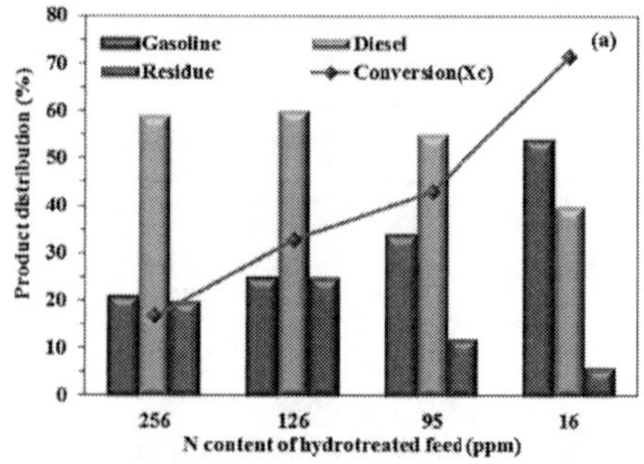


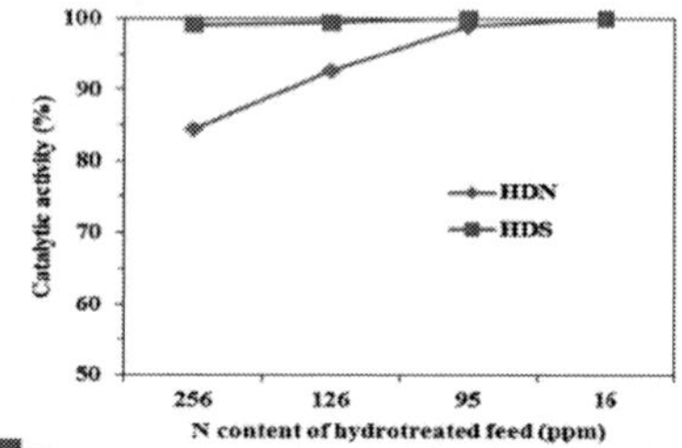


- HDN and HDS are predominant functions rather than HDM during hydrotreating.
- The excellent hyrocracking activity can be exhibited when the nitrogen content in hydrotreated feed was below 100 ppm.



## Transformation of hydrocarbon compounds

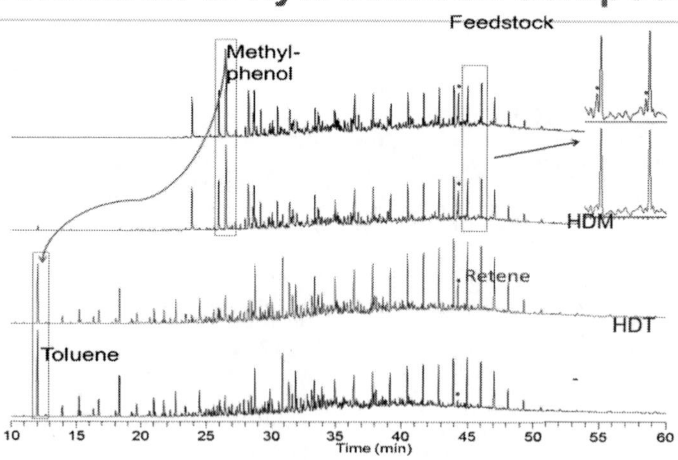


- The distribution of main compounds in coal tar was unaffected.
- The decrease of retene content indicated that hydrogenation of aromatics occurred during every hydrogenation stage.

## Transformation of Sulfur Compounds

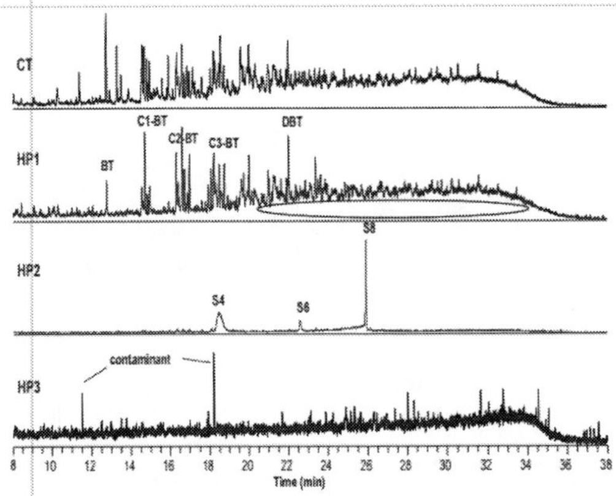


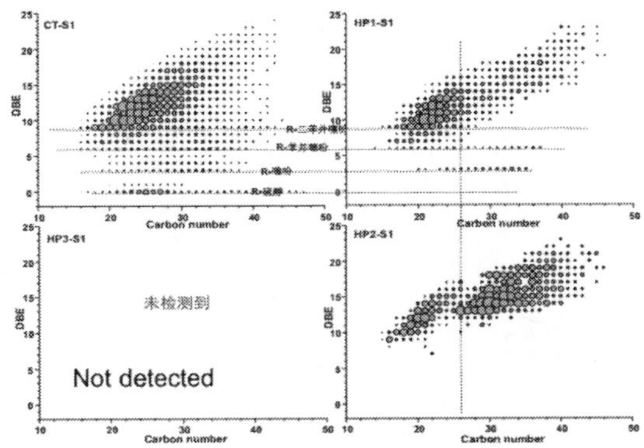


- Coal tar contained long-chain mercaptan and dibenzothiophene-like compounds with high DBE.
- With the hydrogenation degree increasing, sulfur-containing compounds were removed from low to high DBE.



## Transformation of neutral Nitrogen

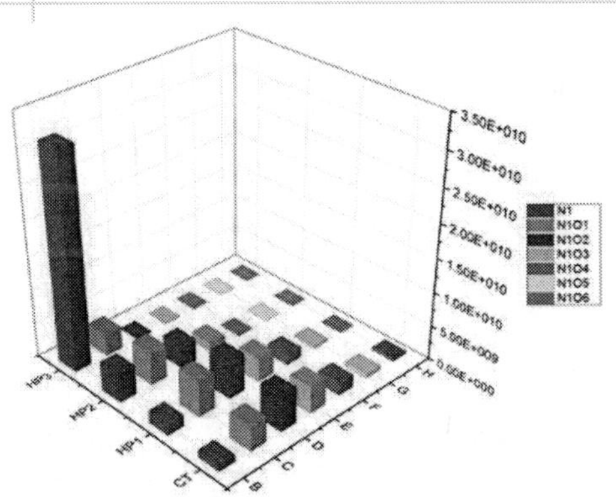


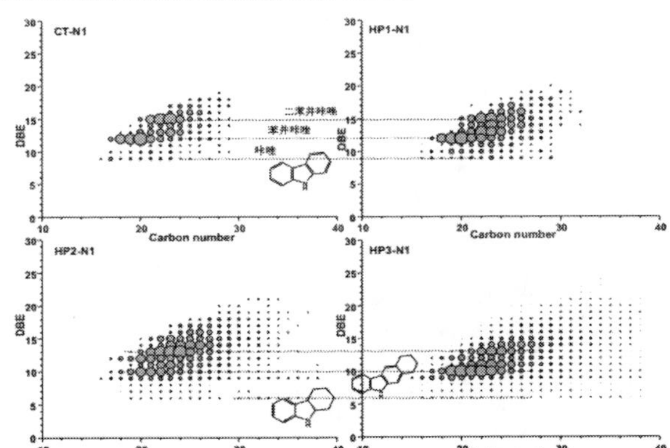


- Before and after hydrogenation, benzo- and Dibenzo-carbazol compounds mainly existed in coal tar, indicating that these species were difficultly converted.

## Transformation of Basic N-containing compounds

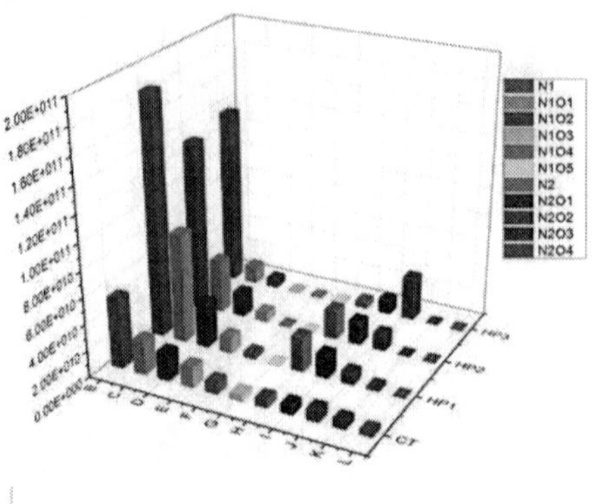


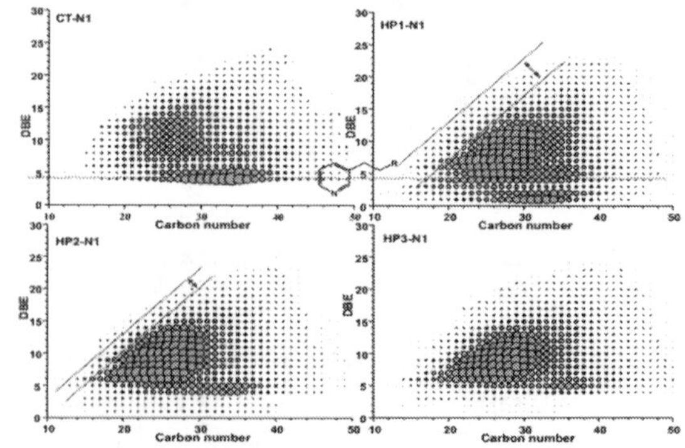


- Basic nitrogen compounds were mainly composed of cyclohydrocarbon and pyridine-type compounds, and 1-3 ring aromatics finally existed with hydrogenation degree.
- Basic nitrogen was much easier to be converted than neutral nitrogen at mild conditions.



## Transformation of Oxygen Compounds

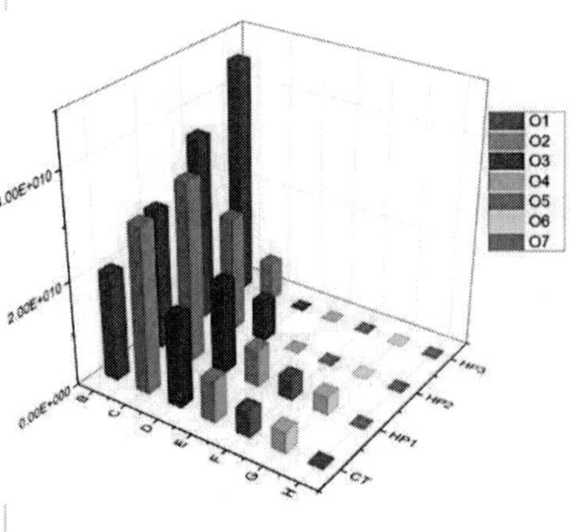


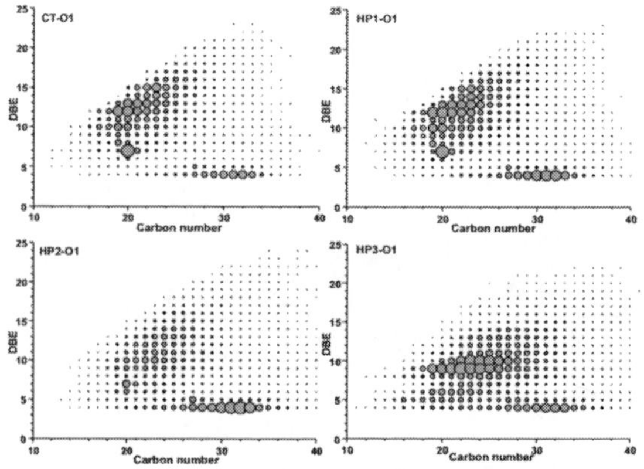


- Oxygen species with high Ox were completed converted after hydrogenation, while low oxygen species such as phenol-type species (O1-2) still existed.

# Characterization of spent HDT catalyst(SEM-EDS)

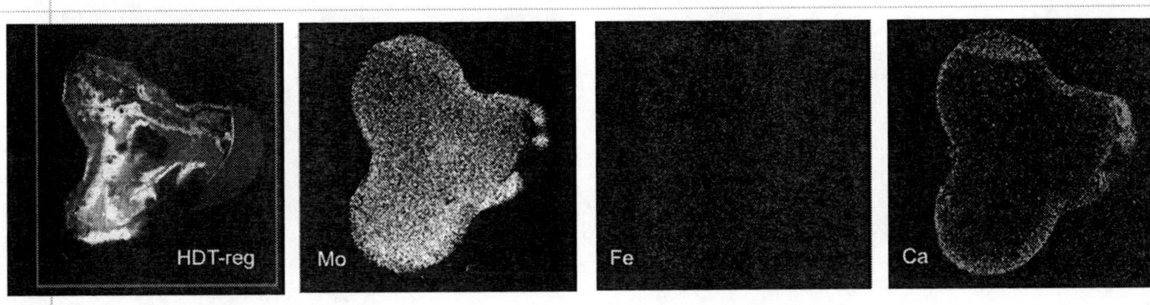


- Mo is homogenously deposited on the catalyst surface, Fe is also homogenously deposited on the spent catalyst surface after HDT reaction.
- Ca is easily deposited and accumulated on the edge of catalyst, much higher than the center of the catalyst.



# Characterization of spent and regenerated catalysts

NMR Spectra

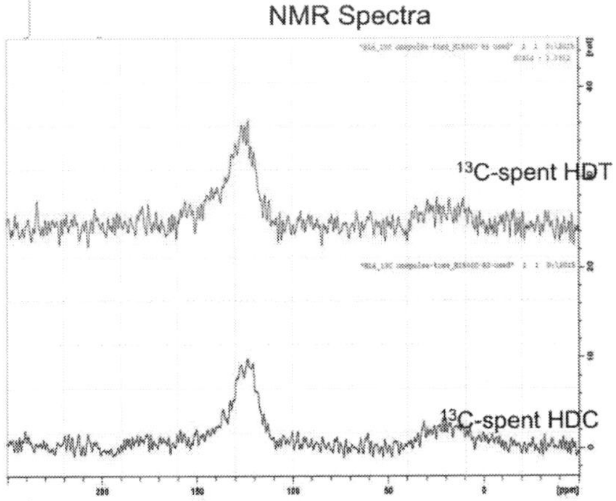


NH3-TPD

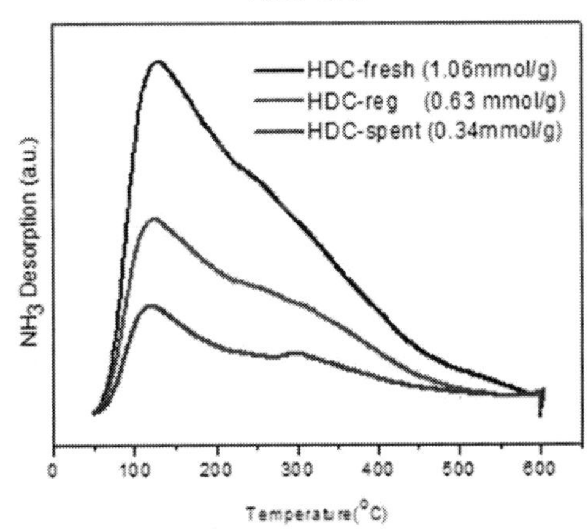


- Carbon deposition on the spent catalyst was present in the form of aromatics and paraffin structure.
- The decrease in the acidity of the HDC catalyst was due to adsorption of nitrogen compounds, decreasing hydrocracking activity.

# Characterization of spent HDC catalysts (STEM)

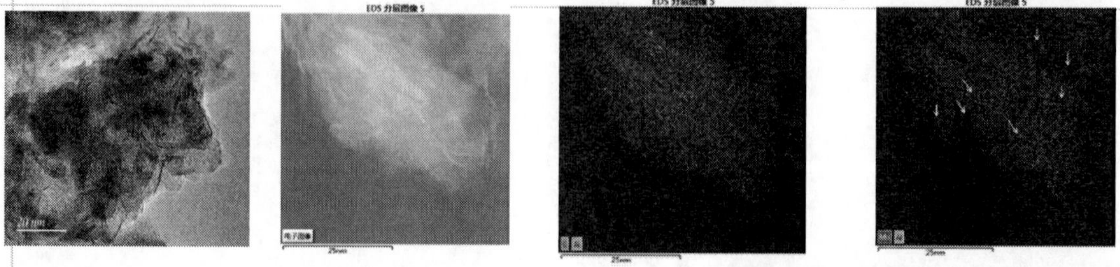

- Carbon were homogenously deposed on the spent HDC catalyst after long-term reaction.
- BET surface and pore volume decreased significantly due to carbon deposition. The micropores were almost blocked, and they can be partially removed after regeneration but some of them also collapsed.

| HDC catalyst | Fresh | Spent | Reg |
|---|---|---|---|
| $S_{BET}$ ($m^2/g$) | 324 | 154 | 237 |
| Pore volume, $cm^3/g$ | 0.39 | 0.24 | 0.36 |
| Pore diameter, nm | 4.8 | 6.6 | 6 |
| T-plot micropore surface area ($m^2/g$) | 194 | 6 | 124 |
| Micropore (<2 nm), % | 22.4 | 12.3 | 10.6 |
| Mesopore (2-50 nm), % | 77.6 | 87.3 | 85.3 |
| Macropore (50 -100nm) , % | 0 | 0.4 | 3.1 |
| Macropore (>100nm) , % | 0 | 0 | 0.9 |



# Summary

- HDM catalyst provides the high capacity of metal removal and asphaltenes conversion with the low removal of sulfur and nitrogen. HDT catalyst exhibited the high activity for the deep removal of sulfur and nitrogen. The product composition is also correlated with the reaction behaviors occurred on the different catalyst surface.
- For the transformation of heter-atom compounds, hydrodesulfurization occurred easily. Basic nitrogen compounds were mainly composed of cyclohydrocarbon and pydine-tpe compounds, and 1-3 ring aromatics finally existed with hydrogenation degree. Basic nitrogen was much easier to be converted than neutral nitrogen at mild conditions.
- The deactivation of spent hydrotreating catalysts were attributed to the carbon and metal deposition and deposited on the catalyst surface. The nitrogen compounds present in the feedstock act as temporary poisons as well as coke precursors for hydrocracking catalysts, resulting in decrease in catalyst activity.

# Acknowledgement


- Analysis Team, China University of Petroleum

  Prof. Quan Shi, Chao Ma, Hongxing Ni
- National High-Tech R&D Program of China (863 project, 2011AA05A202)
- International Science & Technology Cooperation Program of China (China-Australia Joint Research Project, 2013DFG61490)


National Institute of Clean & low-carbon Energy (NICE) www.nicenergy.com NICE Confidential

NATIONAL INSTITUTE OF CLEAN AND-LOW-CARBON ENERGY
北京低碳清洁能源研究所

# Thanks!


中国，北京，昌平区，未来科技城

T: +86-10-57595508 F : +86-10-57339834

www.nicenergy.com

# 2016 International Pittsburgh Coal Conference
# Cape Town, SOUTH AFRICA
# August 8 - 12, 2016

**Coal Science**

## Hazardous particles from coal fires from Santa Catarina, South Brazil

***Diego de Medeiros*[a]*;** *Marcos L.S. Oliveira*[a]; *Frans Waanders*[b]; *Fagner T.C. Sá*[a]; *Silvio R. Taffarel*[a]; *Luis F. O. Silva*[a]
[a] *Laboratory of Environmental Researches and Nanotechnology Development, Centro Universitário La Salle, Mestrado em Avaliação de Impactos Ambientais em Mineração. Victor Barreto, 2288 Centro 92010-000, Canoas, RS, Brazil.*
[b] *School of Chemical and Minerals Engineering North West University (Potchefstroom campus) Potchefstroom 2531, South Africa*

**Abstract**
Coal mining in Santa Catarina, South Brazil, impacted the environment by means of a large number of coal waste piles emplaced over the old mine sites and the adjacent areas of the Criciúma, Urussanga, and Siderópolis. Poor disposal of coal cleaning rejects represents significant environmental concerns due to their potential influence on soils and river sediments, as well as on the surface and groundwater in the surroundings of these zones. In this research advanced analytical techniques were performed to provide an improved understanding of the complex processes related with sulphide-rich coal waste oxidation, spontaneous combustion, sequences of mineral formation, and the transport mechanisms of hazardous elements by neoformed soluble nanominerals and ultrafine particles. The results showed the presence of massive amorphous Si–Al–Fe–Ti–Zr (oxy-) hydroxides and Fe-hydr/oxides with goethite and hematite with various degrees of crystallinity, containing hazardous elements, such as Cr, Hf, Hg, Mo, Ni, Se, Pb, Th, U, Zr, and others. Several of the neoformed nanominerals found in the burned coal rejects are the same as those commonly associated with coal mine drainage, in which oxidation of sulphides plays an important role to environment and human health. This is the first Brazilian coal spontaneous combustion research.

**2016 International Pittsburgh Coal Conference**
**Cape Town, SOUTH AFRICA**
**August 8 – 12, 2016**

**Abstract Submission**

**PROGRAM TOPIC:**
**SUSTAINABILITY AND THE ENVIRONMENT**

**Exact Title of Paper:**

# An investigation on incorporation of a local thermal power station boiler ash as a raw material in clinker manufacture

***Nonhlanhla G. Mguni, Miss, Department of Chemical Engineering, National University of Science and Technology, Box AC 939, Ascot, Bulawayo, Zimbabwe.***
***nonhlanhla.mguni@nust.ac.zw***
***+263772831388***

*Marcia R. Nyika, Miss, Department of Chemical Engineering, National University of Science and Technology, Box AC 939, Ascot, Bulawayo, Zimbabwe*
*marcianyika@gmail.com*

*. Liberty L. Mguni, Mr, Department of Chemical Engineering, National University of Science and Technology, Box AC 939, Ascot, Bulawayo, Zimbabwe*
*liberty.mguni@nust.ac.zw*

**Abstract:**
Coal is the major fuel used for electricity generation in Zimbabwe and a huge contributor of solid waste like boiler ash which presents major disposal challenges. Due to high contents of silica, aluminium oxide and iron oxide, boiler ash may find use in the cement industry. Research was carried out on the possibility of incorporating boiler ash from a local thermal power station as a third raw material in clinker manufacture. The boiler ash was characterised using XRF, bomb calorimeter and proximate analysis. It was observed that the boiler ash contained an average amount of 24% silica, 21% aluminium oxide, and 15% iron oxide which are important materials in clinker manufacture. The boiler ash was also observed to contain high fixed carbon of 28% and had a calorific value of 8.21 kJ/kg. The Mix Solution Region 96 program was used to assess the effects of varying percentages of boiler ash incorporated with other raw materials (limestone and clay). It was observed that the boiler ash can be incorporated replacing up to 14% clay without affecting the clinker quality. For the plant trial 1% boiler ash was used to replace clay instead of 14% due to high carbon content. Carbon forms carbon monoxide in the flue gases during the process which might lead to electric sparks in the electrostatic precipitator. The resulting clinker was of good quality with $C_3S$ (tricalcium silicate) above 53% and high burnability factors. There was an increase in carbon monoxide produced during the plant trial. Using boiler ash also contributed to lowering coal consumption by 10% in the furnace.

*Keywords:* burnability, cement, clinker, bottom ash,

## 1 Introduction

Zimbabwe still heavily depends on coal fired power stations for 40% of its electricity needs. Huge amounts of bottom ash are produced daily i.e one of the smallest power stations produces over 50 tonnes per day [1]. This ash contains traces of heavy metals and hazardous substances and presents problems in disposal. This waste also contains large amounts of silica, alumina and ferrites which are constituents of raw materials for cement production. The use of bottom ash will reduce the cement industry carbon footprint and also conserve the natural resources. Portland cement is produced by high temperature sintering of limestone and clay. Use of calcium oxide bearing waste material such as bottom ash can reduce $CO_2$ emissions during cement production by reducing use of limestone high in $CaCO_3$ by CaO in bottom ash [2].

The use of bottom ash in cement manufacture has been reported by a number of researchers. It has been used in cement at different stages such as before clinkerization as a raw mix and after clinkerization as a cement replacement in mortar and concrete [3,4,5,6]. Bottom ash from different sources have been used in cement production these sources including from coal [7], municipal solid waste incineration [2,8,9] and sugar bagasse ash [7]. It has been reported that incorporation of bottom ash or fly ash showed higher compressive strength [7]. The clinker produced was observed to contain accepted phases and compositions only up to 6% addition of MSWI. Clinker contains four primary phases: $Ca_3SiO_5$ ($C_3S$), $Ca_2SiO_4$ ($C_2S$), $Ca_3Al_2O_6$ ($C_3A$) and $Ca_4Al_2Fe_2O_{10}$ ($C_4AF$). In this work not only was the effect of using bottom ash from a coal firing power station on clinker properties investigated but also its effect on process parameters at a large scale.

## 2. Materials and Methods

### 2.1 Sampling

Random bulk samples of boiler ash were taken from two different points on the bottom ash stockpiles. These samples were homogenized using a riffler then cone and quartered so as to collect a representative sample. Calorific value, proximate analysis and XRF analysis were carried on the bottom ash samples.

### 2.2 Raw material preparation for clinker formation

3 kg of limestone sample, 1 kg of boiler ash and 1 kg of clay were obtained and dried overnight in an oven at $100^{\circ}C$. After drying, each material was crushed in a 2 mm-jaw crusher. 1 kg of each sample was divided into

small masses and each portion milled in the puck mill. All the small portions were combined in a plastic bag and homogenized. A representative 200 g sample of each pulverized/fine material was saved for XRF analysis.

### 2.3 Raw Mix Design

Mix Solution Region 96 program was used to design the required mix. The mass of the coal ash incorporated into the clinker was calculated from the heat consumption of the kiln.

### 2.4 Characterisation

#### 2.4.1 Burnability test method

Nodules were calcined at 1000°C for an hour in a silica dish. After calcining, all nodules were stored in silica dishes in a desiccator waiting for further processing. Nodules were put into the platinum dish and placed in the furnace for 1 hour at 1450°C. The nodules were allowed to cool thoroughly analysed in an XRF.

#### 2.4.2 Loss on Ignition (LOI)

This test is for estimating the amount of unburnt carbon content in bottom ash [10] percentage of the amount of calcinable material. A 1.00g sample was weighed into a platinum crucible of known mass, $W_s$, and ignited in a muffle furnace for 1 hour, cooled and weighed, $W_a$.

**%LOI = $(W_s - W_a)$/weight after ignition x 100**

### 2.5 Plant trial

The ash was then incorporated into the raw mix at plant scale using results of the mix design calculation. Boiler ash was introduced into the process with clay at the milling stage and the process monitored continuously. Particular attention was given to the effect of boiler ash on fuel consumption, clinker production and plant CO levels.

## 3. Results and discussion

### 3.1 Boiler ash characterisation

The boiler ash had high carbon content and high LOI values indicating poor combustion efficiencies of the boiler (Table 3.1). Boiler ash rich in carbon affects the general applicability of concrete mixtures and hence increase the water binder ratio in order to show good workability [11].

Table 3.2 outlines the XRF analysis of boiler ash (B1 and B2) samples, the clinker made without bottom ash samples M and clinker made from formulations of boiler as, C1 and C2. Boiler ash samples, B1 and B2 are rich in silica, alumina and iron, making them suitable replacement for clay as a raw material. The XRF analysis done on C1 and C2 show that the standard deviation (Stdev) from clinker made without boiler ash, M is significantly small. This shows that the addition of boiler ash has no significant effect on the quality of clinker produced. The resulting clinker was of good quality with $C_3S$ (tricalcium silicate) above 53% and higher burnability factors than without ash which contributed to less energy requirements.

**Table 3.1: Boiler ash proximate analysis, calorific value and Loss of Ignition (LOI)**

| | | B1 | B2 |
|---|---|---|---|
| **PROXIMATE ANALYSIS** | Ash % | 63.99 | 74.21 |
| | Volatile matter % | 2.96 | 3.18 |
| | $H_20$ % | 1.29 | 1.24 |
| | Fixed carbon % | 31.76 | 21.37 |
| | Calorific Value, KJ/kg Ash | 10.72 | 6.14 |
| | L.O.I | 34.68 | 22.79 |

**Table 3.2: XRF analysis of the boiler ash and clinker**

| | B1 | B2 | Stdev | M | C1 | Stdev | M | C2 | Stdev |
|---|---|---|---|---|---|---|---|---|---|
| $SiO_2$ | 24 | 27.6 | 2.55 | 21.17 | 20.91 | 0.18 | 20.83 | 20.81 | 0.01 |
| $Al_2O_3$ | 20.72 | 24.16 | 2.43 | 4.25 | 4.54 | 0.21 | 4.59 | 4.58 | 0.01 |
| $Fe_2O_3$ | 14.76 | 18.08 | 2.35 | 3.15 | 3.4 | 0.18 | 3.32 | 3.32 | 0.00 |
| $Mn_2O_3$ | 0.20 | 0.14 | 0.04 | 0.56 | 0.56 | 0.00 | 0.56 | 0.56 | 0.00 |
| $TiO_2$ | 1.18 | 1.39 | 0.15 | 0.32 | 0.32 | 0.00 | 0.32 | 0.32 | 0.00 |
| CaO | 3.15 | 3.71 | 0.40 | 63.83 | 63.56 | 0.19 | 63.96 | 63.99 | 0.02 |
| MgO | 0.14 | 0.15 | 0.01 | 5.13 | 5.08 | 0.04 | 4.72 | 4.72 | 0.00 |
| $P_2O_5$ | 0.42 | 0.24 | 0.13 | 0.03 | 0.03 | 0.00 | 0.03 | 0.03 | 0.00 |
| $SO_3$ | 1.65 | 1.39 | 0.18 | 0.71 | 0.74 | 0.02 | 0.68 | 0.68 | 0.00 |
| Cl | 0.00 | 0.00 | 0.00 | 0.00 | 0.00 | 0.00 | 0.00 | 0.00 | 0.00 |
| $K_2O$ | 0.23 | 0.19 | 0.03 | 0.40 | 0.38 | 0.01 | 0.47 | 0.47 | 0.00 |
| $Na_2O$ | 0.95 | 1.12 | 0.12 | 0.48 | 0.48 | 0.00 | 0.52 | 0.52 | 0.00 |

The measured loss on ignition (LOI) of boiler ash samples B1 and B2 was found to be 34.68% and 22.79%. This approximates to an average of over 28% residual carbon within the ash. The high carbon content of the boiler ash poses the greatest risk, that of potential electrostatic precipitator (ESP) explosion should the carbon be converted to carbon monoxide leading to high concentrations of carbon monoxide in the ESP. With high loads of carbon monoxide in the ESP any spark is likely to trigger a catastrophic explosion in the ESP [12]. Pertinent to this fact this risk assessment was done. Based on this, calculations on the maximum allowable CO was calculated and the greatest dosage of boiler ash was calculated as 1%.

### 3.2 Clinker Formation

The Mix solution 96 program was used to determine boiler ash that can be incooperated without affecting clinker properties. It was observed that 14% boiler ash can be incooperated.

Figure 3.1 shows the change in burnability index of the mix before and after the incorporation of boiler ash into the raw meal. The burnability of a kiln fed is the relative ease or difficulty with which the feed is changed into a clinker in the kiln; that is, the indication of the amount of fuel required to burn the kiln feed into a clinker of good quality [13]. The burnability index is based on the potential clinker products, $C_3S$, $C_4AF$ and $C_3A$. A higher content of $C_3S$ with lower contents of $C_3A$ and $C_4AF$ will cause harder burnability. It was observed in Figure 3.1, that the burnability of C2 which was mixed with B2 was quite high; this proved beneficial as the conversion of alite to belite was successful because the limestone Lime Saturation Factor (LSF) was in the required range. The LSF is the ratio of CaO to three main oxides, $SiO_2$, $Al_2O_3$ and $Fe_2O_3$ [14]. The burnability of B1 was low which was caused mainly by the presence of very little lime for conversion reactions as compared to B2. Boiler ash contributed vastly to the aluminium and iron content for C1 and C2, manufactured from B1 and B2.

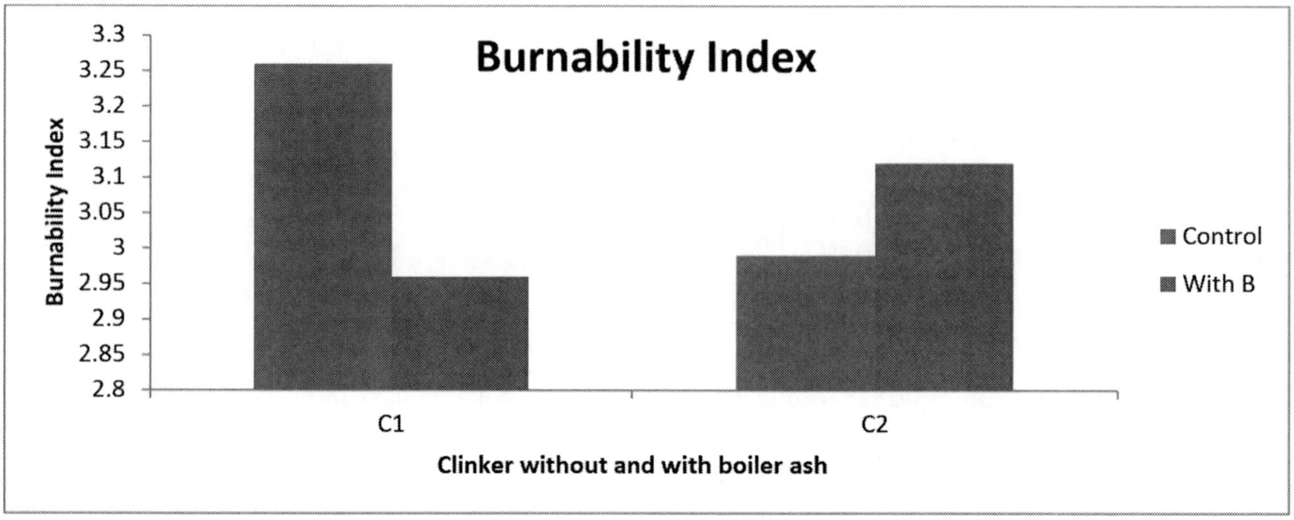


**Figure 3.1: Burnability Index of clinker without ash (control) and with boiler ash (B)**

Figure 3.2 shows the change in the burnability factor before and after the incorporation of boiler ash into the raw meal, there is indeed an increase in the burnability factor after incorporating boiler ash into the mix. All the clinkers with boiler ash showed an increase in burnability (Figure 3.2) of the mix which may be attributed to the fact that boiler ash had already been burnt in the power plant boilers at high temperatures. For that reason less energy could be needed in-order to form the main clinker phases which, as can be seen from Tables 3.2 and 3.3 that all phases are present, implying that boiler ash did not affect the mineral composition.

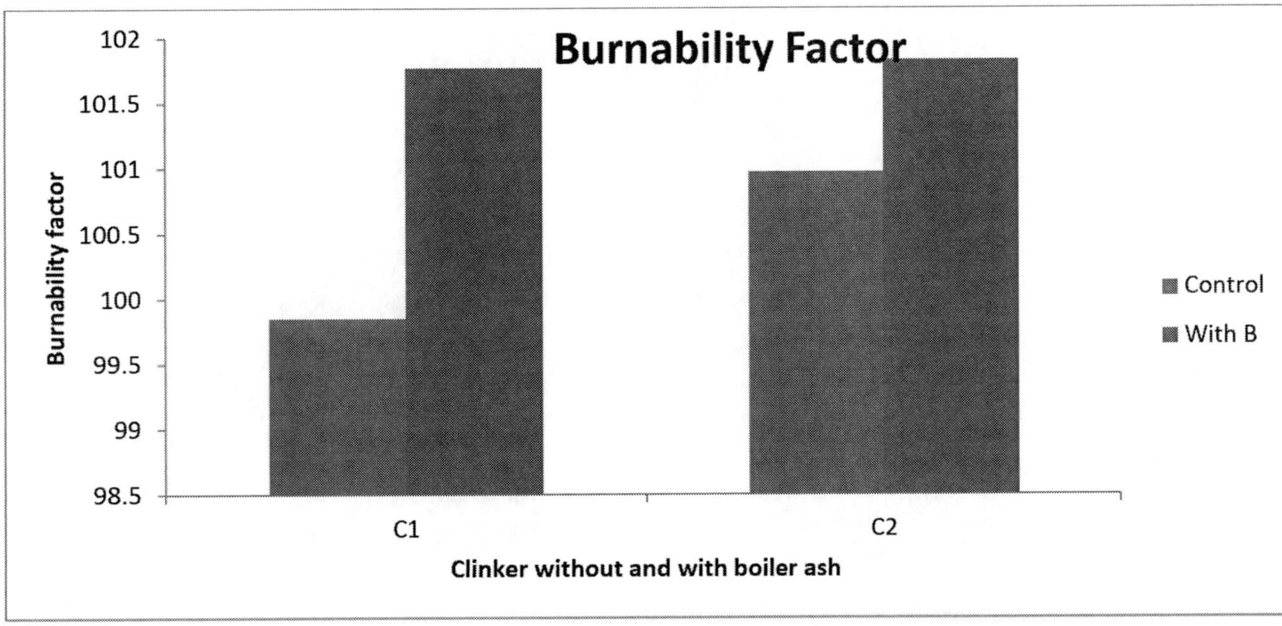


**Figure 3.2: Burnability Factor of clinker without ash (control) and with boiler ash (B)**

Figure 3.3 shows the change in Liquid potential at 1450°C before and after the addition of boiler ash into the raw meal with an average increase from 22.33-23.79%. Liquid potential increased with the addition of boiler ash due to high Al and Fe content compared to limestone and clay. Aluminium and Fe govern the amount of liquid formed and they determine the rate of clinker mineral formation by being the transport medium for reactants. Basically boiler ash acts as a flux, assisting the development of liquid phase and improving its transport properties. The liquid has a critical role in clinker nodulization, as well as clinker minerals development and properties. Without the presence of liquid, the conversion of $C_2S$ and free lime to $C_3S$ at normal clinkering temperatures would be almost impossible. Higher values can be damaging to most refractory bricks in the absence of a stable coating.

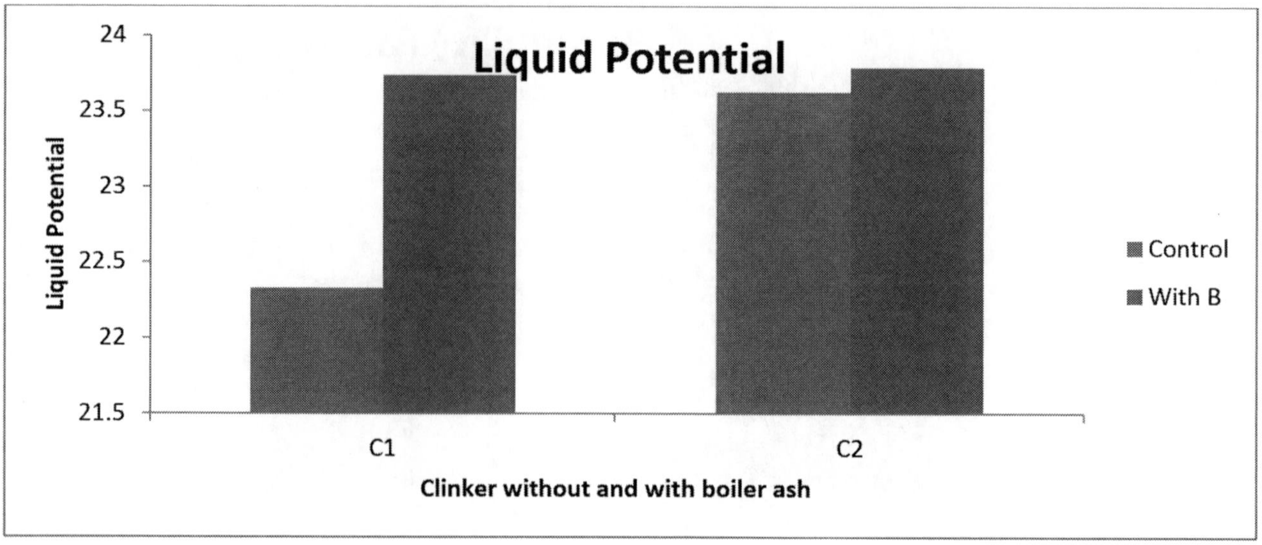


**Figure 3.3: Liquid potential of clinker without ash (control) and with boiler ash (B)**

### 3.3 Plant Trial

The measured loss on ignition (LOI) of boiler ash was found to be 30.7%. This approximates to 28% residual carbon within the ash. In addition to this, $SO_3$ content in the ash is high, B1, 1.69 and B2, 1.39 (Table 3.2) and therefore imposing a higher risk of condensation of sulphur within the colder areas of the kiln, resulting in the formation of undesired rings.

The high carbon content of the boiler ash poses the greatest risk, that of potential electrostatic precipitator (ESP) explosion should the carbon be converted to carbon monoxide leading to high concentrations of carbon monoxide in the ESP. With high loads of carbon monoxide in the ESP any spark is likely to trigger a catastrophic explosion in the ESP. Pertinent to this fact this risk assessment was done and for the plant trial 1% boiler ash was used as stated earlier.

#### 3.3.1 Effect of ash on fuel consumption

Figure 3.4 gives a comparison of the amount of coal used in clinker production with and without the use of boiler ash. 10% drop in coal consumption was observed during the trial period as indicated in Figure 3.1. This confirms that addition of ash to the raw mix results in increased burnability as kiln burning zone temperatures were maintained at 1450°C and yet significantly less fuel was used. The clinker produced during the addition of boiler ash also remained constant.

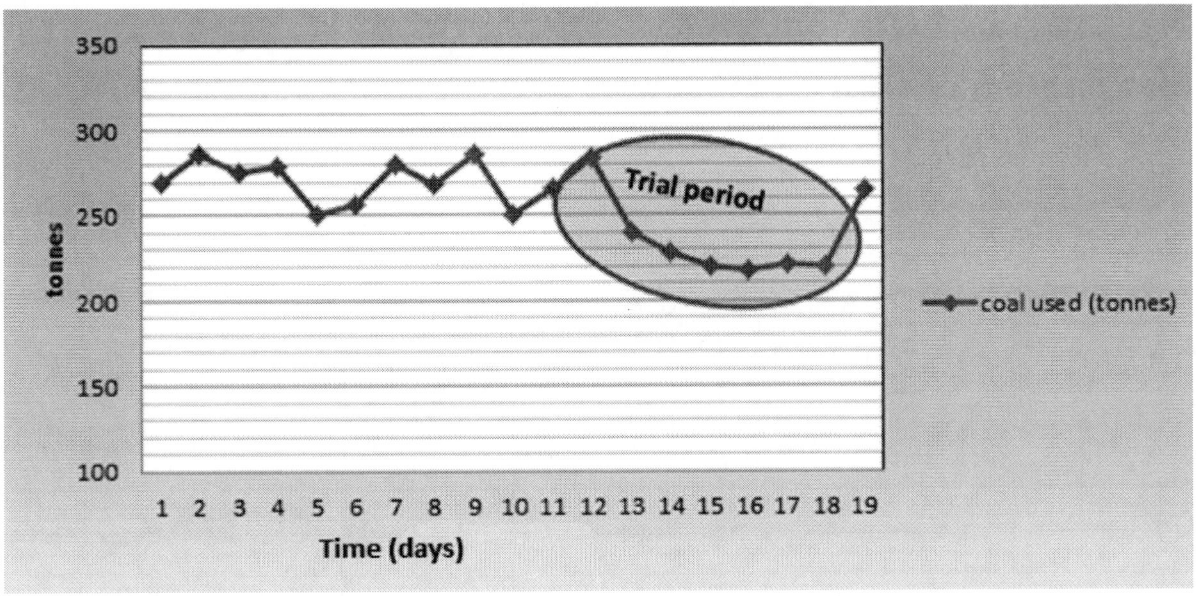


**Figure 3.4: Fuel consumption for a period of 19 days with and without incorporating boiler ash**

### 3.3.2 Effect of ash on clinker production

Figure 3.5 shows clinker production tonnages before and during the trial period. Clinker production remained within the normal range even after addition of ash. It is clear then, that addition of ash has no direct effect on the amount of clinker produced.

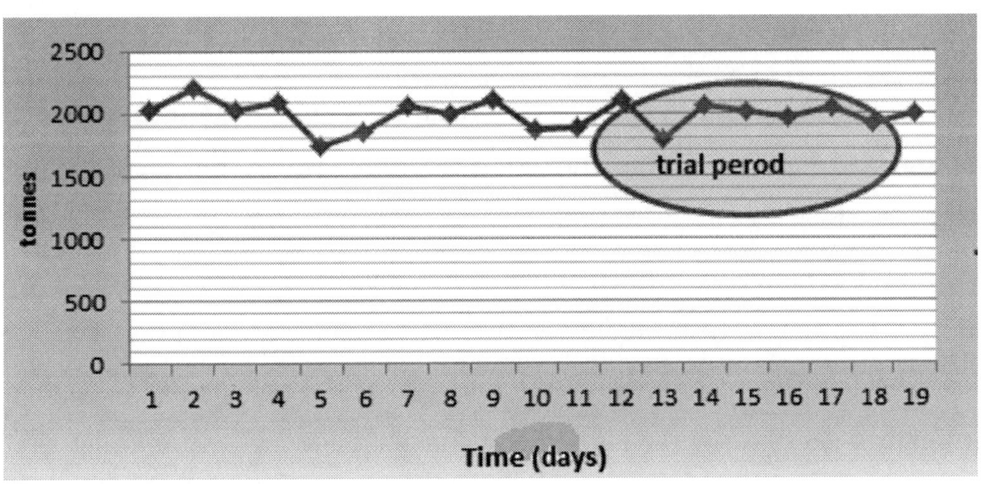


**Figure 3.5: Clinker Production before and during trial**

### 3 3.3 Effect of ash on plant CO

Carbon monoxide was generally high even before the trial as can be seen in Fig 3.6. Addition of ash to the raw did not indicate a significant change in the CO levels.

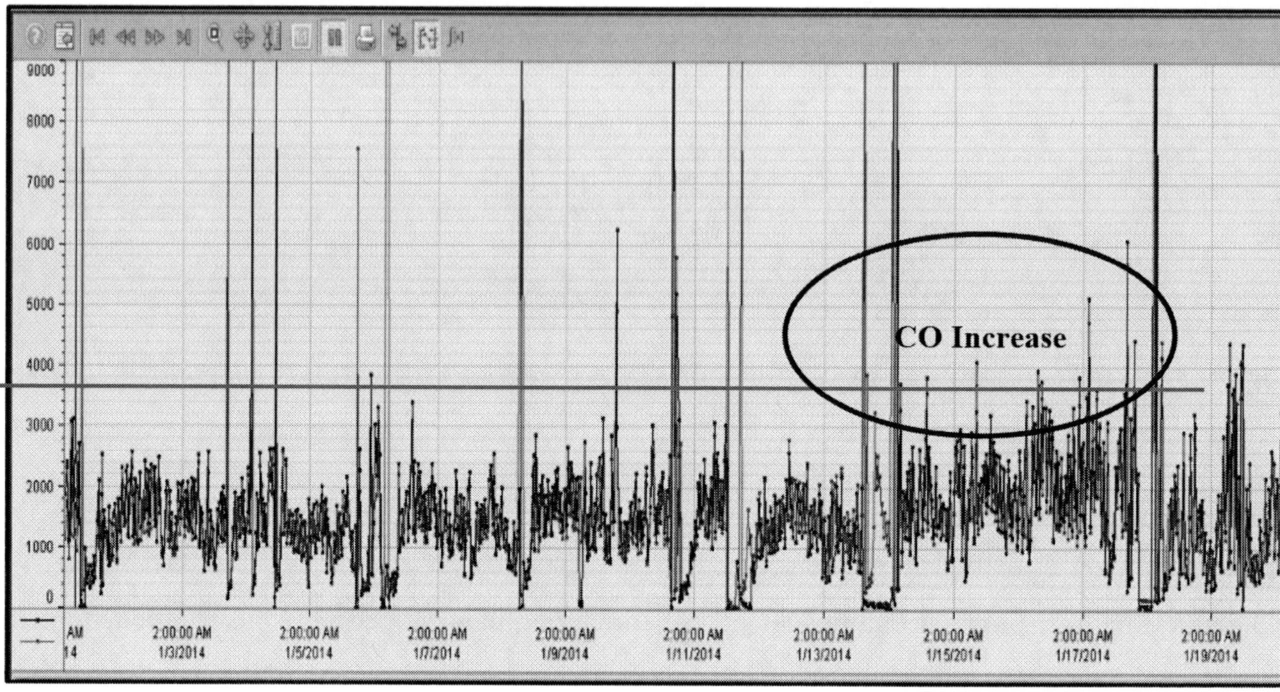


**Figure 3.6: Plant CO before and during trial**

### 4 Conclusions and recommendations

The use of boiler ash in clinker manufacture has a beneficial effect on clinker quality as it increases the liquid phase, which is the transport medium for the clinkering reactions. An overall increase in liquid potential was achieved with a rise from 22.33 – 23.79%. In addition to this, boiler ash incorporation contributed to high $C_3S$, which is a characteristic of high quality clinker. Boiler ash is a good source of Aluminium which improves the granulometry of clinker, resulting in reduced clinker dust. The incooperation of boiler ash also contributed to a 10% decrease in coal consumption. Incorporation of boiler ash in clinker manufacture is therefore advantageous to both the cement industry and the environment. In Zimbabwe, there are lots of boiler ash landfills which are regarded as waste, thus such research could be beneficial to our country so as to reduce the environmental effects of the trace metals in the ash and the amount of land taken up by these landfills.

It is recommended that power stations need to improve their boiler efficiencies so as to combust the coal more effectively resulting in lower carbon content within the ash. The effects of using ash on grinding media and kiln lining could be investigated as well as the effects of trace metals (which still remain entrained in the boiler ash) to the product and the environment.

**References**


1. Baruya, P. and Kessels, J. (2013). *Coal prospects in Botswana, Mozambique, Zambia, Zimbabwe and Namibia.* IEA Clean Coal Centre.

2. Lam C. H. K., Barford J. P and McKay G. (2010). *Utilization of incineration waste ash residues in Portland cement clinker.* Chemical Engineering Transactions, 21, 757-762.

3. Ahmaruzzaman, M. (2010 ). *A review on the utilization of fly ash.* Prog. Energy Combust. Sci., 36: 327-363.

4. Pairsh, R. (2002). *Fly Ash Concrete: Future Foundations?* Southern Illinois University Carbondale. Honors Theses. Paper 141.

5. Rust, D. E. (2008). *Production of Low energy, 100% by-product cement utilizing coal combustion products.* University of Kentucky Masters Theses. Paper 541.

6. . Noor-ul-Amin. (2014). *A multi-directional utilization of different ashes.* RSC Advances 4, 627-669.

7. Kula I., Olgun A., Erdogan Y and Sevinc V. (2001). *Effect of colemanite waste, coal bottom ash and fly ash on properties of cement.* Cement and concrete research, 31, 491-494.

8.Tang, P., Florea, M.V.A., Spiesz, P.R. and Brouwers, H.J.H. (2014). *The application of treated bottom ash in mortar as cement replacement. Proceedings of the EurAsia Waste Management Symposium 2014, 28-30 April 2014, Istanbul, Turkey*, Yildiz Technical University (YTU) and ISTAC Inc, pp. 1077-1082.

9. Pan J. R., Huang C., Kuo J-J and Lin S-H. (2014). *Recycling MSWI bottom and fly ash as raw materials for Portland cement.* International Journal of Concrete Structures and Materials, 8 (2), 173–182.

10. Mohebbi, M., Rajabipour, F. and Scheetz, B. E. (2015). *Reliability of Loss on Ignition (LOI) Test for determining the unburned carbon content in fly ash.* World of Coal Ash (WOCA) Conference in Nashville, TN

11. Johansen, V. and Bhatty, J. I. *Fluxes and mineralizers in clinkering process.* In: J.I. Bhatty, F.M. Miller, and S.H. Kosmatka (eds.), Innovations in Portland cement manufacturing, 2004, Section 3.5, 369-402.

12. Khattak, Z., Ahmad, J., Ali, H. M. and Shah, S. (2013). *Contemporary Dust Control Techniques in Cement Industry, Electrostatic Precipitator-A Case Study.* World Applied Sciences Journal, 22(2), 202-209.

13. Ghosh, S, N. (1991). *Cement and Concrete Science and Technology*, Volume 1. Progress in Cement and Concrete.

14. Banerjea, H. N. (1985). *Technology of Portland Cement and Blended Cements.* Second Edition. Wheeler Publishing.

TECHNISCHE UNIVERSITÄT
BERGAKADEMIE FREIBERG
The University of Resources. Since 1765

IEC EVT

# Environmental and Economic Evaluation of the Production of Organic Platform Chemicals from different Feedstock

**Florian Keller, Christian Wolfersdorf, Bernd Meyer**
**Institute of Energy Process Engineering & Chemical Engineering (IEC),**
**TU Bergakademie Freiberg**

2016 International Pittsburgh Coal Conference, Cape Town
August 8 – 12, 2016

---

## Motivation

IEC EVT

**Motivation for investigation in feedstock alternatives**

- Availability of alternative feedstock
- Import dependancy
- Availability of renewable power
- Political conditions
- Technological advancements

**Feedstock for organic chemical industry in Germany 2013 [VCI 2016]**

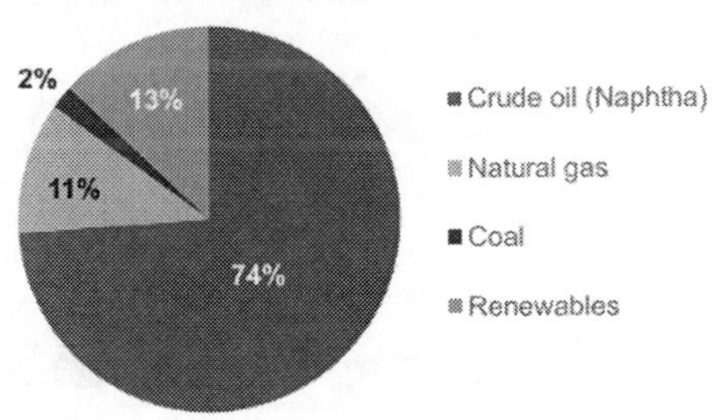


**→ Exemplary case study for production of olefins from different feedstock**

## Structure

1. **Case study framework**
2. Technological assessment
3. Environmental assessment
4. Economic assessment



## Four Olefin Production Pathways

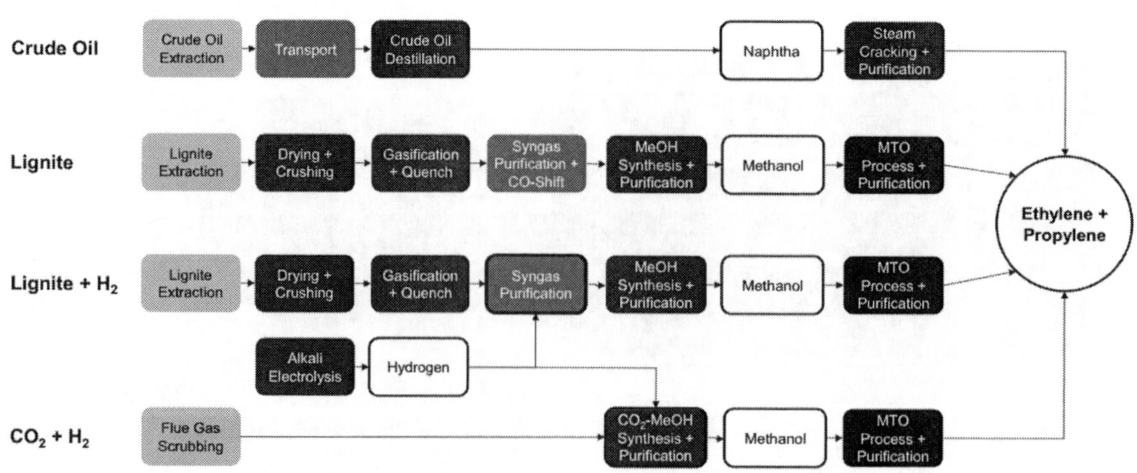

## Framework Conditions

### Technological Evaluation

- Uniform total plant capacity (1 Mio t Olefins / year)
- Polymer-grade olefin production (ethylene > 99.9 %, propylene > 99.5%)
- Excess steam converted to electric power

### Ecological Evaluation

- Life Cycle Assessment, effect based methodology (CML 2001)
- Cradle-to-gate approach, Construction of production facilities not considered
- Mitigation of emissions and benefit of by-products considered ($CO_2$ and power)

### Economic Evaluation

- CAPEX based on scaling of reference facilites (capacity, time)
- OPEX includes fixed and variable operation costs, personnel, fuel and energy costs
- Production cost include side-product revenues (sulfur, power) and financial costs (interest rates)



## Methodology

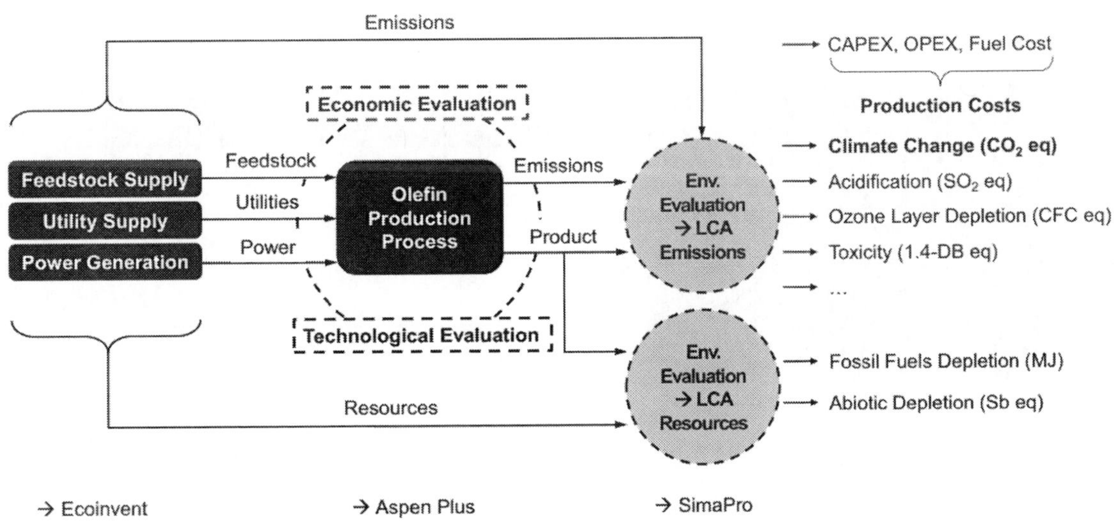

## Structure

1. Case study framework
2. **Technological assessment**
3. Environmental assessment
4. Economic assessment



## Technological Assessment

| | | Naphtha | Lignite | Lignite + $H_2$ | $CO_2$ + $H_2$ |
|---|---|---|---|---|---|
| **Feed demand** | [kg (wf)/ kg olefins] | 1.67 | 9.21 | 2.17 | 6.03 |
| **Process power demand** | [$MJ_{el}$/ kg olefins] | -4.65 | 2.96 | -1.06 | 5.11 |
| **$H_2$ demand** | [kg/ kg olefins] | - | - | 0.32 | 0.59 |
| **$H_2$ power demand** | [$MJ_{el}$/ kg olefins] | - | - | 57.76 | 106.95 |

Structure

1. Case study framework
2. Technological assessment
3. **Environmental assessment**
4. Economic assessment



Power Production Scenarios

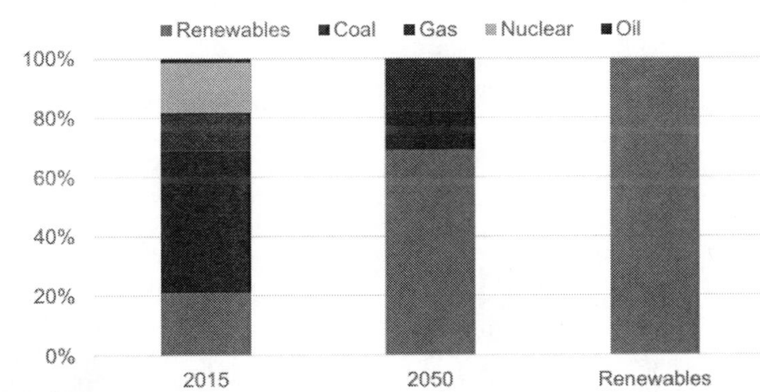


| | 2015 | 2050 | Renewables |
|---|---|---|---|
| **Fossil Fuels Depletion** [MJ/ $MJ_{el}$] | 1.900 | 0.901 | 0.048 |
| **Global Warming Potential** [kg $CO_2$eq/ $MJ_{el}$] | 0.169 | 0.074 | 0.006 |

Sources:
Entwicklung der Energiemärkte– Energiereferenzprognose, Prognos AG, Study 2014;
Power Production Statistic, BDEW 2016

## Global Warming Potential in 2015 Power Scenario

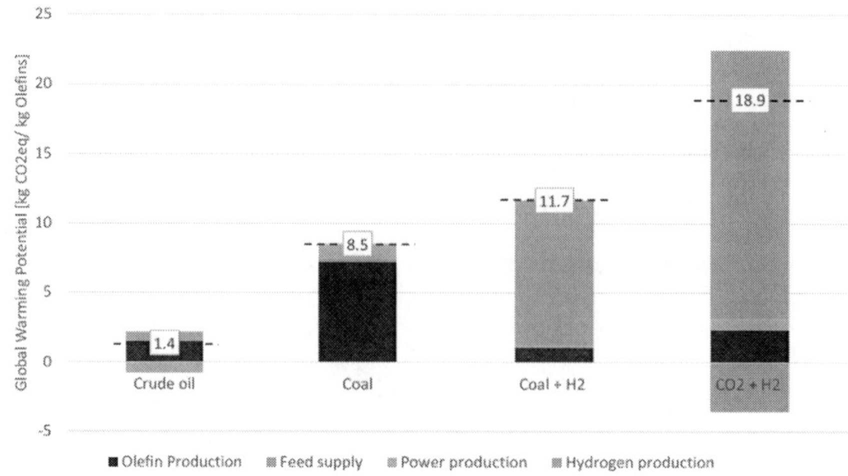




## Global Warming Potential in 2050 Power Scenario

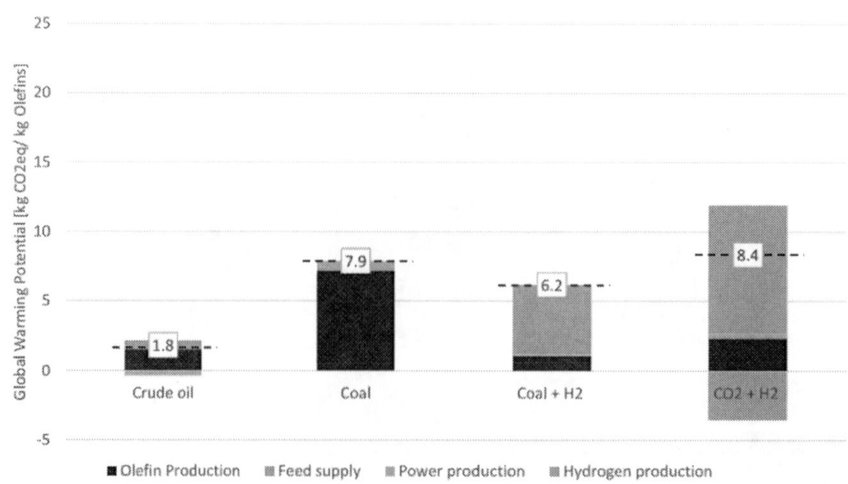

## Global Warming Potential in Renewables Power Scenario

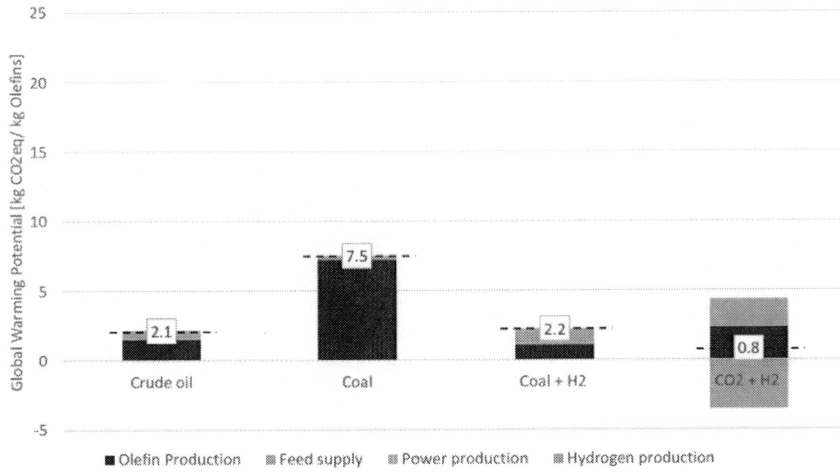




## Structure

1. Case study framework
2. Technological assessment
3. Environmental assessment
4. **Economic assessment**

## Olefin Production Costs current situation

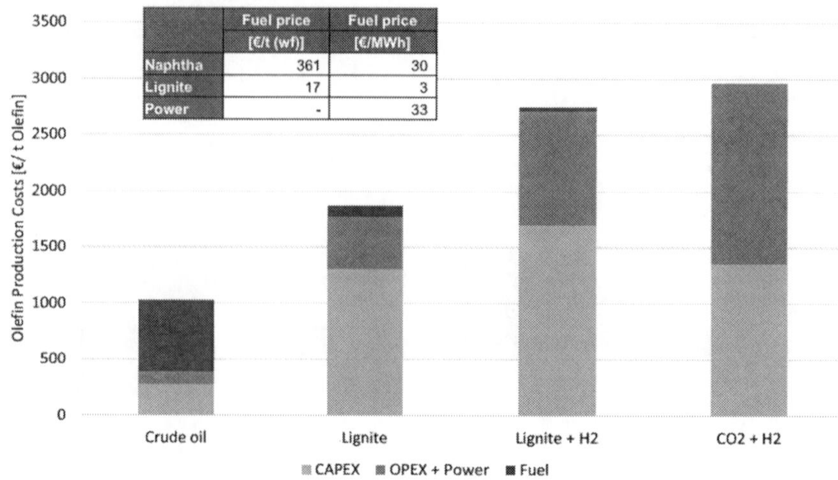


| | Fuel price [€/t (wf)] | Fuel price [€/MWh] |
|---|---|---|
| Naphtha | 361 | 30 |
| Lignite | 17 | 3 |
| Power | - | 33 |

Florian Keller "Environmental and economic Evaluation of the Production of organic Platform Chemicals from different Feedstock", Pittsburgh Coal Conference 2016, Cape Town 15

## Olefin Production Costs alternate scenario

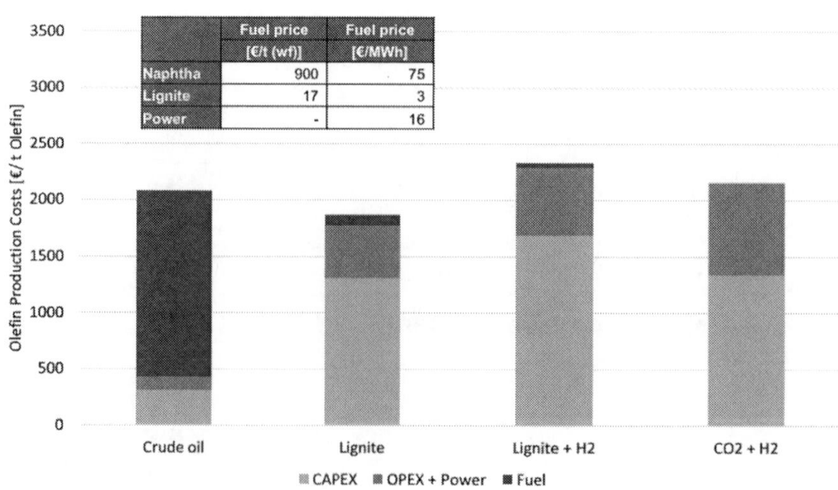


| | Fuel price [€/t (wf)] | Fuel price [€/MWh] |
|---|---|---|
| Naphtha | 900 | 75 |
| Lignite | 17 | 3 |
| Power | - | 16 |

Florian Keller "Environmental and economic Evaluation of the Production of organic Platform Chemicals from different Feedstock", Pittsburgh Coal Conference 2016, Cape Town 16

IEC EVT

**Case Study Olefin Production**

**Today:** → Crude oil is most environmental friendly and most economic feedstock for olefin production

**Tomorrow:** → Coal and $CO_2$ could become viable feedstock options by integration of hydrogen from renewable energies

**But:** → High availability of renewable energy necessary

→ High crude oil price and low renewable energy price necessary

→ more realistic: CAPEX reduction necessary



TECHNISCHE UNIVERSITÄT
BERGAKADEMIE FREIBERG
The University of Resources. Since 1765

IEC EVT

# Thank you for your attention.

Florian Keller
Institute of Energy Process Engineering and Chemical Engineering (IEC)
TU Bergakademie Freiberg
Email: Florian.Keller@iec.tu-freiberg.de
web: www.iec.tu-freiberg.de

## Fossil Fuel Depletion in 2015 Power Scenario

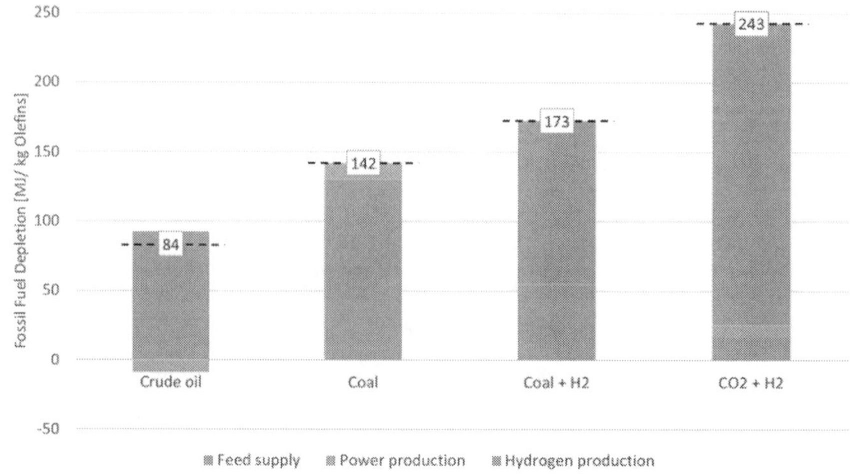


Florian Keller "Environmental and economic Evaluation of the Production of organic Platform Chemicals from different Feedstock", Pittsburgh Coal Conference 2016, Cape Town 19

## Fossil Fuel Depletion in 2050 Power Scenario

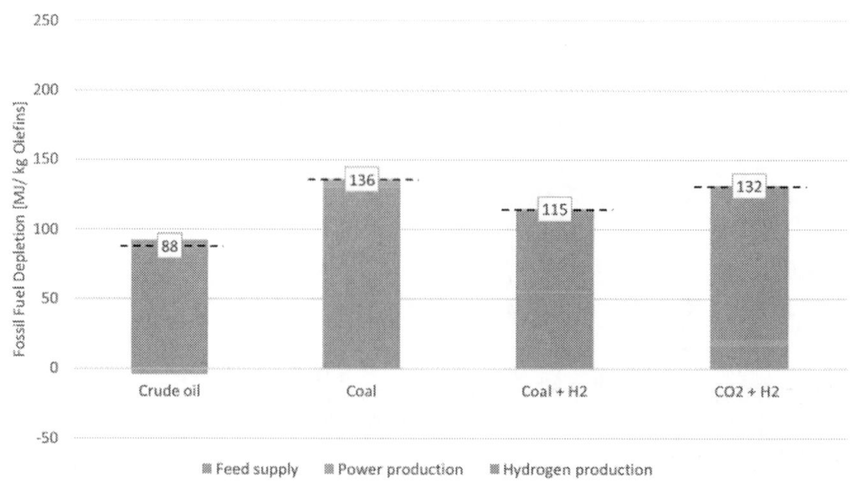


Florian Keller "Environmental and economic Evaluation of the Production of organic Platform Chemicals from different Feedstock", Pittsburgh Coal Conference 2016, Cape Town 20

## Fossil Fuel Depletion in Renewables Power Scenario

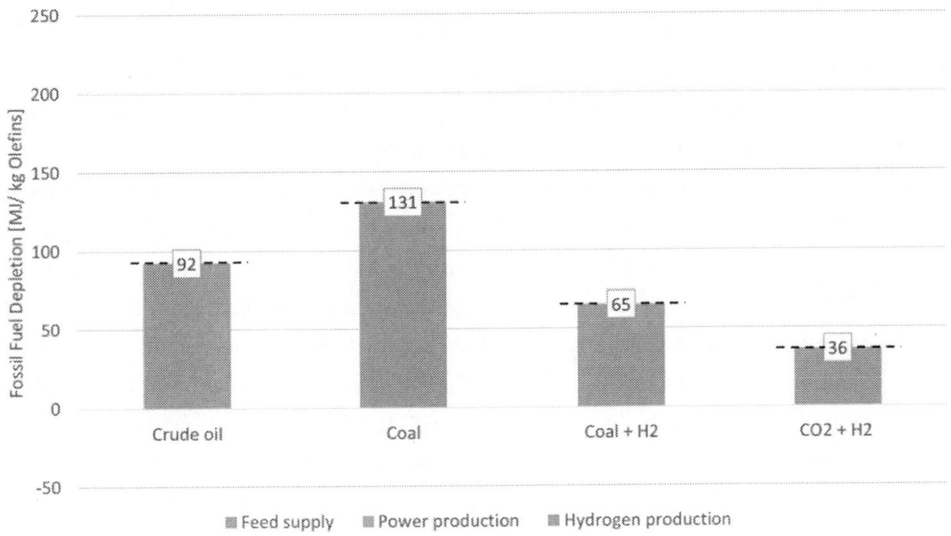


Florian Keller "Environmental and economic Evaluation of the Production of organic Platform Chemicals from different Feedstock", Pittsburgh Coal Conference 2016, Cape Town

21

# Process Simulation and Assessment of Coke Oven Gas to Methanol Integrated with $CO_2$ Recycle

*Min-Hui Gong, Qun Yi, Yi Huang, Guo-sheng Wu, Jie Feng, Wen-Ying Li**

*Key Laboratory of Coal Science and Technology (Taiyuan University of Technology), Ministry of Education and Shanxi province, Taiyuan 030024, PR China*

*Training Base of State Key Laboratory of Coal Science and Technology Jointly Constructed by Shanxi Province and Ministry of Science and Technology, Taiyuan 030024, PR China*

* *Corresponding author. E-mail: ying@tyut.edu.cn*

**Abstract:** COG as a byproduct of coking industry abounds in China, half of it is burned out for heat required of coke oven, the other half is usually combusted or discharged into air which is not only a waste of the hydrogen-rich valuable resource but also causing some environmental problems. Though methanol synthesis from COG is mature and a good way for COG utilization, $H_2/CO$ ratio of nearly 5 is too high for methanol synthesis, to which $H_2/CO$ close to 2 is optimal. Generally, to tackle this issue, carbon-rich gas making subsystems is adopted to supply carbon resource and balance the surplus $H_2$. From this way, it can get more $H_2$ conversion and methanol output, but it will undoubtedly cause system complexity and investment increase. In this work, a newly designed COG to methanol process with $CO_2$ recycle was proposed and simulated by Aspen Plus software. The new system maintains the same merits, and also presents the potential in terms of reduction of system complexity, curtailment of capital investment, and decrease of energy consumption compared with the traditional carbon-supplementary method. Two projects concerning whether H2 is separated from COG or not are presented, and they are marked by project CWOSH (COG without separation $H_2$) and CWSH (COG with separation $H_2$), respectively. In the CWOSH, COG is splits into two parts, one for dry reforming and the other for combustion, and syngas from reforming reactor is directly sent for methanol synthesis. In project CWSH, $H_2$ is firstly separated from the COG, and then COG is divided into two streams likewise in CWOSH. However, the separated $H_2$ mixed with the reformed gas to produce syngas with the proper $H_2/CO$ for methanol synthesis in CWSH. In both projects, the unreacted syngas partially recycles back to react and the other part to be mixed with COG for combustion with oxygen. The heat from combustion is supplied to coking process and reforming reaction. High content of $CO_2$ in flue gas is roughly separated by a cooler, and then recycled to reform with COG, also to coke oven for temperature adjustment. Optimization of the two projects involving element utilization, energy conversion, and economy benefit is given by adjusting variables, like recycle ratio, COG reforming ratio, etc. It is revealed that 3-8% energy saving, about 50% $CO_2$ emission reduction and 5-10% internal rate of return (IRR) increase will be obtained in both projects, versus traditional COG to methanol process. And project CWSH will get more $CO_2$ emission reduction and economic benefit but lower energy saving than CWOSH.

***Keywords:*** coke oven gas; $CO_2$ recycle; methanol; $H_2$ separation; techno-economic

# 1. Introduction

China has the biggest production and consumption amount of coke around the world, accompanied by a large quantity of COG. In 2015, coke production reach to 448 million tons[1], containing $H_2$ (55-60 vol.%), $CH_4$ (23-27 vol.%), CO (5-8 vol.%) and $N_2$ (3-5 vol.%) along with some impurities like $H_2S$, $NH_3$, COS, $CS_2$[2, 3]. Half of it is burnt to supply heat for the combustor in independent coking plants. Direct combustion is not only a waste of the valuable resource but also causing environmental pollution[4]. A better and widely alternative for COG usage is to synthesize methanol[3]. Methanol is a platform chemical, which can be further conversed into more than 20 kinds of following chemicals. Besides, methanol is also a potential liquid fuel[5]. Olah's proposition "methanol economy", may start a new fuel manufacturing era, especially in China and USA[6]. COG to methanol (CTM) is one of the feasible routes to utilize COG properly.

CTM should first involve COG reforming to syngas which is the direct raw material gas for methanol synthesis, besides, syngas with a $(H_2\text{-}CO_2)/(CO+CO_2)$ around 2.0-2.1 is more suitable[6-8], below or above the ratio are bad for methanol synthesis. But the value of COG is around 5.2. All the reforming technologies can change $(H_2\text{-}CO_2)/(CO+CO_2)$ to a lower value. To balance the H/C ratio of the syngas, carbon supplementary is adopted from the view point of element complementary and the gasification subsystem addition is often the case because coal gasification gas has a high content of carbon (CO and $CO_2$). There are several commercial plants that have applied this method to adjust the $(H_2\text{-}CO_2)/(CO+CO_2)$ ratio, COG conversion efficiency and output of methanol are enhanced. Whereas, addition of a subsystem means addition of a serials of units including gasification unit, gas cleaning unit, and so on, which would make the system more complex and increase a high capital investment. There exists a contradiction in the utilization of COG from coking plant that half of it is combusted for coking heat supply, therefore with some $CO_2$ output; the other half for methanol synthesis should supplement carbon for hydrogen-carbon balance, with addition of carbon input. The trade-off between carbon emission and carbon supplementary can be realized by the $CO_2$ recycle based COG to methanol synthesis. Part of COG from coking oven mixed with part of unreacted syngas substitutes half COG from the coking oven to be introduced to combustor for heat supply. Because of pure-oxygen instead of air combustion, the $CO_2$ generated has a high content and can be easily separated from the flue gas. The separated $CO_2$ is partly recycled to reforming unit, reacted with COG to produce syngas, and partly is recycled to the coking combustor as the dilute to adjust the temperature in it.

Dry reforming is selected as the COG conversing technology. Though it has not been industrialized yet with the commercial nickel catalyst because of severe carbon deposition and sintering[9, 10], much research has been taken in laboratory and good catalytic effect has been obtained[11-13]. It is promising that better catalyst and synthesis method will be found in the future. As a technology on one hand to converse greenhouse gas $CH_4$ and $CO_2$[3], dry reforming is receiving higher and higher attention, on the other hand, it is a route to get a syngas with $H_2/CO$ lower than 1 which right meets hydrogen carbon balance in COG reforming for methanol synthesis[2]. And the outcome has been demonstrated in J.M. Bermúdeza etc.'s work[14]. J.M. Bermúdeza etc.[15] and Qian Yu's group[2, 16] have also chosen the technology to converse COG in their proposed conceptual processes, low carbon emission as well as high COG conversion efficiency can be got. Some research works also take $H_2$ separation before COG reforming into consideration[2, 16, 17]. $H_2$ as a product, it is better to remove it from the reactant system for the convenience of $CH_4$ conversion efficiency, however, with a sacrifice of the additional capital investment of pressure swing adsorption unit (PSA). So it is worthy to have an extensive techno-economic analysis to make a comparison between the two scenarios concerning $H_2$ separation or not.

### 2.1 System process description

COG from coke oven is split into two parts, one for dry reforming, the other for combustion. Syngas from reforming reactor is directly sent for methanol synthesis.

The unreacted syngas partially recycles back to react and the other part to be mixed with COG for combustion with oxygen. The heat from combustion is supplied to coking process and reforming reaction. High content of $CO_2$ in flue gas is separated in a flash tank, and then recycled to reform with COG, also to coke oven for temperature adjustment. As for CWSH, $H_2$ separation is completed by PSA. All the hydrogen in COG is separated apart, then mixed with reformed gas from reforming reactor. There is no difference between the routes from the syngas to methanol in the two scenarios. Considering most parts are the same in CWOSH and CWSH, the common sections are depicted as follows.

## 2.2 System boundary declaration

For the convenience of the following assessment of the system, system boundary is of vital to be declared. In this work, the system scale is set at an equivalent production capacity of 2 million tones coke per year. Concerning that 400 $Nm^3$ COG can be obtained with one ton coke production, the molar rate of raw material COG is supposed to be 4509.3 kmol/h. Coke oven is deemed as two independent parts, and the combustor of the coke oven is included within the boundary while the coking chamber is excluded. The boundary encloses COG from the coke oven, COG purification, COG reforming, methanol synthesis, crude methanol rectifying, mixture gas composed of COG and unreacted syngas combustion and necessary electricity and steam input. As for the materials involved in mass balance of the system, COG and oxygen are the raw materials and methanol is the single product, apart from which there is some exhaust gas out from the system. Emphasis on the assumptions that if the LHV of the combustible gas mixture meets the demand of coking process, the coking process will proceed normally, is indispensable for the separated view on the coking chamber and combustor. The boundary of the two scenarios are depicted as Fig. 1 and Fig. 2.

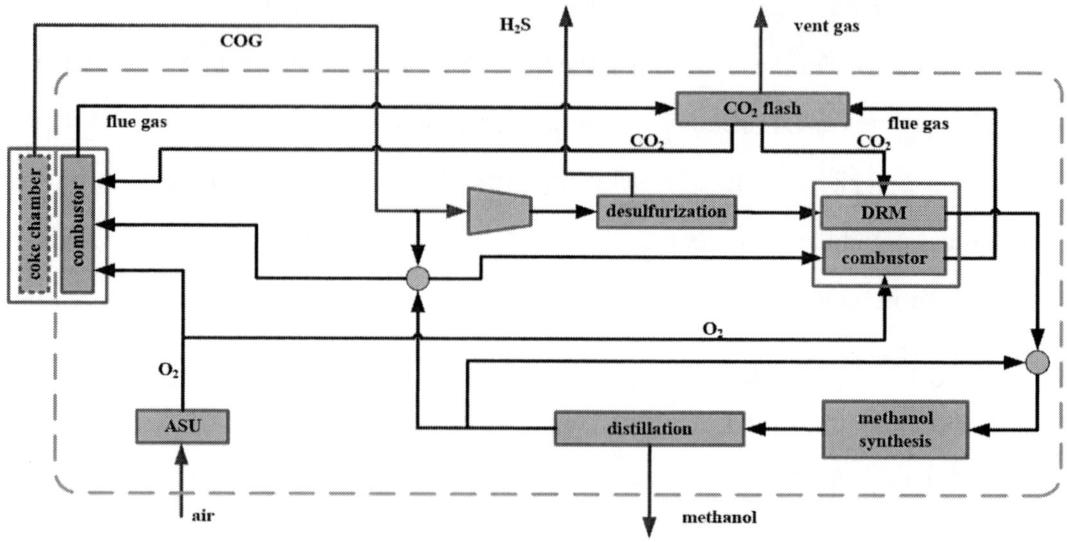


**Fig. 1 Boundary of CTM (CWOSH) process**

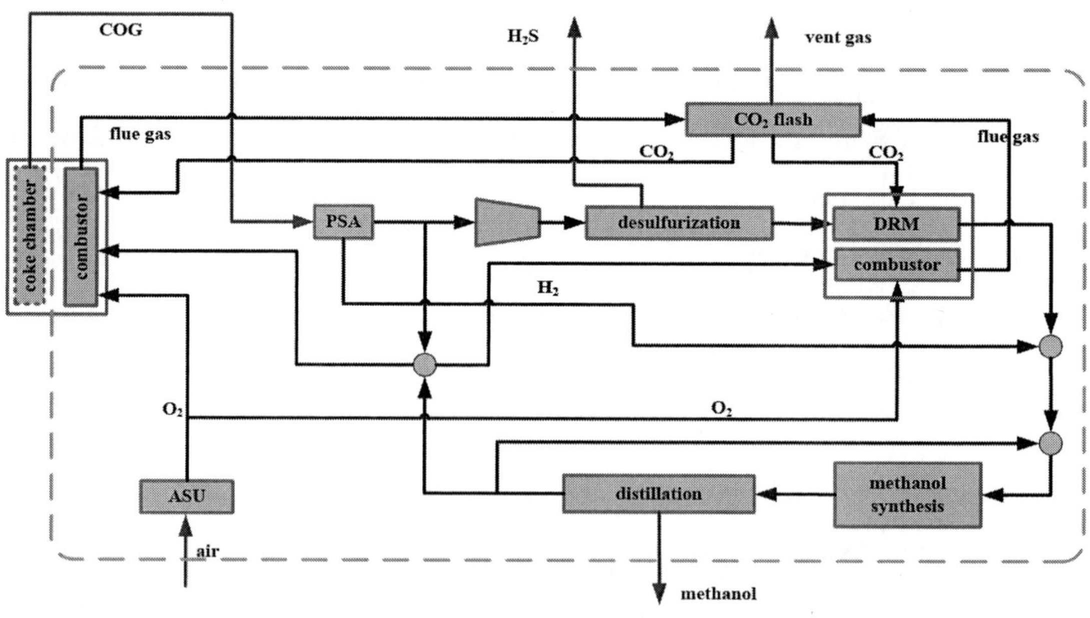


**Fig. 2 Boundary of CTM (CWSH) process**

The whole processes are modeled by Aspen Plus 11.1 which is a useful software for chemical engineering process simulation[18]. The detailed models for each reactor or unit are listed in Table 1. Each model has been verified according to literatures and experimental data[2, 14, 17]

## 3. Techno-economic analysis

In order to have a thorough understanding of the proposed systems, a techno-economic analysis is most widely applied. Some indexes have been selected to measure the performances of the systems. As to the process, COG is more aimed to chemical synthesis rather than combustion, so the key variable is naturally the ratio of COG reforming to combustion ($\lambda$). Its range is from 0.55/0.45-1.00/0. For the sake of ensuring ($H_2$-$CO_2$)/($CO$+$CO_2$) equal to 2.0, $\lambda$ determines the amount of $CO_2$ recycled to reforming and the energy needed of reforming, as well as the remaining heat from the unreacted syngas combustion to supply for coke oven combustor. The temperature of the combustor should be controlled around 1200°C, which accords to the industrious operational parameter. To obtain this aim, the amount of $CO_2$ recycled to the combustor is adjusted.

### 3.1 Methods

#### 3.1.1 $CH_4$+$CO_2$ conversion efficiency

In this work, $CH_4$ and $CO_2$ are main greenhouse gas (GHG), $CH_4$+$CO_2$ conversion efficiency can show the potential of emission reduction of the system. It is expressed as Eq (4):

$$Y_C = \left[\frac{(F_{CH_4}+F_{CO_2})_{in}-(F_{CH_4}+F_{CO_2})_{out}}{(F_{CH_4}+F_{CO_2})_{in}}\right] \times 100\% \qquad (4)$$

Where Yc is $CH_4$+$CO_2$ conversion efficiency, $(F_{CH_4} + F_{CO_2})_{in}$, $(F_{CH_4} + F_{CO_2})_{out}$ denote $CH_4$+$CO_2$ molar flow rate that enter in and out of the reforming reactor, respectively.

#### 3.1.2 Energy and exergy efficiency

Energy efficiency is of importance in evaluating chemical systems from the point of first law of thermodynamics. It is defined as the Eq. (5).

$$\eta_1 = \frac{\Sigma E_o}{\Sigma E_{in}} \times 100\% \tag{5}$$

$\Sigma E_o$ and $\Sigma E_{in}$ denote total energy out and in the system. Energy calculation in this work is based on the LHV of all the material flows. Compression power and electricity consumption is based on the results of the compressor models in the Aspen and estimable data from the equivalent scale reference.

Exergy is a concept based on the First and Second Laws of Thermodynamics, which allows measuring the quality of energy with a quantifiable description. The exergy of a stream is related to a dead state where exergy of all materials is zero, usually the environmental state is taken as the dead state. The exergy can be divided into three parts, physical exergy, chemical exergy, and mixing exergy[19, 20]. The following equation shows the expression of total exergy.

$$EX_{tot} = EX_{phy} + EX_{chem} + EX_{mix} \tag{6}$$

Where $EX_{tot}$ is the total exergy of a material stream, $EX_{phy}$ is the physical exergy of the stream, which caused by the difference from the current state (temperature of T and pressure of P) to the environmental state (temperature of $T_0$ and pressure of $P_0$). $EX_{chem}$ is the chemical exergy that initiated by the difference of pure components distributions between the material stream and environment in the environmental state. $EX_{mix}$ is generated by the change of mixing pure materials to a mixture containing the same components. In this study, $EX_{mix}$ is neglected for its relative low value. $EX_{phy}$ and $EX_{chem}$ can be calculated with the aid of Aspen.

The exergy efficiency is very instructive to evaluate system feasibility, it is calculated as the following equation.

$$\eta_2 = \frac{\Sigma EX_o}{\Sigma EX_{in}} \times 100\% \tag{7}$$

$\Sigma EX_o$ and $\Sigma EX_{in}$ denote total exergy out and in the system.

### 3.1.3 Economic assessment

To put an industrial plant into operation, economic benefit is of the most important to be considered. In this work, total capital investment (TCI), cost of methanol (COM) and internal rate of return (IRR) are taken as the indexes for economic assessment.

TCI is calculated by using equipment factorial method, which takes total equipment cost (TEC) as 100% and other costs are relative with corresponding factors[21, 22]. TCI is the sum of all the factors multiplied by TEC expressed as Eq. (8).

$$TIC = TEC \times (1 + \Sigma RF_i) \quad (8)$$

Where $RF_i$ is the ratio factor of cost i. In this work, according to China's extant COG to methanol economical analysis, $\Sigma RF_i$ is set to be 2.18.

Each equipment's cost is estimated by scaling factor method. The calculating equation is shown as Eq. (9).

$$I_2 = I_1 \times \theta \times (\frac{Q_2}{Q_1})^n \quad (9)$$

Where $I_1$ and $Q_1$ are the equipment cost and corresponding handling scale of the extant equipment. $I_2$ and $Q_2$ are the equipment cost and corresponding handling scale of the estimated equipment. $\theta$ is the domestic factor. n is the scale exponent. Some key equipments and their costs as well as their relative scales are listed in Table 1.

**Table 1 Investment data for main equipment components**

| Unit | Benchmark | Scale $Q_1$ | Domestic-made factor $\theta$ | $I_1$(M$) | Scaling factor ($n$) | ref |
|---|---|---|---|---|---|---|
| ASU | Oxygen supply | 21.3 kg/s | 0.5 | 45.7 | 0.5 | [21] |
| PSA | Hydrogen supply | 21.3 kg/s | 0.5 | 45.7 | 0.74 | [23] |
| DRM | Mixed gas to reforming | 563 $m^3$/h | 0.65 | 1.6 | 0.67 | [24] |
| Methanol synthesis | Syngas input | 10810 mol/s | 0.65 | 20.4 | 0.6 | [16, 25] |
| Distillation Columns | Methanol production | 87 t/h | 0.65 | 7 | 0.67 | [8] |
| Compressor | Work | 10 MW | 0.65 | 6.3 | 0.67 | [26] |

COM is an index that measures total consumption brought about by methanol production[27]. It is calculated as Eq. (10).

$$COM = \frac{TIC \times CRF + F_{O\&M} + C_{fuel} + C_{power} + C_{labor}}{Methanol\ production} \quad (10)$$

Where $F_{O\&M}$, $C_{fuel}$, $C_{power}$, $C_{labor}$ denote fixed operation and maintenance cost, fuel cost, power cost, labor cost respectively. $F_{O\&M}$ is usually calculated by 4% of TIC[28]. CRF is the capital recovery factor and which is expressed as Eq. (11). $i$ and $N$ are the discount rate (12%) and the plant life (20 years).

$$CRF = \frac{i}{1-(1+i)^{-N}} \quad (11)$$

IRR is another robust economic index to assess the feasibility of a project. If IRR > i, it means the project is feasible to operate and vice versa, the bigger the IRR is, the more feasible the project is. It is defined as Eq. (12).

$$\sum_{t=0}^{N} C_t \times (1+IRR)^{-t} = 0 \quad (12)$$

Here, $C_t$ is net cash flow of year $t$. Assumptions are listed in table 2 of some basic parameters to evaluate the economic indexes.

**Table 2 Assumption for economic analysis**

| Parameter | Value |
|---|---|
| plant availability, d/y | 330 |
| depreciation life, y | 10 |
| design time, y | 1 |
| construction life, y | 1 |
| plant life, y | 20 |
| discount ratio, % | 12 |
| depreciation rate, % | 8 |
| tax rate, %[29] | 25 |
| methanol price, US$/t[21] | 396.8 |
| average price of industrial electricity, US$/kWh[21] | 0.11 |

### 3.2 Results and discussion

As mentioned above, λ is a key variable in the system. It affects element flow, energy flow and equipment cost.

#### 3.2.1 Technologic performance

To set $(H_2-CO_2)/(CO+CO_2)$ steady at 2.0, it is adjustable of $CO_2$ recycled to reforming with varied λ. Likewise, to meet the energy supply for coke oven and reformer, the recycle ratio of unreacted gas varies with the change of λ. In two scenarios, $CH_4+CO_2$ conversion efficiency is higher in CWSH (81.7-90.7%) than CWOSH (79.1-79.9%). Under different λ in CWSH, the separated $H_2$ from COG are all mixed with the reformed gas from reformer. To ensure the $(H_2-CO_2)/(CO+CO_2)$ steady under all the inspected λ, $CO_2$ recycled for reforming increases, the $CH_4$ conversion efficiency increases as well, while $CO_2$ conversion efficiency decreases with the increasing of λ. So $CH_4+CO_2$ conversion efficiency represents a trend of uprising first and then downward. $CH_4$ conversion efficiency is more with $H_2$ separation than without $H_2$ separation under the same λ. So in CWSH, the more λ, the less $CH_4$ whose standard enthalpy of combustion is -802.9 kJ/mol, nearly four times of that of CO and $H_2$, left in the reformed gas and unreacted gas. To meet the requirement of combustion heat, the recycle ratio of unreacted gas increases with the increasing of λ. While in CWOSH, under every λ, relative composition of $CH_4$ in the syngas is relatively steady, and so are the $CH_4$ and $CO_2$ conversion efficiency. With more COG reformed, recycled unreacted gas decreases to spare more for combustion. The $CH_4+CO_2$ conversion efficiency, recycle ratio of unreacted gas and $(H_2-CO_2)/(CO+CO_2)$ of two scenarios are shown as Fig. 5 a(i), a(ii). The temperature of coke oven is necessary to be set at ~1200 °C, which is a normal temperature in conventional coke oven combustor. $CO_2$ recycled to coke oven controls the temperature. $CO_2$ recycled to reforming increases with the λ increases. The changing trends of the mentioned above in scenario CWOSH and CWSH are represented as Fig. 5 b(i), b(ii). While more COG or COG depleted $H_2$ sent for reforming, energy and exergy efficiency are less obtained. This may be more electricity consumption of gas compression and air separation. $H_2$ separated scenario has a higher energy and exergy efficiency then the one without $H_2$ separation, it demonstrates that surplus $H_2$ in the reformer causes more energy consumption. Fig. 5 c(i), c(ii) show the energy and exergy efficiency to the λ in two scenarios. From the point of element and energy efficiency, both scenarios are under best operation on the same condition with 55% COG sent for reforming.

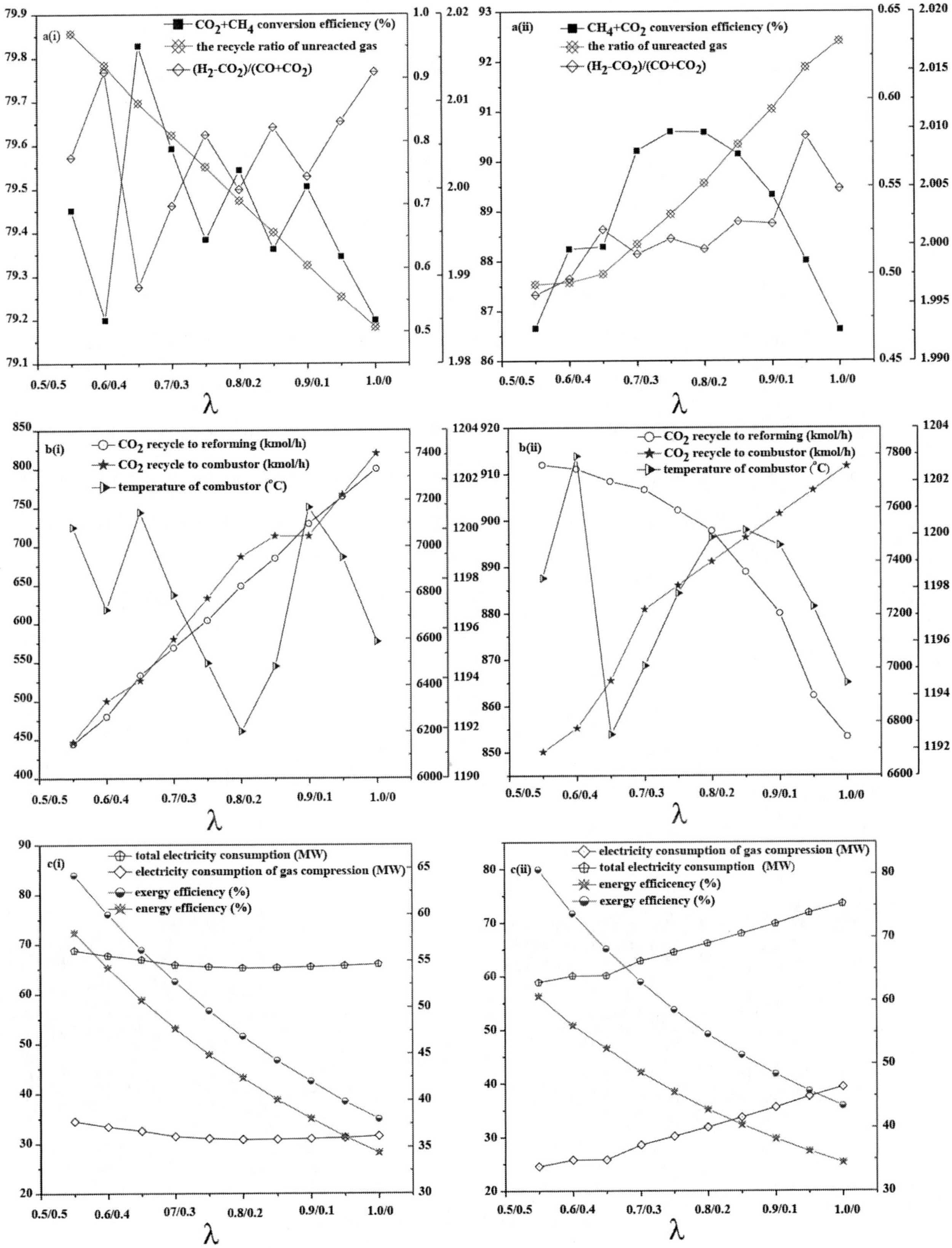


**Fig. 3 Technologic performance**

### 3.2.2 Economic performance

As Fig. 4 show, with more COG introduced to reforming, TCI and COM increase and IRR decreases in two scenarios. The best economic case of CWSH and CWOSH are under the same λ (0.55/0.45). In CWSOH, the lowest TCI and COM are 109.7 M$, 284.7 $/t-methanol with the highest IRR of 28.9%; in CWSH, the values are correspondingly 102.1 M$, 272.7 $/t-methanol and 37.3%. The high $CH_4$ conversion efficiency and low handling capacity of gas volume to reformer make a big economic advantage to CWSH. Though there is additional fixed investment of PSA unit, it does a little discount. The distribution of equipment cost in each scenario is shown as Fig. 7 b(i), b(ii). PSA equipment cost is very low, less than 1% of TEC, while reformer and synthesis reactor costs are lower in CWSH, the amount equal to the cost of PSA. From all the economic indexes, it is evident to conclude that CWSH is better.

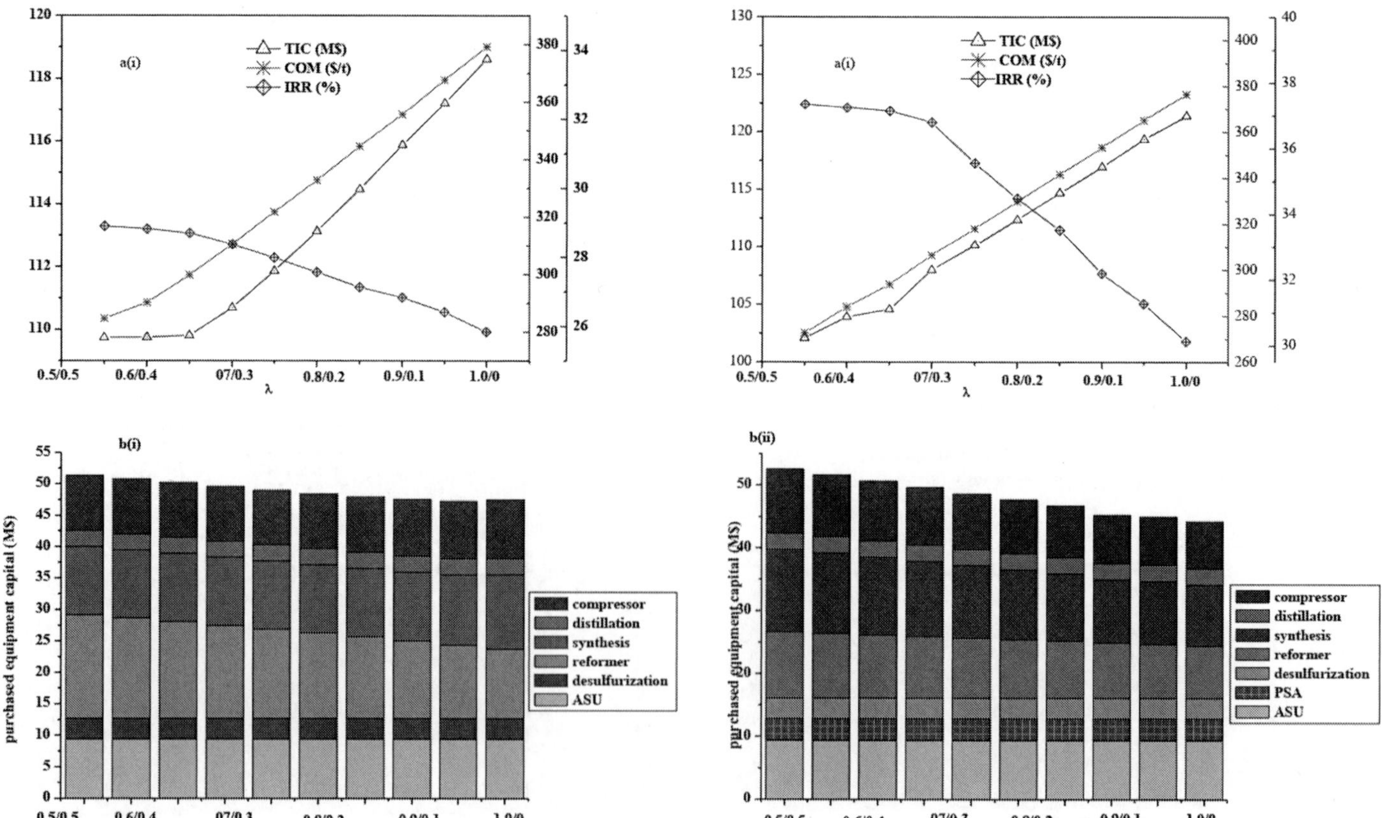


**Fig. 4 Economic performance**

### 3.3 Comparison of similar processes

A thorough comparison has been listed in table 3. In the former work, the comparison between the conventional and carbon-supplementary processes of COG to methanol and the proposed $CO_2$ recycle with tri-reforming technology process. In this work, dry reforming is selected as the technology to utilize $CO_2$ and $CH_4$. By comparison with the two reforming technologies, it is found that the tri-reforming route is a little more productive of methanol than dry reforming; the $H_2$ separated scenario produces more methanol than the one without $H_2$ separation. All the three processes involve ASU, and the handling capacity reaches to 50000 $Nm^3/h$, which generate a high cost though it has been realized localization in China. Dry reforming and tri-reforming may have different configurations but both account for a high (nearly equal to ASU) proportion ( about 25%) of TEC. The TIC and IRR of CWOSH are 29.1%, 28.9% respectively. It is worthy to note that CWSH presents 60.7% and 80.6% of energy and exergy efficiency, followed by the CWOSH of 58.0% and 64.2%, respectively. Put the indirect $CO_2$ emission into the gross $CO_2$ emission of the system, the emissions of $CO_2$ of the two scenarios are 1.56 t/t-methanol in CWOSH and 1.50 t/t-methanol in CWSH. The output of methanol in the two scenarios are $29.4 \times 10^4$ t/a and $29.9 \times 10^4$ t/a. Both have a great enhancement of production scale over the industrious plants. By comparing with the two scenarios, it is instructive when there exists a trade-off between the material conversion efficiency and capital investment addition. If the former has a more positive effect on the outcome, it is wise to invest for increase of material conversion efficiency.

**Table 3 Comparison between different scenarios and previous work**

| Item | CTMCR | CWOSH | CWSH |
|---|---|---|---|
| **Investment** | | | |
| TCI (M$) | 103.2 | 109.7 | 102.1 |
| O&M cost (M$/y) | 4.10 | 4.38 | 4.08 |
| Labor number | 320 | 320 | 330 |
| Labor cost (M$/y) | 2.0 | 2.0 | 2.1 |
| COG cost (M$/y) | 27.9 | 27.9 | 27.9 |
| Electricity cost (M$/y) | 16.9 | 29.4 | 20.9 |
| **Consumption** | | | |
| Steam (t/t-methanol) | balance | balance | balance |
| Electricity (kWh/t-methanol) | 812 | 922.2 | 908.5 |
| Oxygen ($m^3$/t-methanol) | 1513 | 1537 | 1514 |
| COG ($m^3$/t-methanol) | 1600 | 1465 | 1444 |
| **Performance** | | | |
| Methanol production ($10^4$ t/a) | 24.0 | 26.2 | 26.6 |
| $CO_2$ emission (t/t-methanol)* | 0.74 | 1.56 | 1.50 |
| Energy efficiency (%) | 59.5 | 58.0 | 60.7 |
| Exergy efficiency (%) | 68.9 | 64.2 | 80.7 |
| COM ($/t) | 272.0 | 284.7 | 272.7 |
| IRR (%) | 29.1 | 28.9 | 37.3 |

**Note:*** the $CO_2$ emission constitute direct and indirect emission; indirect emission comes from electricity consumption. Usually, 350 g standard coal produces 1 kWh electricity and 917 g $CO_2$ emission.

## 4. Conclusions

The serials of $CO_2$ recycled COG to methanol processes all show a good performance in $CO_2$ emission reduction and methanol output amplification as well as decrease of capital investment, cost of methanol and increase of IRR. Except the reforming technology, the other focus on this work is the concern of $H_2$ separation. Some conclusions are obtained as follows:

- In CWOSH and CWSH scenarios, the optimal ratios of COG for reforming are the same, 55% of COG or COG depleted with $H_2$ is sent for reformer, which accord with our previous work;
- By comparing CWOSH and CWSH, CWSH shows higher energy and exergy efficiency (60.7%>58.0%, 80.6%>64.2%), lower $CO_2$ emission (1.50<1.56 t/t-methanol), more methanol output ($29.9\times10^4 > 29.4\times10^4$ t/a), and higher IRR (37.3%>28.9%). The technologic and economic performances make a good agreement, it is undoubtedly preferable to choose scenario CWSH when considering to establish a plant among the two processes.

**Acknowledgements**

The authors gratefully acknowledge the financial support from the National Natural Science Foundation of China (51404164, 51276120, U1361202).

**Nomenclature**

CTM= COG to methanol

CWHS= COG with $H_2$ separation

CWOHS= COG without $H_2$ separation

$E_{in}$= the total energy input

$E_o$ = the total energy output

$EX_{in}$= exergy of the total fuel input

$EX_o$= exergy of the total fuel output

$EX_{tot}$=total exergy

$EX_{phy}$= physical exergy

$EX_{chem}$= chemical exergy

$EX_{mix}$= exergy change by mixing

$(FCH_4+FCO_2)_{in}$ = $CO_2$ and $CH_4$ inflow rate of the reforming unit

$(FCH_4+FCO_2)_{out}$ = $CO_2$ and $CH_4$ outflow rate of the reforming unit

KCO = adsorption constant of CO, $bar^{-1}$

$KCO_2$= adsorption constant of $CO_2$, $bar^{-1}$

$KH_2$= adsorption constant of $H_2$, $bar^{-1}$

$KH_2O$= adsorption constant of $H_2O$, $bar^{-1}$

$K_{MeOH}$ = adsorption constant of methanol, $bar^{-1}$

$K_{01}$,$K_{02}$,$K_{03}$= thermodynamic equilibrium constant of methanol synthesis reaction (1-3)

Yc = $CO_2$+$CH_4$ conversion efficiency of the system

**Capital Letters**

$C_{fuel}$= COG cost

$C_{labour}$= labor cost

$C_{power}$ = electricity cost

$C_t$= net cash flows of the year t

$F_{O\&M}$= fixed Operation & Maintenance cost

**Lowercase Letters**

d =day

$f_{CO}$= fugacity of CO in gas phase, bar

$fCO_2$ = fugacity of $CO_2$ in gas phase, bar

$fH_2$= fugacity of $H_2$ in gas phase, bar

$fH_2O$ = fugacity of $H_2O$ in gas phase, bar

$f_{MeOH}$ = fugacity of methanol in gas phase, bar

i = the discount rate, %

$k_1$, $k_2$, $k_3$, $k_4$=reaction rate constant of methanol synthesis reaction (1-3), mol/(s•kg•bar)

$r_1$, $r_2$, $r_3$, $r_4$= reaction rate of methanol synthesis reaction (1-3), mol/(s•kg)

n = the expected plant lifetime

y = year

**Greek Letters**

$\lambda$ = the ratio of coke oven gas to reforming

$\eta_1$ = the energy efficiency of system

$\eta_2$ = the exergy efficiency of system

**Acronyms**

Aspen = Advanced System for Process Engineering

ASU= air separation unit

BWRS = Benedict–Webb–Rubin–Starling equation of state

CGG = coal gasified gas

COG = coke oven gas

COM= cost of methanol

Compr model = compressor/turbine

CRF = capital recovery factor

DME = dimethyl ether

DRM= dry reforming

GHG = greenhouse gas

Flash model = flash tank

Heater model = heater/cooler

RGibbs=thermodynamic equilibrium reactor based on Gibbs free energy minimization

Heatx model = two-stream heat exchanger

IRR = internal rate of return

LHV = low heat value

M = million

Mcompr model = multistage compressor/turbine

MeOH = methanol

O&M = operation & maintenance

PCOMCR = production cost of methanol with considering benefits from CO2 tax

PR-BM method= Peng-Robinson equation of state with Boston–Mathias modifications

Radfrac columns = rigorous distillation tower

Rplug = plug flow reactor

PSA= pressure swing adsorption

PSRK= Predictive Redlich-Kwong-Soave

RStoic = stoichiometric reactor

Sep model = multi-outlet component separator

TCI= total capital investment

TEC= total equipment cost

## Reference


[1] NBS. 2016 2016-01-19 [cited 2016 6-12]; Available from: http://www.stats.gov.cn/tjsj/zxfb/201601/t20160119_1306083.html.

[2] Man Yi, Yang Siyu,Qian Yu. Integrated process for synthetic natural gas production from coal and coke-oven gas with high energy efficiency and low emission[J]. Energy Conversion and Management, 2016, 117: 162-170.

[3] Bermúdez José M., Arenillas Ana, Luque Rafael,Menéndez J. Angel. An overview of novel technologies to valorise coke oven gas surplus[J]. Fuel Processing Technology, 2013, 110: 150-159.

[4] Zhang Jiayuan, Zhang Xiaohui, Chen Zhuo,Li Li. Thermodynamic and kinetic model of reforming coke-oven gas with steam[J]. Energy, 2010, 35(7): 3103-3108.

[5] Zhen Xudong,Wang Yang. An overview of methanol as an internal combustion engine fuel[J]. Renewable and Sustainable Energy Reviews, 2015, 52: 477-493.

[6] G.A. Olah A. Goeppert, G.K.S. Prakash, *Beyond Oil and Gas: The Methanol Economy*, 2006, Wiley-VCH.

[7] Xiang Dong, Yang Siyu, Mai Zihao,Qian Yu. Comparative study of coal, natural gas, and coke-oven gas based methanol to olefins processes in China[J]. Computers & Chemical Engineering, 2015, 83: 176-185.

[8] Arteaga-Pérez Luis E., Gómez-Cápiro Oscar, Karelovic Alejandro,Jiménez Romel. A modelling approach to the techno-economics of Biomass-to-SNG/Methanol systems: Standalone vs Integrated topologies[J]. Chemical Engineering Journal, 2016, 286: 663-678.

[9] Mortensen Peter Mølgaard,Dybkjær Ib. Industrial scale experience on steam reforming of CO2-rich gas[J]. Applied Catalysis A: General, 2015, 495: 141-151.

[10] Hou Zhaoyin, Chen Ping, Fang Heliang, Zheng Xiaoming,Yashima Tatsuaki. Production of synthesis gas via methane reforming with CO on noble metals and small amount of noble-(Rh-) promoted Ni catalysts[J]. International Journal of Hydrogen Energy, 2006, 31(5): 555-561.

[11] Bermúdez J. M., Arenillas A.,Menéndez J. A. Syngas from CO2 reforming of coke oven gas: Synergetic effect of activated carbon/Ni–γAl2O3 catalyst[J]. International Journal of Hydrogen Energy, 2011, 36(21): 13361-13368.

[12] Bermúdez J. M., Fidalgo B., Arenillas A.,Menéndez J. A. Dry reforming of coke oven gases over activated carbon to produce syngas for methanol synthesis[J]. Fuel, 2010, 89(10): 2897-2902.

[13] Bermúdez J. M., Fidalgo B., Arenillas A.,Menéndez J. A. CO2 reforming of coke oven gas over a Ni/γAl2O3 catalyst to produce syngas for methanol synthesis[J]. Fuel, 2012, 94: 197-203.

[14] Bermúdez J. M., Arenillas A.,Menéndez J. A. Equilibrium prediction of CO2 reforming of coke oven gas: Suitability for methanol production[J]. Chemical Engineering Science, 2012,

82: 95-103.

[15] Bermúdez J. M., Ferrera-Lorenzo N., Luque S., Arenillas A.,Menéndez J. A. New process for producing methanol from coke oven gas by means of CO2 reforming. Comparison with conventional process[J]. Fuel Processing Technology, 2013, 115: 215-221.

[16] Man Yi, Yang Siyu, Zhang Jun,Qian Yu. Conceptual design of coke-oven gas assisted coal to olefins process for high energy efficiency and low CO2 emission[J]. Applied Energy, 2014, 133: 197-205.

[17] Lin Hu, Jin Hongguang, Gao Lin,Zhang Na. A polygeneration system for methanol and power production based on coke oven gas and coal gas with CO2 recovery[J]. Energy, 2014, 74: 174-180.

[18] Yi Qun, Feng Jie,Li Wen Ying. Optimization and efficiency analysis of polygeneration system with coke-oven gas and coal gasified gas by Aspen Plus[J]. Fuel, 2012, 96: 131-140.

[19] Li Sheng, Jin Hongguang, Gao Lin,Zhang Xiaosong. Exergy analysis and the energy saving mechanism for coal to synthetic/substitute natural gas and power cogeneration system without and with CO2 capture[J]. Applied Energy, 2014, 130: 552-561.

[20] Anvari Simin, Jafarmadar Samad,Khalilarya Shahram. Proposal of a combined heat and power plant hybridized with regeneration organic Rankine cycle: Energy-Exergy evaluation[J]. Energy Conversion and Management, 2016, 122: 357-365.

[21] Xiang Dong, Yang Siyu, Liu Xia, Mai Zihao,Qian Yu. Techno-economic performance of the coal-to-olefins process with CCS[J]. Chemical Engineering Journal, 2014, 240: 45-54.

[22] Xiang Dong, Yang Siyu,Qian Yu. Techno-economic analysis and comparison of coal based olefins processes[J]. Energy Conversion and Management, 2016, 110: 33-41.

[23] Reyes Valle C., Villanueva Perales A. L., Vidal-Barrero F.,Gómez-Barea A. Techno-economic assessment of biomass-to-ethanol by indirect fluidized bed gasification: Impact of reforming technologies and comparison with entrained flow gasification[J]. Applied Energy, 2013, 109: 254-266.

[24] Hao Yanhong, Huang Yi, Gong Minhui, Li Wenying, Feng Jie,Yi Qun. A polygeneration from a dual-gas partial catalytic oxidation coupling with an oxygen-permeable membrane reactor[J]. Energy Conversion and Management, 2015, 106: 466-478.

[25] Petersen Abdul M., Farzad Somayeh,Görgens Johann F. Techno-economic assessment of integrating methanol or Fischer–Tropsch synthesis in a South African sugar mill[J]. Bioresource Technology, 2015, 183: 141-152.

[26] Zhou Li, Hu Shanying, Chen Dingjiang, Li Yourun, Zhu Bing,Jin Yong. Study on Systems Based on Coal and Natural Gas for Producing Dimethyl Ether[J]. Industrial & Engineering Chemistry Research, 2009, 48(8): 4101-4108.

[27] Yi Qun, Fan Yang, Li Wenying,Feng Jie. CO2 Capture and Use in a Novel Coal-Based Polygeneration System[J]. Industrial & Engineering Chemistry Research, 2013, 52(39): 14231-14240.

[28] Li Sheng, Gao Lin, Zhang Xiaosong, Lin Hu,Jin Hongguang. Evaluation of cost reduction potential for a coal based polygeneration system with CO2 capture[J]. Energy, 2012, 45(1): 101-106.

[29] SAT. *China's Tax System*. 2012; Available from: http://www.chinatax.gov.cn/2013/n2925/n2959/c307248/content.html.

# Status of Worldwide CMM and CBM Development, Including Sub-Saharan Africa

**Thirty Third Annual International Pittsburgh Coal Conference**
**Cape Town South Africa**
**August 8th – 12th, 2016**

**Jonathan Kelafant, Advanced Resources International and**
**Felicia Ruiz, U.S. Environmental Protection Agency**

Global Methane Initiative

## Presentation Outline

- Introduction
- U.S.
- Australia
- Canada
- China
- India
- Sub-Saharan Africa
- What's Next for CBM?

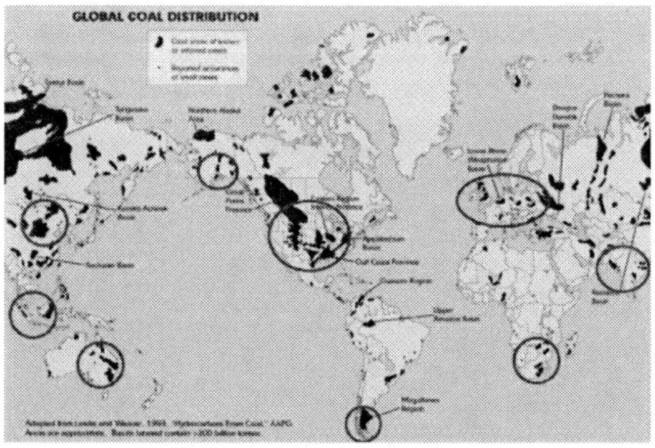


Global Methane Initiative

2

# Introduction

- CBM, which was once the "hot" sector of the natural gas industry in the 1990s and early 2000s, has been eclipsed by the steep rise in shale gas drilling and production.
- Despite a precipitous drop in U.S. CBM activity, CBM drilling and production has been rising steadily in a number of other countries including Australia, China, and India; pilot well testing is underway in several countries including South Africa and Kazakhstan.
- CBM offers several advantages over conventional and shale gas drilling:
  - CBM wells are shallower and lower cost;
  - Technical requirements are simpler which makes conducting projects easier in developing countries with limited oil and gas infrastructure;
  - Decline curves are not as steep as shale wells, resulting in longer reserve life.

Global Methane Initiative

3

# United States

Global Methane Initiative

## In the U.S.

- Over 30 Tcf (0.9 Tcm) of CBM has been produced over the past 25 years;
- Over 400 Tcf (11.4 Tcm) of CBM resources remain to be developed.

### Coalbed Methane Fields, Lower 48 States

Source: EIA

Global Methane Initiative

5

CBM production has fallen 25% over the past 5 years while the number of wells drilled has declined 75% over the same period. This reflects the relatively shallow decline curves for CBM wells and their long-lived reserves. Recompletions also account for a portion of the lower production decline.

### Annual U.S. CBM Well Drilling

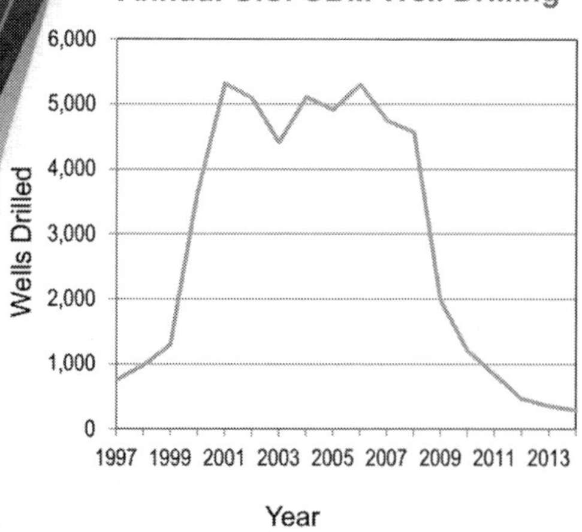


### Annual U.S. CBM Production

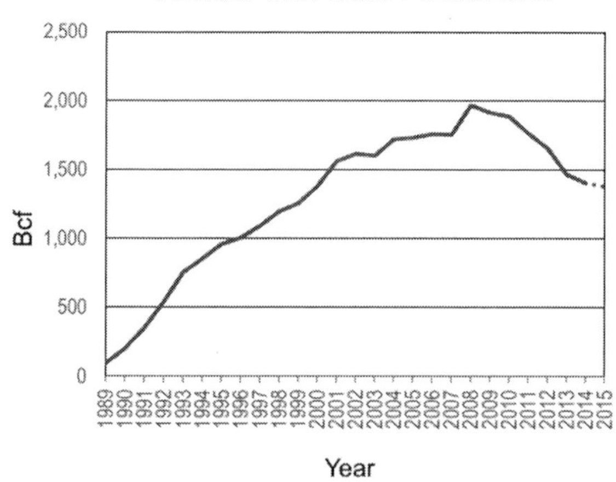


Global Methane Initiative

6

# Australia

Global Methane Initiative

# Australia

### Major Australian Coal Basins

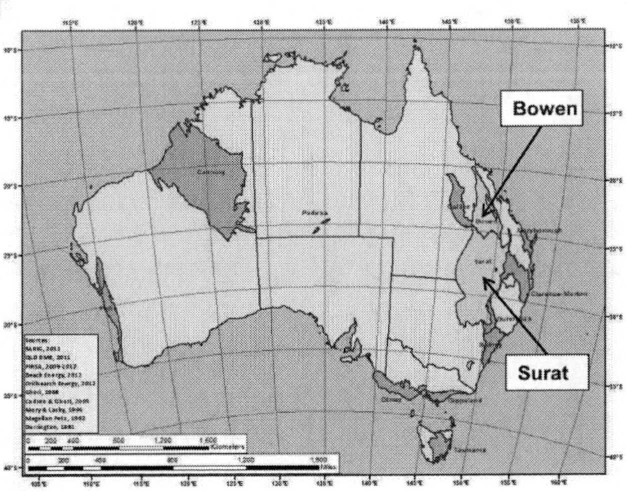


- Australia's CBM resources are estimated to be 203 Tcf.
- 14 basins are actively being explored or producing CBM with LNG driving the activity.
- The Surat and Bowen basins are the two principal producing basins.
- Production of 246 Bcf in 2012 (approximately 13% of total natural gas. production); production is expected to reach 6 Bcfd by 2020.

Global Methane Initiative

- Australia's CBM well count has nearly tripled over the past five years to meet the demands from LNG.
- Major Producers Include BG Group, ConocoPhillips, Origin Energy and Santos.

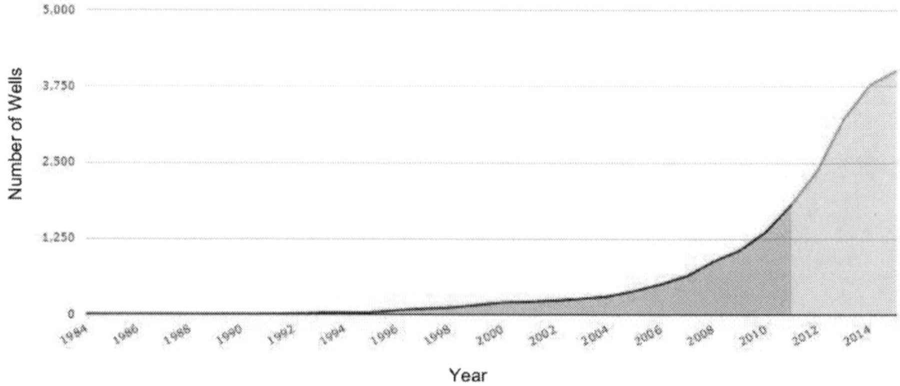


http://www.abc.net.au/news/specials/coal-seam-gas-by-the-numbers/#sources

Global Methane Initiative

9

# Queensland Curtis LNG Project

- Bowen and Surat basin CBM production feeds LNG facilities on nearby Curtis Island.
- Queensland Curtis LNG (QCLNG) is world's first project to turn CBM into LNG.

**QCLNG Facilities**

- The project began construction in 2010. In December 2014, the first quantities of LNG were produced and sold.
- BG Group and CNOOC are partners in the LNG facility. At plateau production, which is expected in mid-2016, QCLNG will be producing enough LNG to load ten vessels per month combined, exporting around eight million tonnes per year.
- The facility will require 375 Bcf per year of natural gas to reach the planned production.

Global Methane Initiative

10

## Santos Gladstone LNG Project

- Developing CBM in the Surat and Bowen Basins to feed an LNG plant on Curtis Island, near Gladstone.
- Santos GLNG is led by Australian company Santos, in partnership with PETRONAS from Malaysia, Total from France, and KOGAS from South Korea.

- Santos GLNG produced and sold its first LNG cargo in October, 2015.
- When fully operational will have the capacity to produce 7.8 million tonnes of LNG each year, or just over 350 Bcf per year.

Global Methane Initiative

11

# Canada

Global Methane Initiative

# Canada

- Canada has an estimated CBM resource base of 801Tcf, according to CSUG (2010).

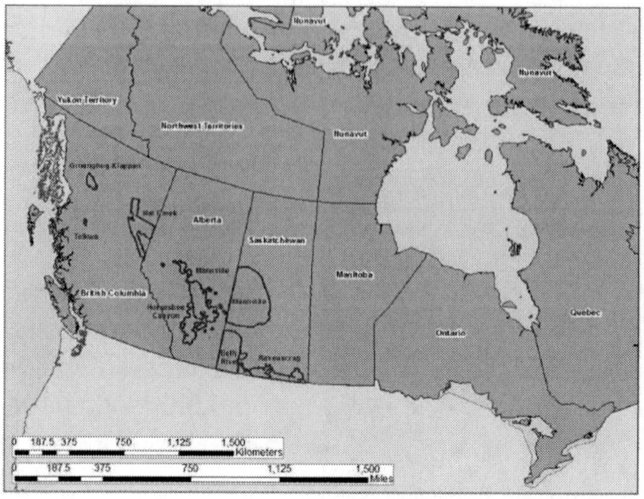


- Most of the CBM resources in Canada occur in Alberta and north eastern British Columbia. Other resource areas have been identified in other parts of British Columbia, Saskatchewan and Nova Scotia.

Global Methane Initiative

13

# Alberta-Canada

- Horseshoe Canyon and Belly River produce little or no water and are referred to as "dry" CBM production. The dry nature of the seams requires nitrogen foam fracturing versus standard slick water hydraulic fractures. Wells are generally vertically drilled to depths of 200 m to 800 m.
- A second set of deeper coal seams, the Mannville coal seams, are also present. These seams are thicker and produce substantial amounts of saline water. Operators tend to develop Mannville coal seams with horizontal wells.

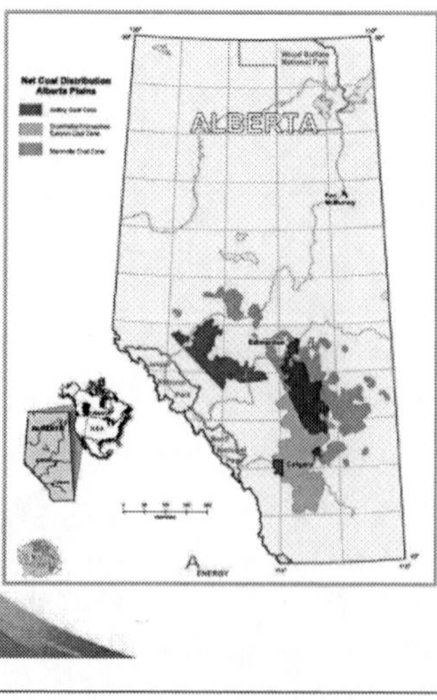


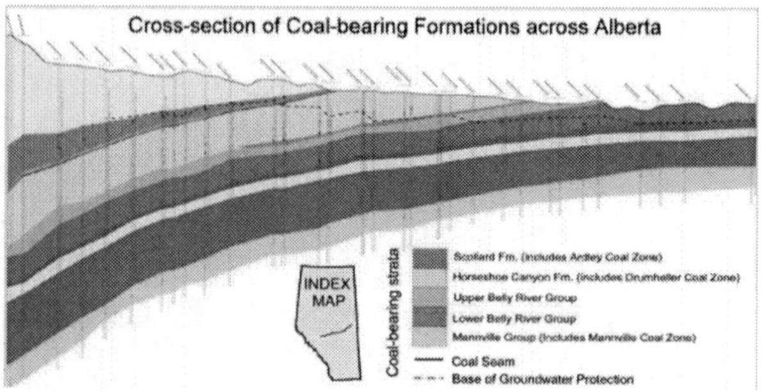


Global Methane Initiative

14

# Canada-British Columbia

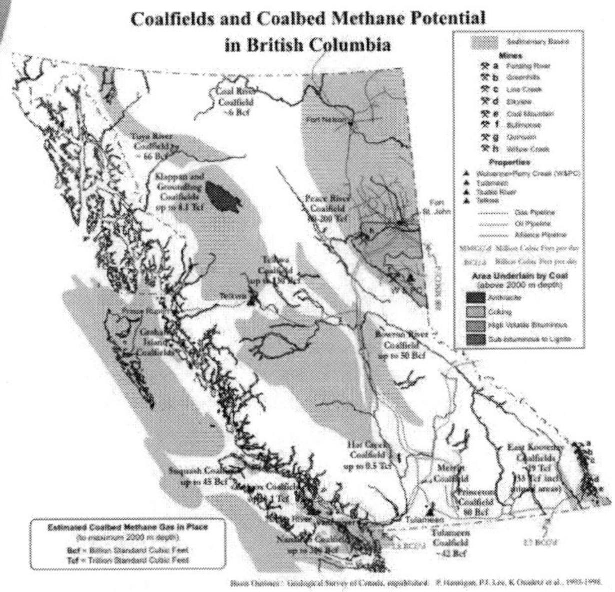


Current estimates place British Columbia's CBM volume between 90-250 trillion cubic feet (Tcf) of gas in coals above a 2,000 meter depth. While only 20 per cent of this gas may be recovered, the marketable volume of 18-50 Tcf is equivalent to 25 to 75 years of gas supply at current production rates (for both domestic and export markets).

Global Methane Initiative

15

# India

Global Methane Initiative

# India

## India's Coalfields

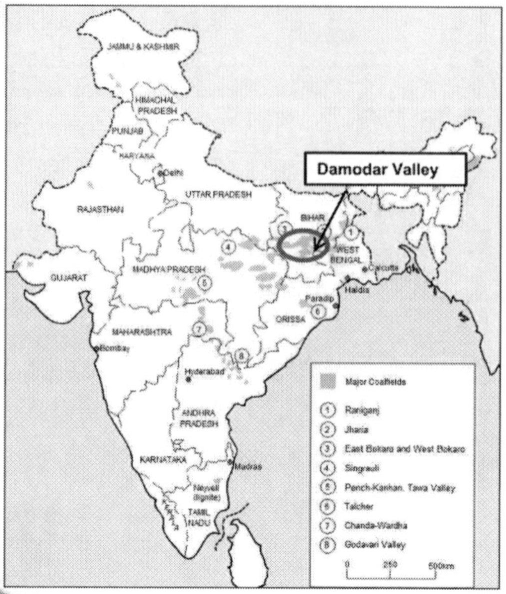


- India's coalfields contain an estimated 120 Tcf of CBM distributed across 44 Coalfields.
- The bulk of the resources are contained in the Damodar Valley Coalfields, the major coal mining province.
- In 1997, the government implemented a new CBM policy to encourage development. Incentives included reduced royalty payments and a sliding production level payment.
- Since 2001, 33 CBM blocks have been awarded on an international tender basis.

Global Methane Initiative

17

# Major Active Projects

Over 600 CBM wells have been drilled in India and this number is expected to double in the next several years as the infrastructure is developed.

| Major Active Projects | | | |
|---|---|---|---|
| **Company** | **Project** | **Status** | **Market** |
| ESSAR Oil | Raniganj East | 230 wells drilled and producing; 200 more wells planned; current production is 35 MMcfd | Matix fertilizer plant |
| Great Eastern Energy | Raniganj South | 150 wells drilled, 100 producing; 150 wells planned; current production 10 MMscfd; Gas Price of $10.00/Mcf | Local industries, steel plants |
| Reliance Industries | Sohagpur | ~170 wells drilled; 200+ wells planned for this year | Pipeline injection |
| Reliance ADA/ Geopetrol | Sohagpur | Series of core wells; 2 production wells; currently planning a 15 well pilot | Power generation |
| Oil and Natural Gas Corp. | Parpatur, Ranuganj, Karanpura | 12-15 pilot wells, limited production | Local usage |

Global Methane Initiative

18

# China

Global Methane Initiative

## CBM Production has been Rising Steadily in China Over the Past Decade

- China has an estimated 1300 Tcf of CBM resources.
- CBM Production Volume was 3.73 billion m³ (361MMcfd) in 2014.
- Jincheng Anthracite Mining Group (JAMC) is the largest CBM producing company in China, producing 1.43 Bm³ in 2014, accounting for 38.3% of the total in China.

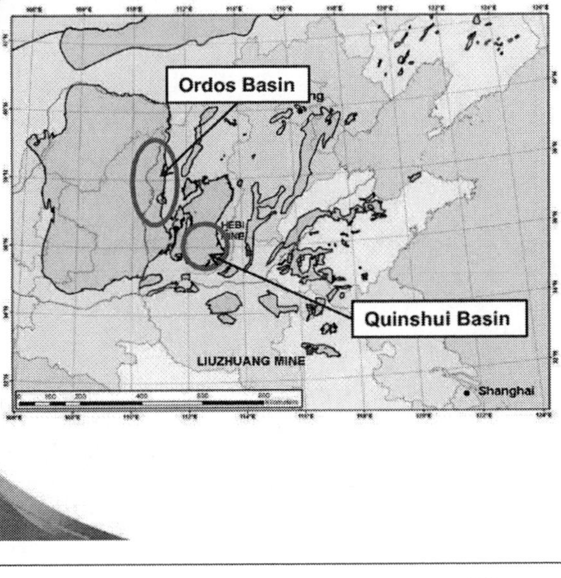


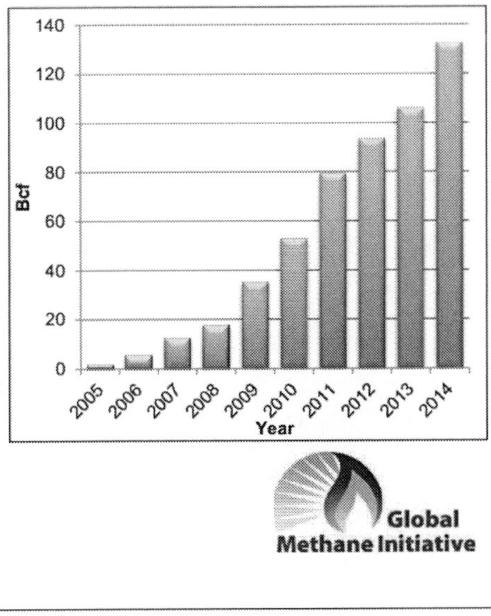


Global Methane Initiative

20

- CBM Exploitation and Development Action and Plan 2020 was published on February 3, 2015, recommending:
  - Building 3-4 bases of CBM development;
  - Integrating CBM recovery and coal mining in the key coal mining areas;
  - The goal is to increase CBM/CMM production to 1,412 Bcf/Year, up from the current 130 Bcf/Year.

Global Methane Initiative

21

# Sub-Saharan Africa

Global Methane Initiative

# Southern Africa

- CBM resources in the Southern African region are estimated to be about 110 Tcf.
- Anglo American has conducted an extensive exploration project in the Waterberg basin of South Africa, including a 5-well pilot project.
- Kalahari Gas and others have drilled over 30 wells in Botswana; development pending power purchase agreements.

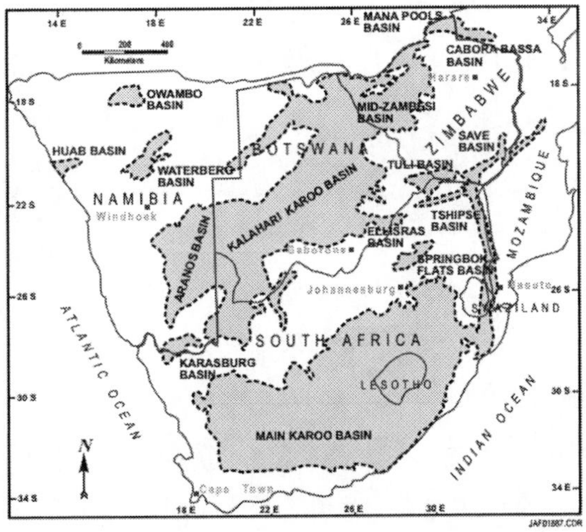


Global Methane Initiative

23

# Southern Africa - Observations

South Africa exploration for CBM focused mainly on the Lephalale and Amersfoort CBM Projects:

## Lephalale CBM Project

- The Lephalale CBM project, conducted by Anglo American, is the most advanced of any in the region; 82 core holes, 30 wells with IFT's, 5-well pilot project.
- Substantially de-risked via extensive drilling and 10-years of pilot wellfield production.
- Key attributes are as follows:
  - No dolerite intrusions are known in this play
  - Basin structural makeup well understood
  - Thick, relatively shallow interbedded coal seams
  - Permeability testing widely performed throughout the basin
  - Baseline water monitoring data collected over 13 years.

## Amersfoort CBM Project

- The Amersfoort CBM Project – in relatively early sate of exploration.
- Key attributes are as follows:
  - Coal extensively intruded by dolerite sills and dykes; structurally complex
  - To date, pilot test wells have flowed gas mainly from sandstone reservoirs
  - Can the project be successfully commercialized solely on the basis of production from sandstone reservoirs?

Global Methane Initiative

24

# Anglo Lephalale Pilot Project

Global Methane Initiative

25

# Southern Africa - Observations

## Botswana

- CBM in Botswana has been investigated by several potential operators including Kalahari Gas and Tlou Energy.
- Deep coal bearing areas are more extensive than in South Africa.
- Key attributes are as follows:
  - Extensively intruded by dolerite dyke swarms and attendant sill offshoots
  - Variable seam thickness but generally less than Waterberg
  - Coal seams are deeper than Waterberg
  - Low gas content and generally low permeability
  - Undersaturated coal seams; suspect no biogenic gas recharge
  - Commercialization of such plays is heavily dependent on the efforts of the operator

## Zimbabwe

- The most prospective CBM area in Zimbabwe is the Mid Zambesi Basin.
- This has mainly been investigated by two operators – AFPenn Exploration and Shangani Energy.
- Key attributes are as follows:
  - No dolerite intrusions are known in this play; prominent faulting exists
  - Areas of good seam thicknesses have been drilled; +30M of clean coal reported
  - Coal seams are deeper than Waterberg
  - Variable gas content has been reported; little permeability testing has been undertaken although several well tests have showed reasonable permeability values (>10 md); gas saturation inadequately studied.

Global Methane Initiative

26

# What's Next for CBM?

Global Methane Initiative

## What's Next for CBM?

- Enhanced CBM Recovery (ECBM)
  - Inject $CO_2$ or $N_2$ to increase methane recovery;
  - Several successful pilot projects in the U.S.;
  - International projects under consideration in China and India;
  - Has the potential to sequester large volumes of $CO_2$
- Improving Hydraulic Fracture Designs
  - Range has increased production 2-3 fold in VA by pumping at higher rates;
  - May open up previously ignored low permeability coal fields like anthracite fields.
- Recompletions
- Horizontal Wells

Global Methane Initiative

28

# Contact Information

Jonathan Kelafant
Advanced Resources International, Inc.
+1 (703) 528-8420,
jkelafant@adv-res.com
www.adv-res.com

Advanced Resources International, Inc.

Felicia Ruiz
Coalbed Methane Outreach Program
U.S. Environmental Protection Agency
+1 (202) 343-9129
ruiz.felicia@epa.gov
www.epa.gov/cmop

U.S. EPA Coalbed Methane OUTREACH PROGRAM

Global Methane Initiative

Global Methane Initiative

29

# Consideration of a Possibility of Enhanced Coalbed Methane Recovery by Injecting Air into Coal Seam

Noritaka Shimatani, Master 1st
Yuichi Sugai, Kyuro Sasaki
Kyushu University, Japan



## Background

ECBM（Enhanced Coalbed Methane Recovery）

$CO_2$-ECBM is one of ECBM technology

- Swelling of coal causes decreasing of permeability and it makes hard to inject additional $CO_2$.
- Capturing, Separating and Transporting $CO_2$ are difficult in terms of economics and technology.

→ Air-ECBM

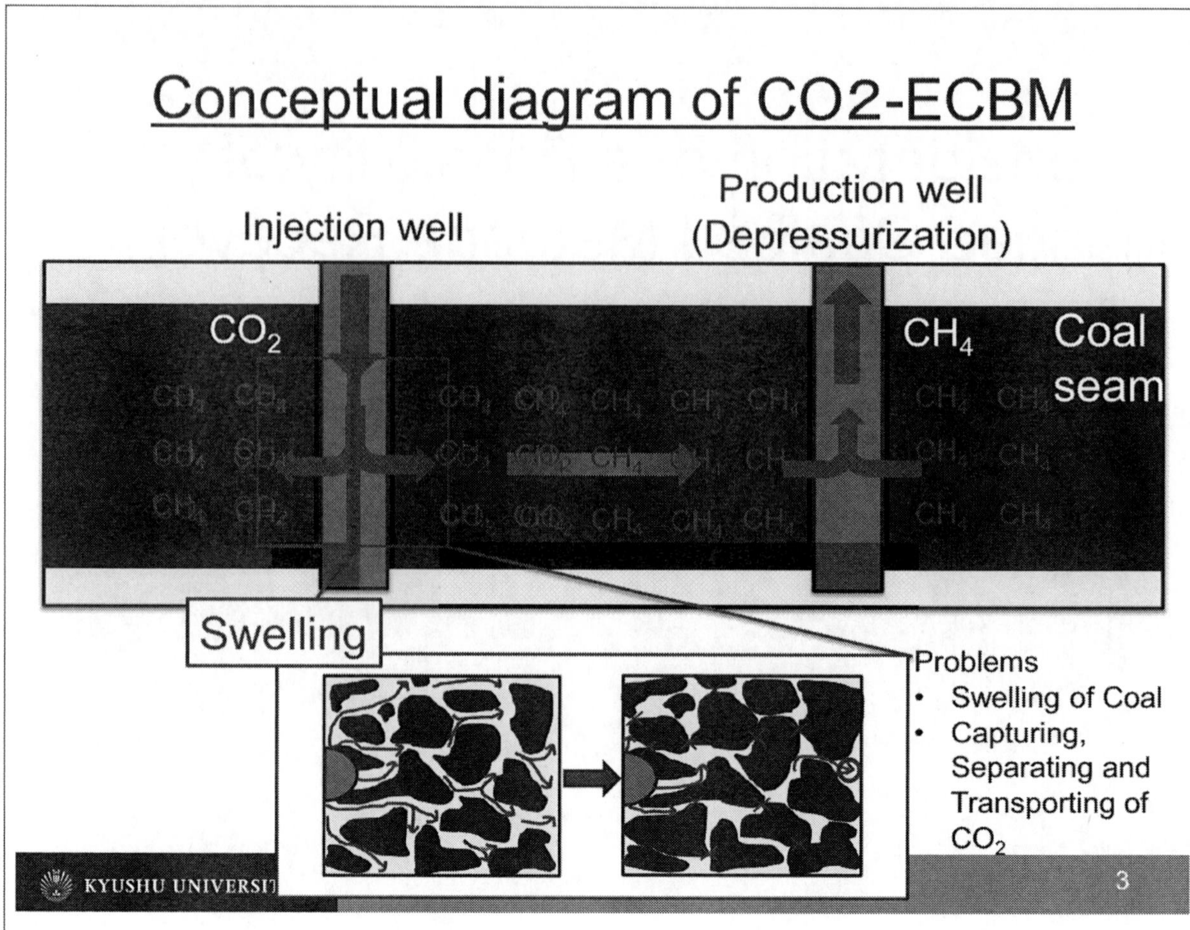
Conceptual diagram of CO2-ECBM
Injection well
Production well
(Depressurization)
$CO_2$
$CH_4$
Coal seam
Swelling
Problems
• Swelling of Coal
• Capturing, Separating and Transporting of $CO_2$
KYUSHU UNIVERSIT
3

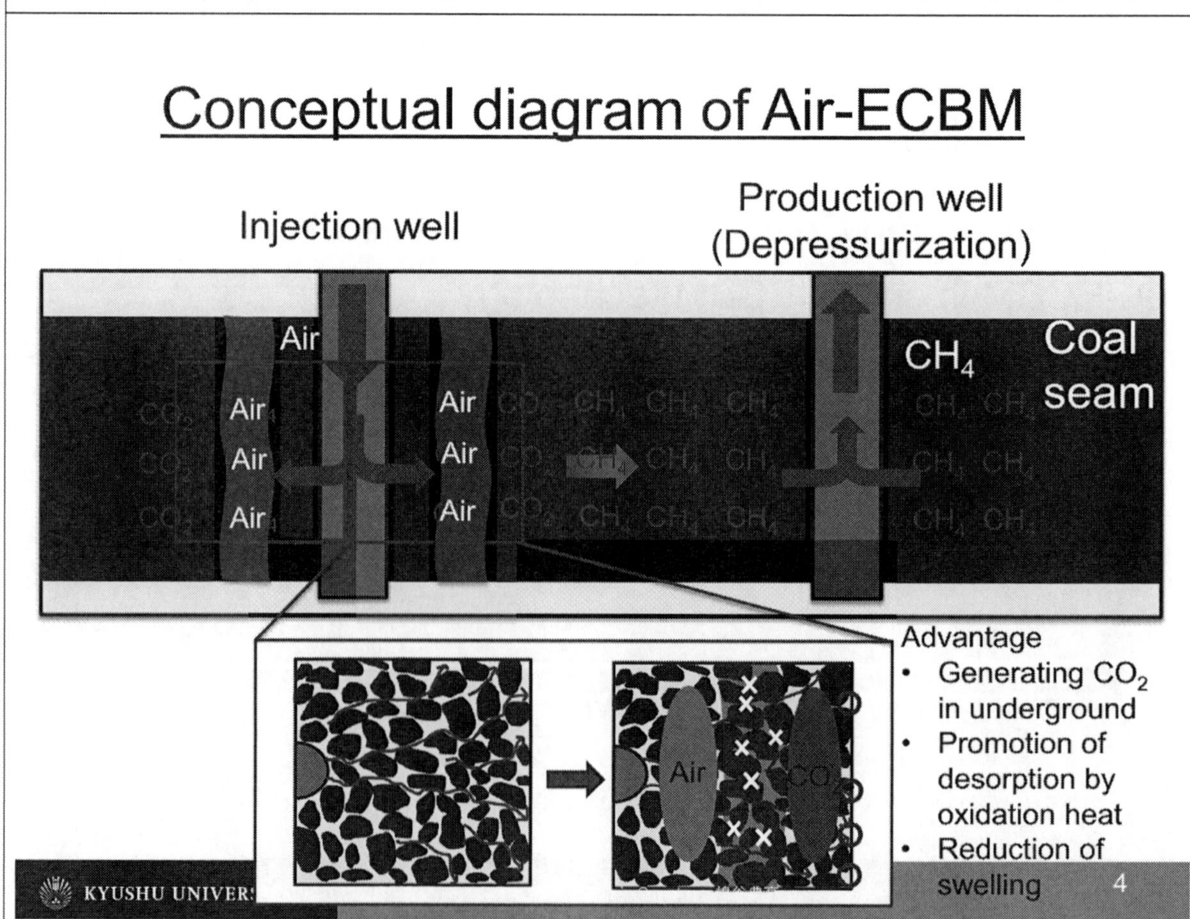
Conceptual diagram of Air-ECBM
Injection well
Production well
(Depressurization)
Air
Air
Air
Air
Air
Air
Air
$CH_4$
Coal seam
Air
Advantage
• Generating $CO_2$ in underground
• Promotion of desorption by oxidation heat
• Reduction of swelling
KYUSHU UNIVER
4

## Objectives

To evaluate the possibility of Air-ECBM, the coal oxidation at low temperature and high pressure condition is measured.



## Experiment Method

- Made powder coal (about 30 μm) and crude coal (about 8 mm)
- Moisture rate was adjusted by vacuum dehydrogenation at 100℃

Proximate analysis

| Lignite | Rate (%) |
|---|---|
| Moisture | 20.1 |
| Volatility | 40.2 |
| Ash | 5.03 |
| Carbon | 44.7 |

Powder coal

Crude coal

# Schematic of Experimental apparatus

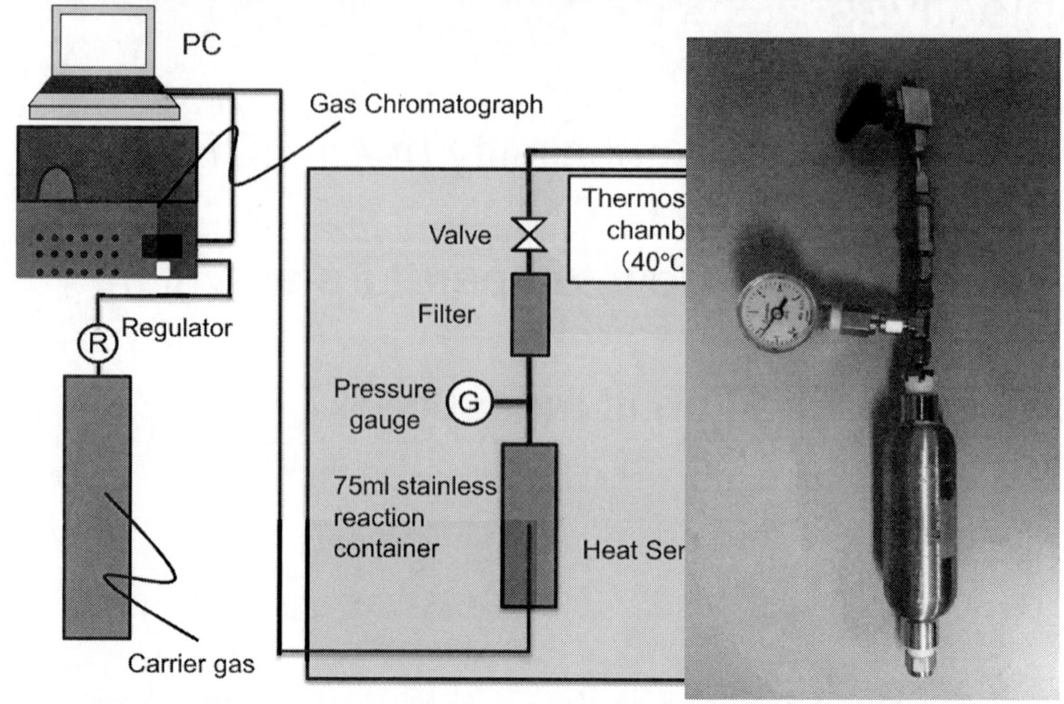




# Experiments

- No.1~5

  The reactivity of coal at high pressure and low temperature was measured.

- No.6

  The oxidation heat at high pressure and low temperature was measured .

## Conditions of experiment

| No. | Injection gas | Particle size (μm) | Moisture rate (%) | Pressure (MPa) | Temperature (°C) |
|---|---|---|---|---|---|
| 1 | Air, $N_2$, $O_2$ | 30 | 20.1 | 4.0, 4.0, 0.8 | 40 |
| 2 | $O_2$ | 30, 8 mm | 20.1 | 1.6 | 40 |
| 3 | $O_2$ | 30 | 5.38, 14.6, 20.1 | 1.2 | 40 |
| 4 | $O_2$ | 30 | 20.1 | 0.4, 0.8, 1.6 | 40 |
| 5 | $O_2$ | 30 | 20.1 | 1.2 | 40, 55, 70 |
| 6 | $N_2$, $O_2$ | 30 | 20.1 | 1.5 | 40 |



# 1. Experiment of $CO_2$ generation from coal in high-pressure Air atmosphere

# $CO_2$ generation by injecting Air and $N_2$

Experimental condition

powder coal (20.1%)

Experimental result

Air, 4.0 MPa

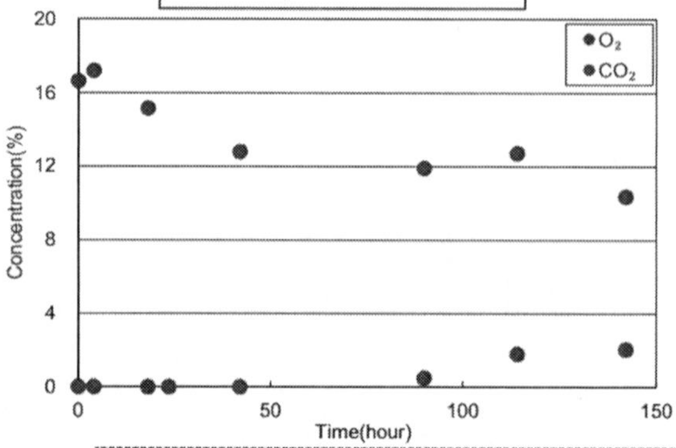


$N_2$, 4.0 MPa

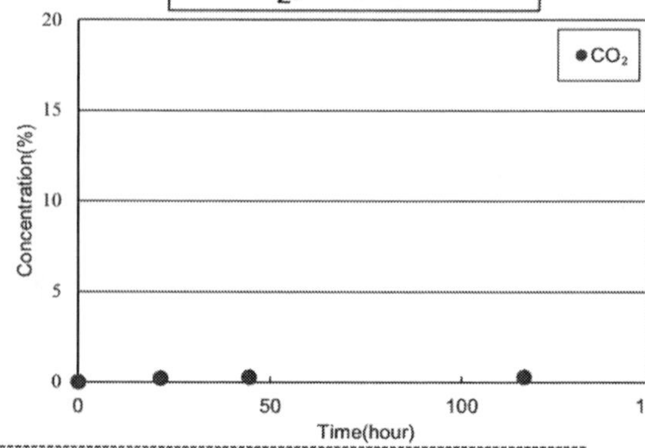


- $CO_2$ was detected by injecting Air
- $CO_2$ was not detected in non $O_2$ atmosphere



# $CO_2$ generation by injecting Air and $O_2$

Experimental condition

Powder coal (20.1%)

Experimental result

Air, 4.0 MPa

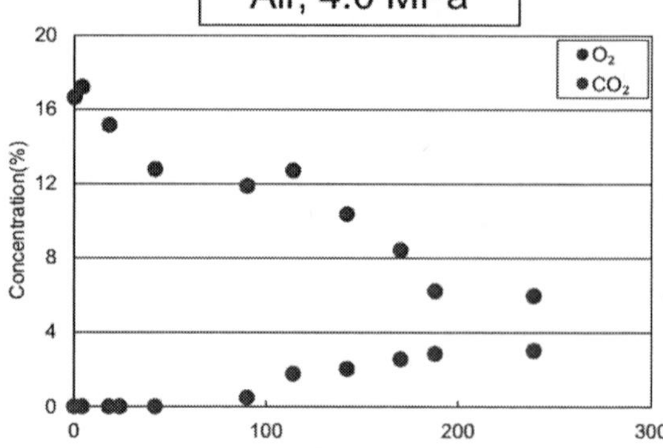


$O_2$, 0.8 MPa

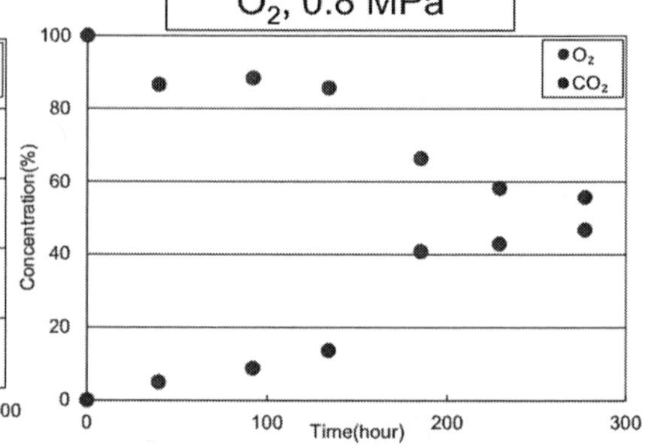


- $N_2$ diminished the adsorption area of $O_2$ and inhibited $CO_2$ generation.

## Adsorption and Desorption at coal surface ( $O_2$, 0.8 MPa)

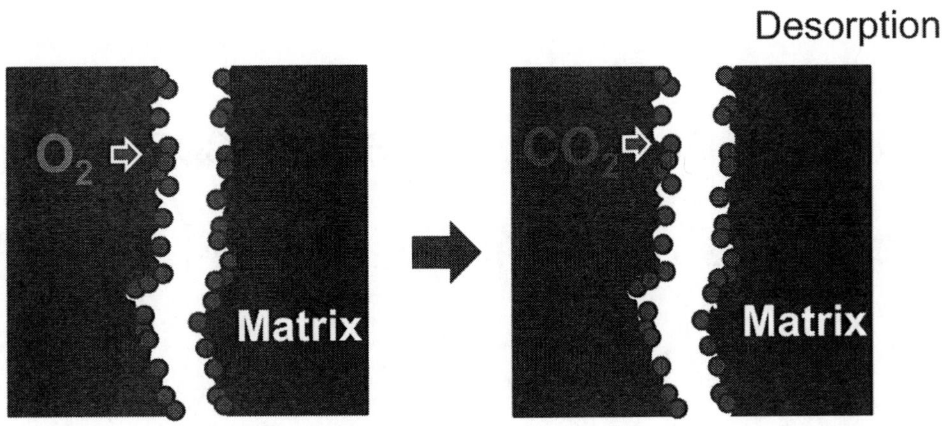


Detected $CO_2$ is equal to 2.5% of adsorbed $O_2$



## Adsorption and Desorption at coal surface ( Air, 4.0 MPa)

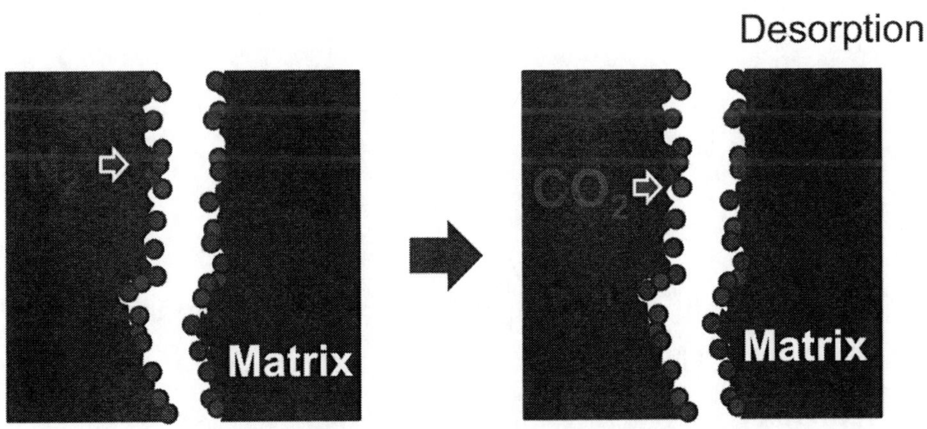


Ratio of adsorption area

$S_{O2} : S_{N2} = 2 : 17$

# 2. Experiment of the effect of coal particle size to the oxidation reaction



## Experimental condition

- Powder coal (20.1%)
- Crude coal (20.1%)
- $O_2$ : 1.6 MPa

## Formula

The quantity of $CO_2$ per unit dry coal mass

$$N = \frac{P \emptyset C}{XW}$$

$N$: The quantity of $CO_2$ per unit dry coal mass (mmol/g)
$P$: Pressure (MPa)
$\emptyset$: Porosity (-)
$C$: Increasing $CO_2$ rate (%)
$X$: 0.335 (MPa/mmol)
$W$: Dry mass (g)

## Experimental result

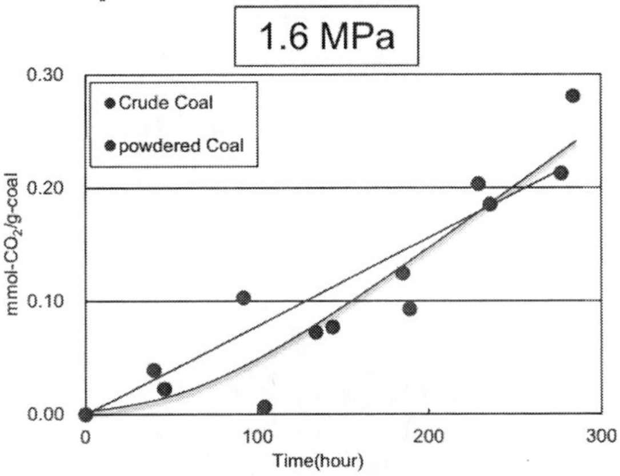


## Internal structure of coal

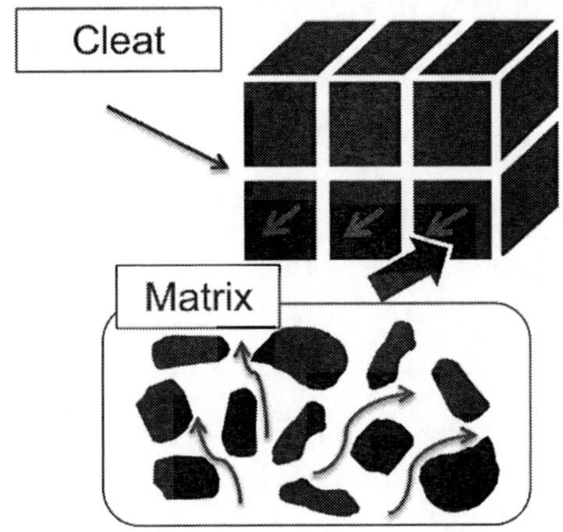


- Generating $CO_2$ by crude coal is later than powder coal
- Increasing $CO_2$ rate of the gas phase after passing through cleat



# 3. Experiment of the effect of coal moisture rate to the oxidation reaction

## Experimental condition

- powder coal
- $O_2$ : 1.2 MPa
- Moisture rate : 20.1, 14.6, 5.38%

## Experimental result

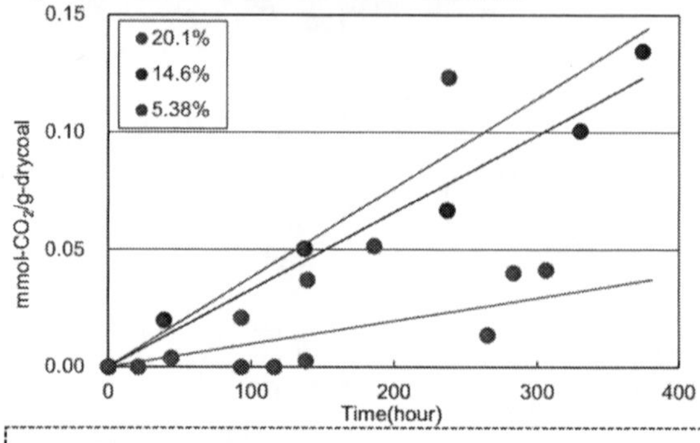


## Coal surface

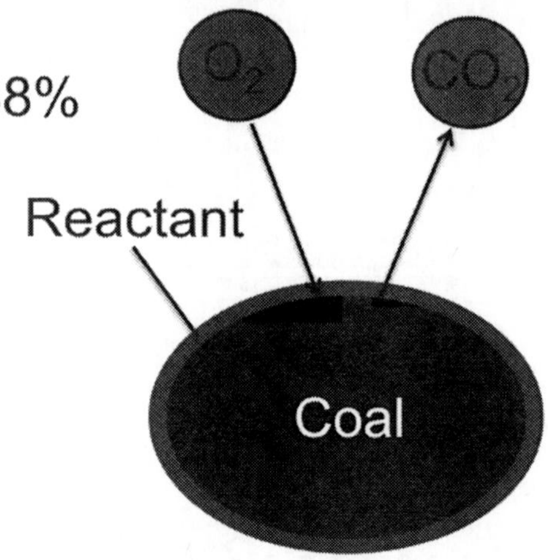


- The quantity of $CO_2$ per unit dry coal mass of the higher moisture rate's coal is larger than the lower coal.
- Reactants from coal to water may involve the oxidation reaction



# 4. Experiment of the relation between $O_2$ partial pressure and the oxidation reaction

## Experimental condition

- powder Coal (20.1%)
- $O_2$ : 0.4, 0.8, 1.6 MPa

## Experimental result

Derivation of reaction rate constant (k)

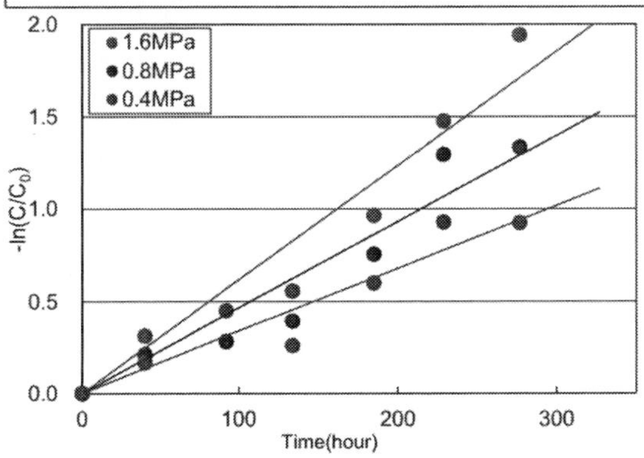


Reaction rate constant and $O_2$ partial pressure

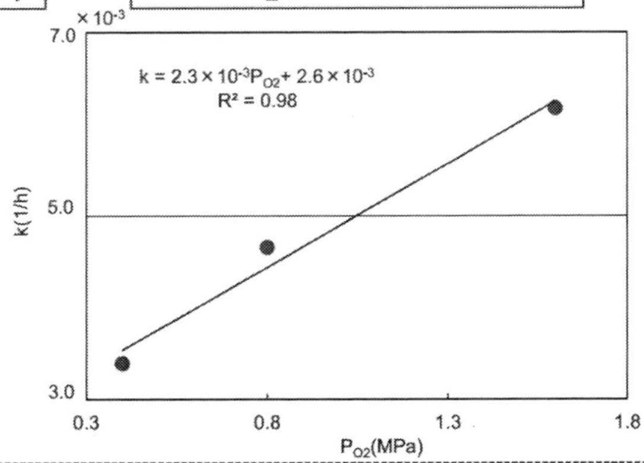


- Reaction rate : High $O_2$ partial pressure is higher
- $k=2.3\times10^{-3}\times P_{O2}+2.6\times10^{-3}$ (1/h, $0.4 \leqq P_{O2} \leqq 1.6$)



# 5. Experiment of the relation between temperature and the oxidation reaction

# Experimental condition

- powder coal (20.1%)
- $O_2$ : 1.2 MPa
- Temperature : 40, 55, 70℃

# Experimental result

Derivation of reaction rate equation

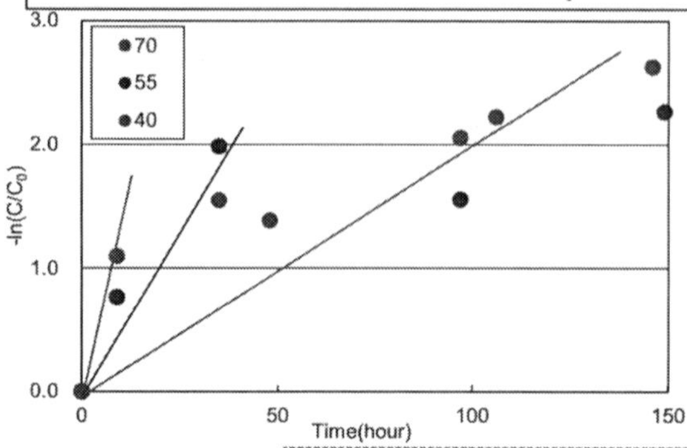


Arrhenius plot

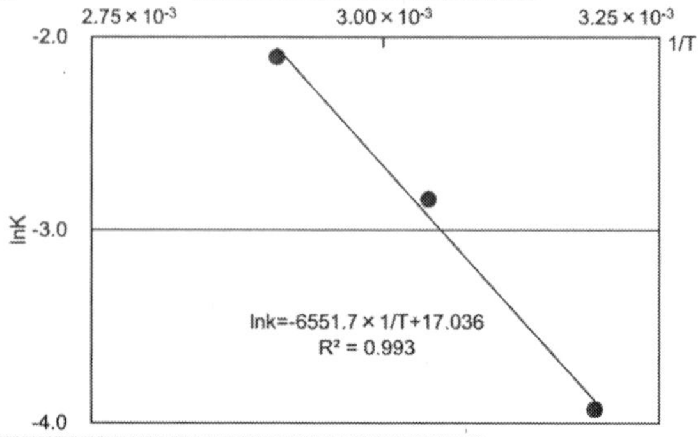


- Activation energy : 54.4 kJ/mol
- Frequency factor : $2.50 \times 10^7$

KYUSHU UNIVERSITY 2016 Pittsburgh Coal Conference in Capetown 嶋谷典高 23

# 6. Experiment of the oxidation heat of coal

KYUSHU UNIVERSITY 2016 Pittsburgh Coal Conference in Capetown 嶋谷典高 24

## Experimental condition

- powder coal (20.1%)
- $N_2$, $O_2$ : 1.5 MPa
- Temperature : 40℃

## Formula

*The quantity of generated heat*

$$Q = \sum \frac{10VS}{X}$$

*Q : The quantity of generated heat (J)*
*X : Sensor sensitivity 1.20 × $10^{-2}$ (mV/W·$m^2$)*
*S : Internal surface area ($m^2$)*
*V : Voltage (mV)*



## Experimental result

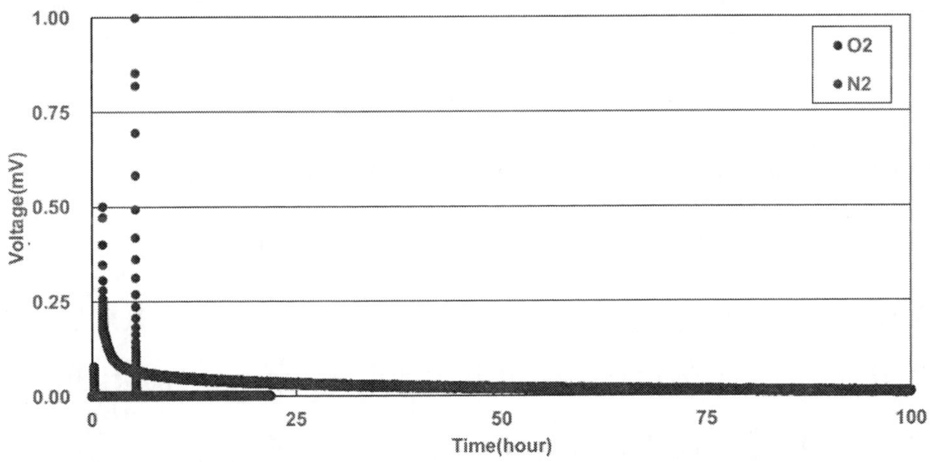


The generated heat
$O_2$ : 175 kJ/g
$N_2$ : 1.36 kJ/g

→ The oxidation heat
174 kJ/g

## Conclusion

- Generation of $CO_2$ by the oxidation reaction at 4.0MPa and 40℃ at air atmosphere was confirmed.
- $CO_2$ is released after desorbing from coal surface.
- The moisture makes reactants and accelerates the oxidation reaction.
- The effect of particle sizes to the oxidation reaction is small.
- Activation energy and frequency factor of oxidation at high pressure and low temperature are derived from the Arrhenius equation.
- The quantity of generated heat by injecting $O_2$ and $N_2$ and oxidation heat were calculated.



## Future plan

- To improve the sensitivity of measuring the oxidation heat.
- To make and improve the simulation model by combing parameters which were measured from this study by using GEM.

Thank you for attention

# The role of the residual bitumen of mature lacustrine shale to the porosity and gas storage capacity:

——A case study of the Triassic Yanchang shale, Ordos Basin, China


Jianfa Chen, Feiran Chen, Zhenxue Jiang, Zhilong Huang Guoheng Liu Fengyang Xiong

油气资源与探测国家重点实验室
State Key Laboratory of Petroleum Resources and Prospecting

State Key Laboratory of Petroleum Resources and Prospecting, China University of Petroleum, Beijing

Fuxing Jiang, Peng Shi,

1905
延长石油
YANCHANG PETROLEUM

Research Institute of Shaanxi Yanchang Petroleum Group Co., Ltd


# 1. Introduction

- **Organic matter in the shale is generally thought to primary control the porosity, pores-structure and gas adsorption capacity of shale;** (Chalmers and Bustin, 2007,; Curtis et al., 2012; Gasparik et al., 2013; Loucks et al., 2009; Rexer et al., 2014; Ross and Bustin, 2007,; Slatt and O'Brien, 2011; Valenza et al., 2013)
- **Positive correlations between the amount of organic matter (TOC) and the pores volume and gas sorption capacity have been observed by many previous works** (Chalmers and Bustin, 2007; Gasparik et al., 2014; Hu et al., 2015; Ross and Bustin,2007, 2009; Strapoc et al., 2010; Weniger et al., 2010; Zhang et al.,2012)

# Introduction

- **Higher mature shale with more micro-nano organic pores could absorb more methane than those of lower maturity under the same geological conditions.** (Bernard et al., 2012; Gasparik et al., 2014; Javie et al., 2007; Ross and Bustin, 2007, 2009)
- **At present, the global shale gas exploration and development mainly focus on the marine shale, such as the Barnett Shale in the U.S. and the Longmaxi Shale in Sichuan Basin, China.**(Biet al., 2014; Gale et al., 2007; Gasparik et al., 2014; Loucks et al., 2012; Rodriguez and Philp, 2010; Slatt and O'Brien, 2011; Tanget al., 2015; Zhang et al., 2012)

# Introduction

- **The Ordos Basin locating in the central part of the North China is the second largest sedimentary basin (covering an area of approximately $32 \times 10^4$ km$^2$ ) in Chin。**
- **The Upper Triassic Yanchang Fm mainly is lacustrine shale in Ordos Basin, of excellent well-developed dark shale and abundant hydrocarbon potential, are a most important oil and gas exploration targets, in China**

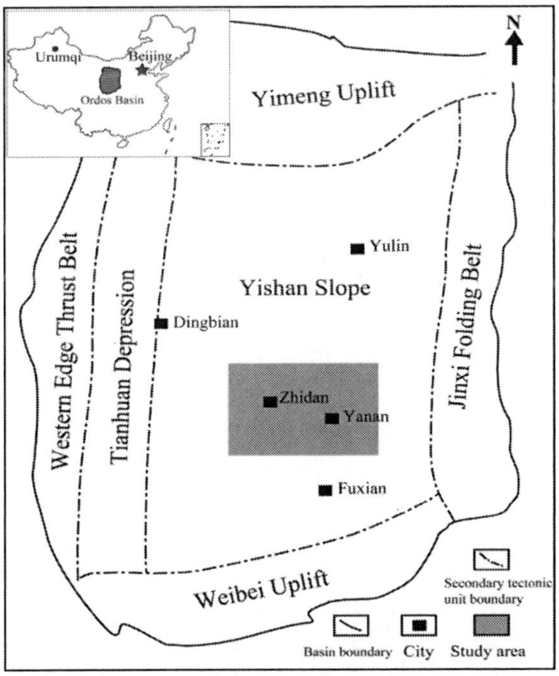


The map of location and general structure of the Ordos basin.

# Introduction

## Geochemical parameters:

- **Vitrinite reflectance(Ro) of YCH Fm. is 0.83%～1.1% in the most part of Ordos basin**
- **TOC of the shale range 0.4 to 12.0%, average is 4.6%.**
- **The shale have very high residual bitumen, especial the shale with higher TOC. The S1 range from 1 to 5 mg/g**
- **This work is to investigate the role of the residual bitumen in the shale to porosity, the pore structure and storage of HC-gas, especially the shale with higher TOC.**

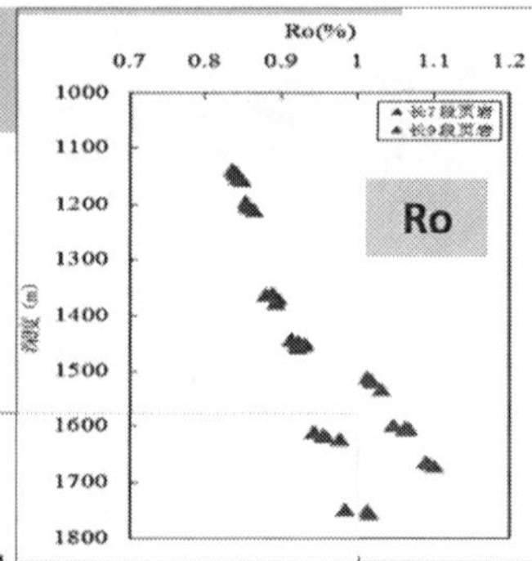


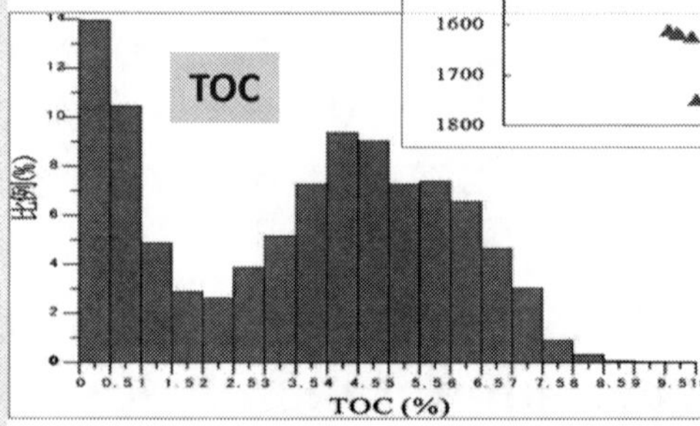


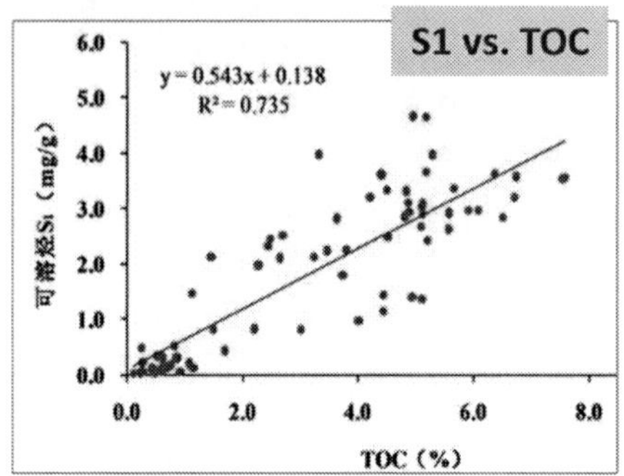


# 2. Samples and experiments

- Three shale core samples were chosen for this work。
- **three samples** ,respectively, are representative of lower，medium，and higher organic matter shale( TOC= 3.4, 7.4, 11.0%).
- **three samples** all are in the mature stage (Ro: 0.89 to 0.87%)

**Table 1 The geochemical parameters of shale samples from Yanchang Fm , Ordos Basin**

| Sample number | lithologic characteristic | TOC（%） | Ro(%) | |
|---|---|---|---|---|
| YY-d01 | Dark shale | 3.4 | 0.87 | |
| YY-d02 | Dark shale | 7.4 | 0.83 | |
| YY-d03 | Dark shale | 11.0 | 0.86 | |

# Samples and experiments

- **Shale samples were crushed to 80 mesh size**
- **Every sample divide in to three groups, one original group, second group extracted by dichloro-methane and the other group extracted by trichloro-methane**
- **Soxhlet extraction:** Soxhlet extractions were carried out for 24h. And after extractions, the **remnant** shale samples were dried in a vacuum at normal atmospheric temperature for 12h.

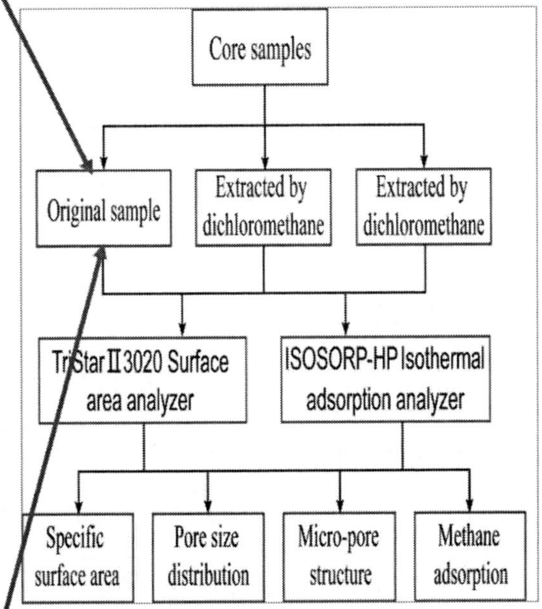


scheme of works process

# Samples and experiments

- **Low pressure $N_2$ isotherm analysis**: Nitrogen sorption experiments were conducted with a Micro-metrics® Tristar II 3020 surface area analyzer
- **High pressure methane sorption:** High pressure methane sorption isotherms (reservoir temperature 52 C; up to 20 MPa) were obtained using FY-KT 1000 isothermal adsorption apparatus。

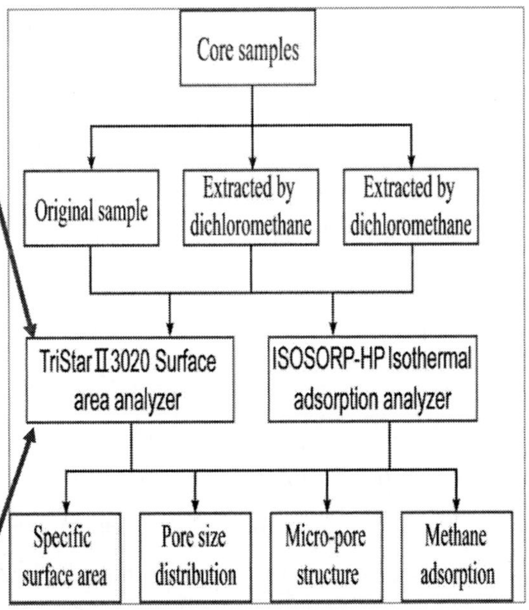


scheme of works process

# 3. Results and discussion

- We can see, from the table, after extraction, PV: (pore volume),and SSA(specific surface area), are significant growth due to the remove of residual bitumen, especially SSA could be over one-magnitude-order.
- It is shows, the higher TOC is, the more pore volume is released after extraction.

Table 2 The PV: (pore volume) and SSA(specific surface area), of shale samples (original and extracted) from Yanchang Fm .

| Sample number | Method of extraced | PV($cm^3/g$) | SSA($m^2/g$) |
|---|---|---|---|
| YY-d01 | Original | 0.0203 | 4.31 |
| | dichloro-methane | 0.0262 | 8.97 |
| | trichloro-methane | 0.0267 | 9.53 |
| YY-d02 | Original | 0.0188 | 4.29 |
| | dichloro-methane | 0.0314 | 18.71 |
| | trichloro-methane | 0.0294 | 19.59 |
| YY-d03 | Original | 0.0099 | 3.48 |
| | dichloro-methane | 0.0195 | 26.48 |
| | trichloro-methane | 0.0189 | 27.38 |

# Results and discussion

The comparison of pore size distributions of pore volume between the original sample and the samples after extraction by dichloromethane and trichloromethane.

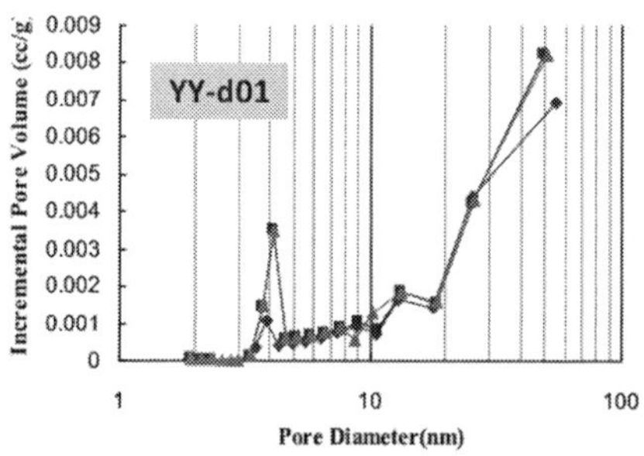


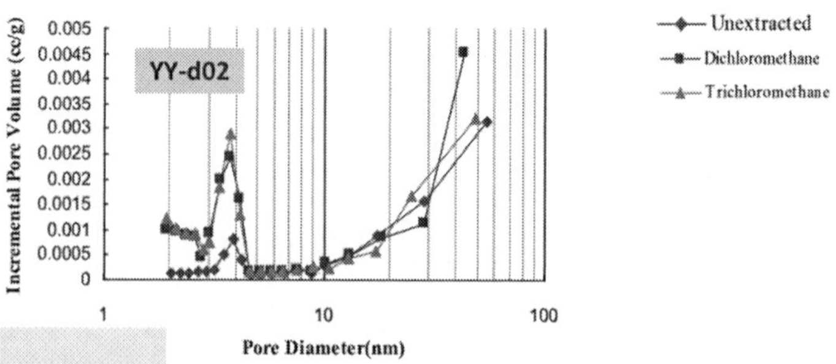


- **Pore size distribution:**

  From the curves of pore size distributions (Figs.), we can see the apparently increase of pore volume range from 2 to 4nm size, especially those higher TOC sample (sample YY-d03),

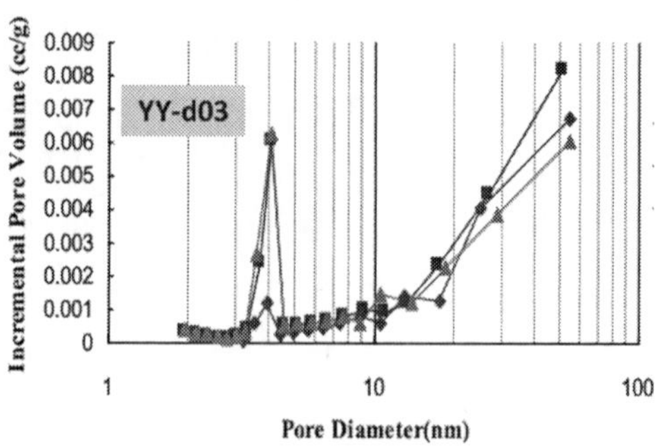

# Results and discussion

- **Specific surface area:**

  It shows, from the Figs, after extraction SSA ( specific surface area) are obvious increase due to the remove of residual bitumen, could be over one-magnitude-order in the higher TOC samples (sample YY-d02,YY-d03),.

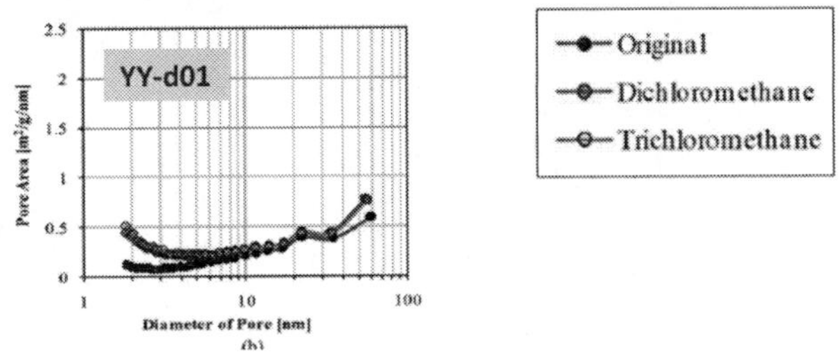


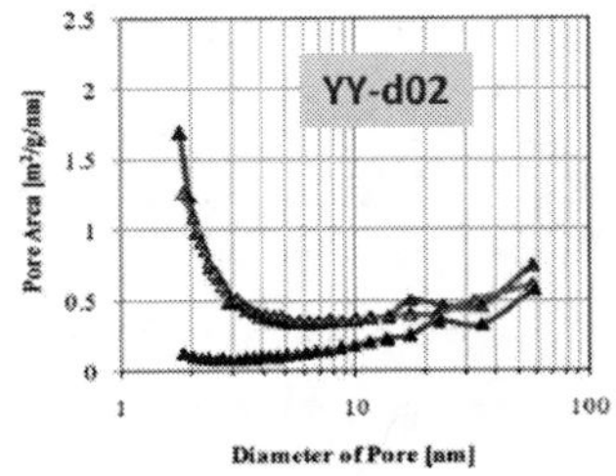


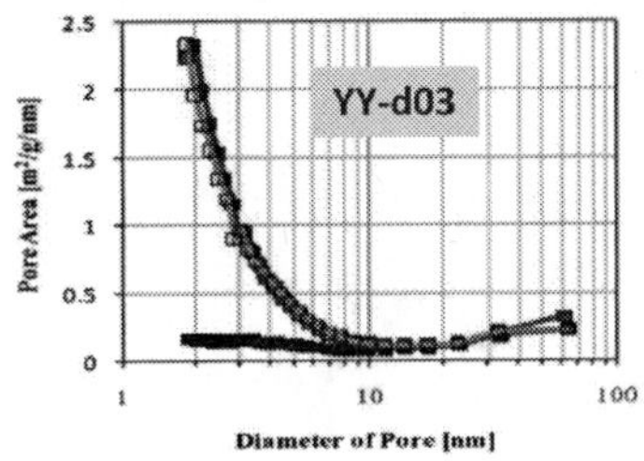


The comparison of pore surface area among the original sample and the samples after extraction

# Results and discussion

- The specific surface area and TOC of shale samples before extraction (Fig. 12) shows a negative correlationship.
- The negative correlationship exists widespread in shale of Yanchang Fm. from Ordos Basin (Fig. 13).
- This may means the abundance of residual bitumen fill the micro-to meso- pores in the shale, because the shale from YCH Fm. is in the mature stage.

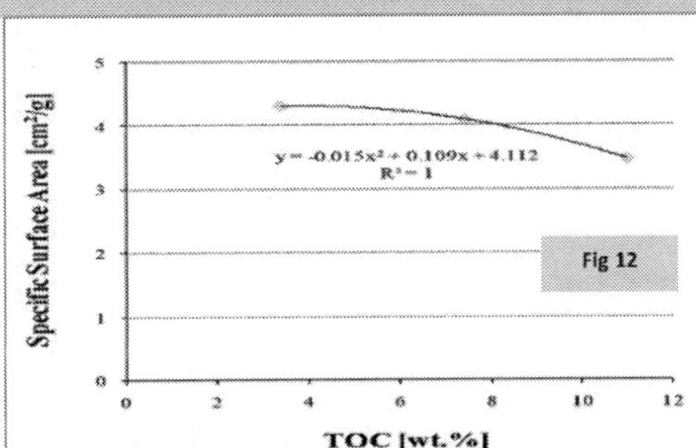


The relationship between TOC and specific surface area of original shale samples.

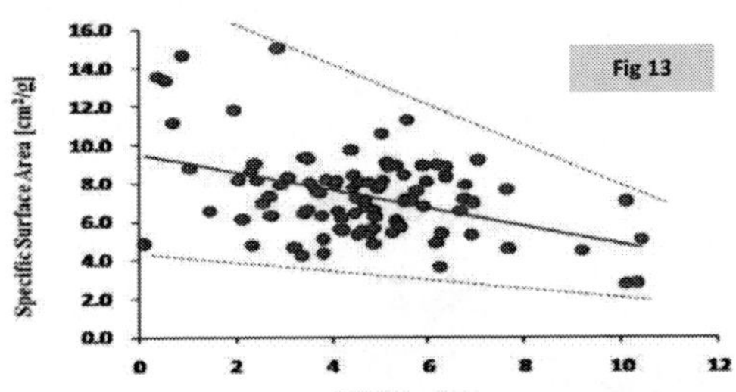


The plot of TOC vs. specific surface area of shale samples from Yanchang Form. In Ordos Basin.

# Results and discussion

- It is shows, comparison the figs, the higher TOC is, the more specific surface area is released after extraction.
- When extracted, the residual bitumen in shale was carried out, and the higher the content of organic matter is, the higher the content of extracted organic matter is, and the greater the growth of specific surface area is.

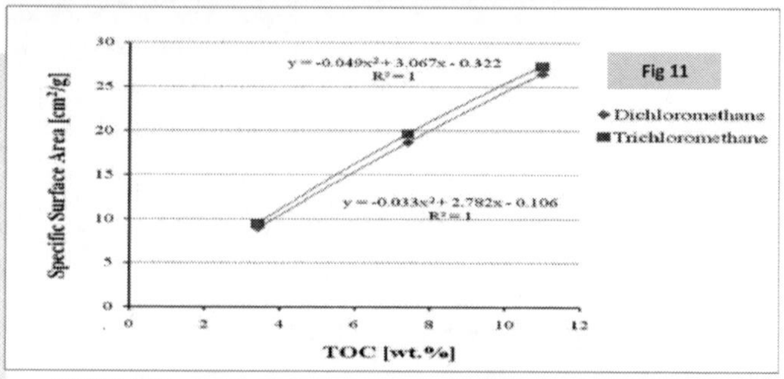


The relationship between TOC and specific surface area of shale samples after extraction.

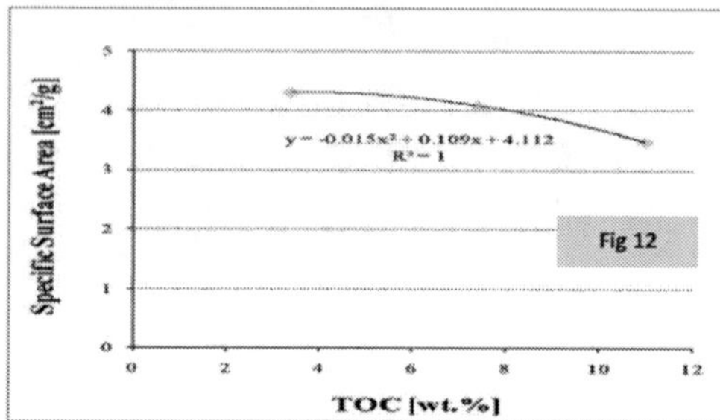


The relationship between TOC and specific surface area of original shale samples.

# Results and discussion

- **Pore type:**

  Nitrogen adsorption /desorption isotherms were applied to investigate pore shapes

  Different adsorption /desorption isotherms means different pore structure, as show in the figs.

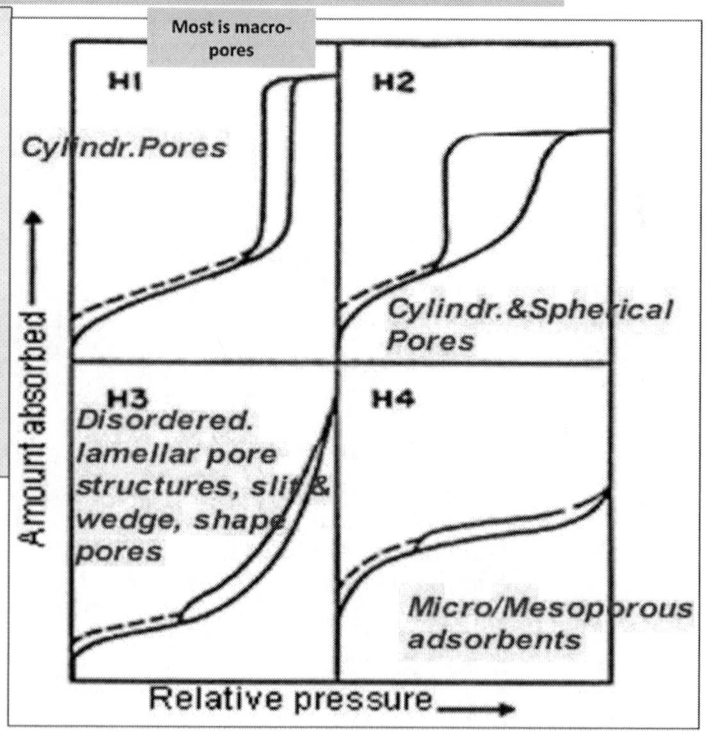


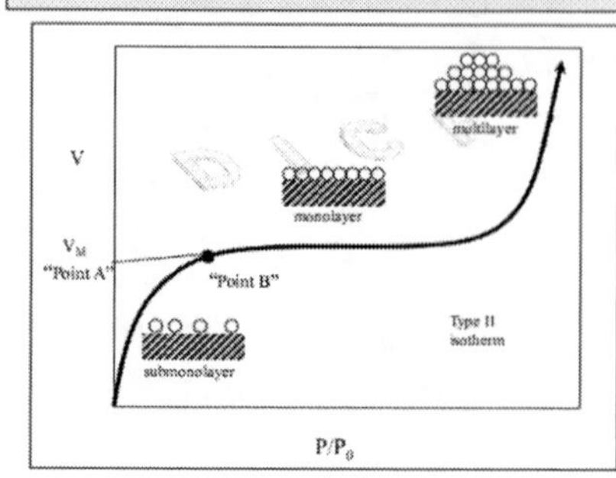

# Results and discussion

- According to nitrogen adsorption /desorption isotherms before and after extraction (Figs. 14), although there are some differences among nitrogen adsorption isotherms, these nitrogen adsorption isotherms almost all present "S" type (type H4). as a whole.
- checking extracted sample against the original, under the lower pressure the volume of adsorption is significant increased, espacially the higher TOC sample (YY-d03), this means the more micro-nano organic pores are released.

Fig 14

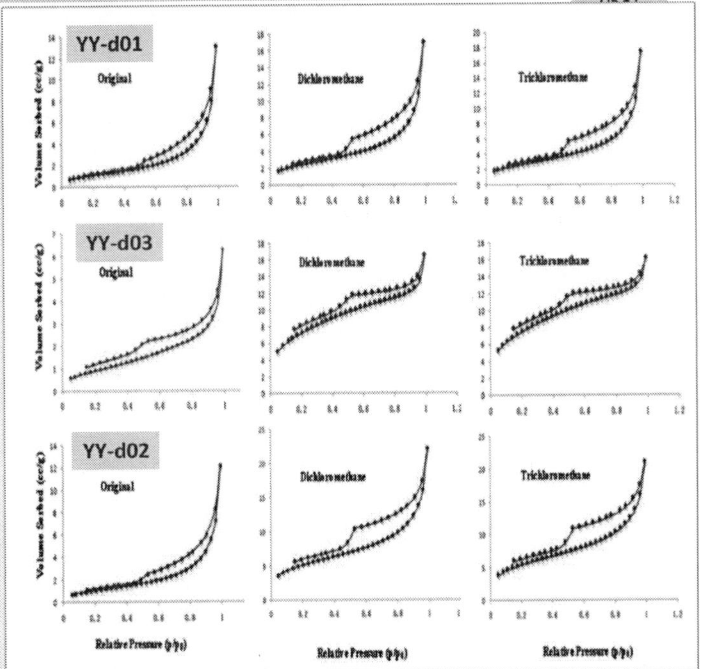


Nitrogen adsorption/desorption isotherms at low temperature (77 K) for shale samples from YCH Fm.

# Results and discussion

- The nitrogen desorption isotherms become steeper than the nitrogen adsorption isotherms at a medium pressure, forming the wide **hysteresis loops**, corresponding to the presence of amorphous pores, such as ink-bottle-shaped pores.
- The results also suggest the well development of micropores and mesopores likely acting as the narrow necks of ink-bottle-shaped pores, which could be conducive for the accumulation of absor- bed gas, however, unfavorable for the seepage of shale gas.

# Conclusions

- For mature shale，  the presence of the residual bitumen dramatically influences the pores of <10 nm, especially the organic-related micropores.
- After extraction, the specific surface area and pore volume greatly increase. The release of a number of micropores is contributed to the great growth of specific surface area.

# Conclusions

- The residual bitumen plays an important role in the pore structure and shale gas storage capacity as it exerts a great influence on the pore volume and specific surface area.

Thanks for Your Attention

THIRTY - THIRD - ANNUAL
INTERNATIONAL PITTSBURGH
COAL CONFERENCE

# Analyses of Adsorbed and Free Gas Production Using Methane Carbon Isotope Fractionation During Canister Desorption of Shale

Ma Yong* Zhong Ningning * Zhejun Pan**

*China University of Petroleum-Beijing
**CSIRO Energy Flagship

*11/8/2016 Cape Town, South Africa*

## Introduction

**Shale gas**

**Adsorbed gas**
+
**Free gas**
+
**Dissolved gas (Rare)**

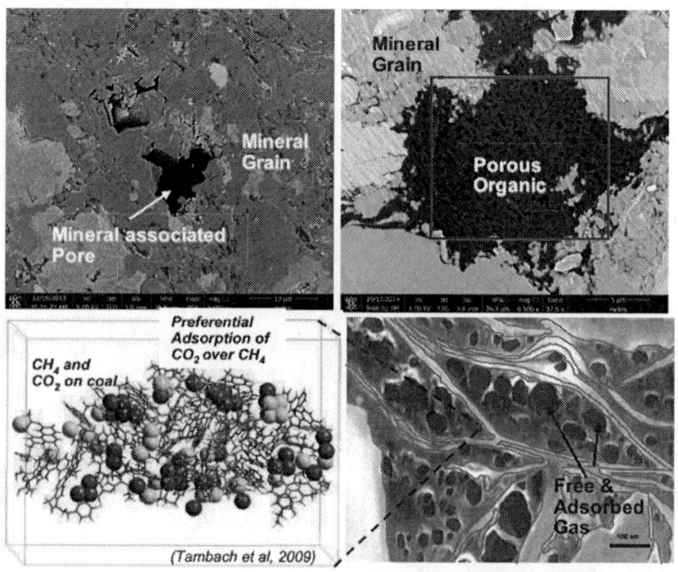


Passey et al. 2009)

Canister Desorption of Shale is a valid method to estimate the gas content and occurrence of shale gas

- ❑ Free gas is crucial to high initial production of shale gas

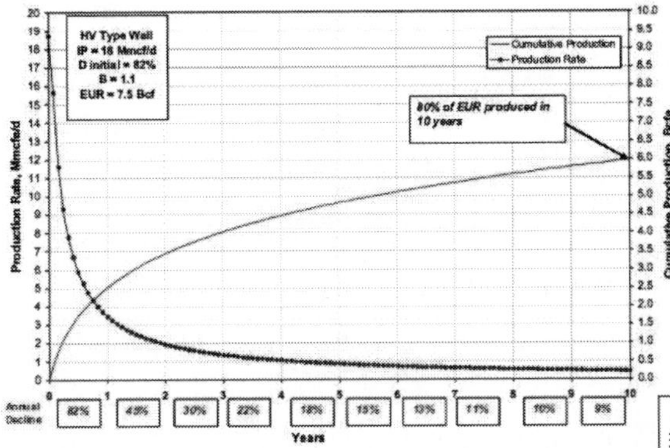


Most of the Production in the First Year (From S. Ellis, 2011. USGS)

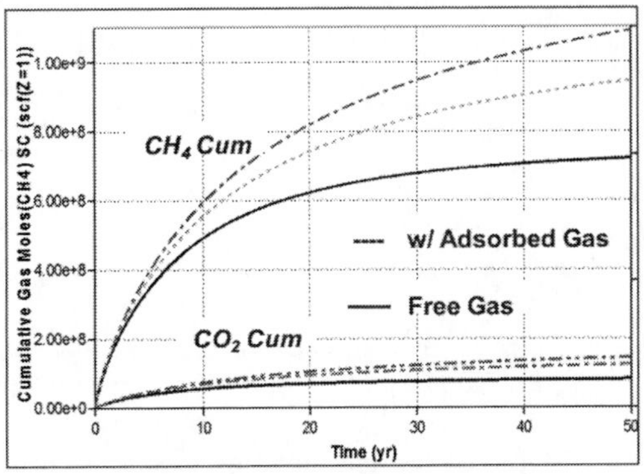


Gas Forecast with 3.3 and 6.2 % TOC, 10% $CO_2$ (From Exxon Mobil)

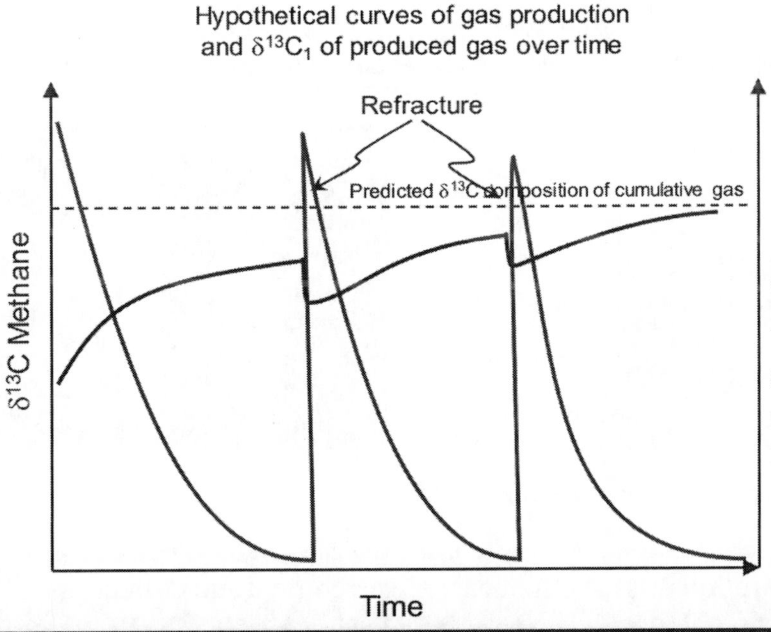


Knowledge of where you are along this curve during production could provide valuable information regarding remaining reserves (S. Ellis, 2011. USGS)

# Sampling and method

Map showing the sampling locations

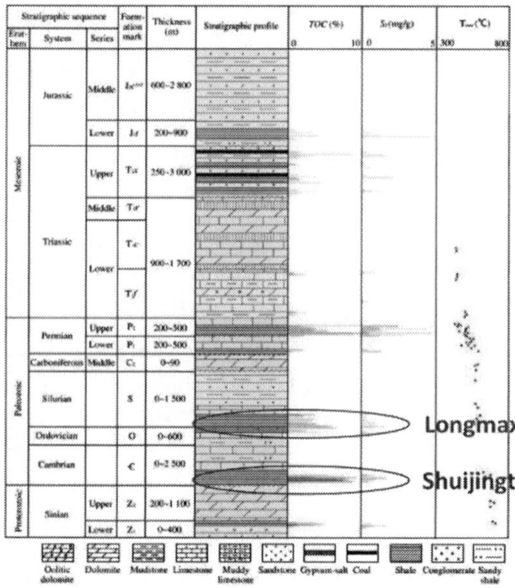


The stratigraphic units in Sichuan Basin (From Dai et al., 2014).

# Sampling and method

Reservoir temperature —<2ml/10hrs→ Elevated temperature (60℃， 90 ℃)

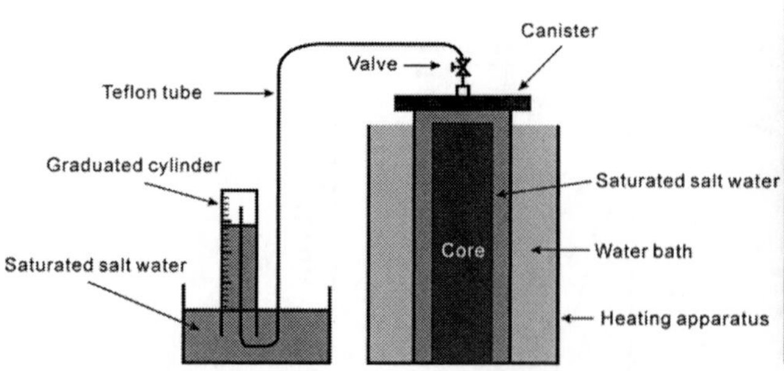


Schematic diagram of desorption experiments

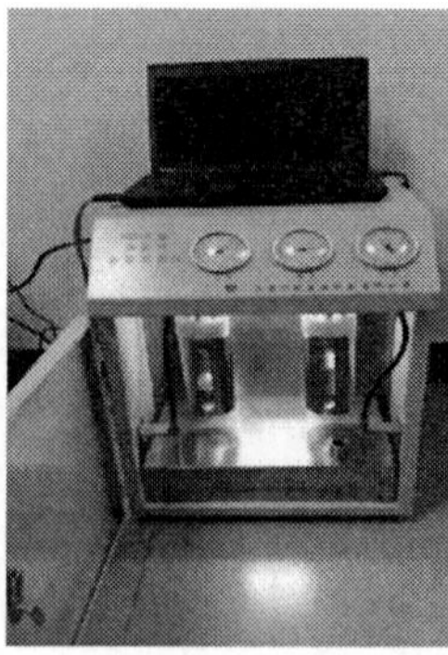

- ❑ Canister desorption was used to measure gas content and collect gas samples
- ❑ Then gas samples were used for gas chromatography (GC) and isotope analyses

# Results

- Longmaxi Shale: about 88% measured gas content was achieved at reservoir temperature.
- Shuijingtuo Shale: only 41%-75% were achieved at reservoir temperature.

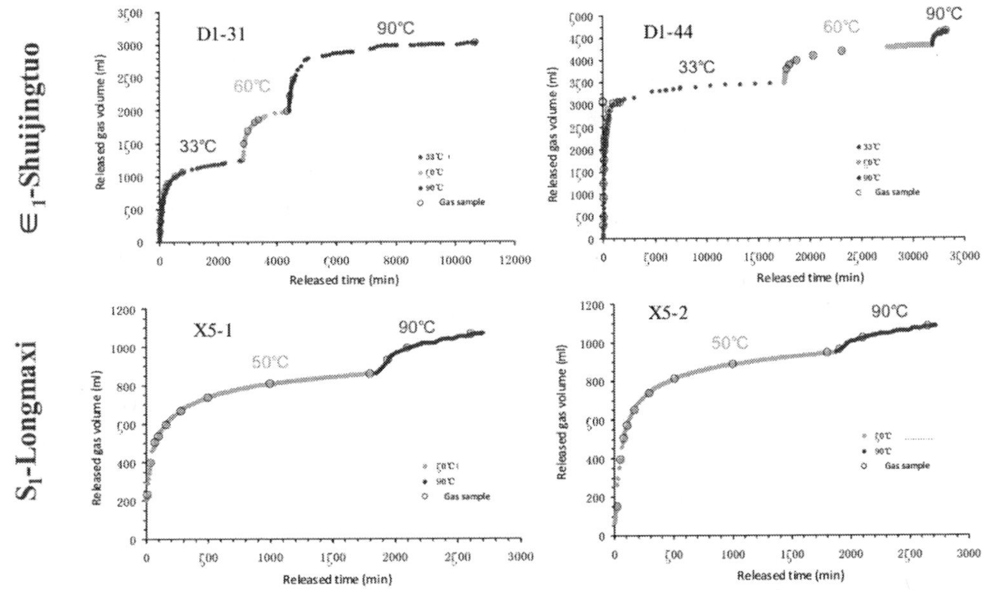


Released gas volumes over time

**GC Results**

- GC analyses of 66 gas samples show that methane is the main component of measured shale gas.
- Low amounts of ethane, carbon dioxide and nitrogen were also detected.

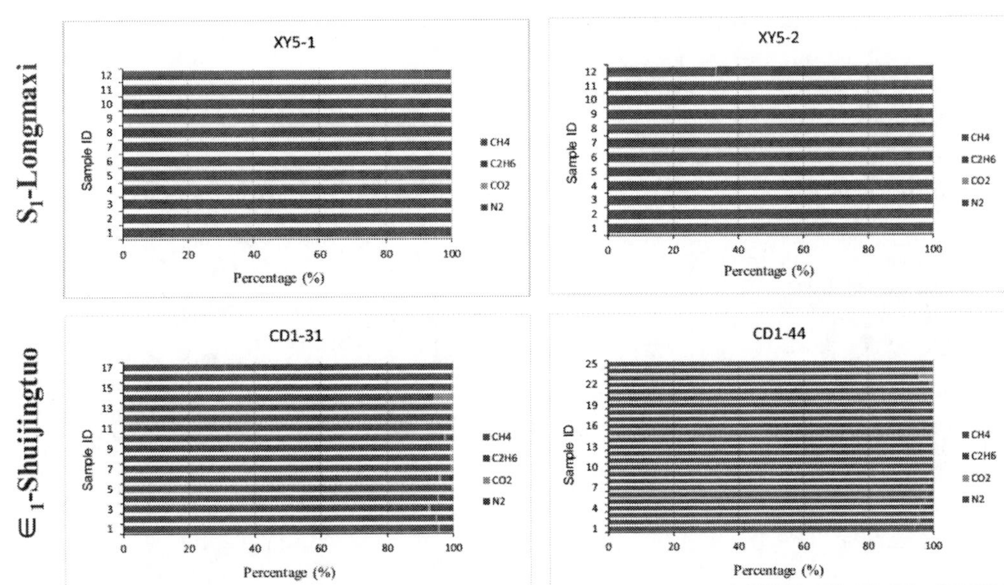

- $S_1$-Longmaxi Shale: $\delta^{13}C$ values of $CH_4$ increase from -37.4‰ to -23.2‰ by 14.4 ‰ after 44 hours desorption
- ∈1-Shuijingtuo Shale: $\delta^{13}C$ values of $CH_4$ increase from -41.9‰ to -34.7‰ by 7.2‰ after 178 hours desorption

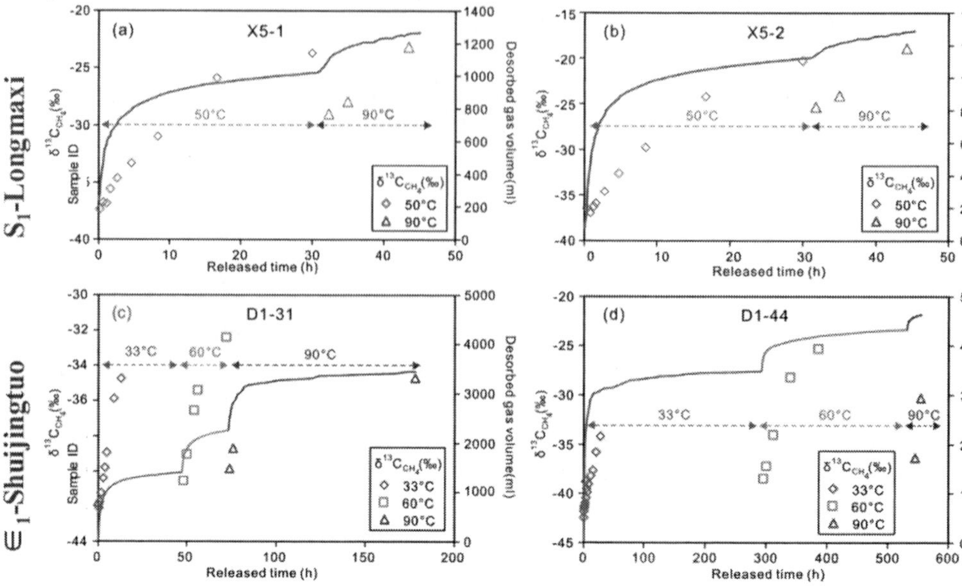


Released gas carbon isotope composition over time

The main phenomena producing isotope fractionations are (Hoefs, 2009):
1. Isotope exchange reactions (e.g. evaporation condensation processes)
2. Kinetic processes that depend primarily on differences in reaction rates of isotopic Molecules (e.g. dissociation reaction and diffusion)

Carbon isotopic fractionation of $CH_4$ during canister desorption of coal or shale were considered to be the difference in diffusivity and adsorbance behavior of gas molecules (Zhang and Krooss, 2001; Strąpoćet al., 2006; Xia and Tang, 2012 )

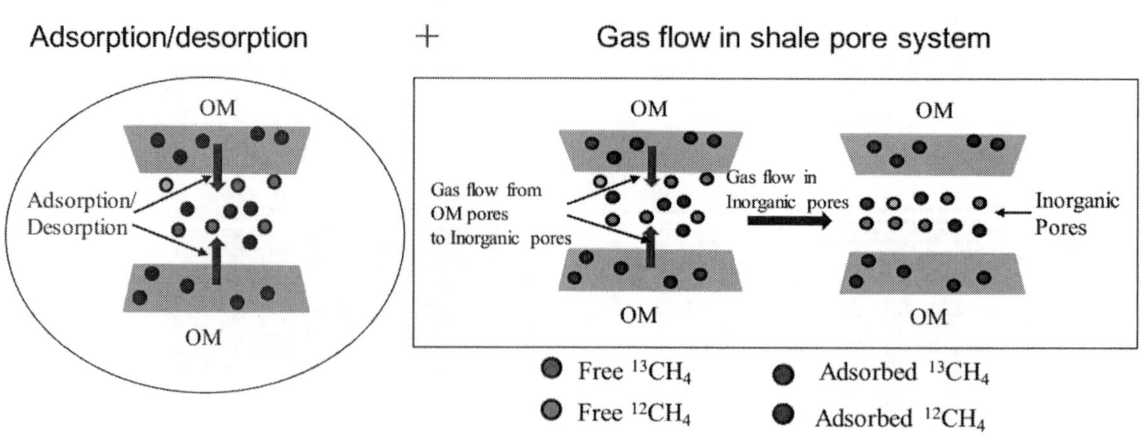

# Multicomponent Gas

Langmuir model:

$$V = \frac{V_L BP}{1+BP}$$

- where V is adsorption volume, $V_L$ is Langmuir volume, B is Langmuir constant, P is pressure.

When extended to mixed-gas adsorption, the extended Langmuir model:

$$V_i = \frac{V_{L,i} B_i P y_i}{1+\sum_j B_j P y_j}$$

- where subscripts i and j mean gas component and y is the gas phase composition.

# Dual porosity

- Adsorbed gas in organic matter, free gas in inorganic pores, gas transfer from organic matter to inorganic:

$$q_{g_i} = f_s \frac{V_m}{\tau_i}\left(\bar{C}_i - C_{eqi}\right)$$

- Gas transport in organic pores:

$$\frac{\partial}{\partial t}\left(\phi\left(\frac{S_g y_i}{B_g}\right)\right) = -\nabla \cdot \left[\frac{k_g k_{rg} y_i}{\mu_g B_g}\left(\nabla p_g + \rho_g g\right)\right] + q_{g_i}$$

# Reservoir simulation – SIMEDWin

- Developed by University of New South Wales and CSIRO, Australia
- Shale Gas/ Coalbed methane production
- Numerical analyses of gas flow in shale and coal cores
- More…

- Multicomponent gas, two phase flow
- Dual or triple porosity
- Adsorption, free gas

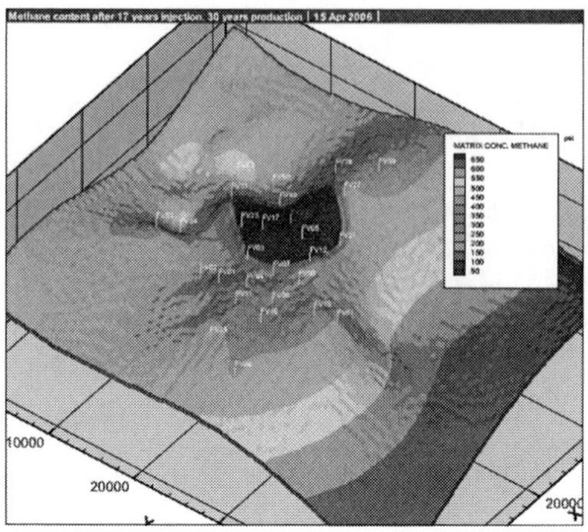


**Table 1.** Relative importance of SIMED-II input parameters

| Parameter | Sensitivity of results to parameter | | | Ability to measure | | |
|---|---|---|---|---|---|---|
| | Low | Middle | High | Low | Middle | High |
| Seam dimensions | | | | | | |
| Roadway dimensions | | | | | | |
| Gas composition | | | | | | |
| Porosity | | | | | | |
| Permeability | | | | | | |
| Langmuir isotherm | | | | | | |
| Desorption time constant | | | | | | |
| Desorption pressure | | | | | | |
| Reservoir temperature | | | | | | |
| Gas content | | | | | | |
| Relative permeability | | | | | | |
| Pore pressure | | | | | | |
| Compressibility | | | | | | |
| Matrix shrinkage | | | | | | |
| Pore pressure | | | | | | |
| Transmissibility | | | | | | |

## Simulation results

According to Zhang and Krooss (2001), Xia and Tang (2012 ) result, we choose the adsorption and diffusivity parameters as follow:

$$K_{13CH4}/K_{12CH4}=1.01$$

$$D_{13CH4}/D_{12CH4}=0.99$$

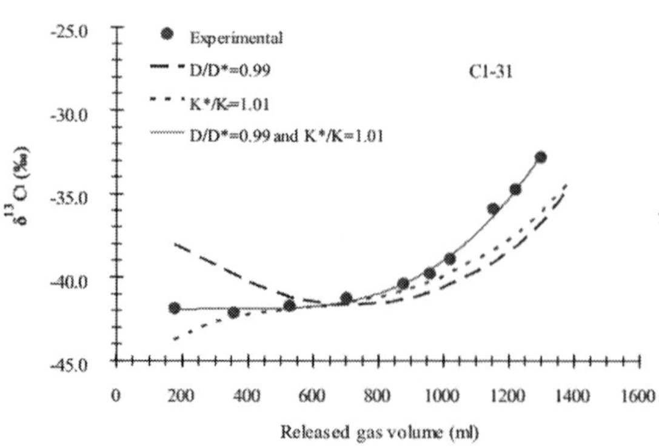


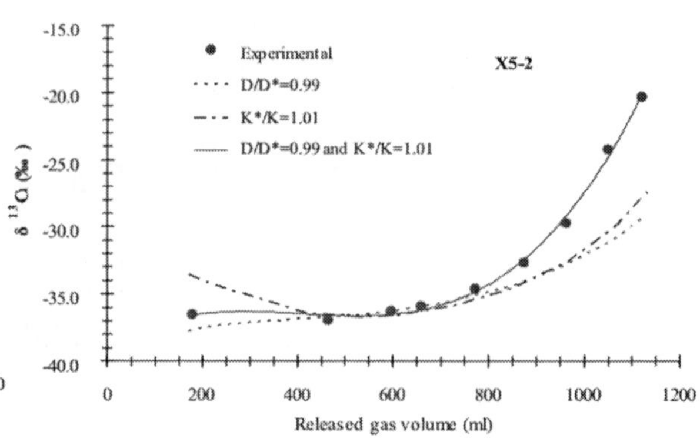

## Simulation results

## $S_1$-Longmaxi Shale

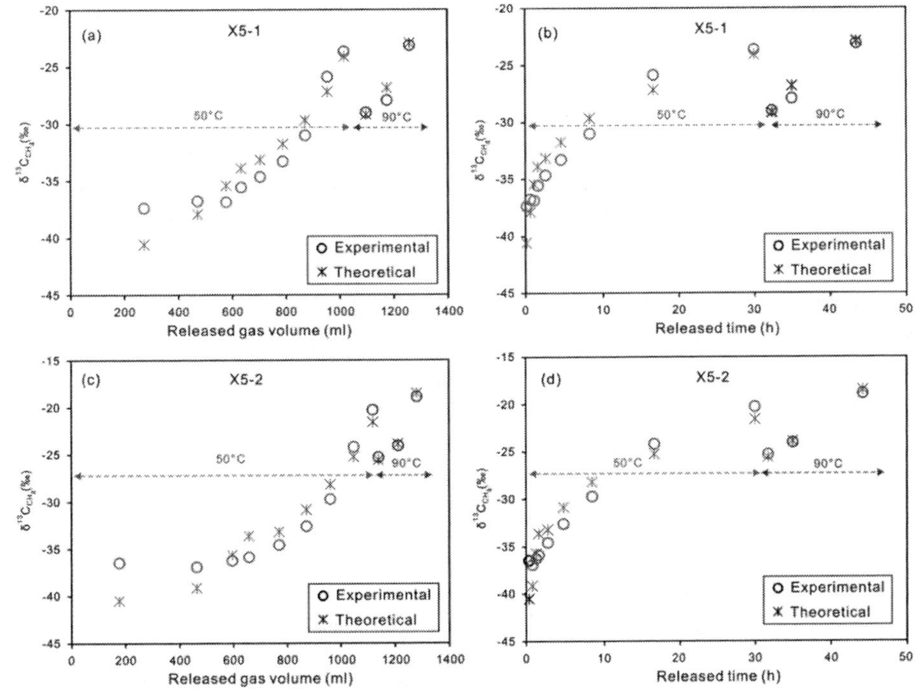


## Simulation results

## $\in_1$-Shuijingtuo Shale

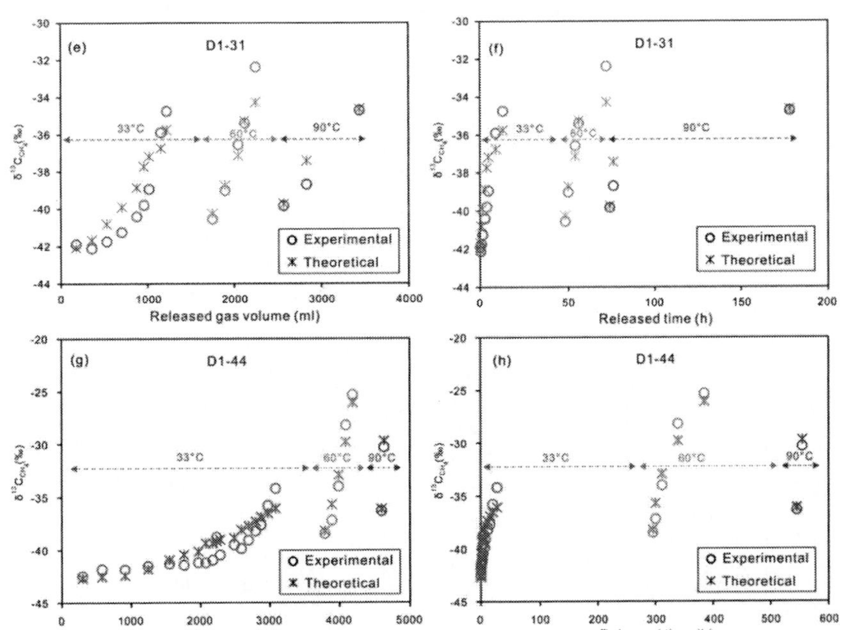

## Free gas and adsorbed gas

**Free gas contributed to the main initial gas production of Longmaxi Shale**

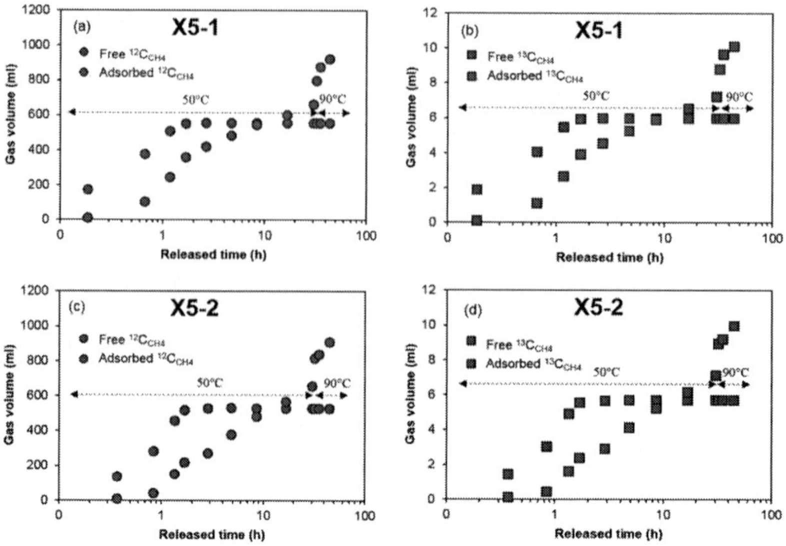


$S_1$-Longmaxi Shale

## Free gas and adsorbed gas

- **The gas production of Shuijingtuo Shale mainly comes from the adsorbed gas**
- **The production rate of Shuijingtuo Shale is much slower compared with the Longmaxi Shale.**

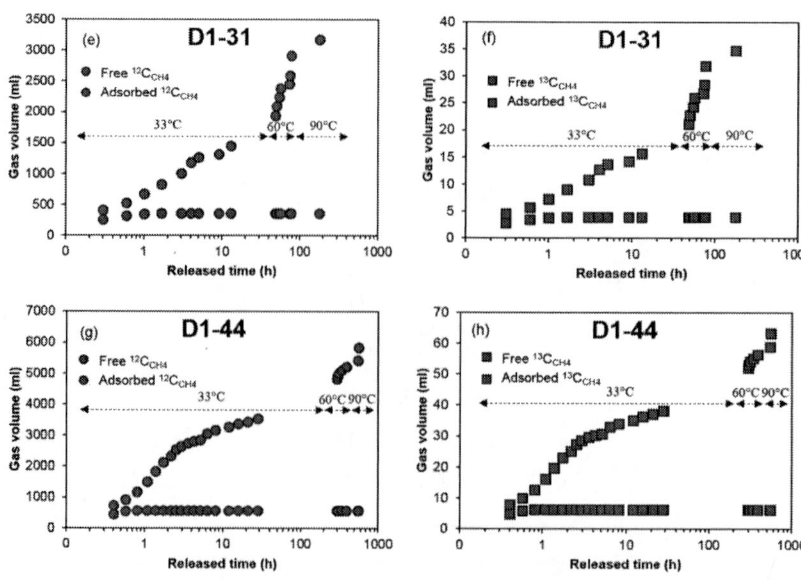


$\in_1$-Shuijingtuo Shale

**Production prediction**

| | initial | | End of RT | | End of Desorption |
|---|---|---|---|---|---|
| • **Percentage of free gas in Longmaxi Shale:** | **85%** | —— | **50%** | —— | **40%** |
| • **Percentage of free gas in Shuijingtuo Shale:** | **40%** | —— | **15%** | —— | **10%** |

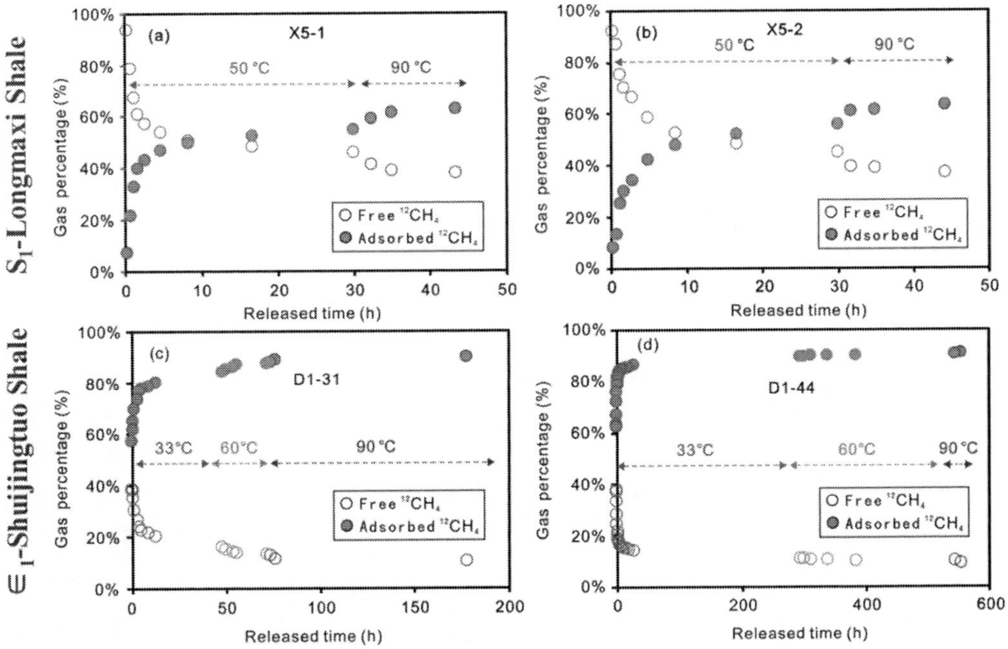


# Conclusions

1. The canister desorption of four shale samples at varied temperature show the changes of released gas volume and carbon isotopic fractionation of methane.
2. The technique of analyzing isotopic composition of produced gas can be used with reservoir simulation for better understanding the shale gas production behaviour in over-matured gas shale especially its long term production behaviour.

Thanks!
Email： mayong0724@126.com
We also want to thank Dr. Xia Xinyu for his assitance in this work!

**THIRTY - THIRD - ANNUAL**
**INTERNATIONAL PITTSBURGH**
# COAL CONFERENCE

# Anisotropic Permeability and its Relationship with Pore/Fracture Structure of the Longmaxi Shales in the Upper Yangtze Region, China

Zhong Ningning*  Ma Yong*  Zhejun Pan**

*China University of Petroleum-Beijing
**CSIRO Energy Flagship

*11/8/2016 Cape Town, South Africa*

## Introduction

- Shale permeability is crucial to high initial production of shale gas
- Shale has strong heterogeneity, thus making shale permeability anisotropic
- High gas yields up to 500 × $10^3$ $m^3$/day have been achieved in the Lower Silurian Longmaxi Shale in the southeastern part of the Upper Yangtze area (Dai et al., 2014), while the reason has not been well studied before.

JY 1HF

Shale outcrop

# Introduction

- Sub-cored plug samples were widely used for determining shale permeability.
- While it may not help to understand the true permeability anisotropy of highly heterogeneous rocks, such as shale (Chen et al., 2014), because the pore/microfracture system in each plug is different.

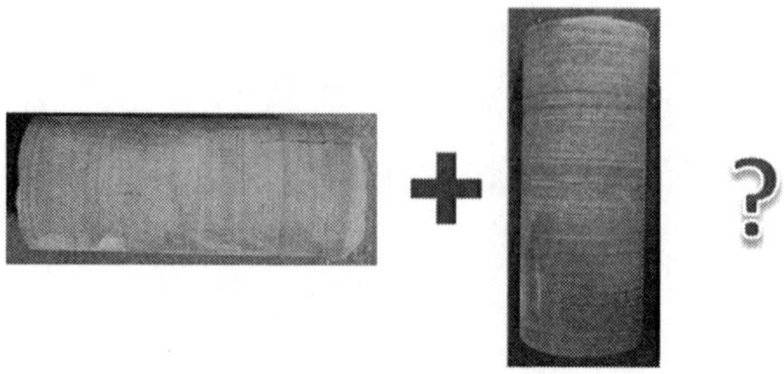

# Experimental method

- Pan et al (2015) reported a new anisotropic permeability measurement method for rocks, using cubic samples with a 3D-printed membrane in a triaxial cell.
- The system can be used to measure permeability in three directions by reorienting the same cubic sample in the triaxial cell each time.

Cubic sample

Membrane

Sample with Membrane

- Main equations used for the permeability:

$$\frac{(P_u - P_d)}{(P_{u,0} - P_{d,0})} = e^{-\alpha t} \quad (1)$$

$$\alpha = \frac{kA\left(P_{u,0} + P_{d,0}\right)}{2\mu L}\left(\frac{1}{V_u} + \frac{1}{V_d}\right) \quad (2)$$

Permeability measurement Rig in CSIRO

# Sampling

- Longmaxi Shale in the southeastern Upper Yangtze Region: Main shale gas exploration and development in China
- Three shale samples in a single well with are selected

**Sample information**

| Sample ID | Depth (m) | TOC (%) |
|---|---|---|
| Y4-60 | 705.4 | 0.95 |
| Y4-98 | 748.3 | 3.72 |
| Y4-103 | 754.2 | 8.49 |

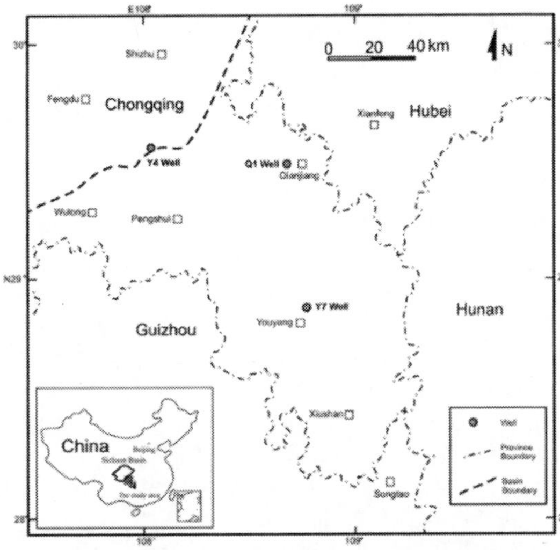

Map showing the sampling locations

Y4-60 Y4-98 Y4-103

Horizontal well site in Fuling

# Results and Discussion

Anisotropic permeability of the shale samples by helium

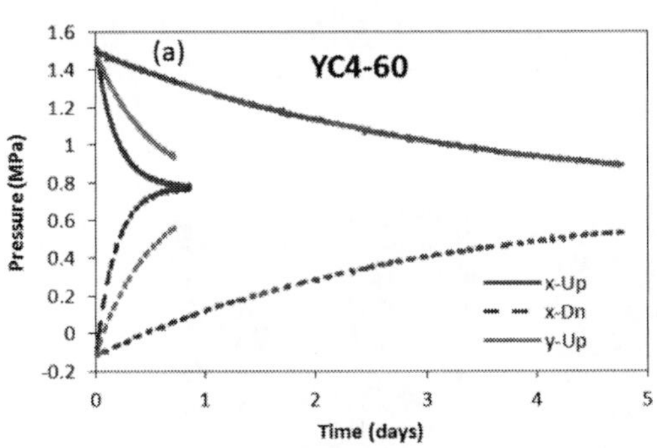


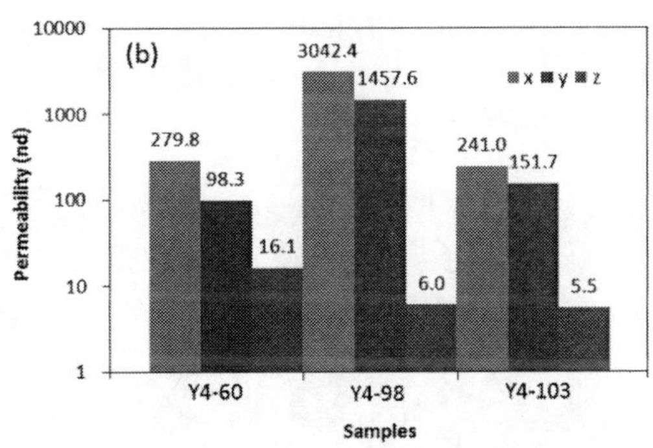


- x-direction (horizontal): 241.0~3042.4 nd
- y-direction (horizontal): 98.3~1457.6 nd
- x-direction/ y-direction: 1.6~2.7
- z-direction (vertical): 5.5~16.1nd
- horizontal permeability to vertical permeability: 6.4~510.5

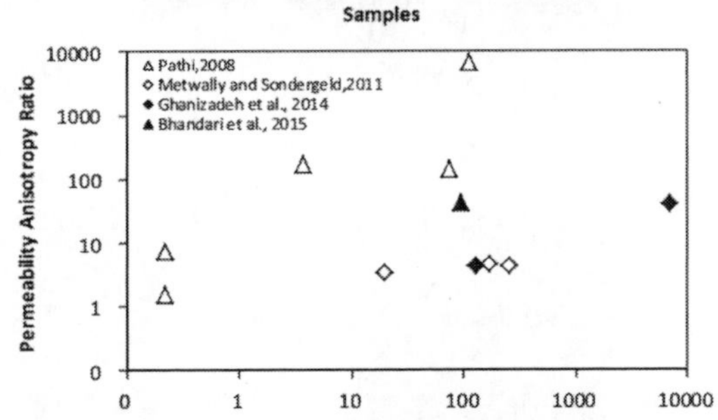

Permeability anisotropy of shales in literature

These results illustrate strong anisotropy between horizontal and vertical permeability

## Results and Discussion

**SEM and SEM Maps Observation**

- Millimeter-sized microfractures along bedding direction are greatly developed
- Micrometer-sized microfractures between mineral grain edges and inside the clay minerals are developed. However, they are non-directional and discontinuous.

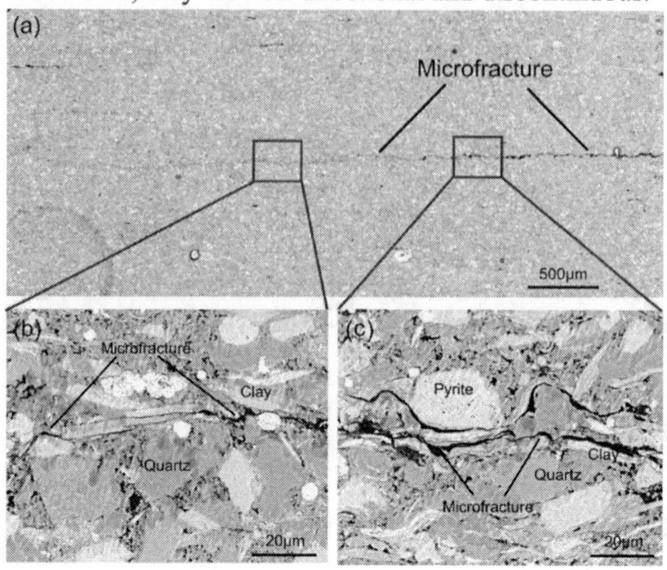


SEM Maps

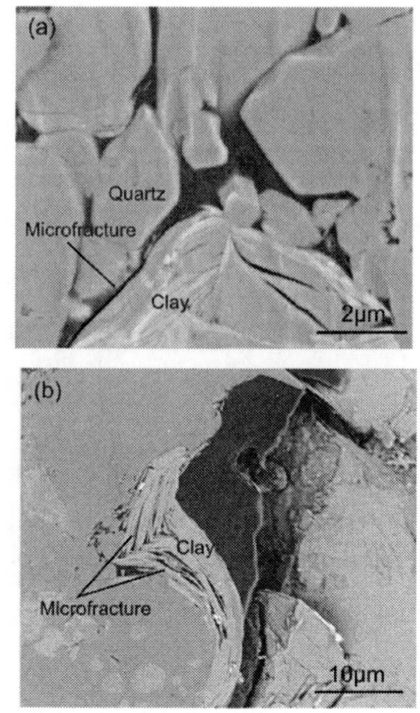


SEM images

## Results and Discussion

**Micro-CT**

- 3D reconstruction of Micro-CT images showing that microfactures along bedding direction are developed in sample Y4-98 and Y4-103 are developed.

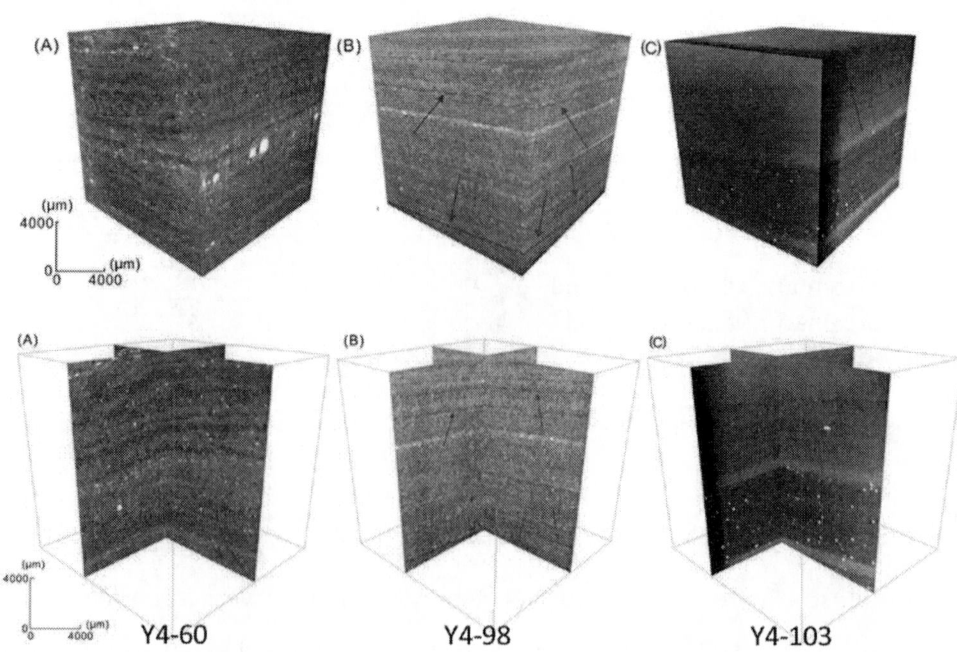

## Results and Discussion

Micro-CT

➢ 2028 slices were equally divided into 15 pieces to manually character the microfractures distribution inside the shales.

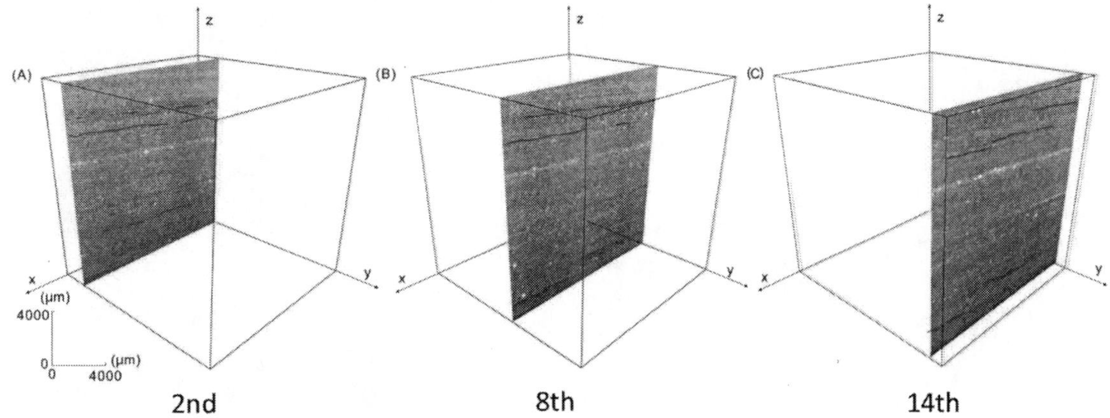


➢ We defined fracture density along one direction in this study:

Fracture density= Fracture numbers/15 pieces

## Results and Discussion

Micro-CT

➢ The permeability has good relationship with fracture density

The relationship with permeability and fracture density

| Sample ID | Direction | Permeability (nd) | Fracture density |
|---|---|---|---|
| Y4-98 | X | 3042.4 | 5.8 |
| | Y | 1457.6 | 4.5 |
| Y4-103 | X | 241.0 | 1.2 |
| | y | 151.7 | 0.8 |

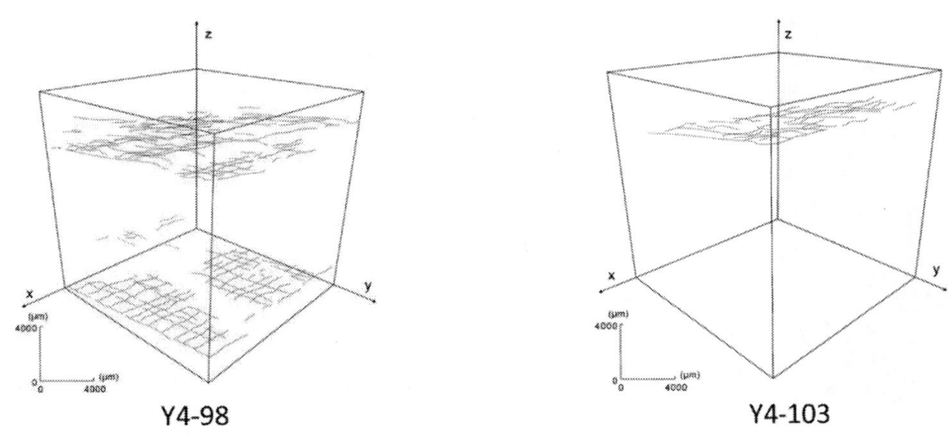

# Results and Discussion

**Origin of the Microfractures**

- The bedding was formed by the difference of deposits, particle sizes, and it was a structural weak plane
- The Lower Silurian formation was a main detachment layer in the eastern Sichuan Basin
- The microfractures might be partly closed or closed under stress in the formation, and became gradually open with the core recovery process due to the stress release

Graptolites in the Longmaxi Shale

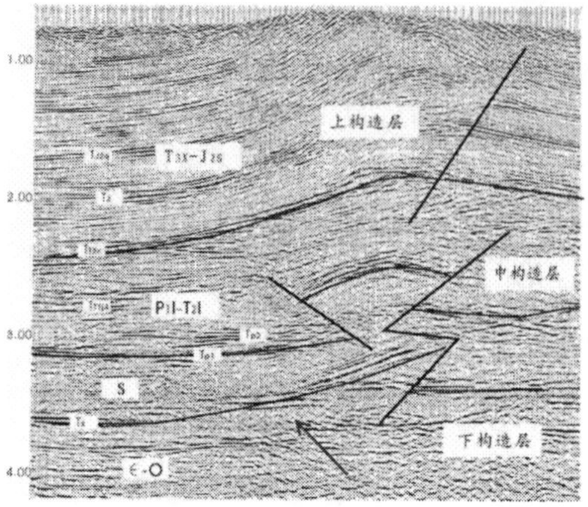


Three detachment layers in the eastern Sichuan Basin. From Hu et al (2004)

The striation in the Longmaxi Shale, 760m, Y4 Well

# Results and Discussion

**Geological controls on shale permeability anisotropy**

- The appearance of microfracture may change permeability variations between shale samples.

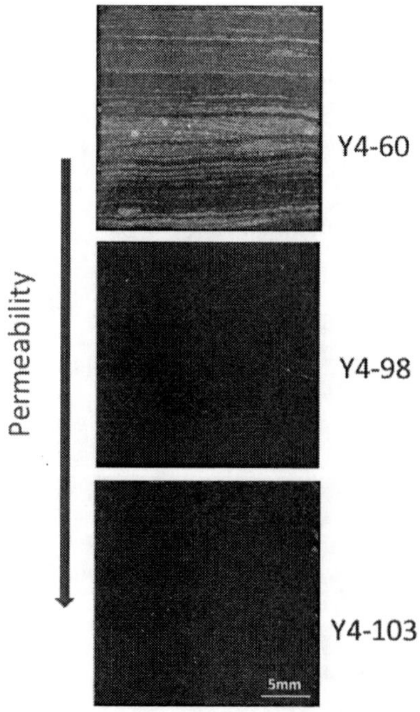


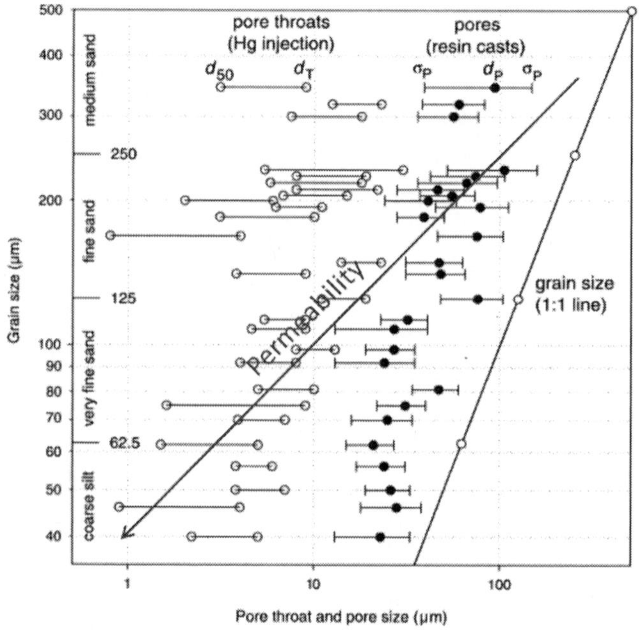


Grain size, pore size, and pore-throat size for sandstone samples (Modified from Wardlaw and Cassan, 1979)

## Results and Discussion

- The appearance of microfracture greatly increase the shale permeability also caused the anisotropic permeability between parallel and vertical to bedding directions.

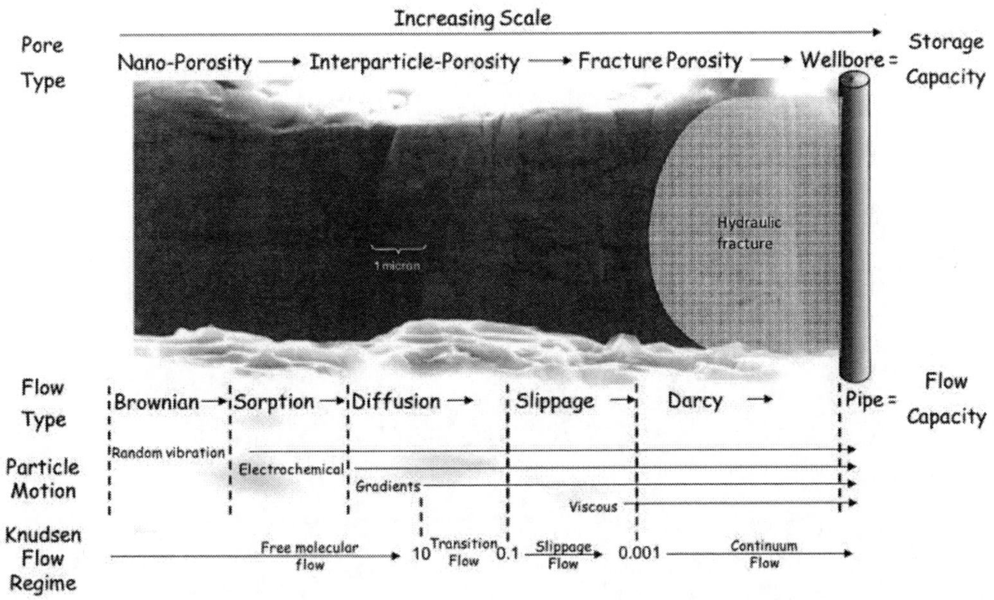


Gas shale storage and flow capacity diagram within a given flow regime (From Sondergeld., 2010)

## Conclusion

- Cubic samples with 3D-printed membrane in triaxial cell is a valid method to determine the anisotropic permeability of tight rocks.
- The three Longmaxi Shale samples all show strong permeability anisotropy between the parallel and perpendicular to bedding directions, varying from 6.4 to 510.5. The permeability anisotropy are mainly influenced by microfractures in the samples.
- The high permeability along bedding direction may contribute to the high initial production of the Longmaxi shale gas.

Ma, Y., Pan Z, Zhong N N, et al. 2016. **Fuel**, 180:106-115

Thank you!
Contact: nnzhongxp@cup.edu.cn

# THE USE OF MICROWAVE ENERGY TO DESUL-PHURIZE S.A COAL

J.H. Potgieter
University of The Witwatersrand & Manchester Metropolitan University

Prof. N. WAGNER
University of Johannesburg
Dr. S. POTGIETER-VERMAAK
Manchester Metropolitan University
Mrs. W. Mohamed
University of the Witwatersrand

## Layout of the presentation

- Microwave desulphurisation - a mad idea?
- How does one actually do it?
- Can it and does it really work?
- OK, so how does one go further?

## General Background

- Gondwana land copals: rich in minerals, relatively difficult to beneficiate and highly variable in rank and organic matter composition
- Sulphur in coals - mainly pyritic, organic and sulphatic. Organic and sulphatic sulphurs are difficult and sometimes impossible to remove
- Microwave applications -extensively used in minerals processing - intense energy flow and penetration depth.
- Desulphurization of coal by microwave energy not new- first patented by Zavitsanos and Golden (1982).
- Dry microwave treatment convert 10 % of the pyrite to phyrrhotite -increase magnetic susceptibly and liberation of iron and sulphide phases during magnetic separation.
- Uslu and Atalay (2003) - 55.1 % sulphur decrease and 21.5 % ash decrease using the same method.

## Previous work (continued)

- Rawson and Rice (1990): Strong caustic solutions (300 g/L NaOH and KOH) and and repeated irradiation - a 70 % total sulphur removal.
- Hayashi *et al.* (1990): Preheated the coal and caustic slurry + microwave treatment for 6 minutes - 90 % of organic sulphur removed.
- Mukherjee and Borthakur (2001): Washed pulverized coal using NaOH solution followed by HCl treatment - 100 % removal of pyritic Sulphur, but only 10 % of the organic sulphur was removed using conventional leaching for 8 hours.
- Yürüm *et al.* (2002): Used hydroiodic acid as a desulphurizing agent with microwaves - complete removal of pyritic sulphur and ±70 % removal of organic sulphur

## S.A. Coal

- Only about 2% sulphur – Gondwana (map on a next slide)
- Sulphur occur in organic + inorganic phase
- Sulphur containing minerals in S.A. coal:
  - Clays
  - Quartz
  - Pyrite (Primary source of sulphur in Coal – not in structure)

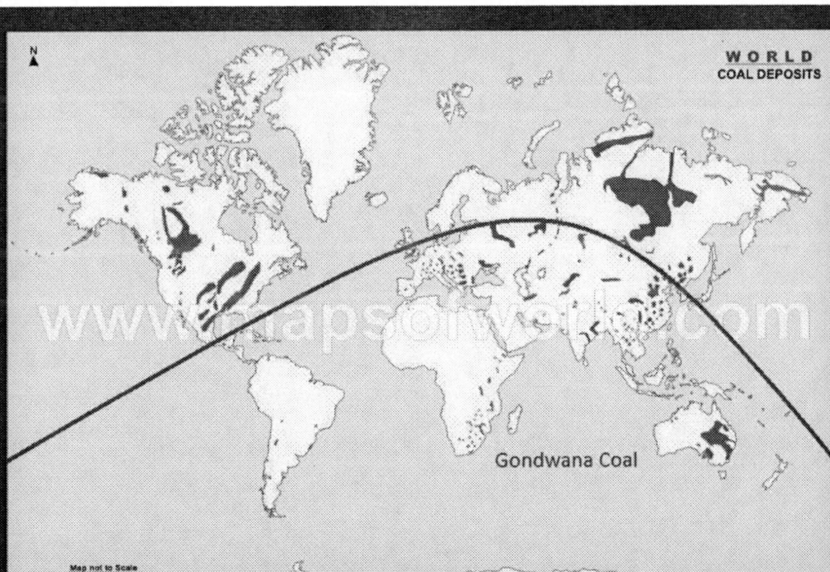

## Aim of the research

- Currently desulphurization done by conventional NaOH leaching
- Microwaves desulphurization could be more effective
- Microwaves = Effective transfer of energy
- Key concerns about sulphur:

1. Corrosion in boilers
2. Quality of coal
3. Pollution

## Experimental Procedure

2 Coal Samples, high (3.3%) and low (0.5%) sulphur concentrations used

Variable power (max 900W) microwave unit

Time and particle sizes varied

Alkali concentrations varied (300 & 400 g/L)

Alkali types varied - solutions used:

1. KOH
2. NaOH
3. NaCl
4. Ca(OH)2

XRD and Raman spectroscopy used for measurements

Promimate analysis and Leco S measurements

## Effect of Microwave Energy (ME) on Coal

- Structure change and Combustion characteristic investigated
- Optimum desulphurization:
  - 650 W Power
  - 10 min exposure time
  - 74 μm and 212 μm particles

## Effect of ME on Coal (continued)

74 μm

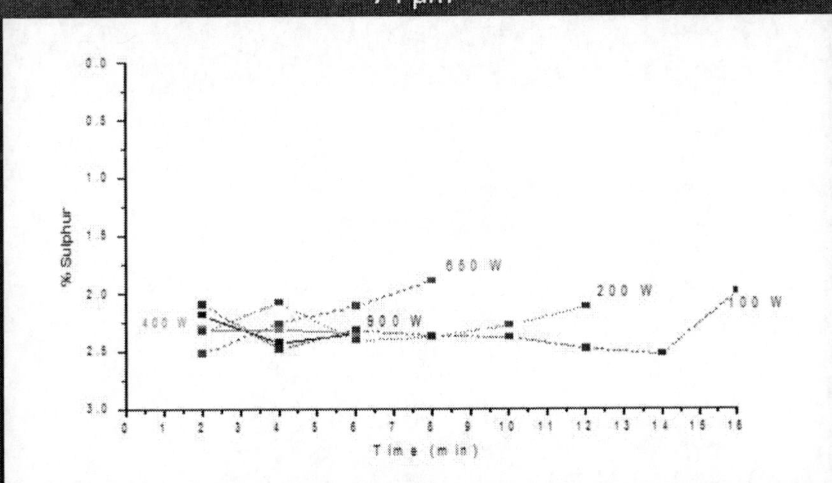

## Effect of ME on Coal (continued)

212 μm

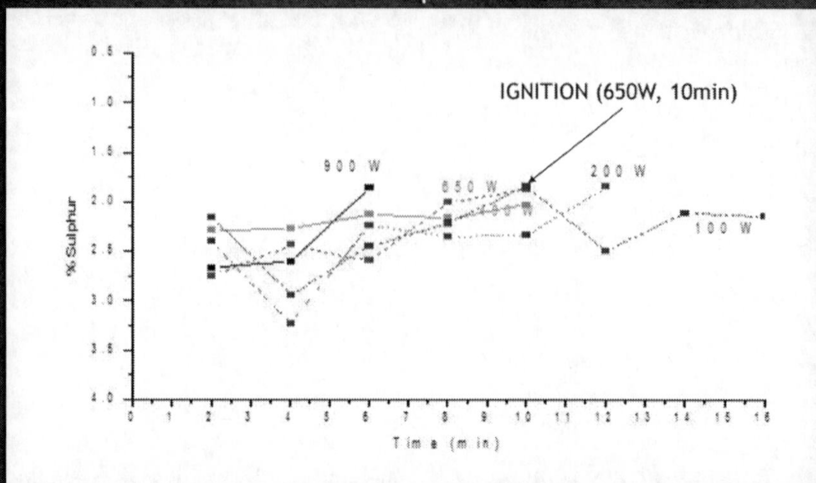


## Effect of ME on Coal (continued)

### Different Particle sizes

300 g/L NaOH, varied time & power

- Desulphurization increases with an increase in exposure time.
- 74 μm coal fraction : 36 % sulphur decrease
- 212 μm size fraction: 6 % sulphur removal
- Calorific value decrease somewhat
- Coal heated for longer than 10 min at 650 W - ignition
- 600 μm size fraction: inconsistent results
- May be due to the pyrite distribution in the coal? Pyrite occurred as large particles, randomly situated, rather than within the coal matrix.

## KOH vs NaOH

- KOH, NaOH halves treatment time (5min) → <u>Energy saving</u>
- KOH better desulphurizing agent on average
- NaOH produces <u>less ash</u> after desulphurization
- Effect of solid to liquid ratio = negligible

## Raman Analysis of Pyrite in Coals

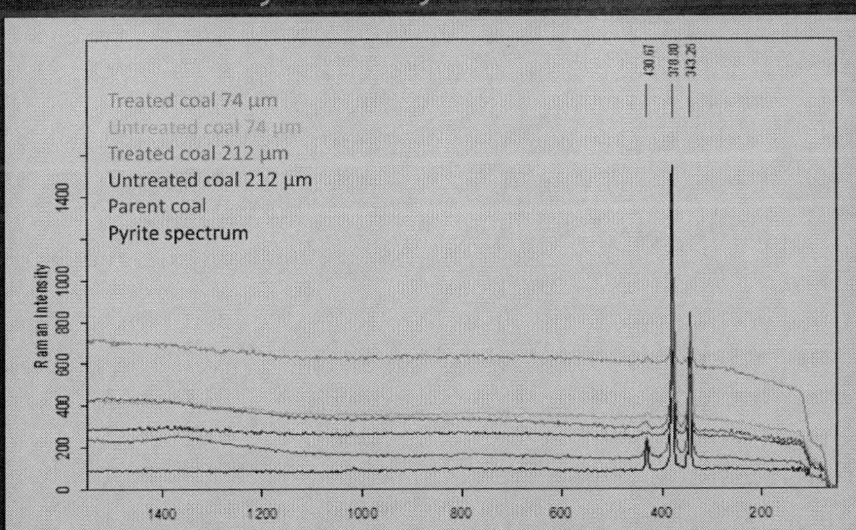

## NaCl vs NaOH

- NaCl lowers max treatment temp + No ignition
- Similar results to using Ca(OH)2
- No difficulty washing coal treated with NaCl
- No sulphur reduction with NaCl
- Noticeable ash reduction with NaCl - further investigation?

## Mixture of NaCl and NaOH (30[l]:70[s])

- NaCl reduces overall temp by forming layer on coal
- NaOH promotes effective leaching of S
- Result = less energy used for effective desulphurization
- 1.37% Sulphur vs 1.30% Sulphur in conventional leaching

## Low Sulphur Coal – did it also work well?

- The low sulphur coal - contained predominantly fine nodules of pyrite
- More difficult to desulphurize.
- Results indicate a 40 % sulphur decrease is possible
- Very similar to 36% obtained for high sulphur coal.

## Summary

- High and Low S coal gave 40% desulphurization
- 60% desulphurization from Rawson and Rice (1990) @500W for 2x60sec intervals
- 30% desulphurization from Araya *et al.* (1981) [Chilean coal] conventional leaching for 8h
- Promising enough to continue?:
- Investigate ME saved vs caloric energy lost
- Overall economics
- Mossbauer to investigate Fe-S structural changes

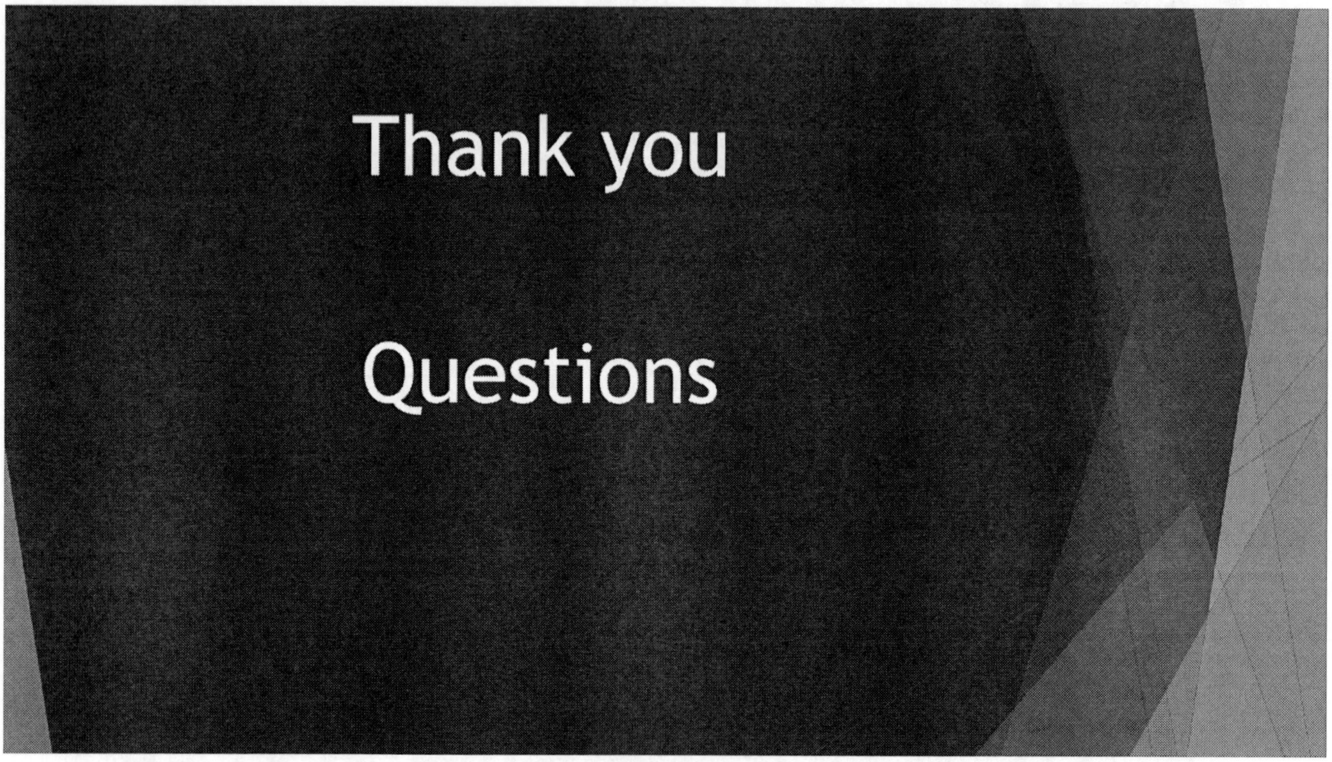
Thank you
Questions

THE Art OF ACCOMPLISHMENT

# Application of Coal Petrography in Understanding the Variation of Coal Composition in the Coalfields of Botswana

**MSc Geology**

**Ndivhuho Nendouvhada**

**Supervisor: Prof. N. Wagner**

**Co-supervisor: Mr A. Golding**

UNIVERSITY OF JOHANNESBURG

---

THE Art OF ACCOMPLISHMENT

## Outline

Introduction - Coal of Botswana

Methodology

Results - South Orapa

Conclusion

UNIVERSITY OF JOHANNESBURG

## INTRODUCTION

- The economy of Botswana – enhanced by Carbon resources
- Vague estimate – 212 billion tonnes coal reserves
  - Sub-bituminous to bituminous rank
  - Contain substantive volume of coal bed methane
- Infrastructure challenges – transport and water availability

UNIVERSITY OF JOHANNESBURG

## INTRODUCTION

- There is only one operating coal mine, Morupule Colliery Ltd, established to supply the adjacent power station with coal for electricity generation.
- Several companies explored the Kalahari-Karoo Basin of Botswana to obtain detailed information about the coal deposits.
  - No coal petrographic information in published literature
  - Lack of petrographic studies impede a full understanding of these deposits

UNIVERSITY OF JOHANNESBURG

# INTRODUCTION

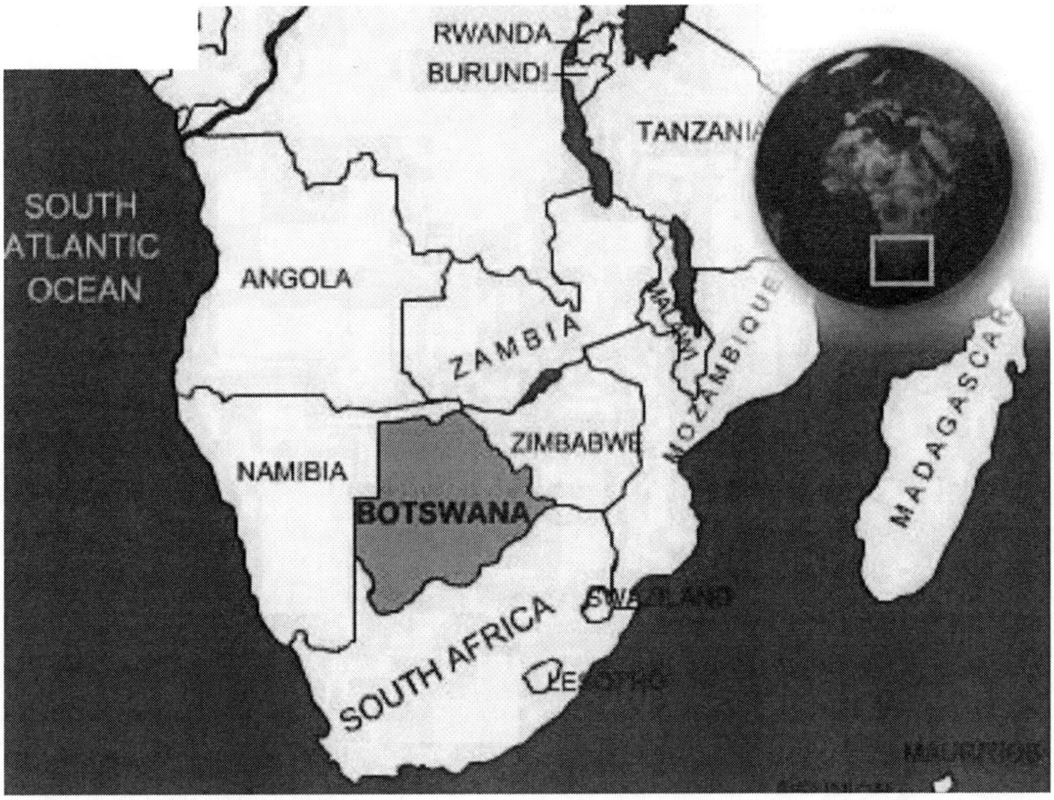


# INTRODUCTION

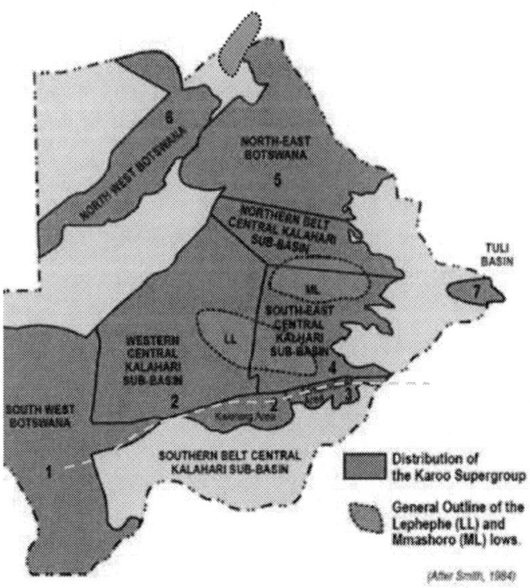


- ❑ **Central Kalahari sub-basin**
  - ➢ Sub-divided into several parts.
  - ➢ Significant basin of the Karoo in Botswana.
- ❑ **Tuli Basin**
  - ➢ Isolated from the Central Kalahari sub-basin.
  - ➢ Karoo Supergroup in Tuli Basin overlies early Precambrian.

UNIVERSITY OF JOHANNESBURG

## INTRODUCTION

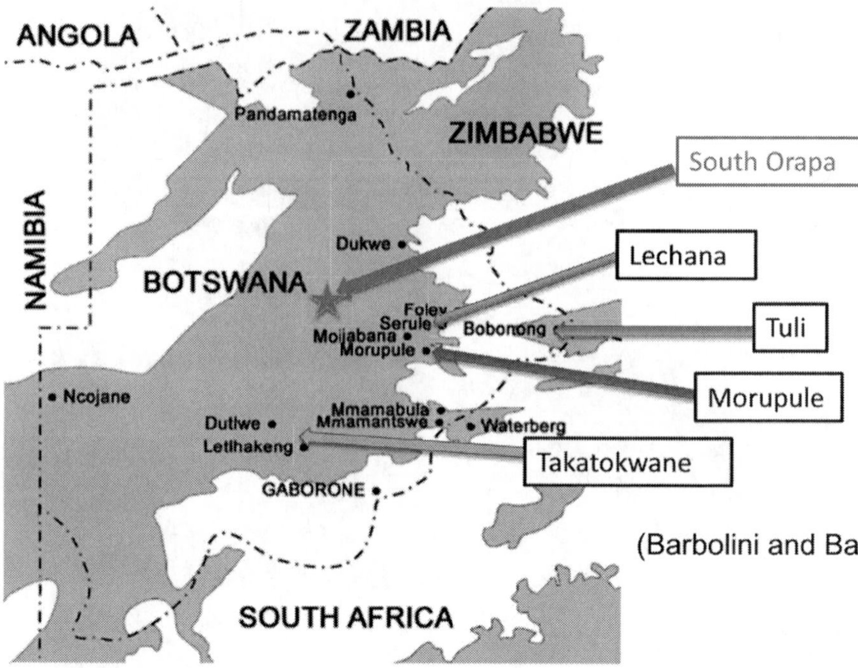


(Barbolini and Bamford, 2014)

UNIVERSITY OF JOHANNESBURG

## AIMS AND OBJECTIVES

- Quantify maceral type, coal rank, microlithotypes, and inorganic components of coal using polished blocks subjected to oil and reflected light.
- Assess the variations in maceral composition and vitrinite reflectance in the South Orapa Coalfield.
- Comparison using proximate, CV and S analyses.
- Data obtained by Petrographic analyses extrapolated to gain understanding of depositional environment.

UNIVERSITY OF JOHANNESBURG

## APPROACH

- ❑ **Bore hole core samples**
- ❑ **Sample preparation**
  - ➢ Crushing and polishing
- ❑ **Petrographic analyses**
  - ➢ Maceral group analysis
  - ➢ Vitrinite reflectance
  - ➢ Microlithotypes
- ❑ **Chemical analyses**
  - ➢ Proximate analysis (TGA)- determination of ash, moisture, volatile and fixed carbon.
  - ➢ CV and S analyses.

UNIVERSITY OF JOHANNESBURG

## RESULTS

### Petrography – Maceral Group Analysis

| | Maceral group analysis | | | |
|---|---|---|---|---|
| | Vitrinite (vol. %) | Inertinite (vol. %) | Liptinite (vol. %) | Mineral matter (vol. %) |
| | Inc.mm | Inc.mm | Inc.mm | |
| Lechana | 11.2-78.0 | 2.9-38.9 | 0-7.3 | 7.8-21.7 |
| South Orapa | 32.6-60.9 | 5.3-32.5 | 0-11.0 | 9.5-54.6 |
| Morupule | 5.3-17.8 | 37.0-84.6 | 0.3-4.0 | 7.9-55.7 |
| Takatokwane | 11.2-78.0 | 2.9-38.9 | 0-7.3 | 11.5-69.9 |

UNIVERSITY OF JOHANNESBURG

## RESULTS

### Petrography – Maceral Group Analysis

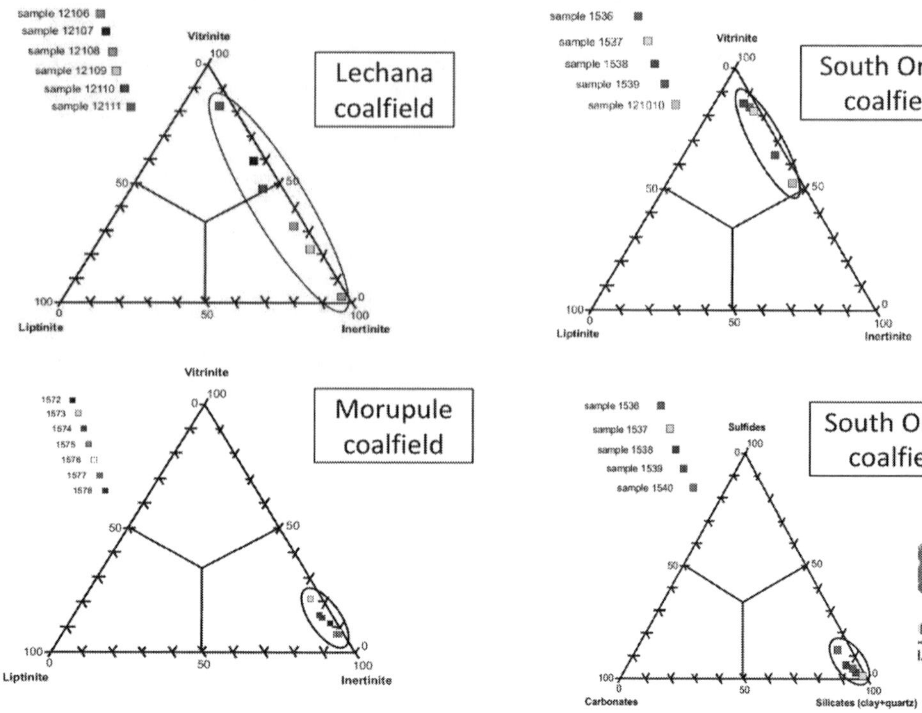


## RESULTS

### Petrography – Maceral Observations

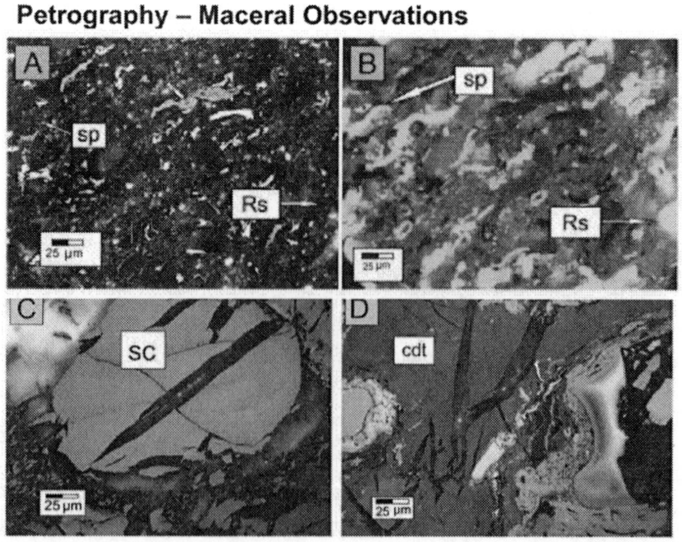


A) Clay associated with liptinites, B) Disintergrated inertinitic fragments, C) Sectretinite associated with carbonate, D) Cracks in the groundmass of vitrinite of vitrinite.

UNIVERSITY OF JOHANNESBURG

# RESULTS

## Petrography – Maceral Observations

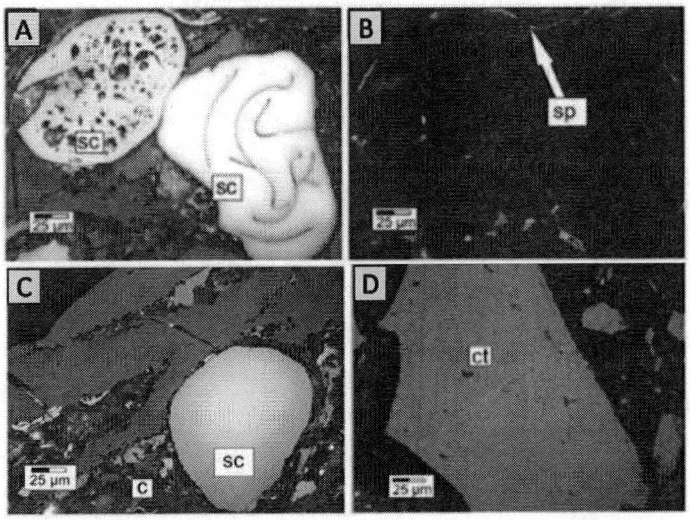


A) & B) various forms of Secretinites, C) Fluorescing liptinite, D) Pale grey collotelinite.

UNIVERSITY OF JOHANNESBURG

# RESULTS

## Petrography- vitrinite reflectance

Lechana- high volatile bituminous rank

| Sample id | % RoVmr | St. deviation |
|---|---|---|
| **12106** | 0.60 | 0.037 |
| **12107** | 0.54 | 0.057 |
| **12108** | 0.0 | 0.0 |
| **12109** | 0.67 | 0.089 |
| **12110** | 0.60 | 0.051 |
| **12111** | 0.57 | 0.058 |

Morupule- high volatile bituminous rank

| Sample id | % RoVmr | St. deviation |
|---|---|---|
| **1572** | 0.67 | 0.097 |
| **1573** | 0.61 | 0.055 |
| **1574** | 0.57 | 0.049 |
| **1575** | 0.60 | 0.10 |
| **1576** | 0.60 | 0.51 |
| **1577** | 0.64 | 0.065 |
| **1578** | 0.59 | 0.066 |

Takatokwane-sub-bituminous-high volatile bituminous rank

| Sample id | % RoVmr | St. deviation |
|---|---|---|
| **1287** | 0.44 | 0.041 |
| **1288** | 0.50 | 0.047 |
| **1289** | 0.46 | 0.043 |
| **1290** | 0.55 | 0.043 |
| **1291** | 0.51 | 0.028 |
| **1292** | 0.61 | 0.056 |
| **1293** | 0.47 | 0.053 |
| **1294** | 0.54 | 0.044 |
| **1298** | 0.49 | 0.048 |

UNIVERSITY OF JOHANNESBURG

# RESULTS

## South Orapa – Petrography – vitrinite reflectance

| Sample id | %RoVmr | Standard deviation | Depth (m) |
|---|---|---|---|
| 1536 | 0.49 | 0.040 | 547.4 - 549.2 |
| 1537 | 0.46 | 0.031 | |
| 1538 | 0.46 | 0.039 | 553.7 - 554.9 |
| 1539 | 1.43 | 0.37 | 648.4 - 650.2 |
| 1540 | 1.58 | 0.045 | |

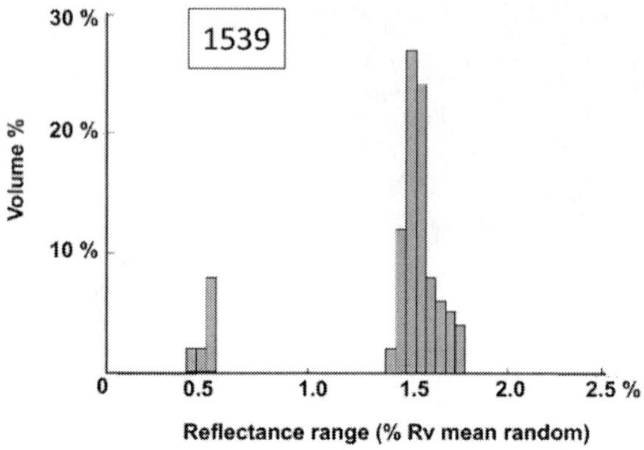


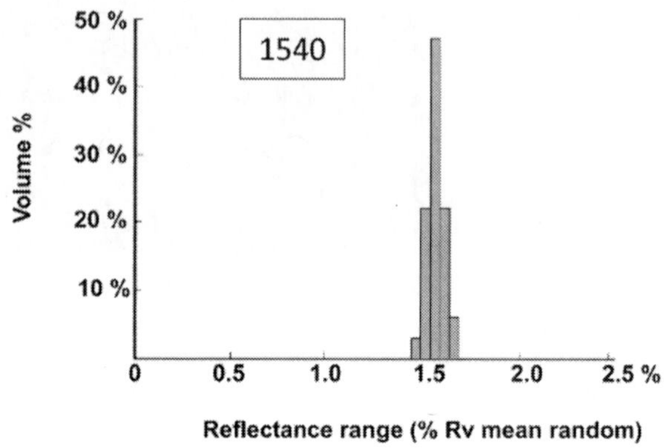


# RESULTS

## South Orapa

Chemical analyses-proximate, CV and S analyses

| | Ash (Wt. %) | Moisture (Wt. %) | Volatile (Wt. %) | Fixed Carbon (Wt. %) | Calorific Values (Wt. %) | Sulphur (Wt. %) |
|---|---|---|---|---|---|---|
| 1536 | 38.62 | 2.28 | 31.73 | 27.37 | 17.07 | 3.98 |
| 1537 | 13.39 | 3.09 | 37.57 | 45.94 | 19.29 | 3.98 |
| 1538 | 43.39 | 1.47 | 28.53 | 26.61 | 15.90 | 4.62 |
| 1539 | 27.54 | 1.31 | 21.55 | 49.59 | 21.06 | 4.74 |
| 1540 | 26.34 | 0.90 | 20.58 | 52.18 | 23.35 | 5.38 |

# RESULTS

## Reconstruction of the depositional environment

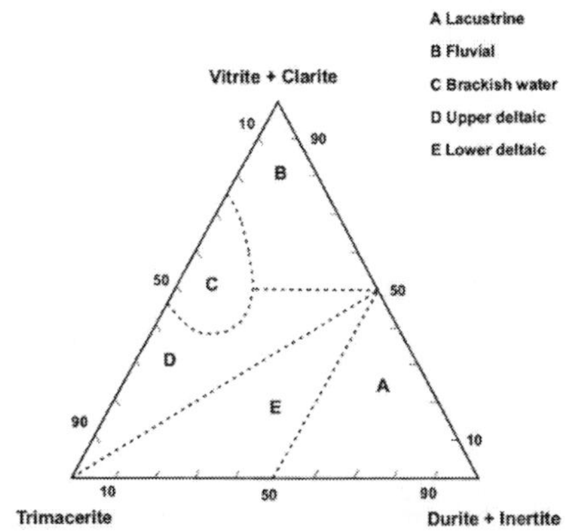


(Smyth, 1984)

$$TPI = \frac{\text{telinite + collinite + fusinite + semifusinite}}{\text{detrovitrinite + macrinite + inertodetrinite}}$$

TPI measures the degree of humification that the maceral precursors has suffered

$$GI = \frac{\text{vitrinite + macrinite}}{\text{semifusinite + fusinite + inertodetrinite}}$$

gelification index relates to the wetness or dryness of peat conditions

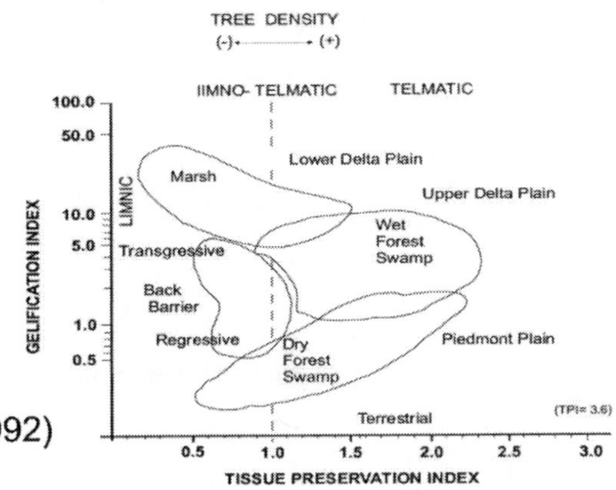


(Diessel, 1992)

# RESULTS

## South Orapa - Reconstruction of the depositional environment

- Fluvial depositional environment
- Limnic-limnotelmatic

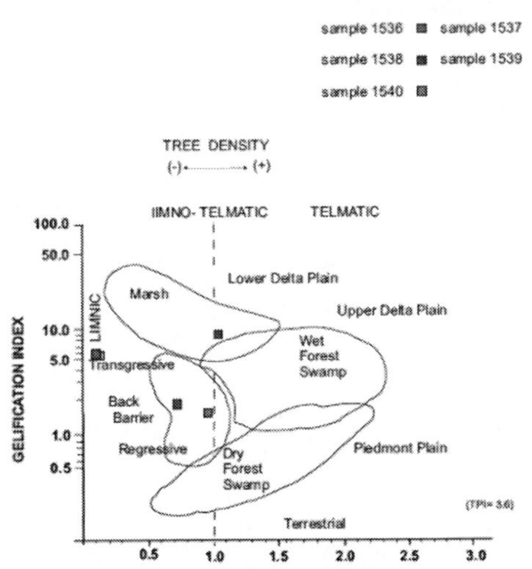


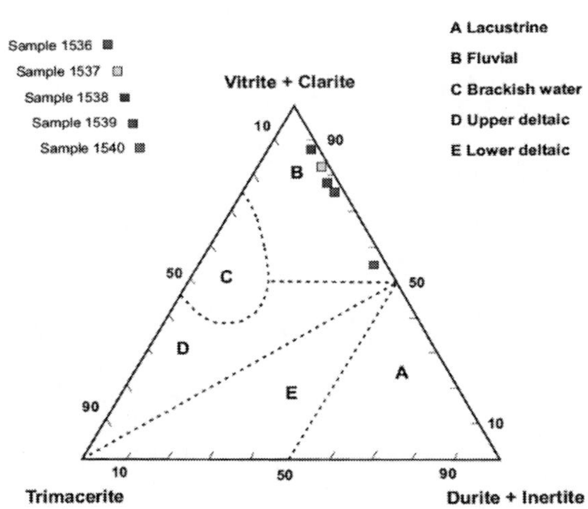


UNIVERSITY OF JOHANNESBURG

# RESULTS

## Morupule - Reconstruction of the depositional environment

- Lacustrine depositional environment

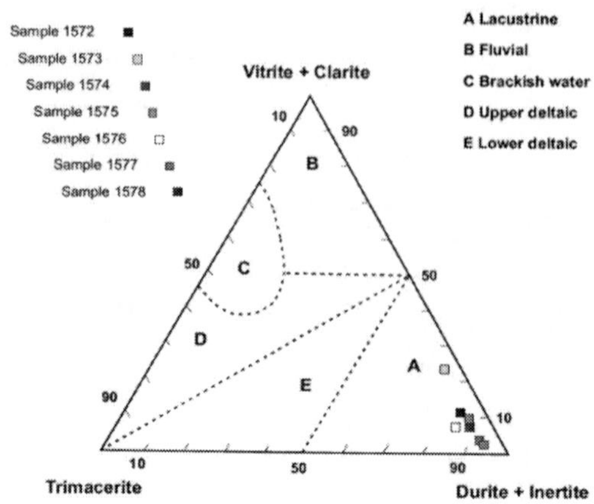


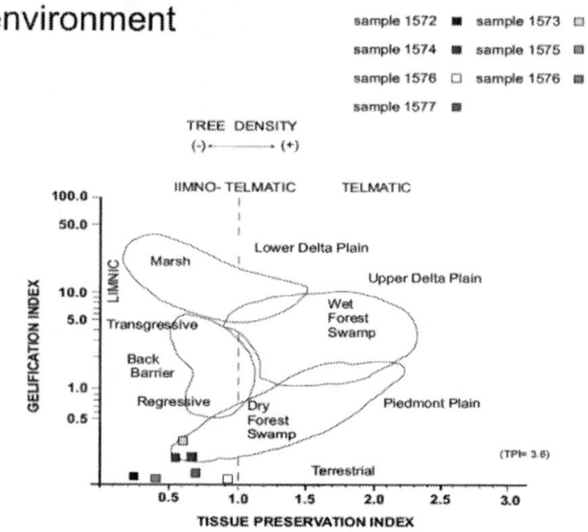


UNIVERSITY OF JOHANNESBURG

# RESULTS

## Tuli - Reconstruction of the depositional environment

- Fluvial depositional environment

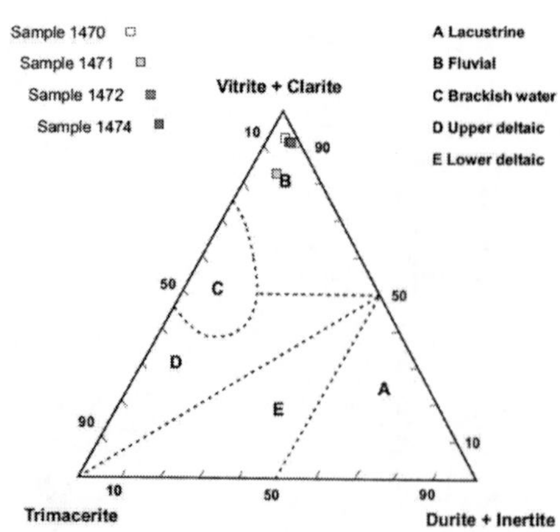


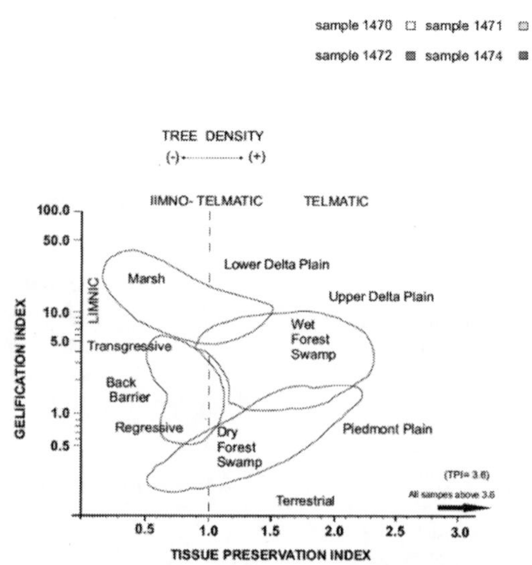


UNIVERSITY OF JOHANNESBURG

## CONCLUSIONS

- Variable maceral compositions
- Sub-bituminous-bituminous rank
- The Botswana coalfields trend from fluvial-lacustrine condition during peat swamp development

UNIVERSITY OF JOHANNESBURG


## Acknowledgements

- Centre of Excellence for Integrated Mineral and Energy Resource Analysis (CIMERA) – Project funding
- Analytika Holdings and Co-supervisor Mr A. Holdings – Provision of samples and additional information on Botswana
- Prof. N Wagner – Supervisor
- Wits Coal Lab and Zama Mthabela for the proximate and CV facilities
- This work is based on the research supported wholly by NRF.


UNIVERSITY OF JOHANNESBURG

THE Art OF ACCOMPLISHMENT

# Thank You

An international university of choice, anchored in Africa, dynamically shaping the future.

UNIVERSITY OF JOHANNESBURG

# Char Gasification Kinetics in Mixtures of $CO_2$ and $H_2O$: Understanding the Roles of Surface Coverage and Catalysis

Daniel Roberts, Mark Kochanek, and David Harris
CSIRO Energy, Brisbane, Australia

ENERGY
www.csiro.au

CSIRO

33rd Annual International Pittsburgh Coal Conference
8-12 August 2016, Cape Town, South Africa

# Context

## Intrinsic Char Reactivity in Gasification

Studies of gasification reaction mechanisms continue to be important as the foundation for modelling and assessment tools for a range of gasification-based applications.

As the requirements of modelling tools increase in their complexity, the ability to accurately reflect coal-specific behaviour in a transportable manner is important.

Intrinsic kinetics are required for this.

This work is part of a wider study into char gasification kinetics, investigating the C-$CO_2$ and C-$H_2O$ reactions in support of coal gasification model development.

D G Roberts, A Y Ilyushechkin and D J Harris. Fuel Processing Technology 94(1) 86-93 (2012).
D G Roberts, A Y Ilyushechkin, A Tremel and D J Harris. Fuel Processing Technology 94(1) 26-33 (2012).

# Introduction

## Char Reactivity in Mixtures

2007

- Paper published [1] providing the first experimental data on intrinsic char reactivity in mixtures of $CO_2$ and $H_2O$ over a relevant range of partial pressures, using chars from bituminous coals
- This work provided data that showed that the presence of one reactant would inhibit the rate of reaction of the other.
- This was shown to be consistent with the accepted Langmuir-Hinshelwood reaction mechanism

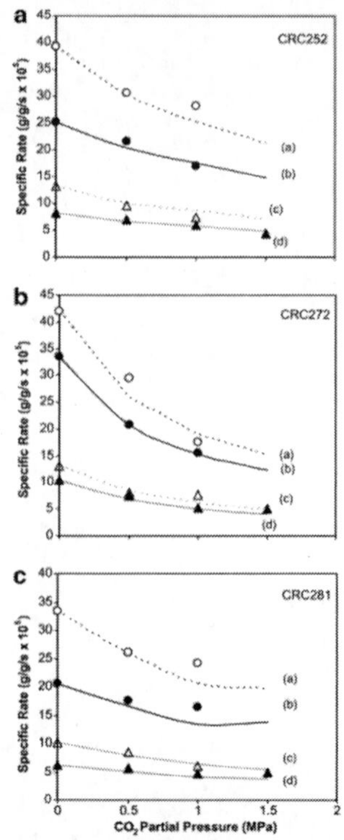


Reaction conditions:

(a) 900°C, 10 bar steam
(b) 900°C, 5 bar steam
(c) 850°C, 10 bar steam
(d) 850°C, 5 bar steam

CRCXXX = unique coal identifier

1. D G Roberts and D J Harris. Fuel **86(17-18)** 2672-2678 (2007).



# Introduction

## Char Reactivity in Mixtures

Since then, there has been considerable activity in the literature with many papers published, many of which suggest two competing outcomes (additive vs competitive).

Fuel Processing Technology

Lignite-char gasification mechanism in mixed atmospheres of steam and $CO_2$ at different pressures

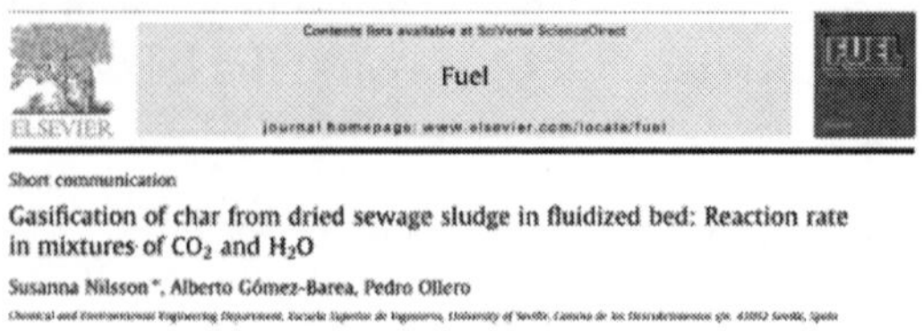

Fuel

Short communication

Gasification of char from dried sewage sludge in fluidized bed: Reaction rate in mixtures of $CO_2$ and $H_2O$

Susanna Nilsson *, Alberto Gómez-Barea, Pedro Ollero

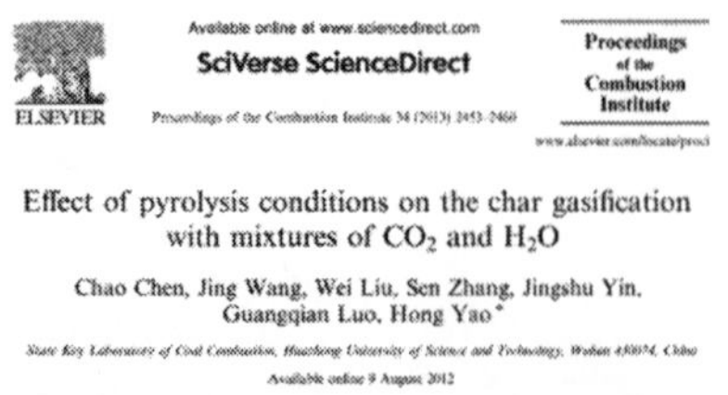

Available online at www.sciencedirect.com

SciVerse ScienceDirect

Proceedings of the Combustion Institute

Effect of pyrolysis conditions on the char gasification with mixtures of $CO_2$ and $H_2O$

Chao Chen, Jing Wang, Wei Liu, Sen Zhang, Jingshu Yin, Guangqian Luo, Hong Yao *

Available online 9 August 2012

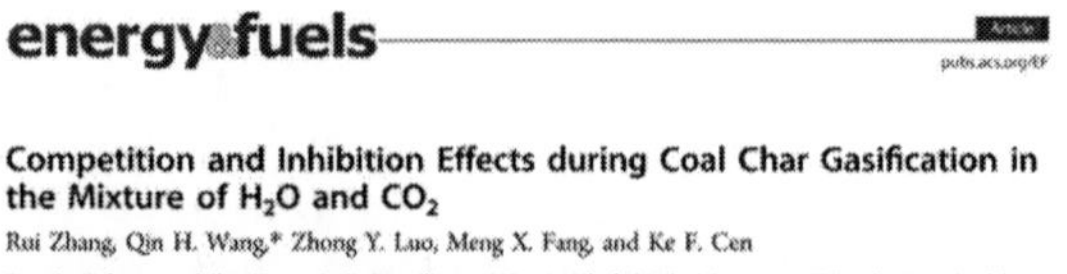

energy&fuels

Competition and Inhibition Effects during Coal Char Gasification in the Mixture of $H_2O$ and $CO_2$

Rui Zhang, Qin H. Wang,* Zhong Y. Luo, Meng X. Fang, and Ke F. Cen

# Introduction

## A spectrum of interactions

More recent work has showed that two competing outcomes are unlikely, but that a spectrum of behaviour all consistent with a single mechanism is probable – certainly for bituminous coals, where catalytic effects are secondary.

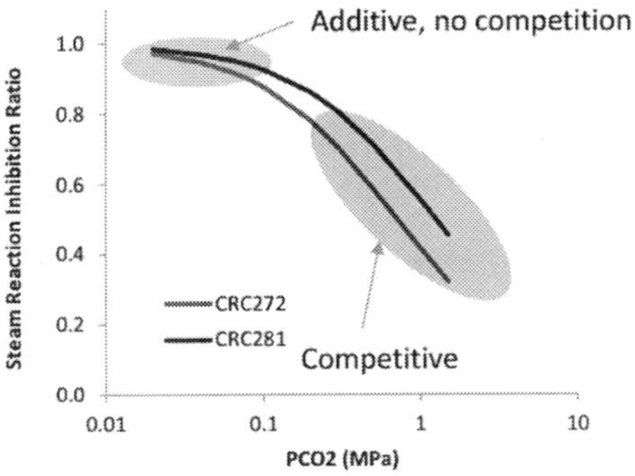


Competition effect is a function of pressure

- Low partial pressures, low surface saturation, additive behaviour
- Surface saturation increases at increasing pressures leading to competitive behaviour

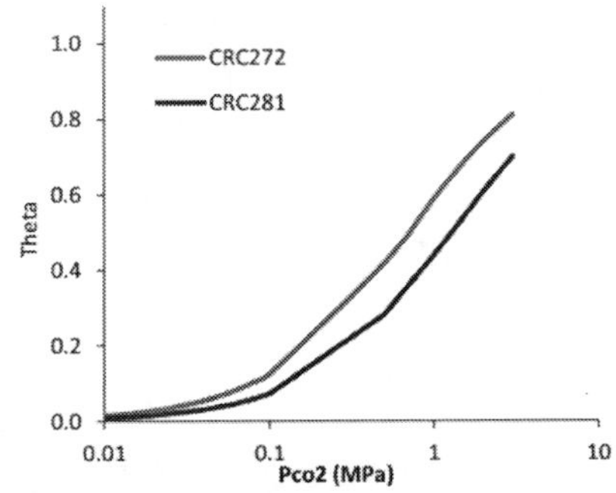


Literature studies suggesting that there are two possible mechanisms – competitive or additive – are probably consistent with the one mechanism



# Impacts of coal type

## Catalytic species complicate kinetic analysis of low rank coals

We know that for chars with high amounts of catalytic material, such as low rank coals, the role of surface is diminished compared with bituminous coals (e.g. [2]).

Hayashi et al have demonstrated this using a LH type of mechanism in parallel with a catalytic reaction mechanism [3].

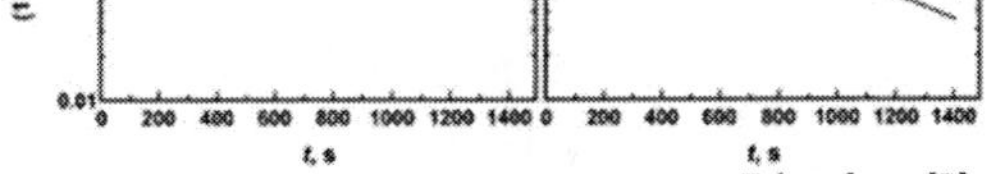


Taken from [2]


2. Kitsuka et al. Energy and Fuels 21 387-394 (2007).
3. M Kajita et al. Energy and Fuels 24 108-116 (2010).

# New data

## Low rank coals (Australian lignite) char

Conversion Profiles

- 5 bar reactants, reacting at 750°C and 800°C
- (low temperatures required to ensure differential, Regime I conditions)

Demonstrate clear non-additive behaviour

- Consistent with our results from bituminous coals

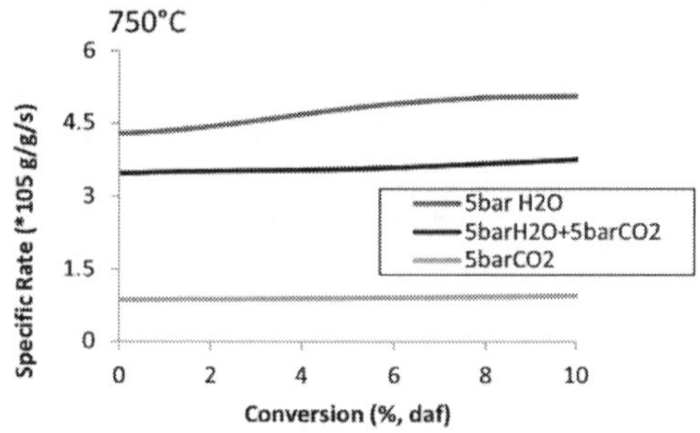


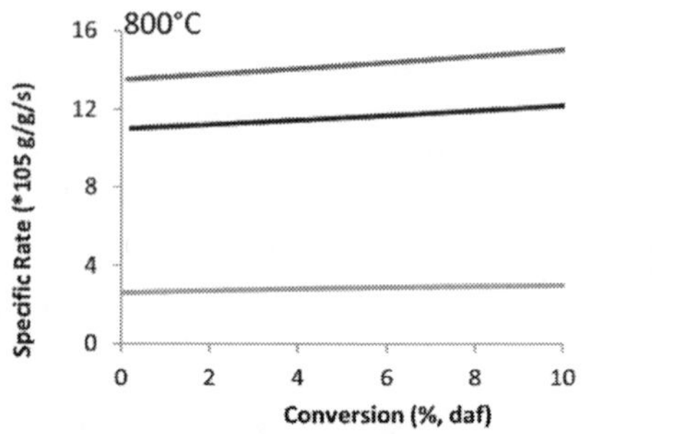




# New data

## Low rank coals (Australian lignite) char

Rate vs $CO_2$ pressure

- Results are qualitatively consistent with data from bituminous coals
- That is, we see $CO_2$ addition decrease the overall rate

Note the pure-$CO_2$ results

- This sample shows a small effect of $CO_2$ partial pressure on reaction rate
- Consistent with low rank coals where catalysis is important (and carbon surface secondary)

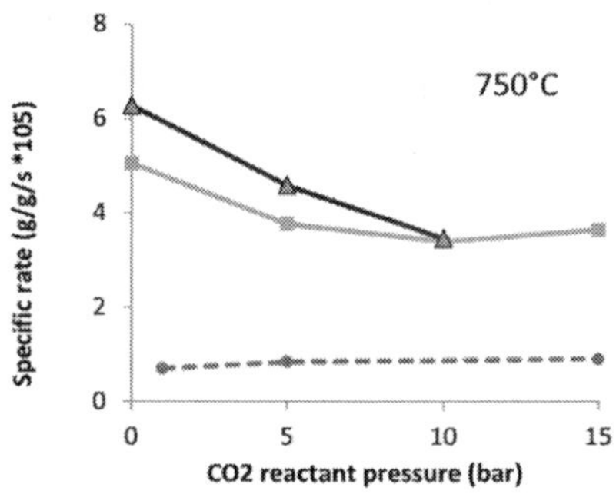


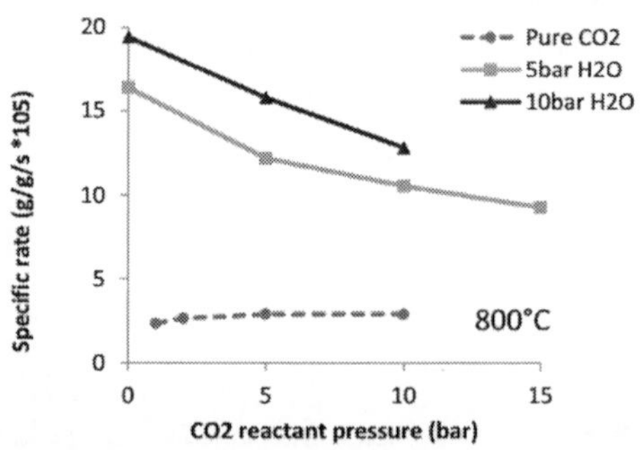

# Reactivity in mixtures

## New data for low-rank coal chars

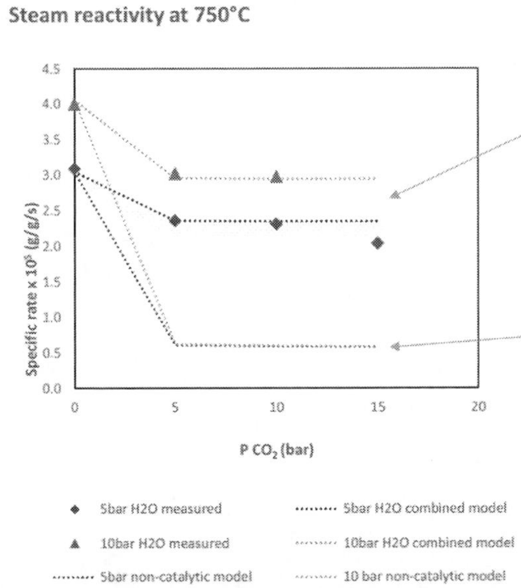


Model where rate is a combination of non-catalytic (LH-based spectrum of competition) and catalytic (generally non-competitive) reaction processes is far more suitable.

Black coal char model relies entirely on surface processes. Leads to significant under-prediction of rates



# Ongoing work

At this early stage, the catalytic component is not quantifiable in the model – this needs to be addressed, building on a considerable amount of work in the literature.

The temperature range for lignites needs to be extended to give a better insight into the wider applicability of the kinetics.

We need to also consider additional low rank coals, with varying catalytic activity driven by a wider range of species.

# Thank you

Daniel Roberts
Research Group Leader, High Efficiency Thermal and Electrochemical Technologies
CSIRO Energy
p: +61 7 3327 4521
e: Daniel.Roberts@csiro.au
w: www.csiro.au/energy

ENERGY FLAGSHIP
www.csiro.au

CSIRO

# Appendix
## Coal Analysis

Proximate analysis (wt%, ar)

| Moisture | Ash | VM | FC |
|---|---|---|---|
| 59.1 | 0.2 | 21.1 | 19.6 |

Ash analysis (wt%)

| SiO2 | Al2O3 | Fe2O3 | K2O | Na2O | CaO | SO3 | P2O5 | TiO | MgO |
|---|---|---|---|---|---|---|---|---|---|
| 17.3 | 10.3 | 14.3 | 1.04 | 11.3 | 6.8 | 26.6 | 0.1 | 0.35 | 12.5 |



CSIRO

# Green Coal development for application in Fixed-Bed Catalytic Gasification.

**John R. Bunt[1,*], Sanette Marx[1], Frans B Waanders[1]**
[1] Coal Research Group, Unit for Energy Systems, School of Chemical and Minerals Engineering, North West-University, Potchefstroom, 2520, South Africa.
** Corresponding author: Tel.: +27 18 299-1376; Fax: +27 18 299-1535;*
*E-mail: john.bunt@nwu.ac.za*

33rd Annual International Pittsburgh Coal Conference (PCC).
August 8 – 12, 2016.

Cape Town, South Africa.

It all starts here ®

NRF RISA
National Research Foundation | Research and Innovation Support and Advancement

NORTH-WEST UNIVERSITY ®
YUNIBESITI YA BOKONE-BOPHIRIMA
NOORDWES-UNIVERSITEIT
POTCHEFSTROOM CAMPUS

## Outline of Presentation

- ❑ Introduction
- ❑ Experimental
- ❑ Results and Discussion
  - ❑ Reactivity
  - ❑ Sulphur capture
- ❑ Next steps
- ❑ Concluding Remarks
- ❑ Acknowledgement

NRF RISA
National Research Foundation | Research and Innovation Support and Advancement

NORTH-WEST UNIVERSITY ®
YUNIBESITI YA BOKONE-BOPHIRIMA
NOORDWES-UNIVERSITEIT

2

## Introduction:

- Of the 255 Mt/a of coal produced in South Africa, 60 Mt/a of coal is discarded as deposits from mining and transport operations because of its perceived poor quality (high ash and sulphur) and volatiles diminish as stockpiles age [1-2].

- Coal is also discarded due to size, i.e. coal fines are generally classified as particles < 500 micron that are separated from the coal during the beneficiation process [3].

- Due to the importance of coal as non-renewable resource, these coal fines can be utilised. Discard coal having a heating value of c.a. 16MJ/kg can be burnt by Eskom, making it a viable energy source; can be beneficiated and added back to washed coarser product where contracts allow; or could be gasified in its current form using fluidised-bed and entrained flow reactor's [4-6], but not in fixed-bed gasifiers (100mm x 5mm).

Storage dam at Twistdraai Colliery for fine discard coal

[1] Statistics SA. 2011. Mineral accounts for South Africa 1980-2008. Discussion document DO405.2. OCLC Number: 743298325. 63 pages.
[2] White Energy Company. 2012. Annual Report 2012. 88 pages.
[3] Wagner N.J. 2008. The characterisation of weathered discard coals and their behaviour during combustion. Fuel 87: 1687-1697.
[4] Bunt JR and Van Nierop P. 1997. Development of a fine coal circuit for Sasol's Twistdraai Colliery. In Proceedings 13th annual International Pittsburgh Coal Conference; Shanxi, Peoples Republic of China.

3

## Introduction:

- Coal utilisation has also led to rising concerns about $CO_2$ emissions causing global warming. The use of biomass is considered to be renewable and assists in reducing $CO_2$ emissions compared with coal, because biomass is suggested to be $CO_2$ neutral with regard to the greenhouse gas balance [7-9].

- Carbonization (also referred to as torrefaction or liquefaction) is a mild thermochemical treatment (200-400°C and 100 bar pressure) that removes moisture and organic acids from the biomass producing a coal like substance (bio-char) as solid product as well as bio-oil and biogas.

- The main aim of liquefaction is to transform a material of low density and energy value into a stable product with high energy density and carbon content ready for use in combustion and gasification processes [10, 11]. Liquefaction has the highest net energy gain for biomass conversion to solid, liquid and gas products compared to pyrolysis and gasification [12].

[5] Statistics SA. 2011. Mineral accounts for South Africa 1980-2008. Discussion document DO405.2. OCLC Number: 743298325. 63 pages.
[6] White Energy Company. 2012. Annual Report 2012. 88 pages.
[7] Usón, S., Valero, A., Correas, L. & Martinez, Á. 2004. Co-Gasification of coal and Biomass in an IGCC Power Plant: Gasifier Modeling. Int.J. Thermodynamics, 7: 165-175.
[8] Zhu, W., Song, W. & Lin, W. 2008. Catalytic gasification of char from co-pyrolysis of coal and biomass. Fuel Processing Technology, 89: 890-896.
[9] Biagini, E., Lippi, F., Petarca, L. & Tognotti, L. 2002. Devolatilisation rate of biomasses and coal-biomass blends; an experimental investigation. Fuel, 81: 1041-1050.
[10] Chen, P., Min, M., Chen, Y., Wang, L., Li, Y., Chen, Q. 2009. Review of the biological and engineering aspects of algae to fuels approach. International Journal of Agricultural and Biological Engineering, 2(4):1-30
[11] Bridgman, T.G., Jone, J.M., Shield, I., Williams, P.J. 2008. Torrefaction of reed canary grass, wheat straw and willow to enhance solid fuel quantities and combustion properties. Fuel, 87:844-856.
[12] Khoo, H.H., Koph, C.Y., Shaik, M.S., Shamatt, P.N. 2013. Bioenergy co-products derived from microalgae biomass via thermochemical conversion – Life cycle energy balance and $CO_2$ emissions. Bioresource Technology, 143:298-307

4

## Introduction:

NRF RISA
National Research Foundation | Research and Innovation Support and Advancement

NORTH-WEST UNIVERSITY
YUNIBESITI YA BOKONE-BOPHIRIMA
NOORDWES-UNIVERSITEIT

- An attractive system for coal gasification is one in which high-sulphur coal is gasified and sulphur in the coal is retained within the gasifier ash thereby eliminating the need for gas clean-up; i.e. the. gas is available for use without first cooling it to preserve the sensible energy; and water vapour is retained in the product gas to make a substantial contribution in the combined-cycle power output.

- These criteria can be met through the use of a calcium-based sorbent such as limestone or dolomite directly in the fluidized-bed gasifier, which acts as both a catalyst for the gasification reactions and captures sulphur as calcium sulphate [13].

- Researchers in the field of chemical kinetics of limestone/dolomite reactions with hydrogen sulphide [14-17] have already verified the potential use of these sorbents for sulphur capture. The reaction of calcined limestone / dolomite is very rapid, and the reaction almost approaches equilibrium. On that basis, it is possible to capture substantial quantities of sulphur and discharge it with the ash [18].

[13] Rehmat, A., Abbasian, M. J., Leppin, D. and Banerjee, D. D., "Reaction of Calcium-Based Sorbents With Sulfur in Coal During Gasification." Paper presented at the International Conference on Coal Science, Eurohal, The Netherlands, 1987.
[14] Chang, E. and Thodos, G., "Complex Nature of the Sulfation Reaction of Limestone and Dolomites," AIChE 3. 30, 3, 1984.
[15] Keairns, D. L., Archer, D. H., Newby, R. A., O'Niell, F. P. and Vidt, E. J., "High Temperature Sulfur Removal System Development Unit," Am. Chem. SOC. Div. Fuel Chem. Prepr. 21, 1976.
[16] Borgwardt, R. H., Roache, N. F. and Brene, K . R . , "Surface Area of Calcium Oxide and Kinetics of Calcium-Sulfide Formation," Envir. Proqr. 2, 2 (1984).
[17] Pell, M., "Reaction of Hydrogen Sulfide With Fully Calcined Dolomite. Ph.D. Thesis, City University of New Yock, 1971.
[18] Jones, F. L. and Patel, J. G., "Performance of Utah Bituminous Coal in the U-GAS Gasifier." Paper presented at the Fifth EPRI Contractors Conference on Coal Gasification, Palo Alto, California, October 1985.

5

## Introduction:

NRF RISA
National Research Foundation | Research and Innovation Support and Advancement

NORTH-WEST UNIVERSITY
YUNIBESITI YA BOKONE-BOPHIRIMA
NOORDWES-UNIVERSITEIT

- Much lab research has been conducted in the field of catalytic gasification, where small coal particles (<1 mm) are gasified ($CO_2$, $H_2O$) in the presence of alkali metals, and have shown that gasification reactivity can be significantly enhanced at temperatures ranging from 800-1000°C. Exxon Research and Engineering Company developed a catalytic coal gasification (CCG) process in the 1970's [19].

- More recent research by Nel *et. al.*, [20] has also shown that the addition of an impregnated $K_2CO_3$ catalyst (1% loading) to 10 mm coal particles of R.O.M. Highveld coal lowered the activation energy during reactivity testing when compared to raw coal. The challenge remains to increase the catalyst loading to large coal particles.

- Botha et al [21] successfully demonstrated that the addition of a catalyst (1%, 3%, 5% $K_2CO_3$ addition) can be incorporated into a discard Highveld fine coal agglomerate mixture (10mm pellet) by physical mixing in order to improve $CO_2$ gasification reactivity measured at high temperatures (900-1000°C). It was concluded that the $CO_2$ gasification reactivity could at least be doubled in this catalysed system when compared to raw coal.

[19] More P, Jagtap N, Kulal A, Dongare M, Umbarkar S. 2012. Effect of support modification by Magnesia on the low temperature catalytic activity and sulphur tolerance of the Ag/Al2O3 system. India. AdMet 2012 Paper No. CM 004. 8 pages.
[20] Nel S., Neomagus H.W.J.P., Bunt J.R., Everson R.C. 2013. Improved reactivity of large coal particles by K2CO3 addition during steam gasification. Fuel Processing Technology 114 (2013): 75-80.
[21] Botha A. 2012. An investigation into fine discard coal agglomeration for use in the fixed-bed gasification process. Journal of Analytical and Applied Pyrolysis, 113 (2015): 723–728.

6

# Introduction:

- From the foregoing it is clear that there is a need to explore the agglomeration of fine discard coal together with bio-char for use in applications such as fixed bed gasification thereby affording a reduction in sulphur emissions and other gaseous pollutants concomitant with increased reactivity rates in a manner that overcomes or at least mitigates the shortcomings associated with conventional fixed-bed gasification operation.

- This paper addresses the development of a *green coal* formulation [22] and discusses preliminary results obtained showing the ability to enhance reaction rate and sulphur capturing propensity during thermal treatment.

[22] International patent pending. Production of a carbonaceous feedstock material from a waste carbon source. 2014/03482.

7

# Experimental: Preparation of Green Coal Pellets

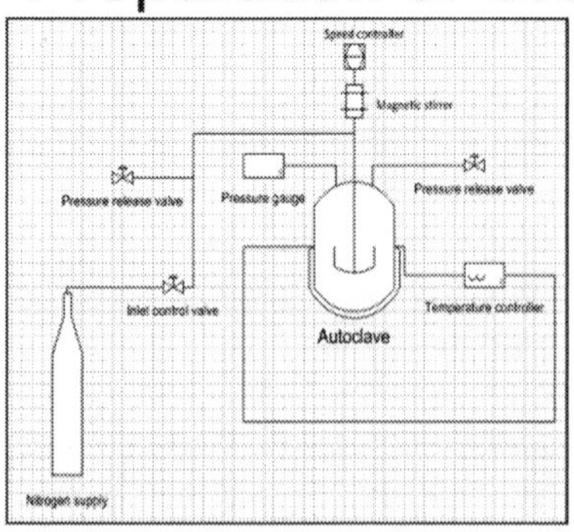


(a) Autoclave setup for the production of bio-char

(b) Sweet sorghum bagasse

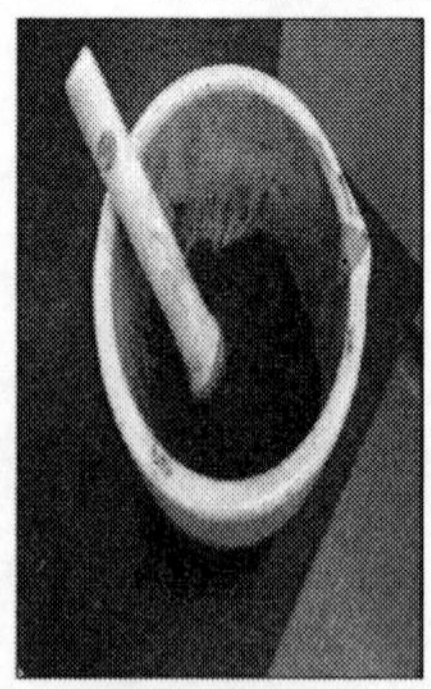

(c) Bio-char residue

(e) Fine discard Coal <212um

(c) Autoclave setup

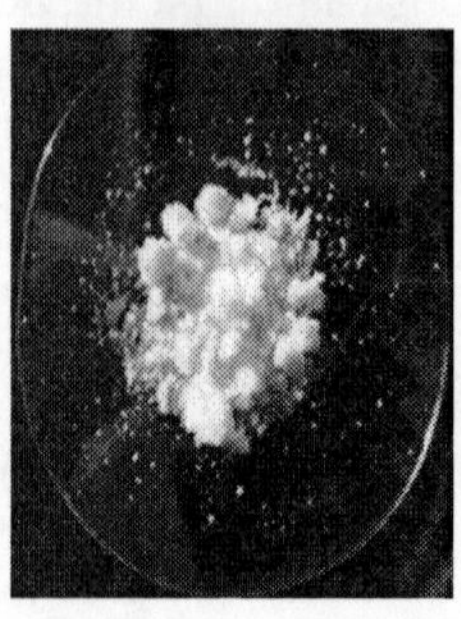

(f) $CaCO_3$

(f) LRX press

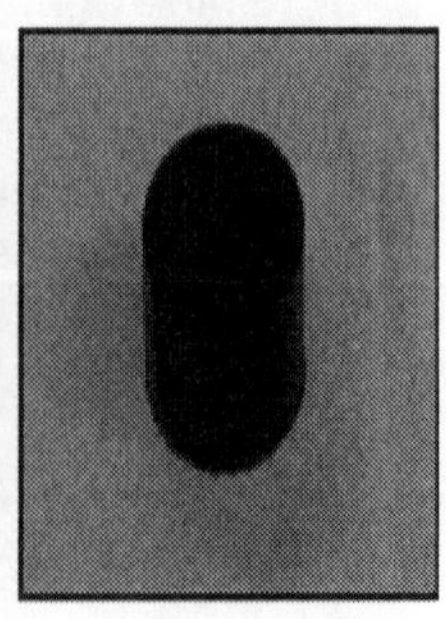

(d) Green coal pellet

8

# Bio-char characteristics

| Procedure | Coal | Bio-char |
|---|---|---|
| Moisture content (wt%) | 5.9 | 1.4 |
| Ash content (wt%) | 29.0 | 5.2 |
| Volatile matter (wt%) | 23.5 | 21.9 |
| Fixed carbon (wt%) | 41.5 | 71.4 |

(a) Proximate analysis results (air dried basis)

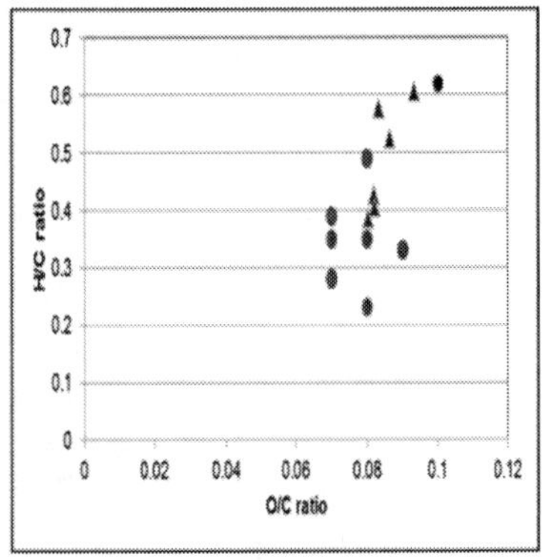


(b) Van Krevelen diagram for sunflower husk based bio-chars from this study compared to that obtained for similar feedstock by other researchers (• - Chars derived from coconut fibre and eucalyptus (Liu *et al.*, 2013), • - Chars derived from paper and wood (Hwang *et al.*, 2012). • - Chars derived from corn stalks (Xiao *et al.*, 2012), • - Coal (Parshetti *et al.*, 2013), ▲ - This study)

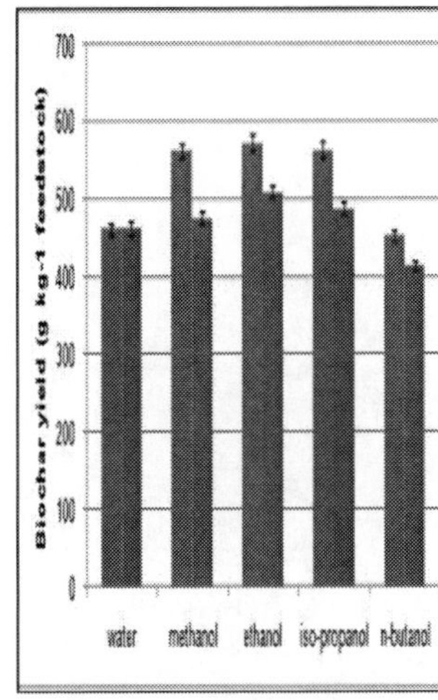


(c) Influence of solvent on bio-char yield in $N_2$ (■) and $CO_2$ (■) atmospheres

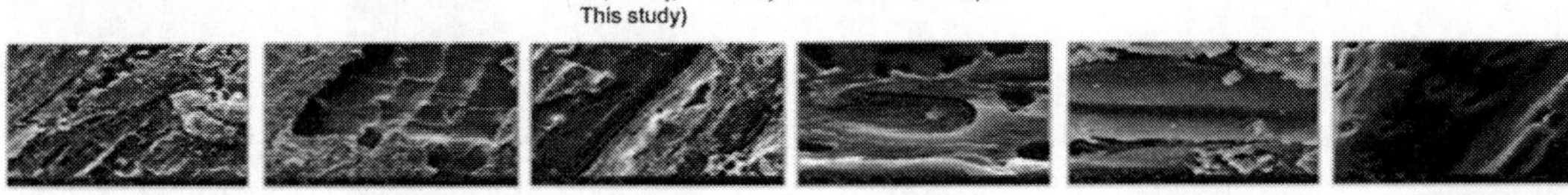


(d) SEM micrographs of biomass and bio-char samples from 240-320 °C in $N_2$ atmosphere

9

# Experimental:
# (a) Reactivity study of Green Coal

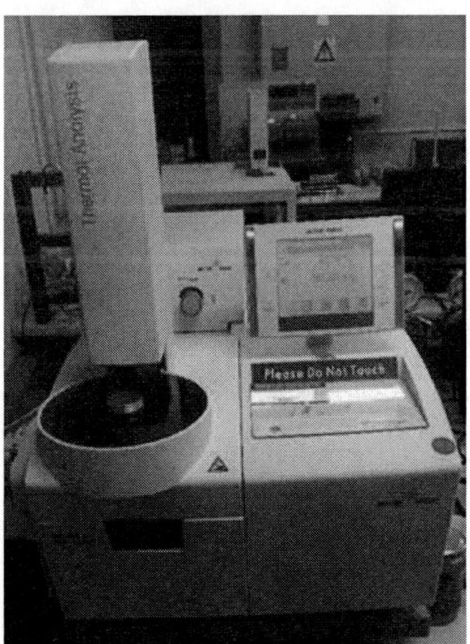


(a) TGA/DSC 1 STAR®

Pellet samples were placed into the TGA and the furnace was heated to 30°C using $N_2$ at a flow rate of 150 ml/min and a heating rate of 50°C/min. This temperature was held constant for 5 minutes.

Pellet samples were then pyrolised at 1000°C, using $N_2$ at a flow rate of 150 ml/min and a heating rate of 50°C/min.

The gasification reactions were carried out at **900°C, 950°C, and 1000°C**, respectively, using $CO_2$ at a flow rate of 150 ml/min. The reaction time was set to 6 hours.

| Pellet Sample variants tested | % |
|---|---|
| Baseline Coal | 100 |
| Bio-char | 100 |
| Coal : Bio-char (Blend) | 75:25 |
| Coal : Bio-char (Blend) | 50:50 |
| Coal : Bio-char (Blend) | 25:75 |
| Coal : Bio-char (Blend) + 1% $CaCO_3$ | 75:25 |
| Coal : Bio-char (Blend) + 3% $CaCO_3$ | 75:25 |
| Coal : Bio-char (Blend) + 5% $CaCO_3$ | 75:25 |

10

# Experimental :
# (b) Sulphur capturing setup

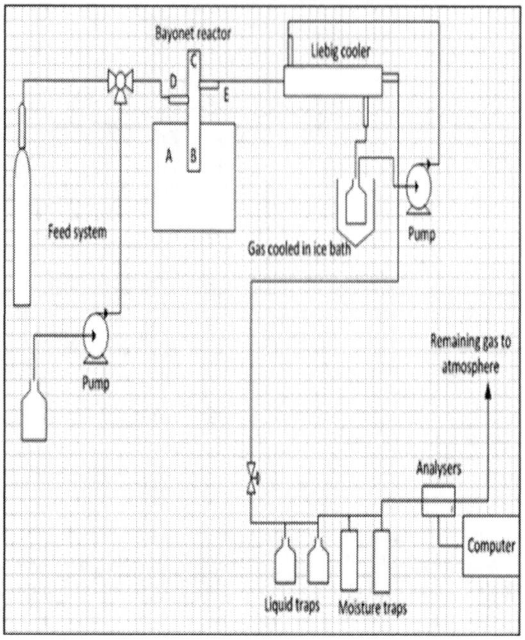


(a) Combustion setup (**Reactions were conduced at 500°C, 600°C, 700°C, 800°C**)

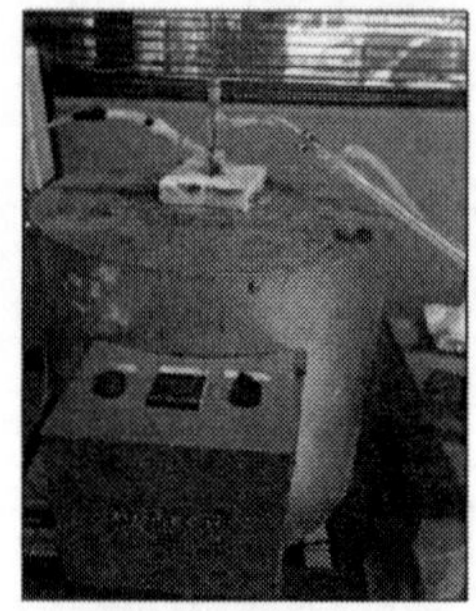

(b) Combustion furnace

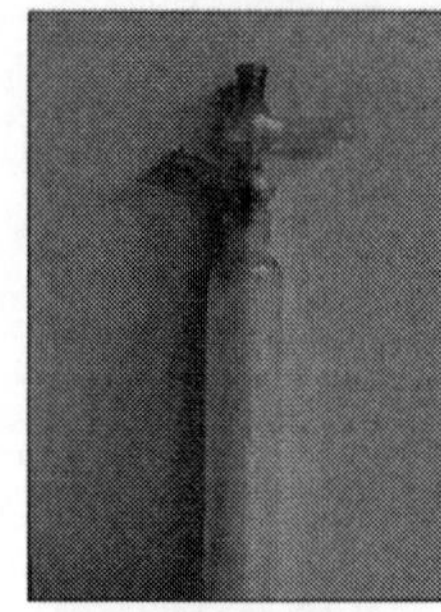

(c) Bayonet reactor

(d) : Mass flow controllers

(e) Gas analyser

11

# Results and Discussion:
# (a) Reactivity results

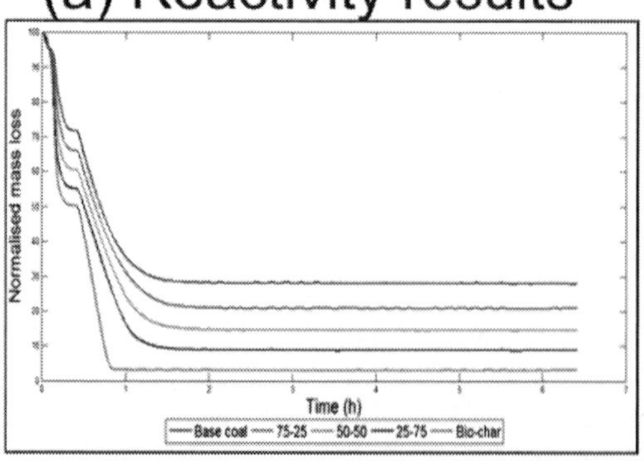


(a) Normalised mass loss curve of base coal, bio-char and blends at 1000°C

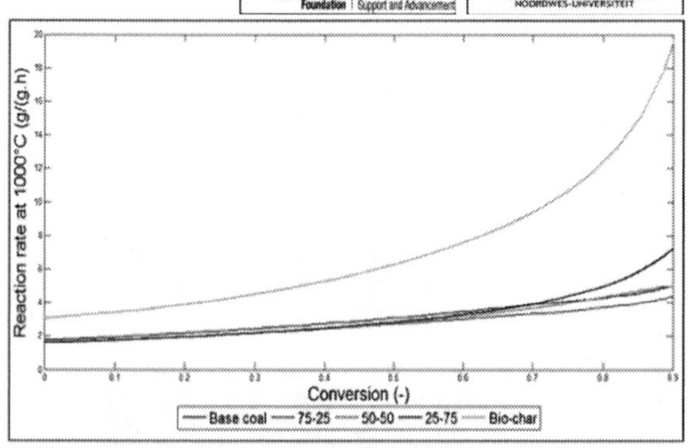


(b) Reaction rate of base coal, bio-char and blends at 1000°C

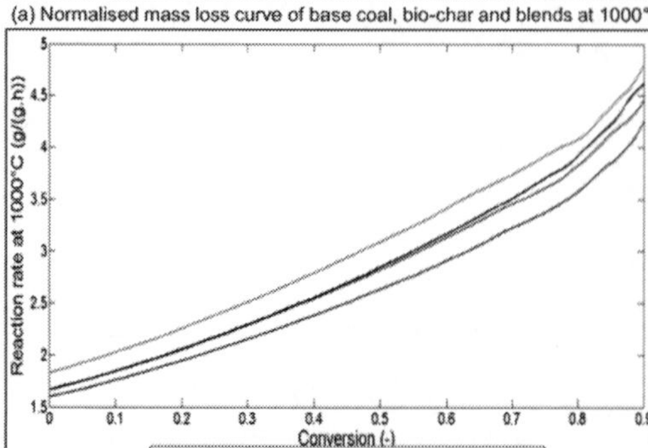


(c) Effect of $CaCO_3$ addition on the reaction rate of 75:25 blend at 1000°C

12

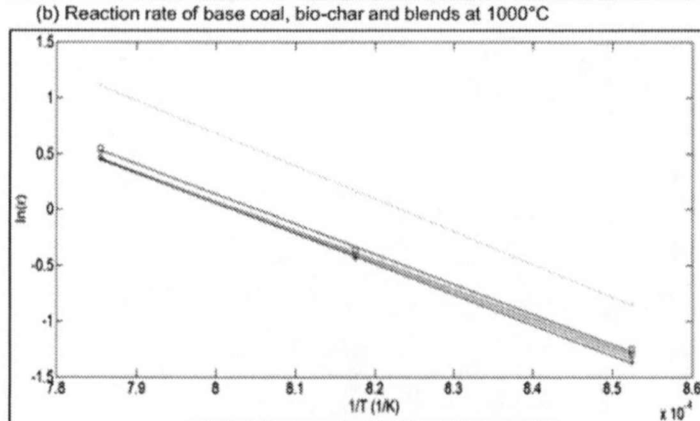


(d) Linearized Arrhenius plot of base coal, bio-char and respective blends

# Results and Discussion:
# (a) Reactivity study Kinetic parameters

| Sample | Activation energy (kJ/mol) | $k_0 \times 10^5$ (1/s) |
|---|---|---|
| Base coal | 224 | 7.02 |
| Bio-char | 244 | 88.39 |
| 75:25* | 219 | 4.04 |
| 50:50* | 221 | 4.88 |
| 25:75* | 227 | 8.91 |
| 75:25 Blend * + 1% $CaCO_3$ | 222 | 5.92 |
| 75:25 Blend * + 3% $CaCO_3$ | 233 | 17.28 |
| 75:25 Blend * + 5% $CaCO_3$ | 222 | 5.59 |

* Ratio of base coal: Bio-char

13

# Results and Discussion:
# (b) Sulphur retention

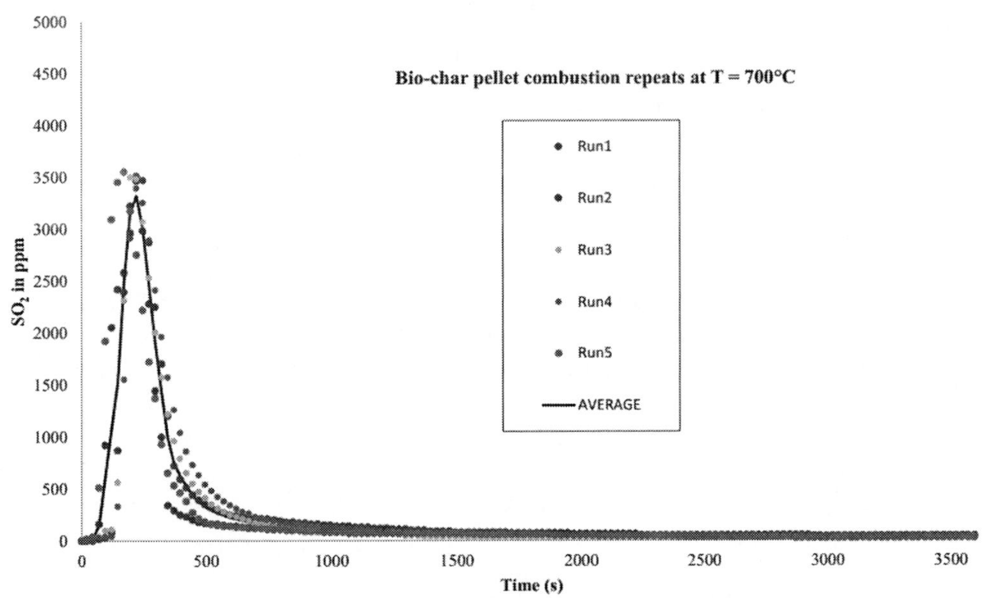


14

# Results and Discussion: (b) Sulphur retention

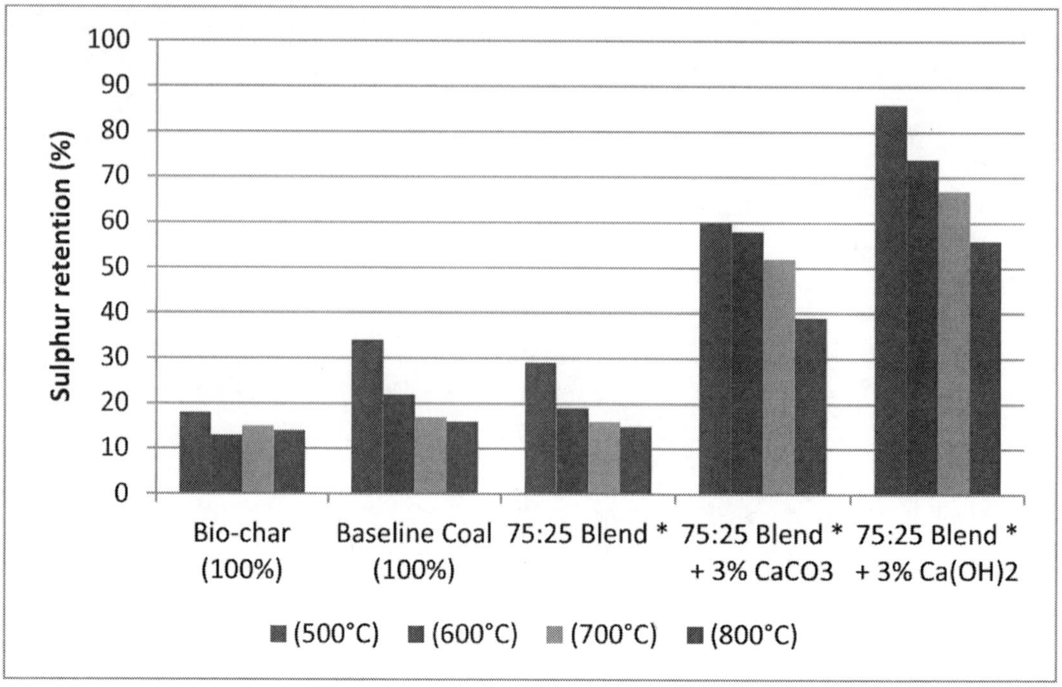


15

# Results and Discussion : Fact-Sage simulation

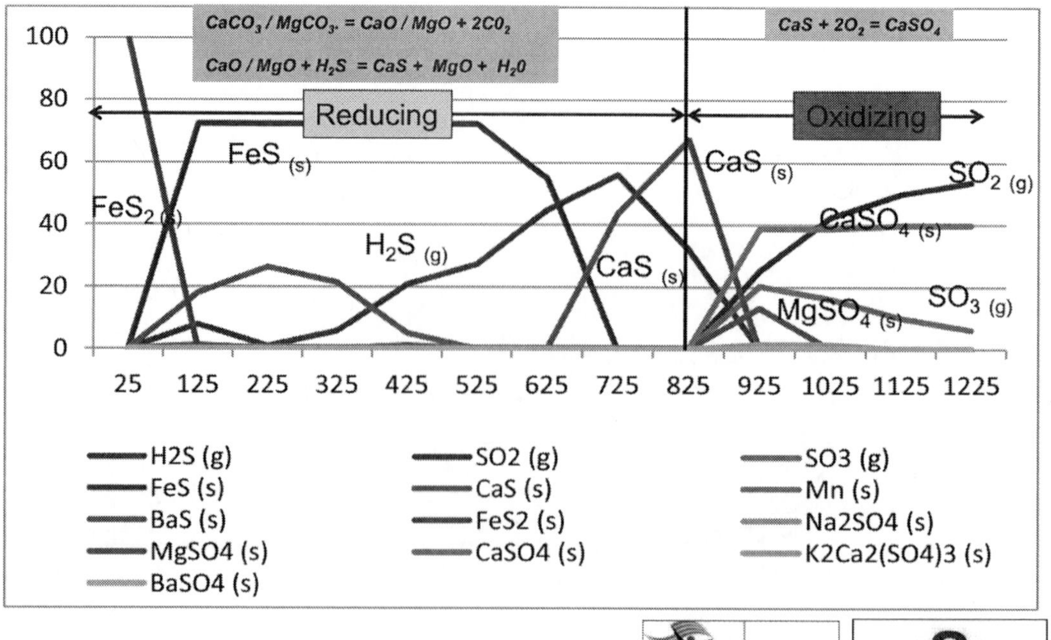


16

# Next Steps:
# NWU / AIST Bilateral Collaboration

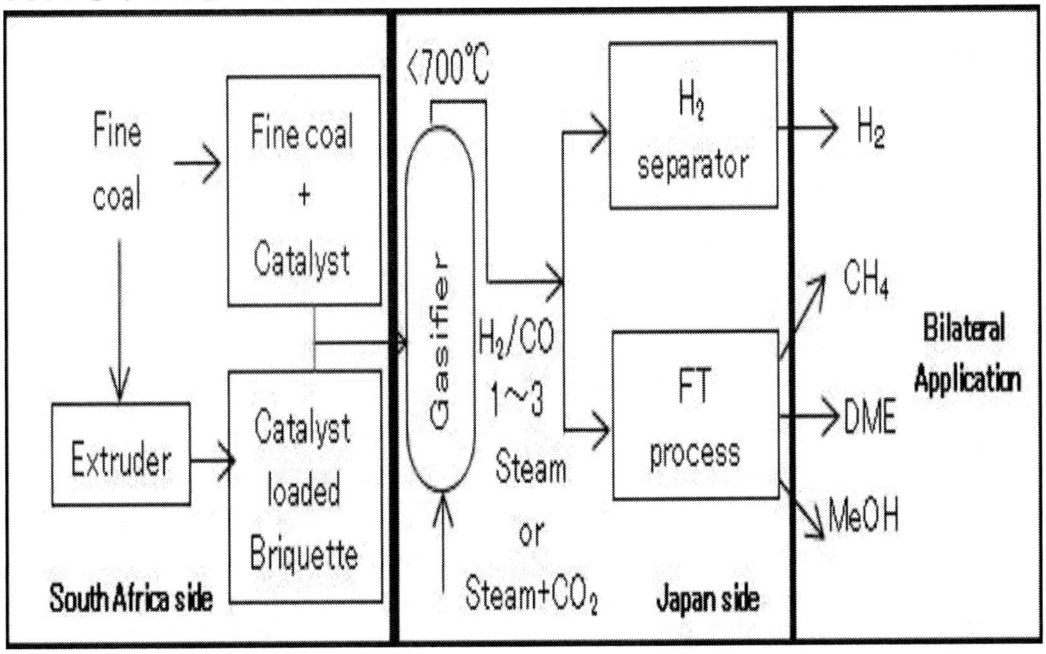


17

# Next steps:

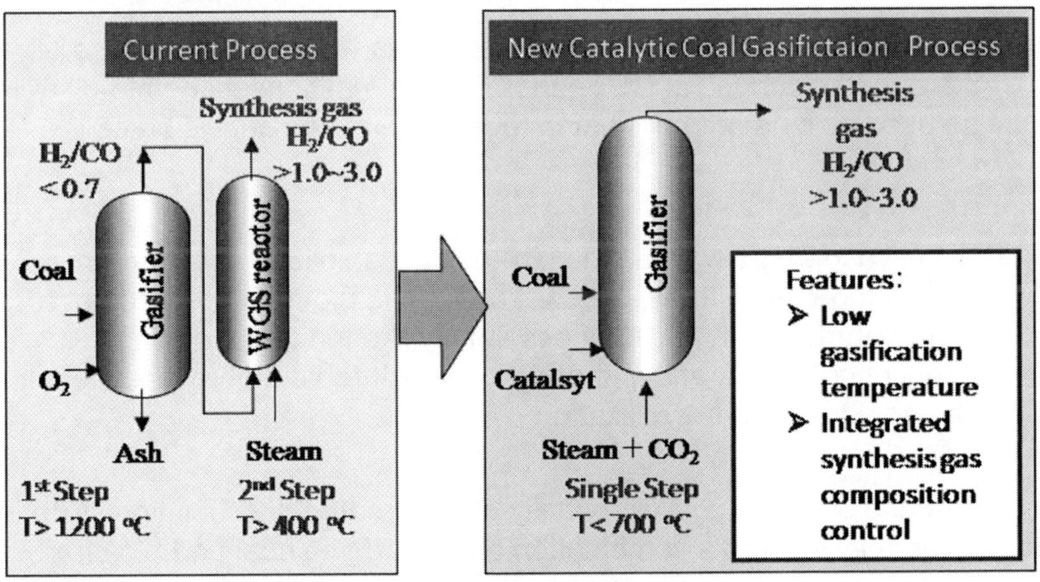


- This process does not need $O_2$ and because of low operating temperature, this process is 10~15% more efficient than the current high temperature gasification process.
- AIST's catalytic coal gasification process has 10~15% higher efficiency than the conventional process and this will lead to 25~30% $CO_2$ reduction.

18

## Concluding Remarks:

- With the *green coal* formulation, the catalyst plays a dual role:
  - (1) Self-scrubbing (in-situ capture) of sulphur during utilization in fixed-bed gasification or any process operating on lump coal,
  - (2) Enhancing the rate limiting $CO_2$ gasification step of the char. *Green coal* thus has the capability of at least doubling the gasification reactivity thus allowing for enhanced throughput possibilities in fixed-bed gasifiers.

- By minimizing the amount of sulphur partitioning to the gas phase during utilization lowers the total cost of the gas clean-up process. *Green Coal* has the potential to reduce up to 50% of the current gaseous sulphur emission during fixed bed gasification, with the balance of the sulphur encapsulated in an insoluble solid phase in the ash assemblage.

19

## Concluding Remarks:

- The co-generation of carbon feed stocks is well known because of the benefits derived in $CO_2$ generation. In most cases the bio-oil will be used as binder if briquetting together with fine coal is being considered.

- In the case of *green coal*, the bio-residue is utilized as a carbon source together with fine discard coal. The bio-char prepared at a temperature of c.a. 300°C is essentially likened to young lignite in terms of coal rank, and is therefore considered a highly reactive carbon source during utilization.

- The use of *green coal* essentially offers an attractive solution since it is formulated from a combination of discard fine coal and bio-char. The bio-char addition to the formulation is considered a renewable energy source, thus carbon credits can be claimed due to the reduction in $CO_2$ production.

20

NRF RISA
National Research Foundation | Research and Innovation Support and Advancement

# Acknowledgements

This work is based on the research financially supported by the South African Research Chairs Initiative of the Department of Science and Technology and National Research Foundation of South Africa.

Any opinion, finding or conclusion or recommendation expressed in this material is that of the author(s) and the NRF does not accept any liability in this regard.

NORTH-WEST UNIVERSITY
YUNIBESITI YA BOKONE-BOPHIRIMA
NOORDWES-UNIVERSITEIT
POTCHEFSTROOM CAMPUS

It all starts here

21

# Thank you

NRF RISA
National Research Foundation | Research and Innovation Support and Advancement

NORTH-WEST UNIVERSITY
YUNIBESITI YA BOKONE-BOPHIRIMA
NOORDWES-UNIVERSITEIT

22

TECHNISCHE UNIVERSITÄT
BERGAKADEMIE FREIBERG
The University of Resources. Since 1765.

Institute of
Energy Process Engineering and
Chemical Engineering

# Investigation of the Boudouard-Reaction at elevated pressure

Sascha Rußig
Pittsburgh Coal Conference


Cape Town, 11.08.2016

## KIVAN – Configuration

**KInetische VersuchsANlage**

- max. operation temperature: 1150 °C
- up to 100 bar
- 4 seperate gas preheaters
- 4 seperate heating zones
- reaction tube: Ø 58 mm x 2.6 m
- inert gases: Ar, $N_2$
- gasifying agents: $CO_2$, steam
- also addition of CO, $H_2$ for investigation of inhibition possible
- max. solids feedrate: 1.5 kg/h

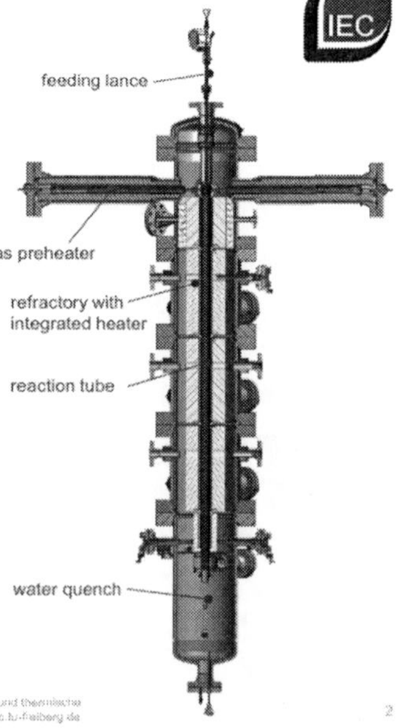

## KIVAN – Special Features

- 12 sampling tubes for product gas along reaction tube
- real time measurement of gas composition with infrared analyser and GC
- 48 temperature measurement points inside reaction tube, distributed along reaction zone
- comparable facilities: PiTER (TUM, Ger), PEFR (CSIRO, Au)

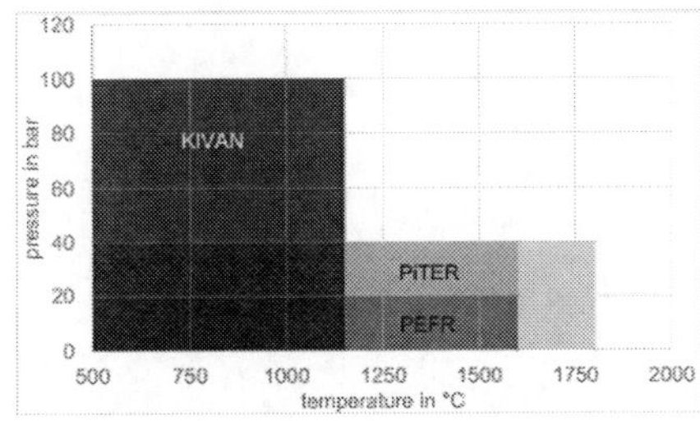

## Experimental Programme

**solid feedstock**

- char from german lignite coal (950 °C)

| proximate analysis in wt%, ad | | | | ultimate analysis in wt%, daf | | | | |
|---|---|---|---|---|---|---|---|---|
| moist | Ash | VM | FC | C | H | N | Sc | O |
| 1.2 | 8.6 | 3.9 | 87.5 | 89.7 | 0.4 | 0.3 | 0.6 | 0.4 |

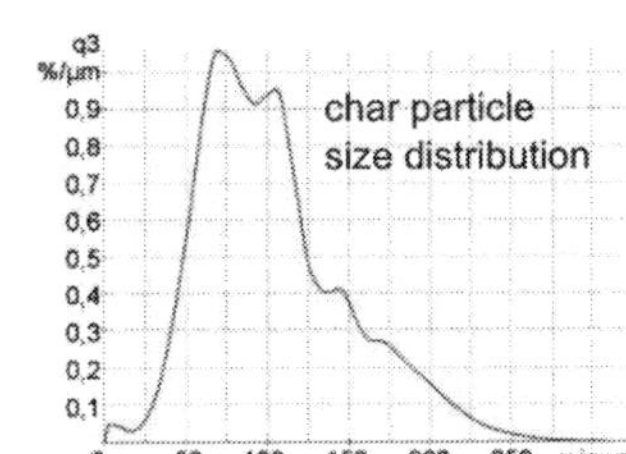


**gasification atmosphere**

- Temperature: 900, 950, 1000, 1050, 1100 °C
- System pressure: 20 bar
- Gasifying agent: $CO_2$, 25, 50, 75 Vol.-%
- Total gas volume flow: 100 l/min STP, 10 l/min i.N. Ar, rest $N_2$
- Goal: constant gas velocity (0.13 m/s)

## Experimental Programme

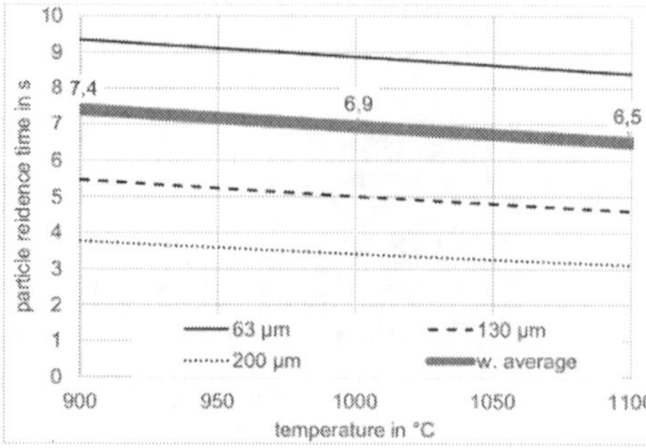


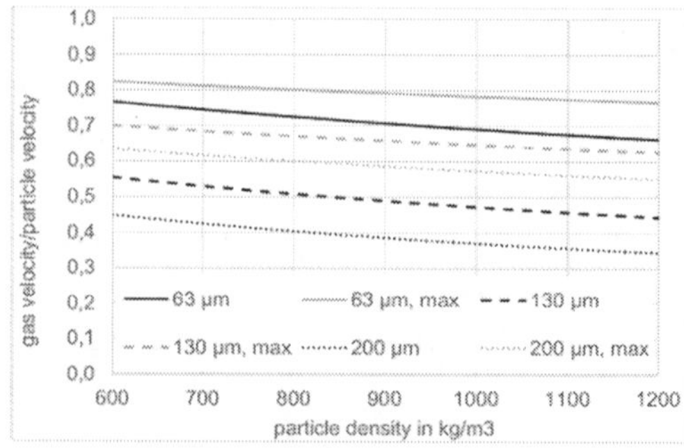


- mean particle residence time 7.4- 6.5 s
- Volume fraction of solids < 0.1 %

- particle velocity > gas velocity
- higher gas velocities to minimize seperating effects possible



## Experimental Results

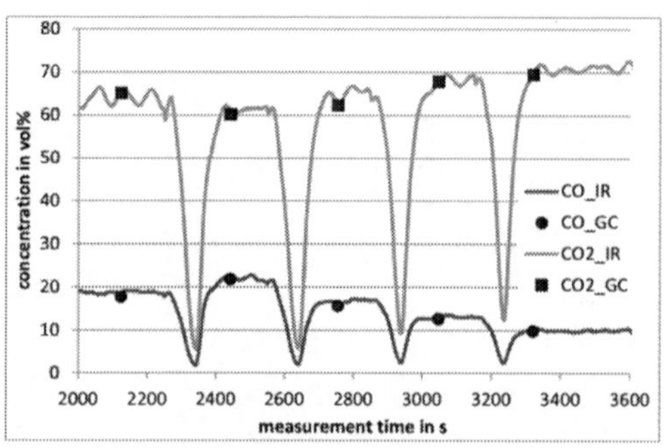


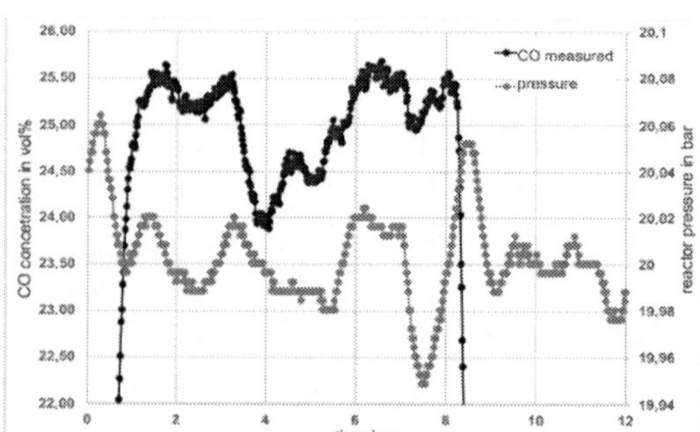


- parallel measurement of product gas composition with GC and IR
- deviation between GC and IR < 1%

- IR enables capturing of process dynamics
- Influence of process parameters on product gas composition can also be determined

## Experimental Results – Approach Temperatures

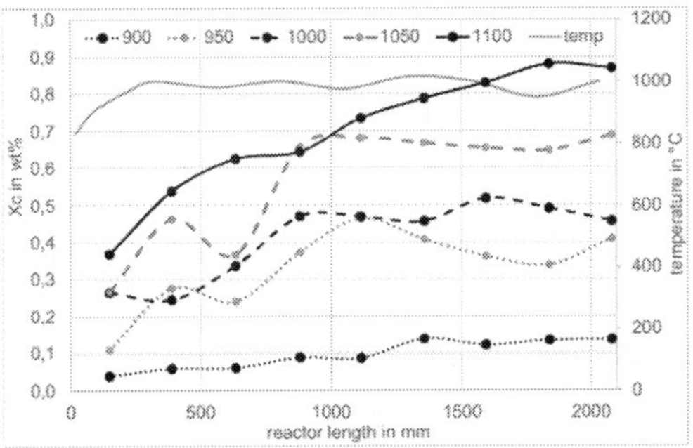


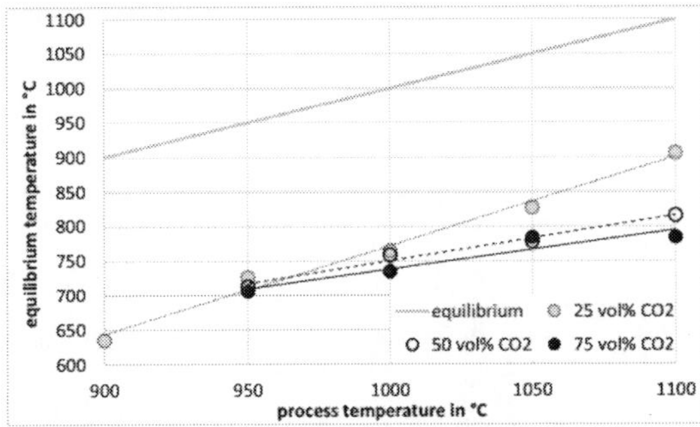


- Xc for 25 vol% $CO_2$
- High carbon conversion levels achieved

- Calculated equilibrium temperatures from measured gas concentrations
- $CO_2$ conversion has to be taken into account in kinetic evaluation



## Evaluation of Experimental Data – modeFRONTIER

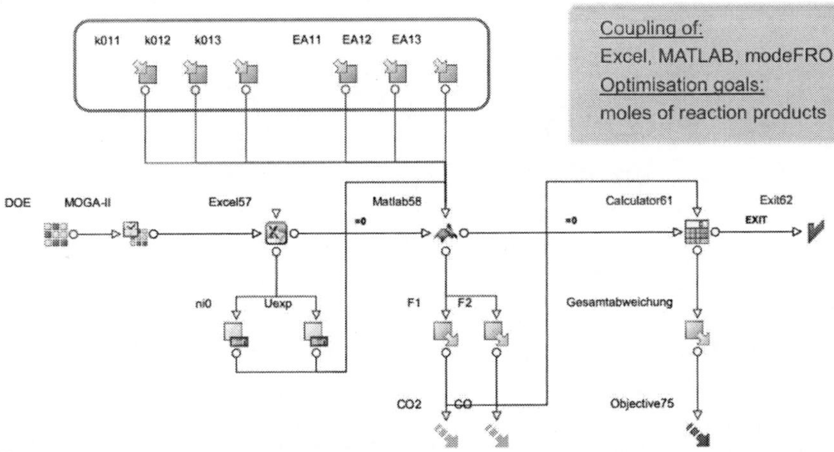


Coupling of:
Excel, MATLAB, modeFRONTIER
Optimisation goals:
moles of reaction products

**-> Effectively represents multidimensional, global and simultaneous regression of experimental results**

## Evaluation of Experimental Data – Sub Models

- Reactor model: $\frac{dl}{dt} = \frac{\dot{V}_{Gas}}{A_R} \cdot \frac{g(\rho_{Koks} - \rho_{Gas})}{18\eta_{Gas}} D_p^2$

- Reaction kinetic: $\frac{dU_{C,CO2}}{dt} = \frac{k_{11} p_{CO2}}{1 + k_{12} p_{CO} + k_{13} p_{CO2}} \cdot f(U_C)$

$$k_{ij} = A_{ij} \cdot exp\left(-\frac{E_{A,ij}}{RT}\right)$$

- Particle model: $f(U_C) = (1 - U_C)\sqrt{1 - \psi \cdot ln(1 - U_C)}\ ;\ \psi = 7{,}5$

- Substance properties: CANTERA-data base

- inclusion of: product gas generation, particle conversion, change of product gas composition

**-> reactor model can easily be extended to acount for aditional gases, reactor geometries, etc.**



## Results of Software-based Evaluation

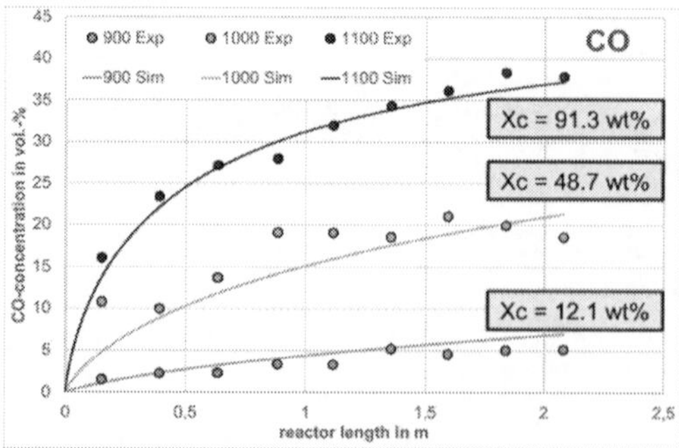


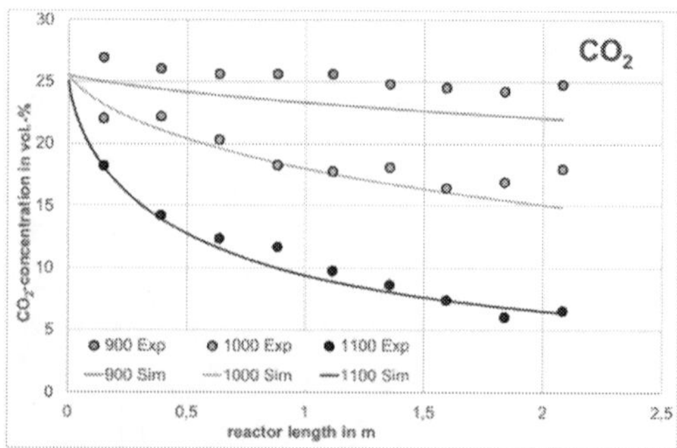


- Good fit can be accomplished with low number of test runs
- Increased number of test runs increases accuracy of model
- Model is able to correctly predict gas concentrations up to high carbon conversion levels

## Results of Software-based Evaluation – L-H-Parameters

| | | KIVAN | Kajitani[1] | | Huang[2] | Umemoto[3] | |
|---|---|---|---|---|---|---|---|
| Regime | | Drop tube | Drop tube | | TGA | Drop tube | |
| Parent Coal | | Lig (Ger) | Lig (Aus) | Lig (PRC) | Lig | HBK (Ind) | SB (PRC) |
| EA11 | kJ/(kg K) | 197 | 251 | 222 | 143 | 332 | 327 |
| EA12 | kJ/(kg K) | -77 | -23 | -23 | -227 | 13,5 | 8,38 |
| EA13 | kJ/(kg K) | -60 | -48,1 | -48,1 | -74 | -72,7 | -96,5 |
| k011 | 1/(bar s) | 9,55E06 | 2,79 | 0,45 | 3,49E04 | 1,33E12 | 3,39E11 |
| k012 | 1/bar | 5,25E-05 | 0,05 | 0,021 | 3,46E-09 | 90,7 | 76,1 |
| k013 | 1/bar | 1,02E-02 | 0,0058 | 0,038 | 1,38E-03 | 0,05 | 1,72E-2 |
| F CO2 | % | 6,7 | - | - | - | - | - |
| F CO | % | 26,3 | - | - | - | - | - |


1: S. Kajitani, N. Suzuki, M. Ashizawa, S. Hara: "CO2 gasification rate analysis of coal char in entrained flow coal gasifier", *Fuel*, Vol. 85, S. 163–169, 2006
2: Z. Huang, J. Zhang, Y. Zhao: "Kinetic studies of char gasification by steam and CO2 in the presence of H2 and CO", *Fuel Processing Technology*, Vol. 91, S. 843–847, 2010
3: Umemoto S et al. Modeling of coal char gasification in coexistence of CO2 and H2O considering sharing of active sites. *Fuel* (2011), doi:10.1016/j.fuel.2011.11.030




## Outlook

- Operation of gasifier at elevated pressure and development of sub models
- Gasification experiments under steam and steam/$CO_2$ atmosphere
- Radioactive particle residence time and residence time distribution measurement

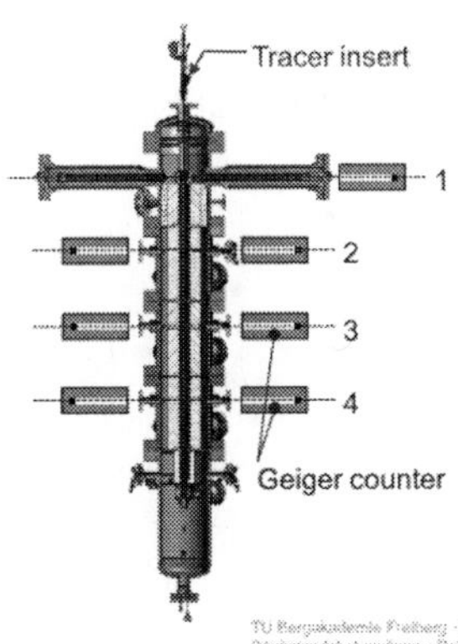


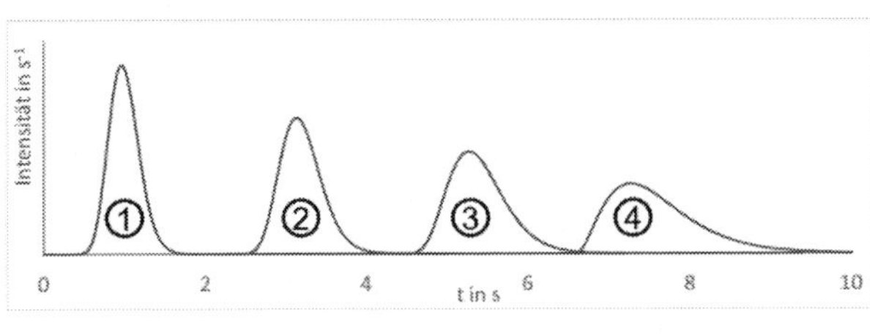

## Acknowledgements

This research has been funded by the German Ministry of Education and Research in the framework of **Virtuhcon** (03Z2FN12) and **HITECOM II** (03Z2F512) projects

Virtuhcon
Virtual High Temperature Conversion

HITECOM
High Temperature Conversion Optical Measurement

Bundesministerium
für Bildung
und Forschung



**Thank you for your attention!**

TU Bergakademie Freiberg · Institut für Energieverfahrenstechnik und Chemieingenieurwesen · Professur für Energieverfahrenstechnik und thermische Rückstandsbehandlung · Reiche Zeche · Fuchsmühlenweg 9 · 09599 Freiberg · Telefon: 03731 39-4511 · Fax: 03731 39-4555 · www.iec.tu-freiberg.de

## Mixing chamber

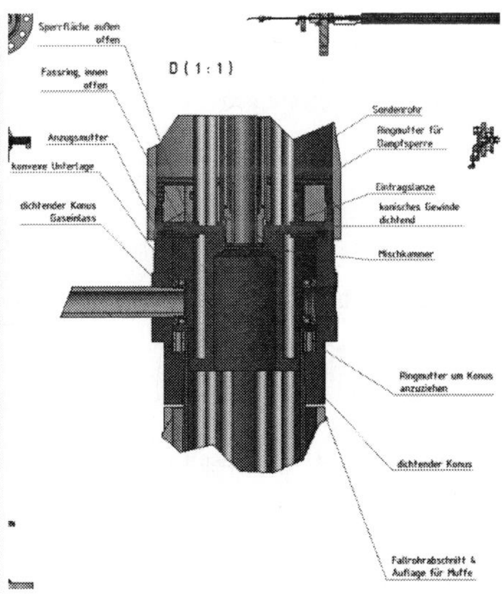

TECHNISCHE UNIVERSITÄT
BERGAKADEMIE FREIBERG
The University of Resources. Since 1765.

Institute of Energy Process Engineering
and Chemical Engineering

# Ash Agglomeration in the Modified Internal Circulating COORVED Coal Gasifier

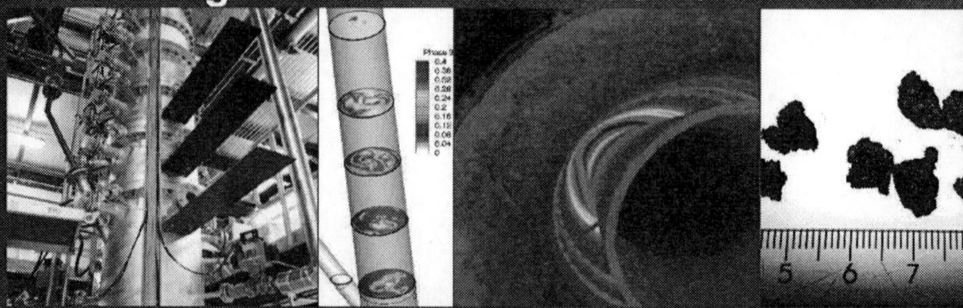

Martin Schurz; Alexander Laugwitz; Steffen Krzack; Bernd Meyer

33rd International Pittsburgh Coal Conference – PCC 2016
8th – 12th August 2016 – Cape Town, South Africa
Session 23: Gasification Technologies: General - 2



## Outline

I. Background & motivation

II. Experimental setup & investigation

III. Results
- Lignite coke gasification
- Gasification of two different lignites with $O_2/CO_2$ and $O_2/H_2O$
- Ash-agglomeration

IV. Summary & future prospects

I. Background & Motivation

## INCI - Gasification Principle

- **In**ternal **Ci**rculation
- Gasification of high ash coal fines
- No slagging
- Ash agglomeration
- Reduced carbon loss via fly ash and bottom product
- Prolonged residence time to increase carbon conversion
- Target outlet temperature 1000-1100°C

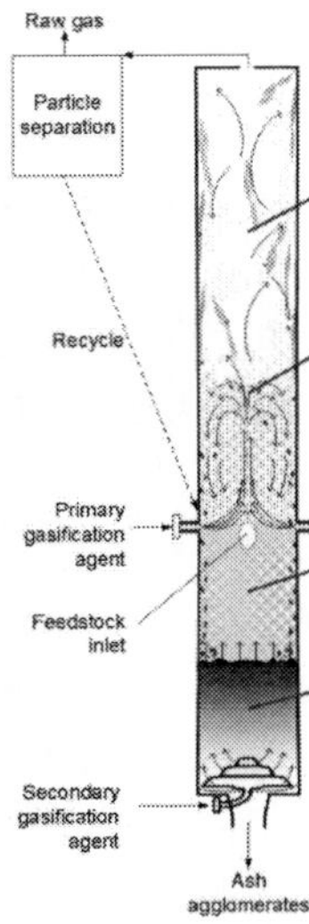


Patent: DE 10 2007 006981 B4 2009.01.29

- CFB flow principle
  - Strand formation
  - Internal circulation
- Jetting fluidized bed
  - Prim. gasific. agent (boxer arrangement)
  - Central flame (T > 2000 °C)
  - **Recirculation cells (internal circulation)**
  - Walls are protected from temperatures
  - **Agglomeration**
- Bubbling fluidized bed
  - Internal separation mechanism
  - Internal heat reuse
- Fixed bed
  - Secondary gasifiaction agent
  - Post gasification (C-content < 5%)
  - Internal heat reuse
  - Determines base fluidization velocity in dense fluidized bed



I. Background & Motivation

## COORVED - Project

- **CO**$_2$ - **R**eduktion durch innovatives **Ve**rgaser**d**esign
- $CO_2$-reduction by innovative gasifier design
- Experimental part (COORVED gasifier)
- Inverse flame research
- Simulation of gasifiers
- Oct. 2009 – Dec. 2016
- Funded by BMWi and Pörner Group

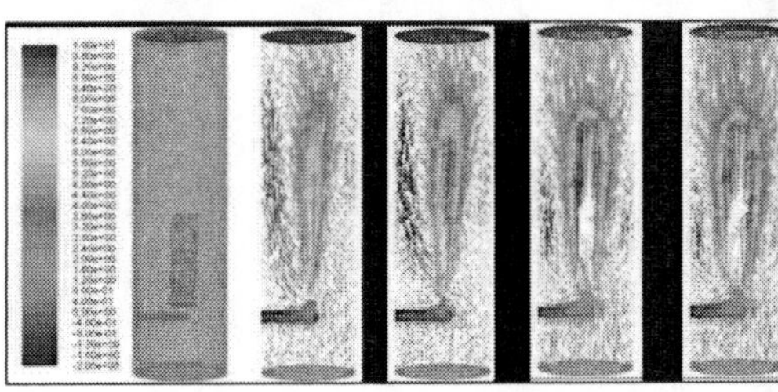

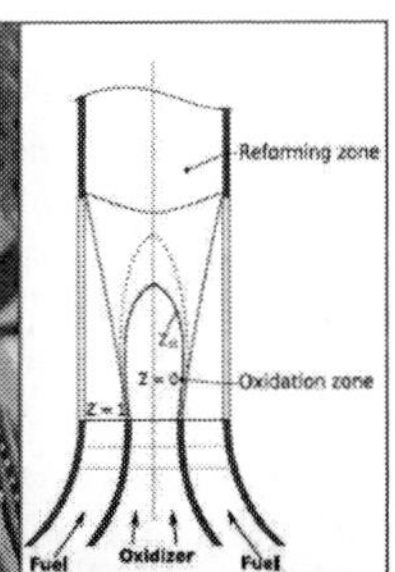

II. Experimental Setup & Investigation

## COORVED lab scale Unit (new Design since 2015)

1 bar(a); ≤ 60 kW
$Ø_i$=0.10-0.30 m; H=6m

44 campaigns
530 operating hours

**Construction**
- Bottom part with metallic reaction tubes
- Central part refractory lined with alumina-based insulating firebricks (preheated by natural gas burner)
- Upper part with ceramic reaction tubes (electrical preheated)

**Measurement equipment**
- More than 30 measurement ports
- temperature, pressure, sampling and optical ports
- Shadowgraphy; flame camera
- Gas-chromatography; FTIR

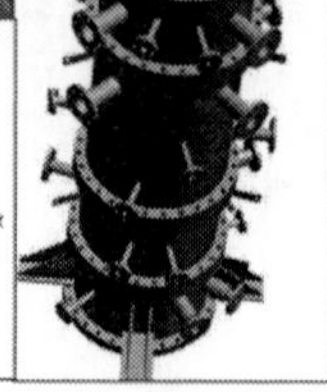

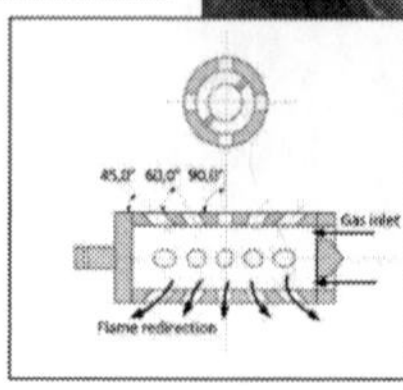


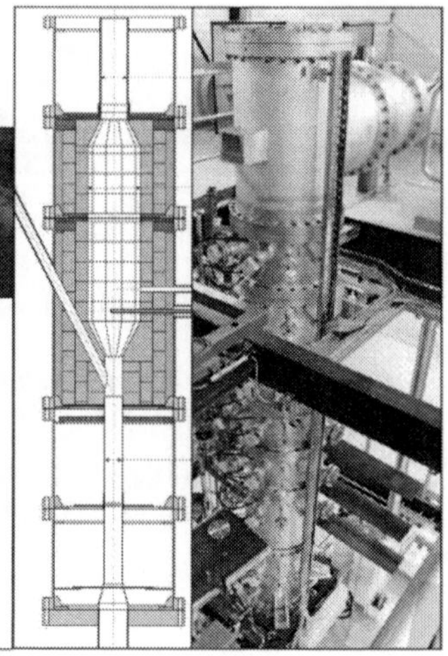



II. Experimental Setup & Investigation

## COORVED – Gasifier: Flow Pattern in Main Reaction Zone

**Spouted bed formation**
- Increasing diameter in the 3rd segment
- Conical widening of 60°
- Strong central fluidization
- Slowly recirculating annular moving bed

**Advantages**
- Increased residence time
- Accumulation of solids in main reaction zone
  → Increased carbon conversion
  → Enforced ash agglomeration
  → Lower wall temperatures
- Stabilized operation and improved controllability

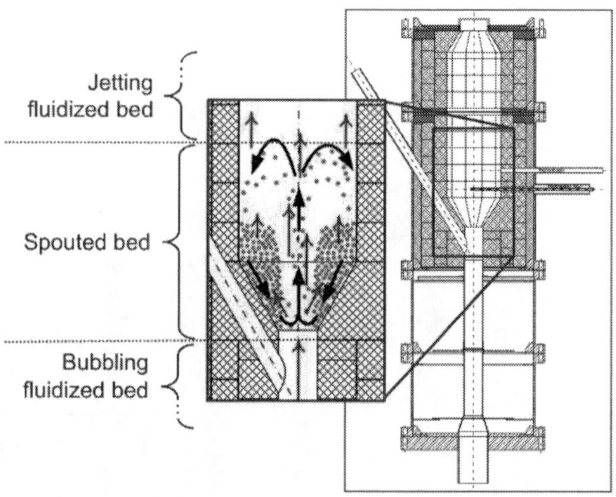

II. Experimental Setup & Investigation

## Steady State Operation in COORVED Reactor

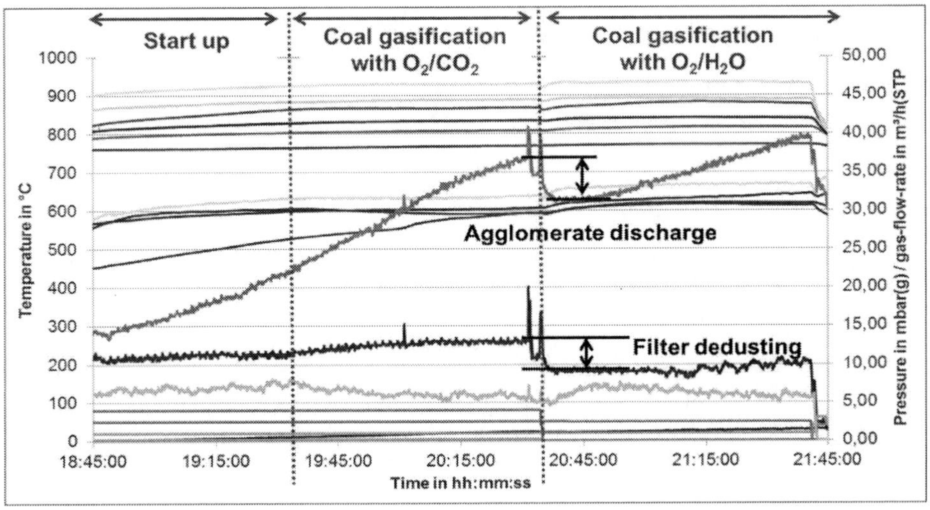




III. Results

## Coke Gasification in COORVED Reactor

**Feedstock characterization**

- Lignite coke (900-950 °C; hearth furnace)
- Powdery coke < 0.4 mm
  ($x_{10}$=13.8 µm; $x_{50}$=79.3 µm; $x_{90}$=239.9µm)
- Proximate analysis:
  - Fixed carbon 87.0 wt.% (wf)
  - Volatile matter 4.3 wt.% (wf)
  - Ash content 8.7 wt.% (wf)
  - Water content 1.1 wt.%

**Coke ash (815°C)**

- Main components (XRF)
  - CaO 31.1 wt.%
  - MgO 14.3 wt.%
  - $Fe_2O_3$ 10.7 wt.%
  - $Na_2O$ 7.8 wt.%
  - $Al_2O_3$ 5.5 wt.%
- Ash melting (red. atmosphere)
  - Init. deform. temp.: 1250°C
  - Fusion temp.: >1394°C

**Achievements**

- Carbon conversion $X_C$ > 93 %
- Cold gas efficiency $\eta_{CGE}$ > 58 %
- Syngas yield $y_{Syn.}$ > 1.35 m³/kg (STP/waf)
- $T_{Wall,max.}$ < Init. deform. temp. of ash

**Limitations**

- Coke with low reactivity
  →Increased oxygen addition (λ↗)
  →Enlarged hot central zone ($T_{avg.}$↗)
  → **Risk of slagging!**
- Missing volatile matter
  → Difficult temperature control
- **No ash agglomeration!**

**Operation with coke-lignite-mixture**

- Addition of lignite →coke-lignite ratio ≈ 50/50

→ **Ash agglomeration!**

III. Results

## Gasification of powdery Lignite (A) in COORVED Reactor

**Feedstock characterization**

- Dried lignite (48 h; 80°C)
- Powdery coal < 0.4 mm
  ($x_{10}$=13.1 µm; $x_{50}$=67.9 µm; $x_{90}$=232.1 µm)
- Proximate analysis:
  - Fixed carbon 42.2 wt.% (wf)
  - Volatile matter 52.1 wt.% (wf)
  - Ash content 5.7 wt.% (wf)
  - Water content 11.4 wt.%

**Lignite ash (815°C)**

- Main components (XRF)
  - $Fe_2O_3$ 26.8 wt.%
  - CaO 25.2 wt.%
  - MgO 9.9 wt.%
  - $SiO_2$ 9.4 wt.%
  - $Al_2O_3$ 4.3 wt.%
- Ash melting (red. atmosphere)
  - Init. deform. temp.: 1300°C
  - Fusion temp.: >1510°C

**Achievements**

**Note for low $\eta_{CGE}$: Heat losses ≈ 27 % of thermal input!**

- Ash agglomeration
- Carbon free bottom product
- Raw gas free of higher hydrocarbons
- $O_2/CO_2$-Operation (λ=0.47-0.53)
  - Carbon conversion $X_C$ > 96 %
  - Cold gas efficiency $\eta_{CGE}$ > 55 %
  - Syngas yield $y_{Syn.}$ > 1.3 m³/kg (STP/waf)
  - **$T_{MRZ}$=1294°C; $T_{Wall,max.}$ = 1146°C → Risk of slagging!**
- $O_2/H_2O$-Operation (λ=0.37-0.48)
  - Carbon conversion $X_C$ > 92 %
  - Cold gas efficiency $\eta_{CGE}$ > 54 %
  - Syngas yield $y_{Syn.}$ ≈ 1.5 m³/kg (STP/waf)
  - $T_{MRZ}$=1200°C; $T_{Wall,max.}$ = 1082°C

**Operation with "ash enriched" lignite**

- Addition of lignite ash (815°C) →Ash content 16.1 wt.% (wf)

→ Enforced ash agglomeration



III. Results

## Exemplary Operation: Lignite A Gasification with $O_2/H_2O$

**Experimental settings**

- Thermal input: 51.6 kW
- Total air ratio ≈ 0.4
- $H_2O/O_2$ ≈ 1.2 kg/m³ (STP)

**Coal ash (815°C)**

- Ash Content: 5.7 wt.% (wf)

**Composition**

- $Fe_2O_3$: 26.8 wt.%
- CaO: 25.2 wt.%
- MgO: 9.9 wt.%
- $SiO_2$: 9.4 wt.%
- $Al_2O_3$: 4.3 wt.%

**Ash melting (red. atmosph.)**

- Init. deform. temp.: 1300°C
- Fusion temp.: >1510°C

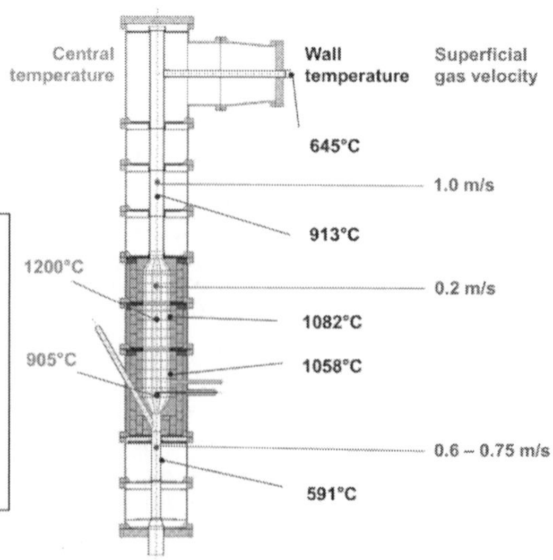


**Raw gas**: 16.8 m³(STP)/h (wf)

- $H_2$: 29.9 vol.%
- CO: 29.9 vol.%
- $CH_4$: 0.1 vol.%
- $CO_2$: 34.2 vol.%
- $N_2$: 5.9 vol.% (Purging gas)
- Gas free of tars and oils!

**Agglomerates**

- Ash Content: > 99 wt.%

**Composition**

- $Fe_2O_3$: 19.5 wt.%
- CaO: 36.0 wt.%
- MgO: 15.3 wt.%
- $SiO_2$: 12.8 wt.%
- $Al_2O_3$: 4.8 wt.%

**Ash melting (red. atmosphere)**

- Init. deform. temp.: 1290°C
- Fusion temp.: 1514°C

III. Results

## Ash Agglomerates of Lignite A

**$O_2/CO_2$-Operation**
- **$T_{MRZ}$=1294°C**
- **Agglomerates of glassy structure**
- **$x(C)_{Agglo.}$ < 1 wt.%**

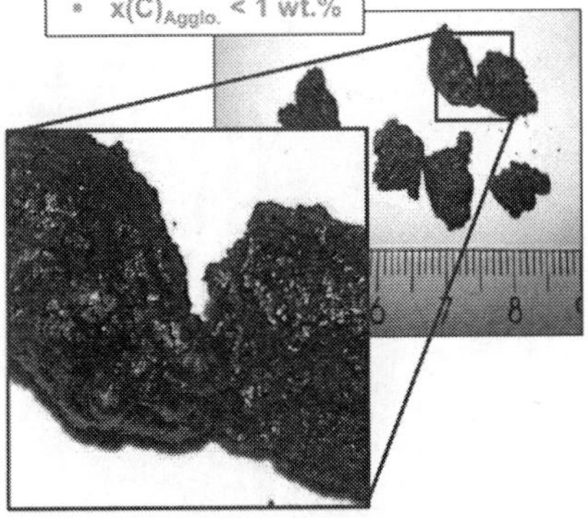

**$O_2/H_2O$-Operation**
- **$T_{MRZ}$=1200°C**
- **Agglomerates of porous structure**
- **$x(C)_{Agglo.}$ < 1 wt.%**

III. Results

## Gasification of granular Lignite (B)

**Feedstock characterization**
- Dried lignite
- Coal size < 2.0 mm
  ($x_{10}$=24.3 µm; $x_{50}$=166.3 µm; $x_{90}$=1242.1 µm)
- Proximate analysis:
  - Fixed carbon 40.6 wt.% (wf)
  - Volatile matter 47.4 wt.% (wf)
  - Ash content 12.0 wt.% (wf)
  - Water content 12.5 wt.%

**Lignite ash (815°C)**
- Main components (XRF)
  - $SiO_2$ 24.5 wt.%
  - CaO 20.4 wt.%
  - $Al_2O_3$ 10.8 wt.%
  - $Fe_2O_3$ 9.1 wt.%
  - MgO 5.6 wt.%
- Ash melting (red. atmosphere)
  - Init. deform. temp.: **1125°C**
  - Fusion temp.: **1230°C**

**Note for low $\eta_{CGE}$:**
**Heat losses ≈ 27 % of thermal input!**

**Achievements**
- **Ash agglomeration**
- **Carbon free bottom product**
- **Raw gas free of higher hydrocarbons**
- $O_2/CO_2$-Operation (λ=0.34-0.38)
  - Carbon conversion $X_C$ ≈ 87…91 %
  - Cold gas efficiency $\eta_{CGE}$ > 55 %
  - Syngas yield $y_{Syn.}$ > 1.13 m³/kg (STP/waf)
- $O_2/H_2O$-Operation (λ=0.35-0.37)
  - Carbon conversion $X_C$ ≈ 90-91 %
  - Cold gas efficiency $\eta_{CGE}$ > 62 %
  - Syngas yield $y_{Syn.}$ ≈ 1.31 m³/kg (STP/waf)

III. Results

## Gasification of Lignite B with $O_2/H_2O$

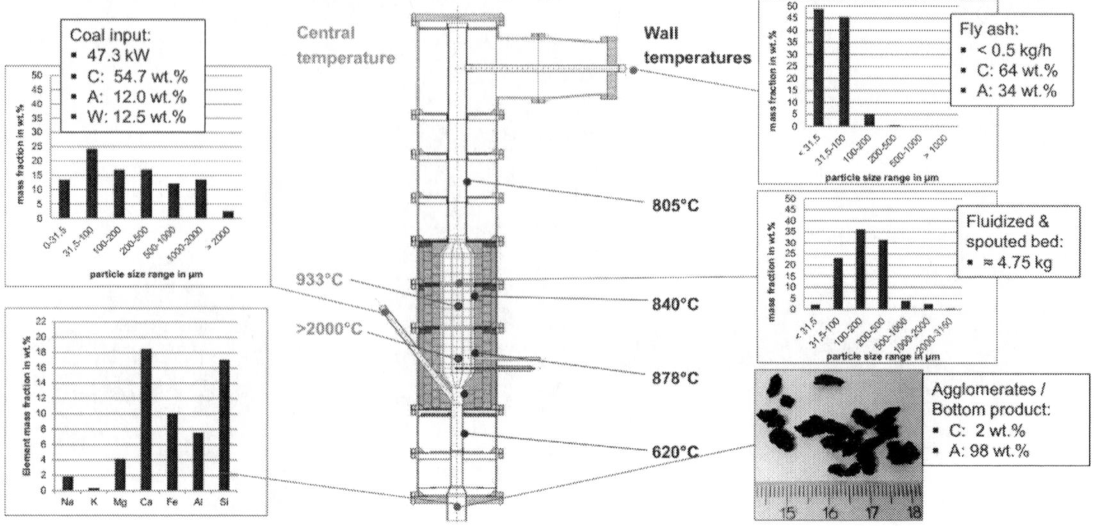




IV. Summary & Future Prospects

## Summary & Future Prospects

**INCI – gasifier COORVED**
- Optimized reactor design
- Improved operability

**Experimental findings**
- Low coke reactivity unfavorable for agglomeration
- **Ash agglomeration in gasification of different lignites and mixtures**
- **Utilization of different coal particle sizes**
- Controlled ash agglomeration in $O_2/CO_2$- and $O_2/H_2O$-atmospheres
- **Efficient agglomerate post-gasification**
- **Tar- and oil-free raw gas**

➡

**INCI – gasifier COORVED**
- Further reactor design optimization
- Changes in moving bed zone (Secondary gas nozzles, pre-heating, diameter)
  →Improved post-gasification for feedstocks with higher percentage of coarse grain and ash
- Varied solids feeding position

**Demanding Feedstocks**
- Lignite with higher ash content
- High ash hard coal

## The End

# Thank you for your attention!

**Funding**

The COORVED project is funded by **German Federal Ministry of Economic Affairs and Energy** (Project ID : **0327865**) and **Pörner Group.**

**Contact Information**

Dipl.-Ing. Martin Schurz

Martin.Schurz@iec.tu-freiberg.de

Tel.: 0049 (0) 3731 – 39 4483

Fax: 0049 (0) 3731 – 39 4555

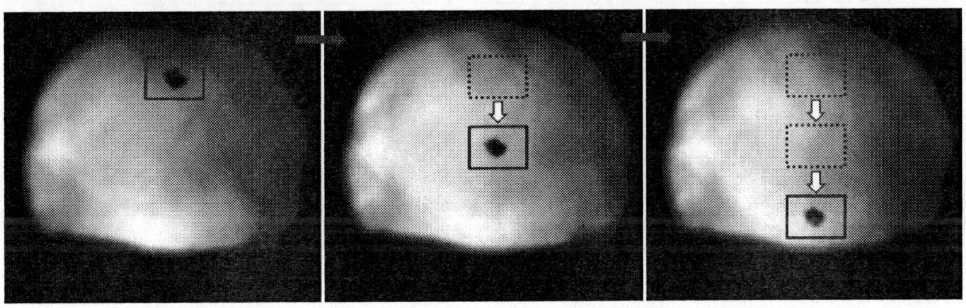

EERC

# SUPERCRITICAL $CO_2$ CYCLES FOR POWER PRODUCTION

Presented at the 2016 International Pittsburgh Coal Conference
Cape Town, South Africa
August 8–12, 2016

Mike Holmes
Director of Energy Systems Development

Critical Challenges. **Practical Solutions.**

# DOE DISCLAIMER

## Acknowledgment

This material is based upon work supported by the U.S. Department of Energy National Energy Technology Laboratory.

## Disclaimer

This presentation was prepared as an account of work sponsored by an agency of the United States Government. Neither the United States Government, nor any agency thereof, nor any of their employees, makes any warranty, express or implied, or assumes any legal liability or responsibility for the accuracy, completeness, or usefulness of any information, apparatus, product, or process disclosed or represents that its use would not infringe privately owned rights. Reference herein to any specific commercial product, process, or service by trade name, trademark, manufacturer, or otherwise does not necessarily constitute or imply its endorsement, recommendation, or favoring by the United States Government or any agency thereof. The views and opinions of authors expressed herein do not necessarily state or reflect those of the United States Government or any agency thereof.

EERC 2 Critical Challenges. **Practical Solutions.**

# NDIC DISCLAIMER

This presentation was prepared by the EERC pursuant to an agreement partially funded by the Industrial Commission of North Dakota, and neither the EERC nor any of its subcontractors nor the North Dakota Industrial Commission nor any person acting on behalf of either:

(A) Makes any warranty or representation, express or implied, with respect to the accuracy, completeness, or usefulness of the information contained in this report or that the use of any information, apparatus, method, or process disclosed in this report may not infringe privately owned rights; or

(B) Assumes any liabilities with respect to the use of, or for damages resulting from the use of, any information, apparatus, method, or process disclosed in this report.

Reference herein to any specific commercial product, process, or service by trade name, trademark, manufacturer, or otherwise does not necessarily constitute or imply its endorsement, recommendation, or favoring by the North Dakota Industrial Commission. The views and opinions of authors expressed herein do not necessarily state or reflect those of the North Dakota Industrial Commission.



# ENERGY & ENVIRONMENTAL RESEARCH CENTER (EERC)

- Nonprofit branch of the University of North Dakota focused on energy and environmental solutions.
- More than 254,000 square feet of state-of-the-art laboratory, demonstration, and office space.

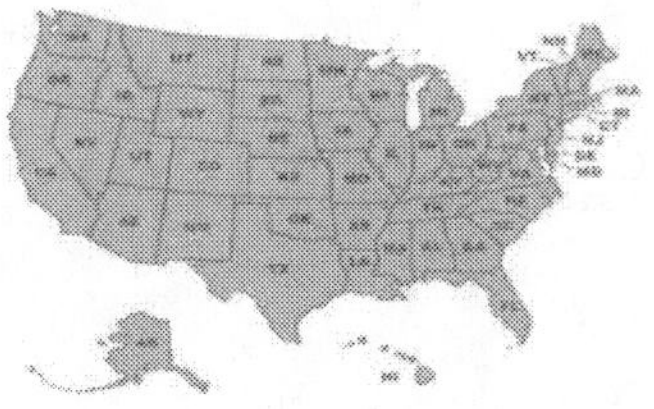

# HISTORY OF SUCCESS

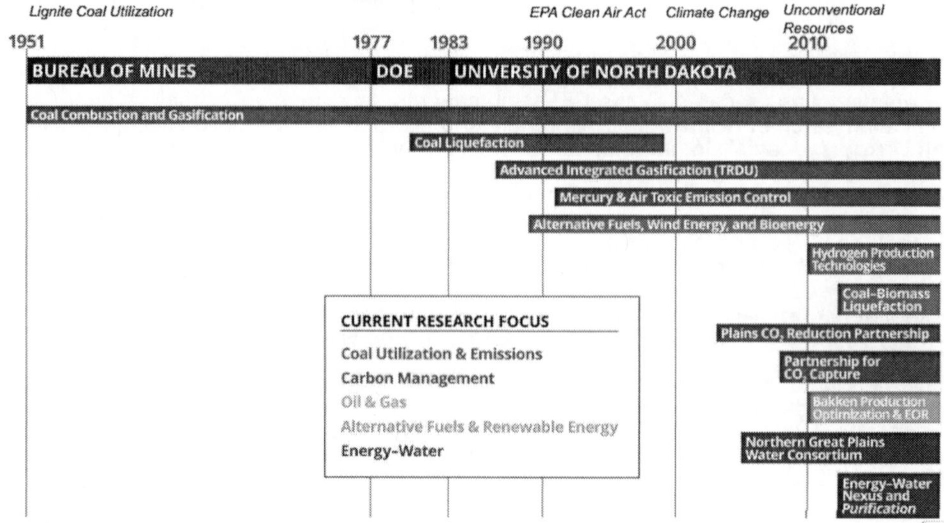


EERC

Critical Challenges. **Practical Solutions.**

# CORE PRIORITIES

- Coal Utilization and Emissions
- Carbon Management
- Oil and Gas
- Alternative Fuels and Renewable Energy
- Energy–Water

EERC

# PRESENTATION OUTLINE

- Technology Drivers
  - Coal Trends and Variables
  - Carbon Management Opportunities
- Next-Generation Power Cycle
  - Supercritical $CO_2$ Cycles for Power Production (Allam Cycle)

EERC

Critical Challenges. **Practical Solutions.**

# PIONEERS AND SYNERGIES

EERC

Critical Challenges. **Practical Solutions.**

# TRENDS

**Comparison of Growth Areas and Emissions, 1970-2014**

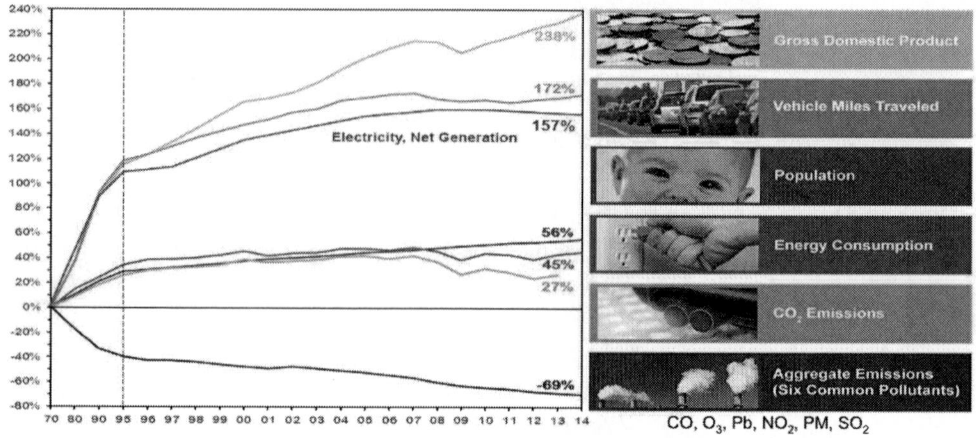


www3.epa.gov/airtrends/aqtrends.html.
www.eia.gov/totalenergy/data/monthly/#electricity.



# COAL INDUSTRY NEEDS

- Coal is challenged by environmental rules.
  - What is needed:
    - ♦ Clean power generation.
    - ♦ Cost-effective carbon capture and storage (CCS).
- The Allam Cycle – a promising technology.
  - High efficiency with intrinsic $CO_2$ capture.
  - $CO_2$ near-ready for enhanced oil recovery (EOR) applications.
- Greatly benefits North Dakota industry.
  - Lignite coal power generation.
  - Oil and gas industry.

# CARBON MANAGEMENT

- Carbon management plans
- Plains $CO_2$ Reduction (PCOR) Partnership
- Partnership for $CO_2$ Capture ($PCO_2C$)

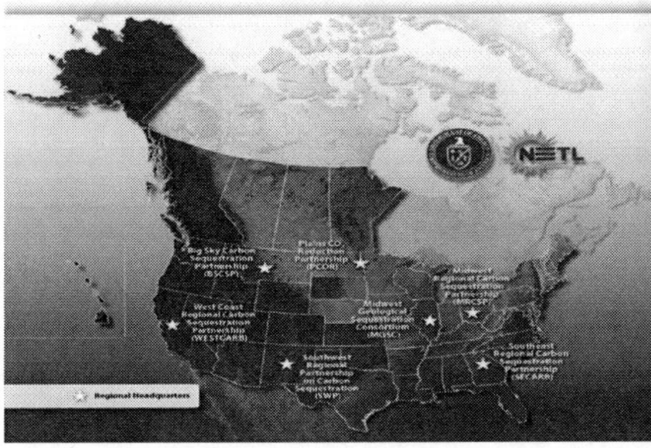

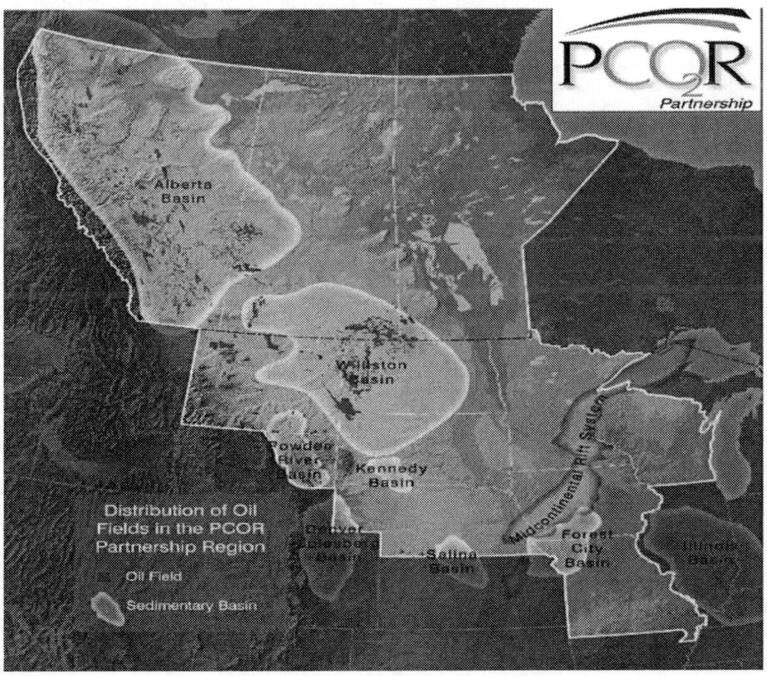


**Oil Fields**

6000+ fields evaluated.

Fields in the Williston, Powder River, Denver–Julesberg, and Alberta Basins were evaluated.

Used two methods:
EOR and volumetric

- **EOR approach:**
  Evaluated ~160 fields.

  **Sequestration capacity = 1 billion tons**
  **Incremental oil >3 billion bbl**
- **Volumetric approach:**
  Thousands of fields, total capacity >10 billion tons.

# BAKKEN $CO_2$ DEMAND FOR NORTH DAKOTA – A 30,000-FT VIEW

- Based on the following:
  - Traditional evaluation techniques
  - North Dakota Industrial Commission (NDIC) original-oil-in-place (OOIP) estimates
  - 4% incremental recovery
  - Net utilization of 5000 and 8000 $ft^3$/bbl
- 2–3.2 Bt of $CO_2$ needed, yielding 4–7 Bbbl of oil.
- North Dakota currently produces ~33 Mtpy of $CO_2$.
- Bakken growth is creating a projected increase in power demand.

EERC UND THE UNIVERSITY OF NORTH DAKOTA

Critical Challenges. **Practical Solutions.**

# WHAT IS THE ALLAM CYCLE?

The Allam Cycle is a supercritical $CO_2$ Brayton cycle:

- Oxy-fueled and direct-fired.
- Recuperates turbine exhaust heat via a recycle stream.
- Can utilize a heat sources in addition to the turbine exhaust.
- Turbine inlet temperature above 800°C (1000°–1200°C optimal)
- Turbine inlet pressure above 80 bar (200–400 bar optimal).
- Exhaust is $CO_2$ with moderate impurities.

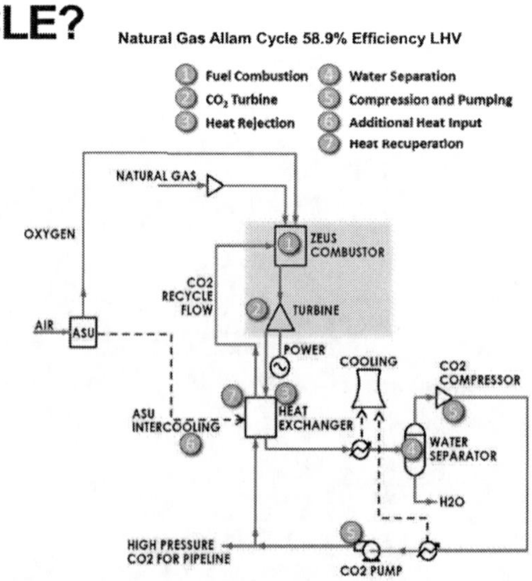


EERC 8RIVERS

Contains the intellectual property of 8 Rivers Capital, LLC and NET Power, LLC

Critical Challenges. **Practical Solutions.**

## CORE NATURAL GAS ALLAM CYCLE IS BEING DEMONSTRATED BY NET POWER

**8 Rivers has partnered with Exelon, CB&I, and Toshiba for the commercialization of the natural gas-fired Allam Cycle via the entity known as NET Power**

- 50-MWth natural gas demonstration plant located in La Porte, TX.
  - Includes all components of the Allam Cycle.
  - Mirrors design of commercial plant to ensure scalability. .

**Plant will undergo full performance evaluation.**
  - Commissioning begins end 2016.
  - Full operations begin in Q2 2017.
  - Will test performance, reliability, controllability, and safety.

**Program is fully funded.**
  - $140 million raised for engineering, construction, and testing.

**300-MWe commercial plant under development.**
  - Pre-FEED study completed on full commercial plant.
  - Beginning FEED and early development work.
  - Toshiba well progressed on commercial turbine design.
  - Working with customers in power, oil and gas industries on development opportunities.

8RIVERS

Contains the intellectual property of 8 Rivers Capital, LLC and NET Power, LLC

## THE ALLAM CYCLE CAN ALSO BE USED WITH COAL

**High efficiency** and **zero emissions** with **existing gasification technology**
Significantly **less integration, complexity, and overall cost** than IGCC.

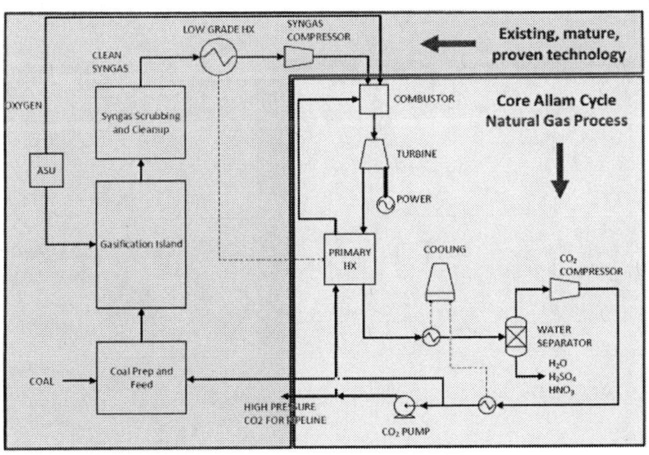


| Efficiency | LHV | HHV |
|---|---|---|
| Gross Turbine Output | 76.3% | 72.5% |
| Coal prep & feed | -0.2% | -0.2% |
| ASU | -10.2% | -9.7% |
| $CO_2$, Syngas Comp. | -9.1% | -8.7% |
| Other Auxiliaries | -6.5% | -6.1% |
| **Net Efficiency** | **50.3%** | **47.8%** |

- High efficiency with **existing gasifier technologies.**
- **Minimal gasifier integration required,** low complexity.
- Process simplification significantly reduces cost vs. IGCC.
- **Zero emissions.** No additional capture or compression equipment needed.

8RIVERS

Contains the intellectual property of 8 Rivers Capital, LLC and NET Power, LLC

# LIGNITE-BASED ALLAM CYCLE

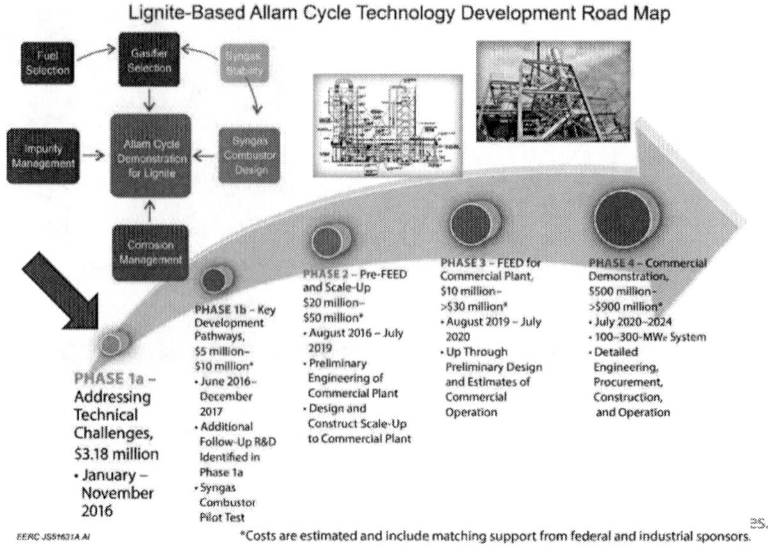


EERC

es. **Practical Solutions.**

# PHASE 1A TECHNICAL CHALLENGE AREAS

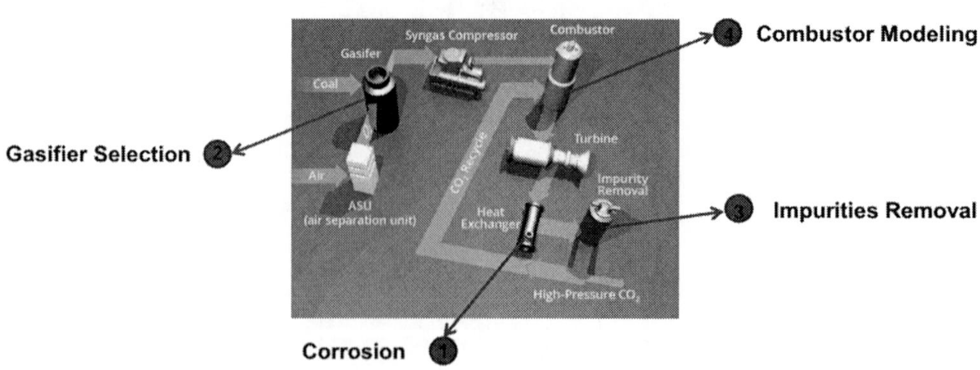


EERC



Critical Challenges. **Practical Solutions.**

# AREA 1: CORROSION

- Work by 8 Rivers and NETL determined that carbon steels were not appropriate.
- Six stainless steels were suggested by 8 Rivers as candidates
- Autoclave tests (1024 hr, 50°C and 90°C) have shown that the selected alloys hold promise.

- Selected Alloys:
  - Alloy 20CB3
  - AL-6XN
  - 347 Stainless
  - 321 Stainless
  - 316L Stainless
  - 304L Stainless

EERC



Critical Challenges. **Practical Solutions.**

# AREA 1: CORROSION

- Little mass loss for all steels.
  - 203BC and AL6X+N best
- Longer test duration and lower pH are currently being evaluated.
  - Up to 3000 hours
  - Lower pH from 3–4 to 1.0 or less

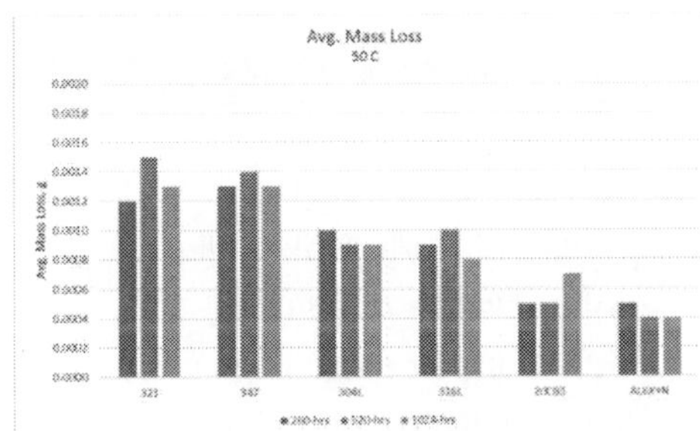


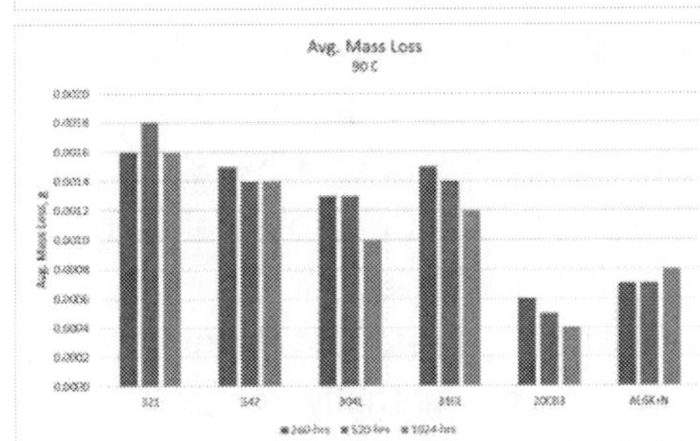


EERC



Critical Challenges. **Practical Solutions.**

## AREA 2: GASIFIER SELECTION

- Selection will impact all areas of the Allam Cycle.
  - Must be compatible with lignite coal.
  - Must be able to integrate with the Allam Cycle.
- Initial selection criteria:
  - Commercial readiness.
  - Applicability to lignite.
  - Operational history with lignite.
  - Thermal integration with Allam Cycle.
  - System reliability.
  - Cost.

EERC 21 Critical Challenges. **Practical Solutions.**

## AREA 2: GASIFIER SELECTION

- Properties of North Dakota lignite
  - High moisture content
  - Friable and nonswelling
  - High reactivity
  - High ash content (6%–12%)
    - ♦ Sodium content (2%–7%).
    - ♦ Calcium content 12%–32%).
    - ♦ Significant Ca and Na are ions bound to the coal.
  - Low-melting slag
  - Similar $T_{250}$ and $T_{cv}$ temperatures
    - ♦ Slag can abruptly freeze.

EERC 22 Critical Challenges. **Practical Solutions.**

## AREA 2: GASIFIER SELECTION

- Initial gasifier selection.
  - 22 gasification systems evaluated.
  - Nine systems seen as most suitable.
- Further selection will be based on:
  - Costs and commercial guarantees.
  - Sodium impacts.
  - Overall system efficiency.

| Gasifier Name | Vendor | Vendor Headquarters | System Type | Feed | Ash |
|---|---|---|---|---|---|
| Lurgi | Air Liquide | Germany | Fixed Bed | Dry | Nonslagging |
| BGL | Envirotherm | Germany | Fixed Bed | Dry | Slagging |
| SE | ECUST/Sinopec | China | Entrained Flow | Dry | Slagging |
| TRIG | KBR/Southern Company | USA | Fluid Bed | Dry | Nonslagging |
| HTW | ThyssenKrupp Uhde | Germany | Fluid Bed | Dry | Nonslagging |
| U-GAS | SES | USA | Fluid Bed | Dry | Agglomerated |
| SFG | Siemens | Germany | Entrained Flow | Dry | Slagging |
| SCGP | Shell | Netherlands | Entrained Flow | Dry | Slagging |
| Prenflo | ThyssenKrupp Uhde | Germany | Entrained Flow | Dry | Slagging |

EERC



Critical Challenges. **Practical Solutions.**

## AREA 3: IMPURITIES REMOVAL

- Choice of precombustion syngas cleanup or postcombustion $CO_2$ cleanup.
  - Syngas cleanup is a proven technology.
  - Postcombustion $CO_2$ cleanup has potential advantages.
- 8 Rivers $DeSNO_x$ $CO_2$ cleanup process:
  - Designed to remove sulfur and nitrogen species.
  - Uses only water in a packed spray column.
  - Produces aqueous sulfuric and nitric acid.

EERC



Critical Challenges. **Practical Solutions.**

## AREA 3: IMPURITIES REMOVAL

- The EERC has completed initial testing of the $DeSNO_x$ process.
  - EERC high-pressure fluidized-bed gasifier used to produce simulated gas stream.
    - ♦ Operated as an oxy-fired combustor firing lignite coal.
    - ♦ High degree of $CO_2$ recycle.
    - ♦ Produced a 95.5% $CO_2$, 1% $O_2$ gas stream with nitrogen and sulfur impurities.
  - $DeSNO_x$ column.
    - ♦ Operated at 30 bar pressure.
    - ♦ Inlet temperature 120°C, outlet temperature 33°C.
    - ♦ Inlet and outlet gas composition and water pH monitored.

EERC 25 Critical Challenges. **Practical Solutions.**

## AREA 3: IMPURITIES REMOVAL

- High $SO_x$ and $NO_x$ removals
  - $SO_x$: >99% removal
  - $NO_x$ ~95% removal

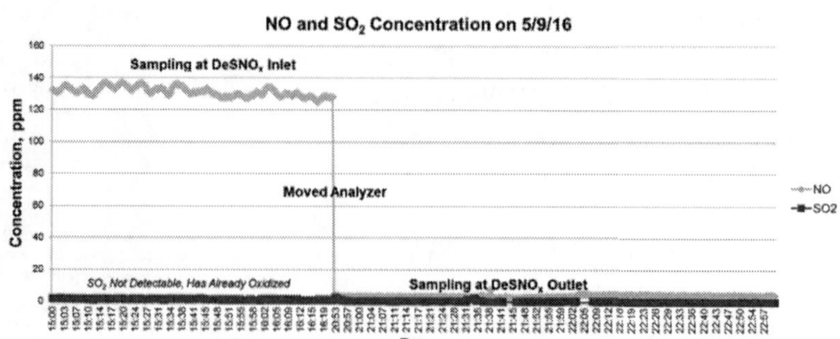


EERC 26 Critical Challenges. **Practical Solutions.**

# AREA 4: SYNGAS COMBUSTION

- ASPEN PLUS modeling was completed by 8 Rivers.
  - Range of syngas compositions modeled for 5-MWth prototype test combustor.
- Design considerations:
  - Swirl-stabilized diffusion flame injector.
  - Primary zone temperature of 3600°F.
  - Equivalence ratio of 0.99.
  - Uniform dilution zone exit temperature of 2100°F.
- Design completed.
  - 5-MWth combustor.
  - Including instrumentation.
  - Validate the combustor.

EERC



Critical Challenges. **Practical Solutions.**

# SUMMARY

- World coal use is growing, while U.S. coal use is losing ground because of regulations, low natural gas prices, policy uncertainties …
- Technology advancements are needed to both meet compliance goals and maintain the current cost of electricity (COE).
- $CO_2$ EOR can be a large factor in economics to help drive carbon capture, utilization, and storage (CCUS) applications.
- Supercritical $CO_2$ cycles offer great potential for next-generation power.
  - High efficiency, low COE, simplified CCUS …
  - Barriers need to be addressed.
  - Accelerated schedules needed.
- Leadership, teamwork, taking advantage of synergies will be key to the future of coal.

EERC



Critical Challenges. **Practical Solutions.**

## LEADERSHIP AND TEAMWORK WILL ENABLE TODAY'S PIONEERS TO DEVELOP TECHNOLOGIES WITH A FOCUS ON SYNERGISTIC DEPLOYMENT

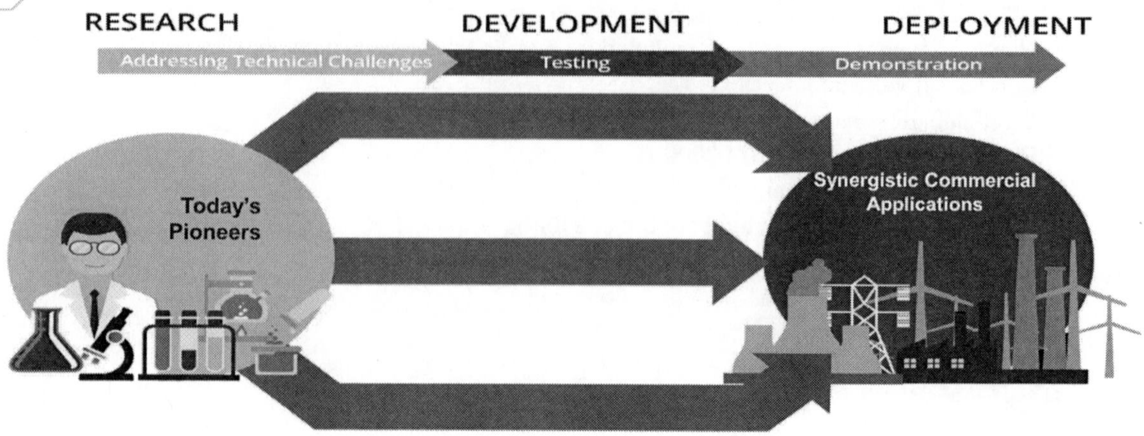


EERC 29 Critical Challenges. **Practical Solutions.**

EERC

# THANK YOU!

Critical Challenges. **Practical Solutions.**

# CONTACT INFORMATION

**Energy & Environmental Research Center**
University of North Dakota
15 North 23rd Street, Stop 9018
Grand Forks, ND 58202-9018

**www.undeerc.org**
701.777.5000 (phone)
701.777.5181 (fax)

**Mike Holmes**
**Director of Energy Systems Development**
701.777.5276
mholmes@undeerc.org

EERC

Critical Challenges. **Practical Solutions.**

Imperial College London

# Integration of Gasification and Fuel Cells: Interaction between the Anode and Contaminants in the Syngas

**Marcos Millan, Esther Lorente, Paul Boldrin, Nigel Brandon**
***Imperial College***

**Pittsburgh Coal Conference, Cape Town, 11th August 2016**

Imperial College London

## Overview of this talk

- Introduction on Gasification-Fuel Cell Integration.
- Effect of Tar on Solid Oxide Fuel Cells.
  - » Model Compounds as Tars.
    - Effect of Steam and Current Density on Suppressing Carbon Formation.
    - Effect of Gas Composition on Carbon Formation.
  - » Comparison between the use of Model Compounds and actual Coal Gasification Tars.
  - » Comparison between the use of different Tar Fractions.
- Summary and Conclusions.

Imperial College
London

# Integration of Gasification and SOFC

## SOFC provide:

- Higher resistance to gas impurities than other types of cell.
- Good matching in temperatures with downdraft and fluidised bed gasifiers.
- Potentially good matching in scales.

**Theoretical studies on performance analysis showed potential CHP efficiencies of around 80%.**

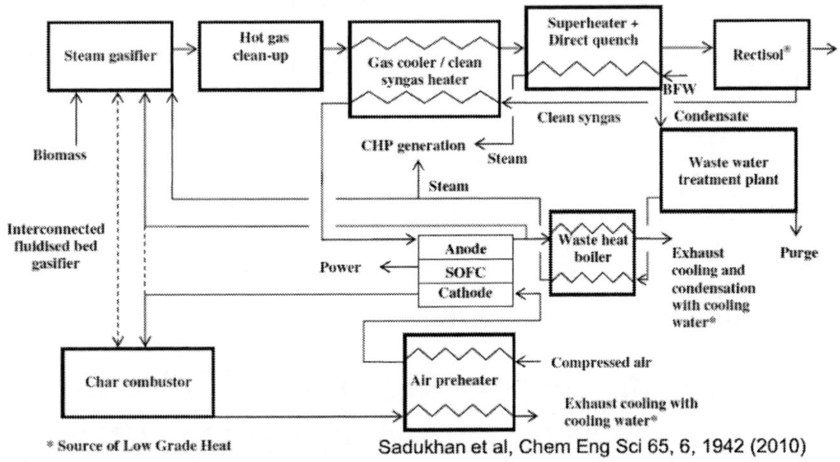


Sadukhan et al, Chem Eng Sci 65, 6, 1942 (2010)

Imperial College
London

# Experimental Set-up

Test were carried out both with a single button cell or a packed bed of anode material

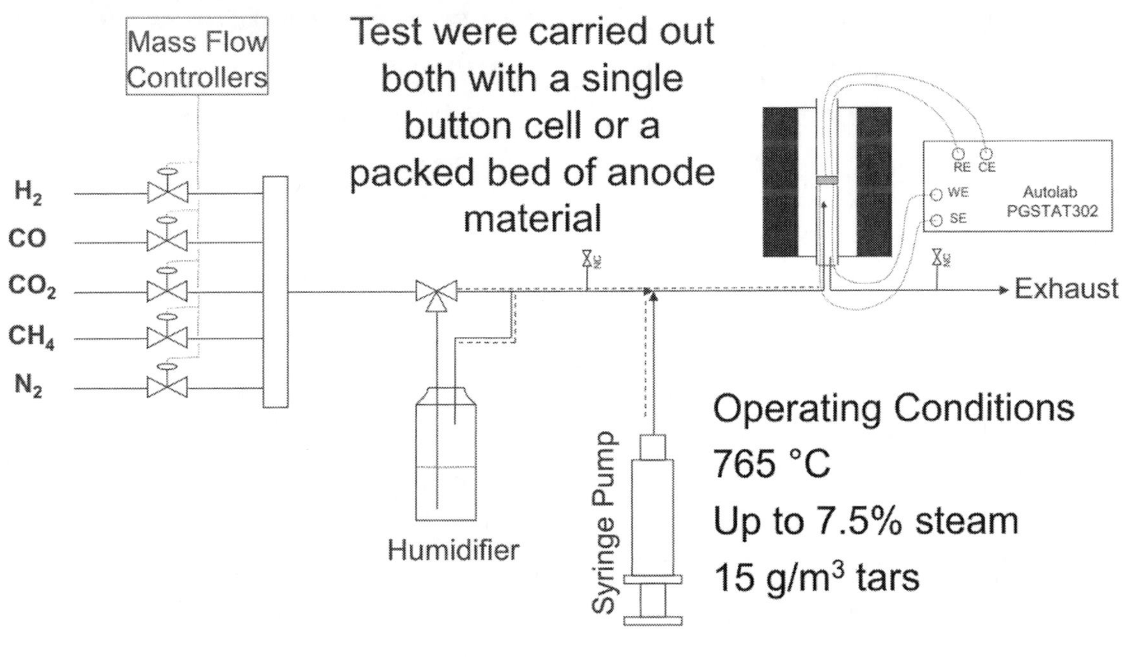


Operating Conditions

765 °C

Up to 7.5% steam

15 g/m$^3$ tars

Imperial College
London

# Carbon Deposition on SOFC Anode Materials

## Thermodynamic Predictions

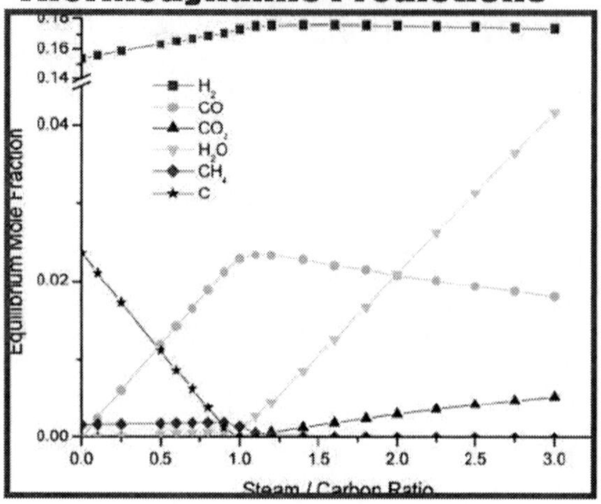


2.5% steam is equivalent to S:C = 1

## Carbon Deposits after 1-hour exposure to benzene

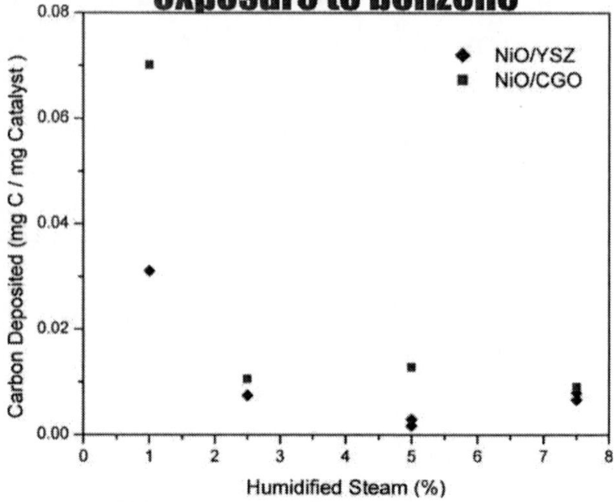


Mermelstein et al, Energy&Fuels, 23, 5042, 2009

**Although reduced, there is still C deposition where it is not thermodynamically favoured**

Imperial College
London

# Effect of Tars and Steam on Anode Performance

## Fuel Cell impedance after 30-min exposure to benzene, dry conditions, open circuit

Mermelstein et al, Chem Eng Sci, 64, 492, 2009

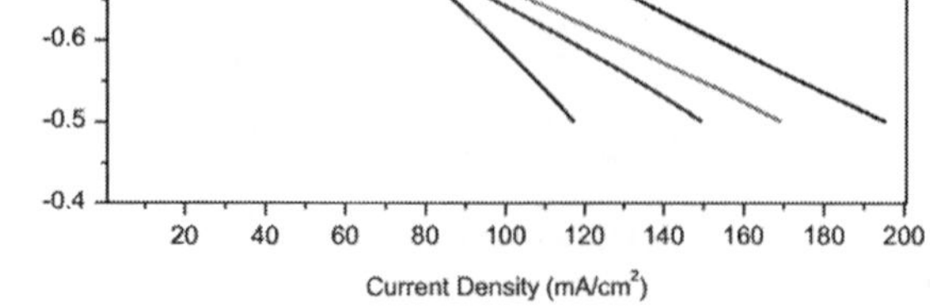


## However, a combination of high S/C ratio and current density may inhibit carbon formation, improving cell performance.

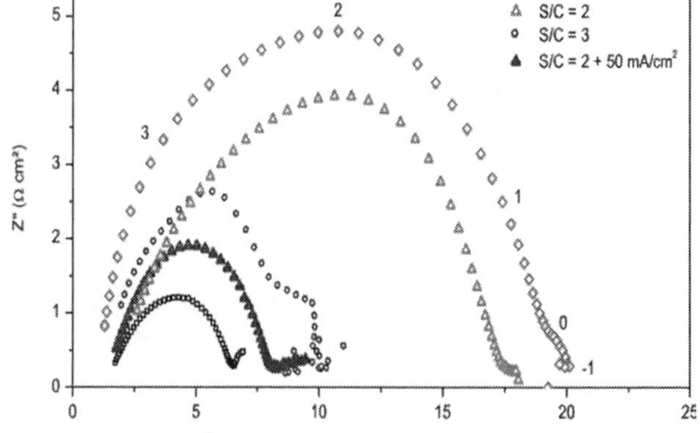


Mermelstein et al, J Power Sources, 195, 1657, 2010

Imperial College
London

# Effect of Tar as part of a Full SynGas Mixture

| Gas Composition | $H_2$ (% vol) | $CO_2$ (% vol) | CO (% vol) | $CH_4$ (% vol) | $H_2O$ (% vol) |
|---|---|---|---|---|---|
| $H_2$ | 15 | | | | 5 |
| $H_2 + CO_2$ | 15 | 10 | | | 5 |
| $H_2 + CO_2 + CO$ | 15 | 10 | 25 | | 5 |
| $H_2 + CO_2 + CO + CH_4$ | 15 | 10 | 25 | 2 | 5 |

Mermelstein et al, J Power Sources, 196, 5027, 2011

- **Complex interactions between gas components were observed.**
- **$CO_2$ led to a decrease in the deposits, due to a combination of dry reforming and competition for Ni active sites.**
- **$CH_4$ led to a large increase in deposits.**

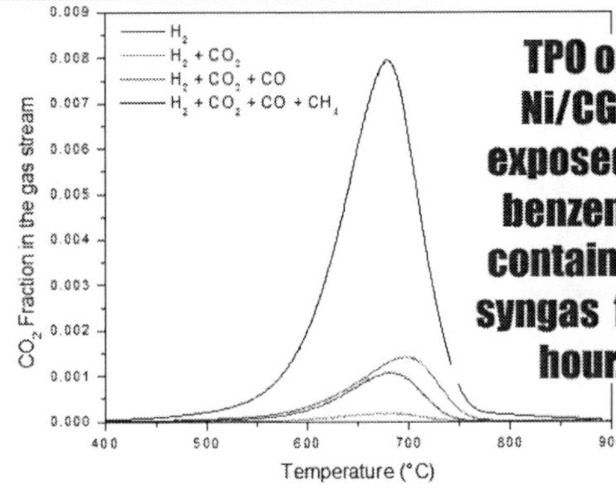


**TPO on Ni/CGO exposed to benzene-containing syngas for 1 hour.**

Imperial College
London

# Influence of Actual Tars on Carbon Deposition

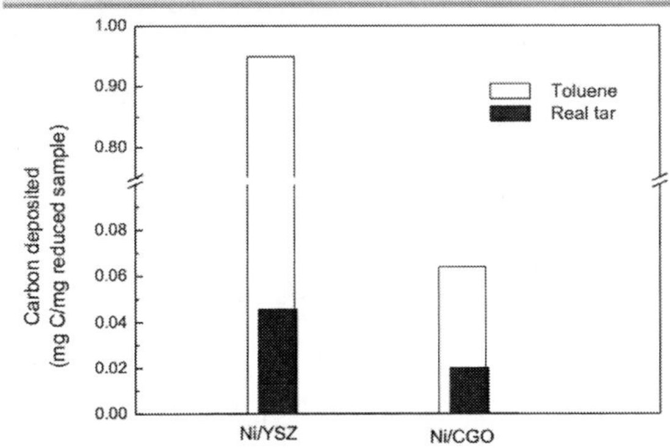


Lorente et al, Int. J. of Hydrogen Energy, 37 (2012), 7271

**Comparison between the use of toluene and actual coal gasification tars**

**Influence of Steam Content on the Carbon Deposited by Gasification Tars**

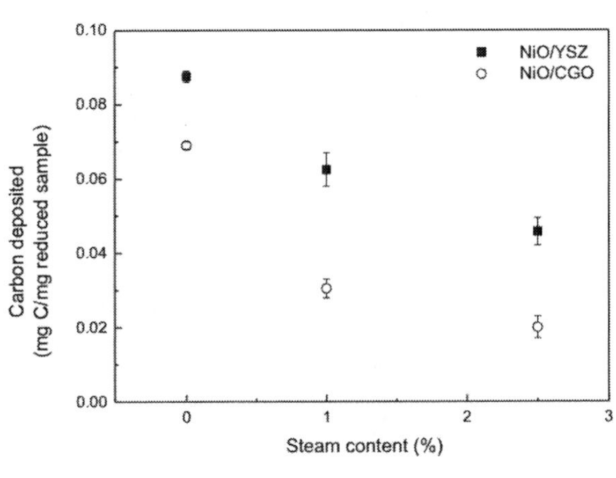

Imperial College
London

## Influence of Tar Fractions on Carbon Deposition

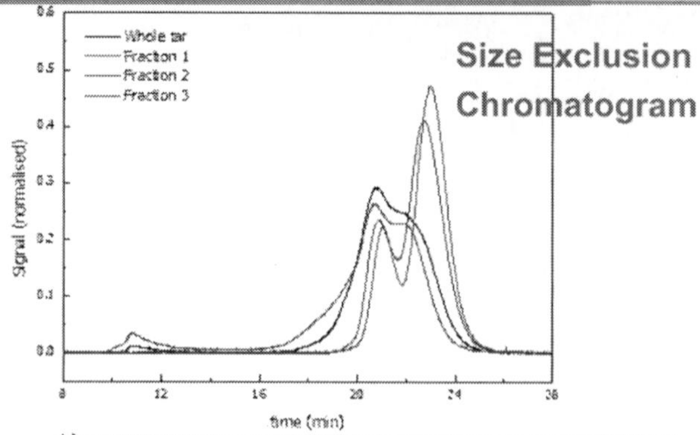


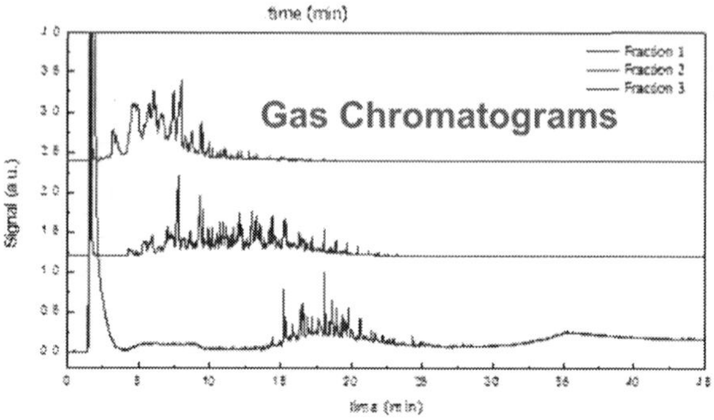


**The tar sample was fractionated by distillation.**

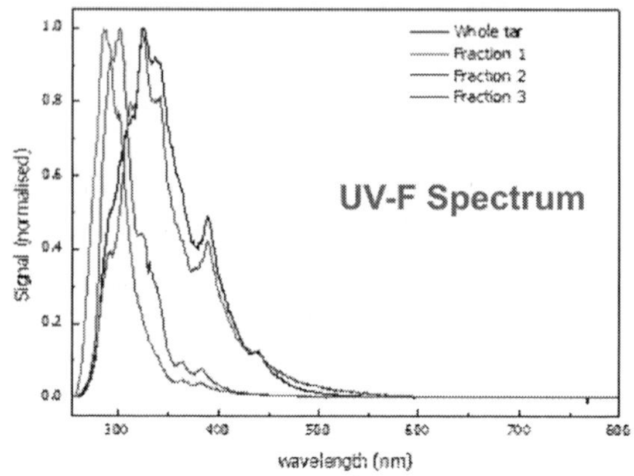


Imperial College
London

## Influence of Tar Fractions on Carbon Deposition

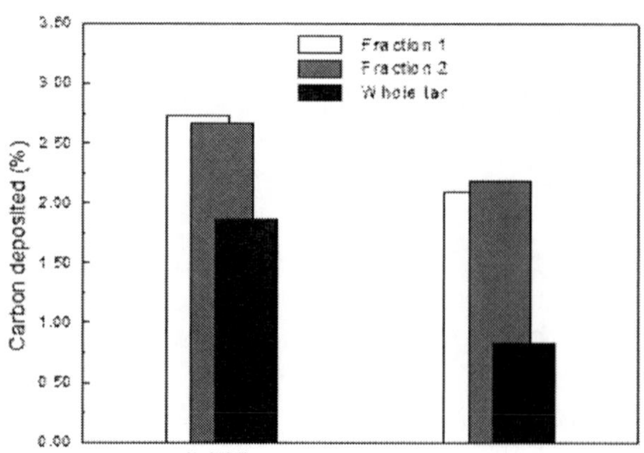


**This behaviour can inform the selection of a gasification technology.**

**A larger formation of deposits resulted from the use of the lighter tar fractions.**

**This is consistent with the trend observed with the use of model compounds and a tar sample.**

Imperial College
London

## Summary and Conclusions

- Fouling due to carbon deposition on the fuel cell anode reduces the performance of the cell.
- The addition of steam above S/C of 1 was found to reduce carbon deposition from tars but not fully suppress coke formation.
- A combination of high S/C=ratio and current density may effectively inhibit carbon formation, maintaining cell performance.
- Benzene and toluene represent a "worst-case scenario" as tar models and light fractions are more likely to produce carbon deposits.

# Direct coal liquefaction of South African vitrinite-rich coal using phenol as solvent


**Zach Sehume[1,*], Christien A. Strydom[1], John R. Bunt[1], Harold H. Schobert[1,2]**

[1]Research focus area for Chemical Resource Beneficiation, North West-University, Potchefstroom, 2520, RSA.
[2]The EMS Energy Institute, Penn State University, University Park, Pennsylvania 16802, USA.


33rd Annual International Pittsburgh Coal Conference (PCC).
August 8 – 12, 2016.

Cape Town, South Africa.

It all starts here

NRF National Research Foundation | RISA Research and Innovation Support and Advancement

crb CHEMICAL RESOURCE BENEFICIATION CHEMIESE HULPBRONVEREDELING

NORTH-WEST UNIVERSITY ®
YUNIBESITI YA BOKONE-BOPHIRIMA
NOORDWES-UNIVERSITEIT
POTCHEFSTROOM CAMPUS

## Presentation Outline

NORTH-WEST UNIVERSITY ®
YUNIBESITI YA BOKONE-BOPHIRIMA
NOORDWES-UNIVERSITEIT
POTCHEFSTROOM CAMPUS

- **Background,**
- **Motivation,**
- **Experimental,**
- **Results and discussion,**
- **Conclusions,**
- **Acknowledgement**

2

# Background



- Interest in direct coal liquefaction (DCL) - intensified by the rapid increase in demand for petroleum
- Targets for products obtained from DCL[1] :
  - Liquid transportation fuels, and
  - Chemicals
- The main challenge of this process is the breaking down of the macromolecular structure of coal, and adding hydrogen to the resultant fragments or free radicals

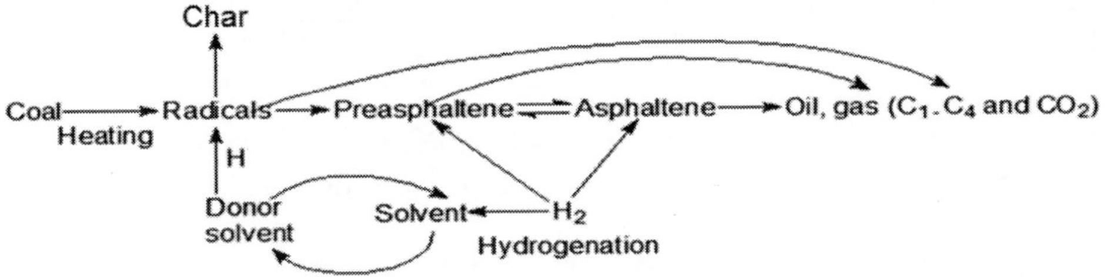


Scheme 1: A general reaction mechanism for coal liquefaction

3

# Background

NORTH-WEST UNIVERSITY
YUNIBESITI YA BOKONE-BOPHIRIMA
NOORDWES-UNIVERSITEIT
POTCHEFSTROOM CAMPUS

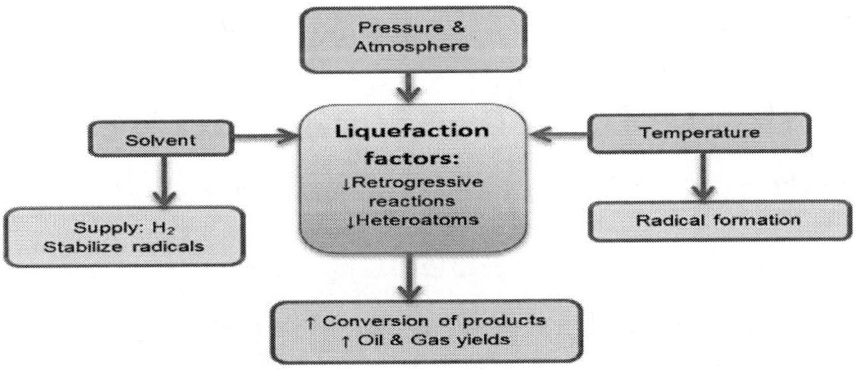


- Severe operation conditions are generally used during DCL[2]:
- ↑ temperature (425 – 550 °C), ↑ pressure (>14 MPa), expensive hydrogen donor solvent (i.e. tetralin), and hydrogenation catalysts
- Products obtained from DCL **are not economically feasible when compared with the refined crude oil→ lower operational conditions for direct coal liquefaction**
- The bituminous coal is preferred for this process because it is “amendable” for liquefaction reactions due to high yield of hydrocarbon products[3]. Most of studies have been done on bituminous coal from other countries.
- **There isn’t much literature available on the behaviour of the South African vitrinite-rich bituminous coal during DCL**

4

# Background

NORTH-WEST UNIVERSITY
YUNIBESITI YA BOKONE-BOPHIRIMA
NOORDWES-UNIVERSITEIT
POTCHEFSTROOM CAMPUS

- 99% phenols production stems from cumene oxidation during petroleum refinement[4]
- Phenol can be an ideal solvent for DCL process?
- Phenolic solvents have been considered to be beneficial for conversion of coal during DCL[3]
- Influences: it enhances the rates of bond cleavage (i.e. various ethers) through hydrogen-bonding interaction between the phenolic species and ethers in the coal[5-6]

❖ **Main interest of this study:** investigation of the influence of phenol as solvent on liquefaction products from vitrinite-rich coal (especially for South African low ash coal) under mild conditions (< 400 °C)

5

# Motivation

NORTH-WEST UNIVERSITY
YUNIBESITI YA BOKONE-BOPHIRIMA
NOORDWES-UNIVERSITEIT
POTCHEFSTROOM CAMPUS

" The desire to liquefy coal into petroleum substitutes, particularly lighter liquid fuel, stems from the abundance of coal and increasing demand for transportation fuels that are currently produced predominantly from the crude petroleum through a series of refining process". Mochida et al.[7]

6

# Aims & Objectives

NORTH-WEST UNIVERSITY
YUNIBESITI YA BOKONE-BOPHIRIMA
NOORDWES-UNIVERSITEIT
POTCHEFSTROOM CAMPUS

- ❖ To convert coal into light liquid products using phenol as solvent, and investigate the influence of this solvent on the products obtained

- ❖ Objectives of this study include:

✓ *Experiments in an autoclave to generate liquefaction products under mild conditions and extraction of liquid products using conventional solvents;*

✓ *Characterisation of the formed liquid products using advanced chemical structural techniques such as proximate, ultimate, GC-MS, NMR ($^{13}C$) and Sim-Dist. GC analyses. Correlation of the results obtained to understand the behaviour of South African bituminous during liquefaction;*

✓ *Characterisation of the formed solid residues using chemical structural techniques such as proximate, ultimate and FTIR analyses. Correlations of the results obtained to determine inter-relationships*

7

# Experimental set-up

NORTH-WEST UNIVERSITY
YUNIBESITI YA BOKONE-BOPHIRIMA
NOORDWES-UNIVERSITEIT
POTCHEFSTROOM CAMPUS

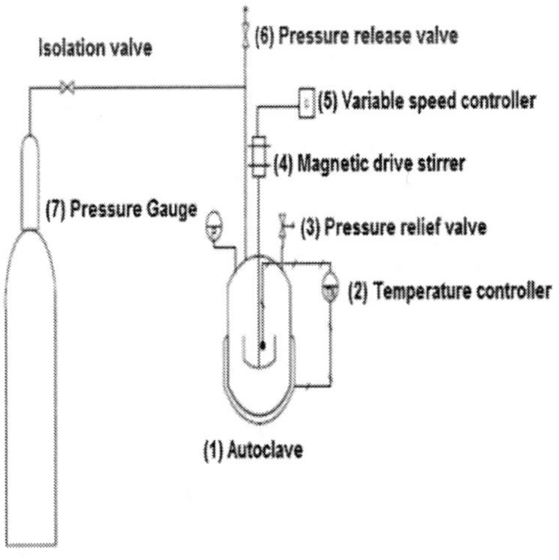


SS316 stainless steel high pressure autoclave system (950 ml)

**Conditions:** Liquefaction temperatures: 300, 320, 340 & 360 °C
**Liquefaction reaction atmosphere**: $N_2$
**Reaction solvent**: commercial phenol as a surrogate solvent (Purity ≥ 89%)
**Coal sample**: vitrinite-rich (washed fraction low ash) Waterberg bituminous coal
**Fixed ratio**: coal to solvent (1:10)

8

# Fractionation scheme for separation of products

NORTH-WEST UNIVERSITY
YUNIBESITI YA BOKONE-BOPHIRIMA
NOORDWES-UNIVERSITEIT
POTCHEFSTROOM CAMPUS

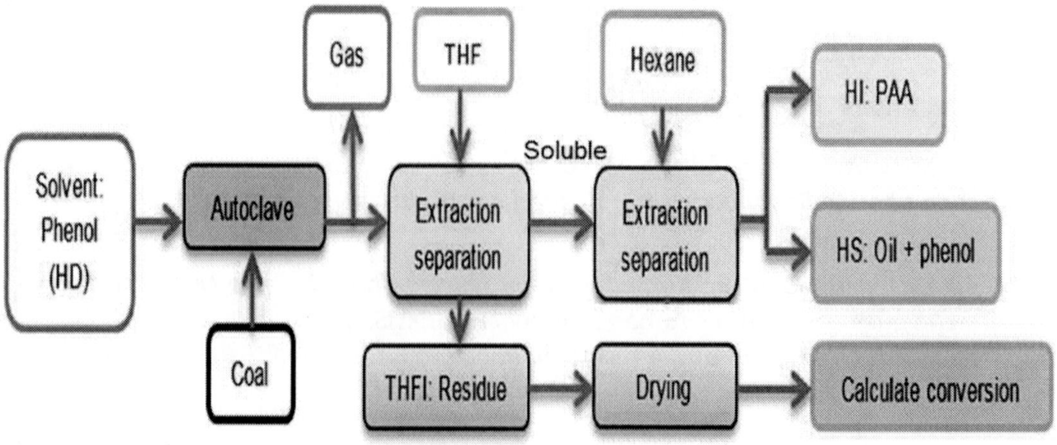


The diagram for liquefaction of coal with phenol

9

# Results and discussion

NORTH-WEST UNIVERSITY
YUNIBESITI YA BOKONE-BOPHIRIMA
NOORDWES-UNIVERSITEIT
POTCHEFSTROOM CAMPUS

| Chemical properties | Standard | Coal |
|---|---|---|
| **Proximate analysis (wt. %, a.d.)** | | |
| Inherent Moisture | ISO11722: 1999 | 3.1 |
| Volatile matter | ISO 562: 2010 | 35.3 |
| Ash | ISO 1171: 2010 | **11.2** |
| Fixed carbon | By difference | 51.8 |
| | | |
| **Ultimate analysis (wt. %, d.a.f)** | ISO 29541: 2010 (Not SANAS Accredited) | |
| Carbon | | 80.9 |
| Hydrogen (H) | | 5.5 |
| Nitrogen (N) | | 1.9 |
| Sulphur (S) | ISO 19579: 2006 | 1.3 |
| Oxygen (O) | By difference | 10.4 |
| | | |
| **Petrography** (m.m.f.) | | |
| Mean random reflectance (%) | | 0.6 |
| Vitrinite content (%) | | 52.8 |
| Total reactive macerals (%) | | 63.0 |

| Product | Chemical Properties | 300 °C | 320 °C | 340 °C | 360 °C |
|---|---|---|---|---|---|
| | **Proximate analysis (wt.%, a.d.)** | | | | |
| | Volatile Matter | 44.7 | 43.0 | 31.7 | 28.5 |
| | Ash | 10.2 | 10.5 | 18.5 | 20.6 |
| | | | | | |
| | **Ultimate analysis (d.a.f.), wt. %** | | | | |
| | % C | 79.1 | 76.4 | 82.9 | 82.1 |
| | % H | 5.1 | 4.7 | 5.1 | 4.8 |
| | % N | 1.6 | 1.6 | 1.9 | 2.0 |
| | % O (by difference) | 14.2 | 17.3 | 10.1 | 11.2 |
| **THFI** | H/C | 0.064 | 0.062 | 0.062 | 0.058 |
| | **Proximate analysis (wt.%, a.d.)** | | | | |
| | Volatile Matter | 98.3 | 98.2 | 98.4 | 98.3 |
| | Ash | 1.1 | 1.2 | 1.3 | 1.4 |
| | | | | | |
| | **Ultimate analysis (wt. %, d.a.f.)** | | | | |
| | % C | 58.3 | 57.5 | 57.2 | 57.2 |
| | % H | 4.8 | 4.8 | 4.8 | 4.8 |
| | % N | 0.2 | 0.3 | 0.2 | 0.2 |
| | % O (by difference) | 36.8 | 37.4 | 37.8 | 37.8 |
| | H/C | 0.081 | 0.084 | 0.084 | 0.084 |
| **HS** | pH | 4.2 | 4.1 | 4.0 | 4.0 |

- volatile matter ↓ with ↑ °C : THFI
- ash content ↑ with ↑ °C: THFI
- HS are characterized by ↑ volatile matter

10

## Results and discussion: Products distribution

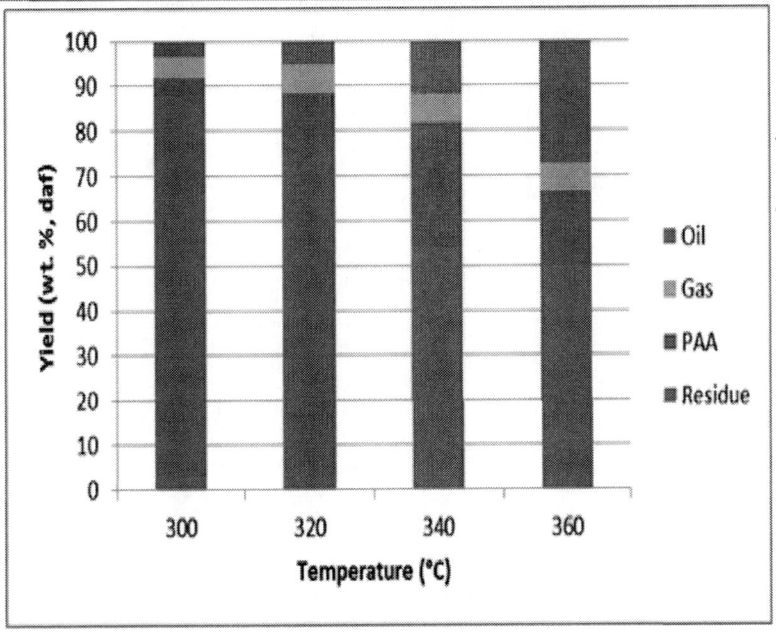


Products distribution of liquefaction under $N_2$ atmosphere

- Liquefaction products ↑ with ↑ in T°C,
- Conversion of 12% at 300 °C was obtained,
- The oil & gas yields were 3.3% & 4.9% respectively,
- Conversion ↑ with rise of T°C & reaching conversion of 50.0% at 360 °C, with oil & gas yield of 27.0% & 6.4% respectively



## Results and discussion: Products distribution

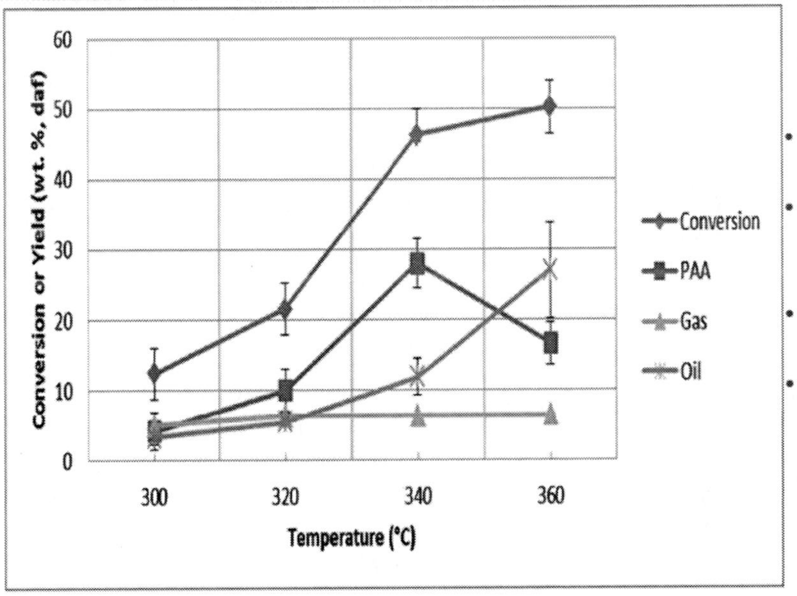


Effect of temperature on products distribution

- Liquefaction products ↑ with ↑ °C
- ↑ T°C can promote the decomposition of coal with solvent to form PAA
- However, at 340 °C PAA started to ↓
- Decomposition of PAA becomes the predominant reaction at this range→ products

# Results and discussion: FTIR

NORTH-WEST UNIVERSITY
YUNIBESITI YA BOKONE-BOPHIRIMA
NOORDWES-UNIVERSITEIT
POTCHEFSTROOM CAMPUS

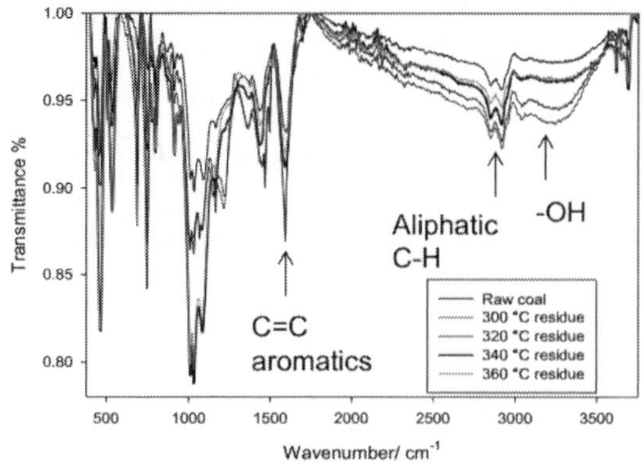


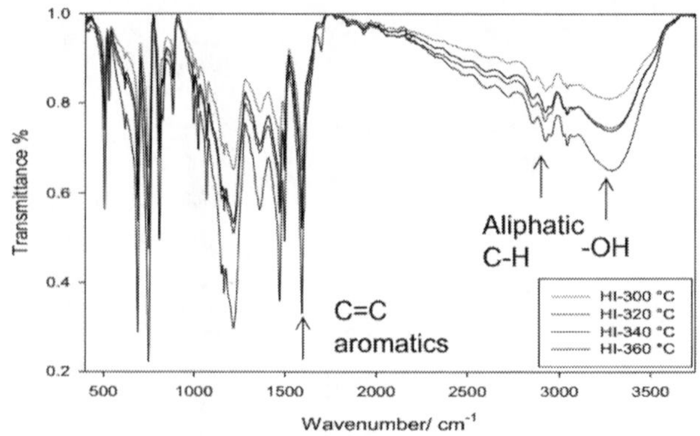


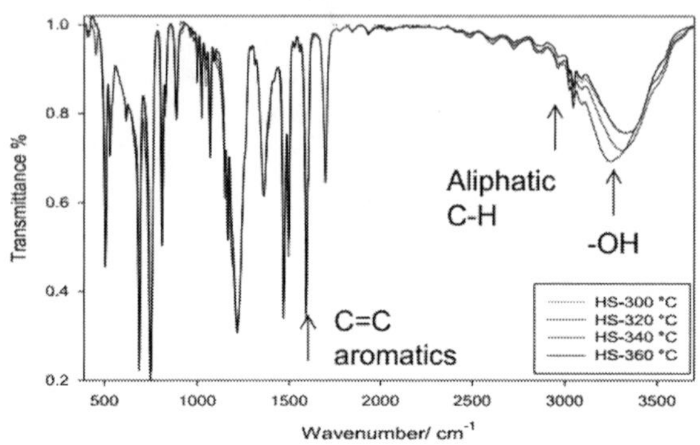


- The functional groups of liquefaction products displayed similar spectra
- Intensity of liquefaction products were more than raw coal
- This suggests that products become more aromatic with ↑ °C
- –OH band may be attributed to phenol and ether

# Results and discussion: GC-MS


NORTH-WEST UNIVERSITY
YUNIBESITI YA BOKONE-BOPHIRIMA
NOORDWES-UNIVERSITEIT
POTCHEFSTROOM CAMPUS


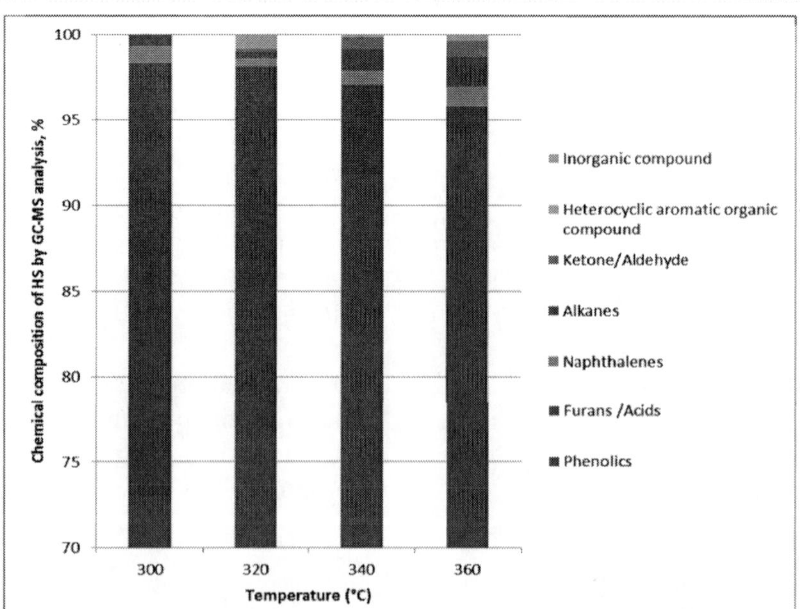


- Oils mainly consisted of phenolics, furans, etc.
- Liquefaction of coal using phenol as solvent:

✓ Significance ↑ of phenolics with ↑ °C
✓ ↑ Alkanes with ↑ °C
✓ Furans/Acids ↓ with ↑ °C which may have → to increased poly-aromatics

Distribution of chemical composition by GC-MS analysis (semi-quantitative) for HS's at selected temperatures

* Some compounds did not separate effectively due to co-elution of peaks

14

# Results and discussion: $^{13}$C-NMR

NORTH-WEST UNIVERSITY
YUNIBESITI YA BOKONE-BOPHIRIMA
NOORDWES-UNIVERSITEIT
POTCHEFSTROOM CAMPUS

| Symbol | Description | Range (ppm) | 300 °C | 320 °C | 340 °C | 360 °C |
|---|---|---|---|---|---|---|
| $C_{ar,ald/ket}$ | **Carbonyl** | 210-205 | 2.63 | 4.12 | 3.68 | 3.22 |
| $C_{ArC}$ | **Catacondensed aromatic carbons, aromatic carbons both with heteroatomic or aromatic substituents** | 167-133 | [illegible] | [illegible] | [illegible] | [illegible] |
| [illegible] | $C_{ar}$-OH | 167-147 | | | | |
| [illegible] | $CH_2$ or CH meta to $C_{ar}$-OH | 147-140 | | | | |
| [illegible] | $C_{ar}$-$CH_3$; $CH_2$ or CH para to $C_{ar}$-OH | 140-133 | | | | |
| $C_{ArP}$ | **Pericondensed aromatic- or protonated aromatic carbons** | **133-110** | | | | |
| [illegible] | Quaternary aromatic carbons; $C_{ar}$-H meta to $C_{ar}$-OH | 133-124 | | | | |
| [illegible] | $C_{ar}$-$CH_3$ para to $C_{ar}$-OH | 124-122 | | | | |
| [illegible] | $C_{ar}$-H para to $C_{ar}$-OH; $C_{ar}$-CH3 ortho to $C_{ar}$-OH | 122-118 | | | | |
| [illegible] | $C_{ar}$-H ortho to $C_{ar}$-OH | 118-110 | | | | |
| $C_{al}$ | **Aliphatic carbons** | **49.3-13** | | | | |
| [illegible] | $CH_2$ and CH carbons a to two aromatic rings (diphenylmethane type) | 49.3-39.5 | | | | |
| [illegible] | $CH_2$ and CH carbons a to two aromatic rings (fluorine type) | 39.5-34.0 | | | | |
| [illegible] | $CH_2$ and CH carbons a to an aromatic ring & b to another (acenaphthene type) | 34-29.5 | 0.1 | 0.01 | 0.09 | 0.01 |
| $CH_2$ | $CH_2$ and CH carbons a or further from an aromatic ring | 29.5-23 | 6.95 | 10.03 | 8.67 | 8.03 |
| $CH_3$ | Methyl carbons | 23-13 | 0.68 | 1.94 | 1.26 | 1.93 |
| Total | | | 99.99 | 100 | 100 | 99.99 |
| $F_a$ | **Fraction of aromatic carbon** | | **0.89** | **0.84** | **0.86** | **0.86** |
| $F_{al}$ | **Fraction of aliphatic carbon** | | **0.08** | **0.12** | **0.10** | **0.10** |

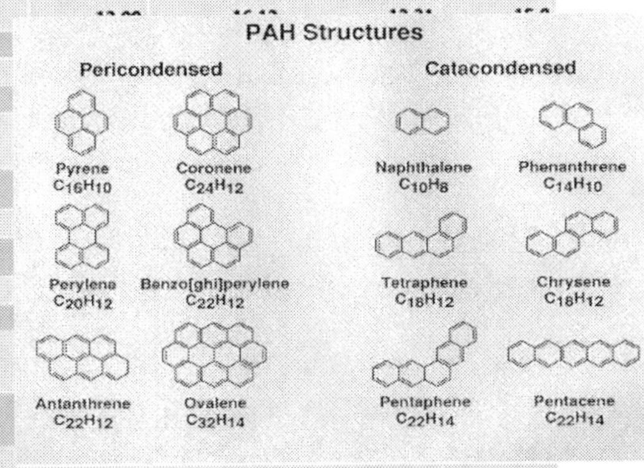


15

# Results and discussion: SIMDIST analysis

NORTH-WEST UNIVERSITY
YUNIBESITI YA BOKONE-BOPHIRIMA
NOORDWES-UNIVERSITEIT
POTCHEFSTROOM CAMPUS

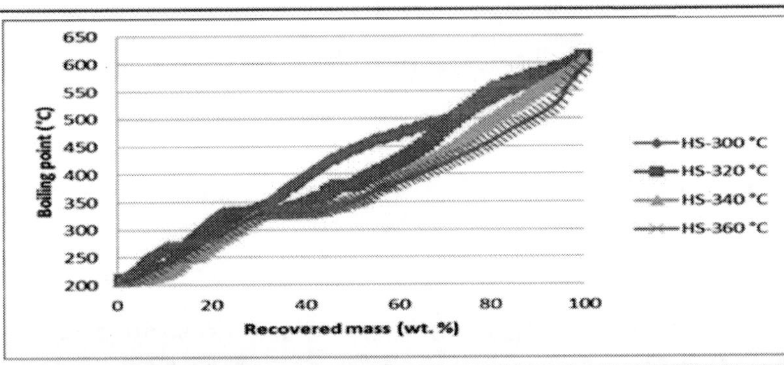


$$WABP = \frac{T_1 + T_2 + T_3 + T_4 + T_5}{5}$$

| Boiling fraction/ definition | Boiling Range a (°C) | Unit | HS-300 °C | HS-320 °C | HS-340 °C | HS-360 °C |
|---|---|---|---|---|---|---|
| Medium naphtha [a] | 79-121 | wt.% | – | – | – | – |
| Heavy naphtha [a] | 121-191 | wt.% | – | – | – | – |
| Kerosene [a] | **91-277** | **wt.%** | **14.0** | **15.0** | **20.0** | **17.0** |
| Distillate fuel oil [a] | **277-343** | wt.% | **16.0** | **23.0** | **26.0** | **30.0** |
| Light vacuum gas oil [a] | **343-455** | **wt.%** | **23.0** | **27.0** | **26.0** | **31.0** |
| Heavy vacuum gas oil [a] | 455-566 | wt.% | 34.0 | 18.0 | 19.0 | 18.0 |
| Residual oil [a] | **>566** | **wt.%** | **12.0** | **16.0** | **8.0** | **3.0** |
| Total | - | wt.% | 99.0 | 99.0 | 99.0 | 99.0 |
| Initial boiling point | - | °C | 209.8 | 208.6 | 208.6 | 209.2 |
| Average boiling point [b] | **100-550** | **°C** | **424.6** | **403.0** | **380.7** | **371.6** |
| Final boiling point | - | °C | 615.4 | 615.6 | 612.4 | 600.6 |

16

# Conclusions

NORTH-WEST UNIVERSITY
YUNIBESITI YA BOKONE-BOPHIRIMA
NOORDWES-UNIVERSITEIT
POTCHEFSTROOM CAMPUS

- Temperature is the dominating factor that determines the distribution of liquefaction products;
- Phenol showed to be a fairly hydrogen donor solvent, as significant upgrading of products started to occur at around 320 °C;
- The conversion, gas and oil yields increased with rising temperatures and reached maximum values of 50.0%, 6.5% & 27% respectively at 360 °C;
- HS products were characterized by larger poly-aromatic compounds with lower aliphatic content;
- The results of this study may provide fundamental understanding on the conversion of a South African coal using phenol as solvent for liquefaction process

17

# Future work

NORTH-WEST UNIVERSITY
YUNIBESITI YA BOKONE-BOPHIRIMA
NOORDWES-UNIVERSITEIT
POTCHEFSTROOM CAMPUS

- The chromatographic analyses methods needs to be evaluated and established;
- Recovery of unreacted phenol without compromising the quality of the oil;
- Determination of the amount of phenol reacted, and also how much $H_2$ was consumed during coal liquefaction;
- Optimise liquefaction process for higher product yields (i.e. hydro-treatment)

18

## Acknowledgements

- North-West University and the South African Research Chairs Initiative of the Department of Science and Technology and National Research Foundation of South Africa for financially supporting the research.
- Any opinion, finding or conclusion or recommendation expressed in this material is that of the author(s) and the NRF does not accept any liability in this regard.

Supervisors

Colleagues

NRF National Research Foundation | RISA Research and Innovation Support and Advancement

CHEMICAL RESOURCE BENEFICIATION CHEMIESE HULPBRONVEREDELING

exxaro SERVICES

# Thank you

It all starts here

NORTH-WEST UNIVERSITY YUNIBESITI YA BOKONE-BOPHIRIMA NOORDWES-UNIVERSITEIT POTCHEFSTROOM CAMPUS

19

## References

NORTH-WEST UNIVERSITY YUNIBESITI YA BOKONE-BOPHIRIMA NOORDWES-UNIVERSITEIT POTCHEFSTROOM CAMPUS


1. Yan, J., Bai, Z., Li, W., Bai, J. 2014. Direct liquefaction of a Chinese brown coal and $CO_2$ gasification of the residues. Fuel, 136: 280-286.

2. Liu, Z., Shi, S., Li, Y. 2010. Coal liquefaction technologies—Development in China and challenges in chemical reaction engineering. *Chemical Engineering Science*, 65(1): 12–17.

3. Akash, B.A. 2013. Thermochemical liquefaction of coal. Int. J. of Thermal & Environmental Engineering, 5: 51-60.Wang, Z., Shui, H., Zhang, D., Gao, J. 2007. A Comparison of FeS, FeS+S and solid superacid catalytic properties for coal hydro-liquefaction. *Fuel,* 86: 835–842.

4. Mantilla, S,V., Manrique, A.M., Gauthier-Maradei, P. 2015. Methodology for extraction of phenolic compounds of bio-oil from agricultural biomass wastes. Waste Biomass Valor, 6:371–383.

5. Kamiya, Y., Sato, H., Yao, T. 1978. Effect of phenolic compounds on liquefaction of coal in the presence of hydrogen-donor solvent. *Fuel,* 57:681.

6. Song et al. 2000. A new process for catalytic liquefaction of coal using dispersed MoS2 catalyst generated in situ with added H2O. *Fuel,* 79: 249–261.

7. Mochida, I., Okuma, O., and Yoon, S-H. 2014. Chemicals from direct coal liquefaction. Chem. Rev., 114: 1637-1672.


20

# EFFECT OF DIRECT ADDITION OF ALCOHOLS TO AN OLIGOMERISATION REACTION SYSTEM

*Mekia Makhubu, Mr., University of Cape Town, Private Bag, Rondebosch, 7701, SOUTH AFRICA, mkhmek001@myuct.ac.za*

*Klaus Moller, Prof., University of Cape Town, Private Bag, Rondebosch, 7701, SOUTH AFRICA, klaus.moller@uct.ac.za*

*Masikana Mdleleni, Prof., PetroSA Synthetic Fuels Innovation Center (PSFIC), University of Western Cape, Private Bag, Bellville, 7535, SOUTH AFRICA, MASIKANA.Mdleleni@petrosa.co.za*

## Abstract:
Conversion of primary and secondary alcohols such as ethanol, propanol, butanol, pentanol and hexanol to gasoline and distillates is one of very attractive concepts. Co-feeding these alcohols to the light olefins in an oligomerisation system may result in an increased production of gasoline and distillates however this is highly dependent on operating conditions, alcohol feed concentration and catalyst activity.

## 1. Introduction
Some of the aforementioned primary and secondary alcohols are by-products of high temperature Fischer-Tropsch (HTFT) synthesis from gas to liquid (GTL) process in which products such as gasoline, diesel and waxes to name a few are produced. The ideology behind dehydration of these alcohols is to further convert generated olefin(s) so as to increase the production of gasoline and distillates.

It is very important to note that methanol was the very first alcohol to be successfully dehydrated and converted to gasoline (MTG) over HZSM-5 by EXXON-MOBIL in 1976, with the first commercial plant operated in New Zealand 10 years later. The MTG was originally designed for the production of gasoline, however, it did not generate any diesel or jet fuel fraction. Thus due to change in the market, it was shut down in 1990 and altered to a combined methanol to olefins (MTO) and Mobil's olefins to gasoline and distillates process devised by EXXON-MOBIL (Dagle et al., 2013).

In MTO process, methanol dehydrates over zeolite (HZSM-5) catalyst to form dimethyl ether which also reacts with the formed (CH2)n pool species to produce products such as ethylene, propylene, and butene (White, 2011) ("see Figure 2 shown in oral presentation") which are oligomerised to higher olefins (Hadi et al., 2014).

Another alcohol of interest is the bio-ethanol which can also be converted to distillates via olefin generation route just like methanol. From ethanol to gasoline (ETG) process, it was proposed by authors such as Gayubo et al. (2001) that ethanol dehydrates to ethylene and water at extremely high temperatures greater than 300 °C at 1 bar. During this dehydration a formed ethylene reacts to form other light olefins such as C3-C4 and then again ethylene reacts with these olefins to generate higher olefins and aromatics necessary for the production of gasoline. This type of the reaction is simply termed auto-catalytic reaction. Methanol also undergoes such a reaction to generate light olefin(s) necessary for the production of gasoline and distillates.

There are different views regarding ethanol dehydration, where authors such as Johansson et al. (2009) suggested that ethanol dehydrates via hydrocarbon pool type mechanism which is more or less the same as the previously mentioned auto-catalyst reaction. But the idea here is that the actual catalytic sites in the zeolite are organic-inorganic hybrids consisting of aromatic species formed during ethanol dehydration at extremely high temperatures that do act as the hydrocarbon pool which after certain period of time split off ethylene, propylene and butene "see Figure 4 shown in oral presentation". Hence, this shows that ethylene needs an aromatic compounds to convert to higher light olefins. With regard to the alcohols higher than ethanol, no hydrocarbon pool type mechanism was observed in the literature. The dehydration of primary alcohols to hydrocarbons over HZSM-5 occurs through two main steps ("these are also shown in Figure 3, Figure 5 and Figure 6 in oral presentation"):

- unimolecular route to produce a respective olefinic product
- bimolecular pathway to produce a respective ether, which further dehydrates to an olefin and water and/alcohol

A choice of the dehydration route is mostly influenced by the temperature as the driving force, especially due to the fact that first route is an endothermic reaction thus it requires more heat to dehydrate any alcohol completely. Whereas the second route is an exothermic reaction with respect to the dehydration of alcohol to ether less heat is required for faster dehydration rate. Theoretically, a second route should diminish with an increasing temperature, hence the first route will be key the reaction in generating light olefins suitable for the production of distillates.

## 2. Proposed dehydration mechanisms and kinetics

Under current oligomerisation reaction system conditions, ethanol is said to dehydrate through the hydrocarbon pool reaction where C3/C4 olefins form a part of the pool ("this pool mechanism is shown in the oral presentation and also see Figure 7"). The kinetic experiments to prove this so-called ethanol hydrocarbon pool are still to be carried out. Nonetheless, the dehydration rate constants used in the model are estimates guided by auto-catalytic reaction of ethanol over HZSM-5. A Gibbs equilibrium reactor ("COCO simulator") with a UNIFAC thermodynamic model was used to study the effect of temperature on dehydration mechanism of alcohols higher than ethanol proposed on the literature. The main goal was to evaluate whether these alcohols will dehydrate to form ether, olefins and water at temperatures greater than 200 °C at 50 bar. Thus it was found that these alcohols dehydrate to olefins and water only ("see dehydration mechanism of 1-propanol to 1-hexanol in oral presentation") provided that in actual oligomerisation reaction system oligomerisation will always take place. It was further found that under the aforementioned conditions, dehydration of alcohols to light olefins and water is irreversible. The same applies to secondary alcohols.

The difference between primary and secondary alcohols dehydration kinetics under the oligomerisation model operating conditions are said to be very insignificant this was confirmed via COCO simulator's Gibbs reactor. The dehydration kinetics for 1-propanol and 1-butanol were sourced from Zhi (2010) and Makarova et al. (1994); John et al. (2015). This is because the kinetic experiments for 1-propanol and 1-butanol dehydration conducted in temperature range of 150-200 °C, 40 bar are still in progress not done yet. Hence 1-pentanol and 1-hexanol kinetic constants were extrapolated from the known kinetics of 1-propanol and 1-butanol. A similar procedure to that of Stauffer and Kranch (1963) to extrapolate kinetic constants other higher alcohols was applied to estimate 1-pentanol and 1-hexanol kinetic constants.

## 3. Effect of alcohols on an oligomerisation system

Increasing feed composition of alcohols slows down the consumption rate of LPG. However as reactor volume and reaction temperature increases more light olefins are formed from the dehydration of primary and secondary alcohols hence after a while the LPG decreases at the faster rate compared to when no alcohols are added into the reactor ("This is shown in Figure 9 in the oral presentation"). Addition of alcohols to the reactor affects the production of gasoline especially for smaller reactor volumes as depicted in Figure 10 ("This figure is shown in oral presentation"). This is can be related to a decrease of cracking reactions due to the addition of alcohols which has lowered down the temperature gradient in a reactor due to endothermic nature of alcohol dehydration reactions ("see Figure 13 in oral presentation"). Nevertheless, the slight decrease in temperature favours an increase in production of diesel and heavies ("see Figure 11 and 12 respectively shown in oral presentation") but for small reactor volumes the impact on diesel production is adverse.

Conversion of C3 and C4 olefins is severely affected for smaller reactor volumes which however improves significantly as the temperature increases ("See Figure 14 and 15 respectively shown in oral presentation"). This is may be due to the reaction of C3 and C4 olefins in the ethanol dehydration pool. Average conversion of primary and secondary alcohols is affected by the increase in feed composition of alcohols which affects the reaction temperature gradient meaning more heat is required to drive dehydration reactions whilst at the same time affecting exothermic reactions/oligomerisation ("see Figure 16 shown in oral presentation").

Co-feeding about 20% of alcohols to the C3/C4 light olefins into an isothermal reactor model does result in an increased generation of higher olefins compared to when only C3/C4 light olefins are added. A prolonged residence time is one of important factors facilitating cracking reactions thus decreasing the production of higher olefins ("see Figure 17 shown in oral presentation"). The similar results are observed when 20% of alcohols are co-fed to the C5/C6 olefins with longer residence times contributing to the cracking reactions which directly impacts production of higher olefins ("see Figure 18 shown in oral presentation").

## 4. Conclusions

The direct addition of primary and secondary alcohols result in an increased production of diesel compared to gasoline. However, it is important to optimize this oligomerisation system in conjunction with the impact of adding high concentration of alcohols which will result in an increased formation of water thus affecting the catalyst activity and production rate as well. Suitable concentrations of alcohols may be determined through the experimental data to validate the oligomerisation reaction system model.

## References

1. Dagle, R. A., Lizarazo-Adarme, J. A., Zhu, Y., Jones, S. B., Hallen, R. T., White, J. F., Palo, D. R., Lebarbier, V. M. C., King, D. L., Gray, M. J., Biddy, M. J., Wang, Y. and Holladay, J. E. (2013), 'Single-Step Syngas-to-Distillates Synthesis via Methanol and Dimethyl Ether Intermediates: Final Report', Pacific Northwest National Laboratory Richland, Washington 99352.

2. Gayubo, A. A. abd Tarrıo, A. M., Aguayo, A. T., Olazar, M. and Bilbao, J. (2001), 'Kinetic Modelling of the Transformation of Aqueous Ethanol into Hydrocarbons on a HZSM-5 Zeolite', Industrial Engineering and Chemistry Resolution 40(16), 3467–3474.

3. Hadi, N., Niaei, A., Nabavi, R., Farzi, A. and Shirazia, N. M. (2014), 'Development of a new kinetic model for methanol to propylene process on mn/h-zsm-5 catalyst', Chemistry and Biochemical Engineering 28(1), 53–63.

4. Johansson, R., Hruby, S. L., Rass-Hansen, J. and Christensen, C. H. (2009), 'The Hydrocarbon Pool in Ethanol-to-Gasoline over HZSM-5 Catalysts', Catalysis Letter pp. 1–6.

5. John, M., Alexopoulos, K., Reyniers, M.-F. and Marin, G. B. (2015), 'Reaction path analysis for 1-butanol dehydration in H-ZSM-5 zeolite: Ab initio and microkinetic modelling', Journal of Catalysis 330, 28–45.

6. Makarova, M. A., Paukshtis, E. A., Thomas, J. M., Williams, C. and Zamaraev, K. I. (1994), 'Dehydration of n-Butanol on Zeolite HZSM-5 and Amorphous', Journal of Catalysis **149**, 36–51

7. Stauffer, J. E. and Kranich, W. L. (1963), 'Kinetics of the catalytic dehydration of primary alcohols', Journal **1**(2), 107–111.

8. White, J. L. (2011), 'Methanol-to-hydrocarbon chemistry: The carbon pool (r)evolution', Catalysis Science & Technology **1**, 1630–1635.

9. Zhi, Y. (2015), Elementary Reactions of 1-Propanol on HZSM-5, MÜNCHEN: ECHNISCHE UNIVERSITÄT MÜNCHEN.

# Environmental assessment and nano-mineralogical characterization of coal, overburden and sediment from Indian coal mining and acid drainage*

**Frans Waanders**, *School of Chemical and Minerals Engineering, North-West University (Potchefstroom campus), Potchefstroom 2531, SOUTH AFRICA*

*Madhulika Dutta[a,b], Jyotilima Saikia[bc], Silvio R. Taffarel[d], Diego de Medeiros[d]; Cesar M. N. L. Cutruneo[d], Luis F.O. Silva[d], Binoy K Saikia[bc]*

*[a]Department of Chemistry, CNB College, Bokakhat-785612, Assam, India, [b]Polymer Petroleum and Coal Chemistry Group, Materials Science and Technology Division, CSIR-North East Institute of Science & Technology, Jorhat-785006, India, [c]Academy of Scientific Innovative research, CSIR-NEIST Campus, Jorhat-785006, India, [d]Laboratory of Environmental Researches and Nanotechnology Development, Centro Universitário La Salle, Mestrado em Avaliação de Impactos Ambientais em Mineração, Victor Barreto, 2288 Centro, 92010–000 Canoas, RS, Brazil*

* *Sent to* **Geoscience Frontiers**

IPCC Aug 2016
Cape Town
South Africa

It all starts here ®

NORTH-WEST UNIVERSITY ®
YUNIBESITI YA BOKONE-BOPHIRIMA
NOORDWES-UNIVERSITEIT
POTCHEFSTROOM CAMPUS

# Outline

- **Introduction**
  - * **Coal mining and the environmental impact**
- **Experimental**
  - * **Materials**
  - * **Geochemical analyses**
  - * **Chemical analyses**
  - * **Spectroscopic analyses**
    **Ion chromatography, XRD, FTIR, Raman, SEM, HR-TEM, Mössbauer**
- **Results and Discussion**
  - * **Characterisation of coals, overburden and sediments**
  - * **Spectroscopic results**
  - * **Water analyses**
- **Final Conclusions**

# Introduction

Out of the almost 1000 billion tons of the world's total coal reserves, 200 billion tons are Indian coal, out of which the North-eastern region of India has a deposit of about 1 billion tons of high sulphur coals

⇩

Distributed unevenly in Assam, Meghalaya, Nagaland and Arunachal Pradesh

**BUT**

The high amounts of sulphur (2-12%) present in the North-eastern coals is found to be in five forms, mainly sulphate, pyritic, organic, elemental and secondary

⇩

This results in significant ecological degradation and exerts a long lasting impact on the landscape, ecosystem and socio-cultural-economic conditions

# Introduction

Pyrite present in the coal and overburden (OB) is leached to form low pH drainage due to amongst others the metabolic activity of some acidophilic bacteria such as *Thiobacillus ferroxidans*

⬇

The oxidation of iron sulphides resulting in the formation of Coal Mining Drainage (CMD) is a significant environmental hazard where the CMD in this study refers to the outflow of acidic water from coal mines and often from abandoned mines where the coal mining activities have exposed the sulphur bearing rocks

Pollutant such as Total Suspended Solids(TSS), Total Dissolved Solids(TDS), oil, grease and heavy metals are found in addition in these coal mining areas.

⬇

As the mining has continued over a long period, huge excavations exist and other environmental problems including air, water, noise pollution and deforestation occur

# Introduction

In the present study the aim is to illustrate the impact of acidic mine water of open cast coal mining in the Ledo and Tirap coal mines (Assam) and throughout the nearby areas.

# Experimental

**Materials**

- The samples that were collected included: (L = Ledo and T = Tirap)
    - Coal (LC-20A,LC-60A, TC-20A, TC-60A),
    - Mine overburden, (LOB-15A,LOB-15B,TOB-15A,TOB-15B),
    - Soil (LS-15A,LS-15B,TS-15A,TS-15B),
    - Sediment (LSE-15A, LSE-15B) and
    - Water samples (LW, TW).
- The samples were collected randomly within a radius of approximately 5 km from the point of mining in all four wind directions.
- The coal, OB, soil and sediment samples were air dried, ground to 0.422 and 0.211 mm size and stored for analysis.
- The water samples were preserved for further analysis by adding 10 ml of 1:1 $HNO_3$ solution.

**Chemical analysis**

- The proximate analyses of coal, soil, sediment and OB samples were carried out with a Proximate Analyzer (Model: TGA 701; Leco Corporation, USA).
- The amount of C, H, N were estimated by using an Elemental Analyzer (Model: Perkin-Elmer 2400) and the total sulphur by 'Sulphur Analyzer' (Leco Corporation, USA) by applying ASTM 2013.

# Experimental

## Spectroscopic analyses

- **XRD**
  Carried out with a X-Ray diffractometer (Rigako, ULTIMA IV), with diffraction data of the samples obtained with starting angle set as 2.00° and stopping angle at 75.00° with a step in angle 0.05°, scanning rate of 1°per minute with target used a Cu- K α (λ =1.7902A°).

- **FTIR**
  Carried out with a Perkin Elmer System 2000 (Model 640B) apparatus in the range of 400-4000 $cm^{-1}$ using KBr pellets with the same amount of dried & ground samples. About 4 - 5 mg of the samples were accurately weighed on a single pan electronic balance (Model: SI-234, Denver instrument) and mixed homogeneously with spectroscopic grade KBr sample in an agate mortar.

- **Raman:**
  Carried out using a LabRam microscope (Horiba John Yvon Ltd.,) fitted with a HeNe laser (λ), with data recording and analysis performed using the proprietary LabSpec software (Horiba Jobin Yvon) run on a Dell Optiplex PC were samples were irradiated with 542 nm excitation wavelength.

# Experimental

- **FE-SEM and HR-TEM analysis**
  The nano-mineralogical analysis of the collected coal and coal mine reject solid samples were performed by Field Emission Scanning Electron Microscope (FE-SEM, Model: FEI Quanta™ 650 FEG), integrated with an Oxford Inca 400 energy dispersive X-ray spectrometer and FEI G2 200 KV High Resolution Transmission Electron Microscope (HR-TEM).

- **Mössbauer spectroscopy**
  A Halder Mössbauer spectrometer, operating in conventional constant acceleration mode, using a proportional counter, filled to 2 atm. with Xe-gas and a $^{57}Co$ γ-ray source was used for the Mössbauer analyses.
  Callibration: α-Fe was used as the reference and a 50 mCi $^{57}Co$ source, plated into an Rh-foil, was used to produce the γ-rays.
  Samples analysed at room temperature and data collected in a multi-channel analyser (MCA) to obtain a spectrum of count rate against source velocity.
  A least-squares fitting program was used and by superimposing Lorentzian line shapes, the isomer shift, quadrupole splitting and/or hyperfine magnetic field of each constituent was determined with reference to the centroid of the spectrum of a standard α-Fe foil at room temperature.
  The amount of each constituent present in a sample was determined from the areas under the relevant peaks.

# Experimental

**Geochemical analysis**

- The water sample analyses included pH, TDS, DO, EC analyses and were carried out at CSIR-NEIST, Jorhat by using a EUTECH-PC700 (pH/mV/Conductivity/°C/°F meter) apparatus.
- The pH values of the solid samples (coal, soil, OB and sediment) were determined by using the same instrument.
- The hardness of the water samples were determined manually by using the ASTM 2010 standard method.

**Ion Chromatographic analysis**

- The concentrations of the major ions including both cations ($NH_4^+$, $Na^+$, $K^+$, $Ca^{+2}$, $Mg^{+2}$) and anions ($F^-$, $Cl^-$, $Br^-$, $SO_4^{-2}$ ) were determined by using Ion chromatography system-Metrohm 882 Compact IC Plus.
- The anions were analysed using 4 X 250mm Metrosep A Supp 5 analytical column with the chemical suppression Sulfuric acid and Metrosep RP2 3.5 guard column.
- Similarly the cations were analysed using a 4 X 150mm Metrosep C-4 analytical column and Metrosep RP2 3.5 guard column.

# Results and discussion

**Chemical analysis**

**Proximate analysis of coal (wt.%)**

| Sample Type | Moist Content | Volatile Matter | Ash Content | Fixed Carbon |
|---|---|---|---|---|
| TC-60A | 2.07 | 40.72 | 2.78 | 54.23 |
| TC-20A | 2.65 | 40.92 | 1.78 | 54.84 |
| LC-60A | 3.12 | 41.27 | 6.68 | 48.93 |
| LC-20A | 5.33 | 35.33 | 19.72 | 39.63 |

**CHNS analysis of coal (wt %)**

| Sample Type | C | H | N | $S_t$ | $S_p$ | $S_s$ | $S_o$ |
|---|---|---|---|---|---|---|---|
| TC-60A | 80.6 | 5.88 | 1.19 | 3.54 | 0.61 | 0.77 | 2.16 |
| TC-20A | 83.3 | 6.14 | 1.35 | 1.94 | 0.35 | 0.57 | 1.02 |
| LC-60A | 76.3 | 5.90 | 1.02 | 2.49 | 0.45 | 0.69 | 1.35 |
| LC-20A | 73.5 | 5.05 | - | 2.94 | 0.49 | 0.71 | 1.74 |

Low moisture and ash content and high volatile content
Relative high S content with organic bound S dominant

**Analysis of OB, soil and sediment (wt %)**

| Sample Type | Moist Content | Volatile Matter | Ash Content | Fixed Carbon | Total S |
|---|---|---|---|---|---|
| LOB-15A | 2.55 | 11.14 | 89.81 | - | 0.24 |
| LOB-15B | 0.82 | 11.13 | 91.72 | - | 0.08 |
| TOB-15A | 2.69 | 12.87 | 86.68 | - | 0.51 |
| TOB-15B | 2.17 | 7.94 | 93.83 | - | 0.06 |
| LS-15A | 0.43 | 11.18 | 92.15 | - | 0.04 |
| LS-15B | 2.13 | 9.71 | 91.96 | - | 0.05 |
| LSE-15A | 3.08 | 11.40 | 89.08 | - | 0.43 |
| LSE-15B | 1.97 | 12.98 | 84.65 | 0.39 | 0.94 |

Mainly mineral matter (quartz) and some S, hematite, pyrite, and marcasite present
➡ speciation obtained from XRD and Mössbauer spectroscopy

# Results and discussion

## XRD

| Sample type | d-values | minerals |
|---|---|---|
| Coal (LC-20A, LC-60A, TC-20A, TC-60A) | 3.3404<br>1.5789 | Quartz<br>Hematite, Quartz |
| Overburden (LOB-15A,LOB-15B, TOB-15A, TOB-15B) | 3.3404<br>1.8252<br>2.2801<br>2.4669<br>1.5789 | Quartz<br>Quartz,Hematite, Pyrite, Marcasite<br>Hematite, Quartz<br>Quartz, pyrite<br>Hematite, quartz |
| Soil & sediment (LS-15A,LS-15B,LSE-15A, LSE-15B) | 4.2654<br>3.3404<br>1.5789<br>2.4469<br>1.8706<br>4.2654 | Quartz<br>Quartz<br>Hematite,Quartz<br>Quartz, pyrite<br>Quartz, Marcasite, Hematite<br>Quartz |

Hematite observed in XRD of C, OB, S and SE
Not observed in MS (sextet, not doublet)
Illite observed in MS of OB, S, SE
Not observed in XRD (low angles)

## Mössbauer

| Sample | Sample | Component | IS mms$^{-1}$ (±0.01) | QS mms$^{-1}$ (±0.01) | H Tesla (±0.3) | Relative intensity (%) |
|---|---|---|---|---|---|---|
| Coal | LC-20A | Pyrite | 0.36 | 0.73 | - | 100 |
| | LC-60A | Pyrite | 0.37 | 0.62 | - | 100 |
| | TC-20A | Pyrite | 0.32 | 0.64 | - | 100 |
| | TC-60A | Pyrite | 0.33 | 0.64 | - | 100 |
| OB | LOB-15A | Pyrite/Marcasite<br>Illite | 0.35<br>1.05 | 0.54<br>2.41 | -<br>- | 61<br>29 |
| | LOB-15B | Pyrite<br>Illite | 0.31<br>1.22 | 0.60<br>1.98 | -<br>- | 8<br>92 |
| | TOB-15A | Jarosite<br>Illite | 0.31<br>1.21 | 0.65<br>1.92 | -<br>- | 68<br>32 |
| | TOB-15B | Pyrite<br>Iron sulphate | 0.34<br>1.27 | 0.60<br>2.40 | -<br>- | 87<br>13 |
| Soil and sediment | LS-15A | Pyrite<br>Illite | 0.32<br>1.22 | 0.57<br>1.84 | -<br>- | 10<br>90 |
| | LS-15B | Pyrite<br>Iron sulphate | 0.34<br>1.12 | 0.67<br>2.65 | -<br>- | 56<br>44 |
| | LSE-15A | Pyrite/Marcasite<br>Iron sulphate | 0.33<br>1.16 | 0.56<br>2.34 | -<br>- | 87<br>22 |
| | LSE-15B | Pyrite<br>Illite | 0.31<br>1.20 | 0.66<br>1.88 | -<br>- | 72<br>28 |

# Results and discussion

## XRD spectra

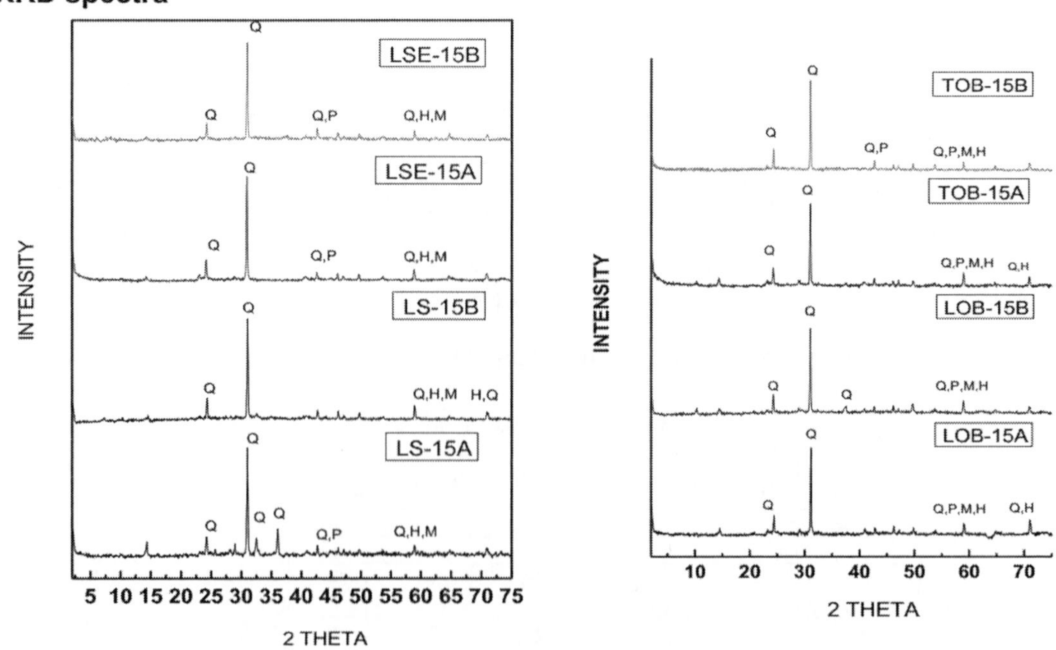


**Figure:** XRD spectra showing only hematite possibility together with pyrite/marcasite

# Results and Discussion

**Mössbauer spectra**

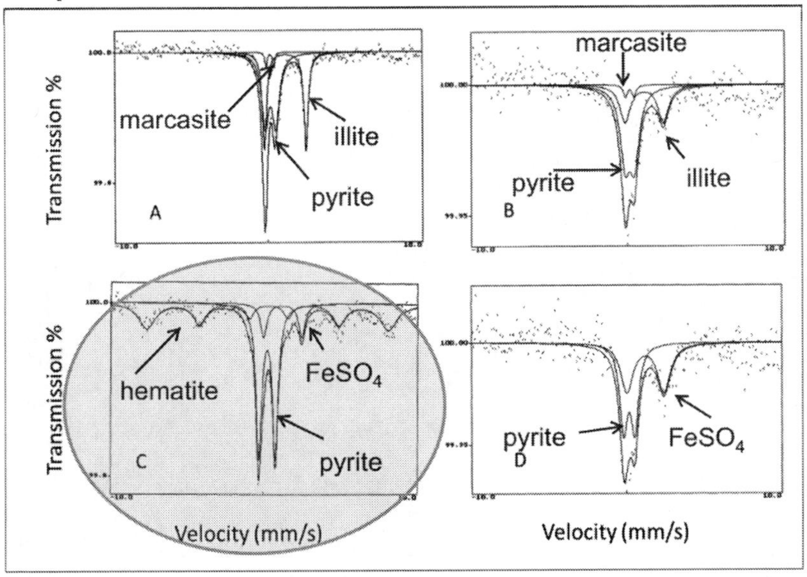


**Figure 1** Mössbauer spectra showing only doublets (A, B and D) present and spectrum showing the hematite sextet in C obtained for another Indian coal study

# Results and discussion

**Observation from FE-SEM analysis**

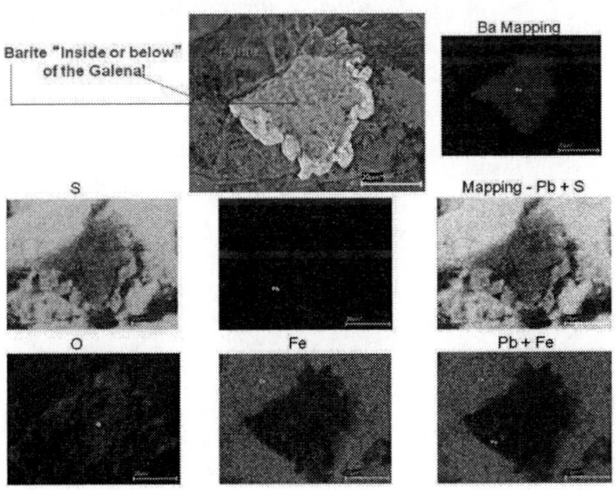


Figure: FE-SEM of coal sample (LC-20A sample)

Figure: Kaolinite (A), NaCl+KCl (B) and gypsum (C) in LOB-15D sample

The sulphate minerals like barite, jarosite, pickeringite analysed by FE-SEM are the primary sources of CMD formation

The overburden materials were found both in amorphous and crystalline forms containing hazardous elements

# Results and discussion

## Observation from HR-TEM analysis

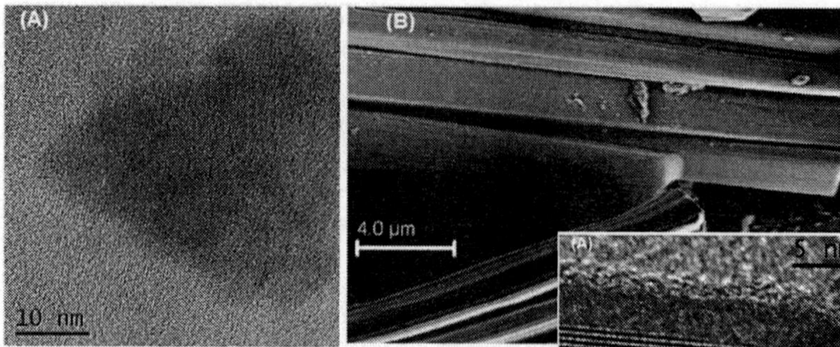


Figure: HR-TEM image of singular nano-hematite (A) and FE-SEM image of pickeringite in LC-15A

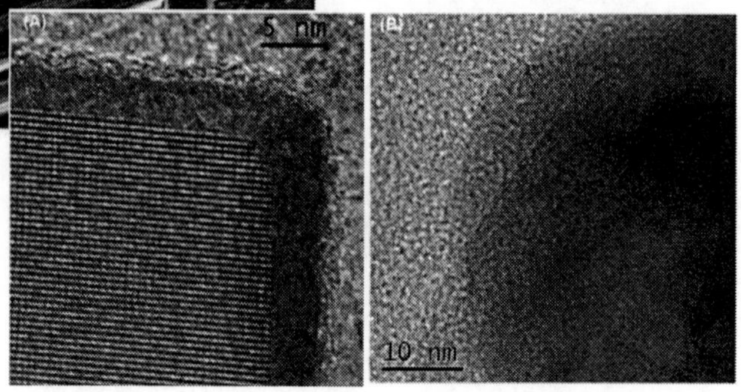


Figure: HR-TEM images of Jarosite (A) and Amorphous and crystalline mixed carbonaceous matter contain hazardous elements (B) (LOB-15F)

From an environmental perspective, one of the major important characteristics of nano-mineral nanoparticles like nano-hematite is their high characteristic surface area and high reactivity

# Results and discussion

## FTIR analysis

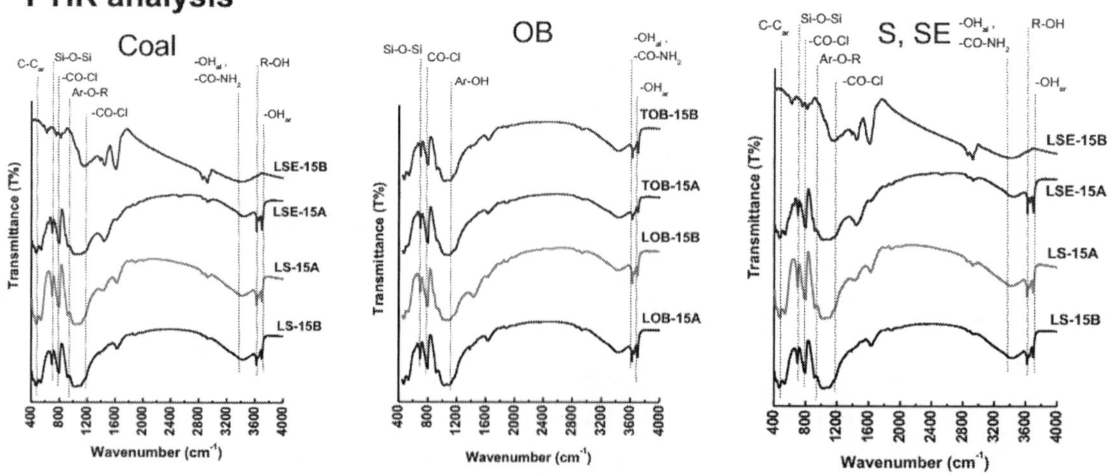


To get the detailed information about the different functional groups as well as other bond stretching bands, FTIR analysis was conducted

- The bands due to stretching vibration of S–S and C–S bonds in primary and secondary thiols were found in the region of 530 and 670 $cm^{-1}$ and are weak.
- The absorption bands due to aliphatic alcohol, ether groups (O-H, C-O-C )and $SO_4^{2-}$ group are found in the region of 1100 - 1205 $cm^{-1}$.
- The aliphatic amino (NH), aromatic C=C, and aliphatic alkyl groups showed the absorption bands within the range of 1440 – 1611 $cm^{-1}$

# Results and discussion

### Raman analysis

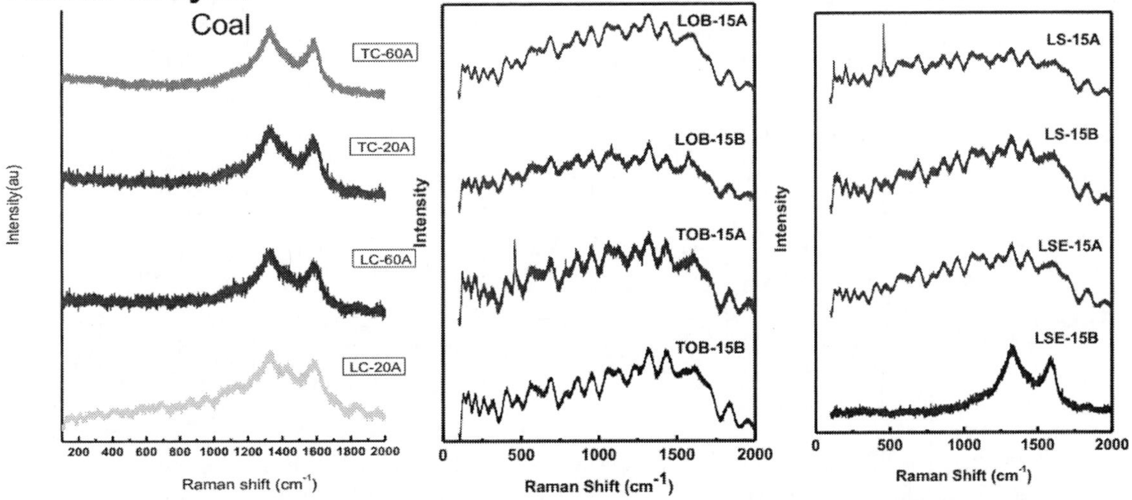


- Raman spectroscopy is used for the characterization of carbon minerals and two distinct peaks were observed at around 1348 and 1589 $cm^{-1}$ indicating a graphite structure
- The band found at 1348$cm^{-1}$ has been found due to defects present in structural units and disorder which is designated as D- band
- The band occurring at 1589 $cm^{-1}$can be interpreted as the peak related to the graphite (or G) band

# Results and discussion

### Geochemical characteristics

| Sample type | Distance from CMD source (km) | pH | Electrical conductivity msem$^{-1}$ | Dissolved oxygen mg/l | Total dissolved solid (ppt) | Total hardness mg/l |
|---|---|---|---|---|---|---|
| LW-15A | source | 4.0 | 1.226 | 0.4 | 0.000627 | 876 |
| LW-15B | 0.5 | 4.1 | 1.234 | 0.2 | 0.000613 | 897 |
| LW-15C | 0.5 | 3.50 | 2.29 | 0.3 | 1.14 | 1276 |
| LW-15D | 1 | 3.30 | 2.49 | 0.8 | 1.18 | 730 |
| LW-15E (Ledopani river) | 2 | 4.4 | 0.936 | 0.5 | 0.000468 | 600 |
| LW-15F (Kachanala stream) | 5 | 5.0 | 0.659 | 0.2 | 0.000328 | 460 |
| TW-15A | 1 | 3.60 | 3.15 | 0.3 | 1.57 | 2350 |
| TW-15B | source | 3.70 | 2.11 | 0.4 | 1.59 | 2470 |
| TW-15C | 0.5 | 3.50 | 5.34 | 0.2 | 2.65 | 1350 |
| TW-15D | 1.5 | 3.90 | 2.11 | 0.3 | 1.06 | 5720 |

pH and EC vary inversely
TDS and EC high and pH low - dissolution of elements as well as ions
Total hardness at Tirap colliery very high compared to Ledo – possible contamination

# Results and discussion

## Ion-Chromatographic analysis

**Cations**

| Cation | LW-15A | LW-15B | LW-15C | LW-15D | LW-15E | LW-15F | TW-15A | TW-15B | TW-15C | TW-15D |
|---|---|---|---|---|---|---|---|---|---|---|
| $Na^+$ | 0.559 | 0.508 | 0.582 | 0.531 | 0.661 | 0.724 | 0.480 | 0.543 | 0.488 | 0.881 |
| $K^+$ | 0.103 | 0.262 | 0.301 | 0.350 | 1.192 | 2.137 | 0.195 | 0.096 | 0.164 | 0.232 |
| $Mg^{+2}$ | 44.913 | 110.25 | 29.291 | 35.015 | 49.781 | 60.784 | 242.81 | 450.05 | 128.04 | 240.44 |
| $NH_4^+$ | - | - | - | - | - | - | - | - | - | - |

**Anions**

| Anion | Lw-15A | LW-15B | LW-15C | LW-15D | LW-15E | LW-15F | TW-15A | TW-15B | TW-15C | TW-15D |
|---|---|---|---|---|---|---|---|---|---|---|
| $F^-$ | - | 0.072 | - | - | - | 0.864 | - | - | - | - |
| $Cl^-$ | 0.268 | 1.689 | 0.741 | 3.384 | 0.495 | 8.846 | 1.788 | 1.067 | 0.675 | 0.571 |
| $Br^-$ | 170.10 | 46.202 | 66.869 | 161.41 | 78.341 | 227.17 | 32.764 | 47.817 | 52.875 | 54.988 |
| $SO_4^{-2}$ | 14.151 | 8.817 | 5.496 | 9.00 | 3.909 | 15.408 | 12.155 | 15.73 | 9.974 | 10.982 |

- The absence of $NH_4^+$ ion confirms the low pH of water bodies near mining area
- The Tirap samples have very high concentration of $Mg^{2+}$, being the 4th common mineral in the coal mine.
- Very high concentration of $Br^-$ in water samples (highest in LW-15F) of the Ledo mine is an indication of contamination of water bodies even at a long distance from the source where the sample was collected from the Kachanalla River, about 5 km from the Ledo colliery.

# Results and discussion

## Mobility of elements in CMD water

- The presence of the possible hazardous elements in the coal, overburden, soil and sediments include element such as S, Cl, Br, F, Si, Al, Fe, Mg, Pb, Ca, As and Ba
- The excavation of the coal exposes the sulphide minerals to atmospheric oxygen and water resulting in the generation of highly acidic mine drainage. Due to its acidic nature, the mine drainage is highly enriched with metals
- Metal concentrations in the CMD from coal mines are controlled by three factors:
  - (a) oxidation rate of pyrite
  - (b) mineralogy of country rock
  - (c) pH of leachate.
- At low pH levels, the toxic (heavy) metals are more soluble and enter into solution and results from secondary reactions between iron sulphates, sulphuric acid and the compounds in nearby clays (kaolinite), sandstones, limestone, sulphides and various organic substances present in the coal mine drainage or streams which explains the existence of Ca, Mg, Al, Na, K, Mn, Pb, As and Ba

# Conclusions

- Relative high S content was found in most coal samples, with organic bound S the dominant species
- The geochemical assessment shows high values of EC, TDS, $SO_4^{-2}$, Ca, Mg, Fe in a number of water samples which make them unsafe for drinking purposes
- Limestone is mainly used for active treatment of CMD, but some problems arise when using limestone as (a) it is less effective for high iron containing water (b) there is some difficulties to get the pH higher than 5 due to its buffering action.
- The formation of gypsum due to limestone addition to the water to increase the pH leads to scaling of pipelines and the equipment used to eradicate the acidity.

↓

For an alternative method in the treatment of CMD water, fly ash zeolites can be used which a low cost material and it is obtained from waste coal fly ash – presently investigated

# Acknowledgments

The financial assistance from CSIR, New Delhi (MLP-6000-WP-III) is duly acknowledged. The authors express their gratitude to the Director of CSIR-NEIST (Jorhat), USTM, and NECL (Margherita) for providing their support and co-operation to the study.

The information presented in this paper is based on the research supported by the South African Research Chairs Initiative (SARChI) of the Department of Science and Technology and National Research Foundation of South Africa (Coal Research Chair Grant No. 86880).
Any opinion, finding or conclusion or recommendation expressed in this material is that of the author(s) and the NRF does not accept any liability in this regard.

Thank you

## Bench-Scale Testing of Stabilized Flue Gas Desulfurization Materials to Treat Acid Mine Drainage


**Tarunjit Singh Butalia[1], Farid Ahmed Malik[2]**

[1]The Ohio State University, USA

[2]FFC University, Pakistan


ccp.osu.edu

THE OHIO STATE UNIVERSITY

COLLEGE OF ENGINEERING

DEPARTMENT OF CIVIL, ENVIRONMENTAL, AND GEODETIC ENGINEERING

## AEP Stingy Run Fly Ash Impoundment

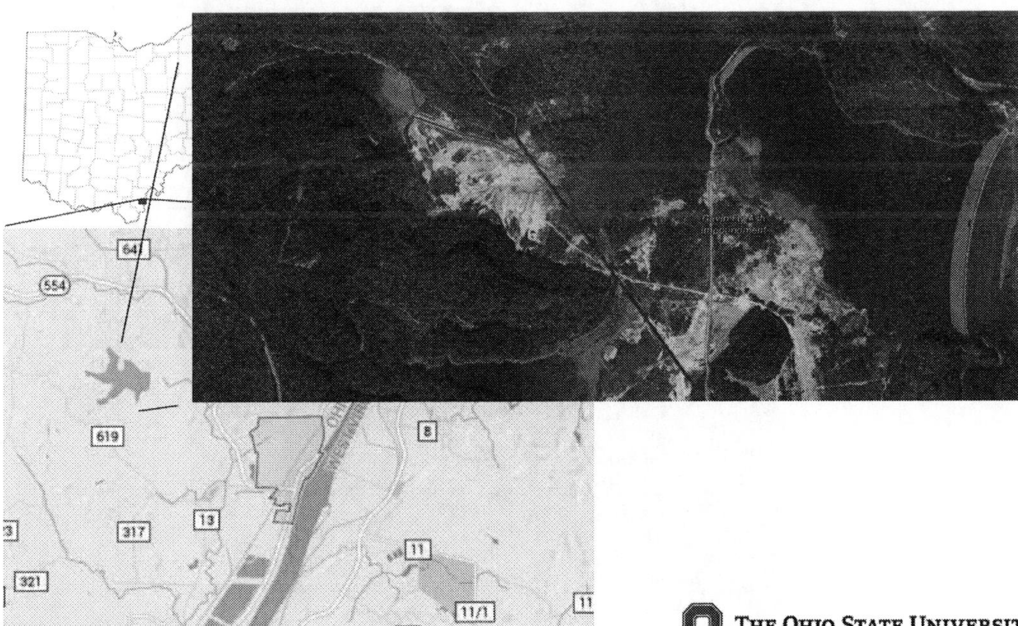


THE OHIO STATE UNIVERSITY

# Acid Mine Drainage

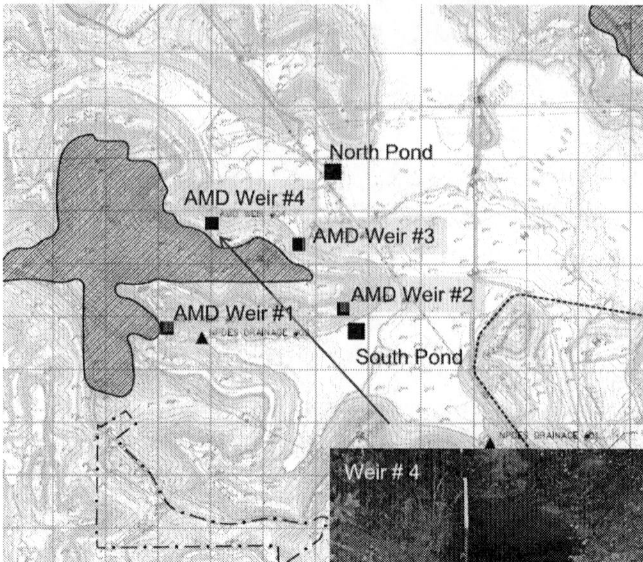


OGS- UGMs GA-031, GA-011

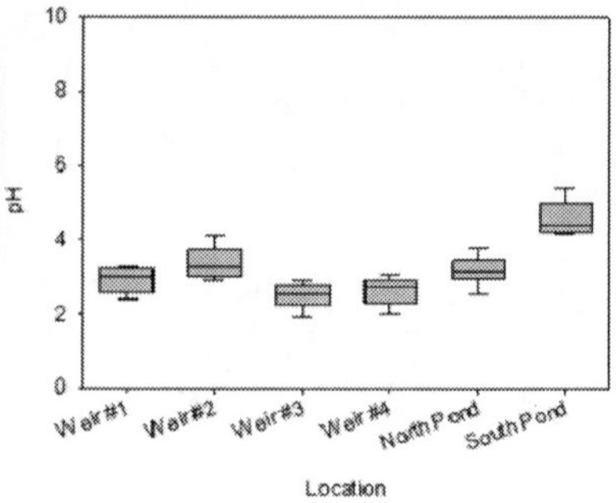


- Acid load to reclamation site was estimated by pH values & flow rates of AMD discharging streams measured by weirs installed along highwall

THE OHIO STATE UNIVERSITY

# Highwalls and Pits Around Stingy Run Impoundment

THE OHIO STATE UNIVERSITY

## Proposed OSU Full-Scale Acid Mine Drainage Remediation Demonstration Project @ AEP Gavin Power Plant

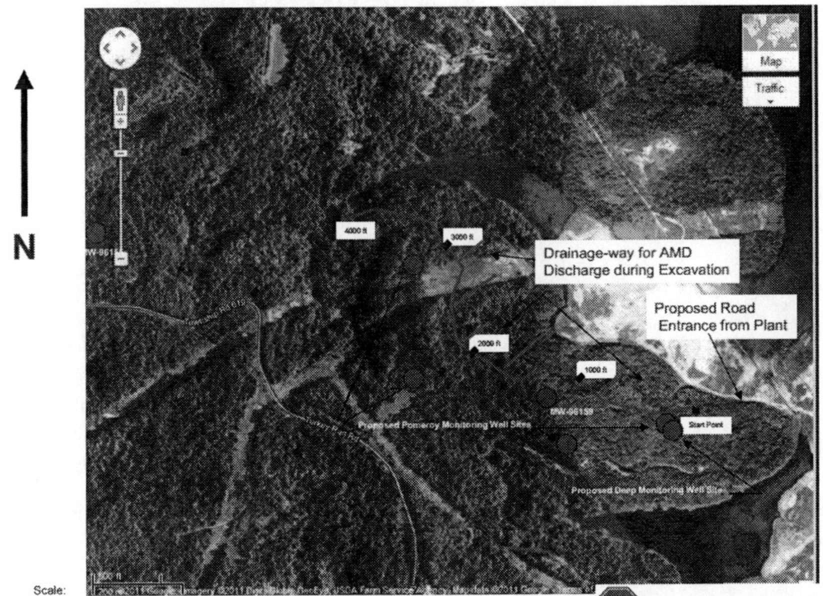


Scale:

Approximate Highwall Location

Approximate Measurement of Distance

THE OHIO STATE UNIVERSITY

## Reclamation using Fixated FGD Material

- **Utilizing large volume of fixated flue gas desulfurization materials**
  - FGD by-product (calcium sulfite) stabilized with fly ash and lime
- **Goals**
  - Encapsulate acid mine drainage (AMD) producing materials
  - Neutralize AMD
  - Re-contour highwalls
- **Approaches**
  - Year I & II: field investigation; laboratory test; bench-scale study; numerical analysis of design approaches; background water monitoring
  - Year III-V: permitting, water quality monitoring, construction of demo project

THE OHIO STATE UNIVERSITY

THE OHIO STATE UNIVERSITY

# Mineral Composition of Gavin Fixated FGD Material

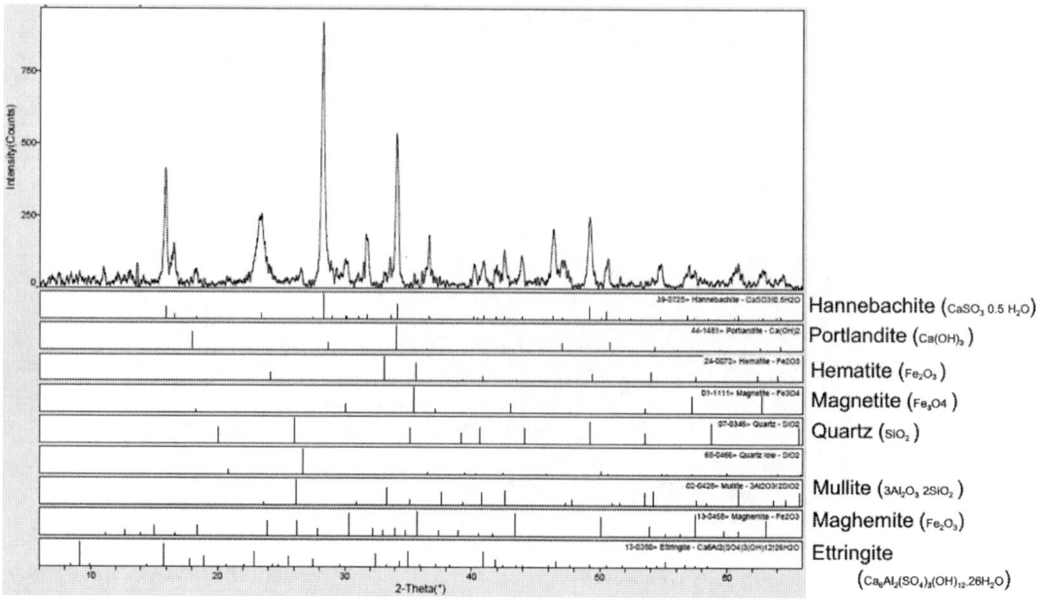


# Mitigating AMD using Fixated FGD Material

- Neutralization Chemical Reactions
  - Portlandite

    $$Ca(OH)_2 + 2H^+ \rightarrow Ca^{2+} + 2H_2O$$

  - Ettringite

    $$Ca_6Al_2(SO_4)_3(OH)_{12}{:}26H_2O + 12H^+ \rightarrow 2Al^{3+} + 6Ca^{2+} + 3SO_4^{-2} + 38H_2O$$

  - Iron oxides

    $$Fe_2O_3 + 6H^+ \rightarrow 2Fe^{3+} + 6H_2O$$

    $$Fe_3O_4 + 8H^+ \rightarrow 2Fe^{3+} + Fe^{2+} + 8H_2O$$

  - Carbonates (calcite or dolomite)

    $$CO_3^{2-} + H^+ \rightarrow HCO_3^- \qquad HCO_3^- + H^+ \rightarrow H_2CO_3$$

- Inundation of nearby underground mine by plugging AMD discharge
  - Low permeability of FGD material
  - Formation of secondary minerals blocking pores

THE OHIO STATE UNIVERSITY

## Laboratory Batch, Column, and Bench-Scale Tests

- Batch Tests
  - Intrinsic pH-dependent leaching property of fixated FGD material under equilibrium conditions (USEPA Method 1313)
  - Leaching of fixated FGD material under different liquid-to-solid (L/S) ratio with acid mine drainage (USEPA Method 1316)
  - Investigating neutralization capacity
  - Establishing geochemical model to explicit/predict geochemical reactions involved in neutralization

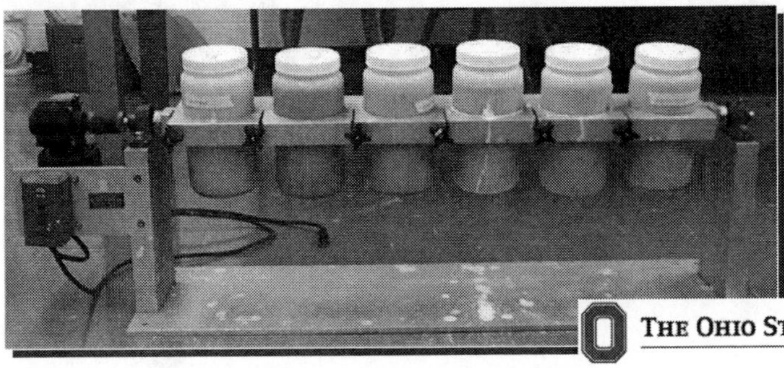

THE OHIO STATE UNIVERSITY

## Laboratory Batch, Column, and Bench-Scale Testing

- Column Tests
  - Simulating AMD neutralization process under similar percolation process as reclamation
  - Two columns with different L/S flow rates
    - Column I: ~1.0 L/S per day
    - Column II: ~2.0 L/S per day
  - Monitoring change of AMD water quality with extended L/S ratio
    - Column I: ~90 (10gallons AMD/lb fixated FGD)
    - Column II: ~160 (20 gallons AMD/lb fixated FGD)

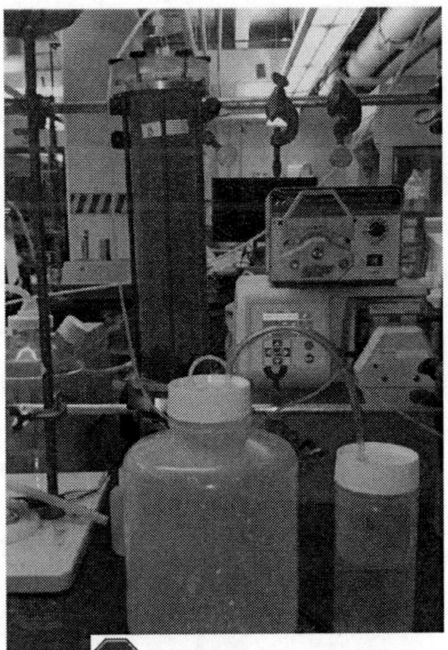

THE OHIO STATE UNIVERSITY

## Laboratory Batch, Column, and Bench-Scale Testing

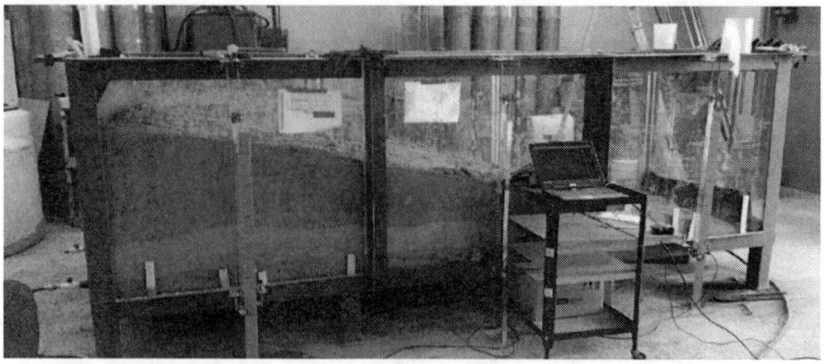

- Bench-Scale Testing
  - Assessment of AMD infiltration in absence and presence of coal drain
  - Calibrate geotechnical and geochemical models representing full-scale demonstration project design

THE OHIO STATE UNIVERSITY

## Bench-Scale Testing @ OSU

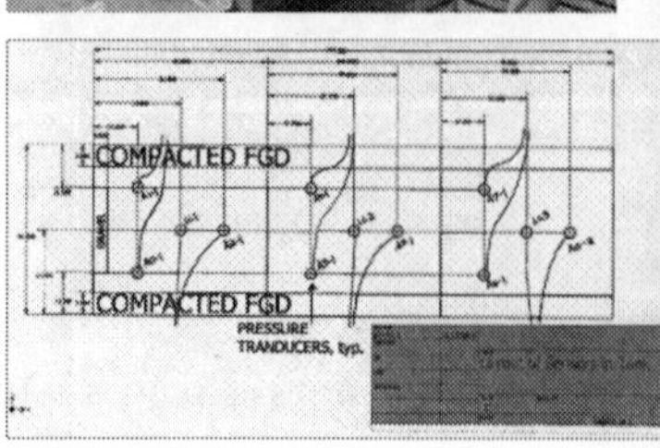

THE OHIO STATE UNIVERSITY

## Bench Scale Model Configurations Tested

| | Parameters | Results |
|---|---|---|
| Test 1 | FGD dumped<br>Water, No Geotextile<br>- Transducers at bottom | Calibrated model for effective permeability ratio (steady state modelling) |
| Test 2 | FGD dumped<br>Water, Geotextile (1 ft long, 1 ft height from bottom)<br>- Transducers above and below geotextile | Flow rates increased significantly |
| Test 3 | FGD lightly compacted<br>AMD, No Geotextile<br>- Transducers at bottom | Calibrated model for effective permeability (transient modelling) for lightly compacted FGD |
| Test 4 | FGD lightly-compacted<br>AMD, Geotextile (1 ft long, 0.5 ft height from bottom)<br>- Transducers at bottom | The presence of geotextile does not decrease the flow rate |
| Test 5 | FGD well-compacted<br>AMD, No Geotextile<br>- Transducers at bottom | Calibrated model for effective permeability (transient modelling) for well-compacted FGD |

THE OHIO STATE UNIVERSITY

## Geotechnical Modeling

### Bench Scale Test 1

- Seep/w was used to predict flow of water through FGD without geotextile
- Calibrated model for effective permeability ratio (V/H permeability ratio = 25) using Steady-State Analysis

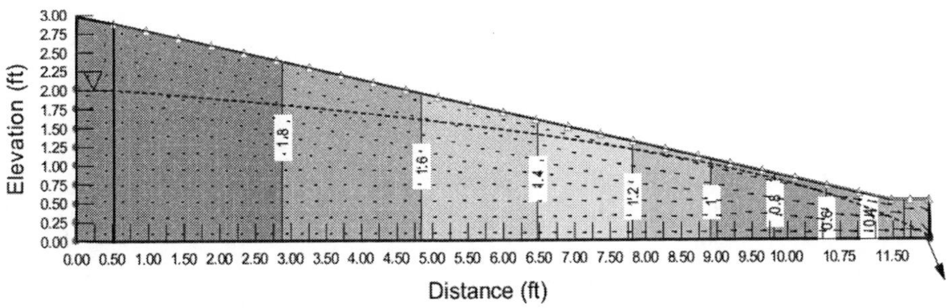


THE OHIO STATE UNIVERSITY

## Geotechnical Modeling

### Bench Scale Test 1: Model vs. Experimental Data

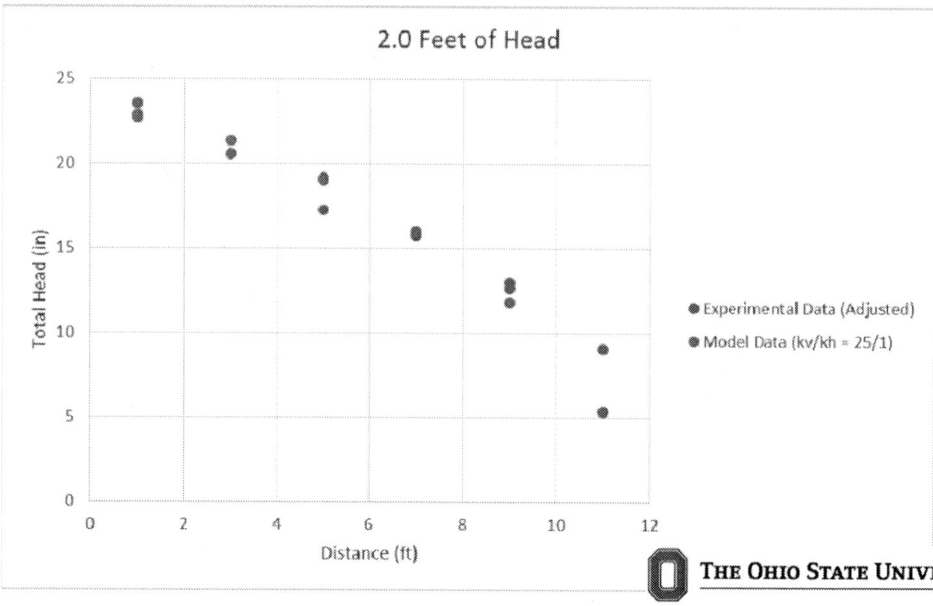


THE OHIO STATE UNIVERSITY

## Geotechnical Modeling

### Bench Scale Test 3

- Seep/w transient modelling to predict flow of AMD through FGD without geotextile
- Performed transient analysis and calibrated model by finding the **effective horizontal permeability** (**$8 \times 10^{-3}$ cm/sec**) to match amount of time needed to reach steady state in experimental data
- Higher end permeability value **($8 \times 10^{-3}$ cm/sec)** identified under loose placement of FGD.

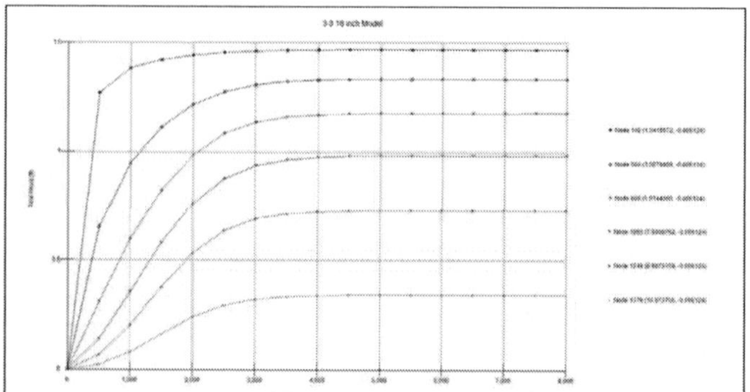

THE OHIO STATE UNIVERSITY

# Geotechnical Modeling

## Seep/W Modeling of Bench Scale Test 3

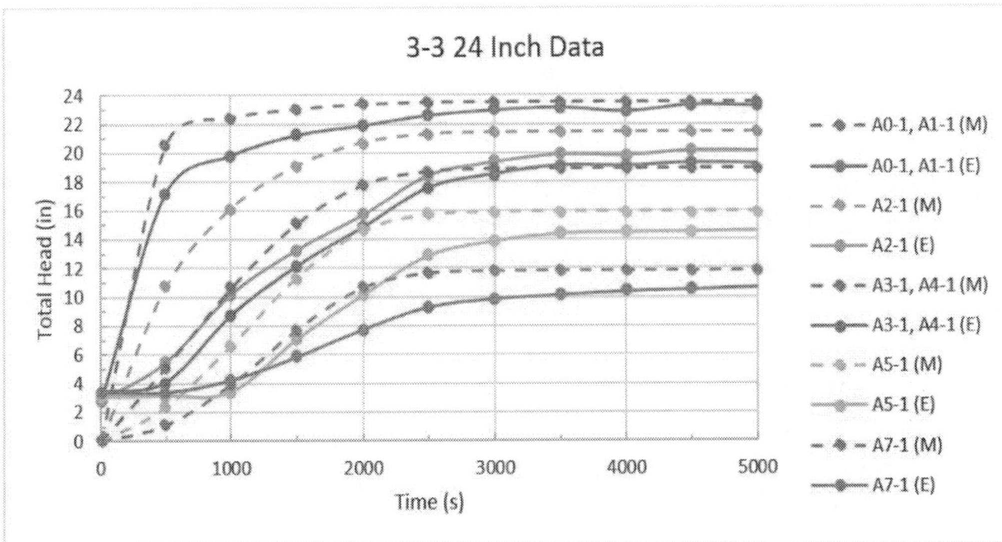


(M) = Model predictions (E) = Experimental observations

# Geotechnical Modeling

## Bench Scale Test 4

- Seep/W model calibrated using same effective horizontal permeability determined in Bench Scale Test 3 **($8 \times 10^{-3}$ cm/sec)**
- Water levels of experimental data were several inches higher than what model predicted

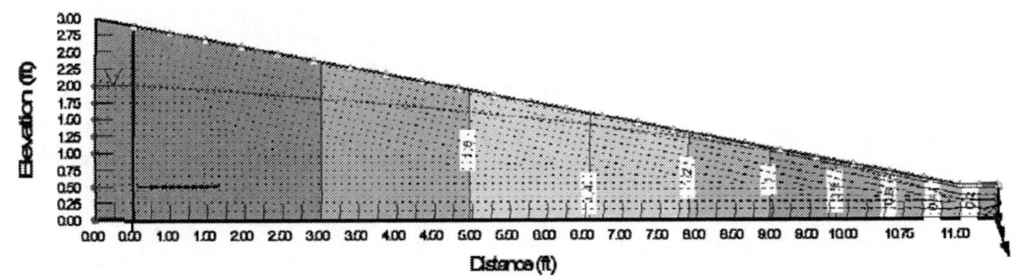

# Geotechnical Modeling

## Bench Scale Test 4

- Central portion (test section) was lightly compacted.
- Geotextile was 1.0 ft long and 0.5 ft from bottom

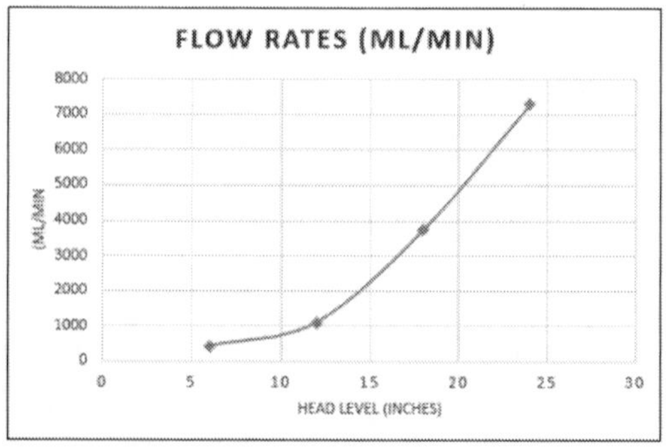


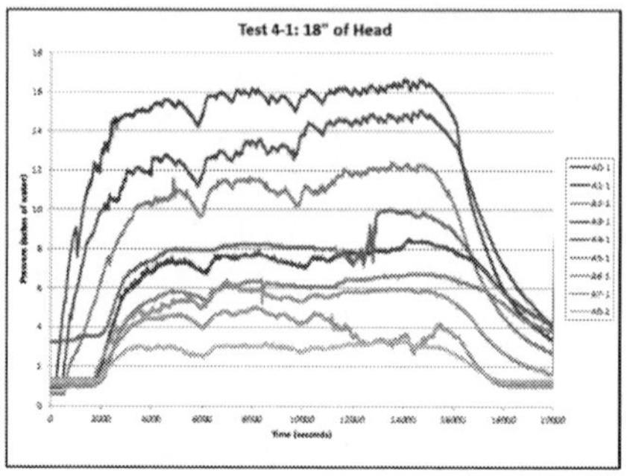


THE OHIO STATE UNIVERSITY

# Geotechnical Modeling

THE OHIO STATE UNIVERSITY

## Seep/W Modeling of Bench Scale Test 4

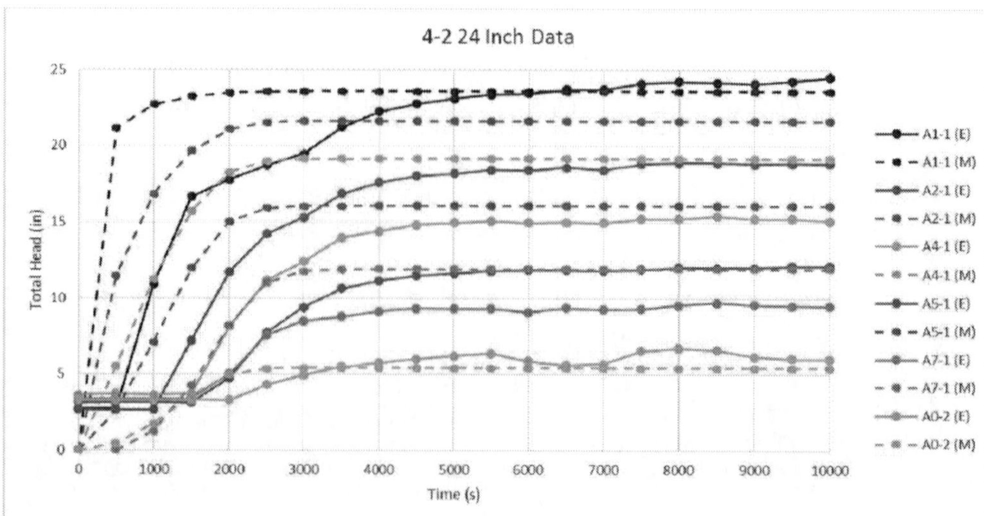


(M) = Model predictions (E) = Experimental observations

## Geotechnical Modeling

### Bench Scale Test 5

- Seep/w transient modelling to predict flow of AMD through well-compacted FGD
- Performed transient analysis and calibrated model by finding the **effective horizontal permeability ($8 \times 10^{-4}$ cm/sec)** to match amount of time needed to reach steady state in experimental data for 12 and 18 inch head
- Permeability value **($8 \times 10^{-4}$ cm/sec)** is smaller for well compacted FGD than for lightly compacted FGD.
- Significantly increased time taken for AMD to reach steady-state

THE OHIO STATE UNIVERSITY

THE OHIO STATE UNIVERSITY

## Geotechnical Modeling

### Seep/W Modeling of Bench Scale Test 5

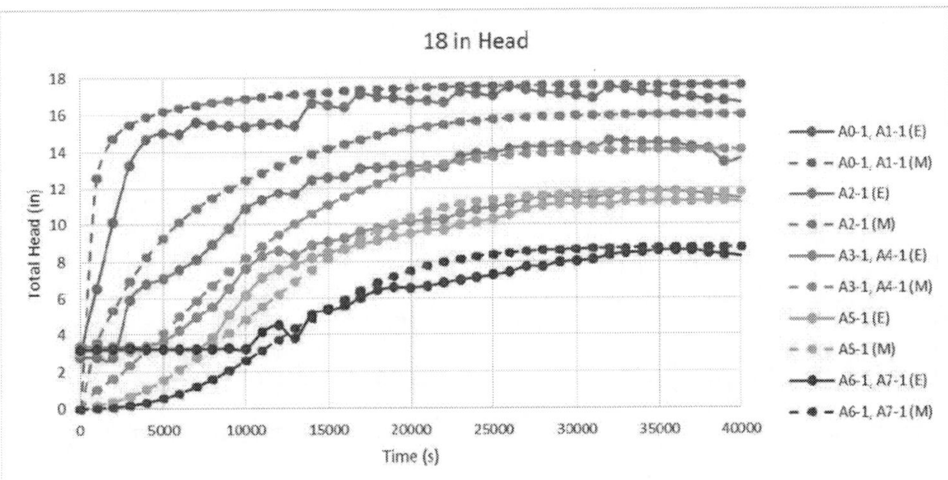


(M) = Model predictions (E) = Experimental observations

## Full-Scale Modeling Work

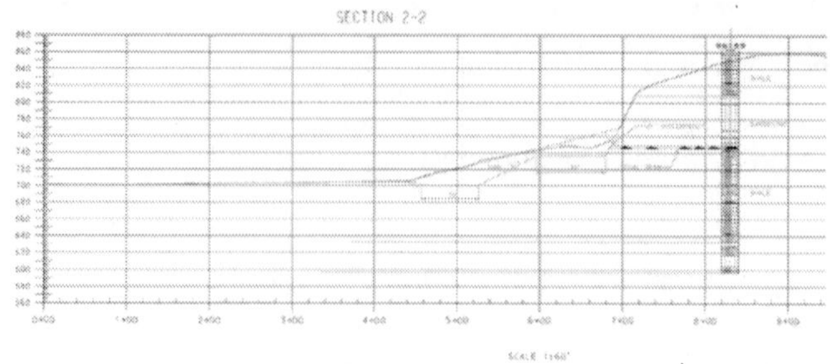


Analyses performed on four different cross-sections representing a region of highwall with proposed FGD embankment at Gavin power plant fly ash pond site

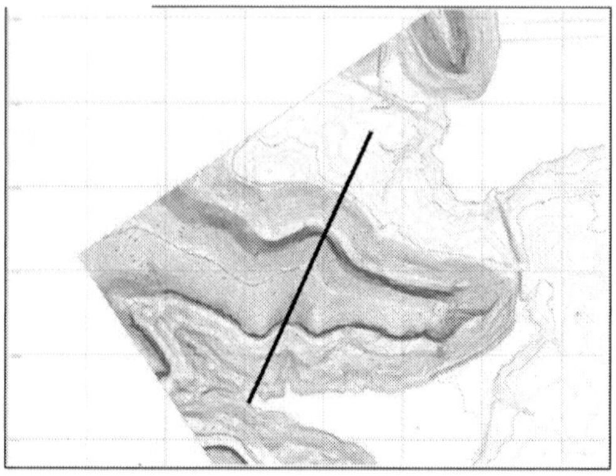

THE OHIO STATE UNIVERSITY

## Full-Scale Geotechnical Modeling: Seepage Analysis

- Evaluated flow paths of AMD through proposed final hillside configuration

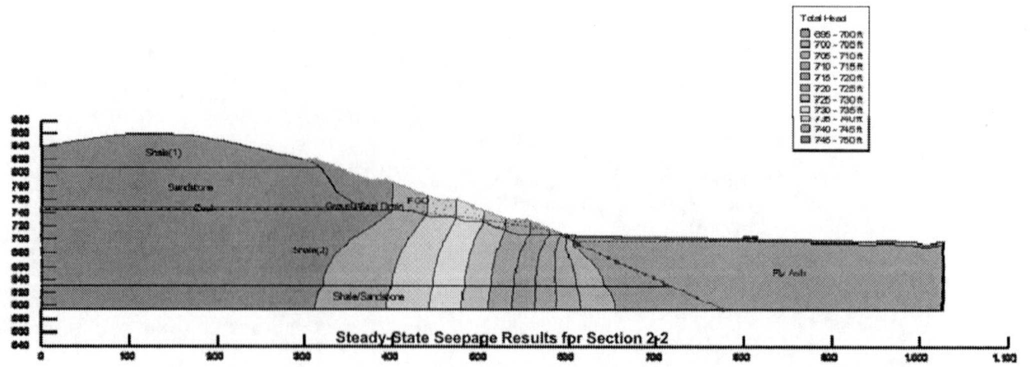


THE OHIO STATE UNIVERSITY

## Full-Scale Geotechnical Modeling: Stability Analyses

- Two stability analyses performed to determine factor of safety of proposed final slope of embankment
- Slope stability analysis
- Seismic pseudo-static analysis examined embankment stability in event of seismic activity

| | Water Elevation (ft) | Factor of Safety (Slope Stability Analysis) | Factor of Safety (Seismic Pseudo-Static Analysis) |
|---|---|---|---|
| Section 1-1 | 748 | 3.88 | 3.20 |
| Section 2-2 | 748 | 3.37 | 2.78 |
| Section 3-3 | 748 | 3.93 | 3.24 |
| Section 4-4 | 748 | 3.53 | 2.89 |

## Full-Scale Geotechnical Modeling: Slope Stability Analysis

- Determined factor of safety of proposed final slope of embankment

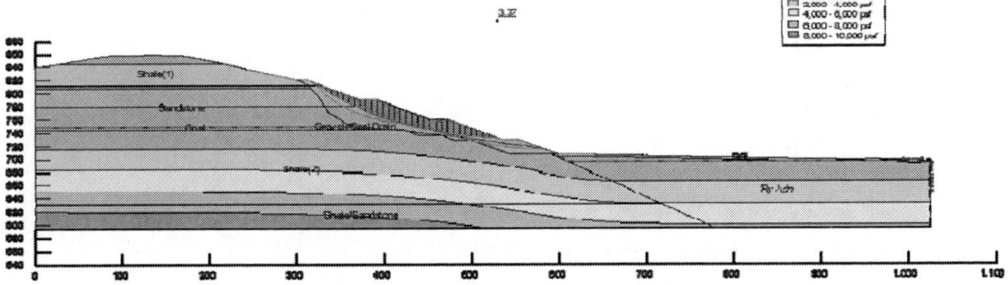


**Slope Stability Results for Section 2-2**

THE OHIO STATE UNIVERSITY

## Full-Scale Geotechnical Modeling: Deformation Analysis

- Evaluated to estimate long-term deformation of proposed embankment at highwall

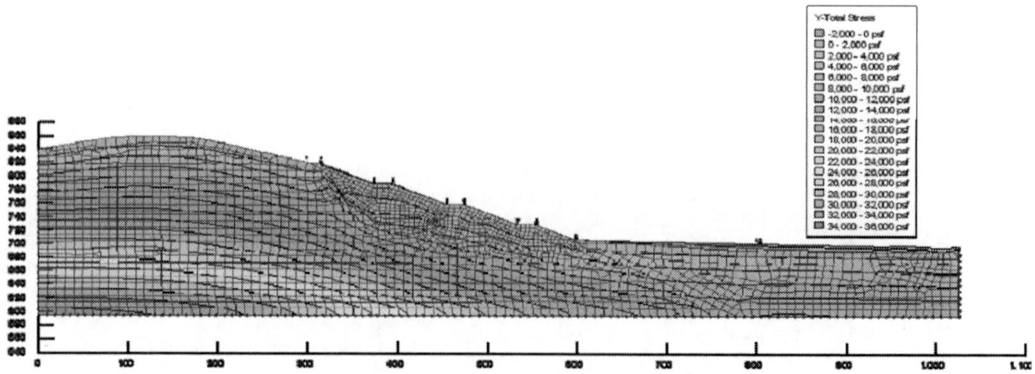


**Load/Deformation Results for Section 2-2**

THE OHIO STATE UNIVERSITY

## Environmental Implications

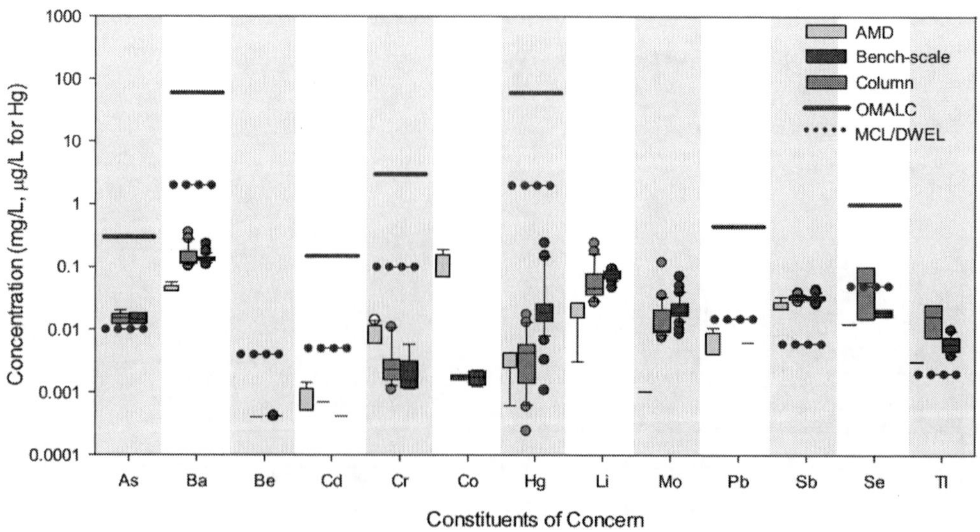


THE OHIO STATE UNIVERSITY

OMALC: Ohio Maximum Acceptable Leaching Concentration
MCL: Maximum Contaminant Level in EPA drinking water Std.
DWEL: Drinking Water Equivalent Level

# Constituents with Elevated Levels

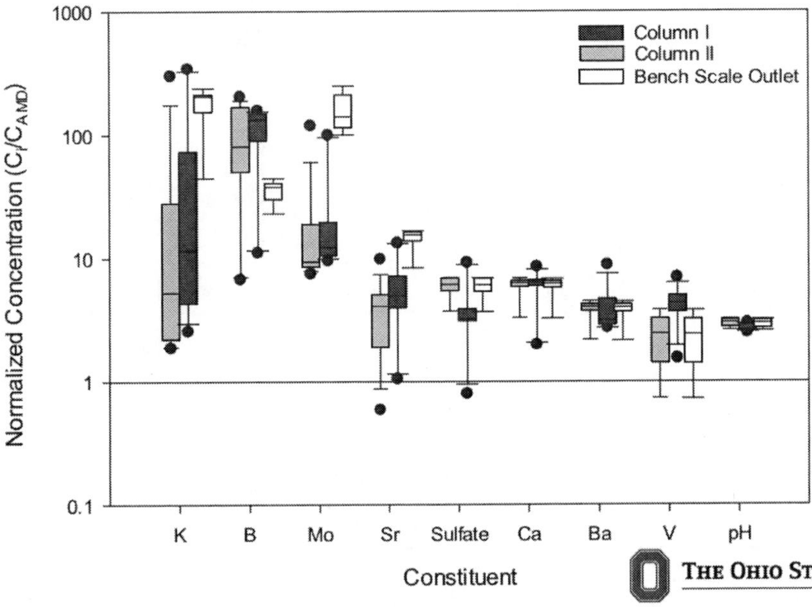


THE OHIO STATE UNIVERSITY

# Elements with Decreased Concentrations

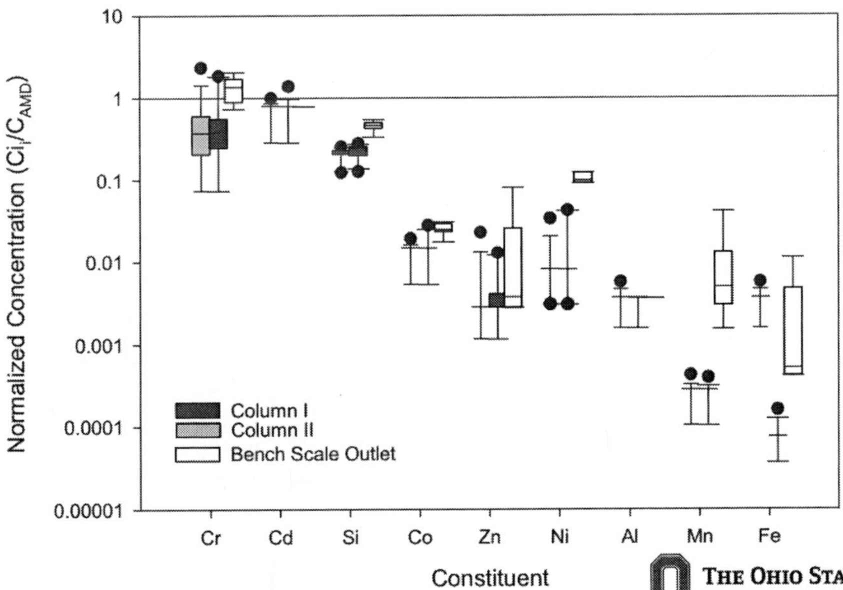


THE OHIO STATE UNIVERSITY

THE OHIO STATE UNIVERSITY

# Elements Showing First Flush Phenomenon

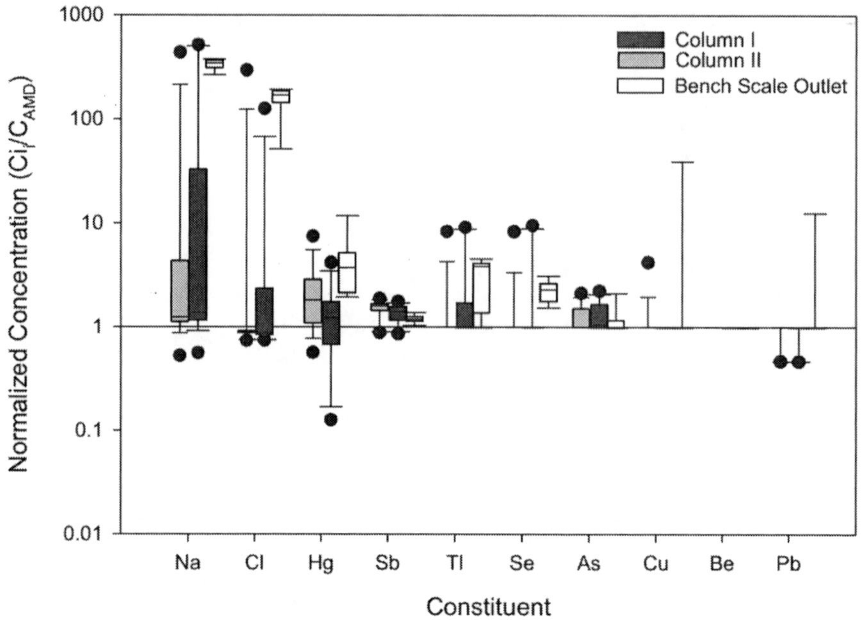


# Neutralization Capacity

- From available neutralization potential analysis

$$Treatment\ Capacity = \frac{available\ neutralization\ potential}{annual\ acid\ load}$$

- From laboratory tests

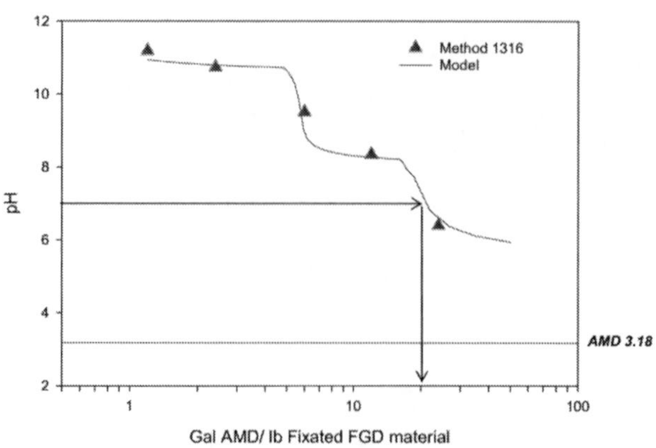


- Estimated under worst case scenario
  - Highest annual acid load
  - 5% of backfilled material in contact with AMD
- ~50 years of treatment capacity

THE OHIO STATE UNIVERSITY

THE OHIO STATE UNIVERSITY

# Geochemical Model

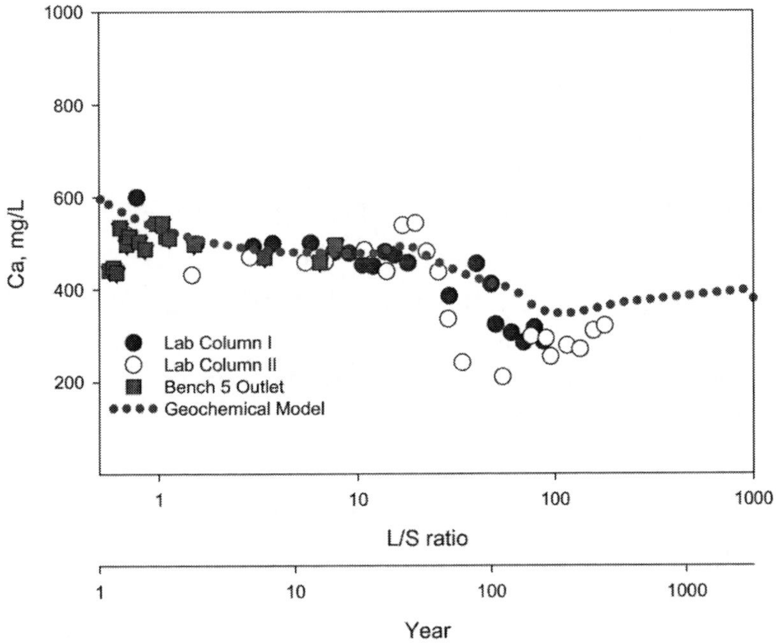


THE OHIO STATE UNIVERSITY

# Summary

- Mitigating AMD using fixated FGD material
  - Based on laboratory leaching study and geochemical model, one pound of Gavin fixated FGD material is able to neutralize approximately 20 gallons of AMD (~160 L/S)
- Bench-Scale modelling
  - Used to calibrate geotechnical models which were used for site design
- Change of AMD water quality
  - Constituents with elevated concentrations
    - pH, sulfate, Ca, K, B, Mo, Ba, Sr, and V
  - Elements with decreased concentrations
    - Fe, Mn, Zn, Cd, Co, Cr, Ni, Si, and Al
  - Elements with first flash phenomenon
    - Na, Cl, Hg, Sb, As, Cu, Se, Tl, Li, and Pb
  - Concentrations of constituents of interest
    - As, B, Pb, Hg, Mo, Se, and Tl exceeded MCL/DWEL during the early stage
    - Sb is constantly higher than MCL
    - All concentrations of COIs are lower than either Ohio Maximum Acceptable Leaching Concentration and/or EPA's Toxicity level

# Future Work

- Geochemical Modeling
  - ☐ establishing models for other major and trace constituents (including transportation and adsorption processes)

- Full-scale demonstration: construction & monitoring
  - ☐ Continuing water quality monitoring
  - ☐ Carrying out statistical and hydrogeochemical analyses of background water quality data
  - ☐ Permit applications submitted. Awaiting approval.
  - ☐ Project site preparation/construction
  - ☐ Backfilling of fixated FGD material to abate AMD
  - ☐ Post-reclamation water monitoring for at least 5 years after completion of construction

THE OHIO STATE UNIVERSITY

COAL COMBUSTION PRODUCTS PROGRAM

# Coal Combustion Products Program

Ohio State's Coal Combustion Products Program focuses on sustainable, high-volume beneficial uses of coal combustion products (CCPs), primarily from sulfur dioxide scrubbing processes, in construction, reclamation, infrastructure rehabilitation, manufacturing and agricultural applications. This program advances the beneficial uses of CCPs from sulfur dioxide scrubbing processes as well as more traditional byproducts, including fly ash, bottom ash, boiler slag and fluidized-bed combustor ash. Re-use of CCPs provides a low-cost raw construction material; extends the life of landfills, and lessens the need for new ones; and helps keep energy production costs in check.

THE OHIO STATE UNIVERSITY

Funded by the Ohio Coal Development Office, Ohio State University, Ohio coal-fired utilities, ash marketers, private businesses and trade and farming organizations, the Coal Combustion Products Program improves and discovers technically sound, environmentally friendly and commercially competitive uses of CCPs in many interdisciplinary sustainable applications.

The program aids the CCP industry through research, education, technology transfer and outreach in its efforts to:

- expand uses in proven areas, such as highway and agricultural applications;
- remove or reduce regulatory and perceptual barriers to use;
- develop new or under-used large-volume market applications, such as mine land reclamation; and
- place greater emphasis on sulfate and sulfite flue gas desulfurization byproducts utilization.

*More than 500 animal feeding pads in more than 12 Ohio counties are made from coal combustion products, including feeding pads at The Wilds in Muskingum County.*

THE OHIO STATE UNIVERSITY

Coal Combustion Products Program
470 Hitchcock Hall
2070 Neil Ave.
Columbus, OH 43210

Program Coordinator: Tarunjit Butalia
Email: butalia.1@osu.edu
Phone: 614-688-3408
To learn more, visit ccp.osu.edu.

2016 International Pittsburgh Coal Conference
Cape Town, SOUTH AFRICA
August 8 – 12, 2016
PROGRAM TOPIC: Sustainability and Environment

# ORBITRAP ANALYSIS OF FULLERENE ON SURFACE SOILS AND RIVER SEDIMENTS FROM SANTA CATARINA (BRAZIL) COAL AREA


*Laboratory of Environmental Researches and Nanotechnology Development - BRAZIL*

***Diego S. de Medeiros****, Marcos L.S. Oliveira, Silvio R. Taffarel,, Luis F.O. Silva, Leonardo K. Botelho*


# Aim of the study

- The fullerene occurrence in soils and sediments and the potential influence of incidental emissions from coal-combustion plants, coal mining and car exhaust have been assessed and characterised for the first time.
- Fullerenes were selected among those with major environmental occurrence such as C60 and C70 or with increasing industrial applications such as [6,6]-phenyl C61 butyric acid methyl ester, [6,6]-thienyl C61 butyric acid methyl ester, C60 pyrrolidine tris-acid ethyl ester and [6,6]-phenyl C71 butyric acid methyl ester fullerene or in the case of N-methylfulleropyrrolidine C60 because it is a common intermediate in the synthesis of other functionalized fullerenes.

# Introduction

**Definition**: A fullerene is a carbon molecule in the form of a hollow sphere, ellipsoid, tube and many other shapes. Spherical fullerenes are called Bucky balls and cylindrical ones are called carbon nanotubes. Fullerenes are similar in structure to graphite, which is composed of stacked graphene sheets of linked hexagonal rings

⇩

The presence of fullerenes in anthropogenic systems has been of big interest since their discovery

Because of their characteristic cage structure they can retain some hazardous elements. Consequently, the questions of the natural occurrence of fullerenes may have substantial environmental and human health implications

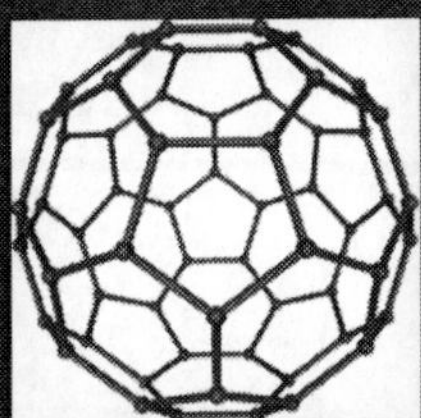

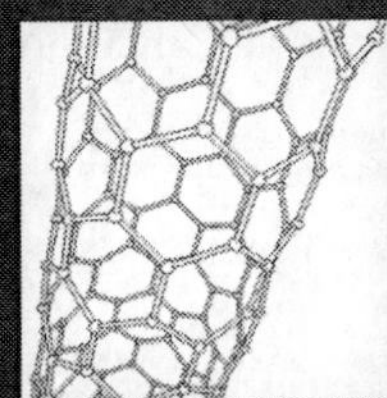

# *Study Area*

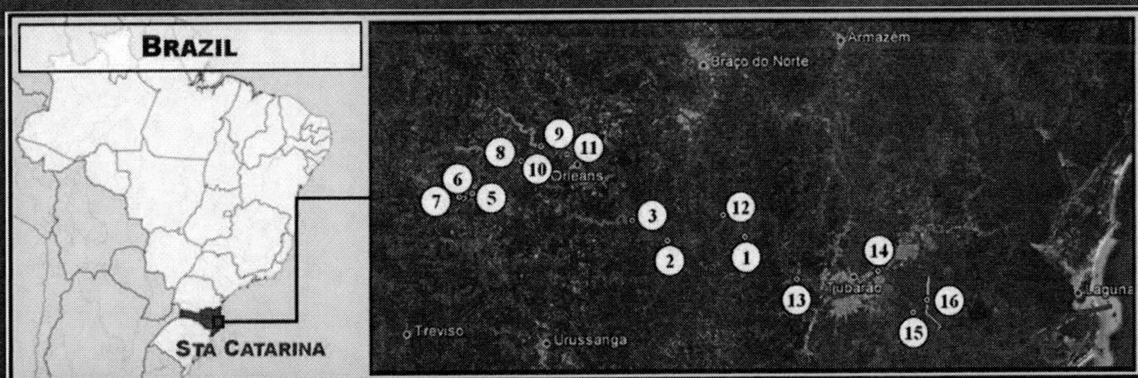


Tubarão River sediments sampling sites locations.

# Sampling collection

- 34 soils, codified from S1 to S34, were from an agricultural area nearby the thermoelectric plant "Complexo Termoelétrico Jorge Lacerda", which is the largest coal-fired thermoelectric power station in Latin America (Godoy et al., 2004).
- 12 soils, codified from S35 to S46, were urban soils collected at the coastal municipalities of Imaruí, Imbituba, Jaguarana and Treze de Maio
- At the collects sites, 15 sediment samples were collected in the Tubarão River and its tributaries.

| Code | Municipality | Latitude | Longitude | Power station distance (km) | Origin |
|---|---|---|---|---|---|
| S1 | Capivari de Baixo | −28.452° | −48.972° | 0.44 | Rice |
| S2 | Capivari de Baixo | −28.449° | −48.969° | 0.43 | Rice |
| S3 | Capivari de Baixo | −28.455° | −48.964° | 0.44 | Rice |
| S4 | Capivari de Baixo | −28.452° | −48.962° | 0.54 | Rice |
| S5 | Capivari de Baixo | −28.453° | −48.967° | 0.00 | Rice |
| S6 | Capivari de Baixo | −28.497° | −48.932° | 6.02 | Rice |
| S7 | Capivari de Baixo | −28.476° | −48.933° | 4.22 | Rice |
| S8 | Capivari de Baixo | −28.456° | −48.922° | 4.45 | Rice |
| S9 | Capivari de Baixo | −28.461° | −48.953° | 1.63 | Rice |
| S10 | Capivari de Baixo | −28.419° | −48.931° | 5.16 | Stevedore |
| S11 | Tubarão | −28.397° | −48.998° | 6.92 | Husbandry |
| S12 | Tubarão | −28.404° | −49.002° | 6.32 | Husbandry |
| S13 | Tubarão | −28.443° | −49.014° | 4.68 | Rice |
| S14 | Tubarão | −28.448° | −48.996° | 2.84 | Rice |
| S15 | Capivari de Baixo | −28.413° | −48.977° | 4.49 | Husbandry |
| S16 | Capivari de Baixo | −28.411° | −48.966° | 4.61 | Husbandry |
| S17 | Tubarão | −28.461° | −48.992° | 2.54 | Publicgym |
| S18 | Tubarão | −28.460° | −48.973° | 1.03 | Stadium |
| S19 | Tubarão | −28.480° | −48.953° | 3.34 | Rice |
| S20 | Laguna | −28.522° | −48.870° | 12.3 | Husbandry |
| S21 | Laguna | −28.534° | −48.871° | 13.1 | Rice |
| S22 | Tubarão | −28.543° | −48.917° | 11.2 | Rice |
| S23 | Tubarão | −28.539° | −48.909° | 11.1 | Husbandry |
| S24 | Tubarão | −28.528° | −48.972° | 8.37 | Rice |
| S25 | Tubarão | −28.501° | −48.951° | 5.56 | Husbandry |
| S26 | Tubarão | −28.486° | −49.021° | 6.40 | Stadium |
| S27 | Tubarão | −28.508° | −49.020° | 8.01 | Cemetery |
| S28 | Tubarão | −28.527° | −49.029° | 10.2 | Rice |
| S29 | Tubarão | −28.545° | −49.045° | 12.7 | Rice |
| S30 | Tubarão | −28.523° | −49.054° | 11.5 | Highway |
| S31 | Jaguaruna | −28.564° | −49.043° | 14.5 | Rice |
| S32 | Jaguaruna | −28.582° | −49.036° | 15.9 | Rice |
| S33 | Jaguaruna | −28.576° | −49.000° | 14.1 | Rice |
| S34 | Jaguaruna | −28.568° | −48.974° | 12.9 | Husbandry |
| S35 | Jaguaruna | −28.544° | −49.001° | 10.6 | Rice |
| S36 | Imarui | −28.208° | −48.738° | 35.3 | Urban |
| S37 | Imarui | −28.125° | −48.788° | 40.4 | Urban |
| S38 | Imarui | −28.097° | −48.774° | 43.9 | Urban |
| S39 | Imarui | −28.081° | −48.735° | 47.2 | Urban |
| S40 | Imbituba | −28.173° | −48.685° | 41.6 | Urban |
| S41 | Imbituba | −28.125° | −48.730° | 43.2 | Urban |
| S42 | Treze de Maio | −28.554° | −49.157° | 21.7 | Urban |
| S43 | Treze de Maio | −28.554° | −49.157° | 21.7 | Urban |
| S44 | Jaguaruna | −28.615° | −49.030° | 19.1 | Urban |
| S45 | Jaguaruna | −28.615° | −49.030° | 19.1 | Urban |
| S46 | Treze de Maio | −28.554° | −49.030° | 21.7 | Urban |

# Sampling Method

- Samples were taken in the upper 5 cm of soil using a metal shovel, wrapped in aluminium foil and transported to the laboratory with dry ice for their preservation.
- In the lab, samples were transferred to glass jars, preserved at −20 ° C and analysed within a week.
  - In order to avoid cross contamination or losses by adsorption, plastic materials were avoided during sampling, transportation and storage.

# *Analytical procedures*

- In this study, an analytical method based on ultrasound-assisted solvent extraction (UAE) with toluene followed by liquid chromatography by a pyrenylpropyl group bonded silica column coupled to high resolution mass spectrometry (HRMS) using an APPI (atmospheric pressure photoionization ionization) source in negative conditions (LC-APPI(−)-HRMS), was used.

# Results and discussion

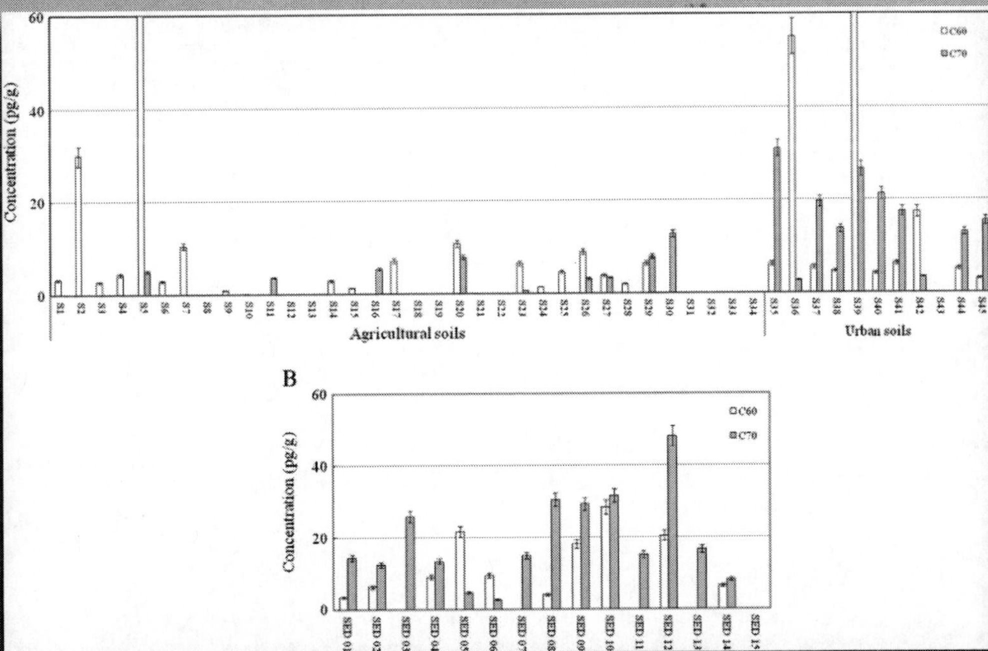


Concentrations of fullerenes (pg/g) in soils (A) and sediments from the Tubarão River (B).

# Results and discussion

- C60 was detected at higher concentrations than C70 fullerene in soils samples. However, the ratio between C60 and C70 was far from being constant. This fact is an indication of the multiple environmental sources of fullerenes.
- As in previous studies (Sanchís et al., 2012), in any of the samples, functionalised fullerenes (engineering-fullerenes) were not detected.
- The absence of functionalized fullerenes may be related to a low industrial production in comparison with accidental combustion sources
- In the river sediments, C70 was predominant and being the fullerene detected at the highest concentrations. C60 and C70 were present in almost all of the samples at concentrations ranging between the MLOD and 48 pg/g and 28 pg/g for C70 and C60, respectively

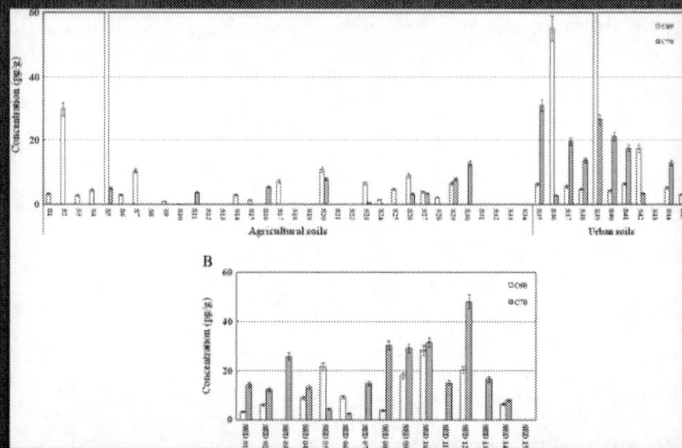


Concentrations of fullerenes (pg/g) in soils (A) and sediments from the Tubarão River (B).

# Conclusion

- The good performance of a pyrenylpropyl group bonded silica column using 100% toluene as mobile phase for the fullerene separation was proved.
- The results using an Atmospheric pressure photoionisation source in negative conditions have shown to be less affected by matrix effects, especially in complex matrices. Hence, in comparison with the performance in previous method based on ESI in negative conditions, an improved sensitivity for the determination of selected fullerene in soils and sediments was achieved. MLOQ down to low 0.1 pg/g level were achieved.
- The accuracy on mass measurements by the use of hybrid quadrupole-Orbitrap mass spectrometer (Q-Exactive Orbitrap, USA) has provided the ultimate confidence for the identification of fullerenes in complex environmental samples. In comparison to previous methods, this new approach in combining the use of a Buckyprep column with toluene in combination with APPI in negative conditions and accurate mass measurements, presented better retention and separation, better sensitivity, less matrix effects and adduct formation in APPI is virtually nonexistent. This approach, has presented MLOQ ranging from 0.062 to 1.3 pg/g in soils and sediments.
- The new analytical procedure was applied to investigation of the occurrence of seven fullerenes in soils and sediments from Brazil. The potential influence of a coal-fired power station, urban contamination from car exhausts and wastewater effluents and coal mining were evaluated as potential sources of contamination. In all types of samples, C60 and C70 fullerenes were detected. Fullerene levels were from a low fg/g up to 154 pg/g. These results contribute to enlarging the currently limited dataset on environmental occurrence of fullerenes.

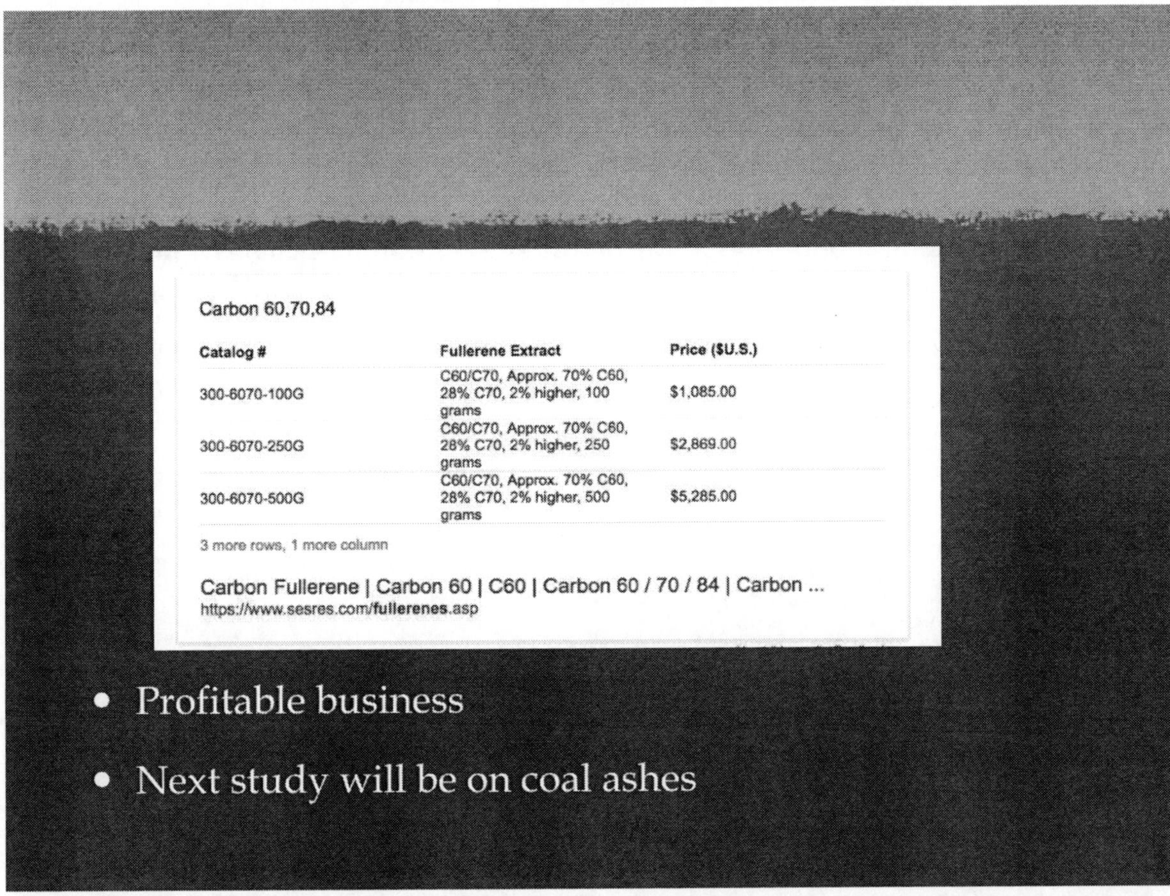
Carbon 60,70,84
Catalog #
Fullerene Extract
Price ($U.S.)
300-6070-100G
C60/C70, Approx. 70% C60, 28% C70, 2% higher, 100 grams
$1,085.00
300-6070-250G
C60/C70, Approx. 70% C60, 28% C70, 2% higher, 250 grams
$2,869.00
300-6070-500G
C60/C70, Approx. 70% C60, 28% C70, 2% higher, 500 grams
$5,285.00
3 more rows, 1 more column
Carbon Fullerene | Carbon 60 | C60 | Carbon 60 / 70 / 84 | Carbon ...
https://www.sesres.com/fullerenes.asp
• Profitable business
• Next study will be on coal ashes

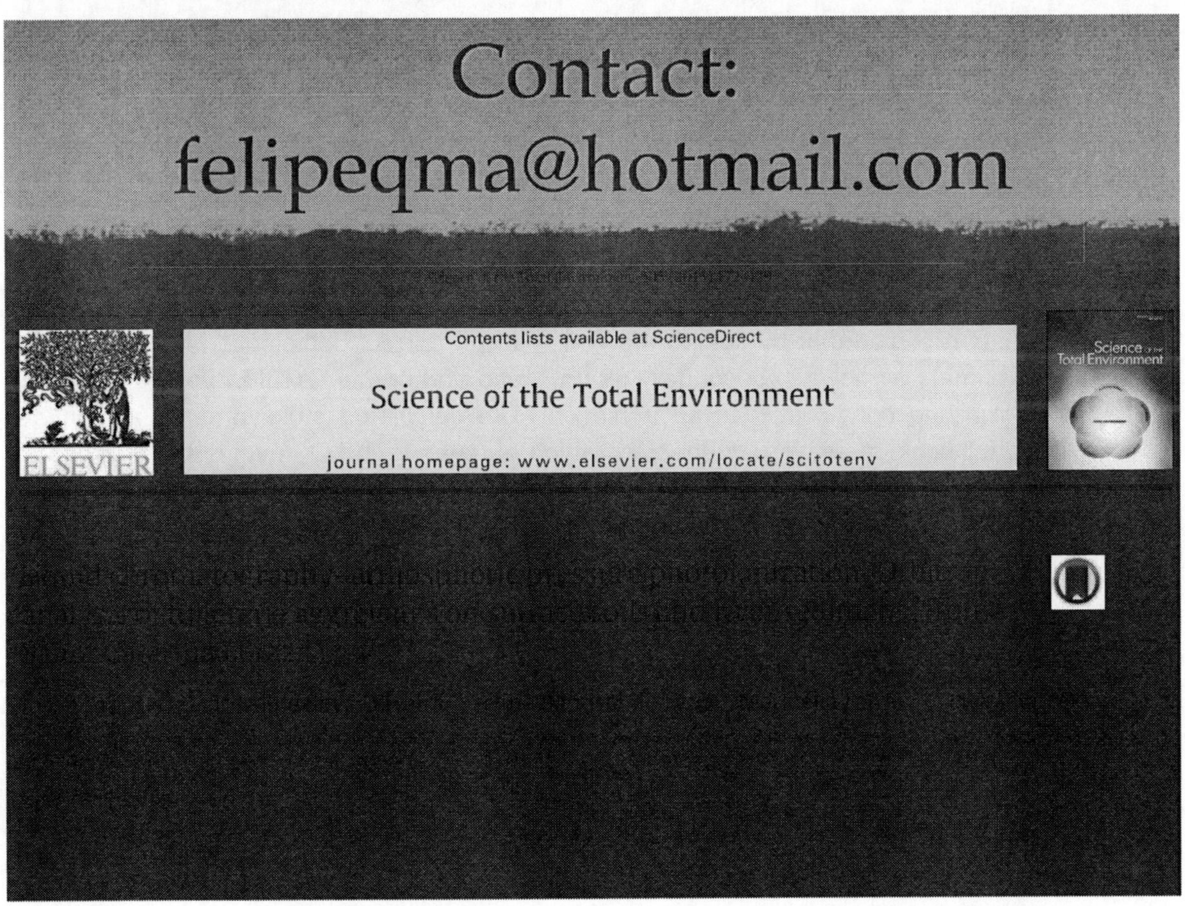
Contact:
felipeqma@hotmail.com
Contents lists available at ScienceDirect
Science of the Total Environment
journal homepage: www.elsevier.com/locate/scitotenv
ELSEVIER
Science of the Total Environment

Second Thirty-Third Annual International Pittsburgh Coal Conference

**Cape Town International Convention Center, Cape Town, SOUTH AFRICA**

**August 8 – 12, 2016**

# Fireside corrosion monitoring, control and prevention in a 600MW$_e$ pulverized fuel boiler

Otilia Mwape[1], Prof. Walter Schmitz[2], Manny de Sousa[3] and Bonny Nyangwa[4]

Combustion Engineering

University of the Witwatersrand

[1] Eskom; Duvha Power Station; mwapevo@eskom.co.za

[2] University of the Witwatersrand; Mechanical Engineering; Walter.Schmitz@wits.ac.za

[3] Eskom; Duvha Power Station; DSousaMJ@eskom.co.za

[4] Eskom; RT&D (Research, Test and Development); NyangwTB@eskom.co.za

## Abstract

Fireside corrosion is the corrosion of tubes due to chemical attack occurring on the radiative section of the boiler. This attack mechanism is one of the problems affecting the efficiency and availability of Eskom (Electricity Supply Commission) coal fired steam generators.

Previous research studies conducted by Research Testing and Development concluded that $H_2S$ (hydrogen sulphide) is the predominant cause of corrosion of the tubes which is prevalent under reducing conditions. The main aim of this research study was to determine the $O_2$ (oxygen) breakpoint when $H_2S$ starts to form under reducing conditions. Initial tests were conducted on one of Eskom's base load power station where flue gas was extracted from the furnace at different excess air levels by means of a water cooled pyro (pyrometer) and using a portable flue gas analyzer to measure all gas species.

Full scale tests revealed that by setting the boiler at the correct optimisation of combustion controls i.e. primary air flows to secondary air flow ratio and primary air flow versus coal flow load lines, oxidizing conditions can be achieved. The calculated stoichiometric air for the boiler where full scale tests were conducted, which equates to the $O_2$ level in the flue gas at the boiler outlet, is 3.3%. During the test the $O_2$ levels were reduced to 2.5% and $H_2S$ was not detected. Due to current limits prescribed by the Fossil Fuel Firing Regulations (FFFR) the boiler could not be reduced beyond 2.5% $O_2$ limit, As a result further tests were carried out at the Eskom PSCTF (pilot scale combustion test facility). Coal utilised at the power station where full scale tests were conducted was combusted and by using similar test method utilised during the full scale tests it was observed that at 0.5% $O_2$ furnace exit conditions, at the sampling port closest to the burner, $H_2S$ was detected and deemed to be $H_2S$ break-point.

**Key words:** Fireside corrosion, sulph'idation, combustion, excess air, oxidizing and reducing conditions.

## 1. Introduction

Fireside corrosion is the thinning of tubes due to chemical attack occurring on the furnace or fireside of heat exchanging surfaces in a fossil fuel-fired furnace [1]. This attack mechanism is one of the problems affecting the efficiency and availability of Eskom coal fired steam generators commonly known as boilers. In server cases fireside corrosion on boiler tube has resulted in unplanned electrical power grid separations and unplanned outage time extensions. However, fireside corrosion is not effectively monitored, controlled and prevented in industrial boilers.

In Eskom the replacement of water-wall tubes below allowable minimum operating wall thickness, during outages, has since been used as the mitigating measure to deal with this phenomenon. The latter mitigating measure has proven to be effective, however, due to the increasing material wastage of water-wall tubes; the practise of tube replacement has become extensive and very costly.

### 1.1. Background

Boiler combustion is the chemical reaction of a fuel and an oxidant for the production of heat and the conversion of chemical species. In conventional power plants the heat is used to produce superheated steam. This steam is produced in the boiler and utilizes to operate steam turbines and auxiliary plant for the generation of electricity.

In the process of combustion boilers are fired with excess air in order to achieve complete combustion. Under ideal combustion conditions, oxidizing conditions are established in the flue gas corresponding to the excess air levels [2]. Oxidising conditions yield an environment that is conducive for favourable boiler combustion conditions, while reducing conditions promote the formation of $H_2S$ (hydrogen sulphide) which causes sulph'idation corrosion on boiler walls [3].

Low alloy or carbon steel waterwall tubes are protected from rapid wastage by a formation of protective iron oxide layer of $Fe_3O_4$. This oxide layer adheres to the tube and is not easily removed by sootblowing or by mechanical or thermal cycling. Mixed oxide/sulphide corrosion scales are produced, which are dense, adherent and protective. The presence of $SO_2$ in the flue gas may increase the porosity and decrease the strength of the scale slightly, but with no major effect on the tube service life. [4]

During combustion most of the carbon burned reacts to form $CO_2$, however small traces of the carbon stay in the intermediary stage of CO. Under sub-stoichiometric or reducing conditions high levels of CO are formed and the sulphur in the coal is partially converted to $H_2S$ at the expense of $SO_2$ as given in the following equations: [4]

$$FeS_2 + H_2 = FeS + H_2S \quad (1)$$
$$S_{(org)} + H_2 = H_2S \quad (2)$$

This $H_2S$ preferentially reacts with the carbon or low-alloy steel to produce iron sulphide (FeS). [4]

$$Fe + H_2S = FeS + H_2 \quad (3)$$

The already formed protective layer ($Fe_3O_4$) on the boiler tube may also be transformed to FeS under the following equation: [4]

$$Fe_3O_4 + 3H_2S = 3FeS + 3H_2O + \frac{1}{2} O_2 \quad (4)$$

Depending on the relative amounts of CO, $CO_2$, $SO_2$, $H_2S$ and other parameters, the scale on the carbon or low-alloy steel may consist of a mixture of magnetite ($Fe_3O_4$) and FeS or, sometimes nearly pure FeS. The strength and adherence of the scale decreases with the amount of FeS present and the permeability and growth rate increases significantly. This results in a significantly increased rate of boiler tube corrosion as shown in Figure 1. [4]

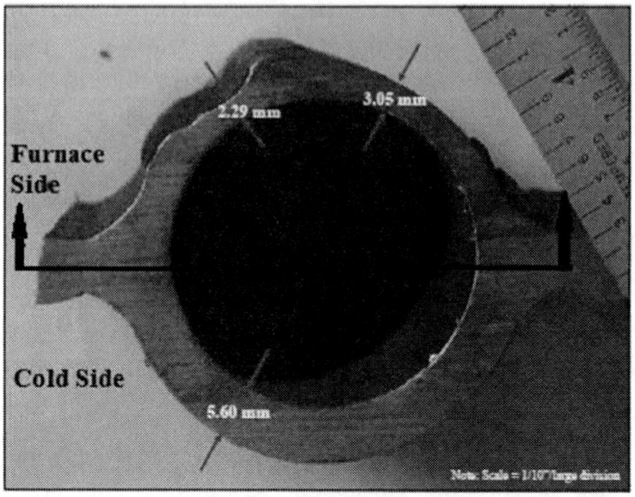


**Figure 1: A cross section of a fireside corroded waterwall tube taken from one of Eskom's PF boilers**
**Source: Dooley B, Water Wall Tube Investigation- 2010**

Fireside corrosion research programs in Eskom were initiated in 1994 due to metal wastage caused by fireside corrosion being identified in certain Eskom boilers [5]. Six of Eskom coal fired power stations were confirmed to be affected by this attack mechanism. Horst Schenk (2003) documented findings from his research that fireside corrosion on Eskom's boilers was limited to furnace water-wall tubes and that the attack mechanism was due to sulph'idation caused by reducing conditions found in the affected areas [5][6]. Subsequent investigation confirmed Schenk's findings.

## 2. Methodology

Experimental tests were initially conducted at MCR (maximum capacity rating) in one of Eskom's 600$MW_e$ (electrical) boiler. It was observed that at a furnace exit $O_2$ concentration of 2.5%, which is a value lower than the 3% recommended value by the Eskom FFFR (Fossil Fuel Fired Regulation), oxidising conditions were still maintained. Reducing the furnace exit $O_2$ concentration below 2.5% would mean operating the plant in an unsafe manner and thus violating regulations. Therefore, for the purpose of concluding the study, Eskom PSCTF (pilot scale combustion test facility) was operated utilizing coal from the Eskom power station where full scale tests were conducted. The sections below details the equipment used for the tests, coal quality of the fuel used and test conditions.

### 2.1. Experimental method

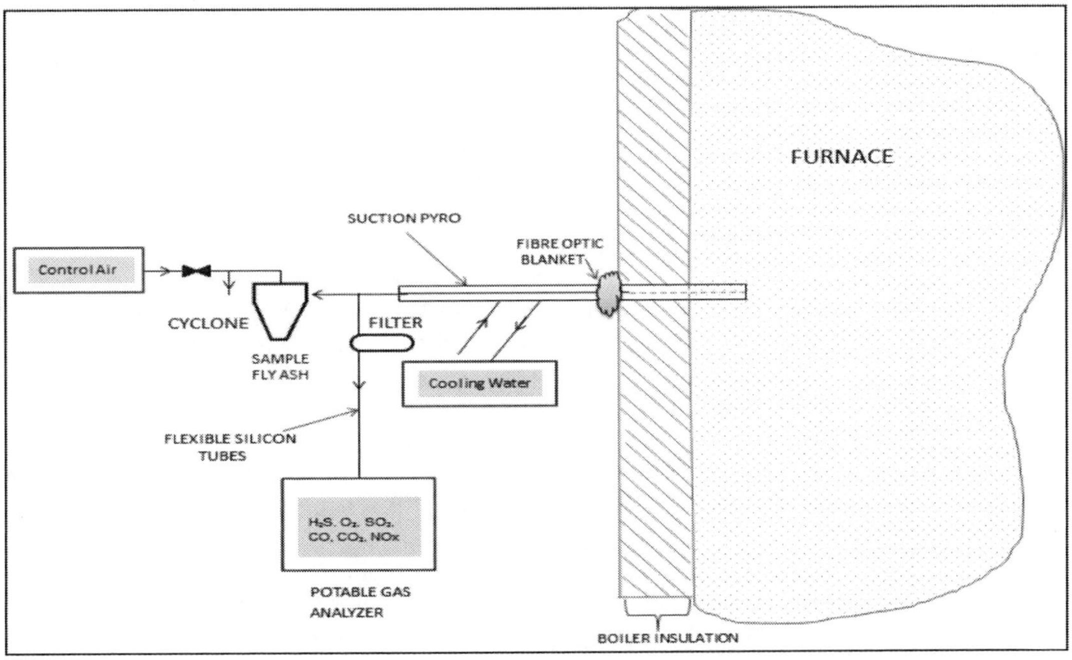


**Figure 2**

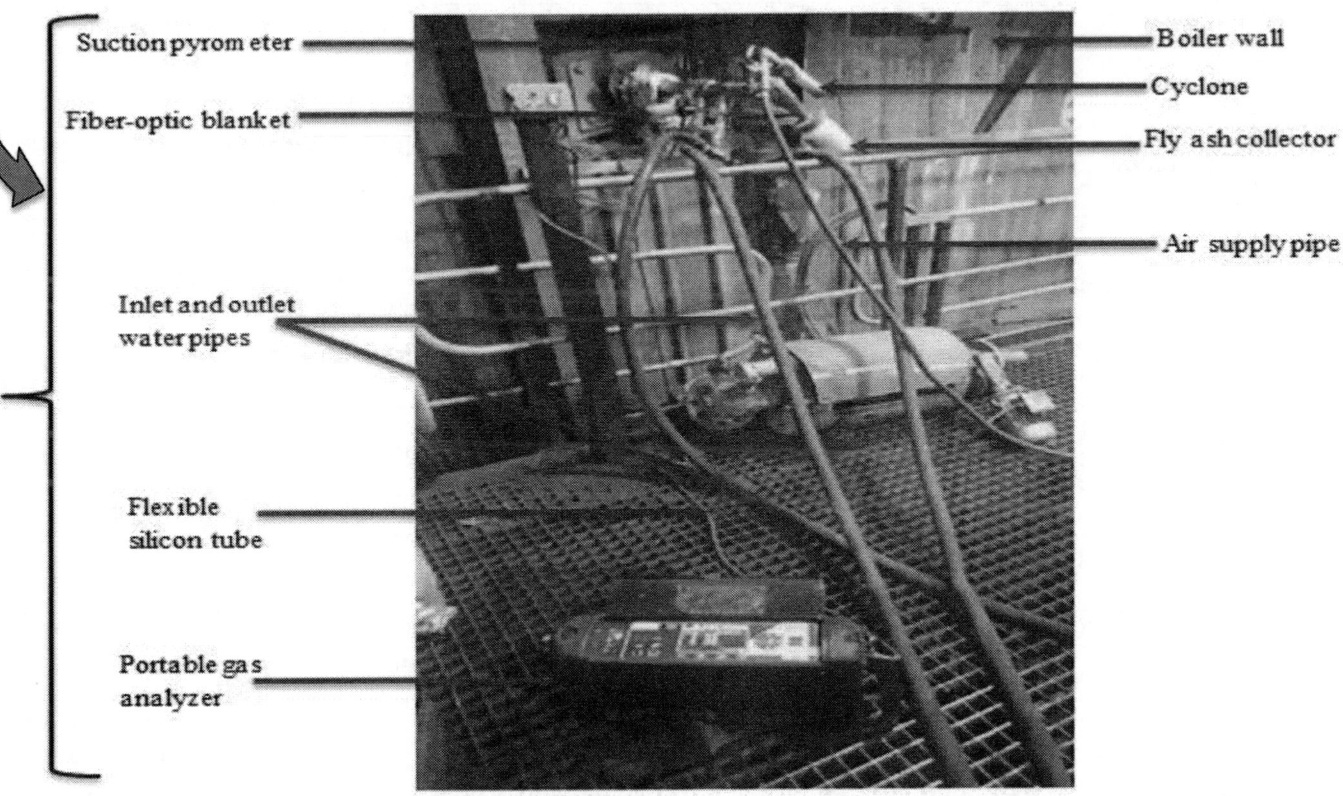


**Figure 3**

**Figure 2 & 3 show the schematic and picture of the assembled equipment used to sample the flue gas in the furnace**

### 2.2. Experimental equipment and material

<table>
<tr><th>Equipment</th><th>Description</th></tr>
<tr><td>Suction pyrometer probe</td><td>1.5 meter IFRF (International Flame Research Foundation) version probe made from 316 low carbon stainless steel. The hot flue gas in the furnace is sucked from an opening in the front end of the probe situated in the envelop pocket inside the cooling water jacket frame.</td></tr>
<tr><td rowspan="2">Cooling water pipes</td><td>Cooling water inlet hose (standard fire hose size – pressure resistant) is connected at the station fire hydrant supply manifold and the other end is connected at the water inlet side of the suction pyro probe. The water is used to cool the suction pyro probe. The cooling water first circulated in the internal tube and then to the external tube.</td></tr>
<tr><td>Cooling water outlet hose (standard fire hose size – pressure resistant) is connected from the water outlet side of the suction pyro probe and the other end to the closest station water drain.</td></tr>
<tr><td>Shut-off valve</td><td>Attached to the outlet of the suction pyro probe and is used to control the cooling water flow rate to maintain sufficient cooling water flow. This valve also allows for the filling of cooling water in the entire external tube before inserting the probe in the furnace to prevent the build-up of bubbles and over-heating.</td></tr>
<tr><td>Air supply pipe</td><td>The one end of the suction hose is connected to the nearest station compressed air supply manifold and the other end is connected to the air ejector via a rotameter.</td></tr>
<tr><td>Rotameter</td><td>The rotameter is connected between the supply air hose and the air ejector and is used for supply air flow indication.</td></tr>
<tr><td>Air ejector</td><td>The ejector uses expansion of compressed air to produce the proper suction rate. The valve on the ejector is opened until the required flue gas suction is achieved. The suction flow rate is dependent upon the slagging probability of the coal.</td></tr>
<tr><td>Fly ash collector</td><td>Fly ash sample bottle is crewed onto the cyclone for collection of fly ash samples extracted along with the flue gas from the furnace.</td></tr>
<tr><td>Fuel filters</td><td>Installed on the flue gas line to trap fly ash particles before entering the gas analyser.</td></tr>
<tr><td>Flexible silicon tube</td><td>2 meter silicon pipe (OD 8mm, ID 6mm) used as a medium through which the sampled flue gas is pump from the probe to the gas analyser.</td></tr>
<tr><td>Optic-fibre blanket</td><td>Certified flame proof material used to cover the furnace access port and prevent air ingress to prevent contamination of the extracted flue gas sample.</td></tr>
<tr><td>Portable gas analyser</td><td>A Lancom IV portable gas analyser was used to measure the flue gas species. The analyser was checked and calibrated on site prior to being utilized for testing. The analyser has a built in vacuum pump. The following gases are measured:
<table>
<tr><th>Gas</th><th>Sensor/measurement</th><th>Range</th><th>Accuracy</th><th>Resolution</th></tr>
<tr><td>$O_2$</td><td>Electrochemical cell</td><td>0-25% vol.</td><td>±1%</td><td>±1% Vol.</td></tr>
</table>
</td></tr>
</table>

| | | | | | |
|---|---|---|---|---|---|
| | CO | Electrochemical cell | 0-5000ppm | ±2% | ±0.1ppm |
| | $CO_2$ | Concentration is calculated | | | |
| | $SO_2$ | Electrochemical cell | 0-2000ppm | ±2% | ±0.1ppm |
| | $H_2S$ | Electrochemical cell | 0-200ppm | ±2% | ±0.1ppm |
| | $NO_2$ | Electrochemical cell | 0-500ppm | ±2% | ±0.1ppm |
| | NO | Electrochemical cell | 0-4000ppm | ±2% | ±0.1ppm |

**Table 1: Summary of the equipment and material used for the test**

### 2.3. Experimental fuel

| Source table | Source field | Quantity name | Unit | Power station A | Power station B |
|---|---|---|---|---|---|
| Coal Proximate | Inherent moisture | Inherent moisture | Mass % | 2.5 | 7.2 |
| Coal Proximate | Ash | Ash | Mass % | 28.1 | 36.5 |
| Coal Proximate | Volatile matter | Volatile matter | Mass % | 20.6 | 19.2 |
| Coal Proximate | Fixed carbon | Fixed carbon | Mass % | 48.8 | 37.1 |
| Fuel Ratio | Fixed carbon/volatile matter | FC/VM | Ratio | 2.37 | 1.93 |
| Coal moisture | Total moisture | Total moisture | Mass % | 6.7 | 11.2 |
| Coal gross CV | Grass CV | Gross calorific value | MJ/kg | 21.76 | 16.15 |
| | | | | | |
| Coal ultimate | Carbon | Carbon | Mass % | 58.82 | 44.51 |
| Coal ultimate | Hydrogen | Hydrogen | Mass % | 2.81 | 2.23 |
| Coal ultimate | Nitrogen | Nitrogen | Mass % | 1.28 | 1.14 |
| Coal ultimate | Total sulphur | Total sulphur | Mass % | 0.69 | 0.62 |
| Coal ultimate | Carbonate (as $CO_2$) | Carbonate (as $CO_2$) | Mass % | 2.85 | 2.26 |

| | | | | | |
|---|---|---|---|---|---|
| Coal ultimate | Oxygen (by difference) | Oxygen (by difference) | Mass % | 2.95 | 5.54 |
| Coal ultimate | Hydrogen / Carbon | Hydrogen / Carbon | Ratio | 0.0478 | 0.0501 |

**Table 2: Properties of the coal utilised for full and pilot scale tests**

### 2.4. Experimental conditions

| Full scale operating parameters | | |
|---|---|---|
| Generated load | | 570 $MW_e$ |
| Total fuel (coal) flow | | 297.86 T/hr |
| Total air flow | | 689.90 kg/s |
| Total secondary air flow | | 486.1 kg/s |
| Total primary air flow | | 146.8 kg/s |
| Total combustion air | | 632.84 kg/s |
| Economiser inlet water flow | | 458.96 kg/s |
| Total steam flow | | 488.35 kg/s |
| Economiser inlet temp | | 242.43°C |
| Economiser outlet temp | | 302.4°C |
| Boiler excess air ($O_2$) | Left hand side | 3.89% |
| | Right hand side | 3.51% |
| Condenser pressure | Cold | 8.8 kPa abs |
| | Hot | 13.2 kPa abs |

**Table 3: Full scale operating parameters**

| PSCTF Burner set-up | |
|---|---|
| Burner load | 1 $MW_{th}$ |
| Burner excess air | 20% |
| Primary air in theoretical air | 0.3 |
| Swirl number | 0 |
| Secondary air ratio | 100% |
| Tertiary air | 0 |

**Table 4: PSCTF set-up for tests**

### 2.5. Full Scale Tests

The boiler used for the trials is of a once through tower design which is split into two sections by a membrane division wall with an opening of approximately 1.4m at the front and rear wall connection. A 3-D model of the boiler is shown below in Figure 4.

Both the front and rear wall have three planes of four lighting-up oil burners. Each burner is fitted with a centrally mounted ignition gun. Full load is achieved with five mills in operation, leaving one mill on stand-by at all times. The burners are located within a common windbox that delivers secondary air heated in the secondary air part of the air heater. The pf (pulverised fuel) is transported from the mills to the boiler by means of primary air. The primary air is also pre-heated to dry the coal during the pulverising process. This air is heated in the primary air part of the air heater and brought to a high pressure by two primary air fans so as to overcome pressure losses in the duct work, air heater, mills, coal supply pipes and burners.

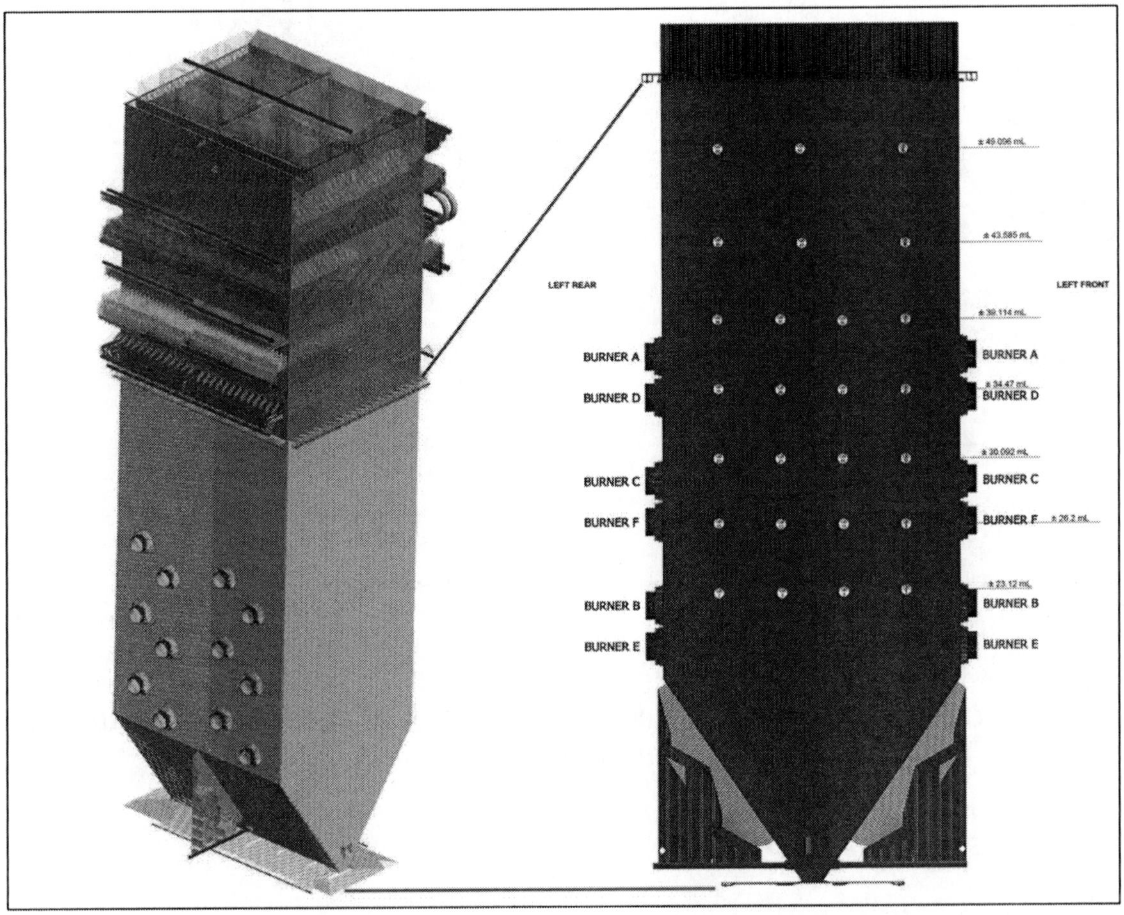


**Figure 4: 3-D model of the boiler where full scale test were conducted with the side view of the boiler furnace projected on the left side**

A corrosion profile as shown in Figure 5 of the boiler, where experimental full scale tests were conducted, was established based on the history of waterwall tube panels replaced due to fireside corrosion during the January 2011 and the July 2014 outages. The boiler corrosion map is indicative that fireside corrosion is experienced in the furnace and is more prominent around the burner regions.

Flue gas samples were extracted by means of a water cooled 1,5m long suction pyrometer from two different locations (Test Point (TP) 1 and TP2) as indicated in Figure 5 at the maximum electrical load achievable. TP1 is within the burner belt region where fireside corrosion is more prominent whereas TP2 is above the burner belt and no waterwall tube thinning have not shown exposure to fireside corrosion during the outage intervals.

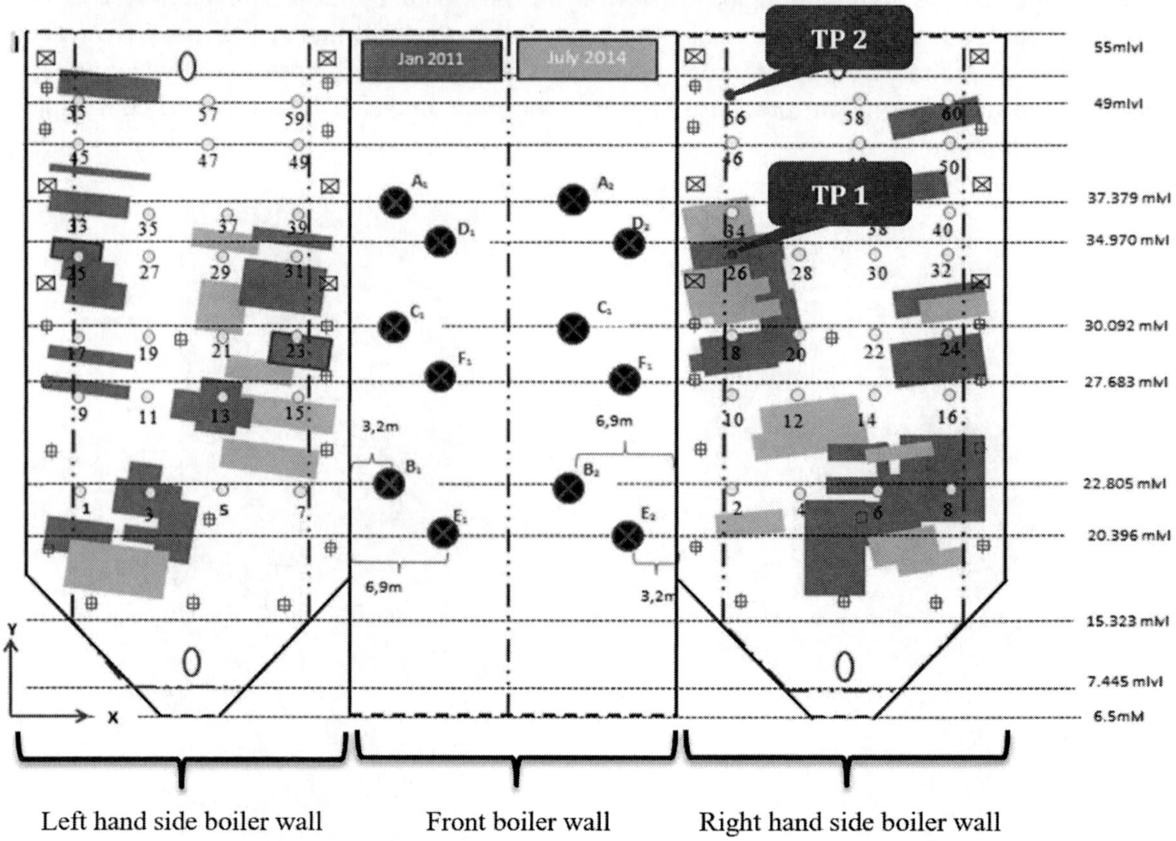


**Figure 5: Corrosion profile of the full scale boiler derived from the January 2011 and July 2014 outages**

## 2.5. Experimental results and discussion

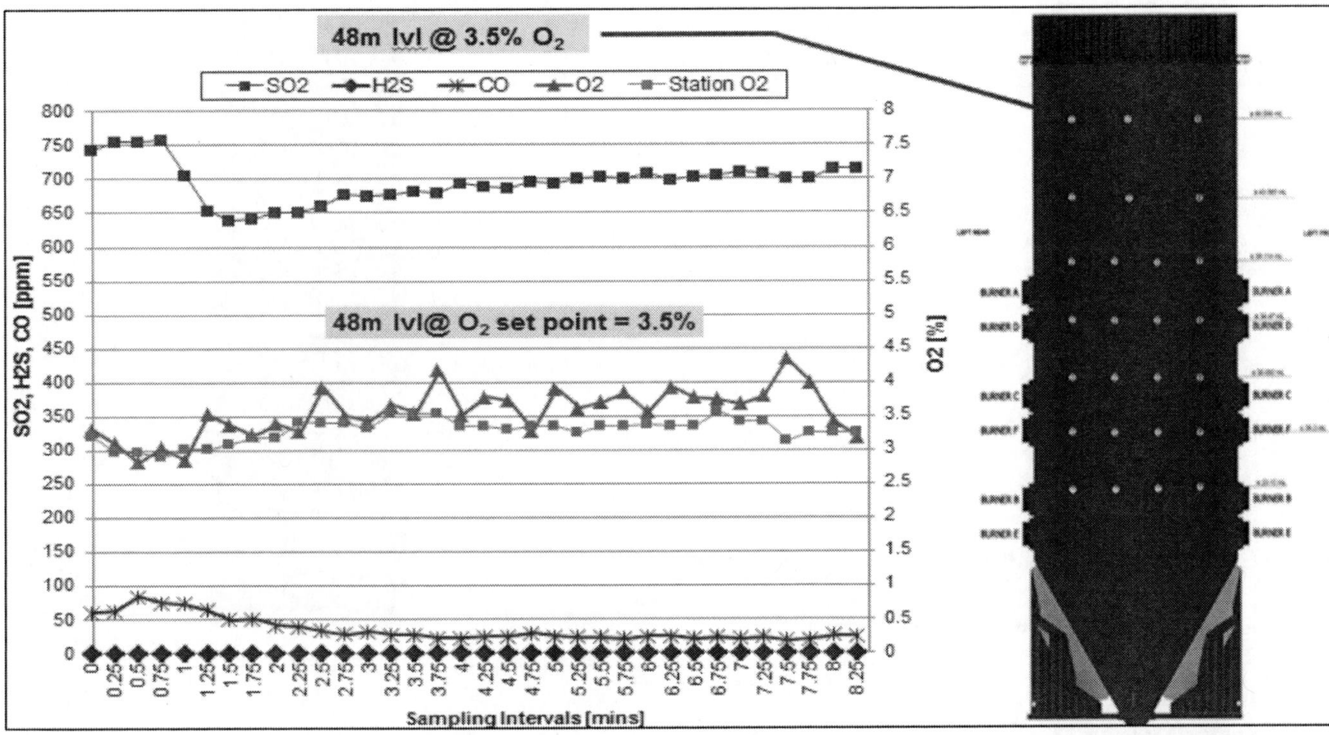


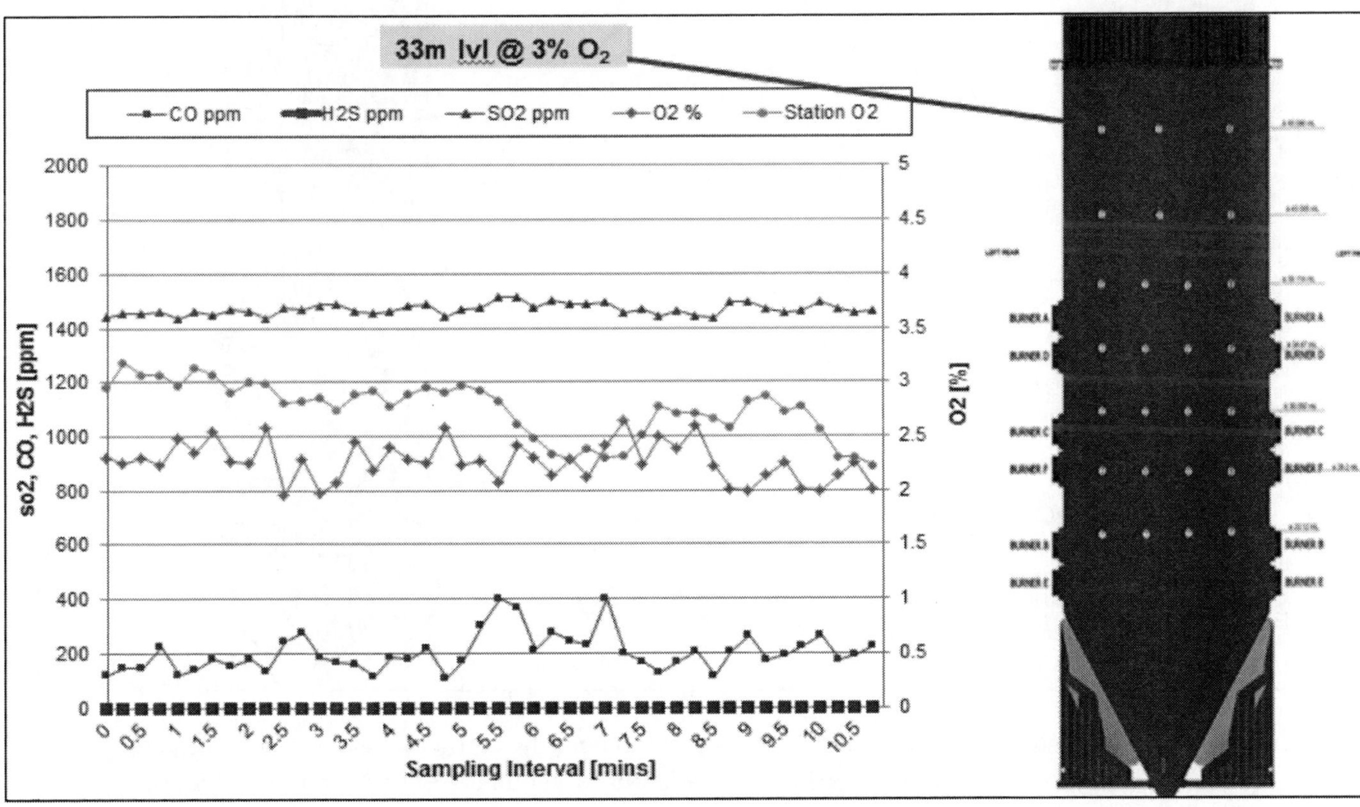

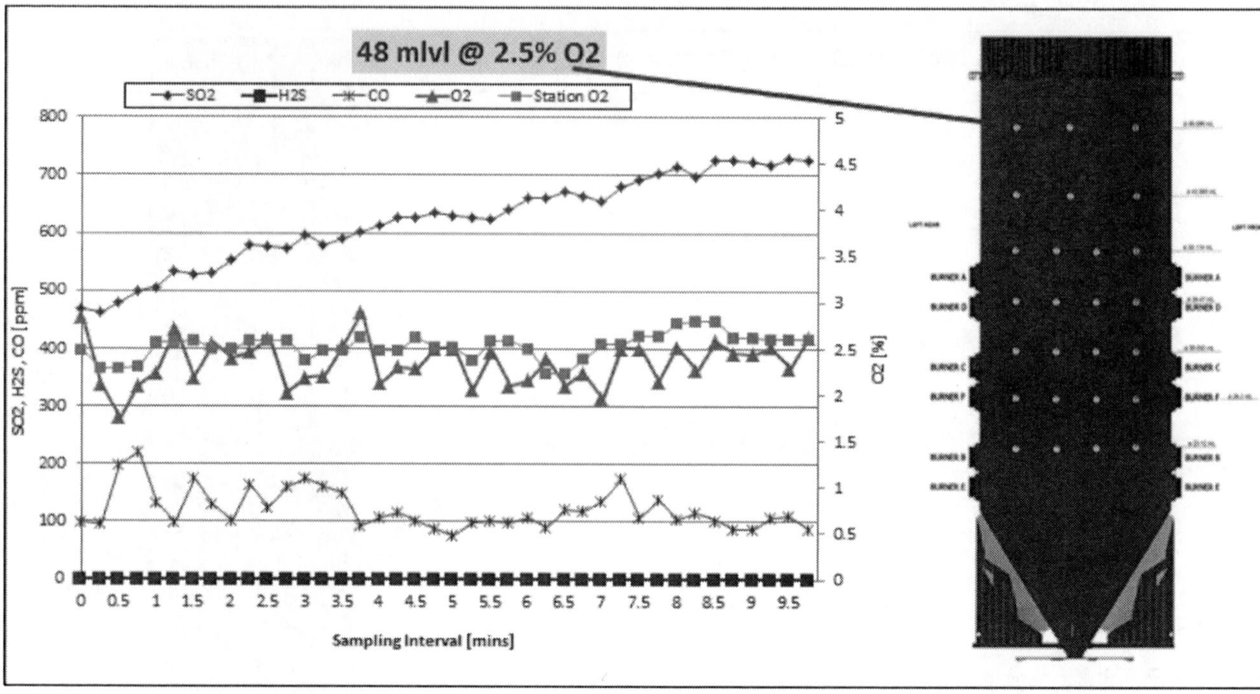


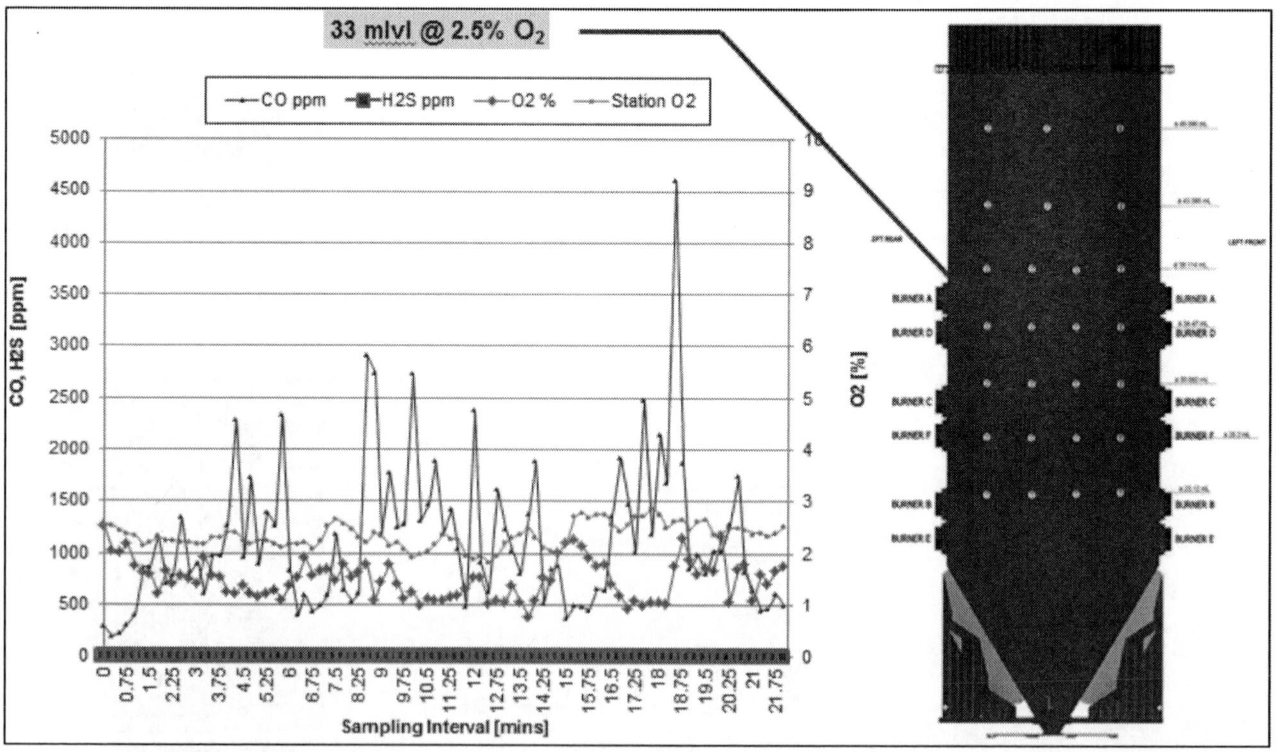


The station $O_2$ in-situ analyser was utilised as a reference, the furnace exit $O_2$ concentration measured using a calibrated Lancom 4 was reduced to simulate reducing atmospheres in and attempt to create corrosive gas species in the furnace. It was observed that at a furnace exit $O_2$ concentration of 2.5%, which is a value lower than the minimum 3% recommended value by the Eskom Fossil Fuel Fired Regulation (FFFR) for the specific boiler, oxidising conditions were still maintained and $H_2S$ did not form. Carbon monoxide at this oxygen concentration spiked ranged from 500-4500ppm. Reducing the furnace exit $O_2$ concentration below 2.5% would mean operating the plant in an unsafe manner and violation the Eskom FFFR.

$$\text{Total air required } (M_{TA}) = Air_{stoich} \times 1.18 \text{ (kg/s)} \quad (5)$$

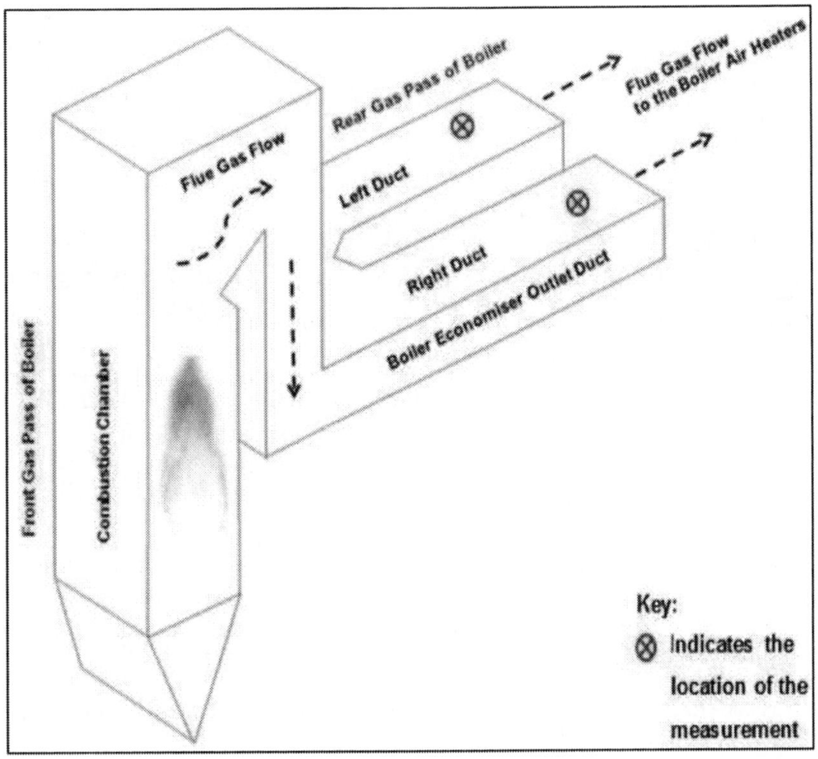


**Figure 6: Boiler indicating the location where station $O_2$ is measured [7]**

### 2.6. Pilot scale tests

Further experimental tests, aimed at concluding this research study, were conducted at the Eskom PSCTF (pilot scale combustion test facility). The Eskom PSCTF is a $1MW_{th}$ (thermal) plant designed with numerous features to aid in the evaluation of fuels used in full scale and industrial boilers. The combustion chamber and the convective section are water cooled and lined with refectory material. The refectory material properties have been chosen to replicate as closely as possible the temperature field of a boiler furnace. The PSCFT has been designed using "residence time scaling". The concept states that a fuel particle entering the PSCTF will experience the same history of gas atmospheres, temperature and particle residence times similar to those of the full scale plant. The furnace is equipped with a single down fired variable swirl burner. Diesel is used as the light-up fuel prior to firing with the coal to be tested. The primary and secondary air is heated in an electrical air heater to temperature values that replicate plant operating conditions. [8]

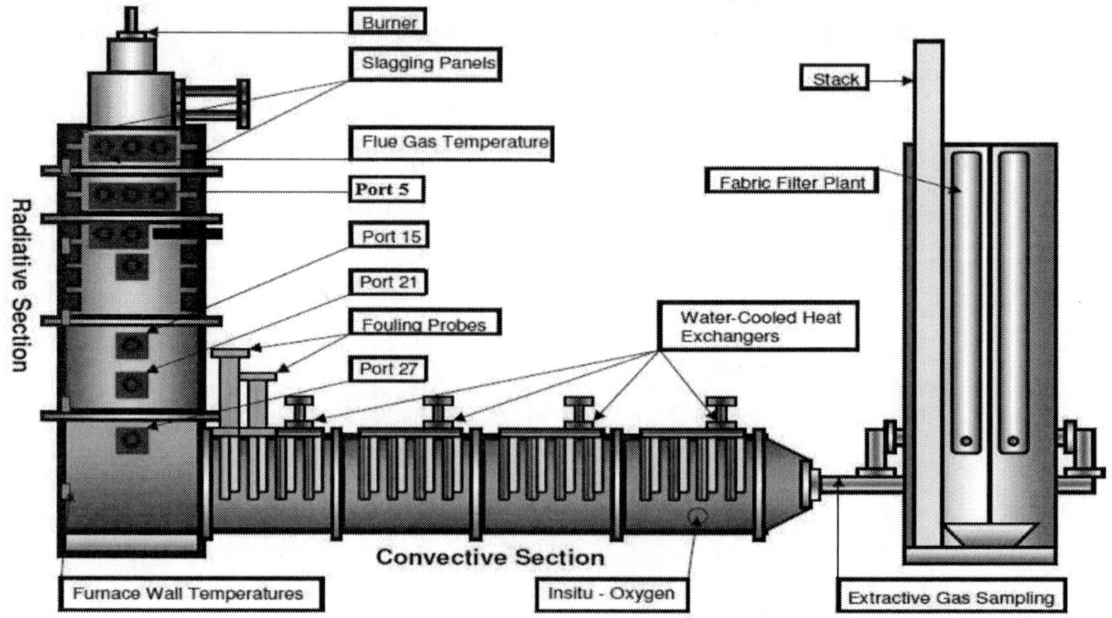


**Figure 6: Schematic diagram of the Eskom PSCTF [7]**

Coal from the Eskom power station where full scale tests were conducted was utilised. Coal properties are shown in table 1. The coal was prepared and pulverised to 75µm which is the typical PF particle size of PF at the power station. Flue gas compositions at various excess air levels were sampled and analysed. Results from the latter analysis were used to determine the break-point of $H_2S$ formation, validate what is stated in literature with regards to $H_2S$ formation.

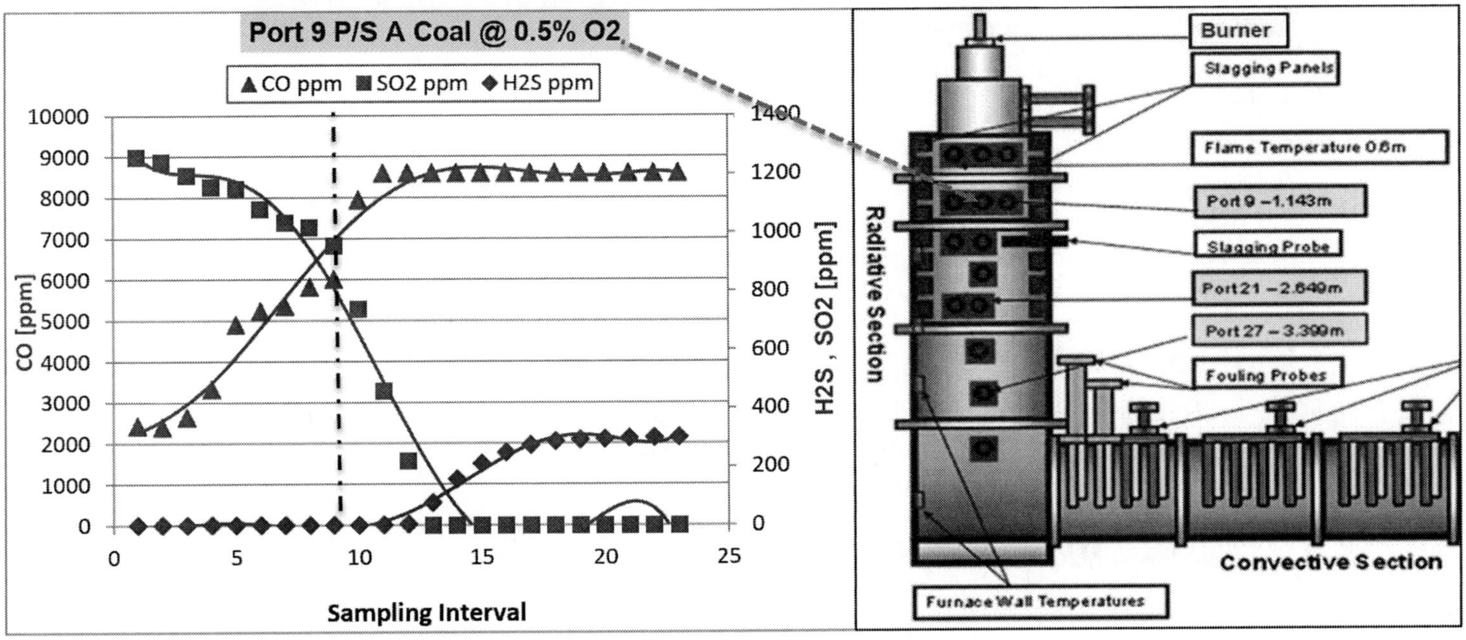


**Figure 7: Concentration distributions of $H_2S$, SO2 and CO at 0.5% $O_2$ recorded at the sampling port closest to the burner**

Referencing the insitu $O_2$ analyser, located down-stream of the convective section of the pilot furnace, $O_2$ concentrations were reduced from 3% with decrements of 0.5% and gas concentrations were sampled from ports 9, 21 and 27 (please refer to the PSCTF schematic diagram above for sampling port locations). Combustion conditions/parameters were maintained for at least 30 minutes before proceeding to the next sampling port.

$H_2S$ was only recorded at 0.5% $O_2$ concentration on port 9 which is the sampling port closest to the burner. As the CO concentration started peaking the $O_2$ concentration decreased and as the CO peaked to 8500ppm $H_2S$ started forming indicating the severity of reducing atmospheres prevalent around the burner area. A similar trend was not depicted when flue samples were analysed from ports 21 and 27 which could be due to ingress air diluting the gas mixture. However, on the contrary it was observed that the concentration of $SO_2$ started increasing as gas concentrations were sampled further away from the burner. This was the exact pattern observed by CRIEPI (Central Research Institute of Electrical Power Industry) in an article published in March 2012 when evaluating the characteristics of $H_2S$ formation in PF coal combustion utilising three bituminous coals of which one was from Witbank South Africa [3].

## 3. Conclusions and recommendations

### 3.1. Conclusions

At a CO breakpoint from 2000 - 5000ppm at 2.5% $O_2$, $H_2S$ did not form at the PSCTF. Similar results were obtained during the full scale tests. However, $H_2S$ breakpoint was achieved at the PSCTF trials when the CO concentration was above 7500ppm at 0.5% $O_2$. Therefore based on results obtained during full scale tests, the lowest $O_2$ limit to prevent the formation of $H_2S$ was 2.5%.

### 3.2. Recommendations

i. To prevent the formation $H_2S$ and minimize CO gas concentrations, Power Station boilers must not be operated below 3% $O_2$ levels as measured in the flue gas duct at boiler outlet or economizer outlet.
ii. Further research is required to establish a relationship between $H_2$ and total sulphur in the coal for the formation of $H_2S$ at lower $O_2$ concentrations.

### Acknowledgements

The first author acknowledges the contribution of Duvha Power Station, Eskom, for the opportunity to participate in the EPPEI programme. Eskom RT&D for the resources made available for this research study. Academic and industrial supervisors for their support and motivation. Eskom combustion consultants for their contribution to this research study. Family and friends for their understanding and encouragement.

### References

[1] J.N. Harb J.N. and E.E. Smith. **Fireside Corrosion in PC-Fired Boilers.** Advanced Combustion Engineering Research Centre and Department of Chemical Engineering, Brigham University, Provo, Utah, USA. 1990.
[2] M. Biarnes, **Combustion booklet-201**. E Intruments International LLC – www.E-Inst.com.
[3] H. Shirai, M. Ikeda, H. Aramaki. **Characteristics of hydrogen sulphide formation in pulverized coal combustion**. Journal publication, 2012.

[4] EPRI: **Boiler and heat Recovery Steam Generator Tube Failures**: Theory and Practise. Volume 2: Water-Touched Tubes. Final Report, March 2007.

[5] H. Schenk. **Flue gas Attack on the Fireside of Boilers**, Eskom, RT&D, Report no TRR/S/95/240, 2003.

[6] K.L. Rodseth. **Fireside Corrosion of Duvha Sidewall Tubes**, Eskom, RT&D, Report no KLR/7173/cm, 1995.

[7] C. Ramsunkar. **Excess air measurement, control and optimization**. Masters of Science Dissertation, University of Witwatersrand, 2014.

**[8] P. Rajoo. The Application of Pilot Scale Coal Evaluation to Full Scale Boilers. Masters of Science Dissertation, University of Witwatersrand, 2010.**

**2016 International Pittsburgh Coal Conference**
**Cape Town, SOUTH AFRICA**
**August 8 – 12, 2016**

**Abstract Submission**

**PROGRAM TOPIC:**
**UNDERGROUND COAL GASIFICATION (UCG)**

**Exact Title of Paper:**
# Can UCG Give Coal a New Lease on Life in 21st Century?

***Michael S. Blinderman,***
***Director, Ergo Exergy Technologies Inc.,***
***465 rue Saint-Jean #1000, Montréal, Québec, CANADA***
***info@ergoexergy.com, Phone 514 499 2093, Fax 514 313 5674***

*Simon S. Maev,*
*President, Laurus Energy, Inc.,*
*PMB #539, 2711 Centerville Road, Suite 300, Wilmington, DE 19808-1645, USA*
*info@laurusenergy.com, Phone 780 862 0134, Fax 780 665 7456*

**Abstract:**
Coal gasification has long been considered a pathway to making cleaner chemicals and energy from coal – a fossil fuel that many in 21st century believe is no longer compatible with sustainable and environmentally responsible development. Aged all but 200 years, conventional coal gasification technologies face problems with the safety, health hazards and environmental stigma of coal mining; handling of ash and other effluents; plant availability and prohibitively high cost. Underground Coal Gasification (UCG) technology safely produces gaseous and liquid hydrocarbons without coal mining, leaves ash deep in the ground and at its best is much less expensive. Low base cost means that, unlike conventional gasification, UCG can afford $CO_2$ sequestration while remaining competitive. Why has not it seen so far a wide global industrial deployment? Is there a chance that broad application of UCG will make wide use of coal environmentally acceptable? What is required of UCG to achieve recognition of investors, regulators and general public?

We answer the above questions by reviewing modern achievements of UCG development and pointing out the aspects that require further work to overcome apparent deficiencies. In doing so, we focus on technical viability of UCG technology and the factors that can make it commercial: scale, conversion efficiency, extraction rates, product quality and consistency. Environmental performance of UCG is presented using examples of recent UCG trials.

The Ergo Exergy's εUCG™ technology is an example of such a commercial technology, which is based on over forty years of operating experience in the Ukraine, Russia and Uzbekistan. Its performance is considered using examples of its application in the UCG projects in Australia, South Africa, New Zealand, the USA and Canada.

The εUCG™ produced syngas at a substantially lower cost than conventional gasification. Performance and economic comparisons are presented for producing SNG, gasoline, diesel fuel, ammonia, urea, methanol, and electricity - from εUCG™ syngas and the syngas of conventional gasification.

Imperial College London

# Life Cycle Sustainability Assessment of Integrated UCG-CCS Systems and Comparison with other Abated Coal Based Power Generation Technologies

Anna Korre
Sevket Durucan, Zenggang Nie, Sebastian Franzsen*

mERG Minerals, Energy and Environmental Engineering Research
Department of Earth Science and Engineering
Royal School of Mines
Prince Consort Road
London, SW7 2AZ, UK

Seamwell INTERNATIONAL Seamwell International Ltd., 3 Shepherd Street, London, W1J 7H, UK

## Outline

- Life Cycle assessment principles
- ICLCA modelling framework
- UCG and power generation LCI model development
  - Underground gasification unit process LCI model
  - Selexol acid gas ($CO_2$ & $H_2S$) removal LCI model
- UCG power generation vs conventional options with and without CCS
  - Comparison with surface gasification
  - Coal to electricity: life cycle energy efficiency
  - Life cycle GHG footprint of electricity

## Life Cycle Assessment

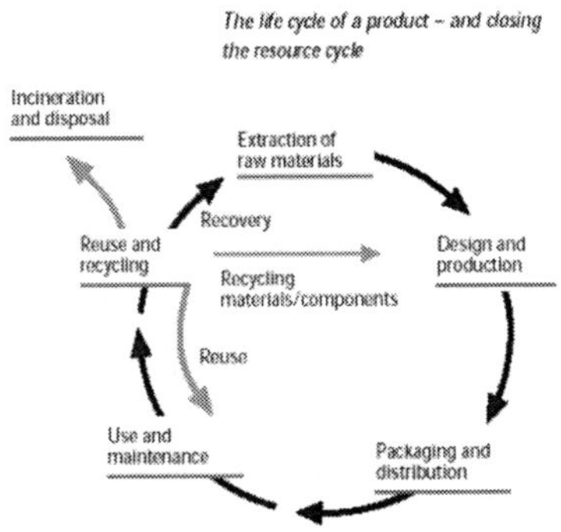


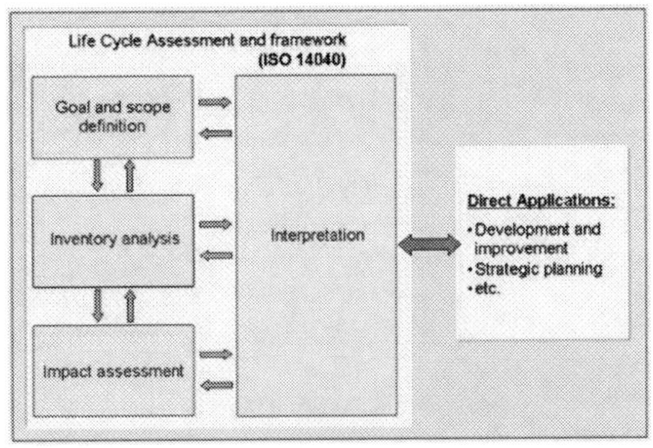


Life Cycle Assessment (LCA) is an objective process to evaluate the environmental burdens associated with a product, process, or activity by identifying energy and materials used and wastes released to the environment, and to evaluate and implement opportunities to affect environmental improvements.





## Environmental Interventions and Economic flows

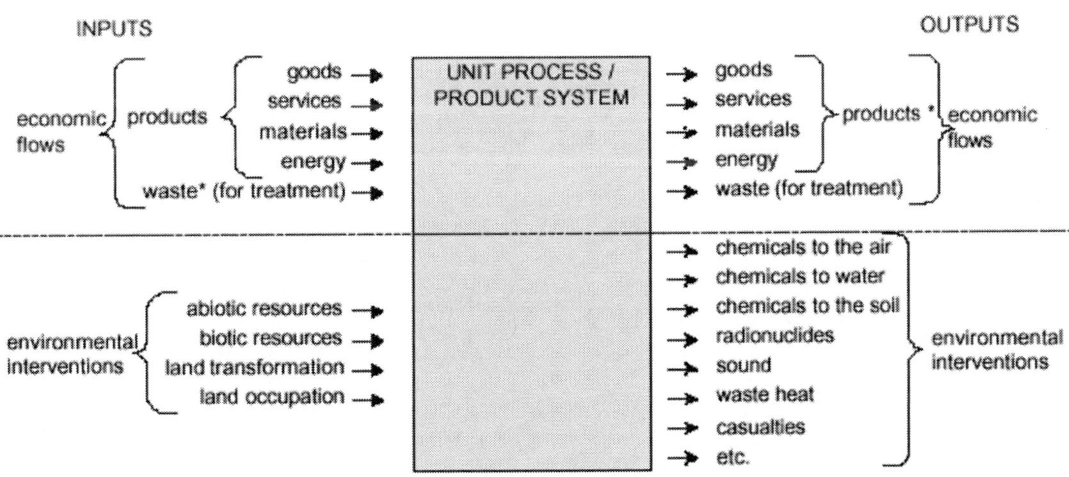


* the functional flows of the process

## Life Cycle Impact Assessment

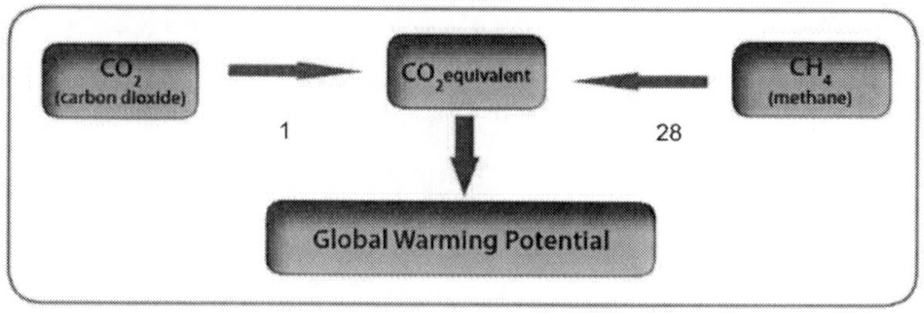


| GWP values and lifetimes from IPCC AR5 (2014) | GWP time horizon | | |
|---|---|---|---|
| | 20 years | **100 years** | Lifetime |
| Carbon dioxide | 1 | **1** | - |
| Methane | 84 | **28** | 12.4 |
| Fossil methane | 85 | **30** | 12.4 |
| Nitrous oxide | 264 | **265** | 121 |

Page 5

## Imperial College's LCA model (ICLCA) of fossil fuel production, transport, power generation value chains

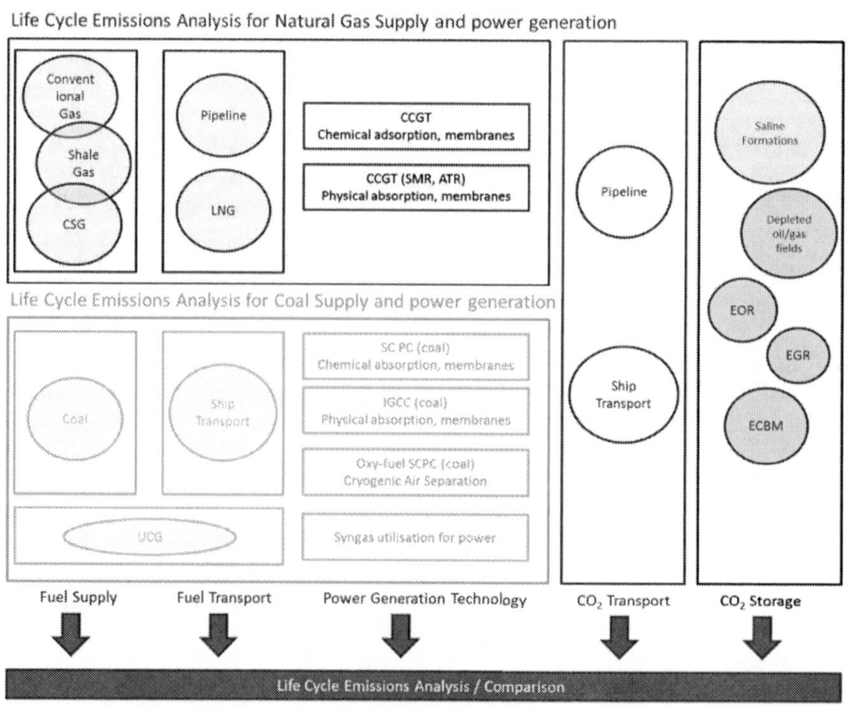




Page 6

## Imperial College's LCA model (ICLCA) of fossil fuel production, transport, power generation value chains

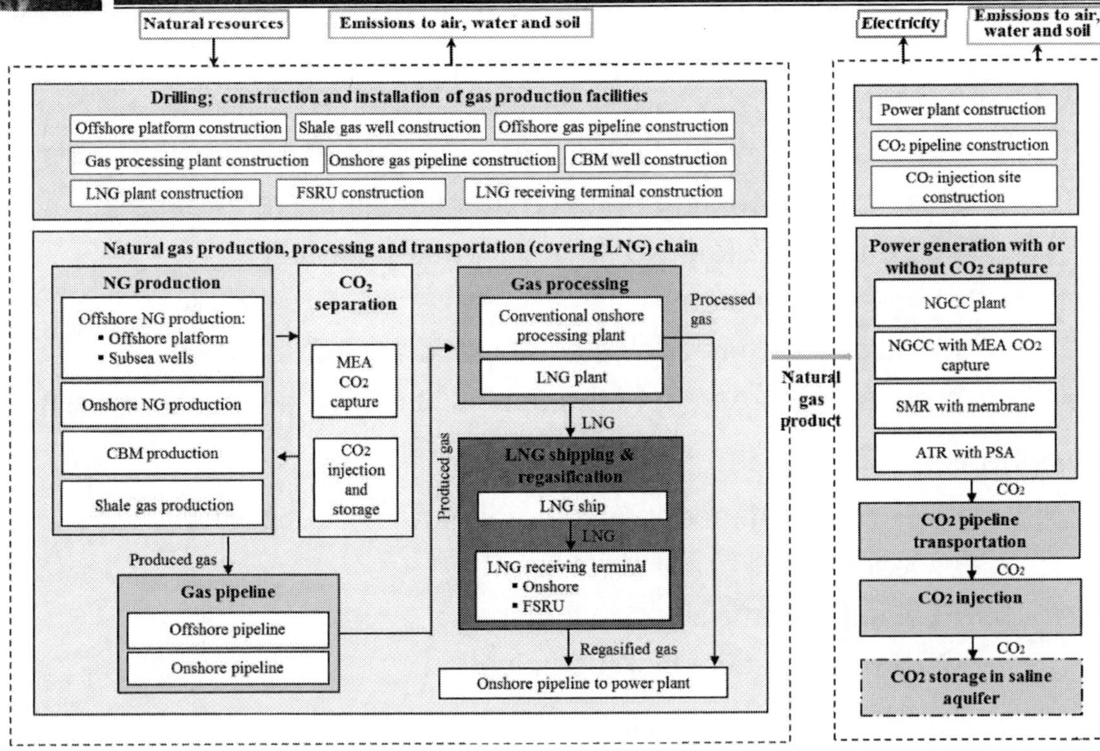


## Approaches for life cycle inventory analysis

- Published emission factors
- Equipment manufacturer emission factors
- Engineering calculations*
- Process simulation or other computer modelling
- Emissions monitoring over a range of conditions and deriving emission factors
- Periodic or continuous monitoring of emissions or parameters for calculating emissions

Improved accuracy

Additional data requirements

Higher cost

* Engineering calculations are based on basic chemical or physical principles of a process, considering operational parameters.

## Imperial College's LCA model (ICLCA) of fossil fuel production, transport, power generation value chains

Key advantages

- Covers conventional / unconventional fossil fuel value chains
- Accurate accounting of fuel/materials supply requirements and emissions (air, water, solid waste) per unit process/subsystem
- Modular structure allows combination of component processes to represent the specifics of any given value chain

Current developments

- Life cycle cost model development and real options assessment
- Fuel supply emissions mitigation
- CCS network evolution optimisation





## Imperial College's LCA model (ICLCA) of fossil fuel production, transport, power generation value chains

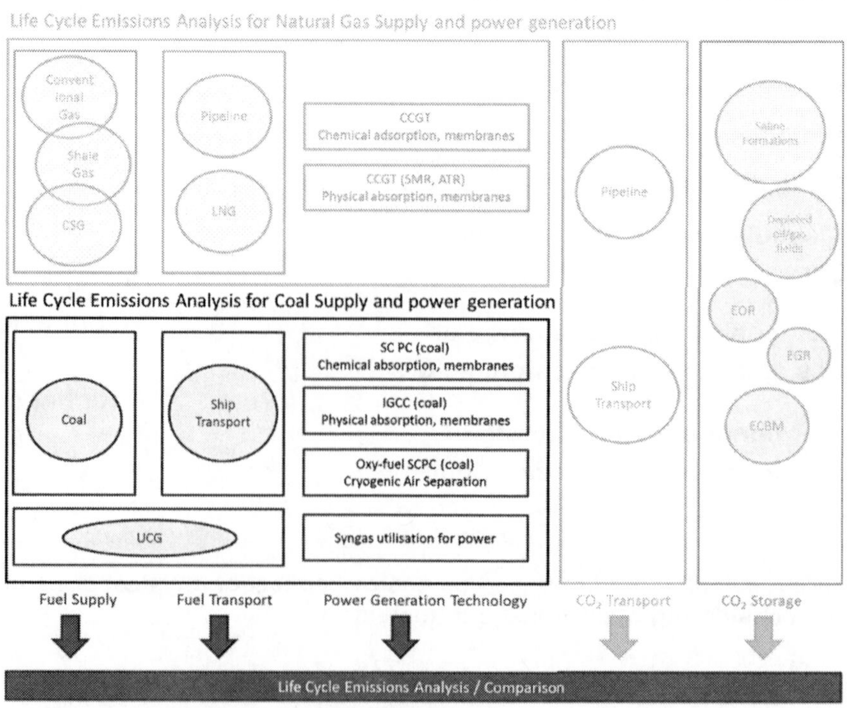

# LCI model development

**Underground coal gasification**

**and**

**power generation with / without CCS**





# Underground coal gasification and power generation

## UCG-CCGT without CCS

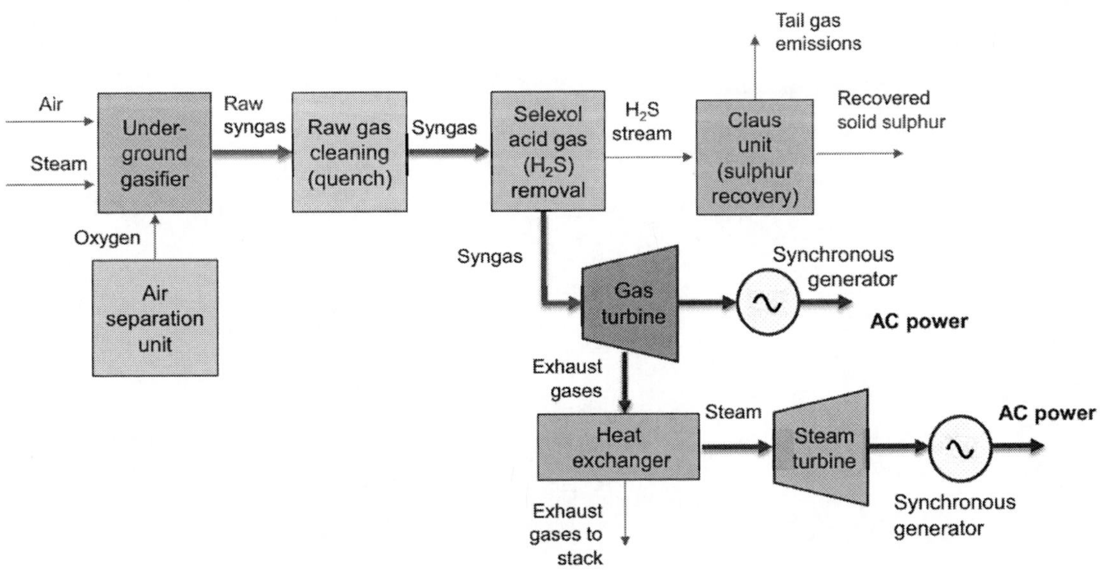

# Underground coal gasification and power generation

## UCG-CCGT with CCS

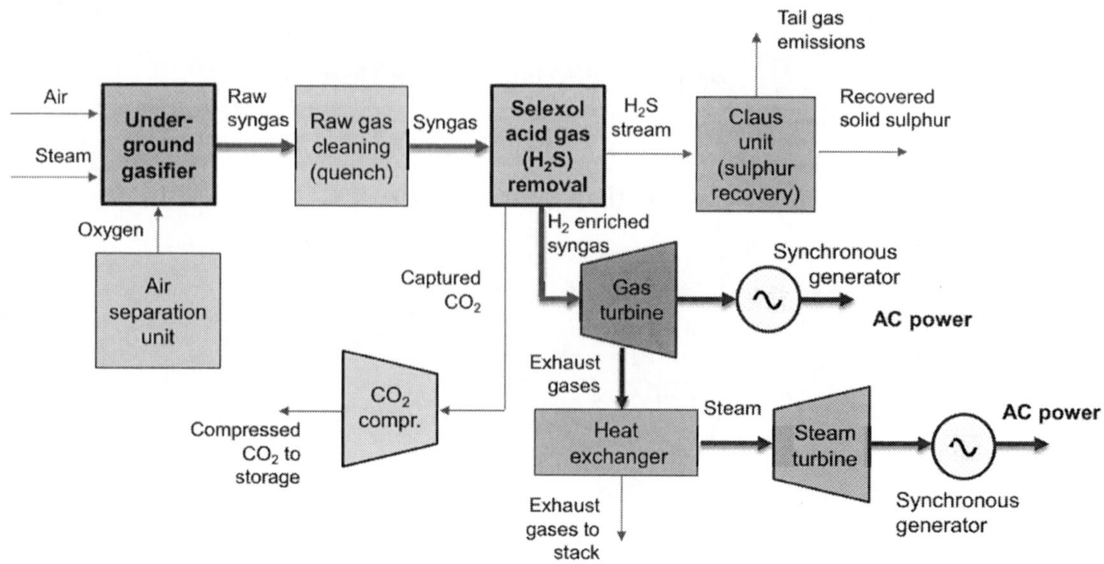






# Underground coal gasifier LCI model

aspentech

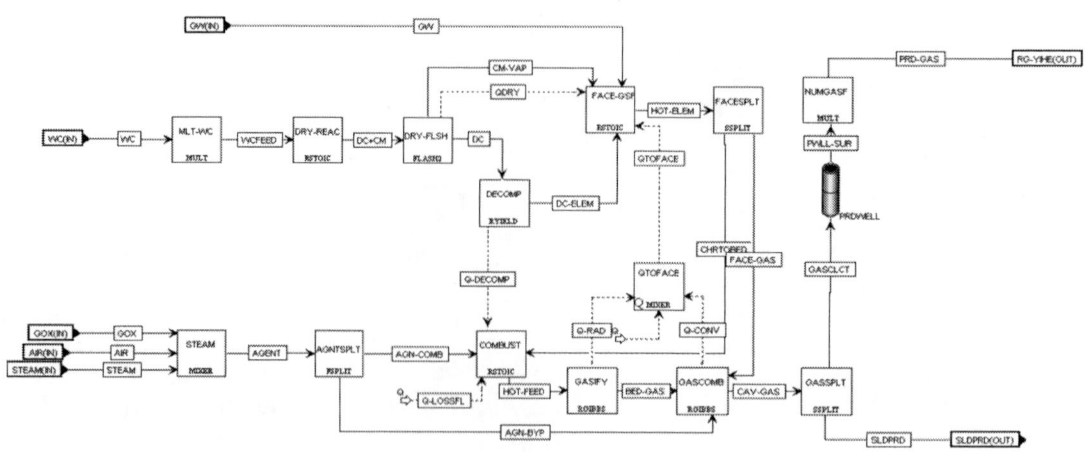

## Selexol acid gas ($CO_2$ and $H_2S$ removal) LCI model

aspentech

**Modified from Field and Brasington (2011)**

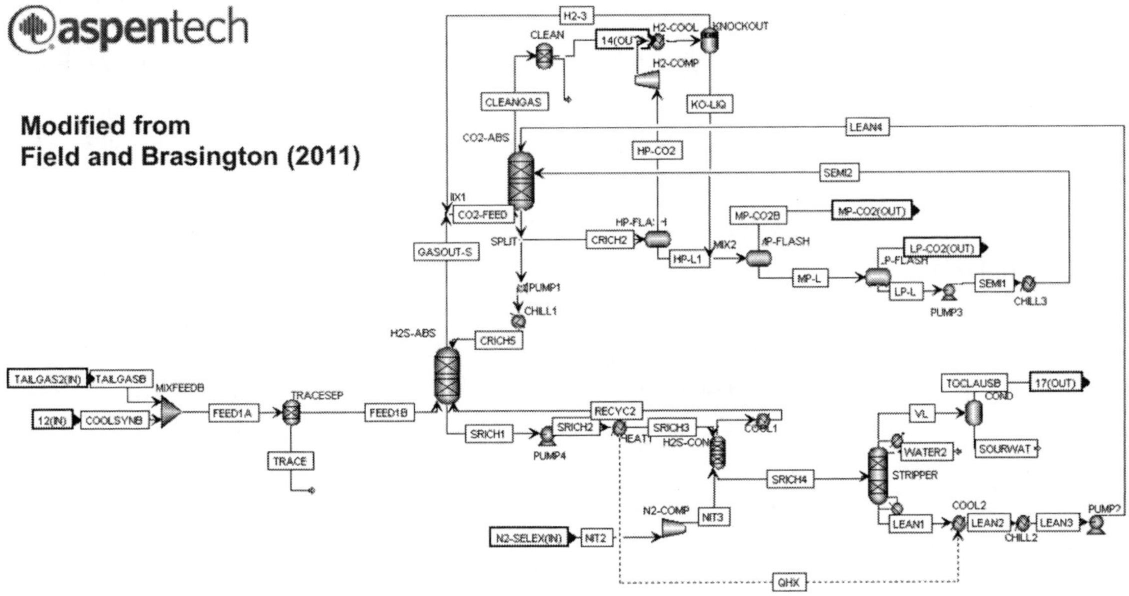






## Gasification parameters (LCI model inputs)

| | |
|---|---|
| **Thermal output:** | 20 – 25 $MW_t$ |
| **Instantaneous coal consumption:** | 6,500 kg/h |
| **Gasification agent:** | enriched air at 35 % wt. dry basis (before steam addition, if required) |
| **Operating pressure:** | 2,000 kPa gauge mode |
| **Ground water influx:** | 5 $kg.m^{-2}.h^{-1}$ |
| **Affected cavity extent:** | 750 $m^3$ or 1,000 tonnes coal consumed |
| **Gasifier roof surface area:** | 150 $m^2$ |
| **Gasifier side wall surface area:** | 100 $m^2$ |
| **Exposed coal surface area:** | 175 $m^2$ (exposed roof rock decreases active cavity extent) |
| **Average coal face (cavity extent) recession rate:** | 0.46 $m.d^{-1}$ |
| **Heat loss into overburden:** | 3% of thermal energy content of coal |
| **Heat loss into underburden:** | 2% of thermal energ |

## Underground coal gasification LCI model

| Coal properties | | Lignite 1 | Lignite 2 | Bituminous Coal |
|---|---|---|---|---|
| **Proximate analysis** | Moisture (wt.% air dried) | 11.13 | 11.6 | 2.89 |
| | Fixed carbon (wt.% dry) | 29.8 | 44 | 67.07 |
| | Volatile matter (wt.% dry) | 63.9 | 39.2 | 30.5 |
| | Ash (wt.% dry) | 6.3 | 16.8 | 2.43 |
| **Ultimate analysis** | Ash (wt.% dry) | 6.19 | 16.8 | 2.43 |
| | Carbon (wt.% dry) | 55.9 | 59.6 | 83.26 |
| | Hydrogen (wt.% dry) | 5.25 | 3.8 | 4.65 |
| | Nitrogen (wt.% dry) | 0.72 | 0.72 | 1.16 |
| | Sulfur (wt.% dry) | 0.74 | 0.28 | 0.21 |
| | Oxygen (wt.% dry) | 31.2 | 18.8 | 8.29 |
| **Sulphur analysis** | Pyritic (wt.% dry) | 0.35 | 0.09 | 0.06 |
| | Sulfate (wt.% dry) | 0.03 | 0.06 | 0.04 |
| | Organic (wt.% dry) | 0.36 | 0.13 | 0.11 |
| **Heating value** | LHV (MJ/kg) | 19.73 | 16.17 | 32.40 |





## Underground coal gasification LCI model

| Gasification results | | Lignite 1 | Lignite 2 | Bituminous Coal |
|---|---|---|---|---|
| **Heating value of product gas** | $MJ/Nm^3$ | 4.55 | 5.51 | 5.29 |
| **Cold gas efficiency*** | % | **61.59** | **81.95** | **60.17** |
| **Thermal energy in product gas** | MW | 21.94 | 24.2 | 35.20 |
| **Radiation heat from bed into coal** | kW | 1,882.4 | 1,881.8 | 2,811.7 |
| **Convection heat from cavity gas to coal** | kW | 722.4 | 516.8 | 378.6 |
| **Heat loss into overburden** | kW | 996.3 | 1,084.7 | 1,709.1 |
| **Heat loss into underburden** | kW | 664.2 | 723.1 | 1,139.4 |
| **Ratio $O_2$ / $Coal_{daf}$** | | 0.75 | 0.72 | 1.00 |
| **Ratio $H_2O$ /$Coal_{daf}$** | | 0.34 | 0.35 | 0.23 |
| **Heating value of product gas** | $MJ/Nm^3$ | 4.55 | 5.51 | 5.29 |

* Cold gas efficiency = heating value of produced raw syngas / heating value of coal gasified

## Underground coal gasification and power generation from 20 panels to supply syngas

| | | UCG without CCS | | | UCG with CCS | | |
|---|---|---|---|---|---|---|---|
| | | Lignite 1 | Lignite 2 | Bituminous Coal | Lignite 1 | Lignite 2 | Bituminous Coal |
| Heating value of consumed coal (20 panel) | MW | 583.92 | 712.47 | 1170 | 583.92 | 712.47 | 1170 |
| Energy used for oxygen production | MW | -10.68 | -11.61 | -13.57 | -10.68 | -11.61 | -13.57 |
| Energy used for oxygen and nitrogen compression | MW | -10.66 | -11.58 | -13.53 | -15.98 | -17.37 | -20.3 |
| Energy used for air compression | MW | -18.06 | -18.48 | -36.42 | -18.06 | -18.48 | -36.42 |
| Energy recovered from water gas shift unit | | | | | 5.86 | -5.67 | 1.52 |
| Energy used for $CO_2$ compression | MW | | | | -13.53 | -13.53 | -18.28 |
| Energy used for acid gas removal | MW | -0.96 | -0.96 | -1.29 | -9.58 | -9.58 | -12.94 |
| Electricity generated from gas turbine | MW | 174.24 | 157.98 | 253.43 | 164.76 | 157.98 | 253.43 |
| Electricity generated from steam turbine | MW | 96.8 | 96.55 | 154.88 | 91.54 | 96.55 | 154.88 |
| **Net electricity output** | **MW** | **230.68** | **211.91** | **343.49** | **194.31** | **178.29** | **308.31** |
| **UCG power generation efficiency*** | **%** | **39.51** | **29.74** | **29.36** | **33.28** | **25.02** | **26.35** |

* electricity generated divided by heating value of coal used.





## Comparison of UCG power generation with surface gasification

aspentech

**Modified from DOE/NETL-2015/1727**

**Shell IGCC with $CO_2$ capture (Case B1B)**

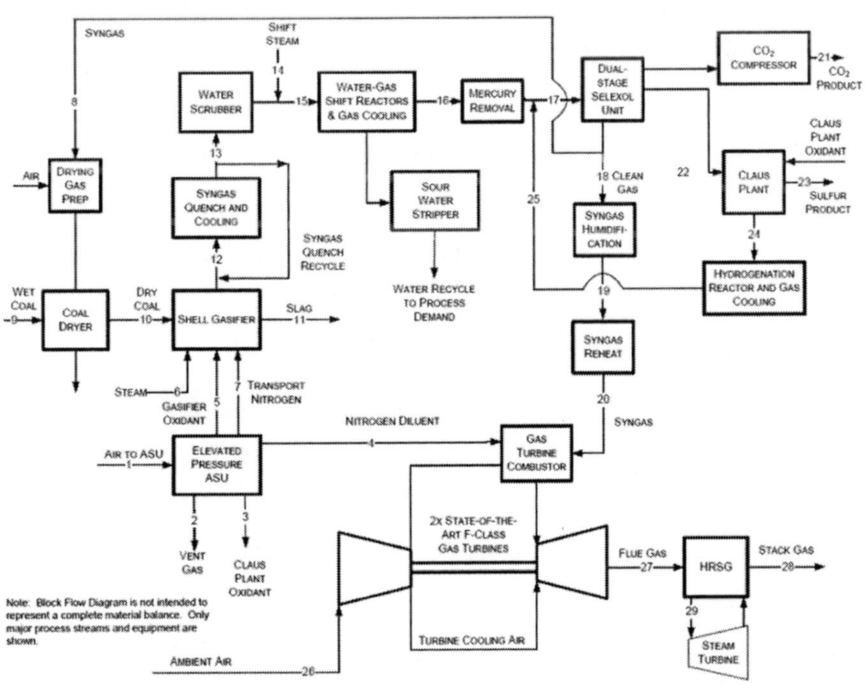

## Comparison of UCG power generation with different surface IGCC plant options

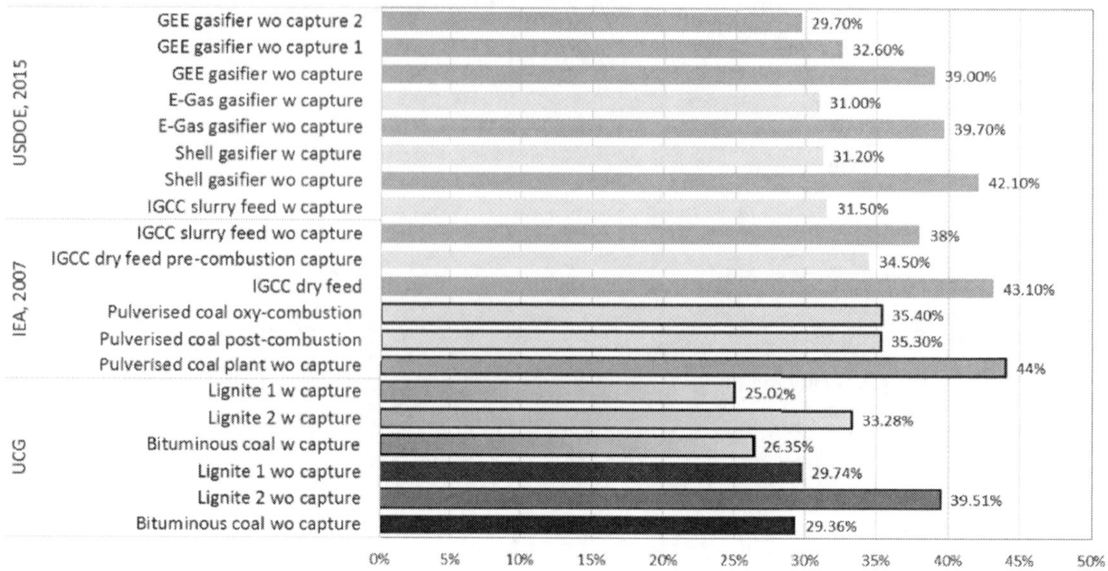






## Coal to electricity: life cycle energy efficiency

- In conventional coal utilisation the life cycle impacts from coal mining, coal transportation and coal preparation need to be added, which further reduce overall coal to electricity efficiency.
  - Considering underground coal mining, 300 km railway transportation and coal preparation, the energy requirement is **158.92 kw** equivalent electricity per tonne coal delivered to the plant (estimated using the ICLCA model).
  - For a power plant with plant energy efficiency of 40%, around **5.18 %** of electricity generated is used for coal mining, transportation and preparation.
- Recovery rate of coal reserve assumed
  - UCG: 50%
  - Underground coal mining: 50%

# Coal to electricity: life cycle energy efficiency

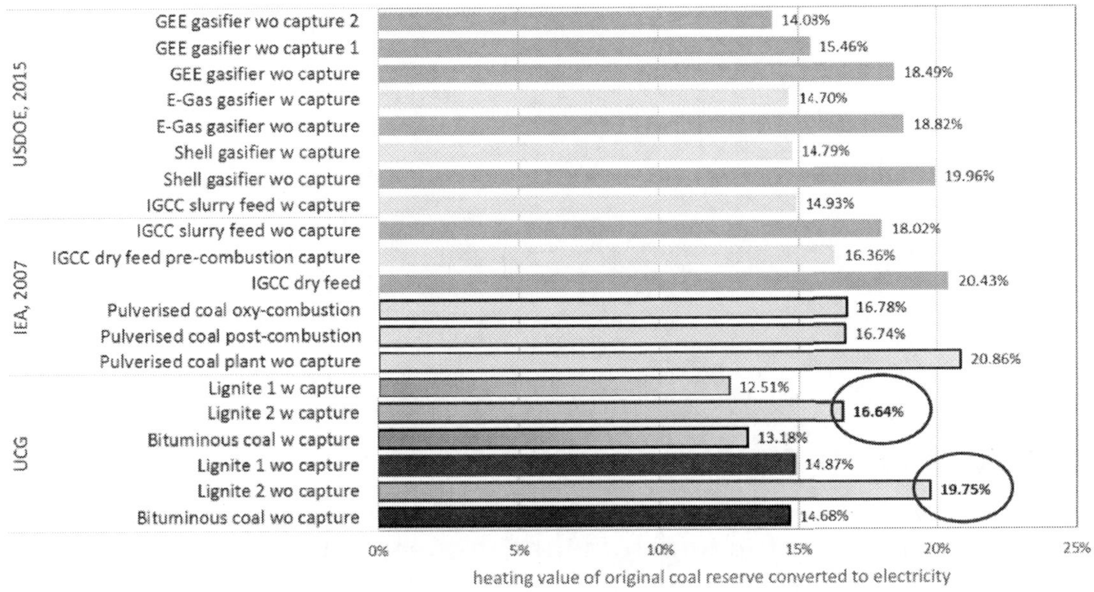






# Life cycle GHG footprint of electricity

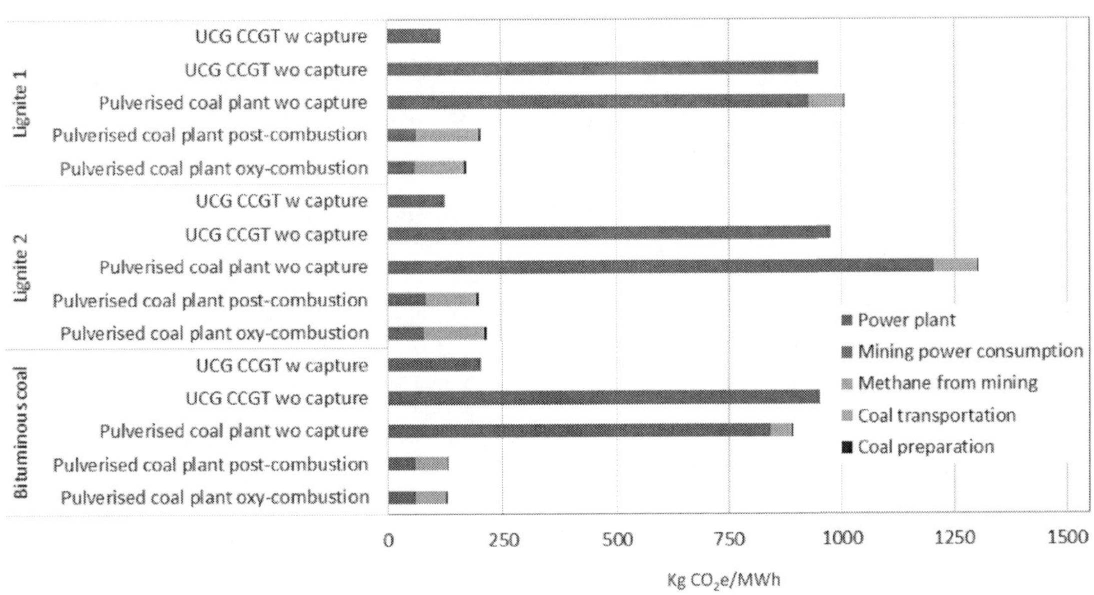

## Acknowledgements

**This project is receiving funding from the European Union's Seventh Framework Programme for research, technological development and demonstration under Grant Agreement No. 608517 ("Technology Options for Coupled Underground Coal Gasification and $CO_2$ Capture and Storage")**

# EX-SITU SIMULATION OF HIGH PRESSURE UNDERGROUND GASIFICATION OF HARD COAL AND LIGNITE

GIG

*Krzysztof Stańczyk, Krzysztof Kapusta, Marian Wiatowski*
*Central Mining Institute, Katowice, POLAND*
*Sevket Durucan, Imperial College London, London, United Kingdom*

2016 International Pittsburgh Coal Conference
Cape Town, SOUTH AFRICA
August 8 – 12, 2016

## Plan of presentation

- Background of the study
- The aim of the research
- Methods
- Results and discussion
- Conclusions

2

## Background

One of the most important factors affecting the chemistry of gasification, and therefore the composition of the product gas from UCG is the operating pressure. Higher operating pressures promote methanation reactions, which lead to the increase in gas calorific value:

$$CO_{(g)} + 3H_{2(g)} \rightarrow CH_{4(g)} + H_2O_{(g)} \qquad \Delta H = -\ 206.2\ MJ/kmol$$

$$C_{(s)} + 2H_{2(g)} \rightarrow CH_{4(g)} \qquad \Delta H = -\ 87{,}5\ MJ/kmol$$

The operating pressure is determined by the seam depth.

3

## Background

Determination of thermodynamic conditions and process parameters for the optimization of CO-rich gas is the other issue.

$$C + CO_2 \rightarrow 2CO \; \Delta H^0_{298} = +\ 159.9\ kJ/mol$$

According to Boudouard equation, the equilibrium point at which an equimolar quantity of the carbon monoxide and carbon dioxide is present, is approximately 700°C and CO content in products gas increases further with temperature. Gasification temperature 750 $^0$C selected for the tests was found as an optimum (and possible) temperature region inside the underground reactor. This selection was made based on the temperature distributions during UCG known from previous ex-situ gasification tests.

4

## The aim of the research

The aim of the study was to evaluate the role of the elevated gasification pressure on the UCG process, depending on the coal rank

5

## Scheme of the ex-situ high pressure UCG installation

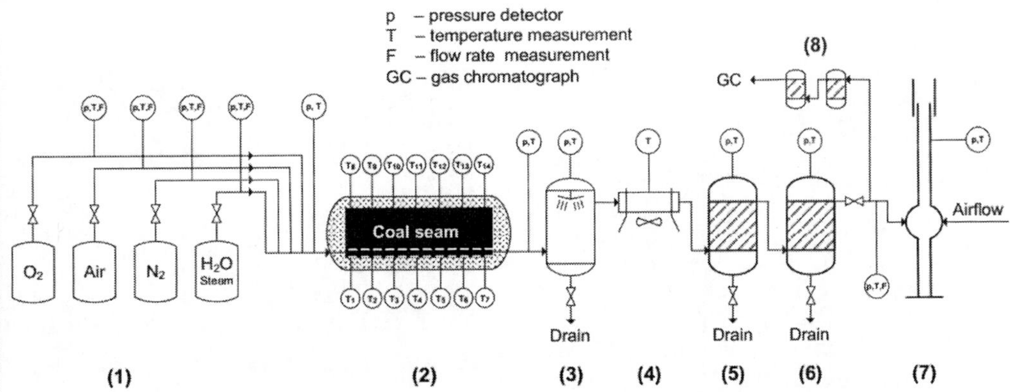


(1)- compressed reagents, (2)- pressure reactor, (3)- wet scrubber for gas cleaning , (4)- air cooler, (5,6)- gas separators , (7)- thermal combustor, (8)- gas treatment module prior to GC analysis

6

## Sections of the high pressure reactor: a) cross-section, b) longitudinal section.

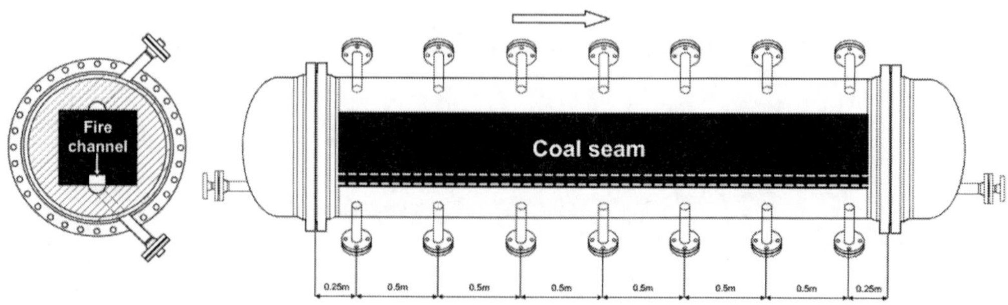


**a** **b**

7

## Experimental stand for the high-pressure simulations of UCG: a) the reaction chamber partially loaded with coal, b) the reactor ready for gasification

**a** **b**

8

## Coal blocks before inserting into the reactor

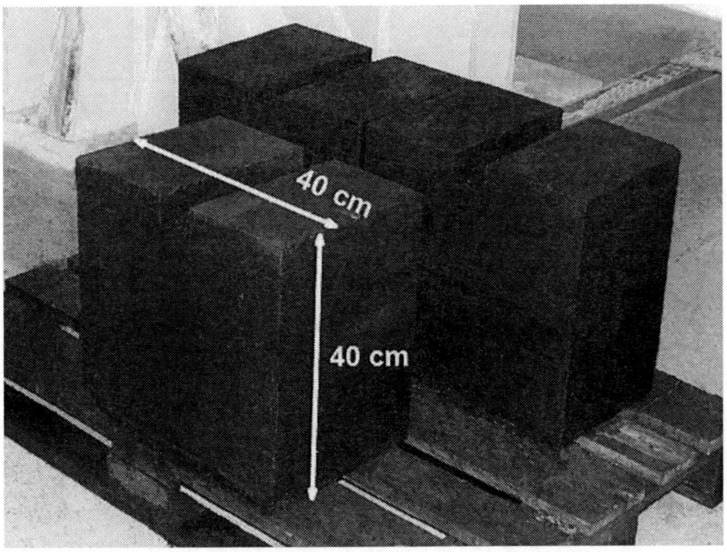


9

## Characteristic of used coal

The coal characterisation included:
•proximate and ultimate analysis,
•sulphur form analysis (speciation),
•ash fusion temperature: sintering temperature, softening temperature, melting temperature, flow (fluid) temperature

•Fischer assay.

10

## Characteristic of used coal;
## **Valenje** (lignite) provided by Premogovnik Valenje (Slovenian)
## **Murcki-Staszic** (hard coal) provided by KHW S.A. (Polish)

| No | Parameter | Coal | |
|---|---|---|---|
| | | Valenje | Murcki-Staszic |
| **As received** | | | |
| 1 | Total moisture $W_t^r$, % | 31.62 | 4.93 |
| 2 | Ash $A_t^r$, % | 4.29 | 4.45 |
| 3 | Volatiles $V^r$, % | 43.67 | 30.16 |
| 4 | Total sulphur $S_t^r$, % | 0.51 | 0.29 |
| 5 | Calorific value $Q_i^r$, kJ/kg | 13 615 | 29443 |
| **Analytical** | | | |
| 6 | Moisture $W^a$, % | 11.13 | 2.37 |
| 7 | Ash $A^a$, % | 5.57 | 4.57 |
| 8 | Volatiles $V^a$, % | 56.76 | 30.97 |
| 9 | Heat of combustion $Q_s^a$, kJ/kg | 19 719 | 31269 |
| 10 | Calorific value $Q_i^a$, kJ/kg | 18 427 | 30300 |
| 11 | Total sulfur $S^a$, % | 0.66 | 0.30 |
| 12 | Carbon $C_t^a$, % | 49.86 | 77.63 |
| 13 | Hydrogen $H_t^a$, % | 4.67 | 4.17 |
| 14 | Nitrogen $N^a$, % | 0.64 | 1.23 |
| 15 | Oxygen $O_d^a$, % | 27.83 | 9.74 |

11

## 10 bar UCG ex-situ experimental simulation (Valenje coal)

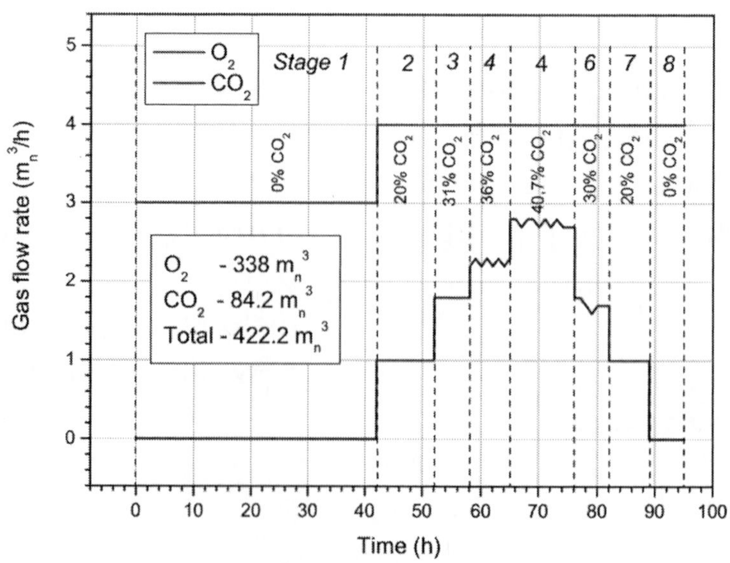


Oxidants supply rate

12

# 10 bar UCG ex-situ experimental simulation (Valenje coal)

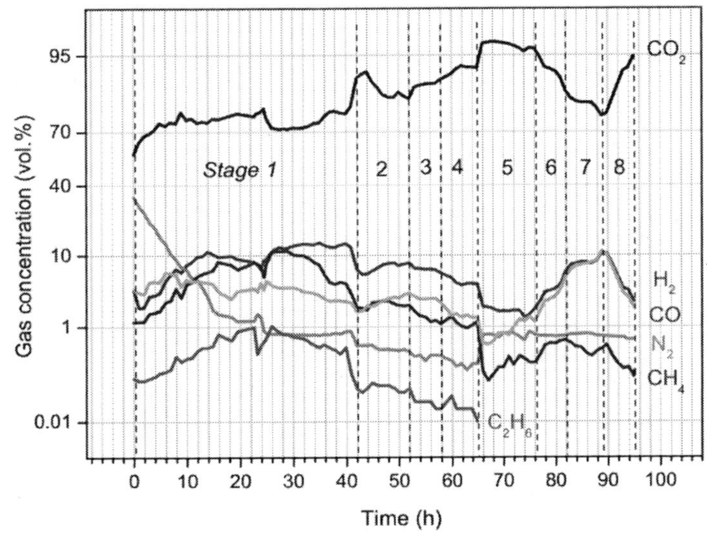


Gas composition



# 40 bar UCG ex-situ experimental simulation (Valenje coal)

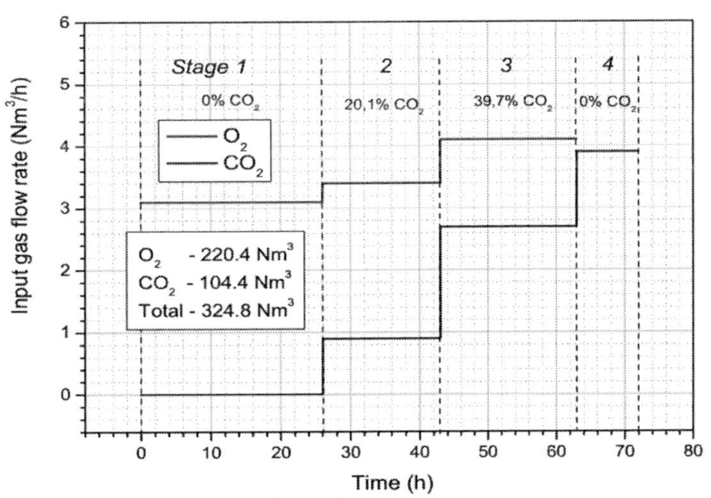


Oxidants supply rate

## 40 bar UCG ex-situ experimental simulation (Valenje coal)

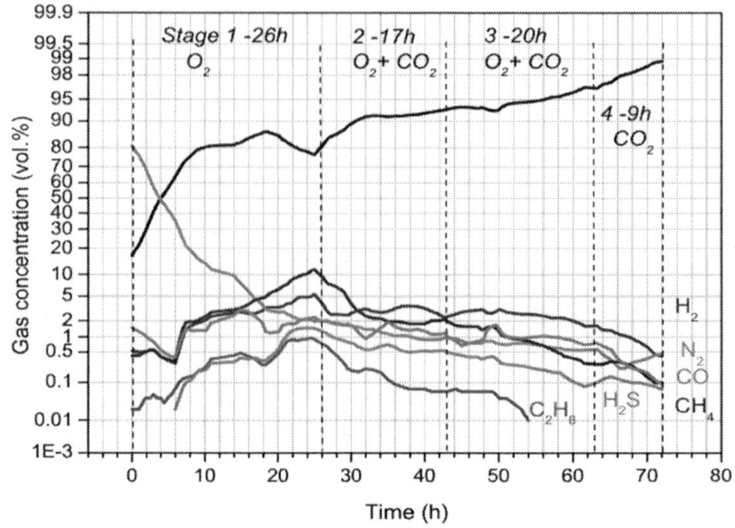


Gas composition

15

## 10 bar UCG ex-situ experimental simulation (Murcki-Staszic coal)

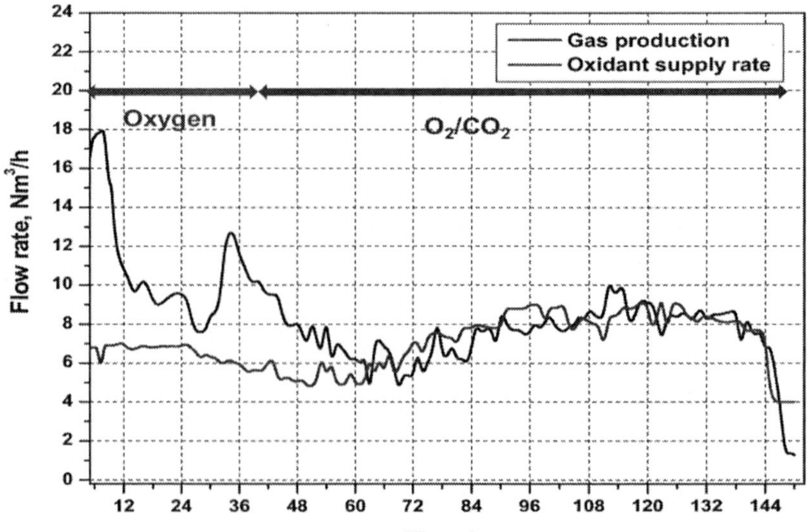


16

## 10 bar UCG ex-situ experimental simulation (Murcki-Staszic coal)

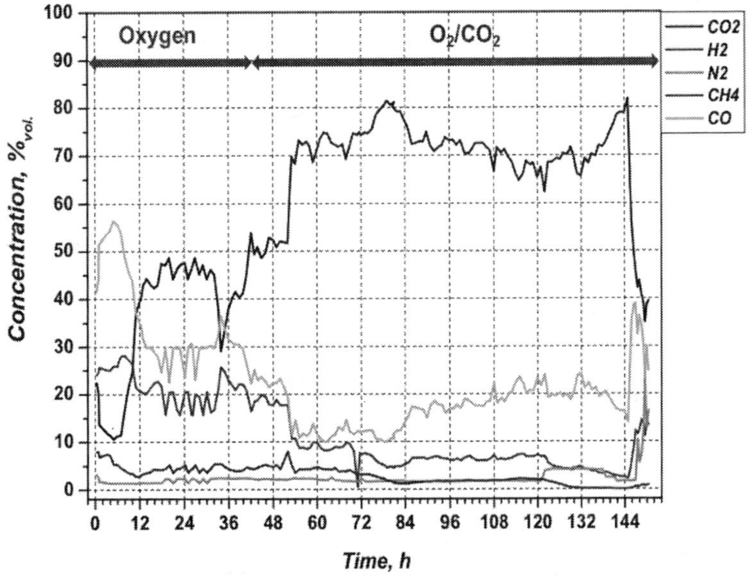


Gas composition

17

## Conclusions

- The experiments conducted proved that the most favourable thermodynamic conditions for the high pressure UCG were obtained using oxygen as gasification reagent. The calorific value of gas produced during the oxygen gasification stage reached 10 MJ/Nm$^3$
- The average gas calorific value dropped from 8.5 MJ/Nm$^3$ in the oxygen-blown stage to 3.8 MJ/Nm$^3$ in the $O_2/CO_2$ gasification stage
- The calorific value of gas obtained from hard coal was higher to compare with gas from lignite however the methan concentration was higher for lignite
- We did not observed significant difference between gas composition and methane concentration under 10 and 40 bars experiments for lignite

18

GIG

**Thank you very much for your attention**

19

# Experimental and Computational Studies on Underground Gasification of Turkish Lignite

Mesut Gür, Nurdil Eskin, Hasancan Okutan[1] , Ahmet Arısoy, Erhan Böke, Ünal Altıntaş, Oğuz Büyükşirin, E. Deniz Canbaz, Olgu Yıldırım

**Istanbul Technical University Mechanical Engineering Department, Istanbul, TURKEY**

**1:Istanbul Technical University Chemical Engineering Department, Istanbul, TURKEY**

**Contact Information: mesutgur@itu.edu.tr, +90 212 293 13 00 / 2476**

**Abstract:**

- Underground coal gasification (UCG) is more advantageous than conventional coal combustion processes, as the first requires lower investment/ handling costs and operates environmentally friendly. Lignite is of utmost importance for the Turkish economy as Turkey has ~15 billion tons of lignite reserves. In this work, we have performed ex-situ reactor experiments to explore the gasification ability of Turkish lignite and to simulate and characterize the UCG process. Preliminary coal gasification tests were carried out with oxygen and steam for Malkara Pirinççesme (Turkey) lignite, which is characterized by high ash contents and low moisture. During the gasification experiments, temperature distribution inside the coal seams was monitored and the syngas outflow rate was measured. Additionally, syngas compositions were analysed using an MRU Varioplus Syngas analyser and an Agilent 7890B gas chromatograph. Results represent that the Turkish lignite sources are suitable for gasification. Experiments were divided into several phase: 1) oxygen phase 2) air- oxygen mixture phase, 3) steam-oxygen mixture phase, 4) air-oxygen-steam mixture phase, The syngas produced in the oxygen gasification stage was characterized by high calorific values of 6-9 MJ/Nm$^3$, carbon monoxide contents of 23-33 vol.%, and hydrogen contents of 20-30 vol.%. Also, computational models of the experimental set-up was built and simulations were performed in ANSYS Fluent to elucidate the optimal parameters required for the high calorific syngas production; such as gasification agent rates, shape of the gasification channel, etc.

## 1.Introduction

Underground coal gasification (UCG) stands out as a logical option to produce energy in Turkey, considering Turkey's large lignite sources and importing country position in terms of energy in global conjuncture. As declared by Turkish General Directorate of Mineral Research and Exploration, Turkey has over 15 billion tons of lignite reserve; however, only 8 percent of these have the high calorific value of 3000 kcal/kg, while 60 percent have less than 1500 kcal/kg. [1, 2, 3].

The main purpose of UCG method is to provide flammable gas (syngas) which mostly comprises $CO_2$, CO, $H_2$ and $CH_4$ [4, 5]. [6] Liu, et al. (2009) has researched the hydrogen production in an ex-situ experiment. They have concluded that $CO_2$ decomposition is increased from 50% to 96%, when the temperature increased from 850°C to 950°C. [7] Daggupati, et al. (2010) manifested the importance of presence cavities by analysing 29 different procedures varying the parameters such as, oxygen feed flow rate, duration of experiment and the distance between gas inlet and outlet points. In another experiment, [8] Stanczyk, et al. (2010) concluded after 54 hours of experimentation that the syngas produced by gasification of lignite with steam in an ex-situ reactor had 7.8 MJ/m$^3$ of calorific value with 46.3% hydrogen content. Also, they stated that gasification with steam is not effective for the used lignite sample with 53% humidity. Another study by [9]. Stanczyk, et al. (2011) proved that the use of air in gasification lowers the quality of the syngas and sudden break-downs in the process occur. On the other hand, they successfully produced a 4.18 MJ/m$^3$ calorific-valued syngas with a composition of 23% $H_2$ and 6.3% CO by adjusting volumetric ratio of oxygen-air mixture (gasification agent) to 2:1. [10] Similar study by Kacur, et al. (2014) has also proven that, a well-established oxygen-air ratio of the gasification agent is a crucial factor in gasification processes. In their experiments, they produced 8.18 MJ/m$^3$ calorically-valued syngas at 1146°C when oxygen-air ratio was volumetrically 1:5 and 8.21 MJ/m$^3$ at 1089°C when oxygen-air ratio was volumetrically 1:3, in contrast to the 2.9 MJ/m$^3$ obtained when the gasification agent was only air.

In addition to these experimental works, several mathematical models have been developed to estimate the performance of the UCG process. A review of mathematical modeling of UCG can be found elsewhere [11, 12, 13].

This paper presents the results of the experimental study performed in an ex-situ gasification reactor which simulates the underground coal gasification of Turkish lignite. In the gasification experiment, oxygen, air, steam and their mixtures were used to find the optimum gasification agent mixture and its composition.

## 2.Experimental

### 2.1 Experimental set up

An experimental set-up was designed and built, which is shown in Figure 1.

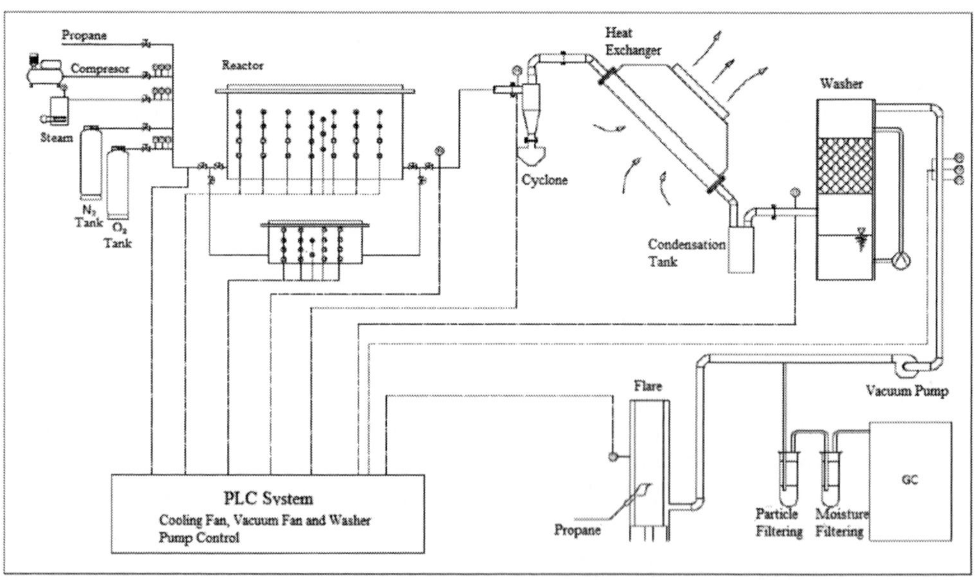


Figure 1: Schematic diagram of semi-pilot scale UCG.

The experimental apparatus was designed to enable arbitrary adjustment of air, steam and oxygen flows, while temperature inside the reactor was monitored. The apparatus consists of gasifying medium inlets with valves and flow regulators, the ex-situ reactors, cyclone, gas cooling system, gas purification system, gas chromatography and gas outlet and flare system

Throughout the whole process, SCADA system has been established to observe the temperature, pressure and flow rate of the system. SCADA system also controls the vacuum pump, heat exchanger, steam generator, compressor and the water pump.

## 2.2 Gasification Reactor

An ex-situ reactor has been built in the shape of a rectangular prism in which the lignite block and the surrounding stratum have been placed. Reactor has the external dimensions 0.65 x 1.0 x 2.00 meter. Refractory concrete has been chosen as an insulating material which is heat-resistant and has low thermal conductivity. A gasification channel was drilled in the bottom of the coal block along the lateral direction. To control the temperature along the gasified coal seam, the experimental unit was equipped with set of 25 thermocouples as shown in Figure 2a. Figure 2b describe the locations of the thermocouples on the reactor. As seen in Figure 2b thermocouples 8, 1216,24,and 28 are placed inside the gasification canal. The other fourteen thermocouples (numbers 2, 6, 7, 8, 10, 11, 12, 13, 14, 15, 19, 22, 23, 27, 38 and 39) have measured the temperatures inside the coal block.

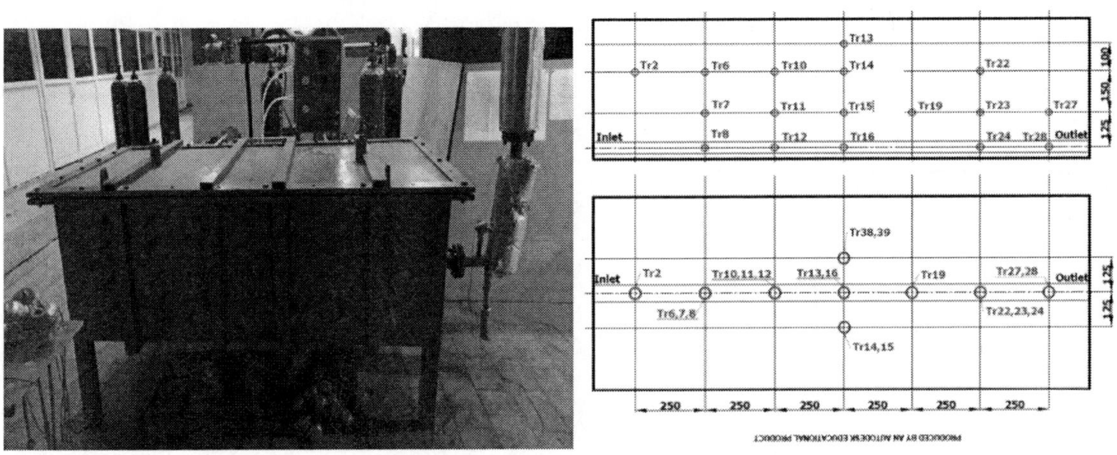


Figure 2: a) A picture of ex situ gasification reactor b) Thermocouple position

## 2.3 Lignite

For the experiments, coal samples have been taken from the third seam in Prinççesme/Malkara lignite site which is managed by Uysal Mining Co. The ultimate, proximate, and calorific value analysis of the coal has been done in TUBITAK-MAM. The analysis of the coal is summarized in Table 1.

Table 1: The analysis data of Malkara/Pirinççeşme lignite.

Table 1: The analysis data of Malkara/Pirinççeşme lignite.

| | Analysis Type | Original Sample | Dry in Air | Dry Sample | Standard |
|---|---|---|---|---|---|
| Approximate Analysis (wt. %) | Moisture | 25.17 | 15.27 | - | ASTM D 7582-2012 |
| | Ash | 17.95 | 20.33 | 23.99 | ASTM D 7582-2012 |
| | Volatile | 28.74 | 32.55 | 38.41 | ASTM D |

| | | | | | |
|---|---|---|---|---|---|
| | Matter | | | | 7582-2012 |
| Heating Value [kcal/kg] | Low Heating Value | 3602 | 4151 | 4998 | ASTM D-5865 |
| | High Heating Value | 3894 | 4409 | 5203 | ASTM D-5865 |

| | Unit | In Dry Sample | Standard |
|---|---|---|---|
| C (Carbon) | % mass. | 57.02 | ASTM D 5373-14 |
| H (Hydrogen) | % mass. | 3.99 | ASTM D 5373-14 |
| N (Nitrogen) | % mass. | 1.44 | ASTM D 5373-14 |
| S (Sulphur) | % mass. | 5.41 | ASTM D 4239-14 |
| Ash | % mass. | 23.99 | ASTM D 7582-12 |
| Oxygen | % mass. | 8.15 | ASTM D 3176-09 |

## 2.3 Experimental Procedure

The selected coal block has the following dimensions and weight: 160cm(L)x50cm(W)x50cm(H), around 650 kg respectively. After that, the coal block has been placed to reactor. Fine coal was used to fill the gap between the coal block and the reactor inner wall. First the stratum and then the isolation material was placed over the coal block. Thereby, the total weight of the coal has reached to around 900 kg.

Even though the reactor has been designed to minimize the heat loss, the coal block was covered by refractory bricks and the reactor cover was closed in order to work under adiabatic condition. .

The experiments were planned to be divided into several phases: 1) ignition phase I of the coal seam, 2) oxygen phase where pure oxygen was supplied to the reaction zone to heat up the coal seam and to accumulate sufficient amount of thermal energy, 3) air- oxygen mixture phase, 4) steam-oxygen mixture phase, 5) air-oxygen-steam mixture phase. The main aim of this experiments was to find the optimal gasification agent flow rate and suitable gasification agent mixture for rich syngas production. The gasification phase of the experiments is shown in Table 2. The main aim of this experiments was to find the optimal gasification agent flow rate and suitable gasification agent mixture for rich syngas production.

Table 2: Gasification phase of the experiments

| Stage | Ignition | Gasification | | | | | | | | | | |
|---|---|---|---|---|---|---|---|---|---|---|---|---|
| Gasification Agent(s) | Air | $O_2$ | Air/$O_2$ | | | $O_2$ | $H_2O/O_2$ | $O_2$ | Air/$O_2$ | Air/$H_2O/O_2$ | Air/$O_2$ | Air/$H_2O/O_2$ |
| Duration [h] | 4 | 2,5 | 8,5 | | | 0,5 | 0,5 | 2 | 1,5 | 1 | 1,5 | 2,5 |
| Ratio | - | - | 3:1 | 1,5:1 | 2:1 | - | 1,5:1 | - | 1:2 | 1:2:2 | 1,5:1 | 1:2:2 |
| Flow Rate (Total) [$m^3$/h] | - | 2-3 | 12 | 8-9 | 10 | 2,5-3 | 7 | 3 | 6 | 9-10 | 5 | 8-9 |

## 4. Results and discussion

In the experiments, all of the planned gasification phases with gasification agents listed in Table 2 were performed. In Figure 3, the picture of the coal block was given before (a) and after (b) the experiment. As clearly shown in Figure 3, after the coal seam in the ex-situ reactor was gasified, a cavity occurred.

a) b)

Figure 3: The view of the block coal a)before and b) after the experiment.

The experiment results will be presented and discussed below considering the several different gasification phases to find the optimum gasification mixture and its composition.

### Oxygen supply phase

After the ignition phase and each time before switching to another gasification agent mixture, only pure oxygen was supplied to the gasification zone to heat the coal seam up to the optimal level of 800-1000°C for gasification. Therefore, the optimal oxygen flow rate was adjusted based on the measurements from the thermocouple placed in the coal seam and simultaneous analysis of the composition of the generated syngas at the reactor outlet. The oxygen flow rate in the experiments lied in the range of 2-3 $m^3$/h. To test the different gasification phases in this experiment, oxygen heating phase was repeated totally 3 times.

The temperature in the coal seam and the reactor canal during the gasification were recorded and plotted in the Figure 4 and 5. From the Figure 4, it is seen that the temperature in the gasification zone in seam reached the highest level of 800-1100 $^0$C, which is the favourable temperature range for gasification.

Figure 4: Temperature in the coal seam

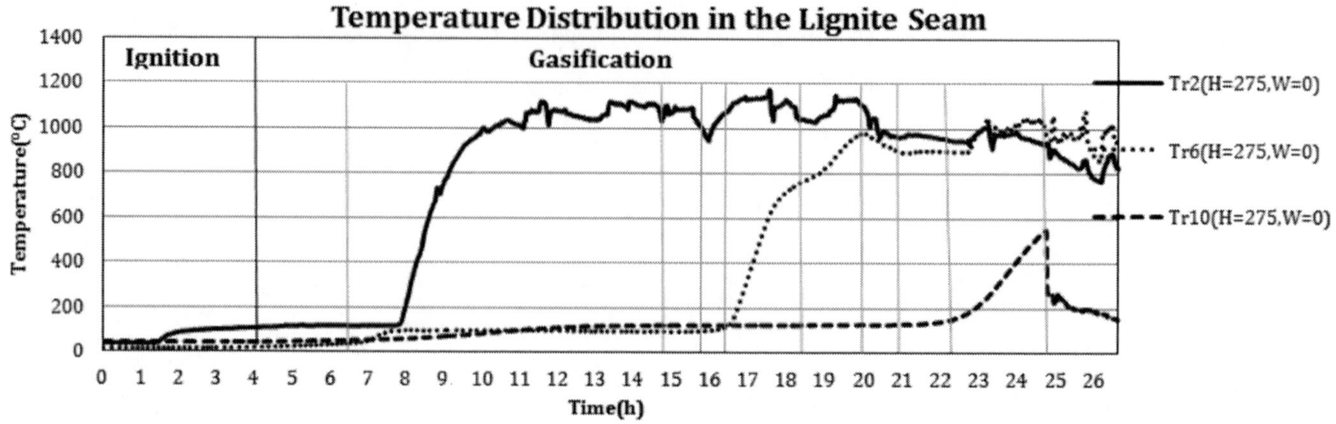


Figure 4: Temperature in the coal seam

In addition, the temperature at several positions horizontally along the coal seam was also measured and drawn in Figure 4. The temperature measurements create a continuous profile in the certain time interval. It means that the gasification zone (cavity) in the coal seam moved forwards as expected. Switching pure oxygen supply after the gasification agent mixture seems to be the best way of heating up the coal seam again.

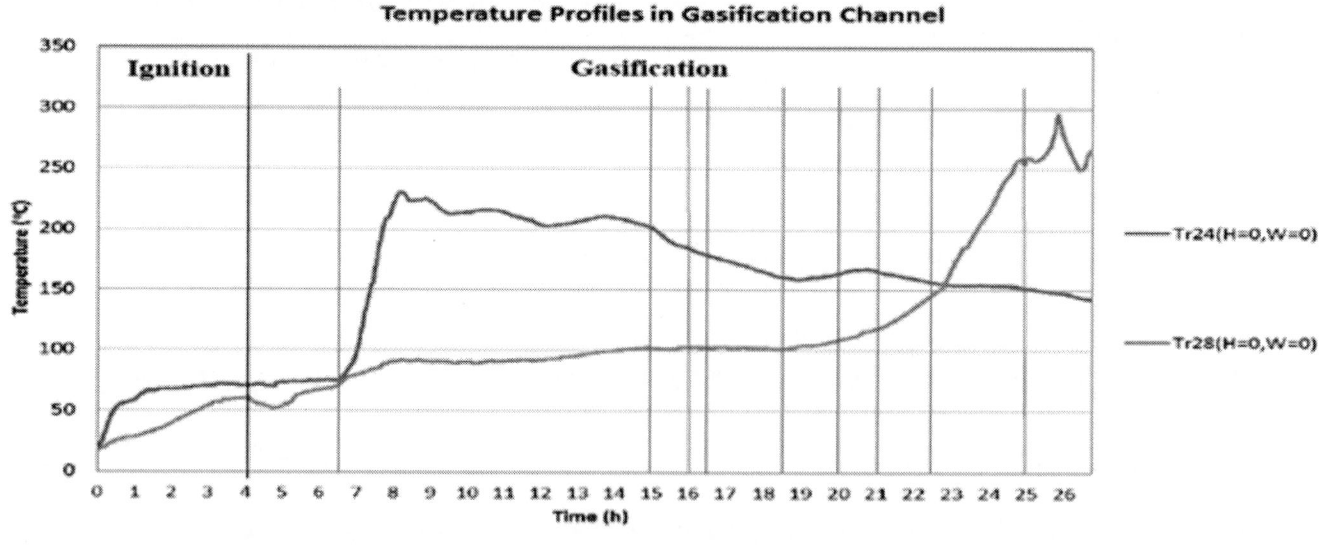


Figure5: Temperature in the gasification canal

The temperature in the gasification was also measured and shown in Figure 5. The relatively lower temperature in gasification canal can be explained by the fact that the produced syngas cools up the canal by losing energy in the gasification zone.

Figure 6 shows the syngas composition measured at the reactor outlet. Generally, the compositions of the components in the produced syngas fluctuated depending on the gasification phase. In the first oxygen phase,

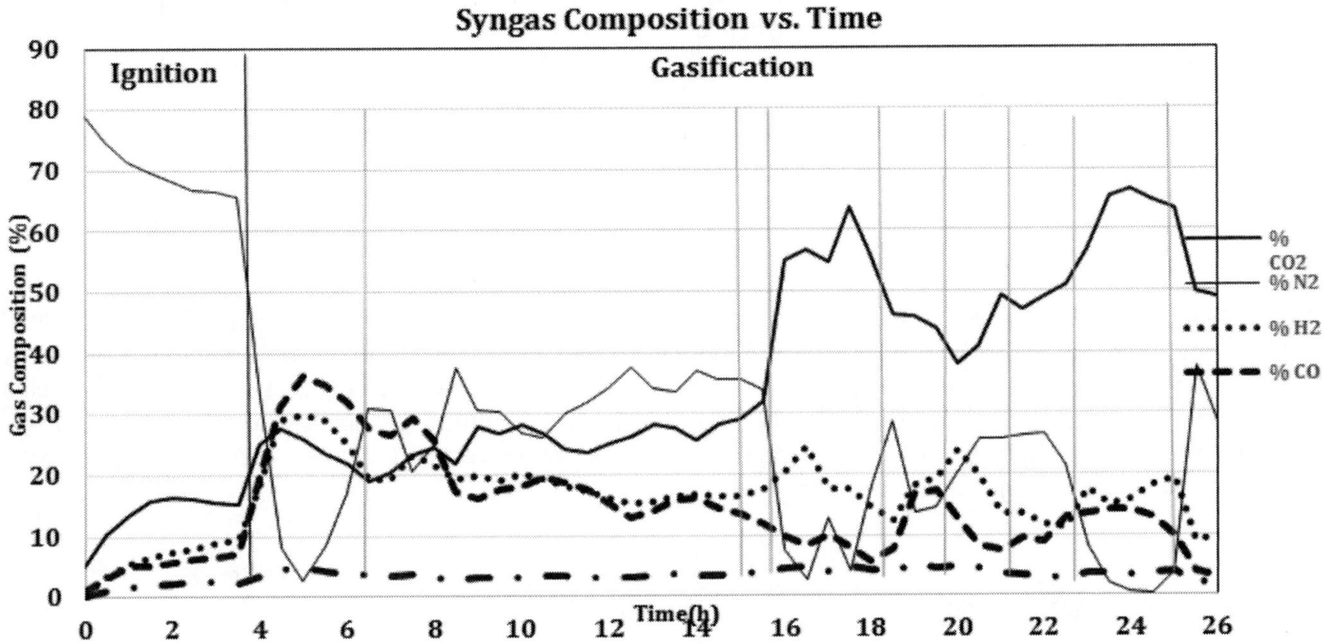


Figure 6: Measured syngas compositions in the several gasification phase

a rich content of $H_2$ (30 vol. %) and CO (35 vol. % ) was produced, however in the further oxygen phase, this value was not achieved as a result of the lower temperature in the coal seam. It is clearly understood that the lignite moisture is used in $H_2$ production reaction. In other words, additional supply of steam is not needed in the gasification reactor, because the seam moisture seems to be sufficient for $H_2$ reaction.

Figure 7 depicts the results of the gasification agent flow rate, the produced syngas flow rate and the heat value of the produced syngas depending on the gasification phase. As seen in Figure 8, the pure oxygen gasification result in the richest syngas generation and the maximum syngas flow rate due to high temperature which speeds up the $H_2$ and CO gasification reactions.

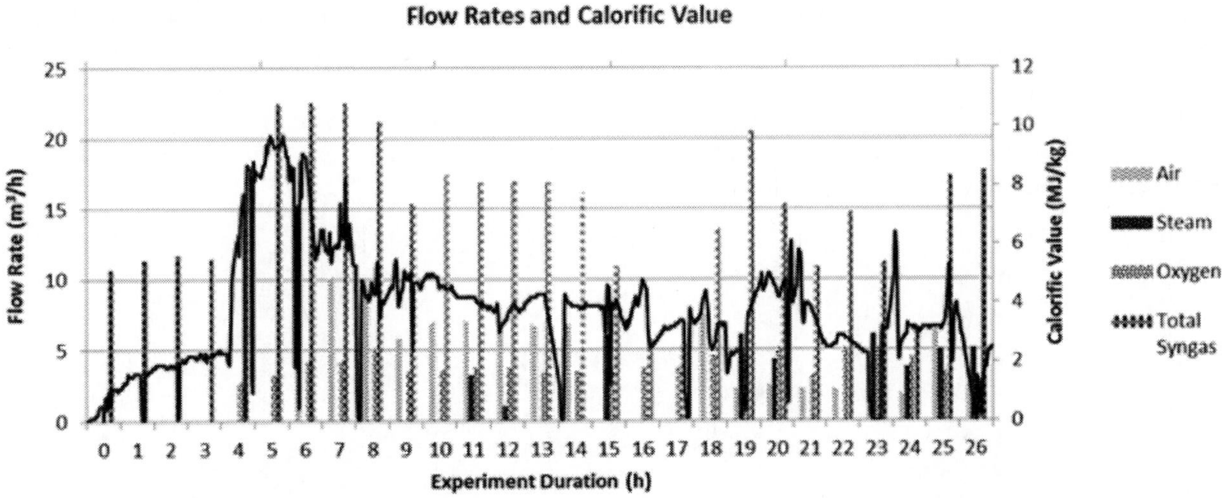


Figure 7: The measured gasification agent flow rate, the produced syngas flow rate and the heat value of the produced syngas depending on the gasification phase

**Air-oxygen (oxygen enriched air) gasification**

During the gasification experiment, air-oxygen gasification was performed totally 3 times using mixture ratios 3:1, 2: 1 und 1.5:1 to find an optimal air-oxygen mixture ratio. The measured temperature in the coal seam, composition of the produced syngas, its calorific value and its flow rate and supplying gasification agent were also presented in the Figure 4, 5, 6 and 7. The experiment results indicate that gasification of lignite seam with an air-oxygen mixture agent improved the syngas quality extremely comparing to pure air gasification. A rich syngas production ($H_2$ 18 vol.%) and CO (20 vol. 35% ) was achieved even with air-oxygen mixture ratio of 1.5:1.

**Steam - oxygen gasification**

Before supplying steam to the gasification reactor, pure oxygen gasification heated up the coal seam to the optimal high temperature (1000 $^0$C). As seen in Figure 4,5,6 and 7, as soon as steam was supplied to the reactor, $H_2$ production increased extremely to 25 vol.% , but the coal seam temperature began to fall down due to steam evaporation, for which steam consume thermal energy . Therefore, an optimal steam-air ratio for gasification was required to hinder the decrease in the seam temperature. From the experiment, it was seen that a steam-oxygen ratio of 2:1 was suitable for such gasification phase.

**Air-steam- oxygen gasification**

An air-steam- oxygen gasification was performed two times with the mixture ratio of 1:2:2, respectively. Under this gasification condition, the seam temperature tended to decrease extremely and the syngas production declined also drastically (Figure 4,5,6 and 7). Therefore, use of such air-steam- oxygen gasification was not recommended.

**CFD Simulation**

Computational models of the experimental set-up was built and simulations were performed in ANSYS Fluent to elucidate the optimal parameters required for the high calorific syngas production; such as gasification agent rates, shape of the gasification channel, etc. First of all, 2D, planar geometry was created and exact inner dimensions of the side view of the ex-situ reactor were used for the calculations. Mesh was generated with quadrilateral shaped cells in ANSYS Meshing. The conservation equations for $O_2$, $CO_2$, CO, $H_2O$, $H_2$, C and coal volatiles from Fluent's database (mv_vol) were discredited and solved. To validate the model, outcomes of the gasification simulation were compared with the results obtained from the gasification experiment with the Turkish lignite.

The first results of the CFD model simulation is shown in Fig 8. There, the $O_2$ consumption and $CO_2$ generation in the reactor by the lignite gasification reactions are clearly seen. As future work, the model will be developed further to simulate all the gasification parameters.

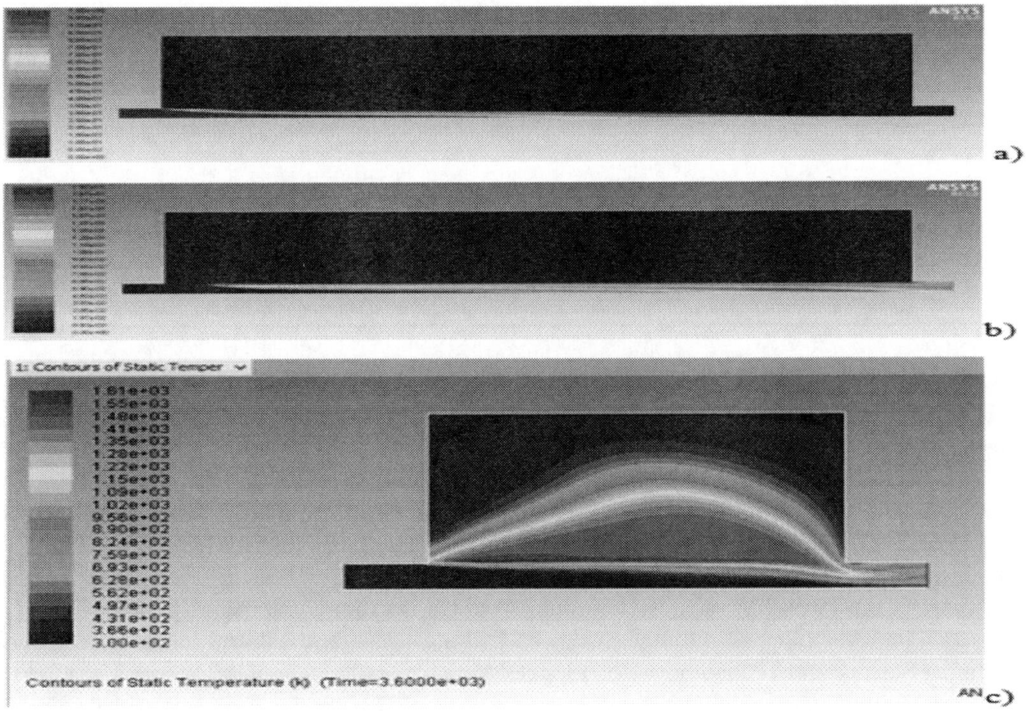


Figure 8: CFD-FLUENT Results: a)$O_2$ and b)$CO_2$ composition and c)seam temperature in rector

## 5. Conclusion

- Underground coal gasification (UCG) with Turkish lignite was simulated in a ex-situ reactor build in Istanbul Technical University Combustion/Gasification Laboratory.
- UCG experiments were carried out at several gasification phases using different gasification agent mixtures; 1) oxygen phase where pure oxygen was supplied to the reaction zone in order to heat up the coal seam and to accumulate sufficient amount of thermal energy, 2) air- oxygen mixture phase, 3) steam-oxygen mixture phase, 4) air-oxygen-steam mixture phase,
- In order to achieve appropriate thermal conditions inside the reactor, a relatively high content of oxygen supply was required because of the high moisture content in the lignite (about 27 vol.%),
- However, it is likely that Turkish lignite with this moisture content would give better results for $H_2$ production.
- In all gasification phases, a gasification mixture agent with a high oxygen content was necessary to sustain the gasification process.
- The experiment results demonstrated that gasification with the application of air and steam rich mixture is difficult because of the rapid decrease in temperature.
- In the gasification experiment,
  A-in the pure oxygen phase, average $H_2$ content in the product gas was 20.9 vol.% and CO was18.3 vol.% and the calorific value was 5.14 MJ/$m^3$.
  B-the optimum air/oxygen ratio was 1.5:1 by volume. Under such conditions, the average $H_2$ content in the product gas was 15.7 vol.% and CO was 14,8 vol.% and the calorific value was 3.87 MJ/$m^3$.
  C- The optimum steam/oxygen ratio was 2:1. The average $H_2$ content in the product gas was 21.3 vol.% and CO 7.6 vol.%. Gas calorific value was 3.60 MJ/$m^3$.
  D-The optimum steam/oxygen/ ratio was 1:2:2. The average $H_2$ content in the product gas was 14.2 vol.% and CO 9.0 vol.%. Gas calorific value was 2.82 MJ/$m^3$
- A mathematical modelling study of gasification coal seam was performed using FLUENT. The first results demonstrated that Fluent model can be used to simulate the seam gasification.

- Further gasification experiments with the vertical wells and also with the lignite coal samples from different locations were planned. Additionally the mathematical modelling studies will continue.


**Acknowledgements**

The works presented in this paper have been performed in the frame of "Development of Underground Coal Gasification Technology For Turkish Lignite" project and supported by TUBİTAK-ARDEB under the contract number 113M038.


## 6. References


[1]http://www.tki.gov.tr/Dosyalar/Dosya/K%C3%B6m%C3%BCrSektorRaporu2009.pdf

[2]Ersin, M. (2006), Türkiye'de Linyit Kömürlerinin Enerji Kaynağı Olarak Önemi, (Yüksek Lisans Tezi), İÜ Sosyal Bilimler Enstitüsü, İstanbul.

[3]Ünalan, G. (2010). Kömür Jeolojisi, Maden Tetkik ve Arama Genel Müdürlüğü, Ankara.

[4]Shimada, S., Ohga, K., Tamari, A and Ishii, E. (1995). Underground Coal Gasificaton-How Much Does It Cost? (Case Study in Japan) MinChem'95, The Fifth Symposium on Mining Chemistry İstanbul, Türkiye.

[5]Glaser, R.R. ve Johnson, L.A. (1986). Physical Simulation of Underground Coal Gasification in an Eastern Bituminous Coal. Proceedings of the Twelfth Annual Underground Coal Gasification Symposium. pp. 408-418.

[6]Liu, S., Wang, Y.Y., Zhao, K. ve Yang, N. (2009). Enhanced-hydrogen gas production through underground gasification of lignite. Mining Science and Technology. Vol.19, pp 389-394.

[7]Daggupati, S., Mandapati, R.N., Mahajani, S.M., Ganesh, A., Mathur, D.K., Sharma, R.K. ve Aghalayam, P. (2010). Laboratory Studies on Combustion Cavity Growth in Lignite Coal Blocks in the Context of Underground Coal Gasification, Energy. Vol.35, pp. 2374-2386.

[8]Stanczyk, K., Smolinski, A., Kapusta, K., Wiatowski, M., Swiadrowski, J., Kotyrba, A. ve Rogut, J. (2010). Dynamic Experimental Simulation of Hydrogen Oriented Underground Gasification of Lignite. Fuel. Vol.89, pp.3307-3314.

[9]Stanczyk, K., Howaniec, N., Smolinski, A., Swiadrowski, J., Kapusta, K., Wiatowski, M., Grabowski, J. ve Rogut, J. (2011). Gasification Of Lignite And Hard Coal With Air And Oxygen Enriched Air In A Pilot Scale Ex-Situ Reactor For Underground Gasification. Fuel. Vol.90, pp. 1953-1962.

[10]Kacur, J., Durdan, M., Laciak, M. ve Flegner, P. (2014). Impact Analysis Of The Oxidant In The Process Of Underground Coal Gasification. Measurement. Vol.51, pp 147-155.

[11]Wachowicz J., Laczny J. M., Iwaszenko S., Janoszek T., Cempa-Balewicz M., (2015). Modelling of Underground Coal Gasification Process Using CFD Methods. Arch. Min. Sci., Vol. 60, No 3, p.663-676

[12]Zogala A., Janoszek T., (2015). CFD Simulations of Influence of Steam in Gasification Agent on Parameters of UCG Process. Journal of Sustainable Mining 14

[13]Shirazi A. S., (2012). CFD Simulation of Underground Coal Gasification, University of Alberta, Department of Chemical and Materials Engineering, MSc Thesis.

# TRACE ELEMENTS ASSOCIATED WITH THE UPPER COAL ZONES OF THE SPRINGBOK FLATS BASIN, SOUTH AFRICA


***Valerie Nxumalo, Ms, Council for Geoscience, 280 Pretoria Street, Silverton, 0184, Pretoria, South Africa; University of Johannesburg, SOUTH AFRICA***
***vnxumalo@geoscience.org.za, +27- 12-8411511***

*Nicola Wagner, Assoc. Prof, University of Johannesburg, PO Box 524 Auckland Park, 2006, Johannesburg, SOUTH AFRICA, nwagner@uj.ac.za , +27-11-5594719*

*Wojciech Przybylowicz, Dr., iThemba LABS, National Research Foundation, P.O. Box 722, Somerset West 7129, SOUTH AFRICA ;AGH University of Science and Technology, Faculty of Physics & Applied Computer Science, al. A. Mickiewicza 30, 30-059 Kraków, POLAND, przybylowicz@tlabs.ac.za, +27-21-8431166*

*Jan Kramers, Prof, University of Johannesburg, PO Box 524 Auckland Park, 2006, Johannesburg, SOUTH AFRICA, jkramers@uj.ac.za , +27-11-5594755*

*Clarisa Vorster, Dr., University of Johannesburg, PO Box 524 Auckland Park, 2006, Johannesburg, SOUTH AFRICA, clarisav@uj.ac.za , +27-11-5594784*

*Bruce Cairncross, Prof, University of Johannesburg, PO Box 524 Auckland Park, 2006, Johannesburg, SOUTH AFRICA, brucec@uj.ac.za , +27-11-559-4713*

*John Paul Hunt, Mr, Council for Geoscience, 280 Pretoria Street, Silverton, 0184, Pretoria, South Africa; University of Johannesburg, SOUTH AFRICA*
*jphunt@geoscience.org.za ,+27- 12-841 1246*



The coal zones of the Springbok Flats Basin, Limpopo Province, South Africa, occur in the upper and lower parts of the Vryheid Formation. They are characterised by intercalations of mudstones, carbonaceous mudstones, and bright coals. The upper coal zones can be subdivided into 3 seams: upper, middle, and lower coal seams, which are separated by mudstones and carbonaceous mudstones. The three seams are well developed in the central sector of the Basin and are characterised by 37 – 62 % coal and 38 – 64% mudstones (carbonaceous and non-carbonaceous). The coals are vitrinite rich (79.7% to 94.1%, mineral matter free basis), dominated by collotelinite, pseudovitrinite, corpogelinite, colledetrinite, telinite, gelinite and vitrodetrinite. Mean random vitrinite reflectance of the coals range between 0.61 and 0.75%, indicating coal rank of medium-rank high volatile bituminous C. The coal seams in the central sector of the Springbok Flats Basin show average moisture content between 1.68 to 2.12%, ash contents of 20.97 to 30.8%, volatile matter average values between 17.72 to 33.07 % (air dried basis) and fixed carbon average values between 40.08 and 50.4%. Average sulphur content of the upper coal seams in the central sector is 2.56% (air dried basis).The average gross calorific value of the coal seams range from 20.39 to 24.36 MJ/Kg (dry basis). The upper coal zones in the central sector of the Springbok Flats Basin are unique in terms of trace elements when compared to the coal zones of other South African coalfields (i.e Witbank, Highveld and Waterberg coalfields) and average global coal Clarke values. The coals are characterised by high amounts of potentially hazardous trace elements such as Zn, V, U, Ni, Mo, Mn, Cu and Co (average values of 46.48 ppm, 80.64 ppm. 106.04 ppm, 39.86 ppm, 26.50 ppm, 1448 ppm, 32.20 ppm and 27.62 ppm respectively). These trace elements are considered to be of environmental interest due to their diverse effects on plants, animals and humans. Proton-Induced X-Ray Emission (PIXE) and proton backscattering spectrometry (BS) in conjunction with Nuclear Microprobe showed that Zn is associated with sphalerite; V and U are finely disseminated in the coal matrix; Mo, Cu and Co are associated with euhedral, subhedral pyrite in the coal samples. Mn is associated with carbonate minerals (i.e. siderite) in coal. The upper coal zones of the Springbok Flats Basin have potential to be mined for electricity generation provided they can be beneficiated (reduction of toxic and hazardous trace elements) and uranium can be extracted (as a by-product).

***Formatting Issue - Best Available Paper/Graphic***

# Organic And Inorganic Characterisation Of Different Coals By Single Particle Analysis

**Sanja Potgieter-Vermaak**

**Maledi, N., Potgieter, J.H., Sparkes, R., Wagner, N.**

Manchester Metropolitan University

PCC 2016: International Pittsburgh Coal Conference 8 – 12 Aug 2016

# Rational - Introduction

Coal behaviour → chemical composition

Coal characterisation focuses on **bulk analysis**, as usage is bulk quantities

What is & why SPA?

Demonstrated in: environmental fields (air pollution, dust storms, climate change) and enables amongst others unambiguous in-vitro behaviour of PM, dust storm effects, etc.

IMPORTANCE FOR COAL:

- SPA can probe organic and inorganic character
- Structural information - organic component: potential for better coal rank classification, reactivity prediction, temperature dependence, etc.
- Structural information - inorganic component provides further detail on mineral content, especially on amorphous / low concentration compounds.
- The combination of the two provide insight in the organic / inorganic association

PREDICTION OF BEHAVIOUR IN APPLICATION

PCC 2016: International Pittsburgh Coal Conference 8 – 12 Aug 2016

***Formatting Issue - Best Available Paper/Graphic***

# Samples

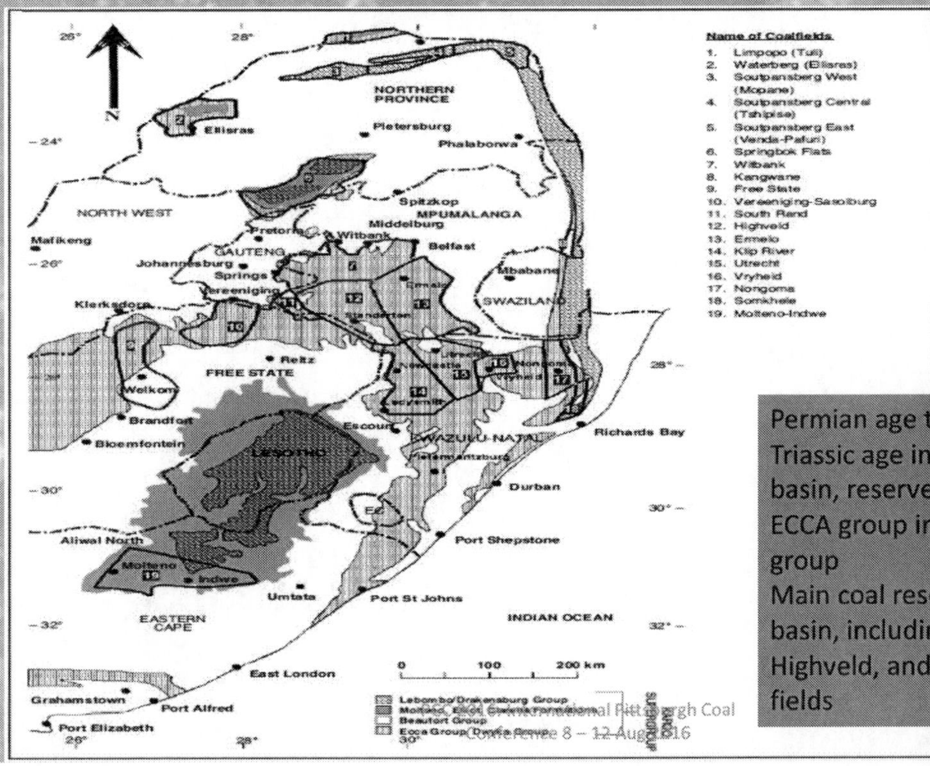


Permian age through to mid-Triassic age in the Gondwana basin, reserves spread over the ECCA group in the Karoo super group

Main coal reserves - central basin, including Witbank, Highveld, and the Ermelo coal fields

International Pittsburgh Coal Conference 8 – 12 August 2016


# Samples


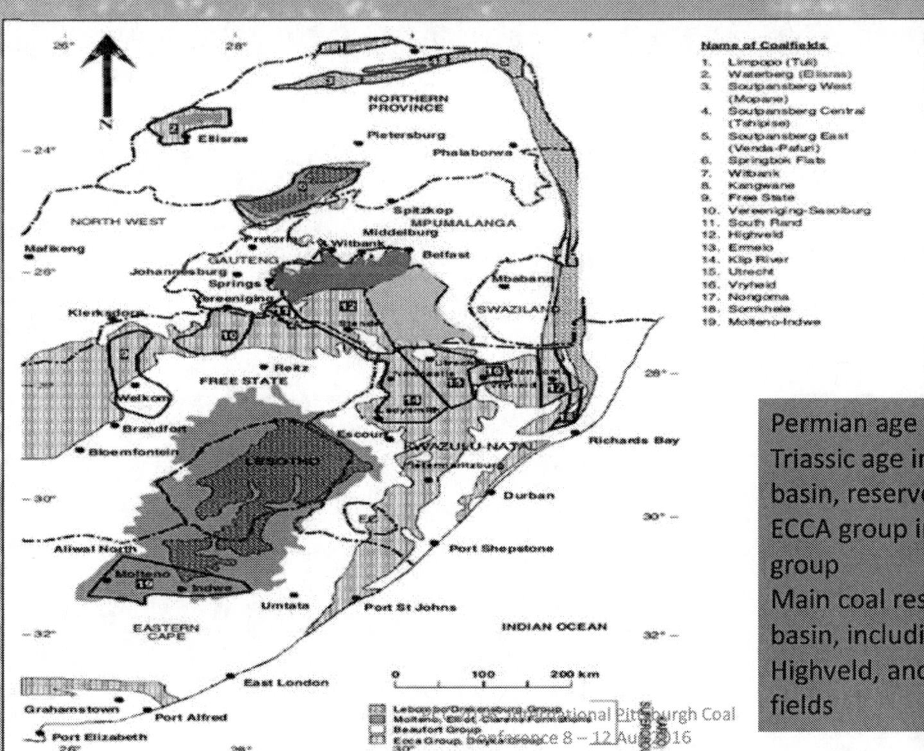


Permian age through to mid-Triassic age in the Gondwana basin, reserves spread over the ECCA group in the Karoo super group

Main coal reserves - central basin, including Witbank, Highveld, and the Ermelo coal fields

International Pittsburgh Coal Conference 8 – 12 August 2016

***Formatting Issue - Best Available Paper/Graphic***

| Coal chemistry | Waterberg | Witbank | Ermelo | Discard |
|---|---|---|---|---|
| **Proximate analysis (mass %)** | | | | |
| Moisture | 2.4 | 4.3 | 4.3 | 2.0 |
| Volatiles | 18.5 | 22.4 | 20.5 | 13.9 |
| Ash | 40.1 | 26.6 | 22.6 | 68.3 |
| Fixed Carbon | 39.0 | 46.7 | 53.6 | 15.8 |
| **Ultimate analysis (mass %)** | | | | |
| C | 48.3 | 53.2 | 60.3 | 17.7 |
| H | 2.9 | 3.4 | 3.7 | 1.8 |
| N | 1.1 | 1.4 | 1.5 | 0.5 |
| O | 10.9 | 13.1 | 14.3 | 8.2 |
| S* | 1.00 | 1.6 | 0.5 | 2.5 |
| C/H | 1.39 | 1.30 | 1.36 | 0.82 |
| **Forms of sulphur** | | | | |
| Pyritic | 0.09 | 0.11 | 0.05 | 0.09 |
| [illegible] | [illegible] | [illegible] | [illegible] | [illegible] |




| Coal chemistry | Waterberg | Witbank | Ermelo | Discard |
|---|---|---|---|---|
| **Proximate analysis (mass %)** | | | | |
| Moisture | 2.4 | 4.3 | 4.3 | 2.0 |
| Volatiles | 18.5 | 22.4 | 20.5 | 13.9 |
| Ash | 40.1 | 26.6 | 22.6 | 68.3 |
| **Fixed Carbon** | **39.0** | **46.7** | **53.6** | **15.8** |
| **Ultimate analysis (mass %)** | | | | |
| **C** | **48.3** | **53.2** | **60.3** | **17.7** |
| H | 2.9 | 3.4 | 3.7 | 1.8 |
| N | 1.1 | 1.4 | 1.5 | 0.5 |
| O | 10.9 | 13.1 | 14.3 | 8.2 |
| S* | 1.00 | 1.6 | 0.5 | 2.5 |
| **C/H** | **1.39** | **1.30** | **1.36** | **0.82** |
| **Forms of sulphur** | | | | |
| Pyritic | 0.09 | 0.11 | 0.05 | 0.09 |
| [illegible] | [illegible] | [illegible] | [illegible] | [illegible] |

PCC 2016: International Pittsburgh Coal Conference 8 – 12 Aug 2016

***Formatting Issue - Best Available Paper/Graphic***

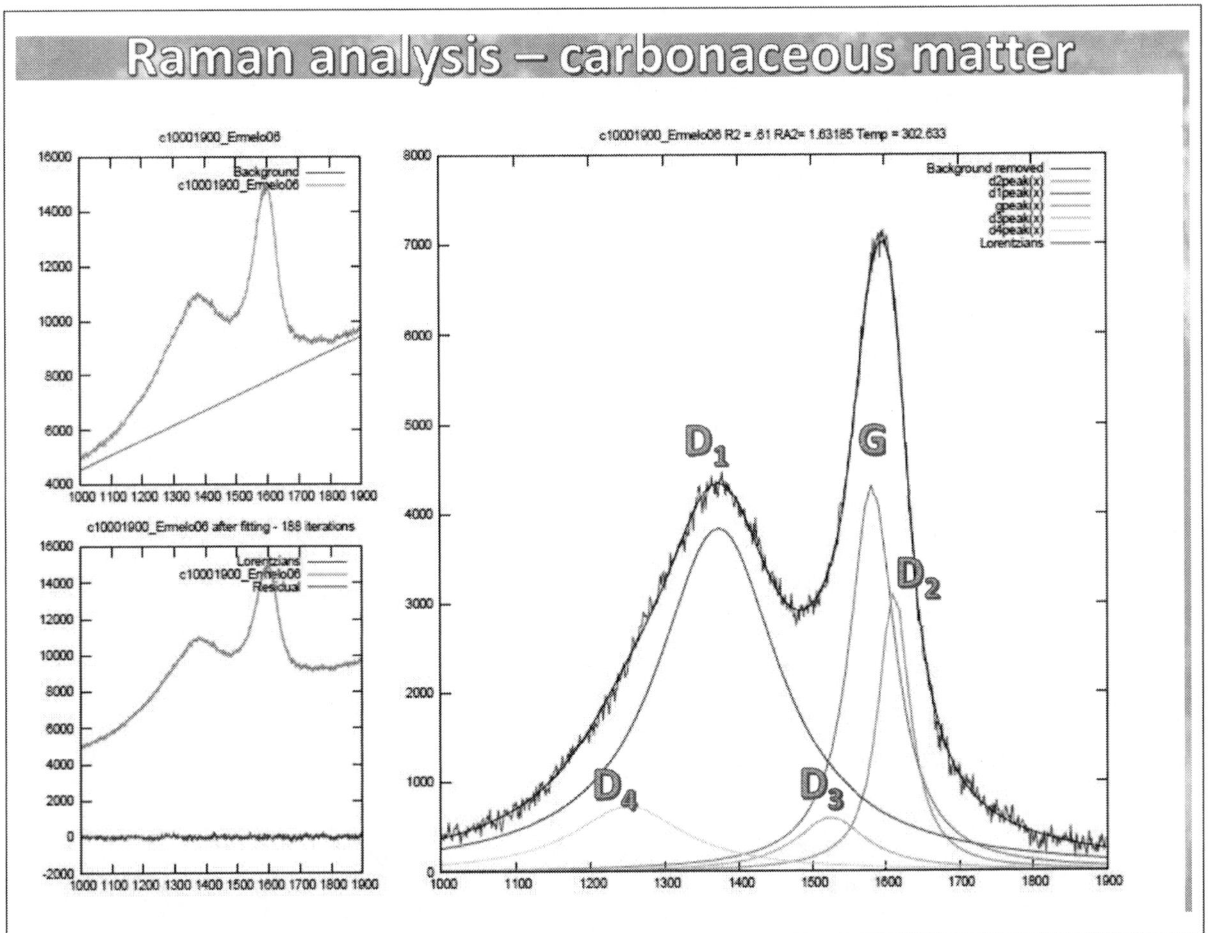


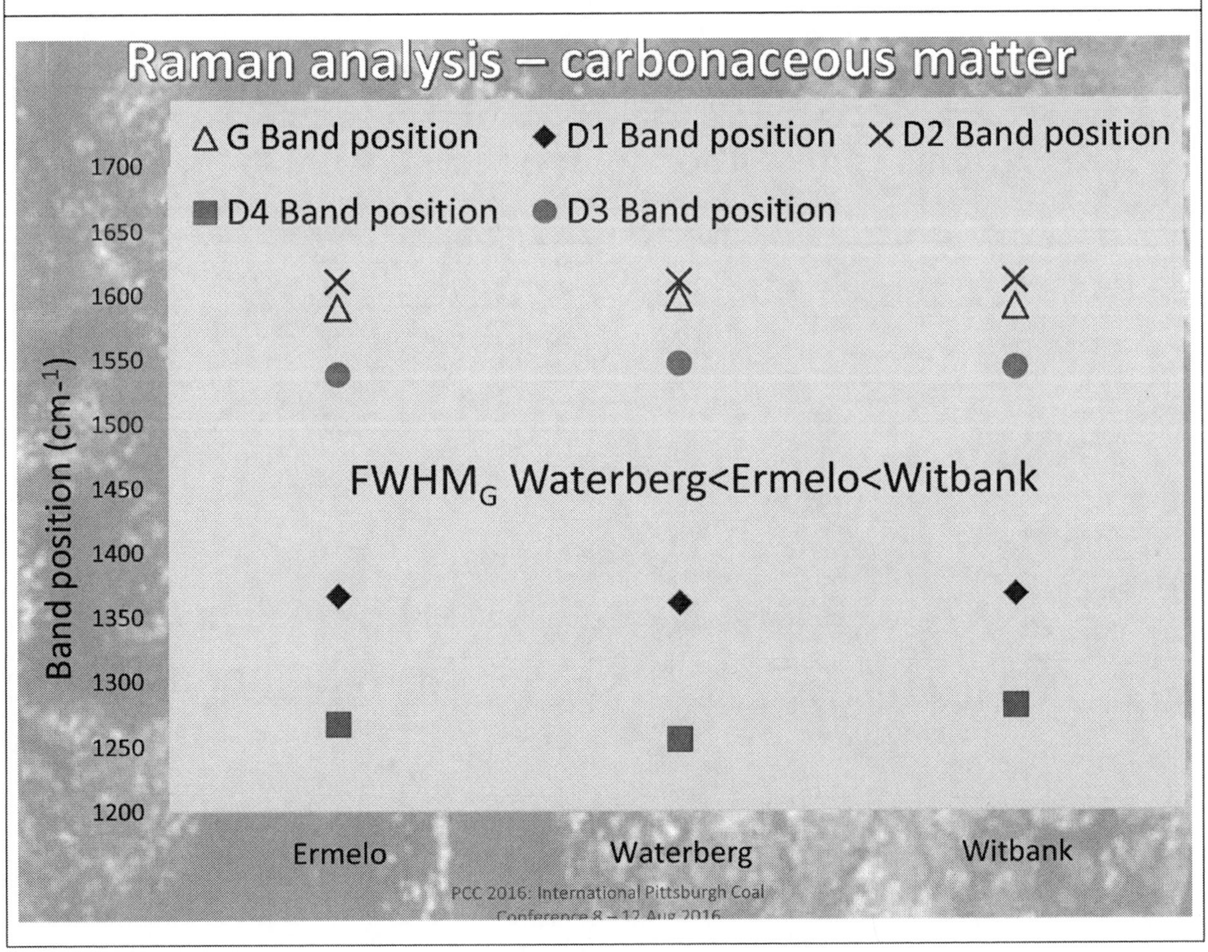

# Petrography

| Mineral (vol%) | Waterberg | Witbank | Ermelo | Discard |
|---|---|---|---|---|
| Clays | 17.4 | 5.3 | 4.8 | 46.6 |
| Sulphides | 5.8 | 2.5 | 0.8 | 4 |
| Carbonates | 3.8 | 5.5 | 1 | 1.2 |
| Quartz | 16.2 | 10.5 | 3.4 | 13.6 |
| Other | 0.4 | 0.2 | 0.2 | 0.2 |
| **Total mineral matter** | **43.6** | **24** | **10.2** | **65.6** |
| **Macerals** | | | | |
| Vitrinite | 2.8 | 15.8 | 9.6 | 1.4 |
| Inertinite | 51.2 | 57.4 | 71.6 | 29 |
| Liptinite | 2.4 | 2.9 | 9.6 | 4 |
| **Total macerals** | **56.4** | **76.1** | **90.8** | **34.4** |
| **Rank Category** | **Medium Rank C** | **Medium Rank C** | **Medium Rank C** | |



# Petrography

| Petrographic composition (vol%) mmf basis | | | | |
|---|---|---|---|---|
| **Maceral Group** | **Waterberg** | **Witbank** | **Ermelo** | **Discard** |
| Vitrinite | 5.0 | 20.8 | 10.7 | 4.10 |
| **Inertinite** | **90.8** | **75.5** | **79.7** | **84.3** |
| Liptinite | 4.3 | 3.8 | 9.60 | 11.6 |

Inertinite-rich with vitrinite bands

***Formatting Issue - Best Available Paper/Graphic***

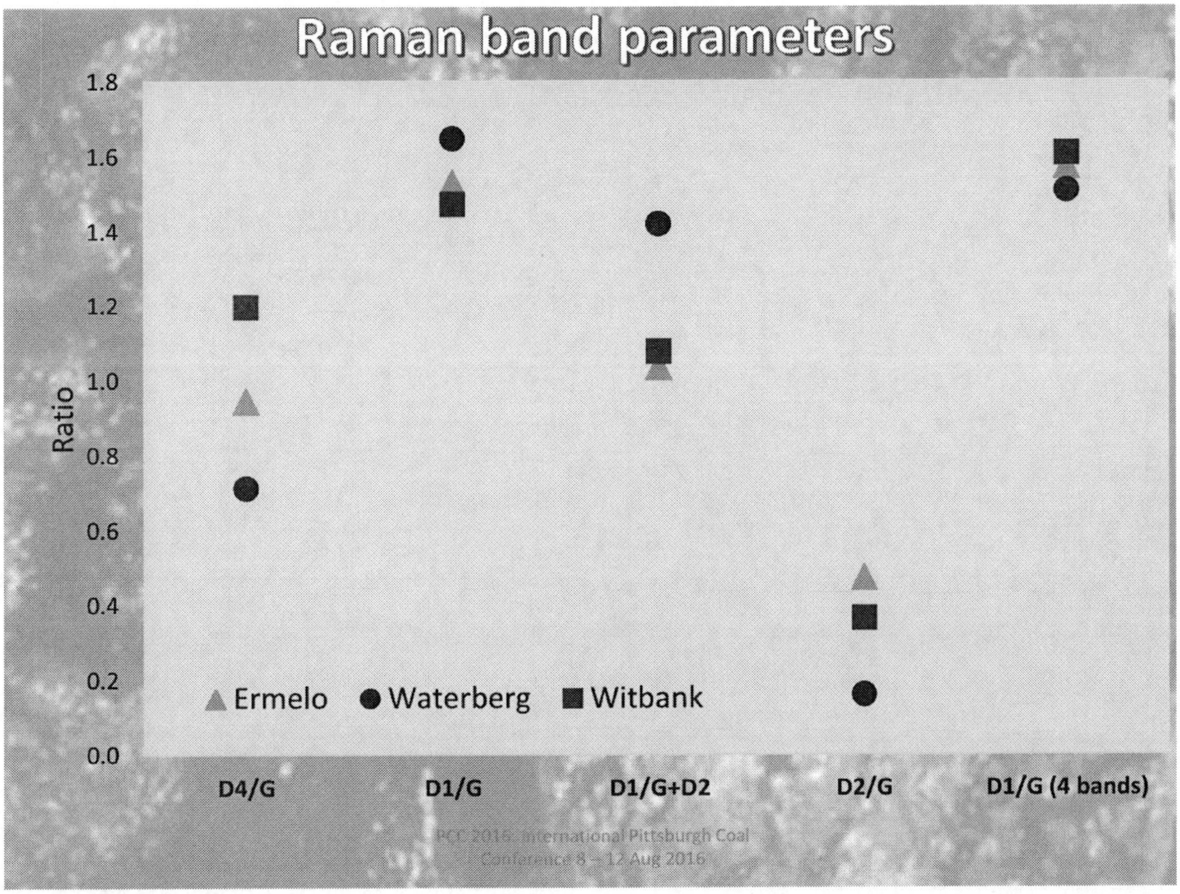


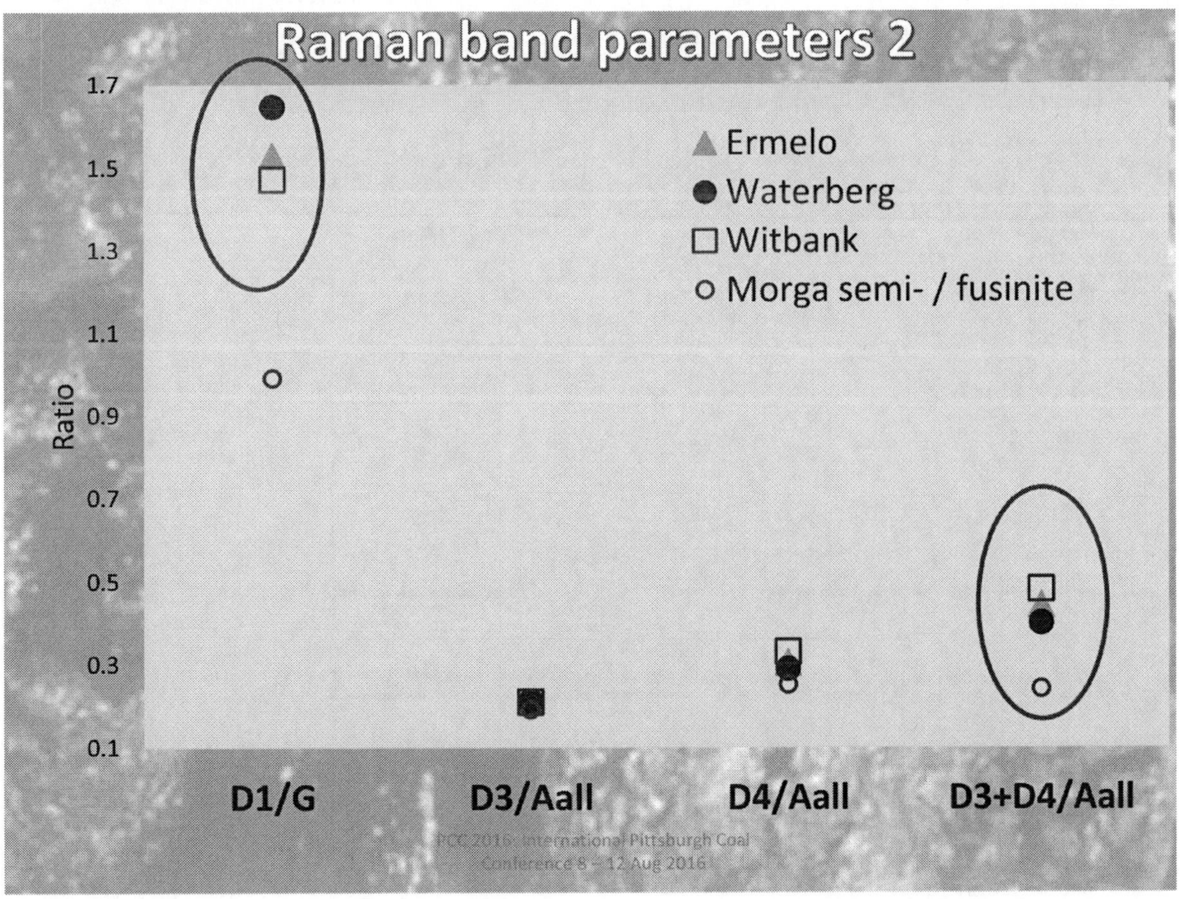

## Coal chemistry

| | Waterberg | Witbank | Ermelo | Discard |
|---|---|---|---|---|
| **Proximate analysis (mass %)** | | | | |
| Moisture | 2.4 | 4.3 | 4.3 | 2.0 |
| Volatiles | 18.5 | 22.4 | 20.5 | 13.9 |
| Ash | 40.1 | 26.6 | 22.6 | 68.3 |
| Fixed Carbon | 39.0 | 46.7 | 53.6 | 15.8 |
| **Ultimate analysis (mass %)** | | | | |
| C | 48.3 | 53.2 | 60.3 | 17.7 |
| H | 2.9 | 3.4 | 3.7 | 1.8 |
| N | 1.1 | 1.4 | 1.5 | 0.5 |
| O | 10.9 | 13.1 | 14.3 | 8.2 |
| S* | 1.00 | 1.6 | 0.5 | 2.5 |
| C/H | 1.39 | 1.30 | 1.36 | 0.82 |
| **Forms of sulphur** | | | | |
| Pyritic | 0.09 | 0.11 | 0.05 | 0.09 |



## Raman & CC-SEM-EDS – inorganic matter

### XRF bulk analysis of the raw coal

| % | Waterberg | Witbank | Ermelo | Discard |
|---|---|---|---|---|
| CaO | 0.5 | 0.8 | 3.5 | 7.5 |
| MgO | 0.2 | 0.5 | 2.0 | 2.4 |
| $K_2O$ | 0.3 | 0.8 | 0.3 | 0.8 |
| $Na_2O$ | 0.2 | 0.2 | 0.3 | 0.7 |
| $Fe_2O_3$ | 2.1 | 6.4 | 3.2 | 5.6 |
| $TiO_2$ | 2.9 | 1.6 | 1.9 | 1.9 |
| ***$Al_2O_3$*** | ***42.0*** | ***28.7*** | ***36.1*** | ***23.8*** |
| ***$SiO_2$*** | ***49.3*** | ***57.9*** | ***45.5*** | ***45.1*** |
| $SO_3$ | 0.1 | 0.6 | 4.0 | 9.9 |
| $P_2O_5$ | 0.4 | 0.2 | 1.0 | 0.9 |
| **TOT** | **97.8** | **97.8** | **98.5** | **97.7** |

Silicates & clays

***Formatting Issue - Best Available Paper/Graphic***

PXRD

| Mineral | Sample Code/ Chem Formula | Waterberg | Witbank | Discard | Ermelo |
|---|---|---|---|---|---|
| Quartz | $SiO_2$ | 6.2 | 23.1 | 29.7 | 1.2 |
| Kaolinite | $Al_2(Si_2O_5)(OH)_4$ | 78.4 | 47.8 | 60.5 | 97.7 |
| Calcite | $CaCO_3$ | | 2.7 | | |
| Aragonite | $CaCO_3$ | | 1.1 | | |
| Dolomite | $CaMg(CO_3)_2$ | 4.2 | 11.4 | | |
| Pyrite | $FeS_2$ | 8.9 | 8.6 | 4.2 | |
| Muscovite | $KAl_3Si_3O_{10}(OH)_2$ | | 1.6 | 5.3 | |
| Illite | $KAl_3Si_3O_{10}(OH)_2$ | | | | |
| Alunite | $KAl_3(SO_4)_2(OH)_6$ | | 1.7 | | |
| Gypsum | $CaSO_4(H_2O)_2$ | | 2 | | |
| Brushite | $CaPO_4(OH)(H_2O)_2$ | 1.4 | | | |
| Gorceixite | $BaAl_3(PO)_4(PO_3OH)(OH)_6$ | 0.9 | | | |
| Anatase | $TiO_2$ | | 0.2 | 0.3 | 1.1 |
| Total | | 100 | 100.2 | 100 | 100 |

SIROQUANT



PXRD

| Mineral | Sample Code/ Chem Formula | Waterberg | Witbank | Discard | Ermelo |
|---|---|---|---|---|---|
| Quartz | $SiO_2$ | 6.2 | 23.1 | 29.7 | 1.2 |
| Kaolinite | $Al_2(Si_2O_5)(OH)_4$ | 78.4 | 47.8 | 60.5 | 97.7 |
| Calcite | $CaCO_3$ | | 2.7 | | |
| Aragonite | $CaCO_3$ | | 1.1 | | |
| Dolomite | $CaMg(CO_3)_2$ | 4.2 | 11.4 | | |
| Pyrite | $FeS_2$ | 8.9 | 8.6 | 4.2 | |
| Muscovite | $KAl_3Si_3O_{10}(OH)_2$ | | 1.6 | 5.3 | |
| Illite | $KAl_3Si_3O_{10}(OH)_2$ | | | | |
| Alunite | $KAl_3(SO_4)_2(OH)_6$ | | 1.7 | | |
| Gypsum | $CaSO_4(H_2O)_2$ | | 2 | | |
| Brushite | $CaPO_4(OH)(H_2O)_2$ | 1.4 | | | |
| Gorceixite | $BaAl_3(PO)_4(PO_3OH)(OH)_6$ | 0.9 | | | |
| Anatase | $TiO_2$ | | 0.2 | 0.3 | 1.1 |
| Total | | 100 | 100.2 | 100 | 100 |

SIROQUANT

PCC 2016: International Pittsburgh Coal Conference 8 – 12 Aug 2016

***Formatting Issue - Best Available Paper/Graphic***

PXRD

| Mineral | Sample Code/ Chem Formula | Waterberg | Witbank | Discard | Ermelo |
|---|---|---|---|---|---|
| Quartz | $SiO_2$ | 6.2 | ***23.1*** | 29.7 | 1.2 |
| Kaolinite | $Al_2(Si_2O_5)(OH)_4$ | 78.4 | ***47.8*** | 60.5 | 97.7 |
| Calcite | $CaCO_3$ | | ***2.7*** | | |
| Aragonite | $CaCO_3$ | | ***1.1*** | | |
| Dolomite | $CaMg(CO_3)_2$ | 4.2 | ***11.4*** | | |
| Pyrite | $FeS_2$ | 8.9 | ***8.6*** | 4.2 | |
| Muscovite | $KAl_3Si_3O_{10}(OH)_2$ | | ***1.6*** | 5.3 | |
| Illite | $KAl_3Si_3O_{10}(OH)_2$ | | | | |
| Alunite | $KAl_3(SO_4)_2(OH)_6$ | | ***1.7*** | | |
| Gypsum | $CaSO_4(H_2O)_2$ | | ***2*** | | |
| Brushite | $CaPO_4(OH)(H_2O)_2$ | 1.4 | | | |
| Gorceixite | $BaAl_3(PO)_4(PO_3OH)(OH)_6$ | 0.9 | | | |
| Anatase | $TiO_2$ | | ***0.2*** | 0.3 | 1.1 |
| Total | | 100 | 100.2 | 100 | 100 |

SIROQUANT



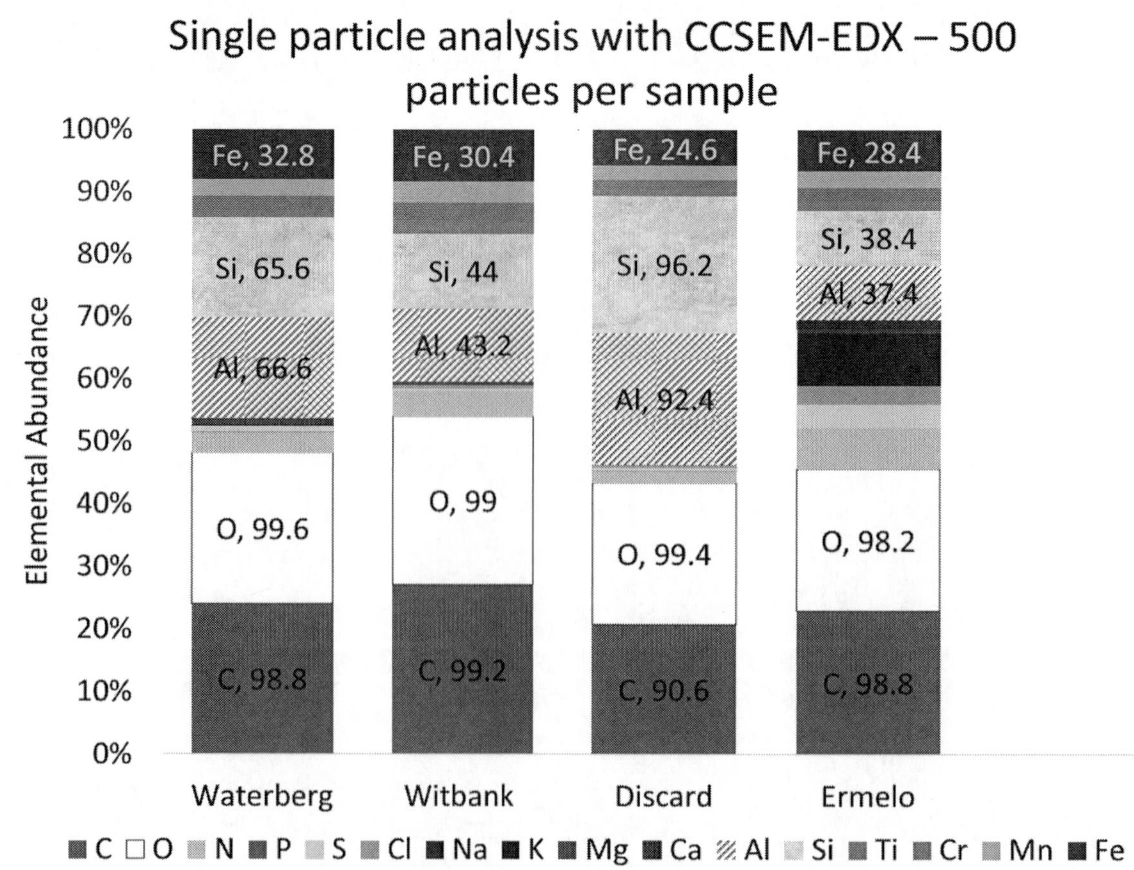

***Formatting Issue - Best Available Paper/Graphic***

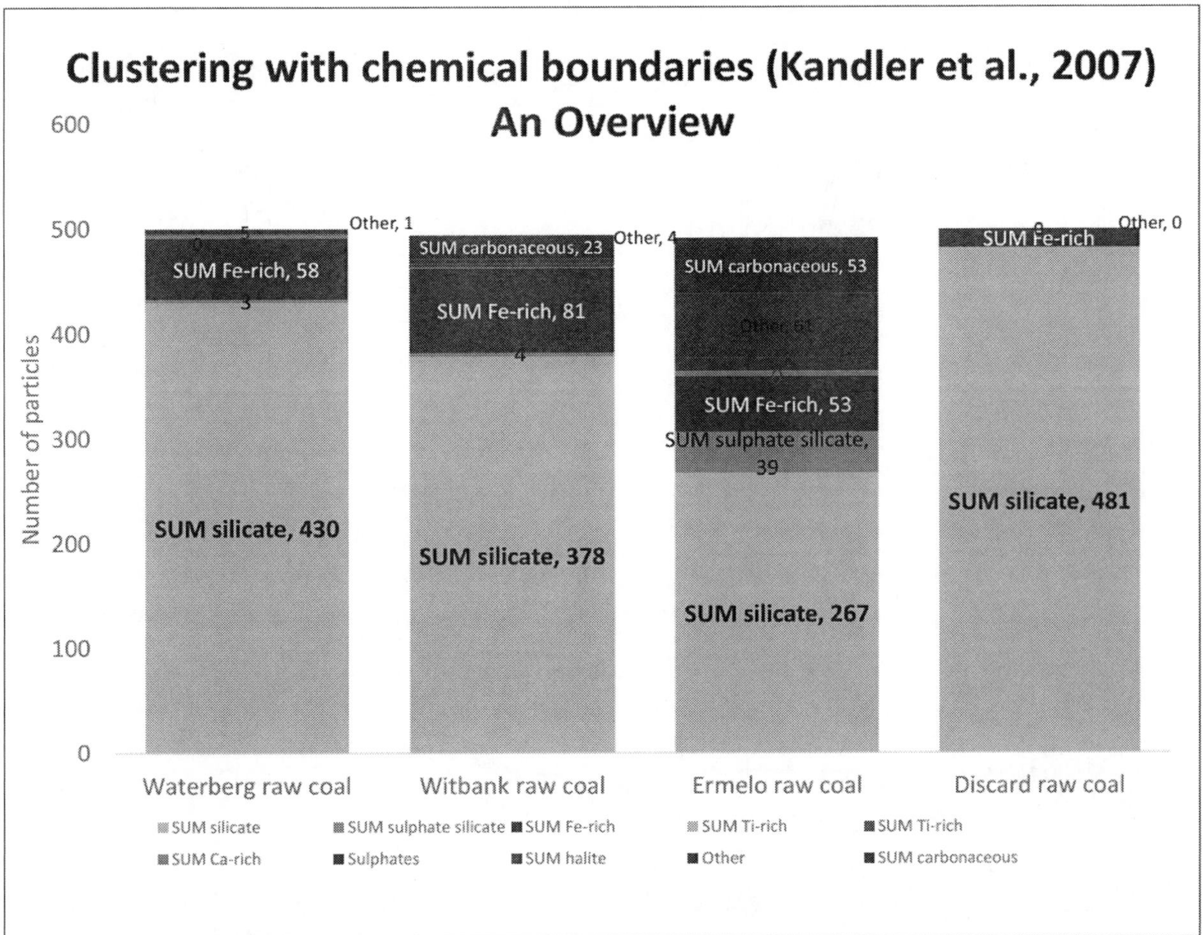


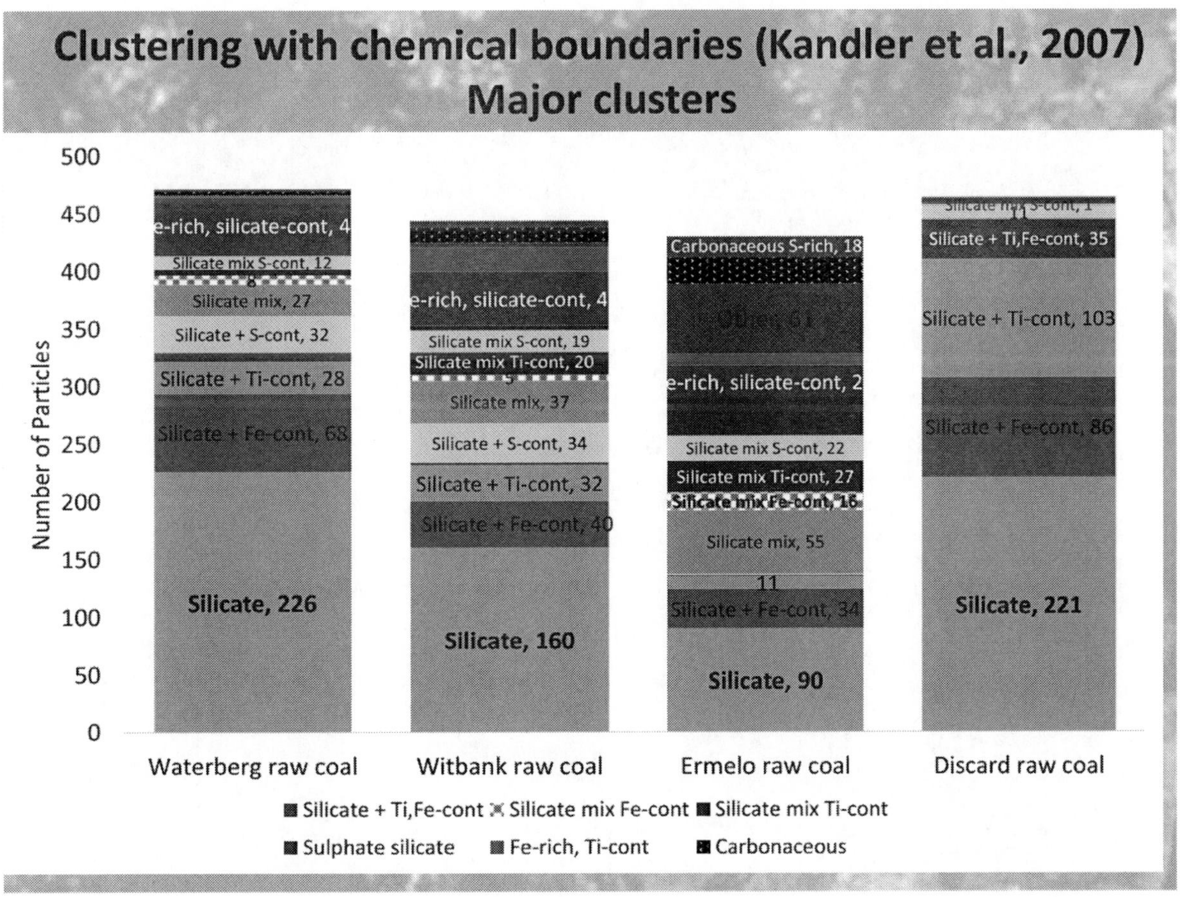

# Micro-Raman Analysis

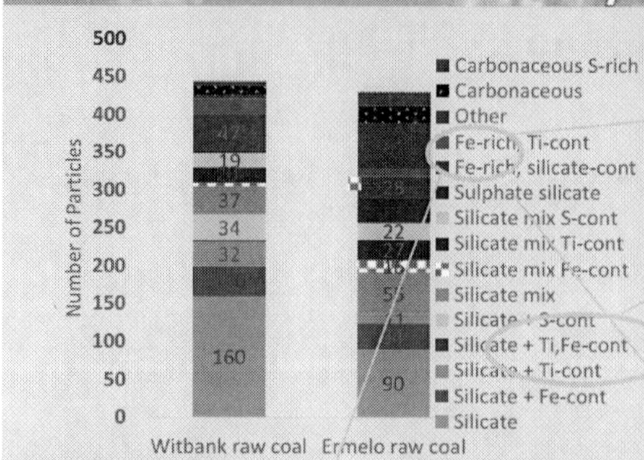


| Ermelo (PXRD) | Ermelo (Raman) | Witbank (PXRD) | Witbank (Raman) |
|---|---|---|---|
| Quartz | Quartz | Quartz | Pyrite |
| Kaolinite | Anatase | Kaolinite | Arsenopyrite |
| Anatase | Rutile | Calcite | Anatase |
| | Pyrite | Aragonite | Rutile |
| | | Dolomite | Jarosite |
| | | Pyrite | Quartz |
| | | Muscovite | Calcite |
| | | Alunite | Aragonite |
| | | Gypsum | Dolomite |
| | | Anatase | Siderite |

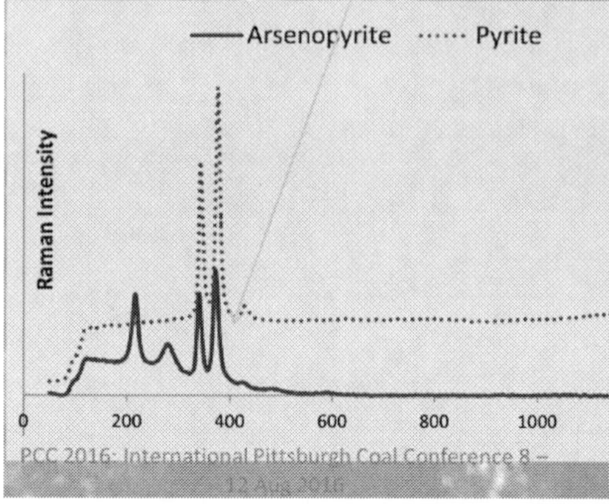


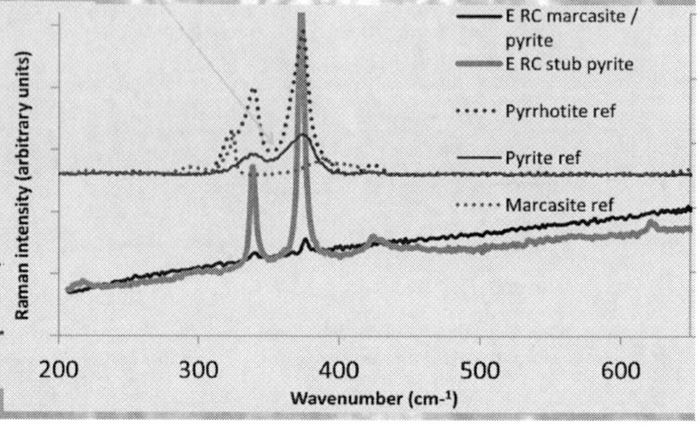


PCC 2016: International Pittsburgh Coal Conference 8 – 12 Aug 2016

# In conclusion:

- Organic and inorganic structure probed and elucidated with the same approach and techniques
- Following Morga (2011) classification, it is suggested that Waterberg will be the least reactive of the 3 coals
- Waterberg seems to have different inorganic/organic association than Witbank and Ermelo
- Inorganic matter in most cases associated with the organic matter on particle level.

PCC 2016: International Pittsburgh Coal Conference 8 – 12 Aug 2016

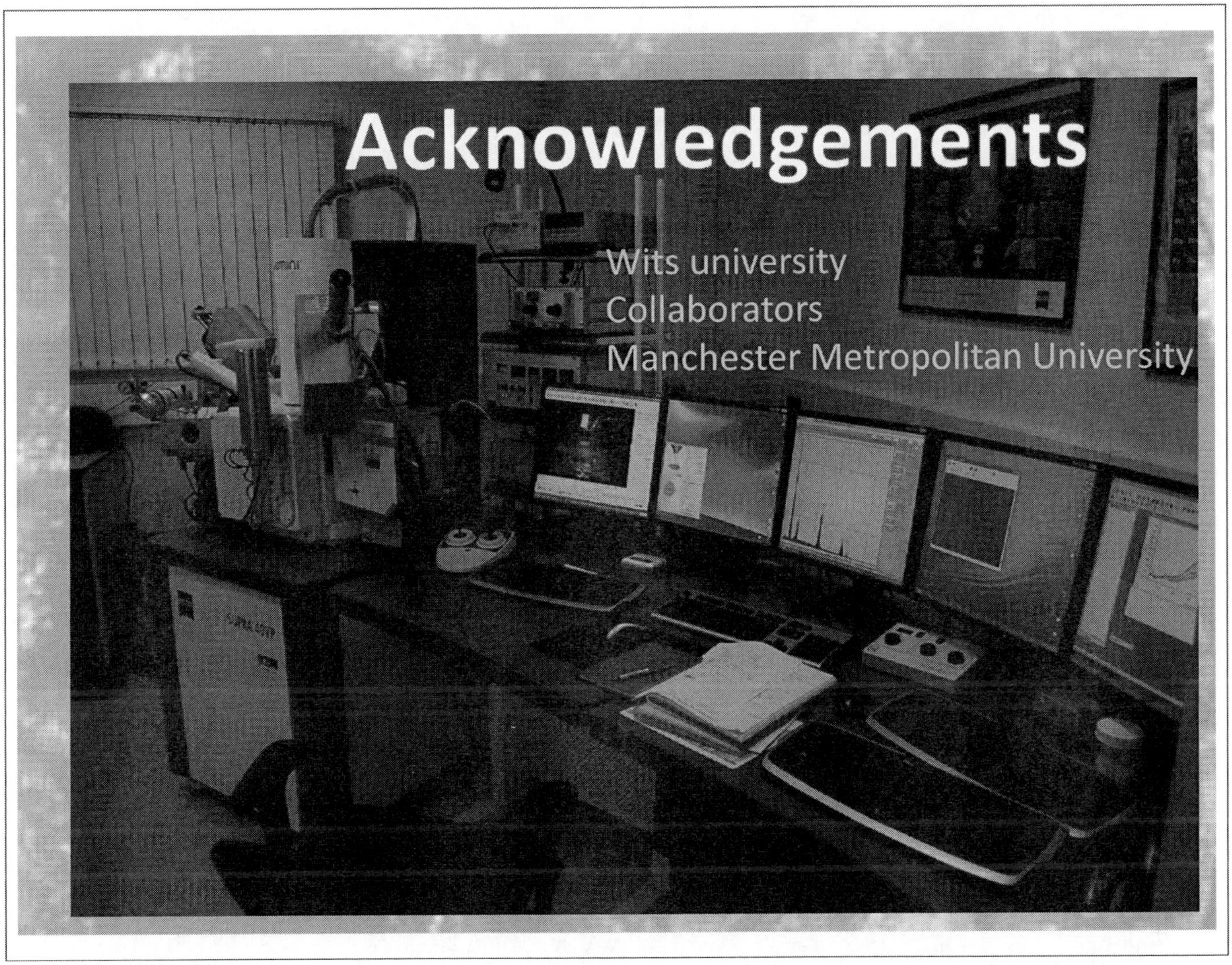
Acknowledgements
Wits university
Collaborators
Manchester Metropolitan University

2016 International Pittsburgh Coal Conference
Cape Town, SOUTH AFRICA
August 8–12, 2016
PROGRAM TOPIC: Coal Science

# COAL CLEANING REJECTS AND SULPHIDE OXIDATION BY FENTON'S REACTION

*Laboratory of Environmental Researches and Nanotechnology Development - BRAZIL*
*Diego S. de Medeiros, Anderson S. Pires, Marcos L.S. Oliveira, Frans Waanders, Bianca D. Lima, Leonardo K. Botelho, Luis F.O. Silva*

# Aim of the study

This study investigated the nature of the coal mining residues (CMRs) and attempts to quantify the extent to which interaction has taken place or could take place between the coal or coal bearing strata and the surrounding environment, by using an accelerated weathering procedure based on the Fenton's reaction, i.e. using $H_2O_2$ as the oxidant to examine experimentally the oxidation of metal sulphide mineral phases associated with Brazilian coals.

- Find a quick, low cost and effective methodology to evaluate the impacts of different coal cleaning rejects.

# **Methods:** Background of the accelerated weathering by Fenton's reaction

- Hydrogen peroxide ($H_2O_2$) has strong oxidizing properties and has been suggested as an appropriate chemical for clean desulphurization treatments of high-sulphur coals. It was determined to be effective in reducing >90% of the pyritic sulphur from high-sulphur coal.

# *Study Area*

Urussanga
Brazil
3
6
12
16

# Sampling Method

- 17 samples used for this study were obtained from Santa Catarina ( The samples were collected directly at source from open air hollows in the region of Lauro Muller city, which impacts in the hydrographic basin of the Tubarao River.
- The samples, gathered from different spots on the most recent stack in the same week, were stored in sample bags weighing between 20 and 25 kg.
- The samples, were homogenized according to norm D2797 and split into batches for the various analyses.
- Samples belong to two groups of materials:
  - CMR-S: Sediments originating from areas in contact with layers of coal, rich in carbonaceous substance (%C = 2.9–5.9) with variable levels of sulphur (%S = 0.2–3.2) and acid reaction.
  - CMR-O: Distinct sediments collected in the influence areas of the mines, but with low levels of carbonaceous matter and sulphur (%S < 0.1) and low acid reaction.

# Analytical procedures

- All samples were acid digested following a two-step digestion method devised to retain volatile elements in coal dissolution;
- Inductively Coupled Plasma Atomic-Emission Spectrometry (ICP-AES)
  - Major trace elements
- Inductively Coupled Plasma Mass Spectrometry (ICP-MS)
  - additional trace elements
- The digestion of the international reference materials (SARM-19) and blanks was also conducted following the same procedure to check the accuracy of the analytical and digestion methods.

# *Analytical procedures*

- Uncertainties were estimated at <3% for most of the elements and around 10% for Cd, Mo and P.
- Mercury analyses were carried out directly on solid samples using LECO AMA 254 gold amalgam atomic absorption spectrometer.
- The occurrence of mineral species was investigated by means of Powder X-ray diffraction (XRD), environmental Scanning Electron Microscope (SEM) coupled with an energy dispersive X-ray microanalysis system (EDX) and Transmission electron microscopy(TEM) for chemical analyses of individual particles.

# Results and discussion

| $mg kg^{-1}$ | CMR-S | CMR-0 | % | CMR-S | CMR-0 |
|---|---|---|---|---|---|
| As | 19.5(1.0) | 8.3(0.5) | Al | 10.23(0.52) | 10.22(0.42) |
| B | 9.8(0.6) | 15.0(1.7) | Ba | 0.12(0.01) | 0,071(0.002) |
| Ba | 538(24) | 952(33) | Ca | 0.53(0.21) | 0.76(0.03) |
| Be | 3.1(0.2) | 4.6(0.1) | Fe | 3.13(0.42) | 2.94(0.17) |
| Bi | 1.5(0.4) | 0.91(0.04) | K | 1.54(0.15) | 1.57(0.04) |
| Cd | 4.5(0.2) | 1.3(0.2) | Mg | 0.27(0.01) | 0.40(0.02) |
| Co | 5.4(0.2) | 11.1(0.5) | Na | 0.12(0.01) | 0.28(0.01) |
| Cr | 48(1) | 53(1) | P | 0.29(0.01) | 0.020(0.003) |
| Cu | 24(1) | 25(2) | Ti | 0.55(0.03) | 0.46(0.09) |
| Ga | 28(1) | 27(1) | | | |
| Ge | 3.8(0.2) | 3.7(0.2) | | | |
| Hf | 5.0(0.3) | 5.4(0.4) | | | |
| La | 37.0(0.2) | 38(1) | | | |
| Li | 89(4) | 81(3) | | | |
| Mn | 51(3) | 235(11) | | | |
| Mo | 32.8(1.3) | 3.4(0.1) | | | |
| Ni | 66(1) | 19.2(0.8) | | | |
| Pb | 71(2) | 45(4) | | | |
| Sb | 1,23(0.05) | 1.03(0.17) | | | |
| Se | 5.4(0.5) | 5.3(0.2) | | | |
| Sn | 57(4) | 5.0(0.1) | | | |
| Sr | 91(7) | 242(66) | | | |
| Ta | 5.1(0.2) | 4.9(0.2) | | | |
| Th | 31(1) | 22.6(0.1) | | | |
| U | 6.5(0.7) | 9.4(0.5) | | | |
| V | 92(2) | 98(2) | | | |
| W | 6.9(0.2) | 7.0(0.3) | | | |
| Y | 22(1) | 24.4(0.7) | | | |
| Zn | 46(2) | 191(11) | | | |
| Zr | 183(9) | 200(11) | | | |

Average range of variation (confidence level at 95%) of the total elemental composition of each metal in the unchanged coal washing residues grouped according to the type of material and the sulphur content, CMR-S and CMR-O respectively.

# Mineralogical characterization

The major minerals in the coal mining residues studied (decreasing order of significance) are quartz (Fig. 2A), kaolinite (Fig. 2C), pyrite (Fig. 2D) and calcite (Fig. 2F) according to the XRD TEM and SEM/EDX investigations. Other minerals, such as illite (Fig.2B), gypsum (Fig. 2E), hematite (Fig. 2G), dolomite, ankerite, chlorite, opal, feldspars, rutile,marcasite and melanterite,were also identified in the CMRs as minor and accessory species.

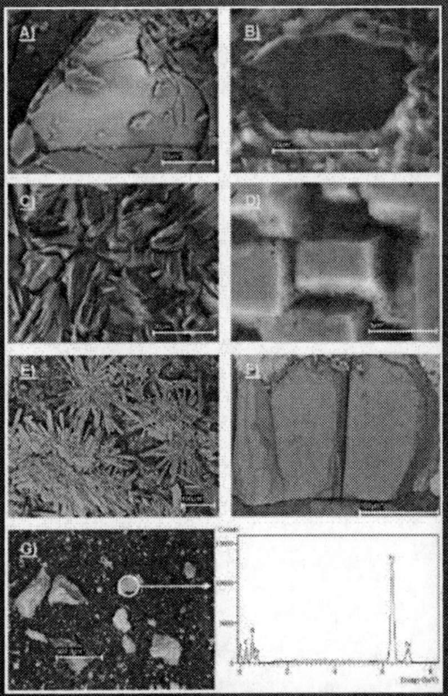


# Mineralogical characterization

- The principal oxide detected in the CMRs analysed was hematite closely mixed with quartz, feldspar and rutile. The hematite probably reflects the oxidation of pyrite in the coal.
- The formation of iron oxides and hydroxides is considered the driving force behind the geochemical cycling of metals and the dissolution of redo sensitive metal oxides or biomineralization products.
- Carbonate minerals are common in the CMRs samples studied. They are represented by calcite and, to lesser extents, dolomite and ankerite.
- Sulphide minerals consist of pyrite, marcasite, pyrrhotite, sphalerite.
- The composition of pyrite surface and the tendency of metal ions from aqueous solution to be adsorbed on themcan, to a considerable extent, also influence the rate of oxidation. This is consistent with the fact the metals like Na, K, Ca, Mg, Al, Ni, Cu, Ag, Pb, Zn, Cd and As, are known to occur on the natural pyrite surface.

# *Mineralogical characterization*

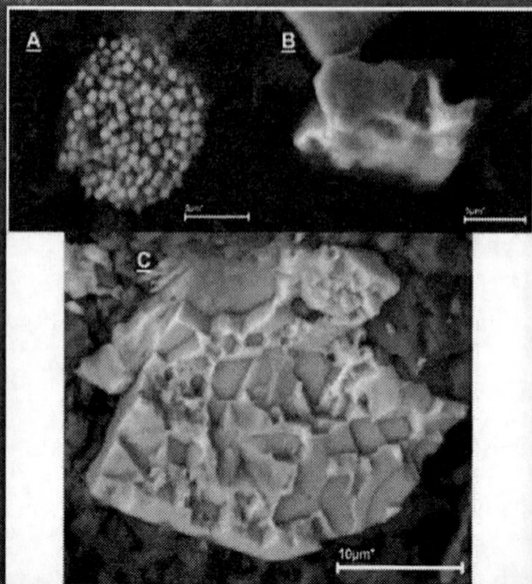


SEM images of (A) pyrite framboids, perhaps the most common form of occurrence of syngenetic pyrite in coal and organic-rich shales; (B) sphalerite grains; and (C) galena fragment.

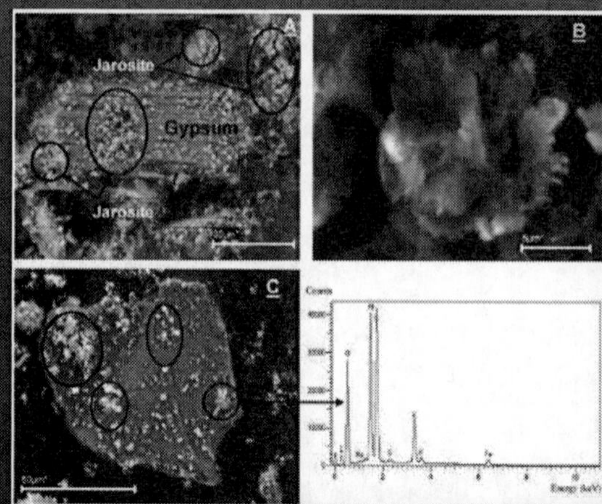


SEM images (A) gypsum crystals in a massive association with new small jarosite crystals formed during the oxidative process by $H_2O_2$ (obtained after the first 3 min of the $H_2O_2$ oxidation); (B) barite; (C) K-feldspar in an association with small jarosite crystals (SEM/EDX).

# Conclusion

- The accelerated weathering procedure based on Fenton's Reaction has shown the release of toxic metals from the sulphide fractions associated with coal residues.
- The experimental results demonstrate the effective oxidation of pyrite, as well as the other metal sulphides, by hydrogen peroxide within a short time period complete after 72 h.
- In terms of relative mobility (% of total metal released during the test), Cu, Co, Cd and Zn appear as metals of high mobility, liberating practically all total content.
- The released concentration of Ni and Mn increases as sulphate increases but a considerable concentration of both metals remains immobile in the solid phases.
- The Fe presentes a relatively lower release, because it suffers processes of further immobilization after pyrite oxidation, even at lower pH values, due to secondary mineral precipitation, such as jarosite; this fact considerably limits its mobility.
- Pb presents a practically null mobility and it does not represent a risk of potential contamination even in samples with high sulphide concentrations due to the possible formation of insoluble lead sulphate.

Contact:
felipeqma@hotmail.com
Journal of Hazardous Materials 186 (2011) 516–525
Contents lists available at ScienceDirect
Journal of Hazardous Materials
journal homepage: www.elsevier.com/locate/jhazmat
ELSEVIER

**2016 International Pittsburgh Coal Conference**
**Cape Town, SOUTH AFRICA**
**August 8 – 12, 2016**

**Abstract Submission**

**PROGRAM TOPIC:**
**7. COAL SCIENCE: TRACE ELEMENTS/EMISSION**

# PETROLEUM COKE-BASED ACTIVATED CARBON MODIFIED BY BINARY METAL OXIDES FOR ELEMENTAL MERCURY REMOVAL IN FLUE GAS


***Presenting Author (Yang Li, Lecturer, State Key Laboratory of Fine Chemicals, Institute of Coal Chemical Engineering, School of Chemical Engineering, Dalian University of Technology, Dalian 116024, P.R. CHINA)***
***Contact Information: yli@dlut.edu.cn, Phone/Fax: 86-411-84986160***
*Co-Author (Ning Zhang, Graduate student, State Key Laboratory of Fine Chemicals, Institute of Coal Chemical Engineering, School of Chemical Engineering, Dalian University of Technology, Dalian 116024, P.R. CHINA)*
*Contact Information: xiongdong@mail.dlut.edu.cn, Phone/Fax: 86-411-84986160*
*Co-Author (Lijun Jin, Associate Professor, State Key Laboratory of Fine Chemicals, Institute of Coal Chemical Engineering, School of Chemical Engineering, Dalian University of Technology, Dalian 116024, P.R. CHINA)*
*Contact Information: ljin@dlut.edu.cn, Phone/Fax: 86-411-84986160*
*Corresponding Author (Haoquan Hu, Professor, State Key Laboratory of Fine Chemicals, Institute of Coal Chemical Engineering, School of Chemical Engineering, Dalian University of Technology, Dalian 116024, P.R. CHINA)*
*Contact Information: hhu@dlut.edu.cn, Phone/Fax: 86-411-84986157*



**Abstract:**
Mercury is considered as one of the most toxic heavy metals because of its persistence and bioaccumulation around the world. The release of mercury and some toxic trace metals/metalloids into the environment have an adverse effect to human health and throughout the ecosystem. Among all the anthropogenic mercury emission sources, coal combustion is the major one. In the face of increasingly serious problems of mercury pollution, it has been an urgent issue to protect environment by reducing mercury emission from coal fired power plants. Activated carbon injection is one of the mercury removal approaches which can remove mercury effectively from flue gas using conventional particulate matter controller. However, application of activated carbon sorbents is limited by huge cost, lack of products for low rank fuels. Therefore, it is important to developing alternative cost-effective sorbents or chemical treatments to promote mercury removal capacity of carbonaceous sorbents. Petroleum coke is a low-priced carbon which can be obtained as a vast sum of byproduct from petroleum refinement. With high carbon content, low volatile and ash content, it is a good precursor for preparing high surface AC. Pyrolyzed high-sulfur petroleum exhibited a mercury removal efficiency of around 50–60% at temperatures around 160-170 °C. However, the pyrolysis temperature had to reach to 1100 °C in $N_2$ atmosphere that could be a heavy burden to apparatus. In this study, mercury adsorption performance of an activated carbon which is derived from petroleum coke (PAC) and modified PAC were evaluated in simulated coal combustion flue gas atmosphere. Potassium hydroxide (KOH) was used as activating agent to obtain PAC by chemical activation. Impregnation method was applied to modify PAC by Mn-Ce binary oxides(MnCe-PAC). The adsorption experiments of vapor-phase elemental mercury were carried out in a laboratory-scale fixed-bed reactor. Specifically, the effects of loading values of metal oxides, temperature (100 - 250 °C) and individual flue gas components ($O_2$, $SO_2$, NO, HCl) on $Hg^0$ removal efficiency were investigated. In addition, thermal stabilities of mercury compounds were studied by using temperature-programmed desorption (TPD). MnCe-PAC showed the best performance at 100 °C, and Ce−Mn binary metal oxides could adsorb more elemental mercury than $MnO_y$-PAC and $CeO_2$-PAC. The samples were characterized by X-ray diffraction (XRD), $N_2$ Adsorption, Fourier Transformed Infrared (FT-IR), X-ray photoelectron spectroscopy (XPS) and temperature programmed desorption (TPD). The results indicated that the presence of 200 ppm $SO_2$ in flue gas significantly promoted the $Hg^0$ adsorption and oxidation over MnCe-PAC. Furthermore, the XPS spectra of $Hg_{4f}$ and Hg-TPD results showed that the form of captured mercury was oxidation state on the samples, which illustrated that catalytic oxidation and chemisorption was the main process for $Hg^0$ removal over sorbent. It indicated that Mn-Ce binary metal oxides are a promising sorbent for the mercury removal from flue gas.The understanding of elemental mercury removal would help developing a cost-effective sorbent by utilizing carbon-rich high sulphur petroleum coke or shot coke with pore activities.

# EFFECT OF ALKALI SPECIES IN AUSTRALIAN COALS ON SLAG COMPOSITION AND VISCOSITY

*Dr Alexander Ilyushechkin, CSIRO, PO Box 883, Kenmore QLD 4069 AUSTRALIA*

*alex.ilyushechkin@csiro.au*

***Dr Daniel Roberts, CSIRO, PO Box 883, Kenmore QLD 4069 AUSTRALIA***

***Daniel.Roberts@csiro.au***

*Dr David Harris, CSIRO, PO Box 883, Kenmore QLD 4069 AUSTRALIA*

*David.Harris@csiro.au*

**Abstract:**

Low rank coals can have high levels of Na and K, which are likely to impact their suitability for use in slagging gasifiers. There are very few studies, however, that allow us to develop predictive models to help with managing slag behaviour. This paper demonstrates the effect of $Na_2O$, $K_2O$ on slag phase composition and viscosity for selected Australian coals that have variable concentrations of alkali oxides in their ashes. For compositions representing Australian black coals, we found that the viscosity and temperature of critical viscosity ($T_{CV}$) can change significantly if the addition of potassium shifts the phase composition of the molten slag to regions with higher or lower liquidus temperatures, or promotes precipitation of crystalline solids. Therefore, a careful analysis of possible effects is required when using coals with high potassium content, which can drive slag composition into a $T_{CV}$ precluded area and hence minimise the gasifier's operational temperature range. For Australian brown coals, our results show that slags with variable sodium content from olivine and cordierite primary phase fields have low liquidus temperatures. The viscosity of these slags at 1200–1600 °C was only ever reduced by an increase of sodium in the liquid phase, and was not affected by solids. For slags from silica and spinel primary phase fields, the increased sodium content reduces both slag viscosity and the formation of solids.

## 1. Introduction

Coal mineral matter plays a significant role in slagging gasifiers, as its characteristics such as ash fusion temperature, slag viscosity and temperature of critical viscosity ($T_{CV}$) are most important in terms of feedstock suitability and gasifier operating conditions. In entrained flow gasification, slag viscosity should generally be within the range of 5–25 Pa·s at operating temperatures of 1200–1500°C to achieve steady slag tapping [1]. The ability to achieve this strongly depends on slag chemical and phase compositions. Silica ($SiO_2$) is the most important component in slag, and the dominant network former. All other oxides can be grouped according to their effect on the silica network: some act as either network formers (increasing the viscosity) or modifiers (decreasing the slag viscosity), while others play less clearly defined roles [2].

Most Australian bituminous coals have mineral matter compositions that can be represented by the four-component system $CaO–SiO_2–Al_2O_3–FeO$. A significant database of ash compositions and slag viscosity has been developed and used to generate models enabling prediction of slag viscosity [1, 4, 5]. Concentrations of other oxides are relatively small, as indicated in Table 1. Most previous studies on Australian coal slags focused on investigating the effects of these 'major' elements in coal ash (Si, Al, Fe and Ca).

Brown coals have different mineral matter compositions from black coal, with significant variations in $SiO_2$, $Al_2O_3$, FeO and CaO. Some brown coal slags also include MgO as a major component. The viscosity of slags representing different groups of chemical compositions based on the major components have been reported [6].

Alkali oxides ($Na_2O$, $K_2O$, $Li_2O$) can all be classified as network modifiers that reduce slag viscosity. In our previous studies, we noticed that even small concentrations (up to 4% by weight) of $K_2O$ can significantly alter slag flow behaviour in some cases [3]. While alkali oxides ($Na_2O$, $K_2O$) are not a major component in ashes, some bituminous coal ashes may have up to 4 wt.% $K_2O$, and $Na_2O$ can reach more than 15 wt.% in brown coal ashes. This can be significant in determining slag formation and flow behaviour, yet there is not a good understanding of the impact of these species on coal performance in entrained flow gasifiers. In this paper, we report the effect of these oxides on slag phase composition and viscosity.

**2. Experimental**

Slag samples with bulk compositions that reproduced ash chemical composition (previously reported in [6]) as well as those obtained in present study) were prepared by mixing laboratory or analytical grade $Al_2O_3$, CaO, $Fe_2O_3$, $K_2CO_3$, MgO, $Na_2CO_3$, $SiO_2$ and $TiO_2$ powders. The mixtures were pelletised and placed in molybdenum (Mo) envelopes pre-melted at 1500–1600°C, slow cooled to 1200–1500°C, and then kept at this temperature for 1 h before being quenched.

The compositions of these slags are listed in Table 1. As the concentrations of MgO, $Na_2O$ and $TiO_2$ in bituminous coals are low, the mixtures representing black coal slags with and without potassium (K) content are based on the four-component system of $Al_2O_3$–CaO–$Fe_2O_3$–$SiO_2$, with fixed $Fe_2O_3$ content (5 wt.%) and variable contents for the other three oxides, making five pairs of comparable samples. The selected samples represent the range of the synthetic slag compositions previously studied for modelling the slag viscosity of Australian black coals.

To investigate the effect of sodium in brown coal slags, the compositions were selected based on our analysis of brown coal ashes, where relatively high Na concentrations were observed. As their compositions vary significantly, they were reproduced in synthetic mixtures of eight major oxides, along with comparable samples where only Na concentration was changed. The pairs of samples are listed in Table 1.

**Table 1: Composition range (in wt.%) of Australian coal ashes and synthetic slags in this study**

| | $SiO_2$ | $Al_2O_3$ | $Fe_2O_3$ | CaO | MgO | $Na_2O$ | $K_2O$ | $TiO_2$ | S/A |
|---|---|---|---|---|---|---|---|---|---|
| *K-series* | | | | | | | | | |
| Black coals | 28–80 | 8–48 | 0.1–32 | 6–12 | 0.1–3 | 0.1–2 | 0.1–4 | 0.1–3 | 1–5 |
| K-1a | 52.3 | 15.2 | 5.9 | 26.6 | – | – | 0 | – | 3.4 |
| K-1b | 50.0 | 14.6 | 5.7 | 25.7 | – | – | 4.0 | – | 3.4 |
| K-2a | 60.0 | 25.0 | 5.0 | 10.0 | – | – | 0 | – | 2.4 |
| K-2b | 57.2 | 23..8 | 5.0 | 10.0 | – | – | 4.0 | – | 2.4 |
| K-3a | 50.0 | 30.0 | 5.0 | 15.0 | – | – | 0 | – | 1.7 |
| K-3b | 47.5 | 28.5 | 5.0 | 15.0 | – | – | 4.0 | – | 1.7 |
| K-4a | 44.9 | 35.1 | 5.0 | 15.0 | – | – | 0 | – | 1.3 |
| K-4b | 42.7 | 33.3 | 5.0 | 15.0 | – | – | 4.0 | – | 1.3 |
| K-5a | 40.0 | 35.0 | 5.0 | 20.0 | – | – | 0 | – | 1.1 |
| K-5b | 37.8 | 33.2 | 5.0 | 20.0 | – | – | 4.0 | – | 1.1 |
| *Na-series* | | | | | | | | | |
| Brown coals | 1–75 | 2–48 | 4–49 | 1–35 | 3–36 | 2–20 | 0–1.2 | 0–1.7 | 0.6–8 |
| N-1a | 79.11 | 10.62 | 3.71 | 1.09 | 2.24 | 1.83 | 0.41 | 0.98 | 7.4 |
| N-1b | 76.46 | 10.27 | 3.59 | 1.06 | 2.16 | 5.12 | 0.40 | 0.95 | 7.4 |
| N-2a | 65.9 | 20.2 | 4.5 | 0.7 | 2.4 | 3.1 | 1.1 | 2.1 | 3.3 |
| N-2b | 63.92 | 19.59 | 4.36 | 0.68 | 2.33 | 6.01 | 1.07 | 2.04 | 3.3 |
| N-3a | 50.60 | 15.93 | 9.12 | 7.66 | 9.74 | 5.00 | 0.60 | 1.34 | 3.2 |
| N-3b | 47.93 | 15.09 | 8.64 | 7.26 | 9.22 | 10.02 | 0.57 | 1.27 | 3.2 |
| N-4a | 59.92 | 21.74 | 4.88 | 1.70 | 3.82 | 4.98 | 1.38 | 1.59 | 2.8 |
| N-4b | 56.78 | 20.60 | 4.62 | 1.61 | 3.62 | 9.96 | 1.31 | 1.51 | 2.8 |
| N-5a | 29.35 | 36.96 | 7.63 | 6.91 | 12.81 | 5.02 | 0.29 | 1.04 | 0.8 |
| N-5b | 27.77 | 34.97 | 7.22 | 6.54 | 12.12 | 10.03 | 0.27 | 0.98 | 0.8 |

Slag viscosity was measured using a Haake high-temperature viscometer using the rotational bob technique described previously [1, 5]. Slags were placed into a Mo crucible and heated to 1650°C in a neutral atmosphere ($N_2$). To ensure the absence of oxygen, the crucible was placed in a sacrificial graphite liner. Viscosity measurements were taken using an Mo rotating sensor. The temperature-cooling step for each set of measurements was 25–50°C for at least 30-min intervals to equilibrate the temperature inside the crucible. At each temperature point, slag viscosity was measured several times to ensure repeatability of the measurements. The average of these measurements was used as the value of viscosity.

Phase compositions of slags were calculated using the FactSage (version 7.1) thermodynamic phase equilibria package with selected FACT53 and FToxid solid solutions and compounds [7]. Most of the predicted phases were confirmed by additional scanning electron microscopy–electron probe microanalysis of quenched slags microstructure.

## 3. Results

### 3.1 Effect of potassium on slag phase composition and viscosity

The effect of K on slag phase equilibria in summarised in Table 2. In most cases, adding K reduced the liquidus temperature, and also shifted slag composition from mullite to feldspar primary phase field. The presence of K in the fully liquid phase generally reduced slag viscosity, while reducing the liquidus temperature also assisted low viscosity. However, the effect of K on slag viscosity is complex and depends on the silica-alumina (S/A) ratio in the bulk composition.

**Table 2: Effect of potassium on primary phase field and phase equilibria**

| Composition | Primary phase field | Effect of potassium (K) |
|---|---|---|
| K-1a | Feldspar | Reduces liquidus temperature |
| K-1b | Feldspar | |
| K-2a | Mullite | Changes primary phase field, reduces liquidus temperature, but promotes feldspar growth at lower temperatures |
| K-2b | Feldspar | |
| K-3a | Feldspar | Reduces liquidus temperature, but promotes feldspar growth at lower temperatures |
| K-3b | Feldspar | |
| K-4a | Mullite | Changes primary phase field, reduces liquidus temperature, but promote feldspar growth |
| K-4b | Feldspar | |
| K-5a | Feldspar | Reduces liquidus temperature |
| K-5b | Feldspar | |

Examples below for slags with different S/A ratios and primary phase fields illustrate the complex manner in which K impacts slag phase composition and slag viscosity. For samples with an S/A ratio of 3.4, the liquidus temperature is decreased by approximately 75°C (Figure 1a-b) with addition of K. However, it did not affect significantly the viscosity-temperature dependence, because both slags mainly have a high-silica liquid phase. For slags with S/A = 1.7, the presence of K slightly reduced the liquidus temperature (Figure 1d-e), yet significantly reduced the viscosity of both liquid slag and slag with solids at temperatures below 1475–1500°C (Figure 1f).

In slags with a low S/A ratio (S/A = 1.1), the addition of K (4 wt.%) significantly reduced liquidus temperatures (Figure 2 a, b). Crystallisation of feldspar ($CaAl_2Si_2O_8$) in the melt without K led to an increase of S/A ratio in the liquid phase. As a result, slag viscosity increased gradually, with solids forming as the temperature decreased from 1500 to 1380°C. The viscosity of the slag with K present was significantly lower at temperatures above the liquidus than viscosity of the slag without K. However, when solids started to precipitate, the growing feldspar ($KAlSi_3O_8$-$CaAl_2Si_2O_8$) consumed more silica than alumina from the liquid phase. This resulted in a sharp viscosity increase (and transformation to non-Newtonian flow behaviour), because solids were in the liquid phase at low and decreasing S/A.

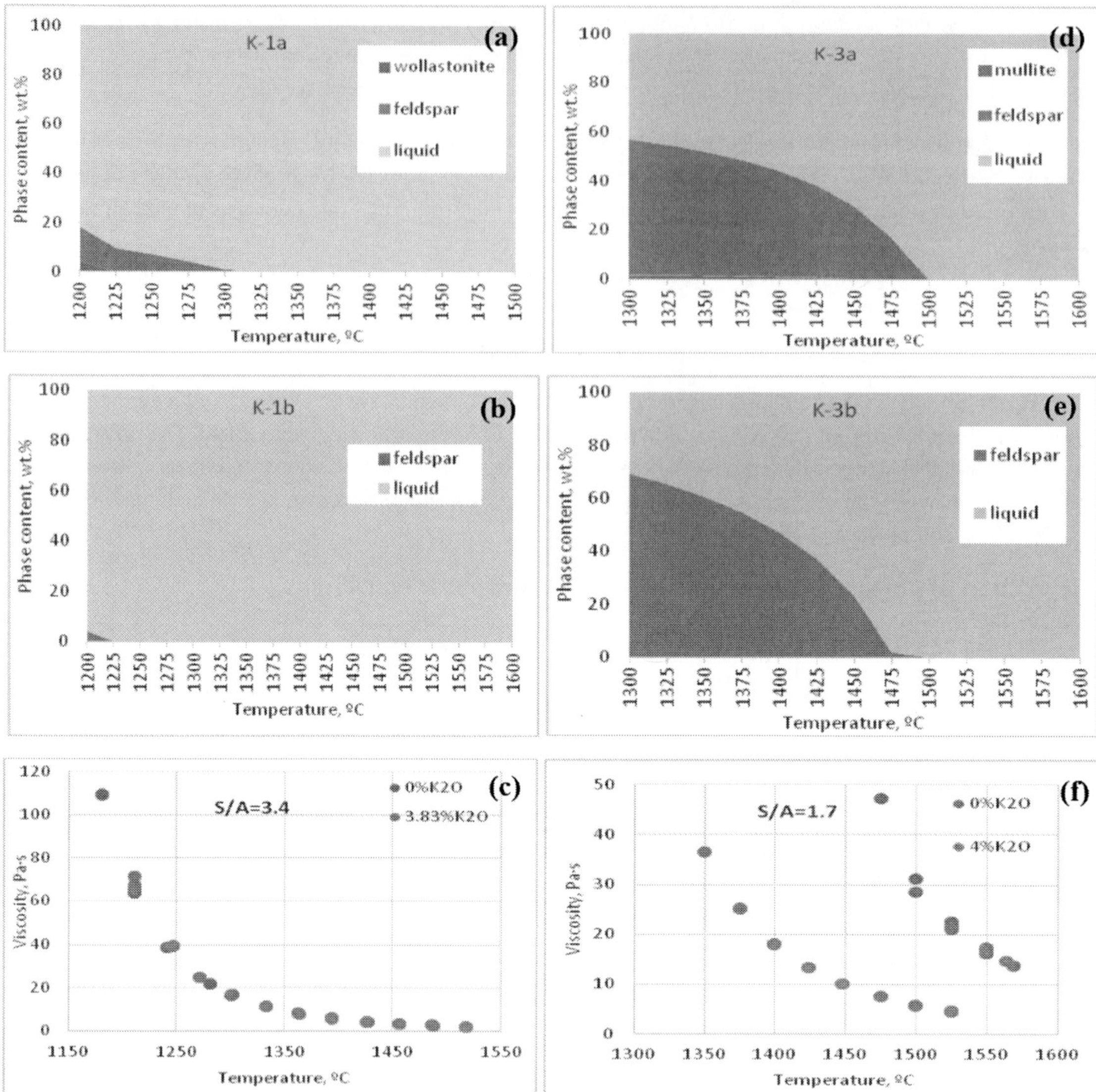


**Figure 1. Phase compositions and viscosity of slags from the feldspar primary phase field with and without potassium; S/A = 3.4 (a–c) and 1.7 (d–f)**

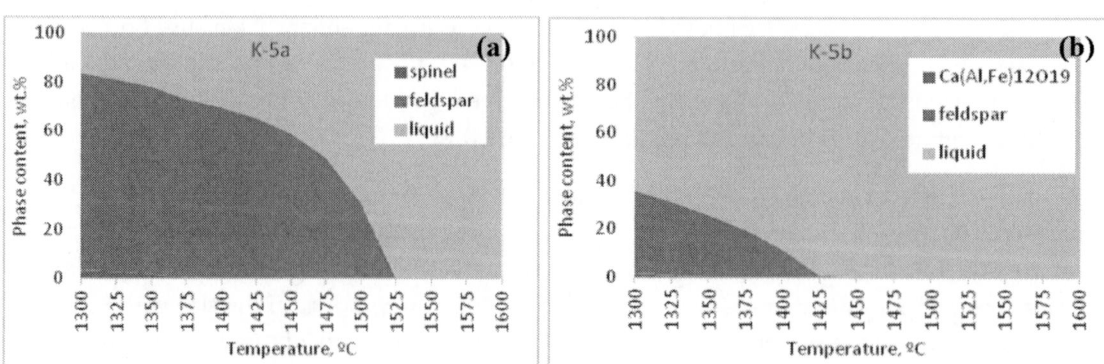

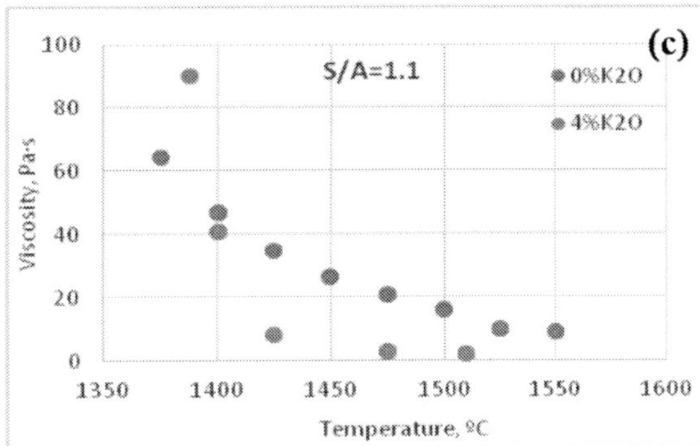


**Figure 2. Phase compositions and viscosity of slags from the feldspar primary phase field (S/A = 1.1) with and without potassium**

In the slags from the mullite primary phase field, the addition of K shifted bulk composition to feldspar primary phase field (Figure 3) and slightly decreased liquidus temperature.

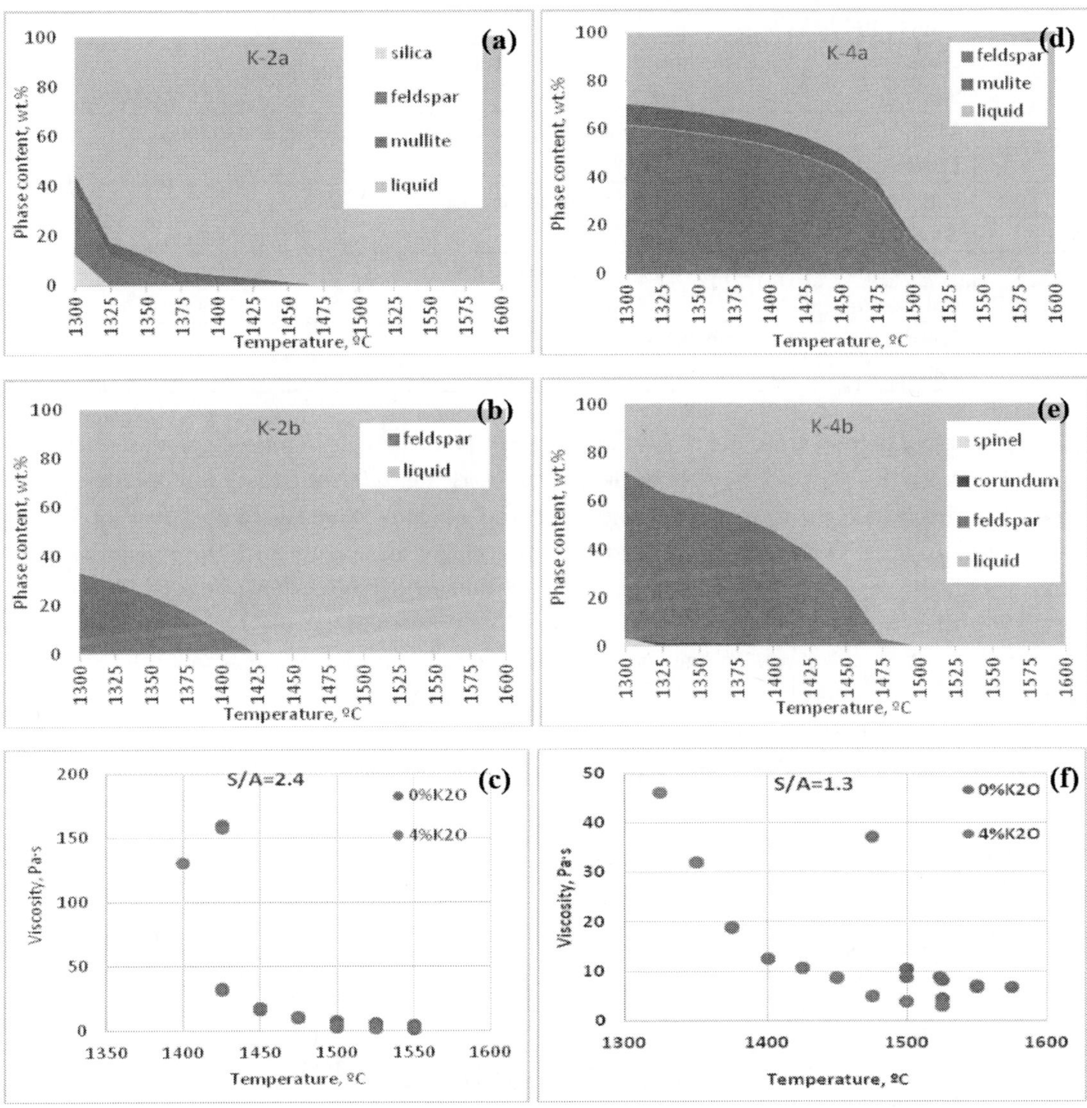


**Figure 3. Phase compositions and viscosity of slags from mullite (without potassium) and feldspar (with potassium) without potassium; S/A = 2.4 (a–c) and 1.3 (d–f)**

Slags with S/A = 2.4 (Figure 3c) did not change viscosity above the liquidus temperatures; however, there was a sharp viscosity increase in the slag without K at temperatures where mullite started to crystallise. A smaller viscosity increase was observed with feldspar crystallisation in the slag with K. For slags with S/A = 1.3, a sharp viscosity increase was observed at ~1500 °C, while the slag with K had a gradual increase in viscosity with growing feldspar at low temperatures (Figure 3d–f).

#### 3.1 Effect of Na on slag phase composition and viscosity

Table 3 summarises the effect of Na on phase equilibria in the slag compositions. Increasing the concentration of Na in bulk composition reduced the liquidus temperature in all slags, except for the N-5a–N5b pair from the spinel primary phase field. However, the concentration of spinel is slightly lower in the slag with higher Na content. For slag from the mullite primary phase field, the increasing Na concentration shifts slag composition to the cordierite primary phase field.

**Table 3: Effect of sodium on slag primary phase field and phase equilibria**

| Composition | Primary phase field | Effect of sodium |
|---|---|---|
| N-1a | Silica | Reduces liquidus temperature |
| N-1b | Silica | |
| N-2a | Mullite | Changes primary phase field, reduces liquidus temperature |
| N-2b | Cordierite | |
| N-3a | Olivine | Reduces liquidus temperature |
| N-3b | Olivine | |
| N-4a | Cordierite | Reduces liquidus temperature |
| N-4b | Cordierite | |
| N-5a | Spinel | Reduces amounts of solids |
| N-5b | Spinel | |

The examples below demonstrate the impact of compositional changes in the slag system on slag viscosity. Slags from the silica primary phase field with a very high S/A ratio (7.4) demonstrate extremely high viscosity (Figure 4a-c), where all measured viscosity values were obtained for fully liquid slag only. Addition of Na to the slag does not affect slag viscosity, because the concentration of silica is very high. Slags N-2a and N-2b (Figure 4d-f) also have high viscosities: despite the moderate S/A ratio, the concentration of network modifiers, such as MgO, CaO and FeO, is very low. A higher concentration of Na in N-2b decreases viscosity in fully liquid slag, because it increases the amount of network modifiers. It also reduces liquidus temperature and shift bulk composition to the cordierite primary phase field; however, slag viscosity is already too high at temperatures above the liquidus, and no data are available at low temperatures.

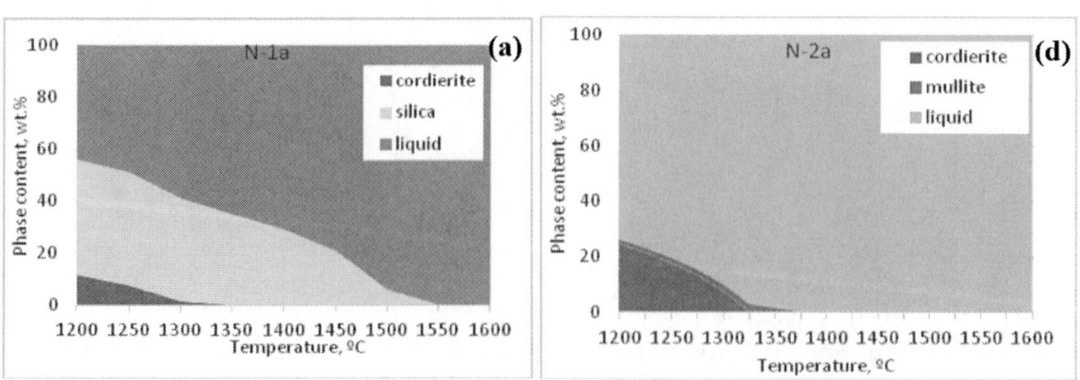

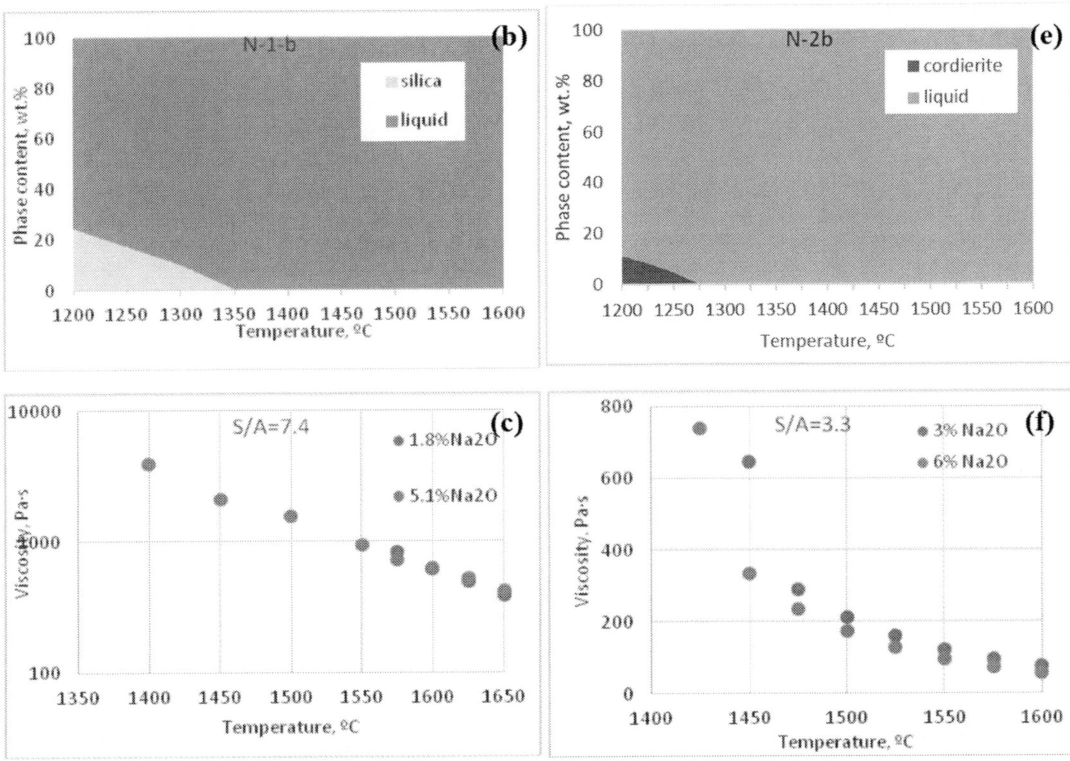


**Figure 4. Phase compositions and viscosity of slags from silica primary phase field (S/A = 7.4) with 1.8 and 5.1 wt.% $Na_2O$, and cordierite primary phase field (S/A = 3.3) with 3 and 6 wt.% $Na_2O$**

Two pairs represent slags from olivine (S/A = 3.2) and cordierite (S/A = 2.8) primary phase fields (Figure 5). All viscosity measurements conducted at temperatures where slags were fully liquid, demonstrated decreasing slag viscosity when the concentration of Na was increased.

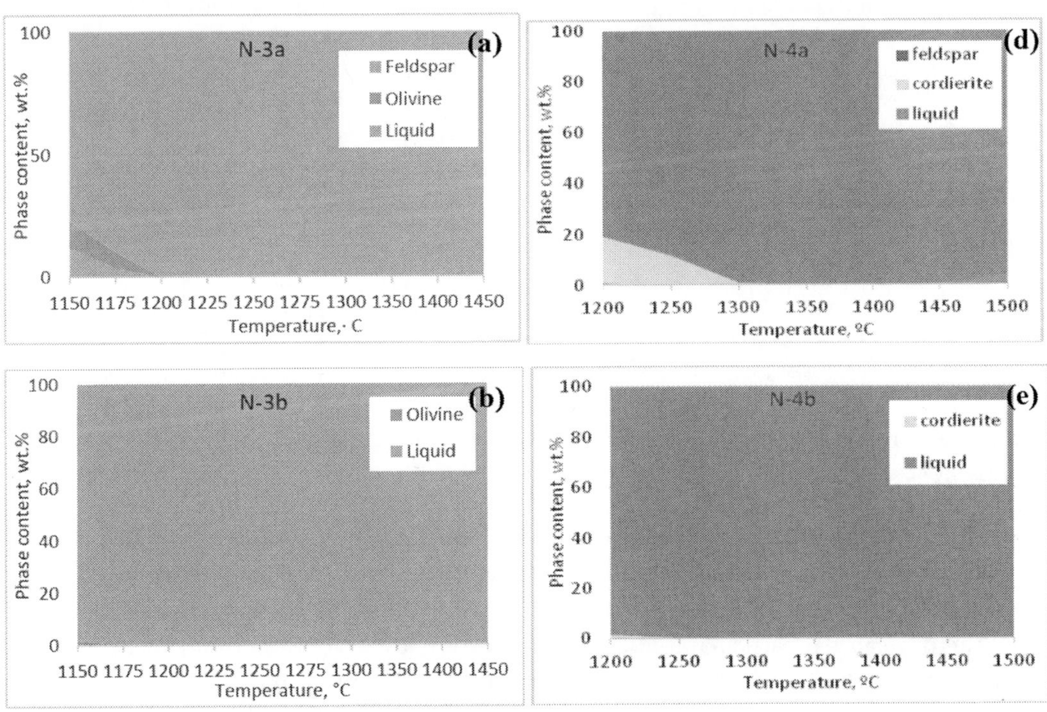

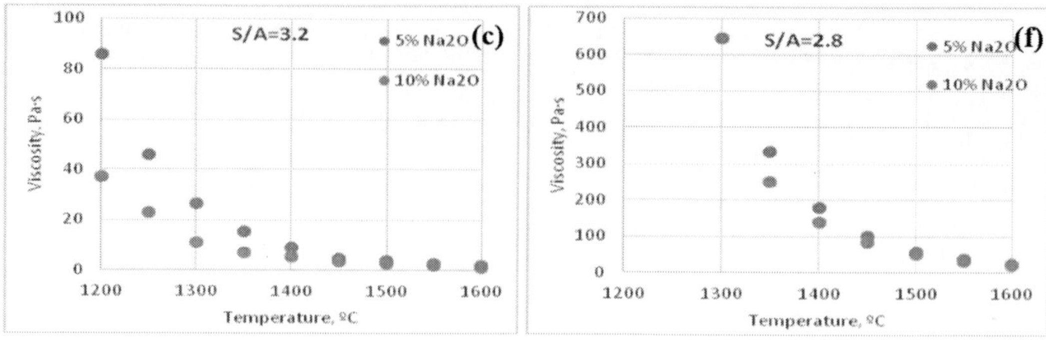


**Figure 5. Phase compositions and viscosity of slags from the olivine and cordierite primary phase fields (S/A = 3.2–3.3) with 5 and 10 wt.% $Na_2O$**

Viscosity of slags from the spinel primary phase field (S/A = 0.8) was determined at the temperatures below liquidus, where slag compositions always have some solids (spinel), as shown in Figure 6. Slag flows demonstrate some deviation from Newtonian flow, due to the low concentration of silica in the melt and the presence of solids. At the same temperature, fewer solids form in the slag with higher Na content, which results in lower viscosity values.

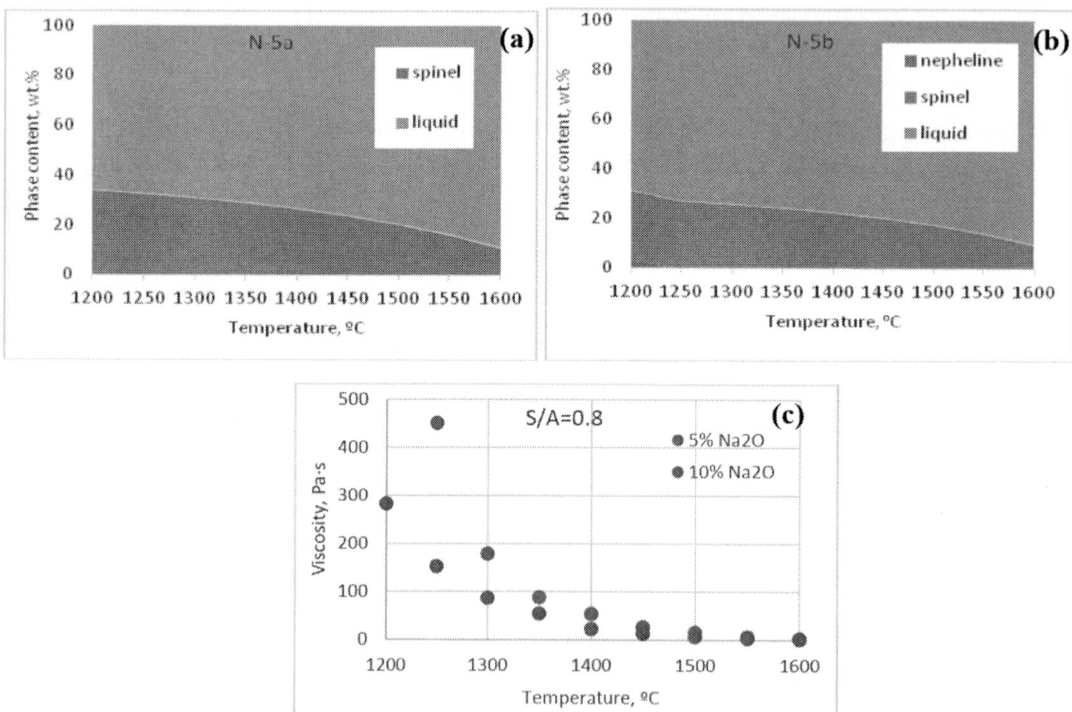


**Figure 6. Phase compositions and viscosity of slags from the spinel primary phase field (S/A = 0.8) with 5 and 10 wt.% $Na_2O$**

The calculated phase compositions and measured slags viscosities for systems with variable Na content demonstrate that presence of Na generally changes slag phase equilibria to lower temperatures, and/or reduces solids content. Both factors decrease slag viscosity. The effect of K is more complicated because it participates in the formation of feldspar, may promote solids formation, and/or may change the primary phase field of slag bulk composition. In some cases, K leads to higher slag viscosity than in slags without K at temperatures below the liquidus.

## 4. Conclusion

We have investigated the effect of alkalis content in mineral matter, representing a range of Australian black coals (focussing on K) and Australian brown coals (focussing on Na).

Potassium has a complex effect on slag behaviour. The addition of K reduces the viscosity of fully liquid slags and reduces liquidus temperatures in the feldspar primary phase field, but promotes feldspar growth. It may also lead to a sharp viscosity increase in slags, due to the large amount of solids that form. Adding K to slags from the mullite primary phase field reduces slag viscosity, and shifts bulk composition to the feldspar primary phase field.

Increasing the concentration of Na in synthetic brown coal slags reduces the phase equilibria (i.e. reduces slag liquidus temperature or solid formation). This generally reduces the slag viscosity.

The effect of alkalis is negligible for slags with a high S/A ratio. In these slags, Na and K do not affect the silica network, because their concentrations are significantly lower than that of silica.

## References


[1] Patterson, J. H.; Hurst, H. J. *Fuel*, **2000**, 79: 1671–1678.

[2] Vargas, S.; Frandsen F.J.; Dam-Johansen K. *Prog. Energ. Combust.*, **2001**, 27: 237–429.

[3] Kinaev, N.N.; Ilyushechkin, A. Y.; Roberts, D. G.; Harris, D. H. *Proceedings ICCS&T,* **2007**

[4] Hurst, H. J.; Novak, F.; Patterson, J. H. *Fuel*, **1999**, 78: 439–444.

[5] Ilyushechkin, A. Y.; Hla, S. S. *Energy Fuels*, **2013**, 27: 3736–3742.

[6] Ilyushechkin, A. Y.; Roberts, D. G. *Fuel Proc. Tech.* **2016,** 147: 47–56.

[7] FactSage. Ecole Polytechnique. Montréal. 2008; http://www.factsage.com/

TECHNISCHE UNIVERSITÄT
BERGAKADEMIE FREIBERG
Die Ressourcenuniversität. Seit 1765

HITECOM ProVirt

# Numerical Study of the HITECOM Reactor for In-Situ Measurement of Single Char Particle Conversion

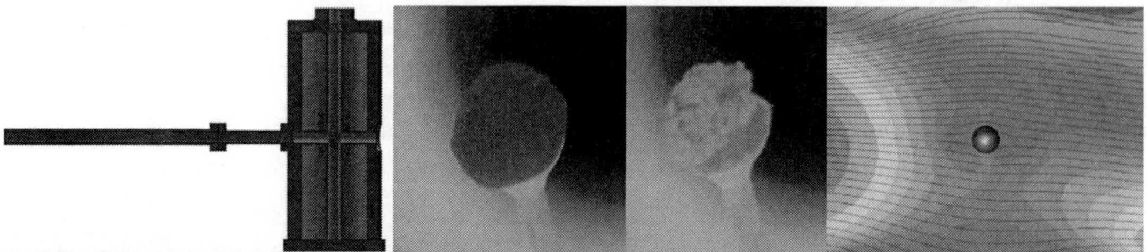

11 August 2016 Cape Town

Fengbo An, Andreas Richter, Felix Küster, Stefan Guhl, Bernd Meyer
Institute of Energy Process Engineering and Chemical Engineering
TU Bergakademie Freiberg, Germany
*fengbo.an@iec.tu-freiberg.de*

HITECOM ProVirt

## Content

- HITECOM reactor
- Motivation of CFD simulation
- Numerical model
- Results
- Conclusion/Outlook

2

## Why HITECOM?

- Fundamental knowledge of processes at single particles necessary for advanced sub model development
- Drop tube and TGA experiments are not able to reflect all process conditions (isolated particles, defined gas flow, high temperature/pressure)
- CFD modelling of isolated reacting particles helps to understand basic physics, but need intensive validation

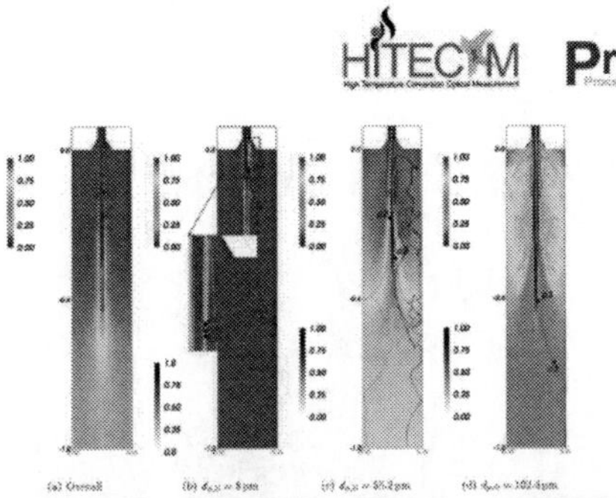

Fig. 1: Particle averaged carbon conversion inside the BYU

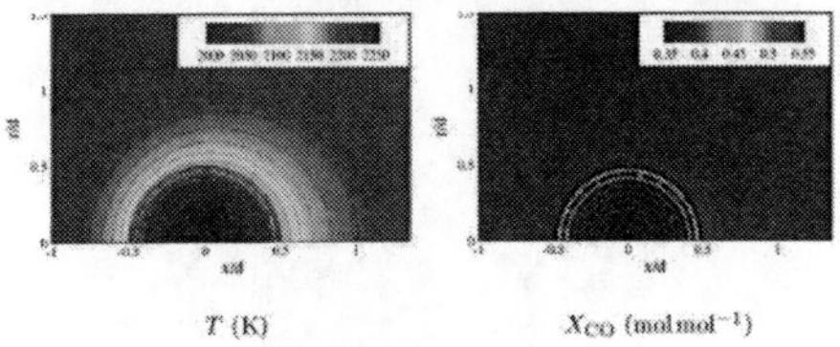


Fig. 2: Temperature and molar fraction of CO

**Fig. 1 and 2 from Richter et al., Detailed analysis of reacting particles in an entrained-flow gasifier, Fuel 144 (2016)**

3

## HITECOM (<u>Hi</u>gh <u>Te</u>mperature <u>C</u>onversion <u>O</u>ptical <u>M</u>easurement)

Fig. 3: HITECOM Reactor

- In-situ measurements of isolated particles at defined pressures, temperatures, and gas flow velocities
- Joint project with Friedrich Schiller University Jena
- Reactor allows for high temperature (up to 1400 °C) and high pressure (up to 40 bar)
- In-situ optical measurement allows for fundamental analysis of char conversion and provides validation data for particle-resolved CFD modelling

4

## HITECOM Reactor

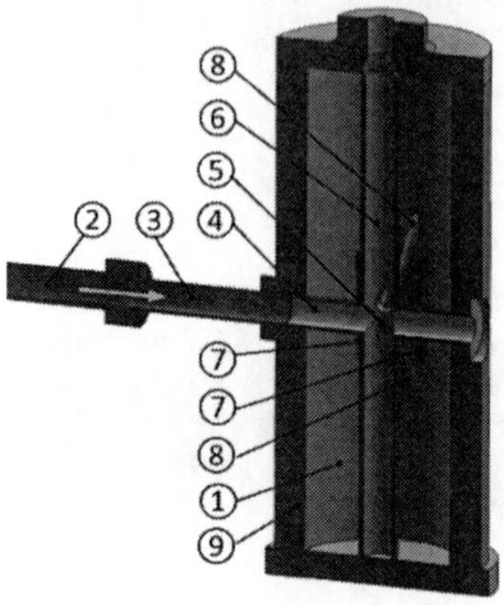

①: Pressure Vessel
②: Gas Preheater
③: Isolation
④: Ceramic Pipe
⑤: Char Particle
⑥: TGA
⑦: Heaters
⑧: Optical Accesses
⑨: Water Cooling

Fig. 4: HITECOM Reactor

- More than 20 temperature measurement points in reactor
- Particle diameter: 0.5...4 mm
- Particle Reynolds number up to 300

5

## HITECOM Reactor

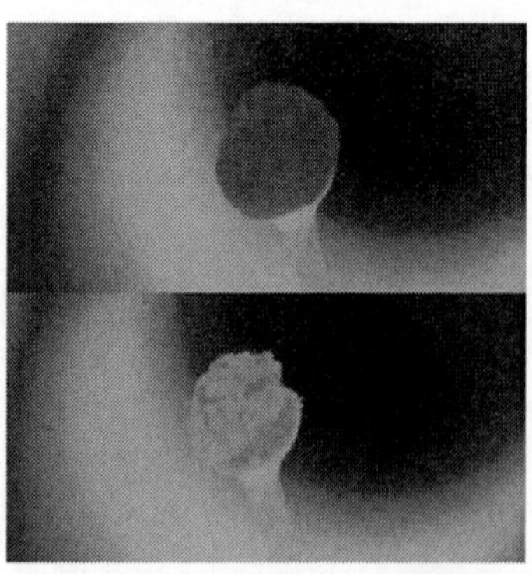

Fig. 5: Single particle reaction in HITECOM

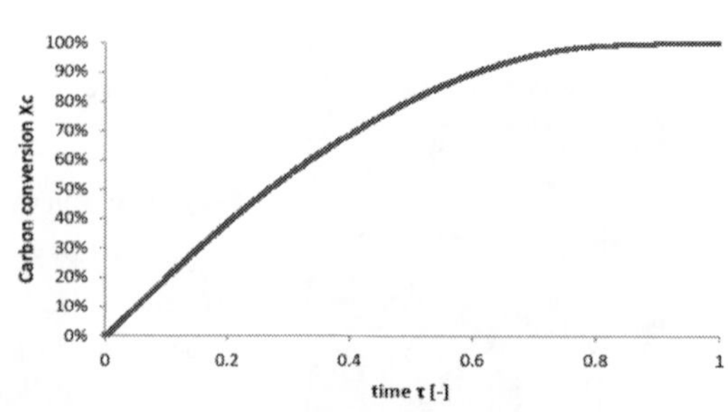


Fig. 6: Carbon conversion of single particle

**Fig. 5 and 6 from F. Küster et al., Experimental investigation of single-particle gasification, International Freiberg Conference (2016)**

6

## Motivation of this work

- Reactor has already gone into operation under purposed operating conditions.
- Kinetics of char particle conversion are measured.
- Reactor supplies really a homogenous flow velocity and temperature around the particle under all operating conditions? → Detailed CFD calculations of the temperature and velocity distribution inside the reactor.

7

Numerical Model

## Calculation case

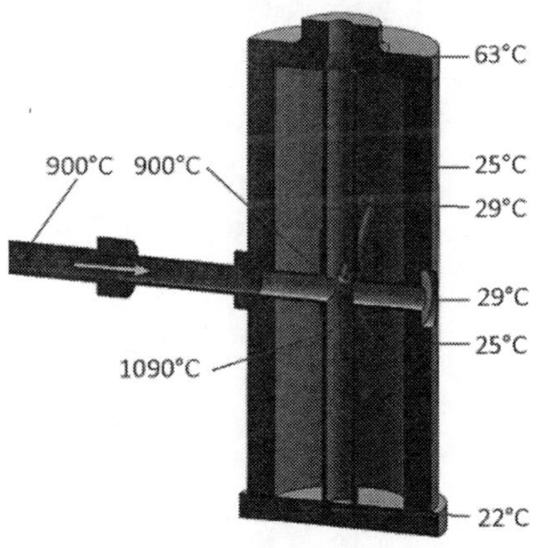


Fig. 6: Measurement points in HITECOM Reactor

- Temperature information from measurement
- Operating pressure 1 and 40 bar
- Particle diameter: 3 and 4 mm
- Particle Reynolds number: 3 and 60
- Gas Reynolds number: 30 and 600

8

HITECOM ProVirt

## Preheater and HITECOM Reactor

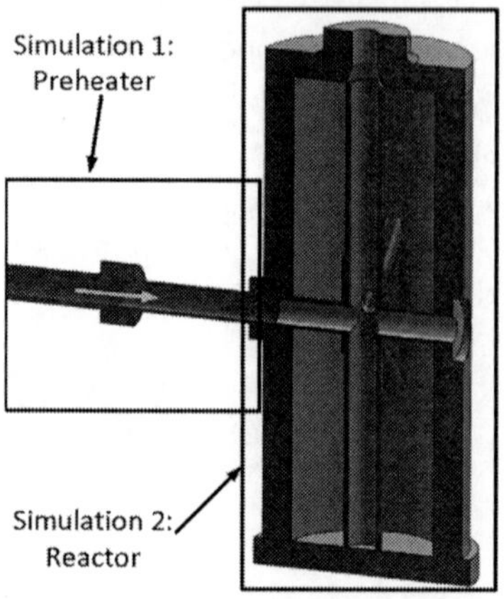


Fig. 7: HITECOM Reactor

- Gas heating process in preheater is also interesting. (Temperature in reactor inlet)
- Two simulation parts are carried out.
- Required calculation time can be reduced.

9



HITECOM ProVirt

## Numerical Model of Preheater

- ANSYS-Fluent 15.0
- 2D-axisymmetric
- steady-state
- buoyancy effects neglected
- Realizable k-ε, scalable wall function
- Species transport without reaction
- Radiation model Discrete Ordinates
- Second order upwind

10

Numerical Model

HITECOM ProVirt

## Calculation Mesh

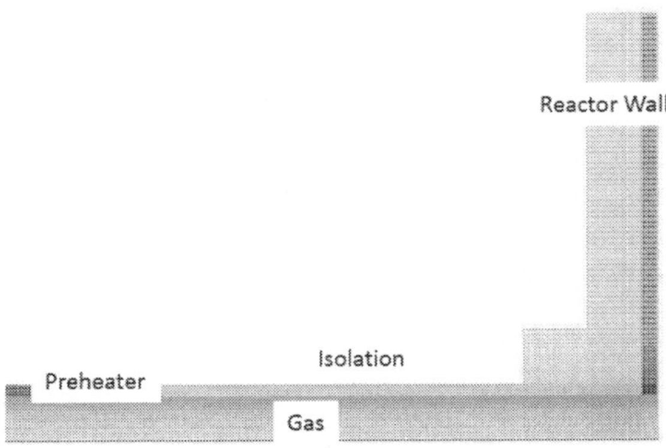


Fig. 7: Calculation mesh Preheater

- Axisymmetric
- 50,000 structured cells
- Heat transfer in solid parts
- Mesh refinement interface

11

HITECOM ProVirt

## Results

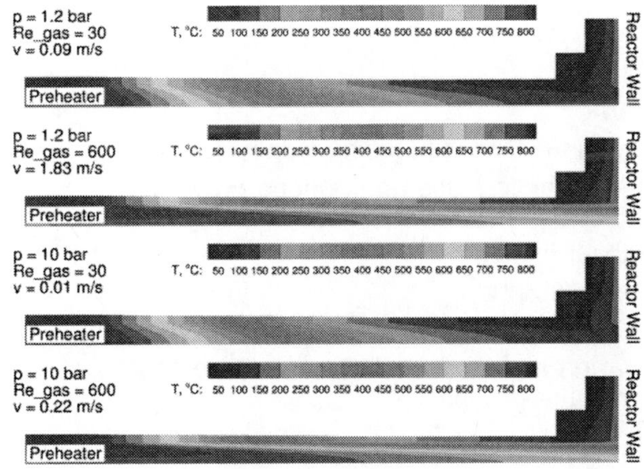


Fig. 8: Temperature distribution in preheater

- Gas is heated to the set temperature through the preheater but cooled through the isolation
- Higher outlet temperature in preheater through high gas velocity

12

## Results for Preheater

| Case | $Re_{gas}$ | Pressure [bar] | $T_{outlet}$ [°C] |
|---|---|---|---|
| vh_1 | 20 | 1.2 | 107 |
| vh_2 | 600 | 1.2 | 651 |
| vh_3 | 20 | 10 | 107 |
| vh_4 | 600 | 10 | 651 |

Tab. 1: Outlet temperature of preheater

- Cases with same Reynolds number have same outlet temperature
- Heating and cooling processes depend only on Re, not on operating pressure

13

## Numerical Model for HITECOM Reactor

- ANSYS-Fluent 15.0
- Steady-state
- Buoyancy effects included
- Realizable k-ε, scalable wall function with full buoyancy effects
- Transport of $CO_2$ inside the reaction tube and transport of Ar in pressure chamber considered
- Radiation model: Discrete Ordinates
- Discretization: Second order upwind

Based on previous study for preheater, calculations are carried out for different velocities at one pressure (1.2 bar).

14

Numerical Model

# Calculation Mesh

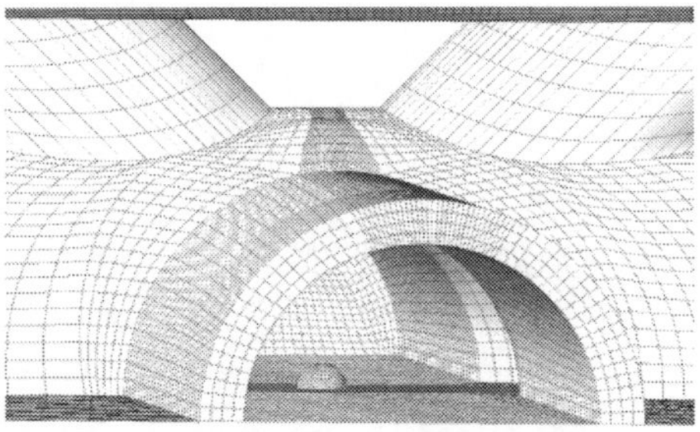
Fig. 9: Calculation mesh HITECOM

- Half of the reactor (Gravity)
- 3 million structured cells
- Heat transfer in solid parts and particle
- Mesh refinement around the particle

15

Numerical Model

# Calculation Mesh

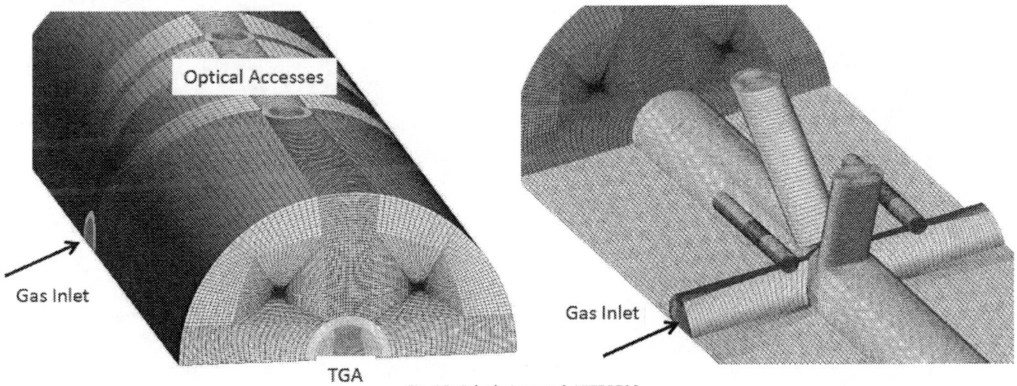

Fig. 10: Calculation mesh HITECOM

16

Numerical Model

HITECOM ProVirt

## Calculation Mesh

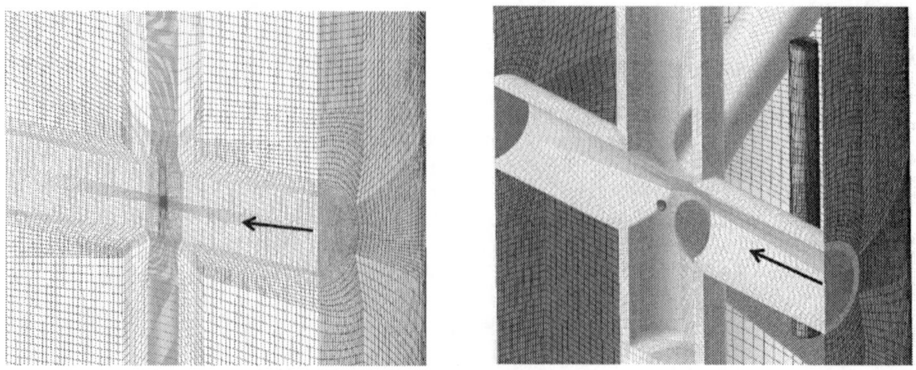

Fig. 11: Calculation mesh HITECOM

17

HITECOM ProVirt

## Temperature Distribution at Reactor Walls

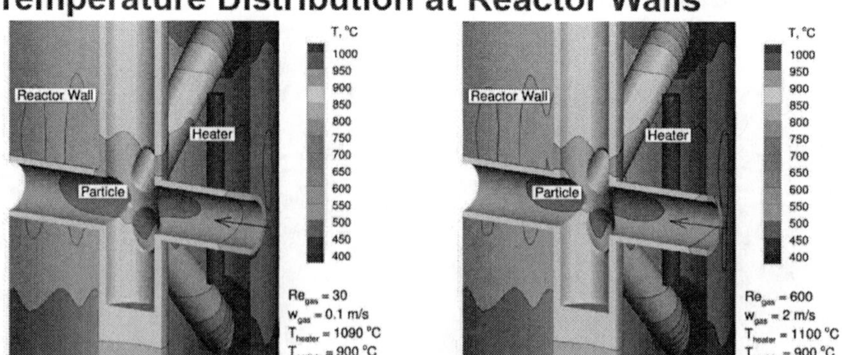


Fig. 12: Temperature distribution on the solid parts

- High temperature at tube mainly through radiative heat transfer from the heater
- Temperatures on gas tube and reactor wall not constant
- Solid around particle reaches the set temperature

18

# Gas Temperature Inside the Reactor

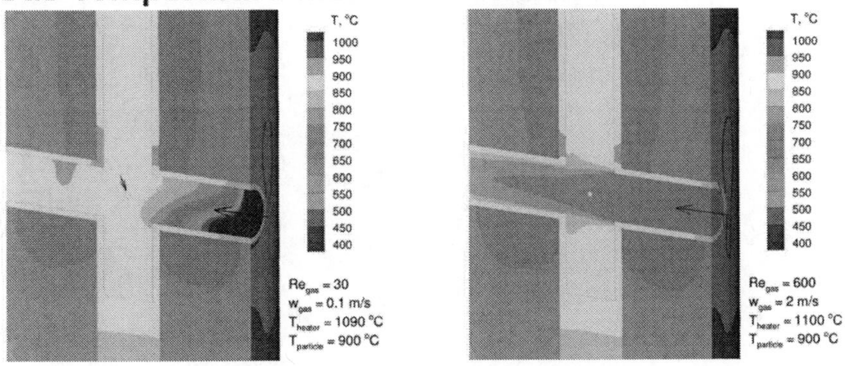


Fig. 13: Gas temperature in HITECOM reactor

- Gas temperature outside the tube (Ar) relatively constant (natural convection)
- Gas inside the tube ($CO_2$) has a lower temperature at inlet and is heated up through radiation and convection
- Particle and gas around the particle features relative constant temperature

19

# Velocity Distribution Inside the Reactor

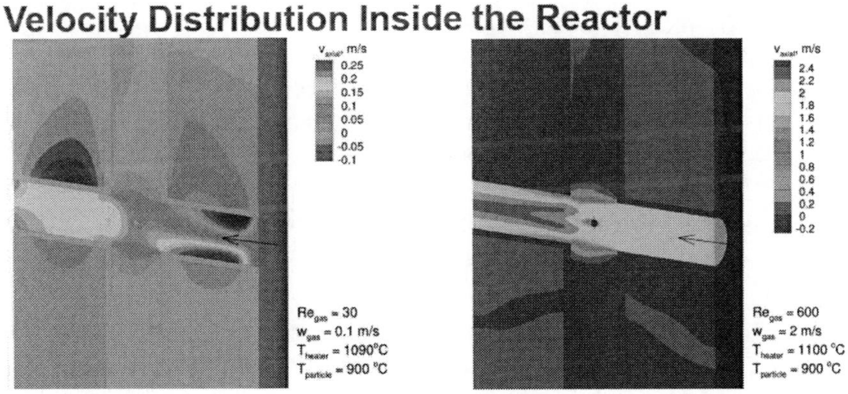


Fig. 14: Axial velocity in HITECOM reactor

- Natural convection in reactor
- Backflow at the beginning of the gas tube through natural convection

20

## Velocity Distribution Around Isolated Particle

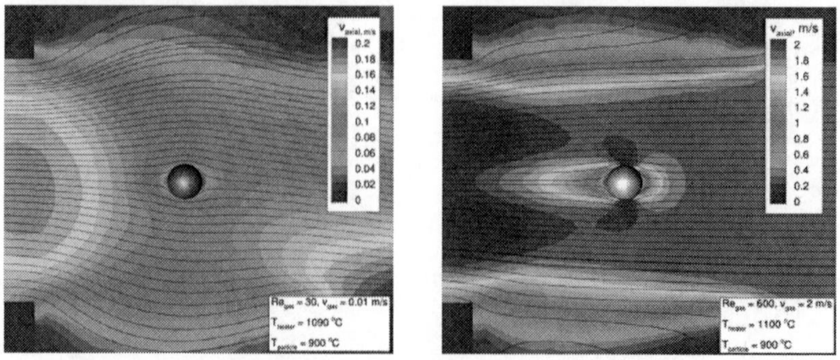

Fig. 15: Gas velocity and stream line near the particle

- Nearly constant axial velocity around the particle

21

## Summary and Outlook

1. CFD studies confirmed the general principal of the HITECOM reactor.
2. Constant temperature and velocity profiles around isolated particles can be achieved for different Reynolds numbers, temperatures and pressures.
3. For different Reynolds numbers, the heat transfer in the reactor differs:
    a. Gas with low Reynolds number would be heated up and cooled down quickly.
    b. Gas with higher Reynolds number has a higher temperature at reactor inlet, but need a higher heater temperature in reactor.
4. As a next step, the HITECOM reactor will be intensively used for single-particle studies on char conversion, kinetic measurements, and the validation of different particle-conversion models.

22

## Acknowledgements

This research has been funded in the framework of HITECOM by the Federal Ministry of Education and Research of Germany (project number 03Z2F512) and in the frame work of ProVirt by the Saxon State Ministry of Science and the Arts and the European Union (project number 100231952).

GEFÖRDERT VOM
Bundesministerium für Bildung und Forschung

ZENTREN FÜR INNOVATIONSKOMPETENZ
UNTERNEHMEN REGION

Europa fördert Sachsen.
ESF

**2016 International Pittsburgh Coal Conference**
**Cape Town, SOUTH AFRICA**
**August 8 – 12, 2016**

**Abstract Submission**

**PROGRAM TOPIC:**
**GASIFICATION TECHNOLOGIES**

**Exact Title of Paper:**

# CFD-BASED ACCELERATED GASIFICATION TECHNOLOGY DEVELOPMENT

***Andreas Richter, TU Bergakademie Freiberg, CIC Virtuhcon, GERMANY***
***Email: a.richter@vtc.tu-freiberg.de, Phone: +49 3731 39 4801, Fax: +49 3731 39-4555***

*Philip Rößger, TU Bergakademie Freiberg, CIC Virtuhcon, GERMANY*
*Philip.Roessger@iec.tu-freiberg.de, Tel. +493731 39 4776, Fax +493731 39 4555*

*Thomas Förster, TU Bergakademie Freiberg, CIC Virtuhcon, GERMANY*
*Thomas.Foerster@iec.tu-freiberg.de, Tel. +493731 39 4239, Fax +493731 39 4555*

*Peter Seifert, TU Bergakademie Freiberg, Chair of Energy Process Engineering and Thermal Waste Treatment, GERMANY*
*Peter.Seifert@iec.tu-freiberg.de, Tel. +493731 39 4552, Fax +493731 39 4555*

*Bernd Meyer, TU Bergakademie Freiberg, Chair of Energy Process Engineering and Thermal Waste Treatment, GERMANY*
*Bernd.Meyer@iec.tu-freiberg.de, Tel. +493731 39 4511, Fax +493731 39 4555*

**Abstract:**

Gasification is a key process to convert carbonaceous materials into basic chemicals. Especially the importance of entrained-flow gasification is increasing, since this technology provides the highest throughput capacities compared to fluidized-bed and fixed-bed gasification processes. Because political and economic frameworks, and therefore the utilized feedstock are changing continuously and more rapidly, an accelerated development of new technology or an optimization of existing technology are needed. CFD provides the opportunity to evolve new technology without the classical, long and cost intensive way from lab-scale to demo- and pilot-scale plants. Basis for an accelerated development are reliable CFD models that are validated against large-scale experiments with conditions close to the industrial process to guarantee the accuracy of the numerical results. In this work, development strategies for new entrained-flow gasifiers are presented. We start from non-catalytic reforming of natural gas and adapt some of the concepts for coal gasification. The CFD model was validated against in-house measurements based on the semi-industrial test plant HP POX at pressures up to 70 bar. This test plant provides a comprehensive set of experimental data, e.g. the gas temperature along the reactor, gas composition and temperature at the outlet, and flame characteristics such as flame shape and length based on an optical probe system.

TECHNISCHE UNIVERSITÄT BERGAKADEMIE FREIBERG
The University of Resources. Since 1765.

IEC EVT

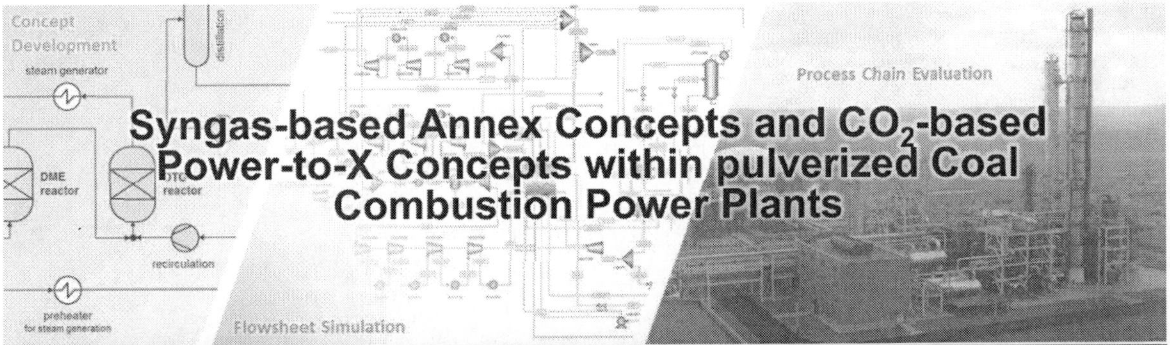


# Syngas-based Annex Concepts and $CO_2$-based Power-to-X Concepts within pulverized Coal Combustion Power Plants

Dipl.-Ing. Christian Wolfersdorf
Institute of Energy Process Engineering & Chemical Engineering, Technische Universität Bergakademie Freiberg
11th August 2016, Cape Town, South Africa

Background › Process Description › Modelling Results › Cost Estimation › Summary

---



IEC EVT

## Electricity generation – Example Germany

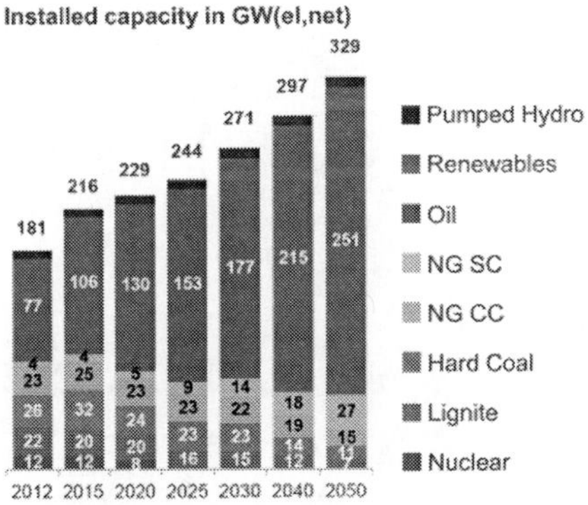


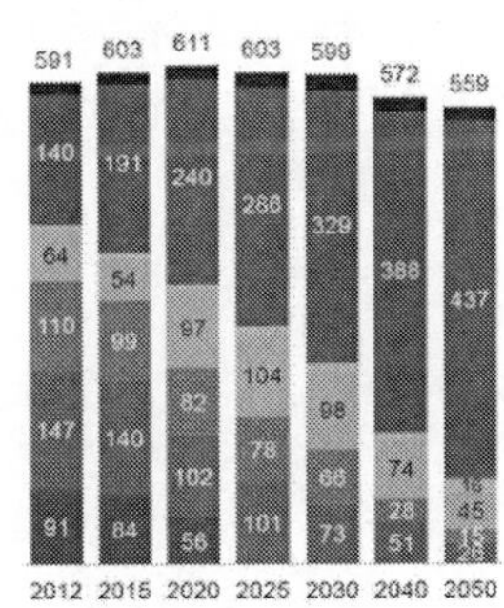


**Excess Electricity in TWh(el)**

| | |
|---|---|
| 2000 | 0.13 |
| 2010 | 0.42 |
| 2020 | 1.1–13 |

Sources:

Energy concept of the Federal Government, Germany, 2010

Jürgensen et al., Biomass and Bioenergy, 2014, 66, 126-132

Wünsch et al., Maßnahmen zur nachhaltigen Integration von Systemen zur gekoppelten Strom- und Wärmebereitstellung in das neue Energieversorgungssystem, Prognos AG, 2013

→ **Challenges:**
- → **Flexibility improvement of fossil-fueled power plants**
- → **Concepts for excess electricity storage in countries with high Renewables share**

2

Background

## Flexibility

Flexibility and part load capability of existing fossil fueled power plants (significant remaining lifetime) are limited

| | | PC Hard Coal | PC Lignite | CC Natural Gas | SC Natural Gas |
|---|---|---|---|---|---|
| Load gradients | % $P_N$ / min | **4** / 6 | **2.5** / 4 | **4** / 8 | **12** / 15 |
| Intervall | % $P_N$ | 40-90 | 50-90 | 40-90 | 40-90 |
| Minimal load | % $P_N$ | **25** / 20 | **50** / 40 | **40** / 30 | **40** / 20 |
| Start-up time | | | | | |
| Hot (< 8 h) | h | **2.5** / 2 | **4** / 2 | **1** / 0.5 | < 0.1 |
| Cold (> 48 h) | h | **5** / 4 | **8** / 6 | **3** / 2 | < 0.1 |

Source: Feldmüller, Siemens Energy, Flexibility of today's power plants from perspective of manufacturers, Germany, 2013

Information:
**State of the Art** / Optimization potential

→ **Challenge for lignite-fired power plants:**
→ **Reducing minimal load to avoid lifetime-consuming start-up, shutdown processes**

3

Process description

## Annex – Coal to Methanol

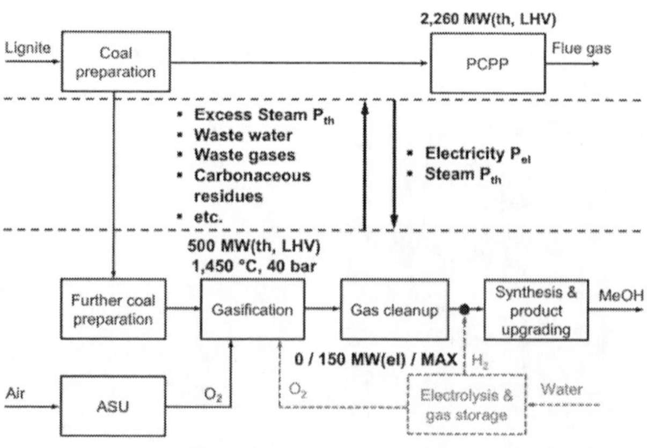


## PTX – $CO_2$ to Methanol

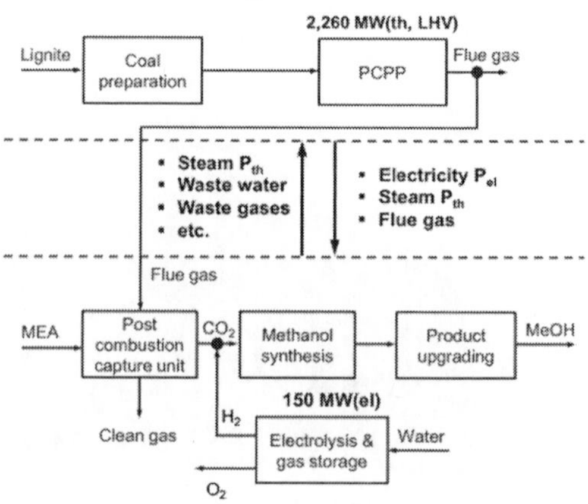


→ **Flexibility improvement of lignite-fired power plant:**
→ **Annex, PTX = power sink → Reduction of net power output, Electricity storage**
→ **Cost reduction by sharing infrastructure and auxiliary equipment**

4

Process description

## Annex & PTX concept configuration

- Case study for different configurations with Methanol production,
- Analysed with flow-sheet simulation tools (Aspen Plus, EBSILON Professional) regarding:
  - Power consumption, steam integration, flexibility improvements, part load operation of PCPP,
  - Efficiency, $CO_2$ emissions, CAPEX, OPEX

| | concepts | | | | loads for electricity prices (cases) | |
|---|---|---|---|---|---|---|
| **Unit** | **Annex 1** | **Annex 2** | **PTX 2** | **Annex 3** | **Low price ↓[2] →** | **High price ↑[3]** |
| **Water Electrolysis** | - | **150 MW(el)** | **150 MW(el)** | > 150 MW(el) | Nominal load | Minimal load |
| **PCPP[1]** | 2,260 MW(th) | 2,260 MW(th) | 2,260 MW(th) | 2,260 MW(th) | Minimal load | Nominal load |
| **Gasifier[1]** | 500 MW(th) | 500 MW(th) | - | 500 MW(th) | Nominal load | Nominal load |
| **ASU** | x | x | - | - | Minimal load | Nominal load |
| **CO-Shift** | x | x | - | - | Nominal load | Nominal load |
| **Synthesis** | x | x | x | x | Nominal load | Nominal load |

[1] LHV-based thermal power input; [2]Reference: 6570 h/a; [3]Reference: 2190 h/a

5

Modelling results

## Efficiency and $CO_2$ emission

| | Annex 1 | Annex 2 | Annex 3 | PTX 2 |
|---|---|---|---|---|
| **Electrolysis capacity** in MW(el) | - | 150 | 780 | 150 |
| **Thermal Input** Coal in MW(th) *1 | 500 | 500 | 500 | - |
| Methanol Output in t/d | 1200 | 1480 | 2740 | 270 |
| Net Power Output (LEP) in % Reference PCPP Minimal load 50 % | 46 | 32 | -26 | 36 |
| **Thermal efficiency** in % *2 | 56,6 | 45,2 | 33,7 | 20,4 |
| **Carbon retention MeOH** in % | 39,0*3 | 48,0*3 | 88,2*3 | 92,0*4 |
| **$CO_2$ emissions** in g/kWh(th,MeOH) electricity from PCPP, 978 g/kWh(el) | 443 | 617 | 930 | 1609 |
| **$CO_2$ emissions** in g/kWh(th,MeOH) renewable electricity, 53 g/kWh(el) | 377 | 276 | 80 | -156 |

*1 LHV-based
*2 Electrolyzer efficiency 67 % (LHV)
*3 based on coal
*4 based on $CO_2$

$$\bar{\eta}_{th} = \frac{\bar{P}_{th,MeOH}}{\bar{P}_{th,coal,Annex} + \left(\bar{P}_{th,el,aux} - \bar{P}_{th,steam}\right)}$$

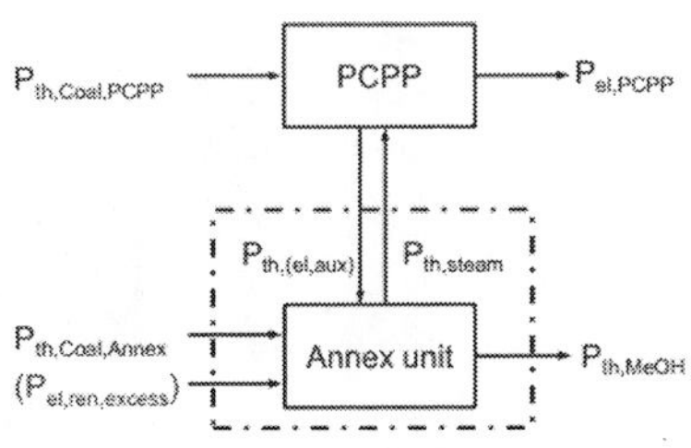


$$\eta_{CR,Annex} = \frac{\dot{n}_{C,MeOH}}{\dot{n}_{C,Coal,Annex}}$$

6

Modelling results

## Efficiency and $CO_2$ emission

| | Annex 1 | Annex 2 | Annex 3 | PTX 2 |
|---|---|---|---|---|
| **Electrolysis capacity** in MW(el) | - | 150 | 780 | 150 |
| **Thermal Input** Coal in MW(th) *1 | 500 | 500 | 500 | - |
| Methanol Output in t/d | 1200 | 1480 | 2740 | 270 |
| Net Power Output (LEP) in %<br>Reference PCPP Minimal load 50 % | 46 | 32 | -26 | 36 |
| **Thermal efficiency** in % *2 | 56,6 | 45,2 | 33,7 | 20,4 |
| **Carbon retention MeOH** in % | 39,0*3 | 48,0*3 | 88,2*3 | 92,0*4 |
| **$CO_2$ emissions** in g/kWh(th,MeOH)<br>electricity from PCPP, 978 g/kWh(el) | 443 | 617 | 930 | 1609 |
| **$CO_2$ emissions** in g/kWh(th,MeOH)<br>renewable electricity, 53 g/kWh(el) | 377 | 276 | 80 | -156 |

*1 LHV-based
*2 Electrolyzer efficiency 67 % (LHV)
*3 based on coal
*4 based on $CO_2$

$$CO + 2H_2 \rightleftarrows CH_3OH$$

$$CO_2 + 3H_2 \rightleftarrows CH_3OH + H_2O$$

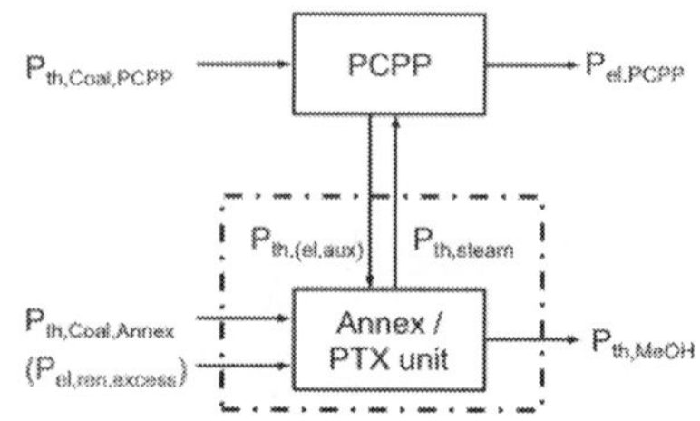


$$\eta_{th} = \frac{\bar{P}_{th,MeOH}}{\bar{P}_{th,coal,Annex} + (\bar{P}_{th,el,aux} - \bar{P}_{th,steam})}$$

$$\eta_{CR,Annex} = \frac{\dot{n}_{C,MeOH}}{\dot{n}_{C,Coal,Annex}}$$

$$\eta_{CR,PTX} = \frac{\dot{n}_{C,MeOH}}{\dot{n}_{C,CO_2}}$$

7

Cost estimation

## Equipment Costs

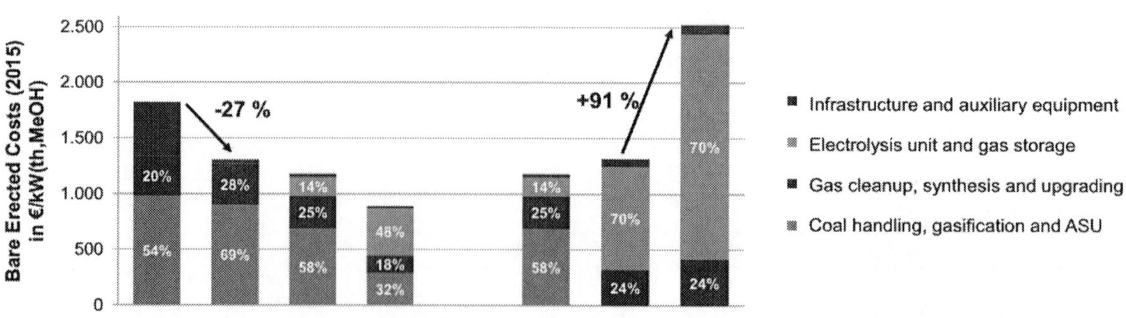


| CTX1 | Annex1 | Annex2 | Annex3 | Annex2 | PTX2 | PTX2 | Concept |
|---|---|---|---|---|---|---|---|
| 1200 | 1200 | 1480 | 2740 | 1480 | 270 | 123 | Methanol production in t/d |
| - | - | 150 | 780 | 150 | 150 | 150 | Electrolysis capacity in MW(el) |
| 5830 | 5830 | 5830 | 5830 | 5830 | 5830 | 2660 | Full load hours Electrolysis in h/a |

8

Cost estimation

## Methanol Costs

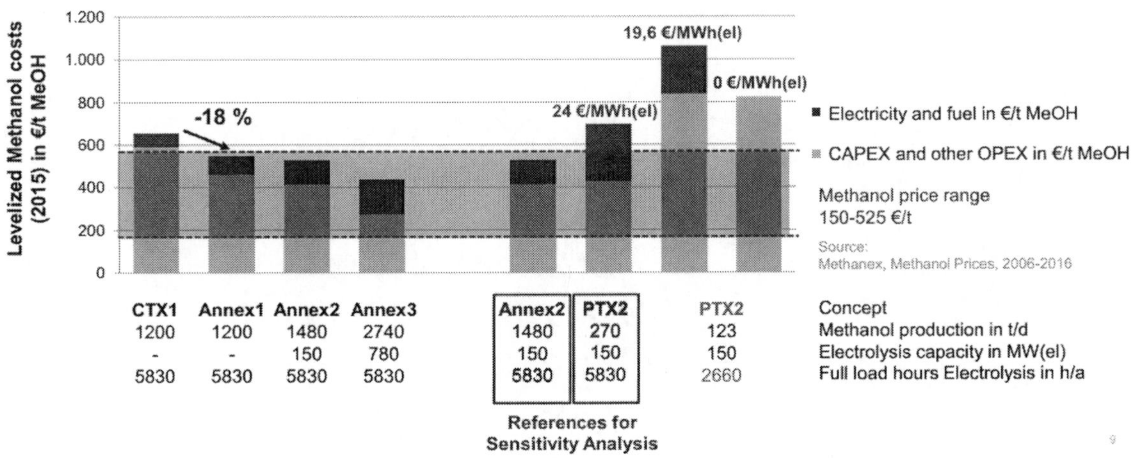


9

Cost Estimation – Sensitivity Analysis – Coal-based Annex

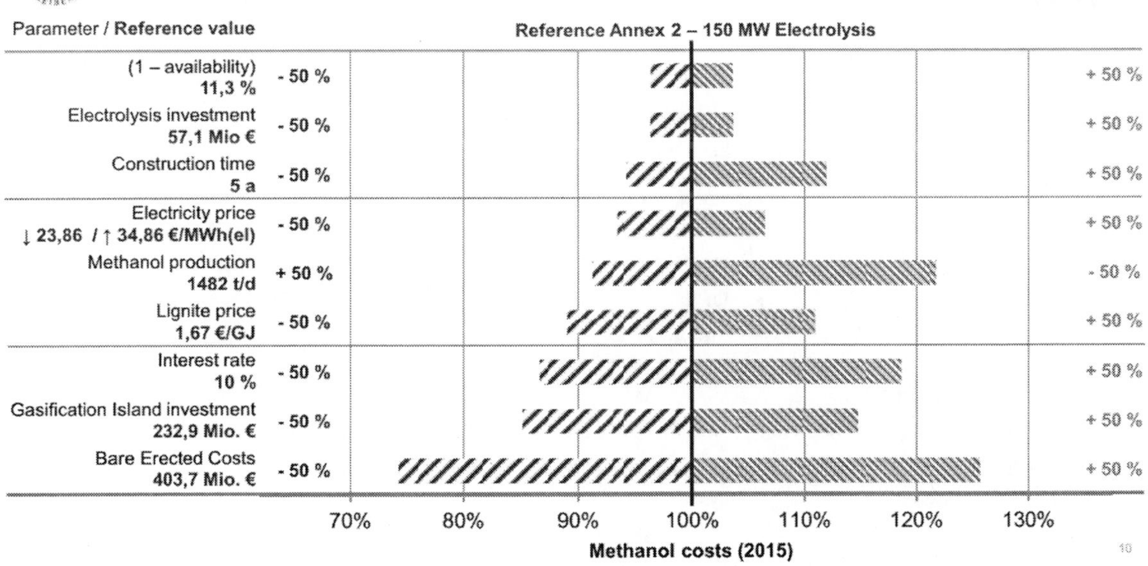


10

Cost Estimation – Sensitivity Analysis – $CO_2$-based PTX

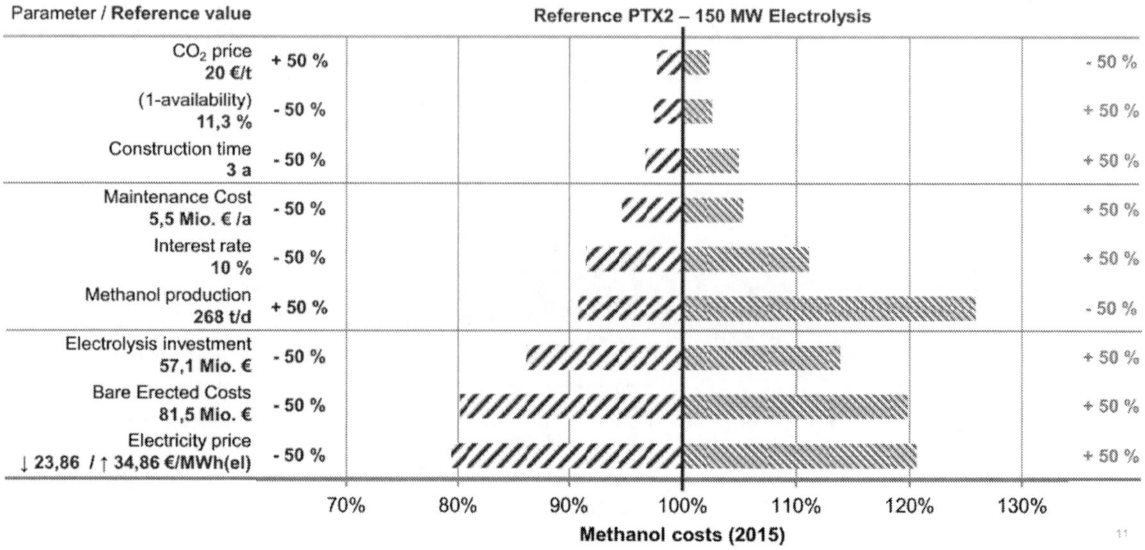


11

Summary

**Annex and PTX concepts** improve flexibility of electricity generation from lignite:

- Power sink → net power output reduction from 50 % (PCPP) to 32 / 36 % (Annex 2 / PTX 2) without lifetime-consuming start-up/shut-down processes
- Electrolysis unit needs high availability for cost reduction

**Annex 1** ↔ **CTX 1** (1200 t/d) without electrolysis unit

- CAPEX ↓ **- 27 %**
- OPEX ↓ **- 8 %**
- Methanol costs ↓ **- 18 %**

**Annex 2** (1480 t/d) ↔ **PTX 2** (268 t/d) 150 MW(el) electrolysis unit

- CAPEX ↓ **- 9 %**
- OPEX ↓ **- 46 %**
- Methanol costs ↓ **- 24 %**

**Electrolysis capacity (Annex)** ↑

- Power sink ↑, $\eta_{CR}$ ↑
- Specific CAPEX ↓
- Total investment costs ↑
- $\bar{\eta}_{th}$ ↓

**Future investigations:**

- Different gasification technologies and synthesis concepts
- Higher value chemical products (MTO, MTG, FT)
- Dynamic modeling

12

## Acknowledgement

Project HotVeGasII: Project number 0327773G
- Participating companies
  - EnBW Kraftwerke AG
  - RWE Power AG, Forschung und Entwicklung
  - Vattenfall Europe Generation AG
  - AIR LIQUIDE Forschung und Entwicklung GmbH
  - Siemens Fuel Gasification Technology GmbH & Co. KG
- Participating research partners:
  - TU München, Institute for Energy Systems
  - TU Bergakademie Freiberg, Institute of Energy Process Engineering and Chemical Engineering
  - Forschungszentrum Jülich GmbH, Institute of Energy and Climate Research (IEK-2)
  - GTT-Technologies, Gesellschaft für Technische Thermochemie und –physik mbH

Project CODY: Project number 03ET7063
- Participating companies
  - E.ON Technologies GmbH / Uniper Technologies GmbH
  - Linde AG (Linde Engineering)
- Participating research partners:
  - TU Bergakademie Freiberg, Institute of Energy Process Engineering and Chemical Engineering, altogether 5 chairs

The results presented in this presentation solely represent the authors view.

Supported by:

Federal Ministry of Economics and Technology

on the basis of a decision by the German Bundestag


13

## THANK YOU FOR YOUR ATTENTION!

For enquiries or further questions, please contact:


**Christian Wolfersdorf**

Email: Christian.Wolfersdorf@iec.tu-freiberg.de
Phone: +49-(0)3731-394512
Fax: +49-(0)3731-394555

Website: www.iec.tu-freiberg.de

14

Back-up

## Coal Analysis

| Ultimate analysis | | |
|---|---|---|
| Ash | wt.-% (dry) | 10.769 |
| Carbon | wt.-% (dry) | 60.000 |
| Hydrogen | wt.-% (dry) | 4.396 |
| Nitrogen | wt.-% (dry) | 0.879 |
| Chlorine | wt.-% (dry) | 0.022 |
| Sulfur | wt.-% (dry) | 1.319 |
| Oxygen | wt.-% (dry) | 22.615 |
| Sum | wt.-% (dry) | 100.000 |

| Heating Value analysis | | |
|---|---|---|
| HHV | MJ/kg (dry) | 23.691 |

| Proximate analysis | | |
|---|---|---|
| Moisture | wt.-% | 12.000 |
| $C_{fix}$ | wt.-% (dry) | 38.242 |
| VM | wt.-% (dry) | 50.989 |
| Ash | wt.-% (dry) | 10.769 |
| Sum | wt.-% (dry) | 100.000 |

15

Back-up

## Electricity generation – Example Germany

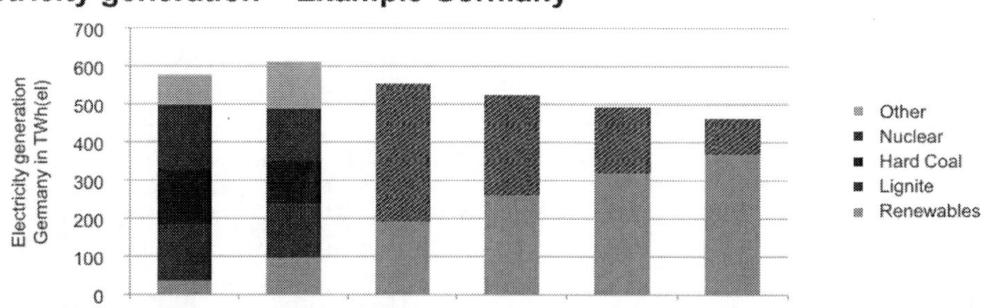


| year | 2000 | 2010 | 2020 | 2030 | 2040 | 2050 |
|---|---|---|---|---|---|---|
| Electrictiy generation | 577 TWh | 611 TWh | 553 TWh | - | - | 461 TWh |
| Renewables Share | 6.2 % | 15.7 % | 35 % | 50 % | 65 % | 80 % |
| Excess Electricity | 0.13 TWh | 0.42 TWh | 1.1–13 TWh | - | - | - |

Source:

Energy concept of the Federal Government, Germany, 2010

Jürgensen et al., Biomass and Bioenergy, 2014, 66, 126-132

→ **Challenges:**
- → **Concepts for excess electricity storage in Germany**
- → **Flexibility improvement of fossil-fueled power plants**

16

Back-up

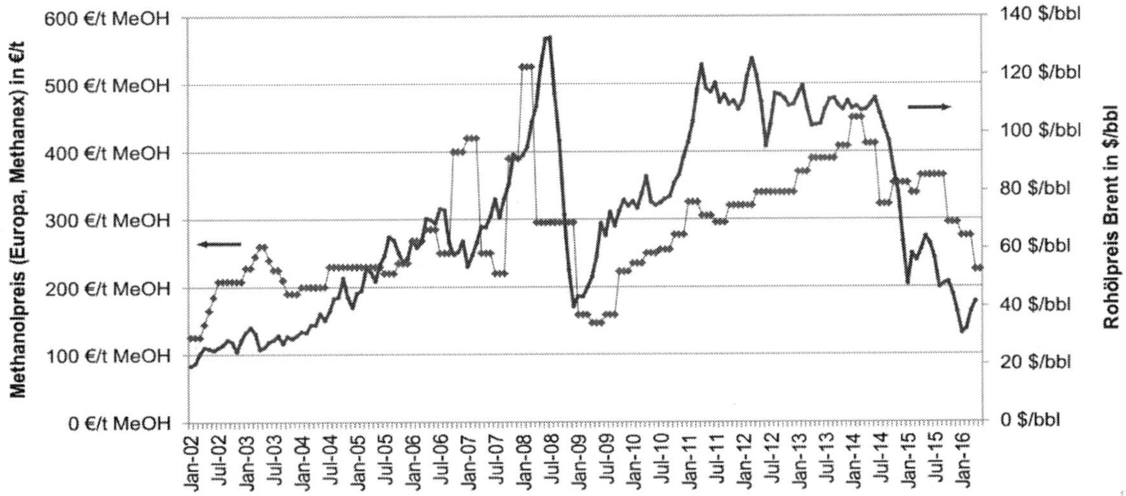


17

Back-up – Electricity prices for „grey Methanol" (power plant electricity)

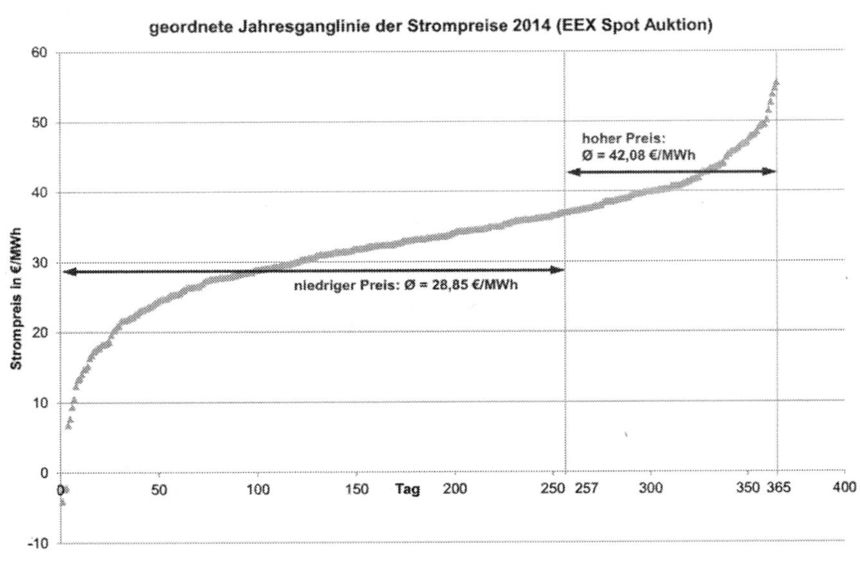


18

## Back-up – Electricity prices for „green Methanol“ (renewable electricity)

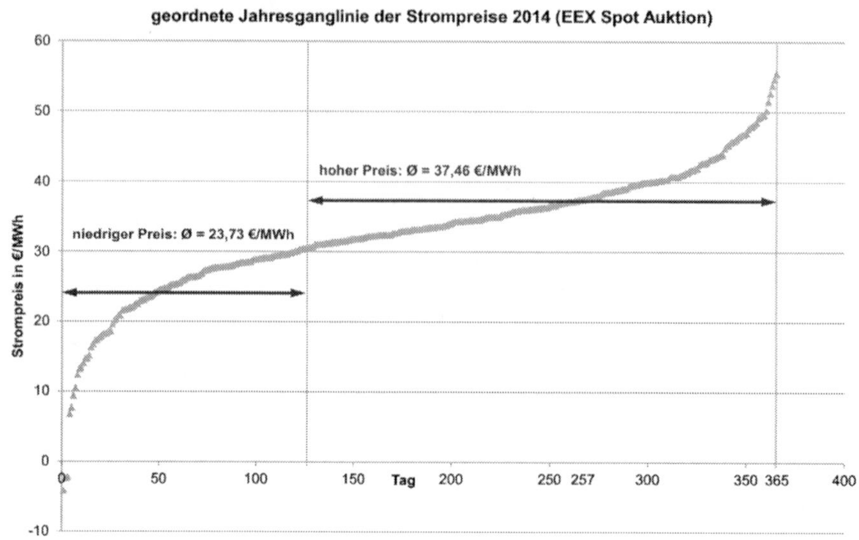


19

## Cost Estimation – Sensitivity Analysis for $CO_2$-based PTX

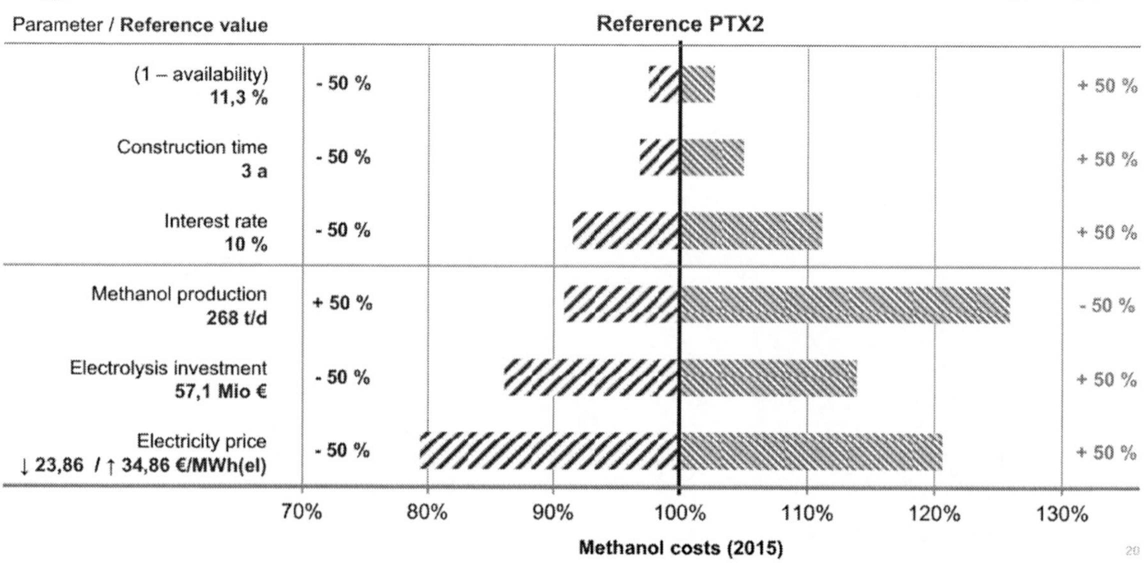


20

Back-up – Methanol costs – Literature

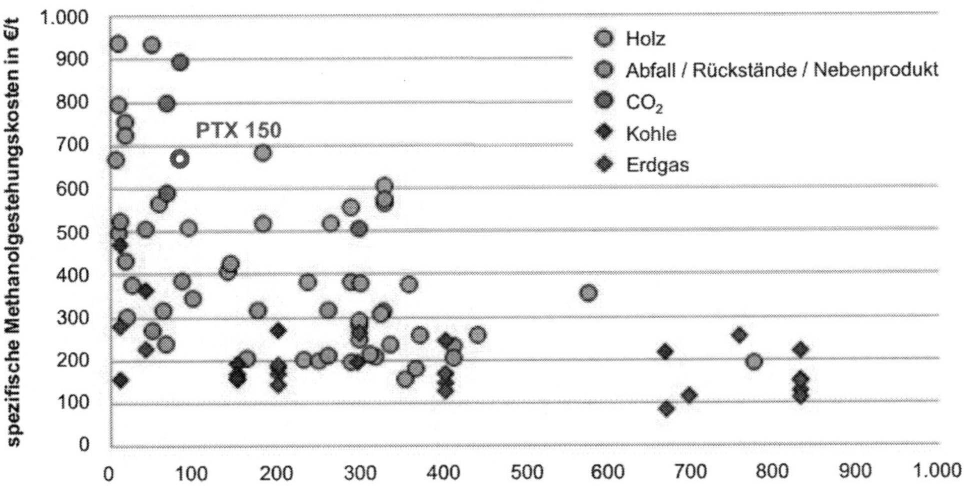


Source: International Renewable Energy Agency (IRENA), Bio-methanol production, 2013 :

21

Back-up – Fluidized bed steam drying

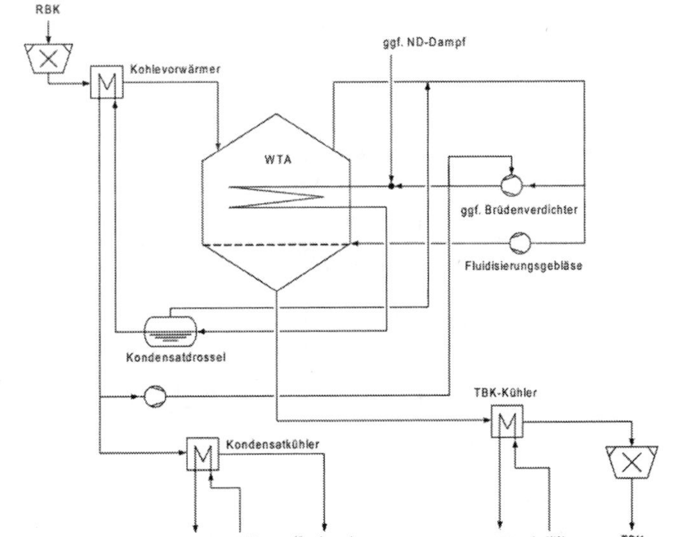


22

## Back-up – Air Separation Unit ASU

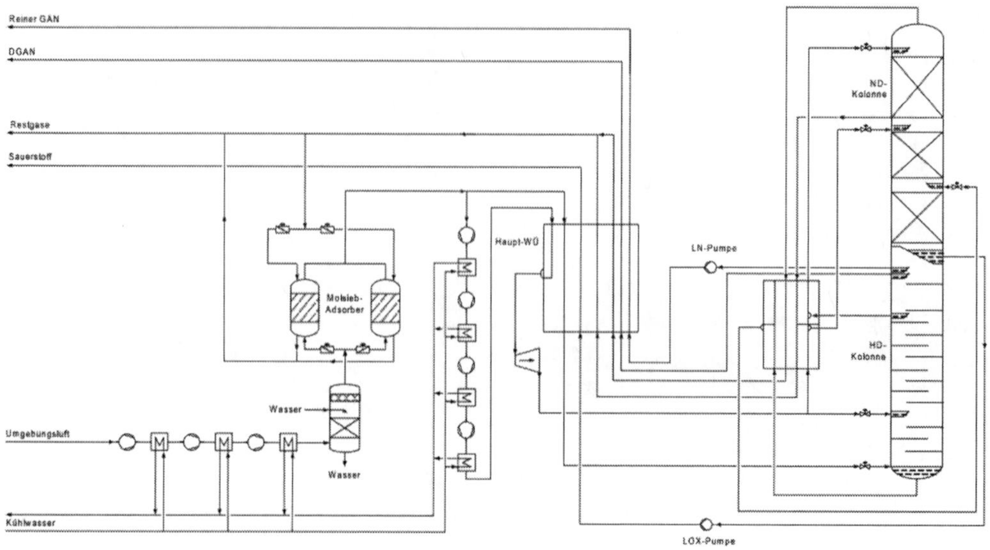


23

## Back-up – Entrained Flow Gasification with Water Quench

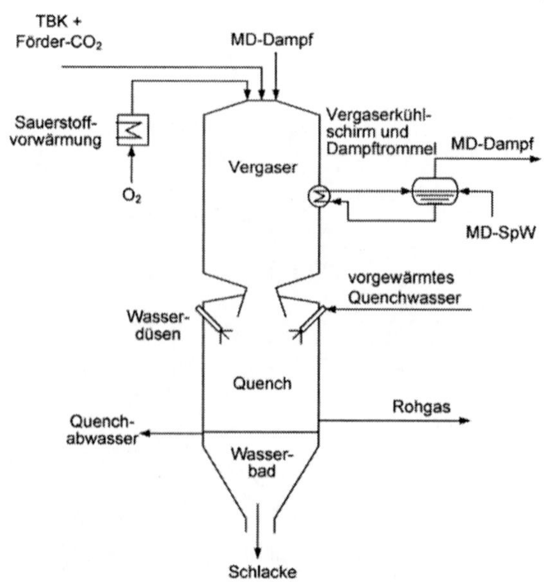


24

## Back-up – Water Scrubbing and Raw Gas Shift Unit

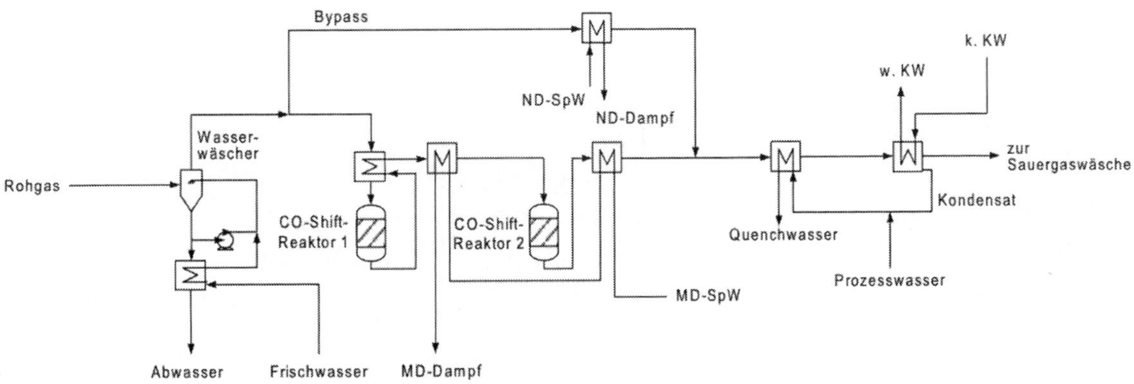




## Back-up – Selective Acid Gas Removal

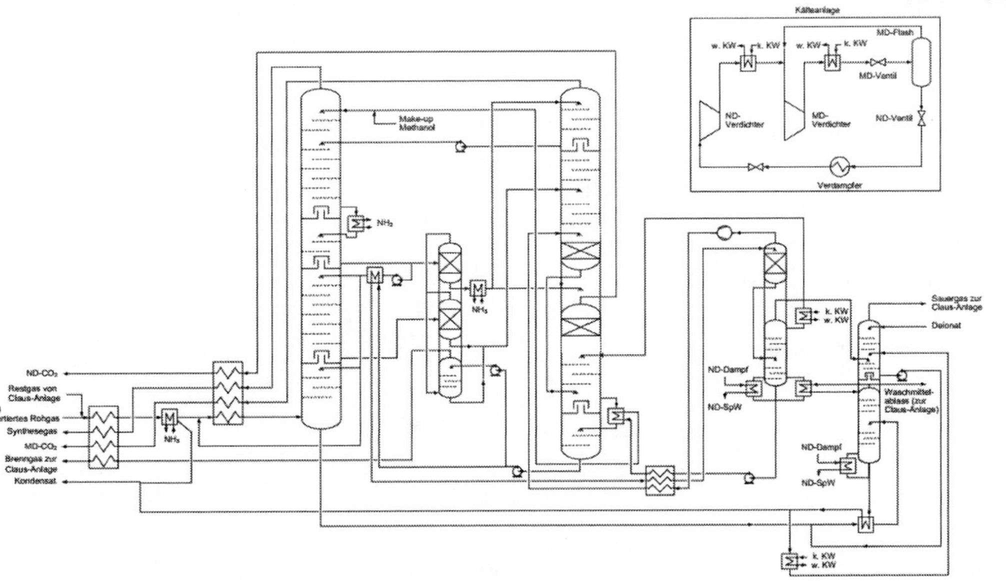

## Back-up – CLAUS Unit

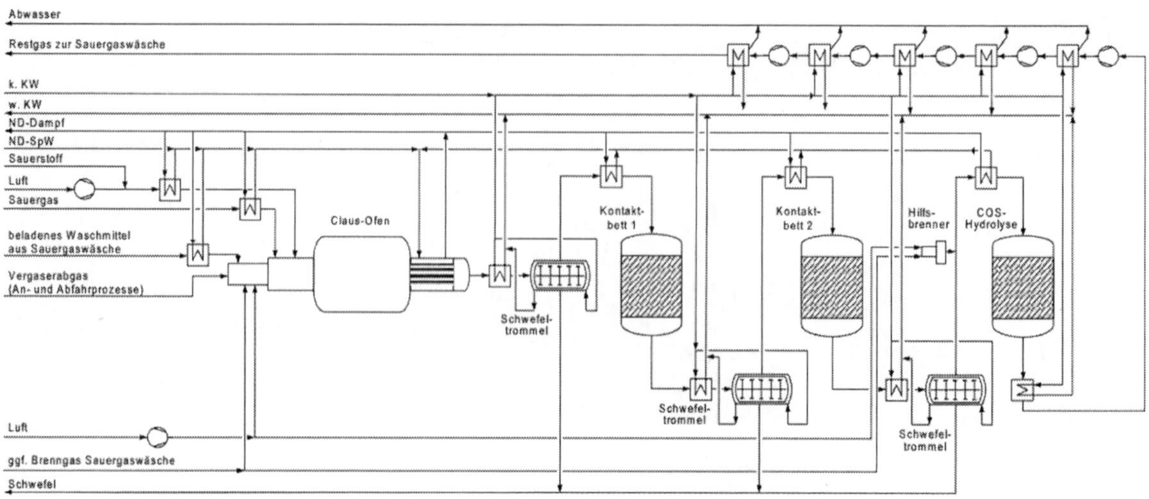


27

## Back-up – Methanol Synthesis

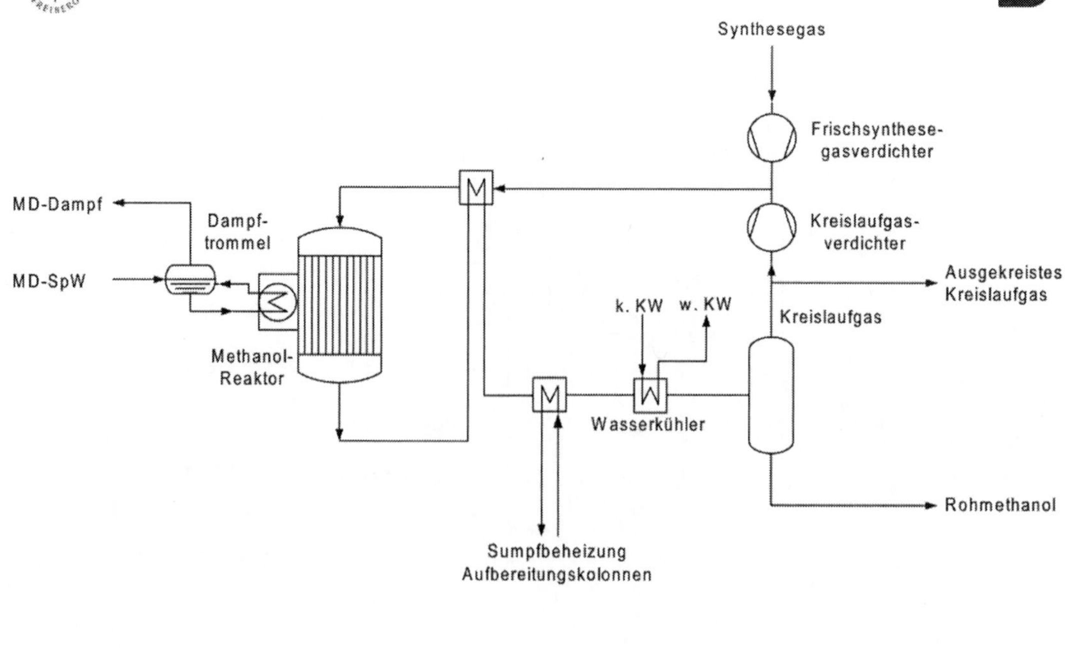


28

## Back-up – Amine Scrubbing (Flue Gas)

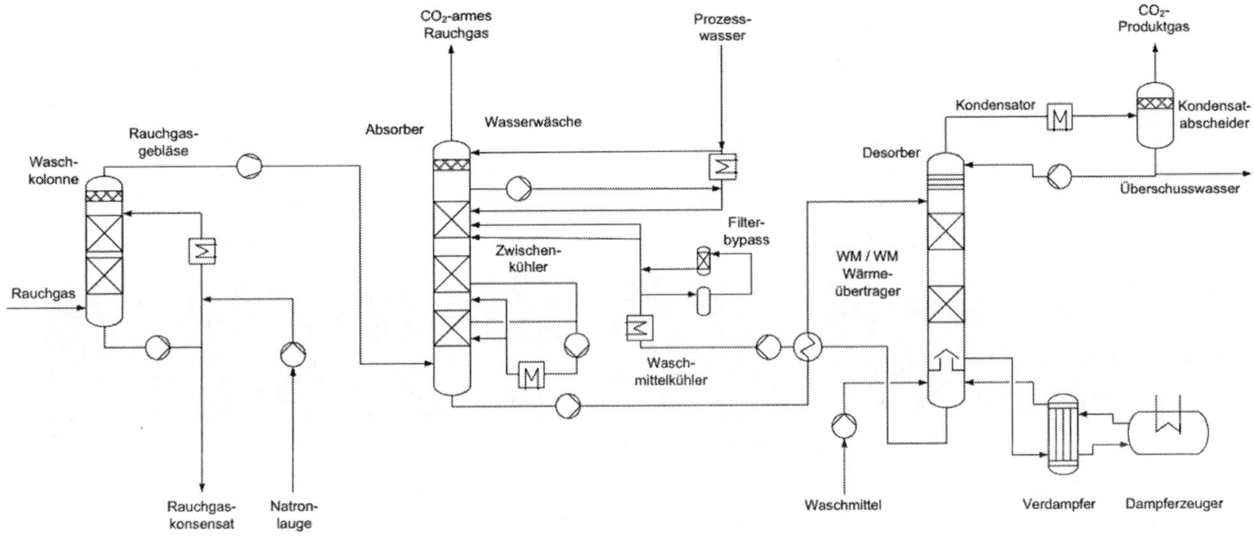


29

THIRTY - THIRD ANNUAL
INTERNATIONAL PITTSBURGH
COAL CONFERENCE
University of Pittsburgh - Swanson School of Engineering

KIER

Session 29: Clean Coal and Gas to Fuels: Catalytic Study - 2

[15:30-15:50] Thursday, August 11, 2016

# Spontaneously Activatable Iron-Based Catalysts in Fischer-Tropsch Synthesis:

## From Lab-Scale Study to Pilot-Scale Demonstration

**Dong Hyun Chun**, Geun Bae Rhim, Ji Chan Park, Ho-Tae Lee, Jung-Il Yang, SungJun Hong, Heon Jung

*Korea Institute of Energy Research, Clean Fuel Laboratory*

---

## Intro: FTS (Fischer-Tropsch Synthesis)

Franz Fischer & Hans Tropsch

A catalytic reaction process to convert syngas ($H_2$+CO) into liquid hydrocarbons

$$nCO + \{2n(+1)\}H_2 \rightarrow C_nH_{2n(+2)} + nH_2O$$

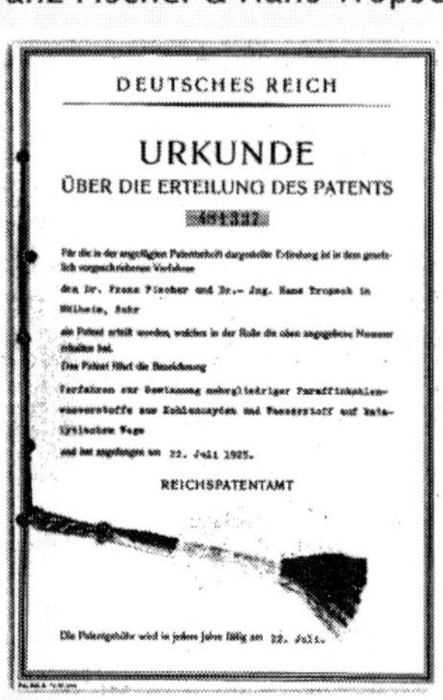

DEUTSCHES REICH

URKUNDE

ÜBER DIE ERTEILUNG DES PATENTS

REICHSPATENTAMT

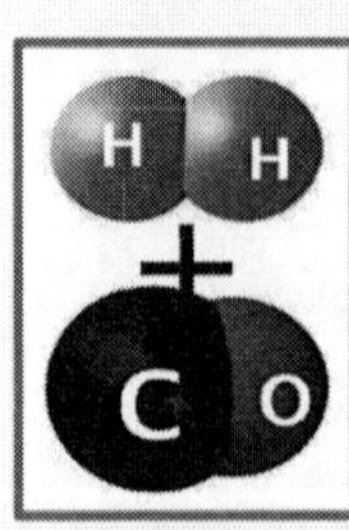


Catalysts
Temperature
Pressure

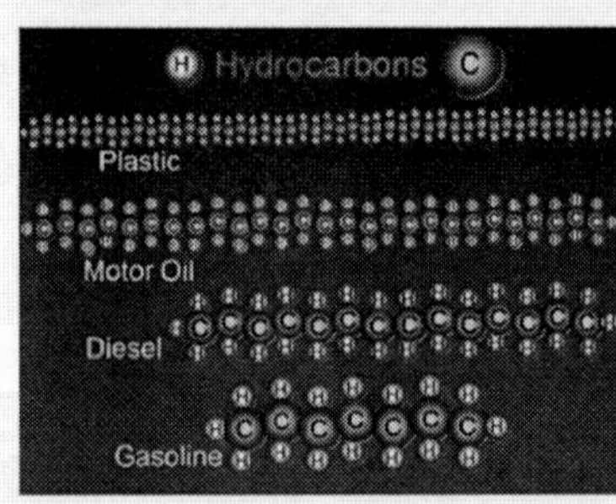


✓ Exothermic
✓ Complicated

Key Technologies: Reactors & Catalysts

# Intro: ICL (Indirect Coal Liquefaction)

A technology for production of synthetic liquid fuels from $H_2$+CO produced via coal gasification

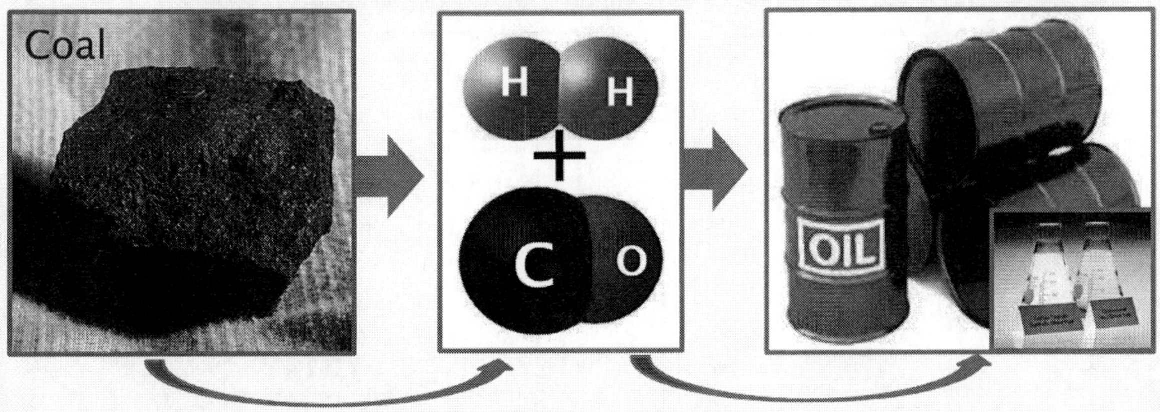


**Gasification + Cleaning** **Fischer–Tropsch**



# Intro: FTS Catalysts & Reactors

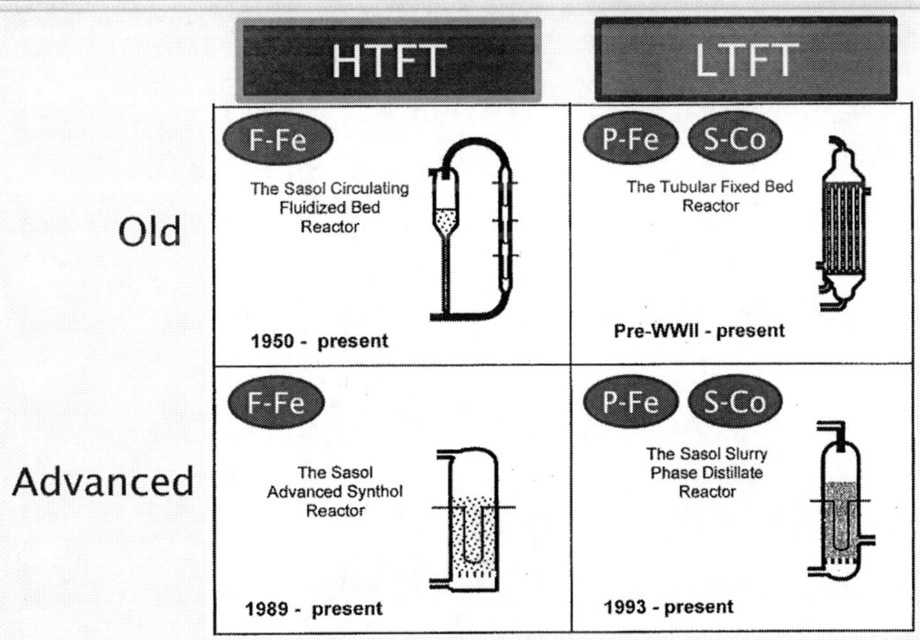


<Figure Source: A.P. Steynberg. M.E. Dry, Fischer-Tropsch Technology, Elsevier, Amsterdam (2004).>

# Intro: KIER FTS Technology

**Low-Temperature Fischer-Tropsch**

Slurry Bubble Column Reactors +
Precipitated Iron-Based Catalysts

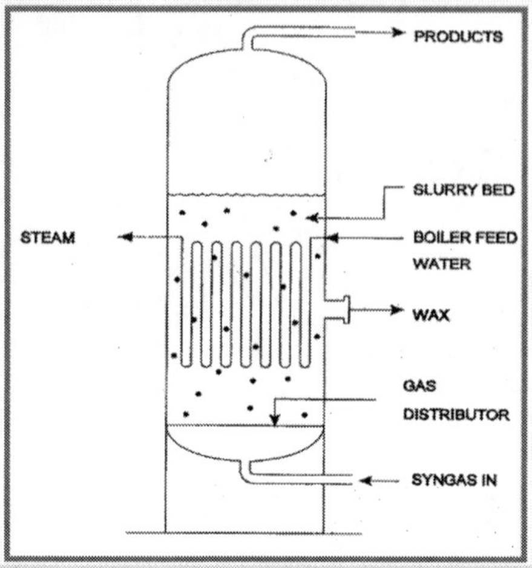


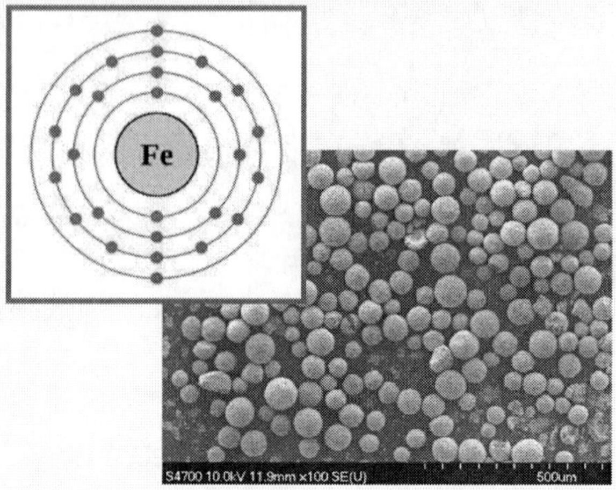






# Intro: Scope of KIER FTS

Hydrodynamic Behavior → **Scale-Up**

Scale-Up Factors

Catalyst

**Lab-Scale**

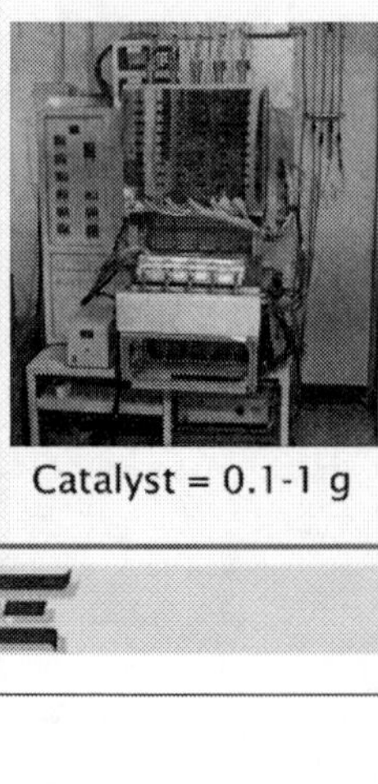

Catalyst = 0.1-1 g

Catalyst = 10-20 g

**Bench-Scale**

Catalyst = 0.5-1 kg

Catalyst = 4-6 kg

**Pilot-Scale**

Catalyst = 30-200 kg

# Intro: KIER FTS History (2007~Present)

- ✓ Our Own Iron-Based Catalysts
- ✓ Pilot-Scale FT Process with Coal Gasifier

# Intro: Co-Precipitation & Spray-Drying

<Process Flow Diagram & Role of Each Process>

- Aqueous Solution of $Fe(NO_3)_3 \cdot 9H_2O$ + Aqueous Solution of $Cu(NO_3)_2 \cdot 2.5H_2O$ → Co-precipitation
- Aqueous Solution of $Na_2CO_3$ → Co-precipitation
- **Co-precipitation**:
  - ✓ Formation of Porous Primary-Particles (Formation of Micro-Pores)
  - ✓ Main Phase: FeOOH
- **Washing**:
  - ✓ Removal of $Na^+$ and $NO_3^-$
- $SiO_2$, $K_2CO_3$ → **Promoter Addition/Mixing**
- **Spray-Drying**:
  - ✓ Removal of $H_2O$ & Catalyst Shaping
- **Calcination**:
  - ✓ Formation of Meso-Pores
    - $2FeOOH \rightarrow Fe_2O_3 + H_2O$
  - ✓ Catalyst Strengthening
  - ✓ Main Phase: Hematite ($\alpha$-$Fe_2O_3$)

**Inactive for FTS**

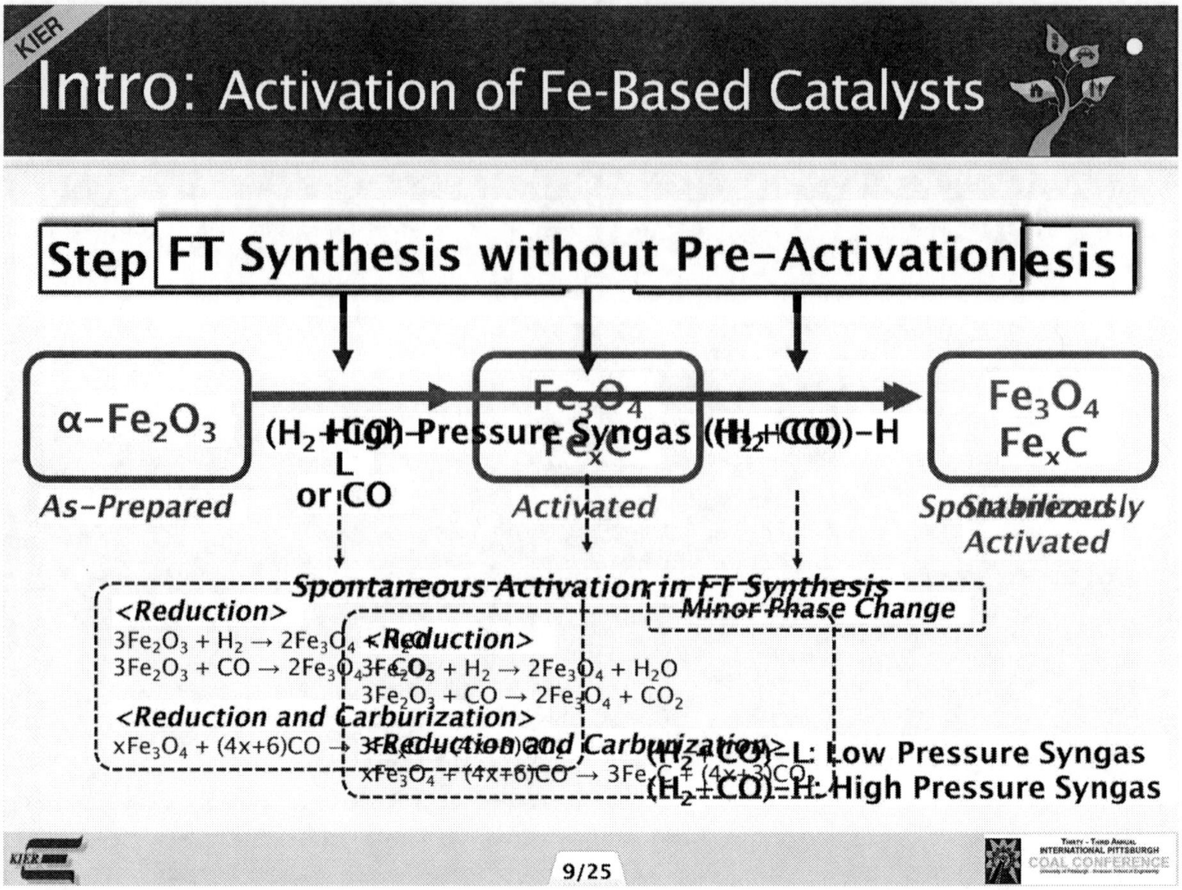
KIER
Intro: Activation of Fe-Based Catalysts
Step
FT Synthesis without Pre-Activation
esis
α-Fe2O3
As-Prepared
(H2+CO)-L
or CO
High-Pressure Syngas
Fe3O4
FexC
Activated
(H2+CO)-H
Fe3O4
FexC
Spontaneously
Stabilized
Activated
Spontaneous Activation in FT Synthesis
Minor Phase Change
<Reduction>
3Fe2O3 + H2 → 2Fe3O4 + H2O
3Fe2O3 + CO → 2Fe3O4 + CO2
<Reduction and Carburization>
xFe3O4 + (4x+6)CO → 3FexC + (4x+3)CO2
(H2+CO)-L: Low Pressure Syngas
(H2+CO)-H: High Pressure Syngas
9/25
Thirty-Third Annual International Pittsburgh Coal Conference

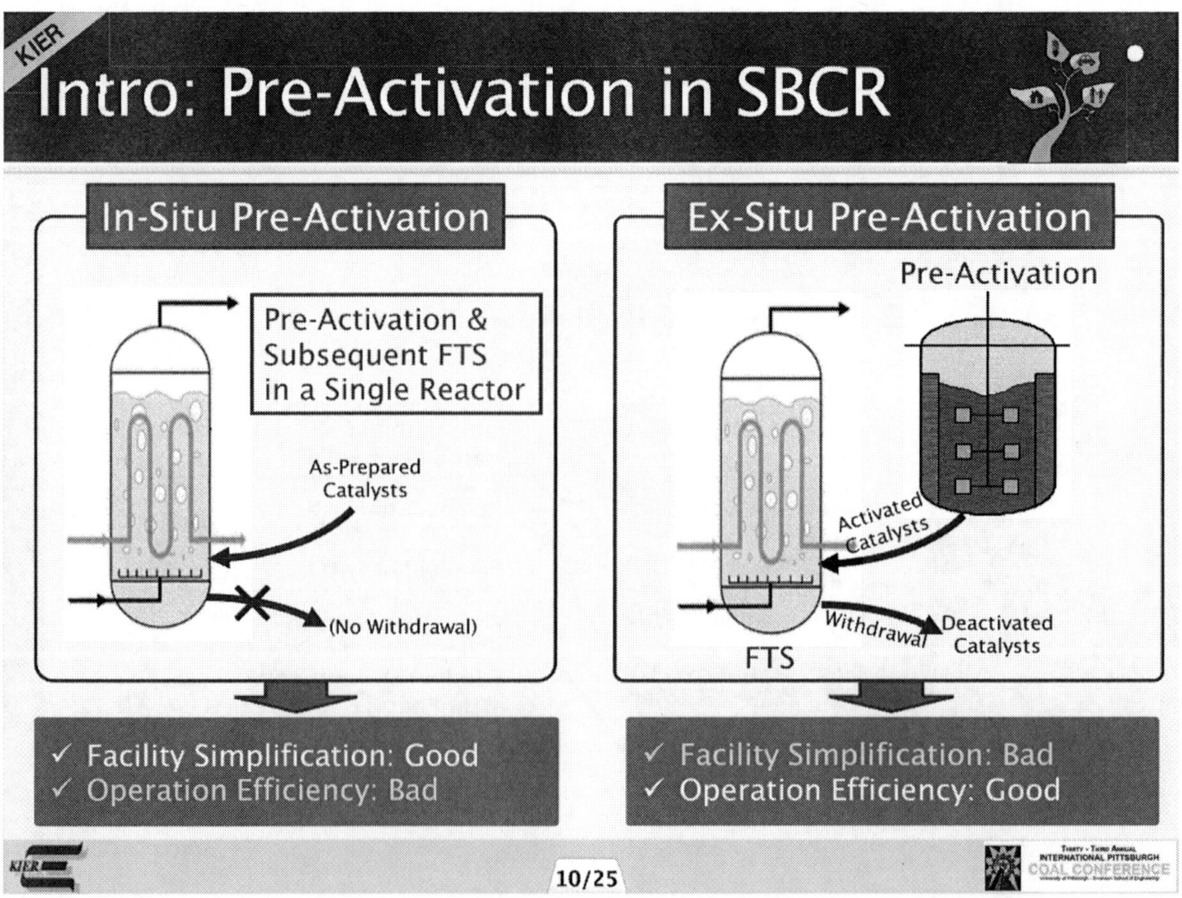
KIER
Intro: Pre-Activation in SBCR
In-Situ Pre-Activation
Pre-Activation &
Subsequent FTS
in a Single Reactor
As-Prepared
Catalysts
(No Withdrawal)
Ex-Situ Pre-Activation
Pre-Activation
Activated
Catalysts
Withdrawal
Deactivated
Catalysts
FTS
✓ Facility Simplification: Good
✓ Operation Efficiency: Bad
✓ Facility Simplification: Bad
✓ Operation Efficiency: Good
10/25
Thirty-Third Annual International Pittsburgh Coal Conference

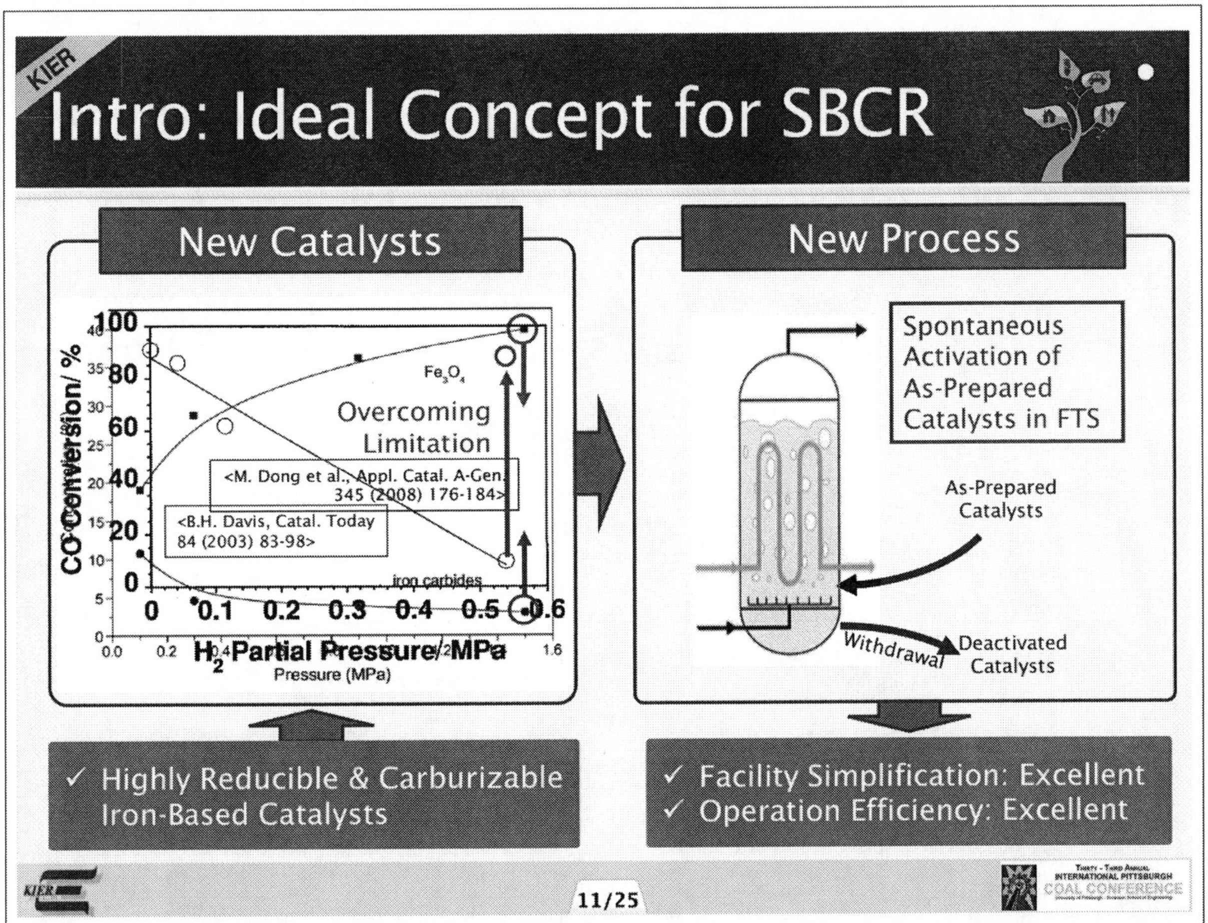
KIER
Intro: Ideal Concept for SBCR
New Catalysts
New Process
100
80
60
40
20
0
0 0.1 0.2 0.3 0.4 0.5 0.6
CO Conversion/ %
H2 Partial Pressure/ MPa
Pressure (MPa)
Fe3O4
iron carbides
Overcoming
Limitation
<M. Dong et al., Appl. Catal. A-Gen. 345 (2008) 176-184>
<B.H. Davis, Catal. Today 84 (2003) 83-98>
Spontaneous Activation of As-Prepared Catalysts in FTS
As-Prepared Catalysts
Withdrawal
Deactivated Catalysts
✓ Highly Reducible & Carburizable Iron-Based Catalysts
✓ Facility Simplification: Excellent
✓ Operation Efficiency: Excellent
INTERNATIONAL PITTSBURGH COAL CONFERENCE
11/25

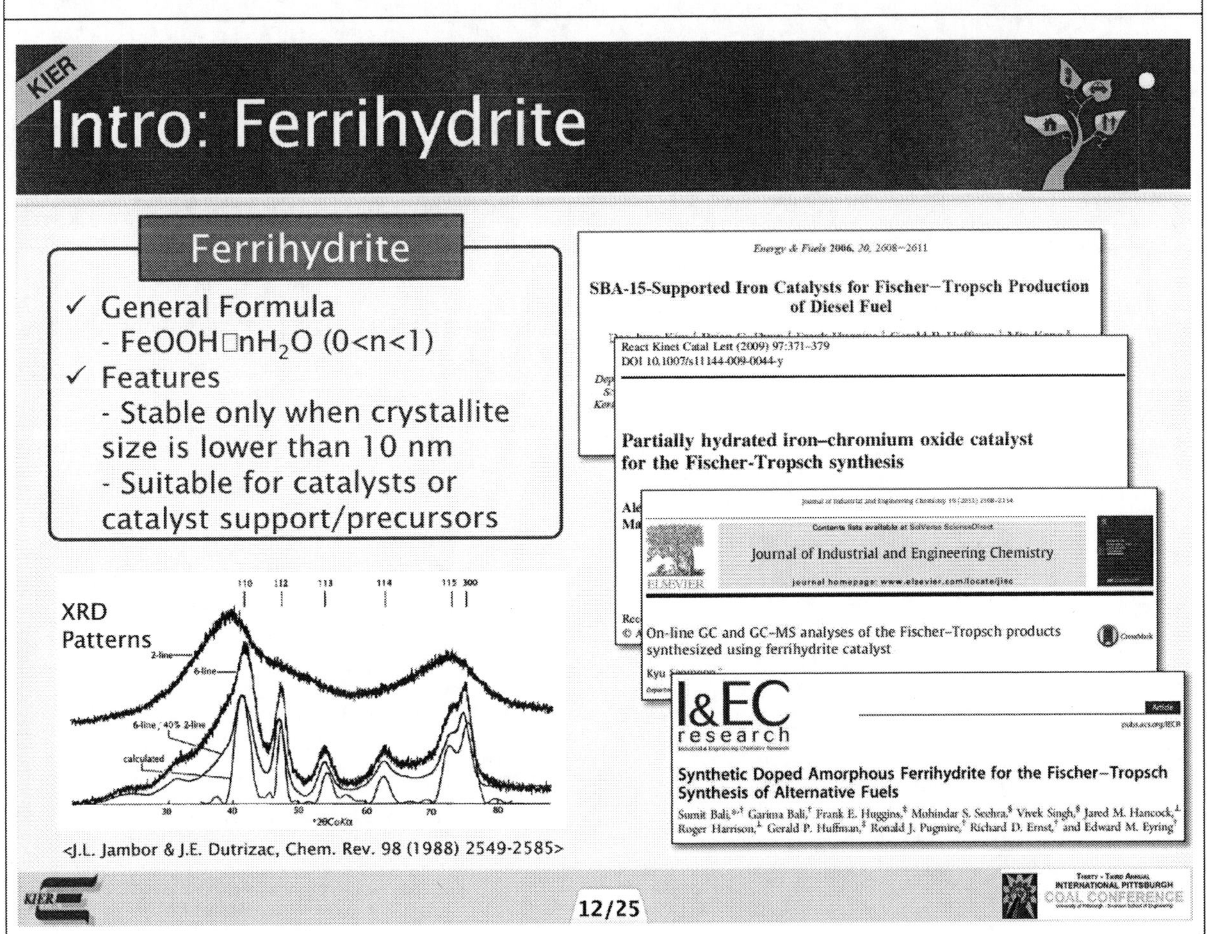
KIER
Intro: Ferrihydrite
Ferrihydrite
✓ General Formula
- FeOOH□nH2O (0<n<1)
✓ Features
- Stable only when crystallite size is lower than 10 nm
- Suitable for catalysts or catalyst support/precursors
XRD Patterns
2-line
6-line
6-line, 40% 2-line
calculated
<J.L. Jambor & J.E. Dutrizac, Chem. Rev. 98 (1988) 2549-2585>
Energy & Fuels 2006, 20, 2608−2611
SBA-15-Supported Iron Catalysts for Fischer−Tropsch Production of Diesel Fuel
React Kinet Catal Lett (2009) 97:371–379
DOI 10.1007/s11144-009-0044-y
Partially hydrated iron–chromium oxide catalyst for the Fischer-Tropsch synthesis
Journal of Industrial and Engineering Chemistry
ELSEVIER
On-line GC and GC-MS analyses of the Fischer–Tropsch products synthesized using ferrihydrite catalyst
I&EC research
Synthetic Doped Amorphous Ferrihydrite for the Fischer−Tropsch Synthesis of Alternative Fuels
Sumit Bali, Garima Bali, Frank E. Huggins, Mohindar S. Seehra, Vivek Singh, Jared M. Hancock, Roger Harrison, Gerald P. Huffman, Ronald J. Pugmire, Richard D. Ernst, and Edward M. Eyring
INTERNATIONAL PITTSBURGH COAL CONFERENCE
12/25

# Intro: In This Study

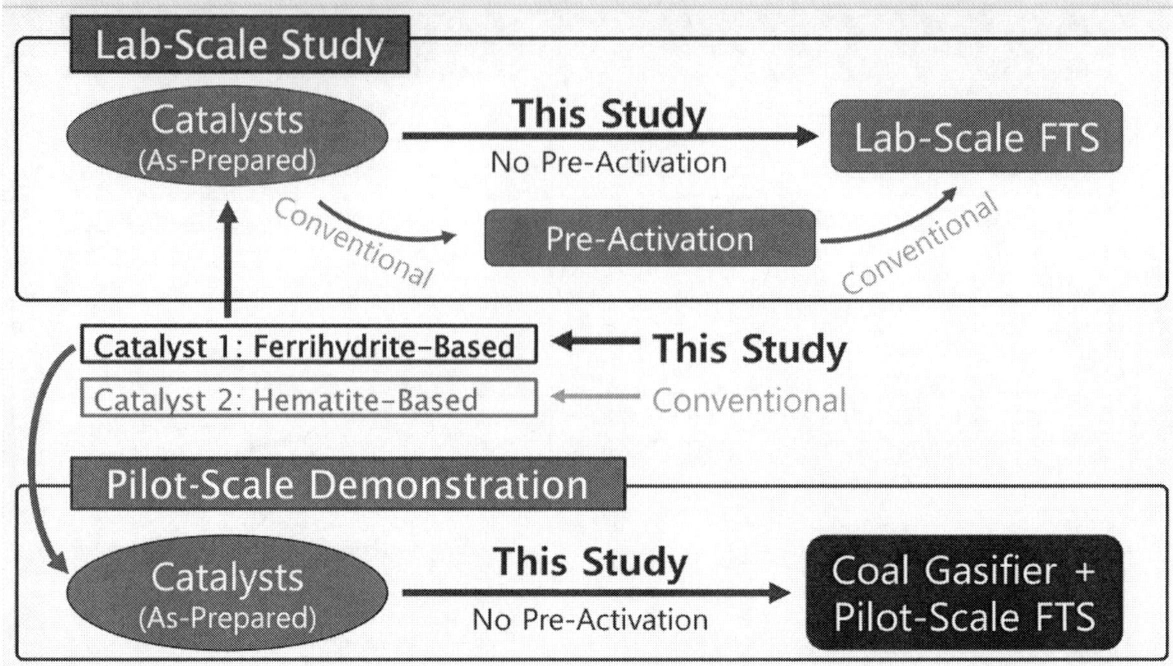




# R&D: Characterization (1)

- **XRD**

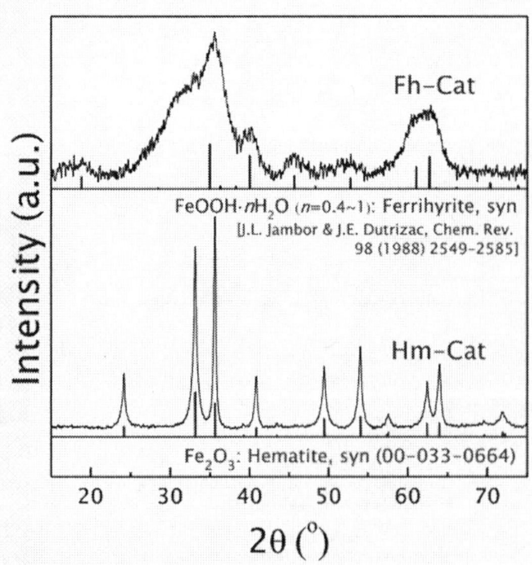


- **Mössbauer**

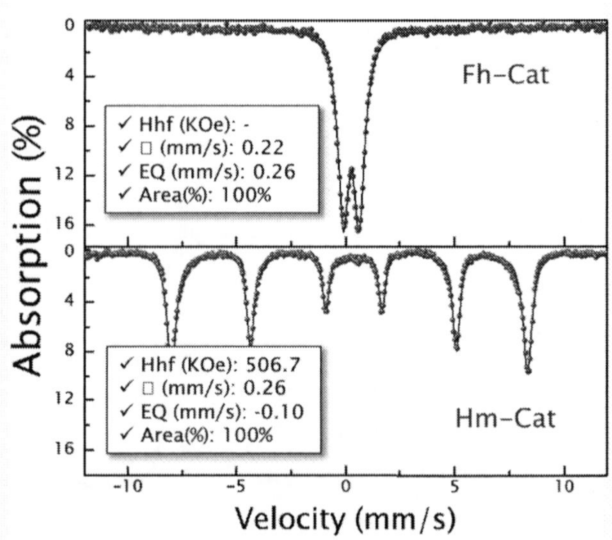


D.H. Chun et al., KR Patent Appl. No. 10-2016-0049574
D.H. Chun et al., PCT Appl. No. PCT/KR2016/004246
D.H. Chun et al., Manuscript in Preparation

# R&D: Characterization (2)

- **TEM (HRTEM)**

**Larger Crystallite**

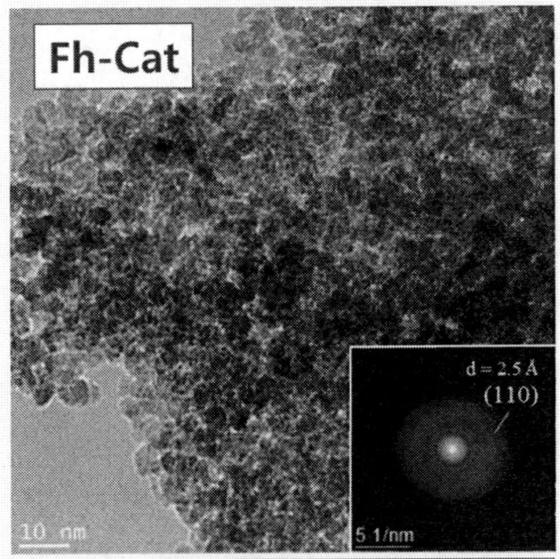


**Smaller Crystallite**

D.H. Chun et al., KR Patent Appl. No. 10-2016-0049574
D.H. Chun et al., PCT Appl. No. PCT/KR2016/004246
D.H. Chun et al., Manuscript in Preparation



# R&D: Reduction/Carburization in TPR

- **$H_2$–TPR**

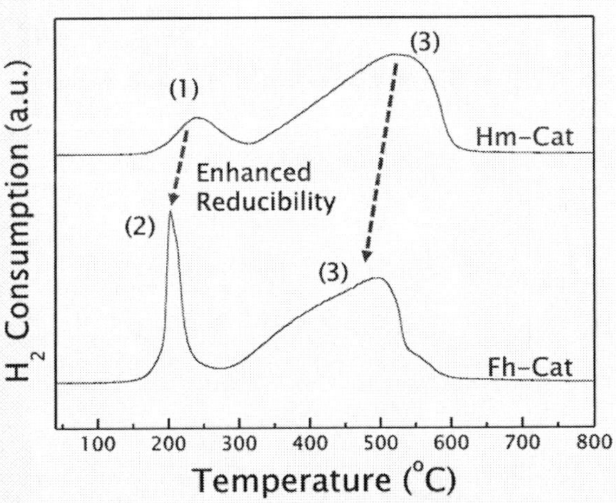


(1) $3Fe_2O_3 + H_2 \rightarrow 2Fe_3O_4 + H_2O$
(2) $2Fe_9O_2(OH)_{23} + 3H_2 \rightarrow 6Fe_3O_4 + 26H_2O$
(3) $Fe_3O_4 + 4H_2 \rightarrow 3Fe + 4H_2O$

- **CO–TPR**

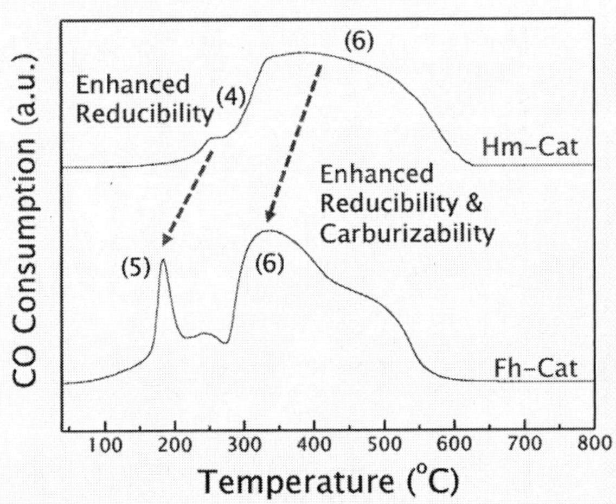


(4) $3Fe_2O_3 + CO \rightarrow 2Fe_3O_4 + CO_2$
(5) $2Fe_9O_2(OH)_{23} + 3CO \rightarrow 6Fe_3O_4 + 23H_2O + 3CO_2$
(6) $xFe_3O_4 + (4x+6)CO \rightarrow 3Fe_xC + (4x+3)CO_2$

D.H. Chun et al., Manuscript in Preparation

# R&D: Reduction/Carburization in FTS

- **Hematite-Based**

✓ GHSV($H_2$+CO) = 2.8 NL/$g_{(cat)}$-h
✓ $H_2$/CO = 1.0
✓ P($H_2$+CO) = 1.5 MPa
✓ **T = 25 → 275 °C**

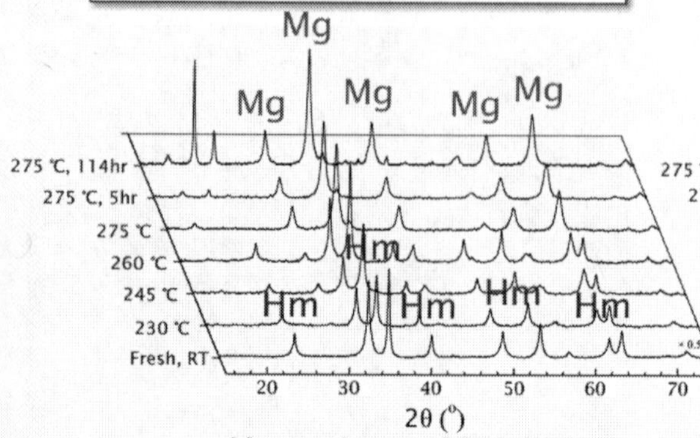


Hematite → Magnetite
($Fe_2O_3$) ($Fe_3O_4$)

- **Ferrihydrite-Based**

✓ GHSV($H_2$+CO) = 2.8 NL/$g_{(cat)}$-h
✓ $H_2$/CO = 1.0
✓ P($H_2$+CO) = 1.5 MPa
✓ **T = 25 → 275 °C**

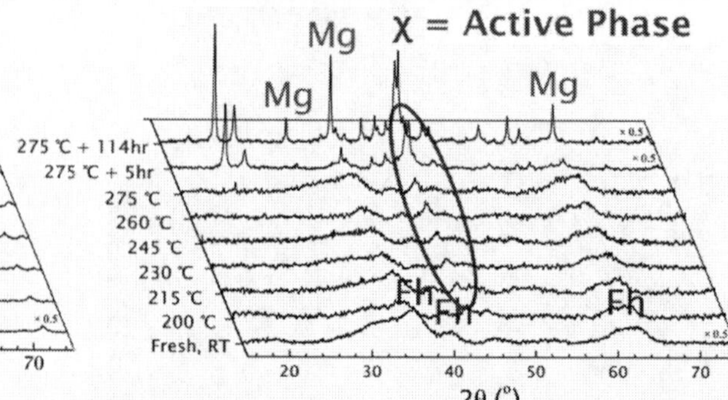


Ferrihydrite → Magnetite + $Fe_{2.5}C$
($Fe_9O_2(OH)_{23}$) ($Fe_3O_4$)

D.H. Chun et al., KR Patent Appl. No. 10-2016-0049574
D.H. Chun et al., PCT Appl. No. PCT/KR2016/004246
D.H. Chun et al., Manuscript in Preparation

17/25

# R&D: Catalyst Performance in Lab (1)

- **With Pre-Activation**

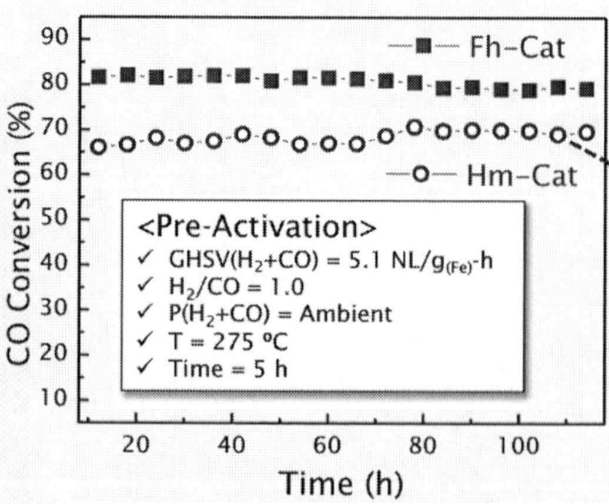


✓ GHSV($H_2$+CO) = 5.1 NL/$g_{(Fe)}$-h
✓ $H_2$/CO = 1.0
✓ P($H_2$+CO) = 1.5 MPa
✓ T = 275 °C

- **Without Pre-Activation**

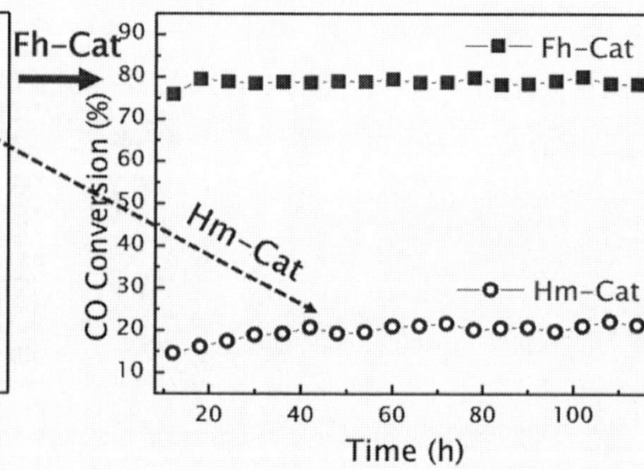


✓ GHSV($H_2$+CO) = 5.1 NL/$g_{(Fe)}$-h
✓ $H_2$/CO = 1.0
✓ P($H_2$+CO) = 1.5 MPa
✓ T = 275 °C

D.H. Chun et al., KR Patent Appl. No. 10-2016-0049574
D.H. Chun et al., PCT Appl. No. PCT/KR2016/004246
D.H. Chun et al., Manuscript in Preparation

18/25

# R&D: Catalyst Performance in Lab (2)

- **Overall FTS Performance**

| Overall Performance (66–114 h) | Pre–Activation | | No Pre–Activation | |
|---|---|---|---|---|
| | Hm–Cat | Fh–Cat | Hm–Cat | Fh–Cat |
| CO Conversion (%) | 69.6 | 79.8 | 20.9 | 78.9 |
| $CO_2$ Selectivity (%) | 45.4 | 45.6 | 36.1 | 46.4 |
| Hydrocarbon Distribution (wt%) | | | | |
| $CH_4$ | 6.98 | 8.22 | 3.92 | 5.96 |
| $C_2$–$C_4$ | 19.7 | 20.5 | 13.5 | 15.8 |
| $C_5$–$C_{11}$ | 19.6 | 19.7 | 11.1 | 15.3 |
| $C_{12}$–$C_{18}$ | 17.0 | 15.9 | 15.2 | 15.6 |
| $C_{19+}$ | 36.7 | 35.7 | 56.4 | 47.3 |
| $C_{5+}$ Productivity (g/g$_{(cat)}$–h) | 0.259 | 0.262 | 0.0857 | 0.296 |

D.H. Chun et al., KR Patent Appl. No. 10-2016-0049574
D.H. Chun et al., PCT Appl. No. PCT/KR2016/004246
D.H. Chun et al., Manuscript in Preparation



# R&D: Pilot-Scale Demonstration (1)

# R&D: Pilot-Scale Demonstration (2)

# R&D: Pilot-Scale Demonstration (3)

- No Pre-Activation (New)

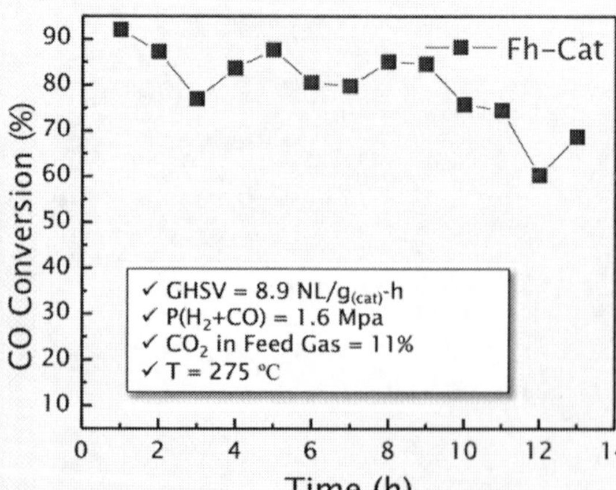


| | Fh-Cat |
|---|---|
| $C_{5+}$ Selectivity (wt%) | 81.4 |
| $C_{5+}$ Productivity (g/$g_{(cat)}$-h) | 0.713 |

Wax Level Increment

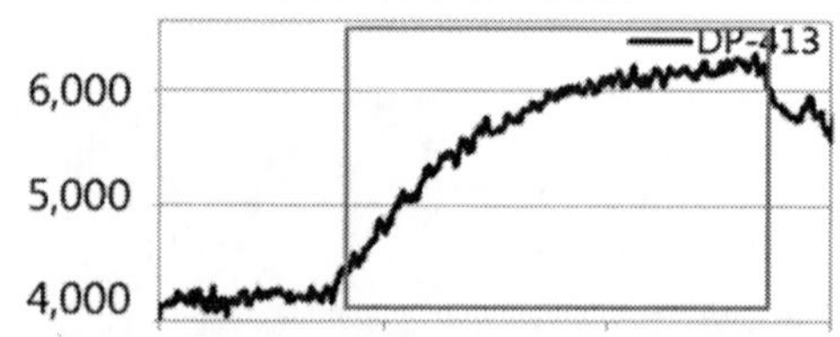


Light Oils

**Scale: 6 bbl/d**

D.H. Chun et al., KR Patent Appl. No. 10-2016-0049574
D.H. Chun et al., PCT Appl. No. PCT/KR2016/004246

KIER

# Conclusion

- ✓ We successfully synthesized ferrihydrite-based catalysts for Fischer-Tropsch synthesis.
- ✓ The ferrihydrite-based catalysts showed greater reducibility and carburizability in $H_2$ or CO atmosphere than the conventional hematite-based catalysts.
- ✓ Even in the reaction condition of Fischer-Tropsch synthesis, the ferrihydrite-based catalysts were spontaneously changed into Hägg-carbide ($Fe_{2.5}C$), an active phase for Fischer-Tropsch synthesis.
- ✓ Even though no activation pre-treatment was carried out, the ferrihydrite-based catalysts showed high catalytic performance for Fischer-Tropsch synthesis.
- ✓ Furthermore, this spontaneous activation holds true for a pilot-scale test (6 bbl/d) in a slurry bubble column reactor.

- ✓ We can conclude that the ferrihydrite-based catalysts are highly advantageous for industrial Fischer-Tropsch processes because the as-prepared catalysts can be used for Fischer-Tropsch synthesis without requiring an extra pre-activation process.



KIER

# Acknowledgment


- ✓ This work was supported in part by the Research and Development Program of the Korea Institute of Energy Research (GP2016-0008/B6-2437) and the National Research and Development Program of the Korea Institute of Energy Technology Evaluation and Planning (NP-2015-0091/B5-3229).

- ✓ We appreciate Prof. Chul Sung Kim and Mr. Jung Tae Lim (Kookmin University, Republic of Korea) for Mössbauer Analyses.


nst National Research Council of Science & Technology

KETEP 한국에너지기술평가원 KOREA INSTITUTE OF ENERGY TECHNOLOGY EVALUATION AND PLANNING

한국에너지기술연구원 KIER KOREA INSTITUTE OF ENERGY RESEARCH

KMU KOOKMIN UNIVERSITY SINCE 1946 국민대학교

KIER
THIRTY - THIRD ANNUAL
INTERNATIONAL PITTSBURGH
COAL CONFERENCE
University of Pittsburgh · Swanson School of Engineering
Session 29: Clean Coal and Gas to Fuels: Catalytic Study - 2
[15:30-15:50] Thursday, August 11, 2016
Thank you!
The KIER, a global energy innovator, does its best in pursuing its mission to invent world-class energy technologies based on open innovation, life-cycle research quality assurance, participatory and open communication. Therefore the KIER will become the best energy technology R&D institute in the world, contributing to the creation of wealth and improvement of quality of life for the people.
Dong Hyun Chun
E-Mail: cdhsl@kier.re.kr

# SINGLE PASS FISCHER-TROPSCH SYNTHESIS OVER IRON-BASED CATALYSTS

Eric van Steen, Michael Claeys, Klaus P. Möller, Doreen Nabaho
*Centre for Catalysis Research, Department of Chemical Engineering, University of Cape Town, Private Bag X3, Rondebosch 7701, South Africa, e-mail: eric.vansteen@uct.ac.za*

**Abstract**
Two iron-based catalysts were tested at various space velocities. It was noted that the silica-containing catalyst was more active at high space velocity, but when the conversion of CO exceeded 85% the alumina-containing catalyst was more active. Furthermore, it was noted that the extent of deactivation was higher with the alumina-containing catalyst than with the silica containing catalyst. This may be attributed to the higher driving force for the formation of iron aluminate under the influence of product water in comparison to the formation of iron silica (the formation of both these compounds may lead to permanent deactivation). The $CO_2$-selectivity increases with increasing CO-conversion up to ca. 60%, at which point the $CO_2$-selectivity remains constant at ca. 25 C-% for the silica containing catalyst and 30 C-% for the alumina containing catalyst. The obtained methane selectivity is low (even at 90% conversion with ca. 5-6C-%), but increases with increasing conversion. At low CO-conversion, the methane selectivity obtained over the silica containing catalyst was lower (ca. 2-3%) compared to the alumina-containing catalyst (ca. 4 C-%). The selectivity for the various classes of organic products can for both catalysts be interpreted in terms of reduced partial pressure of CO and extent of secondary reactions. The obtained data can form a basis for a rationale decision on the extent of reaction required in the Fischer-Tropsch reactor, when operating in a single reactor.

**Keywords:** Fischer-Tropsch, iron, conversion, deactivation, selectivity

## Introduction

The Fischer-Tropsch synthesis is a versatile route to convert carbon-containing feedstock into liquid hydrocarbons [1]. The actual reaction is industrially catalysed using either iron-based or cobalt-based catalysts [2]. The choice of the catalyst is based on criteria, such as catalyst activity, catalyst stability and catalyst selectivity. It has been reported that iron-based Fischer-Tropsch catalysts are more active than cobalt-based catalysts provided the conversion is low [3]; this has been originally attributed to the role of water on the activity on the reaction kinetics over these catalysts [3-6]; however, the strong role of water on the intrinsic reaction kinetics has been disputed [6,7] and the role of water has been attributed to changes in the catalyst [6-10].

The strong decline in the catalytic activity obtained over iron-based catalysts together with a possible runaway in fixed bed reactors [11] has focused the attention for industrial application mainly on relatively low conversions, although reports of iron-based catalysts operating at high level of conversion are plentiful [12-18]. The comparison of activity and selectivity should strictly be performed over at the same level of conversion and thus typically a range of conditions has to be probed [14-15]. Here we compare the effect of the binder in two iron-based catalysts with a view of operating these catalysts at a high CO-conversion for a feed gas with $H_2/CO=2$.

## Experimental

The preparation of the iron-based catalyst with a composition of 100Fe: 5Cu: $5K_2O$: $25SiO_2$ was prepared from a heated (60°C) solution containing iron and copper nitrate (2 g/l Cu and 40 g/l Fe respectively), which was added step-wise to an agitated sodium carbonate solution at 60°C (100 g/l $Na_2CO_3$) until a slurry with a pH of 7 was obtained. The resultant mixture was added stepwise to a $Na_2CO_3$ solution (100 g/L) kept at 60°C in a stirred vessel. The resulting slurry was subsequently filtered and the filtrate washed with warm deionised water (60°C). The washed filtrate was re-slurried with deionised water, to which a potassium silicate solution was added to obtain the desired Si and K loadings. After precipitation of the silica at pH 7 (by addition of a nitric acid solution), the resultant slurry was again filtered and the filtrate added to deionised water to obtain a slurry with 20 wt.-% solids. The slurry was spray dried (d = 83 ± 67 m), after which the dried catalyst was calcined in air at 400°C.

The preparation of the iron-based catalyst with a composition of 100Fe: $6K_2O$: $30Al_2O_3$ was prepared using via co-precipitation from an aqueous iron nitrate and aluminium nitrate solution at its boiling point using ammonium carbonate as the precipitation agent. The catalyst precursor was filtered off and the filter cake was dried in an oven at 100°C for 24hrs and subsequently calcined at 350°C for 16 hrs. Potassium was added by impregnation of the catalyst precursor (d = 240-300 μm) with the appropriate amount of an aqueous $KNO_3$-solution.

The Fischer-Tropsch activity and selectivity were tested in a slurry reactor. The reactor was filled with 280 g of wax, which was molten at 120°C, after which the catalyst was added. The iron-based catalysts (m = 5g) were activated in-situ in synthesis gas ($H_2/CO=2$) at 1 bar and 270°C for 16hrs, after which the reactor was cooled down to the reaction temperature (250°C) and pressurized to 21 bar. The Fischer-Tropsch synthesis was performed using a synthesis gas with $H_2/CO/Ar = 2/1/1$.

The flow rate of the constituents of the synthesis gas was controlled independently by mass flow controllers, after which the gases were mixed and analysed. The synthesis gas mixture was fed into the slurry reactor. The reactor product passed a hot trap maintained at 220°C and a cold trap, which was maintained at room temperature to remove water and alcohols before being sent to vent. A sample of the permanent gases was taken after the cold trap, which was analysed on a 3-channedl micro gas chromatograph Varian CP-4900 equipped with a TCD. Conversion, yield and selectivity were determined by using the inert component in the feed gas as an internal standard. A sample of the organic product was taken after the hot trap gas using the off-line ampoule sampling technique

[19], which were analysed on a Varian CP-3800 equipped with a FID. The yield of organic product compounds was determined using methane as the link component.

**Results and Discussion**

Figure 1 shows the CO-conversion as a function of time on line. The catalysts show some initial deactivation, but settle in rather quickly. The initial deactivation of the silica containing catalyst is evidenced in the period up to 50 hrs on line, where small decreases in the space velocity did not yield a commensurate increase in the activity. After this time period, this catalyst appeared stable. This was further evidenced by the activity obtained at a space velocity of 24 $ml_n$/min/g (after ca. 100 hrs on line), which is between with the activity obtained at a space velocity of 30 $ml_n$/min/g and 12 mln/min/g. A detailed kinetic analysis showed that no deactivation occurred in this time period between 100 and 450 hrs on line. The alumina containing catalyst showed evidence of deactivation over a longer period. The catalyst seemed to deactivate continuously (see period 0-125 hrs on line and period 250-290 hrs on line). The deactivation behaviour of this catalyst could be modelled as a 1$^{st}$ order deactivation process with $k_d = 10^{-3}$ $hr^{-1}$.

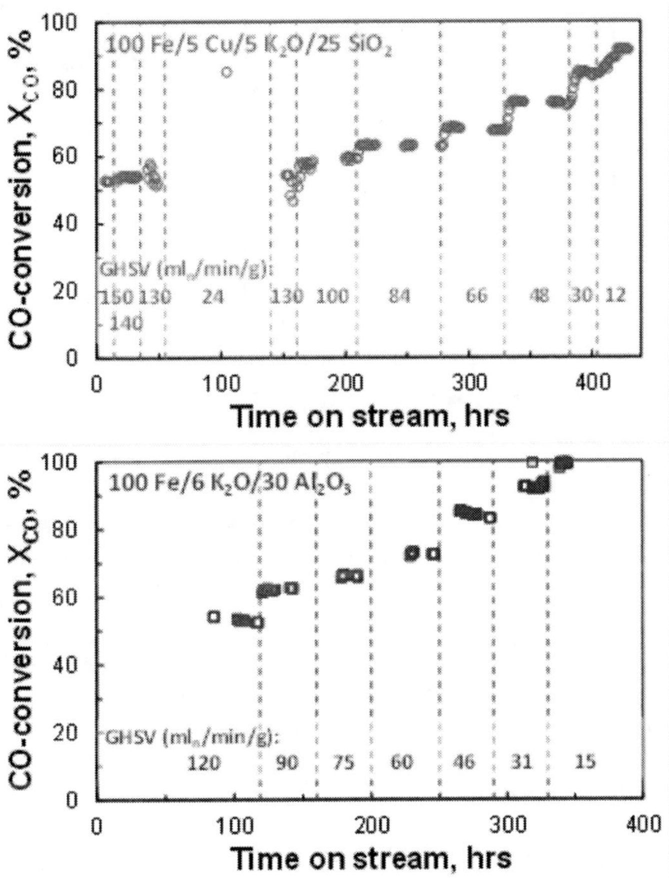


**Figure 1:** Measured conversion as a function of time on line behaviour of silica containing iron-based Fischer-Tropsch catalyst (top) and alumina containing iron-based Fischer-Tropsch catalyst (bottom)

The activity of the alumina containing iron-based Fischer-Tropsch catalyst at a high CO-conversion is higher than that of the silica based catalyst. A kinetic evaluation of the obtained data as a function of the space velocity [5] indicates that this is mainly due to a ca. 5-fold increase in the intrinsic activity rather than to a change in the inhibition factor in the rate expression. The change in the activity may be brought about by a change in the metal-support interactions [20,21].

Iron-based Fischer-Tropsch synthesis does not only produce hydrocarbons, but also $CO_2$. The formation of $CO_2$ is typically seen as a consecutive reaction [22], and the selectivity is thus expected to increase with increasing CO-conversion approaching equilibrium at high levels of CO-conversion (it should be noted that the equilibrium selectivity for the formation of $CO_2$ decreases with increasing CO-conversion due to the change in the $H_2$/CO-ratio in the reactor as a consequence of the $CO_2$-formation). This silica-containing catalyst does show the expected behaviour of increasing $CO_2$-selectivity with increasing CO-conversion. However, this behaviour does not seem to be universal, since the obtained $CO_2$-selectivity over the alumina containing catalyst is almost independent of the CO-conversion in the range between 50 and 98%. This may thus indicate another origin for the formation of $CO_2$, e.g. via the carburization of magnetite in the dynamic iron-based catalyst [8,23-25].

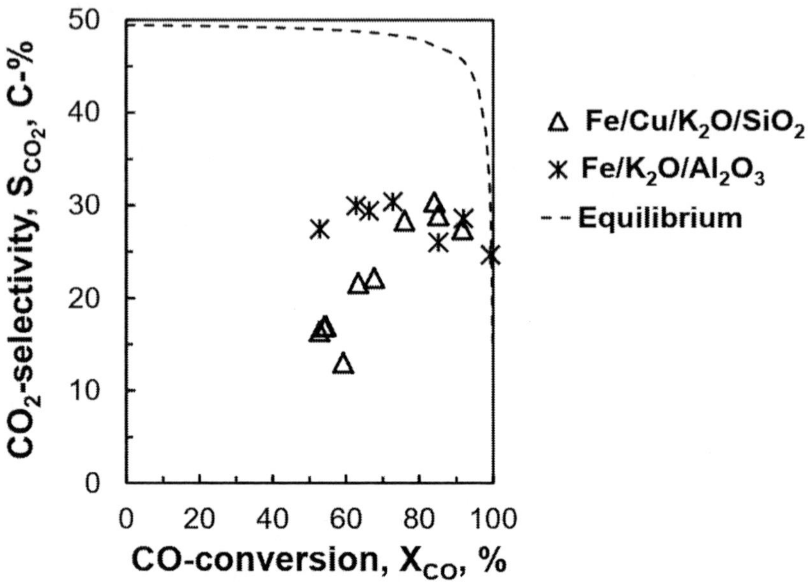


**Figure 2:** $CO_2$-selectivity as a function of CO-conversion.

Figure 3 shows the methane selectivity obtained over these catalysts as a function of the CO-conversion. A low methane selectivity of ca. 2-3C-% is obtained over the Ruhrchemie-based catalyst, 100 Fe/5 Cu/5 $K_2O$/25 $SiO_2$, when the CO-conversion is below 70%. Over the alumina-containing iron-based catalyst, 100 Fe/6 $K_2O$/30 $Al_2O_3$, the methane selectivity up to a CO-conversion level of ca. 85% is ca. 4-5 C-%. At higher levels of CO-conversion, the methane selectivity increases albeit less strongly than over the silica containing iron-based Fischer-Tropsch catalysts. The generally observed increase in the methane selectivity with increasing CO-conversion may be attributed to the change in the $H_2$/CO–ratio in the reactor, which increases to beyond 5 at these conversion levels, lifting the inhibition for methane formation [26-28].

Figure 4 shows the Anderson-Schulz-Flory distribution of the linear hydrocarbons formed over these catalysts. The ASF-plot is characterized by the typical deviations of more methane than expected, less $C_2$, an increase in the chain growth probability with increasing chain length, and the artificial decrease in the rate of formation beyond $C_{12}$ due to retention in the liquid phase [28]. Some interesting difference can be obtained between the Ruhrchemie catalyst and the iron-based catalyst containing alumina, viz. the decrease in the formation of $C_2$ is much more pronounced with the silica-containing iron-based catalyst than with the alumina containing catalyst. This is also evident when looking at the olefin content in the $C_2$-fraction (see Fig. 5), which shows that the olefin content in the C2-fraction over the silica containing Ruhrchemie-catalyst is much lower than that in the alumina containing iron-based catalyst. This may indicate a stronger secondary hydrogenation function on the silica-containing catalyst compared to the alumina containing catalyst.

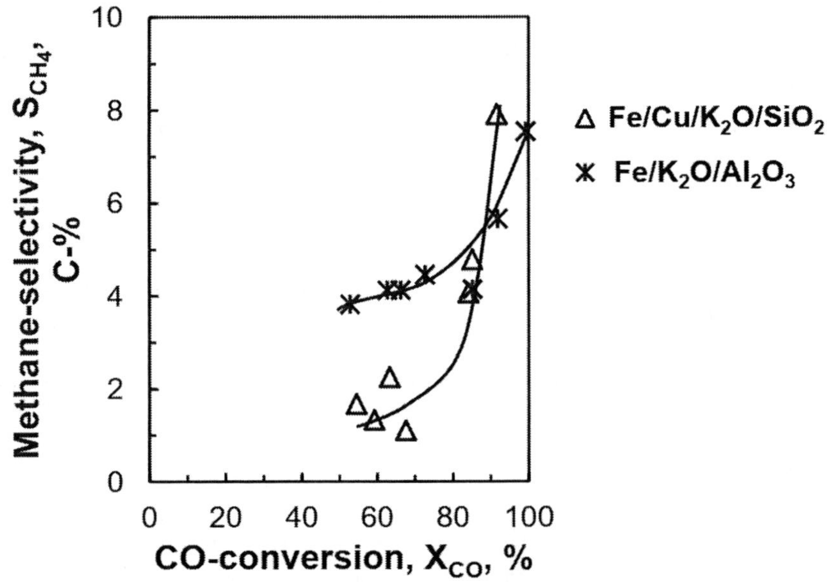


**Figure 3:** Methane selectivity as a function of CO-conversion.

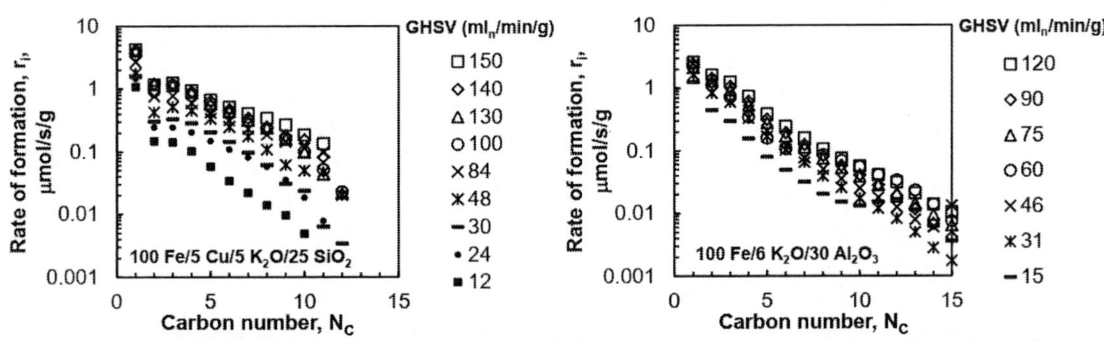


**Figure 4:** ASF-plot for silica containing (left) and alumina containing (right) iron based catalysts.

Figure 5 shows the olefin content in the fraction of linear hydrocarbons as a function of the carbon number. The shape of the curve is characteristic with a low value at $C_2$, a maximum at $C_3/C_4$ and a decrease with increasing carbon number. The low value at $C_2$ has been explained due to the high reactivity of ethylene for secondary reactions (hydrogenation and incorporation) [27]. The maximum typically occurs at the carbon number with the containing less reactive olefin, propene, and decreases with increasing carbon number as a consequence of increased extent of secondary reaction due to the higher concentration of the olefins in the liquid products in the catalyst pores. With decreasing space velocity, i.e. increasing conversion, the characteristic aspects of secondary olefin reactions become more dominant. At the very low space velocity, the maximum in the olefin content has shifted from $C_3$ to $C_4$. This can be ascribed to the formation of internal olefins, which are less reactive than the terminal olefins [26].

A comparison in the olefin content obtained with the Ruhrchemie and the alumina containing iron-based catalyst showed some subtle differences. It was noted the characteristics for secondary olefin hydrogenation (lower olefin content in $C_2$ and a stronger decline in the olefin content with increasing carbon number from $C_3/C_4$ onwards, but in particularly from $C_6$ onward) is more pronounced with the Ruhrchemie catalyst than with the alumina containing iron-based catalyst. Furthermore, it can be noted that the highest olefin content (in the $C_3$-fraction) is obtained with the silica containing Ruhrchemie catalyst. The highest olefin content can be taken as a measure for the primary formation of olefins over these catalyst; it is then prudent to conclude that the binder (possibly in conjunction with the reduction promoter [24,25]) affects the primary formation of olefins in the Fischer-Tropsch synthesis possibly via a surface modification as proposed for silica and titania-based iron catalysts [20,21].

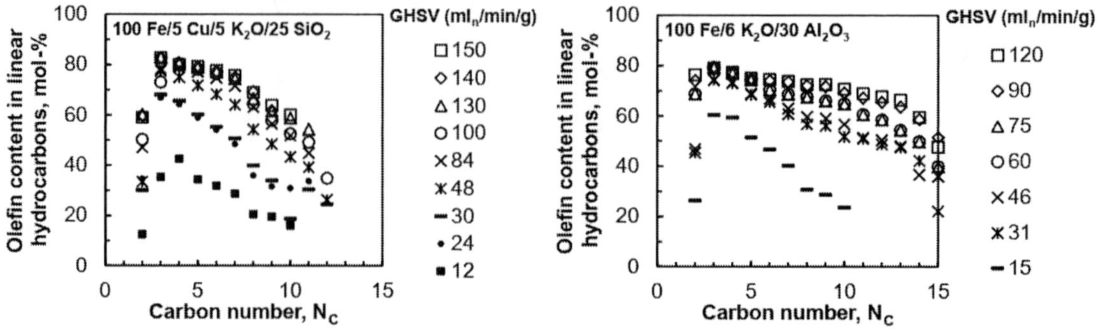


**Figure 5:** Olefin content within the fraction of linear hydrocarbons as a function of carbon number for silica containing (left) and alumina containing (right) iron based catalysts.

Re-adsorption of primarily formed olefins can result in secondary hydrogenation (and thus a lowering of the olefin content), but also in secondary double bond isomerization [26]. Figure 6 shows the 1-olefin content in the fraction of linear olefins as a function of carbon number. Again some interesting and significant differences can be observed between the two catalysts. First of all, it should be noted that even at the highest space velocity, the 1-olefin content is less than 100%. The lowest CO-conversion obtained with these catalysts is 50%; at this level of CO-conversion, the 1-olefin content obtained with the Ruhrchemie catalyst is ca. 91.5%, and with the alumina containing catalyst ca. 96.5%. At these conversion levels the 1-olefin content does not seem to be dependent on the conversion. However, lower levels of CO-conversion need to be probed in order to evaluate the behaviour towards 0% conversion in order to assess the extent to which the Fischer-Tropsch synthesis produces primarily olefins with an internal double bond. It can be further noted that the alumina containing iron-based Fischer-Tropsch catalyst shows a low extent of double bond isomerisation even at low space velocity (high levels of conversion).

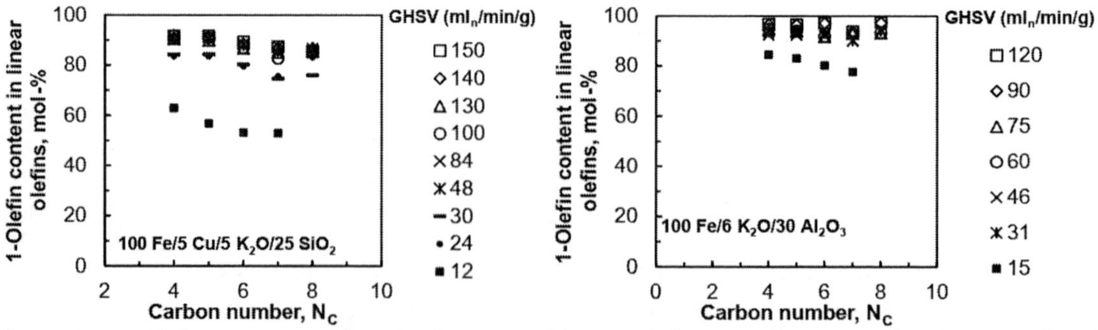


**Figure 6:** 1-Olefin content within the fraction of linear olefins as a function of carbon number for silica containing (left) and alumina containing (right) iron based catalysts.

The Fischer-Tropsch synthesis does not only produce hydrocarbons, but also oxygenates such as alcohols, aldehydes, ketones and carboxylic acid [29]. Figure 7 shows the fraction of oxygen containing product compounds in the fraction of linear product compounds as a function of carbon number. The distribution shows the characteristic dependency on carbon number being low in $C_1$, a maximum at $C_2$, a local minimum at $C_3$ and passing a local maximum at higher carbon numbers [29]. It might be argued that oxygen containing compounds belong to the carbon number group *(n-1)*, since CO-insertion may not lead to chain growth but rather to chain termination [22]. Hence, the oxygen containing compounds with a carbon number of $N_C$ were added to the carbon number fraction *(n-1)* and plotted as a function of the carbon number *($N_C$-1)* (see Fig. 7 bottom). The assignment of oxygen containing compounds to the fraction with a carbon number less one yields a more logical trend, which can be readily explained in terms of increased extent of secondary conversion of oxygen containing compounds with increasing carbon number due to increased solubility within the liquid layer surrounding the catalyst particle. Comparing the two iron-based catalysts, it appears as if at medium space velocity (100-150 $ml_n$/min/g) more oxygen containing compounds are formed when using alumina as a binder compared to the Ruhrchemie catalyst.

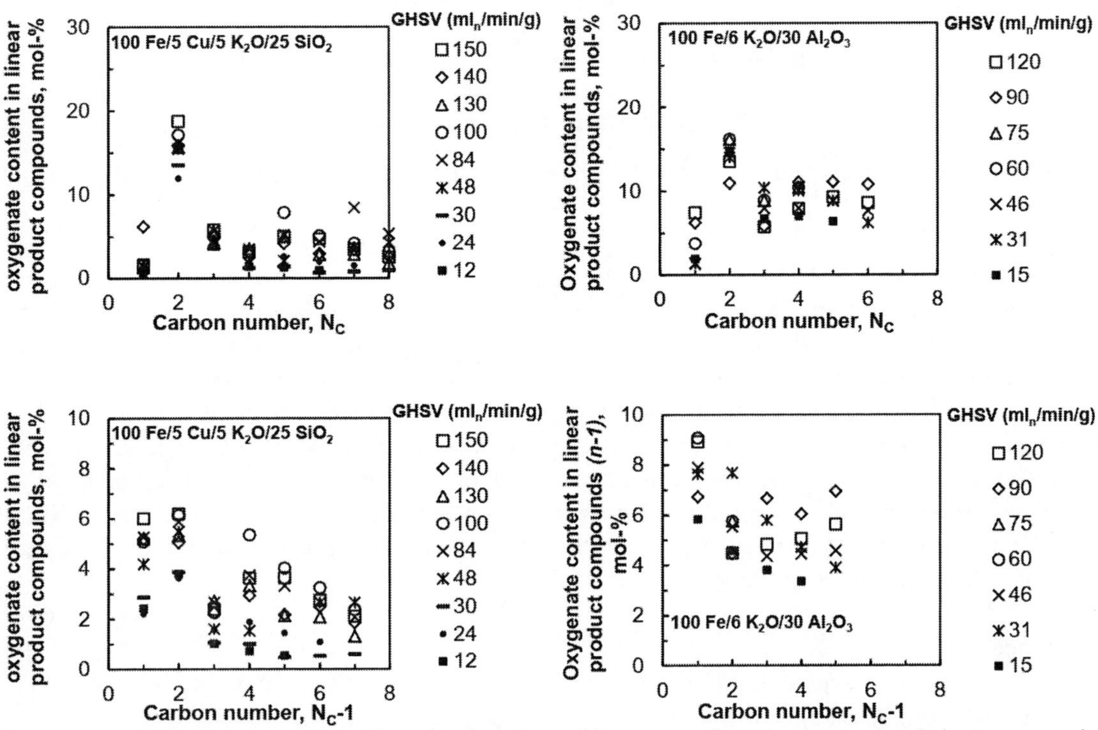


**Figure 7:** Oxygenate content within the fraction of linear product compounds of the same carbon number (top) and within the fraction of oxygenate compounds of carbon number $N_c$ and linear hydrocarbons with a carbon number ($N_C$-1) (bottom) as a function of carbon number for silica containing (left) and alumina containing (right) iron based catalysts.

The Fischer-Tropsch synthesis produces not only linear product compounds, but also branched compounds [26-29]. Figure 8 shows that ratio of branched to linear hydrocarbons in the C5-fraction as a function of the CO-conversion. The ratio shows diametrically opposite trends for the Ruhrchemie catalyst and the alumina containing catalyst. The Ruhrchemie catalyst shows the expected trend of decreasing extent of branching, since the hydrogen to carbon monoxide ratio increases and thus the primary formation of branched compounds is suppressed [26]. It might be argued that the increased extent of branching observed with the alumina containing catalyst is due to the acidity introduced in this catalyst via the binder. However, this catalyst does not show much extent of double bond isomerization, a much more facile reaction, implying that acid catalysed skeletal isomerization is not the origin of the increased formation of branched product compounds.

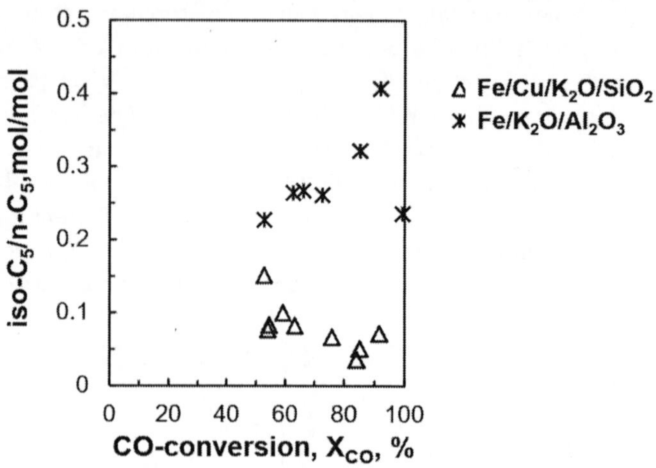


**Figure 8:** Ratio of branched to linear hydrocarbons in the $C_5$-fraction as a function of CO-conversion.

## Conclusions

Iron-based Fischer-Tropsch catalysts can be operated at high levels of CO-conversion, but a fraction of CO is converted to $CO_2$ rather than to hydrocarbons. There are obvious differences in the product selectivity depending on the formulation of iron-based catalysts, which needs to be taken into account when selecting a catalyst for a particular set of operation conditions. It was observed that alumina containing iron-based catalysts produce a more olefinic product with a lower extent of double bond isomerization even at rather high levels of CO-conversion. On the other hand, the Ruhrchemie type of catalyst produces less methane, less branched product compounds, and more long chain hydrocarbons. Hence, the definition of the desired product spectrum is of great importance in selecting the optimum catalyst.

## References

[1] H. Schulz, *Appl. Catal. A: General* **186** (1999), 3-13.
[2] E. van Steen, M. Claeys, *Chem. Eng. Techn.* **31** (2008), 655-666.
[3] P.J. van Berge, R.C. Everson, *Stud. Surf. Sci and Catal.* **107** (1997), 207-212.
[4] E. van Steen, H. Schulz, *Appl. Catal. A: General* **186** (1999),
[5] F.G. Botes, B. van Dyk, C. McGregor, *Ind. Eng. Chem. Res.* **48** (2009), 10439-10447.
[6] F.G. Botes, B. Breman, *Ind. Eng. Chem. Res.* **45** (2006), 7415-7426.
[7] F.G. Botes, *Ind. Eng. Chem. Res.* **48** (2009), 1859-1865.
[8] Z.H. Chonco, L. Lodya, M. Claeys, E. van Steen, *J. Catal.* **307** (2013) 283-294.
[9] E. van Steen, M. Claeys, M.E. Dry, J. van de Loosdrecht, E.L. Viljoen, J.L. Visagie, *J. Phys. Chem. B* **109** (2005), 3575-3577.
[10] M. Claeys, M. Dry, E. van Steen, P. van Berge, S. Booyens, R. Crous, P. van Helden, J. Labuschagne, D. Moodley, A. Saib, *ACS Catal.* **5** (2015), 841-852.
[11] A. Jess, R. Popp, K. Hedden, *Appl. Catal. A: General* **186** (1999), 321-342.
[12] D.B. Bukur, L. Nowicki, X. Lang, Catal. *Today* **24** (1995), 111-119.
[13] R.J. O'Brien, L. Xu, R.L. Spicer, B.H. Davis, *Energy & Fuels* **10** (1996), 921-926.
[14] S. Li, S. Krishnamoorthy, A. Li, G.D. Meitzner, E. Iglesia, *J. Catal.* **206** (2002), 202-217.
[15] R.J. O'Brien, B.H. Davis, *Catal. Lett.* **94** (2004), 1-6.
[16] C.-H. Zhang, Y. Yang, B.-T. Teng, T.-Z. Li, Ho.-Y. Zheng, H.-W. Xiang, Y.-W. Li, *J. Catal.* **237** (2006) 405–415.
[17] H. Wan, B. Wu, C. Zhang, H. Xiang, Y. Li, *J. Mol. Catal. A: Chemical* **283** (2008) 33–42.
[18] H. Suo, S. Wang, C. Zhang, J. Xu, B. Wu, Y. Yang, H. Xiang, Y.-W. Li, *J. Catal.* **286** (2012) 111–123.

[19] H. Schulz, S. Nehren, *Erdöl und Kohle Erdgas, Petrochem.* **39** (1986) 93-94
[20] R.P. Mogorosi, N. Fischer, M. Claeys, E. van Steen, *J. Catal.* **289** (2012), 140-150.
[21] R.P. Mogorosi, M. Claeys, E. van Steen, *Top. Catal.* **57** (2014), 572-581.
[22] R.A. van Santen, I.M. Ciobica, E. van Steen, M.M. Ghouri, *Adv. Catal.* 54 (2011), 127-187.
[23] E. van Steen, M. Claeys, *Chem. Eng. Techn.* 31 (2008), 655-666
[24] Z.H. Chonco, A. Ferreira, L. Lodya, M. Claeys, E. van Steen, *J. Catal.* **308** (2013) 363-373.
[25] Z.H. Chonco, A. Ferreira, L. Lodya, M. Claeys, E. van Steen, *J. Catal.* **307** (2013), 283-294.
[26] H. Schulz, H. Gorre, E. Erich, E. van Steen, *Catal. Lett.* **7** (1990), 157-168.
[27] H. Schulz, E. van Steen, M. Claeys, *Stud. Surf. Sci. and Catal.* **81** (1994), 455-460.
[28] M. Claeys, E. van Steen, *Stud. Surf. Sci. and Catal.* **152** (2004), 601-680.
[29] T. Grobler, M. Claeys, E. van Steen, M. Janse van Vuuren, *Catal. Commun.* **10** (2009), 1674–1680.

中国科学院山西煤炭化学研究所
INSTITUTE OF COAL CHEMISTRY CHINESE ACADEMY OF SCIENCES
ICCCAS

# Low Temperature Methanation in Slurry Reactor over $NiO@SiO_2$ Core-shell Catalysts

Hui Wang, Junfeng Zhang, Qingde Zhang, Yisheng Tan, Yizhuo Han

Aug. 11, 2016

Cape Town

The 33rd International Pittsburgh Coal Conference

## Research Background——Low-temperature methanation

Challenge: easy deactivation

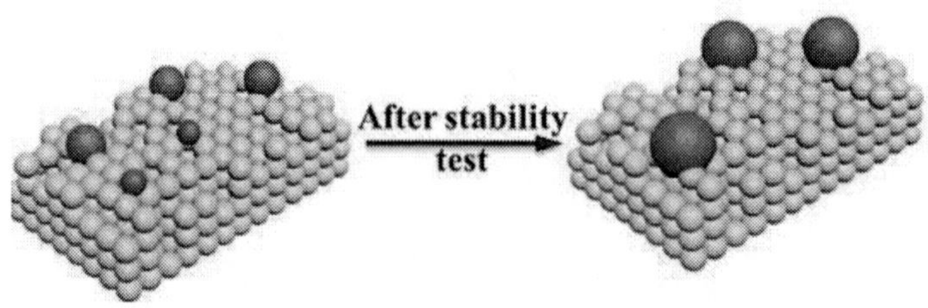


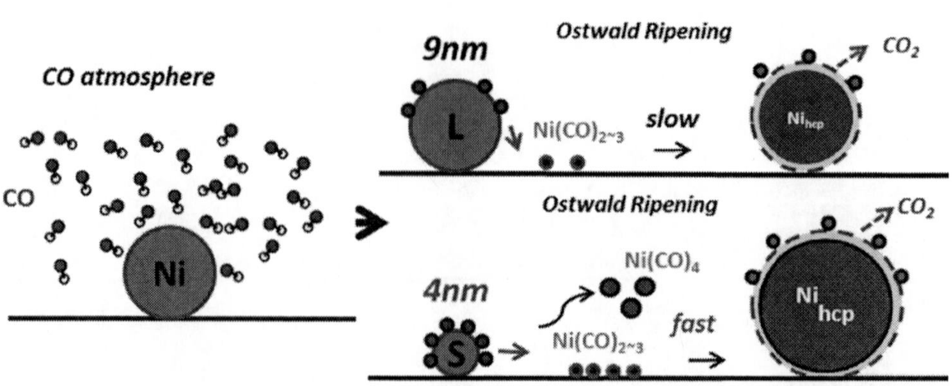

## Research Background——Low-temperature methanation

Solutions :

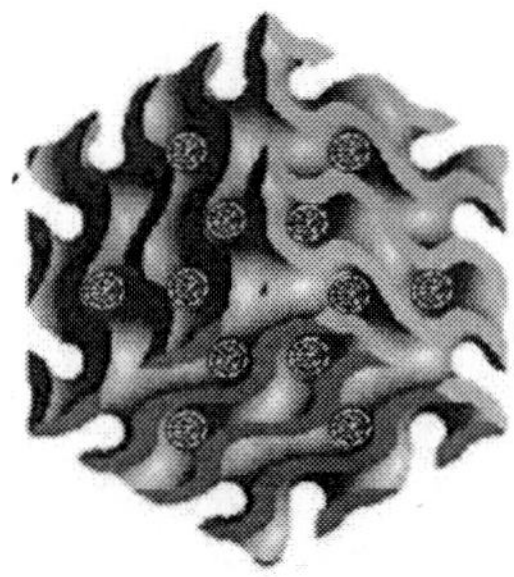

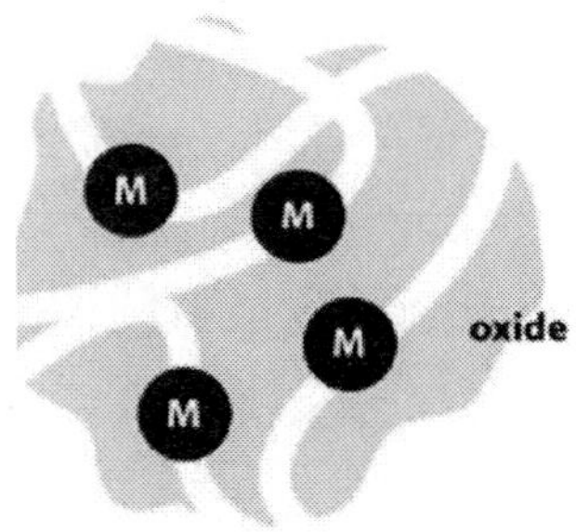


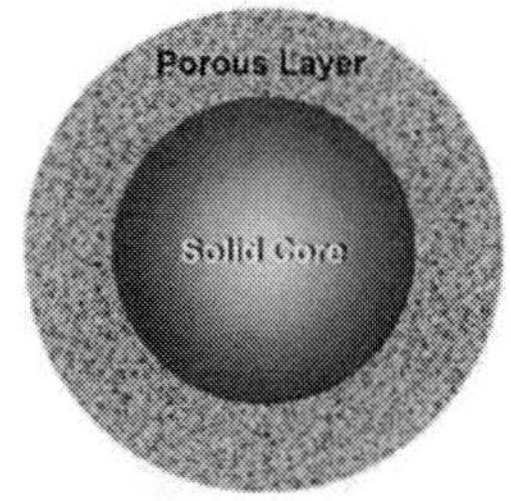


**Embedded mesoporous structure**

**Core-shell**

## Research Background——preparation of core-shell catalysts

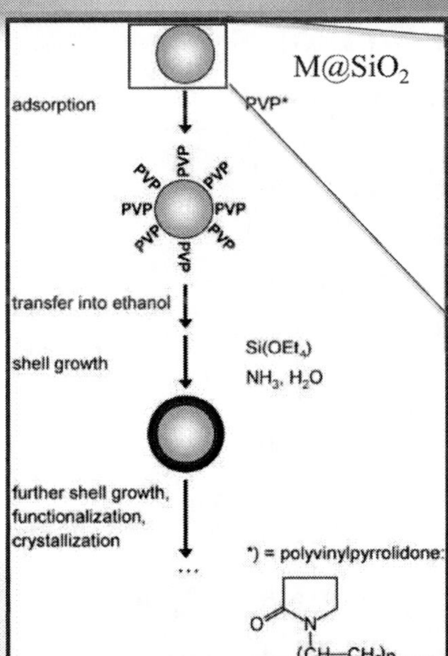


Stöber method

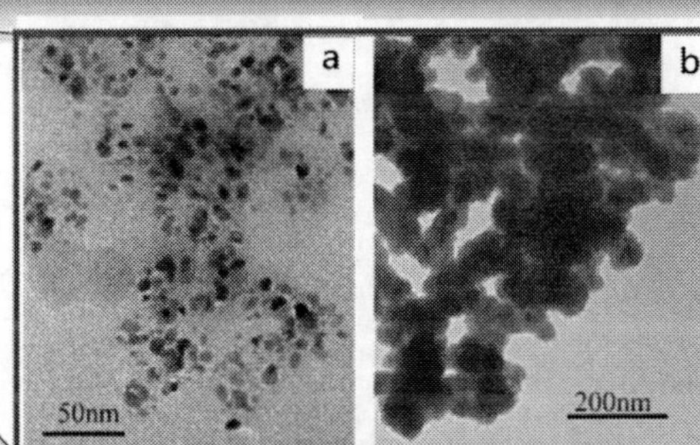


Precipitation -calcination method

Simple, High productivity

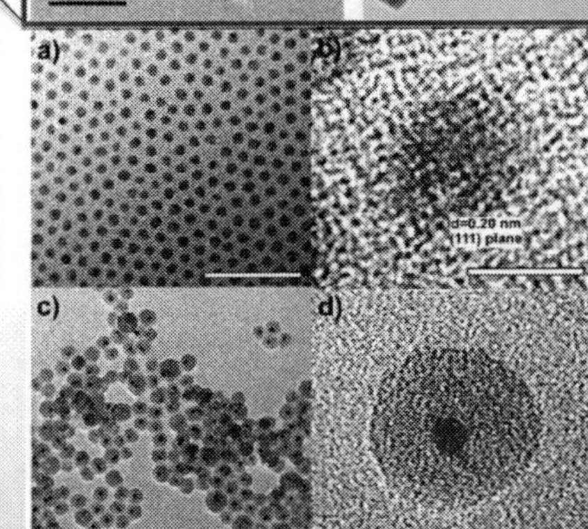


Mono-dispersed NPs：
Thermal decomposition
Solvothermal method
Microemulsion method

Complicated
Low productivity
Regular morphology

NiO NPs was prepared by precipitation-calcination method.

## Reserch Content

1. Preparation and methanation of $NiO@SiO_2$ core-shell catalysts with different core particle sizes.

2. Preparation and methanation of $NiO@SiO_2$ core-shell catalysts with different shell structure and different shell thickness.

## The effect of core particle size on morphology and catalytic performance

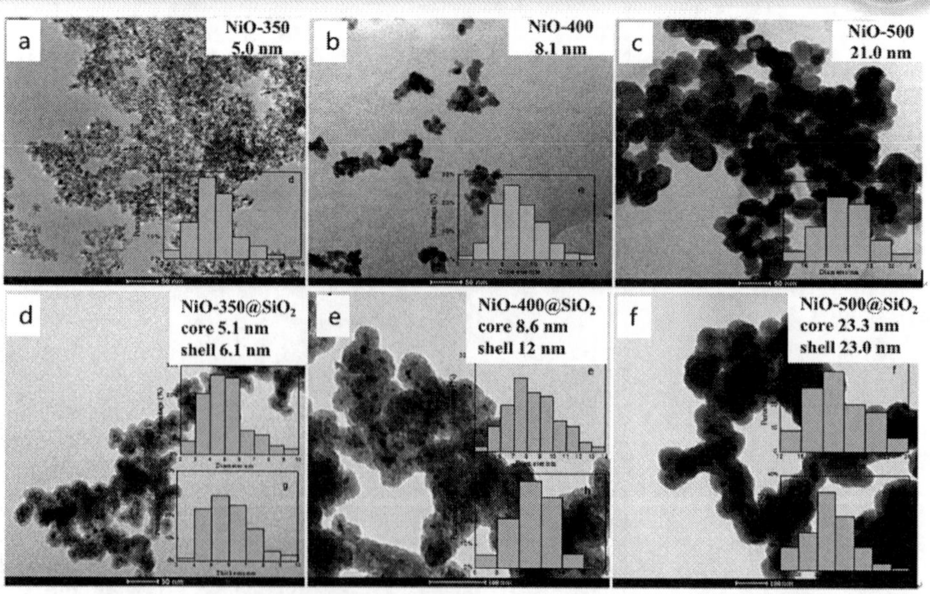


**Figure 1 TEM images of NiO nanoparticles before and after capsulation**

Core particles size doesn't change after calcination; The shell thickness increases with the increase of the core partilce size.

## The effect of core particle size on morphology and catalytic performance

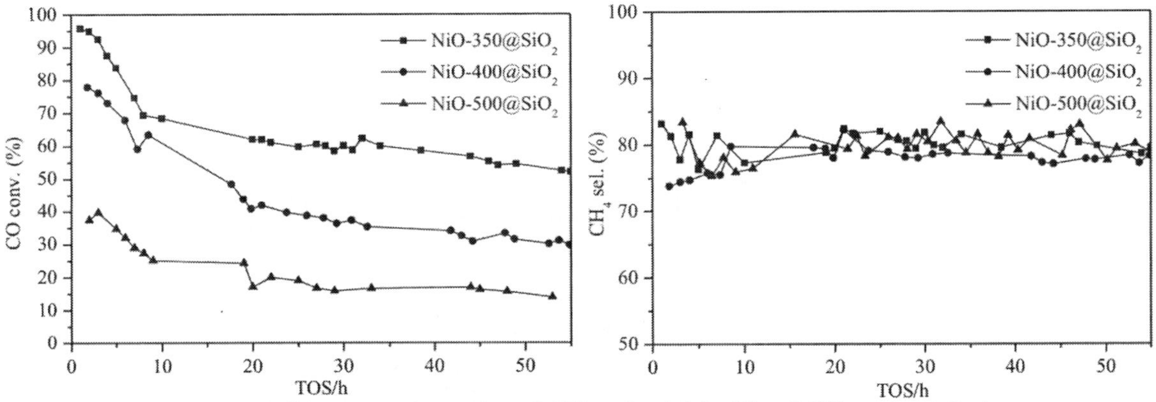


**Figure 2 CO conversions (a) and $CH_4$ selectivities(b) of different catalysts**

【slurry reactor, 10 000 mL/(g·h), 320 ℃, 1.5 MPa】

The core-shell catalysts deactivated rapidly in the initial 20 h and then deactivated slowly.


## The effect of core particle size on morphology and catalytic performance


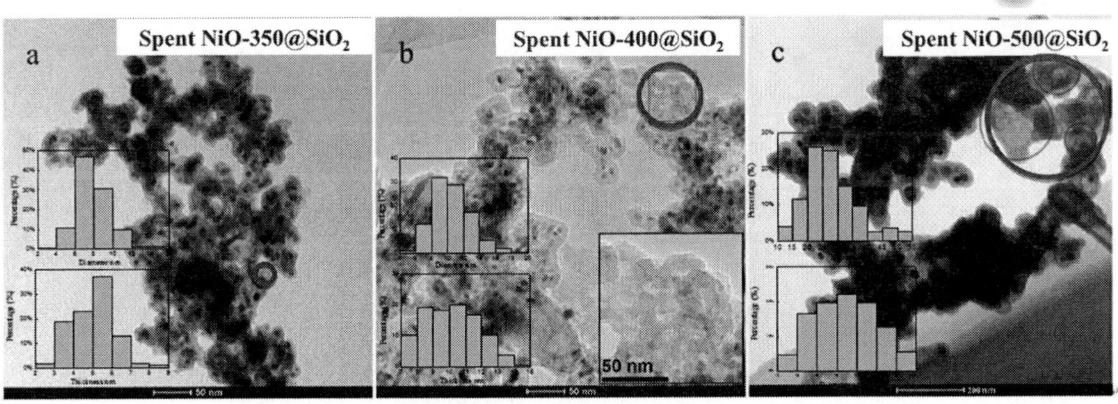


**Figure 3.4 TEM images of Ni@ $SiO_2$ after reaction**

Void shell emerged after reaction, but the Ni particles existed on the outside of shell cannot be found. Ni particles size increased after reaction, while the shell thickness doesn't change after reaction.

## The effect of core particle size on morphology and catalytic performance

**Table 1 The core particles size and shell thickness before and after reaction**

| Sample | | core/nm | shell/nm |
|---|---|---|---|
| NiO-350@$SiO_2$ | fresh | 5.1 | 6.1 |
| | spent | 8.1 | 5.0 |
| NiO-400@$SiO_2$ | fresh | 8.6 | 11.2 |
| | spent | 10.6 | 10.1 |
| NiO-500@$SiO_2$ | fresh | 23.3 | 23.0 |
| | spent | 27.8 | 20.6 |

The core particles size of the three catalysts increased after reation, the shell thickness doesn't change after reation.

## The effect of core particle size on morphology and catalytic performance

**Table 2 Surface atom concentration and Ni content of the catalysts before and after reaction**

| Sample | | XPS | | | | | ICP |
|---|---|---|---|---|---|---|---|
| | | Surface atom concentration/% | | | | Ni/Si | （Ni%） |
| | | Ni | Si | C | O | | |
| NiO-350@$SiO_2$ | fresh | 6.5 | 11.4 | 35.9 | 46.2 | 0.57 | 57.3 |
| | spent | 6.3 | 10.9 | 36.6 | 46.3 | 0.58 | 55.9 |
| NiO-400@$SiO_2$ | fresh | 6.1 | 17.6 | 27.8 | 48.5 | 0.35 | 59.0 |
| | spent | 3.8 | 20.9 | 24.7 | 50.6 | 0.18 | 61.2 |
| NiO-500@$SiO_2$ | fresh | 4.2 | 13.9 | 34.7 | 47.3 | 0.30 | 60.7 |
| | spent | 3.4 | 20.8 | 26.8 | 49.0 | 0.16 | 63.1 |

Ni bulk content of the catalysts doesn't change obviously, indicating no loss of Ni occurred.
The surface Ni content of the catalysts decreased remarkably.

## The effect of core particle size on morphology and catalytic performance

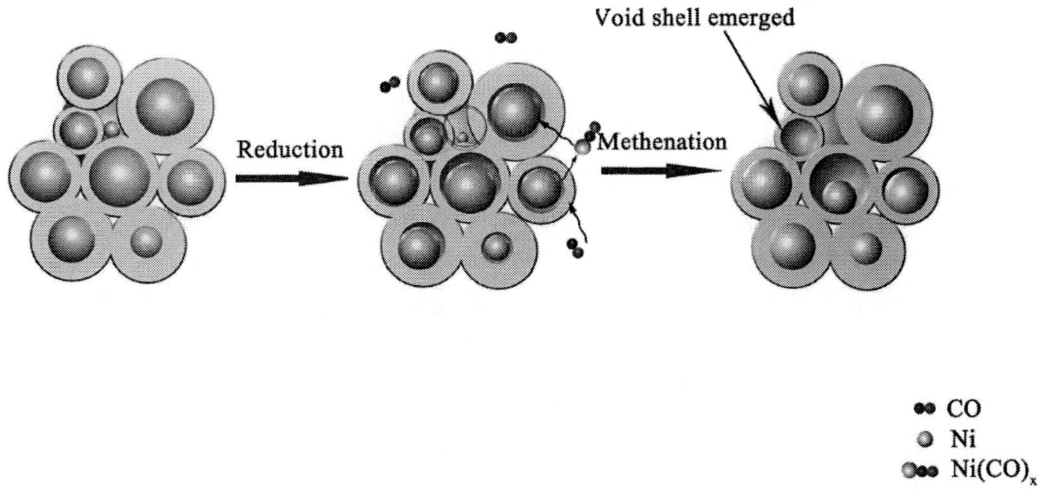


## The effect of core particle size on morphology and catalytic performance

**Table 3 $N_2$ sorption of NiO@$SiO_2$ before and after reaction**

| Sample | | BET surface area ($m^2/g$) | Pore volume ($cm^3/g$) | Pore diameter (nm) |
|---|---|---|---|---|
| NiO-350@$SiO_2$ | fresh | 117.2 | 0.218 | 7.4 |
| | spent | 52.7 | 0.147 | 11.2 |
| NiO-400@$SiO_2$ | fresh | 82.9 | 0.197 | 8.2 |
| | spent | 43.5 | 0.101 | 9.3 |
| NiO-500@$SiO_2$ | fresh | 34.4 | 0.074 | 8.6 |
| | spent | 23.7 | 0.073 | 12.3 |

## Summary

1. The prepared core-shell catalysts with cross-linked shell and multi-core encapsulation were tested. The shell thickness increased with the increase of core particle size.

2. The methanation activity of the catalysts decreased with the increase of core particle size. The catalysts deactivated rapidly at the initial period of 20 h and then deactivated slowly during the following reaction.

3. Void-shell was formed during the reaction because easily-migrated Ni(CO)x species were generated. The nucleation of the Ni(CO)x species outside of the shell is not available due to low supersaturation. Those dissociative Ni(CO)x species spread into inside of shell spontaneously through the pores in shells, and are enriched and decomposed on Ni cores with relatively low surface energy due to their larger size.

4. The increase of core particle size, decrease of BET surface area, pore volume and abatement of mesopores within 3-5 nm in the shell were Attributed to the deactivation of the core-shell catalysts.

## Reserch Content

1. Preparation and methanation of $NiO@SiO_2$ core-shell catalysts with different core particle sizes.

1. Preparation and methanation of $NiO@SiO_2$ core-shell catalysts with different shell structure and different shell thickness.

## The effect of shell structure and thickness on morphology and catalytic performance

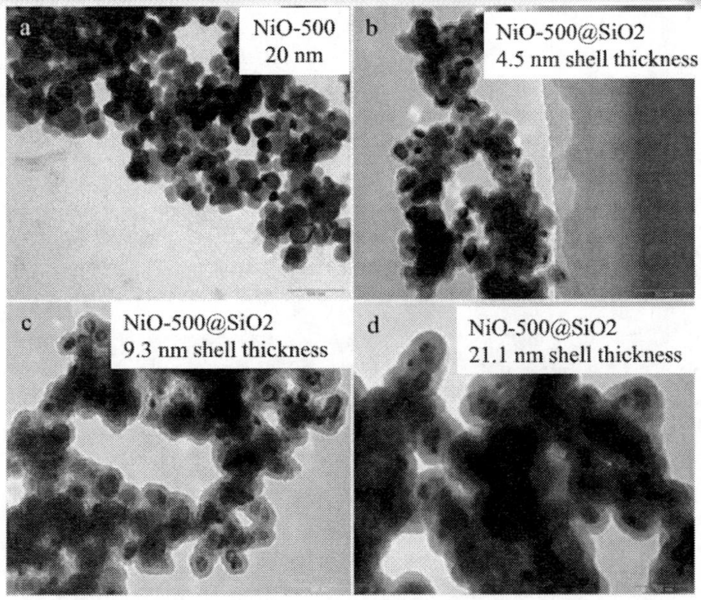


**Figure 5 TEM images of NiO-500 and NiO-500@SiO$_2$ with different shell thickness**

(a)NiO-500 (b) NiO-500@1SiO$_2$ (c)NiO-500@2SiO$_2$ (d) NiO-500@4SiO$_2$


## The effect of shell structure and thickness on morphology and catalytic performance


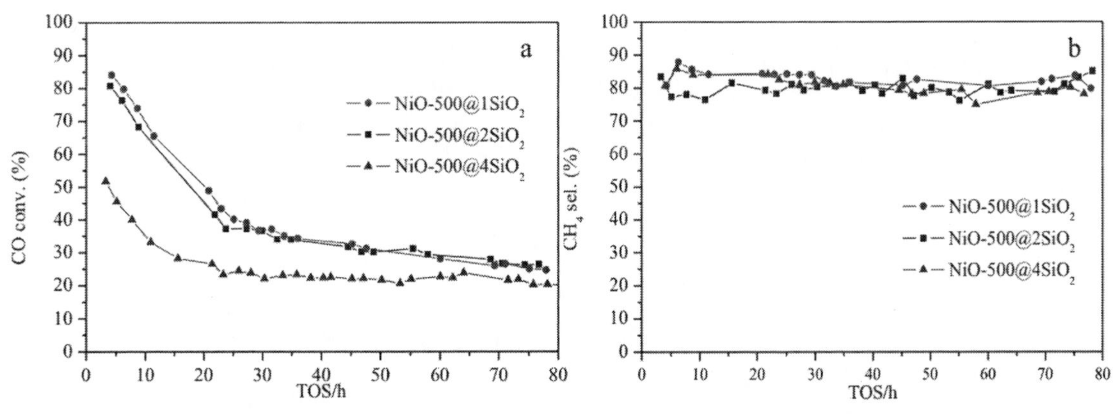


**Figure 6 (a)CO conversion, (b)CH$_4$ selectivity of the catalysts**

【16 000 mL/(g·h), 320 ℃, 1.5 MPa】

## The effect of shell structure and thickness on morphology and catalytic performance

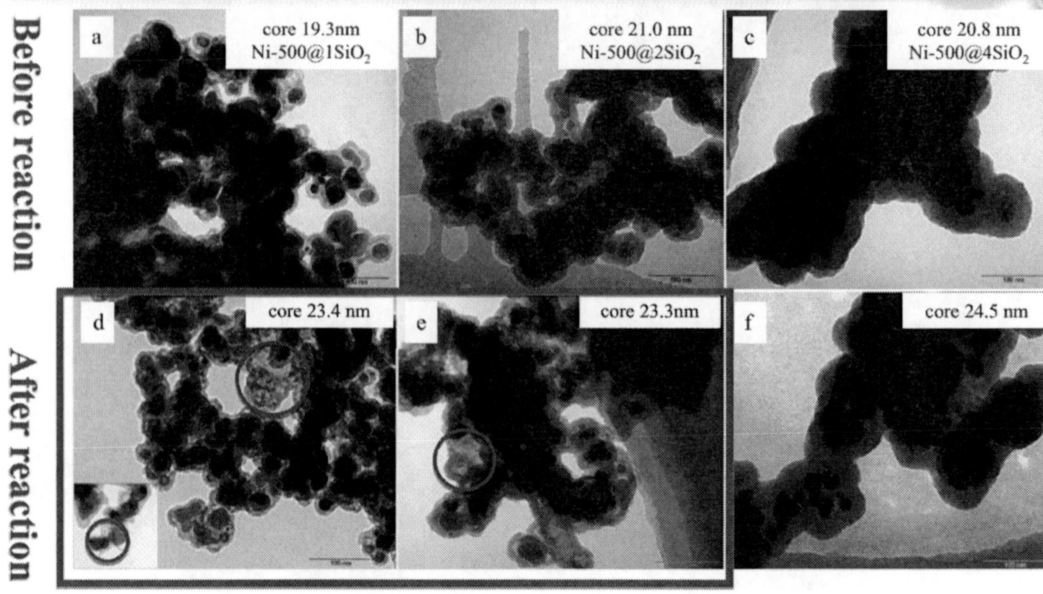


**Figure 7 TEM images of Ni-500@$SiO_2$ with different shell thickness before and after reaction**
**(a) Ni-500@1$SiO_2$, (b)Ni-500@2$SiO_2$, (c)Ni-500@4$SiO_2$**
**(d) spent NiO-500@1$SiO_2$, (e) spent NiO-500@2$SiO_2$, (f) spent NiO-500@4$SiO_2$**

## The effect of shell structure and thickness on morphology and catalytic performance

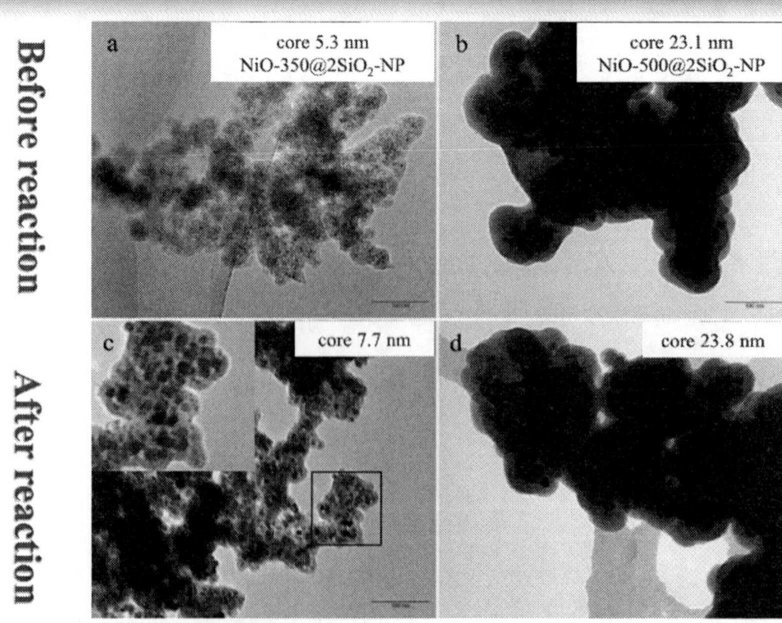


**Figure 8 TEM images of core shell catalyst without the addtion of PVP before and after reaction**
**(a) NiO-350@2$SiO_2$-NP,(b) NiO-500@2$SiO_2$-NP,**
**(c) Used NiO-350@2$SiO_2$-NP, (d) Used NiO-500@2$SiO_2$-NP**

## The effect of shell structure and thickness on morphology and catalytic performance

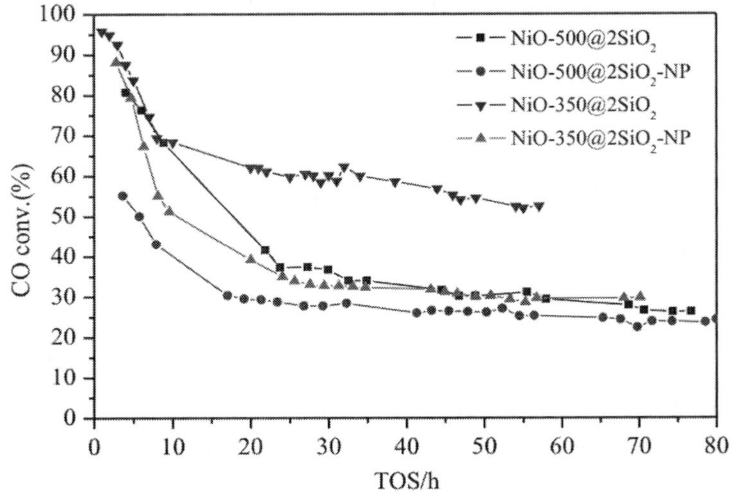


**Figure 9 CO conversion of different catalysts**

Compared with the catalysts prepared with PVP, the stability of catalysts prepared without PVP improved but the methanation activity decreased.

## The effect of shell structure and thickness on morphology and catalytic performance

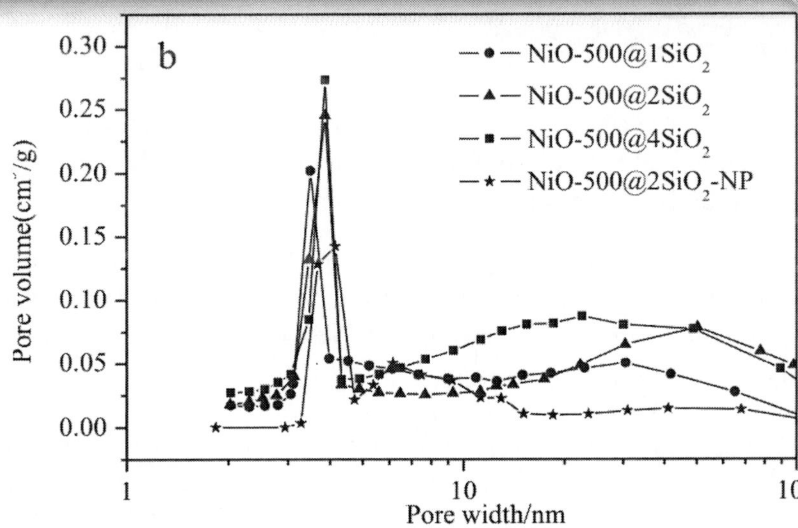


**Figure 10 BJH pore size distribution(b) of the core-shell catalysts before reaction**

Compared with NiO-500@$2SiO_2$ , the mesopores on the shell of NiO-500 @$2SiO_2$-NP decreased significantly.

## The effect of shell structure and thickness on morphology and catalytic performance

**Table 4 Textural properties of the NiO-500@$SiO_2$ catalysts with different shell thickness before and after reaction**

| Sample | BET SurfaceArea ($m^2/g$) | Pore volume ($cm^3/g$) | Pore diameter (nm) |
|---|---|---|---|
| NiO-500@1$SiO_2$ | 32.9 | 0.10 | 11.6 |
| NiO-500@2$SiO_2$ | 34.1 | 0.09 | 10.1 |
| NiO-500@4$SiO_2$ | 37.8 | 0.13 | 13.8 |
| NiO-500@2$SiO_2$-NP | 26.2 | 0.10 | 15.0 |
| Used NiO-500@1$SiO_2$ | 21.7 | 0.06 | 11.6 |
| Used NiO-500@2$SiO_2$ | 30.5 | 0.09 | 11.5 |
| Used NiO-500@4$SiO_2$ | 29.5 | 0.10 | 14.2 |
| Used NiO-500@2$SiO_2$-NP | 17.7 | 0.06 | 12.7 |

The surface area of the catalysts increased with the increase of the shell thickness.The mesopores with 3 nm on the shell increased with the addition of PVP. The surface area of the catalysts decreased after reaction.

## Summary

1. The surface area of the catalysts increased with the increase of the shell thickness.The methanation activity decreased and the stability improved with the increase of shell thickness.

2. The core-shell catalysts prepared without PVP showed severe multi-core encapsulation and shell cross-linking. The application of surfactant PVP during encapsulation promoted the dispersion of NiO nanoparticles and reduced the shell cross-linking.

3. The mesopores with 3 nm pore size in the shell increased with the addition of PVP. The stability of catalysts improved with the decrease of mesopores in the shell.

## Conclusions

1. The core particles size, shell thickness and mesopores have great effects on methanation activity and stability of the core-shell catalysts. The methanation activity decreased with the increase of core particle size. The increase of shell thickness or the decrease of mesopores on the shell can improve the stability of catalysts.

2. Void-shell was formed during the reaction because easily-migrated Ni(CO)x species were generated. The nucleation of the Ni(CO)x species outside of the shell did not occure due to low supersaturation. Those dissociative Ni(CO)x species spread into inside of shell spontaneously through the pores of shells, and are enriched and decomposed on Ni cores with relatively low surface energy due to their larger size.

3. The increase of core particle size, decrease of BET surface area, pore volume and mesopores in the shell were attributed to the deactivation of the core-shell catalysts.

# Thank you for your attention!

# American Electric Power's Experience with Mercury and Air Toxics Standard (MATS) Compliance


Presented by Joseph Bittinger

Contributors: Doug Ritzenthaler, Dan Connor, Dean Bell, Jason Baker


AEP AMERICAN ELECTRIC POWER

# Fun with Acronyms

- HAPs – Hazardous Air Pollutants
- MATS – Mercury & Air Toxics Standard
- PM – Particulate Matter or Total PM
  - FPM - Filterable Particulate Matter
  - CPM –Condensable Particulate ESP
- HCl – Hydrochloric Acid (this is HCL not HCI!)
- FGD – Flue Gas Desulfurization
  - DFGD – Dry FGD
  - WFGD – Wet FGD
  - NID – Dry FGD with J-ducts & pebble lime
- PJFF – Pulse Jet Fabric Filter
  - AKA Baghouse/BH, or just FF
- ACI – Activated Carbon Injection
  - PAC – Powdered Activated Carbon
  - BAC – Brominated Activated Carbon
  - ACI-ESP/FF
- DSI – Dry Sorbent Injection (Trona, Sodium Bicarbonate)
  - SBC – Sodium Bicarbonate
  - $SO_3$ – Sulfur Trioxide
- ESP – Electrostatic Precipitator
  - Hot Side or Cold Side (which side of the air preheater)
- SCR- Selective Catalytic Reduction
  - DeNOx
- PM – Particulate Matter or Total PM
  - FPM - Filterable Particulate Matter
  - CPM –Condensable Particulate ESP
- Heat Input
  - mmBTU – million BTU
  - TBtu – Trillion BTU
- ORP – Oxidizing Reducing Potential

AEP AMERICAN ELECTRIC POWER

# Summary of MATS Regulations

- Proposed legislation is known as "HAPs," "MATS" (Mercury and Air Toxics Standard), "EGU MACT" (Electric Generating Unit Maximum Achievable Control Technology), or "Utility MACT"
- Issued as proposed – March 2011
- Final issuance – December 2011, 40 CFR Part 63
- Imposes new emissions limits on particulate matter, mercury, non-mercury heavy metals, acid gases
- Compliance by 4/15/2015
  - Option to file for 1 year extension

AEP AMERICAN ELECTRIC POWER

# MATS Limits

- Total Particulate Matter – 0.03 lb/mmBtu
  - Surrogate for non-mercury heavy metals
- Hydrogen Chloride – 0.002 lb/mmBtu
  - Surrogate for all acid gases <0.2 lb $SO_2$/mmBTU
- Total Mercury – 1.2 lb/TBtu
  - Applies to all fuels except lignite (less than 8300 Btu/lb)
- Total Mercury – 4.0 lb/TBtu
  - Lignite fuels only (less than 8300 Btu/lb)

AEP AMERICAN ELECTRIC POWER

# Industry Experience & EPRI

Units with SCR & wFGD would be in compliance

- Hg removal is a co-benefit
- SCR oxidizes mercury along with $NO_x$
  - Oxidized mercury is soluble in water, allowing for the removal in the WFGD
  - Stack total expected to be approximately the same as FGD inlet elemental (generally accepted 90% total reduction)

PM is captured in the ESP and WFGD

AEP already capturing acid gas on these units through DSI

AEP AMERICAN ELECTRIC POWER

# Industry Experience & EPRI

Units with cold side ESP only

- ACI and maybe halogen addition for Hg removal
- Need controls for PM and acid gas
- Hot side ESPs ACI does not work
- DFGD also needs ACI to capture Hg
- Ash sales will be eliminated by ACI

AEP AMERICAN ELECTRIC POWER

# Mercury Removal

- Mercury commonly occurs in coal in varying amounts (2-35 lb/TBtu)
- Mercury is vaporized as coal is burned at high temperatures
- Mercury in flue gas leaving boiler is mostly in the elemental form (typically 50+%)
- SCR typically oxidizes mercury <90% in the presence of halogens
- Oxidized mercury can be removed with ACI or wet FGD
- Elemental mercury difficult to remove

AEP AMERICAN ELECTRIC POWER

# Plan for Compliance for units without SCR&FGD

- Test SCR/FGD units and verify compliance.
- Test DSI and ESP for $SO_2$ and acid gas removal
- Test ACI and halogen addition
- Start Engineering & Design on 4 fabric filter additions with ACI injection
- Test/develop, install stack monitors/probes
    - Batch type proven, continuous still being refined
- Retirements

AEP AMERICAN ELECTRIC POWER

# American Electric Power

**AEP IS THE NATION'S LEADING REGULATED ELECTRIC UTILITY**

Provides generation, transmission and / or distribution services to approximately 5.4 million customers in eleven states with headquarters in Columbus, Ohio (NYSE: AEP)

**Our regulated electric assets include:**

- Approximately 32,000 megawatts of generating capacity in 3 RTOs (one of the largest US generation portfolios with a significant cost advantage in many of our market areas)
- Approximately 40,000 circuit miles of transmission lines, including 447 miles of Transco lines, and 2,116 miles of 765kV lines, the backbone of the electric interconnection grid in the Eastern U.S.
- AEP Transco has $1.8B of transmission assets in-service with plans to construct approximately $3B additional through 2017
- Over 223,000 miles of overhead and underground distribution lines

**AEP consistently produces strong financial results:**

- Operating earnings growth of 4% to 6%
- Strong balance sheet including $61 billion of assets
- Cash dividend paid every quarter since 1910 with targeted payout of 60% to 70% of operating earnings

**AEP: THE NEXT PREMIUM REGULATED UTILITY**

**Source: www.AEP.com, 2015 AEP Fact Book**

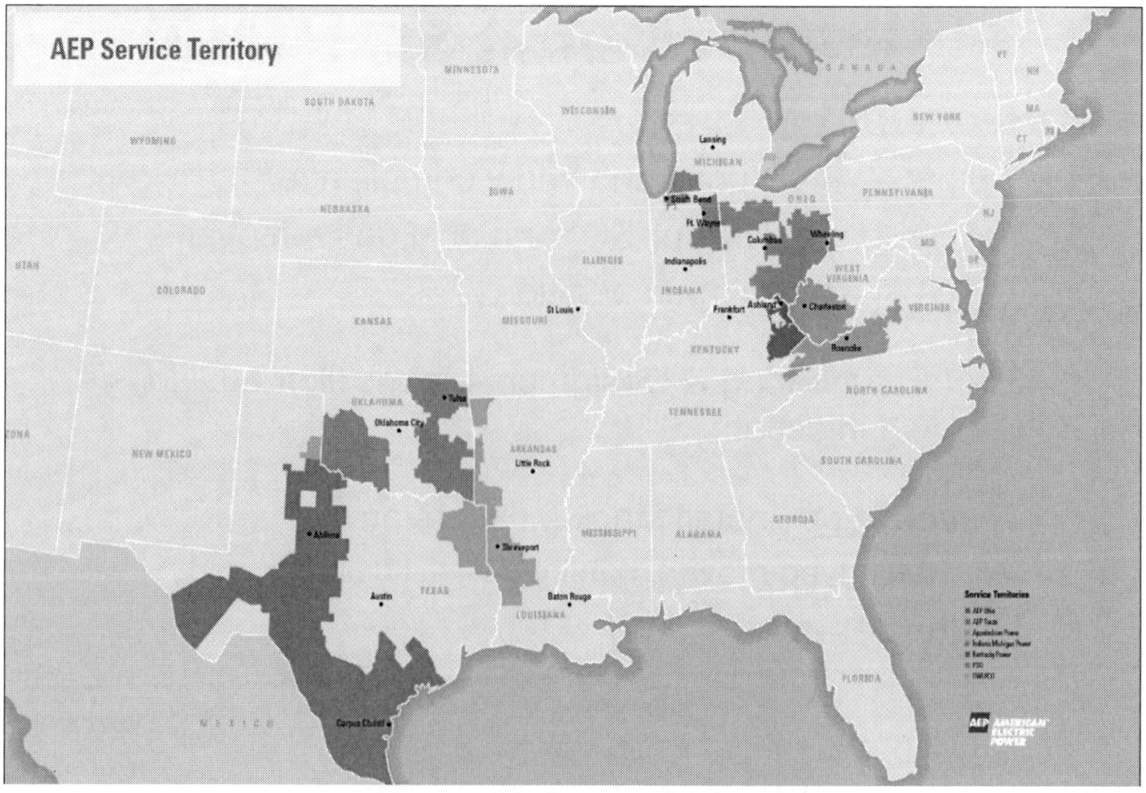

# Transforming our Generating Fleet

## AEP Owned Generating Capacity by Fuel (actual & projected)

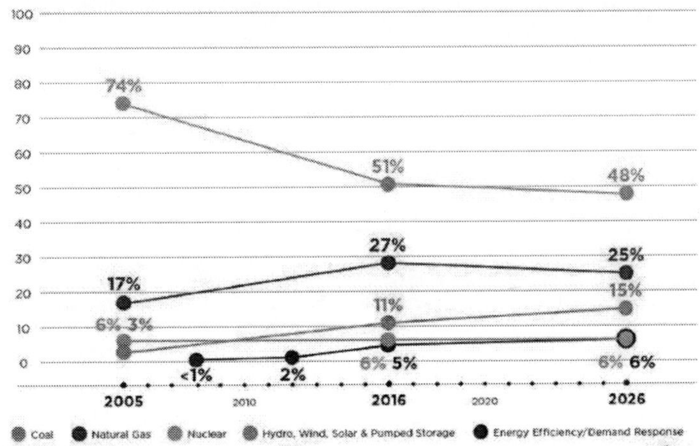


Hydro, Wind, Solar & Pumped Storage ↑ 12%

Natural Gas ↑ 8%

Energy Efficiency/Demand Response ↑ 6%

Coal/Lignite ↓ 26%

20

Source: www.AEP.com, 2015 AEP Fact Book

# Regulated Environmental Investment & Retirements

| Operating Company | Plant | MW | Potential Type of retrofit |
|---|---|---|---|
| APCO | Clinch River 1 | 242 | Refuel with Natural Gas |
| | Clinch River 2 | 242 | Refuel with Natural Gas |
| I&M | Rockport | 2,620 | DSI, SCR |
| KPCO | Big Sandy 1 | 268 | Refuel with Natural Gas |
| PSO | Oklaunion | 102 | ACI, CaBr Injection |
| | Northeastern 3 | 460 | ACI, DSI, Baghouse |
| SWEPCO | Welsh 1 | 528 | ACI, Baghouse |
| | Welsh 3 | 528 | ACI, Baghouse |
| | Pirkey | 580 | ACI, CaBr Injection |
| | Dolet Hills | 256 | ACI, Baghouse |
| | Flint Creek | 264 | FGD, ACI |
| **Total Regulated retrofits =** | | **6,090** | |

ACI – Activated Carbon Injection
DSI – Dry Sorbent Injection
FGD – Flue Gas Desulfurization
SCR – Selective Catalytic Reduction

28

| Operating Company | Plant | MW | Retirement Date |
|---|---|---|---|
| APCO | Glen Lyn 5 | 95 | 2015 |
| | Glen Lyn 6 | 240 | 2015 |
| | Clinch River 3 | 235 | 2015 |
| | Sporn 1 | 150 | 2015 |
| | Sporn 3 | 150 | 2015 |
| | Kanawha River 1 | 200 | 2015 |
| | Kanawha River 2 | 200 | 2015 |
| | Total MW | 1270 | |
| I&M | Tanners Creek 1-4 | 995 | 2015 |
| | Total MW | 995 | |
| KPCO | Big Sandy 2 | 800 | 2015 |
| | Total MW | 800 | |
| SWEPCO | Welsh 2 | 528 | 2016 |
| | Total MW | 528 | |
| PSO | Northeastern 4 | 470 | 2016 |
| | Total MW | 470 | |
| **Total Regulated retirements** | | **4,063** | |

Source: www.AEP.com
2015 AEP Fact Book

# Competitive Environmental Investment & Retirement

| Operating Company | Plant | MW | Type of retrofit |
|---|---|---|---|
| **AEP Generation Resources** | Conesville 5 & 6 | 810 | Gore |
| **TNC** | Oklaunion* | 355 | ACI, CaBr Injection |
| | **Total Competitive retrofits =** | **1,155** | |

*PPA

| Operating Company | Plant | MW | Retired |
|---|---|---|---|
| **AEP Generation Resources** | Beckjord | 53 | Oct 2014 |
| | Conesville 3 | 165 | Dec 2012 |
| | Muskingum River 1-5 | 1,440 | Jun 2015 |
| | Picway 5 | 100 | Jun 2015 |
| | Sporn 2,4 | 300 | Jun 2015 |
| | Sporn 5 | 450 | Dec 2011 |
| | Kammer 1-3 | 630 | Jun 2015 |
| | **Total Competitive retirements =** | **3,138** | |

29

**Source: www.AEP.com**
**2015 AEP Fact Book**

*14*

# $CaBr_2$ with ACI

- Calcium Bromide enhances oxidation of mercury by providing excess halogens.
- ACI captures mercury in the flue gas and is removed by an ESP or baghouse
  - Can only capture oxidized mercury
- Brominated ACI also increases oxidation of mercury, leading to higher capture
- Many tests successful with up to 15 GPH $CaBr_2$ of a 52% solution.

AEP AMERICAN ELECTRIC POWER

# Calcium Bromide Addition

- Two categories for Calcium Bromide addition
  - Refined Fuels – Section 45 Tax Credit
  - Trim mercury oxidation for Hg removal in FGD or with activated carbon injection (ACI)
    - Potential reduction in the amount of ACI

AEP AMERICAN ELECTRIC POWER



# BOP Issues

- EPRI and others in the industry reported numerous corrosion issues with the application of $CaBr_2$ solution to the coal (PRB)
- Affected areas in the process
  - Air Heater Baskets
  - ID Fan
  - Pulverizer
  - FGD

AEP AMERICAN ELECTRIC POWER

# Air Heater Cold End Basket Damage from $CaBr_2$ Condensation

Less than one year of operation

AEP AMERICAN ELECTRIC POWER

# Air Heater Basket Corrosion

Same air heater, basket elements on floor of flue

AEP AMERICAN ELECTRIC POWER

# Lignite ACI Testing

- January - February 2013

(Fabric Filter design under way)

- Lignite fuel
  - Low rank virgin coal
  - MATS Hg limit <4 lb/TBtu
- ACI alone not sufficient to comply with MATS Rule Hg limit
  - From 14 lb/TBtu to 6 lb/TBtu
- ACI coupled with $CaBr_2$
  - Hg <4 lb/TBtu
- Optimize with additive to FGD scrubber

AEP AMERICAN ELECTRIC POWER



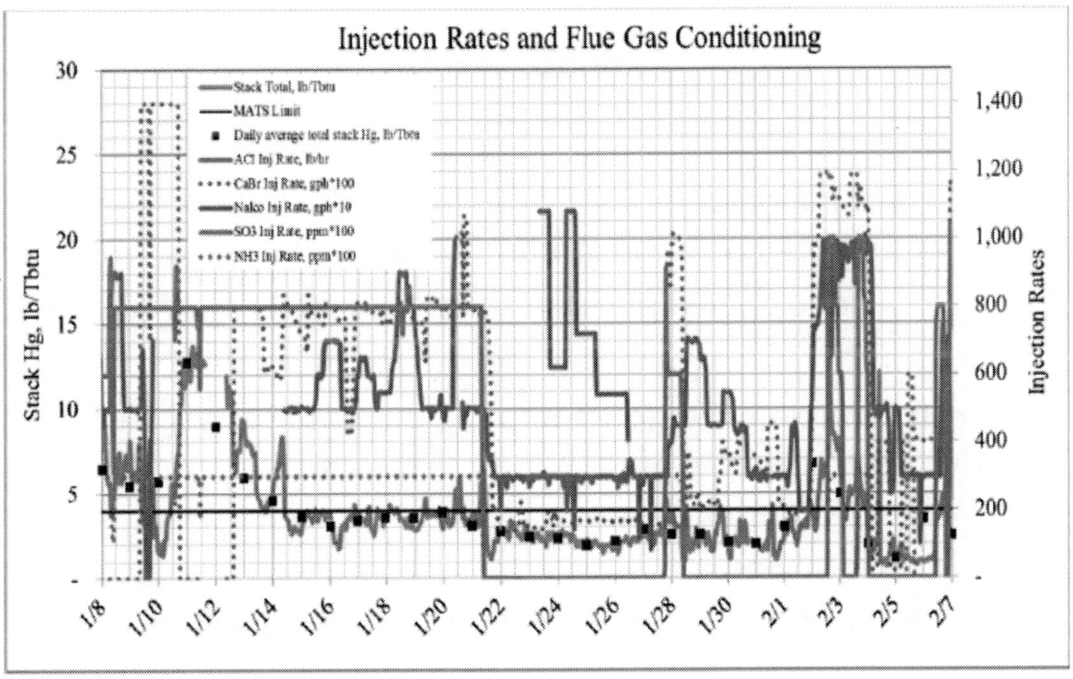


AEP AMERICAN ELECTRIC POWER

# More Testing at AEP

Unit with cold ESP, FGD and no SCR – June 2013

- Eastern bituminous
  - High Sulfur
    - MATS Hg limit <1.2 lb/TBtu
- Brominated ACI
  - High $SO_3$ which renders ACI less effective
  - Feed rate from 200 to 750 lb/hr
- $CaBr_2$ solution fed at 1 – 15 gph
  - Applied to coal at the feeders of the pulverizers

AEP AMERICAN ELECTRIC POWER



# Results

- $CaBr_2$ addition and/or brominated ACI were not able attain MATS Rule Hg compliance
  - Results well above 1.2 lb/TBtu limit

-$CaBr_2$ coupled with low temp Hg ox. catalyst allowed for MATS rule Hg compliance

- Gore more economic vs. $CaBr_2$ and low temp. catalyst
- Oxidation catalyst further from commercial development

AEP AMERICAN ELECTRIC POWER

# Pilot

AEP AMERICAN ELECTRIC POWER

# Mercury Control System

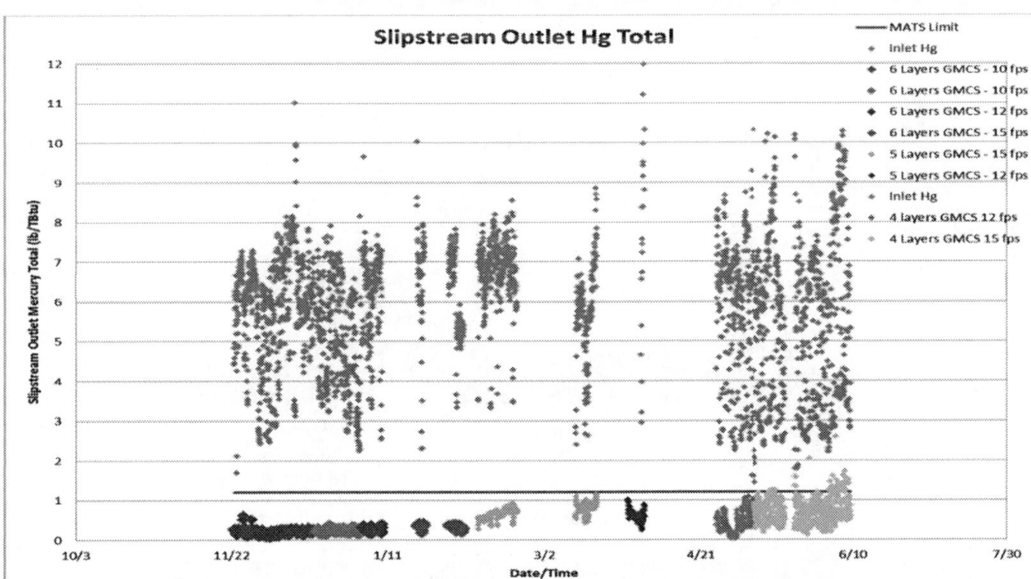


AEP AMERICAN ELECTRIC POWER

# ACI and Fabric Filter Retrofits

Better performance than expected

- Turn Down issues

  Equipment over designed
- Fuel Hg lower than planned
  - Targeted 15 lb/Tbtu in design
  - Delivered 5 lb/Tbtu
- Early design based on field tests, less than optimal distribution

AEP AMERICAN ELECTRIC POWER



# CFD Modeling to Ensure Distribution and Residence Time

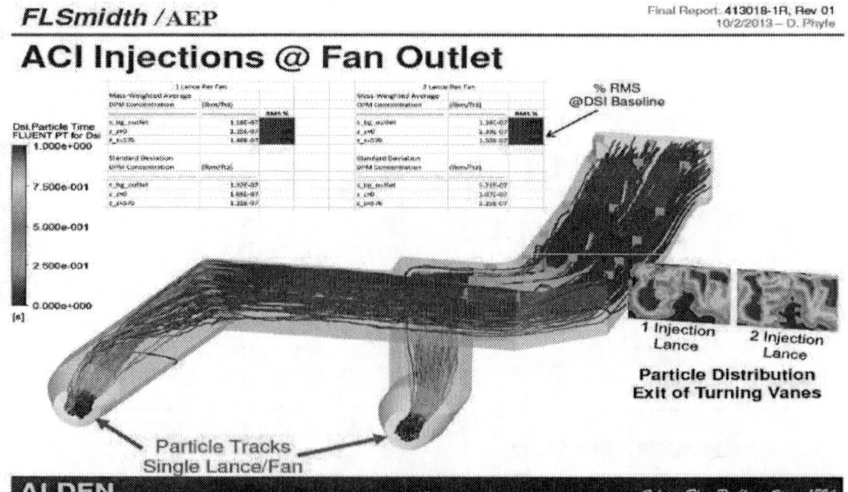


AEP AMERICAN ELECTRIC POWER

# MATS Compliance Better than Expected

- $CaBr_2$ – optimized to reduce amount of application (1/3rd of trials)
- ESPs maintained/ upgraded to minimize $SO_3$ flue gas conditioning
- Continuous Hg monitors were responsive and helpful

AEP AMERICAN ELECTRIC POWER

# BAC Lessons Learned

ACI metering valve corrosion

Figure 12: Corrosion on Non-Drive Side Rotor Shaft

Figure 5: Non-Drive End Seal Air Plugged Pipe Nipple

Figure 17: Corrosion of Non-Drive Side Seal Seat

AEP AMERICAN ELECTRIC POWER

# BAC Lessons Learned

Seal air to the metering valve was not dry air and the moisture and the Bromine reacted with the carbon steel.

The carbon steel storage silos needed to be coated to stop corrosion from the moisture in the fluidizing air reacting with the Bromine in the BAC.

AEP AMERICAN ELECTRIC POWER

# Performance of the SCR/WFDG units

Base Load Coal Units are cycling more than ever from
Low gas prices, low manufacturing energy need, efficiency improvements, renewables…

The SCR / FGD were supposed to be the easy to comply units. Today that is not neccesiliary the case.

AEP AMERICAN ELECTRIC POWER

# Cycling Units Effect Hg Stack Emissions

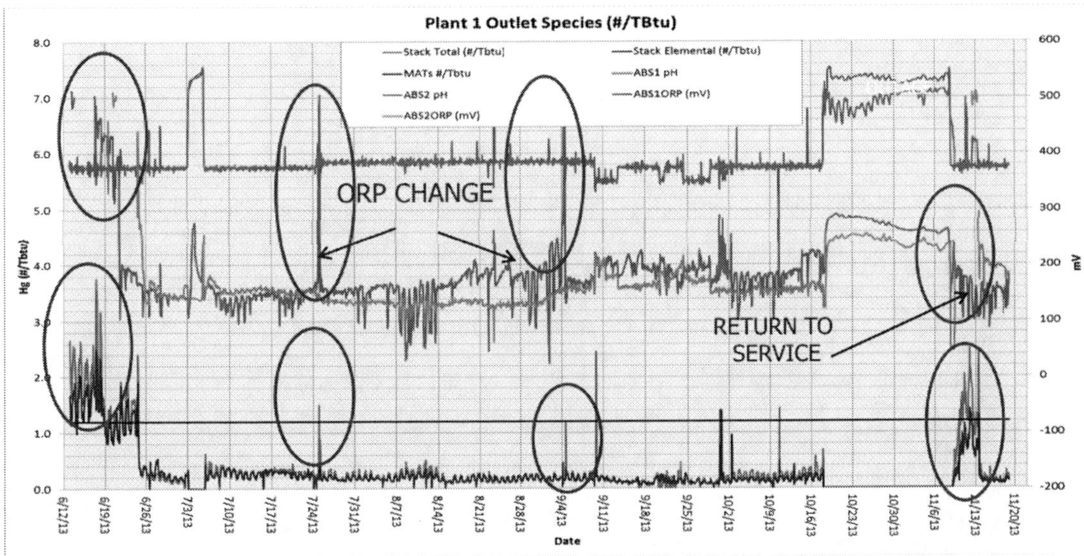


AEP AMERICAN ELECTRIC POWER

32

# Plant B Hg Emissions

## due to cycling

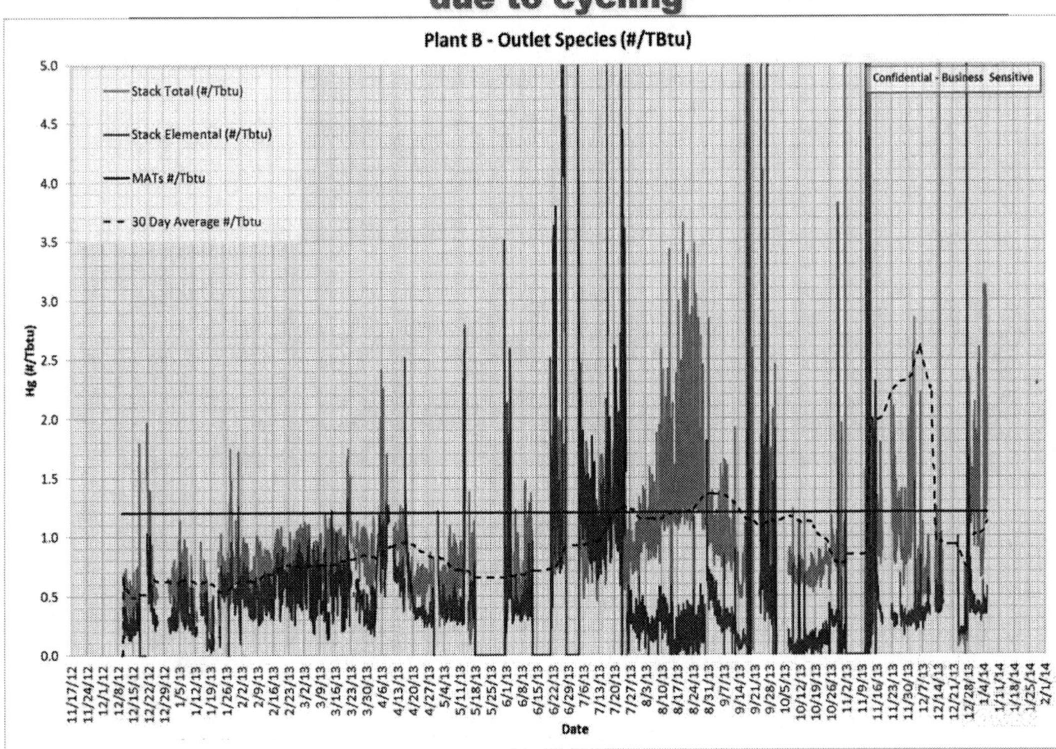

# Plant A Hg Emissions

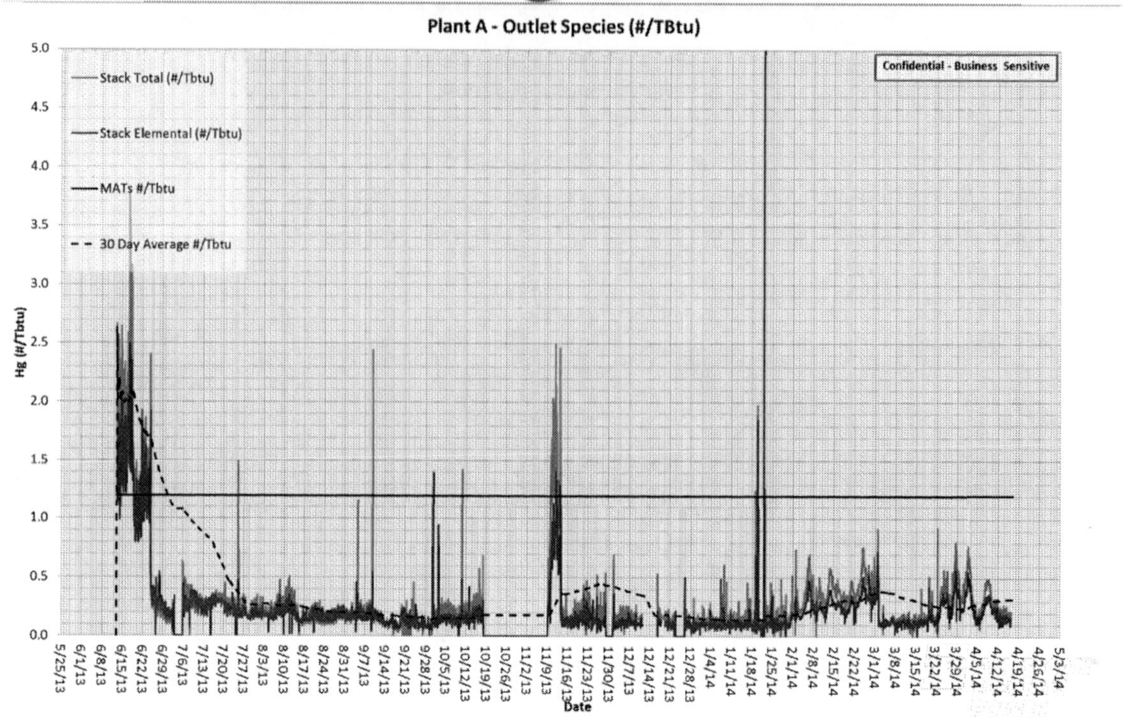


# ORP and Mercury

$HgCl_2$ $CaCl_2 + Hg^o$

100 200 300

ORP

AEP AMERICAN ELECTRIC POWER

# Dealing with ORP Changes

Testing chemical additives for Hg re-emission. These chemicals go in the FGD liquor.

Developing procedures for changing ORP

- Trouble shooting guide
- Modifying blowdown procedures

AEP AMERICAN ELECTRIC POWER

## Questions?

AEP AMERICAN ELECTRIC POWER

# Mercury Collection

- Traps are run in pairs of collection tubes per system (Clean Air & APEX systems)
- Clean Air system is primary system activated after unit is on line (producing electricity to grid)
- APEX system is backup and also collects mercury during start up periods

- Continuous mercury analyzers are used for process feedback but are not compliance devices (Ohio Lumex and Thermo) for AEP

AEP AMERICAN ELECTRIC POWER

# Mercury Traps and Probe

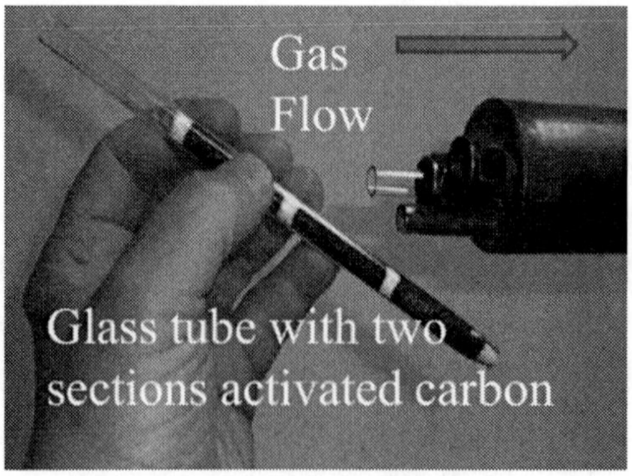


AEP AMERICAN ELECTRIC POWER

# Mercury

- Trap assembly (probe) is pulled. The system records volume and time stamp (start/stop).
- The traps are pulled and serial number recorded to be matched up later with the volume.
- A new set of traps are put in the probe and the probe is restarted.
- The pulled traps are sent out to be analyzed.
- Once analyzed the data in input into the system where the volume and time stamps are used to calculate the µg/dscm.
- This data is then matched to the appropriate PI data to calculate the lbs/tBtu emission values.

AEP AMERICAN ELECTRIC POWER



# Continuous Hg Monitor

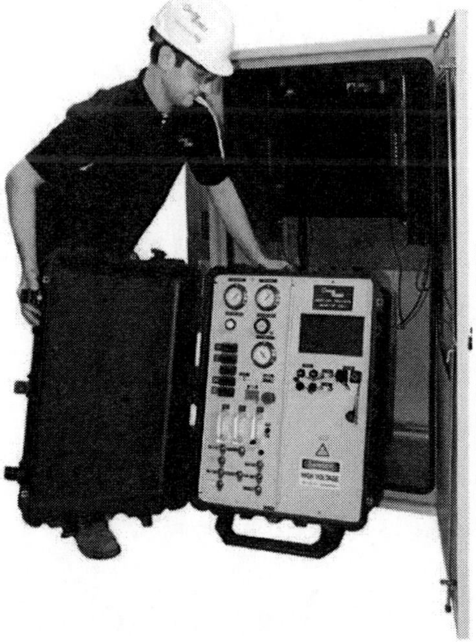

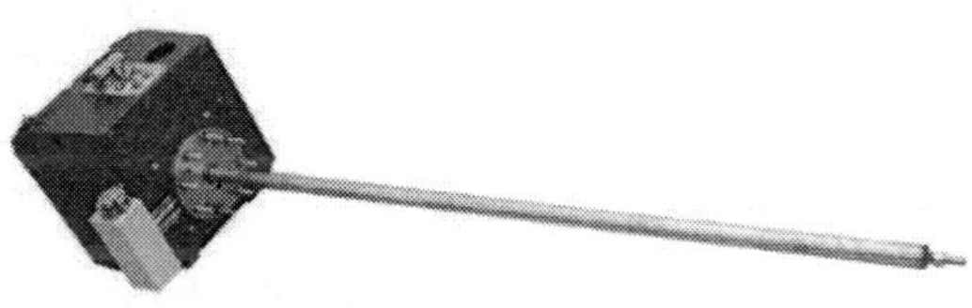

| | |
|---|---|
| Spectrometer | The 915 J features a highly sensitive atomic absorption spectrometer with Zeeman background correction providing interference free measurements. |
| Calibration | Built in Elemental Hg calibrator with low span (2-5 ug/m³). No other calibration source required. |
| Umbilical Lines | Heated line from 10ft to 50ft (others available). |
| Data Acquisition/Communications | Built-in industrial grade computer, real time readings with Excel data file or direct communications with ESC8832 via Modbus TCP. On-Line remote data transfer and analyzer control/calibration capabilities. |

OHIO LUMEX

US★China Energy Center
中美能源中心

2016 International Pittsburgh Coal Conference
Cape Town, SOUTH AFRICA
August 8 – 12, 2016

# The Future of Coal in the United States and China

Qingyun Sun and Richard Bajura,[1]
Rui Zhang and Xianliang Meng,[2]
and Zefu Ye[3]

Presentation by
Francis Lau, Synthesis Energy

(1) West Virginia University
(2) China University of Mining & Technology (Beijing, Xuzhou, resp.)
(3) Shanxi International Energy Group

West Virginia University

US★China Energy Center
中美能源中心

# Presentation Outline

- 1. Coal Production in the United States (U.S.) and China
- 2. Challenges to the U.S. Coal Sector
- 3. Challenges to the Chinese Coal Sector
- 4. Outlook for the Future of Coal in the U.S. and China

2

West Virginia University

US★China Energy Center
中美能源中心

# 1. Coal Production in U.S. and China

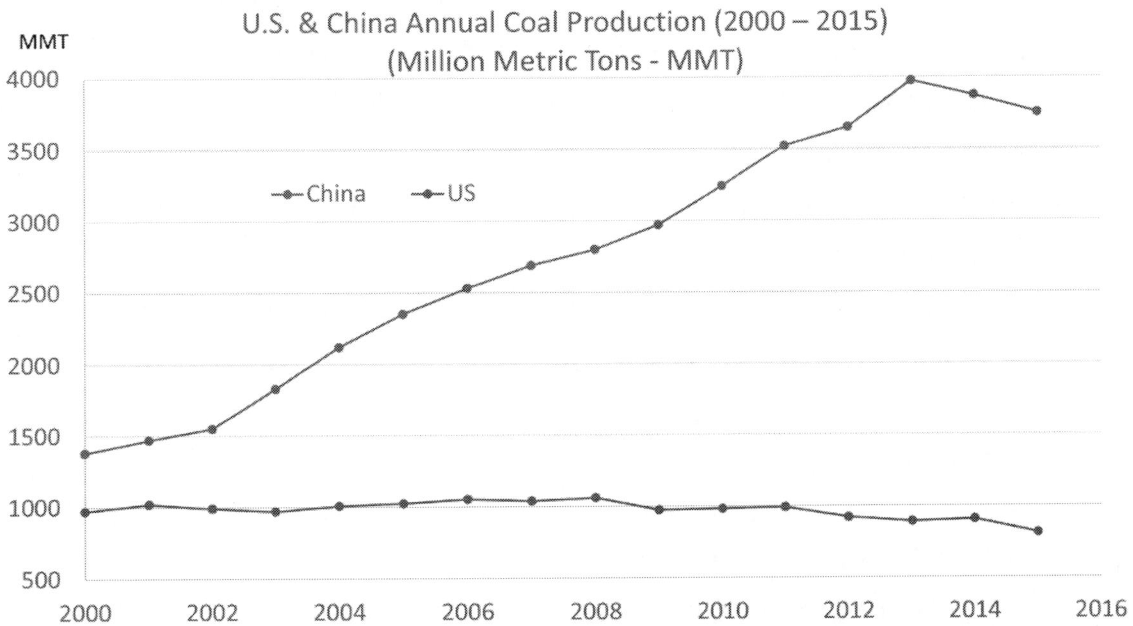


West Virginia University

3

US★China Energy Center
中美能源中心

# Energy Consumption in U.S. and China (2015)

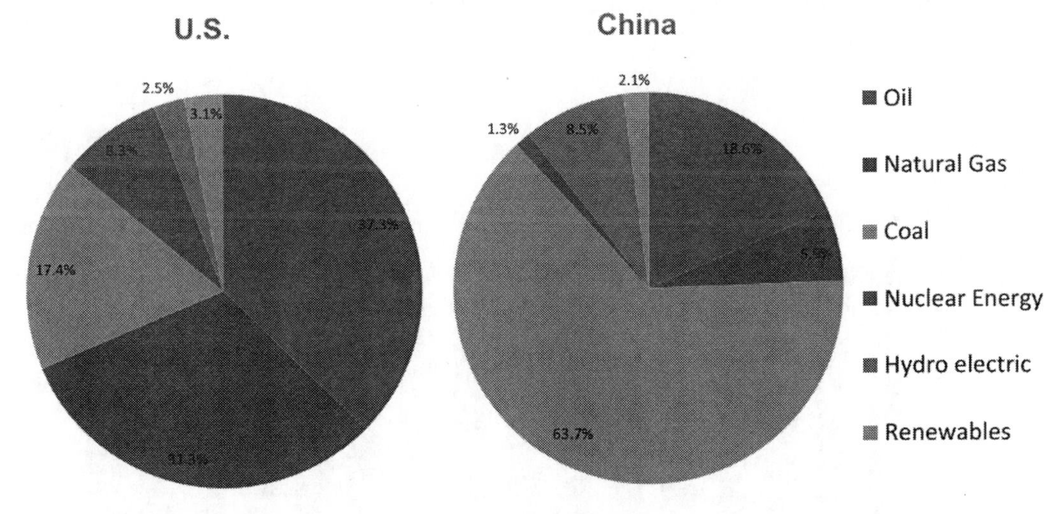


Total: 2280 Mtoe

Total: 3014 Mtoe

Mtoe = Million tons of oil equivalent

West Virginia University

4

US★China Energy Center
中美能源中心

# 2. U.S. Coal Sector Challenges

- Strict environmental regulations increase mining cost
- Both domestic coal demand and prices are falling
  - International prices also fall, reducing trade dollars
- New EPA rules hasten closure of coal-fired power plants
- Shale Gas Revolution advances natural gas use at the expense of coal
  - New power plants use natural gas; few coal-fired plants built
- Increased public and political resistance to the use of coal
- In some areas:
  - reduced coal production →
  - high levels of unemployment →
  - substance abuse →
  - Ghost Town Syndrome in smaller communities

West Virginia University



US★China Energy Center
中美能源中心

# Coal and Energy Consumption in U.S.

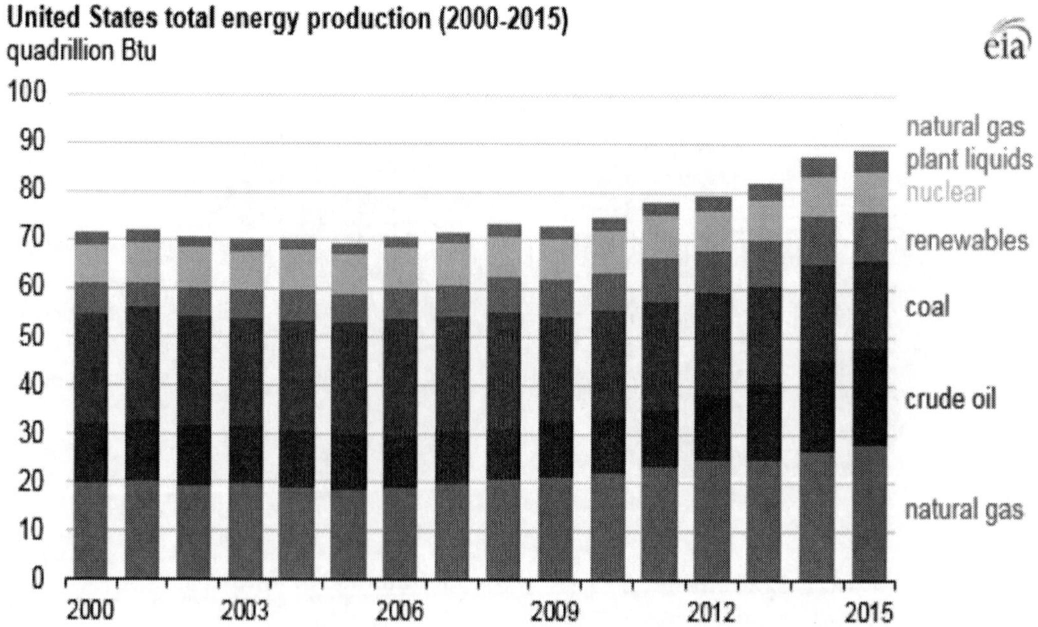


West Virginia University

US★China Energy Center
中美能源中心

# Gas Replaces Coal for Electricity Generation

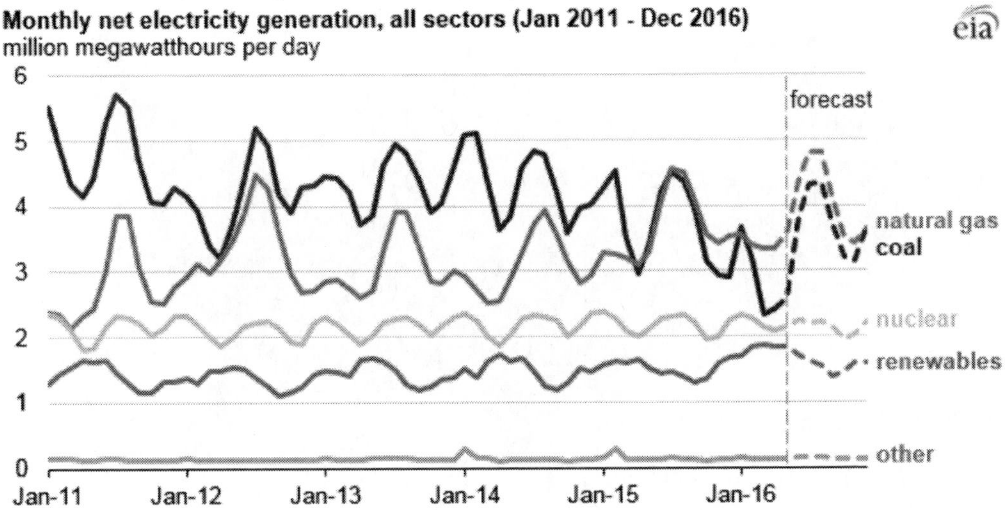


7

West Virginia University

US★China Energy Center
中美能源中心

# EPA New Rules Affect Closure of 280+ Coal-Fired Units

- More than 34 gigawatts (GW) of coal-fired capacity are now scheduled for retirement due to new EPA air pollution regulations
  - over 10% of the U.S. coal-fired generating capacity.
  - more than 32 coal-fired power plants will close and another 36 plants could also be closed.
- New analysis shows 280+ coal-fired generating units across 32 states may be closed in the future.
- Impacts illustrated on following map (Institute of Energy Research)

8

West Virginia University

US★China Energy Center
中美能源中心

## POWER PLANTS TO BE CLOSED BY EPA REGULATION*

- TOTAL MW TAKEN OFFLINE: 34,705 GW
- THAT'S NEARLY 10% OF OUR COAL ENERGY CAPACITY - GONE
- EACH RED DOT MEANS LESS ENERGY FOR THE U.S. ECONOMY

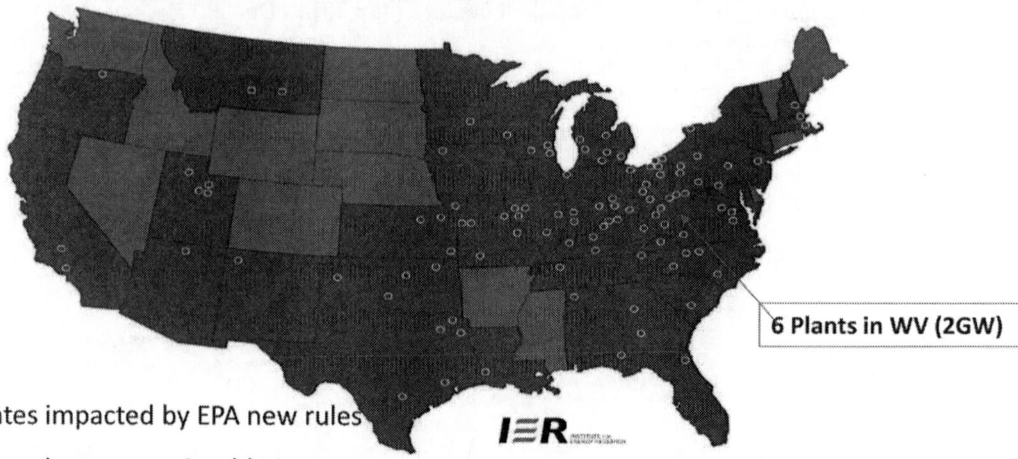


States impacted by EPA new rules

States have no noticeable impacts

*EACH DOT REPRESENTS A POWER PLANT WHERE ONE OR MORE ELECTRIC GENERATING UNITS ARE EXPECTED TO CLOSE. SEE HTTP://INSTITUTEFORENERGYRESEARCH.ORG/EPA-POWERPLANT-CLOSURES

9 West Virginia University

US★China Energy Center
中美能源中心

# 3. Chinese Coal Sector Challenges

- China has excess coal available because of over- production
- Lower economic growth rates lead to shrinking demand
- Renewable and nuclear energy replace coal to reduce air pollution
- Economic downturn hits coal producing areas:
  - Bankruptcy of private coal companies → lost jobs
  - Large state owned coal companies → layoffs & salary reductions

10 West Virginia University

US★China Energy Center
中美能源中心

# China's National Plan for Coal Industry

- ❖ Restrict production to raise prices:
  - Close older and lower efficiency mines
  - Restrict development of new mines
  - Regulate the production of existing mines
  - Restrict or stop use of low quality coal
- ❖ Promote coal conversion and clean coal technology development to improve market and reduce pollution
- ❖ Reform and restructure coal companies
- ❖ Promote coal companies to diversify their business model

11

West Virginia University

US★China Energy Center
中美能源中心

# 4. Future of Coal

**U.S.**

- ❖ Coal remains fuel for power generation, but market share will continue shrinking
- ❖ EPA's new regulations will promote plant closings and cause coal demand to decrease
- ❖ Shale gas development reduces coal use
- ❖ Less costly CCS and CCUS technology will reduce climate change concerns over coal use

**China**

- ❖ Coal remains principal energy source, 4.2 BMt/y coal consumption, 62% of total energy by 2020
- ❖ Current coal business downturn may continue for a short period
- ❖ Clean coal use will be key:
  - Coal conversion
  - Advanced combustion
  - CCUS

12

West Virginia University

US★China Energy Center
中美能源中心

# Major CCS Demonstration Projects in US

## Project Locations & Cost Share

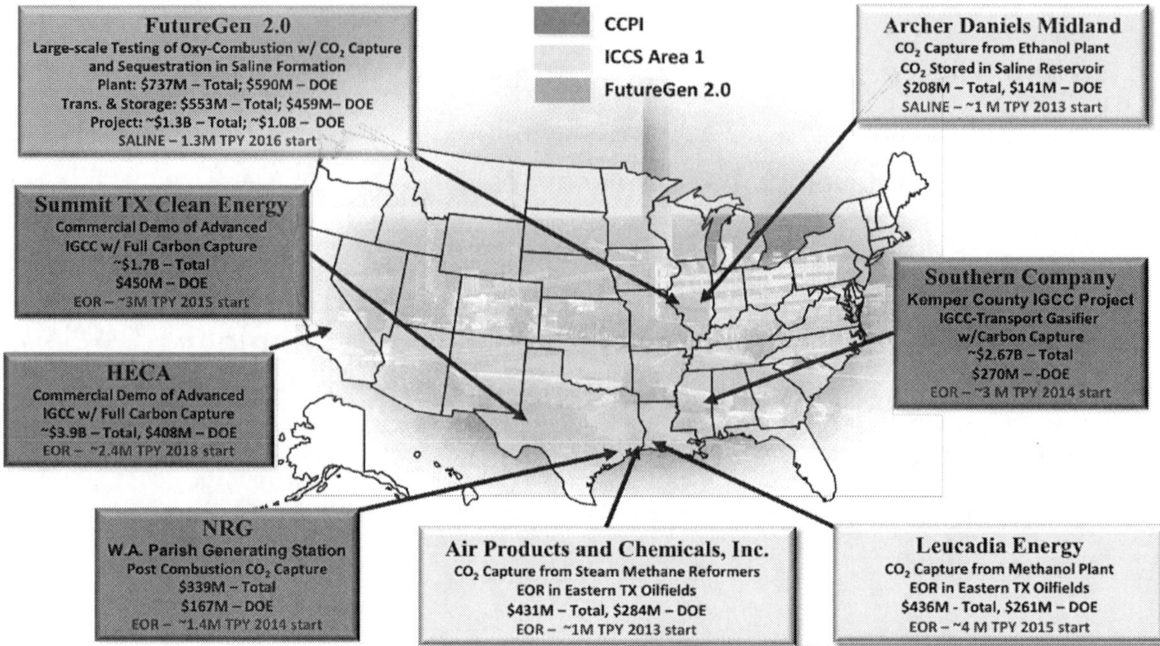


Ref: Office of Fossil Energy, US DOE

West Virginia University

US★China Energy Center
中美能源中心

# Clean Coal and CCUS in China

- Development of a series of CTL and CTO plants
- Advanced power generation (achieve emissions lower than NG power plant [except CO2])
- CCUS development
  - Shenhua DCL and CCS demonstrations
  - Huaneng GreenGen demonstration
  - Huaneng Shidongkou CO2 capture
  - 4 MMt/y CCUS and 300km CO2 pipeline project by Yanchang

Huaneng GreenGen Plant

Shenhua Direct Coal Liquefaction and CCS Demonstration

US★China Energy Center
中美能源中心

# Clean Coal Conversion Development in China

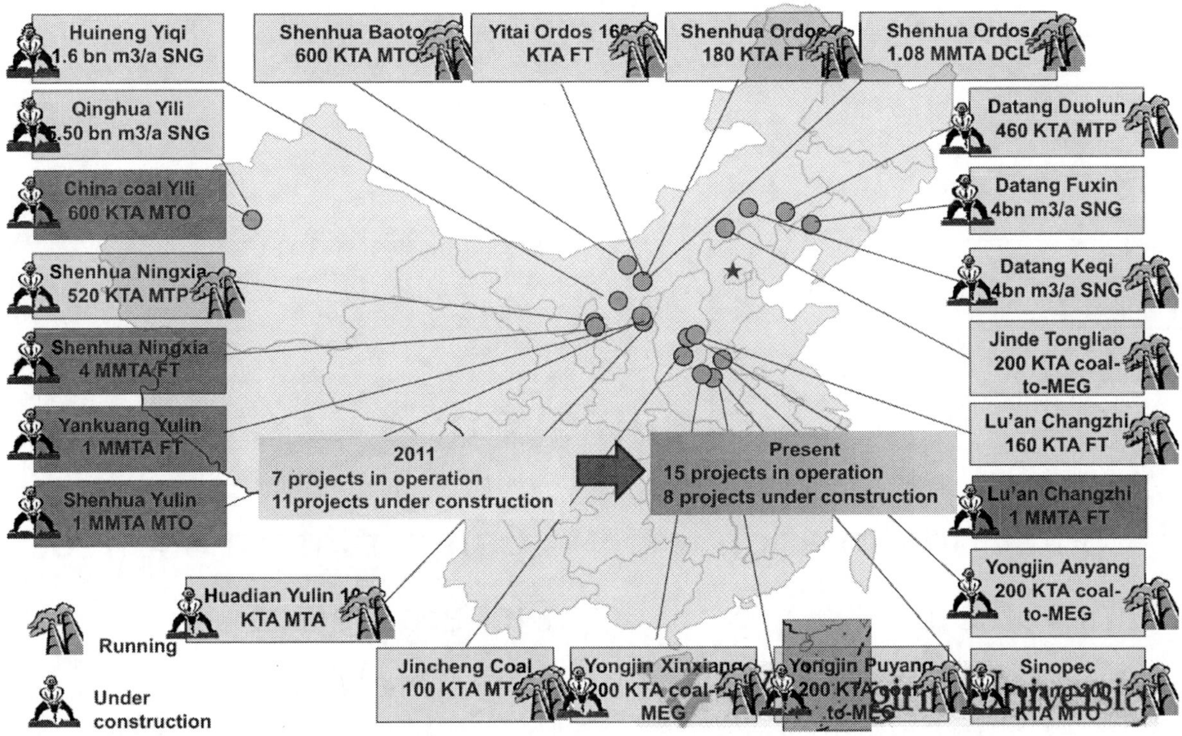


US★China Energy Center
中美能源中心

# Challenges requiring U.S. - China Collaboration on Energy Issues

- Being the two largest energy consumers and CO2 emitters, the U.S. and China are responsible for leadership in emissions reductions and health protection
- As two of largest coal producers and consumers, the U.S. and China can collaborate to help their respective coal industries
- Many mechanisms exist for the U.S. and China to collaborate on coal, energy, and environment:
  - US-China Clean Energy Research Center (CERC) — Advanced Coal Technology Consortium (ACTC)
  - US-China Fossil Energy Protocol signed by DOE and MOST/NEA
  - US China Climate Change Work Group (CCWG)
  - Asia-Pacific Partnership on Clean Development and Climate (APEC)
  - Carbon Sequestration Leadership Forum (CSLF)
  - Coal State Collaboration: West Virginia and Shanxi Strategical Partnership on Energy and Economics

16 West Virginia University

US★China Energy Center
中美能源中心

# Thank you

Contact Information for:

Qingyun Sun

- Associate Director, US-China Energy Center
- Qingyun.Sun@mail.wvu.edu
- (1) 304 293 5545

Contact Information for:

Richard Bajura

- Director, National Research Center for Coal and Energy
- Richard.Bajura@mail.wvu.edu
- (1) 304 293 6034

17

West Virginia University

# CARBON TAX IN SOUTH AFRICA: SUSTAINABILITY CHALLENGES

**L. Landu[1a], R Minnitt[2] and R Falcon[2]**

[1]*Mineral Economic and Strategy Unit (MESU), Council for Mineral Technology (MINTEK), Randburg, South Africa;*
[2] *The University of the Witwatersrand, Braamfontein, Johannesburg,South Africa*
[a]*Corresponding author: Tel.: +27 11709 4457; Email: landul@mintek.co.za*

**Keywords: sustainable development, carbon tax, challenges, mining, policy.**

**Abstract**

The study aims to review the sustainability challenges pertaining to the implementation of carbon tax in the South African mining industry. The primary objective of the South African government in imposing carbon tax is to reduce greenhouse gase emissions (GHG) and address the environmental dimension of sustainability. Amongst the policy' expectations are firstly pricing mechanisms on the environmental and economic damages caused by excessive emissions of GHG. Secondly, the policy aims to drive further investment in green technology projects thereby creating jobs and alleviating poverty in the country. The study will demonstrate that, despite government's good intention in reducing GHG emissions, its subsequent impacts on different sectors of the economy particularly the mining industry is still uncertain. Conversely, the business's view is that carbon tax will inflict destructive socio-economic implications across emitting industries including the mining industry. There is still uncertainty that carbon tax will achieve environmental objectives and succeed to put much emphasis on the prevention of further pollution to the environment in which businesses operate. The government put emphasis on the preservation of environment on one side, the industry on the other side prefers to salvage the socio-economic in lieu of environmental consideration.

The investment in renewable technologies expected to create jobs and alleviate poverty is yet to overcome challenges pertaining to the cost involved in repairing the environmental damages caused by the carbon emissions. Although carbon taxation was designed to promote environmental sustainability, however its implementation in the mining industry may not lead to economic and social sustainability. Sustainable development is an integrated theory structured in three dimensions namely the economic, social and environmental aspects. A policy instrument which is effective in one dimension but detrimental to the other dimensions may be an ineffective choice when considering sustainable development as a whole. If the carbon tax as an instrument is to be considered effective it has to be sustainable from an environmental, economic and social perspectives. In this study, an overview of sustainable development and carbon tax challenges is presented from literature data gathered through books, journals, scientific papers, academic conferences and other relevant publication channels. The literature review will be applied as an approach to develop the topic in an integrated way such that new perspectives and ways of addressing sustainable development and carbon tax challenges would be generated. A three-pronged approach encompassing environmental, economic and social review will be used in discussing and validating the sustainability challenges of carbon tax on South African mining industry. In this paper, the economic challenges that limit sustainability of carbon tax will also be reviewed.

## 1. Introduction

For decades, sustainable development theory has been used to promote the socio-economic and environmental paradigm. In lieu of a carbon emissions target, the literature often focuses on the environmental performance of the carbon tax without looking at the socio-economic performance of the tax. Morris (2013) indicates that from an economic perspective, policymakers should set the price of carbon that is equal to the present value of the environmental and socio-economic damages produced by each additional ton of $CO_2$ emissions. The sustainable development evidence in the current carbon tax design that causes the business society including the mining industry to anticipate more risks than opportunities is still limited. Figure 1 depicts sustainable development limits of carbon tax in the South African context.

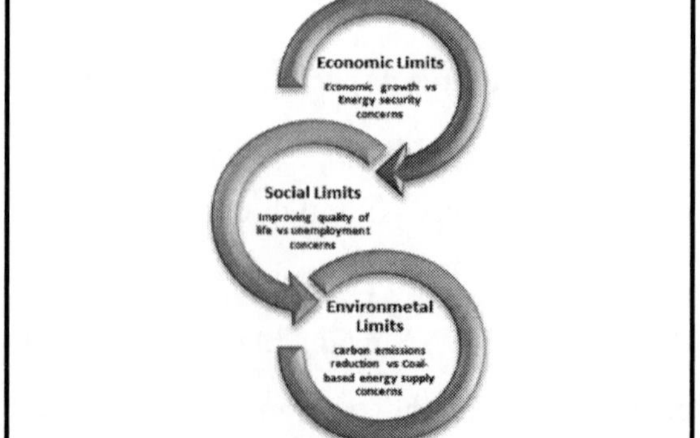


Figure 1: Sustainability Limits of Carbon Tax
Source: Mintek, 2016

The research that demonstrates sustainable development evidence of the carbon tax as an adequate supporting mechanism seems lacking in the global market. Environmentally friendly policy implementation has been examined for years, but new approaches towards its positive socio-economic impacts are continuously researched. Furthermore, the empirical model that considers the carbon tax as the most efficient fiscal solutions for promoting sustainable development is needed. There is a widespread literature in the areas of ecology, and environmental conservation (Böhringer *et al.*, 2009, Jackson, 2010, Golusin *et al.*, 2011), while studies that can establish scientific based relation between ecology, management and socio-economy are still limited. In this regard, most of the scientific studies point to positive environmental impacts of carbon tax without considering in-depth link between socio-economic growth and the environment. Whereas successful environmental policy implementation has the primary goal by influencing not only the environmental conservation policies, but also the economy and the society.

The United Nations Framework Convention on Climate Change (1992) acknowledges that among other important issues that all countries and especially developing countries need to achieve is mostly sustainable social and economic development, changing their orientation towards policies that will only achieve GHG emissions abatement. The carbon policy usually internalises the environmental damage costs of carbon emissions. Following this paradigmatic challenge, this research focuses on the importance of assessing the impacts of carbon tax on the mining industry with the view of incorporating sustainable development approach. The focus is mainly on conceptual issues pertaining to the limited evidence on the economic effects of carbon tax using sustainable development approach. Specifically, this paper evaluates challenges of achieving economical sustainability of carbon taxation. The social and environmental goals of the carbon tax to achieve sustainable development are not covered under the scope of this paper.

## 2. Carbon Tax in South Africa: Current Status

Considering the significance of curbing carbon emissions and anticipating the sustainability benefits that a low carbon based-economy can bring to the country, the South African government has ambitious goals and commitment to achieve GHG emissions reductions of 34% by 2020 and 42% by 2025 (National Treasury, 2013). In response to this commitment, government released the Carbon Tax Bill with the objective of providing clarity to a myriad of issues raised by the different business sectors. The business fraternity is of the primary view that the carbon tax has been designed as a way to increase the tax base in order to growth the economy, while the goal communicated by government is to send the necessary price signals in order to change consumer behaviours as a way of stimulating investor's appetite and local industry to shift towards low carbon technology trajectory (National Ttreasury, 2013). Although, the national government's commitment to revenue recycling measures expected to assist the industry in the transition to the low carbon economy is still uncertain, different role-players in the industry are of the view that the effects on economic growth will be somewhat minimal (Mammatt *et al*, 2011). Establishing a dynamic, sustainable and reliable carbon tax is politically not easy. Policy uncertainty and indecisiveness seems likely to be the rule rather than the exception.Table 1 provide the key sectors of the South African economy viewed as potential casualties of the upcoming carbon tax.The sectors listed below anticiticpate more risks than opportunities that would result to the loss of their competitive advantage in the global market.

Table 1: Key sectors of the South African economy

| Key sectors | Example of companies |
|---|---|
| Energy | Eskom |
| Energy | SASOL coal |
| Mining | Anglo American |
| Cement | Larfarge SA and AfriSam |
| Iron and Steel | Kumba Iron Ore Ltd |
| Manufacturing/Automotive circle | Toyota |
| Paper and pulp | Sappi Ltd |
| Liquid fuels (oil) | SASOL Oil |

Source: Mintek, 2016

Since its inception, the consultation process conducted over the past 5 years, starting from the 2010 discussion paper on carbon tax, the 2013 Carbon Tax Policy Paper to the 2014 Carbon Offsets Paper has been dragging and challenging. Although its valuable contribution to the country's economy, the concept of carbon tax has yet to receive support from a wide range of stakeholders including the mining industry. Figure 2 summarises the legislative evolution of South Africa's carbon tax and shows that the policy was afflicted by uncertainty long before it get to this phase where is being currently debated by different stakeholders at different levels.

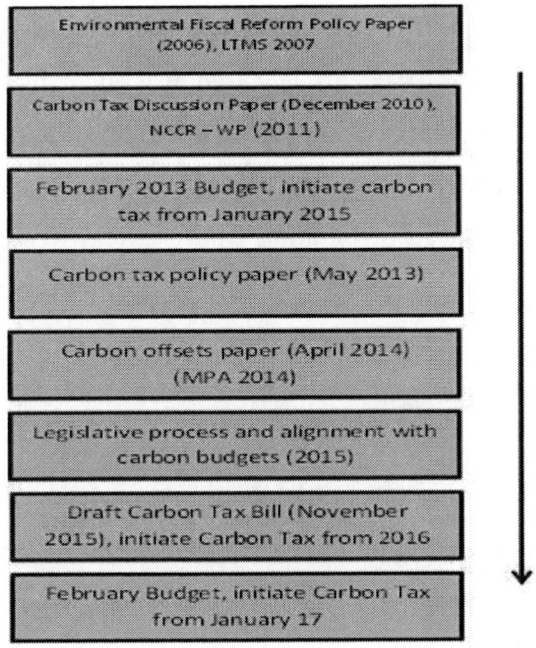


Figure 1: Summary of the Legislative History of Carbon Tax
Source: Adapted from DNA Economics, 2016.

In 2006, the environmental fiscal reform policy paper gave effect to the carbon tax discussion paper in 2010 and introduced it to the public and to businesses including the mining industry. In addition, the paper was exposed to a number of significant changes and comments almost immediately after it came into effect, decreasing the level of certainty in the industry. For example, amongst the received comments were that there was no guidance on the auditing requirements or standards that may be used in the accounting of emissions on which tax will be payable (Promethium, 2013). This, indeed was already a significant issue that directly contradicted one of the original justifications for the tax.

Immediately after the December 2010 carbon tax discussion paper for public comments, came the Carbon Tax Policy paper with the aim to facilitate the transition to a Green Economy released on 2 May 2013. This paper updated the 2010 paper, and became a Carbon Tax Policy Paper. To this effect, the National Treasury planned for the carbon tax to come into effect in early 2014 at a rate of R120 per ton of $CO_2$ on direct emissions expected to increase at 10% per year in the course of the first execution phase of the policy (National Treasury, 2010; National Treasury, 2013).

After the above postponement, the National Treasury initiated the carbon offsets in April 2014 (National Treasury, 2014). The carbon offsets paper was introduced with the aim to complement the policy package that address climate change adversities and protect business including the mining industry. This is a trading system that allows companies, especially emitters to use their carbon offsets in order to mitigate the tax liability of GHG emissions and achieve their carbon budget. However, being amongst the first system of this kind, it was highly suggested that instead of building from scratch, it was recommended to learn from countries that have been successful in the usage of this scheme and experience gained internally in the carbon trading space (Louw *et al,* 2014). The emitters will be able to reduce their carbon tax liability by up to 10% of their actual emissions.

Following these suggestions, the 2014 budget speech by the Minister of Finance again suggested the execution of carbon tax in early January 2015 with the same rate as mentioned above. After all comments have been received in writing from various platforms formed by a wide range of stakeholders including business, Non-Government Organisations, academia, civil society and labour, the national government agreed to initiate an alignment of the design and proposed reduction of GHG emissions. In his speech budget of 2014, the Minister of Finance clearly stated: "To allow for an alignment with desired emissions reduction outcomes (DEROs) being developed by Department of Environmental Affairs and Tourism (DEAT), the implementation of carbon tax was postponed by a year to 2016". This process has been called the legislative process and alignment with carbon budget. Subsequent to this alignment and to allow enough time for consultation of draft legislation, the implementation of carbon tax was delayed to 2016. Despite the continuous postponement of the carbon tax implementation, the popular support for the policy has been fragile except the strong belief from the national government. In the meantime, on the 2$^{nd}$ November 2015, the Draft Carbon Tax Bill was released with an objective to provide another platform set for more comments from various stakeholders. Furthermore, these comments pertain to the design and also some technical and administrative aspects which are still unclear.

The expected comments from the stakeholders consist of environmental and socio-economic impacts of carbon tax including the legal phrasing of the Bill (National Treasury, 2015). In this regard, a revised Bill will then be submitted to the Cabinet of the Minister of Finance for approval prior to tabling it in to Parliament. The Draft Regulations on the carbon offset in terms of the draft carbon tax bill has now been released for public comments ending on 29 July 2016. This document has been developed jointly by the National Treasury, the Department of Energy and the Department of Environmental Affairs relative to Sections 13 and 20 *(b)* of the Draft Carbon Tax Bill and sets out the procedure for the use of carbon offsets by emitters to reduce their carbon tax liability. The legislation that gives effect to this draft regulations has not yet been tabled into Parliament, therefore it is still uncertain whether such draft regulations would actually pass. As a result, there is still a great deal of uncertainty surrounding the future status of the carbon tax. After the compilation and finalisation including the approval of all the above drafts by the national government, the implementation is still delayed to early next year 2017.

### 3. Economic Challenge to the Sustainability of Carbon Tax

In South Africa, proponents of carbon tax have long argued that the economic implications of carbon tax will help the country reduce its emissions while it achieves its normal growth (Winkler *et al*, 2011). The transition to low carbon economy would involve significant structural changes, however, such investment would be in renewable energy and other various clean energy technologies. Liu *et al* (2013) articulated that the implementation of carbon tax in the mining industry would increase energy efficiency practices, promote clean energy sources and process innovation. In addition, the December 2010 carbon tax paper noted that potential advantages are to be achieved from improved energy efficiency and probably the carbon storage technologies. These are actions that could be taken by the country's leading emitters to achieve the goals and commitment of GHG emissions reductions of 34% by 2020 and 42% by 2025 (National Treasury, 2013). The government's view is that these actions would move the country towards an economy that supports an international and agreed target of limiting a global temperature rise to below 2 degree Celsius (UNFCCC, 2015).

The economic aspect of the carbon tax is all about practicalities. The needed economic transformation and transition to low carbon trajectory in the coming years meaning by 2025 is just enormous. Despite the alerting signs of the rapidly emerging carbon tax and climate change adversities, the implementation of low carbon technologies is progressively taking place ,but at a relatively slow pace. The scientists argue that the trajectory towards low carbon technology is feasible and needed in order to reduce GHG emissions and keep the world safe for future generations (Rogelj *et al*, 2016).

Policy-makers would like to balance the economic benefits of the mining industry with the need for the industry to significantly reduce its $CO_2$ emissions in the long-term. Observers from the business industry argue that this intention is probable but not feasible in ten years. The reason is that considering future expansion projects in Asia including projected growth of automobile industry, gold and platinum production in South Africa is forecast to increase implying an increase of energy usage over the years in a medium-term period. Assuming that the additional minerals production would be mined and processed at the same energy intensity as the current production, this would imply additional energy consumption. In this case, the $CO_2$ intensity of energy production processes would be expected to increase. In the same way, diesel and other carbon related sources of energy would increase in coal mining and supply activities. The electricity production in the country is more than 50% carbon intensive, hence the carbon emissions intensity is significantly higher (DEA, 2010). The emissions attributable to electricity represent a larger portion than any other emissions generated from other sources. Until the carbon tax incorporates tangible strategies to mitigate potential impacts on the trade competiveness of the mining industry, the coal, gold and platinum mines may become the biggest casualties of the looming carbon tax.

Furthermore, should the South Africa's gross annual coal exports grow in the years to come, the export volumes and overheads would be expected to increase as well. This would result to a competitive disadvantage for South African iron ore, aluminium and coal producers relative to competitors in China, Brazil and Indonesia (Deloitte, 2016). Consequently, the continuous hope of the South African government which is the introduction of the carbon tax in order to free the country from its dependence on fossil fuels and slowdown the effects on climate change could be affected. In light of the slow pace and cost pertaining to the introduction of energy efficiency technologies including innovation of renewable energy supplies, the achievement geared towards reducing the excessive usage of natural resources becomes unlikely.

Very often and in most of carbon tax policies existing, observers are of the view that the implementing reality regarding ring-fencing of carbon tax expected to be reinvested in order to develop the green economy is still clouded with uncertainty. The fact remains that current available renewable energy technologies are more expensive than the efficient fossil fuel usage and to some extent nuclear technologies. In the South African context, the real increase in electricity prices due to the usage of renewables would mean a further increase of approximately 20% in real electricity prices (AGRISA, 2016). The benefits of renewables have been exaggerated and their external costs in many cases have not been accurately assessed (Stark, *et al*, 2015). In the same order, the existing literatures have found that most energy taxes policies aim to stimulate the use of renewables. However, often the empirical analysis that test support mechanisms for promoting the use of renewables is still lacking (Dombrovski, 2015). It is important to point out that the real economic cost of implementing the renewable energy alternatives has been variable throughout the years as the result of price variability of inputs in the global market.

## 4. Conclusions

The first carbon tax proposal initiated in 1920 by a well-known economist Pigots alluded that a sustainable carbon tax should be revenue neutral, protect the impoverished, the poor, and which allows business and consumers time to adapt, plan, and reinvent processes. Pigots's view is that a sustainable carbon tax should be better than direct action because carbon can be used as a single economic lever capable of propelling all environmental and social impacts into one reform measure for the economy, allowing business capital to be directed towards clean technology. Emissions reductions at the scale envisioned by the national government would necessitate a structural transformation of the economy. However, there are barriers and concerns that can prevent the smooth implementration process of the tax in the country. The challenge is not only to design and implement an effective carbon tax, but also to strike a balance between socio-economic and environmental goals. This balanace can translate to the energy, development, acceptable and achievable emissions reduction targets.

The South African National Treasury designed carbon tax with the primary intention to promote environmental sustainability objective. However, its implementation appeared to be inimical to economic and social sustainability. For this reason, the business society including the mining industry considered as a backbone of the national economy has been sceptical rather than embracing this policy. Unless the carbon tax system clearly demonstrates that environmental goals are not achievable without socio-economic dimensions, its implementation is deemed to encounter more resistance from various fronts of the society.

The introduction of carbon taxation will not yield result if more focus is on environmental sustainability but decreased socio-economic sustainability. Sustainable development is an integrated concept organised in three dimensions namely the economic, social and environmental. A policy instrument which is effective in one dimension but detrimental to another may be considered as ineffective policy choice. Indeed, one of the reasons many developing countries have failed to fully and successfully implement carbon tax is because of its limited sustainability evidence in general and regressive economic effect particularly on the business society including the mining industry. Carbon tax is a commendable instrument and a significant move in the fight against climate change. If the instrument is to be considered effective it must prove to be sustainable from an environmental, economic and social perspective. Furthermore, the instrument should be seen as an opportunity by both sides referring to the decision-makers and the business society as a whole. The effectiveness of the instrument would be measured if both sides see lesser risks by embracing this policy. In addition, sustainable development effectiveness of this instrument would have to be triangulation-based meaning three sides coming together as one to fight one cause which is GHG emissions. The business society including the mining industry needs legislation and regulation compliance to control their GHG emissions. In the same way the government has the responsibility to create healthy and conducive environment for the business society to operate and contribute to socio-economic development of the country. Finally, developing an evidence-based study relating to sustainable development evidence of carbon tax in order to support the carbon tax implementation process and the transition to a green economy is vital.

**References**

AGRISA (2016). AGRISA comments on Eskom application for tariff increase: Regulatory clearing account application – MYPD3 Year 1 (2013/14). Communication to the National Energy Regulator of South Africa, 15 January 2016.
Available at: http://www.growinggreatness.co.za/wp-content/uploads/2016/01/Agri-SA-comments-on-Eskoms-application-for-tariff-increase-2016.pdf. Accessed on 20 May 2016.

Böhringer, C., Mennel, T. P., & Rutherford, T. F. (2009). Technological change and uncertainty in environmental economics. Energy Economics, 31(S1), S1-S3.

Deloitte (2016). Tracking the trends 2016: the top 10 issues mining companies will face in the coming year.

DNA Economics (2016). NBI Public Seminar. Seminar hosted by the National Business Initatives, February 2016.

Dombrovski, R (2015). The Effectiveness of financial and fiscal instruments for promoting sustainable renewable energy technologies. Doctoral Dissertation, University of Ljubljana, Faculty of Economics, Republic of Slovenia. Available at: http://www.cek.ef.uni-lj.si/doktor/dombrovski.pdf. Accessed on 23 May 2016.

Golusin, M., Munitlak-Ivanovic, O., & Teodorovic, N. (2011). The review of the achieved degree of sustainable development in South Eastern Europe – The use of linear regression method. Renewable and Sustainable Energy Reviews, 15(1), 766-772.

Jackson, J. (2010). Promoting Energy Efficiency investments with risk management decision tools. Energy Policy, 38(8), 3865-3873.

Liu, B., & Espinola –Arredondo, A (2013). The Impacts of Environmental Taxes on Firms'Technology and Entry-decisions. Washingtn Sate University Working Paper No. 2, WEashington, USA.

Morris, Adele C. 2013. "The Many Benefits of a Carbon Tax." Washington, DC: Hamilton Project, Brookings Institution. Available at: http://www.brookings.edu/research/papers/2013/02/benefits-of-carbon-tax. Accessed on 11 June 2016.

Mammatt, J., Watt van der, C., Mandy, K., Erasmus, L and Issacs, K (2011). The Case against introducing carbon tax in South Africa. PricewaterCoopers Inc, South Africa.
Robbie Louw, R., Swanepoel, SJ and Immink, H (2014). Potential Alignment between the South African Carbon Offset Trading Scheme and the Carbon Budget Approach. *Promethium, South Africa, September 2014.*

Rogelj, J., Elzen den, M., Hohne, N., Fransen, T., Fekete, H., Winkler, H., Schaeffer, R., Sha, F., Riahi, K and Meinshausen, M (2016). Paris Agreement climate proposals need a boost to keep warming well below 2 °C. *Nature International Weekly Journal of science.*

Robbie Louw, R and Immink, H (2012). Initial Framework for Carbon Offset Opportunities and Verification Options. *Promethium Carbon, South Africa, September 2012.*

Stark, C., Pless, J., Logan, J., Zhou, E and Arent, J (2015). Renewab;e Electrticity: Insights for the Coming Decade. Joint Institute for Strategic Energy Analysis, Technical Report, NREL/TP – 6A50 – 63604, February 2015.

The South African Department of Environmetal (2010). National Climate Change Response Green Paper 2010. DEAT, Notice 1083 of 2010, South Africa.

The South African Department of National Treasury, (2010). Discussion Paper for Public Comment Reducing Greenhouse Gas Emissions: Carbon Tax Option.

The South African Department of National Treasury, (2013). Carbon Tax Policy Paper: Reducing GHG and facilitating the transition to a green economy. Policy paper for public comment.

The South African Department of National Treasury (2014). Carbon Offset Paper. Carbon Offset Paper, April 2014, South Africa.

The South African Department of Treasury (2015). Draft Carbon Tax Bill for public comment. National Treasury, South Africa, 2nd November 2015.

UNFCCC (1992). the Declaration of the United Nations Conference on the Human Environment, adopted at Stockholm on 16 June 1972. United Nations Framework Convention on Climate Change, United Nations 1992.

Winkler, H and Marquard, A (2011). Analysis of Economic Implications of a carbon tax. *Journal of Energy in Southern Africa vol 22 no. 1, Energy Research Center, University of Cape Town.*

Zhu, X., Yujie Lu., and Cui, Q, 2010. Efficiency and Equity Implications of Carbon Tax in the Construction Industry. Engineering Project Organizations Conference (EPOS), Working Paper Proceedings, South Lake Tahoe, CA November 4-7, 2010. Available at: https://elibrary.utas.edu.au/utas/file/fe845ac5-d8cd-4b32-a369 8b5566fa93e6/1/A%20Hutchinson-Ch1-8.pdf. acceesed on 16 June 2016.

THE Art OF ACCOMPLISHMENT
The Application of Coal Quality Database over Space and Time Focussing on Sulphur, Mpumalanga Coalfields
By: Papi Kekana
Supervisor: Prof. N. Wagner
UNIVERSITY OF JOHANNESBURG

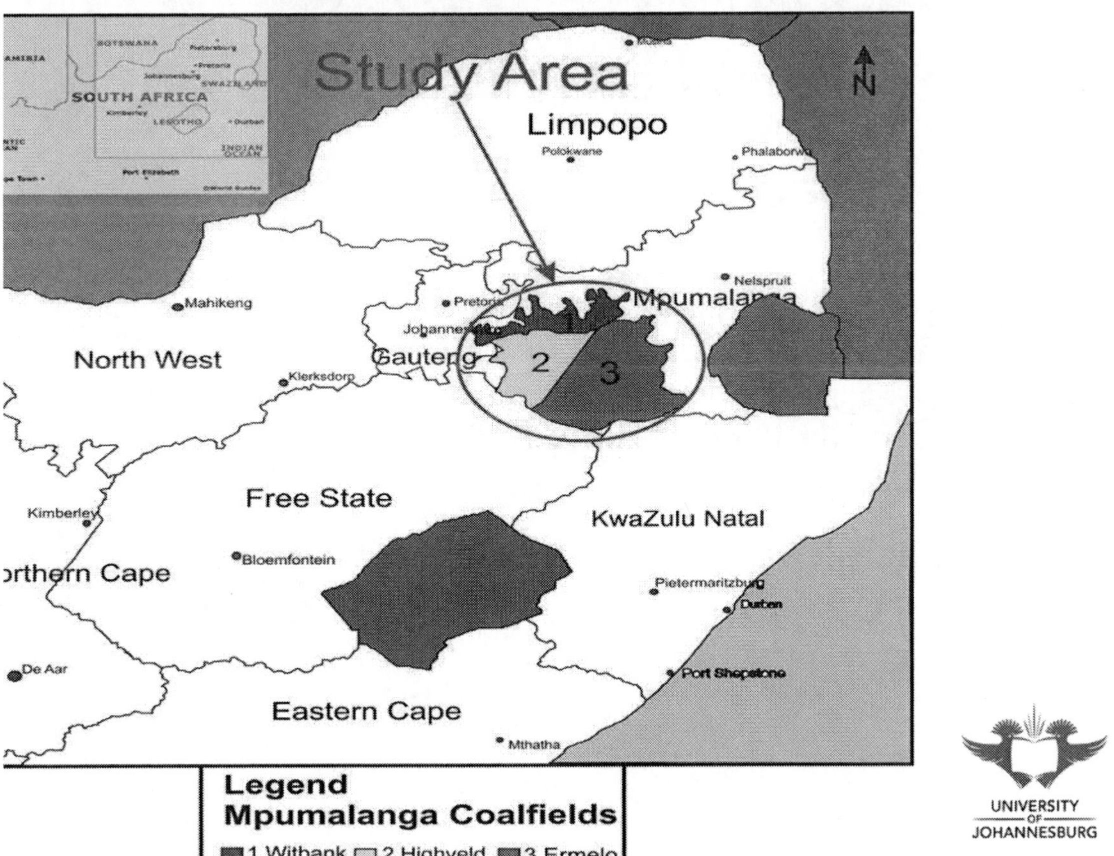
Study Area
N
SOUTH AFRICA
Limpopo
Polokwane
Phalaborwa
Nelspruit
Mpumalanga
Mahikeng
Pretoria
North West
Gauteng
Klerksdorp
1
2
3
Free State
KwaZulu Natal
Kimberley
Bloemfontein
Pietermaritzburg
Durban
Port Shepstone
De Aar
Eastern Cape
Mthatha
Legend
Mpumalanga Coalfields
1 Witbank
2 Highveld
3 Ermelo
UNIVERSITY OF JOHANNESBURG

## Purpose of project

- Consolidate data into a single source.
- Demonstrate coal quality changes over time.
- Explore the influence of changing environmental issues and regulations.
- Use geostatistical methods to demonstrate the changes over time and space.

UNIVERSITY OF JOHANNESBURG

## Presentation Outline

- Data collection
- Data consolidation
- Geostatistical analysis
- Conclusion

UNIVERSITY OF JOHANNESBURG

# Data Collection

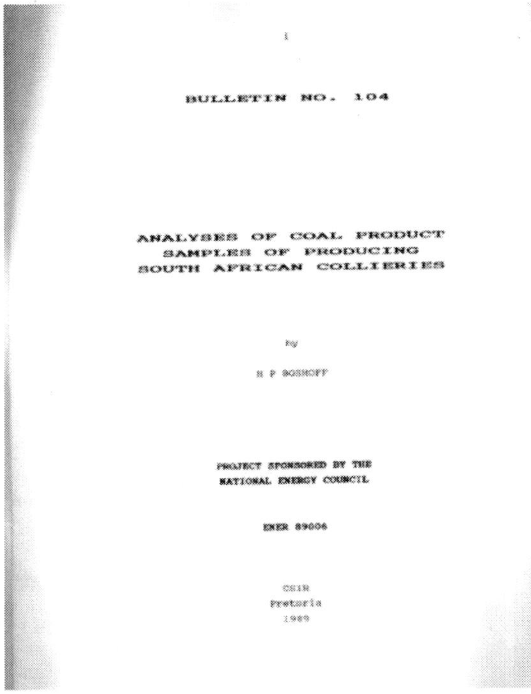

1

BULLETIN NO. 104

ANALYSES OF COAL PRODUCT
SAMPLES OF PRODUCING
SOUTH AFRICAN COLLIERIES

by

H P BOSHOFF

PROJECT SPONSORED BY THE
NATIONAL ENERGY COUNCIL

ENER 89006

CSIR
Pretoria
1989

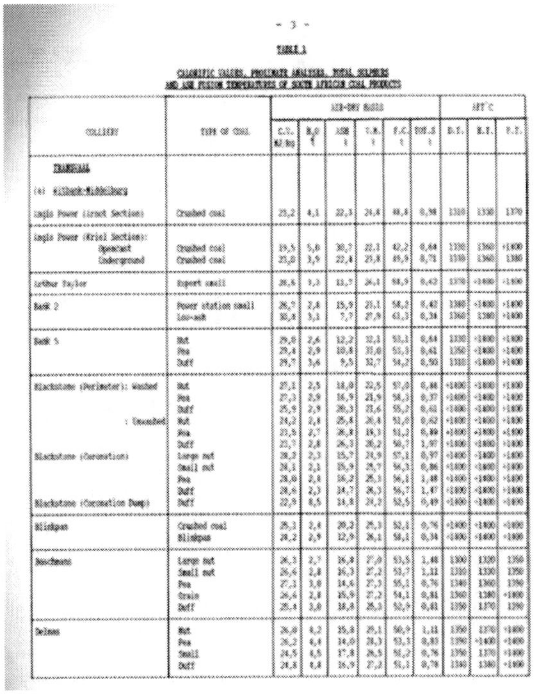

# Database Consolidation

| | 1980 | | | | 1981 | | | | 1982 | | | | 1983 | | | | 1984 | | | | 1985 | | | |
|---|---|---|---|---|---|---|---|---|---|---|---|---|---|---|---|---|---|---|---|---|---|---|---|---|
| | C.V. | Ash | Vols. | T. S. | C.V. | Ash | Vols. | T. S. | C.V. | Ash | Vols. | T. S. | C.V. | Ash | Vols. | T. S. | C.V. | Ash | Vols. | T. S. | C.V. | Ash | Vols. | T. S. |
| Collieries | 44 | 44 | 44 | 44 | 44 | 44 | 44 | 44 | 44 | 44 | 44 | 44 | 44 | 44 | 44 | 44 | 44 | 44 | 44 | 44 | 44 | 44 | 44 | 44 |
| Average | 25.50 | 18.74 | 23.47 | 1.00 | 24.71 | 19.46 | 24.67 | 1.01 | 25.49 | 17.50 | 25.27 | 1.00 | 25.16 | 18.43 | 25.77 | 1.14 | 25.54 | 17.76 | 26.05 | 1.15 | 24.28 | 21.26 | 24.98 | 1.18 |
| Maximum | 28.82 | 39.30 | 35.80 | 2.58 | 29.20 | 33.80 | 36.40 | 2.53 | 30.60 | 33.10 | 36.00 | 2.24 | 29.10 | 27.20 | 37.10 | 3.72 | 30.60 | 27.20 | 34.00 | 3.16 | 30.16 | 37.80 | 35.10 | 2.93 |
| Minimun | 19.60 | 10.50 | 12.50 | 0.34 | 20.31 | 6.40 | 7.40 | 0.29 | 19.30 | 5.20 | 7.70 | 0.31 | 21.36 | 7.20 | 7.50 | 0.42 | 21.30 | 5.30 | 8.50 | 0.20 | 17.69 | 10.40 | 18.30 | 0.20 |
| Mode | 26.4 | 13.4 | 23.5 | 0.50 | 22.7 | 25.6 | 21.9 | 1.01 | 25.8 | 17.5 | 27.9 | 0.86 | 21.96 | 24.6 | 21.3 | 0.90 | 24.7 | 17.5 | 24.9 | 0.86 | 28.1 | 17.5 | 26.8 | 0.62 |
| Skewness | -0.61 | 1.44 | -0.08 | 1.06 | 0.05 | 0.08 | -0.40 | 1.03 | -0.42 | 0.37 | -0.65 | 1.04 | -0.068 | -0.35 | -0.46 | 2.22 | 0.23 | -0.01 | -1.13 | 1.03 | 0.07 | 0.42 | 0.76 | 0.94 |
| Kurtosis | -0.66 | 3.46 | 0.71 | 0.48 | -1.03 | -0.34 | 0.95 | 0.91 | 0.13 | 1.03 | 2.18 | 1.05 | -0.958 | -0.09 | 2.85 | 8.23 | -0.25 | 0.51 | 5.45 | 1.62 | -0.27 | -0.19 | 0.23 | 0.25 |
| Std Dev. | 2.43 | 5.94 | 4.72 | 0.58 | 2.28 | 5.90 | 5.76 | 0.51 | 2.43 | 5.54 | 5.08 | 0.45 | 2.105 | 4.38 | 5.06 | 0.57 | 2.11 | 4.51 | 4.26 | 0.59 | 2.69 | 6.21 | 3.95 | 0.70 |

- TCOA, FRI, CSIR, CGS bulletins
- Last published in 2002
- For time periods: [1986-1990],[1991-1996],[1997-2000],
  [2001-2005],[2006-2010],[2011-2015]

UNIVERSITY OF JOHANNESBURG

## Geostatistical Analysis

- **Micromine**
- **Surfer**
- Histograms
- Scattergrams
- Variography
- Contouring

UNIVERSITY OF JOHANNESBURG

## Total Sulphur (%) (1980 to1985) No.2 Seam Witbank (peas & nuts)

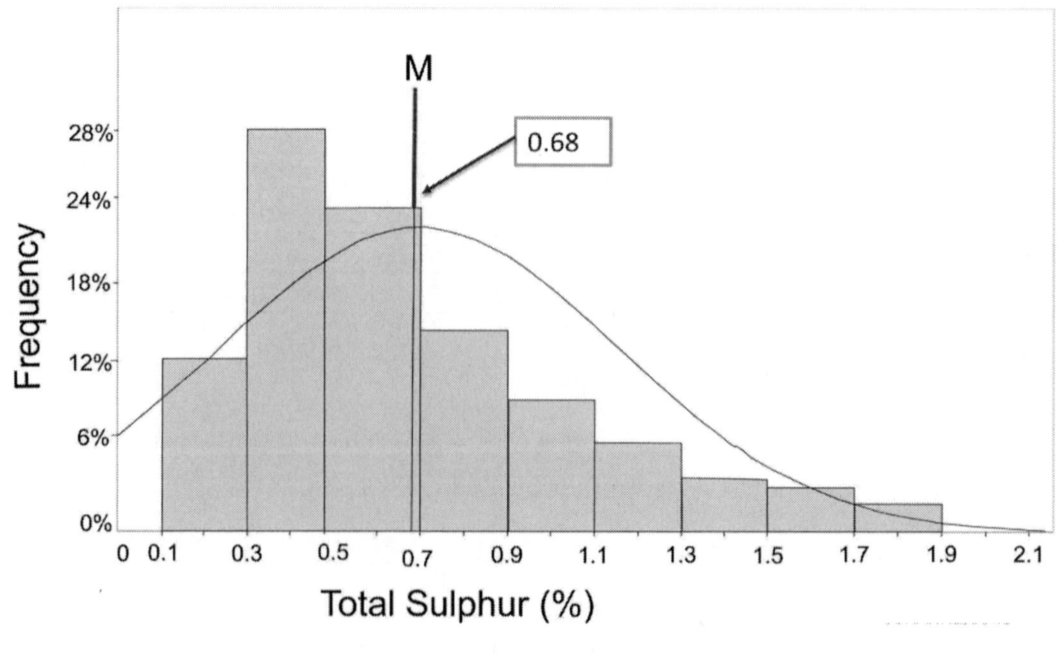

## Total Sulphur (%) (1986 to 1990) No.2 Seam Witbank (peas & nuts)

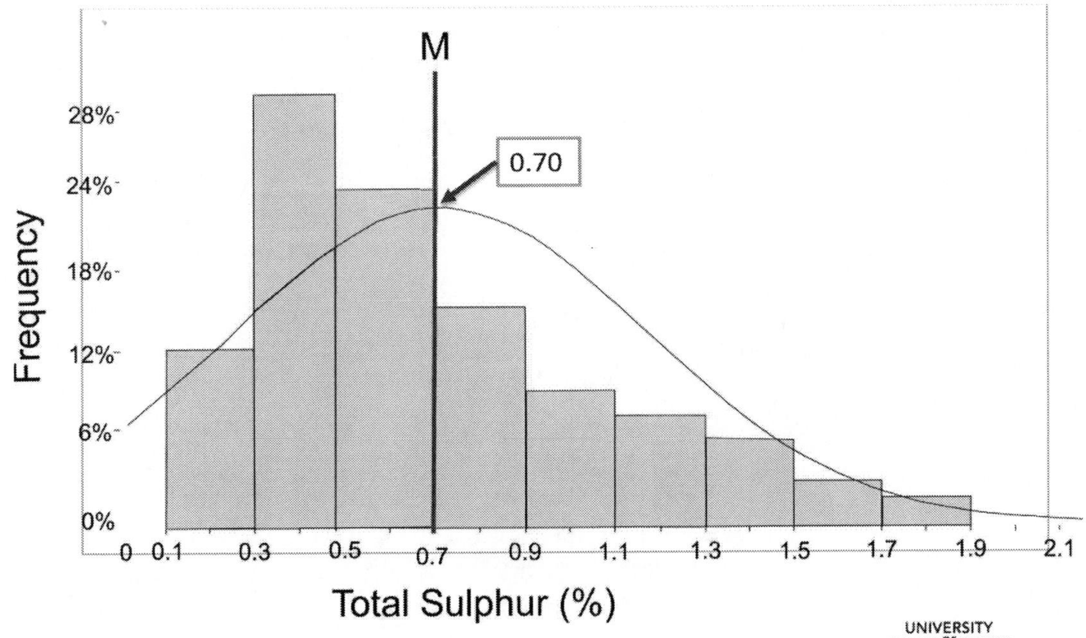


UNIVERSITY OF JOHANNESBURG

## Total Sulphur (%) No.2 Seam 2011 to 2015 (peas & nuts)

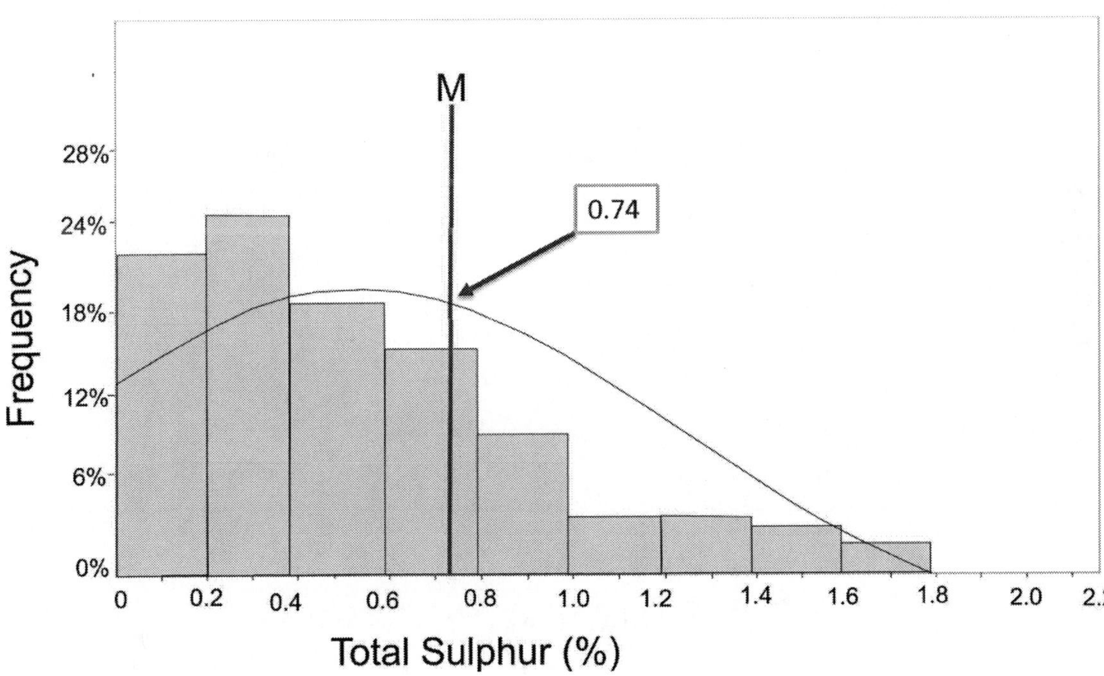

# Total Sulphur (%) Witbank

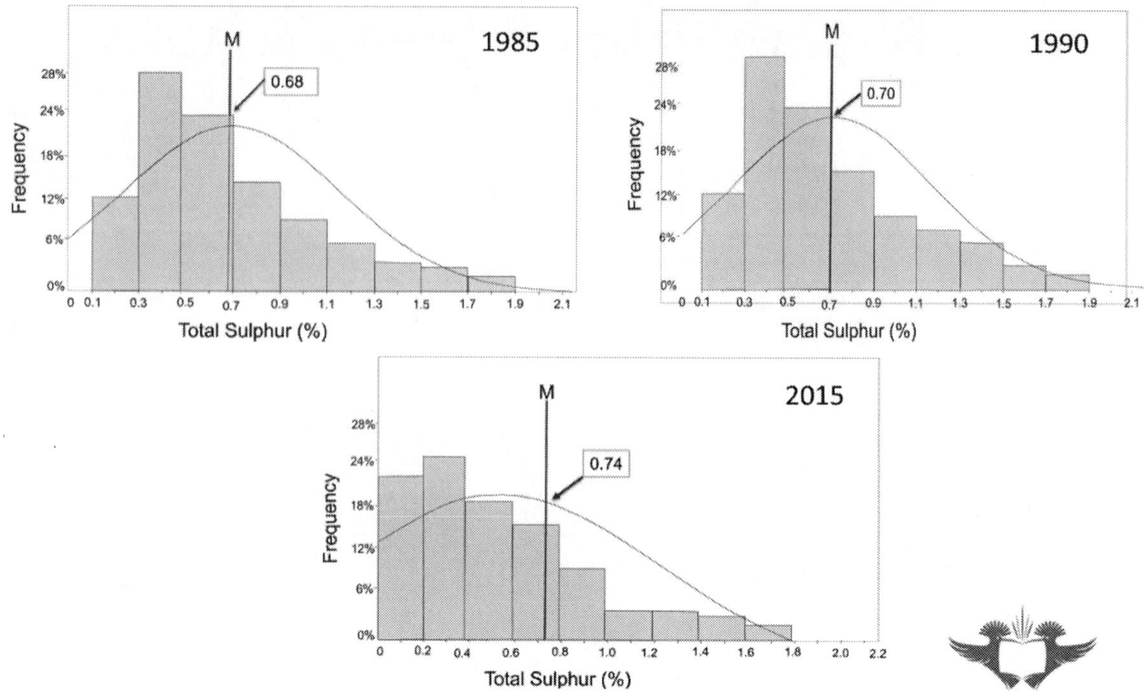


UNIVERSITY OF JOHANNESBURG

# Total Sulphur (%) Witbank 1980 to 2015

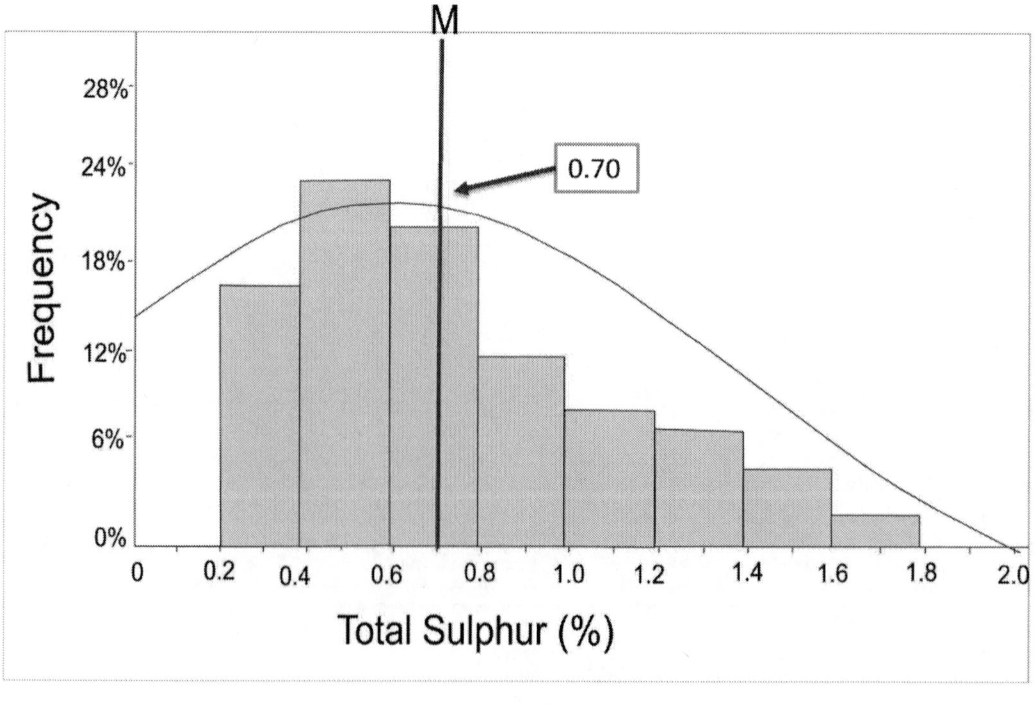

## Total Sulphur (%) Highveld 1980 to 2015

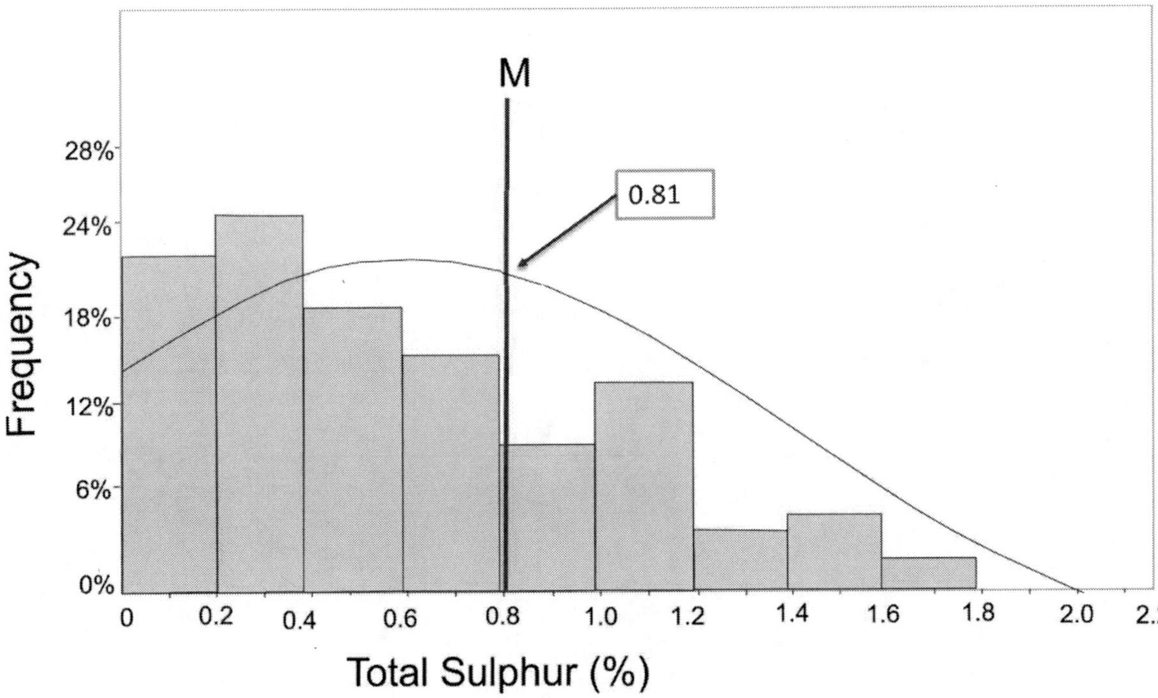


## Total Sulphur (%)Ermelo 1980 to 2015

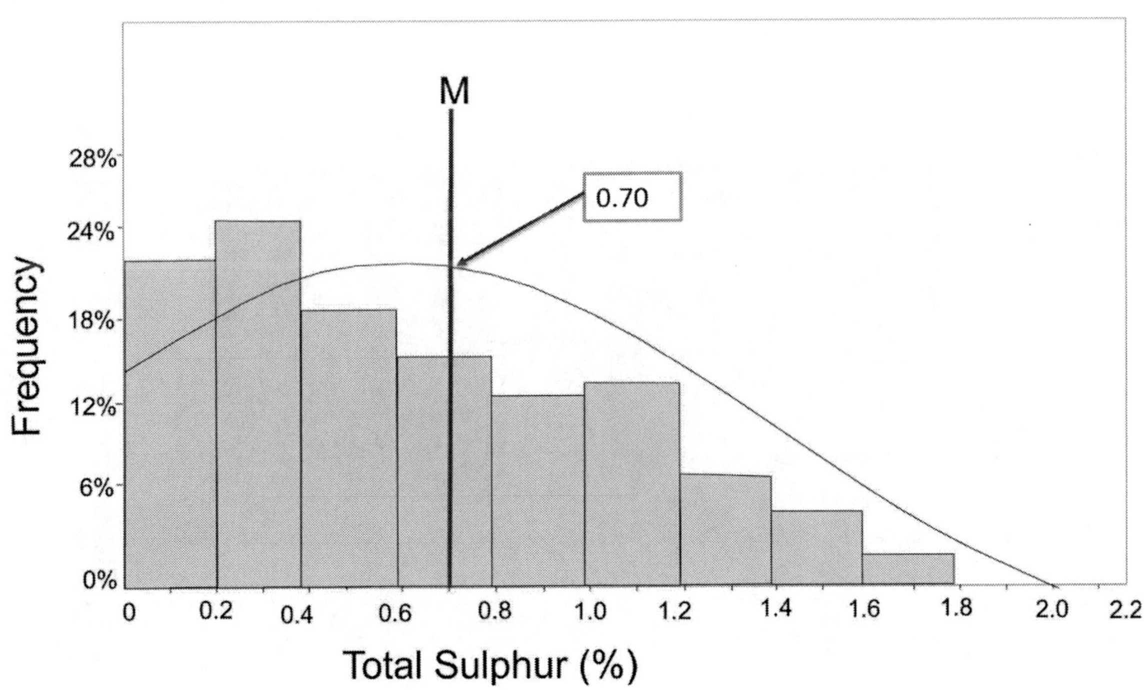

## Total Sulphur (%) Sample localities No.2 Seam

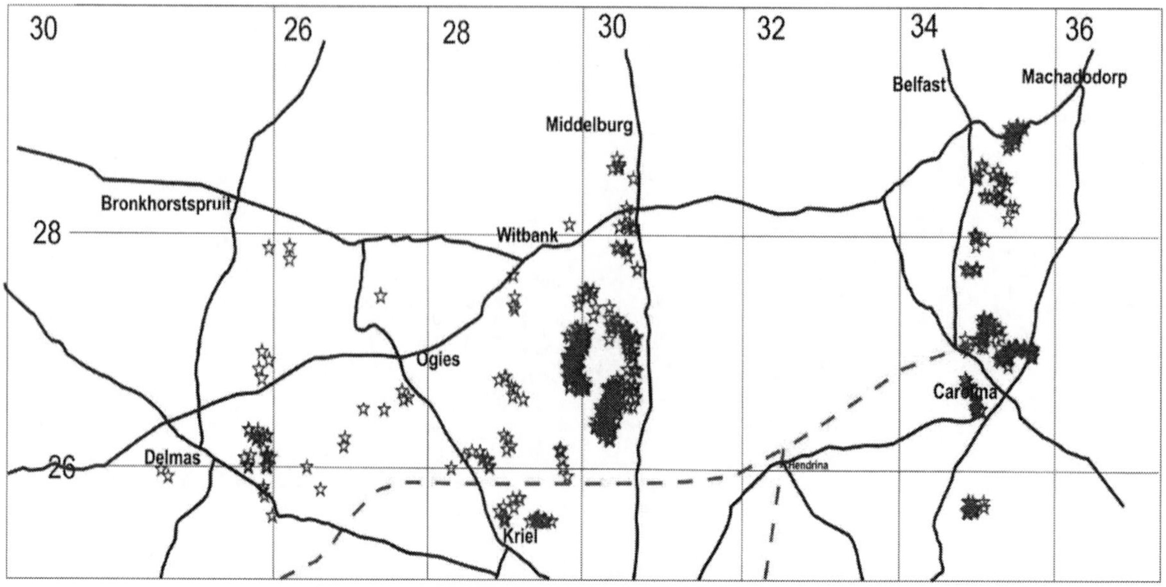


## Total Sulphur (%) No.2 Seam (1980 to 1985)

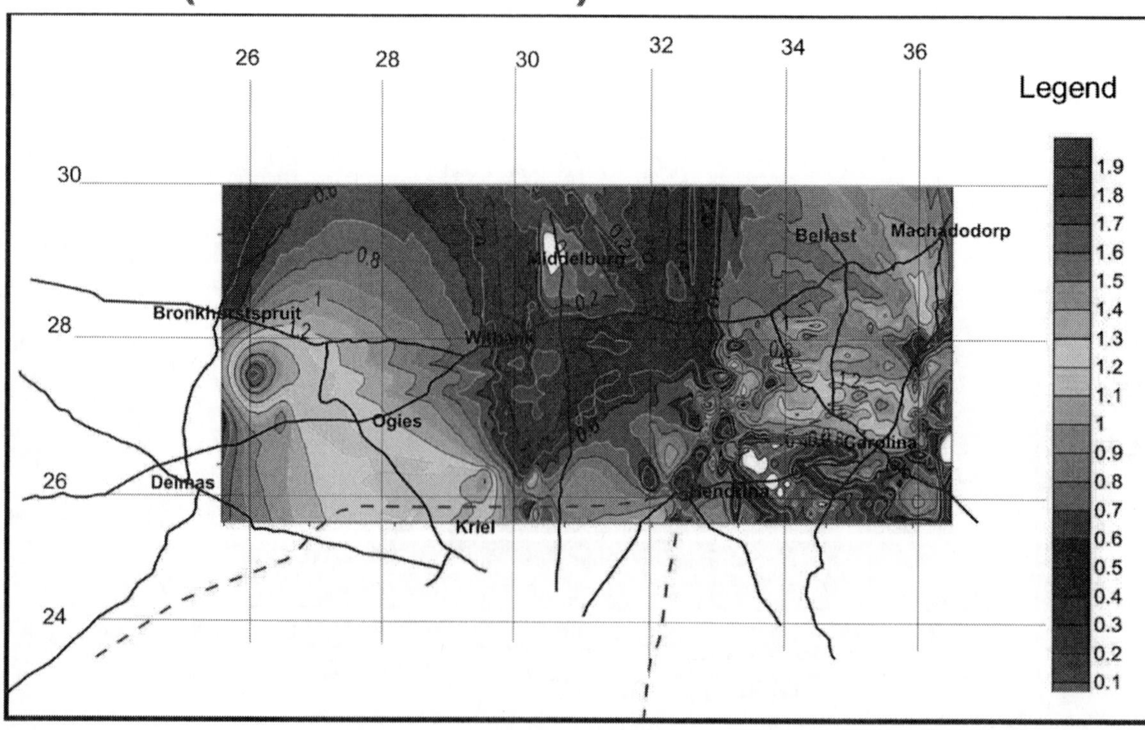

## Total Sulphur (%) No.2 Seam (1986 to 1990)

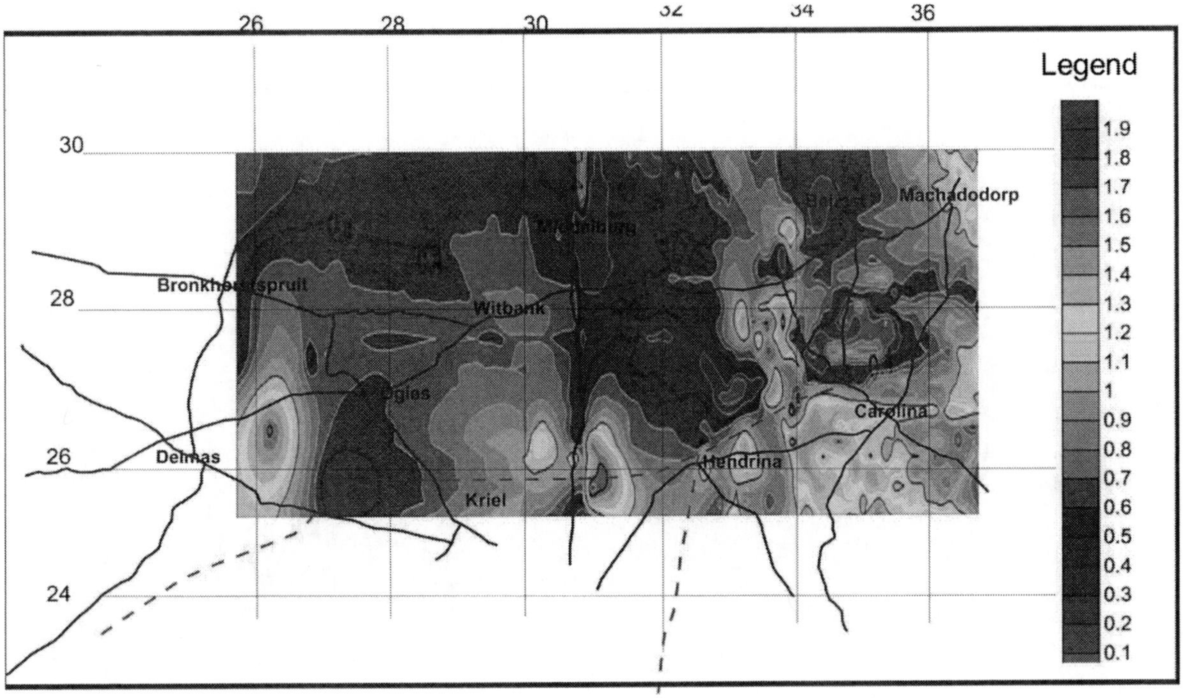


## Total Sulphur (%) No.2 Seam (1996 to 2000)

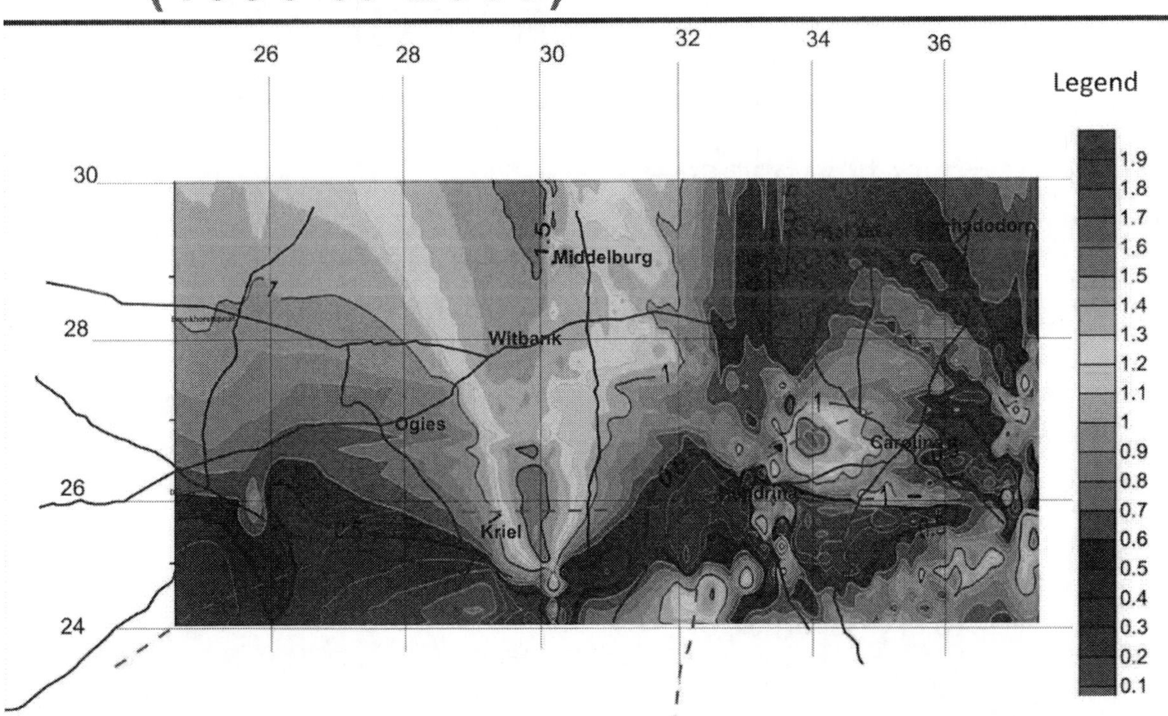

# Total Sulphur (%) No.2 Seam (2011 to 2015)

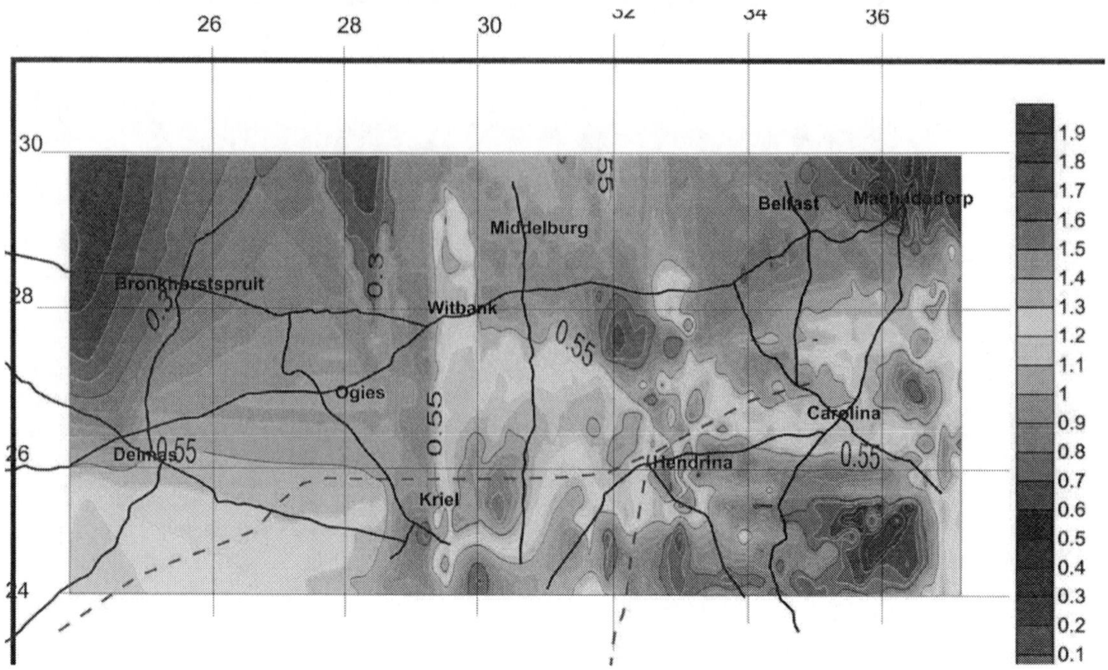


# Conclusion

- Use of Micromine and Surfer software packages helped in demonstrating changes over time and space.
- Witbank area shows a slight increase in sulphur values (1980-2015).
- The increase show coal mining will continue to have a negative impact on the environment.
- Market acceptance of poorer quality coal due to high demand.

UNIVERSITY OF JOHANNESBURG

# Benefits to Industry

- Extracted scientific data from historical databases could be used in decision making by small scale miners.
- A quick reference for one reliable source can be used for future research.
- Promote easy access to information for the general public.

UNIVERSITY OF JOHANNESBURG

# Aknowledgements


- **Coaltech and CIMERA for sponsoring the project.**
- **Council for Geosciences for providing data and assistance.**
- **Gail Gordon for providing data and assistance at the Coal library.**

COALTECH RESEARCH ASSOCIATION NPC

cimera DST-NRF Centre of Excellence for Integrated Mineral and Energy Resource Analysis

Council for Geoscience
Applied Geoscience Solutions

NRF National Research Foundation


UNIVERSITY OF JOHANNESBURG

# Thank You

TECHNISCHE UNIVERSITÄT
BERGAKADEMIE FREIBERG
Die Ressourcenuniversität. Seit 1765.

Zentrum für Innovationskompetenz:
Virtual High Temperature Conversion
Virtuhcon

# KINETIC STUDIES FOR $CO_2$ GASIFICATION OF GERMAN COKE USING THERMOGRAVIMETRIC ANALYZER AT HIGH PRESSURE

**Ziad Abosteif, Stefan Guhl**

**Pittsburgh Coal Conference PCC 2016.**

TU Bergakademie Freiberg - Institute of Energy Process Engineering and Chemical Engineering - 09596 Freiberg - Germany-Tel. +49 3731 394272 - Fax +49 3731 394555
Email Ziad.Abosteif@iec.tu-freiberg.de - Web www.iec.tu-freiberg.de

## Outline

Virtuhcon

1. Objective
2. Introduction
3. Material used
4. Procedure and discussion
   - Set 1:Partial pressure influence
   - Set 2: Absolute pressure influence
   - Set 3: Absolute pressure influence under constant gas velocity
5. Other factors influenced by pressure
6. Summary and outlook

# 1- Objectives and motivation

- Determination of gasfication kinetic parameters for different coals in framework of different projects.

→ Developing kinetic models for coal gasification for later modelling steps (CFD-Modelling for process optimisation and development) .

➢ To find answers :
- Which influence has the partial pressure on the kinetic parameters.
- Which influence has the absolute pressure on the kinetic parameters. Why there are contradicting statements in the litreature.
- Which role plays the volume flow / gas velocity in the gasification experiments using TGA.

# 2- Introduction

## Thermogravimetric analysis

-Measurement of mass loss as a function of the temperature or time dm/dt=f(t,T) Isothermal dm/dt=f(t)
-Most frequently used in research and development.
-Fully automated equipment was used*

Affecting factors**:
-crucible geometry
-heat tracing
-sample pretreatment HOK
-heating rate 20 K/min until T=const
-initial weight 50 mg

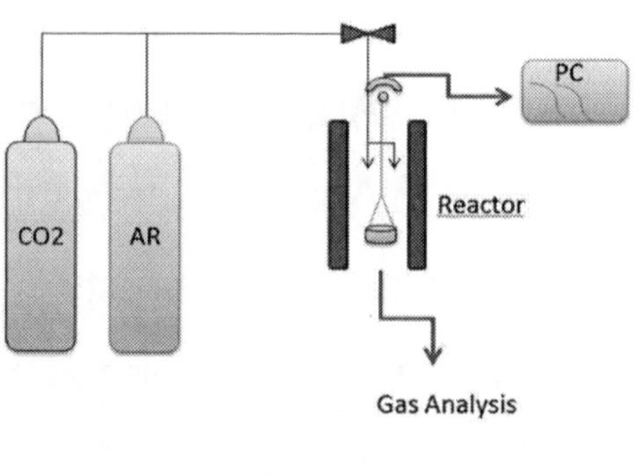


* Rubotherm™
**H.C Gottfried et al. Praxis der Thermischen Analyse von Kunststoffen. 2. Auflage 2003

## 3- Material

Virtuhcon

### Coke Composition

| proximate analysis | | |
|---|---|---|
| moisture | wt.-% (ar) | 0.99 |
| ash | wt.-% (d) | 8.48 |
| volatiles | | 3.96 |
| $C_{fix}$ | | 87.54 |
| Σ | | 100 |
| **ultimate analysis** | | |
| ash | wt.-% (d) | 8.48 |
| C | | 89.82 |
| H | | 0.35 |
| N | | 0.33 |
| $S_C$ | | 0.13 |
| O | | 0.86 |
| Σ | | 100 |

## 3- Material

Virtuhcon

### Particle size distribution

Coke particle size 63 – 200 µm

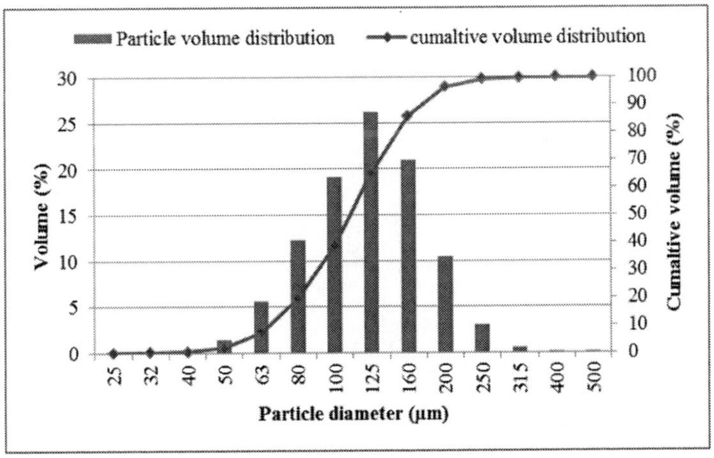

# 3- Material

## Coke properties

| Properties | values | remarks |
|---|---|---|
| Porosity ε | 0.54 | measured |
| initial specific area $S_0$ ( $m^2/g$) | 299.16 | BET analysis |
| initial pore diameter $d_0$ (nm) | 1.2 | Mean value estimated from BET analysis |
| raw density ($g/cm^3$) | 1.0097 | measured |
| true density ($g/cm^3$) | 2.1965 | measured |

*$N_2$ adsorption at 77 K (Micrometrics ASAP 2020) by BET method

# 4-Procedure

Classical evaluation method:

Instantaneous reaction rate:

$$r'' = \frac{\Delta m_C}{\overline{m_C} \cdot \tau}$$

Potential approach

$$r'' = k \cdot p_{CO2}^n$$

logarithmic calculus (determine n)

$$lnr'' = \ln k + n \cdot \ln P$$

Arrhenius-approach

$$k(T) = A \cdot e^{-\frac{E_A}{R \cdot T}}$$

logarithmic calculus

$$\ln k = \ln A - \frac{E_A}{R} \cdot \frac{1}{T}$$

# 4- Procedure

Virtuhcon

## Set 1: partial pressure influence

-Main reaction: C + $CO_2 \rightarrow$ 2CO. $\Delta_R H°$=+159.7 $kJmol^{-1}$
-Coke particle size 63 – 200 µm
-50 mg of sample → "Monolayer"
-Gasflow is a constant 380 ml/min STP during all experiments.
-Total pressure in all experiments 1 bar

| $y_{CO2}$ ( vol%) | T(°C) | | | | | |
|---|---|---|---|---|---|---|
| | 800 | 850 | 900 | 950 | 1000 | 1100 |
| 10 | * | * | * | * | * | * |
| 20 | * | * | * | * | * | * |
| 50 | * | * | * | * | * | * |
| 70 | * | * | * | * | * | * |

# 4- Discussion set 1

Virtuhcon

2. $CO_2$ concentrations influence

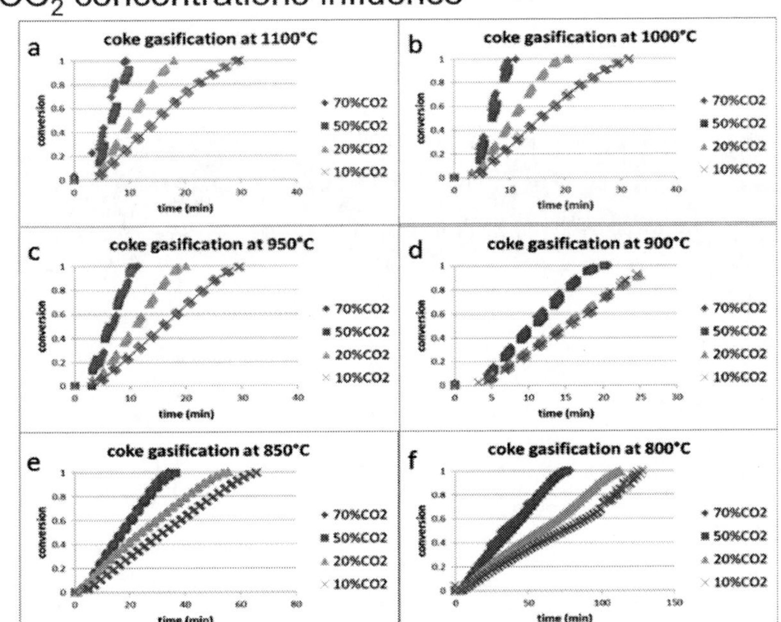


10

# 4-Discussion set 1

Virtuhcon

Activation energy $E_a$ and the pre-exponential factor $A_0$

$$\ln k = -\frac{E_a}{RT} + \ln A_0$$

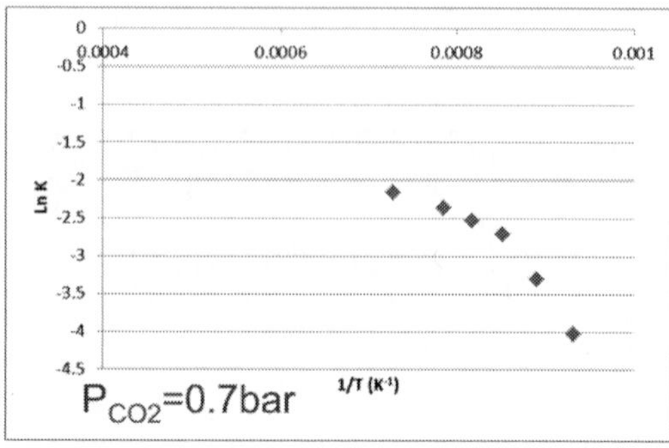


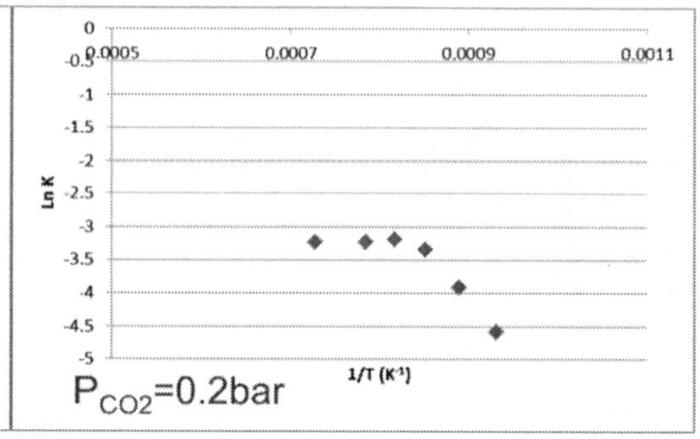


→ Shift of kinetic regime 1 toward higher temperatures

18

# 4- Procedure

Virtuhcon

**Set 2: Absolute pressure influence**

-Gasflow is a constant 260 ml/min STP during all experiments.

| T(°C)<br>$CO_2$(Vol%)<br>(tot.Pressure) | 800 | 850 | 900 | 950 | 1000 |
|---|---|---|---|---|---|
| 100 (1 bar) | * | * | * | * | * |
| 100 (5 bar) | * | * | * | * | * |
| 100 (10 bar) | * | * | * | * | * |

## 4- Disucussion set 2

### 2. Total pressure influence

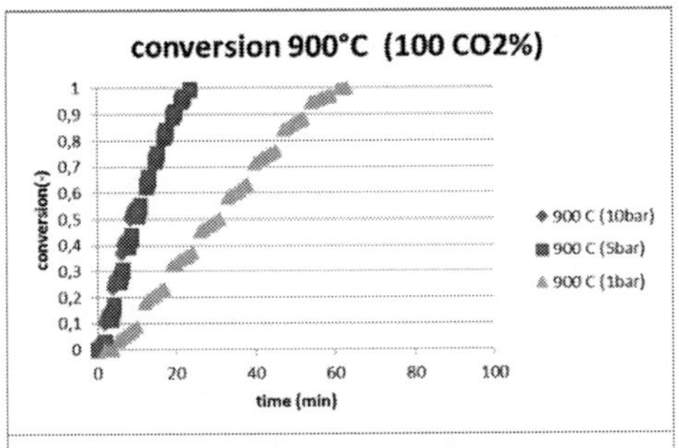


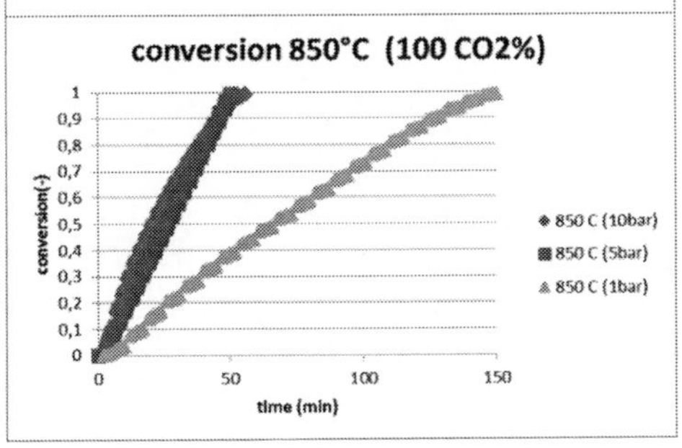


- P↑→ conversion faster till a certain limit.

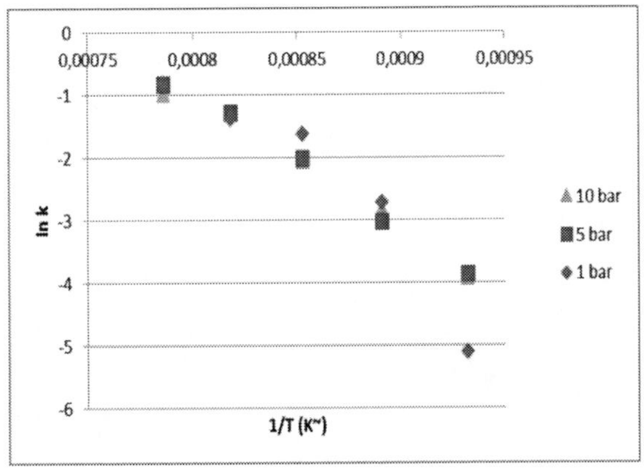


Decreasing flow influence ?

## 4- Procedure

**Set 3: Absolute pressure influence under constant gas velocity**

- **+ set 2 influence of gas velocity at const. pressure**

$V_{CO2\ max}$= 260 ml/min STP

| Total Pressure | Flow CO2 |
|---|---|
| 10 bar | Vmax |
| 5 bar | Vmax/2 |
| 1 bar | Vmax/10 |

| T(°C) / $CO_2$(Vol%) (tot.Pressure) | 800 | 850 | 900 | 950 | 1000 |
|---|---|---|---|---|---|
| 100 (1 bar) | * | * | * | * | * |
| 100 (5 bar) | * | * | * | * | * |
| 100 (10 bar) | * | * | * | * | * |

# 4- Discussion set 3

Virtuhcon

## 2. Total pressure influence

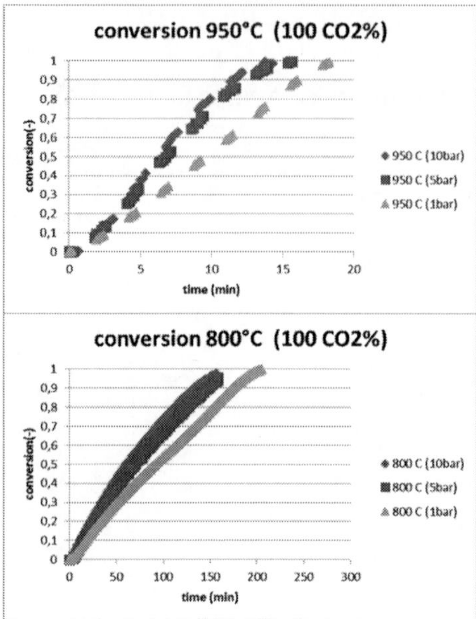


- Pure pressure influence is higher at constant gas velocity

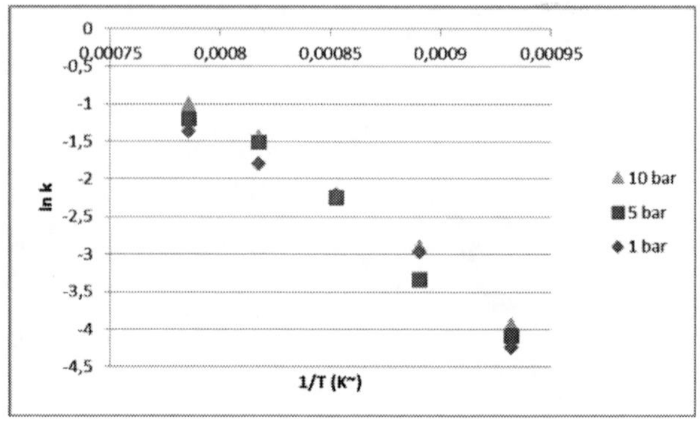


# 4- Discussion set 3

Virtuhcon

## Combination set 2+3

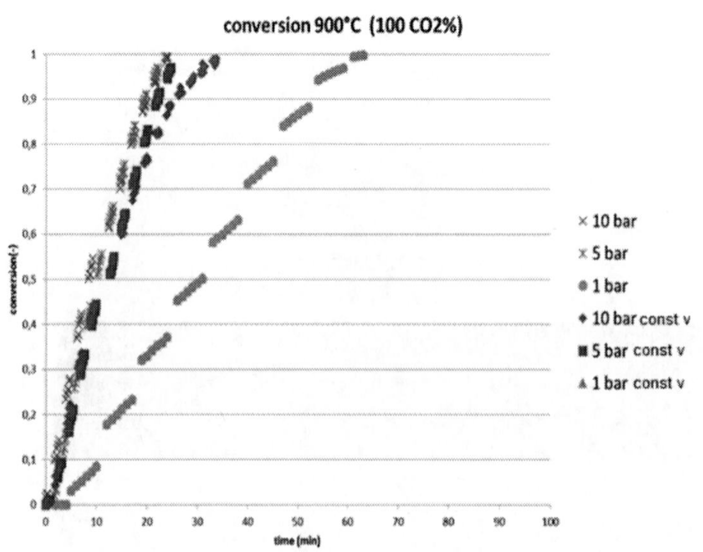


- $\vartheta_g \uparrow$ and P=const→ conversion higher (set2+3) at higher pressures
- $\vartheta_g$ enhance the concentration of the gas on the sample surface in TGA

# 4- Discussion

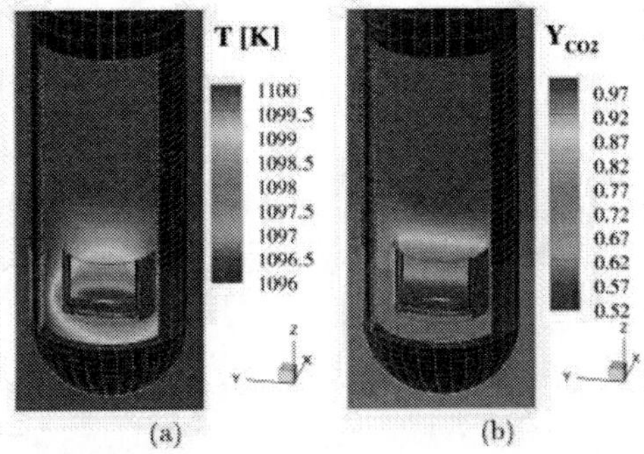


Figure 2: Contour plots of temperature (a) and $CO_2$ mass fraction (b) inside of the TGA crucible based on 3-d simulation with $T_G$= 1100 K

*S. Schulze, P. Nikrityuk, Z. Abosteif, S. Guhl, A. Richter, B. Meyer,* Heat and Mass Transfer within Thermogravimetric Analyser: From Simulation to Improved Estimation of Kinetic Data in Char Gasfication, Submitted to Fuel 2016

# 5-Negative influence of the pressure

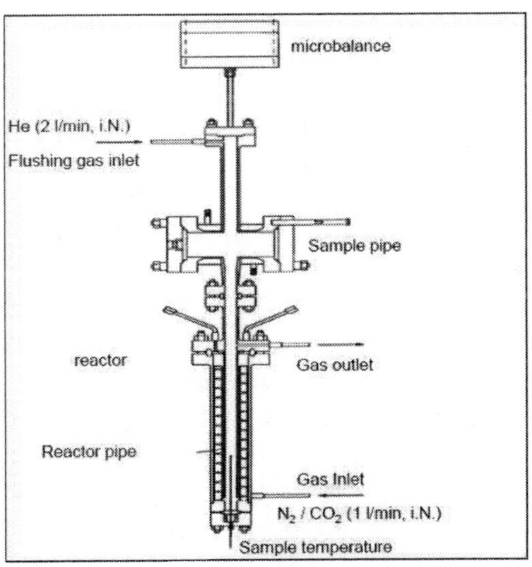


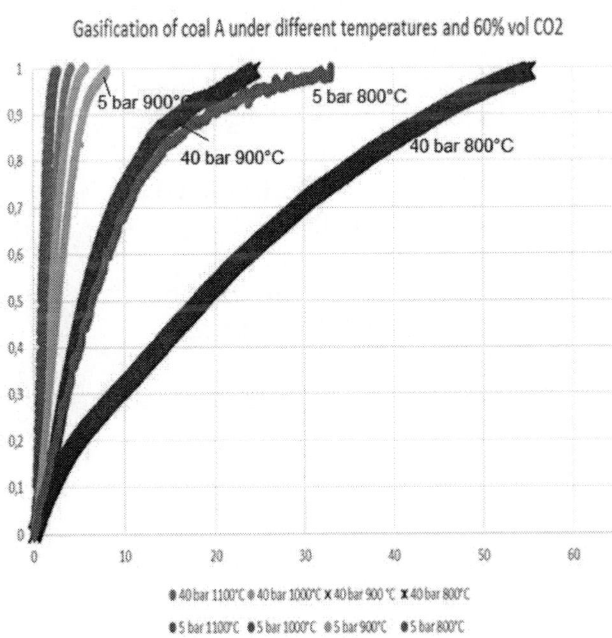

## 5-Other influences from high pressure

swelling/cacking influenced mass transfer, especially at high p in the presented case and lose the shape of sphere. Example coal pyrolysis at 40 bar „drop in“ mode

## 5- Summary and Outlook

-Using TGA for gasification experiments gives fast information about the material, but its application has some limitation;

-Higher partial pressure makes the conversion faster and shifts the kinetic regime toward higher temperature;

-Higher pressure enhances the rate of the conversion, this can be seen at constant gas velocity.

-Increasing the pressure has limitation from the equiment such as max. flow or from the sample such as caking coal ;

-The high pressure would increase the concentration on the surface of the samples.

-Outlook: Optimizing the concentration gradient on the sample surface by using crucible with frit (10 µm), to let the gas flow throw the crucible.

# Acknowledgment

Part of this research has been funded by the Federal Ministry of Education and Research of Germany in the framework of Virtuhcon (Project Number 03Z2FN12).

TU Bergakademie Freiberg
Institute of Energy Process Engineering and Chemical Engineering
09596 Freiberg - Germany
Tel. +493731-394272
Fax +493731-39 4555
Email Ziad.Abosteif@iec.tu-freiberg.de
Web www.iec.tu-freiberg.de

# Effect of particle size on gasification characteristics of different ranked chars under high temperature

**Xin Gong, Xiaopeng Zou,**
**Haifeng Lu, Qinghua Guo**
Institute of Clean Coal Technology
East China University of Science and Technology

# Contents

# Introduction

- Coal is a primary energy source in China.
- Entrained-flow gasification, the leading technology of high efficient and clean utilization of coal resource, has been widely applied in industry.

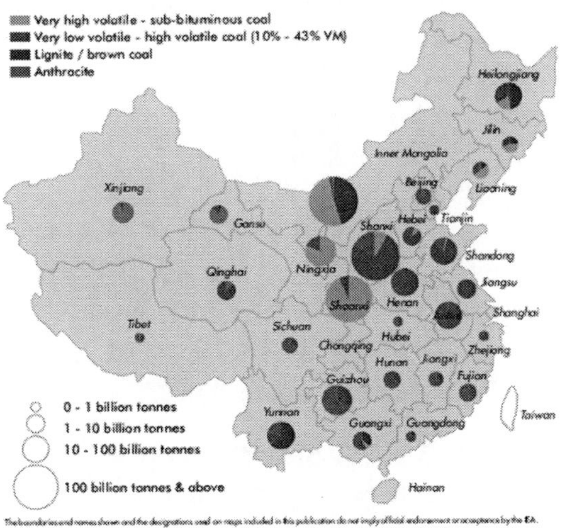


**Fig. 1 China's coal resource**

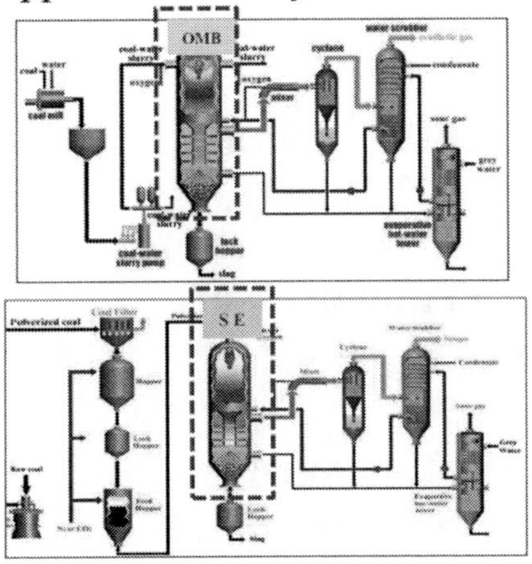


- **High temperature (1300-1600°C)**
- **Short residence time (~6-8s)**

**Fig. 2 Entrained-flow Gasifier**



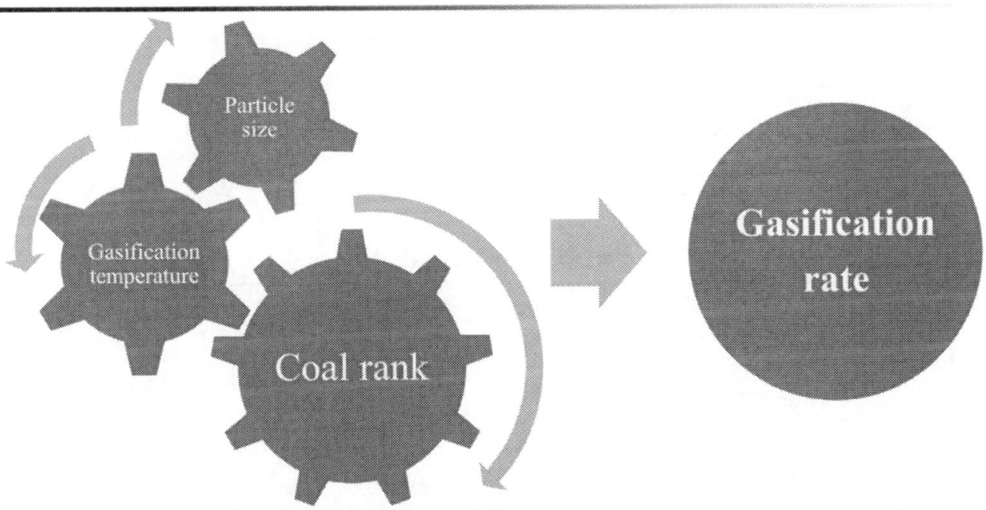


- Gasification rate is affected by coal rank, gasification temperature and particle size etc.
- Effects of gasification temperature and particle size on gasification rate are related to coal rank.

# Experimental set-up

**Experimental apparatus:**
NETZSH STA4493F3
ThermogravimetricAnalyzer (TGA)

**Gasification temperature range:**
From 1000 °C to 1300°C

**Gasification pressure:**
atmospheric pressure

**Gasification agent:**
$CO_2$ (120 mL/min)

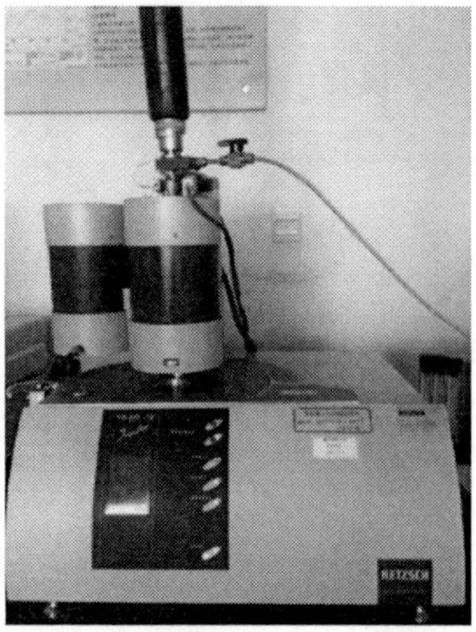

**Fig. 3 NETZSH STA4493F3 Thermogravimetric Analyzer**



# Materials

- Lignite: YN, NM (Indonesia and Inner Mongolia, China).
- Bituminous coal: SF, SH, ZJX (Shaanxi and Anhui, China).
- Anthracite: JC, GZ, GP (Shanxi and Guizhou, China).

**Table 1. Properties of tested coals**

| Sample | Coal rank | Proximate analysis | | | Ultimate analysis | | | | |
|---|---|---|---|---|---|---|---|---|---|
| | | VM | FC | Ash | C | H | N | S | O |
| YN | Lignite | 55.23 | 39.86 | 4.91 | 59.80 | 4.86 | 0.76 | 0.26 | 29.41 |
| NM | | 49.22 | 40.33 | 10.45 | 74.65 | 3.32 | 1.20 | 0.58 | 9.80 |
| SF | Bituminous coal | 35.42 | 58.29 | 6.29 | 79.14 | 2.32 | 1.12 | 0.77 | 10.36 |
| SH | | 32.29 | 60.89 | 6.82 | 77.18 | 3.73 | 1.04 | 0.60 | 10.63 |
| ZJX | | 30.19 | 51.58 | 18.23 | 70.23 | 4.28 | 1.19 | 0.34 | 5.73 |
| JC | Anthracite | 5.35 | 78.32 | 16.33 | 78.76 | 1.08 | 0.91 | 1.97 | 0.95 |
| GZ | | 8.45 | 73.76 | 17.79 | 74.52 | 2.00 | 1.05 | 4.07 | 0.57 |
| GP | | 7.96 | 76.81 | 15.23 | 74.59 | 3.03 | 0.91 | 3.12 | 3.12 |

# Preparation of Samples

- **Samples were sieved into sizes of ~40μm, ~100μm and ~300μm by vibrating screen and pneumatic screen.**
- **Particle size distribution of sieved samples were measured by Malvern Mastersizer 2000.**

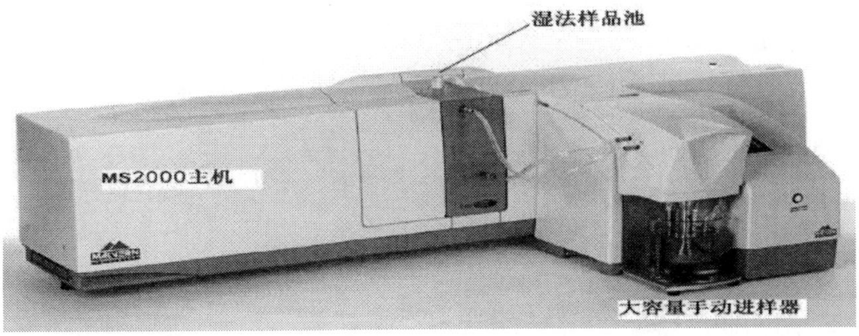


**Fig. 4 Malvern Mastersizer 2000 Particle size Analyzer**



# Particle size distribution of Samples

Fine samples (~40μm) were used to study the effects of coal rank and gasification temperature.

The particle sizes of ~40μm, ~100μm and ~300μm were used to study the effect of particle size on gasification rate.

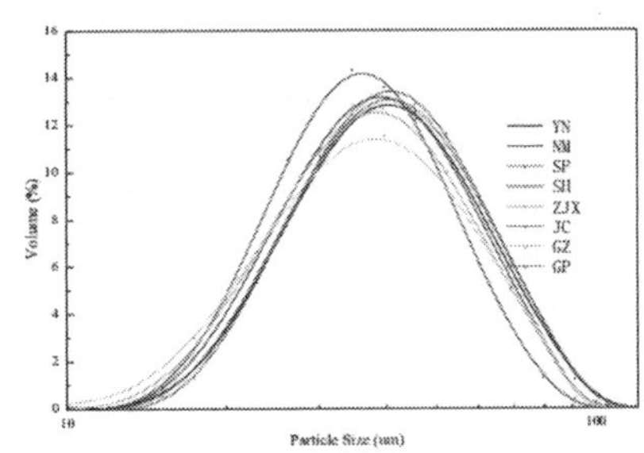


**Fig. 5 Particle size distribution of fine samples**

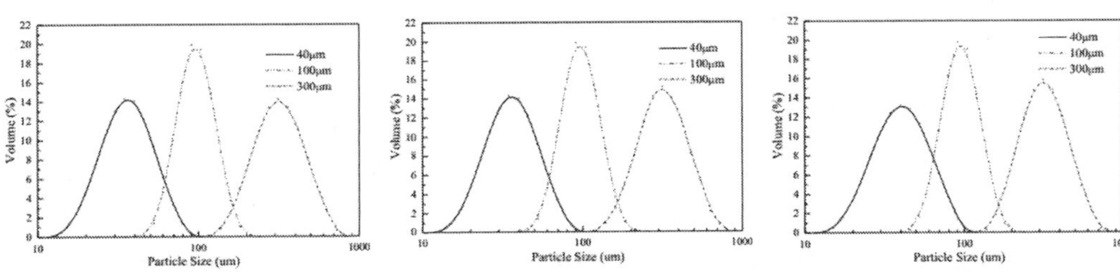


(a) (b) (c)

**Fig. 6 Particle size distribution; (a): NM; (b)ZJX; (c)JC**

Carbon conversion $x$:

$$x = \frac{w_0 - w_t}{w_0 - w_{ash}}$$

$W_{0}$ is the initial weight of char, $w_t$ is the weight at time t and $w_{ash}$ is ash mass in the char.

$R_{0.5}$ to compare gasification reactivity of different samples

$R_{0.95}$ to compare gasification reactivity of different samples at high conversion.

$R_{0.5}$ and $R_{0.95}$:

$$R_{0.5} = {0.5}/{t_{0.5}}$$

$$R_{0.95} = {0.95}/{t_{0.95}}$$

$t_{0.5}$ is reaction time for 50% carbon conversion and $t_{0.95}$ is reaction time for 95% carbon conversion.



# Results and Discussion

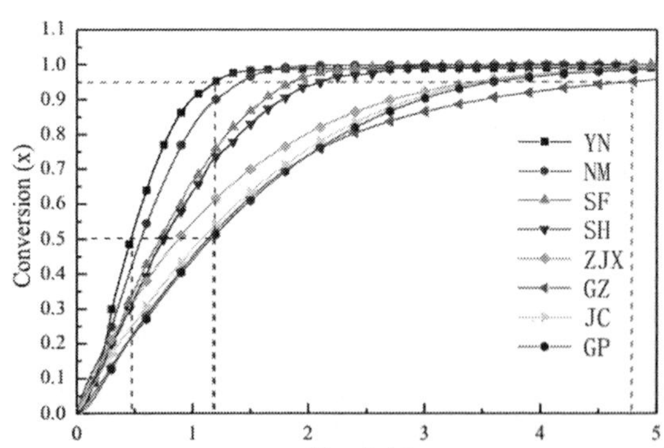


**Fig.7 Carbon conversion $x$ vs. reaction time**

**gasification temperature: 1300°C**

**Table 2 Reactivity indexes of char gasification**

| Sample | $t_{0.5}$ /min | $t_{0.95}$ /min | $R_{0.5}$ /min | $R_{0.95}$ /min | $r_{0.5,R}$* | $r_{0.95,R}$* |
|---|---|---|---|---|---|---|
| YN | 0.47 | 1.19 | 1.06 | 0.80 | 1.00 | 1.00 |
| NM | 0.57 | 1.42 | 0.88 | 0.52 | 1.21 | 1.53 |
| SF | 0.73 | 1.87 | 0.68 | 0.50 | 1.55 | 1.61 |
| SH | 0.76 | 2.12 | 0.66 | 0.44 | 1.62 | 1.82 |
| ZJX | 0.87 | 3.43 | 0.57 | 0.28 | 1.85 | 2.88 |
| GZ | 1.13 | 4.72 | 0.44 | 0.20 | 2.40 | 3.97 |
| JC | 1.16 | 3.60 | 0.43 | 0.26 | 2.47 | 3.03 |
| GP | 1.08 | 3.50 | 0.46 | 0.27 | 2.30 | 2.94 |

$r_{0.5,R}$*and $r_{0.95,R}$*: the ratio of the $R_{0.5}$ and $R_{0.95}$ of YN char sample to the $R_{0.5}$ and $R_{0.95}$ of any char, respectively.

- Low rank coals have higher reactivity than high rank coals.
- Coals with higher volatiles contents have higher reactivity in our study.
- The difference of gasification curves is more significant under high conversion.
- $R_{0.95}$ is a better reactivity index to compare gasification reactivity of different samples at high conversion.

# Effect of gasification temperature on char-$CO_2$ gasification

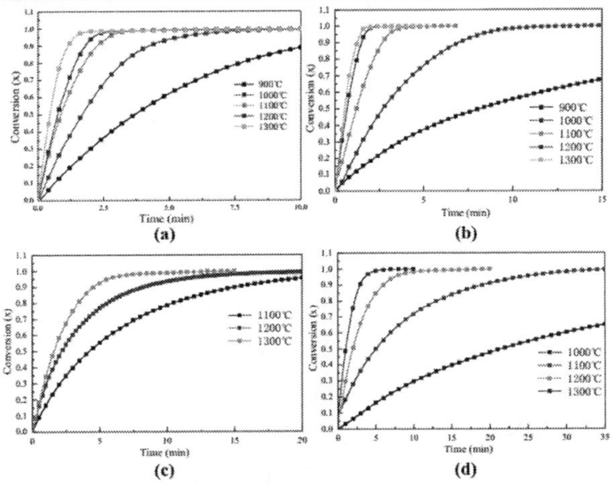


Fig.8 Carbon conversion $x$ vs. reaction time

a:NM; b:SF; c:ZJX; d:JC

**Table 3 Reactivity indexes at different temperature**

| Sample | $t_{0.95}$/min | $R_{0.95}$/min | $r_{0.95,T}^*$ |
|---|---|---|---|
| NM-1000°C | 4.25 | 0.22 | 3.22 |
| NM-1300°C | 1.32 | 0.72 | |
| SF-1000°C | 8.22 | 0.12 | 5.24 |
| SF-1300°C | 1.57 | 0.61 | |
| JC-1000°C | 111.42 | 0.01 | 31.47 |
| JC-1300°C | 3.54 | 0.27 | |

$r_{0.95,T}$*: the ratio of $R_{0.95}$at 1300 °Cto $R_{0.95}$at 1000 °C

- The gasification rates of four chars significantly increase with gasification temperature.
- For low rank coals, such as NM and SF char, the gasification rate increases slowly with temperature at high temperature zone ( >1100°C ).
- Gasification temperature has less effect on low rank coal than that onhigh rank coal.



# Effect of particle size on char-$CO_2$ gasification

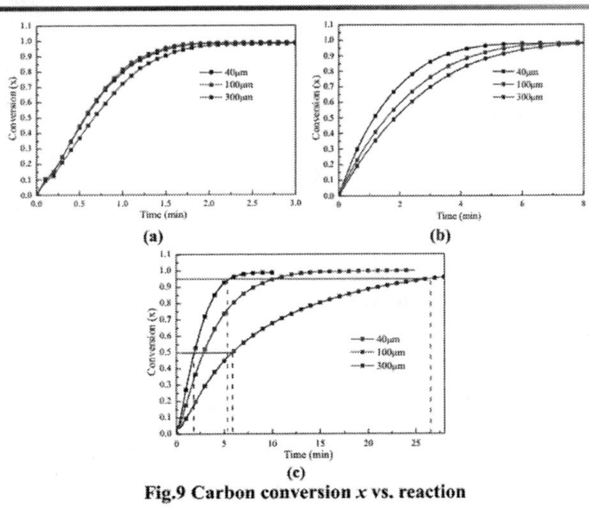


Fig.9 Carbon conversion $x$ vs. reaction

a: NM; b: ZJX; c: JC

**Table 4 Reactivity indexes of chars with different particle sizes**

| Sample | $t_{0.95}$/min | $R_{0.95}$/min | $r_{0.95,P}^*$ |
|---|---|---|---|
| NM-40μm | 1.32 | 0.72 | 1.16 |
| NM-300μm | 1.53 | 0.62 | |
| ZJX-40μm | 3.43 | 0.28 | 1.61 |
| ZJX-300 | 5.51 | 0.17 | |
| JC-40μm | 3.60 | 0.26 | 7.38 |
| JC-300μm | 26.56 | 0.04 | |

$r_{0.95,p}$*: the ratio of $R_{0.95}$ of ~40μm chars to $R_{0.95}$ of ~300μm chars.

- Effect of particle size on char-$CO_2$ gasification is much more significant for high rank coals than that for low rank coals.
- High rank coals with larger particle size are more difficult to reach high conversion.

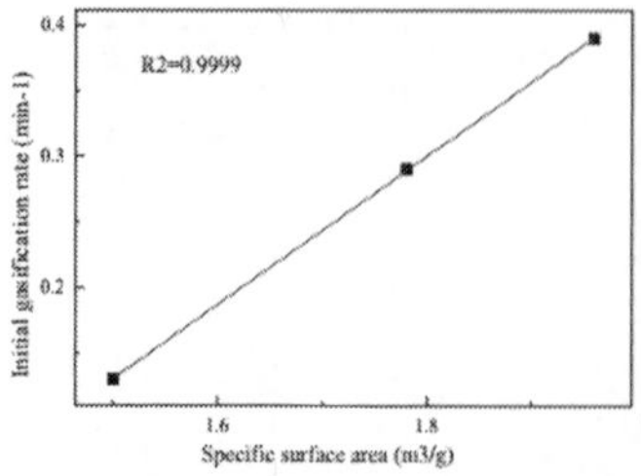


**Table 5 Specific surface area of NM chars and JC chars**

| Sample | 40μm ($m^3/g$) | 100μm ($m^3/g$) | 300μm ($m^3/g$) |
|---|---|---|---|
| NM | 108.5 | 153.9 | 149.2 |
| JC | 1.96 | 1.78 | 1.50 |

**Fig.10 Initial gasification rate vs specific surface area of JC char (Gasification temperature 1300°C)**

- The significant difference of specific surface area between NM char and JC char contributes to the different effects of particle size.
- The specific surface area of JC char is much smaller than that of NM char and the effect of particle size on gasification is much more significant for JC char than that for NM char.
- Good correlation is found between initial gasification rate and specific surface area of JC char.



# Comparison of reactivity indexes

**Table 6 Comparison of different reactivity indexes**

| Sample | $r_{0.5,R}$ | $r_{0.95,R}$ | $r_{0.5,P}$ | $r_{0.95,P}$ | $r_{0.5,T}$ | $r_{0.95,T}$ |
|---|---|---|---|---|---|---|
| NM | 1.21 | 1.53 | 1.27 | 1.16 | 3.76 | 3.22 |
| SF | 1.55 | 1.61 | 1.11 | 1.15 | 4.05 | 5.24 |
| JC | 2.47 | 3.03 | 5.01 | 7.38 | 11.78 | 31.47 |

- Gasification temperature is the most important parameter to control gasification rate.
- Effect of particle size is significant for high rank coals, but negligible for low rank coals.
- Coal rank is not an important factor for gasification at high temperature (≥ 1300°C).

# CONCLUSIONS

- The effects of coal rank, temperature and particle size on char-$CO_2$ gasification were studied at temperatures ranged from 1000°C to 1300°C. Low rank coals had higher reactivity than high rank coals. The difference of char-$CO_2$ gasification rates of different chars was more significant with an increase of the carbon conversion, and high rank coals were more difficult to obtain high conversion.
- The effects of gasification temperature and particle size on char-$CO_2$ gasification is related to coal rank. Effects of temperature and particle size on char-$CO_2$ gasification were both more significant for high rank coal. For high rank coals, the efficiency of industrial entrained-flow gasifier could be improved by increasing gasification temperature and decreasing coal particle size. For low rank coals, the gasification temperature could be lower in order to improve effective gas contents.



***Thanks for your attention !***

**E-mail: gongxin@ecust.edu.cn**

Thirty-Third Annual International
PITTSBURGH COAL CONFERENCE
August 8 - 12, 2016 Cape Town, South Africa

# Experimental study of particle evolution characteristics in an opposed multi-burner (OMB) gasifier


**Presenter: Qing Zhang**

**Co-authors:** Yan Gong, Qinghua Guo, Zhicun Xue, Fuchen Wang, Guangsuo Yu*

*Key Laboratory of Coal Gasification and Energy Chemical Engineering of Ministry of Education;*
*Shanghai Engineering Research Center of Coal Gasification;*
*East China University of Science and Technology (ECUST), Shanghai, China*

## Outline

1. Introduction
2. Experiments
3. Measurement Principles
4. Results and Discussion
5. Conclusions

# 1. Introduction

**large numbers of particles**

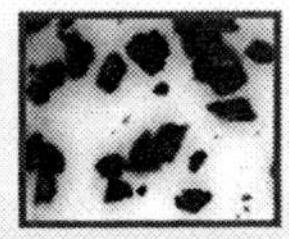

In gasifier/furnace → (particle image) → Research hotspots → **Characteristics:** Ignition, Pyrolysis, Gasification, Fragmentation etc.

The environment is complicated under hot condition

Numerical simulations

Experimental studies: Using particle sampling system → analyzed → Particle size, morphology, compositions etc.

**Particle dynamic behaviors under hot condition ?** → **Visualization techniques**

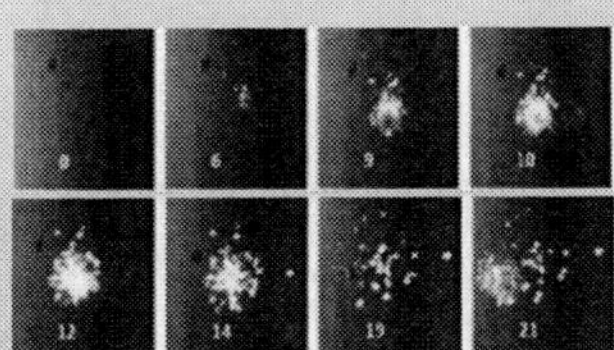

**Combustion behavior of single coal particle [1]**

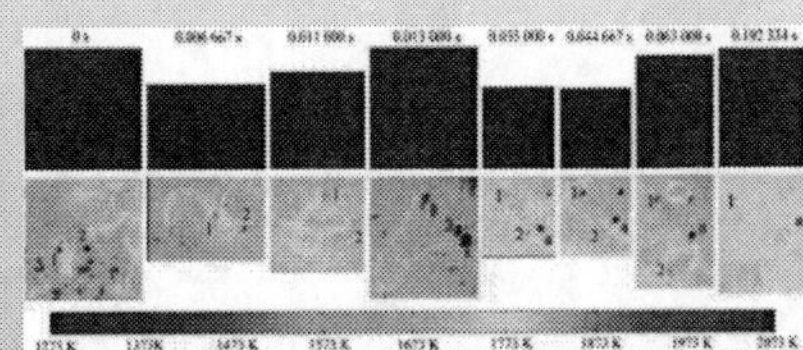

**Low temperature particles with high temperature wake impinging on refractory wall [2]**

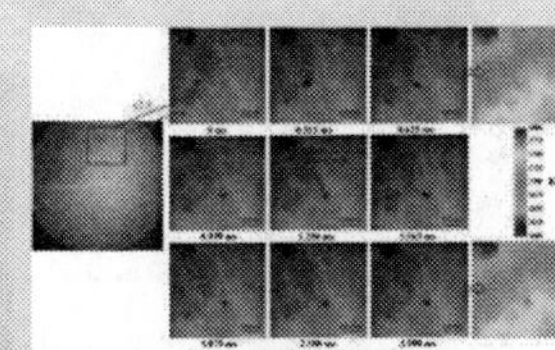

**Low temperature particle impact and adherence to the wall [3]**


[1] Y.A. Levendis, K. Joshi, R. Khatami et al. Combustion behavior in air of single particles from three different coal ranks and from sugarcane bagasse, Combust. Flame 158 (2011) 452-465.

[2] Y. Gong, G.S. Yu, Q.H. Guo et al. Experimental study on particle characteristics in an opposed multi-burner gasifier, Chem. Eng. Sci. 117 (2014) 93-106.

[3] Y. Gong, G.S. Yu, Q.H. Guo, et al. Experimental study of the particle deposition characteristics in an entrained flow gasifier, Chem. Eng. Sci. 138 (2015) 291-302.




# 2. Experiments

*2.1 Experimental setup*

Based on a bench-scale OMB gasifier and the advanced visualization techniques, particle evolution characteristics in the gasifier has been studied.

**Bench-scale OMB gasification platform**

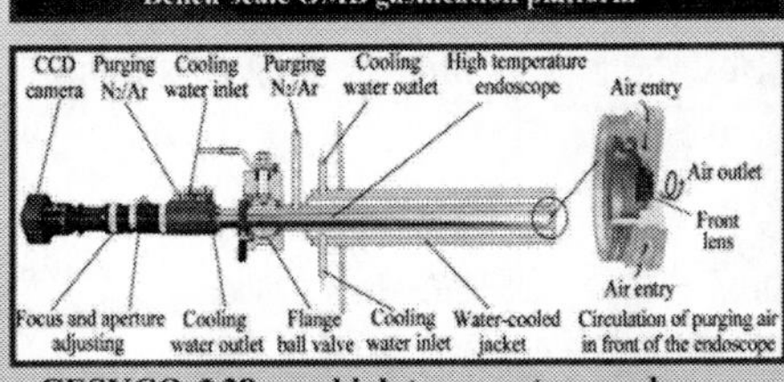


**CESYCO Φ38 mm high temperature endoscope**

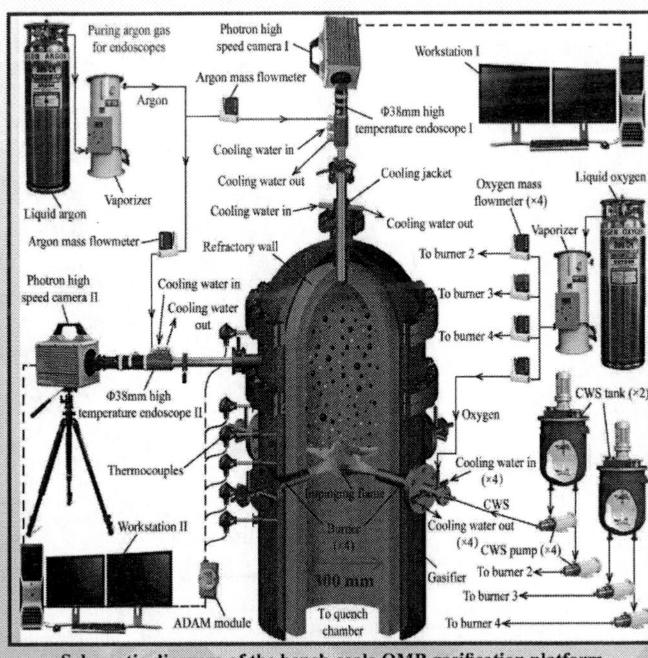


**Schematic diagram of the bench-scale OMB gasification platform**

# 2. Experiments

*2.2 Operating conditions*

**Shenfu bituminous coal**, widely used in the industrial gasifier, is prepared to CWS with water and a few additives as feedstock.

**Table 1 Properties of coal and CWS**

| | | |
|---|---|---|
| Proximate analysis (wt%, dry basis) | Moisture | 4.67 |
| | Volatile matter | 42.66 |
| | Fixed carbon | 45.31 |
| | Ash | 12.03 |
| Ultimate analysis (wt%, dry basis) | Carbon | 72.67 |
| | Hydrogen | 4.38 |
| | Nitrogen | 1.25 |
| | Sulfur | 0.55 |
| | Oxygen | 9.12 |
| Properties of CWS | Average particle size (μm) | 45.8 |
| | Mass concentration | 61% |
| | Viscosity (mPa·s) | 288.5 |
| | Density (kg·m$^{-3}$) | 1132 |

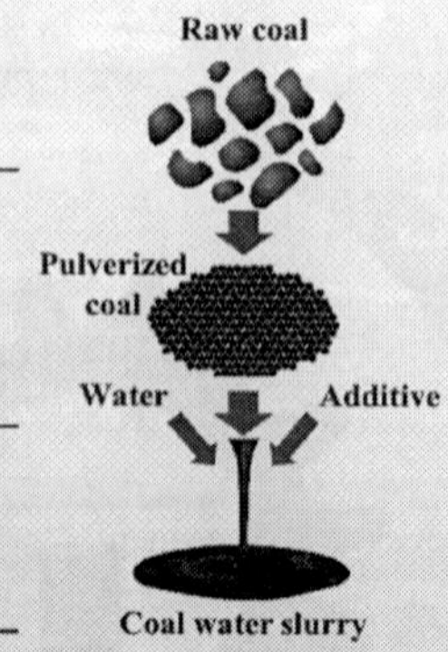


**Table 2 Operating conditions of CWS gasification**

| | CWS flow rate (kg/h) | $O_2$ flow rate (Nm³/h) | O/C (mol/mol) |
|---|---|---|---|
| Burner 1 | 10 | 4.24 | 1.0 |
| Burner 2 | 10 | 4.24 | 1.0 |
| Burner 3 | 10 | 4.24 | 1.0 |
| Burner 4 | 10 | 4.24 | 1.0 |



# 3. Measurement Principles

*3.1 Different focusing plane in the gasifier*

Particle images of different axial space in the gasifier can be obtained by the optical sectioning tomography (OST) method.

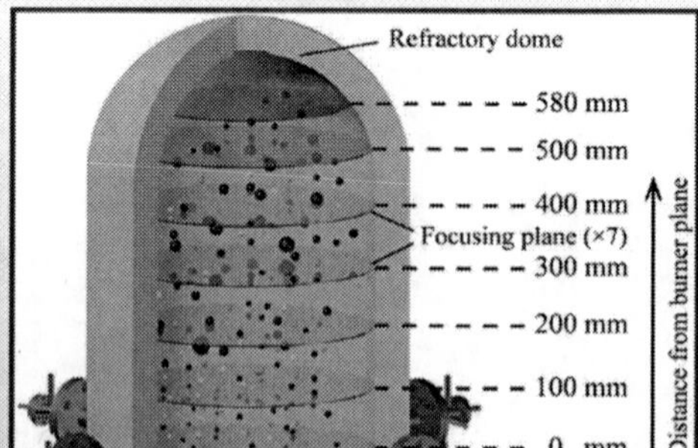


**Schematic of different focusing planes in the gasifier**

This method is only applicable to the space between the burner plane and the refractory dome.

In the axial direction of the gasifier, seven planes are selected as the focusing planes of the imaging system. They are located at the burner plane (0 mm), 100 mm, 200 mm, 300 mm, 400 mm, 500 mm and 580 mm (close to the refractory dome) above the burner plane.

# 3. Measurement Principles

*3.2 Image processing and analysis*

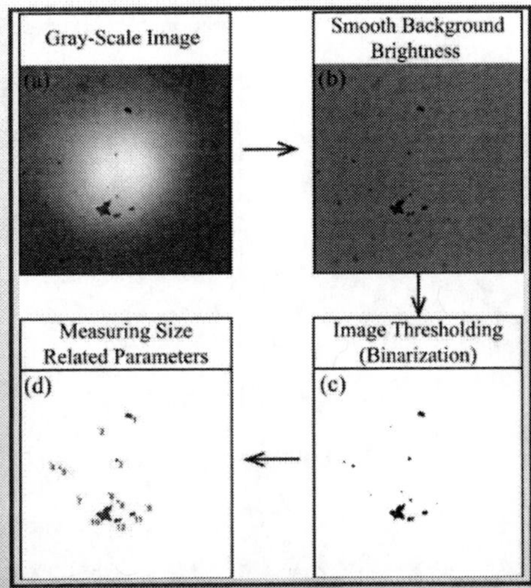


**Schematic for the key steps of the image processing algorithm**

The particle diameter is defined by the equivalent circle diameter, which is the diameter of a circle that has the same area as the particle projection area. Equivalent circle diameter ($D_e$) can be calculated using Eq. (1):

$$D_e = \sqrt{\frac{4A}{\pi}} \cdots\cdots(1)$$

*A* is the particle projection area



# 3. Measurement Principles

*3.3 Calibration of image pixel scale*

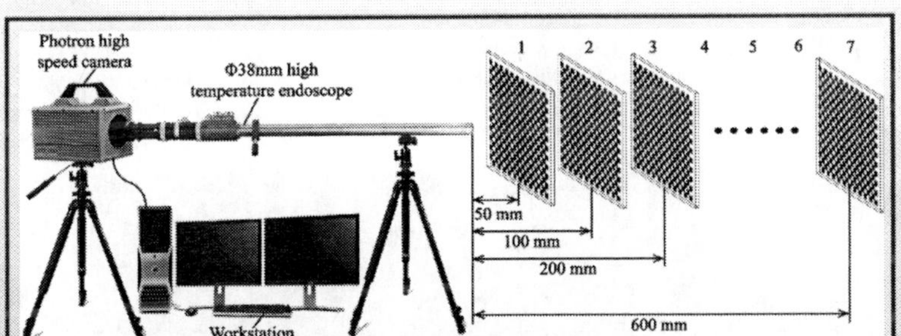


**Schematic of the image pixel scale calibration**

**Table 3 The corresponding relationship between the pixel and the actual size**

| Distance between endoscope and the calibration plate (mm) | Actual distance from the burner plane (mm) | Scale factor (μm/pixel) |
|---|---|---|
| 50 | 580 | 23.793 |
| 100 | 500 | 48.692 |
| 200 | 400 | 98.900 |
| 300 | 300 | 147.514 |
| 400 | 200 | 199.762 |
| 500 | 100 | 248.828 |
| 600 | 0 | 298.325 |

# 4. Results and Discussion

*4.1 Particle type and transformation:* ***The burner plane***

The flow field in the burner plane are divided into three sections: jet zone (JZ), impinging zone (IZ) and recirculation zone (RZ)

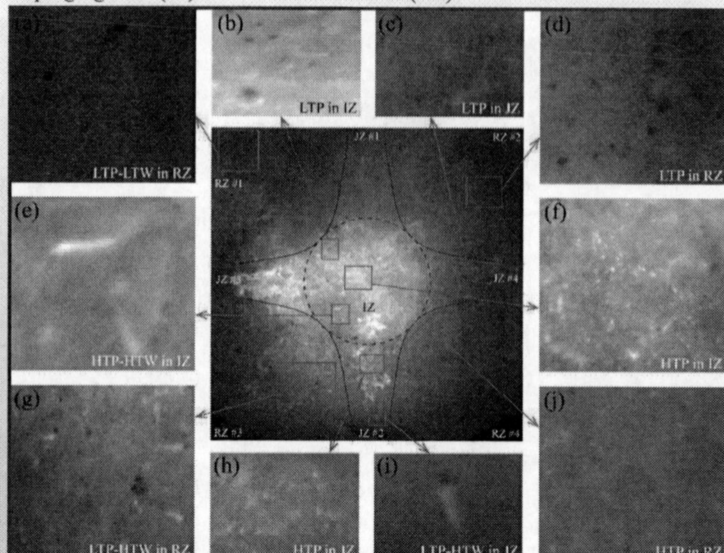


**Different kinds of particles in the burner plane: (a) LTP-LTW in RZ, (b) LTP in IZ, (c) LTP in JZ, (d) LTP in RZ, (e) HTP-HTW in IZ, (f) HTP in IZ, (g) LTP-HTW in RZ, (h) HTP in JZ, (i) LTP-HTW in JZ, (j) HTP in RZ**

LTP:
Low temperature particle without wake

LTP-LTW:
Low temperature particle with low temperature wake

LTP-HTW:
Low temperature particle with high temperature wake

HTP:
High temperature particle without wake

HTP-HTW:
High temperature particle with high temperature wake



# 4. Results and Discussion

*4.1 Particle type and transformation:* ***100 mm above the burner plane***

In this plane, the number of LTP-LTW significantly increased, their particle size is obviously larger than that of LTP-HTW, and the low temperature wake is longer than the high temperature wake.

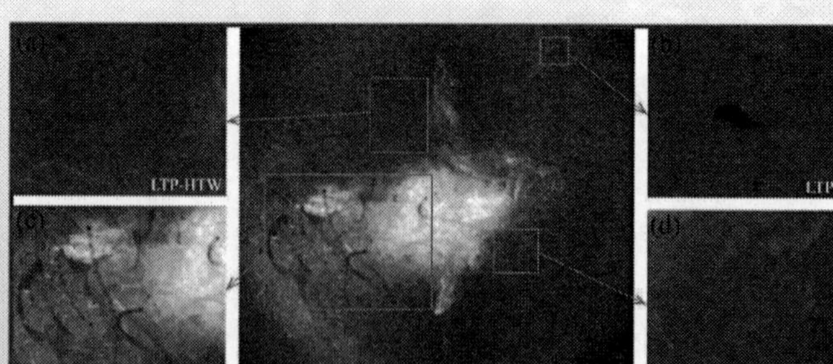


**Different kinds of particles at 100 mm above the burner plane: (a) LTP-HTW, (b) LTP, (c) LTP-LTW, (d) HTP-HTW**

Wakes only appear in pyrolysis stage[1]. The space of 100 mm above the burner plane is the main region of particle pyrolysis.

[1] W.R.Seeker, G.S.Samuelsen, M.P.Heap et al. Thermal decomposition of pulverized coal particles, Symp. (Int.) Combust.18 (1981) 1213-1226.

## 4. Results and Discussion

*4.1 Particle type and transformation: **200 mm and 300 mm above the burner plane***

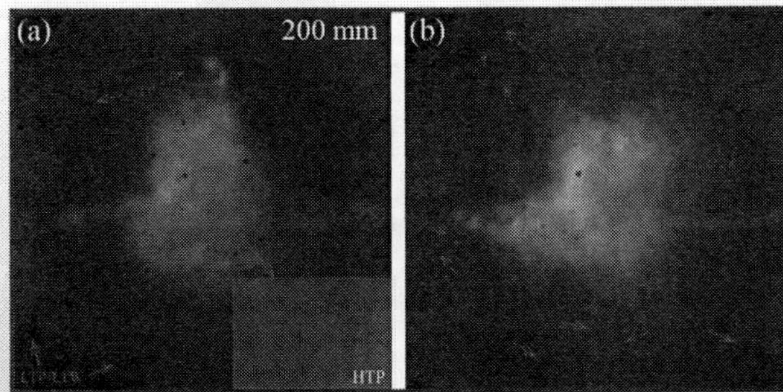

**Different kind of particles at 200 mm above the burner plane. The arrows in (b) refer to the low temperature wake.**

The space around 200 mm and 300 mm above the burner plane is also an important area of particle pyrolysis.

The number of LTP-LTW significantly decreased at 300 mm above the burner plane.

The axial space between 100 mm to 300 mm above the burner plane is the main regions of particle pyrolysis, and the most intense pyrolysis occurs at 100 mm above the burner plane.

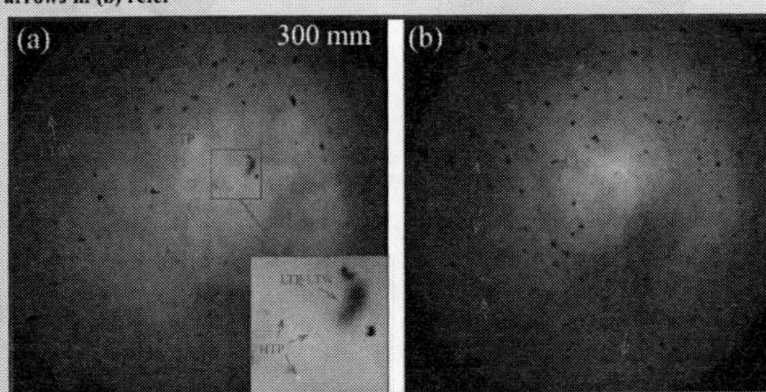

**Different kind of particles at 300 mm above the burner plane. The arrows in (b) refer to the low temperature wake.**



## 4. Results and Discussion

*4.1 Particle type and transformation: **400 mm above the burner plane***

There are plenty of LTP in this plane. The LTP-LTW has a low probability to appear. Nearly all of the particles have completed pyrolysis reactions , and transformed to char or fly ash particles before reaching the space around this plane.

**Different kinds of particles at 400 mm above the burner plane**

A large number of HTP appear in the region near the refractory wall, and the LTP breaks up frequently around this plane.

The HTP in impinging up-flow stream have high speed when forced to retrace after reaching dome. The large kinetic energy and inertia generate large centrifugal force when velocity direction changes, which makes particles tend to move downward close to the refractory wall after retracing.

# 4. Results and Discussion

*4.1 Particle type and transformation:* ***500 mm and 580 mm above the burner plane***

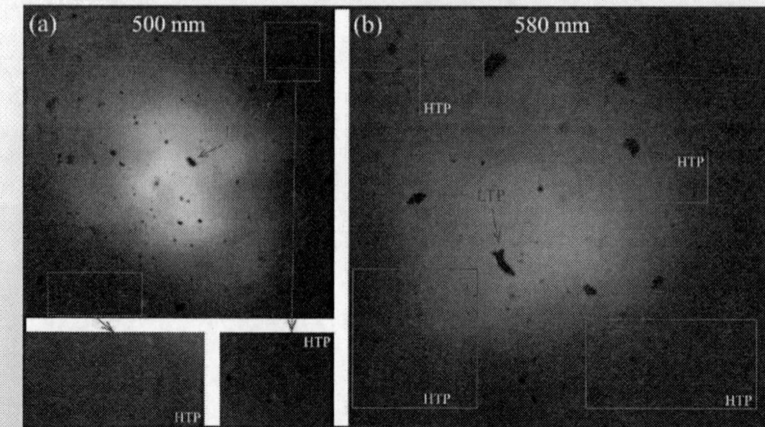


**Different kinds of particles at 500 mm and 580 mm above the burner plane: (a) 500 mm above the burner plane, (b) 580 mm above the burner plane.**

These two planes are close to the refractory dome, particles in which are mainly LTP, and they are almost char particles with low carbon content. In addition, a few HTP can still be found in these two planes.



# 4. Results and Discussion

*4.2 Particle fragmentation characteristics:* ***The burner plane***

The particle breakage in the burner plane is very violent and usually accompanied by sudden bursts which would result in the change of the particle motion status.

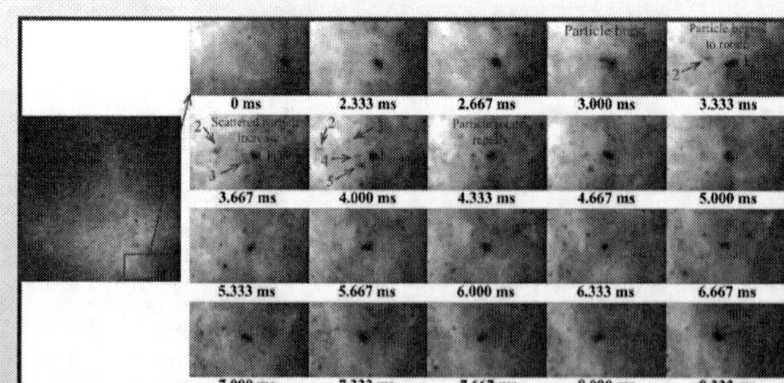

**A particle bursts in the burner plane. The numbers in the figure refer to the scattered particles after bursting**

This particle bursting process is frequently occurred when the particle contacts with the flame, and only occurred in the burner plane.

Particle temperature and thermal stress are increased rapidly, at the same time, the rapid evaporation of residual moisture and the fast devolatilization of volatile matters lead to forming a very high internal pressure in an instant, and then the particle bursts.

# 4. Results and Discussion

*4.2 Particle fragmentation characteristics:* ***100, 200 and 300 mm above the burner plane***

Most of the particles in these regions are in the pyrolysis stage, the fragmentation process is mainly controlled by the pyrolysis behavior.

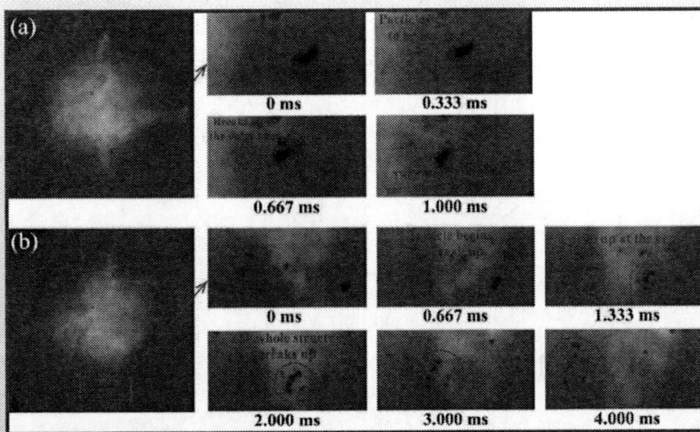


**Typical particle fragmentation process in the space of 100 mm, 200 mm and 300 mm above the burner plane**

The particle fragmentation process can be divided into two categories:

(a) Fragmentation at the outer zone (about 1/3)

(b) Fragmentation at the center (about 2/3)

The fragmentation at the outer zone has a slight effect on the shape and motion state of the mother particle.
The fragmentation at the center completely destroyed the whole structure of the particle.



# 4. Results and Discussion

*4.2 Particle fragmentation characteristics:* ***400, 500 and 580 mm above the burner plane***

**(a) Fragmentation process of a single char particle**

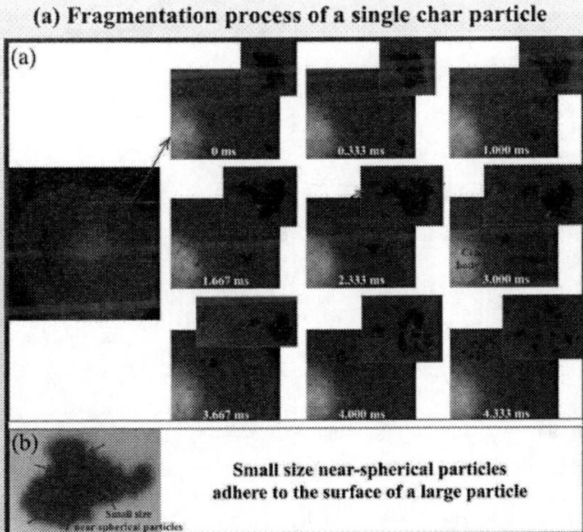


**(b) Small size near-spherical particles adhere to the surface of a large particle**

**(c) Fragmentation process of a particle-group agglomerated by small size near-spherical particles**

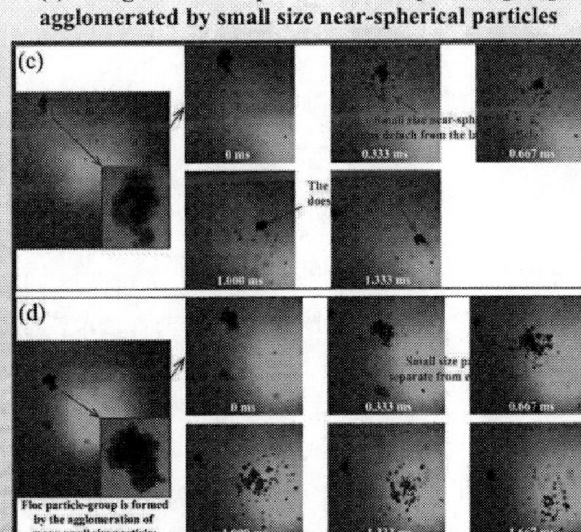


**(d) Fragmentation process of a floc particle-group**

# 4. Results and Discussion

*4.2 Particle fragmentation characteristics:* ***Probability of particle fragmentation***

The probability of particle fragmentation is the ratio of the number of fragmentized particles to the total number of particles.

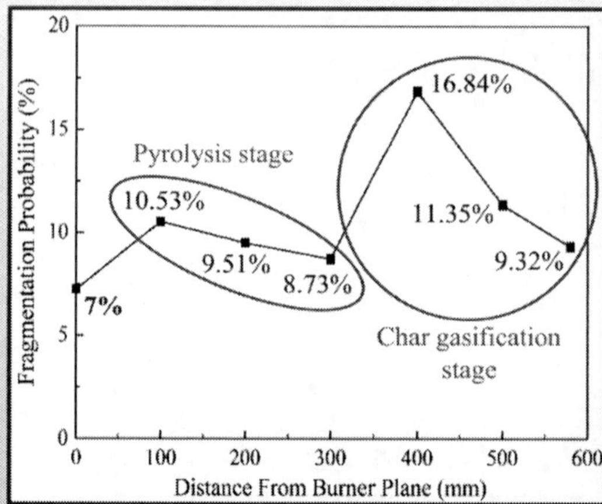


**Probability of particle fragmentation at different focusing plane**

The region with highest particle fragmentation probability is 400 mm above the burner plane (16.84%).

The particle fragmentation has more probability to take place in the char gasification stage than in the pyrolysis stage in the OMB gasifier.



# 4. Results and Discussion

*4.3. Particle size distribution (PSD) characteristics*

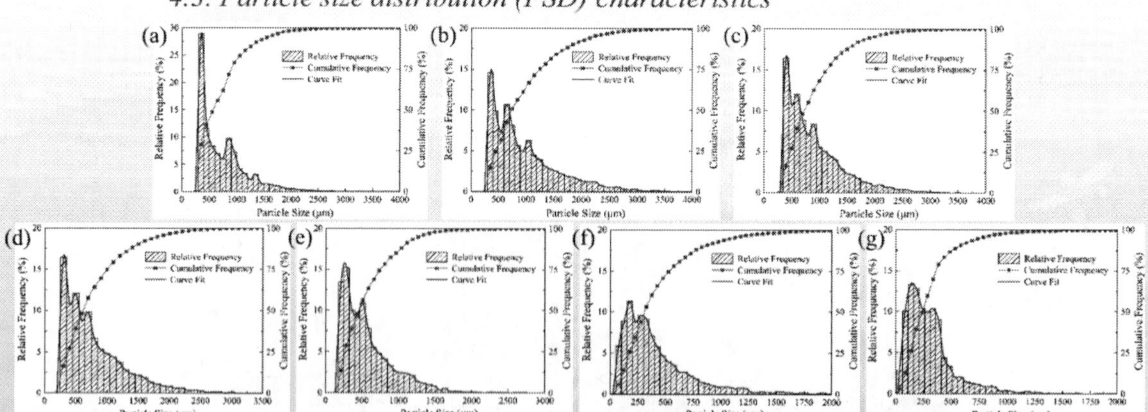


**PSDs of particles at different focusing planes in the gasifier: (a) The burner plane, (b) 100 mm above the burner plane, (c) 200 mm above the burner plane, (d) 300 mm above the burner plane, (e) 400 mm above the burner plane, (f) 500 mm above the burner plane, (g) 580 mm above the burner plane.**

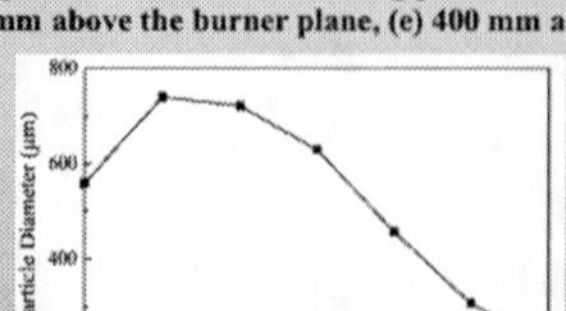


**Median particle diameter of different focusing planes in the gasifier**

Particles at 100 mm above the burner plane have the largest $D_{50}$ and the more close to the refractory dome, the smaller the $D_{50}$ is.

Particle breakage is the main reason to cause particle size decrease.

## 5. Conclusions

- The burner plane contains all of the particle types, most of them are HTP, LTP, HTP-HTW and LTP-HTW. Particle fragmentation process in the burner plane usually accompanied by sudden bursts, the probability of which is relatively low (about 7%).
- The space between 100 mm to 300 mm above the burner plane is the main regions of particle pyrolysis, particle types are mainly the LTP-LTW and the LTP. Particle fragmentation process in these space can be divided into two categories: fragmentation at the outer zone and fragmentation at the center.
- There are plenty of LTP in the space between 400 mm above the burner plane to the refractory dome. The region with highest particle fragmentation probability is 400 mm above the burner plane (16.84%)
- The particle fragmentation has more probability to take place in the char gasification stage than in the pyrolysis stage in the OMB gasifier. The closer to the refractory dome the smaller the $D_{50}$ is.

Page 19 / 20

Thirty-Third Annual International
PITTSBURGH COAL CONFERENCE
August 8 - 12, 2016 Cape Town, South Africa

# *Thank you*


**East China University of Science and Technology (ECUST)**
www.ecust.edu.cn

**Qing Zhang**
Add: No.130 Meilong Road, Shanghai, China
Email: angus_zq@163.com / 020120063@mail.ecust.edu.cn

Page 20 / 20

# CORROSION TESTING OF STEEL FOR PRODUCTION AND INJECTION WELL APPLICATIONS IN A UCG PROCESS

AFRICARY

African Carbon Energy


JC van Dyk[1,3], J Brand[2], FB Waanders[3], AC van Wyk[3] and CJ Kahts[4]

[1]African Carbon Energy, email: johan.vandyk@africary.com
[2]African Carbon Energy, email: johan.brand@africary.com
[3]North-West University, e-mail: frans.waanders@nwu.ac.za
[4]Aqua Alpha Drilling, e-mail: conrad@aquaalpha.com


NORTH-WEST UNIVERSITY YUNIBESITI YA BOKONE-BOPHIRIMA NOORDWES-UNIVERSITEIT

AQUA ALPHA "Directional Drilling Solutions"

33rd International Pittsburgh Coal Conference, Cape Town, 11 August 2016

# OUTLINE OF PRESENTATION

AFRICARY

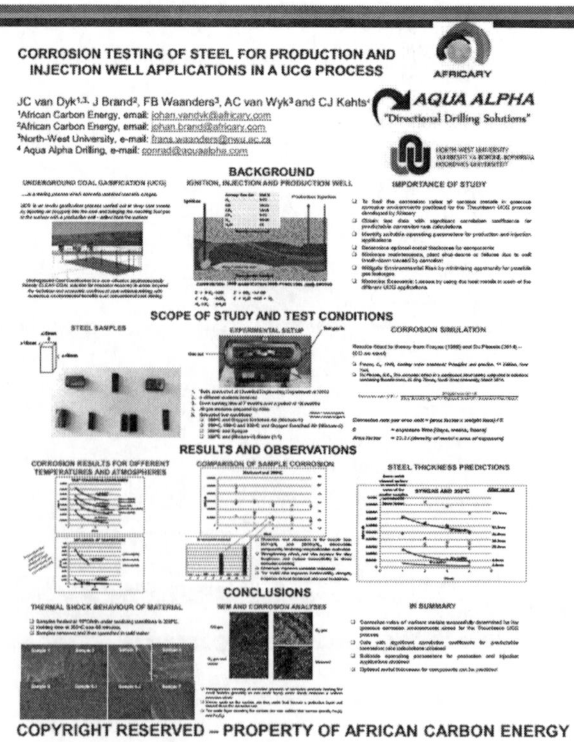

# UNDERGROUND COAL GASIFICATION

AFRICARY

*UCG is an in-situ gasification process carried out in deep coal seams by injecting air (oxygen) into the coal and bringing the resulting fuel gas to the surface with a production well - drilled from the surface*

*Underground Coal Gasification is a cost-effective environmentally friendly CLEAN COAL solution for resource recovery in areas beyond the technical and economic confines of conventional mining with numerous environmental benefits over conventional coal mining*



# IGNITION, INJECTION AND PRODUCTION WELL

AFRICARY

| Average Raw Gas | Mol % |
|---|---|
| $H_2$ | 8-12 |
| CO | 16-24 |
| $CO_2$ | 18-25 |
| $CH_4$ | 8-12 |
| $N_2$ | 30-40 |
| $H_2O$ | <5 |

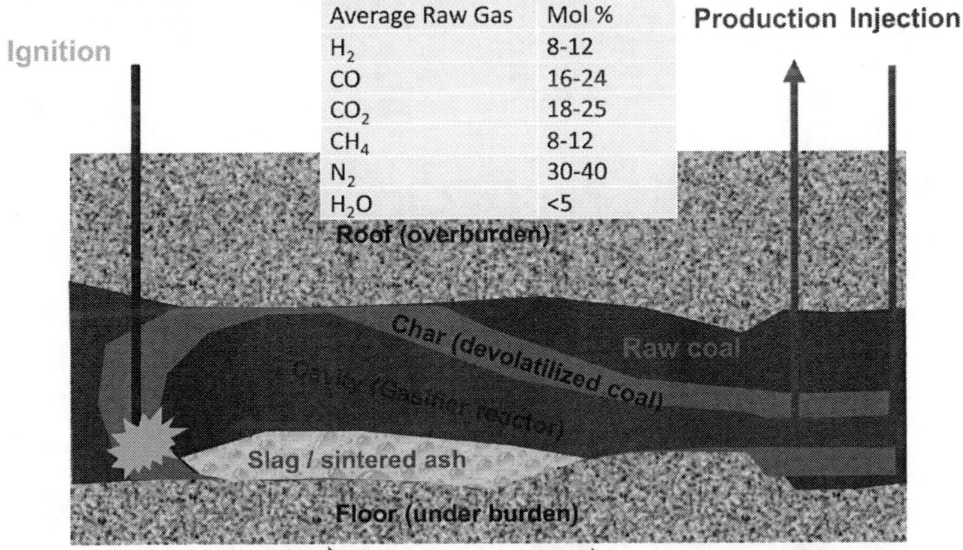


**COMBUSTION** → **GASIFICATION** → **PYROLYSIS** → **DRYING**

$C + \frac{1}{2}O_2 \rightarrow CO$

$C + O_2 \rightarrow CO_2$

$H_2 + O_2 \rightarrow H_2O$

$C + CO_2 \rightarrow 2\,CO$

$C + H_2O \rightarrow CO + H_2$

# IMPORTANCE OF STUDY

AFRICARY

- To **find the corrosion rates of various metals** in gaseous corrosive environments predicted for the Theunissen UCG process developed by Africary
- Obtain test data with significant correlation coefficients for **predictable corrosion rate** calculations
- **Identify suitable operating parameters** for production and injection applications
- **Determine optimal metal thickness** for components
- **Minimize maintenance,** plant shut-downs or failures due to well break-down caused by corrosion
- **Mitigate Environmental Risk** by minimising opportunity for possible gas leakages
- **Minimize Economic Losses** by using the best metals in each of the different UCG applications



# STEEL SAMPLES

AFRICARY

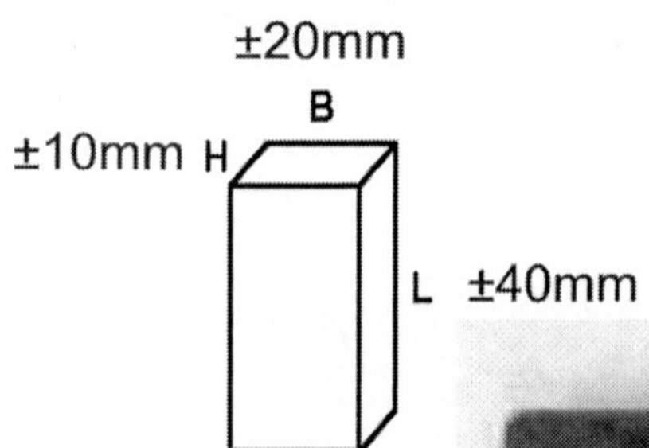


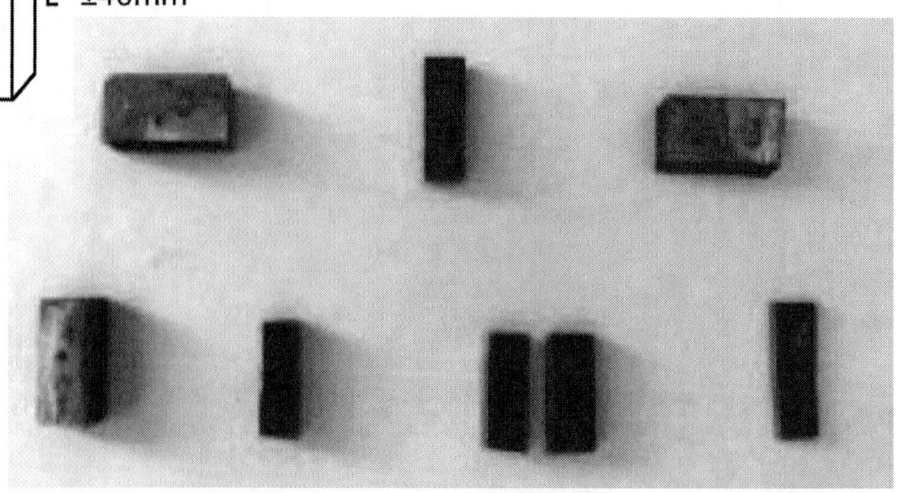

# EXPERIMENTAL SETUP

AFRICARY

1. Tests conducted at Chemical Engineering Department at NWU
2. 4 different students involved
3. Oven running time of 7 months over a period of 18 months
4. All gas mixtures prepared by Afrox
5. Simulated test conditions:
   - ❑ **350°C and Oxygen Enriched Air (Mixture-1)**
   - ❑ **350°C, 650°C and 850°C and Oxygen Enriched Air (Mixture-2)**
   - ❑ **350°C and Syngas**
   - ❑ **350°C and (Mixture-2):Steam (1:1)**

**Mixture 1 = low oxygen %**
**Mixture 2 = high oxygen %**



# EXPERIMENTAL SETUP

AFRICARY

# CORROSION SIMULATION

AFRICARY

## Results fitted to theory from Frayne (1999) and Du Plessis (2014) – (CO on steel)

- *Frayne, C., 1999, Cooling water treatment: Principles and practice, 1st Edition, New York.*
- *Du Plessis, B.C., The corrosion effect in a continuous steel caster, subjected to solutions containing fluordie irons, M. Eng Thesis, North West University, March 2014.*

$$\textit{Corrosion rate (CR)} = \frac{\textit{Weight loss }(g) * K}{\textit{Alloy density }(g.cm^3) * \textit{Exposed Area }(A) * \textit{Exposure time (days)}}$$

***Corrosion rate per area unit = (area factor x weight loss) / E***

***E = exposure time (days, weeks, hours)***

***Area factor = 22.2 / (density of metal x area of exposure)***



# CORROSION RESULTS FOR DIFFERENT TEMPERATURES AND ATMOSPHERES

AFRICARY

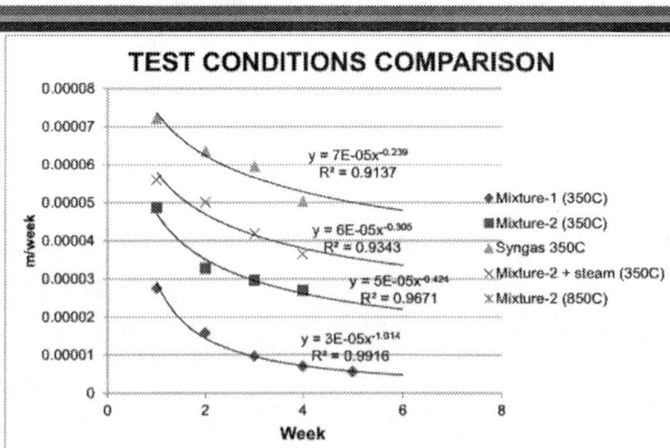


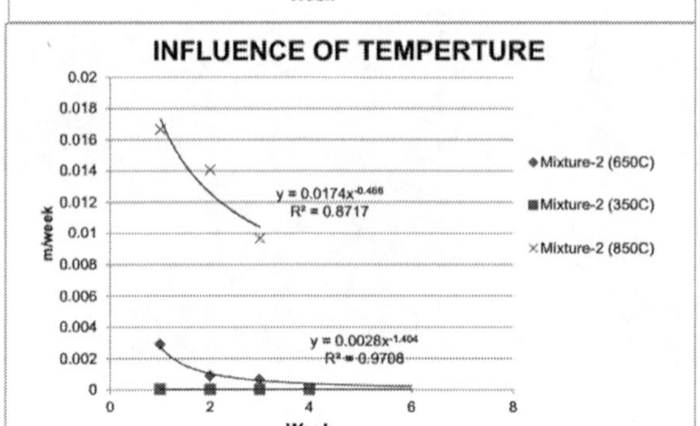


Illustrates the significant effect of high temperature on corrosion rate

## COMPARISON OF SAMPLE CORROSION

AFRICARY

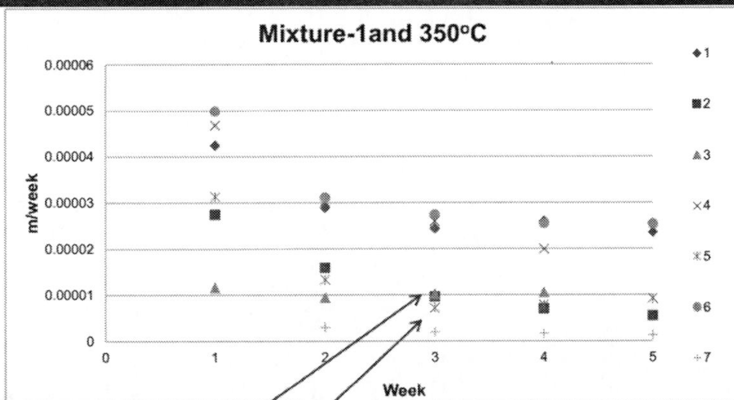


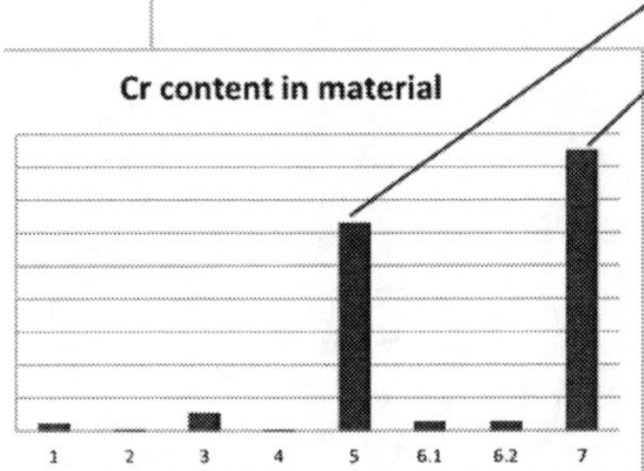


- Chromium and aluminium in the sample form $(CrFe)Al_7$ and $(CrMn)Al_{12}$ intermetallic compounds, hindering recrystallization nucleation
- Strengthening effect, and also improve the alloy toughness and reduce susceptibility to stress corrosion cracking
- Chromium improves corrosion resistance
- The metal also improves hardenability, strength, response to heat treatment and wear resistance.



## STEEL THICKNESS PREDICTIONS

AFRICARY

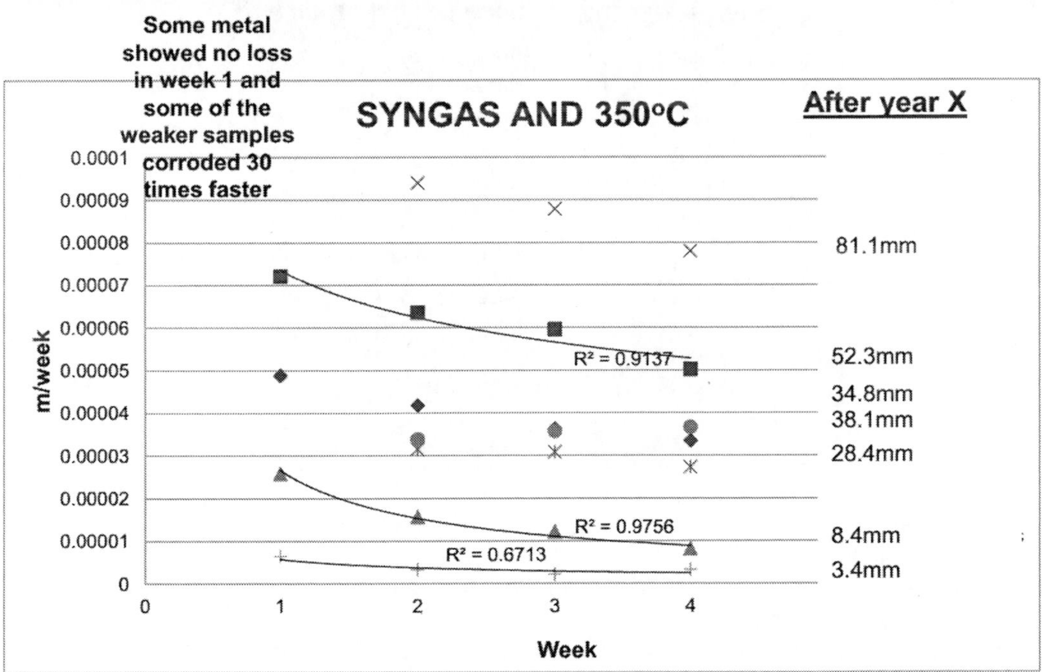

## THERMAL SHOCK BEHAVIOUR OF MATERIAL

AFRICARY

- ❑ Samples heated at 10°C/min under oxidizing conditions to 350°C.
- ❑ Holding time at 350°C was 60 minutes.
- ❑ Samples removed and then quenched in cold water

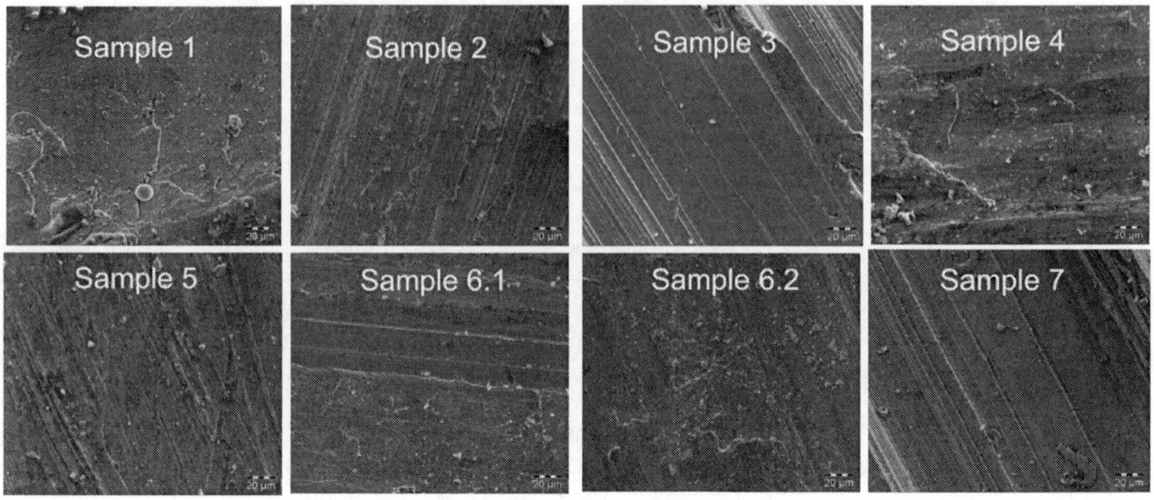




## SEM AND CORROSION ANALYSES

AFRICARY

CO gas

$O_2$ gas

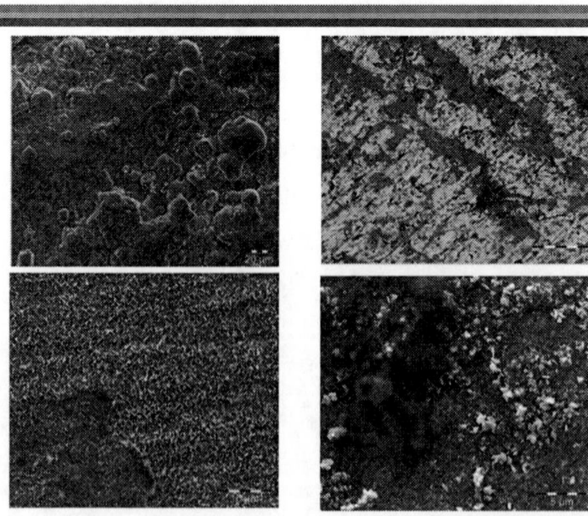

$O_2$ gas and steam

Mixture-2

- ❑ Homogeneous covering of corrosion products of corrosion products looking like small bubbles (possibly an iron oxide layer) which clearly indicates a uniform corrosion attack
- ❑ Darker spots on the surface are iron oxide that formed a protective layer and slowed down the corrosion rate
- ❑ The oxide layer covering the surface are iron oxides that formed (mostly $Fe_2O_3$ and $Fe_3O_4$)

## IN SUMMARY

AFRICARY

- **Corrosion rates of various metals** successfully determined for the gaseous corrosion environments aimed for the Theunissen UCG process
- Data with significant correlation coefficients for **predictable corrosion rate** calculations obtained
- **Suitable operating parameters** for production and injection applications obtained
- **Optimal metal thickness** for components can be predicted



AFRICARY

African Carbon Energy

WWW.AFRICARY.COM

# PROPERTY OF EXTRACT PREPARED BY HYDROTHERMAL EXTRACTION OF BROWN COAL


***Masato Morimoto, PhD, Research Institute of Energy Frontier, National Institute of Advanced Industrial Science and Technology , 16-1 Onogawa Tsukuba 305-8569, JAPAN***
***m.morimoto@aist.go.jp, +81-298610553, +81-298618441***

*Qingxin Zheng, PhD, Research Institute of Energy Frontier, National Institute of Advanced Industrial Science and Technology , 16-1 Onogawa Tsukuba 305-8569, JAPAN*
*qingxin.zheng@aist.go.jp, +81-0298613940, +81-298618441*

*Toshimasa Takanohashi, PhD, Research Institute of Energy Frontier, National Institute of Advanced Industrial Science and Technology , 16-1 Onogawa Tsukuba 305-8569, JAPAN*
*toshi-takanohashi@aist.go.jp, +81-298618038, +81-298618441*


Brown coals and lignites are valuable resources expected as not only fuels but also material feedstocks in this century, because of their worldwide abundant minable reserves. Brown coals contain large amount of water (~60%) in general due to its high oxygen contents, and some of them do not contain a large amount of sulfur, nitrogen, and mineral matters. We have already presented a hydrothermal extraction method that not only removes water from brown coal but also upgrades the coal and extracts low molecular mass compounds simultaneously [1-3]. Recently, we found that the solid extracts obtained by the hydrothermal extraction of brown coals, which were performed using a flow of subcritical water at ~350°C and ~20 MPa, had almost the same morphology: spherical particles of <5 μm in diameter with smooth surfaces (microsphere), as shown in Fig. 1. This microsphere production is very surprising when we consider that this is a one-step product obtained directly from brown coal and that the process is quite simple and uses only water. The method can be applied to various brown coals with carbon yields above 20%, which would meet the requirement for industrial application. The coal organic microsphere we proposed here may be utilized as carriers and/or supporters.

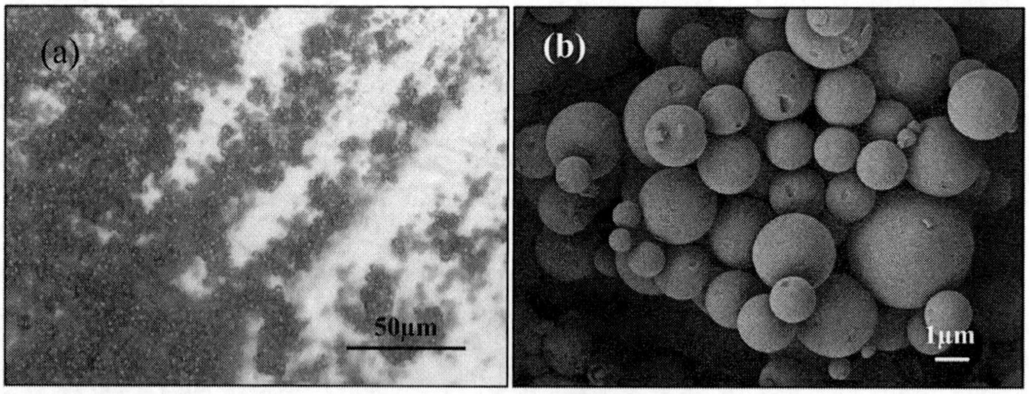


Fig.1 Microscopic (a) and SEM (b) images of coal organic microsphere prepared by hydrothermal extraction of brown coal

Reference


1. Morimoto, M.; Nakagawa, H.; Miura, K. *Energy Fuels* 2009, 23, 4533-4539.
2. Morimoto, M.; Nakagawa, H.; Miura, K. *Energy Fuels* 2010, 24, 3060-3065.
3. Zheng, Q. X.; Morimoto, M.; Sato, H.; Takanohashi, T. *Fuel* 2015, 159, 751-758.

# Process Characteristics Investigation of Shendong Coal Co-pyrolysis through Solid Heat Carrier

*Baofu Li, Xiaohong Li, Wenying Li*, Jie Feng*

*Key Laboratory of Coal Science and Technology (Taiyuan University of Technology), Ministry of Education and Shanxi province, Taiyuan 030024, PR China*

*Training Base of State Key Laboratory of Coal Science and Technology Jointly Constructed by Shanxi Province and Ministry of Science and Technology, Taiyuan 030024, PR China*

*corresponding author. E-mail: ying@tyut.edu.cn

**Abstract:** Pyrolysis of low-rank coal and the coal direct liquefaction residue (DLR) using co-char as solid heat carrier (SHC) can increase tar yield and improve upgrading efficiency. Besides the roles of heat storage and transfer, SHC also have some catalytic activity to regulate tar quality. The interaction of volatiles and SHC led to certain change of the specific surface area and pore structure of char SHC with catalytic active component would promote the stabilization of tar fragments by H radicals in volatiles. Ultimate analysis of the co-char SHC indicants that the content of O decreased slightly, while C and H had a modest increase. Co-pyrolysis with a rapid heating rate through SHC lead to extensive thermal breakdown of macromolecular fragments of coal and DLR, which outweigh the cracking effect of SHC. Simulated experiments of pyrolysis of Shendong coal and DLR with co-char as SHC were carried out to study product distribution and cycling performance. After certain cycles over the range of the conditions employed, the dilution of large amount of char heat carrier could counteract adhesion of DLR softening and melting during pyrolysis.

**Key words:** Solid heat carrier, Low rank coal co-pyrolysis, Volatiles and semi-coke interaction

## 1. Introduction

Low rank coal makes up around 55% of Chinese coal reserves, with the characteristics of high volatile matter, high oxygen content, and high moisture content, which make it difficult for wide utilization[1]. As the core of cascade utilization of low rank coal, pyrolysis is an effective way to get oil under mild conditions[2]. Pyrolysis through solid heat carrier (SHC) can increase the yield of tar and improve the efficiency of upgrading and heat utilization[3, 4]. Besides the roles of heat storage and transfer, SHC can also have some catalytic activity to regulate the quality of tar and gas.

The coal direct liquefaction residue (DLR) is a material with high carbon, high ash, and high sulfur and accounts for about 30% of feed coal[5]. Part of heavy oil can be recovered in pyrolysis process, simultaneously the sulfur iron catalyst and minerals in DLR may also have a catalytic effect on gasification reactivity of the residual char. Introducing DLR into low rank coal pyrolysis process showed inhibitory effect on coal pulverization[6]and an interactive effect to increase tar yield[7]. Co-pyrolysis of coal and DLR with interaction lead to an increased tar yield, yet a relatively high content of the heavy components in tar. As a part of the co-pyrolysis products, co-char SHC can recover the sensible heat and without the investment of solid-solid separation equipment. During SHC pyrolysis process, there existed catalytic cracking as well as thermal cracking of char SHC on volatiles, especially the effects of hot char[8]. Zeolites[9, 10], natural minerals[11-13], char-based catalysts coupled with a packed bed[14, 15] can also upgrade pyrolysis tar , which show cracking effects on primary volatiles for valuable chemicals[16, 17]. Liang[18] studied the process of simulated fluidized bed combustion combined with coal pyrolysis and indicated that most of the gas and tar are released during the first few minutes of the pyrolysis holding time. Deng[19] investigated lignite pyrolysis by Co/olivine heat carrier could increase on tar yield while decrease the heavy tar components. However, due to the adhesive effect of DLR during pyrolysis process, the gasification reactivity of char product might decline to some extent[20]. However, few attention has been devoted to the effect of SHC on tar quality-improving as well as high tar yield in co-pyrolysis process.

In this paper, the catalytic effects of co-char heat carrier on co-pyrolysis of SD coal and DLR were studied in a two-stage reactor, and structural and compositional changes of co-char SHC were analyzed before and after the pyrolysis process. To investigate the relationship between catalytic cracking and thermal history, operational stability of co-pyrolysis process by co-char SHC, the simulated co-pyrolysis cycling experiment of SD and DLR by co-char SHC were also carried out.

## 2. Experimental section

### 2.1. Coal and semi-cokes preparation

Shendong coal (SD) and its direct liquefaction residue (DLR) were used in the pyrolysis experiments. The coal samples were pulverized and sieved to acquire a fraction with the particle size ranging from 0.425 to 0.85 mm. Before the experiments, the samples were dried in a vacuum oven at 70 °C for 10 h. The proximate and ultimate analysis of the samples are listed in Table 1.

Table 1 Proximate analysis and Ultimate analysis of the samples

| Sample | Proximate analysis (wt.%, ad) | | | | Ultimate analysis (wt.%, daf) | | | | |
|---|---|---|---|---|---|---|---|---|---|
| | M | V | A | FC | C | H | O[a] | N | St |
| SD | 6.38 | 6.85 | 29.80 | 56.97 | 80.49 | 4.34 | 13.52 | 1.23 | 0.41 |
| DLR | 0.04 | 18.33 | 32.02 | 49.61 | 81.97 | 4.36 | 9.78 | 1.10 | 2.79 |

[a] By difference.

## 2.2. Apparatus and test procedure

**Apparatus.** A fixed-bed reactor made from quartz was used for the pyrolysis tests, and the preparation of semi-cokes was also carried on it. The schematic diagram of the reactor for semi-cokes preparation and is depicted in Fig. 1. The pyrolysis experimental apparatus mainly consists of fixed bed reactor, the air inlet system, temperature control system and condensation system.

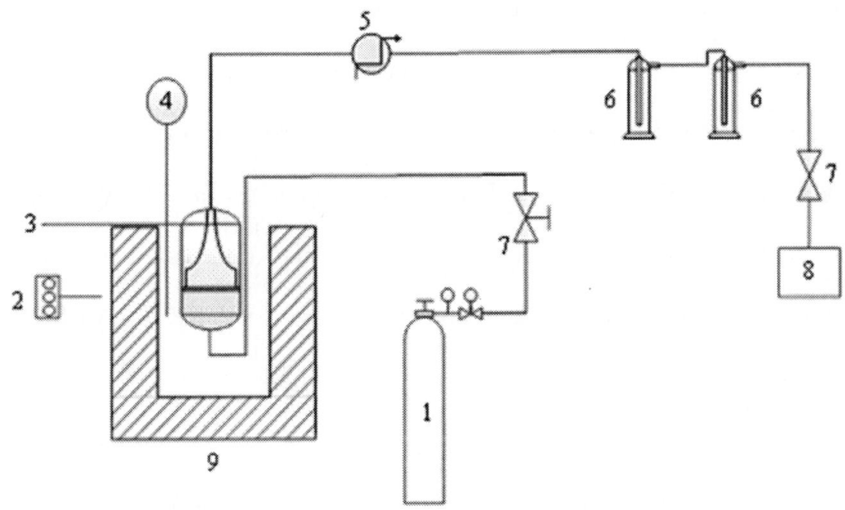


1-$N_2$ cylinder; 2-Temperature controller; 3-Reactor; 4-Thermacouple;
5-Tar condenser; 6-Washing bottle; 7-Valve; 8-Gas collecting bag; 9-Furnace

Fig. 1 Schematic of experimental apparatus for lignite pyrolysis

**Char preparation.** About 20 g SD coal with a 15% mass fraction of DLR were pre-loaded in the reactor with continuous nitrogen flow to exhaust air in the system. The reactor with coal particles was then lifted into the constant temperature zone of the furnace programmed to stay at 650 °C with a heating rate of 10 °C/min. Liquid products in volatiles produced during lignite pyrolysis were condensed in the cold trap with ice-salt bath. Gaseous products were collected in gas sample bag after two-stage wash bottles to absorb uncondensed matters. 30 min residence was given for the complete decomposition of coal particles. Finally the reactor was lifted out of the furnace under nitrogen and cooled down to the room temperature. The char product was collected and used as solid heat carrier in co-pyrolysis of SD coal and DLR tests.

**Experimental procedure and operating conditions.** The furnace was programmed to stay at 800 °C with a heating rate of 10 °C/min. SHC was first put into the reactor after flowing nitrogen had exhausted air in the system. The reactor was then lifted into the constant temperature zone of the furnace to storage heat for 30 min. SD coal and DLR was quickly added into the reactor, with a certain ratio of feed coal to SHC. The reactor was lifted out from the furnace and shaken for 30 s to mix the hot SHC and feed coal, and the reaction time was 5 min. The liquid and gas collection systems were the same in char preparation part.

The catalytic cracking effects of SHC were related to their physical and chemical properties. However, it was difficult to directly separate the newly generated chars from co-char SHC for the possibility of adhesion during pyrolysis. To study the interaction in the co-pyrolysis system, co-char was adopted as SHC were performed on a two-stage quartz reactor and was placed on the up layer, while the feeding material of SD coal and DLR on the down layer. Investigation of catalytic cracking effect of co-char heat carrier on pyrolysis volatiles was also conducted on the two-stage reactor, which allowed the separation of SHC and co-char for further investigation of SHC before and after the pyrolysis process.

A two-stage reactor were used to testify the interactive effect in co-pyrolysis of SD and DLR with co-char heat carrier. The heat carriers were pre-treated at 800 °C for 30 min, and it was considered that second cracking of semi-coke itself was not obvious at this temperature[21]. Quartz sand heat carrier was also used as the control group.SD coal and DLR were pyrolyzed with co-char individually to calculate the theoretical product yield. The difference of tar yield between experimental and theoretical caused by interaction was defined[7] as $\Delta Y_{tar}$,

Where, $\Delta Y_{tar} = Y_{E\text{-}tar} - Y_{C\text{-}tar} = Y_{E\text{-}tar} - (W_{SD} \times Y_{SD\text{-}tar} + W_{DLR} \times Y_{DLR\text{-}tar}) / (W_{SD} + W_{DLR})$

**Data treatment.** Calculation formulas of pyrolysis products yield are as followed:

Char yield: $Y_C = M_C/M \times 100\%$

Tar yield: $Y_T = M_T/M \times 100\%$

Pyrolysis water yield: $Y_W = (M_L - M_T)/M \times 100\%$

Gas yield: $Y_W = 100 - Y_C - Y_T - Y_W$

N-hexane soluble/tar: $f_{N/T} = (M_T - M_{NI})/M_T \times 100\%$

Where, $M$ is weight of feed coal (g); $M_C$ is weight of char (g); $M_T$ is weight of tar; $M_L$ is weight of liquid products (g); $M_{NI}$ is weight of n-hexane insoluble.

**2.3. Analysis and characterization**

The incondensable pyrolysis gases were analyzed by a gas chromatograph (GC-950) equipped with two flame ionization and a thermal conductivity detector. $H_2$, $N_2$, CO, $CO_2$ and hydrocarbons up to $C_5$ were analyzed. Qualitative analysis referred to the retention time of the standard gas, and quantitative method is the normalization of peak area based on correction factors.

The liquid products were first solved in Tetrahydrofuran (THF), and the rotary evaporation instrument was used to separate water and the solvent, the THF soluble was defined as tar. Tar product was then solved into n-hexane to separate the n-hexane soluble and asphaltenes, in which the n-hexane soluble was defined as the light tar, whose content was regarded as the evaluation basis of the quality of tars.

Specific surface area and pore structure characteristics of char samples were analyzed on an $N_2$ physisorption analyzer based on Brunauer-Emmett-Teller (BET) and Barrett-Joyner-Halenda (BJH) methods with a JW-GB instrument. And ultimate analysis of the char heat carrier was conducted with an Elementar vario MACRO cube, with the CHNS and O elemental determinator.

## 3. Results and discussion

**3. 1 Interaction in co-pyrolysis using co-char as SHC**

*3.1.1. Variation between experimental and theoretical product distribution*

As depicted in Fig. 2 and 3, there existed difference between SD coal and DLR pyrolyzed with co-char SHC individually. For DLR, more volatiles produced than SD coal, and high content of $H_2$ but low CO in pyrolysis gas. And the yield of pyrolysis water was much lower than SD. QS SHC was taken as the control group for investigating cracking effect of co-char SHC. Through comparison with co-py/QS, it can be found that tar yield decreased but n-hexane soluble content increased. That the theoretical value of n-hexane soluble content higher than experimental implied interactions between volatiles and co-char SHC.

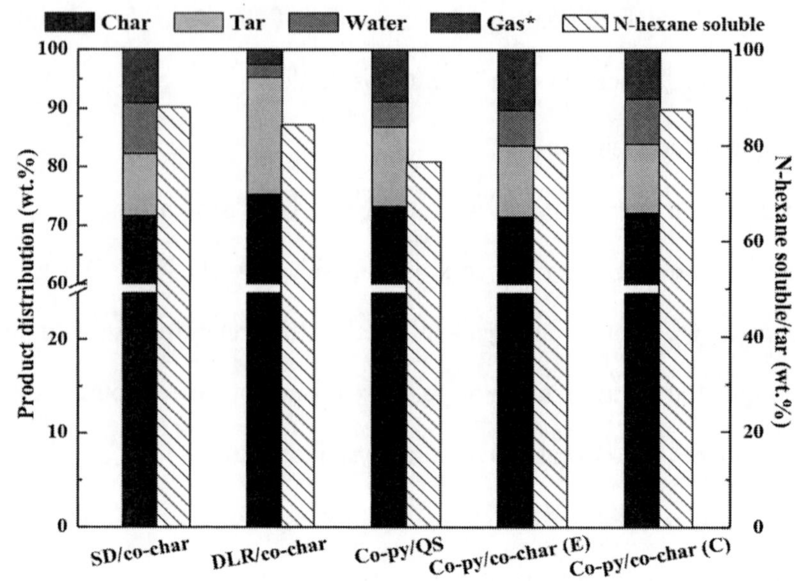


Fig.2 Product distribution of co-pyrolysis of SD coal and DLR with co-char SHC

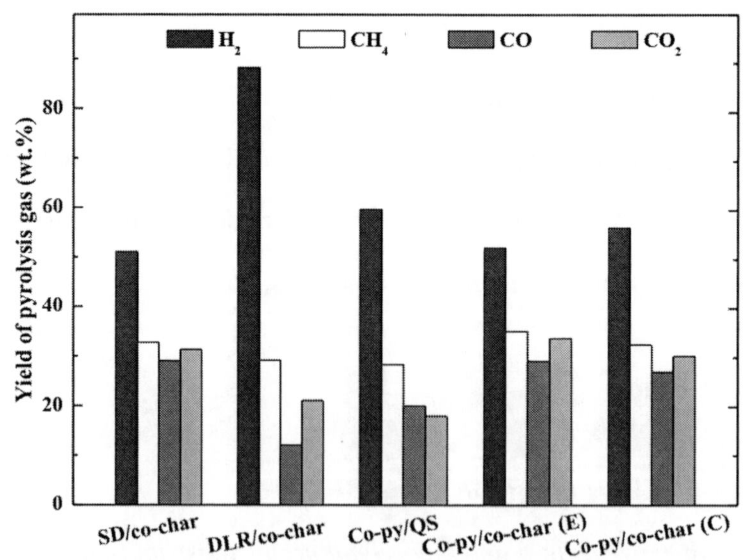


Fig.3 Yield of main pyrolysis gas of SD coal and DLR with co-char SHC

The experimental yield of co-pyrolysis tar was slightly higher than the calculated value, yet the n-hexane soluble fraction dropped. Compared with the Co-py/QS group, the tar yield of co-pyrolysis decreased for the interactions between volatiles and co-char SHC. However, the fraction of n-hexane soluble in co-pyrolysis tar with co-char SHC was higher than QS. Hydrogenation reactions of DLR might occur as it was relatively rich in hydrogen, and the reduced yield of $H_2$ during co-pyrolysis process indicated an improved hydrogen utilization.

There existed cracking effect of co-char on volatile matters during co-pyrolysis process as the yield of char and tar deceased, yet pyrolysis gas increased. Of the main components of co-pyrolysis gas, $H_2$ had dropped while $CH_4$ and COx all increased to certain extent. It was surmised that volatile matters experienced cracking reactions on surface of SHC, which resulted in the decrease of tar yield. Meantime, reforming reactions could took place while volatiles passing through SHC on up-layer, reasons from these two aspects changed the final product distribution.

*3.1.2. The structural and compositional changes of co-char SHC*

. It can be seen from Table 2 that specific area and pore volume of co-char decreased after pyrolysis with different feeding materials, yet the pore diameter increased. These changes might resulted from the physically attached of volatiles released to and/or coke deposited from cracking reactions on co-char. The micropores in co-char were blocked and some pore channels collapsed for larger pores formed.

Table 2 specific surface area and pore structure analysis of co-char SHC

| Samples | $S_{BET}/m^2 \cdot g^{-1}$ | $V_{total\ pore}/cm^3 \cdot g^{-1}$ | Pore diameter /nm |
|---|---|---|---|
| co-char | 16.18 | 0.021 | 2.59 |
| co-char - SD | 3.67 | 0.008 | 2.96 |
| co-char - DLR | 0.93 | 0.003 | 2.93 |
| co-char - co-py | 1.79 | 0.006 | 2.90 |

Ultimate analysis of co-char SHC was conducted, and changes of C, H, and O content were computed and exhibited in Table 3. The C content improved significantly after interacted with volatiles, H had a modest increment but O dropped. The relative change rates of C and H indicated that coke deposited on co-char surface in co-pyrolysis and the possibility of reactions between oxygen groups on co-char SHC and volatiles, yet further study of specific reactions and influence factors needs conducting.

Table 3 Changes of elements content in co-char SHC after pyrolysis

| Samples | Variation of ultimate contents △E (wt. /%, dry basis) | | |
|---|---|---|---|
| | C | H | O |
| co-char - SD | 0.45 | 0.183 | -0.349 |
| co-char - DLR | 2.99 | 0.198 | -0.265 |
| co-char - co-py | 3.47 | 0.2 | -0.128 |

Where, $\Delta E = E_{ij},- E^{0}_{ij}$ $E$, $E^{0}$ represented ultimate analysis of heat carrier before and after pyrolysis test, respectively. And *i* stood for C, H, O, and *j* for pyrolyzed with different material.

### 3.2. Cycling performance of SD coal and DLR co-pyrolysis by co-char SHC

In order to investigate the circulation stability of the process, the mixed pyrolysis char and co-char SHC was cooled to room temperature under nitrogen, and then screened to obtain its particle size distribution, and after that was evenly mixed for the next cycle. The product distribution of the rapid co-pyrolysis of SD coal and DLR with fresh co-char as heat carrier and the results of size distribution were recorded as Cycle-0, followed by a sequence of Cycle-1, Cycle-2... Cycle-10.

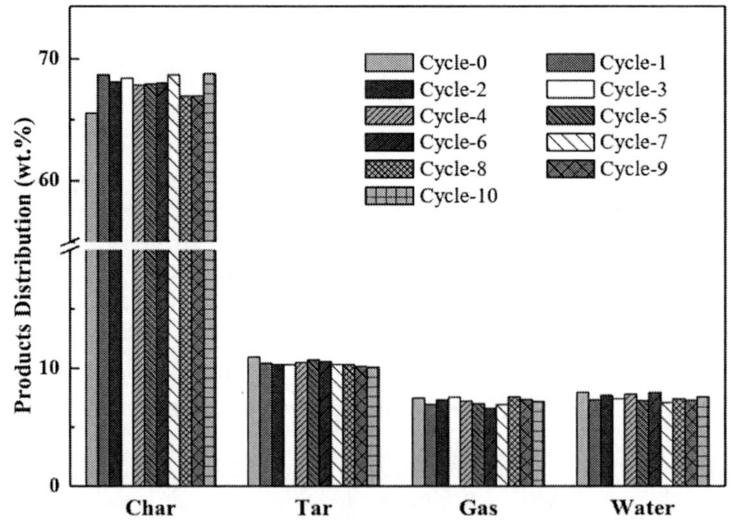


Fig. 4 Products distribution of co-pyrolysis with co-char as SHC in cycling

As shown in Fig. 4, with the increasing number of cycles, the char yield increased about 3% and the tar yield decreased slightly. Tar yield of the 10th cycle dropped less than 1% compared with the initial, and the yield of pyrolysis gas and water fluctuated in a certain range. Overall, the yield of the product was relatively stable.

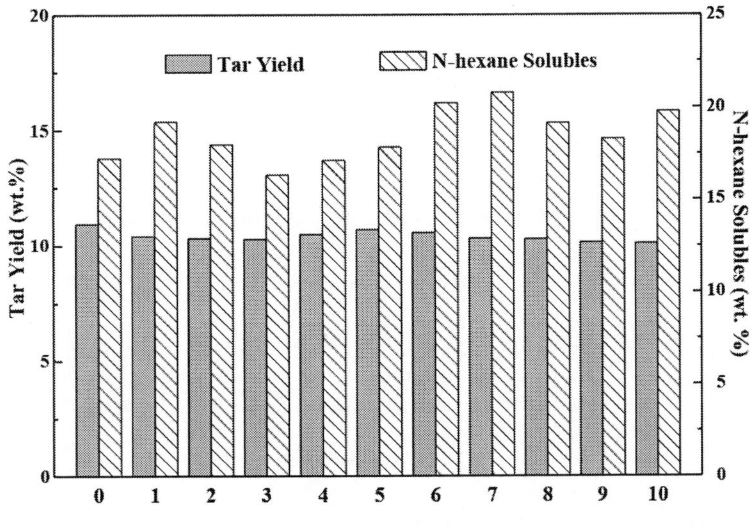


Fig. 5 Tar yield and composition of co-pyrolysis in cycling

According to the Figure 5, total tar yields were approximate during different cycles, but the content of n-hexane soluble exhibited fluctuations. Previous study demonstrated that the catalytic cracking performance of char had great relationship with its pore structure[22]. With the increasing of cycles, the newly produced co-char mixed with the heat carrier. Yet the amount of DLR accumulated gradually in the whole system, which further affected the structure and properties of co-char SHC.

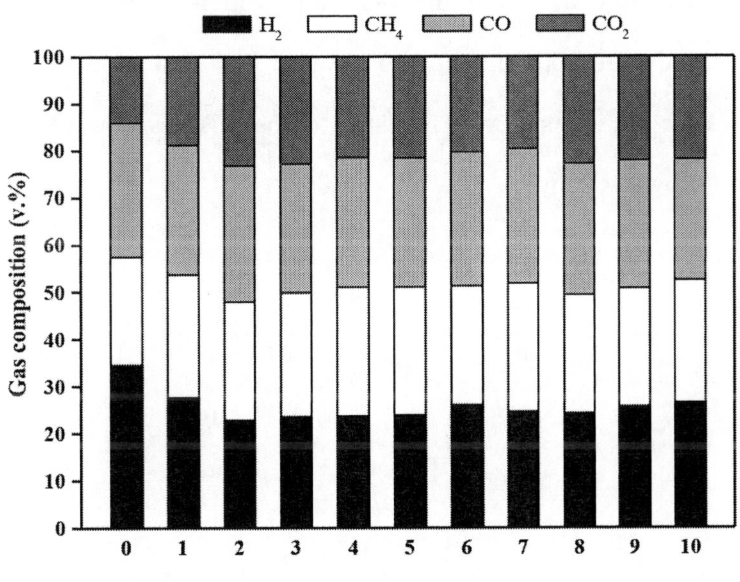


Fig. 6 Gas composition of co-pyrolysis in cycling

The volume percentage of gas in cycling was shown in Fig. 6. The yield of $CH_4$ changed little in the cycling, and the yield of CO and $H_2$ decreased, while the yield of $CO_2$ increased compared with those in cycle-0. From the aspect of circulation, the volume fraction of four main pyrolysis gas tended to be stable gradually after certain times of cycling.

### 3.3. Granular analysis of co-char in cycling

DLR has the property of softening and melting during its pyrolysis. Introducing DLR into the pyrolysis system caused decline of specific surface area and pore volume of co-char, and weakened the cracking activity of char. In rapid pyrolysis of Shendong coal and DLR, bonding among the mixed feeding materials and heat carrier char occurred. This resulted in changes of particle size, heat and mass transfer properties of char heat carrier during the pyrolysis process, which ultimately affected the distribution of pyrolysis products.

After the co-char cooled to room temperature in nitrogen and was taken out from the reactor, granular analysis was carried out through sieves with different number of meshes to obtain the particle size distribution. Compared with the preparation of co-char, the bonding effect of co-char was weakened obviously when the mixing ratio of heat carrier and coal and liquefied residue was 5:1. In co-char SHC preparation, the bonding of char particles to form bulks (char particle size larger than 0.85 mm) accounted for more than 80% of the total yield, which made it difficult to discharge. When adopted as SHC in the cycling experiment, the co-char was first ground to 0.425-0.85 mm. The change regulation of char particle size was investigated on this basis, and bulks of co-char were no longer ground in the followed cycling.

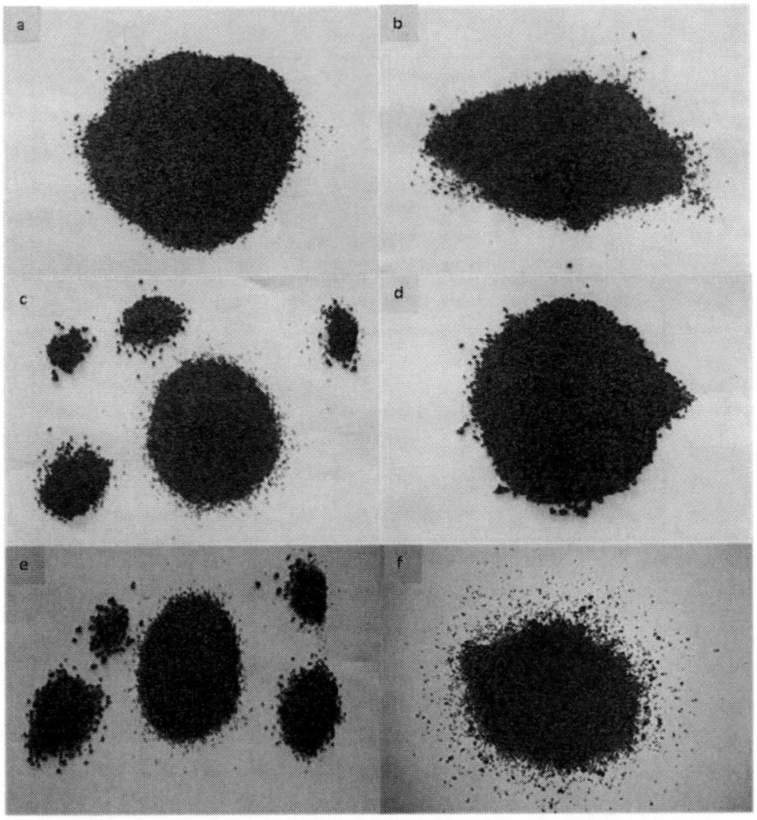


a. Co-char Heat Carrier; b. Cycle-0 Char; c. Cycle-0 Char Sieved;
d. Cycle-9 Char; e. Cycle -9 Char Sieved; f. Cycle-10 Char

Fig. 7 Granular analysis of co-char in cycling

With the increasing of cycling, fraction of char particle size bigger than 2 mm increased gradually, while the proportion of char particle size less than 0.85 mm decreased. From the first several cycles, the influence of pyrolysis reaction on mixed char particle size distribution was larger. The addition of mixed material adhered to heat carrier in the pyrolysis process, leading to the decreasing of mixed char particles less than 0.85 mm. But the granular distribution of mixed char was also influenced by "dilution effect" from the newly generated co-char. Due to the large amount of co-char heat carrier, the direct contact among coal and DLR particles was blocked. Those chars without adhesion added to fraction of mixed-char with less than 0.85 mm. Along with the circulation process went on, the heat storage and transfer characteristics of co-char heat carrier changed due to particle size alteration. And the thermal history of the co-pyrolysis of coal and DLR was affected, and this in turn impacted on the release of pyrolysis products. The new generated semi-coke changed the size of heat carrier on the next cycle to a certain extent.

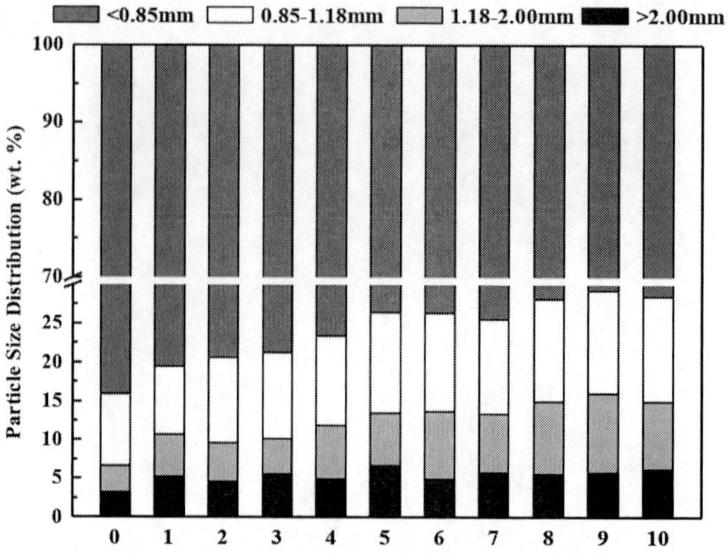


Fig. 8 Particle size distribution of co-char in cycling

Within the range of conditions employed in the experiments, after a certain cycles, there existed a certain degree of offset effect between the bulk formation from char particles bonding and the dilution of new generated semi-coke in the co-pyrolysis process, so that the particle size distribution of the mixed char could in relatively stable.

## 4. Conclusions

Co-pyrolysis of Shendong coal and direct liquefaction residue by co-char solid heat carrier could raise tar yield, meantime, cracking effect of co-char SHC produced more light oil fraction. Physical structure and main elements composition changes of co-char SHC was tested, and it surmised that char reforming/ gasification could play a role in the final product yield. The 10-cycle co-pyrolysis of SD and DLR by co-char SHC experiments were conducted. The product distribution changes and the tendency of mixed char particle size distribution in cycling were analyzed. And it can be testified the stable trend and feasibility of the process.

## Acknowledgements

The authors wish to thank their colleagues who have contributed to the information in this paper. Support was provided under the National High Technology Research and Development Program 863 (2011AA05A202).

## References

[1]. Qu Y, Chu M, Li D, et al. Fragmentation characteristic of lignite during heat upgrading[J]. Journal of China University of Mining & Technology. 2014,43(3):508-13.

[2]. Cen J, Fang M, Wang Q, et al. Development and prospect of coal staged conversion poly-generation technology[J]. Chemical Industry and Engineering Progress. 2010,30(1):88-94.

[3]. Hu G, Gong X, Huang H, et al. Experimental studies on flow and pyrolysis of coal with solid heat carrier in a modified rotating cone reactor[J]. Chemical Engineering & Processing. 2008,47(9–10):1777-85.

[4]. Liang P, Wang Z, Bi J. Simulation of coal pyrolysis by solid heat carrier in a moving-bed pyrolyzer[J]. Fuel. 2008,87(4–5):435-42.

[5]. Li J, Yang J, Liu Z. Pyrolysis behavior of direct coal liquefaction residues[J]. Journal of Fuel Chemistry & Technology. 2010,38(4):385-90.

[6]. Qu Y, Chu M, Shen GD, et al. Inhibitory effect of coal direct liquefaction residue on lignite pulverization during co-pyrolysis[J]. Fuel Processing Technology. 2016.

[7]. Li X, Xue Y, Feng J, et al. Co-pyrolysis of lignite and Shendong coal direct liquefaction residue[J]. Fuel. 2015,144:342-48.

[8]. Liu S, Wang Y, Wu R, et al. Fundamentals of Catalytic Tar Removal over in Situ and ex Situ Chars in Two-Stage Gasification of Coal[J]. Energy & Fuels. 2013,28(1):58-66.

[9]. Li S, Chen J, Hao T, et al. Pyrolysis of Huang Tu Miao coal over faujasite zeolite and supported transition metal catalysts[J]. Journal of Analytical & Applied Pyrolysis. 2013,102(7):161-69.

[10]. Wen Y, Cain E. Catalytic pyrolysis of a coal tar in a fixed-bed reactor[J]. Industrial & Engineering Chemistry Process Design & Development. 1984,23(4):627-37.

[11]. Devi L, Ptasinski KJ, Janssen FJJG, et al. Catalytic decomposition of biomass tars: use of dolomite and untreated olivine[J]. Renewable Energy. 2005,30(4):565-87.

[12]. Yu Q, Brage C, Nordgreen T, et al. Effects of Chinese dolomites on tar cracking in gasification of birch[J]. Fuel. 2009,88(10):1922-26.

[13]. Shamsi A. Catalytic and Thermal Cracking of Coal-Derived Liquid in a Fixed-Bed Reactor[J]. Industrial & Engineering Chemistry Research. 1996,35(35):1251-56.

[14]. Han J, Wang X, Yue J, et al. Catalytic upgrading of coal pyrolysis tar over char-based catalysts[J]. Fuel Processing Technology. 2014,122(6):98–106.

[15]. Wang FJ, Zhang S, Chen ZD, et al. Tar reforming using char as catalyst during pyrolysis and gasification of Shengli brown coal[J]. Journal of Analytical & Applied Pyrolysis. 2013,105(5):269-75.

[16]. Sarıoğlan A. Tar removal on dolomite and steam reforming catalyst: Benzene, toluene and xylene reforming[J]. International Journal of Hydrogen Energy. 2012,37(10):8133-42.

[17]. Hesp WR, Waters PL. Thermal Cracking of Tars and Volatile Matter from Coal Carbonization[J]. Ind Eng Chem, Prod Res Dev. 1970,9(2):194-202.
[18]. Liang P, Wang Z, Bi J. Process characteristics investigation of simulated circulating fluidized bed combustion combined with coal pyrolysis[J]. Fuel Processing Technology. 2007,88(1):23-28.
[19]. Deng J, Li W, Li X, et al. Product distribution of lignite pyrolysis with olivine-based solid heat carrier[J]. Journal of Fuel Chemistry & Technology. 2013,41(08):937-42.
[20]. Li X, Ma J, Xue Y, et al. Properties of semi-coke from co-pyrolysis of lignite and direct liquefaction residue of Shendong coal[J]. Journal of Fuel Chemistry & Technology. 2015,43(11):1281-86.
[21]. Wang X, Han J, Lu J, et al. Catalytic cracking of coal pyrolysis product for oil and gas upgrading over char-based catalysts[J]. Journal of Chemical Industry and Engneering. 2012,63(12):3897-905.
[22]. Li X, Ma J, Li L, et al. Semi-coke as solid heat carrier for low-temperature coal tar upgrading[J]. Fuel Processing Technology. 2016,143:79-85.

# The prediction of pyrolysis products in grading Lignite pyrolysis process

*Lan Yi, Jie Feng, Yuhong Qin, Wenying Li**

*Key Laboratory of Coal Science and Technology (Taiyuan University of Technology), Ministry of Education and Shanxi province, Taiyuan 030024, PR China*

*Training Base of State Key Laboratory of Coal Science and Technology Jointly Constructed by Shanxi Province and Ministry of Science and Technology, Taiyuan 030024, PR China*

* *Corresponding author. E-mail: ying@tyut.edu.cn*

**ABSTRACT:** Pyrolysis is the core unit of lower-rank coal upgrading and refining. A coal pyrolysis model which can be used to predict the product distribution is important to study the grading Lignite pyrolysis process. Based on mass balance in this paper, a pyrolysis model which combines prediction of volatile yields and component analysis of coal tar under different operating temperature is developed. The FG-DVC (Functional group-Depolymerization Vaporization Crosslinking) model is used to quantitatively predict the yields of pyrolysis char, light gases and coal tar. The Van Krevelen diagram which can provide the input parameters of the FG-DVC model by interpolation method is modified to cover more lower-rank coal. A non-linear programming (NLP) model is employed to analyze the component of coal tar, which includes phenols, aromatics, aliphatics, oxy-components and pitchs. The proximate analysis of char and ultimate analysis of coal tar are obtained based on the mass and element conservation of coal pyrolysis unit. The comparison between model prediction and experimental data validates this model both in total volatile yield and in tar yield.

***Key words***: coal pyrolysis; FG-DVC model; component analysis of coal tar

## 1 Introduction

Grading lignite pyrolysis combines gases and light liquid product extracts, as well as electricity and chemicals with heat production to realize the efficient utilization of chemical and thermal energy of the lignite. Many studies[1-3] have focused on the grading lignite pyrolysis on the laboratory, and the results showed that the grading lignite pyrolysis was economically viable and improved yields of high value-added products. Pyrolysis is the core unit in grading lignite pyrolysis. However, the treatment of coal pyrolysis in the process simulation of grading lignite pyrolysis by ASPEN PLUS is either to calculate the related products yields on the basis of coal composition by using the correlative empirical formula[4], or to obtain the products yields by directly putting experimental results into the RYield modul[5]. Because of the large deviation in the empirical formula and the limitation of simulating only a certain temperature's pyrolysis reaction by directly replacing experimental results, a coal pyrolysis model which can accurately predict the product distribution and analyze the tar composition is meaningful for the grading lignite pyrolysis.

Coal pyrolysis models have progressed from simple empirical to network models. The empirical models confirm total volatile release by single-step expressions[6], or by two competitive reactions[7], or by multi-step consecutive reactions[8, 9]. The distributed activation energy model (DAEM) [10] includes standard deviation of the activation energy as an additional parameter and inputs the volatile yields from experimental results. Sommariva et al.[11] have presented a multi-step kinetic pyrolysis model which described coal pyrolysis by three reference coals. With the rapid development of modern instruments, it is possible to study the coal macrostructure by means of Py-FIMS, $^{13}$C-NMR, and TG-FTIR. These techniques accelerate the propose of network pyrolysis models, such as FG-DVC (group-depolymerization, vaporization and cross-linking)[12] model, FLASHCHAIN (distributed-energy chain)[13] model, CPD (chemical percolation model for devolatilization)[14] model, and FD (Fragmentation and diffusion)[15] model. These pyrolysis models focusing on coal structure have been widely used to inspect the pyrolysis of biomass[16, 17], and accurately predict the yields of gas and tar.

Coal tar is an important by-product of coal conversion and the analysis and detection of coal tar is carried out by technological means or by mathematic model. Traditional processing methods prefer to distill the crude tar at first, then separate and refine the distilled fractions step by step, until some light chemical products such as BTX and naphthalene are obtained. Wang et al[18] used GC/MS and NMR to analyze the coal tar compound obtained under different pyrolysis atmosphere. Luo et al[19] characterized the derived chars, tars and gases from coal pyrolysis and analyzed the distribution of seven phenolic compound species in tar by GC/MS. He et al[20] used a non-linear programming (NLP) model to predict the distribution of coal pyrolysis product and tar composition in the process modeling of Lurgi fixed-bed coal gasifier.

This paper presents a lower-rank coal pyrolysis model which underlines the prediction of pyrolysis product and the analysis of tar component. The FG-DVC model is taken advantage of to fulfill the prediction of pyrolysis product and a non-linear programming (NLP) model is established to achieve the analysis of phenols, aromatics, aliphatics, oxy-components and pitchs in coal tar. The model has been validated against experimental measurements and the prediction of product distribution can be directly applied to the process simulation of grading lignite pyrolysis.

# 2 Modeling and analysis of coal pyrolysis

## 2.1 Prediction of coal pyrolysis products

In this section, the FG-DVC model is used here to describe the coal pyrolysis. The model currently predicts volatile yields from the functional group sources in the coal using rank-independent kinetics. The evolution of each functional group into gas and the evolution of molecular aromatic cluster into tar are represented by first-order decay, as shown in Eqs. (1) and (2):

$$\frac{dY_i}{dt} = -k_i Y_i \quad (1)$$

$$\frac{dX}{dt} = -k_x X \quad (2)$$

Where $Y_i$ is the amount of $i$ functional group, $X$ is the amount of tar precursor. Here the reaction rare constant, $k_i$ and $k_x$, are given by

$$k_i = k_{i0} \exp\left(-\frac{E_i}{RT}\right) \quad (4)$$

$$k_x = k_{x0} \exp\left(-\frac{E_x}{RT}\right) \quad (5)$$

The remained functional groups and aromatic clusters constitute the char. Thus the time dependent weight fraction of the char, $W_i(char)$, is

$$W_i(char) = XY_i \quad (6)$$

Before the end of pyrolysis, all light gases and tar come from the so-called char which is different from the product char. The time dependent weight fractions of each gas, $W_i(gas)$, and tar, $W_i(tar)$ are

$$\frac{dW_i(gas)}{dt} = k_i W_i(char) \quad (7)$$

$$\frac{dW_i(tar)}{dt} = k_x W_i(char) \quad (8)$$

To analyze and systemize the above equations, the following formulas can be obtained to get the yields of each gas and tar:

$$W_i(tar) = \int_0^t \left(-\frac{dX}{dt}\right) Y_i dt = (X^o Y_i^0 - XY_i)\frac{k_x}{k_x + k_i} \quad (9)$$

$$\mathrm{W(tar)} = \sum W_i(tar) \quad (10)$$

$$W_i(gas) = \int_0^t k_i(1 - X^o + X)Y_i dt = (1 - X^o)(Y_i^0 - Y_i) + W_i(tar)\frac{k_i}{k_x} \quad (11)$$

Where $X^o$ is the potential tar-forming fraction, $Y_i^0$ is the initial amount of $i$ functional group, and $\mathrm{W(tar)}$ is the total yield of tar. Except of $H_2$, all gases have more than two formation paths, so the yield of a given product is equal to the sum of the yields of that product from each path. As an example, the yield of $CO_2$ is associated with reactions in three paths, and thus can be written as follows:

$$W_i(\mathrm{CO_2}) = W_1(\mathrm{CO_2}) + W_2(\mathrm{CO_2}) + W_3(\mathrm{CO_2}) \quad (12)$$

Table1. The ultimate and proximate analyses of coals

| samples | Ultimate analysis [a] (wt.%) | | | | | Proximate analysis [a] (wt.%) | |
|---|---|---|---|---|---|---|---|
| | C | H | N | S | O[b] | VM | FC |
| Coal A | 74.59 | 5.18 | 1.06 | 0.28 | 18.89 | 46.36 | 53.64 |
| Coal B | 71.13 | 5.04 | 1.68 | 1.38 | 20.77 | 54.49 | 45.51 |

a. daf basis b. by difference

Table 2 Kinetic parameters

| Param. | Gas | Rate equation |
|---|---|---|
| $Y_1^0$ | $CO_2$ extra loose | $k_1 = 5.6E + 17\exp(-(30000 \pm 2000)/T)$ |
| $Y_2^0$ | $CO_2$ loose | $k_2 = 6.5E + 16\exp(-(33850 \pm 1500)/T)$ |
| $Y_3^0$ | $CO_2$ tight | $k_3 = 1.1E + 15\exp(-(38315 \pm 2000)/T)$ |
| $Y_4^0$ | $H_2O$ loose | $k_4 = 2.2E + 18\exp(-(30000 \pm 1500)/T)$ |
| $Y_5^0$ | $H_2O$ tight | $k_5 = 1.7E + 13\exp(-(32700 \pm 1500)/T)$ |
| $Y_6^0$ | CO ether loose | $k_6 = 1.4E + 18\exp(-(40000 \pm 6000)/T)$ |
| $Y_7^0$ | CO ether tight | $k_7 = 1.5E + 15\exp(-(40500 \pm 1500)/T)$ |
| $Y_8^0$ | CO extra tight | $k_8 = 2.0E + 13\exp(-(45500 \pm 1500)/T)$ |
| $Y_9^0$ | CHx aliphatic | $k_9 = 8.4E + 14\exp(-(30000 \pm 1500)/T)$ |
| $Y_{10}^0$ | $CH_4$ extra loose | $k_{10} = 8.4E + 14\exp(-(30000 \pm 1500)/T)$ |
| $Y_{11}^0$ | $CH_4$ loose | $k_{11} = 7.5E + 13\exp(-(30000 \pm 2000)/T)$ |
| $Y_{12}^0$ | $CH_4$ tight | $k_{12} = 3.4E + 11\exp(-(30000 \pm 2000)/T)$ |
| $Y_{13}^0$ | H aromatic | $k_{13} = 1.0E + 14\exp(-(40500 \pm 6000)/T)$ |
| $X^0$ | tar | $k_x = 8.6E + 14\exp(-(27700 \pm 1500)/T)$ |

$X^o$ and $Y_i^0$ are the only two parameters in the FG-DVC model to predict the yields of gas and tar. In the model, Zhao et al[21] have studied the parameters of different coals and defined the coals (library coals) to form a triangular mesh in the Van Krevelen diagram. If an unknown coal is within a triangle formed by three library coals, all the parameters of this unknown coal can be interpolated from those of the three library coals based solely on the elemental composition. The coalification degree of Zhao's study is wide, which covers from lignite to anthracite. However, the main study objects in this paper are the low rank coal, to modify the range of Zhao's Van Krevelen diagram is requisite. By using the same experimental method with Zhao's, Hobbs et al[21] detailedly studied the parameters of the FG-DVC model in the research of fixed bed gasifier. The appropriate coals in Hobbs' study are chosen to modify Zhao's Van Krevelen diagram. Fig.1 is the modified Van Krevelen diagram. Coal 1, 6, 9, 10,11and 12 are from ref. [22], the coal 2, 3, 4, 5, 7, and 8 are from ref. [21].

In this model, individual species evolve from certain functional group, both the activation energy, E, and $k_i$, of different products are considered independent of coal types and operating conditions approximately. All the Kinetic parameters are shown in Table 2, while the proximate and ultimate analyses data for coal A and coal B are listed in Table 1.

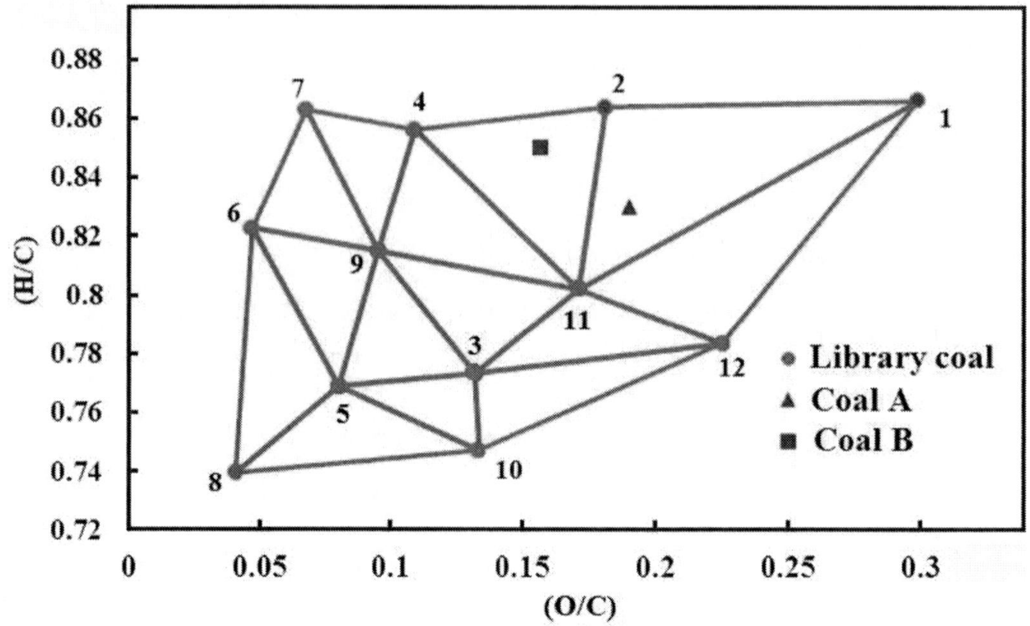


Fig.1 modified Van Krevelen diagram

## 2.2 Analysis of tar component

The composition of low temperature coal tar is complicated and the content of individual component is very low, which results in the difficulty of constituent analysis. Based on boiling point of coal tar, Dai et al.[23, 24] classified coal tar into four lump to implement the study of coal tar hydrogenation. In the thermodynamic and economic analysis of poly-generation system integrating atmospheric pressure coal pyrolysis, Guo et al.[25] choose five model compounds to represent the coal tar. Here based on mass/element conservation, a non-linear programming (NLP) mathematical model is formulated to estimate the composition of coal tar. During the process of analysis, the following assumptions are made:

1. The ashes in coal are treated as inert and they remain in the solid residue;
2. The coal tar does not include the elements of N and S, which is only composed of C, O and H.

### 2.2.1 Determination of proximate and ultimate analyses of solid product

Coal pyrolysis is equal to the process of dehydrogenation, deoxidation and carbon-rich, and the proximate and ultimate analyses of solid product are changed. Through the FG-DVC model, the yields of gas, liquid (water and tar), and char are estimated. Based on mass conservation, the following Eq. (15)~(17) are applied to calculate the contents of fixed carbon (FC), ash and volatile matters (VM) in char.

$$\text{char(ash)} = \frac{coal(ash)}{Y_{char}} \quad (15)$$

$$\text{char(VM)} = \frac{coal(VM) - Y_{\text{gas}} - Y_{liquid}}{Y_{char}} \quad (16)$$

$$\text{char(FC)} = 1 - \text{char(ash)} - \text{char(VM)} \quad (17)$$

Where char(ash), char(VM) and char(FC) are the contents of ash, VM and FC in char, respectively. Coal(ash) and coal(VM) are the contents of ash and VM in coal, respectively. $Y_{\text{gas}}$, $Y_{liquid}$ and $Y_{char}$ are the yields of gas, liquid and char, which are predicted by FG-DVC model.

Proximate and ultimate analyses are two important indexes to understand the characteristics of solid fuel, and a remarkable correlation exists within them. Due to the many experimental items, long test time and complex operations, correlations between the proximate analysis and ultimate analysis have been put forward. Parikh et al.[26] and Shen et al.[27] presented correlations for calculating the elemental composition based on proximate analysis of biomass. Nhuchhen et al.[28] proposed a correlation to estimate the elemental compositions using the proximate analysis of raw and torrefied biomass. Li et al.[29] studied some relevant equations for proximate and ultimate analyses of coal. Pyrolysis is a process of upgrading and accelerating coalification, therefore low temperature pyrolytic char can be regarded as a higher-rank coal. The ultimate analysis of low temperature pyrolytic char can be estimated based on the proximate analysis. Li's correlations will be used in this paper, which are shown:

$$C_{ad} = -0.0588A_{ad} + 0.4773V_{ad} + 0.6131FC_{ad} + 0.8251 \quad (18)$$

$$H_{ad} = 0.0559V_{ad} + 0.05702FC_{ad} + 0.07282 \quad (19)$$

$$N_{ad} = 0.0082V_{ad} + 0.0290 \quad (20)$$

$$S_{ad} = 0.1251A_{ad} - 1.6997 \quad (21)$$

$$O_{ad} = 100 - (C_{ad} + H_{ad} + N_{ad} + S_{ad} + M_{ad} + A_{ad}) \quad (22)$$

Where $C_{ad}$, $H_{ad}$, $N_{ad}$, $S_{ad}$, and $O_{ad}$ are the contents of carbon, hydrogen, nitrogen, sulfur and oxygen in coal on air dry basis. $FC_{ad}$, $V_{ad}$, $A_{ad}$ and $M_{ad}$ are the contents of fixed carbon, volatile matter, ash and moisture in coal on air dry basis.

### 2.2.2 Determination of ultimate analyses of coal tar

The pyrolysis temperature in grading lignite pyrolysis is between 500~650°C, Zhang et al.[30] analyzed the element composition of low temperature and concluded the results shown in Table 3. The contents of carbon, hydrogen and oxygen are between 82~85%, 8.5~10.5% and 4.5~6.1%, respectively. Meanwhile, the pyrolysis unit is a separate flat, the elements of carbon, hydrogen and oxygen in the products of light gas, tar, char and water all abide by the element conservation. The conservation equations, which can be used to calculate the element composition of coal tar, are listed in the following:

$$\mathrm{C(tar)} = 1 - \mathrm{C(char)} - \mathrm{C(gas)} \quad (23)$$

$$\mathrm{H(tar)} = 1 - \mathrm{H(char)} - \mathrm{H(gas + water)} \quad (24)$$

$$\mathrm{O(tar)} = 1 - \mathrm{O(char)} - \mathrm{O(gas + water)} \quad (25)$$

Table 3 Ultimate analysis of low temperature coal tar

| Ultimate composition (wt.%) | | | | | H/C(atom) |
|---|---|---|---|---|---|
| C | H | O | N | S | |
| 82-85 | 8.5-10.5 | 4.5-6.1 | 0.48-0.82 | 0.3-1.5 | 1.24-1.49 |

## 2.3 mathematic model of tar compound

The low-temperature coal tar is made up of tens of thousands of compounds, in which only about five hundred compounds can be identified. However, it is amazing that the content of most compounds is very low. Specifically, the number of compounds whose contents are higher than 1% is only between 10~30.

To study the composition and characteristic of coal tar, choosing appropriate coal tar model compound is a key and difficult problem. In the study of catalytic hydrogenation of coal tar, the tar has been divided into light oil, phenol oil, naphthalene oil, wash oil and asphaltene according to the cut fraction and five model compounds are chosen to represent each kind of oil[23, 24]. In the processing simulation study of coal tar hydrocracking by ASPEN PLUS , the tar is divided into the kinds of heterocyclic compounds, acid compounds, aromatic hydrocarbons and aliphatic hydrocarbons[31]. In the process simulation of coal pyrolysis poly-generation, the model compounds are selected according to the tar cut fraction and the difference of substance species[2], and phenol, hexadecane, quinolone, naphthalene and dibenzothiophene are picked as the model compound, which can basically describe the tar characteristic[1], or to accurately analysis the hydrogen and energy consumption of coal tar hydrogenation unit in poly-generation, as many as possible model compounds are elected[3]. In the study of coal pyrolysis model, the components which have different aliphatic/oxygenic side chains and benzene number are chosen as model compounds[11]. In this paper, the pyrolysis temperature in grading lignite pyrolysis is 500~650 °C, and the product is low temperature coal tar. For the great variety, it is impossible to fulfill the precise quantification of every tar component. Applying tar model compound to describe coal tar characteristics is a feasible method.

Due to that each tar cut fraction is not separated and that the typical component of each cut fraction cannot completely represent the tar characteristics of whole cut fraction, it is not in strict accordance with the material content of each cut fraction when choosing the coal tar model compounds. Being as a whole, the model compounds are picked meeting the following two requires: 1) the kind of coal tar, 2) the boiling spread of the cut fraction. Total 27 tar model compounds are applied in Table 4. And the relative physical property of this model compounds are listed in Table 5.

Table 4 Model compounds of coal tar

| Component | <215°C | 215-320°C | >320°C |
| --- | --- | --- | --- |
| Phenols | phenol | | |
| | Cresol, xylenol, tricresol | | |
| Alkanes<br>Olefins<br>Cycloalkanes | Tridecane, tetradecane, pentadecane, cetane, heptadecane, octodecane, nonadecane, eicosane, heneicosane, docosane, tricosane, lignocerane, pentacosane | | |

| | | | |
|---|---|---|---|
| Aromatics | Toluene, ethylbenzene | Methylnaphthalene, dimethylnaphthalene | dimethylphenanthrene |
| Oxy-components | tetradecanol | trimethylphenanol | |
| pitchs | | | Pyrene, chrysene, wax |

Table 5 Model compounds of tar and its properties

| Components | Molecular formula | Molecular weight | Boiling point | H/C |
|---|---|---|---|---|
| phenol | $C_6H_6O$ | 94 | 182 | 1 |
| Cresol | $C_7H_8O$ | 108 | 191-202 | 1.14 |
| xylenol | $C_8H_{10}O$ | 122 | 225 | 1.25 |
| tricresol | $C_9H_{12}O$ | 136 | 232 | 1.33 |
| Toluene | $C_7H_8$ | 92 | 111 | 1.14 |
| ethylbenzene | $C_8H_{10}$ | 106 | 136 | 1.25 |
| Methylnaphthalene | $C_{11}H_{10}$ | 142 | 240-245 | 0.91 |
| dimethylnaphthalene | $C_{12}H_{12}$ | 156 | 262-266 | 1 |
| dimethylphenanthrene | $C_{16}H_{14}$ | 206 | -- | 0.88 |
| Tridecane | $C_{13}H_{28}$ | 184 | 235 | 2.15 |
| tetradecane | $C_{14}H_{30}$ | 198 | 253 | 2.14 |
| pentadecane | $C_{15}H_{32}$ | 212 | 268 | 2.13 |
| cetane | $C_{16}H_{34}$ | 226 | 287 | 2.13 |
| heptadecane | $C_{17}H_{36}$ | 240 | 292 | 2.12 |
| octodecane | $C_{18}H_{38}$ | 254 | 316 | 2.11 |
| nonadecane | $C_{19}H_{40}$ | 268 | 330 | 2.11 |
| eicosane | $C_{20}H_{42}$ | 282 | 343 | 2.10 |
| heneicosane | $C_{21}H_{44}$ | 296 | 100 | 2.10 |
| docosane | $C_{22}H_{46}$ | 310 | 369 | 2.09 |
| tricosane | $C_{23}H_{48}$ | 324 | 199-200 | 2.09 |
| lignocerane | $C_{24}H_{50}$ | 338 | 391 | 2.08 |

| | | | | |
|---|---|---|---|---|
| pentacosane | $C_{25}H_{52}$ | 352 | 169-170 | 2.08 |
| tetradecanol | $C_{14}H_{30}O$ | 214 | 289 | 2.14 |
| trimethylphenanol | $C_{13}H_{14}O$ | 186 | -- | 1.08 |
| Pyrene | $C_{18}H_{12}$ | 228 | 448 | 0.67 |
| chrysene | $C_{16}H_{10}$ | 202 | 394 | 0.63 |
| Wax(pseudocomponents) | $C_{50}H_{75}O_4$ | 739 | -- | 1.50 |

In this work, a mathematical model is established to describe the coal tar distribution in pyrolysis process in grading lignite process. The following subsections outline this model that gives the overall tar formation reaction.

**Sets**. The set of all atoms $A_{tar,i}$ is defined as

$$a_i \in A_{tar,i} = \{C, H, O\} \qquad i \in [1,3]$$

The set of all tar species produced from the pyrolysis step is given as follows:

$$s_j \in S_{tar,j} = \begin{Bmatrix} C6H6O, C7H8O, C8H10O, C9H12O, C7H8, C8H10, C11H10, C12H12, \\ C16H14, C13H28, C14H30, C15H32, C16H34, C17H36, C18H38, \\ C19H40, C20H42, C21H44, C22H46, C23H48, C24H50, C25H52, \\ C14H30O, C13H14O, C18H12, C16H10, C50H75O4 \end{Bmatrix} \quad j \in [1,27]$$

We define a new index, called ratio, that represents the relationship between certain species involved in the coal pyrolysis process. The set Ration contains these specific relationships as denoted below:

$$\text{Ratio} = \begin{Bmatrix} ratio_1, ratio_2, ratio_3, ratio_4, ratio_5, ratio_6, ratio_7, ratio_8, ratio_9, ratio_{10} \\ ratio_{11}, ratio_{12}, ratio_{13}, ratio_{14}, ratio_{15}, ratio_{16}, ratio_{17}, ratio_{18}, ratio_{19} \\ ratio_{20}, ratio_{21}, ratio_{22}, ratio_{23}, ratio_{24}, ratio_{25}, ratio_{26}, \end{Bmatrix}$$

Where

$$ratio_j = \frac{m^*_{s_j}}{m^*_{s_{j+1}}} \qquad j \in [1,26]$$

$m^*_{s_j}$ represents the mass fraction of individual component, $s_j$, in coal tar.

**Parameters.** The following parameters are defined:

$w_a$: weight fraction of atom a in coal tar sample

$E_{a,s}$: atom weight fraction of atom a in tar species

The composition of the coal tar products varies depending on the coal rank, pyrolysis conditions and other factors. Tables 6 and 7[30] are the distillation and composition analyses of low temperature coal tar in the areas of Shanxi, Inner Mongolian and Shaanxi. And the generic data in the tables are used to calculate the tar formation reaction.

**Variables.** The following variables are defined to simulate the tar formation reaction. Weight fractions are used to simulate the species weight fractions in the tar products. To allow for the possibility that the species composition will not exactly match the data in Tables 7 and 8, slack variables are introduced.

$m_{s_j}$: weight fraction of species $s_j$

$S_{ratio}$: slack variable for species ratio constraints, where ratio ∈ Ratio

**Constraints**. The equation that gives the stoichiometric coefficients of the tar formation reaction is in the following. Equations 26 models the atomic balances during tar formation:

$$m_{tar} * w_a = E_{a,s} * m_{s_j} \quad (26)$$

Where $m_{tar}$ is the tar mass. All atoms in tar are assumed to be converted to tar species. Additional constraints are added based on the expected yields of the tar formation. The following constraints utilize information from Tables 7 and 8 to constrain the ration of tar species.

$$-S_{ratio_j} \leq \frac{m_{s_j}}{m_{s_{j+1}}} - \frac{m *_{s_j}}{m *_{s_{j+1}}} \leq S_{ratio_j} \quad j \in [1,26] \quad (27)$$

$$m_{s_j} \geq 0 \qquad \forall s \in S_{tar}$$

$$S_{ratio} \geq 0 \quad \forall \text{ratio} \in \text{Ratio}$$

**Objective Function.** The composition yield of the tar formation can be estimated by minimizing the slack variables as follows:

$$\min_{m_j, S_{ratio}} \sum_{ratio} S_{ratio} \quad (28)$$

Table 6 Distillation analysis of low temperature coal tar

| Distillation range/°C | Yield/ % | | |
|---|---|---|---|
| | A certain area in Shanxi province | A certain area in Inner Mongolian | A certain area in Shaanxi province |
| <215 | 22.07 | 16.88 | 17.33 |
| 215-320 | 47.07 | 44.99 | 52.30 |

| | | | |
|---|---|---|---|
| >320 | 30.86 | 37.99 | 30.34 |
| Total | 100.00 | 99.86 | 99.97 |

Table 7 Composition analysis of low temperature coal tar

| Components | Yield/ % | | |
|---|---|---|---|
| | A certain area in Shanxi province | A certain area in Inner Mongolian | A certain area in Shaanxi province |
| Hydrocarbons | 21.70 | 30.50 | 21.63 |
| Aromatic | 19.48 | 19.58 | 22.29 |
| Phenols | 23.37 | 17.32 | 19.95 |
| Unknown | 35.45 | 34.02 | 36.13 |
| Total | 100.00 | 100.00 | 100.00 |

# 3 Results and discussion

The results obtained from predicting of coal A and coal B pyrolysis products are provided in Figs. 2 and 3, together with the related experimental data. Fig.2 summarizes the yields of tar and total volatile at different temperatures, from which it is evident that the yield of total volatile increases with increasing pyrolysis temperature. While the tar yield increases until a maximum is reached at about 650 °C for coal A and 800 °C for coal B, and then decreases with further increase in temperature, since at high temperatures the cracking of tar occurs more rapidly than the tar is formed. Fig.3 shows that the yields of key gas products increase with the higher temperatures. As shown in Figs.2 and 3, in which experimental data obtained under the same conditions are also displayed. Except of $CO_2$, all other products' predictions agree well with the measurements, which validate the modified Van Krevelen diagram. The deviation existing in $CO_2$ derives from the neglect of mineral matters decomposition in coal pyrolysis, which can release plenty of $CO_2$ when pyrolysis temperature exceeds 700 °C[32].

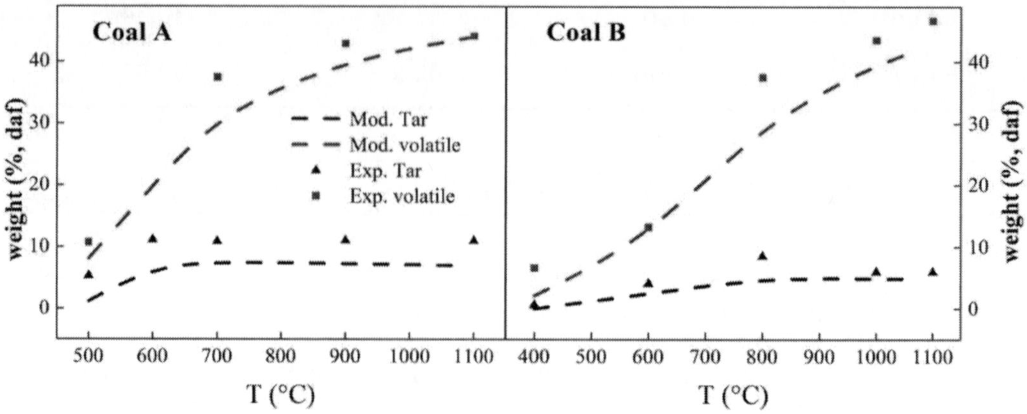


Fig. 2 Yields of tar and volatile as a functional of temperature

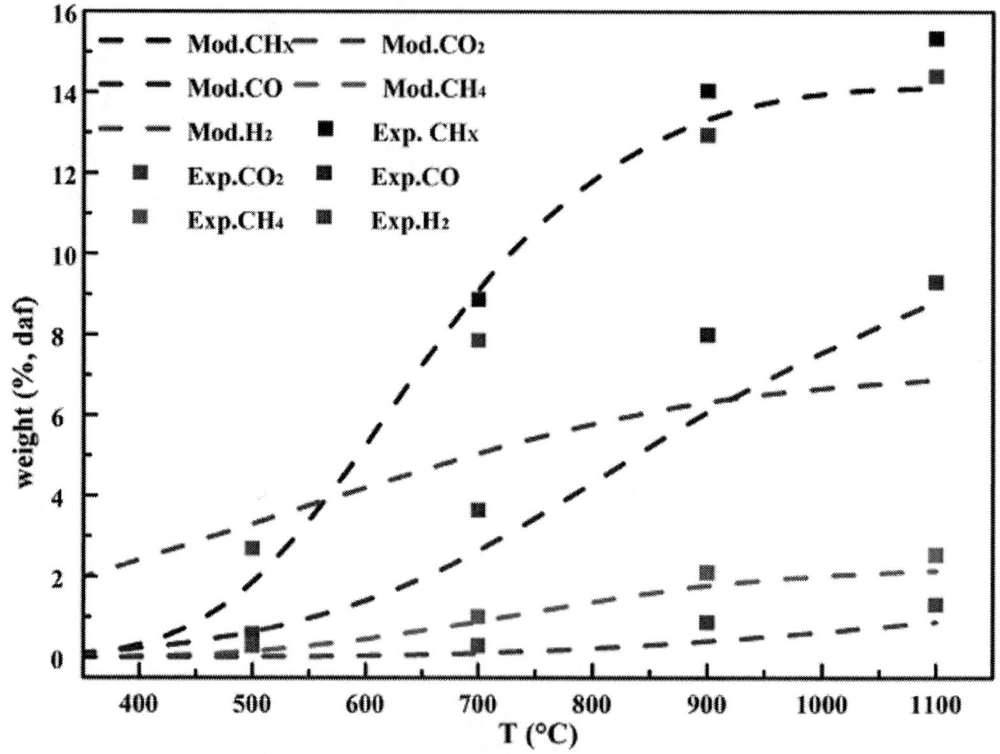


Fig. 3 Gases yields of coal A at different temperatures

The proximate analysis of char calculated by the Eq. (15)-(17) is shown in Table 8. The data in this table illustrate that with the increase of pyrolysis temperature, the content of volatile decreases, while the contents of fixed carbon and ash increase. These calculated results are accordant with the experimental results[5], which indicate that it is feasible to calculate the proximate analysis of char through the mass conservation.

Table 8 Proximate analysis of different temperature char ( dry basis, wt.% )

| | COAL A | | | COAL B | | |
|---|---|---|---|---|---|---|
| | 500°C | 600°C | 700°C | 500°C | 600°C | 700°C |
| VM | 49.11 | 45.46 | 37.96 | 38.95 | 30.75 | 19.80 |
| FC | 44.71 | 47.92 | 54.51 | 54.63 | 61.96 | 71.76 |
| Ash | 6.17 | 6.62 | 7.53 | 6.42 | 7.28 | 8.43 |

The calculated element compositions of different temperature char are listed in Table 9. From this table it can be seen that with the increase of pyrolysis temperature, the contents of carbon, nitrogen and sulfur are gradually increasing, and the contents of oxygen and hydrogen are gradually decreasing. The variation tendency of element contents is consistent with the experimental results.

Table 9 Element composition of different temperature char (dry basis, wt.%)

| | | C | H | O | N | S |
|---|---|---|---|---|---|---|
| COAL A | 500°C | 68.93 | 4.74 | 17.15 | 1.65 | 1.36 |
| | 600°C | 70.04 | 4.59 | 15.53 | 1.77 | 1.46 |
| COAL B | 500°C | 73.50 | 4.75 | 13.96 | 1.08 | 0.28 |
| | 600°C | 76.89 | 4.31 | 9.97 | 1.23 | 0.32 |

The calculated element composition of coal tar at 500°C is shown in Table 10. These values coincide with the literature results in Table 4, which indicates that it is feasible to calculate the tar element composition by the element conservation.

Table 10 Ultimate analysis of 500°C coal tar

| | C | H | O |
|---|---|---|---|
| COAL A | 85.58 | 9.97 | 4.45 |
| COAL B | 85.48 | 9.87 | 4.65 |

Fig.4 is the tar composition distribution of coal A and coal B at 500 °C. The tar composition is divided into five kinds: phenols, aromatics, aliphatics, oxy-components and pitch, whose contents are around 24%, 20%, 30%, 4.5% and 20%, respectively. Cui et al[33] analyzed the composition of low temperature coal tar and discovered that the content of aliphatic hydrocarbon is the highest, up to

33.39%, and the phenols and aromatic hydrocarbons take second place, being 11.68% and 16.86%, respectively. And it includes 4.47% of oxy-compound. The calculated values of tar composition are accordance with the experiment results. Since the proximate and ultimate analyses on air dry basis of coal A and coal B are not very different and the prediction of coal pyrolysis products by FG-DVC model is based on the air dry basis, the tar element composition of coal A and coal B are similar, leading to the similarity of tar composition distributions of these two coals.

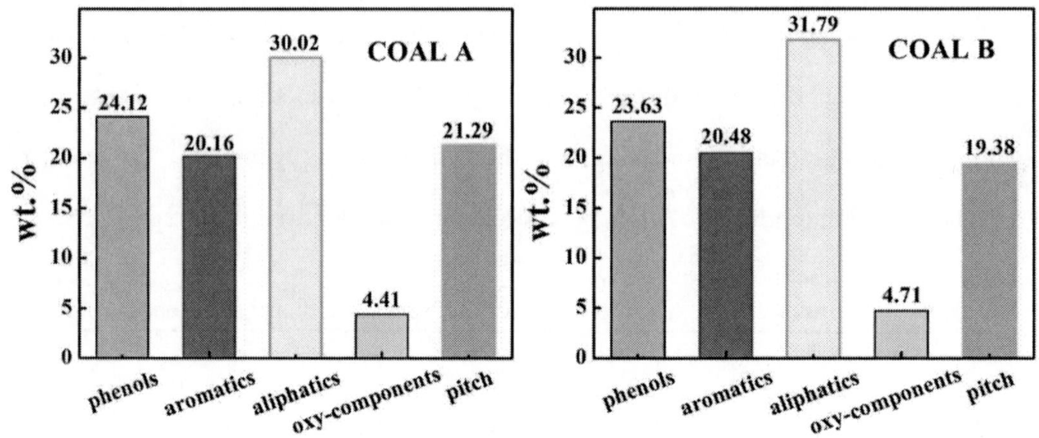


Fig.4 Component distribution of 500°C coal tar

# 4 Conclusion

A model for coal pyrolysis in grading lignite pyrolysis was developed by combining FG-DVC model with NLP model, in which both the kinetic behaviors of coal pyrolysis and component analysis of coal tar were considered simultaneously. Using this model, not only can the yields of key products of coal pyrolysis be predicted, but the component distribution of low temperature coal tar can also be obtained.

A series of simulations using coal A and coal B as the feed stock were carried out at different temperatures. The simulation results show that the modified Van Krevelen diagram can provide the key parameters for the FG-DVC model to predict the pyrolysis products distribution. Based on mass/element conservation of the pyrolysis unit, the proximate analysis of pyrolysis char and the element composition of coal tar can be approximately obtained. By the NLP mathematic model, the tar component distribution of coal A and coal B are parallel, which results from the similar physical characteristics of these two coals. The predicted product distributions of coal pyrolysis can be directly embed into the process simulation of grading lignite pyrolysis, which supplies the fundamental data of pyrolysis unit for this process simulation.

## ACKNOWLEDGEMENTS

The authors gratefully acknowledge the financial support from the Natural Science Foundation of China and National Youth Science Foundation (U1361202, 21206104).

## Reference:

[1] Guo Z, Wang Q, Fang M, et al. Thermodynamic and economic analysis of polygeneration system integrating atmospheric pressure coal pyrolysis technology with circulating fluidized bed power plant[J]. Applied Energy 2014;113:1301-14.
[2] Fan Y, Yi Q, Li F, et al. Tech-economic assessment of a coproduction system integrated with lignite pyrolysis and Fischer–Tropsch synthesis to produce liquid fuels and electricity[J]. Fuel Processing Technology 2015;135:180-6.
[3] He C, You F, Feng X. A novel hybrid feedstock to liquids and electricity process: Process modeling and exergoeconomic life cycle optimization[J]. AIChE Journal 2014;60:3739-53.
[4] Liu B, Yang X, Song W, et al. Process simulation development of coal combustion in a circulating fluidized bed combustor based on Aspen Plus[J]. Energy & Fuels 2011;25:1721-30.
[5] Yi Q, Feng J, Lu B, et al. Energy evaluation for lignite pyrolysis by solid heat carrier coupled with gasification[J]. Energy & Fuels 2013;27:4523-33.
[6] Badzioch S, Hawksley PG. Kinetics of thermal decomposition of pulverized coal particles[J]. Industrial & Engineering Chemistry Process Design and Development 1970;9:521-30.
[7] Kobayashi H, Howard J, Sarofim AF, editors. Coal devolatilization at high temperatures. Symposium (international) on combustion; 1977: Elsevier.
[8] Reidelbach H, Summerfield M, editors. Kinetic-model for coal pyrolysis optimization. ABSTRACTS OF PAPERS OF THE AMERICAN CHEMICAL SOCIETY; 1975: AMER CHEMICAL SOC 1155 16TH ST, NW, WASHINGTON, DC 20036.
[9] Suuberg EM, Peters WA, Howard JB. Product composition and kinetics of lignite pyrolysis[J]. Industrial & engineering chemistry process design and development 1978;17:37-46.
[10] Anthony DB, Howard JB, Hottel HC, et al. Rapid devolatilization and hydrogasification of bituminous coal[J]. Fuel 1976;55:121-8.
[11] Sommariva S, Maffei T, Migliavacca G, et al. A predictive multi-step kinetic model of coal

devolatilization[J]. Fuel 2010;89:318-28.
[12]Solomon PR, Hamblen DG, Carangelo R, et al. General model of coal devolatilization[J]. Energy & Fuels 1988;2:405-22.
[13]Niksa S. FLASHCHAIN theory for rapid coal devolatilization kinetics. 3. Modeling the behavior of various coals[J]. Energy & fuels 1991;5:673-83.
[14]Grant DM, Pugmire RJ, Fletcher TH, et al. Chemical model of coal devolatilization using percolation lattice statistics[J]. Energy & Fuels 1989;3:175-86.
[15]Chen Y, He R. Fragmentation and diffusion model for coal pyrolysis[J]. Journal of analytical and applied Pyrolysis 2011;90:72-9.
[16]Fletcher TH, Pond HR, Webster J, et al. Prediction of tar and light gas during pyrolysis of black liquor and biomass[J]. Energy & Fuels 2012;26:3381-7.
[17]Varhegyi G, Antal MJ, Jakab E, et al. Kinetic modeling of biomass pyrolysis[J]. Journal of analytical and Applied Pyrolysis 1997;42:73-87.
[18]Wang P, Jin L, Liu J, et al. Analysis of coal tar derived from pyrolysis at different atmospheres[J]. Fuel 2013;104:14-21.
[19]Luo K, Zhang C, Zhu S, et al. Tar formation during coal pyrolysis under N 2 and CO 2 atmospheres at elevated pressures[J]. Journal of Analytical and Applied Pyrolysis 2016;118:130-5.
[20]He C, Feng X, Chu KH. Process modeling and thermodynamic analysis of Lurgi fixed-bed coal gasifier in an SNG plant[J]. Applied energy 2013;111:742-57.
[21]YUXIN ZHAO MAS, ROSEMARY BASSILAKIS AND PETER R. SOLOMON. A method of predicting coal devolatilization behavior based on the elemental composition[J]. Advanced Fuel Research 1994:553-60.
[22]Hobbs ML, Radulovic PT, Smoot LD. Modeling fixed - bed coal gasifiers[J]. AIChE Journal 1992;38:681-702.
[23]Dai F, Gao M, Li C, et al. Detailed description of coal tar hydrogenation process using the kinetic lumping approach[J]. Energy & Fuels 2011;25:4878-85.
[24]Dai F, Gong M, Li C, et al. New kinetic model of coal tar hydrogenation process via carbon number component approach[J]. Applied Energy 2015;137:265-72.
[25]Guo Z, Wang Q, Fang M, et al. Simulation of a Lignite - Based Polygeneration System Coproducing Electricity and Tar with Carbon Capture[J]. Chemical Engineering & Technology 2015;38:463-72.
[26]Parikh J, Channiwala S, Ghosal G. A correlation for calculating elemental composition from proximate analysis of biomass materials[J]. Fuel 2007;86:1710-9.
[27]Shen J, Zhu S, Liu X, et al. The prediction of elemental composition of biomass based on proximate analysis[J]. Energy Conversion and Management 2010;51:983-7.
[28]Nhuchhen DR. Prediction of carbon, hydrogen, and oxygen compositions of raw and torrefied biomass using proximate analysis[J]. Fuel 2016;180:348-56.
[29]Li C, Liu D, Yang F. Study on Relevant Equations for Two Kinds of Coal Quality Analysis[J]. coal geology of China 2004;16:6-8.
[30]Zhang J, Liu G. Comprehensive Utilization of Low-temperature coal Tar[J]. coal conversion 2010;33:92-6.
[31]Lei Z, Zongkuan L, Zhaolin G. Simulation of Low-Temperature Coal Tar Hydrocracking in

Supercritical Gasoline[J]. CHINA PETROLEUM PROCESSING & PETROCHEMICAL TECHNOLOGY 2013;15:70-6.
[32]Ko GH, Sanchez DM, Peters WA, et al., editors. Correlations for effects of coal type and pressure on tar yields from rapid devolatilization. Symposium (International) on Combustion; 1989: Elsevier.
[33]Cui XT, Zhang YF, Dong DL, et al., editors. GC-MS Analyzing of Coal Tar of Lignite Briquette from Low Temperature Pyrolysis. Applied Mechanics and Materials; 2014: Trans Tech Publ.

# Influence of Oxygen Group and Coal Structure on the Migration of Biomass Alkalis During Coal/Biomass Co-pyrolysis

*Rong Zhu, Yuncai Song, Laisong Wang, Jie Feng, Wengying Li**

*Key Laboratory of Coal Science and Technology (Taiyuan University of Technology), Ministry of Education and Shanxi Province, Taiyuan 030024, PR China*

*Training Base of State Key Laboratory of Coal Science and Technology Jointly Constructed by Shanxi Province and Ministry of Science and Technology, Taiyuan 030024, PR China*

* *Corresponding author. E-mail: ying@tyut.edu.cn*

**Abstract**: Co-gasification of biomass and coal can not only decrease coal consumption effectively but also reduce gas pollutants especially $CO_2$. Alkali metal K/Na rich in biomass could transfer to the coal char in co-pyrolysis stage and then promote catalytic gasification of coal char in our previous studies. However, the biomass alkalis K/Na how to migrate and how to retain in coal char is still unclear. The raw material was placed in fixed bed reactor to be fast pyrolyzed on separate and co-pyrolyzed experiments. Gasification reactivity characteristics of char samples were investigated in a thermogravimetric analysis. By designing co-pyrolysis of acid coal with biomass and graphite with biomass under 973 K, 1073 K and 1173 K to obtain the effect of the oxygen or carbon structure of coal on the migration of biomass alkalis K/Na. Results showed that co-pyrolysis of biomass and coal was beneficial to enhance the oxygen content in acid coal char (ACC) and which was increased with the pyrolysis temperature. Additionally, the increase of the oxygen content in ACC is advantageous to the migration of alkali metals K/Na of biomass. XRD analysis indicated that the biomass K mainly existed in coal char as amorphous phase. Co-pyrolysis of graphite with biomass results reveals that biomass K/Na combined with carbon structure of graphite, the presence of biomass K/Na enable the conversion rate of graphite char increased by 20%.

***Keywords:*** Biomass alkalis, co-pyrolysis, catalyst, alkalis migration

## 1. Introduction

Biomass has been attracting great attention nowadays as a potentially significant renewable and clean source with $CO_2$-neutral property[1]. Gasification is one of the most promising means in utilizing biomass for producing gaseous fuel[2]. However, the seasonal supply nature and low heating values of biomass remarkably limits its widespread application. Therefore, more and more researcher [3-7] come up with co-thermochemical conversion of biomass and coal which can solve these problems and utilize coal in a cleaner method. In recent years, extensive studies [3, 8, 9] have reported in the literature a synergetic effect in the co-utlization of coal and biomass, especially in co-pyrolysis and co-gasification. This synergy [8, 10, 11] during co-processing is likely to the transferring of active

OH and H radicals from the biomass to the coal as well as the catalytic role of alkalis and alkaline earth metals (AAEMS) from the biomass.

Nevertheless, some disagreement [7, 10, 12, 13] about the existence of synergies between coal and biomass can be found in the literature when using coal/biomass blends. This visible difference might depend on the operating parameters used, such as temperature, type of reactor, biomass blending ratio. Kumabe et al. [13]did not find any synergy in the carbon distribution of products in the mixture of Japanese cedar and Mulia coal, but they observed a slight improvement in the extent of the water-gas shift reaction. Kajitani et al.[14], who performed co-gasification of two bituminous coals and cedar bark using entrained flow gasifiers were gasified with carbon dioxide, the results shows that no any synergy at higher temperature, however, a little improvement of the char gasification reactivity was found because of the catalysis of AAEMS in biomass. Most researcher [1, 3, 15, 16]have reported that the char derived from the pyrolysis of biomass can act as a good catalysis because of its high amount of AAEM catalytic species. Krerkkaiwan et al.[2] studied that co-pyrolysis of coal and two types of biomass was performed in a drop tube fixed-bed reactor, he inferred the potential catalytic effect of AAEMS species in biomass, especially K, appeared to be a potentially significant parameter affecting the observed synergy during the co-pyrolysis. Habibi et al.[17] investigated the $CO_2$ co-gasification of potassium-rich switchgrass with sub-bituminous coal and fluid coke respectively by thermogravimetric experiments, the switchgrass-coal mixtures show a deactivation due to sequestration of the mobile alkalis elements by reaction with aluminosilicate minerals in coal to form inactive $KAlSi_3O_8$, $KAlSiO_4$. Only when the excess alkali is present in the feed mixture to satisfy these deactivation reactions, remaining catalytic activity is evident. Ding [18] , studied the co-gasification of biomass char with coal char and the mechanism responsible for the various interactions, he is considered that the high content of K in biomass char and the remarkable difference of gasification rate between biomass char and coal char was caused synergistic effect during co-gasification and inhibiting effect was attributed to the intimate contact to form of $KAlSiO_4$. In light of this, studying of alkali metal migration and catalytic reaction mechanism during coal and biomass co-gasification, is greatly significant. However, the biomass alkalis how to migrate and how to catalyze in coal particle is still unclearly. Therefore, obtain a clear understanding of the migration and catalytic mechanism of biomass alkalis during co-pyrolysis is indispensable for promoting the co-utlization of biomass and coal.

In the present work, co-pyrolysis characteristics of rice straw (RS) blends with acid coal (AC) have been investigated to obtain a knowledge about the alkalis how to migrate and how to catalyze in coal char particle and which chemical forms of alkalis can be useful in co-gasification of biomass and coal. The method of sequential chemical extraction was employed to study the distribution of different chemical forms of alkalis in ACC produced from different atmosphere. The reactivity of acid coal char was evaluated using a thermogravimetric analyzer.

## 2. Experimental section

### 2.1 Sample

Two raw materials, rice straw (RS) and Jincheng coal (JC) from Shanxi Province in China were chosen as feedstock for co-pyrolysis in this study. Coal and biomass samples were milled and sieved to a particle size ranging from 0.104mm to 0.124mm for the coal particles and less than 0.09mm for the biomass particles. Then the (RS) powder were pressed, ground and sieved into particle sizes of 2.00mm-2.36mm as experimental samples. All samples were dried at 105℃for 24h prior to use. Proximate and elemental analysis results of the samples are shown in Table1. The biomass presents a significantly higher volatile matter content and a lower carbon content than the coal samples.

Table 1 Proximate and ultimate analysis of JC and RS

| Sample | Proximate analysis / wt% | | | | Ultimate analysis / wt%, daf | | | | |
|---|---|---|---|---|---|---|---|---|---|
| | $M_{ad}$ | $A_{ad}$ | $FC_{daf}$ | $V_{daf}$ | C | H | O | N | S |
| JC | 1.07 | 25.42 | 91.66 | 8.34 | 81.22 | 2.569 | 3.274 | 0.82 | 4.734 |
| RS | 4.37 | 14.12 | 19.91 | 80.09 | 47.05 | 6.97 | 49.22 | 1.69 | 0.53 |

Notes: daf--- on the basis of dry and ash free; ad--- on the basis of dry air.

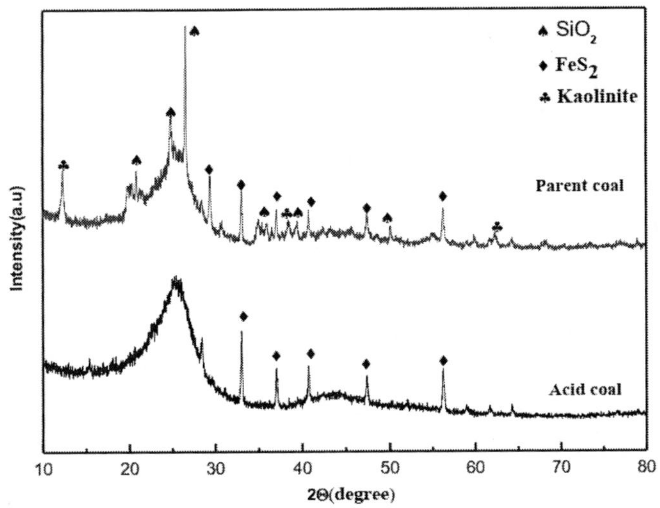


Fig.1 XRD patterns of the minerals distribution in parent coal and acid coal

Some studies [17-19] found that potassium deactivated during gasification due to the formation of potassium aluminosilicate, as the potassium reacted with mineral matter (e.g. kaolinite) in the coal. In this study, picking is used as a pre-treated process targeted on the elimination of Si and Al in parent coal to eliminate the influence of mineral on the migration/fixation of biomass alkalis in coal char. Therefore, the JingCheng coal was washed by the 30% hydrofluoric acid and 15% hydrochloric acid to obtain the sample with very lower Si and Al content. From the Fig.1, it can be seen acid wash mainly to remove the kaolinite and silica in parent coal, but there are a few $FeS_2$ in acid coal.

The inorganic elements in the fuels used in this study were determined by an inductively coupled plasma – mass spectrometer (ICP-MS, Agilent 7700).The samples were dissolved by the mixed acid ($HNO_3$, HF and $H_2O_2$) in a microwave digestion instrument, and the solution was diluted and measured by ICP-MS. The ICP-MS determination use external standard method, the standard curve of K, Na as shown in Fig.2. The content of alkalis K and Na in acid coal (AC) and (RS) samples listed in table 2, the alkalis K/Na content of the biomass sample is clearly higher than that of AC.

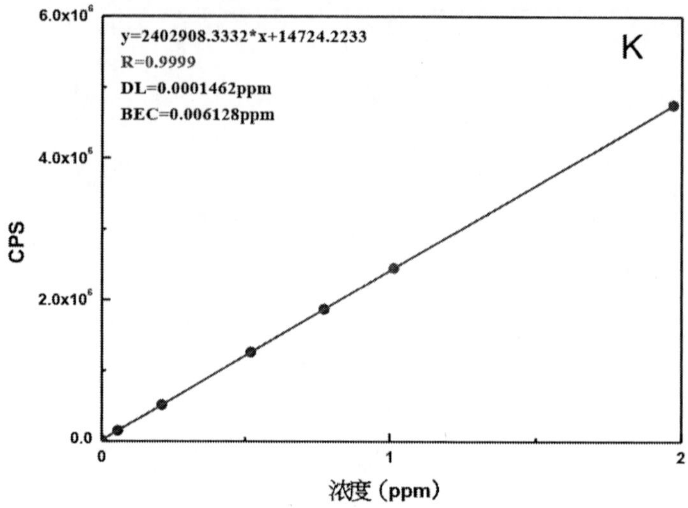


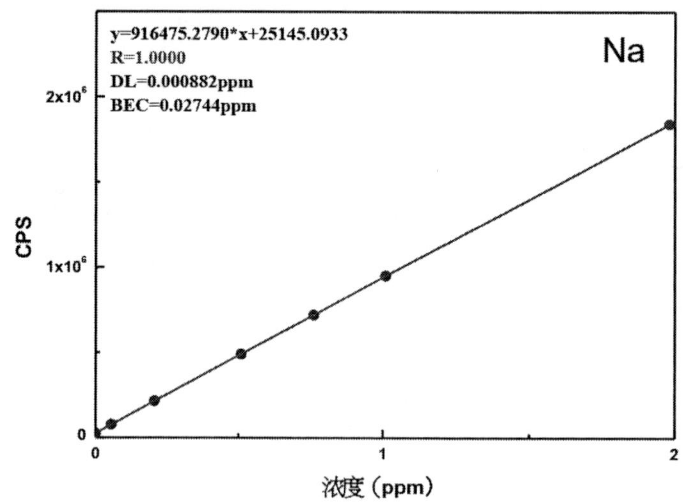


Fig.2 The standard curves for concentration measure of K and Na

Table 2 The content of alkalis K and Na in acid coal (AC) and (RS) samples

| Sample | K(mg/g) | Na(mg/g) |
|---|---|---|
| AC | 0.295 | 0.973 |
| RS | 19.045 | 8.102 |

## 2.2 Experiment Apparatus and Method

This study was taken in the fixed bed reactor which was described in Fig.3.The quartz reactor, which is the main part of the apparatus, was heated by a furnace with a maximum temperature of 1273K. Coal and biomass about 2g respectively are fully blended and then placed on the quartz plate under a flow of pyrolysis gas at atmosphere pressure. The furnace was heated at 6℃/min to the desired temperature. The temperature of this study were 973K, 1073K and 1173K, respectively. When the temperature reached the set values, rapidly rise the reactor furnace until the sample at a constant temperature zone. After the mixed sample charred for 30min at this temperature drop the reactor furnace quickly. After the reactor cooling to room temperature, screen mixed char repeatedly and weight char individually.

Three types of atmospheres were considered here to investigate the effect of atmosphere on the distribution and catalytic of alkalis: 100 mL/min $N_2$, 50 mL/min $N_2$+50 mL/min $CO_2$ and 50 mL/min steam+20 mL/min $O_2$+50 mL/min $N_2$.

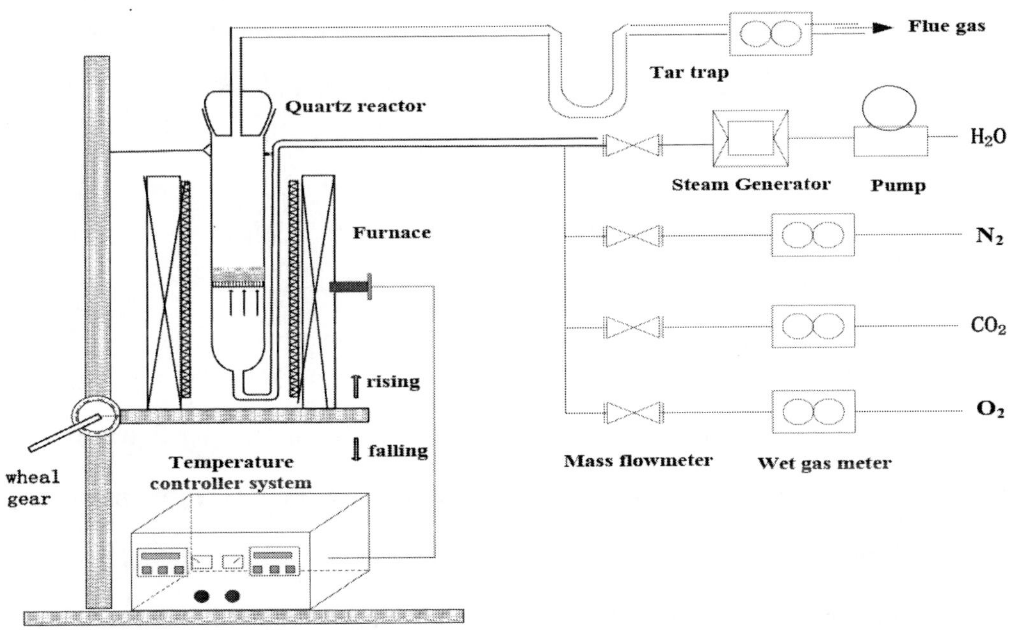


Fig.3 Sketch of the fixed-bed reactor

### 2.3 Characterization and calculations methods

X-ray diffraction (XRD) patterns of the parent coal, acid coal, single AC char and co-pyrolysis AC char were obtained on a Multiflex X-ray diffractometer (Rigaku, D/max-3B, Japan) using Cu/Ka radiation (λ=1.54056Å), a 40 kv tube voltage, a 40 mA tube current, and a scan rate of $4°min^{-1}$.

Surface area measurements were conducted on an adsorption instrument (Beijing JWGB Sci. &Tech. Co.,Ltd), prior to each measurements, the samples were degassed at 210℃for 3h.

Samples of the oxygen content is determined by element analyzer in oxygen mode （VsrioMACROcube, Germany）.

This paper only determined the alkalis content of K, Na of the collecting solid product such as char. The alkali metal content in the gas phase is calculated by the subtraction method in the percentage of weight. Where $m_{co\text{-}bc}$ and $m_{co\text{-}cc}$ is the absolute content of AAEMS in the co-pyrolysis biomass char and co-pyrolysis coal char, $y_{co\text{-}bc}$ and $y_{co\text{-}cc}$ is the amount of char yield. Where $m_{cc}$ and $y_{cc}$ is the absolute amount of AAEMS and char yield in the separate coal char. $m_b$ is the absolute content of AAEMS biomass materials. Where $w_{co\text{-}bc}$, $w_{migration}$, $w_{co\text{-}loss}$ is the percentage of AAEMS retained in co-biochar, migrated to acid coal char and released to gas phase.

$$w_{co-bc} = \frac{m_{co-bc} * y_{co-bc}}{m_b} * 100\% \quad (1)$$

$$w_{migration} = \frac{m_{co-cc} * y_{co-cc} - m_{cc} * y_{cc}}{m_b} * 100\% \quad (2)$$

$$w_{co-loss} = 1 - w_{migration} - w_{co-bc} \quad (3)$$

The gasification reactivity of the char produced from co-pyrolysis and single pyrolysis was measured by using a thermogravimetric analyzer (HCT-1，Beijing Henven Scientific Instrument Factory ). The mass of sample was approximately 11mg. Before TG experiments, a flow of 100ml/min of high purity carbon dioxide was used as sweeping gas. The sample were heated to 1000℃ with a heating rate of 10℃/min at a flow of 80ml/min of carbon dioxide, and kept 30min in set temperature. Then the moment gasification carbon conversion rate ($W$) was calculated by Eq. (4). Where $m_0$ is the initial quality of coal char at the onset of gasification, $m_\infty$ is the final mass, $m_t$ indicates the instantaneous mass of char at time t.

$$W = \frac{m_0 - m_t}{m_0 - m_\infty} * 100\% \quad (4)$$

## 3. Results and discussion

### 3.1 Effect of Reaction Temperature on Co-pyrolysis of RS and AC

It is proven that the temperature significantly influences the release amount of the AAEMS species from biomass. Some studies [20-22] demonstrated that, below 973K, no significant K release was observed during the pyrolysis of the RS in fixed-bed reactor and above 973K, K release increased progressively. Therefore, in this study select (973K, 1073K, 1173K) as pyrolysis temperature to investigate the migration and catalytic principle of biomass alkalis during co-pyrolysis of RS and AC.

#### 3.1.1 Biomass Alkalis Distribution

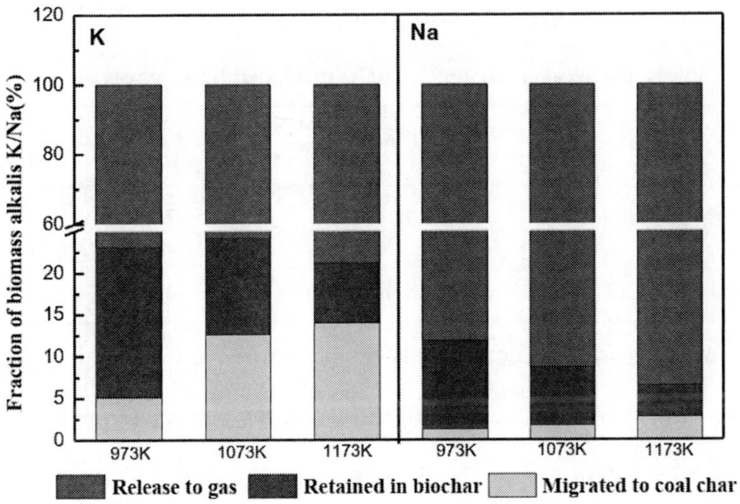


Fig.4 The effects of temperature on the biomass alkalis distribution during co-pyrolysis of RS and AC

Fig.4 shows the distribution of the alkalis content in different temperature during co-pyrolysis of RS and AC. It can be seen that the content of biomass alkalis K and Na were distributed in acid coal char and gas phase increased with the temperature, but the biomass alkalis retained in biochar decreased. This is why inorganic potassium under higher temperature releasing into gas phase due to the vapor pressure increased. Sodium characteristics of pyrolysis are similar with potassium.

Krerkkaiwan [23] speculated in his study that the oxygenated compounds in the original RS-tar condensed over the coal char to originated oxygenated-coke species, which preferentially attract the volatile K interact with the coke to from the phonolate groups (K-O-C) over the char surface. If the oxygen products produced from biomass pyrolysis deposit on the coal char surface will promote the volatile K to migration in coal char, thus the total oxygen content of co-pyrolysis ACC higher than that of single pyrolysis ACC. Hence the total oxygen content of ACC were measured on the Elemental Analyzer. Table 3 shows the oxygen content of the acid coal char produced from different pyrolysis temperature. The oxygen content of all of the co-pyrolysis ACC higher than single pyrolysis ACC, it can be concluded that the oxygen products derived from biomass pyrolysis likely migrate to ACC during co-pyrolysis of RS and AC, and enhance the oxygen content in ACC. The oxygen contents increased with the pyrolysis temperature during co-pyrolysis, combined with Fig. 4 and Table 3 is analyzed that the increase of the oxygen content in ACC is advantageous to the migration of alkali metals K/Na of biomass.

Table 3 The oxygen content of ACC produced from pyrolysis and co-pyrolysis

| Sample | 973K | 1073K | 1173K |
|---|---|---|---|
| Single pyrolysis AAC | 1.968 | 1.978 | 2.064 |
| Co-pyrolysis ACC | 2.328 | 2.541 | 2.837 |

To study which species form of biomass alkalis K/Na in ACC, a pair of single and co- pyrolysis ACC at 1073K were conducted by X-ray diffraction, and their XRD profiles are exhibited in Fig.5. However, the species form of alkalis K/Na did not check in ACC, in other words, the biomass alkalis K/Na migration to ACC exists in the amorphous state.

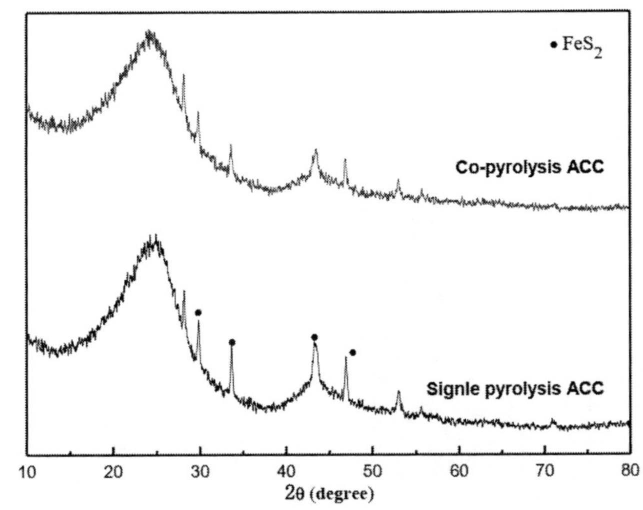


Fig.5 X-ray diffraction patterns of ACC produced from single pyrolysis and co-pyrolysis at 1073K

### 3.1.2 Char Gasification Reactivity

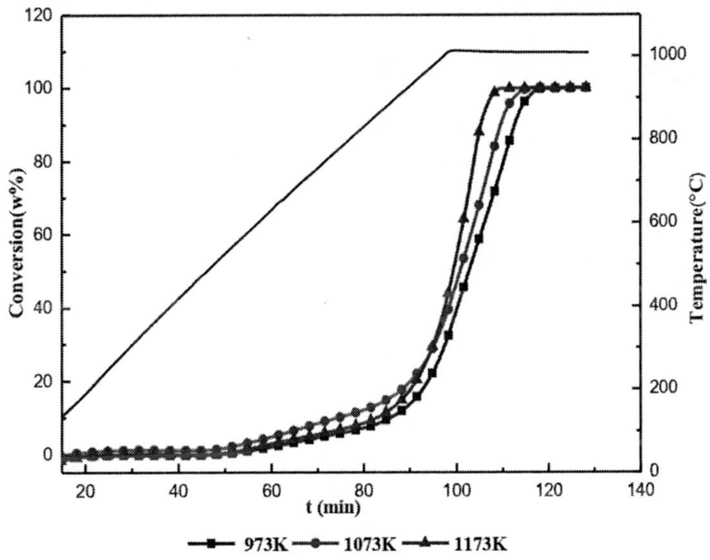


Fig.6 The carbon conversion of ACC during co-pyrolysis at different temperature

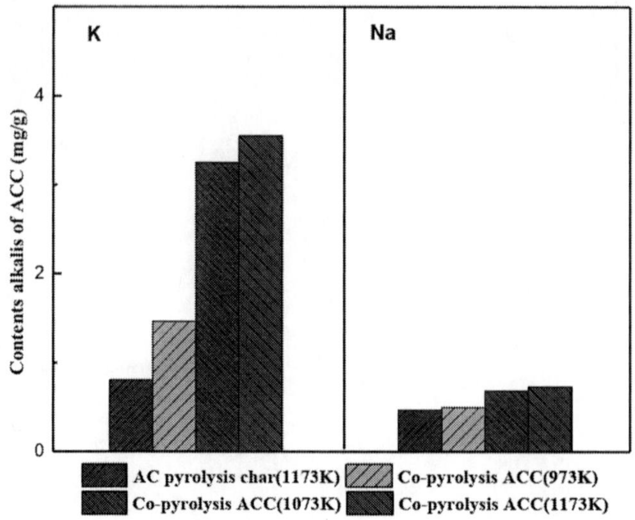


Fig.7 The absolute contents of alkalis in ACC produced from co-pyrolysis at different temperature

Fig.6 presents the carbon conversation *W* versus gasification time and temperature under carbon dioxide atmosphere. It can be seen from the picture the gasification reactivity of ACC increased with the temperature during co-pyrolysis. In the Fig.7, the absolute contents of alkalis in ACC also increased with the pyrolysis temperature, these enhancements are related to the catalytic roles of alkalis K and Na which may be contribute to the gasification reactivity in ACC.

### 3.2 Effect of Reaction Atmosphere on Co-pyrolysis of RS and AC

#### 3.2.1 Chemical Form of the Alkalis

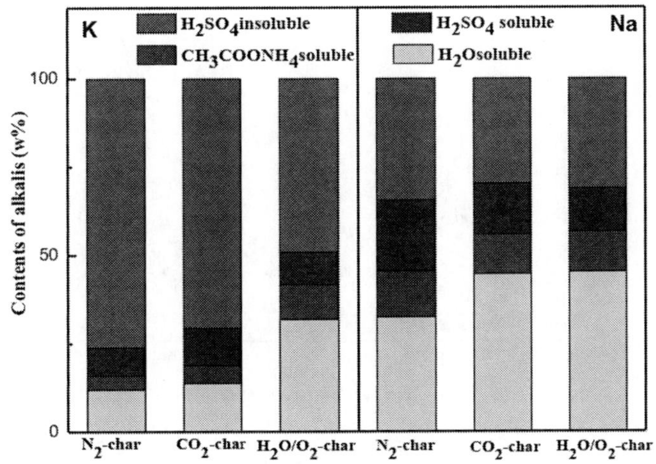


Fig.8 The effects of different co-pyrolysis atmospheres on the chemical forms of alkalis in ACC during co-pyrolysis of RS and AC

The AAEMS species in biomass or low-rank coal mainly exists in four different chemical forms[24-27]: $H_2O$-soluble form (soluble salts), $CH_3COONH_4$-soluble forms (carboxylates), acid-soluble forms (silicates and aluminates) and stable form (covered by carbon and silico-aluminates).The catalytic effects of AAEMS species largely depend on their chemical form[28], $H_2O$-soluble form and $CH_3COONH_4$-soluble forms were reported to have significant catalytic effects during pyrolysis or gasification of brown coal[26]. Therefore, to deeply understand which chemical forms of alkalis K and Na exist in ACC during co-pyrolysis of RS and AC, the distribution and catalytic characteristic of alkalis K and Na during co-pyrolysis at different atmosphere were studied.

Fig.8 indicates the chemical forms of alkalis K and Na in ACC during co-pyrolysis of RS and AC. It can be concluded that alkali K of the $H_2O$-soluble and $CH_3COONH_4$-soluble at the atmosphere of steam and oxygen significant higher than nitrogen and carbon dioxide atmosphere. $H_2SO_4$-soluble alkalis K, which considered to be no catalytic activity due to combined with silicates and aluminates [25], exits in ACC at nitrogen exceed at steam and carbon dioxide atmosphere.

### 3.2.2 Char Gasification Reactivity

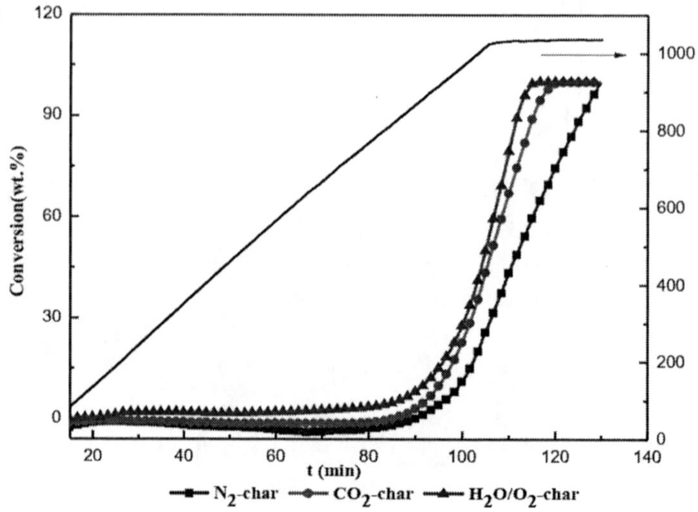


Fig.9 The carbon conversion of ACC during co-pyrolysis at different atmosphere

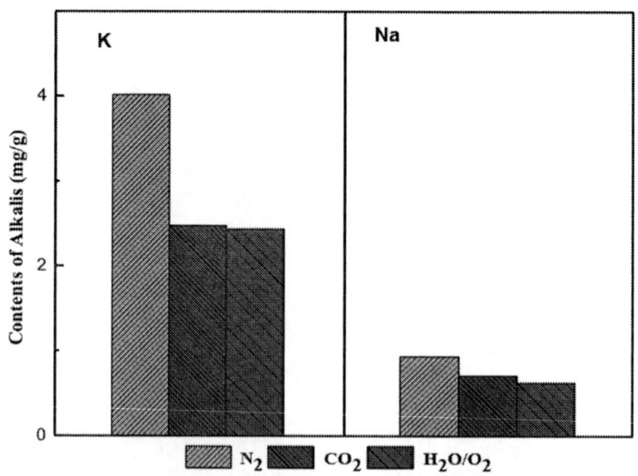


Fig.10 The absolute contents of alkalis in ACC during co-pyrolysis at different atmosphere

Fig.6 shows the carbon conversion of ACC during co-pyrolysis at nitrogen, carbon dioxide and steam atmosphere. The gasification reactivity of ACC produced from steam and carbon dioxide atmosphere higher than nitrogen atmosphere why the higher reactive chemical forms in steam and carbon dioxide atmosphere better than nitrogen, as Fig.8 described, the more the $H_2O$-soluble form and $CH_3COONH_4$-soluble forms exists in ACC ,the more the catalytic effects promote AC gasification. The absolute contents of alkalis in ACC produced from different atmosphere was exhibited in Fig.10, the contents of alkalis K and Na in $N_2$-char distinctly higher than $CO_2$-char and $H_2O/O_2$-char. whereas, the carbon conversion of $CO_2$-char and $H_2O/O_2$-char obviously stronger than $N_2$-char in that the more the $H_2SO_4$-soluble in $N_2$-char no play a actively catalytic role during $CO_2$ gasification. From the BET analysis in table 4, only the $H_2O/O_2$-char had a higher BET surface area and pore volume than $CO_2$-char and $N_2$-char, the result is that $CO_2$-char sample had much lower BET surface area and therefore lower amount of active sites is not conducive to the gasification reaction than $H_2O/O_2$-char. it can be conclude that the large specific surface area and more active form of alkalis in ACC also greatly help to alkalis promote coal char gasification.

Table 4 The pore structure characteristics of ACC produced from co-pyrolysis at different atmosphere

| Sample | Reaction atmosphere | Specific surface area ($m^2/g$) | Pore volume ($cm^3/g$) | Pore diameter (nm) |
|---|---|---|---|---|
| $N_2$-char | nitrogen | 2.394 | 0.005 | 8.009 |
| $CO_2$-char | Carbon dioxide | 2.574 | 0.005 | 7.782 |
| $H_2O/O_2$-char | Steam and oxygen | 37.746 | 0.035 | 3.745 |

### 3.3 Effect of Reaction Temperature Co-pyrolysis of RS and Graphite

The objective of this section was to investigate the effect of pure carbon structure on the migration and catalytic of biomass alkalis during co-pyrolysis. Therefore, the graphite, no mineral and oxygen group were selected as the research materials.

### 3.3.1 Biomass alkalis Distribution

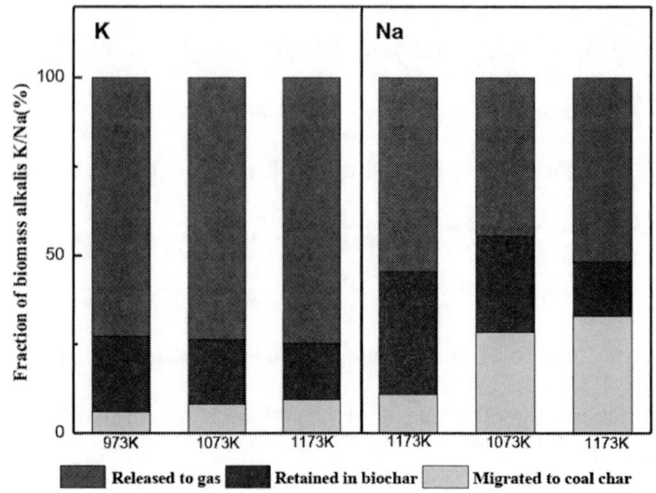


Fig.11 The effect of temperature on the biomass alkalis distribution during co-pyrolysis of RS and graphite

It is seen in Fig.11 that the contents of biomass alkalis K migrate to graphite char rise from 5.94% to 9.46% increase with the temperature. This shows biomass alkalis is likely to combine with the carbon structure during co-utilizing of biomass and coal, but this combination is more difficult.

### 3.3.2 Gasification Reactivity

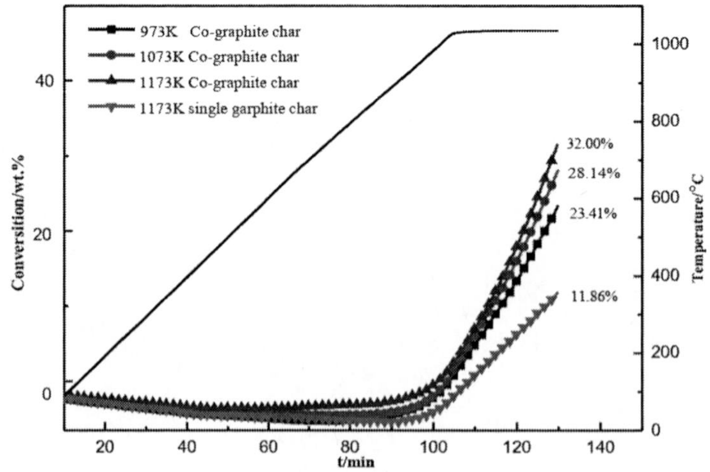


Fig.12 The carbon conversation of graphite char produced from co-pyrolysis at different temperature

It is seen from Fig.12 that increased with the pyrolysis temperature, the carbon conversation of co-pyrolysis graphite char rise from 23.4% to 32.0% , this phenomena could be explain why the biomass alkalis slightly migrate to graphite char in Fig.8. Compared with the gasification reactivity of co-pyrolysis and single graphite char in Fig.12, it can be concluded that the presence of biomass K/Na enable the conversion rate of graphite char increased by 20% ,this is due to the biomass alkalis K/Na combined with carbon structure of graphite during co-pyrolysis process.

## 4. Conclusions

In this study, the migration and catalytic characteristics of biomass alkalis to coal char during co-pyrolysis of rice straw (RS) and acid JingCheng coal (AC) was investigated in a fixed-bed reactor. The prepared pyrolysis char were conducted by thermogravimetric experiments and XRD analysis.

The influence of the temperature on the distribution of biomass alkalis during co-pyrolysis of RS with AC and graphite, and the effect of the atmosphere on the chemical forms distribution and catalytic principle were studied in this study. The main conclusions are summarized as below: (1) The enhancement of the oxygen content in ACC is advantageous to the migration of alkali metals K/Na of biomass during co-pyrolysis of RS and AC. (2) The higher reactivity of $H_2O/O_2$- char implies that reactive chemical forms of alkalis, as well as higher surface area played a significant role in ACC reactivity. (3) The presence of biomass K/Na enable the conversion rate of graphite char increased by 20%, probably due to the biomass alkalis K/Na combined with carbon structure of graphite during co-pyrolysis process.

## ACKNOWLEDGEMENTS

The authors gratefully acknowledge the financial support from the National Natural Science Foundation of China under Grant No.21376158.

## Reference:

[1]. Meng H, Wang S, Chen L, Wu Z, Zhao J. Thermal behavior and the evolution of char structure during co-pyrolysis of platanus wood blends with different rank coals from northern China[J]. Fuel. 2015,158:602-611.

[2]. Krerkkaiwan S, Mueangta S, Thammarat P, Jaisat L, Kuchonthara P. Catalytic Biomass-Derived Tar Decomposition Using Char from the Co-pyrolysis of Coal and Giant Leucaena Wood Biomass[J]. Energy & Fuels. 2015,29:3119-3126.

[3]. Zhu W, Song W, Lin W. Catalytic gasification of char from co-pyrolysis of coal and biomass[J]. Fuel Processing Technology. 2008,89:890-896.

[4]. Zhang Y, Zhang XM, Fan D, Song YC. A promising approach to co-processing calcium-rich coal and an aqueous condensate from biomass carbonization[J]. Fuel. 2014,133:82-88.

[5]. Robert C.Brown QL. Catalytic effects observed during the co-gasification of coal and switchgrass[J]. Biomass and Bioenergy. 2000,18:499-506.

[6]. A-G.Collot YZ, D.R. Co-pyrolysis-and-co-gasification-of-coal-and-biomass-in-bench-scale-fixed-bed-and-fluidised-bed-reactors.[J]. Fuel. 1999,78:667-679.

[7]. K.Sjostrom GC,Q.Yu,C.Brage,C.Rosen. Promoted-reactivity-of-char-in-co-gasification-of-biomass-and-coal-synergies-in-the-thermochemical-process.[J]. Fuel. 1999,78:1189-1191.

[8]. Yang X, Yuan C, Xu J, Zhang W. Co-pyrolysis of Chinese lignite and biomass in a vacuum reactor[J]. Bioresource technology. 2014,173:1-5.

[9]. Ferrara F, Orsini A, Plaisant A, Pettinau A. Pyrolysis of coal, biomass and their blends: Performance assessment by thermogravimetric analysis[J]. Bioresource technology. 2014,171:433-441.

[10]. Xu CF, Hu S, Xiang J, Zhang LQ, Sun LS, Shuai C, et al. Interaction and kinetic analysis for coal and biomass co-gasification by TG-FTIR[J]. Bioresource technology. 2014,154:313-321.

[11]. Krerkkaiwan S, Fushimi C, Yamamoto H, Tsutsumi A, Kuchonthara P. Influences of heating rate during coal char preparation and AAEMs on volatile-char interaction with different sources of biomass volatile[J]. Fuel Processing Technology. 2014,119:10-18.

[12]. Li S, Chen X, Liu A, Wang L, Yu G. Study on co-pyrolysis characteristics of rice straw and Shenfu bituminous coal blends in a fixed bed reactor[J]. Bioresource technology. 2014,155:252-257.

[13]. Kumabe K, Hanaoka T, Fujimoto S, Minowa T, Sakanishi K. Co-gasification of woody biomass and coal with air and steam[J]. Fuel. 2007,86:684-689.

[14]. Kajitani S, Zhang Y, Umemoto S, Ashizawa M, Hara S. Co-gasification Reactivity of Coal and Woody Biomass in High-Temperature Gasification[J]. Energy & Fuels. 2010,24:145-151.

[15]. Liu Q, Zhang YS, Liu Z, Orndorff W, Cao Y, Pan WP. Applications of thermal stepwise reactions on the co-gasification of coal and tobacco stems[J]. Journal of Thermal Analysis and Calorimetry. 2014,116:1205-1212.

[16]. Masnadi MS, Grace JR, Bi XT, Lim CJ, Ellis N, Li YH, et al. From coal towards renewables: Catalytic/synergistic effects during steam co-gasification of switchgrass and coal in a pilot-scale bubbling fluidized bed[J]. Renewable Energy. 2015,83:918-930.

[17]. Habibi R, Kopyscinski J, Masnadi MS, Lam J, Grace JR, Mims CA, et al. Co-gasification of Biomass and Non-biomass Feedstocks: Synergistic and Inhibition Effects of Switchgrass Mixed with Sub-bituminous Coal and Fluid Coke During CO2 Gasification[J]. Energy & Fuels. 2013,27:494-500.

[18]. Ding L, Zhang YQ, Wang ZQ, Huang JJ, Fang YT. Interaction and its induced inhibiting or synergistic effects during co-gasification of coal char and biomass char[J]. Bioresource technology. 2014,173:11-20.

[19]. Gerald P. Huffmant FEH, Ronald W. Shoenberger, James, S. Walker FWLaRBG. Investigation of the structural forms of potassium in coke by electron microscopy and X-ray absorption spectroscopy[J]. Fuel. 1986,65:621-632.

[20]. P. A. Jensen FJF, K. Dam-Johansen,and B. Sander. Experimental Investigation of the Transformation and Release to Gas Phase of Potassium and Chlorine during Straw Pyrolysis[J]. Energy & Fuels 2000,14:1280-1285.

[21]. John G.Olsson UJ, and Jan B,C,Pttersson. Alkali Metal Emission during Pyrolysis of Biomass[J]. Energy & Fuels 1997,11:779-784.

[22]. Tsutomu Okuno NS, Jun-ichiro Hayashi,Chun-Zhu Li,. Primary Release of Alkali and Alkaline Earth Metallic Species during the Pyrolysis of Pulverized Biomass[J]. Energy & Fuels 2005,19:2164-2171.

[23]. Krerkkaiwan S, Fushimi C, Tsutsumi A, Kuchonthara P. Synergetic effect during co-pyrolysis/gasification of biomass and sub-bituminous coal[J]. Fuel Processing Technology. 2013,115:11-18.

[24]. Liu S, Qiao Y, Lu Z, Gui B, Wei M, Yu Y, et al. Release and Transformation of Sodium in Kitchen Waste during Torrefaction[J]. Energy & Fuels. 2014,28:1911-1917.

[25]. Pettersson A, Zevenhoven M, Steenari BM, Amand LE. Application of chemical fractionation methods for characterisation of biofuels, waste derived fuels and CFB co-combustion fly ashes[J]. Fuel. 2008,87:3183-3193.

[26]. Wang Ca, Jin X, Wang Y, Yan Y, Cui J, Liu Y, et al. Release and Transformation of Sodium during Pyrolysis of Zhundong Coals[J]. Energy & Fuels. 2015,29:78-85.

[27]. Xiaofang Wei JH, Tiefeng Liu,Yitian Fang,and Yang Wang. Transformation of Alkali Metals during Pyrolysis and Gasification of a Lignite[J]. Energy & Fuels. 2008,22:1840-1844.

[28]. Katsuyas Sugawara, Yukio Enda HI. Dynamic behavior of trace elements during pyrolysis of coals[J]. Fuel. 2002,81:1439-1443.

Heat release characteristics of the oxidation of aliphatic hydrocarbon groups and its effects on the spontaneous combustion of coal

Tao Xu[1]*, Haihui Xin[2], Deming Wang[2], Jianghong Liu[1], Jinhui Wang[1], Yu Jiao[1], Yutao Kang[1], Xiaobei Yin[1], Qimiao Xie[1], Xianbin, Li[1]

[1]Shanghai Maritime University

[2]China University of Mining and Technology

*Corresponding author. Tel.: +86 02138282517. E-mail address: xutaosmu@163.com.

College of Ocean Science and Engineering, Shanghai Maritime University, No.1550, Haigang Avenue, Pudong New District, Shanghai, 201306, China

ABSTRACT: The heat release characteristics of reactions of -$CH_3$、-$CH_2$- in coal were studied based on organic chemistry and quantum chemistry analysis methods. The structure model of -$CH_3$、-$CH_2$- was established and the reaction processes of the functional groups during spontaneous combustion were analyzed. Using quantum chemistry analysis method, gibbs free energy change and enthalpy change of the reactions were calculated. The results indicate that aliphatic hydrocarbons can react with oxygen and there is a lot of heat of about 400 to 500 kJ/mol released from these reactions. By comparing the heat release characteristics of the reactions of the aliphatic hydrocarbon groups with the oxygen consumption, gas production and temperature rising characteristics of the coal, it is concluded that oxidation of aliphatic hydrocarbons do not occurred in the initial stage of spontaneous combustion of coal and it is the key to determine the spontaneous combustion characteristics of the coal.

KEYWORDS: coal spontaneous combustion; aliphatic hydrocarbon groups; quantum chemistry; heat

## 1. Introduction

The spontaneous combustion of coal is the main hazard in the coal related industry. Great efforts have been done to find out the mechanism of coal spontaneous combustion. Previous studies mainly focused on the crossing-point temperature [1-3], activation energy [4,5], temperature rising [6-8], oxygen consumption [9,10] and gas production properties [10-12], as well as the distribution of the functional groups in raw coal and their variation during spontaneous combustion processes [13,14], attempting to describe the self-heating process of coal and explain why some ranks of coal have higher tendency to spontaneous combustion than the others. Present studies have proposed that the spontaneous combustion of coal is essentially resulted from the accumulation of heat released from reactions of decomposition and oxidation reactions of the active functional groups within the coal [14,15]. Therefore, research towards the heat effects of the active functional groups in coal can lead to a deeper understanding of the spontaneous combustion of coal.

Aliphatic hydrocarbon is key functional group consisting of coal structure [16-19], which is always connected to aromatic hydrocarbons [20-24]. Present studies have demonstrated that aliphatic hydrocarbon is active in the coal structure and reacts with oxygen during the processes of coal spontaneous combustion [25]. To determine the extent to which aliphatic hydrocarbon contributed to the spontaneous combustion of coal, several model of aliphatic hydrocarbon groups in coal were constructed firstly in this study. Equations describing the chemical reactions occurring in each model were subsequently analyzed according to their chemical characteristics, and chemical calculation methods were applied to determine the heat effects associated with each reaction. The results of these calculations were also compared with the temperature, oxygen consumption, and gas production characteristics during the self-heating process of coal. The results of this study will provide useful information concerning the effects of aliphatic hydrocarbon on the spontaneous combustion of coal at low temperatures.

## 2. Model and methods

### *2.1. Types and computational models of aliphatic hydrocarbon*

To study the heat effects associated with aliphatic hydrocarbon in coal, the model of the aliphatic hydrocarbon structure and the chemical reaction processes were analyzed. It has been demonstrated that aromatic functional groups are stable and the coal structure are almost unaffected by the number of aromatic rings [26]. Therefore, single phenyl group was used to model the aromatic rings in this paper, which is represented by Ar. R and R' have been used to represent segments of the aliphatic structure of coal.

The previous study has shown that methyl (-$CH_3$) and methylene (-$CH_2$-) are the main reactant of aliphatic hydrocarbon in coal [27]. -$CH_3$ always located at the end of the aliphatic chain and when it is oxidized, two hydroxyl (-OH) would be produced (shown as reaction model 1). The structure with two –OH is not stable, and dehydration reaction will take place to produce aldehyde (-HC=O) (shown as reaction model 2)

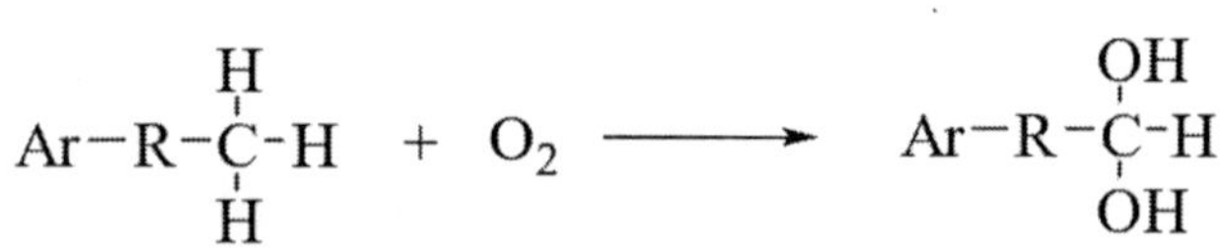

(Model 1)

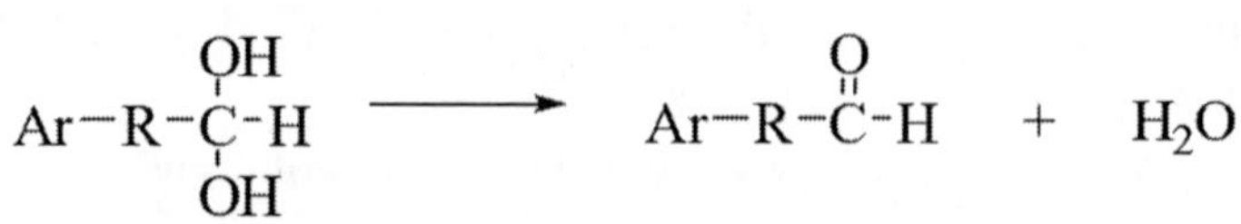

(Model 2)

Oxidation can also take place on -$CH_2$- and the reaction processes can be shown as reaction model 3 to reaction model 5. These reactions will lead to the production of alkyl and olefins ($C_xH_y$), which are the key gas products during the spontaneous combustion of coal.

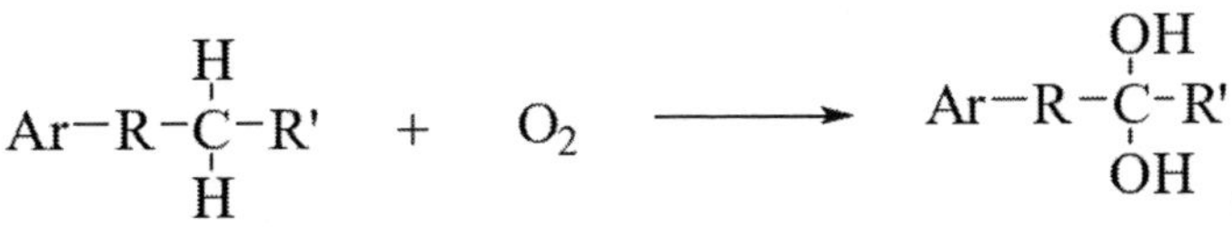
( Model 3)

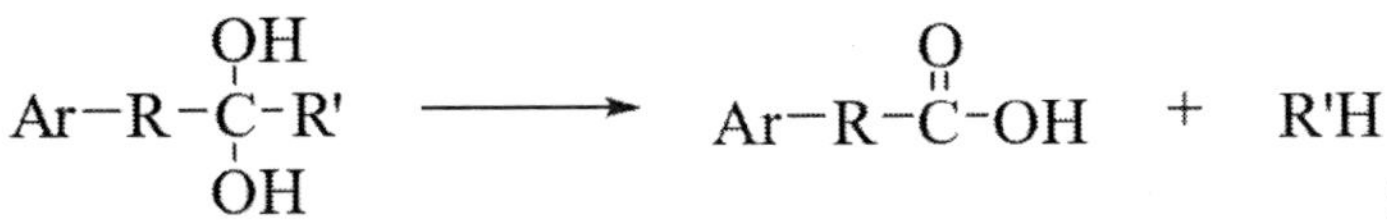
( Model 4)

( Model 5)

Based on the results described above, we have constructed molecular models of the different types of -$CH_3$ and -$CH_2$- described in this paper. The details of these models are shown in Table 1.

Table 1
Molecular model of aliphatic hydrocarbon

| Molecular model | | |
| --- | --- | --- |
| 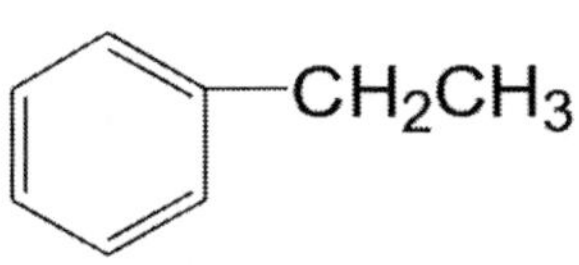 | 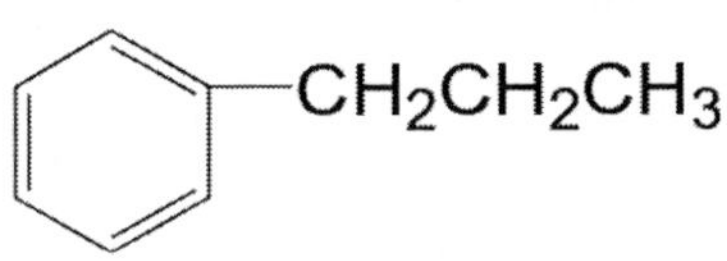 | 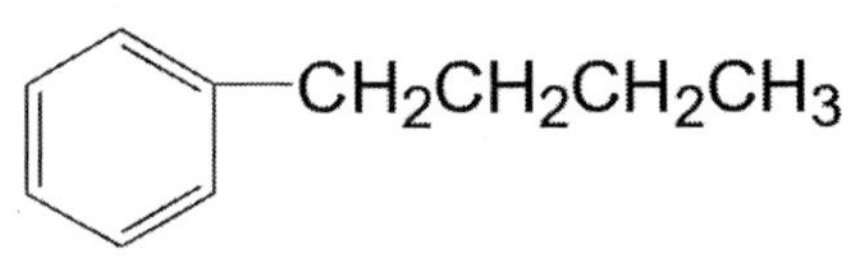 |
| R1 | R2 | R3 |

*2.2. Coal sample*

In this study, brown coal from Beizao coal mine of China (BZ) was selected as the coal samples and its main characteristic parameters are shown in Table 2.

Table 2
Proximate analysis of the coal samples

| Coal sample | $M_{ad}$ | $A_{ad}$ | $V_{ad}$ | $FC_{ad}$ | $Q_{net, ad}$ |
| --- | --- | --- | --- | --- | --- |
| BZ | 25.05 | 2.91 | 30.41 | 41.63 | 22.23 |

*2.3. Methods*

2.3.1. Heat calculation

Gibbs free energy change and enthalpy change are key parameters to determine the heat effect of a reaction process. Positive Gibbs free energy change (ΔG) indicates that the reaction process is spontaneous, or else the reaction can only take place under certain conditions (i.e. high reaction temperature, pressure or catalyst etc.). Positive enthalpy change (ΔH) represents that the reaction process is endothermic while the reaction with negative enthalpy change is exothermic. Therefore, Density Functional Theory was used to calculate the Gibbs free energy (G) and enthalpy (H) of the reactants and products. In this study, Gaussian 09 software package was used to perform the calculation.

2.3.2. Evaluation of the temperature rising properties

Adiabatic method (Fig. 1) was used to evaluate the temperature properties of the coal samples. The coal samples were dried under vacuum condition at 35 °C for approximately 24 h to dry the external water completely [28]. The dried coal samples (80 ± 0.01 g) were put into the adiabatic reaction vessel with the oxygen flowing into the coal samples. The temperature of the oven was maintained to be the temperature of the coal samples, which meant that the temperature rising of the coal was completely due to the heat released from the reaction of the coal.

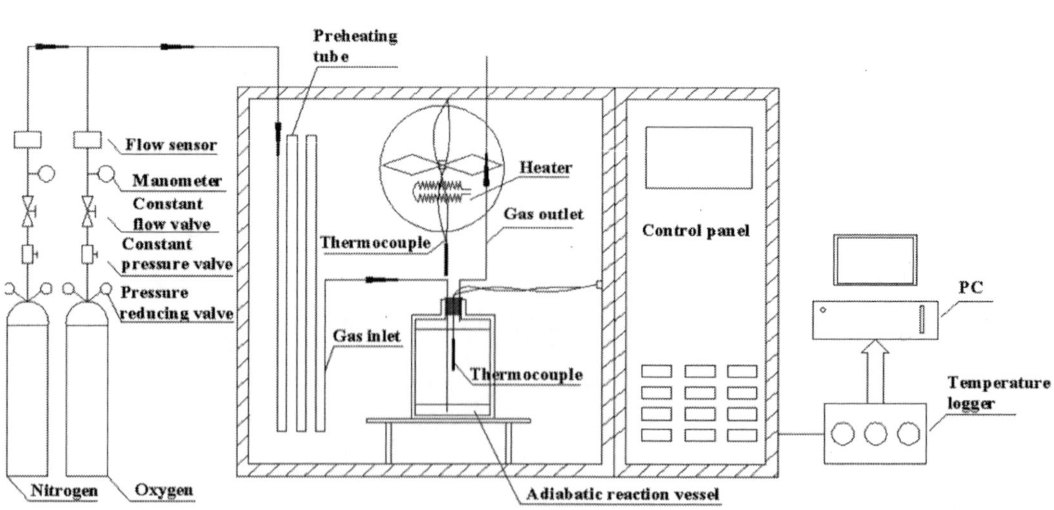


Fig.1. Schematic diagram of the adiabatic testing system

2.3.3. Testing of oxygen consumption and gas production

Gas analysis system (Fig. 2) was used in these experiments. Dried coal samples (50 ± 0.01 g) were packed into the reaction vessel with the dry air (with the oxygen concentration of 20.96% (v/v)) flowing into it at a rate of 50 mL/min, where the oxygen reacted with the coal samples. The rest of the oxygen not participating in the reaction and the gas productions flew out from the outlet, which were measured by gas chromatography.

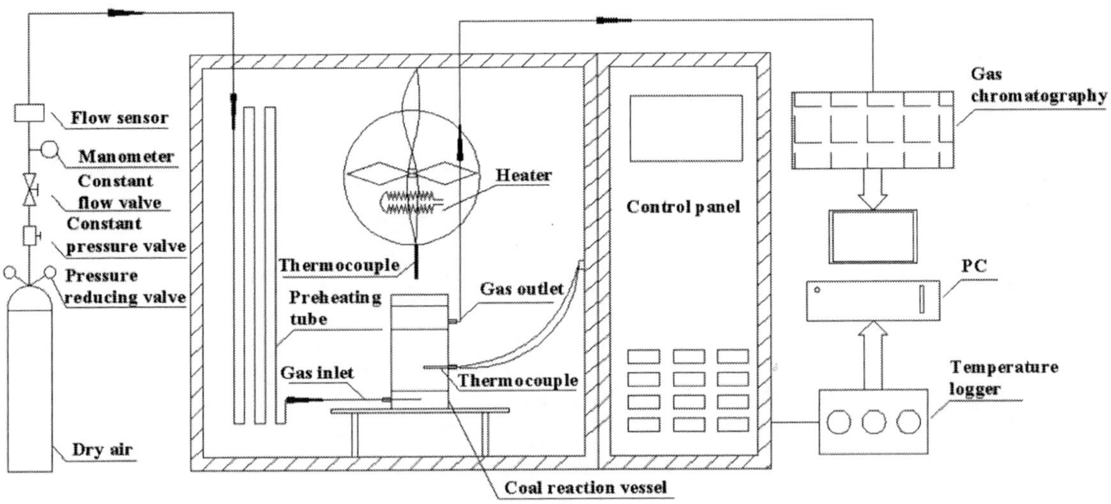


Fig.2. Schematic diagram of the programmable heating system

## 3. Results and discussion

### *3.1. Heat release characteristics of the oxidation of aliphatic hydrocarbon*

To find out the heat release characteristics, G and H of reactants and products were calculated and the results are shown in Table 3. For any given reaction, ΔG = G(products) – G(reactants), and ΔH = H(products) – H(reactants). Therefore, ΔG and ΔH of nine typical reactions were calculated, shown in Table 4.

Reactions 1 to 9 are the oxidation of $-CH_3$ and $-CH_2-$ in three different aliphatic chain length. As ΔG of all the nine reactions are negative, the oxidation of $-CH_3$ and $-CH_2-$ can take place spontaneously and all of the reactions are exothermic, for ΔH of the reactions are all negative too.

Table 3 Optimized structure, gibbs free energy and enthalpy of the reactants and products

| Species | Optimized structure | H (Hartree) | G (Hartree) | Species | Optimized structure | H (Hartree) | G (Hartree) |
|---|---|---|---|---|---|---|---|
| R1 | 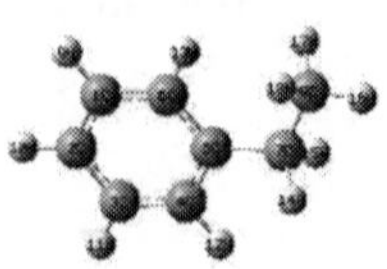 | -310.641475 | -310.679712 | R2 | 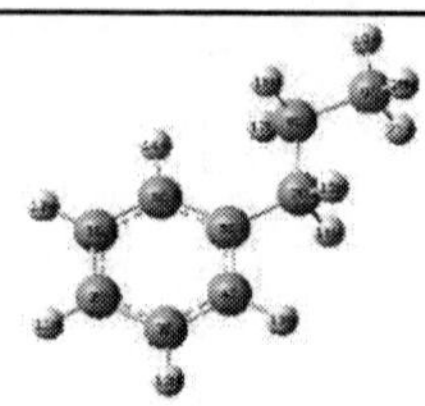 | -349.915263 | -349.956421 |
| R3 |  | -389.189181 | -389.233348 | R4 |  | -150.158973 | -150.181033 |
| R5 | 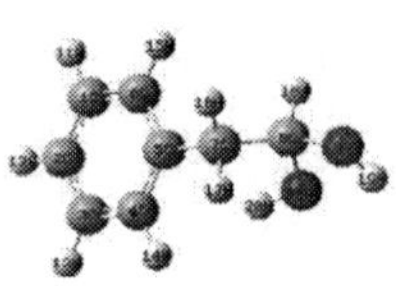 | -460.997755 | -461.039479 | R6 |  | -500.264898 | -500.308352 |

Table 4

Typical reactions and their Δ G and Δ H

| Reaction order | Reaction equation | Δ H (kJ/mol) | Δ G (kJ/mol) |
|---|---|---|---|
| 1 | 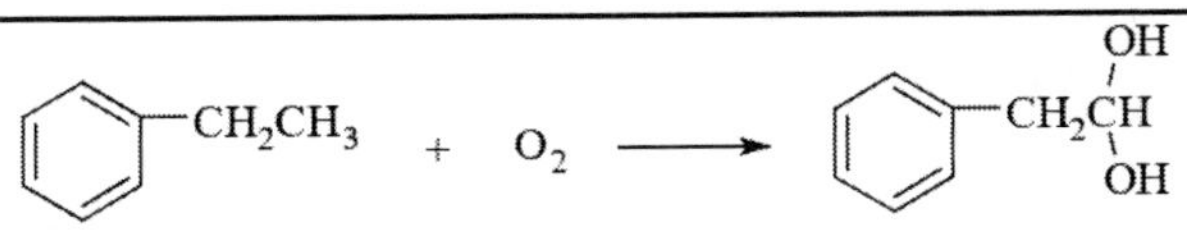 R1 R4 R5 | -465.52 | -469.266 |
| 2 | 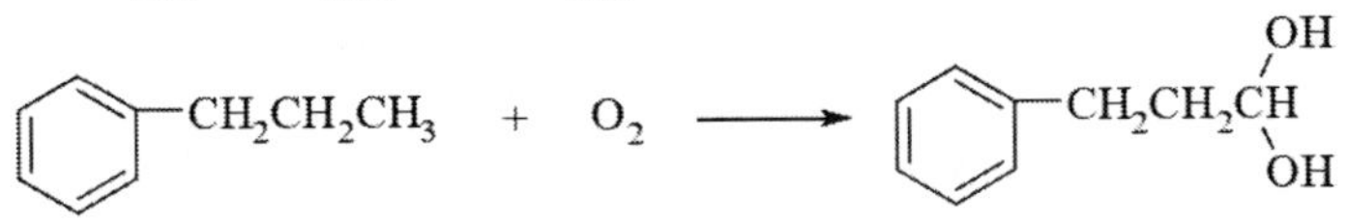 R2 R4 R6 | -500.583 | -448.693 |
| 3 | 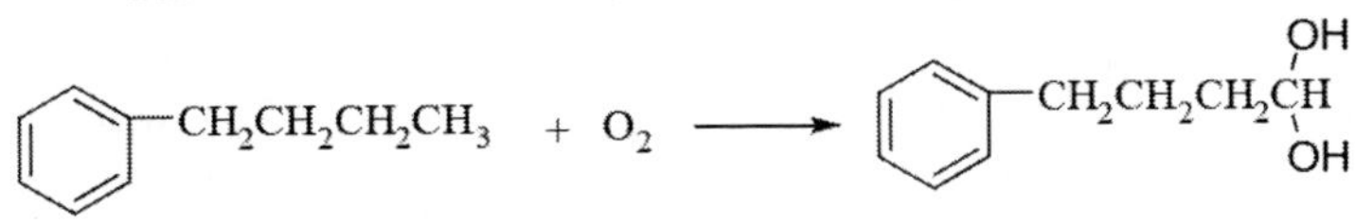 R3 R4 R7 | -500.683 | -448.551 |
| 4 | 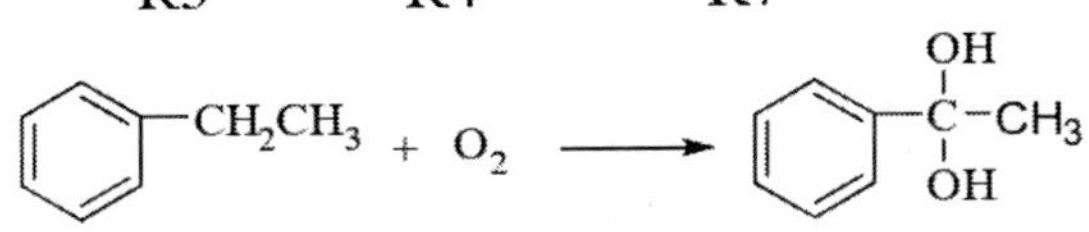 R1 R4 R8 | -534.602 | -482.467 |
| 5 | 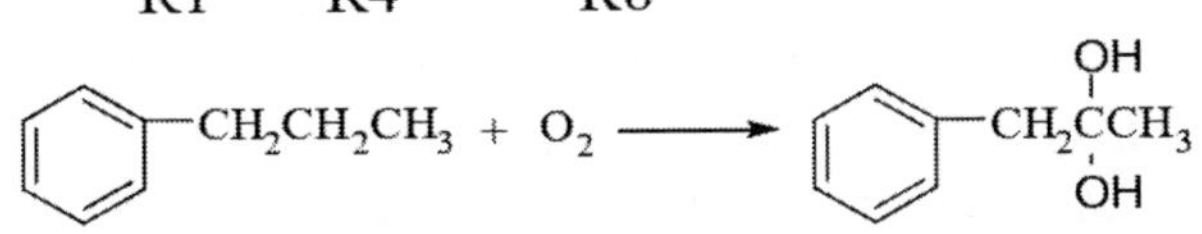 R2 R4 R9 | -542.909 | -490.039 |
| 6 | 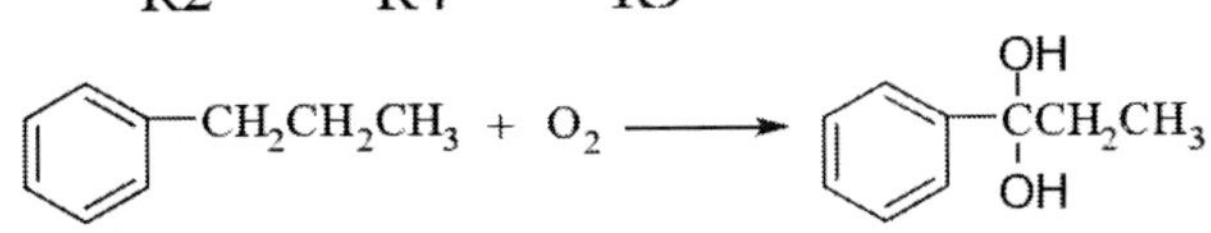 R2 R4 R10 | -528.043 | -474.856 |
| 7 | 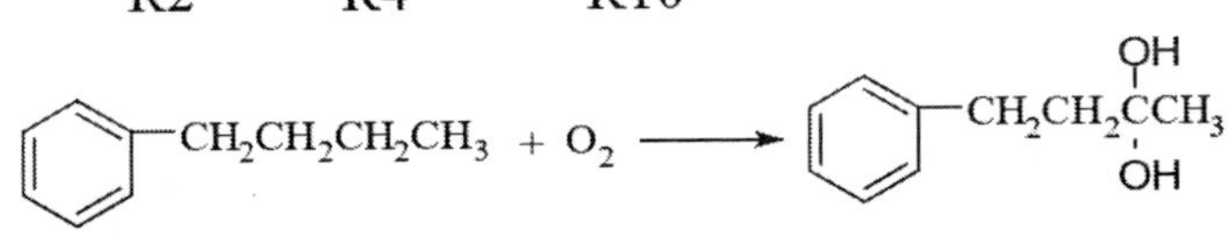 R3 R4 R11 | -525.441 | -527.92 |
| 8 | 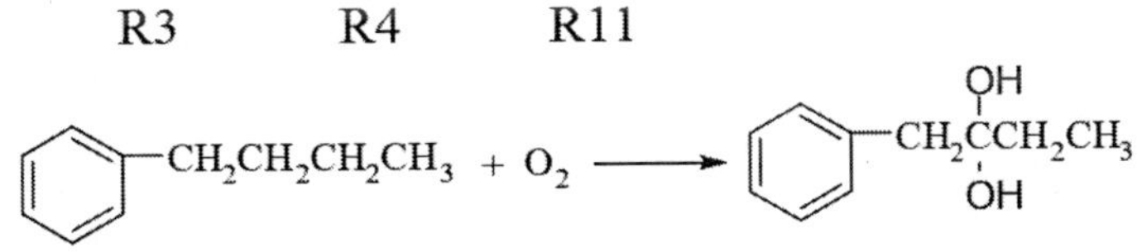 R3 R4 R12 | -525.168 | -476.158 |
| 9 | 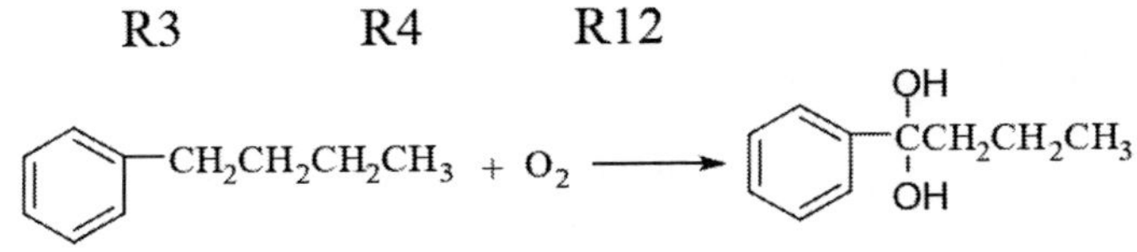 R3 R4 R13 | -525.061 | -479.91 |

Reactions 1 to 3 represent the oxidation of $-CH_3$ in aliphatic chain with two to four carbon atoms. There is a trend that the oxidation of $-CH_3$ in longer aliphatic chain can release slightly more heat than the one in shorter aliphatic chain, for the heat release of $-CH_3$ in aliphatic chain with two, three and four carbon atoms are 465.52 kJ/mol, 500.583 kJ/mol and 500.683 kJ/mol respectively. The heat release of the oxidation of $-CH_2-$ is larger than the oxidation of $-CH_3$. In the same chain length, the oxidation of $-CH_2-$ locates near the aromatic ring can release slightly more heat than the one far away from the aromatic ring. In general, the oxidation of the aliphatic hydrocarbon groups releases much greater heat than the heat release from reaction of the oxygen-containing functional groups [29].

*3.2. The effect of aliphatic hydrocarbon on the spontaneous combustion of coal*

As oxidation takes place on aliphatic hydrocarbon, analyzing the oxygen consumption characteristics can lead to understanding of the reaction properties of aliphatic hydrocarbon in coal. The oxygen concentration of the outlet of the reaction vessel were analyzed every 10 °C and the results are shown in Figure 3.

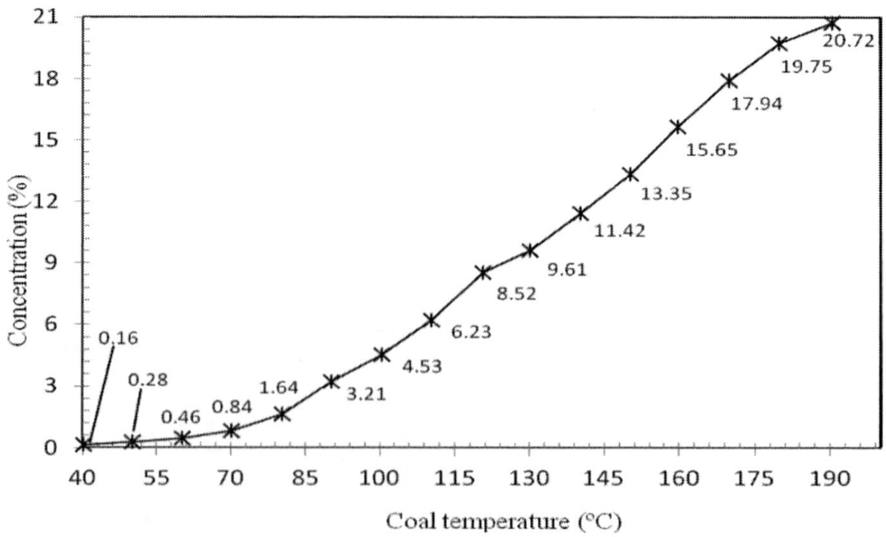


Fig.3. Oxygen consumption of BZ changing with coal temperature

It can be found that for brown coal of BZ, the oxygen consumptions are less than 1% before 70 °C. After 70 °C, the oxygen consumption increases greatly with the increase of the coal temperature. The same oxygen consumption characteristics were found on anthracite [29]. As it has been proved that oxygen is mainly consumed by physical adsorption and reservation in pores and crannies in the low temperature of the coal [10,30], it can be deduced that when the reaction temperature of the coal is low (below about 70 °C for brown coal of BZ), oxidation will not take place, which means aliphatic hydrocarbon groups in coal barely react with oxygen during this period. This conclusion can be also confirmed by the gas production results of $C_xH_y$, shown in Figure 4.

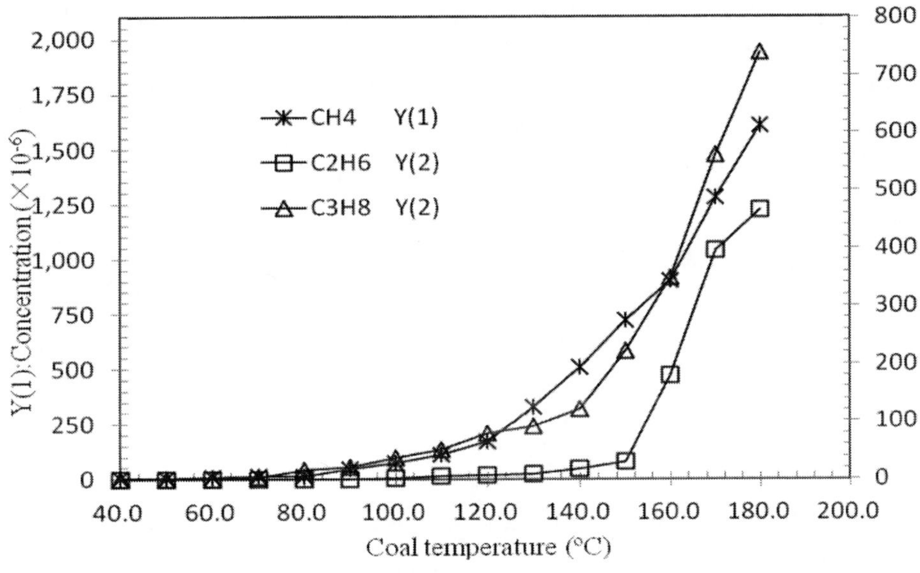


Fig.4. Concentration of the gas products from the reaction of the coal in different coal temperature

Gas products of $C_xH_y$ are from the reaction of aliphatic hydrocarbon groups, which is described in reaction model 4. If R' represents $-CH_3$, gas product of $CH_4$ would generate. Similarly, $C_2H_6$ and $C_3H_8$ would be produced when R' represent $-CH_2CH_3$ and $-CH_2CH_2CH_3$

respectively. From figure 4, $C_2H_6$ and $C_3H_8$ cannot be detected when the temperature of the coal is below 70 °C, which indicates that the structure of R12 and R13 (shown in Table 4) are not generated in this period. For the structure of R12 and R13 is very unstable, reaction model 4 will take place immediately R12 and R13 are generated, leading to the production of $C_2H_6$ and $C_3H_8$. As there were no gas products of $C_2H_6$ and $C_3H_8$ been detected, it can be concluded that when the temperature of the coal below about 70 °C, reaction 8 and 9 (shown in Table 4) do not happen. $CH_4$ was detected from the beginning of the reaction of the coal and the quantity. As it has been widely recognized that $CH_4$ can be adsorbed in raw coal [31, 32], it is normal $CH_4$ shows from the beginning of the experiment even if the oxidation of aliphatic hydrocarbon groups do not take place (reaction 4 to 7).

The temperature rising property of spontaneous combustion of BZ was obtained from adiabatic experiment. The temperature of the coal sample was collected every 20 seconds during the adiabatic process, shown in Figure 5.

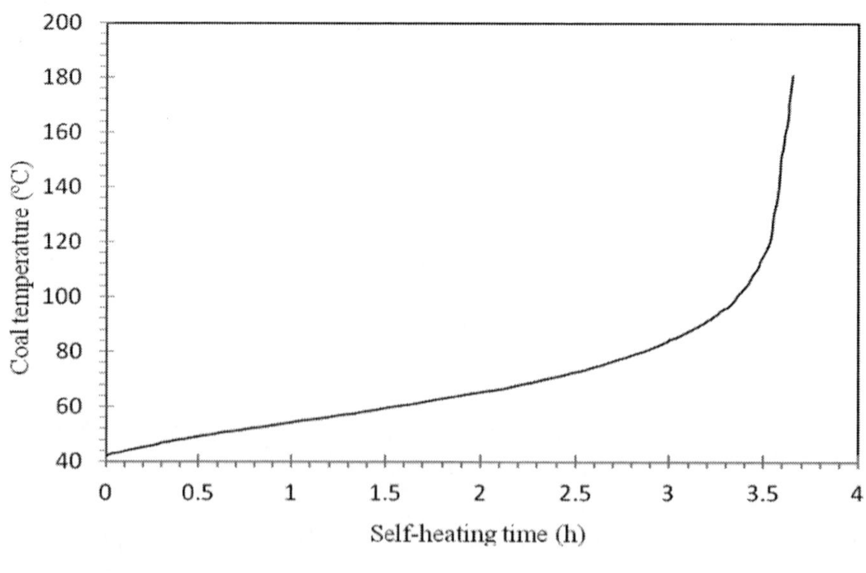


Fig.5. Temperature rising property of BZ during spontaneous combustion process

It can be found that the temperature rising of BZ can be roughly divided into two stages. Below about 100 °C, temperature of the coal sample rises very slow, taking 3.36 hours for BZ to self-heating from 40 °C to 100 °C. Then the temperature grows much more quickly and it only takes 0.33 hours for the coal rising to 180 °C from 100 °C. The results indicate that heat generated below about 100 °C is very small and after that greater of heat is produced. Since heat is produced from the reaction of coal, it can be concluded that above about 100 °C for BZ coal, some reactions with great heat release takes place. Theoretical calculations results (Table 4) show that the oxidation of the aliphatic hydrocarbon groups in coal can release a plenty of heat, which is about 5 to 100 times greater than the reaction of the oxygen functional groups [29]. Accompany with the oxygen consumption, gas production and temperature rising properties, it can be deduced that when the oxidation of the aliphatic hydrocarbon groups happens, great heat will be generated and the temperature of the coal will rise quickly to combust.

## 4. Conclusions

Oxidation always takes place on aliphatic hydrocarbon groups in coal. Gibbs free energy change and enthalpy change of the typical reactions of $-CH_3$ and $-CH_2-$ were calculated. The results indicate that oxidation of $-CH_3$ and $-CH_2-$ is endothermic and the reaction processes are spontaneous. Heat released from the oxidation of $-CH_3$ and $-CH_2-$ is over 500kJ/mol, which is much greater than the reaction of the oxygen functional groups in coal. By analyzing the oxygen consumption, the production of $CH_4$, $C_2H_6$ and $C_3H_8$ and the temperature rising properties and comparing them with the theoretical calculation results, it can be deduced that oxidation of aliphatic hydrocarbon groups is the key to determine the spontaneous combustion characteristics of the coal. The oxidation of the aliphatic hydrocarbon groups always takes place at a relatively higher temperature (over about 70 °C). Once the oxidation has taken place, great heat would be released, leading to the immediate rising of the temperature of the coal. This also give the reason for the cause of the two typical stages of the temperature rising properties during the spontaneous combustion of the coal.

**Acknowledgements**

The authors deeply appreciate the financial support provided by the National Natural Science Foundation of China (grant No. 51304134), Program for Innovative Research Team in Shanghai Maritime University, Shanghai Municipal Education Commission First-class Disciplines Project and Innovation Project of Shanghai Graduate Education (20131129).

**References**

[1] Y.S. Nugroho, A.C. Mcintosh, B.M. Gibbs, Using the crossing point method to assess the self-heating behavior of Indonesian coals, Symposium (International) on Combustion 27 (1998) 2981-2989.

[2] B.B. Beamish, M.A. Barakat, J.D. George, Spontaneous-combustion propensity of New Zealand coals under adiabatic conditions, International Journal of Coal Geology 45 (2001) 217-224.

[3] W. Sujanti, D.K. Zhang, X.D. Chen. Low-temperature oxidation of coal studied using wire-mesh reactors with both steady-state and transientmethods, Combustion and Flame 117 (1999) 646-651.

[4] B.B. Beamish, A. Arisoy. Effect of mineral matter on coal self-heating rate, Fuel 87 (2008), 125-130.

[5] R.R. Martin, J.A. Macphee, M. Workinton, E. Lindsay, Measurement of the activation energy of the low temperature oxidation of coal using secondary ion mass spectrometry. Fuel 68 (1989) 1077-1079.

[6] B.B. Beamish, Comparison of the $R_{70}$ self-heating rate of New Zealand and Australian coals to Suggate rank parameter, International Journal of Coal Geology 64 (2005) 139-144.

[7] T.X. Ren, J.S. Edwards, D. Clarke, Adiabatic oxidation study on the propensity of pulverised coals to spontaneous combustion, Fuel 78 (1999) 1611-1620.

[8] B.B Beamish, M.A. Barakat, J.D. Georg, Adiabatic testing procedures for determining the self-heating propensity of coal and sample ageing effects, Thermochimica Acta 362 (2000) 79-87.

[9] X.Y. Qi, D.M. Wang, X.X. Zhong, Y.L. Xu, Oxygen consumption of coal at low temperatures, Procedia Earth and Planetary Science 1 (2009) 366-370.

[10] H.H. Wang, B.Z. Dlugogorski, E.M. Kennedy, Coal oxidation at low temperatures: oxygen consumption, oxidation products, reaction mechanism and kinetic modeling, Progress in Energy and Combustion Science 29 (2003) 487-513.

[11] J. N. Pone, A.A. Hein, G.B. Stracher, H.J. Annegarn, R.B. Finkleman, D.R. Blake, J.K. McCormack, P. Schroeder, The spontaneous combustion of coal and its by-products in the Witbank and Sasolburg coalfields of South Africa, International Journal of Coal Geology 72 (2007) 124-140.

[12] P. Lu, G.X. Liao, J.H. Sun, P.D. Li, Experimental research on index gas of the coal spontaneous at low-temperature stage, Journal of Loss Prevention in the Process Industries 17 (2004) 243-247.

[13] H.H. Xin, D.M. Wang, X.Y. Qi, G.S. Qi, G.L. Dou, Structural characteristics of coal functional groups using quantum chemistry for quantification of infrared spectra, Fuel Processing Technology 118 (2014) 287-295.

[14] D.M. Wang, X.X. Zhong, J.J. Gu, X.Y. Qi, Changes in active functional groups during low-

temperature oxidation of coal, Mining Science and Technology 20 (2010) 35-40.

[15] X.Y. Qi, D.M. Wang, H.H. Xin, G.S. Qi, In situ FTIR study of real-time changes of active groups during oxygen-free reaction of coal, Energy & fuels 27 (2013) 3130-3136.

[16] T. Shi, J. Deng, X.F. Wang, Z.Y. Wen, Mechanism of spontaneous combustion of coal at initial stage, Journal of Fuel Chemistry and Technology 32 (2004) 652-657. (in Chinese)

[17] Y.Y. Chen, M. Mastalerz, A. Schimmelmann, Characterization of chemical functional groups in macerals across different coal ranks via micro-FTIR spectroscopy, International Journal of Coal Geology 104 (2012) 22-33.

[18] X.G. Sun, The investigation of chemical structure of coal macerals via transmitted-light FT-IR microspectroscopy, Spectrochimica Acta Part A: Molecular and Biomolecular Spectroscopy 62 (2005) 557-564.

[19] L. Wu, Y.M. Zhu, S.B. Chen, Y. Zhou, Research on the structural characteristics of vitrinite in different coal ranks, Fuel 107 (2013) 647-652.

[20] L.M. Ge, H.L. Xue, J.C. Xu, J. Deng, X.H. Zhang, Study on the oxidation mechanism of active groups of coal, Coal Conversion 24 (2001) 23-27. (in Chinese)

[21] A. Marzec, Towards an understanding of the coal structure: a review, Fuel Processing Technology 77-78 (2002) 25-32.

[22] Y. Yürüm, N. Altuntas, Air oxidation of Beypazari lignite at 50 °C, 100 °C and 150°C, Fuel 77 (1998) 1809-1814.

[23] Q.N. Dong, X.Y. Chen, G.Q. Jin, Y.D. Gu, Study of the lignite oxidation at low temperatures by FTIR emission spectroscopy, Journal of Fuel Chemistry and Technology 25 (1997) 333-338. (in Chinese)

[24] I. JoséV, M. Edgar, M. Rafael, FTIR study of the evolution of coal structure during the

coalification process, Organic Geochemistry 24 (1996) 725-735.

[25] W. Lu, Q.T. Hu, Relation between the change rules of coal structures when being oxidized and spontaneous combustion process of coal, Journal of China Coal Society 32 (2007) 939-944. (in Chinese)

[26] T. Shi, X.F. Wang, J. Deng, Z.Y. Wen, The mechanism at the initial stage of the room-temperature oxidation of coal, Combust Flame 140 (2005) 332-345.

[27] T. Xu, D.M. Wang, H.H. Xin, X.Y. Qi, Experimental Study on the Temperature Rising Characteristic of Spontaneous Combustion of Coal, Journal of Mining and Safety Engineering 29 (2012) 575-580. (in Chinese)

[28] T. Xu, D.M. Wang, Q.L. He, The study of the critical moisture content at which coal has the most high tendency to spontaneous combustion, International Journal of Coal Preparation and Utilization 33 (2013) 117-127.

[29] T. Xu, J.H. Liu, Y.T. Kang, Y. Jiao. Heat effect of the oxygen-containing functional groups in coal during spontaneous combustion processes. (under review)

[30] X.Y. Qi, D.M. Wang, X.X. Zhong, J.J. Gu, T. Xu, Characteristics of oxygen consumption of coal at programmed temperatures, Mining Science and Technology 20 (2010) 0372–0377.

[31] S.X. Hao, W. Chu, Q. Jiang, X.P. Yu. Methane adsorption characteristics on coal surface above critical temperature through Dubinin–Astakhov model and Langmuir model, Colloids and Surfaces A: Physicochemical and Engineering Aspects 444 (2014) 104–113.

[32] S. Bhowmik, P. Dutta. Adsorption rate characteristics of methane and $CO_2$ in coal samples from Raniganj and Jharia coalfields of India, International Journal of Coal Geology 113 (2013) 50–59.

# $CO_2$ AND $H_2O$ GASIFICATION UNDER CHEMICALLY AND DIFFUSION CONTROLLED CONDITIONS

**T. KREITZBERG, B. GÖVERT, S. PIELSTICKER and R. KNEER**

*Institute of Heat and Mass Transfer (WSA), RWTH Aachen University*

*Augustinerbach 6, D-52056 Aachen, Germany*

*kreitzberg@wsa.rwth-aachen.de*

**Abstract**

*The heterogeneous char-$CO_2$ and char-$H_2O$ gasification reactions are investigated experimentally in a small-scale fluidized bed reactor. This set-up enables high heat up rates ($\approx 10^4$ K/s), long timescale observation up to several hours, operation over a wide range of particle sizes and accurate control of reaction conditions. Time dependent carbon conversion curves and reaction rates are established from real-time gas product analysis by FTIR-spectrometry through a data analysis procedure. This procedure employs different particle surface-evolution models and accounts for sampling system signal attenuation. In this work, the gasification reactions of biogenic wood-char particles with $CO_2$ (Boudouard reaction) and $H_2O$ (water gas reaction) are investigated under atmospheric pressure. The char particles are gasified at different temperatures (1073 - 1473 K), $CO_2$ and $H_2O$ partial pressures (0.2 - 0.8 bar) and particle sizes (50 - 1120 μm) in order to examine the influence of these parameters on reaction rates. Experiments conducted are divided into two groups: The first set of experiments is operated under conditions where intrinsic kinetics control the observed reaction rate (low temperatures, small particles). For the expression of measured intrinsic chemical kinetics an nth-order approach has been adopted. Activation energies, reaction orders and frequency factors are determined for both investigated reactions. For the second group of experiments, temperatures and particle sizes are increased. This leads to a rising influence of mass transfer processes, limiting the overall reaction rate. The Thiele-analysis is used for the description of reaction rates measured under those conditions. Thereby, the effect of intra particle gas-diffusion on carbon conversion rates is quantified by calculation of effectiveness factors.*

# 1 Introduction

Heterogeneous reaction kinetics of carbon with $CO_2$ (Boudouard reaction) and carbon with $H_2O$ (water gas reaction) are studied by a variety of research groups around the globe. Since these reactions are understood to be the slowest in gasification processes limiting the overall efficiency, precise knowledge about those reaction rates is mandatory for the design and dimensioning of technical apparatuses. In this context, experimentally verified kinetic models of gasification reactions are required for the implementation in pertinent CFD simulations. The biggest share of publications about experimentally investigated gasification kinetics focuses on the determination of reactivities, where the observed rate of reaction is predominantly controlled by the intrinsic chemical reaction (e.g. [1, 2, 3]). This case, where mass transfer of the reactant gas through the particle by diffusive processes is faster than the chemical reaction itself, is denoted as regime I. The boundary conditions for these type of studies demand low temperatures and small particle sizes (typically below 1173 K and 200 µm [4]). As temperature or particle diameter are increased to a characteristic value, the effect of diffusion through the particle pore structure becomes the rate controlling mechanism, affecting observed reactivities significantly. These conditions, where intra particle gas diffusion is slower than the chemical reaction, are regarded as regime II. Neglecting any diffusive resistances, predictions considering only kinetics determined for regime I conditions will overpredict reaction rates for higher temperatures and bigger particles. Literature, and thus available kinetic data, on regime II conditions is scarce in comparison to those targeting regime I conditions [5]. However, in many technical processes, particles most likely will react under regime II conditions, such as in entrained flow gasifiers, where temperatures above 1873 K may occur. Hence, besides the knowledge about intrinsic chemical kinetics a sound understanding about particle reactivities under diffusion controlled conditions is indispensable for a complete kinetic analysis. In this paper, a kinetic study of wood char particles is undertaken regarding $CO_2$ and $H_2O$ gasification under both chemically and diffusion controlled conditions in a small-scale fluidized bed reactor. After establishing intrinsic chemical kinetics for an $n$th-order rate equation additional experiments have been conducted to investigate transition from regime I to II. Therefore, temperature and particle diameters have been varied up to 1473 K respectively 1120 µm to examine the increasing influence of intra particle diffusion. The experimental data then are analyzed by calculating Thiele-moduli and effectiveness factors to obtain an expression, from which reaction rates can be predicted in both regimes. The pore diameter $d_{\text{pore}}$, influencing Knudsen diffusivity, has been interpreted as an adjustable parameter and is determined by fitting the model to experimental data. In this context, the paper is primarily addressing the following questions:

- Is the model developed by *Thiele* [6] generally capable of predicting reactivities in the transition from regime I to II for $CO_2$ and $H_2O$ gasification?

- Can one constant pore diameter be fitted for different a) temperatures b) particle sizes and c) type of reactions (Boudouard, water gas) or are there deviations?

# 2 Experimental

## 2.1 Experimental setup

A small-scale fluidized bed reactor (FBR) has been designed to implement the concept of a well stirred reactor approximating uniform distributions of thermodynamic state variables as well as of reacting species. Small batches of pulverized solid fuel are supplied to a fluidized bed consisting of alumina ($Al_2O_3$) particles while simultaneously analyzing the exhaust gas. The experimental set-up is completed by a gas feeding system, including thermal mass flow controllers and a Fourier transform infrared spectrometer (FTIR) for product gas analysis

### 2.1.1 Feed gas system

The employed feed gas system is adapted depending on the type of experiment conducted (cf. fig. 1): For $CO_2$ gasification experiments, nitrogen and $CO_2$ are fed from separate gas bottles into a mixing vessel. The flow of each component is regulated by Vögtlin red-y thermal mass flow controllers with a precision of $\pm 1.0\,\%$ FS (full scale) and a stability of $\pm 0.2\,\%$ of the nominal value. The resulting $CO_2$ concentration in the mixing vessel is simply adjusted by the ratio of the entering nitrogen and carbon dioxide streams. A different feeding system is used for $H_2O$ gasification experiments: In this case, a controlled nitrogen stream is bubbled through a heated water trap; thus, the $N_2$ gas flow gets enriched with water vapor. Since the temperature of the water trap is adjusted to be higher than the dew point temperature of the target steam concentration, the stream leaving the water dosage unit has a $H_2O$-concentration above the aspired value. To lower the steam concentration, this water dosage unit is followed by a condenser to cool down the nitrogen/steam mixture. In this way, the gas is cooled to the dew point temperature corresponding to the target steam concentration. Excess water is condensed and fed back to the water dosage unit. By these means, a saturated $N_2/H_2O$ gas mixture is obtained which serves as the feed gas for $H_2O$ gasification experiments. All following pipes are heated to at least 373 K to prevent further water condensation.

### 2.1.2 Fluidized bed reactor

The fluidized bed reactor shown in Figure 1 consists of two coaxial ceramic pipes mounted in a stainless steel reactor head. The fluidizing gas is fed in through the reactor head and flows downwards through the annular gap between the two pipes to the bottom of the reactor. The gas enters the inner reactor pipe ($d = 55\,\mathrm{mm}$) through a porous distributor plate made from sintered silica glass and fluidizes the bed. The distributor plate with a typical pore size between 40 and 100 µm creates a homogeneous inflow and is impermeable for the $Al_2O_3$ bed particles. The bed height is approx. 30 mm in non-fluidized state and increases to approx. 70 mm under typical flowing conditions. Details on homogeneity, mixing and fluidization conditions of the bed can be found in [7]. The feed gas flow for the present experiments has been approx. $100\,\mathrm{L_s/h}$. All

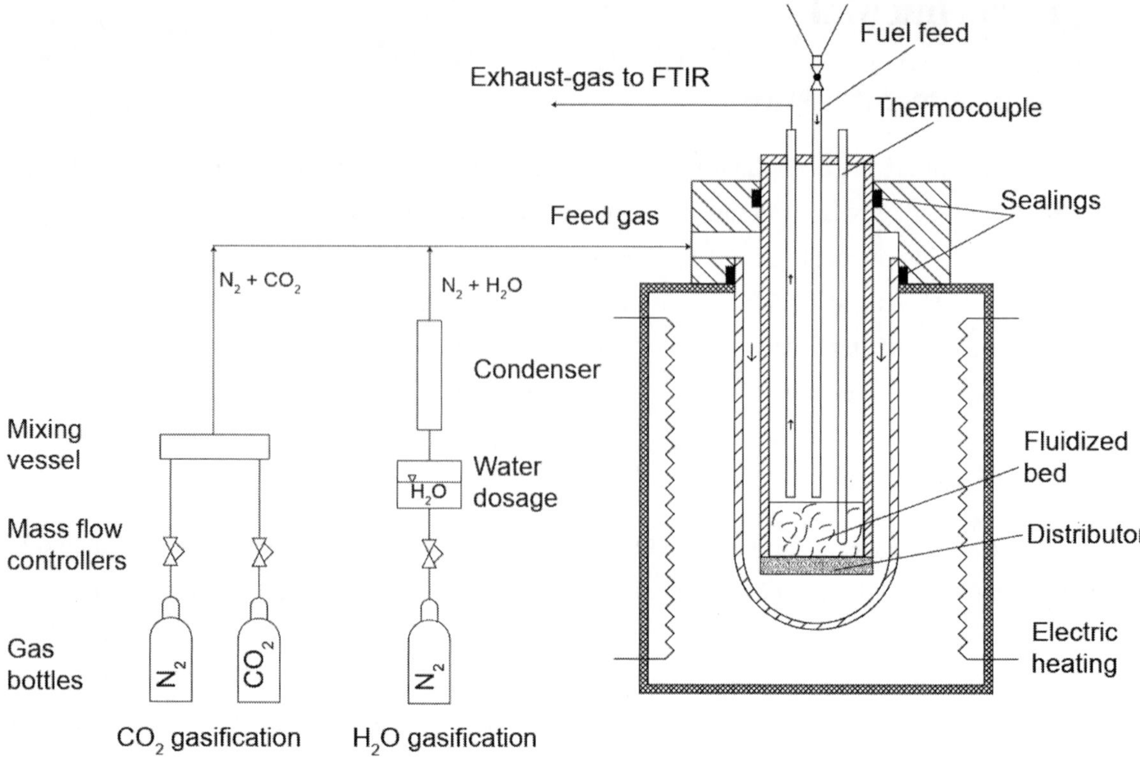


Figure 1: Scheme of the feed gas system and fluidized bed reactor

ceramic parts of the reactor are located in an electrically heated furnace which can be heated up to 1553 K with a stability of ±2 K. The bed temperature is measured separately by an S-type thermocouple placed in a ceramic shielding 50 mm above the distributor plate inside the fluidized bed.

### 2.1.3 Exhaust gas sampling and analysis

Gaseous reaction products are captured just above the fluidized bed by an exhaust pipe. A ceramic filter, which is mounted to the lower pipe end, prevents entrainment of small particles. Driven by a moderate overpressure inside the reactor of typically 10 mbar the exhaust gas is fed into the FTIR-gas analyzer through a sampling line and a filter. A GASMET DX-2000 FTIR spectrometer, measuring in the mid-IR at a wave number range from 600 $cm^{-1}$ to 4200 $cm^{-1}$ is employed for real-time gas analysis, sampling at 0.56 Hz with an accuracy of 2 % of the measured concentration after initial calibration.

## 2.2 Particle preparation and characterization

In this work a biogenic char made from beech-wood chips pyrolyzed in a screw reactor under pure nitrogen atmosphere with a residence time of 5 minutes at 773 K is investigated. The method applied can be considered as an intermediate-pyrolysis with heating rates of ap-

Table 1: Compositions based on ultimate and proximate analysis

| C | H | N | S | O | ash | water | volatiles |
|---|---|---|---|---|---|---|---|
| daf[1] | daf | daf | daf | daf | dry | - | daf |
| 81.2 | 3.3 | 0.3 | 0.0 | 12.1 | 3.1 | 3.6 | 23.3 |

[1] reference state: dry, ash-free

proximately 3 $\mathrm{K\,s^{-1}}$. Detailed information about the screw reactor used is provided by *Tomasi Morgano et al.* [8]. Table 1 gives an overview of the proximate and ultimate analysis results for the fuel: The char mainly consists of carbon (>80 %) with a remaining volatile content of 23.3 %.

After pyrolysis, particles were ground and sieved to nine different fractions (cf. table 2). For the determination of gas-solid reaction kinetics, precise knowledge about the particles' size is crucial. Hence, the different fractions were further characterized by digital imaging and analyzed with respect to shape and size. Images were taken by a NIKON D5100 DSLR camera attached to an OLYMPUS SZ-40 stereo microscope for magnification. Analysis of the images is performed within the software MATLAB: First, the projected surface area and the aspect ratio (length/width) of each registered particle are calculated. As shown in table 2, the aspect ratio of the particles averages around a value of 2, with a slightly decreasing trend from 2.24 (50-63 µm) to 1.9 (1000-1120 µm), indicating a non-spherical particle shape. This finding is further exemplified by four images of different sieve fractions which are illustrated in figure 2. Here, the cylindrical, rod-like shape is clearly revealed, which is typical for biogenic particles [9]. Thus, in the following considerations particles are approximated as ideal cylinders and not as spheres. For the determination of the particles' characteristic diameter, volume and diameter of a cylinder with the same aspect ratio and projected surface area as the detected particle are calculated. Then, an averaged diameter $\bar{d_p}$ weighted by the particles volume is found by summing up the product of individual volumes and diameters, divided by the total volume of all particles (eq. 1). Results of each sieve fraction are presented in table 2. The calculated mean diameters $\bar{d_p}$ generally lie close to the upper mesh size for most of the fractions.

$$\bar{d_p} = \frac{\sum_{j=1}^{N} d_{\mathrm{cyl,j}} \cdot V_{\mathrm{cyl,j}}}{V_{\mathrm{tot}}} \tag{1}$$

## 2.3 Evaluation procedure

The establishment of kinetic parameters is based on the dimensionless carbon conversion $X$, which represents the mass fraction of reacted solid carbon. The carbon conversion curve can be derived experimentally ($X_{\mathrm{exp}}$) from the temporal concentrations of the reactor exhaust gas species discharging the reacted carbon from the reactor – namely $CO_2$ and CO. It can also

Table 2: Sieve fractions, mean diameter, aspect ratio and number of analyzed particles

| Fraction (µm) | Mean diameter $\bar{d}_p$ (µm) | Mean aspect ratio (-) | # Particles |
|---|---|---|---|
| 50-63 | 56.1 | 2.24 | 672 |
| 90-100 | 95.5 | 2.27 | 280 |
| 112-125 | 128 | 1.67 | 151 |
| 125-160 | 145 | 2.04 | 106 |
| 280-315 | 319 | 1.90 | 110 |
| 450-500 | 530 | 1.87 | 53 |
| 600-630 | 655 | 1.89 | 103 |
| 850-900 | 958 | 1.85 | 50 |
| 1000-1120 | 1123 | 1.90 | 52 |

be computed by a model approach ($X_{mod}$), which can be interpreted using the characteristic reaction rate $r_0$ as an adjustable fit-parameter. This characteristic reaction rate is determined iteratively by fitting the model prediction to the experimental data. A detailed overview of all steps of the employed evaluation routine is given in [7].

### Carbon conversion curve from exhaust gas analysis

The calculated conversion $X_{exp}$ of solid carbon $C_{(s)}$ into gaseous products is based on a carbon-species balance around the reactor (eq. 2).

$$\frac{m_C}{M_C}\frac{\partial X_{exp}}{\partial t} = \Delta C \cdot \dot{n}_{out} \tag{2}$$

Where $M_C$ is the molar weight of carbon and $m_C$ represents the total captured carbon mass in the exhaust gas flow. The molar change in carbon due to reaction is symbolized by $\Delta C$ and the molar gas flow leaving the reactor by $\dot{n}_{out}$. Depending on the type of experiment conducted ($CO_2$ or $H_2O$ gasification) the balance yields different expressions for both of these factors, described in the following:

For $CO_2$ gasification the feed gas entering the reactor consists only of carbon dioxide and nitrogen. Inside the reactor, $CO_2$ and solid carbon are converted into CO according to (3). Balancing the feed gas stream and the stream leaving the reactor, equation 4 can be derived.

Boudouard reaction:
$$C_{(s)} + CO_2 \rightarrow 2CO \tag{3}$$

Carbon balance:
$$\frac{\partial X_{exp}}{\partial t} = \frac{M_C}{m_C} \cdot \underbrace{\frac{y_{CO}(t)}{2}}_{\Delta C} \cdot \underbrace{\frac{\frac{\dot{m}_{N2}}{M_{N2}} + \frac{\dot{m}_{CO_2}}{M_{CO_2}}}{1 - 1/2\, y_{CO}(t)}}_{\dot{n}_{out}} \tag{4}$$

In eq. 4, $\dot{m}$ indicates the feed gases' mass flow, divided into the individual species $N_2$ and $CO_2$. The variable $y_{CO}(t)$ denotes the time dependent gas concentration of the reaction product CO,

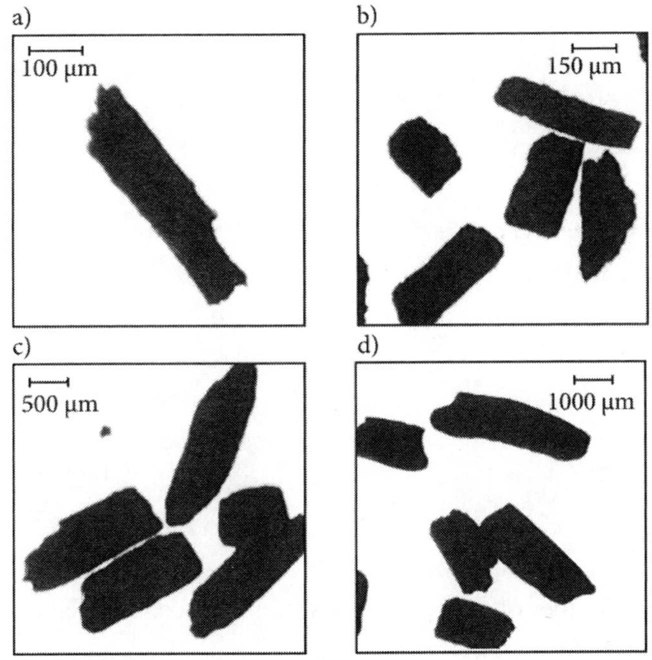


Figure 2: Particle images of different sieve fractions: a) 90-100 µm, b) 160-180 µm, c) 600-630 µm, d) 1000-1120 µm

measured by FTIR. Produced CO enlarges the molar flow leaving the reactor $\dot{n}_{out}$ by the factor $1/(1-1/2 y_{CO}(t))$.

Regarding $H_2O$ gasification, nitrogen and steam are the gases entering the reactor. In this case, solid carbon reacts together with steam forming one molecule of hydrogen and carbon monoxide, respectively (eq. 5). Part of the produced carbon monoxide is further converted to $CO_2$ due to the occurring watergas-shift reaction (eq. 6). As a consequence, both time dependent concentrations of $CO_2$ and CO need to be considered for the carbon balance, yielding eq. 7.

Water gas reaction: $$C_{(s)} + H_2O \rightarrow CO + H_2 \quad (5)$$

Water gas shift reaction: $$CO + H_2O \rightarrow CO_2 + H_2 \quad (6)$$

Carbon balance: $$\frac{\partial X_{exp}}{\partial t} = \frac{M_C}{m_C} \cdot \underbrace{(y_{CO}(t) + y_{CO_2}(t))}_{\Delta C} \cdot \underbrace{\frac{\frac{\dot{m}_{N2}}{M_{N2}} + \frac{\dot{m}_{H2O}}{M_{H2O}}}{1 - y_{CO}(t) - y_{CO_2}(t)}}_{\dot{n}_{out}} \quad (7)$$

Here, $y_{CO_2}$ and $y_{CO_2}$ are molar fractions of $CO_2$ and CO measured by FTIR. In comparison to the feed gas, the stream leaving the reactor is increased by a factor of $1/(1-y_{CO}(t)-y_{CO_2}(t))$.

Figure 3a and 3b exemplify typical concentration profiles of the gaseous species: In fig. 3a, data for a $CO_2$ gasification experiment at 900 °C are shown. Therein, a typical sharp peak in CO concentration can be seen, while the reacting gas ($CO_2$) retains a nearly constant concentration level. In fig. 3b a water gas experiment is illustrated, showing similar trends. In contrast to the previously described experiment, a distinct $CO_2$ peak occurs beside the CO peak. This is due

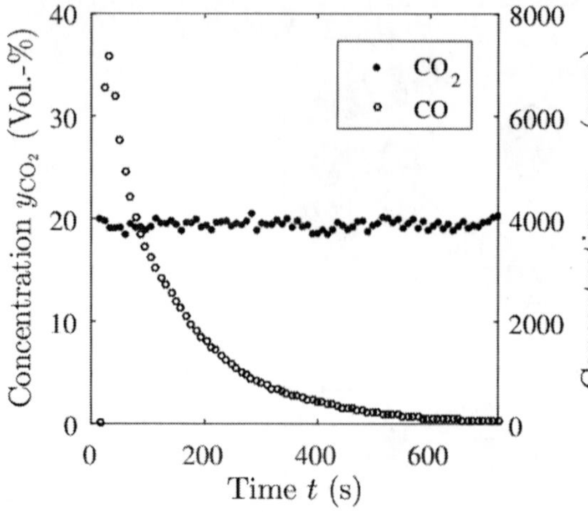

(a)

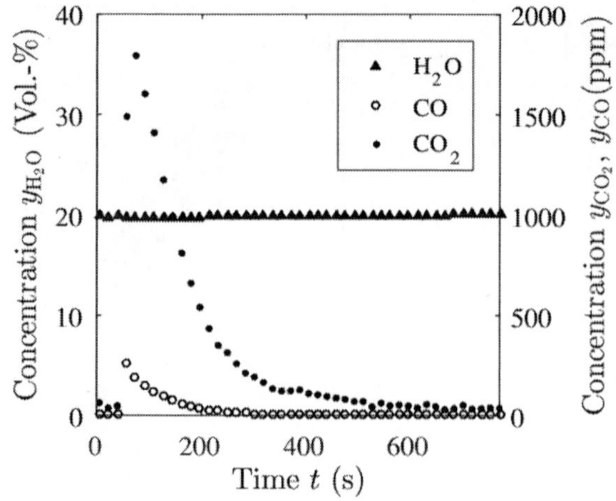

(b)

Figure 3: Typical gas concentration profiles at $T = 1173\,K$: a) $CO_2$ gasification and b) $H_2O$ gasification. Number of data points has been reduced for visualization purposes.

to the water gas shift reaction, converting around 80 % of the carbon monoxide to $CO_2$. The experimental carbon conversion curves $X_{exp}$ are calculated by numerical integration over time of equations 4 (Boudouard) or 7 (Water gas) respectively.

**Determination of characteristic reaction rate**

To determine $r_0$ for each experiment, the calculated model burnout curve $X_{mod}$ is numerically fit to the measured burnout curve $X_{exp}$ by a non-linear least squares regression. As the concentration measured by the FTIR does not equal the concentration directly in the fluidized bed, due to dispersion in the sampling line and mixing in the filter and FTIR gas cell, these effects have to be taken into account. This can be done either by deconvoluting the measured signal, or by convoluting the model curve. As the latter is numerically more stable, the model data given by the chosen model is convoluted. For this purpose, a transfer function $F(t)$ is used. This procedure of convoluting signals, which is based on the Taylor-Aris convection-diffusion transport equation, is described in detail by *Abad et al.* [10]. The transfer function $F(t)$ describes the signal response of a measurement section, in this case consisting of filter, sampling line and gas cell, to a Heaviside step function. Convoluting $X_{mod}$ with $F(t)$ allows to derive a theoretical curve $X_{mod,conv}$, which represents the modeled signal after being passed through sampling line, filter and FTIR gas cell. This procedure of convoluting data is further described in detail in [7]. To establish the characteristic reaction rate, the convoluted burnout curve $X_{mod,conv}$ is fitted to the experimental burnout curve $X_{exp}$. To reduce the influence of the reaction onset and the late conversion phase, only data in the range $0.05 < X_{exp} < 0.90$ is considered for the fit.

### Carbon conversion model selection

In this work, two common approaches describing the time dependent conversion of solid carbon to gaseous products are taken into account: An $m$th-order power law approach (eq. 8) and the random pore model (RPM) from *Bhatia* and *Perlmutter* [11] (eq. 9).

$$m\text{th} - \text{order power law}: \quad \frac{\partial X_{\text{mod}}}{\partial t} = (1 - X_{\text{mod}})^m \cdot r_0 \tag{8}$$

$$\text{Random pore model[11]}: \quad \frac{\partial X_{\text{mod}}}{\partial t} = (1 - X_{\text{mod}})\sqrt{1 - \psi \ln(1 - X_{\text{mod}})} \cdot r_0 \tag{9}$$

In equations 8 and 9, $X_{\text{mod}}$ is the carbon conversion and $t$ time. Function $r_0$ reflects the characteristic rate of reaction, which is used as the fit parameter and contributes to the influence of the reacting gas concentration $y_B$ as well as temperature $T$. Regarding the power law approach, the exponent $m$ can, on the one hand, be interpreted as parameter describing the particles conversion behavior. For instance $m = 2/3$ turns eq. 8 into the shrinking core model (SCM), where reaction is assumed to take place on a thin surface layer of a spherical particle proceeding towards the particle core. $m = 1$ leads to the uniform reaction model (URM), which assums the reaction to occur uniformly distributed throughout the entire particle. On the other hand, $m$ may be understood as a general fit-parameter detached from any physical meaning.

The random pore model is able to describe a changing surface area during conversion due to pore growth and coalescence and features the structural parameter $\psi$. Small values of $\psi$ describe particles with an overlapping and intersecting pore evolution, leading to a reduced effective surface area. Large values describe growing pores, increasing the surface area. [12]

To judge the influence and to find appropriate values of $m$ and $\psi$ respectively, the sum of residues between $X_{\text{exp}}(t)$ and $X_{\text{mod}}(t)$ is calculated according to equation 10 for all Boudouard and Water gas experiments conducted:

$$R = \sum_{j=1}^{\text{end}} \sum_{t=t_0}^{\text{end}} (X_{\text{exp,j}}(t) - X_{\text{mod,j}}(t))^2 \tag{10}$$

The inner summation adds up the residues of each discrete time step of a single gasification experiment (from $t_0$ to the end of the curve fit), while the outer sum accumulates the residues of all experiments conducted (from experiment $j = 1$ to the last experiment). Since the total number of experiments is not equal for Boudouard and water gas, $R$ is normed to the highest value found in the investigated parameter interval to make numbers comparable. Results are illustrated in figures 4a and 4b. The plots depict the variation of parameters $m$ and $\psi$ over an interval from 0.1 to 1, respectively 0 to 8. Figure 4a reveals a local minimum at $m = 1$ for both type of reactions. Comparing the sum of residues reached at $m = 2/3$ (SCM) with the minimal value at $m = 1$ (URM), it can be seen that the curve fits from the shrinking core model show approx. 3 (Boudouard) and 4 times (water gas) higher deviations by reproducing the experimental carbon conversion curves than the uniform reaction model. For the parameter $\psi$ a value of $\psi = 0$ minimizes the sum of residues (cf. 4b) for both reactions. This is in accordance with the findings from the variation of parameter $m$, as the RPM is transformed to the URM for $\psi = 0$ (cf. eq. 9). Subsequently, the uniform reaction model is chosen for estimation of the characteristic reaction rate $r_0$ in this work.

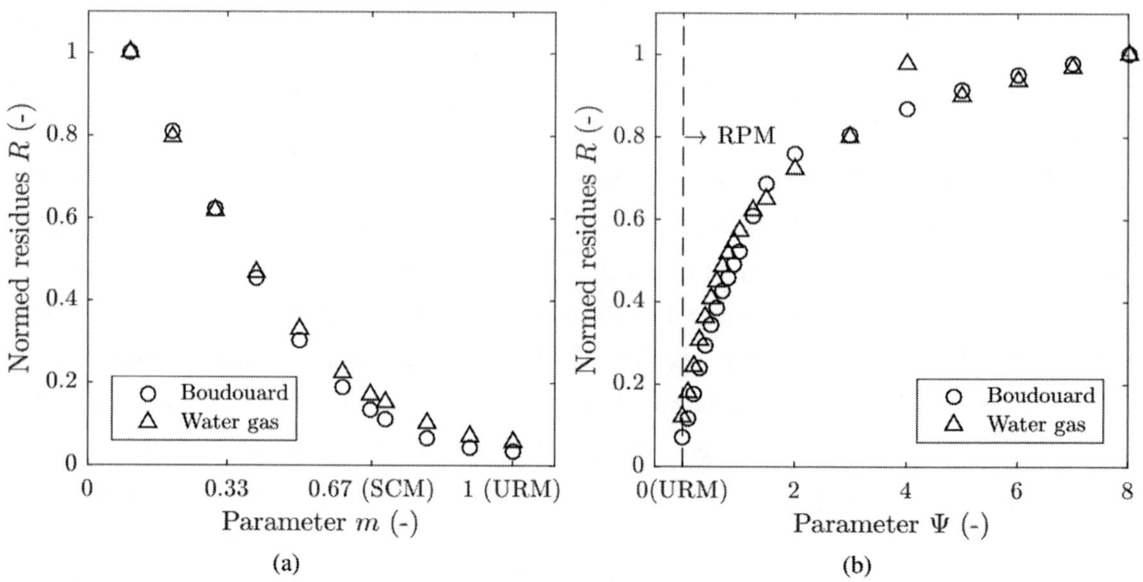


Figure 4: Normed Residues $R$ of curve fits (eq. 10) by varying parameters $m$ (a) and $\psi$ (b)

# 3 Experimental Results

## 3.1 Intrinsic chemical kinetics of $CO_2$ and $H_2O$ gasification

In this section, the determination of intrinsic chemical kinetics for $CO_2$ and $H_2O$ gasification is presented. The sieve fraction used for these experiments has a diameter range from 125 µm to 160 µm, with a mean value of 145 µm (cf. table 2). Experimental data is evaluated according to the procedure described in section 2. Each experiment is repeated at least three times to reduce the impact of random errors. Kinetic parameters of the $n$th-order rate equation (eq. 11) are established for both type of reactions by varying temperature and reacting gas partial pressure. In this rate equation, $A_\mathrm{i}$ is the pre-exponential factor, $E_\mathrm{A,i}$ is the activation energy of either Boudouard or water gas reaction and $\Re$ is the universal gas constant. $p_i$ denotes the reacting gas partial pressure ($CO_2$ or $H_2O$).

$$r_\mathrm{I,i} = A_\mathrm{i} \exp\left(\frac{-E_\mathrm{A,i}}{\Re T}\right) p_\mathrm{i}^{n_\mathrm{i}} \tag{11}$$

First, temperature has been varied at a constant reacting gas partial pressure of 0.2 bar (fig. 5a). Second, reacting gas partial pressure is changed while temperature is held at 900 °C (fig.

5b). Figure 5a shows an Arrhenius plot for the determination of activation energies. Triangles symbolize results for water gas experiments, while circles stand for the Boudouard reaction. The two straight lines are regressions and their slopes are directly proportional ($-E_A/\Re$) to the reactions' activation energy. Linear trends have been observed from 800 °C to 1050 °C for $CO_2$, and from 800 °C to 1000 °C for $H_2O$ gasification. The slope of the water gas experiments is steeper than that of the Boudouard reaction, corresponding to a higher activation energy. More precisely, the activation energy of the water gas reaction is found to be $E_{A,H_2O} = 253\,kJ\,mol^{-1}$, whereas a value of $E_{A,CO_2} = 213\,kJ\,mol^{-1}$ is calculated for the $CO_2$ gasification, which is about 19 % lower. Furthermore, the diagram reveals that for a temperature of approximately 850 °C, both reactions show about the same rate. Above that temperature, the water gas reaction has a higher reactivity, while for lower temperatures the Boudouard reaction is the dominant. The deviation in reactivity increases quickly with rising temperatures due to the difference in activation energies: At 1000 °C a ratio of $r_{H_2O}/r_{CO_2} = 1.75$ can be observed. In figure 5b, reaction orders are estimated by analysis of the slopes of the straight lines again. Here, values of $n_{H_2O} = 0.51$ and $n_{CO_2} = 0.36$ are determined, indicating a greater sensitivity of the $H_2O$ gasification reaction to a change in the reactant gas partial pressure. The determined kinetic parameters are summarized in table 3 and agree well with literature: *Barrio* [1] investigated $CO_2$ gasification of birch char and found a similar activation energy of $215\,kJ\,mol^{-1}$ and an order of reaction of $n_{CO_2} = 0.38$. The same char was also subject to $H_2O$ gasification, where parameters of $E_{A,H_2O} = 237\,kJ\,mol^{-1}$ and $n_{H_2O} = 0.57$ have been obtained.

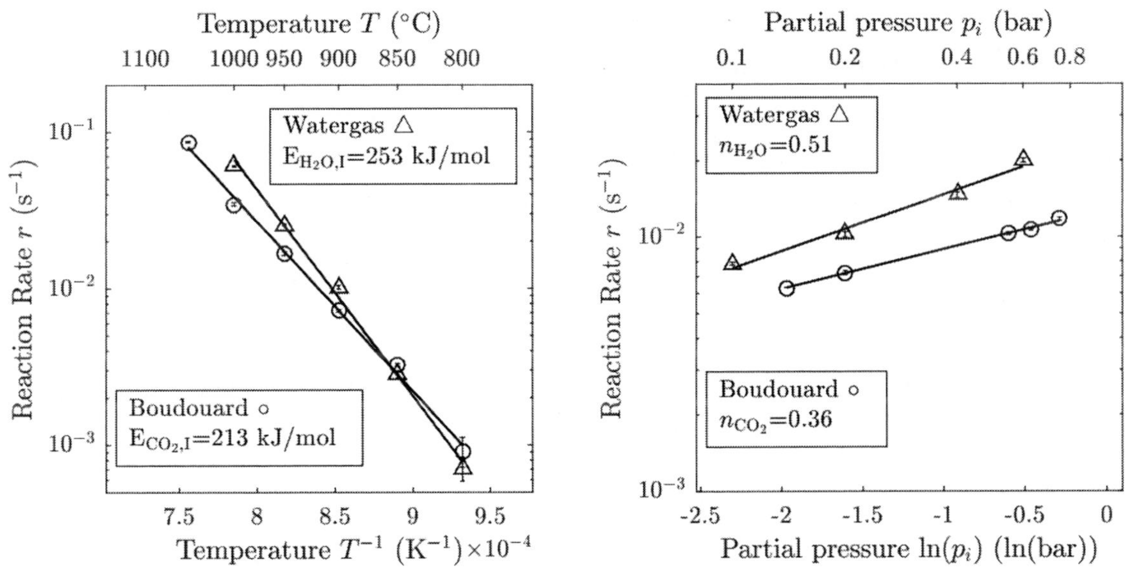


(a) (b)

Figure 5: a) Arrhenius plot for determination of activation energies, b) plot for determination of reaction order

Table 3: Calculated kinetic parameters of the $n$th-order rate equation

| Reaction | $A_0$ $s^{-1}$ $bar^{-n}$ | $E_A$ kJ $mol^{-1}$ | $n$ - |
|---|---|---|---|
| $CO_2$ gasification | $3.83 \cdot 10^7$ | 213 | 0.36 |
| $H_2O$ gasification | $3.57 \cdot 10^9$ | 253 | 0.51 |

## 3.2 Diffusion controlled kinetics of $CO_2$ and $H_2O$ gasification

For both $CO_2$ and $H_2O$ gasification, two sets of experiments have been conducted to investigate transition from regime I to regime II. First, temperature is increased beyond the values shown in figure 5a at constant reactant gas concentration and particle diameter. Next, particle size has been varied at 1000 °C. The experimental results have been analyzed to quantify the influence of pore diffusion on the observed reaction rates. Therefore, the Thiele-modulus $\phi_i$ and effectiveness factor $\eta_i$ are calculated. In the following, subscript i denotes either $CO_2$ for Boudouard or $H_2O$ for water gas considerations. The Thiele modulus $\phi_i$ is a dimensionless number describing the ratio between reaction rate and the rate of diffusion inside the particle. For an $n$th-order reaction $\phi_i$ is defined as (eq. 12, [13]):

$$\phi_i = \frac{\bar{d}_p}{6} \cdot \sqrt{\frac{n+1}{2} \frac{r_{I,i}}{D_{eff,i}} \frac{\rho_C \Re T}{M_C \, p_i}} \tag{12}$$

Where $D_{eff}$ is the effective diffusion coefficient of the reactant gas inside the particle, $\rho_C$ is the apparent density of the char particle and $p_i$ is the partial pressure of species i inside of the reactor. Since the intrinsic chemical reactivity $r_{I,i}$ (either Boudouard or water gas) and the mean char diameter $\bar{d}_p$ have been determined beforehand, the only unknown parameters in this equation remain $\rho_C$ and $D_{eff,i}$. The apparent density has been estimated by mercury porosimetry to $680\,kg\,m^{-3}$. Both, molecular ($D_{m,i}$) and Knudsen ($D_{k,i}$) diffusion contribute to the effective diffusivity, resulting in equation 13. Therein, $\varepsilon$ is the initial char porosity and $\tau$ is the tortuosity. Equation 14 shows the applied expression for the Knudsen diffusion coefficient. $D_{m,i}$ is calculated according to the *Chapman-Enskog* kinetic theory and collision integrals $\Omega_{i,N_2}$ and diameters $\sigma_{i,N_2}$ are tabulated in [14] (eq. 15).

$$D_{eff,i} = \frac{\varepsilon}{\tau} \cdot \frac{1}{1/D_{m,i} + 1/D_{k,i}} \tag{13}$$

$$D_{k,i} = \frac{d_{pore}}{3} \cdot \sqrt{\frac{8 \Re T}{\pi M_i}} \tag{14}$$

$$D_{m,i} = 1.8583 \cdot 10^{-7} \sqrt{T^3 \left( \frac{1}{M_i} + \frac{1}{M_{N_2}} \right)} \frac{1}{p \sigma_{i,N_2}^2 \Omega_{i,N_2}} \tag{15}$$

The Thiele-modulus is then used to calculate the effectiveness factor $\eta_i$ for a particle with spherical geometry, which gives the ratio of reactivity with and without pore diffusion limitation (eq. 16, [13]):

$$\eta_i = \frac{1}{\phi_i}\left(\frac{1}{\tanh(3\phi_i)} - \frac{1}{3\phi_i}\right) \tag{16}$$

By multiplying the effectiveness factor $\eta_i$ with the already determined intrinsic kinetic $r_{I,i}$ (cf. eq. 17), the char reactivity in regime II $r_{II,i}$ can be predicted. At this point it should be noted that strictly speaking equation 16 is only valid for first order reaction kinetics. However, *Hong* [13] introduced a correction factor $f_i$ that allows the application of equation 16 for any reaction order $0 < n < 1$ (cf. 18). With a value for char porosity $\varepsilon = 0.77$ (determined by Hg-porosimetry) and a typical guess for tortuosity of $\tau = 3$ the equation for the effectiveness factor only has the char pore-diameter $d_{pore}$ left as an unknown, which is interpreted as an adjustable fit parameter in the following.

$$r_{II,i} = f_i \cdot \eta_i \cdot r_{I,i} \tag{17}$$

### Variation of temperature

Based on the results shown in figure 5a, additional experiments have been conducted with increased temperatures up to 1200 °C. These added data points do not coincide with the regressions determined in section 3.1, indicated by dashed lines (Figure 6a). Comparing the trend of the dashed lines to the actual data points at high temperatures, it gets obvious that the error emerging by predicting reaction rates for regime II conditions with intrinsic kinetics, is huge: At 1200 °C the rate calculated with intrinsic kinetics is about 7.2 times (watergas) and 2.3 times (Boudouard) higher than the experimentally measured reactivity. Consequently, separate linear regressions have been performed for temperatures from 1050 °C to 1200 °C. Hereof, activation energies in regime II can be determined and are found to be 125 kJ mol$^{-1}$ (water gas) and 112 kJ mol$^{-1}$ (Boudouard). These are approximately 49 % and 53 % of the values under regime I conditions. This reduction of $E_A$ to about one half is typical for the transition from regime I to II [15]. The intersection of both regressions can be used to approximate the specific temperature where the transition of the governing reacting regime occurs: Regarding $CO_2$ gasification values of $T = 1060$°C and for $H_2O$ gasification $T = 1010$°C are estimated. Evidently, the shift of the water gas reaction appears at a lower temperature compared to the Boudouard reaction. This finding can be explained by the higher relative intrinsic reactivity of the water gas reaction in this temperature range. At the temperature of transition itself, the individual reaction rates are $9.8 \times 10^{-2}$ s$^{-1}$ ($CO_2$) and $8.1 \times 10^{-2}$ s$^{-1}$ ($H_2O$).

As pointed out above, the *n*th-order rate equation (eq. 11) in combination with the parameters found in section 3.1 cannot be used to predict reaction rates above a specific temperature.

To overcome this issue, intrinsic kinetics $r_{\mathrm{I,i}}$ are used in combination with equation 16 and 17 to find an appropriate expression for the reactivity $r_{\mathrm{II,i}}$ under regime II conditions . $r_{\mathrm{II,i}}$ is determined by a least squares fit, where the only adjustable parameter is the pore diameter $d_{\mathrm{pore}}$ in the effective diffusivity. Its value is calculated to be $d_{\mathrm{pore}} = 31.4\,\mathrm{nm}$ for the Boudouard and $d_{\mathrm{pore}} = 15.2\,\mathrm{nm}$ for the water gas reaction. The two fitted curves are presented in figure 6b and reproduce the experimental data with a mean deviation of 9 % (water gas) and 7 % (Boudouard), respectively.

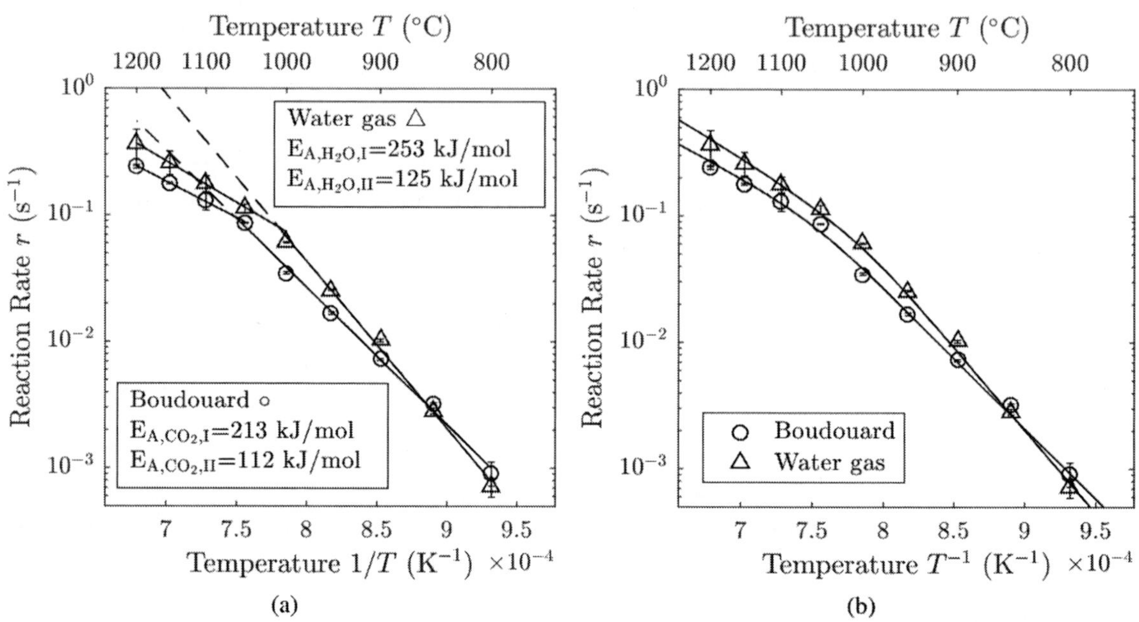


Figure 6: Arrhenius plots for Boudouard and water gas reaction at $p_i = 0.2\,\mathrm{bar}$: a) experimental data points with linear regressions for regime I and II; dashed lines correspond to extended predictions from regime I; b) experimental data points with curve fit (eq. 17) for both regimes

## Variation of particle diameter

Next, a particle diameter variation has been performed. Gasification experiments are carried out at a constant temperature of 1000 °C and a constant reactant gas partial pressure of 0.2 bar. The results of these experiments are presented in figure 7. Therein, reaction rates are plotted against the mean particle diameter $\bar{d}_{\mathrm{p}}$ for both types of reaction. The temperature of 1000 °C has been chosen with respect to the findings from the temperature variation above: the transition from regime I to II was found to be just above this temperature for particles with a mean diameter of 145 µm. Hence, an increase in particle diameter is expected to have an influence on char reactivity. The data points in figure 7 depict a general decrease in the rate of reaction for an increase in particle size. For small particles up to about 100 µm, the change is relatively small, indicating a weak dependency of reactivity on particle diameter. For larger particles, the influence of particle size gets more and more dominant leading to a reactivity decrease. The two curves in the diagrams are fits based on equations 12 to 16. Again, the particle pore diameter has been used as an adjustable parameter to fit experimental data. Here, values of $d_{pore} = 12.9$ nm (Boudouard) and $d_{pore} = 15.5$ nm (water gas) have been obtained.

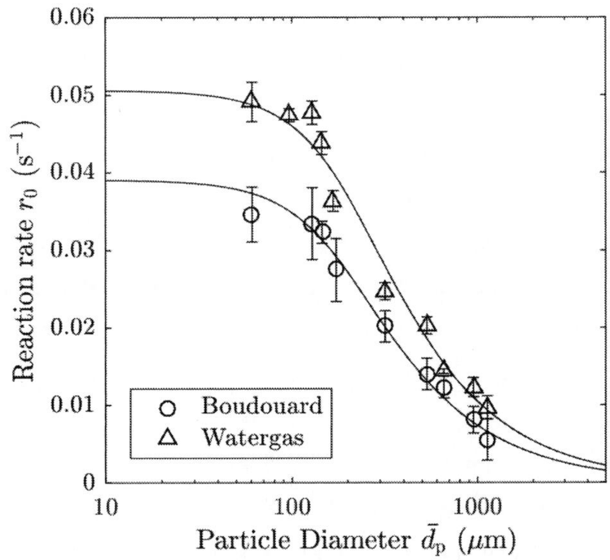


Figure 7: Variation of particle diameter at $T$ =1000 °C and $p_i$ =0.2 bar for $CO_2$ and $H_2O$ gasification

Concluding, it can be stated that for four different sets of experiments, the calculation of effectiveness factors delivers results sufficiently describing char reactivities for regime II conditions. Three of the fits performed, result in very similar values for the adjusted pore diameters $d_{pore}$ (12.9 nm, 15.5 nm and 15.2 nm). A remarkable deviation is found only for the high temperature measurements of the Boudouard reaction, which outnumbers the other values by a factor of approximately 2 ($d_{pore} = 31.4$ nm). Although the particle pore structure is complex and the true pore diameter-range might vary between several orders of magnitude [16], the conducted calculations with only one adjusted diameter revealed good agreement with experiments. To sum up, it has been shown that the model developed by *Thiele* can be used to predict reaction rates under regime I and II conditions. However, the interpretation of the physical meaning of the adjusted parameter $d_{pore}$ requires further investigation to set the obtained values of $d_{pore}$ in a context with the true pore structure of the char. *Kajitani et al.* [17] performed similar calculations for $CO_2$ gasification of a coal char in an entrained flow gasifier and interpreted the determined diameter as an average value. There, values between 0.4 nm to 10 nm were found. In contrast *Guizani et al.* [4] investigated $H_2O$ and $CO_2$ gasification of a wood char in a thermogravimetric analyzer and obtained very different results with pore diameters above 1000 nm. In that work, it has been suggested that the calculated diameters are without any physical meaning, serving only as a fit parameter.

# 4 Conclusion

Gasification reactions of wood char particles with $H_2O$ and $CO_2$ have been investigated experimentally in a small-scale fluidized bed reactor. The char has been sieved to nine different fractions and was analyzed by digital imaging to determine a mean diameter for each class of particles. Experiments have been conducted at ambient pressure and in a temperature range from 800 °C to 1200 °C at 0.2 bar reactant gas partial pressure. An appropriate carbon conversion model has been selected by investigating the influence of the parameters $\psi$ and $m$ on the random pore model and on a power law approach. This analysis revealed the uniform reaction model ($\psi = 0$ or $m = 1$) to have best consistency with the experimental conversion curves. From experiments performed under regime I conditions, kinetic parameters of an $n$th-order rate law were established. Particles with an average diameter of 145 µm were used for these experiments. For this particle size, regime I conditions have been observed up to temperatures of 1010 °C for $H_2O$ gasification and to 1060 °C for $CO_2$ gasification and their activation energies were found to be 213 kJ mol$^{-1}$ and 253 kJ mol$^{-1}$. For temperatures below 850 °C, the Boudouard reaction revealed a higher reactivity, whereas at higher temperatures the water gas reaction is dominant. Order of reactions were determined at 900 °C and account for values of $n = 0.36$ (Boudouard) and $n = 0.51$ (water gas). To further investigate reactivities under intra particle diffusion control, temperature has been increased up to 1200 °C. Under these conditions, activation energies were reduced to about 50 % of the values found at low temperatures. Effectiveness factors and Thiele-moduli have been calculated to obtain an expression that is capable of predicting reactivities under both, regime I and II conditions. The pore diameter, which influences Knudsen diffusivity, was used as a fit parameter. A non-linear least squares regression exhibited values of $d_{pore} = 31.4$ nm (Boudouard) and $d_{pore} = 15.2$ nm (water gas). An additional set of experiments has been carried out, where particle size has been varied. Again, effectiveness factors were calculated to obtain curve fits describing the dependency of reactivity on particle size. There, the pore diameters were found to be $d_{pore} = 12.9$ nm (Boudouard) and $d_{pore} = 15.5$ nm (water gas).

# Acknowledgments

The authors would like to thank the Helmholtz association for funding Thobias Kreitzberg within the framework of the Helmholtz Virtual Institute for Gasification Technology - HVI-GasTech (Grant VH-VI-429). Therein, funding has been provided via the Initiative and Networking Fund of Helmholtz Association. Also, the authors would like to express their gratitude to the German Research Foundation (DFG) for funding Benjamin Gövert and Stefan Pielsticker within the SFB/Transregio 129 "Oxyflame".

# A Additional formula

$$f_{\mathrm{i}} = \left(1 + \frac{\sqrt{1/2}}{2\phi_{\mathrm{i}}^2 + \frac{1}{2\phi_{\mathrm{i}}^2}}\right)^{1/2(1-n_{\mathrm{i}})^2} \quad [13] \tag{18}$$

# Nomenclature

| | | |
|---|---|---|
| **Latin** | | |
| $\Delta C$ | Molar change in carbon | $\mathrm{mol\ mol^{-1}}$ |
| $\dot{m}$ | Mass flux | $\mathrm{kg\ s^{-1}}$ |
| $\dot{n}$ | Molar flux | $\mathrm{mol\ s^{-1}}$ |
| $\mathfrak{R}$ | Universal gas constant | $\mathrm{J\ mol^{-1}\ K^{-1}}$ |
| $A$ | Pre-exponential factor | $\mathrm{s^{-1}\ bar^{-n}}$ |
| $D$ | Diffusion coefficient | $\mathrm{m^2\ s^{-1}}$ |
| $d$ | Diameter | m |
| $E_{\mathrm{A}}$ | Activation energy | $\mathrm{J\ mol^{-1}}$ |
| $F$ | Transfer function | - |
| $f$ | Correction factor for calculation of effectiveness factors | - |
| $M$ | Molar mass | $\mathrm{kg\ mol^{-1}}$ |
| $m$ | Mass | kg |
| $m$ | Order of reaction with respect to solid carbon | - |
| $n$ | Order of reaction with respect to reactant gas | - |
| $p$ | Pressure | Pa |
| $R$ | Residues | - |
| $r$ | Reaction rate | $\mathrm{s^{-1}}$ |
| $T$ | Temperature | K |
| $t$ | Time | s |
| $X$ | Carbon Conversion | - |
| $y$ | Molar fraction | $\mathrm{mol\ mol^{-1}}$ |
| **Greek** | | |
| $\varepsilon$ | Porosity | - |
| $\eta$ | Effectiveness factor | - |
| $\Omega$ | Collision integral | - |
| $\phi$ | Thiele modulus | - |
| $\psi$ | Structural Parameter of the random pore model | - |
| $\rho$ | Density | $\mathrm{kg\ m^{-3}}$ |
| $\sigma$ | Collision diameter | m |
| $\tau$ | Tortuosity | - |

**Subscripts**

| | |
|---|---|
| C | Carbon |
| eff | Effective diffusivity |
| exp | experimental value |
| I | Regime I |
| i | $CO_2$ for Boudouard and $H_2O$ for water gas reaction |
| II | Regime II |
| k | Knudsen diffusivity |
| m | Molecular diffusivity |
| mod | Model prediction |
| out | flux leaving the reactor |
| p | Average particle diameter |
| pore | Particles' (apparent) pore diameter |

# References

[1] Barrio, Experimental investigation of small-scale gasification of wood biomass, Ph.D. thesis, Norwegian University of Science and Technology Trondheim (2002).

[2] H. Risnes, L. H. Sørensen, J. E. Hustad, $CO_2$ reactivity of chars from wheat, spruce and coal, Progress in Thermochemical Biomass Conversion (2008) 61–72.

[3] J. Kramb, J. Konttinen, A. Gómez-Barea, A. Moilanen, K. Umeki, Modeling biomass char gasification kinetics for improving prediction of carbon conversion in a fluidized bed gasifier, Fuel 132 (2014) 107 – 115.

[4] C. Guizani, F. E. Sanz, S. Salvador, Influence of temperature and particle size on the single and mixed atmosphere gasification of biomass char with $H_2O$ and $CO_2$, Fuel Processing Technology 134 (2015) 175 – 188.

[5] D. G. Roberts, E. M. Hodge, D. J. Harris, J. F. Stubington, Kinetics of char gasification with $CO_2$ under regime II conditions: effects of temperature, reactant, and total pressure, Energy & Fuels 24 (10) (2010) 5300–5308.

[6] E. W. Thiele, Relation between catalytic activity and size of particle, Industrial & Engineering Chemistry 31 (7) (1939) 916–920.

[7] H. Haustein, T. Kreitzberg, B. Gövert, A. Massmeyer, R. Kneer, Establishment of kinetic parameters of particle reaction from a well-stirred fluidized bed reactor, Fuel 158 (2015) 263 – 269.

[8] M. Tomasi Morgano, H. Leibold, F. Richter, H. Seifert, Screw pyrolysis with integrated sequential hot gas filtration, Journal of Analytical and Applied Pyrolysis 113 (2015) 216 – 224.

[9] Q. Guo, X. Chen, H. Liu, Experimental research on shape and size distribution of biomass particle, Fuel 94 (2012) 551 – 555.

[10] A. Abad, S. Cardona, J. Torregrosa, F. Lopez, J. Navarro-Laboulais, Flow analysis deconvolution for kinetic information reconstruction, Journal of Mathematical Chemistry 38 (2) (2005) 271–292.

[11] S. Bhatia, D. Perlmutter, A random pore model for fluid-solid reactions: II. diffusion and transport effects, AIChE Journal 27 (2) (1981) 247–254.

[12] S. Bhatia, D. Perlmutter, A random pore model for fluid-solid reactions: I. isothermal, kinetic control, AIChE Journal 26 (3) (1980) 379–386.

[13] J. Hong, W. C. Hecker, , T. H. Fletcher, Improving the accuracy of predicting effectiveness factors for mth order and langmuir rate equations in spherical coordinates, Energy & Fuels 14 (3) (2000) 663–670.

[14] R. Bird, W. Stewart, E. Lightfoot, Transport Phenomena, 2nd Edition, Wiley, 2002.

[15] O. Levenspiel, Chemical reaction engineering, 3rd Edition, Wiley, 1999.

[16] G. Liu, P. Benyon, K. Benfell, G. Bryant, A. Tate, R. Boyd, D. Harris, T. Wall, The porous structure of bituminous coal chars and its influence on combustion and gasification under chemically controlled conditions, Fuel 79 (6) (2000) 617 – 626.

[17] S. Kajitani, N. Suzuki, M. Ashizawa, S. Hara, $CO_2$ gasification rate analysis of coal char in entrained flow coal gasifier, Fuel 85 (2) (2006) 163 – 169, special Issue: The 21st Annual International Pittsburgh Coal Conference.

*The 33rd International Pittsburgh Coal Conference 2016*

# New Approach in Gasifying Reactive Low Rank Coals under Non/Partial Slagging Entrained-Bed Conditions: 20 TPD Pilot Test Results

2016. 8. 11.

Cape Town International Convention Center, South Africa

Yongseung Yun*, Jin-Wook Lee, Seok Woo Chung, Young-Tai Choi[1]
Plant Engineering Center, Institute for Advanced Engineering
[1]Korea Institute of Industrial Technology
Republic of Korea

IAE Institute for Advanced Engineering

IAE

## Background

### ■ Low Rank Coals

- High reactivity, High volatile matter, High moisture content
- Low ash fusion temperatures
- Relatively under-represented in coal utilization

### ■ Construction cost in Korea's case

- Coal IGCC : US$3,800-4,000/$kW_e$
- USC : less than US$2,000/$kW_e$
- Conventional PC(Pulverized Coal) combustion-based power plant : US$1,300-1,400/$kW_e$

### ■ Slagging vs. Non-Slagging → Partial/Non-slagging

- Slagging of coal ash frequently causes operational problems in gasifier. Non-slagging gasification is much more easy to operate. But, a higher possibility in lower carbon conversion (higher entrained fines).
- In large commercial size gasifiers, it would be better to obtain over 99% carbon conversion. Few percent of fines could pose a burden in disposing.
- Captured fines contain 30-60% carbon content, which suggests a possibility as a combustion fuel or other economically useful way.

## Criteria/Results of Fly-ash (Fines) for Economic Usage

◆ Typically, the combustion fly-ash should contain less than 5 wt% remaining carbon to be useful as a cement filler.

➡ However, fines from the one-stage entrained-bed coal gasifier of pilot-scale contain 30-60% remaining carbon. In this case, final fines might be applicable instead as a low-grade fuel if combustibility and other criteria are satisfactory.

◆ For direct use of fines, heavy metal should not leach out to contaminate underground water.

➡ Gasification slags are not leaching any heavy metals. Entrained fines from a non/partial-slagging gasifier were confirmed passing environmental standards.

Source: Y. Yun, "Development of Partial/Non-Slagging Entrained-bed Coal Gasifier to Lower Capital Cost", Gasification Technologies Conference, 2013, Colorado Springs.

IAE

## Target of Development

◆ Main factors of high capital cost in entrained-bed gasifiers :
- Too high operation temperatures for slagging
- Exotic, expensive, 'not-easy-to-get' metal materials
- Too many welding points + Not in mass production
- Safety and Control for explosive, toxic, corrosive syngas

◆ Situation of slag/fly-ash usage :
- Before: fly-ash was dumped → Now: most combustion fly-ashes are sellable as cement filler, etc.
- Large quantities of slag have a limited usage till now.
- Fly-ash at 1,100-1,350°C range → Solve ash leaching problem

◆ Target :
- Gasifier of partial/non-slagging mode of entrained-bed type (operation temperature 1,100-1,350°C)
- Carbon conversion of single pass in gasifier: above 90% (92-96%)
- Fly-ash recycling for >99% carbon conversion, or use as another profitable feedstock
- Easy-to repair and replaceable feature
- 200-500 TPD gasification system with spare gasifier
- Capital cost 1/3-1/4 of current slagging entrained-bed system.

◆ Limitations :
- Not suitable for all available coals (limited in fusibility characteristics)

Source: Y. Yun, "Development of Partial/Non-Slagging Entrained-bed Coal Gasifier to Lower Capital Cost", Gasification Technologies Conference, 2013, Colorado Springs.

## World High- & Low- Rank Coal Amount

- at end 2007, in Millions Tonnes -
(high grade coal: anthracite and bituminous)

### Anthracite, Bituminous

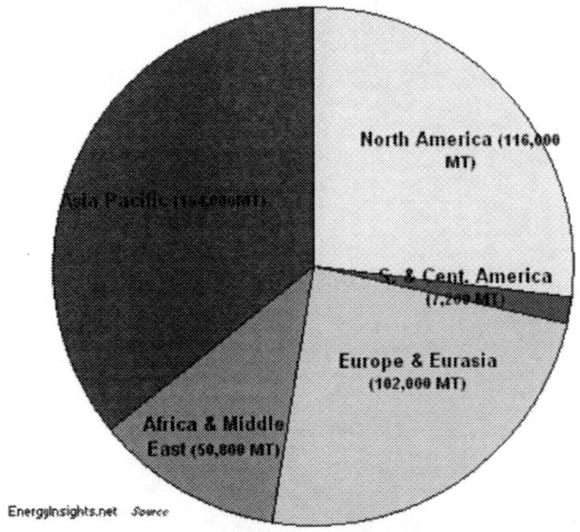


430,000 MT

- at end 2007, in Millions Tonnes -
(low grade coal: sub-bituminous and lignite)

### Subbituminous, Lignite

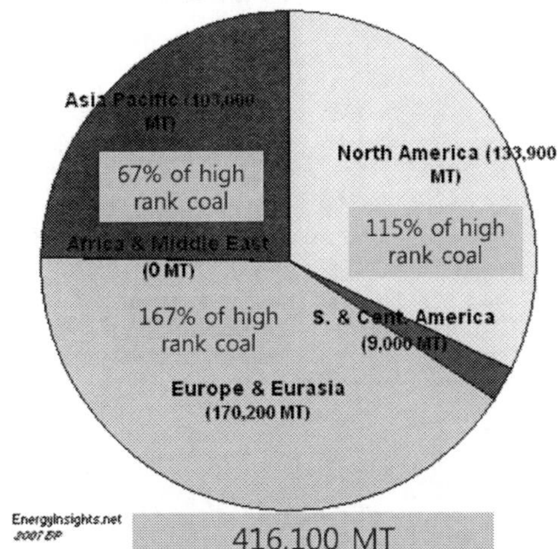


416,100 MT
97% of high rank coals

- Low rank coal is cheaper, but difficult in transport and in conventional coal usage.
- **Gasification is more suitable for low-rank coals.**

IAE

Source: http://www.energyinsights.net/cgi-script/csArticles/articles/000000/000058.htm

## Gasification is More Suitable for Low Rank Coals

- Low ash fusion coals like Indonesian low rank coals → More suitable for IGCC application

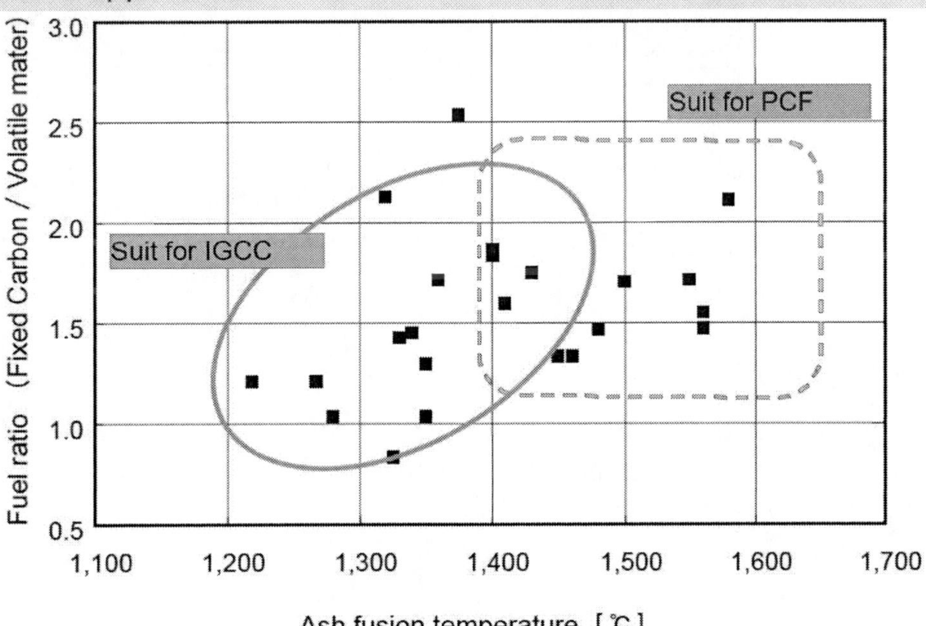


IAE

Source: Tetsuji Asano, 'Progress in Japanese Air-blown IGCC Demonstration Project Update,' Coal Gasification Symposium, Jeju, Korea, 2012/04/26.

# IGCC Demo Plants with Oil Price

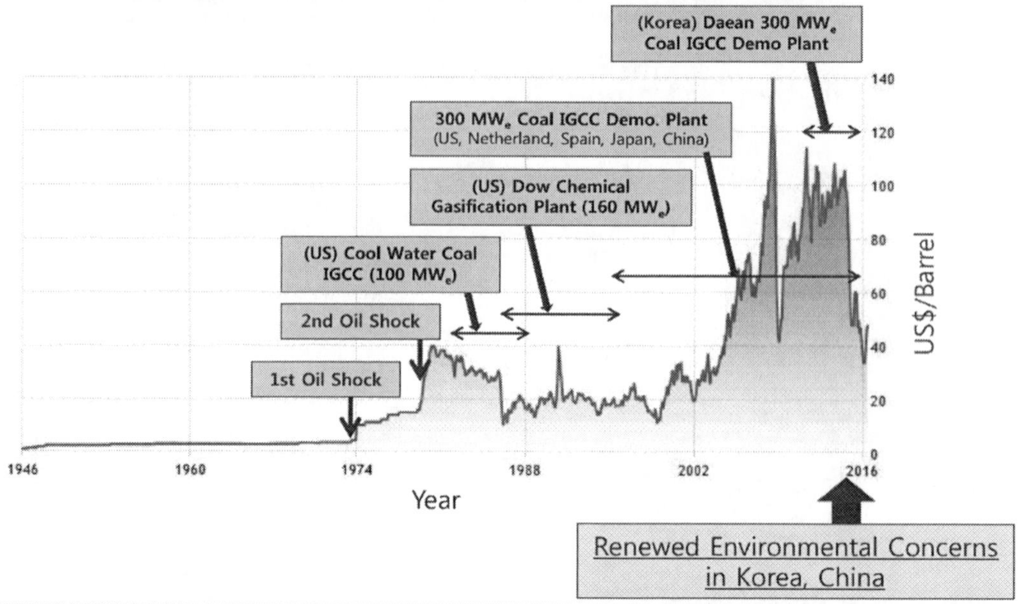


- Driving forces for IGCC installation : Oil price, Environmental concerns
- Can IGCC be upgraded to meet new environmental concerns (PM2.5 particles, etc.) with acceptable price ? → In Korea's case : appears to be YES.

Fig. Source : http://www.tradingeconomics.com/commodity/crude-oil

IAE

# IAE's Coal Gasification Scale-up History

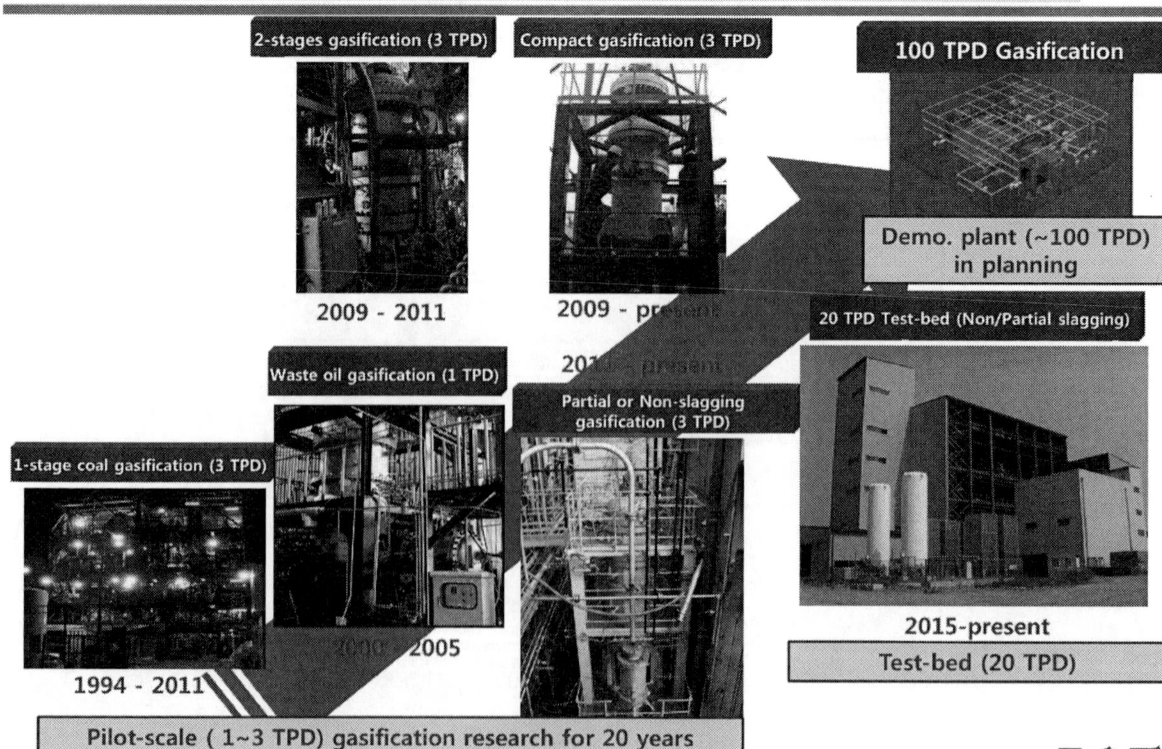


8 IAE

# 2 ton/day(TPD) Non-Slagging Coal Gasification System

-Pressure : 6-20 bar
-Temperature : max. 1,500°C
-Operation from 2011

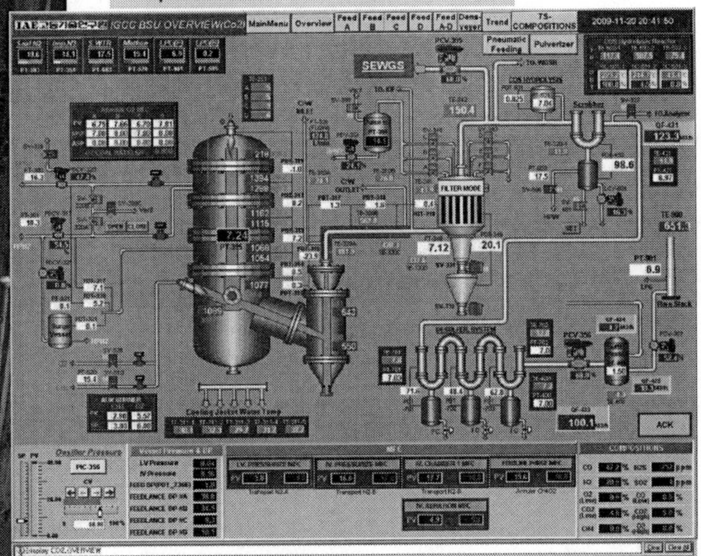


IAE

# 2 TPD testing(2011) : Profiles of Key Operation Parameters

Indonesian ABK coal, Gasifier Temperature: *ca.* 1,250°C

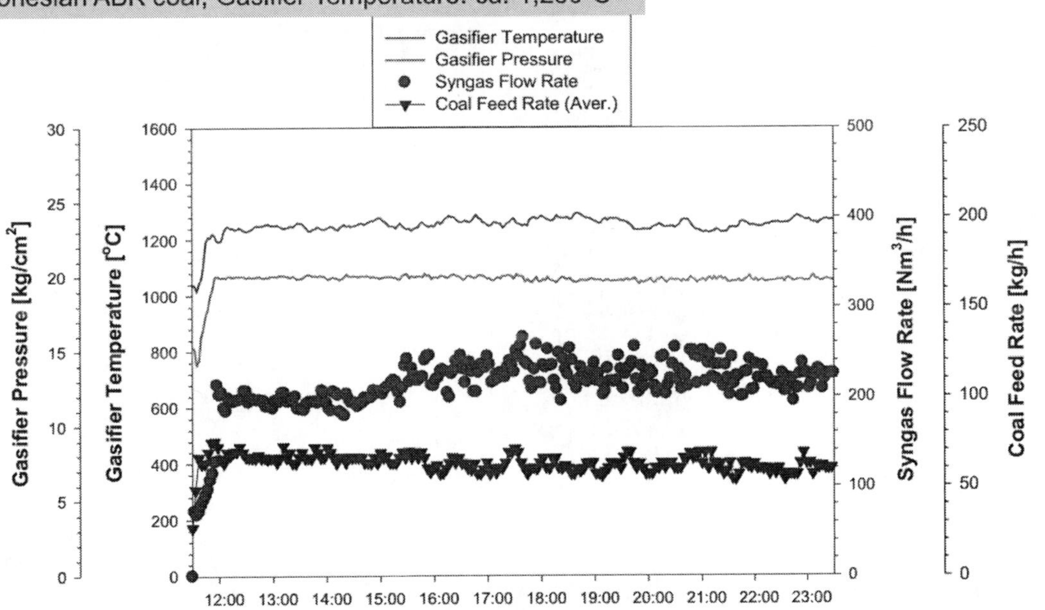


Source: Y. Yun et. al., "Performance Comparison of Non-slagging Entrained-bed Gasifier with Different Ash Melting Temperature Coals," Fifth International Conference on Clean Coal Technologies, CCT2011.

IAE

## 2 TPD Testing(2011) : Syngas Composition

Indonesian ABK coal, Gasifier Temperature: *ca.* 1,250°C

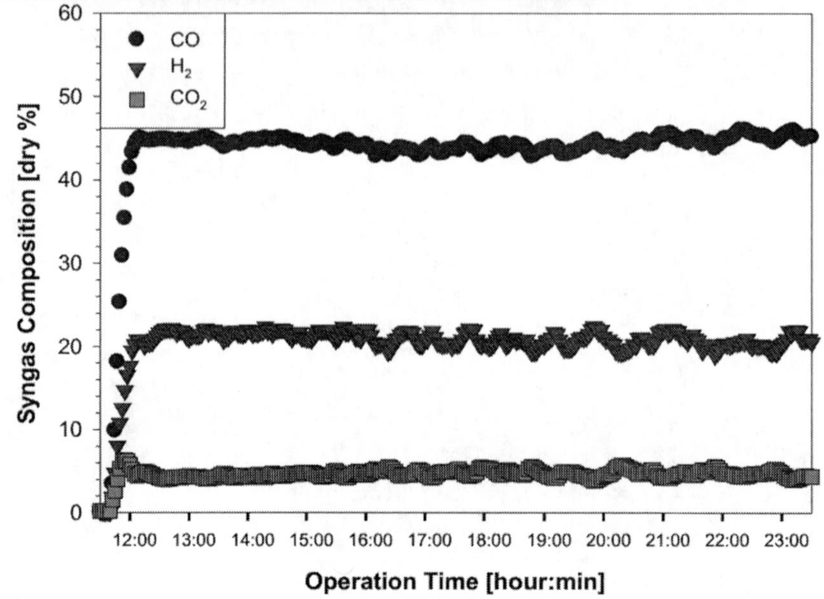


Source: Y. Yun et. al., "Performance Comparison of Non-slagging Entrained-bed Gasifier with Different Ash Melting Temperature Coals," Fifth International Conference on Clean Coal Technologies, CCT2011.

IAE

## 2 TPD Testing: Material Balance
(Down-flow / Non-Slagging, ABK coal)

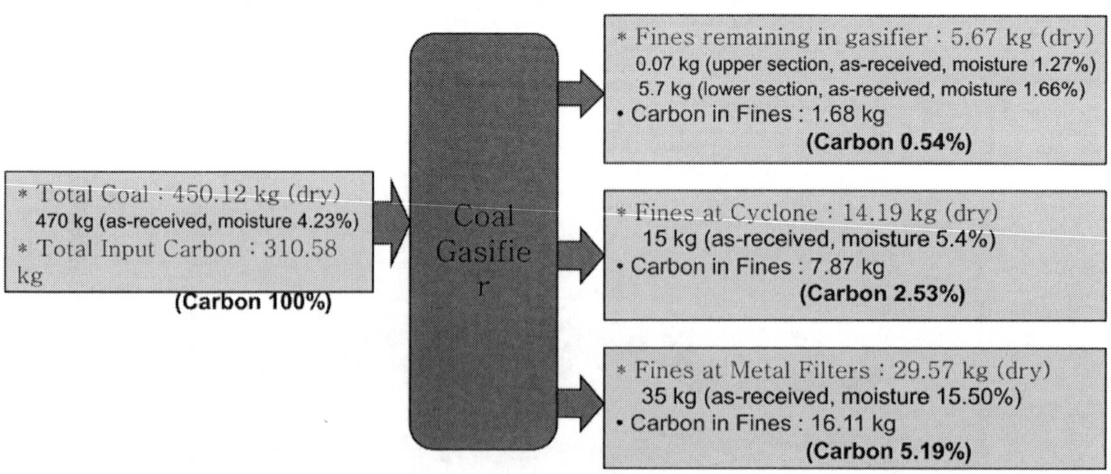


Carbon Conversion (one-pass) : 91.74%
Fines % of Input Coal : 10.98%

→ will increase with cyclic recycling

IAE

Source: Y. Yun, "Development of Partial/Non-Slagging Entrained-bed Coal Gasifier to Lower Capital Cost", Gasification Technologies Conference, 2013, Colorado Springs.

## **2 TPD Testing** : Raw Coal and Fines Captured by Metal Filters
(Down-flow / Non-Slagging)

Raw KPC Coal

Captured Fines by Metal Filters

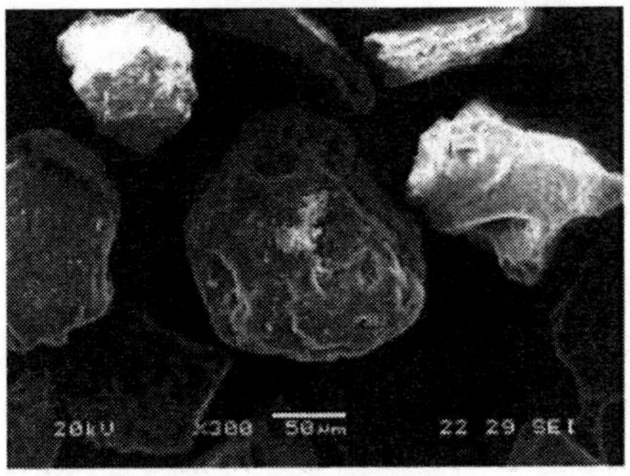

Raw KPC Coal (SEM, x300)

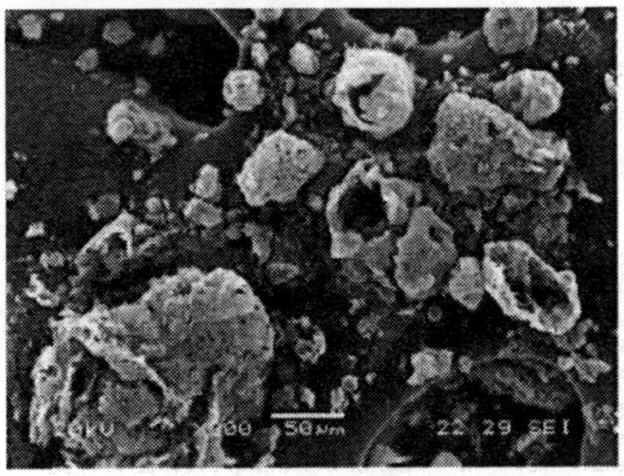

Captured Fines (SEM, x300)

IAE

## **2 TPD Testing** : Analysis Data of Raw Coal / Fines
(Down-flow / Non/Partial-Slagging)

| Sampling Point | | Raw Feeding Coal | Gasifier | | Cyclone | Metal Filters |
|---|---|---|---|---|---|---|
| | | | Upper section | Lower section | | |
| Proximate Analysis (dry basis, wt.%) | Volatiles | 43.36 | 2.04 | 1.35 | 3.16 | 4.37 |
| | Ash | 7.42 | 35.47 | 70.13 | 42.92 | 42.44 |
| | Fixed Carbon | 49.22 | 62.49 | 28.52 | 53.92 | 53.19 |
| Ultimate Analysis (dry basis, wt.%) | C | 69.00 | 62.70 | 29.15 | 55.43 | 54.47 |
| | H | 5.02 | 0.30 | 0.21 | 0.35 | 0.62 |
| | N | 1.34 | 0.34 | 0.06 | 0.39 | 0.15 |
| | S | 0.48 | 0.44 | 0.45 | 0.90 | 1.10 |
| | O (by-difference) | 16.74 | 0.75 | 0.00 | 0.00 | 1.23 |
| | Ash | 7.42 | 35.47 | 70.13 | 42.92 | 42.44 |
| Heating Value (dry basis, kcal/kg) | | 6325.5 | 4984.3 | 1475.8 | 4642.8 | 4237.2 |

- Raw coal : Indonesian ABK subbituminous coal

Source: Y. Yun et. al., "Performance Comparison of Non-slagging Entrained-bed Gasifier with Different Ash Melting Temperature Coals," Fifth International Conference on Clean Coal Technologies, CCT2011.

IAE

# 2 TPD Testing(2012) : Indonesian KPU Coal

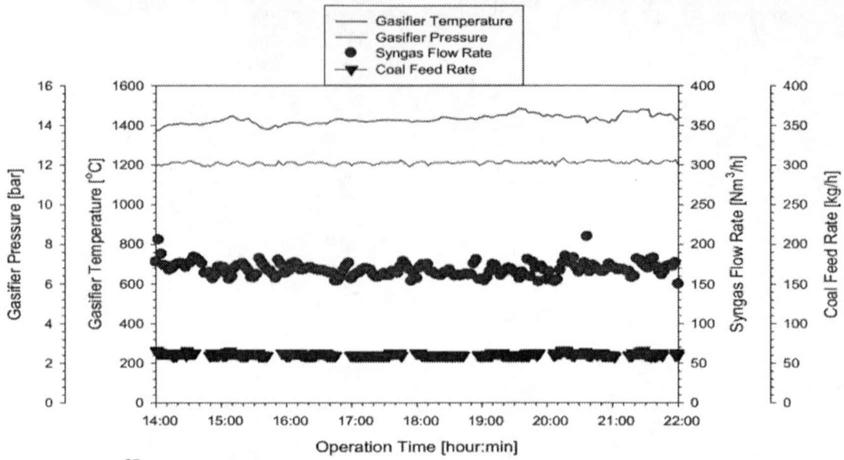


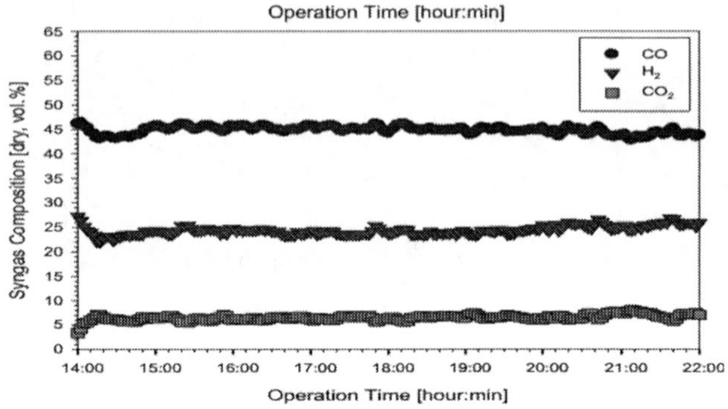


IAE

# 2 TPD Testing : Heavy Metal Content & Leaching of Coal / Fines (Down-flow / Partial/Non-Slagging)

| Heavy Metal | Content of Heavy Metal (mg/kg) | | | | | Leaching of Heavy Metal (mg/L) | | | | |
|---|---|---|---|---|---|---|---|---|---|---|
| | Pulverized Raw Coal (ABK) | Gasifier-Upper | Gasifier-Lower | Cyclone | Metal Filters | Pulverized Raw Coal (ABK) | Gasifier-Upper | Gasifier-Lower | Cyclone | Metal Filters |
| Hg | n.d. | n.d. | n.d. | n.d. | n.d. | n.d. | n.d. | n.d. | n.d. | n.d. |
| Pb | n.d. | n.d. | n.d. | n.d. | 11.9 | n.d. | n.d. | n.d. | n.d. | n.d. |
| Cd | n.d. | n.d. | n.d. | n.d. | n.d. | n.d. | n.d. | n.d. | n.d. | n.d. |
| Cr | n.d. | 72.61 | 411.27 | 178.2 | 945.86 | n.d. | n.d. | n.d. | n.d. | n.d. |
| Cu | n.d. | 10.76 | 33.34 | 22.33 | 28.54 | n.d. | n.d. | n.d. | n.d. | n.d. |
| Fe | 5172.97 | 23073.16 | 39992.1 | 28983.48 | 29755.9 | 0.47 | n.d. | n.d. | n.d. | n.d. |
| Al | 2956.23 | 6924.81 | 6557.13 | 10413.99 | 2860.7 | **1** | 0.62 | n.d. | **10.98** | 0.05 |
| Zn | 0.87 | 14.38 | 9.54 | 3.8 | 76.57 | n.d. | n.d. | n.d. | n.d. | n.d. |
| As | n.d. | n.d. | n.d. | n.d. | n.d. | n.d. | n.d. | n.d. | n.d. | n.d. |
| Sb | n.d. | n.d. | n.d. | n.d. | n.d. | n.d. | n.d. | n.d. | n.d. | n.d. |
| Ti | 205.5 | 1095.78 | 2183.78 | 1414.42 | 1473.09 | n.d. | n.d. | n.d. | n.d. | n.d. |
| Ba | 104.89 | 409.22 | 692.98 | 587.81 | 566.01 | 0.02 | 0.14 | 0.09 | 0.63 | 0.08 |
| K | 273.12 | 1430.34 | 2973.87 | 1943.68 | 2011.23 | 1.41 | 1.42 | 0.92 | 1.89 | 4.03 |
| Ca | 4108.1 | 6499.76 | 8883.22 | 15224.18 | 6910.11 | **8.18** | 116.63 | 24.95 | **73.38** | **59.21** |
| Na | 410.59 | 1694.77 | 3594.83 | 2608 | 2910.28 | 24.08 | 17.13 | 3.16 | 5.58 | 10.5 |
| Mg | 318.01 | 298.49 | 489.47 | 974.67 | 377.25 | **5.96** | 0.06 | 0.16 | 0.22 | **12.09** |
| $Cr^{+6}$ | n.d. | n.d. | n.d. | n.d. | n.d. | n.d. | n.d. | n.d. | 0.006 | 0.007 |
| $CN^-$ | n.d. | 0.008 | 0.004 | 0.013 | n.d. | n.d. | 0.021 | 0.006 | 0.005 | 0.034 |

(n.d. : not detected)

Source: Y. Yun et. al., "Performance Comparison of Non-slagging Entrained-bed Gasifier with Different Ash Melting Temperature Coals." Fifth International Conference on Clean Coal Technologies. CCT2011.

IAE

## 20 TPD Gasification Test Facility (Taean 300MW$_e$ IGCC Site, Korea)

- Non/Partial-slagging entrained-bed type
- Suitable for Low-rank coals (moisture in dried coal : 7-18%)
- Constructed in 2015

IAE

## 20 TPD Coal Gasification Test Results (2015)

- Feeding Rate & Corresponding Syngas Composition

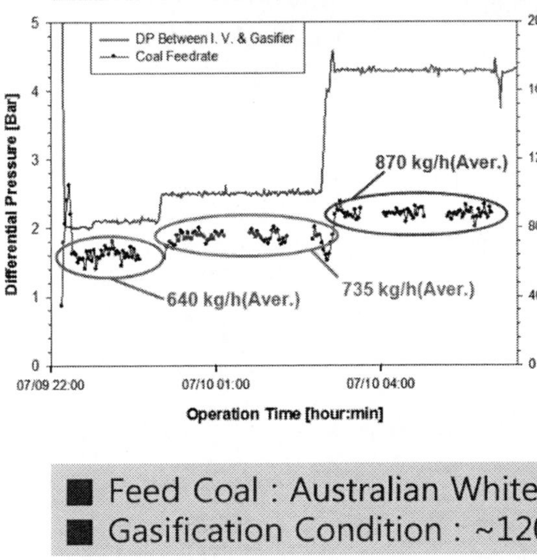


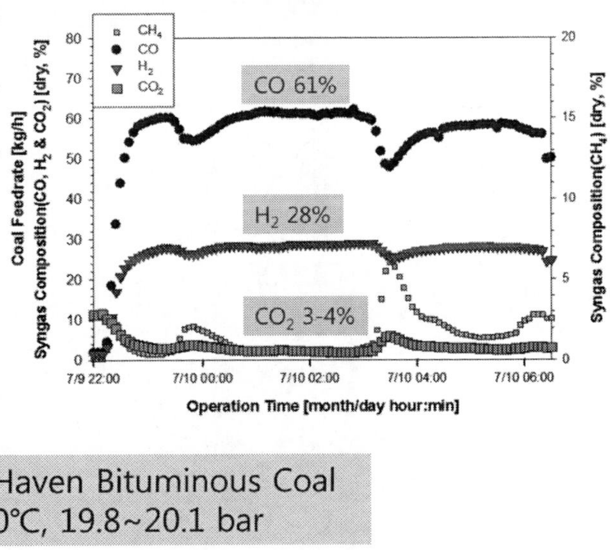


- Feed Coal : Australian White Haven Bituminous Coal
- Gasification Condition : ~1200℃, 19.8~20.1 bar

IAE

# 20 TPD Coal Gasification Test Results (2015)

## ■ Gasification Conditions / Syngas Composition

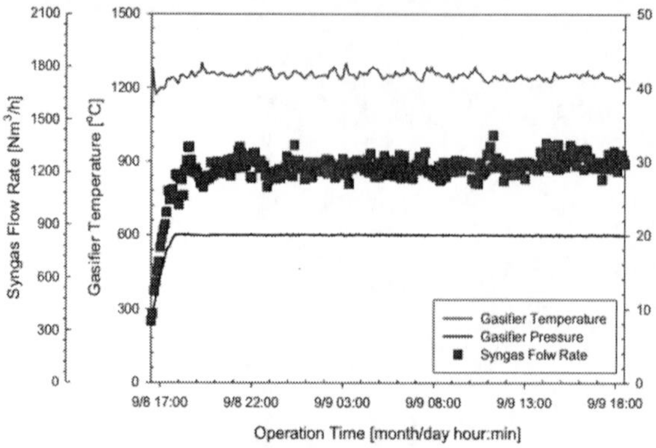


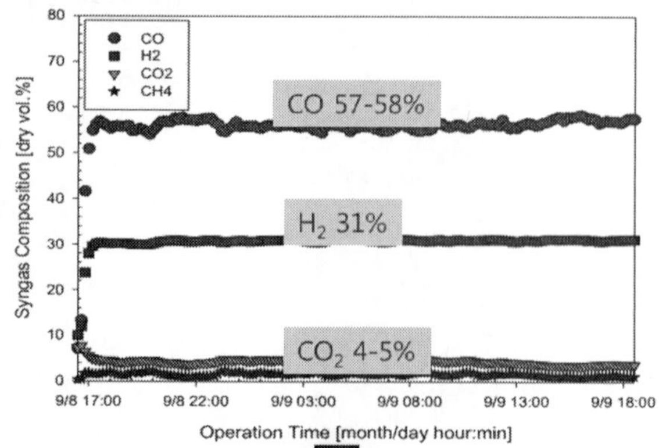


- Carbon conversion : 94%
- Cold gas efficiency : 76%

Lower syngas exit temperature from non/partial slagging gasifier → Lower sensible heat loss → High cce even with low cc.

## ■ Feed Coal

| Dried Feed Coal | | Indonesia KCH Coal |
|---|---|---|
| Proximate Analysis (wt.%) | Moisture | 8.19 |
| | Volatile Matter | 41.91 |
| | Fixed Carbon | 45.30 |
| | Ash | 4.61 |

# Partial Slagging Gasification : Dry-Feeding, Slurry-Feeding

20 ton/day Dry-feeding coal gasification (20 bar)

Australian White Haven coal

2 ton/day Slurry-feeding coal gasification (6.5 bar)

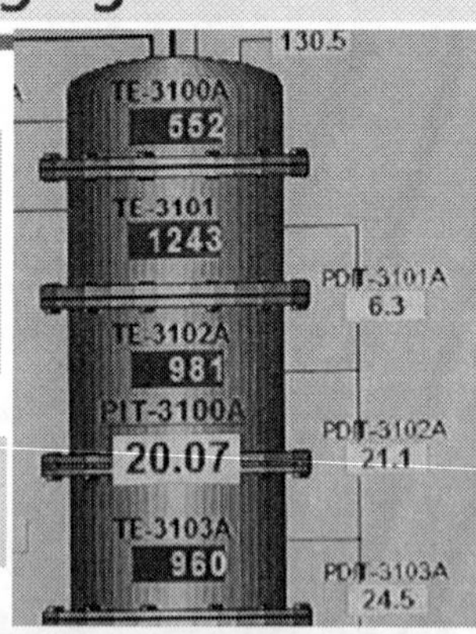


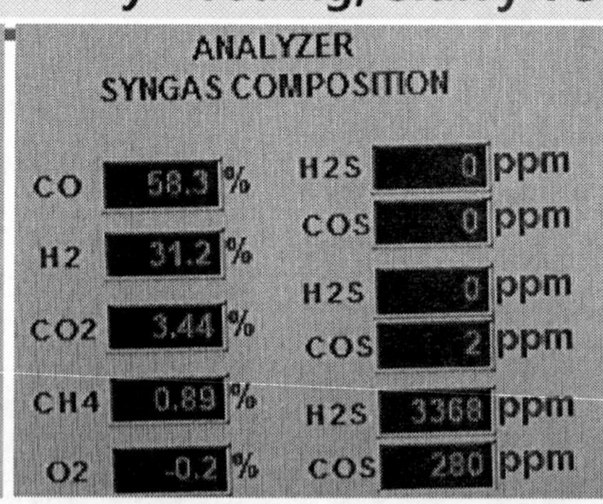

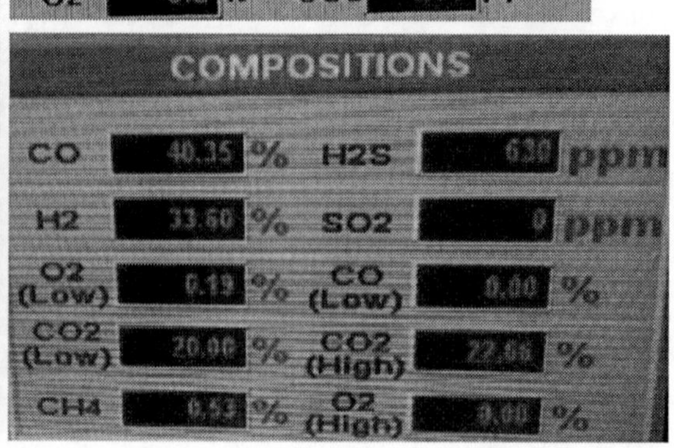


IAE

## 20 TPD Operation(2016): Indonesian KCH-mixed coal

20 TPD Coal Gasification–Syngas Cleaning System Control Panel

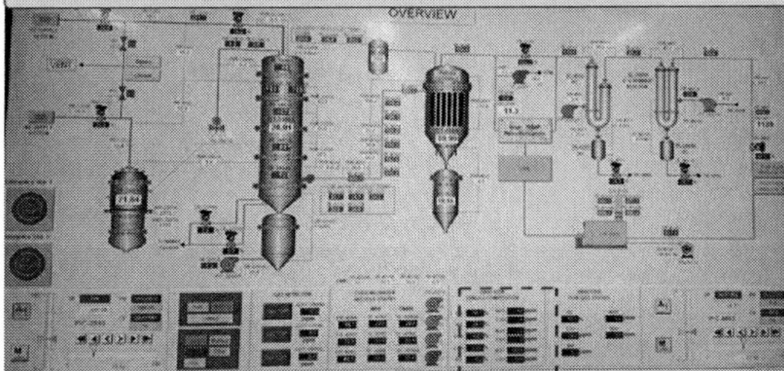


Control Room

Syngas Composition

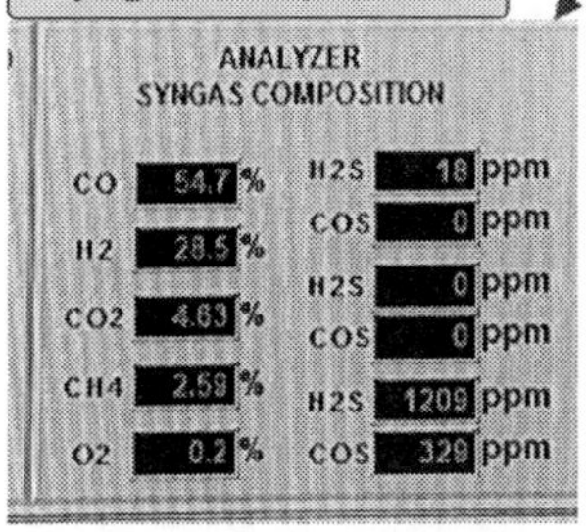


System Monitoring

IAE

## 20 TPD Gasification(2016): Indonesian KCH-mixed coal

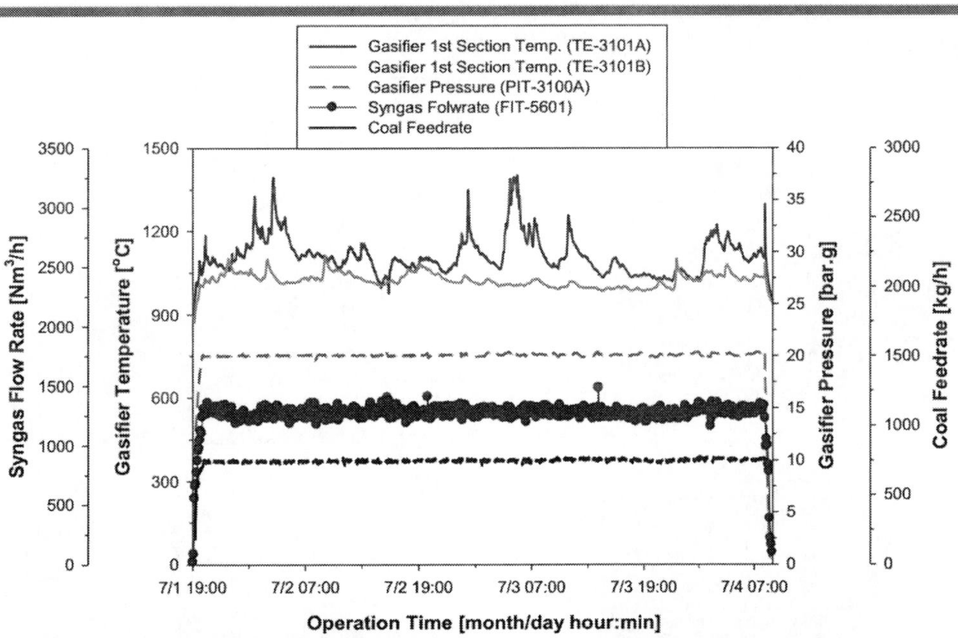


Note: TE-3101A thermocouple measures the temperature around the reaction zone near the coal/oxygen burner → sensitive to reactants mixing, etc.

IAE

## 20 TPD Gasification(2016): Indonesian KCH-mixed coal

### Syngas Composition

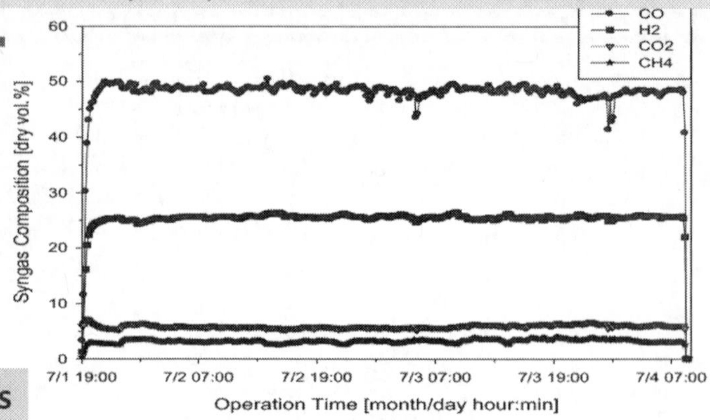


### Conversion Efficiencies

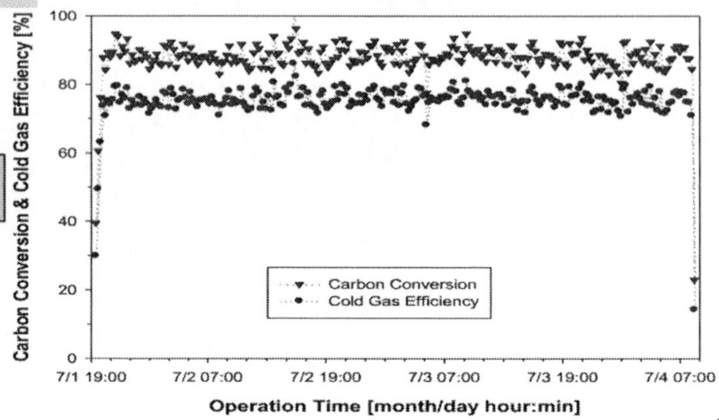


Carbon conversion : avg. 88.3%
Cold gas efficiency : avg. 75.6%

IAE

## 20 TPD Gasification : Slag

- Total produced amount is small.

Slag : Australian White Haven coal (2015/7)

Slag : Indonesian KCH-mixed coal (2016/7)

IAE

## Conclusions

- To compete with USC and other technologies in power generation using coal, much cheaper versions of entrained-bed gasification system appear to be essential.
- Non-/Partial-slagging entrained-bed coal gasification looks promising as a low-cost gasification option, specifically for low rank coals.
- The 20 ton/day scaled-up system has provided a practical data in developing the partial/non-slagging entrained-bed gasification system.
- Carbon content of captured fines in partial/non-slagging entrained-bed gasifier showed 30-60%, with carbon conversion of around 90% in a single-pass through the gasifier. Higher carbon conversion is possible by recycling fines or increasing gasification temperature till 1,400°C.
- Captured fines contain a heating content of 4,200-4,600 kcal/kg, which is higher than the heating value of low-grade coals. Technically, captured fines can be utilized as a combustion fuel.
- Captured fines do not present the possibility of heavy metal leaching according to the current regulation, as confirmed by leaching tests.

IAE

## Thank you for your attention.

Authors thank for the financial supports from KETEP (Korea Institute of Energy Technology Evaluation & Planning) and Korea Western Power Co., Ltd.

IAE

TECHNISCHE UNIVERSITÄT BERGAKADEMIE FREIBERG

The University of Resources. Since 1765.

Institute of Energy Process Engineering and Chemical Engineering

Chair EVT

# Influence of quench water temperature and pH value variations on selected trace components in autothermal gas reforming processes.

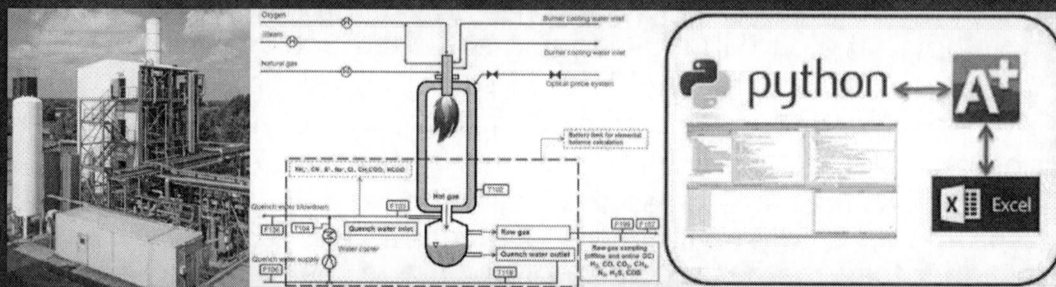


33rd International Pittsburgh Coal Conference August 8 - 12, 2016 Cape Town, South Africa.

Ibrahim Kolawole Muritala

Supervisors: Dr.-Ing. Peter Seifert

Dr.-Ing. Fred Compart

TU Bergakademie Freiberg Institute of Energy Process Engineering and Chemical Engineering Chair of Energy Process Engineering and Thermal Waste Treatment

Reiche Zeche Fuchsmuehlenweg 9 09599 Freiberg Phone: +49 3731 39-4511 Fax: +49 3731 39-4555 www.iec.tu-freiberg.de

## Outline

- Introduction
- Overview of Freiberg HP POX test plant
- HP POX quench water system
- Basis of calculations
- Results of re-calculation of trace components
- Summary and Conclusion

2

## Introduction

### Aims

- The effects of cooling operation on the syngas quality by quench water operations.

### Motivation

- To understand the influence of various trace components on syngas treatment process.
- To elucidate the chemistry behind the formation and depletion of certain trace components and their ions.
- To develop methods that can influence the distribution of these trace components and their ions.

3

## Overview: Freiberg HP POX test plant

**High Pressure Partial Oxidation (HP POX)**

**Process applications:**

- Autothermal catalytic reforming (**ATR**) at pressures up to 70 bar(g)
- Autothermal non-catalytic partial oxidation of natural gas (**GasPOX**) at pressures up to 100 bar(g)
- Autothermal non-catalytic gasification of liquid hydrocarbons (**OilPOX**) at pressures up to 100 bar(g)
- In operation since 2004 with adjustable reactor volume

**Operating parameters:**

- Temperature: up to 1450 °C
- Pressure: up to 100 bar (g)
- Input: up to 500 m³(STP)/h natural gas
- or 500 kg/h liquid feeds
- Syngas output: up to 1500 m³(STP)/h
- Thermal power: up to 5 MW

**Feedstock:**

- Natural gas
- Light and heavy oils
- Heavy residues from crude oil processing

4

## HP POX quench water system

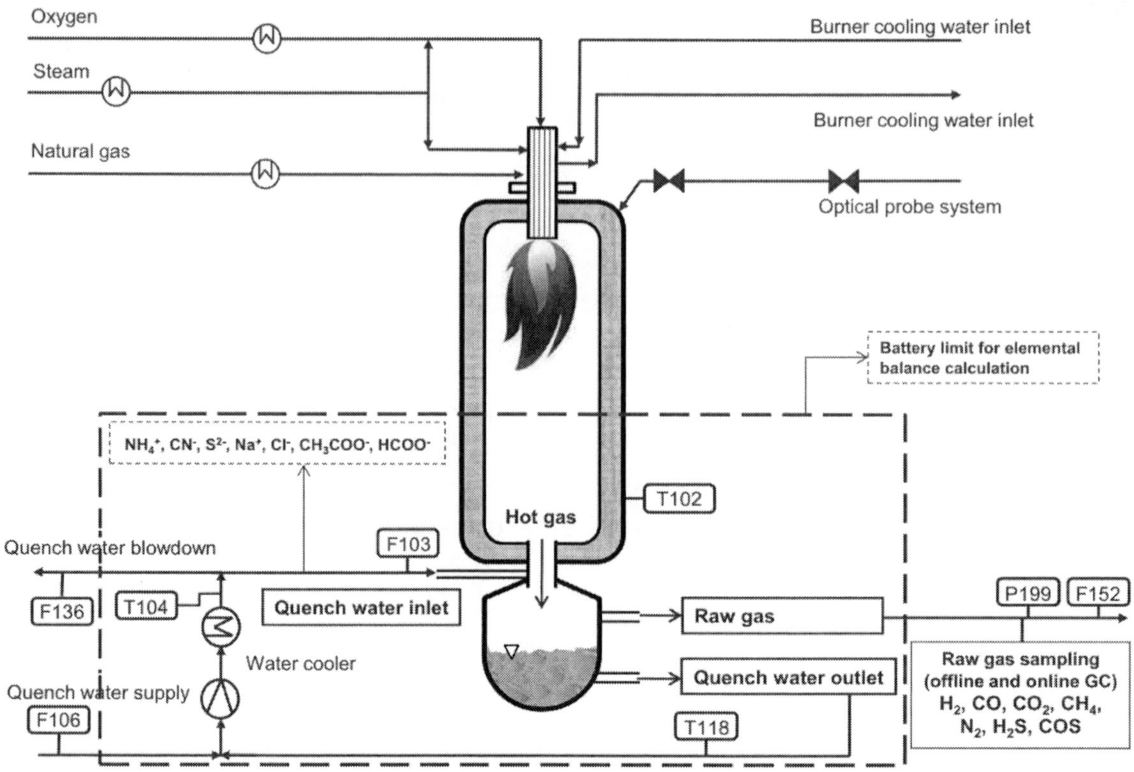


5

## Overview: GasPOX

**GasPOX**

- **Operation pressure: 30 – 100 bar**
- **Outlet temperature up to 1400 °C**
- **Quench water outlet temperature: 155 – 195 °C**
- **34 test points**
- **Test point GasPOX 201 VP1 was chosen for these studies**

**Examples for synthesis gas quality requirement***

| Composition | Methanol synthesis | Fischer-Tropsch synthesis |
|---|---|---|
| CO | 30 - 32 vol % | 30 - 32 vol % |
| $H_2$ | 64 - 68 vol % | 64 - 68 vol % |
| $CO_2$ | 2 - 4 vol % | - |
| $CH_4$ | 4% | 5% |
| Left overs | $H_2O$, $N_2$, Ar, traces | - |
| Total sulfur | < 0.1 mg/m³ (7.00E-6 %) | < 1E-05 vol % |
| Alkali components | < 0.2 mg/m³ | < 1E-08 vol % |
| Dust | < 0.2 mg/m³ | 0 mg/m³ |

*Source: Prof. Dr.-Ing. habil. W. Heschel – Vorlesung Gasaufbereitung

6

## Overview: GasPOX

| Hot gas (GasPOX 201 VP1) | |
|---|---|
| Outlet temperature | 1172 °C |
| Pressure | 71 bar |
| Volumetric flow | 1339 $Nm^3/h$ |
| Mole flow | 59.74 kmol/h |
| Composition | vol % |
| CO | 23.7 |
| $H_2$ | 48.4 |
| $CO_2$ | 4.1 |
| $CH_4$ | 4.6 |
| $H_2O$ | 18.6 |
| $H_2S$ | 1.40E-04 |
| $NH_3$ | 1.60E-02 |
| HCN | 4.30E-03 |
| $N_2$ | 0.6 |
| COS | 1.10E-04 |

- Re-calculated hot gas (synthesis gas composition) and traces

Other trace components and ions in quench water: acetic acid, formic acid, $Cl^-$, $NH_4^+$, $Na^+$, $H_3O^+$, $NH_2COO^-$, $HS^-$, $HCO_3^-$, $OH^-$ , $CO_3^{2-}$,$Cl^-$, $CH_3COO^-$, $HCOO^-$, $CN^-$, $S^{2-}$

7

## Basis of calculations

- The HP POX quench system cycle and operating conditions (pressure, temperature, and flow rates of stream) were measured during the GasPOX 201 VP1.
- Quench water streams and raw gas outlet (under investigation)

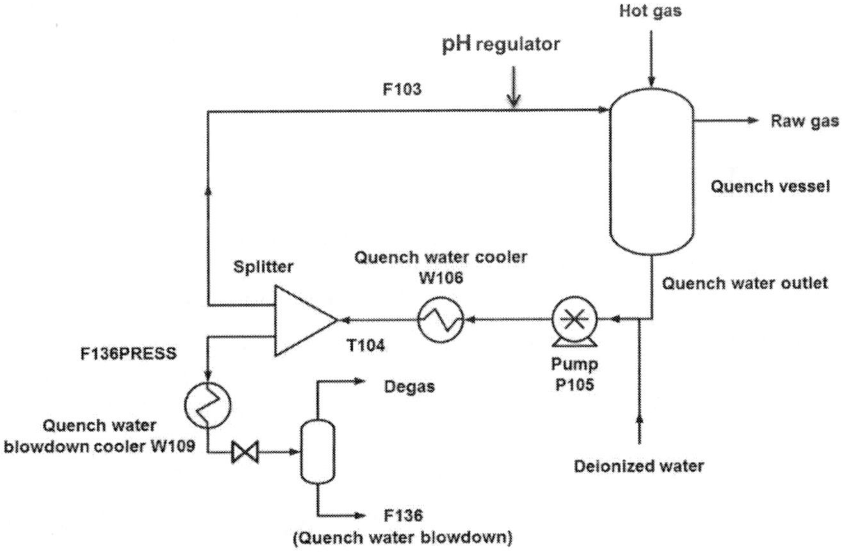


Quench system (simplified scheme)

8

## Basis of calculations

**Operating parameters of the streams measured during the test point GasPOX 201 VP1* and results obtained from the quench water system model in Aspen Plus**.**

| Streams | Operating parameters (** Calculated values, * Measured values) Flow rate (t/h) | Temperature (°C) | Pressure (bar) | Phase |
|---|---|---|---|---|
| Hot gas | 0.80 ** | 1172 * | 71 ** | Gas |
| Raw gas | 0.70 * | 177.4 ** | 71 * | Gas |
| Degas (Quench water blowdown) | 0.0020 ** | 20 ** | 1 * | Gas |
| Quench water outlet | 5.40 ** | 177.4 * | 71 ** | Aqueous |
| Stream T104 (Quench water cooled) | 5.90 ** | 79.9 * | 72 ** | Aqueous |
| Stream F103 (Quench water inlet) | 5.30 * | 79.9 ** | 72 ** | Aqueous |
| Stream F136PRESS (Quench water blowdown) | 0.70 * | 79.9 ** | 72 ** | Aqueous |
| Stream F136 (Quench water blowdown) | 0.70 ** | 20 °C * | 1 * | Aqueous |
| Deionized water | 0.50 ** | 174.4 * | 71 ** | Aqueous |

9

## Basis of calculations

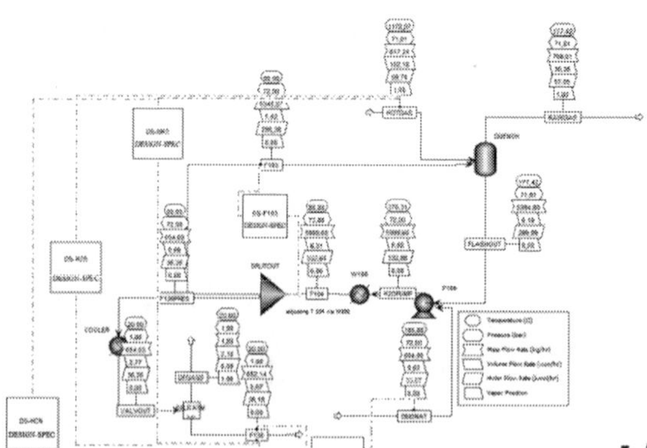

- **Aspen Plus flowsheet for trace component calculation**
- **Property method: ELECNRTL**
- **Electrolyte-NRTL activity coefficient model**

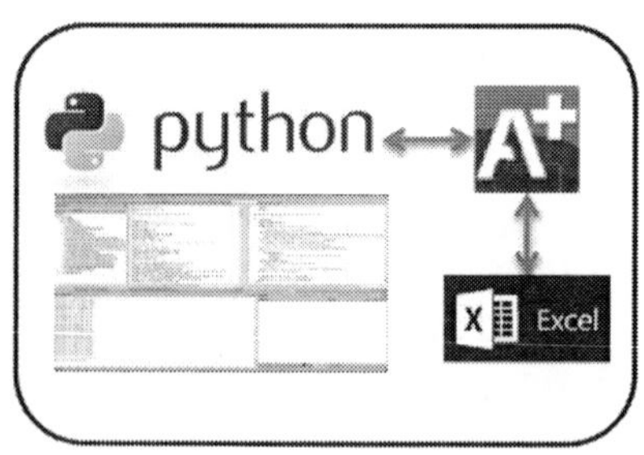


- **Calculation of all 34 test points of the campaign by mass calculation using python – excel – Aspen Plus connectivity (interfaces via python)**
- **Choice of one test point for further consideration**
- **Presentation of species distribution of test point (VP1)**
- **Calculation of species distribution during sensitivity studies originating from this test point**

10

## Main global reactions leading to molecular dissolution of $CO_2$, $NH_3$, $H_2S$ and HCN in water and dissociation of their corresponding molecular dissolved species which were considered in these studies.

| Main components | Reactions | Reaction number (R No.) |
|---|---|---|
| $CO_2$ | | |
| | $CO_2$ (gas phase) ↔ $CO_2$(aq) (molecular dissolved) | 1 |
| | $CO_2$ (aq) + $H_2O$ (l) ↔ $HCO_3^-$(aq) + $H^+$(aq) | 2 |
| | $HCO_3^-$(aq) ↔ $CO_3^{2-}$(aq) + $H^+$(aq) | 3 |
| | $NH_3$ (aq) + $HCO_3^-$ (aq) ↔ $NH_2COO^-$ (aq) + $H_2O$(l) | 4 |
| Over all concentration in liquid phase | $[CO_{2\ total}(aq)] = [CO_2(aq)] + [HCO_3^-(aq)] + [CO_3^{2-}(aq)] + [NH_2COO^-(aq)]$ | [.....] = mol/kg solvent |
| $NH_3$ | | |
| | $NH_3$ (gas phase) ↔ $NH_3$(aq) (molecular dissolved) | 5 |
| | $NH_3$(aq) + $H_2O$(l) ↔ $NH_4^+$(aq) + $OH^-$(aq) | 6 |
| | $NH_3$(aq) + $HCO_3^-$(aq) ↔ $NH_2COO^-$(aq) + $H_2O$(l) | 7 |
| Over all concentration in liquid phase | $[NH_{3\ total}(aq)] = [NH_3(aq)] + [NH_4^+(aq)] + [NH_2COO^-(aq)]$ | [.....] = mol/kg solvent |
| $H_2S$ | | |
| | $H_2S$ (gas phase) ↔ $H_2S$(aq) (molecular dissolved) | 8 |
| | $H_2S$(aq) ↔ $HS^-$(aq) + $H^+$(aq) | 9 |
| | $HS^-$(aq) ↔ $S^{2-}$(aq) + $H^+$(aq) | 10 |
| Over all concentration in liquid phase | $[H_2S_{total}(aq)] = [H_2S(aq)] + [HS^-(aq)] + [S^{2-}(aq)]$ | [.....] = mol/kg solvent |
| HCN | | |
| | HCN (gas phase) ↔ HCN (molecular dissolved) | 11 |
| | HCN(aq) ↔ $CN^-$(aq) + $H^+$(aq) | 12 |
| Over all concentration in liquid phase | $[HCN_{total}(aq)] = [HCN(aq)] + [CN^-(aq)]$ | [.....] = mol/kg solvent |

11

## Quench water temperature variation

- original quench water outlet temperature = 177.4 °C
- pH value = 6.17

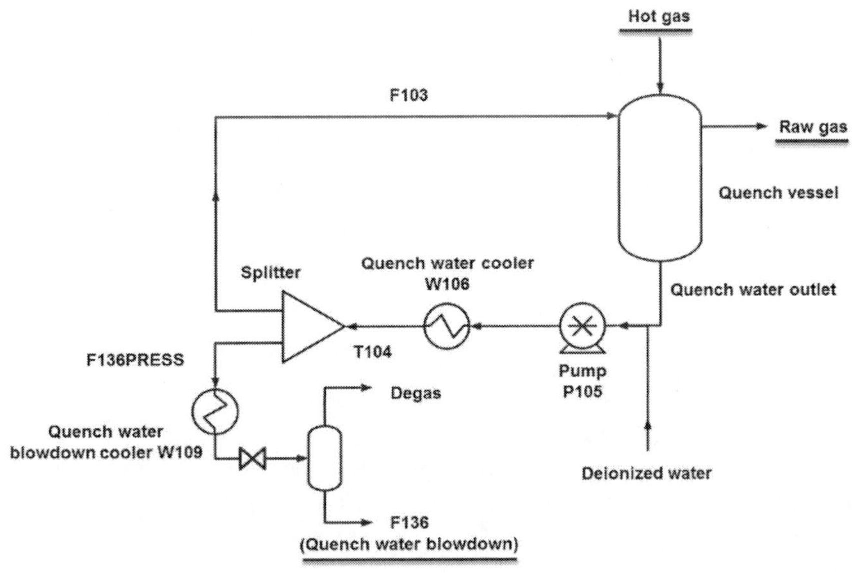


12

## Results of investigation of trace components ($CO_2$, $NH_3$, $H_2S$, HCN)

Sensitivity study: Quench water temperature variation and influence on $CO_2$ distribution

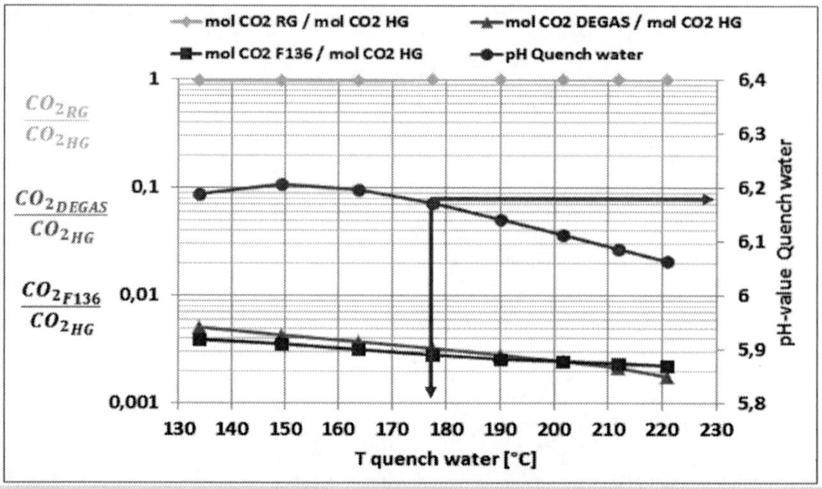


- Almost all $CO_2$ from hot gas remains in raw gas ($CO_{2\ Hot\ gas\ (kmol/h)} \approx CO_{2\ Raw\ gas\ (kmol/h)}$)
- $CO_{2\ F136}$, $CO_{2\ DEGAS}$ decrease with increase in Quench water temperature:

  $CO_2$ 1st dissociaction constant decreases with increase in Quench water temperature → dominant effect
- pH decreases ($K_{H2O}$ increases)

13

## Results of investigation of trace components ($CO_2$, $NH_3$, $H_2S$, HCN)

Sensitivity study: Quench water temperature variation and influence on $NH_3$ distribution

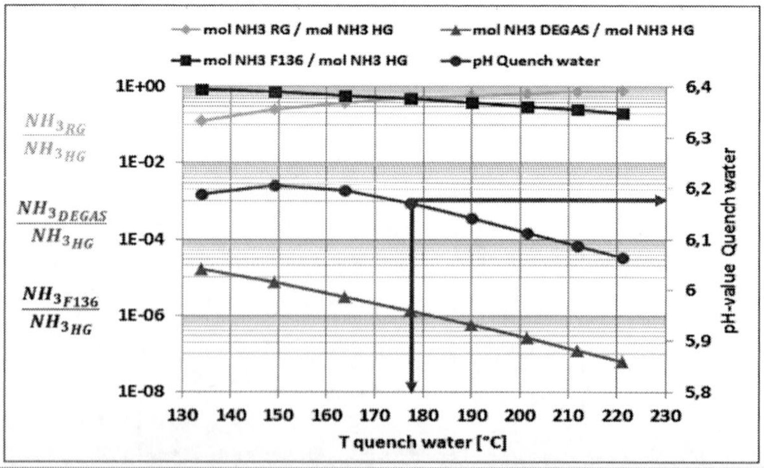


- $NH_{3\ Raw\ gas\ (kmol/h)} < NH_{3\ Hot\ gas\ (kmol/h)}$
- $NH_{3\ Raw\ gas}$ increases with increasing quench water temperature
- Blowdown stream F136: $NH_{3\ F136}$ mostly $NH_4^+$
- $NH_{3\ F136}$, $NH_{3\ DEGAS}$ decreases with increasing quench water temperature: $NH_3$ dissociation constant increases
- $NH_3$ Henry constant increases ($NH_3$ molecular solubility decreases) with increasing Quench water temperature (dominant effect)

14

## Results of investigation of trace components ($CO_2$, $NH_3$, $H_2S$, HCN)

Sensitivity study: Quench water temperature variation and influence on $H_2S$ distribution

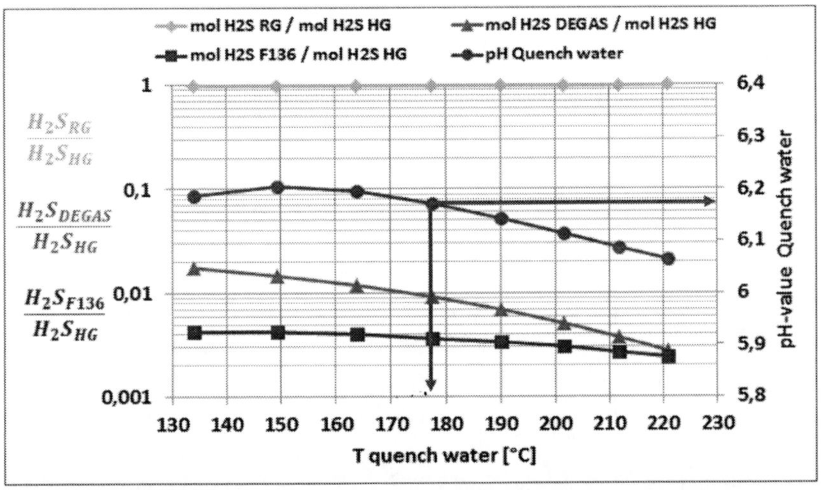


- **Almost all $H_2S$ from Hot gas remains in Raw gas ($H_2S_{\text{Hot gas (kmol/h)}} \approx H_2S_{\text{Raw gas (kmol/h)}}$)**
- **$H_2S_{F136}$ and $H_2S_{DEGAS}$ decrease with increase in Quench water temperature:**
  **$H_2S$ 1st dissociaction constant increases with increase in Quench water temperature → but $H_2S$ Henry constant increases (dominant effect)**

15

## Results of investigation of trace components ($CO_2$, $NH_3$, $H_2S$, HCN)

Sensitivity study: Quench water temperature variation and influence on HCN distribution

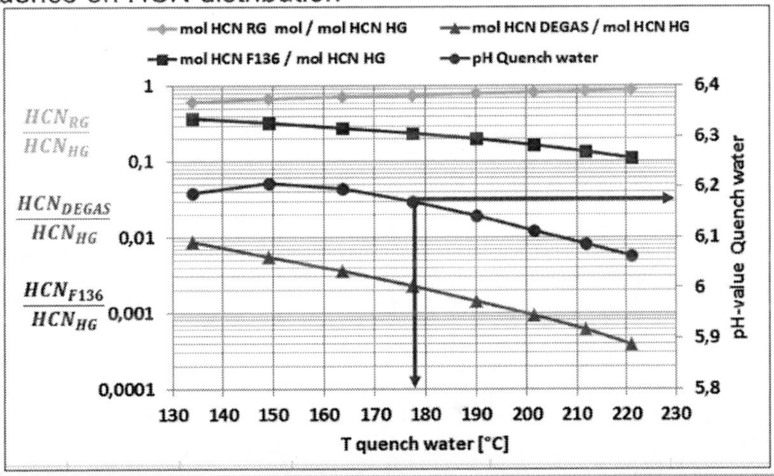


- **As Quench water temperature increases: more HCN from Hot gas remains in Raw gas ($H_2S_{\text{Raw gas (kmol/h)}} / H_2S_{\text{Hot gas (kmol/h)}}$ increases)**
- **$HCN_{F136}$ and $HCN_{DEGAS}$ decrease with increase in Quench water temperature:**
  **HCN dissociaction constant increases with increase in Quench water temperature → but HCN Henry constant increases (dominant effect)**

16

## pH value variation influence

- original quench water pH value = 6.17 at outlet temperature of 177.4 °C
- introducing pH regulator with 20% NaOH as base and 20% $H_2SO_4$ as acid

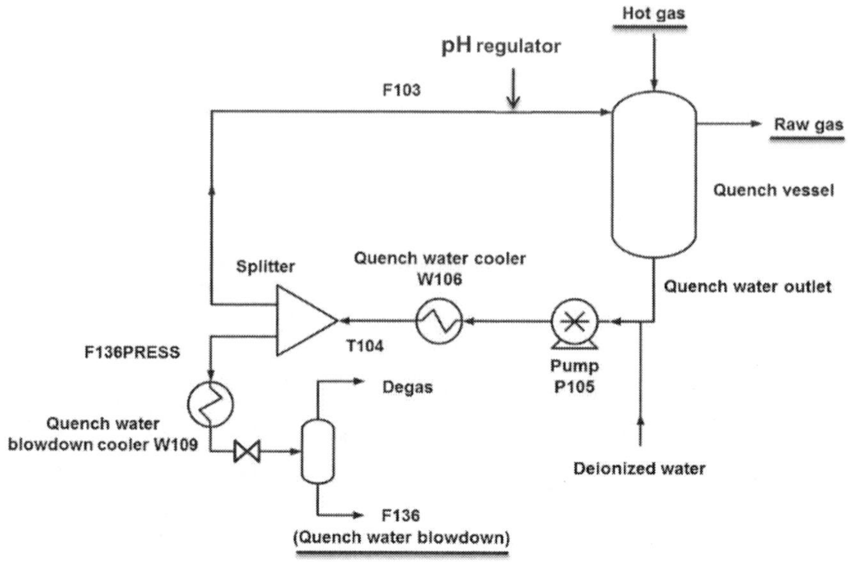


17

Results of investigation of trace components ($CO_2$, $NH_3$, $H_2S$, HCN)

Sensitivity study: pH variation influence on $CO_2$ distribution

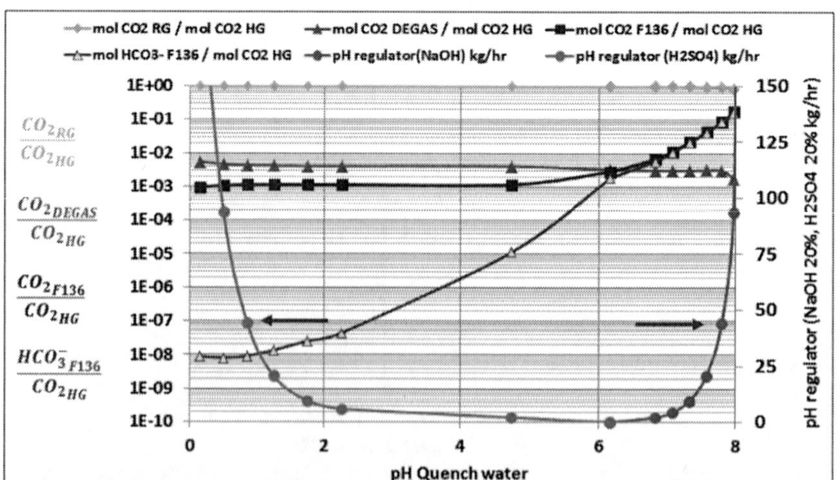


$CO_{2\ \text{Hot gas (kmol/h)}} \approx CO_{2\ \text{Raw gas (kmol/h)}}$ for low pH value

**When pH regulator NaOH is introduced:**

- $CO_{2\ \text{Raw gas}}$ decreases
- Blowdown stream F136: dissolved total $CO_2$ increases as more $CO_2$ goes into dissociation leading to increase in $HCO_3^-$ formation→ $HCO_3^-{}_{\text{F136 (kmol/h)}}$ / total $CO_{2\ \text{F136 (kmol/h)}}$ increases

18

## Results of investigation of trace components ($CO_2$, $NH_3$, $H_2S$, HCN)

### Sensitivity study: pH variation influence on $NH_3$ distribution

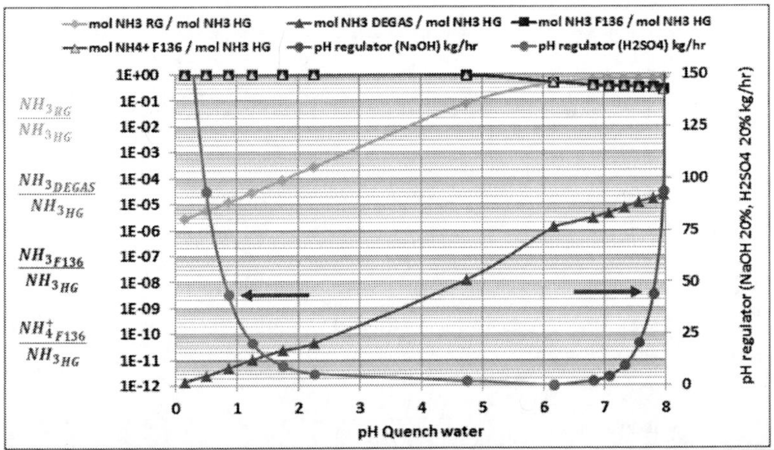


**$NH_{3\ Raw\ gas}$ high, $NH_{3\ Raw\ gas}$ low for low pH value**

**When pH regulator NaOH is introduced:**

- **$NH_{3\ Raw\ gas}$ increases**
- **Blowdown stream F136:**

  **Total $NH_{3\ F136}$ decreases, $NH_4^+{}_{F136}$ / total $NH_{3\ F136}$ decreases**
- **$NH_{3\ DEGAS}$ remains very minute**

19

## Results of investigation of trace components ($CO_2$, $NH_3$, $H_2S$, HCN)

### Sensitivity study: pH variation influence on $H_2S$ distribution

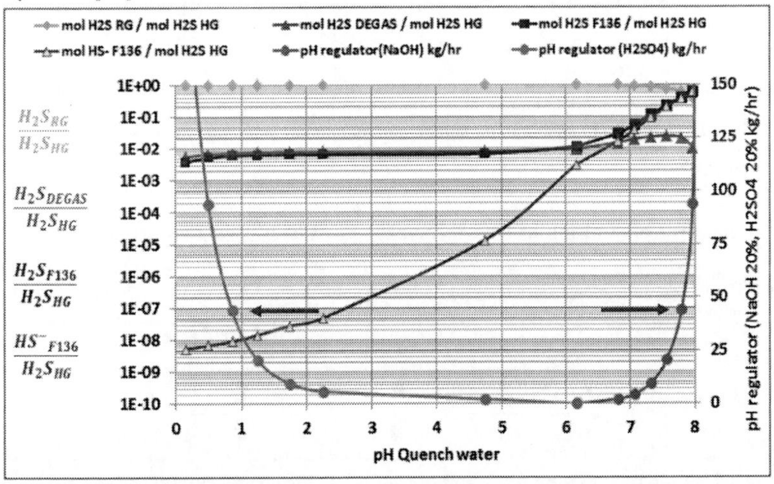


**$H_2S_{\ Hot\ gas} \approx H_2S_{\ Raw\ gas}$ for low pH value**

**When pH regulator NaOH are introduced:**

- **$H_2S_{\ Raw\ gas}$ decreases**
- **Blowdown stream F136: dissolved total $H_2S$ increases as more $H_2S$ goes into dissociation leading to increase in $HS^-$ formation → $HS^-{}_{F136}$ / total $H_2S_{\ F136}$ increases**

20

## Results of investigation of trace components ($CO_2$, $NH_3$, $H_2S$, HCN)

### Sensitivity study: pH variation influence on HCN distribution

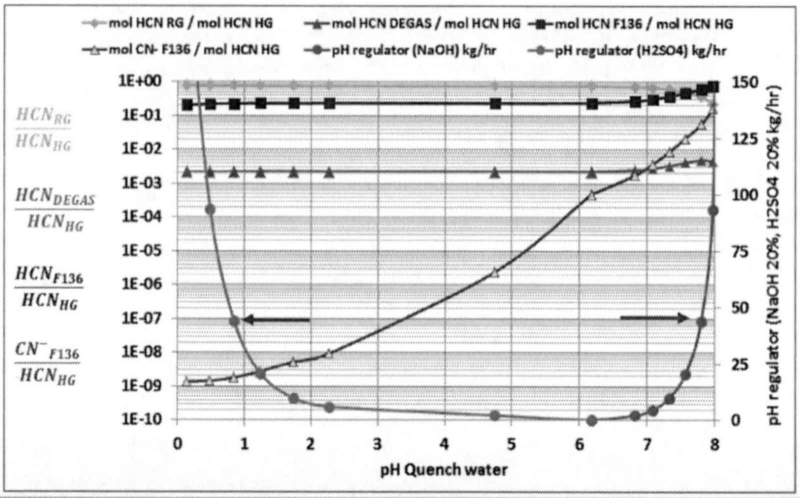


**When pH regulator NaOH are introduced:**

- **$HCN_{\text{Raw gas}}$ decreases**
- **Blowdown stream F136: dissolved total HCN increases as more HCN goes into dissociation leading to increase in $CN^-$ formation → $CN^-_{F136}$ / total $HCN_{F136}$ increases**
  **→ but HCN is the weakest acid (compared with the other acid gases $H_2S$ and $CO_2$)**

21

## Summary and Conclusion

- **The distribution of the different traces between raw gas and quench water is influenced by quench water temperature and quench water pH value variation**
- **Quench water outlet stream contains a lot of molecular dissolved and ionic species with different complex interactions due to temperature dependency with respect to the species Henry´s and dissociation constant.**

**Traces and pH values in quench water (and in blowdown stream F136)**

| | $CO_2$ | $NH_3$ | $H_2S$ | HCN | $pH_{\text{Quench water}}$ |
|---|---|---|---|---|---|
| ↑$T_{\text{Quench water}}$ | ↓ | ↓ | ↓ | ↓ | ↓ |
| ↑pH regulator (20% NaOH) | ↑ | ↓ | ↑ | ↑ | ↑ |
| ↑pH regulator (20% $H_2SO_4$) | ↓ | ↑ | ↓ | ↓ | ↓ |

- **However, this is the first set of steps in the initial sensitivity analysis studies to predict the behaviors of these participating species in their molecular dissolved and ionic forms.**

- **A detail look at other traces such as HCOOH (formic acid) and $CH_3COOH$ (acetic acid) and their behaviors are important in understanding of quench water system (pH-dependent reaction mechanism).**

22

TECHNISCHE UNIVERSITÄT
BERGAKADEMIE FREIBERG
The University of Resources. Since 1765.

Institute of Energy Process Engineering
and Chemical Engineering
Chair EVT

IEC

# Thank you for your attention.

**M.Sc. Ibrahim Kolawole Muritala**

**Email: ibrahim.muritala@iec.tu-freiberg.de**

**www.iec.tu-freiberg.de**


IEC

# Additional information

24

## Dissociation & Henry constant

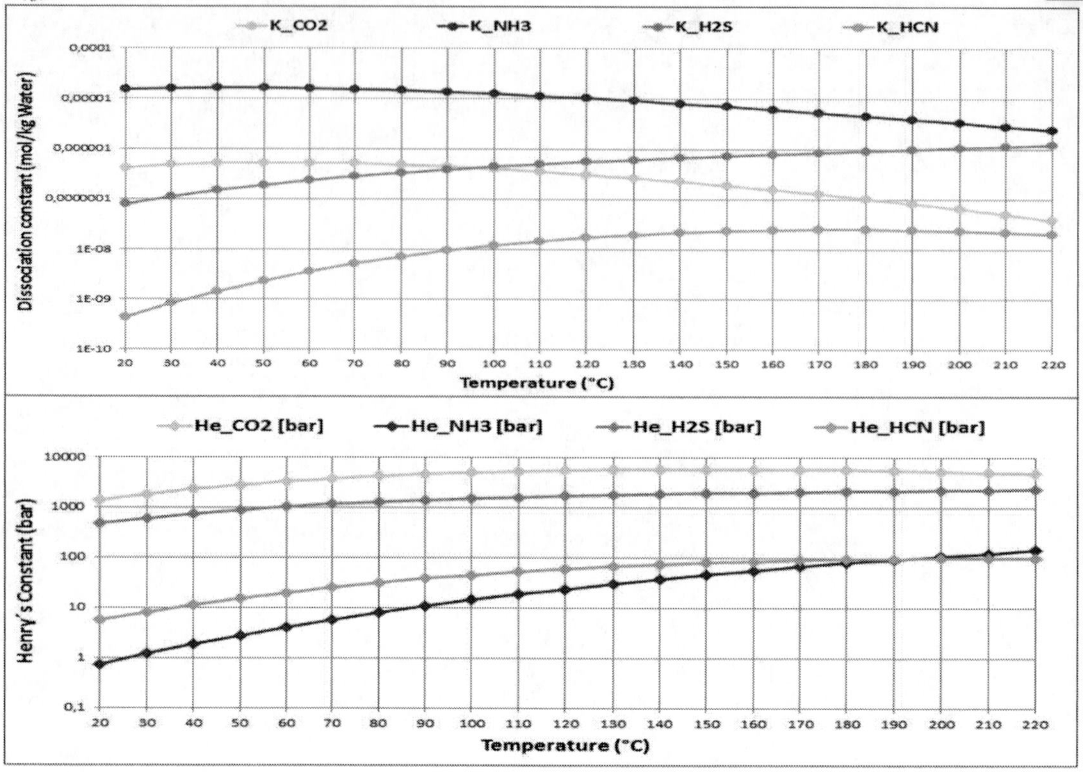


25

## Main global reactions leading to molecular dissolution of $CO_2$, $H_2S$, $NH_3$ and HCN in water and dissociation of their corresponding molecular dissolved species which were considered in these studies.

| Main components | Reactions | Reaction number (R No.) |
|---|---|---|
| $CO_2$ | | |
| | $CO_2$ (gas phase) ↔ $CO_2$(aq) (molecular dissolved) | 1 |
| | $CO_2$ (aq) + $H_2O$ (l) ↔ $HCO_3^-$(aq) + $H^+$(aq) | 2 |
| | $HCO_3^-$(aq) ↔ $CO_3^{2-}$(aq) + $H^+$(aq) | 3 |
| | $NH_3$ (aq) + $HCO_3^-$ (aq) ↔ $NH_2COO^-$ (aq) + $H_2O$(l) | 4 |
| Over all concentration in liquid phase | [$CO_{2\ total}$ (aq)] = [$CO_2$(aq)] + [$HCO_3^-$(aq)] + [$CO_3^{2-}$(aq)] + [$NH_2COO^-$(aq)] | [.....] = mol/kg solvent |
| $NH_3$ | | |
| | $NH_3$ (gas phase) ↔ $NH_3$(aq) (molecular dissolved) | 5 |
| | $NH_3$(aq) + $H_2O$(l) ↔ $NH_4^+$(aq) + $OH^-$(aq) | 6 |
| | $NH_3$(aq) + $HCO_3^-$(aq) ↔ $NH_2COO^-$(aq) + $H_2O$(l) | 7 |
| Over all concentration in liquid phase | [$NH_{3\ total}$ (aq)] = [$NH_3$(aq)] + [$NH_4^+$(aq)] + [$NH_2COO^-$(aq)] | [.....] = mol/kg solvent |
| $H_2S$ | | |
| | $H_2S$ (gas phase) ↔ $H_2S$(aq) (molecular dissolved) | 8 |
| | $H_2S$(aq) ↔ $HS^-$(aq) + $H^+$(aq) | 9 |
| | $HS^-$(aq) ↔ $S^{2-}$(aq) + $H^+$(aq) | 10 |
| Over all concentration in liquid phase | [$H_2S_{\ total}$ (aq)] = [$H_2S$(aq)] + [$HS^-$(aq)] + [$S^{2-}$(aq)] | [.....] = mol/kg solvent |
| HCN | | |
| | HCN (gas phase) ↔ HCN (molecular dissolved) | 11 |
| | HCN(aq) ↔ $CN^-$(aq) + $H^+$(aq) | 12 |
| Over all concentration in liquid phase | [$HCN_{\ total}$ (aq)] = [HCN(aq)] + [$CN^-$(aq)] | [.....] = mol/kg solvent |

26

| Stream names (Temp.,Pressure) | Stream total (t/h) | $CO_{2\ total}$ (kmol/h) | $CO_{2\ total}$ (mmol/l) | $CO_2$ (mmol/l) | $HCO_3^-$ (mmol/l) | $CO_3^{2-}$ (mmol/l) | $NH_2COO^-$ (mmol/l) | pH value |
|---|---|---|---|---|---|---|---|---|
| F103 ( 79.9 °C, 72 bar) | 5.3 | 1.2E-01 | 2.2E-02 | 1.6E-02 (69.7%) | 6.8E-03 (30.3%) | 5.2E-07 (0.002%) | 3.6E-07 (0.002%) | 5.90 |
| QUENCH WATER ( 177.4 °C, 71 bar) | 5.4 | 1.4E-01 | 2.3E-02 | 1.9E-02 (86%) | 3.3E-03 (14%) | 1.7E-07 (0.001%) | 1.5E-06 (0.01%) | 6.17 |
| T104 (79.9 °C, 72 bar) | 5.9 | 1.4E-01 | 2.3E-02 | 1.6E-02 (69.7%) | 6.8E-03 (30.3%) | 5.2E-07 (0.002%) | 3.6E-07 (0.002%) | 5.90 |
| F136PRESS (79.9 °C, 72 bar) | 0.7 | 1.5E-02 | 2.2E-02 | 1.6E-02 (69.7%) | 6.8E-03 (30.3%) | 5.2E-07 (0.002%) | 3.6E-07 (0.002%) | 5.90 |
| F136 (20 °C, 1 bar) | 0.7 | 6.9E-03 | 1.1E-02 | 3.6E-03 (34.1%) | 7.0E-03 (65.9%) | 1.6E-06 (0.015%) | 2.9E-07 (0.0027%) | 6.60 |
| DEGAS (20 °C, 1 bar) | 0.002 | 8.0E-03 | 9.3E-02 (mol/mol) | | | | | |
| HOT GAS (1172 °C, 71 bar) | 0.8 | 2.47 | 4.1E-02 (mol/mol) | $\frac{1}{M_{H_2O}} = \frac{1}{18.015}\frac{kmol}{kg} = 0.056\frac{kmol}{kg} = 56\frac{mol}{kg} = 56000\frac{mmol}{kg} \approx 56000\frac{mmol}{L}$ | | | | |
| RAW GAS (117.4 °C, 71 bar) | 0.7 | 2.45 | 4.3E-02 (mol/mol) | | | | | |

| Stream names (Temp.,Pressure) | Stream total (t/h) | $NH_{3\ total}$ (kmol/h) | $NH_{3\ total}$ (mmol/l) | $NH_3$ (mmol/l) | $NH_4^+$ (mmol/l) | $NH_2COO^-$ (mmol/l) | pH value |
|---|---|---|---|---|---|---|---|
| F103 ( 79.9 °C, 72 bar) | 5.3 | 3.7 E-02 | 6.8E-03 | 7.8E-05 (1.13%) | 6.8E-03 (98.9%) | 3.6E-07 (0.01%) | 5.90 |
| QUENCH WATER ( 177.4 °C, 71 bar) | 5.4 | 4.2E-02 | 6.9E-03 | 3.7E-03 (52.8%) | 3.3E-03 (47.2%) | 1.5E-06 (0.02%) | 6.17 |
| T104 (79.9 °C, 72 bar) | 5.9 | 4.2E-02 | 6.8E-03 | 7.8E-05 (1.13%) | 6.8E-03 (98.5%) | 3.6E-07 (0.005%) | 5.90 |
| F136PRESS (79.9 °C, 72 bar) | 0.7 | 5.0E-03 | 6.8E-03 | 7.8E-05 (1.13%) | 6.8E-03 (98.8%) | 3.6E-07 (0.005%) | 5.90 |
| F136 (20 °C, 1 bar) | 0.7 | 5.0E-03 | 7.0E-03 | 1.1E-05 (0.16%) | 7.0E-03 (99.8%) | 2.9E-07 (0.004%) | 6.60 |
| DEGAS (20 °C, 1 bar) | 0.002 | 1.3E-08 | 1.5E-07 (mol/mol) | | | | |
| HOT GAS (1172 °C, 71 bar) | 0.8 | 9.4E-03 | 1.6E-04 (mol/mol) | $\frac{1}{M_{H_2O}} = \frac{1}{18.015}\frac{kmol}{kg} = 0.056\frac{kmol}{kg} = 56\frac{mol}{kg} = 56000\frac{mmol}{kg} \approx 56000\frac{mmol}{L}$ | | | |
| RAW GAS (117.4 °C, 71 bar) | 0.7 | 4.8E-03 | 8.4E-05 (mol/mol) | | | | |



| Stream names (Temp.,Pressure) | Stream total (t/h) | $H_2S_{\ total}$ (kmol/h) | $H_2S_{\ total}$ (mmol/l) | $H_2S$ (mmol/l) | $HS^-$ (mmol/l) | $S^{2-}$ (mmol/l) | pH value |
|---|---|---|---|---|---|---|---|
| F103 ( 79.9 °C, 72 bar) | 5.3 | 1.4E-05 | 2.6E-06 | 2.1E-06 (79%) | 5.5E-7 (20.9%) | 5.2E-14 (2E-06%) | 5.90 |
| QUENCH WATER ( 177.4 °C, 71 bar) | 5.4 | 1.6E-05 | 2.7E-06 | 1.9E-06 (71.7%) | 7.5E-7 (28.2%) | 2.5E-11 (0.001%) | 6.17 |
| T104 (79.9 °C, 72 bar) | 5.9 | 1.6E-05 | 2.6E-06 | 2.1E-06 (79%) | 5.5E-7 (21%) | 5.3E-14 (2E-06%) | 5.90 |
| F136PRESS (79.9 °C, 72 bar) | 0.7 | 1.7E-06 | 2.6E-06 | 2.1E-06 (79%) | 5.5E-7 (20.9%) | 5.2E-14 (2.1E-06%) | 5.90 |
| F136 (20 °C, 1 bar) | 0.7 | 9.5E-07 | 1.5E-06 | 1.0E-06 (72%) | 4.0E-7 (27.9%) | 1. 4E-15 (9.5E-08%) | 6.60 |
| DEGAS (20 °C, 1 bar) | 0.002 | 8.0E-07 | 9.2E-06 (mol/mol) | | | | |
| HOT GAS (1172 °C, 71 bar) | 0.8 | 8.6E-05 | 1.4E-06 (mol/mol) | $\frac{1}{M_{H_2O}} = \frac{1}{18.015}\frac{kmol}{kg} = 0.056\frac{kmol}{kg} = 56\frac{mol}{kg} = 56000\frac{mmol}{kg} \approx 56000\frac{mmol}{L}$ | | | |
| RAW GAS (117.4 °C, 71 bar) | 0.7 | 8.5E-05 | 1.5E-06 (mol/mol) | | | | |

| Stream names (Temp.,Pressure) | Stream total (t/h) | $HCN_{\ total}$ (kmol/h) | $HCN_{\ total}$ (mmol/l) | HCN (mmol/l) | $CN^-$ (mmol/l) | pH value |
|---|---|---|---|---|---|---|
| F103 ( 79.9 °C, 72 bar) | 5.3 | 5.1E-04 | 9.3E-05 | 9.3E-05 (99.4%) | 5.9E-07 (0.6%) | 5.90 |
| QUENCH WATER ( 177.4 °C, 71 bar) | 5.4 | 5.7E-04 | 9.5E-05 | 9.1E-05 (96%) | 3.7E-06 (4%) | 6.17 |
| T104 (79.9 °C, 72 bar) | 6.0 | 5.7E-04 | 9.3E-05 | 9.3E-05 (99.4%) | 5.9E-07 (0.6%) | 5.90 |
| F136PRESS (79.9 °C, 72 bar) | 0.7 | 6.3E-05 | 9.3E-05 | 9.3E-05 (99.4%) | 5.9E-07 (0.6%) | 5.90 |
| F136 (20 °C, 1 bar) | 0.7 | 6.2E-05 | 9.5E-05 | 9.5E-05 (99.8%) | 1.9E-07 (0.2%) | 6.60 |
| DEGAS (20 °C, 1 bar) | 0.002 | 5.9E-07 | 6.8E-06 (mol/mol) | | | |
| HOT GAS (1172 °C, 71 bar) | 0.8 | 2.6E-04 | 4.3E-06 (mol/mol) | $\frac{1}{M_{H_2O}} = \frac{1}{18.015}\frac{kmol}{kg} = 0.056\frac{kmol}{kg} = 56\frac{mol}{kg} = 56000\frac{mmol}{kg} \approx 56000\frac{mmol}{L}$ | | |
| RAW GAS (117.4 °C, 71 bar) | 0.7 | 2.0E-04 | 3.4E-06 (mol/mol) | | | |

## Dissociation constant

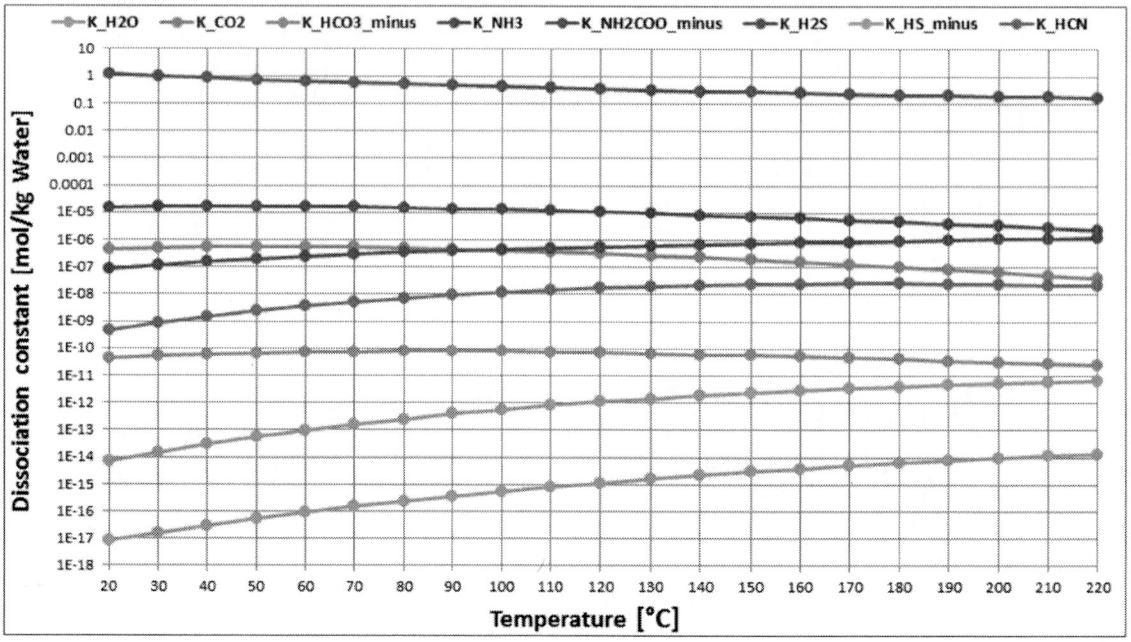


29

## Dissociation constant

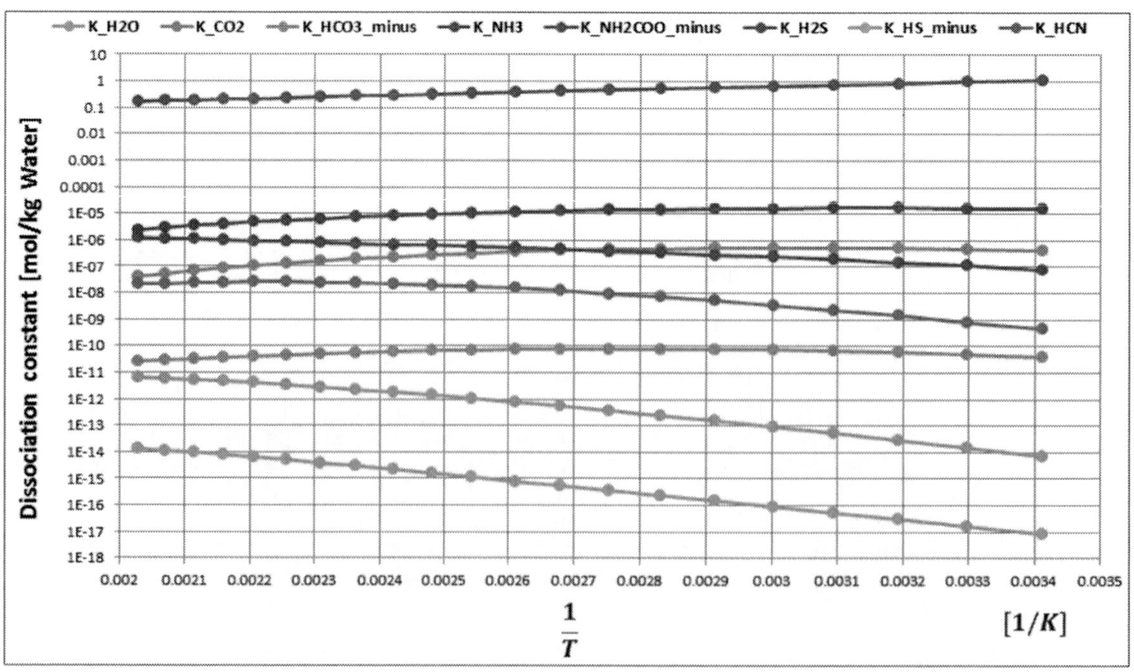


30

## Henry´s constant

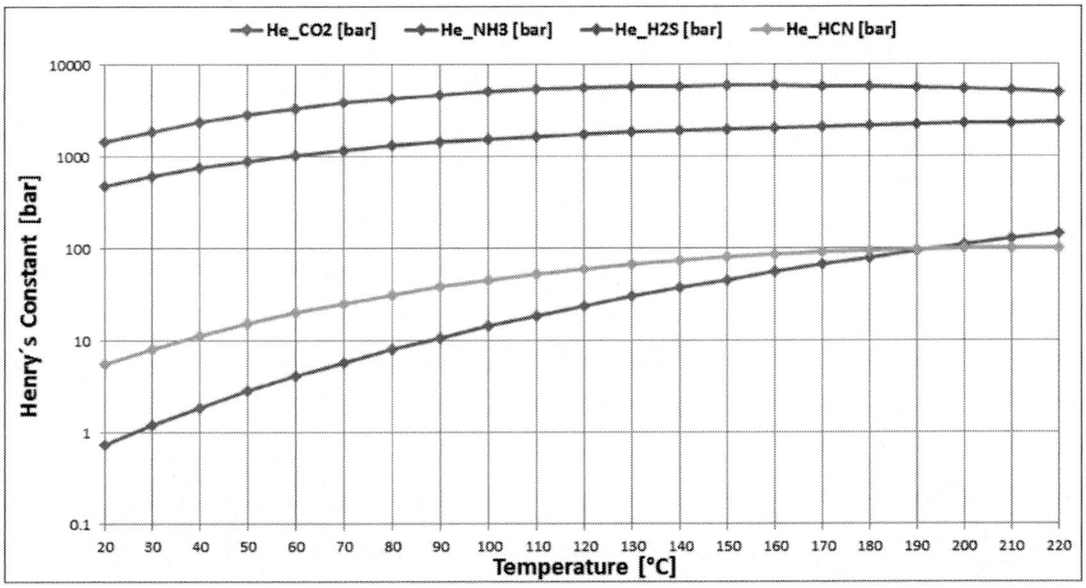


31

## Henry´s constant

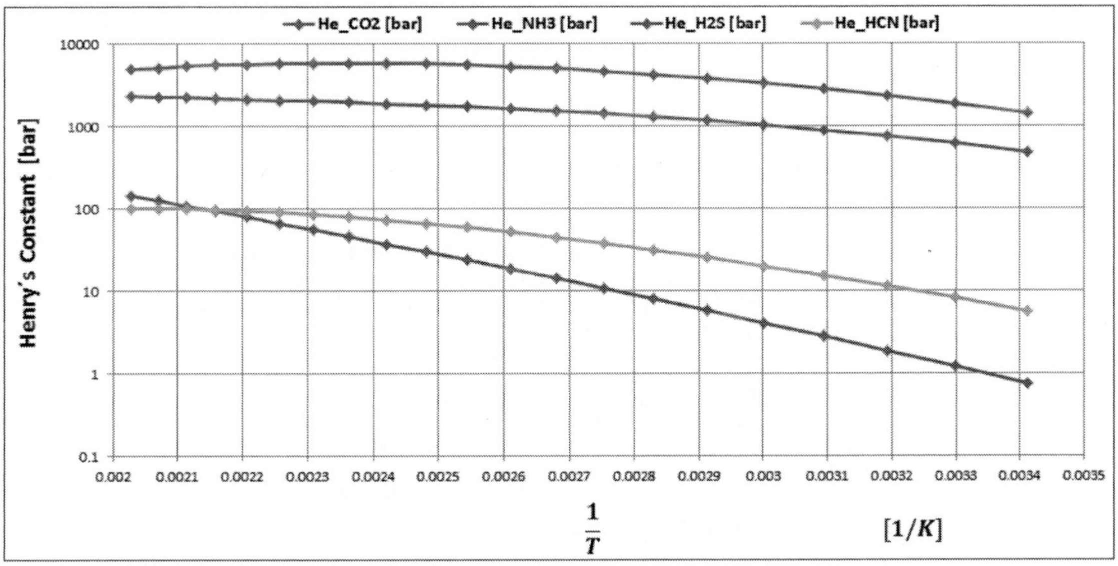


32

# Influences of the Cross Angle on the Granular Jet in Annular Gas Stream with Thick Wall of Inner Channel

**Presenter: Chenchen Fang**
East China University of Science and Technology (ECUST), China

# Content

1. Background
2. Experiments
3. Results and discussion
4. Conclusions

# 1. Background

**Dense gas-solid coaxial jets**

Cold state

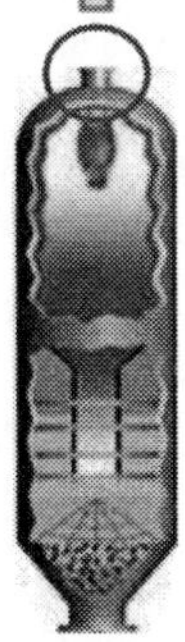

**Gasifier**

**Interactions between gas and particles**

**Particle dispersion**

- **Macro** characteristics
  - Dispersion modes
  - Wavelength, Amplitude, Frequency
  - Undisturbed length, Dispersion length, Dispersion angle
- **Micro** characteristics
  - **Particle motion: Entrainment**

# 1. Background

**Industrial condition**

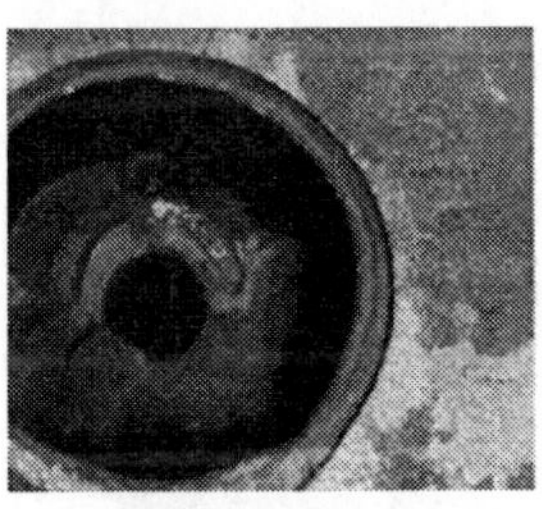

Abrasion

**Entrainment phenomenon**

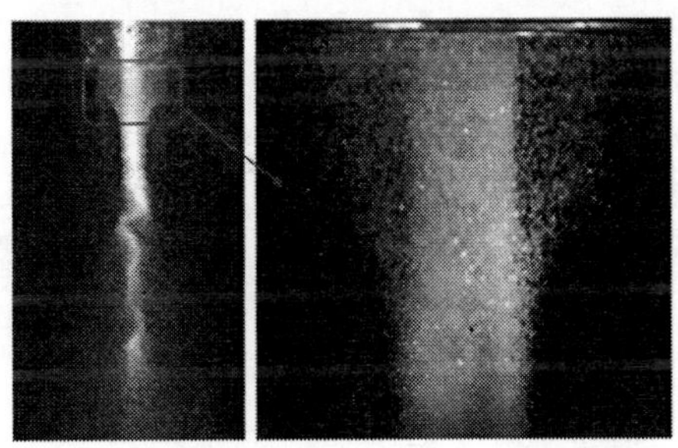

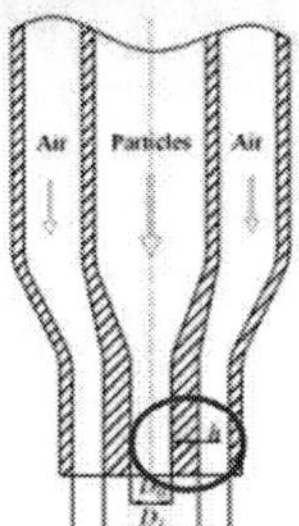


Coaxial nozzle with thick wall

Space effect of thick wall

- Experimental investigation on particle entrainment behaviors near a nozzle in gas–particle coaxial jets. **Powder Technology** 2015, 286: 55-63.
- Influences of the wall thickness on the granular dispersion in a dense gas–solid coaxial jet. **International Journal of Multiphase Flow** 2016, 81: 20-26.

# 1. Background

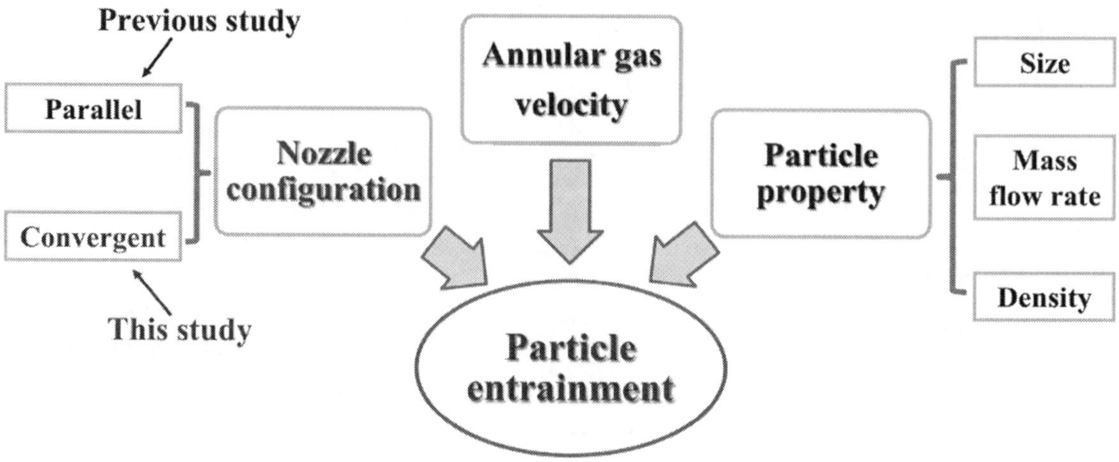


# 2. Experimental setup

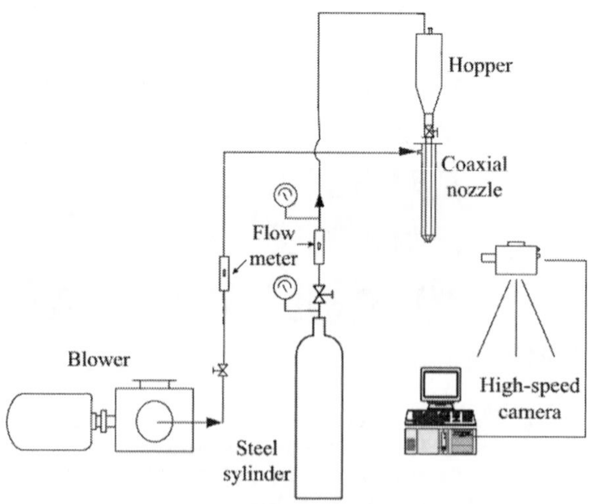


Flow chart of the high-speed photography system

Schematic diagram of the high-speed camera

# 2. Experimental setup

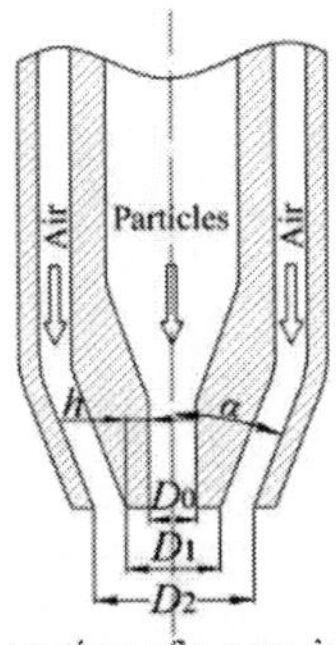


Configuration of a coaxial nozzle

**Table 1.** Geometrical parameters of a coaxial nozzle with the cross angle.

| Nozzle No | $D_0$ (×10$^{-3}$ m) | $D_1$ (×10$^{-3}$ m) | $D_2$ (×10$^{-3}$ m) | $h$ (×10$^{-3}$ m) | $\alpha$ (°) | Area (×10$^{-6}$ mm$^2$) |
|---|---|---|---|---|---|---|
| 1 | 4.1 | 12.1 | 16.9 | 4.0 | 0 | 109.3 |
| 2 | 4.1 | 12.1 | 17.1 | 4.0 | 20 | 109.9 |
| 3 | 4.1 | 12.1 | 17.8 | 4.0 | 40 | 109.9 |

**Table 2.** Experimental parameters of the gas phase and the particle phase.

| Phase | Parameters | |
|---|---|---|
| Gas (air) | Superficial gas velocity, $u_g$ (m/s) | 0-65 |
| | Gas dynamic viscosity, $\mu_g$ (μPa·s) | 17.9 |
| | Gas density, $\rho_g$ (kg/m$^3$) | 1.247 |
| | Reynolds number, $Re_g$ | $0 - 5.6 \times 10^4$ |
| Particle (glass beads) | Particle diameter, $d_p$ (μm) | 77 |
| | Particle density, $\rho_p$ (kg/m$^3$) | 2490 |
| | Particle bulk density, $\rho_{p,b}$ (kg/m$^3$) | 1445 |
| | Particle mass flow rate, $m_p$ (kg/s) | 0.012 |
| | Stokes number, $St$ | 0-241 |

## Features

- **Cross angle $\alpha$**
- **Thick wall $h$**

# 3. Results and discussion

## 3.1 Particle entrainment phenomenon

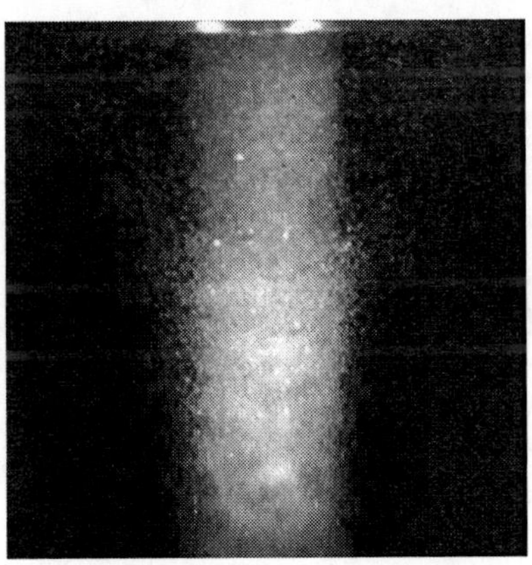

Experimental instantaneous images

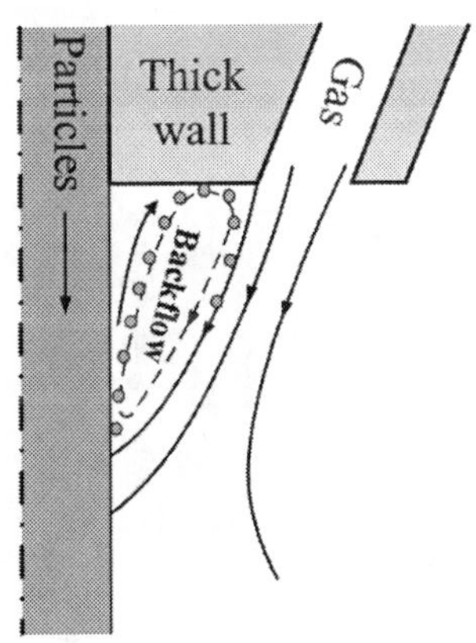


Schematic of the particle entrainment

- **The recirculation occurs at the nozzle exit due to the thick wall of an inner channel.**
- **Particles on the surface of the granular stream are influenced by drag force of the backflow and entrained back to the nozzle wall.**

# 3. Results and discussion

## 3.1 Particle entrainment phenomenon

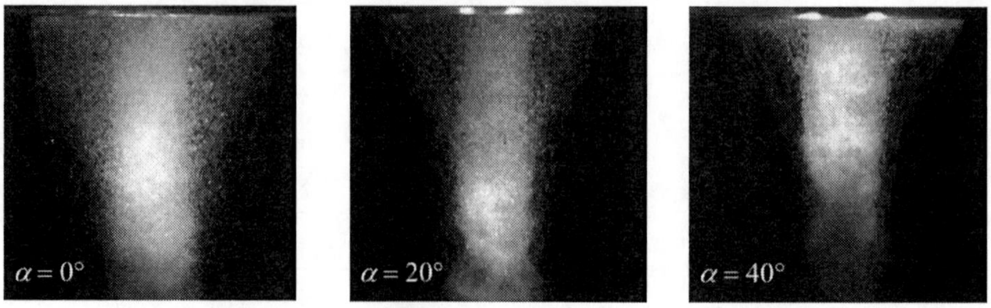


Entrainment phenomena for different cross angles $\alpha$ ($u_g = 50$ m/s)

- **The larger a cross angle is, the less significant the particle entrainment phenomenon is.**
- **The entrained particles are closer to the nozzle wall for the larger cross angle due to the enhancement of the radial component of the annular gas .**

# 3. Results and discussion

## 3.2 Particle entrainment modes

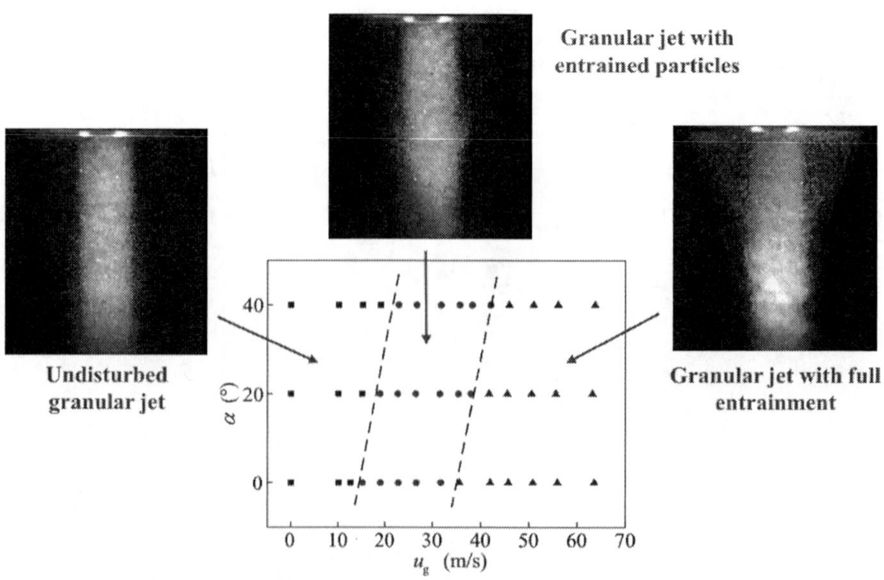


Phase diagram of the entrainment modes for different cross angles $\alpha$

# 3. Results and discussion

### 3.3 Particle entrainment length

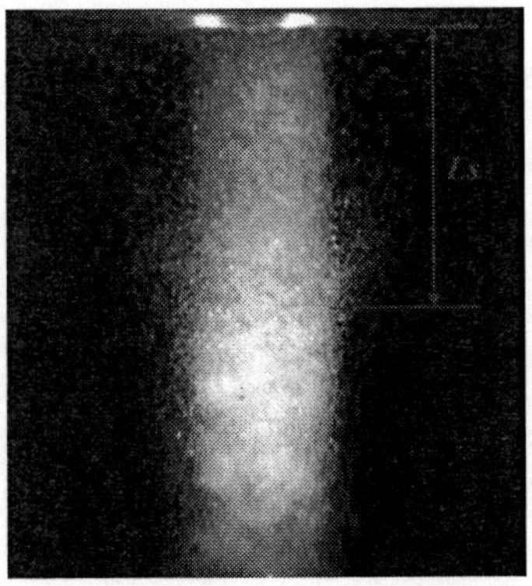


Schematic diagram of the particle entrainment length

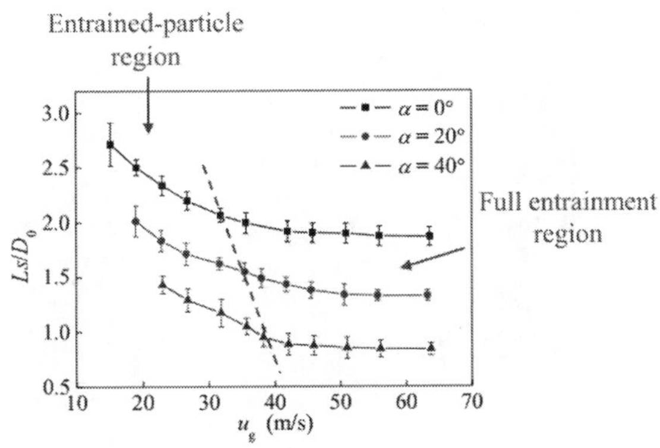


Influence of $u_g$ on $Ls/D_0$ for different cross angles $\alpha$

# 3. Results and discussion

### 3.4 Entrained particle velocity

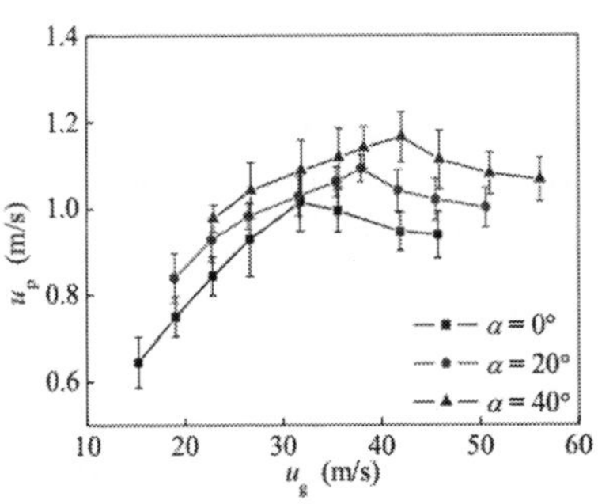


Influence of $u_g$ on $u_p$ for different cross angles $\alpha$

- **An entrained particle velocity first increases and then declines to a constant with an increase of an annular gas velocity.**
- **The larger a cross angle is, the larger an entrained particle velocity is.**

**Cross angle** → (Enhance) → **Turbulence intensity** → (Promote) → **Momentum transfer** → (Increase) → **Particle velocity**

# 3. Results and discussion

## 3.5 Particle entrainment angle

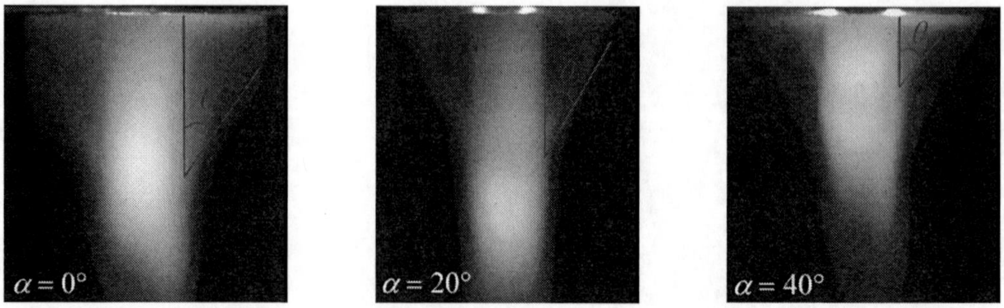


Schematic diagram of the particle entrainment angle for different cross angles $\alpha$

# 3. Results and discussion

## 3.5 Particle entrainment angle

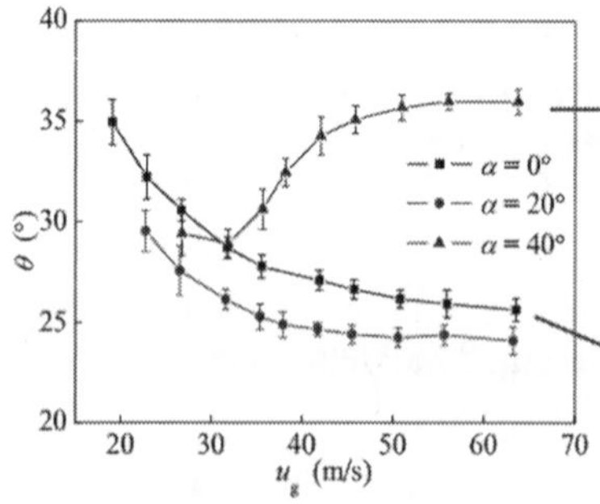


Influence of $u_g$ on $\theta$ for different cross angles $\alpha$

- **For large cross angle, particle entrainment angle increases with an annular gas velocity.**
- **For small cross angles, particle entrainment angle decreases with an increase of an annular gas velocity.**

# 4. Conclusions

- The entrained particles are closer to the nozzle wall with an increase of the cross angle.
- Three entrainment modes of the granular jet emerge in sequence with an increase of annular gas velocity and the phase diagram of the entrainment modes for different cross angles is obtained.
- The particle entrainment length decreases with the increase of the cross angle for the same annular gas velocity.
- The entrained particle velocity increases with the cross angle.
- The particle entrainment angle greatly depends on the cross angle.

Thank you

E-mail:
020120060@ecust.edu.cn
fcc0305@163.com

Klaus Rademann, Thomas Ryll

**ENERGY CHALLENGES**
**METHANE CONVERSION**
**HYDROGEN PRODUCTION**

MICROWAVES

8-12 AUG 2016: Pittsburgh

Cape Town

HUMBOLDT-UNIVERSITÄT ZU BERLIN

# Campus Berlin Adlershof

AZM:
Dr. B. Löchel, Dr. U. Scheunemann

BESSY

Dr. A. Hoell, Dr. D. Tatchev

hmi

Dr. M. Radtke, Dr. H. Riesemeier

BAM

**6000 Students + 4000 Scientists + 6000 Employed**

# Energy Conversion Processes

- Fundamental Aspects of C-H Activation*
- Biomass to Biogas upgrading
- Cellulose and wood to food
- CATALYSIS by Au, Ag, Cu nanoparticles

- Heat fluxes into power (thermoelectrics)
- Light up conversion and down conversion

***Outlook***

- Micro(Nano)structuring for optical applications: data storage, waveguides, photonic crystals (plasmonics) → emitter and waveguides in one glass

- Newest generation of white LEDs: ***Phosphor converted LED*** glasses can be excited with a UV LED (e.g. InGaN)

- Emission color tuning by different dopants (energy transfer to Eu, Sm, Dy, Tb; ZnO)

- Gold nanoparticles in nanolayers accessible to biological molecules: SERS sensor for adenine (cooperation with Prof. Dr. Panne, Dr. Kneipp / BAM)

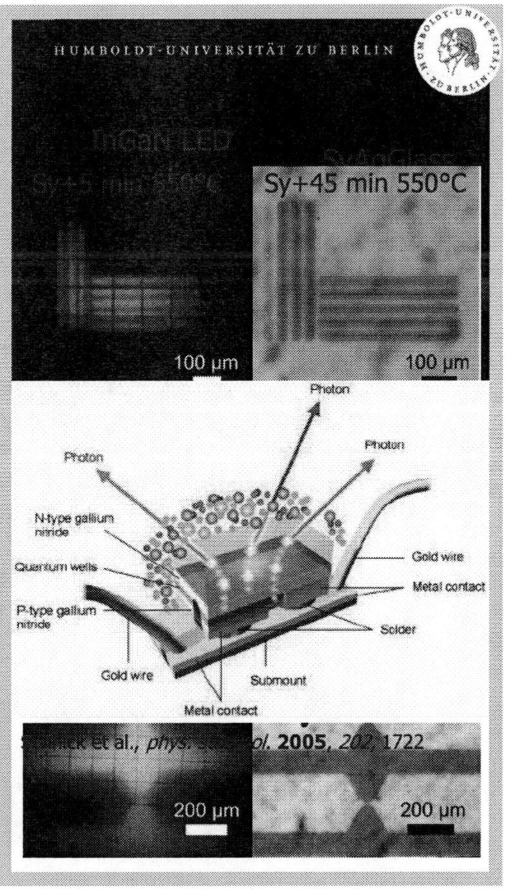

## Glass Nanocomposites, 1st Edition

Synthesis, Properties and Applications

| | |
|---|---|
| Editor(s): | Karmakar & Rademann & Stepanov |
| Expected Release Date: | 11 Jan 2016 |
| Imprint: | William Andrew |
| Print Book ISBN : | 9780323393096 |
| Pages: | 408 |
| Dimensions: | 235 X 191 |

Researchers and engineers in a range of academic fields and industries, such as optoelectronics, semiconductors, photonics, plasmonics and energy harvesting will find this book to be a one-stop reference source that covers all aspects of glass nanocomposite materials

# Co-workers at HU Chemistry

# INVESTMENT DECLARATION

For the time being
it is possible to
Invest as much money
as you like
We make however
no contract on the ROI,
Nor even on the break even point.

Before we look into the future we look into the past

# Science
# 10 August 2007

## Orbital and Millennial Antarctic Climate Variability over the Past 800,000 Years

J. Jouzel,[1*] V. Masson-Delmotte,[1] O. Cattani,[1] G. Dreyfus,[1] S. Falourd,[1] G. Hoffmann,[1] B. Minster,[1] J. Nouet,[1] J. M. Barnola,[2] J. Chappellaz,[2] H. Fischer,[3] J. C. Gallet,[2] S. Johnsen,[4,5] M. Leuenberger,[6] L. Loulergue,[2] D. Luethi,[6] H. Oerter,[3] F. Parrenin,[2] G. Raisbeck,[7] D. Raynaud,[2] A. Schilt,[6] J. Schwander,[6] E. Selmo,[8] R. Souchez,[9] R. Spahni,[6] B. Stauffer,[6] J. P. Steffensen,[2] B. Stenni,[10] T. F. Stocker,[6] J. L. Tison,[9] M. Werner,[11] E. W. Wolff[12]

A high-resolution deuterium profile is now available along the entire European Project for Ice Coring in Antarctica Dome C ice core, extending this climate record back to marine isotope stage 20.2, ~800,000 years ago. Experiments performed with an atmospheric general circulation model including water isotopes support its temperature interpretation. We assessed the general correspondence between Dansgaard-Oeschger events and their smoothed Antarctic counterparts for this Dome C record, which reveals the presence of such features with similar amplitudes during previous glacial periods. We suggest that the interplay between obliquity and precession accounts for the variable intensity of interglacial periods in ice core records.

# CHAPPELLAZ Jerome: Temperature, $CO_2$ and $CH_4$

strong correlations between northern and southern hemispheres

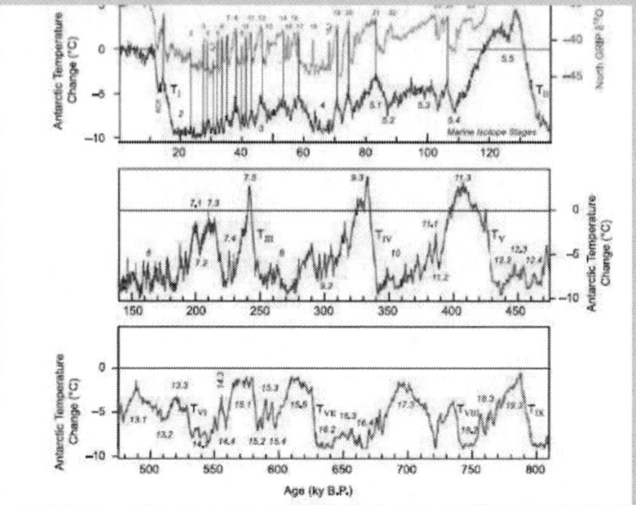

Fig. 2. Dome C temperature anomaly as a function of time over the past 810 ky. Back to 140 ky B.P.

# Methane and Carbon Dioxide

- affected the climate, the ocean, the atmosphere, the earth through the past 800 000 years
- The future within the anthropocen (1500 to 15000) must be taken care of through education and research

# Current Measurements

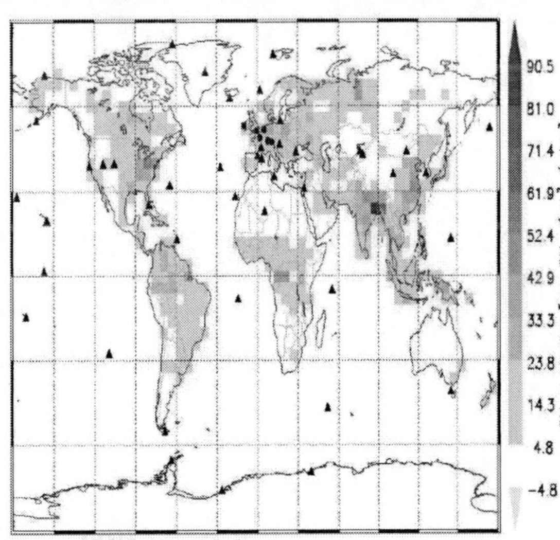


(a) Global true emissions

Methan Distribution on Ground: In-Complete

M. Villani, P. Bergamaschi, M. Krol, J.F. Merinik, F. Deneter, *Atmos. Chem. Phys.*. 10. 5. 1249-1267. **2010**.

# MERLIN for unexplored regions

- MEthane Remote Sensing LIDAR MissioN
- Co-operation CNES & DLR
- Start 2020
- Running time 3 years

MERLIN-Kleinsatellit, http://www.dlr.de/rd/desktopdefault.aspx/tabid-2440/3586_read-31672/

# Methan, Carbon Dioxide

- Methan: concentration in air in time and space
- Methan: concentration throughout the history
- Methan: Concentration in hydrates

# Methane and Biogas UP-Grading

## Biogas upgrading

Typical composition of Biogas

| Matter | % |
|---|---|
| Methane, $CH_4$ | 50–75 |
| Carbon dioxide, $CO_2$ | 25–50 |
| Nitrogen, $N_2$ | 0–10 |
| Hydrogen, $H_2$ | 0–1 |
| Hydrogen sulfide, $H_2S$ | 0–3 |
| Oxygen, $O_2$ | 0–2 |

# DRY REFORMING of METHANE

## Why the dry reforming of methane (DRM)?

$$CH_4 + CO_2 \longrightarrow 2CO + 2H_2$$

$\Delta H^{\circ}_{298} = 247\,kJ/mol$ $\Delta G^{\circ} = -8545 - 7.84T$

**Advantages:**

- Greenhouse gases as feedstock
- $CH_4$: natural gas or methanization
- Syngas production
- $H_2/CO$ ratio suitable for GTL processes

**Disadvantages:**

- High temperature required (T > 600°C)
- Rapid deactivation of the catalyst

# Syngas is of Central Importance

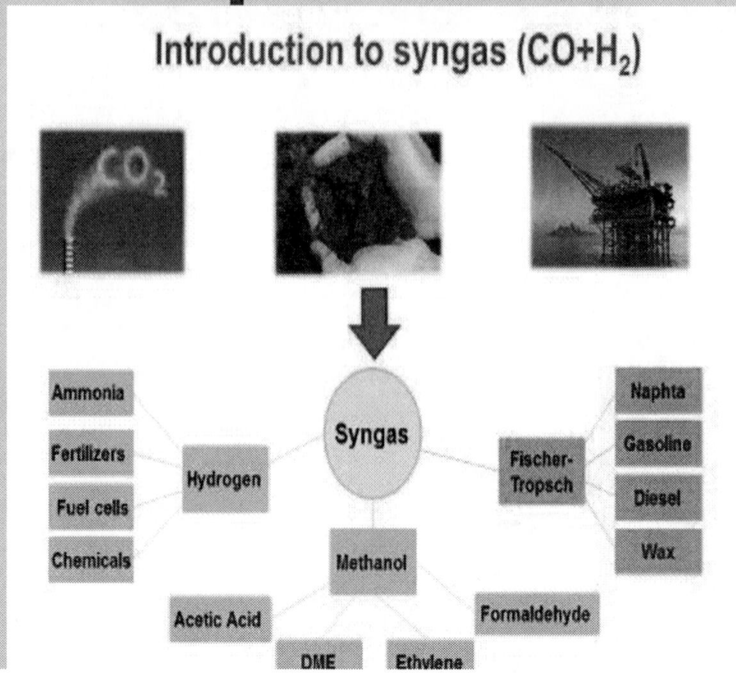


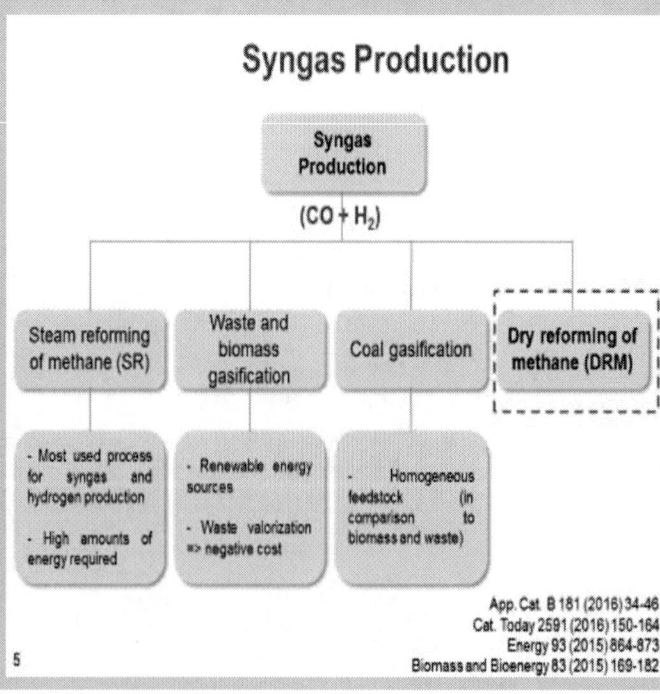

## DRM: Catalyst deactivation

**Main reaction:** DRM: $CH_4 + CO_2 \leftrightarrow 2H_2 + 2CO$ $\Delta H^{\circ}_{298} = 247 kJ/mol$

**Side reactions:**

Boudouard reaction: $2CO \leftrightarrow C + CO_2$ $\Delta H^{\circ}_{298} = -172 kJ/mol$

Methane cracking: $CH_4 \leftrightarrow C + 2H_2$ $\Delta H^{\circ}_{298} = 75 kJ/mol$

Clogging of active sites

Reverse water-gas shift: $CO_2 + H_2 \leftrightarrow H_2O + CO$ $\Delta H^{\circ}_{298} = 41 kJ/mol$

Carbon gasification: $H_2O + C_{(s)} \leftrightarrow CO + H_2$ $\Delta H^{\circ}_{298} = 131 kJ/mol$

↑ Basicity of catalyst/support ↑ $CO_2$ adsorption ↑ removal of coke[1]

Mg, K, Ca...

Objectives : - Study the influence of basicity of support on catalyst activity/stability
- Impact of the carbon deposit on the catalyst efficiency
- Investigation of the role of the water formed during the reaction

7

Int. J. Hyd. Ener. 37 (2012)

## Gas-Analysis GC

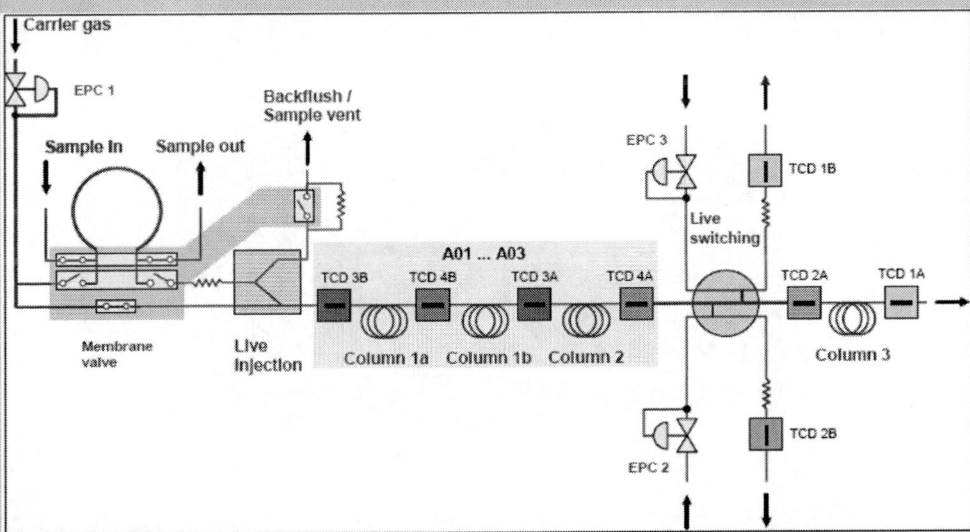

# Reactor Size: 300m³/h

# Microwave Generator

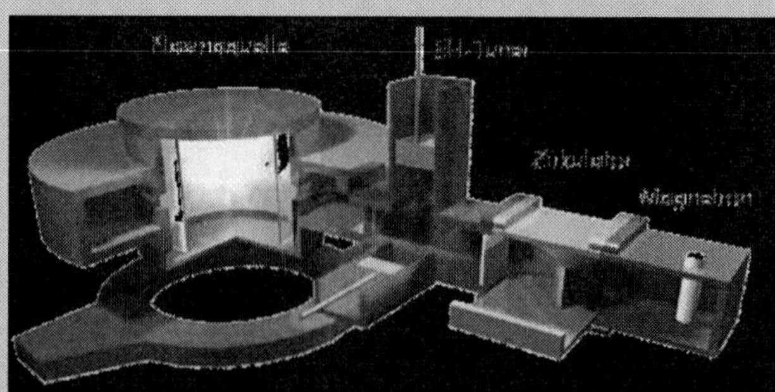

| year | process | reactor | conversion rate | max. (norm.) acetylen yield | selectivity $C_2H_2 : C_mH_n$ |
|---|---|---|---|---|---|
| **2014** | **MW (cw)** | **<100 kW** | **>99 %** | **99 %** | **>200:1** |
| 2010 | MW (cw) | 400 W | 93,7 % | 88,6% | 9:1 |
| 2002 | MW (pulsed) | 16 – 81 W | 90 % | 80 % | 4:1 |
| 1996 | MW | 50 - 250 W | >85 % | 90 % | 9:1 |
| 2002 | HFPP | 11KV / 8kHz | 23,5 % | 92 % | 9:1 |
| 1962 | DC arc | 8 MW | 70,5 % | 72,9 % | approx. 3,5:1 |

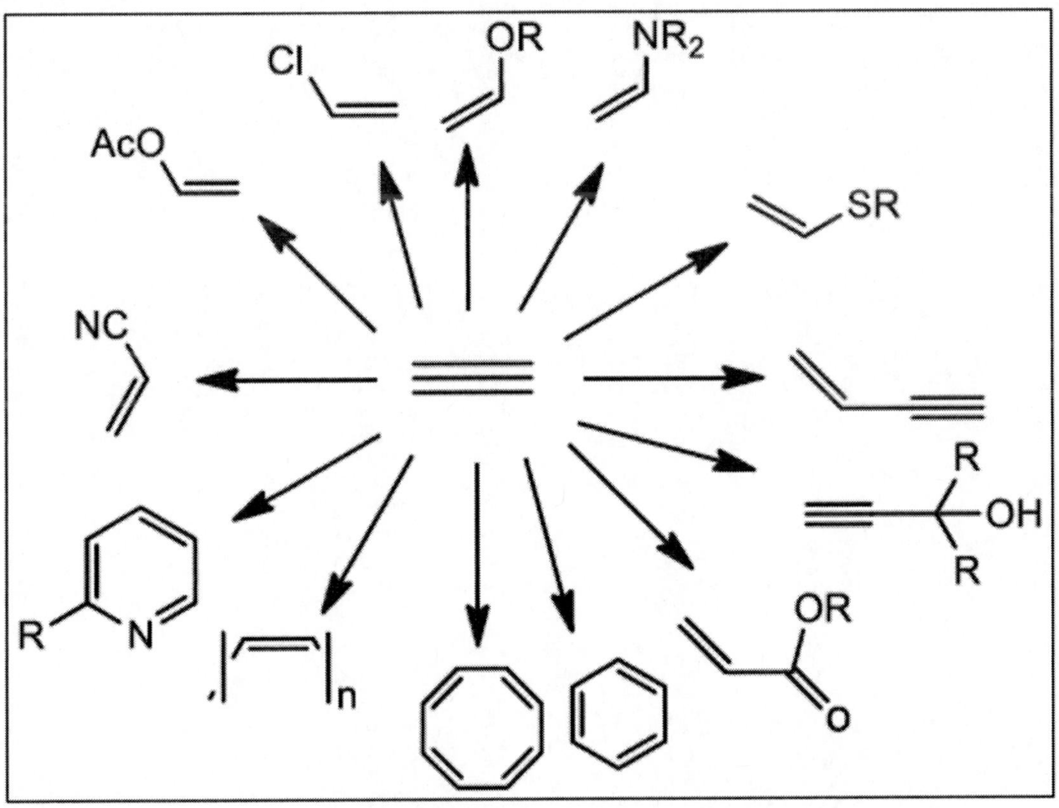

## Summary

- Methane and carbon dioxide must be treated together
- Catalytic conversion processes
- Chemical thermal equilibrium at 1000 K for methane to acetylene: 5%
- Microwave assisted: 80-90 %
- Under up-scaling conditions: 100%

## Summary

- Methane and Carbon Dioxide
- MW assisted conversion processes can be highly selective and effective

# Mechanism

$CH_4 \rightleftharpoons CH_3\cdot + H\cdot$ : $\Delta H = 421$ kJ/mol

$H\cdot + H\cdot \rightleftharpoons H_2$ : $\Delta H = -436$ kJ/mol

$CH_4 + H\cdot \rightleftharpoons CH_3\cdot + H_2$ : $\Delta H = -15$ kJ/mol

# Thank you

# RESULTS: GOLD (13 nm) and Bimetallic Au/Pd nanoclusters

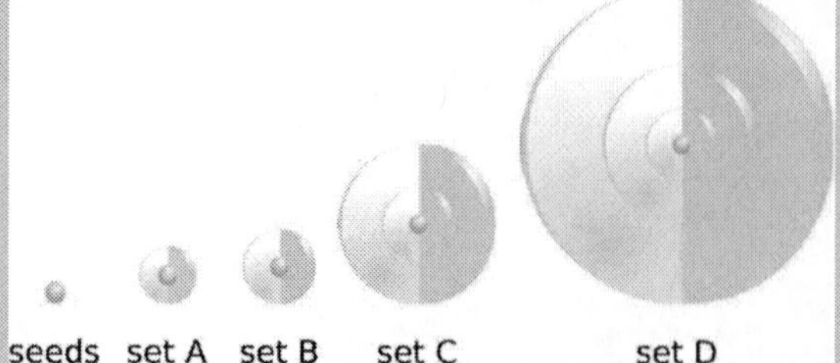


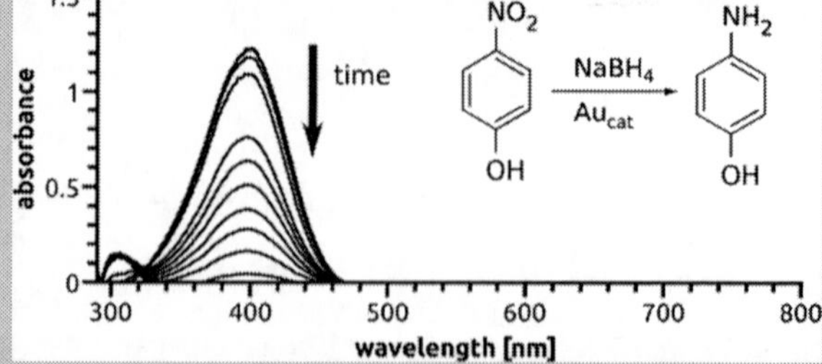


R. Fenger, E. Fertitta, H. Kirmse, A. F. Thünemann and K. Rademann

**"Size dependent catalysis with CTAB-stabilized gold nanoparticles"** *Phys. Chem. Chem. Phys.*, **14**, 9343 (2012).

## Au/Pd nanoclusters

HZB Helmholtz Zentrum Berlin

### Catalytically highly active bimetallic Gold/Palladium nanoclusters: Relationship between nanostructure and catalytic activity

**Samples:** Samples with different Au and Pd amounts were prepared.

**Catalysis:** UV-VIS spectroscopy was used to study the catalytic activity of those colloidal particles in solution.

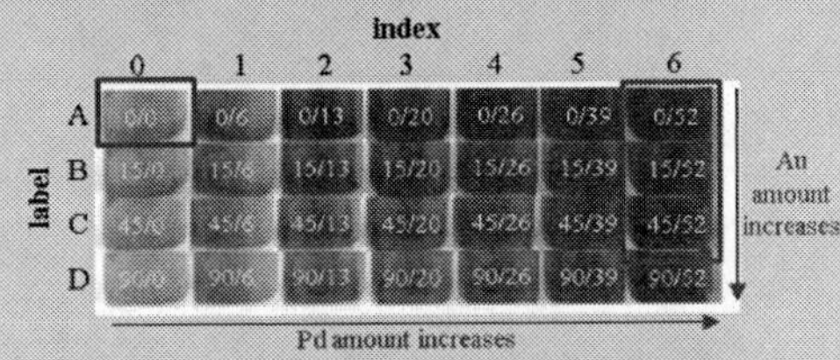


**Question:** Structure of the particles: spherically shaped or core-shell ?

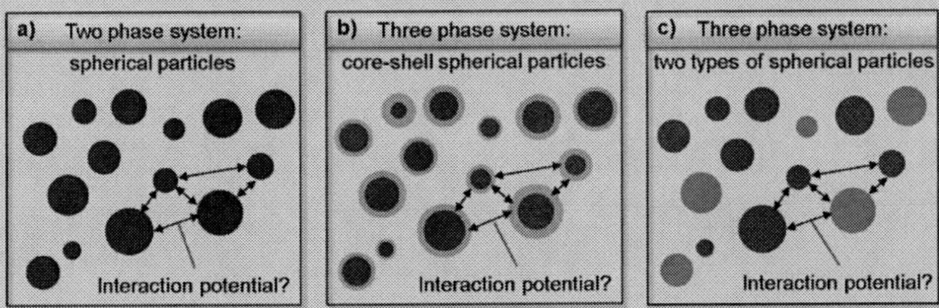


Combined ASAXS at the Au-$L_3$ and Pd-K X-ray absorption allows determination of the nanostructure of the catalytically active nanoclusters!

## Au/Pd nanoclusters

HZB Helmholtz Zentrum Berlin

### Proposed Mechanism

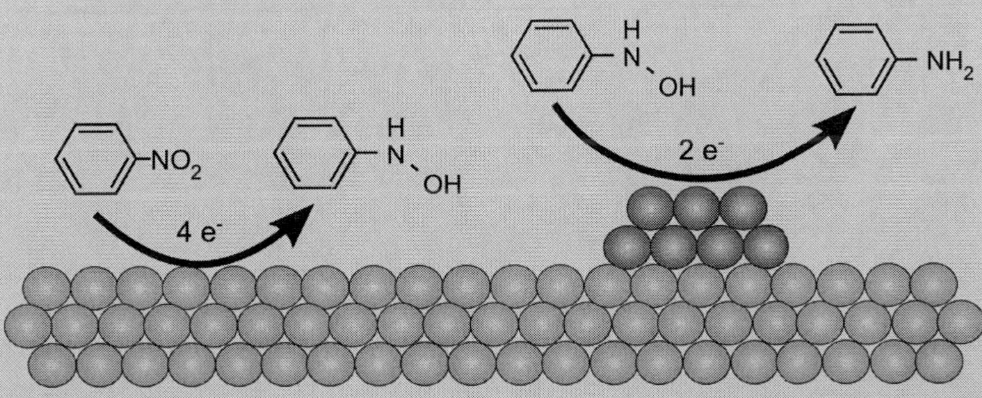


### [4 + 2] Mechanism

# Size Effects and Outlook

- Gold in soda lime silicate glass can be grown and stabilized beginning from the ion [Au(III)) via Au(I) ] all the way up to atoms, dimers, trimers, clusters, nanoparticles in the sub-nanometer range:
- Strong red luminescence for dimers: >60% **clusters below 20 atoms.**

ON CHOLIN-
CHLORIDE/UREA

VIKRAM RAGHUWANSHI
ZEITSCHRIFT FÜR
PHYSIKALISCHE CHEMIE
2015

Cu/Ag/Au in PLA

PLA Stool
Biodegradable

Dipl.-Ing. Jörg Petri
(Designer-Preis)

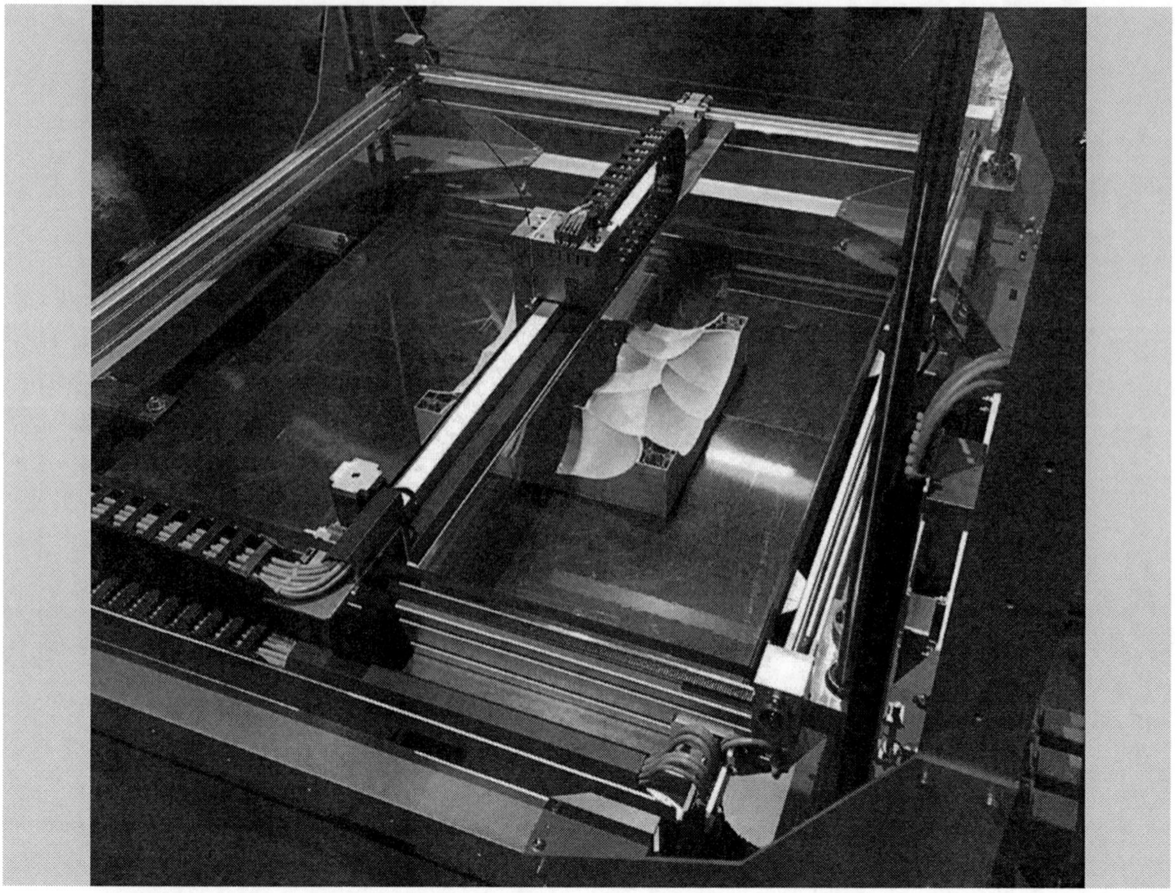

PLA:
AS A STOOL
OR AS A TABLE
IS VERY STABLE

# THE DEVELOPMENT OF HIGHLY ANISOTROPIC COKE FROM AN ALIPHATIC WAXY OIL RESIDUE

The Thirty-Third Annual Pittsburgh Coal Conference

August 8 - 12, 2016 - Cape Town, SOUTH AFRICA

Prof. John G Clark

## Lecture content

- Process overview
- Delayed coking and calcining
- Needle coke overview
- Introduction
- Microstructure and XRD of high ash Waxy Oil coke
- Experimental protocol
- Molecular composition of Waxy Oil and thermally treated variants
- Waxy Oil modified microstructure
- Proposed reaction mechanism
- Mesophase development and coalescence
- Conclusions
- Acknowledgements
- References

## Sasol Secunda site

sasol frontiers

Maintaining momentum 3

## Process overview

sasol frontiers

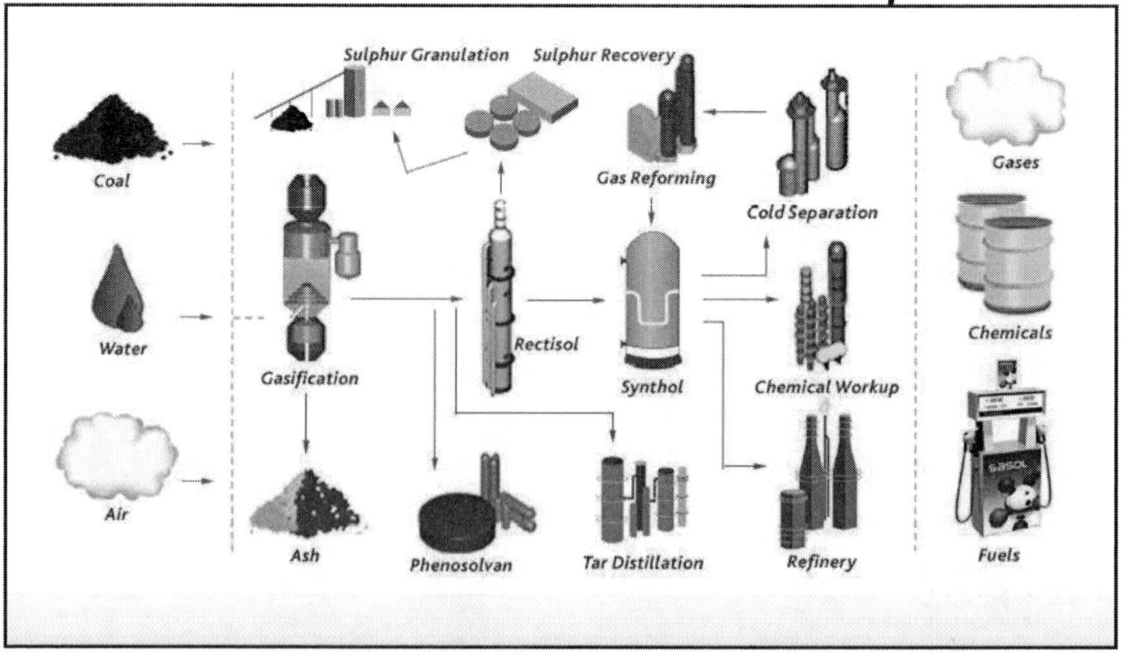


Maintaining momentum 4

## Delayed coking

sasol frontiers

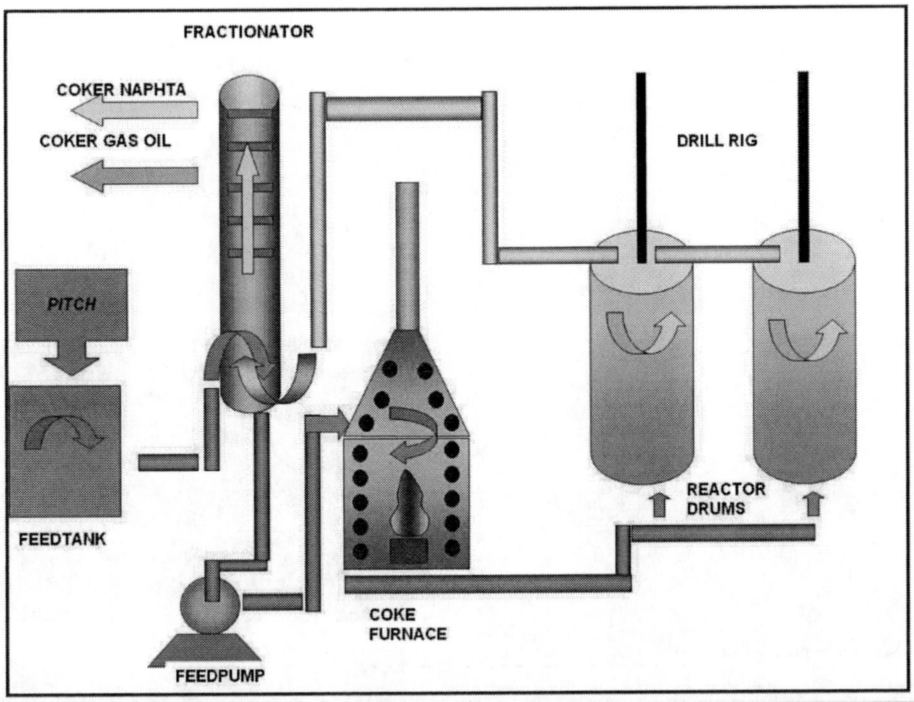




## Calcination

sasol frontiers

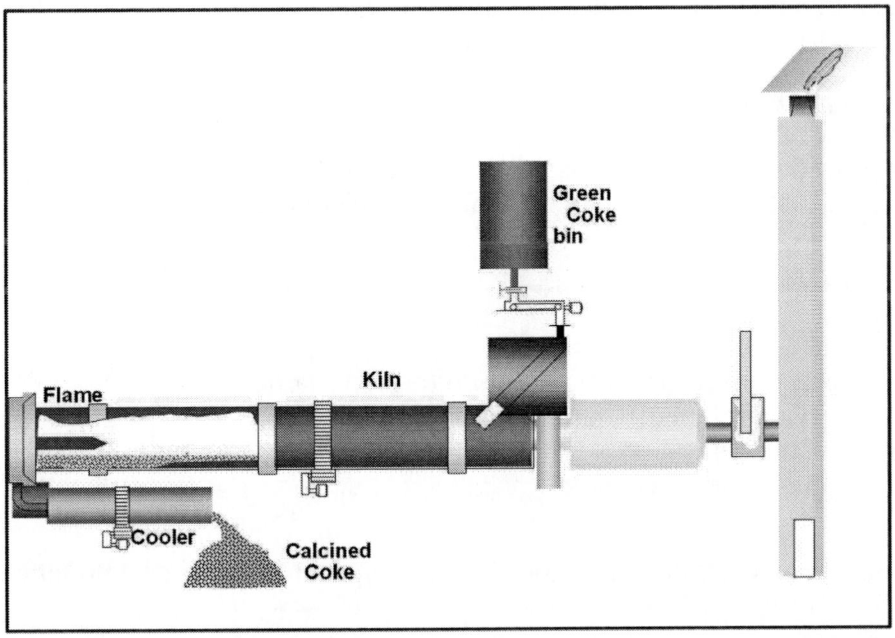

## Electrodes in the steel industry

sasol frontiers

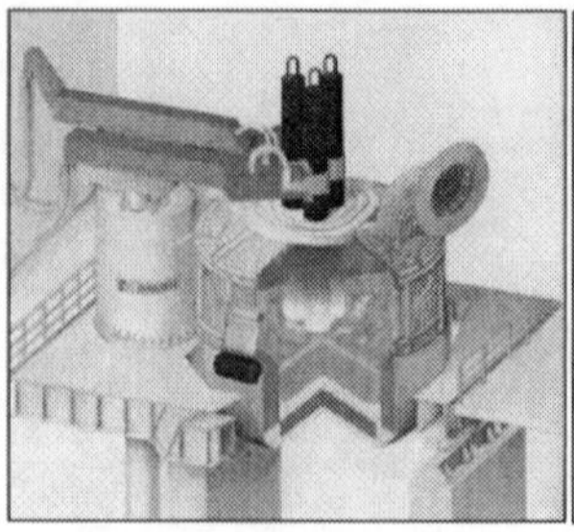

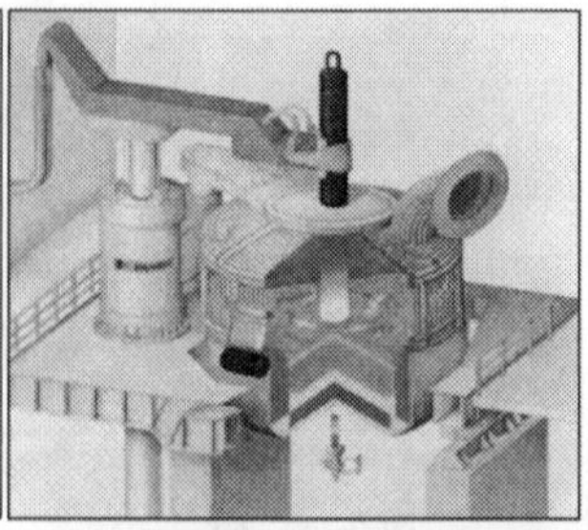

## Needle coke overview

sasol frontiers

- Requirements for a needle coke precursor:
  - High aromaticity
  - Low ash content
  - Low concentration of stable heteroatoms (N,S)
  - Low asphaltenes content (dominating microstructure)
- Requirements for needle coke:
  - Highly anisotropic microstructure (preferred orientation)
  - Low Coefficient of Thermal Expansion (CTE)
  - Low "puffing" ratio
  - High calcined coke yield
- Given the limitations with respect to feed of coal-tar pitch or petroleum decant oil, a new feedstock would prove extremely beneficial

## Introduction

- The Sasol Synthol reaction produces hydrocarbons from synthesis gas (hydrogen and carbon monoxide); the distillation of which produces a heavy residue product namely Waxy Oil
- Waxy Oil is predominantly composed of long normal and multi-alkylated alkanes making it a reactive carbonisation precursor ×
- Waxy Oil is periodically affected by higher entrained catalyst concentration building up during delayed coking and calcination causing multiphase graphitisation ×
- Waxy Oil has a comparatively low green coke yield ×
- Waxy Oil has a comparatively high oxygenate content ×
- As Waxy Oil is devoid of other stable heteroatom species, the calcined coke contains negatable nitrogen and sulphur contents ✓



## Microstructure and XRD of calcined Waxy Oil coke

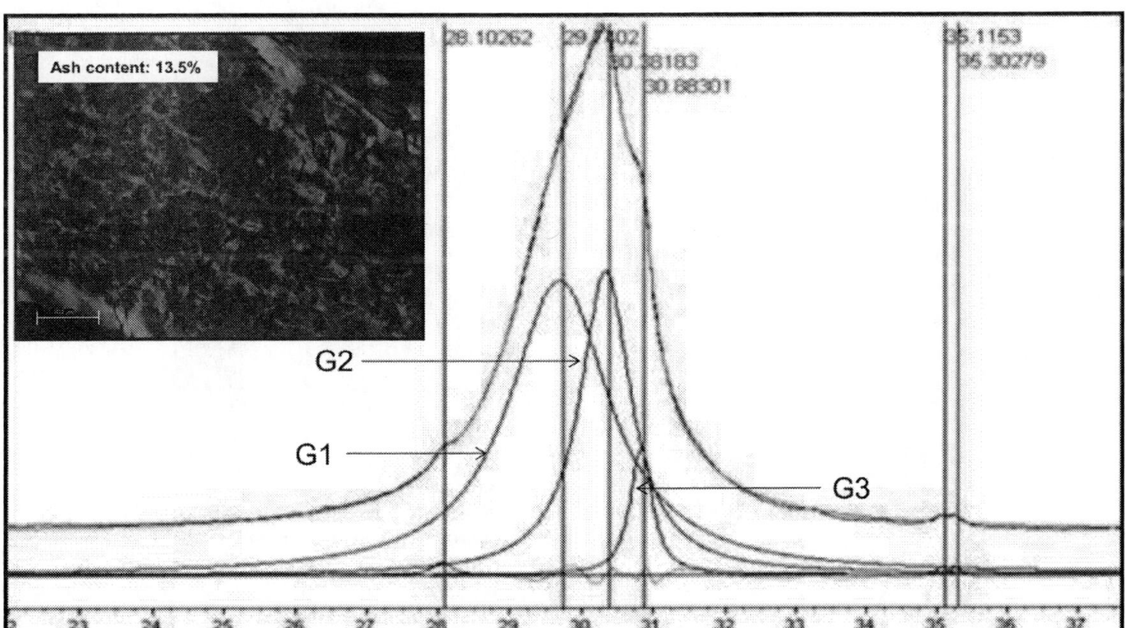

# Experimental protocol

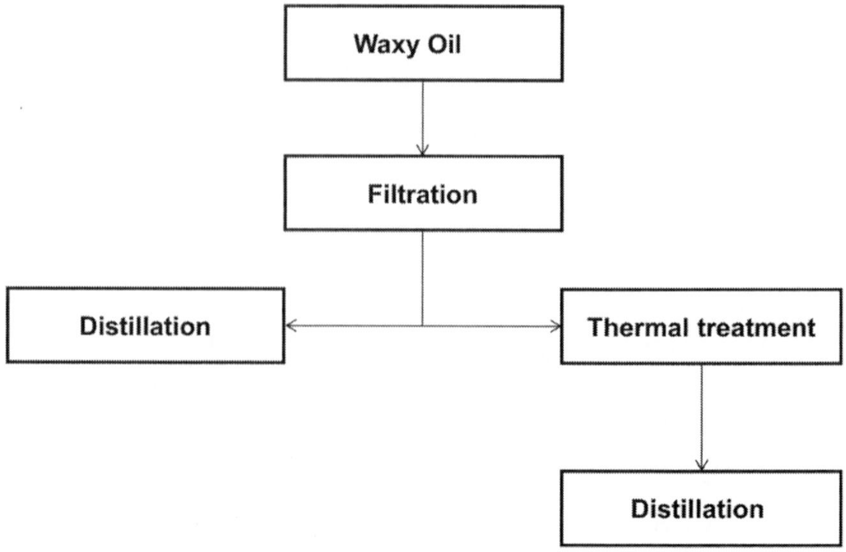


Maintaining momentum 11

# Composition of Waxy Oil

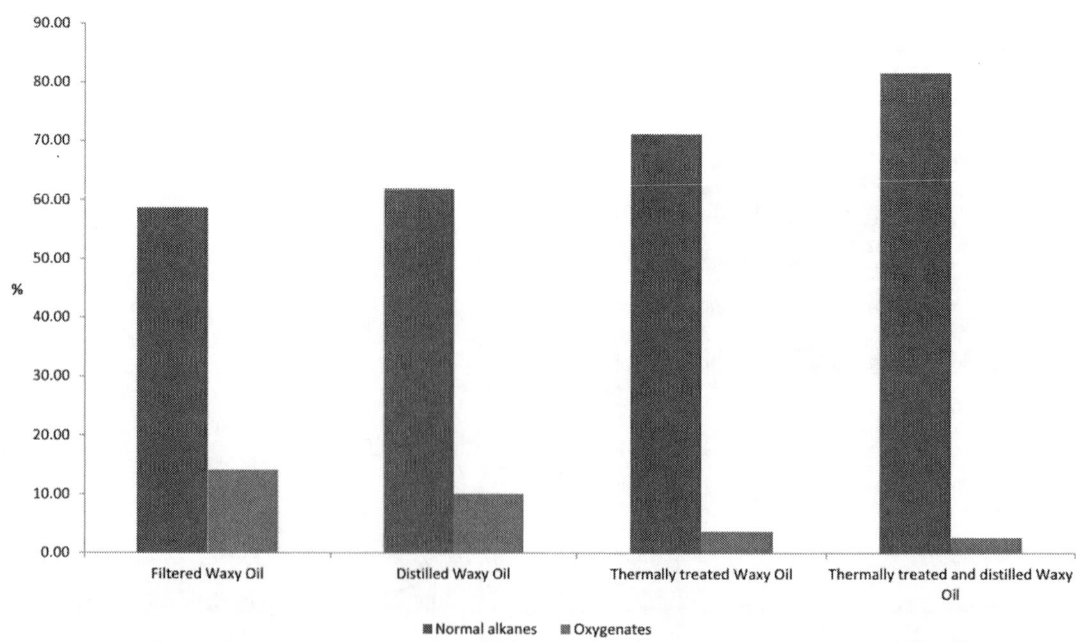


Maintaining momentum 12

# Influence of Waxy Oil modification

sasol frontiers

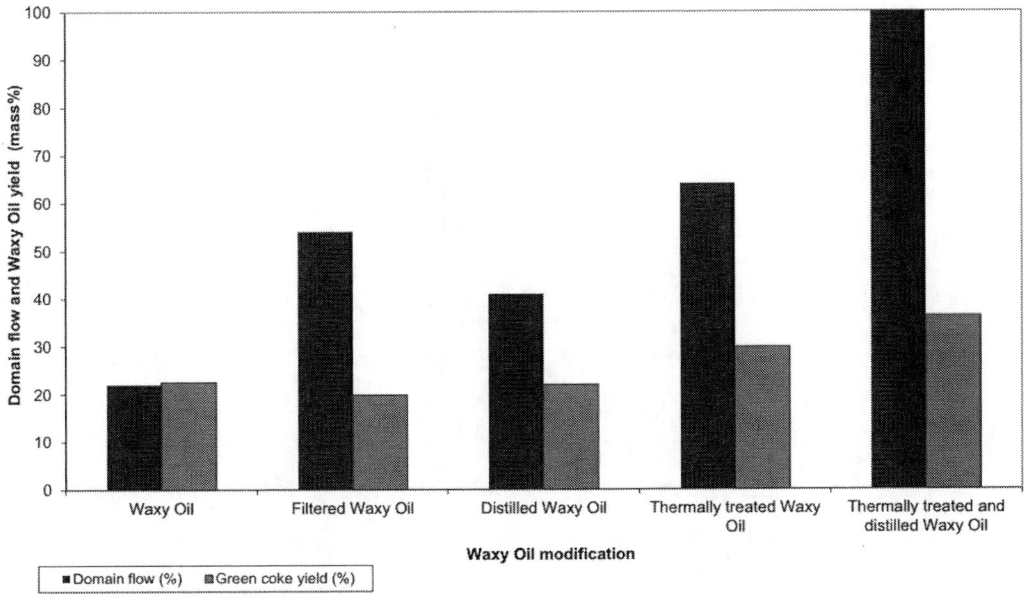




# Proposed carbonisation mechanism of Waxy Oil

sasol frontiers

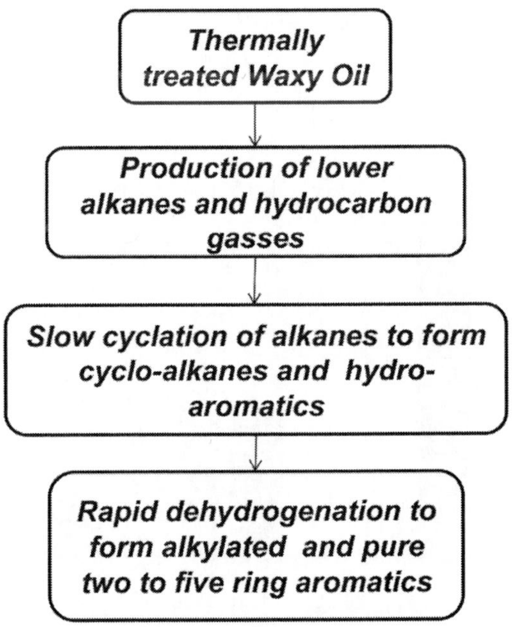

## The influence of hydro-aromatics during carbonisation

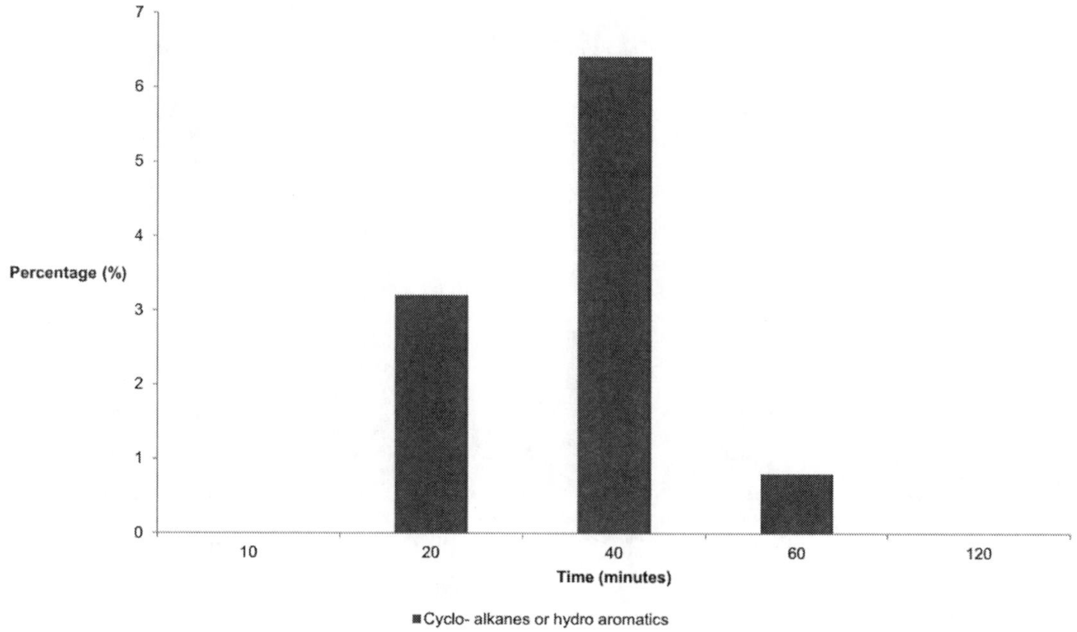


Maintaining momentum 15

## Variation of the organic composition of Waxy Oil as a function of carbonisation duration

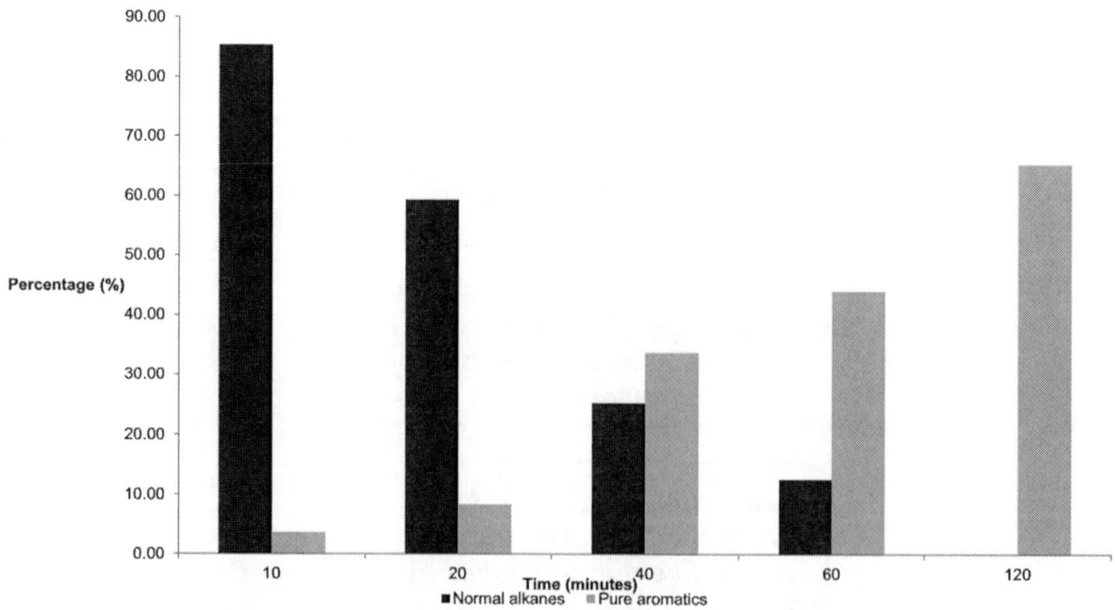


Maintaining momentum 16

## Incipient mesophase growth from isotropic pitch

sasol frontiers

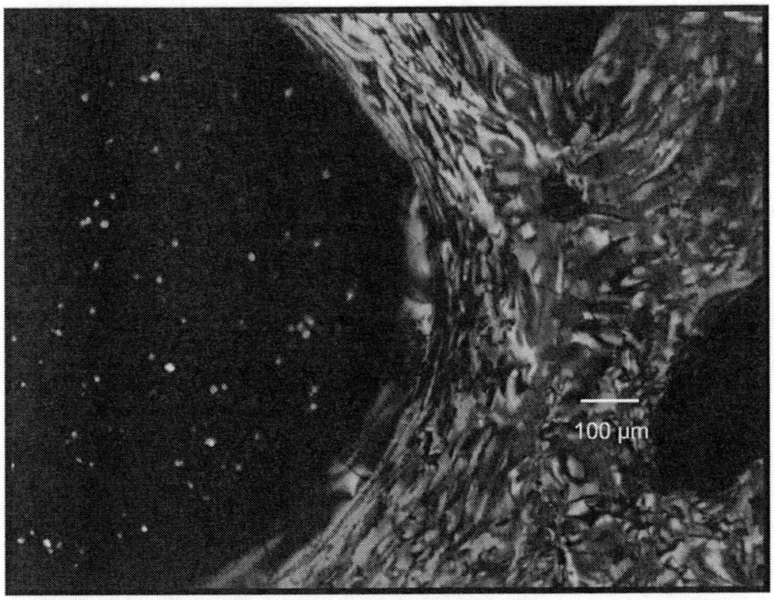




## Mesophase coalescence

sasol frontiers

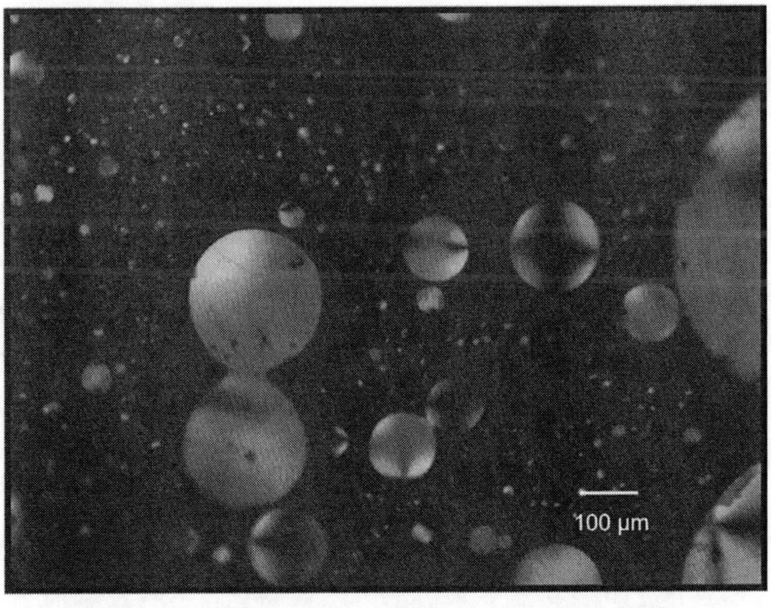

## Mesophase coalescence with carbon microstructure

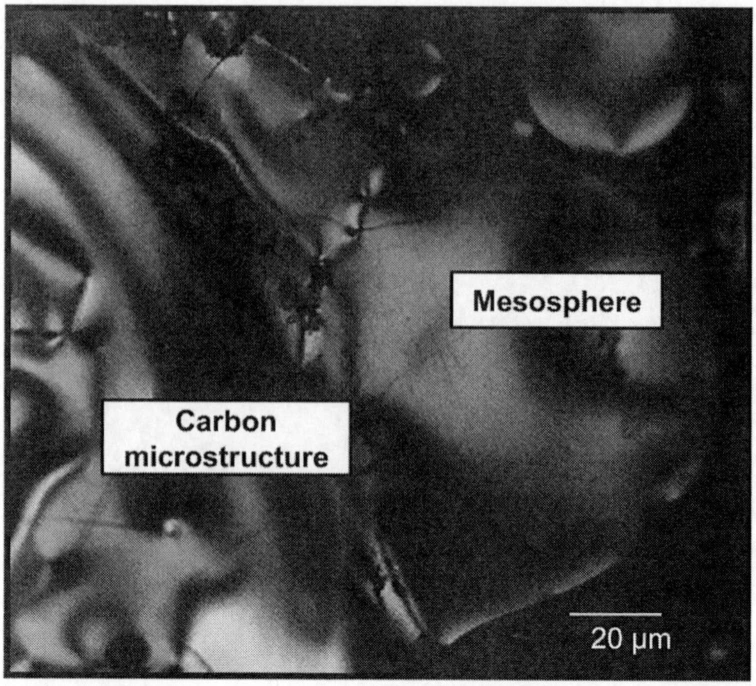




## Microstructure of Waxy Oil coke

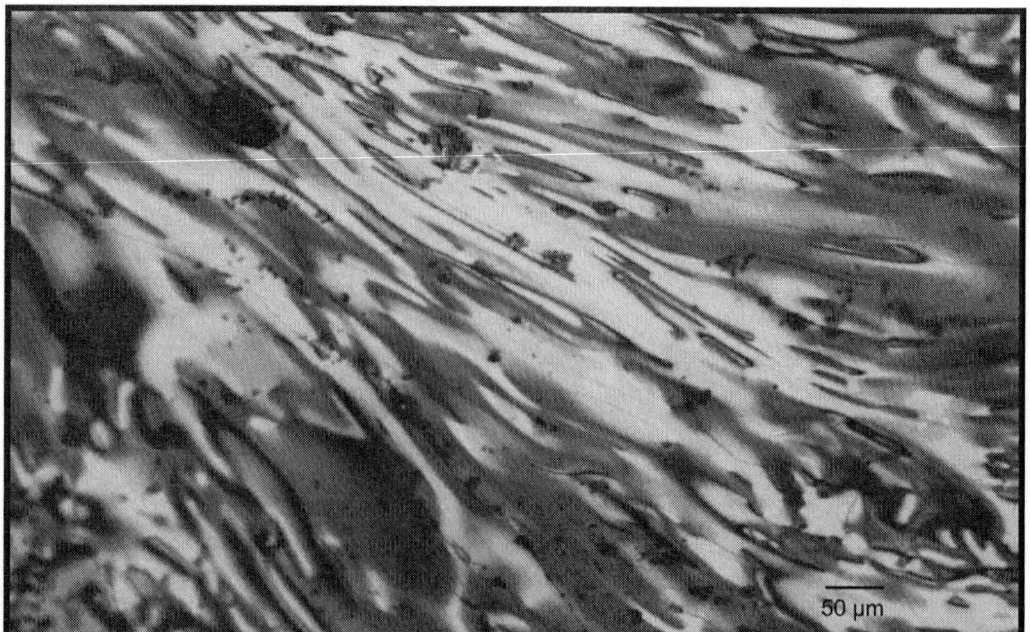

## Conclusions

sasol
frontiers

- The catalyst concentration in Waxy Oil is detrimental to the characteristics of the coke in terms of the ash content, oxidative polymerisation, catalytic graphitisation, carbon dioxide reactivity, microstructure and air reactivity

- Thermal treatment followed by distillation is capable of substantially increasing the concentration of thermally stable normal alkanes while increasing the fraction of pre-carbonisation distillates

- "Static" carbonisation provides a simple method to consolidate the molecular composition of modified Waxy Oils with the degree of anisotropy of the green coke from the bottom of a longitudinal coke section

- The carbonisation mechanism of thermally treated Waxy Oil involves initial production of lighter hydrocarbon gasses and lower alkanes. This is followed by a slow cyclation and rapid dehydrogenation step to form alkylated and pure aromatics

Maintaining momentum 21

## Acknowledgements

sasol
frontiers

- Sasol Performance Chemicals

- University of Pretoria

- This work is based upon research supported by the South African Research Chairs Initiative of the Department of Science and Technology and the National Research Foundation. Any opinion, findings and conclusions or recommendations expressed in this material are those of the authors and therefore the NRF and DST do not accept any liability with regard thereto

- INSTITUTO NACIONAL DEL CARBON (INCAR)

Maintaining momentum 22

## References and bibliography


1. Mochida, I; Qing, Y & Korai, Y 1990, 'A study of the carbonization of ethylene tar pitch and needle coke formation', *Fuel*, vol. 69, pp. 667-671
2. Jenkins, R; Nandi, S & Walker, P 1973, 'Reactivity of heat-treated coals in air at 500 °C', *Fuel*, vol. 52, pp. 288-293
3. Walker, P; Matsumoto, S; Hanzawa, T; Muira, T & Ismail, M 1983, 'Catalysis of gasification of coal-derived cokes and chars', *Fuel*, vol. 62, pp. 140-149
4. Wang, Y; Korai, Y; Mochida, I; Nagayama, K; Hatano, H & Fukuda, F 2001, 'Modification of synthetic mesophase pitch with iron oxide, $Fe_2O_3$', *Carbon*, vol. 39, pp.1627-1634
5. Obara, T; Yokono, T; Sanada, Y & Marsh, H 1985, 'Carbonization behaviour of pitch in the presence of inert material', *Fuel*, vol. 64, no. 7, pp. 995- 998
6. Guillen, M; Dominguez, A; Iglesias, M; Fuente, E & Blanco, CB 1996, 'Analysis of coal-tar pitch: relations between thermal behavior and composition', *Fuel*, vol. 75, no. 9, pp. 1101-1107
7. Domine, F; Bounaceur, R; Scacchi, G; Marquaire, PM; Dessort, D; Pradier, B & Brevart, O 2002, 'Up to what temperature is petroleum stable? New insights from a 5200free radical reactions', Organic Geochemistry, vol. 33, pp.1487-1499




## Forward-looking statements

Forward-looking statements: Sasol may, in this document, make certain statements that are not historical facts and relate to analyses and other information which are based on forecasts of future results and estimates of amounts not yet determinable. These statements may also relate to our future prospects, developments and business strategies. Examples of such forward-looking statements include, but are not limited to, statements regarding exchange rate fluctuations, volume growth, increases in market share, total shareholder return and cost reductions. Words such as "believe", "anticipate", "expect", "intend", "seek", "will", "plan", "could", "may", "endeavour" and "project" and similar expressions are intended to identify such forward-looking statements, but are not the exclusive means of identifying such statements. By their very nature, forward-looking statements involve inherent risks and uncertainties, both general and specific, and there are risks that the predictions, forecasts, projections and other forward-looking statements will not be achieved. If one or more of these risks materialise, or should underlying assumptions prove incorrect, our actual results may differ materially from those anticipated. You should understand that a number of important factors could cause actual results to differ materially from the plans, objectives, expectations, estimates and intentions expressed in such forward-looking statements. These factors are discussed more fully in our most recent annual report under the Securities Exchange Act of 1934 on Form 20-F filed on 9 October 2015 and in other filings with the United States Securities and Exchange Commission. The list of factors discussed therein is not exhaustive; when relying on forward-looking statements to make investment decisions, you should carefully consider both these factors and other uncertainties and events. Forward-looking statements apply only as of the date on which they are made, and we do not undertake any obligation to update or revise any of them, whether as a result of new information, future events or otherwise.

# The conversion of FT olefins: a multiphase reactor simulation study


Klaus Möller, Bongani Mthombeni-Möller,
Cyril Knottenbelt, Masikana Mdleleni



Acknowledgements: PetroSA and University of Cape Town

## Introduction

### Outline

- Background on COD
- Objectives
- Kinetics
- Thermo-chemical properties
- Thermo-physical properties
- Reactor model
- Results
  - Chemical equilibrium
  - Effect of lumping
  - Thermodynamics
  - Multi-phase operation
- Conclusions

**Gas-to-liquids process**

## Introduction
## Summary of olefin to distillate technology

### COD and MOGD

- 200-300°C, 30-60atm
- fixed bed/trickle bed, adiabatic
- ZSM5 type catalysts
- Olefinic feeds, typically C3–C6
- Products in the gasoline/diesel range depending on conditions
- mostly olefins but also cyclic olefins, paraffins and aromatics depending on conditions.

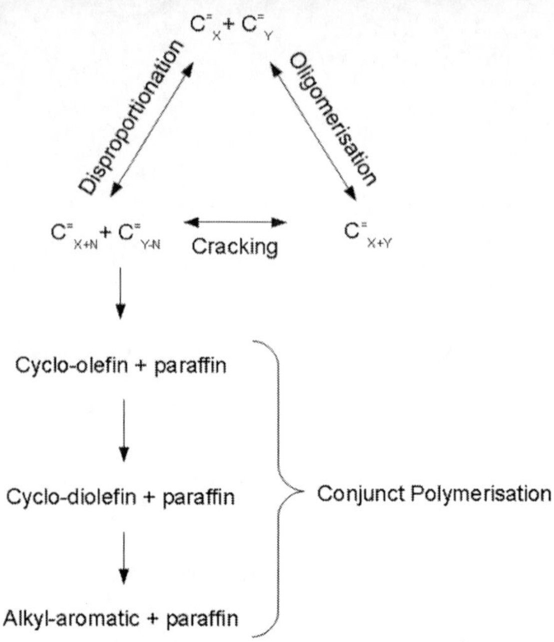


Process Modelling & Optimisation Group

UNIVERSITY OF CAPE TOWN
IYUNIVESITHI YASEKAPA • UNIVERSITEIT VAN KAAPSTAD

## Introduction
## Process modelling and kinetics

### State of the art

- Quann et al., 1988 and Quann & Krambeck,1991, "gas phase" kinetics, thermodynamics and reactor models based on CN (ZSM5)
- Paratello et al., 1999, Kinetics, propene, fugacity based (Silica-Alumina)
- Shahrouzi et al., 2008, Single events, 4 isomer groups, limited to C12.

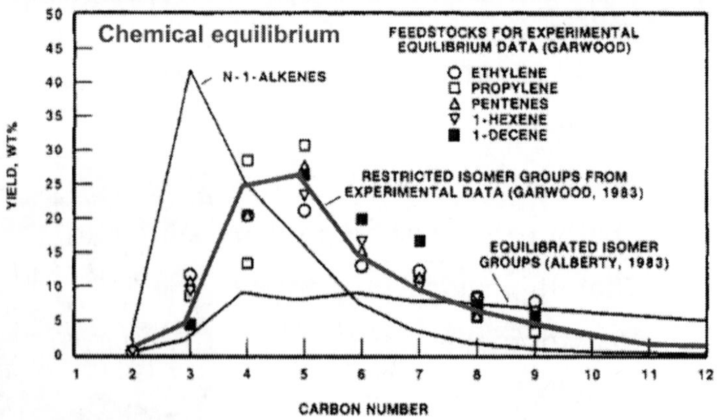


Kinetic data

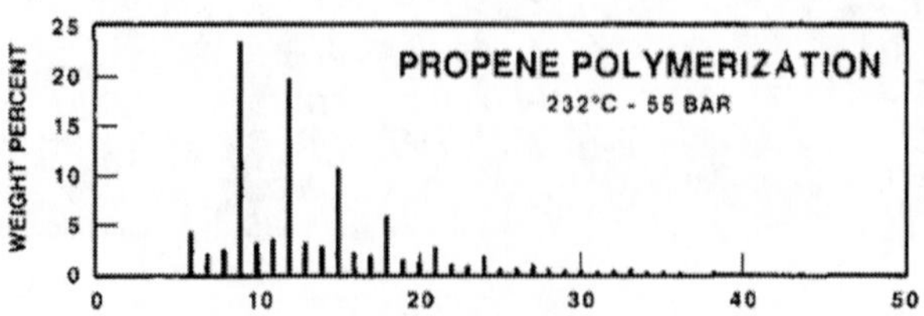


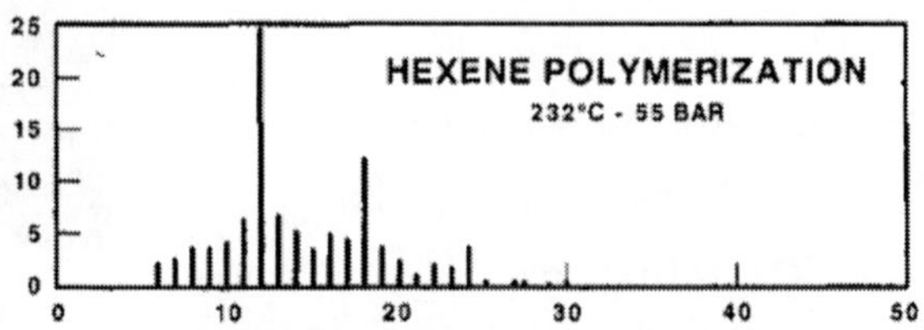


Process Modelling & Optimisation Group

UNIVERSITY OF CAPE TOWN
IYUNIVESITHI YASEKAPA • UNIVERSITEIT VAN KAAPSTAD

# Objectives

**Develop an adiabatic reactor model that can accept any olefinic feed stock and predict reactivity and product distribution (cetane and octane?)**

**Kinetics:**

- **Oligomerisation, cracking, isomerisation**
- **Hybrid, between single events and lumped modelling**
- **Carbon numbers up to C40, 4 distinct structural isomers**
- **Variation of reactivity with CN embedded in approach**

**Thermodynamics:**

- **Account for chemical and phase equilibrium**

**Reactor:**

- **Adiabatic fixed bed**

Table 1. Numbers $N_I$ of alkene isomers including stereoisomers

| $N_c$ | $N_I$ |
|---|---|
| 2 | 1 |
| 3 | 1 |
| 4 | 4 |
| 5 | 6 |
| 6 | 18 |
| 7 | 42 |
| 8 | 118 |
| 9 | 314 |
| 10 | 895 |
| 15 | 185,310 |
| 20 | 46,244,031 |
| 25 | $12.70 \times 10^9$ |



# Kinetic model development "Elementary" reaction steps

**Oligomerisation-cracking:**

- **oligmerisation increases degree of branching by 1 (limit 3)**
- **species form sequencially**

$$C_3^= + C_6^= \rightleftarrows 1C_9^=$$
$$C_3^= + 1C_6^= \rightleftarrows 2C_9^=$$
$$C_3^= + 2C_6^= \rightleftarrows 3C_9^=$$

$C_3^= + C_6^= \rightleftarrows 1C_9^=$ followed by $1C_9^= \rightleftarrows 2C_9^=$

$C_6^= \rightleftarrows 1C_6^=$ followed by $C_3^= + 1C_6^= \rightleftarrows 2C_9^=$.

**Isomerisation:**

- **increases/decreases the degree of branching**

$$C_9^= \rightleftarrows 1C_9^= \rightleftarrows 2C_9^= \rightleftarrows 3C_9^=$$

**The rules for ZSM5:**

- **every second carbon atom, maximum 3 methyl branches**

**Rate constants:**

- **Two for now, k1-oligomerisation, k2-isomerisation**

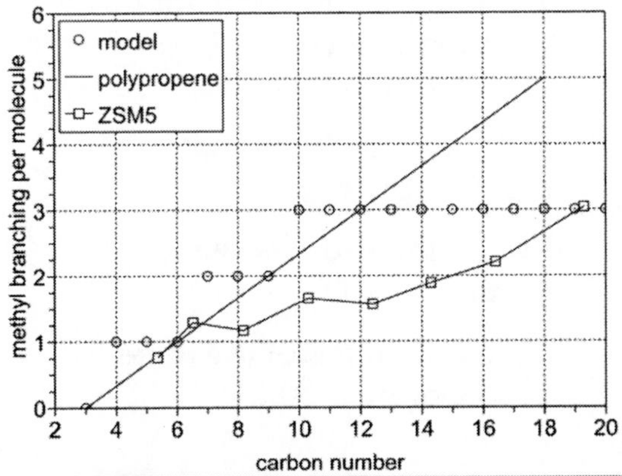

## Kinetic model development
## Chain truncation dependence

**Why C40 to predict diesel at C16**

- **CN truncation causes wrong distributions**

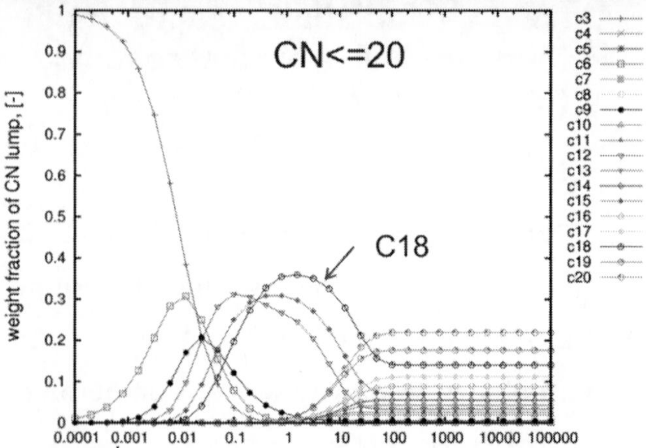


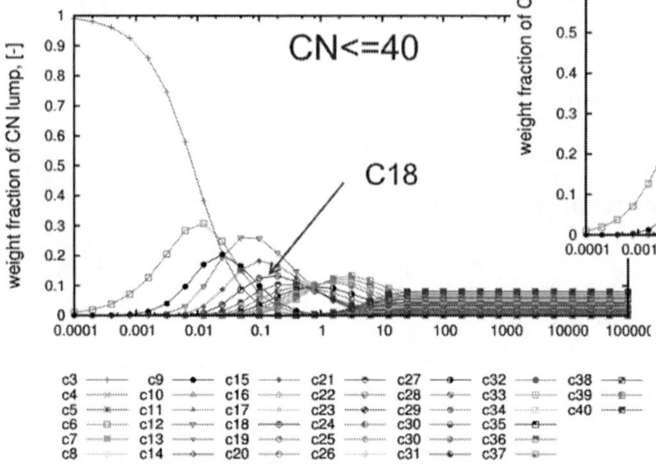


**To predict up to C20 need simulation up to C40 and need to ensure that C40<0.1%**



## Chemical Thermodynamics
## Ideal gas thermo-chemical constants

**ΔGf(T), ΔHf(T), ΔSf(T), Cp(T) up to C40**

- **Bensons Method combined with Alberty's approach for equilibrated isomer groups**
- **Consider all possible mono-branched alkane/olefin structures to C12/C13**
- **4 isomers with alternate methyl spacing**
- **Prediction reduced isomer groups ΔGf(T)**
- **Consequences of the ideal gas reference state**

**e.g. C10**

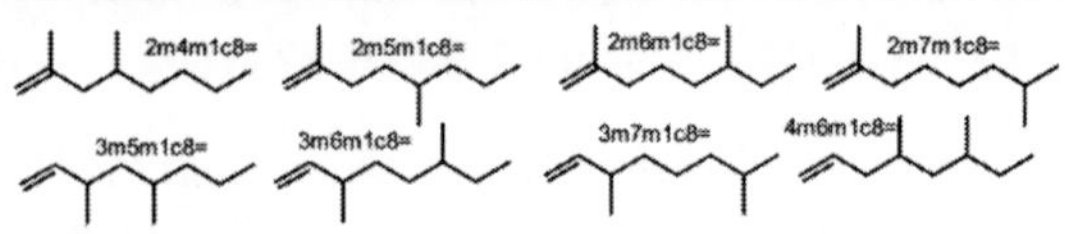


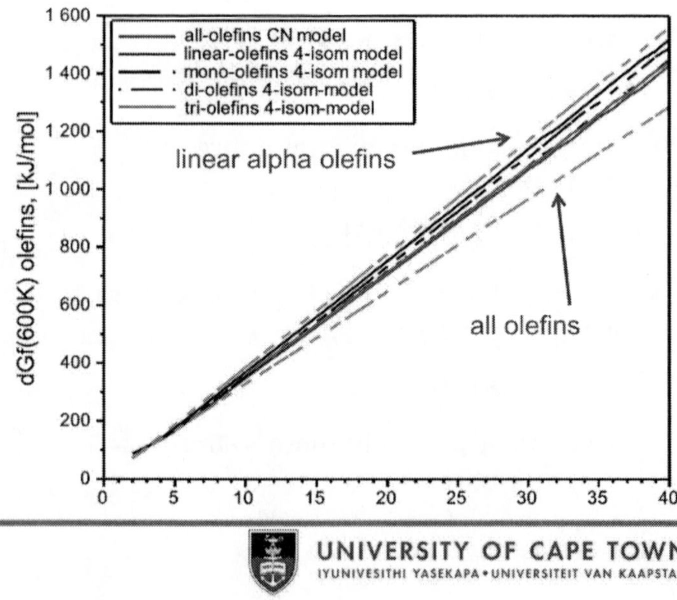

# Physical thermodynamics thermo-physical property relations

Tc

Pc

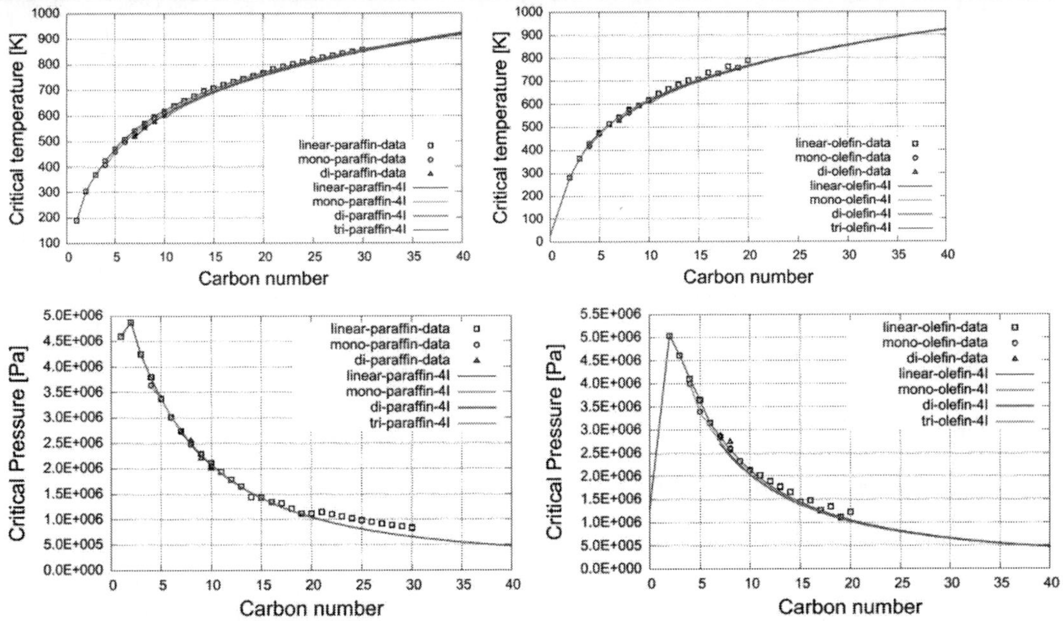


(a) paraffins (b) olefins



# Development of the reactor model Ref state: ideal gas

## Rate equations

- **activity** $\boxed{A_i + A_j \rightleftarrows A_{i+j},}$ $r_i^{olig} = k_{olig}\left(a_i a_j - \frac{a_{i+j}}{K_{i,j}^{olig}}\right)$ $\boxed{A_i \rightleftarrows A_j,}$ $r_i^{isom} = k_{isom}\left(a_i - \frac{a_j}{K_{i,j}^{isom}}\right)$
- **gas phase** $r_i^{olig} = k_{olig}\phi_i^v\phi_j^v P_t^2\left(y_i y_j - \frac{y_{i+j}}{P\frac{\phi_i^v\phi_j^v}{\phi_{i+j}^v}K_{i,j}^{olig}}\right)$ $r_i^{isom} = k_{isom}\phi_i^v P_t\left(y_i - \frac{y_j}{\frac{\phi_i^v}{\phi_j^v}K_{i,j}^{isom}}\right)$
- **liquid phase** $r_i^{olig} = k_{olig}\phi_i^l\phi_j^l P^2\left(x_i x_j - \frac{x_{i+j}}{P\frac{\phi_i^l\phi_j^l}{\phi_{i+j}^l}K_{i,j}^{olig}}\right)$ $r_i^{isom} = k_{isom}\phi_i^l P\left(x_i - \frac{x_j}{\frac{\phi_i^l}{\phi_j^l}K_{i,j}^{isom}}\right)$
- **ideal liquid** $r_i^{olig} = k_{olig}\boxed{p_i^{sat}p_j^{sat}}\left(x_i x_j - \frac{x_{i+j}}{\boxed{\frac{p_i^{sat}p_j^{sat}}{p_{i+j}^{sat}}}K_{i,j}^{olig}}\right)$ $r_i^{isom} = k_{isom}\boxed{p_i^{sat}}\left(x_i - \frac{x_j}{\boxed{\frac{p_i^{sat}}{p_j^{sat}}}K_{i,j}^{isom}}\right)$

## Design equations

- **mass & energy** $\frac{dF_i}{dV} = R_i$ $\left(\sum F_i H_{f,i}\right)_{V_r} - \left(\sum F_i H_{f,i}\right)_{V_{r,0}} + \int_{V_{r,0}}^{V_r} Q_{loss}\, dV = 0$

## Reactor model results: Chemical Equilibrium

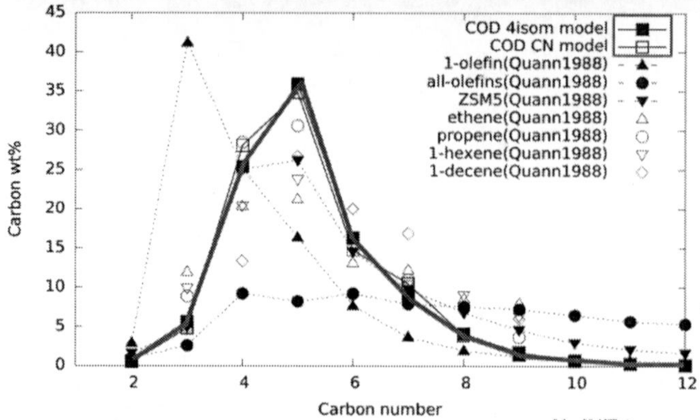


**Prediction of chemical equilibrium based on carbon number lumps and 4-isomer lumps**

**Additional information:**

- **The structural isomer distribution at equilibrium**

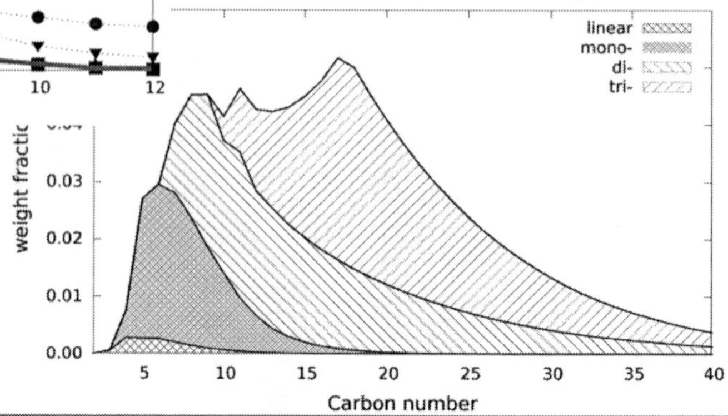


Process Modelling & Optimisation Group

UNIVERSITY OF CAPE TOWN
IYUNIVESITHI YASEKAPA • UNIVERSITEIT VAN KAAPSTAD

## Reactor model results: Degree of branching

**Since the structure is known, simply count the methyl branches**

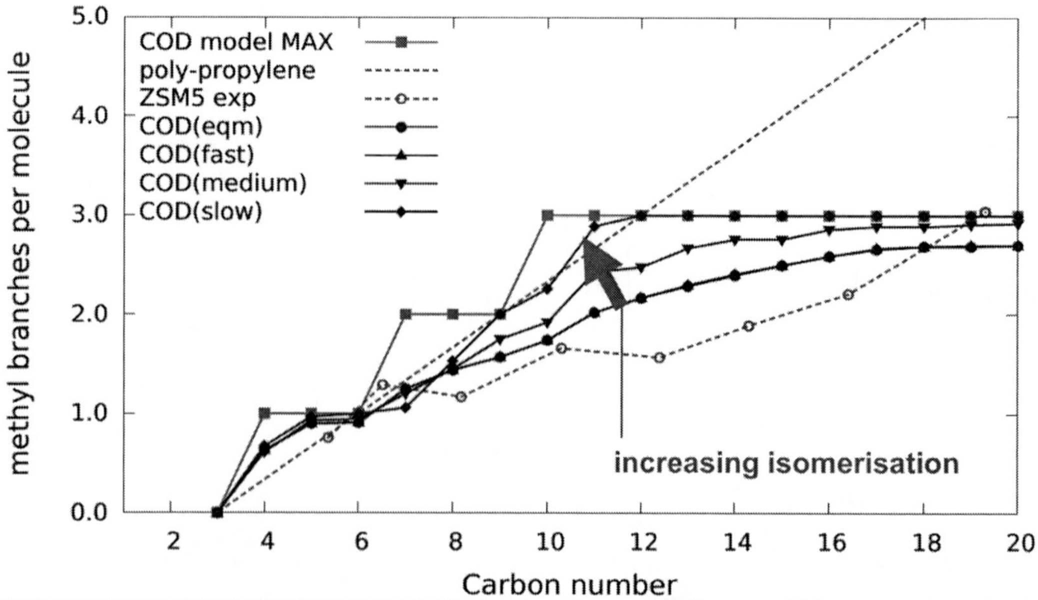


Process Modelling & Optimisation Group

UNIVERSITY OF CAPE TOWN
IYUNIVESITHI YASEKAPA • UNIVERSITEIT VAN KAAPSTAD

## Reactor model results: influence of chemical lumping (feed:C3=)

| group name | LPG | Gasoline (G) | Diesel (D) | Heavies |
|---|---|---|---|---|
| olefin CN range | C2-C4 | C5-C9 | C10-C20 | C21-C40 |

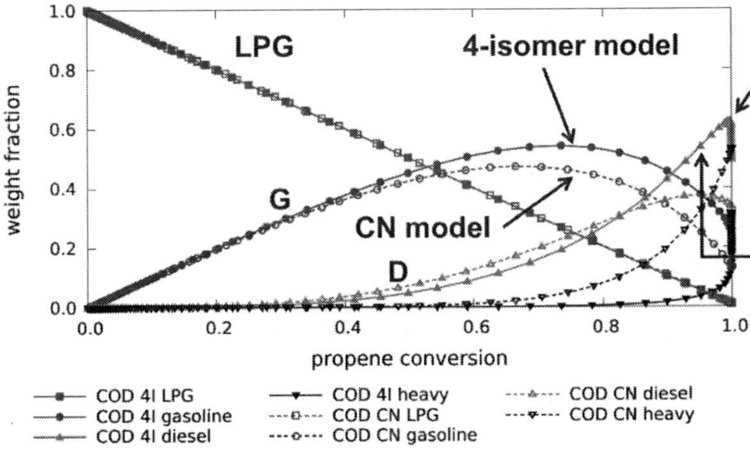


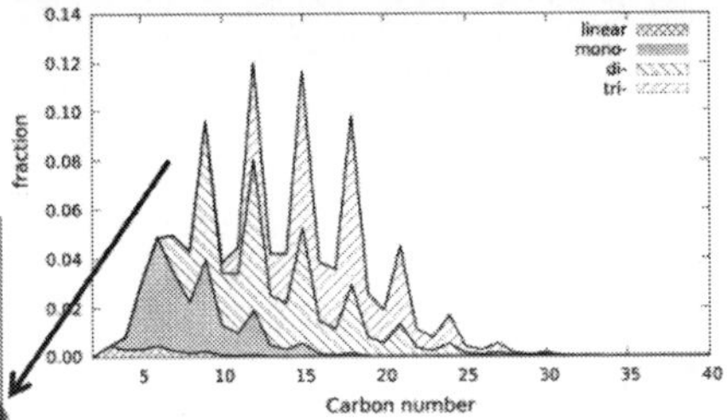


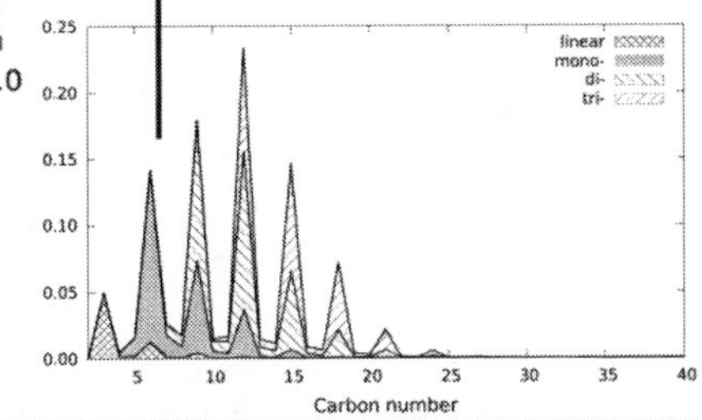


**note the different selectivity of the CN model**



## Reactor model results: gas phase vs liquid phase (feed: C3=)

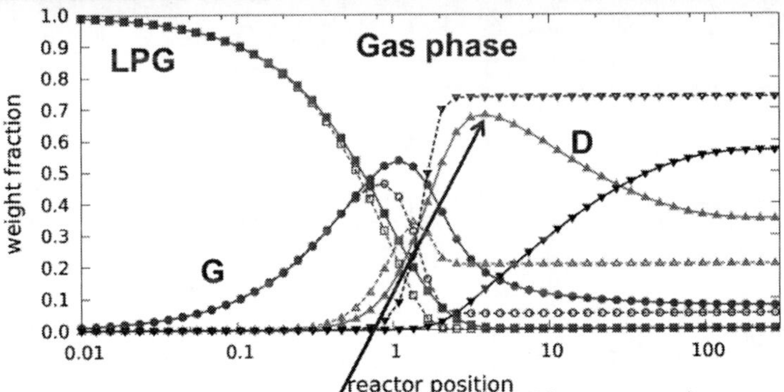


**liquid fugacity coefficients depend strongly on carbon number**

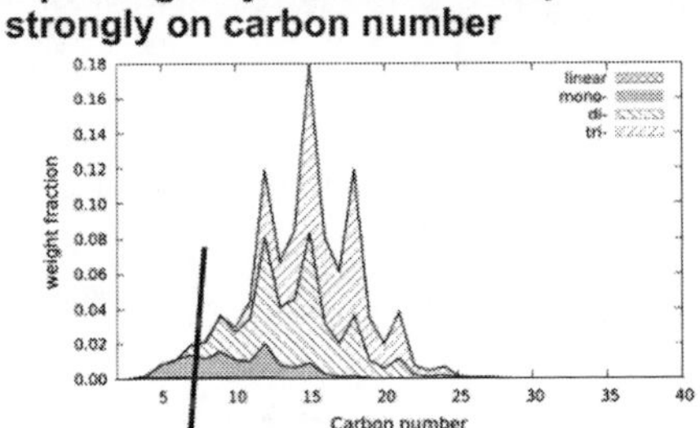


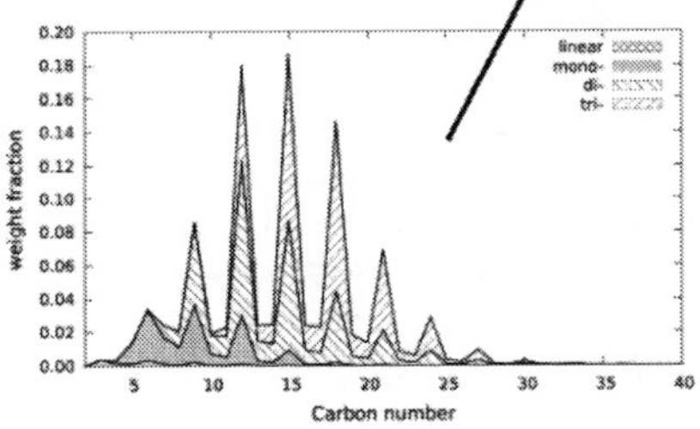


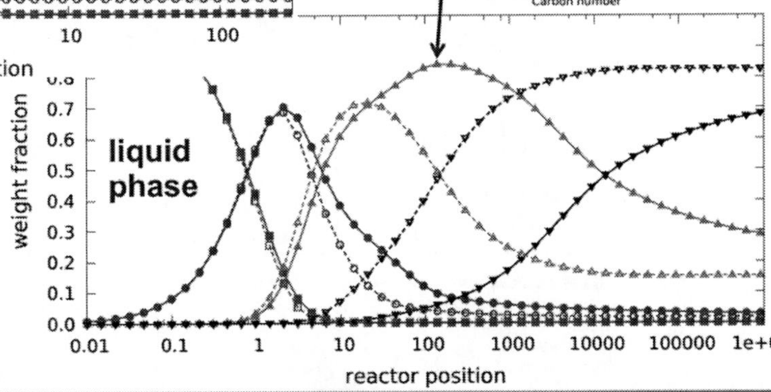

# Reactor model results:
# Adiabatic operation: effect of lumping

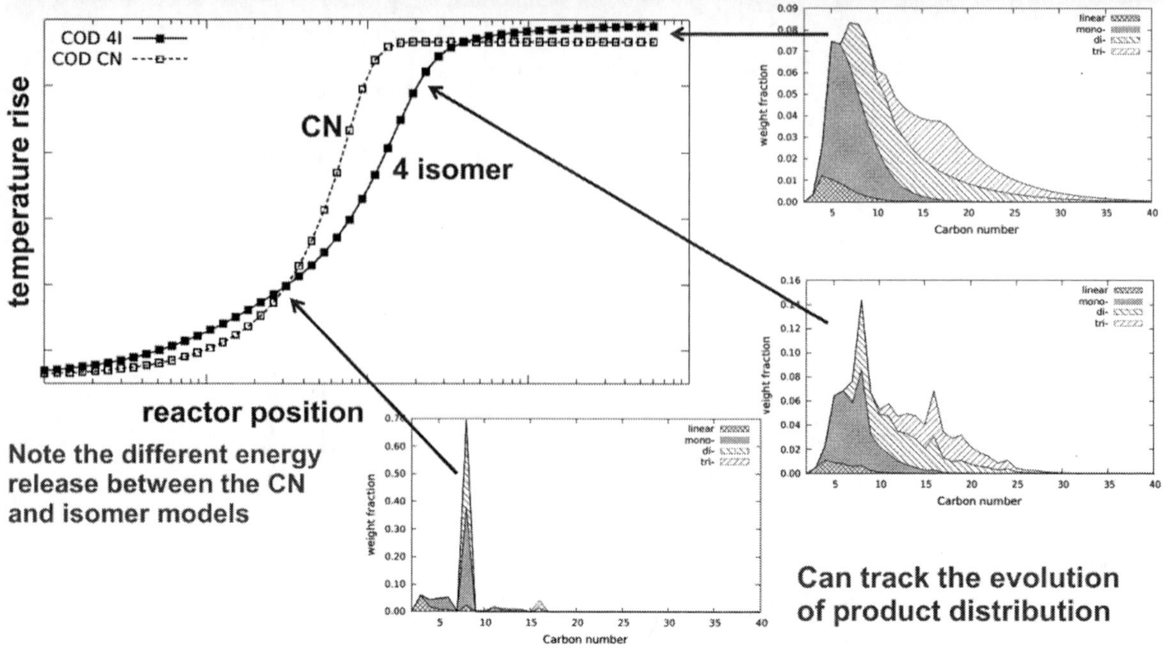


**Note the different energy release between the CN and isomer models**

**Can track the evolution of product distribution**



# Phase thermodynamics and VLE:
# The decision tree

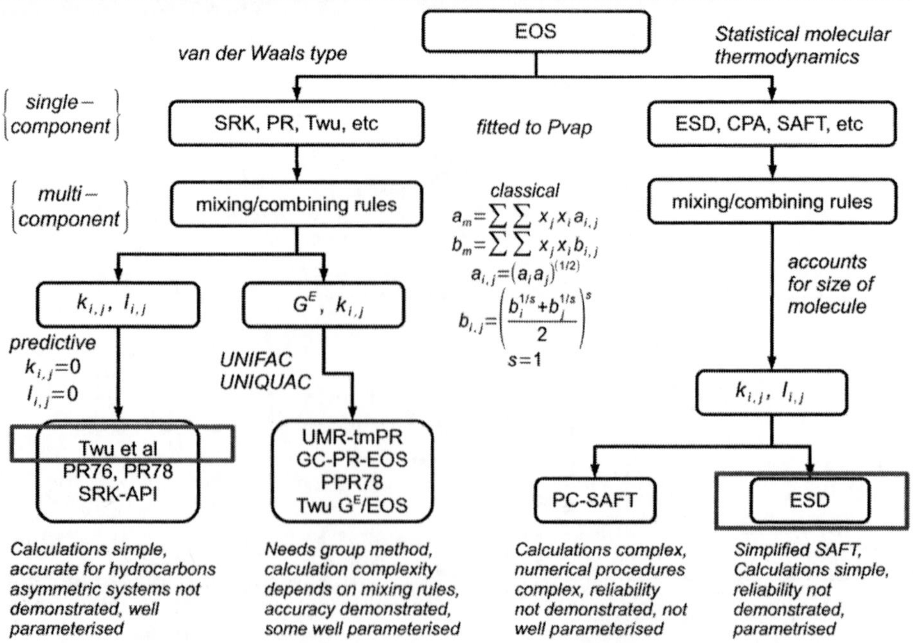

## Phase thermodynamics and VLE: behaviour of Super-critical species

**Comparing alpha functions used in equations of state at a temperature of 523 K**

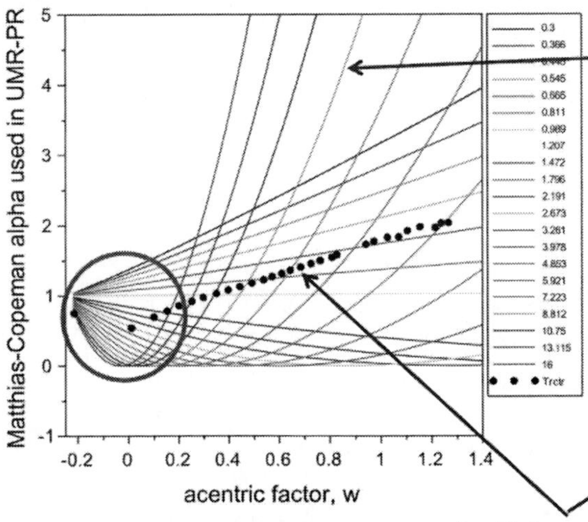


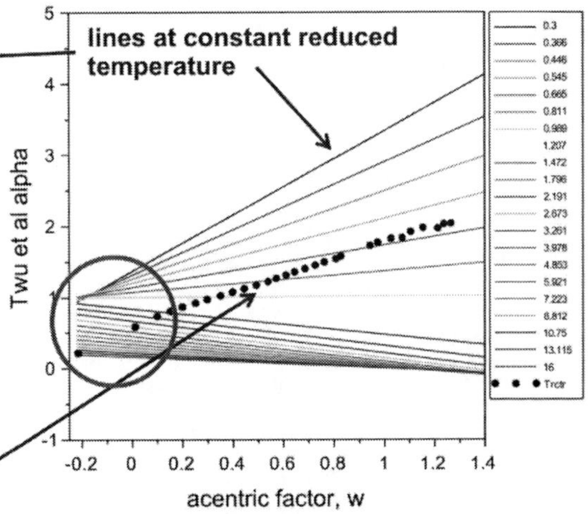


**super-critical data not accounted for (most EOS's)**

- **Alpha values for $H_2$, C1 – C20 at reactor conditions**

**Special effort made to develop alpha functions for supercritical species (Twu et al.)**

Process Modelling & Optimisation Group

UNIVERSITY OF CAPE TOWN
IYUNIVESITHI YASEKAPA • UNIVERSITEIT VAN KAAPSTAD

## Phase thermodynamics and VLE: Elliott, Suresh and Donohue (ESD) EOS

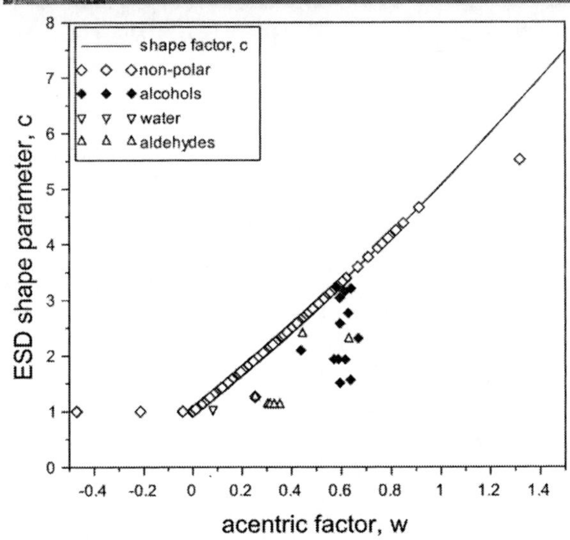


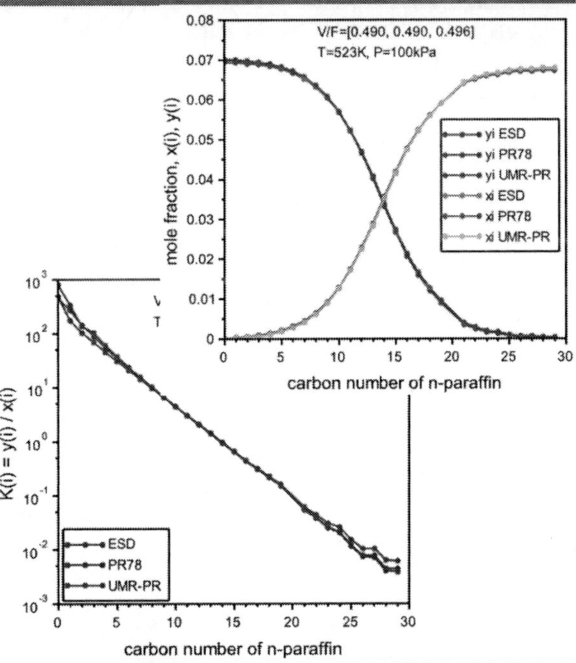


- **Explicitly takes into account the molecular size and polarity**
- **Does not have an alpha function**

Process Modelling & Optimisation Group

UNIVERSITY OF CAPE TOWN
IYUNIVESITHI YASEKAPA • UNIVERSITEIT VAN KAAPSTAD

## Results: Two-phase operation 25atm (feed: C3...C6,C8,H2)

**isothermal** — V/L phase split — **adiabatic**

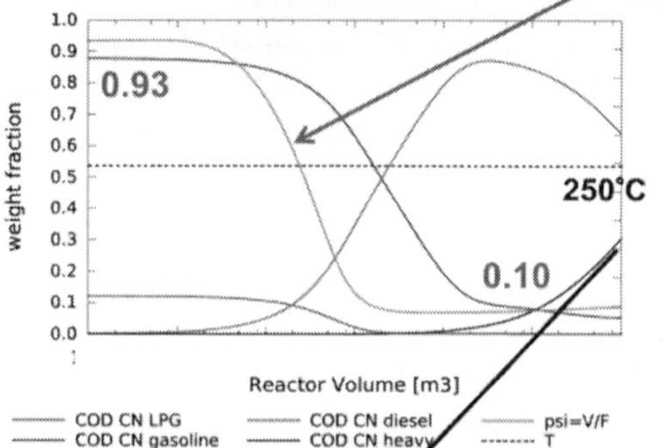


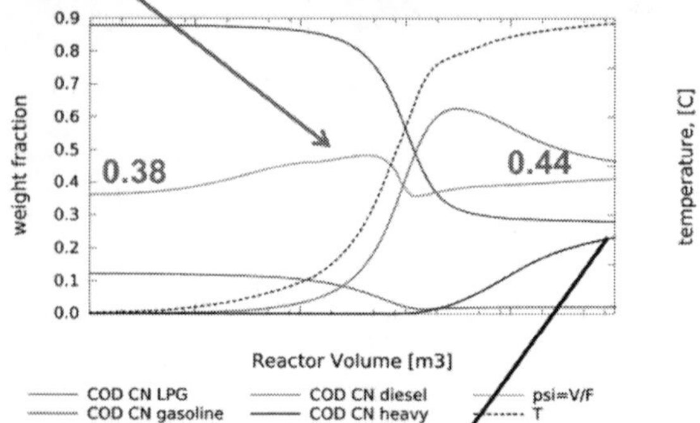


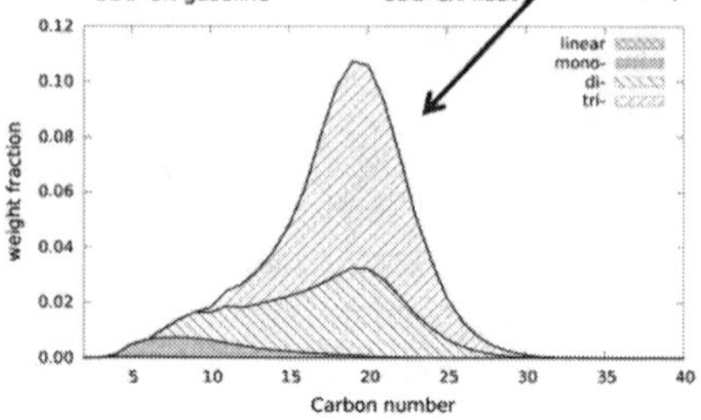


- Able to simulate multiphase operation with 280 species
- Difference in product spectrum due to temperature

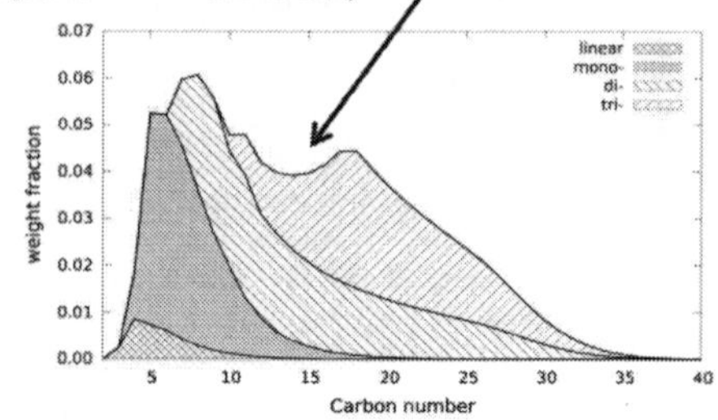


Process Modelling & Optimisation Group

UNIVERSITY OF CAPE TOWN
IYUNIVESITHI YASEKAPA • UNIVERSITEIT VAN KAAPSTAD

## Results: Two-phase operation Liquid phase acceleration (2 atm)

**vapour phase CN model** — V/L phase split — **two phase CN model**

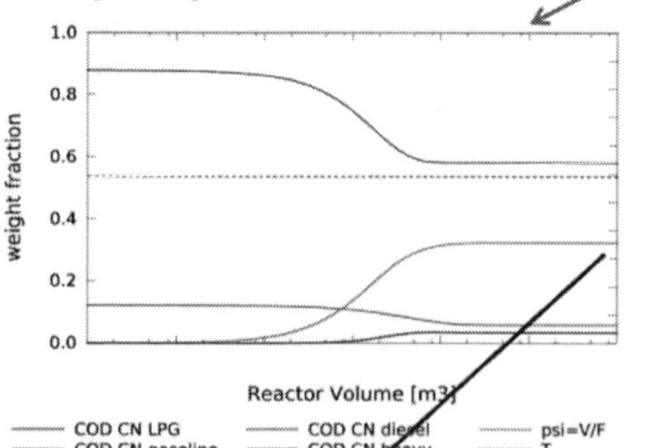


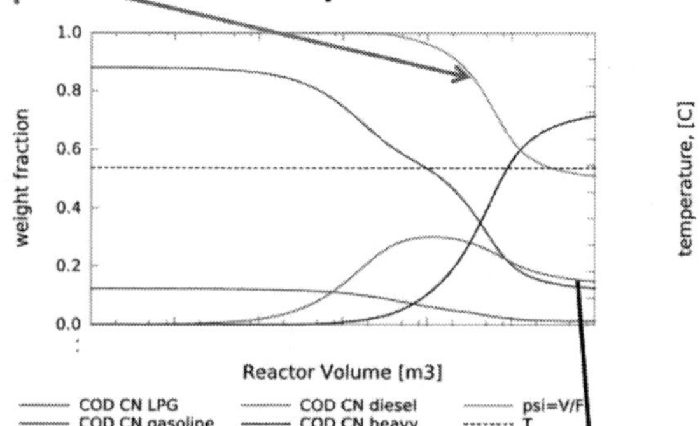


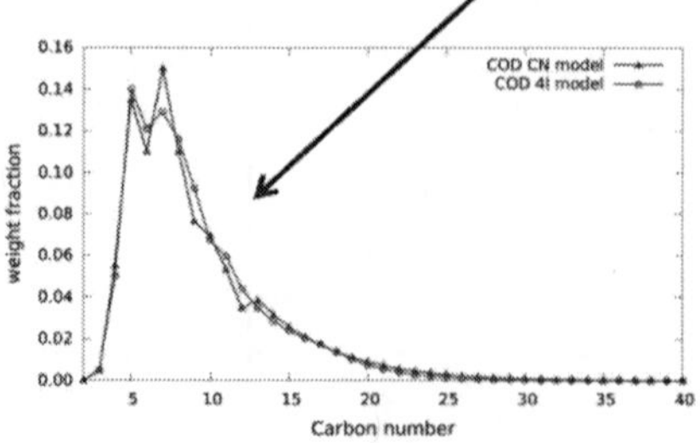


- **Vapour only:** reaction ends at vapour phase equilibrium
- **Two-phase:** formation of liquid phase accelerates reaction to heavies

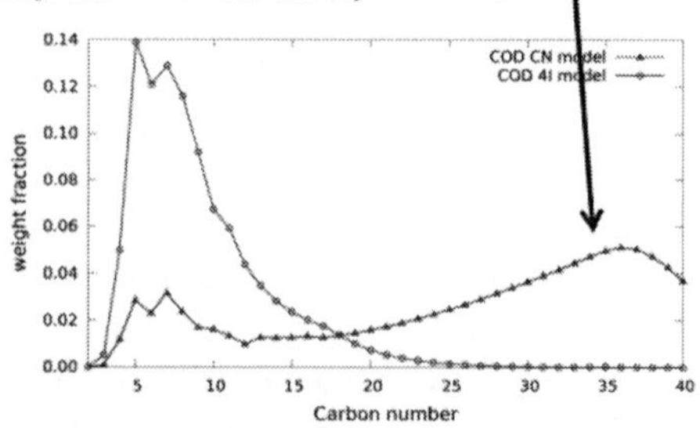


Process Modelling & Optimisation Group

UNIVERSITY OF CAPE TOWN
IYUNIVESITHI YASEKAPA • UNIVERSITEIT VAN KAAPSTAD

## Results: Two-phase operation Phase boundary crossing

**feed: C3=, 50 atm, 250°C**

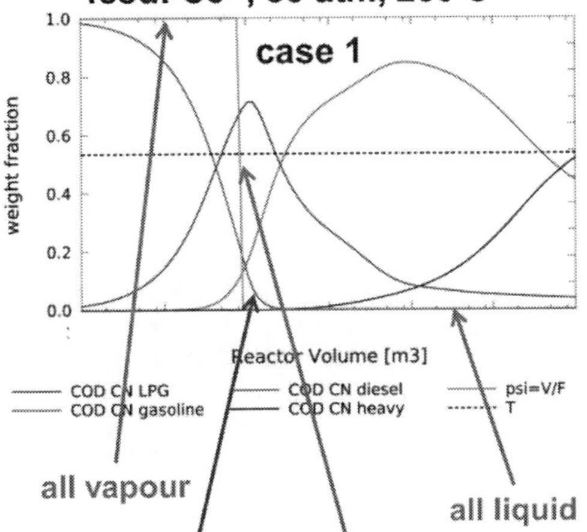


**2 phase region:**

- Insufficient super critical C3= to sustain vapour phase
- phase envelope small

- Diesel formation, liquid only
- Rising temperature produces 2nd phase, presence of H2

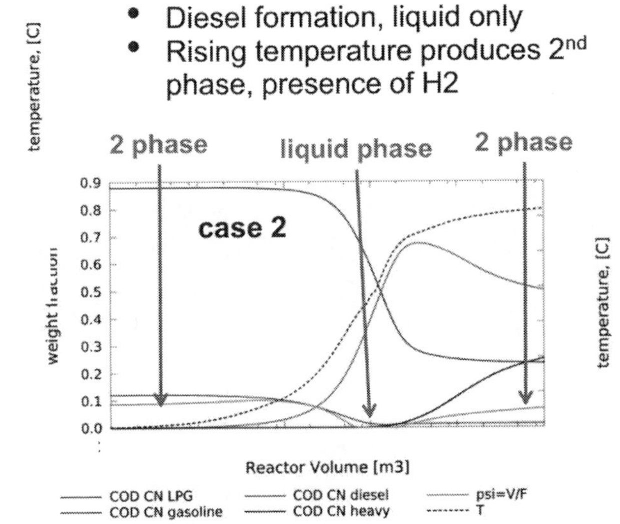


**feed: H2,C3...C6,C8, 40 atm, adiabatic**

Process Modelling & Optimisation Group

UNIVERSITY OF CAPE TOWN
IYUNIVESITHI YASEKAPA • UNIVERSITEIT VAN KAAPSTAD

## Results: Two-phase operation Micro-reactor data – model comparison

**Micro-reactor data**

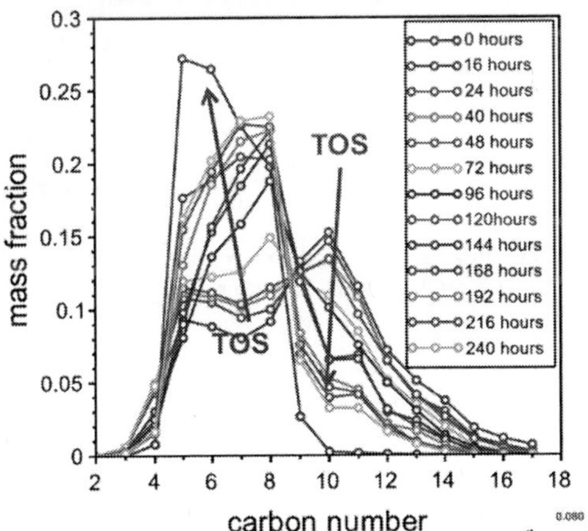


- deactivation with time on stream
- maximum "conversion" limited by WHSV

- **deactivation ≡ reducing reactor size**
- **comparison promising**

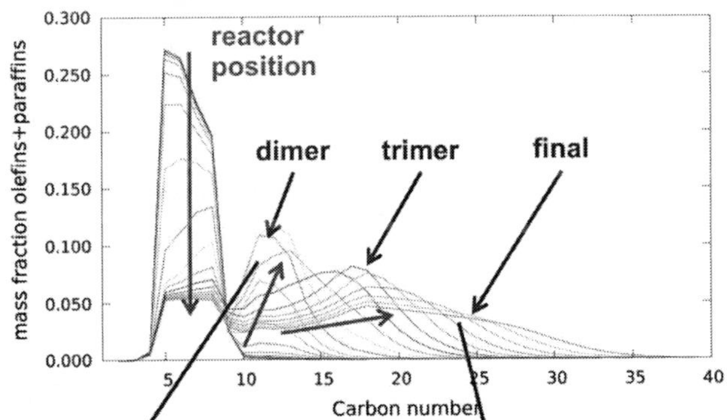


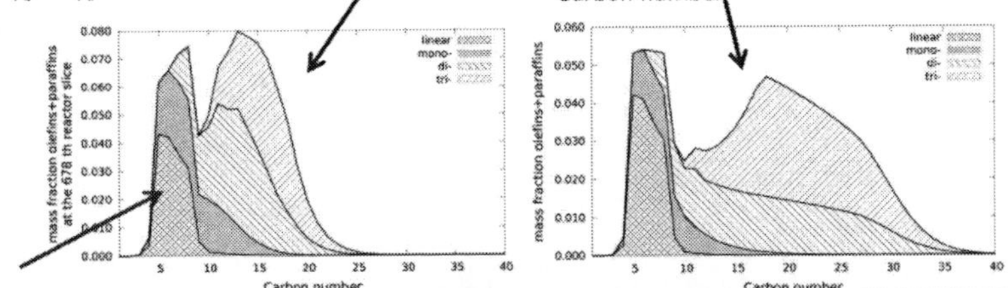


**paraffin content in feed**

Process Modelling & Optimisation Group

UNIVERSITY OF CAPE TOWN
IYUNIVESITHI YASEKAPA • UNIVERSITEIT VAN KAAPSTAD

# Concluding remarks

1. Model can predict both kinetic and equilibrium and distinguish between oligomerisation, cracking and isomerisation.
2. Computing times order of seconds to minutes (≈300 species, 1000's of reactions)
3. Demonstrate the effects of CN lumping
4. Tracking of carbon number and carbon structure with conversion
5. Multi-phase operation, seamless crossing of the phase boundaries

**Thank you for your attention!**

Process Modelling & Optimisation Group

UNIVERSITY OF CAPE TOWN
IYUNIVESITHI YASEKAPA • UNIVERSITEIT VAN KAAPSTAD

Imperial College London

# Coal to Liquids by Oxidative Cracking in Near-Critical and Super-critical Water

**Marcos Millan, Pedro Arcelus-Arrillaga, Klaus Hellgardt**
***Department of Chemical Engineering, Imperial College***

**Pittsburgh Coal Conference, Cape Town, 12th August 2016**

Imperial College London

## Introduction: Conceptual Framework for Oxidative Cracking

- **The aim is to maximise the liquid fraction of the products.**
- **Oxidation attacks preferentially the central ring in phenanthrene-like structure.**
- **The oxidised ring is weakened.**
- **Subsequent cracking improves the selectivity to liquid products.**

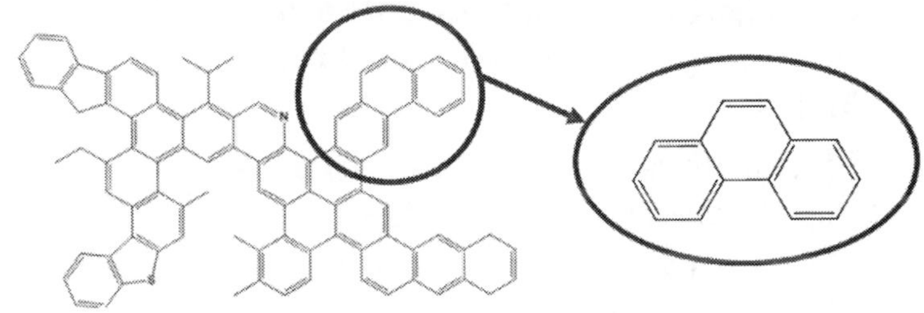

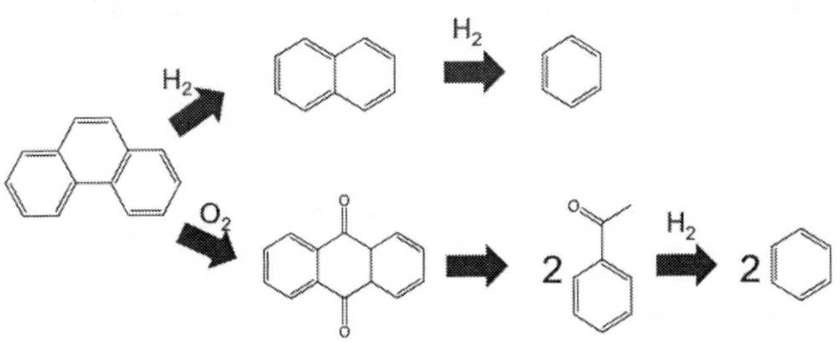

Imperial College London

# Introduction: Advantages of SCW for Oxidative Cracking

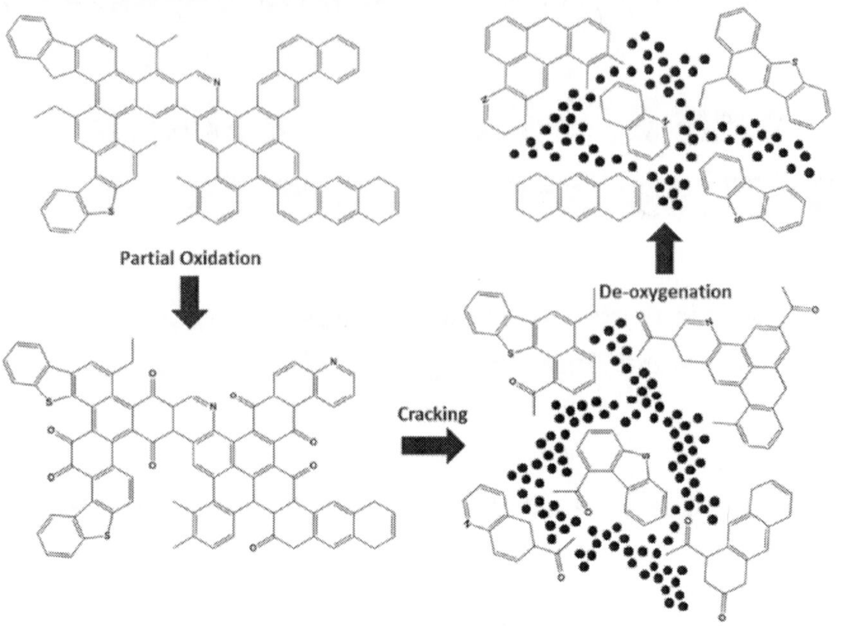


**Supercritical Water provides:**

- Good solvent properties.
- Dispersion of heavy hydrocarbons.
- Stabilisation of products by "cage effect".
- Reduced mass transfer limitations.
- Complete gas miscibility.

Imperial College London

# Experimental Setup

- **5 mL Batch Reactor capable of operating at up to 300 bar and 500 °C.**
- **Phenanthrene was selected as a model compound for PAH structures.**

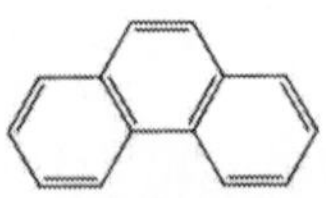


| Reaction Conditions | |
|---|---|
| Temperature (°C) | 400, 425 and 450 |
| Pressure (bar) | 230 |
| $O/O_{stoich}$ | 0.20 |
| Time (min) | 0, 30, 60 and 90 |

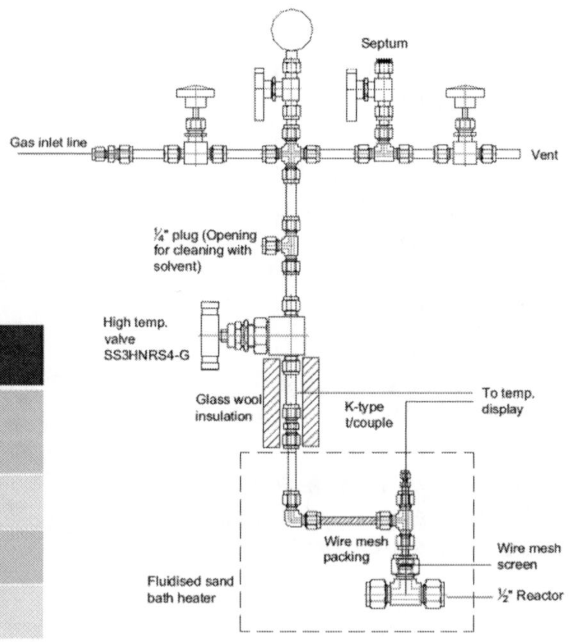

Imperial College
London

## Phenanthrene Conversion and Product Yields

**Reaction Conditions: 230 bar, $O/O_{stoich}$ ratio = 0.2**

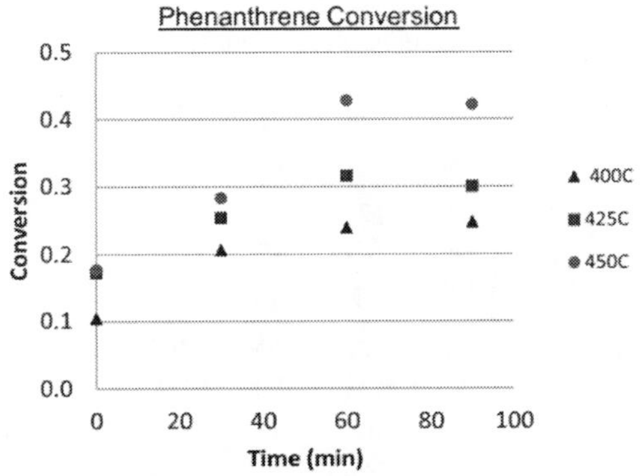


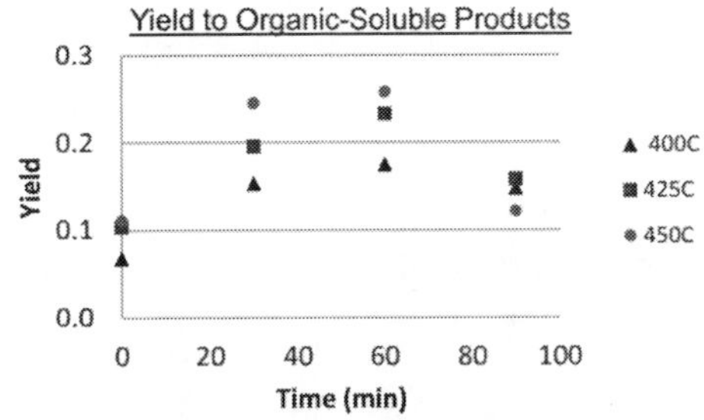


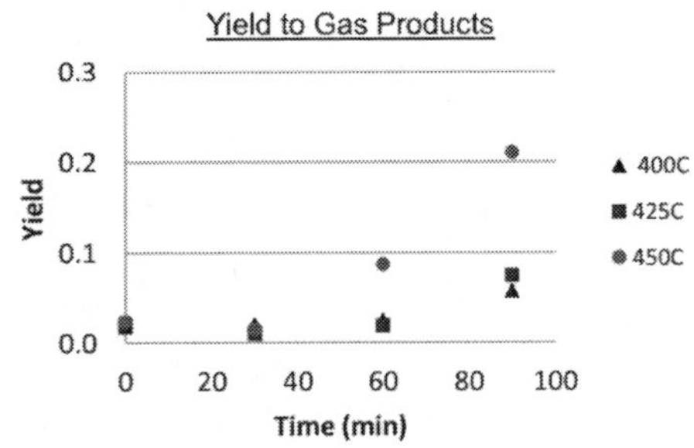


- **Phenanthrene reacts quickly with the oxidant and is stable in SCW in the absence of oxidant.**
- **Oxygenated intermediates react in SCW leading eventually to gas.**
- **Low yield to coke (< 8%) was observed at all conditions.**

Imperial College
London

## Product Distributions Organic-soluble Fraction

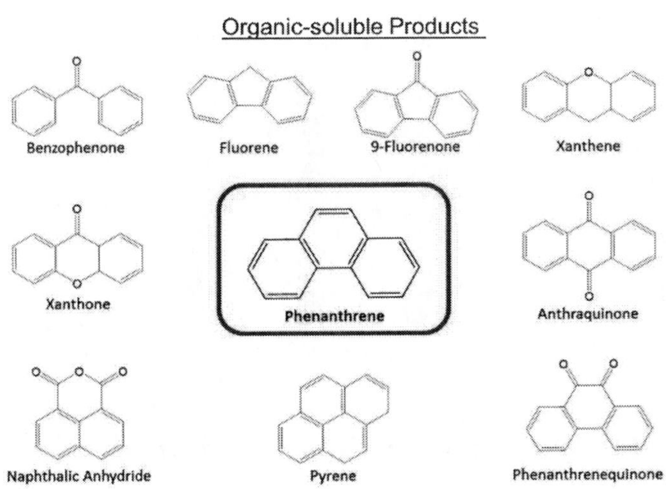


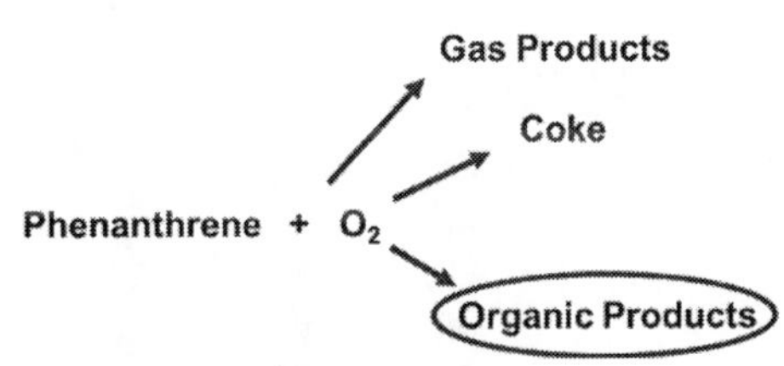


**Mainly oxygenated compounds from attack in the central ring were observed**

Imperial College
London

# Product Distributions
# Organic-soluble Fraction

**Reaction Conditions: 425 °C; 230 bar; O/Ostoich ratio = 0.2**

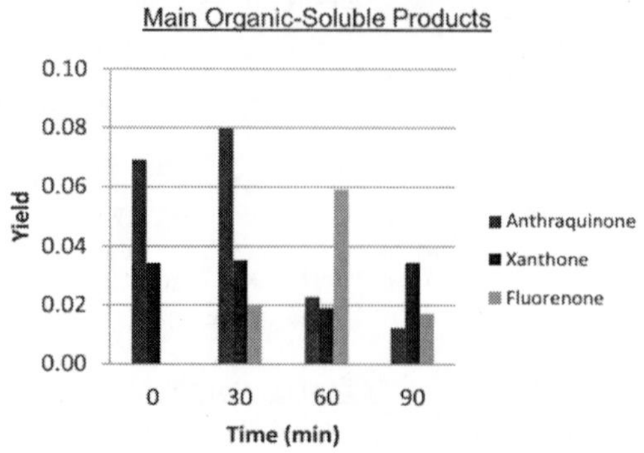


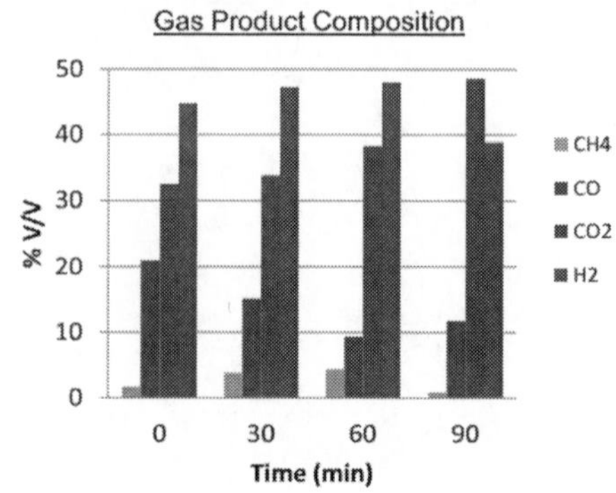


- **Main intermediate products of reaction:**

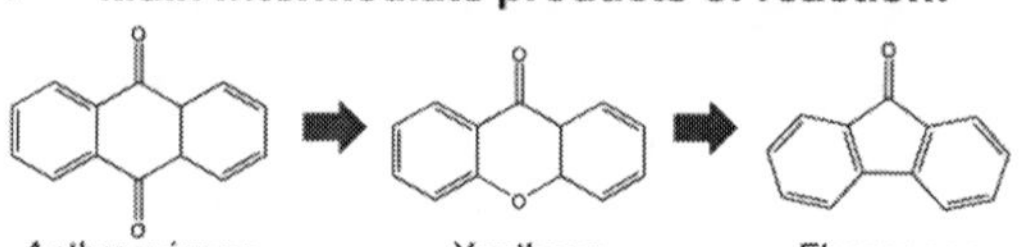

Anthraquinone Xanthone Fluorenone

- **$H_2$-rich gas fraction.**
- **$H_2$ can react in-situ for hydrogenation and de-oxygenation.**

Imperial College
London

# O/C Ratio Evolution – De-oxygenation

**Reaction Conditions: 425 °C; 230 bar; O/Ostoich ratio=0.2**

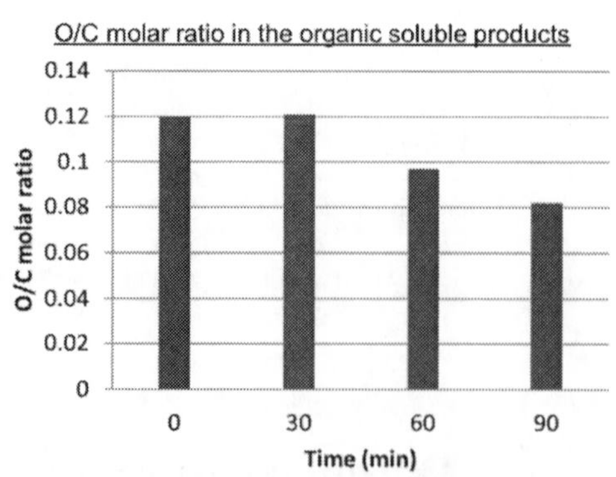


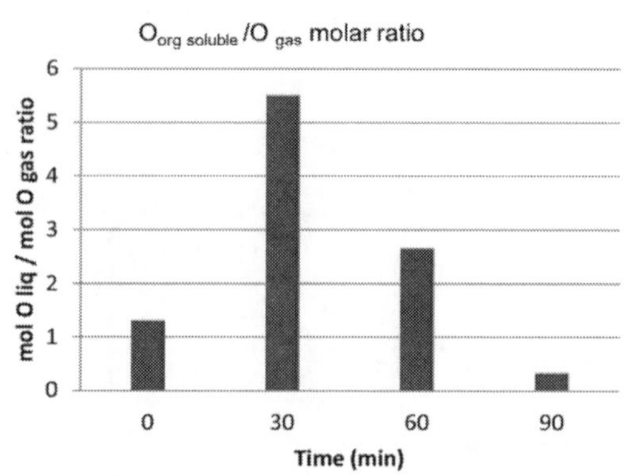


- **Oxygen is incorporated to the organic soluble fraction at early reaction stages.**
- **At longer reaction times intermediate products undergo deoxygenation in SCW with gas release.**

Imperial College
London

# Simplified Potential Reaction Pathway

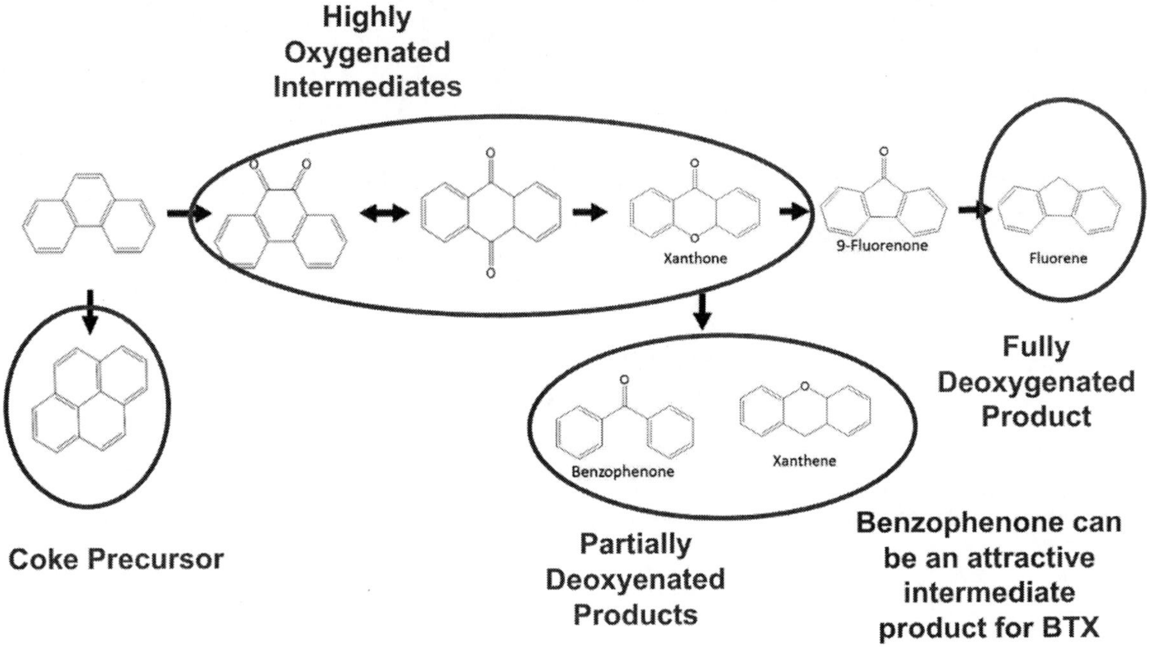


Imperial College
London

# Conclusions

- Partial oxidation of polycyclic aromatic structures proceeds through the central rings.
- Initial oxidation is quick, leading to significant conversions during heat-up time.
- Oxygenated intermediates continue to react in SCW alone leading to gas and a decrease in O/C ratio in the liquid fraction.
- PAH are non reactive in the absence of reactive oxygen species.
- A $H_2$-rich gas fraction is obtained.
- Coke formation is relatively small.

Ashcaan Khazali
khzash001@myuct.ac.za
Prof. Klaus Möller
klaus.moller@uct.ac.za
Prof. Eric van Steen
eric.vansteen@uct.ac.za


# Optimization of Configurational Alternatives of a Gas To Liquids Process based on Fischer-Tropsch Technology


**University of Cape Town**
**Department of Chemical Engineering**


UNIVERSITY OF CAPE TOWN
IYUNIVESITHI YASEKAPA · UNIVERSITEIT VAN KAAPSTAD

Process Modeling and Optimization Group

## Configurations: ASU-ATR-RC

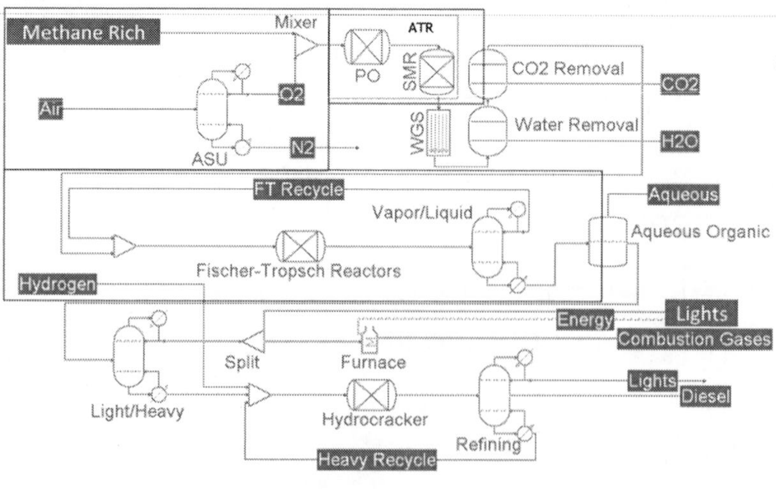


*Fig.* 1: *ASU − ATR − RC*

Acronyms:

- ASU- Air Separation Unit
- ATR – Autothermal Reformer
- SMR – Steam Methane Reformer
- RC – Recycle
- OT – Once through


UNIVERSITY OF CAPE TOWN
IYUNIVESITHI YASEKAPA · UNIVERSITEIT VAN KAAPSTAD

Process Modeling and Optimization Group

## Objectives and Key Questions

Objectives:

- Identify promising configurations from literature study
- Build a thermodynamic database up to C200
- Development of the flowsheet in COCO/Scilab
- Model analyses

Key Questions:

- What is the optimal value of α that will maximize the diesel yield?
- What are the key parameters
- How do they affect diesel yield

UNIVERSITY OF CAPE TOWN
IYUNIVESITHI YASEKAPA · UNIVERSITEIT VAN KAAPSTAD
Process Modeling and Optimization Group

## Tools

cape open to cape open simulation environment

SciLab

Open source software for numerical computation

- COFE - graphical user interface to chemical flowsheeting
- TEA - COCO's thermodynamics for engineering applications

ChemSep is a column simulator for distillation, absorption, and extraction operations - PCD Manager

amsterCHEM

SciLab CAPE-OPEN Unit Operation
SciLab CAPE-OPEN Thermo Import

UNIVERSITY OF CAPE TOWN
IYUNIVESITHI YASEKAPA · UNIVERSITEIT VAN KAAPSTAD
Process Modeling and Optimization Group

## Configurations: AIR/ASU-ATR/SMR-OT/RC

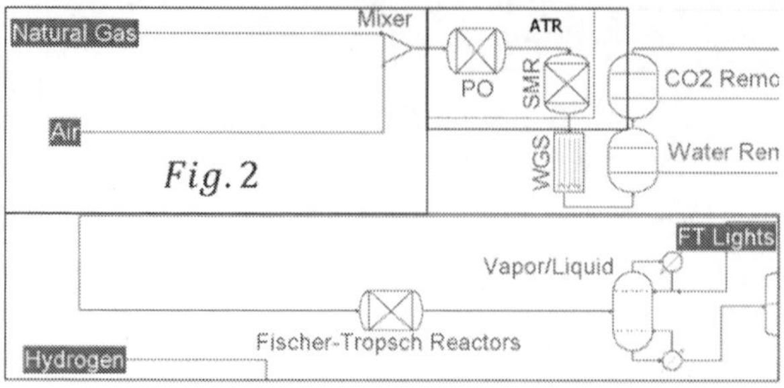


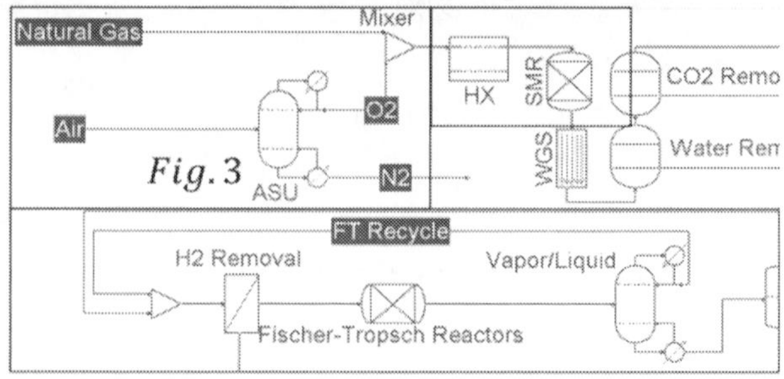


$Fig. 2: AIR - ATR - OT$

$Fig. 3: ASU - SMR - RC$

$Fig. 4: AIR - SMR - OT$

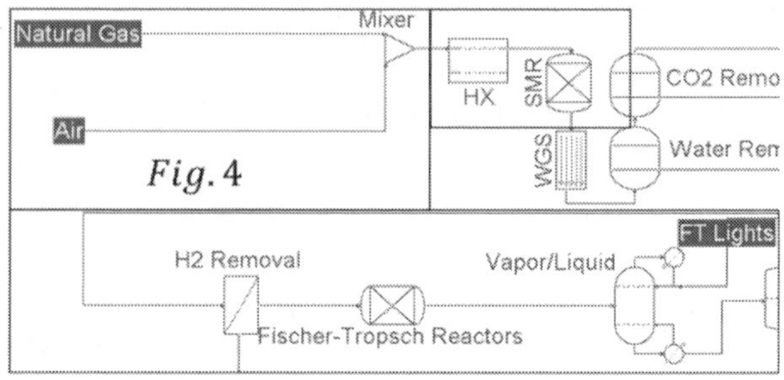


## Methods: Thermophysical Property Estimation

Constantinou and Gani, (1994) Group contribution methods

First order groups:

$$f(X) = \sum_i N_i C_i \quad (1)$$

| αOlefins |
|---|
| $CH_2 = CH \text{----} (CH_2)_n \text{--} CH_3$ |

| nParaffins |
|---|
| $CH_3 - CH_2 \text{----} (CH_2)_n \text{--} CH_3$ |

Critical Temperature:

$$exp\left(\frac{T_c}{t_{c0}}\right) = \sum_i N_i t_{c1i} \quad (2)$$

Critical Pressure:

$$(P_c - p_{c1})^{-0.5} = \sum_i N_i p_{c1i} \quad (3)$$

Critical Volume, Normal Boiling Point, Standard Gibbs Energy of Formation, Standard Enthalpy of Vaporization, Standard Enthalpy of Formation.

UNIVERSITY OF CAPE TOWN
IYUNIVESITHI YASEKAPA • UNIVERSITEIT VAN KAAPSTAD

Process Modeling and Optimization Group

## Methods: Thermophysical Property Estimation

Marano and Holder, (1997) T independent properties

$$Y = Y_{\infty} - \Delta Y_0 exp(-\beta(n \pm n_0)^{\gamma}) \quad (4)$$

Marano and Holder, (1997) T dependent properties

$$Y = Y_{\infty,0}(T) - \Delta Y_{\infty}(T)(n - n_0) - \Delta Y_0(T) exp(-\beta(n \pm n_0)^{\gamma}) \quad (5)$$

Constraints:

$$\Delta Y_0(paraffins) = \Delta Y_0(olefins) \quad (6)$$

$$\Delta Y_{\infty,0}(p) = \Delta Y_{\infty,0}(o) \quad (7)$$

$$\Delta Y_{\infty}(p) = \Delta Y_{\infty}(o) \quad (8)$$

$$\beta(p) = \beta(o) \quad (9)$$

$$\gamma(p) = \gamma(o) \quad (10)$$

UNIVERSITY OF CAPE TOWN
IYUNIVESITHI YASEKAPA • UNIVERSITEIT VAN KAAPSTAD

Process Modeling and Optimization Group

## Results: Thermophysical Property Estimation

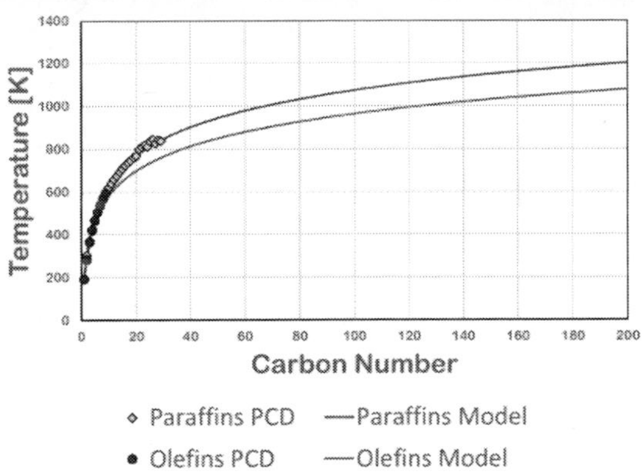


$Fig. 5: Critical\ Temperature\ C1 - C200$

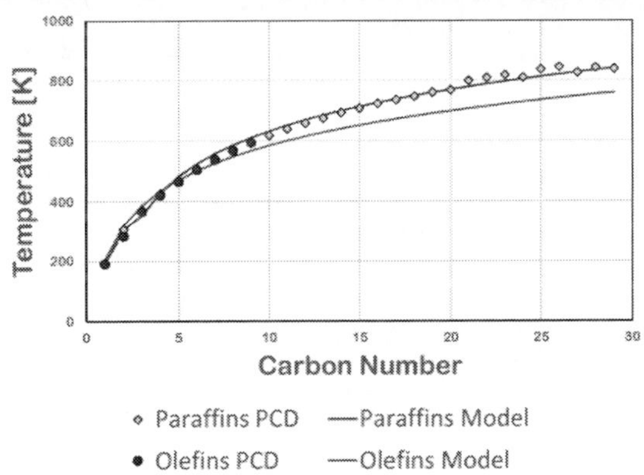


$Fig. 6: Critical\ Temperature\ C1 - C30$

UNIVERSITY OF CAPE TOWN
IYUNIVESITHI YASEKAPA • UNIVERSITEIT VAN KAAPSTAD

Process Modeling and Optimization Group

## Reactor Models: Fischer-Tropsch Synthesis

$$CO + 3H_2 \leftrightarrow CH_4 + H_2O \quad (18)$$

**nParaffins**

$CH_3 — CH_2 ---- (CH_2)_n -- CH_3$

$$(2n+1)H_2 + nCO \leftrightarrow C_nH_{2n+2} + nH_2O \quad (19)$$

**αOlefins**

$CH_2 = CH ----- (CH_2)_n -- CH_3$

$$(2n)H_2 + nCO \leftrightarrow C_nH_{2n} + nH_2O \quad (20)$$

Iglesia et al., (1993).

$$r_{CO} = \frac{(1.331 \times 10^{-9})P_{H_2}^{0.6}P_{CO}^{0.65}}{1 + (3.3 \times 10^{-5})P_{CO}} \quad (21)$$

$$r_{CH_4} = \frac{(7.334 \times 10^{-10})P_{H_2}P_{CO}^{0.05}}{1 + (3.3 \times 10^{-5})P_{CO}} \quad (22)$$

UNIVERSITY OF CAPE TOWN
IYUNIVESITHI YASEKAPA • UNIVERSITEIT VAN KAAPSTAD

Process Modeling and Optimization Group

## Reactor Models: Fischer-Tropsch Synthesis

Kruit et al., (2013). Correlation for α

$$\alpha = \frac{1}{1 + k_a\left(\frac{C_{H_2}}{C_{CO}}\right)^{\beta} exp\left\{\frac{\Delta E_a}{R}\left(\frac{1}{498.15} - \frac{1}{T}\right)\right\}} \quad (23)$$

Flory, (1936). ASF product distribution

$$w_n = n(1-\alpha)^2\alpha^{n-1} \quad (24)$$

Shi and Davis, (2005). Paraffin to Olefin mass ratio

$$\gamma_i = k_1 exp\{-k_2\} \quad (25)$$

$$k_1 = 1 \quad k_2 = 0.15$$

Process Modeling and Optimization Group

## Results: FTS ASF Distribution

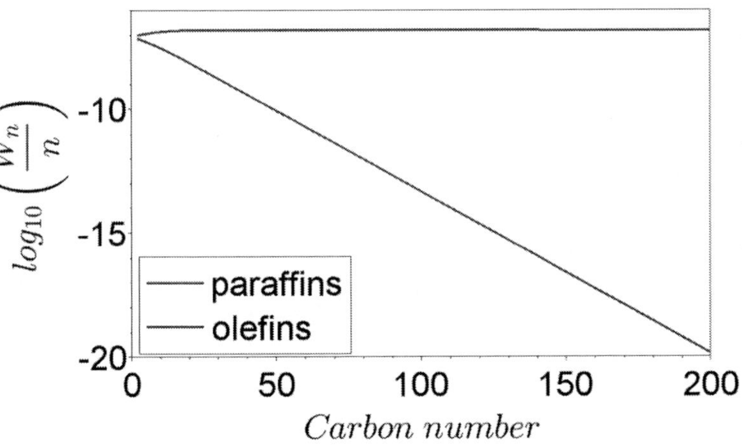


$Fig. 7: ASF\ distribution,$
$180^oC,$
$22bar,$
$H_2: CO = 2,$
$50\ m^3,$
$k2 = 0.15$
$X = 30\%$



## Results: FTS Flow Profiles

Shi and Davis, (2005) Paraffin to Olefin mass ratio

$$\gamma_i = k_1 exp\{-k_2\} \quad (25)$$

$$k_1 = 1 \quad k_2 = 0.15$$

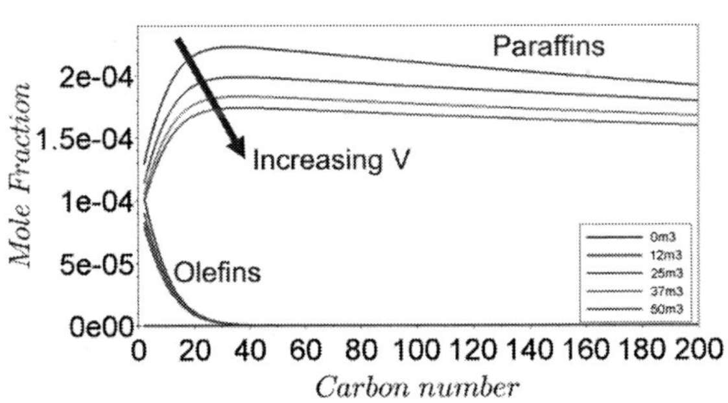


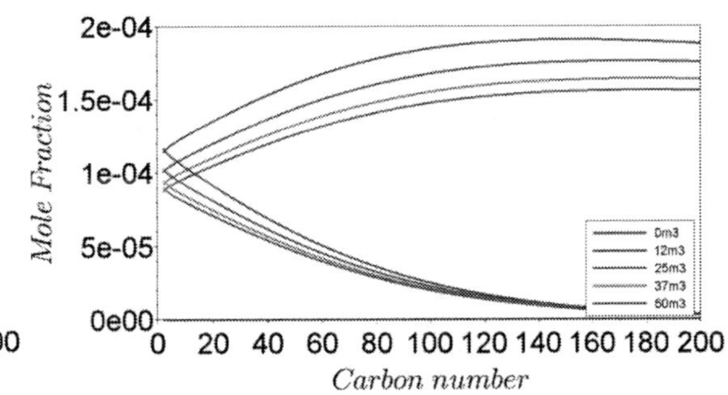


$Fig. 8: FTS, 180^oC, 22bar,$
$H_2: CO = 2,$
$k2 = 0.15$

$Fig. 9: FTS, 180^oC, 22bar,$
$H_2: CO = 2,$
$k2 = 0.02$

## Results: FTS Temperature & $H_2$:CO Ratio

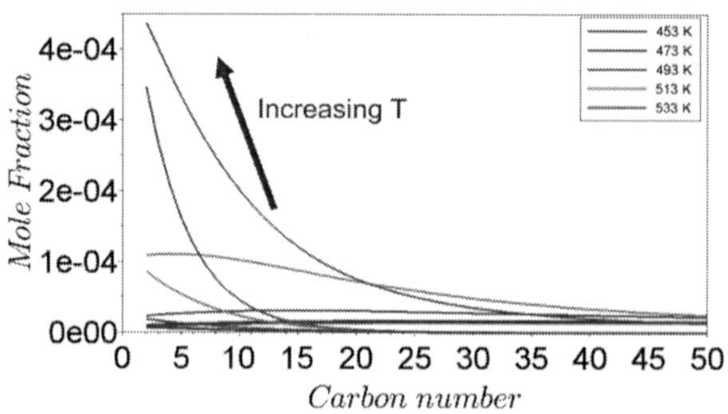


$Fig. 10: FTS, 22bar,$
$H_2: CO = 2,$
$X = \sim 30\%, k2 = 0.15$

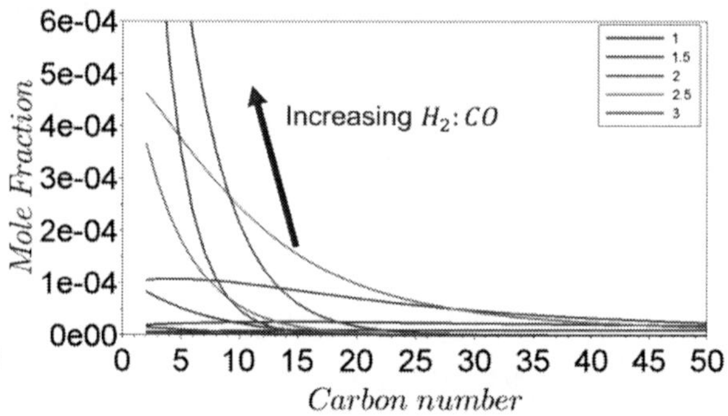


$Fig. 11: FTS, 180^o, 22bar,$
$X = \sim 30\%, k2 = 0.15$

UNIVERSITY OF CAPE TOWN
IYUNIVESITHI YASEKAPA • UNIVERSITEIT VAN KAAPSTAD

Process Modeling and Optimization Group

## Results: FTS Pressure & $P_{N2}$

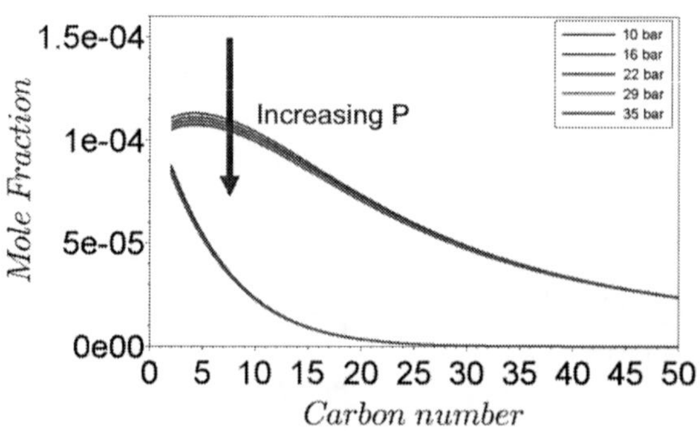


$Fig. 12: FTS, 180^oC, 22bar,$
$H_2: CO = 2,$
$50\ m^3, k2 = 0.15$

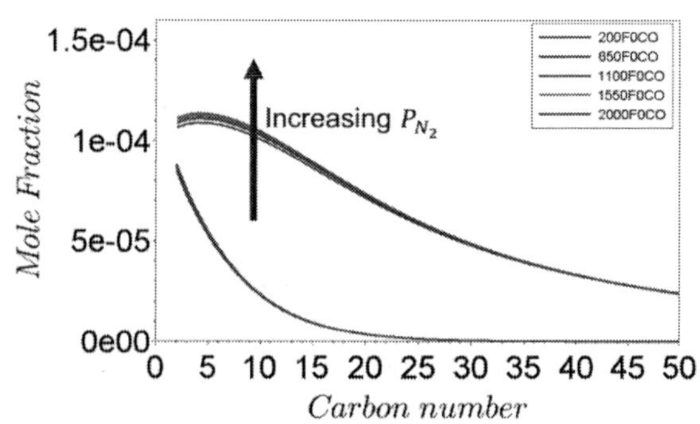


$Fig. 13: FTS, 22bar,$
$H_2: CO = 2,$
$50\ m^3, k2 = 0.15$

UNIVERSITY OF CAPE TOWN
IYUNIVESITHI YASEKAPA • UNIVERSITEIT VAN KAAPSTAD

Process Modeling and Optimization Group

## Results: FTS Recycle

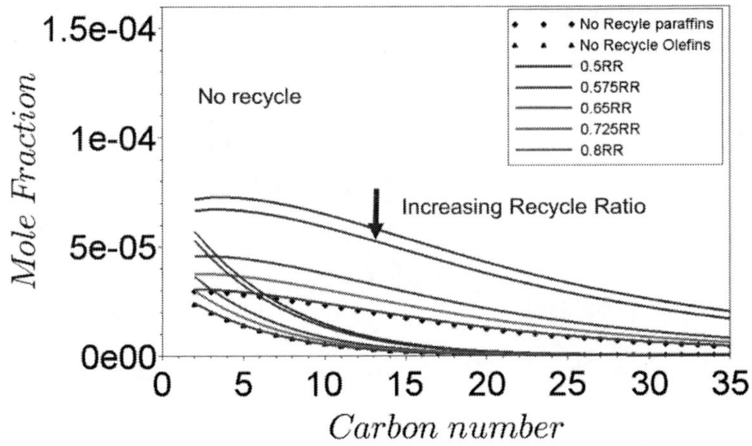


*Fig.* 14: *FTS*, $180^oC$, $22bar$, $H_2$: $CO = 2$, $50\ m^3$, $k2 = 0.15$, Xnr= ~30%, Xr= 90%



## Reactor Models: Hydrocracking

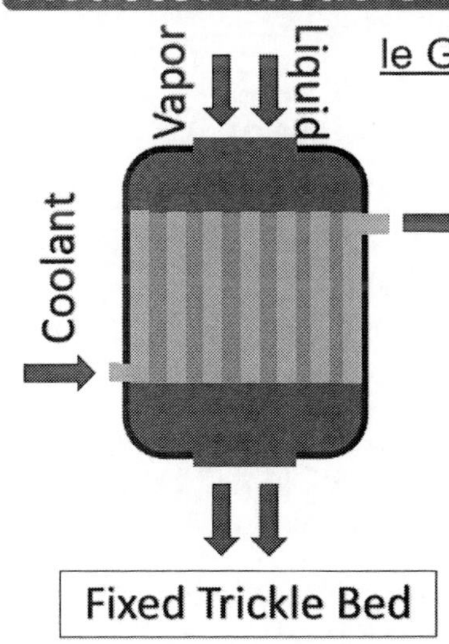


le Grange, (2009) & Möller et al., (2009)

$$r_3 = (k_{B1} + k_{B2} + 2k_C)\sum_{7}^{n} C_i + 2k_C C_6 \quad (11)$$

$$j = (4,5)\ r_j \quad (12)$$

$$= 2(k_A + k_{B1} + k_{B2} + 2k_C)\sum_{i=j+4}^{n} C_i$$

$$+ (k_{B1} + k_{B2} + 2k_C)C_{j+3}$$

$$etc \quad r_6, r_7, r_8\ to\ r_{(n-4)}, r_{(n-3)}, r_{(n-2)}\ to\ r_n \quad (13)$$

# Reactor Models: Hydrocracking

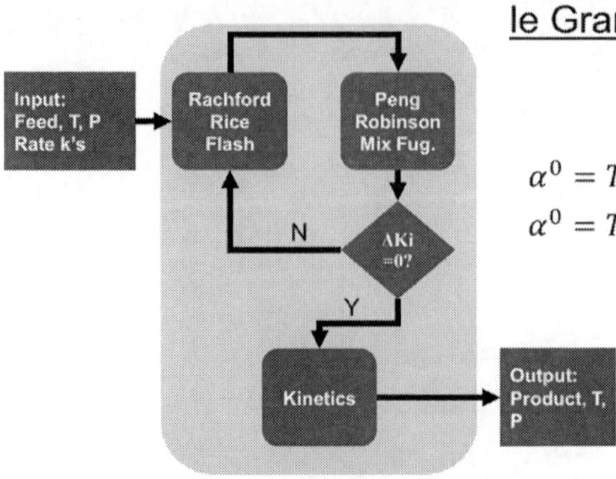


Hydrocracker ODE Solver

le Grange, (2009) & Möller et al., (2009)

Twu et al, (1995)

$$\alpha(\omega, T_r) = \alpha^0 + \omega(\alpha^1 - \alpha^0) \quad (14)$$

$$\alpha^0 = T_r^{-0.171813} exp\{0.125282(1 - T_r^{1.77634})\} \quad (15)$$

$$\alpha^0 = T_r^{-0.607352} exp\{0.511614(1 - T_r^{2.20517})\} \quad (16)$$

Rachford Rice Flash

$$\sum_{i=1}^{N} \frac{(K_i - 1)z_i}{\varphi(K_i - 1) + 1} = 0 \quad (17)$$



# Results: Hydrocracker Parameter Estimation

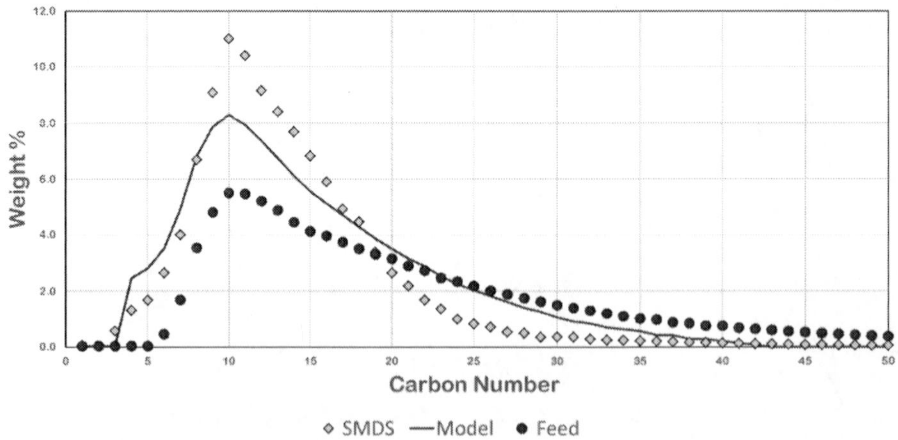


*Fig.* 15:Hydrocracker Model vs SMDS Heavy and SMDS Feed, 300°C, 30bar and H2:Hydrocarbons of 10:1 (molar)

## Results: Hydrocracker-FTS Alpha

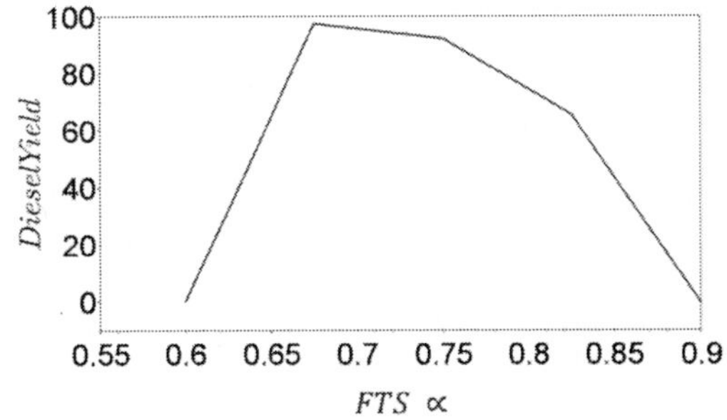


$Fig. 16: Diesel\ Yield\ vs\ FTS\ \propto, 573K, 3MPa,$ H2:Hydrocarbons of 10:1 (molar)

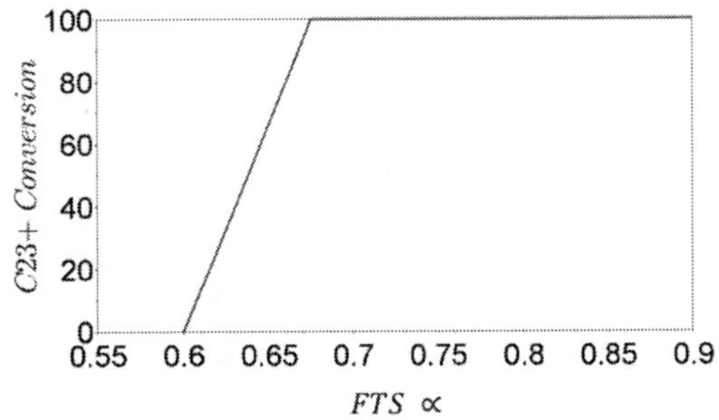


$Fig. 17: C23 + conversion\ vs\ FTS\ \propto, 573K, 3MPa,$ H2:Hydrocarbons of 10:1 (molar)



## Reactor Models: Steam Reforming

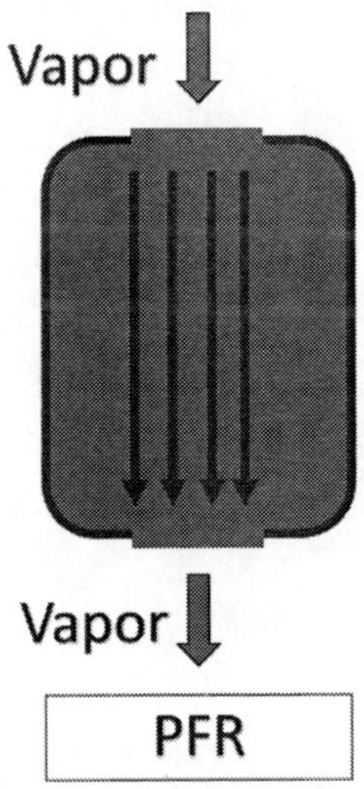


Rowshanzamir et al., (2009) & Xu and Froment, (1989)

$$CH_4 + H_2O \leftrightarrow CO + 3H_2 \quad (26)$$

$$CO + H_2O \leftrightarrow CO_2 + H_2 \quad (27)$$

$$CH_4 + 2H_2O \leftrightarrow CO_2 + 4H_2 \quad (28)$$

$$R_1 = k_1 p_{H_2}^{-2.5}\left(p_{CH_4}p_{H_2O} - p_{H_2}^3 p_{CO} K_{e1}^{-1}\right)(DEN)^{-2} \quad (29)$$

$$R_2 = k_2 p_{H_2}^{-1}\left(p_{CO}p_{H_2O} - p_{H_2} p_{CO_2} K_{e2}^{-1}\right)(DEN)^{-2} \quad (30)$$

$$R_3 = k_3 p_{H_2}^{-2.5}\left(p_{CH_4}p_{H_2O} - p_{H_2}^4 p_{CO_2} K_{e3}^{-1}\right)(DEN)^{-2} \quad (31)$$

$$DEN = 1 + K_{CO}p_{CO} + K_{H_2}p_{H_2} + K_{CH_4}p_{CH_4} + K_{H_2O}p_{H_2O}p_{H_2}^{-1} \quad (32)$$

## Results: ATR Temperature and Pressure

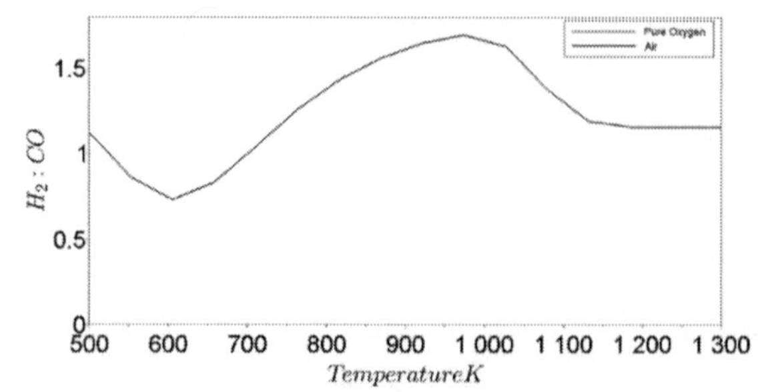


*Fig.* 18: $H_2$: *CO vs T*, 22*bar*, *1MPa*

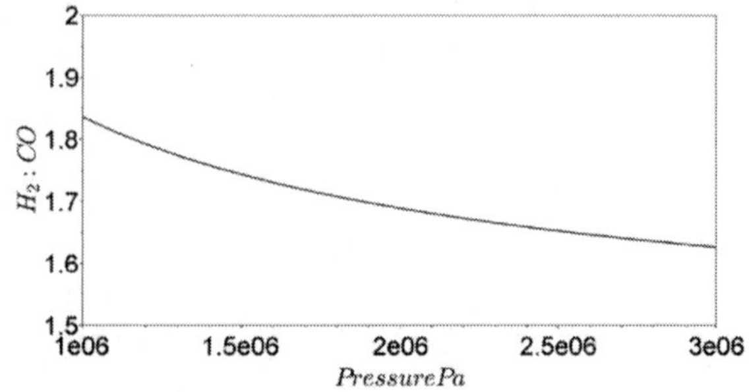


*Fig.* 19: $H_2$: *CO vs P*, 975*K*, *1MPa*

UNIVERSITY OF CAPE TOWN
IYUNIVESITHI YASEKAPA · UNIVERSITEIT VAN KAAPSTAD

Process Modeling and Optimization Group

## Results: ATR CO2 & Partial Oxidation Conversion

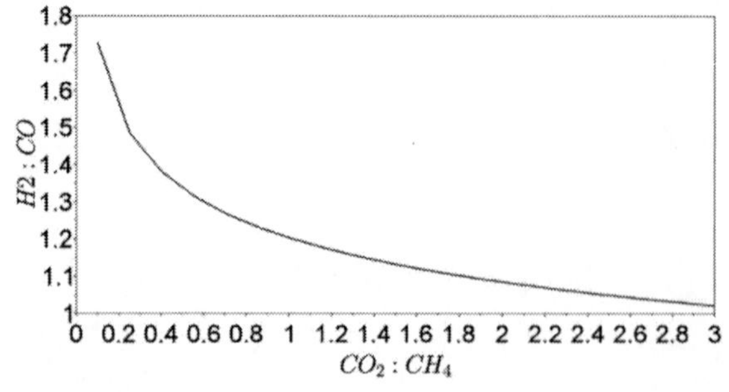


*Fig.* 20: *FTS*, $180^oC$, 22*bar*
$H_2: CO = 2$,
$50\ m^3, k2 = 0.15$

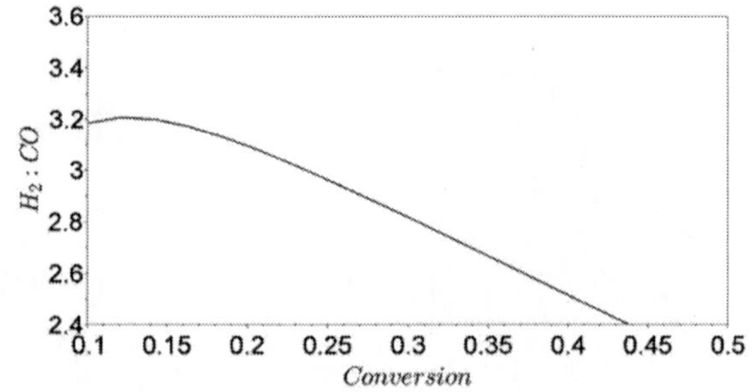


*Fig.* 21: *FTS*, 22*bar*,
$H_2: CO = 2$,
$50\ m^3, k2 = 0.15$

UNIVERSITY OF CAPE TOWN
IYUNIVESITHI YASEKAPA · UNIVERSITEIT VAN KAAPSTAD

Process Modeling and Optimization Group

## Preliminary Conclusions

1. The optimal value of alpha lies between 0.68 and 0.75.
2. Nitrogen will have little effect on conversion provided the partial pressures of hydrogen and carbon monoxide are unchanged.
3. Fischer-Tropsch T and $H_2$:CO will have the strongest effect on diesel Yield.
4. Reformer T, P and $CO_2$ have the greatest effect on H2:CO ratio

UNIVERSITY OF CAPE TOWN
IYUNIVESITHI YASEKAPA • UNIVERSITEIT VAN KAAPSTAD

Process Modeling and Optimization Group

## Outlook

1. Couple paraffin and olefin property estimations
2. Implement FTS elementary reactor model
3. Develop FBR model for FTS
4. Develop SBCR model for FTS
5. Find suitable model for PO
6. Energy balance
7. Unit Sizing
8. Unit Costing
9. Test more configurations
10. Plantwide optimization

UNIVERSITY OF CAPE TOWN
IYUNIVESITHI YASEKAPA • UNIVERSITEIT VAN KAAPSTAD

Process Modeling and Optimization Group

## References


- Constantinou, Leonidas and Rafiqul Gani (1994). "New group contribution method for estimating properties of pure compounds". In: AIChE Journal 40.10, pp. 1697–1710.
- Flory, Paul J (1936). "Molecular size distribution in linear condensation polymers1". In: Journal of the American Chemical Society 58.10, pp. 1877–1885.
- Iglesia, Enrique, Sebastian C Reyes, and Stuart L Soled (1993). "Reaction-transport selectivity models and the design of Fischer-Tropsch catalysts". In: CHEMICAL INDUSTRIES-NEW YORK-MARCEL DEKKER-,pp. 199–199.
- Kruit, Katrina D et al. (2013). "Selectivity of the Fischer–Tropsch process: deviations from single alpha product distribution explained by gradients in process conditions". In: Catalysis Science & Technology 3.9, pp. 2210–2213.
- Le Grange, Philip (2009). "Models for the hydrocracking of Fischer-Tropsch derived waxes". MA thesis. University of Cape Town.
- Marano, John J and Gerald D Holder (1997). "A general equation for correlating the thermophysical properties of n-paraffins, n-olefins, and other homologous series. 3. Asymptotic behaviour correlations for thermal and transport properties". In: Industrial & engineering chemistry research 36.6, pp. 2399–2408.


UNIVERSITY OF CAPE TOWN
IYUNIVESITHI YASEKAPA • UNIVERSITEIT VAN KAAPSTAD

Process Modeling and Optimization Group

## References


- Moller, Klaus, Philip le Grange, and Carlo Accolla (2009). "A Two-Phase Reactor Model for the Hydrocracking of Fischer- Tropsch-Derived Wax". In: Industrial & Engineering Chemistry Research 48.8, pp. 3791–3801.
- Rowshanzamir, S, MH Eikani, et al. (2009). "Autothermal reforming of methane to synthesis gas: Modeling and simulation". In: international journal of hydrogen energy 34.3, pp. 1292–1300.
- Shi, Buchang and Burtron H Davis (2005). "Fischer–Tropsch synthesis: the paraffin to olefin ratio as a function of carbon number". In: Catalysis today 106.1, pp. 129–131.
- Twu, Chorng H, John E Coon, and John R Cunningham (1995). "A new generalized alpha function for a cubic equation of state Part 1. Peng-Robinson equation". In: Fluid Phase Equilibria 105.1, pp. 49–59.
- Xu, Jianguo and Gilbert F Froment (1989). "Methane steam reforming, methanation and water-gas shift: I. Intrinsic kinetics". In: AIChE Journal 35.1, pp. 88–96.



UNIVERSITY OF CAPE TOWN
IYUNIVESITHI YASEKAPA • UNIVERSITEIT VAN KAAPSTAD

Process Modeling and Optimization Group

2016 International Pittsburgh Coal Conference
Cape Town, SOUTH AFRICA
August 8 – 12, 2016

**PROGRAM TOPIC:**
**GASIFICATION SCIENCE AND MODELING**

# SYNGAS PRODUCTION FROM BRAZILIAN COAL CHARS IN THE PRESENCE OF STEAM


***Humberto Jorge José, Dr., Department of Chemical Engineering and Food Engineering, Federal University of Santa Catarina, 88040-900, Florianópolis-SC, BRAZIL***
***humberto.jose@ufsc.br, Phone/ Fax +55 48 3721 4074***

*Gabriela Carvalho Collazzo, Dr., Department of Chemical Engineering, Federal University of Santa Maria, 97105-900, Santa Maria -RS, BRAZIL.*
*gabriela.collazzo@ufsm.br, Phone/ Fax + 55 55 3220 8841*

*Michele Di Domenico, Dr., Department of Chemical Engineering, Federal University of Technology – Paraná, 85601-970, Francisco Beltrão/PR, BRAZIL.*
*micheled@utfpr.edu.br, Phone/ Fax+55 46 3520 2649*

*Regina de Fátima P. M. Moreira, Dr., Department of Chemical Engineering and Food Engineering, Federal University of Santa Catarina, 88040-900, Florianópolis-SC, BRAZIL.*
*regina@enq.ufsc.br, Phone/ Fax+55 48 3721 4074*



**Abstract:**
The steam gasification of two Brazilian coal chars, classified as sub-bituminous (CMRS-02) and bituminous (CMSC-02), were investigated using a pressurized thermogravimetric analyzer (PTGA, model Dyntherm-HP-ST, Rubotherm Company) in the temperature range of 1073-1173 K and pressure range of 1-20 bar. The study area lay within the Leão-Butiá coalfield (M2 seams – sample CMRS-02) and South Santa Catarina coalfield (Barro Branco seams – sample CMSC-02), both located in the Paraná Basin. To evaluate the coals and chars, the physical and chemical properties were determined by proximate and ultimate analysis, calorific value, grindability index, free swelling index and mineralogy. For the gasification experiments, firstly, the samples were pyrolysed in a quartz tubular reactor under nitrogen and then gasified under steam in a pressurized thermogravimetric analyzer (PTGA). The operating variables used were in agreement with differential reactor and free of diffusion limitations. Gasification reactions were conducted using the isothermal method and each experiment was performed in duplicate in order to test the repeatability of the results. The gas produced during the gasification reactions was detected using a gas chromatograph (model 2014, Shimadzu) connected to the PTGA output. From the results of carbon conversion as function of reaction time it was observed that the curves are strongly dependent on the specific coal properties; char CMRS-02 showed a higher conversion rate, which is probably related to the coal rank. Thus, the sub-bituminous coal was more reactive than the bituminous coal. This result can be ascribed to the vitrinite and liptinite contents, since higher values of these macerals are known to increase the intrinsic reaction rate. Char CMRS-02 and char CMSC-02 presented, respectively, vitrinite contents of 50.8% and 34.5%. Higher vitrinite and liptinite contents also indicate higher H/C ratios, which can also influence the reaction rate. In this work, the effect of both temperature and total system pressure on the gasification rate was stronger for the bituminous sample than for the sub-bituminous one. Finally, from the analysis of gasification product gas, it was observed that hydrogen, carbon monoxide and carbon dioxide were produced in significant amounts and an increase in the reaction temperature clearly increased the concentration of all gases and no $CH_4$ was detected. The $H_2/CO$ molar ratio was found to be > 2 for both coal chars, indicating that the produced syngas from Brazilian coal gasification is promising for the Fischer-Tropsch synthesis.


## 1. Introduction

Coal is one of the most commonly used fossil fuels which currently meets approximately 40% of the global electricity demand, while also covering one quarter of the global energy needs. However, these figures barely reflect the importance of coal as an energy source, considering that only 15 to 20% of the known coal resources are suitable for conventional raw material extraction through surface and underground mining [1], and it is expected to be applied as the energy resource for over 110 years [2]. However, the huge consumption of coal over the past few decades has caused serious environmental impact locally and globally.

The clean and efficient use of coal by means of gasification is considered to be a promising alternative to ensure the continued demand for this fossil fuel in the energy markets [3]. Coal gasification is more advantageous because of the conversion for coal into a combustible gas, making it a more efficient and sustainable technology for the generation of energy and chemical production, as well as a predominant replacement for conventional fossil fuels utilization [4, 5]. In the chemical processing of coal gasification, the first problem to be addressed is creating a large-scale coal gasification facility that is clean, feasible and highly efficient [6]. At the same time, it should meet economic and environmental demands [7]. The selection of coal gasification technique depends on several factors, including the scale of the enterprise, the type of original coal, technique maturity, cost and consumption, as well as pollution and treatment [8].

In a coal gasification process is entrained in a stream of steam/carrier gas, and a series gas-phase reactions result in a syngas product composed predominantly of $H_2$ and CO with fuel-dependent concentrations of $CO_2$, $CH_4$, $H_2O$, and $H_2S$. Some by-products of the coal gasification process include tars and gas- and solid-phase inorganic species. Any evolved tar species represent a reduction in process efficiency in terms of overall conversion to syngas as carbon and hydrogen are bound in the tars and therefore unavailable for conversion to the desirable products, $H_2$ and CO.

Syngas can then be directly converted to energy through combustion or to a variety of fuels: (i) liquid hydrocarbons from methanol, (ii) liquid hydrocarbons through Fischer-Tropsch synthesis, and (iii) synthetic natural gas (SNG) [9, 10].The selective conversion of syngas to liquid hydrocarbons or SNG requires, however, specific ratios of $H_2$ to CO in the amount of 2 or 3, respectively [10]. Obtaining those explicit values is not a straightforward task as the yields of produced non condensable gases depend on several factors such as raw material composition and operating conditions of the converting process.

According to the latest Brazilian Mineral Yearbook [11] the Brazilian coal reserves are around 32 billion tons and, as in other countries, one of the most significant uses of this resource is power generation. There are many studies on steam coal gasification [12-17] but little is known about the gasification of Brazilian coal [18, 19], especially at high pressure.

In this context, the aim of this work was to investigate the gasification of two Brazilian coals, classified as sub-bituminous (CMRS-02) and bituminous (CMSC-02), using steam as the gasifying agent in the temperature range of 1073-1173 K and pressure range of 1-20 bar.

## 2. Experimental

### 2.1. Coal and char preparation

The study area lay within the Leão-Butiá coalfield (M2 seams) and South Santa Catarina coalfield (Barro Branco seams), both located in the Paraná Basin. The samples tested were CMRS-02 (sub-bituminous coal from Leão-Butiá coalfield) and CMSC-02 (bituminous coal from South Santa Catarina coalfield). These two coal types were formed from different geological processes.

The samples were ground, sieved to a particle size of 500-1000μm and stored in nitrogen atmosphere. The samples were then pyrolyzed in a quartz tubular reactor under nitrogen (400 ml $min^{-1}$ and purity 99.996%) using a heating rate of 190 K $min^{-1}$ up to 1173 K and holding for 5 min. The resulting char was ground, sieved to a particle size of < 100μm and stored in nitrogen atmosphere.

### 2.2 Coal and char characterization

To evaluate the coals and chars, the physical and chemical properties were determined. The proximate and ultimate analysis, calorific value, grindability index, free swelling index and mineralogy data are given in Table 1. The proximate analysis was performed according to the standard method ASTM E-1131 [20] using a thermogravimetric analyzer (DTG-60, Shimadzu). The ultimate analysis and mineralogical characterization were carried out in an elemental analyzer (Vario Macro) and diffractometer (Siemens 5000), respectively. To determine the coal and char porous structures, an ASAP 2020 (Micromeritics) equipment was employed (see Table 2). The data on the petrographic composition of the samples are given in Table 2 and their classification in terms of rank was obtained according to the International Standard [21].

Table 1. Coal and char characterization.

| | CMRS-02 | | CMSC-02 | |
|---|---|---|---|---|
| | **Coal** | **Char** | **Coal** | **Char** |
| **Proximate analysis (wt. %, db[1])** | | | | |
| Ash | 32.0 | 41.8 | 33.6 | 42.6 |
| Volatile matter | 31.0 | 6.9 | 24.4 | 6.7 |
| Fixed carbon | 37.0 | 51.3 | 42.0 | 50.7 |
| **Ultimate analysis (wt. %, daf[2])** | | | | |
| Carbon | 54.5 | 51.8 | 56.2 | 50.1 |
| Hydrogen | 4.3 | 0.8 | 4.5 | 0.5 |
| Nitrogen | 1.0 | 2.1 | 1.1 | 0.9 |
| Sulfur | 1.1 | 0.7 | 1.5 | 1.1 |
| Oxygen | 7.1 | 2.8 | 3.1 | 4.8 |
| **Heating value** | | | | |
| **Higher heating value ($MJ\cdot kg^{-1}$)** | 21,3 | - | 23,5 | - |
| **Lower heating value ($MJ.kg^{-1}$)** | 20,3 | - | 22,5 | - |
| **Grindability index** | 50 | - | 59 | - |
| **Free Swelling index** | 0 | - | 1 | - |
| **Mineralogy** | | | | |
| Illite | - | - | x[3] | - |
| Kaolinite | x | - | x | - |
| Gypsum | - | - | x | - |
| Quartz | x | - | x | - |
| Calcite | x | - | x | - |
| **Pore structure** | | | | |
| **Area ($m^2\cdot g^{-1}$)** | | | | |
| BET surface area | - | 18.22 | - | 3.08 |
| **Volume ($cm^3\cdot g^{-1}$)** | | | | |
| Single point adsorption total pore volume | - | 0.022 | - | 0.011 |
| **Pore size (Å)** | | | | |
| Adsorption average pore diameter | - | 48.48 | - | 150.12 |

[1]db: dry basis; [2]daf: dry ash-free basis; [3]x: indicates the presence of the mineral in the sample.

Table 2. Petrographic composition of coal samples.

| | CMRS-02 | CMSC-02 |
|---|---|---|
| **Macerals (Vol. %)** | | |
| Collotelinite | 30.2 | 4.8 |
| Desmocolinite | 19.0 | 27.2 |
| Vitrodetrinite | 1.6 | 2.6 |
| **Total vitrinite** | **50.8** | **34.6** |
| Cutinite | 0.4 | 1.2 |
| Sporinite | 5.0 | 4.8 |
| Resinite | 0.2 | 0.2 |
| Liptodetrinite | 1.6 | 0.0 |
| **Total liptinite** | **7.2** | **6.2** |
| Fusinite | 14.6 | 25.0 |
| Semifusinite | 6.4 | 3.6 |
| Intertodetrinite | 8.2 | 10.6 |
| Secretinite | 0.2 | 0.0 |
| Macrinite | 0.6 | 0.0 |
| **Total inertinite** | **30.0** | **39.2** |
| Carbonate | 0.0 | 0.0 |
| Clay mineral | 10.0 | 17.4 |
| Pyrite | 0.6 | 1.4 |
| Quartz | 1.4 | 1.2 |
| **Total Mineral Matter** | **12.0** | **20.0** |
| **Reflectance of vitrinite (Vol. %)** | | |
| Min[1] | 0.375 | 0.734 |
| Max[2] | 0.609 | 1.097 |
| MV[3] | 0.447 | 0.911 |
| SD[4] | 0.050 | 0.086 |
| N[5] | 100 | 100 |
| Rank | Sub. B[6] | HVB A[7] |

[1]minimum value of vitrinite reflectance, [2]maximum value of vitrinite reflectance, [3]average value of vitrinite reflectance, [4]standard deviation, [5]number of vitrinite reflectance measurements, [6]Sub-bituminous, [7]high volatile bituminous.

### 2.3 Kinetic measurements

The gasification reactions were carried out in a pressurized thermogravimetric analyzer (PTGA, model Dyntherm-HP-ST, Rubotherm Company) as schematically shown in Fig. 1. The equipment allows experiments to be carried out at temperatures up to 1373 K within the range of 1-40 bar. The PTGA has a magnetic, non-contact, mass measurement system, which allows one to tare and calibrate the balance during the experiment, this being extremely important for ensuring measurement accuracy. As confirmed by carrying out runs with and without (blank) the sample, the system corrects the effects of buoyancy.

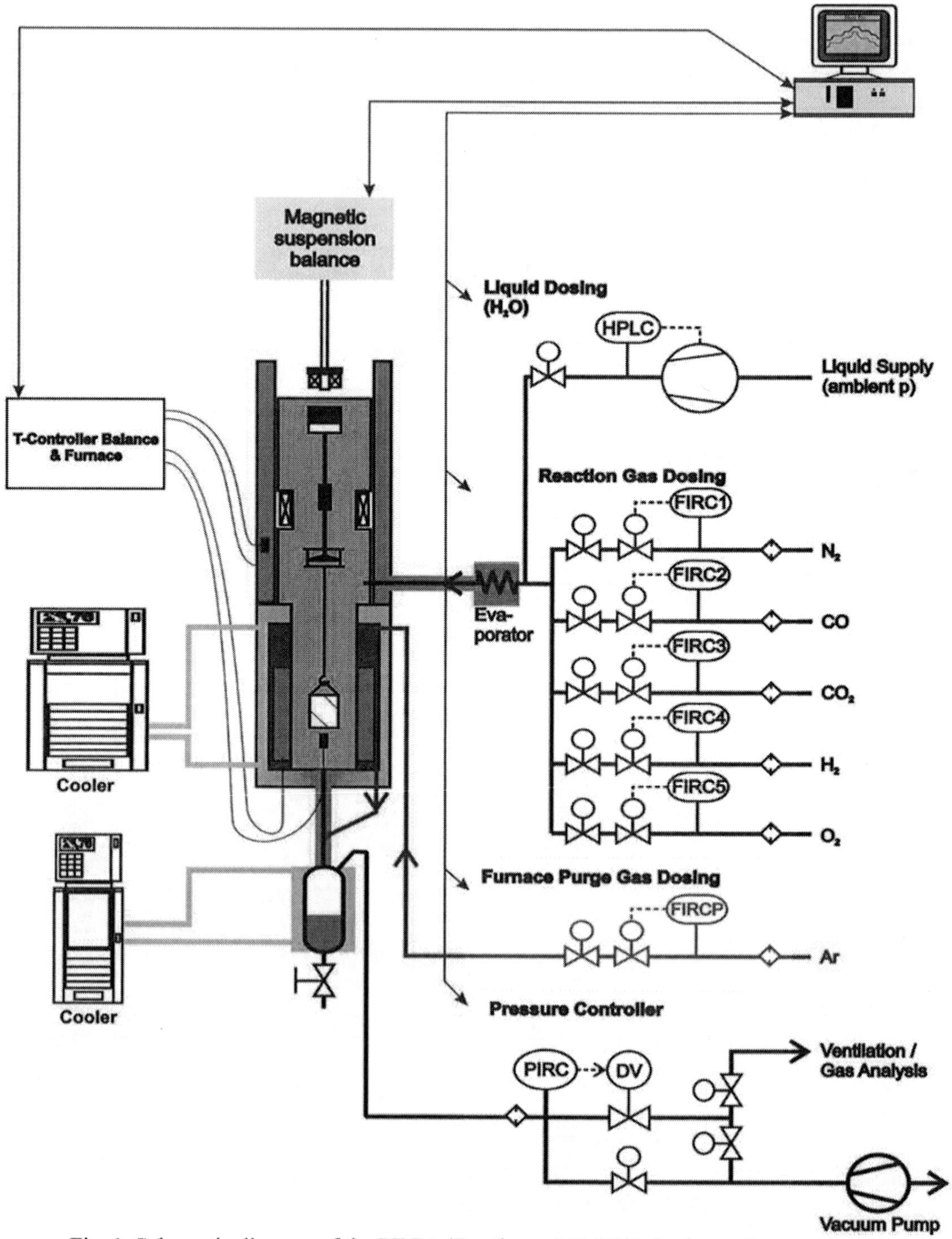

Fig. 1. Schematic diagram of the PTGA (Dyntherm-HP-ST/Rubotherm Company).

The effects of particle size, total sample mass and gas flow on the diffusion limitations were pre-tested. It was found that when the particle size was smaller than 100 μm, the sample mass was 100 mg and the total gas flow was 300 mL $min^{-1}$ of argon/steam mixture (270 $ml_N$ $min^{-1}$ of argon/ 30 $ml_N$ $min^{-1}$ of steam) these effects could be neglected. The operating variables used were in agreement with differential reactor conditions, where the concentration of product gases was < 1 %. In this study the effectiveness factors were calculated according to [22], and the values were found to be close to unity, indicating that there are no diffusion limitations to the external mass transfer or pore diffusion, which confirm the existence of kinetic controlled regimes.

Approximately 100 mg of char was placed in an alumina crucible (15 mm diameter x 10 mm height) forming a thin layer. Initially, the system was purged with argon (100 mL $min^{-1}$ and purity 99.999%). Subsequently, the temperature was increased at 20 K·$min^{-1}$ up to the desired reaction temperature and then the argon was replaced with an argon/steam mixture (300 $ml_N$ $min^{-1}$ and 10 vol.% steam). Prior to reaching the sample, the argon/water mixture flows through an electrically heated evaporator at 473 K to vaporize the water. Gasification reactions were conducted using the isothermal method (1173 K, 1148 K, 1123 K, 1073 K) and the weight loss was recorded at intervals of 10.0 s with computerized digital scales. Each experiment was performed in duplicate in order to test the reproducibility of the results.

### 2.4 Gas analysis

The gas produced during the gasification reactions was detected using a gas chromatograph (GC-2014-TCD/FID with methanizer, Shimadzu) connected to the PTGA output. The gas collected also contained the purge gas (argon) from the system. The products were injected automatically at 483 K, and then separated in a Carboxen 1000 column (308 K for 5 min, 20 K $min^{-1}$ up to 483 K which was held for 15 min, using argon as the carrier gas). $H_2$ was quantified in a thermal conductivity detector (TCD) and CO, $CO_2$ and $CH_4$ were directed to the methanizer (at 688 K) and quantified in a flame ionization detector (FID).

## 3. Results and discussion

### 3.1 Effect of temperature on char conversion

Fig. 2 shows the results for the carbon conversion for the two chars as a function of reaction time for the reactions carried out at 1073-1173 K. It can be observed that in this temperature range the profiles of the $X$ vs. $t$ curves are strongly dependent on the specific coal properties, the Char CMRS-02 showing a higher conversion rate, which is probably related to the coal rank. Also, it can be observed that the temperature effect on the reaction rate was stronger for Char CMSC-02. It is also evident that the gasification reactions of Brazilian coals are strongly dependent on the temperature.

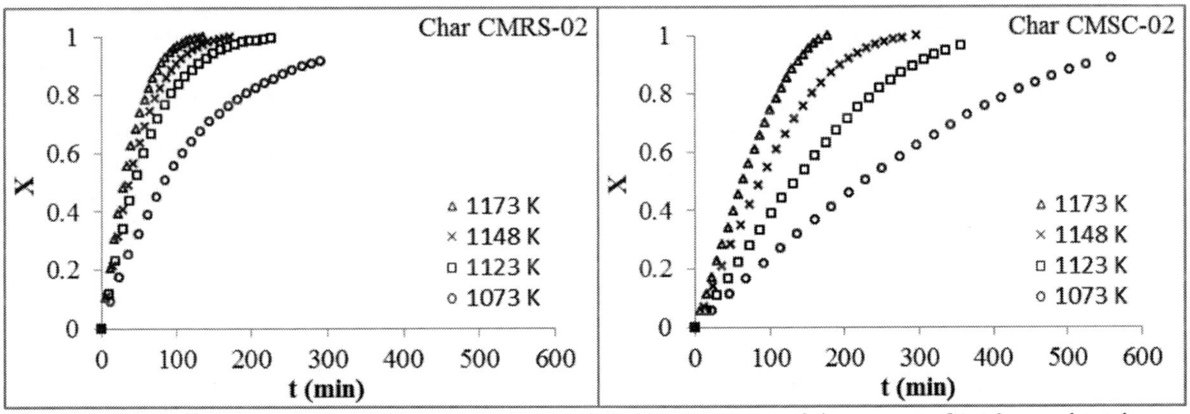

Fig. 2. Temperature effect on conversion (X) versus reaction time (t) at steam partial pressure of 0.1 bar and total system pressure of 1 bar.

Based on these results, Char CMRS-02 (sub-bituminous) was more reactive than Char CMSC-02 (bituminous), which can be ascribed to the vitrinite and liptinite content, since a higher content of these macerals increases the intrinsic reaction rate. In relation to the vitrinite content, it can be noted that Char CMRS-02 has a higher value (50.8 %) than Char CMSC-02 (34.5 %). Consequently, the sample with a higher vitrinite and liptinite contents has a higher H/C which influences the intrinsic reaction rate. Messenböck et al. [23] in their research with vitrinite and liptinite-rich coals found that a higher content of these macerals increases the intrinsic reaction rate in steam gasification. Although these coals have similar carbon contents they have different ranks, i.e., the nature of their geological formation differs, the coal with lower rank being more reactive. This result is consistent with the findings of Walker and coworkers. Walker [24] and coworkers have published data on the reaction rate of chars during gasification with air [25], $CO_2$ [26], $H_2$ [27] and $H_2O$ [28]. They concluded that the reaction rate of the char generally increases with a decrease in the rank of the parent coal in the presence of an oxidizing gas atmosphere. The higher the rank the greater the degree of coalification of the coal and the carbon aromaticity, i.e., the greater the fraction of carbon in aromatic rings will be. Hence, the reaction of carbon with the gasifying agent is hindered, decreasing the reaction rate [29]. Another observation in the difference between the reaction rate of the coal chars is about the microporous area (shown in Table 2) once the Char CMRS-02 have a higher than Char MSC-02. On comparing the properties of coal chars in relation to reaction rate, the most significant in terms of the reaction rate is the rank. As it can be seen in the results for the characterization, the coal chars have similar physical and chemical properties, the most significant difference being in terms of the coal rank.

### 3.2 Effect of total system pressure on char conversion

Fig. 3 shows the results for the effect of total system pressure on char conversion for the two chars at 1173 K and 10 vol.% steam.

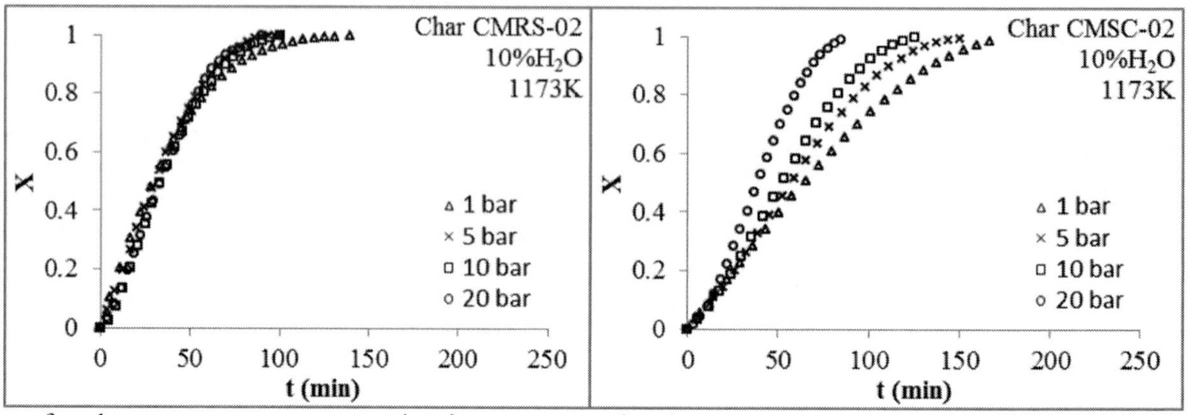


Fig. 3. Effect of total system pressure on conversion data versus reaction time at reaction temperature of 1173 K and steam partial pressure of 0.10 bar.

In this study, the effect of the total system pressure on carbon conversion was stronger for the bituminous sample than for the sub-bituminous one. As in the case of the temperature effect, the effect of the total system pressure was more significant for Char CMSC-02. For the CMSC-02, the pressure dependency follows a linear trend with angular coefficient of 3.9 x $10^{-4}$ $min^{-1}$ $bar^{-1}$ in the pressure range 1-20 bar. The pressure effect on the intrinsic reaction rate constant can vary with the coal char rank. The pressure dependence of steam gasification reactions was also investigated by Mühlen et al. [30] using a thermobalance at pressures up to 70 atm. The results showed a linear increase in the reaction rate at lower pressures for all gasifying reactant gases. Sha et al. [31] measured the gasification rates of seven coal chars varying in rank from lignite to bituminous, using a pressurized thermobalance. It was observed that the gasification rates increased with increasing pressure up to 10 atm and leveled off at higher pressures. Goyal et al. [32] studied the gasification kinetics of chars prepared from bituminous coals, with steam, steam-hydrogen and synthesis gas mixtures in a temperature range of 1198-1311 K and a pressure range of 4-28 bar. The authors observed that the total pressure had a negligible effect on the reaction rate.

### 3.3 Product gas

The products of the gasification reactions were analyzed at 1073-1173 K and the results obtained are shown in Fig. 4. It can be observed that hydrogen, carbon monoxide and carbon dioxide were produced in significant amounts and an increase in the reaction temperature clearly increased the concentration of all gases analyzed. No $CH_4$ was detected. It can be observed in Fig. 4 that the hydrogen and carbon monoxide production rate is higher for Char CMRS-02. According to Cao et al. [6], a syngas with a stoichiometric ratio ($H_2$/CO) of 2 (or ranging from 1 to 2) is generally used for the production of syngas-based chemicals. The $H_2$/CO ratio was found to be > 2 for both chars. This is a significant result because this is the ratio required value for the Fischer-Tropsch synthesis to produced liquid fuels.

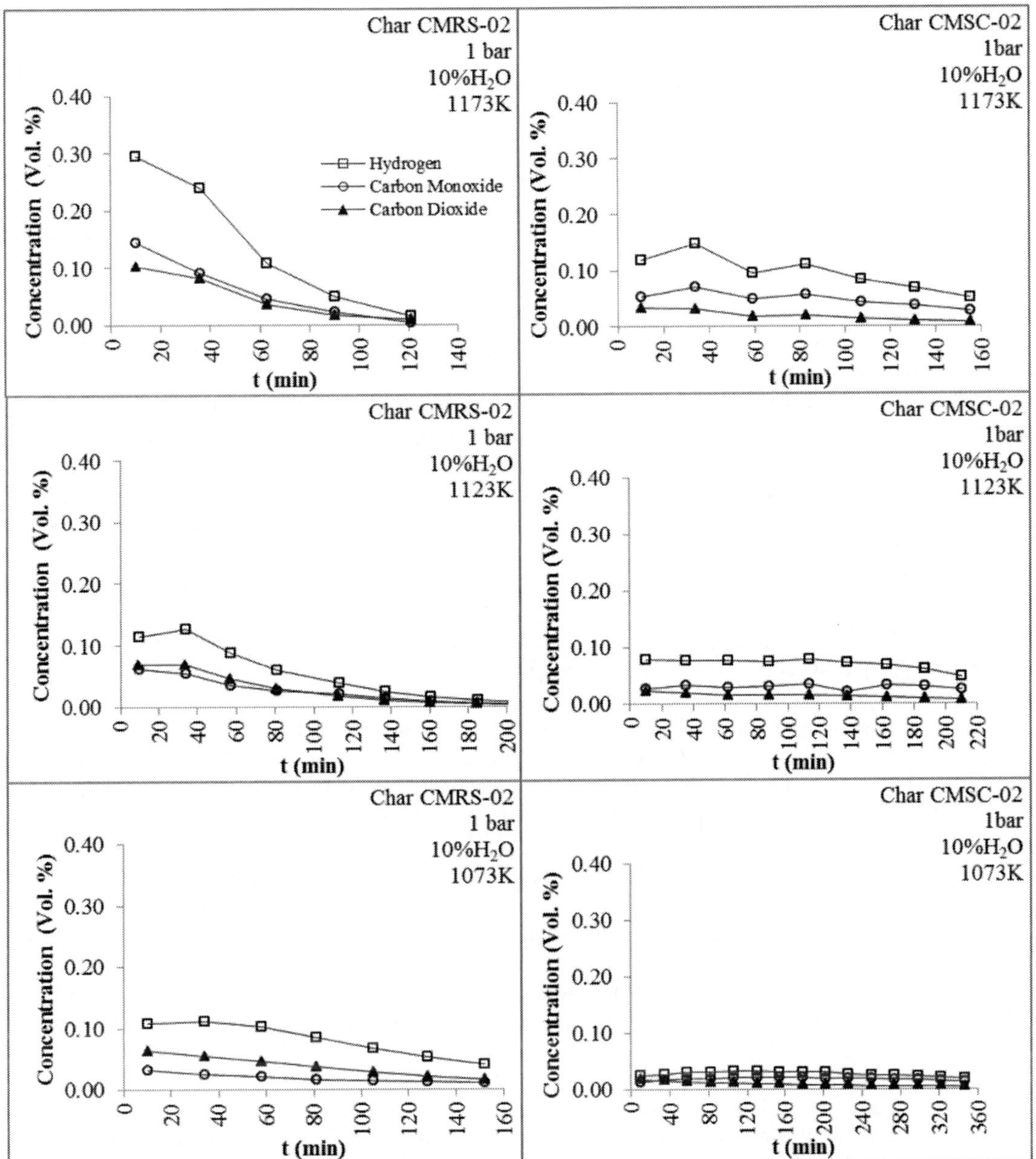

Fig. 4. The composition of the gaseous products of steam gasification reactions.

## 4. Conclusions

The gasification kinetics of Brazilian coals was investigated in a PTGA in the temperature range of 1073-1173 K and pressure range of 1-20 bar. The samples showed similar carbon contents, different ranks and, consequently different vitrinite and liptinite contents. CMRS-02 (sub-bituminous) presented a higher vitrinite and liptinite content than CMSC-02 (bituminous). Sample CMRS-02 was more reactive than sample CMSC-02, indicating that the reaction rate increases with a decrease in the rank of the parent coal. In this study, the effect of temperature and total pressure on the gasification rate was stronger for the bituminous sample when compared to the sub-bituminous one. The main gases formed were $H_2$ and CO (with smaller amounts of $CO_2$), and the $H_2$/CO molar ratio (> 2) indicates that the syngas produced is promising for Fischer-Tropsch synthesis.

## 5. Acknowledgements

The authors would like to acknowledge the Brazilian agencies CNPq and CAPES for financial support and the Mineral Processing Laboratory of the Federal University of Rio Grande do Sul, Brazil, for the preparation of the coal sample.

## 6. References


[1] World Energy Council, Coal, 2013. in: World Energy Resources: 2013 Survey, pp. 1-32.
[2] BP, 2014. Statistical review of world energy (accessed 15.05.14).
[3] Fermoso J, Gil M V, Arias B, Plaza M G, Pevida C, Pis J J, Rubiera F, 2010. Application of response surface methodology to assess the combined effect of operating variables on high-pressure coal gasification for $H_2$-rich gas production. Int J Hydrogen Energ 35, 1191-1204.
[4] Pereira E G, da Silva J N, de Oliveria J L, Machado C S, 2012. Sustainable energy: A review of gasification technologies. Renew Sust Energ Rev 16, 4753-4762.
[5] Taba L E , Irfan M F, Daud W A M W, Chakrabarti, M H, 2012. The effect of temperature on various parameters in coal, biomass and CO-gasification: A review. Renew Sust Energ Rev 16, 5584-5596.
[6] Loha C, Gu S, de Wilde J, Mahanta P, Chatterjee P K, 2014. Advances in mathematical modeling of fluidixed bed gasification. Renew Sust Energ Rev 40, 688-715.
[7] Xie K C, Li W Y, Zhao W, 2010. Coal chemical industry and its sustainbale development in China. Energy 35, 4349-4355.
[8] Tan Q L, Wen Z G, 2016. Goal and technology path of $CO_2$ mitigation in China's cement industry: from the perspective of co-benefit. J Clean Prod 114, 299-313.
[9] Baliban R C, Elia J A, Weekman V, Floudas C A, 2012. Process synthesis of hybrid coal, biomass, and natural gas to liquids via Fischer-Tropsch synthesis, ZSM-5 catalytic conversion, methanol synthesis,methanol-to-gasoline, andmethanolto-olefins/distillate technologies. Comput Chem Eng 47, 29–56.
[10] Boateng A A, Hicks K B, Vogel K P, 2006. Pyrolysis of switchgrass (Panicum virgatum) harvested at several stages of Maturity. J Anal Appl Pyrols 75, 55–64.
[11] Brazilian Mineral Yearbook – National Departament of Mineral Production 2010 (accessed 01.06.11).
[12] Cetin E, Gupta R, Moghtaderi B, 2005. Effect of pyrolysis pressure and heating rate on radiata pine char structure and apparent gasification reactivity. Fuel 84,1328-1334.
[13] Takarada T, Tamai Y, Tomita A, 1985. Reactivities of 34 coals under steam gasification. Fuel 64, 1438-1442.
[14] van Heek K H, Muhlen H-J, 1987. Effect of coal and char properties on gasification. Fuel Process Technol 15,113-133.
[15] Shufen L, Xinyan X, 1993. Gasification reactivity of three Chinese coal chars with steam at elevated pressure. Fuel 72,1351-1353.
[16] Ahn D H, Gibbs B, Ko K H, Kim J J, 2001. Gasification Kinetics of an Indonesian sub-bituminous coal-char with $CO_2$ at elevated pressure. Fuel 80,1651-1658.
[17] Tremel A, Haselsteiner T, Kunze C, Spliethoff H, 2012. Experimental investigation of high temperature and high pressure coal gasification. Appl Energ 92,279-285.
[18] José, H J, 1989. Zur Reaktivität von Koksen aus Santa Catarina-Steinkohle. Brasilien, bei der Vergasung mit Wasserdampf und Kohlendioxid. Thesis, RWTH Aachen, Germany.
[19] Schmal M, Monteiro J L F, Castellan J L, 1982. Kinetics of Coal Gasification. Ind Eng Chem Process Des Dev 21, 256-266.
[20] ASTM E 1131-03 - American Society for Testing and Material: Standard Test Method for Compositional Analysis by Thermogravimetry. 2004.
[21] International Standard 7405/5. Methods for the petrographic analysis of bituminous coal and anthracite - Part 5: Method of determining microscopically the reflectance of vitrinite, 1984.
[22] Mehta B N, Aris R, 1971. Communications on the theory of diffusion and reaction - VII. The isothermal nth order reaction. Chem Eng Sci 26,1699-1712.
[23] Messenböck R C, Paterson N P, Dugwell D R, Kandiyoti R, 2000. Factors governing reactivity in low temperature coal gasification. Part 1. An attempt to correlate results from a suite of coals with experiments on maceral concentrates. Fuel 79,109-121.
[24] Walker P L Jr, 1981. Structure of coals and their conversion to gaseous fuels. Fuel 60, 801-802.
[25] Jenkins R G, Nandi S P, Walker P L Jr, 1973. Reactivity of heat-treated at 500°C. Fuel 52, 288-293.
[26] Hippo E., Walker P L Jr, 1975. Reactivity of heat-treated coals in carbon dioxide at 900°C. Fuel 54, 245-248.
[27] Tomita A, Mahajan O P, Walker P L Jr, 1977. Reactivity of heat-treated coals in Hydrogen. Fuel 56, 137-144.
[28] Linares-Solano A, Mahajan O P, Walker P L Jr, 1979. Reactivity of heat-treated coals in steam. Fuel 58, 327- 332.
[29] Miknis F P, 1995. Composition, Geochemistry and Conversion of Oil Shales. NATO ASI Series C, 455, Kluwer, Dordrech.
[30] Mühlen H J, van Heek K H, Jüntgen H, 1985. Kinetic studies of steam gasification of char in the presence of $H_2$, $CO_2$ and CO. Fuel 64, 944-949.
[31] Sha X Z , Y G, Cao J, Yang Y M, Ren D Q, 1990. Effects of operating pressure on coal gasification. Fuel Chen 69, 656-659.
[32] Goyal A, Zabransky R F, Rehmat A, 1989. Gasification Kinetics of Western Kentucky Bituminous Coal Char. Ind Eng Chem Res 28, 1767-1778.
[33] Cao Y, Gao Z, Jin J, Zhou H, Cohron M, Zhao H, Liu H, Pan W, 2008. Synthesis Gas Production with an Adjustable $H_2$/CO Ratio through the Coal Gasification Process: Effects of Coal Ranks and Methane Addition. Energ Fuel 22, 1720-1730.

**2016 International Pittsburgh Coal Conference**
**Cape Town, SOUTH AFRICA**
**August 8 – 12, 2016**

**Abstract Submission**

**PROGRAM TOPIC:**
**GASIFICATION SCIENCE AND MODELING**

**Exact Title of Paper:**

# EVALUATION OF $CO_2$ GASIFICATION REACTIVITY OF BIOMASS AND COAL CHARS BY PHYSICOCHEMICAL PROPERTIES

***Zhiqing Wang, State Key Laboratory of Coal Conversion, Institute of Coal Chemistry, Chinese Academy of Sciences, Taiyuan, Shanxi 030001, People's Republic of China, People's Republic of China***
***Emai: qcumt@sxicc.ac.cnx***

*Jiejie Huang, State Key Laboratory of Coal Conversion, Institute of Coal Chemistry, Chinese Academy of Sciences, Taiyuan, Shanxi 030001, People's Republic of China, People's Republic of China*
*Yitian Fang, State Key Laboratory of Coal Conversion, Institute of Coal Chemistry, Chinese Academy of Sciences, Taiyuan, Shanxi 030001, People's Republic of China, People's Republic of China*
*Jiantao Zhao, State Key Laboratory of Coal Conversion, Institute of Coal Chemistry, Chinese Academy of Sciences, Taiyuan, Shanxi 030001, People's Republic of China, People's Republic of China*
*Libo Dong, State Key Laboratory of Coal Conversion, Institute of Coal Chemistry, Chinese Academy of Sciences, Taiyuan, Shanxi 030001, People's Republic of China, People's Republic of China]*

**Abstract:**

The isothermal and non-isothermal $CO_2$ gasification of different biomass charss and different coal rank char were conducted by thermogravimetric analyzer (TGA), and the reactivity index and peak temperature were obtained to express the gasification reactivity. In addition, the physicochemical properties of these chars, such as the alkali index, Brunauer-Emmett-Teller surface area ($S_{BET}$), micropore area ($S_{mic}$), carbon crystalline structure were examined by inductively coupled plasma spectrometer (ICP), $N_2$ adsorption and X-ray diffraction (XRD). The $CO_2$ chemisorption was also measured by TGA and the total, strong and weak chemisorbed volumes ($C_{total}$, $C_{str}$ and $C_{wea}$) were obtained. All of these properties were used to evaluate the $CO_2$ gasification reactivity of these chars. The results show that the gasification reactivity of chars decreases with increasing coal rank. Comparing with the other properties, the $C_{total}$ and $C_{str}$ are the best parameters for predicting the gasification reactivity, and the linear correlation coefficients between $C_{total}$ /$C_{str}$ and gasification reactivity can reach up to 0.90 or even 0.98. It suggests that the $CO_2$ chemisorption can be used to evaluate the gasification reactivity of different coal rank chars. And this evaluation is valid when it is extended to biomass chars, thus the $CO_2$ chemisorption may be a common and useful method for gasification evaluation.

China University of Mining Technology (CUMT)

# The CUMT International Research Centre for Underground Coal Gasification (UCG)

Dr Cliff Mallett, Director CUMT UCG Research Centre

Dr. Shirong Ge, President, CUMT

Dr Jim Zhang, Vice-Director, CUMT UCG Research Centre

PCC August 2016

# China Energy Consumption

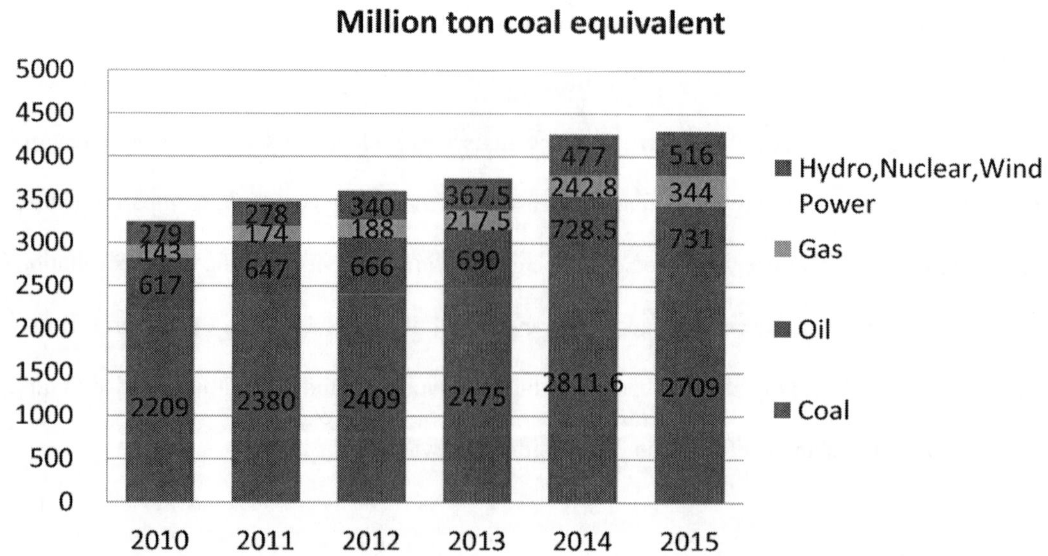


2

# China Coal Distribution

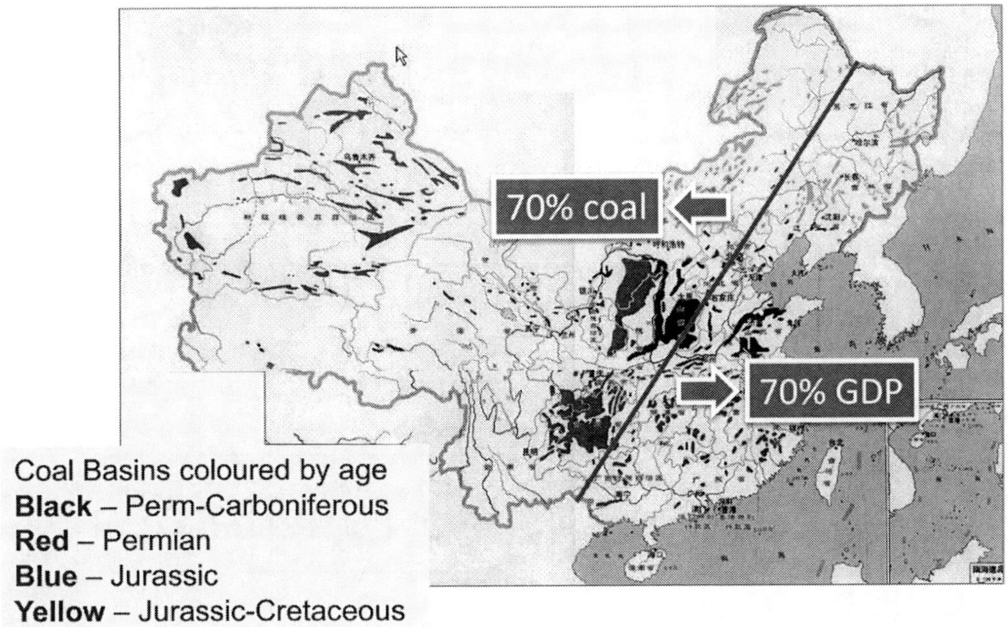


3

# Revised Trend of Coal Output in China

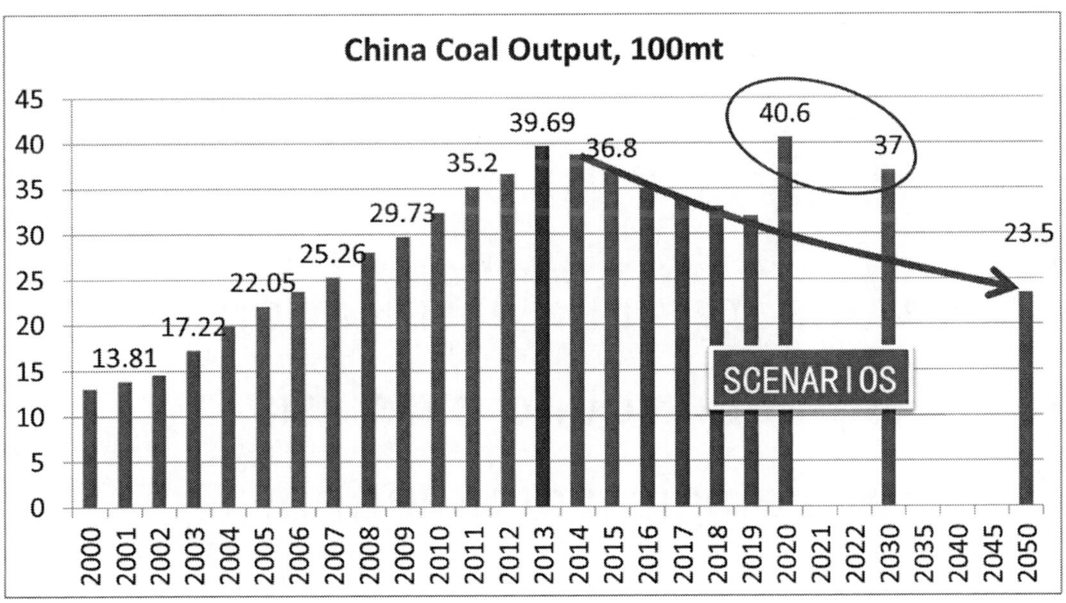


4

# Evolution of Coal Mining & Use

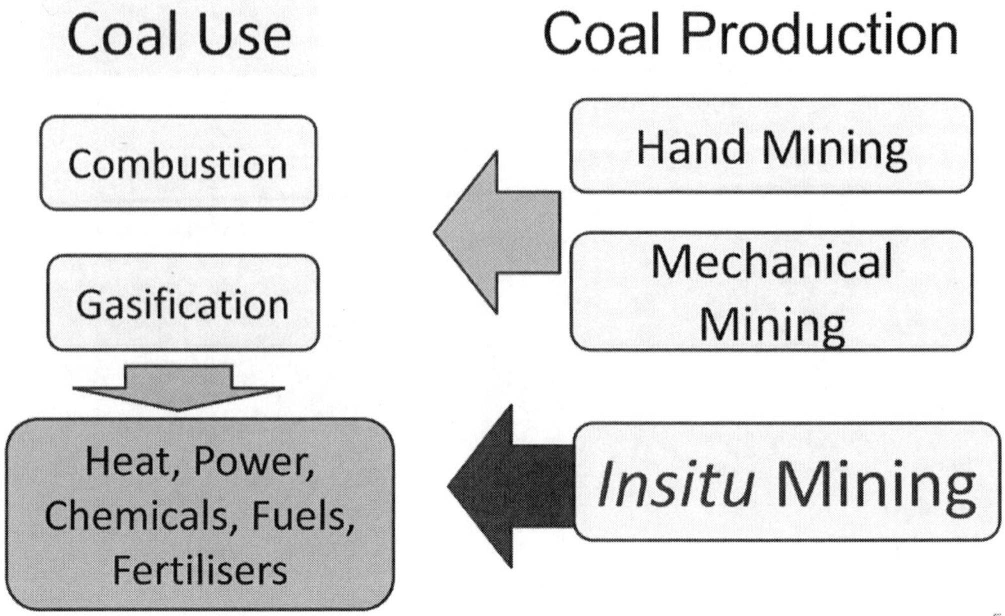


5

# *Insitu* Mining of Coal

Coal *insitu* mining is a method to control the chemical reaction of coal seam, and then extract the energy and chemicals of the raw coal. Includes:

- underground gasification
- underground pyrolysis
- biological dissolution.

6

# Fathers of UCG – Chemical Engineers

1868
Conceived the
UCG concept

**Karl William Siemens, 1823～1883**

1888
Published paper
on coal to gas
underground

**D. I. Mendeleev, 1834-1907**

1904
Conducted
gasification
trials at
Durham, UK.

**William Ramsay, 1852-1916**

1947
Published paper
on replacement
of miners with
UCG

**Dawei YU (Jing HUANG), 1912-1958**

7

# UCG Fundamentals

- **Coal gasification reactions**
- Gasifier cavity is a pressure vessel
- Progressive movement of gasification face
- Oxidant injection & product removal
- No escape of by-products

8

# UCG Fundamentals

- Coal gasification reactions
- Gasifier cavity is a pressure vessel
- Progressive movement of gasification face
- Oxidant injection & product removal
- No escape of by-products

9

# Interactions with UCG

P. Pei et al. / Energy Conversion and Management 113(2016) 95–103

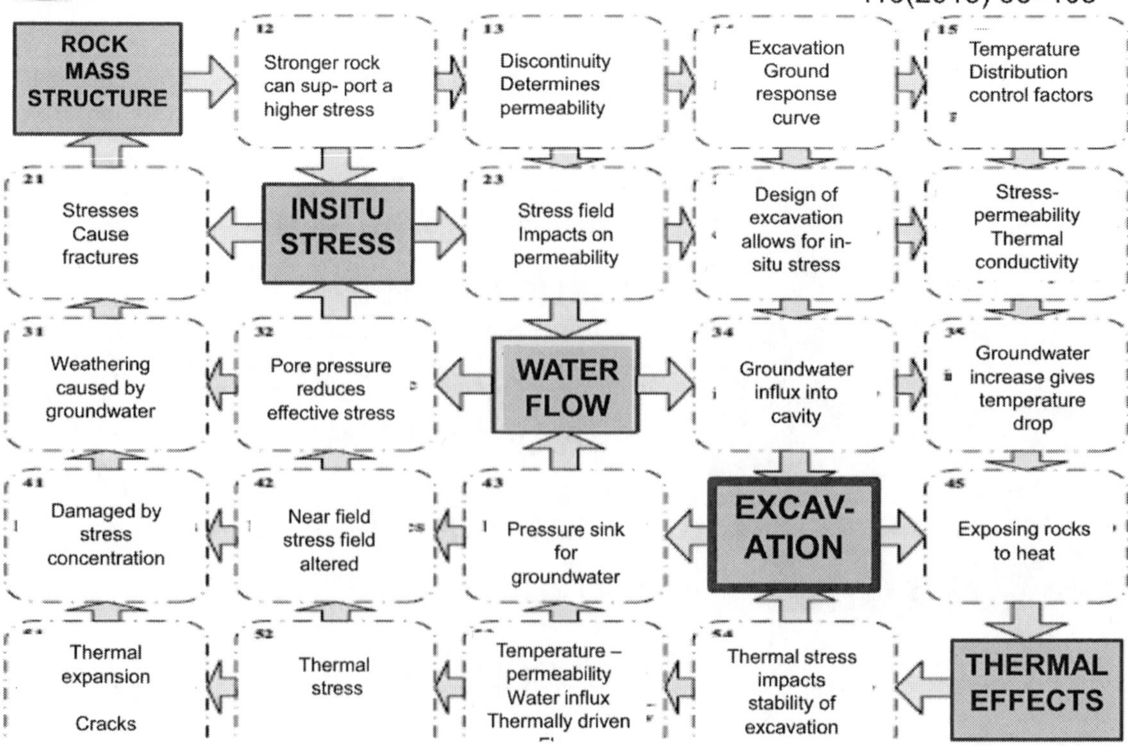

# Advantage of Energy Efficiency

| | Gas Output $Nm^3/kg$ | Coal Energy MJ/kg | Syngas Energy $MJ/Nm^3$ | Energy Used kWh/GJ | CO2 Emissions kg/GJ |
|---|---|---|---|---|---|
| LURGI Gasifier | 1.91 | 22.3 | 11.69 | 3.09 | 2.70 |
| Shell Gasifier | 1.47 | 16.3 | 11.07 | 2.70 | 2.36 |
| UCG | 1.97 | 18.3 | 9.27 | 1.68 | 1.47 |
| | No1 | No2 | No3 | No1 | No1 |

11

# UCG Tests in China, 1958-1959

20 UCG test sites have been built up in China, 11 UCG sites were designed by CUMT.

| Year | Location | Kinds of Coal | Gasifier | Syngas | Average Heating Value ($MJ/m^3$) | Syngas Output (x10k $m^3/d$) |
|---|---|---|---|---|---|---|
| 1958 | Datong | No-Caking Coal | Borehole | Air-blown | 3.81 | |
| 1958 | Jilin | Gas-Coal | Borehole | Air-blown | 3.50 | |
| 1958 | Hegang | Gas-Coal | Borehole | Air-blown | 5.54 | |
| 1959 | Anhui | | Tunnel | Air-blown | | 0.15 |

12

# 1959, Anhui UCG Trial

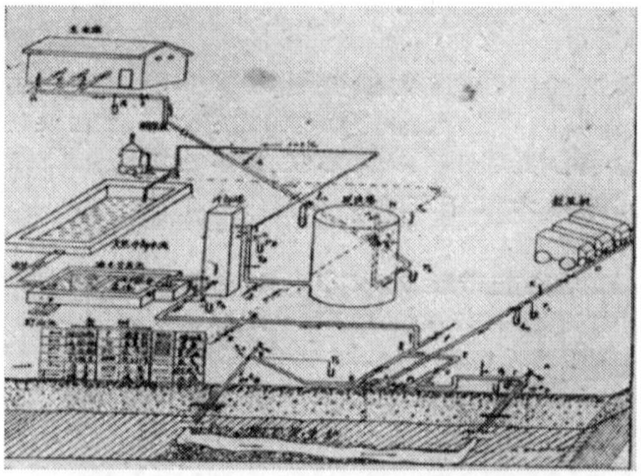

| 日期 | | | | CO | | H₂ | | $C_nH_m$ | | $CH_4$ | | $CO_2$ % | $O_2$ % | 热值 | 鼓风量 |
|---|---|---|---|---|---|---|---|---|---|---|---|---|---|---|---|
| 年 | 月 | 日 | 时 | 含量 % | 热值 | 含量 % | 热值 | 含量 % | 热值 | 含量 % | 热值 | 含量 | 含量 | 大卡/标准米³ | 米³/时 |
| 59 | 12 | 15 | 11 | 7.55 | 230 | 14.7 | 380 | 1 | 150 | 1.5 | 128 | 9.55 | 1.85 | 888 | 550 |
| 59 | 12 | 17 | 19 | 5.1 | 155 | 10 | 258 | 0.3 | 45 | 1.5 | 128 | 11.2 | 3.45 | 586 | 800 |
| 58 | 12 | 19 | 9 | 29 | 88.5 | 8.55 | 221 | 0.25 | 37.5 | 1.8 | 162 | 14.2 | 0.95 | 509 | 520 |
| 59 | 12 | 26 | 11 | 78 | 258 | 5.9 | 152 | 2.0 | 500 | 1.6 | 137 | 10 | 3.9 | 847 | 720 |
| 59 | 12 | 26 | 23 | 9.6 | 293 | 18.3 | 472 | 1.0 | 150 | 1.4 | 120 | 12.7 | 0.6 | 1085 | 730 |
| 59 | 12 | 29 | 23 | 12.1 | 358 | 22.4 | 578 | 0.5 | 75 | 1.4 | 120 | 13 | 1.4 | 1.41 | 1190 |

# UCG tests in China, 1987-2000

| Year | Location | Kinds of Coal | Gasifier | Syngas | Average Heating Value ($MJ/m^3$) | Syngas Output (x10k $m^3/d$) | Cost ($RMB/m^3$) |
|---|---|---|---|---|---|---|---|
| 1987 | Xuzhou-1 | Gas-Coal | Borehole | Air-Blown | 3.53 | 0.03 | |
| 1994 | Xuzhou-2 | Gas-Coal | Yuli-Model | Steam-Blown | 13.7 | 3.6 | 0.088 |
| 1996 | Tangshan | Gas-Coal | Yuli-Model | Steam-Blown | 12.74 | 11 | 0.19 |
| 1997 | Heilongjiang | Long-Flame Coal | Tunnel | Air-Blown | 5 | 9 | |
| 1998 | Hebi | Lean-Coal | Tunnel | Air-Blown | 4.82 | 7 | 0.063 |
| 1998 | Yima | Long-Flame Coal | Tunnel | Air-Blown | 5.24 | Coal Consumption 3470t | 0.009 |
| 2000 | Xinmi | Lean-Coal | Tunnel | Rich-Oxygen Blown | 7.92 | 6 | |
| 2000 | Xinwen | Gas-Coal | Yuli-Model | Steam-Blown | 14.7 | 6 | 0.20 |

14

# 1987, Xuzhou UCG Trial

**Professor YU Li, CUMT**

**UCG Process** – Long Channel, Large Cross-Section and Two Stage

15

# 1996, Tangshan UCG Trial

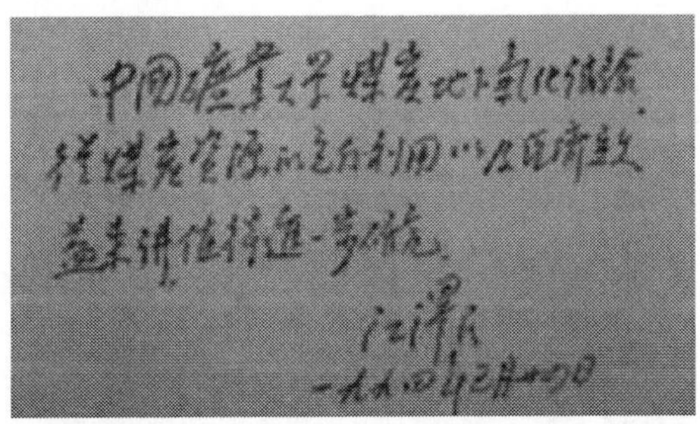

表 0-6 两阶段工艺煤气组分及热值(唐山刘庄煤矿工业性试验)

| 1996 年 | | 煤气组分 (%) | | | | | 煤气热值 | 煤气流量 |
|---|---|---|---|---|---|---|---|---|
| 月 | 日 | $H_2$ | CO | $CH_4$ | $CO_2$ | $N_2$ | $MJ/m^3$ | $m^3/h$ |
| 6 | 21 | 40.68 | 28.02 | 7.84 | 5.41 | 17.97 | 11.88 | 1963 |
| 6 | 23 | 39.25 | 21.28 | 9.50 | 1.13 | 28.84 | 11.46 | 2121 |
| 6 | 26 | 48.98 | 5.02 | 13.65 | 22.61 | 9.74 | 12.26 | 2315 |
| 7 | 5 | 52.12 | 14.48 | 12.40 | 1.77 | 19.21 | 13.37 | 2139 |
| 7 | 6 | 43.57 | 15.68 | 11.02 | 5.01 | 22.81 | 11.89 | 2287 |
| 7 | 8 | 51.53 | 16.96 | 14.66 | 8.47 | 7.33 | 14.51 | 2936 |
| 7 | 9 | 49.11 | 13.21 | 14.11 | 16.82 | 6.75 | 13.51 | 2871 |
| 7 | 10 | 39.25 | 28.25 | 17.15 | 5.10 | 10.00 | 15.42 | 2656 |
| 7 | 11 | 47.14 | 13.36 | 12.38 | 20.48 | 6.65 | 12.59 | 2263 |
| 7 | 12 | 55.94 | 12.69 | 14.59 | 9.43 | 7.34 | 14.50 | 2435 |
| 7 | 15 | 46.69 | 14.45 | 10.27 | 23.55 | 5.04 | 11.83 | 2345 |
| 7 | 16 | 47.73 | 9.09 | 15.73 | 26.12 | 1.33 | 13.45 | 2233 |
| 7 | 17 | 65.16 | 8.29 | 8.89 | 14.06 | 3.60 | 12.83 | 2315 |
| 7 | 18 | 60.21 | 10.39 | 12.29 | 14.21 | 2.90 | 13.82 | 2216 |
| 7 | 24 | 47.91 | 16.63 | 12.04 | 18.17 | 5.21 | 12.97 | 2462 |
| 7 | 25 | 53.01 | 24.77 | 7.23 | 10.46 | 4.53 | 12.74 | 2346 |
| 7 | 26 | 41.58 | 11.67 | 7.46 | 28.93 | 10.35 | 9.71 | 2989 |
| 7 | 27 | 71.68 | 11.38 | 13.03 | 2.07 | 1.83 | 15.70 | 2324 |
| 7 | 28 | 59.96 | 4.91 | 9.38 | 22.97 | 2.78 | 11.93 | 2523 |
| 7 | 29 | 52.00 | 11.24 | 8.65 | 21.83 | 6.27 | 11.45 | 2430 |
| 7 | 31 | 36.77 | 8.83 | 8.54 | 39.31 | 6.55 | 9.17 | 2750 |

16

# UCG tests in China, 2001-2016

| Year | Location | Kinds of Coal | Gasifier | Syngas | Average Heating Value (MJ/m$^3$) | Syngas Output (x10k m$^3$/d) |
|---|---|---|---|---|---|---|
| 2001 | Xinwen | Gas-Coal | Yuli-Model | Semi-Steam Blown | 8.37 | 2 |
| 2001 | Feicheng | Gas-Coal | Yuli-Model | Air-Blown | 6.24 | 3.5 |
| 2001 | Xiyang | Anthracite | Yuli-Model | Air-Blown | 5.68 | 4 |
| 2001 | Panzhihua | Lean-Coal | Tunnel | Air-Blown | 4.2 | 8 |
| 2002 | Xinwen-1 | Gas-Coal | Yuli-Model | Air-Blown | 5.2 | 6 |
| 2003 | Xinwen-2 | Gas-Coal | Tunnel | Air-Blown | | 15 |
| 2005 | Zhongliangshan | Cokeing-Coal | Tunnel | Steam-Blown | 11.5 | 5 |
| 2007 | Inner Mongolia | Lignite | Borehole | Steam-Blown | 16.7 | 15 |
| 2010 | Huating | No-Caking Coal | Tunnel | Mixed | 4.77 | 16 |
| 2016 | Panjiangsha | Coking-Coal | Tunnel | Steam-Blown | 9 | 18 |

17

# 2005, Zhongliangshan UCG Trial

Since 2000, CUMT UCG Process

***Underground Guide Control Chemical Mining Technology***

Zhongliangshan, Gansu and Guizhou Trial and Industrial Demonstration

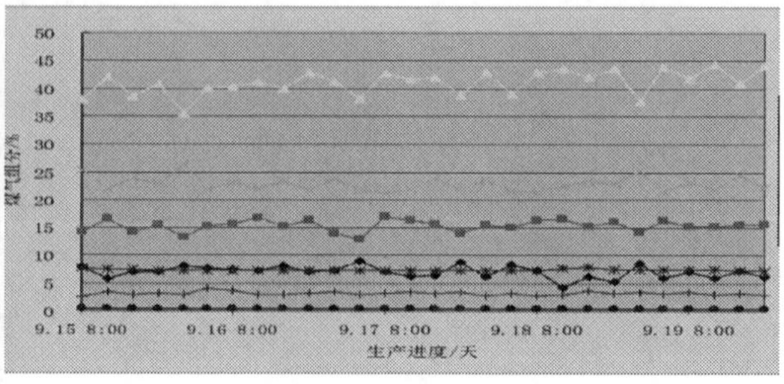


18

# UCG Research Cluster

In January 2012, the largest UCG industrial plant was founded in Pingding, Shanxi.

In July 2015, CUMT, Yilong Energy Co. and Carbon Energy Company joined in cooperation.

19

# China Energy Innovation Action (2016-2030) (NDRC & NEA)

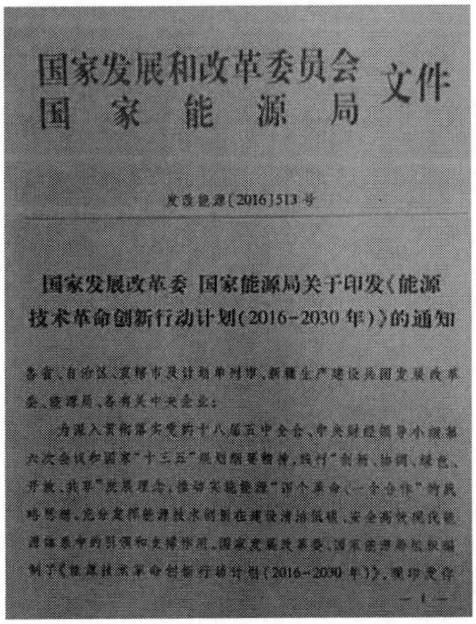

国家发展和改革委员会
国　家　能　源　局　文件

发改能源[2016]513号

国家发展改革委 国家能源局关于印发《能源技术革命创新行动计划(2016-2030年)》的通知

各省、自治区、直辖市及计划单列市、新疆生产建设兵团发展改革委、能源局、各有关中央企业：

[illegible]

— 1 —

1. The No-harm Coal Mining Technological Innovation
   - Improved the efficiency of coal exploitation and intelligent level
   - High efficiency tunnel construction & high speed excavating, intelligent mining panel, high energy recovery for special coal seam, *Underground Coal Gasification*, associated resources comprehensive exploitation and utilization technology, etc.
   - Key mining area implementation unmanned panel, the national coal mining mechanization achieved above 95%.

20

# CUMT International UCG Research Centre

Opening ceremony March 2016

**FOCUS**

***UCG Syngas Production and use, environment***

***UCG panel construction*** - drilling, excavation, well completion
***Enhanced gasification designs*** - simulations and demonstrations
***Syngas synthesis*** - to fuels and chemicals
***Clean low cost power*** – UCG+IGCC
***UCG and emission reduction*** – UCG+CCS+CCU
***Pollution reduction*** through UCG

21

# Activities of the Centre

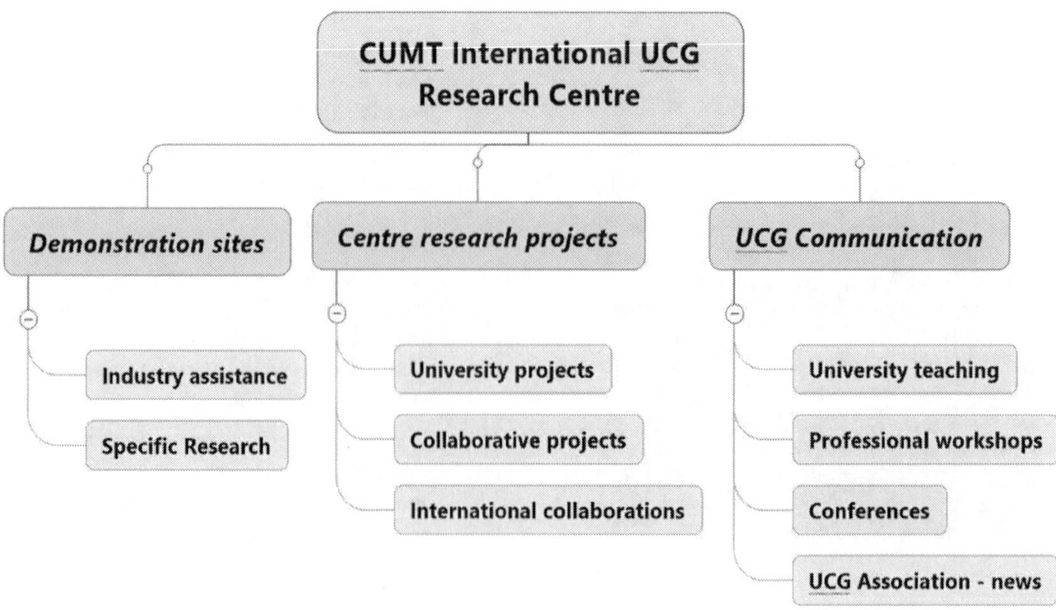

# Thank You.

23

# Online Detection and Retention of Trace Contaminants in Reducing Gasification Atmosphere

*Philip Edinger* *[a,b], *Mohamed Tarik* [b], *Christian Ludwig**[a,b]

[a] École Polytechnique Fédérale de Lausanne (EPFL), ENAC-IIE, 1015 Lausanne (Switzerland)

[b] Paul Scherrer Institut (PSI), ENE, 5232 Villigen PSI (Switzerland)

* corresponding authors:

Philip Edinger, e-mail: philip.edinger@alumni.ethz.ch

Prof. Christian Ludwig, e-mail: christian.ludwig@epfl.ch

ABSTRACT

Gasification of coal creates a syngas which can be used for the production of electricity or chemicals. Besides other undesired by-products, the raw syngas exiting the gasifier, contains various trace elements. Among them are mercury (as $Hg^0$), selenium (as $H_2Se$) and arsenic (as $AsH_3$) and P (as $PH_3$). All of them tend to remain in the gas phase and can thereby pose a threat to down-stream equipment as well as to the environment. An option to retain these contaminants are sorbent materials. A profound evaluation of their sorption performance requires reliable analytical tools with high temporal resolution and detection limits at parts per billion by volume (ppbv) levels. This study therefore addresses the development and successful employment of an online analytical method, based on the direct introduction of a simplified flue gas ($H_2$, $H_2O$, Ar) into an ICP-MS. Applying matrix matched calibration, the individually calculated detection limits are < 2 ppbv for $H_2Se$ and $PH_3$, < 0.1 ppbv for $AsH_3$ and < 0.01 ppbv for Hg, thus being comparable to current offline methods. These low detection limits and the high temporal resolution (≈ 1 Hz) enable the use of this method as a quick and robust tool for the lab- and bench-scale performance evaluation of sorbent materials. The capture of $H_2Se$ by a $Zn/Al_2O_3$ sorbent at temperatures from 150 to 350°C was chosen as a case study. The results show that the support itself interacts with $H_2Se$ at dry reaction conditions. However, as small amounts of steam are added to the gas stream, the sorption capacity strongly decreases, indicating competitive adsorption of $H_2Se$ and $H_2O$ on the support. For experiments with $Zn/Al_2O_3$ sorbents, sorption of $H_2Se$ was observed up to steam contents of 10 vol% and temperatures up to 350°C with initial $H_2Se$ concentrations below the method detection limit.

# 1. Introduction

During gasification of coal, trace elements in the fuel are volatized to different extents. Thermodynamic studies show that elements which are most volatile and mainly remain in the gas phase are Hg, Br, Cl, F, B, Se, I and As. [1-3] Of these elements As, B, Hg and Se were categorized by the Environmental Protection Agency (EPA) as elements of major concern due to their high toxicity to biological systems.[4] Of these elements of major concern, As and Se can also negatively affect downstream equipment.[5] As (in form of $AsH_3$) can poison catalysts for methanol production [6], other synthesis catalysts [7] and lead to degradation of SOFCs.[8-10] Selenium is known to cause degradation of SOFCs.[8] For this reason there is a growing interest to quantify their concentration in gasification environments.

This study aims at establishing an experimental setup that allows for the online detection of $AsH_3$, $H_2Se$, Hg and $PH_3$ in a simplified reducing gas atmosphere containing $H_2$, $H_2O$ and Ar. An ICP-MS was chosen as measurement device. In combination with matrix matched calibration, this allows for a highly time resolved study of sorbent materials even at concentrations in the low ppbv range. $Zn/Al_2O_3$ was chosen as material to investigate the sorption of $H_2Se$. The aim of this study is therefore twofold. The first step is to evaluate the ICP-MS with direct gas introduction as a tool for quantification of As, Hg, P and Se in reducing gas atmosphere. In the second step, this method is applied to investigate the Se capture performance of $Zn/Al_2O_3$ as a sorbent at warm gas cleanup conditions.

## 2. Matrix Effects

ICP measurement in the gas phase has to consider its composition. Already small amounts of $H_2$ and $H_2O$ are known to change the plasma conditions and thereby the analytical performance of an ICP-MS. The addition of $H_2$ to ICP-MS plasmas has been studied by several authors. Guillong and Heinrich [11] added small amounts of $H_2$ (0-14 ml/min) to the helium carrier gas flow of a LA-ICP-MS dry plasma. Under optimized ICP-MS operating parameters and $H_2$ flowrates, a general increase in sensitivity (for $^{77}$Se by a factor 3) was observed. Limit of detections could generally be decreased. However, in case of $^{77}$Se, a two-fold increase due to polyatomic interferences was observed. In the work of Ebdon, et al. [12], the addition of 5 vol% $H_2$ to the nebulizer gas of a wet plasma caused a significant increase in polyatomic interferences based on Ar but decreased interferences from metal oxides. Louie and Soo [13] found a slight increase in $^{78}$Se signal when adding small amounts of $H_2$ (< 10 ml/min) to a wet plasma, to be followed by a continuous decline in signal when adding higher amounts of $H_2$, resulting in a 50% signal decrease when 50 ml/min $H_2$ were added. Under optimum conditions ($H_2$ flow of 20 ml/min), the authors qualitatively reported a minor increase in S/B ratio for most elements. Hutton, et al. [14] observed an increase of the $^{127}$I signal by 70% when adding 7.5 ml/min $H_2$ to the central channel of a dry plasma. The increase was ascribed to an increased energy transfer between bulk plasma and the central channel due to the high thermal conductivity of $H_2$. A noted side effect of $H_2$ addition was the increase in reflected power (30 W for 7.5 ml/min $H_2$). Liu, et al. [15] investigated the effect of small additions of water vapour (5.1 Nml/min) to the central gas flow in LA-ICP-MS. For a shielded torch (as used in the current work), they found an increase in sensitivity for most investigated elements (for $^{82}$Se by a factor 2.5) and a change in spatial profiles of the ion distribution upon addition of $H_2O$. Alary and Salin [16] added small amounts of water (4.1 Nml/min) to the carrier gas flow between an ETV and an ICP-MS and observed changes to the signals of the measured elements (Zn, Pb, Cu, As, Sb).

# 3. Material and Methods

## 3.1. Experimental Apparatus

### 3.1.1. Analytical Method

Experiments were performed using an Agilent 7700x ICP-MS spectrometer. Gas sampling was done continuously by sending it to the plasma torch by means of a quartz adapter shown in Figure 1. Four different sets of ICP-MS operating conditions were used as listed in Table 1. Helium was used in the octopole reaction system (ORS) to lower polyatomic interferences.

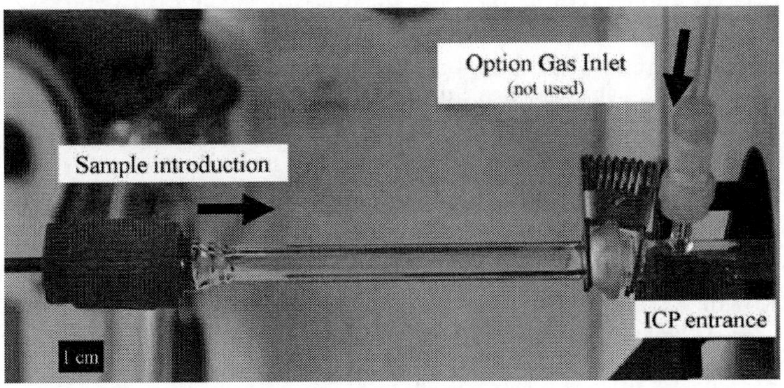


*Figure 1 – In-house constructed quartz adapter (similar to the connector tube of the used ICP instrument), directly connected to the plasma torch*

*Table 1 - ICP-MS operating conditions*

<table>
<tr><th></th><th>Matrix effects</th><th>$H_2Se$</th><th>$PH_3$, $AsH_3$</th><th>Hg</th></tr>
<tr><td colspan="5">ICP operating conditions</td></tr>
<tr><td>RF power (W)</td><td colspan="4">1350</td></tr>
<tr><td>Reflected RF power (W)</td><td>< 60</td><td colspan="3">< 3</td></tr>
<tr><td>Sample Depth (mm)</td><td colspan="2">10</td><td colspan="2">8</td></tr>
<tr><td>Plasma gas flow rate (Nl/min)</td><td colspan="4">15</td></tr>
<tr><td>Carrier Gas (Nml/min)</td><td>900</td><td>1100</td><td colspan="2">1000</td></tr>
<tr><td>He flow (Nml/min)</td><td>1.5</td><td colspan="3">2.0</td></tr>
<tr><td colspan="5">Mass spectrometer acquisition settings</td></tr>
<tr><td>Detector</td><td colspan="4">pulse mode</td></tr>
<tr><td>Scan mode</td><td colspan="4">time resolved, peak hopping</td></tr>
<tr><td>Settling + scan time (ms)</td><td colspan="4">10</td></tr>
<tr><td>Integration time</td><td colspan="4">0.2 s / m/z</td></tr>
<tr><td>Recorded isotopes</td><td>$^{13}C$, $^{76}Se$, $^{77}Se$, $^{78}Se$, $^{82}Se$</td><td>$^{13}C$, $^{66}Zn$, $^{68}Zn$, $^{76}Se$, $^{77}Se$, $^{78}Se$, $^{82}Se$</td><td>$^{13}C$, $^{31}P$, $^{35}Cl$, $^{75}As$, $^{124}Xe$</td><td>$^{124}Xe$, $^{199}Hg$, $^{200}Hg$, $^{201}Hg$, $^{202}Hg$</td></tr>
</table>

3.1.2. Experimental Setups

Two setups were used to perform experiments. Figure 2 (a) is a schematic representation of the setup used to perform calibration and sorption experiments with $H_2Se$, $AsH_3$ and $PH_3$ as gas contaminants. The setup encompasses a set of mass flow controllers (MFC, EL-Flow, Bronkhorst) to deliver defined amounts of gases. Argon (quality 6.0), $H_2$ (4.6) and $O_2$ (4.8) were taken in pure form from gas cylinders. Predefined gas mixtures of $H_2Se$ (106 ppmv $H_2Se$ in $H_2$) or $AsH_3$ and $PH_3$ (48 ppmv $AsH_3$, 55 ppmv in $H_2$) were used to feed the contaminants to a three way valve, sending it either to a gas scrubber (two impingner bottles of 0.7 M $HNO_3$) or to the reactor. In order to minimize pressure pulsations in the experimental setup, steam was generated by the controlled reaction of $H_2$ and $O_2$ (ratio $H_2/O_2 > 6$) at 250°C over a Pt catalyst. [17] Tubing in contact with the analytes was $SiO_2$ coated (SilcoNert 2000, SilcoTek) and heated to 160°C to avoid adsorption effects. The gas flow entering the reactor was kept at 100 Nml/min. After the reactor the gas flow was diluted by Ar to give total carrier flow rates between 900-1100 Nml/min, depending on the experimental set (see Table 1). In this work concentrations on a molar basis refer to those encountered in the reactor.

The second experimental setup used for Hg calibration experiments is depicted in Figure 2 (b). It shares many of the features of setup 1, but has been modified to include an Hg generation device. Steam generation was omitted in this setup. A constant Ar flow of 37.5 Nml/min passes along an Hg permeation tube (Vici). The $Hg^0$ permeation rate is temperature dependent and controlled by means of a water bath. The $Hg^0$ containing gas flow is either directed to a gas scrubber or towards an empty quartz reactor. The latter stream is diluted by $H_2$, Ar and an Xe/Ar mixture, where the Xe acts as an internal standard.

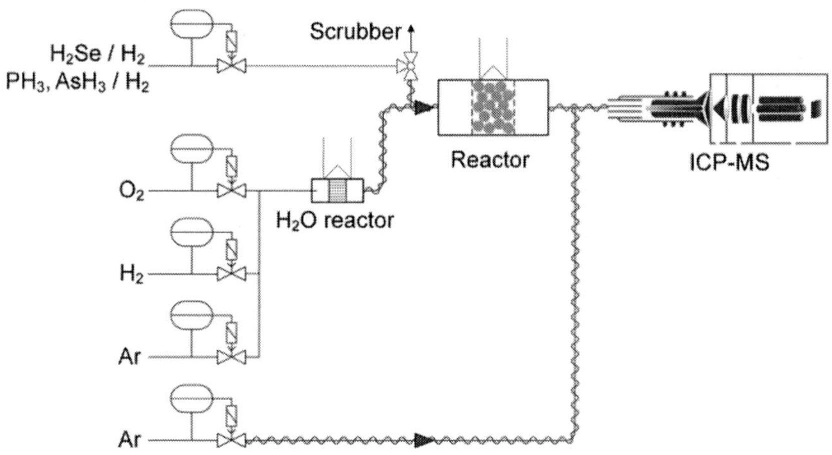


(a)

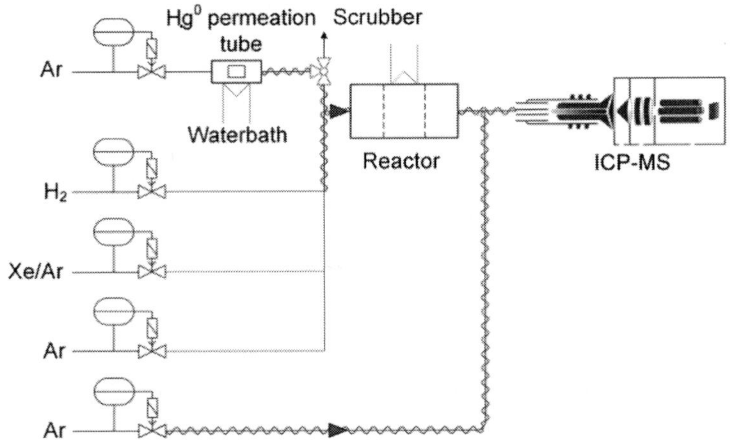


(b)

*Figure 2 – Schematic representations of experimental setups. (a) Setup 1 was used to perform calibration experiments for contaminants $AsH_3$, $PH_3$ and $H_2Se$ and sorption experiments including $H_2Se$; (b) Setup 2 was used to perform calibration experiments involving Hg.*

## 3.2. Sorbent Material

$H_2Se$ sorption experiments were performed with an in-house prepared sorbent consisting of 1 wt% Zn supported on commercial $\gamma$-$Al_2O_3$ (Puralox Nwa-155, Sasol Germany GmbH). The sorbent was prepared by a wet impregnation technique. $\gamma$-$Al_2O_3$ was sieved (sieve 125-160 μm) and subsequently degassed in Ar atmosphere at 300°C for 3 h. Impregnation was performed by soaking the support in a $Zn(NO_3)_2$ (analytical grade, Sigma Aldrich) solution at room temperature for 3 h. The impregnated support was then dried at 80°C, 25 mbar for 1 h and calcined in air at 500°C for 7 h.

## 3.3. Experimental Procedures

### 3.3.1. Matrix Effects

In order to evaluate effects and possible interactions of the two factors $H_2$ and $H_2O$ on the measurement of the $^{78}Se$ isotope, a two-level full factorial design ($2^2$) was performed.[18] The factors were varied within boundaries, representative for gasification processes, with $H_2$ concentrations ranging from 20 to 40 vol% and the steam content ranging from 0 to 5 vol%. A geometric representation of the experimental design with the corresponding gas flows is shown in Figure 3. Each factor combination, noted by numbers 1-4, in this plot consists of both a blank run and a run containing 2.25 ppmv of $H_2Se$. For each run, replicates with more than 100 data points were acquired on isotope $^{78}Se$ after signal stabilisation.

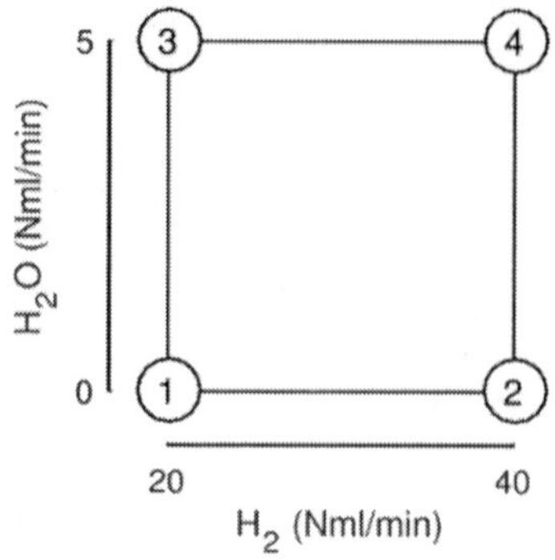


*Figure 3 - $2^2$ factorial design with the gas concentrations of $H_2$ and $H_2O$ as factors. The run numbers are indicated in the corners.*

### 3.3.2. Sorption Experiments

Before sorption experiments a matrix-matched calibration was performed using an empty reactor. Afterwards a quartz reactor (2 mm inner diameter), containing the sorbent bed (20.5 mg) was placed in the oven (MTF 12/25A, Carbolite Limited) and flushed with the $H_2Se$-free experimental matrix for 30 min. All sorption experiments have H2 concentrations of 40 vol%. Sorption experiments were started by adding 4.77 ppmv $H_2Se$ to the gas feed.

The removal efficiency $\eta_{H_2Se}$ of $H_2Se$ is calculated according to Eq (1), where $y_{Se,in}$ and $y_{Se,out}$ refer to the $H_2Se$ gas concentrations at the reactor inlet and outlet, respectively.

$$\eta_{H_2Se} = \frac{y_{Se,in} - y_{Se,out}}{y_{Se,in}} \tag{1}$$

The sorbent capacity $q_{Se}$ up to the time $t_{Bt}$ where the exit gas concentration exceeds 500 ppbv (representing $\eta_{H_2Se}$= 0.9) is calculated in Eq. (2). In this equation $M_{Se}$ refers to the molecular weight of Se, $m_{bed}$ the sorbent bed mass, $\dot{V}_{rx}$ the total gas flow through the reactor and $V_m$ the molar volume.

$$q_{Se} = \frac{M_{Se}\,\dot{V}_{rx}}{m_{bed}\,V_m}\int_0^{t_{Bt}} \left(y_{Se,in} - y_{Se,out}\right) dt \tag{2}$$

# 4. Results and Discussion

## 4.1. Calibration Results

The response time characteristics of an analytical method become of importance when transient processes such as sorption processes are studied. In order to investigate these characteristics, step response and wash-out experiments involving $AsH_3$, $H_2Se$, Hg and $PH_3$ were performed. A signal increase of species $AsH_3$, $H_2Se$ and $PH_3$ is detected 2 to 8 s after 2.4 to 2.7 ppmv of analyte were added to the inlet gas stream. 98% of the stable signal is reached after 13 to 20 s, depending on analyte. The washout behavior was analyzed by stopping the analyte feed at a certain point in time and observe the signal response. All three species show an initially steep decrease in signal, followed by significant tailing. While this is of no direct consequence when measuring breakthrough curves it can affect the time in-between measurements as detection limits are increased due to the increased background levels. In case of the performed experiments, detection limits of 4 to 7 ppbv were measured, 5 min after the contaminant addition was stopped.

Experiments involving $Hg^0$ were performed at different concentration (8 ppbv) levels in setup 2. The response time is similar to those of the other analytes, with a stable signal being reached after 40 s. When $Hg^0$ addition is stopped, a 99% decrease is observed after 1min, and the limit of detection after 5 min is 0.07 ppbv.

Calibration curves for all four analytes are shown in Figure 4. The calibration results are summarized in Table 2. The coefficients of determination for all species are >0.998, indicating a good linear response. The detection limits for Se and P are below 2 ppbv, while those of As (0.1 ppbv) and Hg (0.002 ppbv) are significantly lower.

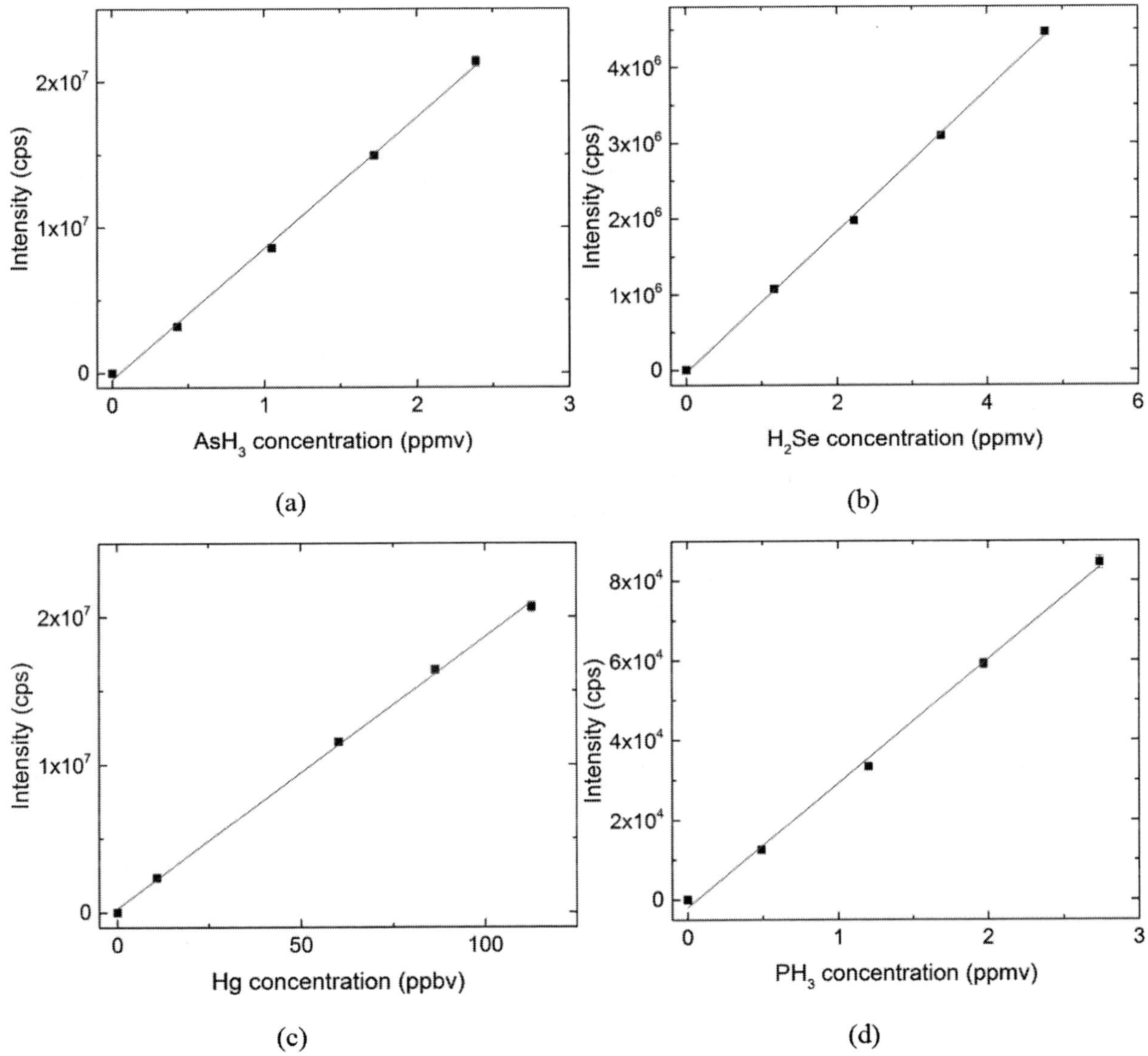


*Figure 4 - Calibration curves for (a) $AsH_3$ measured on isotope $^{75}As$; (b) $H_2Se$ measured on isotope $^{78}Se$; (c) $PH_3$ measured on isotope $^{31}P$; (d) Hg measured on isotope $^{202}Hg$.*

*Table 2 - Calibration results*

| | $^{76}$**Se** | $^{77}$**Se** | $^{78}$**Se** | $^{82}$**Se** | $^{199}$**Hg** | $^{200}$**Hg** | $^{201}$**Hg** | $^{202}$**Hg** | $^{75}$**As** | $^{31}$**P** |
|---|---|---|---|---|---|---|---|---|---|---|
| $R^2$ | 0.9992 | 0.9990 | 0.9990 | 0.9992 | 0.9992 | 0.9992 | 0.9991 | 0.9990 | 0.9978 | 0.9982 |
| LOD (ppbv) | 1.36 | 1.15 | 0.74 | 1.03 | 0.003 | 0.003 | 0.004 | 0.002 | 0.1 | 1.6 |
| Abundance (mol/mol %) | 9.4 | 7.6 | 23.8 | 8.7 | 16.9 | 23.1 | 13.2 | 29.9 | 100 | 100 |
| Interferences [19] | $^{40}Ar\,^{36}Ar^+$, $^{38}Ar\,^{38}Ar^+$ | $^{36}Ar\,^{40}Ar\,^{1}H^+$, $^{38}Ar_2\,^{1}H^+$ | $^{40}Ar\,^{38}Ar^+$ | $^{40}Ar_2\,^{1}H_2^+$ | | | | | $^{36}Ar^{38}Ar^{1}H^+$ | $^{14}N^{16}O^{1}H^+$, $^{15}N^{15}N^{1}H^+$, $^{15}N^{16}O^+$, $^{14}N^{17}O^+$, $^{13}C^{18}O^+$, $^{12}C^{18}O^{1}H^+$ |

#### 4.1.1. Matrix Effects

Even though a change of the gas composition in the current case is not expected during single experiments, an understanding of their qualitative influence on the plasma is of interest. The investigation of these effects will reveal the necessity or not of using a 100% matrix matched calibration and will evaluate the accuracy/reliability of the used setup and the calibration procedure. Therefore $H_2$ and $H_2O$ were varied as two factors in a $2^2$ full factorial design to obtain an estimate of the main and interaction effects and to get a general understanding of their significance. Based on the performed experiments, the main effects $E_{H_2}$ and $E_{H_2O}$ of $H_2$ and steam can be calculated according to Eq. (3) and Eq. (4). The interaction $E_{H_2,H_2O}$ of these two compounds is calculated according to Eq. (5). In these equations the response $\gamma$ can represent the blank signal or the sensitivity obtained from a 2-point calibration.

$$E_{H_2} = \frac{(\gamma_2 + \gamma_4) - (\gamma_1 + \gamma_3)}{2} \tag{3}$$

$$E_{H_2O} = \frac{(\gamma_3 + \gamma_4) - (\gamma_1 + \gamma_2)}{2} \tag{4}$$

$$E_{H_2,H_2O} = \frac{(\gamma_1 + \gamma_4) - (\gamma_2 + \gamma_3)}{2} \tag{5}$$

For calculation of the variance, runs were not genuinely repeated but rather a pooled estimate of error averaged from the data points acquired in each run was calculated. The computed error therefore represents the analytical rather than the experimental run-to-run variance and is likely to be smaller than the latter.

The calculated absolute values of the effects on the sensitivity and blank signal are plotted as Pareto charts of effects in Figure 5 (a) and Figure 5 (b), respectively, in order to find the most influential factors. The red line in the plot represents the 95% confidence interval, giving an estimate of the significance of the individual effects. In Figure 5 (a), all bars show a negative effect on the sensitivity. However, only the main effect $E_{H2}$ caused by a change in $H_2$ concentration, is significant in the varied range, showing a decrease of 32% in sensitivity compared to the constant effect. Increase of the $H_2$ concentration also resulted in a significant increase of reflected power up to 60 W but no plasma instabilities were observed. An optimization of the matching network was not performed. The effect of the factors on the blank signal is shown in Figure 5 (b). Again, only the change in $H_2$ concentration has a significant, negative, effect on the signal, decreasing it by 17%, compared to the constant effect. The effect of water or the interaction thereof are negligible.

While in literature, the use of mixed-gas plasmas tends to improve figures of merit concerning Se analysis, this work cannot confirm these observations for the given experimental settings. Two facts might contribute to these findings. Previous studies investigated mixed-gas plasmas with the intention of improving figures of merit. This typically involved optimization of the plasma parameters that can deviate significantly from all-Ar plasmas. However, when measuring a transient process, where the gas composition is subject to change, does not allow for continuous parameter optimization and calibration. Performing no optimization, therefore can, as observed in the current study, lead to adverse effects concerning the analysis of Se. Another difference lies in the introduced sample. Studies typically feature aerosol introduction where the evaporation of droplets and the decomposition of particles plays a central role. In this case however, the introduced sample is a gas. Knowing that the curvature of the response might not be linear, limits the range of conclusions that can be drawn. In case we assume not too much of a non-linearity, the obtained results imply that changes in steam content between 0 to 0.56 vol% at the ICP-MS inlet do not affect the quantification of selenium on $^{78}Se$. The influence of $H_2$ on the other hand is significant which results in increased detection limits.

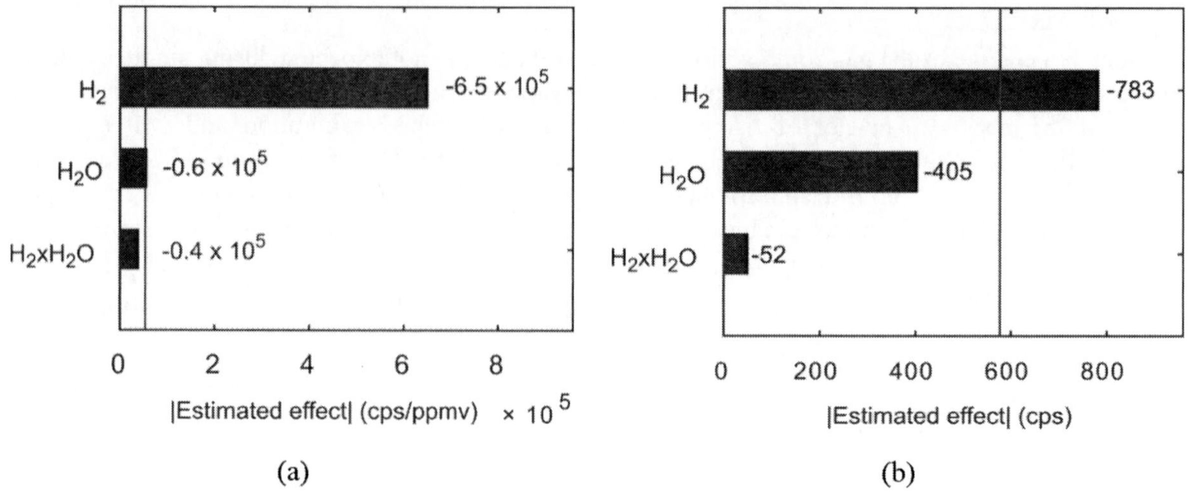


*Figure 5 - Pareto charts of effects (a) Sensitivity; (b) Blank signal.*

## 4.2. $H_2Se$ Sorption Experiments

An overview of experimental conditions and obtained Se capacities is given in Figure 6.

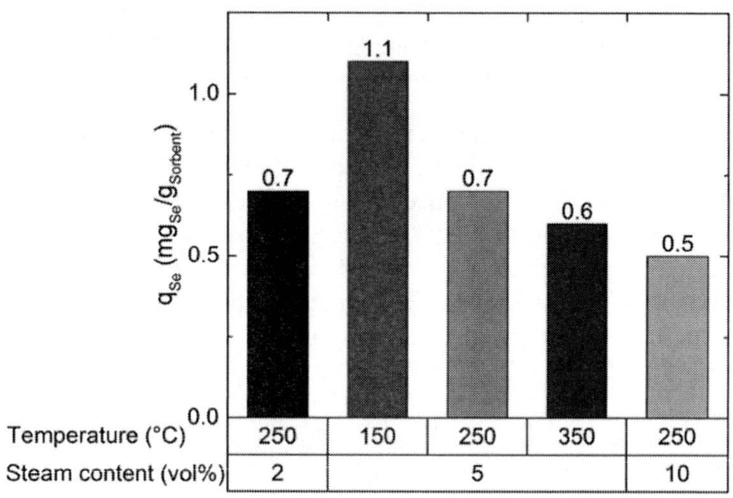


*Figure 6 – Breakthrough capacities of $Zn/Al_2O_3$ sorbent for different experimental conditions*

### 4.2.1. Effect of $H_2O$ Content

Because most gasification processes contain steam, its effect on the sorption behavior is of interest. Figure 7 (a) depicts the $H_2Se$ breakthrough curve of the $Zn/Al_2O_3$ sorbent in a gas atmosphere containing no steam. A 100% $H_2Se$ removal efficiency is observed for 30 min, afterwards the efficiency drops to ≈98% until the measurement is stopped after 3 h. The stoichiometric breakthrough time, based on a stoichiometric reaction of ZnO and $H_2Se$ according to Eq. (6) is surpassed after 130 min.

$$ZnO + H_2Se \leftrightarrow ZnSe + H_2O \quad (6)$$

This indicates an interaction of the support in the $H_2Se$ removal as it has previously been observed at lower temperatures.[20] Indeed, additional experiments involving the sole $\gamma$-$Al_2O_3$ support exhibit capacities of $\approx 1 mg_{Se}/g_{Sorbent}$ in dry atmosphere. However, as steam (5 vol%) is added to the gas composition, this capacity reduces drastically (<2 $\mu g_{Se}/g_{Sorbent}$). This indicates competitive adsorption of $H_2O$ and $H_2Se$ on $\gamma$-$Al_2O_3$.

Figure 7 (b) shows breakthrough curves for steam contents of 2, 5 and 10 vol% at 250°C. All three curves share a similar profile with high initial $H_2Se$ removal rates that decrease rapidly after 4 min and become less steep after ≈15 min at removal efficiencies of 20-30%. Breakthrough curves for experiments with 2 and 5 vol% steam content are almost identical. A further increase in steam content to 10 vol% results in an earlier breakthrough time (6 min *versus* 8 min) Compared to the expected stoichiometric breakthrough time of 130 min, the observed ones are short, indicating a small conversion of the sorbent. This is concurrent with previous work on the sorption of $H_2Se$ on supported Zn sorbents ($Zn_2TiO_4$).[21]

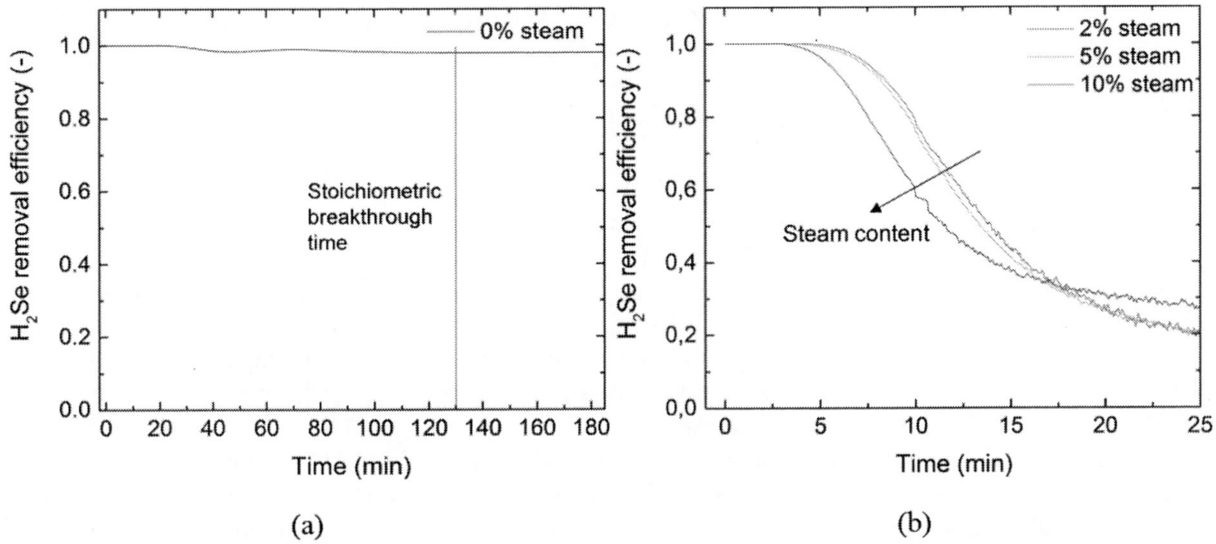


*Figure 7 - (a) Breakthrough curve for 0 vol% steam content (250°C); (b) Breakthrough curves for steam contents of 2, 5 and 10 vol% (250°C).*

## 4.2.2. Effect of Temperature

Figure 8 plots the $H_2Se$ removal efficiency of the $Zn/Al_2O_3$ sorbent for three temperatures 150, 250 and 350°C and 5 vol% steam loading. All curves show high initial removal rates followed by a decrease in removal efficiency after 5-10 min which starts to level off after 15-25 min. The breakthrough at 250 and 350°C occurs after 7 and 8 min, respectively. When the temperature is further decreased to 150°C the breakthrough time decreases to 14 min. Based on reaction kinetics and mass transfer considerations, an increase in temperature should effect an increase in conversion leading to longer breakthrough times. This is not observed. The thermodynamic equilibrium constant of reaction in Eq. (6) increased by two orders of magnitude ($10^{12} \rightarrow 10^{8}$) as the temperature is increased from 150 to 350°C.

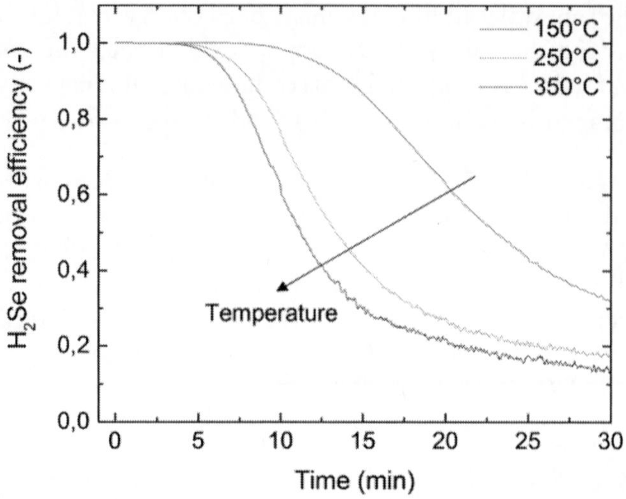


*Figure 8 - Breakthrough curves for temperatures 150, 250 and 350°C (5 vol% steam)*

# 5. Conclusions

In this study, a new analytical method for the online detection of coal gasification contaminants involving $AsH_3$, $H_2Se$, Hg and $PH_3$ has been developed. Using a direct gas inlet on an ICP-MS in combination with matrix matched calibration, these contaminants can reliably be detected with high temporal resolution (≈1 Hz). Detection limits of $H_2Se$ and $PH_3$ were found to be <2 ppbv while those of $AsH_3$ with 0.1 ppbv and Hg with 0.002 ppbv were lower. An investigation of matrix effects of $H_2$ and $H_2O$ content in the carrier gas showed that, in the range of varied parameters, $H_2O$ had a negligible and $H_2$ a negative effect on figures of merit. A case study investigated the capture of $H_2Se$ by a $Zn/Al_2O_3$ sorbent at temperatures from 150 to 350°C. The results show that the support itself interacts with $H_2Se$ at dry reaction conditions. However, as small amounts of steam are added to the gas stream, the sorption capacity strongly decreases, indicating competitive adsorption of $H_2Se$ and $H_2O$ on the support. For experiments with $Zn/Al_2O_3$ sorbents, sorption of $H_2Se$ was observed up to steam contents of 10 vol% and temperatures up to 350°C with initial $H_2Se$ concentrations below the method detection limit.

REFERENCES


1. Froment, K.; Defoort, F.; Bertrand, C.; Seiler, J. M.; Berjonneau, J.; Poirier, J., Thermodynamic equilibrium calculations of the volatilization and condensation of inorganics during wood gasification. *Fuel* **2013,** 107, 269-281.
2. Porbatzki, D.; Stemmler, M.; Mueller, M., Release of inorganic trace elements during gasification of wood, straw, and miscanthus. *Biomass Bioenergy* **2011,** 35, S79-S86.
3. Diaz-Somoano, M.; Martinez-Tarazona, M. R., Trace element evaporation during coal gasification based on a thermodynamic equilibrium calculation approach. *Fuel* **2003,** 82, (2), 137-145.
4. United States Environmental Protection Agency, National Emission Standards for Hazardous Air Pollutants From Coal- and Oil-Fired Electric Utility Steam Generating Units and Standards of Performance for Fossil-Fuel-Fired Electric Utility, Industrial-Commercial-Institutional, and Small Industrial-Commercial-Institutional Steam Generating Units; Final Rule. **2012,** 77, (32), 9304-9513.
5. Higman, C.; Tam, S., Advances in Coal Gasification, Hydrogenation, and Gas Treating for the Production of Chemicals and Fuels. *Chemical Reviews* **2014,** 114, (3), 1673-1708.
6. Quinn, R.; Mebrahtu, T.; Dahl, T. A.; Lucrezi, F. A.; Toseland, B. A., The role of arsine in the deactivation of methanol synthesis catalysts. *Appl. Catal., A* **2004,** 264, (1), 103-109.
7. Nielsen, B.; Villadsen, J., Poisoning of nickel catalysts by arsenic. *Applied Catalysis* **1984,** 11, (1), 123-138.
8. Cayan, F. N.; Zhi, M. J.; Pakalapati, S. R.; Celik, I.; Wu, N. Q.; Gemmen, R., Effects of coal syngas impurities on anodes of solid oxide fuel cells. *Journal of Power Sources* **2008,** 185, (2), 595-602.
9. Trembly, J. P.; Gemmen, R. S.; Bayless, D. J., The effect of coal syngas containing AsH3 on the performance of SOFCs: Investigations into the effect of operational temperature, current density and AsH3 concentration. *Journal of Power Sources* **2007,** 171, (2), 818-825.
10. Bao, J.; Krishnan, G. N.; Jayaweera, P.; Lau, K.-H.; Sanjurjo, A., Effect of various coal contaminants on the performance of solid oxide fuel cells: Part II. ppm and sub-ppm level testing. *Journal of Power Sources* **2009,** 193, (2), 617-624.
11. Guillong, M.; Heinrich, C. A., Sensitivity enhancement in laser ablation ICP-MS using small amounts of hydrogen in the carrier gas. *Journal of Analytical Atomic Spectrometry* **2007,** 22, (12), 1488-1494.
12. Ebdon, L.; Ford, M. J.; Goodall, P.; Hill, S. J., Hydrogen Addition to the Nebulizer Gas for the Removal of Polyatomic Ion Interferences in Inductively Coupled Plasma Mass Spectrometry. *Microchemical Journal* **1993,** 48, (3), 246-258.
13. Louie, H.; Soo, S. Y.-P., Use of nitrogen and hydrogen in inductively coupled plasma mass spectrometry. *Journal of Analytical Atomic Spectrometry* **1992,** 7, (3), 557-564.
14. Hutton, R. C.; Bridenne, M.; Coffre, E.; Marot, Y.; Simondet, F., Investigations into the direct analysis of semiconductor grade gases by inductively coupled plasma mass spectrometry. *Journal of Analytical Atomic Spectrometry* **1990,** 5, (6), 463-466.
15. Liu, S. H.; Hu, Z. C.; Gunther, D.; Ye, Y. H.; Liu, Y. S.; Gao, S.; Hu, S. H., Signal enhancement in laser ablation inductively coupled plasma-mass spectrometry using water and/or ethanol vapor in combination with a shielded torch. *Journal of Analytical Atomic Spectrometry* **2014,** 29, (3), 536-544.

16. Alary, J. F.; Salin, E. D., Quantitation of water and plasma diagnosis for electrothermal vaporization-inductively coupled plasma mass spectrometry: The use of argon and argide polyatomics as probing species. *Spectrochimica Acta Part B-Atomic Spectroscopy* **1998,** 53, (12), 1705-1721.
17. Peitz, D.; Bernhard, A.; Elsener, M.; Kroecher, O., Laboratory test reactor for the investigation of liquid reducing agents in the selective catalytic reduction of NOx. *Review of Scientific Instruments* **2011,** 82, (8), 1-8.
18. Box, G. E. P.; Hunter, J. S.; Hunter, W. G., *Statistics for experimenters design, innovation, and discovery*. 2nd ed.; Wiley-Interscience: Hoboken, N.J., 2005; p 633.
19. May, T. W.; Wiedmeyer, R. H., A table of polyatomic interferences in ICP-MS. *Atomic Spectroscopy* **1998,** 19, (5), 150-155.
20. Watanabe, T.; Suzuki, T.; Kinomura, N., Adsorption of diborane and hydrogen selenide on porous alumina and silica. *Journal of Materials Science: Materials in Electronics* **1998,** 9, (2), 127-132.
21. Diaz-Somoano, M.; Lopez-Anton, M. A.; Martinez-Tarazona, M. R., Retention of arsenic and selenium during hot gas desulfurization using metal oxide sorbents. *Energy Fuels* **2004,** 18, (5), 1238-1242.

Thirty Third Annual International

PITTSBURGH COAL CONFERENCE

# INTERACTION BETWEEN RICE STRAW AND JINCHENG ANTHRACITE SINGLE PARTICLE DURING ASHING PROCESS


***Zibing Zhao, Qingqing Han, Chengcheng Zhu, Yuhong Qin*, Jie Feng, Wenying Li***


***Cape Town, South Africa***
***August 8-12, 2016***


***Taiyuan University of Technology***


## Background

Table 1 Requirements for ash chemically of coal gasification technology

| Gasifier | Fixed bed | | Fluidized bed | | Entrained bed | |
|---|---|---|---|---|---|---|
| **Representative Technologies** | BGL | Lurgi | Winkler<br>HTW<br>CFB | U-gas<br>KRW<br>ICC | Texaco<br>EGAS | Shell<br>GSP<br>HT-L |
| **Feeding method** | Lump | Lump | Crushed | Crushed | Slurry | Pulverized |
| **Slag discharge** | ***Liquid*** | Solid | Solid | Ash agglomerating | ***Liquid*** | ***Liquid*** |
| Ash fusion temperature (°C) | FT<1500 | FT>1050 | FT>1050 | ST>1100 | FT<1350 | - |

**2/15**

# Background

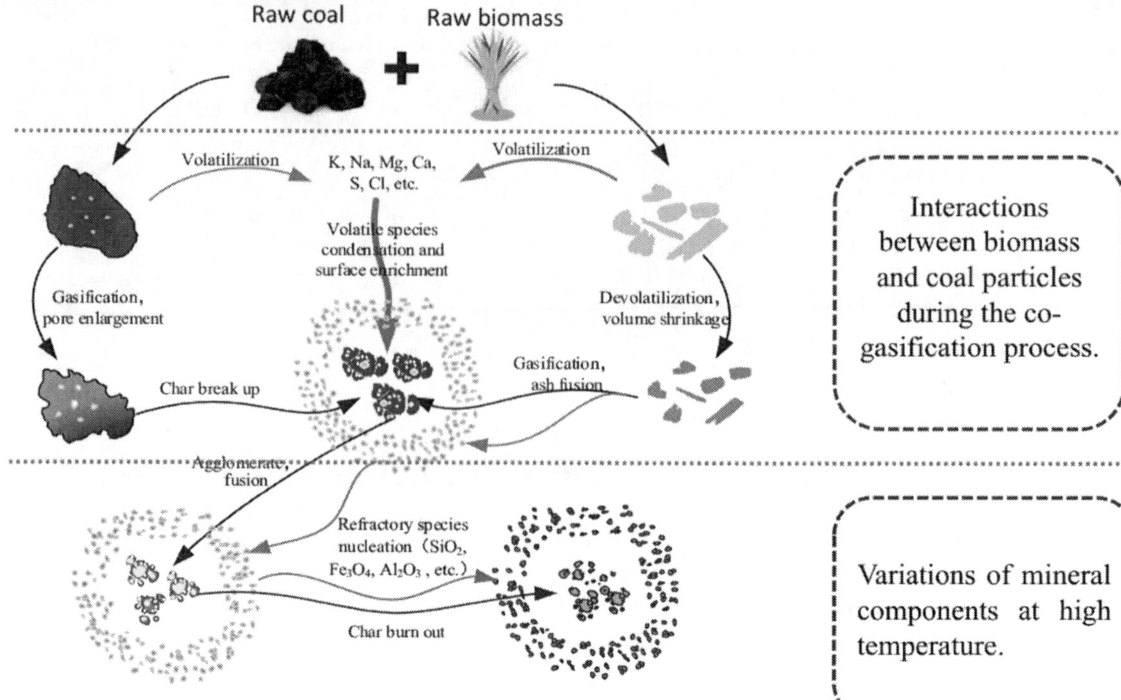


Fig.1 Ash formation and transformation mechanisms during co-gasification of coal and biomass

Interactions between biomass and coal particles during the co-gasification process.

Variations of mineral components at high temperature.

3/15

# Experimental design

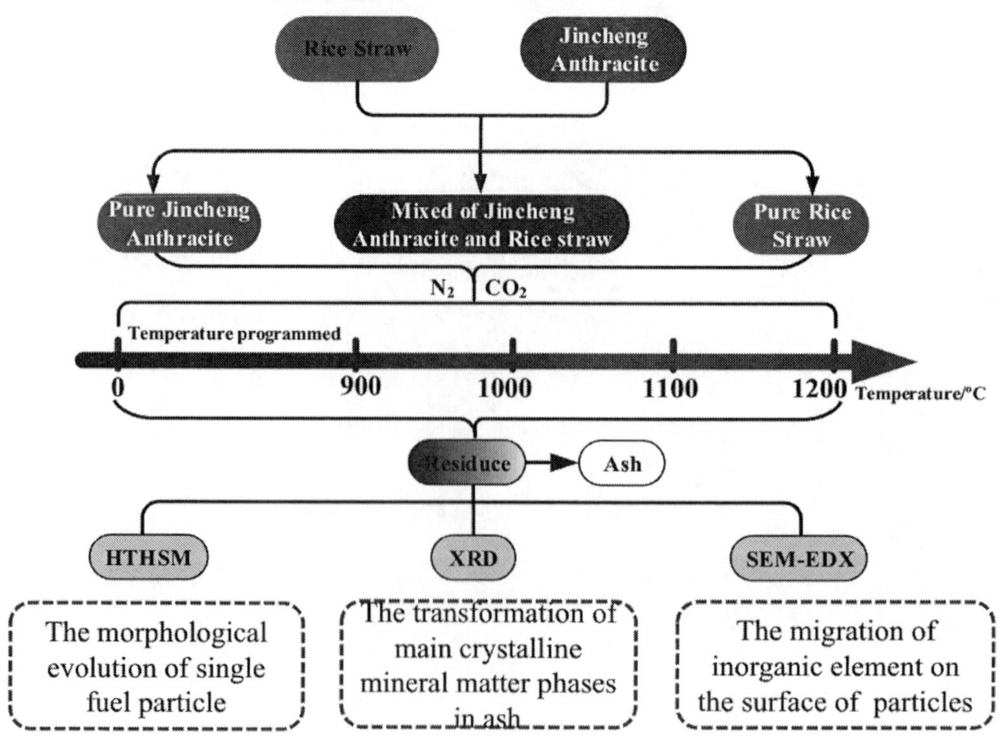


HTHSM: High temperature hot stage microscope

4/15

# *Experimental*

# Sample

Table 2 Proximate and ultimate analysis of Jincheng anthracite and rice straw

| Feedstock | Proximate analysis w/% | | | | Ultimate analysis $w_{daf}$/% | | | | |
|---|---|---|---|---|---|---|---|---|---|
| | $M_{ad}$ | $A_d$ | $FC_{daf}$ | $V_{daf}$ | C | H | O* | N | S |
| Jincheng Coal | 0.95 | 21.4 | 89.79 | 10.21 | 87.94 | 3.22 | 5.32 | 1.10 | 2.42 |
| Rice straw | 5.83 | 15.79 | 14.65 | 85.35 | 45.91 | 2.88 | 48.50 | 0.93 | 1.78 |

ad-air dried basis, d- dried basis, daf-dry ash free basis, *by difference

Table 3 Chemical compositions* of ash from Jincheng anthracite and rice straw

| Sample | Chemical composition w/% | | | | | | | | |
|---|---|---|---|---|---|---|---|---|---|
| | $SiO_2$ | $Al_2O_3$ | $TiO_2$ | $Fe_2O_3$ | CaO | MgO | $K_2O$ | $Na_2O$ | $SO_3$ |
| Jincheng ash | 49.37 | 31.94 | 1.46 | 11.10 | 2.91 | 1.12 | 0.80 | 0.57 | 0.73 |
| Rice straw ash | 42.68 | 5.03 | 1.56 | 2.74 | 13.37 | 6.76 | 11.45 | 5.52 | 10.89 |

*:The Chemical compositions* of ash were determined according the Chinese Standard GB/T 1574-2007

Table 4 AFTs* of Jincheng anthracite and rice straw

| | Temperature t/°C | | | |
|---|---|---|---|---|
| | DT | ST | HT | FT |
| Jincheng ash | >1500 | | | |
| Rice straw ash | 1090 | 1127 | 1130 | 1133 |

*: The fusion temperatures of the ash were determined according the Chinese Standard GB/T 219-2008.

# Experimental setup 1

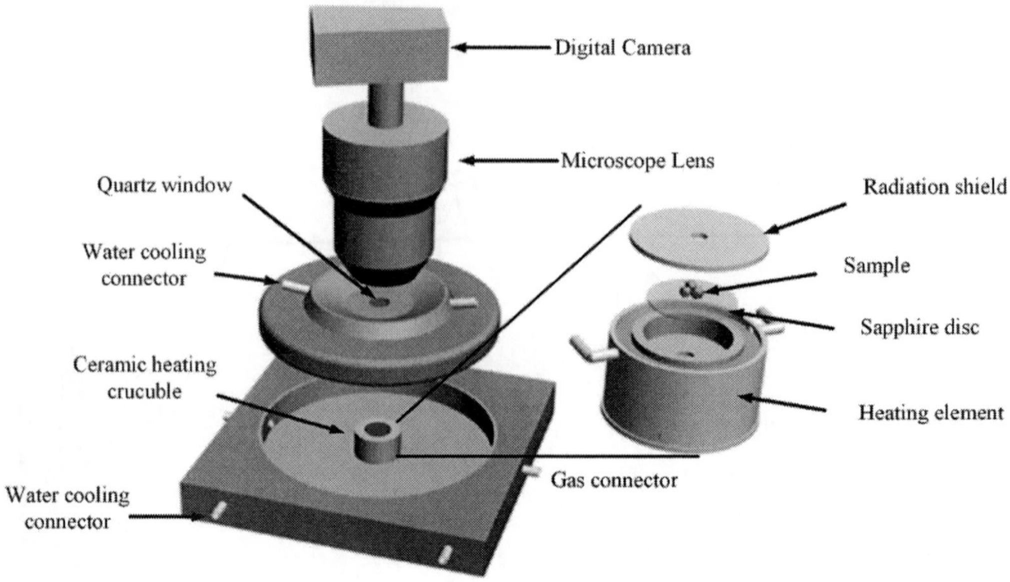


Fig. 2. The diagram of the heating stage microscope

7/15

# Experimental setup 2

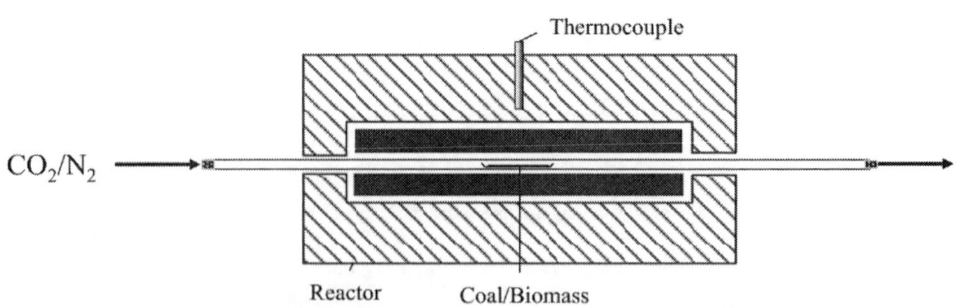


***Preliminary test*:**

***Particle size: 0.106-0.125 mm ; $CO_2/N_2$ flow rate 200 mL/min; and sample mass 160-240 mg.***

8/15

# Results and Discussion

## Morphology of the sample

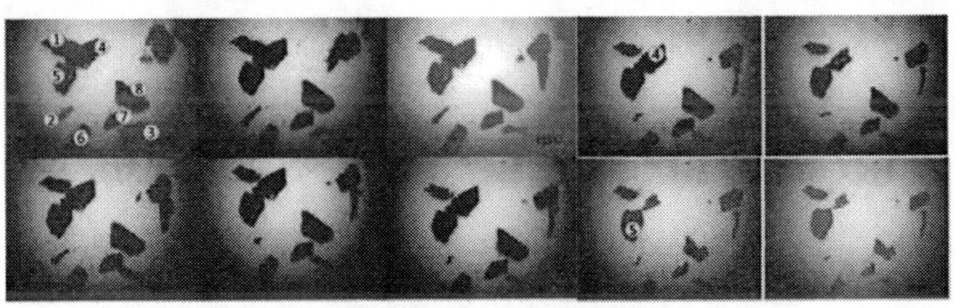

Fig. 3. The morphology transformation during the first heating stage from 100-1100 ° C (100×)
(①—④ Biomass particle ⑤—⑧ Coal particle)

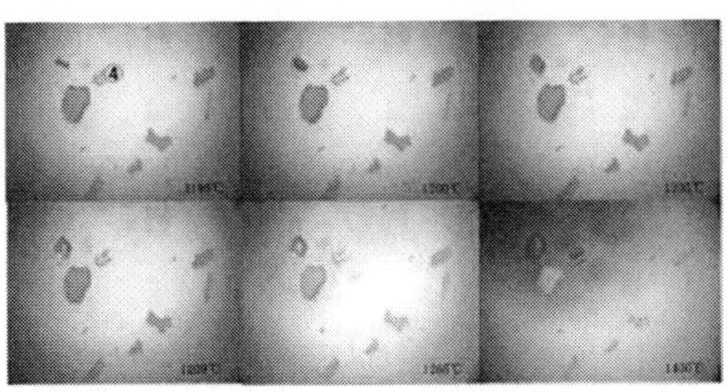

Fig. 4. The morphology transformation during the second heating stage from 1100-1400 °C (100×)

The Jincheng anthracite particles and rice straw particles affected each other during the ashing process.

# Mineral matter transformation

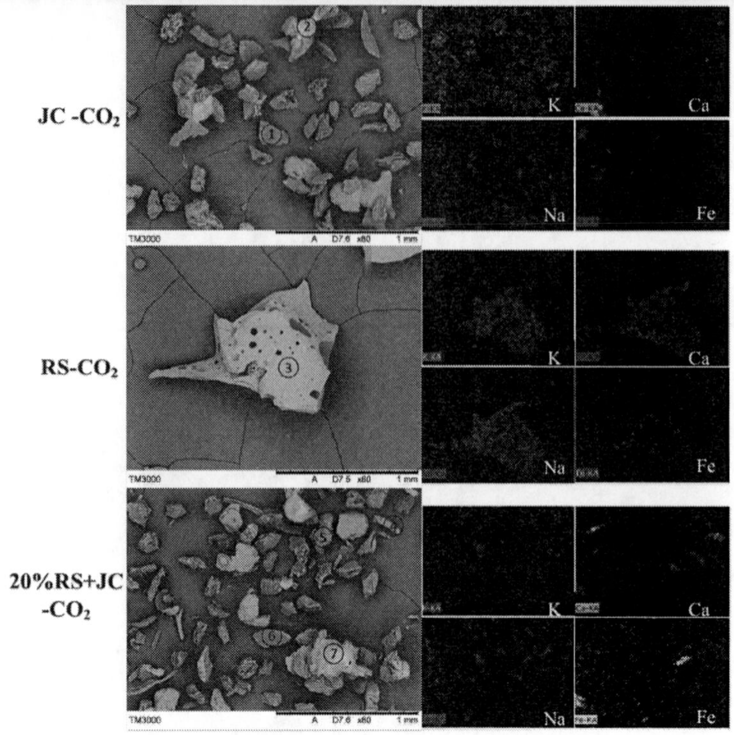


The Jincheng anthracite particles preferred to adsorb the alkali and alkali earth metals from rice straw.

Fig. 5. Morphology of residues at 1100°C



# Mineral matter transformation

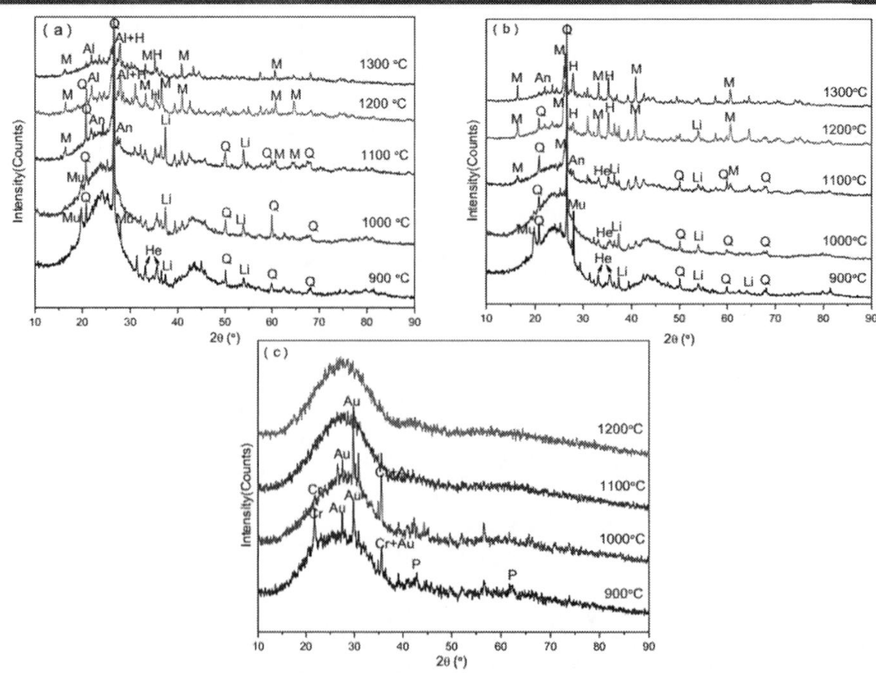


Fig.6. XRD patterns of samples at different heat-treatment temperatures at $CO_2$ atmosphere

(a): Mixture (b) Jincheng coal (c): Rice straw; Q-Quartz($SiO_2$); He-Hematite($Fe_2O_3$); Mu-Muscovite($KAl_2(AlSi_3O_{10})(OH)_2$); Li-Lime(CaO); An-Anorthite($CaO\cdot Al_2O_3\cdot 2SiO_2$); Al-Albite(Na(Si3Al)O8; M-Mullite ($Al_6Si_2O_{13}$); P-Periclase (MgO); Cr-Cristobalite($SiO_2$); Au-Augite($Ca(Mg,Fe)Si_2O_6$) He-Hercynite($FeO\cdot 2Al_2O_3$)

# Conclusions

1. The inorganic elements of rice straw under $CO_2$ atmosphere more easily transferred onto the surface of Jincheng anthracite than under $N_2$ atmosphere during co-thermal chemical process.
2. The Jincheng anthracite particles could absorb the alkali and alkali earth metal emitted by devolatilization process of rice straw.
3. The potassium-containing and sodium-containing mineral matter was produced through the addition of rice straw during co-thermal chemical process.

13/15

# Acknowledgements

***The authors would like to thank the financial support from the National Youth Science Foundation of China (21206104, 51406129) and National Natural Science Foundation of China (21376158).***

14/15

Thanks for your attention !

# Catalytic Gasification with Carbon Dioxide Characteristics of Alkali and Alkaline Earth Metals Catalysts in the Fixed-bed Reactor

Yifei WANG*, Jilin Li, Guangsuo Yu, Fuchen Wang

Key Laboratory of Coal Gasification and Energy Chemical Engineering of Ministry of Education, East China University of Science and Technology, Shanghai 200237, China

## Abstract

Low temperature gasification is a talent technology to reduce the energy consumption and improve the heat value of gasification products. Catalytic gasification has advantage in accelerating reaction rate, and the catalytic gasification technology has gained great attention, alkali and alkaline earth metals have proved as effective catalysts for the catalytic gasification reaction. Effects of temperature, catalyst types and catalyst loadings on carbon dioxide gasification reaction were investigated on a fixed-bed reactor in this paper. Catalyst dispersion effect, BET specific surface area and crystallite constituents of Jinyou coal char with different loadings of K and Ca were also analyzed by energy-dispersive X-ray analyzer (EDX), $N_2$ adsorption at 77K and X-ray diffraction (XRD). Results showed that Potassium carbonate ($K_2CO_3$), as a catalyst, could improve gasification rate significantly, when gasification temperature was above 850℃. Cation species of the catalyst could influence its catalytic activity, and the main effects of the catalysts were the alkali and alkaline earth metal elements, the catalytic effect order was $K_2CO_3$>KOH>$Na_2CO_3$>NaOH>$Ca(NO_3)_2$. The reactivity of coal char increased continuously with the catalyst loadings, and showed a loading saturation level (LSL) which was 10% for K. K and Ca added to Jinyou coal could decrease its BET specific surface area, compared to the raw char. Moreover, K and Ca added to Jinyou coal could change its crystallite constituent during the process of coking and gasification, which indicated that some K and Ca reacted with minerals in the Jinyou coal char during the process of coking and gasification, resulting deactivation of some K and Ca. $SiO_2$ was the main mineral constituent in the raw char and its gasification residue. During the coking process, the soluble K and Ca would reacted with the minerals in Jinyou coal to produce insoluble salt of $KAlSiO_4$, $MgSiO_3$ and $Ca(OH)_2$ and $Ca_2Al_2SiO_7$, respectively. During the gasification of $CO_2$ process, some soluble K in the char became into insoluble salt of $K_2MgSi_3O_8$, and CaO diffraction peak appeared in its gasification residue, which indicated that some soluble K and Ca deactivated during the $CO_2$ gasification process.

**Keywords:** $CO_2$ gasification reactivity; catalytic reaction; carbon conversion

*Corresponding author. Tel:86-021-64252522, Fax:86-021-64251312

E-mail address: wangyf@ecust.edu.cn

## 1 Introduction

Recently, greenhouse effect has resulted in the rise of the global temperature. Many national governments pay attention to $CO_2$ emissions and low carbon economy. By using $CO_2$ reacting with coal char is an effective way to realize $CO_2$ emissions and coal clean using. There are many literatures about coal gasified by $CO_2$ at a high temperature, but many problems will exist, such as difficulty in cooling and purifying the gas at a high temperature [1]. To further lower the temperature of coal gasification, catalytic gasification has advantage in accelerating reaction rate, reducing the energy consumption and directing gasification products. Thus, the catalytic gasification technology has gained great attention [2-4]. And alkali and alkaline earth metals being as effective catalysts have become a consensus [5-10].

However, catalyst is usually added to the coal with the pretreatment of acid pickled and ash removed, focusing on the effect of catalyst types on the gasification reactivity. And this pretreatment will change coal structure and coal gasification reactivity. Therefore, it could not reveal the essence of coal reactivity [11,12]. Some progress of catalytic gasification of coal char by $CO_2$ has been achieved, such as Huang et al. found that Na and Ca added to the coal char could improve the formation of crystal structure by XRD [13], however, there are few research reports on the effect of catalyst on the coal char's BET specific surface area and crystallite constituent simultaneously. Therefore, the effects of catalyst on the coal char's BET specific surface area and crystallite constituent of the Jinyou coal char with different loadings of K and Ca become the focus in this paper, to further study characteristic of catalyst reacting with the mineral matter in the coal.

## 2 Experimental

2.1 Raw material and method of catalyst loading

Shanghai coking plant char and Jinyou coal were used as samples. The composition analysis was shown in table 1. KOH, $K_2CO_3$, $Na_2CO_3$, NaOH and Ca $(NO_3)_2.4H_2O$ were catalysts. The amount of catalyst loading was according to the ratio of metal element weight and coal weight. The way of catalyst loading was by immersion. Specifically, the raw material and catalyst were put into water and stirred by a glass rod. And kept it static for 48h and then dried the coal at 105℃ for 4 h.

2.2 Gasification process

The gasification process flow diagram was shown in Fig.1. It was composed of quartz tube and corundum casing tube. The length and diameter of the quartz tube were 1000mm and 28mm, respectively. And the length and diameter of corundum casing tube were 200mm and 25mm, respectively. The reactor was heated by electric furnace. A thermocouple of S model was placed in the outer wall of quartz tube, showing the factual internal temperature of furnace. $CO_2$ and $N_2$ gases controlled by mass flowmeters were as gasifying agent and carrier gas, respectively. Raw material was placed in the middle part of corundum, and the corundum was placed in the middle of the quartz tube. The gasification pressure was in atmospheric. The carrier gas of $N_2$ flowed into reactor when the furnace was heated at a 50 °C/min heating rate. When the temperature reached its fixed temperature, keep it for 15mins. Then the gasifying agent of $CO_2$ was flowed into the reactor. After product gas flowed out the reactor, it was purified by tar filter. Then, the composition of the product gas was analyzed by Gas Chromatographic of Agilent 7890. The heated equipment was turned off when the gasification was over. The carrier gas of $N_2$ did not flow into the reactor until the internal temperature dropped to room temperature.

2.3 Experimental parameters

Carbon conversion: $$x_c = \frac{\sum_{t=0}^{t} y_{CO} v_t \times 12 \times T_0}{2 \times 22.4 \times m \times w_0 \times T_t} \times 100\% \quad (1)$$

Reaction rate: $r = \frac{dx}{dt}$ (2)

Specific gasification rate: $M = \frac{dx/dt}{1-x}$ (3)

## 3 Results and discussion

### 3.1 Gasification reactivity of coal char from Shanghai coking plant (SCP)

#### 3.1.1 Effect of temperature on the gasification reactivity

As to Boudouard reaction, an increase in reaction temperature would result in an increase in carbon conversion and gasification rate, because this reaction was controlled by chemical reaction process at our experimental conditions. In this experiment, $K_2CO_3$ was as catalyst, the K loading was 5%. The effects of gasification temperature on carbon conversion and gasification reaction rate were showed in Fig.2 and Fig.3, respectively. Fig.2 showed that the gasification reaction of coal char happened at a very slow rate when the reaction temperature was 750°C. The carbon conversion rate was only 10.3% after a 60 minutes reaction time. The carbon conversion increased by 36% and 23.7% as reaction temperature increased from 750 °C to 850°C and 850°C to 950°C, respectively. It indicated that increasing the same temperature under the condition of different gasification temperatures would result in different accelerating degrees of the gasification reaction. Only the gasification temperature being higher than 850°C was the catalysis of $K_2CO_3$ obvious. And the catalytic gasification reaction was insignificantly below 750°C, this implying that a change in the temperature between 750°C and 850°C significantly affected the catalysis of $K_2CO_3$. It should be mentioned that even at high gasification temperature, the total carbon conversion was below 100%. The missing fraction of carbon might partly be occluded in the residual ash, to which the atmospheric gasifying reactant was hardly accessible and partly from miscellaneous miner carbon-containing products (e.g., COS etc). The increase of temperature was favorable for this catalytic gasification reaction, which was as the same as the report of Hiroyuki Iwaki [14]. In addition, Fig.3 illustrated that reaction rate increased with an increase of reaction temperature. The reaction rate showed a decreasing trend with the increase of reaction time. Before the reaction time reached 30 minutes, the instant gasification rate decreased significantly as the reaction time increased. However, the instant gasification rate hardly changed as the reaction time increased, when the reaction time was more than 30 minutes. The reason for this was that the effective carbon in solid-phase carbon declined, resulting in a continuous decrease of the concentration of solid reactant in the solid-phase and solid area contacting with gas reactant [15].

For a further understanding of influence of potassium carbonate on coal char gasification, the relations between reaction rate and carbon conversion of char-$CO_2$ gasification at different gasification temperatures in the range 750-950℃ were proposed. Coal char particles had abundant cylindrical pores with different sizes, the inner surface of the pores served as reaction surface of gasification [16]. In this paper, the random pore model was used to fit the kinetic data, which is shown as follows:

$$dx/dt = P_1 * (1-x)\sqrt{1-P_2 \ln(1-x)} \tag{4}$$

Where $P_1$ is the initial reaction rate of gasification; $P_2$ is the pore structure parameter of chars, and $x$ refers to carbon conversion.

Coefficient of determination ($R^2$) of fitted curves were above 0.9, indicated that the random pore model was reasonable. Through the fitting results, the initial reaction rate of coal char gasification at experiment temperatures were received, and the average activation energy of this temperature was available on the basis of these reaction rate data. Experiments of non-catalyst coal char gasification were also carried out with gasification temperature higher than 850℃, because non-catalyst coal char gasification needed higher gasification temperature and CO in the outlet gas could not be detected if gasification temperature was below 800℃. The activation energy of coal char-$CO_2$ gasification under different conditions was shown in table 2.

In experiment temperature range, the activation energy of non-catalyst sample was 193.05 $kJ\cdot mol^{-1}$, while after loading potassium carbonate into coal sample, the activation energy of char-$CO_2$ gasification decreased 42.17 $kJ\cdot mol^{-1}$. It indicated that by reducing the activation energy of gasification reaction, the addition of potassium carbonate could accelerate the reaction obviously when gasification temperature is 750-950℃. Moreover, with the increase of gasification temperature, the initial rate of coal char gasification reaction increased, which indicated that the char-$CO_2$ gasification reaction in the temperature is controlled by chemical reaction.

3.1.2 Effect of catalyst types on the gasification reactivity

Since K, Na and Ca had different reaction activities due to different reaction mechanism, catalyst type was an essential factor to be considered. The effect of $K_2CO_3$, KOH, $Na_2CO_3$, NaOH and Ca $(NO_3)_2.4H_2O$ on the gasification reactivity were investigated. Moreover, gasification reactivity of coal char without catalyst was also tested. The results were showed in Fig.4, Fig.5 and Fig.6.

Fig.4 illustrated that these catalysts had catalytic effect and the catalytic effect of K and Na were obvious. The catalytic order was $K_2CO_3$>KOH>$Na_2CO_3$>NaOH>Ca $(NO_3)_2.4H_2O$. There was no big difference between $K_2CO_3$ and KOH, neither was between NaOH and $Na_2CO_3$. However, the catalytic effect of Ca $(NO_3)_2.4H_2O$ was not obvious. The result showed that catalytic effect of potassium was higher than sodium, and alkali metal was higher than alkaline earth metal. The main effects of the' catalysts were the alkali and alkaline earth metal elements. In addition, the non-catalytic gasification reaction was happened at very slow rate. The total carbon conversion was only 27%. The catalysis effect of $K_2CO_3$ was the most obvious. In addition, R-t curve in Fig.5 showed that the gasification reaction rate decreased as the reaction time increased. However, M-t curve in Fig. 6 showed the specific reaction rate with a catalyst of K or Na reached a peak at a reaction time of 2 minutes while specific reaction rate with a catalyst of Ca was as the same trend of reaction rate in R-t curve in Fig. 5. This illustrated that the catalyst of K or Na was a compulsory additive if the gasification reaction with carbon dioxide was happened significantly at a temperature below 950℃.

3.1.3 Effect of catalyst adding amount on the gasification reactivity

As was pointed out by a previous experiment on the carbon dioxide gasification, the catalytic effect of $K_2CO_3$ was the most obvious. In order to investigate different adding amount of $K_2CO_3$ on the gasification reactivity, the various K adding amount of 3%, 5%, 8%, 10%, 12% and 15% were added to the carbon dioxide gasification. And the variations of carbon conversion with reaction time at 950℃ in all six char samples with different K adding amount were measured as was showed in Fig.7. And the gasification reaction rate was plotted in Fig.8.

As was showed in Fig.7, a LSL (loading saturation level) of K was about 10%. When K adding amount was lower than the LSL, there was a continuous increase in carbon conversion rate as an increase of K adding amount. And the curves of variation of carbon conversion with reaction time with K adding amount of 12% and 15% almost overlapped with the curve of K adding amount of 10%. And the reaction rate showed the same decrease trend with the increase of reaction time as it was illustrated in Fig. 8. Therefore, it further explained the LSL of K was 10%.

### 3.2 Study on Jinyou coal char

#### 3.2.1 The preparation of Jinyou coal char

Jinyou coal was placed in the center of the reaction tube in the fixed-bed reactor. The carrier gas of $N_2$ whose volume flow was 1000 mL/min flowed into reactor when the furnace was heated at a heating rate at 50℃/min. When the temperature reached 950℃, keep it for 30 minutes. The Jinyou coal was added different amount of $K_2CO_3$ and Ca $(NO_3)_2.4H_2O$. The char preparation condition was the same for different samples. For it satisfied the condition of gasification characteristics compared when the preparation conditions of char were the same [17]. The char added 5%K, 10%K, 5%Ca, 10%Ca and non-catalyst were marked as K-5-char, K-10-char, Ca-5-char, Ca-10-char and raw-char. Their proximate and ultimate analyses were showed in Table 3.

#### 3.2.2 The gasification reactivity of Jinyou coal char

The gasification condition was the same as the coal char produced from Shanghai coking plant. Fig.9 and Fig.10 showed the carbon conversion of different catalysts in varying loadings versus the reaction time. It was illustrated that the alkali metal K and the alkaline earth metal Ca had a catalytic effect on Jinyou coal. Moreover, the catalytic effect of K was higher than Ca with the same loading. The carbon conversion of Jinyou coal increased with an increase of catalyst loading. The catalytic effect of K was once again proved more obvious than Ca.

3.2.3 EDX analysis of Jinyou coal char

In order to investigate the distribution of catalyst in the coal char surface, EDX was used to analyze the char samples. Fig.11 showed the X-ray spectrum of the chars. It was showed that the surface of raw-char mainly contained Al, Si and Fe, which mainly came from the minerals in the Jinyou coal. And the peaks of Al, Si and Fe were very low. As for K-5-char, the surface mainly contained K and little Ca. And Ca mainly came from the minerals in coal. While for Ca-5-char, the surface mainly contains Ca and a small amount of Al and Si. The result showed that the added catalyst of K and Ca to the Jinyou coal mainly distributed on the coal surface.

3.2.4 Effect of catalyst on Jinyou char's BET specific surface area

In order to analyze the effect of catalysts on the Jinyou coal char' s pore structure, the BET specific surface area of the Jinyou coal char with different loadings of K and Ca were examined by an $N_2$ adsorption at 77K. The result was showed in Table 4. The alkali metal K and alkaline earth metal Ca had a decrease effect on BET specific surface area of Jinyou coal char, indicating that a substantial portion of catalyst was seeped into the pores. In addition, the BET specific surface area of the chars decreased with an increase of K and Ca adding amount. And this effect caused by the adding of K was more obvious than Ca. Though adding the catalyst to the coal resulting in the decrease the specific surface area of the coal char, it still increased the gasification rate. This was because the catalyst improved the gasification reactivity of char was mainly due to that it increased the number of activation center and decreased the reaction activation energy. It also explained in Hamiltontl's [18] report.

3.2.5 Effect of catalyst on crystallite constituent in Jinyou coal char

In order to investigate the catalyst on crystallite constituent in Jinyou coal char, a further study on the interaction of the catalyst with minerals in coal was important. Fig.12 showed the XRD spectra of the chars and their gasification residue. The first spectrum in Fig.12 illustrated $SiO_2$ was the main mineral constituent in the raw char with non-catalyst and its gasification residue, for the raw was at a low carbon conversion of 11% in 40 minutes reaction time. The second spectrum in Fig.12 was the XRD spectrum of K-10-char and its gasification residue. The K-10-char, compared to raw char, peaks of $KAlSiO_4$ and $MgSiO_3$ were appeared except $SiO_2$. It indicated that some K was turned into K-insoluble salt by reacting with the coal minerals during the process of coking. The carbon conversion of K-10-char with $CO_2$ was 83.5%. And $K_2MgSi_3O_8$ diffraction peak in its gasification residue became higher and $SiO_2$ diffraction peak became weaker compared with K-10-char. It indicated that during the $CO_2$ gasification process, some K in the char was turned into $K_2MgSi_3O_8$ without catalysis through the reaction with the mineral matters in the char. It showed some soluble K occurred deactivation during the gasification of $CO_2$. The third spectrum in Fig.12 was the XRD spectrum of Ca-10-char and its gasification residue. Compared to the raw char, the weak peaks of $Ca(OH)_2$ and calcium aluminum melilite $Ca_2Al_2SiO_7$ appeared except $SiO_2$. The carbon conversion of Ca -10-char with $CO_2$ was 27.4%. CaO diffraction peak was appeared in its gasification residue. $Ca_2Al_2SiO_7$ diffraction peak got higher and $Ca(OH)_2$ got weaker compared with Ca-10-char. It also showed some soluble Ca occurred deactivation through the reaction with the mineral matters in the char during the gasification process.

**4 Conclusions**

1 Potassium carbonate, as a catalyst, could improve gasification rate significantly, when gasification temperature was above 850℃.

2 Cation species of the catalyst could influence its catalytic activity. The main effects of the catalysts were the alkali and alkaline earth metal elements. The catalytic effect order was $K_2CO_3$>KOH>$Na_2CO_3$>NaOH> $Ca(NO_3)_2$. The reactivity of coal char increased continuously with catalyst loadings, and showed a loading

saturation level (LSL) which was 10 % for K.

3 The catalyst adding to the Jinyou coal, after being coked, was mainly scattered on the coal's surface and seeped into the pores. The catalyst would lessen BET specific surface area of the coal char. And the effect of K on decreasing the BET specific surface area of the coal char was more obvious than Ca.

4 $SiO_2$ was the main mineral constituent in the raw char and its gasification residue. During the coking process, the soluble K and Ca would reacted with the minerals in Jinyou coal to produce insoluble salt of $KAlSiO_4$, $MgSiO_3$ and $Ca(OH)_2$ and $Ca_2Al_2SiO_7$, respectively. During the gasification of $CO_2$ process, some soluble K in the char became into insoluble salt of $K_2MgSi_3O_8$, and CaO diffraction peak appeared in its gasification residue, which indicated that some soluble K and Ca deactivated during the $CO_2$ gasification process.

**Acknowledgements**

This study is sponsored by the National "863" Scientific Research Program (Contract No. 2008AA05Z310).

**Symbols used**

| | | |
|---|---|---|
| $m$ | [g] | weight of coal char |
| $T_0$ | [K] | temperature in the normal condition |
| $T_t$ | [K] | outlet gas temperature |
| $v_t$ | [L] | the outlet gas accumulated flow in the adjacent sampling period |
| $w_0$ | [-] | carbon quality percent in the element analysis |
| $y_{CO}$ | [-] | CO volume concentration at t moment |
| $P_1$ | [-] | initial reaction rate of gasification |
| $P_2$ | [-] | pore structure parameter of chars |
| $x$ | [%] | carbon conversion |

**References**


[1] J. J. Huang, Y. T. Fang, Y. Wang. *Chem. Tec.* **2002**, 30(5), 385-391.

[2] D. Mckee, C. Spiro, P. Kosky, E. Lamby. *Fuel.* **1983**, 62, 217-220.

[3] W. D. Mckee. *Fuel.* **1983**, 62(2), 170-175.

[4] S. Li, Y. Cheng. *Fuel.* **1995**, 74(3), 456-458.

[5] M. Quyn, H. Wu, J. Hayashi, C. Li. *Fuel Process Technol.* **2005**, 86(12-13), 1241-1251.

[6] X. J. Li, C. Z. Li. *Fuel.* **2006**, 85(10-11), 1518-1525.

[7] J. H. Zhou, J. P. Kuang, Z. J. Zhou. *Proceedings of the CSEE.* **2006**, 26(12), 149-155(in Chinese).

[8] S. Atul, Y. Yeboah, A. Godavarty. *Fuel.* **2003**, 82(3), 305-317.

[9] Q. L. Sun, W. Li, H. K. Chen. *Fuel.* **2004**, 83 (13), 1787-1793.

[10] R. Struis, C. Scala, S. Stucki, R. Prins. *Chem Eng Sci.* **2002**, 57(17), 3593-3602.

[11] Y. Onal, K. Ceylan. *Fuel.* **1995**, 74(7), 972-977.

[12] A. Lem, Y. Zhuo, G. P. Reed. *Fuel.* **2002**, 81(3), 315-326.

[13] Y. Q. Huang, X. L. Yin, C. Z. Wu. *Biotechnol Adv.* **2009**, 27(5), 568-572.

[14] H. Iwaki, S. F. Ye, H. Katagiri. *Appl Catal.* **2004**, 270(1), 237-243.

[15] B. C. Zhu. Chemical Reaction Engineering. Chemical Industry Press, Beijing **2007**(in Chinese).

[16] B. B. Beamish, K. J. Shaw, J. Newman. *Fuel Processing Technology*, 1998, 53(3): 243-252.

[17] J. S. Dennis, S. A. Scott, A. N. Hayhurst. *J Energy Inst.* **2006**, 79(3), 187-190.

[18] R. T. Hamilton, D. A. Sams, F. Shadman. *Fuel.* **1984**, 63(7), 1008-1012.

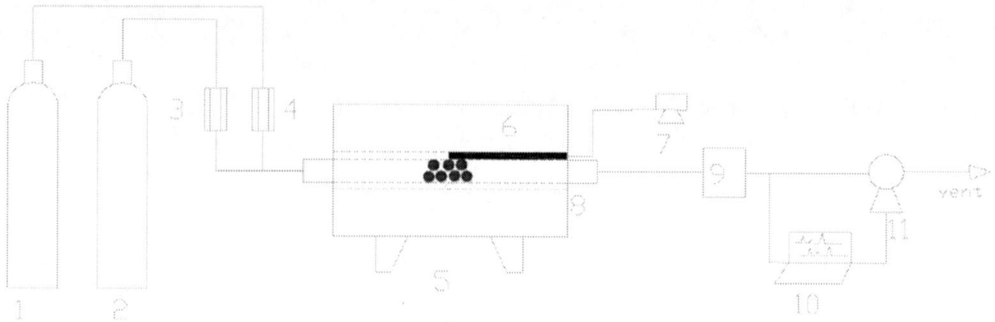


Fig.1 Schematic diagram of the fixed-bed reactor system

1-nitrogen 2-carbon dioxide 3,4-mass flowmeter 5-fixed-bed reactor 6-thermocouple 7-temperature controller 8-reaction tube 9-tar filter 10-gas chromatography 11-wet flowmeter

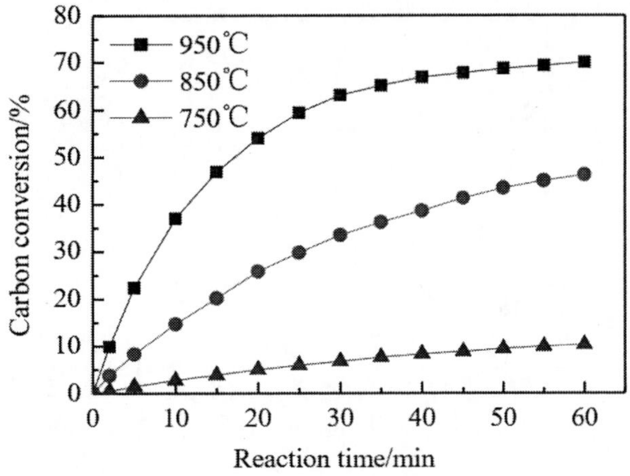


Fig.2 Carbon conversion vs. reaction temperature

(Catalyst $K_2CO_3$, K added amount 5%, Coal char from SCP)

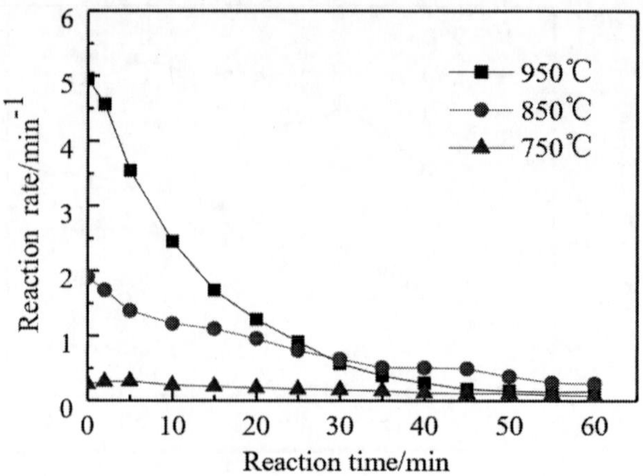


Fig.3 Reaction rate vs. reaction temperature

(Catalyst $K_2CO_3$, K added amount 5%, Coal char from SCP)

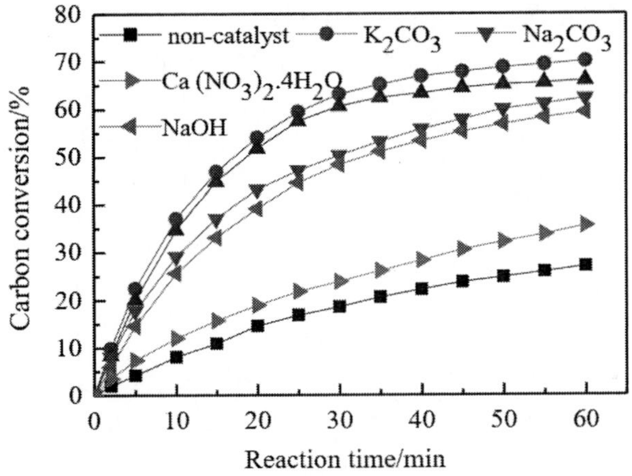


Fig.4 Carbon conversion vs. reaction time

(Catalyst: $K_2CO_3$, KOH, $Na_2CO_3$, NaOH and Ca $(NO_3)_2.4H_2O$, K, Na and Ca added amount 5% respectively,

Coal char from SCP)

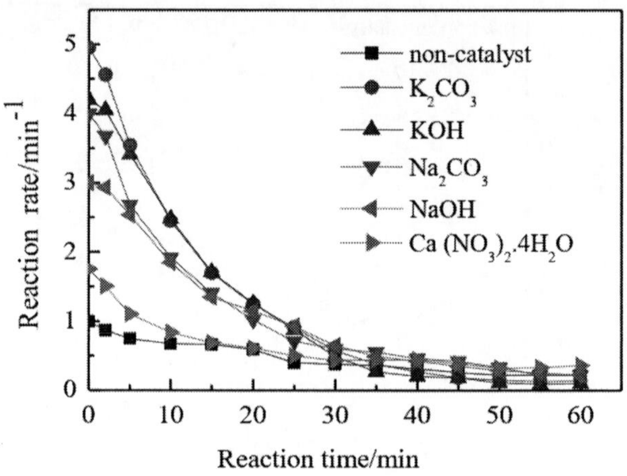


Fig.5 Reaction rate vs. reaction time

(Catalyst: $K_2CO_3$, KOH, $Na_2CO_3$, NaOH and Ca $(NO_3)_2.4H_2O$, K, Na and Ca added amount 5% respectively, Coal char from SCP)

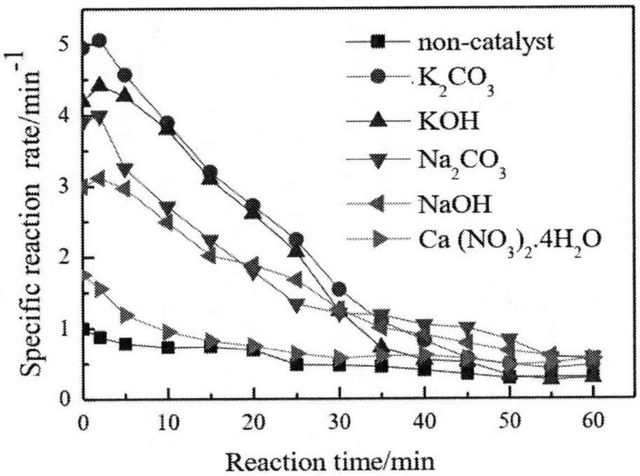


Fig.6 Specific reaction rate vs. reaction time

(Catalyst: $K_2CO_3$, KOH, $Na_2CO_3$, NaOH and $Ca (NO_3)_2.4H_2O$, K, Na and Ca added amount 5% respectively, Coal char from SCP)

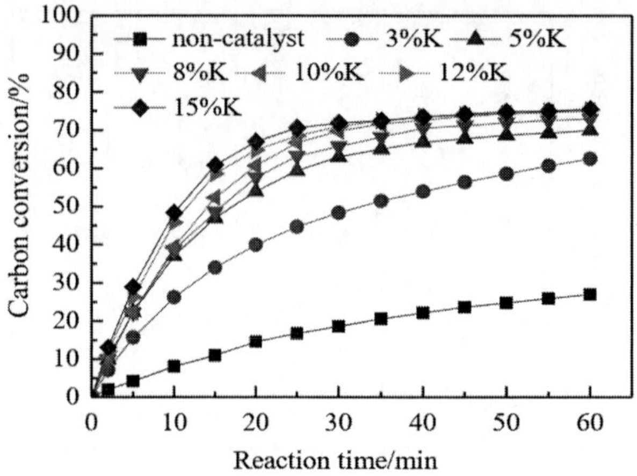


Fig.7 Carbon conversion vs. reaction time

(Catalyst $K_2CO_3$, K added amount: 3%, 5%, 8%, 10%, 12%and 15%, Coal char from SCP)

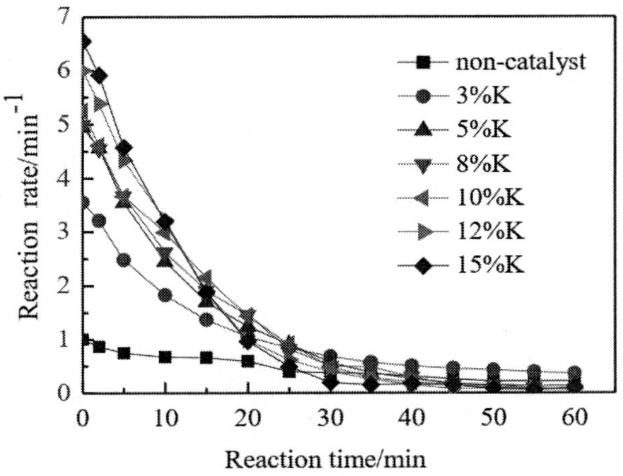


Fig.8 Reaction rate vs. reaction time

(Catalyst $K_2CO_3$, K added amount: 3%, 5%, 8%, 10%, 12%and 15%, Coal char from SCP)

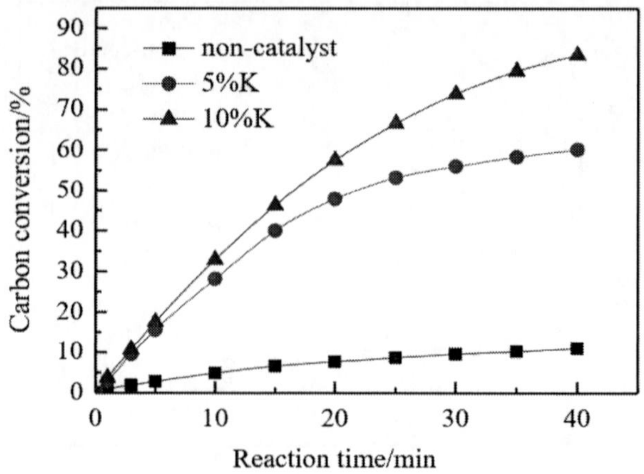


Fig.9 Effect of K loading on the carbon conversion

(Catalyst $K_2CO_3$, K added amount 5% and 10%, Jinyou coal char)

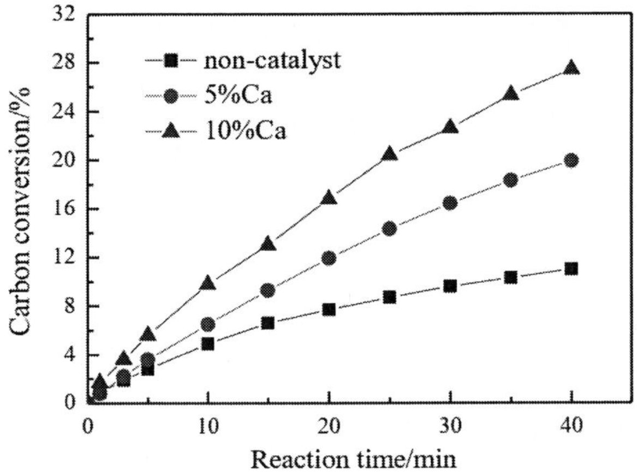


Fig.10 Effect of Ca loading on the carbon conversion

(Catalyst Ca $(NO_3)_2.4H_2O$, Ca added amount 5% and 10%, Jinyou coal char)

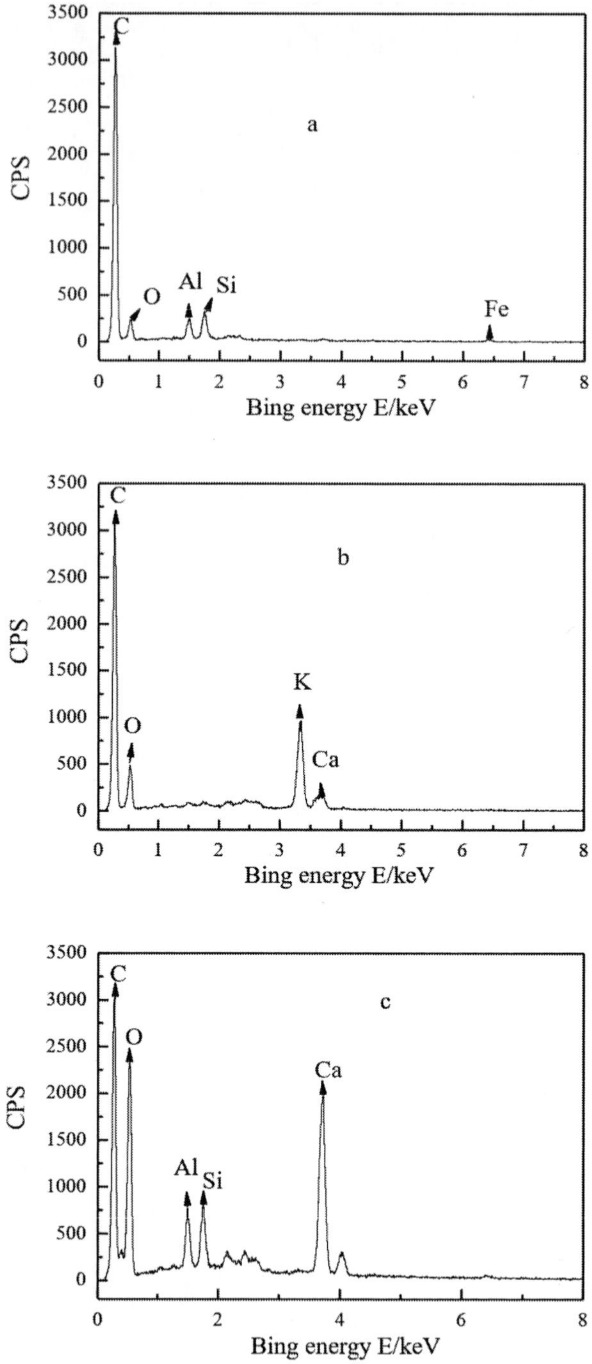


a) raw-char b) K-5-char c) Ca-5-char

Fig.11 EDX spectra of the surface of Jinyou coal char

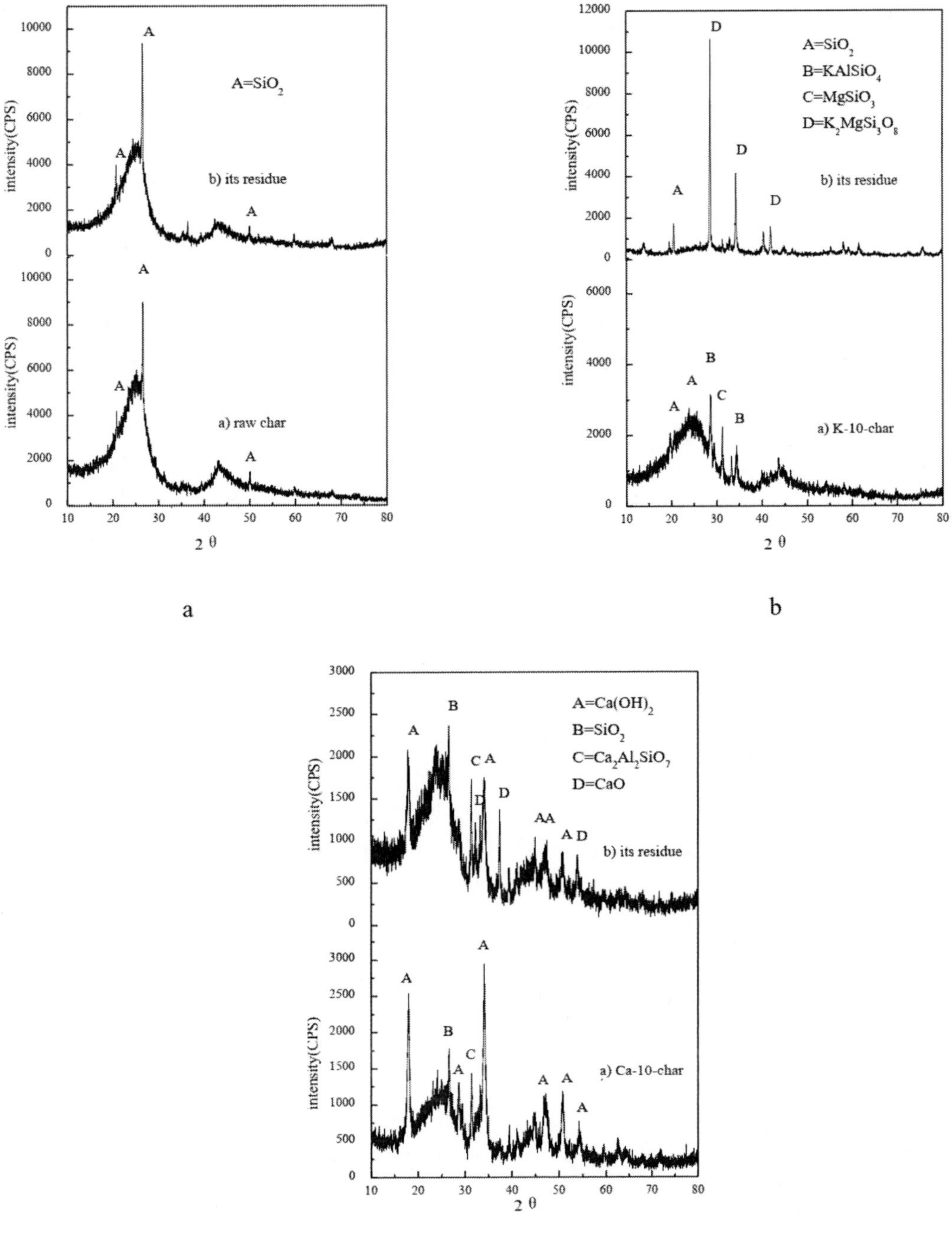


c

a) raw-char b) K-10-char c) Ca-10-char

Fig.12 XRD spectra of Jinyou coal char and it residue

Table 1 Proximate and ultimate analyses of coal samples

| Sample | Proximate analysis $w_{ad}$ / % | | | | Ultimate analysis $w_{ad}$ / % | | | | |
|---|---|---|---|---|---|---|---|---|---|
| | M | A | V | FC | C | H | N | S | O |
| Coal char | 1.89 | 7.07 | 14.08 | 76.96 | 85.96 | 2.261 | 0.961 | 0.856 | 1.002 |
| Jinyou Coal | 0.83 | 24.93 | 15.03 | 59.21 | 68.66 | 2.552 | 1.031 | 1.163 | 0.834 |

Table 2 Ea of coal char-$CO_2$ gasification under different conditions

| Sample | Ea / $kJ.mol^{-1}$ | temperature / ℃ |
|---|---|---|
| 5%K | 150.88 | 750-950 |
| non-catalyst | 193.05 | 850-950 |

(Catalyst $K_2CO_3$, K added amount 5%, Coal char from SCP)

Table 3 Proximate and ultimate analyses of Jinyou coal char samples

| Sample | Proximate analysis $w_{ad}$ / % | | | | Ultimate analysis $w_{ad}$ / % | | | | |
|---|---|---|---|---|---|---|---|---|---|
| | M | A | V | FC | C | H | N | S | O |
| raw char | 3.61 | 15 | 13.43 | 67.96 | 76.17 | 2.022 | 0.93 | 0.795 | 1.473 |
| K-5-char | 1.5 | 29.21 | 11.99 | 57.3 | 64.36 | 1.780 | 0.817 | 0.815 | 1.518 |
| K-10-char | 1.06 | 22.51 | 8.87 | 67.56 | 71.43 | 1.838 | 0.896 | 0.722 | 1.544 |
| Ca-5-char | 0.99 | 22.21 | 14.09 | 62.71 | 72.99 | 1.789 | 1.008 | 0.881 | 0.132 |
| Ca-10-char | 0.98 | 24.67 | 14.41 | 59.94 | 69.28 | 1.885 | 0.915 | 0.941 | 1.329 |

Table 4 BET surface area of Jinyou coal char samples

| Name | Raw char | K-5-char | K-10-char | Ca-5-char | Ca-10-char |
|---|---|---|---|---|---|
| BET surface area/$m^2.g^{-1}$ | 0.9013 | 0.2197 | 0.0562 | 0.6944 | 0.2893 |

# Oxygen Co-gasification of Coal and rice straw in Drop Tube Furnace

East China University of Science and Technology

Xueli Chen; Jianye Tang; Qinghua Guo; Guangsuo Yu

PCC

Thirty Third Annual International
PITTSBURGH COAL CONFERENCE

## CONTENT

# INTRODUCTION

Hydroenergy

Solar Energy

Geothermal Energy

Wind Energy Source

Biomass Energy

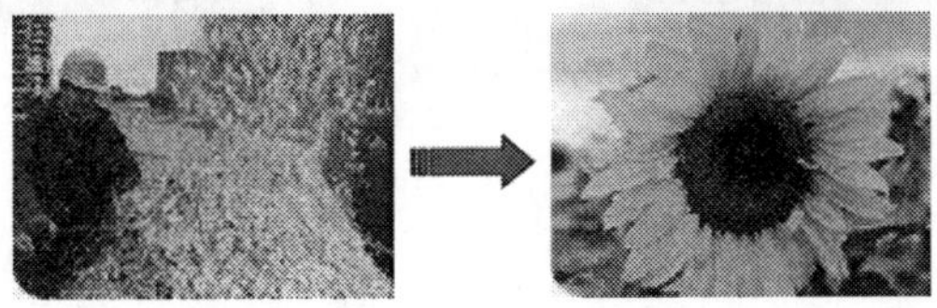

Among the renewable energies, biomass is expected to be one of the most important in the near future.

# INTRODUCTION

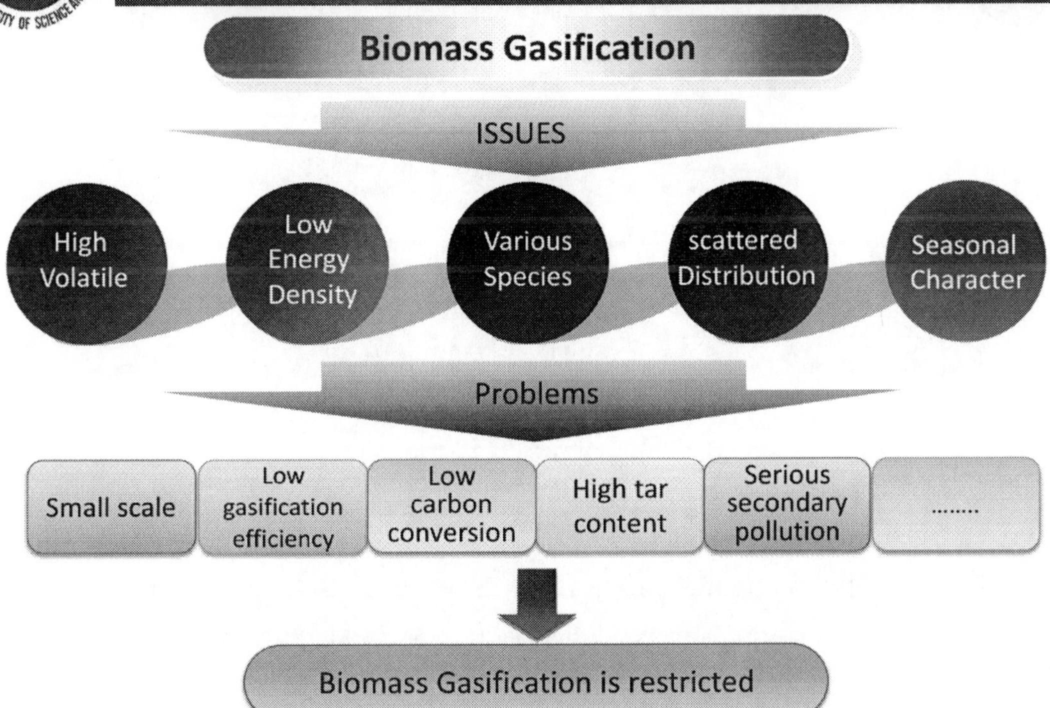

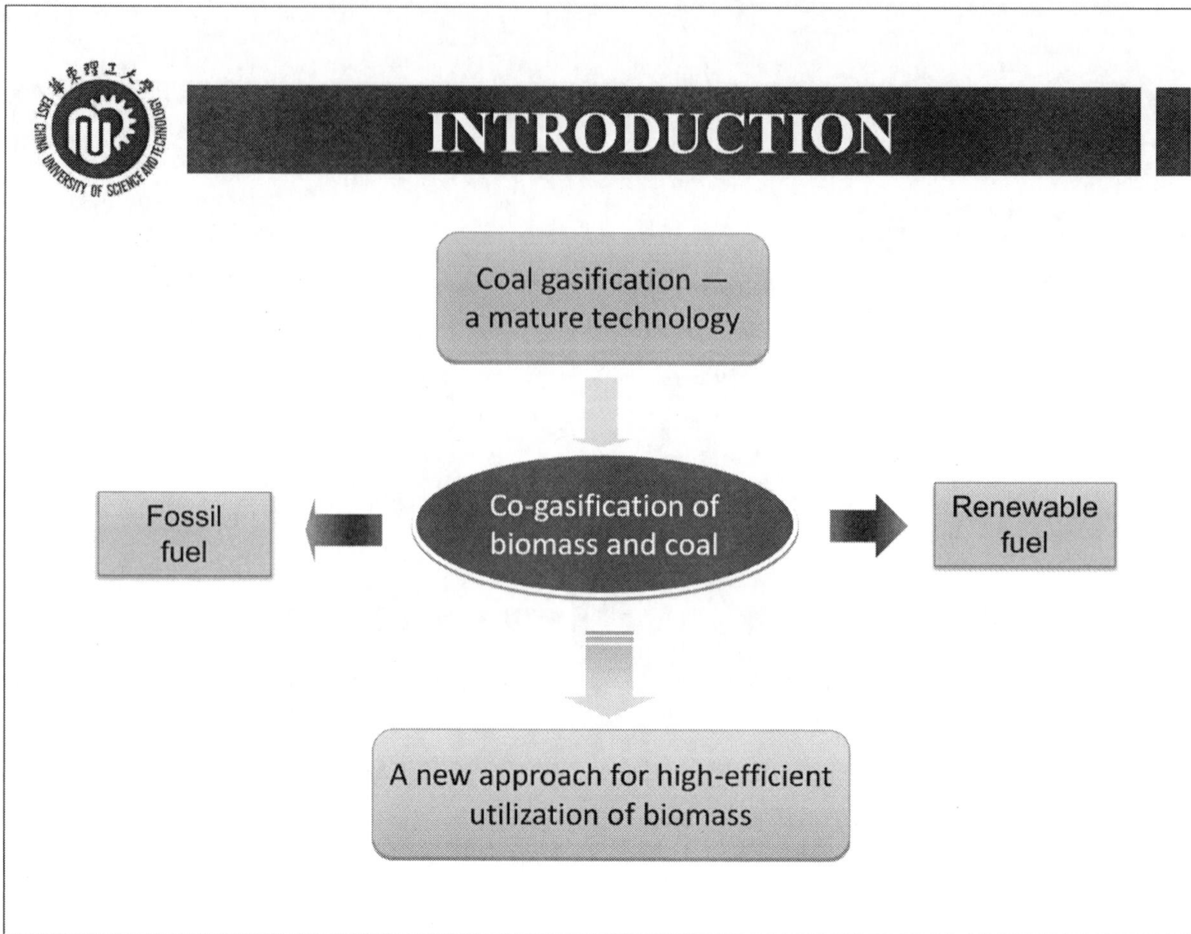
EAST CHINA UNIVERSITY OF SCIENCE AND TECHNOLOGY
INTRODUCTION
Coal gasification — a mature technology
Fossil fuel
Co-gasification of biomass and coal
Renewable fuel
A new approach for high-efficient utilization of biomass

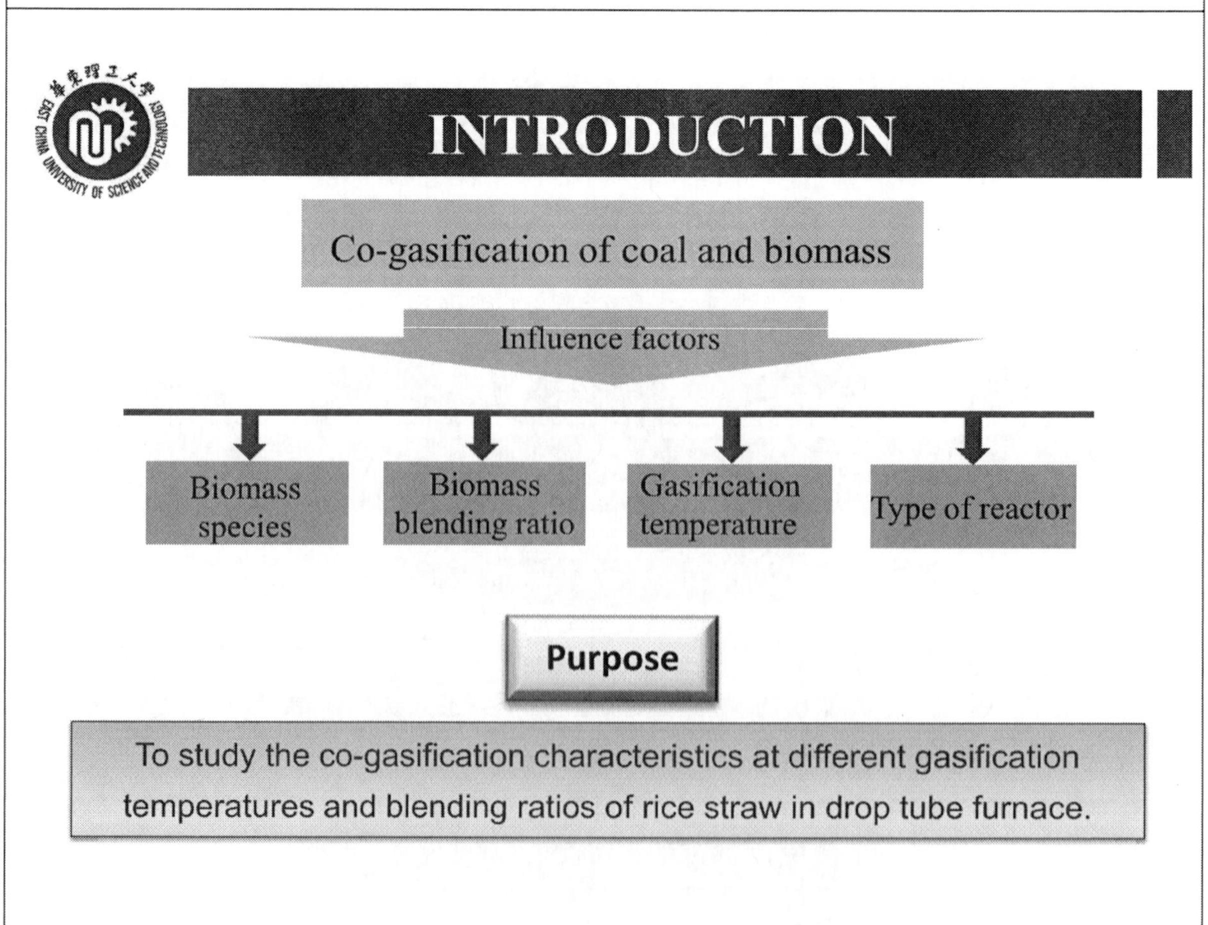
EAST CHINA UNIVERSITY OF SCIENCE AND TECHNOLOGY
INTRODUCTION
Co-gasification of coal and biomass
Influence factors
Biomass species
Biomass blending ratio
Gasification temperature
Type of reactor
Purpose
To study the co-gasification characteristics at different gasification temperatures and blending ratios of rice straw in drop tube furnace.

# EXPERIMENT

## Raw materials

Materials ← Coal → From China

Materials ← Rice straw → From suburb of Shanghai in China

Table 1. Proximate and ultimate analyses of samples on a dry basis

| Samples | Proximate analysis(*wt.*%) | | | Ultimate analysis(*wt.*%) | | | | |
|---|---|---|---|---|---|---|---|---|
| | A | V | FC | $C_{daf}$ | $H_{daf}$ | $O_{daf}$ | $N_{daf}$ | $S_{daf}$ |
| Coal | 11.97 | 10.98 | 77.05 | 94.05 | 0.78 | 1.28 | 2.36 | 1.53 |
| Rice straw | 10.21 | 80.11 | 9.68 | 49.85 | 2.90 | 45.45 | 1.58 | 0.22 |

# EXPERIMENT

## Experiment Condition

Blending ratio of rice straw:
0, 20, 50, 80 and 100 wt.%.
Feeding rate of raw material:
2.0 – 20.0 g/min.
Total flow rate of $O_2$ and $N_2$:
5 L/min.
The value of O/C:
0.8.
Gasification temperature:
1200 – 1400 °C

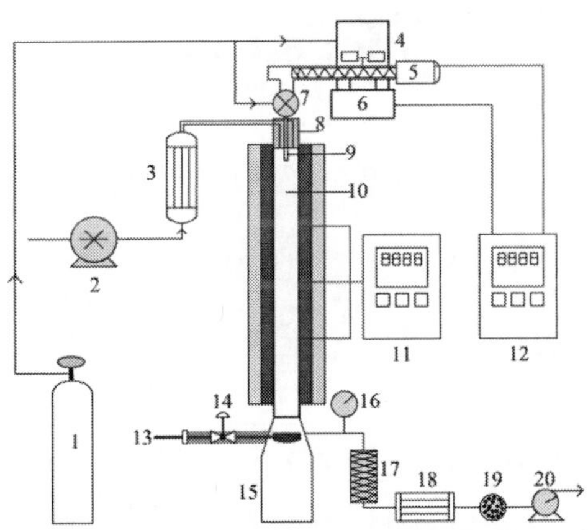

Fig.1 Schematic diagram of the drop tube furnace

1: $N_2$ cylinder; 2:double-type axial piston pump; 3: steam generator; 4: hopper; 5: screw feeder; 6: hopper weigher; 7: gas lock; 8: electrical preheater; 9: feeding probe; 10:corundum tube reactor; 11: temperature controller; 12: weighing controller; 13: char sampler ladle; 14: ball valve; 15: char collector; 16: pressure gage; 17: filter; 18: cooler; 19: dry box; 20: gas flowmeter

# RESULTS & DISCUSSION

## 1. Coal and rice straw separate gasification characteristics

### 1.1 Gas yield and composition

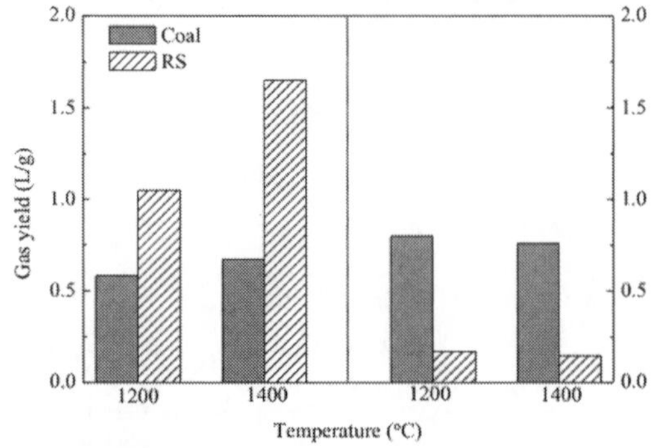


Fig.2 Total gas yield and solid yield of coal and RS gasification

Table 2 Gas components of coal and RS gasification

| Gas composition (%) | $CO_2$ | $H_2$ | $CH_4$ | CO |
|---|---|---|---|---|
| RS-1200 | 7.0 | 50.5 | 1.8 | 40.7 |
| RS-1400 | 0.0 | 58.6 | 0.0 | 41.4 |
| Coal-1200 | 46.4 | 22.8 | 0.0 | 30.8 |
| Coal-1400 | 27.5 | 26.0 | 0.0 | 46.5 |

# RESULTS & DISCUSSION

### 1.2 Proximate and ultimate analyses of solid products

Table 3 Proximate and ultimate analyses of raw samples and solid products

| **Samples** | **Proximate analysis(wt.%)** | | | **Ultimate analysis(wt.%)** | | | |
|---|---|---|---|---|---|---|---|
| | **Ad** | **Vd** | **FCd** | **Cd** | **Hd** | **Nd** | **Sd** |
| **RS** | **9.97** | **77.56** | **12.47** | **45.36** | **4.88** | **0.90** | **0.26** |
| **RS-1200** | **59.15** | **14.43** | **26.42** | **34.93** | **0.29** | **0.65** | **0.38** |
| **RS-1400** | **68.72** | **9.36** | **21.92** | **27.61** | **0.20** | **0.38** | **0.20** |
| **Coal** | **11.97** | **10.98** | **77.05** | **94.05** | **0.78** | **2.36** | **1.53** |
| **Coal-1200** | **14.96** | **4.44** | **80.60** | **85.54** | **0.46** | **1.43** | **1.04** |
| **Coal-1400** | **15.76** | **2.16** | **82.08** | **84.18** | **0.178** | **0.94** | **0.87** |

# RESULTS & DISCUSSION

## 1.3 Apparent morphology of solid products

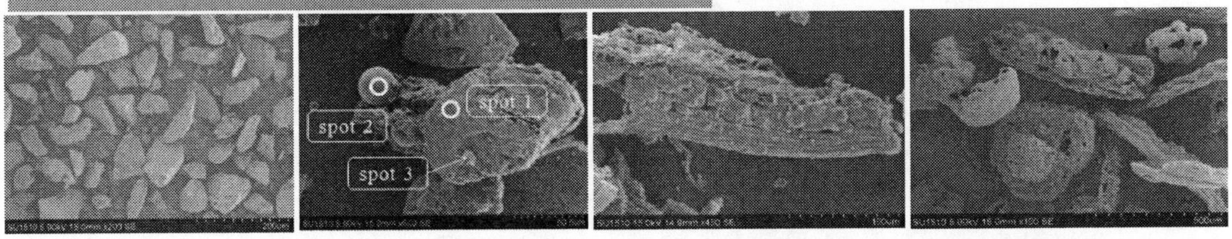


(1) 1200 ℃ (2) 1400 ℃ (3) 1200 ℃ (4) 1400 ℃

Fig.3 SEM image of typical coal solid and ash at different gasification temperature (1)、(2) – Coal solid products; (3)、(4) – RS solid products.

- ◆ Coal gasification solid products are generally compact and irregular. The temperature has no obvious influence.
- ◆ There is some melting part on the particle surface at 1400 ℃.
- ◆ Solid particles hold the regular straw particle morphology at 1200 ℃. When the temperature increases up to 1400℃, some straw particles soften and curl , further become melting state.


# RESULTS & DISCUSSION


## Apparent morphology of solid products

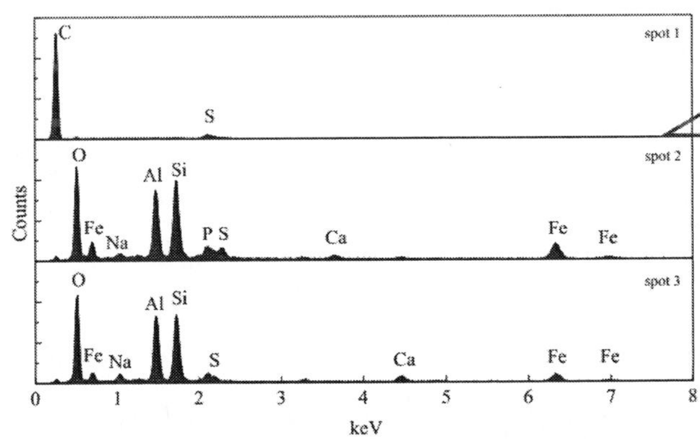


Fig.4 Elemental spectrum of coal ash particles

- ◆ Globular melting part is relatively rich in Fe and Ca besides Si and Al.

# RESULTS & DISCUSSION

## 2. Oxygen co-gasification behaviors of coal and rice straw

### 2.1 Gas yield and component

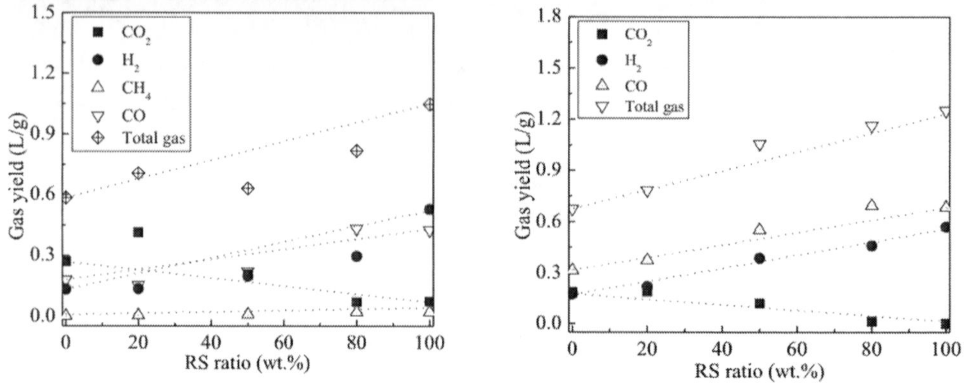


1200 ° C    1400 ° C

Fig.5 Gas yield during oxygen co-gasification of coal and RS with different RS ratio

◆ There was obvious interaction between coal and RS and strength of interaction effect increased with the increase of RS ratio.


# RESULTS & DISCUSSION


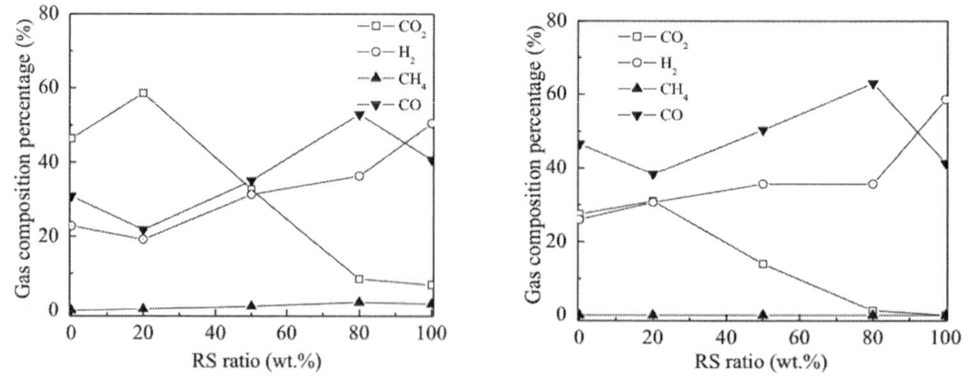


1200 °C    1400 °C

Fig.6 Effects of RS ratio on gas component during oxygen co-gasification of coal and RS

# RESULTS & DISCUSSION

## 2.2 Solid yield

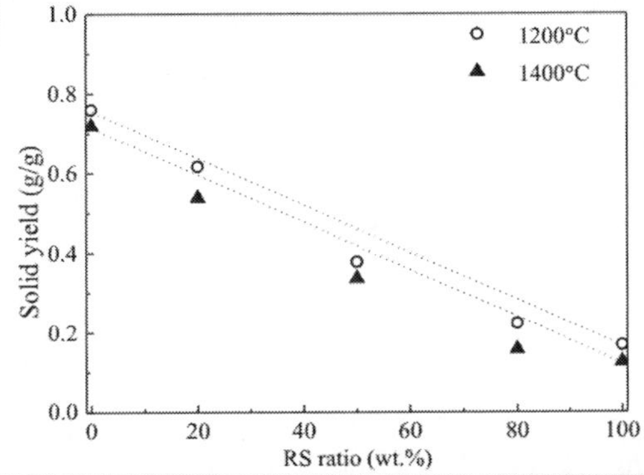


Fig.7 Ash yield during oxygen co-gasification of coal and RS

- ◆ Experimental values of solid yields of all different blending ratios are slightly lower than weighted average of yield during separate gasification of coal and RS. It may be correlation with the catalysis effect of alkalis in RS.
- ◆ The addition of RS is beneficial to the gasification reaction of chars, boosting the conversion of coke.

# RESULTS & DISCUSSION

## 2.3 Transformation of C and H

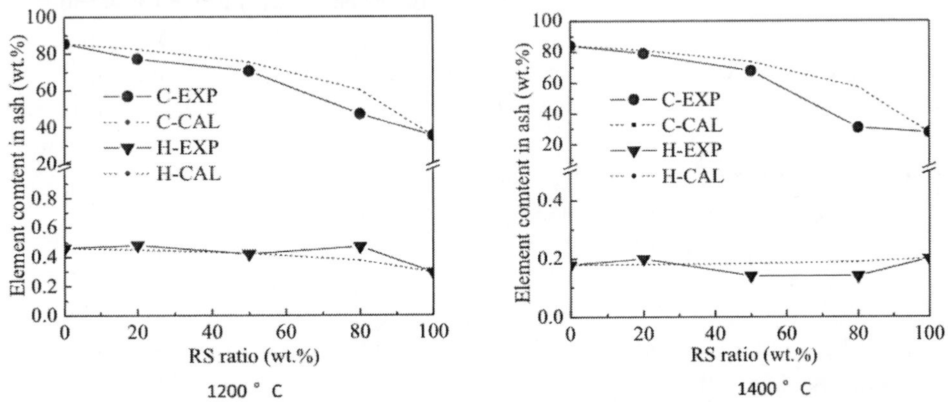


Fig.8 Content of C and H in solid from oxygen co-gasification of coal and RS

- ◆ Co-gasification of coal and RS promotes the carbon conversion .
- ◆ Coal would interact with RS during oxygen co-gasification.

# RESULTS & DISCUSSION

## 2.4 Apparent morphology of solid products

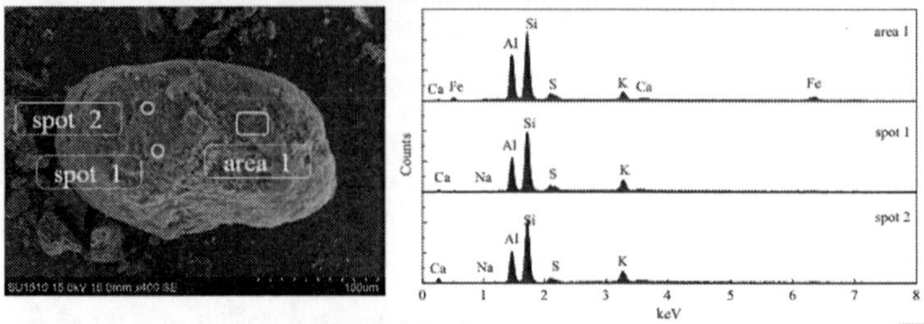


(a) Particle I

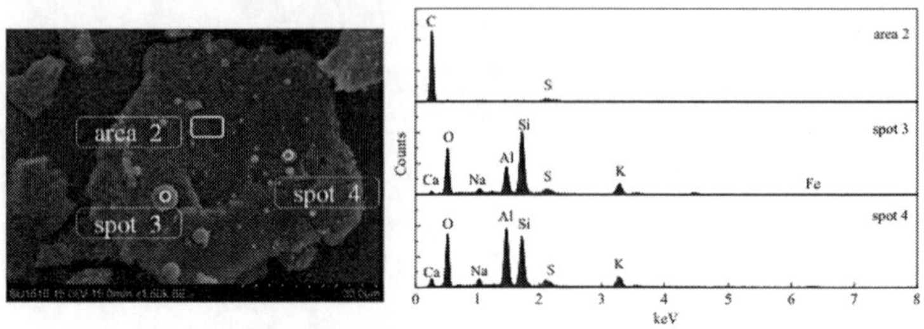


(b) Particle II

Fig.9 SEM-EDX analyses of typical oxygen co-gasification solid at 1400℃

- ◆ The ash of RS and coal may interact with each other and form spherical ash. It may inhibit or promote gasification reactions.
- ◆ The condensation of released K from RS on the surface of coal ash could promote the fusion process.

# CONCLUSION

01

**Coal and straw separate gasification**

- ◆ The gas yield increases and solid yield decreases with increasing temperature.
- ◆ The syngas from rice straw gasification is mainly composed of $H_2$ and CO, while that of coal gasification is mainly composed of $H_2$, $CO_2$ and CO.
- ◆ For both coal and rice straw, volatile matter and fixed carbon of solid products decrease.
- ◆ The contents of C, H, N and S decrease with increasing temperature and the content of C obviously decreases for rice straw.
- ◆ The shape and structure of rice straw during gasification will change.

02

**Oxygen co-gasification**

- ◆ At 1200 °C , CO and $H_2$ are inhibited and $CO_2$ is promoted when the blending ratio is 20% and 50%. However, when the blending ratio is 80%, $H_2$ and $CO_2$ are inhibited.
- ◆ At 1400 °C, $H_2$ is inhibited and $CO_2$ is promoted when the blending ratio is 20%, and CO is promoted when the blending ratio is 50% and 80%.
- ◆ The interaction occurs during the co-gasification of coal and rice straw. The fused ash particles are formed, which is rich of Si, Al, K and Ca et al.

Thank you
for your attention!

**2016 International Pittsburgh Coal Conference**
**Cape Town, SOUTH AFRICA**
**August 8 – 12, 2016**

**Abstract Submission**

**PROGRAM TOPIC: NOVEL COMBUSTION AND CYCLE TECHNOLOGIES**

# INFLUENCE OF COMBUSTION TEMPERATURE AND OXYGEN CONCENTRATION IN COMBUSTION MEDIUM ON EMISSIONS OF $NO_X$, $N_2O$ AND $SO_2$.

***Assoc. Prof. Karel Svoboda, Ph.D.***
***Institute of Chemical Process Fundamentals, Academy of Science of the Czech Republic, Rozvojová 135, 165 02 Praha 6-Suchdol, Czech Republic.***
***Contact Information: svoboda@icpf.cas.cz***

*Jaroslav Moško[a,b], Tomáš Durda[a,b], Assoc. Prof. Michael Pohořelý, Ph.D.[a,b], Boleslav Zach[a,b], Michal Šyc Ph.D.[a], Aneta Krausová[a,b], Michal Jeremiáš, Ph.D.[a]*
*[a]Institute of Chemical Process Fundamentals, Academy of Science of the Czech Republic, Rozvojová 135, 165 02 Praha 6-Suchdol, Czech Republic.*
*[b]Department of Power Engineering, Institute of Chemical Technology Prague, Technická 5, 166 28 Praha 6, Czech Republic.*
*Contact Information: pohorely@icpf.cas.cz, pohorelm@vscht.cz*

**Abstract:**
Sewage sludge incineration is the only suitable way for disposal of large volumes of sewage sludge containing POPs and some heavy metals. Sewage sludge combustion processes reduce the volume of such waste and cause destruction of organic contaminants such as POPs, detergent residues, pharmaceuticals and various compounds with endocrine-disrupting potential. Due to higher phosphorus content, the ash from sewage sludge combustion can be potentially used for phosphorus recovery. The more volatile heavy metals and their compounds (with exception of mercury) can be concentrated in fly ash. Oxy-fuel combustion is a technology for simplifying $CO_2$ sequestration from the flue gas by using nearly pure oxygen instead of air for the combustion together with flue gas recirculation contribution to higher $CO_2$ content in inlet gas and flue gas. In order to control the oxygen concentration in inlet gas, combustion temperature, and formation of $NO_x$, part of the flue gas is recycled to combustion chamber. Fluidized bed technology offers also significant advantage to the combustion process due to its flexibility to burn various types of fuel with low and fluctuating heating value. Our paper brings information about investigation of the effects of combustion (fluidized bed) temperature and concentration of oxygen in inlet combustion media on the emissions of $NO_x$, $N_2O$ and $SO_2$ in FB incineration of sewage sludge at temperatures 750–930 °C. Concentrations of oxygen in the inlet gas during experiments were 21, 25, 28 and 30 vol. %. Experiments were carried out in a bubbling fluidized bed (BFB) reactor with inner diameter of 9.36 cm and height of 130 cm above the gas distributor. To simulate oxy-fuel BFB combustion, a mixture of oxygen and carbon dioxide as a gaseous combustion medium was used. Dry sewage sludge (SS) from mesophilic stabilization (moisture content 9 mass % and ash content 44 mass % in dry matter) was used for experiments. Size of prepared particles from dry sewage sludge was between 0.5 and 2 mm. The particles were fed into the FB reactor by gas-driven slide feeder with two fuel containers (quasi-continual feeding). The SS particles were introduced to the very bottom of the FB in a cooled tube (inner diameter of 9 mm) by pneumatic transport. $CO_2$ was used as the gaseous medium for the pneumo-transport. The ash from SS incineration was used as primary fluidized bed material. Main components of the primary fluidized bed material were $SiO_2$, $P_2O_5$, $Al_2O_3$, CaO and $Fe_2O_3$, size of primary fluidized bed material particles ranged from 0.71 to 1.6 mm. Minimum fluidizing velocity of the primary fluidized bed material was experimentally measured and calculated to have a value of 30 cm/s at 25 °C and 101,325 kPa. The experimental results have shown that an increase in combustion temperature leads to an increase in $NO_x$ and to a decrease in $N_2O$ emissions in a temperature range 750–930 °C. The lowest $SO_2$ emissions were measured at the temperature interval from 790 to 820 °C in all experiments with different oxygen concentration in inlet gas. It was found that when the oxygen concentration in inlet gas increased, the emissions of $SO_2$ increased while emissions of $NO_x$ slightly decreased. Experimental results suggest that flue gas from oxy-fuel BFB combustion of dry, stabilized sewage sludge is suitable for $CO_2$ sequestration (CCS process) after further purification and compression. Furthermore, the bottom ash from the fluidized bed combustion of SS can be considered as potential source for phosphorus recovery due to high phosphorus content (with about 80 % of phosphorus content from the sludge). Based on our results, emissions of nitrous oxides and $SO_2$ are higher at temperature 750–930 °C during oxy-fuel BFB combustion of sewage sludge when compared to combustion by $O_2/N_2$ at the same operating conditions. The oxy-fuel combustion or co-combustion processes are, on the other hand, generally handicapped by additional costs for production of pure or highly concentrated oxygen.

33rd Annual International Pittsburgh Coal Conference, Cape Town (South Africa), August 8-12, 2016

# Methanol synthesis by catalytic hydrogenation and photoelectrochemical reduction of $CO_2$

Alberto Pettinau, Mauro Mureddu, Francesca Ferrara

apettinau@sotacarbo.it m.mureddu@sotacarbo.it f.ferrara@sotacarbo.it

Sotacarbo S.p.A., Grande Miniera di Serbariu, 09013 Carbonia, ITALY

SOTACARBO

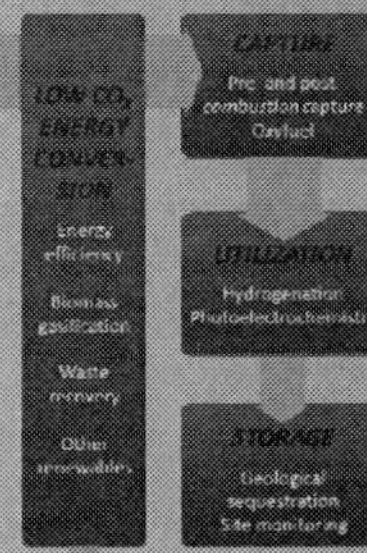


## Abstract

The increasing attention toward climate changes and the recent strategic policies to stabilize and reduce the amount of $CO_2$ emission is promoting research in the field of carbon capture, storage and more recently utilization technologies, whereby $CO_2$ is captured from industrial flue gas, reused if possible or permanently stored in geological formations.

This work presents a general overview of the researches on $CO_2$ utilization by conversion into liquid fuels (mainly methanol), highlighting the research infrastructure and the main experimental results.

$CO_2$ hydrogenation - catalysts: Cu/ZnO/ZrO$_2$ catalysts have been synthetized as nanopowders by a sol-gel self-combustion method. Characterization by:

- XRD (X-ray diffraction)
- TEM (Transmission Electron Microscopy)
- TGA (ThermoGravimetric Analysis)
- CHN/S (Carbon, Hydrogen, Nitrogen, Sulphur) analysis
- TPR (Temperature Programmed Reduction)
- $N_2$ physisorption

$CO_2$ hydrogenation – pilot plant: The catalysts will be finally characterized in a fully automated pilot unit equipped with two different fixed-bed reactors which can operate up to 100 bar and 700 °C.

$CO_2$ photoelectrochemical (PEC) reduction: "green" technology for $CO_2$ conversion driven by solar energy and water (like in natural photosynthesis). The possibility to use a photoanode-driven PEC cell in which a metal has been used as cathode, while the anode material is a nanotubular structure of $TiO_2$, is investigated.

## Catalysts for $CO_2$ hydrogenation

Unsupported catalysts (Cu/ZnO/ZrO$_2$)

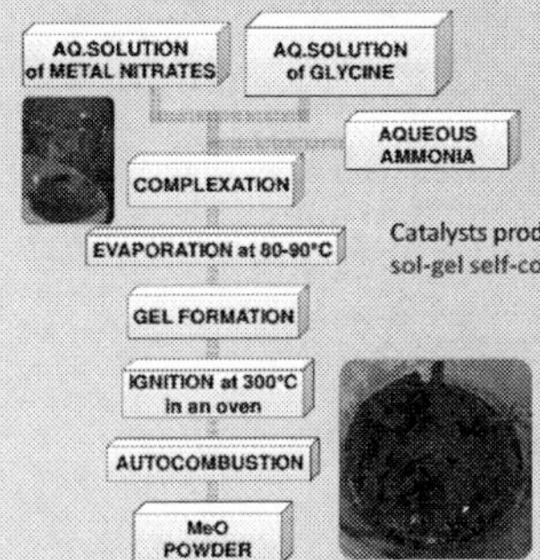


Catalysts production by the innovative sol-gel self-combustion method

Supported catalysts (Cu/ZnO@SBA-15)

Nanocomposites fabrication procedure

Impregnation-sol-gel self-combustion approach

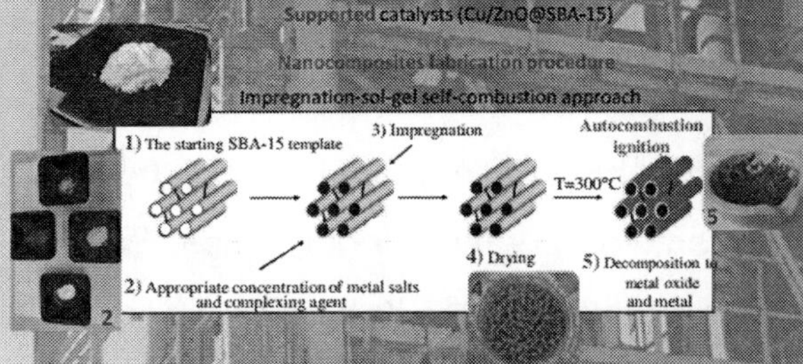


SBA-15 mesostructured silica

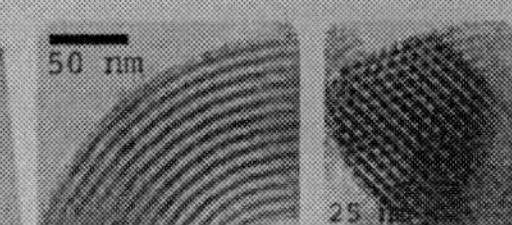


Materials characterization

| Sample | $S_{BET}$ ($m^2/g$) | $D_p$ (nm) | $V_p$ ($cm^3/g$) |
|---|---|---|---|
| SBA-15_T140 | 500 | 8.2 | 1.23 |
| CZS_20_1_T300 | 383 | 8.2 | 1.025 |
| CZS_20_2.3_T300 | 393 | 8.1 | 1.0649 |
| CZS_35_2.3_T300 | 312 | 8.3 | 0.846 |

## $CO_2$-to-Liquids pilot unit

Gas/Liquid/Liquid separator detail

Schematic diagram of the synthesis section for liquid fuels production

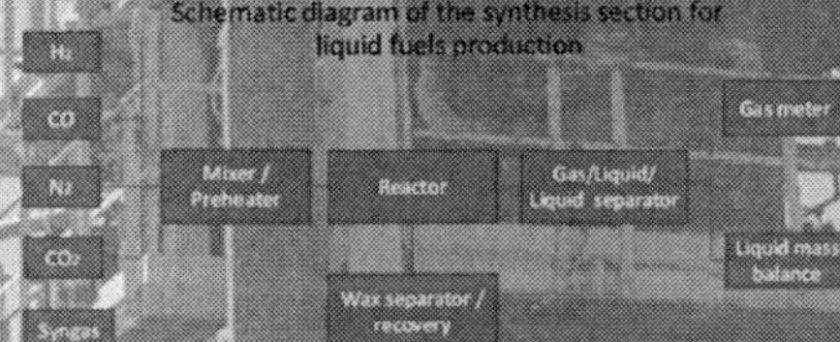

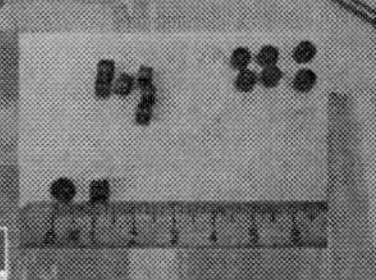

Tubular reactors specifications

| | | |
|---|---|---|
| Material | Hastelloy C | Hastelloy C |
| Internal diameter (mm) | 9.1 | 23.8 |
| Max. temperature at atm. pressure (°C) | 1080 | 1080 |
| Max. temperature at 100 bar (°C) | 700 | 700 |
| Max volume of catalyst at isothermal conditions ($cm^3$) | 3.3 | 21.9 |

## Photoelectrochemical reduction of $CO_2$

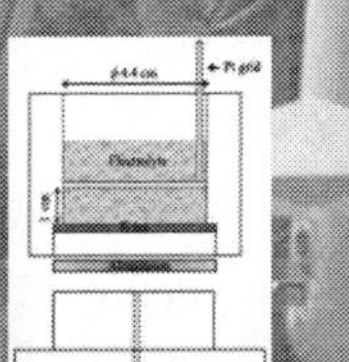

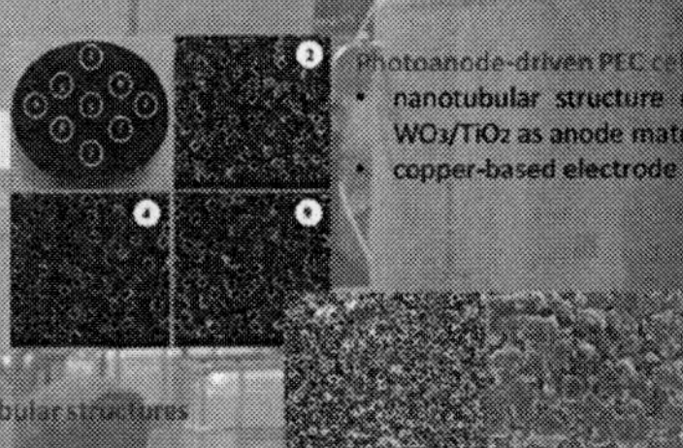

Photoanode-driven PEC cell:

- nanotubular structure of $TiO_2$ and $WO_3/TiO_2$ as anode material
- copper-based electrode as cathode

Synthesis of nanotubular structures

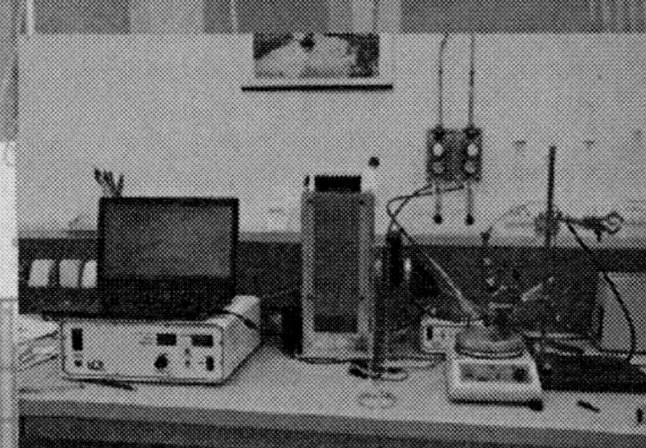

*Acknowledgements*

This work has been done within the "Centre of Excellence on Clean energy" project (CUP: D82I13000250001), funded by the Regional Government of Sardinia.

Many thanks to:

University of Cagliari, Italy

- Prof. Elisabetta Rombi
- Prof. Simonetta Palmas

SOTACARBO SpA, Grande miniera Serbariu, 09013 Carbonia, ITALY

# Effect of Internals Configurations and Sparger Design on the Local Hydrodynamics in Slurry Bubble Column Reactors Operating under Fischer-Tropsch Conditions

**Omar Basha and Badie I. Morsi**

Department of Chemical and Petroleum Engineering, University of Pittsburgh, Pittsburgh, PA 15261, USA

## Abstract

A three-phase, three-dimensional multi-Eulerian Computational Fluid Dynamics (CFD) model, which has been developed in ANSYS, Fluent v. 14.5 and experimentally validated [1, 2], was used to predict the effects of gas spargers on the local hydrodynamics (gas holdup, Suter mean bubble diameters and backmixing) in a pilot-scale (0.3-m ID, 3-m height) Slurry Bubble Column Reactor (SBCR). The CFD model was also used to predict the effects of internals on the local hydrodynamics in a large-scale conceptual (1-m ID, 10-m height) SBCR. The simulations were carried out in the two reactors operating under typical Fischer-Tropsch (F-T) process conditions (pressure = 20.5 bar, temperature = 443 K, catalyst concentration = 11 vol% and inlet superficial gas velocity = 0.20 m/s) without chemical reactions.

The CFD simulation was conducted for the pilot-scale SBCR provided with two different gas spargers (6-arms spider and concentric-ring), having downward- and upward-facing orifices. No internals were used in this reactor in order to solely highlight the effects of gas spargers. The simulation results showed that the spargers with downward-facing orifices were more effective in solids dispersion and led to lower Sauter mean bubble diameters than those found with the spargers provided with upward-facing orifices. Also, concentric-ring spargers resulted in larger Sauter bubble diameter values than those obtained with the 6-arms spiders.

The CFD simulation was also carried out for the conceptual SBCR provided with different internals configurations in parallel and bundled arrangements. The reactor was provided with 3 concentric-rings sparger, having 611 identical orifices with 7-mm ID. The simulation results indicated that the use of parallel internals led to smaller average bubble size, and faster as well as larger liquid recirculations, which were developed much earlier, when compared with those obtained with bundled internals. Also, internals bundled configurations with smaller internals-reactor wall clearance resulted in stronger liquid recirculations than those observed for bundled configurations with wider internals-reactor wall clearance.

## 1. Introduction and Background

Gas spargers are an essential component in SBCRs as they can control the bubbles size as well as distribution and consequently the gas holdup, gas-liquid interfacial area, local hydrodynamics and the overall performance. Some examples of gas spargers include perforated plates, porous plates, porous membranes, rings type distributors and multiple-arms spargers [3]. Although significant work has been done investigating bubbles formation at the orifice [2], the effect of sparger design on the performance of bubble columns and SBCRs has received little attention.

An efficient sparger design should aim at minimizing the pressure drop across the sparger while maintaining a uniform distribution of gas bubbles at the orifices, which could represent significant savings in operating costs and energy requirements. Generally, when the diameters of SBCR are large than 1-m, multi-arms spiders or concentric-rings spargers are preferred over perforated plates [4]. Also, it was reported that the effects of sparger type on the gas holdup and bubble diameter were more significant at low superficial gas velocities (0.1 m/sec), however, such effects become insignificant at greater superficial gas velocities as in the churn-turbulent flow regime [5].

Reactor internals are also an integral part of the design and optimization of numerous industrial applications. Different internals, including rings, inverse-cone tubes, and horizontal ring-type baffles were used in circulating fluidized beds (CFBs) to improve the radial solids distribution [6]. Fixed and floating internals were used in fast fluidized-beds to reduce the agglomerate size in cohesive powders for fluidization improvement [7]. Porous draft tubes, open-sided draft tubes and non-porous draft tubes were used in conical spouted-beds to increase the gas flow rate from the spout into the annulus and enhance heat transfer [8, 9]. Also, different internals were employed in SBCRs to maintain a near isothermal operation in highly exothermic processes.

Generally, heat transfer rate in multiphase systems is controlled by many variables, including heat transfer area, heat transfer coefficients, superficial gas velocity, liquid circulation velocity, local and overall phase holdups, bubble velocities, bubble sizes, and bubble frequency. In SBCRs, good mixing and high heat capacity of the slurry-phase allow for excellent temperature control, making it ideal for the F-T synthesis process, which is highly exothermic. Kolbel and Ralek [10] reported a temperature difference of less than 1 °C over the total length of a pilot-scale SBCR (1.29-m ID, 8-m height) used for F-T synthesis at 268 °C. The cooling surface requirements in SBCRs have been reported to be less than a quarter of those used in a multitubular fixed-bed reactors, due to the improved heat transfer film coefficients in addition to the higher permissible temperature difference between the reactant and coolant [11, 12].

In this work, a three-phase, three-dimensional multi-Eulerian Computational Fluid Dynamics (CFD) model, which has been developed in ANSYS and experimentally validated [1, 2], was used to investigate the effect of gas spargers on the local hydrodynamics (local gas holdup, Suter mean bubble diameters, gas and liquid velocity vectors, mixing intensity and liquid backflow) in a pilot-scale (0.3-m ID, 3-m height) SBCR. The CFD model was also used to investigate the effect of internals on the local hydrodynamics in a large-scale conceptual (1-m ID, 10-m height) SBCR.

## 2. CFD Model Formulation

The developed CFD model is a 3-D, three-phase, multi-fluid Eulerian model. The kinetic theory of granular flow (KTGF) was used to represent the solids behavior in the system, and the model developed by Gidaspow [13] was used to describe the solid viscosity. The bubbles induced turbulence was considered through the introduction of two source terms $S_k$ and $S_\varepsilon$, into the $k$-$\varepsilon$ equation to account for the covariance of the velocities of the liquid and dispersed phases, in addition to the relative and drift velocity effects. The bubble population balance was included to represent the bubble size distribution. The model also incorporated the breakup and coalescence models by Lou and Svendsen [14] and Prince and Blanch [15], respectively. A first order implicit time stepping was used to advance the solution in time; and due to the strong non-linear nature of the model, relaxation coefficients by Patankar [16] were introduced in the momentum conservation holdup equations.

The drag and turbulence model parameters used were optimized by comparison against experimental data by Yu and Kim [17], Chen et al. [18] and Sehabiague et al. [19]. The models of Ishii and Zuber [20], Schiller and Naumann [21] and Wen and Yu [22] were employed to represent the gas-liquid, gas-solid and liquid-solid drag, respectively, along with the RNG $k$-$\varepsilon$ turbulence model. Also, the lift coefficient was represented using the correlation by Tomiyama et al. [23]. A summary of the governing equations and constitutive relations used in the model are shown in Table 1, and the momentum exchange terms in addition to interphase coefficient expressions are reported by Basha et al. [1].

The simulations were carried out using ANSYS Fluent v14.5. An upwind Quadratic Upstream Interpolation for Convective Kinematics (QUICK) scheme was used to solve the model equations. This is because the upwind QUICK is superior than the second order upwind scheme since upwind QUICK has less dispersion and dampens the high frequency components much less.

**Table 1: Summary of the governing equations and constitutive relations used in the model**

| | |
|---|---|
| **Mass conservation** | $\frac{\partial(\rho_k\alpha_k)}{\partial t} + \nabla(\rho_k\alpha_k u_k) = -\dot{m}_{kn}$ (1) |
| **Momentum conservation** | $\frac{\partial(\rho_k\alpha_k u_k)}{\partial t} + \nabla(\rho_k\alpha_k u_k u_k) = -\alpha_k \cdot \nabla P_k + \nabla(\alpha_k\tau_k) + \alpha_k\rho_k g + u_{kn}\dot{m}_{kn} + M_k$ (2) |
| **Stress tensor** | $\tau_{k=G,L} = \alpha_k\mu_{k,eff}\left(\nabla u_k + (\nabla u_k)^T - \frac{2}{3}I(\nabla \cdot u_k)\right)$ (3) |
| **Effective viscosity of liquid phase** | $\mu_{eff} = \mu_{L,M} + \mu_{L,T} + \mu_{L,B}.$ (4)<br>$\mu_{L,T} = C_{\mu,B}\rho_L(\frac{k^2}{\varepsilon}),$ with $C_{\mu,B} = 0.6$ (5) |
| **Effective viscosity of the gas phase** | $\mu_{eff} = \mu_{G,M} + \mu_{G,T}.;\ \mu_{G,T} = \frac{\rho_L}{\rho_G}(\mu_{L,T})$ (6) |
| **Solid phase representation** | $\tau_S = \alpha_S\mu_S(\nabla u_S + (\nabla u_S)^T) - P_s I + \alpha_S\mu_S\left(\lambda_S - \frac{2}{3}\right)I(\nabla \cdot u_S)$ (7)<br>$\lambda_S = \frac{4}{3}\alpha_S^2\rho_S d_p g_0(1+e_p)\sqrt{\frac{\Theta}{\pi}}$ and $P_s = \rho_S\alpha_S\Theta + 2g_0\rho_s\alpha_S^2\Theta(1+e_p)$ (8)<br>$g_0 = 0.6\left[1 - \left(\frac{\alpha_S}{\alpha_{S,max}}\right)^{1/3}\right]^{-1}$ and $\Theta = \frac{d_p^2}{30(1-e_p)}(\nabla u_S + (\nabla u_S)^T)$ (9)<br>$\mu_S = \frac{4}{5}\alpha_s^2\rho_s d_p g_0(1+e_p)\sqrt{\frac{\Theta}{\pi}} + \frac{2\frac{5\sqrt{\pi}}{96}\rho_s d_p\sqrt{\Theta}}{g_0(1+e_p)} \cdot \left[1 + \frac{4}{5}\alpha_s g_0(1+e_p)\right]^2$ (10) |
| **Turbulence model (RNG *k*-ε model)** | $\frac{\partial(\rho k)}{\partial t} + \nabla(\rho k u) = \nabla(Pr_{eff,k}\mu_{eff}\nabla k) + G_k + G_b - \rho\varepsilon - Y_M + S_k$ (11)<br>$\frac{\partial(\rho\varepsilon)}{\partial t} + \nabla(\rho\varepsilon u) = \nabla(Pr_{eff,\varepsilon}\mu_{eff}\nabla\varepsilon) + C_{\varepsilon 1}\frac{\varepsilon}{k}(G_k + C_{\varepsilon 3}G_b) - C_{\varepsilon 2}\rho\frac{\varepsilon^2}{k} - R_\varepsilon + S_\varepsilon$ (12) |
| **Bubble induced turbulence** | $S_k = (\frac{\rho_G}{\rho_G + C_{VM}\rho_G})\alpha_G(\frac{3\,C_D}{4\,d_B})\lvert u_G - u_L\rvert(q_{LG} - 2k_L + u_d \cdot u_r),\quad S_\varepsilon = C_{\varepsilon 3}\frac{\varepsilon}{k}S_k$ (13)<br>$u_r = [u_G - u_L] - u_d,\quad u_d = -D_{12}^t\left[\frac{\Delta\alpha_2}{\alpha_2} - \frac{\Delta\alpha_1}{\alpha_1}\right]$ (14)<br>$D_{12}^t = \frac{1}{3}\tau_{12}^t q_{LG} \cdot I,\quad Where\ I\ is\ a\ 3 \times 3\ identity\ matrix$ (15) |
| **Bubble population balance** | $\frac{\partial n_i}{\partial t} + \nabla(u_G n_i) = \left(\sum_J R_J\right)_i = B_C + B_B - D_C - D_B$ (16)<br>$f_{BV} = \frac{v_I}{v} = \frac{d_I^3}{d} = \frac{d_I^3}{d_I^3 + d_{II}^3}$ (17)<br>$\Omega_B(v: vf_{BV}) = 0.923(1-\alpha_G)n\left(\frac{\varepsilon}{d^2}\right)^{1/3}\int_{\xi_{min}}^{1}\frac{(1+\xi)^2}{\xi^{11/3}}exp\left(-\frac{12c_f\sigma}{\beta\rho_L\varepsilon^{2/3}d^{5/3}\xi^{11/3}}\right)d\xi$ (18)<br>$B_B = \sum\Omega_B(v_j: v_i)n_i;\ D_B = \Omega_i n_i;\ where\ \Omega_i = \sum_{m=1}^N\Omega_{mi}$ (19)<br>$u_T = \frac{r_{ij}^{2/3}}{\varepsilon^{1/3}};\quad \tau_{ij} = \frac{r_{ij}^{2/3}}{\varepsilon^{1/3}};\quad r_{ij} = \frac{2r_i r_j}{(r_i + r_j)};\quad t_{ij} = \left(\frac{r_{ij}^3\rho_l}{16\sigma}\right)^{0.5} ln\left(\frac{h_0}{h_f}\right)$ (20)<br>$B_C = \frac{1}{2}\sum_{k=1}^i\sum_{j=1}^i\eta_{jki}\chi_{ij}n_i n_j;\ D_C = \sum_{m=1}^N\chi_{ij}n_i n_j$<br>Where $\eta_{jki} = \begin{cases}(v_j + v_k) - v_{i-1}/(v_i - v_{i-1}) & if \quad v_{i-1} < v_j + v_k < v_i \\ v_{i+1} - (v_j + v_k)/(v_{i+1} - v_i) & if \quad v_i < v_j + v_k < v_{i+1} \\ 0 & if \quad neither\end{cases}$ (21)<br>$\chi_{ij} = \frac{\pi}{4}[d_i + d_j]^2(u_{Ti}^2 + u_{Tj}^2)^{0.5}exp\left(-\frac{t_{ij}}{\tau_{ij}}\right)$ |

- *k* indicates the phase (L for liquid, G for gas and S for solid)

## 3. Effect of Sparger Design on Local Hydrodynamics in the Pilot-scale SBCR

The effect of two different gas spargers (6-arms spider and concentric-rings) on the gas holdup, bubble size and liquid-phase recirculation in the pilot-scale SBCR were investigated using our CFD model in the absence of F-T reactions. The spider-type sparger used consists of 6 arms, each arm has 6 orifices of 5 mm ID on each side and on the bottom, totaling 18 orifices. Thus, there are a total of 108 orifices on the 6-arms spider gas sparger. It should be noted that there are no orifices at the top of the arms so that the solid particles could not plug them and the gas downward jets should be able to lift any solid particles, which might settle at the bottom flange of the reactor. The gas sparger is screwed onto a 0.0254 m ID pipe to a height of 0.102 m from the top of the bottom flange. Also, its maximum height from the bottom of the reactor is about 0.152 m. The concentric-ring sparger consists of three concentric-rings with 30, 15 and 10 orifices per ring, respectively, totaling 55 identical orifices. Each orifice has an ID of 7 mm. The geometry and numerical mesh details of the spider and concentric-ring spargers are shown in Figure 1. It should be noted that no internals were used in the pilot-scale SBCR in order to solely highlight the effect of sparger type.

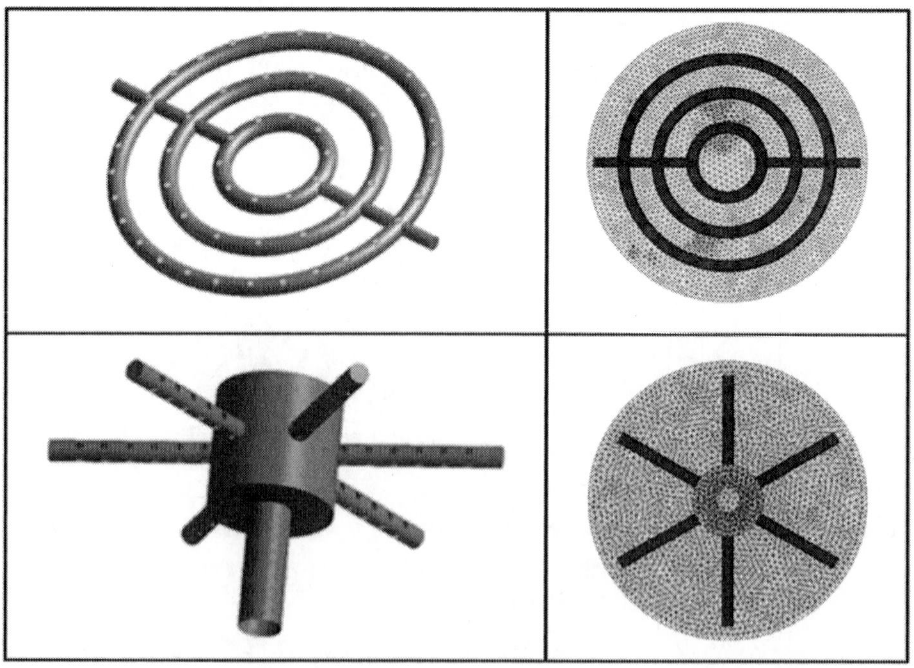

**Figure 1: Constructed geometry and mesh of the concentric-rings (Top) and 6-arms spider (Bottom) spargers used**

The spatial variations of the local gas holdup in the SBCR give rise to the local pressure variations, which induce liquid recirculations inside the reactor. These liquid recirculations are crucial as they govern the rate of mixing, affecting the heat and mass transfer characteristics. Our CFD model predictions illustrated in Figures 2 and 3 for the spider and concentric-rings spargers, respectively, show several snapshots of the gas holdup inside the reactor at different times of 1, 20, 40 and 60 s. These data were obtained using an inlet superficial gas velocity of 0.20 m/s at pressure of 20.5 bar, temperature of 443 K and catalyst concentration of 11 vol%. Under these operating conditions, the overall gas holdup in the SBCR were 0.42, and 0.36 for the spider and concentric-rings, respectively.

As can be seen in Figures 2 and 3, the gas holdups obtained using the two spargers vary with the axial and radial positions in the SBCR; and the radial gas holdup exhibits a non-linear behavior at any axial position as illustrated by the gradual variation in the color shades of the gas holdup from the center to the wall of the reactor. It should be pointed out that the maximum gas holdup value can be seen near the center of the reactor before it gradually decreases along the radii with various profiles in the axial direction.

The effect of the sparger orifice orientation on the overall gas holdup and the axial solids dispersion was investigated using our CFD model. The 6-arms spider sparger similar to that shown in Figure 1, provided with upward-facing orifices in addition to the side-orifices and the concentric-rings sparger shown in Figure 1 with only downward-facing orifices were used in this investigation. The effect of the sparger type and orifices orientation on the overall gas holdup, axial solid loadings and axial bubble size variation were investigated using three different superficial gas velocities of 0.1, 0.2 and 0.3 m/s at pressure of 20.5 bar, temperature of 443 K, and catalyst concentration of 11 vol%.

In general, a sparger with downward-facing orifices is expected to be more effective in solids dispersion in a SBCR when compared with a similar sparger provided with only upward-facing orifices. It is argued that the gas jets from downward-facing orifices provide the necessary agitation to facilitate solids dispersion. Also, the sparger used definitely determines the bubble sizes observed in the reactor, hence small orifices enable the formation of small gas bubbles.

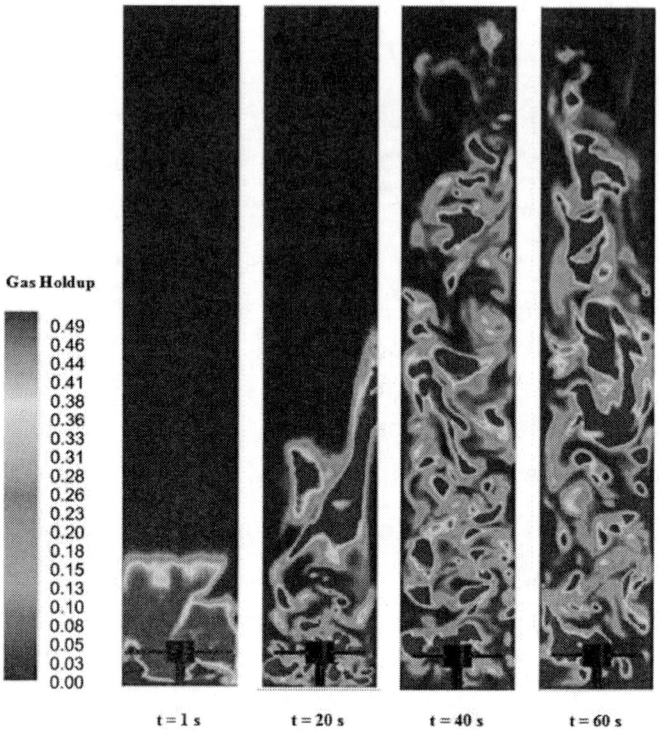


**Figure 2: Snapshots of gas holdup contours at different times for the spider sparger**

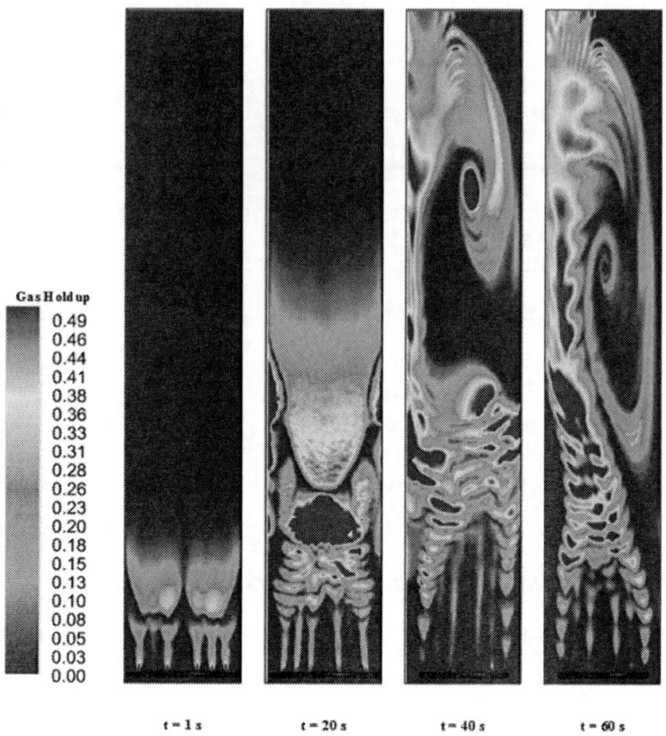


**Figure 3: Snapshots of gas holdup contours at different times for the concentric-rings sparger**

Figure 4 (a) shows the effect of the 6-arms sparger orifice orientations on the overall gas holdup at different superficial gas velocities. As can be observed in this figure, both the upward and downward 6-arms spider orifice orientations give higher gas holdup values when compared with those of the concentric-rings spargers. The 6-arms spider with upward-facing orifices resulted in the highest gas holdups. However, at lower superficial gas velocities ($U_g < 0.15$ m/s), the concentric-rings sparger with downward-facing orifices gives higher gas holdup values than those using a concentric-rings sparger with upward-facing orifices.

Figure 4 (b) shows the effect of the sparger orifices orientations on the Sauter mean bubble diameter ($d_{32}$) at different superficial gas velocities. As can be observed in this figure, the concentric-rings spargers resulted in greater $d_{32}$ when compared with those of the 6-arms spider, which can be attributed to the larger orifice size (7 mm ID) for the concentric-rings sparger versus 5 mm ID for the 6-arms spider. The effect of the superficial gas velocity on the $d_{32}$ obtained with the 6-arms spider was insignificant, which is in agreement with the experimental data [24]. The effect of superficial gas velocity on the $d_{32}$, however, was more significant for the concentric-rings provided with upward-facing orifices as a result of the turbulent eddies created, which enhanced the bubbles coalescence. This is because bubbles coalescence and breakup mechanisms is primarily governed by the turbulent eddies, which in turn, are controlled by the liquid and gas velocities [25]. Nonetheless, the bubbles breakup is governed by the turbulent stresses, which are controlled by the pressure variations within the fluid and subsequently are related to the physical property gradients within the continuous-phase [15].

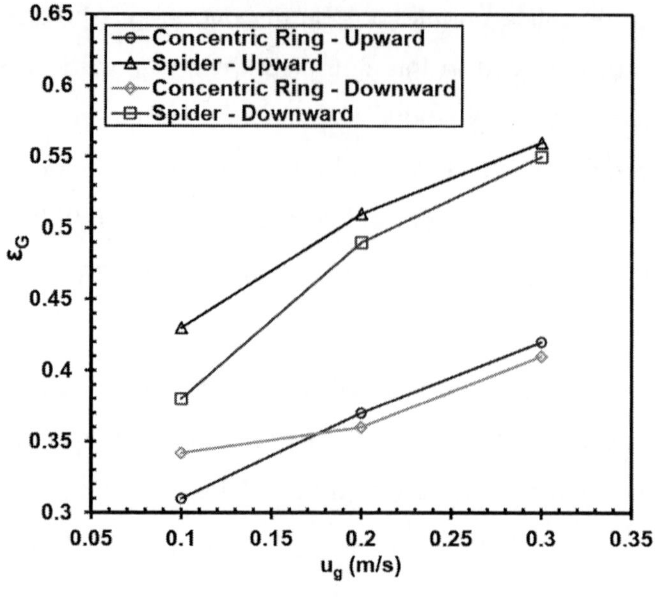


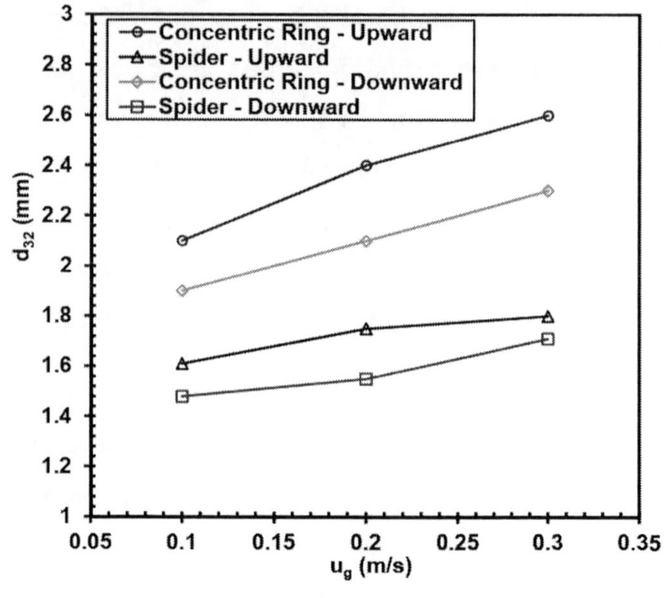


**Figure 4: Effect of sparger type and nozzles orientation on the (a) gas holdup and (b) Sauter mean bubble diameter at different superficial gas velocities**

Figure 5 (a) to (c) show the effects of the sparger orifices orientations on the axial solid concentration profiles at superficial gas velocities of 0.1, 0.2 and 0.3 m/s, respectively. As can be seen in these figures, the spargers with downward-facing orifices give a more uniform axial solids distribution, whereas, those with upward-facing orifices result in a solid accumulation near the bottom of the reactor. Actually, high solid concentrations (up to 17.2 vol%) can be seen at the bottom of the reactor at superficial gas velocity 0.2 m/s, leading to lower solid concentrations above the sparger for the spargers with upward-facing orifices.

It is important to note that the concentric-rings sparger with upward-facing orifices resulted in higher solid concentrations at the bottom of the reactor when compared with those of the 6-arms spider with upward-facing orifices. This behavior could be due to the fact that the 6-arms spider has also lateral nozzles on both sides of each arms, which helped in enhancing mixing around this sparger.

The effect of sparger orifices orientation on the axial solids distribution profile at the ratio of an axial position (z) to the total height (H), (z/H) greater than 0.2 appear to decrease with increasing the superficial gas velocity. Generally, the spargers with downward-facing orifices were more effective in solids distribution when compared with those with upward-facing orifices. This is due to the increased mixing provided by the gas jets from the downward-facing orifices, which increased the solid particles distribution.

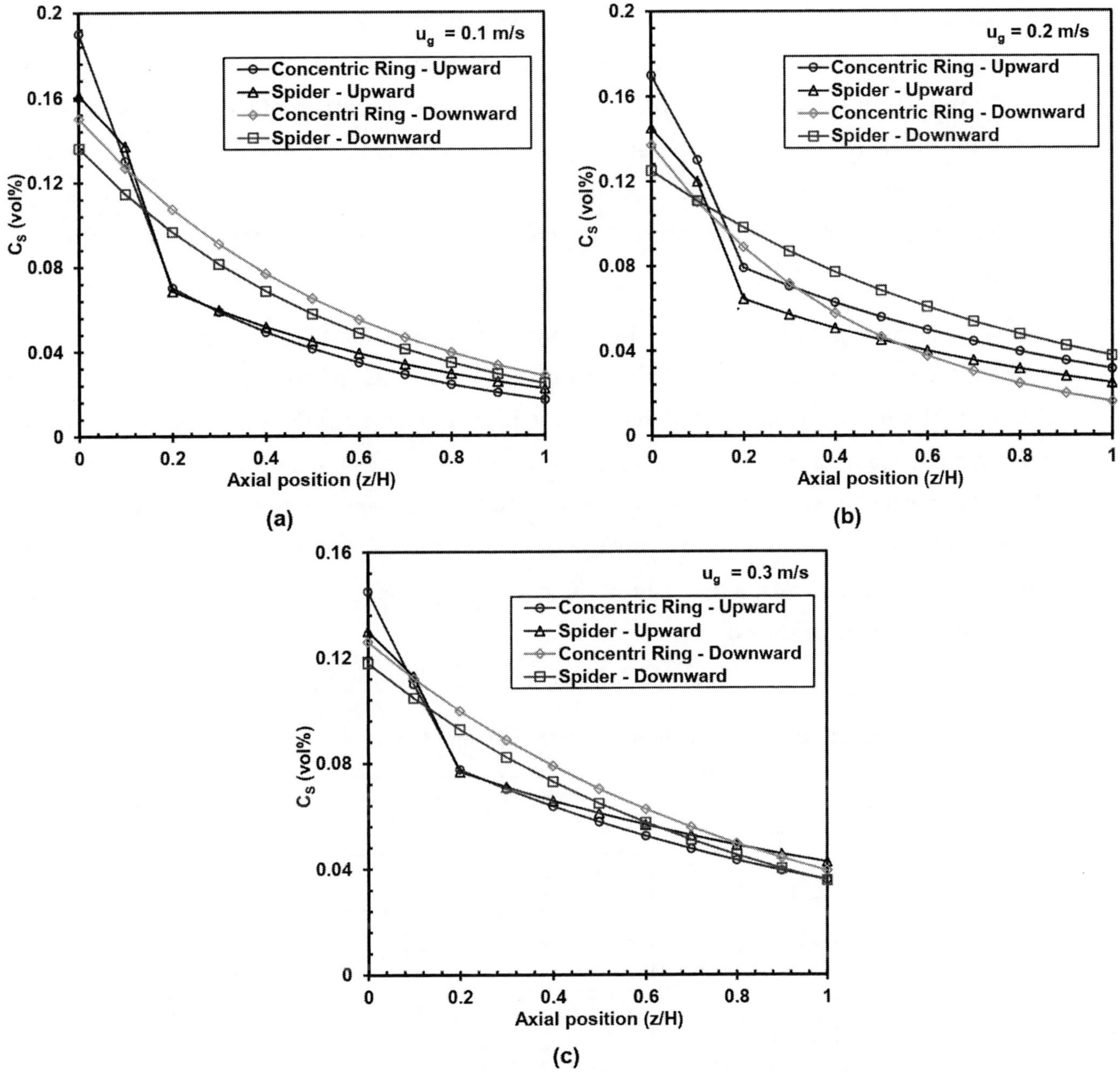


**Figure 5: Effect of sparger orifice orientations on the axial solid concentration profile at different superficial gas velocities**

## 4. Effect of Internals Configurations on the Local Hydrodynamics in a Large-scale Conceptual SBCR

The effect of internals configurations on the liquid axial velocity and gas holdup was predicted using our CFD model in three different parallel configurations inside the conceptual large-scale SBCR (1-m ID, 10-m height). The areal coverage and volume fraction of the internals are shown in Table 2. In all the cases investigated, the internals extended from the top of the reactor up to 86% of the reactor height. The gas was introduced using a concentric-ring sparger, consisting of 3 concentric-rings with 333, 167 and 111 orifices, totaling 611 identical orifices, each with an ID of 7-mm.

The simulations were carried out in the absence of F-T reactions under pressure of 20.5 bar, temperature of 443 K and catalyst concentration of 11 vol. % at an inlet superficial gas velocity of 0.20 m/s. The meshing was built in the same way as described by Basha et al. [1]. In order to prevent numerical problems, the calculations were carried out with a 0.05 s increment for the first 1000 time steps, increased to 0.1 s for the next 1000 time steps, then increased to 0.25 s until the end of the simulation time. The simulated physical time was for 300 s, at which the statistical quasi-steady state was attained, i.e., once the converging time averages of the local holdups and axial velocities were stabilized.

**Table 2: Details of the parallel internals used in the Conceptual SBCR**

| Parallel Arrangement | | |
|---|---|---|
| P-1 | P-2 | P-3 |
| 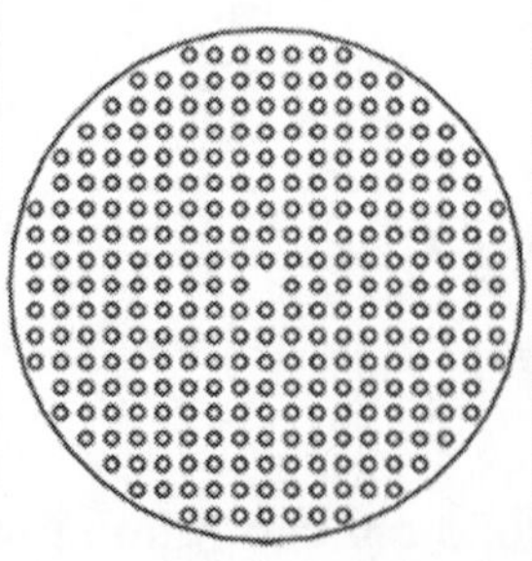 | 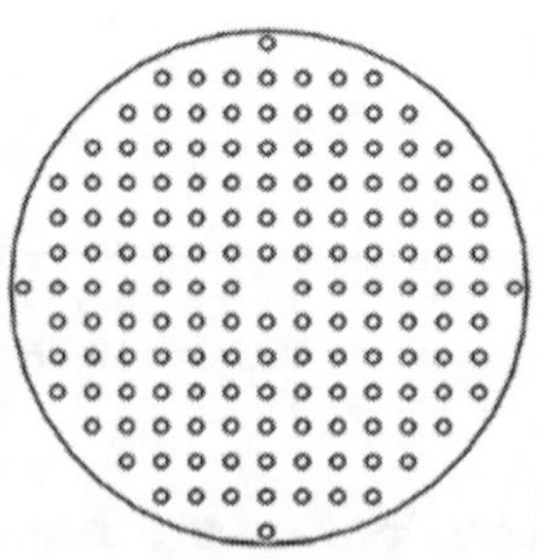 | 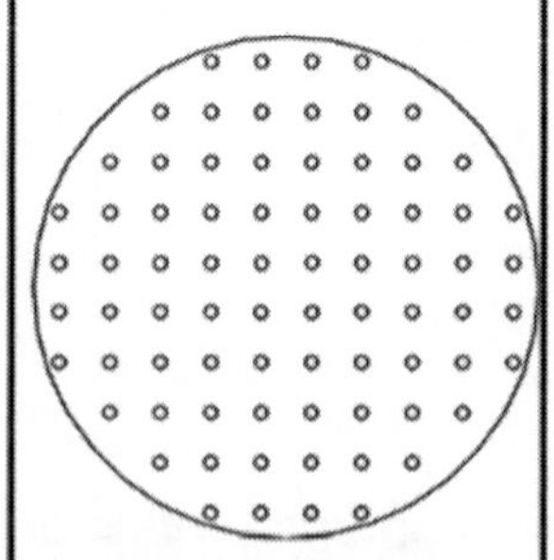 |
| 19% CSA Coverage<br>16.34% Volume coverage<br>293 – 1 in internal pipes<br>1 m –ID, 10 m high SBCR | 10% CSA Coverage<br>8.6% Volume Coverage<br>155 – 1 in internal pipes<br>1m –ID, 10 m high SBCR | 5% CSA Coverage<br>4.3% Volume Coverage<br>77 – 1 in internal pipes<br>1m –ID, 10 m high SBCR |

Figure 6 shows the axial liquid velocity contours for cases P-1 through P-3. As can be seen in this Figure, a strong liquid recirculations are exhibited at a much earlier time, marked by smaller scale liquid circulations throughout the reactor, which eventually grow into large scale recirculation cells with positive axial velocity near the reactor center and strong backmixing near the reactor wall at steady-state conditions. Similar flow structure and behavior is also exhibited for cases P-1 and P-2, in which the liquid recirculations develop at a much faster rate, and eventually result in large liquid recirculations throughout the reactor. This behavior is primarily due to liquid funneling because of the presence of internals in the reactor, which significantly affect the flow structure, resulting in large, fast and more turbulent liquid recirculations.

Figure 7 shows the presence of internals increases the gas holdup throughout the reactor at different height/dimeter ratios (Z/D), which is in agreement with earlier findings [26]. Also, in all cases at z/D value of 2.5, the radial gas holdup profiles exhibit higher values near the reactor center and lower values near the reactor wall, which eventually flatten along the reactor height, with more distinctly flatten gas holdup profiles at higher z/D values of 5 and 7.5.

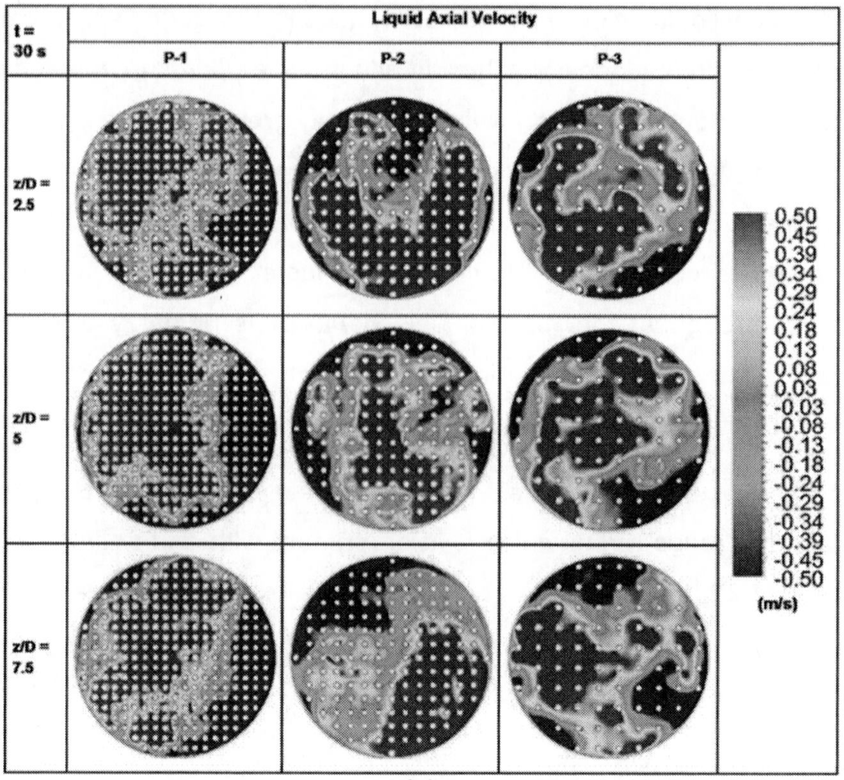


(a)

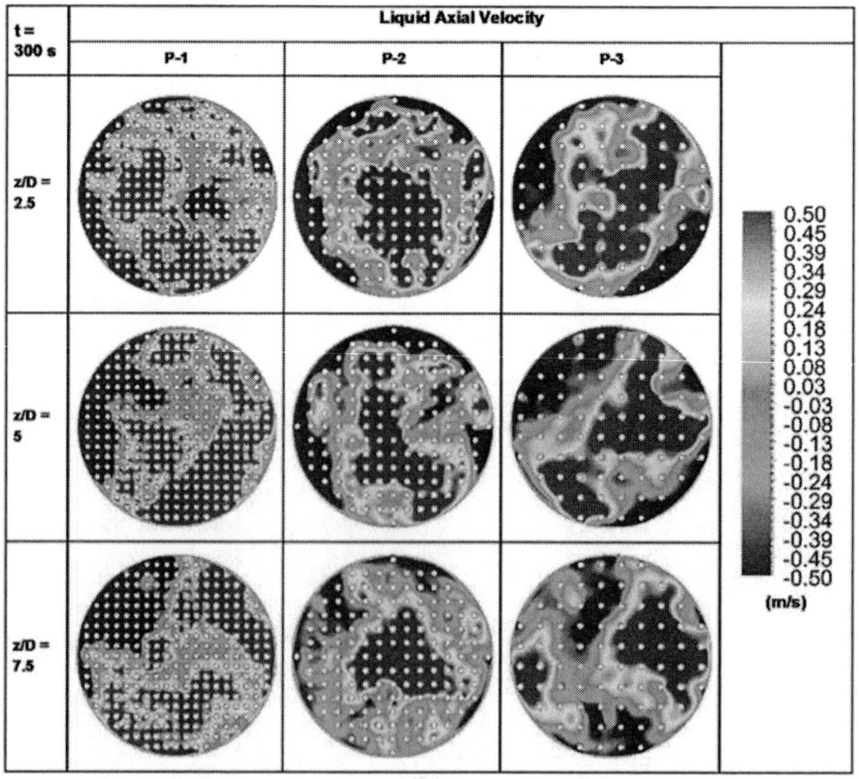


(b)

**Figure 6: Liquid axial velocity cross-sectional contours at different heights for the parallel internals configurations at (a) 30 s and (b) 300 s**

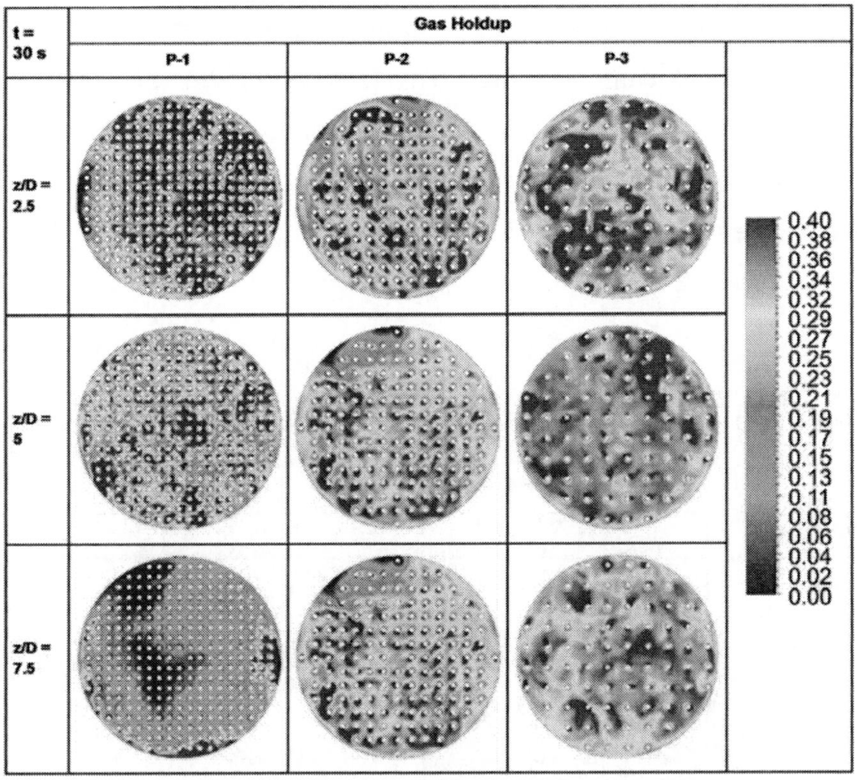


(a)

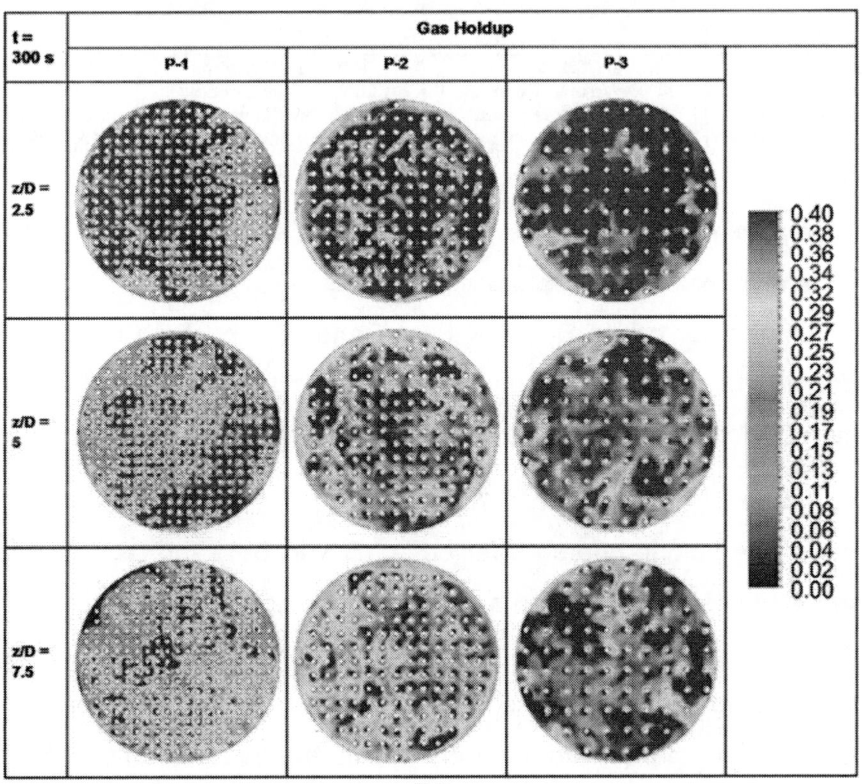


(b)

**Figure 7: Gas holdup cross-sectional contours at different heights for the parallel internals configurations at (a) 30 s and (b) 300 s**

In addition, four different bundled configurations in the conceptual reactor with 1, 3, 4 and 5 bundles, as shown in Table 3, were simulated. This was conducted to investigate the effect of the reactor geometry on the local hydrodynamics in the presence of bundled internals, and the impact of using a large number of internals inside the reactor. Again, simulations were carried out under pressure of 20.5 bar, temperature of 443 K and catalyst concentration of 11 vol. % at an inlet superficial gas velocity of 0.20 m/s.

**Table 3: Details of the bundled internal used in the Conceptual SBCR**

| Bundled Arrangement | | | |
|---|---|---|---|
| **B-1** | **B-2** | **B-3** | **B-4** |
| 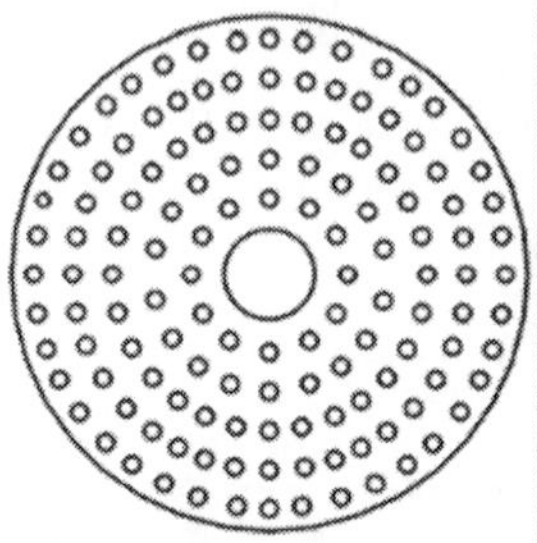 | 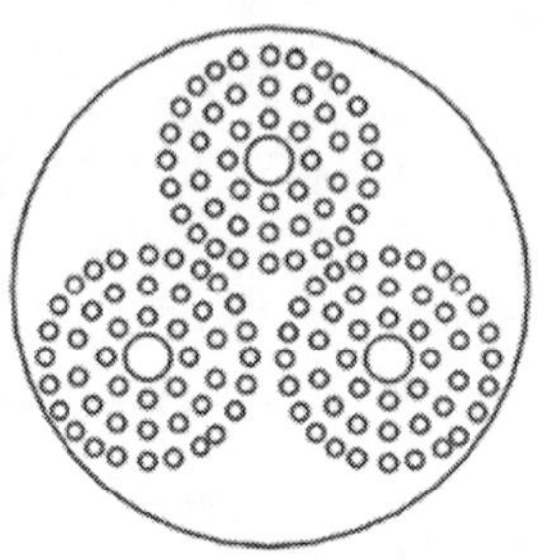 | 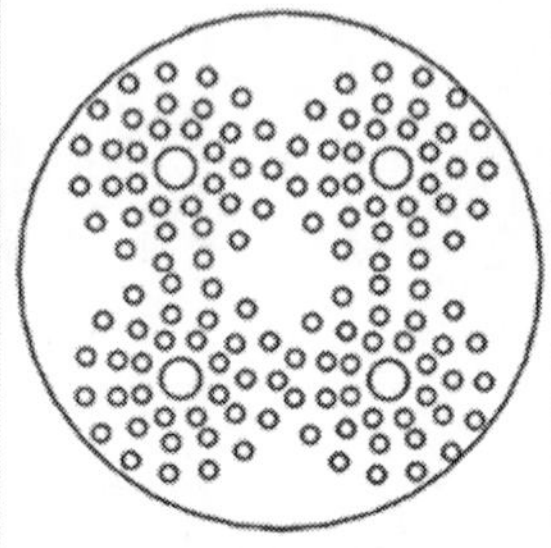 | 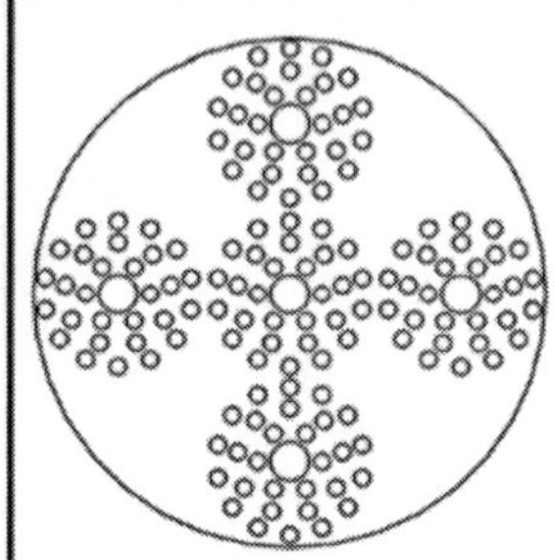 |
| 16.7% CSA Coverage<br>14.3% Volume Coverage<br>1 - 7 in bundle center<br>134 - 1.25 in internal pipes | 16.3% CSA Coverage<br>14.0% Volume Coverage<br>3, 3.5 in bundle centers<br>138, 1.25 in internal pipes | 16% CSA Coverage<br>13.8% Volume Coverage<br>4, 3 in bundle centers<br>136, 1.25 in internal pipes | 17.5% CSA Coverage<br>15.1% Volume Coverage<br>5, 3 in bundle centers<br>145, 1.25 in internal pipes |

Figure 8 shows the axial liquid velocity contours for cases B-1 through B-4; and as can be seen strong liquid recirculations are exhibited for cases B-1 and B-2, whereas weaker liquid recirculations are shown for cases B-3 and B-4. Generally, backmixing zones in the bundled arrangements seem to be affected by the wall clearance zones, especially in the configurations, which have large wall clearances, such as B-2 and B-4, where unique backmixing profiles can be observed.

Figure 9 shows the gas holdup contours for cases B-1 to B-4; and as can be observed for all cases, the presence of internals results in smoother radial gas holdup profiles, however, the local gas holdup values were lower than those exhibited for the parallel configurations (Figure 7). This behavior could be related to the increase of the Sauter mean bubbles size in the presence of bundled internals when compared with that obtained in the presence of parallel internals.

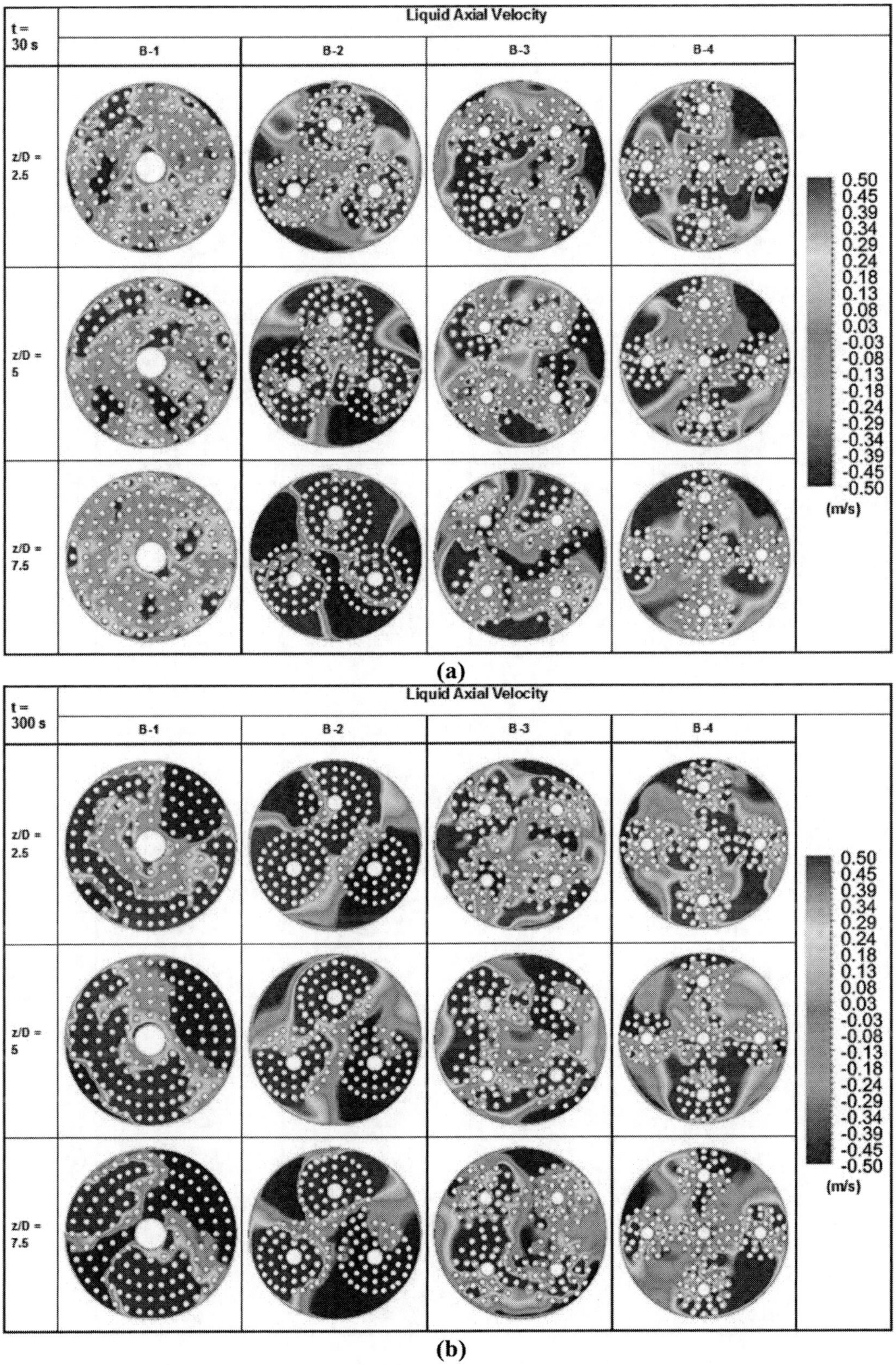


**Figure 8: Liquid axial velocity cross-sectional contours at different heights for the bundled internals configurations at 30 and 300 s (1-m ID, 10-m high SBCR)**

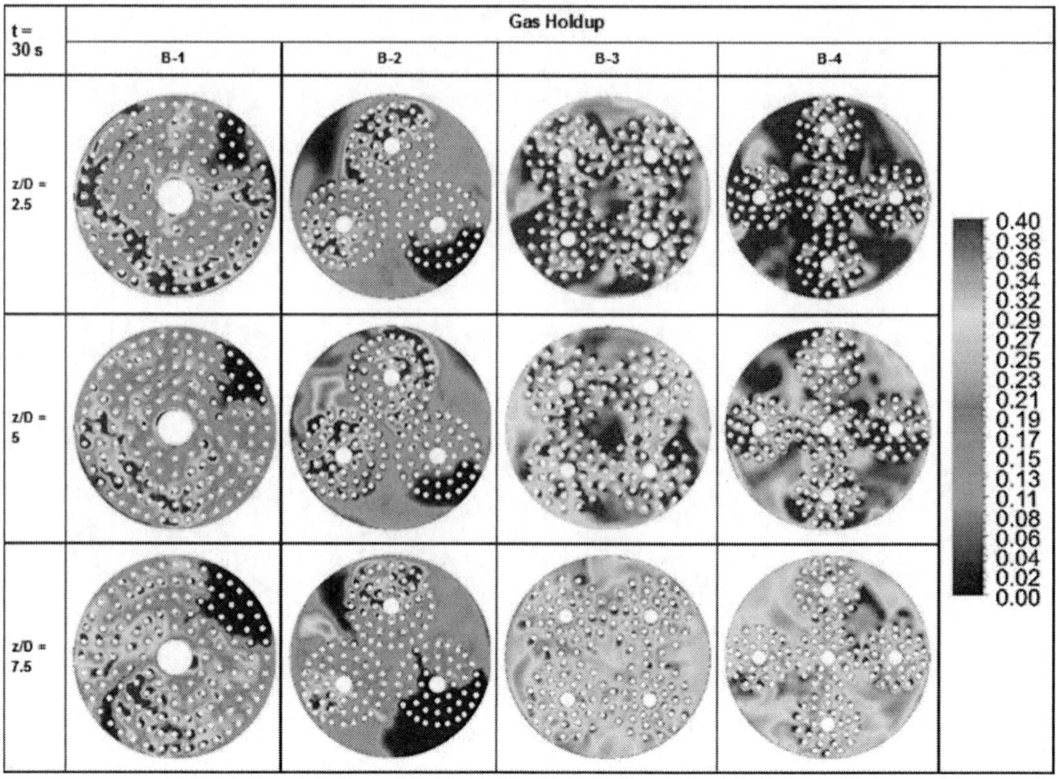


(a)

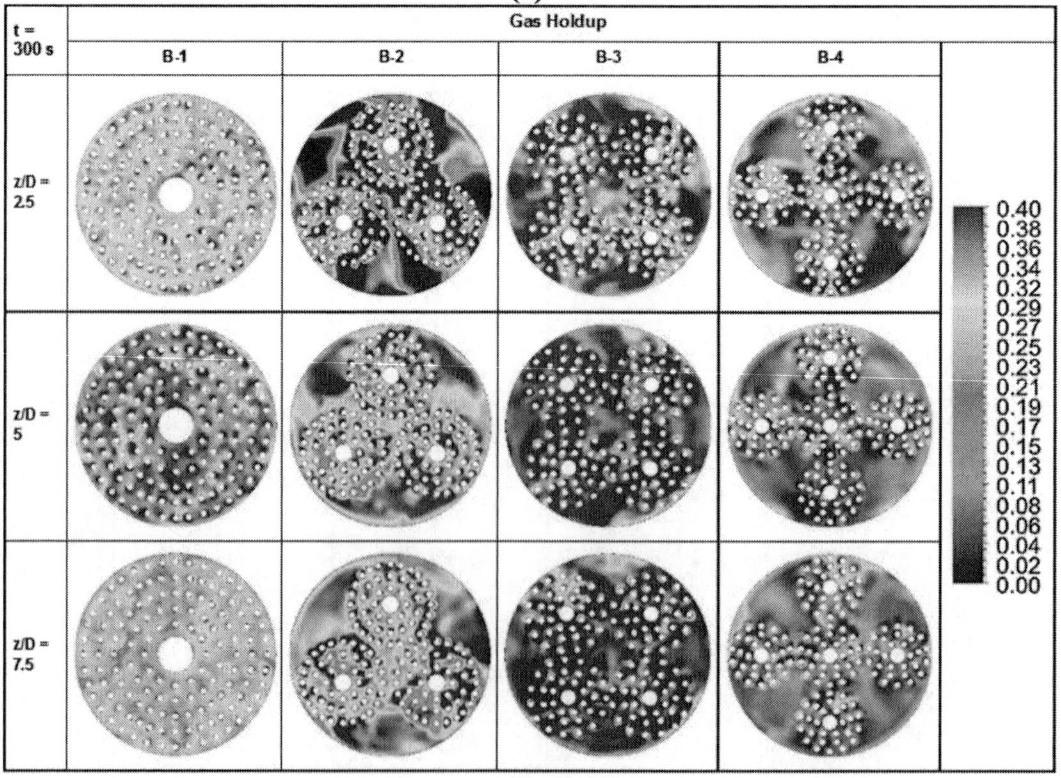


(b)

**Figure 9: Gas holdup cross-sectional contours at different heights for the bundled internals configurations at 30 and 300 s (1-m ID, 10-m high SBCR)**

## 5. Concluding remarks

The CFD model was used to predict the effects of using 6-arm spider and concentric-ring gas spargers, with upward and downward-facing orifices, on the local hydrodynamics (gas holdup, Sauter mean bubble diameter, and axial solid concentration profiles) in the pilot-scale (0.3-m ID, 3-m height) SBCR, operated under typical F-T conditions without chemical reactions. The CFD simulation results showed that the spargers with downward-facing orifices were more effective in solids dispersion than those with upward-facing orifices. This was due to the agitations provided by the gas jets from the downward-facing orifices, which enhanced the solid particles distribution. Also, the spargers with upward-facing orifices resulted in larger Sauter bubble diameters when compared with those from spargers with downward-facing orifices. In addition, whether provided with upward-or downward-facing orifices, the concentric-ring spargers resulted in larger Sauter bubble diameters when compared with those obtained with the 6-arms spider. The gas sparger design had a significant effect on the local gas holdup within the reactor, and for both spargers the maximum gas holdup was exhibited at the center of the reactor. Moreover, the 6-arms spider with upward-facing orifices resulted in the highest gas holdups, whereas at lower superficial gas velocities ($U_g < 0.15$ m/s), the concentric-rings sparger with downward-facing orifices gives higher gas holdup values than those using a concentric-rings sparger with upward-facing orifices.

The CFD model was also used to predict the effects of using parallel and bundled internals, with different configurations, on the local hydrodynamics in the conceptual SBCR (1-m ID, 10 m high) SBCR operating under F-T conditions without chemical reactions. The reactor was provided with a concentric-ring sparger, consisting of 3 concentric-rings with 333, 167 and 111 orifices, totaling 611 identical orifices, each with an ID of 7 mm. The CFD simulation results showed that the use of internals significantly enhanced the liquid funneling in the reactor. Faster and larger liquid recirculations were exhibited at a much earlier time with parallel internals when compared with those obtained using the bundled internals. Stronger liquid recirculations were found when using B-1 and B-2 bundled internals than those for when using B-3 and B-4. Also, unique backmixing profiles were observed in the presences of B-2 and B-4, which exhibited large internals-wall clearance than those with B-1 and B-3. Furthermore, the used of bundled internals exhibited smoother radial gas holdup profiles, with lower local gas holdup values, when compared with those obtained with parallel internals. This behavior could be related to the increase of the average bubble sizes obtained in the presence of the bundled internals when compared with those when using the parallel internals.

# 6. Nomenclature

| | | | |
|---|---|---|---|
| $a$ | Gas-liquid interfacial area per unit liquid volume ($m^{-1}$) | $n$ | Bubble number density ($m^{-3}$) |
| $B_B$ | Birth rate of bubbles due to the turbulence induced breakup | $P$ | Pressure (Pa) |
| $B_C$ | Birth rate of bubbles due to the turbulence induced coalescence | $P_v$ | Liquid-phase saturation pressure (bar) |
| $C_D$ | Drag Coefficient | $r$ | Bubble radius (m) |
| $C_P$ | Concentration of solid particles in the slurry phase (kg. $m^{-3}$) | $Re$ | Reynolds number |
| $C_V$ | Volumetric concentration of solid particles in the slurry phase (vol%) | $t$ | Time (s) |
| $d_C$ | Column diameter (m) | $t_{ij}$ | Coalescence time (s) |
| $d_p$ | Particle diameter (m) | $T$ | Temperature (K) |
| $d_{32}$ | Sauter mean gas bubble diameter (m) | $u_b$ | Gas bubble rise velocity ($m \cdot s^{-1}$) |
| $D_B$ | Death rate of bubbles due to the turbulence induced breakup | $u_g$ | Superficial gas velocity ($m \cdot s^{-1}$) |
| $D_C$ | Death rate of bubbles due to the turbulence induced coalescence | $u_k$ | The velocity of phase $k$ |
| $e_p$ | Coefficient of restitution | $u_{kn}$ | Relative velocity between phases k and n |
| $g$ | Acceleration due to gravity, $m/s^2$ | $u'$ | Root-mean-square of the turbulent velocity fluctuations ($m \cdot s^{-1}$) |
| $k$ | Turbulent kinetic energy (J/mol) | U | Mean velocity ($m \cdot s^{-1}$) |
| $k_L a$ | Volumetric liquid-side mass transfer coefficient ($s^{-1}$) | $v$ | Linear flow velocity (m/s) |
| $Mw$ | Molecular weight ($kg \cdot kmol^{-1}$) | $\vec{v}_{eff,rad}$ | Radial momentum transfer coefficient ($m^2/s$) |
| $\dot{m}_{kn}$ | Mass transfer rate between phases k and n due to interfacial mass transfer | $X_W$ | Foaming Coefficient of the liquid-phase (See Behkish et al. [27]) |

**<u>Greek Letters</u>**

| | | | |
|---|---|---|---|
| $\alpha$ | Volume fraction of each phase | $\upsilon$ | Local velocity of the dispersed phase |
| $\varepsilon$ | Void fraction (-) | $\rho$ | Density ($kg/m^3$) |
| $\eta$ | Kinematic viscosity ($m^2/s$) | $\sigma$ | Surface tension (N/m) |
| $\theta$ | ($m^2/s^2$) | $\tau$ | Viscous stress tensor (Pa) |
| $\lambda_S$ | Solids bulk viscosity ($kg/m \cdot s$) | $\tau_{rz}$ | Reynolds shear stress |
| $\kappa$ | Coefficient of bulk viscosity | $\vec{\phi}$ | Body source per unit mass |
| $\mu$ | Viscosity ($kg/m \cdot s$) | $\chi$ | Coalescence rate of bubbles ($m^{-3}$ $s^{-1}$) |
| $\mu_b$ | Relative apparent bed viscosity | $\Omega$ | Break-up rate of bubbles ($m^{-3}$ $s^{-1}$) |
| $\mu_T$ | Turbulent or eddy viscosity | | |

**<u>Subscripts</u>**

| | | | |
|---|---|---|---|
| $g$ | Gas | $s$ | Solid |
| $k$ | Phase | $w$ | Wall |
| $l$ | Liquid | | |

# 7. Acknowledgement

The Authors would like to thank the National Institute of Clean-and-low-carbon Energy (NICE), China for the financial support of this research.

## 8. References


[1] M. Basha Omar, L. Weng, Z. Men, and I. Morsi Badie, "CFD Modeling with Experimental Validation of the Internal Hydrodynamics in a Pilot-Scale Slurry Bubble Column Reactor," in *International Journal of Chemical Reactor Engineering* vol. 0, ed, 2016.

[2] O. M. Basha, L. Sehabiague, A. Abdel-Wahab, and B. I. Morsi, "Fischer–Tropsch Synthesis in Slurry Bubble Column Reactors: Experimental Investigations and Modeling–A Review," *International Journal of Chemical Reactor Engineering,* vol. 13, pp. 201-288, 2015.

[3] N. Kantarci, F. Borak, and K. O. Ulgen, "Bubble column reactors," *Process Biochemistry,* vol. 40, pp. 2263-2283, 2005.

[4] A. V. Kulkarni and J. B. Joshi, "Design and selection of sparger for bubble column reactor. Part I: Performance of different spargers," *Chemical Engineering Research and Design,* vol. 89, pp. 1972-1985, 10// 2011.

[5] B. A. Toseland, "Engineering Development of Slurry Bubble Column Reactor (SBCR) Technology," *2000.,* 2000.

[6] N. Balasubramanian, C. Srinivasakannan, and C. Ahmed Basha, "Drying kinetics in the riser of circulating fluidized bed with internals," *Drying Technology,* vol. 25, pp. 1595-1599, 2007.

[7] H. Tong and H. Li, "Floating internals in fast bed of cohesive particles," *Powder Technology,* vol. 190, pp. 401-409, 2009.

[8] J. Makibar, A. R. Fernandez-Akarregi, L. Díaz, G. Lopez, and M. Olazar, "Pilot scale conical spouted bed pyrolysis reactor: Draft tube selection and hydrodynamic performance," *Powder Technology,* vol. 219, pp. 49-58, 2012.

[9] M. Olazar, G. Lopez, H. Altzibar, M. Amutio, and J. Bilbao, "Drying of biomass in a conical spouted bed with different types of internal devices," *Drying Technology,* vol. 30, pp. 207-216, 2012.

[10] H. Kölbel and M. Ralek, "The Fischer-Tropsch synthesis in the liquid phase," *Catalysis Reviews Science and Engineering,* vol. 21, pp. 225-274, 1980.

[11] H. Koppers, "Rheinpreussen-Koppers Liquid-Phase Process of Fischer-Tropsch Synthesis," *Chem. Age India,* vol. 12, p. 7, 1961.

[12] J. Fox, III, "Fischer-Tropsch reactor selection," *Catalysis letters,* vol. 7, pp. 281-292, 1990/01/01 1990.

[13] D. Gidaspow, *Multiphase flow and fluidization: continuum and kinetic theory descriptions.* San Diego: Academic Press, 1994.

[14] H. F. Svendsen, H. A. Jakobsen, and R. Torvik, "Local flow structures in internal loop and bubble column reactors," *Chemical Engineering Science,* vol. 47, pp. 3297-3304, 1992.

[15] M. J. Prince and H. W. Blanch, "Bubble coalescence and break-up in air-sparged bubble columns," *AIChE Journal,* vol. 36, pp. 1485-1499, 1990.

[16] S. Patankar, *Numerical Heat Transfer and Fluid Flow.* New York: McGraw-Hill Book Company, 1980.

[17] Y. H. Yu and S. D. Kim, "Bubble characteristics in the radial direction of three-phase fluidized beds," *AIChE Journal,* vol. 34, pp. 2069-2072, 1988.

[18] J. Chen, F. Li, S. Degaleesan, P. Gupta, M. H. Al-Dahhan, M. P. Dudukovic, *et al.*, "Fluid dynamic parameters in bubble columns with internals," *Chemical Engineering Science,* vol. 54, pp. 2187-2197, 1999.

[19] L. Sehabiague and B. I. Morsi, "Hydrodynamic and Mass Transfer Characteristics in a Large-Scale Slurry Bubble Column Reactor for Gas Mixtures in Actual Fischer–Tropsch Cuts," *International Journal of Chemical Reactor Engineering,* vol. 11, pp. 83-102, 2013.

[20] M. Ishii and K. Mishima, "Two-fluid model and hydrodynamic constitutive relations," *Nuclear Engineering and Design,* vol. 82, pp. 107-126, 1984.

[21] L. Schiller and Z. Naumann, "A drag coefficient correlation," *Zeitung Ver. Deutsch. Ing.,* vol. 77, pp. 318-320, 1935.

[22] C. Y. Wen and Y. H. Yu, "Mechanics of Fluidization," *Chem Eng Prog Symp Ser,* vol. 62, pp. 100-111, 1966.

[23] A. Tomiyama, H. Tamai, I. Zun, and S. Hosokawa, "Transverse migration of single bubbles in simple shear flows," *Chemical Engineering Science,* vol. 57, pp. 1849-1858, 2002.

[24] L. Sehabiague, O. M. Basha, S. Zhansheng, H. Jia, Y. Hong, L. Weng, *et al.*, "Assessing the Performance of an Industrial SBCR for Fischer-Tropsch Synthesis: Experimental and Modeling," *AIChE J.,* vol. In Press, 2015.

[25] H. Luo and H. F. Svendsen, "Turbulent circulation in bubble columns from eddy viscosity distributions of single-phase pipe flow," *The Canadian Journal of Chemical Engineering,* vol. 69, pp. 1389-1394, 1991.

[26] K. Bernemann, "On the hydrodynamics and mixing of the liquid phase in bubble columns with longitudinal tube bundles," PhD Thesis, University of Dortmund, 1989.

[27] A. Behkish, R. Lemoine, L. Sehabiague, R. Oukaci, and B. I. Morsi, "Gas holdup and bubble size behavior in a large-scale slurry bubble column reactor operating with an organic liquid under elevated pressures and temperatures," *Chemical Engineering Journal,* vol. 128, pp. 69-84, 2007.

**2016 International Pittsburgh Coal Conference**
**Cape Town, SOUTH AFRICA**
**August 8 – 12, 2016**

**Abstract Submission**

**PROGRAM TOPIC:**
**4. CLEAN COAL AND GAS TO FUELS: COAL-TO-LIQUIDS FUEL, CTL**

**Exact Title of Paper:**
# Isotope Analysis for Understanding the Hydrogen Transfer Mechanism in Direct Liquefaction of Bulianta coal

*Ben Niu, Ph.D. candidate, State Key Laboratory of Fine Chemicals, Institute of Coal Chemical Engineering, School of Chemical Engineering, Dalian University of Technology, Dalian 116024, China*
*Contact Information: Email: wsniuben@163.com*
*Zhiwei Shi, Ph.D. candidate, State Key Laboratory of Fine Chemicals, Institute of Coal Chemical Engineering, School of Chemical Engineering, Dalian University of Technology, Dalian 116024, China*
*Contact Information: Email: szw19891121@163.com*
***Haoquan Hu*, Professor, State Key Laboratory of Fine Chemicals, Institute of Coal Chemical Engineering, School of Chemical Engineering, Dalian University of Technology, Dalian 116024, China***
***Contact Information: Email: hhu@dlut.edu.cn, Tel/Fax: +86-411-84986157***
*Lijun Jin, Ph.D, Institute of Coal Chemical Engineering, School of Chemical Engineering, Dalian University of Technology, Dalian 116024, China*
*Contact Information: Email: ljin@dlut.edu.cn, Tel/Fax: +86-411-84986160*
*Yang Li, Ph.D, Institute of Coal Chemical Engineering, School of Chemical Engineering, Dalian University of Technology, Dalian 116024, China*
*Contact Information: Email: yli@dlut.edu.cn, Tel/Fax: +86-411-84986160*

**Abstract:**
Deuterium has been used to trace the pathways hydrogen in Bulianta coal liquefaction with or without Fe-based catalyst. The recovered solvent, asphaltene plus preasphaltene and residue were analyzed by $^2$H-NMR or IR-MS. The $^2$H-NMR spectroscopy of the recovered tetralin after liquefaction reaction shows that Fe-based catalyst can inhibit hydrogen transfer of tetralin in the $H_\alpha$ position. Results of the analysis of $^2$H content in asphaltene plus preasphaltene by IR-MS shows that the $^2$H content with Fe-based catalyst is higher than that without catalyst when using tetralin-$^2H_4$ as liquefaction solvent in $N_2$ atmosphere, indicating that Fe-based catalyst can accelerate transfer of hydrogen in tetralin to asphaltene plus preasphaltene. However, using tetralin-$^2H_4$ as liquefaction solvent in $^2H_2$ atmosphere, $^2$H content with Fe-based catalyst is similar to that without catalyst. This indicates that the role of Fe-based catalyst is to promote the hydrocracking of asphaltene plus preasphaltene. When using tetralin-$^2H_4$ as liquefaction solvent in the absence of catalyst, $^2$H content in $^2H_2$ atmosphere is obviously higher than that in $N_2$ atmosphere, however, when using tetralin-$^2H_4$ as liquefaction solvent and in the presence of Fe-based catalyst, $^2$H content is almost the same in $^2H_2$ atmosphere and $N_2$ atmosphere. These results indicate that the predominant pathway of hydrogen transfer to coal is from molecular hydrogen to solvent and then from solvent to coal in the absence of a catalyst, and in the presence of Fe-based catalyst, the predominant pathway of hydrogen transfer to coal is from molecular hydrogen to coal.

**2016 International Pittsburgh Coal Conference**
**Cape Town, SOUTH AFRICA**
**August 8 – 12, 2016**

**Abstract Submission**

## CLEAN COAL AND GAS TO FUELS: COAL-TO-LIQUIDS FUELS, CTL

# Design ZSM-22 Crystal Morphology with Urea and its catalytic property

***Presenting Author (Debao Li, Institute of Coal Chemistry Chinese Academy of Science, 27 South Taoyuan Road, Taiyuan, Shanxi, P.R.China)***
***Contact Information: Email:dbli@sxicc.ac.cn***
***Phone:86351-4121877***

*Co-Authors (Pengyu Niu, Litao, Jia, Bo Hou Institute of Coal Chemistry Chinese Academy of Science, 27 South Taoyuan Road, Taiyuan, Shanxi, P.R.China))*

**Abstract:**

**Tuning the morphology of zeolites to achieve special properties remains a challenge. Zeolite growth modifiers, which are molecules absorbing on select crystal faces, have been used to tailor the morphology of zeolites. Xiao et al have used urea to alter morphology of TS-1 successfully. Recently, we have moderated the length-to-diameter aspect ratio of ZSM-22 crystals by adding urea to the synthesis mixture. The conventional ZSM-22 zeolite was synthesized following the recipe reported in the literature[4]. The composition of synthesis mixture was $Al_2O_3/13KOH/27NH_2(CH_2)_6NH_2/91SiO_2/3670H_2O$. Urea was chosen as the zeolite growth modifier. The urea was added in a molar ratio of 0.1urea/1 $SiO_2$. The synthesis mixtures were heated in autoclaves at 160℃ for 36h under stirring at about 42 rpm in an oven. After crystallization, the samples were filtered, washed with deionized water, and then dried overnight.**

**The XRD patterns of ZSM-22 and urea-ZSM-22 samples are shown in Fig.1, which are in good agreement with TON standard diffraction data without any crystal impurities. Adding urea into the starting synthesis mixture don't affect the crystallization of ZSM-22. SEM images of ZSM-22 and urea-ZSM-22 are also investigated. The conventional ZSM-22 sample shows needle-shaped crystals. Notably, when used urea in the synthesis process, the length-to-diameter aspect ratio of ZSM-22 becomes more high than the conventional ZSM-22. Urea have an important effect on ZSM-22 aspect ratio.**

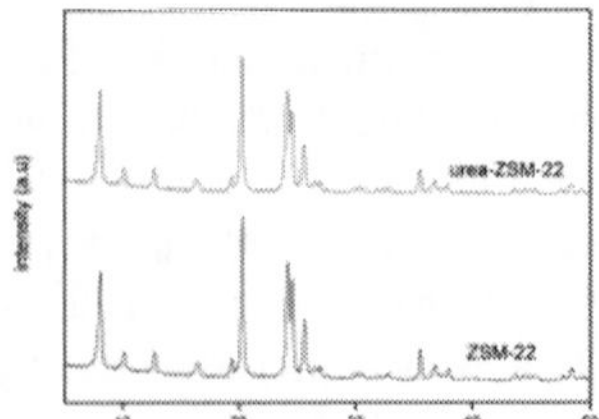


**Fig.1 Powder XRD patterns of ZSM-22**

**In summary, urea as zeolite growth modifier plays a great role in tailoring the morphology of ZSM-22. ZSM-22 crystals with high length-to-diameter aspect ratio have been obtained successfully by adding urea in the starting synthesis mixture. The role of urea in morphology control of ZSM-22 could be that urea may adsorb on the surface of ZSM-22 crystals or change the pH during the crystallization process. Tailoring ZSM-22 crystal morphology has great influence on catalytic properties in hydroisomerization of n-alkanes.**

# 2016 International Pittsburgh Coal Conference
# Cape Town, SOUTH AFRICA
# August 8 - 12, 2016

**Sustainability and Environment**

**Geochemical Mobility of Potential Hazardous Elements of Coal Cleaning Rejects**

***Diego de Medeiros***[a]; *Marcos L.S. Oliveira*[a]; *Frans Waanders*[b]; *Bianca D. Lima*[a]; *Luis F. O. Silva*[a]
[a] *Laboratory of Environmental Researches and Nanotechnology Development, Centro Universitário La Salle, Mestrado em Avaliação de Impactos Ambientais em Mineração. Victor Barreto, 2288 Centro 92010-000, Canoas, RS, Brazil.*
[b] *School of Chemical and Minerals Engineering North West University (Potchefstroom campus) Potchefstroom 2531, South Africa*

**Abstract**
The geochemical characteristics of coal cleaning rejects (CCR) in Santa Catarina State, Brazil, were investigated. Around 3.5 million ton/year of coal waste are dumped in Santa Catarina State. Coal beneficiation by froth flotation results in large amounts of CCR composed of coaly and mineral matter, the latter characterised by the occurrence of sulphide minerals and a broad array of leachable elements. The total and leachable contents of more than 60 elements were analysed. Atmospheric exposure promotes sulphide oxidation that releases substantial sulphate loads as well as $Ca^{2+}$, $K^{+}$, $Mg^{2+}$, $Cl^{-}$ and $Al^{3+}$. The metals with the most severe discharges were Zn, Cu, Mn, Co, Ni and Cd. Most trace pollutants in the CCR displayed a marked pH-dependent solubility, being immobile in near-neutral samples. The results highlight the complex interactions among mineral matter solubility, pH and the leaching of potentially hazardous elements.

# 2016 International Pittsburgh Coal Conference
# Cape Town, SOUTH AFRICA
# August 8 - 12, 2016

**Sustainability and Environment**

## Occurrence of the hazardous elements in agricultural soils surrounding a coal power plant complex from Santa Catarina (Brazil)

***Diego de Medeiros***[a]; *Frans Waanders*[b]; *Marcos L.S. Oliveira*[a]; *Fagner T.C. Sá*[a]; *Luis F. O. Silva*[a]

[a] *Laboratory of Environmental Researches and Nanotechnology Development, Centro Universitário La Salle, Mestrado em Avaliação de Impactos Ambientais em Mineração. Victor Barreto, 2288 Centro 92010-000, Canoas, RS, Brazil.*

[b] *School of Chemical and Minerals Engineering North West University (Potchefstroom campus) Potchefstroom 2531, South Africa*

**Abstract**

Hazard element contamination coming from coal power plants is something obvious, but when this contamination is accompanied by other contamination sources, such as, urban, coal mining and farming activities the study gets complicated. This is the case of an area comprised in the southern part of Santa Catarina state (Brazil) with the largest private power plant generator. After the elemental analysis of 41 agricultural soils collected in an extensive area around the thermoelectric (from 0 to 47 km), the high presence of As, Co, Cr, Cu, Fe, Mn, Mo, Pb, Sb, Sn, Tl, V and Zn was found in some specific areas around the power plant. Nevertheless, as the NWAC (Normalized-and-Weighted Average Concentration) confirmed, only soils from one site were classified as of very high concern due to the presence of potential toxic elements. This site was located within the sedimentation basin of the power plant. The spatial distribution obtained by kriging in combination with the analysis of the data by Principal Component Analysis (PCA) revealed three important hotspots in the area according to soil uses and geographic localization: the thermoelectric, its area of influence due to volatile compound deposition, and the area comprised between two urban areas. Farming practice turn out to be an important factor too for the quantity of hazard element stored in soils.

# A Density Functional Theory Study on Acetic Acid Decomposition on Ni(111)

*Yanxiong Ran, Zhenyi Du, Jie Feng, Wenying Li**

*Key Laboratory of Coal Science and Technology (Taiyuan University of Technology), Ministry of Education and Shanxi province, Taiyuan 030024, PR China*

*Training Base of State Key Laboratory of Coal Science and Technology Jointly Constructed by Shanxi Province and Ministry of Science and Technology, Taiyuan 030024, PR China*

* *Corresponding author. E-mail: ying@tyut.edu.cn*

**Abstract:** Catalytic steam reforming of bio-oil is a potential process to convert renewable biomass feedstocks into hydrogen. However, carbon deposition on the catalyst is the major problem encountered in the steam reforming process. To shed light on the coke formation mechanism, acetic acid ($CH_3COOH$), the smallest organic molecule that contains C-O, C-C, C-H, O-H and C=O bonds, is selected as the model compound of the oxygenates in bio-oil. Thus, density functional theory is applied to investigate the decomposition mechanism of acetic acid on the Ni (111) surface. To gain a good knowledge of the high selectivity towards hydrogen and a comprehensive interpretation of coke formation mechanism, all of the possible decomposition pathways of acetic acid are considered, especially for the dehydrogenation pathways and C-C bond scission reactions. The stable adsorption geometry configuration and adsorption energies of different species and the important intermediates are determined. Based on our calculations, the most favorable decomposition pathway on Ni (111) surface is $CH_3COOH^* \rightarrow CH_3COO^* \rightarrow CH3CO^* \rightarrow CH_2CO^* \rightarrow CH_2^* \rightarrow CH^*$, and the decomposition of acetyl ($CH_3CO^*$) via C-H bond cleavage to form $CH_2CO$ is the rate-determining step with a high activation barrier of 1.33 eV. Meanwhile, the carbon deposits formation results from the accumulation of $CH^*$ species, which is mostly from the dehydrogenation of $CH_2^*$. Moreover, the formation pathways and initial dissociation of acetone intermediates is also investigated.

***Key words***: Acetic acid; Steam reforming; Carbon deposit; Density functional theory; Nickel

## 1. Introduction

In recent years, the extensive use of fossil fuels has led to a sharp increase in greenhouse gas emissions and a rugged environment, therefore, biomass, as a sole renewable carbon source, is an alternative energy to ease the crisis of fossil fuel and improve energy efficiency. Converting biomass into hydrogen is a promising way to make full use of biomass, as hydrogen can be more efficiently utilized in internal combustion engines and fuel cells without pollution and biomass is a carbon neutral energy. At present, the vast majority of hydrogen is produced through methane steam reforming (SR), while hydrogen derived from biomass undergoes two steps, biomass pyrolysis to generate bio-oil as much as 70%-85%[1, 2] and then via bio-oil steam reforming to produce hydrogen. The advantage of bio-oil is that the heating value (16-19 MJ/Kg) is much higher than biomass and bio-oil is easier to reserve and transport, thus being a promising alternative energy. However, due to the high water content (15-30%), relatively strong acidity (pH=2.5) and high oxygen content (35-40%) of bio-oil[3], it is directly restricted to use as the conventional power fuels. Hence, bio-oil steam reforming for hydrogen provides for a feasible way. Concerning the feasibility of bio-oil steam reforming process, Eleni Heracleous et al[4] carried out that $CO_2$ emission of biomass derived hydrogen process is 50% less than methane SR. Ekaterini Ch. Vagia et al[5] pointed out the total energy consumption between bio-oil SR and methane were nearly the same. More importantly, the energy efficiency of bio-oil SR is as high as 84%[6]. Moreover, it can take full advantage of the water in bio-oil in situ during bio-oil SR. In a word, bio-oil SR is feasible with promising prospect, and the aim of bio-oil SR is to improve hydrogen yield.

However, carbon deposition [7-11] that results in deactivation of catalyst is the main problem in the experimental study of bio-oil SR. Thus, developing a catalyst that resists carbon deposit and improves the hydrogen yield with high catalytic activity and stability has become a heated research in recent years. To achieve these goals, it is worth illustrating the mechanism of catalytic reactions, and a model compound of bio-oil is needed to represent bio-oil. Bio-oil consists of more than 100 oxygenated compounds, including acids, ketones, alcohols, aldehydes, phenolic derivatives, and sugars. Here, acetic acid (HOAc), the major compound in bio-oil, account for 32%[12, 13], is selected as the model compound of bio-oil. Furthermore, it is the smallest molecule that contains abundant bonds C-O, C-C, C-H, O-H and C=O as well as the same C/O ratio as C6 sugars ($C_6H_{12}O_6$). Ni-based catalyst[10, 11, 14-17] has been widely used in bio-oil SR and HOAc SR processes, because Ni catalyst shows an excellent activity on bond breaking performance of C-C, C-H and O-H of oxygenates[18, 19]. Meanwhile, Ni (111) surface is considered as one of the most stable and catalytic active surfaces and has been studied by many researchers [20-22], hence, Ni (111) is identified as the final catalytic active surface.

Similarly, carbon deposition is the major problem during HOAc SR. In current experimental studies, the understanding of coke is limited to the deduction based on the experimental results. It is still lacking in some theoretical evidence closely linked to carbon deposition, such as, the source of coke, the mechanism of coke formation and the carbon deposition precursor. Feng Cheng et al [23] and Rong-rong Hu et al[24] carried out that acetone produced during HOAc SR was a carbon deposition precursor that caused catalytic deactivation. T.M.C. Hoang et al [25] considered that coke formation resulted from two aspects, further decomposition of acetic acid and acetone formation. Therefore, it is necessary to explore the mechanism of HOAc decomposition and acetone formation. Shurong Wang et al [26] studied the mechanism of HOAc decomposition on Co(111) with an optimal pathway of $CH_3COOH^* \rightarrow CH_3COO^* \rightarrow CH_3CO^* \rightarrow CH_3^* + CO^*$. Xinbao Li et al [13] has systematically studied the mechanism of HOAc decomposition on Co stepped surface, including the acetone formation reactions by DFT calculations. However, a completely decomposition of HOAc on Ni (111) surface is still lacking, especially for acetone formation part. Therefore, a density functional theory (DFT) calculations is applied to the decomposition of HOAc on Ni (111) surface, and all of possible dissociation pathways are taken into account to identify the optimal decomposition pathway and rate-determining step, more importantly, to identify the carbon deposition precursor, moreover, the mechanism of acetone formation and the initial dissociation reactions are also investigated. The purpose of this work is to shed insight on the origin of coke formation and serve as the further coke elimination study and the design of catalyst.

## 2. Calculational methods

Density functional theory (DFT) calculations were performed using Dmol3 program package in Materials Studio 5.5, in which a double numerical polarization (DNP) basis set was used[27]. The Perde-Burke-Ernzerhof (PBE) within the generalized gradient approximation (GGA)[28] was applied to describe the exchange-correlation functional. As Ni possesses the magnetic property and the gas phase species derived from acetic acid cintain unpaired electrons, spin polarization was considered during the Ni surface reactions and the gas phase radical calculations. The criterion of total energy, maximum displacement, and maximum force tolerances were converged at the value of $2\times10^{-5}$ hartree (Ha), $5\times10^{-3}$ Å, and $4\times10^{-3}$ Ha/Å, respectively. The Methfessel-Paxton method[29] was applied in the calculations with a Femi smearing width of 0.001 Ha and the total energies were extrapolated to 0 K.

For surface model, a three-layer, p (4×4) periodic slab unit cell of Ni (111) surface was established to represent the actual nickel-based catalyst. The bottom layer was constrained, while the top two layers were free to move. Also, a 20 Å vacuum thickness was used to avoid the interaction of adsorbates and periodic slab in the z direction. The Brillouin zone was sampled by a 3×3×1 Monkhorst-Pack k-piont mesh. Transition state (TS) configuration was located by a complete LST/QST method[30]. Adsorption energy $E_{ad}$ was defined as the following equation:

$$E_{ad} = E_{adsorbate/slab} - E_{adsorbate} - E_{slab}$$

where $E_{adsorbate/slab}$ represents the total energy of adsorbate and slab, $E_{adsorbate}$ refers to the energy of gas-phase radical species, and $E_{slab}$ means the energy of clean Ni(111) surface. According to the definition, a negative $E_{ad}$ represents an exothermic process.

## 3. Results and discussion

### 3.1 Surface adsorption on Ni (111)

Before the elementary reaction occurs, the adsorbates must be adsorbed on the catalyst surface, and it is a foundation for the catalytic reactions to identify the most stable adsorption configuration of the reactant and intermediates on Ni (111) surface. Due to the multiatomic system property of acetic acid, all possible intermediates derived from different bond dissociation of acetic acid and subsequent dissociation species are considered. The most stable adsorption geometries and the adsorption energy of the intermediates adsorbed on Ni (111) surface are shown in Fig. 1 and Table 1.

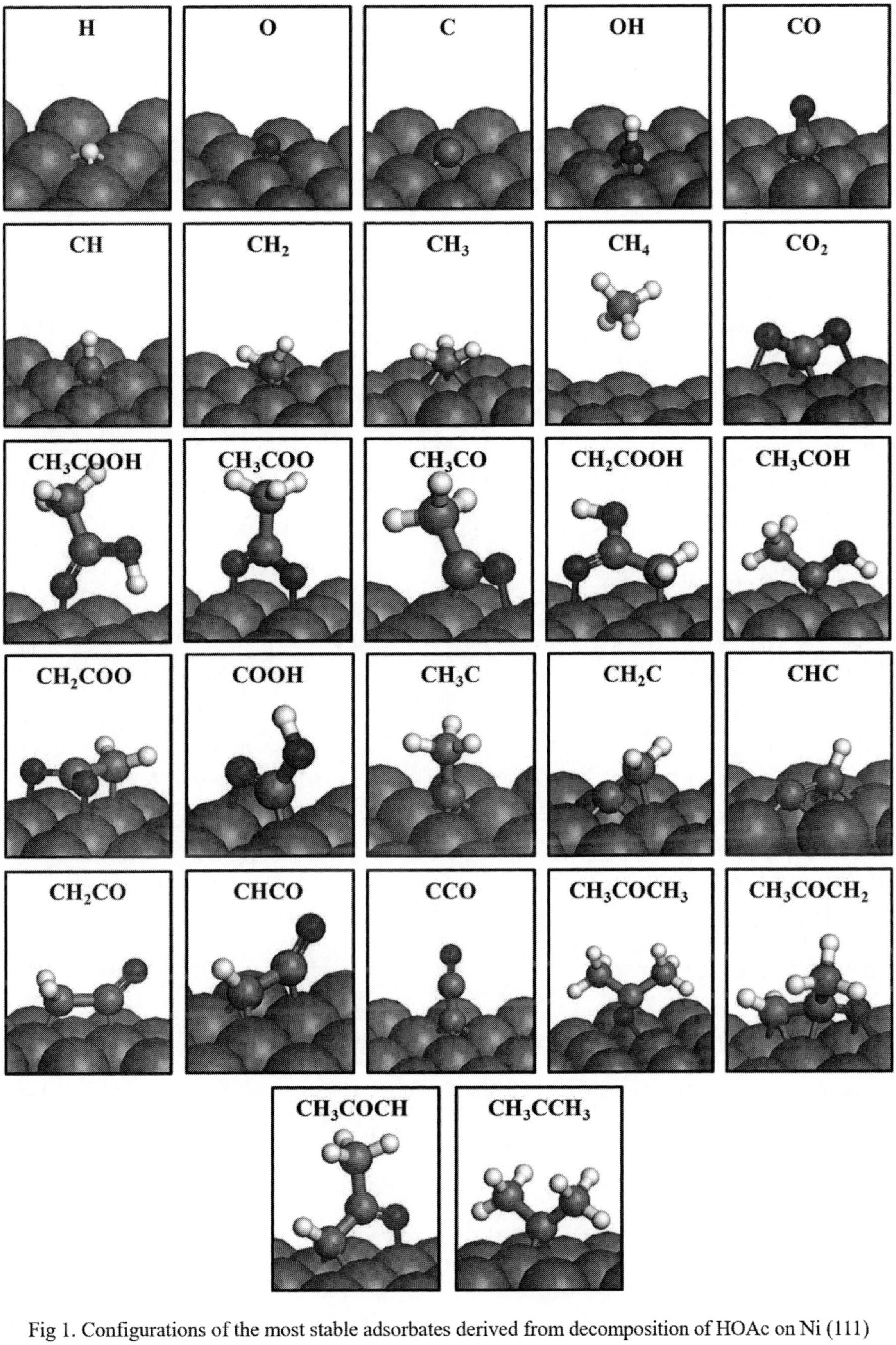


Fig 1. Configurations of the most stable adsorbates derived from decomposition of HOAc on Ni (111)

Table 1. Adsorption energy Ead/eV and configuration parameters of the most stable adsorbates on Ni (111)

| Species | Adsorption site | $d_{O\text{-}Ni}$/Å | $d_{C\text{-}Ni}$/Å | $E_{ad}$/eV |
|---|---|---|---|---|
| H | H-fcc | | | -2.86 |
| O | O-fcc | 1.86 | | -5.72 |
| C | C-fcc | 1.77 | | -6.85 |
| OH | O-fcc | 2.00 | | -3.22 |
| CO | C-fcc | | 1.95 | -1.85 |
| CH | C-fcc | | 1.84 | -6.44 |
| $CH_2$ | C-fcc | | 1.91/1.91/1.98 | -4.10 |
| $CH_3$ | C-fcc | | 2.11/2.12/2.13 | -1.98 |
| $CH_4$ | weak adsorption | | | -0.06 |
| $CO_2$ | C-bri, O-top | 2.14/2.16 | 2.00/2.03 | 0.25 |
| $CH_3COOH$ | $O_\alpha$-top | 1.77 | | -0.41 |
| $CH_3COO$ | $O_\alpha$ and $O_\beta$-top | 1.97/1.98 | | -3.01 |
| $CH_3CO$ | $C_\alpha$ and O-top | 2.03 | 1.84 | -2.10 |
| $CH_2COOH$ | $O_\alpha$ and $C_\beta$-top | 2.05 | 2.01 | -1.75 |
| $CH_3COH$ | $C_\alpha$-bri | | 1.97 | -2.57 |
| $CH_2COO$ | $O_\alpha$, $O_\beta$ and $C_\beta$-top | 1.95/1.94 | 2.01 | -1.96 |
| COOH | $O_\alpha$ and C-top | 2.07 | 1.85 | -2.35 |
| $CH_3C$ | $C_\alpha$-fcc | | 1.85 | -5.66 |
| $CH_2C$ | $C_\alpha$-fcc, $C_\beta$-top | | $1.85^\alpha/1.85^\alpha/1.85^\alpha/2.17^\beta$ | -4.22 |
| CHC | $C_\alpha$-fcc, $C_\beta$-hcp | | $1.82^\alpha/1.94^\alpha/1.94^\alpha/2.07^\beta$ | -5.40 |
| $CH_2CO$ | $C_\alpha$ and $O_\beta$-top | | $1.87^\alpha/1.98^\beta$ | -0.87 |
| CHCO | $C_\alpha$-top, $C_\beta$-bri | | $1.91^\alpha/1.94^\alpha/1.96^\beta$ | -3.02 |
| CCO | $C_\beta$-fcc | | 1.91 | -5.28 |
| $CH_3COCH_3$ | C-top, O-bri | 2.03/2.03 | 2.00 | -0.17 |
| $CH_3COCH_2$ | $O_\alpha$ and $C_\beta$-bri, $C_\alpha$-top | 2.04/2.09 | $2.04^\alpha/2.07^\beta/2.18^\beta$ | -2.03 |
| $CH_3COCH$ | $O_\alpha$-top, $C_\beta$-bri | 1.96 | 1.97/1.97 | -4.32 |
| $CH_3CCH_3$ | $C_\alpha$-bri | | 1.94/1.94 | -3.27 |

**H*, O* and C***

H*, O* and C*, as the single atomic species generated from the HOAc dissociation, are most stably adsorbed on the fcc site of Ni(111) surface, and the corresponding binding energies are -2.86 eV, -5.72 eV, and -6.85 eV, respectively. The computational bond lengths of adsorbed H, O, C atom with Ni atom are 1.70 Å, 1.86 Å, and 1.77 Å, which is consistent with previous reports[27, 31].

**OH***

The hydroxyl produced via dehydroxylation of HOAc binds through O atom at the hcp site of Ni (111) forming a O-Ni bond distance of 2.00 Å. The adsorption process is greatly exothermic by

3.22 eV, which is in good agreement with reported results[32].

**CO***

CO* prefers to bind at the fcc site with the linear CO molecule straightly perpendicular to the surface. The three C-Ni bond lengths are all 1.95 Å, and the adsorption energy is calculated to be 1.85 eV, which is close to the reported value of 1.89 eV[32].

**$CH_x$* (x=1-3)**

As the C-C bond cleavage products of HOAc and some C2 intermediates, $CH_x$* (x=1-3) tend to stably adsorb with C atom at the fcc site of Ni (111). With the increase of unsaturated C atom, the adsorption energies increase gradually, following the order: $CH_3$ (-1.98 eV) < $CH_2$ (-4.10 eV) < CH (-6.44 eV), the high binding energies indicate that they are difficult to desorb from the surface. Meanwhile, the bond distances between C and Ni atoms are also gradually shortened with the sequence: $CH_3$ (2.11, 2.12, 2.13 Å) > $CH_2$ (1.91, 1.91, 1.98 Å) > CH (1.84, 1.84, 1.84 Å). This calculated adsorption configuration is in line with $CH_x$* species from $CH_4$ dissociation [22].

**$CH_4$***

On account of its closed-shell property, $CH_4$* shows an extremely weak adsorption on the surface with a high value of -0.06 eV. The nearest distance between C and Ni atom is 3.87 Å, also, the C-H bond lengths of adsorbed $CH_4$ stay the same as gas-phase species, indicating methane shows a typical physisorption[22].

**$CO_2$***

$CO_2$* preferentially interacts with Ni slab through C atom combining with a bridge site formed by two adjacent Ni atoms and two O atoms binding with two top positions individually, forming a bent structure with an O-C-O angle of 135.26°, yet, the C-Ni bond lengths are 2.00, 2.03 Å, while the O-Ni are 2.14 Å and 2.16 Å, respectively. The binding energy for this configuration presents a endothermic chemisorption of 0.25 eV, which is in accordance with published result[33].

**$CH_3COOH$***

Acetic acid, as the initial reactant, is of significant importance to identify the most stable adsorption configuration of HOAc on Ni (111) surface. For the reason that the lone pair electrons of oxygen atom is easy to interact with Ni metal, there are two adsorption sites in acetic acid molecule:

carbonyl oxygen $O_\alpha$ and hydroxyl oxygen $O_\beta$. The calculated result proves that the surface adsorption of $O_\alpha$ is more stable than $O_\beta$. As seen in Fig 1, HOAc preferentially adsorbs on the slab via $O_\alpha$ linked to the top site with a distance of 1.77 Å O-Ni bond, also, the methyl group points out of the surface. HOAc shows a relatively weak adsorption energy of 0.41 eV due to its closed shell structure.

**$CH_3COO$***

Acetate species $CH_3COO$* produced by hydroxyl dehydrogenation of acetic acid is adsorbed by two O atoms combining with the top sites individually, resulting in two O-Ni bond lengths of 1.97 Å and 1.98 Å, and the whole structure is vertical to the surface with an O-C-O angle of 125.41° formed. This adsorption process is remarkably exothermic by 3.01 eV, indicating that acetate is relatively stable on Ni(111) surface and thermodynamically favorable.

**$CH_3CO$***

CH3CO*, generated via deoxygenation of HOAc, adsorbs on the top sites of Ni (111) through bond valence unsaturated carbonyl $C_\alpha$ and O atoms. The interaction between C and Ni is stronger than O and Ni, due to the shorter bond distance of C-Ni (1.84 Å) than O-Ni (2.03 Å). The binding energy is calculated to be -2.10 eV.

**$CH_2COOH$***

As the source of dehydrogenation of methyl group in acetic acid, $CH_2COOH$* prefers to adsorb on the top site with unsaturated $C_\beta$ forming a single bond to satisfy its stable valence bond structure, yet, the carbonyl oxygen $O_\alpha$ is also adsorbed on the top site for its lone pair electrons. The adsorption energy of this geometry is found to be -1.75 eV.

**$CH_3COH$***

$CH_3COH$*, derived from C=O bond cleavage of HOAc, is more likely to interact with Ni slab for the unsaturated $C_\alpha$ atom, therefore, $C_\alpha$ is located on the bridge site formed by two adjacent Ni atoms to satisfy the valence structure of C atom, thus making it stable on the surface with a binding energy of 2.57 eV. The hydroxyl group points to the surface as that of acetic acid.

**$CH_2COO$***

As shown in Fig 1, the most stable adsorption configuration of $CH_2COO$* is that Oα, $O_\beta$ and $C_\beta$

are situated on the top sites of neighbouring three Ni atoms, and these three single bonds are exactly in line with their own stable bond structure with the bond distances of 1.95 Å, 1.94 Å and 2.01 Å, respectively, forming a paralled molecular plane to the surface. This stable configuration shows an exthermic property by 1.96 eV.

**COOH***

As a radical species of C-C bond breakage of HOAc, COOH* is chemically combined with Ni (111) surface via C and carbonyl $O_\alpha$ lying on the top sites. The hydroxy points out of the surface and O-C-O bond angle is 117.62°. The binding energy for this geometry is calculated to be -2.35 eV.

**$CH_3C$***

Due to the extremely valence linkage unsaturated and severely lacking in electron of $C_\alpha$ in $CH_3C$*, $C_\alpha$ is preferentially adsorbed on the fcc site of three Ni atoms and these three single bonds just meet the four valence bonds of C atom, while the $C_\beta$ atom has been saturated and has no direct interaction with the surface. Therefore, a configuration of upright $CH_3C$* with C-C bond perpendicular to the surface is identified as the final adsorption state. The adsorption process is highly exothermic by 5.66 eV, implying that $CH_3C$* is stably adsorbed and difficult to desorb from the Ni (111) surface.

**$CH_2C$* and CHC***

$CH_2C$* and CHC* species are all the products of continuous dehydrogenation of $CH_3C$*. As seen in Fig 1, both of the two species have similar adsorption geometry with $C_\alpha$ all lies on the fcc site, while $C_\beta$ lies on top ($CH_2C$*) and hcp (CHC*) sites. It is thermodynamically favourable for $CH_2C$* and CHC* adsorption, along with strong heat release, and CHC* (-5.40eV) is more stable than $CH_2C$* (-4.22eV) by 1.18 eV.

**$CH_2CO$*, CHCO* and CCO***

$CH_2CO$*, CHCO* and CCO* are the source of serial dehydrogenation of $CH_3CO$*. Ketene ($CH_2CO$*), is one of the most important intermediates during the HOAc steam reforming and can be observed with experimental methods. To satisfy their own full valence bond, $C_\beta$ atoms are located on top ($CH_2CO$*), bri (CHCO*) and fcc (CCO*) sites. As shown in Fig 1, the configuration of $CH_2CO$* is similar as CHCO*, but CCO* presents a completely different structure, with its C-C-O linear structure vertically adsorbed on Ni (111) surface. The computational binding energies increase with the increase of the degree of unsaturated $C_\beta$ atom, in the order that, -0.87 eV (CH2CO) < -3.02 eV (CHCO) < -5.28 eV (CCO).

**$CH_3COCH_3$***

Acetone, detected as a major reaction intermediate during HOAc SR, has been considered as one of the sources of carbon deposition. The $C_\alpha$ and O atoms separately interact with top and bri sites of Ni (111) surface, resulting in a configuration of C-C-C plane perpendicular to the surface and a C-C-C angle of 112.8°. The bond distances between C, O atoms and Ni atoms are 2.00 Å and 2.03 Å, respectively. Acetone can be easily desorbed from the surface for its low binding energy (-0.17 eV).

**$CH_3COCH_2$***

$CH_3COCH_2$* tends to adsorb on the surface with unsaturated carbonyl O, $C_\alpha$ and methylene $C_\beta$ binding on bri, top and bri sites. The species is inclined to the surface, and the calculated binding energy is found to be -2.03 eV.

**$CH_3COCH$***

$CH_3COCH$* is a product of acetone dehydrogenation, a configuration of combined Cβ and O atoms with bri and top sites separately is confirmed as the final adsorption state. This radical is approximately perpendicular to the surface with methyl group pointing out of the slab. The adsorption process is strongly exothermic by 4.32 eV, implying its stable adsorption and not easy to desorb from Ni (111).

**$CH_3COCH_3$***

As a product of carbonyl C=O bond cleavage from acetone, $CH_3COCH_3$* prefers to bind on bri site with $C_\alpha$, forming two C-Ni bond lengths of 1.94 Å. Two methyl groups symmetrically distributed on both sides of $C_\alpha$ and the plane formed by a 108.33° angle of C-C-C is vertical to Ni (111) surface. The calculated binding energy of this optimized structure is found to be -3.27 eV.

### 3.2 Elementary reactions

This section focuses on all possible elementary reaction pathways involved in the decomposition of acetic acid, and has systematically investigated about the mechanism of HOAc dissociation and the formation pathways and preliminary dissociation reactions of acetone. The activation energies $E_a$/eV and reaction enthalpy changes ΔH/eV for each elementary reaction on Ni (111) surface are listed in Table 2, and the corresponding configurations of transition states for each reaction are shown in Fig 2.

Table 2. Activation energies $E_a$/eV and reaction enthalpy changes ΔH/eV for each elementary reaction on Ni (111) surface.

| Elementary reactions | | TS | $E_a$/eV | ΔH/eV |
|---|---|---|---|---|
| **$CH_3COOH$ decomposition** | | | | |
| R1 | $CH_3COOH^* \rightarrow CH_3COO^* + H^*$ | TS1 | 0.45 | -0.43 |
| R2 | $CH_3COOH^* \rightarrow CH_3CO^* + OH^*$ | TS2 | 1.26 | 0.30 |
| R3 | $CH_3COOH^* \rightarrow CH_3COH^* + O^*$ | TS3 | 1.95 | 0.16 |
| R4 | $CH_3COOH^* \rightarrow CH_3^* + COOH^*$ | TS4 | 3.99 | -0.03 |
| R5 | $CH_3COOH^* \rightarrow CH_2COOH^* + H^*$ | TS5 | 1.42 | 0.52 |
| **$CH_3COO$ decomposition** | | | | |
| R6 | $CH_3COO^* \rightarrow CH_3CO^* + O^*$ | TS6 | 1.29 | 0.59 |
| R7 | $CH_3COO^* \rightarrow CH_3^* + CO_2^*$ | TS7 | 3.22 | 0.97 |
| **$CH_3CO$ decomposition** | | | | |
| R8 | $CH_3CO^* \rightarrow CH_2CO^* + H^*$ | TS8 | 1.33 | 0.49 |
| R9 | $CH_3CO^* \rightarrow CH_3C^* + O^*$ | TS9 | 1.33 | -0.21 |
| R10 | $CH_3CO^* \rightarrow CH_3^* + CO^*$ | TS10 | 1.73 | -0.67 |
| **$CH_2CO$ decomposition** | | | | |
| R11 | $CH_2CO^* \rightarrow CH_2^* + CO^*$ | TS11 | 0.49 | -0.83 |
| R12 | $CH_2CO^* \rightarrow CH_2C^* + O^*$ | TS12 | 0.44 | -0.25 |
| R13 | $CH_2CO^* \rightarrow CHCO^* + H^*$ | TS13 | 0.50 | -0.16 |
| **$CH_3C$ decomposition** | | | | |
| R14 | $CH_3C^* \rightarrow CH_2C^* + H^*$ | TS14 | 1.17 | 0.36 |
| R15 | $CH_3C^* \rightarrow CH_3^* + C^*$ | TS15 | 2.13 | 1.28 |
| **$CH_2C$ decomposition** | | | | |
| R16 | $CH_2C^* \rightarrow CH_2^* + C^*$ | TS16 | 2.35 | 1.16 |
| R17 | $CH_2C^* \rightarrow CHC^* + H^*$ | TS17 | 1.35 | 0.17 |
| **CHCO decomposition** | | | | |
| R18 | $CHCO^* \rightarrow CH^* + CO^*$ | TS18 | 0.42 | -0.93 |
| R19 | $CHCO^* \rightarrow CHC^* + O^*$ | TS19 | 1.93 | 0.38 |
| R20 | $CHCO^* \rightarrow CCO^* + H^*$ | TS20 | 0.83 | -0.01 |
| **$CH_x$(x=0-4) formation and decomposition** | | | | |
| R21 | $CH_4^* \rightarrow CH_3^* + H^*$ | TS21 | 1.45 | 0.19 |
| R22 | $CH_3^* + H^* \rightarrow CH_4^*$ | | 1.26 | -0.19 |
| R23 | $CH_3^* \rightarrow CH_2^* + H^*$ | TS22 | 0.79 | 0.16 |
| R24 | $CH_2^* + H^* \rightarrow CH_3^*$ | | 0.63 | -0.16 |

| | | | | |
|---|---|---|---|---|
| R25 | $CH_2^* \rightarrow CH^* + H^*$ | TS23 | 0.35 | -0.31 |
| R26 | $CH^* + H^* \rightarrow CH_2^*$ | | 0.66 | 0.31 |
| R27 | $CH^* \rightarrow C^* + H^*$ | TS24 | 1.39 | 0.53 |
| R28 | $C^* + H^* \rightarrow CH^*$ | | 0.86 | -0.53 |
| **$CH_3COCH_3$ formation** | | | | |
| R29 | $CH_3CO^* + CH_3^* \rightarrow CH_3COCH_3^*$ | TS25 | 1.75 | -0.12 |
| R30 | $CH_3CO^* + CH^* \rightarrow CH_3COCH^*$ | TS26 | 0.72 | -0.12 |
| R31 | $CH_3COCH^* + H^* \rightarrow CH_3COCH_2^*$ | TS27 | 0.48 | -0.14 |
| R32 | $CH_3COCH_2^* + H^* \rightarrow CH_3COCH_3^*$ | TS28 | 1.21 | 0.00 |
| **$CH_3COCH_3$ decomposition** | | | | |
| R33 | $CH_3COCH_3^* \rightarrow CH_3CO^* + CH_3^*$ | TS25 | 1.87 | 0.12 |
| R34 | $CH_3COCH_3^* \rightarrow CH_3COCH_2^* + H^*$ | TS28 | 1.29 | -0.27 |
| R35 | $CH_3COCH_3^* \rightarrow CH_3CCH_3^* + O^*$ | TS29 | 1.22 | 0.06 |

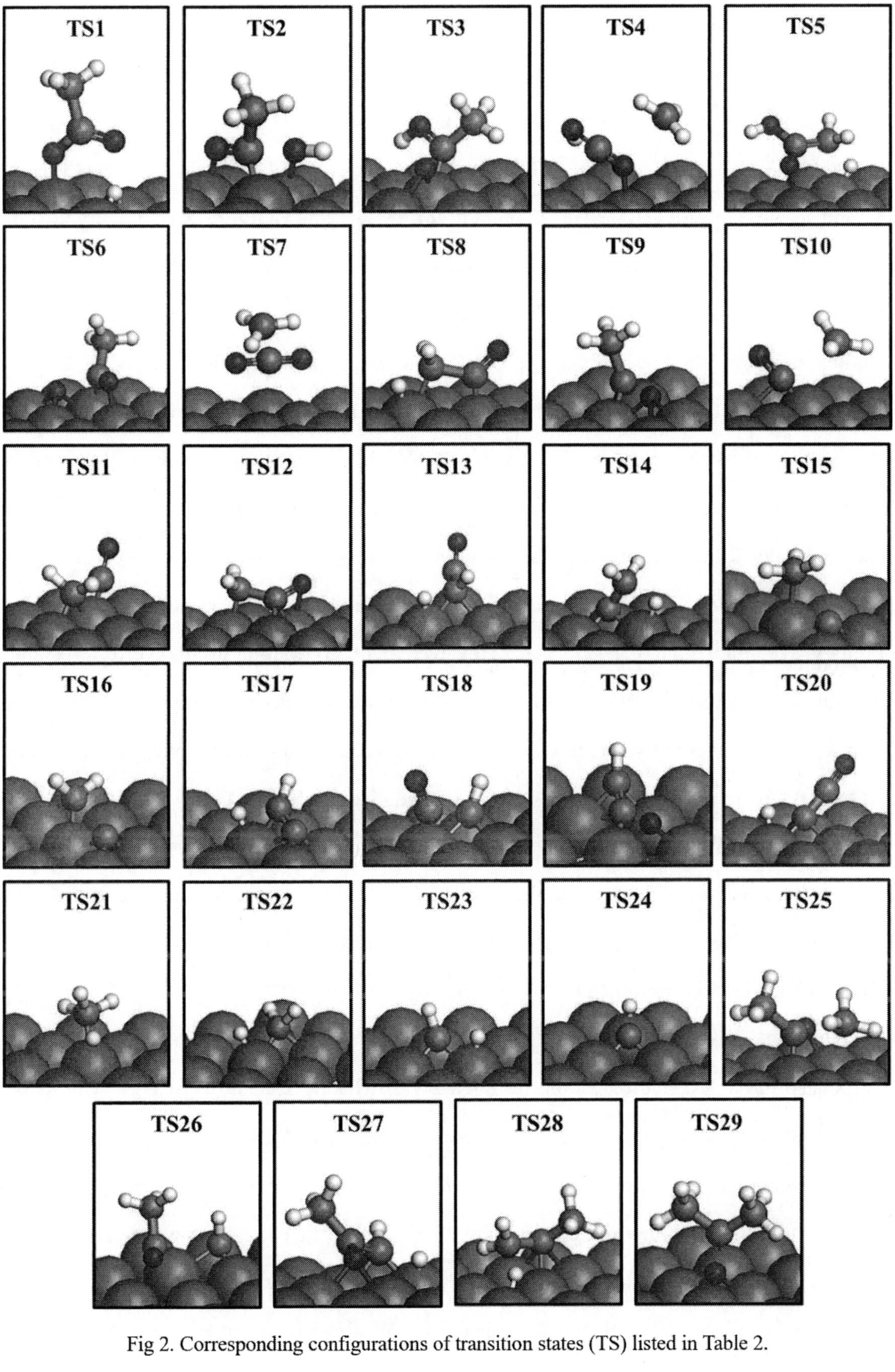


Fig 2. Corresponding configurations of transition states (TS) listed in Table 2.

**$CH_3COOH$* decomposition**

According to the different chemical bonds cleavage in acetic acid molecule, it can be divided into five possible cases for the initial decomposition of reactant: 1) dehydrogenation of hydroxy to form acetate $CH_3COO$* (R1), 2) $CH_3CO$* formation via dehydroxylation reaction (R2), 3) carbonyl C=O bond scission reaction (R3), 4) C-C bond to form $CH_3$* and COOH* (R4), 5) dehydrogenation from methyl to form $CH_2COOH$* (R5). Among these five pathways, the C-C bond dissociation reaction R3 is considered to be impossible to occur for its extremely high activation energy of 3.99 eV, indicating there is a significant obstacle on the dynamics. The activation barriers of dehydroxylation reaction R2, deoxidization from carbonyl R3 and dehydrogenation from methyl R5 are all exceed 1 eV with the value of 1.26 eV, 1.95 eV and 1.42 eV, respectively, making it difficult to take place for the relatively large kinetic block. At the same time, these three reactions have the similar thermodynamic properties with exothermal values of 0.30 eV, 0.16 eV and 0.52 eV, respectively. Therefore, the R2, R3 and R5 are considered as the less possible pathways for the preliminary decomposition of acetic acid. However, dehydrogenation from hydroxy reaction R1 reveals the lowest activation energy of 0.45 eV with an enthalpy change of -0.43 eV, indicating that whether kinetics or thermodynamics is favorable for R1. In addition, the bond length of breaking O-H (1.75 Å) in transition state TS1 is longer than that of HOAc (1.01 Å) by 0.74 Å. After dissociation, the H atom binds on the fcc site. Hence, dehydrogenation from hydroxyl reaction R1 to generate acetate species $CH_3COO$* is the optimal path with the activation energy of 0.45 eV for HOAc initial decomposition on Ni (111) surface. The further decomposition of $CH_3COO$* is thus explored on Ni (111) followed below.

**$CH_3COO$* decomposition**

Produced through dehydrogenation from hyroxyl of HOAc, acetate ($CH_3COO$*) can be broken down via C-C bond cleavage reaction R7 to generate methyl $CH_3$* and carbon dioxide $CO_2$*, which requires to overcome a high energy barrier of 3.22 eV, that is, the kinetics significantly restricted this path, meanwhile, the endothermic energy of 0.97 eV means that it is against the reaction at low temperature. Compared to R7, the reaction CH3CO* formation R6 is more likely to happen with the activation barrier (1.29 eV) and enthalpy change (0.59 eV) are lower by 1.93 eV and 0.38 eV, respectively, but there are still some obstacles in the kinetics and thermodynamics. Also, another pathway of dehydrogenation from methyl in acetate is not taken into account, which is attributed to the far distance (4.25 Å) between methyl and the slab and big steric hindrance for the reaction. Thus, R6 is the only path for $CH_3COO^*$ dissociation to form $CH_3CO^*$, which will immediately undergo the subsequent dissociation.

**$CH_3CO^*$ decomposition**

For further decomposition of $CH_3CO^*$ derived from acetate, there exist three different reaction pathways: C-H bond scission to form ketene ($CH_2CO^*$) R8, carbonyl C=O bond cleavage reaction R9 and C-C bond breakage R10. As can be seen from Table 2, the activation barriers for R8-10 are over 1 eV, that is, 1.33 eV, 1.33 eV and 1.73 eV, respectively, making it not so easy to occur, also, it can be deduced that $CH_3CO^*$ species is more abundant on Ni(111) surface. In comparison with R8 and R9, R10 is thermodynamically dominant, while it is 0.40 eV higher than R8 and R9 on dynamics and it is obviously hampered to some degree. Despite that R9 (-0.21 eV) is more thermodynamically favorable than R8 (0.49 eV), both have the same activation energy. Therefore, R8 and R9 belong to parallel competitive reactions, and both need to be taken into account for the subsequent deep dissociation. In the configurations of TS8 and TS9, the breaking C-H and C-O bond distances are measured to be 1.41 Å and 1.75 Å.

**$CH_2CO^*$ decomposition**

Ketene, one of the most important intermediates during HOAc steam reforming, can be detected as a final product in the experiment. Some researchers infer that $CH_2CO^*$ derived from intramolecular dehydration of HOAc ($CH_3COOH \rightarrow CH_2CO + H2O$). In this paper, dehydrogenation from methyl of $CH_3CO^*$ is the major source to form ketene with an activation energy of 1.33 eV. On one hand, the key intermediate $CH_2CO^*$ can directly desorb from catalytic surface as a product owing to its relatively low adsorption energy (-0.87 eV). On the other hand, according to different types of bond scission, ketene can be decomposed via three paths: bond breakage of C-C (R11), C-O (R12) and C-H (R13), and the calculated activation barriers are 0.49 eV, 0.44 eV and 0.50 eV, separately. Apparently, R11 –R13 are all kinetically favored and considerably easy to take place, meanwhile, the reaction processes accompanied different degrees of heat release of 0.83 eV, 0.25 eV and 0.16 eV. Therefore, these three reaction pathways are all thermodynamically and kinetically favorable, and the slight differences in activation energies make them coexistent during the dissociation of ketene. The rate-determining steps of succeeding decomposition of $CH_2CO^*$ depends on further cracking of easily formed $CH_2^*$, $CH_2C^*$ and $CHCO^*$.

**$CH_3C^*$ decomposition**

$CH_3C^*$ is produced via another paralleled $CH_3CO^*$ dissociation R9 with an activation barrier of 1.33 eV. $CH_3C^*$ can be broken down through C-H bond scission R14 to form $CH_2C^*$ and C-C bond scission R15 to form $CH_2^*$ and $C^*$. Based on our calculations, the activation barrier of C-C bond scission reaction (2.13 eV) is higher than that of C-H bond scission reaction (1.17 eV) by 0.96 eV, implying R 15 is kinetically restricted. Consequently, there is no doubt that the optimized decomposition pathway of $CH_3C^*$ is dehydrogenation of methyl to form $CH_2C^*$ species, and the activation energy and thermal change are 1.17 eV and 0.36 eV, separately. Even so, there are still some dynamic and thermodynamic obstacle for R14, which is in consistent with the similar configurations between TS14 (Fig 2) and the reactant $CH_3C^*$ (Fig 1). Additionally, $CH_3C^*$ can hardly desorb from surface due to its extremely high binding energy (-5.66 eV). Therefore, $CH_3C^*$ fragment is most likely to be a precursor of carbon deposition. In order to confirm this, it is quite necessary to study the further decomposition of $CH_2C^*$.

**$CH_2C^*$ decomposition**

$CH_2C^*$, a free radical species derived from optimal dissociation path of $CH_3C^*$, may undergo subsequent C-C bond cleavage reaction R16 and C-H bond cleavage R17, forming the products of $CH_2^*$ and $CHC^*$ fragment. R17 is most likely to occur for its relatively low activation barrier (1.35 eV) compared with that of R16 (2.35 eV), most importantly, both reactions are thermodynamically unfavorable with strong endothermic of R16 (1.16 eV) and moderate endothermic of R17 (0.17 eV). Hence, C-H bond scission reaction R17 is more dynamically and thermodynamically prefer than C-C bond scission reaction R16, and $CHC^*$ fragment undoubtedly be the ultimate product for $CH_2C^*$ decomposition. However, the optimal decomposition of $CH_2C^*$ is harder than that of $CH_3C^*$ by 0.18 eV, and most importantly, $CH_3CO^* \rightarrow CH_2CO^*$ path is much easier than $CH_3CO^* \rightarrow CH_3C^*$ path to occur, thus, the further decomposition of $CHC^*$ is not considered below.

**$CHCO^*$ decomposition**

$CHCO^*$ can be effortlessly obtained via C-H bond breakage R13 (0.50 eV) of $CH_2CO^*$. There are three possible pathways for further dissociation of $CHCO^*$: C-O bond cleavage R18, C-O bond cleavage R19, and C-H bond cleavage R20, generating $CH^*$, $CHC^*$, and $CCO^*$, respectively. Except for R19 (1.93 eV), there seems to be no obvious obstacle for R18 and R20 on dynamics, as the activation energies are less than 1 eV. Meanwhile, C-C bond cleavage R18 (0.42 eV) is 0.41 eV preferred than C-H bond cleavage R20 (0.83 eV), and releases more exothermic heat than R20 by 0.92 eV. In sum, C-C bond breakage of $CHCO^*$ to form $CH^*$ and $CO^*$ is kinetically and thermodynamically preferred, thus dominating the $CHCO^*$ decomposition pathways.

**$CH_x^*$ (x=0-4) formation and dissociation**

$CH_x^*$ (x=0-3) is produced along with decomposition of acetic acid on Ni(111) surface, particularly derived from C-C bond scission reactions, and the current result in this paper shows that C-C bond breakage from $CH_2CO^*$ and $CHCO^*$ to generate $CH_2^*$ and $CH^*$ is the primary source of $CH_x^*$. In general, the carbon deposition precursor is often attributed to $CH_x^*$ (x=0-3) species that leads to deactivate on catalyst. For the aim of determining the coke precursor species during HOAc decomposition on Ni (111) surface, in this paper, the formation and dissociation of $CH_x^*$ (x=0-4) has been systematically examined, and the forward and reverse reactions are presented in Table 2 (R21-28). The calculated result shows that the activation energies of continuous dehydrogenation of methane reactions R21, R23, R25 and R27 are 1.45 eV, 0.79 eV, 0.35 eV and 1.39 eV, respectively, which is agree well with the reported result[20] 1.34 eV, 0.68 eV, 0.27 eV and 1.40 eV, separately. The result indicates that $CH_4^*$ is stable are difficult to dissociate, while $CH_2^*$ (0.79 eV) and $CH^*$ (0.35 eV) can be effortlessly generated. For further dehydrogenation, it is less likely for $CH^*$ to dissociate into pyrolytic carbon in consideration of its high activation barrier (1.39 eV). Therefore, $CH^*$ is the most abundant species in decomposition of $CH_4^*$. When compared with formation and consumption reactions of $CH^*$, it is quite clear that $CH^*$ decomposition to form pyrolytic carbon R27 (1.39 eV) occurs much slower than the reverse reaction R28 to form $CH^*$ (0.86 eV), similarly, the activation barrier of $CH^*$ hydrogenation to form $CH_2^*$ R26 is 0.31 eV higher than that of $CH_2^*$ dehydrogenation to generate $CH^*$ R25. At the same time, $CH^*$ fragment can hardly desorb from Ni

(111) surface for its high adsorption energy of -6.44 eV. Under these evidence, it can be undoubtedly concluded that CH* species is enriched as the carbon deposition precursor on Ni (111) surface during the HOAc decomposition, and the reason for the coke formation is ascribed to the migration and further polycondensation of CH* species.

**Acetone formation**

Acetone is not only one of the most important intermediates during acetic acid steam reforming, but also acknowledged as the source of carbon deposition, therefore, it is of great necessity to carry out the formation mechanism of acetone. To date, some researchers reached a consensus that $CH_3COCH_3$* is directly formed via $CH_3CO$* combined with methyl $CH_3$* (R29), even though, an indirect acetone formation route has been put forward in this paper, that is, two enriched species on Ni (111), $CH_3CO$* and CH* fragments, can be firstly combined to form $CH_3COCH$* (R30), afterwards, $CH_3COCH$* undergoes sequential hydrogenation reactions (R31 and R32) and ultimately form acetone.

From Table 2, we can see that compared with the activation energy of direct acetone formation R29 (1.75 eV), the activation barriers of indirect acetone formation reactions R30 (0.72 eV), R31 (0.48 eV) and R32 (1.21 eV) are much lower, what is more important is that the two abundant species are more likely to interact with each other and the fact proves that $CH_3COCH$* can be readily formed, which confirms the feasibility of the indirect route for acetone formation. Moreover, the $CH_3$* radical prefers to react by dehydrogenation to form CH* species on Ni (111) rather than stably adsorbed and enriched on the slab, thus inhibiting the direct acetone formation reaction to some extent. In conclusion, the formation of acetone is kinetically favorable via indirect route rather than direct route.

**Acetone initial decomposition**

For the reason why acetone is one of sources of coke formation, some researchers[25] hold the view that acetone was condensed to generate the carboatomic rings that result in coke formation and eventually deactivate the catalyst. In this paper, considering the huge calculated amount, the complete decomposition and condensation of acetone is not carried out. Here, the initial decomposition of $CH_3COCH_3$* is provided. Based on different bond cleavage in acetone, it can be divided into three possible pathways: 1) C-C bond scission reaction R33, 2) C-H bond scission from methyl reaction R34, 3) C-O bond scission from carbonyl reaction R35. All of these three pathways represent a high activation energies more than 1 eV with 1.87 eV, 1.29 eV and 1.22 eV, respectively, indicating that the dissociation of acetone on Ni catalyst is dynamically unfavorable. Compared with the dissociation reactions, acetone is more likely to desorb from the surface into the gas phase and to be detected in the experiment as $CH_3COCH_3$*, which is slightly adsorbed on Ni (111) surface with only 0.17 eV.

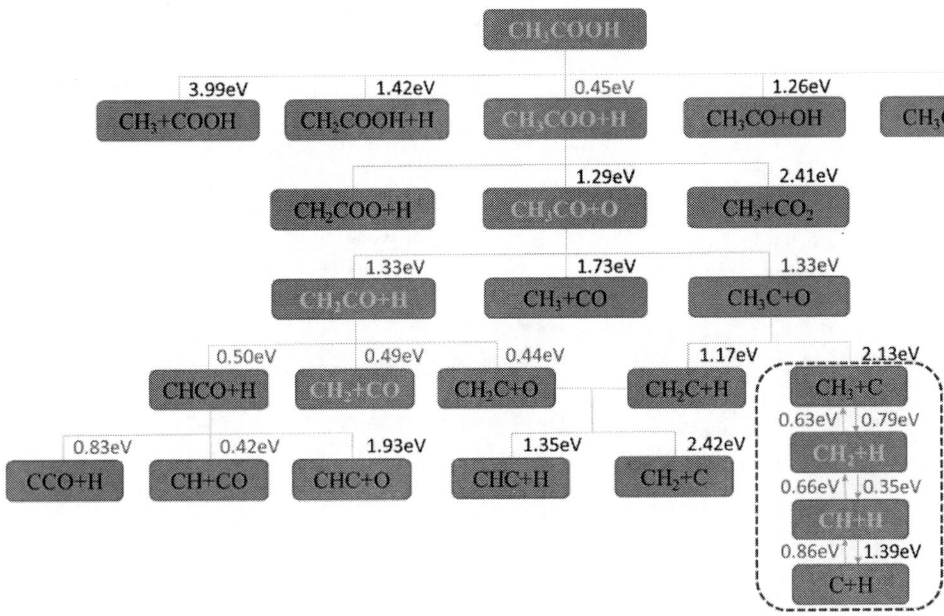


Fig 3. The overall possible decomposition pathways of HOAc on Ni (111) surface. The purple dash line involved is $CH_x$ (x=0-3) formation and decomposition. The optimal decomposition paths are in orange. The activation energies below 1 eV are in green. The rate-determining steps are in red.

To sum up the whole HOAc decomposition on Ni(111) surface, the complete possible decomposition pathways are presented in Fig 3. It is obvious that the dissociation reactions can be divided into to pathways, ketene ($CH_2CO$*) formation path P1: $CH_3COOH$*→$CH_3COO$*→$CH_3CO$*→$CH_2CO$*→$CH_2$*→CH* and $CH_3C$* formation path P2: $CH_3COOH$*→$CH_3COO$*→$CH_3CO$*→$CH_3C$*→$CH_2C$*→CHC*. From the potential energy graph for the possible pathways of HOAc decomposition (Fig 4), it is straightforwardly stated that ketene formation and $CH_3C$* formation by $CH_3CO$* are compatible with the same activation energies, but $CH_3C$* path is thermodynamically favored. Even so, as proceeding description, a serious of subsequent dissociation of $CH_2CO$* is kinetically and thermodynamically preferred than $CH_3C^*$. Therefore, the most favorable pathway for HOAc decomposition on Ni (111) surface is the ketene formation path P1, and the detected ketene in the experiment provides an evidence for the optimal pathway, but the formation of ketene from $CH_3CO$ is a bottleneck during the whole process. Meanwhile, CH* species contributes to the carbon deposition precursor and enriches for further polymerization to form coke.

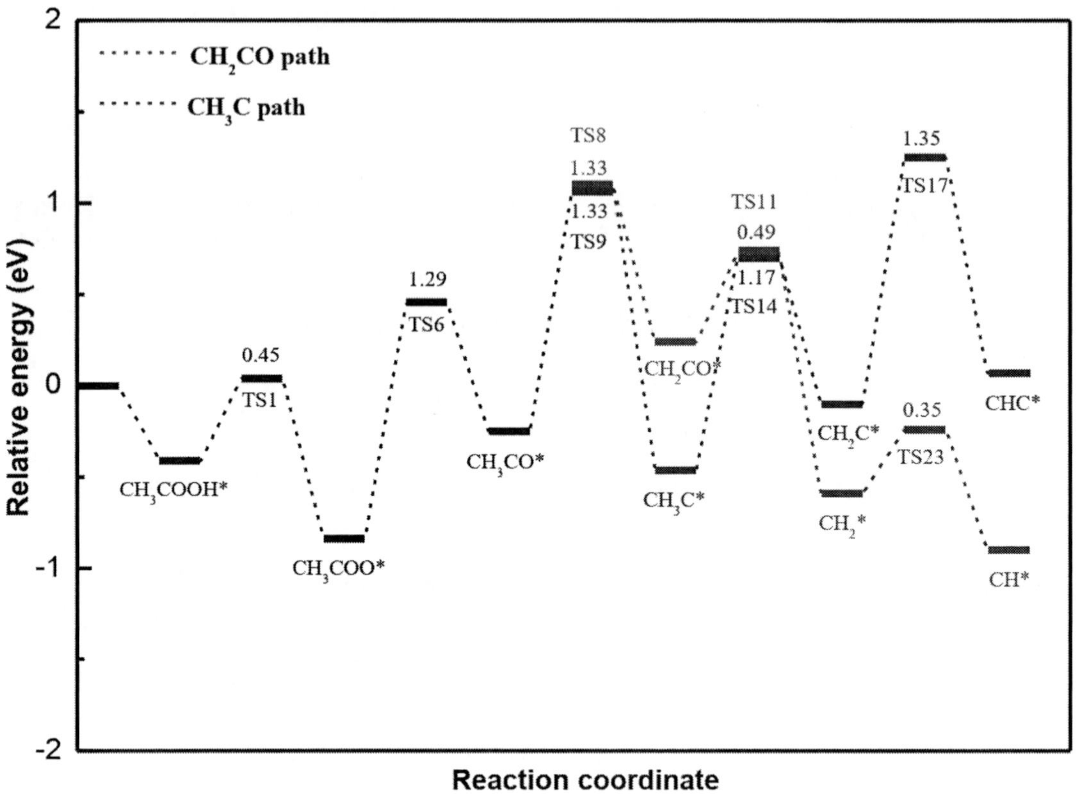


Fig 4. The potential energy graph for the possible pathways of HOAc decomposition on Ni (111) surface.

## 4. Conclusions

In this paper, acetic acid is selected to simulate the property of bio-oil derived from the renewable biomass. To illustrate the mechanism of carbon deposition formation and high hydrogen yield during bio-oil steam reforming, we have thoroughly investigated the HOAc decomposition mechanism on Ni(111) surface with all of the possible reaction pathways are considered. The calculated binding energies, activation barriers, enthalpy changes and transition state configurations are systematically involved to describe the reaction routes. The HOAc decomposition pathway starts from initial O-H bond scission with an activation energy of 0.45 eV, and the generated acetate $CH_3COO^*$ undergoes the further deoxygenation to form $CH_3CO^*$. The subsequent $CH_3CO^*$ dissociation reactions can be divided into two paths, ketene formation path and $CH_3C^*$ formation path, both have the same and the highest kinetic barriers of 1.33 eV, but $CH_3C^*$ formation path is thermodynamically favorable, while ketene formation path is not. Compared with subsequent decomposition of $CH_3C^*$ and ketene, it is dramatically proved that ketene dissociation is dynamically and thermodynamically preferred. Therefore, the most favorable decomposition pathway of HOAc on Ni(111) surface is that $CH_3COOH^* \rightarrow CH_3COO^* \rightarrow CH_3CO^* \rightarrow CH_2CO^* \rightarrow CH_2^* \rightarrow CH^*$. The rate-determining step is ketene formation from $CH_3CO^*$ and $CH^*$ species is identified as the major carbon deposition precursor. What's more, the mechanism of acetone formation and the primary dissociation has also explored. The result shows that the formation of acetone is kinetically favorable via the combination of enriched $CH_3CO^*$ and $CH^*$ species and succeeding hydrogenation reactions rather than directly combined by $CH_3CO^*$ and $CH_3^*$. It can be also deduced that the formation of $CH_3C^*$ and acetone causes the coke formation.

## ACKNOWLEDGEMENTS

The work was supported by the National Natural Science Foundation of China (51406129); the Program for the Top Young Academic Leaders of Higher Learning Institutions of Shanxi (154010148-s); the Research Project Supported by Shanxi Scholarship Council of China (2015-046); and the Research Project Supported by Key Laboratory of Renewable Energy, Chinese Academy of Sciences (Y607k31001).

## Reference

[1] W.N.R.W. Isahak, M.W.M. Hisham, M.A. Yarmo, T.-y. Yun Hin, A review on bio-oil production from biomass by using pyrolysis method, Renewable and Sustainable Energy Reviews, 16 (2012) 5910-5923.

[2] H.B. Goyal, D. Seal, R.C. Saxena, Bio-fuels from thermochemical conversion of renewable resources: A review, Renewable and Sustainable Energy Reviews, 12 (2008) 504-517.

[3] S. Ayalur Chattanathan, S. Adhikari, N. Abdoulmoumine, A review on current status of hydrogen production from

bio-oil, Renewable and Sustainable Energy Reviews, 16 (2012) 2366-2372.

[4] E. Heracleous, Well-to-Wheels analysis of hydrogen production from bio-oil reforming for use in internal combustion engines, International Journal of Hydrogen Energy, 36 (2011) 11501-11511.

[5] E. Vagia, A. Lemonidou, Thermodynamic analysis of hydrogen production via steam reforming of selected components of aqueous bio-oil fraction, International Journal of Hydrogen Energy, 32 (2007) 212-223.

[6] Y. Zhang, T.R. Brown, G. Hu, R.C. Brown, Comparative techno-economic analysis of biohydrogen production via bio-oil gasification and bio-oil reforming, Biomass and Bioenergy, 51 (2013) 99-108.

[7] J. Remón, J.A. Medrano, F. Bimbela, L. García, J. Arauzo, Ni/Al–Mg–O solids modified with Co or Cu for the catalytic steam reforming of bio-oil, Applied Catalysis B: Environmental, 132-133 (2013) 433-444.

[8] F. Seyedeyn Azad, J. Abedi, E. Salehi, T. Harding, Production of hydrogen via steam reforming of bio-oil over Ni-based catalysts: Effect of support, Chemical Engineering Journal, 180 (2012) 145-150.

[9] R. Trane, S. Dahl, M.S. Skjøth-Rasmussen, A.D. Jensen, Catalytic steam reforming of bio-oil, International Journal of Hydrogen Energy, 37 (2012) 6447-6472.

[10] R. Trane-Restrup, A.D. Jensen, Steam reforming of cyclic model compounds of bio-oil over Ni-based catalysts: Product distribution and carbon formation, Applied Catalysis B: Environmental, 165 (2015) 117-127.

[11] B. Valle, B. Aramburu, A. Remiro, J. Bilbao, A.G. Gayubo, Effect of calcination/reduction conditions of Ni/La2O3–αAl2O3 catalyst on its activity and stability for hydrogen production by steam reforming of raw bio-oil/ethanol, Applied Catalysis B: Environmental, 147 (2014) 402-410.

[12] K. Takanabe, K. Aika, K. Seshan, L. Lefferts, Sustainable hydrogen from bio-oil—Steam reforming of acetic acid as a model oxygenate, Journal of Catalysis, 227 (2004) 101-108.

[13] X. Li, S. Wang, Y. Zhu, G. Yang, P. Zheng, DFT study of bio-oil decomposition mechanism on a Co stepped surface: Acetic acid as a model compound, International Journal of Hydrogen Energy, 40 (2015) 330-339.

[14] L. An, C. Dong, Y. Yang, J. Zhang, L. He, The influence of Ni loading on coke formation in steam reforming of acetic acid, Renewable Energy, 36 (2011) 930-935.

[15] E.C. Vagia, A.A. Lemonidou, Investigations on the properties of ceria–zirconia-supported Ni and Rh catalysts and their performance in acetic acid steam reforming, Journal of Catalysis, 269 (2010) 388-396.

[16] R. González-Gil, I. Chamorro-Burgos, C. Herrera, M.A. Larrubia, M. Laborde, F. Mariño, L.J. Alemany, Production of hydrogen by catalytic steam reforming of oxygenated model compounds on Ni-modified supported catalysts. Simulation and experimental study, International Journal of Hydrogen Energy, 40 (2015) 11217-11227.

[17] B. Valle, A. Remiro, A.T. Aguayo, J. Bilbao, A.G. Gayubo, Catalysts of Ni/α-Al2O3 and Ni/La2O3-αAl2O3 for hydrogen production by steam reforming of bio-oil aqueous fraction with pyrolytic lignin retention, International Journal of Hydrogen Energy, 38 (2013) 1307-1318.

[18] S. Li, C. Zhang, P. Zhang, G. Wu, X. Ma, J. Gong, On the origin of reactivity of steam reforming of ethylene glycol on supported Ni catalysts, Physical chemistry chemical physics : PCCP, 14 (2012) 4066-4069.

[19] H. Xie, Q. Yu, X. Yao, W. Duan, Z. Zuo, Q. Qin, Hydrogen production via steam reforming of bio-oil model compounds over supported nickel catalysts, Journal of Energy Chemistry, 24 (2015) 299-308.

[20] D.W. Blaylock, T. Ogura, W.H. Green, G.J. Beran, Computational investigation of thermochemistry and kinetics of steam methane reforming on Ni (111) under realistic conditions, The Journal of Physical Chemistry C, 113 (2009) 4898-4908.

[21] C. Dong, L. An, Y. Yang, J. Zhang, A Density Functional Investigation on C Atom Adsorbed on Ni (111) Surface in the Steam Reforming of Acetic Acid, in: Power and Energy Engineering Conference (APPEEC), 2010 Asia-Pacific, IEEE, 2010, pp. 1-4.

[22] K. Li, C. He, M. Jiao, Y. Wang, Z. Wu, A first-principles study on the role of hydrogen in early stage of graphene growth during the CH4 dissociation on Cu(111) and Ni(111) surfaces, Carbon, 74 (2014) 255-265.

[23] F. Cheng, V. Dupont, Nickel catalyst auto-reduction during steam reforming of bio-oil model compound acetic acid, International Journal of Hydrogen Energy, 38 (2013) 15160-15172.

[24] R.-r. Hu, C.-f. Yan, X.-x. Zheng, H. Liu, Z.-y. Zhou, Carbon deposition on Ni/ZrO2–CeO2 catalyst during steam reforming of acetic acid, International Journal of Hydrogen Energy, 38 (2013) 6033-6038.

[25] T.M.C. Hoang, B. Geerdink, J.M. Sturm, L. Lefferts, K. Seshan, Steam reforming of acetic acid – A major component in the volatiles formed during gasification of humin, Applied Catalysis B: Environmental, 163 (2015) 74-82.

[26] S. Wang, X. Li, L. Guo, Z. Luo, Experimental research on acetic acid steam reforming over Co–Fe catalysts and subsequent density functional theory studies, International Journal of Hydrogen Energy, 37 (2012) 11122-11131.

[27] J. Ren, H. Guo, J. Yang, Z. Qin, J. Lin, Z. Li, Insights into the mechanisms of CO2 methanation on Ni(111) surfaces by density functional theory, Applied Surface Science, 351 (2015) 504-516.

[28] R. Zhang, L. Song, H. Liu, B. Wang, The interaction mechanism of CO2 with CH3 and H on Cu (111) surface in synthesis of acetic acid from CH4/CO2: A DFT study, Applied Catalysis A: General, 443-444 (2012) 50-58.

[29] S. Lin, J. Huang, X. Gao, X. Ye, H. Guo, Theoretical Insight into the Reaction Mechanism of Ethanol Steam Reforming on Co(0001), The Journal of Physical Chemistry C, (2015) 150126155826007.

[30] H.-p. Zhang, X.-g. Luo, H.-t. Song, X.-y. Lin, X. Lu, Y. Tang, DFT study of adsorption and dissociation behavior of H2S on Fe-doped graphene, Applied Surface Science, 317 (2014) 511-516.

[31] R.C. Catapan, A.A.M. Oliveira, Y. Chen, D.G. Vlachos, DFT Study of the Water–Gas Shift Reaction and Coke Formation on Ni(111) and Ni(211) Surfaces, The Journal of Physical Chemistry C, 116 (2012) 20281-20291.

[32] A. Mohsenzadeh, T. Richards, K. Bolton, DFT study of the water gas shift reaction on Ni(111), Ni(100) and Ni(110) surfaces, Surface Science, 644 (2016) 53-63.

[33] S.-G. Wang, D.-B. Cao, Y.-W. Li, J. Wang, H. Jiao, Chemisorption of CO2 on nickel surfaces, The Journal of Physical Chemistry B, 109 (2005) 18956-18963.

**2016 International Pittsburgh Coal Conference**
**Cape Town, SOUTH AFRICA**
**August 8 – 12, 2016**

**Abstract Submission**

**(CARBON MANAGEMENT)**

# PRETREATMENT OF FLUE GAS USING POLYMERIC MEMBRANE FOR $CO_2$ CAPTURE

***Soohyun Choi, Senior Technologist, Korea Institute of Energy Research, 152 Gajeongro, Yuseong-gu, Daejeon, Republic of KOREA***
***hyeon@kier.re.kr, +82428603193, Fax +82428603134***

*Ilhyun Baek, Principal Researcher, Korea Institute of Energy Research, 152 Gajeongro, Yuseong-gu, Daejeon, Republic of KOREA*
*ihbaek@kier.re.kr, +82428603648, Fax +82428603134*

**Abstract:**
The carbon dioxide, the greatest contributor to the global warming, has been mostly produced in the power generation sector so it seems like a foremost requirement for us to come up with a new power generation technology in an attempt to pro-actively respond to global warming. Therefore, it is critical to develop cost effective technologies to reduce carbon dioxide processing cost. One very promising approach to reducing carbon dioxide emissions is carbon dioxide capture. However, if the promise of this approach is to be realized, capture costs will have to be reduced through membrane technology. In this work the removal of oxygen from flue gas by combustion using polymer membrane for carbon dioxide capture have been studied. The composite membrane used in the present study the morphology was examined by scanning electron microscope to understand and correlate morphology with the performance of membrane. The effects of parameters involved in the spinning of PES hollow fiber and PDMS layer deposition, and further meso-tetraphenylporphyrin cobalt complex selective layer coating as resulting composite membranes were investigated.

The prepared composite membrane showed good facilitated oxygen transport, especially for $O_2/N_2$ gas pairs, it reversed the permeaselectivity and gave the selectivity 1.24. It is expected to be developed further for oxygen removal from flue gas in the pretreatment process for carbon dioxide capture. On the other hand, the activation energies of $CO_2$ and $N_2$ have positive values, which mean that they are dominated by the diffusion coefficient under the operating conditions. Generally, an increase in temperature could supply more energy, increase the mobility of the gas penetrants and enhance the mobility of the polymeric chain, leading to an increase of the penetrate diffusion coefficient. This was because the increase in $CO_2$ diffusivity compensated the decrease of solubility parameter under the experimental conditions. In contrast to the activation energy of $N_2$, that of $CO_2$ decreased with pressure. In the mixture gas $CO_2/N_2$ experimental results demonstrated that the permeance of $CO_2$ is higher than that of $N_2$, because the kinetic diameter of $CO_2$ is smaller than $N_2$. The composite membrane shows relatively good $N_2/CO_2$ selectivity in consideration of operating conditions with lower pressure. Compared to flat sheet membranes, hollow fiber membrane has the advantages of supporting itself, and of having a large membrane area per unit volume. Therefore, hollow fiber membranes are more acceptable for the application to industrial processes.

This membrane showed good facilitated oxygen transport from flue gas, especially for $O_2/N_2/CO_2$ gas pairs, it reversed the selectivity compared with existing data released. It is expected to be developed further for oxygen removal from flue gas in the carbon dioxide capture as well as reducing carbon dioxide capture cost due to the pretreatment of pollutant using cost-effective membrane system.

# Experimental study on $CO_2$ capture from simulated flue gas by using aqueous ammonia with external magnetic field

**Qi Zhang, Jiayi Wu, Ping Lu**

(School of Energy and Mechanical Engineering, Nanjing Normal University, Nanjing 210042, Jiangsu, P. R. China)

**Abstract:** A new ammonia-based $CO_2$ capture process is proposed in the paper. The process uses magnetically gas-solid-liquid bed as the reactor in which nano-ferromagnetic particles suspend in aqueous ammonia solution under the influence of external magnetic field (EMF). A continuous of $CO_2$ passes through the reactor, absorbed by aqueous ammonia. Experiments in a laboratory-scale apparatus were carried out to investigate the roles of the ferromagnetic particles and the applied magnetic field in $CO_2$ absorption. Results shows that EMF and nanoparticles (such as $Fe_3O_4$) can significantly improve the performance of $CO_2$ absorption. The highest $CO_2$ absorption efficiency with EMF and nanoparticles reaches 99.40%, more than 7% compared to that without EMF and nanoparticles. With EMF and nanoparticles, the $CO_2$ absorption efficiency increase significantly when using low concentration aqueous ammonia (5%~8%), large flux of simulated flue gas (3.5 L/min) and low absorption temperature (22-36°C). The promoting mechanism of EMF and nano-$Fe_3O_4$ on $CO_2$ absorption may be the enhancements of contact efficiency and mass transfer among gas-liquid-solid three phases, as well as the change of thermo-physical property of aqueous ammonia leading to the increase of the solubility of aqueous ammonia and the decrease of ammonia slip.

**Keywords:** Magnetically gas-solid-liquid bed; aqueous ammonia-based $CO_2$ absorption, gas-liquid mass transfer

## 1. Introduction

Nowadays rising energy production is associated with increasingly greenhouse gases emissions. Innovative carbon neutral technologies will be required to enable the power sector to meet the global demand for electricity, while controlling $CO_2$ emissions and thus reducing the impact on global warming[1,2].

Flue gas post-combustion $CO_2$ capture in coal-fired power plants consists of absorption, adsorption, membrane separation, chilling ammonia, etc. Considered with $CO_2$ in flue gas with low partial pressure, huge gas flow rate as well as complex co-existing components, the chemical absorption is the most widely used method. Conventional $CO_2$ absorbents for chemical absorption are alkanolamines, while ammonia solution has been proposed as an alternative absorbent. The ammonia solution process needs less recycling energy, and the heat of the absorption reaction is one-third that of the MEA process. Sensible heat of ammonia-based separation(25~80℃) is also lower than that of the current MEA process(55~115℃). Furthermore the cost of the ammonia process is one-tenth that of the MEA process. There are no degradation and corrosion problems in ammonia process, while MEA has a high corrosion rate for the equipment and amine is easily degradable by acid gases. The ammonia solution process can remove $SO_2$, $CO_2$ and $NO_x$ from flue gas simultaneously [3].

**Correspondence:** P. Lu (luping@njnu.edu.cn), School of energy and mechanical engineering, Nanjing Normal University, Nanjing 210042, Jiangsu, P.R. China.

The ammonia-based $CO_2$ capture has been reported in two configurations based on the operating temperature of the absorber. The absorption temperature of $ECO_2$ is above 20℃, while the absorption temperature of CAP (Chilled Ammonia Process) is 2~10 ℃. The successful commercialization of ammonia-based $CO_2$ capture lies on the technical and economic feasibility optimization. New means should be explored to strengthen the absorption, lower the energy consumption of the regeneration step, and resolve some technical issues including ammonia slip, precipitation and slurry formation.[4]

In the paper, magnetically gas-solid-liquid bed is used as the reactor in a new ammonia-based $CO_2$ absorption. The ferromagnetic particles suspend in aqueous ammonia solution under the influence of external magnetic field. A continuous of $CO_2$ passes through the reactor, absorbed by the aqueous ammonia. The magnetically gas-solid-liquid bed has the primary advantage of high contact efficiency of different phases, which makes the $CO_2$ absorption reaction more efficient. Moreover, nano-particles can enhance the mass transfer between gas and liquid, and the external magnetic field can increase the dissolution rate and the solubility of aqueous ammonia.

The paper investigates the roles of the ferromagnetic particles and the external magnetic field in $CO_2$ absorption. Experiments are carried out in a laboratory-scale semi-batch apparatus to measure $SO_2$ removal efficiency. The roles of ferromagnetic particles and external magnetic field are revealed by comparing $SO_2$ removal efficiency.[5-8]

## 2. Experimental

A schematic diagram of the experimental apparatus is displayed in Fig 1.The experiment is conducted in a self-designed system, which consists mainly of the simulated flue gas generator, the reactor, the flue gas reprocessing and the measure equipment. The reactor in a thermostat water bath is a magnetically gas-solid-liquid bed with gas inlet and outlet tubes.

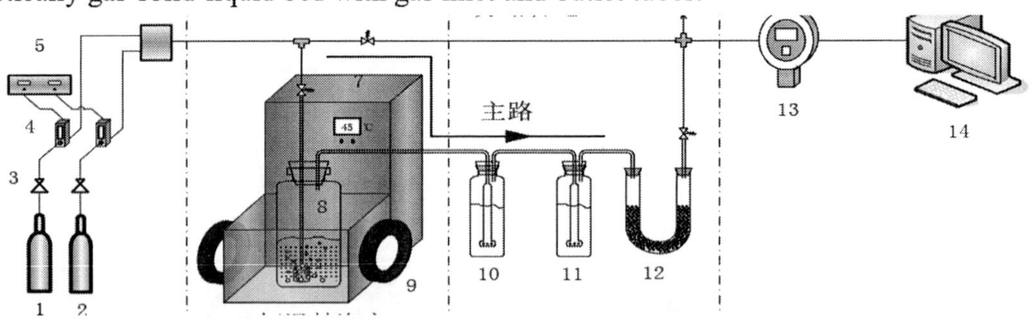


Fig 1. schematic diagram of the experimental apparatus

The external magnetic field is generated by the electromagnets. An electromagnet, a ring with 155mm in outer diameter, 80 mm in inner diameter and 20 mm in thickness, can generate a magnetic field up to 150mT. Four arrangements have been made for the electromagnets, which are shown in Fig 2. The EMF(a), (b), (c) and (d) correspond to the horizontal magnetic field with two electromagnets, the vertical magnetic field with an electromagnet above the reactor, the vertical magnetic field with an electromagnet below the reactor, and the vertical magnetic field with two electromagnets respectively.

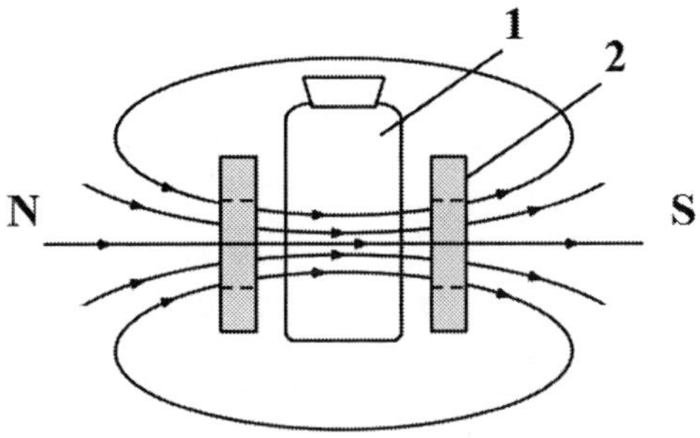


(a) Horizontal (two electromagnets)

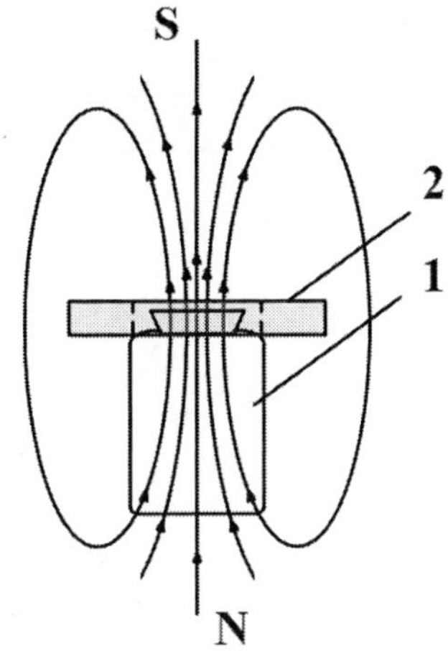


(b) Vertical (electromagnet above the reactor)

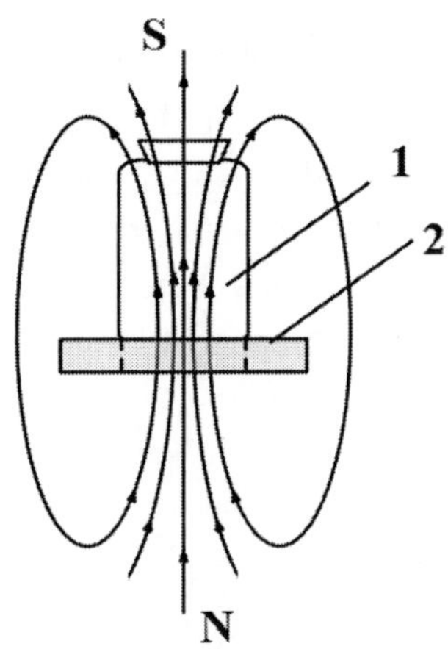


(c) Vertical (electromagnet below the reactor)

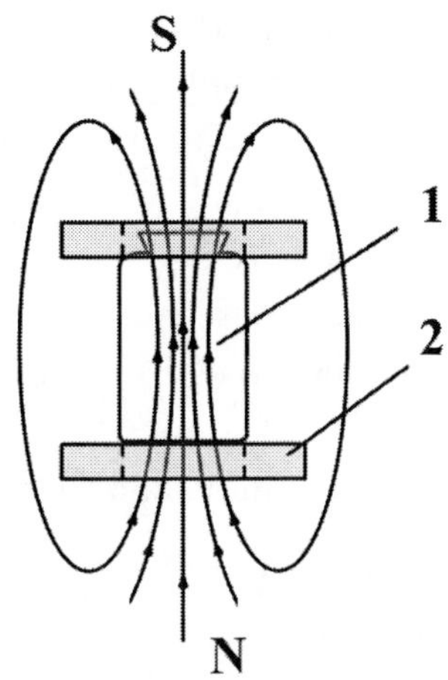


(d) Vertical (two electromagnets)

Fig 2. schematic diagram of the external magnetic fields

Three ferromagnetic particles, $Fe_3O_4$、 $Fe_2O_3$ and Fe, are used in the experiments. The properties of them are listed in Table1.

Table 1 The properties of the particles

| Material | $Fe_3O_4$ | $Fe_2O_3$ | Fe |
|---|---|---|---|
| Diameter | 30nm | 1μm | 10μm |
| Purity | ⩾98% | ⩾98% | ⩾98% |
| Magnetism | strong | weak | strong |
| Reaction activity | stable | stable | reductibility |
| Color | black | red | silvery |
| Appearance |  |  | 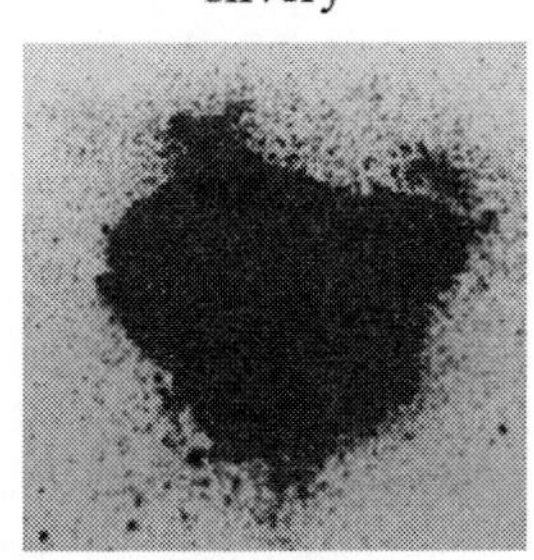 |

The reaction temperature is controlled by the thermostat water bath. Inlet and outlet $CO_2$

concentration are measured by a gas analyzer (FIX550-$CO_2$-G-IR). Compositions of the ferromagnetic particles are measured by an X-ray diffraction analyzer (D/max2500/PC).
The range of experimental conditions used in the work is listed in Table 2

Table 2 Experimental conditions

| Solution volume ($V_{NH3}$) | Ammonia concentration ($C_{NH3}$) | Inlet $CO_2$ concentration ($\varphi_{CO2,in}$) | Inlet flue gas flux ($Q_{in}$) | Reaction temperature ($T_a$) | Magnetic field intensity ($H_{EMF}$) | Particle concentration ($F_d$) |
|---|---|---|---|---|---|---|
| mL | Wt% | % | L·min$^{-1}$ | ℃ | mT | g/L |
| 500 | 5 | 20 | 2.0 | 20 | 0-16 | 0-5 |

$N_2$ and $CO_2$ firstly pass through the mass flowmeters which are used to measure and adjust their feeding fluxes separately, and then mix to form simulated flue gas. Meanwhile, the ammonia solution is mixed with water to form aqueous ammonia solution with certain ammonia concentration. A quantity of particles is put into the solution, and the electromagnets are fixed outside the reactor. $CO_2$ reacts with aqueous ammonia in the reactor under the influence of ferromagnetic particles and external magnetic field. Concentrated sulfuric acid is used to absorb escaped ammonia and $CaCl_2$ powder is used to remove water from the cleaned flue gas.

In this study, $CO_2$ removal efficiency is defined as

$$\eta = \frac{\varphi_{CO_2,in} - \varphi_{CO_2,out}}{\varphi_{CO_2,in}(1-\varphi_{CO_2,out})} \times 100\% \qquad (1)$$

Where $\varphi_{in}$ and $\varphi_{out}$ are the inlet $CO_2$ concentration and outlet $CO_2$ concentration, respectively.

# 3. Results and discussion

## 3.1 Enhancement effects of EMF and ferromagnetic particles

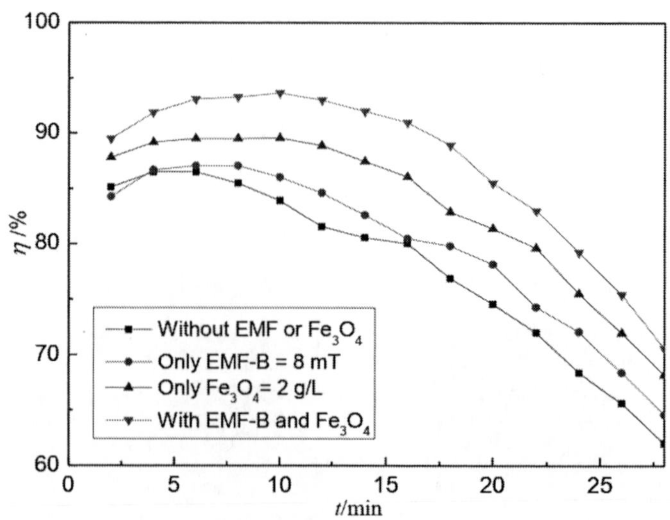


Fig.3 The enhancement effects of EMF and ferromagnetic particles

The $CO_2$ removal efficiencies shown in Fig3 have been measured under the conditions listed in Table2. As shown in Fig3, the efficiency with 2g/L $Fe_3O_4$ and 8mT EMF-B is maximally 9% higher than that without ferromagnetic particles and EMF, the removal efficiencies decrease with time, and the differences in efficiency become smaller with time.

When the EMF is used alone, $CO_2$ removal efficiency increases slightly. The magnetic field can increase the solubility of aqueous ammonia, which leads to faster reaction between aqueous ammonia and $CO_2$, higher $CO_2$ absorption rate, and then higher $CO_2$ removal efficiency. As the intensity of the EMF is weak, the effect of the EMF alone is weak too.

When the nanoparticles $Fe_3O_4$ is used alone, $CO_2$ removal efficiency increases obviously. The nanoparticles can increase the contact efficiency of gas and liquid phase, and enhance the absorption of $CO_2$ consequently.

When ferromagnetic particles $Fe_3O_4$ and EMF are used together, the enhancement effect is the most obvious. Firstly, the use of ferromagnetic particles can highly enhance the intensity of magnetic field in the reactor. Secondly, the EMF can control the move of the ferromagnetic particles, which leads to the uniform dispersion of ferromagnetic particles in the reactor, and the contact of gas and liquid phase becoming more efficiently.

### 3.2 Effects of the arrangement of the EMF

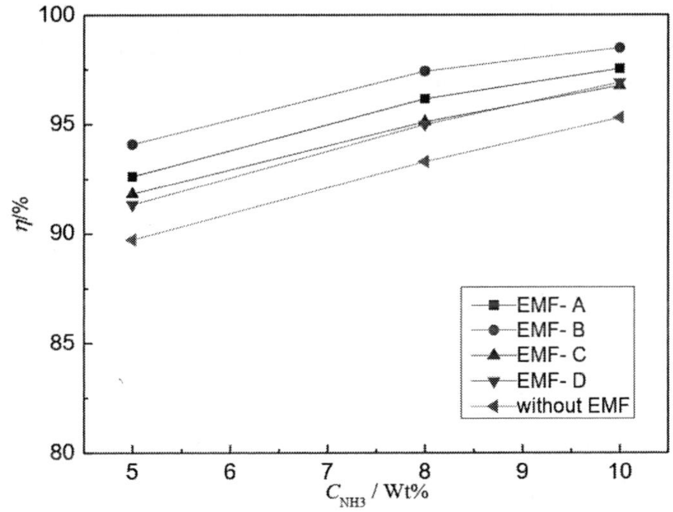


Fig 4 Effects of applied field arrangement on removal efficiency

As shown in Fig4, the arrangement of EMF-B refer to the highest $CO_2$ removal efficiency. When EMF-B is used, the magnetic force of particles is upward, while the gravity of particles is downward. Two force cancels out, leading the dispersion of particles in the reactor uniform.

### 3.3 Effects of different ferromagnetic particles

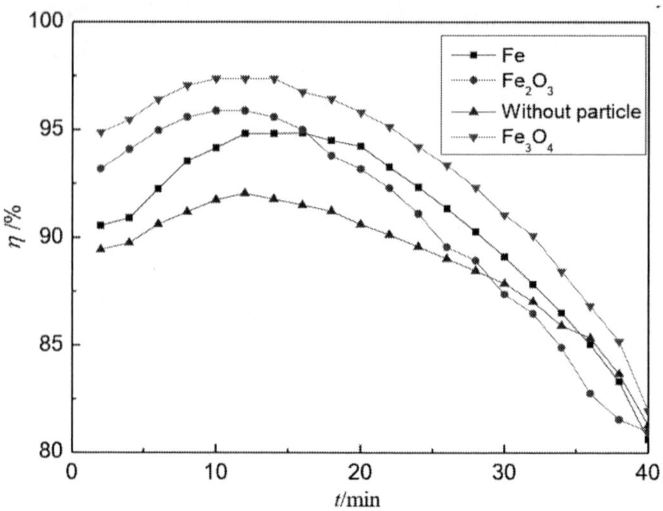


Fig 5 Effects of different ferromagnetic particles on removal efficiency

As shown in Fig5, $Fe_3O_4$ refer to the highest $CO_2$ removal efficiency. As the magnetism of $Fe_3O_4$ is the strongest and the diameter of $Fe_3O_4$ is the finest, the roles of $Fe_3O_4$ on $CO_2$ removal efficiency is the most obvious.

## 4. Conclusion

Experimental results shows that EMF and nanoparticles (such as $Fe_3O_4$) can significantly improve the performance of $CO_2$ absorption. The highest $CO_2$ absorption efficiency with EMF and nanoparticles reaches 99.40%, more than 7% compared to that without EMF and nanoparticles. With EMF and nanoparticles, the $CO_2$ absorption efficiency increase significantly when using low concentration aqueous ammonia (5%~8%), large flux of simulated flue gas (3.5 L/min) and low absorption temperature (22-36°C). The promoting mechanism of EMF and nano-$Fe_3O_4$ on $CO_2$ absorption may be the enhancements of contact efficiency and mass transfer among gas-liquid-solid three phases, as well as the change of thermo-physical property of aqueous ammonia leading to the increase of the solubility of aqueous ammonia and the decrease of ammonia slip.

## References

[1] Oliver JE. Intergovernmental Panel in Climate Change (IPCC) [J]. Encyclopedia of Energy Natural Resource & Environmental Economics, 2013, 26(D14): 48-56.

[2] Iizumi T, Luo JJ, Challinor AJ, et al. Impacts of El Nino Southern Oscillation on the global yields of major crops. [J]. Nature Communications, 2014, 5: 3712-3712.

[3] Mclarnon CR, Duncan JL. Testing of Ammonia Based $CO_2$ Capture with Multi-Pollutant Control Technology[J]. Energy Procedia, 2009, 4(1): 1027-1034.

[4] Telikapalli V, Kozak F, Francois J, et al. CCS with the Alstom chilled ammonia process development program–Field pilot results[J]. Energy Procedia, 2011, 4(1): 273-281.

[5] Salasperegrin M, Carrascomarin F, Lopezgarzon F, et al. Additions and Corrections - Adsorption of $CO_2$ on Activated Carbons from Diluted Ambient Environments.[J]. Energy & Fuels, 1994, 8(1): 239-243.

[6] Yong Z, Vera Mata A, Rodrigues AE. Adsorption of Carbon Dioxide onto Hydrotalcite-like

Compounds (HTlcs) at High Temperatures[J]. Ind.eng.chem.res, 2000, 40(1): 204-209.

[7] Siriwardane RV, Shen MS, Fisher EP, et al. Adsorption of $CO_2$ on Zeolites at Moderate Temperatures[J]. Energy Fuels, 2005, 19(3): 1153-1159.

[8] Yang L, Du K, Niu XF, et al. Experimental study on enhancement of ammonia–water falling film absorption by adding nano-particles [J]. International Journal of Refrigeration, 2011, 34(3): 640-647.

Acknowledgement: The study was supported by Project BK20130905 supported by NSF of Jiangsu Province of China.

**2016 International Pittsburgh Coal Conference**
**Cape Town, SOUTH AFRICA**
**August 8 - 12, 2016**

**Coal Science**

# Coal cleaning residues and Fe-ultrafine Particles and Fe-minerals Consequences

***Diego de Medeiros***[a]; *Frans Waanders*[b]; *Felipe B. de Leão*[a]; *Fagner T.C. de Sá*[a]; *Luis F. O. Silva*[a]
[a] *Laboratory of Environmental Researches and Nanotechnology Development, Centro Universitário La Salle, Mestrado em Avaliação de Impactos Ambientais em Mineração. Victor Barreto, 2288 Centro 92010-000, Canoas, RS, Brazil.*
[b] *School of Chemical and Minerals Engineering North West University (Potchefstroom campus) Potchefstroom 2531, South Africa*

**Abstract**
In the present investigation, a study was undertaken to understand the origin of Fe-minerals presents in Brazilian coal mining and to understand the environmental implication and the chemical heterogeneity in the study area. Coal cleaning residue samples rich in clays, quartz, sulphides, carbonates, sulphates, etc. were sampled from Lauro Muller, Urussanga, Treviso, Siderópolis, and Criciúma cities in the Santa Catarina State and a total of 19 samples were collected and Mössbauer, XRD, SEM/EDX, and TEM analyses were conducted on the samples. The major Fe-minerals identified are represented by the major minerals chlorite, hematite, illite, and pyrite, while the minor minerals include ankerite, chalcopyrite, goethite, hematite, jarosite, maghemite, magnetite, marcasite, melanterite, natrojarosite, oligonite, pyrrhotite, rozenite, schwertmannite, siderite, and sideronatrile. Pyrite is relatively abundant in some cases, making up to around 10% of the mineral matter in several samples. The sulphates minerals such as jarosite and others, probably represent oxidation products of pyrite, developed during exposure or storage.

**2016 International Pittsburgh Coal Conference**
**Cape Town, SOUTH AFRICA**
**August 8 - 12, 2016**

**Coal Science**

**Zinc occurrence in coal power plant burning combinations of coal and tires**

***Diego de Medeiros*[a]; *Fernanda R. Silveira*[a]; *Frans Waanders*[b]; *James C. Hower*[c]; *Silvio R. Taffarel*[a]; *Luis F. O. Silva*[a]
[a] *Laboratory of Environmental Researches and Nanotechnology Development, Centro Universitário La Salle, Mestrado em Avaliação de Impactos Ambientais em Mineração. Victor Barreto, 2288 Centro 92010-000, Canoas, RS, Brazil.*
[b] *School of Chemical and Minerals Engineering North West University (Potchefstroom campus) Potchefstroom 2531, South Africa*
[c] *University of Kentucky Center for Applied Energy Research, Lexington, KY, 40511, USA.*

**Abstract**
Fly ash from the cyclone-boiler co-combustion of high-S, high volatile bituminous coal and tire-derived fuel (tdf) was studied using a variety of chemical, optical, and microbeam techniques. Fly ash, dominated by Al-Si glass with lesser amounts of coal derived carbons, Fe-spinels, and tire-derived carbons, has Zn concentrations ranging from 2200 ppm (1st ESP row) to 6900 ppm Zn (3rd ESP row). Zinc occurs in Zn-rich nanoparticles in the Al-Si glass phases and as ZnO in amorphous and crystalline nanominerals, Fe- and Zn-sulphides, Pb-Al-Fe sulphates, and Zn sulphates. Iron-rich, Al- and Ti-bearing spinels contain accessory $Zn^{2+}$, $Cr^{3+}$, $Mn^{2+}$, and $Pb^{2+}$. Fe-sulphates and phosphates nanoparticles incorporate As, Cr, V, Ni, and Zn. Fullerenes were not detected in this fly ash, potentially due to the higher temperature of combustion in the cyclone boiler. Zinc was detected by XPS, but the low binding energies mitigated against the determination of the speciation of the element.

# Hard Carbon from Coal Heavy Organics as Anode Material for Li-ion Batteries

Ying Zhou*, Yulong Li, Ruifeng Liu, Jun Guan, Peng Wan, Lei He, Weiran Cao, Dongju Ma,

*(School of Chemical Engineering, Dalian University of Technology, Dalian, Liaoning, 116024, E-mail: zhouying.dlut@dlut.edu.cn)*

**Abstract**

A hard carbon (HC) material was prepared from heavy organics of coal direct liquefaction by a simple stabilization -carbonization two-step method. Structural analysis shows that the HC is poorly crystallized and contains some graphite-like micro-crystallites. The effects of heat treatment temperature varying from 500 to 900 °C on electrochemical performance were systemically studied. HC-700 obtained at 700 °C as a negativematerial for lithium-ion batteries exhibits a reversible capacity of about 503 $mAhg^{-1}$ at 100 $mAg^{-1}$ with a corresponding coulombic efficiency of 68.8%. After 700 cycles, the reversible capacity of 380 $mAhg^{-1}$ and 189 $mAhg^{-1}$ is still achieved at current density of 500 $mAg^{-1}$ and 5 $Ag^{-1}$. Moreover, the 10%-HCG-700 exhibits superb capacity and stability over long-term cycling because of the multidimensional pathways served by the conductive graphene networks to facilitate the transport of electrons and lithium ions in the bulk electrode. It still delivers 520 $mAhg^{-1}$ and 200 $mAhg^{-1}$ at current density of 500 $mAg^{-1}$ and 10 $Ag^{-1}$ after 700 cycles.

*Keywords : Heavy Organics of coal direct liquefaction; Hard carbon ;Li-ion batteries*

## 1. INTRODUCTION

Lithium-ion batteries have been widely used for the portable/consumer electric markets on account of its various advantages. Various forms of carbonaceous materials are widely used as anode materials in Li-ion batteries, since their high thermal and chemical stability, environmental friendliness, abundant source and low cost[1-4]. Nowdays graphite is most commonly used an anode material because of its high initial coulombic efficiency and low electrode potential relative to lithium metal. However, since low theoretical capacity (372$mAhg^{-1}$) and bad rate performance, the graphitized carbon anode material is far from meeting the requirements for high energy/power density especially in electric vehicles[5]. In this regard, the researchers hope to improve the electrochemical performance of anode materials by developing new nanostructure carbon materials. Hard carbon, as a novel anode material, has received considerable attention for its high capacity (740mAh/g), good rate performance, low cost and easy to large-scale production[6-8]. On the other hand, a novel two-dimensional graphitic carbon, graphene, has drawn special attention due to its its outstanding electrical conductivity, superior mechanical flexibility, large specific surface area, and high thermal/chemical stability[1, 9-13].

Gerenally, there are a lot of kinds of hard carbon, for example resin carbons (such as phenolic resin, epoxy resin, etc.), organic polymer pyrolytic carbons (such as PFA, PVC, PAN, etc.) , biomass derived carbons (such as peanut shells, potato starch, glucose, etc.) and the heavy organic matters derivesd carbons (coal tar pitch, petroleum asphalt), etc[6, 8, 14]. Among them, the heavy oraganic matters have been accounted as promising hard carbon precursors because of their low cost and extensive sources. The coal liquefaction residues (CLR) is a waste of coal liquefaction process and hinders the large-scale application of coal liquefaction technology. It's very important to develop high value-added utilization of CLR technology. CLR rich in asphaltene and preasphaltene has a lot of advantages like high degree of aromaticity, good reactivity, high carbon residue. So it is accounted as promising hard carbon precursor.

In this study, hard carbon (HC) material is prepared from the coal direct liquefaction residue heavy organic matters by a simple crosslinking-solidifying and carbonization two-step method. The composite of HC/G is synthesized by carbonizing a mixture of GO and S-AR, and carbonization. We investigated the influence of pyrolysis temperature (TP) and graphene on structural characterization and electrochemical performance of HC.

## 2. EXPERIMENTAL

*2.1 Synthesis of hard carbon (HC)*

Hard carbon (HC) was prepared by a two-step heat treatment consisting of stabilization and carbonization. In a typical run, the mixture of the extract (30g) from Shenhua coal direct liquid residue and Benzenedimethanol (9 g) were melt in the nitrogen atmosphere at 140℃. 6 g of 20% sulfuric acid was added into the melt mixture dropwise and kept vigorously stirring for 2h at 160℃, the product was labeled as AR. The as-obtained AR was set in a quartz tube reactor, heated in an Ar flow to 300℃ for 8h to convert the AR to thermosetting S-AR. Hard carbon was then easily obtained by carbonizing the S-AR at 500℃, 700℃, 800℃ and 900℃ for 3 h under a flow of Ar, and the products were labeled as HC-500, HC-700, and HC-800, HC-900, respectively.

*2.2 Synthesis of hard carbon/graphene (HC/G) composite*

Graphene Oxide (GO) was prepared from graphite powder through a modified Hummers' method[15]. Firstly, S-AR and GO with a designated mass ratio of 10:1 are added into ethanol solution and sonicated for 0.5 h to disperse GO sheets and S-AR powders uniformly. Then some polyvinyl pyrrolidone (PVP-k30), which can help the GO sheets coating on the surface of S-AR, is added into the suspension, and intensely stirred for 5h at room temperature. The composite of S-AR/GO is obtained by distillation and drying overnight in an oven at 80℃. Finally, the composite of hard carbon and graphene (HC/G) is then easily obtained by carbonizing the S-AR/G at 700℃ for 3 h under a flow of Ar, and the products were labeled as 10%-HCG-700.

*2.3 Materials characterization*

Field-emission scanning electron microscopy (FE-SEM, FEI NOVA NanoSEM 450, operated at 3 kV) and transmission electron microscopy(TEM, FEITF30, operated at 300 kV) are used to examine morphologies and microstructures of the samples. X-ray diffraction (XRD) patterns are recorded on D/MAX-2400 using Cu Kα (=1.5406) radiation over the range of 5°to 80°angles (2θ) at room temperature to examine the crystallite structures of samples. Raman analysis was studied on an inVia Raman Microscope (RENISHAW) with a 532nm$^{-1}$ laser beam. Pore-structure is measured with ASAP 2020 (Micromeritics, USA). Before sorption measurement, the samples are degassed at 250℃ for 5h. The specific surface areas (SSA) of the samples is calculated according to the Brunauer–Emmett–Teller (BET) equation within a relative pressure range of 0.05~0.20. The pore size distributions (PSD) of the samples are computed using density functional theory (DFT) model.

*2.4 Electrochemical measurements*

The electrochemical measurements are carried out with LIR2016 coin cells using pure lithium foil as the counter and reference electrodes at room temperature. The working electrodes consist of active materials, carbon black and polyvinylidene difluoride (PVDF) in a weight ratio of 8:1:1. First, the composite dissolve in NMP to make into uniform slurry. The slurry is then spread uniformly onto a copper foil and dried in a vacuum oven at 120℃ for 12 h. The solution of 1 M LiPF6 in a 50:50 (wt./wt.) mixture of ethylene carbonate (EC) and diethyl carbonate (DEC) is used as the electrolyte. The Cell assembly was conducted in an Ar-filled glove box at room temperature. The cells are galvanostatically charged and discharged at various current densities of 100 mAg$^{-1}$ to 10 Ag$^{-1}$ between 0.01 and 3.0 V (vs. Li/Li$^+$) on the LAND CT2000 battery testing system at room temperature. The cyclic voltammograms (CVs) between 0.01~3.0V at a scan rate of 0.1 mV s$^{-1}$ and the electrochemical impedance spectroscopy (EIS) measurements employing the AC amplitude of 5 mV over the frequency range of 100 kHz to 0.01kHz are obtained at the CHI660D electrochemical station.

## 3. RESULTS AND DISCUSSIONS

*3.1 Characterizations of morphology and structure*

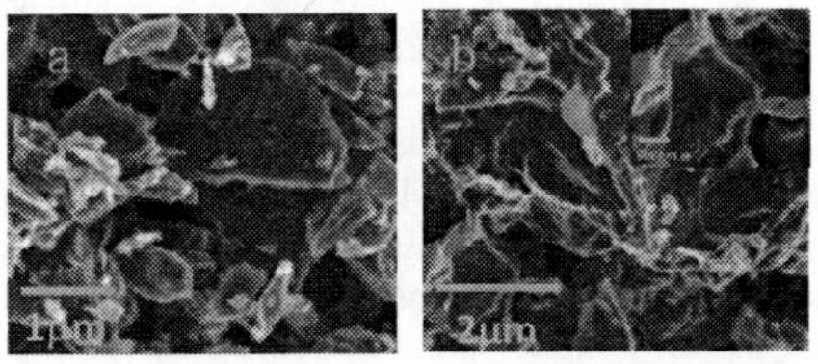


Fig1. SEM images of a)HC-700 and b)10%-HCG-700

The microstructure and morphology of the samples were examined by scanning electron microscopy (SEM). Fig.1 shows the field emission scanning electron microscopy (FESEM) images for the HC-700 (Fig. 1a), 10%HC/G-700 (Fig. 1b). As shown in Fig. 1a, the sample HC-700 show a non-uniform morphology with an average particle size about 5μm. Apparently, the surface of HC particles is smooth and non-porosity is observed. Fig. 1b shows that the surface of HC particles is wrapped by rGO sheets uniformly. The high magnification images show that wrinkled rGO nanosheets are distribute and inter connected in interstice of HC particles, which provides pathways to facilitate the transport of electrons and lithium ions in the negative electrode.

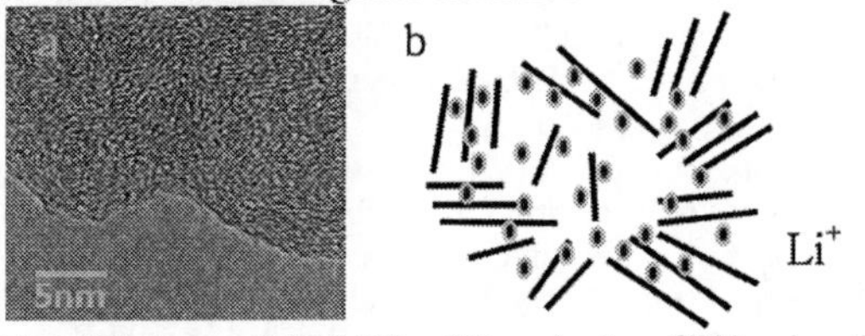


Fig2. TEM image of a)HC-700 and b) mechanism of lithium insertion in HC

The TEM images in Fig. 2 reveal the microstructure of the samples. As shown in Fig. 2a, HC-700 is poorly crystallized and the short-range parallel stacking of small and curved graphene layers gives rise to disordered nanoscale porosity, which can provide as the active sites for lithium ions. This poorly ordered turbostratic structure can be described using the falling cards model. Fig2. b shows the possible lithium storage mechanism. As shown in Fig2. b, $Li^+$ can be intercalated between graphene layers or adsorb at the edge or on the both surface of the graphite microcrystallites of the HCs.

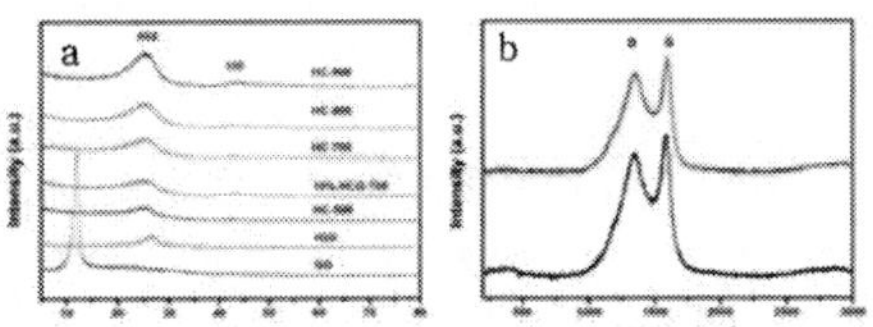


Fig3. a) XRD patterns and b) Raman spectra of the HC and HCG composite

Table1. Parameters of XRD spectra and Raman spectra.

| Sample | $d_{002}$(nm) | Lc(nm) | La(nm) | N | R | $I_D/I_G$ |
|---|---|---|---|---|---|---|
| HC-500 | 0.354 | 1.695 | 4.13 | 5.84 | 4.73 | 0.639 |
| HC-700 | 0.355 | 1.279 | 2.69 | 4.6 | 3.52 | 0.872 |
| HC-800 | 0.352 | 1.285 | 4.68 | 4.7 | 4.69 | 0.957 |
| HC-900 | 0.354 | 1.562 | 5.11 | 5.4 | 6.61 | 0.995 |

Raman spectroscopy and X-ray diffraction (XRD) are carried out to further investigate the variation in microstructures of HC/G composite and the HC prepared by different thermal treatment. The X-ray diffraction patterns of the samples are displayed in Fig. 3a, which are typical of carbonaceous materials with an amorphous-like structure. As shown in XRD patterns, two broad peaks around 25°and 43°are observed, corresponding to the 002 diffraction and the 100 or 101 diffraction peaks of graphite. The d-spacing between 002 plane of HC-700, calculated by Bragg equation, is 0.356nm, which is larger than that in graphite. The big interlayer spacing of $d_{002}$ is favorable for storage and transport between graphene layers, especially under high current density. Fig. 3a shows that the diffraction patterns become sharpen and move to slightly higher angles with increasing heat temperature. Notably, this trend is quite weak because of its cross-linked microstructure holds back graphite crystallite growing up. The GO exhibit only one peak centered at ca. 12°, corresponding to a basal spacing of 0.75 nm. Owing to the introduction of various oxygen containing functionalities (e.g., carboxyl, hydroxyl, or epoxy) between layers, the basal spacing was greatly expanded. Interestingly, in contrast to the broad peak around 26° of the rGO pattern, it suggest that the GO transformed into graphene successfully by thermal treatment under Ar. As shown in Fig. 3b, Roman spectra of all the samples have two characteristic bands around 1350 $cm^{-1}$ (D-band) and 1580 $cm^{-1}$ (G-band), corresponding to the vibrational mode of a tetrahedral sp3 configuration and sp2-bonded carbon respectively. The intensity ratio of the D-band over the G-band ($I_D/I_G$) used to quantify the disorder degree of the carbon. According to Fig. 3b, the ratio of $I_D/I_G$ decreases slightly from 0.872 to 0.858 with the addition of graphene. A 2D-band(~2700 $cm^{-1}$) is not observed because the 2D-band signals of graphene are relatively too weak to be detected. The 2D-band isassociated with a second-order zone boundary phonon mode for graphite or graphene .

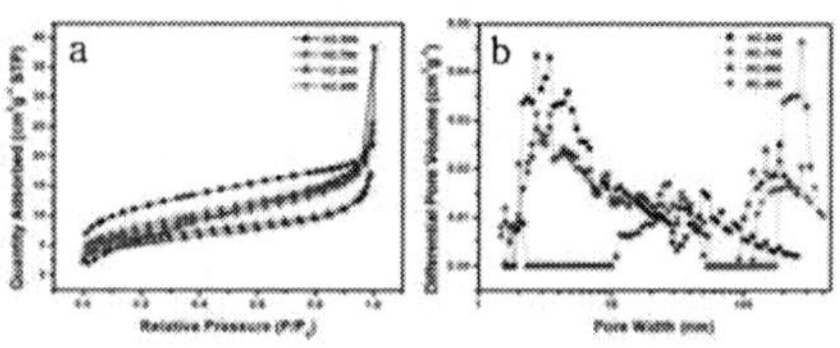

Fig4. a) Nitrogen adsorption isotherm and b)pore size distribution of HC

Table2. Textural properties of HC

| Sample | $S_{BET}$ ($m^2/g$) | $V_{tot}$ ($cm^3/g$) | $V_{mso}$ ($cm^3/g$) | $V_{mso}/V_{tot}$ (%) |
|---|---|---|---|---|
| HC-500 | 27.33 | 0.032 | 0.028 | 87.5 |
| HC-700 | 27.3 | 0.034 | 0.024 | 70.6 |
| HC-800 | 18.8 | 0.026 | 0.018 | 69.2 |
| HC-900 | 25.71 | 0.033 | 0.023 | 69.7 |

Nitrogen physisorption measurements were carried out at 77 K to analyze the textural characteristics of the HC obtained at several temperatures: 500, 700, 800 and 900℃. As shown in Fig.4a, the isotherms of the HC obtained under different temperature give rise to a type II curve which is characteristic of mesoporous and macropore materials. The Brunauer-Emmett-Teller (BET) surface area is around 20 $m^2g^{-1}$, and the pore size distribution derived from DFT model is shown in Fig.4b, it can be seen that HC obtained under different temperature all possess a wide pore size distribution. The corresponding data is summarized in table 2. In the peculiar hierarchical structure, mesopores and macropores in the nanoparticles can promote the electrolyte diffusion into the internal electrode materials and provide excellent transport pathways to shorten the diffusion distance of Li-ions, which may enhance the electrochemical performance

*3.2 Electrochemical characterization*

The as-prepared HC and 10%-HCG-700 was investigated as the anode material for lithium ion battery and its electrochemical performance is presented in Fig 5.

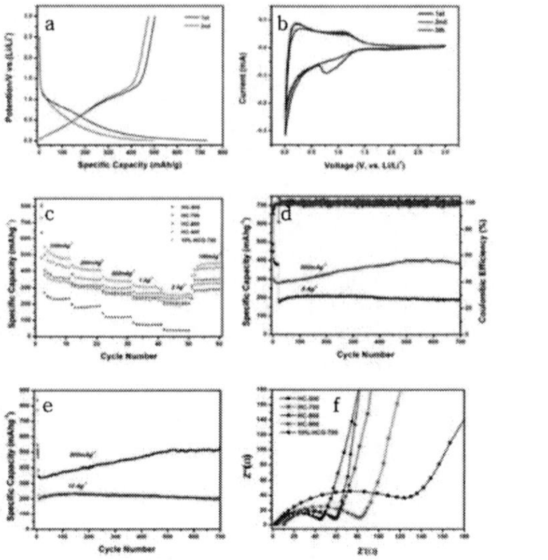

Fig.5 Electrochemical performances of HC and 10%-HCG-700: a) charge/discharge curves of HC-700 at a current density of 100 $mAg^{-1}$, b) cyclic voltammograms for HC-700 at a scan rate of 0.01 $mVs^{-1}$, c) rate performance of the as-prepared HC under different temperature and 10%-HCG-700, d) cycle performance of HC-700 at acurrent density of 500$mAg^{-1}$ and 5$Ag^{-1}$, e) cycle performance of 10%-HCG-700 at acurrent density of 500$mAg^{-1}$ and 10$Ag^{-1}$ and f) Nyquist plots of HC and 10%-HCG-700 over the frequency range from 100 kHz to 0.01 Hz.

The galvanostatic charge/discharge profiles of the HC-700 during the first and second cycles at a current density of 100 $mAg^{-1}$ are given in Fig. 5a. In the first cycle, the HC-700 has a discharge capacity of 731 $mAhg^{-1}$ and a charge capacity of 503 $mAhg^{-1}$, with a corresponding coulombic efficiency of 68.8%. No obvious plateau or staging phenomenon was observed in the first charge curve, indicating a low degree of graphitization in the HC-700. As shown in Figure 5b, the HC-700 shows typical cyclic voltammetry (CV) curves of hard carbon. An apparent irreversible region can be seen from 0.6 to 1.25 V in the first cycle, which is originated from the electrolyte decomposition on the electrode/electrolyte surface and irreversible lithium insertion into potentially unique positions, such as cavities or sites in the vicinity of residual hydrogen atoms in the carbon material[16].

Fig. 5c compares the rate performance of HC prepared under different temperature and 10%-HCG-700 composite. In the case of the HC-500 and HC-900 samples, HC-500 exhibited poor rate capability (delivering only ~38 $mAhg^{-1}$ at 2 $Ag^{-1}$), which may be associated with bad electrical conductivity. HC-700 displayed superior specific capacity at different current density from 100 $mAg^{-1}$ (450 $mAhg^{-1}$), larger than the theoretical specific capacity 372$mAhg^{-1}$ of graphite, to 2 $Ag^{-1}$ (240 $mAhg^{-1}$), which may be related with both the good electrical conductivity and suitable microstructure. When the current rate decreased from 2 $Ag^{-1}$ to 100 $mAg^{-1}$ after the 50th rate cycle, a capacity of 430 mA $hg^{-1}$ was still recoverable. The effect of heat treatment temperature on electrochemical characteristics varies depending on the combined effect of the electric conductivity and the amount of intercalation sites available. And the electric conductivity and the amount of intercalation sites available are two important parameters for anode materials. The graphitization proceeds with increasing heat treatment temperature, while the amount of intercalation sites available decreases and the lithiumion diffusion in the layered structure becomes difficult compared with the diffusion in the turbostratic disordered structure. Furthermore, comparing to HC-700, 10%-HCG-700 delievers the better rate performance. Because the conductive graphene networks can serve as multidimensional pathways to facilitate the transport of electrons and lithium ions in the bulk electrode.

Figure 5d illustrates the cycle performance of the HC-700 at a current density of 100 $mAg^{-1}$ for the first several cycles and then 500 $mAg^{-1}$ and 5 $Ag^{-1}$for subsequent cycles. After 700 cycles, there is no obvious sign of fade in capacity, and a capacity of 380 $mAhg^{-1}$ and 189 $mAhg^{-1}$ is still retained. As shown in fig 5e, 10%-HCG-700 exhibits superb capacity and stability over long-term cycling. It still delivers 520 $mAhg^{-1}$ and 200 $mAhg^{-1}$ at current density of 500 $mAg^{-1}$ and 10 $Ag^{-1}$ after 700 cycles. As shown in fig. 5d and e, the capacity increases slightly with the cycle number increasing. This capacity-rise phenomenon may be derived from the delayed infiltration of the electrolyte into the hard carbon materials in previous cycles and the generation of structural defects to provide more active sites for $Li^+$ diffusion and storage[4].

Fig 5f shows the Nyquist plots of the asprepared HC and 10%-HCG-700. The Nyquist plots for the samples display a depressed semicircle in the high frequency range and followed by linear part toward lower frequency. From Fig. 5f, it can be seen that 10%-HCG-700 present smaller charge transfer resistance than those of pure hard carbon materials. It implies that the graphene network could effectively enhance the electrical conductivity and reduce the contact and charge-transfer resistances in the HCG composite electrode.

## 4. CONCLUSION

In summery, heavy organics of coal direct liquefaction -based hard carbon is prepared by a simple stabilization -carbonization method. Structural characterization indicates that the obtained hard carbon is poorly crystallized and contains some contacted graphite-like micro-crystallites. Because of the conductive graphene networks serving as multidimensional pathways, the 10% HC/G exhibits large capacity, excellent rate performance and superb cycle performance as the anode material in lithium ion batteries.

## ACKNOWLEDGEMENTS

The work was supported by the National Natural Science Foundation of China (Grant No. 21276045, No. 21576047, No.U1510204).

## References

[1] Liu X., Wu Y., Yang Z., et al. Nitrogen-doped 3D macroporous graphene frameworks as anode for high performance lithium-ion batteries[J]. Journal of Power Sources, 2015,293: 799-805.

[2] Sun D., Yang J. and Yan X. Hierarchically porous and nitrogen, sulfur-codoped graphene-like microspheres as a high capacity anode for lithium ion batteries[J]. Chemical Communications, 2015,51: 2134-2137.

[3] Wang F., Song R., Song H., et al. Simple synthesis of novel hierarchical porous carbon microspheres and their application to rechargeable lithium-ion batteries[J]. Carbon, 2015,81: 314-321.

[4] Wang S., Xiao C., Xing Y., et al. Carbon nanofibers/nanosheets hybrid derived from cornstalks as a sustainable anode for Li-ion batteries[J]. Journal of Materials Chemistry A, 2015,3: 6742-6746.

[5] Xing W. B. and Dahn J. R. Study of irreversible capacities for Li insertion in hard and graphitic carbons[J]. Journal of the Electrochemical Society, 1997,144: 1195-1201.

[6] Lv W., Wen F., Xiang J., et al. Peanut shell derived hard carbon as ultralong cycling anodes for lithium and sodium batteries[J]. Electrochimica Acta, 2015,176: 533-541.

[7] Yinghu L., Xue J. S., Tao Z., et al. Mechanism of lithium insertion in hard carbons prepared by pyrolysis of epoxy resins[J]. Carbon, 1996,34: 193-200.

[8] Ni J., Huang Y. and Gao L. A high-performance hard carbon for Li-ion batteries and supercapacitors application[J]. Journal of Power Sources, 2013,223: 306-311.

[9] Yu C., Meng X. T., Song X. D., et al. Graphene-mediated highly-dispersed MoS2 nanosheets with enhanced triiodide reduction activity for dye-sensitized solar cells[J]. Carbon, 2016,100: 474-483.

[10] Zhang M., Sun Z. H., Zhang T. F., et al. Excellent cycling stability with high SnO2 loading on a three-dimensional graphene network for lithium ion batteries[J]. Carbon, 2016,102: 32-38.

[11] Dou Y. H., Xu J. T., Ruan B. Y., et al. Atomic Layer-by-Layer Co3O4/Graphene Composite for High Performance Lithium-Ion Batteries[J]. Advanced Energy Materials, 2016,6:

[12] Wang X. P., Lv L. X., Cheng Z. H., et al. High-Density Monolith of N-Doped Holey Graphene for Ultrahigh Volumetric Capacity of Li-Ion Batteries[J]. Advanced Energy Materials, 2016,6:

[13] Chen K. F., Song S. Y., Liu F., et al. Structural design of graphene for use in electrochemical energy storage devices[J]. Chemical Society Reviews, 2015,44: 6230-6257.

[14] Guo Z., Wang C., Chen M., et al. Hard Carbon Derived from Coal Tar Pitch for Use as the Anode Material in Lithium Ion Batteries[J]. International Journal of Electrochemical Science, 2013,8: 2702-2709.

[15] Zhao J. P., Pei S. F., Ren W. C., et al. Efficient Preparation of Large-Area Graphene Oxide Sheets for Transparent Conductive Films[J]. Acs Nano, 2010,4: 5245-5252.

[16] Park C. W., Yoon S. H., Lee S. I., et al. Li+ storage sites in non-graphitizable carbons prepared from methylnaphthalene-derived isotropic pitches[J]. Carbon, 2000,38: 995-1001.

# The relationship between oxygen transfer and pyrolysis conditions

*Chaopeng Wang, Huijun Huang, Xiaohong Li, Jie Feng*, Wenying Li**
*Key Laboratory of Coal Science and Technology (Taiyuan University of Technology), Ministry of Education and Shanxi province, Taiyuan 030024, PR China*
*Training Base of State Key Laboratory of Coal Science and Technology Jointly Constructed by Shanxi Province and Ministry of Science and Technology, Taiyuan 030024, PR China*

* *Corresponding author. E-mail:* fengjie@tyut.edu.cn; ying@tyut.edu.cn

**Abtract**: Oxygen is an important factor that contributes to the complex structure of coal, and its content and form influence coal pyrolysis properties. In the process of pyrolysis, different pyrolysis conditions will make products (volatile and char) distribution and constitute different. In order to explore the relationship between oxygen transferring and pyrolysis conditions, Hulunbeir lignite was chosen to pyrolyze in the fixed bed reactor. The oxygen amount of solid product in different heating rates was measured by elemental analyzer. The results show that the oxygen gradually transfers from coal to volatile with the increase of temperature. At 650°C and 1.18-2.00mm, at the heating rate of 15°C /min, 60°C /min, and 300°C /min, there were 17.31%, 17.79%, and 27.73% of oxygen transferred to solid products respectively. Rapid heating rate helps more oxygen transfer to char and tar while block the production of gas-O and water. Particle size has a sight effect on oxygen transfer to solid products under high heating rate. Coal will suffer stronger power when the heating rate is faster, resulting in more free radicals forming and releasing in short time. The change of process conditions will affect elemental distribution of pyrolysis products, which means oxygen transformation is not independent, but connected with the quality and quantity of hydrogen radical in pyrolysis.

**Keywords**: oxygen transfer; lignite pyrolysis; temperature; heating rate; particle size

## 1. Introduction

Lignite accounts for more than 16% of total coal reserves in China. However, its own characteristics (three high and two low) limit its use efficiently. Grading lignite pyrolysis technology is a more efficient and cleaner way for lignite conversion. Pyrolysis is an initial unit for this technology, it can extract the light components under relatively mild conditions, producing gas, tar and char can be used for subsequent utilization processing, to improve the efficiency of lignite[1]. Lignite, with high oxygen content, the oxygen-containing functional groups influence its pyrolysis conversion processes and utilization of the downstream products directly. It has important academic significance and potential application value to study the oxygen migration characteristics and influencing factors in pyrolysis process for the oriented conversion of oxygen under different technical conditions. Pyrolysis characteristics of lignite is related to its structure and composition, the presence of oxygen contributes to lignite's complex molecular structure[2], its content and the existing form have a tremendous impact on lignite's characteristic, so that directly affect the the reaction behavior among pyrolysis, liquefaction and gasifaction. The effect of oxygen on the pyrolysis process is mainly derived from the consumption of hydrogen, oxygen and hydrogen combine to generate pyrolysis water, consuming large amounts of hydrogen. The effective pyrolysis products yield decreased and the quality deteriorates, reducing the element utilization. Thus the pyrolysis process migration of oxygen need to be figured out, explore the main factors affecting the migration of oxygen, and then regulate migration path of oxygen, so that more hydrogen and oxygen come into the organic component and realised efficient use in oxygen and hydrogen elements of lignite.

In the pyrolysis process, a portion of oxygen in coal release in the form of gas (mainly CO, CO2), tar and pyrolysis water, the rest will remain in semi-coke[3, 4]. Liu[5] investigated the migration path of oxygen of three kinds of Chinese coal during pyrolysis and found that more than 70% of the oxygen transported to the gas product, mainly CO and CO2, but it did not consider oxygen in the water in the study. Luik[6] studied the migration of oxygen at different pyrolysis operating conditions using pine bark and found that most of oxygen come into the water. Their research also showed that the migration of oxygen is closely related to operating conditions in the pyrolysis process. Feng[7] studied the pyrolysis process of Baiyinhua lignite and found that oxygen conversion during pyrolysis have a great relationship with the temperature. When the temperature is 400 ° C, the pyrolysis of organic oxygen in coal was mainly in the form of water and pyrolysis gases, pyrolysis tar had rich oxygen substances (especially phenols) before 500 ° C. However, the current literature have been lack of migration of oxygen under different pyrolysis operating conditions.

In this paper, Hulunbeier lignite has been selected for the study, by examining how the pyrolysis temperature and heating rate effect the migration of oxygen, to explore the main factors affecting the migration of oxygen, to provide theoretical guidance for the rational use of lignite under different pyrolysis process oxygen elements.

## 2. Experimental

### 2.1 coal sample

Pyrolysis experiments were carried out with Hulun Buir lignite (HL) . Proximate analysis and ultimate analysis of the samples have already been given in Table 1. Samples were groud, sieved to the particle size range of 1.18-2.00mm, dried 10h at 110°C, and sored until required.

Table 1 Proximate analysis and Ultimate analysis of the samples

| Sample | Proximate analysis (wt.%, ad) | | | | Ultimate analysis (wt.%, daf) | | | | |
|---|---|---|---|---|---|---|---|---|---|
| | M | A | V | FC* | C | H | O | N | S |
| HL | 11.44 | 13.11 | 33.08 | 42.37 | 68.65 | 4.97 | 28.69 | 0.97 | 0.32 |

* By difference

## 2.2 pyrolysis experiments

Pyrolysis experiments were carried out in a fixed bed reactor. Details of the reactor configuration have already been given in figure 1. The experimental apparatus includes four parts: a quartz reactor, gas supply systems, heating systems and condensing temperature, the product collection system.

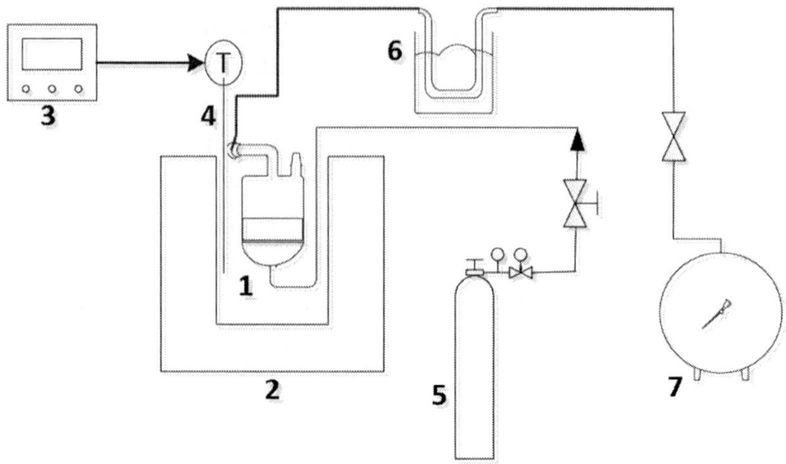


Figure 1 Schematic of the pyrolysis apparatus

1- fixed-bed reactor; 2-electrical oven; 3-temperature display; 4-thermocouple; 5-$N_2$ cylinder; 6-ice-salt bath; 7-wet gas flow meter

It was divided into three modes to get different heating rate, as Figure 2. Mode 1 (M1 for short) utilized Temperature Programmed to heat up the sample in the rate of 15°C /min. Isothermal decomposition were carried out in mode 2 (M2 for short). Quartz sand was selected as heat carrier due to its good capability of heat storage, heat transfer and non-catalytic cracking, the blend ratio of heat carrier and coal was 3: 1 (mass ratio) in mode 3 (M3 for short).

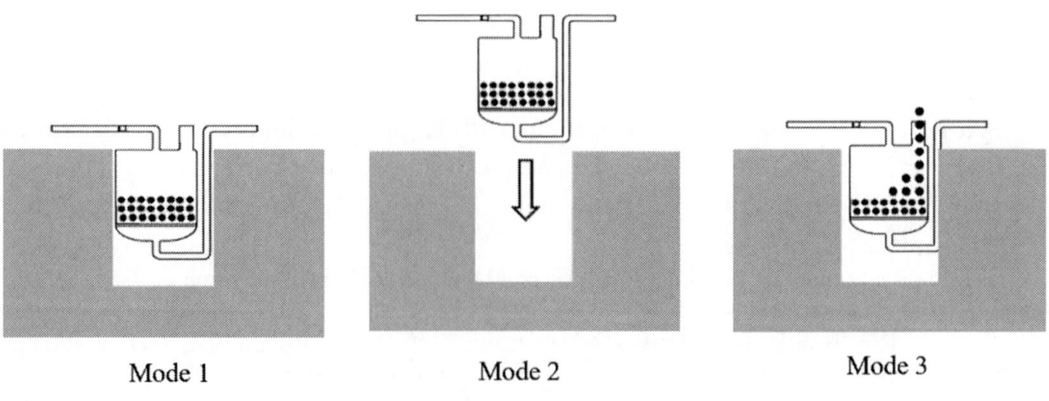


Figure 2 Different modes of pyrolysis

2.3 Calculation pyrolysis product distribution parameters of oxygen

$$O_i = \frac{M_i \omega_i}{M \times \omega_M} \times 100\%$$

$$O_{gas} = \frac{M_{CO_2} \omega_{CO_2} + M_{CO} \omega_{CO}}{M \times \omega_M} \times 100\%$$

Where $\omega_i$ is the mass fraction of oxygen in the pyrolysis products of i, $M_{CO_2}$、$M_{CO}$ are the quality of pyrolysis gas ($CO_2$ and CO), $M_i$ is the quality of pyrolysis products, $M$ is the quality of the sample.

2.4 Analysis of pyrolysis products

(1) Analysis of the gaseous products: pyrolysis gas is detected by Shanghai Haixin gas chromatography (GC-950), the main pyrolysis gases include H2, CH4, CO, and CO2, using area normalization method to calculate volume fraction.

(2) The Elemental Analysis of products: Elemental composition of pyrolysis products use Elementar Vario MICRO cube element analyzer made in German to test.

## 3. Results and discussion

3.1 Product yields varying with the pyrolysis temperature

As an important factor in pyrolysis technology, temperature has profound influence on product distribution. In this paper, the migration of oxygen in the process of pyrolysis is mainly studied. The organic oxygen in coal is mainly in the form of oxygen functional groups, so it is necessary to study the transformation rule of oxygen functional groups in pyrolysis process. The experiment used section analysis of isothermal pyrolysis of lignite arranging from 250°C (carboxyl group is the most flexible of oxygen-containing functional groups on the thermal stability and it will decompose at 250°C) to 700°C. Figure 3 shows the trend about the yield of the pyrolysis products with the pyrolysis temperature.

As the figure 3 shows, the pyrolysis degree of coal increased with the pyrolysis temperature gradually, while the yield of solid products decreased. It hardly generate any tar before 300°C, only a little water and gas were produced. At this stage, lignite was heated, made dehydration and degassing. After 350°C, lignite occured depolymerization and decomposition reaction, lots of volatile (pyrolysis tar and gas) released at this stage. The period of 400 °C-550 °C generated large sums of pyrolysis tar, while a small amount of tar was generated at 550 °C -650 °C. The yield of tar reached top of 6.3% at 650 °C. With higher temperature, the yield of tar has decreased slightly, but pyrolysis water and gas still increased slowly. This is mainly due to the second reactions (including cracking and condensation reactions) of volatile aggravating with the increasing of pyrolysis temperature. Solid product yields declined with pyrolysis temperature rising[8], secondary degassing process (polycondensation reaction) was dominant after 650 °C which weakened the tendency to slow down.

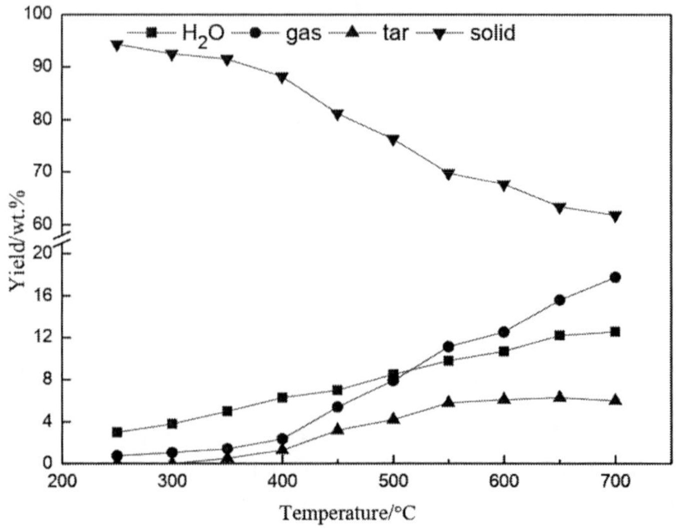


Fig 3 Product yields varying with the pyrolysis temperature

## 3.2 Oxygen transferring varying with the pyrolysis temperature

In order to explore the migration of oxygen during the pyrolysis process further, the distribution characteristics of oxygen in the pyrolysis products at different pyrolysis temperatures were investigated.

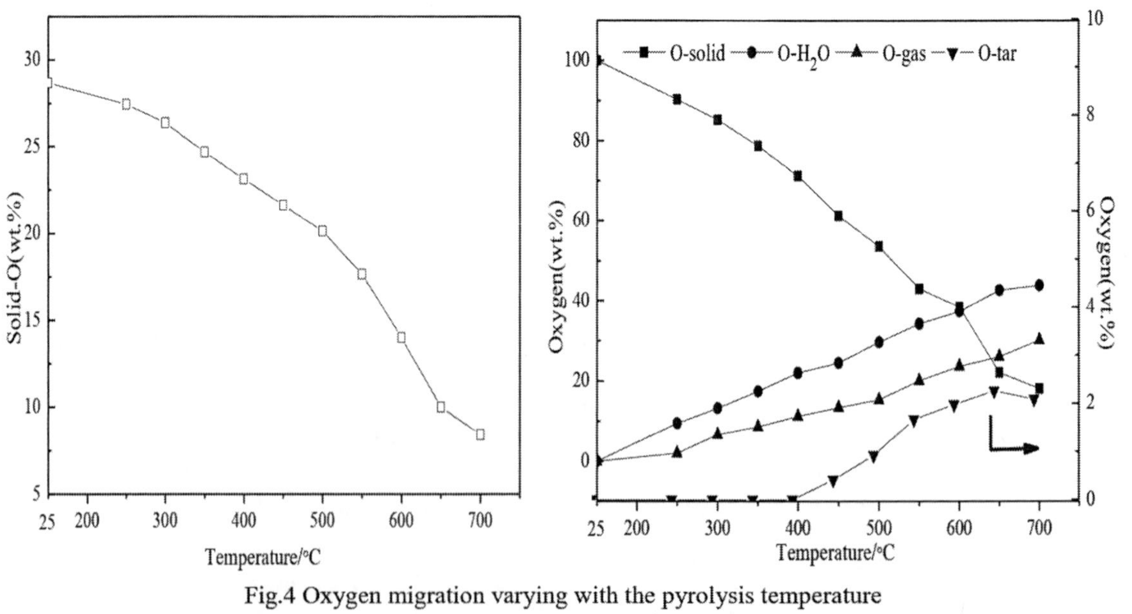


Fig.4 Oxygen migration varying with the pyrolysis temperature

As shown in Figure 4, the left revealed the relationship between pyrolysis temperatures and oxygen percentage in solid product by elemental analyzer, With the increasing of pyrolysis temperature, the oxygen content in the solid products gradually decreased. The oxygen content in solid product decreased rapidly from 600 ° C to 650 ° C, the trend slowed down after 650 °C. Eventually, oxygen content in the solid product became 8.41% paralled with 28.69% of oxygen in coal. In other words, 70.71% of the oxygen in coal transported to the volatiles (including tar, pyrolysis water and pyrolysis gas (CO, CO2)) at 700 °C. Oxygen distribution parameters in the pyrolysis products can be known in the right. At 250 °C, there are 9.48% and 2.02% of oxygen which migrate to volatiles in the form of pyrolysis water and gas, coming from the desorption of adsorbed gas by pore and decomposion of part of carboxyl groups. With the pyrolysis temperature increasing, the ratio of oxygen transported to the pyrolysis water and pyrolysis gas rised, Oxygen content migrating to water pyrolysis slowed trend after 650 °C. There were 43.84% and 30.17% respectively of oxygen transported to pyrolysis water and gas when pyrolysis temperature of 700 °C, only 18.12% of oxygen remained in the solid product. Only 3.19% of oxygen migrated to the tar at 450 °C and there were almost no oxygen migrating to the tar under 450 °C. At the temperatures below 450 °C, the amount of tar generated little, and it were mainly pyrolysis gas and water (0.5% of pyrolysis tar produced at 350 °C and the tar yield was 1.3% at 400 °C). At this stage, The cleavag of weak bridge bonds (such as -CH2-, CH2-CH2-, and -CH2-O-.) and some functional groups (mainly the weaker aliphatic side chain and oxygen-containing functional groups) occurred and produced light tar excluding the oxygen-containing substances. With the pyrolysis temperature rising, the oxygen migrating to tar showed the tendency of going up firstly and going down secondly. It reached peak value at 650°C, similar to the tendency of pyrolysis tar. Above 600 °C, the pyrolysis temperature was higher, the secondary reaction volatile was more drastic. Radical fragments occurred secondary cracking and polycondensation intensified, so tar yield decreased, acid production in tar reduced[9]. Oxygen mainly exist in form of oxygen-containing phenolic substances in low temperature pyrolysis tar. Phenolic hydroxyl group will crack at 700 °C, accompany by decomposion, dealkylation and dehydroxylation reactions, so the yield of phenols reduced, content of oxygen in tar reduced, too[10, 11].

### 3.3 oxygen migration of different sizes of samples

This paper selected three particle sizes of samples (1.18-2.00 mm, 0.425-0.85 mm and 0.25-0.425 mm) to do isothermal pyrolysis experiments (final pyrolysis temperature is 650 °C), exploring the effects of different particle sizes on oxygen migration during pyrolysis process.

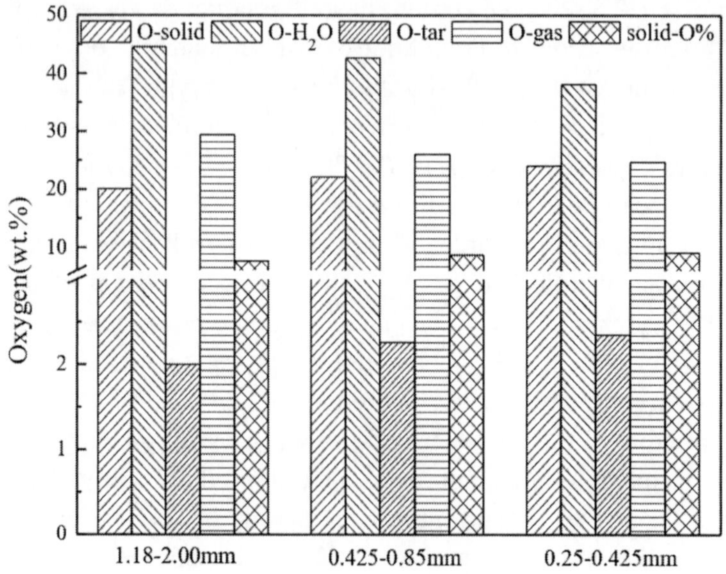

Fig. 5 Oxygen migration paths with different particle sizes

Figure 5 shows that, with decreasing of particle size, the percentage of oxygen in the solid product showed a gradual increment trend (respectively 7.67%, 8.70%, and 9.12%), the smaller particle promote more oxygen to remain in the solid product during the pyrolysis process. However, the ratio of oxygen migrated to pyrolysis tar does not have a clear change with particle size (oxygen migration to coal tar was 2.06%, 2.26 %, and 2.35%). The distribution parameters of three particle sizes of coal showed the same trend, under the experimental conditions, most of the oxygen will migrate to volatile products (mainly pyrolysis gas and water oxygen). Small size is good for reducing Heat transfer block, cross-linking reactions decrease, oxygen of pyrolysis gas and water become less.

## 3.4 Oxygen transferring varying with different heating rate

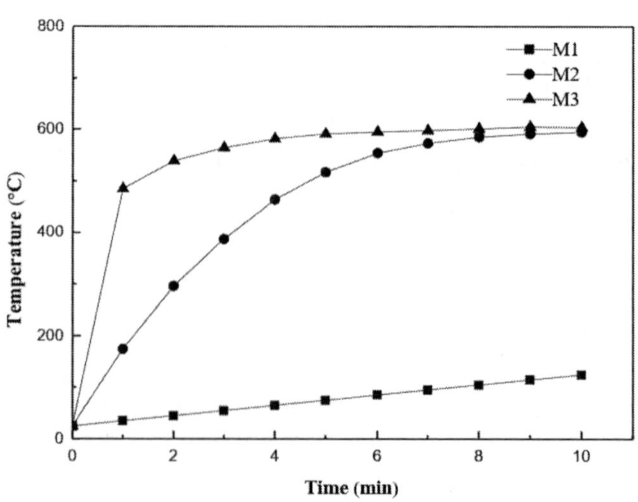

Fig 5 The relationship between temperature and time

The heating process of the sample in three different models were measured by thermocouple and obtained the data of the relationship between temperature and time. According to the figure 5, the heating rate of three pyrolysis modes were respectively 15°C /min, 60°C /min, and 300°C /min.

Kandiyoti[12] pointed out that the volatiles of coal pyrolysis is relatively unstable and volatile liquidity will receive certain restrictions, secondary reactions of volatile occur very rapidly. This study used a fixed bed quartz reactor, it needs to undergo a certain time for volatile from generation to leave the reactor, so secondary reactions of volatile will occur in the reactor, the different heating rates will affect the degree of secondary reactions in the reactor.

Pyrolysis products distribution under three heating rate shown in Figure 6-7. The two figures both show that the pyrolysis product yields of three kinds of heating rates followed a similar trend: With increasing of heating rate, the oxygen migration to pyrolysis gas and pyrolysis water reduced, more oxygen remaining in the solid product, while the ratio of oxygen entering the pyrolysis tar also increased. When coal particle size is 1.18-2.00 mm, 27.73% of oxygen retained in the solid product in mode 3, being 10.42% and 9.94% more than mode 1 and mode 2. When coal particle size is 0.425-0.85 mm, 28.09% of oxygen remained in the solid product, being 7.98% and 5.97% more than mode 1 and mode 2.

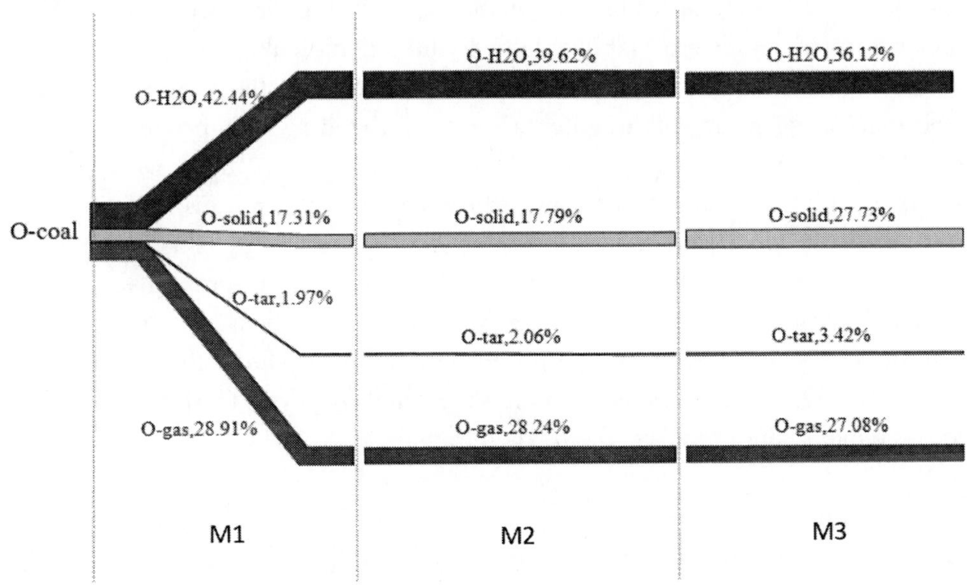


Fig. 6 Allocation of oxygen in products with different heating rates (1.18-2.00mm in particle size)

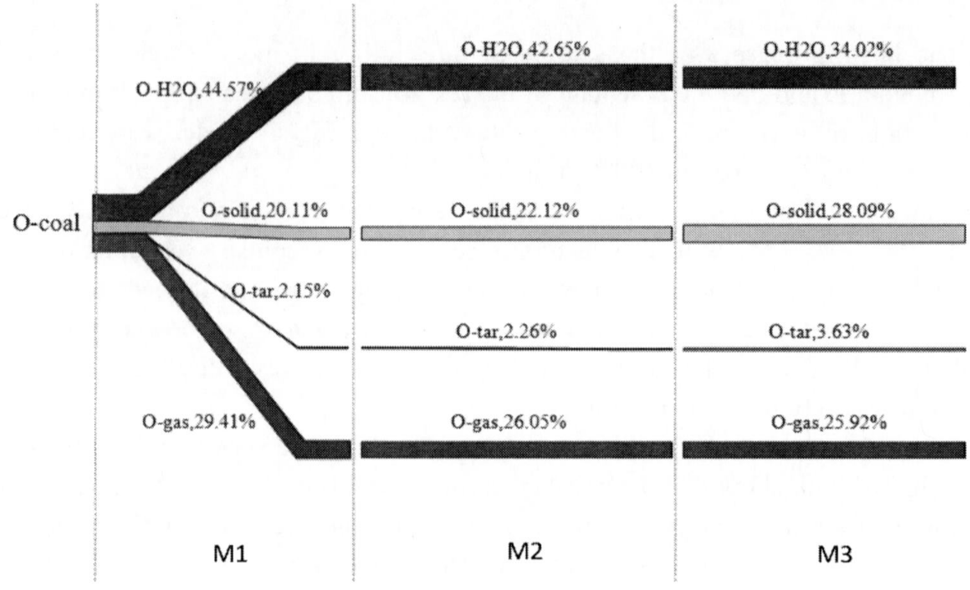


Fig. 7 Allocation of oxygen in products with different heating rates(0.425-0.85mm in particle size)

Heating rate was conducive to the oxygen remaining in the solid product.The heating rate is correlative with heat transfer. When the sample gets a fast heating rate, its structure would obtain enough energy to support rapid cracking. The time to achieve the desired temperature when the bond is broken was shorten, more free radicals formed and were stabilized by hydrogen radicals fragments, it weakened condensation among the radical fragments. So it reduced the formation of pyrolytic water, more oxygen radicals fragments were stabilized in solid product and pyrolysis tar.

At the same time, the impact of coal particle size on oxygen transport can be noticed, especially in mode 1 and mode 2. Under low heating rates, small particle size is helpful to remain more oxygen in the solid product, the migration to tar increases slightly. When the samples were heated with solid heat carrier, particle size become no more clear to oxygen migration to the solid product, the proportion were respectively, 27.73% and 28.09%. In the same final pyrolysis temperature, the heating rate is the main factor to affect the oxygen transfer.

## 4. conclusion

Temperature is a main factor affecting the migration of oxygen. When pyrolysis temperature is higher, the proportion of oxygen migrating to the product will increase in the form of pyrolysis water and pyrolysis gas, the migration to tar oxygen will reach a maximum at 650°C (2.26% of oxygen in coal). Particle size is only valid for low heating rate. In the same final pyrolysis temperature, the heating rate is crucial to affect the oxygen transportation.

## References


[1] Z. Wang, Z. Bai, L. Wen, H. Chen, B. Li, Quantitative study on cross-linking reactions of oxygen groups during liquefaction of lignite by a new model system, Fuel Processing Technology, 91 (2010) 410-413.

[2] Z. Zhu, Coal Chemistry, Shanghai Science and Technology Press, 1984.

[3] J.A. Macphee, J.P. Charland, L. Giroux, Application of TG–FTIR to the determination of organic oxygen and its speciation in the Argonne premium coal samples, Fuel Processing Technology, 87 (2006) 335-341.

[4] H.N.S. Schafer, Pyrolysis of brown coals. 4. Relation between acidic groups and evolved carbon oxides, Fuel, 59 (1980) 302–304.

[5] S. Liu, The Fate of Organic Oxygen During Coal Pyrolysis, Energy Sources, 25 (2003) 479-488.

[6] H. Luik, I. Johannes, V. Palu, L. Luik, K. Kruusement, Transformations of biomass internal oxygen at varied pyrolysis conditions, Journal of Analytical & Applied Pyrolysis, 79 (2007) 121-127.

[7] X.B. Feng, J.P. Cao, X.Y. Zhao, C. Song, T.L. Liu, J.X. Wang, X. Fan, X.Y. Wei, Organic oxygen transformation during pyrolysis of Baiyinhua lignite, Journal of Analytical & Applied Pyrolysis, 117 (2015) 106-115.

[8] P. Dong, G. Chen, X. Zeng, M. Chu, S. Gao, G. Xu, Evolution of Inherent Oxygen in Solid Fuels during Pyrolysis, Energy & Fuels, 29 (2015).

[9] W.C. Xu, A. Tomita, The effects of temperature and residence time on the secondary reactions of volatiles from coal pyrolysis, Fuel Processing Technology, 21 (1989) 25-37.

[10] M. Siskin, T. Aczel, Pyrolysis studies on the structure of ethers and phenols in coal, Fuel, 62 (1983) 1321-1326.

[11] H. Wachowska, W. Pawlak, Effect of cleavage of ether linkages on physicochemical properties of coals, Fuel, 56 (1977) 422-426.

[12] J.R. Gibbins, R. Kandiyoti, The effect of variations in time-temperature history on product distribution from coal pyrolysis, Fuel, 68 (1989) 895-903.

# AUTHOR INDEX

# AUTHOR INDEX

# AUTHOR INDEX

# AUTHOR INDEX

# AUTHOR INDEX

**International Pittsburgh Coal Conference**
University of Pittsburgh
Swanson School of Engineering
3700 O'Hara St.
940 Benedum Hall
Pittsburgh, Pennsylvania 15261
USA

ISBN 978-1-5108-5366-9

CURRAN ASSOCIATES INC.
proceedings.com